CONSTRUCTION MATERIALS

CONSTRUCTION MATERIALS

TYPES, USES, AND APPLICATIONS

Caleb Hornbostel, D.P.L.G., R.A.

A WILEY-INTERSCIENCE PUBLICATION

JOHN WILEY & SONS, New York • Chichester • Brisbane • Toronto

Library of Congress Cataloging in Publication Data:
Hornbostel, Caleb.
Construction materials.

"A Wiley-Interscience publication."
1. Building materials. I. Title.

TA403.H59 624′.18 78-6278
ISBN 0-471-40940-5

Printed in the United States of America

10 9 8 7 6 5 4 3 2 1

TO
THE DEPARTMENT OF ARCHITECTURE
COLLEGE OF ENGINEERING TECHNOLOGY
TEMPLE UNIVERSITY

PREFACE

This book represents the correlation of knowledge scattered piecemeal throughout construction, architectural, engineering, chemical and other diversified technical books, journals, and manufacturers' promotional and advertising literature into one volume for use in the entire field of construction. It is meant to serve architects and engineers, both professional and student, contractors and subcontractors, estimators, specification writers, purchasing agents, etc. The format of the book is alphabetical, like an encyclopedia, and materials are discussed under the following subheadings: "Physical and Chemical Properties," "Types and Uses," "Application," "Condensed Checklist," "Conditions Favorable" and "Conditions Unfavorable" to the use of the material, and "History and Manufacture."

The basic purpose of this book is to give a comprehensive overview of construction materials in general and not to provide a complete treatise on any specific construction material.

The evolution of this book was caused by three circumstances. The first was the task I undertook of developing a logical indexing system for architectural materials for use in reading and evaluating catalogues and pamphlets. I immediately realized how much basic information was missing and how many superficial data were given. The second was a series of happenings that occurred in my professional practice. I used new materials in the construction of buildings, and some very unfortunate experiences developed only because the specific information necessary to avoid the trouble was not available in a form that could be simply understood. The third circumstance arose from my teaching experiences, where I found no basic reference book or text covering the materials used in construction.

Since all materials for construction are derived from only 92 natural elements, I therefore used the elements as the building blocks for construction materials and as starting points to which all other data could be anchored.

In the entire field of construction there exist major problems of terminology and semantics, which are going to be further complicated with the shift to metric. In many cases no one can be sure what another means by a word such as "plastic," "paint," or "plank." For example, "board" and "plank" originally designated specific kinds of lumber. Now these words are meaningless.

Therefore, it was decided that the simplest terminology possible would be used throughout the book. The word "plank" has been eliminated, and very clear definitions are given for "plastic," "sheet," "flooring," "decking," etc.

To simplify data on the physical and chemical properties of materials, only information pertinent to the construction field has been given, such as specific gravity, coefficient of expansion, and compressive and tensile strength. For each element, the symbol, atomic number, specific gravity, melting and boiling points, tensile strength, and coefficient of expansion are given.

The entire construction field probably will be one of the last to convert completely to the metric system because of the multitude of different types of materials that must be used in constructing any building. For this reason I have given both the metric and the currently used sizes, dimensions, etc., in almost all parts of the book. I have also in most cases used decimals, such as 6.25 ft instead of 6′3″ and 0.125 in. instead of $\frac{1}{8}''$, because the entire metric system is based on decimals. In all cases where I have made direct conversions to metric, I have generally used three or four decimal places so that, when the dimensions of a material are converted to metric, the user of this book can easily make the corrections within the text. From my experience during the 4 years I lived and studied in Europe, I found that it took me almost 2 years to think completely in metric. First I started with conversions from feet and inches to meters, centimeters, and millimeters, then graduated to more complex conversions, and finally was able not to convert at all but to think in metric.

There are two basic approaches for using this book. Active practitioners in any aspect of construction would use a sort of "backward" approach. They would go directly to a basic material, such as glass, pick out the type of glass desired, such as tempered glass, and immediately read "Types and Uses" and "Application." Then they might backtrack to the basic data about glass, its physical and chemical properties, and, if interested, learn what and how special properties are given to glass. The students, on the other hand, would use a "forward" approach. They could begin with the basic elements and materials, their properties, components, history and manufacture. Then they could work their way up to "Types and Uses" and proceed to the fabricated forms of these materials, such as sheet, brick, or insulating materials, and their applications. Finally they could read about fundamental parts of buildings, such as flooring, windows, and roofing.

Throughout the book there is frequent mention of specifications, standards, codes, and other requirements, yet ASTM standards, for example, are not quoted or referred to by number. The major reasons for not including ASTM and other specifying agencies' code systems and numbers were (*a*) these organizations are constantly retesting familiar materials and testing new ones, thus causing continual shifting and changing of old code systems and numbers, and adding new code systems and numbers, and (*b*) most importantly, all specifications, standards, codes, etc., will be completely revised with the changeover to the metric system.

Every effort has been made to achieve the original objectives set for this book, but the task is almost impossible. The book represents my own ideas concerning materials for construction, based on my professional practice and teaching, amplified by the data I gathered, and modified by countless specialists who helped by reading, correcting, and criticizing parts of the manuscript. I shall welcome any further constructive criticism, suggestions, and additional information that will help to make this book more valuable to those for whom it was written.

CALEB HORNBOSTEL

Philadelphia, Pennsylvania
June 1978

ACKNOWLEDGMENTS

The sources of information for this book are so varied, broad, and extensive that it is impossible to give credit to all who contributed information, data, suggestions, and criticisms. Instead I must content myself with generalities.

For supplying me with authoritative data I am deeply indebted to three groups: (1) the manufacturers, (2) the many associations and institutes devoted to representing various professional and manufacturing groups, and (3) all the various departments of the United States Government.

The manufacturers very generously not only contributed printed literature of all sorts, but also made available their technical, research, sales, and product development personnel for reading pertinent sections of the manuscript to check for accuracy, up-to-dateness, and completeness of information from their point of view, as well as to offer other suggestions.

Among individuals who specifically supplied data and suggestions I wish to sincerely thank Lloyd R. Cutler of the Miracle Adhesives Corporation, J. H. Gross of the United States Steel Corporation, R. Pinner of the Lazy Laboratories Limited, Peter J. McQuillin of the PPG Industries, J. J. Lenzotti of The Sherwin-Williams Company, T. J. Jones of Wheeling Corrugating Company, and William N. Wilson of The National Association of Architectural Metal Manufacturers. I especially want to thank and praise John Coken, who revised the original illustrations and made the new ones so graphically perfect that it is impossible to distinguish one from the other or to know where changes were made. Last but not least I thank wholeheartedly Ida Norden and Francesca Di Cosmo for struggling with my handwriting and typing the manuscript changes, and my wife, Barbara, who not only played the role of student to read and clarify the text and illustrations, but also helped to edit and proofread the entire manuscript.

One of the pleasures of transforming the manuscript into its final book form was working with the editorial staff of John Wiley & Sons. They were encouraging, cooperative, and creative at every step of the way.

C. H.

A

ABRASIVES

Physical and Chemical Properties

An abrasive is any hard, sharp material that wears away a softer, less resistant surface when the two are rubbed together. The use for which an abrasive is intended determines its form. It may be a solid such as a sharpening stone, grinding wheel, or pumice block; it may be coated on a backing of another material; or it may be a loose powder. The hardness, toughness or brittleness, and the fracture or shape of the mineral after it has been crushed to a given grade, are the determining characteristics of abrasives.

Hardness. Hardness is usually expressed in terms of the Mohs scale, which rates the diamond as 10, corundum as 9, and talc as 1. Actually this scale is merely comparative, not absolute, as the gap between 10 and 9 is even greater than that between 9 and 1.

Particle Size. Particle size is given as a grit number which is based upon screen mesh sizes (*see* Table *A1*). Coarse grain is 12 to 24 grit; very fine grain is 150 to 240 grit. Grinding flours range in grain size from 280 to 600 mesh.

Types and Uses

The natural abrasives are quartz, sand, flint, corundum, emery, and the industrial gem stones, diamond and garnet. The artificial abrasives are silicon carbide; aluminum oxide; miscellaneous mineral abrasives such as magnesium oxide, tin oxide, iron oxide, and cerium oxide, all used mainly for polishing; and boron carbide and tungsten carbide for extreme hardness. Soft or "mild" abrasives used in powder form for polishing glass and silver are rouge, talc, and chalk; for polishing metals, they are pumice, diatomite, opal, silica flour, whiting, putty powder, and china clay.

Sanding Machines. In construction work various portable sanding machines are used to finish wood, cork, terrazzo, concrete, cement, marble, and certain types of stone floor. Portable hand-sanding machines are used for woodwork and painted finishes; special kinds exist for areas, such as terrazzo cove bases, that are

Table A1 Grit numbers and other designations for the commonly used abrasives

	Grit number in mesh sizes[a]		Type	
General description	Silicon carbide	Aluminum oxide and garnet	Flint	Emery
Very fine	600			
	500			
	400	400 (10/0)		
	360			
	320	320 (9/0)		
	280	280 (8/0)		
	240	240 (7/0)		
	220	220 (6/0)		
			Extra fine	
Fine	180	180 (5/0)		
	150	150 (4/0)		Fine
	120	120 (3/0)	Fine	
Medium	100	100 (2/0)		Medium
	80	80 (0)	Medium	
	60	60 ($\frac{1}{2}$)		Coarse
Coarse	50	50 (1)		
	40	40 ($1\frac{1}{2}$)	Coarse	
	36	36 (2)		Very coarse
Very coarse	30	30 ($2\frac{1}{2}$)	Extra coarse	
	24	24 (3)		
	20	20 ($3\frac{1}{2}$)		
	16	16 (4)		
	12			

[a]The numbers in parentheses are arbitrary grading designations used before adoption of screen mesh sizes.

Table A2 Characteristics and uses of abrasive materials

Material	Type	Hardness (Mohs)	Toughness (percent)	Type of particle	Uses
Flint	Flint quartz from New Hampshire and Maryland	6.8–7.0	20	Sharp crystals	Ordinary sandpaper for carpentry, leather, paint
Garnet	Almandite, rhodolite	7.5–8.5	60	Sharp crystals	Sanding in shop-applied paint work; sanding and finishing of wood and furniture; terrazzo
Emery	40% iron oxide, 60% corundum (natural aluminum oxide)	8.5–9.0	80	Round and blocky grain	Mainly for polishing; minor use in finishing metals
Aluminum oxide	Electric furnace	9.4	75	Chunky, very hard grain	Woodworking, particularly furniture; metalworking industry; cutting and tooling of heavier metals; finishing terrazzo
Silicon carbide	Electric furnace	9.6	55	Very long, sharp crystals	Stone, glass, plastics; soft and ductile metals such as soft brass, pure aluminum, soft bronze, lead castings; surface finishing of lacquers, enamels, etc.

inaccessible to the large machines used for floor finishing. Most of these machines use coated abrasives or, less often, solid shapes made of the same abrasive materials. (*See* Table *A2*.)

Scouring Abrasives. Scouring abrasives are natural sand grains or pulverized quartz employed in scouring compounds and soaps, buffing compounds, and metal polishes. Federal specifications require that all abrasive grains used in grit cake soap and scouring compounds shall pass a No. 100 screen. All grains for scouring compounds for marble floors must pass a No. 100 screen, and 95% of these must pass a No. 200 screen. For ceramic tile floors, 90% must pass a No. 60 screen. Very fine air-floated quartz is employed in metal polishes, and in this case all grains must pass a 325-mesh screen. Ground glass is still regularly marketed as an abrasive for use in scouring compounds and in match-head compositions.

Coated Abrasives. This term refers to abrasives coated onto a backing of another material, the most familiar form to most readers. The abrasive coating may vary in type, particle size, and manner of coating. It may be closed, that is, with the abrasive grains completely covering the surface of the backing; or it may be open, with the grains separated, leaving bare spaces of controlled area on the backing. Either type may be applied by electrocoating, which causes the particles to become aligned in the direction of flow of the electrical force and to become embedded with sharp end uppermost and with equal spacing.

Backings can vary in strength. They may be paper, cloth, or a combination of both; for extremely heavy conditions a backing of heavy vulcanized fiber can be used. The adhesives or bonding agents are hide glues and synthetic resins. Hide glues are suitable only for dry sanding, and they can be greatly hardened and toughened by the addition of a finely pulverized mineral. Synthetic resins are designed for wet sanding and for special conditions.

Grinding Wheels. Grinding wheels were introduced into the West in 1825. They came from India, where they had been made for centuries by binding natural corundum grains with gum resin or shellac. Vitrified wheels, used for heavy work, are made by running the mixture into molds and, when dry, subjecting to intense heat. The silicate process consists of tamping the material into molds with a silicate binder and then baking in an oven. The silicate bond releases the grains more easily than does the vitrified type. Synthetic resins, such as the phenol resins, are used for bonding when greater strength is required than is obtainable with the silicate, but there is less openness of grain than with the vitrified. A binder of shellac is used on wheels for light work and for finishing. Rubber is used for high speed on fine finishing of surfaces.

History and Manufacture

Sand, rubbed on with a piece of flexible hide, was probably the first abrasive used by man—how long ago, we do not know. The first true abrasive was emery, called *shamir* in the Old Testament and used to shape and sharpen metal implements of that time. In the 13th century the Chinese glued crushed sea shells onto parchment to make the first coated abrasive on record. Over two centuries ago the Swiss began to produce an abrasive of crushed glass, coated onto a paper backing. About a century ago glass was replaced by flint, and today flint paper and emery cloth are still the most widely known abrasives. However, these have been largely replaced in industry by products developed toward the end of the 19th century. Garnet, the newest of the natural abrasives, appeared first in 1878 and proved to be much harder and sharper than flint. The two electric furnace abrasives, silicon carbide and aluminum oxide, were discovered next. These and further developments in backings and especially in adhesives have produced a wide range of abrasives for specialized industrial uses.

ACCESSORIES

In this book the term "accessories" refers to all the miscellaneous items (some of which are also categorized as rough hardware) used to anchor, stiffen, cover and strengthen any materials that are being put together. Each category of related items is described under a major group heading such as Anchors, Nails, Inserts, or Screws. All items needed for the installation of a material are described in connection with the material for which they are specifically intended. These accessories are always included in the "Applications" section of a given material; for instance, the section dealing with burned brick lists accessories for brick masonry.

In architectural terminology, "accessories" can also refer to an entirely different category generally known as "bathroom accessories." These include medicine cabinets, towel bars, and holders for toothbrushes, water glasses, soap, and toilet paper. All these items, with the exception of medicine cabinets, are made either of fired clay similar to the ceramic wall-tile material or of various metals with a chromium, brass, bronze, silver, or gold finish. Figure *A1* shows where such items should be located as a rule.

ACETATE

An acetate is a salt of acetic acid. Acetate is also the official name for a fiber, formerly called acetate rayon or acetate silk, made from regenerated cellulose by the acetate process. Partially acetylated fibers or resins dissolve in acetone and are thermoplastic. Fully acetylated fibers, insoluble in acetone and having a higher melting point, are called Arnel.

Basic copper acetate is known as verdigris and is the coloring ingredient of one type of patina on that metal.

ACETIC ACID

Acetic acid (CH_3COOH) is an organic acid which occurs naturally in plant juices and is most familiar in its dilute form as vinegar. Glacial acetic acid (also called acetic anhydride) is the term for the pure concentrated compound. Acetic acid can be obtained by fermentation from ethyl alcohol, by distillation of wood, and on a commercial scale by synthesis from calcium carbide made in the electric furnace from limestone and coal.

Acetic acid itself is used for cleaning and etching certain metals, as a flux in soldering, and in pickling steel; as a bleach and stain remover; and as an ingredient and solvent in the manufacture of rubber, plastics, cellulose acetate fibers, dyes, and photographic film.

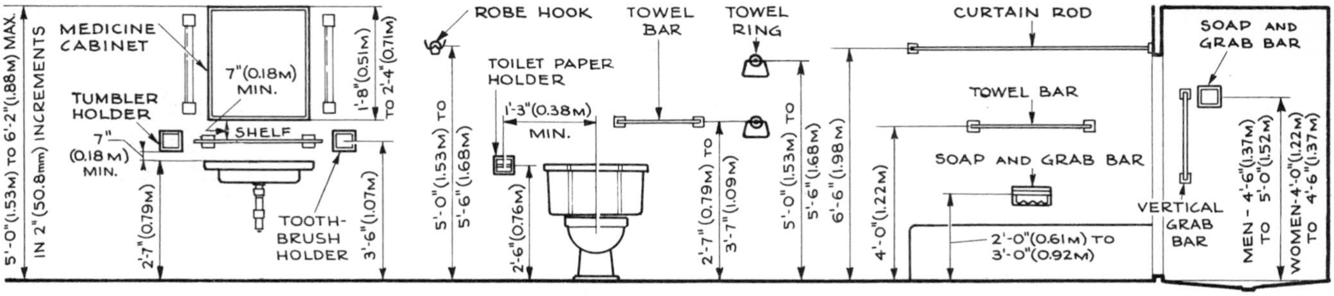

Figure A1 General location of bathroom accessories.

ACETONE

Acetone is a colorless, fragrant, inflammable, volatile liquid which is miscible in all proportions with water, alcohol, and ether. Therefore it is an important solvent for lacquers, paints, adhesives, acetylene, nitrocellulose, celluloid, cellulose acetate, and many gums, resins, and fats. This is its best known use in construction.

ACETYLENE

Acetylene is a toxic, highly flammable, colorless gas (C_2H_2) which contains over 90% carbon. It is produced on a large scale for chemical and fuel uses and is sold compressed (dissolved in acetone to make it nonexplosive) in steel cylinders.

In the construction field we are familiar with acetylene as an ingredient in the welding and cutting of metals. For welding it is mixed with oxygen and develops a temperature of 6332°F (3500°C). It can also be used for general illumination and is an important ingredient in the manufacture of many synthetic products in everyday use, such as neoprene, vinyl, and acetate fibers for textiles.

ACIDS, BASES, AND SALTS

Acids, bases, and salts may be generally defined in terms of the characteristics given in Table *A3* (*see also* "History and Manufacture"). The few acids and bases that have any direct application in construction materials are listed in Tables *A4* and *A5*.

Table A3 General characteristics of acids, bases, and salts

Acid	Base
Tastes sour in a water solution	Tastes bitter in a water solution
Reacts with some metals to liberate hydrogen	Slippery to touch in a water solution
Changes blue litmus to red	Changes red litmus to blue
Neutralizes bases to form salts	Neutralizes acids to form salts
Ionizes in water to yield hydrogen ions	Ionizes in water to yield hydroxide
Salt	
When an acid and a base neutralize each other, their individual characteristics are destroyed and the compound produced is a salt	

Table A4 Some common acids used in construction

Acid	Formula	General uses
Acetic	CH_3COOH	Manufacture of plastics, white lead, resins, rubber
Boric	H_3BO_3	Manufacture of glass, ceramics; fireproofing agent
Butyric	$C_2H_5CH_2CO_2H$	Manufacture of plastics, resins
Hydrochloric	HCl	Metal pickling; cleaning ceramic tile, brick, structural clay tile, stone
Hydrofluoric	HF	Etching glass
Lactic	$CH_3CHOHCO_2H$	Manufacture of synthetic resins
Nitric	HNO_3	Etching and pickling metals
Oxalic	$(HO_2C \cdot CO_2H)2H_2O$	Rust and stain removal; bleaching
Phosphoric	H_3PO_4	Rustproofing metals
Sulfuric	H_2SO_4	Pickling steel

Table A5 Some common bases used in construction

Base	Formula	General use
Ammonia	HN_3	Manufacture of plastics, nitric acid, refrigerants
Calcium hydroxide (lime)	$Ca(OH)_2$	Cements; manufacture of glass, water softeners
Potassium	K_2CO_3	Manufacture of glass
Sodium carbonate (soda ash)	Na_2CO_3	Manufacture of glass, ceramics, cleaners, water softeners
Sodium hydroxide (caustic soda)	$NaOH$	Manufacture of paper, rubber, textiles, plastics

History and Manufacture

The ancients knew only one acid, vinegar, which is a dilute impure solution of acetic acid in water. Actually the word "acid" comes from the Greek word for sour and vinegar. Prussic acid also was used but without knowledge of its nature except that it was a very effective poison. Bases, or alkalis, were similarly used in their natural forms long before their chemistry was known.

Today acids, bases, and salts are manufactured on a vast scale to serve the needs of all major industries and represent a large segment of our economy. Sulfuric acid alone is produced at the rate of about 13 million tons per year, a tonnage exceeding that of any metal produced except iron.

Theory of Acids and Bases. Many theories about what constitutes an acid and a base were developed and in turn abandoned as more knowledge was obtained. In 1778 Lavoisier established oxygen as the acid-producing element. Then, in 1808, the researches of Davy, Gay-Lussac, and Thenard seemed to establish beyond doubt (at least until 1923) that hydrogen, not oxygen, was the element essential for acidic properties. In 1923 Brönsted and Lowry, each working alone, simultaneously introduced new definitions for acid and base, namely, any compound that can transfer a proton to any other compound is an acid, and the compound that accepts the proton is a base. These new definitions not only simplify acid-base reactions but also include nonaqueous solutions and enlarge the number of compounds that can be considered as acids and bases. Another definition, devised by Lewis, is that the word "acid" includes all substances that, during a chemical reaction, become attached to an unshared pair of electrons in some other molecule. According to these theories, hydrogen is no longer essential to acidity. At present the Brönsted-Lowry definitions are used and questionable substances are called Lewis acids.

ACOUSTICS AND ACOUSTICAL MATERIALS

Acoustics is the science of sound. Most simply defined, sound is a wave traveling in air and perceived by the human ear. Noise, increasingly a problem in construction, is unwanted sound. A sound wave causes very small changes in atmospheric pressure above and below the static pressure. The average deviation in atmospheric pressure is called the sound pressure. In addition to pressure, sound waves have frequency, i.e., the number of times per second that the sound pressure alternates above and below the ambient atmospheric pressure. Each complete alternation is called a cycle, and frequency is expressed in hertz units, denoting the number of cycles per second. The change in frequency from one musical note to another is known as the pitch. The higher the frequency, the higher is the pitch.

Since the human ear can detect atmospheric pressures over a range of 1 million to 1, it was necessary to devise a compact scale in order to measure and calculate audible sound. The decibel is the logarithmic unit of measure of sound pressure or power. Zero on the decibel scale corresponds to a standardized reference pressure (0.0002 microbar), or sound power (10^{-12} watt).

Sound waves travel through the air at a constant speed of 1125 ft/s (342.9 m/s). In a theater at room temperature, approximately 70°F (21.11°C), it takes 0.10 second for sound to travel from the stage to the last row of seats (*see* Table *A6*).

A more scientific definition of sound—a propagated elastic disturbance in a material medium—brings vibra-

Table A6 Sound pressure levels of everyday sounds and noises

Type of noise	Range (decibels)	Type of environment
Very faint	0–20	Rustle of leaves; whisper; soundproof room; threshold of audibility
Faint	20–40	Quiet home; private office; average auditorium; quiet conversation; babbling brook
Moderate	40–60	Average office; noisy house; average conversation; quiet radio, television, hi-fi
Loud	60–80	Noisy office; average factory; average radio, television, hi-fi; loud conversation; children's playground
Very loud	80–100	Accelerating motorcycle; police whistle; noisy factory; New Orleans jazz band; very loud, angry conversation; loud street noises; very loud hi-fi
Deafening	100–120	Nearby riveter or compressed air hammer; boiler factory; thunder; artillery; air raid siren

tion and the control of vibration in a building within the scope of acoustics in construction (*see also* Vibration Control).

Types and Uses

Acoustical materials can be classified into three categories: (1) fabrics and textiles; (2) rigid, hard materials with holes, slots, or perforations of various types, backed with a soft, sound-absorbing material; and (3) artificial or natural materials with soft or porous surfaces. The last group, materials which are soft and porous in character, may be prefabricated into tiles, sheets, and other easily handled forms, or they may be installed on the job, for example, as hung baffles, swinging panels, or decorative panels on the walls.

All these materials control sound by absorption. Tables *A7* and *A8* give general data on size, color variations, textures, and maintenance problems for each type. Table *A9* shows the comparative sound absorption properties of typical construction materials.

Truly soundproof construction requires complete understanding of the physics of sound and the acoustic properties of materials, both areas of specialization beyond the scope of this book.

Table A7 Typical acoustical materials used for walls and ceiling of buildings

Material	Advantages	Disadvantages	Major uses
Aluminum, perforated[a] (*see* Aluminum Sheet and Strip)	Permanent, rigid, durable; acoustic properties can be controlled by quantity and sizes of holes and sound absorption value of pads used behind it; wide range of colors when porcelain enameled	Requires special supporting and attaching systems for ceilings; requires acoustic pads	Ceilings of buildings where acoustic control is important
Asbestos, perforated[a] (*see* Asbestos-Cement Board and Sheet)	Permanent, rigid, durable; withstands rough usage; acoustic properties can be controlled by quantity and size of holes and sound absorption value of pads used behind it; easily painted; fire resistant	Available only in cement color; acoustic pads are required; fire resistance controlled by type of pad used and method of support	Walls and ceilings of buildings where acoustic control is important
Clay tile, structural, perforated (*see* Clay Tile, Structural)	Permanent, durable; withstands rough usage; wide selection of colors; fire resistant; load-bearing	Limited acoustic value; can be used only for walls; acoustic pad within structural tile becomes permanent part of building	Walls of buildings where acoustic treatment is important, generally corridors
Concrete, lightweight (*see* Concrete Decking and Structural Units	Permanent, durable; roof or floor decking gives ceilings with acoustic value; fire resistant	Limited acoustic value; cannot be painted and repainted; can be used only for ceilings; no color selection	Ceilings of buildings where some acoustic treatment is important
Concrete block (*see* Concrete Block)	Permanent; durable; withstands rough usage; fire resistant; load-bearing	Can be used only for walls; acoustic pads, when used, become permanent part of building	Walls and partitions of building where acoustic treatment is important; corridors, side walls, auditoriums, theaters, etc.
Cork (*see* Cork)	Limited acoustic value; easily installed; flame resistant	As acoustic values increase it becomes softer; for ceiling tile the cork is granular in form with binders and other materials; limited color selection	Flooring; bulletin, tack, and similar boards; ceilings (when in tile form)
Glass fibers (*see* Glass Fibers; Mineral Wool)	Good acoustic value; available with finely perforated plastic coating; flame resistant; easily installed	Soft; will not withstand rough usage	Ceilings of buildings where acoustic treatment is relatively important

Table A7 Typical acoustical materials used for walls and ceiling of buildings (*continued*)

Material	Advantages	Disadvantages	Major uses
Mineral wool, sprayed (*see* Mineral Wool)	Good acoustic value; relatively easy to install by spraying on surface; fire resistant	Very soft; will not withstand rough usage; cannot be painted and repainted; no color selection	Ceilings of buildings where acoustic control is important
Mineral wool tile[a] (*see* Mineral Wool)	Limited acoustic value; easily installed; flame resistant	Soft; will not withstand rough usage; cannot be painted and repainted; limited color selection	Ceilings of buildings where acoustic treatment is relatively important
Paper[a] (*see* Fibers; Paper and Paper Pulp Products)	Limited acoustic value; easily installed; flame resistant	Soft; will not withstand rough usage; cannot be painted and repainted; limited color selection	Ceilings of buildings where acoustic treatment is relatively important
Paper, perforated[a] (*see* Fibers; Paper Pulp Sheets and Rigidized Sheets)	Durable, thin; withstands relatively rough usage; acoustic properties can be controlled by quantity and size of holes and acoustic value of pads behind it; easily painted; available in prefinished baked enamel in a wide range of colors; flame resistant	Acoustic pads are necessary; requires careful support as it will sag and buckle	Walls and ceilings of buildings where acoustic control is important
Perforated tempered hardboard[a] (*see* Wood Hardboard)	Permanent; rigid; durable; acoustic properties can be controlled by quantity and sizes of holes and sound absorption value of pads used	Requires special supporting and attaching systems for both walls and ceilings; fire resistant depending on pads and supporting systems	Walls requiring acoustic treatment
Plastic foam (*see* Plastics)	Good acoustic value; easily installed; flame resistant	Soft, will not withstand rough usage; cannot be painted; limited color selection	Ceilings of buildings where acoustic treatment is important
Stainless steel, perforated (*see* Stainless Steel Sheet, Strip, and Plate)	Permanent, durable; acoustic properties can be controlled by quantity and size of holes and acoustic value of pad behind it	Requires special supporting and attaching systems; requires acoustic pads; fire resistance controlled by type of pad used and method of support; no color selection	Ceiling of buildings where acoustic control is important
Steel, perforated[a] (*see* Steel Sheet and Strip)	Permanent, durable, rigid; acoustic properties can be controlled by quantity and size of holes and acoustic value of pads behind it; easily painted; wide selection of colors in porcelain enamel type	Requires special supports and attaching systems; requires acoustic pads; fire resistance controlled by type of pad used and method of support	Ceilings of buildings where acoustic control is important
Textiles (*see* Textiles)	Good acoustic value when draped; available perforated to be applied over padding; flame resistant	Require periodic cleaning; realtively easily damaged; have limited life	Walls of buildings where acoustic treatment and control are important
Textiles, carpet	Good acoustic properties; easily installed; can be used on floors, walls, or ceilings; flame resistant; wide variety of colors, textures, designs, and thickness	Require periodic cleaning; have limited life	Floors, not only for acoustic treatment but also for comfort and decorative purposes; on walls and ceilings for special types of acoustic treatment

[a]Many of these acoustic materials are prefinished and require careful handling during installation.

Table A8 Thickness, surface appearance, and noise reduction coefficients[a] of acoustical materials used in construction

Type of material	Surface appearance	Thickness in.	Thickness mm	NRC[a,b]
Paper pulp materials	Regularly perforated	$\frac{1}{2}$	12.70	0.60–0.80
		$\frac{3}{4}$	19.05	0.65–0.85
	Randomly perforated	$\frac{1}{2}$	12.70	0.45–0.65
		$\frac{3}{4}$	19.05	0.60–0.80
	Textured, fissured, or simulated fissured, finely perforated	$\frac{1}{2}$	12.70	0.40–0.60
		$\frac{9}{16}$	14.29	0.40–0.55
		$\frac{3}{4}$	19.05	0.45–0.75
	Lay-in panels	$\frac{1}{2}$	12.70	0.40–0.60
		$\frac{5}{8}$	15.88	0.60–0.70
		1	25.40	0.35–0.45
		$1\frac{1}{2}$	38.10	0.40–0.50
		2	50.80	0.45–0.55
Mineral wool materials	Perforated	$\frac{1}{2}$	12.70	0.45–0.70
		$\frac{5}{8}$	15.88	0.55–0.85
		$\frac{3}{4}$	19.05	0.60–0.85
		$\frac{7}{8}$	22.23	0.45–0.65
	Fissured	$\frac{1}{2}$	12.70	0.50–0.65
		$\frac{5}{8}$	15.88	0.55–0.85
		$\frac{3}{4}$	19.05	0.50–0.90
	Textured, finely perforated, or or smooth	$\frac{1}{2}$	12.70	0.45–0.70
		$\frac{5}{8}$	15.88	0.50–0.80
		$\frac{3}{4}$	19.05	0.55–0.85
	Lay-in panels	$\frac{1}{2}$	12.70	0.55–0.70
		$\frac{5}{8}$	15.88	0.45–0.90
		$\frac{3}{4}$	19.05	0.25–0.95
		1	25.40	0.65–0.95
		$1\frac{1}{2}$	38.10	0.75–0.85
		3	76.20	0.75–0.85
		$3\frac{3}{8}$	85.73	0.70–0.80
	Rated as part of fire-resistive assemblies; perforated, fissured, textured tile	$\frac{5}{8}$	15.88	0.50–0.75
		$\frac{3}{4}$	19.05	0.60–0.85
	Rated as part of fire-assemblies; perforated, fissured, textured lay-in panels	$\frac{1}{2}$	12.70	0.55–0.65
		$\frac{5}{8}$	15.88	0.45–0.70
		$\frac{3}{4}$	19.05	0.50–0.65
		1	25.40	0.80–0.95
Metal pans with mineral fiber pads or blankets	Perforated	$1\frac{9}{16}$	39.69	0.70–1.10
	Perforated; rated as part of fire-resistive assemblies	$1\frac{9}{16}$	39.69	0.85–1.10
		$2\frac{13}{16}$	71.44	0.85–0.95
Asbestos cement panels with mineral fiber pads	Perforated	$1\frac{3}{16}$	30.16	0.55–0.65
		$1\frac{11}{16}$	42.86	0.65–0.75
		$2\frac{3}{16}$	55.56	0.70–0.85
	Lay-in perforated panels with paper sound-absorbing element glued to back	$\frac{3}{16}$	4.76	0.60–0.80
Glass	Cellular glass with perforations	2	50.8	0.55–0.65
Textile	Fabric-faced, quilted mineral wool blanket, flameproofed	1	25.4	0.65–0.75
		2	50.8	0.55–0.65
Plastic	Translucent plastic lay-in panels	1	25.4	0.65–0.75

[a] The various methods of installing acoustical material affect the NRC values. When an installation system is selected, a much closer NRC value can be obtained.
[b] The range is from 250 to 2000 Hz.

Application

Acoustics is now a very highly specialized science, and when the design and construction involve an area in which perfect reproduction of musical sound is a key factor—for example, auditoriums, theaters, television and radio studios—it is necessary to work with specialists in this field. Even noise control in factories, restaurants, and large office areas should be planned with acoustical engineers and manufacturers of acoustical materials.

Table A9 Noise reduction coefficients for typical construction materials

Type of material	Surface or texture	Method of application	Noise reduction coefficient
Brick (burned, cement, sand-lime, etc.)	Smooth to rough, painted	–	0.01–0.02
	Smooth to rough, unpainted	–	0.02–0.05
	Glazed	–	0.01–0.02
Ceramic veneer	Glazed	–	0.01–0.02
Clay tile, structural	Glazed	–	0.01–0.02
	Glazed, perforated with built-in acoustic pad	–	0.50–0.70
Concrete	Floated, steel troweled	–	0.01–0.02
Concrete block	Coarse	–	0.36–0.25
	Painted	–	0.10–0.08
	Acoustical	–	0.50–0.70
Cork resilient flooring	Smooth	With adhesive to concrete	0.02–0.02
Glass	Smooth		0.18–0.02
Linoleum	Smooth	With adhesive to concrete	0.03–0.05
Plaster	Smooth, hard	On clay tile or brick	0.01–0.05
		On lath	0.02–0.03
	Rough	On lath	0.04–0.06
Plastic resilient flooring	Smooth	With adhesive on concrete	0.02–0.02
Rubber resilient flooring	Smooth	With adhesive on concrete	0.02–0.02
Stone	Smooth	–	0.01–0.02
Terrazzo	Smooth, polished	–	0.01–0.0
Textiles	Carpet (heavy)	On concrete	0.02–0.65
		With 40-oz (1143-g) underlay	0.08–0.73
	Drapes (medium)	10 oz/yd^2 (339.1 g/m^2) hung straight	0.03–0.35
	Drapes (medium)	1 oz/yd^2 (339.1 g/m^2) draped to half area	0.07–0.60
	Drapes (heavy)	18 oz/yd^2 (610.38 g/m^2) draped to half area	0.14–0.65
Wood	Smooth	Applied as paneling	0.28–0.11
Wood	Strip flooring	–	0.15–0.07
	Parquet or thin block on concrete	–	0.04–0.07

In construction the major problems in relation to acoustics are (1) to determine what types of material are available and the maintenance problems presented by these materials; (2) to become familiar with the colors, textures, patterns, and other design features of these materials; (3) to design with and to install these materials after the correct type has been selected to answer the particular acoustical problems involved; and (4) to decide how to handle the areas that receive rough usage. Acoustical materials are available in a wide variety of environmental coordinated systems that are integrated with lighting, heating and air conditioning, and sprinkler systems. Sound inside a room can usually be controlled by sound-absorbing material (this does not include sound radiation resulting from vibrations). Outside noise, however, can be controlled only by proper construction of partitions. Walls transfer sound largely by acting as vibrating diaphragms; only if massive enough, can they be adequate sound insulators. Usually lightweight layered partitions with air spaces 4 in. (10.16 cm) or more thick between them acting as acoustic filters are better. Doors and windows (glass) transmit more sound than walls do and must be double if soundproofing treatment is desired. Floors are more difficult to insulate than walls because they are part of the basic construction of the building and, therefore, any direct contact with a floor such as the tapping of high heels, the dropping of a heavy object, or vacuum cleaning sets up vibrations that immediately transmit these sounds to the room below. Carpeting and/or applying a dense material on the ceiling below, including acoustical treatment, are the general methods used to overcome this problem.

Acoustical materials vary in thickness from $\frac{1}{2}$ in. (12.7 mm) to $\frac{3}{4}$ in. (19.05 mm) for 12 in. × 12 in. (304.8 mm × 304.8 mm) acoustical tile, and from $\frac{1}{2}$ in. (12.7 mm) to 1 in. (25.4 mm) for the 24 in × 48 in. (60.96 cm × 121.92 cm) lay-in type of acoustical panels. The minimum thickness is $1\frac{1}{16}$ in. (26.99 mm) for 12 in. × 24 in. (30.48 cm × 60.96 cm) perforated metal pans with fiber pads or blankets, and for 24 in. × 24 in. (30.48 cm × 30.48 cm) perforated asbestos cement or perforated tempered Prestwood.

The noise reduction coefficient (NRC) of an acoustical material is the arithmetic average of its absorption coefficients at 250, 500, 1000, and 2000 Hertz inclusive. These noise reduction coefficients vary considerably in relation to the various methods of installation, namely, (1) mechanical suspension systems; (2) fire retardant time design rated systems; (3) environmental coordination systems; (4) adhesive application; and (5) nailing, screwing, or stapling to wood strips. Depending on the installation system used, the NRC can vary from 0.40 to 1.00 for similar materials.

Condensed Checklist

1. When designing any auditorium, theater, cinema, concert hall, or opera house, always consult with acoustical engineers in regard to design, shape, and acoustical treatment.
2. Always check local, municipal, and state codes and the codes of the fire underwriters, insurance companies, and federal government (army, navy, etc.) for fire retarding limitations and requirements for the installation of acoustical ceiling treatments.
3. Always check with manufacturers for the various coordinated systems that are available for the building to be constructed.
4. Always check with manufacturers not only about the various types of acoustical materials, textures, designs, and installation systems that are available but particularly about the design patterns, as these are changed constantly and certain ones are discontinued.
5. Always check the noise reduction coefficients (NRCs) of the various acoustical materials in relation to the areas to be acoustically treated and the method of installation.
6. Always check with the manufacturers of acoustical materials as to the best method of treating an acoustical problem, or consult an acoustical engineer.
7. When installing partitions between acoustically treated areas, always check that these partitions are continuous from floor to underside of the floor construction of the floor above.

History

Before 1914 in construction the problems of acoustics were considered only in regard to theaters, concert halls, opera houses, and the like. At that time, most of these structures were designed in the classical style; therefore there were enough sound-controlling elements in the form of ornamentation, columns, moldings, and other projections to enable many of these buildings to have fairly good acoustics. (There are also notable examples of buildings with bad acoustics.)

The first serious work in this country on applying the science of acoustics to buildings was done by Sabine in 1895. About 1920 in Europe, where architecture had already broken away from classicism, studies were made in relation to acoustics in theaters and opera houses, and many such structures were designed in the contemporary idiom, applying this knowledge. Today we are conscious of the problem of acoustics in the home, office, and factory. In almost every type of building noise control is considered necessary, and methods of answering acoustical problems must be found.

ADHESIVES

Physical and Chemical Properties

In this book adhesives include all the various substances (pastes, glues, pyroxylin adhesives, rubber cements, latex cements, special cements of chlorinated rubber, natural rubber, and synthetic rubber, natural resins and synthetic resins, and mucilage) that perform a similar function, that of holding materials together by surface attachment. Today adhesives are directly related in their chemistry to plastics, rubbers, paints, and other materials now included under organic coatings.

An adhesive may be defined as a material for sticking, adhering, or bonding the surface of one material to another surface of the same material or another material.

Pastes are water solutions of starches or dextrins, sometimes mixed with gums, resins, or glue to add strength to their bonding power.

Glues are water solutions of animal gelatin. Hide and bone glues are available as dry flake, and fish glue is available in liquid form.

Mucilage is obtained from linseed or other seeds and from water-soluble gums. It is a light adhesive for paper.

Categories. Adhesives may be subdivided into five categories: (1) rubbers, or elastomers, (2) resins, (3) rubber resins, (4) synthetic resins, and (5) water-base adhesives, including animal, fish, and vegetable glues.

Setting and Bonding. The curing (time of setting), temperature of setting (room temperature or high applied temperature), and method of bonding (on contact, with clamps, or under high pressure) are important data which must be known before an adhesive can be chosen and/or evaluated for a particular end use.

Strength. Adhesives are strongest in tensile and shear stresses and weakest in cleavage and peeling stresses. The reaction to moisture and the resistance to peeling or stripping determine the value and lifetime of any adhesive. Even if theoretically no water can ever reach the adhesive or the joint, often during construction or later moisture does penetrate. What then happens to the adhesive and the strength of the joint is a subject for concern. *See* Plastics; Rubber; Mortar; Wood; Wood Integrants (Plywood); Paint; Painting.

Commercial Forms. Adhesives can often be tailormade to specific materials and conditions of use. They are available as powders, liquids, solids, and films and in any of these forms combined with catalysts.

Types and Uses

Today it is a good general rule always to specify the adhesive that is recommended by the manufacturer of the materials to be joined.

Adhesives are finding ever-increasing use in the field of construction for joining porous, rigid, or impervious materials together; for installing stuck-ups, as shown in Figure *A2*; and for prefabricated or preassembled units

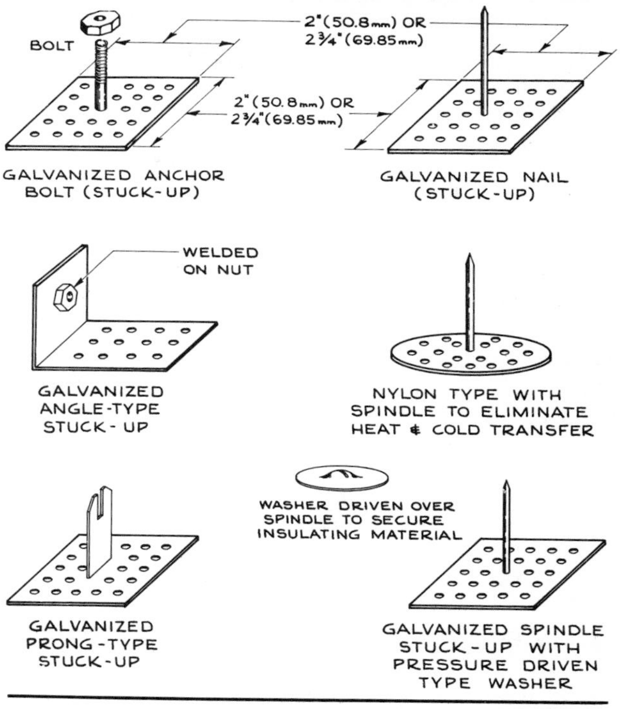

Figure A2 Different types of stuck-ups for adhesive application.

of various types such as skylights, curtain walls, pregrouted ceramic tile panels, skin-type structural units, laminated lumber, and preassembled masonry panels. (*See* Concrete, Precast; Concrete, Reinforced; Concrete, Prestressed; Clay Tile: Floor and Wall Tile; Brick; Stone; Wood *and all other* Wood *sections*; Plastics; Rubbers.)

Tables *A10* through *A12* list the commonly used adhesives by type, describing general characteristics, curing, setting temperatures, and major use for each one. Asphalt adhesives are also covered under Asphalt *and* Asphalt Coatings.

Thermosetting Resins. The thermosetting resins, urea-formaldehyde, melamine, acid-catalyzed phenolic, resorcinol, phenolic, and phenol-resorcinol, are described from the viewpoint of construction applications not here, but under particular materials that require thermosetting resins and thermoplastic resins.

Water-Base Adhesives. Miscellaneous water-base adhesives, including animal and vegetable glues, are described in general terms in Table *A13.* Starch, dextrin and soybean are further discussed in relation to their use in joining wood under Wood, Joining. Magnesium oxychloride, another water-base adhesive, is discussed in detail under Cement.

Table A10 Typical rubber adhesives

Type of rubber or elastomer	Form and method of application	Characteristics	Methods of curing; temperature; adhesion time	Major use
Natural rubber	Liquid gasoline or naphtha solvents; applied to both surfaces	Good moisture resistance; good strength; weakens on exposure to sunlight; fair aging properties; loses strength at 160° F (71.11° C)	Evaporation of solvent; room temperature; adheres immediately	General purpose; interior uses
Rubber latex and dispersion	Liquid water dispersion; applied to both surfaces	Good moisture resistance; good strength; is damaged by freezing	Evaporation of water; room temperature; adheres immediately	General purpose; interior uses
Reclaimed rubber	Liquid to heavy mastic; hydrocarbon solvents; applied to one surface	Good moisture resistance; good strength; weakens on exposure to sunlight; loses strength at high temperatures	Evaporation of solvent; room temperature; adheres immediately	General purpose; interior uses; felt to metal; rubber products to metal
Butadiene–styrene rubber	Liquid solvent or latex; applied to both surfaces	Good moisture resistance; fair strength; poor resistance to solvents and oils	Evaporation of solvent; room temperature; adheres within minutes	Laminating metal foil to paper and plastic films to other materials; interior uses
Neoprene	Liquid to heavy mastic, also in two parts; applied to both surfaces	Good strength; excellent aging properties; good resistance to sunlight; good resistance to moisture	Evaporation of solvent; room temperature; adheres within minutes	Applying plastic laminates; interior uses
Buna-N (acrylonitrile-butadiene rubber)	Liquid; applied to both surfaces	Good strength; good moisture resistance; ages well and resists mold growth; loses strength at 160° F (71.11° C)	Free from tackiness when dry and can be reactivated with heat, solvents or coat of adhesive; room temperature; adheres within minutes	Most versatile general purpose adhesive; interior uses

Table A11 Typical resins, pyroxylin cements, oleoresins, and asphalt adhesives

Type of resin, pyroxylin cement, oleoresin, or asphalt adhesive	Form and method of application	Characteristics	Methods of curing; temperature; adhesion time	Major use
Cellulose nitrate	Liquid; applied to one or both surfaces	Good moisture resistance; ages fairly well but discolors in sunlight	Evaporation of solvent; room temperature; within minutes	Known as household cement; interior uses; bonding thermoplastics
Cellulose acetate	Liquid; applied to one or both surfaces	Fair moisture resistance; good aging	Evaporation of solvent; room temperature; within minutes	Interior uses; for cellulose acetate plastics
Ethyl cellulose	Liquid or hot melts; applied to one or both surfaces	Good for porous materials; good strength; good resistance to oil, mold, and fungi; mediocre moisture resistance	Evaporation of solvent; room temperature; within minutes	Interior uses; for porous materials like paper and textiles
Polyvinyl acetate	Liquid solvents or water emulsion; applied to both surfaces	Fair moisture resistance; good strength; good resistance to sunlight; excellent resistance to oils, gasoline, and grease	Evaporation of solvent; baking at elevated temperatures; minimum 20 min or longer	Interior uses; for metal foil to paper and for porous materials like paper and textiles

Table A11 Typical resins, pyroxylin cements, oleoresins, and asphalt adhesives (*continued*)

Type of resin, pyroxylin cement, oleoresin, or asphalt adhesive	Form and method of application	Characteristics	Methods of curing; temperature; adhesion time	Major use
Polyvinyl alcohol	Liquid water emulsion; applied to both surfaces	Fair moisture resistance; good strength; good resistance to oils and grease	Evaporation of solvent; room temperature; within minutes	Interior uses; for porous materials like paper and textiles
Vinyl-vinylidene	Liquid; applied to both surfaces	Good strength; good resistance to oils, grease, and hydrocarbon and chlorinated solvents; fine moisture resistance	Evaporation of solvent; room temperature; within minutes	Interior and exterior uses; where excellent solvent and water resistance is required along with a transparent film
Vinyl butyral	Liquid; applied to both surfaces	Good strength; excellent resistance to sunlight; transparent; relatively poor moisture resistance; excellent adhesion to glass	Evaporation of solvent; room temperature and higher; within minutes	Laminating glass
Acrylics	Liquid solvents, water emulsion, or two-part combinations; applied to both surfaces	Good strength; excellent resistance to sunlight; transparent; good moisture resistance	Evaporation of solvent and catalysts; room temperature and higher; within minutes	Bonding of acrylic plastics, such as plastic skylights
Shellac	Liquid with alcohol solvent and hot melts; applied to both surfaces	Good strength; good resistance to moisture, oil, grease, and hydrocarbon solvents	Evaporation of solvent; room temperature; within minutes	Cork assemblies
Manila gum	Liquid with alcohol solvent; applied to one surface	Good strength; fine moisture resistance	Evaporation of solvent; room temperature; within minutes	For resilient flooring
Limed rosin	Heavy mastic; applied to one surface	Good strength; fair resistance to oil, grease, and moisture	Evaporation of solvent; room temperature; within minutes	For acoustic tile and fiberboard
Oleoresin	Heavy mastic; applied to one surface	Good strength; fair moisture resistance; weak resistance to common solvents	Evaporation of solvent; room temperature; within minutes	For acoustic tile and fiberboard, caulking compounds
Asphalt	Liquid solvents, cut-backs, water emulsion, and hot melts; applied to one or both surfaces	Good for little or no stress; weak at slightly raised temperatures; becomes brittle at 20 to 30° F (−2.16 to −1.11° C); excellent water and alkali resistance; ages well; poor resistance to oil, grease, and common solvents	Evaporation of solvents; room temperature; relatively a few minutes to 30 min or longer	Laminating papers, treated paper, foils, felts, and installation of asphalt resilient flooring

Miscellaneous Types. A pressure-sensitive adhesive is generally a mixture of phenolic resin and nitrile rubber in a solvent. Another type consists of silicone resins in a solvent or water-soluble vinyl resin powder in a water solvent. Wash-away adhesives are acrylic or other low melting thermoplastic resins. They are used for holding materials for grinding, polishing, or similar operations and can be removed either with a solvent or by heating.

Application

In construction adhesives are divided by usage into two categories: (1) those applied at the site of construction, and (2) those applied under controlled conditions in shops, mills, etc. When adhesives are used for on-site joining of materials, both the adhesive manufacturers and the manufacturers of the materials to be joined

Table A12 Typical synthetic resin, rubber-resin, and resin blend adhesives

Type of resin, rubber-resin, or resin blend adhesive	Form and method of application	Characteristics	Method of curing; temperature; adhesion time	Major use
Polyester	Liquid monomeric styrene solution converted to film with catalysts at time of use; applied to both surfaces	Resistant to moisture, chemicals, heat, and weathering	Catalysts; room temperature or 200–220° F (93.33–104.44° C); adheres in minutes	Exterior and interior uses; bonding polyester laminates
Epoxy	Liquid and heavy paste (catalyst to be added when used) and also powder and stick form for high-temperature application; applied on one or both surfaces	Good strength; resistant to oil, gasoline, benzene, alcohols, and acetone; fair moisture resistance	Chemical reaction; room temperature or 250–500° F (121–260° C); 30 min to 24 hr	Exterior and interior uses; joining dissimilar materials; sandwich panels; wood-to-metal and glass-to-metal seals
Furan	Liquid porfuryl alcohol resin (combined with catalyst when used); applied to one or both surfaces	High strength; resistant to chemicals	Chemical reaction	Chemical-resistant cements
Latex cements	Liquid synthetic rubber in solvent; applied to both surfaces	Good bond, waterproof; deteriorates on exposure unless cured	Evaporation of solvent	Bonding paper, leather, and textiles
Neoprene-resin	Liquid (solvents are toluene, methyl ethyl ketone, or a combination of solvents) and also in tape form; applied to both surfaces if liquid, at joint if tape form	High strength; resistant to water	Evaporation of solvent; room temperature or baked at 180° F (82.22° C) and cured at 325° F (162.78° C) with pressures of 50–500 lbf/in.^{2}; 15 to 30 min	Metal-to-metal bonding for structural purposes; bonding different metals
Polyurethane	Liquid or heavy paste; applied to one or both surfaces	High strength (with elasticity); resistant to water and chemicals	Evaporation of solvent	Bonding dissimilar materials
Buna-N-resins	Liquid, extruded, and tape form; applied to both surfaces if liquid, applied at joint if extruded or tape form	High strength; good fatigue properties; good resistance to shock; good moisture resistance	Evaporation of solvent; with 325°F (162.78° C) and pressure; 20 min (also at higher temperatures and pressures of 50–250 lbf/in.^{2})	Bonding rubber to metal
Rubber cement	Liquid natural rubber in solvent; applied to both surfaces	Good bond, waterproof; deteriorates on exposure as rubber is uncured	Evaporation of solvent	Bonding paper
Vinyl formal-phenolic	Liquid, applied to both surfaces; tape, applied at joint; two-part combination: liquid phenolic resin applied to both surfaces and powdered vinyl formal sprinkled over the liquid	Higher strength; good fatigue properties; good resistance to shock; high resistance to moisture and weathering	Evaporation of solvent; then baked at 150–180° F (65.56–82.22° C) and cured at 240–300° F (115.56–148.89° C) with pressures of 50–500 lbf/in.^{2}; from 15 min to 1 h	Bonding metal-to-metal or paper cellular cores for sandwich panels; bonding metal to plywood
Vinyl butyral-phenolic	Liquid with combination solvents, applied to both surfaces; tape and coated paper tape, applied at joint	Properties similar to those of vinyl formal-phenolic but not as strong or tough	Same type of curing as vinyl formal-phenolic adhesives	Bonding plastic laminates to other materials, bonding metal to resin-treated cellular cores

Table A12 Typical synthetic resin, rubber-resin, and resin blend adhesives (*continued*)

Type of resin, rubber-resin, or resin blend adhesive	Form and method of application	Characteristics	Method of curing; temperature; adhesion time	Major use
Nylon-phenolic	Liquid; applied to both surfaces	Properties similar to those of vinyl formal-phenolic but not as good moisture resistance	Same type of curing as vinyl formal-phenolic adhesives	Same uses as vinyl formal-phenolic adhesives
Thiokol	One part is compounded liquid rubber and the other part the setting agent; mixed when used; sets into a rubbery solid	Good strength; resistant to oil, grease, and solvents; fine resistance to weathering; good aging characteristics	Chemical action curing; room temperature; from 4 to 7 days	Sealing applications; transparent bonding; plastic to glass and rubber to metal

Table A13 Miscellaneous water-base adhesives

Type of water-base adhesive	Form and method of application	Characteristics	Method of curing; temperature; adhesion time	Major use
Vegetable starch and dextrine	Dry powder mixed with water when used, also prepared liquids; applied to one or both surfaces	Lower in strength than other adhesives; light color and nonstaining; not resistant to mold and moisture	Evaporation of water; adheres almost immediately (not more than several minutes)	Bonding paper; applying wallpapers
Vegetable soybean and zein[a]	(same as above)			Interior plywoods, bonding paper, applying wallpapers
Animal hide glue	Dry flake or powder mixed with water; applied to one or both surfaces by hand or machine at elevated temperatures	Good strength; poor water resistance; susceptible to mold	Evaporation of water; adheres before glue cools and gels (within minutes)	Interior uses; woodworking industries; furniture; bonding paper and textiles; as a binder
Animal bone glue	Dry powder mixed with cold water; applied to one or both surfaces	Good strength; poor water resistance, reactivated by water	Evaporation of water; room temperature; adheres almost immediately	Gummed paper tape
Animal fish glue	Dry powder mixed with cold water; applied to both surfaces	Good strength; poor resistance to water, mold, and fungi	Evaporation of water; room temperature; adheres almost immediately	Interior uses; furniture; woodworking requiring a cold-type adhesive
Animal blood albumen	Dry powder mixed with water; applied to one or both surfaces	Good strength; fair resistance to moisture and heat when cured at 180° F (82.22° C)	Evaporation of water; room temperature or at 180° F (82.22° C); adheres within minutes	Bonding paper; textiles; cork
Animal casein	Dry powder mixed with water for use, also liquid household-type adhesive	Good strength; fair resistance to moisture	Evaporation of water; room temperature; adheres within minutes	Interior uses; woodworking; plywoods; furniture
Animal casein-latex	Liquid; applied to both surfaces	Good strength; ages well; poor moisture resistance	Evaporation of solvent; room temperature; adheres almost immediately	Interior uses; bonding linoleum and laminated plastics to wood or metal

[a]A protein found in corn.

should be consulted, first to obtain the proper adhesive and then to obtain full and accurate data on the method of application and curing. These data should always be included in the specifications.

For prefabricated units or millwork made under controlled conditions, it is necessary always to specify shop drawings and require that they show the methods of joining and describe the adhesives used.

When selecting an adhesive for a particular end use, the following question should be considered: will the adhesive be required to be rigid or flexible; to resist shear, twisting, torsion, shock, impact, bending, static load, and vibration; and to withstand water, heat, and cold? For on-site applications, not only all the foregoing properties should be checked, but also climatic conditions, such as hot, warm, dry, cold, wet, and windy, must be considered and actually checked by inspection. Other considerations include working conditions during application, time required for setting, and whether bracing or clamps are necessary. For shop or mill application, all the above conditions should be considered during application, and also whether the end product will be installed on the exterior or interior.

History and Manufacture

The first adhesives used by men were probably natural asphalt and gums from various trees. The first manufactured adhesive known in history was the animal glue used by the early Egyptians. Animal glue, asphalt, and gums continued to be the main adhesives until gypsum and cement (portland) were discovered and used to hold materials in place. (In current terminology, however, gypsum and cement are not considered adhesives.)

The search for a water resistant adhesive led to the discovery of casein adhesives about 1900, and from then on the development of adhesives advanced by leaps and bounds. Gradually, after the discovery of vulcanization of rubber in 1839 and with the development of celluloid after World War I, it became increasingly difficult to define adhesives, rubbers, plastics, and even paints, and to differentiate between them.

ADMIXTURES

Physical and Chemical Properties

Admixtures are substances that are added to mortar, stucco, cement plaster, and concrete to produce specific results. They may or may not cause a chemical reaction within the mortar, stucco, cement plaster, or concrete, but usually a chemical reaction does occur. They can be divided into three categories: those for (1) mixing into concrete, (2) surface application or finish, and (3) mixing into mortar, stucco, and cement plasters.

Concrete admixtures include accelerators, retarders, finely divided powders, plasticizing agents, air-entraining agents, waterproofing compounds, and color pigments.

Surface applications or finishes for concrete consist of hardeners, color pigments, special aggregates, sealers, abrasive materials, waterproofing agents, and fillers and patchers.

Mortar and exterior plaster admixtures consist of accelerators, plasticizing agents, waterproofing agents, and color pigments.

Commercial Forms. In practice, admixtures come in powder, paste, and liquid form, and are usually patented and sold under trademark names. In many cases various admixtures are compounded to provide specific, controlled results in concrete and mortar. Almost all concrete and some mortars manufactured in the United States contain at least one admixture.

Types and Uses

Admixtures suitable for concrete, mortar, stucco, and exterior plasters are shown in Table *A14*, and those for

Table A14 Admixtures for concrete and mortar generally used in construction

Type of admixture	Effect on concrete or mortar	Ingredients generally used	Method of of adding	Advantages	Disadvantages	Major use in concrete and mortar
Accelerators	Speeds up hydration of the cement	Calcium chloride	Max. 2 lb/94-lb bag of cement (0.7464 kg/ 35.081 kg)	Speeds up setting time; develops strength earlier; lowers freezing point of water by 3° F (−16.11° C); increases heat due to hydration	Increases expansion and contraction; reduces resistance to sulfate attack; increases efflorescence and the corrosion of high-tension steels	In cold weather to speed up setting time and strength; to reduce length of time for protection; used for both concrete and mortar
Air-entraining agents	Introduces minute air bubbles throughout concrete	Rosin, beef tallow, stearates, vinsol resin, lauryl sodium sulfate, foaming agents	Usually $\frac{1}{4}$–$1\frac{1}{2}$ oz/100 lb cement (0.0078–0.0467 kg/37.32 kg cement)	Increases plasticity and cohesiveness; reduces bleeding; greatly increases resistance to freezing and thawing	Requires careful control; may require more frequent slump tests; causes some loss of strength	For all concrete exposed to freezing, thawing, and salt application; mostly for concrete

Table A14 Admixtures for concrete and mortar generally used in construction (*continued*)

Type of admixture	Effect on concrete or mortar	Ingredients generally used	Method of of adding	Advantages	Disadvantages	Major use in concrete and mortar
Color pigments	Either integral or as a topping color	For red, yellow, brown, and black, oxides of iron; for green, chromium oxide 98% pure, and cobalt blue 98% pure	Not more than 10% by weight of the cement, generally 3–6 lb/per bag of cement (1.1196–2.2392 kg per bag)	Permanent colors for concrete floors; for exterior, use only colors that do not fade	Reduces strength with increasing quantity of colors; increases water requirements, resulting in increased drying shrinkage	Colored floors, walks, and terraces; for both mortar and concrete
Latex (non-re-emulsifiable bonding type)	Improves adhesion and increases both tensile and flexural strength	Organic polymer-type latex and air-detraining agents (do not use air-entraining cements)	Generally 4 gal to 94-lb bag of cement (15.14 liter to 35.081 kg of cement)	Increases water retention, adhesion to substrates, tensile strength, and resistance to freezing and thawing	More difficult to finish; a steel-trowelled finish should be avoided	Flash coats, toppings, leveling of courses, and patching
Inert, finely divided powders	Corrects gradation of aggregates deficient in fines	Powdered glass, sand (silica), slate flour, stone dust, lime	As per manufacturer's directions	Corrects deficiency in fines for coarse aggregates; improves workability	Increases water requirements and drying shrinkage; decreases strength in rich mixes	To improve workability; mostly for concrete
Plasticizing agents (correctly called water-reducing agents)	Lowers water-cement ratio; lubricates solid particles in mix (aggregates)	Polyhydroxylated polymers, lignosulfonates, or hydroxylated carboxylic acids with calcium chloride or some other accelerator in some formulation	As per manufacturer's directions	Reduces water content; increases workability and plasticity	May slow hydration and decrease early strength	To make mix more workable and plastic; mostly for mortar
Pozzolanic, finely divided powders	Reacts with free lime during hydration of cement to form cementitious materials	Volcanic ashes, fly ash (residue from burning coal), calcined shale and clay, siliceous materials, natural cements, some slags	As per manufacturer's directions	Controls alkali-aggregate reaction; improves workability; reduces heat generation and expansion and contraction; increases strength after 28 days; increases resistance to sulfate attack; may increase impermeability	May cause excessive drying shrinkage; reduces durability; reduces early strength	To control alkali-aggregate reaction; to increase resistance to sulfate attack; mostly for concrete
Retarders	Slows up hydration of the cement	Zinc oxide, calcium lignosulfonate, derivatives of adipic acid	As per manufacturer's directions	Slows up setting; reduces heat due to hydration; reduces expansion and contraction	Some loss of early strength; requires careful control; may require more frequent slump tests	For very hot weather, and massive concrete; mostly for concrete
Waterproofing (permeability-reducing) compounds	Reduces capillary attraction of the voids in concrete or mortar	Stearic acid or its compounds, mainly calcium stearate; asphalt emulsions	0.1–4.0% of weight of cement	Decreases water absorption of concrete or mortar if no hydrostatic pressure exists	Reduces strength; does not make concrete waterproof	For slabs and basement or cellar foundation walls and other concrete in contact with earth; for both mortar and concrete

surfacing concrete in Table *A15*. Actually those for surface application or the finishing of concrete are admixtures only in the broad sense of the word. In fact, they are surface treatments added to concrete work already installed.

Included in the term "admixtures" are a few substances that are patching and grouting cements and two substances which are added to various admixtures to counteract the effects of other admixtures (*see* Table *A16 for details*).

Table A15 Admixtures used for surface treatments of concrete

Type of surface	Ingredient generally used	Method of application	Major use
Abrasive	Abrasive aggregate	Mixed and applied as surface treatment to freshly floated concrete floors	Floor surfaces where nonslip-type surface is necessary
Colored	Metallic oxides	Mixed integrally with mortar or concrete or surface-applied to freshly floated concrete floors	To add color to concrete or mortar
Hardeners	Usually zinc or magnesium fluosilicates	Mixed integrally with concrete or surface-applied after concrete floor is cured	To harden surface
Heavy wear	Metallic aggregate, usually iron particles	Mixed and applied as surface treatment to freshly floated concrete floors	Floor surfaces where hardness, durability, and wear resistance are necessary
Nonsparking	Metallic aggregate, usually iron particles	Mixed, electrically grounded, and applied as surface treatment to freshly floated concrete floors	Floor surfaces where static electricity is a hazard; areas where highly explosive or flammable materials are used or stored

Table A16 Special admixtures for concrete and mortar

Type of admixture	Ingredient generally used	Method of application	Major use
Grouting and patching	Finely ground iron particles or nonmetallic noncorrosive additives for strength, for shrinkage compensation, and as hardening elements	When applied to correctly prepared surfaces, to obtain a high-strength nonshrinking material that provides a long-lasting installation	To set steel, columns, etc., onto concrete or masonry walls; also heavy machinery installations
Metallic waterproofing	Powdered iron particles with oxidizing material	Applied into cracks, holes, and onto surface of concrete	To waterproof concrete after water penetration develops
Reducing efflorescence	Barium carbonate	Generally mixed with cement in manufacturing or mixed with mortar admixtures	To reduce development of efflorescence in brickwork
Reducing laitance	Natural siliceous materials	Generally mixed with concrete or mixed with admixture	To reduce development of laitance

In the construction field, admixtures are used to meet the requirements and demands for special types of concrete and mortar. Admixtures have become the means by which concrete and mortar performance can be closely tailored to meet specific job requirements.

Application

Condensed Checklist

1. Make sure that the concrete or mortar design mixes include all the types of concrete or mortar to be used on a job, and specify the type and quantity of admixture or admixtures to be used.
2. Make sure that the admixture used is the correct one for the specific result desired.
3. Make sure that for a given specific result the admixture will not cause other weaknesses that will be harmful to the concrete in its end use.
4. When an admixture is to be used, make sure that the method and proportions of its use are adequately controlled and additional tests carried out.
5. If a substitution is submitted by a contractor, check very thoroughly that the substituted admixture is equal to the one called for in the specifications.

Conditions Favorable to the Use of Admixtures

1. Where resistance to sulfate attack is necessary.
2. Where freezing and thawing will be encountered on the site where the building is to be erected.
3. Where quick setting is important, particularly during freezing weather, to shorten the time needed for protection.
4. Where slow setting is important, particularly during very hot, dry weather and for massive concrete work.
5. Where the capillary attraction of concrete for water must be halted.
6. Where the workability and plasticity of mortar are important.

Conditions Unfavorable to the Use of Admixtures

1. As the only method to protect concrete or mortar from freezing.
2. As a method to save on cement.
3. As a sure method of making concrete waterproof.
4. As a means to completely prevent efflorescence.
5. As a cure-all for concrete.

History and Manufacture

Egyptians and Greeks used color pigments intelligently in their concrete work. Oxblood, which the Romans added to concrete, made an excellent air-entraining agent, and various types of urine were also used. All this knowledge was lost with the fall of the Roman Empire, and admixtures did not reappear until after portland cement was developed in 1824 and reinforced concrete in 1868 stimulated research into concrete mixtures. But even as late as 1920 admixtures were a minor item, although calcium chloride, stearates, and color pigments were in use. Farmers put "Gold Dust" soap powder as a waterproofing agent into their concrete, and the stearates in this soap did a good job of water-repelling. An important development in 1938 was the discovery that small amounts of well-dispersed, entrained air not only improved the workability of concrete but also increased its resistance to freezing and thawing. Since then admixtures have been widely promoted as a means to rid concrete of all its ills.

AGGREGATES

Physical and Chemical Properties

An aggregate is any hard, inert material composed of fragments in a wide gradational range of sizes, which is mixed with a cementing material to form concrete or the like. The term also may refer to any mass of similar mineral fragments which are used, with or without a binder, in many ways, including some that involve physical and chemical alteration of the aggregate material itself.

Aggregates should be clean, sound, tough, durable, and uniform in quality. They should also be free of soft, friable, thin, or laminated fragments and deleterious substances such as alkali, oil, coal, humus, or other organic matter.

Specific gravity, color, compressive strength, and roundness of particles are important in some instances. Pore characteristics and roughness of surface affect the bond between cement and aggregate. Pore structure, that is, the size, number, and continuity pattern of void spaces, also affects water absorption, permeability, and apparent specific gravity.

The chemical activity of impurities such as volcanic glass, chats, gypsum, and pyrite (iron sulfide ore) that occur in mineral aggregates with the alkalis in portland cement is a factor to be considered, as oxidation and hydration of certain minerals may result in crystal growth and formation of sulfuric acid, with possible undesirable consequences.

Types and Uses

Aggregates may be classed into two groups: (1) natural materials such as sand, gravel, crushed stone, and pumice; and (2) artificial materials, produced by crushing blast furnace slag or burning and crushing clays or shales. The second group includes most of the lightweight aggregates.

Fine Aggregates. Two basic grades of aggregates are used in construction work, fine and coarse. Fine aggregates consist of particles from 0.02 to $\frac{1}{4}$ in. (0.508 to 6.35 mm) in diameter (*see* Sand). They should have a fairly regular shape and should not include flat, elongated particles. The fineness modulus should be not less than 2.3 or greater than 3.1 (*see* Table *A17*). The fine aggregates most commonly used are natural sand and finely crushed stone, slag, and cinders (*see* Table *A18*).

Table A17 Grading requirements for fine aggregates

Size of sieve		Percent of aggregate passing
$\frac{3}{8}$ in.	9.53 mm	100
No. 4	4.76 mm	95–100
No. 16	1.19 mm	45–80
No. 50	297.00 microns	10–30
No. 100	149.00 microns	2–10

Table A18 Fineness modulus[a] for fine aggregates

Size of sieve		Percent passing	Percent passing (cumulative)
No. 4	4.76 mm	98	2
No. 8	2.38 mm	85	15
No. 16	1.19 mm	65	35
No. 30	595.00 microns	45	55
No. 50	297.00 microns	21	79
No. 100	149.00 microns	3	97

[a]The fineness modulus is obtained by dividing the total (cumulative) percent retained by 100. Thus, adding the figures in the last column gives a total of 283; dividing 283 by 100 gives a fineness modulus of 2.83.

Table A19 Grading requirements for coarse aggregates

Designated size of aggregate	Size of sieve openings (square openings) in inches and millimeters										
	4 in. 101.6 mm	$3\frac{1}{2}$ in. 88.9 mm	3 in. 76.2 mm	$2\frac{1}{2}$ in. 63.5 mm	2 in. 50.8 mm	$1\frac{1}{2}$ in. 38.1 mm	1 in. 25.4 mm	$\frac{3}{4}$ in. 19.05 mm	$\frac{1}{2}$ in. 12.7 mm	$\frac{3}{8}$ in. 9.6 mm	No. 4 4.76 mm
	Percentage of aggregates passing										
$3\frac{1}{2}$ to $1\frac{1}{2}$ in. (88.9 to 43.1 mm)	100	90–100		25–60		0–15		0–5			
$2\frac{1}{2}$ to $1\frac{1}{2}$ in. (63.5 to 43.1 mm)			100	90–100	35–70	0–15		0–5			
2 to 1 in. (50.8 to 25.4 mm)				100	95–100	35–70	0–15		0–5		
2 in. to No. 4 (50.8 to 4.76 mm				100	95–100		35–70		10–30		0–5
$1\frac{1}{2}$ to $\frac{3}{4}$ in. (43.1 to 19.5 mm)					100	90–100	20–55	0–15		0–5	
$1\frac{1}{2}$ to No. 4 (43.1 to 4.76 mm)					100	95–100		35–70		10–30	0–5
1 in. to No. 4 (25.4 to 4.76 mm)						100	95–100		25–60		0–10
$\frac{3}{4}$ in. to No. 4 (19.05 to 4.76 mm)							100	90–100		20–55	0–10
$\frac{1}{2}$ in. to No. 4 (12.7 to 4.76 mm)								100	90–100	40–70	0–15

Coarse Aggregates. Coarse aggregates consist of particles $\frac{1}{4}$ in. (6.35 mm) in diameter and over. Gravel, crushed rock, and blast furnace slag are generally used. In some areas, chats are sold as aggregate. The term is a local one for waste from zinc ore mining in Oklahoma and Missouri. (*See* Table *A19*.)

Special Aggregates. Certain special aggregates (*see* Table *A20*) are often added to impart specific properties to the finished product.

Trends. Commercial developments and research programs indicate the following trends in aggregates:

1. A sawdust-diatomite-clay mixture as a possible source of lightweight aggregate.
2. A mixture of perlite and finely ground diatomaceous earth to form an aggregate for concrete, claimed to reduce the stratification of aggregate and cement that often occurs in a regular perlite-cement mix.
3. Increased utilization of clays, shales, slates, and phosphate stones that "bloat" or expand at incipient fusion to form lightweight aggregates. These have the advantage over other lightweight aggregates in that they yield high-strength concretes comparable to regular concrete.
4. Research toward development of a variety of synthetic binders for molding sands.
5. Steam-treating or fusing mixtures of ground silica and lime to produce an aggregate.
6. Research toward the use of waste glass, such as bottles and containers, as an aggregate for concrete and asphalt concrete.

AIR

Air is composed mainly of oxygen and nitrogen with small amounts of carbon dioxide, hydrogen, and the rare gases argon, neon, helium, krypton, and xenon. In the construction field the principal interest in air involves the effect of oxygen, water vapor, sulfur dioxide, salt (in marine atmospheres), and possible industrial fumes on construction materials. (*See* Table *A21*.)

Table A20 General uses of special aggregates

Kinds of aggregate	Application
Asbestos	High temperature
Kyanite and sillimanites (aluminum silicates)	High temperature
Zircon	High temperature
Diatomite	Lightweight concrete
Expanded clay, shale, and slate	Lightweight concrete
Expanded slag	Lightweight concrete
Perlite	Lightweight concrete
Pumice	Lightweight concrete
Vermiculite	Lightweight concrete
Iron	Very dense concrete
Barite	Resistance to gamma rays; heavyweight concrete
Limonite ore (brown iron ore)	Concrete for nuclear shielding
Aluminum oxide	Abrasive aggregates
Emery	Abrasive aggregates
Silicon carbide	Abrasive aggregates
Slag (air-cooled)	Abrasive aggregates
Marble, granite	Colored, decorative aggregate for terrazzo and surfacing

From the ecological viewpoint in construction the environmental control systems and equipment have to be carefully checked in regard to air pollution. Any industrial types of structure where manufacturing or chemical processes are in operation must also be checked against air pollution standards.

AIR STRUCTURES

Two basic principles control the design of air structures. (1) At sea level the pressure of the earth's atmosphere due to gravity is 15 lb/in.^{2}(10 546 kg/m^2); by slightly increasing this pressure inside a sealed form, the form is inflated and maintains this stabilized shape so long as the inside pressure is maintained. (2) The shape of the inflated form is generally ovoid or spherical and can be designed to be structurally rigid. These two fundamental

Table A21 Composition of air

N	O	Ar	CO_2	H	Ne	He	Kr	Xe
				(percent of content)				
78.03	20.99	0.94	0.03	0.01	0.00123	0.0004	0.00005	0.000006

types of air structures—one in which the basic shape is ovoid or spherical, and the other in which units of the air structure when inflated become structural supporting members—are shown in Figure *A3*.

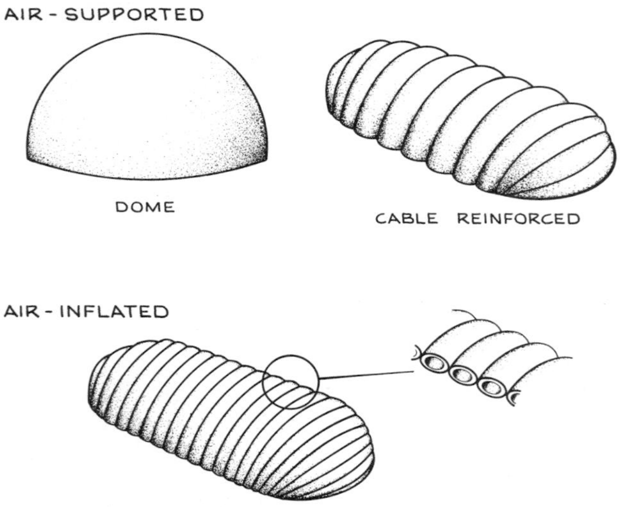

Figure A3 Typical forms for air structures.

Air structures are made of plastic textiles or reinforced films, nylons, vinyls, Teflon, and other similar material (*see* Plastics; Textiles). It is beyond the scope of this book to discuss the engineering technology required to design an air structure, involving as it does the problems of aerodynamics in general and air locks in particular.

ALKALI

In industry the word "alkali" refers to compounds that are soluble and that behave actively as a base (*see* Acids, Bases, and Salts). In chemistry, it refers to soluble hydroxides of the alkali metals and of ammonia. Originally the term was applied to ashes of plants from which potassium carbonate (potash) and sodium carbonate (soda ash) were leached.

ALLOTROPY

Allotropy (polymorphism) is a characteristic or ability of certain compounds and elements to exist in at least two distinct crystalline forms (atomic arrangements) which are differentiated by their outward crystal form, optical properties, volume, etc. The most familiar example is carbon, which may be found native as a diamond or as graphite.

ALLOYS

Physical and Chemical Properties

An alloy is defined by the American Society for Testing Materials as "a substance having metallic properties, consisting of two or more metallic elements or of metallic and nonmetallic elements, which are miscible with each other when molten, and have not separated into distinct layers when solid." Taken literally, this definition means that all commercial metals are really alloys, since none is available 100% pure. In practice, however, metals are divided into two main groups: commercially pure metals (metallic elements with controlled amounts of impurities) and alloys.

The properties of an alloy depend on but also differ materially from those of its constituents. For example, two soft, malleable metals such as copper and tin combine to form a hard, much less plastic alloy, bronze. The characteristics of an alloy also depend on the proportions and arrangement of the constituents. All metals and alloys in the solid state are composed of crystals, and it is the chemical composition, size, and shape of these crystals as well as the interrelation of planes within the individual crystal itself, and of the crystals to each other within the larger lattice or physical pattern, that determine what proportions and arrangements are possible.

The crystals form as the molten alloy changes from liquid to solid, and the action which occurs during this solidification, down to the least detail of conditions and materials present, determines the characteristics of the end product. Physical processes applied to the alloy, such as working or heat treatment, also affect its properties.

Types and Uses

Alloys are classified industrially into two main divisions: ferrous (iron-base alloys) and nonferrous. Many ferrous alloys are termed steel, whereas the rust-resisting steels are generally grouped as stainless steels. Nonferrous alloys are further classified industrially in terms of the major metal present, for example, copper-base alloys, tin-base alloys, high nickel-copper alloys. Aluminum-base alloys and magnesium-base alloys are also known as "light" (lightweight) alloys and are increasingly important in construction.

Metallurgic Classification. In metallurgy alloys are further classified as binary, ternary, quaternary, and multimetal systems, depending on the number of metallic elements present. Study of the constitution and crystal structure of alloys shows that three types of constituents are possible in these systems: pure or substantially pure metals; solid solutions of metals; and chemical compounds made up of two metals or a metal and a nonmetal.

Comparing Alloys. At present almost all metals used in construction are alloys. Therefore, while not expected to be a metallurgist, one should have a general idea of both the industrial and scientific classifications of metals and alloys. One should also know why a certain alloy is stronger or more corrosion resistant or more ductile than another, roughly similar alloy.

The data in this book have been organized for this purpose. For example, if one wishes to use bronze, one can find available types by looking under Bronze. There one will learn, for example, that phosphor bronze contains tin, which gives it good corrosion resistance, whereas architectural bronze, actually a brass containing no tin, is less resistant to corrosion. On the other hand, if one follows through by reading the Tin section, one learns that tin also makes an alloy very hard and hence difficult to work and affects its color. Using this method, one can finally determine which alloy will have the right characteristics for a particular design requirement.

History and Manufacture

The Far East appears to be the main source of early knowledge on methods of purifying, combining, and working metals. The early bronzes, Phoenician tin alloys, and Chinese *paktong* (nickel silver) are the earliest known alloys and were probably made by a simple, direct smelting together of ores containing these metals.

Industrial Alloying Techniques. In industrial work, alloys are usually made by melting one metal, usually the least volatile one, and dissolving the other metals or master alloys in it. Alternatively, two or more metals may be melted separately and mixed. Alloys may also be formed (1) by mixing metal powders and heating to a temperature below the melting point, (2) by direct electrodeposition, or (3) by a more recently developed technique, powder metallurgy.

Powder Metallurgy. In powder metallurgy, massive materials or shaped objects are manufactured by using the substance in powdered form under special pressures and temperatures. Thus metals can be combined with nonmetals normally incompatible in the usual smelting or other alloying techniques to form special alloys, for example, alloys that are capable of withstanding tremendous temperatures.

Specific Processes. The exact process of alloying used depends on the end product desired and varies more or less with each one. Since there are currently more than 5000 alloy compositions in use, a more detailed discussion of methods is impractical here but can be found under individual metal headings (*see* Brass; Bronze; *etc.*).

ALUM

Alum is a clear to white crystalline potassium aluminum sulfate with a specific gravity of 1.757. It melts in its water of crystallization at 197.0°F (92°C). It is available in lump, nut, pea, powdered, and ground form. In the construction field its main uses are for water purification and in Keene's cement to control setting time. In the art field it is used as a wet wash to remove any oil or grease from fine mounted drawing paper.

ALUMINUM

Physical and Chemical Properties

Symbol: Al
Atomic number: 13
Specific gravity: 2.6989 (20°C, 68°F)
Melting point: 660.37°C, 1220.67°F
Boiling point: 2467°C, 4422.6°F
Linear coefficient of thermal expansion: 0.0000128/°F (0.0000231/°C)
Tensile strength: 7000 lbf/in.^{2}, 48.265 MN/m^2 (for 99.996% pure aluminum)

Aluminum is a soft, nonmagnetic silvery metal characterized by its light weight (one-third that of iron, brass, or copper), low melting point, high thermal and electrical conductivity (surpassed only by silver and copper), and moderately high coefficient of expansion.

Strength. The strength of aluminum depends on its composition and on the thermal and mechanical treatment to which it has been subjected. Its strength can be increased by alloying and proper working to as high as 100,000 lbf/in.^{2} (689.5 MN/m^2).

Solubility of Other Metals. Silicon, copper, iron, zinc, tin, manganese, and magnesium dissolve readily in aluminum. Sodium and potassium are practically insoluble in it, and titanium, vanadium, boron, nickel, and chromium have low solubilities. Hydrogen is the only gas soluble in aluminum to any extent.

Corrosion Resistance. Aluminum combines readily with oxygen and is made corrosion resistant by the transparent film of aluminum oxide that quickly forms and is relatively inert to further chemical action. Aluminum is readily attacked by alkalis and hydrochloric acid and slowly attacked by dilute acids. It is inert to sulfur. In direct contact with metals other than zinc, cadmium, magnesium, and nonmagnetic stainless steel, aluminum is subject to certain types of galvanic action and should therefore be electrically insulated from other metals.

Workability. Aluminum is easily worked and can be hot or cold rolled, extruded, forged, pressed, drawn, molded, stamped, bent, and shaped. It can be riveted, bolted, welded, brazed, and soldered.

Commercial Forms. Aluminum is available in ingot and pig form. These are fabricated into bar, rod, bus bar, wire, cable, tube, conduit, pipe and fittings, plate, sheet, foil, and powder form, as well as into structural shapes, castings, forgings, and extrusions. Aluminum can be obtained 99.996% pure (99.9999% in laboratories); commercial aluminum is generally 99.0 to 99.6% pure (*see* Table *A22*).

Table A22 Composition of commercially pure aluminum

Al	Si	Fe	Cu	Mn	Mg	Zn	Ti
(percent of content)							
99.5	0.2	0.2 max.	0.04 max.	0.03 max.	0.03 max.	0.03 max.	0.03 max.

Types and Uses

Aluminum has a wide range of uses in compound form and metallic form. Table *A23* indicates only generally where and how it is used in construction and lists its major nonconstruction applications. Asterisks indicate the use of metallic aluminum in pure or alloy form.

Table A23 Uses of aluminum

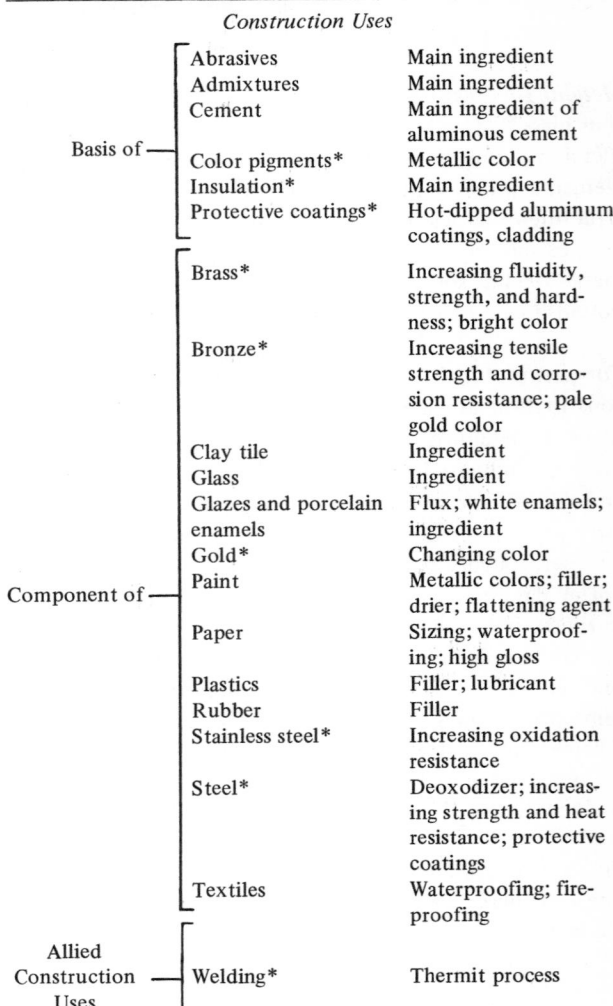

Construction Uses

Basis of	Abrasives	Main ingredient
	Admixtures	Main ingredient
	Cement	Main ingredient of aluminous cement
	Color pigments*	Metallic color
	Insulation*	Main ingredient
	Protective coatings*	Hot-dipped aluminum coatings, cladding
Component of	Brass*	Increasing fluidity, strength, and hardness; bright color
	Bronze*	Increasing tensile strength and corrosion resistance; pale gold color
	Clay tile	Ingredient
	Glass	Ingredient
	Glazes and porcelain enamels	Flux; white enamels; ingredient
	Gold*	Changing color
	Paint	Metallic colors; filler; drier; flattening agent
	Paper	Sizing; waterproofing; high gloss
	Plastics	Filler; lubricant
	Rubber	Filler
	Stainless steel*	Increasing oxidation resistance
	Steel*	Deoxodizer; increasing strength and heat resistance; protective coatings
	Textiles	Waterproofing; fireproofing
Allied Construction Uses	Welding*	Thermit process

Nonconstruction Uses

Aircraft, automotive and transportation industries,* communication equipment,* electric power construction,* electronics,* industrial refractories, jewelry,* lubricants, machinery and equipment,* radio,* television,* X-ray filters*

Alloys important in the construction field are discussed in detail under Aluminum Alloys, and mill-produced and fabricated aluminum products under the specific headings that follow Aluminum Alloys.

History and Manufacture

Aluminum is one of the newest of the common metals, despite the fact that it is the most abundant element in the earth's crust. However, its naturally occurring compounds were used as astringents and mordants as long ago as 500 B.C. One which the Romans called *alumen* was a natural potassium aluminum sulfate. About A.D. 1200 these mineral salts were purified into crystalline alum, and in the 16th century alum, $Al_2(SO_4)_3$, was produced from clay. In 1809 Davy proved what had been suspected earlier, that alum had a metallic base, by producing an aluminum-iron alloy through electrolysis of fused alumina in a hydrogen atmosphere. Dissolving this alloy yielded aluminum oxide. Davy's suggested name for this element, alumium (from the Latin *alumen*), has been adapted to aluminum in North America and modified to aluminium in England, Europe, and other countries. In 1825 Oersted produced the first metallic aluminum, but only as powder. In 1845 Woehler managed to transform the powder into particles, from which he discovered its outstanding physical properties. In 1852 aluminum was more precious than gold, costing \$545 per pound. Napoleon III ordered it made into forks and spoons used only by honored guests at state banquets where other guests used gold and silver service. In 1854 St. Claire Deveille and Von Bunsen simultaneously but independently isolated a 96 to 97% pure aluminum by using sodium instead of potassium reduction. Deveille displayed several large bars of it at the 1855 Paris Exposition. The following year saw the first architectural use of aluminum, in the tip of the Washington Monument.

In 1886 the aluminum industry was born. Hall in the United States and Héroult in France simultaneously discovered the modern electrolytic method of aluminum production. (A curious sidelight on the history of aluminum is that Hall and Héroult not only were born in the same year, but also both made their discovery at the age of 23 and both died in 1914). The price of aluminum dropped from \$11.33 per pound in 1885 to 57 cents per pound in 1892.

Its earlier use was for household utensils. Then lighter-than-air craft were developed; the earliest structural use of alloys of the duralumin (copper) type was in the Zeppelins built by the Germans. Development of alloys for the aviation industry between World Wars I and II led to increased use in other fields until aluminum has become a major construction material in the transportation industry (including automobiles), the chemical industry, the electrical industry, and architecture.

Aluminum began to appear in construction about 1926, and its use has steadily increased since then for spandrels, windows, doors, ornamental railings and grill work, roofing and siding, store fronts, fencing, reflective insulation, and miscellaneous hardware. About 1954 curtain wall construction became widespread, and buildings were covered with panels made of aluminum sheet, backed with light concrete, in patterned, anodized, and enameled finishes.

Refining Procedures. Today aluminum is refined by a two-stage method: the first stage, the Bayer process, yields almost pure alumina (Al_2O_3) from bauxite ore; the second stage, the Hall-Héroult process, reduces the oxide to metallic aluminum more than 99% pure. The two steps are shown as a continuous process on the flow chart in Figure *A4*.

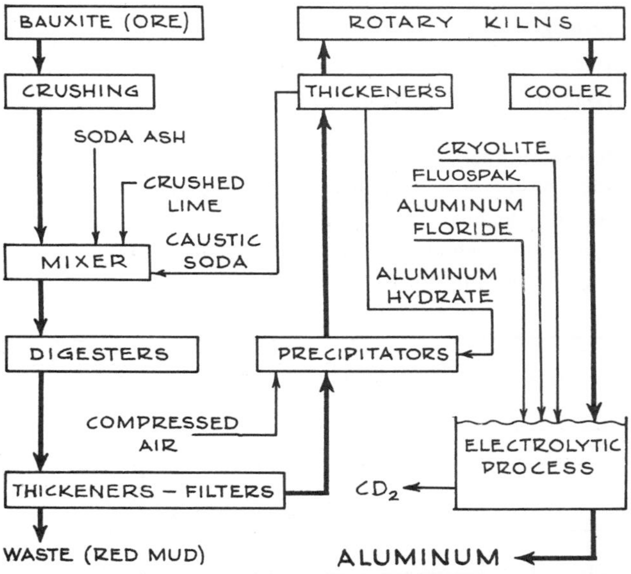

Figure A4 Dryer and Hall-Héroult process of aluminum production.

Bayer Process. In the Bayer process the bauxite ore is first crushed to a fine powder and then mixed with soda ash (Na_2CO_3), crushed lime (CaO), and hot water to exact proportions. The Na_2CO_3 and CaO form caustic soda (NaOH). This mixture is pumped into digester tanks; there, after mixing under high pressure, steam is injected to accelerate the reaction and agitators keep churning the mixture to bring all the materials into close contact. The caustic soda dissolves the alumina to form sodium aluminate ($Na_2Al_2O_4$ or $Na_2O \cdot Al_2O_3$), and the insoluble impurities form a residue, called red mud, which contains iron oxide, titanium dioxide, and silica. After digestion the aluminum-containing solution is passed through pressure-reducing tanks and then into filter presses (shown in the flowchart as one unit called thickeners-filters). Here the red mud is removed; and the sodium aluminate solution passes into precipitators, where aluminum hydrate, $Al_2O_3 \cdot 3H_2O$ or $Al(OH)_3$, obtained from a later step, is used to seed the solution. Agitation with compressed air and subsequent slow cooling causes the aluminum hydrate particles (the seed) to grow in size until all the sodium aluminate is precipitated as aluminum hydrate. This mixture is pumped into thickeners (filters); these separate the aluminum hydrate from the caustic soda solution, which is fed back to the digester tanks for reuse. The aluminum hydrate is then calcined in rotary kilns at 2000°F (1099.3°C) to convert it to alumina (Al_2O_3), which is then ready for the second stage.

Hall-Héroult Process. In the second stage, the Hall-Héroult process, alumina is dissolved in molten cryolite (Na_3AlF_6) and electrolyzed by direct current. A carbon-lined steel box acts as the cathode, and carbon rods or blocks immersed in the molten bath are the anodes. Electrolytic action breaks down the Al_2O_3 into oxygen, which reacts with the anode to form carbon dioxide, and aluminum, which is deposited on the cathode. Since molten aluminum is heavier than molten cryolite at the operating temperature of 1800°F (981.67°C), it accumulates on the bottom. From time to time additional alumina, cryolite, and aluminum fluoride (to neutralize the small amount of soda present as an impurity in the alumina) are added, and the molten aluminum is drawn off and cast into pigs or passed through alloying furnaces and cast as ingots. These ingots are now ready for further processing, alloying, and fabrication into various end products, as shown in the flowchart in Figure *A5*.

ALUMINUM ALLOYS

Physical and Chemical Properties

In construction practically all fabricated forms of aluminum are used, that is, rod, bar, extrusion, casting, sheet, strip, etc. However, these products are not fabricated from aluminum in the purest obtainable form. The 99.996% pure metal, which has superior corrosion

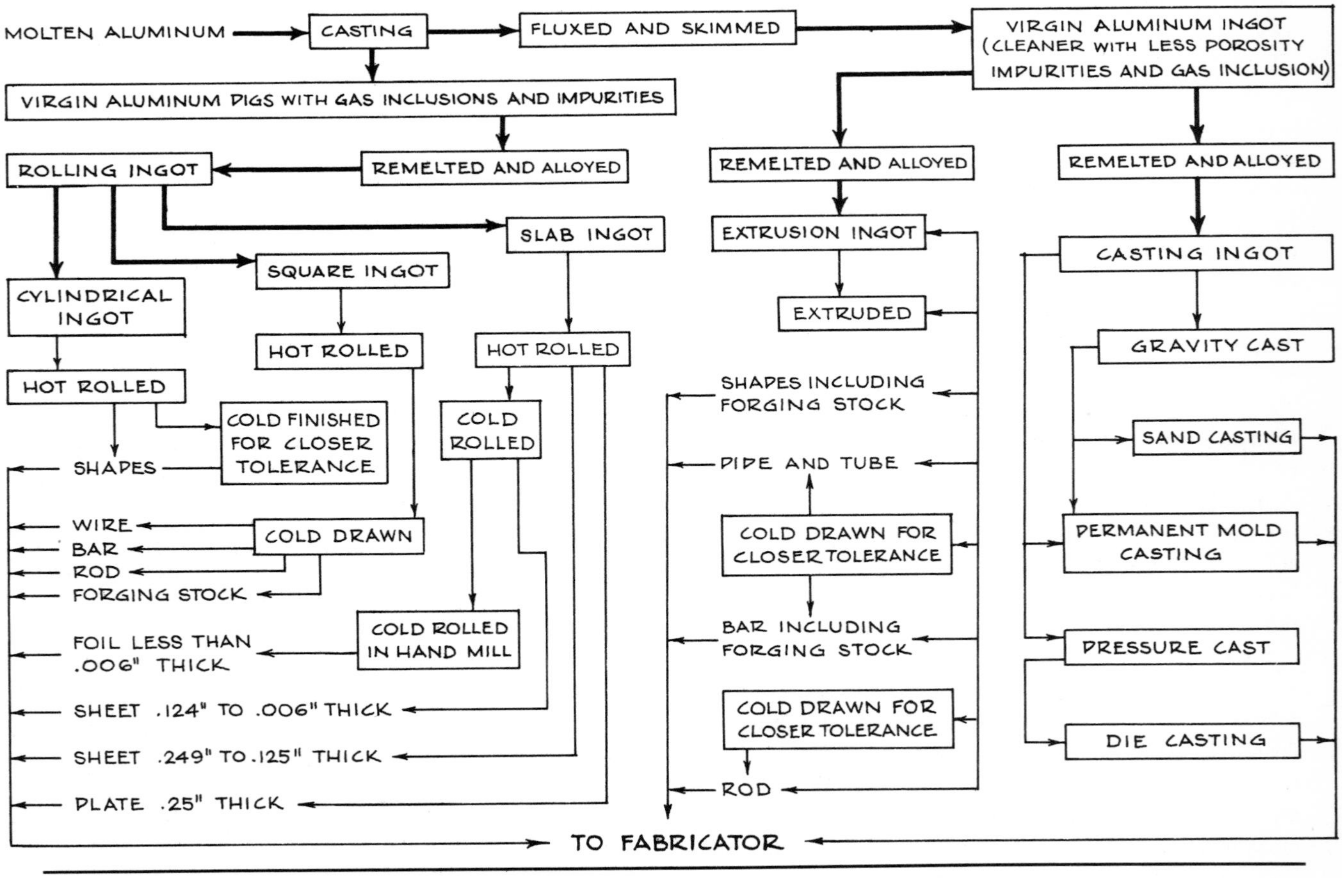

Figure A5 Processing of aluminum into mill products for fabricators.

resistance, is at present used commercially in small quantities, mainly for trim, for jewelry, as a catalyst, and in applications where high electrical conductivity is required.

Commercial Aluminum. Strictly speaking, commercial aluminum, generally 99.0 to 99.6% pure, is an alloy of pure aluminum with small amounts of iron and silicon that increase its strength but decrease its resistance to corrosion. Iron is generally considered an undesirable impurity and is held to a very low percentage, whereas silicon may be either an impurity or a major alloying element, depending on the characteristics desired in the aluminum alloy.

Alloying Elements. Other major alloying elements are copper, manganese, magnesium, zinc, chromium and nickel. Tin, cadmium, lead, bismuth, and cobalt may be used in special alloys. Beryllium, boron, lithium, titanium, vanadium, and zirconium are also used on occasion for special purposes, for example, beryllium (under 0.15%) for grain refinement, and lithium for superior mechanical strength at moderately elevated temperatures.

Types and Uses

Aluminum alloys are divided into two major types, casting alloys and wrought alloys. Casting alloys are those in which the metal is cast in its final form. Wrought alloys are those in which the cast metal is worked mechanically by a process such as rolling, extruding, forging, or drawing.

Classification of Casting Alloys. Casting alloys are classified according to a system of designation devised by the Aluminum Association, based on a four-digit system similar to the system for wrought alloys and described as follows:

- 1xx.x Commercially pure aluminum
- 2xx.x Copper as the major alloying element
- 3xx.x Manganese as the major alloying element
- 4xx.x Silicon as the major alloying element
- 5xx.x Magnesium as the major alloying element
- 6xx.x Magnesium and silicon combined as the major alloying elements
- 7xx.x Zinc as the major alloying element
- 8xx.x Other elements
- 9xx.x Unused series

The first digit indicates the alloy group, the second two digits identify the aluminum alloy or indicate the aluminum purity, and the last digit, which is separated by a decimal point, indicates the product form, namely, casting or ingot. For example, 1xx.0 indicates casting, and 1xx.1 indicates ingot. (*See* Table *A24*.)

Classification of Wrought Aluminum and Aluminum Alloys. Wrought aluminum and aluminum alloys are classified according to a system of designation devised by the Aluminum Association, based on a four-digit system schematically shown and briefly described as follows. The first digit represents the major constituent of the alloy:

- 1xxx Commercially pure aluminum
- 2xxx Copper as the major alloying element
- 3xxx Manganese as the major alloying element
- 4xxx Silicon as the major alloying element
- 5xxx Magnesium as the major alloying element
- 6xxx Magnesium and silicon combined as the major alloying elements
- 7xxx Zinc as the major alloying element
- 8xxx Other elements
- 9xxx Unused series

The second digit represents either modifications of impurity limits in the 1xxx series, or modifications of

Table A24 Common aluminum casting alloys used in construction

Alloy number and temper	Al	Si	Fe	Cu	Mn	Mg	Zn	Ti	Others (total)	Form	Major use
			(percent of content)								
B443.0-F	Remainder	4.5–6.0	0.8	0.15	0.35	0.05	0.35	0.25	0.15	Castings, sand casting	Panels, lettering, hardware, grilles, posts, sculpture
514.0-F	Remainder	0.35	0.5	0.15	0.35	3.5–4.5	0.15	0.25	0.15	Casting, sand casting	Lettering, posts
356.0-T6	Remainder	6.5–7.5	0.6	0.25	0.35	0.2–0.4	0.35	0.25	0.15	Casting	Hardware, railings, supports

Table A25 Example of high-purity wrought aluminum alloy of the 1xxx series

Alloy number	Al	Fe	Si	Cu
		(percent of content)		
1175	99.75	Fe + Si 0.15 (max)		0.10

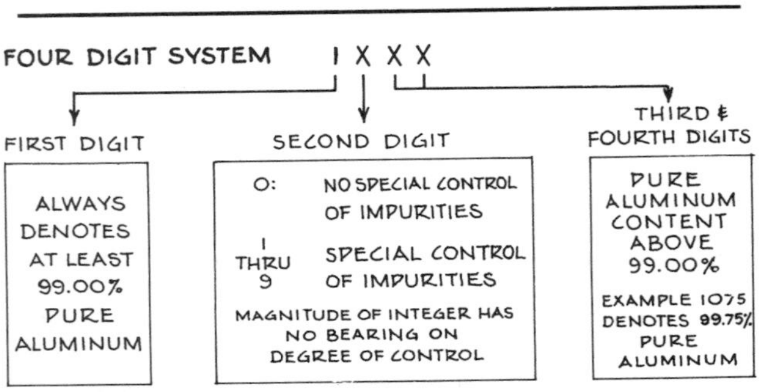

Figure A6 Four-digit system for wrought aluminum and aluminum alloys (1xxx series).

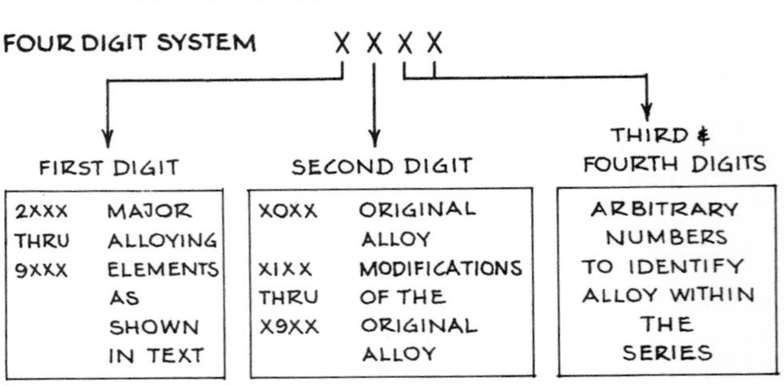

Figure A7 Four-digit system for wrought aluminum and aluminum alloys (2xxx series).

Table A26 Example of wrought aluminum alloy of the 2xxx series

Alloy number	Al	Fe	Si	Cu	Mn	Mg	Others	
			(percent of content)				Each	Total
2017	Remainder	1.0	0.50	4.0[a]	0.7	0.6	0.05	0.15

[a]Main alloying element.

the alloy in the 2xxx through 9xxx series. The last two digits either indicate aluminum purity over 99% in the 1xxx series or, in the 2xxx to 9xxx series, further identify the alloy by its old commercial designation. Thus alloy 2014 is the former commercial alloy 14S.

For example, alloy 1075 represents an aluminum alloy containing 0.75% more aluminum than the minimum of 99.0% required and subject to specific control of impurity limits (as shown by the zero as second digit). Alloy 1175 has the same aluminum content but is subject to modified impurity control. (*See* Figure *A6* and Table *A25*.)

Alloy 2017 belongs to the copper series and is the alloy as originally developed (as indicated by the zero as second digit) in the 2X17 group of copper alloys. Designations 2217 through 2917 are available to represent future commercially accepted and manufactured modifications of the original alloy designated as 2017. (*See* Figure *A7* and Table *A26*.)

For the compositions and forms of common wrought alloys in construction use, see Table *A27*.

Temper. The temper of alloys is indicated by letters and numbers as shown in Table *A28*. Such a designation follows the alloy number and is separated from it by a dash. It should be noted that certain temper designations apply only to wrought products, and others only to cast products. F, O, and H designate alloys hardened by cold work, usually but not necessarily non-heat-treatable. W and T designate heat-treatable alloys.

Some wrought alloys respond to heat treatment, whereas others do not. The latter, known as non-heat-treatable alloys, are those in which mechanical properties are determined by the alloying elements and the amount of cold work present. The properties obtained by cold work are destroyed by subsequent heating above 392°F (200°C) and can be restored only by additional cold work. Heat-treatable alloys are those in which the mechanical properties are improved by heat treatment. Specifically, strength is increased with little sacrifice of ductility. Also, heat-treatable alloys can be heat-treated again after annealing to restore their original characteristics.

Table A27 Common wrought aluminum alloys in construction

Alloy number and temper	Composition (maximum allowable percentages)											Form used	Major use
	Al	Fe	Si	Cu	Mn	Mg	Cr	Zn	Ti	Others Each	Others Total		
1100-H16[a]	99.00	1.0	Fe + Si	0.2	0.05			0.10		0.05	0.15	Sheet	Ducts
1235[b]	99.35	0.65	Fe + Si	0.05								Foil	Insulation, vapor barrier
2014-T6	Remainder	0.7	1.2	5.0^c	1.2	0.8	0.1	0.25	0.15	0.05	0.15	Rolled shapes, extrusions	Structural members
2017-T4	Remainder	0.7	0.8	4.5^c	1.0	0.8	0.1	0.25		0.05	0.15	Screw machine stock	Screws
2024-T4	Remainder	0.5	0.5	4.9^c	0.9	1.8	0.1	0.25		0.05	0.15	Wire Screw machine stock	Rivets Screws
3003-O	Remainder	0.7	0.6	0.2	1.5^c			0.10		0.05	0.15	Sheet	Concealed and exposed flashing
3003-H14	Remainder	0.7	0.6	0.2	1.5			0.10		0.05	0.15	Sheet	Siding and shingles, gutters and leaders, wall and acoustic tile, etc.
3003-H18	Remainder	0.7	0.6	0.2	1.5			0.10		0.05	0.15	Sheet	Hardware
5050-O	Remainder	0.7	0.4	0.2	0.1	1.8^c	0.1	0.25		0.05	0.15	Sheet	Hardware
5052-H18	Remainder	Fe + Si, 0.45		0.1	0.1	2.8^c	3.5	0.10		0.05	0.15	Sheet	Venetian blinds
5052-H38	Remainder	Fe + Si, 0.45		0.1	0.1	2.8^c	3.5	0.10		0.05	0.15	Sheet	Weatherstripping, screens
6061-T13	Remainder	0.7	0.8^c	0.4	0.15	1.2^c	0.35	0.25	0.15	0.05	0.15	Extrusion	Pipe railings
6061-T6	Remainder	0.7	0.8^c	0.4	0.15	1.2^c	0.35	0.25	0.15	0.05	0.15	Wire, extrusion, bar, forging, pipe, sheet, rolled shapes	Structural members, gratings, treads and steps, windows
6061-T913	Remainder	0.7	0.8^c	0.4	0.15	1.2^c	0.35	0.25	0.15	0.05	0.15	Wire	Nails
6063-T4	Remainder	0.35	0.6^c	0.1	0.10	0.9^c	0.10	0.10	0.10	0.05	0.15	Tubing	Poles (light, flag, etc.)
6063-T42	Remainder	0.35	0.6^c	0.1	0.10	0.9^c	0.10	0.10	0.10	0.05	0.15	Extrusion	Exposed flashing
6063-T5	Remainder	0.35	0.6^c	0.1	0.10	0.9^c	0.10	0.10	0.10	0.05	0.15	Extrusion	Door and window frames, trim, moldings, sills, stools, thresholds, etc.; hardware, exposed flashing, grilles and louvers, etc.
6063-T6	Remainder	0.35	0.6^c	0.1	0.10	0.9^c	0.10	0.10	0.10	0.05	0.15	Extrusion	Structural members, door and window frames, etc.; nails, rigidized sheets
7075-T6	Remainder	0.5	0.4	2.0	0.3	2.9	0.35	6.10^c	0.2	0.15		Sheet, plate, extrusion, bar, bar, rod, wire	Structural members

[a]Not original alloy, as second digit is not 0 but 1, which shows modifications, and last two digits (00) indicate 99.00% pure aluminum.
[b]Not original alloy, as second digit is 2, which shows modifications, and last two digits (35) indicate 99.35% pure aluminum.
[c]Denotes major alloying elements keyed to first digit.

Table A28 Temper designations and characteristics of aluminum alloys

Temper	Characteristics
F	As fabricated (Products produced from ingots without any subsequent controlled amount of cold work. Different lots of the same material show reasonably uniform properties, but these properties are not guaranteed and may vary considerably.)
O	Fully annealed and recrystallized (wrought products only)
H	Strain hardened
H1[a]	Plus one or more digits, strain hardened only
H2[a]	Plus one or more digits, strain hardened and then partially annealed
H3[a]	Plus one or more digits, strain hardened and then stabilized
W	Solution heat-treated, unstable temper
T[b]	Treated to produce stable tempers other than F, O, or H
T2	Annealed (cast products only)
T3	Solution heat-treated, then cold-worked
T4	Solution heat-treated and naturally aged to a substantially stable solution
T5	Artificially aged only (precipitation heat-treated)
T6	Solution heat-treated, then artificially aged
T7	Solution heat-treated, then stabilized
T8	Solution heat-treated, cold-worked, then artificially aged
T9	Solution heat-treated, artificially aged, then cold-worked
T10	Artificially aged, then cold-worked

[a]Second digit indicates the final degree of strain hardening as follows: H X 2 = $\frac{1}{4}$ hard, H X 4 = $\frac{1}{2}$ hard, H X 6 = $\frac{3}{4}$ hard, H X 8 = fully hard.
[b]The T designation is always followed by one or more digits. The first digit indicates variation in basic treatment as described. Deliberate variations to produce other characteristics are indicated by adding one or more digits to the basic designation.

In non-heat-treatable alloys, the addition of alloying elements and work hardening (cold work) are responsible for improved mechanical properties, whereas in heat-treatable alloys the addition of alloying elements, subsequent heat treatment, and possibly cold working may be combined to achieve the desired properties for a given application. If cold working is the final mill processing of a heat-treated alloy, maximum properties are reached.

A three-digit H temper designation has been assigned for wrought products for all alloys and for alloys containing over a nominal 4% magnesium as follows:

For All Alloys

H111 Products strain-hardened less than the amount required for a controlled H11 temper
H112 Products acquiring some temper from shaping processes; not having special control over the amount of strain-hardening or thermal treatment, but subject to mechanical property limits

For Alloys Containing over 4% Magnesium

H311 Products strain-hardened less than the amount required for a controlled H31 temper
H321 Products strain-hardened less than the amount required for a controlled H32 temper
H32 and H345 Products specially fabricated to have acceptable resistance to stress cracking

There is also a three-digit H temper designation for patterned or embossed sheet (*see* Table *A29*).

Table A29 Three-digit "H" temper designations for patterned or embossed sheet

Alloy from which fabricated	Patterned or embossed sheet
O	H114
H11, H21, and H31	H124, H224, and H324
H12, H22, and H32	H134, H234, and H334
H13, H23, and H33	H144, H244, and H344
H14, H24, and H34	H154, H254, and H354
H15, H25, and H35	H164, H264, and H364
H16, H26, and H36	H174, H274, and H374
H17, H27, and H37	H184, H284, and H384
H18, H28, and H38	H194, H294, and H394
H19, H29, and H39	H195, H295, and H395

These specific additional digits have been assigned for stress-relieved tempers of wrought products:

H51 Stress-relieved by stretching
T510 No further straightening after stretching
T511 Minor straightening after stretching to comply with standard tolerances

T52 Stress-relieved by compressing

T54 Stress-relieved by combined stretching and compression

The following temper designations have been assigned for wrought products heat-treated from 0 or F temper to demonstrate response to heat treatment:

T42 Solution heat-treated from 0 or F temper to demonstrate response to heat treatment and naturally aged

T62 Solution heat-treated from 0 or F temper to demonstrate response to heat treatment and artificially aged

Controlling Castings. Castings vary greatly in their physical properties, depending not only on the composition of the alloy but also on the casting condition. Thin sections generally show greater strength than thick ones. Heat-treated alloys may possess greater strength and ductility than non-heat-treated alloys. Castings in permanent (metal) molds are stronger by 5000 to 10,000 $lbf/in.^2$ (34.48 to 68.95 MN/m^2) than similar sand castings. Prolonged heat treatment at a temperature approaching that of solution, followed by quenching in hot or cold water or oil and then by natural aging (at room temperature) or artificial aging (at the boiling point of water or higher), these are the important steps in the preparation of castings that control their final characteristics.

Controlling Wrought Aluminum. In the production of wrought aluminum, not only control of the alloy constituents but also rather elaborate mechanical arrangements are required to produce ingots satisfactory for working. Annealing of the cold-worked material produces maximum ductility but minimum strength. Cold working (rolling, forging, and drawing through dies at ordinary temperatures) hardens the material, increasing both tensile and yield strength.

Cladding. "Alclad," a term applied to certain aluminum products, refers to the protective coating (cladding) applied, primarily for corrosion resistance, to thin sheets of an alloy whose corrosion resistance has been decreased by the constituents added to give strength and other desirable characteristics. A clad surface of pure aluminum on both sides, totaling about 10% of the sheet thickness, retains the corrosion resistance of aluminum and utilizes the high strength of the core. Cladding also improves the appearance of the alloy. This thin integral cladding usually consists of pure aluminum, magnesium silicide, or zinc alloys with or without manganese. Specific examples are heat-treatable alloys of the copper (2xxx) and zinc (7xxx) series clad with high-purity aluminum, low magnesium-silicon alloy, or a 1% zinc alloy. Alloys 3003, 3004, 6061, and 7075 are generally used as core material for cladding.

Effect of Alloying Elements. The effects that the various alloying elements, as well as the impurities, introduce can be generally stated.

Aluminum-Copper-Magnesium-Manganese Group (2xxx). Copper was one of the first metals to be alloyed with aluminum (*see* Duralumin). It is still the principal alloying substance of a series of important aluminum alloys. A copper content of 4.0 to 5.5% imparts the strength and hardness that characterize alloys in this group (for example, 2014, 2017, 2024, and 2025). However, as the strength and hardness of an alloy increase, its ductility and workability decrease. Any significant amount of copper as an alloying element reduces the natural corrosion resistance of aluminum. Consequently, these wrought copper-aluminum alloys are frequently used in the form of clad sheet or plate. Alloys in this group are classified as heat-treatable.

Aluminum-Manganese Group (3xxx). Manganese increases the strength to a moderate degree with only slight reduction in ductility. Alloys in this group are general-purpose alloys for applications where moderate strength is acceptable. Workability, corrosion resistance, and brightness are only slightly less than in alloy 1100. The best known alloy in this group is commercial 3003, which contains 1.25% manganese. This alloy and its successor, 5005, anodize in both plain and colored finishes very satisfactorily. Alloys in this group are common, or non-heat-treatable.

Aluminum-Silicon Group (4xxx). These alloys have the best corrosion resistance, particularly to marine environments. They have very good resistance to weathering and chemical attack and can be used even in aggressive locations.

Aluminum-Magnesium Group (5xxx). Magnesium in tenths of 1% increases the tensile strength of aluminum alloys. Alloys containing 1% or more of this element work-harden much faster during forming than does either 1100 or 3003. Higher magnesium content alloys such as 5052 at 2.5% have excellent resistance to fatigue stresses. The magnesium contents of 5050, 5052, 5056, etc., have a major influence on their excellent corrosion resistance, especially in marine environments. The high-strength weldable common alloys depend on magnesium for good weldability characteristics and good corrosion resistance. Alloys in this group are common, or non-heat-treatable.

Aluminum-Magnesium-Silicon Group (6xxx). This group of heat-treatable alloys containing magnesium silicide as a hardening constituent is finding increasing applications. Of slightly lower strength than the aluminim-copper-magnesium-manganese (2xxx) group, these alloys are more readily formed and fabricated and have a higher corrosion resistance. A characteristic of the group is a relatively stable T4 temper and good formability in this temper. Formed parts may be given a low-temperature aging treatment to develop T6 temper without distortion. Alloy 6061, an example of this group, also has excellent fusion weldability properties.

Aluminum-Magnesium-Zinc Group (7xxx). Zinc improves the machining characteristics of aluminum alloys. High mechanical properties, such as endurance strength and yield strength, are characteristic of this group, which is in the heat-treatable classification. These alloys are the strongest yet developed; they are more difficult to work and more expensive. The principal alloy in this category is 7075.

Nickel: The addition of nickel contributes increased hardness to aluminum alloys at elevated temperatures. It is used mainly for alloys suitable for permanent mold castings.

Other Metals. Bismuth, cadmium, lead, and tin are all usually considered impurities. Small amounts of one or more of these metals are sometimes added to aluminum alloy 2011 to improve machinability. Appreciable amounts result in decreased mechanical properties and a greater susceptibility to corrosive attack.

Applications

Joining. Joining is an important detail in any construction use of aluminum and its alloys. Riveting and oxy-gas (acetylene) welding with a suitable flux are the older methods, and inert-gas shielded arc welding is a newer development which has become common practice. Electric spot and seam welding are especially useful resistance welding processes for alclad sheet. Brazing is a common practice for joining light sections such as window sections. When soldering is used, the solder must be of the tin-zinc type and the soldered joints must be protected against moisture. The advantages and limitations of each method in relation to the alloy or alloys to be joined should be fully investigated (*see also* Adhesives; Brazing; Soldering; Welding).

Oxy-gas or acetylene welding depends on an active flux whose residue is a potential source of corrosion that should be thoroughly removed. This process is inexpensive and readily available. It is used most widely to join thin gauges of aluminum and to repair welding when other processes are not available.

Brazing also uses an active flux that must be completely removed. It produces a joint with good appearance and with a strength comparable to that of an oxy-gas welded joint, but with much less tendency toward distortion. It is particularly advantageous for joining thin to thicker sections in sheet, extrusions, tubing, and complicated assemblies with relatively inaccessible parts.

Resistance welding methods for aluminum include spot, seam, and flash welding. Resistance welding is applicable to all aluminum alloys but particularly to the heat-treatable alloys, which are difficult to weld by the fusion processes. Spot and seam welding are in general use for the effective joining of aluminum where the thickness of the thinnest piece is not over $\frac{1}{8}$ in. (3.175 mm). Flash welding can be used to join aluminum and its alloys in the form of plate, tubing, extrusions, wire, rod, and bar having a cross-sectional area up to approximately 2 in.^{2} (12.9 cm^2). It can also be used to join aluminum to copper and other metals.

Inert-gas shielded arc processes require no flux and produce joints that have a clean, bright appearance and require little if any finishing. The argon-tungsten arc welding process is used to produce X-ray quality welds in materials $\frac{1}{64}$ to $\frac{3}{8}$ in. (0.5 to 9.5 mm) thick. It is suitable for manual welding in all positions. Thicker material is better welded by the inert metal arc process, using a consumable aluminum electrode (filler wire). This partially automatic process is used extensively on manual welding applications involving thick materials, although its usual range includes material as thin as 0.032 in. (0.8 mm). Inert metal arc welding can be used only when the addition of filler is permissible.

Soldering is used for aluminum in a wide range of industrial applications in which dry conditions prevail or the joint can be protected adequately. Soldered aluminum joints are susceptible to galvanic corrosion in outdoor and corrosive environments because solders in current use are electropositive with respect to aluminum. Also, in flow soldering the flux is a potential source of corrosion and its removal is recommended.

Adhesive bonding is another method for joining aluminum to aluminum or to other metals and materials. Adhesive bonding is used extensively in aircraft fabrication, construction of aluminum-faced building panels, and lightweight honeycomb structural panels. The design of joints for adhesive bonding requires care to avoid localized stresses and to keep joint deflection to a minimum. Adhesive bonding is usually confined to the thinner gauges of aluminum, since other methods of joining are preferred for the thicker gauges. Many types of adhesive are available, but those based on phenol-formaldehyde and epoxy resins appear to be the most suitable ones at present.

Aluminum Alloy Nomenclature in Foreign Countries

The foreign producers of wrought aluminum alloys employ different nomenclatures from those used in this country. Table *A30* gives the designations for the United States, Canada, France, Germany, Great Britain, Italy, Spain, Switzerland, and the International Organization for Standardization.

Table A30 United States and foreign designations for wrought aluminum alloys

U.S.A. AA[a]	Canada CSA[a]	France NF[a]	Germany DIN[a]	Great Britain BS	Great Britain DTD	Italy UNI	Spain UNE	Switzerland USM	International Organization for Standardization ISO
EC		A5/L	E-Al99.5 3.0257	1E				Al99.5E	
1100	990C	A45							Al99.0Cu
2011	CB60		AlCuBiPb 3.1655						
2014	CS41N	A-U4SG	AlCuSiMn 3.1255			P-AlCu4.4-SiMnMg	L-313		AlCu4SiMg
Alclad 2014	CS41N Alclad					P-AlCu4.4-SiMnMg placc.			
2017	CM41	A-U4G	AlCuMg1 3.1325	H14 5L.36 L.87	150A	P-AlCu4-MgMn		Al3.5Cu-0.5Mg	AlCu4MgSi
2018	CN42								
2024	CG42	A-U4G1	AlCuMg2 3.1355	L.97 L.98	5090	P-AlCu4.5-MgMn	L-314	Al4Cu1.2-Mg	AlCu4Mg1
Alclad 2024	CG42 Alclad				5100	P-AlCu4.5-MgMn placc.			
2025	CS41P								
2117	CG30	A-U2G	AlCuMg0.5 3.1305	L.86		P-AlCu2.5-MgSi			AlCu2Mg
2218		A-U4N		6L.25			L-315	Al-Cu-Ni	
2618		A-U2GN			717 724 731A 745 5014 5084				
3003	MC10	A-M1							AlMn1Cu
3004		A-M1G							
4032	SG121	A-S12UN			324A	P-AlSi12-MgCuNi			
4043	S5		AlSi5 3.2245	N21					

Table A30 United States and foreign designations for wrought aluminum alloys (*continued*)

U.S.A. AA[a]	Canada CSA[a]	France NF[a]	Germany DIN[a]	Great Britain BS	Great Britain DTD	Italy UNI	Spain UNE	Switzerland USM	International organization for standardization ISO
5005		A-G0.6							AlMg1
5050		A-G1		3L.44		P-AlMg1.5		Al-1.5Mg	AlMg1.5
5052	GR20			2L.55		P-AlMg2.5			AlMg2.5
				2L.56					
				L.80					
				L.81					
5056	GM50R			N.6					AlMg5
				21.58					
5083	GM41		AlMg4.5Mn	N8					AlMg4.5Mn
			3.3547						
5086		A-G4MC							AlMg4
5154									AlMg3.5
5356	GM50P								
5454	GM31N								AlMg3Mn
5657						P-AlMg0.9			
6053	GS11P								
6061	GS11N			H20					AlMg1SiCu
6063	GS10	A-GS	AlMgSi0.5	H9	372B	P-AlSi0.4Mg			AlMgSi
			3.3206						
6101		A-GS/L	E-AlMgSi0.5			P-AlSi0.5Mg		Al-Mg-Si	
			3.3207	91E					
6151	SG11P								
7075	ZG62	A-Z5GU	AlZnMg-Cu1.5	L.95			L-371	Al-Zn-Mg-Cu	AlZn6MgCu
			3.4365	L.96					
Alclad 7075	ZG62 Alclad							Al-Zn-Mg-Cu pl.	

[a]AA: Aluminum Association; CSA: Canadian Standards Association; NF: Normes Françaises; DIN: Deutsche Industrie-Norm; BS: British Standard; DTD: Directorate of Technical Development; UNI: Unificazione Nazionale Italiana; UNE: Una Norma Espanola; USM: Verein Schweizerischer Maschinenindustrieller.

ALUMINUM, CORRUGATED

Physical and Chemical Properties

Corrugated aluminum is rigidized sheet fabricated of special aluminum alloys specifically developed for this purpose. It usually consists of an aluminum alloy core of one type clad with another, highly corrosion-resistant aluminum alloy, but it may also be all of one alloy.

Corrugated aluminum has a high insulating value and is about one-sixth lighter than similar materials. It is silvery in color, strong, and does not stain adjoining materials. It has another advantage in that it can be lapped, thus eliminating many joint problems and at the same time forming a weather seal.

Corrosion Resistance. Corrugated aluminum is resistant to the action of industrial gases and fumes and to salt air. It reacts, however, with dissimilar metals, green or damp wood, certain wood preservatives, lime mortar, concrete, and other masonry materials and must be insulated from contact with these materials.

Commercial Forms. Corrugated aluminum is available in plain mill and embossed finishes. Both highly reflective and low-gloss surfaces, as well as a limited number of colors in either baked enamel or an organic coating, are available.

Types and Uses

Corrugated aluminum is available in two generally used types, commercial-industrial and standard, and in two special types, curved and perforated.

Industrial Corrugated Aluminum. This type is used primarily in heavy construction for roofing and siding and some flashing (*see* Figure *A8*).

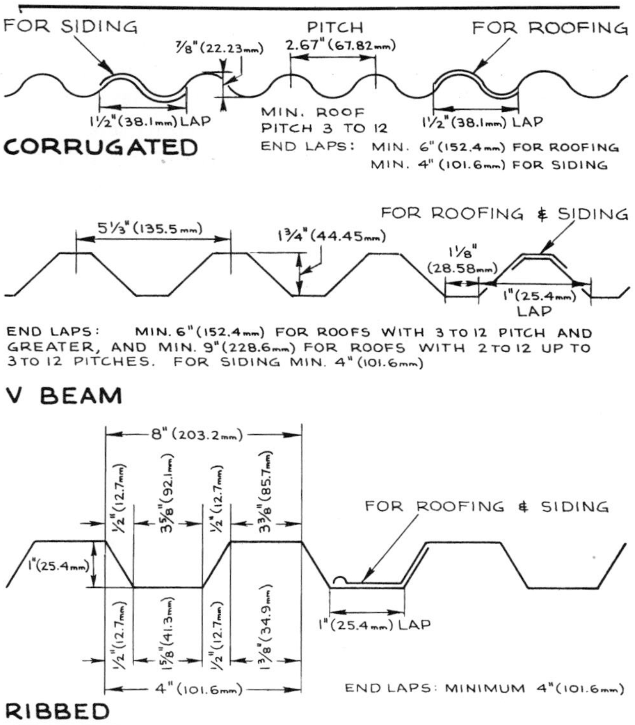

Figure A8 Industrial corrugated aluminum.

Sheet thickness: 0.024, 0.032, and 0.040 in. (0.61, 0.813, and 1.016 mm).

Width: $33\frac{3}{4}$, 35, 43, $47\frac{1}{8}$, and $48\frac{1}{3}$ ft (0.853, 0.888, 1.197, and 2.223 m).

Length: 5 to $14\frac{1}{2}$ ft (1.524 to 4.420 m) in 6-in. (152.4 mm) increments.

Depth of corrugations: $\frac{7}{8}$ in. (22.23 mm).

Commercial-industrial corrugated aluminum also includes a ribbed sheet with the following dimensions:

Sheet thickness: 0.032, 0.040, and 0.050 in. (0.813, 1.016, and 1.270 mm).

Width: $33\frac{1}{2}$ and $41\frac{1}{2}$ in. (0.846 and 1.049 m).

Length: 5 to 30 ft (1.524 to 9.144 m) in 6-in. (152.4-mm) increments.

Both industrial types are available in plain mill or embossed surface finishes, including double embossed, which gives a low-gloss finish, and in natural aluminum and a limited number of colors in either baked enamel or an organic coating (*see* Figure *A9*).

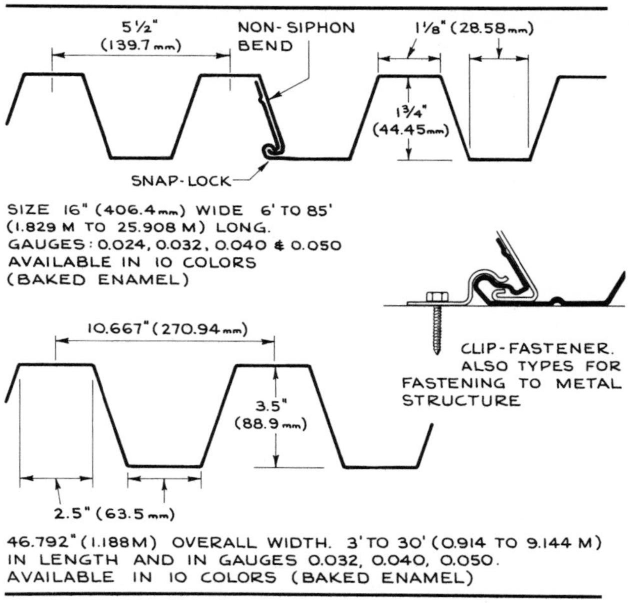

Figure A9 Colored baked enamel corrugated aluminum.

Standard Corrugated Aluminum. This type is mainly used in farm construction work for roofing and siding and some flashing. It is available in plain mill or embossed surface finishes and in a natural aluminum, green, red, gold, and white color (*see* Figure *A10*).

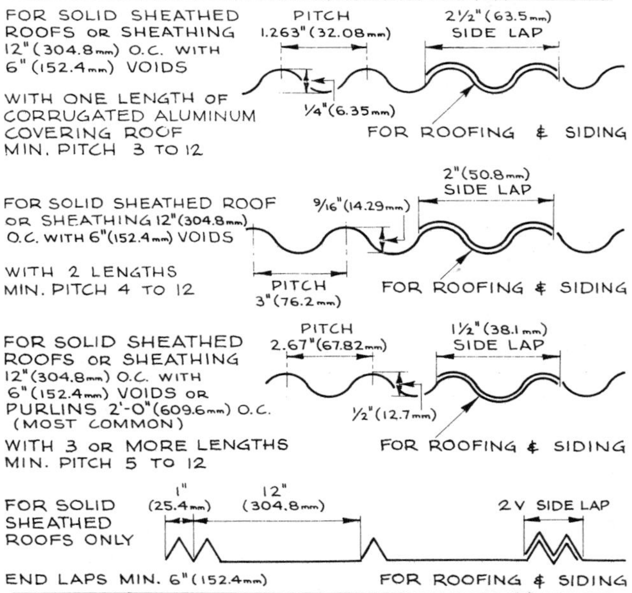

Figure A10 Standard corrugated aluminum.

Sheet thickness: 0.019, 0.024, and 0.027 in. (0.473, 0.610, and 0.686 mm).

Width: 26 and $51\frac{1}{3}$ in. (0.660 and 1.303 m).

Length: 6 to $14\frac{1}{2}$ ft (1.829 to 4.240 m) in 6-in. (152.4-mm) increments.
Depth of corrugations: $\frac{1}{4}$, $\frac{9}{16}$, and $\frac{1}{2}$ in. (6.35, 14.29, and 12.70 mm).

This category also includes a V-crimp type of the same thicknesses, lengths, and widths, but available only in a $\frac{1}{2}$-in. (12.7-mm) depth of corrugations.

Curved Corrugated Aluminum. This type finds use in curved roofing, siding, and flashing applications where the curve would otherwise be perpendicular to the corrugations (*see* Figure *A11*).

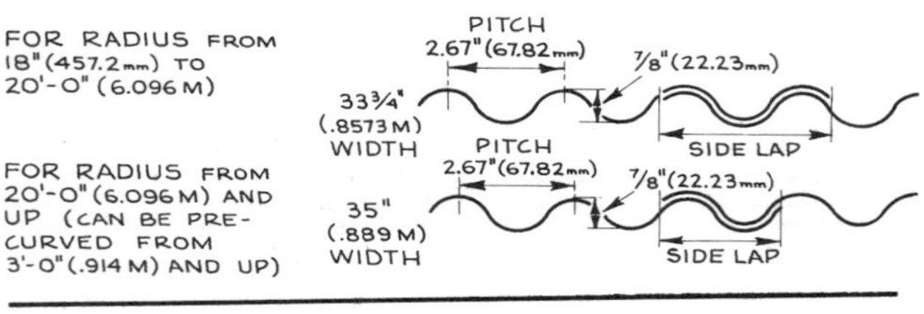

Figure A11 Curved corrugated aluminum.

Sheet thickness: 0.024 and 0.032 in. (0.610 and 0.813 mm).
Width: $33\frac{3}{4}$ and 35 in. (0.853 and 0.888 m).
Length: 5 to $14\frac{1}{2}$ ft (1.524 to 4.42 m) in 6-in. (152.4-mm) increments.
Depth of corrugations: $\frac{7}{8}$ in. (22.23 mm).

Perforated Corrugated Aluminum. This may be corrugated or ribbed and is used mainly for acoustic treatment of interior walls and ceilings.

Sheet thickness: 0.024 in. (0.610 mm).
Width: $33\frac{3}{4}$ in. (0.853 m).
Length: 5 to $14\frac{1}{2}$ ft (1.524 to 4.42 m) in 6-in. (152.4-mm) increments.
Depth of corrugations: $\frac{7}{8}$ in. (22.23 mm).
Open surface amounts to approximately 14% of the total area.

Application

Load Considerations. Maximum roof spacing without support (purlin spacing) for uniform roof load in pounds per square foot (kilograms per square meter) is based on the thickness of the material and the type and temper of the aluminum alloy used by the manufacturer. This same type of variation among manufacturers' products holds true for siding and roofing in relation to wind loads. When designing with corrugated aluminum, one should check the limitations of the material in question, and specifications should be written without an "or equal" clause.

Preventing Corrosion. Galvanic action is possible with metals other than nonmagnetic stainless steel, cadmium, zinc, and nickel bronze. Aluminum can be isolated from such metals with a suitable paint or physically separated by a joint filled with a suitable mastic or plastic membrane.

Chemical action between aluminum and lime mortar, concrete, or other masonry materials can be prevented by coating the aluminum with a heavy-bodied bituminous or other suitable organic coating. Chemical reaction between aluminum and green or constantly damp wood can be prevented by painting the wood with two coats of aluminum paint or by using a suitable membrane. Fastening devices, special nails, and other hardware and accessories are manufactured for use with both standard and commercial-industrial corrugated aluminum (*see* Figure *A12* for typical examples).

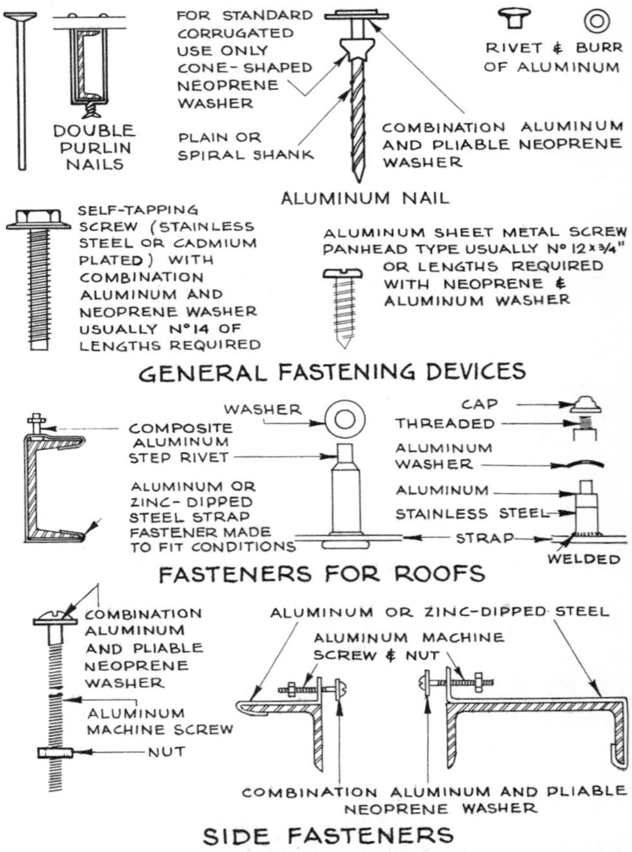

Figure A12 Typical fastening devices for industrial, standard, and colored baked enamel corrugated aluminum used for roofing and siding.

Table A31 Fastener spacing for industrial corrugated aluminum

Type of industrial corrugated aluminum	Supports			Side laps		End laps	
	Roofing	Siding	Ends	Roofing	Siding	Roofing	Siding
	Fastener spacing						
	in. (mm)	in. (mm)	in. (mm)	in. (mm)	in. (mm)	in. (mm)	in. (mm)
Corrugated	$10\frac{2}{3}$ (270.76)	$10\frac{2}{3}$ (270.76)	$10\frac{2}{3}$ (270.76)	12 (304.8)	12 (304.8)	$10\frac{2}{3}$ (270.76)	$10\frac{2}{3}$ (270.76)
V-beam ribbed	$10\frac{2}{3}$ (270.76)	$10\frac{2}{3}$ (270.76)	$5\frac{1}{3}$ (135.28)	12 (304.8)	12 (304.8)	$5\frac{1}{3}$ (135.28)	$5\frac{1}{3}$ (135.28)
		12 (304.80)	8 (203.20)		12 (304.8)		8 (203.20)
Curved	$10\frac{2}{3}$[a] (270.76)	$10\frac{2}{3}$[a] (270.76)	$10\frac{2}{3}$[a] (270.76)	12 (304.8)		$10\frac{2}{3}$ (270.76)	

[a]For extreme wind conditions, 8-in. spacing is recommended.

Note: Fasteners can be installed in either the top or bottom of the corrugations, usually at the bottom for roofing and at the top for siding.

Table A32 Fastener spacing for standard corrugated aluminum, based on solid or spaced sheathing

Type of standard corrugated aluminum	Supports			Side laps		End laps		Supporting nails
	Roofing	Siding	Ends	Roofing	Siding	Roofing	Siding	
	Fastener spacing							
	in. (mm)	in. (mm)	in. (mm)	in. (mm)	in. (mm)	in. (mm)	in. (mm)	
$\frac{1}{2}$-in. (12.7-mm) thick	$10\frac{2}{3}$ (270.76)	$10\frac{2}{3}$ (270.76)	$4\frac{7}{8}$ (123.83)	12 (304.8)	12 (304.8)	$5\frac{1}{3}$ (135.28)	$5\frac{1}{3}$ (135.28)	Roofing nails with washer
$\frac{9}{16}$-in. (14.29-mm) thick	12 (304.8)	12 (304.8)	6 (152.4)	12 (304.8)	12 (304.8)	12 (304.8)	12 (304.8)	Roofing nails with washer
V-crimp	12 (304.8)	12 (304.8)	6 (152.4)	12 (304.8)	12 (304.8)	12 (304.8)	12 (304.8)	Roofing nails with washer

For fastening commercial-industrial corrugated aluminum side laps and end laps and flashing to other pieces of aluminum, No. 12 × $\frac{3}{4}$ in. (No. 304.8 × 6.35 mm) pan-head-type sheet metal screws should be used. For siding, the spacing is 12 in. (304.8 mm) o.c. for ribbed and $10\frac{2}{3}$ in. (270.76 mm) o.c. for corrugated; for roofing, $10\frac{2}{3}$ in. (270.76 mm) o.c. under normal conditions and 8 in. (203.2 mm) o.c. for high-stress situations. For roofs the screws should be installed at the high point of corrugations, and for siding they may be installed at either the high or low point. (*See* Table *A31*.)

For fastening standard-type corrugated aluminum, roofing and siding nails made of aluminum should be used. They should be installed at the high point of the corrugations, and a pliable neoprene washer should be placed under the nailhead. (*See* Table *A32*.)

Accessories. Several stock accessories are available for use with corrugated aluminum, including ridge and hip rolls, end wall flashing pieces, valley shapes, and closure strips (*see* Figure *A13*).

Condensed Checklist

1. Roof spans: When designing with corrugated sheet aluminum, all data should be carefully checked with manufacturers' recommendations.
2. Contacting materials: A careful check should be made for the possibility of galvanic or chemical action between contacting materials, and protective measures should be taken.
3. Accessories and fastening devices: All should be of a compatible metal or material, and of the correct type, properly installed.

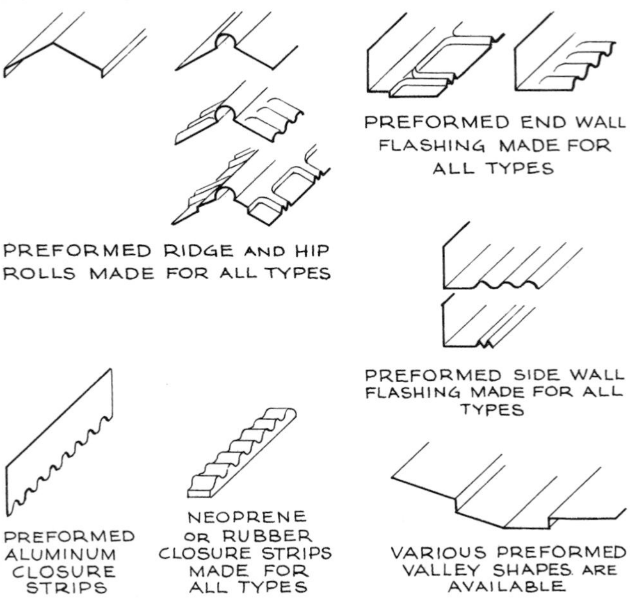

Figure A13 Typical stock accessories for industrial, standard and colored baked enamel corrugated aluminum.

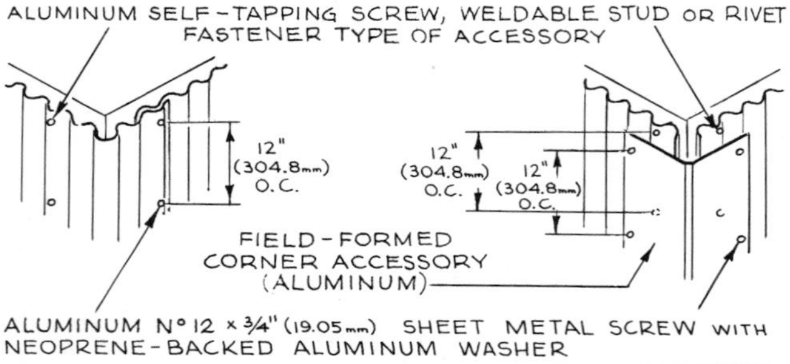

Figure A14 Details of external corners for siding.

4. Treatment of external corners for industrial-type corrugated aluminum siding (*see* Figure *A14*).
5. Method of lapping at sides and ends for both roofing and siding (*see* Figure *A15*).
6. Treatment of roof intersections at ridge and hip (*see* Figure *A16*).
7. Treatment of valleys and gable ends for corrugated aluminum roofs (*see* Figure *A17*).
8. Treatment of corrugated aluminum at sills and at joint with roof (*see* Figure *A18*).
9. Treatment of openings for windows and doors: At head and sill a closure accessory is generally used; at jambs, the siding is flattened and secured as shown in Figure *A19*.

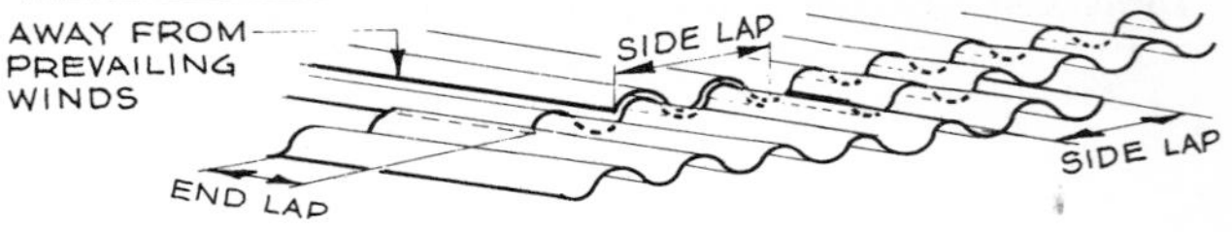

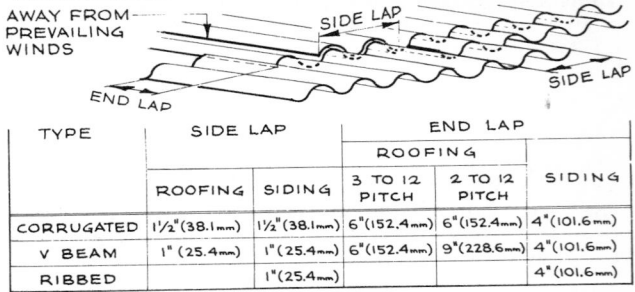

TYPE	SIDE LAP		END LAP		
			ROOFING		
	ROOFING	SIDING	3 TO 12 PITCH	2 TO 12 PITCH	SIDING
CORRUGATED	1½" (38.1mm)	1½" (38.1mm)	6" (152.4mm)	6" (152.4mm)	4" (101.6mm)
V BEAM	1" (25.4mm)	1" (25.4mm)	6" (152.4mm)	9" (228.6mm)	4" (101.6mm)
RIBBED		1" (25.4mm)			4" (101.6mm)

Figure A15 Side lapping and end lapping for roofing and siding industrial-type corrugated aluminum.

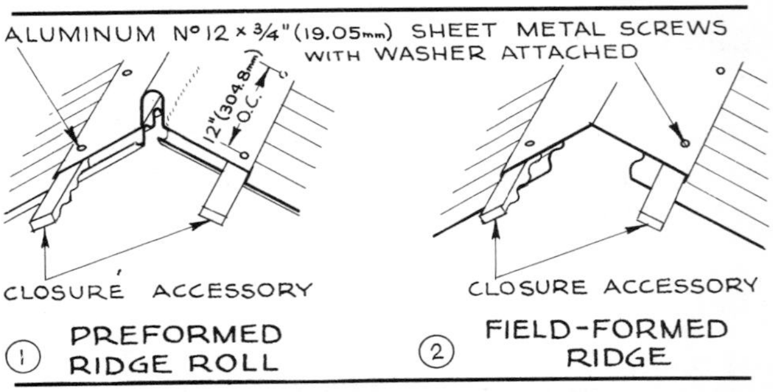

Figure A16 Details of preformed (1) and field-formed (2) roof ridges and hips.

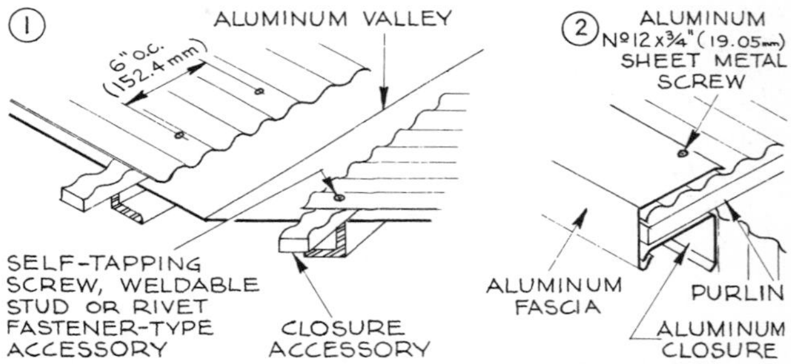

Figure A17 Details of valley (1) and gable end (2) using corrugated aluminum roofing.

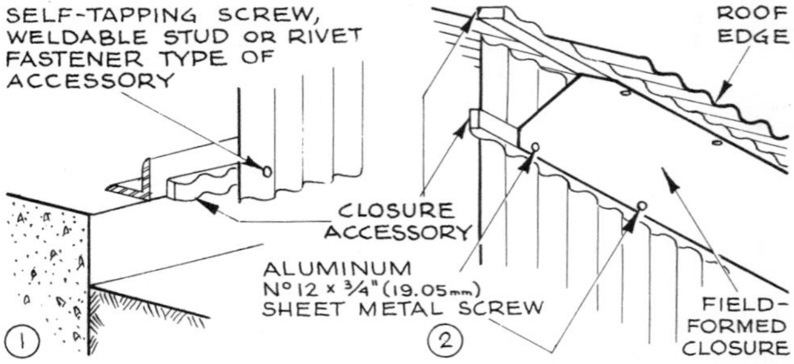

Figure A18 Details of corrugated aluminum siding at sill (1) and at roof edge and roof-siding joint (2).

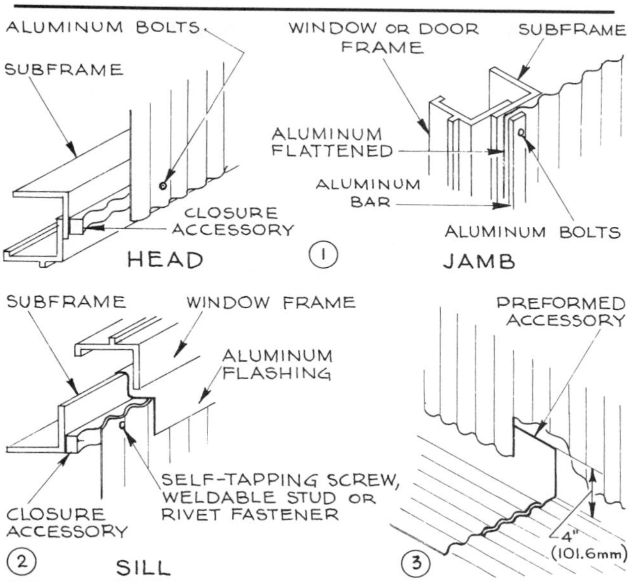

Figure A19 Details of window and door (1, 2) and at roof openings (3).

Conditions Favorable to the Use of Corrugated Aluminum

1. Where maintenance must be kept at a minimum.
2. Where a lightweight material that will reduce roof loading and hasten erection is desired.
3. Where a highly reflective surface is desired.
4. Where the added reflective insulation value of corrugated aluminum is helpful.
5. Where a corrosion-resistant material which will not stain adjoining materials is required.
6. Where a fire-resistant material is desired. Municipal, state, and local codes should be checked, and also the fire rating codes of the Fire Underwriters, insurance companies, labor departments, and the federal government (Army, Navy, etc.).

Conditions Unfavorable to the Use of Corrugated Aluminum

1. Where spans larger than 10 ft (3.048 m) for V-beams and 7 ft (2.134 m) for heavy industrial work are required.
2. Where direct contact with materials causing galvanic action or chemical action is unavoidable.
3. On curved surfaces where the curve is perpendicular to the corrugations unless special curved corrugated aluminum is available.

History and Manufacture

Corrugated aluminum appeared before World War II but was not available on a commercial scale until about 1945. It is fabricated by passing flat aluminum sheet between two corrugated rollers, the rollers resembling in cross section large gears, the teeth of which form the corrugations. The dimensions and shape of the corrugations are determined by those of the roller used and the temper of the metal. The shapes commonly used are corrugated, V-beam, and ribbed. V-beam and ribbed shapes are formed on brake press or roll form equipment. The corrugated (sine wave) shape is formed on drum corrugaters or roll form equipment.

ALUMINUM DOORS

Physical and Chemical Properties

Aluminum doors are controlled by standards set by the Architectural Aluminum Manufacturers Association. Doors are divided into two qualities, one for commercial and one for residential use. These standards control the types of alloys to be used, methods of joining, resistance to wind, types of cladding alloys, fasteners, hardware, and weatherstripping for exterior doors. The glass to be installed in any aluminum door should be tempered plate or float or sheet glass, and the thickness is controlled by federal specifications.

Aluminum doors are generally fabricated from extrusions, rolled shapes, sheet and strip made of aluminum alloys that meet strict quality standards such as Nos. 6063-T5, 6063-T6, and 6061-T6. Doors made of aluminum are lightweight, corrosion resistant, rotproof, and termiteproof. Also, they will not swell, split, or warp.

Since doors receive hard usage and the bare surface of aluminum is relatively soft, any of the finishes that improve the resistance of the surface to staining and abrasion is desirable. These would include electrochemical, mechanical, baked enamel, vinyl and organic coatings, and laminating. Anodizing is the finish most commonly used. (*See* Aluminum Finishes.)

Preventing Galvanic Action. As the possibility of galvanic action exists, aluminum doors and all related metal accessories and attachments subject to such interaction must be insulated from all common metals other than nonmagnetic stainless steel, cadmium, zinc, or other noncorrosive materials compatible with aluminum. The same precaution applies to hardware, weatherstripping, screens, thresholds, mullions, subframes, flashing, and trim.

Strength Considerations. The relation of alloy characteristics to extrusion shapes is important, as strength in aluminum alloys is not in direct relation to size or weight of the metal but is the result of alloy composition and methods of metal working.

Fire-Resistant Doors. Aluminum has a relatively low melting point, and for this reason labeled fire doors formerly could not be made of aluminum. However, with the development of fire-resistant inert materials, labeled aluminum fire doors are available with specific hour ratings.

Types and Uses

The types of doors available in aluminum are summarized under the section Doors. Specifically, they include entrance, screen, storm, labeled fire doors, garage, and interior doors of many types, including roll-up, in a wide range of dimensions. Table *A33* gives the more common sizes.

Special types like hangar, telescoping, bi-parting, and revolving doors are also available, but they are usually designed to meet specific job conditions such as wind pressure, speed of operation, and weight.

Store Fronts. Store fronts utilizing aluminum doors in design combinations with windows and other types of large glass display areas are widely used. Innumerable aluminum accessories are available for such installations in many types and shapes for bracing, strengthening and covering of joints, mullions, frames, etc. (*See* Figures *A20* and *A21*.)

A check should always be made of recent manufacturers' data for possible new types of store front shapes, since these are constantly being developed in the aluminum industry. A check should also be made of length limitations, methods of joining, finishes, and type of alloy from which the shapes are fabricated (generally from Nos. 6063-T5 and 6063-T6).

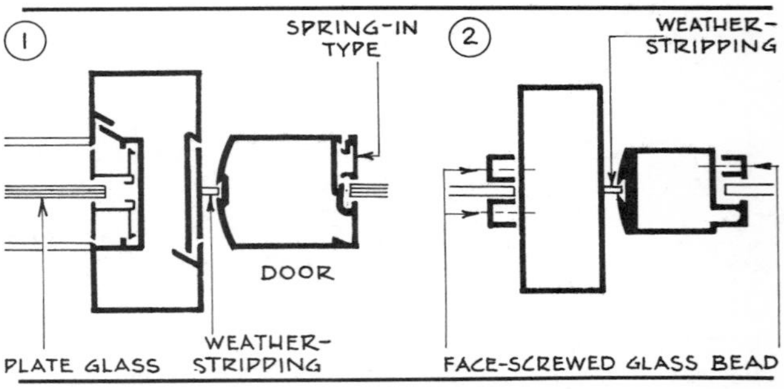

Figure A20 **Typical concealed (1) and exposed (2) anchoring devices for aluminum doors and fixed glass.**

Figure A21 **Typical concealed (1) and exposed (2) anchoring devices for fixed glass.**

Table A33 **Common sizes of aluminum doors**

	Entrance doors		Garage[b] doors		Roll-up[b] doors, mesh or solid	Screen doors	Storm doors	Interior doors	Sliding doors	Labeled fire doors
	Single	Double	Resi-dential	Commer-cial						
Size[a]	ft (m)	ft (m)	ft (m)	ft (m)	ft (m)	ft (m)	ft (m)	ft (m)	ft (m)	ft (m)
Width	2.5–3.5 (0.762–1.067)	5–7 (1.524–2.134)	8 (2.438) 9 (2.743) 15 (4.572) 16 (4.877)	8–20 (2.438–6.096)	Up to 25 (Up to 7.62)	2.67–3.33 (0.813–1.016)	2.67–3.33 (0.813–1.016)	2.5–3.5 (0.762–1.067)	6–20 (1.829–6.096	2.5–3.5 (0.762–1.067)
Height	7 (2.134)	7 (2.134)	6 (1.829) 7 (2.134)	8.5–14.5 (2.591–4.42)	Up to 25 (Up to 7.62)	6.67 (2.032) 7.00 (2.134)	6.67 (2.032) 7.00 (2.134)	6.67 (2.032) 7.00 (2.134)	6.75 (2.057) 8.00 (2.438)	6.67–7.00 (2.032–2.134)

[a] Sizes refer to door opening sizes, i.e., door plus clearance.
[b] Can be motor-operated by switch, radio, electric eye, or other means.

The current practice is to consider the entrance a package unit, consisting of door(s), frame, transom, side-lights, and the attached hardware, even to the extent of including an electric operating mechanism if desired. This trend, with the encouragement of both the construction industry and the manufacturer, goes further to integrate side-lights, show windows, and even the entire curtain wall system, all of which are related through characteristics of design, function, and installation. (*See* Figures *A22* and *A23*.)

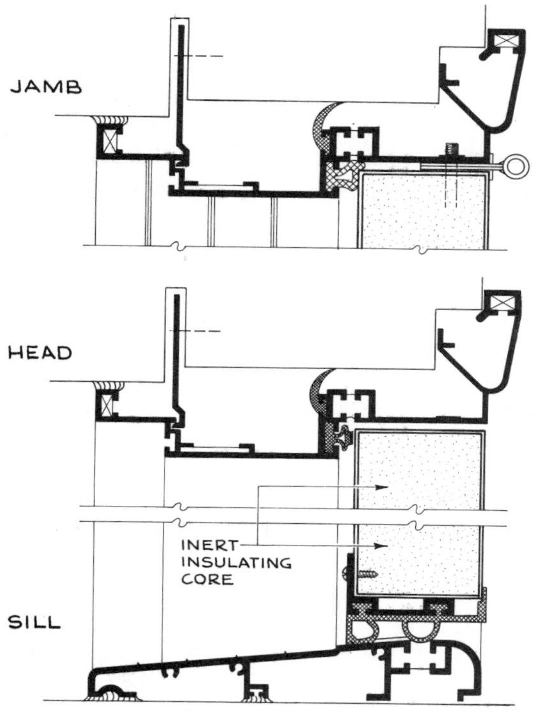

Figure A22 Typical aluminum entrance door and frame.

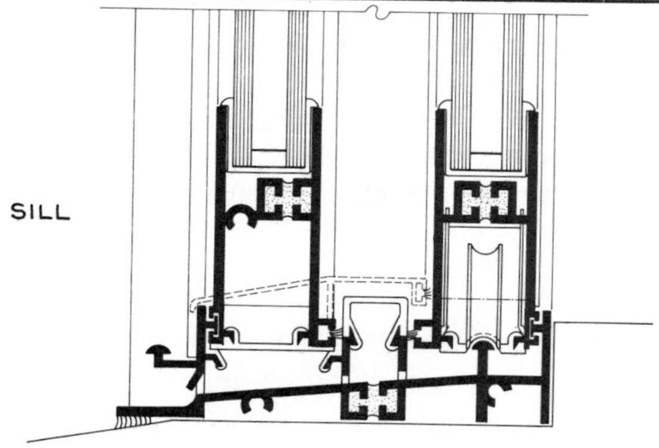

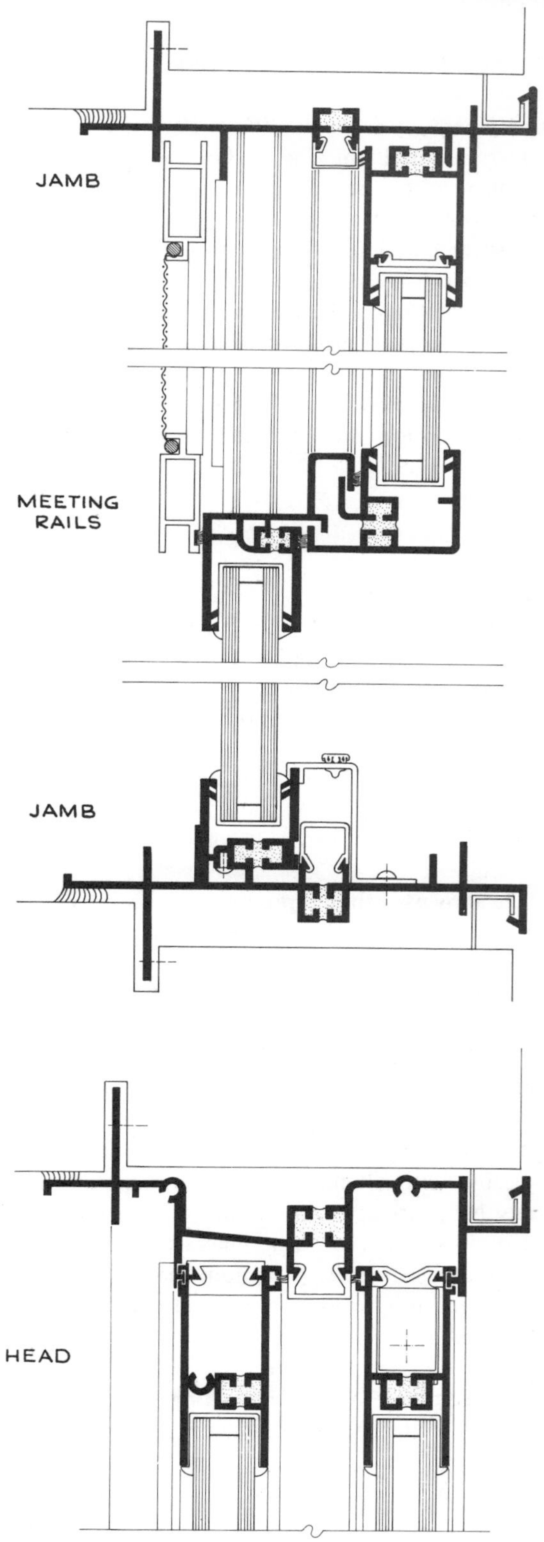

Figure A23 (*above and right*) Typical aluminum sliding glass doors.

Application

Hardware. Hardware made of aluminum is available for use with all types of aluminum doors and may be purchased separately from the doors. The hardware is fabricated from aluminum alloys. Hardware made of other metal also can be used for aluminum doors if the metal is compatible with aluminum or is insulated from it to avoid galvanic action. Effective insulation is provided by organic coatings, by plastic coatings, by laminating, or by metal plating with aluminum, cadmium, zinc, or other noncorrosive materials that are compatible with aluminum.

Fastening accessories such as screws, nuts, and bolts are usually fabricated from aluminum alloys.

Condensed Checklist

1. Fire rating: The doors selected should comply with local, municipal, and state fire rating codes and also with those of the Fire Underwriters, insurance companies, and federal government (Army, Navy, etc.) Aluminum as a metal has too low a melting point to meet codes written for steel doors. Therefore various core materials with aluminum skins must be used to meet code requirements. The glazed narrow-stile aluminum door generally used today carries no rating.
2. The alloy used should provide the necessary strength and corrosion resistance for the extrusions selected and for the conditions of use.
3. The finish selected should be suitable for the conditions of use.
4. The door selected should be suited to the type of glazing (single or double) desired.
5. To prevent galvanic action, all metals with which the door is likely to come in contact should be compatible with aluminum. If the contacting metal is anything other than nonmagnetic stainless steel, cadmium, zinc, or other metals compatible with aluminum, then the aluminum should be insulated. This precaution applies to all components in relation to installation.
6. Protective coatings: The doors should be delivered with a suitable protective coating against lime, mortar, and other materials which may drop, spatter, or otherwise come in contact with the doors and frames during construction.
7. Caulking and glazing compounds should be of the correct type for aluminum. Where dry-set glazing is used, the lifetime of the glazing materials should be established for the conditions of climate, usage, and maintenance practices.
8. Moving hardware parts or door members made of aluminum should be isolated from direct contact or friction with each other; otherwise the protective coating (oxide) will be constantly rubbed off.
9. When support is required for structural loads, steel members must be used within the aluminum framework as it is difficult to achieve support with aluminum framing members.
10. When gaskets are used for installing glass, it is always advisable to check that the color of these gaskets matches or blends with the color finish of the aluminum door(s) and frame, trim, etc.
11. The glass for installation should always be tempered plate, float, or sheet glass (known as safety glass). If installed in labeled fire doors, wire glass must be used and must meet maximum size limits as set by the codes that control the area where the building is to be constructed.

Conditions Favorable to the Use of Aluminum Doors

1. Where a lightweight, attractive, corrosion-resistant material is required for doors, particularly at entrances.
2. Where maintenance must be kept at a minimum.
3. Where initial painting is not desired.

Conditions Unfavorable to the Use of Aluminum Doors

1. Where fire resistance or a fire rating is required which cannot be met with the available labeled aluminum fire doors.
2. Where a color other than the natural silvery aluminum color is required, and none of the available aluminum finishes or other type of finish is suitable or desirable.

History and Manufacture

Aluminum doors were available before World War II, but only since 1945 has their use become widespread, especially for store fronts (in conjunction with large glazed areas for display) and in office, commercial, and industrial buildings. Almost from the beginning, the package concept—that is, door, frame, and integral hardware—prevailed among the manufacturers of aluminum doors. Aluminum doors are usually made of glass framed in extrusions. The choice of alloy, the fabrication of suitable extrusions, and the assembling of parts are described fully under Aluminum Windows. Doors made of aluminum sheet or strip applied to frames constructed from rolled shapes or extrusions have been largely discontinued.

ALUMINUM FINISHES

Physical and Chemical Properties

Although aluminum has inherently good corrosion resistance, aluminum and aluminum alloy products can be and often are given a wide range of finishes for decorative or protective purposes or both. There are basically six kinds: (1) mechanical finishes, (2) chemical treatments, (3) electrochemical finishes, including anodizing, (4) electroplating, (5) porcelain or vitreous enamels, and (6) paints, or organic coatings, including lacquers and enamels.

See also Metals: Aluminizing; Metals: Chemical Finishes; Metals: Coatings; Metals: Mechanical Finishes; Metals: Sprayed Metal Coatings.

Chemical Finishes. These include (1) cleaning of the surface without affecting the metal, (2) a clean, matte-textured (etched) surface, (3) a smooth, bright finish, and (4) a chemical conversion of the surface to receive applied coatings.

Nonetch cleaning consists of degreasing by vapors of chlorinated solvents or cleaning with hydrocarbon solvents. Chemical cleaning by spraying with or dipping into inhibited chemicals is another method.

Matte finishes are described as fine, medium, and coarse. Fine matte is obtained by using a mild alkali solution, either trisodium phosphate or sodium carbonate, and a solution of hot chromic sulfuric acid or ammonium bifluoride. Medium matte is obtained by using caustic soda (sodium hydroxide), which gives a silvery white finish. Coarse matte is obtained by using a solution containing sodium fluoride, sodium hydroxide, or proprietary compounds.

Bright finishes are used for reflectors, for mirrors, and in light fixtures. A highly specular finish is obtained by preliminary buffing, followed by chemical brightening or electropolishing. Diffuse bright is obtained by adding caustic etch, followed by chemical brightening.

Conversion coatings utilize acid-chromate-fluoride-phosphate, generally proprietary solutions, to produce a clear or greenish color which is suitable as a final finish. Another type of acid-chromate-fluoride solution, also usually proprietary, produces a clear or yellowish color. Alkaline chromate solutions, also generally proprietary, produce a gray color which is suitable as a final finish.

Electrochemical (or Electrolytic) Finishes. Commonly referred to as anodized finishes, these finishes are based on the specific ability of aluminum to develop a protective coating of oxide on its surface. Generally the anodizing solutions are oxalic, sulfuric, and chromic acid. The coating formed may be transparent or opaque. It is hard, yet when colored finishes are desired, it is porous enough to absorb dyes until the final hot water treatment, which seals the surface by increasing its volume. The final treatment may utilize, instead of hot water, sodium chromate or dichromate, sodium silicate, and nickel or cobalt acetate. The resulting anodized coating can be and usually is given a further protective sealing coating by one of two means: (1) saturating the porous protective oxide film with wax or grease; or (2) covering it with a clear coating, usually a transparent, colorless synthetic resin lacquer. The problems of quality anodizing are discussed under "Application."

Electroplating. Aluminum can be covered with a protective or decorative film of another metal, usually by electrodeposition (*see* Table *A34*).

Table A34 Major uses of electroplating metals for aluminum

Type of plating on aluminum	Pretreatment needed	Characteristics and general uses
Copper	Zincating	Most common preplate; also used in electrical field and where soft soldering of joints is required
Chromium	Zincating; copper, nickel, or brass	Extra abrasion resistance and highly polished surfaces
Brass	Zinc	Decorative color effects ranging from light yellow to deep brown; otherwise similar to copper
Nickel	Copper preferable; zinc immersion or anodizing with phosphide electrolyte	Best preplate for bright chromium plating
Black nickel	Dipping in sulfuric acid	Decorative finish (background color) with limited resistance to exposure

Table A34 Major uses of electroplating metals for aluminum (*continued*)

Type of plating on aluminum	Pretreatment needed	Characteristics and general uses
Tin	Zinc, copper, or brass	Applications requiring joining by soldering
Zinc	Zincating	Anodic to aluminum and therefore protective under corrosive conditions
Cadmium	Zincating	Almost neutral to aluminum and therefore protective against galvanic action
Silver	Zincating	Decreasing electrical contact resistance

The coating should be complete and unbroken (except for coatings of nickel or other metals that are cathodic to aluminum in the everyday environment). Otherwise there will be galvanic action in which the aluminum will be sacrificed for the protection of the plating metal. Electroplated finishes are not generally utilized in construction but are used for hardware, various types of fasteners, and factory-produced components.

Porcelain or Vitreous Enamel. This finish forms a hard, resistant surface. It is available in a broad color range that creates a different feeling in that the colors are glassy, whereas anodic color is metallic in nature. A porcelain enamel finish is controlled by the melting point of the particular aluminum alloy to be covered. Its use on aluminum has been made possible by the development of low-melting glazes (920 to 1050°F, or 493.3 to 565.6°C) with a high coefficient of expansion compatible with that of the aluminum base metal. Conversion coating treatments are generally required before application of the glaze. The relatively high temperature used for this final baking softens the metal to a degree, depending on the alloy composition, and alters the characteristics of the metal, leaving it with a lower temper than it had before baking. The baking may cause distortion of thin gauges.

Organic Coatings. Paint, lacquer, and enamel can be applied as finishes to aluminum surfaces which have been prepared by one of the processes suitable for pretreatment. The degree of protection depends on the type of paint, enamel, or lacquer used (*see* Paints; Painting; Plastics).

Special Sealing Compounds for Aluminum Assemblies. Special sealants are used in joints between different parts of an assembly as an additional precaution when severely corrosive conditions may be encountered in service. Sealants are always necessary when dissimilar metals or materials are in contact with aluminum in the presence of moisture. Sealers should be specified to protect the faying, or mating, surfaces in all areas where moisture may be trapped. Sealing compounds currently used include zinc chromate impregnated tapes, and aluminum pigmented pastes. The caulking or sealing compound used should be determined by the nature of the crevice or joint.

Application

Anodizing Techniques. Quality anodizing has become a complicated and highly specialized science. The production of a hard, uniform, nonporous film of equal thickness is no mean task. Among the many variables that must be controlled during the anodizing process are (1) concentration of the electrolyte, (2) temperature of the electrolyte, (3) time in the bath, (4) current density and proper use of rectifiers, (5) agitation of the bath, and (6) preanodizing preparations and sealing of the film after anodizing.

There are several commonly used anodizing processes, each based on a different acid, namely, sulfuric, chromic, oxalic, phosphoric, and boric.

Colored anodized finishes for aluminum can be obtained with organic and inorganic dyes by three methods: (1) by impregnating the anodic coating with dyes or pigments, (2) by electrolytically depositing pigments in the coating, and (3) by using alloys and processes that produce integral color in the coating.

Slight variations in the chemical composition of a given lot of aluminum can cause variations in hue, especially if sheet and extrusions are used together in construction. Studies are being continuously made by both the aluminum producers and the anodizers on the matter of achieving color match.

Of the colors used with the anodic treatments, architectural gold has proved to be one of the most stable from the standpoint of fade resistance. Other satisfactory colors are blue, brown, and black. Various mechanical finishes and anodizing treatments can be combined to produce a spectrum of color effects.

Condensed Checklist

1. The type of aluminum finish used should be one which will fulfill all the necessary requirements, particularly if used on the exterior.

2. The type of aluminum alloy used must be suitable both for the construction purpose and for the type of finish desired.
3. Local equipment should be investigated. Tanks in the area must be of an adequate size and shape to fulfill the job requirements.
4. Protection from damage must be ensured. The finish surfaces must be protected during transit, storage, and erection.
5. The thickness of aluminum oxide finishes controls the degree of resistance to corrosion and abrasion and should always be correlated to the end use.
6. The correct type of cleaning agent must be chosen, as many acids and alkalis will attack aluminum finishes. Advice on this point should also be given to the owner(s).
7. Accessories and components should be of an aluminum alloy that can receive the same finishes as the major parts.
8. Color matching and color fastness should be carefully controlled. In most architectural applications, color blending, whether of natural oxide films or dyed anodic films, is very important. Therefore uniform surface preparation is a prerequisite wherever color matching or blending is a primary problem.

History and Manufacture

Mechanical finishing, painting, and electroplating of metals were known long before the introduction of metallic aluminum on a commercial scale. Since the late 1920s and especially during and after World War II, experiments with known finishes and the development of specific treatments for aluminum were undertaken. These resulted in a wide range of kinds and colors of finishes.

Most of the chemical and electrolytic methods are protected by patents or employ proprietary solutions and should not be used without permission of the patentee.

ALUMINUM FOIL

Physical and Chemical Properties

Aluminum foil is rolled to a thickness of 0.005 in. (0.127 mm) or less; it can be rolled as thin as 0.00017 in. (0.0043 mm). Above 0.005 in. (0.127 mm) it is technically considered to be sheet. The maximum available width is 6 ft (1.829 m), and the maximum length 3000 ft (914.4 m). Foil is fabricated from fairly high-purity aluminum alloys, Nos. 1035-O, 1145, 1160, and 1235 being the most commonly used as they have both good corrosion resistance and good working qualities.

Aluminum foil is available in thicknesses ranging from 0.00017 to 0.00314 in. (0.00432 to 0.07976 mm) and from gauges No. 40 (equal to 0.00314 in. or 0.00432 mm) to No. 34 (0.00634 in. or 0.16104 mm).

Vapor Permeability. Aluminum foil is nearly vapor-proof, except at joints. The 0.00035-in. (0.00889-mm) foil used on reflective insulation has some vapor permeability because of minute perforations which exist in normal production. The joints can be made waterproof if properly sealed with aluminum tape.

Thermal Reflectivity. Aluminum foil has the highest thermal reflectivity (95 to 98%) of available materials. This high value holds true only for a mirror-bright finish. For example, if the reflectivity is 80%, this will decrease the overall value of the reflective air space by 25%.

Finishes. Aluminum foil is available with a bright or dull finish, in a wide variety of embossed patterns, and in as many as 33 colors, including gold and silver. It can be covered with protective coatings, combined with plastic films, laminated, or mounted onto paper of all types and weights and onto sheets and boards.

Types and Uses

In construction aluminum foil is used for thermal insulation; cigarette-proofing for plastic laminates; vapor barriers, and added insulation to paper pulp decking and boards; protective finish on plywood, hardboard, and particle board; and decorative finishes on various types of panels and boards.

Thermal insulation and vapor barriers are manufactured in several types. The most common is foil solidly laminated to a flat paper backing with asphalt or an aqueous adhesive (*see* Table *A35*). Another widely used

Table A35 Typical laminations of aluminum foil and paper for insulation

	Weight of paper		Width		Area per roll	
Foil	lb	kg	in.	cm	ft^2	m^2
One or both sides	40	18.144	25	63.50	250	23.225
			36	91.44	500	46.450
One or both sides	80	36.288	25	63.50	250	23.225
			33	83.82		
			36	91.44		

type consists of alternate layers of paper and foil, and the third general type consists of layers of aluminum foil. In the last two types the layers are divided by air spaces, which add to the overall insulating value. Foil can be laminated in a similar manner and for similar purposes to cellophane, acetate, polyethylene, polyester, and other types of plastic films.

For special types of refrigeration and for areas where a constant temperature must be maintained, aluminum foil installed in several rigid layers separated by trapped air spaces can be used. A check should always be made with manufacturers and specialists in this field before this type of insulated area is designed.

Foil can also serve as a surface finish material when laminated to various sheet and board materials. In this form it also supplies additional insulation value to the sheet or board material.

Nonconstruction Uses. Aluminum foil is used also for packaging, wrapping, labeling, electrical and electronic equipment, air filters, and printing. Special inks and gravure methods have been developed for printing on aluminum.

Application

Condensed Checklist

1. The insulation values derived from the number of reflective surfaces should always be checked. An air space should always be provided next to each reflective surface. This air space should be a minimum of $\frac{1}{2}$ in. (12.7 mm) wide, but is most efficient at $\frac{3}{4}$ in. (19.05 mm).
2. Any breaks occurring in transit or handling impair the value of foil as insulation or a vapor barrier.
3. For single-layer insulation, a material having foil on both sides should be used to divide the framing spaces. This will provide two or more fully insulating spaces and therefore increase the total insulating value.
4. Aluminum foil should be installed so that the first layer of foil will be close to the warm side of construction, where it is also most effective as a vapor barrier.
5. When plaster is to be used as the interior surface, aluminum foil should always be recessed so as to avoid direct contact with the wet plaster or cement.
6. To prevent galvanic action, aluminum foil should be installed with aluminum nails, staples, rods, channels, etc., and should be isolated from any metals except those that are compatible with aluminum (*see* Metals).
7. The type of framing and construction must be correlated with the type of foil used. The width of the foil should be checked against the spacing of studs, furring, etc.
8. Sizes of areas to be covered should be considered, as too many small areas will cause difficulty. Window and door openings, in particular, should be checked as they can be difficult to seal.
9. Sealing of joints must be ensured. When foil is used as a vapor barrier, all joints should be sealed with an adhesive tape suitable for aluminum. An adequate vapor barrier can be obtained if the joints are lapped well and secured.
10. When the foil is used as a laminated surface finish, all joints should be sealed with the tapes specially made for this type of work.
11. When installing aluminum, one should not only use aluminum nails, screws, staples, rods, bars, channels, and angles to eliminate the possibility of galvanic action, but should also make sure that these installation materials are similarly isolated from noncompatible materials.

Conditions Favorable to the Use of Aluminum Foil

1. As a vapor barrier in humid atmospheres.
2. Where a permanent, corrosion-resistant material is required.

Conditions Unfavorable to the Use of Aluminum Foil

1. Where the areas to be covered for insulation or vapor barriers are small or very broken up.
2. Where winds are constant, as they will cause aluminum foil to vibrate and oscillate.
3. As insulation when aluminum foil is placed in direct contact between two layers of conductive materials. The foil material remains effective as a vapor barrier, but no insulating value can be assumed.

History and Manufacture

Aluminum foil had been introduced as a new insulating material before 1940. During and after World War II, however, its uses diversified widely until it has largely replaced lead and tin foil in many of their applications.

Aluminum foil is manufactured by hot-rolling appropriate aluminum alloy billets to a thickness of $\frac{1}{8}$ in. (3.175 mm); the sheets are then cold-rolled in a foil-rolling mill to the thickness desired and are annealed frequently between rollings to overcome the hardening effects of working the cold metal. The bright finish is obtained during this cold-rolling process. Pack-rolling may be necessary on the thinner gauges and results in one side being dull and the other bright.

ALUMINUM MESH AND WIRE CLOTH

Physical and Chemical Properties

The bulk of aluminum mesh and wire cloth is made of wire fabricated from aluminum alloys that are strong, lightweight, corrosion resistant, and nonstaining. A special type of insect screen is made from stamped sheet. The specific alloys used depend on the end use of the mesh and wire cloth.

Aluminum mesh and wire cloth are fabricated from various aluminum alloys and alclad alloys in a wide variety of tempers to meet the requirements for all the various areas where mesh and wire cloth are used in construction.

The expanded decorative mesh made from aluminum sheet is available in a wide variety of designs and patterns, as well as in numerous finishes (*see* Mesh and Wire Cloth).

Types and Uses

Construction uses of aluminum mesh and wire cloth fall into several categories: protective guards and grilles; various decorative applications of mesh and screen; insect screening; and fencing, particularly chain link fencing.

Decorative mesh and wire cloth for guards, grilles, and screening are made in a wide choice of mesh patterns, wire crimps, and mesh sizes.

Insect Screens. Insect screens are usually made from No. 12 square-woven wire mesh cloth fabricated from 0.029 in. (0.737 mm) aluminum wire. The woven wire mesh comes in rolls 50 and 100 ft (15.24 and 30.48 m) long, with a maximum width of 6 ft (1.829 m). Stamped sheet aluminum, specially slit and formed into tiny louvers, is also used but to a lesser extent. The stamped sheet comes in rolls 50 ft (15.24 m) long and from 1.5 to 4 ft (0.457 to 1.219 m) wide in 6-in. (152.4-mm) increments.

Fencing. Fencing other than chain link fencing is made from stock aluminum mesh and wire cloth. Chain link fencing is made in a woven diamond mesh pattern of a special type used for exterior enclosures, fencing, and protective barriers. It is generally in two weights: aluminum wire gauge 0.192 in. (4.877 mm) and 0.148 in. (3.759 mm). It comes in rolls 50, 100, and 150 ft (15.24, 30.48, and 45.72 m) long and 3 to 10 ft (0.914 to 3.048 m) wide in 12-in. (304.8-mm) increments.

For additional information, tables, and illustrations pertaining to types, uses, and applications of aluminum mesh and wire cloth, *see* all headings under Mesh and Wire Cloth.

Application

Condensed Checklist

1. The alloy used for fabricating the mesh and wire cloth should be the correct one for the particular end use.
2. The finish should be of the type, color, and texture desired, especially for decorative screens.
3. To prevent galvanic action, direct contact with metals not compatible with aluminum must be avoided. Insulation must be provided if necessary.
4. Correct accessories must be used. Frames, supports, and accessories such as screws, nuts, and bolts should be made from aluminum or a compatible metal, and protected from incompatible materials.
5. The sizes and shapes designated for guards, grilles, etc., should be based on available stock widths.
6. Design limitations should be investigated. For decorative designs with mesh and wire cloth, a check should be made with the manufacturer to determine the availability of materials and the limitations on size, shaping, welding, design, and finishing.
7. The strength of frame and mesh should be correlated to the end use. For example, in insect screening a heavier mesh should be used where heavy wear is likely to occur.
8. For chain link fencing, a check should be made of the spacing between supports, of the width of openings for gates, and of the type of edges desired (barbed or knuckled).

Conditions Favorable to the Use of Aluminum Mesh and Wire Cloth

1. Where a strong, lightweight, corrosion-resistant, and nonstaining material is required.
2. Where the design calls for a decorative color finish for which neither paint nor enamel nor the natural color of the metal is desired.

Conditions Unfavorable to the Use of Aluminum Mesh and Wire Cloth

1. Where direct contact with metals that will cause galvanic action is unavoidable.
2. Where there will be direct contact with alkalis or acids as, for instance, in cement, mortar, and other masonry materials.

History and Manufacture

Mesh and wire cloth began to be fabricated from aluminum alloys during the general expansion of the aluminum industry after 1945. Use of this material has grown steadily. The same manufacturing processes are used as for wire and mesh of other metals (*see also* Mesh and Wire Cloth; Wire and Wire Rope).

ALUMINUM, ORNAMENTAL

Many kinds of rods, bars, pipes, railing, fittings, flanges, letters, plaques, and special shapes are manufactured as stock items from appropriate aluminum alloys for use in ornamental designs of railings, grilles, letters, and fences (*see* Figures *A24* and *A25*).

In these applications the fact that the aluminum alloys can be anodized, plated, coated, and colored, as well as finished by various mechanical types of surface treatment, is as important as the fact that they are lightweight, corrosion resistant, and easily worked (*see* Aluminum Finishes).

Porcelain enamel coatings in a wide range of colors, as well as a wide variety of colored or clear plastic resin coatings, are available for application on ornamental aluminum units.

FORM	ALLOY	PROCESS
RAILINGS	6063 - T5	EXTRUDED
PIPE	6063 - T832	EXTRUDED
TUBE	6063 - T5	EXTRUDED
ROD	6063 - T5	ROLLED EXTRUDED
BAR	6063 - T5	ROLLED
FITTING FOR PIPE RAILINGS	B 214	CAST
RAILING FITTINGS	F 214	CAST
FLOOR AND ANGLE FLANGES FOR PIPE RAILINGS	6061 - T6	FORGED
WALL FLANGE FOR PIPE RAILINGS	356 - T6	CAST

Figure A24 Ornamental forms of aluminum alloys.

Types and Uses

Ornamental aluminum shapes are available in many simple, stock forms and also in specially designed co-

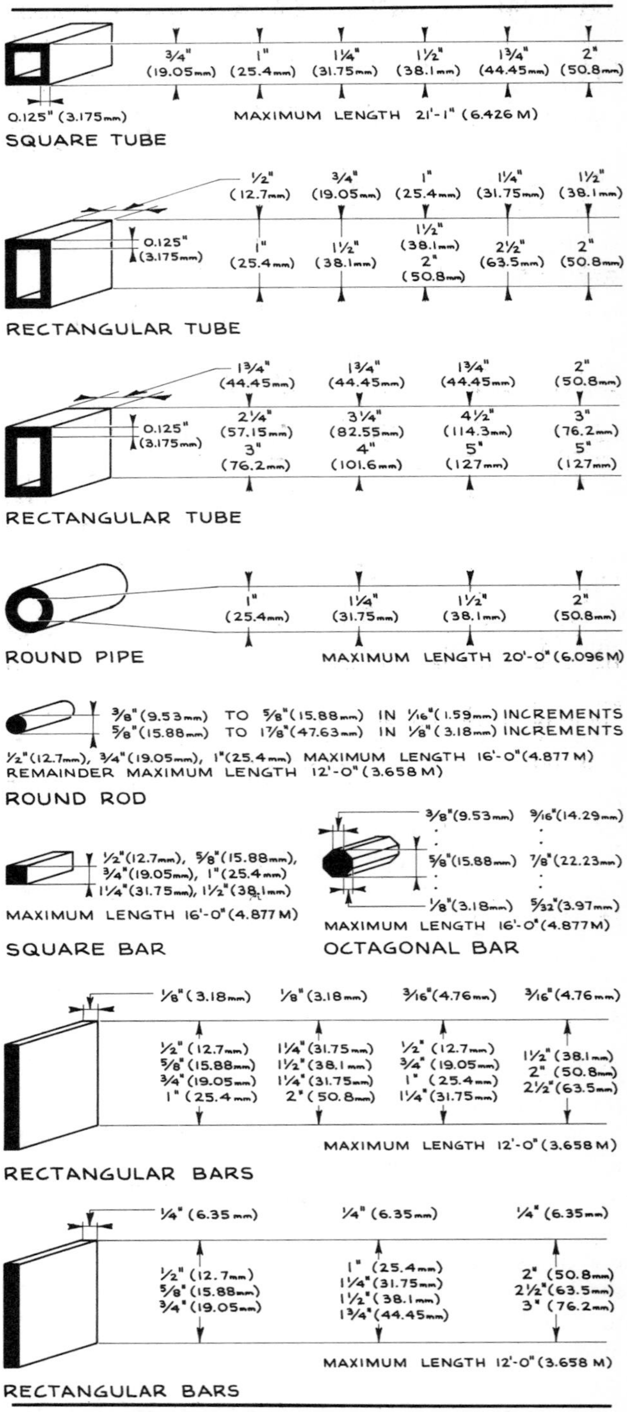

Figure A25 Standard dimensions of aluminum tubes, pipes, rods, and bars.

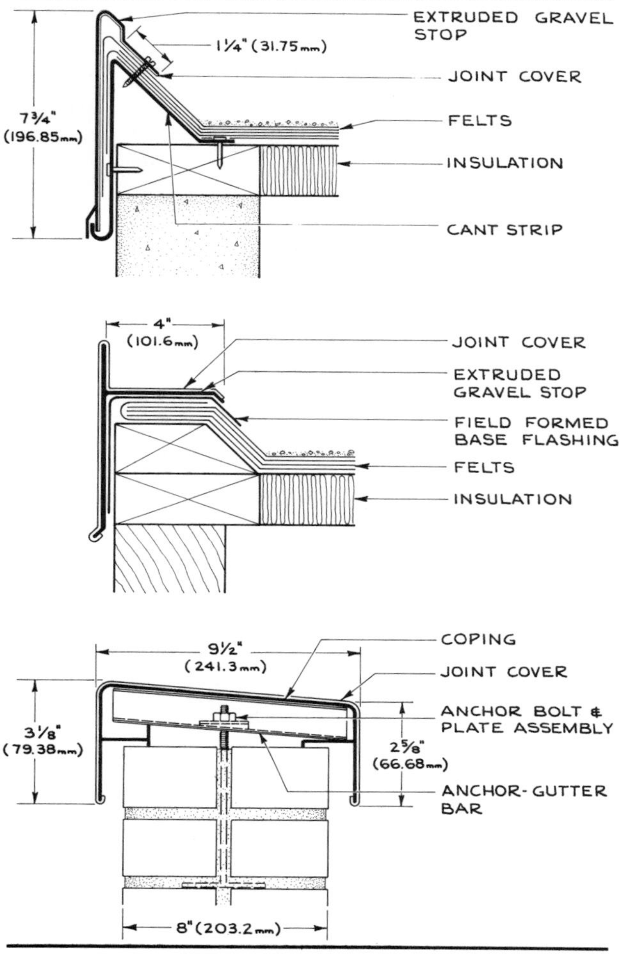

Figure A26 Cast aluminum gravel stops and coping.

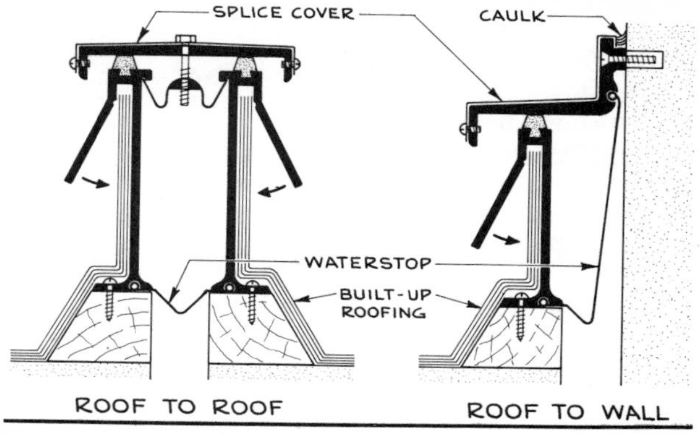

Figure A27 Cast aluminum expansion joint covers for roofing.

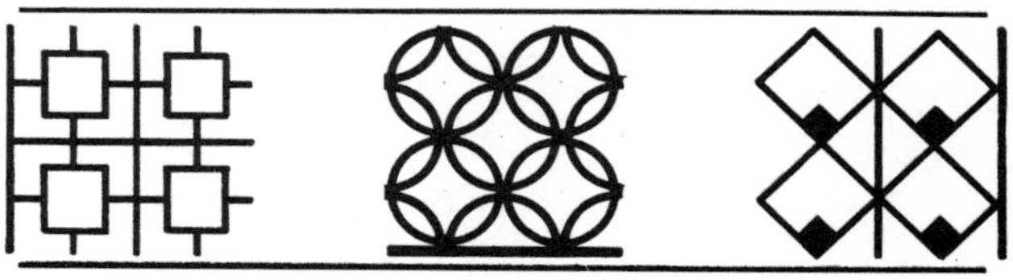

Figure A28 Typical examples of decorative aluminum screens.

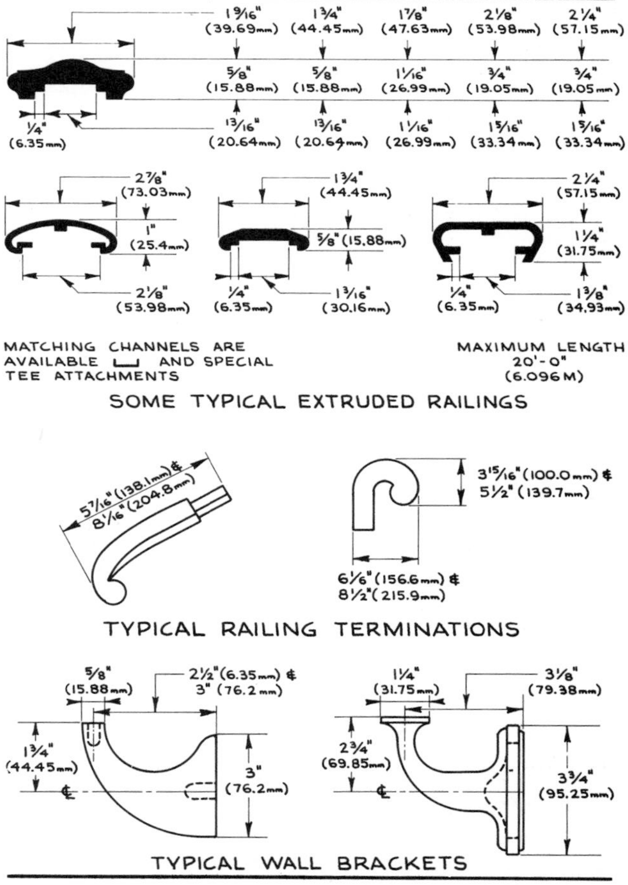

Figure A29 Typical cast and extruded aluminum railing shapes, brackets, and terminations.

ordinated systems, based in some cases on patented devices. Certain types of expanded aluminum sheet also are manufactured primarily for decorative purposes. There are also many cast aluminum ornamental designs (*see* Figures *A26* and *A27*). Thus this metal presents a wide scope for design in construction.

In designing ornamental railings, grilles, and fences of aluminum, the exact type of alloy, fastening devices, method of welding, dimensions, and finishes should be shown and specified. (*See* Figures *A28* to *A32*.)

New types and designs are constantly being developed for extruded railings, including termination pieces, scrolls, brackets, and attachments. High-strength adhesives are used both for joining ornamental units by fabricators and for on-site assemblage. Therefore it is a good general rule to check with both the fabricator of components and the metalwork manufacturers to obtain the latest data on shapes and finishes before beginning the design.

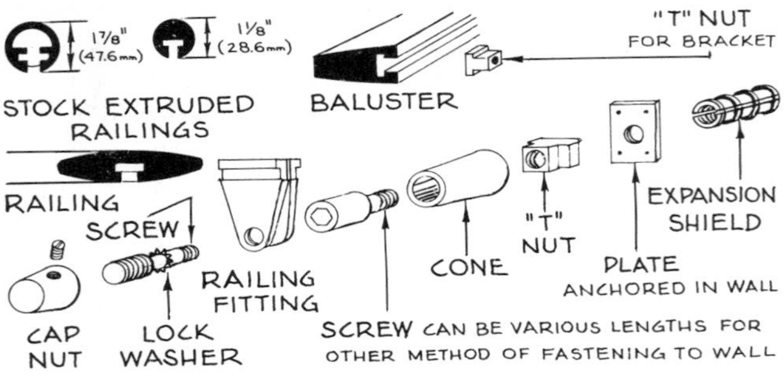

Figure A30 Typical examples of a patented type of aluminum railing.

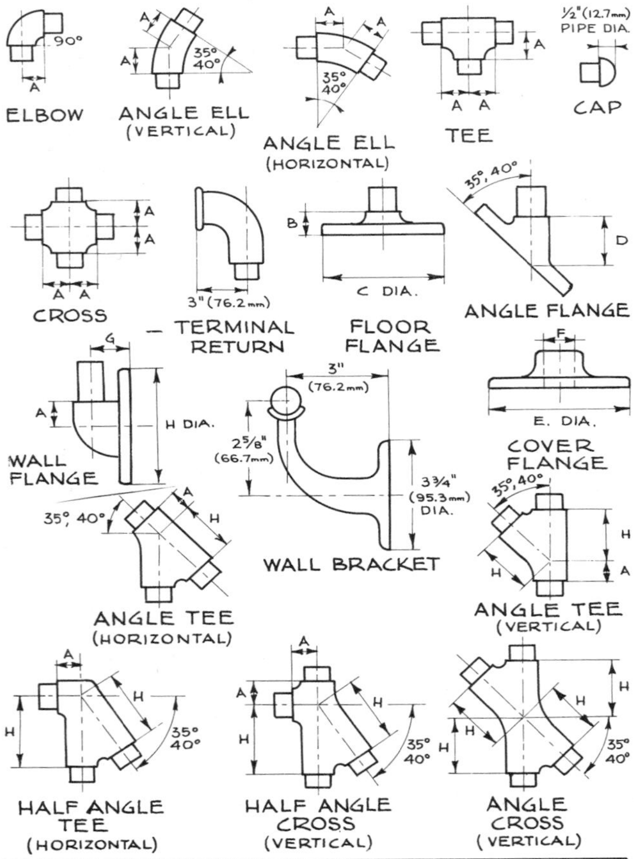

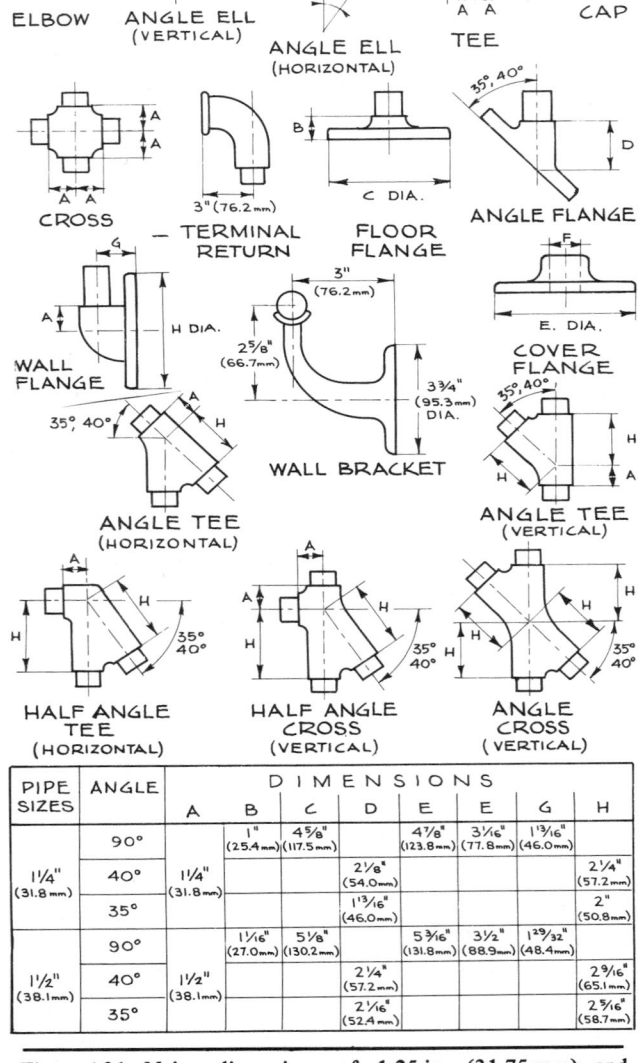

PIPE SIZES	ANGLE	DIMENSIONS A	B	C	D	E	E	G	H
1¼" (31.8mm)	90°	1¼" (31.8mm)	1" (25.4mm)	4⅝" (117.5mm)		4⅞" (123.8mm)	3 1/16" (77.8mm)	1 13/16" (46.0mm)	
	40°				2⅛" (54.0mm)				2¼" (57.2mm)
	35°				1 13/16" (46.0mm)				2" (50.8mm)
1½" (38.1mm)	90°	1½" (38.1mm)	1 1/16" (27.0mm)	5⅛" (130.2mm)		5 3/16" (131.8mm)	3½" (88.9mm)	1 29/32" (48.4mm)	
	40°				2¼" (57.2mm)				2 9/16" (65.1mm)
	35°				2 1/16" (52.4mm)				2 5/16" (58.7mm)

Figure A31 Major dimensions of 1.25-in. (31.75-mm) and 1.50-in. (38.10-mm) aluminum pipe railings.

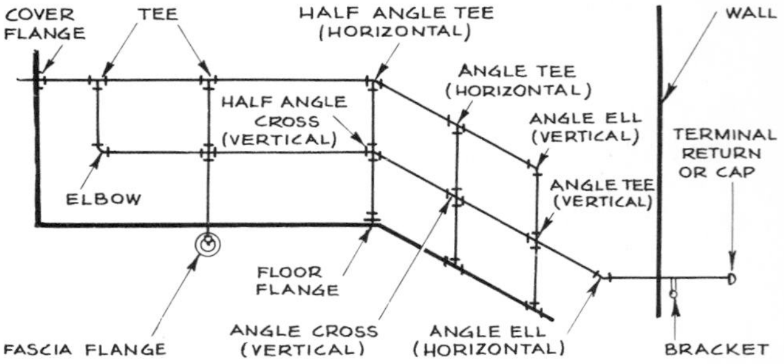

Figure A32 Location of fittings for aluminum pipe railings.

Application

Condensed Checklist

1. To prevent galvanic action, aluminum should not come into direct contact with metals that are not compatible. It is advisable to check with manufacturers or fabricators for the compatibility of metals and for methods of anchoring or assembling aluminum materials.
2. Adequate anchorage must be provided. Floor, wall, and fascia flanges should have secure anchorage to which they can be firmly attached.
3. Self-tapping screws, wood screws, and bolts should be made of alloy No. 2024-T4 or a compatible metal. The use of Phillips-type heads prevents slipping of the screwdriver and thus will avoid scratches on the finish surface.
4. Shop drawings should always be specified for any ornamental aluminum designs.
5. Before actual manufacture of the ornamental items is begun, a field check should be made (1) by the architect or designer to ensure that all necessary anchorage has been installed properly, and (2) by the fabricator to obtain all on-site, actual dimensions necessary for the design submitted so that the fabricated product, when installed, will correspond exactly with the requirements as shown on the contract documents.
6. All welding should, if possible, be performed in the shop; inert gas shielded arc-welding should always be specified.
7. A check should always be made of the permanence of the surface finishes to be used. For anodized or plated finishes a check should be made of the length, width, and depth of the plating baths available locally, as this will control the type of finish and size of design units.
8. When using high strength adhesives, it is advisable

always to check with manufacturers and fabricators to obtain the correct type and methods of application.

9. When using a porcelain enamel coating, a check should always be made with manufacturers and fabricators for available colors and limitations.
10. When using plastic resin coatings, one should always check with aluminum manufacturers, plastic resin manufacturers, and fabricators to obtain the correct type of resin, color, and method of application.

Conditions Favorable to the Use of Ornamental Aluminum

1. Where a lightweight, corrosion-resistant, easily worked metal is required.
2. Where a decorative or flush type of railing, fence, or grille is required.
3. Where a colored metal is desired (aluminum is available in a spectrum of effects produced by combinations of mechanical finishes, anodic treatments, porcelain enamels, and plastic-resin clear or colored coatings).
4. Where special designs are required and the quantity to be made will justify the manufacture of new extrusions. The manufacturer should be consulted to determine the quantity of new extrusions required to match the cost of stock extrusions.

Conditions Unfavorable to the Use of Ornamental Aluminum

1. Where the sizes and designs cannot be made from existing stock shapes and the quantity needed is insufficient to justify the cost of new dies.
2. Where welding or another metalworking procedure is necessary after the aluminum has already received a surface finish or coating that would be damaged by this procedure.

ALUMINUM OXIDE

Aluminum oxide exists in several crystalline forms, many of them found in nature (e.g., corundum and gem stones such as the ruby and sapphire) and all characterized by their hardness (*see* Abrasives). Aluminum oxide is also commercially made and is important as an abrasive and refractory material.

Naturally occurring aluminum oxide in hydrated form is a major constituent of bauxite ore and is the ultimate source of the world's aluminum. It is generally referred to as alumina in technical literature.

The hydrates of aluminum oxide and various aluminates are important mineral constituents of many rocks, of clay, and of other materials used in construction either directly or indirectly as major ingredients of materials such as cement, concrete, brick, clay tile, or glass.

Hydrated alumina is widely used both in ceramic bodies and in glazes and as a filler in plastic and rubber ware.

Impure bauxite ore is frequently used as a component of aluminous-type cement, and for electric insulators, high-intensity lamps, extrusion dies, and numerous applications in the electronic field.

ALUMINUM PANELS, SANDWICH PANELS, AND CURTAIN WALLS

The terms "panel," "sandwich panel," and "curtain wall" are not synonymous.

Panels. Aluminum panels are prefabricated units which are generally manufactured on modular and nonmodular window-width dimensions for the exterior of buildings and generally in 2, 3, and 4 ft (0.610, 0.914, and 1.419 m) widths for interior partitions, dividers, and enclosures.

Panels for the exteriors of buildings primarily consist of an aluminum exterior facing, which may be an aluminum casting, an extrusion, or sheet material which has been pressed, stamped, or formed into specially designed shapes. This facing may either have the natural aluminum finish or be colored by the various specialized processes used for finishing aluminum (*see* Aluminum Finishes). The facing is strengthened and stiffened by the aluminum frame to which it is attached.

The panel also consists of the following components: vapor barrier, condensation drains and lead-offs, insulation, and an interior finish. This last may be any of the following: (1) the exposed back of the insulation; (2) an aluminum sheet given any aluminum finish desired; (3) a backing ready to receive a job-applied finish; or (4) a factory-applied laminate finish.

Sandwich Panels. A sandwich panel, on the other hand comprises a system of construction, termed skin construction, that has been used for many years in the aviation industry. A cellular core of aluminum, foam insulation, or other material has a skin of aluminum applied and bonded to both sides, thereby forming a unified whole in which all the components work as one. Thus a true sandwich panel is a structure and should be designed for the particular situation in which it is to be used (*see* Figure *A33*).

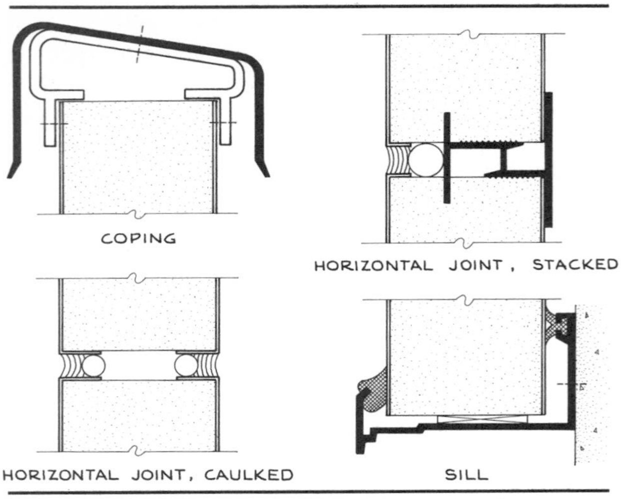

Figure A33 Typical types of aluminum panel wall systems.

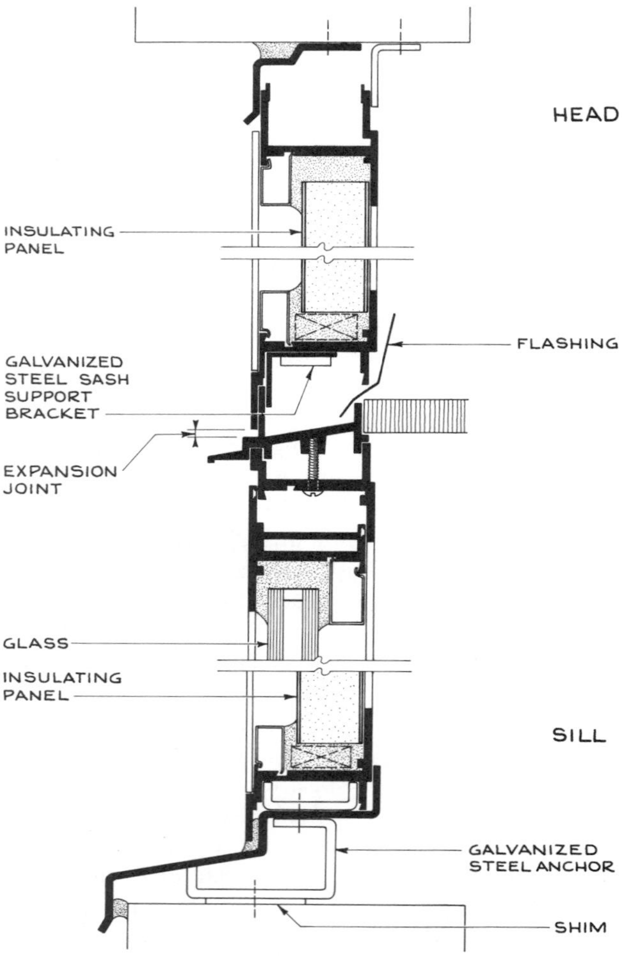

Figure A34 Typical aluminum curtain wall with windows and spandrel panels.

Curtain Walls. Aluminum curtain walls originally were made up of aluminum windows, a fixed type of solid or patterned aluminum filler panel, and mullions to attach to structural framing and masonry walls, with each part considered as a separate unit. Now curtain walls are a complete system including windows, panels, mullions, trim, insulation, exterior and interior finish, condensation channels and drips, etc. (*see* Figure *A34*).

The rapid developments made in adhesives, particularly the epoxys, has simplified the fabrication of panels and sandwich panels and has expanded the uses of these types of panels in construction, particularly as floors, roofs, supporting walls, partitions, and spandrels.

Types and Uses

Aluminum panels, curtain wall systems, and sandwich panels are available in a very wide variety of types, shapes, sizes, forms, and systems. They are manufactured together with all the necessary covers, mullions, flashing, and accessories for fastening and anchoring.

When designing with these panels, curtain walls, and sandwich panels, one should always check the manufacturers' latest data on sizes, shapes, textures, colors, etc.

Application

Condensed Checklist

1. A check should be made that there is sufficient insulation, vapor barrier, and condensation drainage within the panel.
2. The aluminum alloy used in the panel should be suitable for the particular job.
3. Permanence of the finish under the particular job conditions must be determined in advance.
4. Distortion or warping must be avoided. A sufficiently heavy gauge should be chosen to avoid the possibility of distortion or warping of the exterior and interior finishing faces. These facing panels should be cast, formed, pressed, stamped, or strengthened and stiffened in such a way that the aluminum face does not deform or distort in its flat plane.
5. Calculations and provisions for expansion should be made for the individual panels and for the entire framework supporting the panels, windows, and other units.
6. Sealing of joints is a major problem in panel construction systems. It is therefore important always to check the latest data on sealants, gaskets, tapes, etc., not only for new products but also for performance data on those already in use.

7. One should always check what types of anodized finishes, organic coatings, and baked enamel finishes are available, and in what range of colors.
8. When selecting a curtain wall system, one should always check what alloys are used for the various parts of the wall system because, if an anodized finish is selected, only certain alloys will produce the same anodized color.
9. Whenever panels, curtain walls, or sandwich panels are to be installed in a system, it is necessary always to specify that a full-size unit be tested under all conditions that may be encountered in the end use of the structure.

Conditions Favorable to the Use of Aluminum Panels, Sandwich Panels, and Curtain Walls

1. Where a strong, corrosion-resistant, lightweight material is required.
2. Where colors or metallic finishes available for aluminum are desired.
3. Where a complete system with metallic or colored finishes for enclosing a structure is required.
4. Where design requirements call for enclosure units finished on interior and exterior surfaces, available in lengths over 20 ft (6.096 m), which are rigid, are semisupporting, and have high thermal insulation values.

Condition Unfavorable to the Use of Aluminum Panels, Sandwich Panels, and Curtain Walls

Where galvanic action cannot be avoided.

History and Manufacture

Design concepts based on prefabricated exterior walls were conceived in the early 1920s but were not realized until the late 1940s. Commercial construction actually started in the 1950s. The first use of such panels was for large office buildings where the quantity of identical units required was sufficient to make the prefabrication economically sound.

ALUMINUM POWDER

Physical and Chemical Properties

Aluminum powder is made from pure aluminum in both flake and granular form. The former consists of aluminum flakes about five-millionths of an inch in thickness that vary in diameter from particles barely fine enough to pass through a 400-mesh or a 100-mesh sieve down to those of submicron dimensions. The granular form consists of small, spheroidal-shaped pieces of pure aluminum. Both types are standardized as to percentages retained on a 325-mesh sieve. They are stable at ordinary temperatures but should be kept dry and stored at 60 to 80°F (15.6 to 26.7°C). The flake types are available in both dry and paste form.

Types and Uses

The largest use of aluminum flake powder is as a metallic pigment in many types of paints for the protection of concrete, masonry, steel, wood, roofing, etc. Its value in these paints is due to the flake orientation that occurs in the paint film. The flakes tend to float to the surface and form a series of layers of aluminum metal separated by the paint vehicle; these metal layers prevent penetration and breakdown of the paint by ultraviolet rays of the sun.

Aluminum flake powder also is used as a pigment in inks, as a coating for textiles, in caulking compounds, and in plastics. In most of these, its main function is a decorative one.

Some of the basically nonpigment uses of aluminum powder, both flaked and granular, are in metallurgy, powder metallurgy, rocket and missile propellants, and ceramics.

Aluminum powder is also used to make aerated concrete and in metalwork for thermit-type reactions (*see* Welding).

History and Manufacture

Aluminum powder first appeared in 1890 and was quickly adopted by the bronze powder industry—hence the name, "aluminum bronze powder," by which it is commonly known. Aluminum powders are currently produced by stamping, by atomizing, and by ball-milling processes.

Stamping Process. In the stamping process either aluminum foil or granular aluminum is used. Both forms are mechanically hammered to achieve the required thickness, lubricated to separate the flakes, and screened for correct particle sizes; in many cases the flakes are subsequently polished.

Atomizing Process. In the atomizing process pure aluminum is first melted and then forced through a small orifice, where it is broken up into spray form by a stream of air. As it hardens, it forms granules which are passed into collecting tanks, where they are selectively screened and graded.

Ball-Milling Process. The wet ball-mill process converts granular aluminum powder to a flake form which is used for the manufacture of pastes as well as powder. Atomized aluminum powder is placed in a steel drum together with steel balls, a lubricant to keep the flakes separated, and a suitable liquid to act as a carrier. As the drum is revolved, the steel balls hammer the aluminum powder into the required form. After this stage of processing, it is filtered, and this filter cake is adjusted to the proper paste form or is dried thoroughly to provide powder.

The dry ball-mill process omits the liquid carrier and makes the powder form directly.

ALUMINUM SHEET AND STRIP

Physical and Chemical Properties

Aluminum sheet and strip are fabricated from alloys that vary in composition and temper. The one most frequently used in construction work is 3003-H14, in which manganese is the major alloying element. Its physical properties are as follows: specific gravity, 2.73; melting point, 1190°F (643.33°C); tensile strength, 22,000 lbf/in.^{2} (151.7 MN/m^2); coefficient of expansion, 0.0000129/°F (0.000024/°C). If strength greater than that of No. 3003 is required, suitable alloy products are readily available.

Sheet and strip aluminum in this and other construction alloys is nonsparking, nonstaining, and highly corrosion resistant under most atmospheric conditions.

Workability. Sheet and strip aluminum can be punched, rolled, formed, machined, welded, brazed, and soldered.

Commercial Forms. Sheet and strip aluminum is available in coils and in rectangular sheets of various dimensions and gauges. It may be flat or rigidized in a wide variety of treatments—for example, corrugated, fluted, ribbed (*see* Aluminum, Corrugated)—and may be given a mill, bright, or embossed finish.

Types and Uses

Alloy No. 3003-H14 is the one most widely used in construction. From it are manufactured such items as air ducts, awnings, louvers, grilles, copings, cornices, fascias, gutters and leaders, moldings, mullions, window sills and stools, panels and sandwich panels, perforated acoustic sheet, prefabricated interior partitions, roof shingles, flat and rigidized sheet and strip for siding and roofing, garage doors, shutters, screen doors, columns and caps, wall tiles, switch plates, termite shields, terrazzo divider strips, trim, letters, and similar applications.

Alloy No. 5005 in all tempers is used if the product must have a high-quality anodically applied film, plus strength comparable to that of No. 3003.

Other alloys and tempers are the following:

No. 1100-O and No. 1100-H16 for ductwork.
No. 3003-O for general flashing.
No. 3003-H18 and No. 5050-O for hardware.
No. 5052-H38 for venetian blinds.
No. 5052-H38 for weatherstripping.
No. 6061-T6 for signs and treadplates.
Alclad No. 4043 and No. 1235 for panels and sandwich panels.

For data on aluminum sheet and strip used in construction, *see* Tables *A36* and *A37*.

Table A36 Sizes of aluminum alloy sheet generally used in construction

Alloy designation	Flat sheet: Width ft	m	Length ft	m	Coiled sheet width ft	m
1100-0	2	0.610	6	1.829	1, 1.5, and 2	0.305, 0.457, and 0.610
	3	0.914	8	2.438	1.17 and 1.67	0.356 and 0.508
	4	1.219	12	3.650	1.33	0.406
					3 and 4	0.914 and 1.219
1100-H14	2	0.610	6	1.829		
	3	0.914	8	2.438		
	3	0.914	10	3.048		
	4	1.219	12	3.658		
3003-0	2	0.610	6	1.829		
	3	0.914	8	2.438		
	4	1.219	12	3.658		
3003-H14	2	0.610	6	1.829	3	0.914
	3	0.914	8	2.438	4	1.219
	3	0.914	10	3.048	1	0.305
	4	1.219	8	2.438	1.33	0.406
	4	1.219	10	3.048	1.5 and 2	0.457 and 0.610
	4	1.219	12	3.658		
	5	1.524	12	3.658		
3003-H14 alclad	3	0.914	10	3.048	1.17 and 1.67	0.356 and 0.508
	3	0.914	12	3.658		
	4	1.219	10	3.048		

Table A37 Construction applications of aluminum sheet and strip

Type of use	Width ft	Width m	Length ft	Length m	Thickness in.	Thickness mm
			Flat sheet			
Flashing	1.5	0.457	4	1.219	0.019	0.483
Roofing	2	0.610	6	1.829	0.032	0.813
	3	0.914	8	2.438		
	4	1.219	12	3.658		
Shingles	1.21	0.369	0.67	0.203	0.025	0.635
Siding	2	0.610	8	2.438	0.025	0.635
	2	0.610	10	3.048		
	2	0.610	12	3.658		
Gutters	0.415	0.127	20	6.096	0.032	0.813
Leaders	Varies		12	3.658	0.020	0.508
Wall tile	0.250	0.076	0.50	0.152	0.025	0.635
	0.415	0.127	0.83	0.254		
	Square wall tile					
	0.333	0.102			0.025	0.635
	0.374	0.115				
	0.415	0.127				
	0.500	0.152				
	0.833	0.254				
			Coiled sheet			
Flashing	1	0.305	5	1.524	0.019	0.483
	1.67	0.508	5	1.524		
	2.33	0.711	5	1.524		
	0.50	0.152	100	30.480	0.024	0.610
	1.67	0.508	50	15.240	0.024	0.610
Roofing	2	0.610	50	15.240	0.032	0.813
	3	0.914	50	15.240	0.040	1.016
	4	1.219	50	15.240	0.040	1.016

Application

Accessories. Many accessories made of aluminum are available for installing aluminum sheet and strip roofing, and flashing, shingles, and siding. These include reglets, drips, ridge rolls, pipe flashing, corner closures, gutters, and leaders.

Aluminum sheet and strip are fabricated into many accessories for use with other materials such as laminates, linoleum and other resilient flooring, wood, and glass. These accessories include glazing beads, mullions, edging strips, and trim.

Expansion Joints. Expansion joints are generally necessary when working with aluminum sheet and strip except in the case of flashing and roofing made with other than flat seams, where the normal joints provide for expansion. Otherwise, in all cases of long horizontal runs—for example, caps, copings, fascias, and gutters—expansion joints are absolutely necessary. To avoid trouble later on, the allowance should be generous rather than minimal. Installation every 24 ft (7.468 m) is a good general rule, therefore, for expansion joints.

Condensed Checklist

1. Sheet and strip should be bent to rounded angles, never sharply.
2. For fastening, all nails should be of aluminum. Other fastening devices should be of aluminum or a compatible metal such as nonmagnetic stainless steel, nickel bronze, or a metal plated with zinc or cadmium. The alloy used should in any case be checked for possible galvanic action.
3. All soldering, brazing, and welding methods and materials should be suitable for the aluminum alloy being used.
4. Sheet and strip should not come into direct contact with metals not compatible with aluminum. They should be protected from direct contact with wet or intermittently wet cement, concrete, or mortar by the use of a bituminous or other suitable coating. Sheet or strip used for flashing should also be given a bituminous or other suitable coating before installation.
5. For roofing, the methods of joining sheet and strip by standing, batten, and single- or double-lock seams should be as detailed in Figure *A35*.

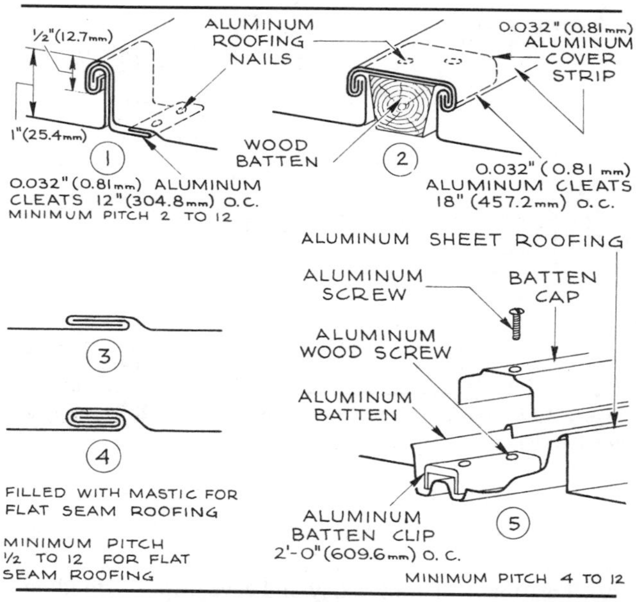

Figure A35 Details of standing (1), batten (2, 3), single-lock (4), and double-lock (5) seams.

6. Treatment of cross seams and expansion joints for roofing and flashing should be as detailed in Figure *A36*. A good general rule is to place an expansion joint every 24 ft (7.468 m) and cross seams every 6.83 ft (2.083 m).

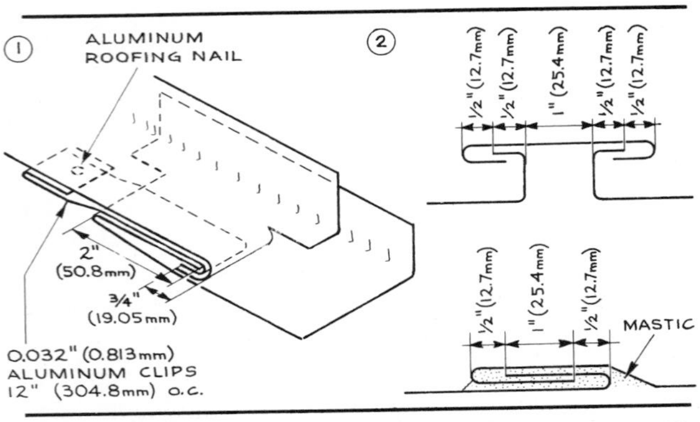

Figure A36 Details of cross seams (1) and expansion joints (2) for aluminum roofing and siding.

7. Treatment of roof intersections at ridges and hips with standing or batten seams should be as detailed in Figure *A37*.

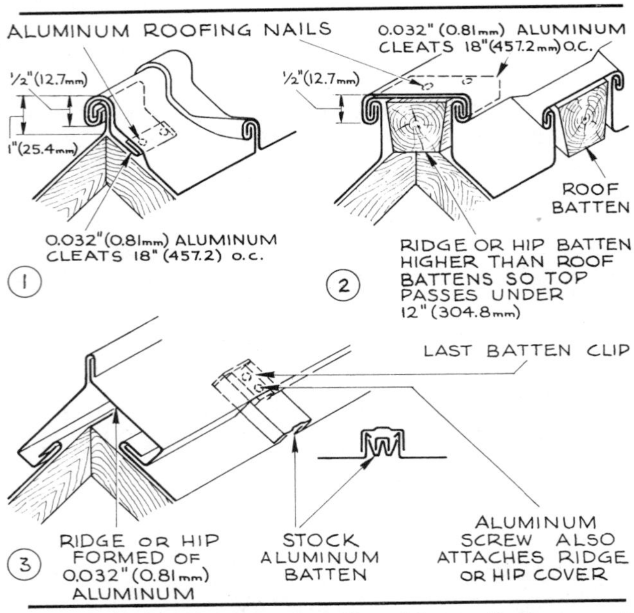

Figure A37 Details of standing (1) and batten (2, 3) seams at hip and ridge of roof.

8. Treatment of standing and batten seams at roof gable ends and edges should be as detailed in Figure *A38*.
9. Treatment of roof valley and eave drip for standing, batten, and flat seams should be as detailed in Figure *A39*.

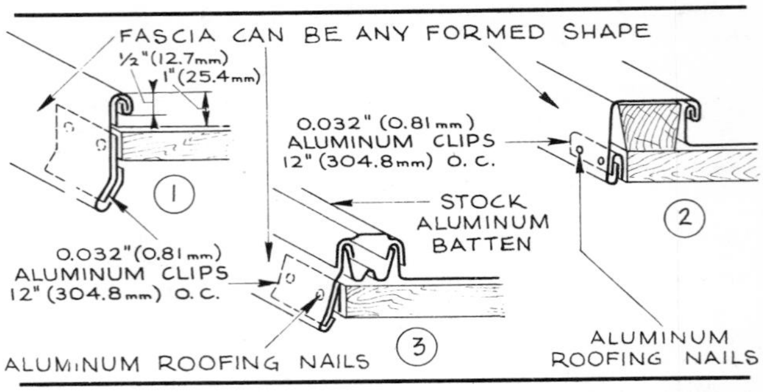

Figure A38 Details of standing (1) and batten (2, 3) seams at gable roof ends.

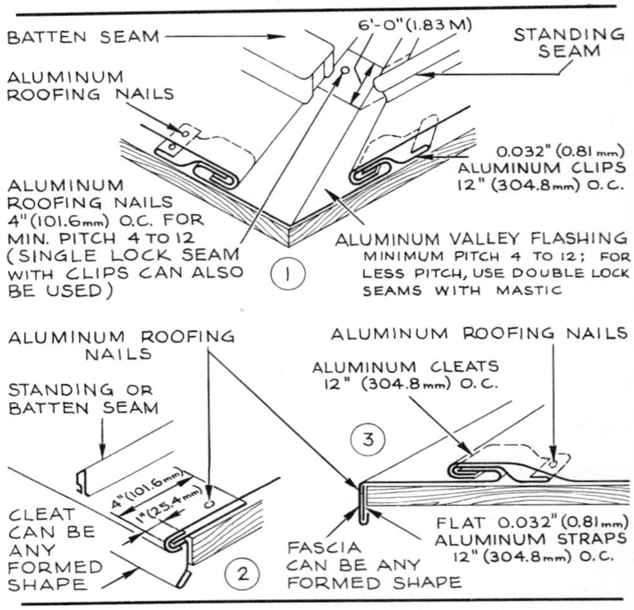

Figure A39 Details of standing and batten seams at valley (1), eave drip (2), and flat seams at eave drip (3).

10. Shingles: For roofs with aluminum shingles, a minimum pitch of 4 to 12 should be used. The shingles should be fastened only with aluminum roofing nails. (*See* Figure *A40*.)

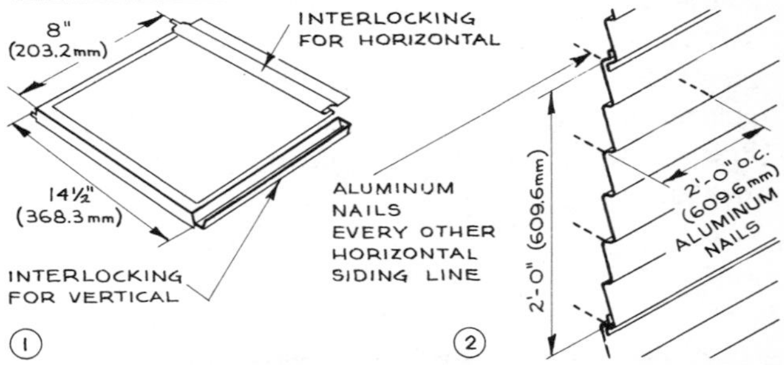

Figure A40 Details of roofing shingles (1) and siding (2) made of aluminum sheet and strip.

11. Siding: Outside and inside corners of siding should always be checked, and the special corner accessories designed for aluminum sheet and strip should be used.
12. Wall tiles: All tilework should be installed with the adhesives specifically recommended for this type of material. The base and cap should always be of another material.
13. The type of trim that is available should always be checked for aluminum panels, siding, and shingles (*see* Figures *A41* and *A42*).
14. When using aluminum gutters and leaders, one should always check methods of attachment, pitch, sizes, shapes, and colors and finishes.
15. When using aluminum columns of classical design, one should always check the manufacturer for current data on sizes, types, and finishes.

Conditions Favorable to the Use of Aluminum Sheet and Strip

1. Where a nonsparking, corrosion-resistant, lightweight material that will not stain adjoining materials is required.
2. For roofing where strong vertical lines are desired (standing or batten seams).
3. For roofing where a highly reflective, partially insulating material is required.
4. For roofing and siding where a comparatively fire-resistant material is desired. Municipal, state, and local codes should be checked; also the fire rating codes of the Fire Underwriters, insurance companies, labor departments, and federal government (Army, Navy, etc.).
5. For shingles or siding where a lightweight, permanent material is desired.
6. For wall tile where clear, bright, or metallic colors are desired and a material that can be bent and easily cut for fitting is required.
7. For permanent, noncorrosive gutters and leaders where ease of maintenance is desired.
8. For permanent, noncorrosive plain or patterned shingles and siding where a wide variety of colors is available for selection.

Conditions Unfavorable to the Use of Aluminum Sheet and Strip

1. Where a high fire resistance or a specific fire rating is required.
2. Where direct contact with materials causing galvanic action or chemical action is unavoidable.

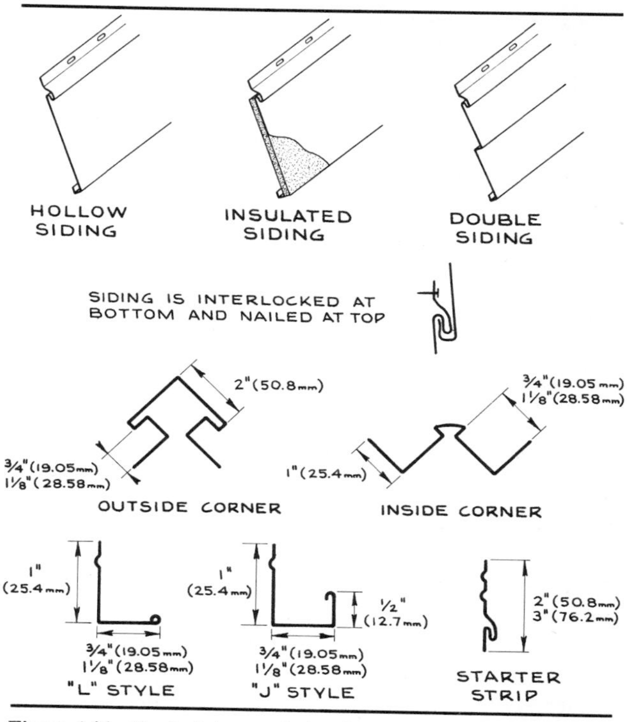

Figure A41 Typical types of aluminum siding panels and trim.

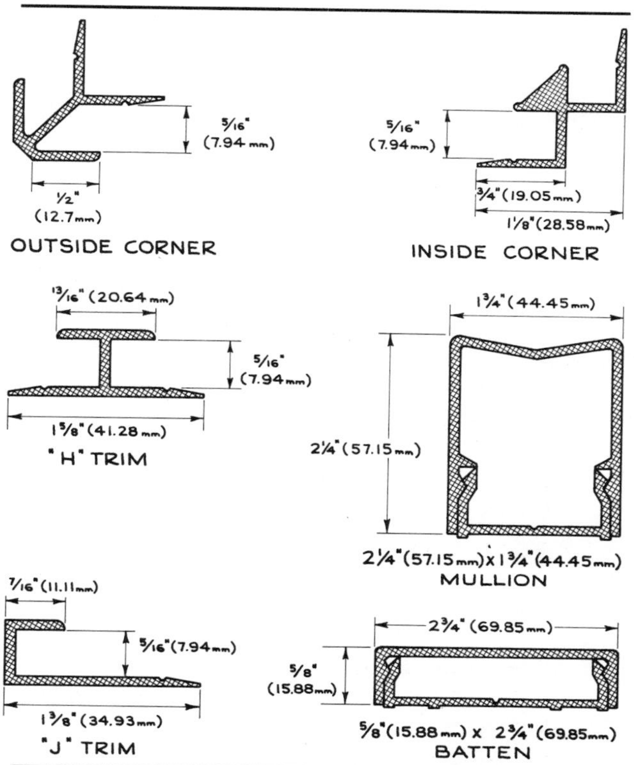

Figure A42 Extruded aluminum trim for aluminum siding and panels.

History and Manufacture

Rolling was one of the first fabrication methods applied to aluminum. Sheet is rolled from slab-shaped ingots weighing 4000 lb or more. The heated slabs are sent through powerful rollers for a series of passes to give the first breakdown. The elongated slab is then removed from this "hot mill" and taken to other mills, where it is cold-rolled down to the desired thickness and temper. Cold rolling gives a better finish and increases strength and hardness. For very thin sheets it may be necessary to anneal between final rollings to counteract hardness.

ALUMINUM, STRUCTURAL

Physical and Chemical Properties

When aluminum is used as a structural material, several important factors enter into the design considerations, all of them arising from its physical and chemical characteristics. Aluminum can be extruded; therefore a structural shape can be produced economically to meet the specific structural design requirements. Very corrosion-resistant aluminum alloys are available; therefore no painting is required, and the thickness of sections can be reduced since a safety margin is almost never necessary to cover loss of strength due to corrosion.

Aluminum is a very lightweight material, and aluminum girders and columns show increased efficiency with large bay spacings. The fact that the modulus of elasticity of aluminum alloys is lower than that of steel, however, means that buckling is a possibility and should always be checked; for example, in the flange and web, the compressive stress has to be checked for local buckling, and for the same reason the shear stress in the web must be checked.

In view of the recent advent of aluminum as a structural material and the current rapid expansion of knowledge in this field, the architect who proposes to design structurally with aluminum should undertake much more research than can come within the scope of this book and should consult a structural engineer.

Types and Uses

Alloy Requirements. The alloys most commonly used for aluminum structural shapes are 6061-T6, 6063-T5, 6063-T6, 2014-T6, and 2024-T4. High strength No. 7075-T6 (high in zinc) also finds certain structural applications. The 6xxx series is characterized by high corrosion resistance and moderate strength, and the 2xxx series by high strength and limited corrosion resistance. When designing structurally with aluminum, strength, cost, and corrosion resistance should always be considered in choosing the type of alloy. Furthermore, it is important not only to check the alloy for conditions of use but also to determine what other materials will be used, what their galvanic actions in contact with aluminum will be, and whether they contain acids or alkalis harmful to aluminum.

Typical Structural Shapes. The typical structural shapes shown in Figures *A43* to *A51* are generally made from the 2xxx and 6xxx alloys previously listed. Lengths vary from 12 to 25 ft (3.658 to 7.620 m) and must always be checked with manufacturers for general shape, alloy type, method of manufacture (extruded or rolled), and exact dimensional and shape limitations.

Joining Methods. Joining aluminum can present problems, especially on the construction site. For field work riveting is possibly still the most common method, although welding, with the advent of inert gas shielded-arc processes, no longer requires inordinate skill. Welding under controlled shop conditions is relatively easy and is used extensively for prefabricated items. A working knowledge of the various alloys and types of welding best suited for the circumstance is of course desirable (*see* Aluminum Alloys).

Pipe. Aluminum pipe can be used for structural supports. It is available in various sizes according to the American National Standard Schedule Numbers. A check should always be made as to whether the pipe (or column) should be filled with concrete, be filled with concrete and reinforcing, or be left hollow to take the required load. Table *A38* gives available dimensions, weights, and alloys.

Plate and Gratings. Aluminum plate, available in $\frac{1}{4}$ to 1 in. (6.35 to 25.4 mm) thicknesses and in various patterns, is used as a self-supporting flat surface for stair treads, access covers, and the like. Aluminum gratings are another structural item used in much the same way as aluminum plate and for similar special purposes.

History and Manufacture

Aluminum was first used as a structural material in commercial aviation, where light weight was the controlling factor, resulting in skin construction and sandwich panel construction. As a structural material for architecture aluminum is just in its infancy, and up to the 1950's only a few bridges, airplane hangars, and special structures had actually been built with it. It is obvious that in the future aluminum will become as basic a structural material as steel and reinforced concrete are today. At present, as happens with most new materials, it is regarded as a substitute for steel, and therefore the tendency has been to adopt or adapt currently used steel forms and design shapes. As the use of aluminum increases, there is no question that a completely new set of engineering techniques and structural shapes will evolve in the same manner that an intrinsic structural design has been developed for reinforced concrete.

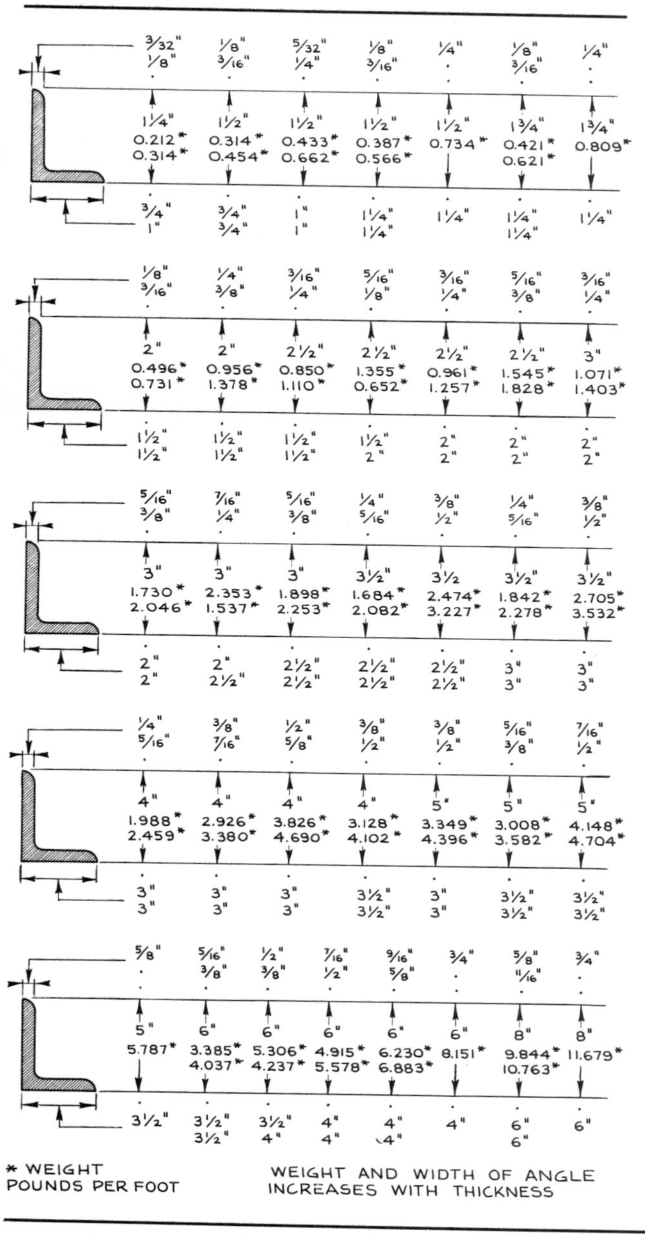

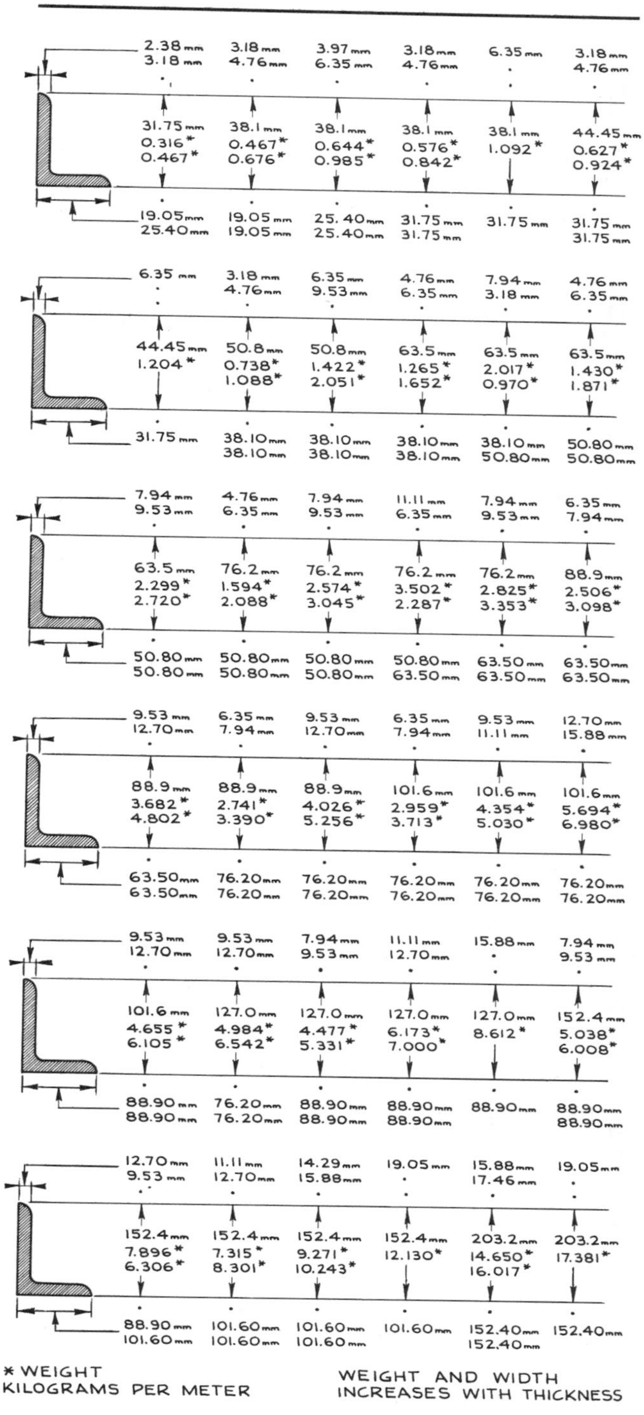

Figure A43 Structural shapes in aluminum: unequal-leg angles.

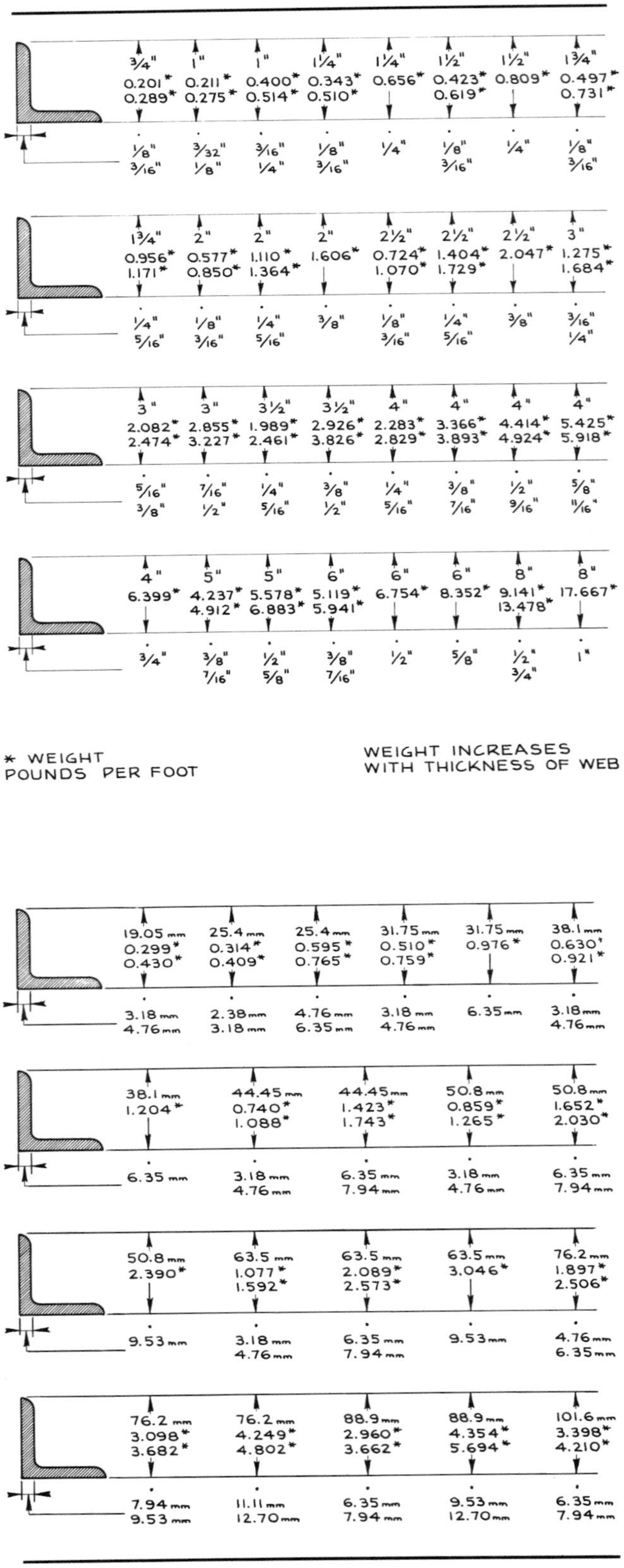

Figure A44 Structural shapes in aluminum: equal-leg angles.

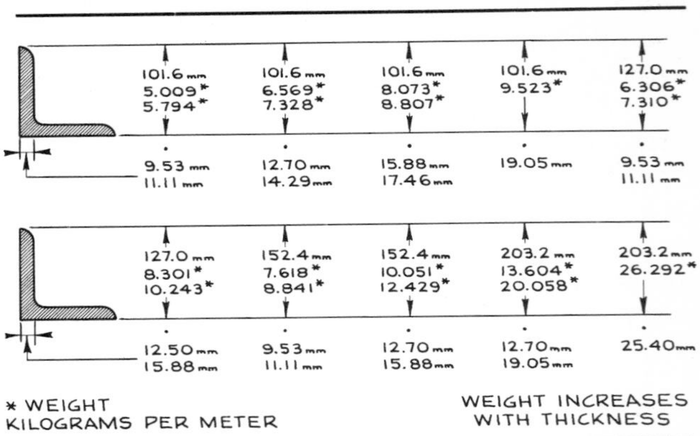

Figure A44 Structural shapes in aluminum: equal-leg angles (*continued*).

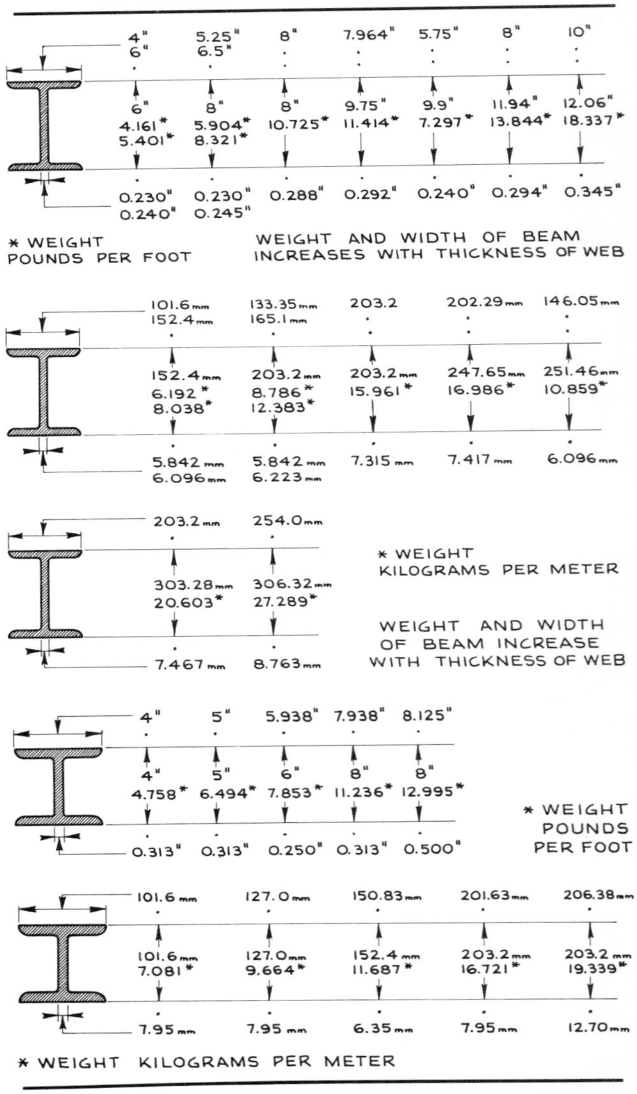

Figure A45 Structural shapes in aluminum: H-beams.

2.50"	2.50"	3.00"	3.00"	3.50"	4.00"	4.00"
3"	3"	4"	4"	5"	6"	6"
1.637*	2.030*	2.311*	2.793*	3.700*	4.030*	4.692*
0.13"	0.15"	0.15"	0.17"	0.19"	0.19"	0.21"

4.50"	5.00"	5.00"	5.50"	6.00"	6.00"	7.00"	7.00"
7"	8"	8"	9"	10"	10"	12"	12"
5.800*	6.181*	7.023*	8.361*	8.646*	10.286*	11.672*	14.292*
0.23"	0.23"	0.25"	0.27"	0.25"	0.29"	0.29"	0.31"

* WEIGHT POUNDS PER FOOT

63.5 mm	63.5 mm	76.2 mm	76.2 mm	88.9 mm
76.2 mm	76.2 mm	101.6 mm	101.6 mm	127 mm
2.436*	3.021*	3.439*	4.157*	5.506*
3.302 mm	3.810 mm	3.810 mm	4.318 mm	4.826 mm

101.6 mm	101.6 mm	139.7 mm	127 mm	127 mm
152.4 mm	152.4 mm	177.8 mm	203.2 mm	203.2 mm
5.997*	6.983*	8.632*	9.199*	10.452*
4.826 mm	5.334 mm	5.842 mm	5.842 mm	6.350 mm

139.7 mm	152.4 mm	152.4 mm	177.8 mm	177.8 mm
228.6 mm	254.0 mm	254.0 mm	304.8 mm	304.8 mm
12.443*	12.867*	15.308*	17.370*	21.269*
6.858 mm	6.350 mm	7.366 mm	7.366 mm	7.874 mm

* WEIGHT KILOGRAMS PER METER

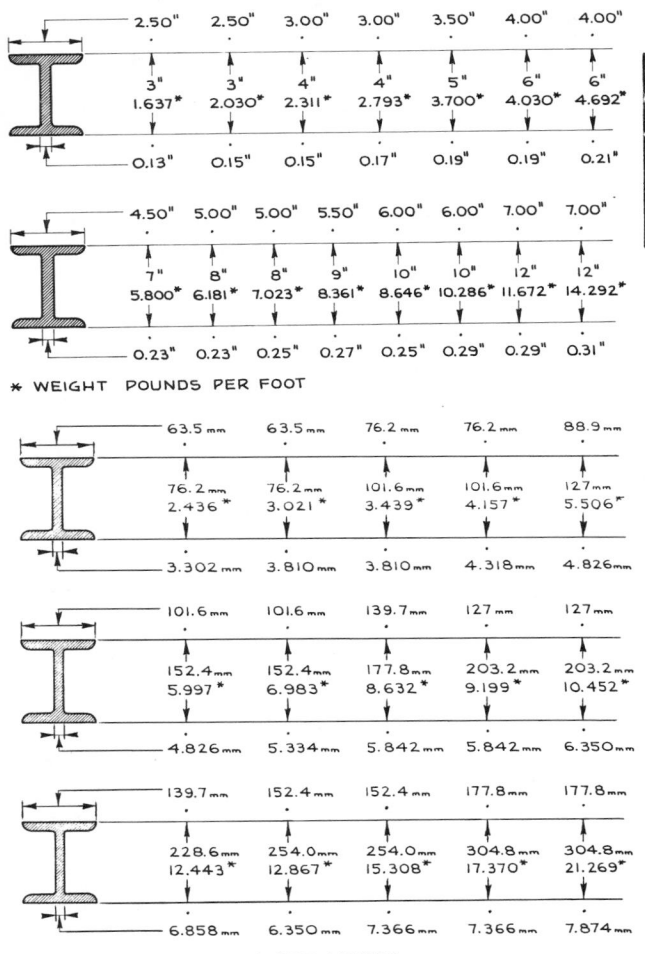

Figure A46 Structural shapes in aluminum: I-beams.

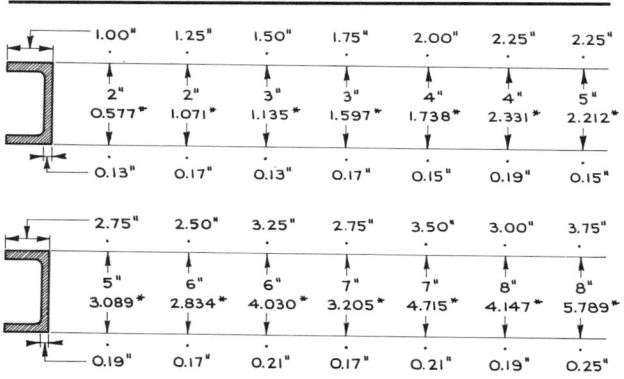

1.00"	1.25"	1.50"	1.75"	2.00"	2.25"	2.25"
2"	2"	3"	3"	4"	4"	5"
0.577*	1.071*	1.135*	1.597*	1.738*	2.331*	2.212*
0.13"	0.17"	0.13"	0.17"	0.15"	0.19"	0.15"

2.75"	2.50"	3.25"	2.75"	3.50"	3.00"	3.75"
5"	6"	6"	7"	7"	8"	8"
3.089*	2.834*	4.030*	3.205*	4.715*	4.147*	5.789*
0.19"	0.17"	0.21"	0.17"	0.21"	0.19"	0.25"

Figure A47 Structural shapes in aluminum: channels.

3.25"	4.00"	3.50"	4.25"	4.00"	5.00"
9"	9"	10"	10"	12"	12"
4.983*	6.970*	6.136*	8.360*	8.274*	11.822*
0.23"	0.29"	0.25"	0.31"	0.29"	0.35"

* WEIGHT POUNDS PER FOOT

25.4 mm	31.75 mm	38.1 mm	44.45 mm	50.8 mm	57.15 mm	57.15 mm
50.8 mm	50.8 mm	76.2 mm	76.2 mm	101.6 mm	101.6 mm	127.0 mm
0.859*	1.594*	1.689*	2.377*	2.586*	3.469*	3.292*
3.302 mm	4.318 mm	3.302 mm	4.318 mm	3.810 mm	4.826 mm	3.810 mm

69.85 mm	63.50 mm	82.55 mm	69.85 mm	88.90 mm	76.20 mm	95.25 mm
127.0 mm	152.4 mm	152.4 mm	177.8 mm	177.8 mm	203.2 mm	203.2 mm
4.597*	4.218*	5.997*	4.770*	7.017*	6.172*	8.615*
4.826 mm	4.318 mm	5.334 mm	4.318 mm	5.334 mm	4.826 mm	6.350 mm

82.55 mm	101.6 mm	88.9 mm	107.95 mm	101.6 mm	127.0 mm
228.6 mm	228.6 mm	254.0 mm	254.0 mm	304.8 mm	304.8 mm
7.416*	10.373*	9.132*	12.441*	12.313*	17.594*
5.842 mm	7.366 mm	6.350 mm	7.874 mm	7.366 mm	8.890 mm

* WEIGHT KILOGRAMS PER METER

Figure A47 Structural shapes in aluminum: channels (*continued*).

2 11/16"	2 11/16"	3 1/16"	3 1/8"	3 3/16"	3 1/4"	3 5/16"
3"	3"	4"	4 1/16"	4 1/8"	5"	5 1/16"
2.333*	3.381*	2.848*	3.575*	4.318*	6.192*	4.813*
0.250"	0.375"	0.250"	0.312"	0.375"	0.500"	0.375"

* WEIGHT POUNDS PER FOOT

68.26 mm	68.26 mm	77.79 mm	79.38 mm	80.96 mm
76.2 mm	76.2 mm	101.6 mm	103.19 mm	104.78 mm
3.472*	5.032*	4.238*	5.320*	6.426*
6.35 mm	9.53 mm	6.35 mm	7.92 mm	9.53 mm

82.55 mm	84.14 mm
127.0 mm	128.59 mm
9.215*	7.163*
12.70 mm	9.53 mm

* WEIGHT KILOGRAMS PER METER

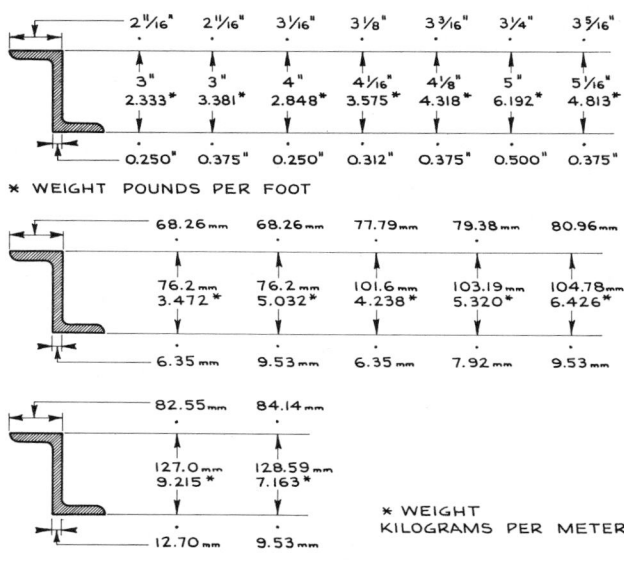

Figure A48 Structural shapes in aluminum: Z's.

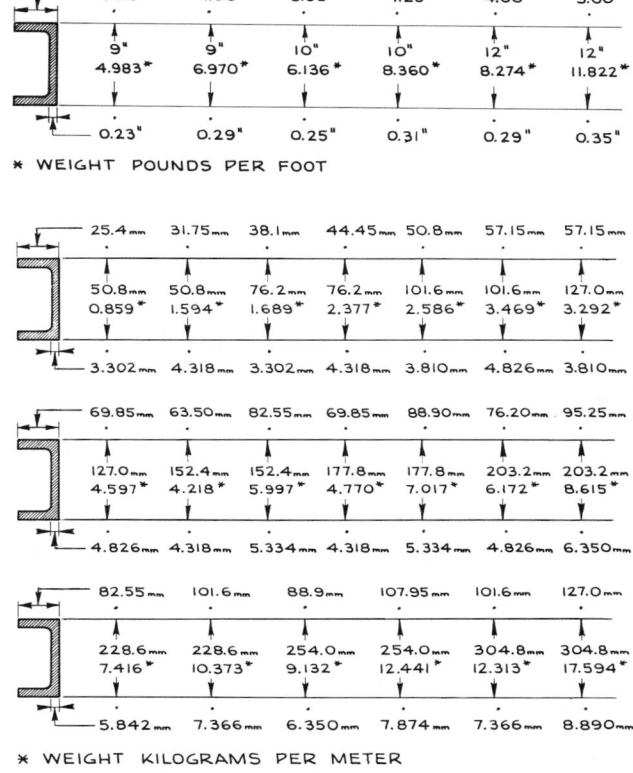

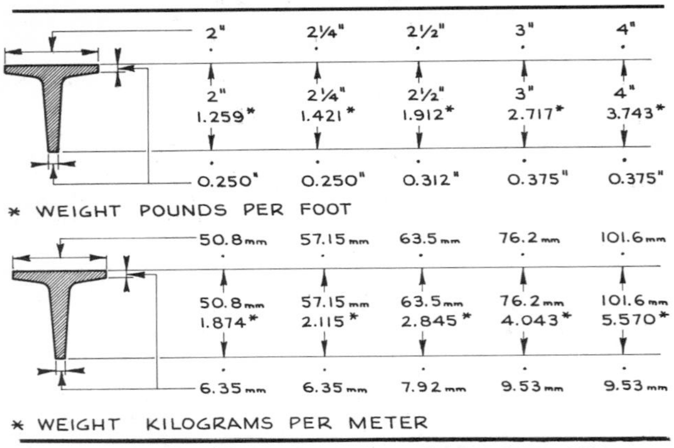

Figure A49 Structural shapes in aluminum: T's.

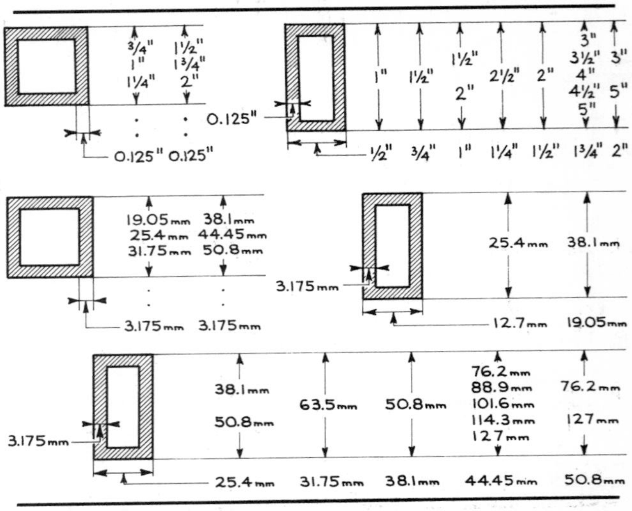

Figure A51 Structural shapes in aluminum: squares and rectangles.

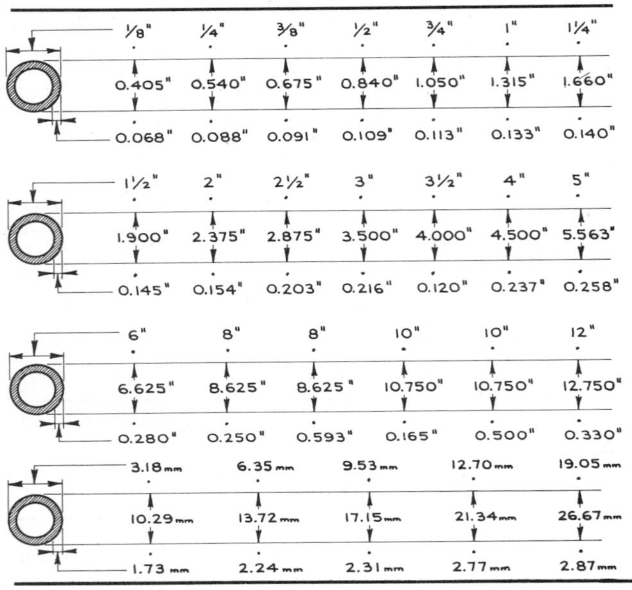

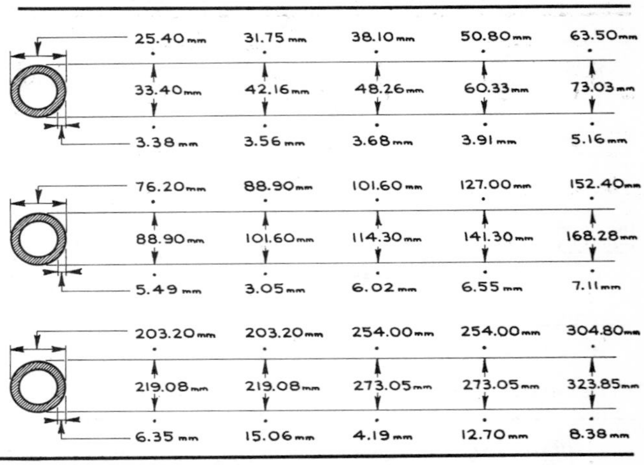

Figure A50 Structural shapes in aluminum: pipe.

Table A38 Typical dimensions, schedule numbers and weights of aluminum pipe

Size		Length		Schedule numbers	Outside diameter		Inside diameter		Wall thickness		Weight	
in.	mm	ft	m		in.	mm	in.	mm	in.	mm	lb/ft	kg/m
⅛	3.175	12	3.658	40	0.405	10.287	0.269	6.833	0.068	1.727	0.085	0.127
				80	0.405	10.287	0.215	5.461	0.095	2.413	0.109	0.162
¼	6.350	12	3.658	40	0.540	13.716	0.364	9.246	0.088	2.235	0.147	0.219
				80	0.540	13.716	0.302	7.671	0.119	3.023	0.185	0.275
⅜	9.525	12	3.658	40	0.675	17.145	0.493	12.522	0.091	2.311	0.196	0.292
				80	0.675	17.145	0.423	10.744	0.126	3.200	0.256	0.382
½	12.700	20	6.096	5	0.840	21.336	0.710	18.034	0.065	1.651	0.186	0.277
				10	0.840	21.336	0.674	17.120	0.083	2.108	0.232	0.347
				40	0.840	21.336	0.622	15.799	0.109	2.759	0.294	0.438
				80	0.840	21.336	0.546	13.868	0.147	3.834	0.376	0.560
				160	0.840	21.336	0.466	11.834	0.187	4.750	0.451	0.671

Table A38 Typical dimensions, schedule numbers and weights of aluminum pipe (*continued*)

Size		Length		Schedule numbers	Outside diameter		Inside diameter		Wall thickness		Weight	
in.	mm	ft	m		in.	mm	in.	mm	in.	mm	lb/ft	kg/m
$\frac{3}{4}$	19.050	20	6.096	5	1.050	26.670	0.920	23.368	0.065	1.651	0.237	0.353
				10	1.050	26.670	0.884	22.454	0.083	2.108	0.297	0.441
				40	1.050	26.670	0.824	20.930	0.113	2.870	0.391	0.582
				80	1.050	26.670	0.742	18.847	0.154	3.912	0.510	0.799
				160	1.050	26.670	0.614	15.596	0.218	5.537	0.670	0.997
1	25.400	20	6.096	5	1.315	33.401	1.185	30.099	0.065	1.651	0.300	0.446
				10	1.315	33.401	1.097	27.864	0.109	2.769	0.486	0.723
				40	1.315	33.401	1.049	26.645	0.133	3.378	0.581	0.865
				80	1.315	33.401	0.957	24.308	0.179	4.547	0.751	1.118
				160	1.315	33.401	0.815	20.701	0.250	6.350	0.984	1.464
$1\frac{1}{4}$	31.750	20	6.096	5	1.660	42.164	1.530	38.862	0.065	1.651	0.383	0.570
				10	1.660	42.164	1.442	36.627	0.109	2.769	0.625	0.930
				40	1.660	42.164	1.380	35.052	0.140	3.556	0.786	1.170
				80	1.660	42.164	1.278	32.461	0.191	3.851	1.037	1.543
				160	1.660	42.164	1.160	29.464	0.250	6.350	1.302	1.938
$1\frac{1}{2}$	38.100	20	6.096	5	1.900	48.260	1.770	44.958	0.065	1.651	0.441	0.656
				10	1.900	48.260	1.682	41.351	0.109	2.769	0.721	1.073
				40	1.900	48.260	1.610	40.894	0.145	3.683	0.940	1.400
				80	1.900	48.260	1.500	38.100	0.200	5.080	1.256	1.869
				160	1.900	48.260	1.338	33.985	0.281	7.137	1.681	2.602
2				5	2.375	60.325	2.245	57.023	0.065	1.651	0.555	0.876
				10	2.375	60.325	2.157	54.788	0.109	2.769	0.913	1.359
				40	2.375	60.325	2.067	52.502	0.154	3.912	1.264	1.881
				80	2.375	60.325	1.939	49.251	0.218	5.538	1.737	2.585
				160	2.375	60.325	1.689	43.901	0.343	8.712	2.575	3.832
$2\frac{1}{2}$	63.500	20	6.096	5	2.875	73.02	2.709	68.809	0.083	2.108	0.856	1.274
				10	2.875	73.02	2.635	66.929	0.120	3.048	1.221	1.817
				40	2.875	73.02	2.469	62.713	0.203	5.156	2.004	2.982
				80	2.875	73.02	2.323	59.004	0.276	7.010	2.650	3.944
				160	2.875	73.02	2.125	53.975	0.375	9.525	3.464	5.155
3	76.200	20	6.096	5	3.500	88.90	3.334	84.684	0.083	2.108	1.048	1.560
				10	3.500	88.90	3.260	82.804	0.120	3.048	1.498	2.229
				40	3.500	88.90	3.068	77.927	0.216	5.486	2.621	3.901
				80	3.500	88.90	2.900	73.660	0.300	7.620	3.547	5.279
				160	3.500	88.90	2.676	67.970	0.437	11.110	4.945	7.359
$3\frac{1}{2}$	88.900	20	6.096	5	4.000	101.60	3.834	97.384	0.083	2.108	1.201	1.787
				10	4.000	101.60	3.760	95.504	0.120	3.048	1.720	2.560
				40	4.000	101.60	3.548	90.119	0.226	5.740	3.151	4.689
				80	4.000	101.60	3.364	85.446	0.318	8.077	4.326	6.438
4	101.600	20	6.096	5	4.500	114.30	4.334	110.084	0.083	2.108	1.354	2.015
				10	4.500	114.30	4.260	108.204	0.120	3.048	1.942	2.890
				40	4.500	114.30	4.026	102.260	0.237	6.020	3.733	5.555
				80	4.500	114.30	3.826	97.325	0.337	8.560	5.183	7.713
				120	4.500	114.30	3.626	92.100	0.437	11.110	6.560	9.764
				160	4.500	114.30	3.438	87.325	0.531	13.487	7.786	11.583

Table A38 Typical dimensions, schedule numbers and weights of aluminum pipe (*continued*)

Size		Length		Schedule numbers	Outside diameter		Inside diameter		Wall thickness		Weight	
in.	mm	ft	m		in.	mm	in.	mm	in.	mm	lb/ft	kg/m
5	127.000	20	6.096	5	5.536	140.61	5.345	135.763	0.109	2.769	2.196	3.268
				10	5.536	140.61	5.397	134.543	0.134	3.404	2.688	4.000
				40	5.536	140.61	5.047	128.194	0.258	6.553	5.057	7.526
				80	5.536	140.61	4.813	122.250	0.375	9.525	7.188	10.697
				120	5.536	140.61	4.563	115.900	0.500	12.700	9.353	13.919
				160	5.536	140.61	4.313	109.550	0.625	15.875	11.400	16.965
6	152.400	20	6.096	5	6.625	168.28	6.407	162.738	0.109	2.769	2.624	3.905
				10	6.625	168.28	6.357	161.468	0.134	3.404	3.213	4.782
				40	6.625	168.28	6.065	154.051	0.280	7.112	6.564	9.769
				80	6.625	168.28	5.761	146.329	0.432	10.973	9.884	14.709
				120	6.625	168.28	5.501	139.725	0.562	14.275	12.590	18.736
				160	6.625	168.28	5.189	131.801	0.718	18.237	15.670	23.320
8	203.200	20	6.096	5	8.625	219.08	8.407	213.538	0.095	2.413	3.429	5.104
				10	8.625	219.08	8.329	211.567	0.130	3.302	4.635	6.903
				20	8.625	219.08	8.125	206.375	0.219	5.563	7.735	11.511
				30	8.625	219.08	8.071	205.003	0.242	6.147	8.543	12.714
				40	8.625	219.08	7.981	202.717	0.282	7.163	9.878	14.600
				60	8.625	219.08	7.813	198.450	0.355	9.017	12.330	18.350
				80	8.625	219.08	7.625	193.675	0.438	11.125	15.010	22.348
				100	8.625	219.08	7.439	188.951	0.519	13.183	17.600	26.192
				120	8.625	219.08	7.189	182.601	0.628	15.951	20.970	31.208
				140	8.625	219.08	7.001	177.825	0.711	18.059	23.440	34.882
				160	8.625	219.08	6.813	173.050	0.793	20.142	25.840	38.455
10	254.000	20	6.096	5	10.750	273.050	10.482	266.243	0.134	3.404	5.256	7.821
				10	10.750	273.050	10.420	264.668	0.165	4.191	6.453	9.603
				20	10.750	273.050	10.250	260.350	0.250	6.350	9.698	14.333
				30	10.750	273.050	10.136	258.454	0.307	7.798	11.840	17.620
				40	10.750	273.050	10.020	254.508	0.365	9.271	14.000	20.835
				60	10.750	273.050	9.750	247.550	0.500	12.700	18.930	28.172
				80	10.750	273.050	9.564	242.926	0.593	15.062	22.250	33.112
				100	10.750	273.050	9.314	236.576	0.718	18.237	26.610	39.601
12	304.800	20	6.096	5	12.750	323.850	12.438	315.925	0.156	3.962	7.268	10.816
				10	12.750	323.850	12.390	314.706	0.180	4.572	8.359	12.440
				20	12.750	323.850	12.250	311.150	0.250	6.350	11.550	16.589
				30	12.750	323.850	12.090	307.080	0.330	8.382	15.140	22.531
				40	12.750	323.850	11.938	303.225	0.406	10.312	18.520	27.561
				60	12.750	323.850	11.750	298.450	0.500	12.700	22.630	33.678
				80	12.750	323.850	11.376	288.950	0.687	17.450	30.620	45.569

ALUMINUM, STRUCTURAL: LIGHT-GAUGE SHAPES

Physical and Chemical Properties

Light-gauge shapes are formed from flat aluminum sheet, strip, and plate. Various extruded sections are used for sills, fascias, door trim, and other parts of light-gauge structural systems. These shapes are generally made from aluminum alloy 6061 of various tempers, the most common being 6061-TG with a tensile strength of 42,000 lbf/in.^{2} (289.59 MN/m^2). Aluminum light-gauge shapes meet mechanical property requirements and vary in thickness from 0.0875 to 0.24 in. (1.59 to 6.35 mm).

Types and Uses

These aluminum light-gauge shapes can be used for spans up to 20 ft (6.096 m), using the heavier thickness with 12-in. (304.8-mm) spacing. They are available in thicknesses ranging from $2\frac{1}{2}$ to 6 in. (38.1 to 152.4 mm), and even thicker. Joists vary in depth from 4 to 14 in. (101.6 to 255.6 mm). Joists are interchangeable with studs where greater thickness is required and/or a greater load must be supported. These aluminum light-gauge shape systems are used for residential, motel, multiple housing, and commercial construction, replacing wood stud and masonry construction (*see* Figure *A52*).

Application

When choosing a light-gauge aluminum framing system, it is always advisable to check with the manufacturers for the latest data on this type of construction. These shapes can be installed directly at the site of construction or can be prefabricated off site or on site in a temporary enclosure. When this type of system is chosen, the selection and method of installing all interior and exterior materials should be checked. These shapes are also used for constructing prefabricated storage sheds, garages, and hothouses.

Accessories. Accessories for this type of construction include bridging, bolts, nuts, screws, and anchors, as well as devices for fastening units together, such as clips, interlocking devices, self-tapping screws, and the like. It is always necessary to check that all accessories are of materials compatible with aluminum to prevent galvanic action.

Condensed Checklist

1. Complete shop drawings, details, and erection drawings, to be supplied by the manufacturer, must be specified.
2. The type of floor and roofing materials must be determined and safe loads must be calculated, be-

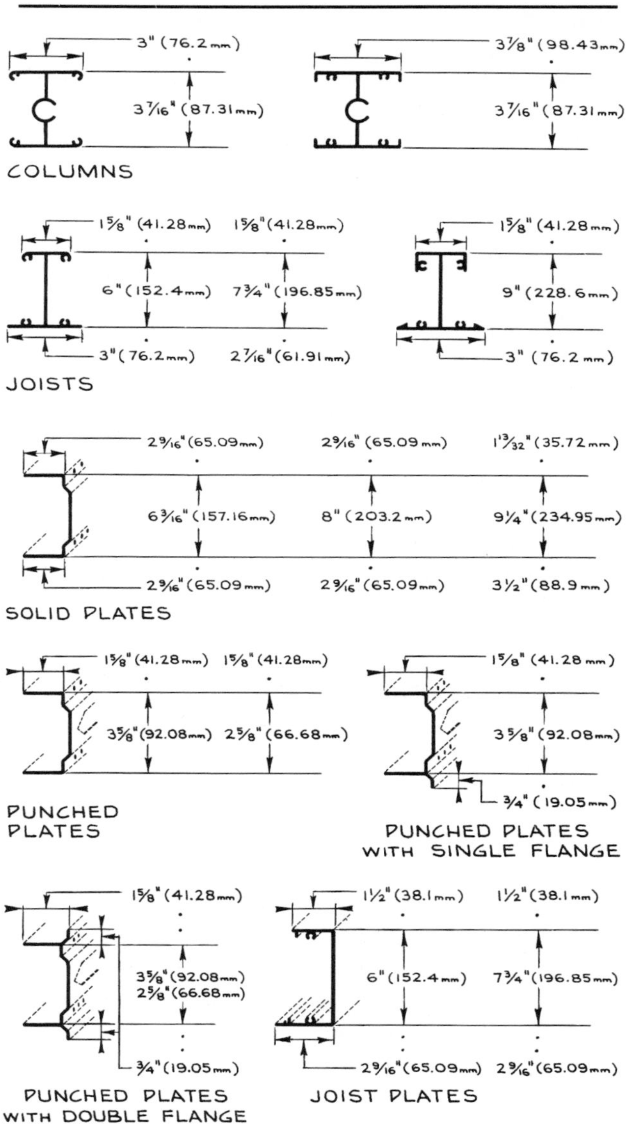

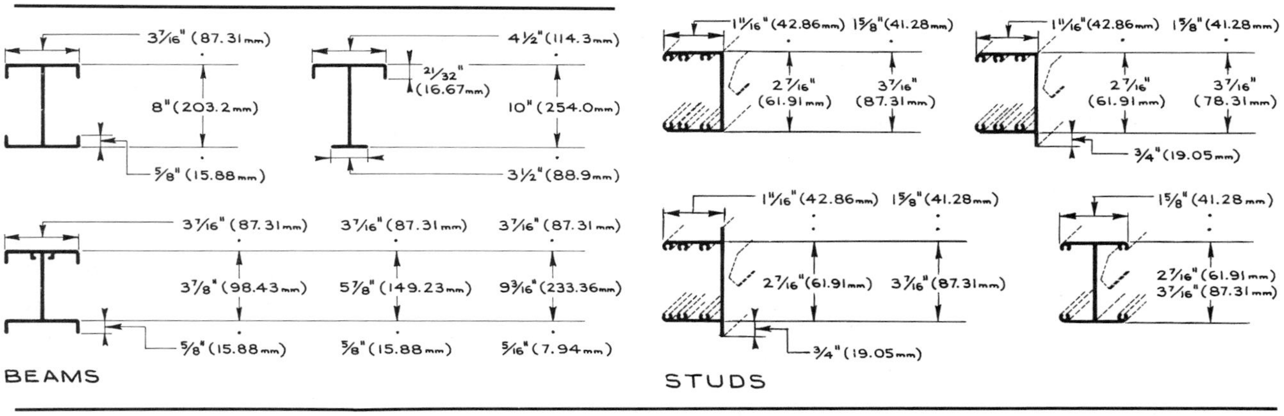

Figure A52 Types of light-gauge aluminum sections and studs.

cause these will control the size, depth, and shape, as well as the spacing, of the light gauge shapes.

3. The types of finish materials for floors, walls, and ceilings must be selected for their suitability for application to this type of structural system. The manufacturers should be consulted for the best methods of applying their materials.
4. A check should be made regarding the type of shape to be used, whether it should be solid or punched to accommodate piping, ducts, wiring, etc.
5. The method of anchoring these systems to the foundations and at the joint between wall and roof should be checked.
6. Locations of any openings for stairs, chimneys, vents, ducts, etc., should be checked in relation to the type of framing.
7. With roof overhangs, balconies, and cantilevers of any type, an important consideration is that light-gauge aluminum shapes can cantilever in one direction, but only with difficulty in two directions, within the same plane.
8. Local, municipal, and state codes should be checked, as well as the fire rating codes of insurance companies, Fire Underwriters, and any other governing agencies having jurisdiction in the area of construction.
9. The possibility of galvanic action due to any metals or materials that are not compatible with aluminum should be checked.
10. The ability of manufacturers and/or erectors to supply all necessary components for the structure should be ascertained.
11. The types of self-tapping screws for connecting light-gauge aluminum units together and also for installing the materials to be applied to the aluminum frame should be determined, and a check made that the correct types are used.
12. When off-site prefabrication is selected as the method of construction, one should check which manufacturers in the area where the proposed construction is to be erected are available for this type of construction.
13. All environmental control systems should be checked for their installation requirements in a light-gauge aluminum construction. Included should be the electrical, plumbing, heating, ventilation, and air-conditioning systems.
14. One should always check the type of tools, machines, and equipment required for light-gauge aluminum construction.
15. On multistory buildings or complicated commercial structures, one should always consult with a structural engineer.

Conditions Favorable to the Use of Light-Gauge Aluminum Shapes

1. Where the design has clear spans not exceeding 30 ft (9.144 m).
2. Where fire resistance is an important consideration in designing the structural system.
3. Where a strong, rigid, rotproof, termiteproof type of construction is required.
4. Where the type of construction requires a prefabricated system.
5. Where interior spaces must be subdivided with floor-to-ceiling partitions.
6. For interior partitions where electrical service and piping are required.

Conditions Unfavorable to the Use of Light-Gauge Aluminum Shapes

1. Where the locality is not near manufacturers or erectors of this type of construction system.
2. Where the structure requires a better than 4-hour fire-resistance rating.

ALUMINUM WINDOWS

Physical and Chemical Properties

Windows fabricated from aluminum alloys are lightweight yet strong. They do not stain adjacent materials. Because they are made of metal, they are rotproof and termiteproof, and will not swell, split, or warp. Also, because aluminum is very easily extruded, it can be formed into small, strong shapes which can be made as complex as necessary to provide for all the special requirements of condensation, weatherstripping, and glazing within the form of the extrusion.

Corrosion Resistance. Variations in the alloys used for the same extrusion can alter the strength and corrosion resistance of the finished product. Resistance to both corrosion and abrasion can be improved by applying an appropriate electrochemical, vinyl-coated, or baked enamel finish.

Relation of Alloys to Extrusions. Alloys 6063-T5, 6063-T6, and 6061-T6 are generally used for window extrusions. The specific alloys in current use may change, but it remains important to know the comparative properties of the alloys being used and the comparative size and efficiency of the extrusions.

Aluminum windows are required to meet rigid standards established by the Architectural Aluminum Manufacturers Association. These standards cover structural strength, water penetration, wind loads, air infiltration, and alloys, hardware, trim, etc.

Types and Uses

Aluminum window units are made in modular and nonmodular sizes for residential, commercial, monumental, and specialized types of construction. Rigid standards called "quality-approved" have been set up for the industry by the Architectural Aluminum Manufacturers Association, and certain classification designations (listed in Table *A39*) indicate that these standards have been complied with by the manufacturer of the window unit showing this designation. Figures *A53* to *A65* show some of the typical shapes and sizes of windows available in aluminum.

In addition to the types of windows illustrated, there are also vertically pivoted, utility, top hung, and continuous windows, not to mention the new types constantly being introduced by the various manufacturers. Among them are windows with a 2-in. (25.4-mm) air space between the outer and the inner window panes in which narrow-slat venetian blinds with remote control can be installed; replacement windows which can be installed in existing buildings just by removing the worn-out sash; and windows with various methods of removing frame and glass for easy cleaning.

Table A39 "Quality-approved" designations of the Architectural Aluminum Manufacturers Association

Type of window	Classification designation: Residential	Commercial	Monumental
Double-hung and single-hung	DH-B1	DH-A2	DH-A3 and A4
Casement	C-B1	C-A2	C-A3
Projected	P-B1	P-A2	P-A3
Awning	A-B1	A-A2	
Horizontal sliding	HS-B1 and B2	HS-A2	HS-A3
Jalousie	J-B1		
Jal-awning	JA-B1		
Vertical sliding	VS-B1		
Top-hinged		TH-A2	TH-A3
Vertically pivoted		VP-A2	VP-A3

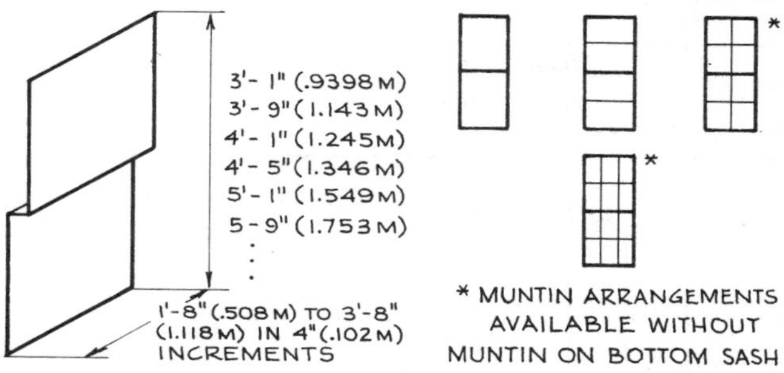

Figure A53 Sizes of double-hung aluminum windows.

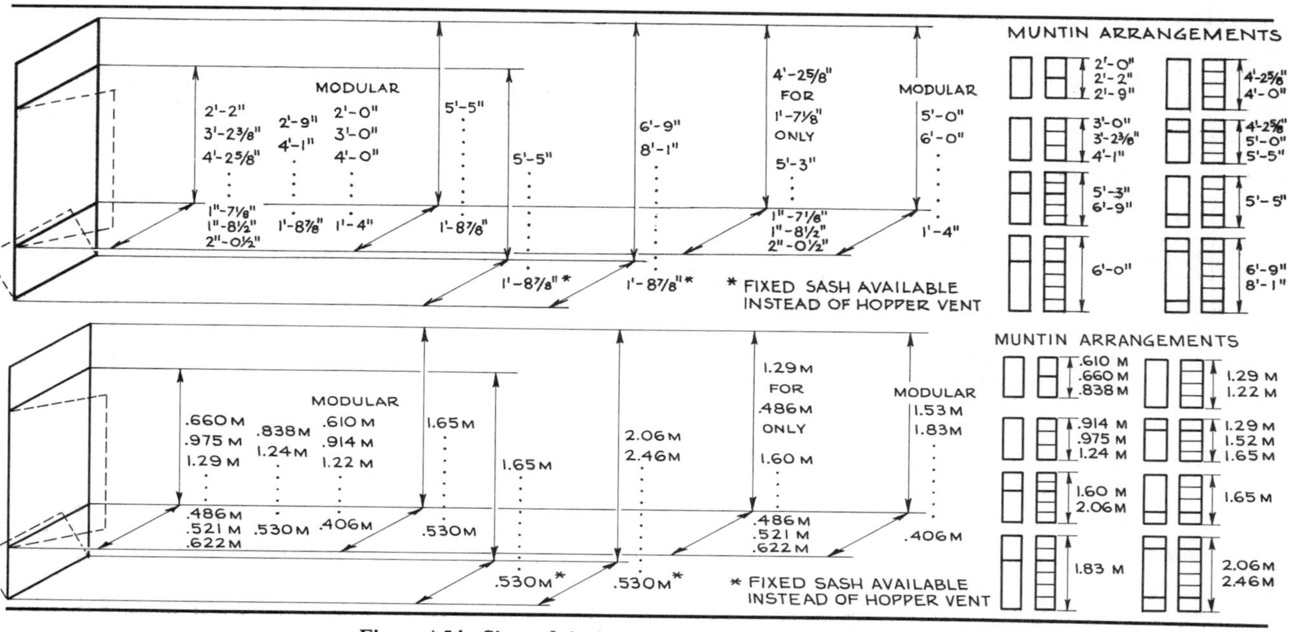

Figure A54 Sizes of single-casement aluminum windows.

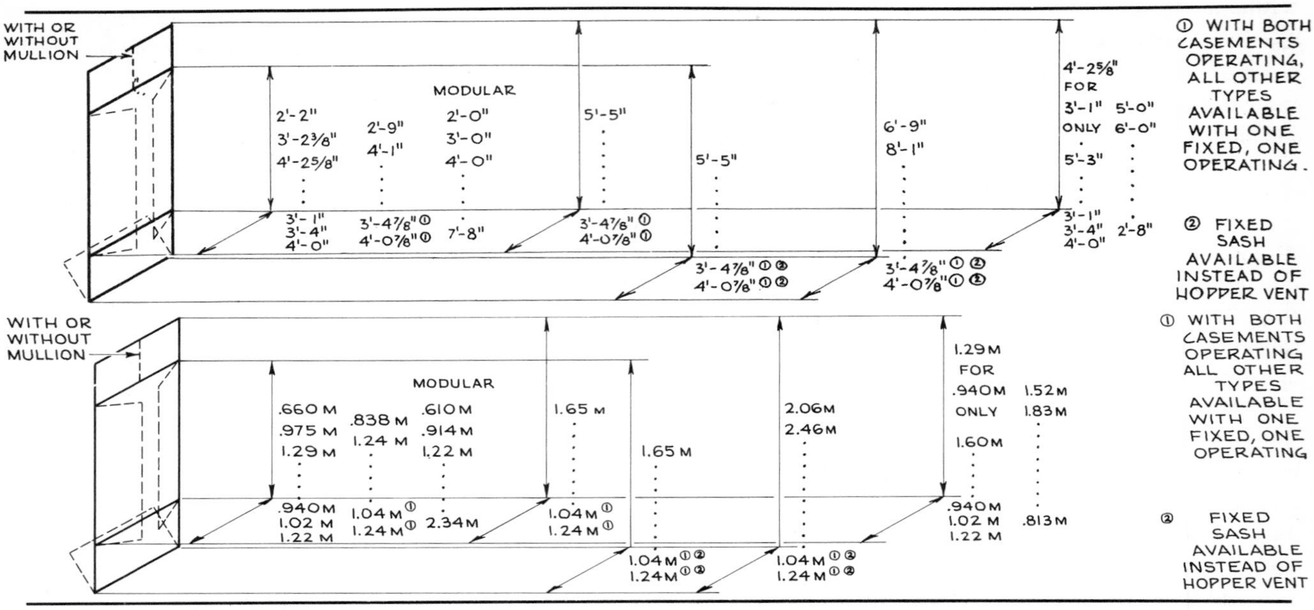

Figure A55 Sizes of double-casement aluminum windows with same horizontal mullions as single-casement windows.

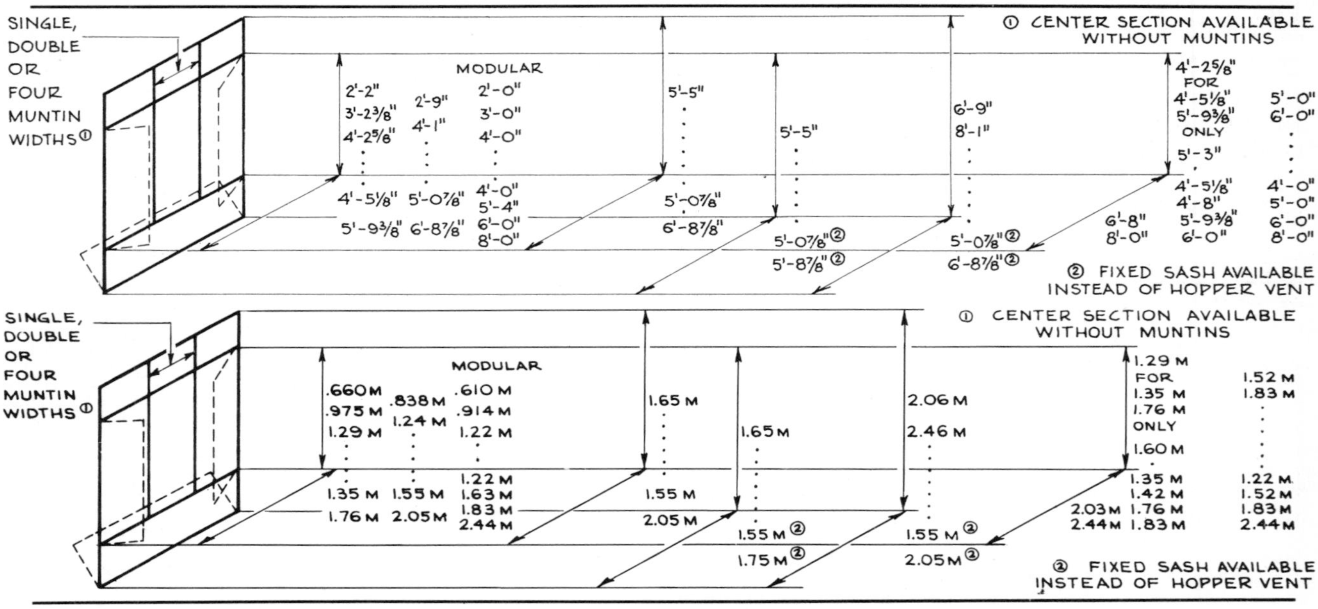

Figure A56 Sizes of double-casement aluminum windows with fixed center section with same horizontal mullions as single-casement windows.

In selecting an aluminum window, a check should be made of the manufacturers' latest data.

Store Fronts. Store fronts, where the design includes doors, glass display areas, and large expanses of fixed glass related into effective combinations, represent a specialized field of the aluminum window industry. Many shapes and methods of installation have been developed for just this purpose (*see* Aluminum Doors).

Custom Shapes. Today extrusion dies are relatively inexpensive, and it is economically feasible to manufacture special shapes for a particular job requirement if it is large enough, for example, a project like the World

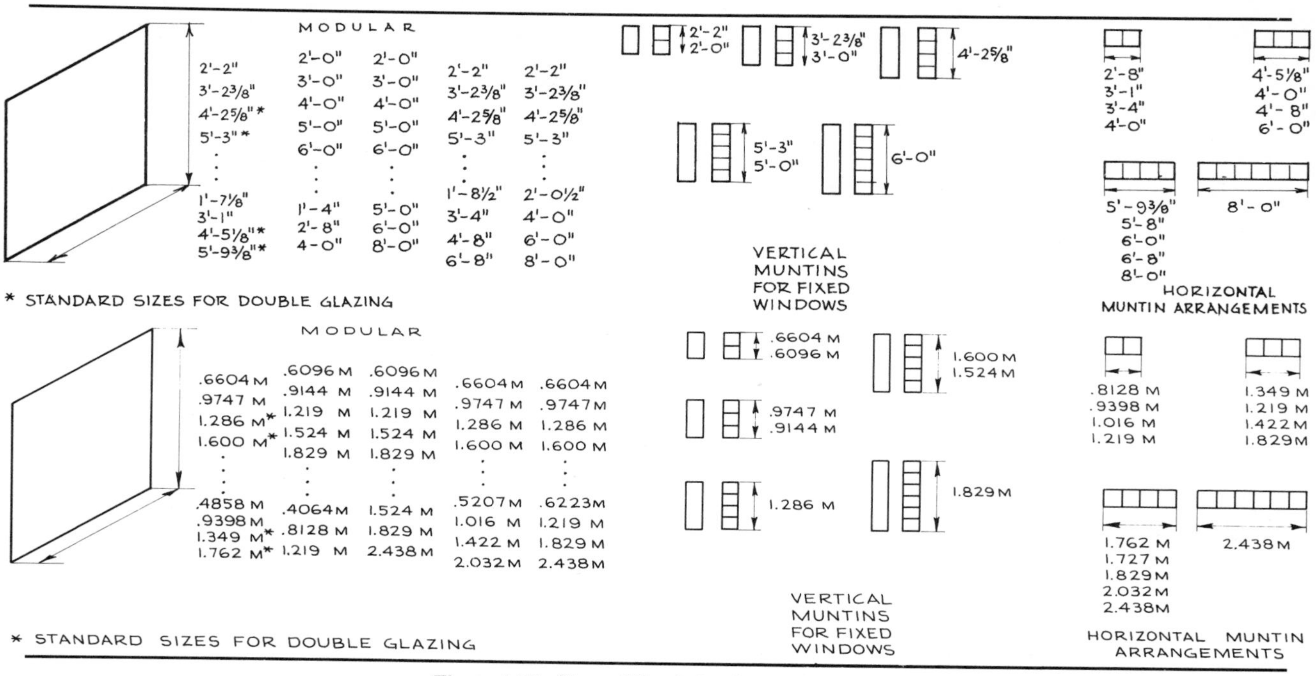

Figure A57 Sizes of fixed aluminum windows.

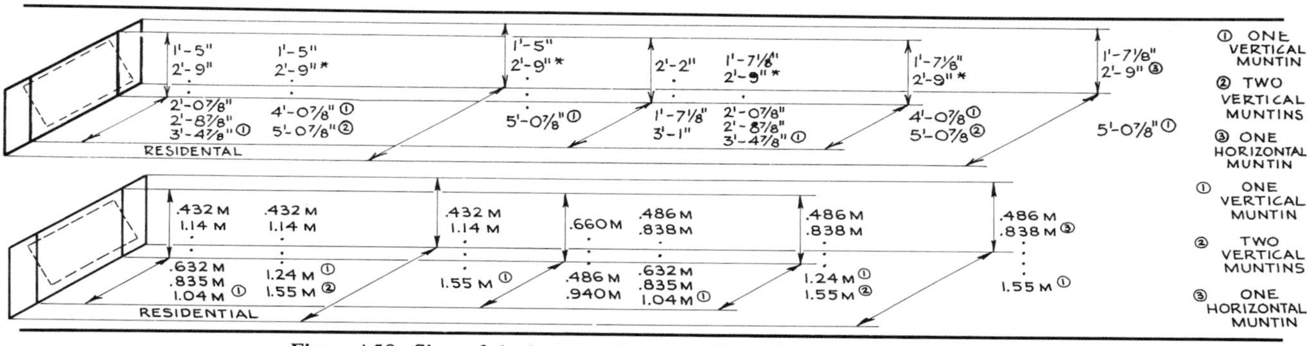

Figure A58 Sizes of single out-projected and in-projected aluminum windows.

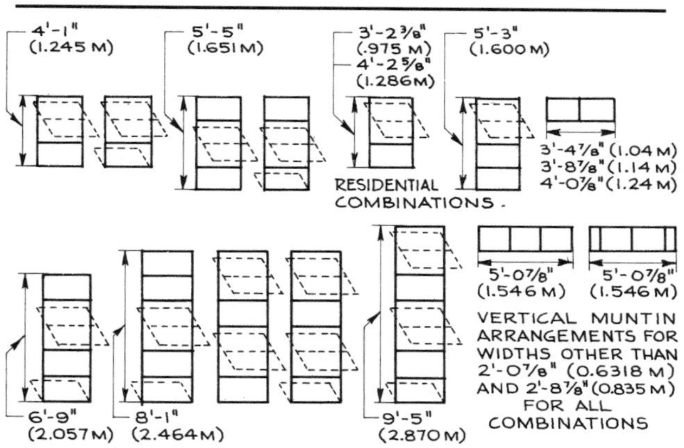

Figure A59 Heights of combination out-projected, in-projected, and fixed sections of aluminum windows.

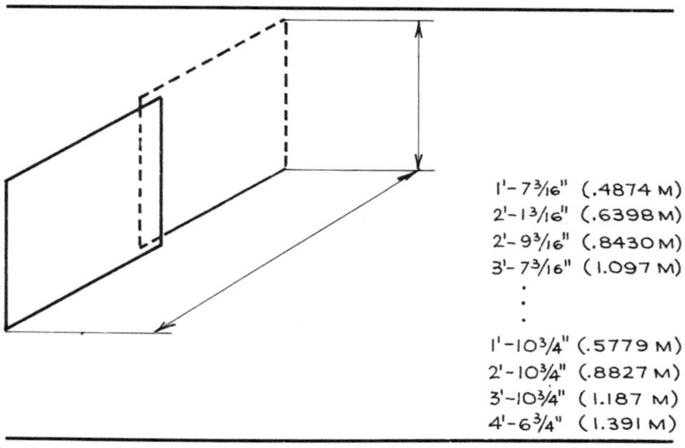

Figure A60 Sizes of residential sliding aluminum windows available with two sliding sashes or one sliding and one fixed sash.

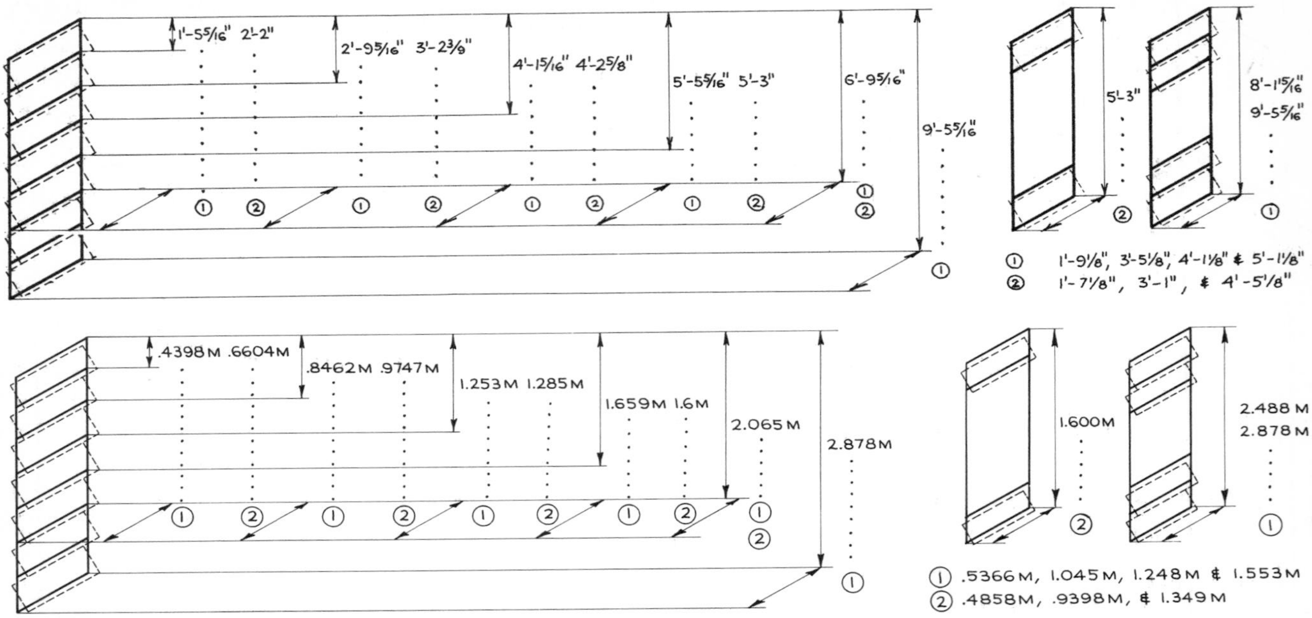

Figure A61 Sizes of aluminum awning windows.

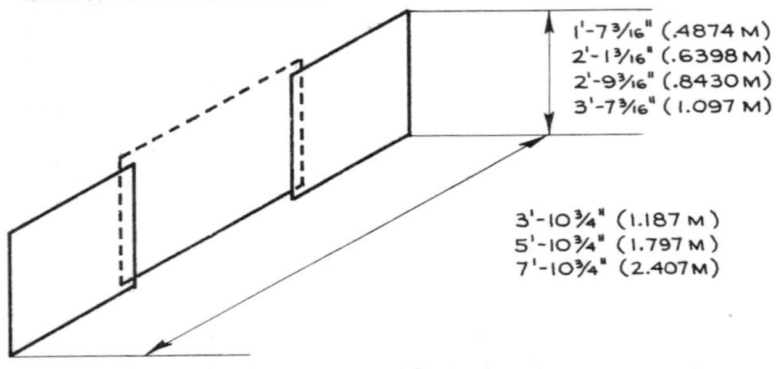

Figure A62 Sizes of residential sliding aluminum windows with fixed center section.

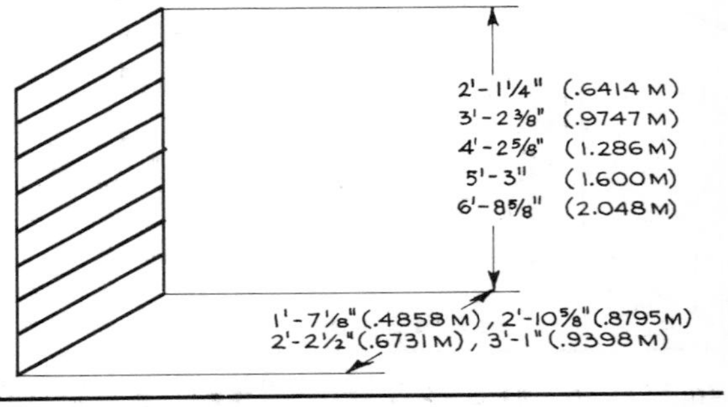

Figure A63 Sizes of jalousie aluminum windows (size of individual louvers has been standardized).

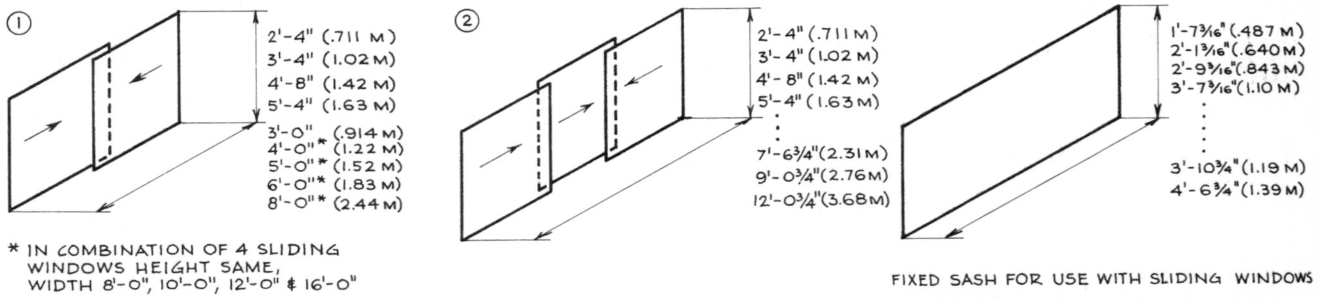

Figure A64 Sizes of sliding (1) and triple sliding (2) aluminum windows with rollers at bottom.

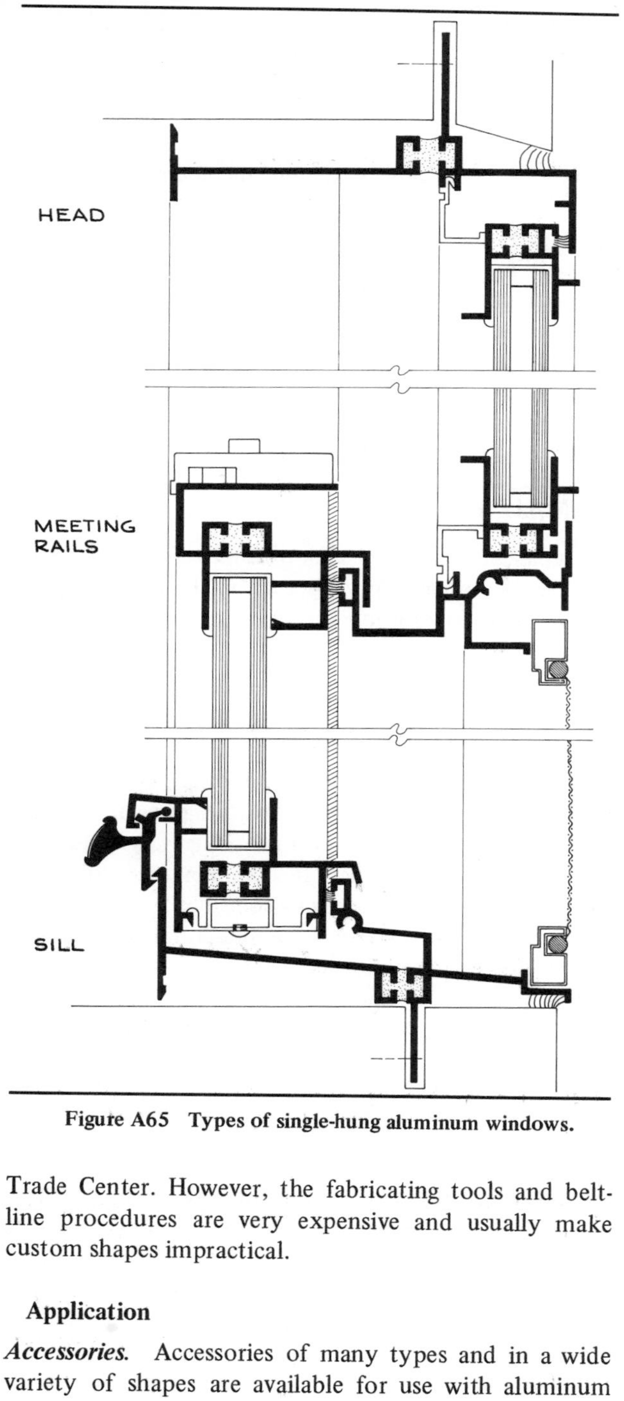

Figure A65 Types of single-hung aluminum windows.

Trade Center. However, the fabricating tools and belt-line procedures are very expensive and usually make custom shapes impractical.

Application

Accessories. Accessories of many types and in a wide variety of shapes are available for use with aluminum windows. These include screens, storm windows, sub-frames, sills, stools, weatherstripping, flashing, trim, glazing beads, mullions, muntins, screws, nuts and bolts, anchors, and miscellaneous shapes for bracing, strengthening and covering joints, mullions, frames, and the like. These accessories must all be of a material compatible with aluminum so that galvanic action cannot occur or the material will not attack aluminum chemically.

Hardware. Aluminum hardware for use with all types of aluminum windows is available and generally comes with the windows. Hardware for aluminum windows can also be made of or plated with other metals, but unless the metal is compatible with aluminum, insulation of the metal is necessary to avoid galvanic action. Effective insulation is provided by protective paints, coatings, or plating with aluminum, nickel, cadmium, or zinc.

Accessories and hardware are fabricated from aluminum alloy No. 5050-O (for items cut from sheet), from Nos. 34 and 356 (for cast items), and from No. 6063-T5 (for extruded items). Fastening accessories such as screws, nuts, and bolts are generally fabricated from Nos. 2911-T3 and 2017-T4.

Condensed Checklist

1. The windows selected should be "quality approved."
2. The type of aluminum alloy used must provide the necessary strength and corrosion resistance.
3. When designing for double glazing in windows, the type of window should be checked to make sure that it can take dual glazing.
4. All metals with which the window is likely to come in contact should be compatible with aluminum. If the contacting metal is anything other than non-magnetic stainless steel, cadmium, zinc, or nickel bronze, the aluminum should be insulated to avoid galvanic action.
5. Windows should be delivered with a protective coating against lime, mortar, and other materials which may drop, spatter, or otherwise come in contact with them during construction.
6. The caulking and glazing compounds, gaskets, shims, etc., used must be the correct types for aluminum.
7. Moving hardware parts or window members made of aluminum must be isolated from direct contact with each other. Otherwise, the protective oxide or other finish will wear off at points of moving contact.
8. The type of finish applied should be suitable for the conditions of use. If color applied by anodizing is desired, the finish should be checked for durability and wearing characteristics.
9. When replacement windows are selected, the specifications must require that field measurements be taken and shop drawings be submitted.

10. When aluminum windows having baked enamel finish or treated with an organic coating are selected, a check should be made with the manufacturers for data concerning the length of weathering life of the coating, the colors available, and the correct methods of installation.
11. When selecting the type of aluminum windows for a building, a check should be made of the windows in relation to available options–single glass or insulating glass, methods of installing glass from either the inside or the outside, and methods and ease of maintenance.

Conditions Favorable to the Use of Aluminum Windows

1. Where a strong, lightweight, corrosion-resistant material is required for windows.
2. Where maintenance must be kept to a minimum.
3. Where initial painting is not desired.
4. Where narrow muntins, mullions, and frames are desired for design purposes.
5. Where permanently colored window frame, sash, and window trim are required.
6. Where an existing building requires the replacement of existing windows, trim, etc.
7. Where storm windows and doors are required.

Conditions Unfavorable to the Use of Aluminum Windows

1. Where windows are to help support structural loads. This is possible with aluminum but much more difficult than with other metals.
2. Where fire resistance is important.

History and Manufacture

Although aluminum windows were being manufactured before World War II, thay have become commonplace only since 1945. This is due in part to the technological advances made during the war years and in part to the popularity of large glass areas for store fronts. This trend led to the designing of many specialized items for framing and anchoring doors and windows which were applicable to other types of construction.

Aluminum windows are made from extrusions which are fabricated by heating cast cylindrical billets of the appropriate alloy to a plastic condition at 600 to 800°F (315.56 to 426.56°C) and forcing them under pressure through a steel die with the required shape of opening. The billets range from 4 to 6 in. (101.6 to 152.4 mm) in diameter; pressures are as high as 5500 lbf/in.^{2} (8492 MN/m^2). Extrusions are made in long lengths which are straightened by stretching if necessary, cut into shorter lengths, and put into curing ovens. The windows are made up by assembling the fabricated frame parts in jigs where the sections are either welded or attached with screw fasteners; stiffening members are then installed, holes are drilled, and weatherstripping, hardware, and other accessories are attached, all in a continuous belt-line system.

The trend is increasingly toward package units consisting of window with frame and complete accessories; window and frame as part of a curtain wall system; and various combinations that simplify construction (*see* Aluminum Doors).

AMALGAM

An amalgam is defined as any combination or mixture of diverse elements. In construction the amalgams used are various alloys of mercury with other metals such as tin or silver.

AMMONIA

Ammonia (NH_3) is one of the important compounds of nitrogen. It is a colorless gas with a pungent odor. It liquifies easily under pressure and is very soluble in water, alcohol, and ether. In common usage the word "ammonia" refers to aqueous solutions of the gas. The only applications related to architectural materials are in refrigeration, textiles, ceramics, the manufacture of rayon and rubber, the fireproofing of wood, and the nitriding of steel. Ammonia is a commonly used cleaning agent.

ANCHORS

Anchors are made of various metals, including iron, steel, aluminum, bronze, copper, and stainless steel. In choosing a metal a check should be made of the possibility of electrolytic action between it and any metal with which it may come in contact. There are two types of anchors.

1. Anchors that are installed as an integral part of the basic construction (*see* Figure *A66*). Such anchors are used primarily either to tie building units, usually masonry, into a unified, stronger whole, or to provide a means whereby the other structural or finish materials can be attached and secured to the basic structure. In wood construction they are one of the means of anchoring together structural members (*see* Figure *A67*).

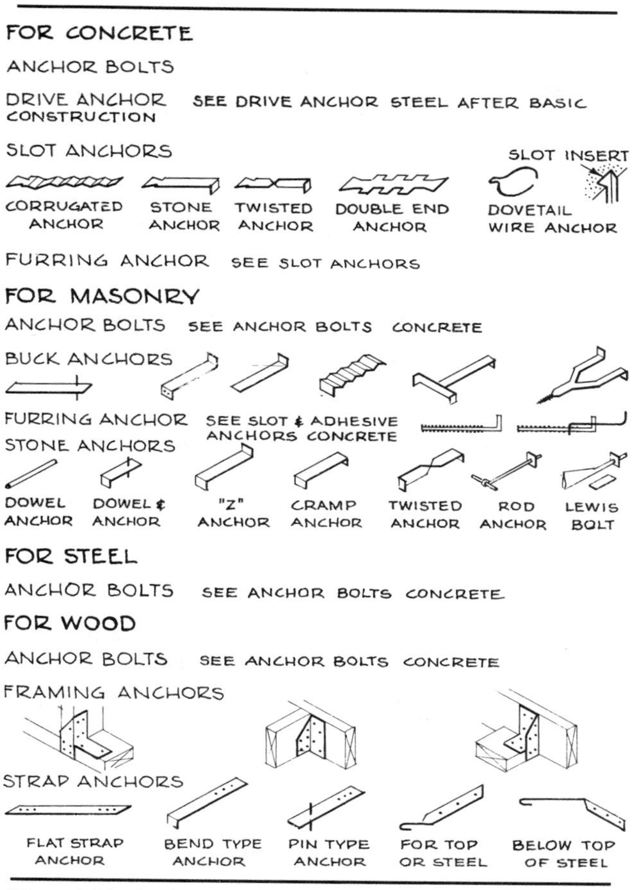

Figure A66 Anchors installed as integral part of basic construction.

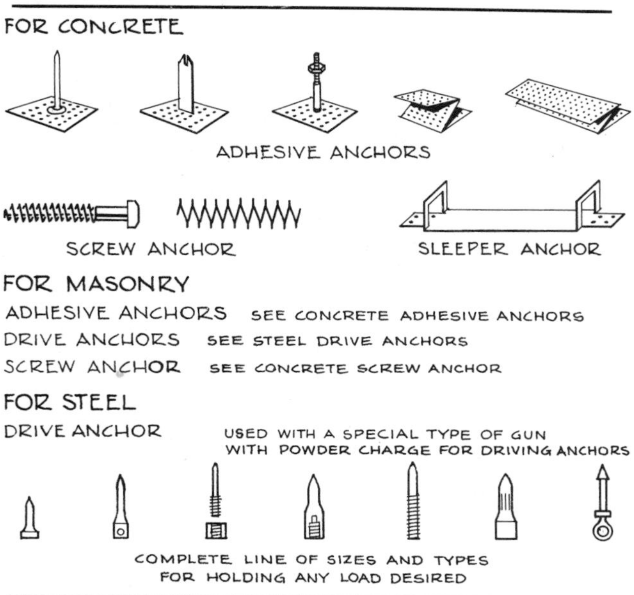

Figure A68 Anchors installed after basic construction.

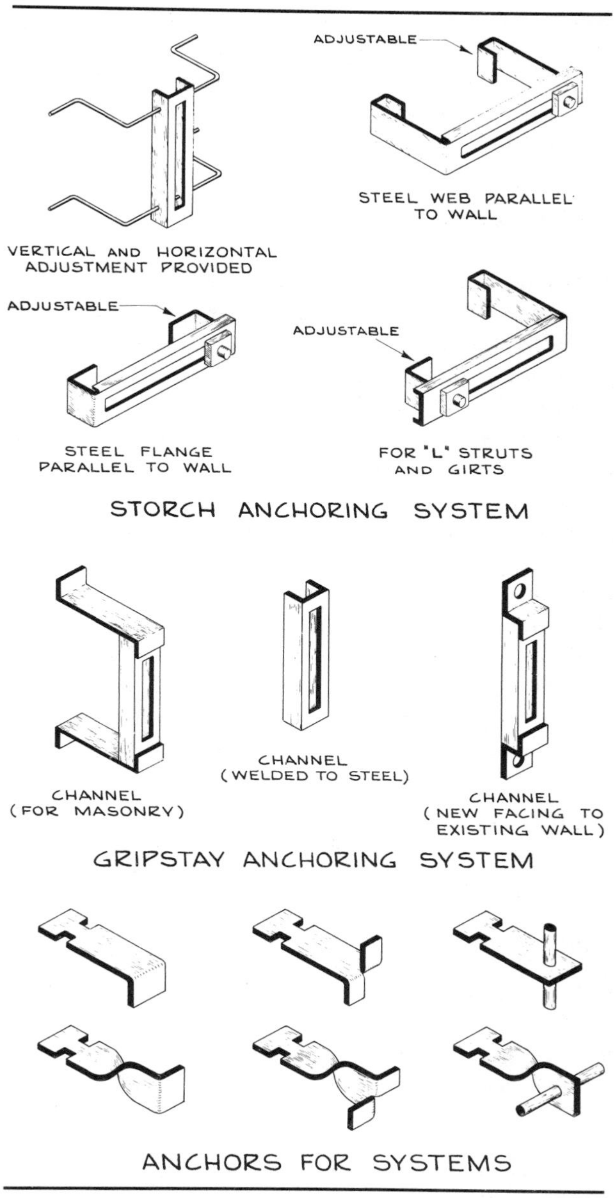

Figure A67 Types of anchoring systems.

2. Anchors that are installed after the basic construction is completed (*see* Figure *A68*). These have the same purpose as the preceding type but must somehow be attached to a completed material. The usual means of attachment are adhesive or cementing materials; the drilling of holes and insertion of some material or device that will expand in the hole, thereby giving the desired anchorage; or, in the case of wood members, nailing something to which other materials can be attached. A more recent device is the use of a driving or penetrating force (i.e., a gun-type mechanism using an

explosive to attain penetration) to create an anchorage to which other materials can be attached.

Other types of anchors are illustrated as part of the accessories used with specific materials. (*See* Screws, Nuts and Bolts, and Related Devices.)

ANNEALING

Annealing is a process of heating followed by cooling. Its purpose may be to eliminate the effects of cold working, to relieve internal stresses, or to improve electrical, magnetic, or other properties of metal, glass, and other materials as they are heated to above the critical or recrystallization temperature and then cooled to produce a recrystallized grain structure with the desired characteristics. Heating and gradual cooling are required for steel and glass but not for copper and brass.

Steel is annealed to reduce hardness, improve machinability, facilitate cold working, and obtain desired physical, mechanical, and other properties.

Copper is annealed for cold working to obtain the optimum combination of ductility and strength.

Annealing glass produces heat-strengthened, structural, and tempered types of glass.

ANTIMONY

Physical and Chemical Properties

Symbol: Sb
Atomic number: 51
Specific gravity: 6.691 (20°C, 68°F)
Melting point: 630.74°C, 1167.33°F
Boiling point: 1750°C, 3182°F
Ultimate tensile strength: 1500 lbf/in.^{2}, (10.34 MN/m^2)

Antimony is a shiny, silvery-white, hard, brittle, crystalline metal that is neither malleable nor ductile. Its chemical properties resemble those of arsenic. At ordinary temperatures it oxidizes slightly and on heating oxidizes readily.

Commercial Forms. Refined antimony is called regulus and is available in slab, lump, and powder form (*see* Table *A40*).

Table A40 Typical analysis of standard refined antimony

Sb	Cu	Fe	Pb	As	Zn	S	Co, Ni
(percent of content)							
99.6	0.046	0.004	0.102	0.092	0.034	0.086	0.028

Types and Uses

Antimony in compound form has several important uses directly related to architectural materials. Its uses in metallic alloy form are noted by asterisks in Table *A41*.

Table A41 Uses of antimony

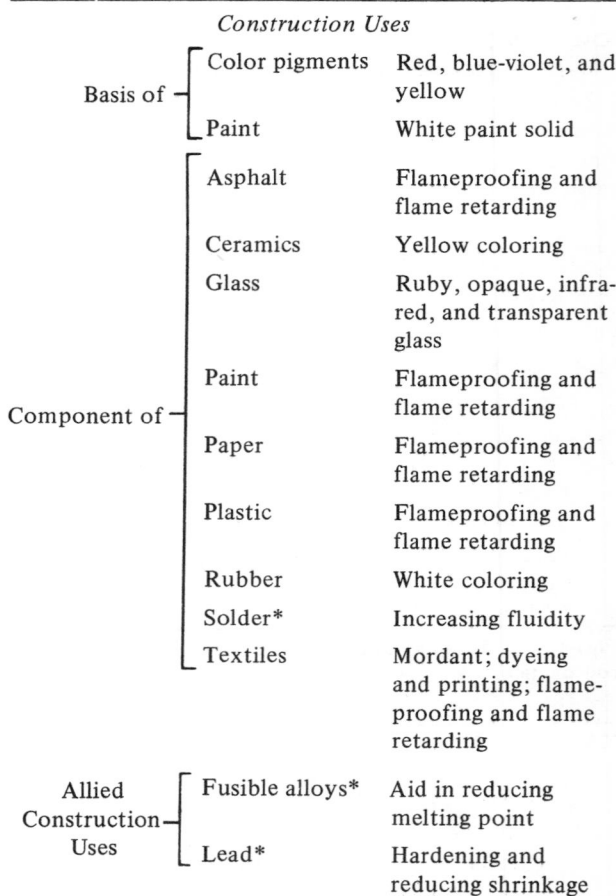

Construction Uses		
Basis of	Color pigments	Red, blue-violet, and yellow
	Paint	White paint solid
Component of	Asphalt	Flameproofing and flame retarding
	Ceramics	Yellow coloring
	Glass	Ruby, opaque, infrared, and transparent glass
	Paint	Flameproofing and flame retarding
	Paper	Flameproofing and flame retarding
	Plastic	Flameproofing and flame retarding
	Rubber	White coloring
	Solder*	Increasing fluidity
	Textiles	Mordant; dyeing and printing; flameproofing and flame retarding
Allied Construction Uses	Fusible alloys*	Aid in reducing melting point
	Lead*	Hardening and reducing shrinkage
Nonconstruction Uses		
Pewter and britannia metal,* type casting*		
Possible Future Uses		
Electroplating of metals for high resistance to tarnish and corrosion*		

Antimony oxide (pigment grade) is a pure white paint solid used in paint, enamel, and lacquer. It is incorporated in these with other solids to improve durability, opacity, suspension, and flow, or to reduce excessive chalking or flooding tendencies. There are two grades: one for highest quality paints, the other for clear, brilliant paints.

Flameproofing Preparations. Flameproofing preparations containing antimony are increasingly important in the growing plastics field. A peculiar chemical phenomenon occurs when flame is applied to an organic substance containing both antimony oxide and chlorine. Just what does occur is not yet clearly defined. However, unquestionably the combination of these two ingredients produces a compound which prevents the spreading of flame.

History and Manufacture

Antimony was obtained as a metal by the Chaldeans as long ago as 4000 B.C.

Both the older and recent methods for extracting antimony are based on the ease with which antimony oxide is volatilized and the oxides and sulfides are reduced to metal.

There are four methods for obtaining antimony from ores: volatilization, blast furnace, reverberatory furnace, and an electrolytic method from which a very high purity metal is obtained (*see* Figure *A69*).

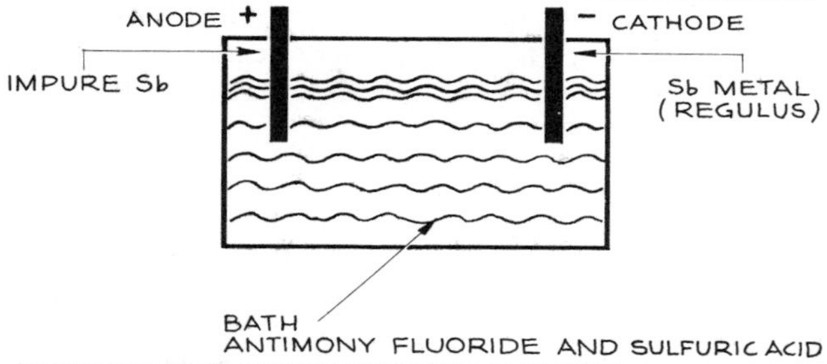

Figure A69 Electrolytic refining of antimony.

ARGON

Physical and Chemical Properties

Symbol: Ar
Atomic number: 18
Freezing point: -189.2°C, -308.56°F
Boiling point: -185.7°C, -302.26°F

Argon is a colorless, inert gas that does not combine chemically with any other element. It glows with a purple light when electric current passes through it.

Types and Uses

Argon is used in industry as an inert gas shield in arc welding, for filling electric lamps, and with neon in "neon" lights. In metallurgy, it is used as a blanket in titanium refining and for flushing molten metals to eliminate porosity in casting.

History and Manufacture

Argon was discovered in 1894 by Lord Rayleigh and Sir William Ramsay and was the first of the inert gases to be isolated. It is obtained from the fractional distillation of air.

ARSENIC

Physical and Chemical Properties

Symbol: As
Atomic number: 33
Specific gravity: 5.72
Melting point: 817°C, 1502.6°F

Arsenic is a silvery-gray, extremely brittle, crystalline metal which turns black when exposed to air. Because it is extremely poisonous and volatile even at the temperature of boiling water, arsenic is unfit for most purposes for which the common metals are used. As a rule it is considered a deleterious element in alloys. When present, it generally tends to cause brittleness and to lower the melting point.

Commercial Forms. Arsenic as a metal is available in lump, powder, or alloy form. An analysis of refined arsenic trioxide is given in Table *A42*.

Table A42 Analysis of refined arsenic trioxide

As_2O_2	Pb	Cu	Fe	Al_2O_3	Sb_2O_3
		(percent of content)			
99.78	0.0022	0.0001	0.0012	0.0001	0.2124

Types and Uses

Arsenic is used mainly in compound form, the bulk of it for insecticides. Its few uses in metallic form are noted by asterisks in Table *A43*.

History and Manufacture

In early Greek writings arsenic minerals are referred to as *sandarach*, *arsenicum*, and *auripigmentum*. The term

Table A43 Uses of arsenic

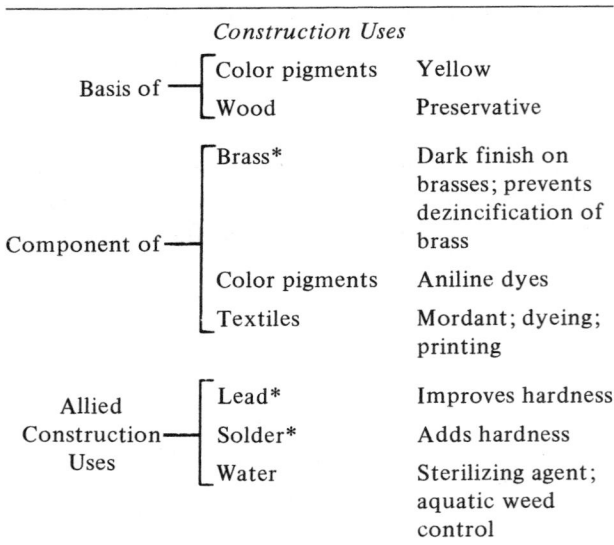

	Construction Uses	
Basis of	Color pigments	Yellow
	Wood	Preservative
Component of	Brass*	Dark finish on brasses; prevents dezincification of brass
	Color pigments	Aniline dyes
	Textiles	Mordant; dyeing; printing
Allied Construction Uses	Lead*	Improves hardness
	Solder*	Adds hardness
	Water	Sterilizing agent; aquatic weed control

"arsenic" may be derived from the Greek word for valiant or bold, alluding to the great energy with which it acts on other metals, or from the Arabic *arsa naki* or the Persian *zimuk* or *zime*. Strangely, early writers mentioned several medicinal properties of arsenical preparations but failed to note the toxicity of white arsenic (arsenious oxide).

Most arsenic is recovered in this form as a by-product of the refining of copper, lead, and other ores. Since the oxide of arsenic is volatile, it is easily purified by sublimation. Metallic arsenic is obtained from arsenious oxide and coke by reduction.

ASBESTOS

Physical and Chemical Properties

Asbestos is the name given to a group of fibrous silicate minerals with a characteristic fibrous structure that can be separated into individual flexible filaments. These fibers are roughly comparable to wool, cotton, or silk fibers with two differences. First, asbestos fiber bundles can be divided into finer fibers, ultimately (theoretically) to the size of a single molecule. Cotton, wool, and silk fibers, on the other hand, are constant in diameter and indivisible. The other most valuable difference is that asbestos is nonflammable.

The nonflammability of asbestos should not be confused with refractoriness, or the ability to withstand very high temperatures. Above 752°F (400°C) asbestos begins to deteriorate from loss of water of hydration, and between 1100°F (593.34°C) and 1400°F (760°C) all molecular water is evaporated with an accompanying loss of strength and embrittlement of the fibers.

Other important properties of asbestos are resistance to heat, moisture, corrosion, and many chemicals. Asbestos is valuable as a heat insulator not because of a low thermal conductivity (it is not low) but because of its fibrous structure.

Fiber characteristics and chemical composition differ considerably with each variety of asbestos. Only six are commercially significant. Chrysotile, the most plentiful and important, is a fibrous form of the mineral serpentine. Crocidolite, amosite, anthophyllite, tremolite, and actinolite are all amphiboles.

Chrysotile. Chrysotile asbestos constitutes the bulk (95%) of the world's production. It is a hydrated magnesium silicate which may contain minor quantities of iron, manganese, nickel, or aluminum. In general, chrysotile is more constant and dependable in quality than are other varieties of asbestos. Chrysotile fibers are the longest and vary in length from an average of $\frac{1}{16}$ to $\frac{3}{4}$ in. (1.59 to 19.05 mm) to as long as 12 in. (304.8 mm) in rare specimens. The fibers are very flexible and have a high tensile strength (150,000 to 440,000 lbf/in.^{2}, or 1034.25 to 3033.8 MN/m^2) and a high resistance to heat.

Crocidolite. Crocidolite, also called blue asbestos, is a complex silicate of iron and sodium. It has a lower resistance to heat and is more difficult to spin than chrysotile, but its tensile strength is high (100,000 to 300,000 lbf/in.^{2}, or 689.5 to 2068.5 MN/m^2), and it has superior resistance to attack by acids.

Amosite. Amosite is an iron silicate. It has good tensile strength and better resistance to heat than chrysotile or crocidolite. The fibers are flexible and long but harsh and relatively difficult to spin. The other varieties are little used.

Table *A44* summarizes the properties of the most commonly used types of asbestos.

Types and Uses

On the basis of use, asbestos falls into two principal classes, spinning and nonspinning fibers. The grading system used for Canadian asbestos divides the fibers into nine major groups. The first two, "crude" asbestos, are hand-selected and hand-cobbed material in its native or unfiberized form. The others are milled asbestos.

Table A44 Physical and chemical properties of types of asbestos

	Chrysotile		Crocidolite		Amosite	
Property	lbf/in.^{2}	MN/m^2	lbf/in.^{2}	MN/m^2	lbf/in.^{2}	MN/m^2
Tensile strength[a]	150,000–500,000	1034.25–3447.5	300,000–550,000	2068.5–3792.25	200,000–350,000	1379.0–2413.25
Flexibility	Very good		Good		Poor	
Acid resistance	Poor		Good		Good	
Texture	Soft		Harsh		Coarse	
Heat resistance	Good		Good (melts earlier)		Good	
Percent of weight loss when exposed to 400°F (204.4°C) for 2 h	0.30		0.08		0.23	
Percent of weight loss when exposed to 1800°F (986.1°C) for 2 h	13.77		0.77[b]		1.53	

[a]1 lbf/in.^{2} = 0.006895 MN/m^2.
[b]Iron changing in weight because of oxidation.

Group 1 (Crude No. 1): fiber greater than $\frac{3}{4}$ in. (19.05 mm) in length. It should be silky and have enough tensile strength to be made into yarn, tape, cloth, carded fiber, and other textiles.

Group 2 (Crude No. 2): generally fiber that has not been milled and has a length of $\frac{3}{8}$ to $\frac{3}{4}$ in. (9.53 to 19.05 mm). It must have good tensile strength.

Group 3: milled fiber suitable for textile yarn; also used for asbestos papers and millboards.

Group 4: including fibers suitable for the manufacture of asbestos cement products such as pipe, shingles, siding, corrugated sheets, boards, and asbestos papers.

Group 5: blended with Group 4 for the manufacture of asbestos products as listed for Group 4, but also for disk brake pads and brake blocks.

Group 6: used in conjunction with Group 4 for the manufacture of asbestos products as listed for Group 4, but also for asbestos roofing felts and papers, and roof coatings.

Group 7: fiber known as shorts and a subgroup known as floats, used in the manufacture of resilient floor tiles and roll goods, and in paper, paints, coatings, adhesives, spackling cements, and a wide variety of plastic products.

Groups 8 and 9: known as sand and gravel, and as stone, respectively. They find little commercial use.

Most textile fibers are taken off in the initial milling stages and packaged for processing at the textile plant.

Asbestos paper, textiles, and the principal asbestos cement products used in construction are discussed separately.

History and Manufacture

Asbestos was known and woven into textiles more than 2000 years ago. The Greek word *asbestos* means "inextinguishable" or "unquenchable," which is directly opposite to the incombustible character of the material. This might be explained by Plutarch's reference to the perpetual asbestos lamp wicks used by the Vestal Virgins. The ancient Chinese and Egyptians are said to have woven asbestos into mats.

Commercial Development. Asbestos was first developed on a commercial scale in Italy as a result of studies and experiments sponsored in about 1808 by an Italian noblewoman. These led to the successful manufacture of thread, fabric, and paper incorporating asbestos.

In 1860 asbestos was discovered in Quebec, Canada, and by 1907 several deposits were being exploited. It was not until 1915, however, that asbestos became a major industry as a result of the growth of the automotive industry and its need for brake linings, as well as a growing demand for heat insulators in industrial construction. For the distribution of asbestos ore throughout the world, see Figure *A70*. Asbestos is mined by both open-pit and underground methods. The ore yields about 5% of usable fibers.

Textile Processing. Only about 4% of all asbestos fibers extracted is of a grade suitable for processing into textiles. Mill fibers are obtained by crushing the rock until the asbestos is freed and then removing the fiber from the rock by screening and air separation. Most textile

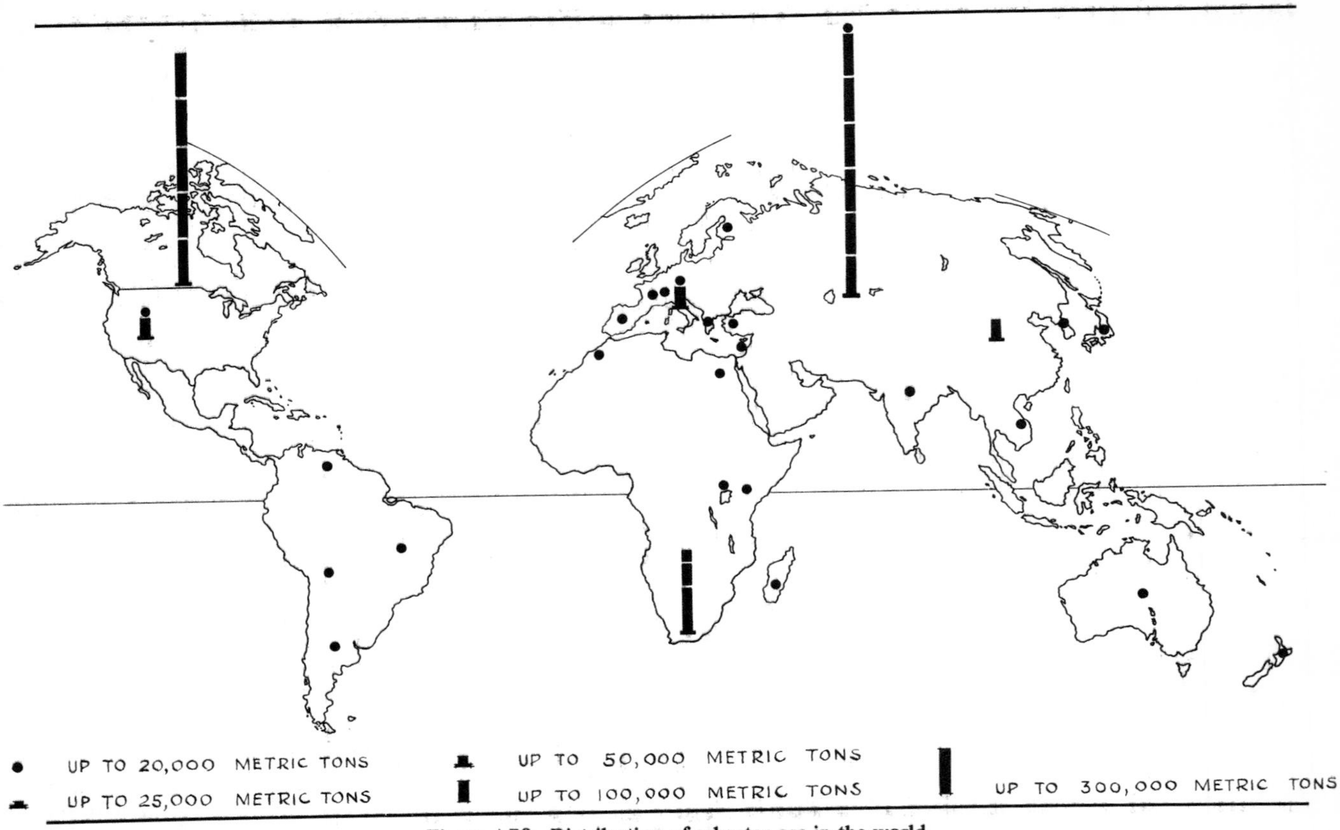

Figure A70 Distribution of asbestos ore in the world.

fibers are taken off in the initial milling stages and packaged for processing at the textile plant.

ASBESTOS-CEMENT BOARD AND SHEET

Physical and Chemical Properties

Asbestos-cement sheet and board are made of portland cement and grade 4, 5, and 6 asbestos fibers, mainly chrysotile, which act as reinforcement. Like all asbestos products, they are fire resistant, strong, water resistant, permanent, resistant to alkalis and acids, and impervious to rot, mold, fungi, insects, and rodents. They are low in tensile strength and shock resistance.

Commercial Forms. Sheet and board are available in the natural cement color, colored with integral mineral colors, and with an acrylic 3-coat finish in a wide variety of colors; with a smooth and shiny or textured surface; and perforated or plain. The most common sizes are 2 ft × 2 ft (0.610 m × 0.610 m) and 4 ft × 8 ft (1.219 m × 2.438 m). Thickness ranges from $\frac{1}{8}$ to $\frac{1}{2}$ in. (3.175 to 12.7 mm).

Types and Uses

Asbestos-cement sheets and boards are manufactured in stock thickness and sizes and in a limited number of textures and in a variety of colors, as shown in Table *A45*. Asbestos-cement sheets are available in two types that might be described as general purpose materials:

1. Type F (flexible): for exterior and interior use where high strength and density, smooth surface, flexibility, and low moisture absorption are needed. This type does not have to be drilled for nailing in thicknesses up to $\frac{1}{4}$ in. (6.35 mm), but predrilling is necessary if nails or screws are located near edges.

2. Type U (utility): for exterior and interior use, with sufficient strength for general utility and construction purposes where flexibility, density, smoothness, and moisture absorption are not essential. This also does not require predrilling in thicknesses up to $\frac{1}{4}$ in. (6.35 mm), unless nails or screws are located near edges.

Table A45 Types of asbestos-cement sheet and board

Type	Texture	Color	Sizes: Thickness in. (mm)	Sizes: Width ft (m)	Sizes: Length ft (m)	Weight lb/ft^2 (kg/m^2)	Nail holes	Major uses
Plain sheet	Smooth	Natural cement color, light gray	$\frac{1}{8}$ (3.18) $\frac{3}{16}$ (4.76) $\frac{1}{4}$ (6.35)	4 (1.219)	4 (1.219) 8 (2.438) 10 (3.048) 12 (3.658)	1.225 (5.981) 1.750 (8.543) 2.350 (11.423)	Not punched; nails, screws, bolts, etc., should be $\frac{3}{8}$ in. (9.53 mm) min. from edge, and 2 in. (50.8 mm) min. from corners	Interior or exterior walls, ceilings, and soffits; for fire resistance in integral areas; general utility
Colored sheet (3-coat acrylic color finish)	Smooth	Wide range of colors and accent colors	$\frac{1}{8}$ (3.18) $\frac{3}{16}$ (4.76) $\frac{1}{4}$ (6.35)	2 (0.610) 4 (1.219)	4 (1.219) 4 (1.219) 8 (2.438)	1.25 (6.1025) 1.80 (8.7876) 2.40 (11.7168)	Not punched; nails, screws, bolts, etc., should be $\frac{3}{8}$ in. (9.53 mm) min. from edge, and 2 in. (50.8 mm) min. from corners	Interior or exterior walls, panels, ceilings, and soffits; for interior partitions, dividers, etc.
Colored sheet (integral)	Striated	Gray, blue, green, brown, red, and yellow; pastel colors; and cement color	$\frac{3}{16}$ (3.18)	2.16 (0.658)	8 (2.438)	2.08 (10.1546)	$\frac{1}{8}$ in. (3.18-mm) nail holes prepunched along side and down middle	Exterior siding
Perforated sheet	Smooth with holes	Natural cement color, light gray	$\frac{3}{16}$ (3.18)	>2[a] (0.606) 2 (0.610)	>2[a] (0.606) >4[a] (1.215) 2 (0.610) 4 (1.219)	0.533 (2.6021)	Perforations used for application	Acoustic treatment
Board[b]	Smooth	Natural cement color	$\frac{5}{16}$ (7.94) $\frac{3}{8}$ (9.53) $\frac{1}{2}$ (12.70) $\frac{5}{8}$ (15.88) $\frac{3}{4}$ (19.05) 1 (25.40)	4 (1.219)	8 (2.438)	2.90 (14.158) 3.45 (16.843) 4.70 (22.945) 5.95 (29.048) 7.05 (34.418) 9.40 (45.891)	Not punched; nails, screws, bolts, etc., should be $\frac{3}{4}$ in. (19.05 mm) min. from edges, and holes should be drilled first	Exterior industrial roofing and siding; interior areas requiring fire resistance; laboratory countertops and fume hoods
Colored board (3-coat acrylic finish)	Smooth	Wide range of colors and accent colors	$\frac{3}{8}$ (9.53) $\frac{1}{2}$ (12.7)	2 (0.610) 4 (1.219)	4 (1.219) 4 (1.219) 8 (2.438)	3.45 (11.423) 4.70 (22.945)	Not punched; nails, screws, bolts, etc., should be $\frac{3}{8}$ in. (9.53 mm) min. from edge, and 2 in. (50.8 mm) from corners	Interior or exterior walls, ceilings, and soffits; for interior partitions, dividers, etc.

Table A45 Types of asbestos-cement sheet and board (*continued*)

Type	Texture	Color	Sizes: Thickness in. (mm)	Width ft (m)	Length ft (m)	Weight lb/ft^2 (kg/m^2)	Nail holes	Major uses
Colored board (integral)	Smooth	Green, gray, brown, and white	$\frac{1}{4}$ (6.35)	4 (1.219)	8 (2.438)	2.35 (11.423)	Should be drilled first	Laboratory tables and equipment and areas open to chemicals
			$\frac{3}{8}$ (9.53)			3.45 (16.843)		
			$\frac{1}{2}$ (12.70)			4.70 (22.945)		
			$\frac{3}{4}$ (19.05)			7.05 (34.418)		
			1 (25.40)			9.40 (45.891)		
			$1\frac{1}{4}$ (31.75)			11.75 (57.365)		
Colored chalkboard	Smooth	Green, brown, and gray	$\frac{3}{16}$ (4.76)	3.5 (1.067)	4 (1.219)	2.08 (10.1546)	Installed in standard chalkboard trim systems	Chalkboards
			$\frac{1}{4}$ (6.35)	4 (1.219)	6 (1.829)	2.40 (11.7168)		
					8 (2.438)			

[a]Units are available in 1′-11$\frac{7}{8}$″ (1.9896-ft) widths and in 1′-11$\frac{7}{8}$″ (1.9896-ft) and 3′-11$\frac{7}{8}$″ (3.9896-ft) lengths. These dimensions allow for expansion and contraction when installed on 2-ft centers without a possible "bubbling" into a quilted effect.
[b]Also available in other larger sizes up to 2″ (50.8 mm) thick.

Application

Condensed Checklist

1. The type, texture, and size of asbestos-cement sheet or board selected must always meet the requirements of its end use. (*See* Table *A45*.)
2. Local, municipal, and state codes and also codes of the Fire Underwriters, insurance companies, labor departments, and federal government (Army, Navy, etc.) should be checked for fire resistance ratings and limitations.
3. In installing asbestos-cement sheets and boards, the correct type of nail must be used for the type of asbestos-cement sheet or board used (*see* Table *A46*). Manufacturers of asbestos-cement boards should be consulted for proper methods of fastening and correct sizes and composition of fastening devices. The type and spacing of fastenings will vary with the end use.
4. Predrilling is always necessary for thicknesses over $\frac{1}{4}$ in. (6.35 mm) and is advisable for all thicknesses as a general rule. If the asbestos-cement sheet or board selected is not prepunched for fastening, predrilling should be required by specifications, and the exact recommendations of the manufacturer must be followed.
5. Asbestos board cannot be bent except on special order. Table *A47* lists minimum bending radii.
6. For siding, the treatments of external corners, vertical and horizontal joints, at roof, at grade, and at openings must always be checked (*see* Figures *A71* and *A72*).
7. When perforated asbestos is used for acoustic treatment, specialists in this field and also the manufacturers should be consulted.
8. When asbestos-cement sheet or board is used in laboratories and production or manufacturing areas where chemical and chemical fumes are present, manufacturers should be consulted for the correct type of asbestos-cement material in relation to the specific conditions of end use.

Conditions Favorable to the Use of Asbestos-Cement Sheet and Board

1. Where a fire-resistant, durable material that is also rot-, insect-, and rodentproof is desirable for finishing exterior or interior walls and ceilings.
2. Where a fire-resistant, durable, strong material is desired for acoustic treatment.
3. Where a material resistant to chemicals and corrosive fumes is needed for laboratories, manufacturing, and processing areas.
4. Where a wide variety of colors is required.
5. Where chalkboards are required.

Table A46 Types of nails for various thicknesses of asbestos-cement sheet and board

Type and composition of nail	Head	For $\frac{1}{8}$ in. (3.18 mm) Length in.	mm	Gauge	For $\frac{3}{16}$ in. (4.76 mm) Length in.	mm	Gauge	For $\frac{1}{4}$ in. (6.35 mm) Length in.	mm	Gauge	Major use
Common wire nail	Flat	$1\frac{1}{4}$	31.75	14	$1\frac{1}{2}$	38.1	12	$2\frac{1}{2}$	63.5	10	Interior walls and ceilings
Roofing aluminum or hot-galvanized steel	Flat or button	$1\frac{1}{4}$ 1	31.75 25.4	10 12	$1\frac{1}{4}$ 1	31.75 25.4	10 12	$1\frac{1}{4}$ 1	31.75 25.4	10 12	
Shingle hot-galvanized steel	Flat	$1\frac{1}{2}$	38.1	12	$1\frac{1}{2}$	38.1	12				Exterior and interior walls and ceilings
Spiral, annular, or helical noncorroding hardened steel	Casing, lead, or neoprene washer	1	25.4	14	$1\frac{1}{2}$	38.1	14	$1\frac{1}{2}$	38.1	14	
Special aluminum	Casing or flat	$1\frac{1}{8}$	28.58	12	$1\frac{1}{2}$	38.1	11	$1\frac{1}{4}$	31.75	10	Exterior and interior walls and ceilings; aluminum basing not recommended for exterior walls and ceilings
Roofing hot-galvanized steel	Casing or flat	$1\frac{1}{4}$	31.75	10	$1\frac{1}{4}$	31.75	10	$1\frac{1}{4}$	31.75	10	Exterior walls

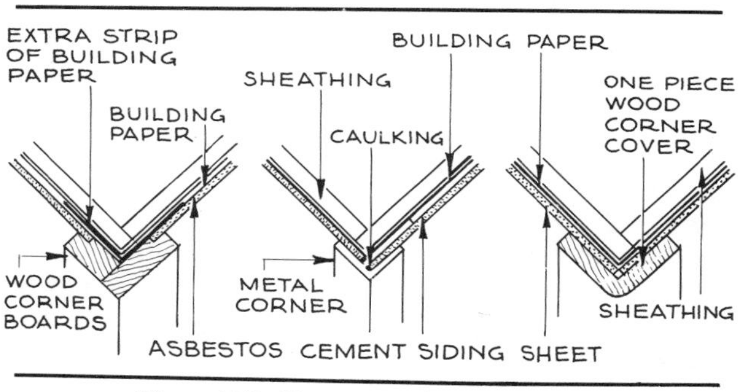

Figure A71 Treatment of external corners.

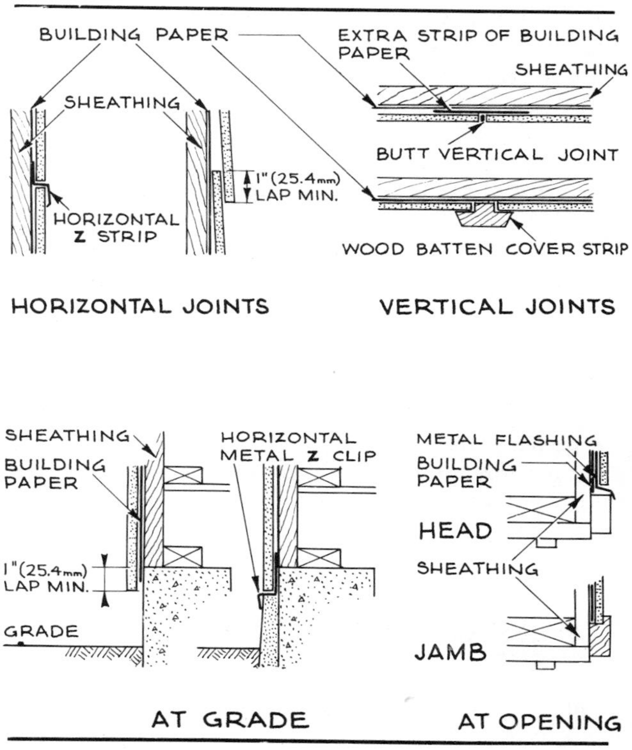

Figure A72 Treatment of horizontal and vertical joints, at grade and at opening.

Table A47 Minimum bending radii for asbestos-cement sheet and board

Thickness of asbestos cement in.	mm	Minimum bending radius for long direction of sheet ft–in.	m	Minimum bending radius for short direction of sheet ft–in.	m
$\frac{1}{8}$	3.18	2–6	0.762	3–0	0.914
$\frac{3}{16}$[a]	4.76[a]	3–0	0.914	4–6	1.346
		3–7	1.092	5–3	1.600
$\frac{1}{4}$[a]	6.35[a]	5–3	1.600	7–0	2.134
		5–6	1.676	6–0	1.829

[a]Check with manufacturer for minimum radii.

Condition Unfavorable to the Use of Asbestos-Cement Sheet and Board

In areas where resistance to shock is necessary.

History and Manufacture

The basic process for manufacturing asbestos cement was developed by Hatschek in Austria in 1900. Asbestos cement was first manufactured in the United States in 1905, but the color additive process was not developed until immediately after World War II.

Hatschek Process. The manufacturing process consists of a thorough dry mixing of carefully selected cement, silica, and the asbestos fibers. This dry mix is then conveyed to a wetting tank and from the wetting tank to an attrition mill, which gives the wet mix (called the slurry) a smooth consistency. From here it is fed to the modern improved version of the Hatschek machine.

The Hatschek machine consists of several large vats in each of which is a screen-covered revolving cylinder. As the cylinder turns, the wet mix clings to the screen. This clinging wet mix is forced onto an endless woolen belt by a roll (called a cooch). From the endless belt, it is transferred to an accumulator roll, where it is built up in layers. When the required thickness has been reached, the sheet is stripped, still wet, from the accumulator roll, and is conveyed to an oven for partial drying.

Various processes are then used to obtain the different types of asbestos sheet or board materials. For textured surfaces, the sheet is pressure rolled and embossed; for smooth surfaces it is smooth rolled; and for more dense products it is hydraulically pressed. These sheets or boards are cut to size, either rough or final dimensions, and at this point they can also be punched for nail holes or perforations. They are cured by stacking or autoclaving. They can be treated with a 3-coat acrylic resin to obtain a wide variety of colors, and integral colors can be added to the wet mix to provide a limited series of pastel colors.

The entire process is a completely controlled operation, and all products are continually inspected from start to final packing.

ASBESTOS-CEMENT CORRUGATED SHEET

Physical and Chemical Properties

Corrugated asbestos is made of the same ingredients—shorter grades of asbestos fiber and portland cement—as other asbestos-cement sheet products. By virtue of its corrugations it can withstand heavy service conditions. Another advantage is that it can be lapped, thus making a weatherseal and eliminating many joint problems. (*See* Table *A48.*)

Corrugated asbestos is resistant to many acid fumes and corrosive smoke; it will not rot, rust, or burn; it is not affected by salt air or extremes in heat and cold; and it is vermin- and termiteproof.

Types and Uses

Corrugated asbestos is available in two types, standard and lightweight. Both are used as roofing and siding material. Minimum roof pitch for both types is 3 to 12. Side laps are one corrugation for both types; end laps are 6 in. (152.4 mm).

Curved sheets of corrugated asbestos are also available; minimum lengthwise radius is 5 ft (1.524 m), and minimum crosswise radius is 4 ft (1.219 m).

Application

Corrugated asbestos-cement sheet, because of its rigidity, can also be used for all sorts of partitions, division walls, decorative panels, etc., on the interior of buildings. The problem of the treatment of the corrugations wherever they meet a flat surface is usually answered by stock filler strips, but for a more reliable, positive sealing action the recommended practice now is to use a durable plastic sealing compound in combination with the closure strip.

The standard handling procedure for storage of corrugated asbestos is on a series of 2-by-4's (50.8 by 101.6 mm), 18 in. (457.2 mm) on center, set at right angles to the corrugations.

Condensed Checklist

1. Predrilling of all holes is necessary in corrugated asbestos-cement sheet.
2. Pliable plastic (neoprene) or lead washers or lead-head bolts must be used for attachment.
3. External and internal corners must be correctly detailed (*see* Figure *A73*).
4. Sill and roof edge joints must be correctly detailed (*see* Figure *A74*).
5. Lapping horizontally and vertically for walls and roofs must be correctly done (*see* Figure *A75*).
6. The correct type of side wall fasteners must be used. A large variety is available for different conditions (*see* Figure *A75*).

Table A48 Dimensions, weight, and unsupported span of corrugated asbestos

Dimensions and properties	Standard	Lightweight
Length	6 in. to 12 ft (152.4 to 3657.6 mm) in 6-in. (152.4-mm) increments	2 in. to 12 ft (50.8 to 3657.6 mm) in 1-in. (25.4-mm) increments
Thickness (nominal)	$\frac{3}{8}$ in. (9.53 mm)	$\frac{3}{16}$ in. (4.76 mm)
Width	42 in. (1066.8 mm)	42 in. (1066.8 mm)
Number of corrugations	10	10
Tolerances:		
width	$+0-\frac{1}{16}$ in. (1.59 mm) off each edge corrugation	$+0-\frac{1}{16}$ in. (1.59 mm) off each edge corrugation
length	$\pm\frac{3}{16}$ in. (4.76 mm)	$\pm\frac{3}{16}$ in. (4.76 mm)
squareness	$\frac{1}{64}$ in. per linear foot (0.4 mm per 304.8 mm)	$\frac{1}{64}$ in. per linear foot (0.4 mm per 304.8 mm)
Temperature limit:		
continuous	600°F (315.56°C)	600°F (315.56°C)
intermittent	1000°F (537.78°C)	1000°F (537.78°C)
Thermal conductivity or K factor: Btu in./ft^2 · h · °F (0.1442 W/m · °C)	4.5	4.5
Maximum roof spacing without support	4.5 ft (1.372 m)	2.5 ft (0.762 m)
Maximum wall spacing without support	5.75 ft (1.753 m)	3.5 ft (1.067 m)
Weight, lb/ft^2 (kg/m^2)	3.8 to 4.1 (18.55 to 20.02)	1.75 to 2.10 (8.54 to 10.25)

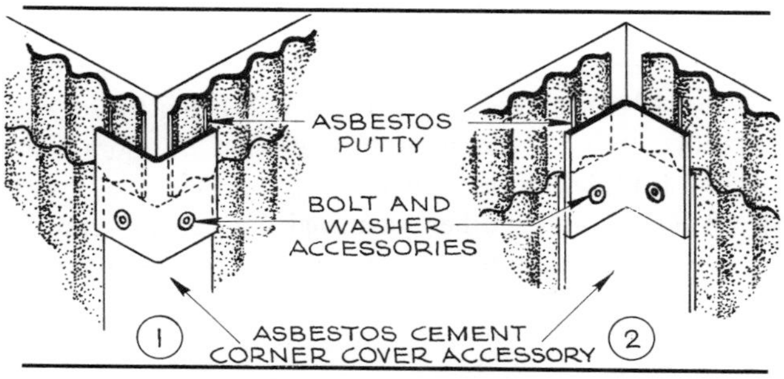

Figure A73 Details of external (1) and internal (2) corners.

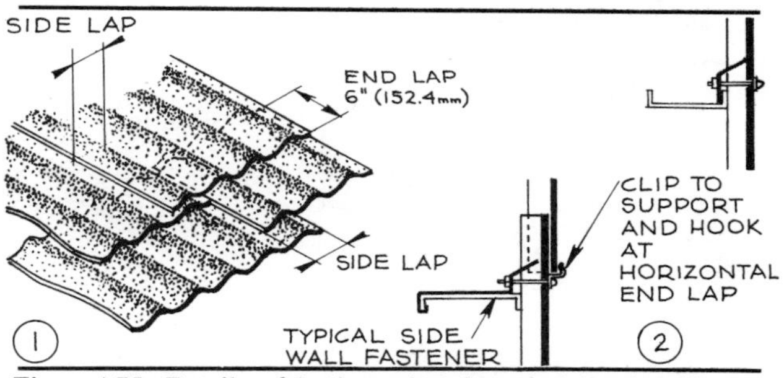

Figure A75 Details of horizontal and vertical lapping (1) and types of side wall fasteners (2).

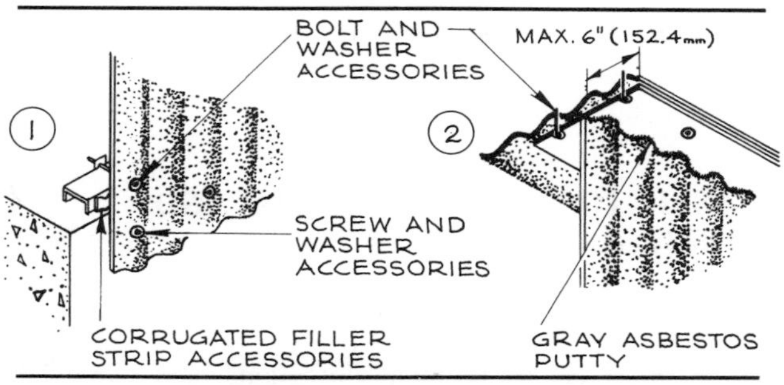

Figure A74 Details of sill (1) and at roof edge (2).

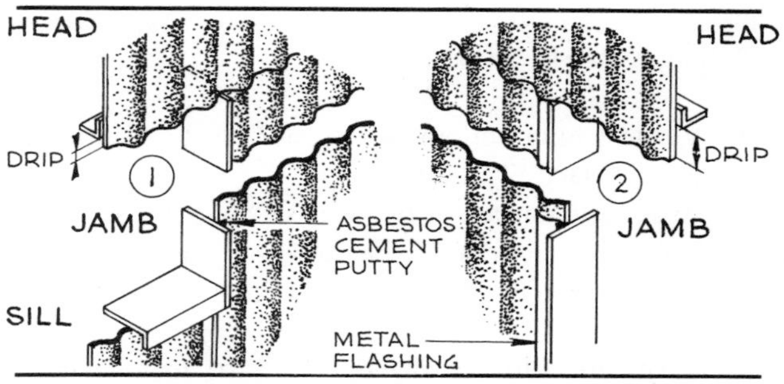

Figure A76 Details of openings at windows (1) and doors (2).

7. At head and sill of windows and doors, stock filler strips are generally used. At the jamb, the corrugated asbestos may be butted; any space may be filled with asbestos cement putty, or flashing may be used (*see* Figure *A 76*).
8. Roof intersections at ridge or hip and at valley must be correctly detailed (*see* Figure *A 77*).

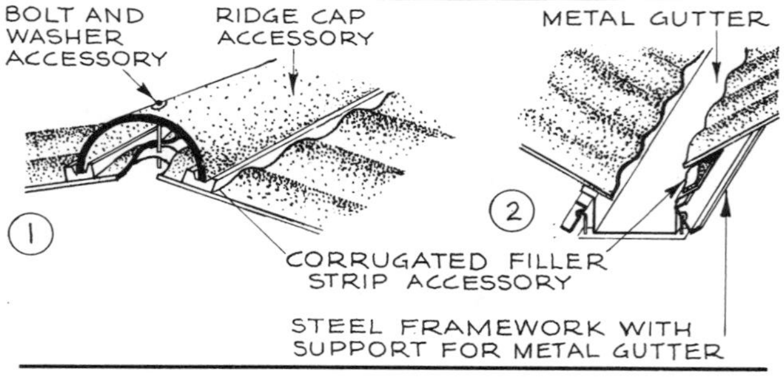

Figure A77 Details of ridge or hip (1) and of valley (2).

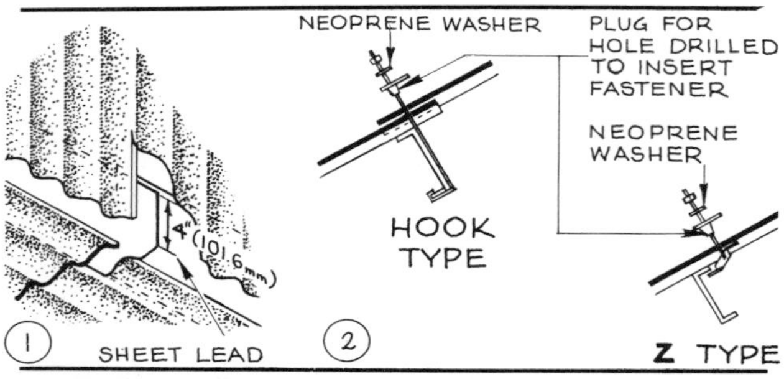

Figure A78 Vertical intersection in the roof (1) and typical top-applied roof fasteners (2).

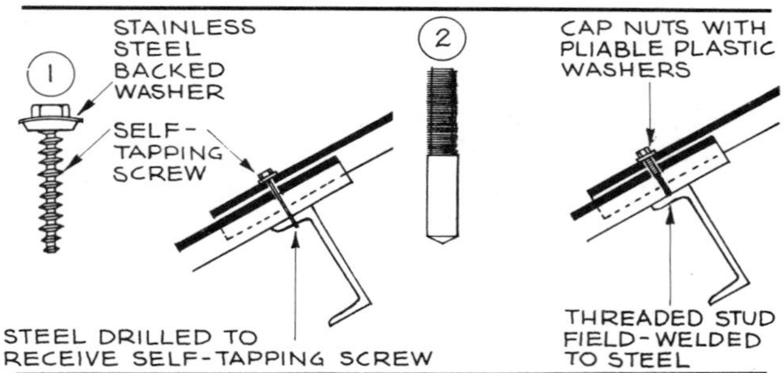

Figure A79 Self-tapping screws (1) and welded steel fasteners (2) for corrugated asbestos roofs.

9. Vertical intersections in the roof must be correctly detailed (*see* Figure *A 78*).
10. Roof fasteners are of two types. One is applied from the top (exterior) of the roof, and the other from underneath (or the interior of) the roof (*see* Figures *A 78* and *A 79*).
11. Accessories for corrugated asbestos include clips, washers, and bolts; self-tapping screws; special studs to be welded; and fillers, caps, and angles (*see* Figure *A 80*).

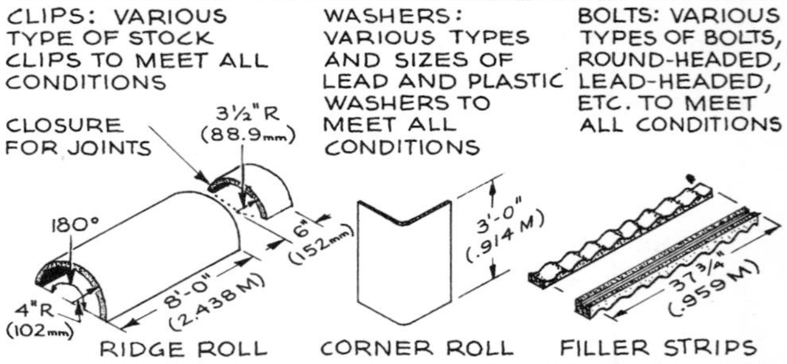

Figure A80 Typical accessories for corrugated asbestos.

Conditions Favorable to the Use of Corrugated Asbestos Cement

1. Where the surface is likely to receive rough usage.
2. Where maintenance must be kept to a minimum.
3. Where fire resistance is required, using suitable accessories and back-up materials. Municipal, state, and local codes, and also the fire rating codes of the Fire Underwriters, insurance companies, labor departments, and federal government (Army, Navy, etc.), should be checked for any restrictions.

Conditions Unfavorable to the Use of Corrugated Asbestos Cement

1. Without predrilling for attachment of accessories.
2. Where impact shock is likely to occur.
3. On curved surfaces with a radius of less than 4 ft (1.219 m) crosswise or 5 ft (1.524 m) lengthwise.
4. For roofing if an unsupported span of more than 4.5 ft (1.372 m) is demanded.
5. Where a color other than natural cement gray is desired but minimum maintenance is also required.

History and Manufacture

Corrugated asbestos was first manufactured in 1905 by a process similar to that used for asbestos sheets, siding, and shingles. For many years it was used only for industrial construction, but now it is being employed effectively in all types of buildings.

ASBESTOS-CEMENT PANELS AND INTEGRANTS

Physical and Chemical Properties

Integrants. Asbestos-cement integrants and sandwich panels consist of a core faced on both sides with $\frac{1}{8}$ in.

Table A49 Sizes, weights, insulating R values, and major uses of asbestos-cement integrants and sandwich panels

Standard sizes						Tolerances							
Thickness		Width		Length		Thickness		Width and length		Weight		Thermal insulation:	
in.	mm	ft	m	ft	m	in.	mm	in.	mm	lb/ft^2	kg/m^2	R factor	Major uses
$\frac{11}{16}$	17.46									3.4	16.62	2.22	Exterior finish
$\frac{3}{4}$	19.05	The values below apply for all sizes								3.2	15.62	2.64	and curtain
1	25.4			6,	1.83,	$\pm\frac{1}{16}$	±1.59	$\pm\frac{1}{8}$	±3.18	2.7–3.4	13.18–16.60	3.84–2.44	wall construc-
$1\frac{1}{8}$	28.58	4	1.22	8,	2.44,					2.7–3.6	13.18–17.59	4.35–2.70	tion; roof deck-
$1\frac{9}{16}$ [a]	39.69			10,	3.05,					2.8–6.5	13.67–31.73	5.88–2.78	ing; interior
2	50.8			and	and					2.8–5.4	13.67–26.36	7.69–4.35	walls and
$2\frac{1}{2}$ [b]	63.5			12	3.66					9.3	44.40	4.00	partitions

[a] 1-Hour fire rating.
[b] 2-Hour fire rating.

(3.18 mm) insulating rigid plastic-type board, paper pulp, or fiber pulp board, either a single board or several laminated boards or noncombustible cores, or noncombustible and insulating cores of expanded perlite. Paper or fiber pulp cores are treated, usually with asphalt, for resistance to mold rot, fungi, insects, and rodents, and are thereby made moisture resistant also. All adhesives are waterproof. Rigid plastic-foam board is self-extinguishing and is not only moistureproof but also resistant to mold, rot, fungi, insects, and rodents. Asbestos-cement integrants and sandwich panels that have noncombustible cores are available with fire ratings as high as 2 hours.

Panels. Asbestos panels are available in thicknesses of $\frac{1}{4}$, $\frac{1}{2}$, and $\frac{3}{4}$ in. (6.35, 12.7, and 19.05 mm) and in a limited number of colors and textures.

Types and Uses

Integrants. Asbestos-cement integrants and sandwich panels with $\frac{1}{8}$ in. (3.18 mm) asbestos-cement sheets as surface finish are available in standard thicknesses and sizes, shown in Table *A49*. Standard weights, insulation values, and major uses are also given in the table. Sheets with thicker surfaces, for example, $\frac{3}{16}$ or $\frac{1}{4}$ in. (4.76 or 6.35 mm), can be obtained on special order. All are available in the natural gray color of asbestos cement, or primed for field painting, or with an acrylic 3-coat finish in a wide range of stable colors.

All asbestos-cement integrants are edge-sealed before shipment from the factory to prevent moisture absorption. If any cutting is done, the new edges should be sealed with a rubber base paint or sealing compound as recommended by the asbestos-cement manufacturer.

Panels. Asbestos-cement panels are available in thicknesses of $\frac{3}{16}$ in. (4.76 mm) up to $\frac{3}{4}$ in. (19.05 mm) and in a limited number of colors and textures, as shown in Table *A50.* They are used for exterior facings of buildings, chalkboards, laboratory countertops, interior partitions, dividers, etc. (*see* Figures *A81* and *A82*).

These integrants, sandwich panels, and panels are finding many uses such as permanent form materials for concrete, decorative exterior or interior screens, solid railings, curtain walls, and exterior facing materials.

Exterior Panels. Asbestos panels for the exterior are not manufactured as prefabricated units at present. The panels are usually assembled and built up on site, using asbestos-cement integrants to serve as thermal insulation and inside wall surface, with corrugated asbestos or another type of surfacing material, generally metal, on the outside, or exterior, wall surface.

Interior Panels. Prefabricated asbestos panels for use as movable partitions to subdivide interior areas are available in a wide variety of types. These prefabricated panels have specialized systems for the treatment of corners, intersections, fixed glass panels, and doors, and for fastening to floors and ceilings. They also are made with various fire-resistance ratings.

Sandwich Panels. Sandwich panels are available in limited sizes: 4 ft (1.219 m) widths, and lengths up to 12 ft (3.658 m); thicknesses range from $\frac{11}{16}$ to $2\frac{1}{2}$ in. (17.46 to 63.5 mm). The weights vary from 3.4 to 5.4 lb/ft^2 (16.60 to 26.36 kg/m^2).

Sandwich panels are used for fire-resistant areas and partitions.

Table A50 Sizes, weights, and major uses of asbestos-cement panels

Standard sizes						Weight		
Thickness		Width		Length				
in.	mm	ft	m	ft	m	lb/ft²	kg/m²	Major use
$\frac{3}{16}$	4.76	$3\frac{1}{2}$	1.07	4, 6, 8	1.22, 1.83, 2.44	1.6	7.81	Chalkboards[a]
$\frac{1}{4}$	6.35	4	1.22	4, 6, 8	1.22, 1.83, 2.44	2.0	9.76	Chalkboards[a]
$\frac{1}{2}$	12.70	4	1.22	8	2.44	5.0	24.40	Exterior facings[b]
$\frac{3}{4}$	19.05	4	1.22	8	2.44	7.5	36.62	Exterior facings[b]
$\frac{1}{4}$	6.35	4	1.22	8	2.44	2.0	9.76	Exterior facings[c]
$\frac{1}{2}$	12.70	4	1.22	8	2.44	5.0	24.40	Exterior facings[c]
$\frac{3}{4}$	19.05	4	1.22	8	2.44	7.5	36.62	Exterior facings[c]
$\frac{1}{4}$	6.35	4	1.22	8	2.44	2.0	9.76	Laboratory countertops[c]
$\frac{1}{2}$	12.70	4	1.22	8	2.44	5.0	24.40	Laboratory countertops[c]
$\frac{3}{4}$	19.05	4	1.22	8	2.44	7.5	36.62	Laboratory countertops[c]
1	25.40	4	1.22	8	2.44	10.0	48.82	Laboratory countertops[c]
$1\frac{1}{4}$	31.75	4	1.22	8	2.44	12.0	58.58	Laboratory countertops[c]

[a]Available in gray, green, and brown.
[b]Textured to resemble stone.
[c]Available in seven pastel colors.

Application

Condensed Checklist

1. When asbestos integrants, sandwich panels, and panels are used, the relative humidity of the area in which they will be installed should always be checked. When the relative humidity is less than 40%, no precautions are necessary. If it ranges between 40 and 60%, all surfaces should be painted with a protective coat of the type recommended by the manufacturer of the integrants, sandwich panels, and panels. They should not be used in areas where the relative humidity is higher than 60%.
2. Local, municipal and state codes and also codes of the Fire Underwriters, insurance companies, labor departments, and federal government (Army, Navy, etc.) should be checked for fire-resistance ratings and restrictions on the use of asbestos integrants, sandwich panels, and panels.
3. Thermal insulation values of asbestos integrants, sandwich panels, and panels should be checked in relation to the geographic area and orientation in which they will be used to make sure that they meet insulation requirements.
4. Painting: Surfaces of asbestos integrants, sandwich panels, and panels must be thoroughly clean of dust and dirt before being painted. Oil and grease spots must be removed with carbon tetrachloride. Asbestos cement and paint manufacturers should always be consulted to determine the correct type of paint to meet the end-use requirements.
5. All asbestos integrants, sandwich panels, and panels must be predrilled with holes of the correct diameter for the type of fastening device selected.

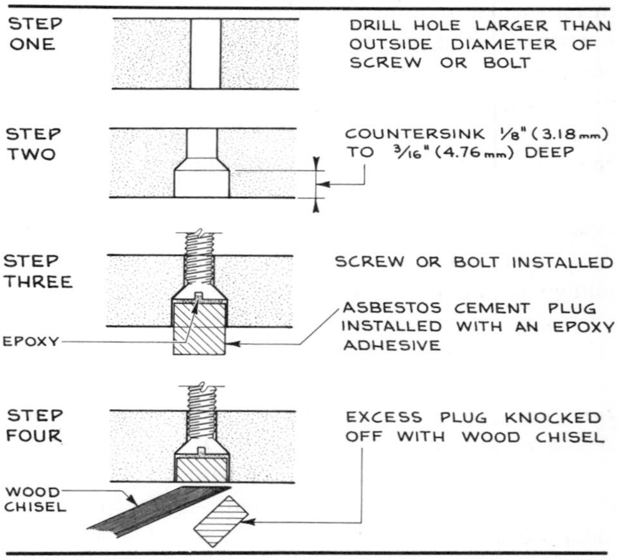

Figure A81 Methods of attachment for asbestos-cement integrants and boards.

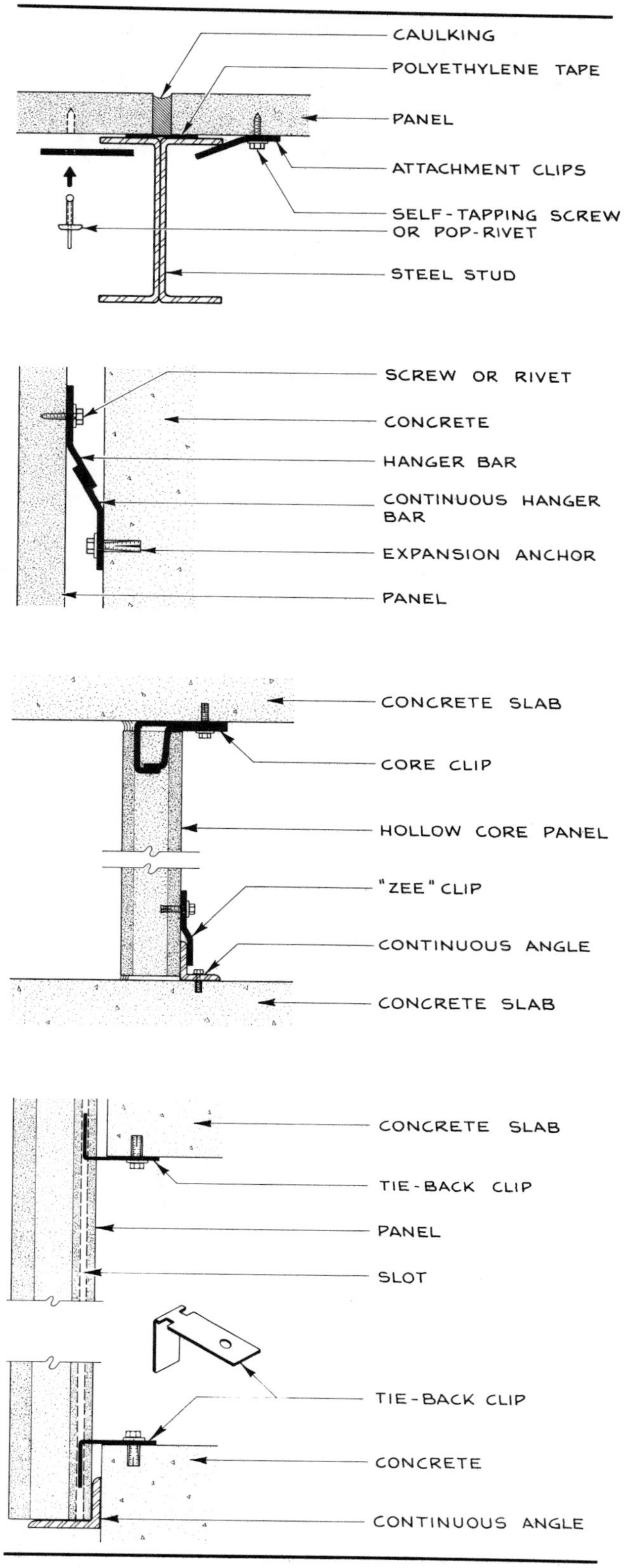

Figure A82 Connections and joining of asbestos-cement panels.

6. For integrants, the type of fastening device to be used for exterior walls should always be checked in relation to structure and material, whether wood or steel (*see* Figure *A83*).

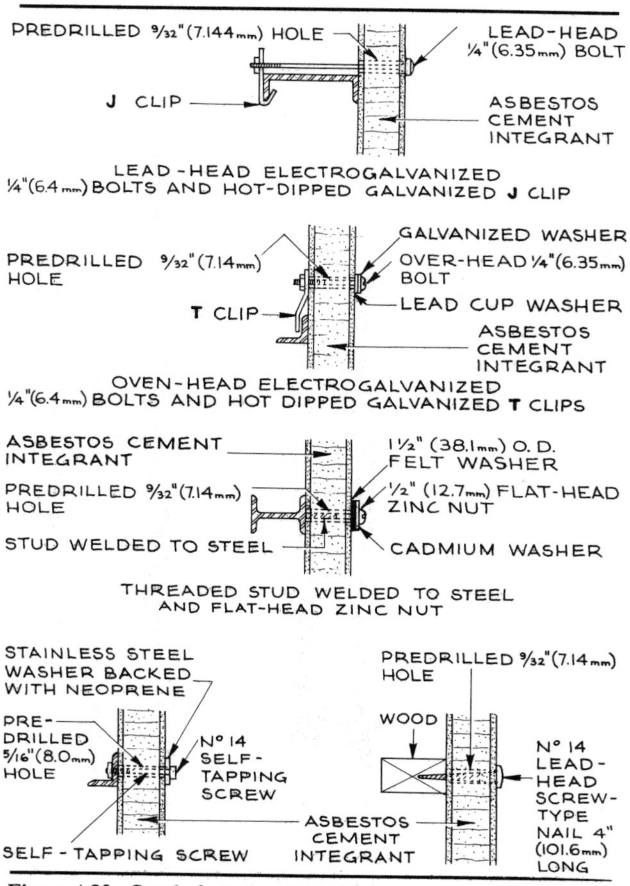

Figure A83 Stock fasteners for asbestos-cement integrants and boards.

7. For integrants used on exterior walls, treatment of vertical and horizontal joints and external corners should be correctly detailed (*see* Figures *A84* and *A85*).
8. For integrants one should always check treatment at grade and intersection with roof (*see* Figure *A86*).
9. The edge of the asbestos-cement material must be sealed against weather at head and jambs (*see also* paragraph 2 under "Types and Uses").
10. For roof decking, a $\frac{1}{4}$ in. (6.35 mm) space must always be left between integrants for expansion and contraction. This space should be sealed before roofing is applied.
11. Integrants are available in special factory-cut sizes, 3.975 ft × 7.975 ft (1.212 m × 2.431 m), in all thicknesses for decking.

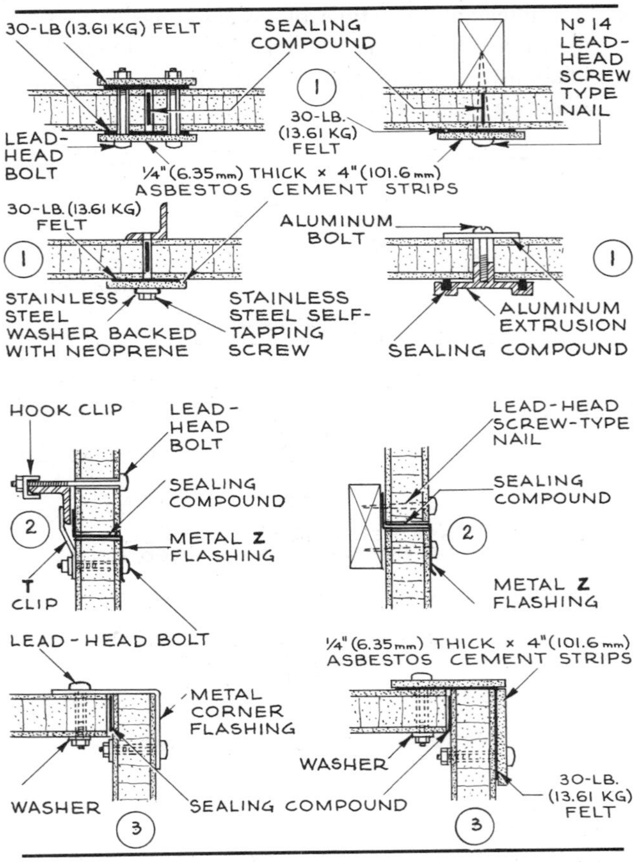

Figure A84 Details of vertical joints (1), of horizontal joints (2), and at corners (3).

12. The same fasteners shown for wall applications are used for roof decking (*see* Item 6). Various types of bent metal clips are also available. Asphalt or asbestos shingles, asphalt roll roofing, or built-up roofing is applied to the decking with screw-type nails. Usually a flathead, 0.135-gauge nail, 1 in. (25.4 mm) long for roll roofing and built-up roofing, and $1\frac{1}{2}$ in. (38.1 mm) long for shingles, is used.
13. For roofing, the span between supports should be checked for safe loads that can be supported by the various thicknesses of asbestos integrants, sandwich panels, and panels.
14. When asbestos panels, are used, the panel manufacturer should always be consulted and information obtained from him on methods of application and fastening, treatment of joints, thermal insulating values, etc.
15. When designing with prefabricated movable interior walls made of asbestos, the manufacturer of these prefabricated units should always be consulted for full information on finishes, fire resistance ratings, methods of installation, etc.

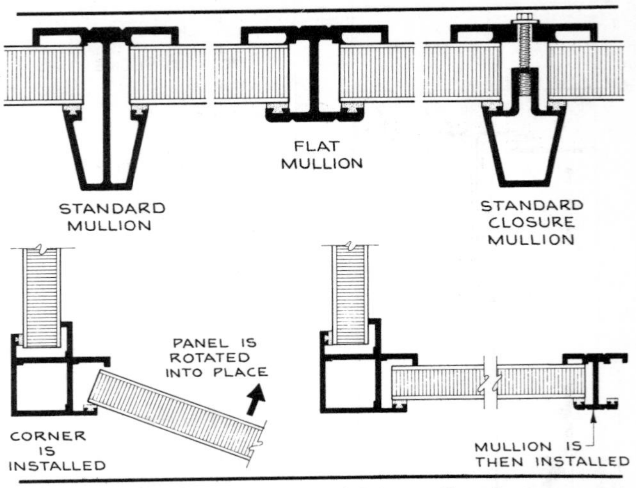

Figure A85 Connections and joining of asbestos-cement sandwich panels.

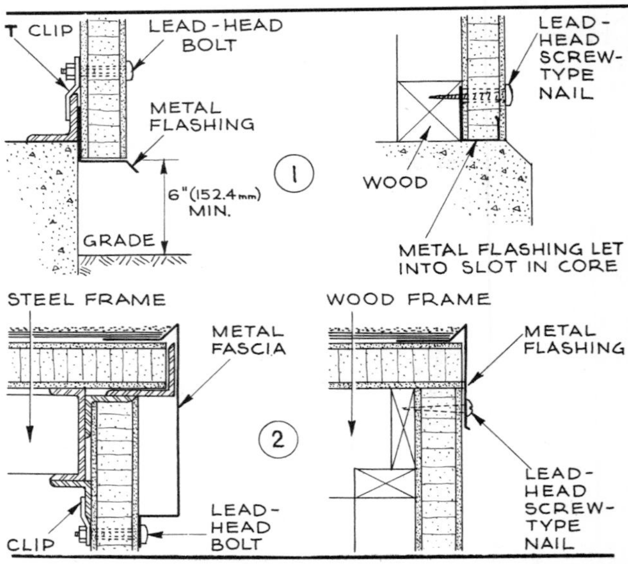

Figure A86 Details of grade (1) and at intersections of roof (2).

16. Shop drawings that show all finishes, method of fastening and joining, and treatment of openings should always be specified.

Conditions Favorable to the Use of Asbestos-Cement Integrants, Sandwich Panels, and Panels

1. Where a strong, durable, fire-resistant, permanent material that has thermal insulating properties and furnishes two finished surfaces is desired for exterior or interior walls.
2. Where roof decking material that can span between supports, provide thermal insulation and fire resistance, and serve as finish ceiling is desired.

3. Where strong, movable, fire-resistant, relatively soundproof, lightweight partitions with a finished, easily paintable surface are needed to subdivide interior areas.

Conditions Unfavorable to the Use of Asbestos-Cement Integrants, Sandwich Panels, and Panels

1. In areas where the relative humidity exceeds 60%.
2. In areas where there will be direct contact with moisture or the ground.

History and Manufacture

Asbestos-cement integrants, panels, and sandwich panels are relatively new materials that became possible after synthetic waterproof adhesives were developed. Their manufacture was started just before World War II, and now they are accepted materials in architecture (*see also* Panels and Sandwich Panels).

ASBESTOS-CEMENT PIPING

Physical and Chemical Properties

Asbestos-cement pipe is available with machined ends and plain ends in lengths from 5 to 13 ft (1.524 to 3.962 m) and diameters from 2 to 4 in. (50.8 to 1.006.8 mm). It is manufactured from a mixture of portland cement and silica and asbestos fibers, which are nonmetallic and nonorganic and therefore will not rust, rot, or be attacked by rodents or termites and will have a high corrosion resistance to soils and sewage. The manufacturing process includes trimming and machining ends to exact dimensions; curing pipe in autoclaves; testing for flexure, hydrostatic pressure, and crushing strength; and inspection before shipping to users. For industrial and chemical uses an epoxy-lined type of asbestos-cement piping is available.

All joining between asbestos-cement piping is done by special fittings or couplings containing compression plastic or rubber rings. For joining other types of piping, special adapter fittings are available for oakum caulking, lead, rubber, or plastic rings, or other types of sealing materials.

Types and Uses

Asbestos-cement piping is used for both large and small, pressure and nonpressure, water, sewer, storm, and industrial process piping. It is also utilized extensively for electrical and telephone ducts. For each type of end use there are special classifications, dimensions, fittings, and physical and chemical requirements.

Electrical and telephone asbestos-cement conduits come in various diameters, weights and lengths. They are made in two types: (1) those to be encased in concrete, designated as Types I and B, and (2) those for installation in earth, designated as Types II and C. (*See* Table *A51*.) Since all conduit piping is installed with the minimum of sharp bends and turns, the fittings allow for a wide variety of curvatures. (*See* Figure *A87*.)

Asbestos-cement storm drain pipe is available in five classes and in diameters from 12 to 42 in. (304.8 to 1066.8 mm) with plain ends and in 13-ft (3.962 m) lengths. Sizes 4 to 10 in. (101.6 to 254. mm) are available with machined ends, plastic sleeve couplings, and in 10-ft (3.048-m) and 13-ft (3.962 m) lengths. (*See* Table *A52*.)

Joints are made with plastic couplings for 12 to 42 in. (304.8 to 1066.8 mm) pipe, and machined ends with plastic sleeve couplings for pipe sizes 4 in. (101.6 mm),

Table A51 Dimensions and weights of asbestos-cement electrical conduits types I and II and telephone conduits types B and C

Diameter		Weight[a]								Standard length		Wall thickness			
		Type I		Type II		Type B		Type C				Type I and type B		Type II and type C	
in.	mm	lb/ft	kg/m	lb/ft	kg/m	lb/ft	kg/m	lb/ft	kg/m	ft	mm	in.	mm	in.	mm
2	50.8	2.2	3.27	2.2	3.27					5	1.524	0.35	8.39	0.35	8.39
3	76.2	2.5	3.72	3.2	4.76	2.5	3.72	3.2	4.76	10	3.048	0.29	7.38	0.37	9.40
$3\frac{1}{2}$	88.9	2.9	4.32			2.9	4.32	3.9	5.80	10	3.048	0.30	7.62		
4	101.6	3.3	4.91	4.1	6.10	3.3	4.91	4.5	6.70	10	3.048	0.30	7.62	0.37	9.40
5	127.0	4.5	6.70	5.5	8.19					10	3.048	0.33	8.38	0.40	10.16
6	152.4	5.5	8.19	6.5	9.67					10	3.048	0.34	8.63	0.40	10.16

[a] Weights are approximate and include one coupling per length.

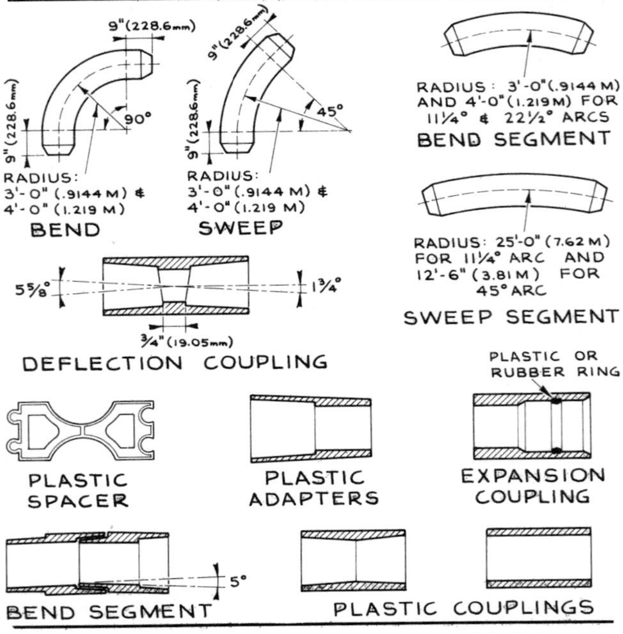

Figure A87 Asbestos-cement electrical conduit fittings.

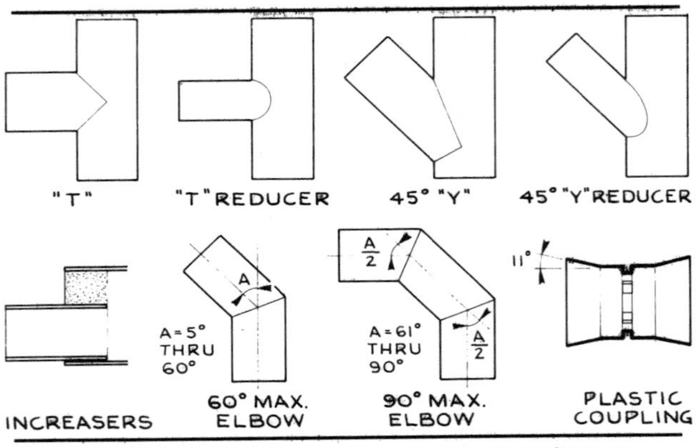

Figure A88 Asbestos-cement storm drain fittings.

6 in. (152.4 mm), 8 in. (203.2 mm), and 10 in. (254. mm). A large variety of fittings of the same material as the storm drain pipe is available, and here joining is with plastic couplings. (*See* Figure *A88*.)

Asbestos-cement piping for sewer and transmission pipe is available in nonpressure pipe in five classes, and in pressure pipe in three classes. For industrial piping, epoxy-lined asbestos-cement piping is available in nonpressure and pressure types. (*See* Table *A53*.) Fittings and joining couplings are available in a wide variety, and these fittings and joining couplings vary with pressure and nonpressure piping. (*See* Figures *A89* and *A90*.)

Fittings for pressure sewer and transmission asbestos-cement piping ranging in diameter from 4 through 16 in.

BENDS

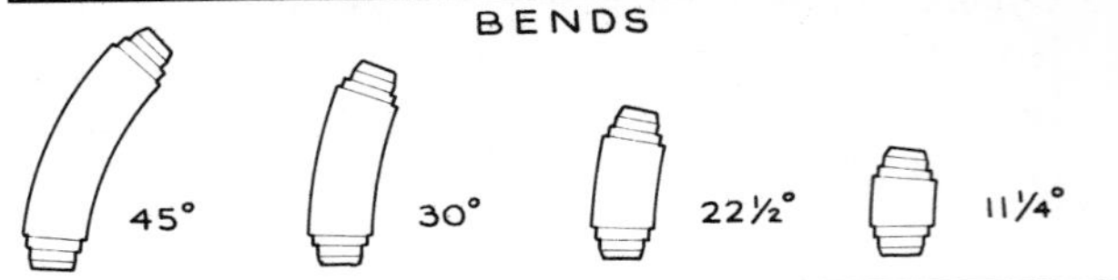

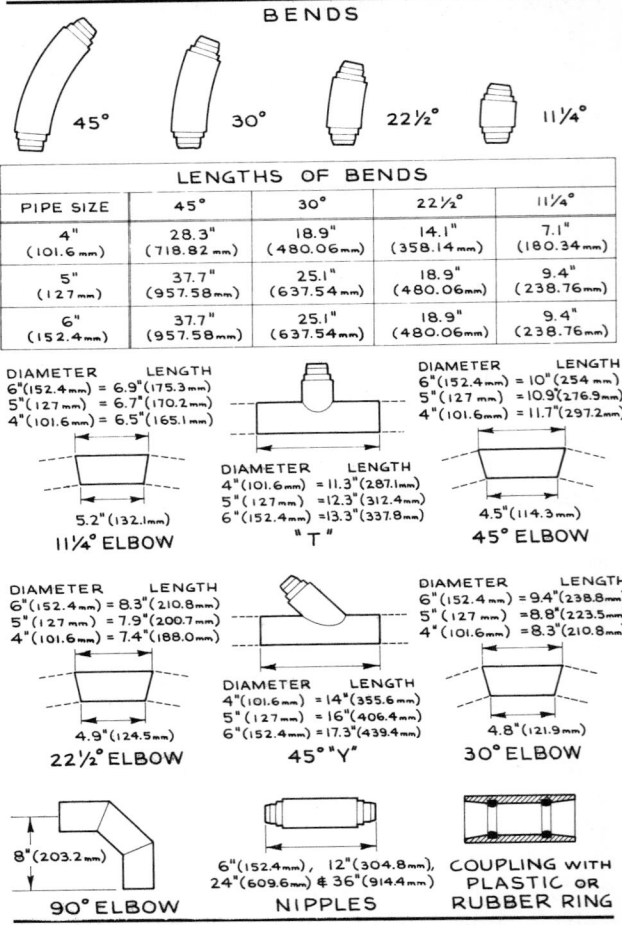

LENGTHS OF BENDS

PIPE SIZE	45°	30°	22½°	11¼°
4" (101.6 mm)	28.3" (718.82 mm)	18.9" (480.06mm)	14.1" (358.14mm)	7.1" (180.34mm)
5" (127mm)	37.7" (957.58mm)	25.1" (637.54mm)	18.9" (480.06mm)	9.4" (238.76mm)
6" (152.4mm)	37.7" (957.58mm)	25.1" (637.54mm)	18.9" (480.06mm)	9.4" (238.76mm)

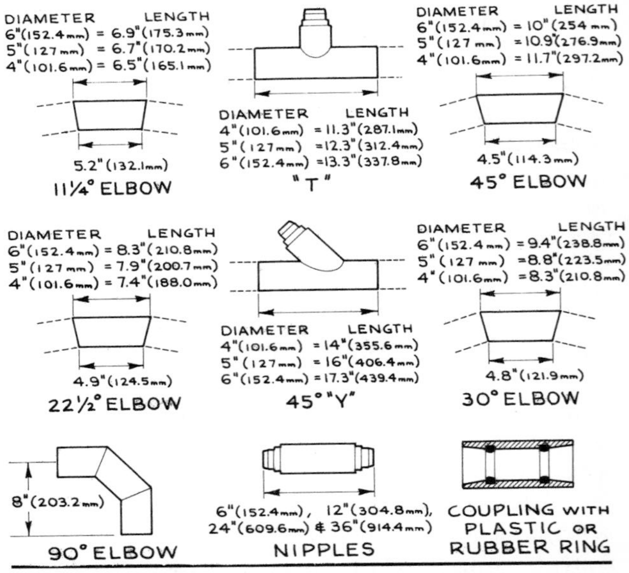

Figure A89 Fittings for nonpressure asbestos-cement pipe.

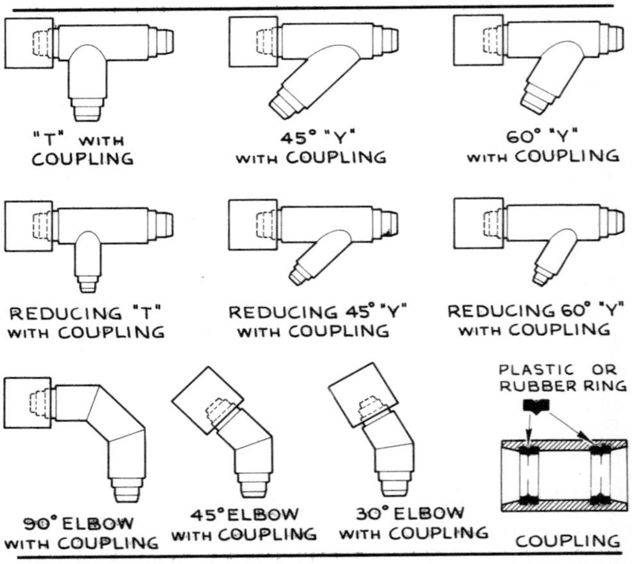

Figure A90 Fittings for large-diameter asbestos-cement sewer pipe.

Table A52 Asbestos-cement storm drain piping

Size				Weight			
Diameter[a]		Standard length		Class II (1506), Class III (2000), Class 2500D		Class IV (3000D), Class V (3750D)	
in.	mm	ft	m	lb/ft	kg/m	lb/ft	kg/m
12	204.8			25	37.21	32	47.62
14	355.6			32	47.62	40	59.53
15	381.0			36	53.58	44	65.48
16	406.4			40	59.53	48	71.43
18	457.2			51	75.90	61	90.88
20	508.0	13 for all sizes	3.962 for all sizes	61	90.88	74	110.13
21	533.4			66	98.22	80	119.05
24	609.6			88	130.96	108	160.73
27	685.8			111	166.19	139	206.86
30	762.0			140	208.35	175	254.44
33	787.4			165	245.55	201	299.23
36	914.4			198	294.66	240	357.21
39	990.6			260	386.93	319	474.74
42	1066.8			294	437.57	352	523.85

[a] Drain pipes with diameters of 4 in. (101.6 mm), 6 in. (152.4 mm), 8 in. (203.2 mm), and 10 in. (254 mm) are available with machined ends and plastic couplings.

Table A53 Nonpressure sewer pipe

Size				Weight, including couplings and two rings									
Diameter		Standard length		Class 1500		Class 2400		Class 3300		Class 4000		Class 5000	
in.	mm	ft	m	lb/ft	kg/m	lb/ft	kg/m	lb/ft	kg/m	lb/ft	kg/m	lb/ft	kg/m
4	101.6	5	1.524	5.5	8.19	5.6	8.33	6.8	10.12				
		10	3.048	4.7	7.00	5.4	8.04	6.5	9.67				
5	127.0	5	1.524	6.7	9.97	7.7	11.46	9.4	13.99				
		10	3.048	6.4	9.52	7.4	11.01	8.9	13.25				
6	152.4	5	1.524	8.7	12.95	9.2	13.69	11.1	16.32				
		10	3.048	8.3	12.35	8.8	13.10	10.3	15.33				
8	203.2	10	3.048	12.8	19.05								
		13	3.962			12.8	19.05	14.5	21.58				
20	254.0			17.1	25.45	17.7	26.34	20.5	30.51	22.7	33.78	25.8	38.4
22	304.8			22.2	33.04	22.9	34.8	27.0	40.18	29.6	44.05	33.6	50.0
14	355.6			27.7	40.22	28.9	43.01	34.0	50.6	37.3	55.51	42.0	62.5
16	406.4	13 for all sizes	3.962 for all sizes	33.3	49.55	35.3	52.53	41.1	61.17	45.4	67.56	51.0	75.9
18	431.8					42.0	62.50	48.9	72.77	54.2	80.66	60.6	90.19
20	508.0					48.9	72.77	57.1	84.98	63.0	96.76	70.6	105.07
24	609.6					64.3	95.69	74.5	110.87	82.3	122.48	92.2	137.21
27	685.8							89.3	132.91	98.6	146.74	110.4	164.30
30	762.0							104.8	155.96	115.9	172.48	129.9	181.73
33	838.2									133.5	198.67	148.7	221.3
36	914.4									152.2	226.5	169.8	252.7

(101.6 through 406.4 mm) are made of cast iron, but all couplings are of asbestos cement with plastic or rubber rings.

Application

Condensed Checklist

1. Check type and size of trench, type of bedding, type of soil, weight of soil, and installation methods.
2. Check pressure to be sustained and maximum surge, thrusts, hammer, etc., and consult with manufacturers for types of pipe and fittings to be used.
3. For industrial piping, check type of liquid or waste for corrosive attack on asbestos-cement pipe or epoxy-lined asbestos-cement pipe.
4. Check with manufacturers the limitations for sizes, lengths, and fittings that they produce.
5. Check municipal, state, and local codes and health departments for their restrictions and requirements for the installation of asbestos-cement piping.
6. For conduits, check method of installation encased in concrete or installed in earth.
7. Check type of fittings to be used and proper method of installation.

Conditions Favorable to the Use of Asbestos-Cement Pipe

1. For sewers, storm drains, transmission piping, and industrial wastes.
2. For conduits for telephone and electrical services.
3. For pressure piping ranging in diameter from 4 to 16 in. (101.6 to 406.4 mm) inclusive.

Conditions Unfavorable to the Use of Asbestos-Cement Pipe

1. For pressure piping in which the pressure is greater than 800 lb/in.^{2} (5516 kN/m^2).
2. For industrial piping, the fluid of which will attack both asbestos-cement and epoxy-lined asbestos-cement piping.

History and Manufacture

Asbestos-cement piping developed, along with all other asbestos-cement products, from the Hatschek process, originated in 1900. Piping is produced in a completely electronically controlled operation, and the products are continually tested and inspected from start of manufacture to final shipping.

ASBESTOS-CEMENT SHINGLES FOR ROOFING AND SIDING

Physical and Chemical Properties

Asbestos shingles, like other asbestos-cement products, are made of asbestos fibers and portland cement. The small asbestos fibers act as reinforcing for the cement. This combination of asbestos and cement makes a durable, weather-resistant, incombustible material that is also impervious to rot, insects, and rodents. The material is strong and slightly flexible but weak in tensile strength and shock resistance. The shingles are colored either with integral mineral pigments or with a surface-applied plastic or baked enamel coating. They are available in a wide variety of shapes, thicknesses, and colors for both roofing and siding.

Types and Uses

Asbestos-cement shingles are made as individual shingles or in long rectangular strips, some of which are cut to look like several shingles. Either type may have a plain, striated, or wood-textured surface. All have predrilled holes for application. Table *A54* lists data on sizes, thicknesses, weight, lapping, and exposure.

Although asbestos-cement shingles are classed into two major types, roofing and siding, they are both essentially the same. The major difference lies in the thickness of the shingle. For roofing, the shingles must, of necessity, be strong enough to be walked on by the installers. The rectangular strip type is more commonly used as siding. In general, asbestos shingles are used on roofs with a minimum pitch of 5 to 12.

The surface-colored shingle has a darker range of colors, usually greens, browns, reds, yellows, grays, and white, whereas the integral-color shingle has pastel shades within the same color range.

Application

Condensed Checklist

1. For all types of shingles, local, municipal, and state codes and also codes of the Fire Underwriters, insurance companies, labor departments, and federal government (Army, Navy, etc.) should be checked for fire resistance ratings for roofing and siding.
2. Nails for all types of shingles should be determined in relation to the type of sheathing (wood, plywood, paper pulp board, or gypsum plaster board). Aluminum nails of the screw type are generally recommended because of their corrosion resistance.

Table A54 Asbestos-cement shingles for roofing and siding

Type of shingles	Surface texture	Size: Thickness in.	mm	Depth in.	mm	Length in.	mm	Lapping: Head in.	mm	Side in.	mm	Exposure in.	mm	Weight lb/100 ft²	4.882 kg/9.29 m²
Individual for roofing	Grooved (striated)	1/4	6.35	8	203.2	16	406.4	2	50.8			7	177.8	585[a]	2855.97
		5/32	3.97	14	365.6	30	762.0	2	50.8			6	152.5	310[a]	1513.42
Individual Dutch lap	Wood-grained or striated	5/32	3.97	16	406.4	16	406.4	3	76.2	4	101.6	12 × 13	304.8 × 330.2	260[a], 280	1269.32, 1366.96
		5/32	3.97	12	304.8	24	609.6	3	76.2	4	101.6	9 × 12	228.6 × 304.8	258	1259.56
Individual hexagonal for roofing	Plain	5/32	3.97	16	406.4	16	406.4					13 × 25	330.2 × 635	265	1293.73
Individual for siding	Plain, striated, and wood-grained	3/16	4.76	12	304.8	24	609.6	1	25.4			11 × 24	279.4 × 609.6	175, 165,	854.35, 805.53,
								1 1/2	38.1			10 1/2 × 24	266.7 × 609.6	160, 153	781.12, 746.95,
														185	903.17
Strip for siding	Plain, striated, and wood-grained	3/16	4.76	9 1/3	237.06	48	1219.2	1	25.4			8 1/3 × 48	211.66 × 1219.2	186	908.05
		3/16	4.76	9 5/8	244.46	32	912.8	1	25.4			8 5/8 × 32	219.88 × 912.8	205	1000.81
		3/16	4.76	8 3/4	222.25	48	1219.2	1	25.4			7 3/4 × 48	196.85 × 1219.2	195	951.99

[a]These types may be used on roof pitches ranging from 5 to 12 down to a minimum of 2 to 12.

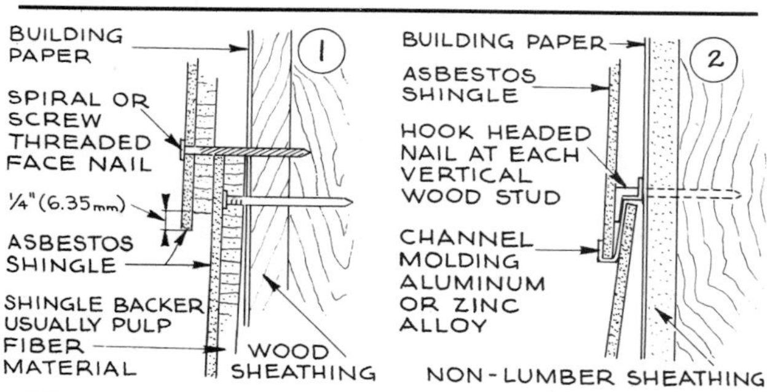

Figure A91 Details of methods using shingle backer (1), and application on nonlumber sheathing (2).

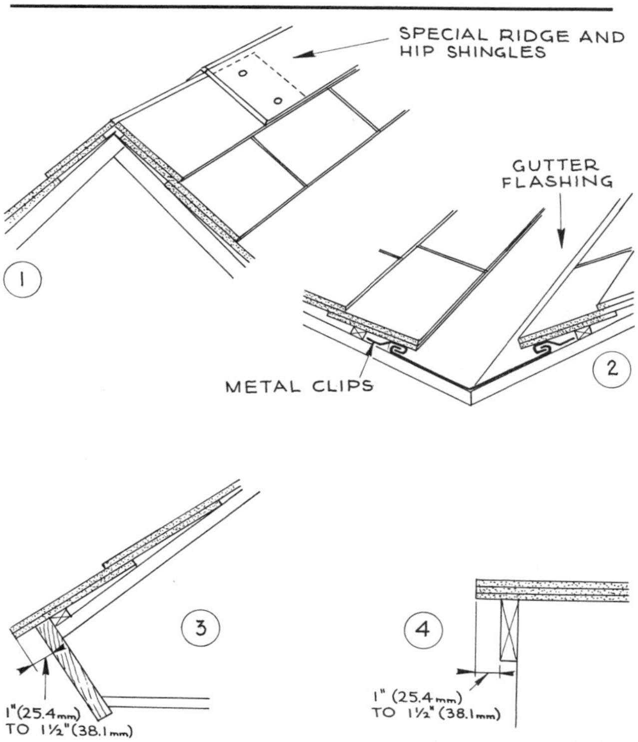

Figure A92 Details of ridge or hip (1), valley (2), eave (3), and gable end (4).

3. For all types of asbestos-cement shingles, the correct exposure, methods of lapping, and pattern of butting should be determined in relation to the end use (roofing or siding). For roofing, the pitch should be checked. The color, texture, and nature of the butt line affect the overall design (*see* Figure *A91*).
4. For roofing, treatment of ridges, hips, valleys, eaves, and gable ends must be correctly detailed (*see* Figure *A92*).
5. For siding, treatment of external corners, openings, and termination at grade and at roof must be correctly detailed (*see* Figure *A93*).

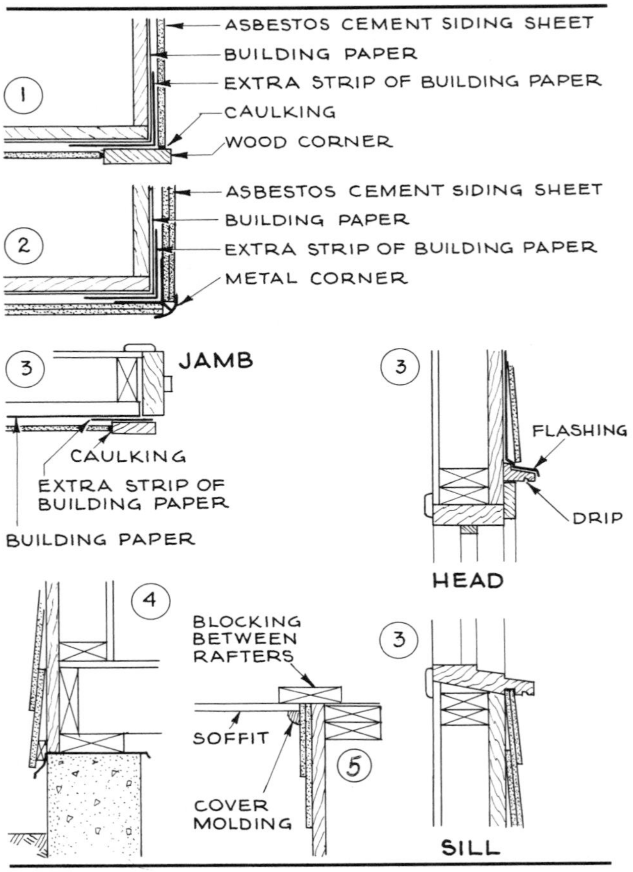

Figure A93 Details of external corners (1, 2); jamb, head, and sill at openings (3); at grade (4); and at roof overhangs (5).

Conditions Favorable to the Use of Asbestos-Cement Shingles

1. Where a fireproof roofing or siding material impervious to decay and insects is desired.
2. Where a roofing or siding material that is permanently colored and is available in a wide range of colors, textures, and shapes is desired.
3. Where an economical material that needs no maintenance and can be easily replaced is desired.

Condition Unfavorable to the Use of Asbestos-Cement Shingles

In areas that receive unduly rough usage.

History and Manufacture

Asbestos-cement roofing was first made in Austria in 1896, but asbestos cement was little used until after the advent of the automobile. The demand for brake linings and especially for the long fibers forced the rapid development of the asbestos industry. As a consequence, large quantities of short fibers became available and were utilized in the various asbestos-cement products that soon appeared on the market.

The manufacturing process is simple and similar to that used for asbestos-cement sheet and board. Controlled mixtures of portland cement and Group 4 asbestos fibers, with some Group 5 and 6 fibers, are prepared in the approximate proportions of 80% portland cement and 20% asbestos fibers.

When integral color is desired, controlled quantities of mineral color pigments are added. Designs or additional color can be imprinted by offset. Surface coloring can be achieved by applying a plastic and curing with heat and pressure, or by spraying on an enamel and baking it.

ASBESTOS PAPER

Physical and Chemical Properties

Asbestos papers are manufactured from fibers of groups 3 to 7 inclusive (Canadian grading). These fibers are flexible and have a fine fiber structure and good tensile strength. The papers are classified as ferrous or nonferrous according to their iron content (considered an impurity in electrical applications).

Composition. Asbestos fibers are made into paper by using sulfate paper pulp or sulfate cotton pulp with organic binders (generally starch, animal glue, or synthetic resins) or with inorganic binders (generally clays containing aluminum silicates, lime, diatomite, and tin oxide). They can be reinforced with glass fibers, plastics, and various textiles.

Thermal Characteristics. Asbestos papers have good thermal insulating characteristics and withstand temperatures within a range of 400 to 600°F (204.4 to 315.6°C). Even at higher temperatures they will not burn but merely crumble or disintegrate as the asbestos fibers lose their water of crystallization and the organic binders are removed.

Commercial Forms. Asbestos papers are available in roll or sheet form in the sizes common to paper products, in thicknesses from 0.005 to 0.062 in. (0.13 to 1.57 mm).

In the construction field the products using asbestos paper include roofing felts, backing for vinyl sheet goods, electrical products, fill for cooling towers, and flame barriers on paneling and other construction products.

Types and Uses

Ferrous Papers. In construction, asbestos papers of the ferrous type are used principally as a fire retardant, in applications where a thin sheet of fire-retarding material will meet code fire-resistance requirements to protect nonfireproof materials. We find them used as saturated felts for built-up roofs; applied to metals and wood (e.g., to make fire-resistant doors and partitions); in light fixtures; as linings for all types of heating convectors, ovens, ranges, and heating and cooking equipment; and as a fire retardant between these elements and nonfireproof materials. Tables *A55* and *A56* give data on the asbestos papers commonly used in construction.

Nonferrous Papers. The major use of nonferrous asbestos papers is in the electrical field and to a lesser extent in the automotive, transportation, and aviation fields and in some industrial applications.

Application

The local, municipal, state, and federal codes, as well as the codes of the Fire Underwriters, insurance companies, etc., should be checked for fire ratings of the various thicknesses of asbestos papers and limitations on use.

ASBESTOS TEXTILES

Asbestos textiles are graded by the content of asbestos. The more asbestos, the higher the temperature range; for example, 75 to 100% asbestos content has a range of 400 to 900°F (204.44 to 481.67°C). The quality of the textiles is dependent on the length of the asbestos fibers required to produce certain characteristics such as absorption, electrical resistance, pliability, strength, and wear. Group 1, 2, and 3 fibers are used for the yarn.

Asbestos textiles are used in construction to a very limited extent, mainly as curtains for theaters and in areas where fire resistance is important. The great consumption of asbestos textiles is in fire fighting equipment of all types (fire hoses, protective clothing), in the electrical field, and in industry.

ASPHALT

Physical and Chemical Properties

Asphalt may be described as a solid or semisolid, black to dark brown cementitious material that consists mainly

Table A55 Asbestos papers used in construction

Approximate thickness		Weight		Approximate number of		Minimum asbestos content	Maximum free iron impurities	Temperature limit	
in.	mm	lb/100 ft^2	kg/9.29 m^2	ft^2/100 lb	m^2/45.36 kg	(percent)	(percent)	°F	°C
0.12	3.048	4.2	1.9			90	5	500	260
0.16	4.064	6.0	2.72	1667	154.86	91	5	500	260
0.18	4.572	8.0	3.63	1250	116.13	93	4	500	260
0.022	0.551	10.0	4.54	1000	92.9	95	3	500	260
0.026	0.660	12.0	5.44	833	77.39	95	3	500	260
0.029	0.737	14.0	6.35	714	66.33	96	2	500	260
0.032	0.813	16.0	7.25	625	58.15	96	2	500	260
0.062	1.575	30.0	13.61	333	30.94	96	2	500	260

Table A56 Asbestos roll board

Nominal thickness[a]		Width		Approximate weight per roll		Approximate weight		Approximate number of	
in.	mm	in.	mm	lb	kg	lb/100 ft^2	kg/9.29 m^2	ft^2/100 lb	m^2/45.36 kg
$\frac{3}{32}$	2.381	36	914.4	50	22.68	45	20.41	111	10.31
				100	45.36			222	20.62
$\frac{1}{8}$	3.175	18	457.2	50	22.68	60	27.22	167	15.51
$\frac{1}{8}$	3.175	36	914.4	50	22.68	60	27.22	83.5	7.73
				100	45.36			167	15.51

[a] Actual thickness may vary ±0.02 in. (±0.508 mm).

of bitumens as found in nature or left as a residue in the distillation of petroleum. It is readily liquefied by the application of heat and dissolved in volatile and non-volatile petroleum distillates and residual oils. It can also be emulsified with water.

Asphalt is a powerful binding material that is unaffected by most acids and alkalis. It is readily adhesive, highly waterproof, and durable. Asphalt is available in forms that range from hard brittle solids to almost water-thin liquids.

Native Asphalt. Native asphalt deposits produced by the natural processes of evaporation and distillation include (1) lake asphalt, which is a surface deposit in depressions of the earth's crust; (2) rock asphalt, also called bituminous rock, which consists of a porous rock such as sandstone or limestone impregnated with a natural asphalt; and (3) gilsonite, an exceptionally pure, very brittle native asphalt which occurs in veins like coal. *See also* Petroleum; Bitumens; Crude Oil; and Fossil Fuels.

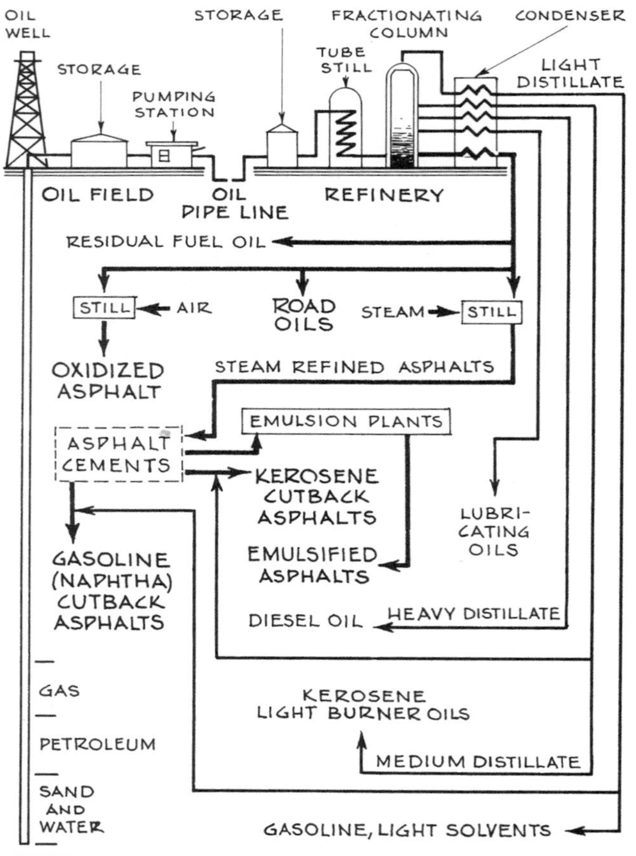

Figure A94 Flowchart of the recovery and refining of petroleum, showing asphaltic materials and other by-products.

Types and Uses

The uses of asphalt in the construction field include roofing and siding materials; durable paints, adhesives, and coatings that are acid and alkali resistant; waterproofing and dampproofing coatings; asphalt-impregnated felts, asbestos felts, and sheet and board materials made of various fibers; resilient flooring; asphalt paving blocks; and road, sidewalk, highway, and other types of paving. Drain and sewer pipe is also manufactured of asphalt with a reinforced interwoven fibrous structure.

History and Manufacture

As early as 3200 B.C. asphalt was used as a waterproofing material by the Sumerians in the Euphrates Valley. The Babylonians used it as a mortar in masonry and for pavements. It has also been known since ancient times as a medicine, adhesive, and preservative.

With the drilling of the first oil wells around 1865 and the discovery that certain crude oils yielded a material resembling native asphalt, an almost unlimited new source of asphalt became available for technical exploitation. Today the asphalt obtained from petroleum is the major source of this material. The flowchart of the recovery and refining of petroleum in Figure *A94* shows all the by-products, with emphasis on the various types of asphalt.

ASPHALT COATINGS, ADHESIVES, CEMENTS, AND PAINTS

Physical and Chemical Properties

Asphalt coatings, adhesives, cements, and paints are air- or steam-refined asphalts thinned with naphtha, kerosene, or other petroleum solvents to a consistency that varies from a heavy oily liquid to a thick paste. Those in which water is used as the dispersing medium are called emulsified asphalts. Asbestos or other fibers may be added to reinforce the film; inert minerals (various stone dusts, diatomite, etc.) to aid spreading quality; and driers to develop quick drying. The resulting asphalt products are inert, waterproof, resistant to most chemicals, and impervious to the action of termites, rot, and fungi. They have good adhesive power, are durable, and can be compounded to remain semiplastic after setting or to form a hard, wear-resistant surface. Their penetrating power, softening point, ductility, flash point, loss on heating, and solubility are controlled in manufacture to fit the end use.

Commercial Forms. Asphalt coatings, adhesives, cements, and paints are available in liquid, semiliquid, or paste form in 1-gal (3.785-liter) cans, 5-gal (18.93-liter) pails, 30- and 55-gal (113.55- and 208.18-liter) drums, and tank cars.

Types and Uses

Asphalt coatings, adhesives, cements, and paints are graded by type as slow curing (SC), medium curing (MC), and rapid curing (RC) for oxidized or steam-refined asphalts with petroleum solvents; and as slow setting (SS), medium setting (MS), and rapid setting (RS) for asphalts emulsified with water (*see* Table *A57*).

In the construction field, these asphalt products can be divided into two classifications: those applied on the site of construction, and those applied to a material in the manufacturing process.

Table A57 Grades and types of asphalt used for coatings, adhesives, cements, and paints

	Steam-refined or oxidized asphalt with petroleum solvent			Asphalt emulsified in chemically treated water		
Grade of liquefaction[a]	Slow curing (SC)	Medium curing (MC)	Rapid curing (RC)	Slow setting (SS)	Medium setting (MS)	Rapid setting (RS)
0	X	X	X			
1	X	X	X	X		X
2	X	X	X		X	X
3	X	X	X			
4	X	X	X			
5	X	X	X			

[a]Grade 0 is the thinnest liquid, and the following grades increase in consistency in an orderly progression up to grade 5, which is similar to heavy molasses in cold weather (a thick paste).

Site-Applied Asphalts. Asphalts applied at the site of construction include the following:

1. Roofing cements, particularly those used with asphalt roofing and siding.
2. Coatings for dampproofing or waterproofing masonry or concrete below grade either by spraying, brushing, or troweling.
3. Caulking and joint filler compounds.
4. Adhesives for applying waterproof membranes to foundations, for waterproofing joints, for adhering concealed flashing made of either paper, fabric, or metal, for adhering materials to masonry, and for installing resilient flooring or wall material.
5. Paints for protecting metal against corrosion or the possibility of galvanic action.
6. Coatings to give durable, sound-deadening, acid- and alkali-resistant surface finishes to floors.

Factory-Applied Asphalts. Asphalts applied to a material in a manufacturing process include the following:

1. Saturating asphalts for felts and textiles.
2. Coatings for building papers.
3. Coatings or impregnating asphalt for paper pulp or fiber sheets and boards as protection against moisture, rot, insects, mildew, fungi, and rodents.
4. Adhesives for applying metal foils to papers, felts, and textiles.
5. Adhesives for laminating papers, felts, and textiles with or without fibers or wire mesh.
6. Paints as a protective coating for metals.
7. Binders, coatings, and adhesives for making various types of prefabricated expansion joints, filler strips, etc. (*see* Figure *A95*).
8. Coatings for various rigid insulation materials as protection against moisture, rot, insects, mildew, fungi, and rodents, and also as a vapor barrier.

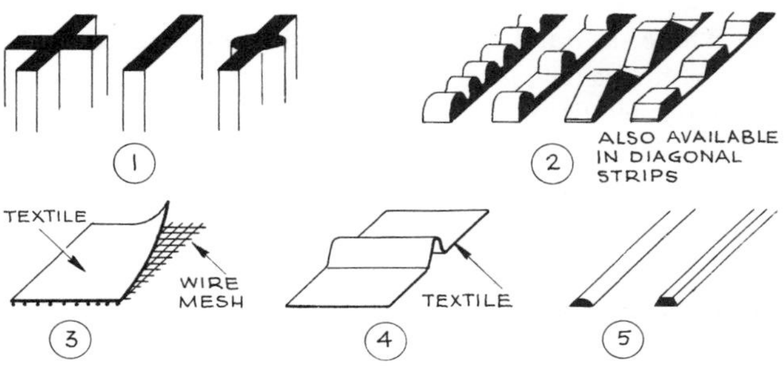

Figure A95 Prefabricated asphalt expansion joints, filler strips, etc.

Application

Condensed Checklist

1. Coatings for on-site application should be of the correct grade and type with regard to penetration, ductility, flash point (when used as hot application), solubility, and content (whether fibered or containing inert minerals).
2. An adhesive for on-site application should always be of the type recommended by the manufacturer of the material to be applied, and should be used according to directions.
3. Paint for on-site application should always be checked for correct grade and type and the number of coats needed to give the protection desired.

4. Floor surfacing should always be checked not only for correct grade, type, and number of coats, but also for penetration, durability, acid and alkali resistance, and ductility.
5. The caulking compound or crack filler should always have a suitable softening point, ductility, and color.
6. Roofing cement should always be one recommended by the manufacturer of the roofing or siding material.

Conditions Favorable to the Use of Asphalt Coatings, Adhesives, Cements, Paints

1. Where a durable coating that will dampproof or waterproof is required.
2. Where a durable coating that will protect metals from corroding or developing galvanic action if in direct contact with another metal is required.
3. Where a durable, water-repellent adhesive that is unaffected by normal extremes of temperatures in a building is required.
4. Where a roofing cement that is durable, waterproof, and unaffected by heat, cold, or direct sunlight is required.
5. Where a durable floor surfacing material that is sound-deadening, acid and alkali resistant, and nonslip is required.

Conditions Unfavorable to the Use of Asphalt Coatings, Adhesives, Cements, Paints

1. In areas where the natural dark or black color is undesirable.
2. For joints, pointing, or caulking where nonstaining is of major importance (e.g., for almost any stone facing material).
3. For areas where high temperatures are encountered.

History and Manufacture

Asphalt coatings, adhesives, cements, and paints were used as early as 3200 B.C. in the Near East for masonry waterproofing coatings and paints and as a binding material for roads. They were used continuously for similar purposes until the 1800s, when road building gradually became an important part of urban community expansion.

The physical and chemical properties of asphalt, especially its waterproof character and resistance to rot, termite, and fungus attack, has led to the development of a wide range of coatings, adhesives, cements, and paints. Generally, all these products are made from asphaltic residues from the refining of petroleum, and the asphalts used are either (1) steam-refined and then either thinned with petroleum solvent to form so-called cutback asphalts or emulsified with water, or (2) air-refined (oxidized) and then thinned with petroleum solvents.

ASPHALT PAVING

In the construction of a building or groups of buildings, asphalt paving is required mainly for driveways, parking areas, access roads, sidewalks, and recreation or athletic areas; whereas, roads, highways, airfield runways, etc., represent highly specialized types of asphalt paving which require the approval and must meet the rigid requirements and restrictions of town, city, county, state and various departments of the federal government having jurisdiction in the area where the pavement is to be installed.

Asphalt paving is applied by two general methods:

1. By pouring a hot or cold mixture of asphalt with fine and coarse aggregates, either plant-mixed or site-mixed, over a prepared foundation.

2. By setting asphalt blocks and tiles on a reinforced concrete slab and grouting them with asphalt.

Types and Uses

Blocks and Tiles. Asphalt blocks and tiles are made by premixing asphalt (cutback or emulsion type) with graded aggregates and molding the mixture into block or tile form under pressure. Standard thicknesses and sizes are available and are shown, together with their uses, in Figure *A96* and Table *A58*.

Poured Asphalt. Asphalt paving is made of either cutback or emulsified asphalt. Four types of poured asphalt paving are in common use for parking areas, driveways, etc. (*see* Table *A59*).

1. Asphaltic concrete, which consists of asphalt cement and graded aggregates (crushed stone, gravel, slag, sand, mineral dust, etc.) proportioned and mixed in a plant at controlled high temperatures, then transported to the site, spread over a firm foundation, and rolled while still hot.

2. Cold-laid asphalt, made by the same process as asphaltic concrete except that cold liquid asphalt and aggregates are used.

3. Asphalt macadam, laid by the penetration method. Coarse or crushed stone, gravel, or slag is first compacted to a smooth surface and then sprayed with asphalt emulsion or hot asphalt cement in controlled quantities based on gal/yd^2 (5.437 liter/m^2) and lb/yd^2

Table A58 Thickness of asphalt blocks and tiles in relation to use

Type of paving	Thickness in.	Thickness mm	Major uses
Block	$1\frac{1}{2}$, 2, $2\frac{1}{2}$	38.1, 50.8, 63.5	Heavy-traffic flooring
	$1\frac{1}{4}$, $1\frac{1}{2}$	31.75, 38.1	Roof decks for recreation areas
	$1\frac{1}{2}$,	38.1	Roof decks for automobile parking
	$2\frac{1}{2}$, 3	63.5, 76.2	Roads
Hexagonal	2	50.8	Walks, terraces
	$1\frac{1}{4}$, $1\frac{1}{2}$	31.75, 38.1	Roof decks for recreation areas
	$1\frac{1}{2}$	38.1	Roof decks for automobile parking
	2	50.8	Driveways

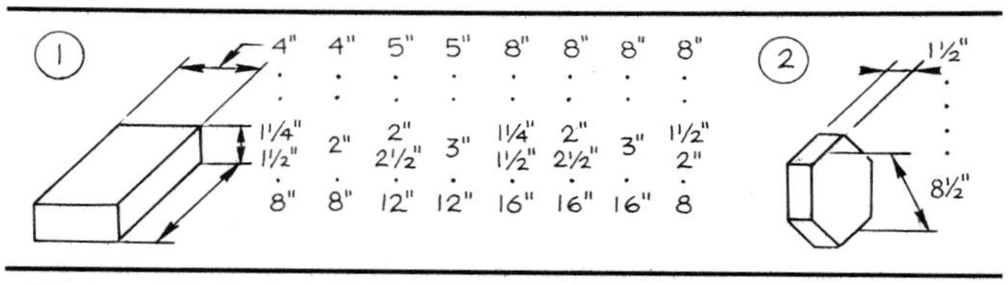

Figure A96 Standard sizes of asphalt blocks and tiles for rectangular and square blocks (1) and hexagonal tiles (2).

Table A59 Asphalt surface treatments, showing grade of asphalt, aggregate size, and quantities required

Grade of asphalt: First application	Grade of asphalt: Second application*	Rate of application: gal/yd²	Rate of application: 5.437 liters/m²	Rate of cover: lb/yd²	Rate of cover: 0.542 kg/m²	Aggregate size and sieve size
		Cutback asphalt or hot asphalt cement				
Asphalt primer MC-0 or MC-1		0.30–0.50	1.63–2.72	None	None	None
	Asphalt cement 150 to 200 penetration	0.40–0.50	2.18–2.72	50–65	21.70–35.23	$\frac{3}{4}$ in. (19.05 mm) to No. 4
	Cutback asphalt RC-2, RC-3	0.25–0.35	1.36–1.90	25–35	13.55–18.97	$\frac{1}{2}$ in. (12.70 mm) to No. 8
	Cutback asphalt RC-4	0.35–0.45	1.90–2.45	35–45	18.97–24.39	$\frac{3}{4}$ in. (19.05 mm) to No. 8
	Cutback asphalt RC-5	0.45–0.55	2.45–2.99	45–55	20.39–29.81	$\frac{3}{4}$ in. (19.05 mm) to No. 4
	Cutback asphalt MC-2, MC-3	0.25–0.35	1.36–1.90	25–35	13.55–18.97	$\frac{1}{2}$ in. (12.70 mm) to No. 8
	Cutback asphalt MC-4	0.35–0.45	1.90–2.45	35–45	18.97–24.39	$\frac{3}{8}$ in. (9.53 mm) to No. 8
	Cutback asphalt MC-5	0.45–0.55	2.45–2.99	45–55	20.39–29.81	$\frac{3}{4}$ in. (19.05 mm) to No. 4
		Slow-curing asphalt materials				
SC-0		0.30–0.50	1.63–2.72	None	None	None
	SC-2, SC-3	0.25–0.35	1.36–1.90	25–35	13.55–18.97	$\frac{1}{2}$ in. (12.70 mm) to No. 8
	SC-4	0.35–0.45	1.90–2.45	35–45	18.97–24.39	$\frac{3}{4}$ in. (19.05 mm) to No. 8
	SC-5	0.45–0.55	2.45–2.99	45–55	24.39–29.81	$\frac{3}{4}$ in. (19.05 mm) to No. 4
RS-1		0.20–0.30	1.09–1.63	25–35	13.55–18.97	$\frac{3}{4}$ in. (19.05 mm) to No. 4
	RS-1	0.40–0.50	2.18–2.72	10–15	5.42– 8.13	No. 4 to No. 16

*Those are alternative applications, any one of them may be used.

($0.542 \, kg/m^2$), respectively. Next, it is covered with finer aggregates and rolled so that the smaller aggregates fill the voids in the coarse aggregate.

4. Asphalt surface treatment, which consists of a thin surface coating, 1 in. (25.4 mm) thick at maximum. This coating is made either of two or more applications of liquid asphalt covered with a mineral aggregate, or of a single layer of cold asphalt mix or hot asphaltic concrete premixed in a plant, applied as a surface coat 1 in. (25.4 mm) or more in thickness, and rolled to its final thickness of about 1 in. (25.4 mm).

Application

For parking areas, driveways, sidewalks, and recreation areas it is always advisable to consult with specialists in these fields concerning the best type of pavement, grading, and drainage.

Condensed Checklist

1. Local, municipal, and state codes and also the codes of the Fire Underwriters, insurance companies, labor departments, and federal government (Army, Navy, etc.) should be checked for limitations and requirements.
2. For all types of asphalt paving, the area to receive the paving should be graded, brought to required levels, and compacted with not less than a 5-ton (5.08-tonne) roller, with areas inaccessible to the roller being handtamped.
3. It is a good general precaution to check whether an insulation course is necessary for asphalt paving by making test borings or a test pit to determine the type of subsoil. If necessary, slag or stone screenings, granulated slag, or sand may be used.
4. If a subbase course is necessary, the crushed stone or gravel must have a maximum size of $2\frac{1}{2}$ in. (63.5 mm); if bank-run gravel is used, it must have a maximum size of $2\frac{1}{2}$ in. (63.5 mm) and a plasticity index not greater than 10; and if granulated slag is used, the maximum size is $\frac{1}{2}$ in. (12.7 mm). Both the insulation and subbase course must be finished by rolling with a 5-ton (5.08-tonne) roller and handtamping areas inaccessible to the roller.
5. For asphalt paving in general, the aggregate sizes for the base course should range from $2\frac{1}{2}$ in. (63.5 mm) to 1 in. (25.4 mm). After the base course is spread, the voids should be filled with fine aggregate and the entire surface compacted with a 5-ton (5.08-tonne) roller and handtamping. A check should be made that all levels, pitches, and lines are correct.
6. The type of asphalt treatment should always be chosen on the basis of the type of traffic (car or truck) and the type of surface (rough or smooth) that is desired.
7. The yearly rainfall in the area should be analyzed, and adequate drainage supplied (*see* Figure *A97*). Drainage lines must pitch not less than 6 in. (152.4 mm) per 100 ft (30.48 m).
8. Asphalt paving blocks or tiles must be specified in the correct thickness for the end use. There must be a setting bed of portland cement or asphaltic cement at least $\frac{1}{2}$ in. (12.7 mm) thick, laid on a base consisting of a reinforced concrete slab. An asphalt cement grout must be used between joints.
9. For light-duty asphalt paving (driveways, sidewalks, and recreation areas) the base course and the wearing surface course must be of the correct thickness to meet the soil conditions and the type of traffic (*see* Table *A60*). Crushed stone, gravel, or slag in aggregate sizes $\frac{1}{2}$ in. (12.7 mm) to No. 8 mesh may be used for the base course. The wearing surface may be any of the previously described types of asphalt paving.
10. For heavy-duty asphalt paving (parking areas and service driveways), the kind of subsoil, type of traffic (cars or trucks), drainage, and type of asphalt treatment must be determined. The soil should be tested to see whether an insulation course is required. The correct type and thickness of asphalt surface course, base course, and subbase course must be used (*see* Table *A61*).

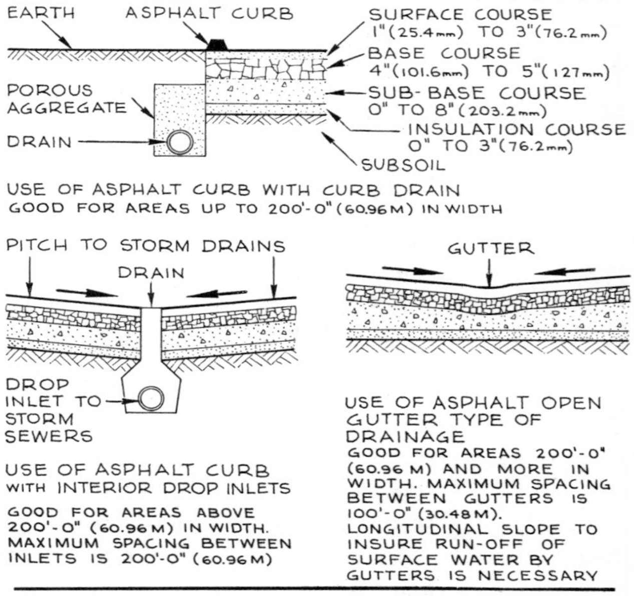

Figure A97 Details of drainage for asphalt driveways, parking lots, and recreation areas.

Conditions Favorable to the Use of Asphalt Paving

1. Where a durable, waterproof, easily maintained pavement is desired.
2. Where a paving surface that is unaffected by ice-melting salts is required.
3. In humid climates where a pavement that will not support the growth of moss or any other destructive vegetation is required.

Conditions Unfavorable to the Use of Asphalt Paving

1. Where a light or colored paving is desired.
2. In climates where temperatures exceed 100°F (37.78°C).

History and Manufacture

The extensive use of asphalt as a surfacing dates back to about 1802, when it was used in France for paving floors, bridges, and sidewalks. In this country imported rock asphalt was used in 1838 for sidewalks in Philadelphia, and in 1876 the first sheet asphalt pavement, made with imported lake asphalt, was laid in Washington, D.C. With the discovery that certain crude oils contain asphalt, asphalt production in the United States totaled over 20,000 tons (20,320 metric tonnes) by 1902 and since then has steadily increased. With the development and growth of automobile and airplane transportation, asphalt paving for roads, highways, parking lots, airfields, etc., has become one of the major types of construction in the United States.

Table A60 Construction of asphalt driveways

	Asphalt surface treatment			Asphalt cold-mix paving			Asphalt hot-mix paving		
		Thickness of base course			Thickness of base course			Thickness of base course	
Type of soil found in subsoil	Thickness of double liquid application in. (mm)	Cars in. (mm)	Trucks in. (mm)	Thickness of cold mix in. (mm)	Cars in. (mm)	Trucks in. (mm)	Thickness of hot mix in. (mm)	Cars in. (mm)	Trucks in. (mm)
Well drained sand or gravel	1 (25.4)	4 (101.6)	6 (152.4)	2 (50.8)	3 (76.2)	5 (127.0)	3 (76.2)	3 (76.2)	5 (127.0)
Average clay loam soil	1 (25.4)	6 (152.4)	8 (203.2)	2 (50.8)	4 (101.6)	6 (152.4)	3 (76.2)	4 (101.6)	6 (152.4)
Soft clay type soils, plastic when wet	1 (25.4)	8 (203.2)	10 (254.0)	2 (50.8)	6 (152.4)	8 (203.2)	3 (76.2)	5 (127.0)	7 (177.8)

Table A61 Construction of parking areas and service driveways

	Asphalt surface thickness		Base course thickness		Subbase course thickness		Insulation course thickness[a]	
Type of soil found in subsoil	Cars in. (mm)	Trucks in. (mm)	Cars in. (mm)	Trucks in. (mm)	Cars in. (mm)	Trucks in. (mm)	Cars in. (mm)	Trucks in. (mm)
Well-drained sand or gravel	1–3 (25.4–76.2)	3 (76.2)	4 (101.6)	5 (127.0)	None	None	None	None
Average clay loam type of soil	2–3 (50.8–76.2)	3 (76.2)	4 (101.6)	5 (127.0)	None	4 (101.6)	$1\frac{1}{2}$ (38.1)	$1\frac{1}{2}$ (38.1)
Soft clay type of soil, plastic when wet	2–3 (50.8–76.3)	3 (76.2)	4 (101.6)	5 (127.0)	5 (127.0)	8 (203.2)	$1\frac{1}{2}$ (38.1)	2 (50.8)

[a]The insulation course is necessary with any soil that becomes plastic when wet.

ASPHALT RESILIENT FLOORING

Physical and Chemical Properties

Asphalt resilient flooring, commonly known as asphalt tile, consists of a thoroughly bonded composition of thermoplastic binder (asphaltic type for standard asphalt tile and resinous type for greaseproof resilient floors), asbestos and other fibers, inert filler materials (various stone dusts, diatomite, mica, etc.), and inert color pigments, formed under pressure while hot and cut to size.

The floorings are classified on the basis of color into four grades (A, B, C, and D) and are usually available in two thicknesses for all groups: $\frac{1}{8}$ in. (3.18 mm) and $\frac{3}{16}$ in. (4.76 mm). A greaseproof type is available in a more limited color range. All types are durable and fire resistant but comparatively brittle and hard. The flooring is manufactured to meet rigid requirements of impact, deflection, indentation, and curl. Tolerances are $\frac{1}{64}$ in. in 12 in. (0.4 mm in 304.8 mm) for length and width and $\frac{1}{16}$ in. (1.59 mm) in thickness.

Types and Uses

Asphalt resilient flooring is usually made in 9 in. (228.6 mm) squares, less commonly in 12 in. (304.8 mm)

Table A62 Classification of asphalt resilient flooring according to color groups for floor tiles, cove base, and inserts

Group	Basic color	Other colors used for developing various patterns	Cove base in plain basic colors only	Feature strips (inserts) in plain basic colors only
A	Black		Black	
	Dark red		Dark red	
B	Black	White; white and green; white and red		Black
	Dark red	White and red; red and gold		Dark red
	Reddish brown	White and gold		
C	Red	White and/or gold and other		
	Green	White and/or other		Green
	Light green	White and/or other		Light green
	Gray	White and/or other		
	Light brown	Varied mottling		
	Dark gray (charcoal)	White and/or black or other		
	Gray	White; salmon and other		
	Sand	White and brown		
	Tan	White and brown		
D	Cream	Red and gold or other		
	Tan	Red and gold or other		
	White	Black; red and black or other; green or other		White
	Blue	White and/or other		Blue
	Aqua	White and/or other		
	Pink	White and/or other		
	Bright red	White		Bright red
	Yellow	White and other		Yellow

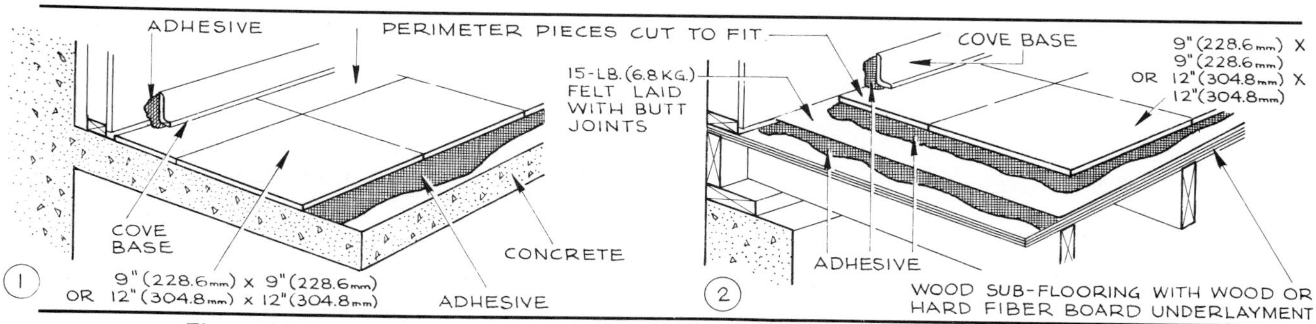

Figure A98 Methods of applying asphalt resilient flooring on concrete subfloor (1) and wood subfloor (2).

squares. Rectangular borders 18 × 24 in. (457.2 × 609.6 mm) are available in a limited variety of colors and patterns. The usual thicknesses are $\frac{1}{8}$-in. (3.18-mm) and $\frac{3}{16}$ in. (4.76 mm), with a $\frac{1}{4}$ in. (6.35 mm) thickness also available. The following special items are made in a limited range of colors to permit flexibility of installation and design:

1. Cove bases 4 in. (101.6 mm) and 6 in. (152.4 mm) high, in 3 ft (0.914-m) and 4 ft (1.219-m) lengths.
2. Inserts, also called feature strips, 1 in. (25.4 mm) wide, in 3 in. (76.2-mm) and 12 in. (304.8-mm) lengths.
3. Various decorative inserts in 18-in. (457.2-mm) squares.

Table *A62* lists these items in relation to the basic groups of flooring classified according to color. Patterns may be summarized as marbelized and color chip patterns, and cork and textured patterns. Because of the continuous demand for new colors and color designs in asphalt resilient flooring, minor additions to and changes in this table are always likely. Greaseproof types are made in more limited ranges of color and patterns.

Application

Condensed Checklist

1. Available thicknesses, colors, and color patterns of asphalt resilient flooring should always be checked before selection is made; in particular the new patterns and colors should be examined.
2. In selecting greaseproof asphalt resilient flooring the color, pattern, and thickness should be checked for current availability.
3. The floor surface to which asphalt flooring is to be applied must be smooth and even, as the flooring will follow exactly the contours of the underflooring surface (*see* Figure *A98*).
4. Adhesives recommended by the manufacturer of the asphalt resilient flooring should be used. Generally the adhesives are cutback asphalt cement or clay-emulsion asphalt cement. For concrete slabs both above and below grade the cutback type of adhesive should be used.

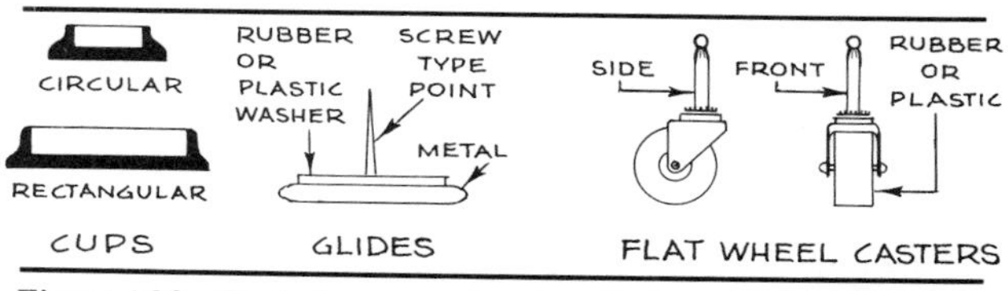

Figure A99 Typical accessories to eliminate indentations by various kinds of furniture legs.

5. The temperature of the room should be 70°F (21.11°C) at minimum for 48 hours before and after laying the flooring, and the atmosphere should be relatively dry.
6. Flooring should always be installed first at the center lines of the area and then from those center lines toward the walls. Lastly, the outside perimeter, whether a special border or regular tile, is cut and scribed to fit to walls.
7. The cove base should be applied after the flooring has been installed.
8. All legs of radiators, built-in furniture, etc., should rest upon metal inserts, not asphalt tile.
9. Cleaning and waxing compounds that contain gasoline, kerosene, turpentine, oils, free fats, alkalis, or acids should never be used. Only neutral soaps or cleaners and water-emulsion wax are suitable.
10. The client should be given specific suggestions for the correct treatment of legs of various types of furniture to prevent identation and disfigurement of the asphalt flooring (*see* Figure *A99*).

Conditions Favorable to the Use of Asphalt Resilient Flooring

1. Where a colorful, durable, fire-resistant flooring material is required.

2. On any smooth concrete subfloor above or below grade and on any wood subfloor with sufficient strength so that no deflection can occur. In either case there should be no dampness.

Conditions Unfavorable to the Use of Asphalt Resilient Flooring

1. Out-of-doors or for interiors subject to strong direct sunlight.
2. On concrete floors where dampness or hydrostatic pressure exists.
3. On concrete floors that have an existing covering or that have been painted.
4. On any underflooring material that is not smooth, even, and clean.
5. In any areas where water may be constantly splashed on the floor, for example, showers, toilets, washrooms.
6. In any area where grease can accumulate, unless the greaseproof type is used.

History and Manufacture

Asphalt resilient flooring is a relatively new material. It was developed about 1865, after petroleum was found to be a new source for asphalt. Asphalt in plastic form, however, was used for flooring by the ancients.

The invention of linoleum in 1860 led naturally to further research and experiment to develop a precast flooring material with asphalt as a binder. A successful product appeared on the market shortly after linoleum did, but the original inventor of what we called asphalt tile is not known. The process evolved consisted of mixing asphalt with various rubbers and gums as a binder plus inert fillers and pigments and then forming the mixture by pressure and heat into a sheet material which was cut into smaller sizes. Today the process is the same except for the addition of asbestos and other fibers, the use of better asphaltic binders, and the development of a wide variety of colors and patterns.

ASPHALT ROOFING AND SIDING

Physical and Chemical Properties

All asphalt roofing and siding products are manufactured of six basic materials:

1. Felt, made from rags, wood, and other cellulose fibers by a method similar to that for making paper (*see also* Fibers). This felt is saturated with asphalt and then used as a base for asphalt shingles and siding and for built-up roofing, and as underlayment for finished surface roofing and siding materials.

2. Asphalt, obtained from the fractional distillation of petroleum (crude oil). This is used as a saturant for the roofing felts; it is combined with stabilizers to make blended coatings; it is applied as a base for fine surfacing or colored granules and as an adhesive for the layers of saturated felt used in built-up roofing. Also, it is the basic ingredient of many roof cements, quick-setting cements, and asphalt coatings for general use.

3. Stabilizers, generally silica, marble, sandstone, slate dust, talc, micaceous materials, dolomite (limestone), and traprock. These are ground into finely divided particles and combined with the asphalt to control its hardness, elasticity, adhesion, and weatherability.

4. Fine surfacing materials, generally talc, mica, and micaceous talc, finely ground and used to prevent the various asphalt materials from sticking together when packed, shipped, and stored.

5. Colored granules, generally natural slate, marble, granite, various other natural stones, or slag, which have been crushed, screened, and graded to sizes. These may be either natural colored or pigmented and fired to produce permanent colors.

6. Gravel, slag, marble chips, and to a lesser extent granite and limestone, crushed and screened to exact size. These are used as protective topping for built-up roofing and are applied to the last coat of asphalt. *See* Figure *A100*.

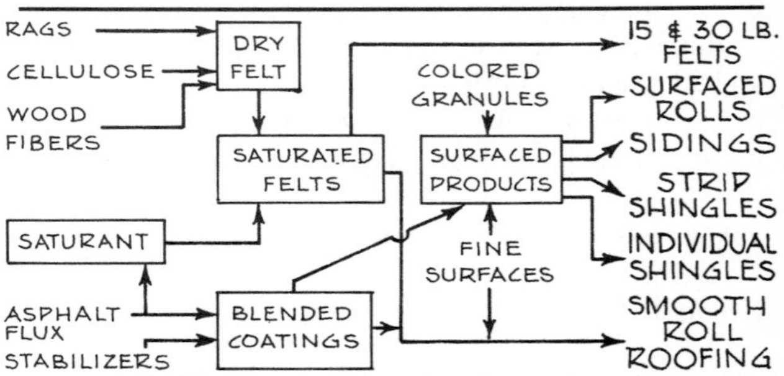

Figure A100 Flowchart of the fabrication of asphalt roofing and siding.

Commercial Forms. Asphalt roofing and siding materials are available in a wide range of colors and textures, in single units, in strips, and in rolls.

Types and Uses

Asphalt roofing and siding materials are identified on the basis of weight per square of applied roofing, a square equaling 100 ft^2 (9.29 m^2) of roof surface, or an area 10 ft by 10 ft (9.29 m by 9.29 m) (*see* Table *A63*).

Table A63 General types of asphalt roofing and siding

Type of roofing or siding	Weight per square 100 ft² (9.2903 m²)[a] lb	(kg)	Size of roll or strip: Length ft	Length (m)	Width ft	Width (m)	Side or end lap in.	(mm)	Head lap in.	(mm)	Exposure in.	(mm)	Uses
Saturated felt in rolls	15, 30	(73.23, 146.46)	144, 72	(43.89, 21.95)	3	(0.9144)	4	(101.6)	2	(50.8)	34	(863.6)	Underlayment and built-up roofs
Roll roofing with smooth surface	45, 55, 65	(219.69, 268.51, 317.33)	36	(10.97)	3	(0.9144)	4	(101.6)	2	(50.8)	34	(863.6)	Underlayment
Roll roofing with mineral surface	90, 93 95	(439.38, 454.03 463.79)	36	(10.97)	3	(0.9144)	4, 6, 6	(101.6, 152.4, 152.4)	2, 3, 4	(50.8, 76.2, 101.6)	34, 33, 32	(863.6, 838.2, 812.8)	Roofs
Roll roofing with selvage	144,[b] 140[c]	(693.01, 638.48)	36	(10.97)	3	(0.9144)			19	(482.6)	17	(431.8)	Roofs
Roll roofing with patterned edge	105	(512.61)	42, 48	(12.80, 14.63)	3, 2.67	(0.9144, 0.8138)			2	(50.8)	16, 14	(406.4, 355.6)	Roofs
Shingles, single rectangular Dutch lap, laid horizontally	162	(790.88)	1.33	(0.409)	1.0	(0.3048)	3	(76.2)	2	(50.8)	10	(254.0)	Roofs
Shingles, single rectangular giant American	325	(1586.25)	1.33	(0.409)	1.0	(0.3048)			6	(152.4)	5	(127.0)	Roofs
Shingles, single hexagonal staple-down or lock-down type	135, 138	(659.07, 673.72)	1.33	(0.409)	1.33	(0.409)	$2\frac{1}{2}$	(63.5)					Roofs
Strip shingle, three-tab square butt	210, 262, 300	(1025.22, 1279.08, 1464.60)	3	(0.9144)	1.0	(0.3048)			2, 4, 4	(50.8, 101.6, 101.6)	5, 4, 4	(127.0, 101.6, 101.6)	Roofs and siding
Strip shingle, two- and three-tab hexagonal	165	(815.29)	3	(0.9144)	0.94[d]	(0.2865)			2	(50.8)	$4\frac{2}{3}$	(118.62)	Roofs and siding
Roll siding, textured	105	(512.61)	43	(13.106)	2.58[e]	(0.9864)			$1\frac{5}{8}$	(41.28)	$4\frac{7}{8}$	(123.83)	Siding

[a]Always check manufacturers for weight in lb/square.
[b]Applied with cold asphalt.
[c]Applied with hot asphalt or tar.
[d]0.94 ft = 11 in.
[e]2.58 ft = 31 in.

Asphalt-saturated felts are used as building papers, as moisture and wind barriers, and as underlayments for various types of roofing materials. (For a fuller description of building, sheathing, and underlayment papers, *see* Paper and Paper Pulp Products).

Built-up Roofing. Built-up asphalt roofing consists of alternate layers of hot asphalt cement and asphalt-saturated felts. These layers are called 3-ply, 5-ply, etc., according to the number of layers of asphalt-saturated felt. As a rule, the ply designation does not include the bottom layer of unsaturated felt, which is generally applied first, except on concrete. The finish surface consists of slag or various types of stone chips. This type of roofing is used for roof surfaces with a pitch not greater than 3 to 12.

Roof Cements. Asphalt roof cements consist of either asphalt-water emulsions, which can be thinned with water, or asphalt thinned with petroleum solvents. Both types contain asbestos fiber. They are used as adhesives to hold down asphalt shingles, siding and flashing, and generally for sealing joints (*see* Asphalt Coatings; Adhesives; *etc.*). Asphalt roof cements are also made with quick-drying solvents which make them more adhesive and quick-setting. This type is used for applying asphalt rolled roofing and for sealing down tabs of strip shingles. It should not be exposed to the weather.

Application

Condensed Checklist

1. Local, municipal and state codes and the codes of the Fire Underwriters, insurance companies, labor departments, and federal government (Army, Navy, etc.) should be checked for fire ratings and other requirements.
2. The roof pitch also controls the type of asphalt roofing which can be used (*see* Figure *A101*).
3. Asphalt roofing or siding materials should never be installed in temperatures below approximately 45°F (7.22°C) without taking the precaution of heating them prior to installation. This includes all roof cementing materials.
4. The correct type of nail and method of fastening the asphalt roofing and siding must be used. Nails should be large-headed, sharp-pointed aluminum or hot galvanized steel nails with barbed or deformed shanks. Generally the heads are $\frac{3}{8}$ to $\frac{7}{16}$ in. (9.53 to 11.11 mm) in diameter, and the nails are made of 11- or 12-gauge wire. They should be long enough to penetrate into the wood sheathing for $\frac{3}{4}$ in. (19.05 mm). A general rule for shank length is 1 in. (25.4 mm) for roll roofing, $1\frac{1}{4}$ in. (31.75 mm) for strip or individual shingles, and $1\frac{3}{4}$ in. (44.45 mm) for installation over old material. If staples and special fastening devices for attaching to gypsum or other types of fibrous materials are used, the directions recommended by the manufacturers of the asphalt roofing and siding should always be followed.
5. For asphalt built-up roofing, the roof pitch should not be greater than 3 to 12. One should always check the actual number of layers, the weight of saturated felt, the type of slag or stone chips for surface finish, and the method of installation in relation to the guarantee of the manufacturer (*see* Figure *A102*).
6. Roll roofing should be installed with concealed nails as a rule, with the joints parallel or perpendicular to the eaves. Installation methods and type of roll roofing should conform to roof pitch and fire-resistance requirements (*see* Figure *A103*).
7. Treatment of roll roofing at ridges, hips, valleys, eaves, gable ends, and flashing must be correctly detailed, as shown in Figures *A104* and *A105*.
8. Asphalt strip shingles and individual shingles must be of the correct type and weight per square in relation to roof pitch and fire-resistance requirements. Methods of application should be as shown in Figures *A106* to *A109*.

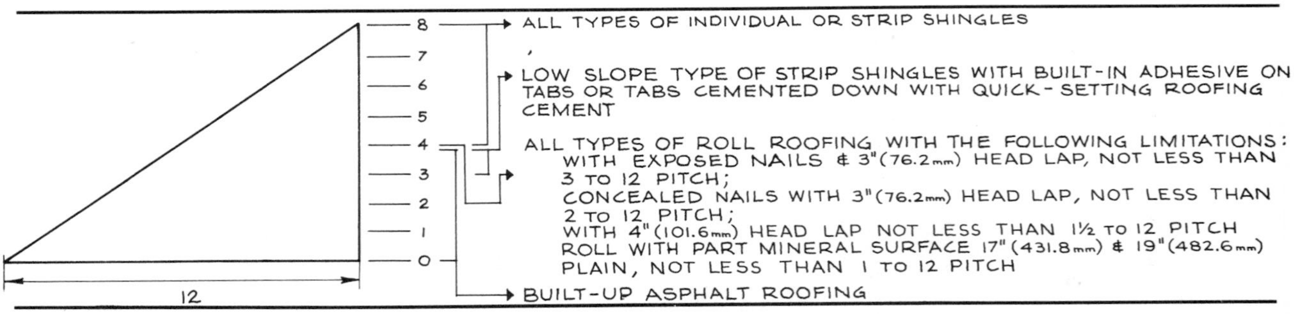

Figure A101 Roof pitch in relation to type of asphalt roofing.

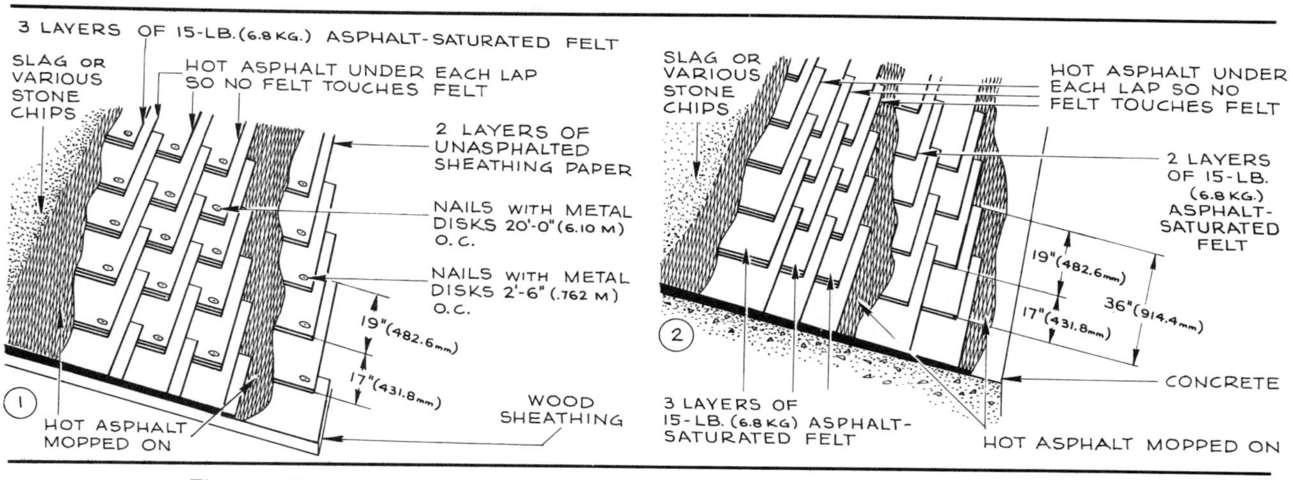

Figure A102 Typical asphalt built-up roofs, showing details of three-ply built-up roof on wood with 10-year guarantee (1) and five-ply built-up roof on concrete with 20-year guarantee (2).

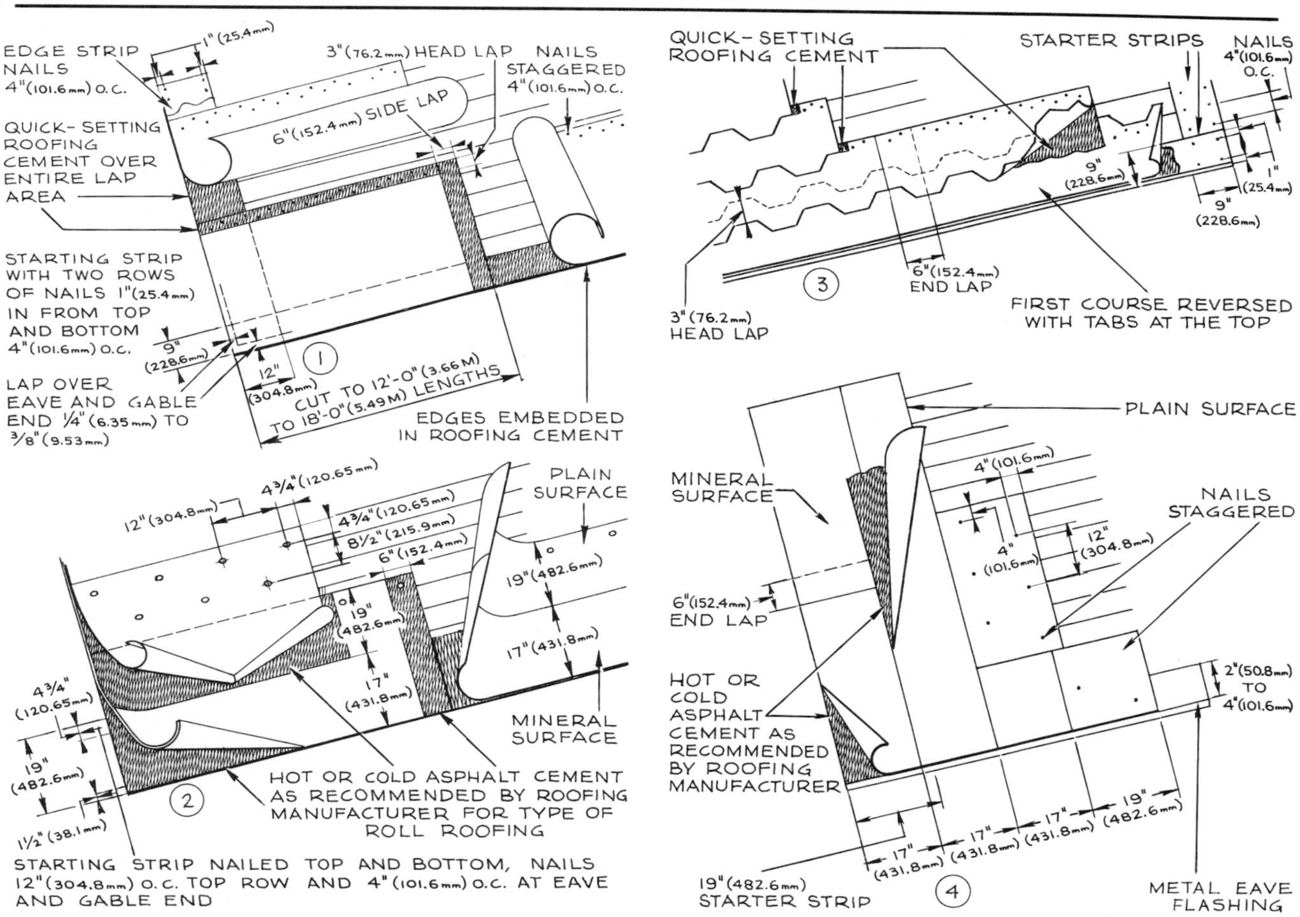

Figure A103 Application of asphalt roll roofing, showing concealed nailing of roll roofing (1); mineral-surfaced roll roofing with selvage, attached with concealed nails (2); mineral-surfaced roll roofing with patterned edge (3); and roll roofing laid vertically (4).

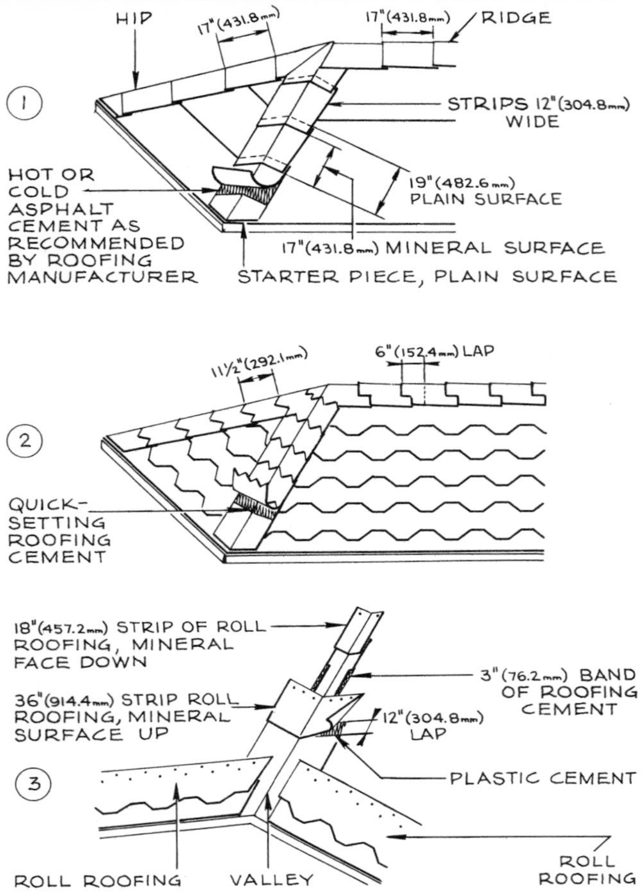

Figure A104 Details of roll roofing at ridge (1, 2) and at valleys (3).

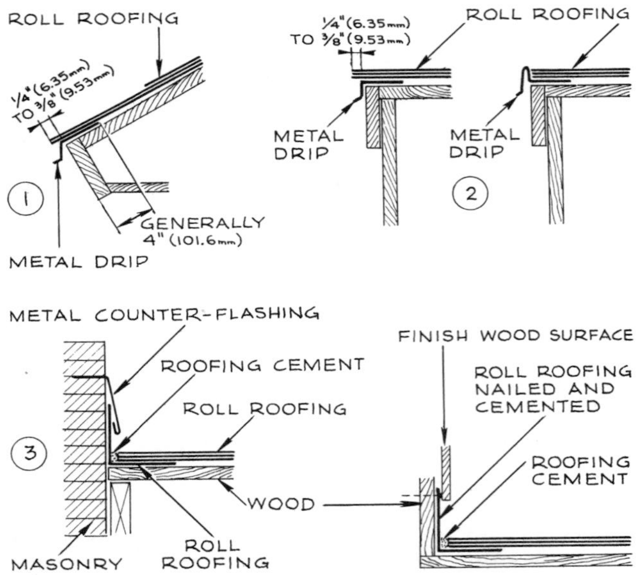

Figure A105 Details of roll roofing at eaves (1) and gable ends (2) and methods of flashing (3).

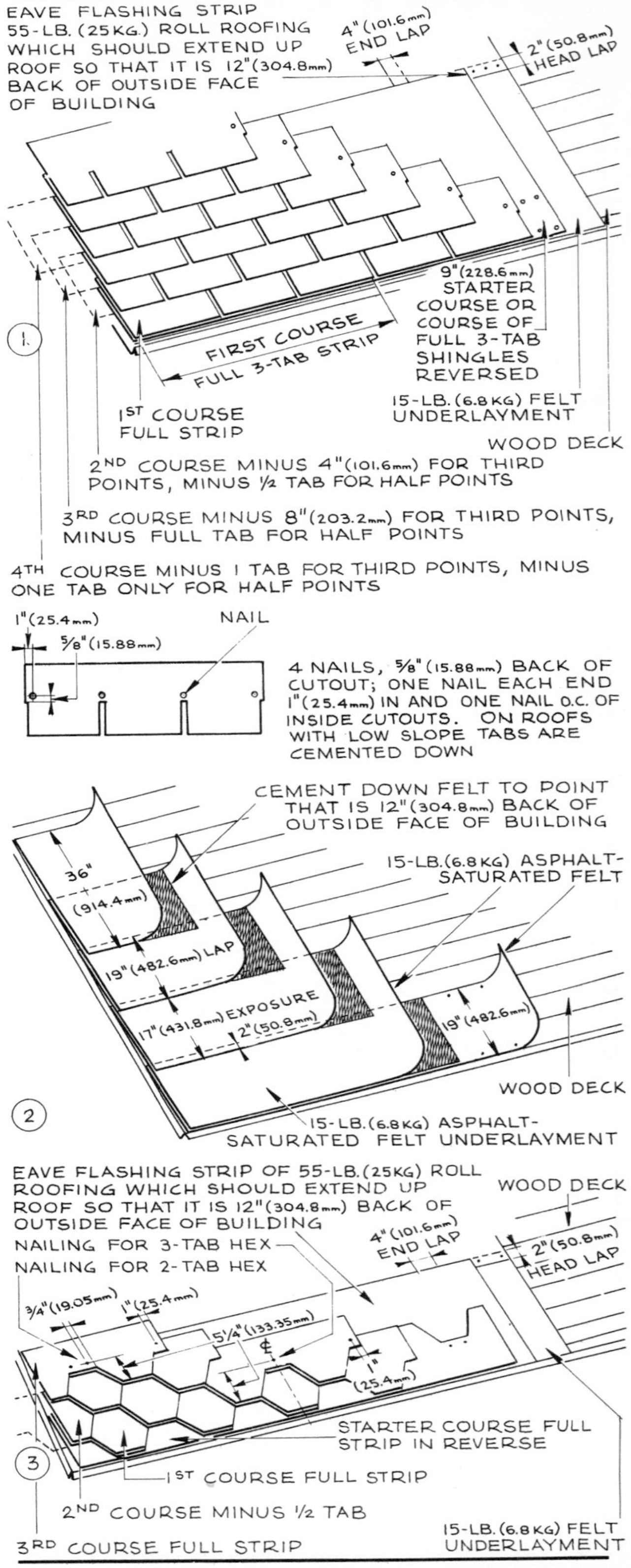

Figure A106 Installation of asphalt strip shingles: three-tab butt-edge strip shingles (1), underlayment for low, sloping roofs (2), and two- and three-tab hexagonal strip shingles (3).

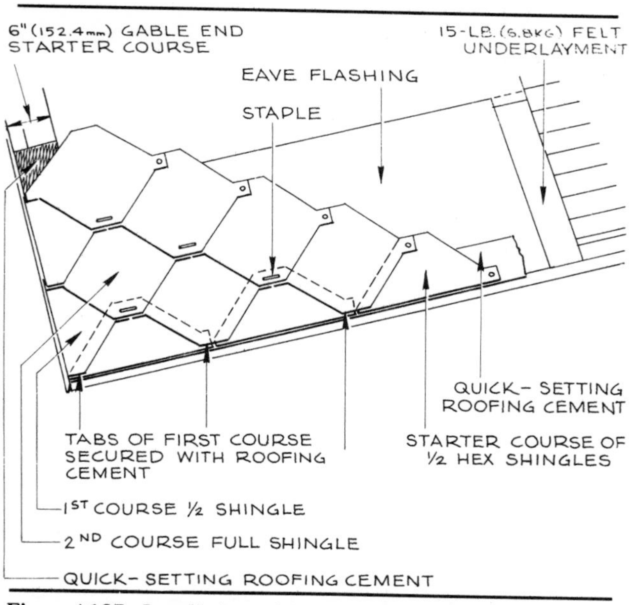

Figure A107 Installation of individual hexagonal staple-down shingles.

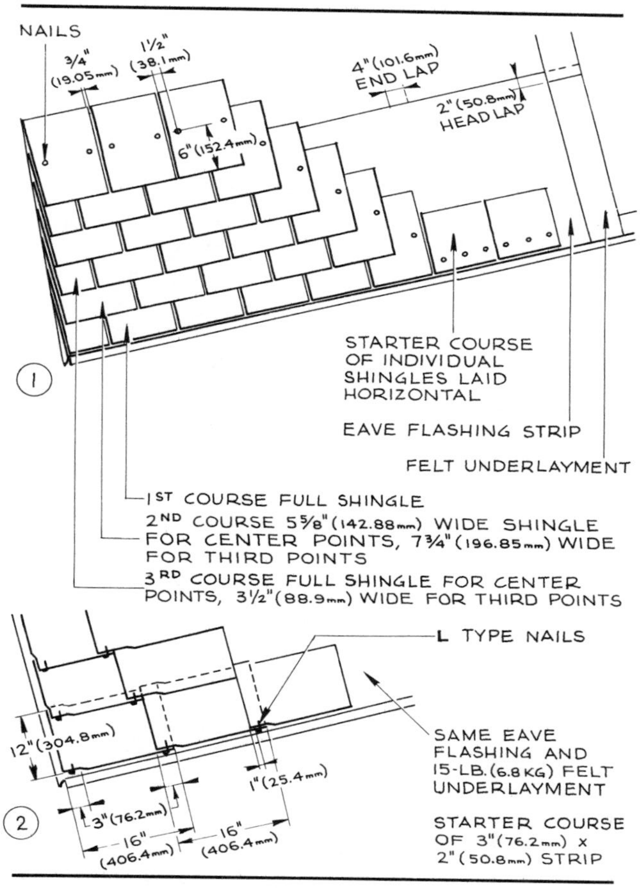

Figure A108 Installation of giant individual shingles showing shingles laid o.c. of shingles below (1) and Dutch lap (2).

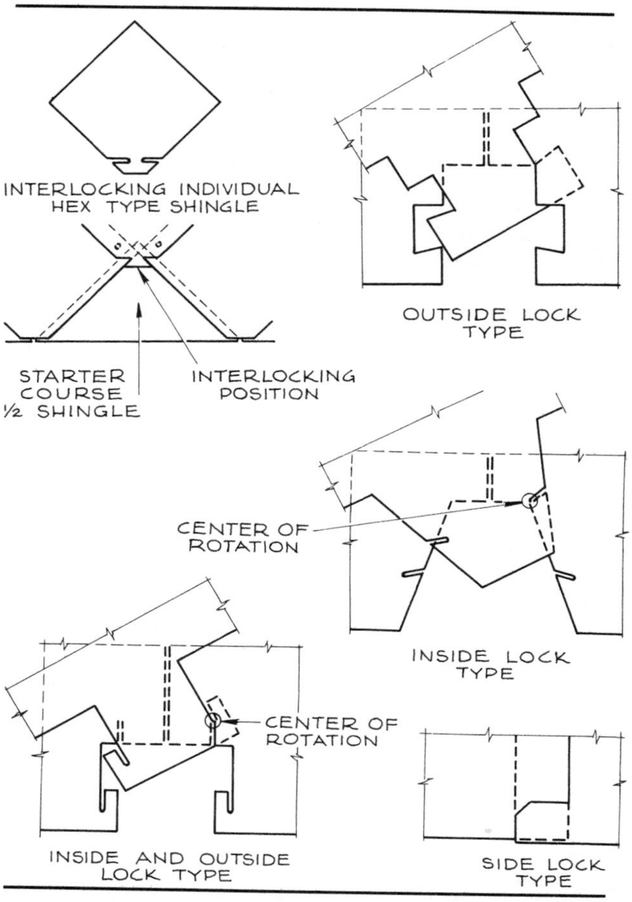

Figure A109 Methods of laying lock-type individual shingles.

9. Treatment of strip and individual shingles at ridges, hips, valleys, eaves, gable ends, and flashing must be correctly detailed (*see* Figures *A110* and *A111*) for the specific asphalt roofing product being used.
10. Asphalt siding must be of the correct type and weight to meet required fire-resistance ratings.
11. Treatment of siding at inside and outside corners, at windows and doors, and at flashing must be correctly detailed (*see* Figure *A112*).

History and Manufacture

The history of asphalt roofing and siding began in the United States in 1892, when a chemist by the name of William Griscom developed a single-ply ready-to-lay roll of asphalt roofing. From this beginning the manufacture of asphalt roofing and siding has steadily grown to the point where about 90% of all roofing applied during the last few decades has been asphalt roofing material.

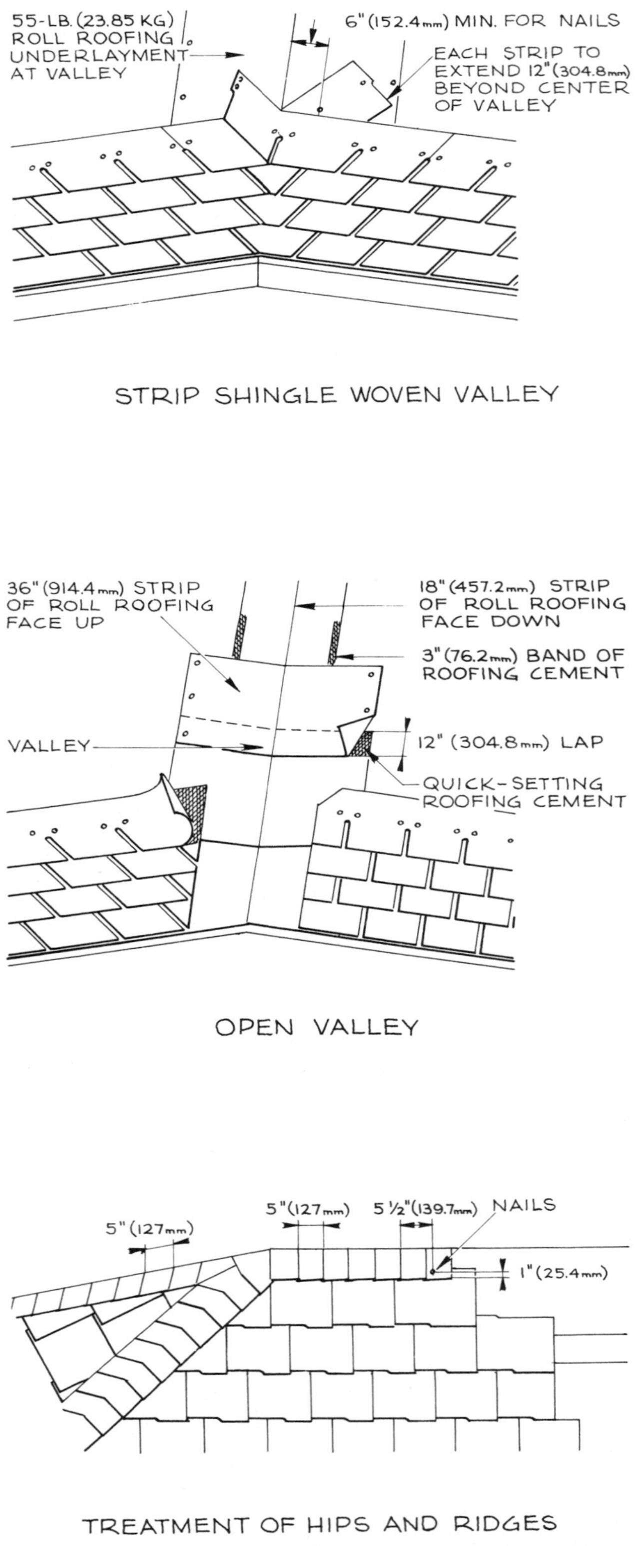

Figure A110 Details of strip shingles at valleys, ridges, and hips.

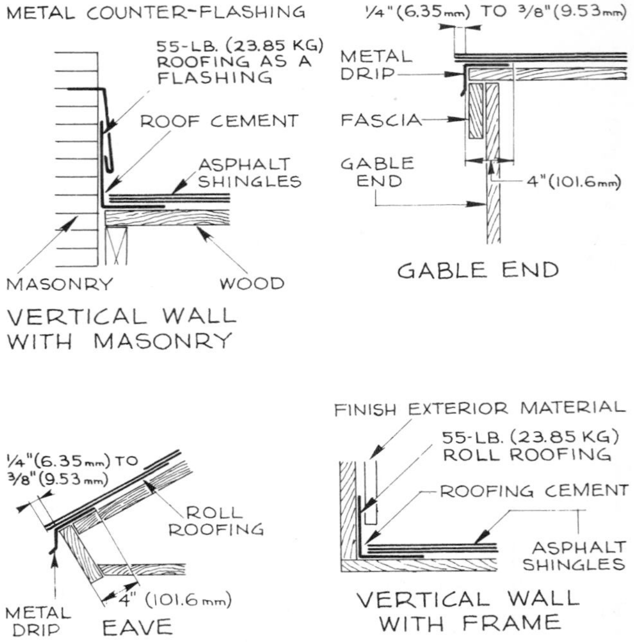

Figure A111 Installation of strip and individual shingles at a vertical masonry wall, gable end, eave, and vertical frame wall.

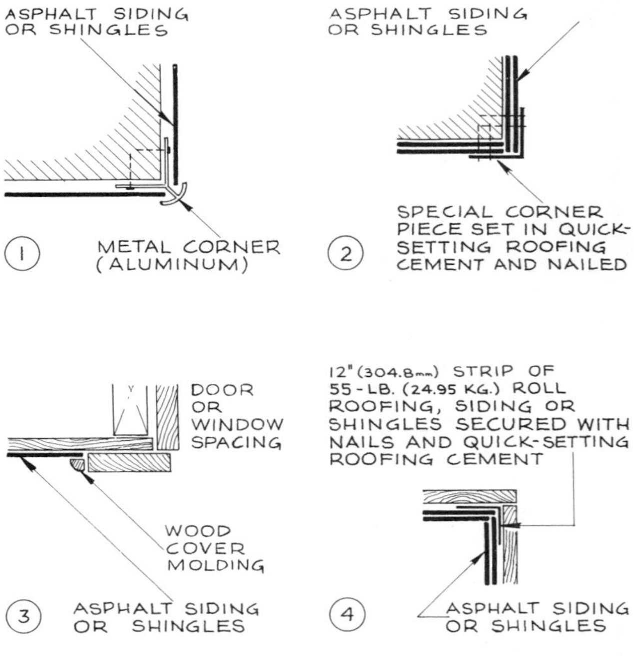

Figure A112 Details for asphalt siding at outside corners (1, 2), at jambs (3), and at inside corners (4).

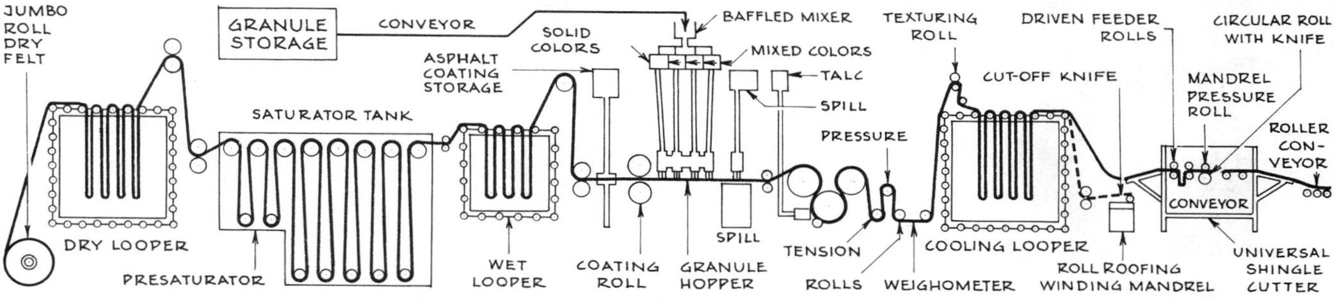

Figure A113 Schematic production line for asphalt roofing and siding products.

A schematic production line for asphalt roofing and siding products is shown in Figure *A113*.

Preparation of Felt. The basis of all asphalt roofing and siding is a rag or asbestos felt. The saturation of this felt is the most important part of the process. The felt, in large rolls, is unwound into a storage bin or reservoir, where the felt is looped over a series of rollers that always maintain a reserve of felt so that, when the process starts, there is always a sufficient supply should imperfections have to be cut out or a new roll of felt installed. The felt first goes through a presaturating process which eliminates any moisture; then it is saturated as completely as possible with hot asphalt. On leaving the hot asphalt the felt travels into an area called the wet looper, in which any excess asphalt forced out of the felt by shrinkage is drawn back into the felt as it cools, thus giving better saturation. At this point, felt to be used for built-up roofing, dampproofing, and waterproofing goes directly to the cooling and winding room for final processing, without the other steps for coating and adding granules shown in the flowchart.

Asphalt Coating. For rolled roofing, shingles, and siding the next step is to apply an asphalt coating to both sides, the amount being controlled by closing or opening the coating rollers. For smooth felt roofing, talc or mica is applied to the underside by spreading and pressing through a press roll. If granules are to be applied, they are added through hoppers and spread thickly on the hot asphalt coating. The roofing or siding then passes through water-cooled drum-type rollers and press rollers, which cool it and embed the granules to the desired depth. If a textured or embossed finish is desired, the felt at this point passes through rollers that form the desired pattern on the surface.

Cooling and Cutting. The final procedure is to allow the material to cool so that it can be rolled, cut, and packed without danger to the finished product. Material for roll roofing, built-up roof products, and dampproofing or waterproofing felts is taken from the cooling area directly to winders, which measure the length desired and cut it off into rolls that are now ready for packing, storage, and shipment. If shingles are being made, the material passes between a cutting roller (bottom) and a pressure roller (top), which cut it through from the smooth side. As the strips of shingles or siding separate, they are stacked, packed, and made ready for shipment.

AUTOCLAVE, AUTOCLAVING

In construction, "autoclaving" refers to a manufacturing process that utilizes a large steel chamber (an autoclave) into which fabricated materials are placed and subjected to high-pressure steam for specific periods of time to establish curing for concrete-type materials. *See* Asbestos-Cement Piping; Concrete Block; Concrete Piping; Concrete Decking.

B

BARIUM

Physical and Chemical Properties

Symbol: Ba
Atomic number: 56
Specific gravity: 3.5 (20°C, 68°F)
Melting point: 725°C, 1337°F
Boiling point: 1640°C, 2984°F

Barium is a silver-white, slightly lustrous, somewhat malleable metal belonging to the alkaline earth metals (others are calcium, strontium, radium). Metallic barium is available in lump or rod form.

Types and Uses

Natural barium sulfate, or barite (*see* Table *B1*), and purified barium sulfate have many direct uses in materials used in or closely allied to construction, as shown in Table *B2*. The uses of barium in metallic form, indicated by asterisks, are very minor; chiefly, it is used as a deoxidizer of copper.

When atomic power plants are commercially constructed, barite as an aggregate in concrete shields may become important.

Table B1 Composition of natural barium sulfate (barite)

$BaSO_4$	SiO_2	Al_2O_3	Fe_2O_3
	(percent of content)		
98.2	1.5	0.15	0.15

History and Manufacture

Barite was first used in the United States as a filler in paint. Lithopone, a white paint solid formed of zinc sulfide and barium sulfate, was first produced in the United States in 1892. Since 1926 barium has become increasingly important as a weighting agent in well-drilling mud.

Table B2 Uses of barium

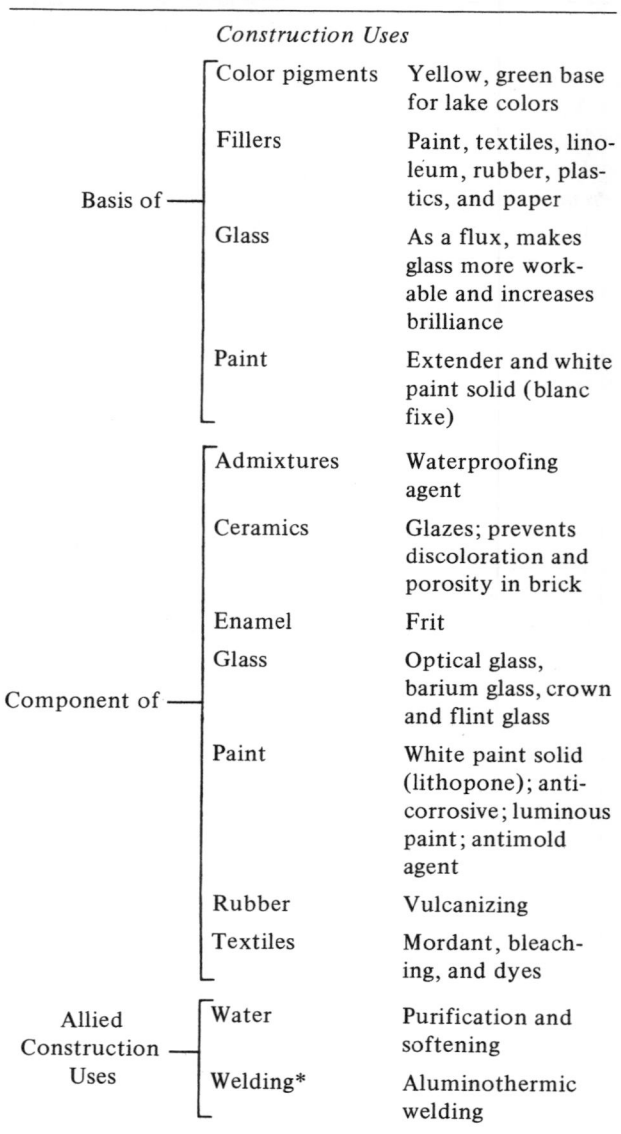

	Construction Uses	
Basis of	Color pigments	Yellow, green base for lake colors
	Fillers	Paint, textiles, linoleum, rubber, plastics, and paper
	Glass	As a flux, makes glass more workable and increases brilliance
	Paint	Extender and white paint solid (blanc fixe)
Component of	Admixtures	Waterproofing agent
	Ceramics	Glazes; prevents discoloration and porosity in brick
	Enamel	Frit
	Glass	Optical glass, barium glass, crown and flint glass
	Paint	White paint solid (lithopone); anticorrosive; luminous paint; antimold agent
	Rubber	Vulcanizing
	Textiles	Mordant, bleaching, and dyes
Allied Construction Uses	Water	Purification and softening
	Welding*	Aluminothermic welding
	Possible Future Uses	
Aggregate in concrete to shield against radiation, and in asphalt paving with synthetic rubber for density and durability.		

Barite is usually processed by grinding and flotation. To obtain barium metal, fused barium chloride is electrolyzed.

BASE

Physical and Chemical Properties

In chemical terminology a base is any compound which is capable of accepting a proton (hydrogen ion) that is transferred from another compound. *See also* Salt; Acids, Bases, and Salts.

Some of the bases that have direct application in construction materials are listed in Table *B3*.

Table B3 Some common bases used in construction

Base	Formula
Ammonia	NH_3
Calcium oxide	CaO
Sodium carbonate	Na_2CO_3

History and Manufacture

The ancients used many bases in their natural forms long before their chemistry was known. In 1923 Brönsted and Lowry, working completely on their own, introduced a new definition for a base, which is known as the Brönsted-Lowry theory. Another concept, developed by G. N. Lewis, is that a base is anything which has an unshared pair of electrons.

BERYLLIUM

Physical and Chemical Properties

Symbol: Be
Atomic number: 4
Specific gravity: 1.84 (20°C, 68°F)
Melting point: 1278°C, 2332°F
Coefficient of expansion: 0.0000123 in./°F, 0.0002214 mm/°C

Beryllium is a hard, exceptionally lightweight, gray-white metal, corrosion resistant at ordinary temperatures; chemically it is related to aluminum. Beryllium is the only light metal other than titanium which has both good strength and a high melting point. It has the additional properties of good electrical conductivity, high modulus of elasticity, exceptional transmission of X-rays, and transmission of sound at very high velocities. Its ability to act as a source, moderator, and reflector of neutrons has made it very valuable in the nuclear energy field.

Toxicity. The mists, dusts, and fumes of beryllium and its soluble compounds are toxic and may present a serious health hazard, but beryllium and its alloys in solid form are not toxic.

Workability. The very pure form of berryllium (*see* Table *B4*) can be worked with care and by special means; it can be forged, rolled, and extruded (at controlled temperatures with the metal encased in a protective jacket), cast by vacuum casting, machined, punched, and sheared.

Table B4 Composition of pure beryllium metal

Be	Al	Fe	Mg	Cu	Mn	Si	C and other impurities
			(percent of content)				
99.5	0.10	0.10	0.05	0.05	0.05	0.03	0.02

Commercial Forms. Beryllium is available in block, rod, sheet, pebble, metal disk, foil, and powder form and in fabricated shapes.

Types and Uses

Beryllium has relatively small but steadily increasing uses in its compound forms, particularly the oxide. Its uses in metallic form are indicated by asterisks in Table *B5* (p. 114).

Beryllium-Copper Alloys. Alloys of copper and beryllium are unsurpassed in their ability to withstand fatigue, wear, and corrosion and in their electrical conductivity at high temperatures. They are easily cast and reproduce fine detail. Nonmagnetic and nonsparking alloys are also available. These and other alloys containing beryllium can be worked in a soft state and then, by simple heat treatment, strengthened and hardened to an accurately controllable degree.

Master Alloys. Master beryllium alloys of magnesium-aluminum, aluminum, nickel, copper, iron, or cobalt are always used to introduce beryllium into the respective metals.

Table B5 Uses of beryllium

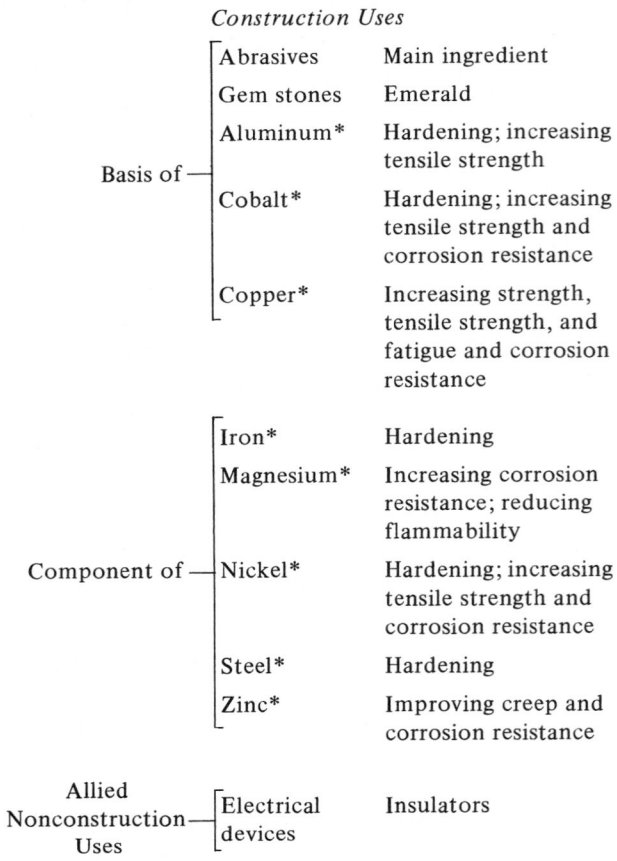

	Construction Uses	
Basis of	Abrasives	Main ingredient
	Gem stones	Emerald
	Aluminum*	Hardening; increasing tensile strength
	Cobalt*	Hardening; increasing tensile strength and corrosion resistance
	Copper*	Increasing strength, tensile strength, and fatigue and corrosion resistance
Component of	Iron*	Hardening
	Magnesium*	Increasing corrosion resistance; reducing flammability
	Nickel*	Hardening; increasing tensile strength and corrosion resistance
	Steel*	Hardening
	Zinc*	Improving creep and corrosion resistance
Allied Nonconstruction Uses	Electrical devices	Insulators
	Nonconstruction Uses	
Powder metallurgy,* special mirrors,* X-ray tubes, windows*		

Beryllium Oxide. Beryllium oxide is a remarkable refractory which already plays a significant role in high-temperature problems. Its melting point is 4685°F, and its specific gravity 3.02. It has strength, hardness, high electrical resistance, high thermal conductivity, and chemical inertness.

Commercial Disadvantages. Wider commercial application of beryllium is retarded by several difficulties. The only commercial source at present is hand-cobbed beryl ore, although low-grade deposits are available; its toxicity, although controllable, presents occupational hazards; and it is difficult to produce beryllium that is pure enough to have the degree of ductility needed for it to be easily worked (one of the major problems researchers are trying to solve).

History and Manufacture

In 1797 Vauquelin isolated the element beryllium from beryl; it was originally called glucinum because of the sweet taste of its compounds. In 1828 Wöhler and Bussy obtained the first metallic beryllium. In 1899 Lebeau produced beryllium-copper alloys by the reduction of beryllium oxide with carbon in the presence of copper.

Commercial development of beryllium in the United States began in 1916, and by the early 1930s both metallic beryllium and beryllium alloys were commercially produced. The principal ore from which beryllium is obtained is beryl ($3BeO$, Al_2O_3, $6SiO_2$). The commercial process mainly used for obtaining the purified metal is based on the discovery that beryl ore completely melted and then quenched in cold water becomes soluble in sulfuric acid. According to one method, this solution is treated with ammonium sulfate to eliminate the aluminum, and then treated with sodium hydroxide and other reagents to convert the remaining beryllium sulfate into a hydroxide, which in turn is heated and ignited to obtain beryllium oxide (BeO).

Several methods may be employed to obtain the commercial metal. One of these is the reduction of beryllium fluoride, obtained from beryllium oxide, with magnesium metal.

BISMUTH

Physical and Chemical Properties

Symbol: Bi
Atomic number: 83
Specific gravity: 9.743 (20°C, 68°F)
Melting point: 271.3°C, 520.34°F
Boiling point: 1560°C, 2840°F
Ultimate tensile strength (special ductile bismuth): 2500 lbf/in.^{2}, 17.24 MN/m^2

Bismuth is a heavy, grayish white, crystalline metal with a high luster and a pinkish tinge. It is one of the few metals that expands during solidification. It is the most diamagnetic of all metals (*see* Magnets and Magnetism).

Commercial Forms. Ordinary bismuth is 99.99% pure and is too brittle to roll, draw, or extrude. It is available in bar, lump, stick, shot, and powder form. Special ductile bismuth, 99.9975% pure, is available in wire, rod, strip, sheet, foil, and special shapes. A typical analysis of standard refined bismuth is shown in Table *B6*.

Table B6 Typical analysis of standard refined bismuth

Bi	Zn	Fe	Ag, Cu, Pb, Sb	As
	(percent of content)			
99.9989	0.0004	0.0003	0.0001 each	0.0

Uses

Bismuth in metallic form is increasingly used in alloys; these applications are indicated by asterisks in Table *B7*. The major alloy use is for fusible alloys. Bismuth as a liquid metal has also been used as a coolant and for shielding atomic reactors.

Possible future uses of bismuth in the electronic field include thermoelectric elements such as cooling junctions for air conditioning and refrigerating devices. A number of these have already been built on pilot model scale.

Table B7 Uses of bismuth

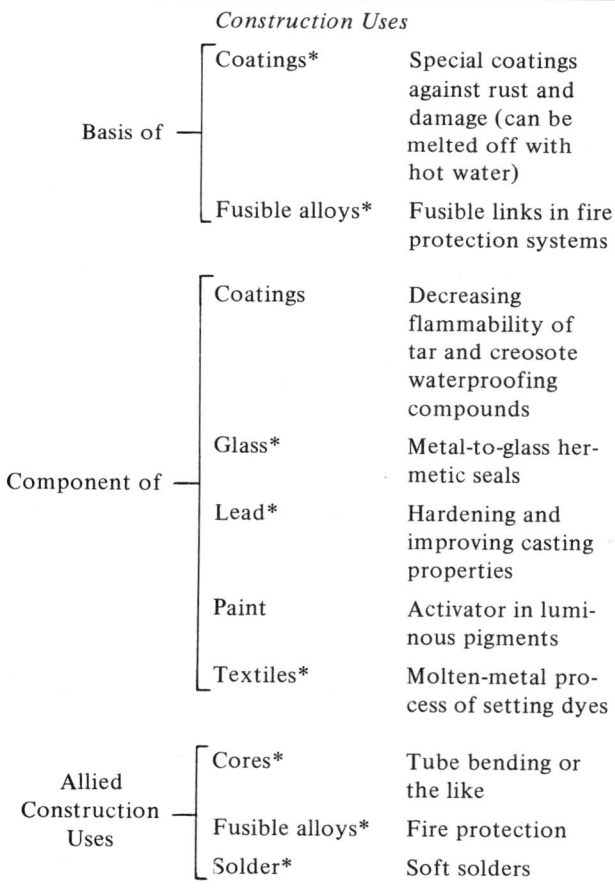

	Construction Uses	
Basis of	Coatings*	Special coatings against rust and damage (can be melted off with hot water)
	Fusible alloys*	Fusible links in fire protection systems
Component of	Coatings	Decreasing flammability of tar and creosote waterproofing compounds
	Glass*	Metal-to-glass hermetic seals
	Lead*	Hardening and improving casting properties
	Paint	Activator in luminous pigments
	Textiles*	Molten-metal process of setting dyes
Allied Construction Uses	Cores*	Tube bending or the like
	Fusible alloys*	Fire protection
	Solder*	Soft solders
	Possible Future Uses	
Permanent magnets*		

History and Manufacture

Bismuth has been known since the Middle Ages. Its name probably came from the German *Wismut*. In 1750 bismuth was identified as a metallic element. Most bismuth is recovered as a by-product of the refining of lead, copper and sometimes tin. After removal of precious metals, this by-product, containing only bismuth, lead and zinc, is heated to about 932°F and treated with chloride. After these have been removed, the molten metal is treated and cast as 99.99% refined bismuth.

BITUMENS

In general usage "bitumen" refers to natural hydrocarbons (*see* Table *B8*). Bitumens as a group are flammable, insoluble in water, but largely soluble in carbon disulfide. Commercially, the term applies to solid and semisolid hydrocarbons and excludes natural gas. All asphalts, asphaltites, and asphaltic pyrobitumens, except shale oil, have been derived originally in nature from liquid petroleum by evaporation or metamorphism. Similar products are obtained from distillation of crude oil and coal.

Table B8 Bitumens

Type	Description
Liquid petroleum	Paraffin base, mixed base, and asphaltic base
Native mineral waxes	Ozocerite (mineral wax) and montan wax
Native asphalts	Vary from a pure solid, bitumen to an impure bitumen containing sand, clay, etc.
Asphaltites	Gilsonite (natural black bitumen), glance pitch (bitumen with luster), and grahamite

Bitumens are widely used in industry and commerce; in the construction field we are familiar with them as asphalt and tar. (*See* Asphalt; Coal; Crude Oil; Fossil Fuel; Petroleum; Tar.)

BOARD

The word "board" was originally part of the terminology for lumber and wood construction. It is now much more

Table B9 Types of board

Material	Advantages	Disadvantages	Major Uses
Asbestos cement (*see* Asbestos, Corrugated; Asbestos-Cement Sheet and Board)	Permanent; durable; fire resistant; can be painted	Brittle; cement color or limited color range; must be predrilled for attachment; nails and joints exposed	Fire-resistant partitions
Glass (*see* Glass, Corrugated)	Fire resistant if wired; durable	Requires special attaching devices; limited in size	Skylights, decorative partitions, and screens
Paper pulp (*see* Paper and Paper Pulp Products)	Easily applied; good insulation value; flameproof; some acoustic value; can be painted	Easily damaged; joints exposed; very limited choice of colors	Interior partitions; sheathing for residential buildings
Plastic (*see* Plastics; Plastics: Surfacing, Siding, and Panels)	Easily applied; wide color range; durable; semitransparent or opaque; very wide variety of patterns and designs	Limited in sizes; requires special attaching devices	Decorative partitions, skylights, and screens
Wood (*see* Wood Particle Board, Siding and Paneling)	Easily applied; permanent; durable; easily painted or stained	Requires painting or staining periodically; limited in width	Exterior siding; interior paneling

loosely used to describe almost any piece of thin material other than metal.

To reduce confusion in terminology, a board in this book is explicitly defined as a flat or corrugated unit made of one material throughout. It is greater than $\frac{1}{4}$ in. (6.35 mm) in thickness; it is rigid and semisupporting. For materials that do not fall under this definition, *see* Decking; Integrant; Sheet. Table *B9* lists various types of board and materials from which they are made.

BONDERIZING

Bonderizing is a trademarked process for furnishing a corrosion-resistant phosphate coating on steel, aluminum, and zinc and their alloys. This coating, by virtue of its porous nature, serves as an excellent base for the application of oil, paint, or shellac to these metals. *See also* Nitriding; Parkerizing.

BORON

Physical and Chemical Properties

Symbol: B
Atomic number: 5
Specific gravity: 2.34 (crystals)
Melting point: 2300°C, 4172°F
Boiling point: 2550°C, 4622°F

Boron is now classed as a semimetallic element. In its purest forms it can be either a very hard, crystalline black powder or semilustrous, large particles.

Electrical Properties. The electrical properties of boron are unique in that it is feebly conductive at room temperature but becomes increasingly conductive as temperature rises (*see* Table *B10*). As little as 0.1% of carbon dissolved in boron will increase its conductivity many times; if 7 to 8% of carbon is present, the conductivity of boron is almost that of carbon.

Workability. Boron is brittle and cannot be worked at room temperature, nor can it be fabricated by melting and casting. By powder metallurgy techniques at about 3632°F (2000°C) it can be formed and molded into various shapes. It can also be bonded with plastics.

Commercial Forms. Boron is available in crystalline and in powder form in a wide range of purities. For composition of high-purity boron, *see* Table *B11*.

Types and Uses

Boron plays an important role in several commonly used construction materials, especially in alloys of aluminum and copper. The wide range of uses of boron compounds and semimetallic boron, indicated by asterisks, is shown in Table *B12*.

Table B10 Changes in electrical resistance of boron with rising temperatures

	80.6°F (27°C)	212°F (100°C)	338°F (170°C)	608°F (310°C)	968°F (520°C)	1112°F (600°C)
Electrical resistance (ohms)	775,000	66,000	7700	180	7	4

Table B11 Composition of high-purity boron powder

B	C	Fe	Impurities
	(percent of content)		
99.70	0.05	0.15	0.10

Table B12 Uses of boron

Construction Uses

	Material	Use
Basis of	Abrasives	Main ingredient
	Adhesives	Main ingredient
	Color pigment	Green, blue
Component of	Aluminum*	Improving strength without reducing conductivity; grain refining; producing ductile alloys with good physical properties
	Brass*	Degasifier and grain refiner
	Brick	Glazes
	Bronze*	Grain refiner
	Cement	High-grade cement for high-polish finish
	Chromium*	Surface facing for cutting tools
	Coatings and protective coatings	Fire resistance
	Copper*	Removing gaseous impurities; high conductivity alloys
	Glass	Optical and special glasses; heat resistant glass; soda-lime-silica glasses
	Glazes and porcelain enamels	Fluxing agent
	Insulation	Glass fiber ingredient
	Iron*	Inhibiting graphitization
	Manganese*	Improving strength; cleaning and deoxidizing

Construction Uses (Continued)

	Material	Use
Component of	Nickel*	Surface facing for cutting tools
	Paints	Plasticizer and adhesion additive to latex paints; fire-resistant additive
	Paper	Surface glaze; fire resistance
	Plastics	Fire retardant
	Steel*	Degasifying; deoxidizing; case hardening; substitute for molybdenum
	Structural clay tile	Glazes
	Terracotta	Glazes
	Textiles	Mordant; bleaching; fireproofing
	Tile	Glazes
	Titanium*	Improving strength; cleaning; deoxidizing
	Vanadium*	Improving strength; cleaning; deoxidizing
	Wood	Fire retardant; hardener for soft woods; aiding resistance to friction, bacteria, weather
Allied Construction Uses	Lamps	Fluorescence
	Solders*	Fluxes
	Water	Water softener
	Welding*	Fluxes

Nonconstruction Uses

Detergents, disinfectants, fertilizers, insecticides, photography, refractories, soil sterilizing, tanning, weed killer

Ferroboron. Ferroboron, an alloy containing 18 to 25% boron, is used for boron steels and nonferrous metals. A manganese-boron alloy is used for steels and a 2% boron-copper alloy for high-conductivity copper alloys and for brasses and bronzes.

Boron 10. Isotope boron 10 has the ability to absorb neutrons without yielding harmful radiation. This is an important advantage over other strong neutron absorbers, which emit hard gamma radiation. Boron 10 is therefore being used increasingly for nuclear reactor shields and control mechanisms.

A mixture of boron, potassium permanganate, and lead oxide or antimony oxide ignites on shock or friction but without detonation.

History and Manufacture

Boron has been known as a mineral for at least 400 years in the form of borax, boric acid, boracic acid, sassolite, and tincal. Boron itself first appeared in 1808, when Gay-Lussac, Thenard, and Davy prepared the free element as a black powder. Davy coined the name boron, which is analogous with carbon. In 1909 Weintraub developed a process to produce pure boron.

The well-known boron mineral, borax, got its name from the Persian *burah*, meaning white. From 1882 to 1925 the minerals for the production of borax in the United States were brought from Death Valley by the famous 20-mule teams. In 1913 a new mineral deposit was discovered at Kramer, California, which eventually caused the abandonment of the Death Valley operation.

Electrolytic Production. There are several processes for producing boron by electrolysis. The Cooper cell shown in Figure *B1* is for production (by patented processes) of high-purity boron on an industrial scale from boron oxide or from potassium fluoborate in a fused alkali metal chloride bath. In the Cooper patented processes the fused salt mixture is first prepared, and then, by electrolysis, boron is deposited on the cathode. By special grading and purification, a 99.7% pure boron is obtained.

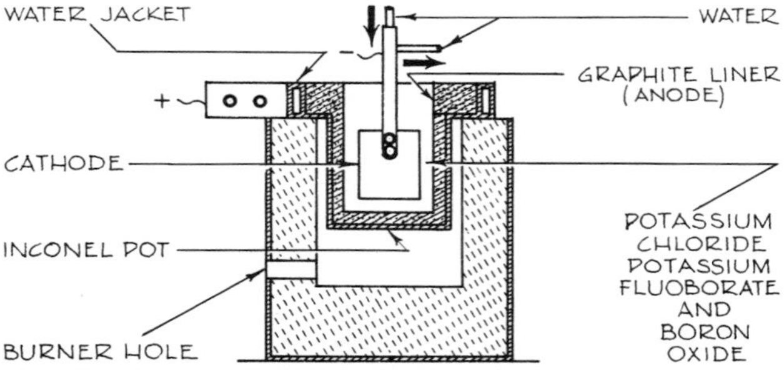

Figure B1 Cooper electrolytic cell for producing boron.

BRASS

Physical and Chemical Properties

Brass is fundamentally an alloy of copper and zinc with small quantities of other elements sometimes added to give special qualities. The copper-zinc proportions may vary from 95% copper and 5% zinc to 55% copper and 45% zinc. Brass varies in color from lemon yellow to deep golden-brown according to the amount of zinc: 10% zinc for bronze color; 15% zinc for golden; 20 to 38% zinc for yellow; and above 45% zinc for silvery-white. As a class, brass alloys are less hard and strong than steels (iron-base alloys) but are superior in workability and resistance to corrosion.

Many of the copper-based alloys known as brasses are used for both wrought and cast forms. The wrought alloys are copper-zinc, copper-zinc-tin, and copper-zinc-lead alloys. The cast alloys are copper-tin-zinc, copper-tin-zinc-lead, manganese bronze, leaded manganese bronzes, and copper-zinc-silicon alloys.

Classification. Brasses can be classified according to color (red, yellow), zinc content (high, low), metallographic constituents (Alpha, Beta), and other metals present (leaded, arsenical). The compositions are given in Table *B13*.

Alpha brasses are ductile and can be both cold worked without annealing and hot worked. Corrosion resistance and electrical and thermal conductivity are fair, these properties decreasing with a decrease in copper content. Alpha brasses have a wide range of color. Copper-base alloys containing more than 64% copper are known as Alpha brasses.

Alpha-Beta brasses are copper-base alloys containing from 64% down to 55% copper, and containing between 36% and 45% zinc. They have comparatively high tensile strength and hardness, fairly low melting points, and relatively poor corrosion resistance and electrical conductivity. They can be easily hot worked and within limits can be cold worked without annealing. Their color range is important because as the copper content decreases, they become less red and at 58% copper the color practically matches that of commercial bronze; therefore complicated shapes may be made by hot extrusion and used with sheets of commercial bronze.

Leaded brasses are those to which lead is added to increase machinability. They may be either Alpha or Alpha-Beta brasses and are especially suited for free-

Table B13 Types of brass and their compositions

Alloy Number	Name	Major alloying elements (percent)								Total other elements[a]
		Cu	Zn	Pb[a]	Fe[a]	Sn[a]	As[a]	Sb[a]	P[a]	
210	Guilding 95%	95.0	5.0							0.10
220	Commercial bronze 90%	90.0	10.0	0.05	0.05					0.10
226	Jewelry bronze $87\frac{1}{2}$%	87.5	12.5	0.05	0.05					0.15
230	Red brass 85%	85.0	15.0	0.05	0.05					0.15
240	Low brass 80%	80.0	20.0	0.05	0.05					0.15
260	Cartridge brass 70%	70.0	30.0	0.07	0.05					0.15
268	Yellow brass 66%	66.0	34.0	0.15	0.05					0.15
270	Yellow brass 65%	65.0	35.0	0.10	0.07					0.20
280	Muntz metal 60%	60.0	40.0	0.30	0.07					0.20
314	Leaded commercial bronze	89.0	9.1	2.50	0.10	0.70				0.50
335	Low-leaded brass	65.0	34.5	0.80	0.10					0.50
340	Medium-leaded brass $64\frac{1}{2}$%	65.0	34.0	1.40	0.10					0.50
349	Brass	62.2	37.5	0.50	0.10					0.50
350	Medium-leaded brass 62%	62.5	36.4	1.40	0.10					0.50
377	Forging brass	60.0	36.0	2.50	0.30					0.50
385	Architectural bronze	57.0	40.0	3.80	0.35					0.50
464	Naval brass uninhibited	60.0	39.7	0.20	0.10	1.0				0.10
465	Naval brass arsenical	60.0	39.7	0.20	0.10	1.0	0.10			0.10
466	Naval brass antimonial	60.0	39.7	0.20	0.10	1.0		0.10		0.10
467	Naval brass phosphorized	60.0	39.7	0.20	0.10	1.0			0.10	0.10
482	Naval brass medium-leaded	60.5	38.0	1.0	0.10	1.0				0.10
485	Naval brass high-leaded	60.0	37.5	2.2	0.10	1.0				0.10

[a]Values given represent maximum content.

machining purposes. They can be easily hot-worked (however, the metal must be supported mechanically) and cold-worked within limits, but they are not as strong, corrosion resistant, or hard as the Alpha and Beta brasses.

Tin brasses (also called naval brasses) are used for chemical, steam power plant, and marine equipment.

Galvanic Action. All brasses react with other metals. When brass is used in direct contact with any other metal, a careful check should be made of its position on the galvanic series. Brass should not come into direct contact with iron, steel or stainless steel, aluminum, zinc, or magnesium if there is an electrolyte present or the possibility of one forming at the point of contact.

Dezincification. Corrosive attack of a peculiar type, called dezincification, occurs in brasses. It is most prevalent in some brasses with more than 15 to 20% zinc. It can be stopped by using inhibitors in the simple two-metal alloys, but in multiple-type alloys inhibitors can only retard it. This corrosive action occurs when acids and other strongly conducting solutions are present. What actually happens is that the copper-zinc alloy is dissolved, the copper is redeposited electrochemically, and the zinc remains in solution or forms a scale. The corrosion occurs in two ways: in one, corrosion pits form and are filled with redeposited copper; in the second, the entire surface acquires a redeposited layer of copper.

Workability. All brasses can be soldered, brazed, welded, and polished.

Commercial Forms. Brasses are available in bar, sheet, strip, rod, tube, and powder form and as special shapes and castings. Gauge equivalents for sheet, strip and wire are given in Table *G3* (*see* Gauges).

Types and Uses

In the construction field, brasses are used for doors, windows, door and window frames, and ornamental metalwork such as railings, trim, and grilles. Cost, type of alloy, and color are the controlling factors in selection. The three brasses commonly used for construction purposes are two so-called bronzes (commercial bronze, architectural bronze) and Muntz metal (*see* Table *B14*).

Table B14 Types and properties of brasses for the construction field

Name	Composition (percent) Cu	Zn	Pb	Color	Machining	Cold working	Other properties[a]	Weldability
Architectural bronze	56.5	41.25	2.25	Bronze	Good	Very poor	Excellent forging and free-machining properties	Poor
Commercial bronze	90.0	10.0		Bronze	Poor	Excellent	Very ductile	Gas, carbon arc, metal arc
Muntz metal	60.0	40.0		Light yellow	Good	Fair	High strength combined with low ductility	Gas, carbon arc, metal arc; spot and seam welding for thin sheets

[a]All can be easily hot worked, soldered and polished.

The brasses are also used extensively for finish hardware, plating of hardware, and other miscellaneous accessories such as screws, nuts and bolts, anchors, weatherstripping, and ties.

The bulk of brass produced is used in plumbing, heating, and air conditioning, and in industrial machinery and equipment, where the special properties of brasses are required.

History and Manufacture

In ancient times many brasses were produced accidentally and were really bronzes because copper ores often contain tin. It was not until the ancients recognized the difference between zinc and tin that brass and bronze were differentiated. The Romans knew of an alloy of copper and zinc which they called *aurichalcum*. One of the earliest examples of Roman brass is a coin which contains 17.3% zinc and is dated 20 B.C. Brass was produced in quantity in the Low Countries of Europe about A.D. 300. Around A.D. 1000 Cologne was the center of brass working in Europe, with the towns of Huy and Dinant as two of the main producers. In 1500 in England, Queen Elizabeth granted a patent and exclusive rights for working calamine (zinc ore) and making brass. Until 1850 brass was made by heating copper shot with zinc ores (calamines) and charcoal. The partially brassed shot was then melted, mixed, and cast. The process of mixing the actual metals of copper and zinc was patented in England in 1781 by Emerson. Today the same process is used, and a large variety of compositions, including other metals, has been developed to give desired properties.

BRAZING

Physical and Chemical Properties

Brazing may be defined as a type of soldering in which the operating temperatures are higher (but lower than in welding) and in which stronger and higher melting alloys are used to fill the joints, which consequently are stronger than ordinary soldered joints (*see* Figure *B2* and compare with Figures *S3* and *W6*).

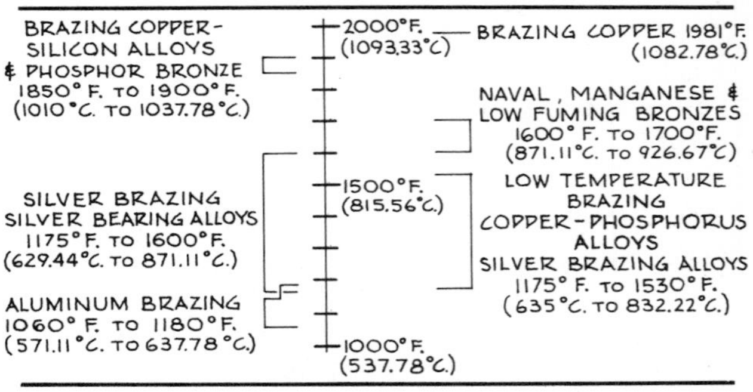

Figure B2 Temperature ranges for brazing.

The bond is obtained by alloying between the brazing material and the surface of the joined metals. Brazing is generally used where the shape and position of the joint or the composition of the metal or metals are not adaptable to welding. In brazing, the type of metal to be joined, the brazing material, and their color are equally important because galvanic action, strength of the joint, and matching of colors play a significant part in the finished product.

Brazing materials vary considerably with the type of

metal to be joined, but all are nonferrous. The various brazing materials are prepared by melting and mixing together the metallic ingredients according to fixed and controlled proportions.

Types and Uses

There are five major methods of brazing: torch brazing, furnace brazing, induction brazing, resistance brazing, and dip brazing.

Torch Brazing. In this method the flame of a gas torch is used to heat the metals and brazing materials. The gas for the torch may be tank gas or public utility Gas for the torch may be tank gas or public utility gas.

Furnace Brazing. This method is used principally when the parts to be joined can be assembled prior to the brazing operation. The entire assembly, including brazing alloy and flux set correctly in place, is raised to brazing temperature in a furnace; in many of the newer procedures, this is done in a controlled atmosphere, for example, in nitrogen or hydrogen.

Induction Brazing. In this method heat is developed by electricity in a heating coil and induced into the parts to be brazed (which do not become part of the electrical circuit).

Resistance Brazing. Here the parts to be brazed are held between two electrodes, and proper pressure and electric current are applied. Heat is developed from the resistance to the flow of electric current through the electrodes and the joint to be brazed. In this method the brazing alloy is preplaced. A conductive flux must be specially chosen, as most fluxes are nonconductive.

Dip Brazing. In one method the metals to be joined and the brazing alloy are dipped into a bath of molten salt, which acts as both flux and source of heat. In the other method the metals to be joined are dipped into a bath of molten salt, which acts as both flux and source of heat.

Other Methods. Three other, less commonly used methods of brazing are (1) the twin-carbon arc method, in which the carbon electrodes are separated to develop a flaming arc by means of an electric current; (2) block brazing, in which preheated, large metal blocks supply heat; and (3) flow brazing, in which the molten brazing alloy is poured onto the joint.

Brazing Materials. Brazing materials fall into six major types: aluminum-silicon, copper-phosphorus, silver, copper and copper-zinc, magnesium, and heat-resistant alloys. Each type is particularly suited to a certain group of metals.

Aluminum-silicon brazing alloys are meant specifically for brazing aluminum (*see* Table *B15*). The flux used must also be suitable for aluminum. The maximum permissible content of impurities is 0.15%.

Copper-phosphorus brazing alloys are used to braze copper and copper alloys (*see* Table *C16*). A flux suitable for copper alloys is always used except when brazing copper, in which case the brazing material is self-fluxing. The maximum permissible content of impurities is 0.15%.

Silver brazing alloys are used for brazing all ferrous and nonferrous metals except aluminum, magnesium, titanium, zinc, and lead (*see* Table *B17*). A suitable flux must always be used except when a vacuum or inert gas shield is utilized in the brazing process. All of these brazing alloys are available in sheet, strip, wire,

Table B15 Aluminum-silicon brazing alloys

Composition (percent)										Melting range °F (°C)		
Basic type alloy		Other metals										
Al	Si	Cu	Fe[a]	Zn[a]	Mg[a]	Mn[a]	Cr[a]	Ti[a]	Form	Solid	Liquid	General method of brazing
Remainder	4.0– 6.0	0.30	0.8	0.1	0.05	0.05		0.2	Wire	1150 (621.11)	1185 (640.56)	Furnace and dip
Remainder	6.8– 8.2	0.25	0.8	0.2					Brazing sheet	1120 (604.44)	1140 (615.56)	Furnace and dip
Remainder	3.3– 4.7	3.30	0.8	0.2	0.15	0.15	0.15		Wire			Furnace and dip
Remainder	11.0–13.0	0.30	0.8	0.2	0.10	0.15			Wire	1090 (587.78)	1185 (640.56)	Torch

[a]Maximum percentage allowable.

Table B16 Copper-phosphorus brazing alloys

Composition (percent)				Melting range °F (°C)		
Basic type alloy		Other				
Cu	P	Ag	Form	Solid	Liquid	General method of brazing
Remainder	4.75–5.25		Strip	1450 (777.78)	1700 (926.67)	Resistance and furnace
Remainder	6.75–7.5	4.75– 5.25	Wire, rod, powder	1350 (732.22)	1550 (843.33)	Torch, furnace, resistance, induction, dip
Remainder	6.0 –6.5	4.75– 5.25	Wire, rod, powder	1300 (707.44)	1550 (843.33)	
Remainder	6.75–7.8	5.75– 6.25	Wire, rod, powder	1300 (707.44)	1500 (815.56)	
Remainder	4.75–5.25	14.5 –15.5	Wire, rod, powder	1300 (707.44)	1500 (815.56)	

Table B17 Silver brazing alloys

Composition (percent)						Melting range °F (°C)			
Ag	Cu	Zn	Cd	Ni	Sn	Solid	Liquid	General method of brazing	Major use
44–46	14–16	14–18	23–25			1145 (618.33)	1400 (760)	Torch, furnace, induction, dip, resistance	General purpose (light yellow color)
34–36	25–27	19–23	17–19			1295 (701.67)	1500 (843.33)		
39–41	29–31	26–30		1.5–2.5		1435 (779.44)	1650 (893.89)		General purpose (whitish yellow color)
55–57	21–23	15–19			4.5–5.5	1205 (646.11)	1400 (760)	Furnace	For white metals, stainless steel, etc. (white color)

rod, and powder form, and the permissible content of impurities is 0.15%.

Copper and copper-zinc brazing alloys are used for brazing the metals listed in the last column of Table *B18*. A flux suitable for use with the particular metals to be brazed is usually required. In furnace brazing with a hydrogen atmosphere the copper alloy does not require a flux. All these brazing alloys are used with torch, furnace, induction, resistance, and dip brazing methods and also with the rarely used twin-carbon arc, block, and flow methods.

Magnesium brazing alloys are used for brazing magnesium and magnesium-base alloys (*see* Table *B19*). If the furnace brazing method is used, beryllium is added to prevent ignition.

Heat-resistant brazing alloys are used for brazing stainless steel, high-nickel alloys, carbon steels, and low-alloy steels (*see* Table *B20*). When the nickel brazing alloy is used, the metal to be brazed can be heated to 2000°F (1093.33°C), whereas with the silver brazing alloy it is heated only to 900°F (427.22°C). A flux is always used in torch and induction brazing; furnace brazing with a special atmosphere requires no flux.

BRICK

Physical and Chemical Properties

Brick may be defined as a small building unit, solid or cored not in excess of 25%. It is commonly a rectangular block composed of inorganic, nonmetallic substances of mineral origin and hardened by heat or chemical action.

Categories. Brick can be divided on this basis into two main categories:

Table B18 Copper and copper-zinc brazing alloys

Composition (percent)											Maximum percentage of impurities	Form	Melting range °F (°C)		Major use
Zn	Cu	Sn	Fe	Ni	P	Ag	Mn[a]	Pb[a]	Al[a]	Si[a]			Solid	Liquid	
	99.9				0.075[a]			0.02	0.01		0.1	Wire, round rod, strip	2000 (1093.33)	2100 (1148.89)	Steel, nickel, copper-nickel alloys
Remainder	58–62							0.05	0.01		0.5	Wire, round rod, strip	1670 (910)	1750 (954.44)	Carbon steel, copper, copper alloys, nickel, nickel alloys, stainless steel
Remainder	57[b]	0.5–1.0						0.05	0.01		0.5	Wire, round rod, strip	1670 (910)	1750 (954.44)	
Remainder	56[b]	1.1[a]	0.25–1.25	1.0[a]			1.0	0.05	0.01	0.25	0.5	Wire, round rod, strip	1670 (910)	1750 (954.44)	
Remainder	50–55	0.1[a]						0.5				Grains, lump	1600 (871.11)	1700 (926.67)	
Remainder	50–53	3.5–4.5						0.5				Grains, lump	1620 (882.22)	1700 (926.67)	
Remainder	46–50		0.25[a]	9–11				0.05	0.005	0.15	0.5	Wire, round rod	1720 (932.78)	1900 (1037.78)	
Remainder	46–48			10–11	0.2–0.5	0.3–1.0				0.15	0.1	Wire, round rod, powder	1690 (921.11)	1800 (982.22)	

[a]Maximum percentage allowable.
[b]Minimum percentage allowable.

Table B19 Magnesium brazing alloys

Composition (percent)							Maximum percentage of impurities	Form	Melting range °F (°C)		General method of brazing
Mg	Al	Mn	Zn	Si[a]	Cu[a]	Ni[a]			Solid	Liquid	
Remainder	8.3–9.7	0.1 min.	1.7–2.3	0.3	0.05	0.01	0.03	Round rod, wire	2000 (1093.33)	2150 (1176.67)	Furnace, torch, induction

[a]Maximum percentage allowable.

Table B20 Heat-resistant brazing alloys

Composition (percent)					Maximum percentage of allowable Fe + Si + C	Maximum percentage of impurities	Form	Melting range °F (°C)		General methods of brazing
Ni	Cr	B	Ag	Mn				Solid	Liquid	
65–75	13–20	2.75–4.75			0.1	0.5	Wire, strip	2000 (1093.33)	2150 (1176.67)	Furnace, torch, induction
			84–86	14–16	–	0.15	Wire, strip	1780 (971.11)	2100 (1148.89)	Furnace, torch, induction

1. Those made of clay burned or fired to hardness. This group represents a major branch of the ceramics industry, the burned clay products for construction purposes. Burned brick includes common (building) brick, facing brick, glazed facing brick, fire brick, flooring brick, paving brick, refractory brick, acid brick, and sewer brick.

2. Those made of cementitious materials that harden by chemical action. This group includes sand-lime brick and cement brick.

A third category of brick, adobe brick, is the simple forerunner of burned brick. It is dried out and hardened in the sun.

Surface Qualities. The following finishes are common to all types of brick:

1. Smooth finish: the surface is plane as formed by the die during manufacture.
2. Scored finish: the surface is grooved as it comes from the die; this increases bond for mortar, plaster, or stucco.
3. Combed finish: the surface is altered by parallel scratches or scarfs to produce a desired texture or to increase bond for mortar, plaster, or stucco.
4. Roughened finish: the surface is entirely broken by wire cutting, etc., to provide a desired texture or to increase bond for mortar, plaster, or stucco.

Texture is the surface quality apart from the color. It includes the texture of the clay body of the brick, which can range from fine through medium to coarse. Other textures that can be introduced in the manufacturing process include marked, rugged, barked, stippled, and hammered.

BRICK, ADOBE

Physical and Chemical Properties

Adobe brick is made of a calcareous, sandy clay or any alluvial desert clay with good plastic properties that dries to a hard, uniform mass. It can best be used in arid or semiarid climates where the clay is easily available and the very good insulating properties of the brick can be used to advantage.

Types and Uses

Adobe bricks are generally 3 to 5 in. (76.2 to 127 mm) high, 10 to 12 in. (254 to 304.8 mm) wide, and 14 to 20 in. (1355.6 to 508 mm) long.

Application

Condensed Checklist

1. Adobe brick should always be set on a waterproof (usually concrete) foundation to prevent capillary action and consequent disintegration.
2. It should be laid with adobe mortar.
3. Adobe brick masonry should be given a finish coat, either of lime and cement plaster or adobe.
4. It should always be protected from moisture penetration, usually by a large roof overhang.
5. Building height should preferably be one story and not over three or four stories.

History and Manufacture

The word "adobe" comes from the Arab word *atob*, meaning sun-dried brick. Its use, or that of clays with similar properties, dates back many hundreds of years. The ancient sites of this type of brick in the Western Hemisphere are found from the United States to Peru. The adobe brick of the Pueblo Indians is perhaps the best known of these. The use of a mold for shaping the brick was probably introduced into Spain from Africa and brought over to the United States by the Spanish conquistadors in the 16th century.

The methods of making adobe brick have not changed appreciably since early times. They consist in wetting a suitable soil and letting it stand for 1 day to become homogeneous. Straw or other fiber is added to prevent shrinkage cracks during the curing process. The whole is then mixed with a hoe, and the mass is worked by machine or by hand into the right consistency for molding. The material is placed in the mold and allowed to dry for 2 weeks.

BRICK, BURNED

Physical and Chemical Properties

Hard burned brick is a clay product fired at high temperatures to near-vitrification, which produces low absorption and high compressive strength. Soft burned brick is fired at lower temperatures and has relatively high absorption and low compressive strength. Vitrification occurs when the temperature of firing is sufficient to fuse all grains and close all pores, thus making the mass impervious.

The specific gravity of burned brick ranges from 2.6 to 2.8, and its weight per unit volume ranges from 0.060 to 0.082 lb/in.^{3} (1.661 to 2.240 g/cm^3), the average being 0.071 lb/in.^{3} (1.965 g/cm^3), which is equivalent to 123 lb/ft^3 (1970.46 g/m^3).

Constituents. The clay from which burned brick is made is technically known as a hydrated silicate of alumina ($Al_2O_3 \cdot 2SiO_2 \cdot 2H_2O$). It also contains varying amounts of the oxides of iron, calcium, magnesium, potassium, sodium, titanium, and sulfur.

Surface Qualities. The natural color range of burned brick is mainly in the reds, buffs, and creams.

The surface features of burned brick may include, besides those common to all brick (i.e., smooth, scored, combed, roughened), the following finishes:

1. Natural finish: the surface is unglazed or uncoated, burned to the natural color of the material forming the body of the brick.

2. Ceramic glazes: these are compounded glassy coatings fused to the brick. Ceramic color glaze is a surface covering of fire-bonded, colored ceramic glaze with a satin or gloss finish. Ceramic clear glaze is a surface covering of fire-bonded, translucent or tinted ceramic glaze with a lustrous finish. Nonlustrous glaze is a surface covering of fire-bonded ceramic glaze with a satin-like or matte finish. Salt glaze is a surface covering of lustrous glaze formed by chemical reaction of vapors of salts or chemicals during burning. (*See* Glazes and Porcelain Enamels for fuller description.)

Textures in burned brick are the same as those for brick in general but also include spotted, water-struck, and sand-struck textures. "Water-struck" means that the molds are wetted with water to stop the clay from sticking to them. "Sand-struck" means that the molds are sprinkled with sand, which acts as a separator and also gives a texture to the brick.

Types and Uses

The types of brick used in construction are building, facing, hollow, glazed facing, and fire brick.

Table *B21* lists sizes of modular and nonmodular building, facing, glazed facing, and hollow brick. Table *B22* lists grade requirements for brick exposures for building (*see* Figure *B3*). Table *B23* shows designation by compressive strength.

Building Brick. Building brick is divided into three grades as follows:

Grade SW: for rigorous exposures with heavy rainfall and freezing conditions.
Grade MW: for exposures with average moisture and minor freezing conditions.
Grade NW: for exposures with minimum moisture and freezing conditions (*see* Table *B24*).

Facing Brick. Facing brick is divided into two grades similar to SW and MW for building brick and three other types based on factors affecting the appearance of the finished wall, as follows:

Type FBX: exterior and interior walls and partitions; high degree of mechanical perfection, narrow color range, and minimum size variation per unit.

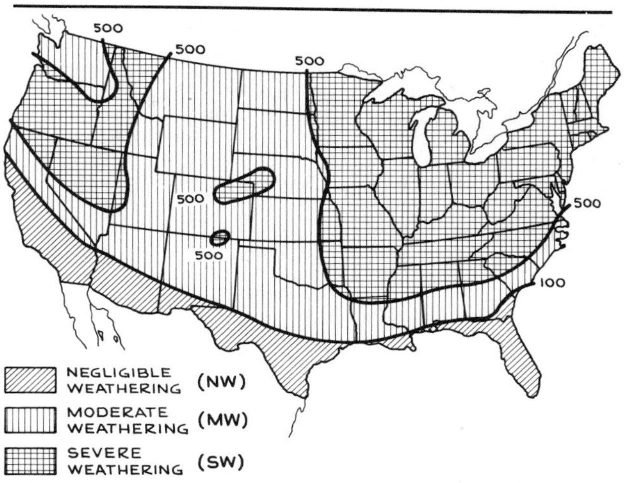

Figure B3 Weathering indices for building, facing, and hollow brick, based on the freezing-cycle days and the winter rainfall.

Type FBS: exterior and interior walls and partitions; wide range of color and greater size variation per unit.
Type FBA: exterior and interior walls and partitions; nonuniformity in size, color and texture per unit.
See Tables *B25*, *B26*, and *B27* for requirements.

Building brick and facing brick are both available solid, cored, or frogged within the following limitations: the net cross-sectional area of cored brick in any plane parallel to the bearing surface shall be at least 75% of the gross cross-sectional area measured in the same plane, and no part of any core shall be less than $\frac{3}{4}$ in. (19.05 mm) from the edge of the brick; one bearing face of a brick may have a recess or panel (frog) with a maximum depth of $\frac{3}{8}$ in. (9.53 mm), and no part of this frog shall be a minimum of $\frac{3}{4}$ in. (19.05 mm) from the edge of the brick.

Hollow Brick. Hollow brick is divided into two grades similar to SW and MW for building brick and into four other types based on factors affecting the appearance of the finish wall, as follows:

Type HBX: exterior and interior walls and partitions; high degree of mechanical perfection, narrow color range, and minimum size variations per unit.
Type HBS: exterior and interior walls and partitions; wide range of color and greater size variations per unit.
Type HBB: exterior and interior walls and partitions; nonuniformity in size, color, and texture per unit.
Type HBA: hollow brick selected and manufactured to produce characteristic design effects resulting from nonuniformity in size, color, and texture of the individual units.

Table B21 Sizes of burned brick

Type of brick	Manufactured dimensions for face dimensions in wall: Height in.	Height mm	Length in.	Length mm	Thickness in.	Thickness mm	Joints in.	Joints mm	Coursing (C): No. of C	Coursing in.	Coursing mm
Nonmodular											
3-in.[a]	$2\frac{5}{8}$	66.68	$9\frac{5}{8}$	244.48	3	72.6	$\frac{1}{2}$	12.7	1	$3\frac{1}{8}$	79.38
(76.2-mm)							$\frac{3}{8}$	9.53	1	3	76.2
	$2\frac{3}{4}$	69.85	$9\frac{3}{4}$	297.65	3	72.6	$\frac{1}{2}$	12.7	1	$3\frac{1}{4}$	82.55
							$\frac{3}{8}$	9.53	1	$3\frac{1}{8}$	79.38
Standard	$2\frac{1}{4}$	57.15	8	203.2	$3\frac{3}{4}$[b]	95.25	$\frac{1}{2}$	12.7	1	$2\frac{3}{4}$	69.85
							$\frac{3}{8}$	9.53	1	$2\frac{5}{8}$	66.86
Oversize	$2\frac{3}{4}$	69.85	8	203.2	$3\frac{3}{4}$[b]	95.25	$\frac{1}{2}$	12.7	1	$3\frac{1}{4}$	82.55
							$\frac{3}{8}$	9.53	1	$3\frac{1}{8}$	79.38
Modular[c]											
Standard modular			$7\frac{5}{8}$	193.68	$3\frac{5}{8}$	92.08					
	$2\frac{1}{4}$	57.15	$7\frac{1}{2}$	190.50	$3\frac{1}{2}$	88.90	$\frac{1}{2}$	12.7	3	8	203.2
Norman			$11\frac{5}{8}$	295.28	$5\frac{5}{8}$	142.88					
SCR brick[d]			$11\frac{1}{2}$	292.10	$5\frac{1}{2}$	139.70					
Economy 8			$7\frac{5}{8}$	193.68	$3\frac{5}{8}$	92.08					
(Jumbo closure)			$7\frac{1}{2}$	190.50	$3\frac{1}{2}$	88.90					
Economy 12					$3\frac{5}{8}$	92.08					
(Jumbo utility)	$3\frac{5}{8}$	92.08			$3\frac{1}{2}$	88.90	$\frac{3}{8}$	9.53	1	4	101.6
	$3\frac{1}{2}$	88.90					$\frac{1}{2}$	12.7			
6-in. (152.4-mm)			$11\frac{5}{8}$	295.28	$5\frac{5}{8}$	142.88					
Jumbo			$11\frac{1}{2}$	292.10	$5\frac{1}{2}$	139.70					
8-in. (152.4-mm)					$7\frac{5}{8}$	193.68					
Jumbo					$7\frac{1}{2}$	190.50					
Engineer			$7\frac{5}{8}$	193.68	$3\frac{5}{8}$	92.08					
			$7\frac{1}{2}$	190.50	$3\frac{1}{2}$	88.90					
Norwegian	$2\frac{13}{16}$	71.44					$\frac{3}{8}$	9.53	5	16	407
	$2\frac{11}{16}$	68.26					$\frac{1}{2}$	12.7			
			$11\frac{5}{8}$	295.28							
6-in. (152.4-mm)			$11\frac{1}{2}$	292.10	$5\frac{5}{8}$	142.88					
Norwegian					$5\frac{1}{2}$	139.70					
Double			$7\frac{5}{8}$	193.68							
	$4\frac{15}{16}$	125.41	$7\frac{1}{2}$	190.50	$3\frac{5}{8}$	92.08	$\frac{3}{8}$	9.53	3	16	407
Triple	$4\frac{13}{16}$	119.06	$11\frac{5}{8}$	295.28	$3\frac{1}{2}$	88.90	$\frac{1}{2}$	12.7			
			$11\frac{1}{2}$	292.10							
Roman	$1\frac{5}{8}$	41.28	$11\frac{5}{8}$	295.28	$3\frac{5}{8}$	92.08	$\frac{3}{8}$	9.53	2	4	101.6
	$1\frac{1}{2}$	38.10	$11\frac{1}{2}$	292.10	$3\frac{1}{2}$	88.90	$\frac{1}{2}$	12.7			
Hollow brick	$7\frac{5}{8}$	193.68	$7\frac{5}{8}$	193.68	$3\frac{5}{8}$	92.08	$\frac{3}{8}$	9.53	1	8	203.2
	$11\frac{5}{8}$	295.28	$11\frac{5}{8}$	295.28	$3\frac{5}{8}$	92.08			1	12	304.8

[a] The 3-in. (76.2-mm) brick is produced under other designations: Kingsize, Big John, Jumbo, Scotsman, and Spartan.
[b] The thickness on standard and oversize nonmodular brick varies from $3\frac{1}{2}$ in. (88.90 mm) to $3\frac{3}{4}$ in. (95.25 mm). Therefore, if other than running bond is to be used, it is necessary to check with the manufacturer of the brick to be used.
[c] Available in solid units or, in a number of cases, as hollow brick.
[d] Registered U.S. Patent Office, SCPI.

The net cross-sectional area of hollow brick in any plane parallel to the bearing surface shall be at least 60% of the gross cross-sectional area measured in the same plane. Cored-shell hollow brick shall have a shell of a minimum thickness of $1\frac{1}{2}$ in. (38.1 mm). Cores greater than 1 in.^{2} (645.2 mm^2) in cored shells shall not be less than $\frac{1}{2}$ in. (12.7 mm) from any edge. Cores not greater than 1 in.^{2} (645.2 mm^2) in shells cored no more than 35% shall not be less than $\frac{3}{8}$ in. (9.53 mm) from any edge. Double-shell hollow brick with inner and outer shells not less than $\frac{1}{2}$ in. (12.7 mm) may have cells no greater than $\frac{5}{8}$ in. (15.88 mm) in width or 5 in. (127.0 mm) in length between the inner and outer shell. Table *B28* shows tolerances on dimensions and distortions of the different types of hollow brick.

Table B22 Grade requirements based on brick exposure for building, facing, and hollow brick

Exposure	Weathering index[a] for building brick,[b] facing brick, and hollow brick: Less than 100	100–500	500 and greater
In vertical surfaces	MW	SW	SW
In contact with earth or not in contact with earth	MW	MW	MW
In other than vertical surfaces in contact with earth	SW	SW	SW
Not in contact with earth	MW	SW	SW

[a]Weathering index is based on the product of the annual number of freezing cycle days and the average winter rainfall in inches.
[b]Brick not exposed to weather may be grade NW.

Glazed Facing Brick. Glazed facing brick is manufactured cored or uncored in two grades and two types, as follows:

Grade S (select): for use with comparatively narrow mortar joints.
Grade SS (select sized or ground edge): for use where variations of face dimension must be very small.
Type I (single-faced units): for general use where only one finished face will be exposed and where one finished end will be exposed.
Type II (two-faced units): for use where two opposite finished faces will be exposed (seldom used, as it would be a special order for glazed facing brick).

The compressive strength of glazed facing brick is the same as that of grade SW for facing brick. The glazed surface shall be fused to the body of the brick, shall be

Table B23 Designation by compressive strength for facing brick, building brick, and hollow brick

Designation by compressive strength		Minimum compressive strength (gross area): Average of five bricks		Individual bricks	
lbf/in.^{2}	MN/m^2 [a]	lbf/in.^{2}	MN/m^2 [a]	lbf/in.^{2}	MN/m^2 [a]
2,500	17.238	2,500	17.238	2,200	15.169
4,500	31.028	4,500	31.028	4,000	27.580
6,000	41.370	6,000	41.670	5,800	39.991
8,000	55.160	8,000	55.160	7,000	48.265
10,000	68.950	10,000	68.950	8,800	60.676
12,000	82.740	12,000	82.740	10,600	73.087
14,000	96.530	14,000	96.530	12,300	84.809

[a]Meganewtons per meter squared; 1 lbf/in.^{2} = 0.006895 MN/m^2.

Table B24 Physical requirements for building brick (common brick), either solid or cored

Grade of brick	Minimum compressive strength (brick flatwise): Average of five bricks, lbf/in.^{2}	Average of five bricks, MN/m^2 [b]	Individual bricks, lbf/in.^{2}	Individual bricks, MN/m^2 [b]	Maximum water absorption after 5 hours of boiling (percent): Average of five bricks	Individual bricks	Maximum saturation coefficient[a]: Average of five bricks	Individual bricks
SW	3000	20.685	2500	17.238	17.0	20.0	0.78	0.80
MW	2500	17.238	2200	15.169	22.0	25.0	0.88	0.90
NW	1500	10.343	1250	8.619	No limit	No limit	No limit	No limit

[a]The saturation coefficient is the ratio of absorption after 24-hour submersion in cold water to that after 5-hour submersion in boiling water.
[b]Meganewtons per meter squared; 1 lbf/in.^{2} = 0.006895 MN/m^2.

Table B25 Physical requirements for facing brick (solid or cored) and hollow brick

Grade of brick	Minimum compressive strength (brick flatwise; hollow brick in bearing position) Gross area				Maximum water absorption after 5 hours of boiling (percent)		Maximum saturation coefficient[a]	
	Average of five bricks		Individual bricks					
	lbf/in.^{2}	MN/m^2 [b]	lbf/in.^{2}	MN/m^2 [b]	Average of five bricks	Individual bricks	Average of five bricks	Individual bricks
SW	3000	20.685	2500	17.238	17.0	20.0	0.78	0.80
MW	2500	17.238	2200	15.169	22.0	25.0	0.88	0.90

[a]The saturation coefficient is the ratio of absorption after 24-hour submersion in cold water to that after 5-hour submersion in boiling water.
[b]Meganewtons per meter squared; 1 lbf/in.^{2} = 0.006895 MN/m^2.

Table B26 Permissible extent of chippage for facing brick, solid or cored

Type	Maximum permissible extent of chippage of finish face or face into the surface				Percentage of shipment that may be allowed chippage over permissible maximum				
	Edge		Corner		Percent allowable	Edge		Corner	
	in.	mm	in.	mm		in.	mm	in.	mm
FBX	$\frac{1}{8}$	3.175	$\frac{1}{4}$	6.35	5	$\frac{1}{4}$	6.35	$\frac{3}{8}$	9.525
FBS (smooth)	$\frac{1}{4}$	6.35	$\frac{3}{8}$	9.525	10	$\frac{5}{16}$	7.938	$\frac{1}{2}$	12.7
FBS (rough)	$\frac{5}{16}$	7.938	$\frac{1}{2}$	12.7	15	$\frac{7}{16}$	11.113	$\frac{3}{4}$	19.05
FBA	As specified by the manufacturer								

Table B27 Tolerances on dimensions and distortion of facing brick, solid or cored

Dimensions	Maximum permissible variation from specified dimensions (plus or minus)				Maximum permissible distortion			
	Type FBX		Type FBS		Type FBX		Type FBS	
	in.	mm	in.	mm	in.	mm	in.	mm
3 in. (76.2 mm) and under	$\frac{1}{16}$	1.588	$\frac{3}{32}$	2.381				
Over 3 to 4 in. inclusive (76.2 to 101.6 mm)	$\frac{3}{32}$	2.381	$\frac{1}{8}$	3.175				
Over 4 to 6 in. inclusive (101.6 to 152.4 mm)	$\frac{1}{8}$	3.175	$\frac{3}{16}$	4.763				
Over 6 to 8 in. inclusive (152.4 to 203.2 mm)	$\frac{5}{32}$	3.969	$\frac{1}{4}$	6.35				
Over 8 to 12 in. inclusive (203.2 to 304.8 mm)	$\frac{7}{32}$	5.556	$\frac{5}{16}$	7.938	$\frac{3}{32}$	2.831	$\frac{1}{8}$	3.175
Over 12 to 16 in. inclusive (304.8 to 406.4 mm)	$\frac{9}{32}$	7.144	$\frac{3}{8}$	9.525	$\frac{1}{8}$	3.115	$\frac{5}{32}$	3.969
8 in. inclusive (203.2 mm) and under					$\frac{1}{16}$	1.588	$\frac{3}{32}$	2.831

Table B28 Permissible tolerances on dimensions and distortion of face dimensions

	Type HBX				Types HBS and HBB			
	Maximum permissible variation		Maximum permissible distortion		Maximum permissible variation		Maximum permissible distortion	
Dimensions	in.	mm	in.	mm	in.	mm	in.	mm
3 in. (76.2 mm) and under	$\frac{1}{16}$	1.59			$\frac{3}{32}$	2.38		
Over 3 to 4 in. inclusive (76.2 to 101.6 mm)	$\frac{3}{32}$	2.38			$\frac{1}{8}$	3.18		
Over 4 to 6 in. inclusive (101.6 to 152.4 mm)	$\frac{1}{8}$	3.18			$\frac{3}{16}$	4.76		
Over 6 to 8 in. inclusive (132.4 to 203.2 mm)	$\frac{5}{32}$	3.97			$\frac{1}{4}$	6.35		
8 in. (203.2 mm) and under			$\frac{1}{16}$	1.59			$\frac{3}{32}$	2.38
Over 8 to 12 in. inclusive (203.2 to 304.8 mm)	$\frac{7}{32}$	5.56	$\frac{3}{32}$	2.38	$\frac{5}{16}$	7.94	$\frac{1}{8}$	3.18
Over 12 to 16 in. inclusive (304.8 to 406.4 mm)	$\frac{9}{32}$	7.14	$\frac{1}{8}$	3.18	$\frac{3}{8}$	9.53	$\frac{5}{32}$	3.97

colorfast and resistant to crazing, and shall have a hardness and abrasion resistance above 5 on the Mohs hardness scale. When ceramic glazed brick is required for exterior use, the manufacturers should be consulted for the suitable grade and type.

Fire Brick. Fire brick (refractory brick) made of fire clay is used for lining the firebox of fireplaces, barbecues and boilers. It usually ranges from white to dark beige in color.

Special Brick Shapes. The most commonly used shape is the bullnose, but there are many other standard shapes for caps, sills, watertables, special corners, etc. Special brick shapes, which have to be made to fit the particular job, are obtainable for flat and round arches, radial brick, and round columns. When special brick shapes are to be used, the manufacturer of the building brick selected for the job should be consulted to make sure that these shapes are part of his line and to find out whether there will be color variations due to separate burning of the special shapes. (*See* Figure *B4*.)

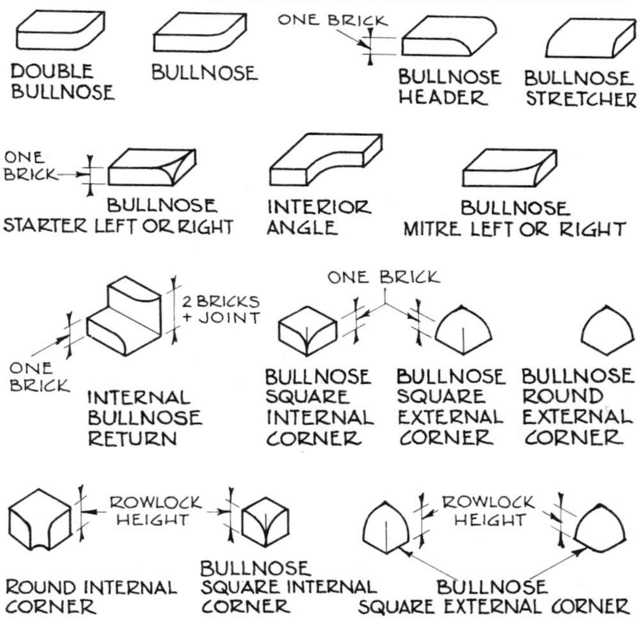

Figure B4 Special brick shapes.

Other Bricks. Prefabricated panels and preassembled brick units using high-bond mortar are accepted building components in the construction field.

Brick veneer 1 in. (25.4 mm) thick, $2\frac{1}{2}$ in. (63.5 mm) deep and 8 in. (203.2 mm) or $11\frac{1}{2}$ in. (292.1 mm) long, is available. It can be applied with nails, adhesive, or mortar to a back-up surface.

Flooring brick is an acid-resistant, dense, hard brick generally used in industrial construction. Flooring brick is manufactured in four types to be used as follows:

Type T: where a high degree of resistance to thermal and mechanical shock is required, but low absorption is not required.
Type H: where resistance to chemicals and thermal shock are factors, but low absorption is not required.
Type M: where low absorption is required. These bricks normally have limited mechanical shock resistance but have high abrasion resistance.
Type L: where minimal absorption and a high degree of chemical resistance are required. These bricks normally have limited mechanical shock resistance, but have high abrasion resistance.

Paving brick is a very dense, hard brick made from fire clay, semifire clay, or shale. It is available in thicknesses ranging from $\frac{1}{2}$ to $2\frac{1}{2}$ in. (12.7 to 63.5 mm), in widths ranging from $3\frac{3}{8}$ to 4 in. (85.73 to 101.6 mm), and in lengths ranging from $7\frac{1}{2}$ to $11\frac{3}{4}$ in. (190.5 to 298.45 mm). Paving brick is also available in squares of 4, 12, and 16 in. (101.6, 304.8, and 406.4 mm), and in hexagons of 6, 8, and 12 in. (152.4, 203.2, and 304.8 mm).

Refractory brick is used primarily where high temperatures are encountered. It is also used for resistance to abrasion, corrosion, pressure, and rapid changes in temperatures. Refractory brick is graded according to fusion temperature, porosity, spalling strength, resistance to rapid temperature changes, thermal conductivity, and heat capacity. Commonly used types of refractory brick are alumina brick, chrome brick, magnesite brick, and silica brick. Because of the large variety available, it is advisable to check the particular conditions and then choose the refractory brick that will best meet them. In this type of construction, it is always best to consult specialists.

Acid brick is a machine-made, uncored brick designed primarily for use in the chemical and allied industries. It is always laid with chemically resistant mortar.

Sewer brick is a clay or shale brick used for sewerage, industrial waste, and storm water conduits. The three grades of this brick are determined by compressive strength and water absorption rates.

Application

Suction or Absorption Rate of Brick. This rate has an important effect on the adhesion of brick and mortar because if the brick absorbs water too quickly from the mortar, the mortar will set too soon and adhesion will be poor. All brick with absorption rates in excess of 0.7 oz (19.85 g) per minute should be wetted sufficiently so that the rate of absorption does not exceed this amount. An exception is brick to be used with grouted masonry, for which the rate of absorption when laid may be 1.4 oz (39.69 g) per minute.

A rough but effective test for determining whether units will give improved adhesion by wetting consists in drawing a circle 1 in. (25.4 mm) in diameter on the surface of the unit which will be in contact with the mortar and placing 20 drops of water inside the circle. If the time taken for the water to be absorbed exceeds $1\frac{1}{2}$ minutes, the unit need not be wetted; if the absorption time is less than $1\frac{1}{2}$ minutes, wetting is recommended.

Strength of Brick Masonry Walls. The strength of brick masonry walls depends on mortar strength and the individual strength of masonry units (*see* Table *B29*).

Mortar and Mortar Joints. Mortar types recommended for brick masonry are as follows:
Type M: a high-strength mortar suitable for general use and recommended specifically where maximum compressive strength is required or where masonry will be below grade and in contact with earth.

Table B29 Compressive strength of brick masonry walls[a]

Compressive strength of masonry walls		Assumed compressive strength of brick masonry											
		Without inspection						With inspection					
		Type M		Type S		Type N		Type M		Type S		Type N	
lbf/in.^{2}	MN/m^2	lbf/in.^{2}	MN/m^2	lbf/in.^{2}	MN/m^2	lbf/in.^{2}	MN/m^2	lbf/in.^{2}	MN/m^2	lbf/in.^{2}	MN/m^2	lbf/in.^{2}	MN/m^2
14,000+	96.53+	3070	21.17	2600	17.19	2140	14.74	4600	31.72	3900	26.89	3200	22.06
12,000	82.74	2670	18.40	2270	16.64	1870	12.89	4000	27.58	3400	23.44	2800	19.30
10,000	68.95	2270	16.64	1930	13.31	1600	11.03	3400	23.44	2900	20.00	2400	16.55
8,000	55.16	1870	12.89	1600	11.03	1340	9.24	2800	19.30	2400	16.55	2000	13.79
6,000	41.37	1470	10.14	1270	8.76	1070	7.38	2200	15.16	1900	13.10	1600	11.03
4,000	27.58	1070	7.38	930	6.41	800	5.52	1600	11.03	1400	9.65	1200	8.27
2,000	13.79	670	4.62	600	4.14	530	3.65	1000	6.89	900	6.21	800	5.52

[a]From Building Code Requirements for Engineered Brick Masonry.

Type S: suitable for general use and recommended specifically where high lateral strength of masonry is desired.

Type N: a medium-strength mortar suitable for general use in exposed masonry above grade, recommended specifically where high compressive and/or lateral strengths are not required.

Type D: a low-strength mortar suitable for non-load-bearing walls of solid units, interior non-load-bearing partitions of hollow units, and load-bearing walls of solid units in which the axial compressive strength developed does not exceed 100 lbf/in.^{2} (0.69 MN/m^2). Type D is also suitable where exterior walls will not be subject to severe weathering. For all types of mortars *see* Mortars.

Pointing mortar: prehydrated type N.

In cavity walls, type S mortar should be used where winds are over 80 mph (128.72 km/h) and type N otherwise. With reinforced brick masonry, type S should be used.

Mortar joints should be watertight and should shed water (i.e., the water should drain off). Joints that do not shed water should be used only where watertightness is not a serious problem or where precautions against leakage are introduced into the brickwork. The commonly used methods of making mortar joints are shown in Figure *B5*.

Standard joint thicknesses for brickwork are $\frac{1}{4}$ in. (6.35 mm) for glazed brick, $\frac{3}{8}$ or $\frac{1}{2}$ in. (9.53 or 12.7 mm) for facing brick, and $\frac{1}{2}$ in. (12.7 mm) for building brick.

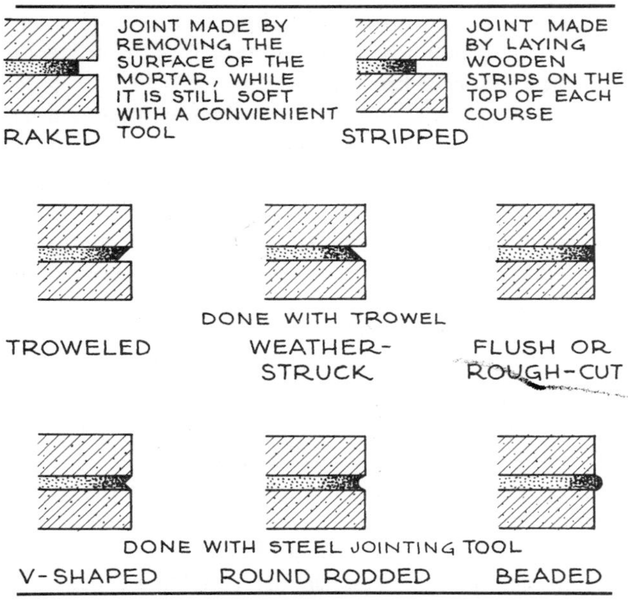

Figure B5 Mortar joints.

Bonding. The method of placing bricks so that the entire wall becomes one solid mass throughout its length and breadth is called bonding. All brickwork should be bonded either by headers, soldiers, special bonding brick shapes or by metal ties, anchors, clips, etc. (*see* Figure *B6*). The fundamental systems of bonding are English bond, Flemish bond, and bonding by metal ties. All bonding patterns are derived from these three types. (*See* Figures *B7 and B13*.)

Efflorescence. Efflorescence is a whitish powder of crystallization on brick masonry walls caused by water-soluble salts deposited on the surface upon evaporation of water. Efflorescence will appear if there are soluble salts in the wall materials and moisture to carry these salts to the surface. To overcome efflorescence, it is

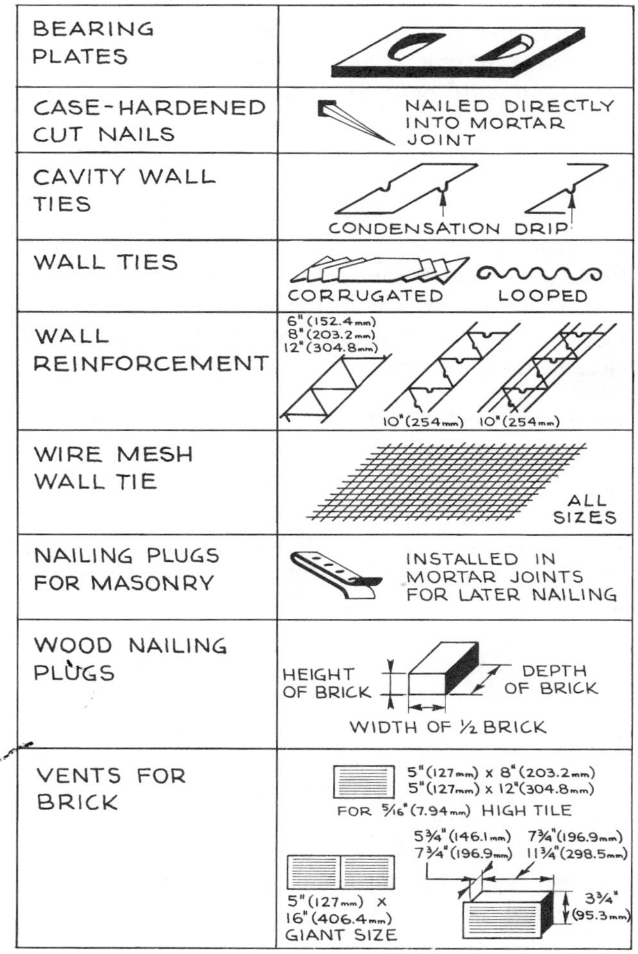

Figure B6 Accessories for brickwork.

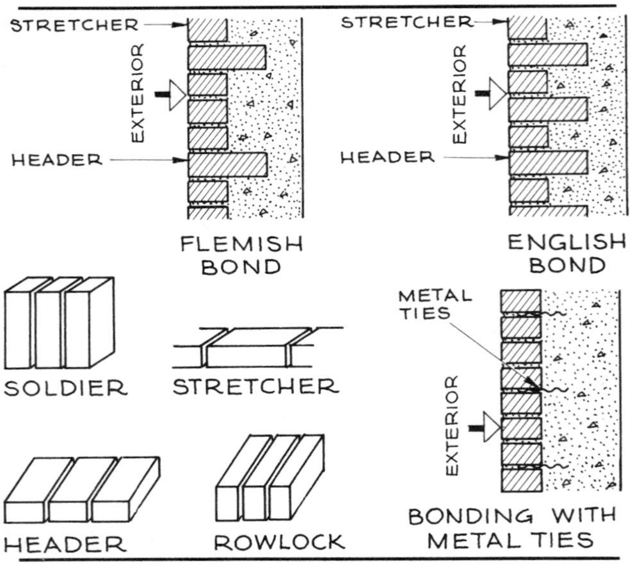

Figure B7 Bonding for brick masonry.

necessary to check types of brick, quantity and quality of water used, type of mortar, and particularly the type of admixture (if used). There is no known sure method of stopping efflorescence. Brick which will not contribute to efflorescence is available in all parts of the United States.

Cleaning of brickwork is generally done with brushes and clean water or with a weak solution of muriatic (hydrochloric) acid followed by a clear water rinse. It is always advisable to consult the manufacturer(s) of brick selected for the job to ascertain the best methods of cleaning the brickwork.

Reinforced Brick Masonry. This refers to brick masonry in which steel reinforcement is embedded and so placed that the masonry will have greatly increased resistance to forces that produce tensile, shearing, and compressive stresses. The principles of reinforced brick masonry design are the same as those commonly accepted for reinforced concrete, and similar formulas may be used.

Since reinforced masonry is designed to resist bending as well as compression, it is essential that all joints in the masonry be completely filled. The recommended method for doing this is to fill all interior joints with grout, which is obtained by adding sufficient water to the mortar to give it a fluid consistency.

Reinforced brick has an advantage over reinforced concrete in that it needs no forms for vertical walls or columns. One of its most general uses is for lintels (*see* Figure *B8*), but it is being utilized increasingly in locations where earthquakes, cyclones, or hurricanes are problems and also in the industrial field.

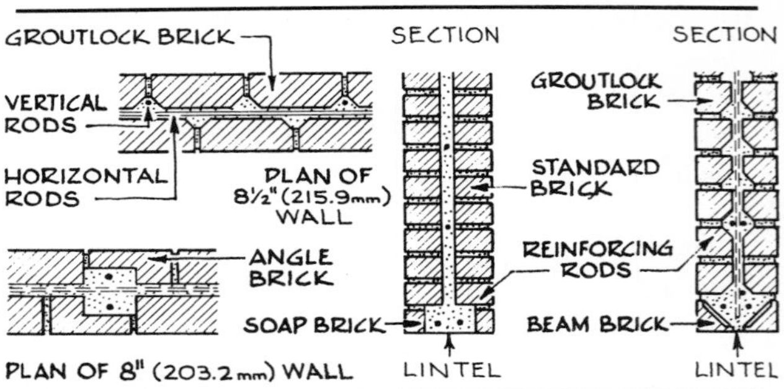

Figure B8 Details of reinforced brick walls and lintels.

The development of high-bond mortar, which develops not only compressive strength but also tensile strength (*see* Mortars), has led to the use of large prefabricated and preassembled brick units $3\frac{5}{8}$ in. (92.09 mm) thick which can be transported to the construction site, lifted by crane, and attached as exterior curtain walls to the building structure.

Most companies now make the brick shapes necessary for reinforced brickwork; they come in standard and modular sizes (*see* Figure *B9*).

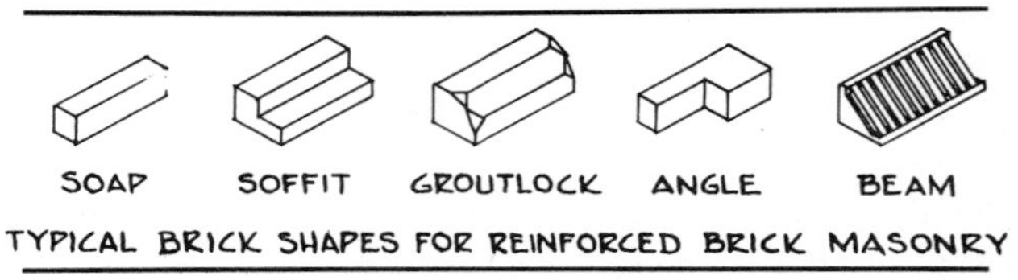

Figure B9 Typical shapes for reinforced brick masonry.

Condensed Checklist

1. The type of brick, its color, texture, and composition should be checked, as well as the necessity for any special brick shapes.
2. Height, length, and thickness limitations should be checked against local, state, and federal codes.
3. Dimensions of both horizontal and vertical brick coursing should always be calculated when designing with brick, either standard or modular. Figures *B10* and *B11* show these vertical and horizontal dimensions.
4. The locations of electrical conduits and outlets, pipes, etc., must be checked, and allowances made for them.
5. The wetness of the brick when laid and items 6, 7, and 8 should be as discussed previously.
6. Type of mortar and mortar joints.
7. Bonding.
8. Efflorescence and cleaning.
9. In the United States all major building codes have requirements related to masonry construction dur-

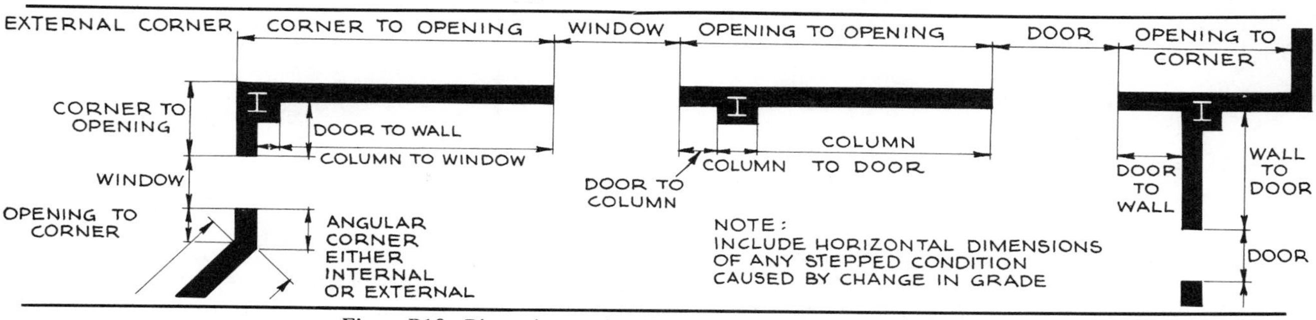

Figure B10 Dimensions to be calculated for horizontal brick coursing.

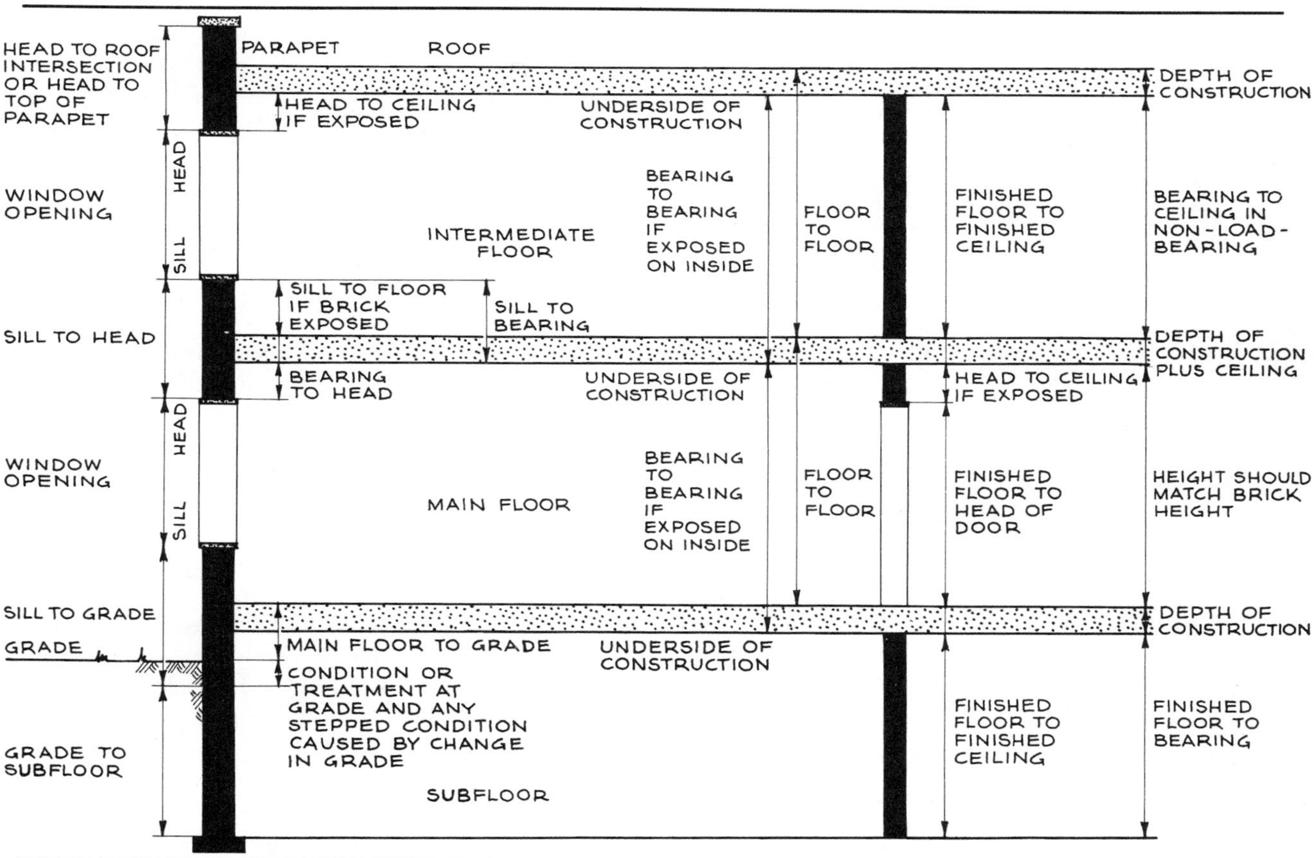

Figure B11 Dimensions to be calculated for vertical brick coursing.

ing cold weather. Codes in the area where construction is to be done should be checked to ensure correct procedures during cold-weather construction.

10. Corners (*see* Figure *B12*).
11. Conditions at grade and roof (*see* Figure *B13*). This illustration shows typical treatment of various types of masonry walls at grade and roof.
12. In any openings in brick masonry the points always to be considered are the through-wall flashing at head, pan flashing at sill, and dampproofing at jamb (*see* Figure *B14*).
13. In all expansion joints a complete opening must be made so that the two parts of the wall are definitely isolated (*see* Figure *B15*).
14. Brickwork should always be protected from adverse weather conditions at the end of each day's work.

Conditions Favorable to the Use of Brick

1. For load-bearing designs, where compressive strength is important.

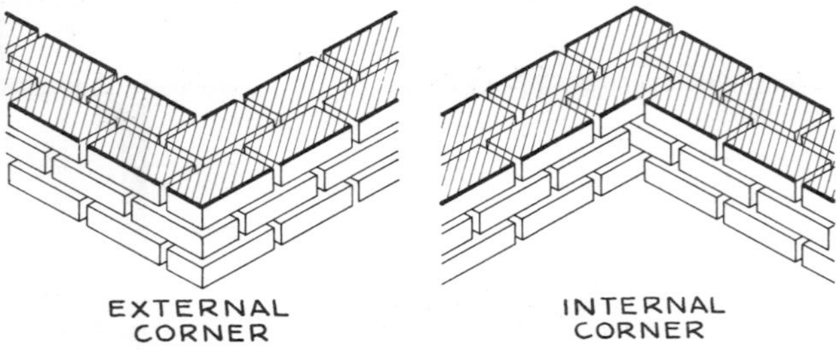

Figure B12 Treatment of corners in brickwork.

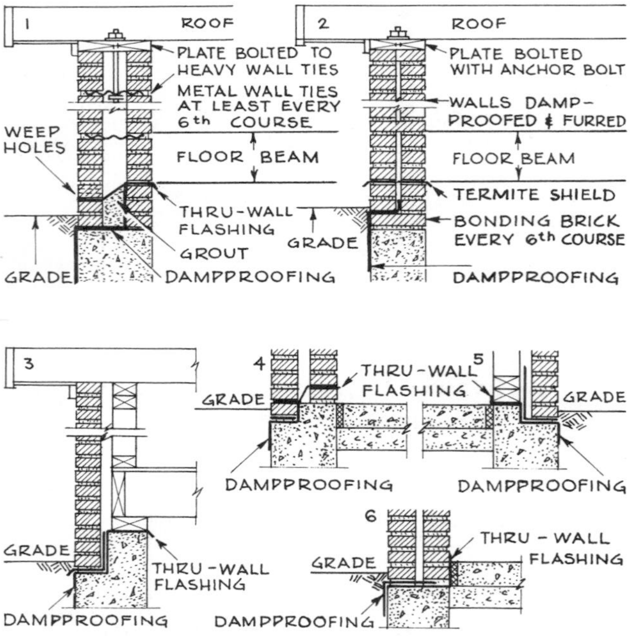

Figure B13 Treatment of conditions at grade and roof.

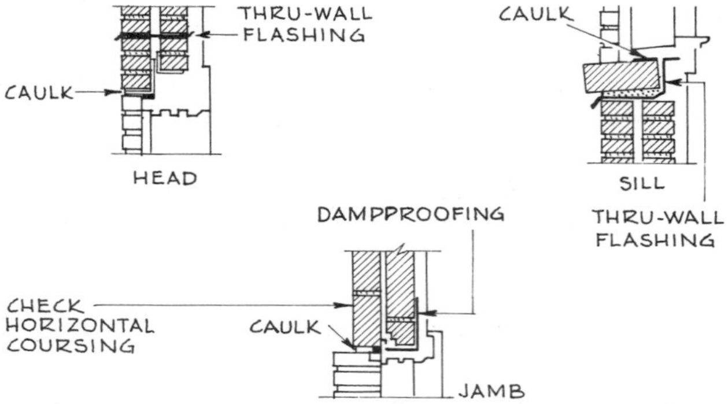

Figure B14 Treatment of conditions at openings.

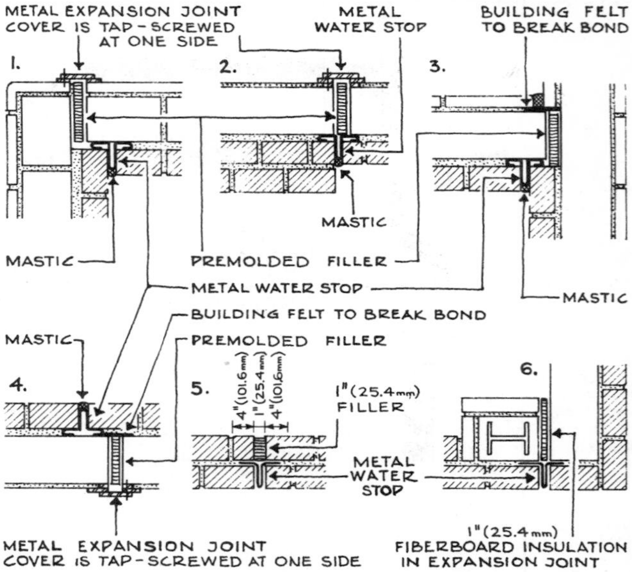

Figure B15 Details of expansion joints.

2. Where maintenance must be kept to a minimum and the surface is likely to receive rough usage.
3. Where fire resistance is required (*see* Table *B30*). Local and state codes should be checked and also the fire rating codes of the Fire Underwriters, insurance companies, labor departments, and federal government (Army, Navy, etc.).
4. Where the color and texture of brick are desirable design factors.
5. Where low sound transmission and sound control are important. *See* Table *B31*; *see also* Acoustics and Acoustical Materials; Vibration Control.
6. Where energy conservation is important. *See* Table *B31*, Figures *B16* and *B17*, *and* Insulation and Insulating Materials.

Conditions Unfavorable to the Use of Brick

1. Where severe tensile or lateral strength cannot be obtained with reinforced brick construction or with high-bond mortar.
2. Where a severe water condition exists below grade and a severe hydrostatic force is present.
3. Where very small radii are important or necessary.

History and Manufacture

Brick is one of the oldest manufactured building materials, made perhaps as early as 12,000 years ago. The Egyptians and Babylonians made a brick, and later, at the beginning of the Christian era, the Romans and

Table B30 Fire resistance of brick masonry walls

Fire rating	Wall thickness in.	Wall thickness mm	Material	Construction
4-hour	8	203.2	Brick, 75% solid	Noncombustible or no members framed in
	8	203.2	Brick and load-bearing tile	Noncombustible or no members framed in, plaster both sides
	10	254.0	Brick cavity, 75% solid	Noncombustible or no members framed in
	12	304.8	Brick, 75% solid	Combustible members framed in
	12	304.8	Brick and load-bearing tile	Noncombustible or no members framed in
	8	203.2	Hollow brick, 71% solid	Noncombustible or no members framed in, plaster both sides
	8	203.2	Hollow brick, 60% solid	Noncombustible or no members framed in, cells filled in with perlite or similar loose-fill insulation
3-hour	8	203.2	Solid brick, 75% solid	Combustible members framed in, plaster both sides
	8	203.2	Hollow brick, 71% solid	Noncombustible members or no members framed in
	8	203.2	Hollow brick, 71% solid	Combustible members framed in, plaster both sides
	8	203.2	Brick and load-bearing tile	Noncombustible members or no members framed in
	8	203.2	Brick cavity, 75% solid	Noncombustible members or no members framed in
	16	406.4	Brick and load-bearing tile	Combustible members framed in
2-hour	4	101.6	Solid brick, 75% solid	Noncombustible members or no members framed in
	6	152.4	SCR brick, 75% solid	Noncombustible members or no members framed in
	8	203.2	Solid brick, 75% solid	Combustible members framed in
	8	203.2	Hollow brick, 72% solid	Combustible members framed in
	10	254.0	Brick cavity, 75% solid	Combustible members framed in; wall filled solidly at combustible members
	12	304.8	Brick and load-bearing tile	Combustible members framed in
1-hour	4	101.6	Solid brick, 75% solid	Noncombustible members or no members framed in
	8	203.2	Brick and load-bearing tile	Combustible members framed in

Chinese made and used a brick that seems almost modern. After the fall of the Roman empire, the art was lost in Europe until the middle of the 13th century. The manufacture of brick in the United States began about the middle of the 17th century, and the 19th century saw the beginning of mechanical production.

Three main processes of brick manufacture are used in the United States (*see* Figure *B18*). In the first, the stiff-mud process, the clay contains just sufficient moisture (12 to 15%) and plasticity to be extruded through a die. In the second, the soft-mud process, the clay is too wet (20 to 30% moisture) to be forced through a die and hence must be molded. In the third, the dry-press process, the clay is in a nearly dry state (7 to 10% moisture) and is molded into shape under high pressure. A fourth process, used to only a limited degree, is the stiff-plastic one, which is intermediate between the stiff-mud and the dry-press processes.

Table B31 Resistance factors and *U* factors of brick masonry walls

Type of wall	*U* Factor	$W/m^2 \cdot °C$	*R* Factor	Reciprocal of $W/m^2 \cdot °C$
Solid walls				
6-in. (152.4-mm) wall[a]	0.68	3.861	1.47	0.259
8-in. (203.2-mm) wall	0.48	2.725	2.08	0.367
4 in. (101.6 mm) facing brick				
4 in. (101.6 mm) building brick				
12-in. (304.8-mm) wall	0.35	1.987	2.86	0.503
4 in. (101.4 mm) facing brick				
8 in. (203.2 mm) building brick				
8-in. (203.2-mm) building brick wall	0.41	2.328	2.44	0.429
8-in. (203.2-mm) wall	0.41	2.328	2.44	0.429
4 in. (101.6 mm) facing brick				
4 in. (101.6 mm) load-bearing tile				
10-in. (254.0-mm) wall	0.35	1.987	2.86	0.503
4 in. (101.6 mm) facing brick				
6 in. (152.4 mm) load-bearing tile				
12-in. (304.8-mm) wall	0.31	1.760	3.23	0.568
4 in. (101.6 mm) facing brick				
8 in. (203.2 mm) load-bearing tile				
Cavity walls[b]				
10-in. (254.0-mm) wall	0.33	1.874	3.03	0.533
4 in. (101.6 mm) facing brick				
4 in. (101.6 mm) building brick				
10-in. (254.0-mm) wall	0.30	1.703	3.33	0.587
4 in. (101.6 mm) facing brick				
4 in. (101.6 mm) load-bearing tile				
10-in. (254.0 mm) wall	0.27	1.533	3.70	0.652
4 in. (101.6 mm) building brick				
4 in. (101.6 mm) load-bearing tile				
10-in. (254.0-mm) wall	0.25	1.430	4.00	0.699
4 in. (101.6 mm) load-bearing tile				
4 in. (101.6 mm) load-bearing tile				

[a]Based on high-density facing brick.
[b]Nominal dimensions: 4 in. (101.6 mm) for inner and outer wythes and a 2-in. (50.8-mm) cavity.

Table B32 Strength and absorption requirements

	Compressive strength[a] (minimum) of concrete brick tested flatwise Average gross area				Water absorption (maximum) based on average of three bricks with oven-dry weight of concrete					
	Average of three concrete bricks		Individual concrete brick		Light weight[b]		Medium weight[b]		Normal weight[b]	
Grade	$lbf/in.^2$	MN/m^2	$lbf/in.^2$	MN/m^2	lb/ft	kg/m	lb/ft	kg/m	lb/ft	kg/m
N-I N-II	3500	24.13	3000	20.69	15	240.30	13	208.26	10	160.20
S-I S-II	2500	17.24	2000	13.79	18	288.32	15	240.30	13	208.26

[a]Metric equivalent: 1 $lbf/in.^2 = 0.006895\ MN/m^2$.
[b]Light weight is less than 105 lb/ft^3 (1682 kg/m^3); medium weight, less than 125 to 105 lb/ft^3 (2003 to 1682 kg/m^3); normal weight, 125 lb/ft^3 (2003 kg/m^3) or more.

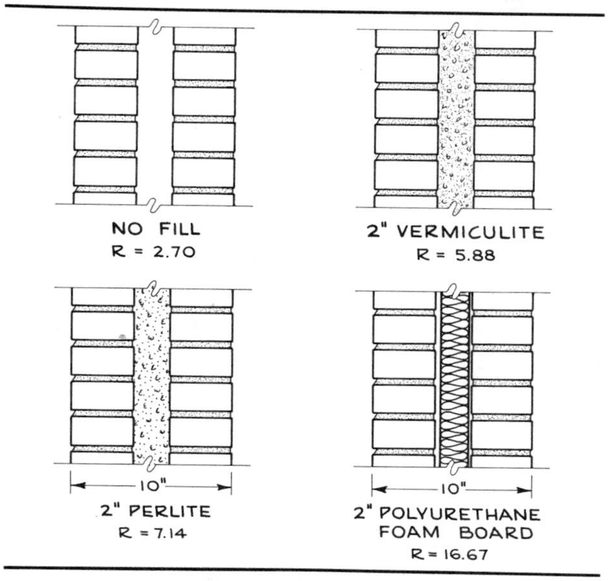

Figure B16 **The difference in "*R*" factors for cavity walls without insulation and cavity walls with various types of insulating materials.**

BRICK, CEMENT

Physical and Chemical Properties

Cement (or concrete) brick is made from a controlled mixture of portland cement and suitable aggregates such as sand, crushed stone, gravel, cinders, burned clay or shale, or blast-furnace slag. Its strength requirements are shown in Table *B32*.

Various colors can be produced in cement brick by the introduction of mineral oxide pigments.

Types and Uses

There are two types of cement brick:

Grade N: for use as veneer and facing units for exterior walls and in applications where high strength and resistance to severe frost action and moisture penetration are required.

Grade S: for general use where moderate strength and resistance to moisture penetration are required.

These grades are subdivided into two types: Type I, which is moisture-controlled, and Type II, which is non-moisture-controlled. Grading is indicated thus: N-I, N-II, and S-I, S-II. *See* Table *B33*.

Cement brick is available in the same stock sizes as burned brick.

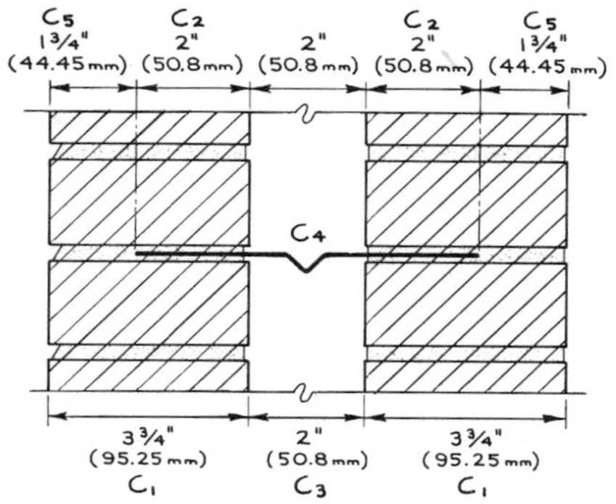

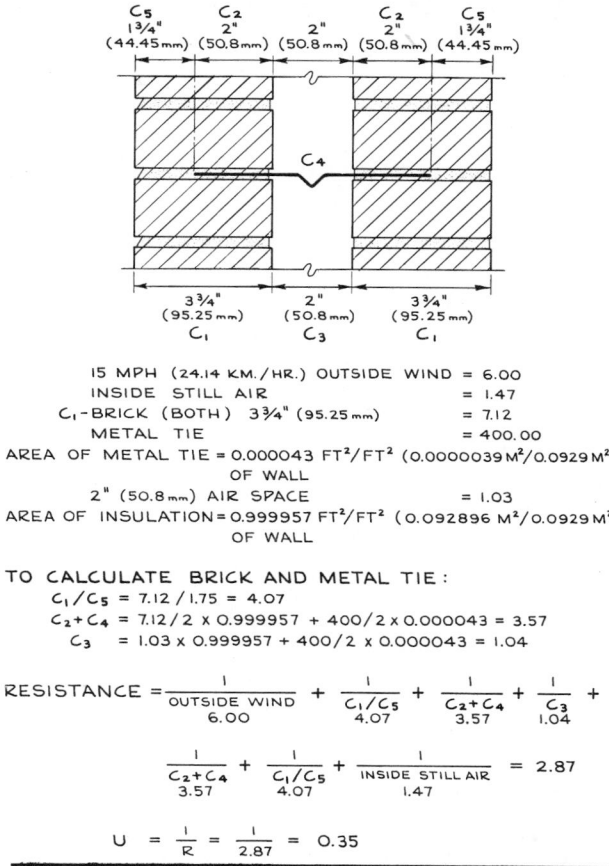

15 MPH (24.14 KM./HR.) OUTSIDE WIND = 6.00
INSIDE STILL AIR = 1.47
C_1-BRICK (BOTH) 3¾" (95.25 mm) = 7.12
METAL TIE = 400.00
AREA OF METAL TIE = 0.000043 FT²/FT² (0.0000039 M²/0.0929 M²) OF WALL
2" (50.8 mm) AIR SPACE = 1.03
AREA OF INSULATION = 0.999957 FT²/FT² (0.092896 M²/0.0929 M²) OF WALL

TO CALCULATE BRICK AND METAL TIE:

$C_1/C_5 = 7.12 / 1.75 = 4.07$

$C_2 + C_4 = 7.12/2 \times 0.999957 + 400/2 \times 0.000043 = 3.57$

$C_3 = 1.03 \times 0.999957 + 400/2 \times 0.000043 = 1.04$

$$\text{RESISTANCE} = \frac{1}{\frac{\text{OUTSIDE WIND}}{6.00}} + \frac{1}{\frac{C_1/C_5}{4.07}} + \frac{1}{\frac{C_2+C_4}{3.57}} + \frac{1}{\frac{C_3}{1.04}} + \frac{1}{\frac{C_2+C_4}{3.57}} + \frac{1}{\frac{C_1/C_5}{4.07}} + \frac{1}{\frac{\text{INSIDE STILL AIR}}{1.47}} = 2.87$$

$$U = \frac{1}{R} = \frac{1}{2.87} = 0.35$$

Figure B17 **Method of calculating the "R" factor for brick cavity wall and cavity wall ties.**

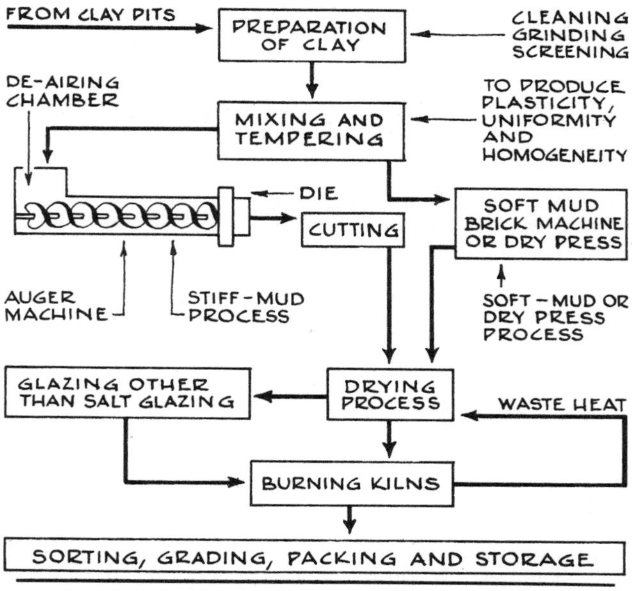

Figure B18 **Flowchart of brick manufacture.**

Table B33 Moisture content requirements for type I concrete brick

Grade and type of concrete brick	Linear shrinkage (percent)	Moisture content (maximum) based on percent of total absorption (average of three concrete bricks) — Conditions[a] at job site or point of use: Humid	Intermediate	Arid
N-I	0.03 or less	45	40	35
	Over 0.03 to 0.045	40	35	30
S-I	0.045 to 0.065 max.	35	30	25

[a]Humid: average annual relative humidity above 75%; intermediate: average annual relative humidity 50 to 75%; arid: average annual relative humidity less than 50%.

Application

Condensed Checklist

1. Dimensions of both horizontal and vertical brick coursing.
2. Type of mortar and mortar joints.
3. Wetness of the brick when laid.
4. Temperature during laying. Freezing temperatures should be avoided if possible.
5. Bonding: type, when, and how.
6. Corners.
7. Conditions at grade, at roof, and at openings.
8. Expansion joints.
9. Waterproof covering of brickwork at the end of each day's work.
10. Wall height, length, and thickness limitations. These should be checked against local, state, and federal codes.
11. Methods of dampproofing for cement brick walls.
12. Locations of electrical conduits, outlets, pipes, etc., must be checked, and allowances made for them.

Most of these problems are discussed in considerable detail under Brick, Burned.

Conditions Favorable to the Use of Cement Brick

1. Where compressive strength is important.
2. Where fire resistance is required. State and local codes should be checked and also the codes of the Fire Underwriters, insurance companies, and federal government (Army, Navy, etc.).
3. Where the color and texture of cement brick are desirable design factors.

Conditions Unfavorable to the Use of Cement Brick

1. Where tensile and lateral strength are important (unless reinforced cement brick masonry is used).
2. Where a watertight condition is required below grade.
3. Where small radii are important or necessary.

History and Manufacture

The history of cement brick is roughly parallel to that of sand-lime brick. The brick is manufactured by either the wet (steam curing) or the dry (air curing) process.

BRICK, SAND-LIME

Physical and Chemical Properties

Sand-lime brick is composed of a sand and lime mixture instead of the more usual clay. Its compressive strength averages between 2500 and 3000 lbf/in.^{2} (17.23 and 20.68 MN/m^2), and its modulus of rupture is about 450 lbf/in.^{2} (3.10 MN/m^2).

Sand-lime brick is pearl-gray in color, although other colors can be produced by introducing mineral oxides. It hardens with age, and it has good frost and acid resistance and better fire resistance than burned brick (common building brick). Its absorption rate is between 7 and 10% over 48 hours. Sand-lime brickwork is easily washed and shows no efflorescence.

Types and Uses

The grading for sand-lime brick is the same as that for burned brick, but a check should always be made with the manufacturer concerned in order to determine whether his brick meets physical requirements.

A good test for sand-lime brick is to set a brick in water for 1 hour. If moisture rises over $\frac{1}{2}$ in. (12.7 mm), the brick is not first class; if it rises 2 in. (50.8 mm), the brick should not be used as a facing brick; if it rises 3 in. (76.2 mm), the brick is suitable for back-up and interior work only.

Application

Calculating Wall Strength. The strength of sand-lime brick masonry walls may be calculated on the basis of type of workmanship in the same way as for burned brick masonry walls.

Recommended Mortars. The mortars recommended for sand-lime brick masonry are types N and S, as described under Brick, Burned.

Condensed Checklist

1. Dimensions of both horizontal and vertical brick coursing.
2. Type of mortar and mortar joints.
3. Wetness of the brick when laid.
4. Temperature during laying: Freezing temperatures should be avoided if possible.
5. Bonding: type, when, and how.
6. Treatment of corners.
7. Conditions at grade, at roof, and at openings.
8. Expansion joints.
9. Waterproof covering of brickwork at the end of each day's work.
10. Wall height, length, and thickness limitations: These should be checked with local, state, and federal codes.
11. Locations of electrical conduits, outlets, pipes, etc., should be checked, and allowances made for them.
12. Methods of dampproofing sand-lime brick walls should be checked.

For a detailed discussion, *see* Brick, Burned.

Conditions Favorable to the Use of Sand-Lime Brick

1. Where compressive strength is considered important.
2. Where fire resistance is required. State and local codes should be checked and also the fire rating codes of the Fire Underwriters, insurance companies, labor departments, and federal government (Army, Navy, etc.).
3. Where the color and texture of sand-lime brick are desirable design factors.

Conditions Unfavorable to the Use of Sand-Lime Brick

1. Where tensile or lateral strength is important (use reinforced brick masonry instead).
2. Where a below-grade watertight condition is required in construction.
3. Where small radii are important or required.

History and Manufacture

Ancient sand-lime brick was made of lime mortar molded into brick form and hardened by exposure to air. In 1875 in Germany it was discovered that by use of heat and pressure the brick could be made in a few hours. It was first produced in the United States in about 1900.

Sand-lime brick is manufactured from a mixture of 5 to 10% hydrated lime and pure silica sand. It is made by the semidry method in presses and hardened in autoclaves with live steam for 24 hours. These bricks can then be cut, carved, or sandblasted to the desired finish.

BROMINE

Physical and Chemical Properties

Symbol: Br
Atomic number: 35
Specific gravity: 3.12 (20°C, 68°F) liquid
Melting point: −7.2°C, −44.64°F
Boiling point: 58.78°C, 163.4°F

Bromine is a deep red liquid at ordinary temperatures. Its chemical properties are similar to those of iodine and chlorine.

Types and Uses

Elemental bromine has no construction uses. Its only applications with any relation to construction are in compound form as a fire extinguisher fluid and as a desiccant in air conditioning. Silver bromide, a light-sensitive compound, is an important constituent of photographic film.

History and Manufacture

Bromine was isolated by Balard, a French chemist, in 1826. Its name is derived from the Greek word meaning "stench." Commercial quantities of bromine are extracted from seawater, well brines, and salt bitterns.

BRONZE

Physical and Chemical Properties

True bronze, historically speaking, is an alloy of copper and tin which varies only slightly from a 90% copper-

Table B34 General composition of bronzes and bronze-type alloys

Alloy number	Type	Cu	Sn	Zn	Al	Si	P	Pe	Fe	Co	Ni	Mn	Others
							(Percent of content)						
613	Aluminum bronze	92.7	0.5[a]		8.0[a]				3.5[a]		0.5[a]	0.5[a]	
614	Aluminum bronze	91.0		0.2[a]	8.0[a]		0.015[a]	0.01[a]	3.5[a]			1.0[a]	
638	Aluminum bronze, silicon-cobalt	95.0		0.5[a]	3.1[a]	2.1[a]		0.05[a]	0.05[a]	0.55[a]	0.10[a]	0.10[a]	
642	Aluminum bronze, silicon	91.2	0.2[a]	0.5[a]	7.6[a]	2.2[a]		0.05[a]	0.3[a]		0.25[a]	0.10[a]	0.15[a] As
	Aluminum bronze powder						Actually aluminum powder						
385	Architectural bronze						A leaded-type brass						
	Bell metal	78.0	22.0										
	Bronze powder						A brass of the Alpha type						
220	Commercial bronze						A brass of the Alpha type						
655	High-silicon bronze	97.0		1.5[a]		3.8[a]		0.05[a]	0.80[a]		0.6[a]	1.5[a]	
226	Jewelry bronze						A brass						
651	Low-silicon bronze	98.5		1.5[a]		2.0[a]		0.05[a]	0.80[a]			0.7[a]	
280	Muntz metal						A brass of the Beta type						
510	Phosphor bronze	94.8	5.8[a]	0.3[a]			0.35[a]	0.05[a]	0.10[a]				
511	Phosphor bronze	95.6	4.9[a]	0.3[a]			0.35[a]	0.05[a]	0.10[a]				
521	Phosphor bronze	92.0	9.0[a]	0.2[a]			0.35[a]	0.05[a]	0.10[a]				
	Statuary bronze	90.0	10.0										

[a]Maximum percent.

10% tin composition. This bronze is a rich golden-brown metal, originally worked by forging and particularly suited for casting since it is corrosion resistant, dense, and hard enough to take an impression of a mold of any delicacy whatsoever.

It is important in the construction field to realize at once that the term "bronze" is no longer used in this limited sense. In commercial practice the terms "brass" and "bronze" may be used without much regard for their original meanings. The term "bronze" now usually has a prefix and indicates alloys of copper with silicon, manganese, aluminum, and other elements with or without zinc, for example, silicon bronze. A few brasses are known as bronzes because they have the characteristic bronze color, for example, commercial bronze. (*See* Table *B34*.)

Chemical Reactivity. When bronze is used in direct contact with any other metal, the same precautions hold as for brass. Bronze should not come into direct contact with iron, steel, stainless steel, aluminum, zinc, or magnesium if conditions are such as to promote galvanic action.

Workability. Bronze-type alloys can be rolled and extruded as well as forged and cast.

Types and Uses

Of the so-called bronzes a few are true bronzes. These are statuary bronze, bell metal, and the phosphor bronzes. Many of the copper-based alloys known as bronzes are used for both wrought and cast forms. The wrought alloys are copper-tin-phosphorus, copper-tin-lead-phosphorus, copper-aluminum, and copper-silicon alloys. The cast alloys are copper-tin, copper-tin-lead, copper-tin-nickel, and copper-aluminum alloys. As for the others, architectural bronze is really a leaded brass, and commercial bronze is one of the more commonly used brasses (90% copper and 10% zinc).

Statuary Bronze. When statuary bronze is used for statues, architectural decoration, and ornamental metalwork, careful attention should be paid to color, alloying ingredients, and type of casting. Statuary bronze should always contain tin. Color can vary from a reddish metallic color in bronze containing over 90% copper to an orange-yellow in bronze containing less than 90% copper. The strongest alloy with the best working characteristics contains 10% tin.

When choosing the type of extrusion, forging, or casting for use in construction, three major items must be taken into consideration: cost, accuracy of reproduction, and quantity of reproductions to be made.

Hardware and Weatherstripping. Rough hardware accessories of many kinds are made of wrought or cast bronze. Among these are screws, nuts and bolts, washers, anchors, ties, and the like for attaching copper, brass, or bronze metals and for various types of masonry work. Bronze is also used for weatherstripping as it is tough, strong, and corrosion resistant, and has spring qualities with good fatigue resistance. Gauge equivalents for sheet and strip are given in Table *G3* (*see* Gauges).

History and Manufacture

The history of man evolves in a cycle of development from stone to copper to bronze to iron, with the third of these stages called the Bronze Age. The oldest piece of bronze discovered is from Egypt and dates back to about 3700 B.C. From Egypt the use of bronze spread rapidly at about the following dates: to Crete in 3000 B.C., to Sicily in 2500 B.C., to Europe in 2000 B.C., and to Britain in 1800 B.C. In China the working of bronze had developed into a fine art as early as the Shang Dynasty, about 1765 B.C. In America bronze was developed and used by the early Maya and Inca about the beginning of the Christian era. The history of the early cast bronzes has come down to us in paintings and written descriptions. An account of the casting of the doors for the temple of Karnak in Egypt is an example of such writings. The ancient classical bronze consisted of 67 to 95% copper with tin and some zinc, lead, or silver.

As an architectural material bronze has a long history. Bronze tiles were relatively common on the most monumental buildings of the Roman Empire. At present, however, true bronze is rarely used because of cost and the heavy commercial demand for tin for other purposes. The production of "bronze" today consists of mixing the copper with various other metals to the desired proportions.

BTU (BRITISH THERMAL UNIT)

The quantity of heat required to raise the temperature of 1 lb (0.4536 kg) of water 1°F (0.5556°C) at sea level. The Btu is equivalent to 1.055 kilojoules or 1055 joules. *See also C* Factor; *U* Factor; *K* Factor; *R* Factor; Insulation and Insulating Materials.

BULKHEAD

In general construction and engineering, a bulkhead is a structure built to prevent earth from sliding into an excavation, or else a wall to shut off water, pressure, fire, etc.

In architectural construction the term includes outside cellar entrances and areaways and also any structure above floor or on a roof used as a means of covering the head of a stair, elevator shaft, or the like. There exist stock parts or complete units for building such structures, usually made of metal or of wood covered and combined with metal. (*See* Figure *B19*.)

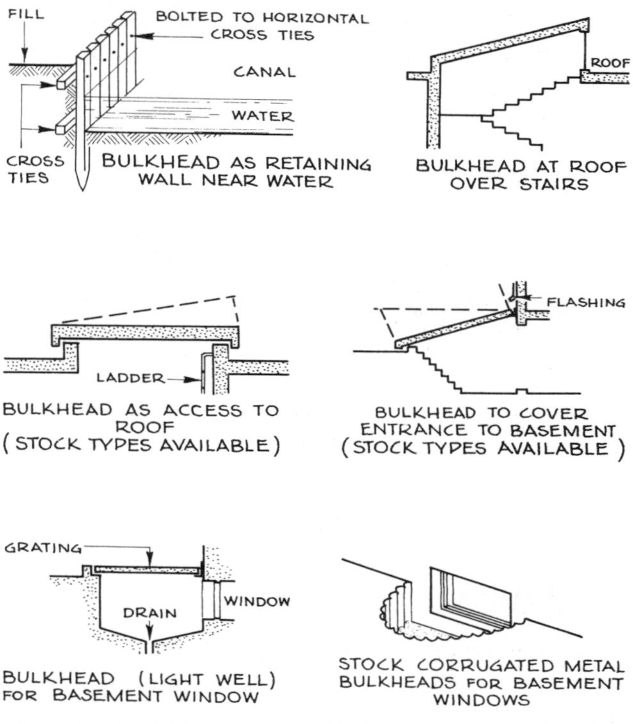

Figure B19 Types of bulkheads.

C

CABLE STRUCTURES

The name itself describes the principle of cable structure—cables suspended between supporting elements (*see* Figure *C1*). The suspension bridge was the first type of cable structure. It was not until the various synthetic materials such as reinforced films, sheet, textiles, and fabrics were developed that cable structure became a practical reality in building construction. Cable construction design is generally used to cover very large areas and can be combined with the air structure design principle (*see* Air Structures; Tent Structures; Plastics; Wire and Wire Rope). It is not within the scope of this book to cover the engineering technology involved in the design of a cable structure.

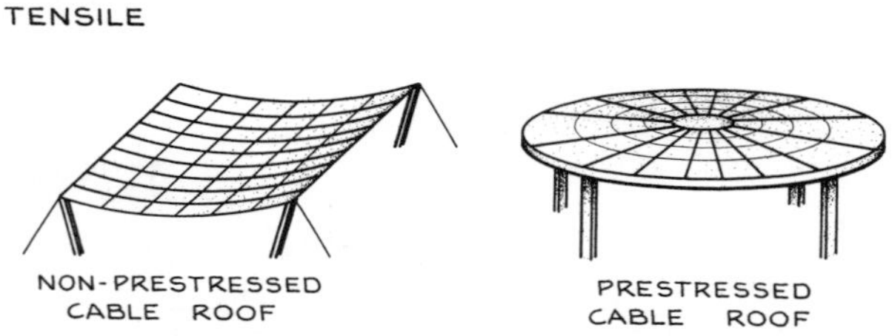

Figure C1 Typical cable structures.

CADMIUM

Physical and Chemical Properties

Symbol: Cd
Atomic number: 48
Specific gravity: 8.65 (20°C, 68°F)
Melting point: 320.9°C, 609.62°F
Boiling point: 765°C, 1409°F

Cadmium is a soft, blue-white, ductile, malleable metal, similar to zinc in many respects. Cadmium is not attacked by dry air; in moist air a superficial film forms which protects the metal from further corrosion. It is attacked by some acids but not by alkalis. Cadmium compounds, solutions, and fumes are highly poisonous.

Workability. Cadmium can be rolled, hammered, drawn, soldered, and cast. When bent, it gives the same characteristic cry as tin.

Commercial Forms. Cadmium is available in cast sheet, slab, stick, and ball form. Electrolytic and commercial cadmium are both 99.9% pure.

Types and Uses

Cadmium compounds have many uses, particularly as color pigments; the uses of cadmium in metallic form are indicated by asterisks in Table *C1*.

Mercury and cadmium form an amalgam which is used as a metallic cement for filling holes in metal when the use of heat is not desired.

Cadmium Coatings. Cadmium coatings on iron, steel, and copper alloy base metals have a better corrosion resistance than zinc. They are usually electrodeposited. A cadmium plating of 0.0003 in. (0.00762 mm) is equal to a zinc coating of 0.001 in. (0.0254 mm), is silvery-white, and is harder than tin. It is easily soldered, retains its luster, and is ductile. In fact, electroplating represents a major use of this metal. One reason for this is the compatibility of cadmium and aluminum with regard to galvanic action.

History and Manufacture

Cadmium was discovered in 1917 by Stromeyer from a zinc impurity. Almost all cadmium is obtained as a by-product in the smelting and refining of zinc, copper, and lead ores.

Table C1 Uses of cadmium

Construction Uses		
Basis of	Color pigments	Red and yellow colors
Component of	Copper*	Corrosion-resistant coatings (hard, silvery-white)
	Fusible alloys*	Lowering melting temperatures
	Glass	Red and yellow colors; special mirrors
	Glazes and porcelain enamels	Iridescence; red and yellow colors
	Iron*	Corrosion-resistant coatings (hard, silvery-white)
	Magnesium	Improving casting and extruding properties
	Mercury	Heat-resistant and alkali-fast color pigments
	Paint	White paint solid; red and yellow colors; fluorescence
	Paper	Red and yellow colors
Construction Uses (Continued)		
Component of	Plastics	Stabilizing against light and heat
	Coatings* and protective coatings*	Plating
	Rubber	Aiding abrasion resistance; red and yellow colors
	Steel*	Corrosion-resistant coatings (hard, silvery-white)
	Textiles	Dyeing; printing; red and yellow colors
Allied Construction Uses	Lamps*	Incandescent filaments; fluorescence
	Solders*	Lowering melting temperature; aluminum solder; substitute for tin
Nonconstruction Uses		
Lithography, metallurgy,* photography, X-ray screens*		
Possible Future Uses		
Solar generator for heat and energy		

CAISSON

In general construction, caissons are boxlike structures similar to pontoons, used where water-bearing or soft soils are encountered. They provide a method of constructing foundations below the water level. Caissons are usually made of timber, steel, or concrete, alone or in combination, and become an integral part of the permanent structure.

CALCIUM

Physical and Chemical Properties

Symbol: Ca
Atomic number: 20
Specific gravity: 1.55 (20°C, 68°F)
Melting point: 842°C, 1547.6°F
Boiling point: 1487°C, 2708°F

Calcium is a moderately soft, white metal which is the fifth most abundant element in the earth's crust, found widely distributed in compound and mineral form. Calcium tarnishes in air, and this tarnish protects the metal from further attack.

Workability. Calcium can be machined, turned, drilled, extruded, drawn, pressed, hammered, and cast by special methods.

Commercial Forms. Calcium can be obtained 99.6% pure in lump, chip, slug, rod, and stick form.

Types and Uses

Many calcium compounds are important ingredients in materials used in construction, namely, in cement, lime, paint, glass, ceramics, and accelerator-type admixtures. The uses of calcium in metallic form are indicated by asterisks in Table *C2*.

A salt of molybdenum, calcium molybdate, is used in place of ferromolybdenum to add molybdenum to steel. Alloys of calcium and silicon with small percentages of iron, aluminum, and manganese are used as deoxidizing agents and desulfurizers in the production of high-grade steels and cast irons.

Radioactive calcium 45 is used in water purification, surface wetting, and other surface phenomena, and in the diffusion of calcium in glass.

Table C2 Uses of calcium

	Construction Uses	
Basis of	Abrasives	Main ingredient of polishing powders
	Adhesives	Main ingredient
	Admixtures and additives	Accelerator for cement and concrete
	Chalk	Main ingredient
	Color pigment	Yellow
	Lime	Main ingredient
	Paint	Whitewash; calcimine; white paint solid
	Putty	Main ingredient
	Whiting	Main ingredient
	Aluminum*	Alloying and modifying agent
Component of	Asbestos	Ingredient
	Cast iron*	Deoxidizer; desulfurizer; decarbonizer; hardener
	Cement	Ingredient
	Ceramics	Ingredient
	Chromium*	Reducing agent; decarbonizer; desulfurizer; deoxidizer
	Coatings and protective coatings	Waterproofing
	Copper*	Alloying agent, deoxidizer
	Diatomite	Ingredient
	Feldspar	Ingredient
	Fluospar	Ingredient
	Glass	Ingredient
	Glazes and porcelain enamels	Ingredient
	Construction Uses (Continued)	
Component of	Iron*	Deoxidizer; desulfurizer; decarbonizer; hardener
	Lead*	Debismuthizer; hardener
	Magnesium*	Modifying agent; fire retarding
	Nickel*	Decarbonizer; desulfurizer; deoxidizer
	Paint	Filler; extender; fireproof driers; emulsifier; stabilizer; luminosity
	Paper	Dissolving lignin in pulp; preservative, filler; coatings; flameproofing; sizing
	Plastics	Stabilizer
	Rubber	Filler; hardener
	Steel*	Deoxidizer; desulfurizer; decarbonizer; hardener; flux
	Textiles	Mordant; flameproofing
	Tin*	Reducing agent; decarbonizer; desulfurizer; deoxidizer
	Uranium*	Reducing agent
	Wood	Filler; preservative
	Zirconium*	Reducing agent
Allied Construction Uses	Lamps	Phosphor for tubes
	Water	Softener
	Nonconstruction Uses	
	Road treatment, photography, X-ray photography	

History and Manufacture

Calcium metal is obtained by three methods: (1) electrolysis of fused calcium chloride; (2) the use of aluminum as a reducing agent with lime; and (3) recovery from the sludge of the electrolytic production of sodium metal from fused salt.

CANVAS

Canvas is a heavy, coarse, tightly woven fabric originally made of cotton, hemp, or flax and used for tents and sails. Now synthetic fibers are woven into the same kind of cloth, which also has applications in construction for roofing, decks, and terraces. Figure *C2* shows details for canvas roofing. *See also* Textiles.

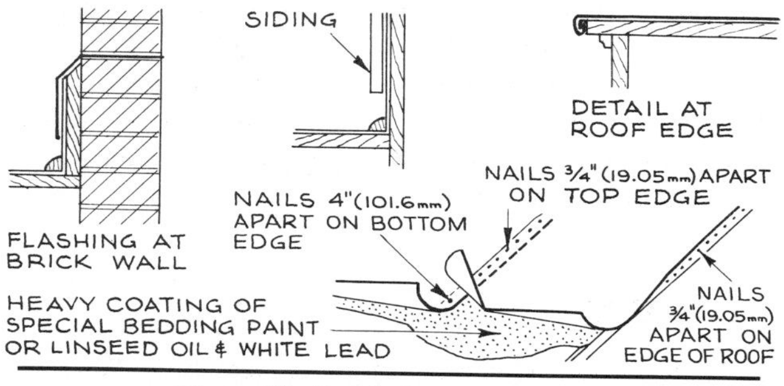

Figure C2 Details for canvas roofing.

CARBON

Physical and Chemical Properties

Symbol: C
Atomic number: 6
Specific gravity: 3.513 (25°C, 77°F) gem diamond
Melting point: 3550°C, 6422°F
Boiling point: 4827°C, 8720.6°F

Carbon is a nonmetallic element which exists in two crystalline allotropic forms, diamond and graphite, and in numerous forms of varying purity such as charcoal, coke, carbon black, lampblack, and activated carbon, all of which can be grouped as black or amorphous carbon.

Carbon is stable at ordinary temperatures and insoluble in common solvents. It is soluble in some molten metals (e.g., iron) from which it crystallizes as graphite; with others (e.g., silicon, tungsten) it forms carbides, many of which are characterized by extreme hardness.

A characteristic of the carbon atom is its great tendency to unite with other carbon atoms to form chains or rings. The study of these carbon chain and ring compounds, half a million of which have been identified and many more of which undoubtedly are possible, comprises the vast, complex field of organic chemistry.

Types and Uses

Since the uses of carbon involve the entire field of organic chemistry, a comprehensive chart is impossible. Table *C3* covers only those few uses of carbon in elemental form where carbon is the main ingredient of a material used in architecture or has a major effect on the characteristics of such a material.

Table C3 Uses of carbon

	Construction Uses	
Basis of	Abrasives	Main ingredient (diamond)
	Color pigments	Black
Component of	Cast iron	0.2% carbon and more
	Iron	Control of carbon percentages for cast and wrought iron, steel, and stainless steel
	Rubber	Filler and strengthener in synthetic and natural rubbers
	Steel	Less than 0.2% carbon
	Wrought iron	0.01 to 0.02% carbon

Diamonds. The diamond, a transparent, colorless to black crystal, is the hardest known substance, rated at 10 on the Mohs scale (*see* Abrasives). Its specific gravity is 3.52. It is a poor conductor of heat and electricity and is not affected by even the strongest oxidizing agents.

Graphite. Graphite, considered the most stable form of carbon, consists of shiny, gray-black, flat crystals which are soft if tested along the plane of cleavage but can be almost as hard as the diamond if tested perpendicular to the plane of cleavage. Its specific gravity is 1.8 to 2.34. Its thermal conductivity is higher than that of many metals, and it also has good electrical conductivity.

Both graphite and the amorphous carbons are attacked by strong oxidizing agents. The amorphous forms tend toward a crystalline structure and are industrially differentiated according to source and method of manufacture.

Carbon Blacks. The amorphous carbons used as color pigments are called carbon blacks and are made by thermal decomposition of natural gas, liquid hydrocarbons, or vaporizable portions of any other carbonaceous material. Carbon blacks vary in intensity of color according to particle size; the smaller the particle, the blacker the pigment.

Carbon blacks are not to be confused with animal, vegetable, or mineral blacks, which are chars. Examples of chars are coke, charcoal, bone black, and ivory black. When ground to 325-mesh size, these are also used to a limited extent as pigments.

Carbon Fibers. Carbon fibers are available as a continuous filament 0.002 in. (0.0508 mm) in diameter, with a tensile strength of 200,000 lbf/in.^{2} (1379 MN/m^2). These fibers retain their dimensional stability to temperatures as high as 5700°F (3148.89°C). The various types of carbon fibers are at present used in industry for filters, mats, special insulators, and high-temperature fabrics. In time, carbon fibers will find applications in the construction field as yarns, ropes, reinforcing materials, and other uses based on their high tensile strength and dimensional stability.

History and Manufacture

Carbon is one of the elements essential to life. It not only is necessary to plant and animal existence but also is considered by some to be the catalyst in the reaction by which the sun is thought to derive its energy, that is, through the formation of helium from hydrogen by a succession of nuclear changes. All forms of carbon are found both free in nature and in combined form. Graphite and diamonds are also artificially produced on an industrial scale.

CASTING

In the construction field, casting refers to two processes: one is the art of making molds of and casting art objects or ornaments, and the other is a method of working or fabricating metal in the foundry, called founding. Since by definition all casting requires a mold, the term "casting" can also be used to cover glass blowing and casting, molding of plastics, and powder metallurgy.

Art Casting. In casting as an art, there are three methods of obtaining reproductions of sculpture, ornaments, and decorations: (1) the *cire perdue*, or lost wax method, suitable for one reproduction; (2) gelatin molds, for several reproductions only; and (3) metal molds, or sand molds, for quantity reproductions. Usually, large-quantity reproduction in metal passes into the category of founding by virtue of the technical problems it poses and the equipment it requires. The building in which this type of casting is done is called a foundry.

Founding. In founding, the problems are (1) to find suitable materials for making the molds into which to pour the molten metal, usually at a very high temperature; and (2) to find methods of allowing the displaced air and the gases created to escape freely. Another important problem is brought about by the fact that most metals shrink upon cooling. In large castings this shrinkage is overcome by designing the castings with openings in the upper part and adding more metal to replace that lost by shrinkage. In another method, the coefficients of contraction being known, the molds are made larger to counteract the shrinkage of the final casting.

Almost all metal casting molds are made in two pieces. This always leaves on the finished product a meeting line which has to be ground and filed off by mechanical metal-finishing procedures.

CATALYST

A catalyst is a substance that by its mere presence alters the velocity of a reaction, and that may be recovered unaltered in nature or amount at the end of the reaction.

CAULKING

Caulking may be defined as the method of filling with an elastic compound all the small crevices, holes, and joints between different materials that cannot be sealed by any other means. For example, in a joint between wood and brick on the exterior of a building, no matter how closely the wood is fitted to the brick, there is always an area or crevice through which rain, moisture, wind, and dust can penetrate. To make this joint tight, a nonstaining, pliable, adherent material must be forced into the opening, thereby sealing it.

Physical and Chemical Properties

A caulking compound should have a long lifetime. In addition, and equally important, it should have elasticity and remain adherent, pliable, and watertight. It should not harden, shrink, or crack in cold weather or soften and ooze when heated by the sun. It should resist minor settling and vibration in the building without cracking.

It should be paintable and not bleed through or discolor the paint.

Constituents. Caulking materials may consist of various combinations of specially treated oils, resins, synthetic rubbers, plastics, inert pigments, volatile solvents, and other ingredients not always identified, since many compositions are sold under trademark names.

Asphalt products can be used for caulking but are usually limited to concealed areas because of their color and because they may stain or bleed through other materials. *See* Plastics; Rubber; Asphalt.

Types and Uses

Caulking compounds are often used for pointing up masonry and making expansion joints. Nonstaining types are especially important for use with marble, granite, limestone, and other stone veneers. Caulking materials are available in colors also.

Application

When cracks or spaces are very large or deep, oakum or a repp yarn should be placed in the joint to fill up the space to about $\frac{1}{4}$ to $\frac{3}{4}$ in. (6.35 to 19.05 mm) from the face of the joint before the caulking is applied.

The type of material to which caulking is to be applied should be checked to determine whether the materials in contact with the caulking will absorb too much oil out of the caulking material itself, thus causing it to dry out, shrink, crack, and lose its sealing action in the joint. Sealing with shellac or varnish prior to caulking can remedy this situation. (Other precautions and details of application are specifically listed under sections discussing the material being caulked or requiring an expansion joint.)

CEILINGS

Ceilings, in this book, refer to the finished surface materials of any overhead surfaces within a building and semienclosed spaces on the exterior of a building.

Table *C4* summarizes the various types of ceiling finish materials and serves as a cross reference to the materials from which they are made.

Table C4 Typical materials used for ceilings

Materials	Advantages	Disadvantages	Major uses
Aluminum, perforated (*see* Aluminum Sheet and Strip)	Permanent; good acoustic value; unlimited preapplied colored and porcelain enamel finish surface; lightweight	Special supporting and attaching systems are required; acoustic pads must be installed behind aluminum	Areas where acoustic treatment and permanence are important
Asbestos cement, perforated (*see* Asbestos-Cement Board and Sheet)	Permanent; good acoustic value; easily painted	Brittle; cement color; nails and joints are exposed; acoustic pads are required above asbestos material	Areas where acoustic treatment is important
Cement plaster (*see* Cement; Plaster and Plastering)	Permanent; smooth or rough texture; easily painted	Metal lath and furring are required for support; expansion joints required for areas over 20 ft (6.096 m) in either direction	Exterior ceilings
Clay tile (*see* Clay Floor and Wall Tile)	Permanent; waterproof; very wide range of colors	No acoustic value; usually limited to small areas	Bathrooms, showers, and other areas where moisture is present
Concrete (*see* Concrete; Concrete Decking; Concrete, Reinforced)	Permanent; fireproof; easily painted	No acoustic value; electrical conduit and other mechanical pipes or ducts must be integrated within structure or left exposed	Ceilings that are part of structural system

Table C4 Typical materials used for ceilings (*continued*)

Materials	Advantages	Disadvantages	Major uses
Concrete, artificial stone	Permanent; wide variety of colors and textures	No acoustic value; special supporting and attaching devices required; heavy	Decorative
Glass (*see* Glass Fibers)	Good acoustic value; fire resistant	Soft; special surface treatment necessary to stop attraction of dirt	Areas where acoustic treatment is important
Mineral wool	Good acoustic value; fire resistant	Soft; limited colors	Areas where acoustic treatment is important
Paper pulp (*see* Paper and Paper Pulp Products; Paper Pulp Sheets and Rigidized Sheets; *see also* Fibers)	Good acoustic value; flame resistant	Soft; limited colors	Areas where acoustic treatment is important
Plaster (*see* Plaster and Plastering)	Smooth, hard finish; easily painted; durable	Tendency to crack; requires lathing before application except on concrete; no acoustic value; requires expansion joints for area over 20 ft (6.096 m) in either direction	Ceilings where acoustic treatment is unimportant
Plaster, acoustic (*see* Plaster and Plastering)	Good acoustic value	Soft; requires lathing before application except on concrete	Areas where acoustic treatment is important
Plaster integrants (*see* Plaster Integrants and Sheets)	Easily applied in areas where it can be nailed; easily painted	Joints and nails are difficult to conceal; no acoustic value	Residential construction
Plastic (*see* Plastics: Sheet; Plastics: Film; Plastics: Foam)	Translucent; some acoustic value	Requires special supporting and attaching systems; can deform	Luminous ceilings
Plastic, luminous (*see* Plastics)	Gives a total luminous ceiling, easily maintained	Requires extra-great depth between light source and hung luminous ceiling	Commercial-type areas; stores, banks, lobbies, etc.
Stainless steel, perforated (*see* Stainless Steel Sheet, Strip, and Plate)	Permanent; good acoustic value	Requires special supporting and attaching systems; requires acoustic pads above stainless steel material	Areas where acoustic treatment is important
Steel, perforated (*see* Steel Sheet and Strip)	Permanent; good acoustic value; unlimited preapplied colors in porcelain enamel finish surface	Requires special supporting and attaching systems; requires acoustic pads above steel ceiling material	Areas where acoustic treatment is important
Stone (*see* Stone, Granite; Stone, Marble)	Permanent; wide variety of colors and patterns	Requires special attaching and supporting devices; no acoustic value	Entrance vestibules and lobbies

Table C4 Typical materials used for ceilings (*continued*)

Materials	Advantages	Disadvantages	Major uses
Wood (*see* Wood Integrants)	Permanent; easily stained and painted; easily applied where it can be nailed	No acoustic value; joints exposed	Residential construction
Wood (*see* Siding and Paneling)	Permanent; easily applied where it can be nailed; easily stained and painted	Limited in widths; joints exposed	Residential construction

CEMENT

Physical and Chemical Properties

The word "cement" in its broad meaning includes any cementitious material that is able either to unite portions of substances not in themselves adhesive into a cohesive whole, or to cement nonadhesive materials together. Cement in common usage refers to portland cement and is so used in this book. Other types of cement—Keene's, oxychloric, natural cement—are described under "Types and Uses."

Constituents of Portland Cement. Portland cement is a closely controlled chemical combination of argillaceous materials (silica, alumina) and calcareous materials (lime) with iron oxide and small amounts of other ingredients, to which gypsum is added in the final grinding process to regulate the setting time of the cement. Lime (CaO) and silica (SiO_2) make up about 85% of the mass. (*See* Table *C5*.)

Raw Materials. Various raw materials may be used as the source of these ingredients:

Cement rock:	Argillaceous limestones (calcium carbonate)
Limestone:	Calcium carbonate and some magnesia, silica, alumina, and iron
Oyster and coquina shells, chalk:*	Calcium carbonate
Marl:	Natural mixtures of clay with calcium carbonate and magnesium carbonate
Clay and shale:†	Aluminum slicates, clay, mica, quartz, and other minerals
Slag:	Lime, alumina, magnesia, manganese, and phosphorus
Sand and sandstone:‡	Silicon dioxide (silica)
Gypsum:	Hydrated calcium sulfate
Iron ore, iron dust:§	Ferric oxide

*Alkali waste, marble.
†Diaspore, copper slag, fly ash, aluminum ore refuse, staurolite, granodiorite, and kaolin.
‡Traprock, calcium silicate, quartzite, and fuller's earth.
§Iron calcine, iron pyrite, iron sinters, and iron oxide.

Other possible materials include diatomite, fluorspar, pumice, flue dust, pitch, red mud and rock, hydrated lime, tufa, cinders, and sludge. Grinding aids, air-entraining compounds, calcium chloride, and other admixtures for imparting special properties may also be added during manufacture. For the typical chemical compositions of the main raw ingredients of portland cement, *see* Table *C6*.

Compound Compositions. The essential components of portland cement are the oxides of calcium silicon, aluminum, and iron. During grinding and subsequent burn-

Table C5 Typical chemical composition of portland cement

Lime (CaO)	Silica (SiO_2)	Alumina (Al_2O_3)	Ferric oxide (Fe_2O_3)	Magnesia (MgO)	Rutile (TiO_2)	Sodium oxide (Na_2O)	Potassium oxide (K_2O)	Sulfuric anhydride[a] (SO_3)	Loss in kiln gases
(percent of content)									
60–66	19–25	3–8	1–5	0–5	0.24	0.54	0.64	2.3–4.5	0.3

[a] SO_3 limits depend on the type of portland cement and on whether sulfate resistance is involved.

Table C6 Typical chemical composition of main raw ingredients of portland cement

	Composition (percent)[a]					
Raw material	Silica (SiO_2)	Alumina (Al_2O_3)	Lime (CaO)	Magnesia (MgO)	Carbonic anhydride (CO_2, water, alkalis)	Ferric oxide (Fe_2O_3)
Cement rock	12.66	3.92	43.26	1.30	36.97	1.50
Limestone, oyster shells	1.16	0.33	54.82	0.28	43.33	0.08
Marl	13.10	3.98	44.58	0.48	36.14	1.72
Clay	58.78	18.42	0.52	1.90	12.78	7.60
Shale	60.20	19.42	0.40	1.46	10.28	8.24

[a]Also contains 0.39% of sulfuric anhydride (SO_3).

ing of the raw materials into cement clinker, these oxides combine into four principal compounds the exact formulation of which is still subject to extensive research. These compound compositions are expressed as complicated chemical formulas in the quaternary system (e.g., $CaO{\cdot}Al_2O_3{\cdot}SiO_2{\cdot}F_2O_3$), which for practical purposes can be abbreviated and written as C_2S, C_3S, C_3A, C_4AF as shown in Table *C7.* These compounds are important to the speed of setting of the cement, to its heat during setting, and its resistance to alkali waters and soils.

Since the same proportions of these oxides can combine to form cements with quite different characteristics, it is not the relative amount of each oxide but the relative proportion of each compound that determines the characteristics of the various types of cement (*see* discussion under "Types and Uses").

Dicalcium and tricalcium silicates (C_2S and C_3S) are the most essential and useful; they control most of the strength-developing characteristics of cement. The sum of the percentages of these two in the various types of cement ranges from 70 to 80. The dicalcium silicates have a lower heat of hydration than the tricalcium silicates.

Tricalcium aluminate (C_3A) is the most active in heat generation and is responsible for most of the undesirable qualities in cement. A high content can increase volume changes and influence the formation of cracks, as well as lower the sulfate resistance of the cement. During manufacture, the tricalcium aluminate is removed as an impurity with iron oxide, which converts it to tetracalcium aluminoferrite (C_4AF), a compound with less heat of hydration but also with less cementing value.

Table C7 Compound compositions within portland cement

Dicalcium silicate $2CaO{\cdot}SiO_2$ (C_2S)	Tricalcium silicate $3CaO{\cdot}SiO_2$ (C_3S)	Tricalcium aluminate $3CaO{\cdot}Al_2O_3$ (C_3A)	Tetracalcium aluminoferrite $4CaO{\cdot}Al_2O_3{\cdot}Fe_2O_3$ (C_4AF)
(percent of content)			
32	50	9.0	9.0

Free or uncombined lime (CaO) *and magnesia* (MgO) are thought to be the chief causes of unsoundness in cement. If present in considerable amounts, these oxides, which remain undehydrated for a long time at ordinary temperatures, may eventually cause expansion and disruption of concrete.

Commercial Requirements. Portland cement is controlled in its manufacture to meet definite requirements established by various government departments (Army, Navy, etc.) and other official specifying agencies (e.g., The American Society for Testing Materials). In general, these requirements fall into five categories.

1. *Chemical limits.* These are shown in Table *C8.*
2. *Soundness.* This refers to the ability of a hardened cement paste to retain its volume after setting. Most specifications limit the magnesia content and the autoclave expansion. The latter is determined by the autoclave expansion test. Since 1943, with the adoption of this test by the American Society for Testing Materials, practically no cases of abnormal expansion attributed to unsound cement have occurred.
3. *Time of setting.* This must be not less than 45 minutes (Vicat needle) or 60 minutes (Gillmore needle) and not more than 10 hours.

Table C8 Major chemical limits of portland cement

Controls	Loss in kiln gases	Residue contained	Sulfuric anhydride[a] (SO_3)	Magnesia (MgO)
Limit (percent)	3.00	0.85	3.00	5.00
Tolerance (percent)		0.15	0.10	0.40

[a]Limit depends on type of portland cement.

Table C9 Types of portland cement

Type	Name	Limits	Compound composition (percent)							
			C_2S	C_3S	C_3A	C_4AF	Calcium sulfate ($CaSO_4$)	Free calcium oxide (lime) (CaO)	Magnesium oxide (mgO)	Fineness[a] (cm^2/g)
I, IA	Normal portland; air-entraining materials added during manufacture	Maximum	43	54	14	10	3.3	1.5	3.8	
		Minimum	22	29	9	6	2.2	0.0	0.7	1800
		Typical	24	50	11	8	2.8	0.8	2.4	
II, IIA	Moderate portland; air-entraining materials added during manufacture	Maximum	46	50	9	18	3.3	1.8	4.4	
		Minimum	22	29	3	10	1.9	0.1	1.5	1800
		Typical	33	42	3	13	2.9	0.6	3.0	
III, IIIA	High-early-strength portland; air-entraining materials added during manufacture	Maximum	38	70	17	10	4.6	4.2	4.8	
		Minimum	0	34	7	6	2.2	0.1	1.0	2600
		Typical	13	60	9	8	3.9	1.3	2.6	
IV	Low-heat portland	Maximum	61	33	8	18	4.0	0.3	4.1	
		Minimum	41	10	3	16	2.5	0.0	1.0	1900
		Typical	50	26	5	12	3.2	0.3	2.7	
V	Sulfate-resisting portland	Maximum	49	55	6	9	3.1	0.6	2.3	
		Minimum	27	35	4	5	2.7	0.1	0.7	1900
		Typical	40	40	4	9	2.9	0.4	1.6	

[a]Wagner turbidimeter test.

4. *Fineness.* Cements are now being ground to micron size. Greater fineness increases the rate at which cement hydrates and thus accelerates strength development, particularly during the first 7 days. Fineness may also be measured in terms of surface area per gram. *See* last column in Table *C9.*

5. *Compressive strength.* This is determined by tests of standard 2-in. (50.8-mm) mortar cubes. Table *C10* shows the effects of various types of portland cement on the relative strength of concrete, using "standard" sand.

All these standards are amended periodically and correlated with general specifications for concrete work (*see* Concrete).

Packaging, Labeling. Packaging, marking or labeling, and storage methods for cement are also controlled by specific requirements. Most portland cement is transported in bulk by railroad, truck, ship, or barge. Pneumatic loading and unloading of the transporting vehicle is almost standard procedure for handling bulk cement. A lesser means of transporting portland cement is in paper bags: the U.S. paper bag weighs 94 lb (42.64 kg), whereas the Canadian paper bag weighs 80 lb (36.29 kg).

Table C10 Relative strength of concrete using the various types of portland cement

Type	Name	Relative compressive strength based on percent of strength of concrete using Type I normal portland cement			
		1 day	7 days	28 days	3 months
I, IA	Normal with air-entraining materials	100	100	100	100
II, IIA	Moderate with air-entraining materials	75	85	90	100
III, IIIA	High-early-strength with air-entraining materials	190	120	110	100
IV	Low-heat	55	55	75	100
V	Sulfate-resisting	65	75	85	100

Types and Uses

Portland Cements. Eight types of portland cement are available and widely used in the construction field (*see* Table *C9*). Each type has definite characteristics which make it the preferred cement for a specific set of job conditions, as follows:

Type I, Normal: for general overall use.

Type IA: Normal Type I portland cement with air-entraining materials interground with the clinker during manufacture.

Type II, Moderate: for applications where slow setting and less heat are required, and particularly in massive concrete structures such as large retaining walls, piers, and abutments, where excessive heat will cause cracking.

Type IIA: Moderate Type II portland cement with air-entraining materials interground with the clinker during manufacture.

Type III, High Early Strength: for applications where quick-setting early strength is important and more heat must be developed to offset low temperatures.

Type IIIA: High Early Strength Type III portland cement with air-entraining materials interground with the clinker during manufacture.

Type IV, Low Heat: for applications where very slow setting is desired and very little heat should be generated. This is used particularly in very massive concrete where cracking due to heat must be avoided.

Type V, Sulfate-resisting: for applications where alkaline water and soils that will attack other types of portland cement are present.

Portland cement Types IA, IIA, and IIIA correspond in composition to Types I, II, and III but produce concrete with improved resistance to freeze-thaw action and to scaling caused by chemicals applied for snow and ice removal. The concrete produced is also more plastic, fluid, and workable. The air-entraining materials impart to the concrete minute, well-distributed, and completely separate air bubbles.

As Table *C9* shows, the differences between the various types of portland cement are the result of differences in their compound compositions. Type III is high in tricalcium silicate; Type IV is high in dicalcium silicate and tetracalcium aluminoferrite (C_4AF); and Type V is high in dicalcium and tricalcium silicates and low in C_3A and C_4AF.

There are four other types of portland cement:

1. *White portland cement,* which is manufactured from select raw materials and contains negligible amounts of iron and manganese oxides. Its major uses in construction are for prefabricated curtain wall and facing panels, terrazzo, stucco, cement paint, various types of stonework, ceramic veneer, tile grout, and applications where colored concrete or mortar is required.

2. *Waterproof white portland cement,* which is manufactured by adding a small amount of calcium, aluminum, or other stearate to portland cement clinker during the final grinding. It is available in either white or gray. Its major use is in various types of stonework and preassembled or prefabricated white concrete units.

3. *Portland pozzolanic cement,* which is manufactured by intergrinding portland cement clinker with a pozzolan, or by blending portland cement or portland blast-furnace slag cement and a pozzolan. A pozzolan is a siliceous material which will react with lime in the presence of water. There are four types of portland-pozzolanic cement: *Type IP*, *Type IP-A* with air-entraining additive, *Type P*, and *Type PA* with air-entraining additive. Their major use is for large hydraulic structures such as bridges, piers, dams, and canal locks.

4. *Portland blast-furnace slag cement*, in which granulated blast-furnace slag of select quality is either interground with the portland cement clinker or blended with portland cement. There are two types: *Type IS*, and *Type IS-A* with air-entraining additive. If moderate heat of hydration is required, the suffix (MH) may be added and the type indicated as IS(MH) or IS-A(MH). If moderate sulfate resistance is required, the suffix (MS) may be added thus: Type IS(MS) and Type IS-A(MS). Also, both suffixes may be added: Type IS(MS)(MH) and Type IS-A(MS)(MH). The rate of strength development and the compressive strength of Type IS cement are about the same as those of Type I portland cement.

Construction Uses of Portland Cements. The two major uses of portland cement in construction are (1) all types of concrete construction and prefabricated concrete materials, and (2) mortar for masonry materials.

Cement also is used for making asbestos-cement products, cement brick, cement pipe, terrazzo, artificial stone, as a basis of paints for concrete and masonry materials, and as a plaster for exterior surfaces. It finds many minor applications such as grouting, pointing, floor surface finishes, and rough fill for leveling.

Other Cements. A number of cements other than portland cement find use in the field of construction.

Aluminous cement, made in much the same way as portland cement, utilizes bauxite ($Al_2O_3 \cdot 2H_2O$), the ore from which aluminum is made, as the major raw material. This cement is a quick-setting one that in 24 hours will have the strength of standard portland cement which has set for 28 days. It is used for roads and areas where this property of very quick setting is important.

It is also resistant to salt action and is used for severe exposures. A specialized application is the insulation of furnaces where high temperatures are encountered. A typical composition is 40% aluminum oxide, 40% lime, 15% iron oxide, and 5% silicon and magnesium oxides.

Plastic cements are manufactured by adding plasticizing agents, up to 12% of total volume, to Type I Normal or Type II Moderate portland cement during manufacture. Their major uses are for mortar, plaster, and stucco.

Expansive cement is a hydraulic cement with a setting time that can be controlled from approximately 1 to 2 minutes to about 1 hour, with correspondingly rapid early-strength development. It is a modified portland cement with set-control and early-strength development components incorporated during manufacture. The expansion of this cement occurs during the early hardening period after setting.

Expansive cement is available in three types, *K*, *M* and *S*, and the expansive properties of the three types can be varied over a considerable range. When expansion is restrained, these cements can be used to compensate for the effects of volume decrease in concrete due to shrinkage; they also can be used to induce tensile stress in reinforcement for prestressing purposes.

Masonry cements are prepared mixtures of portland cement with various other ingredients such as hydrated lime, granulated slag, silica, clay, diatomite, and finely pulverized limestone. Small additions of calcium stearate, petroleum, colloidal clays, and other admixtures are sometimes included. The ingredients and proportions vary widely and usually are patented or kept secret. These cements make a more workable and plastic mortar for masonry materials. They are available in two types, masonry mortar Types I and II, either white or gray in color, and also waterproof or nonwaterproof.

Natural cements are made of natural raw materials, found mixed in the correct proportions, which need only grinding and burning in a kiln to produce a cement. Weights are marked on the package or bag.

Pozzolanic cement (not to be confused with portland pozzolanic cement, already discussed) is a cement made of lime mortar and a pozzolanic material. Lime in the presence of water combines with silica in the active state to form calcium silicate similar to that in portland cement. Various natural or artificial materials contain active silica, among them pozzolan (volcanic ash), granulated slag, and pumice. This pozzolanic cement is produced by grinding together lime and the material with the active silica until they are completely intermixed. This cement has a strength almost equal to that of portland cement.

Keene's cement is made by heating gypsum to about 230°F (110°C) to make plaster of paris, which is then placed in a solution of alum, dried, reheated to 932°F (500°C), and ground to a powder. Keene's cement makes a dense, hard type of interior plaster used where moisture and rough treatment prevail.

Oxychloric (*Sorel*) *cement*, more accurately termed magnesium oxychloride cement, is based on Sorel's discovery in 1853 that zinc chloride solution, when mixed with zinc oxide, forms an oxychloride that makes a very hard white cement. Today oxychloric cement is made from magnesium oxide which is freshly burned (calcined) and ground to a fine powder. After it is mixed with carborundum, wood sawdust, sand, or other material, and a solution of magnesium chloride has been added, the entire mass is intimately mixed, put into molds placed into position, and floated (troweled) off. This cement has a certain resilience and is used for finish floors, in making artificial stone and marble, and as a binder in making grindstones. Oxychloric cement must be fresh (not older than 6 months), carefully stored, and used correctly. Depending on its end use, four groups of ingredients are combined in its formulation: inert fillers (marble flour, talc); fibrous fillers (asbestos, wood flour, sawdust), which are good for marble floors; inert aggregates (sand, crushed stone); and inorganic alkali-resistant pigments.

Sulfur cement is made by mixing melted sulfur with brick dust or other mineral filler and is used as a mortar or to anchor metal into stone or masonry.

Mixtures of animal, vegetable, and mineral substances used to cement other materials are discussed under Adhesives; for bituminous cement compounds, *see* various Asphalt headings *and also* Tar.

The use of various types of plastic such as epoxies and saran polymers to produce high compressive and tensile strength in mortars is fully discussed under Mortar. The use of these or other, similar types of additives with portland cements or other types of cement for producing stronger concrete has not as yet become common practice in the field of concrete construction.

Application

The two major applications of cement, concrete and mortar, are discussed at length under specific headings, as are the major prefabricated cement products. The following precautions hold true for all uses of cement.

Condensed Checklist

1. Type of portland cement in relation to site: The site and its location should be checked to determine the

type of portland cement best suited for the particular end use.
2. Type of water and subsoil: A check should be made to determine whether there are subsoil waters; if there are, a careful analysis of these waters should be made, particularly of their alkaline and sulfur contents. The subsoil should also be checked in a like manner to determine what type of cement should be used.
3. Climate and the time of the year that the cement materials are to be used: Temperature, humidity, and weather conditions affect both the type of cement and the conditions of use.
4. Cement for mortars: The cement should be of a type recommended for the particular masonry work being done.
5. Type of concrete construction design: This will also dictate the type of cement to be used.
6. With preassembled or prefabricated concrete units, the finish surfaces will determine to a great extent the type of cement to be used.

History and Manufacture

The search for cementitious materials started when man built his first stone walls and looked for something to bind them together. The Babylonians developed and used clay, and the Egyptians discovered gypsum plaster, which they used as a binder. The Romans developed a cement, and their *caementum*, meaning rough, unhewn stone or chips of marble from which a kind of mortar was made, is the root of our wood "cement." Roman cement was made by mixing slaked lime with pozzolana (volcanic ash), which was named after Pozzuoli, located near Mt. Vesuvius, where the ash was found. This Roman cement also hardened under water. In the Americas, the Maya and Aztec Indians used a type of hydraulic cement made from limestone deposits.

With the fall of the Roman Empire the art of cement making was lost, and for several centuries the only cements were those made from naturally occurring mixtures of lime and clay. Gradually improvements were introduced; in 1756 Smeaton, an Englishman, rediscovered hydraulic cement; and in 1769, in the United States, priests and Indians at San Diego, California, made a cement from local limestone deposits by techniques learned from the Mexican Indians, who had long used a form of hydraulic cement.

Invention of Portland Cement. In 1824 Aspdin, an English bricklayer and mason, invented and patented portland cement, a predetermined and carefully proportioned chemical combination of lime, silica, iron oxide, and alumina, which he named after the Isle of Portland in the English Channel. Aspdin's contribution lay in his careful proportioning of limestone and clay and in his processing method, which consisted of pulverizing the mixture, burning it to a clinker, and then grinding the clinker into the finished product, portland cement.

Portland cement was gradually accepted throughout the world, replacing cements made from the burned, naturally occurring mixtures of lime and clay which varied widely in their properties. But this came about slowly and largely as a result of a general increase in cement production, stimulated in the United States by the construction of canals, which started about 1818. The first portland cement was not produced until 1871, and by 1879 only 82,000 barrels had been made. In 1972 production was 84.6 million short tons (85.95 tonnes).

At first all cements, portland and natural, were made in vertical kilns, which had to be cooled after each burning. In 1885, in England, Ransome patented the horizontal rotating kiln, in which, because it was slightly tilted, the material moved gradually from one end to the other. In 1902 Edison introduced the 150-ft–long horizontal kiln, and today kilns are as long as 720 ft (219.46-m).

Commercial Production. Portland cement is manufactured by two methods, the "wet process" and the "dry process." In the wet process, the raw materials are ground and mixed with water and fed to the kilns as a moist slurry; in the dry process, the raw materials are ground and mixed and fed into the kilns in a dry state. In the horizontal, slightly tilted kilns lined with refractory material, a blast of flame at the lower end heats the wet or dry raw materials to approximately 2700°F (1482.22°C). As the raw materials pass through the kiln, they unite and form a new substance with its own chemical and physical characteristics, which makes clinker about the size of marbles. This red-hot clinker is discharged from the lower end of the kiln and is cooled. After cooling, it can be stockpiled or immediately ground with added gypsum until it is so fine that 90% will pass through a mesh with 32,400 meshes/in.^{2} (6.452 cm^2). Figure *C3* shows both the wet and dry processes.

There have been several improvements in the manufacture of cement: (1) longer kilns, using waste heat from the clinker cooler kiln and the preheater to dry materials during grinding; (2) preheater systems; and (3) used by the Japanese, a flash furnace which provides about 90% calcination of raw material feed prior to entering the kiln, thereby increasing the capacity of the kiln 100%.

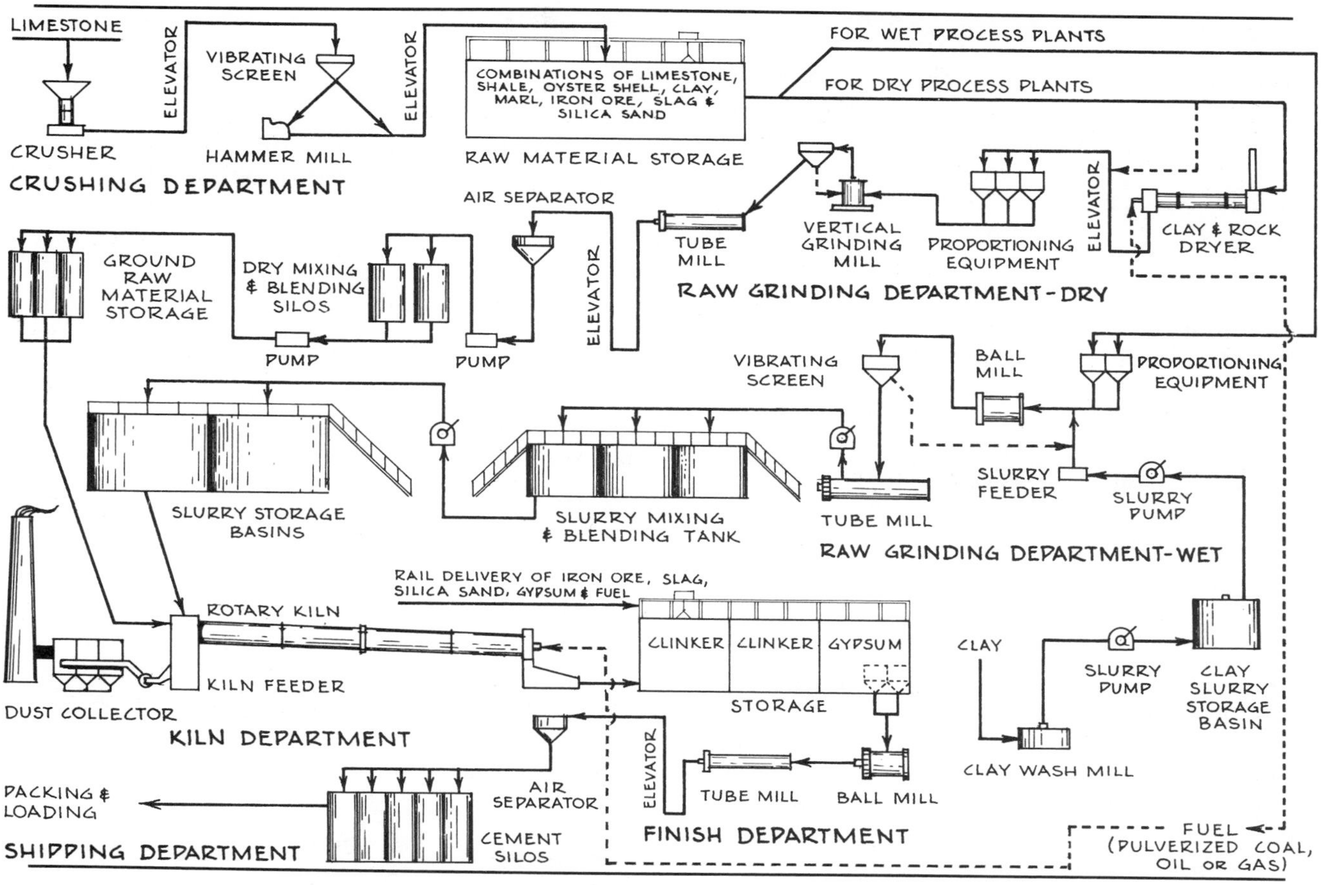

Figure C3 Production of portland cement, showing both the wet and the dry process.

Originally all cement was shipped in barrels, and it is interesting to note that the trademarks or brands of the older manufacturers are circular in design to fit the top of a barrel.

CERAMICS

The word "ceramics" originally referred to a very limited range of products made from natural earths that had been exposed to high temperatures. Currently ceramics are defined as inorganic, nonmetallic materials or products that have been subjected to heat treatment and are usually serviceable through high-temperature processing and use. They are hard and brittle, able to withstand compression, resistant to abrasion, and able to sustain large compressive loads. Therefore the field of ceramics covers a widely diversified group of materials which include not only china and porcelain, but also many materials used in construction. Some of these are abrasives, glass, glazes and porcelain enamels, cements, limes, plasters, and structural clay products such as brick, tile, conduit, sewer pipe, and ceramic veneer (terracotta). Cermets, electrical insulators, and electronic and nuclear ceramics represent newer products for specialized non-construction uses.

The current and continuing developments in ceramics were brought about by the demand for special materials in the areas of aerospace, electronics, and nuclear energy. This scientific research has negated the earlier, limited definition of ceramics and has led to Parker's statement; "The brittleness of ceramic materials is not an inherent property common to all nonmetallic inorganic compounds. . . . Certain solids can be made ductile merely by changing the surface condition."

The following is a list of some of these new ceramic materials that at present have no application in construction but may eventually find uses there also.

Type	*Present Use*
Ceramic fuel elements	Enriched uranium dioxide, fuel element for nuclear electric power plants

Type	*Present Use*
Laser material	Extension of the range of frequencies of the laser beam
Piezoelectric materials	Active elements for sonar and ultrasonic devices
Magnetic ceramics	Computer memory cores, telecommunication systems, small direct-current motors
Cermets (metal and ceramics)	Jet engines, brake shoe linings, oxidation-resistant products
Refractory ceramics	Manufacturing equipment requiring extremely high temperatures, such as 3900°C (7030°F)
Dry film lubricants	For temperatures where conventional lubricants cannot function

C FACTOR

Thermal conductance (*C* factor) is the measurement of heat flow through any single material regardless of thickness from surface to surface. The *C* factor is measured in British thermal units (Btu) per square foot per hour per degree F (or in the metric unit of watts per square meter per degree C). The *C* factor can be said to be related to the *K* factor (thermal conductivity) in that the latter measures heat flow through 1 in. (25.4 mm) of the material under the same conditions.

The *C* factor, then, is a measure of the rate of heat flow for the actual thickness of a material 1 ft^2 (0.0929 m^2) in area at a temperature difference of 1°F (0.5556°C). If the *K* factor of a homogeneous material is known, the *C* factor can be determined by dividing the *K* factor by the thickness. For example, if the *C* factor for an 8-in. (203.2-mm) common sand-gravel hollow concrete block is 0.90, then the *K* factor is 0.90 divided by 8, or 0.1125.

See also Insulation and Insulating Materials.

CHAIN

Chain is a series of links generally made of iron, wrought iron, steel, stainless steel, brass, bronze, nickel, aluminum, or other metals or alloys. Large chain is made from bars, medium or small chain is made from wire, and some types of weldless chain are made from sheet material. Chains that are to withstand loads should be tested, usually by a load equal to one-half of the average ultimate tensile strength. When safety is important, the chain is tested to destruction.

Types and Uses

Chain is available in all the standard gauges of wire, sheet, and bar for the particular metal of which it is made. It is further classified according to the number of links per foot, weight per 100 ft, and tensile strength in pounds. Chain made of bar or wire may be either welded or nonwelded. The more common types are shown in Figure *C4*.

In construction, chains are used mostly for railings or barriers and occasionally for decoration. Fine chains are a part of many hardware items.

Industry uses great quantities of chain in heavy equipment, hoisting engines, and in transportation, especially railroads.

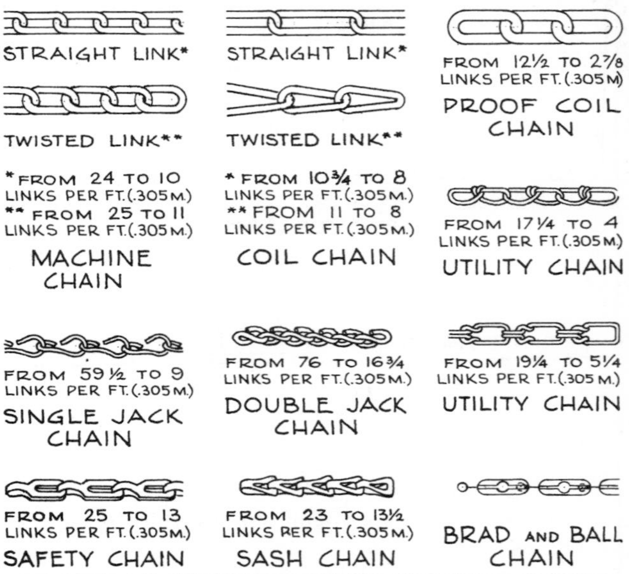

Figure C4 Common types of chain.

History and Manufacture

The history of chain follows that of the metals, as it is one of the simplest forms into which a metal can be worked. Historical records tell of slaves chained to their oars in fighting ships; chain mail armor, which reached a high peak of development in both the East and the West; and reinforcing chains around the dome of St. Peter's, which are a well-known architectural curiosity. All these early chains were handmade, usually from bronze or wrought iron. Today almost all chain is machine-made except for large chains, which still require hand-hammering for the weld.

CHIMNEYS

A chimney may be defined as the shaft by which the hot, used air and fumes from fireplaces, boilers, incinerators, and industrial processes are eliminated from a building. In today's construction, apart from large industrial stacks, most chimneys consist of a clay tile flue lining surrounded by a protective layer of masonry, serving the heating system alone or in combination with fireplaces. In light construction, an asbestos-cement pipe or prefabricated chimney formed of either asbestos cement or a vitreous enameled metal surrounded by insulation can be used (*see* Figure *C5*).

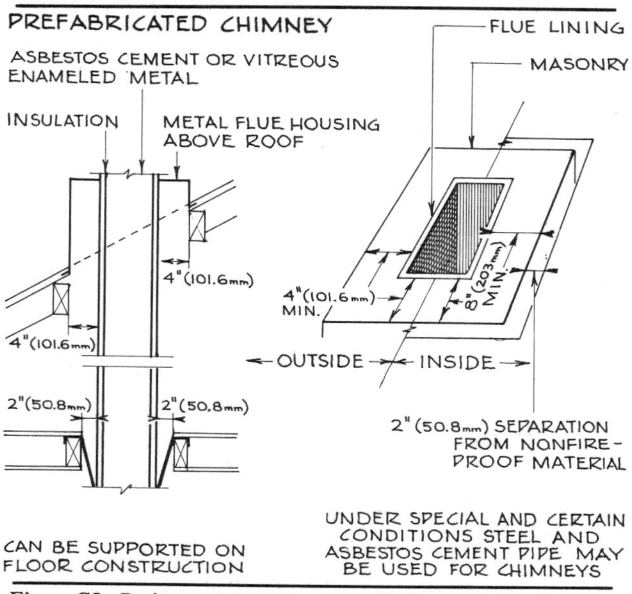

Figure C5 Basic components and limitations of a typical prefabricated chimney for residential use.

Fire Prevention Measures. Chimneys carry enough heat at times for their materials to reach a very high temperature; also, they can catch fire from the burning of excess soot that has collected in their interior passage and develop dangerously high temperatures. To obviate this fire hazard, chimneys should always be completely isolated from nonfireproof materials by an air space of 2 in. (50.8 mm)

Building code requirements in most localities set the thickness of the masonry from flue lining to outside surface on the exterior of the building at a minimum of 4 in. (101.6 mm) and the thickness from flue lining to outside surface on the interior of the building at a minimum of 8 in. (203.2 mm).

Draft. The most important point about a chimney is its draft, which in turn is the product of the size (area) of the flue and the height of the chimney. The height should always extend at least 2.5 ft (0.762 m) above the highest point of the building. The size of the flue is controlled by the manufacturer in the case of incinerators and various types of heating units and by the size of the opening in the case of fireplaces.

In many chimneys other than residential, fans are added to induce a forced draft. The advantage of the forced draft is that it cuts down on the height necessary to develop a sufficiently strong natural draft.

Industrial Chimneys. Industrial chimneys represent an engineering problem. Their size and other requirements are usually beyond the scope of the ordinary type of clay tile flue and chimney construction. Special refractory materials are needed, and the choice of material (commonly in brick form), type of mortar, and design of chimney are controlled by the type and volume of gases and chemical fumes carried by the chimney. Smoke and air pollution control may also enter into the picture.

CHLORINE

Physical and Chemical Properties

Symbol: Cl
Atomic number: 17
Specific gravity: 1.56 (−33.6°C, −28.48°F)
Freezing point: −100.98°C, −149.76°F
Boiling point: −34.6°C, −30.28°F

Chlorine is a heavy, poisonous, greenish yellow gas or a clear amber liquid with an irritating odor. It is soluble in water and alkalis and chemically very active. Chlorine is commercially available in liquid or gas form.

Types and Uses

Chlorine has few uses in relation to finished materials. It would be hard to enumerate all its industrial applications, however, particularly in compound form. It is widely used as a bleach for paper and textiles, as a disinfectant, and as a raw ingredient in the preparation of many plastics, nylon, synthetic rubber, and other organic compounds. In Table *C11* an asterisk indicates the use of elemental chlorine as such.

History and Manufacture

Chlorine never occurs in nature uncombined; as sodium chloride (salt) it is very abundant and an essential

Table C11 Uses of chlorine

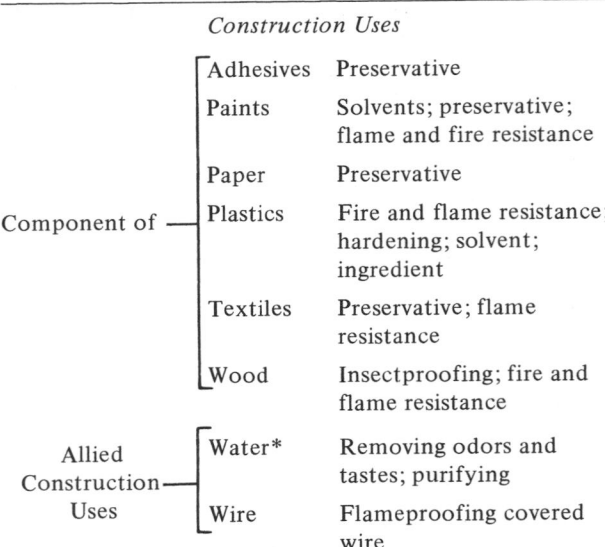

	Construction Uses	
Component of	Adhesives	Preservative
	Paints	Solvents; preservative; flame and fire resistance
	Paper	Preservative
	Plastics	Fire and flame resistance; hardening; solvent; ingredient
	Textiles	Preservative; flame resistance
	Wood	Insectproofing; fire and flame resistance
Allied Construction Uses	Water*	Removing odors and tastes; purifying
	Wire	Flameproofing covered wire
	Nonconstruction Uses	
Electroplating, gold-plating, manufacture of paper and silicones, metallurgy, photography.		

constituent of life. It was first prepared in 1774 by Scheele from hydrochloric (muriatic) acid and the mineral pyrolusite (MnO_2). In 1810 Davy established it as a new element and named it after the Greek work *chlorus*, meaning "grass-green." With the discovery of chlorine, Lavoisier's theory that all acids contain oxygen was disproved and discarded.

Almost from the time of its discovery it has been manufactured on a large scale for industry. The electrolytic method is now the only commercial source of chlorine.

CHROMIUM

Physical and Chemical Properties

Symbol: Cr
Atomic number: 24
Specific gravity: 7.19 (20°C, 68°F)
Melting point: 1890°C, 3434°F
Boiling point: 2482°C, 4499.6°F

Chromium is a steel-white metal which takes a brilliant polish and is harder than cobalt or nickel. It is nonmagnetic at ordinary temperatures but becomes magnetic at -13°F (-25°C). It does not tarnish in air, resists oxidizing agents, and is soluble in acids (except nitric) and in strong alkalis.

Table C12 Composition of high-purity commercial chromium

Cr	Al	Fe	Si	C	S	H_2O	O
			(percent of content)				
99.41	0.03	0.27	0.10	0.03	0.02	0.08	0.06

Workability. Pure metallic chromium is ductile, but as now commercially available it is brittle. Chromium can be cast and hot-forged but, because of trace impurities that cause brittleness, cannot be cold-worked.

Commercial Forms. Chromium is available in flake, lump, and powder form. High-purity chromium is available but not on a large commercial scale (*see* Table *C12*).

Types and Uses

Chromium in compound form is widely used, primarily as a pigment. Its uses in metallic form are noted by asterisks in Table *C13*.

Ferrochromiums. The principal use of chromium is as an alloying ingredient in ferrous and nonferrous metallurgy. In cast iron, alloy steel, and stainless steel production, ferrochromiums of various types are used because of their low cost compared to chromium metal. Recently a low-carbon ferrochromium produced by reducing chromic oxide with silicon was introduced and has replaced other ferrochromiums in stainless steel production. It reduces melting time and costs and improves the quality of the steel.

Plating. Chromium plating is the most commonly encountered usage of this metal in construction. It gives a thin, hard, bright, wear-resistant surface which sheds water when highly polished.

The metals that can be plated with chromium include aluminum, copper, iron, magnesium, nickel, titanium, zinc, and their alloys. The chromium is electrodeposited as a thin layer of pure metal. It usually contains microscopic pinholes or cracks which permit the atmosphere or other corroding agents to affect the underlying metal. To be highly protective it should be 0.001 in. (0.0254 mm) thick. Usually it is applied as a decorative, bright, tarnish-free finish 0.00001 to 0.00002 in. (0.000254 to 0.000508 mm) thick. Its protective value is increased by and to a great measure dependent on the

Table C13 Uses of chromium

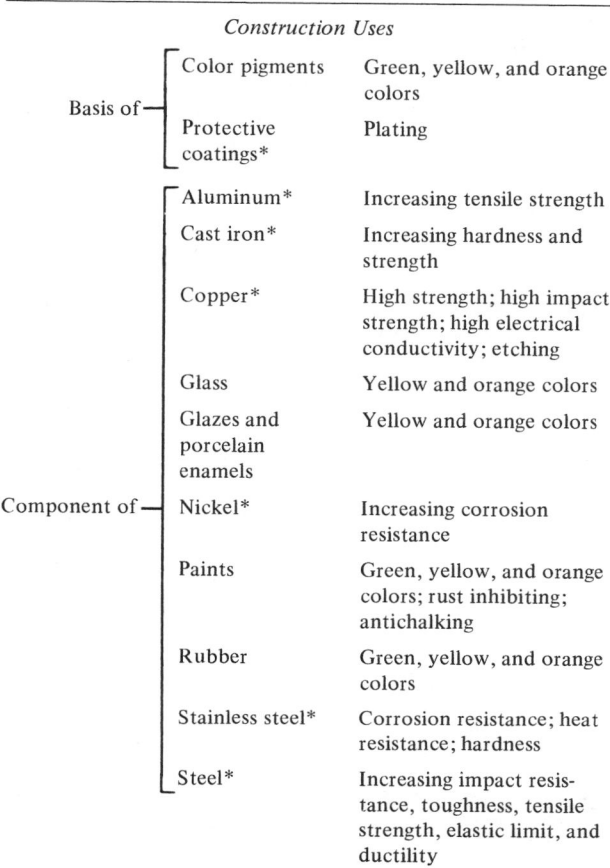

	Construction Uses	
Basis of	Color pigments	Green, yellow, and orange colors
	Protective coatings*	Plating
Component of	Aluminum*	Increasing tensile strength
	Cast iron*	Increasing hardness and strength
	Copper*	High strength; high impact strength; high electrical conductivity; etching
	Glass	Yellow and orange colors
	Glazes and porcelain enamels	Yellow and orange colors
	Nickel*	Increasing corrosion resistance
	Paints	Green, yellow, and orange colors; rust inhibiting; antichalking
	Rubber	Green, yellow, and orange colors
	Stainless steel*	Corrosion resistance; heat resistance; hardness
	Steel*	Increasing impact resistance, toughness, tensile strength, elastic limit, and ductility
	Nonconstruction Uses	
Photographic processes, refractories		

undercoating of copper, nickel, or both (*see* Nickel Plating).

Other Forms. Chromite ore is used in making bricks, cements, and other chromium refractories for lining industrial furnaces.

Chromium oxide is a green pigment used in the paint industry and for roof granules.

History and Manufacture

In 1798 Vauquelin and Klaproth discovered chromium simultaneously and independently. It is named for the Greek word meaning color because of the varied colors of its compounds. In 1859 Wöhler first isolated it in its metallic form. Because of the difficulties of obtaining it by electrodeposition, chromium remained a curiosity until Goldschmidt developed the aluminothermic or thermite process of producing it in 1895. In this process powdered aluminum and chromium oxide are mixed and then ignited. The heat of reaction is explosive and sufficient to produce chromium metal.

The use of chromium as ferrochromium in cast iron, alloy steel, and stainless steels began about 1912 and has grown steadily in importance. The principal chromium ore is chromite ($FeCr_2O_4$ or $FeO \cdot Cr_2O_3$).

Commercial Production. There are several methods for producing chromium.

1. Reduction of chromite ore by carbon in a submerged-arc electric furnace to produce ferrochromium, which consists of 65% chromium.
2. Electrolysis of chromium solutions.
3. Reduction of the oxide by the thermite process.
4. Reduction of chromic oxide by silicon under a lime slag in a tilting electric furnace to produce 99% chromium.

CLAY

Physical and Chemical Properties

Clay may be defined in simplest terms as a material that is plastic when wet and hard when fired; this is the characteristic that makes it commercially valuable in the ceramics industries. Clay consists of finely divided minerals, chiefly aluminum silicates, which have a sheetlike crystal structure. Clays differ in composition, characteristics, and degree of purity. Most clays including ceramic clays, clay shales, mudstones, glacial clays, oceanic clays, red clay, blue clay, and blue mud, are the result of weathering or alteration of aluminous and silicate rocks. The major part of any clay is composed of the following clay minerals: the kaolinite group, the montmorillonite group, and potash clay, or illite. Clay may contain other minerals, predominantly quartz, but including also calcite, limonite, gypsum, and mica (muscovite).

Types and Uses

There are four major classifications of clay, listed here with their main uses (*see also* Table *C14*).

1. *China clay* or kaolin, used for paper, rubber, refractories, and finer grades of pottery.
2. *Ball clay*, used for pottery, ceramic tile, and ceramic veneer (terra cotta) to increase plasticity.
3. *Fire clay*, used for refractory materials, brick, ceramic tile, and structural tile.
4. *Miscellaneous clays*, used for brick, structural clay tile, ceramic veneer, ceramic tile, and cement.

Table C14 Uses of clay

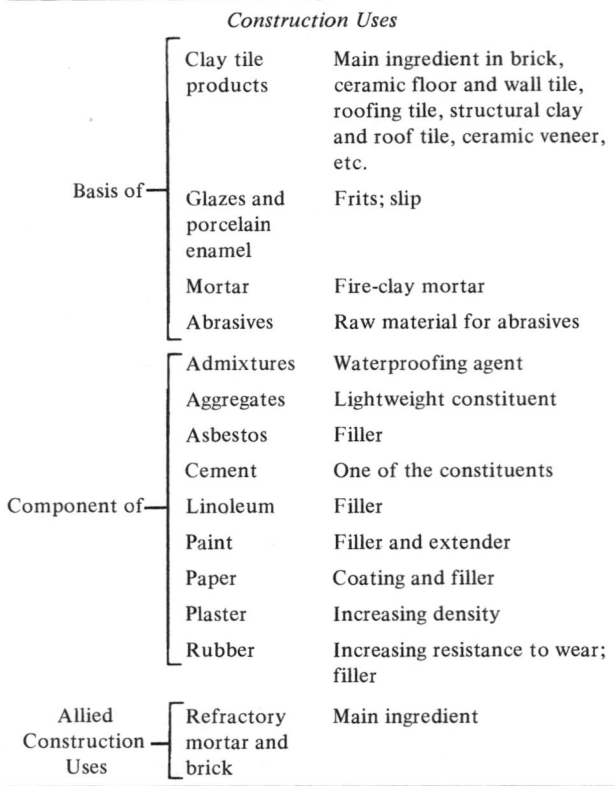

	Construction Uses	
Basis of	Clay tile products	Main ingredient in brick, ceramic floor and wall tile, roofing tile, structural clay and roof tile, ceramic veneer, etc.
	Glazes and porcelain enamel	Frits; slip
	Mortar	Fire-clay mortar
	Abrasives	Raw material for abrasives
Component of	Admixtures	Waterproofing agent
	Aggregates	Lightweight constituent
	Asbestos	Filler
	Cement	One of the constituents
	Linoleum	Filler
	Paint	Filler and extender
	Paper	Coating and filler
	Plaster	Increasing density
	Rubber	Increasing resistance to wear; filler
Allied Construction Uses	Refractory mortar and brick	Main ingredient

History and Manufacture

Fired clay objects are usually found in the ruins of every ancient civilization, and it is safe to say that the use of clay antedates recorded history. The widespread occurrence of clay and the simplicity of its manufacture probably account for the appearance of fired clay building materials at such an early period.

Today clay not only is the raw ingredient for brick, structural clay tile, ceramic veneer, ceramic floor and wall tile, and ceramics in general but also has become one of the major mineral products used in industry. In the United States almost every state is a producer of some type of clay, which is normally mined by power shovel, dragline, shale planer, scraper, or various other mechanical methods.

CLAY TILE: CERAMIC VENEER (TERRA COTTA)

Physical and Chemical Properties

Ceramic veneer may be defined as a molded clay unit, decorative or plain, whose properties are similar to those of burned brick. The clays from which it is formed should be dense-burning and contain no soluble salts; they should have low shrinkage and freedom from warpage. Tolerance limits for any dimension are $\pm\frac{1}{16}$ in. (±1.59 mm). A good test of quality for ceramic veneer is its tone when struck, which should be bell-like.

Weight. The average weight of hand-molded ceramic veneer is 70 lb/ft^3 (1121.4 kg/m^3). Today, 90% of ceramic veneer is machine-extruded and weighs as follows: 1 in. (25.4 mm) ceramic veneer (adhesion type), 11 lb/ft^2 (57.7 kg/m^2); 2 in. (50.8 mm) anchor-type, 16 lb/ft^2 (78.1 kg/m^2).

Color. The color range of natural, unglazed ceramic veneer is usually dull ocher to red. It can vary according to the composition of the clay and the temperature of firing.

Finishes. The surface features of ceramic veneer include the following finishes:

1. Smooth finish: plane surface as formed by the die in manufacturing.
2. Scored finish: surface grooved as it comes from the die. This increases the bond for mortar, plaster, or stucco.
3. Combed finish: surface altered by parallel scratches to produce a desired texture or to increase the bond for mortar, plaster, or stucco.
4. Roughened finish: surface entirely broken by wire cutting, etc., to provide a desired texture or to increase the bond for mortar, plaster, or stucco.
5. Ceramic glaze: a compounded, transparent, hard coating fused to ceramic veneer.
6. Nonlustrous glaze: surface covered by a fire-bonded ceramic glaze with satin or matte finish.
7. Ceramic color glaze: a surface coating of bonded ceramic glaze, either solid (one color) or mottled (two or more colors evenly distributed and somewhat blended), with a satin or gloss finish. A very wide range of colors is possible in glazes.
8. Polychrome finish: two or more colors applied separately to specific areas and burned separately for each color.

Texture refers to surface qualities other than color and includes the surface finishes just listed and any elaboration thereof. Extruded ceramic veneer is available in smooth, beveled, fluted, and scored surface texture. In handmade types, unlimited varieties of textures are available. Local manufacturers should be consulted to determine possible textures and colors.

Types and Uses

Extruded Ceramic Veneer. Extruded ceramic veneer, made by the stiff-mud process, includes (1) adhesion-type or ceramic veneer, which is not more than $1\frac{1}{4}$ in. (31.75 mm) thick and does not exceed 30 in. (762 mm) in any one dimension or 540 in.^{2} (0.35 m^2) in superficial area; and (2) anchor type, which is not less than $1\frac{1}{4}$ in. (31.75 mm) thick (*see* Figure *C6*).

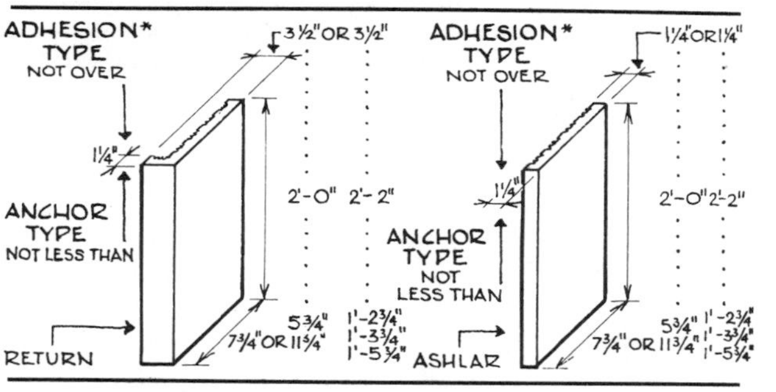

Figure C6 Standard sizes (not allowing for mortar joint) of adhesion-type and anchor-type extruded ceramic veneer. Asterisk indicates that 30 in. (76.2 cm) in any one dimension is not to be exceeded, or 540 in.^{2} (3483.86 cm^2) in superficial area.

Handmade Ceramic Veneer. Handmade ceramic veneer may be of three types: closed back, open back, and solid slab (anchor type). Sizes may range from less than 2 to 12 ft^2 (0.1858 to 1.1148 m^2) in surface area. Larger sizes can be obtained, but surface distortion then becomes a problem. (*See* Figure *C7*.)

The shells and webs of handmade ceramic veneer must be properly proportioned and able to resist expansion and contraction stresses when they are burned.

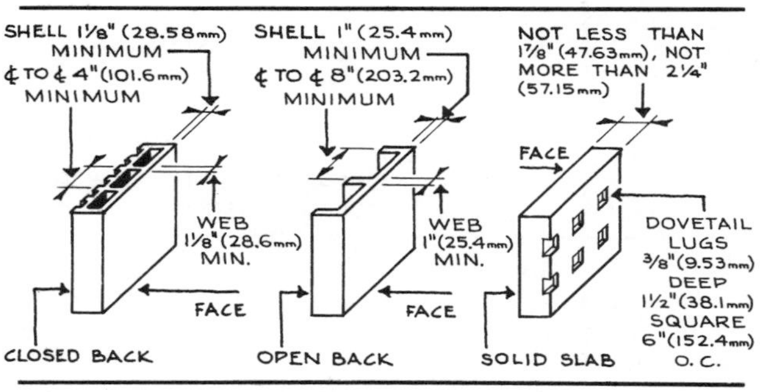

Figure C7 Types of handmade ceramic veneer and limitations of shells and webs.

Ceramic Veneer Panel. A prefabricated ceramic veneer panel, consisting of a ceramic veneer facing with a lightweight concrete backing (including setting clips, etc.) and reinforcing, is available for curtain wall construction (*see* Figure *C8*). This wall unit, with a total thickness of $2\frac{1}{2}$ in. (63.5 mm), is capable of carrying a load of 20,000 lb/ft (29,764 kg/m).

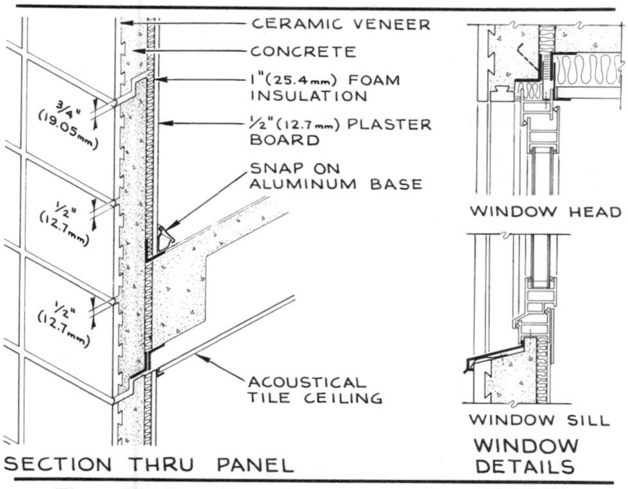

Figure C8 Precast concrete panel with ceramic veneer.

Ornamental Uses. Sculptural reproductions can be obtained in ceramic veneer, either plain or in polychrome colors. This requires the use of the original sculptural piece, from which a handmade reproduction is made.

Decorative, perforated ceramic veneer units are now available for use as screens, grilles, and facades in architectural work. They are made in a large range of colors, sizes, and designs. Although any design is possible, it is advisable to check with the local manufacturer or supplier for current stock items and their availability.

Application

Extruded adhesion-type ceramic veneer may be cemented to a masonry or concrete wall or to exterior plaster, provided that the mortar bond is sufficient to withstand a shearing stress of 50 lbf/in.^{2} (0.343 MN/m^2) after curing for 28 days. Installation includes the following steps. Pieces must be soaked in clean water for 1 hour or more just prior to installation. At the beginning of each day all walls must be drenched with clean water and again drenched no sooner than 1 hour before setting. Just prior to application of mortar coats, the wall and entire back of the ceramic veneer about to be set are given a brush coat of neat portland cement and water. Immediately after, half of the mortar coat should be spread on the wall and the other half on the back of the ceramic veneer, which should then be tapped into place on the wall so as to fill all voids completely and leave no air pockets. The thickness of the mortar coat should average $\frac{3}{4}$ in. (19.05 mm), but a slight excess of mortar should be used so that, when it is tapped into place, the excess will be forced out of the joints.

Joints. Joints in ceramic veneer installations are an important detail. Mortar used should consist of 1 part portland cement by volume, 0.5 part high-lime putty, and 0.5 part clean sand, to which is added 1 quart of ammonium stearate or its equivalent. (*See* Table *C15*.) Mortar joints are usually $\frac{1}{4}$ in. (6.35 mm) thick, struck flush. The usual admixture for mortar for adhesion-type ceramic veneer is ammonium stearate or its equivalent.

Expansion joints vary in number with the size of the ceramic veneer units used. When the size of the ceramic veneer has been determined, the manufacturer should be required to indicate the locations of expansion joints on his setting drawings. Expansion joints are filled under pressure with a nonstaining, elastic pointing compound to within $\frac{1}{2}$ in. (12.7 mm) of the face and then pointed with pointing mortar.

Table C15 Types of mortar for ceramic veneer

Ceramic veneer	Proportions of mortar mix	Use
Extruded (anchor and adhesion)	$1-\frac{1}{2}-4\frac{1}{2}^{a}$	Setting and pointing
	$1-0-1^{a}$	Grout, grout filler
	$1-0-5^{b}$	Grout, grout filler
	$1-\frac{1}{2}-4^{a}$	Setting and pointing
Handmade (closed back, open back, and solid slab)	$1-0-7^{c}$	Grout filler
	$1-\frac{1}{2}-4\frac{1}{2}^{a}$	Pointing

[a] Sand.
[b] Pea gravel passing $\frac{3}{8}$-in. (9.53-mm) sieve.
[c] Sand and top gravel.

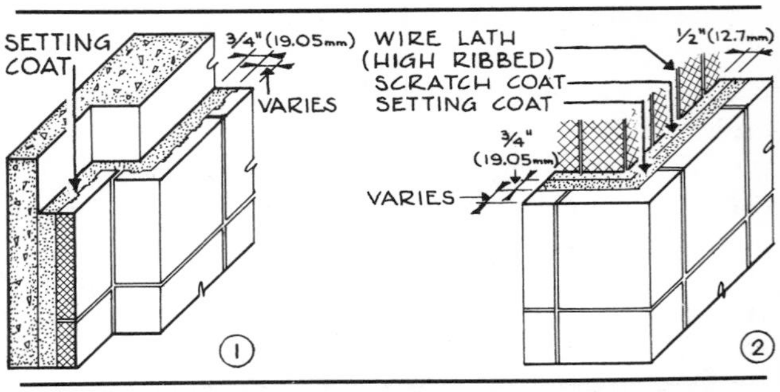

Figure C9 Construction details for extruded adhesion-type ceramic veneer on masonry or concrete walls (1) and on wood or metal walls (2).

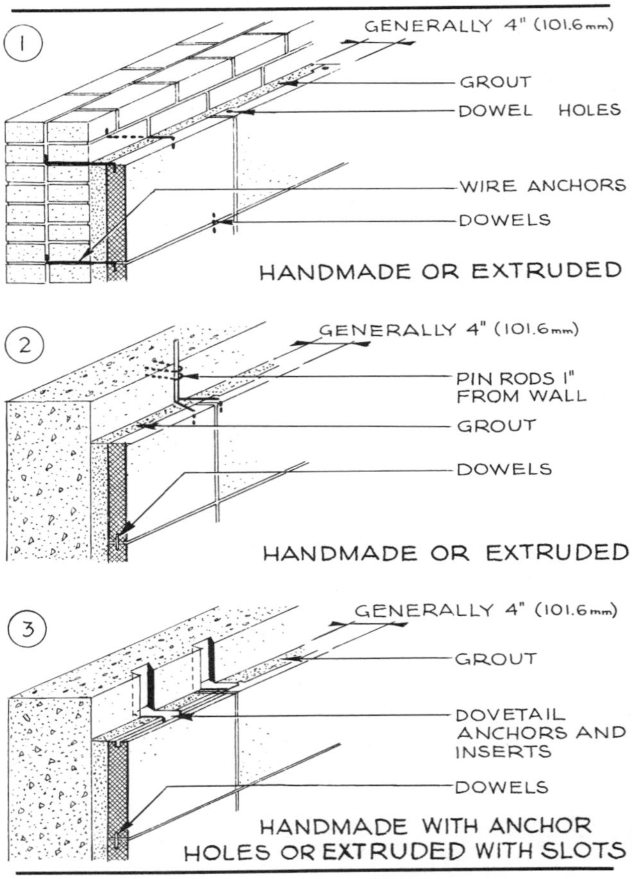

Figure C10 Construction details for extruded anchor-type or handmade ceramic veneer on brick or masonry walls (1) and on concrete walls (2), showing method for alteration to existing buildings (3).

Installation Temperatures. Freezing weather should be avoided when installing ceramic veneer. No work should be done unless adequate means are provided for maintaining a temperature above 32°F (0°C) for 48 hours after lay-up.

Cleaning. Ceramic veneer work is usually cleaned with brushes and clean water or with a weak solution of muriatic (hydrochloric) acid, followed by a thorough rinsing with clear water.

Condensed Checklist

1. Specifications: Setting drawings that show exact details of placement of individual units and a setting number for each piece of ceramic veneer should be called for in the specifications.
2. Type of ceramic veneer and method of application: Figures *C9* and *C10* show construction details for each type.
3. Type of mortar.

Table C16 Maximum facial distortion and required number of anchors for ceramic veneer

Face area		Maximum permissible distortion		Minimum number of anchors
ft^2	m^2	in.	mm	
1 and under	0.0929 and under	$\frac{1}{16}$	1.5875	None
Over 1 to 2	Over 0.0929 to 0.1858	$\frac{1}{8}$	3.1750	2
From 2 to 4	From 0.0929 to 0.3716	$\frac{1}{8}$	3.1750	3
From 4 to 12	From 0.3716 to 1.1148	$\frac{3}{16}$	4.7625	4
From 12 to 20	From 1.1148 to 1.8580	$\frac{3}{16}$ and over	4.7625 and over	6
Over 20	Over 1.8580	$\frac{3}{16}$ and over	4.7625 and over	One anchor for each 3 ft^2 (0.2787 m^2) of area

4. Joints, including type of mortar joint and location and type of expansion joints. An allowance of $\frac{1}{4}$ in. (6.35 mm) for mortar joints should be made in dimensioning.
5. Wetness of the ceramic veneer and of the surface to which it is to be applied.
6. Anchorage: A check should be made of the location and type of anchorage and also of metal-to-metal contacts to detect any possibility of electrolytic action.
7. Facial distortions and dimensions (*see* Table *C16*).
8. Temperature during setting.
9. Special shapes at openings.

Conditions Favorable to the Use of Ceramic Veneer

1. Where large units are a design consideration.
2. Where a wide choice of color and texture ranges is desired.
3. Where the surface is likely to receive rough treatment.
4. Where maintenance must be kept to a minimum.
5. Where a special surface design is desired.
6. Where fire resistance is required. Municipal, local, and state codes should be checked and also the fire rating codes of the Fire Underwriters, insurance companies, labor departments, and federal government (Army, Navy, etc.).

Conditions Unfavorable to the Use of Ceramic Veneer

1. As a structural material, unless specifically designed, tested, and approved for such an application.
2. Where a watertight condition is required, unless properly put up with all precautions, that is, solid back-up, solid panels, and solid joints.
3. Where a high coefficient of thermal insulation or sound absorption is required.

History and Manufacture

In Italian "terra cotta" literally means cooked or baked earth. Terra cotta was used throughout the ancient world for pottery and sculpture. It was probably first used extensively as a construction material by the Egyptians, Greeks, and Etruscans. The Romans also used terra cotta as a substitute for stone in construction and developed its ornamental use in buildings. After several centuries of disuse, it was revived as a construction material in the 13th century, first in Italy and Germany and then throughout Europe. At the end of the 16th century its popularity once more declined until the 19th century, when again it was widely used for architectural ornament.

Ceramic Veneer. Today, terra cotta is known as ceramic veneer. It is manufactured primarily by the de-aired stiff-mud extrusion process similar to that used for brick. Some is still pressed in molds by hand. In the stiff-mud process the clay contains just sufficient moisture (12 to 15%) and plasticity to be extruded through a die. Handmade ceramic veneer requires that the clay be pressed into the mold in thin layers and that it be thoroughly dry before burning.

To ensure straightness and lack of warpage, all ashlar is face-planed after drying and before glazing. After drying, the glaze or glazes are then applied and the whole fired to high temperatures of about 2200°F (1204.44°C). All ceramic veneer is sized by grinding and cutting after firing so that no dimension varies by more than $\frac{1}{16}$ in. (1.59 mm) over or under that specified. All beds, ends, and backs must be free from glaze.

CLAY TILE: DRAIN AND VITRIFIED CLAY PIPE

Physical and Chemical Properties

The clays used for clay drain tile and vitrified clay pipe and fittings are similar to those used for burned brick. Both drain tile and vitrified clay pipe are designated by their inside diameters.

Clay drain tile is usually round, unglazed, and either solid or perforated below the horizontal axis (*see* Figures *C11* and *C12*). It is generally manufactured with straight ends and in three classes: Standard, Extra Quality, and Heavy Duty; perforated drain tile has a fourth class, Extra Strength.

Vitrified clay pipe and fittings are round, glazed or unglazed, and able to withstand attack by acids, alkalis, bases, detergents, etc., as well as bacterial action and root infiltration. Vitrified clay pipe and fittings are generally manufactured with bell and spigot and in two classes: Extra Strength and Standard Strength (*see* Figure *C13*).

Types and Uses

Clay drain tile is manufactured in the following sizes: 3 to 24 in. (76.2 to 609.6 mm) in diameter, and in 1-, 2-, and 3-ft (304.8-, 609.6-, and 914.4-mm) lengths. Perforated drain tile is available in diameters of 3 to 8 in. (76.2 to 203.2 mm) and in lengths of 12 and 24 in. (304.8 and 609.6 mm). A wide variety of fittings is available (*see* Figure *C11*).

The major uses of clay drain tile in construction are for foundation drains, storm water leaders, and gutters for subsoil distribution to dry wells, subsoil sewage disposal drains, and drainage under concrete floor slabs. Clay drain tile is also used for drainage in the construction of highways, athletic fields, and recreational areas. Clay drain tile is usually laid with open joints (*see* Figure *C14*), and perforated drain tile with either open joints or couplings.

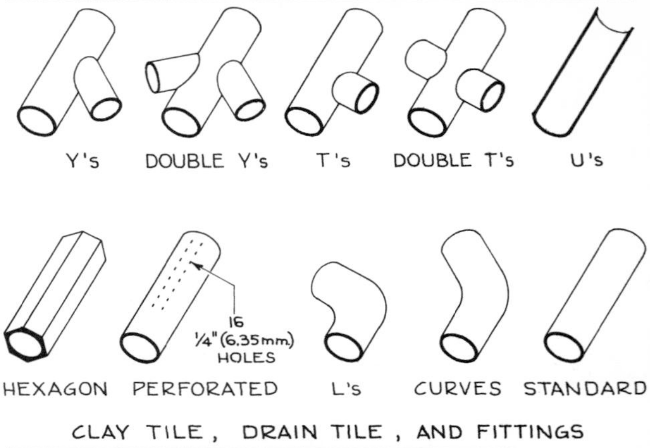

Figure C11 Clay drain tile and fittings.

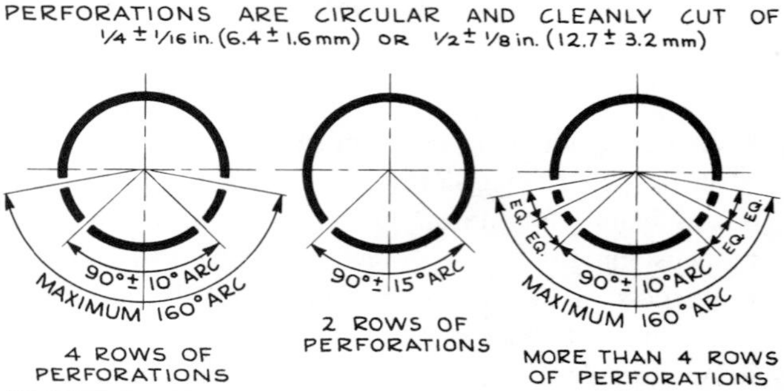

Figure C12 Spacing of perforations for perforated drain pipe.

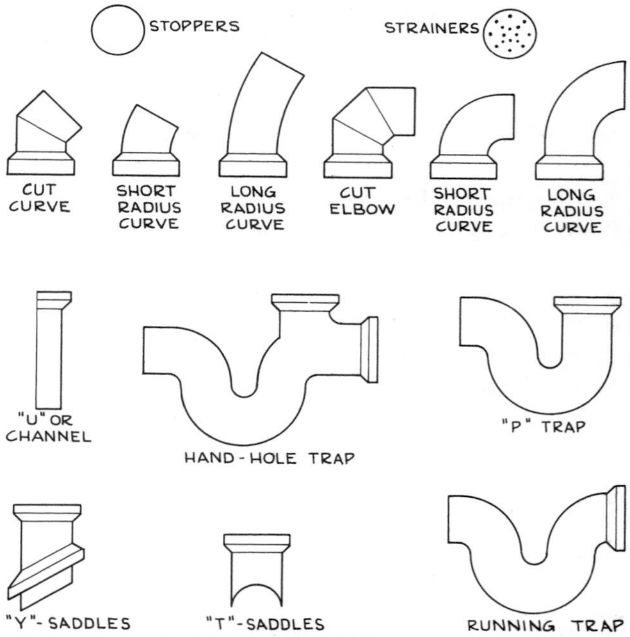

Figure C13 Bell and spigot vitrified pipe and fittings.

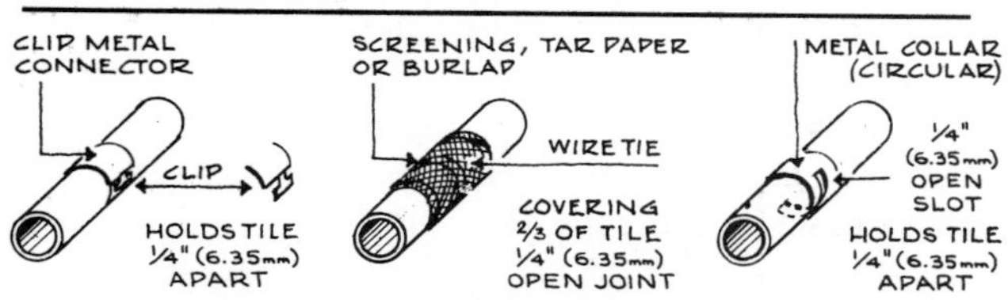

Figure C14 Details of clay drain tile joints.

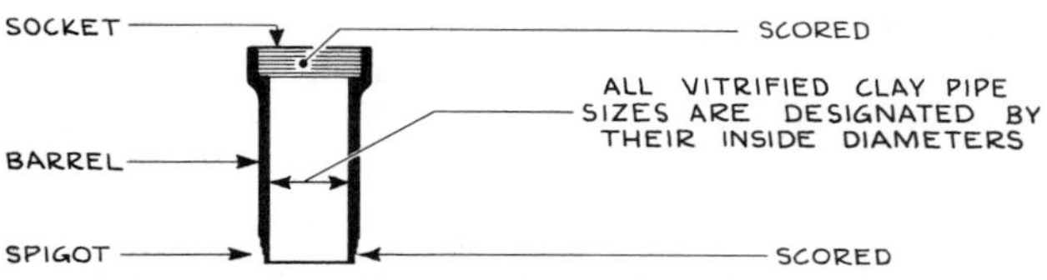

Figure C15 Nomenclature of bell and spigot vitrified clay pipe.

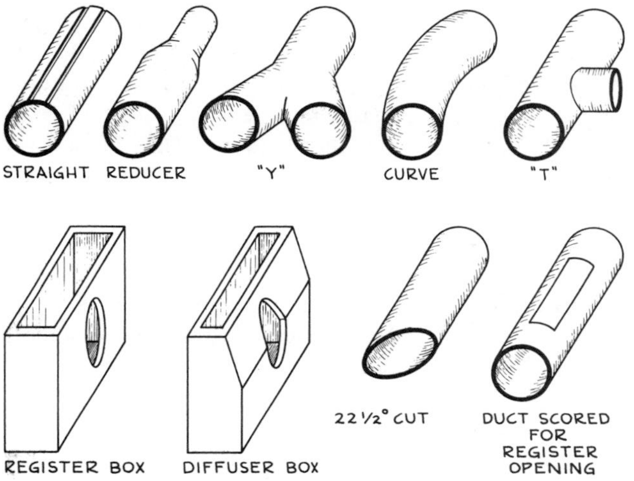

ALL DUCTS AND FITTINGS ARE MADE FOR 6" (152.4mm), 8" (203.2mm), 10" (254mm), 12" (304.8mm), 15" (381mm), 18" (457.2mm), 20" (508mm), AND 24" (609.6mm) DIAMETER DUCTS. STRAIGHT DUCTS COME IN 6" (152.4mm), 9" (228.6mm), 12" (304.8mm), 18" (457.2mm), 24" (609.6mm), AND 36" (914.4mm) LENGTHS.

Figure C16 Vitrified clay duct and fittings for underfloor heating and cooling systems.

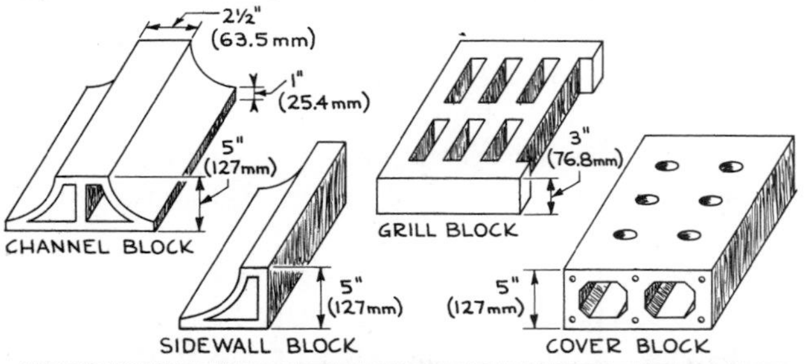

Figure C17 Clay tile underground system for sewage treatment.

Vitrified clay pipe is usually manufactured in bell and spigot with plain end available, and in the following sizes: 4 to 42 in. (101.6 mm to 1.07 m) in diameter and in various lengths up to 8 ft (2.44 m). Perforated pipe is available 4 to 24 in. (101.6 to 609.6 mm) in diameter, and in 2- and 3-ft (609.6- and 914.4-mm) lengths. Stock fittings are available in a large variety of sizes and shapes (*see* Figures *C13* and *C15*).

The major uses of vitrified clay pipe in construction are for sanitary and storm sewer systems; drainage of streets, highways, airports, parks, and playgrounds; and building sewers and drains. Vitrified clay pipe is also used for heating ducts in slab-type buildings, using special polyvinyl chloride tape with a pressure-sensitive adhesive for installation (*see* Figure *C16*). Vitrified clay pipe, when installed, is usually laid with compression joints or with oakum and cement or bituminous compounds to fill in the joints.

There are special tiles and pipe for sewage treatment plants (*see* Figure *C17*) and special products for meter boxes, grease traps, etc. (*see* Figure *C18*).

Application

Condensed Checklist

1. Pitch for vitrified clay pipe for sanitary and storm systems should be always checked against local, municipal, and state codes and health departments.
2. Pitch for clay drain tile should be a minimum of 3 in. (76.2 mm) and a maximum of 4 in. (101.6 mm) per 100 lin. ft (30.48 m).

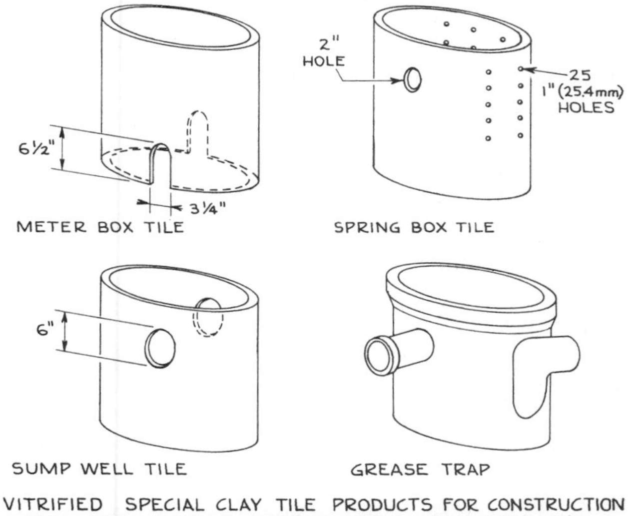

Figure C18 Special clay tile products for construction.

3. Clay drain tile with open joints should have the joints covered or treated by other methods to prevent displacement and clogging (*see* Figure *C21*).
4. Clay perforated drain tile should be installed with compression couplings field assembled in accordance with manufacturers' directions.
5. Vitrified clay pipe should be made leakproof with oakum and cement or bituminous compounds or with compression joints which consist of gaskets molded to both bell and spigot shapes.
6. Support of clay tile drain pipe should be as detailed in Figure *C19*.
7. Support for vitrified clay pipe should be checked against trench width, thickness of bedding material, required pitch, load to be supported, and method for backfilling. The trench should always be free of water.

Figure C19 Details of support for clay tile drain pipe.

History and Manufacture

Clay tile drain pipes of various shapes, many of them cylindrical, have been discovered in excavations dating back to before 4000 B.C. Present-day clay drain tile and its modern development, vitrified clay pipe, are both produced by machines similar to those for brick and structural clay tile products. But the machines in this case are assembled vertically because the clay column must flow down instead of horizontally to form the bell and the other, more intricately shaped parts such as joints, increasers, and decreasers.

CLAY TILE: FLOOR AND WALL TILE (CERAMIC TILE)

Physical and Chemical Properties

The designation "tile" here refers only to relatively small surfacing units, glazed or unglazed, conventionally flat, made from clay or a mixture of clay and other ceramic materials, and fired according to various processes.

Tiles differ principally in (1) composition of the tile body; (2) surface finish, that is, glazed or unglazed; (3) process of manufacture, that is, whether the dust-press or plastic method is used in shaping the body of the tile (the appearance of the tile usually denotes the process used); and (4) the degree of vitrification or fusion of the tile body after firing, as indicated by the extent to which it absorbs moisture.

Vitrification. Vitrification is also a measure of the density of tile bodies and is generally classified into four degrees: nonvitreous, semivitreous, vitreous, and impervious.

Nonvitreous tiles have a degree of density that permits moisture absorption of more than 7% of the weight of the tile but does not prevent the product from having a high degree of strength. Such tiles often have sufficient strength even when their absorption is as high as 18%. The moisture absorption of nonvitreous tiles facilitates installation because of their ready adherence to mortar.

Semivitreous tiles have a degree of density that limits moisture absorption to 3 to 7% of the weight of the tile.

Vitreous tiles have a moisture absorption of 0.5 to 3% and a body density which prevents any penetration of dirt that cannot be easily removed.

Impervious tiles are the hardest. Their moisture absorption is less than 0.5% and they are readily cleansed of stains and dirt. Few tiles have this degree of vitrification, and usually they are manufactured by special order.

Composition. Tiles are made of compounded and of natural clay bodies. Those made of compounded bodies contain three principal constituents—the plastic, the filler, and the flux or solvent—all of which are carefully proportioned and mixed before shaping the tiles.

Plastics are chiefly clays having high bonding power and some fluxing ability. The so-called ball clays are universally used as plastics in tile manufacture, not only because of their high plasticity and bonding power but also because of their white or creamy color when fired.

Fillers may be nonplastics and partial nonplastics. Flint or finely pulverized silica is the leading nonplastic filler used in the compounding of tile bodies. It reduces shrinkage in drying and firing and imparts to the body a certain rigidity that prevents deformation under heat.

Kaolins or china clays act to some extent as fillers in tile bodies and are the most important partial nonplastics employed in tile manufacture. Because of variations in their physical and chemical properties, kaolins of two types are generally blended in order to utilize the desirable qualities of each.

Talc and pyrophyllite have recently come into wide use as filler ingredients of wall tile bodies. Among the outstanding advantages they impart are lower firing temperatures, less shrinkage, and low moisture expansion, which greatly lessens the crazing or cracking of glazes.

The flux most widely used in floor and wall tiles is the mineral feldspar, which melts under intense heat and fuses the more heat-resistant ingredients into a solid mass. It is sold to the ceramic trade in finely pulverized form. The potash type of feldspar, which contains over 9% potassium oxide, is the one generally used.

Proportions of Constituents. Compositions of tile bodies vary considerably, depending on the properties and proportions of clay, flint, and feldspar used in compounding them. For nonvitreous wall tiles, clay usually constitutes about 50 to 60% of the mixture, flint about 30 to 35%, and feldspar about 10 to 15%. A blend of ball clay and of one or more kaolins is generally used, the kaolins usually exceeding the ball clay by a small percentage.

In the newer wall tile compositions using talc or pyrophyllite, the incorporation of these materials has been largely at the expense of feldspar and flint. Some mixes require as low as 10%, whereas others may contain over 75% of talc or pyrophyllite.

For ceramic mosaic tile of the porcelain type, the proportions of clay, flint, and feldspar often approximate the following: clay (ball clay and kaolin), 30 to 35%; flint, 10 to 15%; and feldspar, 45 to 50%. The large feldspar content promotes vitrification by increasing the formation of a glassy matrix in the fired ware.

Finish. Unglazed tiles are composed of the same ingredients throughout and derive their color and texture from the materials of which the body is made.

Glazed tiles have a glassy surface of ceramic materials fused upon their face to give them a decorative appearance and to make the surface impervious to moisture. Glazes are produced in a large variety of colors and shades, ranging from pure white to jet black. Glaze finishes are of two general classes in regard to their light-reflecting properties: (1) bright glazes, which have a highly polished surface and reflect an image clearly; and (2) matte glazes, which do not clearly reflect an image or are entirely without sheen. All degrees of semilustrous or satinlike finish may be produced between the two extremes of reflection and nonreflection. In addition, glazes may have a plain, textured, polychrome, mottled, stippled, or rippled surface.

Crystalline glazes are a distinctive type which is costly to produce. These glazes are characterized by a texture featuring crystals of various sizes, shapes, and colors. The finish varies from bright to matte on the same piece of tile, the crystals differing from the surrounding glaze.

Glazes for floor and wall tiles are also made in different degrees of hardness in accordance with their ultimate use. For example, a soft glaze may be satisfactory in a purely decorative use, whereas a hard, dull glaze is necessary when the tile is subjected to the abrasive action of foot traffic. The harder glazes must generally be fired at temperatures considerably above those required for soft glazes.

Standardization of Grades. Standardization was initiated when the U.S. Bureau of Standards suggested grading, and in 1930 two designations, standard and seconds, were established for white and colored glazed tiles and for ceramic mosaic tiles. No grades were established for faience and quarry tiles or for floor tiles other than ceramic mosaics, their wide variation making any classification impossible.

Warpage and blemishes on the surface of tile are primary factors in grading, for there is no difference in the wearing and sanitary qualities of different grades. Tile of "standard" grade must not have a warpage of more than 0.4%, that is, 0.024 in. (0.61 mm) in the 6-in. (152.4-mm) length or 0.017 in. (0.43 mm) in the $4\frac{1}{4}$-in. (107.95-mm) length; also, it must be free from defects such as specks and blots (face stains), visible at a distance of more than 3 ft (0.9144 m). "Seconds" may have blemishes and defects not permissible in the higher grade but must be free from blots and biscuit cracks (any body fracture visible on both the face and back of tiles). Standard grade tiles may include up to 5% of these seconds.

Types and Uses

The leading types of clay floor and wall tile are the following: (1) glazed wall tile, (2) mosaic tile, (3) quarry tile, and (4) paver tile (*see* Table *C17*).

Sizes are coordinated to permit combinations of different sizes into patterns and of different types of tile (*see* Figure *C20*).

Trim tiles are variously shaped units used for bases, caps, corners, moldings, angles, etc., of tilework. In contrast to flat tiles, which are sold on a square-foot basis, trim tiles are sold by the piece. Practically every tile job, regardless of size, requires some trim tiles that correspond to the flat tiles with which they are installed.

Glazed Wall Tile. Glazed wall tiles are made with an impervious glaze over a nonvitreous body and are used

Table C17 Types of clay tile floor and wall tile

Type of tile	Physical characteristics	Disadvantages	Major use	Standard sizes: Thickness		Width		Length	
				in.	mm	in.	mm	in.	mm
Glazed wall tile[a]	Nonvitreous body, matte or bright glaze	Not suitable for floors or areas subject to freezing	Interior walls	$\frac{5}{16}$	7.94	$4\frac{1}{4}$	107.95	$4\frac{1}{4}$	107.95
						$4\frac{1}{2}$	107.95	6	152.40
						6	152.40	6	152.40
	Nonvitreous body, crystalline glaze	Not suitable for areas subject to freezing	Interior walls and floor areas subject to rough usage and countertops						
	Vitreous body, matte or bright glaze	Not suitable for floors	Interior and exterior areas subject to freezing						
Mosaic tile	Porcelain body, unglazed	Disadvantages imposed by method of installation and type of grout	Interior and exterior walls and floors; major use is for bathrooms, kitchens, and swimming pools	$\frac{1}{4}$	6.35	1	25.4	1	25.4
						1	25.4	2	50.8
						2	50.8	2	50.8
	Porcelain body, glazed	Not suitable for floors	Interior and exterior walls						
	Natural clay body, unglazed	Requires a frostproof body when used on the exterior	Interior and exterior walls and floors						
	Natural clay body, glazed	Requires a frostproof body when used on the exterior	Interior and exterior walls						
	Conductive	Requires special installation	Areas where danger of static electricity is present						
Quarry tile	Natural clays and shales, unglazed, impervious	None	Interior and exterior moderate to heavy-duty floors	$\frac{1}{2}$	12.7	3	76.2	3	76.2
						4	101.6	4	101.6
						3	76.2	6	152.4
						6	152.4	6	152.4
						4	101.6	8	203.2
				$\frac{3}{4}$	19.05	6	152.4	6	152.4
						$2\frac{1}{2}$	63.5	8	203.2
						4	101.6	8	203.2
						6	152.4	9	228.6
						9	228.6	9	228.6
Paver tile	Unglazed vitreous, semi-vitreous, and impervious	None	Interior and exterior heavy-duty floors	$\frac{3}{8}$	9.53	4	101.6	4	101.6
				$\frac{1}{2}$	12.7	4	101.6	4	101.6
						6	152.4	6	152.4
						4	101.6	8	205.2

[a]Glazed wall tiles are available in a variety of shapes other than the standard sizes.

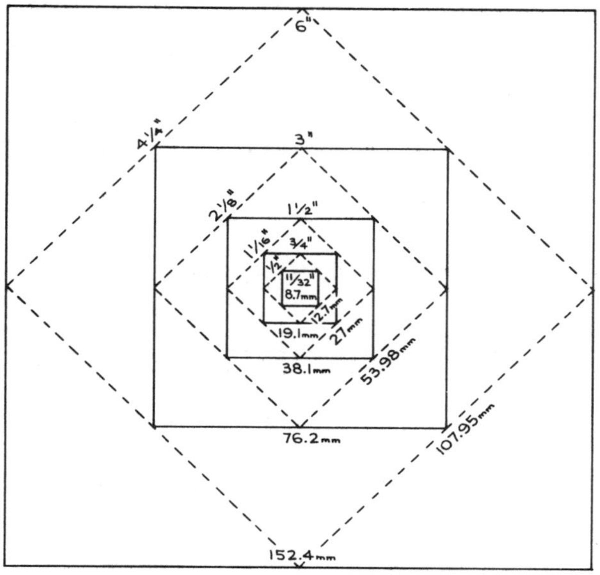

Figure C20 Basis of sizes for common square tile of less than 36 in.

generally on the interior. They are available with an impervious (vitreous) body for exterior use where they are exposed to moisture and freezing. Glazed wall tiles are manufactured with plain or scored surfaces, in bright, matte and crystalline finishes, and in hand-decorated and sculptured designs. They are suitable for flooring in areas subject to moderate foot traffic. Most wall tile is made with square edge or cushioned edge. Bathroom accessories are available in matching colors. (*See* Figures *C21*, *C22*, *C23*, and *C24*.)

Mosaic Tiles. Mosaic tiles are tiles less than 6 in.^{2} (38.712 cm^2) in facial area, and are available glazed or unglazed. Ceramic mosaic tiles are made in two distinct types, the porcelain type and the natural clay or shale type. The former is made by the dust-press process from a carefully proportioned blend of ceramic materials, and the latter is made largely from natural clays or shales by either of the two basic processes, the plastic method being the more common. Unglazed tiles have a wide range of natural earth colors that extend throughout the body of the tile. Slip-resistant and conductive tiles are produced with special additives incorporated. These tiles all may have a plain, mottled, textured or flashed surface. (*See* Figures *C25* and *C26*.)

To facilitate installation, ceramic mosaic tiles are usually mounted at the factory on sheets of paper about 2 ft^2 (0.1858 m^2) in area, the individual tile units being spaced so as to allow for the insertion of cement between them when the paper is removed and the face of the tiles is exposed.

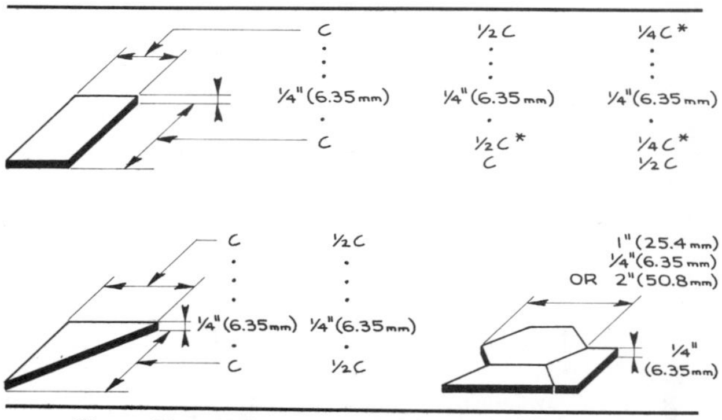

Figure C21 Sizes and shapes of glazed interior tile.

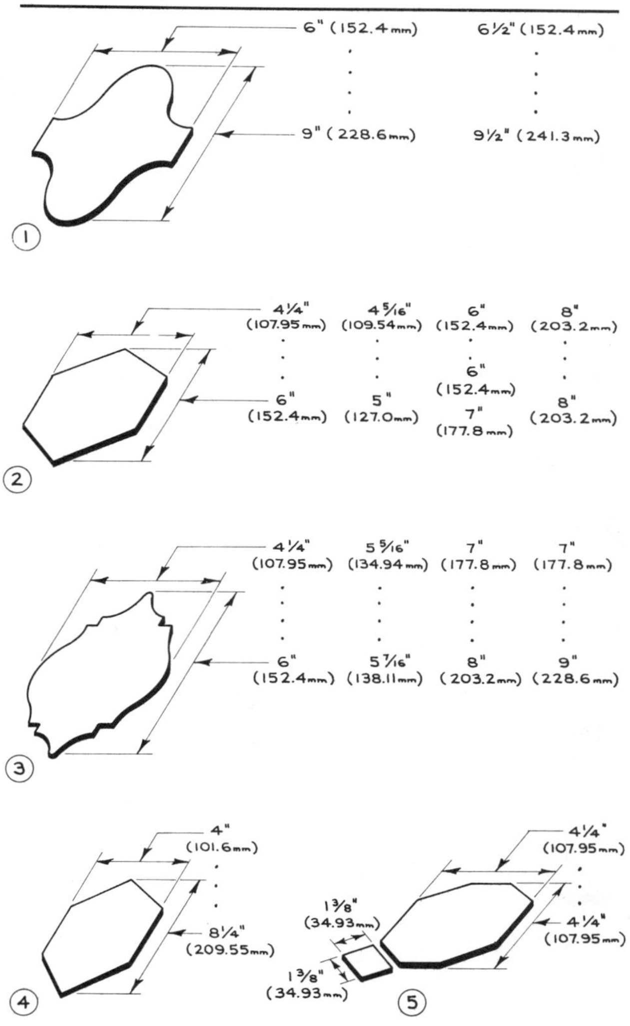

NUMBERS 2 & 3 - GLAZED FLOOR AND WALL TILE AND QUARRY TILE
NUMBER 4 - QUARRY TILE
NUMBERS 1 & 5 - GLAZED FLOOR TILE

Figure C22 Types of glazed floor, wall, and quarry tile: glazed floor tile (1, 5); glazed floor and wall tile (2, 3); and quarry tile (2, 3, 4).

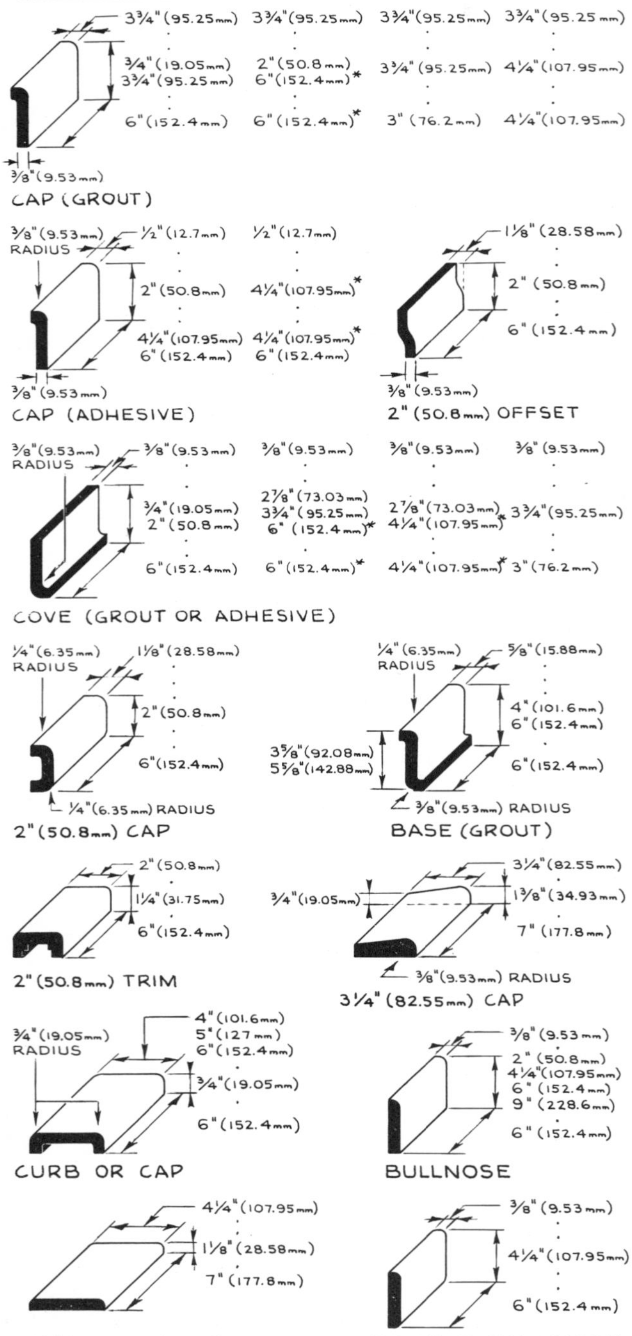

Figure C23 Trimmers for glazed interior tile.

Unglazed Porcelain Tile. Frostproof unglazed porcelain tile is made from refined ceramic materials and is a vitreous (impervious) tile that is dense, smooth, highly stain-resistant, and wear-resistant; its surface is characterized by clear, luminous colors or granular blends. Slip-

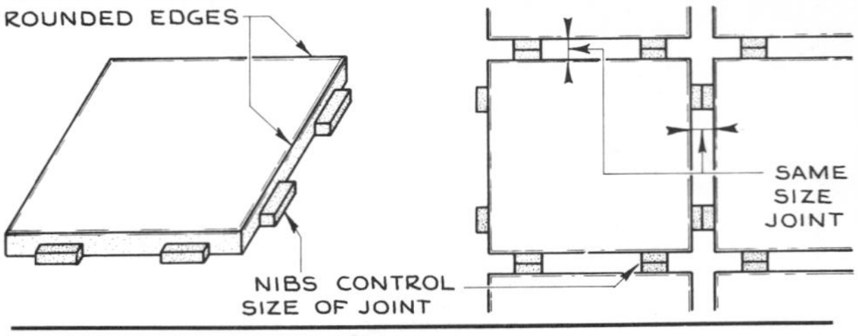

Figure C24 Typical wall tile with nibs that control the size of joints.

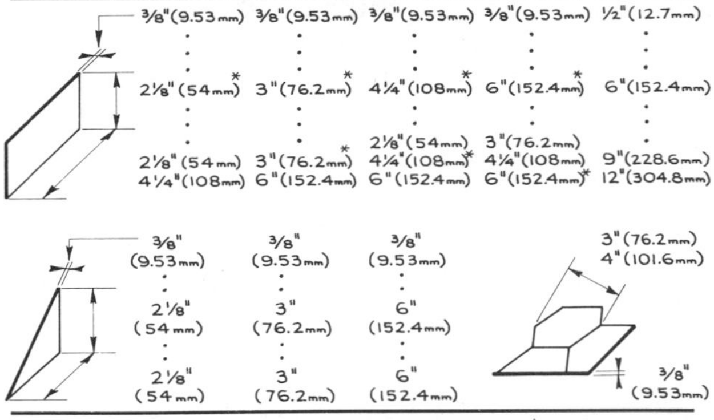

Figure C25 Sizes and shapes of unglazed mosaic tile.

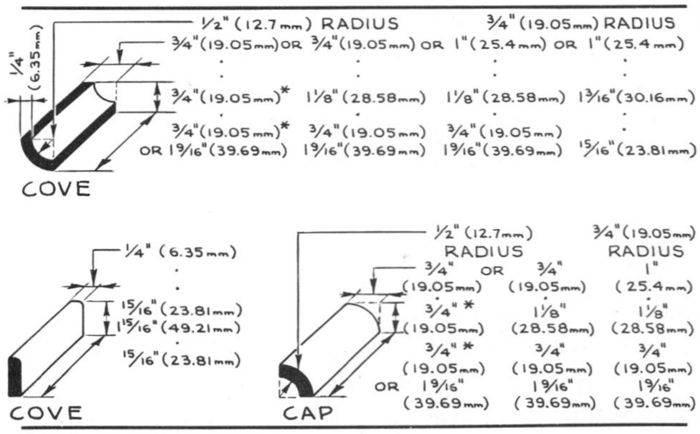

Figure C26 Trimmers for unglazed mosaic tile.

resistant 1-in. (25.4-mm) squares in limited colors are available.

Glazed Porcelain Tile. Frostproof glazed porcelain tile has the same body as unglazed porcelain tile with various glazes applied, including transparent ceramic, textured, metallic, and special decorative glazes.

Unglazed Natural Clay Tile. Frostproof and non-frostproof unglazed natural clay tile is made from unwashed clays and has a dense, abrasion-resistant body with a rugged, slightly textured surface. Slip-resistant

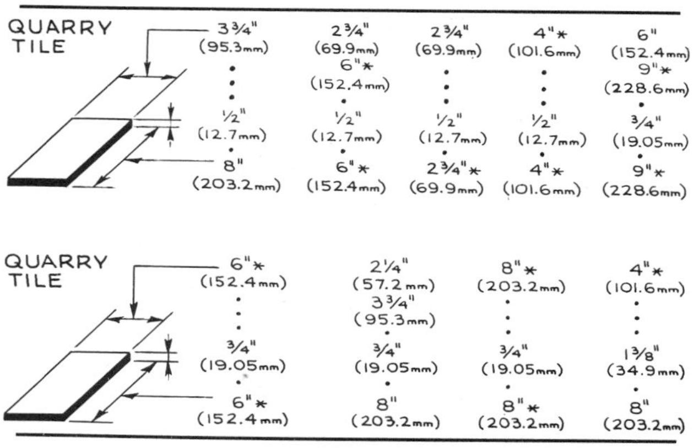

Figure C27 Sizes and shapes of quarry tile.

1-in. (25.4-mm) squares are available in a limited range of colors.

Glazed Natural Clay Tile. Frostproof and nonfrost-proof glazed natural clay tile has the same body as unglazed natural clay tile with various glazes applied, including transparent ceramic, metallic, and special decorative glazes.

Quarry Tile. Quarry tile is unglazed floor tile made from natural clays or shales by a plastic method similar to the machine extrusion methods employed in making plastic (stiff-mud) face brick. Quarry tiles are a very durable flooring material, being impervious to moisture, stains, and dirt, and resistant to abrasion, freezing, and thawing. They are generally sold in three grades: select, standard, and seconds or commercial. (*See* Figures *C27* to *C29*.)

Pavers. Pavers are standard-size unglazed tiles, resembling ceramic mosaic tiles in composition and physical characteristics but usually having a facial area of 6 in.^{2} (645.2 mm^2) or more. Because of their greater size, these tiles are generally not pasted onto paper but are laid out individually. Like the porcelain type of ceramic mosaic tile, dust-pressed porcelain pavers are either impervious or vitreous. When the natural clay paver is plastic-made, it is either vitreous or semivitreous; but when made by the dust-press method, it is sometimes impervious. All pavers, regardless of body density, are weatherproof and are especially suitable for heavy floor service. (*See* Figure *C30*.)

Application

Clay floor and wall tile can be applied by a wide variety of mortars, adhesives, and methods, including portland cement mortar, dry-set mortar, latex portland cement mortar, epoxy mortar, epoxy adhesive, furan mortar, organic adhesive, and pressure-sensitive adhesive methods. Pregrouted sheets and panels of wall tile are also available ready to be applied directly to the wall with adhesives.

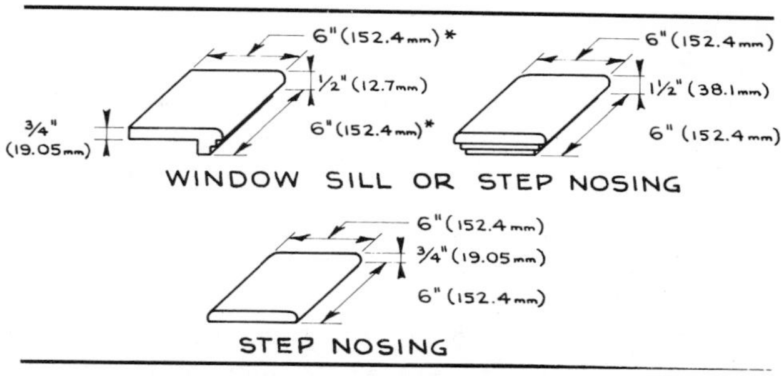

Figure C28 Special shapes of quarry tile.

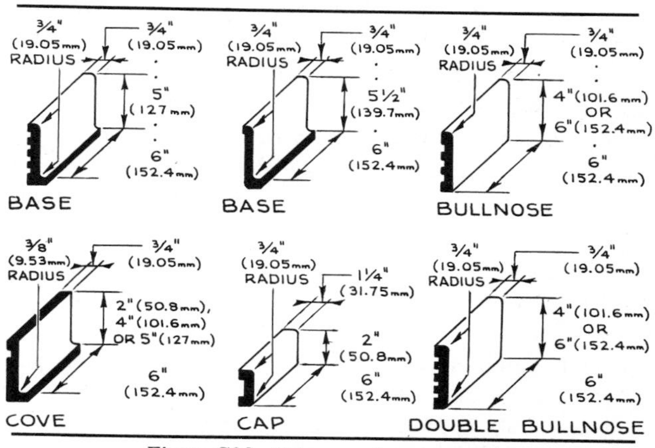

Figure C29 Trimmers for quarry tile.

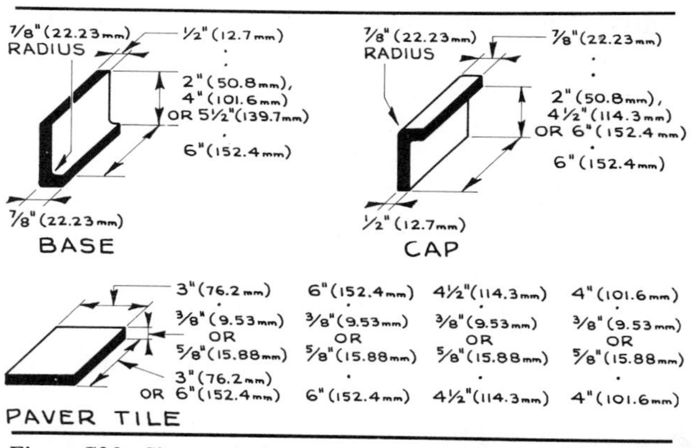

Figure C30 Sizes and shapes of paver tile, including trimmers.

Portland cement mortar is a mixture of portland cement, sand, hydrated lime, and water. The mixture varies depending on whether the mortar is to be used for a leveling bed or a setting bed, and whether it is to be applied on a vertical or horizontal surface. The coat that applies or holds the tile is a mixture of portland cement and water and is called the bond coat. A portland cement bed should always be reinforced with metal lath whenever it is backed with a cleavage membrane or open stud framing or furring. (*See* Figures *C31* to *C33*.) Suitable backings are brick, concrete block, structural clay tile, concrete, foam insulation board, gypsum wallboard, and gypsum plaster or plaster block. Portland cement mortar is used for both interior and exterior applications.

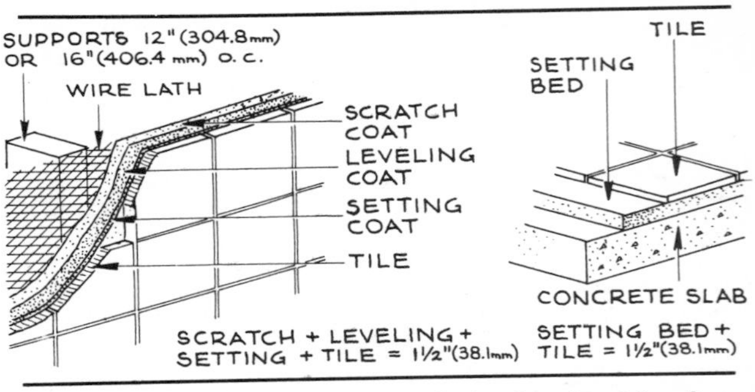

Figure C31 Number of coats for vertical and horizontal surfaces for cement mortar application of tile.

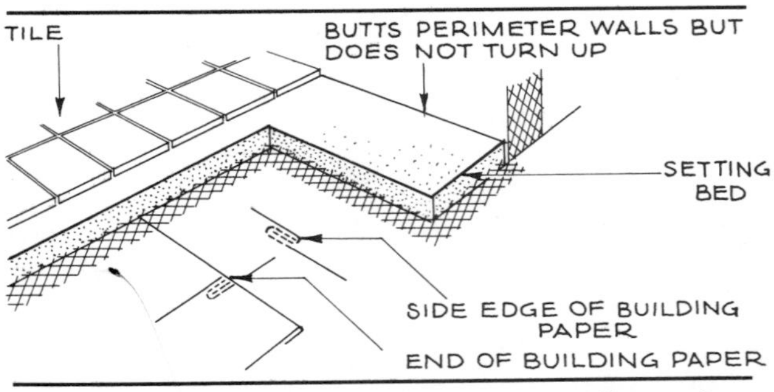

Figure C32 Cleavage planes on smooth structural floor surfaces.

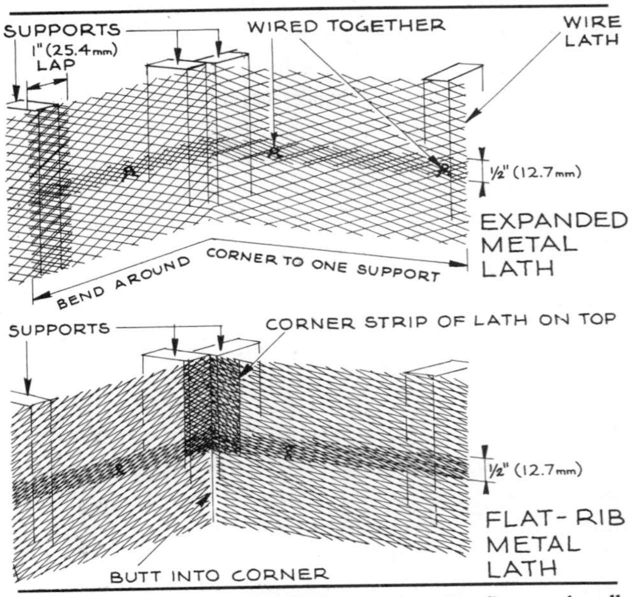

Figure C33 Application of metal lath for clay floor and wall tile installed with cement mortar.

Dry-set mortar is a mixture of portland cement, sand, and resinous additives. Dry-set mortar has impact resistance, is nonflammable, and is suitable for both interior and exterior applications. It is available as an unsanded mortar or as a factory-sanded mortar that needs only the addition of water. It is limited for trueing or leveling to a thickness of about $\frac{1}{4}$ in. (6.35 mm). Suitable backings are brick, concrete block, concrete, foam insulation board, gypsum wallboard, and cured portland cement mortar.

Latex-portland cement mortar is a mixture of portland cement, sand, and a latex additive. The uses of this type of mortar are similar to those for dry-set mortar. It is generally applied in a coat $\frac{1}{8}$ to $\frac{1}{4}$ in. (3.18 to 6.35 mm) thick.

Epoxy mortar is a two-part mortar consisting of an epoxy resin and a hardener. Epoxy mortar is used where chemical resistance and/or high bond strength are required. As a rule, it is applied in a layer $\frac{1}{16}$ to $\frac{1}{8}$ in. (1.59 to 3.18 mm) thick. Suitable backings are concrete, plywood, steel plate, and chipboard.

Epoxy adhesive is similar to epoxy mortar in that it consists of an epoxy resin and a hardener. It is generally used for thin-set application of tile on walls, floors, and counters where high bond strength and ease of application are desirable.

Furan mortar is a two-part mortar system consisting of a furan resin and a hardener. It is used where chemical resistance is of prime importance. Suitable backings are concrete and steel plate.

Organic adhesives are ready to use and harden by evaporation. They are applied in layers $\frac{1}{16}$ in. (1.59 mm) thick and spread with a toothed trowel. Suitable backings are concrete, gypsum wallboard, cement-asbestos board, plywood, chipboard, and gypsum plaster.

Grouting materials for clay wall and floor tile are manufactured and prepared in a large variety of types (*see* Table *C18*).

Installation systems for floor and wall tile are being simplified by the tile manufacturers' use of high-bond-strength and resilient grouts and the production of pregrouted sheets of floor and wall tile. In general, four systems are available in pregrouted sheets: (1) sheets of wall tile in $17\frac{1}{4}$-in. (438.15-mm) squares; (2) sheets of wall tile in 24-in. (609.6-mm) squares, with internal corner strips 24 in. (609.6 mm) long for showers; (3)

Table C18 Grouting materials for clay tile floor and wall tile

Type of grout	Physical characteristics	Disadvantages	Advantages	Major use
Commercial cement[a]	Prepared mix of portland cement and other ingredients; white, available in colors	Requires presoaking of tile and damp-curing of grouted surfaces	Suitable for interior and exterior installation	For installing wall tile
Sand-portland cement[a]	Field-mix of portland cement, fine sand, water, and sometimes hydrated lime	Requires presoaking of tile and damp-curing of grouted surfaces	Suitable for interior and exterior installations	For installing floor and wall tile
Dry-wall[a]	Prepared mix of portland cement and additives that provide water retention	Dampening of tile surface prior to grouting with extremely dry conditions	Does not require presoaking and damp-curing of grouted surfaces; suitable for interior and exterior installations	For installing wall tile
Latex-portland cement[a]	Prepared mix of commercial cement, sand-portland, or dry-wall plus latex additives	Dampening of tile surface prior to grouting with extremely dry conditions	Does not require presoaking and damp-curing of grouted surfaces; suitable for interior and exterior installations; increases resilience of grout	For installing floor and wall tile
Mastic	Ready-to-use one-part formulation	Not suitable for exterior use or areas in continuous contact with water; not suitable for heavy-duty floors	Does not require presoaking and damp-curing of grouted surfaces; stain-resistant to mildew growth	For installing floor and wall tile
Epoxy	Ready-to-use two-part system consisting of epoxy resin and hardener	Requires special skills for proper installation; has limited pot life.	Does not require presoaking and damp-curing of grouted surfaces; highly resistant to chemical staining; high bond strength; suitable for interior and exterior installations	For installing floor and wall tile
Furan resin	Ready-to-use two-part system consisting of furan resin and hardener	Requires special skills for proper installation; has limited pot life	Does not require presoaking and damp-curing of grouted surfaces; has extremely high chemical resistance; suitable for interior and exterior installation	For installing floor and wall tile

[a]Requires addition of water at the job site.

sheets of wall tile in sizes to fit different tub sizes and various heights of tilework above the tub, including internal corner strips and tub side trim; and (4) mosaic floor tile in sheets usually 24 in. (609.6 mm) square.

Another system for installing floor tile is the use of pressure-sensitive sheets 24 in. (609.6 mm) square or individual floor tile which can be applied directly to the substrate. These are suitable for light traffic areas and can be applied to concrete, plywood, and hardwood floors.

Condensed Checklist

1. Terminology: One should always make sure that specifications are written with the correct official terminology for the various clay floor and wall tiles.
2. Grade of tile to be used (*see* "Types and Uses" and "Standardization of Grades").
3. Samples: It is necessary to specify that samples of tile and trim pieces be submitted for approval.
4. Large-scale details: Elevations, plans, and sections of areas where floor and wall tile are to be used must be drawn at large scale to show tile spacing. This spacing, based on the size of tile selected, must fit heights and widths for walls and widths and lengths for floors so that no small thin slivers of tile need be used. Trimmers for base, caps, facing, and openings must also be calculated so that their dimensions and spacing are correct in relation to floor and wall tiles and also meet other dimension limitations.
5. Countertops: When tile is used for countertops, the method of treating sink or lavatory openings and

the treatment at intersections, at walls, and at edges must be correctly detailed (*see* Figure *C34*).

6. Joints: One should always check that the correct width is used for the various clay floor and wall tiles (*see* Table *C19*).

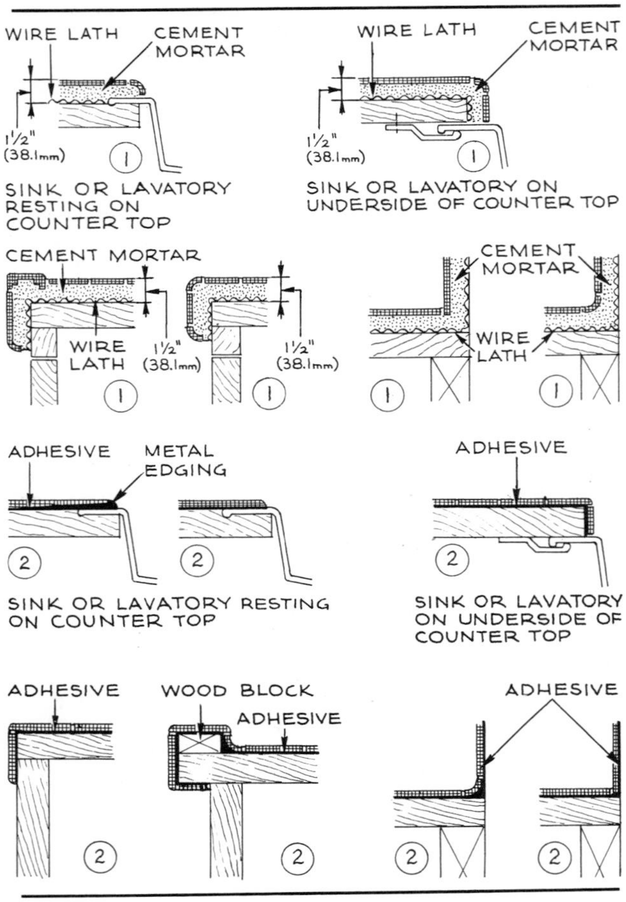

7. Temperature during installation: Clay floor and wall tile should not be installed in temperatures below 50°F (10°C) by either setting method. With cement mortar, low temperatures will alter and damage the mortar; and with adhesives which have volatile solvents, a temporary type of heating (salamanders) may cause explosions.

8. Grouting, pointing, and caulking materials: For clay floor and wall tile, one of three general types of material is commonly used: (a) trademarked or patented mixtures; (b) grouting, pointing, and caulking material for special end-use requirements such as resistance to acids or alkalis, and electrical conductivity; and (c) portland cement. In each case the manufacturer of clay floor and wall tile should be consulted in order to select the correct grouting,

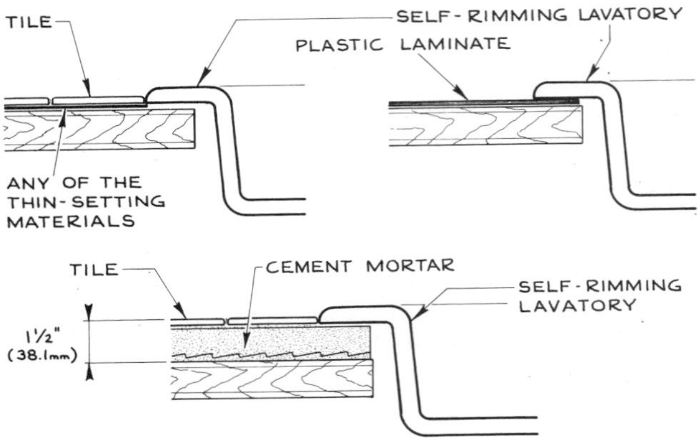

Figure C34 Details of ceramic tile countertops installed by the cement mortar method, by the dry-set latex-portland cement method, by organic mastic products, by epoxy and furan tile-setting products, and with self-rimming lavatories.

Table C19 Width of joints for clay floor and wall tile

Size limitations	Types of tile and form in which delivered to job	Width of joints: Minimum in.	Minimum mm	Maximum in.	Maximum mm
$2\frac{3}{16}$-in. (55.6-mm) or smaller square	Glazed and unglazed, mounted on paper	$\frac{1}{16}$	1.59	$\frac{1}{8}$	3.18
$2\frac{3}{16}$-in. (55.6-mm) and larger square	Glazed and unglazed, mounted on paper	$\frac{1}{16}$	1.59	$\frac{1}{4}$	6.35
$2\frac{3}{16}$- to $4\frac{1}{4}$-in. (55.6- to 107.9-mm) square	Unglazed, unmounted.	$\frac{1}{8}$	3.18	$\frac{1}{4}$	6.35
3-in. (76.2-mm) and larger square	Unglazed, unmounted	$\frac{1}{4}$	6.35	$\frac{3}{4}$	19.65
6-in. (152.4-mm) and larger square	Glazed, unmounted	$\frac{1}{16}$	1.59	$\frac{1}{4}$	6.35
All sizes	Quarry and paver, unmounted	$\frac{3}{8}$	9.53	$\frac{3}{4}$	19.65

Table C20 Sieve analysis of sand for setting beds and for pointing of clay floor and wall tile

	Percentage passing sieve					
Type of mortar	No. 4	No. 8	No. 16	No. 30	No. 50	No. 100
Mortar for setting beds	100	95–100	60–85	35–60	15–30	0–5
Pointing mortar				100		0–5

Table C21 Clay floor and wall tile mortar mixes by volume

	Mortar mix by volume for the various coats for vertical surfaces			Mortar mix by volume for horizontal surfaces
Mortar ingredient	Scratch	Leveling[a]	Setting bed[a]	Setting bed
Portland cement	1 part	1 part	1 part	1 part
Hydrated lime[b]	20%	$1\frac{1}{2}$ parts	$\frac{1}{2}$–1 part	10% (max.)
Sand	4 parts	4–7 parts	4–7 parts	4–7 parts

[a] Before application of these coats, the preceding coat should be thoroughly wetted.
[b] *See* Mortar; Lime.

pointing, and caulking material. In general, for wide joints the grouting and pointing materials are applied with a tool and for narrow joints they are flowed into the joint.

See Tables *C20* and *C21*.

9. Cleaning: Upon completion of tilework, one should always make sure that cement or adhesive is removed from the surface by dilute solutions of acid or other means recommended by the tile manufacturer for the installation method used.
10. Setting and curing: After pointing and grouting of the floor tile has been completed, exits must be closed to traffic and protected until tilework sets. For the cement mortar setting method, the surface should be covered with bitumen-saturated felt lapped 4 in. (101.6 mm) in both directions and sealed so that moisture does not escape for a full 3 days. For the various thin-set methods, the curing should be done as recommended by the manufacturer of the adhesive.
11. Color fastness of mortar pigments: If colored cement mortar or colored trade-name products are used, the mineral pigment must not stain the type of clay floor or wall tile selected.
12. Expansion and contraction: Expansion and contraction of the structure which supports the tile flooring may cause cracks in the setting bed and open joints in the tile, especially in large floor areas. One should always check with a structural engineer for large areas to be covered with tile to see whether cleavage planes are necessary (*see* Figures *C32* and *C35*).

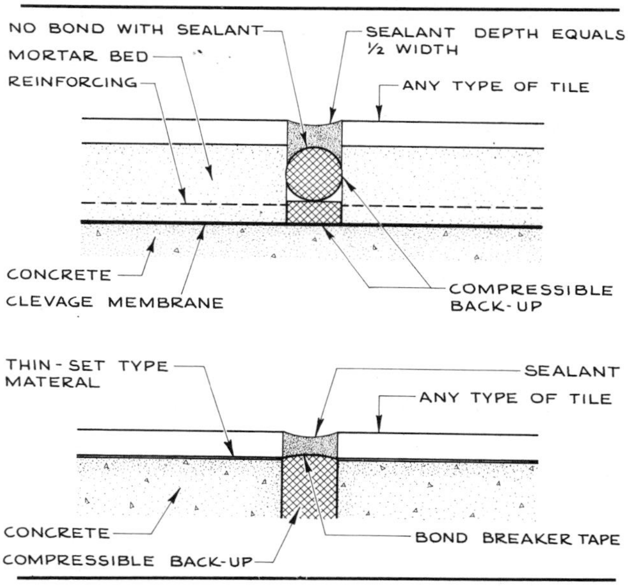

Figure C35 Treatments of expansion and contraction joints in clay floor tile installations.

Building paper should be a bitumen-saturated felt. This paper is used in frame construction over a wood subfloor as a cleavage plane in the same manner except that its ends and edges are lapped 3 in. (76.2 mm) and it is secured to the subfloor. For the various thin-set mortar or adhesive methods, always check with the manufacturer for the requirements for expansion and contraction joints. (*See* Figure *C35*.)

13. Installation of accessories: Bathroom accessories must be installed by the proper method. For matching tile accessories, the locations are established in relation to tile size. For the recessed type, the locations must not interfere with piping. For metal surface-mounted accessories, once the locations are established, blocking must be installed behind tile to which the accessories are secured. For the recessed metal type, holes are left or rough boxes installed behind tile into which the accessories can be secured. These openings or boxes must be established according to dimension and tile size so that they fit the tile pattern and also do not interfere with piping.
14. Cement mortar application method: One should always check the proper mortar mix (*see* Table *C21*), type of lath, and number of coats required for wall and floor application. The leveling coat can often be eliminated if the surface to receive clay floor and wall tile is smooth, level, and plumb (*see* Figures *C31* to *C33*). Cement is generally standard gray portland cement. White cement is used when colors are to be added. High early strength cement is used in periods of cool weather for faster setting. Waterproofed cement is used for areas subject to continuous wetting and for exterior work. Air-entrainment improves the workability of the mortar and is desirable for exterior work.
15. Lath for cement mortar application: Wire lath for supports spaced 16 in. (406.4 mm) o.c. can be any of the following: flat-rib metal lath, wire, or sheet lath. For supports 12 in. (304.8 mm) o.c., any of the lath listed can be used, as well as flat expanded metal lath and wire lath. (*See* Figure *C33*.)
16. Fill: Where floor levels are to be raised by fill to meet grades, the fill must consist of aggregate graded from fine to coarse—generally within the limits of $\frac{1}{4}$ in. (6.35 mm) to a ring size one-half the thickness of fill—in the following proportions by volume: 1 part portland cement, $2\frac{1}{2}$ parts sand, and 5 parts coarse aggregate.
17. Cement mortar application: For cement mortar application, mounted (paper-backed) tile must not be soaked in water beforehand; unmounted tile should be soaked in water, but no free moisture should remain on the back of tiles when being set. No fire clay, cinders, or slag are to be used in any cementitious mixture.
18. Dry-wet mortar application: One should always check with the manufacturers of both tile and dry-set mortar to be sure that the correct backing, type of tile, and method of grouting and pointing are used.
19. Organic adhesive setting method: The manufacturers of both tile and adhesive should be consulted to assure the correct combination of materials. When the adhesive has been selected, always check: (a) the proper method of applying tile, that is, dry or presoaked in water, (b) setting and curing time, (c) method of pointing and grouting, and (d) preparation of surfaces to receive tile. For the adhesive setting method, the thickness and shape of tile and tile trimmers to be used should always be specified.
20. Various types of thin-set methods: When these methods as described under "Types and Uses" are specified, the manufacturers of both the thin-set material and the tile should be consulted to make sure that the correct backing, type of tile, and method and type of grouting and pointing material are used.
21. Pregrouted sheets of clay floor and wall tile: When specifying any pregrouted materials, one should always check with manufacturers of clay floor and wall tile for size of sheets, trim, closures and corner strips, type of back-up, bathroom accessories, and type of thin-set material to be used.
22. Pressure-sensitive systems: When these are used, the backing, type of tile, and type of grout and pointing material must be checked to make sure that they are the correct kinds.
23. With all methods of setting clay floor and wall tile, one should always check backing, type of tile, and type of grout and pointing materials to make sure that all these materials are compatible when combined.

Conditions Favorable to the Use of Clay Floor and Wall Tile

1. Where a permanent, durable, waterproof, easily maintained, decorative, and colorful wall surface, floor, or ceiling is needed on the exterior or interior.

Conditions Unfavorable to the Use of Clay Floor and Wall Tile

1. In areas where acoustic requirements are of utmost importance.
2. In areas that will be subject to severe shock or mechanical damage.

History and Manufacture

Floor and wall tiles of fired clay are one of the oldest building materials known to man and are essentially the same today as those discovered by archeologists in the Nile Valley and Tigris-Euphrates Basin. Now, as then, they are used for their durability, sanitary qualities, and decorative appearance, on both exterior and interior surfaces, in residences and in public and commercial buildings. Clay tiles were an important material in Egyptian, Babylonian, Cretan, Greek, and Roman architecture. After the fall of the Roman Empire the Saracens continued to develop the manufacture and use of clay tiles. The metallic luster surface is credited to the Persian ceramists.

In the early 12th century tile was again made in Italy, and by the 14th century the town of Faenza, which is credited with developing what is now known as faience tile, had become the center of Italian production. Germany and Austria in about the 12th century were also producing a wall tile used primarily for the exterior finish of stoves. In the 15th century Holland started production of delft tile, characterized by a figure or landscape design in blue or violet-brown. In the 17th century the English developed a method of printing scenes on tile from copper plates. Once the dust-press technique appeared, tiles could be mass-produced.

In the United States in 1867 Samuel Keys started production of floor and wall tiles in Pittsburgh, and in a few years plants were established in Ohio, Indiana, and Massachusetts. Mosaic tiles were introduced shortly before 1900, and by 1937 there were 52 manufacturers of floor and wall tiles.

Dust-Press Process. Today floor and wall tiles are manufactured by dust-press and plastic methods. Dust-pressed tiles are shaped in steel dies by applying heavy pressure to the damp ceramic mix while it is in finely pulverized form. This mix contains only enough moisture to cause the particles to cohere under pressure. The dust-press method of production gives greater mechanical precision and a more regular appearance to the tiles than do other methods. (*See* Figure *C36*.)

Plastic Process. Plastic-made tiles are shaped from clay rendered plastic by mixing with sufficient water. They are made either by hand-molding or by extrusion from an auger machine. When shaped by machine, the extruded ribbon of clay is cut into the desired sizes as it emerges from the die. Most types made by the plastic method vary slightly from the true geometric forms and therefore have a more handmade appearance than do dust-pressed tiles.

CLAY TILE FLUE LININGS

Flue linings made of clay, commonly (but incorrectly) called flue tiles, provide a smooth surface in chimneys that discourages the accumulation of soot and protects

Figure C36 Flowchart of the dust-press process.

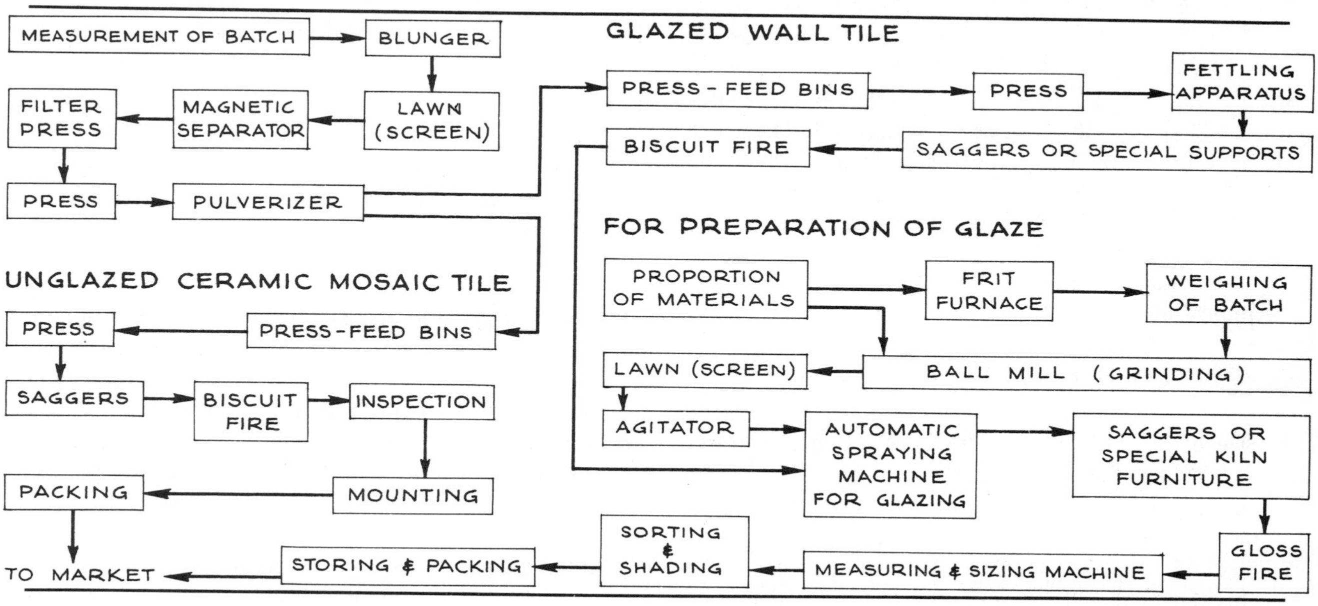

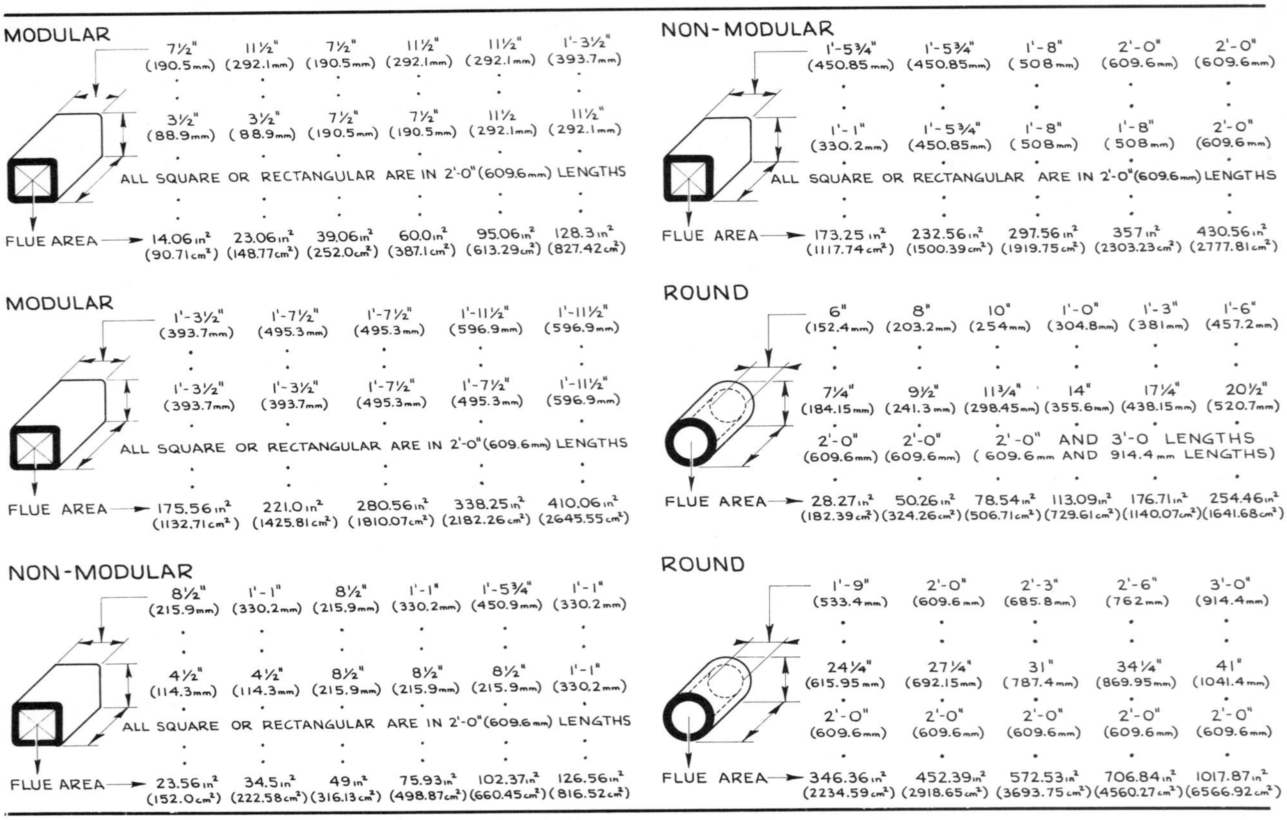

Figure C37 Types and dimensions of flue tiles.

the surrounding masonry. The fire clays used for flue linings are often the same as those used for burned brick. Flue linings are usually, but not necessarily, salt glazed. They are available in modular and nonmodular sizes in 2-ft (0.6096-m) lengths for square and rectangular shapes and 2- and 3-ft (0.6096- and 0.9144-m) lengths for round shapes (*see* Figures *C37* and *C38*). The inside cross-sectional area is the controlling factor in choosing the size of flue for chimneys of fireplaces, boilers, incinerators, etc. (*See also* Chimneys; Fireplaces.)

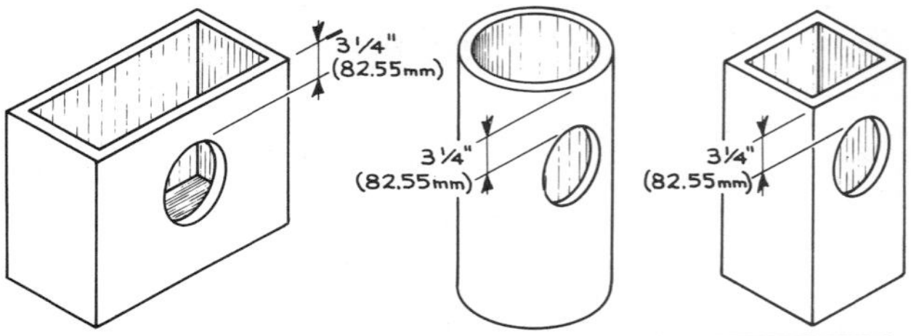

Figure C38 Clay tile flue linings preholed for breaching connections.

Application

Condensed Checklist

1. Size of flue: Recommendations of the manufacturer of the heating unit, incinerator, etc., should be checked for the correct size of flue to be used on that particular job.
2. Size of fireplace opening: This should always be checked as it will determine the size (cross-sectional area) of flue to be used (*see* Fireplaces).
3. Smoothness of joints: The flue lining joints should fit exactly since a smooth inner surface is absolutely necessary for a good draft. For the same reason the angle of the flue must not be less than 60°.
4. Fire resistance: Local, state, and municipal codes should be checked for fire resistance ratings, and also the codes of the Fire Underwriters, insurance companies, labor departments, and federal government (Army, Navy, etc.) in relation to the size, thickness, and necessary amount of masonry surrounding the flue.

History and Manufacture

Clay flue linings are a modern development of the drain tile and are produced by the same machines that manufacture brick and structural clay tile.

CLAY TILE ROOFING

Physical Properties

The clays used for roofing tile are the same as those used for brick. The color of unglazed roofing tile can range from orange-yellow to dark red, and glazed finishes in polychrome and various solid colors are also obtainable. Minimum roof pitch is 3 to 12 for interlocking types of roofing tile, 5 to 12 for Norman, and 4 to 12 for roll-pattern type roofing. For the various types of clay tile roofing *see* Figure *C39*.

Cement Roofing Tiles. Cement roofing tiles are made in shapes similar to clay roofing tiles. Generally, they should be used in areas where the climate is dry and temperature never severely low. Their color is limited in

Figure C39 Types of roofing tile.

TYPE		COLORS	LENGTH IN.	LENGTH mm	WIDTH IN.	WIDTH mm	EXPOSURE LENGTH IN.	EXPOSURE LENGTH mm	EXPOSURE WIDTH IN.	EXPOSURE WIDTH mm	WEIGHT LB/FT²	WEIGHT KG/M²
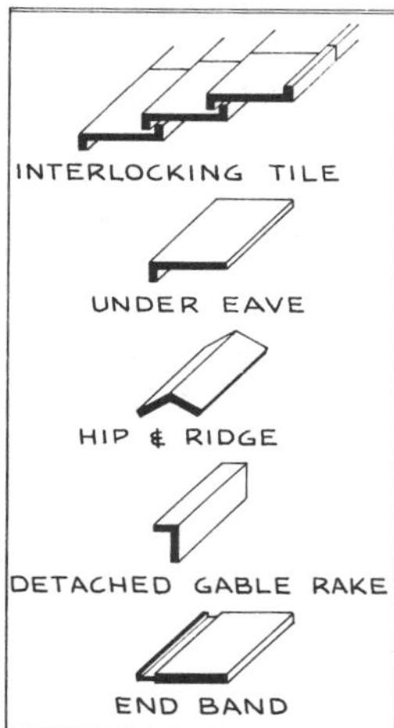 INTERLOCKING TILE; UNDER EAVE; HIP & RIDGE; DETACHED GABLE RAKE; END BAND	CLASSIC	RUSSET RED, MEDITERRANEAN BLUE, DESERT SAND, ANTIQUE GRAY, FROST WHITE, SLATE GREEN, LIGHT RED, BROOKVILLE GREEN										
	WILLIAMSBURG	COLONIAL GRAY, FORREST GREEN										
	LANAI	SUNSET RED, HAWAIIAN GOLD, BEACH BROWN, LAVA BLACK	14	355.6	9	228.6	11	279.4	8¼	209.55	8	39.056
	EARLY AMERICAN	CONCORD RED, CAPE COD GREY, LEXINGTON GREEN										
	BARK TILE SHAKES	AGED CEDAR										
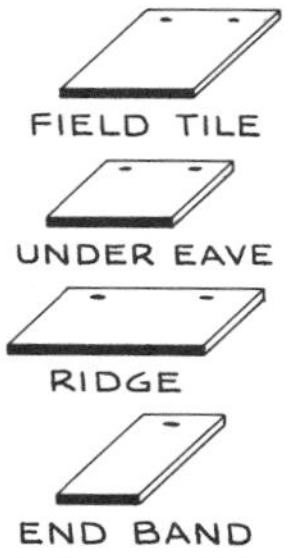 FIELD TILE; UNDER EAVE; RIDGE; END BAND	NORMAN	COLOR - BLEND OF MEDIUM-DARK GREEN, SMOKED RED, GRAYS, BLACK AND IN-BETWEEN TONES	15	381.0	7	177.8	6½	165.1	7	177.8	16	78.11
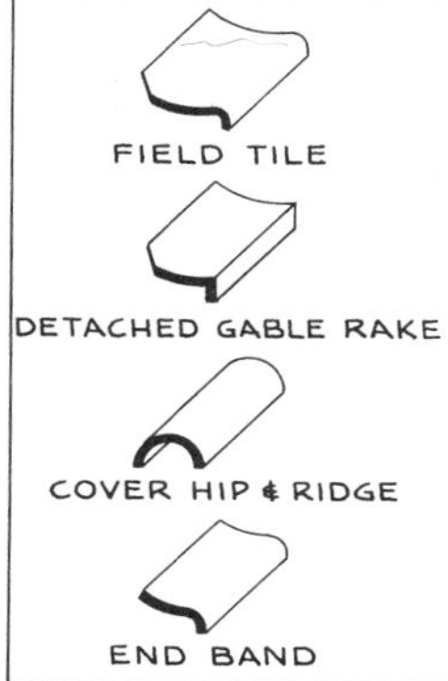 FIELD TILE; DETACHED GABLE RAKE; COVER HIP & RIDGE; END BAND	SCANDIA	LAVA BLACK, BEACH BROWN, HAWAIIAN GOLD, MEDITERRANEAN BLUE, NORDIC RED, BROOKVILLE GREEN	14¼	301.95	10	254.0	11¼	285.75	8¼	209.55	8.25	40.28

TYPE		COLORS	LENGTH		WIDTH		EXPOSURE				WEIGHT	
							LENGTH		WIDTH			
			IN.	mm	IN.	mm	IN.	mm	IN.	mm	LB/FT²	KG/M²
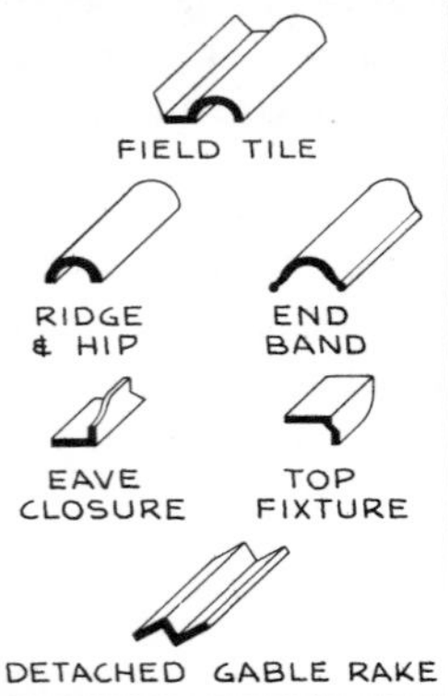	SPANISH AND SPECIAL SPANISH	NATURAL RED, MEDITERRANEAN BLUE, GRANADA RED, BROOKVILLE GREEN	13¼	336.55	9¾	247.65	10¼	260.35	8¼	209.55	9	43.94
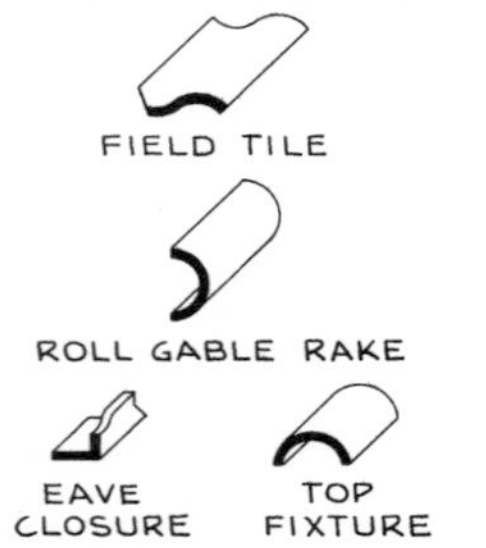	CASA RANCHO	LIGHT RED, FIREFLASHED RED	18	457.2	12	304.8	15½	393.7	10	254.0	9	43.94
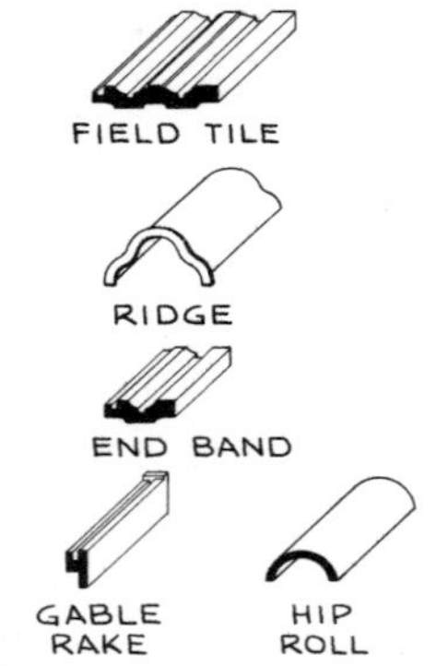	FRENCH	MEDITERRANEAN BLUE, ROYAL RED, BROOKVILLE GREEN	16¼	412.75	9	228.6	13⅜	339.73	8⅛	206.38	9.35	45.65
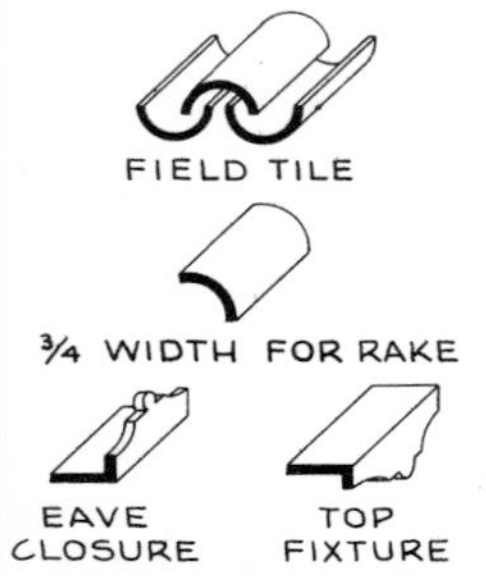	MISSION	LIGHT RED, RANCH RED, DARK RED, GRANADA RED, MEDITERRANEAN BLUE, BARCELONA BUFF, SANTIAGO ROSE	14¼ 18	301.95 457.2	11½	292.1	11¼ 15	285.75 381.0			12.5 12.2	61.03 59.56

Figure C39 **Types of roofing tile (*continued*).**

Table C22 Roof dimensions from ridge to a 2-inch projection at eave in roofing tile courses

	Straight Mission Barrel												Early American, Classic, Lanai, Williamsburg					
	$14\frac{1}{4}$-inch		18-inch		Spanish		French		Norman		Scandia				Caso Rancho		Bark Tile Shakes	
Course	in.	m	in.	m	in.	m	in.	m	in.	m	in.	m	in.	m	in.	m	in.	m
1	$12\frac{1}{4}$	0.311	16	0.406	$11\frac{1}{4}$	0.286	14	0.356	13	0.330	$12\frac{1}{4}$	0.311	12	0.305	16	0.406	14	0.356
2	$23\frac{1}{2}$	0.597	31	0.787	$21\frac{1}{2}$	0.546	$27\frac{3}{8}$	0.695	$19\frac{1}{2}$	0.495	$23\frac{1}{2}$	0.597	23	0.584	$31\frac{1}{2}$	0.800	$27\frac{1}{4}$	0.695
3	$34\frac{3}{4}$	0.883	46	1.168	$31\frac{3}{4}$	0.906	$40\frac{3}{4}$	1.035	26	0.660	$34\frac{3}{4}$	0.883	34	0.864	47	1.194	$40\frac{1}{2}$	1.029
4	46	1.168	61	1.549	42	1.067	$54\frac{1}{8}$	1.376	$32\frac{1}{2}$	0.826	46	1.168	45	1.143	$62\frac{1}{2}$	1.588	$53\frac{3}{4}$	1.365
5	$57\frac{1}{4}$	1.454	76	1.930	$52\frac{1}{4}$	1.327	$67\frac{1}{2}$	1.715	39	0.991	$57\frac{1}{4}$	1.454	56	1.422	78	1.981	67	1.702
6	$68\frac{1}{2}$	1.740	91	2.311	$62\frac{1}{2}$	1.588	$80\frac{7}{8}$	2.060	$45\frac{1}{2}$	1.156	$68\frac{1}{2}$	1.740	67	1.702	$93\frac{1}{2}$	2.375	$80\frac{1}{4}$	2.038
7	$79\frac{3}{4}$	2.026	106	2.692	$72\frac{3}{4}$	1.848	$94\frac{1}{4}$	2.394	52	1.321	$79\frac{3}{4}$	2.026	78	1.991	109	2.769	$93\frac{1}{2}$	2.375
8	91	2.311	121	3.073	83	2.108	$107\frac{5}{8}$	2.736	$58\frac{1}{2}$	1.486	91	2.311	89	2.261	$124\frac{1}{2}$	3.163	$106\frac{3}{4}$	2.705
9	$102\frac{1}{4}$	2.597	136	3.434	$93\frac{1}{4}$	2.368	121	3.073	65	1.651	$102\frac{1}{4}$	2.697	100	2.540	140	3.556	120	3.048
10	$113\frac{1}{2}$	2.883	151	3.835	$103\frac{1}{2}$	2.629	$134\frac{3}{8}$	3.719	$71\frac{1}{2}$	1.816	$113\frac{1}{2}$	2.883	111	2.819	$155\frac{1}{2}$	3.696	$133\frac{1}{4}$	3.384
11	$124\frac{3}{4}$	3.169	166	4.216	$113\frac{3}{4}$	2.889	$147\frac{3}{4}$	3.753	78	1.981	$124\frac{3}{4}$	3.169	122	3.099	171	4.343	$146\frac{1}{2}$	3.721
12	136	3.454	191	4.851	124	3.150	$161\frac{1}{8}$	4.093	$84\frac{1}{2}$	2.147	136	3.434	133	3.378	$196\frac{1}{2}$	4.737	$159\frac{3}{4}$	4.058

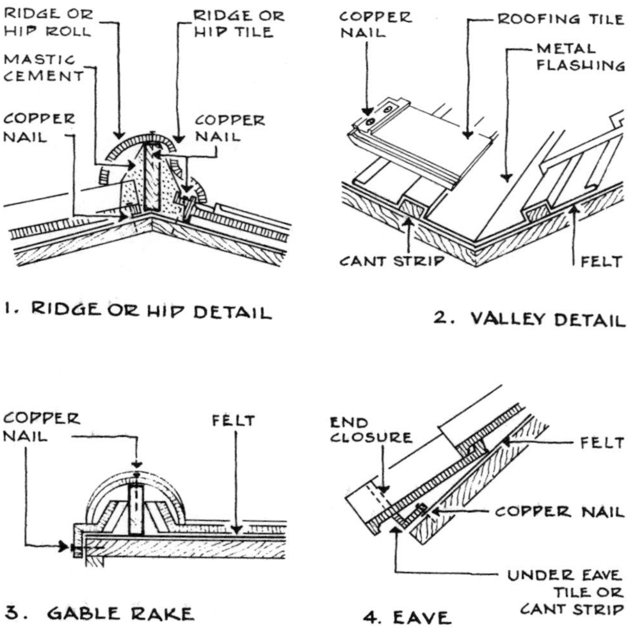

Figure C40 Details at ridge or hip (1), valley (2), gable rake (3), and eave (4).

range and subdued in hue. Such tiles are less desirable than clay roofing tile but otherwise are similar in their application.

Industrial cement roofing tile is 2 ft (0.610 m) wide, 4 ft (1.219 m) (and more) long, $1\frac{1}{2}$ in. (38.1 mm) thick, and reinforced. The ends are formed to overlap roof purlins, thus eliminating any fastening devices.

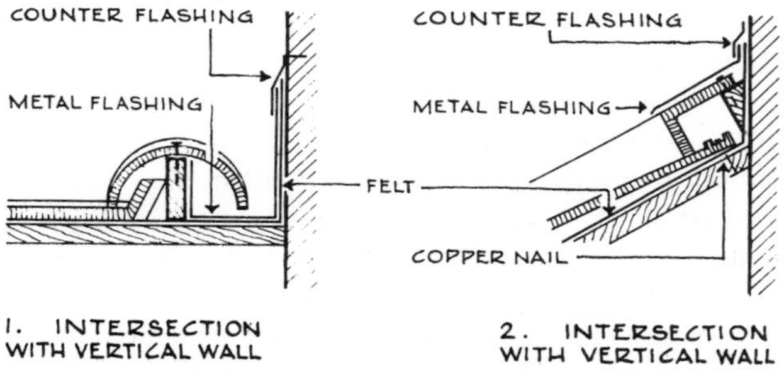

Figure C41 Details of flashing at intersections with vertical walls.

The length of the roof (ridge to eave) should be carefully calculated when designing for clay tile roofing because the length exposure of each roof tile is a fixed measurement. Table *C22* gives roof dimensions in terms of tile courses.

Application

Condensed Checklist

1. Pitch of roof and support to carry weight (total load) of the tile roofing.
2. Treatment of ridges, hips, valleys, gables, and eaves for tile roofing (*see* Figure *C40*).
3. Use of copper or aluminum nails or wire for fastening.
4. Flashing at intersection of flat or pitched roof with vertical wall (*see* Figure *C41*).
5. Gutter details.

Conditions Favorable to the Use of Tile Roofing

1. Where an almost permanent type of roof covering is desired.
2. Where a nearly fireproof roofing material is desired.
3. For a strong roof texture.
4. Where a permanent roof color is desired, and a range of color from subtle hues to brilliant colors is needed.

Conditions Unfavorable to the Use of Tile Roofing

1. Where light roof construction is essential.
2. Where cost is a factor.

History and Manufacture

The Greeks were the first known users of tile roofing, and the earliest discovered examples show such a high degree of development that it must be assumed that they used roofing tiles at an even earlier date. The Romans used the S-shaped tile, and the Greeks used both the flat and curved tile with joint cover. The ancient Chinese and Japanese fashioned beautiful roof tiles glazed in a wide range of colors and shades, including black, although the most common was a bright yellow. Since earliest times roofing tile has retained substantially the same form, but the methods of manufacture have been improved for machine and large-scale production. The soft-mud and dry-press processes used for manufacturing burned brick are also used to make tile roofing.

CLAY TILE, STRUCTURAL

Physical Properties

Structural clay tile may be defined as a hollow, or cored, burned-clay masonry unit with parallel cells either in the vertical or the horizontal direction. The clays used are the same as for brick. The two basic categories of structural clay tile—load-bearing and non-load-bearing—differ in their characteristics.

Non-load-bearing structural clay tile is subdivided into four categories: non-load-bearing partition title, furring tile, fireproofing tile, and screen tile. Load-bearing structural clay tile is subdivided into three categories: load-bearing wall tile, structural facing tile, and ceramic glazed structural facing tile.

There are also two types of solid masonry units, defined as follows: (1) a masonry unit whose net cross-sectional area in every plane parallel to the bearing surface is 75% or more of its gross cross-sectional area measured in the same plane; and (2) a masonry unit whose net cross-sectional area in every plane parallel to the bearing surface is less than 75% of its gross cross-sectional area measured in the same plane.

Average Weight. The average weight of load-bearing structural clay tile walls is 60 lb/ft^3 (961.2 kg/m^3) and that of non-load-bearing structural clay tile partitions is 50 lb/ft^3 (801.0 kg/m^3).

Finishes. The surface of structural clay tile may be given one of the following finishes, either as it is extruded through the die or soon thereafter.

1. Smooth finish: plane surface as formed by the die with no further changes.
2. Scored finish: surface grooved as it comes from the die to provide better bond for mortar, plaster, or stucco.
3. Combed finish: surface altered by parallel scratches to produce a desired texture or to increase bond for mortar, plaster, or stucco.
4. Roughened finish: surface entirely broken by wire cutting, etc., to provide a desired texture or to increase bond for mortar, plaster, or stucco.

Textures of the clay itself in structural clay tile range from fine through medium to coarse. Some of the various special textures other than the finishes already described include vertically or horizontally marked, rugged, barked, stippled, hammered, and spotted. Perforations to help absorb sound may also be called a texture. The surface may be either glazed or left natural.

Natural finishes depend on the natural color of the clay forming the body and the degree and manner of burning. The color range of natural unglazed structural clay tile is mainly (actually 99%) in the reds, buffs and creams.

Glazed finishes include salt glazes, clear glazes with either a shiny or matte (nonlustrous) finish, and color glazes in solid, mottled, speckled, or spotted colors, also with shiny or matte finish.

Shapes. The shape of structural clay tile is controlled by the die through which the clay is extruded, and the ease with which different designs can be produced has led to a wide variety of sizes and patterns.

Types and Uses

Each of the two basic categories of structural clay tile is made in several types. Load-bearing tile is subdivided into (1) wall tile, including back-up tile that carries part or all of the total weight of the wall; (2) facing tile; and (3) glazed facing tile. Non-load-bearing tile is subdivided into furring and partition tile, including non-load-bearing back-up tile, fireproofing tile, and screen tile.

Load-Bearing Structural Clay Wall Tile. This type includes (1) wall tile for the construction of exposed or faced load-bearing walls, which is designed to carry the total load, including facing of stucco, plaster, stone, or other material, and (2) back-up tile for construction of combination walls of brick or other masonry, in which the total load is supported by both the facing and backing. (*See* Tables *C23* to *C26 and* Figure *C42*.) Both categories of structural clay load-bearing wall tile are classed into two grades as follows:

Grade LBX: for general masonry work, for masonry exposed to weathering, and as a base for stucco.

Grade LB: for masonry not exposed to frost action or for exposed masonry protected with a facing of 3 in. (7.62 cm) or more of stone, brick, ceramic veneer, etc.

Load-bearing Structural Clay Facing Tile. Facing tile is divided into two classes, standard and special duty, both of which are based on the thickness of the face shells. (*See* Tables *C27 to C30 and* Figures *C43 and C44*.) Each of these classes is graded on the basis of factors that determine appearance as follows:

Type FTX: for exterior and interior walls and partitions. These tiles must show a high degree of mechanical perfection, a narrow range of color variation and minimum variations in face dimensions of units. They are used where tiles of low absorption, easy cleaning, and resistance to staining are required.

Type FTS: for exterior and interior walls.

Table C23 Fire-resistance rating of load-bearing structural clay wall tile

Type of wall construction	Wall thickness[b] 4-hour		3-hour		2-hour		1-hour	
			U = units,[c] C = cells[c]					
COMBUSTIBLE MATERIAL FRAMED IN WALL:[a]								
Unplastered	16 in. (406.4 mm)	2U 4C	16 in. (406.4 mm)	2U 4C	12 in. (304.8 mm)	3C	12 in. (304.8 mm)	[d] 3C
Plastered one side with $\frac{5}{8}$ in. (15.88 mm) of 1:3 gypsum and sand plaster	16 in. (406.4 mm)	2U 4C	12 in. (304.8 mm)	3C	12 in. (304.8 mm)	2C	8 in. (203.2 mm)	2C
Both sides plastered as above	16 in. (406.4 mm)	2U 4C	12 in. (304.8 mm)	3C	12 in. (304.8 mm)	[d] 2C	8 in. (203.2 mm)	2C
NONCOMBUSTIBLE MATERIAL FRAMED IN WALL:								
Unplastered	12 in. (304.8 mm)	2U 4C	12 in. (304.8 mm)	2U 3C	12 in. (304.8 mm)	3C	8 in. (203.2 mm)	2C
Plastered one side with $\frac{5}{8}$ in. (15.88 mm) of 1:3 gypsum and sand plaster	12 in. (304.8 mm)	2U 3C	12 in. (304.8 mm)	3C	8 in. (203.2 mm)	3C	8 in. (203.2 mm)	2C
Both sides plastered as above	12 in. (304.8 mm)	2U 3C	8 in. (203.2 mm)	3C	8 in. (203.2 mm)	3C	8 in. (203.2 mm)	2C

[a] Ratings for walls with combustible members framed into the wall apply for framed-in members not over 4 in. (101.6 mm).

[b] Thickness given does not include that of any plaster specified.

[c] Ratings of load-bearing wall tile depend in certain cases on the number of cells and units in the wall thickness. These are shown in the table along with the total thickness, in inches and millimeters, of the wall; for example, 2U represents 2 units and 4C represents 4 cells in the wall thickness.

[d] An 8-in. (203.2-mm) tile wall may be used for this rating if hollow spaces near combustible members are filled with fire-resistant materials for the full thickness of the wall and for 4 in. (101.6 mm) or more above, below, and between the combustible members.

Table C24 Physical requirements of load-bearing clay wall tile

Grade	Maximum water absorption[a] after 1 hour in boiling water (percent): Average of five tests	Maximum water absorption: Individual	Minimum compressive strength based on gross area[b] — End construction[c]: Average of five tests, lbf/in.^{2}	End, Average of five tests, MN/m^2	End, Individual, lbf/in.^{2}	End, Individual, MN/m^2	Side construction[c]: Average of five tests, lbf/in.^{2}	Side, Average of five tests, MN/m^2	Side, Individual, lbf/in.^{2}	Side, Individual, MN/m^2
LBX	16	19	1400	9.653	1000	6.895	700	4.827	500	3.448
LB	25	28	1000	6.895	700	4.827	700	4.827	500	3.448

[a] The range in percentage absorption for tile delivered to any one job shall not be more than 12.
[b] The gross area of a unit shall be determined by multiplying the horizontal face dimension of the unit as placed in the wall by its thickness.
[c] 1 lbf/in.^{2} = 0.006895 MN/m^2.

Table C25 Nominal modular sizes of load-bearing structural clay wall tile, including mortar joints

Thickness in.	Thickness mm	As back-up, Face dimensions in wall: Height in.	Height mm	Length in.	Length mm	As finished wall, Face dimensions in wall: Height in.	Height mm	Length in.	Length mm
4	101.6	$2\frac{2}{3}$	67.74	8 or 12	203.2 or 304.8				
		$5\frac{1}{3}$	135.47	12[a]	304.8	$5\frac{1}{3}$	135.47	12	304.8
		8		8 or 12	203.2 or 304.8	8	203.2	8 or 12	203.2 or 304.8
		$10\frac{2}{3}$		12	304.8				
						12	304.8	12	304.8
6	152.4	$5\frac{1}{3}$	135.47	12	304.8	$5\frac{1}{3}$	135.47	12	304.8
		8	203.2	12[a]	304.8				
		$10\frac{2}{3}$	270.94	12	304.8				
						12	304.8	12	304.8
8	203.2	$5\frac{1}{3}$	135.47	12	304.8	$5\frac{1}{3}$	135.47	12	304.8
						6	152.4	12	304.8
		8	203.2	8 or 12[a]	203.2 or 304.8	8	203.2	8, 12, or 16	203.2, 304.8, or 406.4
		$10\frac{2}{3}$	270.94	12	304.8				
						12	304.8	12	304.8
10	254.0					8	203.2	12	304.8
						12	304.8	12	304.8
12	304.8					12	304.8	12	304.8

[a] Includes header and stretcher units.

Load-Bearing Ceramic Glazed Structural Clay Facing Tile. Ceramic glazed structural clay facing tile is divided into two grades and two types as follows:

Grade S (select): for use with comparatively narrow mortar joints.

Grade SS (select sized or ground edge): for use where variations of face dimensions must be very small.

Type I: for general use where only one finished face will be exposed.

Type II: for use where the two opposite finished faces will be exposed.

Table C26 Minimum number of cells in the direction of wall thickness for load-bearing structural wall tile

	Nominal horizontal thickness of wall tile as laid in wall				
Cells[a]	4 in. (101.6 mm)	6 in. (152.4 mm)	8 in. (203.2 mm)	10 in. (254.0 mm)	12 in. (304.8 mm)
Minimum number of cells in direction of wall thickness	1	1	2	2	3

[a]Cells are hollow spaces enclosed within the perimeter of the exterior shells and having a minimum dimension of not less than $\frac{1}{2}$ in. (12.7 mm) and a cross-sectional area of not less than 1 in.^{2} (654.2 mm^2).

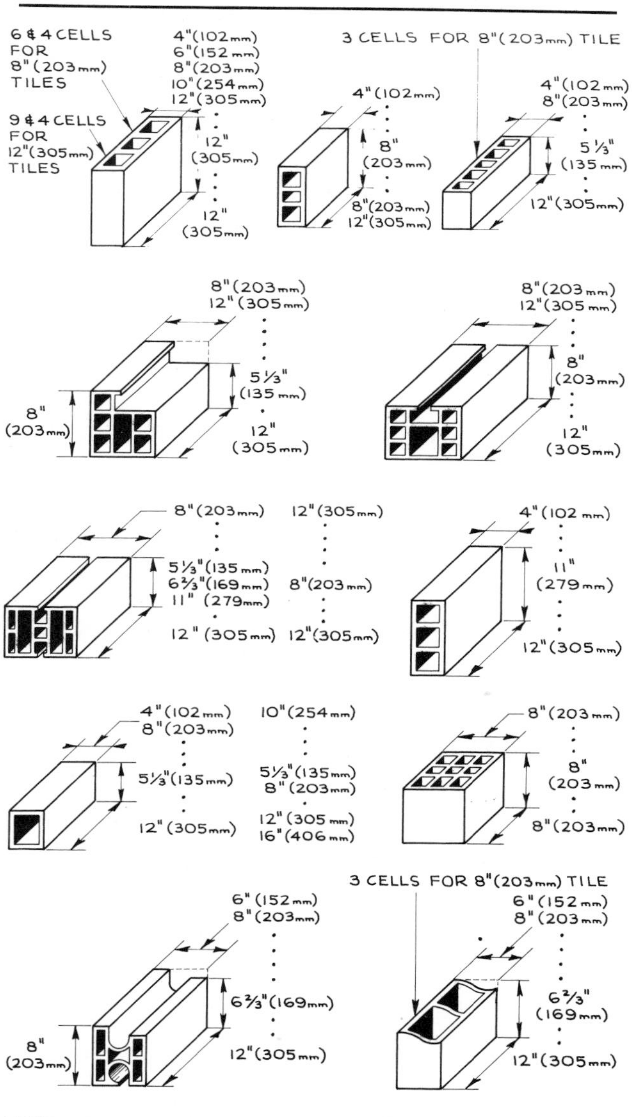

NOTE: LARGER SIZES HAVE FINGER SLOT FOR EASIER HANDLING

Figure C42 Standard shapes and sizes of load-bearing structural clay wall tile.

Table C27 Maximum water absorption rate for load-bearing structural clay facing tile

	Maximum water absorption (percent)			
	After 24-hour submersion in cold water		After 1-hour submersion in boiling water	
Type	Average	Single unit	Average	Single unit
FTX	7	9	9	11
FTS	13	16	16	19

Ceramic glazed structural clay facing tile is available in the same shapes and sizes, with the same water absorption rate, as load-bearing structural clay facing tile. There are other shapes available in addition to those illustrated in Figures *C43* and *C44*. Requirements for compressive strength are shown in Table *C31*. Sizes and permissible variations in sizes and distortion are shown in Tables *C32* to *C34*. The finish is tested for imperviousness, opacity, hardness, and chemical, fading, crazing and abrasion resistance. Mottled, stippled, and smooth textures and colors in matte or glossy finish are available. A check should always be made to ensure that the supplying manufacturer has or can provide the required special shapes. Also, the possibility of color variations resulting from separate burning of the various special shapes should be investigated.

Non-Load-Bearing Structural Partition and Furring Tile. Such tile has one grade only. The maximum absorption rate is 28%, and the maximum absorption range for one job is 12%. Modular sizes of partition and furring tile have not been established throughout, although in some areas manufacturers do make them. (*See* Tables *C35* and *C36 and* Figure *C45*.)

Partition tile is designed for the construction of non-load-bearing interior partitions or for backing up non-load-bearing combination walls.

Table C28 Compressive strength of structural clay facing tile, based on gross area

Class	Minimum compressive strength[a]							
	End construction				Side construction			
	Minimum average of five tests		Individual minimum		Minimum average of five tests		Individual minimum	
	lbf/in.^{2}	MN/m^2	lbf/in.^{2}	MN/m^2	lbf/in.^{2}	MN/m^2	lbf/in.^{2}	MN/m^2
Standard	1400	9.653	1000	6.895	700	4.827	500	3.448
Special duty	2500	17.238	2000	13.790	1200	8.274	1000	6.895

[a] 1 lbf/in.^{2} = 0.006825 MN/m^2.

Table C29 Minimum number of cells in direction of wall thickness for load-bearing structural facing tile

Cells[a]	Nominal horizontal thickness of wall tile as laid in wall				
	4 in. (101.6 mm)	6 in. (152.4 mm)	8 in. (203.2 mm)	10 in. (254.0 mm)	12 in. (304.8 mm)
Minimum number of cells in direction of wall thickness	1	2	2	3	3

[a] Cells are hollow spaces enclosed within the perimeter of the exterior shells and having a minimum dimension of not less than $\frac{1}{2}$ in. (12.7 mm) and a cross-sectional area of not less than 1 in.^{2} (654.2 mm^2).

Table C30 Nominal modular sizes of load-bearing structural facing tile, including mortar joints

Thickness		Face dimension in wall			
		Height		Length	
in.	mm	in.	mm	in.	mm
2, 4, 6, or 8	50.8, 101.6, 152.0, or 203.2	4	101.6	8 or 12	203.2 or 304.8
		$5\frac{1}{2}$	135.47	8 or 12	203.2 or 304.8
		6	152.0	12	304.8
		8	203.2	12	304.8
2 or 4	50.8 or 101.6	8	203.2	16	406.4

Table C31 Compressive strength of ceramic glazed structural facing tile[a]

Direction of coring or cells	Minimum average of five tests[b]		Individual minimum[b]	
	lbf/in.^{2}	MN/m^2	lbf/in.^{2}	MN/m^2
Horizontal	2000	13.790	1500	10.343
Vertical	3000	20.685	2500	17.238

[a] Special duty units are available where higher compressive strengths are required.

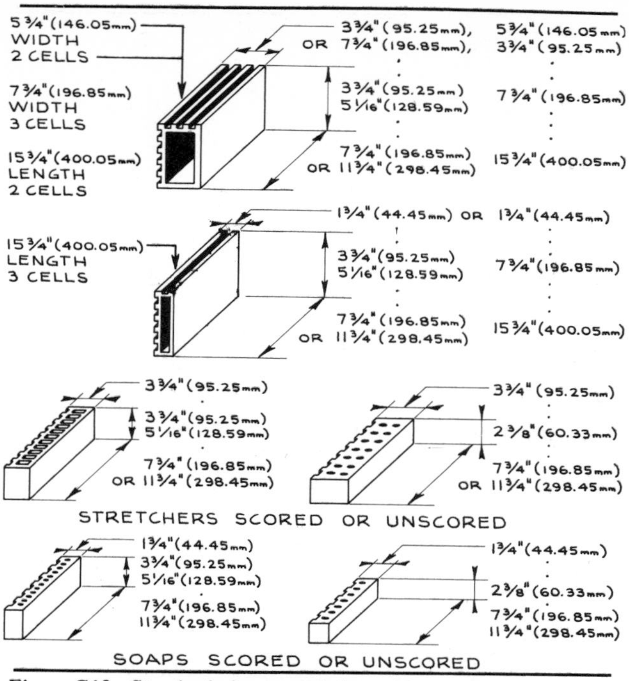

Figure C43 Standard shapes and sizes of load-bearing structural clay facing tile, excluding mortar joints.

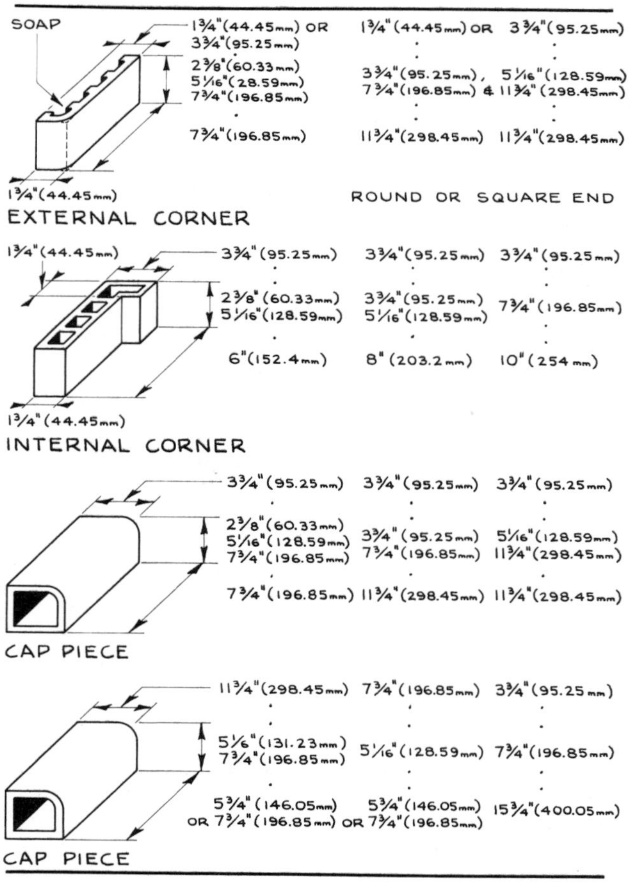

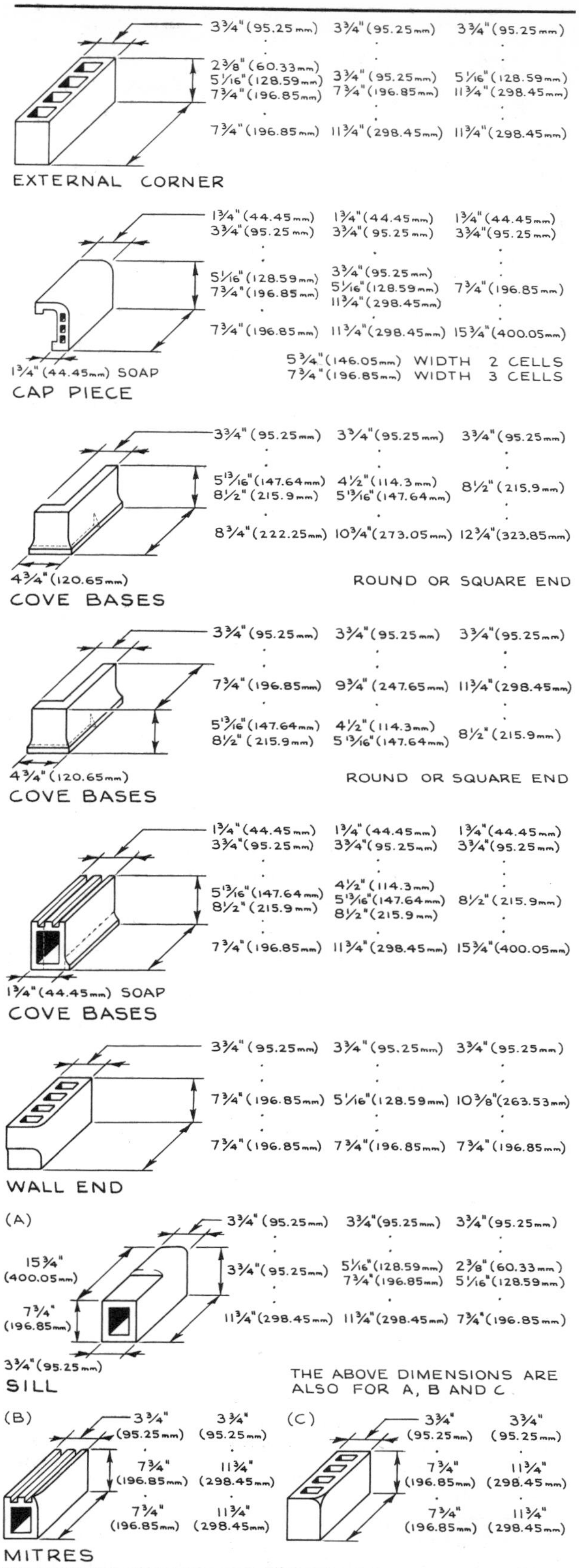

Figure C44 Special shapes and sizes of load-bearing structural clay facing tile other than stretchers and soaps.

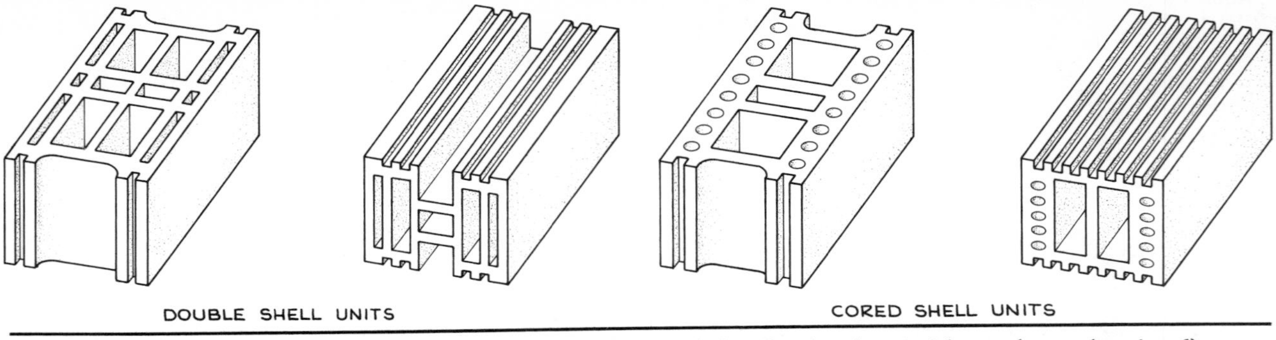

Figure C44 Special shapes and sizes of load-bearing structural clay facing tile other than stretchers and soaps (*continued*).

Table C32 Standard sizes of type I and type II ceramic glazed structural clay facing tile

Type	Face dimensions: Height in.	Height mm	Length in.	Length mm	Thickness in.	Thickness mm
Type I[a] and Type II[b]	$2\frac{3}{8}$	60.39	$7\frac{3}{4}$	196.85	$1\frac{3}{4}$ or $3\frac{3}{4}$	44.45 or 95.25
	$5\frac{1}{16}$	128.59	$7\frac{3}{4}$	196.85	$1\frac{3}{4}$, $3\frac{3}{4}$, $5\frac{3}{4}$, or $7\frac{3}{4}$	44.45, 95.25, 146.05, or 196.85
	$3\frac{3}{4}$	95.25	$11\frac{3}{4}$	298.45	$1\frac{3}{4}$, $3\frac{3}{4}$, $5\frac{3}{4}$, or $7\frac{3}{4}$	
	$5\frac{1}{16}$	128.59	$11\frac{3}{4}$	298.45		
	$5\frac{3}{4}$	146.05	$11\frac{3}{4}$	298.45		
	$7\frac{3}{4}$	196.85	$15\frac{3}{4}$	400.05	$1\frac{3}{4}$ or $3\frac{3}{4}$	44.45 or 95.25

[a] Standard sizes available from all manufacturers.
[b] Type II tiles are standard only in the $3\frac{3}{4}$-in. (95.25-mm) and $5\frac{3}{4}$-in. (146.05-mm) heights from all manufacturers.

Table C33 Permissible variations in face dimensions and depth dimensions

Face dimension (height or length)		Maximum difference between dimension of any unit and the specified dimension: If larger		If smaller		Depth dimension[a] (wall thickness)		Maximum difference between dimension of any unit and the specified dimension: If larger		If smaller	
Grade S units						**Type I single-faced units**					
in.	mm	in.	mm	in.	mm	in.	mm	in.	mm	in.	mm
$2\frac{3}{4}$	69.85	$\frac{1}{16}$	1.59	$\frac{3}{32}$	2.38	$1\frac{3}{4}$	44.45	$\frac{1}{8}$	3.18	$\frac{1}{8}$	3.18
$3\frac{3}{4}$	95.25					$3\frac{3}{4}$	95.25			$\frac{3}{16}$	4.76
$5\frac{1}{16}$	128.59					$5\frac{3}{4}$	146.05			$\frac{1}{4}$	6.35
$5\frac{3}{4}$	146.05					$7\frac{3}{4}$	196.85			$\frac{5}{16}$	7.94
$7\frac{3}{4}$	196.85			$\frac{1}{8}$	3.18						
$11\frac{3}{4}$	298.45			$\frac{5}{32}$	3.97						
Grade SS units[b]						**Type II two-faced units**					
$7\frac{3}{4}$	196.85	$\frac{1}{16}$	1.59	$\frac{1}{16}$	1.59	$3\frac{3}{4}$	95.95	$\frac{1}{8}$	3.18	$\frac{1}{8}$	3.18
$15\frac{3}{4}$	400.05					$5\frac{3}{4}$	146.05				

[a] No bed depth greater than $3\frac{3}{4}$ in. (95.25 mm) is made in trim shades.
[b] Other sizes are available conforming to the provisions of Grade SS.

Table C34 Permissible distortion

Face dimension				Maximum permissible distortion	
Height		Length			
in.	mm	in.	mm	in.	mm
$2\frac{3}{8}$	60.33	$7\frac{3}{4}$	196.85	$\frac{1}{16}$	1.59
$5\frac{1}{10}$	128.59	$7\frac{3}{4}$	196.85		
$3\frac{3}{4}$	95.25	$11\frac{3}{4}$	298.45		
$5\frac{1}{16}$	128.59	$11\frac{3}{4}$	298.45		
$5\frac{3}{4}$	146.05	$11\frac{3}{4}$	298.45		
$7\frac{3}{4}$	196.85	$11\frac{3}{4}$	298.45	$\frac{5}{32}$	3.97
$7\frac{3}{4}$	196.85	$15\frac{3}{4}$	400.05	$\frac{3}{32}$	2.38

Table C35 Number of cells to size of partition and furring tile

Dimensions						Minimum number of cells	
Thickness		Height		Length		In unit	In direction of wall thickness
in.	mm	in.	mm	in.	mm		
Partition tile							
2	50.8					3	1
3	76.2					3	1
4	101.6					3	1
6	152.4	12	304.8	12	304.8	3	1
6	152.4					4	2
8	203.2					4	2
10	254.0					4	2
12	304.8					4	2
Furring tile							
$1\frac{1}{2}$	38.1	12	304.8	12	304.8	3	
2	50.8					3	

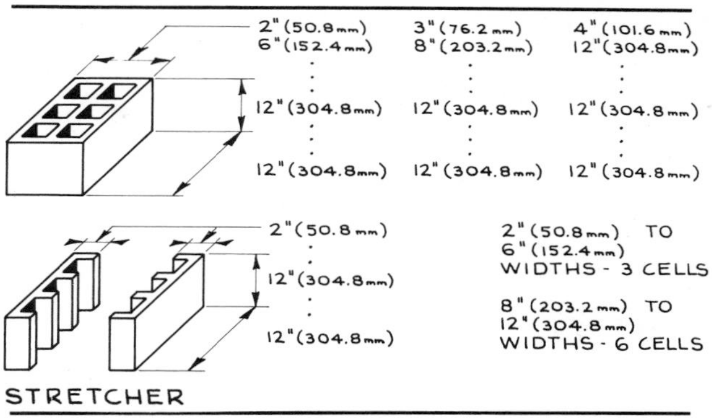

Figure C45 Shapes and sizes of structural clay partition and furring tile.

Furring tile is used for lining the inside of walls to provide a plaster base and an air space between plaster and wall. When furring tile is set against masonry walls or composite walls of solid masonry units, it should be properly fastened or attached.

Fireproofing Tile. This is a particular type of non-load-bearing clay tile used to protect structural members such as steel girders, beams, and columns against fire. Tile manufacturers are equipped to meet almost any set of conditions that might arise in this type of tilework. (*See* Figure *C46*.)

Non-Load-Bearing Structural Clay Screen Tile. There are three grades and two types of screen tile:

Grade SE: high resistance to weathering, freezing, and thawing.
Grade ME: moderate resistance to weathering.
Grade NE: for interior use only.

Table C36 Height and length limitations of non-load-bearing structural clay partition and furring tile

Partition tile				Furring tile			
				24-in. (609.6-mm) vertical and horizontal tie spacing (maximum)		16-in. (406.4-mm) vertical and 24-in. (609.6-mm) horizontal tie spacing (maximum)	
Thickness and type of tile		Maximum unsupported height		Maximum unsupported height			
in.	mm	ft	m	ft	m	ft	m
2 split	50.8			Up to 14	Up to 4.267	14–35	4.267–10.668
2 hollow	50.8	9^a	2.743	9–14	2.743–4.267	14–35	4.267–10.668
3 hollow	76.2	12	3.658	12–18	3.658–5.486	18–35	5.486–10.668
4 hollow	101.6	15	4.572	15–22	4.572–6.708	22–35	6.708–10.668
6 hollow	152.4	20	6.096				
8 hollow	203.2	25	7.620				

[a] Not over 6 ft (1.829 m) in length.

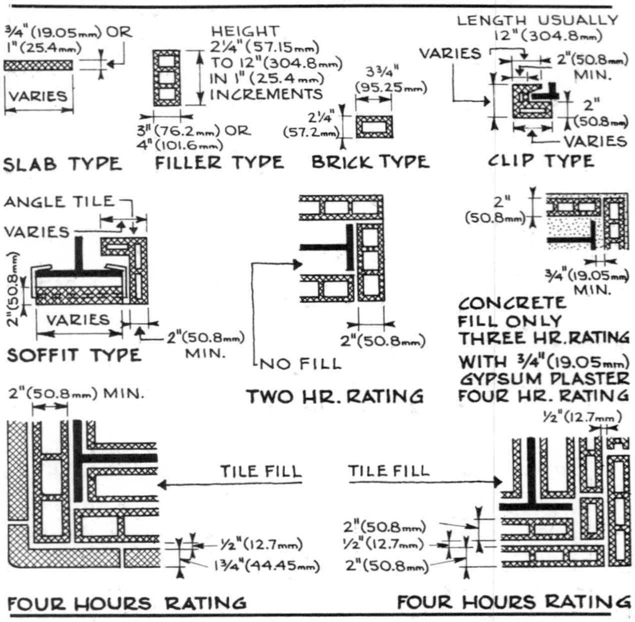

Figure C46 Shapes of structural clay tile for fireproofing beams, girders, and columns.

Type STX: high degree of mechanical perfection and minimum size variation.

Type STA: larger degree of size variation.

Screen tile is available in a large variety of patterns, sizes, and shapes and in a limited number of colors. The surfaces may be smooth, scored, combed, or roughened. The physical requirements and dimensional variations are shown in Tables *C37* and *C38*.

Table C37 Physical requirements for non-load-bearing structural clay screen tile

Grade	Maximum water absorption during 1 hour of boiling (percent): Average of five tests	Individual
SE	10	12
ME	14	16
NE	20	24

Table C38 Permissible dimensionable variations for non-load-bearing structural clay screen tile

Specified dimensions		Maximum permissible variations from specified dimensions (plus or minus)			
		Type STX		Type STA	
in.	mm	in.	mm	in.	mm
Up to and including 4	Up to and including 101.6	$\frac{3}{32}$	2.38	$\frac{1}{8}$	3.18
Over 4 to and including 6	Over 101.6 to and including 152.4	$\frac{1}{8}$	3.18	$\frac{3}{16}$	4.76
Over 6 to and including 8	Over 152.4 to and including 203.2	$\frac{5}{32}$	3.97	$\frac{1}{4}$	6.35
Over 8 to and including 12	Over 203.2 to and including 304.8	$\frac{7}{32}$	5.56	$\frac{5}{16}$	7.94
Over 12	Over 304.8	$\frac{9}{32}$	7.15	$\frac{3}{8}$	9.53

Application

Water Absorption Rate. The suction or absorption rate of structural clay tile has an important effect on the adhesion of tile and mortar, because if the tile absorbs water too quickly from the mortar, the latter will set too soon and adhesion will be poor. All structural clay tile with absorption rates in excess of 0.7 oz (19.85 g)/min should be wetted sufficiently so that the rate of absorption does not exceed this amount. Tile laid with grouted masonry may have an absorption rate of 1.4 oz (39.69 g)/min. A rough but effective test for determining whether units will give improved adhesion by wetting consists in drawing a circle 1 in. (25.4 mm) in diameter on the surface of the unit which will be in contact with the mortar and placing 20 drops of water inside the circle. If the time taken for the water to be absorbed exceeds $1\frac{1}{2}$ minute, the unit need not be wetted; if less than $1\frac{1}{2}$ minute, wetting is recommended.

Mortars. The following type of mortar is recommended for all types of structural clay tile: mortar controlled by property specifications in which (1) the types of cementitious material, lime, aggregate, and water are controlled; (2) the time of mixing is limited to at least 3 minutes; and (3) the mortar is required to be placed in final position within $2\frac{1}{2}$ hours after mixing. The compressive strength shall be tested from three 2-in. (50.9-mm) cubes and shall meet the requirements shown in Table *C39*. Mortar controlled by proportion specifications is shown in Table *C40*. (*See also* Aggregates; Cement; Mortar.)

Mortar joints are usually $\frac{1}{4}$ in. (6.35 mm) for glazed facing tile, and $\frac{3}{8}$ or $\frac{1}{2}$ in. (9.53 or 12.7 mm) for facing tile and for wall tile. When $\frac{1}{4}$-in. (6.35-mm) joints are used with structural clay tile, all the aggregate should pass a No. 16 sieve.

Bonding. A method of placing units so that they overlap and bind the entire wall into a solid mass throughout its length and breadth is necessary in all structural tilework. Either special bonding shapes or metal ties, anchors, clips, etc., are used. (*See* Figures *C47* and *C48*.)

Special Shapes. Some of the many special shapes of structural clay tile for particular purposes are illustrated in Figures *C48* and *C49*.

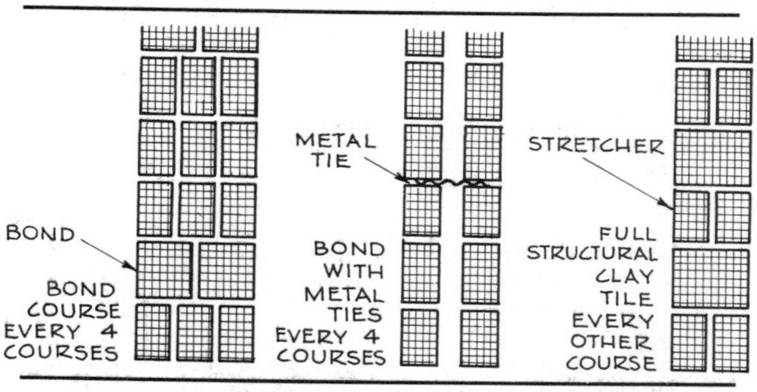

Figure C47 Bonding of structural clay tile.

Condensed Checklist

1. Check whether modular or standard sizes are available in the locality where the building is to be constructed.
2. Check limitations on height, length, and thickness of walls with local, municipal, and federal codes.

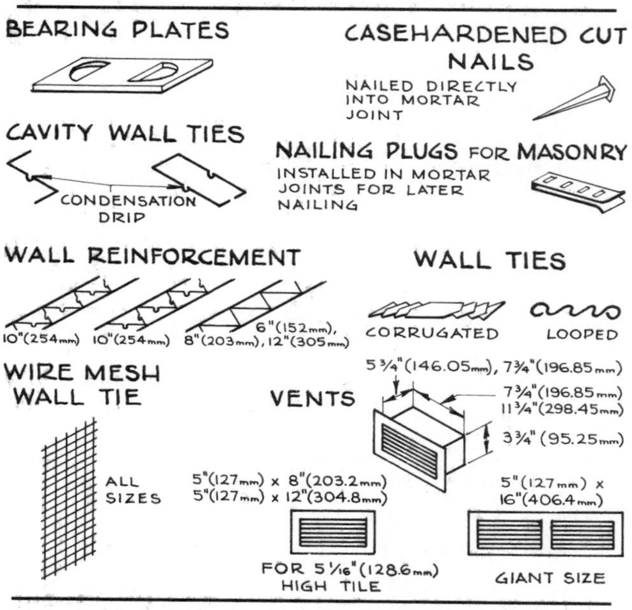

Figure C48 Accessories for structural clay tile.

Table C39 Compressive strength of mortar types for property specifications

Mortar type[a]	Average compressive strength[b] at 28 days lbf/in.^{2}	MN/m^2
M	2500	17.24
S	1800	12.41
N	750	5.17
O	350	2.41
K	75	0.52

[a] Formerly these types were designated as A-1, A-2, B, C, and D, respectively.
[b] 1 lbf/in.^{2} = 0.006895 MN/m^2.

Table C40 Mortar proportions of volume for proportion specifications

Mortar type[a]	Parts by volume of portland cement or portland blast-furnace slag cement	Parts by volume of masonry cement	Parts by volume of lime putty or hydrated lime	Aggregate measured in a damp, loose condition
M	1 1	1 None	None $\frac{1}{4}$	Not less than $2\frac{1}{4}$ and not more than 3 times the sum of the volumes of cements and lime used
S	$\frac{1}{2}$ 1	1 None	None Over $\frac{1}{4}$ to $\frac{1}{2}$	
N	None 1	1 None	None Over $\frac{1}{2}$ to $1\frac{1}{2}$	
O	None 1	1 None	None Over $1\frac{1}{4}$ to $2\frac{1}{2}$	
K	1	None	Over $2\frac{1}{2}$ to 4	

[a] Formerly the above types were designated as A-1, A-2, B, C, and D, respectively.

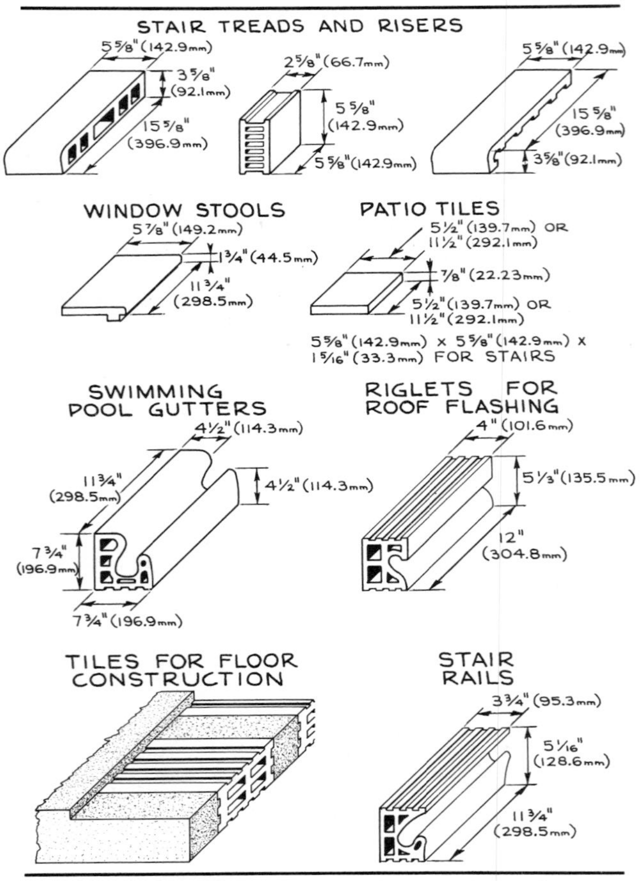

Figure C49 Special shapes of structural clay tile.

3. Type of mortar and mortar joint.
4. Type and number of expansion joints.
5. Wetness of the tile when laid (*see* Brick).
6. Avoid freezing temperatures below 40°F (4.45°C) in laying tile masonry unless suitable means are provided to heat the completed work and thus protect it from freezing.
7. Check electrical conduit, outlets, pipes, etc., and make allowances for their installation.
8. Carefully calculate dimensions of both horizontal and vertical coursing in relation to the dimensions of the tile units to be used so that no cutting of tile on the job will be necessary. There should be no need for small slivers to fit either horizontally or vertically. (*See* Figures *C50* and *C51*.)
9. Bonding (*see* Figure *C47*).
10. Treatment of corners (*see* Figure *C52*).
11. Conditions at grade and roof (*see* Figure *C53*).
12. Conditions at floor and ceiling for interior partitions. The type of base is controlled by the type of finish floor used (*see* Figure *C54*).
13. Conditions at openings (*see* Figure *C55*).
14. Check fire-resistance regulations of governing codes for the type of tile that can be used.
15. Always check the design of a screen wall for necessary expansion and pressure-relieving joints and for strength and compressive strength.
16. Always specify the length and the height in courses for the sample wall to be built on site and the color range allowed for the type of structural tile selected.
17. When using acoustical-type structural tile, always check the noise reduction coefficients.
18. Always check with the manufacturers the colorfastness, hardness, and abrasion resistance of the structural clay facing tile selected for the job.
19. When using high-strength mortars with structural clay tile, always check with the manufacturers of both the structural tile and the high-strength mortar and with state, municipal, and local codes for requirements regarding high-strength mortars and limitations on their use.

Conditions Favorable to the Use of Structural Clay Tile

1. Where compressive strength is important (all load-bearing tile).
2. Where the surface is likely to receive rough treatment (facing or glazed facing tile).
3. Where maintenance must be kept to a minimum (facing or glazed facing tile).
4. Where a high degree of imperviousness and chemical resistance is required (facing or glazed facing tile).
5. Where fire resistance is required (fireproofing, facing and glazed facing tile). State, municipal, and local codes should be checked and also the fire rating codes of the Fire Underwriters, insurance companies, labor departments, and federal government (Army, Navy, etc.).
6. Where any of the available colors and textures is desired for design purposes (facing and glazed facing tile).
7. Where a plaster finish is desired on partitions (partition tile) or on furred walls (furring tile).
8. On the interior of masonry walls where an integral interior finish (facing or glazed facing tile) or a plaster finish (wall tile) is desired.
9. Where an integral finish is desired on one or both sides of a partition (facing or two-surfaced glazed facing tile).
10. Where a decorative screen wall is important.
11. Where fireproofing is required for structural steel, stairways, shafts, etc. Always check state, municipal, and local codes for fire rating of various types of partition, furring, and fireproofing structural clay tile.

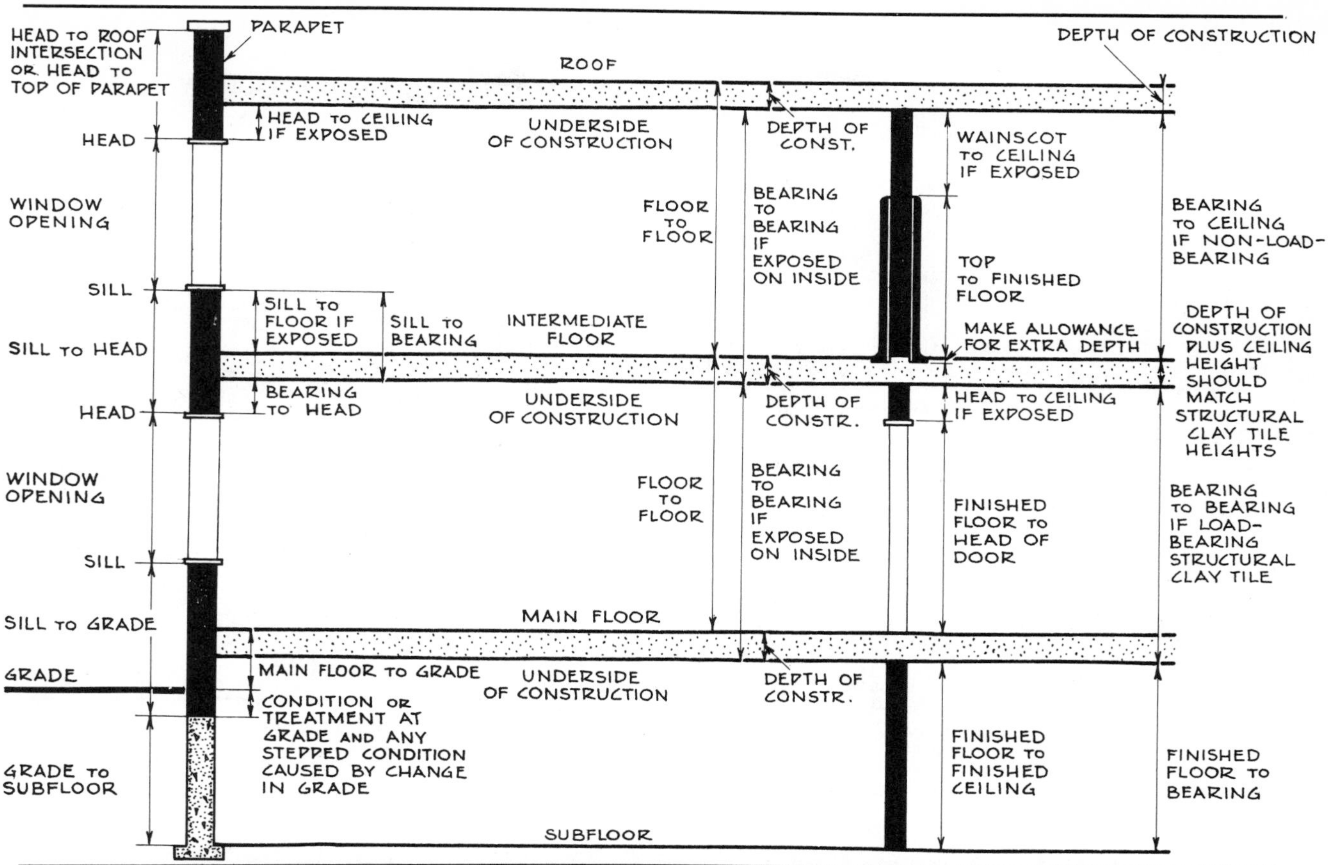

Figure C50 Dimensions of vertical coursing for load-bearing and non-load-bearing structural clay tile.

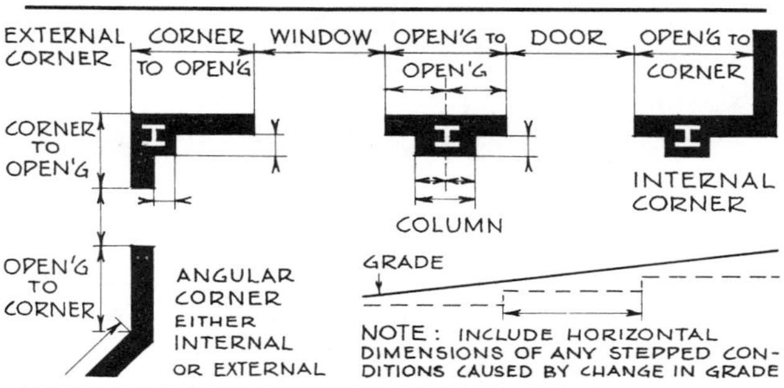

Figure C51 Dimensions of horizontal coursing for load-bearing and non-load-bearing structural clay tile.

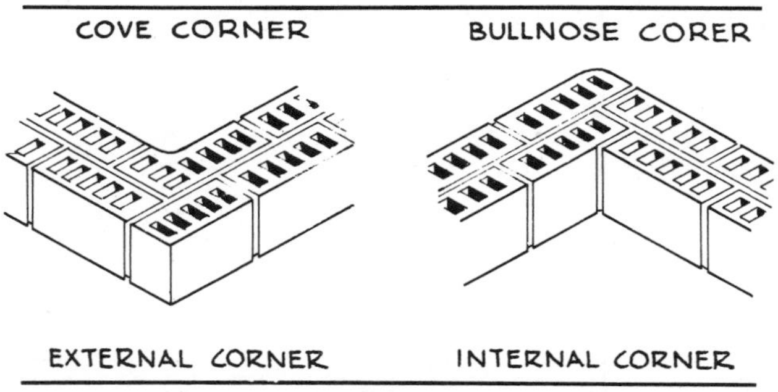

Figure C52 Treatment of corners for structural clay tile.

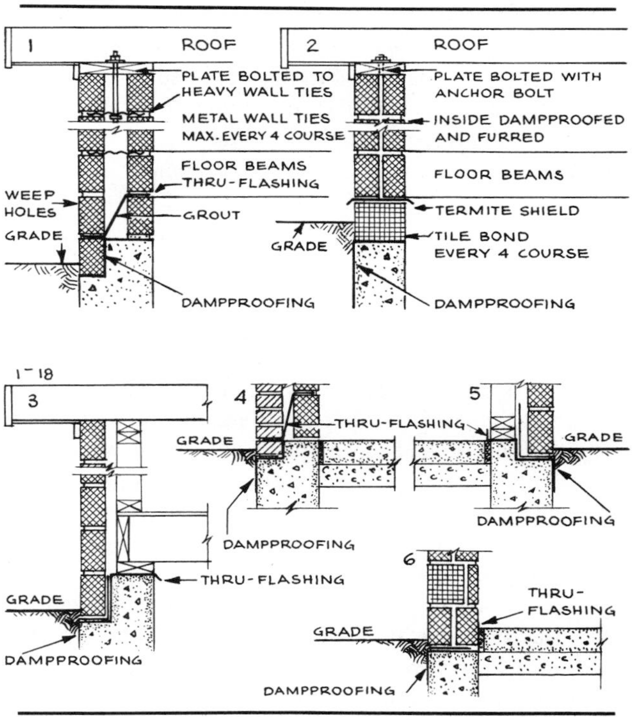

Figure C53 Conditions at grade and roof for structural clay tile.

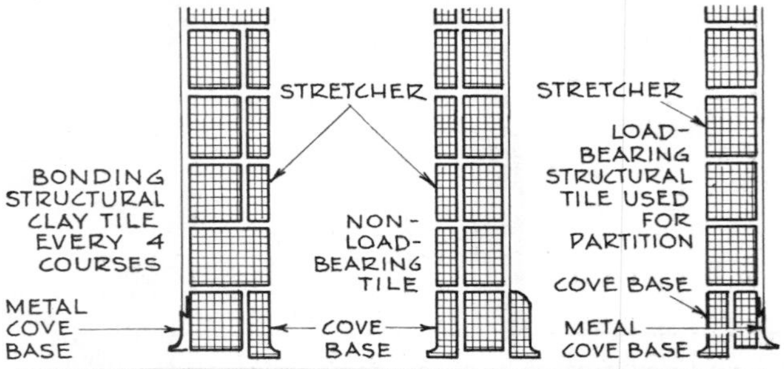

Figure C54 Conditions at floor for structural clay tile.

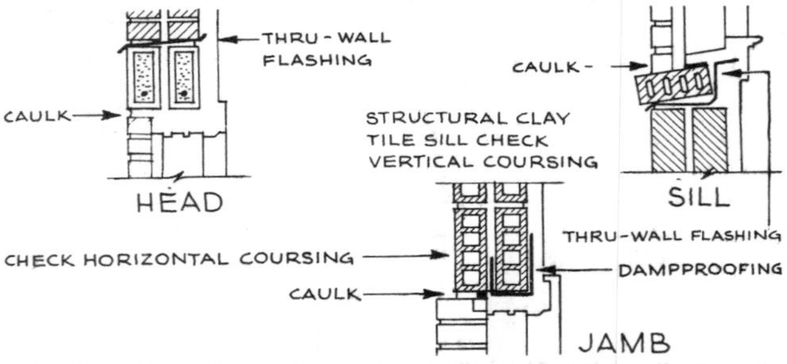

Figure C55 Conditions at openings for structural clay tile.

Conditions Unfavorable to the Use of Structural Clay Tile

1. Where tensile or lateral strength is important, unless reinforced structural clay tile masonry is used.
2. Where a watertight condition is required.
3. Where a high coefficient of insulation or sound absorption is required.
4. Where small radii are important or required.

History and Manufacture

Structural clay tile, unlike brick, ceramic veneer, and other clay products, is a machine-made product of recent origin. It was first produced in the United States in 1875 in New Jersey. From 1884 to 1887 Ohio, Illinois, and Indiana also started production, and now, since clays for this type of tile are plentiful, almost every state contributes to total production and supplies local demand.

The number of shapes and sizes produced by the industry reached its maximum about 1903. Since then, with the growing acceptance of modular coordination, the structural clay tile industry has been in the process of standardizing special shapes in addition to the standard ones. The manufacturing process is fundamentally the same as that used for brick.

CLIMATE AND CLIMATOLOGY

In the construction field, climate and climatology have become increasingly important factors in selecting the orientation of a building or buildings and the type of construction, including materials, roofing, drainage,

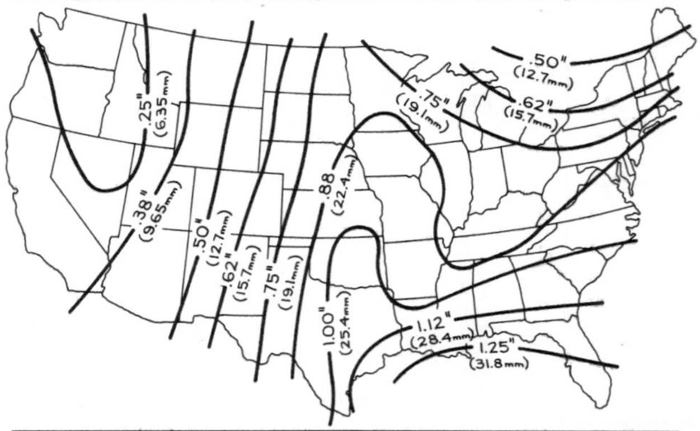

Figure C56 Fifteen-minute rainfall to be expected in 2 years.

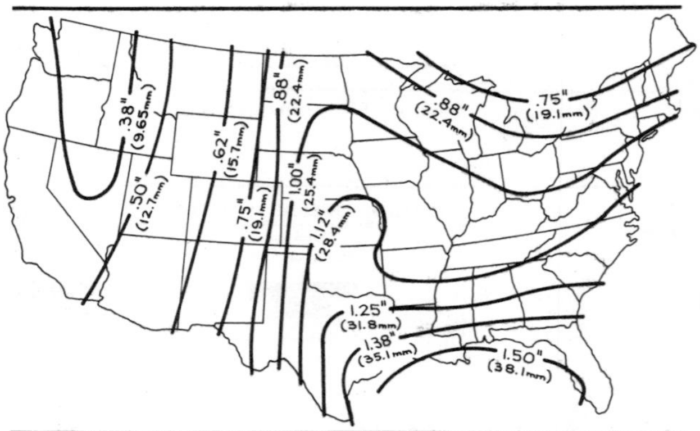

Figure C57 Fifteen-minute rainfall to be expected in 5 years.

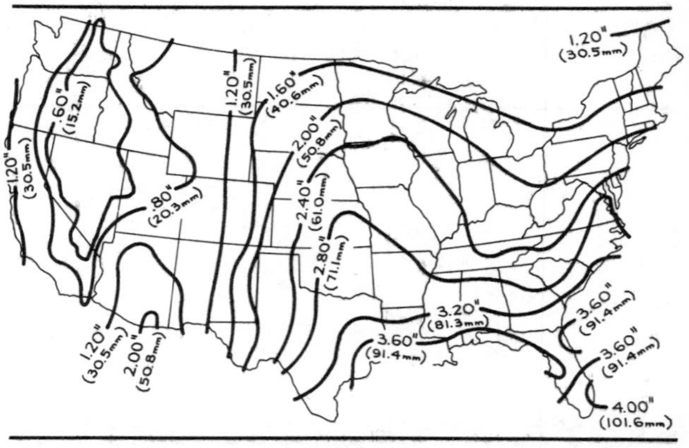

Figure C58 Thirty-minute rainfall to be expected once in 2 years.

Table C41 Proximate characteristics of types of coal

Type of coal	Moisture (percent)	Volatile matter (proximate percent)	Fixed carbon	Ash	S	H (ultimate percent)	C	N	O	Heating value[a] Btu/lb	kJ/kg
Anthracite	4.4	4.8	81.8	9.0	0.6	3.4	79.8	1.0	6.2	13,130	30 540
Semianthracite	2.8	11.9	75.2	10.1	2.2	3.7	78.3	1.7	4.0	13,360	31 075
Low-volatile bituminous	2.3	19.6	65.8	12.3	3.1	4.5	74.5	1.4	4.2	13,220	30 751
Medium-volatile bituminous	3.1	23.4	63.6	9.9	0.8	4.9	76.7	1.5	6.2	13,530	31 470
High-volatile A bituminous	3.2	36.8	56.4	3.6	0.6	5.6	79.4	1.6	9.2	14,090	32 773
High-volatile B bituminous	5.9	43.8	46.5	3.8	3.0	5.7	72.2	1.3	14.0	13,150	30 587
High-volatile C bituminous	14.8	33.3	39.9	12.0	2.5	5.8	58.8	1.0	19.9	10,550	24 589
Subbituminous coal rank A	13.9	34.4	41.0	10.9	0.6	6.2	57.5	1.4	23.4	10,330	24 028
Subbituminous coal rank B	22.2	32.2	40.3	4.3	0.5	6.9	53.9	1.0	33.4	9.610	22 353
Subbituminous coal rank C	25.8	31.1	38.2	4.7	0.3	6.3	50.0	1.6	38.1	8,580	19 957
Lignite	36.8	27.8	30.2	5.2	0.4	6.9	41.2	0.7	45.6	6,960	16 189

[a] 1 Btu per pound = 2.326 kilojoule per kilogram.

grading, selection of lawn and planting materials, and the type of environmental controls to be installed.

The construction field encompasses not only the United States but also the entire world, as architects, engineers, planners, construction firms, etc., are now designing and building structures of all types in all parts of the world and in all kinds of climatic conditions.

In almost all cases data concerning climate are available from county, city, state, or federal agencies, as well as specialized private organizations. Climate and climatology include temperature; humidity; amounts of rain, snow, and ice; duration of sunshine; amounts of clouds; wind direction and speed; weather phenomena such as fog, frost, rain, thunderstorms, hurricanes, and cyclones; changes in barometric pressure; evaporation; condition of soil; and atmospheric pollution.

The subject is beyond the scope of this book, but for certain basic data concerning the United States *see* Wood Preservatives; Insulation and Insulating Materials; *and* Figures *C56*, *C57*, and *C58*.

COAGULANT

A coagulant is a material or substance that causes a liquidlike material to thicken (*see* Paint).

COAL

Coal is a combustible material consisting of elemental carbon with various amounts of hydrocarbons, complex organic compounds, and inorganic materials. The varieties of coal are: lignitic, bituminous, subbituminous, anthracite, and semianthracite coals (*see* Table *C41*). Most of the coal mined is used as fuel to produce electricity, steam for heating, and as fuel in industry.

Coal is converted into solid, liquid and gaseous products which are used as fuels, materials for the chemical and construction industries, and for roads, highways, and turnpikes. One ton (1.016 metric tonne) of coal yields 8.8 gallons (33.31 liters) of coal tar. *See* Tables *C42* and *C43*. There are over 25,000 chemical by-products of coal.

Coal which is made into coke, an ingredient used in blast furnaces to produce steel, gives off coal gas. This gas is used directly for heating in open hearth furnaces. *See* Petroleum; Natural Gas; Bitumens; Petrochemicals; Fossil Fuels; Asphalt; Tar.

COBALT

Physical and Chemical Properties

Symbol: Co
Atomic number: 27

Table C42 Main products from coal and the destructive distillation of coal

Coal and coal derivatives	Use
COAL	
Anthracite, semianthracite, bituminous, subbituminous, and lignite	Commercial and industrial heating; production of electric power
CARBONIZATION OF COAL	
Acetylene	Welding
Carborundum	Abrasives
Coke	Manufacture of steel
Coal gas	Industrial heat
Water gas	Heating and cooking
DISTILLATION OF COAL	
Soft tar pitch	Paving; coatings; building papers and felts; waterproofing and dampproofing coatings; impregnating fiber boards and pipes
Medium-soft tar pitch	Built-up roofing; building papers and felts; protective metal coatings; expansion joints and joint fillers
Hard tar pitch	Coatings
COKING AND DISTILLATION	
Raw material	Synthetic rubbers, resins, and fibers; plastics; dyes, color pigments; solvents and resins for paint

Table C43 Uses of coal derivatives

	Construction Uses	
Basis of	Abrasives	Main ingredient
	Acetylene	Welding
	Adhesives	Main ingredient
	Caulking	Main ingredient
	Paving	Roads, driveways, and runways
	Preservative	Creosote
	Protective coatings	Waterproofing; protecting ferrous materials against rust
	Roofing	Built-up type of roofing
	Waterproofing paints	Main ingredient
Component of	Fibers	Ingredient
	Fiber pipe and board	Impregnating and waterproofing
	Expansion joint and joint filler	Ingredient
	Paint	Waterproofing and coatings
	Paint casein	Vehicle
	Paper and felt	Coating and impregnating
	Rubber	Ingredient
	Textiles	Impregnating and waterproofing fabric flashing
	Wood	Preservative
Allied Construction Uses	Plumbing	Coating pipes
	Electrical	Coatings

Specific gravity: 8.9 (20°C, 68°F)
Melting point: 1495°C, 2723°F
Boiling point: 2870°C, 5198°F
Coefficient of expansion: 0.0000068 in./°F, 0.003109 mm/°C
Tensile strength: 34,400 lbf/in.^{2}, 237.188 MN/m^2 (as cast)

Cobalt is a hard, magnetic, malleable, somewhat brittle, silver-white metal, similar to nickel and iron in appearance and chemical properties. It is stable in air and water but is attacked by certain gases and the common acids. Cobalt metal dust is injurious to human beings.

Cobalt has the ability to dissolve large amounts of chromium, tungsten, carbon, nickel, and iron, resulting in alloys in which the special properties of each of these metals can be utilized. Cobalt alloyed with iron, nickel, and other metals imparts exceptional magnetic properties to the resulting material.

Commercial Forms. Cobalt is available in rondel, shot, anode, and powder form; a new form, electrolytic cobalt, is also now on the market. Commercial cobalt varies from 98 to 99.8% purity (*see* Table *C44*).

Table C44 Composition of electrolytic (cathode) cobalt

Co	Ni	Fe	Si	C	Cu	S
		(percent of content)				
99.5	0.45	0.03	0.01	0.03	0.01	0.003

Table C45 Uses of cobalt

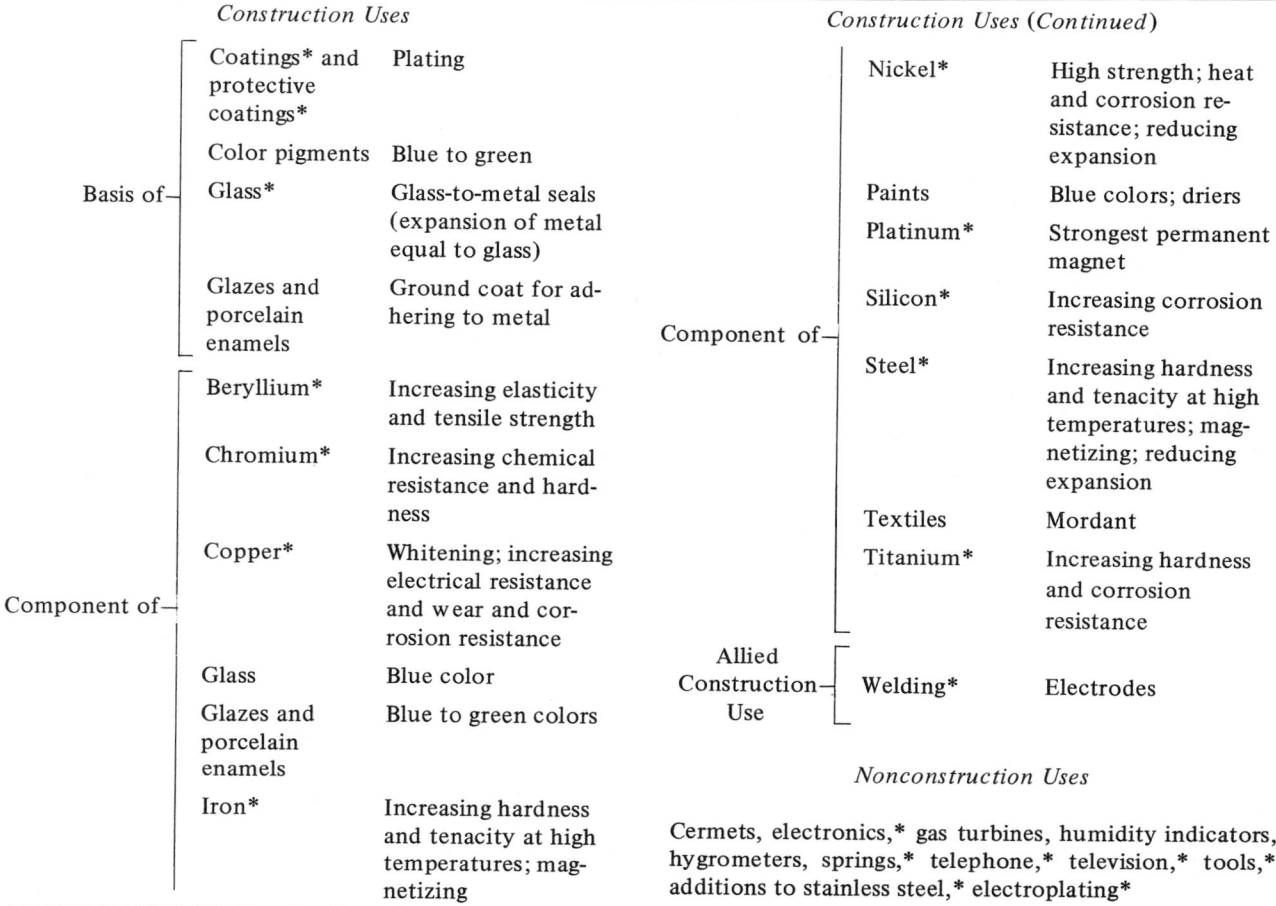

Construction Uses		
Basis of	Coatings* and protective coatings*	Plating
	Color pigments	Blue to green
	Glass*	Glass-to-metal seals (expansion of metal equal to glass)
	Glazes and porcelain enamels	Ground coat for adhering to metal
Component of	Beryllium*	Increasing elasticity and tensile strength
	Chromium*	Increasing chemical resistance and hardness
	Copper*	Whitening; increasing electrical resistance and wear and corrosion resistance
	Glass	Blue color
	Glazes and porcelain enamels	Blue to green colors
	Iron*	Increasing hardness and tenacity at high temperatures; magnetizing
Construction Uses (Continued)		
Component of	Nickel*	High strength; heat and corrosion resistance; reducing expansion
	Paints	Blue colors; driers
	Platinum*	Strongest permanent magnet
	Silicon*	Increasing corrosion resistance
	Steel*	Increasing hardness and tenacity at high temperatures; magnetizing; reducing expansion
	Textiles	Mordant
	Titanium*	Increasing hardness and corrosion resistance
Allied Construction Use	Welding*	Electrodes
Nonconstruction Uses		
Cermets, electronics,* gas turbines, humidity indicators, hygrometers, springs,* telephone,* television,* tools,* additions to stainless steel,* electroplating*		

Types and Uses

Until 1914 almost all cobalt was used in the ceramic industry. Now the major use of cobalt is in magnetic, high-temperature alloys, corrosion-resistant alloys, and hard facing where corrosion and abrasion resistance are combined. Cobalt compounds are very widely used to color and pigment glass and other ceramic products and as a drier in paints. In Table *C45* the uses of cobalt in metallic form are indicated by asterisks.

Magnetic alloys of interest to the construction field are found in all electrical devices and are increasingly used in finish and rough hardware. The high-temperature alloys are promising but have not yet found application in the architectural field. The only direct use for cobalt metal itself is for plating, which promises to become a much more important application than at present.

Plating. Cobalt-nickel plating is bright, ductile, and adherent. Phosphorus, added as an alloy to cobalt or nickel used for plating, makes it hard, corrosion resistant, and bright. A plating of cobalt-tungsten is extremely hard, its hardness under heat approaching that of the stellites.

Stellites. In 1899 Haynes developed a cobalt-chromium alloy which resisted chemical fumes and remained hard even at red heat. By 1908 he had developed a cutting alloy whose performance equaled that of tempered steel. Tungsten, molybdenum, and carbon were added to this cobalt-chromium base to produce cutting tools superior to high-speed steel tools. Today these cobalt-chromium-tungsten alloys, called stellites, are important in industry.

Radioactive Cobalt. Radioactive cobalt (cobalt 60) is used for radiographic inspection in testing weldings and castings, for measuring paint thickness, for determining wearing qualities of floor wax, and for locating buried telephone and electrical conduits.

History and Manufacture

Persian glass beads colored with cobalt date back to about 2250 B.C. Egyptian and Babylonian pottery of about 1450 B.C. and early Roman and Venetian blue glass also were colored with cobalt. Recently discovered Egyptian glass statuettes of about 2680 B.C. proved to have a cobalt-base blue coloring agent.

The chemistry and metallurgy of cobalt started about A.D. 1500. Its name originated in Germany. Certain ores from the Harz Mountains, when roasted, yielded only dangerous fumes instead of copper and were called *Kobold*, or "goblin" in German, by the miners. In 1742 Brandt prepared an impure metallic cobalt and was the first to describe the properties of the metal. About 1780 Bergman established it as an element.

Cobalt is a coproduct of the metallurgy of ores containing arsenic, iron, nickel, copper, manganese, and silver, those containing copper and nickel being practically the entire source. The production method for cobalt depends on the type of ore used and the products desired.

The largest producer in the United States uses ore from the Blackbird Creek District, Idaho. The ore is beneficiated to yield copper concentrate and a cobalt concentrate containing 17.5% cobalt, 20% iron, 24% arsenic, 1% nickel, 0.50% copper, and 5% insolubles. Further processing of this is shown in Figure *C59*.

Blast Furnace Process. The oldest method of obtaining cobalt is the blast furnace process of reducing ores to cobalt oxide. Here the ores are smelted to a speiss, which is ground and roasted to eliminate the arsenic. This roasted speiss is dissolved in acid, the copper is precipitated out, and then the iron and remaining arsenic are removed from solution by adding limestone and blowing air through the solution. The cobalt and nickel remaining in solution are precipitated out as hydroxides. Roasting the cobalt hydroxide converts it into cobalt oxide, which is then sold commercially in that form or is reduced to metal.

Electric Furnace Process. The electric furnace method of production yields cobalt and copper, both in crude alloy form. The furnace is charged with high-cobalt ores, concentrates from copper ores with lower cobalt content, and slag from the copper refinery which is part of this process, together with lime as the flux and coke as conductor and reducing agent. Within the electric furnace a slag is first run off; then silicon is added to reduce the cobalt, copper, and some iron to the metallic state. This produces a light, crude cobalt alloy (white alloy), which is run off at the top, and a heavier, crude copper alloy (red alloy), which is run off at the bottom and refined for copper. The gravity separation is effected by keeping the silicon in the white alloy at about 2.5%. However, very little cobalt is produced by this method.

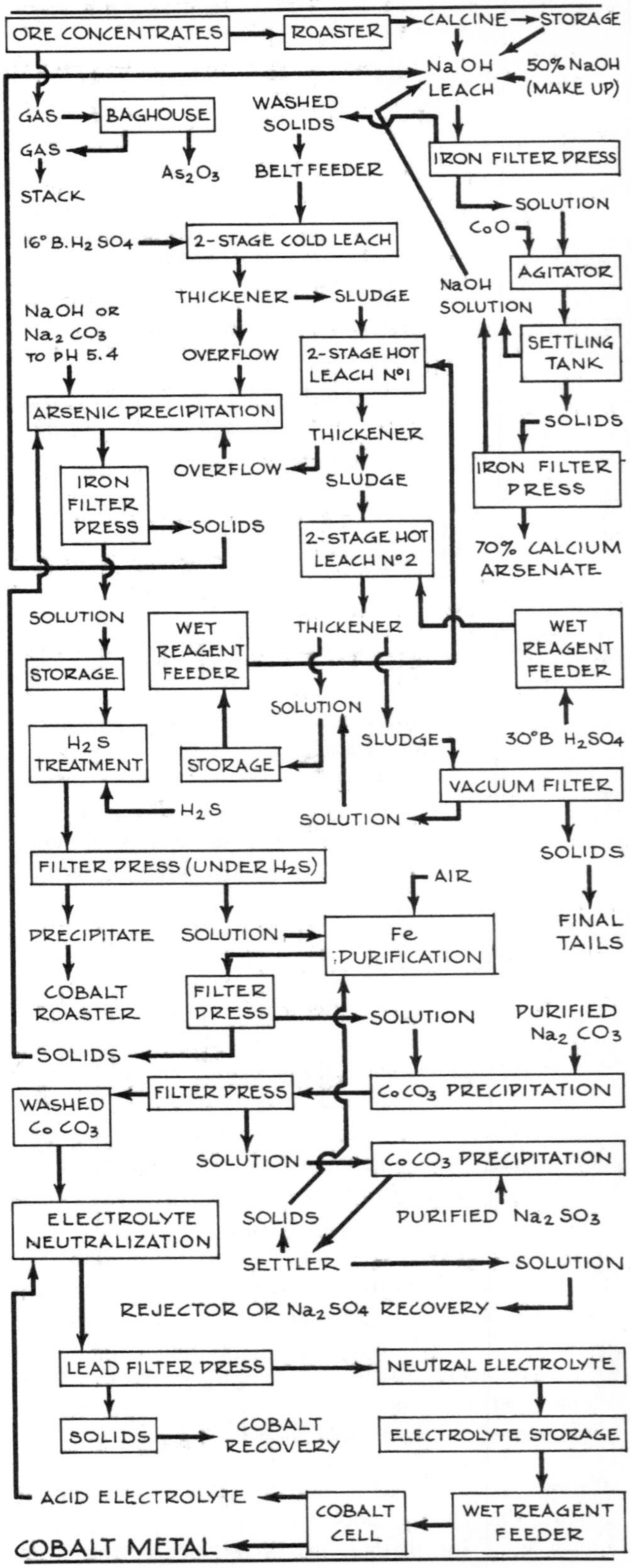

Figure C59 Flowchart of cobalt production from cobaltite ore.

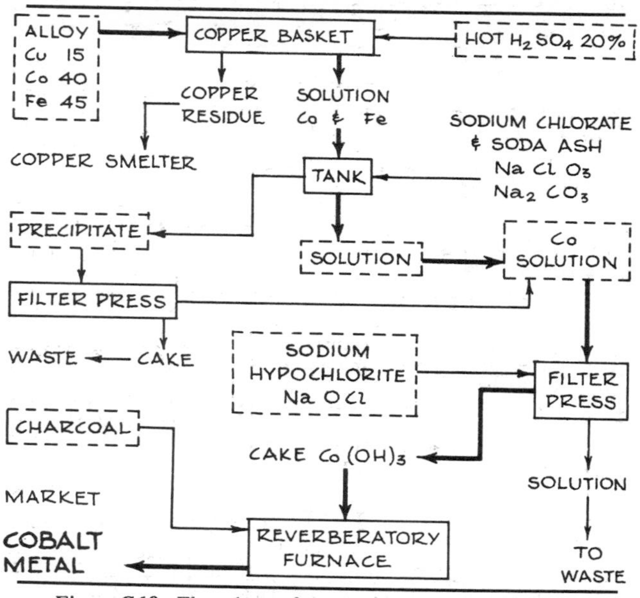

Figure C60 Flowchart of processing crude cobalt alloy.

Pure Cobalt. To obtain pure cobalt, the crude alloy is first dissolved in sulfuric acid and the copper is precipitated out. The iron and cobalt solution is treated with sodium chlorate to oxidize the iron, which is precipitated with lime solution. The iron hydrate is removed by filter press, and the cobalt-bearing solution is treated with sodium hypochlorite to precipitate cobalt hydrate. This hydrate is mixed with dextrin, water, and aerosol, pressed into rondelles, and then reduced with charcoal in a reverberatory-type furnace at 1922°F (1050°C) to metallic cobalt. (*See* Figure *C60*.)

COEFFICIENT OF EXPANSION

The linear coefficient of expansion of a material is the rate at which the unit length changes with an increase of 1° of temperature. *See* Table *C46*.

COLOR PIGMENTS

Physical and Chemical Properties

The pigments discussed here are the finely divided insoluble white or colored powders used to impart color to paints, enamels, floor coverings, rubber and plastic compositions, roofing granules, and other similar construction materials.

The important characteristics of a pigment are its light absorption, refractive index and particle size, and its resistance to heat, light, and chemical agents. The light absorption, refractive index, and particle size of a pigment control its opacity or hiding power in a paint or other mixture. The resistance of a pigment to heat, light, particularly sunlight, and chemical agents determines its practicality for a given end use (*see* Paint for general discussion of relation of color pigment to paint).

Light Absorption. Pigments ordinarily impart color by selective absorption of light. The predominant light wavelength reflected by the coloring material gives its hue or color designation. Value, or brightness, is the percentage of light reflected. Jet black reflects practically no light; brilliant white reflects practically 100% of the total spectrum. Chroma, or saturation, is the percentage of reflected light that is colored. It gives the exact tint (strength or weakness) of a hue.

Refractive Index. In white pigments the refractive index determines the opacity and covering or hiding power of paints. Particle size being equal, when the difference between the refractive index of the pigment and the refractive index of the vehicle is large, the paint exhibits high covering power; the less the difference between them, the closer the paint approaches transparency.

Particle Size. The optimum particle size for pigments is a diameter just less than the wavelength of visible light. According to paint manufacturers, the optimum size is from 0.15 to 0.4 μ (1 micron equals 0.001 mm or 0.0004 in.). If the particles are too small, light is easily diffracted around them.

Types and Uses

Inorganic Pigments. Mineral or inorganic color pigments are substances that occur in nature and that are now chemically prepared. These so-called earth colors are the most stable and widely applicable.

Organic Pigments. Organic pigments are available in brighter and more varied colors but are generally inferior to inorganic pigments in light fastness, heat resistance, and tendency to dissolve or bleed in oils and solvents. "Lakes" are insoluble organic pigments which consist of a soluble organic coloring agent combined with a metallic oxide that holds the color. "Toners" are organic color pigments which contain neither inorganic color pigment nor carrying base.

For common color pigments, *see* Table *C47* (*see also* Antimony; Barium; Lithopone; Titanium Dioxide; Zinc Oxide; Zinc Sulfide.) For metallic color pigments, *see* Table *C48* (*see also* Aluminum Powder).

Table C46 Coefficients of linear expansion

Material	Coefficient of expansion for 100°		Expansion for 100° F/10 ft (37.78°C/3.048 m)		
			in.		
	°C	°F	Fractions	Decimals	mm
Aluminum	0.00231	0.00128	$\frac{5}{32}$	0.156	3.969
Brass	0.00188	0.00104	$\frac{1}{8}$	0.125	3.175
Brick masonry	0.00055	0.00031	$\frac{1}{32}$	0.0313	0.794
Bronze	0.00181	0.00101	$\frac{1}{8}$	0.125	3.175
Clay tile structure	0.00059	0.00033	$\frac{1}{32}$	0.0313	0.794
Concrete	0.00143	0.00079	$\frac{3}{32}$	0.0938	2.381
Concrete structure	0.00098	0.00055	$\frac{1}{16}$	0.0625	1.588
Copper	0.00168	0.00093	$\frac{7}{64}$	0.0156	2.778
Glass sheet	0.00085	0.00047	$\frac{3}{64}$	0.0469	1.191
Granite	0.00084	0.00047	$\frac{3}{64}$	0.0469	1.191
Iron, cast	0.00106	0.00059	$\frac{1}{16}$	0.0625	1.588
Iron, wrought	0.00120	0.00067	$\frac{5}{64}$	0.0781	1.984
Lead	0.00286	0.00159	$\frac{3}{16}$	0.1875	4.763
Limestone	0.00080	0.00044	$\frac{3}{64}$	0.0469	1.191
Magnesium	0.00290	0.00143	$\frac{11}{64}$	0.1719	4.366
Marble	0.00100	0.00056	$\frac{1}{16}$	0.0625	1.588
Monel	0.00136	0.00075	$\frac{3}{32}$	0.0938	2.381
Plaster	0.00166	0.00092	$\frac{7}{64}$	0.0156	2.778
Stainless steel	0.00173	0.00096	$\frac{7}{64}$	0.0156	2.778
Steel	0.00132	0.00073	$\frac{3}{32}$	0.0938	2.381
Steel, galvanized	0.00116	0.00065	$\frac{5}{64}$	0.0781	1.984
Terne	0.00116	0.00065	$\frac{5}{64}$	0.0781	1.984
Titanium	0.00085	0.00047	$\frac{3}{64}$	0.0469	1.191
Wood, fir	0.00037	0.00021	$\frac{1}{64}$	0.0156	0.397
Wood, oak	0.00049	0.00027	$\frac{1}{32}$	0.0313	0.794
Wood, pine	0.00054	0.00030	$\frac{1}{32}$	0.0313	0.794
Zinc	0.00311	0.00173	$\frac{12}{64}$	0.2031	5.159

Extender Pigments. Extender pigments are a group of mineral substances used as fillers. They are characterized by low refractive indices. Many of them are white but produce virtually no opacity in an oil-paint film. In properly formulated paints these extender pigments may be used to confer certain special properties, for example, improved brushing characteristics and storage qualities. They are essential components of priming and sealing paints and of flat (matte) or lusterless finishes.

The following is a list of the materials commonly used for this purpose:

Barium sulfate: natural barites, precipitated barites such as blanc fixe and permanent white.
Calcium carbonate: whiting, chalk.
Calcium sulfate: gypsum, terra alba, plaster of paris.
Kaolin: china clay, bentonite, and other clays.
Magnesium carbonate: magnesian limestones.
Magnesium silicate: talc, soapstone, asbestine.
Metallic soaps: aluminum stearates and palmitates.
Silica: quartz, chemically prepared silica, diatomaceous earth.
Miscellaneous: mica, pumice.

These are also called "inert" pigments because they do not react chemically with oils or other pigments. They are sometimes referred to as reinforcing pigments because some of them add strength or other structural features to a paint film (*see* Paint, Table *P2*).

Table C47 Common color pigments

Color	Permanence (resistance to light and chemical action)	Color pigments: Name	Color pigments: Percentages and materials contained
Black	Excellent	Black synthetic iron oxide	Ferroso-ferric oxide $Fe_3O_4(FeO, Fe_2O_3)$ by chemical action
Black: blue gray	Excellent	Lampblack	Carbon from burned oils, tars
Black: brownish gray	Excellent	Ivory black (bone black, drop black)	Carbon from charred bones
Black: reddish gray	Excellent	Carbon black	Carbon from burning natural gas under low oxidation
Blue	Excellent	Phthalocyanine blue	Copper phthalocyanine (complex organic color)
Blue: brilliant	Fades on exterior exposure	Ultramarine blue	Double silicate of sodium and aluminum with sodium sulfide
Blue: greenish	Excellent (used mainly in artists' colors)	Cobalt blue	Combination of aluminum oxide and cobalt oxide
Blue: reddish	Fades on exterior exposure (do not use where there are alkalis or acids)	Prussian blue (iron blue, milori blue, Chinese blue)	Ferric ferrocyanide
Brown: dark	Excellent	Burnt umber	Iron oxide (42%) from raw umber calcined at low heat
Brown: dull greenish	Excellent	Raw umber	Natural hydrated iron oxide (37% min.) with siliceous base
Brown: rich red	Excellent	Burnt Sienna	Iron oxide (40% min.) made by heating raw Sienna
Green: brilliant bluish	Excellent	Phthalocyanine green	Chlorinated copper phthalocyanine, related to phthalocyanine blue
Green: brilliant bluish	Excellent	Hydrated chromium oxide (viridian, emeraude green)	Hydrated chromic oxide
Green: brilliant yellow to bluish	Good (do not use where there are alkalis)	Pure chrome green	Precipitated mixture of lead chromate and Prussian blue
Green: dull bluish	Excellent (most permanent green)	Chromium oxide green	Chromic oxide (97% min.)
Orange: brilliant	Darkens on exterior exposure	Molybdated orange	Combination of lead chromate and lead molybdate
Orange: dark brilliant	Excellent (do not use where there are alkalis)	Dark chrome orange	Precipitated lead chromate (55% min.)
Orange: light brilliant	Excellent (do not use where there are alkalis)	Light chrome orange	Precipitated lead chromate (60% min.)
Red	Excellent	Venetian red	Iron oxide (20 to 50%) on a calcium sulfate base
Red	Excellent	Ocher	Hydrated iron oxide (17% min.) with siliceous base
Red: bright yellowish to deep purplish	Excellent	Red iron oxides (light red, Indian red, Tuscan red)	Red iron oxide (95 to 98%)
Red: brilliant	Excellent (bleeds in linseed oil)	Toluidine red (neutral red)	Insoluble azo compound (aniline color)

Table C47 Common color pigments (*continued*)

Color	Permanence (resistance to light and chemical action)	Color pigments: Name	Color pigments: Percentages and materials contained
Red: light	Fades on exterior exposure, bleeds in lacquer	Lithol red	Tobias acid and β-naphthol combined into lakes and toners; capable of considerable variation in hue
Red: light	Moderate (fades on exterior exposure, bleeds)	Para red	Azo compound (oldest organic red color for paint)
Red: very light	Fades on exterior exposure	Chlorinated para red	Variation of para red
Reds: light to deep	Excellent (poor resistance to acids)	Cadmium reds	Cadmium sulfide and barium sulfate
Red: orange	Fades on exterior exposure	American vermilion	Lead chromate
Slate gray	Excellent	Blue lead	Basic lead sulfate-blue
White	Excellent	Zinc white	Zinc oxide (98% min.)
White	Excellent	Zinc sulfide	Zinc sulfide (97% min.)
White	Excellent	Zinc, titanated	Zinc sulfide, barium sulfate, and titanium dioxide
White	Excellent	Lithopone, titanated	Zinc sulfide, barium sulfate, and titanium dioxide
White	Excellent	Antimony trioxide	Antimony trioxide (98% min.)
White	Excellent	Titanium dioxide (anatase or rutile)	Unextended titanium dioxide (94% min.)
White	Excellent	Lead silicate	Basic lead silicate with 45% of lead as metal
White	Good	Titanium barium	Titanium dioxide (28% min.) and barium sulfate
White	Good	Titanium calcium	Titanium dioxide (28% min.) and anhydrous calcium sulfate
White	Good (turns gray with any sulfur fumes)	White lead (basic carbonate)	Lead carbonate (62 to 75%) and lead hydroxide
White	Good (use only on exterior)	Lithopone	Zinc sulfide (26%) and barium sulfate
White	Good	Zinc sulfide magnesium	Zinc sulfide (45%) and magnesium ($3MgSiO_4$, $5H_2O$) with or without micaceous silicate
White	Good	Zinc sulfide barium	Zinc sulfide (45%) and barium sulfate (55%)
White	Good	Zirconium oxide	Zirconium dioxide
Yellow: deep brilliant	Tends to darken on exterior exposure (do not use where there are alkalis)	Medium chrome yellow	Precipitated lead chromate (93% min.)
Yellow: dark brownish	Excellent	Raw sienna	Natural hydrated iron oxide (37% min.) with manganese oxides in siliceous base
Yellow: dull	Excellent	Yellow ocher	Hydrated iron oxide on siliceous base
Yellow: greenish	Excellent (do not use where there are acids)	Cadmium yellow	Cadmium sulfide and barium sulfate
Yellow: lemon	Darkens on exterior exposure	Barium chromate	Precipitated barium chromate

Table C47 Common color pigments (*continued*)

Color	Permanence (resistance to light and chemical action)	Color pigments: Name	Color pigments: Percentages and materials contained
Yellow: lemon	Tends to darken on exterior exposure (do not use where there are alkalis)	Lemon chrome yellow	Precipitated lead chromate (68% min.)
Yellow: light	Tends to darken on exterior exposure (do not use where there are alkalis)	Primrose chrome yellow	Precipitated lead chromate (50% min.)
Yellow: light	Excellent	Hydrated yellow iron oxide	Hydrated iron oxide (97% min.) prepared by chemical action
Yellow: light	Excellent	Strontium chromate	Strontium chromate

Table C48 Metallic color pigments

Name	Material	Color
Aluminum powder	Fine aluminum flakes ((99.9% Al)	Bluish white
Copper powder	Fine copper powder (98.3% Cu) for antifouling paints	Copper tint
Zinc dust	Fine zinc powder (Zn 94% min.)	Dull gray
Pale gold bronze	Fine bronze flakes (95% Cu, 5% Zn)	Reddish gold
Rich gold bronze	Fine bronze flakes (70% Cu, 30% Zn)	Pale gold

Application

From the standpoint of harmony and design, color is both a physiological and a psychological sensation. It is not inherent in pigments, dyes, or other materials but is the effect of light rays reflected to the eye by the material. Different materials absorb and reflect different light waves and thus appear differently colored. The type of illumination also affects the color as seen by the eye. Therefore color should never be considered independently of light.

Although theoretically each wavelength of light represents a different color, the human eye can perceive only a narrow range of light wavelengths (from 0.000035 cm for violet to 0.000070 cm for red). Within these limits, however, under proper illumination it is possible to detect with the eye exceedingly slight color differences, the number of distinguishable colors being estimated at 10,000,000 (U. S. Bureau of Standards).

Duplication of Colors. Colors or hues vary slightly with different batches of paints, dye, etc. For this reason products that must be matched exactly in hue are usually finished from the same batch or lot. The necessity for duplicating colors within narrow limits in mass-produced materials has led to the development of several color systems. The Munsell is the most successful one. It is based on hue, value, and chroma, which correspond to the basic physical parameters of dominant wavelength, reflectance, and purity.

Visibility. Visibility at a distance varies with different colors. Red can be seen and recognized at long distances, whereas blue can be seen at only short distances. The comparative scale of visibility of colors at a distance is as follows: red, green, white, yellow, blue. Legibility, however, varies also with the background. Black on yellow is more legible than black on white. Green or red or blue on white is more legible than black on white.

Safety Color Code. There is a safety color code which specifies the application of colors for uniformly designating and marking physical hazards and the location of fire, safety, and protective equipment (*see* Table *C49*). There is also a color code for piping (*see* Table *C50*).

Esthetics of Color. Harmony of color or tone design is a complicated art that uses the color relationship to convey a pleasing emotional reaction. The esthetics of color and its effective use in architecture are beyond the scope of this book. Briefly, a rich color is a hue at its fullest intensity. A warm color is one in which red-orange predominates. In general, warm colors are considered pleasing or exciting, while cold colors are restful to the senses.

Table C49 Safety color code

Basic color	Physical hazards or equipment identified	Suggested areas of use
Red	(1) Fire protection equipment and apparatus	(1) Fire alarm boxes, sand and water buckets, extinguishers, hydrants, hose locations, pipe lines and valves of sprinkling systems and all other lines carrying fire protection or control materials
	(2) To indicate "danger"	(2) Containers of flammable liquids, barriers at temporary obstructions, and possible hazards on temporary construction
	(3) To indicate "stop"	(3) Emergency stop bars on hazardous machinery, stop buttons for electrical switches for emergency stopping
Orange	(1) Dangerous parts of moving machinery	(1) Inside of movable guards; exposed edges of rollers, gears, and the like
	(2) Pipelines carrying dangerous materials	(2) Pipelines carrying hot water or corrosive, flammable, explosive, or poisonous substances
Yellow	(1) To designate "caution"	(1) Exposed and unguarded edges of platforms, pits, and wells; fixtures or equipment suspended from ceilings, etc., which extend into normal operating areas; hand rails, guards, top and bottom treads of stairways where caution is needed; inside covers of switch and fuse boxes; lips on horizontally closing elevator doors; doorways, low beams and pipes; pillars, posts, or columns; caution signs
	(2) Pipelines carrying dangerous materials	(2) Pipelines carrying hot water or corrosive, flammable, explosive, or poisonous substances
Green	(1) To designate "safety"	(1) First-aid dispensaries, first-aid kits, etc.; safety bulletin boards, safety showers, safety instruction signs, safety starting buttons
	(2) Pipelines carrying safe materials	(2) Pipelines carrying cold water, brine, air, etc.
Blue	(1) To warn against movement or use of equipment being worked on during construction	(1) Electrical equipment, elevators, ladders and other equipment, and parts of scaffolding
	(2) Pipelines carrying electric conduit	(2) Pipelines carrying electric conduit
Purple	(1) Radiation hazards	(1) Areas where radiation hazards are present; containers of radioactive materials
Black, white, or combination thereof	(1) To designate housekeeping and traffic markings	(1) Dead ends of aisles or passageways, direction lines for stairways, direction signs, location of refuse cans (used in solid white, solid black, single color striping, alternate stripes of black and white, or black and white checkers)

Table C50 Color code for piping

Color	Classification	Where used
Red	Class F	Sprinkler piping, mains, and risers
Orange or yellow	Class D	Dangerous materials; materials easily ignited or explosive such as fuel oil, gasoline, or naphtha; corrosive and toxic chemicals such as acids, alkalis, and hydrogen sulfide; materials at high temperatures and pressures such as steam, high-pressure water, and air
Green or achromatic colors such as white, black, gray, or aluminum	Class S	Safe materials involving little or no hazard to life or property; materials at low pressures and temperatures which are neither toxic nor poisonous and will not produce fire or explosions
Bright blue	Class P	Materials piped for the express purpose of lessening the hazards of dangerous materials; all protective materials other than fire protection

CONCRETE

Physical and Chemical Properties

Concrete is a mixture of sand, gravel, crushed rock or other aggregate held together by a hardened paste of cement and water. This mixture, when properly proportioned, is at first a plastic mass that can be cast or molded into a predetermined size and shape. Upon hydration of the cement by the water, concrete becomes stonelike in strength, hardness, and durability. It is differentiated from other cement-water-aggregate mixtures on the basis of aggregate size. When cement is mixed with water and a fine aggregate of less than $\frac{1}{4}$ in. (6.35 mm), it is known as mortar, stucco, and cement plaster. When a large aggregate more than $\frac{1}{4}$ in. (6.35 mm) in diameter is added to cement, fine aggregate, and water, the product is concrete.

Characteristics of concrete can vary through a wide range, depending on the characteristics of the ingredients and the proportions of the mix. The techniques used for mixing, placing, finishing, and curing can also affect the quality of the concrete. Figure *C61* summarizes these relationships.

Generally, cement paste constitutes between 25 and 40% of the total volume of concrete, and the volume of cement is between 7 and 15% (375 and 750 lb/yd^3, or 222.49 to 444.98 kg/m^3).

Aggregate constitutes between 40 and 60% of the volume of the concrete, and in air-entrained concrete the air contents range up to about 8% of the volume of the concrete. Figure *C62* shows compressive strength with various water-cement ratios for air-entrained and non-air-entrained concrete.

The unit of measure for concrete is the cubic foot (equivalent to 0.02832 m^3). A sack of portland cement containing 94 lb (42.6384 kg) equals 1 ft^3. Fine and coarse aggregate is measured by loose volume. Water is measured by the gallon (1 gal = 8 lb = 4.546 liters = 3.6288 kg).

Concrete weighs from 110 to 155 lb/ft^3 (1762.2 to 2483.1 kg/m^3). Weights of specific types of concrete are as follows:

Construction (general) concrete: 140 to 150 lb/ft^3 (2242.8 to 2403.0 kg/m^3)
Gravel or limestone concrete: 142 to 148 lb/ft^3 (2272.84 to 2370.96 kg/m^3)
Traprock concrete: 148 to 155 lb/ft^3 (2370.96 to 2483.10 kg/m^3)

Concrete can weigh as little as 35 lb/ft^3 (560.7 kg/m^3); special lightweight aggregate is used.

General Types and Uses

Concrete construction, utilizing plain, reinforced, and prestressed concrete depending on the particular end use, plays a major role in the construction field. It is used for piling, footings, foundation and retaining walls, structural members, floors, walls and roofs, and paved areas such as roads, highways, parking lots, sidewalks, and driveways. Concrete is also prefabricated into various types of block, decking, columns, beams, girders, artificial stone, roofing and flooring tile, etc.

Structural methods such as thin-shell, tilt-up, prefabricated, preformed, and lift-slab construction produce construction forms and design effects unique to this material.

Other major types of concrete construction include highways, landing fields, dams, docks, breakwaters, bridges, boats, tanks, and pipe.

Types of Concrete Mixtures

Water-Cement-Aggregate Proportions. The durability of concrete exposed to weather, its potential strength, and its other properties depend on the amount of water in proportion to the amount of cement and the amount of entrained air. The water-cement ratio controls the binding power of the paste which coats and surrounds the aggregates and, upon hardening, holds the entire mass together. The actual amount of water required thoroughly to hydrate the cement, that is, develop its binding power, is very small in comparison to the water needed to develop the workable plastic consistency of the concrete. (*See* Table *C51*.)

The strength of concrete remains the same for a given water-cement ratio of the paste irrespective of the amount of aggregate embedded in the paste. But the quantity of aggregate that can be mixed with a fixed amount of cement paste with a chosen water-cement ratio depends on the size and grading of the sand and coarse aggregate and on their proportions to each other. The larger the aggregate, the greater the volume of cement paste required. However, for each type of use there is an optimum size and proportioning of the sizes of aggregate. Figure *C63* shows the relationship of size of aggregate to quantity of water.

Tables *C52* to *C55* show, respectively, the effect that increased quantities of mixing water have on the strength of concrete, the comparative water absorption of different types of aggregate, and the relative quantity of water carried by the different types of aggregate. All these values must be calculated and correlated to arrive at a correct water-cement ratio in relation to the compressive strength of the concrete. The amount of

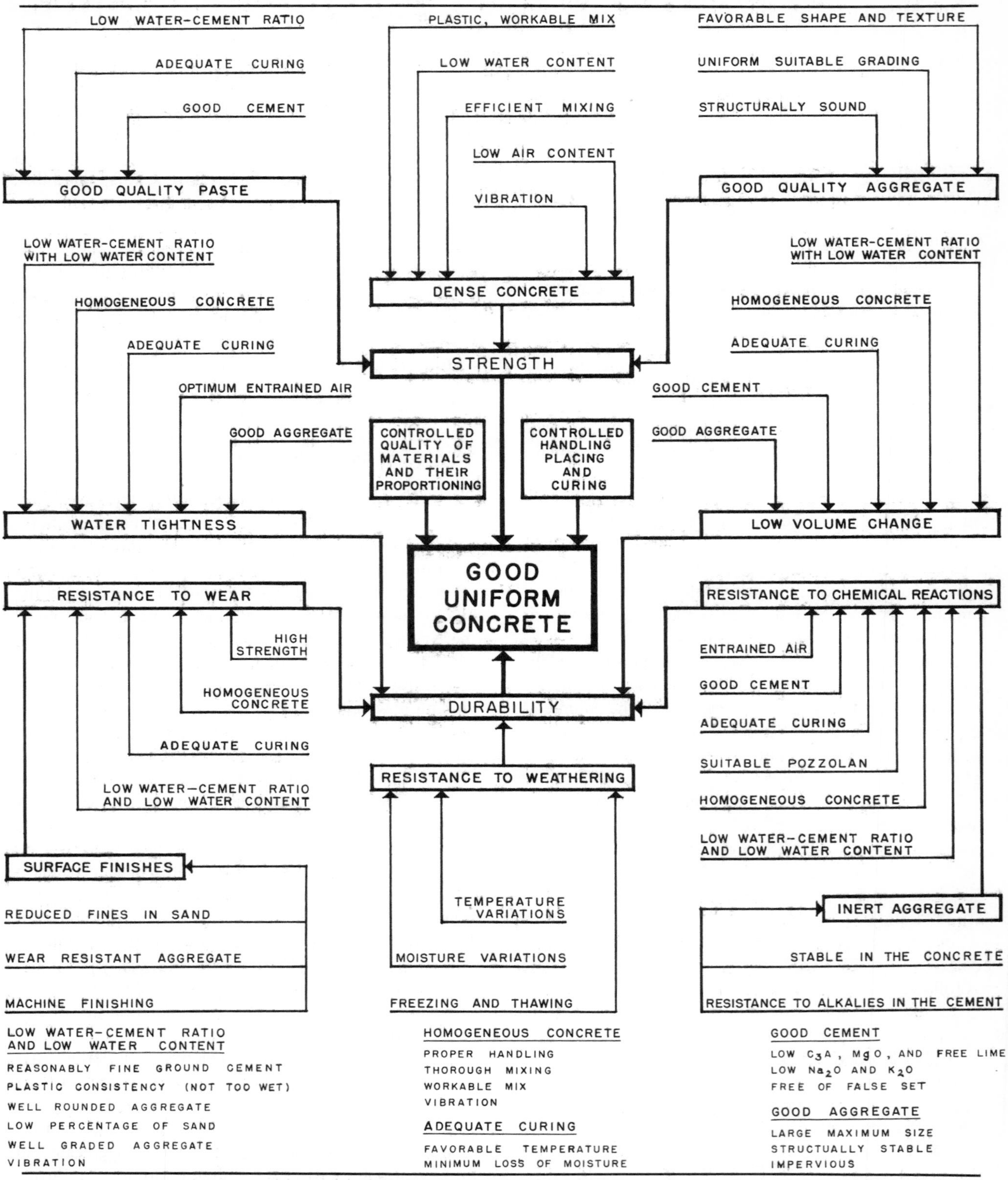
LOW WATER-CEMENT RATIO
ADEQUATE CURING
GOOD CEMENT
PLASTIC, WORKABLE MIX
LOW WATER CONTENT
EFFICIENT MIXING
LOW AIR CONTENT
VIBRATION
FAVORABLE SHAPE AND TEXTURE
UNIFORM SUITABLE GRADING
STRUCTURALLY SOUND
GOOD QUALITY PASTE
GOOD QUALITY AGGREGATE
LOW WATER-CEMENT RATIO WITH LOW WATER CONTENT
HOMOGENEOUS CONCRETE
ADEQUATE CURING
OPTIMUM ENTRAINED AIR
GOOD AGGREGATE
DENSE CONCRETE
STRENGTH
CONTROLLED QUALITY OF MATERIALS AND THEIR PROPORTIONING
CONTROLLED HANDLING PLACING AND CURING
LOW WATER-CEMENT RATIO WITH LOW WATER CONTENT
HOMOGENEOUS CONCRETE
ADEQUATE CURING
GOOD CEMENT
GOOD AGGREGATE
WATER TIGHTNESS
LOW VOLUME CHANGE
GOOD UNIFORM CONCRETE
RESISTANCE TO WEAR
HIGH STRENGTH
HOMOGENEOUS CONCRETE
ADEQUATE CURING
LOW WATER—CEMENT RATIO AND LOW WATER CONTENT
DURABILITY
RESISTANCE TO WEATHERING
RESISTANCE TO CHEMICAL REACTIONS
ENTRAINED AIR
GOOD CEMENT
ADEQUATE CURING
SUITABLE POZZOLAN
HOMOGENEOUS CONCRETE
LOW WATER–CEMENT RATIO AND LOW WATER CONTENT
SURFACE FINISHES
REDUCED FINES IN SAND
WEAR RESISTANT AGGREGATE
MACHINE FINISHING
TEMPERATURE VARIATIONS
MOISTURE VARIATIONS
FREEZING AND THAWING
INERT AGGREGATE
STABLE IN THE CONCRETE
RESISTANCE TO ALKALIES IN THE CEMENT
LOW WATER-CEMENT RATIO AND LOW WATER CONTENT
REASONABLY FINE GROUND CEMENT
PLASTIC CONSISTENCY (NOT TOO WET)
WELL ROUNDED AGGREGATE
LOW PERCENTAGE OF SAND
WELL GRADED AGGREGATE
VIBRATION
HOMOGENEOUS CONCRETE
PROPER HANDLING
THOROUGH MIXING
WORKABLE MIX
VIBRATION
ADEQUATE CURING
FAVORABLE TEMPERATURE
MINIMUM LOSS OF MOISTURE
GOOD CEMENT
LOW C_3A, MgO, AND FREE LIME
LOW Na_2O AND K_2O
FREE OF FALSE SET
GOOD AGGREGATE
LARGE MAXIMUM SIZE
STRUCTUALLY STABLE
IMPERVIOUS

Figure C61 Characteristics of good uniform concrete, their interrelationship, and the factors that control them.

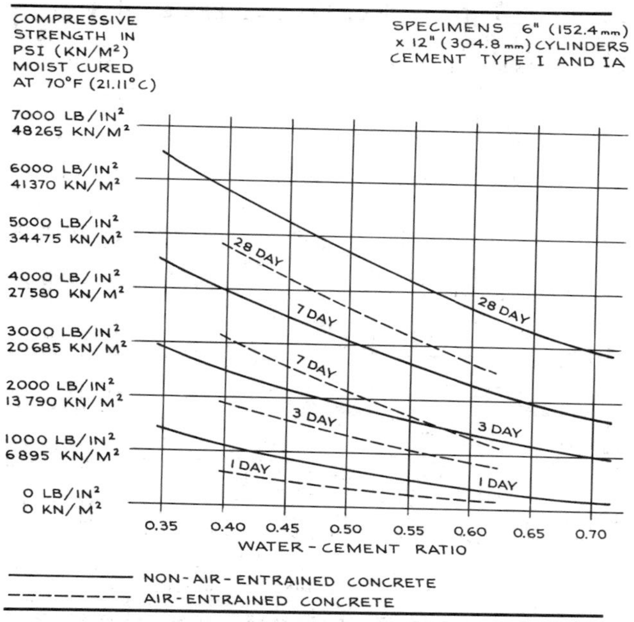

Figure C62 Compressive strength with various water-cement ratios for air-entrained and non-air-entrained concrete.

absorbed water and water carried by the aggregates should be deducted from the total amount of mixing water specified by the chosen water-cement ratio.

In the construction field, concrete is generally specified on the basis of the compressive strength it develops in 7 and 28 days. These specifications also require that trial mixes be made and tested to arrive at the correct water-cement ratio, the correct proportioning of fine and coarse aggregate, the minimum compressive strength developed, and the workability of the concrete for the given end use.

The slump test is used in the field to control water content. In the slump test, a stock metal cone which has had its top cut off so that it forms a frustum is placed on a flat surface, filled with the prepared concrete mix and rodded with a $\frac{5}{8}$ in. (15.88 mm) rod, 24 in. (609.6 mm) long. When it is completely filled, the cone is carefully

Table C51 Advisable maximum permissible water-cement ratios for various construction uses and exposures

Exposure[a]						
Wide range of temperatures or freezing and thawing conditions (air-entrained concrete only)			Mild temperatures barely below freezing, or rainy or arid conditions (air-entrained or non-air-entrained concrete)			
Exposed in air	In fresh water	In seawater or in contact with sulfates[b]	Exposed in air	In fresh water	In seawater or in contact with sulfates[b]	Type of construction
0.49	0.44	0.40	0.53	0.49	0.40	Thin sections, reinforced pipe and piles
0.44	0.44	0.40	0.49	0.49	0.44	Bridge decks
0.49			0.53	0.49		Thin sections with less than 1 in. (25.4 mm) of concrete cover over concrete
0.53	0.49	0.44	c	0.53	0.44	Retaining walls, piers, girders, beams with moderate-size sections, and exterior portions of heavy mass sections
	0.44	0.44		0.44	0.44	Concrete deposited under water by tremie
0.53			c			Concrete slabs on grade
0.49				0.53		Pavements
c			c			Concrete protected from weather, interiors of buildings, and concrete below grade
0.53			c			Concrete that will be protected from freezing and thawing within 2 or 3 years

[a] Air-entrained concrete should always be used under freezing and thawing conditions.
[b] Soil or ground water containing sulfate concentrations of more than 0.2%.
[c] Water-cement ratio should be selected on the basis of strength and workability requirements.

Table C52 Effect of the quantity of mixing water on the strength of concrete

Quantity of water to sack of cement			Approximate compressive strength at 28 days[a]	
gal	liter	Water-cement ratio by weight	lb/in.^{2}	kN/m^2
$8\frac{1}{4}$	37.42	1.10	1500	10 343
$7\frac{1}{2}$	34.02	1.00	1750	12 066
7	31.75	0.966	2000	13 790
$6\frac{3}{4}$	30.62	0.90	2250	15 514
$6\frac{1}{2}$	29.48	0.866	2500	17 238
6	27.22	0.80	2750	18 961
$5\frac{1}{4}$	23.81	0.70	3000	20 685
$4\frac{1}{4}$	19.28	0.60	4000	27 580
$3\frac{3}{4}$	17.01	0.50	5000	34 475
3	13.61	0.40	6000	41 370

[a] 1 lbf/in.^{2} = 0.006895 MN/m^2

Table C53 Absorption of water by the different types of aggregate

Type of aggregate	Absorption of water (percent by weight)
Average sand	1.00
Pebbles and crushed rock	1.00
Traprock and granite	0.50
Porous sandstone	7.00

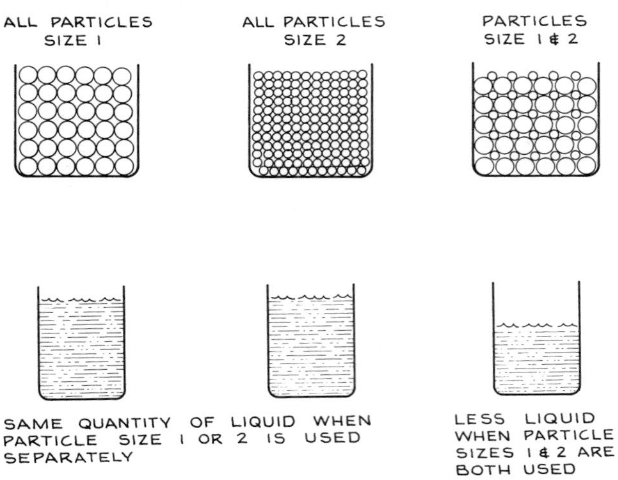

Figure C63 Aggregate (particle size) in relation to water ratio.

Table C54 Water carried by aggregate

Type and condition of aggregate	Water carried by aggregate gal/ft^3	kg/m^3
Sand, very wet	0.75–1.00	12.02–16.02
Sand, moderately wet	0.50	8.0
Sand, moist	0.25	4.0
Gravel (pebbles), moist	0.25	4.0
Crushed rock, moist	0.25	4.0

Table C55 Slump test requirements for various types of concrete construction

Slump[a] Maximum in.	mm	Minimum in.	mm	Major uses
5	127.0	2	50.8	Reinforced footings and foundation walls
4	101.6	1	25.4	Plain footings and nonreinforced walls
6	152.4	3	76.2	Reinforced slabs, beams, walls, and columns
3	76.2	2	50.8	Pavements
3	76.2	1	25.4	Massive concrete

[a] The slump is generally designated in the approved-test concrete mixes. For volume-mixed concrete, the above should always be followed.

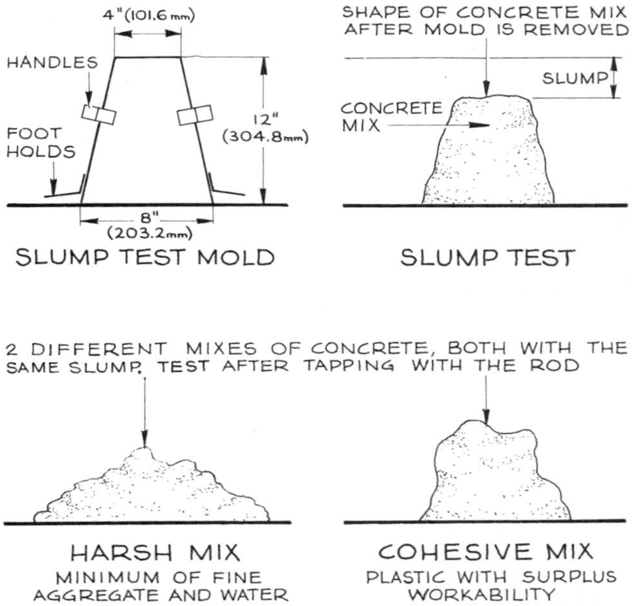

Figure C64 Slump test for concrete mixes.

Table C56 Accepted maximum aggregate size for various types of concrete construction[a]

Minimum dimension of section of concrete to be poured		Maximum size of aggregate[b] based on square screen openings					
		Reinforced walls, beams and columns		Heavy reinforced concrete slabs		Lightly reinforced or plain concrete slabs	
in.	mm	in.	mm	in.	mm	in.	mm
5 or less	127 or less			$\frac{3}{4}$–$1\frac{1}{2}$	19.05–38.1	$\frac{3}{4}$–$1\frac{1}{2}$	19.05–38.1
6–11	152.4–279.4	$\frac{3}{4}$–$1\frac{1}{2}$	19.05–38.1	$1\frac{1}{2}$	38.1	$1\frac{1}{2}$–3	38.1–76.2
12–29	204.8–736.6	$1\frac{1}{2}$–3	38.1–76.2	3	76.2	3–6	76.2–152.4
30 or more	762 or more	$1\frac{1}{2}$–3	38.1–76.2	3	76.2	6	152.4

[a] Aggregate size should always be checked in relation to the spacing of reinforcement rods, bars, etc., and to the size of reinforcing mesh.
[b] For pumping concrete, the aggregate size is controlled by the height of pumping, air entrainment, and reinforcement and mesh spacing.

removed and placed adjacent to the now-slumped pile of concrete. The distance from the top of the metal cone to the top of the slumped concrete measures water content and the consistency, or fluidity, of the mix. If the slumped concrete is now tapped with the rod, some indication of its workability is also obtained. (*See* Figure *C64*.)

A harsh mix is efficient for slabs, pavements, or mass concrete where the lowest possible water-cement ratio is desirable. This mix is feasible for such use only because it can be consolidated by vibration; it is completely unfit for complicated reinforced concrete work. A cohesive mix is the correct one for the latter use but is unfit for the former applications. Table *C55* shows the slump in inches for various types of concrete in relation to their uses.

Fine and coarse aggregate are proportioned according to the fluidity, consistency, and workability that are required in the concrete for its end use (*see* Table *C56*). Sand is the usual fine aggregate, and many tables and formulas for concrete mixes are based on the fineness modulus of sand. Table *C57* shows a typical sand analysis and computation of the fineness modulus.

Data for a typical concrete mixture are given in Table *C58*, and Table *C59* shows how the values for typical concrete must be adjusted for other conditions.

Table *C60*, which gives the relative proportions of cement and coarse and fine aggregate in relation to maximum aggregate size, can be used for formulating job-mixed concrete for small jobs where more elaborate calculations and testing procedures are impractical.

Table C57 Example of fine aggregate fineness modulus[a]

Sieve size: Number	Sieve size: mm or microns	Percentage passing	Percentage retained (cumulative)
4	4.76 mm	98	2
8	2.38 mm	90	10
16	1.19 mm	60	40
30	595 microns	45	55
50	297 microns	20	80
100	149 microns	2	98
			100

Fineness modulus = 285 ÷ 100 = 2.85

[a] The fineness modulus for fine aggregate should be not less than 2.3 or more than 3.1.

Type of Cement, Aggregate, and Admixture in Relation to Mixes. The various types of concrete are controlled not only by the water-cement ratio and the proportions of fine and coarse aggregates but also by the type of cement, the type of aggregate, the presence of entrained air, admixtures, and the various kinds of reinforcing.

The types of portland cement that are particularly adapted to produce a specific kind of concrete are shown in Tables *C61* and *C62*.

Lightweight aggregates of various types may be used to control the weight, thermal insulating, and nailing characteristics of concrete (*see* Table *C63*).

When lightweight aggregates are used, it is necessary to test the unit weight. Also, slump tests should be made more frequently. In general, the slump should be less than 3 in. (76.2 mm); for sawdust concrete the slump should be from 1 to 2 in. (25.4 to 50.8 mm). The cement and water content should be adjusted to compensate for variations in the properties and condition of the aggregate. During curing, the concrete should be kept well wetted.

Table C58 Data for a typical concrete, giving air and water contents and relative proportions and size of aggregate

Maximum size of coarse aggregate		Unit weight of coarse aggregate for unit volume of concrete (percent)	Approximate amount of entrapped air (percent)	Average water content*		Amount of sand in total aggregate by solid volume[a] (percent)
in.	mm			lb/yd^3	kg/m^3	
$\frac{3}{8}$	9.53	41	3.0	352	208.85	61
$\frac{1}{2}$	12.70	52	2.5	336	199.36	53
$\frac{3}{4}$	19.05	62	2.0	316	187.49	45
1	25.4	67	1.5	300	178.00	41
$1\frac{1}{2}$	38.1	73	1.0	280	166.13	36
2	50.8	76	0.5	266	157.83	33
3	76.2	81	0.3	242	143.59	31
6	152.4	87	0.2	210	124.60	28

[a]Tables based on concrete with natural sands (with fineness modulus of 2.75), average coarse aggregate, and a slump of 3 to 4 in. (76.2 to 101.6 mm) at mixer.

Table C59 Adjustment of values for typical concrete foı other conditions

Change in conditions	Effect on values: Unit water content	Effect on values: Sand (percent)	Effect on values: Coarse aggregate (percent)	Change in conditions	Effect on values: Unit water content	Effect on values: Sand (percent)	Effect on values: Coarse aggregate (percent)
0.1 increase or decrease in fineness modulus of sand		±0.5	±1.0	1% increase or decrease in air content	±3%	±0.5 to 1.0	
1 in. (25.4 mm) increase or decrease in slump	±3%			For less workable concrete (pavements, mass concrete, etc.)	-8 lb	-3	+6

Table C60 Relative proportions for job-mixed concrete for small concrete jobs

Maximum size of coarse aggregate		Designation of mixes[a]	Bags of cement per yd^3 (m^3) of concrete	Weight of aggregate per bag of cement batch: Sand[b]		Weight of aggregate per bag of cement batch: Gravel or crushed rock	
in.	mm			lb	kg	lb	kg
$\frac{1}{2}$	12.7	1	7.0	245	111.13	170	77.11
		2	6.9	235	106.60	190	86.18
		3	6.8	235	106.60	205	92.99
$\frac{3}{4}$	19.05	1	6.6	235	106.60	225	102.06
		2	6.4	235	106.60	245	111.13
		3	6.3	225	102.06	265	120.20
1	25.4	1	6.4	235	106.60	245	111.13
		2	6.2	225	102.06	275	124.74
		3	6.1	215	97.52	290	131.54

Table C60 Relative proportions for job-mixed concrete for small concrete jobs (*continued*)

Maximum size of coarse aggregate		Designation of mixes[a]	Bags of cement per yd^3 (m^3) of concrete	Weight of aggregate per bag of cement batch			
				Sand[b]		Gravel or crushed rock	
in.	mm			lb	kg	lb	kg
$1\frac{1}{2}$	38.1	1	6.0	235	106.60	290	131.54
		2	5.8	225	102.06	320	145.15
		3	5.7	215	97.52	345	156.49
2	50.8	1	5.7	235	106.60	330	149.69
		2	5.6	225	102.06	360	163.30
		3	5.4	215	97.52	380	172.37

[a]Procedure is as follows. First select maximum-size aggregate, use mix No. 2, and add just enough water to produce sufficiently workable consistency. If mix is undersanded, use mix No. 1; if mix is oversanded, use mix No. 3.

[b]Weights shown are for dry sand. For damp sand, add 10 lb (4.536 kg); for wet sand, add 20 lb (9.072 kg).

Table C61 Type of cement in relation to type of concrete

Portland cement	Type of concrete	Major use
Type I, Normal[a]	Normal standard-type concrete	For general construction purposes
Type IA, air-entraining	Standard with air entraining, more workability, and resistance to freezing and thawing	For general construction purposes
Portland blast-furnace slag IS	Standard-type concrete	For general construction purposes
Portland blast-furnace slag, air-entraining IS-A	Standard-type concrete with more workability and resistance to freezing and thawing	For general construction purposes
Type II, Moderate[a]	Slower setting, lower heat generation, and smaller volume change than Types I and I-A; develops strength in 28 days	For general construction purposes and where exposed to moderate sulfate action
Type IIA, air-entraining	Same as concrete using Type II moderate cement, but with more workability and resistance to freezing and thawing	For general construction purposes and where exposed to moderate sulfate action
Type III, High-Early-Strength[a]	Rapid setting, higher heat generation (which helps offset freezing), some volume change, develops strength in 7 days	For construction where rapid development of strength is essential
Type IIIA, air-entraining	Same as concrete using Type III high-early-strength, but with more workability and resistance to freezing and thawing	For construction where rapid development of strength is essential
Type IV, Low Heat of Hydration[a]	Slow setting, low heat generation, small volume change, good strength with age	For massive concrete construction
Type V, Sulfate-Resisting[a]	High resistance to sulfate attack, fairly low heat generation, high strength with age	Where there is ground water or soil that has sulfates
Portland-pozzolan P and PIP	A hydraulic concrete	For large hydraulic structures
Portland-pozzolan, air-entraining P-A and IP-A	A hydraulic concrete with air entraining	For large hydraulic structures

[a]These are CSA (Canadian Standards Association) designations for the five types designated as I, II, III, IV, and V by the ASTM.

Table C62 Heat generated during first 7 days for portland cement

Type of portland cement	Percent based on 100% for Type I, Normal
Type I, Normal	100
Type II, Moderate	80–85
Type III, High-Early-Strength	Up to 150
Type IV, Low Heat of Hydration	40–60
Type V, Sulfate-Resisting	60–75

All types of lightweight concrete are subject to high shrinkage except those made with expanded shales, clays, and scoria. The strength of lightweight concrete varies considerably. Relatively high strength is obtained with expanded shale and clay; intermediate strength with pumice, scoria, and expanded slag; and low strength with diatomite, perlite, and vermiculite. However, the insulating properties of the last group are better than those of the first group. But even crushed shale and clay concrete has four times the insulating value of ordinary concrete.

Air-entraining agents are soaplike resinous or fatty materials either interground with portland cement during manufacture or added to concrete mixes as admixtures. These agents produce millions of tiny air bubbles throughout the concrete mix. Best results are obtained when the air content is between 3 and 6%. Table *C64* shows the desirable content of entrained air in relation to aggregate size for concrete subject to freezing. Figure *C65* illustrates a simple pocket-sized air indicator to check the air content of freshly mixed concrete. This test takes only a few minutes to do. This method of testing is not a substitute for the more accurate pressure and volumetric methods.

Air-Entrained Concrete. This type of concrete was developed in the 1930s for road building to combat the problem of freezing and thawing and the use of salt, which causes normal dense concrete to deteriorate. This concrete is now used for all types of roads, airfield landing strips, and other construction subject to freezing and thawing and salt. Not only does entrained air make concrete resistant to these destructive influences but it also improves the workability and cohesiveness; and when concrete is placed in forms, the entrained air helps to keep the ingredients from separating so that there is less tendency for the water to rise to the top of fresh concrete.

Entrained air bubbles are extremely small in size, varying in diameter from about 0.001 to 0.003 in.

Table C63 Lightweight aggregate for making lightweight, nailable, and special concrete

Type of lightweight aggregate	Characteristics and process of production	Approximate weight of concrete	
		lb/ft³	kg/m³
Cinders	Average combustible content 35% by weight of dry-mixed aggregate; sulfides less than 0.45%; sulfates less than 1.0%	85	1361.70
		(with natural sand added for workability)	
		110–115	1762.20–1842.30
Expanded slag	Blast-furnace slag heated and expanded by various applications of water	75–110	1201.50–1726.20
Expanded shale or clay	Shale or clay heated to fusion point, expanded by entrapped gases; weight 40–70 lb/ft^3 (640.80–1121.40 kg/m^3)	75–110	1201.50–1726.20
Natural pumice (volcanic glass)	Crushed and screened	90–100	1441.80–1602.00
Natural scoria (volcanic rock)	Crushed and screened	90–110	1441.80–1726.20
Natural volcanic cinders	Crushed and screened	90–110	1441.80–1726.20
Natural diatomite	Crushed and screened	90–110	1441.80–1726.20
Perlite and vermiculite	Heated quickly and literally exploded; high thermal-insulating and fire-resisting properties	50–80	801.00–1281.60
Sawdust	Pine sawdust should pass $\frac{1}{4}$-in. (6.35-mm) screen but not all pass a No. 16 screen; produces fine, nailable concrete	50–80	801.00–1281.60

Table C64 Relation of aggregate size to entrained air[a]

Minimum size of coarse aggregate		Air content (percent of volume)[b]
in.	mm	
$\frac{3}{8}$ to $\frac{1}{2}$	9.53 to 12.7	7 ± 1
$\frac{3}{4}$ to 1	19.05 to 25.4	6 ± 1
$1\frac{1}{2}$, 2, or $2\frac{1}{2}$	38.1, 50.8, or 63.5	5 ± 1

[a]For structural lightweight concrete, add 2% to the values as an allowance for the entrapped air in the aggregates. A $\pm 1\frac{1}{2}$% range is permissible.
[b]Approximately 3% should be the air content of the mortar fraction of the concrete.

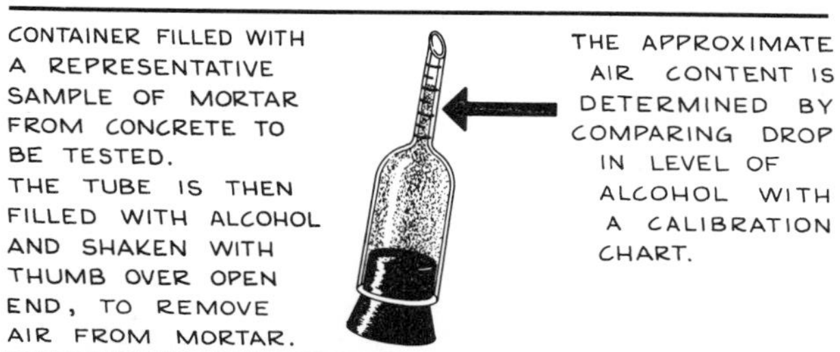
CONTAINER FILLED WITH A REPRESENTATIVE SAMPLE OF MORTAR FROM CONCRETE TO BE TESTED.
THE TUBE IS THEN FILLED WITH ALCOHOL AND SHAKEN WITH THUMB OVER OPEN END, TO REMOVE AIR FROM MORTAR.
THE APPROXIMATE AIR CONTENT IS DETERMINED BY COMPARING DROP IN LEVEL OF ALCOHOL WITH A CALIBRATION CHART.

Figure C65 Air indicator for testing air entrainment in concrete.

(0.0254 to 0.0762 mm). One cubic yard (0.7646 m^3) of air-entrained concrete may contain as many as 300 to 500 billion bubbles having an air content between 4 and 6% by volume and $1\frac{1}{2}$ in. (38.1 mm) maximum size of aggregate.

Mixing action is the most important factor in the production of entrained air in concrete. Uniform distribution of entrained air voids in concrete requires that adequate mixing be maintained at all times for ready-mixed concrete. "Void" is defined as the total volume of air plus water plus entrained or entrapped air. Strength of air-entrained concrete depends principally on the voids-cement ratio. The gradation of aggregate and the cement content of a mix also have important effects on the air content of both air-entrained and non-air-entrained concrete. (*See* Table *C65*.)

Admixtures for concrete are prepared formulations which are either trademarked or patented or whose ingredients are kept secret. They are added to concrete mixes to alter certain characteristics, for example, to accelerate or retard setting, to increase water repellency, to aid workability and plasticity, to add color, to increase surfate resistance, or to harden (*see* Admixtures). The use of admixtures with concrete requires closer control of the mix and more slump tests. It is always necessary to check the final characteristics of the resulting concrete against end-use requirements to evaluate correctly the advantages and disadvantages obtained by using admixtures.

Reinforcing for concrete is usually steel because the two materials have an almost identical expansion and because concrete, though strong in compression, is weak in tension, whereas steel is exceptionally strong in tension. A combination of the two therefore produces an ideal material for construction. Reinforced concrete is fully discussed under its own heading.

When concrete work is to be installed during cold weather conditions, the concrete should be heated by heating the water, sand, coarse aggregate, and cement. Table *C66* shows the temperature requirements for concrete work during cold weather, including minimum temperatures after concrete is poured and maximum drop in temperature within 24 hours after the end of protection.

Special Cement Mixtures. Several special cement mixtures are important in concrete work.

Grouting cement mortar is used for filling reglets, for installing machinery, and for setting steel on masonry or concrete. Mortar for these purposes must not shrink and must permanently hold its original volume. This type of mortar can be made by three methods: (1) prolonged mixing of ordinary mortar; (2) use of special (high silica) cements; and (3) addition of aluminum powder, which reacts to produce hydrogen gas throughout the mortar, thereby causing separation of the water (bleeding) and thus stopping shrinkage. This also expands the mortar so that it completely fills the space in which it is confined.

Antibacterial cement is used to produce a special finish on concrete floors in food processing plants where bacteria that break down concrete may develop. When this type of concrete is used, the manufacturer's directions must be followed exactly.

Soil-cement refers to a concrete made by compacting a mixture of cement, water, and soil. Actually the soil replaces the fine and coarse aggregates. Soil-cement is generally used as a base for a bituminous surface.

Concrete fill is used for many purposes such as building up floor levels, building up fill under lockers, closets, etc., between sleepers, stair pans, and all areas where fill is necessary. Concrete fill requires little compressive strength and in many of these applications is lightweight. Its exact weight and strength are usually specified.

Heavyweight concrete is used where human beings and equipment must be shielded from the harmful effects of X-ray, gamma ray, and neutron radiation. Examples are X-ray equipment in cancer hospitals and atomic energy power plants.

Table C65 Approximate mixing water requirements for different slumps and maximum sizes of aggregates

Maximum size of aggregate		Air-entraining concrete						
		Recommended average total air content (percent)[a]	Slump					
			1–2 in. 25.4–50.8 mm		3–4 in. 76.2–101.6 mm		5–6 in. 127.0–152.4 mm	
			Water: lb/yd³ (0.5933 kg) of concrete[b]					
in.	mm		lb	kg	lb	kg	lb	kg
$\frac{3}{8}$	9.53	7.5	310	183.92	340	201.72	360	213.59
$\frac{1}{2}$	12.70	7.5	300	177.99	325	192.82	340	201.72
$\frac{3}{4}$	19.05	6.0	275	163.16	300	177.99	315	186.89
1	25.40	6.0	260	154.26	285	169.09	300	177.99
$1\frac{1}{2}$	38.10	5.0	240	142.39	265	157.23	285	169.09
2	50.80	5.0	225	133.49	250	148.33	265	157.23
3	76.20	4.0	210	123.92	235	139.43		
6	152.40	3.0	185	109.76	200	118.66		

[a]Plus or minus 1%.
[b]These quantities of water are for use in computing cement factors for total batches.

Application

Temperature. The ultimate strength of concrete mixed and cured at high temperatures is lower than that of concrete mixed and cured at 70°F (21.11°C). It requires more water, shows an increased tendency to cracking, and has a greater shrinkage. The temperature can be lowered by mixing with cold water, even adding ice if necessary; by cooling the coarse aggregate with cold water or cold air blasts; and by working at night. Cement in bags should not be hot from standing in the sun.

For cold weather there are minimal temperature requirements for concrete work, as shown in Table *C66*. In freezing weather, calcium chloride is generally used to speed up setting but not when there is any possibility of sulfate attack. For very severe freezing conditions, the forms and surface of the concrete should be protected by covering with insulation. The water and aggregate for mixing concrete can be heated, but only within specified limits and under controlled conditions.

Tests. Mixed concrete is tested for consistency, temperature, air content, and unit weight; concrete test cylinders, usually 6 in. (152.4 mm) in diameter and 12 in. (304.8 mm) long, are made for compression tests. For concrete work of large scope, specifications should include provisions for making job-site slump tests.

Test cylinders should be taken from each pouring, or one test for each 100 yd³ (76.46 m³). These cylinders should be tested by an independent testing laboratory for 7- and 28-day compressive strength. On large jobs or those where pozzolanic materials are used, 90-, 180-, and 365-day tests should also be made for every tenth testing sample and one cylinder per week for 90-day strength.

Transportation. Ready-mixed concrete should not be in transit from plant to job over a distance greater than 3 to 4 miles.

Placing. When concrete is being poured, it should be installed in layers of 15 to 20 in. (381 to 508 mm) for mass concrete and 12 to 20 in. (304.8 to 508 mm) for structural concrete. Each layer should be soft when a new layer is poured upon it. Concrete should never be allowed to drop freely over 5 ft (1.524 m) for unexposed work and over 3 ft (0.914 m) for exposed work.

Vibration. When vibrators are used, they should be inserted vertically at points 18 to 30 in. (457.2 to 762 mm) apart for periods ranging from 5 to 15 seconds. Vibrators should have 7000 rpm for heads 4 in. (101.6 mm) or less in diameter and 6000 rpm for diameters larger than 4 in. (101.6 mm).

Pumped Concrete. Entrained air acts like a lubricant to concrete mixes, giving them great plasticity. This plasticity plus the introduction of air-entrained concrete into the construction field naturally led to the pumping of concrete from ready-mix trucks directly to place of

Table C65 Approximate mixing water requirements for different slumps and maximum sizes of aggregates (*continued*)

Maximum size of aggregate (in.)	Maximum size of aggregate (mm)	Approximate amount of entrapped air (percent)	Non-air-entrained concrete — Slump: 1–2 in. 25.4–50.8 mm — Water: lb/yd³ (0.5933 kg) of concrete[b] — lb	kg	3–4 in. 76.2–101.6 mm — lb	kg	5–6 in. 127.0–152.4 mm — lb	kg
$\frac{3}{8}$	9.53	3.0	350	207.66	385	228.42	410	243.25
$\frac{1}{2}$	12.70	2.5	335	198.76	365	216.56	385	228.42
$\frac{3}{4}$	19.05	2.0	310	183.92	340	201.72	360	213.59
1	25.40	1.5	300	177.99	325	192.82	340	201.72
$1\frac{1}{2}$	38.10	1.0	275	163.16	300	177.99	315	186.89
2	50.80	0.5	260	154.26	285	169.09	300	177.99
3	76.20	0.3	240	142.39	265	157.23		
6	152.40	0.2	210	123.92	235	139.43		

Table C66 Temperature requirements for concrete work during cold weather

Type of concrete	Minimum temperature of concrete as mixed[a]: Above 30°F (1.11°C) °F	°C	0 to 30°F (−18 to 1.11°C) °F	°C	Below 0°F (−18°C) °F	°C	Minimum temperature of concrete after placing for first 72 hours °F	°C	Maximum gradual drop in temperature in 24 hours after end of protection °F	°C
Thin concrete members (walls, beams, slabs, etc.)	55–60	12.78–15.56	60–65	15.56–18.33	65–70	18.33–21.11	50–55	10.00–12.78	40–50	4.44–10.00
Mass-type concrete (foundations, footings, dams, etc.)	45–50	7.22–10.00	50–55	10.00–12.78	55–60	12.78–15.56	40–45	4.44–7.22	20–30	−6.67 to 1.11

[a]These temperatures are obtained by heating the cement, sand, aggregates, and water.

pouring. The height to be pumped dictates the type and size of the pumping equipment and the concrete mix. The concrete mix must meet all the structural requirements and the pumping limitations such as friction, size of pipe, speed of pumping, and different pumping heights.

***Underwater Installation*.** Concrete should not be placed underwater unless special precautions are taken. There should be no running water where concrete is to be placed. Concrete should be fed continuously to the bottom by a pipe whose lower end should always be embedded in newly placed concrete. Other specialized methods may be used. For this type of installation, concrete should not be puddled or vibrated but should be left undisturbed. (*See* Figure *C66*.)

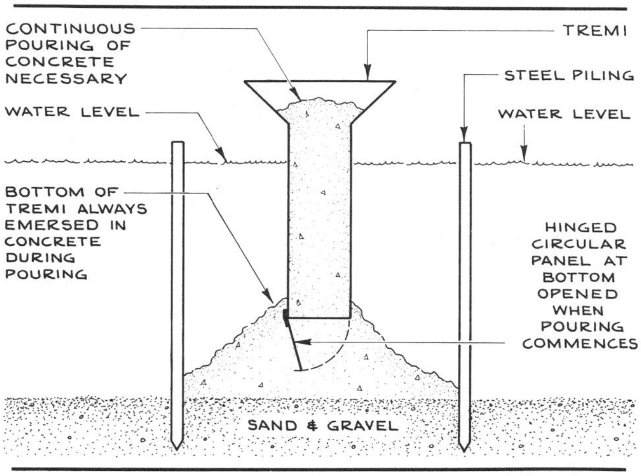

Figure C66 Tremi for installing concrete under water.

Curing. If concrete is allowed to dry out too quickly, complete hydration of the cement will be prevented. There are several methods of controlling curing: (1) keeping the concrete surfaces moist either by spraying water periodically or by covering with earth, sand, or burlap that is kept moist; (2) covering with special types of paper that keep moisture from evaporating; (3) using special sealing compounds; and (4) in cold weather, using steam. Concrete forms should not remain any longer than necessary, as they will not keep the concrete sufficiently moist for curing.

Forms. Forms may be defined as the molds into which concrete is placed, that support and hold it until it sets sufficiently to perform the function for which it has been designed. In many cases forms for reinforced concrete are so special that their design is handled by the structural engineer. In many states, by law or code requirements, the forms must be designed by a structural engineer.

Forms for general types of concrete work may be either completely prefabricated units or job-assembled prefabricated units made of steel, wood, and, more recently, reinforced paper (for cylindrical forms). These forms and all the accessories for supporting, tying, bracing, etc., are stock items available throughout the United States. Forms are also made of special materials that serve as a form and also become the finish material of a wall or ceiling. Materials commonly used for such permanent forms are artificial stone, ceramic veneer, asbestos cement, metals, and plastics (still experimental).

Forms can be used to create pattern and texture in concrete, since the concrete will reproduce a negative impression of the texture and pattern of the material of which the form is made or with which it is lined.

The forms for any large building in general must be designed by a structural engineer. A check should be made of all codes and with all agencies and departments that have jurisdiction over concrete construction.

Reinforcing. Reinforcing should be clean and free from mud, oil, paint, or loose dried mortar. Loose rust or mill scale will be automatically removed in handling and placing. Any reinforcing that projects above concrete should not be carelessly handled for at least 7 days, lest the bond be impaired.

Expansion. The coefficient of linear expansion of concrete is 6.5 millionths (0.0000065) per degree Fahrenheit (0.00000117 per degree Celsius) of temperature change. In large buildings expansion joints must be installed, and in many cases reinforcing rods (temperature rods) are installed to counteract expansion within concrete to stop minor cracking.

Construction Joints. In concrete work it is often necessary to pour concrete in sections (usually not over 80 ft, or 24.384 m, in a horizontal direction), as only so much concrete can be poured in 1 day. The methods of closing off these sections are called construction joints. These joints should be detailed by a structural engineer, especially when reinforced concrete is used. The adjoining section of concrete should not be installed until 48 hours have elapsed.

Slots, Inserts, Hangers, and Other Accessories. Before concrete is poured into the forms, all slots, inserts, hangers, anchors, sleeves, etc., should be installed in their correct locations.

Finishes. Concrete finishes may be roughly divided into two groups: (1) finishes for vertical and ceiling surfaces, applied to concrete already set; and (2) finishes for floor surfaces, applied while the concrete is still plastic and workable.

Walls and Ceilings. These may be given any of the following typical finishes:

Rough finish—This requires only that, after forms are removed, pockets be filled, rough edges and projections be removed, and bellies be trimmed off to an even plane.

Smooth finish—Plywood forms or special form linings, carefully installed with tight joints, must be used. After the forms are removed, joint marks should be smoothed out, blemishes removed, and the surface made smooth and unmarred. Tolerances, offsets, and projections must be not more than $\frac{1}{16}$ in. (1.59 mm) and variations in level not more than $\frac{3}{8}$ in. (9.53 mm) in 10 ft (3.048 m).

Rubbed finish—Concrete should not be finally set (i.e., should still be "green") when this type of finish is applied. First, projections, offsets, and blemishes are removed and any damage is repaired. Next, the surface is thoroughly wetted in sections and rubbed with cement or carborundum bricks and water until it is uniformly smooth. The rubbed area is then washed clean, and work proceeds on the next section until the entire surface receives a rubbed finish.

Design finish—Special pieces are installed in the forms to create a pattern on concrete when the form is removed. After the forms are removed the surface receives the same treatment as is used for smooth finish.

Plaster, stucco, or cement plaster finish—Concrete that is to receive these finishes must be provided with some sort of bonding surface. For example, flexible dovetail inserts may be applied to the forms to create bonding cavities.

Floor Surfaces. These may be finished as follows:

Wood float finish—The concrete aggregate is forced below the surface; the surface is then leveled with a straight wood screed, and, while the concrete is still not firmly set, it is given a float finish with straight, flat, large wood trowels (called floats).

Steel-troweled finish—After the concrete aggregate is forced below the surface, the surface is leveled with a straight wood screed and given a wood float finish. Before the concrete finally sets, the entire surface is steel-troweled.

Textured finishes—A wide variety of textured finishes are obtained by using various shapes, colors, and depth of exposure of aggregates. One method used for precast concrete panels involves previously treating the form with a retarder. When the panel is removed from the form, the surface paste can be brushed away. Other methods involve removal of surface cement by sand-blasting, bush hammering, or grinding. For sidewalks and various types of paving, the method used is to apply various types of aggregate to the leveled surface of the slab and then embed the aggregates flush with the level surface of the slab. Before the concrete has completely set, the surface concrete can be brushed away, exposing the aggregates.

Colored finishes—There are two methods of obtaining colored finishes for concrete slabs, precast panels and prefabricated panels, blocks, slabs, and other forms: (1) adding color pigments to the concrete mixture, and (2) applying dry colored material to the floated slab, precast panel or prefabricated unit by the dry-shake method. In general, white portland cement will give cleaner, brighter colors than will normal gray cements. When color pigments are added to the concrete mix, the amount of colored pigment should not be more than 10% of the weight of the cement.

Integral cement finish—While the concrete is still green but surface water is gone, the surface is leveled with a straight wood screed. Then a finish coat, 1 in. (25.4 mm) thick, is applied. The mix consists of 1 part portland cement, 1 part fine aggregate which must pass a $\frac{1}{4}$ in. (6.35 mm) mesh, 2 parts coarse aggregate ($\frac{1}{8}$ to $\frac{3}{8}$ in., or 3.18 to 9.35 mm in size), and 5 gal (22.73 liters) water. This finish coat is leveled with a wood screed, given a wood float finish, and then steel-troweled smooth. As soon as the surface rings when tapped with a trowel, a second steel-troweling is given.

Separately applied cement finishes—For these the bare concrete surface must be left rough, having been only leveled and screeded. These separate finishes are designated as nonslip, heavy-duty, colored, etc. Chemical coatings for hardening concrete, paints, synthetic resins, enamels, and other types of applied finishes are discussed under Admixtures; Painting of Concrete; etc.

Special Casting Methods. There are several special methods for casting concrete.

"*Intrusion prepacked concrete*" refers to a patented method of placing concrete which forces a cement mortar into the voids remaining in a compacted mass of coarse aggregate.

"*Vacuum-processed concrete*" refers to a patented treatment of concrete in which a vacuum is applied to the surfaces of the concrete immediately after the concrete is placed. The vacuum removes water from the concrete adjacent to the surface and also removes any air bubbles. Its main advantage is that it speeds up production of a durable strong concrete. The process is particularly useful in manufacturing precast concrete materials.

Pneumatically applied cement mortar is a mixture of portland cement, sand, and water actually shot into place by compressed air. It can be applied to any surface regardless of its shape or slope. The effectiveness of this process depends on the materials to which it is applied. As is natural on vertical surfaces, overhangs, and ceilings, and at corners, a large percentage of the cement mortar bounces from the surface and is wasted.

The Max True machine is used for a method that shoots concrete, not just mortar, into place.

Condensed Checklist

1. When designing with concrete, one should check the locality for the availability of ready-mixed concrete and the distance from plant to site.
2. Soil and underground water should be checked for possible sulfate attack.
3. Reinforced concrete should always be designed by a structural engineer, who should also design the forms.
4. Location and quantity of expansion joints must be determined. These joints should be specially designed to meet all conditions (exterior and interior) at floors, walls, roofs, windows, doors, partitions, etc. They must be watertight yet able to move freely.
5. Specifications should indicate the type of cement, sand, aggregate, admixture, and water to be used. The compressive strength of the concrete must be specified, and trial batches must be made and then

Table C67 Example of a trial batch of concrete

Material	Type of concrete: Lightweight fill 1500 lbf/in.^{2} (10.35 MN/m^2)	Controlled 2500 lbf/in.^{2} (17.24 MN/m^2)	Pump 2500 lbf/in.^{2} (17.24 MN/m^2)
Portland cement 1 yd^3 (0.764 m^3) dry weights	500 lb (226.80 kg)	495 lb (224.53 kg)	591 lb (268.08 kg)
Sand	625 lb (285.50 kg)	1275 lb (578.34 kg)	1445 lb (655.45 kg)
Coarse aggregate	725 lb (328.86 kg) Nytrolite	1850 lb (839.16 kg)	1550 lb (703.08 kg)
Water	40.9 gal (185.93 liters)	36.9 gal (167.75 liters)	38.9 gal (176.84 liters)
Air-entraining agent	5.3 oz (150.49 cm^3)	5.3 oz (150.49 cm^3)	5.3 oz (150.49 cm^3)
Water-cement ratio	7.7	7.01	6.19
Lightweight fines	450 lb (204.12 kg)	450 lb (204.12 kg)	450 lb (204.12 kg)
Weight: ft^3 (m^3)	97.8 ft^3 (2.27 m^3)	145.5 ft^3 (3.38 m^3)	144.4 ft^3 (3.35 m^3)
Slump	5 ± 1 in. (127 ± 25.4 mm)	5 ± 1 in. (127 ± 25.4 mm)	5 ± 1 in. (127 ± 25.4 mm)

approved by the structural engineer. For each type of concrete, two specimens are required for 7- and 28-day testing for each 150 yd^3 (114.69 m^3). A sample of a trial batch as it would come from a testing laboratory is shown in Table *C67*.

6. Local, municipal, and state codes, as well as federal (Army, Navy, etc.) codes, should be checked for strength requirements, reinforcing, sizes, and other limitations, as well as fire resistance ratings for the concrete work.
7. The time sequence in relation to the season of the year when construction operations are to begin and the duration of time necessary for the concrete work to be completed must be checked to ensure that adequate precautions can be taken for freezing weather or dry hot weather, as the case may be.
8. One should always check that all inserts, slots, hangers, anchors, sleeves, supports, etc., have been installed in the forms in their correct locations, and that those to be installed after pouring (while concrete is green) are available and their locations identified. This is particularly necessary for anchor bolts where templates are to be used.
9. There must be adequate inspection at the construction site to check forms, reinforcing, slump tests, handling, pouring, and placing of concrete; and to ensure proper curing, form removal, etc.
10. For pumping concrete, a structural engineer should always be consulted for the trial mixes that must be made and approved including the type of pumping equipment.
11. A check should always be made that form detailing and drawings meet local, municipal, state, and federal government codes.
12. When special design or surface textures are required, specification should provide that full-size samples be poured and cured at the site, shop, or factory and that they be approved before concrete work is commenced. Local, municipal, state, and federal codes should be checked as to requirements for the design of concrete form work.

Conditions Favorable to the Use of Concrete

1. For all footings and foundation walls, including most residential work, and especially in areas where there is a subsurface water condition.
2. For the structural frame of buildings with design and structural requirements that make it economical to use concrete in preference to other materials.

3. For floor areas directly on earth.
4. For exterior walks, walls, roadways, steps, etc., where freezing and thawing occur.
5. Where fire-retarding or fireproofing materials are necessary.
6. In localities where steel is not readily available.
7. For precast, prefabricated, and preformed panels, curtain wall panels, and structural members, both vertical and horizontal.
8. For thin shell domes, irregular shaped forms, and combination cable and concrete structures.
9. For prefabricated prestressed concrete beams, hollow slabs, and single and double T's.

Conditions Unfavorable to the Use of Concrete

1. Where the design calls for very small or thin structural members, unless the structural design is specifically planned for thin shell construction.
2. Where weight is an important factor.
3. For excessively long spans between supporting members unless the structural design is specifically planned for prestressed concrete.

History and Manufacture

The Romans were the first to build with what we know as concrete, and our word "concrete" comes from the Latin *concretus*, meaning "growing together." Perhaps the best commentary on concrete is found in the specifications attributed to the Roman architect Gaius Franciscus for concrete work presumably designed by him: "The concrete on this project shall be in accordance with the best principles of design, most of which are well known but few of which are ever observed." The binding material the Romans used was lime, to which they added pozzolana, a volcanic ash obtained from a small town near Mount Vesuvius. The result was a cement that hardened under water. With this hydraulic cement they constructed aqueducts, bridges, foundations, and other massive buildings, all characterized by extreme durability. Generally these structures were faced with brick or stone. With the fall of Rome, the art of building with concrete was lost until the 18th century. During this period lime in some form was the only binding material available, other than gypsum, until Aspdin patented portland cement in 1824.

Modern Developments. Before the development of reinforced concrete construction during 1850 to 1880, concrete was placed nearly dry and compacted with heavy tampers. Thereafter, until about 1914, concrete was made very wet and was literally poured into place, resulting in weak concrete with poor durability.

Investigations were undertaken to improve quality, first by controlling mix proportions to obtain uniform concrete which had good workability, durability, and strength. As a result, Abrams formulated the water-cement ratio law, which showed the importance of restricting this ratio to the lowest value consistent with the required workability of concrete for a particular job.

Further research led to a series of developments that have made concrete a major material for modern construction.

It was found that vibrating the concrete eliminated the necessity of sloppy or watery mixes. The appearance of high-early-strength, low-heat, and sulfate-resisting cements increased the versatility of concrete. The introduction of pozzolanic materials into concrete improved its characteristics and permitted sound concrete to be made with aggregates that were otherwise unsuitable. About 1938 it was discovered that small amounts of well-dispersed entrained air not only improved the workability of concrete but also tremendously increased its resistance to freezing and thawing.

Where once it was thought that the securing of a maximum of solid substance gave all the desirable properties to concrete, it is now recognized that the most dense concrete is not necessarily the most durable.

Improvements were made in the manner in which concrete is mixed. Today all concrete is mixed by weight instead of by volume, as was done formerly, and the various aggregates are separated into two or more sizes, thus minimizing segregation during handling.

Concrete, which was once considered a simple mixture of cement, sand, coarse aggregate, and water haphazardly mixed and placed, now is recognized to be a carefully controlled mixture that combines other substances (admixtures) as needed to obtain the optimum in quality and economy for any use. The technical development of prestressed concrete has made possible new, more efficient construction methods for architectural and engineering projects.

CONCRETE BLOCK

Physical and Chemical Properties

Concrete block is a masonry unit, usually with single or multiple hollows but also available solid, made of the following ingredients: water, portland cement, blended cements, and various types of aggregate such as sand, gravel, crushed stone, air-cooled slag, coal cinders, ex-

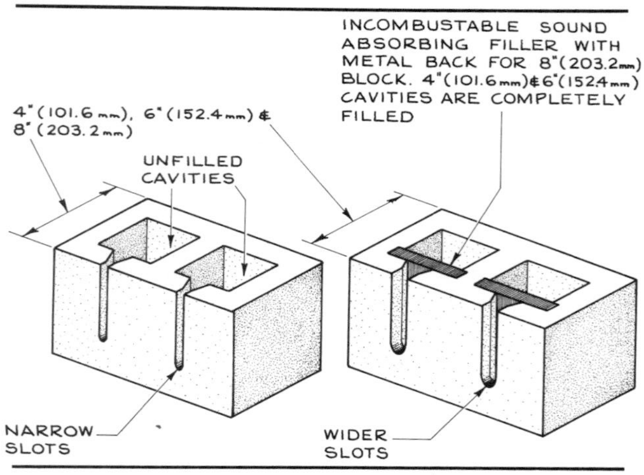

Figure C67 Acoustical types of concrete block.

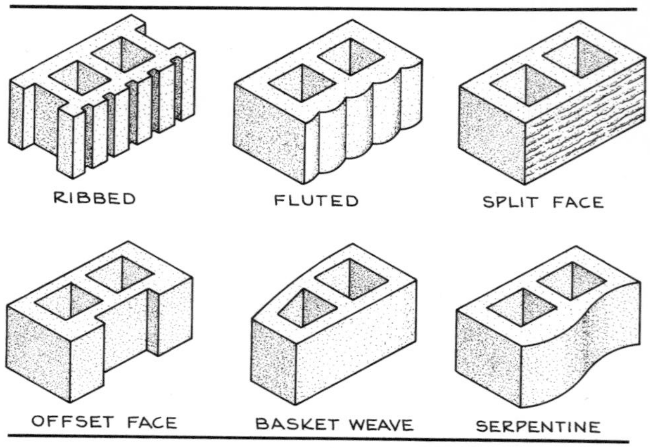

Figure C68 Typical examples of concrete block with various surface textures.

panded shale or clay, expanded slag, volcanic cinders (pozzolan), pumice, and scoria (refuse obtained from the reduction of ores and from the smelting of metals). The term "concrete block" was formerly limited to hollow masonry units made with aggregates such as sand, gravel, and crushed stone, but today the term covers all types of concrete block, including solid units, made with any of the various kinds of aggregate. It includes units with applied glazed surface(s), various pierced designs,

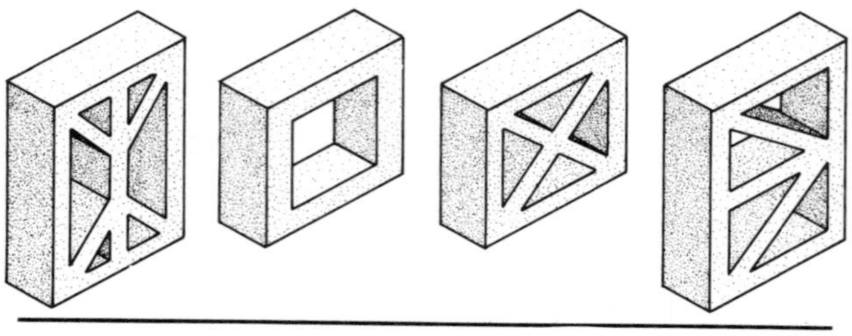

Figure C69 Typical examples of pierced concrete block for decorative screen walls.

Table C68 Physical requirements for concrete block

Type of concrete block	Grade desig- nation	Compressive strength[c] Minimum of average gross area: Average of three units, lb/in.^{2}	MN/m^2	Individual unit, lb/in.^{2}	MN/m^2	Maximum water absorption lb/ft^3 (kg/m^3) Average of three units with oven-dry weight of concrete in lb/ft^3 (kg/m^3): Lightweight, Less than 85 lb/ft^3 (1367.7 kg/m^3), lb/ft^3	kg/m^3	Lightweight, Less than 105 lb/ft^3 (1682.1 kg/m^3), lb/ft^3	kg/m^3	Medium weight, Less than 105–125 lb/ft^3 (1682.1–2002.5 kg/m^3), lb/ft^3	kg/m^3	Normal weight, 125 lb/ft^3 (2002.5 kg/m^3 or more), lb/ft^3	kg/m^3
Hollow load-bearing	N-I N-II	1000	6.90	800	5.52	20	320.4	18	168.2	15	240.3	13	208.26
	S-I[b] S-II[b]	700	4.43	600	4.14								
Hollow non-load-bearing	I II	350	2.35	300	2.07	20	320.4	18	168.2	15	240.3	13	208.26
Solid load-bearing	N-I N-II	1800	12.41	1500	10.34	20	320.4	18	168.2	15	240.3	13	208.26
	S-I[b] S-II[b]	1200	8.27	1000	6.90								

[a]Protective coatings should be applied on exterior face when below grade and when required on exterior face above grade.
[b]Limited to use for above-grade exterior walls with weather-protective coatings and for interior walls.
[c]1 lbf/in.^{2} = 0.006895 MN/m^2

acoustical pierced types, and a wide variety of surface textures, as shown in Figures *C67*, *C68*, and *C69*. It is most commonly 8 × 8 × 16 in. (203.2 × 203.2 × 406.8 mm) in size and meets specified requirements covering size, type, weight, moisture content, compressive strength, and other special characteristics (*see* Tables *C68*, *C69*, and *C70*).

Table C69 Moisture-content requirements for type I units

Type of concrete block	Linear shrinkage (percent)	Moisture content Maximum percentage of total absorption (average of three units) Humidity[a] conditions at job site or point of use		
		Humid	Intermediate	Arid
Solid load-bearing	0.03 or less	45	40	35
Hollow load-bearing	0.03–0.045	40	35	30
Hollow non-load-bearing	0.045–0.065	35	30	25

[a]Humid; average relative humidity less than 50%; intermediate: average relative humidity 50–75%; arid: average relative humidity less than 50%.

Effect of Aggregate. The kind of aggregate used affects the color, texture, weight, heat transmission, and other characteristics of the concrete block (*see* Table *C71*).

Thermal Values. Thermal values for the various types of concrete block wall are given in Table *C72* in terms of resistance (R) values.

Types and Uses

There are three types of concrete block: (1) hollow load-bearing, (2) hollow non-load-bearing, and (3) solid load-bearing. The load-bearing types are available in two grades: Grade N for general use such as exterior walls above and below grade that may or may not be exposed to moisture penetration or the weather, and for back-up and interior walls; Grade S for above-grade exterior walls with weather-protective coating, and for interior walls. These grades are subdivided into two types: Type I moisture-controlled units N-I and S-I, and Type II non-moisture-controlled units N-II and S-II.

Concrete block is made in various sizes and shapes and in both modular and nonmodular dimensions. (*See* Figures *C70* to *C75* and Table *C68*.)

Lightweight Blocks With Colored Finishes. Various types of concrete block are manufactured with permanent colored surface finishes applied to one or both sides and are available in a wide range of colors and tex-

Table C70 Minimum thicknesses of face shells and webs of concrete block

Type of concrete block	Nominal width of units (minimum)		Face-shell thickness (minimum)		Web thickness (average of web measurements on three units at thinnest points)	
	in.	mm	in.	mm	in.	mm
Hollow load-bearing	3, 4	76.20, 101.60	$\frac{3}{4}$	19.05	$\frac{3}{4}$	19.05
	6	152.40	1	25.40	1	25.40
	8	203.20	$1\frac{1}{4}$	31.75	1	25.40
	10	254.00	$1\frac{3}{8}$	34.93	$1\frac{1}{8}$	28.58
			$1\frac{1}{4}$[a]	31.75		
	12	304.80	$1\frac{1}{2}$	38.10	$1\frac{1}{8}$	28.50
			$1\frac{1}{4}$[a]	31.75		
Hollow non-load-bearing	3, 4	76.20, 101.60				
	6	152.40				
	8	203.20	$\frac{1}{2}$	12.7		
	10	254.00				
	12	304.80				

[a]This face-shell thickness is applicable where allowable design load is reduced in proportion to the reduction in thickness from basic face-shell thickness shown.

Table C71 Effect of aggregate on color, texture, weight, and other characteristics of concrete block

Type of aggregate	Concrete block: Color	Texture	Weight per unit: lb.	kg	Characteristics	Coefficient of heat transmission (R) for block alone
Sand and gravel	Light gray	Fine	38–43	17.24–19.51	High compressive strength; low absorption; dense and durable	8 in. (203.2 mm) = 1.70 12 in. (304.8 mm) = 2.04
Cinder	Dark gray	Medium to coarse	26–33	11.79–14.97	Good strength; high insulating and fire-resistant properties; good acoustic properties; nailable	8 in. (203.2 mm) = 2.70 12 in. (304.8 mm) = 2.86
Shale	Steel gray or light gray	Fine, medium, and coarse	26–33	11.79–14.97	High strength; high insulating and fire-resistant properties; good acoustic properties; durable and nailable	8 in. (203.2 mm) = 3.03 12 in. (304.8 mm) = 3.11

Table C72 Resistance (R) values for various types of concrete block wall

Type of concrete block or wall[a]	Width of block: in.	mm	Concrete block without plaster R	$\frac{1}{2}$ in. (12.7 mm) plaster directly on concrete block R	$\frac{1}{2}$ in. (12.7 mm) plaster on $\frac{3}{8}$-in. (9.53-mm) plaster lathing board furred $\frac{3}{4}$ in. (19.05 mm) R	$\frac{1}{2}$ in. (12.7 mm) plaster on $\frac{1}{2}$-in. (12.7-mm) insulation board furred $\frac{3}{4}$ in. (19.05 mm) R
Sand and gravel aggregate	8	203.2	1.89	2.04	3.23	4.55
	12	304.8	2.04	2.22	3.33	4.55
Cinder aggregate	8	203.2	2.70	2.85	4.00	5.26
	12	304.8	2.85	3.03	4.17	5.26
Shale, clay, or slag aggregate	8	203.2	3.03	3.13	4.35	5.26
	12	304.8	3.13	3.23	4.35	5.26
10-in. (254-mm) cavity wall of two 4-in. (101.6-mm) blocks of sand and gravel aggregate			2.94	3.03	4.17	5.26
10-in. (254-mm) cavity wall of two 4-in. (101.6-mm) blocks of clay or slag aggregate			3.85	4.17	5.26	6.67

[a] All concrete block is hollow type; exterior walls have two coats of portland cement, and all calculations for R values are based on an exterior wind velocity of 15 mph (24.135 km/h).

tures. The glazed-type surface is a factory-applied finish consisting of a resinous binder of the thermosetting type with glass (silica) sand and color pigments or colored granules. This type of concrete block must have resistance to various chemicals and substances as shown in Table *C73*, and also be resistant to pencil, magic marker, carbon paper, and lanolin. The glazed-type surface should have a flame-spread classification no higher than 25 without evidence of continued progressive combustion and a smoke density rating no higher than 50. This type of concrete block can be used wherever a colored, washable, smooth, durable interior finish surface is required. The commonly available standard sizes and shapes are illustrated in Figure *C76*. When this type of block is used, all horizontal and vertical dimensions should be calculated and planned so that very small cut pieces will not be necessary, especially at openings and the bases of walls.

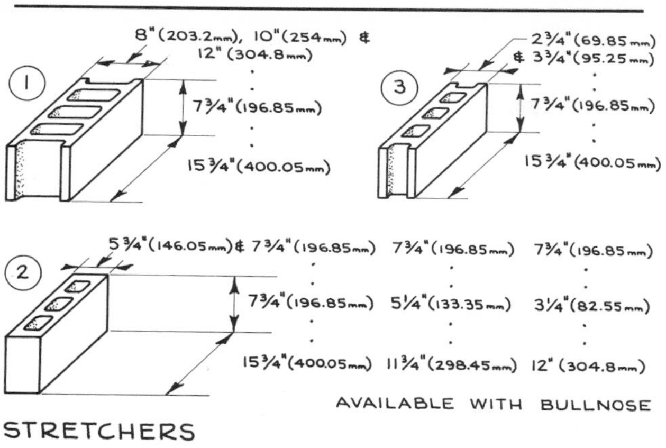

Figure C70 Modular sizes of load-bearing (1) and non-load-bearing (2, 3) concrete block.

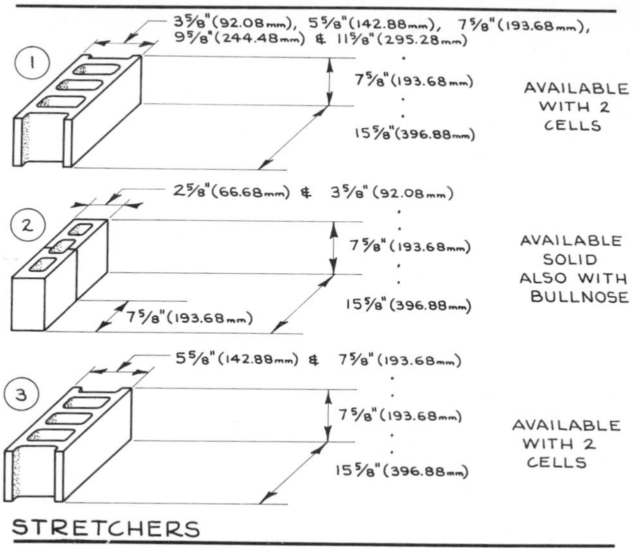

Figure C72 Nonmodular sizes of load-bearing (1) and non-load-bearing (2, 3) concrete block.

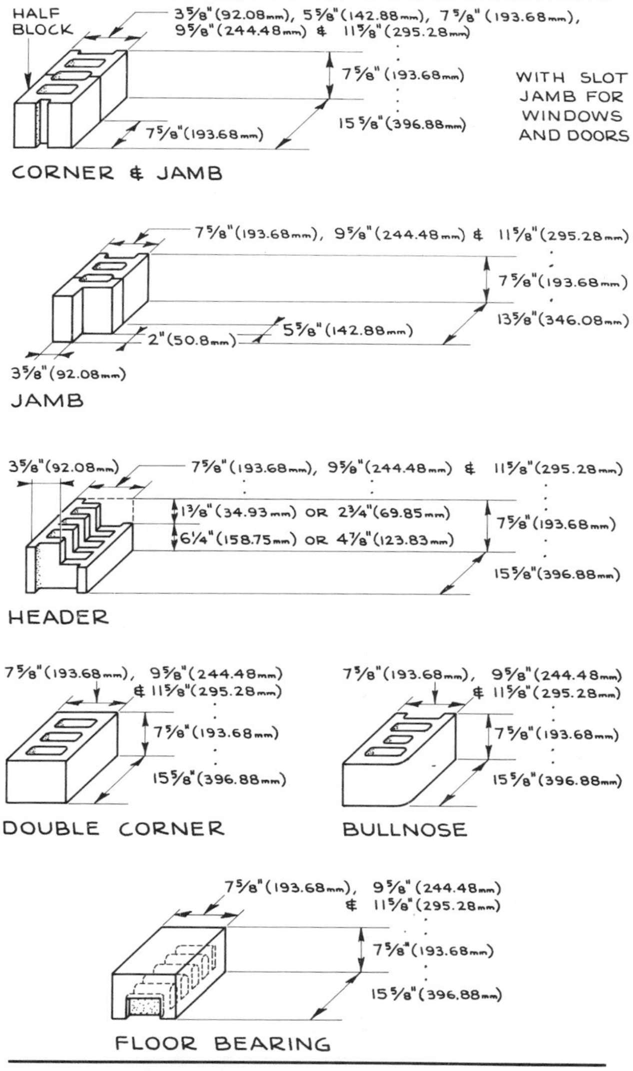

Figure C71 Typical modular sizes of special-shape concrete block.

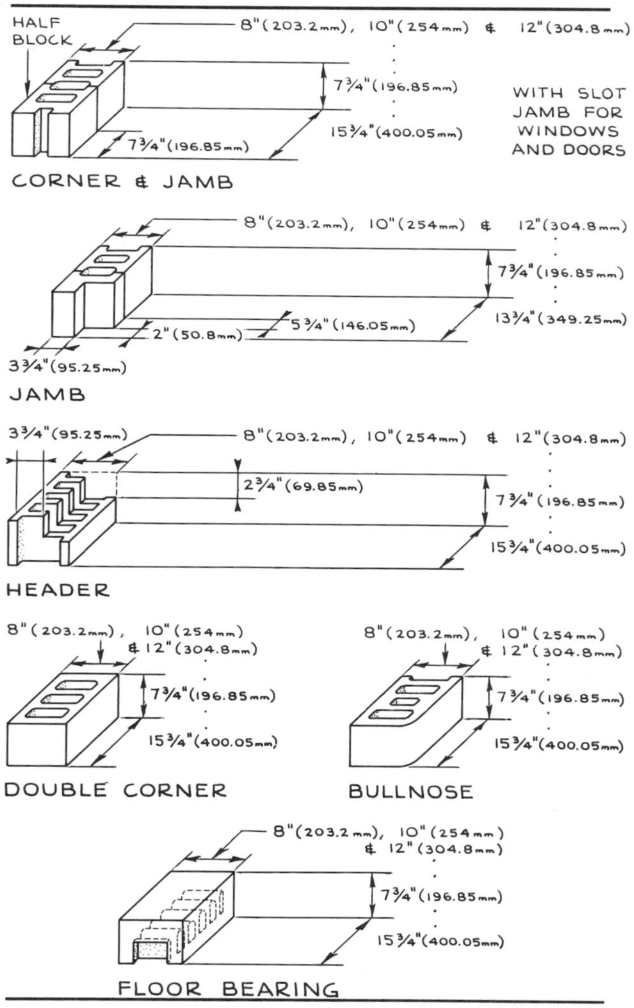

Figure C73 Typical nonmodular sizes of special-shape concrete block.

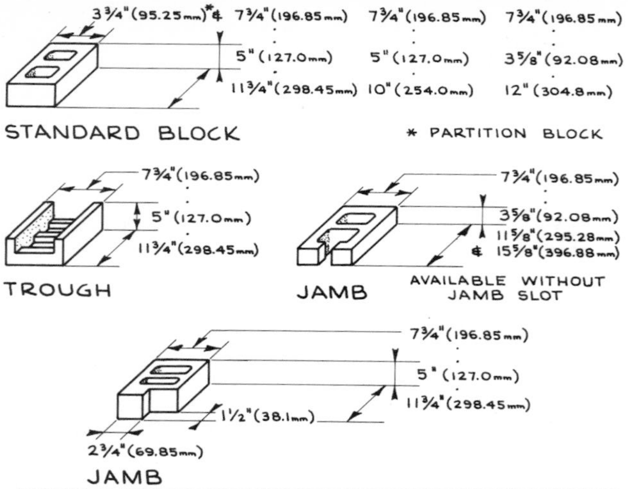

Figure C74 Nonmodular concrete block used in the southeast.

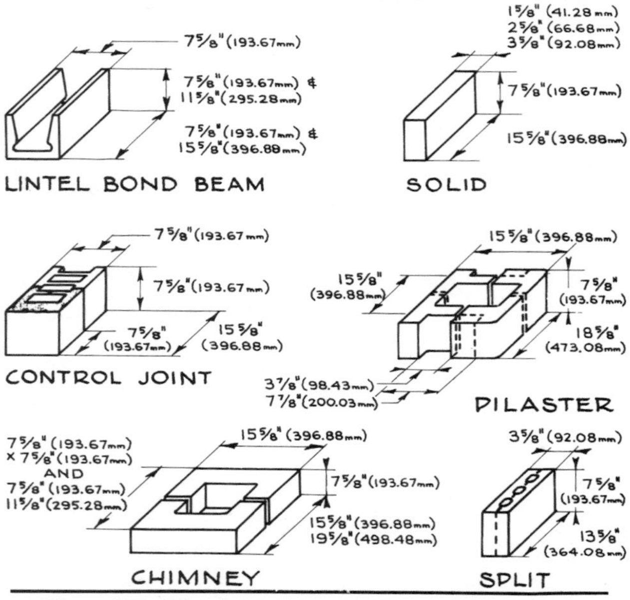

Figure C75 Miscellaneous special concrete block shapes.

Table C73 Chemicals to which glazed faced concrete block must be resistant

Chemical	Duration of application of chemical (hours)
Acetic acid	24
Potassium hydroxide	3
Trisodium phosphate	24
Hydrogen peroxide	24
Household detergents	24
Vegetable oil	24
Blue-black ink	1
Ethyl alcohol (industrial, denatured 95%)	3
Hydrochloric acid	3

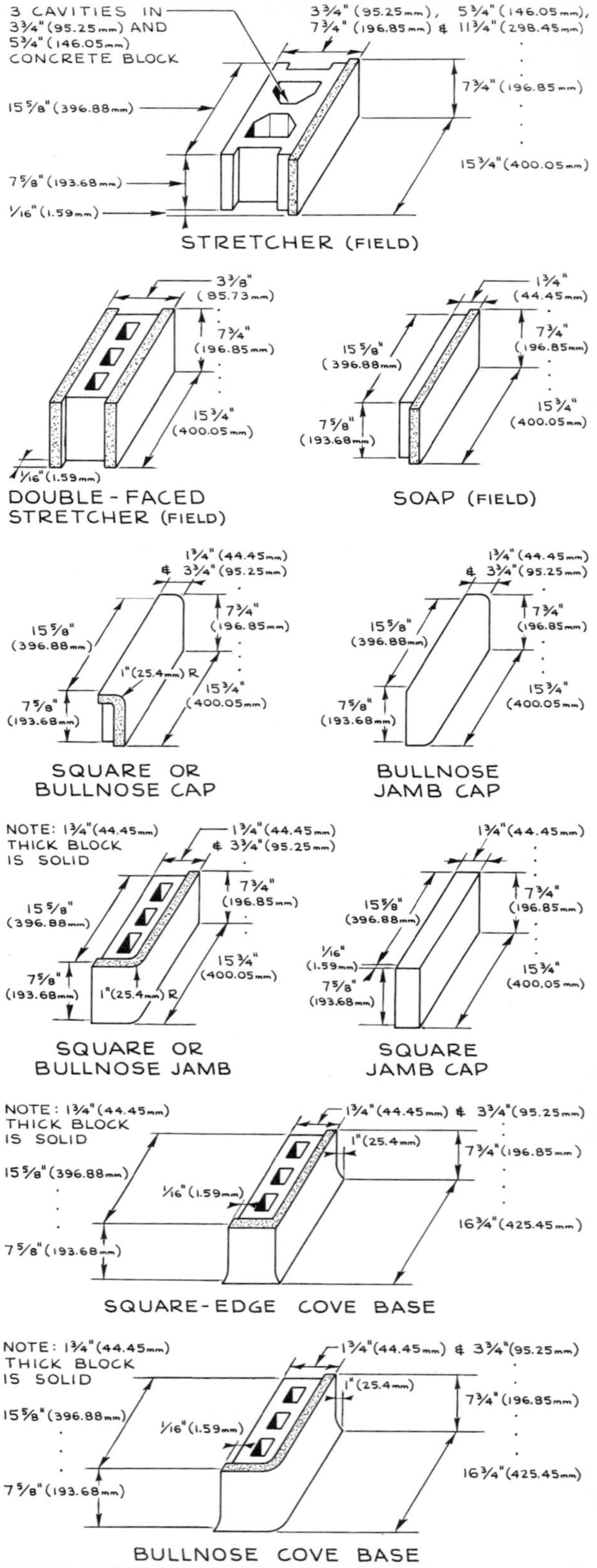

Figure C76 Typical shapes and dimensions of lightweight concrete block with a smooth, durable, colored surface finish.

Application

Accessories. Accessories for concrete block work include reinforcing, anchors, and fillers for construction joints, as shown in Figures *C77* and *C78.* Most of the reinforcing for concrete block work consists of reinforcing rods deformed and welded into special shapes and usually galvanized. When detailing the construction of concrete block walls, the architect should determine and indicate the correct size of rods, the amount of added strength developed in the walls, and the placement of the reinforcing rods (number of courses between rods).

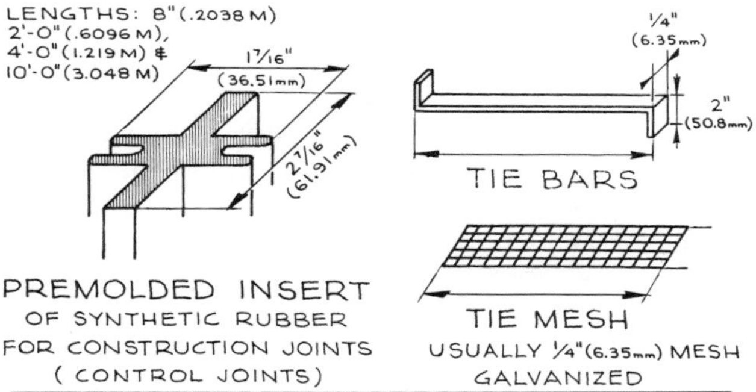

Figure C77 Reinforcing and accessories for concrete block work.

Mortars. Mortars for concrete block work are designated either by property specifications, where the mortar is based on the properties of the materials and the properties of water retention and compressive strength, or by proportion specifications. (*See* Tables *C74* and *C75*; *also see* Mortars.)

Length and Height Restrictions. There are important limitations to the height and length of concrete block walls. The maximum distance between lateral supports of both vertical and horizontal walls is 12 ft (3.658 m) for 8-in. (203.2-mm) block, 18 ft (5.468 m) for 12-in. (304.8-mm) block, and 11.66 ft (3.556 m) for 10-in. (254.0-mm) cavity walls. Limitations for vertical dimensions are 12 ft (3.65 m) for 8-in. (203.2-mm) block, and 35 ft (10.66 m) for 12-in. (304.8-mm) block. The size of block must increase by 4 in. (101.6 mm) for every succeeding 35 ft (10.668 m) of wall height. The size of concrete block walls for various types of multistoried buildings are given in Tables *C76* and *C77.*

Joints. Joints for concrete block vary in type as shown in Figure *C79.* Raked and extruded joints should not be used for the exterior except in dry climates. Jointing tools consist of a bar at least 22 in. (558.8 mm) long for horizontal joints of the V and concave type, and a small

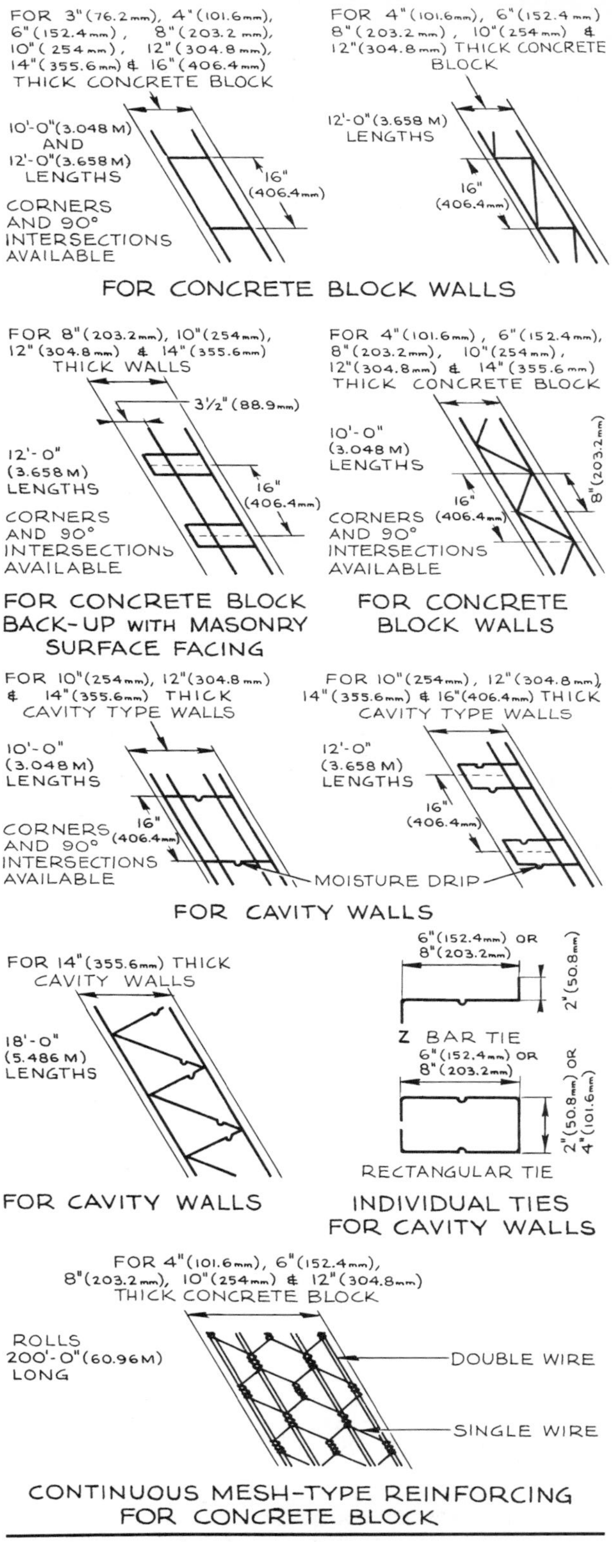

Figure C78 Reinforcing and accessories for concrete block work.

S-shaped tool for vertical joints. Concave-type joints are made with a $\frac{5}{8}$-in. (15.88-mm) round bar, and V-type joints with a $\frac{1}{2}$-in. (12.7-mm) square bar, for both horizontal and vertical tooling. Flush joints are tooled, then refilled with mortar, and when the mortar has stiffened, rubbed flush. A raked joint requires a special tool to finish the joint after sufficient mortar has been raked out. Weatherstruck joints are made with the normal small trowel.

Table C74 Compressive strength for mortar types as required by property specifications

Mortar type[a]	Average compressive strength[b] at 28 days	
	lbf/in.^{2}	MN/m^2
M	2500	17.24
S	1800	12.41
N	750	5.17
O	350	2.41
K	75	0.52

[a]Before 1954 mortar-type designations were A-1, A-2, B, C, and D.
[b]lbf/in.^{2} = 0.006895 MN/m^2.

Condensed Checklist

1. The type of concrete block and mortar should meet the requirements set by height, structural conditions, loads and end use of the design.
2. The type of joint to be used in various areas of the building and the type of tool for forming the joints should be specified.
3. Concrete block should always be laid with a full bed of mortar.
4. A check should be made that all the special shapes for corners, jambs, lintels, etc., are installed.
5. Both expansion joints and construction joints (control joints) must be installed. Control joints relieve stresses within concrete block walls and thus allow movement without cracking of the masonry (*see* Figure *C80*). Expansion joints for concrete block

Table C75 Mortar proportions by volume[a] as required by proportion specifications

Mortar type	Parts by volume of portland cement, cement, or portland blast-furnace slag cement	Parts by volume of masonry cement	Parts by volume of hydrated lime or lime putty	Aggregate measured in a damp, loose condition
M	1	1	—	Not less than $2\frac{1}{4}$ and not more than 3 times the sum of the volumes of the cements and lime used
	1	—	$\frac{1}{4}$	
S	$\frac{1}{2}$	1	—	
	1	—	Over $\frac{1}{4}$ to $\frac{1}{2}$	
N	—	1	—	
	1	—	Over $\frac{1}{4}$ to $\frac{1}{2}$	
O	—	1	—	
	1	—	Over $\frac{1}{4}$ to $\frac{1}{2}$	
K	1	—	Over $\frac{1}{4}$ to $\frac{1}{2}$	

[a]To convert proportions by volume to weight proportions, multiply the unit volumes by the following weights: portland cement, 94 lb; hydrated lime or lime putty, 40 lb; sand (damp and loose), 85 lb; masonry cement, weight as printed on bag.

Table C76 Generally applicable thicknesses of concrete block for single- and multistoried buildings other than residences

Number of stories	Basement		First story		Second story		Third story		Fourth story	
	in.	mm	in.	mm	in.	mm	in.	mm	in.	mm
1	12 or 8[a]	304.8 or 203.2[a]	12 or 8	304.8 or 203.2[a]						
2	12	304.8	12	304.8	12 or 8[a]	304.8 or 203.2[a]				
3	12	304.8	12	304.8	12	304.8	12 or 8	304.8 or 203.2		
4	16 or 12[a]	506.4 or 304.8[a]	16 or 12[a]	506.4 or 304.8[a]	12	304.8	12	304.8	12	304.8

[a]These sizes should be checked for height and live, dead, and other loads, as well as for requirements of local, municipal, state, and federal codes.

Table C77 Generally applicable thicknesses for cavity walls for single and multistoried residential buildings

Number of stories	Basement		First story		Second story		Third story	
	in.	mm	in.	mm	in.	mm	in.	mm
1	12	304.8	10	254.0				
2	12	304.8	10	254.0	10	254.0		
3	12	304.8	12	304.8	10[a]	254.0	10	254.0

[a]Cavity wall 10 in. (254.0 mm) thick is limited to 25 ft (7.620 m) in height. Check with local, municipal, state, and federal codes.

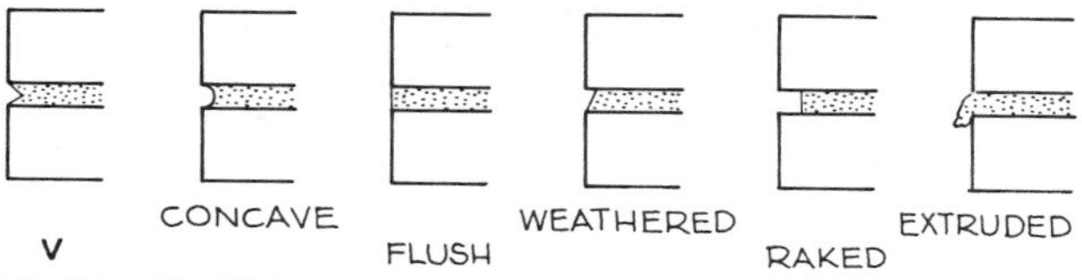

Figure C79 Various types of joints for concrete block work.

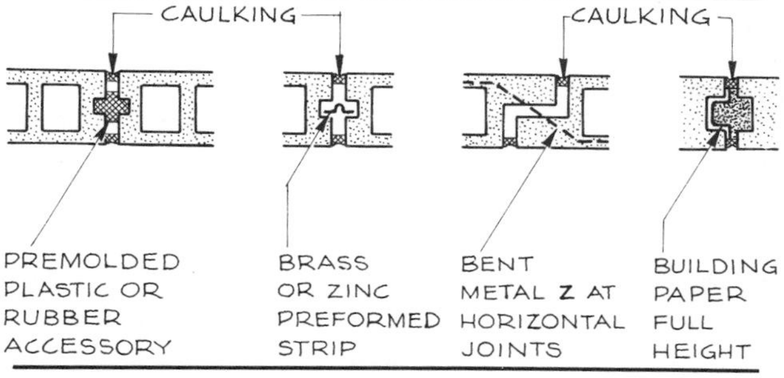

Figure C80 Control of construction joints for concrete block work.

work are the same as those for brick and other masonry work.

6. Vertical coursing must be correctly planned for heads of windows, doors, and window sills; coursing should also be checked for ceiling heights, floor-to-floor height, and structural bearing. For 6′-10″ (2.083 m) and 7′-0″ (2.134 m) openings there is a $4\frac{7}{8}$″ (123.83 mm) starter course which brings the horizontal joint at 7′-$1\frac{1}{4}$″ (2.165 m).
7. A check should be made of local suppliers to see that they have in stock all the special pieces for corners, jambs, lintels, etc., required for the job.
8. Local, state, municipal, and state codes and also codes of the Fire Underwriters, insurance companies, labor departments, and federal government (Army, Navy, etc.) should be checked for requirements and limitations.
9. Concrete block should be delivered dry and cured. Block stored on site should be kept covered, and the top of concrete block work, after a day's work, should also be covered. Concrete block must always be dry.
10. When concrete block is to be used for exterior walls, the type of treatment needed to make it weathertight should be determined.

Conditions Favorable to the Use of Concrete Block

1. Where the design requires a fire-resistant, strong, relatively light, load-bearing or non-load-bearing masonry wall.
2. Where a masonry back-up material is required for brick, stone, or other type of facing material.
3. As a strong, lightweight, fire-resistant masonry partition, either load-bearing or non-load-bearing.
4. As a strong, relatively lightweight, fire-resistant, load-bearing or non-load-bearing exterior wall in dry climates.

Conditions Unfavorable to the Use of Concrete Block

1. Where weathertight exterior walls that will not need further maintenance to keep them weathertight are required.
2. Where there will be extremely rough usage.

History and Manufacture

During the reign of Caligula in A.D. 37 to 41, in the region of present-day Naples, the Romans used a type of concrete block that was allowed to harden before being laid. Concrete block of the now-familiar hollow type is a rather recent invention. However, solid blocks made of a cement and some kind of aggregate date back to ancient times, and their early history parallels that of sand-lime brick. Today most concrete block is manufactured by machine-mixing the ingredients, pouring the mix into molds, and curing the block by air-drying. Some manufacturers use a wet steam-and-pressure curing process that can produce concrete block in a few hours.

CONCRETE DECKING AND STRUCTURAL UNITS

Concrete decking can be defined as prefabricated reinforced concrete slabs that will span between supporting beams, walls, or partitions to form floors or roofs. These slabs are made of general construction concrete, lightweight concrete, and nailable concrete with fiber, ordinary steel, or special prestressed steel reinforcing. Depending on the type of concrete used, such decking can serve as thermal and acoustic insulation as well as finish ceiling surface.

Concrete structural units are comparable to structural steel. They consist of prefabricated structural beams and girders, made of concrete reinforced with steel or prestressed steel, which act as structural supporting members for reinforced concrete slabs or other type of decking for floors and roofs. Both concrete decking and structural units are controlled in manufacturing to meet exact requirements and rigid tests.

Decking is available in sizes ranging from 15 to 32 in. (38.1 to 812.8 mm) in width, 4 to 10 ft (1.219 to 3.048 m) in length, and 2 to 4 in. (50.8 to 101.6 mm) in thickness.

Structural units are available in a limited number of stock sizes but can be obtained in almost any size to meet load and span requirements. Their design and use constitutes a highly technical engineering problem.

Types and Uses

Concrete decking can be divided into two main types: (1) decking that is merely laid upon and attached to intermediate beams, which are the structural support, and (2) decking that is both structural support and decking in one. Typical examples of each kind are shown in Figures *C81* and *C82*.

A series of prestressed concrete beams and girders that are designed primarily for use with concrete decking is shown in Figures *C83*, *C84*, and *C85*.

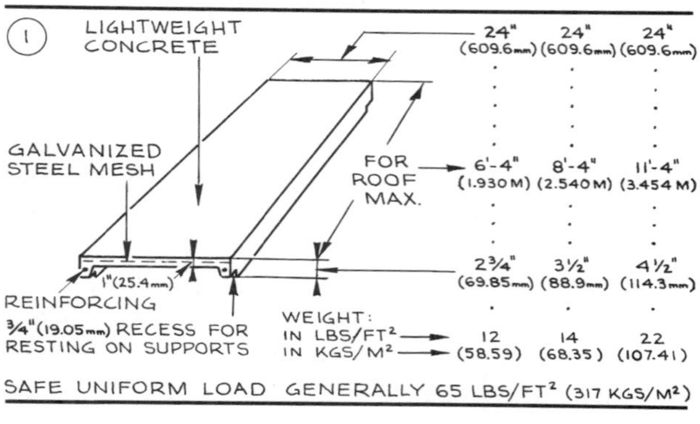

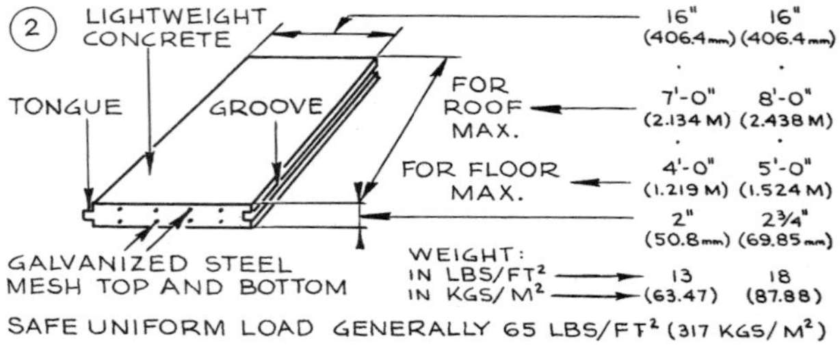

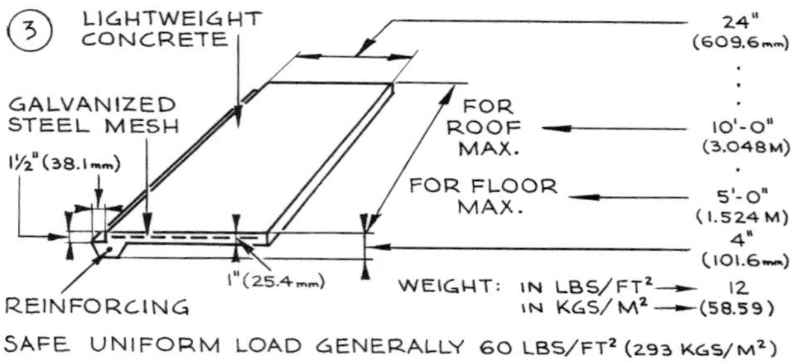

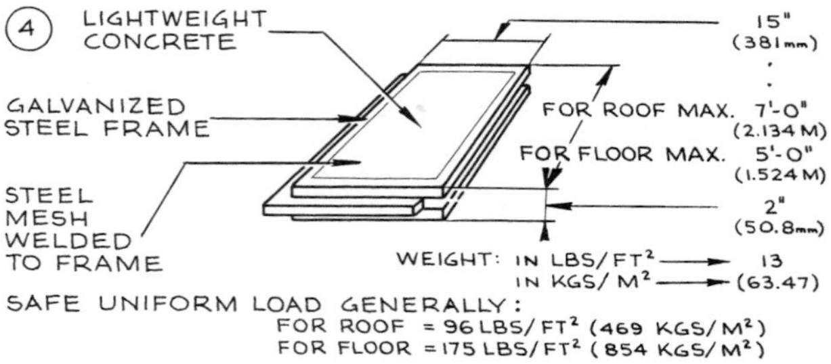

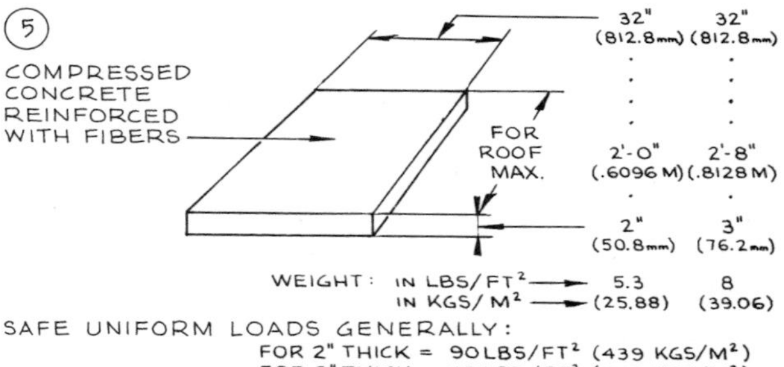

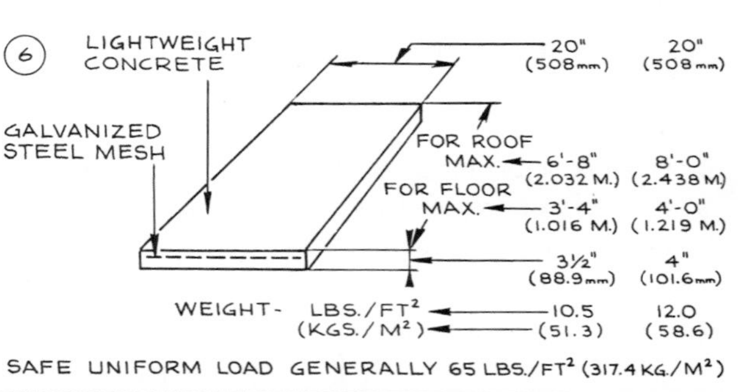

Figure C81 Typical examples of decking to be installed on intermediate structural supports: channel (1), tongue and groove (2), lap (3), steel frame (4), reinforced with fiber (5) and flat (6).

DEPTH OF DECKING	MAXIMUM SPANS FOR ROOFS	MAXIMUM SPANS FOR FLOORS	SPAN LIMITATIONS IN 2'-0" (609.6mm) INCREMENTS
8" (203.2mm)	30'-0" (9.144 M.)	26'-0" (7.925 M.)	10'-0" TO 30'-0" (3.048 M. TO 9.144 M.)
6" (152.4mm)	24'-0" (7.315 M.)	22'-0" (6.706 M.)	10'-0" TO 24'-0" (3.048 M. TO 7.315 M.)

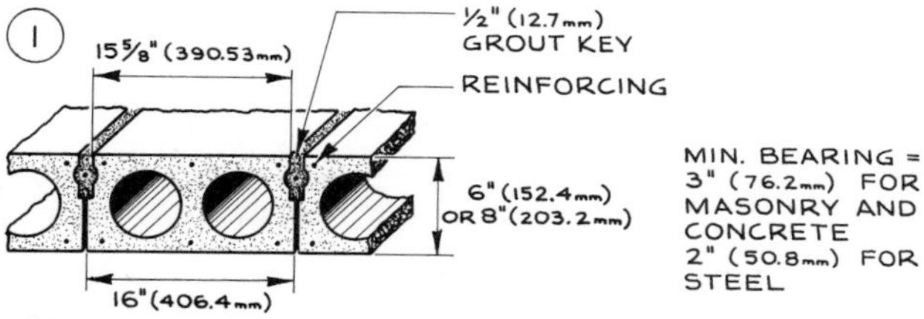

8" (203.2mm) BLOCK HAS 2 HOLES 6⅛" (155.58mm) DIAMETER
6" (152.4mm) BLOCK HAS 3 HOLES 4⅛" (104.78mm) DIAMETER

DEPTH OF DECKING	SPAN LIMITATIONS IN 2'-0" (609.6mm) INCREMENTS		
4" (101.6mm)	8'-0" (2.438 M.)	TO	16'-0" (4.877 M.)
6" (152.4mm)	8'-0" (2.438 M.)	TO	24'-0" (7.315 M.)
8" (203.2mm)	12'-0" (3.658 M.)	TO	32'-0" (9.754 M.)

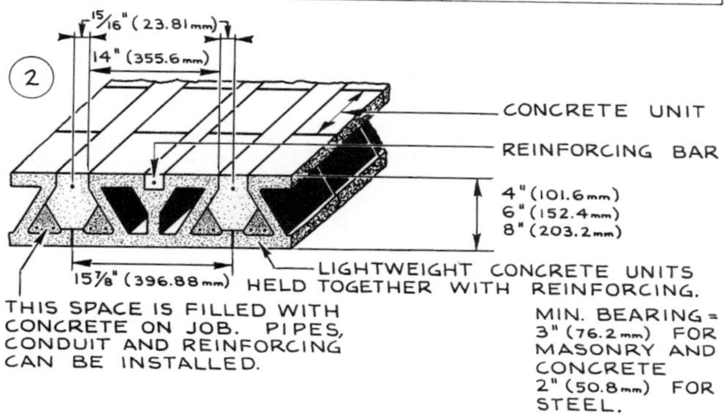

DEPTH OF DECKING	SPAN LIMITATIONS IN 2'-0" (609.6mm) INCREMENTS		
4" (101.6mm)	8'-0" (2.438 M.)	TO	16'-0" (4.877 M.)
6" (152.4mm)	8'-0" (2.438 M.)	TO	20'-0" (6.096 M.)
8" (203.2mm)	14'-0" (1.219 M.)	TO	26'-0" (7.925 M.)

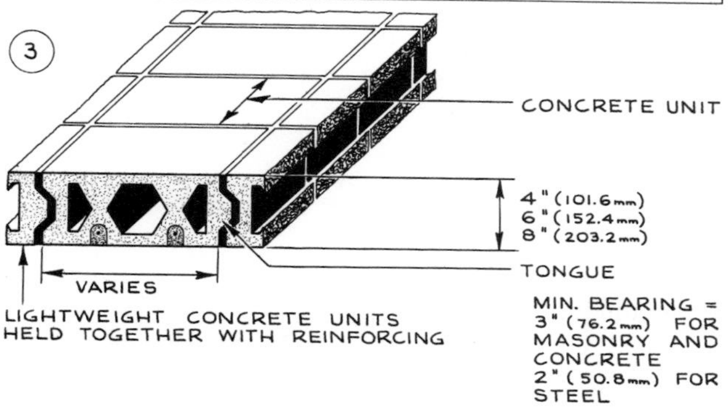

Figure C82 Typical examples of decking that is both structural and structural support in one: unit cast in one piece for entire length (1), concrete units forming needed length and held together with reinforcing (2), tongue-and-groove units forming needed length and held together with reinforcing (3), deep channel type (4), and T-section type (5).

DEPTH OF DECKING	SPAN LIMITATIONS IN 1'-0" (304.8mm) INCREMENTS		
6" (152.4mm)	10'-0" (3.048 M.)	TO	20'-0" (6.096 M.)
8" (203.2mm)	10'-0" (3.048 M.)	TO	24'-0" (7.315 M.)
9" (228.6mm)	10'-0" (3.048 M.)	TO	26'-0" (7.925 M.)
12" (304.8mm)	10'-0" (3.048 M.)	TO	32'-0" (9.754 M.)

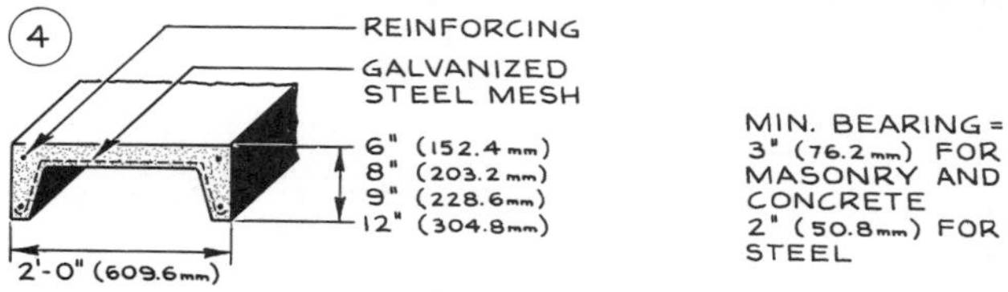

SIZE OF REINFORCING AND DEPTH OF DECKING CONTROLLED BY LOAD TO BE SUPPORTED AND SPAN

DEPTH OF DECKING	SPAN LIMITATIONS IN 1'-0" (304.8mm) INCREMENTS		
12" (304.8mm)	20'-0" (6.096 M.)	TO	33'-0" (10.058 M.)
14" (355.6mm)	20'-0" (6.096 M.)	TO	35'-0" (10.668 M.)
16" (406.4mm)	20'-0" (6.096 M.)	TO	37'-0" (11.278 M.)
18" (457.2mm)	25'-0" (7.620 M.)	TO	45'-0" (13.716 M.)

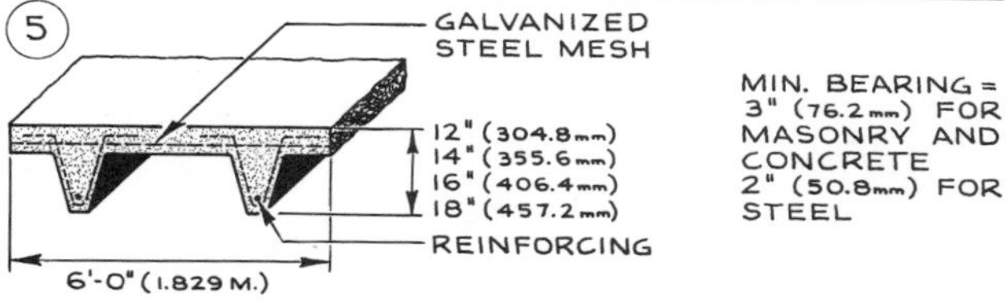

SIZE OF REINFORCING AND DEPTH OF DECKING CONTROLLED BY LOAD TO BE SUPPORTED AND SPAN

Application

Condensed Checklist

1. The locality should be checked to see whether concrete decking and structural units of the type desired are available.
2. Local, municipal, and state codes and also codes of the Fire Underwriters, insurance companies, labor departments, and federal government (Army, Navy, etc.) should be checked for structural limitations and fire resistance ratings.
3. The concrete structural beams, girders, joists, etc., must fit the structural, economic and design conditions of the building.
4. The type of decking selected for the design must meet load requirements and insulation codes and be suited to the type of finish ceiling desired for the buildings.
5. Methods of lighting, design of lighting fixtures, and installation of conduits must all be taken into account and provisions made for their installation.
6. Vapor barriers may be necessary on roofs.

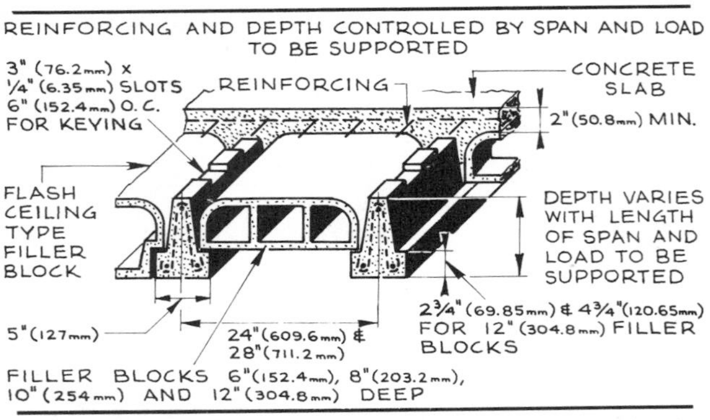

Figure C83 Typical examples of prefabricated concrete beams and filler blocks for concrete decking.

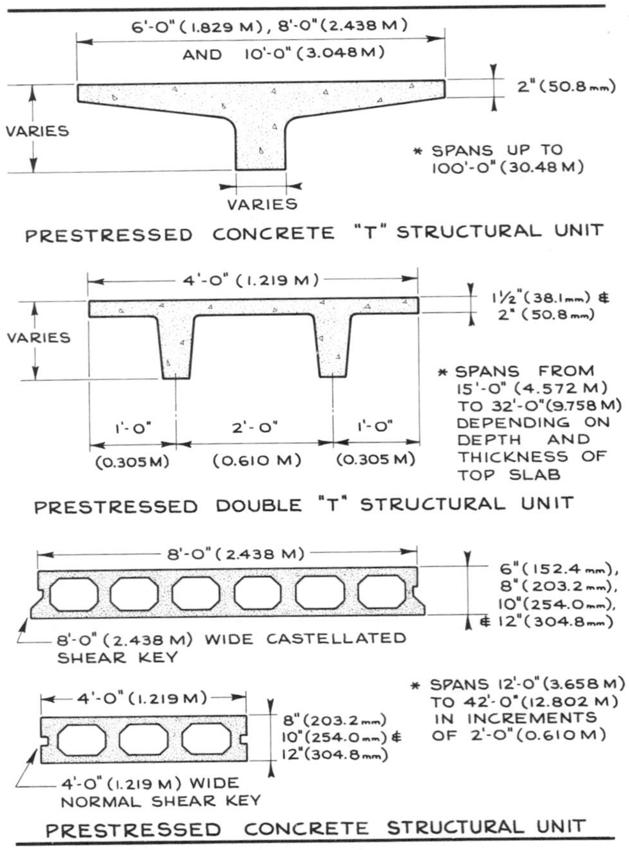

Figure C84 Typical structural units for floors and roofs.

7. The type and method of attachment must be checked (*see* Figure *C86*).
8. One should experiment with various spans and support spacings to arrive at an optimal spacing with the lightest possible supporting members (*see* Figure *C87*). Selecting the decking with the maximum possible span between supports does not always give the overall minimum tonnage of supporting steel. The range of support spacings for each type of decking in relation to the weight of supporting steel (both the total number and weight per support) should be calculated to find the best combination of span of decking and weight of steel. The optimal solution for a given distance to be spanned by the supports often falls between the minimum and maximum spacing of supports for the decking.
9. Openings for stairs, shafts, ducts, etc., in decking, all of which represent a design problem, must be correctly detailed.

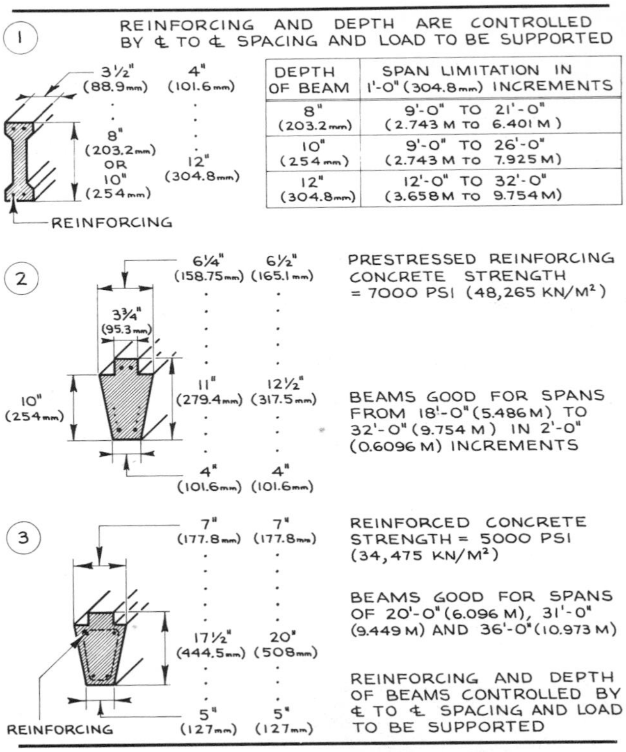

Figure C85 Precast beams for concrete decking (1), and precast beams with top projection (2, 3).

Conditions Favorable to the Use of Concrete Decking and Structural Members

1. Where fire-resistant or fireproof construction is required.
2. Where concrete decking and structural members are readily available in the locality.

Conditions Unfavorable to the Use of Concrete Decking and Structural Members

1. Where the design is multistoried and wind bracing is important, as this system of floor and roof construction gives practically no diagonal bracing.

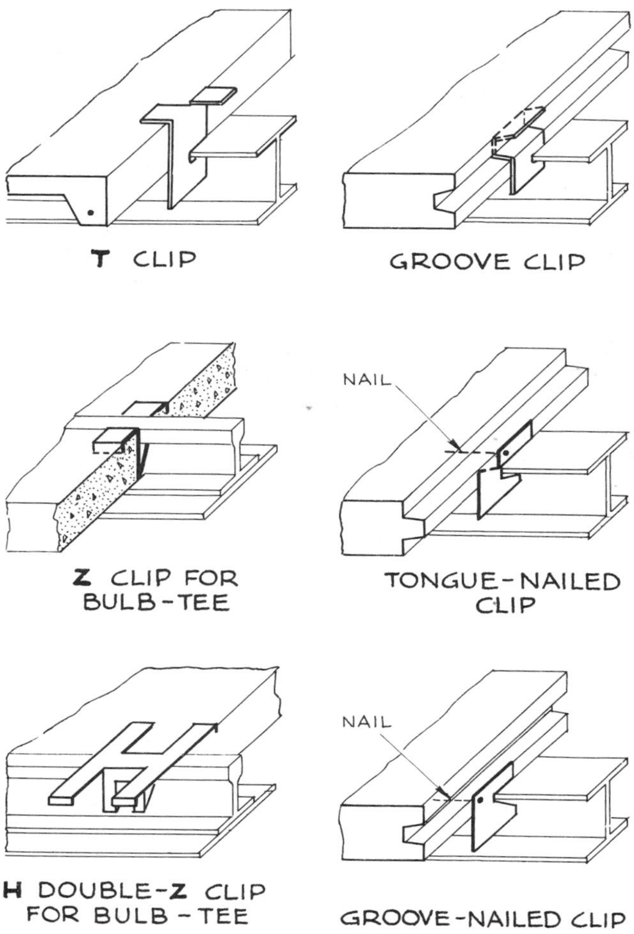

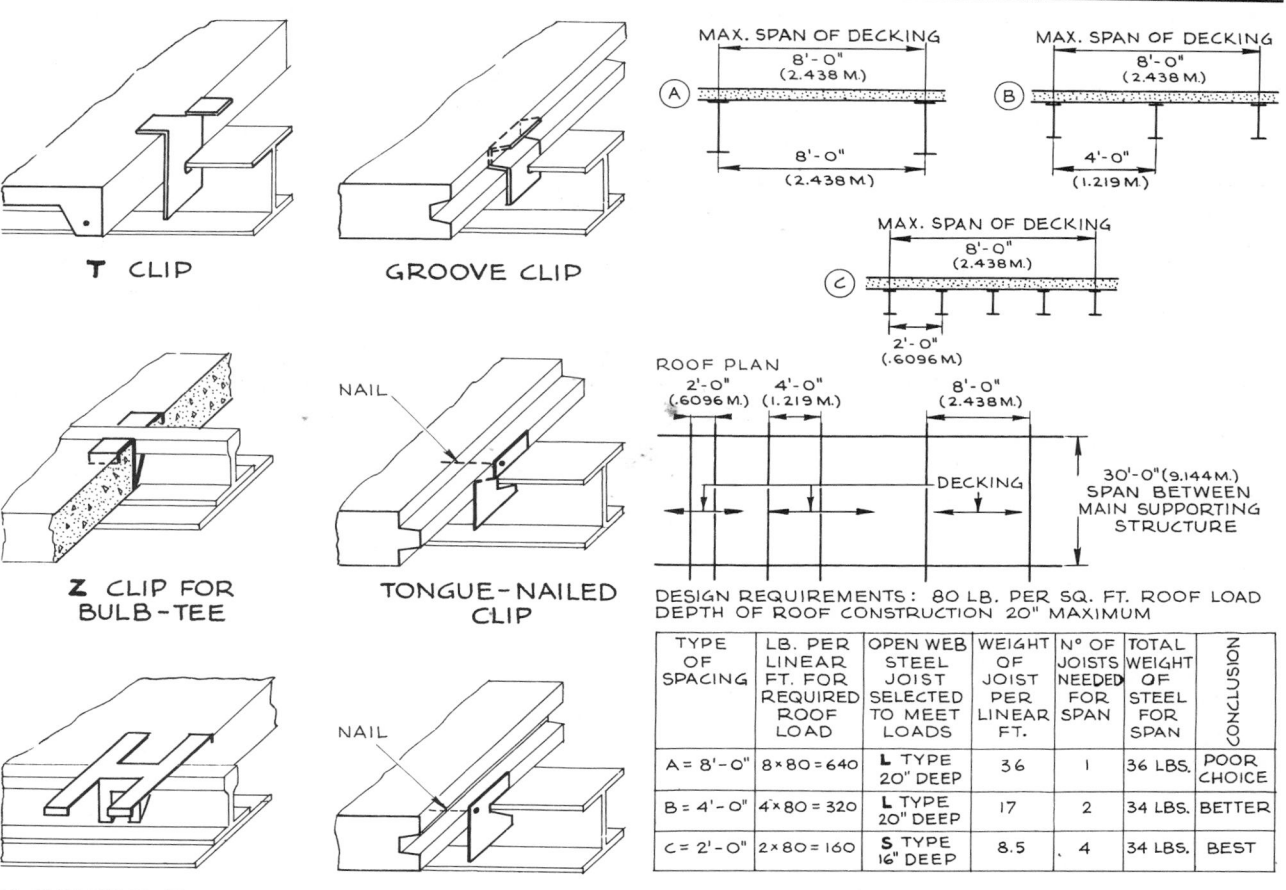

Figure C86 Typical methods of attachment for concrete decking.

2. For buildings in which large numbers of chases, shafts, ducts, and piping must pierce the floor or roof.

History and Manufacture

Prefabricated reinforced concrete was first used for structural members or decking soon after the discovery of reinforced concrete in 1868 and prestressed concrete in 1888, but it was not widely accepted as a material for construction or used to any extent until after World War I. From then on, the use of concrete decking and structural members steadily increased, particularly during and after World War II. Today, because of their light weight, ease of installation, and control in manufacture, they represent a facet of concrete construction throughout the world.

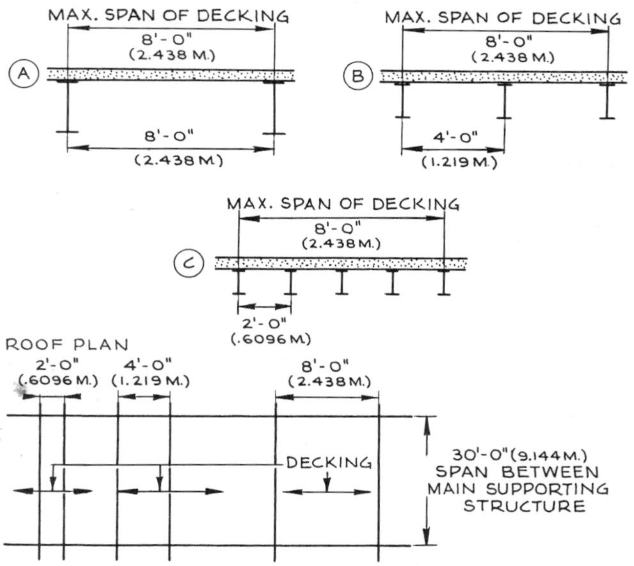

DESIGN REQUIREMENTS: 80 LB. PER SQ. FT. ROOF LOAD
DEPTH OF ROOF CONSTRUCTION 20" MAXIMUM

TYPE OF SPACING	LB. PER LINEAR FT. FOR REQUIRED ROOF LOAD	OPEN WEB STEEL JOIST SELECTED TO MEET LOADS	WEIGHT OF JOIST PER LINEAR FT.	N° OF JOISTS NEEDED FOR SPAN	TOTAL WEIGHT OF STEEL FOR SPAN	CONCLUSION
A = 8'-0"	8 × 80 = 640	L TYPE 20" DEEP	36	1	36 LBS.	POOR CHOICE
B = 4'-0"	4 × 80 = 320	L TYPE 20" DEEP	17	2	34 LBS.	BETTER
C = 2'-0"	2 × 80 = 160	S TYPE 16" DEEP	8.5	4	34 LBS.	BEST

DESIGN REQUIREMENTS: 390.6 KGS PER M² ROOF LOAD
DEPTH OF ROOF CONSTRUCTION 508 mm MAXIMUM

TYPE OF SPACING	KGS. PER LINEAR M. FOR REQUIRED ROOF LOAD	OPEN WEB STEEL JOIST SELECTED TO MEET LOADS	WEIGHT OF JOIST PER LINEAR METER	N° OF JOISTS NEEDED FOR SPAN	TOTAL WEIGHT OF STEEL FOR SPAN	CONCLUSION
A = 2.438 M.	2.438 M. x 390.56 KG. = 952.19 KG.	L TYPE 508 mm DEEP	53.58 KG.	1	53.58 KG.	POOR CHOICE
B = 1.219 M.	1.219 M. x 390.56 KG. = 476.1 KG.	L TYPE 508 mm DEEP	25.30 KG.	2	50.6 KG.	GOOD CHOICE
C = 0.610 M.	0.610 M. x 390.56 KG. = 238.2 KG.	S TYPE 406.4 mm DEEP	12.65 KG	4	50.6 KG.	BEST CHOICE

Figure C87 Relation of span to supporting members.

CONCRETE PIPE

Physical and Chemical Properties

Concrete pipe is available in bell and spigot, either nonreinforced or reinforced, and in lengths varying from 4 to 6 ft (1.22 to 1.83 m), and in diameters of 4 to 16 in. (101.6 to 406.4 mm) for unreinforced pipe, and 12 to

144 in. (0.3048 to 3.6576 m) for reinforced. Reinforced concrete pipe is available with lift eyes or holes.

All joining is done either with prefabricated gaskets, mortar, or asphaltic cement. The cement used is a portland blast-furnace slag or portland pozzolan cement; in no case should the proportion of portland cement in the concrete be less than 564 lb/yd^3 (33 kg/m^3) of concrete.

Both types of concrete pipe and fittings are cured by one of three methods: steam curing, water curing, and membrane curing, as long as the specified strength requirements at 28 days or less are developed.

All types and classes of concrete pipe are subject to tests of strength, permeability, absorption, and hydrostatic properties (*see* Figure *C88*).

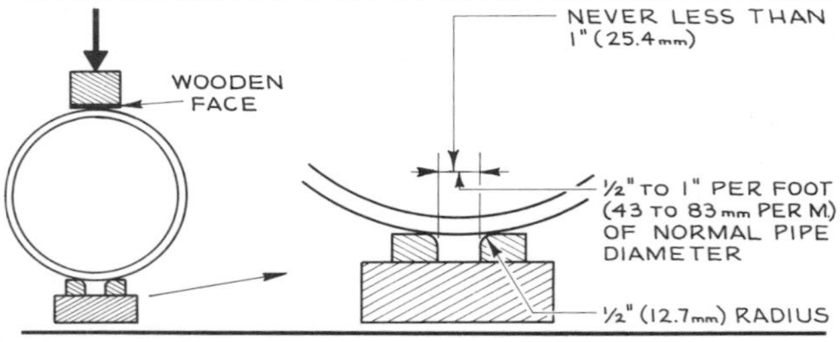

Figure C88 Three-edge bearing method for testing concrete pipe.

Types and Uses

Nonreinforced concrete pipe must meet strict physical and dimensional requirements and tolerances (*see* Table *C78*). It is manufactured in three classes, and each class meets strict physical and dimensional requirements (*see* Table *C79*). Special shapes and fittings such as Y's, T's, L's, and adapters are manufactured in the three classes and meet the physical and dimensional requirements of these classes. All nonreinforced pipe must have the following identification clearly marked on each piece of pipe or fitting: class, date of manufacture, name or trademark of manufacturer, and plant identification.

Reinforced concrete pipe meets not only the strict physical and dimensional requirements already mentioned, but also concrete strength, circular reinforcement, and elliptical reinforcement requirements. It is manufactured in five classes, and each class must meet specifically designated strict physical and dimensional requirements. There are also strict requirements for the reinforcing steel, depending on whether it is for one line of circular reinforcing, for two lines of circular reinforcing, or for elliptical reinforcing. These include location, size, joining, and lapping. All reinforced concrete pipe must have the following identification clearly

Table C78 Tolerances for concrete pipe

Type of concrete pipe	Internal diameter	Wall thickness	Length	Length of two opposite sides	Straightness
		Dimensions shall not vary more than			
Nonreinforced pipe	±1% + $\frac{1}{8}$ in. (3.2 mm) to the nearest $\frac{1}{8}$ in. from design diameter	−1.5% or $\frac{1}{8}$ in. (3.2 mm), whichever is smaller from design diameter	−$\frac{1}{2}$ in. (−12.7 mm) from design length	$\frac{1}{4}$ in. (6.4 mm) or 2%, whichever is larger	$\frac{1}{8}$ in./ft (0.1 mm/mm) of length in alignment
Reinforced pipe[a]	12–24 in. (304.8–609.6 mm), ±1.5% from design diameter; 27 in. (685.8 mm) and larger, ±1% or $\frac{3}{8}$ in. (9.6 mm), whichever is greater, from design diameter	5% or $\frac{3}{16}$ in. (4.8 mm), whichever is greater	$\frac{1}{8}$ in./ft (0.01 mm/mm), with a maximum of $\frac{1}{2}$ in. (13 mm) in any length of pipe	$\frac{1}{8}$ in./ft (0.01 mm/mm), with a maximum of $\frac{5}{8}$ in. (16 mm) in any length of pipe except where beveled, and pipe for laying on curves is specified	

[a]Variation in position of reinforcement shall be ±10% of wall thickness or ±$\frac{1}{2}$ in. (13 mm), whichever is greater.

Table C79 Physical and dimensional requirements[a] for nonreinforced concrete pipe

Internal diameter of pipe		Minimum thickness of wall		Minimum strength three-edge bearing		Minimum thickness of wall		Minimum strength three-edge bearing		Minimum thickness of wall		Minimum strength three-edge bearing	
in.	mm	in.	mm	lb/linear ft	kN/linear m	in.	mm	lb/linear ft	kN/linear m	in.	mm	lb/linear ft	kN/linear m
4	100	$\frac{5}{8}$	15.9			$\frac{3}{4}$	29.0			$\frac{3}{4}$	19.0		
6	150			1500	21.9			2000	29.2	$\frac{7}{8}$	22.2	2400	35.0
8	200	$\frac{3}{4}$	19.0			$\frac{7}{8}$	22.2			$1\frac{1}{8}$	28.6		
10	250	$\frac{7}{8}$	22.2	1600	23.3	1	25.4			$1\frac{1}{4}$	31.8		
12	310	1	25.4	1800	26.3	$1\frac{3}{8}$	34.9	2250	32.8	$1\frac{3}{4}$	44.5	2600	37.9
15	380	$1\frac{1}{4}$	31.8	2000	29.2	$1\frac{5}{8}$	41.3	2600	37.9	$1\frac{7}{8}$	47.6	2900	42.2
18	460	$1\frac{1}{2}$	38.1	2200	32.1	2	50.8	3000	43.8	$2\frac{1}{4}$	57.2	3300	48.1
21	530	$1\frac{3}{4}$	44.5	2400	35.0	$2\frac{1}{4}$	57.2	3300	48.1	$2\frac{3}{4}$	69.9	3950	56.2
24	610	$2\frac{1}{8}$	54.0	2600	37.9	3	76.2	3600	52.5	$3\frac{3}{4}$	95.3	4400	64.2
27	690	$3\frac{1}{4}$	82.6	2800	40.9	4	102.0	3950	57.7	4	102.0	4600	67.2
30	760	$3\frac{1}{2}$	88.9	3000	43.8	$4\frac{1}{4}$	108.0	4300	62.8	$4\frac{1}{4}$	108.0	4750	69.4
33	840	$3\frac{3}{4}$	85.2	3150	46.0	$4\frac{1}{2}$	114.0	4400	64.2	$4\frac{1}{2}$	114.0	4875	71.2
36	910	4	102.0	3300	48.1	$4\frac{3}{4}$	121.0	4500	65.7	$4\frac{3}{4}$	121.0	5000	73.0

[a] *See* Table C78 for tolerances.

marked on each piece of pipe: class, date of manufacture, name or trademark of manufacturer; and on one end of each pipe there must be marked the elliptical or quadrant reinforcement on the inside and outside of opposite walls, along the minor axis of the ellliptical reinforcing or along the vertical axis for quadrant reinforcing.

In the construction field, concrete pipe is used for sewers, storm drains, combination sewers, culverts, industrial wastes, and diverting small streams and brooks.

Application

Condensed Checklist

1. Check the type of soil, the weight of soil, installation methods, and bedding requirements.
2. Check with manufacturers for their limitations on the sizes and lengths of pipe and fittings that they produce.
3. For industrial wastes, always check the type of wastes and their possible effects on concrete pipe.
4. Check with municipal, state, and local codes and health departments for their restrictions and requirements for the installation of concrete pipe.
5. Check the type of joints to be used for the installation of concrete piping.

Conditions Favorable to the Use of Concrete Pipe

1. For sewers, storm drains, culverts, combination sewers, and industrial wastes.
2. For diverting small streams and brooks.

Conditions Unfavorable to the Use of Concrete Pipe

1. For industrial wastes that attack and break down concrete.

History and Manufacture

Concrete pipe is the outgrowth of the development of reinforced concrete. The real growth in the manufacture of nonreinforced and reinforced concrete pipe occurred after 1945, and its development and manufacture have continued since this date.

CONCRETE, PRECAST (ARTIFICIAL OR CAST STONE)

Physical and Chemical Properties

Precast concrete, sometimes called cast stone or artifical stone, is a masonry material made of reinforced concrete whose facing surface has a decorative value. It may be either smooth, polished, or textured; and colored or uncolored. It may incorporate granite, quartz, and vitreous ceramic material in aggregate or chip form; and it may have a slightly exposed aggregate or a largely exposed aggregate texture, depending on the aggregate size, shape, and color. Precast concrete is available in stock sizes (within certain limitations in square-foot area). Reinforcing, molds, anchor bolts, inserts, slots for anchoring, holes, etc., have been carefully integrated so that the units can be mass-produced economically. Because this material is a manufactured product, special sizes and shapes can be made and the characteristics can be controlled by the choice of material. For example, units can be made of lightweight insulating concrete, or with built-in insulation, or with a backing of rigid insulation. Precast concrete walls are available in an unlimited variation of sizes, shapes, textures, and design, depending on what one wishes to select for the exterior design of the building (*see* Figure *C89*). The plain or colored concrete type is available in many special shapes. (*See* Table *C80*.)

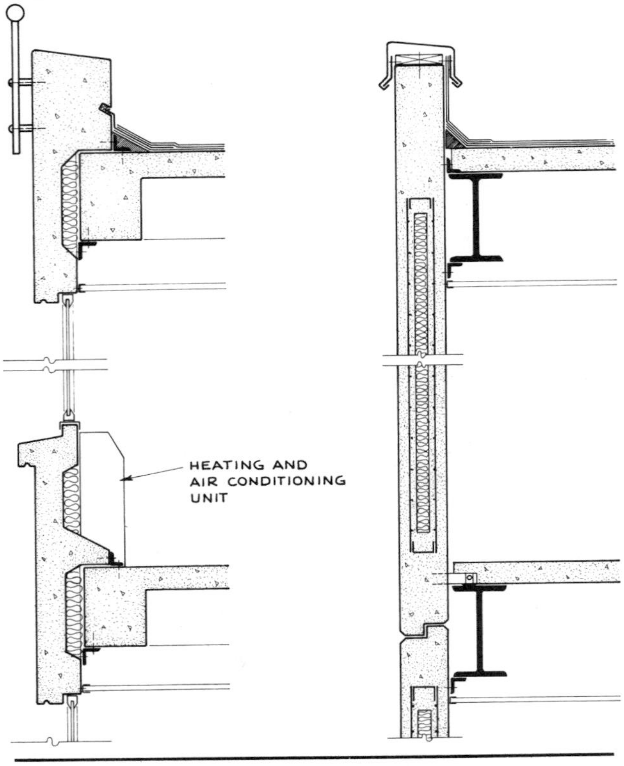

Figure C89 Typical precast concrete curtain wall.

Types and Uses

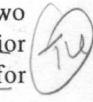

Precast concrete products may be divided into two general categories: (1) facing for exterior or interior walls of buildings, and (2) miscellaneous items for masonry work such as stair treads and copings.

Facing. The precast concrete used for facing buildings usually has a textured or polished surface made with granite, quartz, and vitreous ceramic materials of a specific color selected for the job. It may have various types of exposed aggregate textures and colors. This facing can be anchored both to existing structures and to new structures, or it may be used as forms for either concrete or reinforced concrete construction, as shown in Figure *C90*. There is almost no limit to possible methods of attachment for artificial stone because the manufacturing process permits any of the various devices to be installed before the artificial stone is cast and finally processed.

The facing units can be made with lightweight insulating concrete, built-in rigid insulation, rigid insulating back-up, or reinforced concrete and prestressed concrete, when precast concrete is used structurally (*see* Figure *C91*).

Table C80 Typical sizes of precast concrete facings

Type of finish	Standard thickness		Maximum size limitations: Width		Length		Face area		Minimum compressive strength[a]		Color
	in.	mm	ft	m	ft	m	ft^2	m^2	$lbf/in.^2$	MN/m^2	
Smooth, polished	$2\frac{1}{2}$	63.5	4	1.219	5	1.624	12–15	1.12–1.40	7500	51.71	Gray to black
	$1\frac{1}{2}$	38.1	4	1.219	4	1.219	16	1.49			
Textured, incised or flush	2	50.8	4	1.219	5	1.624	60	5.58	7500	51.71	Gray to black

[a] $1\ lbf/in.^2 = 0.006895\ MN/m^2$

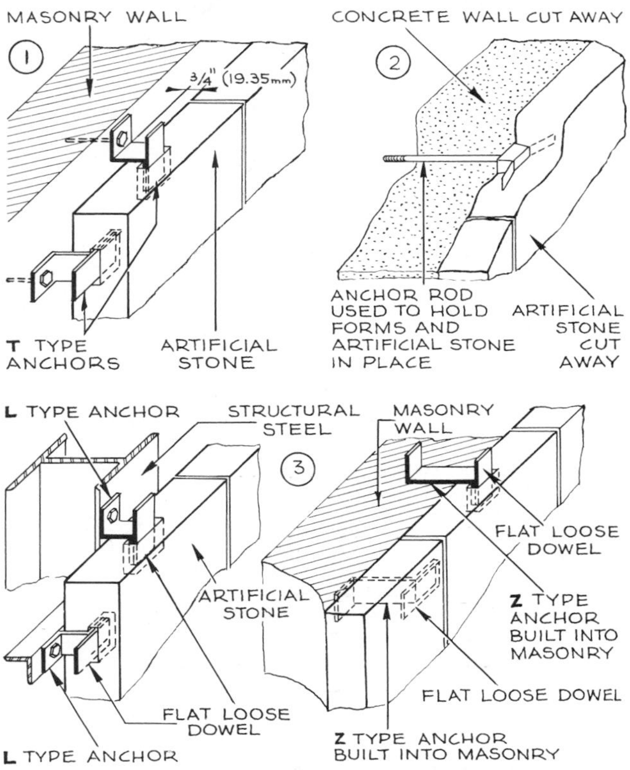

Figure C90 Attachment of precast concrete facing, showing details for existing structures (1), new structures (2), and use as forms for concrete work (3).

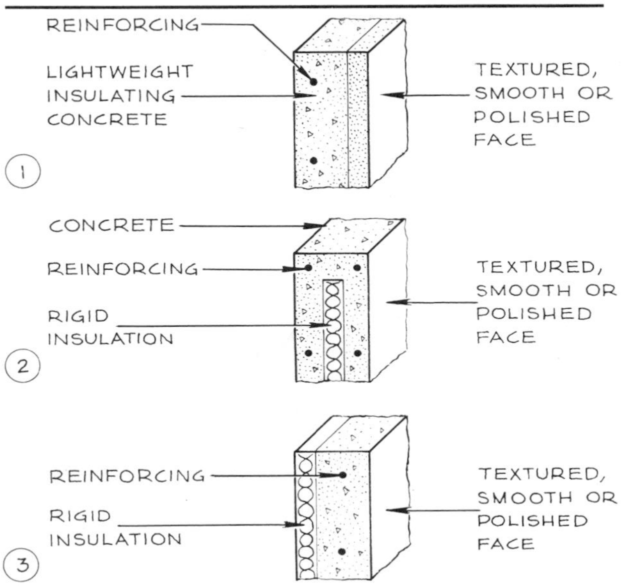

Figure C91 Precast concrete facing with insulation provided by insulating-type concrete (1), built in (2) and as back-up panel (3).

Miscellaneous Items. The miscellaneous artifical stone items used for masonry work are generally made of plain or colored concrete, but where used in connection with the facing material, they can be made to match the color and texture of the facing (*see* Figure *C92*). The items are manufactured to exact sizes and dimensions set by the architectural working drawings and the shop drawings which always accompany them.

For lintels, information regarding the sizes of reinforcing and type of concrete is supplied to the manufacturer by the structural engineer.

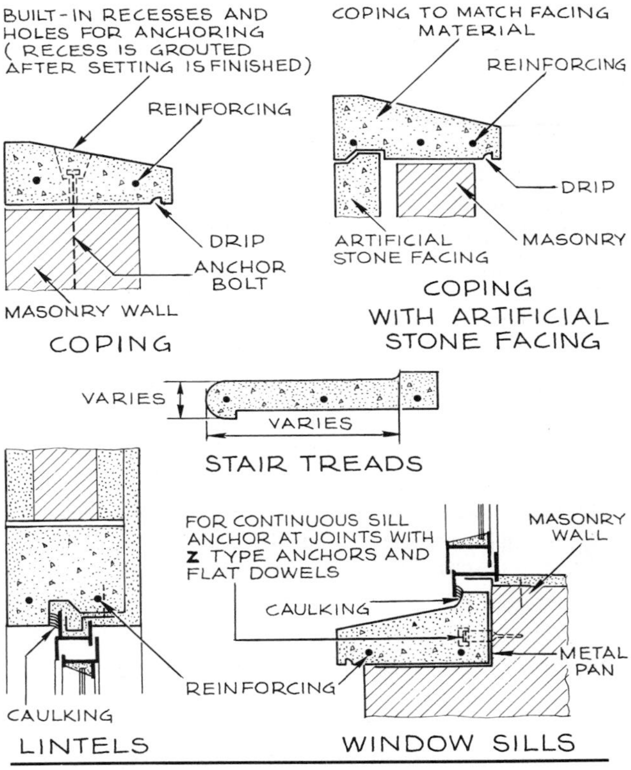

Figure C92 Typical precast concrete products for copings, stair treads, lintels, and window sills.

Application

Condensed Checklist

1. The locality should be checked to see whether precast concrete facing or a similar product is available.
2. Local, municipal, and state codes and also codes of the Fire Underwriters, insurance companies, labor departments, and federal government (Army, Navy, etc.) should be checked for structural requirements and for fire resistance ratings.

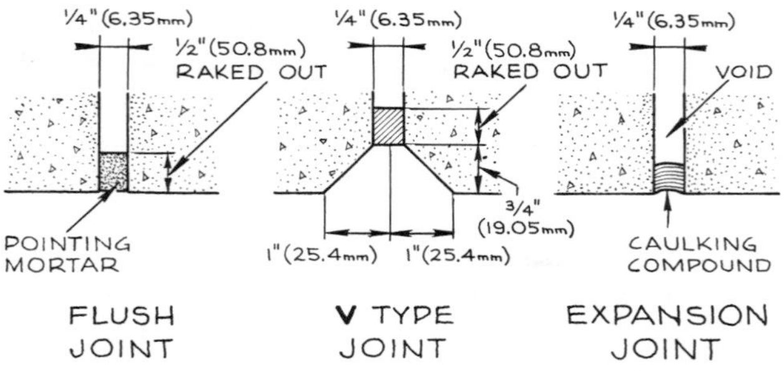

Figure C93 Details of joints for precast concrete, showing flush (1), V-type (2), and expansion (3) joints.

3. In the construction field it is always required that shop drawings be made and samples be submitted. Final material as delivered must match the sample.
4. All dimensions, joints (both vertical and horizontal), and conditions at openings must be worked out to fit the design.
5. Type and treatment of joints and location of expansion joints must be detailed (*see* Figure *C93*).
6. Pointing mortar should consist of 1 part white portland cement, $\frac{1}{4}$ part hydrated lime, and 3 parts fine sand.
7. Treatment of all external corners must be carefully considered. Typical detailing methods are shown in Figure *C94*.
8. When precast concrete is used as forms for concrete, it must be correctly anchored, all joints must be plugged with fillers, and where four corners meet, a temporary bolt must be installed to keep the pieces of facing accurately aligned.
9. When precast concrete curtain walls or facing are used with steel construction, methods of anchoring must be correctly detailed and calculations for the facing must include wind load. Metals that will not cause galvanic corrosion must be used.
10. In freezing weather, precast concrete curtain walls or facing can be installed to steel or concrete, but pointing and grouting must not be done in such temperatures. Precast concrete used in conjunction with masonry or concrete must not be installed in freezing weather.
11. Precast concrete must always be cleaned only with detergents and water. Acid or alkaline solutions must never be used.
12. Manufacturers of precast concrete curtain walls and facings should be consulted regarding the best method of accomplishing design requirements, insulation, methods of securing to structure, and limitations on widths and lengths.

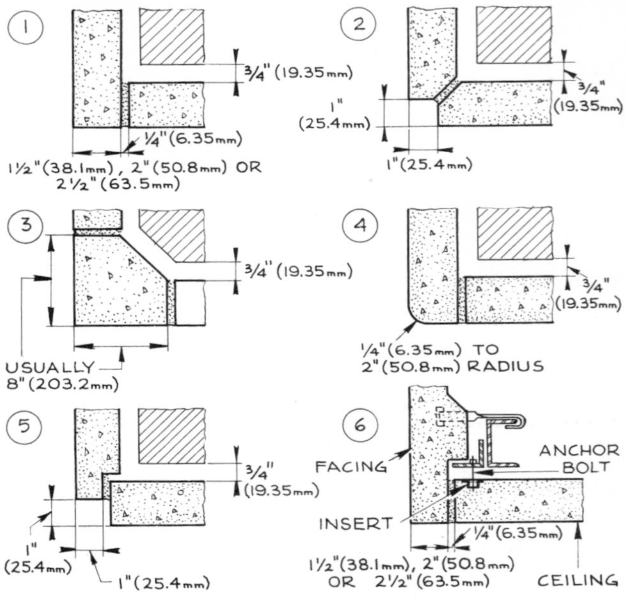

Figure C94 Details of external joints for precast concrete, showing thickness of facing (1), quirk joint (2), filler block (3), bullnose (4), offset (5), and opening with ceiling (6).

13. Samples of finishes, joints, and full-size units should be specified whenever large quantities of precast concrete panels are to be installed.
14. In specific cases, it should be specified that full-size mock-ups, including windows, gaskets, and joints, be tested under conditions similar to those of the area of installation, including wind, rain, freezing and thawing conditions, and temperature changes due to orientation.

Conditions Favorable to the Use of Precast Concrete

1. Where masonry materials cannot be obtained in the desired shape or in the lengths required to support the loads.
2. Where a durable, permanent material that has a wide range of textures and colors is needed.
3. Where a masonry material is needed that can be manufactured with built-in anchoring devices to meet flexible design and dimension requirements.
4. Where the design requires the precast concrete curtain wall or exterior wall facing to become a structural supporting member of the building.

Conditions Unfavorable to the Use of Precast Concrete

1. As a load-bearing facing material, except in cases where the structural design requires the precast concrete to become structural.

2. In areas that are remote from the manufacturing source.
3. Where prestressed types of concrete curtain wall or exterior facing of a building require structural forces that can be obtained only by using prestressed concrete.

History and Manufacture

The use of precast concrete parallels the history of cement brick. After reinforced concrete was invented in 1968, techniques for making precast concrete were developed. The early products were used for stair treads, coping, lintels, window sills, and areas where it was difficult to build with stone or to obtain long lengths of stone. Also, the natural cement was similar in color and texture to many natural stones. Later it was used for casting sculpture and ornament for the exterior. After World War I, when nonfading limeproof colors were developed, it was used for large-scale decorative treatment on buildings. Just before World War II new surface effects were achieved by incorporating stone aggregates and chips of the desired color and texture and then grinding and polishing them to create the first real artificial stone, in contrast to the early versions, which were actually tamped cast concrete. Today almost any texture or color can be obtained by variation of the color, size, and projection of the aggregates.

Precast concrete is manufactured by first making an accurate mold of the shape to be cast, including anchor slots, anchors, inserts, and holes; reinforcing is also accurately placed in the mold. The correct proportions of cement, aggregates, pigment, and water are mixed and placed in the mold. If the surface is to be given a polished finish, the mold is not completely filled, and a mixture of cement and usually a mixture of granite aggregates, screened and proportioned by size and color, are used to completely fill up the mold. After the concrete has set, this surface is mechanically ground and polished.

When a textured surface is desired, the mold is filled to the exact height necessary to receive the type and size of surface aggregates desired, thus producing a finished surface that is either incised or roughly flush.

Exposed-type aggregate surfaces require that a concrete with retarders be used when aggregates are applied either at the bottom or at the top of the form. When concrete is partially set, the surface with the aggregates to be exposed is either brushed slightly or severely, or sand-blasted to obtain the required aggregate exposure.

CONCRETE, PRESTRESSED

The prestressing of reinforced concrete is a means whereby part of the tensile stress (bending) that develops when normal loads are applied can be counteracted by prestressing (stretching) the reinforcing steel within the concrete and thereby compressing the concrete. By this method the cross-sectional area, or size, of the reinforcing steel can be reduced, and the concrete members can become smaller, span greater distances between supports, and support greater loads.

Types and Uses

There are two methods for making prestressed concrete, namely, pretensioning and posttensioning.

Pretensioning. In the method called pretensioning, the reinforcing steel is first prestressed and then the concrete is poured. When the concrete has developed strength, the stress in the steel is released, and the bond between the steel and the concrete causes the concrete to be compressed.

Posttensioning. In the posttensioning method tubes, conduits, or channels are inserted in the concrete where reinforcing steel is required. After the concrete is adequately cured, steel reinforcement is inserted in the tubes or channels, stretched to the proper tension, and anchored at the ends. The steel remains in tension, and the concrete is in compression. In construction, combinations of posttensioned and pretensioned members are used extensively, as are combinations with poured-in-place reinforced concrete. Posttensioning and pretensioning are both commonly used.

Structural Forms. Prestressed concrete design may be applied to almost all types of standard reinforced concrete construction. Structures commonly made of prestressed concrete are beams, slabs, columns, decking, panels, curtain walls, building facings, structural roofing units, and girders, all of which are prefabricated (*see* Figures *C95* and *C96*). Prestressed concrete is also used for circular concrete tanks and pipe, usually made with a winding tendon (*see* Concrete Pipe; Concrete, Reinforced; Steel for Concrete; Wire).

Compressive Strength Requirements. Concrete for prestressed concrete must develop compressive strength usually greater than that of AA-type concrete, which has a strength of 3750 lbf/in.^{2} (25.8 MN/m^2) in 28 days.

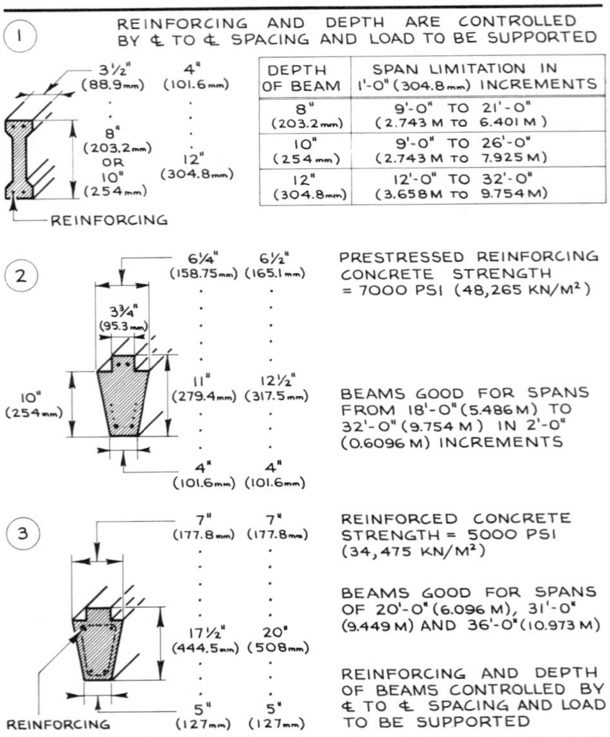

DEPTH OF BEAM	SPAN LIMITATION IN 1'-0" (304.8mm) INCREMENTS
8" (203.2mm)	9'-0" TO 21'-0" (2.743 M TO 6.401 M)
10" (254mm)	9'-0" TO 26'-0" (2.743 M TO 7.925 M)
12" (304.8mm)	12'-0" TO 32'-0" (3.658 M TO 9.754 M)

Figure C95 Typical example of prefabricated prestressed concrete beam for roof or floor construction.

Reinforcing. The reinforcing is usually wire, strand, bar, or rope made of heat-treated steel. In decking, lightweight prefabricated beams, girders, etc., a high-strength, stress-relieved wire is sometimes used (*see* Figures *C97* and *C98*).

Wire rope is of the seven-wire type, stress-relieved by continuous heat treatment. Table *C81* gives the mechanical requirements for wire rope used for prestressed reinforced concrete. (See page 240.)

The designing of prestressed concrete for structures is highly technical and should always be done by a structural engineer, even when using prefabricated prestressed concrete units.

Application

Condensed Checklist

1. For prefabricated units, data should be obtained on the type of concrete, the type of reinforcing steel, the method of prestressing, the standards met, and the tests made.
2. Span load limitations and the method of securing prefabricated units should always be checked.

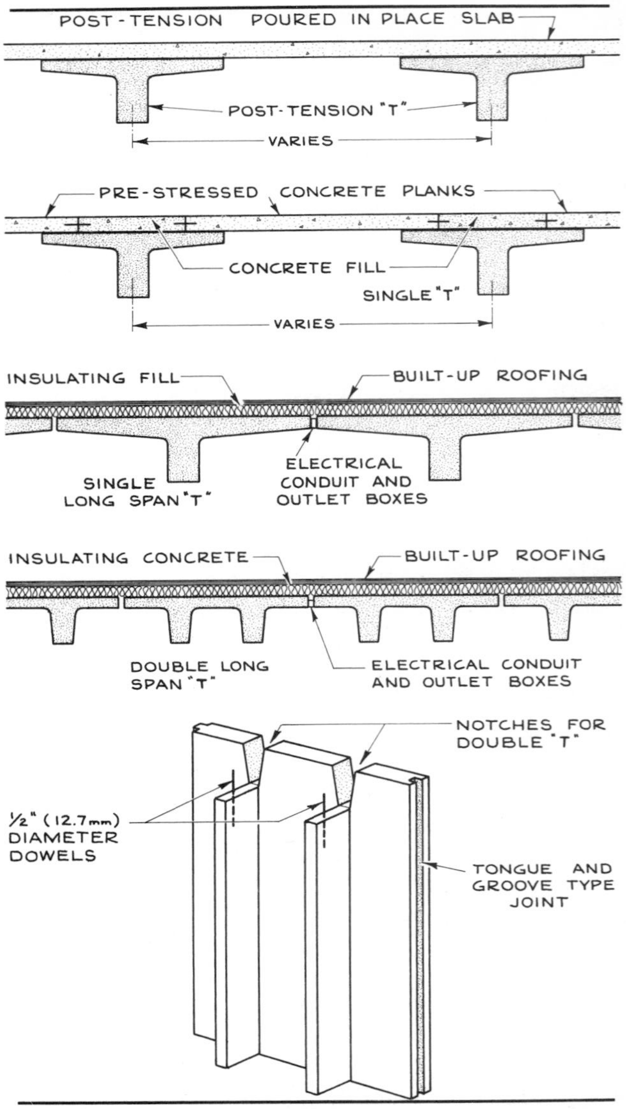

Figure C96 Examples of various prestressed concrete construction units.

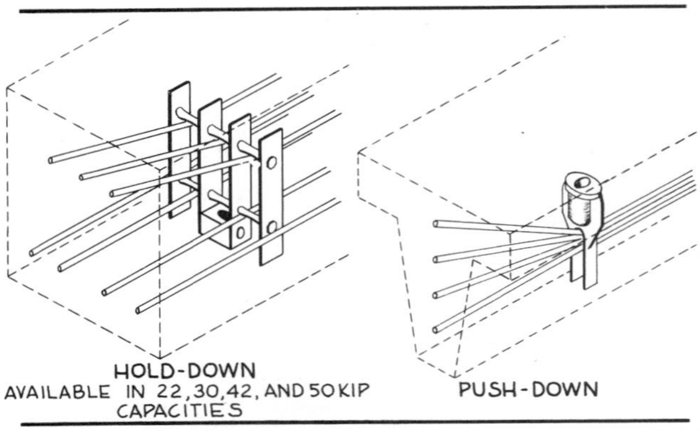

Figure C97 Hold-down and push-up form ties for prestressed concrete.

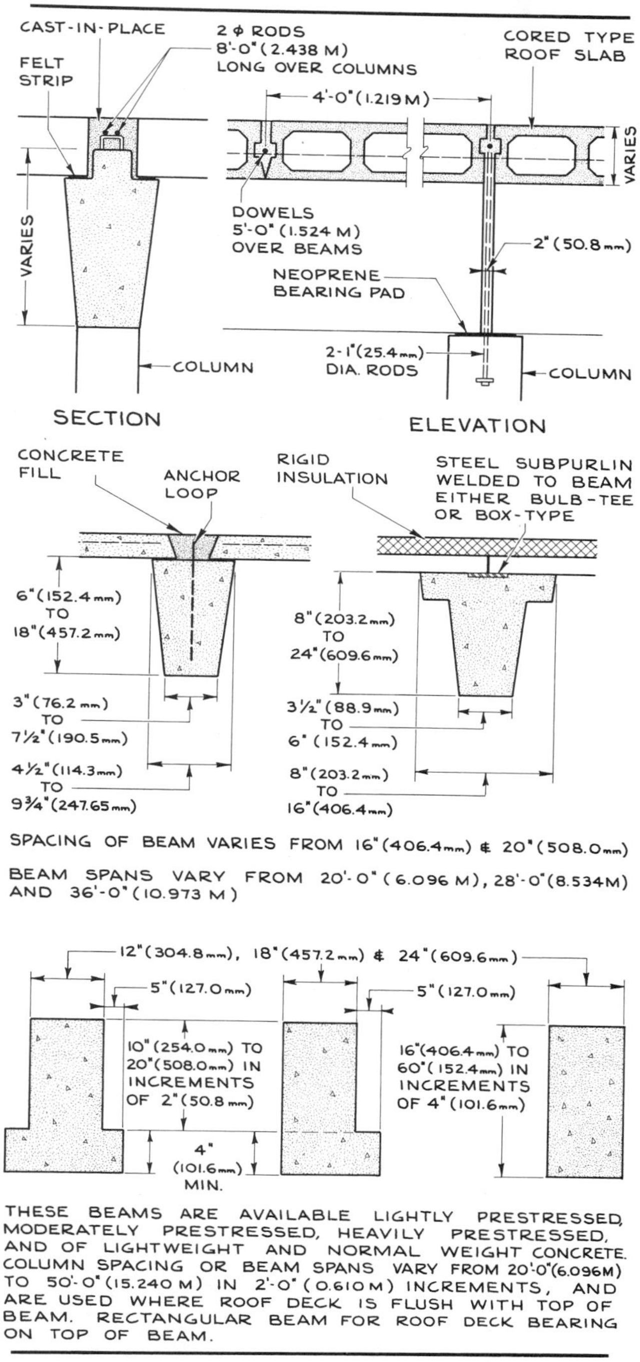

Figure C98 Types of prestressed concrete beams with various types of flooring and decking.

3. Limitations and fire resistance requirements of local, municipal, and state codes, and also codes of the Fire Underwriters, insurance companies, labor departments, and federal government (Army, Navy, etc.) should be checked.
4. The locality should always be checked for the availability of knowledge, equipment, and experienced contractors for prestressed work. One should specify all necessary controls, tests, and requirements.
5. One should always consult with structural engineers when planning to use prestressed or poststressed concrete or combinations of the two, or combinations with reinforced concrete.
6. Shop drawings and samples of color, texture, and patterns should always be specified to be submitted. It is suggested that a full-size mock-up also be built for approval.
7. Manufacturers should always be consulted for methods of attachment, joining, and structural characteristics necessary to meet the building's exterior design factors.

Conditions Favorable to the Use of Prestressed Concrete

1. Where spans and loads cannot be adequately designed in reinforced concrete.
2. For decking, beams, girders, and other prefabricated units where greater spans and loads with thinner, stronger, and in some cases lighter members are required.
3. For construction where fire resistance is an important factor.
4. For simple, easily erected exterior facings for buildings.
5. Where large, clear spans are required for wall-bearing construction and where steel or concrete column systems are contemplated.
6. Where the exterior facing of a structure is not only to act as exterior facing but also to be part of the structural system.

Conditions Unfavorable to the Use of Prestressed Concrete

1. Where steel, other metals, or other materials will do the job more simply, quickly, and economically.
2. Where loads are light and spans are short.

History and Manufacture

Although the early experiments were made in the United States and the first patent for prestressed concrete was issued in 1888 to P. H. Jackson of San Francisco, it was in Europe that this type of construction was developed. French, German, and Danish engineers tried various ways of prestressing concrete, but none was successful until the 1930s, when, in France, Freyssinet used high-

Table C81 Requirements for wire rope used for prestressed concrete

Nominal dimensions							
Diameter of wire rope		Area of steel in wire rope		Weight of wire rope		Breaking strength of wire rope	
in.	mm	in.^{2}	mm^2	lb/1000 ft	kg/304.8 m	lb	kg
$\frac{1}{4}$	6.35	0.036	23.23	122	55.34	9,000	4 082.40
$\frac{5}{16}$	7.94	0.058	37.42	198	89.81	14,500	6 577.20
$\frac{3}{8}$	9.53	0.080	51.62	274	124.29	20,000	9 072.00
$\frac{7}{16}$	11.11	0.109	70.33	373	169.19	27,000	12 247.20
$\frac{1}{2}$	12.70	0.144	92.91	494	224.08	36,000	16 329.60

strength steel wire. After that, many systems of prestressed concrete construction were developed and used extensively throughout Europe, especially during and after World War II. This stimulated interest in prestressed concrete in the United States, but it was not until 1950 that the first prestressed concrete bridge was completed in this country. Since then, prestressed concrete has found its major use here in prefabricated materials where it can be shop-controlled and tested. As experimentation continued and knowledge of this material grew, prestressed concrete began to be used at the construction site. Lift slab construction with precast or cast-on-the-job columns is an example. Prestressed concrete in all its various forms and combinations is an accepted material in the entire field of construction.

CONCRETE, REINFORCED

Concrete is very strong in compressive strength but weak in tensile strength, whereas steel is very strong in tensile strength and relatively weak in compressive strength. Reinforced concrete is concrete combined with steel in such a way that the compressive strength of the concrete and the tensile strength of the steel are used to best advantage.

Reinforced concrete work is in reality a manufacturing process carried on at the site of construction, in a plant, or both. Its characteristics vary with those of the concrete and the reinforcing. Prestressing the concrete permits an even wider range of structural characteristics.

(*See also* Concrete; Concrete, Prestressed; Mesh and Wire Cloth; Steel for Concrete; Steel Mesh and Wire Cloth.)

Types and Uses

Reinforced concrete is very widely used in the construction field for foundations, structural framing, floors, roofs and walls, various precast beams and decking, precast concrete, decoration, sculpture, walks, roads, etc. It is used as a fire retardant for steel and in combination with steel for floors. Reinforced concrete can be molded into almost any form or shape desired and is particularly adaptable to thin shell construction, plastic shapes, and prestressed concrete construction in general.

Lift Slab System. The lift slab system requires special equipment and expertise and is an accepted method of concrete construction (*see* Figure *C99*). The slabs are

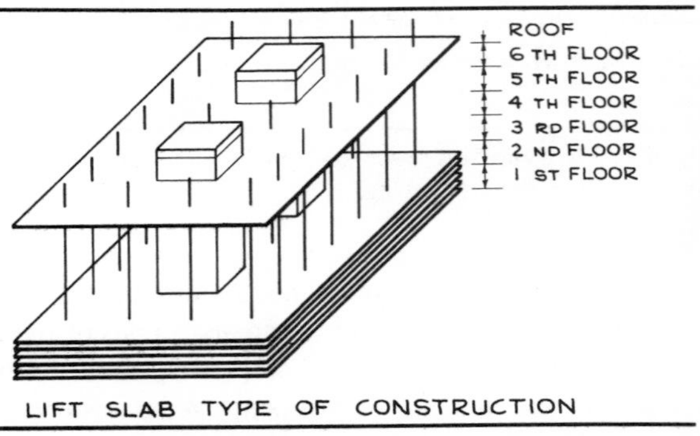

Figure C99 Lift-slab concrete construction system.

Table C81 Requirements for wire rope used for prestressed concrete (*continued*)

Yield strength				Differences in diameter of center wire and outside wires		Length per reel of wire rope	
Initial load		Minimum load at 1% extension					
lb	kg	lb	kg	in.	mm	ft	m
900	408.24	7,650	3 470.14	0.001	0.0254	25,000	7620
1450	657.72	12,300	5 579.28	0.0015	0.0381	15,000	4572
2000	907.20	17,000	7 711.20	0.002	0.0508	15,000 or 10,000	4572 or 3048
2700	1224.72	23,000	10 432.80	0.0025	0.0635	12,000, 10,000, or 8,000	3657.6, 3048, or 2438.4
3600	1632.96	30,600	13 880.76	0.003	0.0762	9,000 or 6000	2738.2 or 1828.80

cast on grade, one on top of the other, and then are raised to the various floor heights by jacks. The slabs can be prestressed or reinforced flat slabs. The environmental systems become more economical because all the electric conduits, chases and shafts for plumbing, and heating and air-conditioning requirements are installed on grade, thereby eliminating the usual floor-by-floor installation and coordination. All the roughing for plumbing, both vertical and horizontal ducts, and vertical electric conduits and panel boxes can be prefabricated and then installed when the floor slabs are hoisted into their permanent position.

One of its most advantageous design factors is that, within a given thickness, beams can project in two directions as cantilevers without the need for one beam to be below the other, as is the case with wood or steel.

Tilt-up System. Another accepted concrete construction method is the tilt-up system. There are two basic methods: load-bearing and non-load-bearing tilt-up systems. Generally, the tilt-up panels are cast on the building floor slab, which must be level and smoothly troweled. The floor slab should be treated with a bond-breaking agent where tilt-up panels are to be formed, so that there is no mechanical bond and there can be a trouble-free lift from the floor slab. Figures *C100* and *C101* show various methods of joining tilt-up panels and assembling them to a structural system or to a cast-in-place column, which can be designed for load-bearing and non-load-bearing tilt-up panels. Tilt-up panels are reinforced with bars or wire mesh and can be prestressed. The exterior surface finish can be smooth, textured, exposed aggregate, or decorative.

Depending on the height of the tilt-up panels, the location and number of cast-in-place pick-up points for lifting should be specified by a structural engineer. In addition, all reinforcement, methods of attachment, and methods of attaching other materials to the load-bearing or non-load-bearing tilt-up panels and structural systems should be checked by the structural engineer. (*See* Figure *C102*.)

In designs where very deep beams are required, the slab may be made the same thickness as the deep beams and made lightweight by installing large paper tubes. This method allows for a completely smooth, flat, and level ceiling.

Figure *C103* illustrates the various types of reinforced concrete and indicates the direction of the reinforcing by arrows.

The designing of concrete structures is beyond the scope of this book. The engineering principles involved and their practical application require close collaboration with the structural engineer. However, a thorough general knowledge of the advantages, disadvantages, flexibility, and limitations of this type of construction is a necessity.

Tests. For structures, the type of cement and aggregates, the water-cement ratio, entrained air, the type of admixtures, the mixing and preparation, and the compressive strength developed in 7 and 28 days are all tested and must meet the requirements of the end use

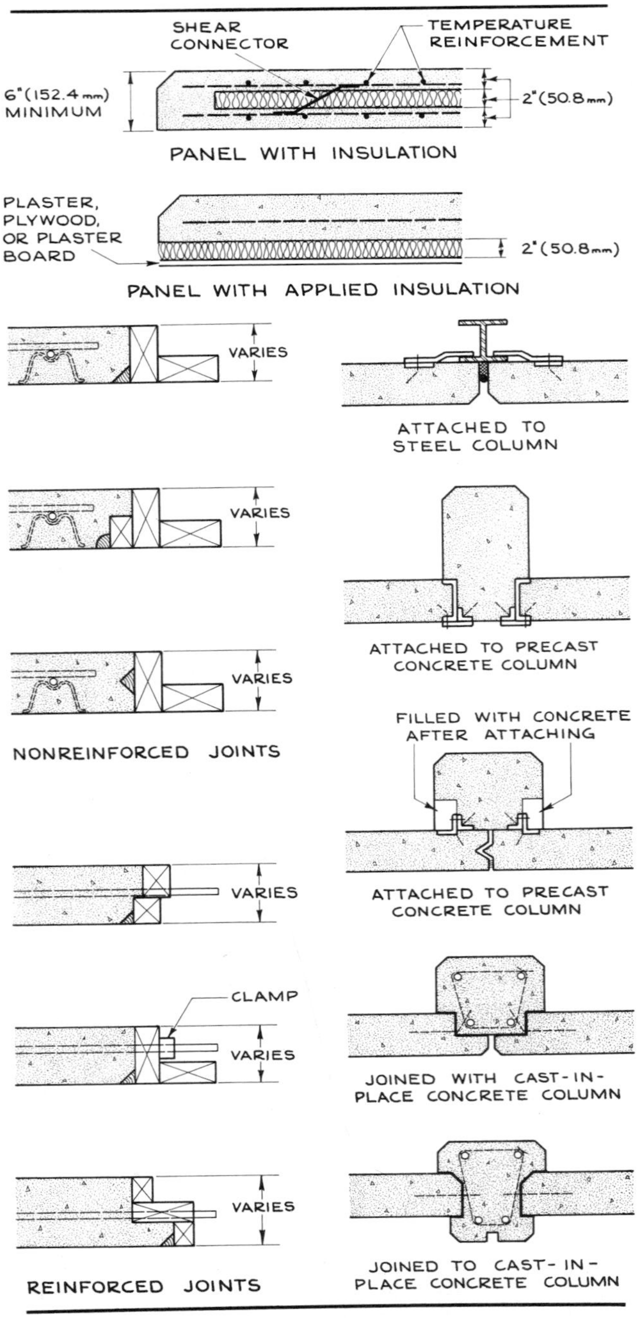

Figure C100 Tilt-up concrete construction systems.

before any concrete is installed, whether the concrete is site-installed or prefabricated. Slump tests and entrained-air tests for on-site poured concrete are made before placing the concrete; compression tests using 6 × 12 in. (15.24 × 30.48 cm) cylinders also have to be taken for each section of concrete that is installed. These cylinders are tested for 7- and 28-day strength.

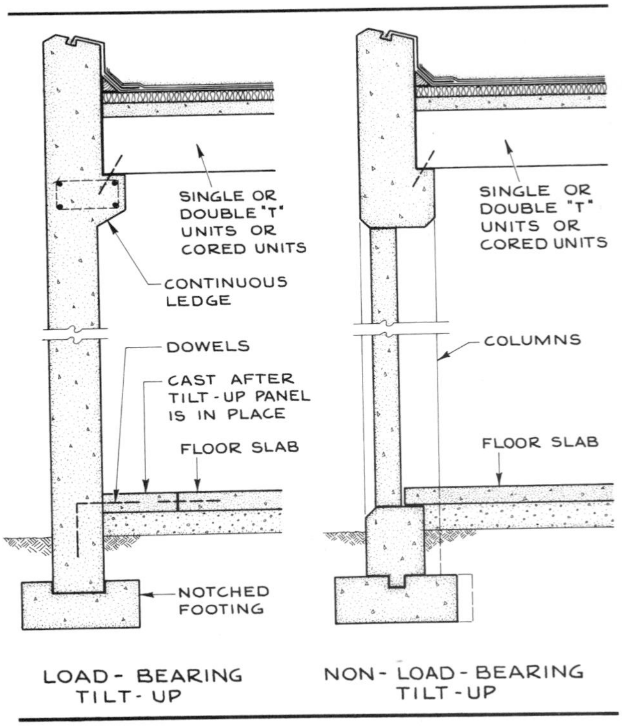

Figure C101 Typical examples of load-bearing and non-load-bearing tilt-up concrete panels.

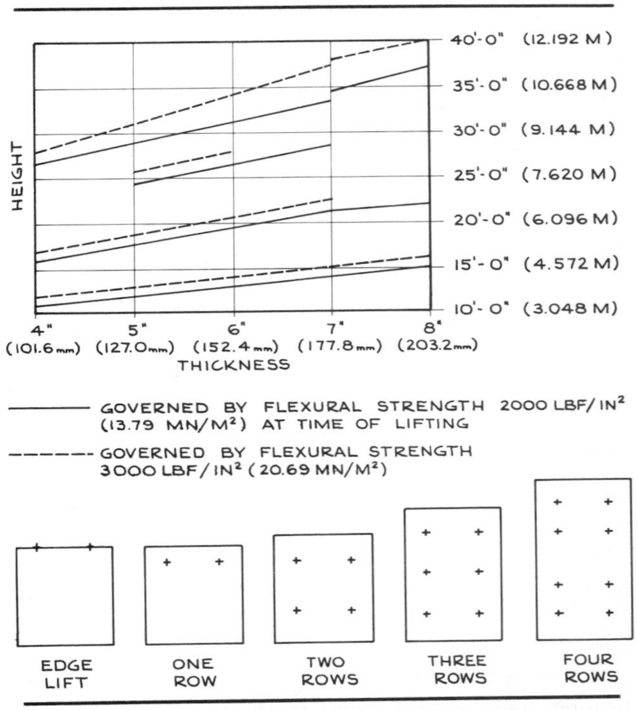

Figure C102 Typical locations of crane pick-up points for tilt-up panel systems.

Water-Cement Ratio. The water-cement ratio for the concrete mix is affected both by climate and by the exposure of the concrete to climate (*see* Table *C82*).

Class of Concrete. Reinforced concrete is generally specified by lbf/in.^{2} (MN/m^2) and approval of trial mixes. This method determines the optimum proportions. These trial mixes are made in relatively small batches in a laboratory and tested for 7- and 28-day strength.

Compressive Strength. For special structural conditions, where greater compressive strength may be required, the concrete can be specified by the minimum strength required and the water-cement ratio. Table *C83* shows the range of compressive strength that can be obtained with various water-cement ratios and includes strengths for which air-entrained concrete may be used. When using concrete with greater strength, great care should be taken, and placing, vibrating, and other details of installation must be specially controlled so that these strengths will be developed.

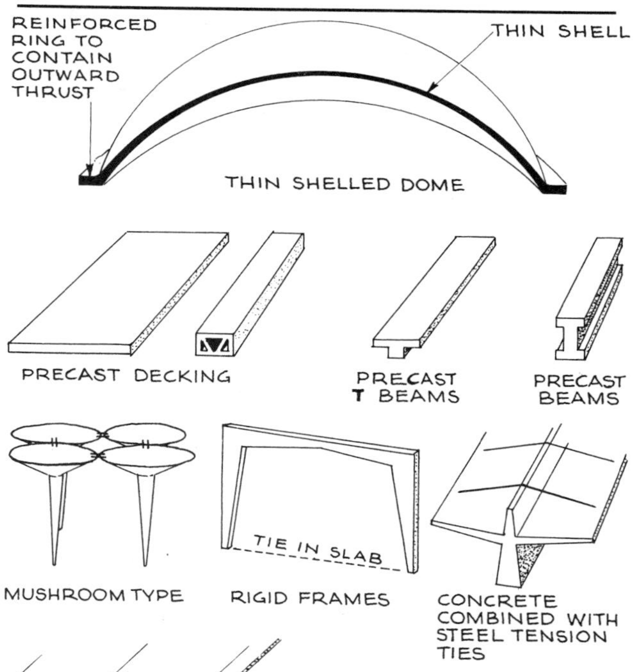

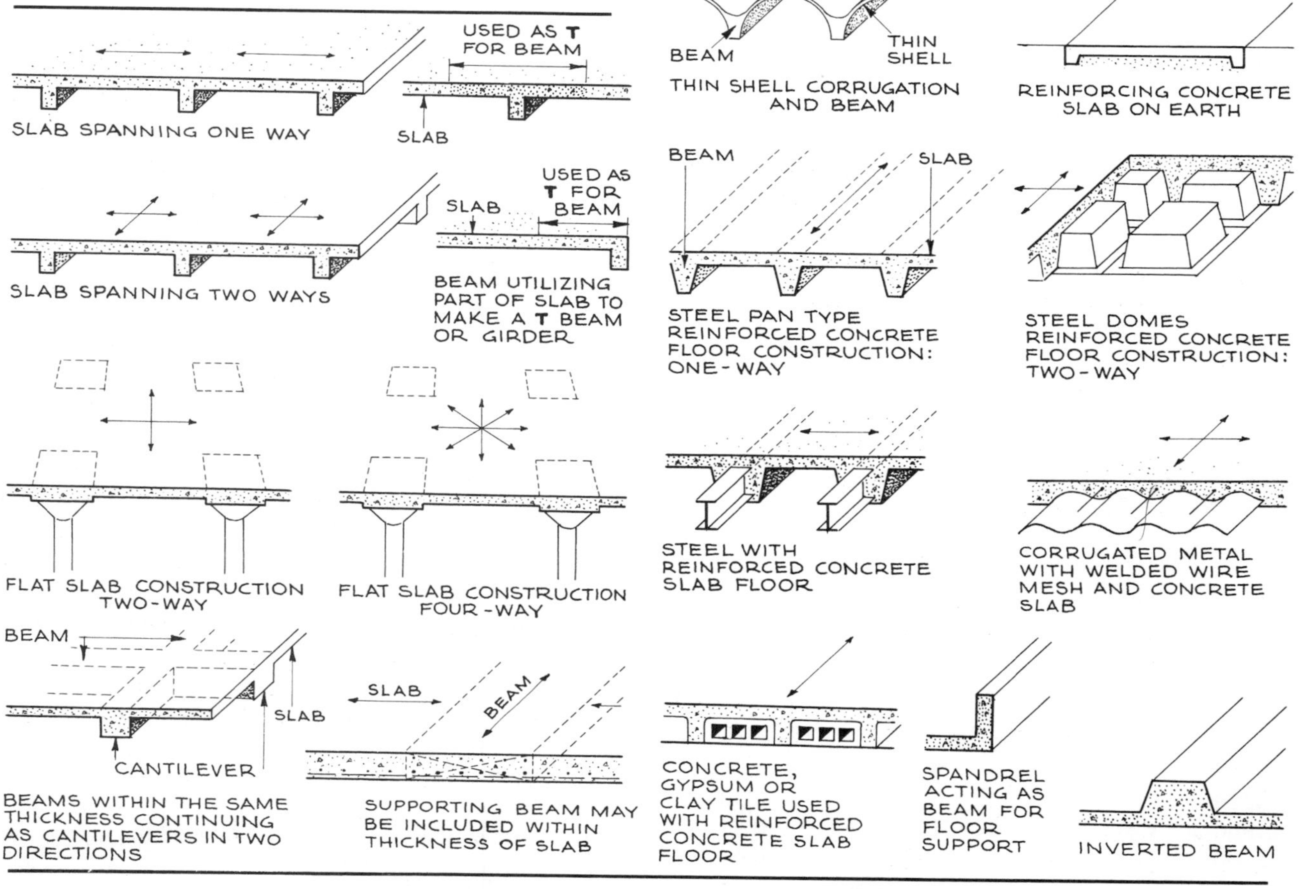

Figure C103 Reinforced concrete construction.

Table C82 Water-cement ratio for reinforced concrete in relation to climate and exposure

Water-cement ratio by weight[a]		
Severe climate (freezing and thawing)	Mild climate (no freezing and thawing)	Exposure of concrete as placed for end use
0.45	0.55	Exterior walls, beams, floors, columns, etc.; foundation walls above grade; concrete exposed to varying water levels and spray
0.50	0.55	Interior walls, beams, floors, columns, etc., which can be protected during construction
0.58	0.58	Walls, beams, columns, footings, floors, etc., which are below grade or are protected from weather[b]
0.45	0.50	Walls, beams, columns, footings, floors, etc., which are subject to sulfate attack[c]
0.45	0.45	Concrete to be placed under water

[a] All water-cement ratios can vary by ±0.02.
[b] If concrete may be subject to severe freezing and thawing during construction, the water-cement ratio should be reduced by 0.05.
[c] Calcium chloride should not be used because it weakens the sulfate resistance of sulfate-resistant cements and admixtures.

Table C83 Compressive strengths required for reinforced concrete and water-cement ratios

Minimum compressive strength at 28 days[a]				
Normal concrete		Air-entrained concrete		
$lbf/in.^2$	MN/m^2	$lbf/in.^2$	MN/m^2	Water-cement ratio by weight
3400	23.44	2700	18.62	0.60
3800	26.20	3100	21.38	0.55
4300	29.65	3500	24.13	0.50
4900	33.79	3900	26.89	0.45
5400	37.23	4300	29.65	0.40

[a] 1 $lbf/in.^2$ = 0.006895 MN/m^2.

Types of Reinforcing Steel. Reinforcing steel is generally of two types, bars and mesh. The bars may be plain or deformed, and their surface may have various types of projections for better bonding to the concrete. They are called billet-steel bars or railsteel or axle steel bars. Billet-steel bars are made to meet fixed chemical compositions. Physical requirements for reinforcing bars are shown in Table *C84*; for fuller information *see* Concrete; Steel for Concrete.

The steel wire used as reinforcing or for making welded reinforcing fabric is fabricated from the same steel that is used for bars and is cold-drawn from hot-rolled billets. The nominal diameters generally used in construction range from 0.080 to 0.625 in. (2.032 to 15.95 mm). Table *C85* gives requirements for reinforcing wire.

Application

Condensed Checklist

1. All tests for concrete must be made, and the concrete mix must meet the design conditions, before placing of concrete is begun.
2. All local, state, and municipal codes and the codes of the Fire Underwriters, insurance companies, labor departments, and federal government (Army, Navy, etc.) should be checked for requirements and limitations and also for fire ratings.
3. Forms must be correctly designed and installed before concrete is placed (*see* Concrete).
4. All reinforcing must be of the correct size, correctly installed and correctly held and tied according to structural requirements.
5. All inserts, hangers, ties, sleeves, anchors, chases, shafts, etc., must be in their assigned locations, and all allowances made, before concrete is placed.
6. Anchor bolts for steel bearing plates, etc., must be available before the concrete is poured. If templets are to be used for accurate placement of anchor bolts in freshly poured concrete, the templets must be also available before the concrete is poured. Special care and supervision are necessary to ensure that anchor bolt locations are marked exactly and that the bolts are correctly located.
7. All construction and expansion joints must be correctly located. Expansion joints must be left completely open with no reinforcing in them.
8. All precautions for meeting freezing weather must be ready for immediate use.
9. Air-entrained concrete should be tested by slump tests and an entrained air test indicator before concrete is placed.

Table C84 Strength requirements for steel reinforcing bars

Type of steel	Size number	Grade	Minimum tensile strength[a] lbf/in.²	Minimum tensile strength[a] MN/m²	Minimum yield[a] lbf/in.²	Minimum yield[a] MN/m²	Cold bend test[b] (180° unless otherwise noted)
Billet steel	3 to and including 18	40, 60, and 75	70,000 to 100,000	482.65 to 689.50	40,000 to 75,000	275.80 to 517.125	Under No. 6: 4d No. 6 and larger: 5d Nos. 7 and 8: 6d Nos. 14 and 18: 10d – 90°
Rail steel	3 to and including 11	50 and 60	80,000 and 90,000	551.60 and 620.55	50,000 and 60,000	344.75 and 413.70	Under No. 9: 6d Nos. 9 and 10: 8d No. 11: 8d – 90°
Axle steel	3 to and including 11	40 and 60	70,000 and 90,000	482.65 and 620.55	40,000 and 60,000	275.80 and 413.7	Under No. 6: 4d No. 6 and larger: 5d Nos. 7 and 8: 6d Nos. 9, 10, and 11: 8d

[a] 1 lbf/in.^{2} = 0.006895 MN/m^2.
[b] In column below, d = nominal diameter of size of reinforcing bar.

Table C85 Typical wire requirements for reinforcing or making welded wire mesh

Nominal diameter of wire in.	Nominal diameter of wire mm	Minimum tensile strength[a] lbf/in.²	Minimum tensile strength[a] MN/m²	Minimum yield[a] lbf/in.²	Minimum yield[a] MN/m²	Minimum reduction in area (percent)	Bend test 0.30 and under	Bend test Over 30
0.08–0.628	2.032–15.95	80,000	551.60	100,000	689.50	30	Bend around pin of wire diameter	Bend around pin of double wire diameter

[a] 1 lbf/in.^{2} = 0.006895 MN/m^2.
[b] Wire for welded wire fabric must have a minimum tensile strength of 70,000 lbf/in.^{2} (482.65 MN/m^2). Material that tests a tensile strength of more than 100,000 lbf/in.^{2} (689.50 MN/m^2) must have a minimum reduction in area of 25%.

10. Concrete should be inspected immediately after forms are removed to see that there is no separation of aggregate and no bad air holes or voids; that forms did not shift, bend, or move during the placing of the concrete; and that no damage has developed because of freezing.
11. Footings must be placed on virgin soil, not on fill.
12. Steel to be covered with concrete must be clean of mud, dirt, and also any waste matter.
13. All electrical conduit, pipes, etc., that are to be buried in the concrete should be checked to see that they have been installed as required.
14. All changes in levels of floors—which represent allowances for various types of floor finishes, mats, door openings, plumbing fixtures, steps, etc.—must be installed as specified.
15. Tilt-up panels should be checked for methods of joining, exterior surface treatments, placing of reinforcement, and any inserts, angles, channels, and plates for methods of attachment.
16. The method of erection, location of area for casting, and crane pick-up points should be checked. (*See* Figure *C102*.)
17. One should always check that the first floor slab is smooth, level, and treated with a good bond-breaking agent or material. Succeeding slabs should also be treated with a bond-breaking agent or material.
18. Lift-slab construction requires careful checking of all reinforcement, electrical conduits, heating, ventilating and air-conditioning openings, pipe chases, stair openings, etc., to make sure that all are accurately located as each successive slab is poured.

Conditions Favorable to the Use of Reinforced Concrete

1. In areas where construction material must be located directly on earth.
2. In areas where steel is not readily available.
3. Where fireproof construction or fire resistance is important.
4. For foundations, especially if there is a subsoil water condition.
5. For circular, spherical, or amorphous forms or any shape other than rectangular.

6. Where weight is not an important design factor.
7. Where heavy loads are to be supported, and widths of spans and bays are not important design factors.
8. Where a lift-slab type of construction has been selected as the method of constructing a building.
9. Where a tilt-up type of construction is required.
10. Where precast or cast-on-site building facings have been selected.
11. Where a textured, exposed aggregate, smooth, decorated exterior facing of a building is required.

Conditions Unfavorable to the Use of Reinforced Concrete

1. In construction where spans and allowances for depths of supports become so great that reinforced concrete is not feasible or economical.
2. Where supporting columns for the lower floors of a building should not, for design reasons, be heavy or large.

History and Manufacture

The Romans were the first to reinforce concrete; they simply inserted bronze bars in various areas to increase strength but had no knowledge of reinforced concrete as we understand it today. After the fall of the Roman Empire, this elementary knowledge of reinforced concrete was lost until W. B. Wilkinson received a patent in 1854 in England and therefore is now considered the inventor of reinforced concrete. Soon after, reinforced concrete was used for buildings, and the first recorded reinforced concrete building in the United States was erected by Ward in 1875 in Port Chester, New York. From this beginning a completely new field of engineering has developed and continues to expand.

CONVERSION TABLES

Table C86 Linear measure

	Inches	Feet	Yards	Rods	Millimeters	Centimeters	Meters	Kilometers
1 inch					25.4	2.54		
1 foot	12				304.8	30.48	0.304 8	
1 yard	36	3			914.0	91.44	0.914 4	
1 rod	198	16.5	5.5		20 292.0	202.92	2.029 2	
1 chain	792	66	22	4		20 111.60	20.116 8	0.002 016
1 furlong	7,920	660	220			20 116.80	201.168	0.021 168
1 mile[a]	63,360	5280	1760			160 934.40	1609.344	1.609 344

[a] 1 mile = 8 furlongs.

Table C87 Land measure

	Square feet	Square yards	Square rods	Square chains	Acres	Section	Square meters	Hectares
1 square rod	90.75	30.75					25.292 85	0.002 529
1 square chain	4,356.00	484.00	48.0				404.685 47	0.040 468
1 acre	43,560.00	4840.00	160.00	10			4 046.854 68	0.404 680
$\frac{1}{4}$ section					160	$\frac{1}{4}$	647 488.000 0	64.748 800
1 square mile		3097.00	102,400.00		640	1	1 789 986.995 20	258.995 520

Table C88 Cubic measure

	Feet long	Feet wide	Feet high	Cubic inches	Cubic feet	Cubic centimeters	Cubic meters
1 cubic inch						16.387	
1 cubic foot				1,728		283.168	0.028 316 8
1 cubic yard				46,656	27	7 645.536	0.764 553 6
1 cord (foot)				27,648	16	4 530.688	0.453 068 8
1 cord (wood)	4	8	4	221,184	128	36 245.504	3.624 550 4
1 perch (stone)	16.5	1.5	1	41,904	24.25	6 866.824	0.686 682 4

Table C89 Surface measure

	Square inches	Square feet	Square centimeters	Square meters
1 square inch			6.451 60	
1 square foot	144		929.030 40	0.092 903
1 square yard		9	8 361.273 60	0.836 127
1 square (roofing)		100	92 903.040 00	9.290 304

Table C90 Circular and angular measure

	Seconds (″)	Minutes (′)	Degrees (°)
1 minute	60		
1 degree		60	1
1 circle			360

Table C91 Feet converted into meters

Feet ⟶	0	1	2	3	4	5	6	7	8	9
0	. . . . . .	0.304 8	0.609 6	0.914 4	1.219 2	1.524 0	1.828 8	2.133 6	2.438 4	2.743 2
10	3.047 9	3.352 7	3.657 5	3.962 3	4.267 1	4.571 9	4.876 7	5.181 5	5.486 3	5.791 1
20	6.095 9	6.400 7	6.705 5	7.010 2	7.315 1	7.619 9	7.924 7	8.229 5	8.534 3	8.839 0
30	9.143 8	9.448 6	9.753 4	10.058 2	10.363 0	10.667 8	10.972 6	11.277 4	11.582 2	11.887 0
40	12.191 8	12.496 6	12.801 4	13.106 2	13.411 0	13.715 8	14.020 5	14.325 3	14.630 1	14.934 9
50	15.239 7	15.544 5	15.849 3	16.154 1	16.458 9	16.763 7	17.068 5	17.373 3	17.678 1	17.982 8
60	18.287 7	18.592 5	18.897 3	19.202 0	19.506 8	19.811 6	20.116 4	20.421 2	20.726 0	21.030 8
70	21.335 6	21.640 4	21.945 2	22.250 0	22.554 8	22.859 6	23.164 4	23.469 2	23.774 0	24.078 8
80	24.383 6	24.688 4	24.993 1	25.297 9	25.602 7	25.907 5	26.212 3	26.517 1	26.821 9	27.126 7
90	27.431 5	27.736 3	28.041 1	28.345 9	28.650 7	28.955 5	29.260 3	29.565 1	29.869 9	30.174 7

Table C92 Dry measure

	Pints	Quarts	Gallons	Pecks	Cubic inches	Liters
1 pint						0.550 61
1 quart	2				67.2	1.101 221
1 peck		8	2		537.6	8.809 77
1 bushel			8	4	2150.42	35.239 1
1 barrel					7056.00	115.627

Table C93 Liquid measure

	Gills	Pints	Quarts	Gallons	Ounces	Cubic inches	Liters	Cubic centimeters
1 ounce						1.805		2.957 35
1 gill					4	7.21875		11.829 4
1 pint	4				16	28.875	0.473 176	
1 quart	8	2			32	57.75	0.946 353	
1 gallon	32	8	4		128	231.00	3.785 41	
1 barrel				42	5376	9702.00	158.987	

Table C94 Fraction of an inch, decimals of an inch, and millimeters

Fraction of an inch	Decimal of an inch	Millimeters	Fraction of an inch	Decimal of an inch	Millimeters
$\frac{1}{32}$	0.03125	0.793 75	$\frac{17}{32}$	0.53125	13.493 78
$\frac{1}{16}$	0.0625	1.587 50	$\frac{9}{16}$	0.5625	14.287 53
$\frac{3}{32}$	0.09375	2.381 25	$\frac{19}{32}$	0.59375	15.081 28
$\frac{1}{8}$	0.125	3.175 01	$\frac{5}{8}$	0.625	15.875 03
$\frac{5}{32}$	0.15625	3.968 76	$\frac{21}{32}$	0.65625	16.668 78
$\frac{3}{16}$	0.1875	4.762 51	$\frac{11}{16}$	0.6875	17.462 53
$\frac{7}{32}$	0.21875	5.556 26	$\frac{23}{32}$	0.71875	18.256 29
$\frac{1}{4}$	0.25	6.350 01	$\frac{3}{4}$	0.75	19.050 04
$\frac{9}{32}$	0.28125	7.143 76	$\frac{25}{32}$	0.78125	19.843 79
$\frac{5}{16}$	0.3125	7.937 52	$\frac{13}{16}$	0.8125	20.637 54
$\frac{11}{32}$	0.34375	8.731 27	$\frac{27}{32}$	0.84375	21.431 29
$\frac{3}{8}$	0.375	9.525 02	$\frac{7}{8}$	0.875	22.225 04
$\frac{13}{32}$	0.40625	10.318 77	$\frac{29}{32}$	0.90625	23.018 80
$\frac{7}{16}$	0.4375	11.112 52	$\frac{15}{16}$	0.9375	23.812 55
$\frac{15}{32}$	0.46875	11.906 27	$\frac{31}{32}$	0.96875	24.606 30
$\frac{1}{2}$	0.5	12.700 3	1	1.0	25.400 65

Table C95 Miscellaneous conversion factors

1 kip (1000 lb)	= 453.59 kg	1 lumen/ft^2	= 10.76 lux
1 ton	= 1.016 tonne	1 lux	= 1 lumen/m^2
lb/ft	= 1.4882 kg/m	1 pound force	= 4.448 N (newton)
lb/in.	= 17.858 kg/m	1 lbf/ft	= 14.59 N/m
lb/ft^2	= 4.882 kg/m^2	1 lbf/in.^{2}	= 0.1751 N/mm
lb/in.^{2}	= 703.07 kg/m^2	1 kip (1000 lbf)	= 4.448 kN (kilonewton)
1 mile/gal	= 0.354 km/liter	1 lbf/ft^2	= 47.88 N/m^2 [a]
lb/ft^3	= 16.02 kg/m^3	1 meganewton	= 1 N/mm^2
lb/in.^{3}	= 27.68/cm^3	1 lbf/in.^{2} (psi)	= 0.006895 MN/m^2 [b]
1 mile/h	= 1.609 km/h	1°F	= $\frac{5}{9}$ (°F - 32) °C
1 ft/min	= 0.3048 m/min	ft/s · s = ft/s^2	= 0.3048 m/s^2
1 ft/s	= 0.3048 m/s	Btu/lb	= 2.326 kJ/kg
1 Btu	= 1055 J (joule)	Btu/ft^3	= 37.26 kJ/m^3
Btu/h	= 0.2931 W (watt)	Btu/gal	= 232.6 J/liter
"U" value in 1 Btu/ft^2 · h · °F	= 5.678 W/m^2 · °C	*K* value in Btu in./ft^2 · h · °F	= 0.1442 W/m · °C
1 ton, refrigeration	= 3519 W	1 mil	= 0.0254 mm
1 horsepower	= 745.7 W	1 short ton	= 907.185 kg
lumen	= lumen	1 long ton	= 1016.047 kg
1 foot-candle	= 10.76 lux	1 calorie	= 4.18 J

[a] One ton force per square foot = 107.3 kN/m^2; one ton force per square inch = 15.44 MN/m^2 = 15.44 MPa (megapascal).
[b] One pascal (Pa) = 1 N/m^2, and 1 MPa = 1 MN/m^2. Many manufacturers are using megapascal (MPa) when converting psi: 20,000 psi = 137.9 MPa.

COPPER

Physical and Chemical Properties

Symbol: Cu
Atomic number: 29
Specific gravity: 8.96 (20°C, 27.78°F)
Melting point: 1083°C, 1981.4°F
Boiling point: 2595°C, 4703°F
Coefficient of expansion: 0.0000094/°F
Tensile strength: 30,000 lb/in.^{2}, (206.85 MN/m^2) (in soft condition)

Copper is a ductile, malleable, nonmagnetic metal with a characteristic bright reddish brown color. It has the highest electrical and thermal conductivity of any substance except silver, its electrical conductivity being 94% that of silver. Copper forms useful alloys, has enough strength for minor structural work, and is easily worked. *See* Copper Alloys.

Corrosion Resistance. Copper is attacked by alkalis and many of the common acids (all of its salts are highly toxic). It is highly resistant to corrosion by air and salt water. On exposure it soon reacts to form a surface layer of an insoluble green salt, usually considered to be copper carbonate, which retards further corrosion; this green color on copper is known as its patina. For the galvanic action of copper, see "Application."

Workability. Copper can be cast, drawn, extruded, hot- and cold-worked, spun, hammered, punched, welded, brazed, and soldered. Soft copper can be made stronger by cold working but can then be softened again by annealing. It does not lend itself to other heat treatment.

Commercial Forms. Commercial copper is available in several types which are graded by chemical composition and electrical conductivity, by refining methods, and by fabrication procedures (*see* Tables *C96* and *C97*). It is available in ingot, billet, cathode, slab, rod, tube, pipe, bar, shot, cake, sheet, strip, foil, wire, and powder form.

Table C96 Typical composition of electrolytically refined tough-pitch copper

Copper[a]	Oxygen (percent of content)	Other impurities
99.945	0.045	0.01

[a] Silver counted as copper.

Formation of Patina. The patina of copper refers to the green or greenish gray coating which develops naturally over a period of time on copper exposed to the atmosphere and which protects it from further corrosion. It cannot develop in pure air but is formed only where there are traces of certain impurities. Although the patina has been called a basic carbonate, actual tests show that on roofs and in places where there are traces of sulfurous fumes in the atmosphere, it consists mostly of basic copper sulfate. It has been proved that carbon dioxide, always present in air, has little influence on the formation of patina. A copper chloride patina forms mostly on objects lying in water or damp earth, and also in sea air free from industrial pollution. It takes about 8 to 10 years, depending on location and atmospheric conditions, for a copper patina to become fully developed; in about 70 years it becomes mineralized (similar to the original copper ore) and impervious to further action. This natural patina is not to be confused with "verdigris."

Verdigris, although similar in color, is formed by a different chemical reaction and is of a different composition. Verdigris is a basic copper acetate which can form naturally in conjunction with the acetic acid produced by decaying organic matter. Development of artificial patina on copper has been attempted by various means, but to date none has proved entirely satisfactory.

Effects of Impurities. Various impurities play an important part in the properties and therefore the use of

Table C97 Minimum purity for various types of copper

Lake coppers		Electrolytic copper: Cathode	Fire-refined copper		
Low resistance	High resistance		Oxygen free	Cast	Wrought
99.9% Cu[a]	99.9% Cu[b]	99.9% Cu[b]	99.9% Cu[a]	99.7% Cu[a]	99.88% Cu[a]

[a] Silver counted as copper.
[b] Silver and arsenic counted as copper.

copper. Oxygen in the small amounts normal to the tough-pitch type of copper does not significantly affect its electrical conductivity but does reduce ductility; in increased quantity it tends to raise tensile strength as well as reduce ductility (not generally encountered except possibly in castings). Residual traces of deoxidants such as phosphorus decrease the electrical conductivity; but because they serve to eliminate the oxygen, they cause marked improvement in ductility and malleability. Sulfur, selenium, and tellurium adversely affect the mechanical properties of copper, although up to 1% of selenium and tellurium increases machinability. Bismuth interferes with hot rolling. Antimony, in amounts of 0.5% and over, hardens copper, decreases ductility, and lowers conductivity. Arsenic, in amounts up to 0.6%, hardens and strengthens but reduces conductivity. Silver is usually present in copper in small quantities and has a great effect on the recrystallization. Iron in amounts up to 2% hardens and strengthens without affecting ductility but reduces conductivity. Lead is added in amounts from 0.005 to 1.0% to aid machinability. Gases have great effects on the physical properties of copper. Oxygen in large amounts reduces the conductivity and forms oxides. Carbon monoxide and hydrogen can make castings unsound if oxygen is also present. In casting, control of the gases becomes a very important, complex problem.

Types and Uses

As copper is one of the best electrical conductors, it finds tremendous use in the entire electrical field, from very fine wire to bus bars in hydroelectric power installations. Because of its high thermal conductivity, it also finds wide use in the plumbing, heating, and air conditioning fields, all beyond the scope of this book.

Copper has many uses in compound form. Its uses in metallic form are indicated by asterisks in Table *C98*. Pure copper finds uses in construction as sheet and strip for roofing and flashing, as foil, and as mesh and wire cloth. Copper alloys are also important as construction materials. Copper is alloyed with tin, zinc, and nickel

Table C98 Uses of copper

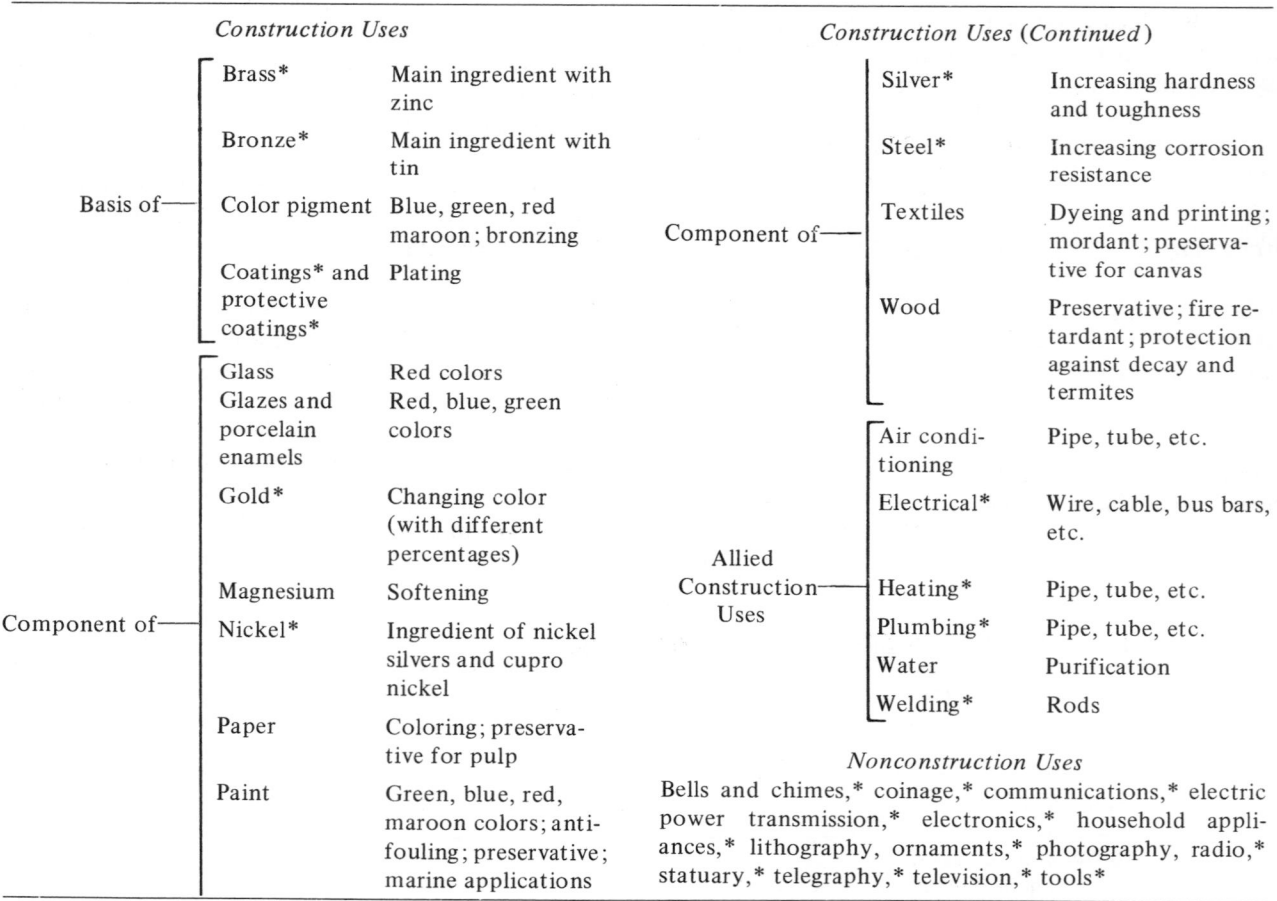

Construction Uses		
Basis of	Brass*	Main ingredient with zinc
	Bronze*	Main ingredient with tin
	Color pigment	Blue, green, red maroon; bronzing
	Coatings* and protective coatings*	Plating
Component of	Glass	Red colors
	Glazes and porcelain enamels	Red, blue, green colors
	Gold*	Changing color (with different percentages)
	Magnesium	Softening
	Nickel*	Ingredient of nickel silvers and cupro nickel
	Paper	Coloring; preservative for pulp
	Paint	Green, blue, red, maroon colors; antifouling; preservative; marine applications
	Silver*	Increasing hardness and toughness
	Steel*	Increasing corrosion resistance
	Textiles	Dyeing and printing; mordant; preservative for canvas
	Wood	Preservative; fire retardant; protection against decay and termites
Allied Construction Uses	Air conditioning	Pipe, tube, etc.
	Electrical*	Wire, cable, bus bars, etc.
	Heating*	Pipe, tube, etc.
	Plumbing*	Pipe, tube, etc.
	Water	Purification
	Welding*	Rods
Nonconstruction Uses		
Bells and chimes,* coinage,* communications,* electric power transmission,* electronics,* household appliances,* lithography, ornaments,* photography, radio,* statuary,* telegraphy,* television,* tools*		

to make bronzes, brasses, and nickel silvers (*see* specific headings). It is alloyed with steel to improve the corrosion resistance of the base metal sheets of terneplate and of ducts and culverts made of that material.

Copper Plating. Copper plating is the method of applying a thin coat of copper by electrolytic action to a base metal. Almost all copper plating is intended as a base for plating with another metal.

Application

Galvanic Action. The galvanic action of copper must be considered when copper is used in construction. Copper is low in the galvanic series. When in contact with many of the common construction materials and in the presence of an electrolyte, it will corrode these materials near the area of contact. The copper itself, being cathode, will not corrode. Therefore a careful check should always be made of the methods of attachment, support, and securing into place. Copper should never be used when sulfur is present; for example, copper in contact with cinders will be destroyed by corrosive attack fairly soon.

Accessories. Accessories made of copper are important in construction applications. Among these are rough hardware items such as nails, screws, washers, ties, anchors, nuts and bolts, and also items such as louvers and vents, gutters and leaders, and screening. For flashing, there are various standard prefabricated accessories for pipe, cap and through-wall flashing, expansion joints, fascias, etc.

History and Manufacture

Copper is one of the most useful and therefore important nonferrous metals. Its history is the history of man's advance from the Stone to the Bronze Age. It occurs in its pure metallic state in some parts of the world and may have been used as long as 20,000 years ago. Archeological findings show that Neolithic man had copper artifacts by about 8000 B.C. By 3500 B.C. copper ornaments, tools, and utensils were in common use among the Egyptian, Aegean, and Near Eastern peoples, and by about 3200 B.C. bronze had been discovered in Egypt. By 2200 B.C. the use of copper and bronze became common throughout the Mediterranean area. A parallel development occurred in China and India. Copper was known and used in China in 2500 B.C., and beautifully worked bronzes of about 1766 B.C. have been found. In Central America and Peru, the stone-copper-bronze cycle evolved at the beginning of the Christian era, but much more slowly.

The island of Cyprus produced large quantities of copper about 3000 B.C. and became so important that as a result it was conquered and controlled successively by Egyptians, Assyrians, Phoenicians, Greeks, Persians, and Romans. It was the major supply of copper for the Romans, who at first called this material *aes cyprium.* The shortened form *cyprium* was later corrupted to *cuprum*, which in English became "copper." The chemical symbol, Cu, is from *cuprum*.

After the fall of the Roman Empire the production of copper declined almost completely until the Middle Ages.

Flotation Process. Copper is produced in this country largely from low-grade sulfide ores, which require concentration before smelting and refining. The flowchart in Figure *C104* of concentration by flotation shows the process generally used. It comprises crushing, grinding, flotation, thickening, and filtering. The ore is first milled to free the precious metals from rock and gangue by crushing and screening. It is further ground in a ball mill and then classified by gravity, using water; next it is mixed with flotation reagents in a conditioner and passed to the primary and secondary flotation cells. The waste (tailings) from the flotation cells is passed to thickeners and then to storage dams, and the water is reclaimed. The concentrates are again ground and classified and passed on to another set of flotation cells, then to thickeners, and finally to a filter, after which they are ready for smelting.

Smelting and Refining. The flowchart in Figure *C105* of the smelting and refining of sulfide ores shows how complex a process this is. The U.S. Bureau of Mines describes it as follows:

"The ores and concentrates are first smelted, with or without roasting, in reverberatory furnaces. Copper matte and slag are produced. The liquid-copper matte (a copper-iron sulfide containing the precious metals and some impurities) is treated in a copper converter, where siliceous flux is added and air is blown through the molten bath. The iron sulfide is oxidized by the air blast to ferrous oxide, which unites with the silica, producing a ferrous silicate slag, which is poured off. The copper sulfide reacts with oxygen to yield metallic copper and sulfur dioxide. Some volatile impurities are also eliminated in the converter as gaseous products and some are slagged off; others remain with the copper, which at this stage is called 'blister' copper. The name 'blister' comes from the rough surface caused by the

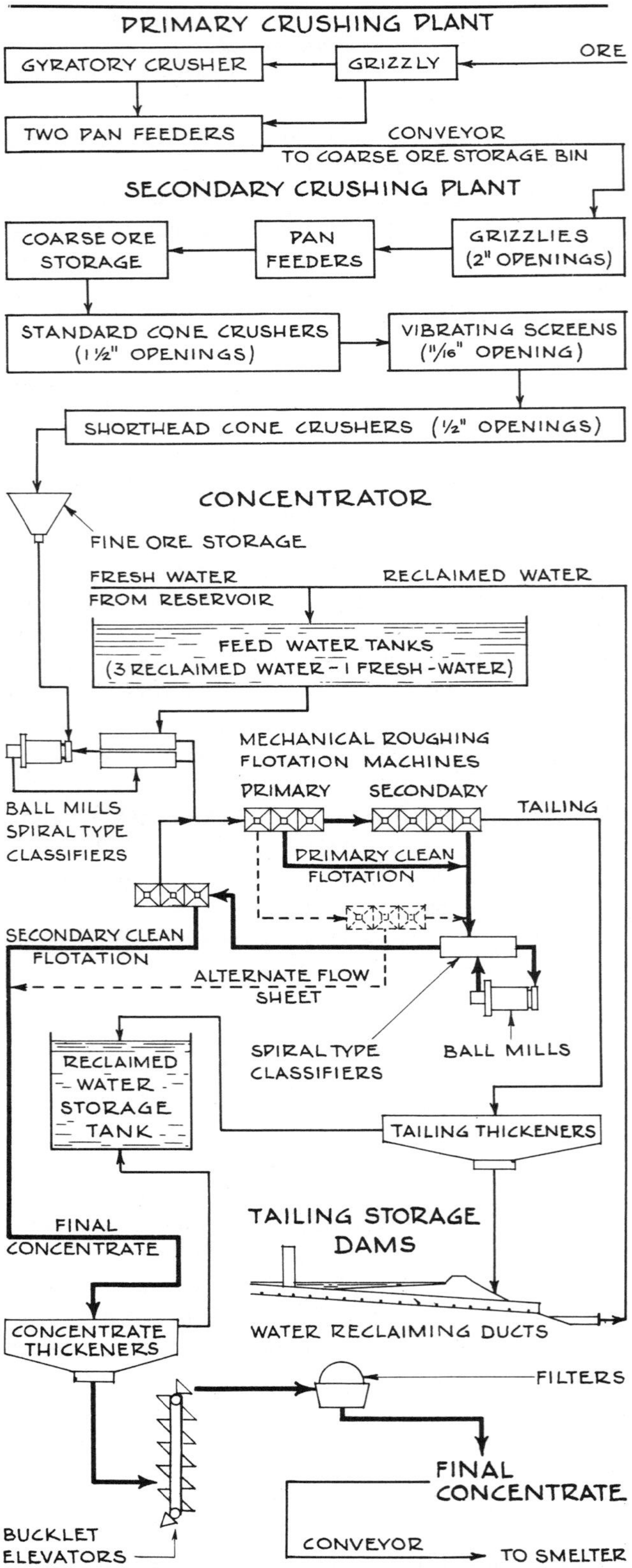

Figure C104 Flowchart of copper concentration by flotation.

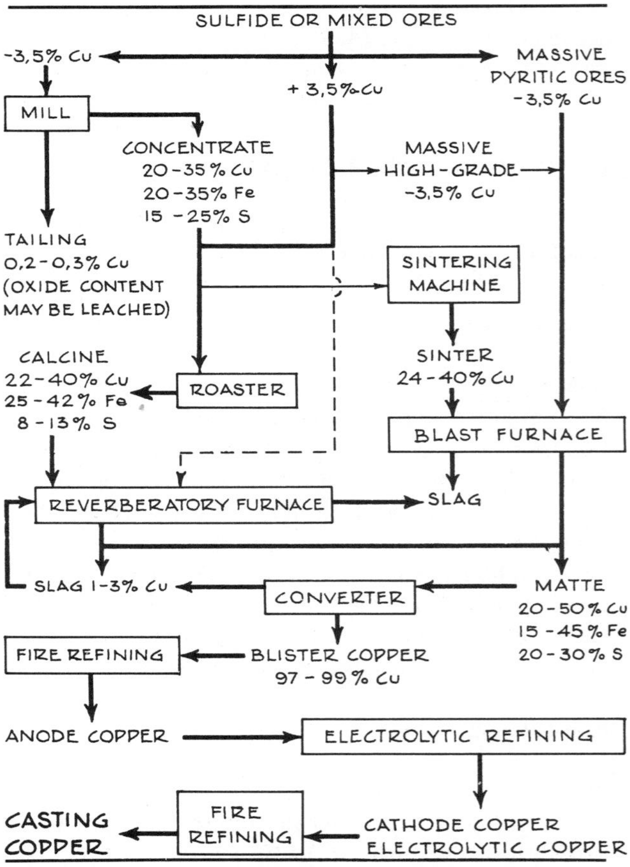

Figure C105 Copper production by the smelting and refining of sulfide ores.

escaping gases of the cooling copper castings. Blister copper cannot be used as such, as it still is too impure. It may be transferred from the converter to small holding or casting furnaces, where it can be cast into blister cakes or anodes for shipment to the refinery; otherwise, it may be transferred molten to reverberatory furnaces for fire refining before casting into commercial intermediate shapes, such as cakes, billets, or ingots. Most copper enters consumption channels in electrolytically refined form for electrical and other uses."

Electrolytic Refining. In electrolytic refining the blister copper anodes are spaced alternately between sheets of refined copper, which serve as the cathodes. This assembly of cathodes and anodes is placed in a lead-lined tank filled with a water solution of 4% copper and 16% sulfuric acid. As the electric current passes through the tank, the cast anodes dissolve and pure copper is deposited on the cathode sheets. The mud and slimes containing precipitated impurities are removed and treated to recover selenium, tellurium, sulfur, gold, silver, and some copper. The cathodes are removed and

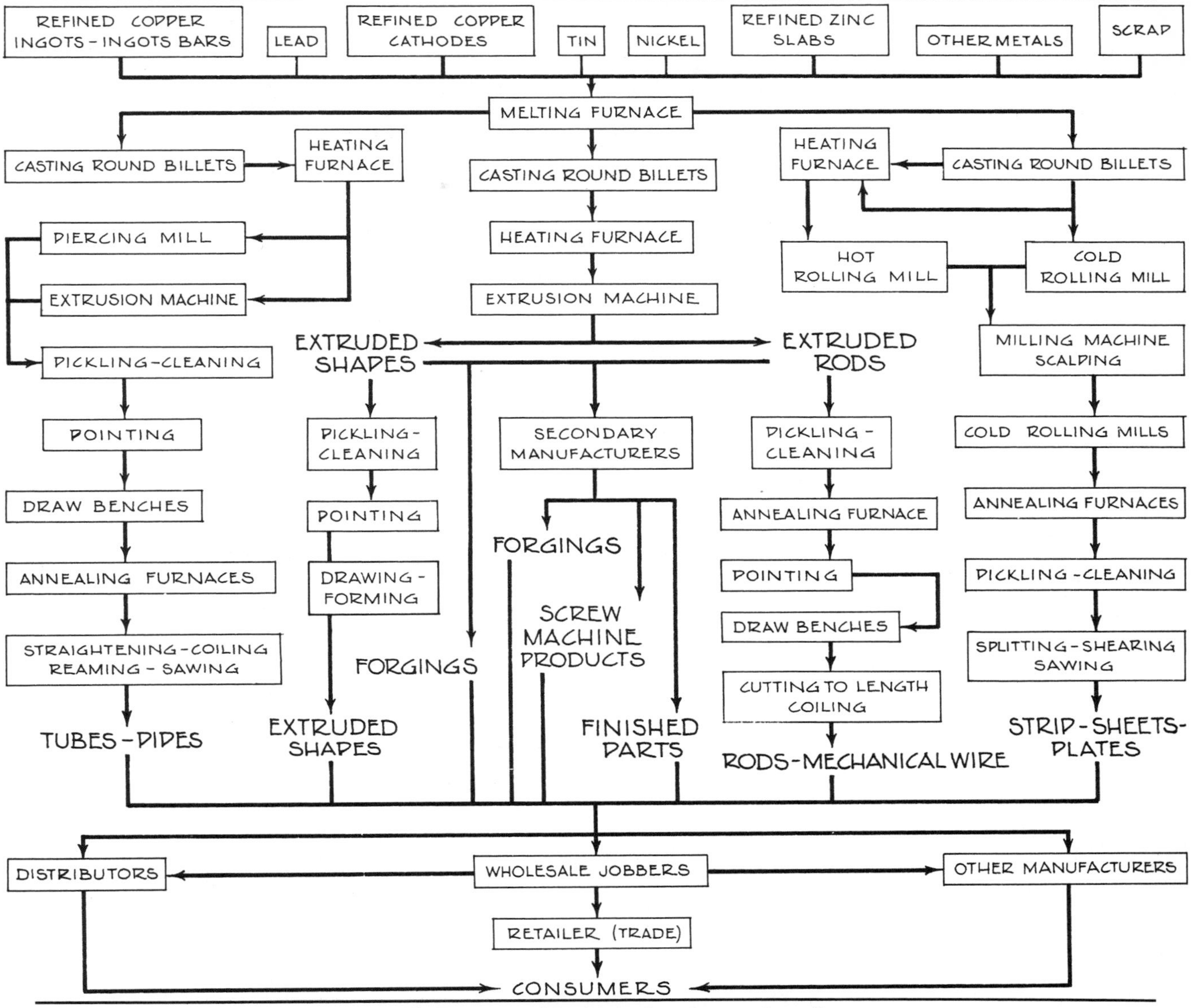

Figure C106 Flowchart of the fabrication of copper-base alloys.

are marketed as cathode or electrolytic copper. They can be fire-refined further to remove any remaining sulfur and to control the oxygen content for casting purposes.

Fabrication of Electrolytic Copper. This purified copper, still known as electrolytic copper, is available in the following refinery shapes: wire bar, cake, billet, ingot, and ingot bar. These refinery shapes are subsequently fabricated into wire, tube, pipe, rod, special shapes, and flat products in pure or alloy form as shown on the flowchart in Figure *C106.* These products can be further shaped, hot- and cold-worked, spun, hammered, punched, welded, brazed, and soldered into consumer products.

Secondary Copper. The production of secondary copper is an important branch of the copper industry, amounting to about 40% of the new copper output. The bulk of this secondary copper is recovered from the brass and bronze makers, brass mills, and scrap refiners.

COPPER ALLOYS

The accepted alloy designation system for wrought and cast copper and copper alloys, as administered by the

Table C99 Wrought copper alloys used in the construction field

Alloy number	Type	General uses
HIGH-COPPER ALLOYS		
110	Electrolytic tough pitch	Gutters, leaders, flashing, roofing, screens, nails, rivets, and tacks
125, 127, 228, 129, 130	Fire-refined tough pitch with silver	Gutters, leaders, flashing, roofing, and screens
145	Tellurium-bearing	Soldering and welding tips, solders, resistance welding and spot welding tips
147	Sulfur-bearing	
150	Zirconium-bearing	
155	with magnesium	
105	with cadmium and tin	
170, 172, 173, 175	Beryllium copper	Fasteners, washers, and miscellaneous rough hardware
182, 184, 185	Chromium copper	Soldering and welding tips, solders, resistance-welding and spot-welding tips
189	with tin and silicon	Welding rod and wire
190, 191	with nickel and phosphorus	Nuts, bolts, screws, and nails
BRASSES		
210	Guilding copper	Base for vitreous enamel
220	Commercial bronze	Grilles, screens, rivets, screws, weatherstripping, and miscellaneous finish hardware
226	Jewelry bronze	Angles, channels, chain, and fasteners
230	Red brass	Trim, weatherstripping, fasteners, and finish hardware
240	Low brass	
260	Cartridge brass	Grilles, chains, fasteners, rivets, screws, and miscellaneous finish hardware
268, 270	Yellow brass	
280	Muntz metal	Sheet and strip, trim, and large nuts and bolts
314	Leaded commercial bronze	Screws and miscellaneous rough hardware
316	Leaded commercial bronze, nickel-bearing	Nuts, screws, fasteners, and miscellaneous rough hardware
335	Low-leaded brass	Hinges, nuts, bolts, rivets, and screws
340, 350	Medium-leaded brass	
349	with some lead 0.3	Finish hardware
385	Architectural bronze	Architectural extrusions, store front trim, hinges, and finish hardware
464, 465,	Naval brass	Nuts, bolts, rivets, and marine rough and finish hardware
466, 467, 482, 485	Naval brass, medium-leaded	
BRONZES		
518, 521	Phosphor bronze	Fasteners, washers, clips, wire brushes, and miscellaneous rough hardware
613, 614	Aluminum bronze	Nuts and bolts
638	Bronze with aluminum silicon, and cobalt	Glass sealing and porcelain enameling
642	Bronze with aluminum and silicon	Nuts and bolts, screws, and marine rough and finish hardware
651	Low-silicon bronze	
655	High-silicon bronze	
NICKEL SILVERS		
745	Nickel–silver 65–10	Rivets, screws, fasteners, trim, and rough and finish hardware
752	Nickel–silver 65–18	
757	Nickel–silver 65–12	

Copper Development Association, is based on a number system in which wrought alloys are designated by numbers from 100 through 799, and cast alloys by numbers from 800 through 999. Within these two categories, the following coppers and copper alloys have been grouped accordingly.

Coppers—Minimum copper content 99.3%.

High-copper alloys—Wrought alloys with copper contents of more than 96% but less than 99.3%. Cast alloys with copper contents in excess of 94%. Silver may be added for special properties.

Brasses—Alloys in which zinc is the principal alloying element, or copper-base alloys containing an appreciable amount of zinc. *Wrought alloys:* copper-zinc, copper-zinc-lead (leaded brasses), and copper-zinc-tin (tin brasses). *Cast alloys:* copper-tin-zinc (red, semi-red, and yellow brasses), manganese bronze (high-strength yellow brasses), leaded manganese bronze (leaded high-strength yellow brasses), and copper-zinc-silicon (silicon brasses and bronzes).

Bronzes—Copper alloys in which the major alloying element is one other than zinc or nickel, or copper-base alloys in which alloying elements, other than zinc, are present in sufficient amounts to impart special characteristics and to be predominant over the effect of zinc in the alloy. *Wrought alloys:* copper-tin-phosphorus (phosphorus bronzes), copper-tin-lead-phosphorus (leaded phosphorus bronzes), copper-aluminum (aluminum bronzes), and copper-silicon (silicon bronzes). *Cast alloys:* copper-tin (tin bronzes), copper-tin-lead (leaded and high-leaded tin bronzes), copper-tin-nickel (nickel-tin bronzes), and aluminum (aluminum bronzes).

Copper nickels—Alloys in which nickel is the principal alloying element.

Nickel silvers—Copper-nickel-zinc alloys.

Leaded coppers—Casting alloys with 20% or more of lead.

Special alloys—Alloys whose chemical compositions do not fall into any of the aforementioned categories.

A unified numbering system for all commercial metals and alloys is being developed. The designations for copper alloys consist of adding the prefix "C" and the suffix "00" to the present Standard Designation System.

Table *C99* lists the copper alloys used in construction.

COPPER FOIL

Copper foil is made in wide, thin sheets of unlimited length by electrodeposition. The foil is made in eight weights, designated by weight in ounces per square foot: 0.5 and 1 through 7 oz in 1-oz gradations (14.175 g and 28.35 through 198.45 g per 0.093 m^2 in 28.35-g gradations). Widths range from 6 to 64-in. (0.1524 to 1.627 m). For most building purposes foil is laminated with high-grade building papers, saturated fabrics, or asphaltic compounds. Its greatest use is for concealed flashing. It is advisable not to use copper foil in exposed conditions because of its thinness and strong, distinctive color.

COPPER MESH AND WIRE CLOTH

Physical and Chemical Properties

Copper mesh and wire cloth are made of wire fabricated from electrolytic, tough-pitch copper or from alloys of copper, usually brass or bronze. This wire is woven into a corrosion-resistant, tough mesh that varies in color from reddish to dark brown. After exposure to weather, mesh made of copper turns green, the brass type loses its luster, and the bronze takes on a greenish brown color. Usually copper or copper-alloy mesh screens are initially painted with a thinned varnish to retard this color change.

Types and Uses

Copper mesh and wire cloth are available in a range of meshes from No. 2 to No. 100, the size of the mesh being based on the number of openings per linear inch. Copper mesh and wire cloth are used in architecture for screening, grilles, detention guards, flameproof vents, etc. (*See* Table *C100*.)

Table C100 Copper meshes and wire cloth

Mesh size	Gauge of wire[a]	Percentage of open area
2	16	76.4
4	18	65.9
6	20	62.7
8	22	60.2
10	23	56.3
12	24	51.8
14	25	51.0
16	26	50.7
18	27	48.3
20	28	46.2
24	31	45.8
30	32	40.8

[a] In circular mils.

Screening. Two types of woven wire mesh cloth are used for screening:

1. Rectangular 18 X 14 mesh, which is mesh with 18 openings per inch (25.4 mm) of width and 14 openings per inch (25.4 mm) of length.
2. Square 18 X 18 mesh.

The standard widths of woven wire mesh cloth rolls used for screening are 2 to 3.5 ft in 2-in. increments (0.6096 to 1.0668 m in 50.8-mm increments), and 4, 4.5, 5, and 6 ft (1.2192, 1.3716, 1.524, and 1.8288 m). All rolls are a maximum length of 100 ft (30.48 m).

The finer meshes of copper are made with a rectangular type of opening, and the smallest wires used are gauges 34 to 40. In using copper mesh other than screening, a check should be made with the manufacturer regarding the type and size recommended for the design purpose concerned.

Weaves. Copper mesh is made with square or rectangular openings in a variety of weaves. The simplest is a plain weave, where each wire is woven over one and under the next wire; in crimped weave, the wires are locked into position; and in double crimp weave, the wires are corrugated as well as locked into position.

History and Manufacture

Wire mesh first appeared in the middle of the 19th century. Between 1877 and 1909 many types and forms were developed, and around 1890 looms for the weaving of mesh were put into operation. At present most meshes come in rolls of definite widths and 100-ft (30.48 m) lengths, in a large variety of mesh sizes.

COPPER SHEET AND STRIP

Physical and Chemical Properties

Copper sheet and strip are formed of tough-pitch copper: specific gravity 8.89; melting point 1981°F (1.082.78°C); tensile strength 36,000 lbf per in.^{2} (248.22 MN per m^2); coefficient of expansion 0.0000094 °F (0.556 per °C). The sheet and strip are yellowish red in color, strong, and corrosion resistant. When exposed, they become coated with a green protective layer of copper carbonate (or sulfate), which stains adjoining materials (*see* Copper for discussion of patina).

Types and Uses

Copper sheet and strip are available as either hot-rolled soft copper (known as soft copper) or cold-rolled copper (known as cornice temper). They are also available with a lead coating of 15 lb (6.804 kg) per 100 ft^2 (9.29 m^2) of copper ($7\frac{1}{2}$ lb or 2.402 kg on each side of the copper). (*See* Table *C101*.)

Application

For roofing both sheet and strip are used, but for flashing only strip is normally used.

Accessories. Accessories for copper sheet and strip are generally of copper or a brass or bronze alloyed especially for the purpose. Included are such items as nails, screws,

Table C101 Sizes and weights of copper sheet and strip

	Weight per ft^2 (929.03 cm^2)		Width		Length		Minimum thickness	
Form	oz	g	in.	mm	ft	m	in.	mm
Sheet	16[a]	453.6	24, 30, 36	609.6, 762, 914.4	8, 10	2.44, 3.05	0.0190	0.4826
	20[a]	567.0	24, 30, 36	609.6, 762, 914.4	8	2.44	0.0245	0.6223
	24	680.4	24, 30, 36	609.6, 762, 914.4	8	2.44	0.0300	0.7620
	32	907.2	24, 30, 36	609.6, 762, 914.4	8	2.44	0.0405	1.0287
Strip	16[a]	453.6	10, 12, 14, 15	254, 304.8, 255.6, 381	8	2.44	0.0190	0.4826
	16[a]	453.6	20	508	8,10	2.44, 3.05	0.0190	0.4826
	20[a]	567.0	20	508	8	2.44	0.0245	0.6223
	24	680.4	20	508	8	2.44	0.0300	0.7620
	32	907.2	20	508	8	2.44	0.0405	1.0287
Roll	16[a]	453.6	6, 7, 8, 10, 12, 14, 16, 18, 20	152.4, 177.8, 203.2, 254, 304.8, 255.6, 406.4, 457.2, 508	Varying lengths		0.0190	0.4826
Parallel-edge strip	16[a]	453.6	6–9.5	152.4–241.3	10, 10.33	3.05, 3.15	0.0190	0.4826

[a]All 16-oz (453.6-g) to 20-oz (567-g) sheet and strip copper is available lead-coated.

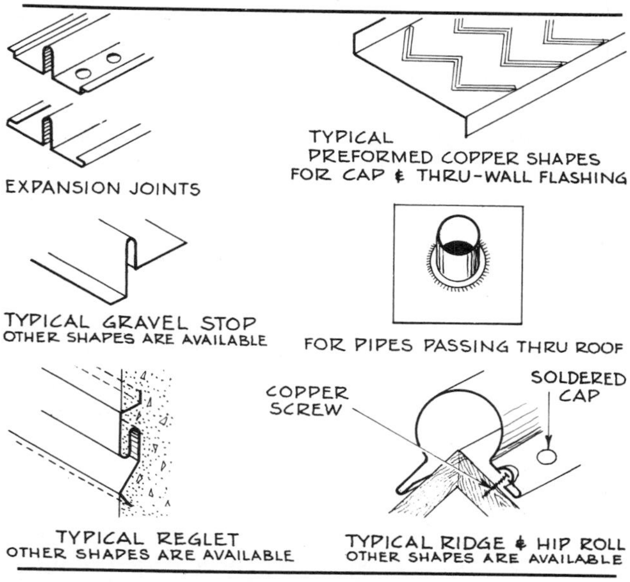

Figure C107 Prefabricated copper shapes for flashing and roofing.

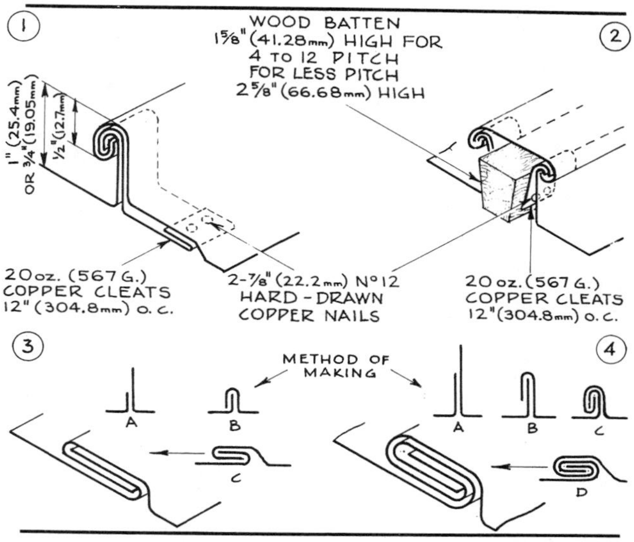

Figure C108 Details of standing (1), batten (2), single-lock (3), and double-lock (4) seams.

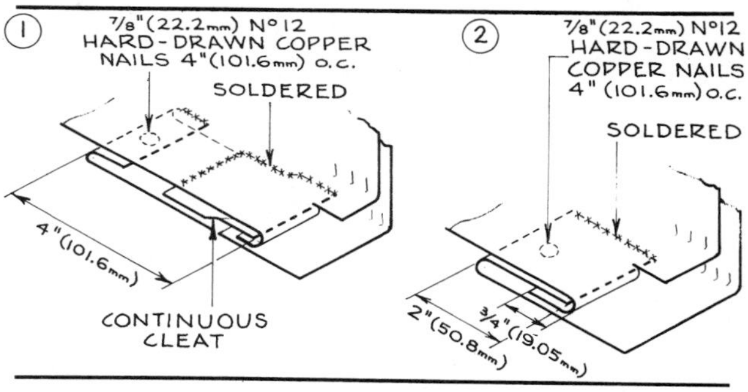

Figure C109 Details of cross seams for low-pitch (1) and steep-pitch (2) standing and batten seam roofing.

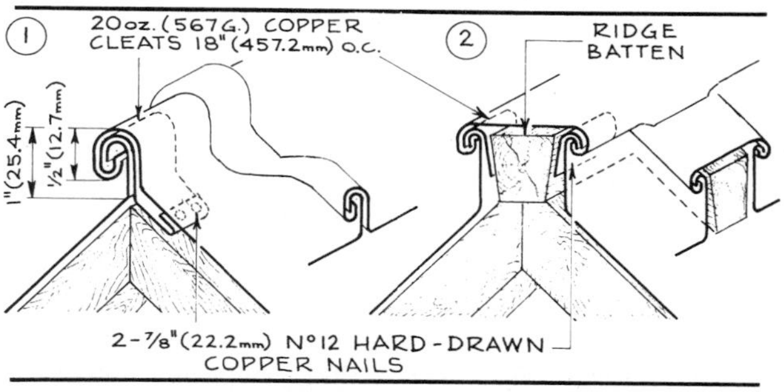

Figure C110 Details of standing seams (1) and batten seams (2) at ridge or hip.

washers, nuts and bolts, rods, bars, and small angles. There are also many prefabricated shapes made of copper for general use in flashing and roofing (*see* Figure *C107*). The expansion joints and similar items have even wider general applications. The same precautions must be observed for both roofing and flashing, whether of ordinary copper or lead-coated copper.

Condensed Checklist

1. Copper sheet and strip should be bent only to rounded angles, never sharply.
2. All fasteners should be of copper; bronze may be used, but the alloy ingredients should be checked for possible galvanic action.
3. All nails should have large, flat heads and sharp points.
4. Solders used for joining, particularly on flat seam roofing, should be 50% lead and 50% tin with a resin flux; joints to be soldered should be pretinned and cleaned immediately after soldering.
5. Copper must not come into direct contact with aluminum, steel, stainless steel, zinc, or magnesium, as it will corrode these metals.
6. Asphalt felt roofing paper must be covered with rosin paper before the copper is applied.
7. Areas of copper flashing in contact with built-up roofing must be precoated with pitch.
8. Treatment of standing, batten, single-lock, and double-lock seams for joining roofing sheets and strips (*see* Figure *C108*).
9. Type of cross seam for roofs with low pitch (3:12 up to 6:12) and steep pitch (6:12 and over) (*see* Figure *C109*).
10. Treatment of standing or batten seams at ridges or hips of roof intersections; in these seams the ridge batten is built high in order to enclose the top of the roof batten (*see* Figure *C110*).

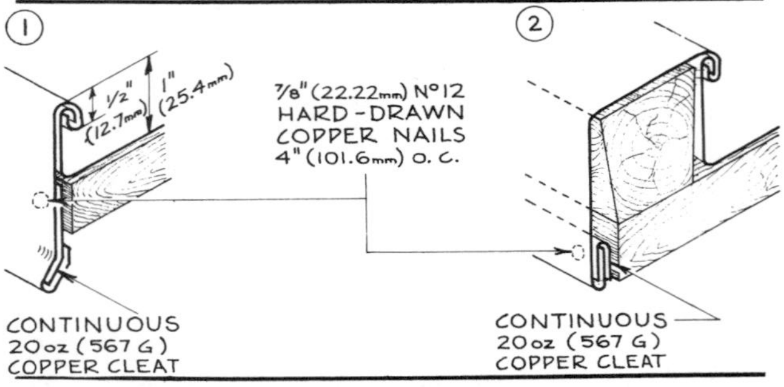

Figure C111 Details of standing seams (1) and batten seams (2) at gable ends.

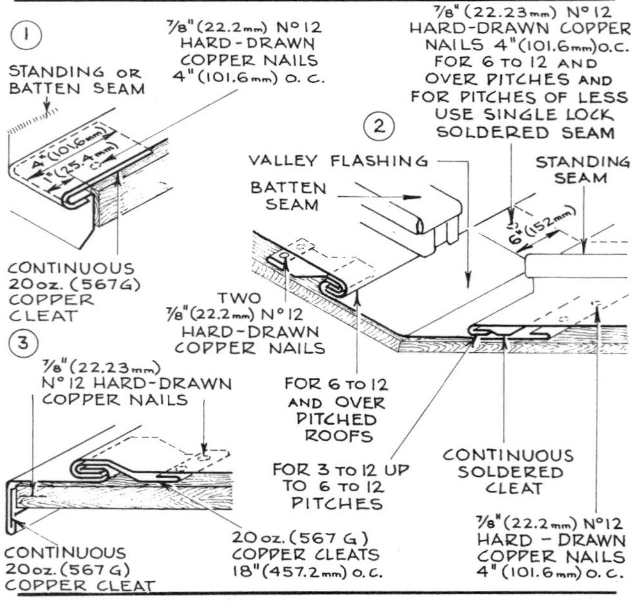

Figure C112 Details of standing and batten seams at eave drip (1) and valley (2), and of flat seam of eave drip (3).

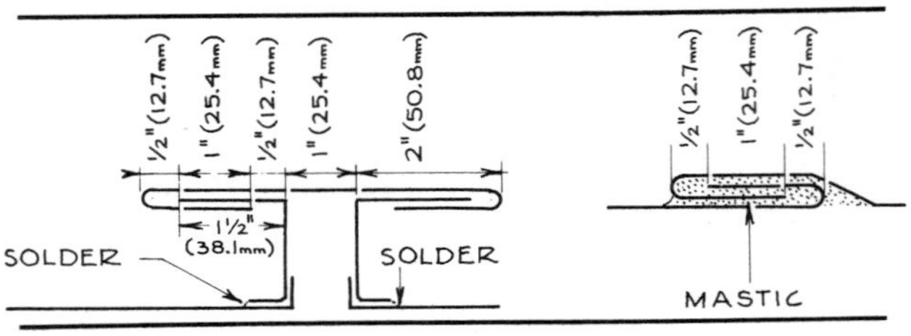

Figure C113 Details of expansion joints.

11. Treatment of standing or batten seams at roof gable ends (*see* Figure *C111*).
12. Treatment of standing and batten seams at eave drip and valley, and use of flat seams at eave drip (*see* Figure *C112*).
13. Roofing lengths of 8 ft (2.44 m) should be used wherever possible so as to avoid the need for expansion joints. When expansion joints beocme necessary, that is, in lengths over 8 ft (2.44 m), they should be installed every 12 ft (3.66 m) at a minimum (*see* Figure *C113*).
14. For flat seam roofing, 16 × 18 in. (406.4 × 457.2 mm) sheets of 20-oz (567-g) copper should be used and expansion joints placed at 30 ft (9.144 m) intervals in both directions.
15. Expansion rates of copper sheet and strip are comparatively great. The sheet and strip should preferably be held with clips to allow for this factor, but when nailing is necessary this should be done only at one end of the sheet or strip. The use of maximum lengths of 8 ft (2.44 m) eliminates the needs for expansion joints, but, in the case of gutter and leader connections, the bottom width of the gutter controls the maximum distance between expansion joints. The wider the gutter, the less is the distance between joints. Stock gutters made of 16-oz (453.6-g) copper are available in 4 and 6 in. 101.6 and 152.4 mm) widths. With these, a good general rule is to allow expansion joints 12 ft (3.66 m) from leaders and every 24 ft (7.315 m) on straight runs. If special gutters and leaders are designed, a check should be made with the manufacturers regarding the number and placing of expansion joints.
16. To prevent discoloration by copper sheet or strip, drainage must be thorough and complete. All drainage water must be removed in copper lead-offs (gutters, leaders, drains, etc.).

Conditions Favorable to the Use of Copper Sheet and Strip

1. Where a permanent, strong, corrosion-resistant material is required for roofing and flashing, particularly in sea air.
2. Where strong vertical lines are required on a roof.
3. Where a flashing material is required that will adhere to concrete, mortar, or plaster and will not be corroded by them.
4. Where a fire-resistant material is desired. Municipal, state, and local codes should be checked and also the fire rating codes of the Fire Underwriters, insurance companies, labor departments, and federal government (Army, Navy, etc.).

Conditions Unfavorable to the Use of Copper Sheet and Strip

1. Where direct contact between copper and aluminum, steel, stainless steel, zinc, or magnesium is unavoidable.
2. Where a nonstaining material is required.

3. Where a color other than the characteristic light green patina of aged copper is desired. *Note:* The copper can be covered with a clear protective coating, but a check should always be made of the length of time this coating will remain protective under the particular conditions of use.
4. In warm climates, as the light green color is not highly reflective.

History and Manufacture

The origins of the use of copper in sheet and strip form and for roofing are unknown. Copper sheet and strip were first manufactured by hammering and first produced on a commercial basis by this method about 1500. About 1750 the rolling mill technique was invented, and ever since sheet and strip have been manufactured by hot and cold rolling. The first sheet copper was produced in the United States by Paul Revere in 1801.

Hot and Cold Rolling. Today all sheet and strip copper is first hot-rolled (known as the breakdown); then the sheet is cold- or hot-rolled to finish thickness, whereas the strip is always cold-rolled to finish thickness. In the preliminary or "breakdown" hot-rolling process, the copper is heated to between 1200 and 1700°F (648.89 and 926.67°C). This provides strain relief, final temper (hardness), and sufficient temperature for hot rolling and rerolling. After heating, the copper is passed through the rolls 11 or 13 times, reducing the original thickness by 90% (a slab $4\frac{5}{8}$-in. or 117.45-mm thick is reduced to 0.2-in. or 5.08-mm thickness). The copper is then hot- or cold-rolled to finish thickness. Between cold rollings it is necessary to reheat the copper, which otherwise becomes harder and less ductile. Most copper rolling mills have edge rollers to control the width and to prevent cracking at the edges. Oil is used in all the rolling processes to cool surfaces and act as a cushion between the rollers and the copper, ensuring a uniform pressure over the entire surface.

Pickling and Finishing. The sheet and strip end-products are usually pickled in a continuous process immediately after the cold rolling. The pickling bath consists of a 5 to 10% sulfuric acid solution at room temperatures up to 125°F (51.67°C). After pickling, the copper sheet and strip are finished by slitting, cutting with shears or saws, and flattening by roller or stretch levelers.

COPPER WIRE

Physical and Chemical Properties

Copper wire is generally fabricated of standard annealed copper made from electrolytic, tough-pitch copper. Special coppers and copper alloys are used for special electrical conditions. All copper wire that is to be rubber-coated is always given a thin, protective plating of tin to prevent corrosion by the sulfur present in rubber. Copper wire to be coated with various synthetics is always coated first with materials that prevent corrosion from substances in the synthetics which can chemically attack copper.

With the growing sophistication of environmental controls in the construction industry, there is increased use of all categories of copper and copper-alloy electrical conductors, namely, wire, cable, bus bars, rods, and the like.

Types and Uses

The largest quantity of copper and copper-alloy wire is used in the electrical field. In construction, wire is woven into mesh and wire cloth for screens, grilles, and protective barriers of various types. The standard gauges of copper are based on those used for steel wire and are given in circular mils [a circular mil is the area of a circle 1 mil, or 0.001 in. (0.0254 mm), in diameter].

History and Manufacture

Copper wire was originally handmade by cutting and hammering, as noted in the Bible. Later it was drawn by hand through stone dies and still later through iron dies. About 1350 water power was first used for drawing; from 1769 steam power was used until electrical power replaced it entirely in industry. The real growth of the copper wire industry began with the discovery of the electric lamp by Edison in 1878. Since then, consumption of copper wire has increased steadily.

Copper wire is manufactured by hot rolling or by hot extrusion of billets into rods about $\frac{1}{4}$ in. (6.35 mm) in diameter, which are then cold-drawn through various stages into wire form. Most copper-alloy wire is drawn from extruded rods. After the wire is drawn to finish size, it receives various types of coverings according to its use. Thus wire may be enameled or tinned; it may be insulated with rubber, paper, plastic, glass fiber, silk, or cotton.

CORK

Physical and Chemical Properties

Cork is the thick spongy bark of an evergreen oak (*Quercus suber*), which grows in southern Europe and along the northern coast of Africa. The outer layer of bark grows thicker each year and gradually becomes a thick, soft, homogeneous mass possessing a peculiar combination of commercially valuable properties.

Cork is characterized by buoyancy, elasticity, low thermal conductivity, and a high coefficient of friction. It is chemically inert and has a relatively high degree of imperviousness to penetration by either air or water. It can be highly compressed vertically without horizontal, or lateral, spread. One of the lightest of solid materials, it has specific gravity of 0.15 to 0.25. It begins to char at 250°F (121.11°C) but ignites only in contact with flame.

Cell Structure. Many of these distinctive properties result from the peculiar structure of the cork cell which comprises the structural unit of the bark. The walls of each cell are very thick and heavily impregnated with a fatty substance that makes them essentially impervious to air and water. A 1-in. (1639 mm^3) cube of cork contains approximately 200,000,000 of these tiny air-filled cells, so that more than 50% of its volume is captive air.

Commercial Forms. Cork is available in bulk form cut into sheets, boards, and blocks (lagging), and in granulated form, graded to size and including flour ground so fine that it floats in air.

Types and Uses

Natural and composition cork products have many direct applications in architecture. Sheet and board are used for bulletin boards, thermal and sound insulation, vibration control, and resilient-type flooring. Cork block (lagging) is widely used for thermal and sound insulation, for example, in refrigerator rooms, piping, ducts, and areas where precise acoustic or thermal control is necessary. A main use of cork in flour form is as an ingredient in linoleum. Ground cork is mixed with wet clay to form fire bricks, in which small insulating air spaces are left as

Table C102 Treatment and major uses of the various forms of cork

Form of cork used	Treatment	Form of material	Major use in construction	Other uses
Ground	Pressure and heat, using natural resins in cork	Sheet, board, and tile	Acoustic ceiling treatments, especially in areas where moisture is present (swimming pools, shower rooms)	Refrigerated rooms, pipes, ducts, etc.
Granular	Pressure and heat, using binders, animal or vegetable glue, glycerin, or synthetic resins	Sheet, board, resilient flooring, and acoustic tile	Bulletin boards, acoustic ceiling treatments, flooring	Refrigerated rooms, pipes, vibration control, gaskets, etc.

Table C103 Sizes, colors, textures, and major uses of cork materials in construction

	Dimensions							
	Thickness		Width		Length			
Type of cork material	in.	mm or cm	in. and ft	cm or m	in. and ft	cm or m	Color and texture	Major use
Bulletin boards in sheet form	$\frac{1}{4}$	6.35 mm	4′-0″	1.219 m	100′-0″	30.48 m	Natural cork colors or pigmented, usually in light and dark green, red, blue, and gray	Bulletin boards in all types of buildings
	$\frac{1}{4}$	6.35 mm	6′-7″	2.007 m	91′-0″	27.74 m		
	$\frac{1}{8}$, $\frac{1}{4}$, and $\frac{1}{2}$	3.20 mm, 6.35 mm, and 12.7 mm	4′-0″	1.219 m	16′-0″	4.88 m		

Table C103 Sizes, colors, textures, and major uses of cork materials in construction (*continued*)

Type of cork material	Dimensions: Thickness, in.	Thickness, mm or cm	Width, in. and ft	Width, cm or m	Length, in. and ft	Length, cm or m	Color and texture	Major use
Resilient flooring	3/32,	2.4 mm,					Light, medium, and	Floors and walls except
	1/8,	3.2 mm,	6"	15.24 cm	6"	15.24 cm	dark natural cork	on subflooring of con-
	3/16,	4.8 mm,			12"	30.48 cm	colors, and pig-	crete slabs on grade or
	5/16;	8.4 mm;	9"	22.86 cm	9"	22.86 cm	mented in light	below grade on earth
	also	also	12"	30.48 cm	12"	30.48 cm	colors	
	1/4,	6.35 mm,			24"	60.96 cm		
	1/2	12.7 mm						
	made to order							
Resilient flooring,			9"	22.86 cm	9"	22.86 cm	Light, medium, and	Floors and walls except
plastic-surfaced	1/8	3.2 mm			12"	30.48 cm	dark, patterned in	on subflooring of con-
	and	and	12"	30.48 cm	and	and	natural cork colors,	crete slabs on grade or
	3/16	4.8 mm			24"	60.96 cm	and pigmented in	below grade on earth
			36"	91.44 cm	36"	91.44 cm	light colors	
Thermal insulation	1,	2.54 cm,	12",	30.48 cm,			Light color of cork	Roof insulation and
sheet and board	1 1/2,	3.81 cm,						refrigerated rooms
	2,	5.08 cm,	1'-6",	45.72 cm,				
	3,	7.62 cm,			36"	91.44 cm		
	4,	10.16 cm	2'-0"	60.96 cm				
	and	and	and	and				
	6	15.24 cm	3'-0"	91.44 cm				
Acoustic tile	1/4	6.35 mm	5 3/4"	14.61 cm	11 1/2"	29.21 cm	Painted, usually white	Ceiling and wall acous-
	1 1/2	3.81 cm	11 1/2"	29.21 cm				tic treatment
	3/8	9.6 mm	2'-0"	60.96 cm	12'-0"	3.66 m	Brick-patterned in	Wall acoustic treatment
							various colors	

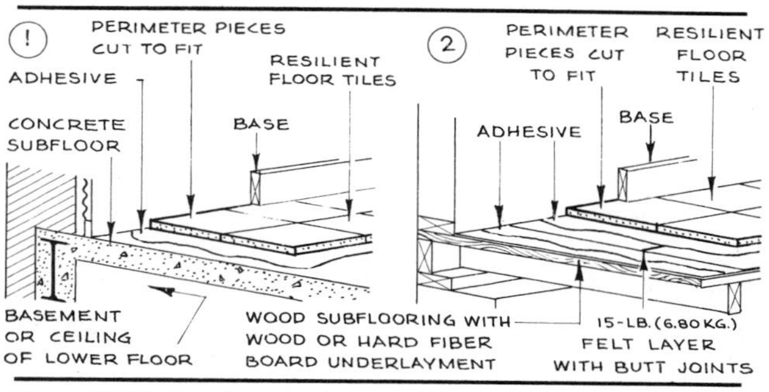

Figure C114 Details of installation of cork flooring on concrete slab (1) and wood subflooring (2).

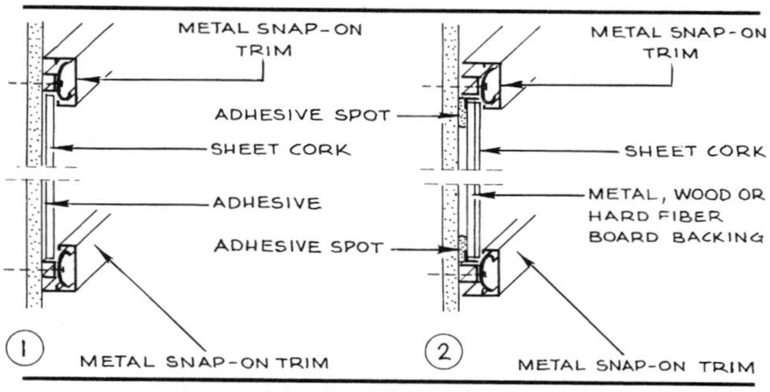

Figure C115 Installation details for cork bulletin boards.

the cork burns out. Nonconstruction uses of cork include gaskets, cork stoppers, cork disks for stoppers made of other materials, polishing wheels, soles and heels of shoes, floats, buoys, and life preservers. It also finds wide use in industrial equipment and in the manufacture of musical instruments. (*See* Tables *C102* and *C103*.)

Application

Condensed Checklist

1. Wall or floor surfaces on which cork is to be installed should be smooth, clean, dry, and free from cracks and warping. Figures *C114* and *C115* show typical installation details.

2. The adhesive recommended by the manufacturer of the cork material should be used according to the manufacturer's directions.
3. Rooms in which cork materials are being installed should remain at 70°F (21.11°C) minimum for 24 hours before and after installation.
4. All legs of radiators, built-in furniture, etc., should rest upon metal inserts, not the cork flooring material (*see* Figure *C116*).

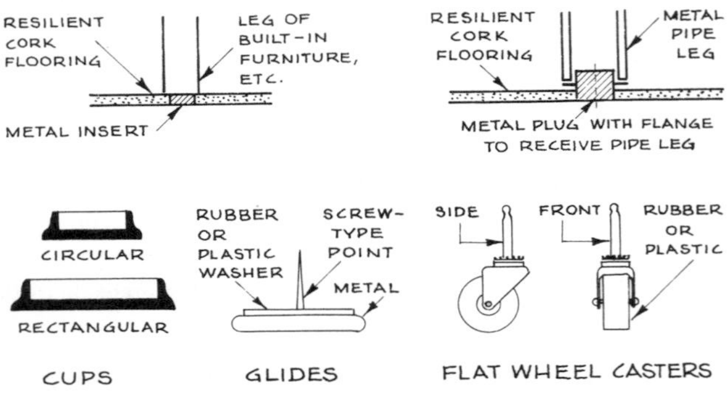

Figure C116 Method of eliminating indentation of cork flooring by furniture, convectors, etc.

5. Cork materials should be cleaned only with neutral soaps and water, and waxes must be of the water-emulsion type.
6. Local, municipal, and state codes and also the codes of the Fire Underwriters, insurance companies, labor departments, and federal government (Army, Navy, etc.) should be checked for limitations and fire resistance ratings.

Conditions Favorable to the Use of Cork Materials

1. Where the design requires resilient flooring that is durable, easy to maintain, and easy to walk upon, and has high sound-absorbing values.
2. In any location where a permanent bulletin board is required, as cork can withstand continual punching by thumb tacks.
3. In areas where acoustic treatment for walls or ceilings is necessary. Always check the acoustic values of the various cork materials, and consult specialists in this field.
4. Where thermal insulation for refrigerated areas, doors, and specialized cool rooms is required. Consult specialists in this field.
5. For machinery vibration control. Always consult specialists in this field.

Conditions Unfavorable to the Use of Cork Materials

1. As resilient flooring in areas of very heavy traffic or on concrete slab subflooring poured on earth either at grade or below grade.
2. As thermal insulation subject to high heat.
3. Wherever high fire resistance is required.

History and Manufacturer

Stripping or "Harvesting." Cork trees can be stripped first when they are 15 to 20 years old and once every 8 to 10 years thereafter. The first yield, called virgin cork, is rough and woody in texture and is used principally for the manufacture of cork insulation and isolation products. Each subsequent yield through the fifth or sixth improves in quality but is still considered low grade (known as refugo) and is used similarly to virgin cork. Thereafter the tree produces a stable, high-quality cork for 150 years or more.

Processing. Natural cork must be processed for most of its uses. First it is seasoned, then flattened out, and dried. After trimming and grading, it is marketed for solid cork products such as stoppers, floats, and other marine products. What remains is used for various cork composition products.

CORROSION

The deterioration of a metal by electrochemical or chemical action, resulting from exposure to weather, moisture, chemicals, or other agents or media. *See* Metals.

CORUNDUM

Corundum is a hard, crystalline mineral composed of aluminum oxide (Al_2O_3). It has a high specific gravity (4.0) and melting point (3660°F or 2015.56°C). The finer transparent, colored varieties are valued as the gem stones ruby and sapphire. Impure massive forms are known as emery and are used as polishing agents. Corundum has been artificially produced both in large crystals as a gem stone and in finer granular, crystalline forms which are widely used as abrasives. The latter are often called aluminum oxide to differentiate the synthetic from the naturally occurring form. *See also* Aluminum Oxide; Abrasives.

CREOSOTE

Creosote is an oily, yellowish to dark greenish brown liquid which is clear at 71.6°F (22°C) or higher. It is soluble in alcohol and in benzene, has a distinct odor, and is poisonous. Creosote is obtained mainly as a by-product from coal which has been coked for blast furnace operation in steel-making; it is separated by fractional distillation of coal tar. It is also a product of wood tar distillation.

Construction Uses

The main use of creosote in the construction field is as a wood preservative. Wood is treated with creosote if it can be attacked by termites, marine borers, etc., and where it will come in direct contact with the earth, such as telephone, power line, and lighting poles; wood supports for exterior fencing; bulkheads for canals, slips, and docks; railroad ties; wood sills and supports in houses where termites and the possibility of alternate wetting and drying occur; and many other similar applications.

Wood treated with creosote stains to a darkish brown color and grows darker with age. The period of effective preservation depends on the amount of creosote applied or forced into the wood. Wood well saturated with creosote applied under pressure will be preserved for as long as 30 years (*see* Wood Preservatives).

CRUDE OIL

A natural hydrocarbon mixture found beneath the earth's surface either on land or under the seas. *See* Bitumens; Fossil Fuels; Petroleum.

CRYOLITE

Cryolite is a transparent, crystalline compound of sodium, fluorine, and aluminum. It has a vitreous to greasy luster and fuses easily. In its natural mineral form it occurs in commercial quantities only in Greenland. It is also produced synthetically on an industrial scale. Its main uses are as a source of aluminum (as the electrolyte in electrolytic production or as a flux), as a flux in ceramics (glass, enamel), in electrical insulation, and as a binder for abrasives. *See also* Fluorine; Aluminum; Glazes and Porcelain Enamels.

CURTAIN WALLS

In this book "curtain wall construction" refers to the systems used to install exterior prefabricated walls on buildings, using panels, window units, integrants, sandwich panels, grid systems, and mullions, both horizontal and vertical. Figure *C117* shows some typical curtain wall systems. *See also* Panels and Sandwich Panels; Integrants; Windows; Aluminum; Stainless Steel; Steel; Asbestos; Concrete; Stone; Plastics.

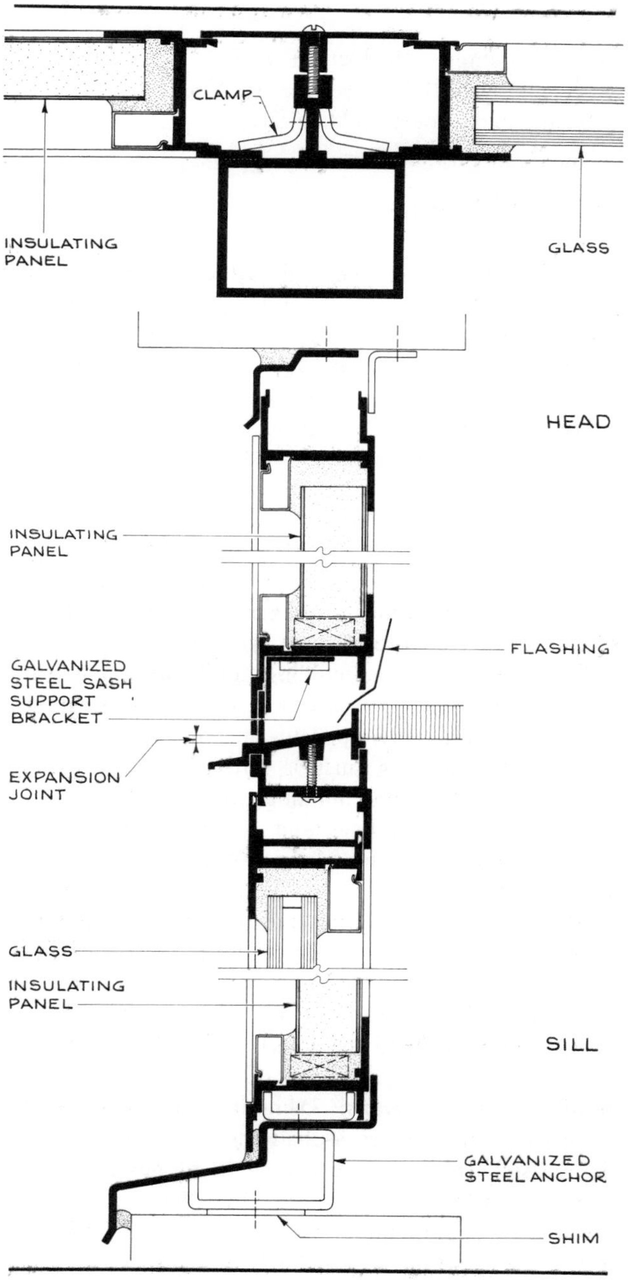

Figure C117 Typical curtain wall system.

D

DAMPENING

In construction, "dampening" refers to checking or deadening vibration. *See* Lead; Vibration Control; Acoustics and Acoustical Materials.

DAMPPROOFING

Dampproofing may be defined as the method of stopping dampness from the earth and a certain amount of surface water caused by rains from penetrating into areas at grade and below ground level. The term usually applies to protection against moisture and intermittent wetting and may be distinguished from waterproofing in that it does not involve constant hydrostatic pressures. (*See also* Flashing; Waterproofing.)

Waterproofing versus Dampproofing. One of the problems in construction is deciding when to dampproof and when to waterproof. For example, an architect may specify only dampproofing on the basis of site conditions and the fact that buildings adjoining or near the new construction have had no water problems; yet excavation work may open up an underground spring and then the building must be waterproofed instead of dampproofed. For large buildings for which test borings are required and underground conditions at adjacent buildings can be accurately determined, one usually knows beforehand what to expect. But for smaller projects where test borings are not the rule and the neighboring buildings are often quite far away, we must sometimes guess and meet problems as they arise. Actually no one knows what will be encountered below the surface until the excavations have been completed.

Methods and Materials. Dampproofing is generally achieved by applying an asphalt-base paint or coating to the outside surface of below-grade foundation walls. Usually two coats are applied so that the total exterior surfaces of the foundation walls are sealed with this water-repellent type of coating. A second method is to apply to the outside surface of the foundation walls a coat of dense cement plaster generally about $\frac{1}{2}$ in. (12.7 mm) thick, well pressed into the wall surface. Both methods will stop any ground moisture or surface water caused by rains from entering the building (*see* Figures *D1* and *D2*). Figure *D3* shows several methods of dampproofing masonry walls above grade.

Other substances are available for dampproofing below grade, and new ones are constantly being developed and introduced for this purpose, for example, silicones and plastics. When using any of these, one should always check the guaranteed effective lifetime of the material, the conditions of testing to arrive at this figure, and the resistance of the material to termites, chemical reaction with the soil, and attack by microorganisms in the soil.

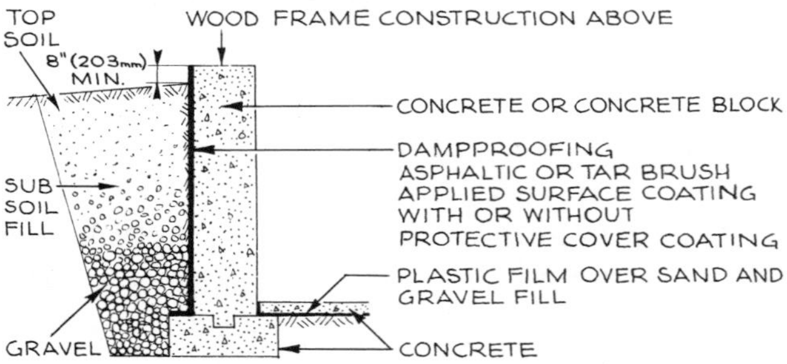

Figure D1 Dampproofing of foundation walls below grade with wood frame construction above.

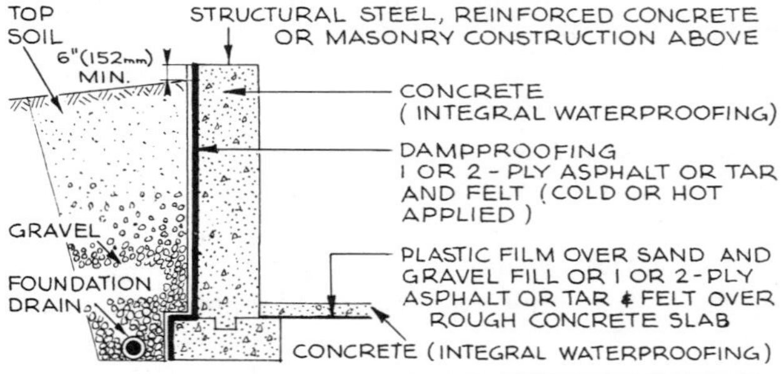

Figure D2 Dampproofing of foundation walls below grade with steel, concrete, or masonry construction above.

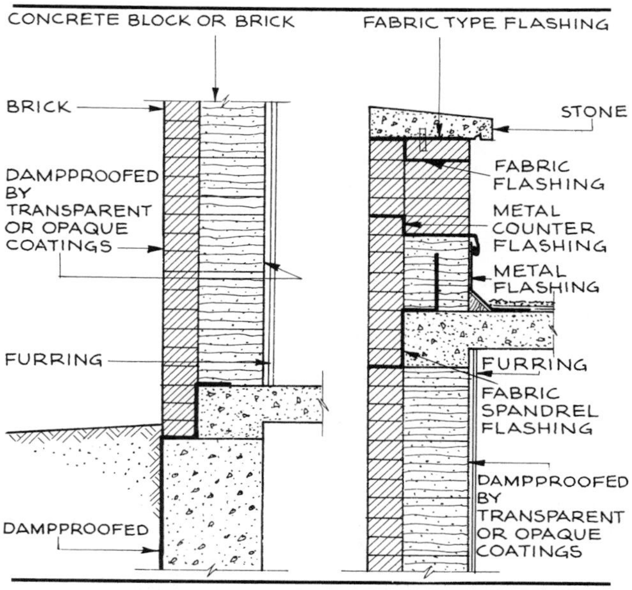

Figure D3 Dampproofing of walls above grade.

Foundation Drainage. In general, it is always good practice to install foundation drainage that empties into dry wells when dampproofing, as this will eliminate the possibility of water pressure building up on the outside.

DECKING

To clarify terminology, the word "plank" has been eliminated from this book and the word "decking" has been substituted for all such materials. "Plank" originally referred to a piece of sawed lumber at least 2 in. (50.8 mm) thick (i.e., a thick board) to be used for barn, factory, and similar heavy-duty flooring. Today the only correct use of the original word is in "plank and beam" construction that is wholly wood.

In this book decking is defined as a self-supporting flooring or roofing unit that is able to span between structural members. It does not include materials that are part of a structural system, such as reinforced concrete or flat-arch clay tile. Table *D1* lists various types of decking and materials from which they are made.

Decking is manufactured in two basic types, roof and floor decking. Roof decking may have a reinforced concrete slab on top of the decking, which is only a form for the concrete; or roof decking may consist of only an applied rigid insulation, and in this case the decking must support dead load, live load, and up-lift loads.

Floor and roof decking, in general, includes light-gauge steel panels formed with corrugations, flutes, ribs, or cellular construction of continuous longitudinal cells.

Table D1 Typical kinds of decking used in construction

Material	Advantages	Disadvantages	Major use
Asbestos (*See* Asbestos Panels and Integrants)	Good insulation value; fire resistance; provides finished ceiling	Limited spanning; limited to areas where there are no severe freezing temperatures; type of roofing material used is limited by thickness and nail-holding power of decking; electrical conduit and piping cannot be concealed within decking	Roof decking for residential construction
Concrete (*see* Concrete Decking and Structural Units)	Good fire resistance; provides finished ceiling; can span up to 7 ft (2.134 m)	No insulation value; electrical conduit and piping cannot be concealed within decking	General roof and floor decking
Concrete, lightweight (*see* Concrete Decking and Structural Units)	Insulation value; fire resistance; provides finished ceiling; some acoustic value	Limited in spanning; electrical conduit and piping cannot be concealed within decking	General roof decking
Concrete, structural units (*see* Concrete Decking and Structural Units)	Insulation value; good fire resistance; some types can span up to 30 ft (9.144 m); some types make allowance for concealing electrical conduit and piping; provides finished ceiling	Structure must be designed to fit type of decking used; electrical conduit and piping must be designed to fit into type of decking used	General roof and floor decking

Table D1 Typical kinds of decking used in construction (*continued*)

Material	Advantages	Disadvantages	Major use
Paper (*see* Paper and Paper Pulp Products; Paper Pulp Decking)	Good insulation value; flame resistance; provides finished ceiling; acoustic value in some types	Limited spanning; limited to areas where severe freezing temperatures do not occur; type of roofing material limited by thickness and nail-holding power of decking used; electrical conduit and piping cannot be concealed within decking	Roof decking for residential construction
Plaster (*see* Plaster Decking)	Insulation value; good fire resistance; provides finished ceiling; electrical conduit can be concealed within decking; installation possible at freezing temperatures	Limited in spanning; special framing system with forms is required as it is poured on the job, not prefabricated; piping cannot be concealed within decking	General roof decking
Steel (*see* Steel, Corrugated)	Fire resistance; provides finished ceiling; some types make allowance for concealing electrical conduit and piping; wide spanning up to 20 ft (6.096 m)	No insulation value; methods of attachment vary with type of decking; barrier between steel and roofing material is necessary	General roof and floor decking
Wood (*see* Wood; Wood Integrants (Plywood); Wood, Structural)	Some insulation value; provides finished ceiling	Limited in spanning; limited flame resistance; electrical conduit and piping cannot be concealed within decking	Roof decking for residential construction

Roof decking will span from 5 to 10 ft (1.524 to 3.048 m) between supports, depending on the type of decking; with multiple supports the bearing capacity increases. Metal roof decking is made of 18, 20, and 22 gauge steel and has a prime coat. This prime coat is only for protection until insulation and roofing are installed.

Floor decking is used as a form for thin reinforced concrete slabs and/or as a structural system and a subflooring. When using this type of structural system, one should always check with the manufacturers of floor decking as to availability, span limitations, depths, method of applying finish flooring, and especially, how to include and coordinate the pipes, ducts, and conduit for the plumbing, heating-ventilating-air conditioning, and electrical work.

Floor decking is made of steel and is manufactured in various gauges, lengths, and depths. It is available in lengths to span from 5 to 30 ft (1.524 to 9.144 m) and greater, and in widths from 18 to 36 in. (0.457 m to 0.914 m). When this type of structural flooring system is selected, a structural engineer should always be consulted.

DIATOMITE

Diatomite, or diatomaceous earth, erroneously called infusorial earth, is actually not earthy but an opaline mineral composed principally of silica and consisting chiefly of the fossil remains of aquatic plant organisms known as diatoms. Kieselguhr is the German term formerly used in this country. Tripoli is a variety of crystalline diatomite. The purest varieties of diatomite are chalklike in appearance; impurities may cause greenish, brown, or gray color.

Diatomite is soft, porous, friable, very light in weight, and able to withstand high temperatures. When dry, its apparent specific gravity (as mined and used) is given as 0.15 to 0.45. The true specific gravity of the mineral ranges from 1.9 to 2.35. Thus it can be used to give bulk without adding appreciable weight.

Diatomite is insoluble in acids, except hydrofluoric acid, and it is soluble in alkalis. It is able to absorb $1\frac{1}{2}$ to 4 times its own weight in water. Its closed cellular structure makes it a poor conductor of sound, heat, and electricity.

Table D2 Uses of diatomite

	Construction Uses	
Basis of	Abrasives	Mild abrasive and polishing agent
Component of	Adhesives	Filtration
	Aggregates	Thermal and sound insulating
	Asbestos	Bonded with asbestos fiber into boards
	Asphalt products	Filler
	Insulation	Heat- and sound-absorbing aggregates, cement, powder, sawed or molded blocks and bricks
	Plaster	Filler
	Rubber	Filler
	Textiles	Filler, ingredient of textile fireproofing agents

Commercial Forms. Diatomite is available as a powder and as a calcined aggregate; combined with a clay binder, it may also be obtained as high-temperature refractory brick.

Uses

Diatomite is extensively used as a filler in paints, paper and paper pulp products, plaster and plaster products, plastics, and rubbers. In paints it also functions as a flatting agent. Its other uses are listed in Table *D2*.

DOORS

Doors may be broadly defined as a means of closing off areas or entrances or exits in buildings. In designing and specifying any type of door, every aspect of the end use to which it will be put must be analyzed and checked. For example, for entrances one must know how many people are going to pass in and out at any given time and how much control of this volume is necessary or desirable; for closets, how much of the closet area should be made accessible; for fire exits, what the code restrictions are and which doors will meet these requirements; for a garage, what types and number of cars or other vehicles are to be accommodated. These same considerations hold true for doors for fire barriers, loading platforms, hangars, garages, industrial and manufacturing buildings, and any type of construction where areas must be separated, closed off, or isolated.

Table *D3* shows the different forms and types of doors available and the general characteristics and possible variations for each type. Figures *D4* and *D5* show various types of doors for industrial, commercial, and

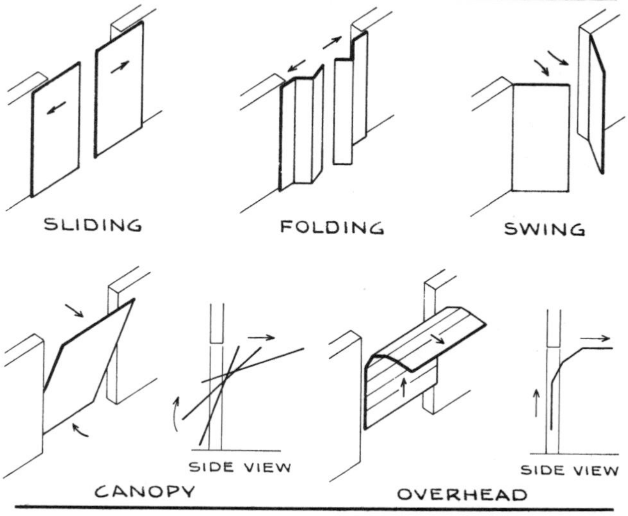

Figure D4 Types of doors for industrial, commercial, and manufacturing construction.

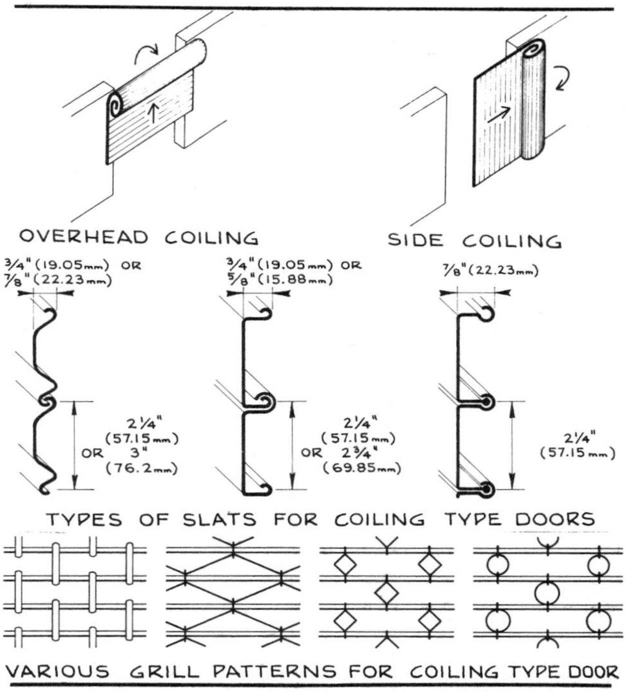

Figure D5 Types of overhead and side-coiling industrial, commercial, and manufacturing doors.

Table D3 Types and characteristics of doors

TYPE	MATERIAL	FORM	OPERATION	OPERATING HARDWARE	WEATHERSTRIP
SWING	ALUMINUM, BRONZE, GLASS, STAINLESS STEEL, STEEL AND WOOD	1 2 3 4 5 6 7 8 9 10 11 12	MANUAL MECHANICAL ELECTRICAL	SIDE HINGES PIVOTS	WITH OR WITHOUT
BY-PASSING CLOSED OPEN	ALUMINUM, BRONZE, STAINLESS STEEL, STEEL AND WOOD	2 3 7 8	MANUAL MECHANICAL ELECTRICAL	ROLLERS SLIDES	WITH OR WITHOUT
FOLDING CENTER PIVOT HINGED SIDE PLASTIC AND FABRIC	ALUMINUM, BRONZE, STAINLESS STEEL, STEEL, WOOD, FABRIC AND PLASTIC	2 3 6 7 8 9 10	MANUAL MECHANICAL ELECTRICAL	ROLLERS SLIDES	
OVERHEAD FOLD-UP ROLLING SWING-UP	ALUMINUM, BRONZE, STAINLESS STEEL, STEEL AND WOOD	2 8 11 12	MANUAL MECHANICAL ELECTRICAL	ROLLERS SLIDES	WITH OR WITHOUT
REVOLVING	ALUMINUM, WOOD, BRONZE, GLASS, STAINLESS STEEL, STEEL, NICKEL-SILVER	2 3 4 8	MANUAL PANIC PROOF ELECTRICAL	PIVOTS	WITH OR WITHOUT
SLIDING ROLL	ALUMINUM, BRONZE, STAINLESS STEEL, STEEL AND WOOD	2 3 4 6 7 8 9 10	MANUAL MECHANICAL ELECTRICAL	ROLLERS SLIDES	WITH OR WITHOUT

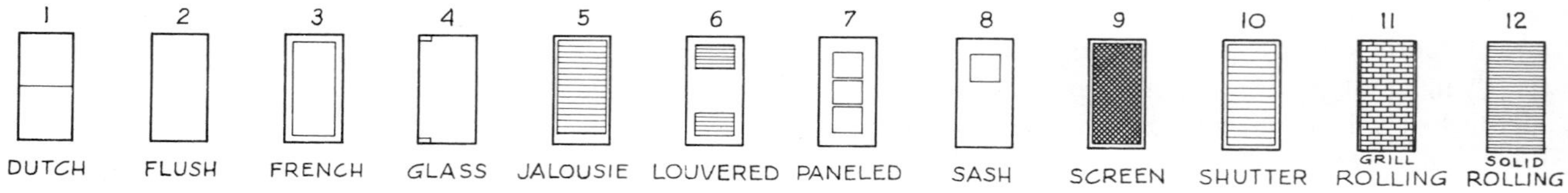

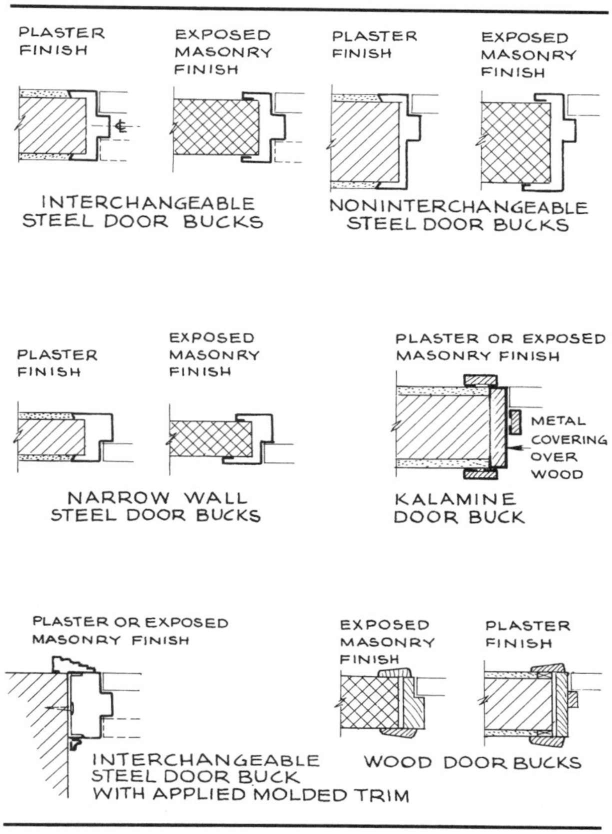

Figure D6 Typical details of interior door bucks.

manufacturing construction. When the types of doors have been selected for a building, more specific information can be found under the appropriate general heading (e.g., Aluminum; Bronze; Glass; Wood; Hardware) about the basic material of which the door and its components are made, and under specific headings (e.g., Aluminum Doors; Wood Windows; Wood Doors) the available sizes and finishes, the methods of application, and the proper precautions to observe in choice and use.

All doors must fit into a frame known as a "buck" (*see* Figures *D6* and *D7*); specific details are shown for various types of interior and exterior door bucks. For further information, *see* Aluminum; Steel; Wood; and Stainless Steel.

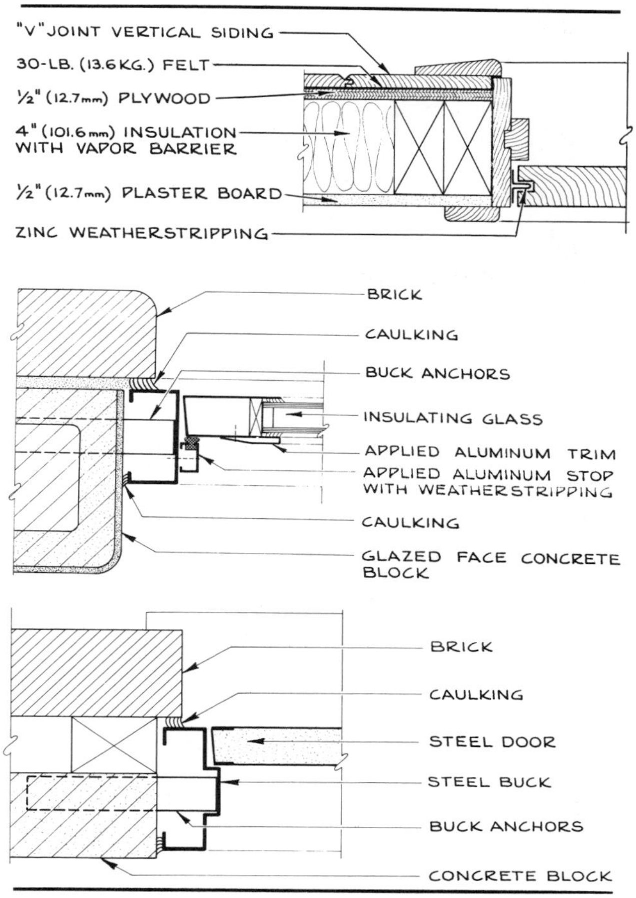

Figure D7 Typical details of exterior aluminum, steel, and wood door bucks.

DURALUMIN

Duralumin was originally the trademarked name for a wrought aluminum-copper alloy containing sufficient magnesium to allow heat treatment and a slight amount of manganese for hardening. The aircraft and metalwork industries use the term "duralumin-type alloy" for any wrought aluminum alloy containing 3 to 4.5% copper, 0.4 to 1.0% magnesium, 0 to 0.7% manganese, and the remainder aluminum with or without very small amounts of iron and silicon. (*See* Aluminum Alloys.)

E

ELEMENTS

Elements can no longer be simply defined as the ultimate constituents of matter. As far as the construction field is concerned, the chemical elements are of interest only because by analogy they comprise the building blocks of which the materials used in construction are made. For example, glass consists of silicon, sodium, calcium, oxygen, and small amounts of other elements. Elements which are classed as metals are of direct concern in construction and are quite fully discussed both as a general class (*see* Metals) and as individual metals and alloys. Of the 103 known elements, 88 are metallic and 15 are nonmetallic; 23 are radioactive. All the elements are listed in Table *E1* in which nonmetals are indicated by a single asterisk.

Table E1 List of the elements[a]

Element	Symbol	Atomic number	Physical state	Element	Symbol	Atomic number	Physical state
Actinium	Ac	89	Metal	Gadolinium	Gd	64	Metal
Aluminum	Al	13	Metal	Gallium	Ga	31	Metal
Americium	Am	95	Metal	Germanium	Ge	32	Metal
Antimony	Sb	51	Metal	Gold	Au	79	Metal
Argon*	Ar	18	Gas	Hafnium	Hf	72	Metal
Arsenic	As	33	Metal	Helium*	He	2	Gas
Astatine*	At	85	Gas	Holmium	Ho	67	Metal
Barium	Ba	56	Metal	Hydrogen	H	1	Metal
Berkelium	Bk	97	Metal	Indium	In	49	Metal
Beryllium	Be	4	Metal	Iodine*	I	53	Solid
Bismuth	Bi	83	Metal	Iridium	Ir	77	Metal
Boron	B	5	Metal	Iron[b]	Fe	26	Metal
Bromine*	Br	35	Liquid	Krypton*	Kr	36	Gas
Cadmium	Cd	48	Metal	Lanthanum	La	57	Metal
Calcium	Ca	20	Metal	Lawrencium	Lw	103	Metal
Californium	Cf	98	Metal	Lead[b]	Pd	82	Metal
Carbon*	C	6	Metal	Lithium	Li	3	Metal
Cerium	Ce	58	Metal	Lutetium	Lu	71	Metal
Cesium	Cs	55	Metal	Magnesium	Mg	12	Metal
Chlorine*	Cl	17	Gas	Manganese	Mn	25	Metal
Chromium	Cr	24	Metal	Mendelevium	Md	101	Metal
Cobalt	Co	27	Metal	Mercury	Hg	80	Metal
Copper[b]	Cu	29	Metal	Molybdenum	Mo	42	Metal
Curium	Cm	96	Metal	Neodymium	Nd	60	Metal
Dysprosium	Dy	66	Metal	Neon*	Ne	10	Gas
Erbium	Er	68	Metal	Neptunium	Np	93	Metal
Einsteinium	Es	99	Metal	Nickel	Ni	28	Metal
Europium	Eu	63	Metal	Niobium[c]	Nb	41	Metal
Fermium	Fm	100	Metal	Nitrogen*	N	7	Gas
Fluorine*	F	9	Gas	Nobelium	No	102	Metal
Francium	Fa	87	Metal	Osmium	Os	76	Metal

Table E1 List of the elements[a] (*continued*)

Element	Symbol	Atomic number	Physical state	Element	Symbol	Atomic number	Physical state
Oxygen*	O	8	Gas	Sodium	Na	11	Metal
Palladium	Pd	46	Metal	Strontium	Sr	38	Metal
Phosphorus*	P	15	Solid	Sulfur	S	16	Solid or liquid
Platinum	Pl	78	Metal				
Plutonium	Pu	94	Metal	Tantalum	Ta	73	Metal
Polonium	Po	84	Metal	Technetium	Tc	43	Metal
Potassium	K	19	Metal	Tellurium	Te	52	Metal
Praseodymium	Pr	59	Metal	Terbium	Tb	65	Metal
Promethium	Pm	61	Metal	Thallium	Tl	81	Metal
Protactinium	Pa	91	Metal	Thorium	Th	90	Metal
Radium	Ra	88	Metal	Thulium	Tm	69	Metal
Radon*	Rn	86	Gas	Tin[b]	Sn	50	Metal
Rhenium	Re	75	Metal	Titanium	Ti	22	Metal
Rhodium	Rh	45	Metal	Tungsten[b]	W	74	Metal
Rubidium	Rb	37	Metal	Uranium	U	92	Metal
Ruthenium	Ru	44	Metal	Vanadium	V	23	Metal
Samarium	Sm	62	Metal	Xenon*	Xe	54	Gas
Scandium	Sc	21	Metal	Ytterbium	Yb	70	Metal
Selenium	Se	34	Metal	Yttrium	Y	39	Metal
Silicon	Si	14	Metal	Zinc	Zn	30	Metal
Silver[b]	Ag	47	Metal	Zirconium	Zr	40	Metal

[a]Nonmetals are indicated by asterisks.
[b]For copper (cuprum), iron (ferrum), lead (plumbum), silver (argentum), tin (stannum), and tungsten (wolfram), the names in parentheses are used when forming compound names from these elements; for example, ferrate, *not* ironate, and wolfrate, *not* tungstenate.
[c]In 1949 the International Union of Chemistry adopted "niobium" as the official name, replacing "columbium."

EMERY

Emery is a natural mixture of corundum with magnetite or hematite and spinel. The aggregate has a gray to black color, a specific gravity varying from 3.7 to 4.3, and a hardness of 8 on the Mohs scale. Although synthetic abrasives have replaced emery in many of its earlier uses, emery is still used in construction as an abrasive and polishing material in the form of emery wheels, disks, paper, and cloth. (*See* Abrasives.)

ENAMEL

The word "enamel" originally referred to a vitreous glaze or combination of glazes applied as a decorative surface finish to metals. Its meaning was then extended to include objects decorated with this type of material. In recent years the word has been used to describe almost any hard, glassy-looking surface finish on almost any type of material, especially after "enamel" became the accepted name for a certain category of paint. Similarly, loose usage of the word "porcelain" has contributed to the general confusion in terminology.

Strictly speaking, then, enamel may be one of two categories of surface coating, defined as follows:

Porcelain enamel is generally an opaque, vitreous surface finish fused at high temperatures and used on metals.

Enamel is a type of oil- or resin-base paint that contains a binder which forms a glossy or flat film on exposure to air and that possesses the ability to level off brush strokes (*see* Paint).

In this book the emphasis is on application of enamel and porcelain enamel to materials for construction, and not on any of the art forms or decorative aspects of enamelwork.

EXPANSION AND EXPANSION JOINTS

Expansion as temperature increases and a corresponding contraction as temperature decreases occur in the various materials used in construction and play a very important part in the construction of buildings. A given length of material, which must inevitably move a certain distance in response to seasonal temperature changes, will suffer from cracks, distortion, breakage, malfunctioning, or other unforeseen developments if it does not have room to move. Therefore it is vitally important in designing buildings of any size or complexity that one calculate the range of expansion and contraction for each material and plan for expansion joints throughout the building to relieve these pressures and prevent the damage that would otherwise occur. For example, a reinforced concrete building 300 ft (91.44 m) long will expand 2.34 in. (59.44 mm), or 1.17 in. (29.72 mm) at each end. If the building has a wing at one end, this 1.17 in. (29.72 mm) expansion has to be taken up by expansion joints.

Situations Requiring Joints. Expansion joints can become a very difficult design problem in large buildings where each separate wing must be isolated. Such wings must become completely independent, movable units by means of an opening that cuts through the entire building. Another difficulty is that expansion joints at the roof, exterior walls, ceiling, floors, partitions, etc., must be watertight, weathertight, and soundproof, yet flexible enough to give and take with the expansion and contraction. These joints can become exceedingly complex, particularly those that must withstand expansion not in just two directions but in four.

Expansion and contraction also affect many minor parts of a building, for example, a railing of any length. Even though the building itself may not need an expansion joint, a railing made of metal may require one because of the fact that the metal expands more than masonry. Wall panels (prefabricated or preassembled), curtain walls, and sandwich panels in particular have to be calculated for expansion and contraction in relation to their framework. Other materials causing installation difficulties are glass block, large pieces of plate glass, various types of pipe, floor slabs in relation to foundation walls, and sidewalks, driveways, and slabs on the exterior of the building. In general, it is always good practice to install too many expansion joints rather than too few.

Expansion joints for various materials are fully covered under specific headings, for example, Copper Sheet and Strip; Stone. Figures *E1* through *E10* show details of expansion joints for major units of a building. In general, four-way expansion joints for floors and at slabs on grade are similar to two-way expansion joints except at intersections of two expansion joints.

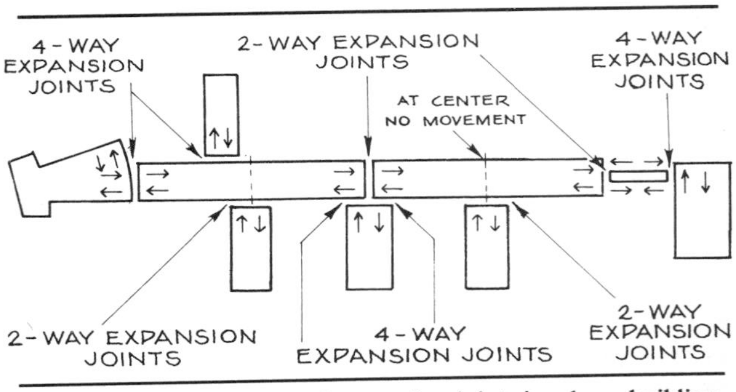

Figure E1 Various types of expansion joints in a large building.

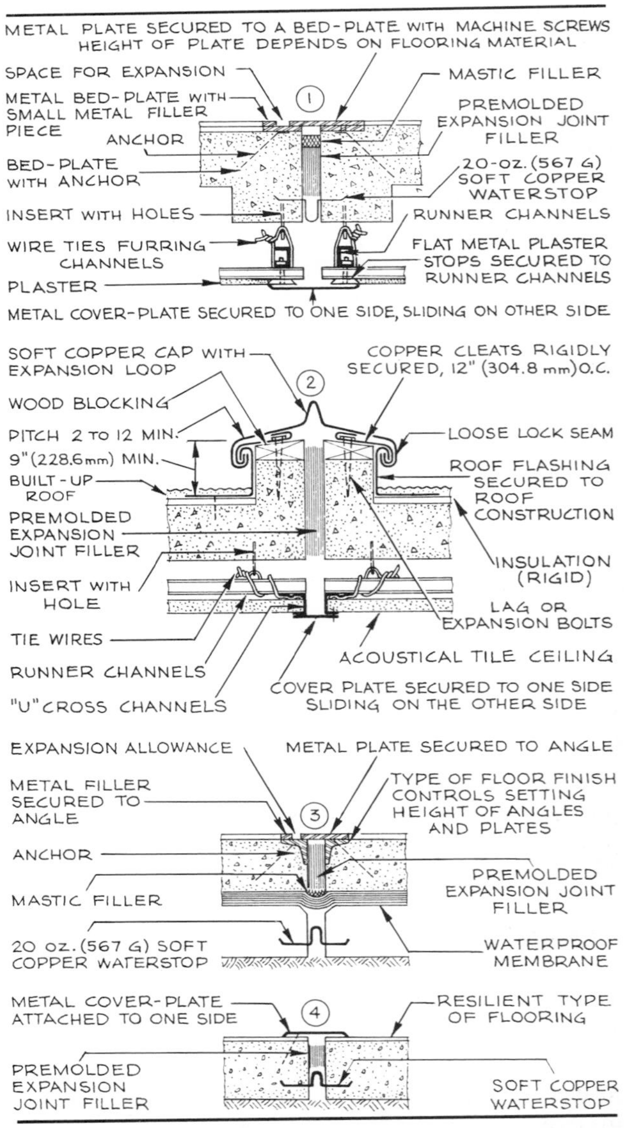

Figure E2 Details of two-way expansion joint: through floor with hung ceiling below (1), through roof with hung acoustical tile ceiling (2), through waterproof slab on grade (3), and through slab on grade (4).

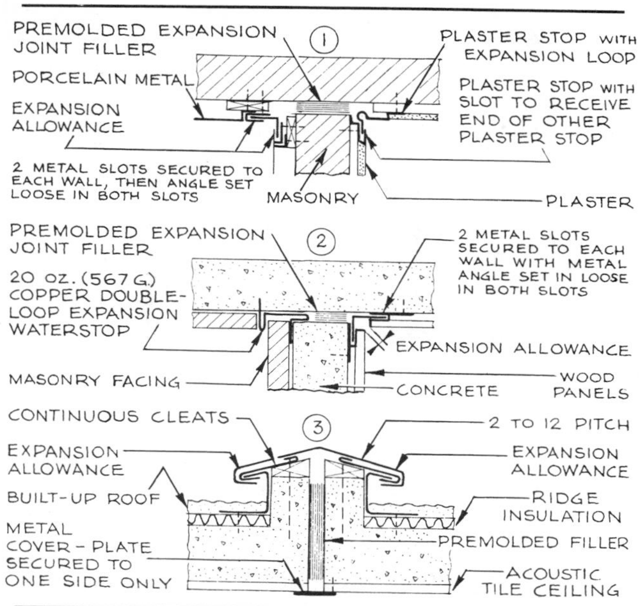

Figure E3 Details of four-way expansion joint: exterior corner with plaster on interior (1), exterior internal corner with double-loop expansion joint with wood panels on interior (2), and through roof with acoustical ceiling below (3).

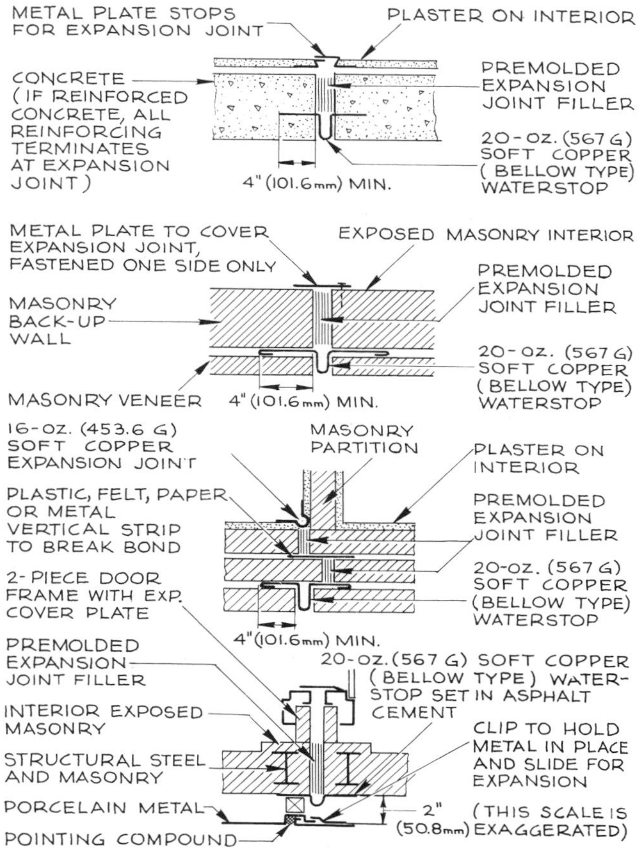

Figure E4 Details of two-way expansion joints in straight walls, showing various interior conditions that may be encountered.

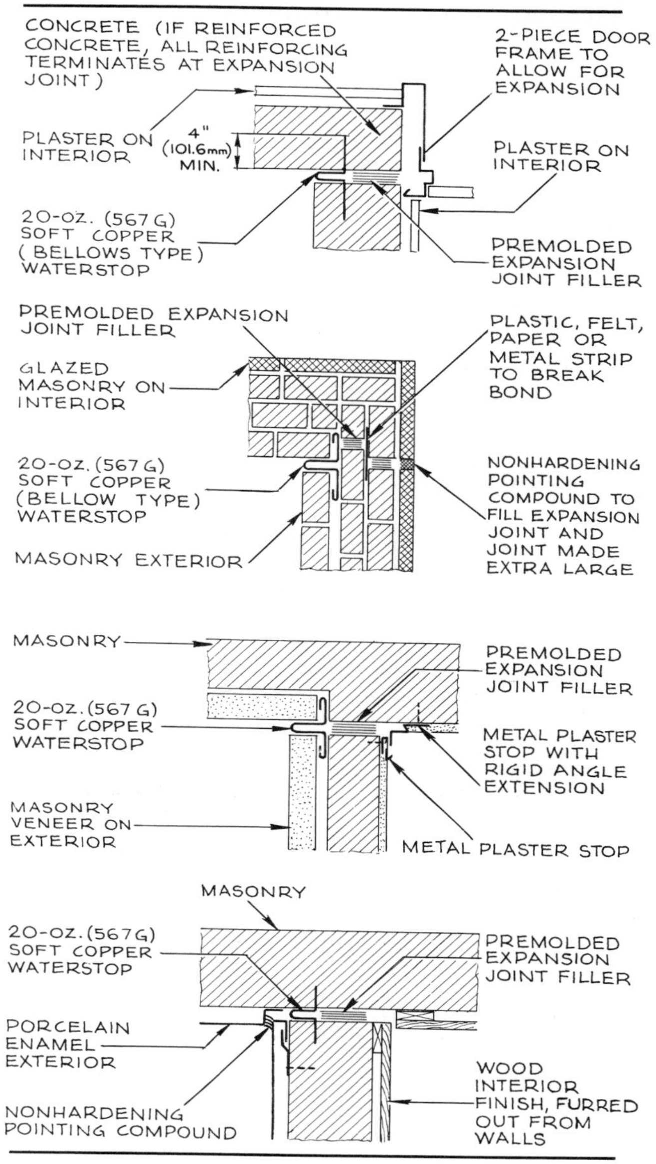

Figure E5 Details of two-way expansion joints at exterior internal corners, showing various internal conditions that may be encountered.

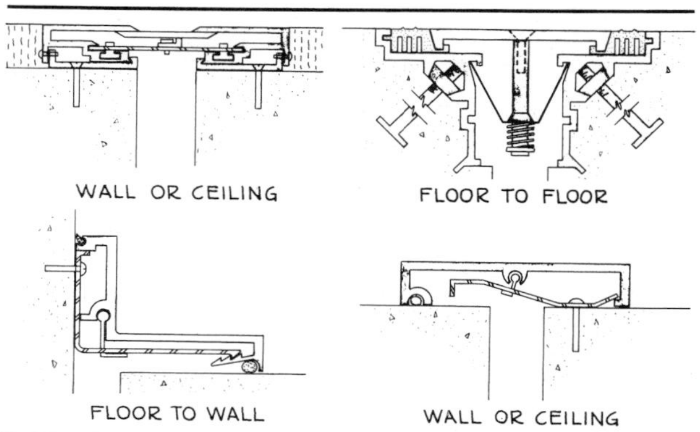

Figure E6 Typical interior expansion joints.

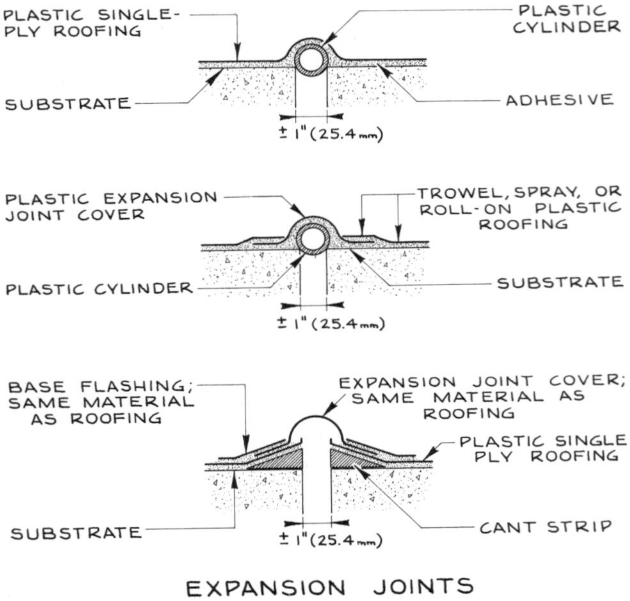

Figure E7 Expansion joints for plastic-type roofing.

WIDTH IN.	WIDTH mm.	THICK. IN.	THICK. mm.	BULB IN.	BULB mm.	WEB WIDTH IN.	WEB WIDTH mm.	MAX. JOINT MOVEMENT IN.	MAX. JOINT MOVEMENT mm.	WEIGHT LBS. PER 100'	WEIGHT KGS. PER 30.5M	STANDARD LENGTHS FT.	STANDARD LENGTHS M.
4*	101.6	3/16	4.76	7/16	11.11	1 3/8	34.9	1/4	6.35	50	22.7	100	30.48
6*	152.4	3/16	4.76	7/16	11.11	2	50.8	1/2	12.70	74	33.6	100	30.48
6	152.4	3/16	4.76	1 1/8	28.6	2 3/4	69.9	1	25.40	98	44.5	50	15.24
6*	152.4	3/8	9.53	5/8	15.9	2	50.8	1/2	12.70	140	63.5	50	15.24
9	228.6	3/16	4.76	7/16	11.11	2 3/4	69.9	1	25.40	130	59.0	50	15.24
9*	228.6	3/8	9.53	5/8	15.9	2 3/4	69.9	1	25.40	220	99.8	50	15.24
9	228.6	1/2	12.70	3/4	19.1	2 3/4	69.9	1	25.40	252	114.3	50	15.24
9	228.6	3/8	9.53	1 1/2	38.1	3 1/4	82.6	1 3/4	44.45	244	110.7	50	15.24

4	101.6	3/16	4.76	3/8	9.53					47	21.3	100	30.48
5	127.0	3/16	4.76	1/2	12.7					70	31.8	100	30.48
6	152.4	3/16	4.76	3/4	19.1					108	49.0	100	30.48
6	152.4	3/8	9.53	3/4	19.1					150	68.0	50	15.24

* THESE BULB TYPES ARE ALSO AVAILABLE WITH SPLIT FLANGE

Figure E8 Typical vinyl waterstops.

Figure E9 Extruded vinyl expansion joint cover for plastic type or built-up type of roofing.

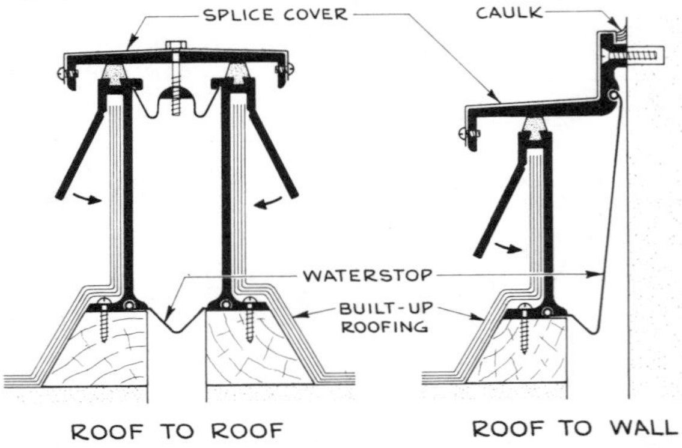

Figure E10 Extruded aluminum expansion joint covers for roofing.

Coefficient of Linear Expansion. The amount of expansion is usually expressed for each material used in construction in terms of increase per unit length per degree rise in temperature of the material. This is called the coefficient of linear expansion.

The coefficient of linear expansion as found in engineering, architectural, and construction data is expressed either in terms of increase per inch of length per degree Fahrenheit or increase per unit length per degree Celsius. Table *E2* gives the coefficients of expansion according to both systems.

For construction purposes, the range in exterior temperature does not exceed 100°F (37.78°C) except in very special localities. For example, in Death Valley, Nevada, the temperature range is approximately 160°F (71.11°C), and in certain areas of Alaska the seasonal range is about 140°F (60°C). Also, expansion is rarely figured closer than $\frac{1}{32}$ in. (0.00254 mm), in contrast to engineering calculations, which may be figured to $\frac{1}{10{,}000}$ in. (0.00254 mm) and smaller. In the construction field the coefficients of linear expansion in Fahrenheit are accurate enough. If more precise figures are necessary, the calculations must be checked by using the coefficients of expansion in Celsius. If very accurate figures for specialized detailing are necessary, the actual type of alloy used must be determined and its specific coefficient of expansion, including the temperature limitations for this value, must be obtained.

Application

To calculate linear expansion by using the thermal coefficient in degrees Fahrenheit, multiply the length of material *in inches* first by 100 (°F temperature range) and then by the coefficient of expansion (in °F). For

Table E2 Coefficients of linear expansion of construction materials

Material	Coefficient of expansion per inch per change in 1°F[a]	Coefficient of expansion per unit length per change in 1°C[a]	Material	Coefficient of expansion per inch per change in 1°F[a]	Coefficient of expansion per unit length per change in 1°C[a]
Aluminum (wrought)	0.0000128	0.0000229	Lead	0.0000159	0.000028
			Limestone	0.0000038	0.000008
Brass	0.0000047	0.000019	Magnesium	0.0000143	0.000029
Brick	0.0000031		Monel alloy	0.0000078	0.000014
Bronze	0.0000101	0.000021	Marble	0.0000056	0.0000117
Carbon steel	0.0000067	0.000011	Plaster	0.0000092	
Cast iron	–	0.000010	Stainless steel	0.0000096	0.00002
Clay tile (structural)	0.0000033	–	Titanium	–	0.0000085
			Wood (parallel to grain)	0.000003	0.0000071
Concrete	0.0000065	–			
Copper	0.0000098	0.000017			
Glass	0.0000047	0.00008	Wood (across grain)	0.0000189	0.000042
Granite	0.0000040	0.000079			
Iron	0.0000068	0.0000117			

[a]Based on types of materials and metal alloys used in construction.

example, for a wrought aluminum pipe railing 150 ft long, the expansion would be

$$150 \times 12 \times 100 \times 0.0000128.$$

This equals a total expansion of 2.3 in. (58.4 mm).

To calculate the expansion by using the thermal coefficient given in degrees Celsius, multiply the length of material *in inches* first by 55.78 (temperature range in °C equivalent to 0 to 100°F temperature range) and then by the coefficient of expansion (in °C). For example, for the wrought aluminum pipe railing mentioned, the expansion would be

$$150 \times 12 \times 55.78 \times 0.0000229.$$

This equals a total expansion of 2.3 in. (58.4 mm).

F

FASTENERS

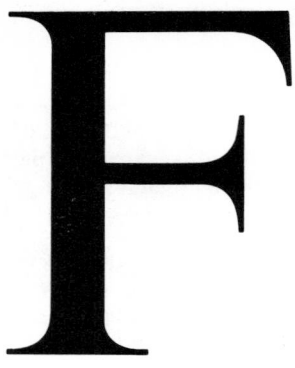

"Fastener" is an overall term covering many types of rough hardware used in construction. A fastener may be defined as a device used to join various types of construction materials together. Included are screws, nuts, bolts, rivets, pins, washers, anchors, and ties. These are used with wood, masonry, metal, stone, and plastics.

Fasteners are made for interior and exterior installations and are corrosion- or non-corrosion-resistant. It is always necessary to check the fasteners selected so that no galvanic action can occur.

See the following headings: Anchors; Hangers; Hardware; Hardware: Rough; Nails; Rivets; Ties; Screws, Nuts and Bolts, and Related Devices.

FELDSPAR

Feldspar is the general name for a group of aluminum silicates of potassium and sodium. Calcium and barium may also be present. In color, feldspars vary from white through pale shades of yellow, red, or green (sometimes, but less commonly, dark) with a vitreous luster. Specific gravity varies between 2.55 and 2.75. Feldspars are the common mineral in igneous rocks.

Commercial feldspars are intergrowths of at least two species of feldspar. After mining, feldspar is processed by flotation, dry grinding, or hand-cobbing. The trend in various flotation methods is to utilize fully the various minerals in the deposit and to improve grading by chemical content and size. Grades are based on silicon dioxide content, potassium-sodium ratio, iron content, and fineness of grinding.

The bulk of feldspar is used as a flux and vitrifying agent in pottery, vitreous or porcelain enamels, ceramic ware including tile, and glass including plate glass and structural block.

Among its other uses, feldspar is a bond for abrasive wheels, an abrasive in cleaning compounds, and a component of cement, concrete, insulating compositions, and tarred roofing materials.

See also Clay Floor and Wall Tile; Glass and Glazing; Glazes and Porcelain Enamels.

FERROUS METALS

"Ferrous" comes from the Latin word *ferrum*, meaning "iron," and includes all iron-based alloys, cast and wrought iron, steel, stainless steels, steels in concrete, supersteels, and weathering steels. (*See* Iron; Metals; Stainless Steels; Steel.)

FIBER BOARDS

In the terminology of this book fiber boards are any flat construction material made from paper pulp, various types of fiber and other materials, and the unregenerated waste from wood preparation and finishing (in the sawmill). (*See* Fibers; Paper and Paper Pulp Products.)

FIBERS

Physical and Chemical Properties

Fibers can be classified into three groups: (1) natural fibers obtained from vegetable, mineral, or animal sources, (2) regenerated fibers (made into better form) synthesized from naturally occurring polymeric materials such as cellulose, protein, or glass, and (3) synthetic fibers produced from man-made chemical polymers. Fibers can also be grouped into organic and inorganic fibers. The latter include asbestos (mineral), glass (silicates), fine wire (metal), and slag (fused waste silicates transformed into wool). (*See* Table *F1*.)

Durability. Generally speaking, all animal fibers are attacked by alkalis, whereas all others are resistant to such attack. Asbestos, glass, and acetated and synthetic types of fiber are generally immune to attack by molds, microorganisms, mildew, and insects; all others, although

Table F1 Fibers used in construction

End use	Natural			Regenerated (made into better form)			Synthetic	Metal
	Vegetable	Animal	Mineral	Cellulose	Protein	Glass		
Paper and paper pulp materials	X	X	X			X		
Textiles	X	X	X	X	X	X	X	X
Brushes	X	X					X	X
Stuffing, padding	X						X	
Plastics						X		

they may be naturally resistant to one or more of these deteriorating agents, must be treated to be resistant to all types of attack. Asbestos, glass, and some synthetic fibers do not burn; all others must be treated for flame or fire resistance.

Properties Affecting End Use. Physical and mechanical properties that affect or control end use include the following:

Tenacity: strength of a fiber in terms of unit of weight (grams per denier equals gram weight of 9000 meters of the fiber).

Tensile strength: strength in units of cross-sectional area (tensile strength in lbf/in.^{2} equals tenacity in grams per denier times 12,800 times specific gravity).

Stiffness: resistance of the fiber to deformation.

Extendibility: elongation at break.

Toughness: amount of work required to rupture fabric.

Elastic recovery: degree of ability to return to original length after being stretched 5%.

Fibers are classified as monofilament, continuous monofilament, or staple in physical character. All fiber types may be spun and twisted into yarns, but generally, long fibers are made into yarn and short fibers into felts and felted materials.

Yarns. The characteristics of continuous fiber yarns depend on strength (which is equal to the sum of the individual fibers that make up the yarn), tenacity, number of fibers, and twist.

In stable fiber yarns, the characteristics are based on strength (equal to about one-half of the strength of the fibers that make up the yarn, as this depends on the frictional hold of the fibers among themselves), staple length of the fibers, tenacity, and degree of crimp and twist.

Designation of Yarns. Yarns are designated by counts, that is, so many yards per pound of weight, and by numbers based on the count used for the particular fiber. The count of cotton is based on 840 yd to the pound (768.096 m to 0.4536 kg), and a No. 10 cotton would indicate a yarn that contained 8400 yd to the pound (7680.96 m to 0.4536 kg). Linen yarns count 300 yd to the pound (274.32 m to 0.4536 kg); for example, No. 10 runs 3000 yd to the pound (2743.2 m to 0.4536 kg). Rayon yarns use the same count as cotton yarns. Silk is based on the International Denier (a 500-m-length weight of 0.05 g); thus 20 denier would indicate a silk yarn that weighs 20 times 0.05 g, or 1 g per 500 m. Silk yarns also use the same count as cotton.

Types and Uses

The type of fiber selected for a given end use depends on its structural characteristics, length, form, uniformity, strength, tenacity, elasticity, resistance to chemical attack, colorability, and adaptability to the necessary processing techniques (*see* Table *F2*). Fibers used in paper and paper pulp products are discussed separately.

Yarns are made into cord, rope, and textiles. For cord and rope, the main requirements are strength and durability. Except for oakum, which is loose fiber obtained by untwisting and picking apart old ropes and which is used for caulking seams, there are no direct architectural applications for rope.

Textiles. Textiles, either knitted or woven, require yarns made from fibers that have resistance to the substances used in the manufacturing process and in the

Table F2 Characteristics, classifications, and major uses of fibers in construction

Fiber	Classification	Staple length in. (mm)	Composition	Fire resistance	Effects: Sunlight	Effects: Age	Major use
Cotton	Vegetable	$\frac{1}{2}$–$2\frac{1}{2}$ (12.7–63.5)	Cellulose	Burns	Loss of strength, yellows	Virtually none	Textiles, cordage, plastic laminates, filler for plastics
Hemp		40–80 (1016–2032)	Cellulose and lignocellulose	Burns	Loss of strength, yellows	Virtually none	Canvas for marine areas, cordage, rope
Jute		10–60 (254–1524)	Lignocellulose	Burns	Loss of strength, yellows	Virtually none	Canvas, cordage, rope, reinforcing for building paper
Kapok		$\frac{1}{4}$–$1\frac{1}{4}$ (6.35–31.75)	Lignocellulose	Burns	Loss of strength, yellows	Virtually none	Stuffing, padding
Palmyra		9–18 (228.6–457.2)	Lignocellulose	Burns	Loss of strength, yellows	Virtually none	High-grade brushes
Ramie		15–25 (381–635)	Cellulose	Burns	Loss of strength, yellows	Virtually none	Canvas, fabrics for exterior use, cordage, rope
Sisal		20–36 (508–914.4)	Lignocellulose	Burns	Loss of strength, yellows	Virtually none	Reinforcing for building paper, cordage, rope
Wool	Animal	$1\frac{1}{2}$–15 (38.1–381)	Protein	Difficult to burn	Loss of strength, yellows	Loss of strength, yellows	Textiles, rugs, insulation, soundproofing
Silk		(400–1300 yd) (365.76–1188.72 m)	Protein	Burns	Loss of strength, yellows	Loss of strength, yellows	Textiles
Asbestos	Mineral	$\frac{3}{8}$ or more (9.6 or more)	Hydrated silicates of magnesium	Will not burn	None	None	Asbestos-cement materials, textiles, paper
Acetate	Regenerated (made into better form)	$\frac{15}{16}$–7[a] (23.8–177.8)	Cellulose acetate	Slow burning	Loss of strength	Loss of strength	Textiles
Casein		$\frac{1}{2}$–6[a] (12.7–152.4)	Protein	Slow burning	Loss of strength	None	Textiles
Glass		Up to 18[a] (Up to 457.2)	Glass	Will not burn	None	None	Thermal and sound insulation, textiles, plastics
Soybean		$\frac{1}{2}$–6[a] (12.7–152.4)	Protein	Slow burning	Loss of strength, yellows	None	Textiles
Viscous rayon		1–8[a] (25.4–203.2)	Regenerated cellulose	Burns rapidly	Loss of strength, yellows	Yellows	Textiles, cordage
Polyacrylonitrile (Orlon)	Synthetic polymers	[a]	Polyacrylonitrile	Burns slowly	Very resistant	None	Textiles, plastic laminates
Polyamide (nylon)		1–5[a] (25.4–127)	Polyamide	Will not burn	Loss of strength	None	Textiles, plastic laminates
Polyester (Dacron)		$1\frac{1}{2}$–$2\frac{1}{2}$[a] (38.1–63.5)	Polyethylene terephthalate	Burns slowly	None	None	Textiles, plastic laminates
Polyethylene		[a]	Polymerized ethylene	Burns slowly	Loss of strength	None	Textiles
Polyvinyl chloride and vinyl acetate (Vinvon)		[a]	Vinyl chloride and vinyl acetate	Will not burn	Not attacked	None	Textiles

[a]Obtainable in continuous lengths.

subsequent washing, cleaning, and dyeing operations. Construction uses of textiles are largely limited to decoration, canvas for roofing, and various types of cloth flashing (*see* Textiles). Table *F3* summarizes the fire retardants commonly used for textile fibers.

Felts. Felt is another category of fibrous material or product that is widely used in industry and construction. The fibers of felts are interlocked by pressure. As a rule, only felts made of wool or wool plus other fibers are classified as textiles.

Table F3 Fire retardants commonly used for textile fibers

Fire retardant	Application	Aftertreatment	Advantages	Disadvantages
Pigmented type (mixture of antimony oxide and chlorinated compound)	Immersed in solution rolled, dried	Silicate solution to retard afterglow	Good flame resistance; durable under severe commercial dry-cleaning and laundering	Decreases porosity; increases stiffness; increases weight; metallic oxide enhances afterglow
Pigmented type (antimony and titanium oxides)	Oxides precipitated onto fibers from hydrochloric acid solution of chlorides of antimony and titanium; neutralized by sodium carbonate	Silicate solution to retard afterglow	Good flame resistance; durable under severe commercial dry-cleaning and laundering	Slight decrease in porosity; slight increase in stiffness and weight; definite afterglow
Pigmented type (mixture of zinc oxide and carbonate plus boric acid and a chlorinated compound)	Impregnated with mixture in double bath		Good flame and afterglow resistance; loses resistance with severe dry cleaning and laundering	Increases weight
Resin type (melamine polyphosphate)	Applied from water solution, then dried	Heat-treated to polymerize resin	Good flame and afterglow resistance	Durable for only several home launderings and 25 dry, wet, or solvent dry cleanings
Resin type (polymerized bromoallyl phosphate)	Applied from solvent solution	Heat-treated to polymerize bromoallyl phosphate; scoured	Excellent flame and afterglow resistance; durable under severe commercial launderings	Very difficult application
Chemically attached (urea phosphate)	Cellulose reacted with mixture of ammonium phosphate and urea or other nitrogen organic compound; heat-cured		Good flame and afterglow resistance; fine distribution	Up to 50% loss of strength; destroyed in alkaline laundering
Chemically attached (organic nitrogen compound replaces urea in above compound)			Highly fire resistant; durable to mild laundering	Semidurable; loss in strength
Chemically attached (cellulose etherified to include amino ether group)	Cellulose reacted with 2-aminoethyl sulfuric acid in caustic solution by heat cure		Excellent flame and afterglow resistance; durable in ordinary laundering solutions	Slight loss in strength; destroyed by strong alkaline and salt solutions; small increase in weight

History and Manufacture

The use of fibers for cord, rope, and clothing and as building materials was one of man's earliest discoveries. Weaving and knitting are prehistoric. We can assume that spinning the fibers into yarn was the first step, followed by development of some sort of hand loom for weaving the yarn into textiles. In 1733 Kay's invention of the fly shuttle simplified and began the mechanization of the hand loom, and in 1799 the first papermaking machine was developed. These two events marked the birth of two major modern industries, textiles and paper products, both utilizing fibers in tremendous quantities.

Although the modern power loom actually had its beginning in 1661, it was not until 1803 that the completely automatic loom was established. It is interesting to note that in 1661 a loom was made in Europe that could operate day and night and weave four to six webs without human aid. However, government authorities suppressed it and did away with the inventor.

Fibers are processed by first cleaning them of foreign matter and then forming them into a continuous strand (a step called carding). In the next step (drawing) this strand is rolled to stretch it and reduce the diameter, and finally it is spun into yarn. The more twists to the inch, the stronger is the yarn. The yarn is usually combed to eliminate all foreign matter. The yarns are made into cloth or fabric by weaving, knitting, or felting. (For treatment of fibers for paper and paper pulp products, *see* Paper.)

FIREPLACES

Until the energy crisis, the fireplace in many parts of the country was used primarily as an ornament, as a focal point of a room or to establish a mood. Heating systems in the United States are so efficient and available that the fireplace had been relegated to a very minor role. In any case, whether the fireplace is an important part of the building or a minor accessory, the architect, mason, and contractor still need to know its vital components, particularly if the fireplace is to be specially designed (*see* Figure *F1*). Each must also know what materials to use in its construction and what the effects of heat will be on them and other contacting materials. The same precautions must be observed as for chimneys. Masonry or concrete fireplaces can be made more efficient by slanting the side walls; by having them open into two rooms; by enclosing them on two sides or on one side, or having them completely open on all sides; and by installing metal hoods. (*See* Figure *F2*.)

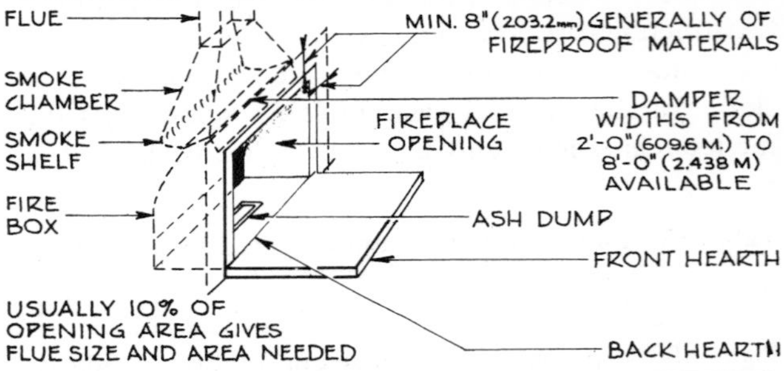

Figure F1 Requirements for individual, special architectural design of fireplaces.

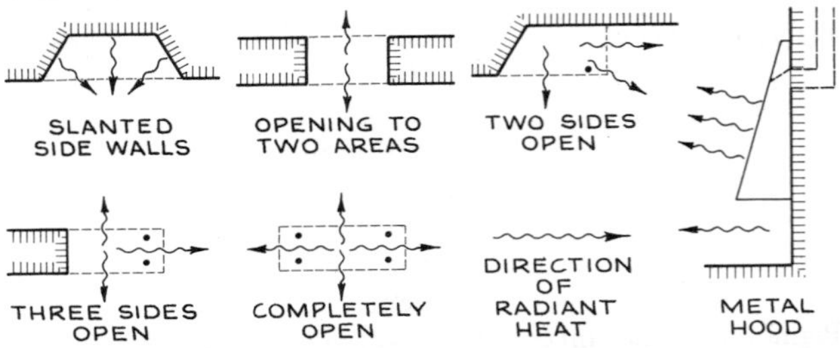

Figure F2 Various types of fireplaces constructed of masonry or concrete.

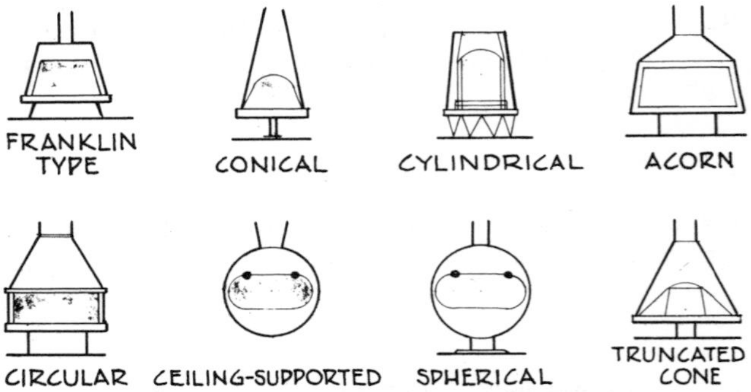

Figure F3 Typical free-standing prefabricated fireplaces.

Prefabricated Fireplaces. Prefabricated fireplaces or, more accurately, fire-boxes, are not at all new. Many of the designs currently available are derived from the basic invention of Benjamin Franklin, the Franklin stove, based on the fact that fire contained within a metal box (usually iron) heats the box and in turn the air in all directions. (*See* Figure *F3*.)

Package Units. For residences, camps, ski lodges, summer homes, and miscellaneous other buildings where the fireplace also functions as the heating system, there are on the market several types of package units for the conventional wall-type fireplace which are planned to increase its efficiency as a heating unit. In these the fireplace proper (the part containing the fire) consists of a metal box-within-a-box with provisions made for intake of cold air into the enclosed air space and for exhaust by gravity of the heated air into the room (*see* Figure *F4*).

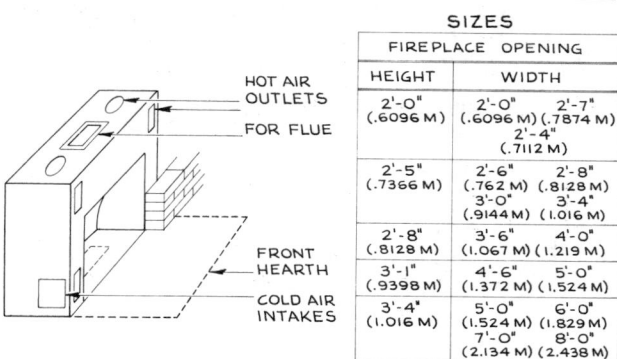

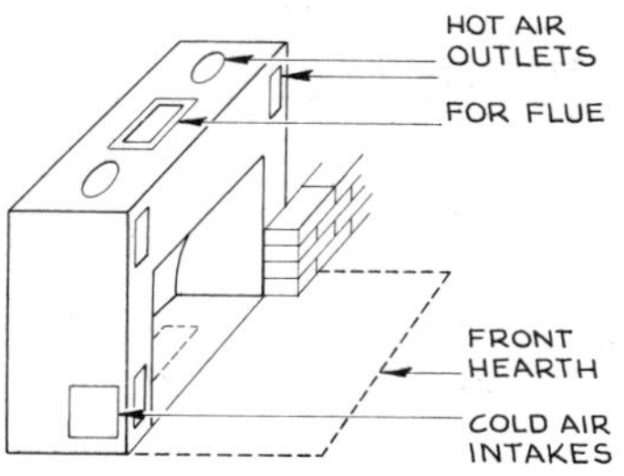

SIZES

FIREPLACE OPENING	
HEIGHT	WIDTH
2'-0" (.6096 M)	2'-0" (.6096 M) 2'-7" (.7874 M) 2'-4" (.7112 M)
2'-5" (.7366 M)	2'-6" (.762 M) 2'-8" (.8128 M) 3'-0" (.9144 M) 3'-4" (1.016 M)
2'-8" (.8128 M)	3'-6" (1.067 M) 4'-0" (1.219 M)
3'-1" (.9398 M)	4'-6" (1.372 M) 5'-0" (1.524 M)
3'-4" (1.016 M)	5'-0" (1.524 M) 6'-0" (1.829 M) 7'-0" (2.134 M) 8'-0" (2.438 M)

PREFABRICATED HOT AIR HEATING UNIT FOR FIREPLACES

Figure F4 Typical prefabricated hot-air heating unit for fireplaces.

Design Considerations. The fireplace has again become an important heating element as our oil, coal, gas, and other raw materials become increasingly scarce and other sources of energy are being developed. The free-standing prefabricated fireplaces and the prefabricated hot-air heating units for fireplaces are the most efficient heating units. Asbestos-cement pipe, prefabricated chimneys, etc., have made the installation of fireplaces much less complex. (*See* Chimneys; Clay Tile Flue Linings; Asbestos-Cement Pipe.)

FIREPROOFING

The word "fireproofing" is actually misleading because almost all materials suffer damage from fire. For example, when steel is heated sufficiently, it loses its strength and buckles or collapses. In general practice, therefore, a completely fireproof building would be impossible to construct since it would have to be made primarily of reinforced concrete and could not contain any material that would in any way catch fire. Thus fireproofing in fact refers to methods of (1) retarding the spread of fire (fire control), (2) lengthening the time that a material can withstand high temperatures (fire resistance), or (3) reducing the flammability of materials (flameproofing).

The purpose in all cases is to provide enough time, allowing a margin of safety, for all occupants of a burning building to be evacuated before the structure weakens sufficiently to become dangerous. Again taking steel as an example, covering a steel structural member with a masonry material, gypsum plaster, concrete, or fireproofing clay tile will increase the time it takes for the member to collapse from the effects of high temperatures. Wood, an easily combustible material, can be chemically treated so that it smolders rather than burns, or it can be covered with a metal that retards its destruction.

Specific techniques of flameproofing paper pulp products and textiles woven of nonmineral fibers have been developed and are especially important in public buildings. Asbestos and glass fibers are intrinsically flameproof substances (meaning they do not catch fire or support combustion) that are now widely used in the form of textiles, sheets, and boards.

Other methods of fire retarding include the installation of automatic sprinkler systems controlled by special metals called fusible alloys, which melt at a given temperature (usually 155°F or 68°C) and allow water to flow out under pressure onto the fire. Instead of water, various special fire-extinguishing materials can be used.

Fire Codes. Requirements for safety measures and regulation of the kinds of materials to be used in buildings exist in most countries. In the United States these requirements are to be found in the codes of the National Board of Fire Underwriters, various insurance companies, building departments and fire marshals on the local, municipal, and state level, and various branches of the federal government (Army, Navy, Air Force).

See also Fibers; Paints; Paper and Paper Pulp Products; Textiles; Wood; and material headings for specific fire-retardant chemicals and treatments.

FLASHING

Flashing may be described as the application of a material to seal and protect the joints that appear wherever different parts of a structure or different materials are brought together. Thus flashing is a major problem in construction. It is used where a roof meets vertical walls, where a building itself meets the earth (at grade), and wherever one material goes through, or pierces, another material. It also includes joints and other devices which provide for expansion and contraction of materials.

There are two types of flashing, concealed and exposed. The two differ in the type of material used and the methods of application.

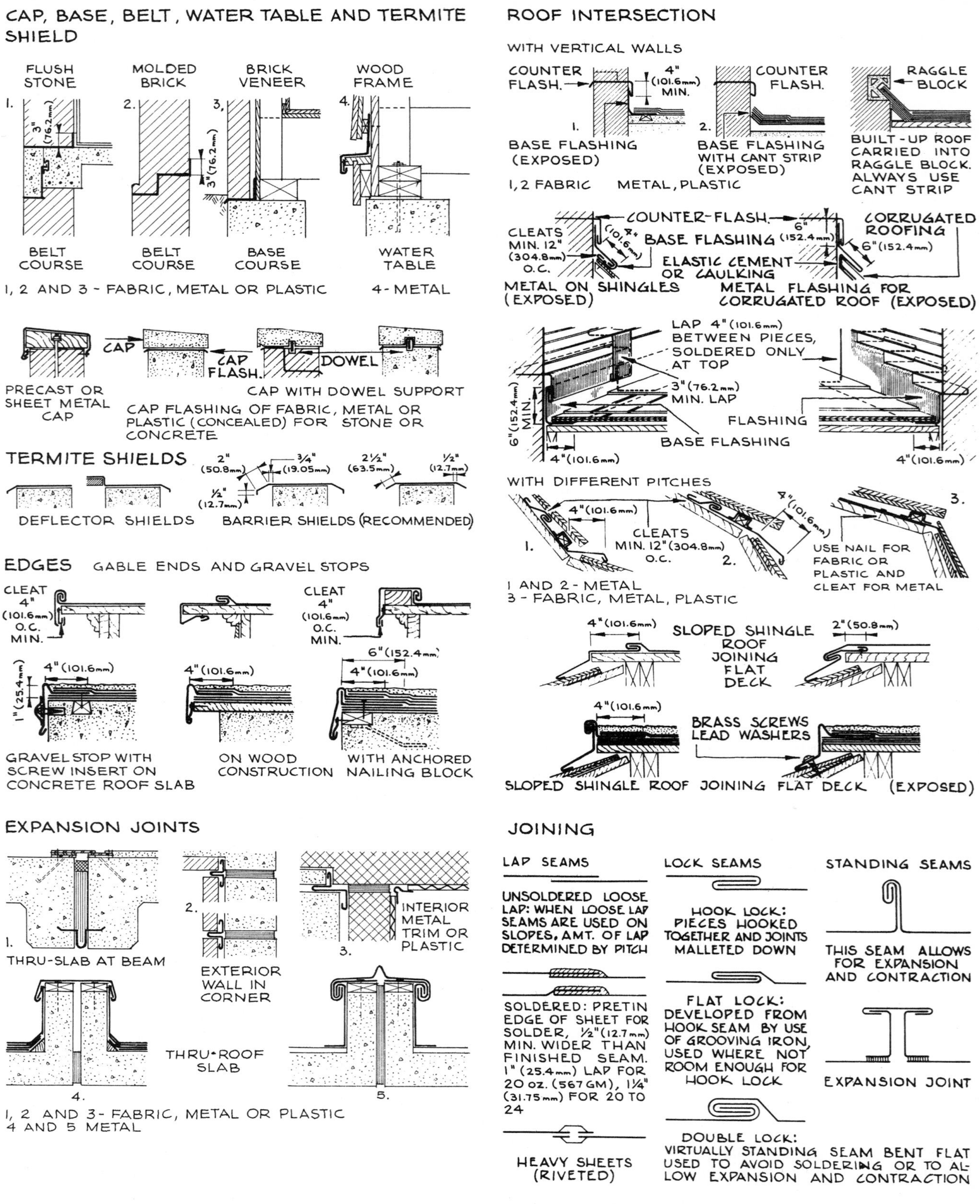

Figure F5 Details of concealed and exposed flashing.

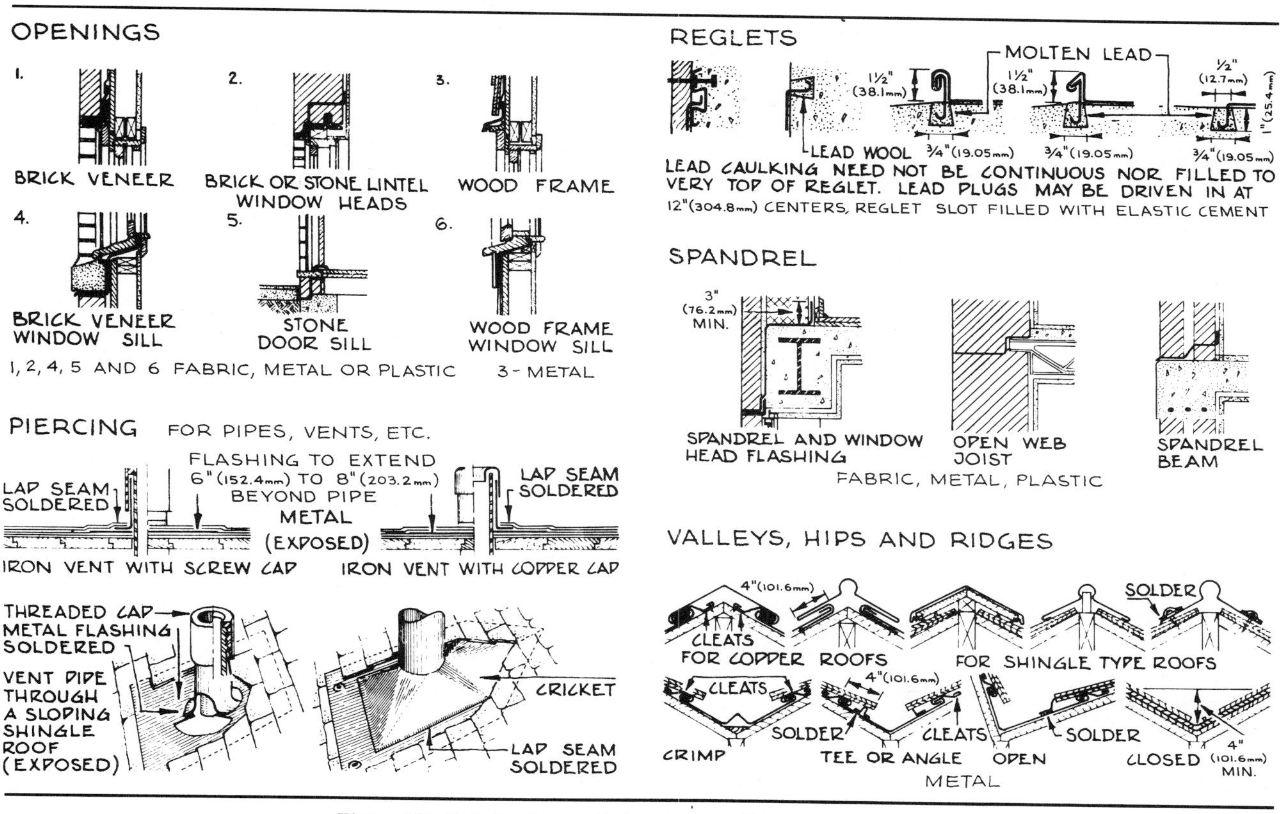

Figure F5 Details of concealed and exposed flashing (*continued*).

Concealed Flashings. With concealed flashings, which are invisible on the exterior or interior of the building, the only problem is to determine what material will do the job of protecting the joint and then to apply it properly. For this purpose, a thin metal sheet or foil, a fabric, a plastic, or various combinations of these materials may be used, depending on climate and structural requirements. Papers and fabrics may be sealed or saturated with protective substances such as asphalt or plastic materials.

Exposed Flashings. Where the joints are exposed to view, the problem is complicated by the fact that the flashing affects the design of the building. Not only must it withstand exposure and other conditions of use, but also it must be so chosen that it will not adversely affect the esthetic value of the structure over a period of time. This immediately limits the choice of materials almost entirely to the metals and plastics. The metal or plastic must conform to requirements of color, texture, and resistance to exposure. It must not stain adjoining materials or react with them; it must hold its form; if its color changes, this change must not detract from the appearance of the building; its maintenance must not become a problem.

Figure *F5* shows typical details for each category of flashing (concealed and exposed), indicating various methods of flashing according to location and construction of the joints. Figures *F6* and *F7* show additional details for specific situations.

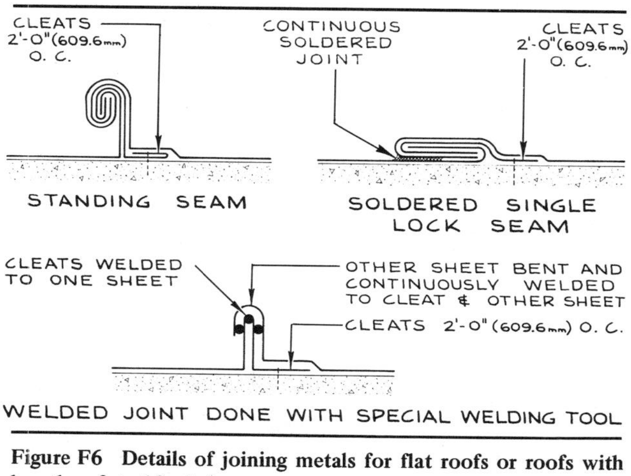

Figure F6 Details of joining metals for flat roofs or roofs with less than 2 to 12 pitch.

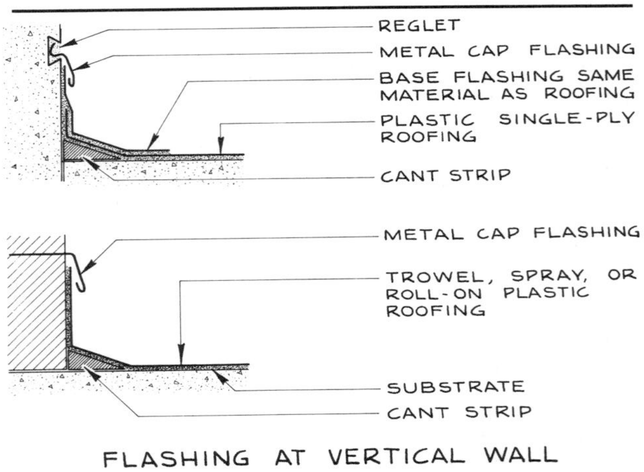

Figure F7 Flashing for vertical intersections with plastic roofing.

FLOORS

Floors, in the terminology of this book, are the finished surfaces within a building upon which people walk, furniture and specialized equipment are placed, and rolling equipment is moved. Table *F4* lists the various types of finish flooring and the materials from which they are made.

There are five basic criteria for selecting a flooring material: (1) the substrate upon which the flooring is to be applied must be of a type suitable to receive the flooring selected; (2) the expansion and contraction of the substrate and of the flooring material itself must be considered in the selection of a flooring material; (3) various types of flooring require different depths for the substrate, and therefore requirements for changes in

Table F4 Typical flooring materials used in construction

Materials	Advantages	Disadvantages	Major use
Aluminum (*see* Aluminum Sheet and Strip)	Durable		Utility stairs and platforms
Asphalt (*see* Asphalt Resilient Flooring)	Easily installed; fairly large color selection; relatively hard	Will show any defects in the surface of the material on which it is applied; will indent where legs of furniture come in contact with it; colors are not brilliant; cannot be used where moisture is present	General utility flooring
Brick (*see* Brick, Burned)	Durable; waterproof; fireproof	Hard; rough and uneven because of joints; difficult to clean	Decorative
Carpet (*see* Textiles; Plastics)	Resilient; insulation and acoustic values; wide variety of colors and patterns	Limited durability, some types will stain; not cigarette-, cigar-, or pipe tobacco-proof	Residences, apartments, hotels and motels, and office areas
Clay tile (*see* Clay Tile: Floor and Wall Tile)	Durable; waterproof; wide variety of colors; easily cleaned	Matte finish; hard	Areas where moisture is present; areas where heavy wear is expected
Concrete (*see* Concrete; Concrete, Reinforced)	Durable; can be painted or treated	Hard; will give off dust if not treated; difficult to clean	Utility flooring
Concrete, treated (*see* Admixtures)	Nonconductive, nondusting, and nonslip; very wear resistant	Requires various special materials and methods of application	Industrial and manufacturing areas, and areas where nonsparking is required
Cork (*see* Cork and Plastics)	Durable; resilient; nonslip; has insulation and acoustic values; easily cleaned	Limited colors; will stain	Areas where resilience and nonslip qualities are important
Linoleum (*see* Linoleum)	Durable; resilient; rather large variety of colors and patterns; easily cleaned	Will indent where legs of furniture come in contact with it	Areas where resilience is important

Table F4 Typical flooring materials used in construction (*continued*)

Materials	Advantages	Disadvantages	Major use
Plastic (vinyl) (*see* Plastics: Flooring)	Durable; resilient; very wide variety of colors and patterns; easily cleaned	Except for thick types, will show any defects in the surface of the material on which it is applied	General utility flooring
Plastic floor coatings (*see* Plastics; Paint; Painting)	Durable; easily cleaned; waterproof	Requires carefully controlled mixing and application; limited colors	Finished flooring surface on concrete
Rubber (*see* Rubber)	Durable; resilient; wide variety of colors and patterns; easily cleaned	Slippery when wet	General utility flooring
Seamless (*see* Terrazzo; Plastics)	Requires no joints; durable; resilient (to varying extents)	Requires special materials and methods of application	General utility flooring
Steel (*see* Steel; Steel Sheet, Strip, and Plate)	Durable	Requires painting	Utility stairs and platforms; mezzanine floors in mechanical and industrial equipment areas
Stone (*see* Stone; Stone, Granite; Stone, Marble; Stone, Slate)	Durable; easily cleaned	Hard; can stain; limited colors	Areas where heavy wear is expected
Terrazzo (all types) (*see* Terrazzo)	Durable; easily cleaned	Hard; can stain; limited colors	Areas where heavy wear is expected
Wood (*see* Wood Flooring)	Durable; relatively easily cleaned	Limited colors; can stain; hard	Residential construction; multiple dwelling buildings

levels of the substrate must be checked if various different types of flooring are selected; (4) where different types of flooring materials meet, the treatment of this joint must be checked; and (5) one must always determine what the maintenance requirements are for the flooring or floorings selected for a building, and then inform the owner of these maintenance requirements.

FLUORINE

Physical and Chemical Properties

Symbol: F
Atomic number: 9
Specific gravity: 1.108 (−188.14°C, −306.65°F) liquid
Freezing point: −219.62°C, −363.32°F
Melting point: −188.14°C, −306.65°F

Fluorine is a pale, greenish yellow, poisonous, and corrosive gas. It is the most reactive and electronegative nonmetallic element known; it attacks glass, metals, and even asbestos. With other elements it forms compounds highly resistant to chemical and physical attack.

Types and Uses

Fluorine compounds (fluorspar and cryolite) are essential in the manufacture of steel and other metallurgical procedures, in the production of hydrofluoric acid for the chemical industry (a growing use), and in the manufacture of glass and other ceramics. Table *F5* gives the composition of ceramic-grade fluorspar. Cryolite serves as the electrolyte in the production of aluminum. Fluosilicic acid and fluosilicates are used as hardening agents in concrete. Table *F6* summarizes these uses.

Fluoridation. Fluoridation is the addition of fluorine (1 part per million) to public drinking water supplies to reduce tooth decay.

Fluorocarbon Compounds. Fluorocarbon compounds are increasingly important in current technology because of their great stability and inertness. They are formed when fluorine replaces hydrogen in organic compounds. Examples are Freon, Genetron, and fluorocarbon plastics.

Table F5 Composition of ceramic-grade fluorspar

CaF_2	SiO_2	$CaCO_2$	Other elements
(percent of content)			
95	2.5	1.5	1.0

Table F6 Uses of fluorine

	Construction Uses	
Basis of	Abrasives	Filler and bonding agent
Component of	Brick	Additive to clays
	Ceramics	Fluxing and increasing hardness
	Concrete	Hardener
	Enamels	Coatings for steel and iron
	Glass	Flint and opal; etching and polishing glass
	Insulation	Glass fiber
	Paint	Filler and paint solid
	Plaster	Hardener
	Plastics	Fluorocarbons
	Steel	Flux and pickling
	Textiles	Fire retarding with antimony compounds
	Wood	Preventing decay
Allied Construction Uses	Fire protection	Ingredient of fire extinguishers
	Refrigerants	Freon, Genetron, etc.
	Welding	Electrode coatings

History and Manufacture

Fluorine was first prepared by the French chemist Moissan in 1886. Modern processing of fluorspar (the mineral fluorite) varies from simple hand sorting, washing, screening, and gravity separation to complex methods such as sink-float and froth flotation or a combination of these methods.

FLUX

A flux is a substance used in metallurgy, ceramics, soldering, brazing, and welding, generally to aid fluidity. In metallurgy it also serves to carry off impurities in the form of gas or slag by means of oxidation or decomposition. In soldering, brazing, and welding its primary purpose is to clean the surfaces of the metals to be joined although it also aids fluidity. In welding it may, in addition, carry the inert gas shield.

FOOTINGS

Footings are the fundamental supporting elements of a building to which the total weight of the building is distributed by the foundation walls and columns. These footings must be calculated to support the load upon whatever type of soil is found on the site. The footings must always lie below the frost line; otherwise the alternate expansion and contraction of the earth during freezing and thawing may heave the footings, damage the foundations and, if severe enough, even damage the superstructure. Figures *F8* and *F9* show the various types of footings.

There are several instances when footings can be eliminated entirely. The basement or first floor can be treated as the footing and used to support the load of the building. In warm climates, it is possible to utilize the floor (slab) of the building as a footing. These are only two examples.

Three complications are possible in the construction of footings. First, in order to obtain the required size of footing on a property line, the footing must be cantilevered over to the next footing (*see* Figure *F9* detail). Footings on soils with low load capacities represent a second type of problem. An example would be construction of a building in New Orleans on the silt of the river

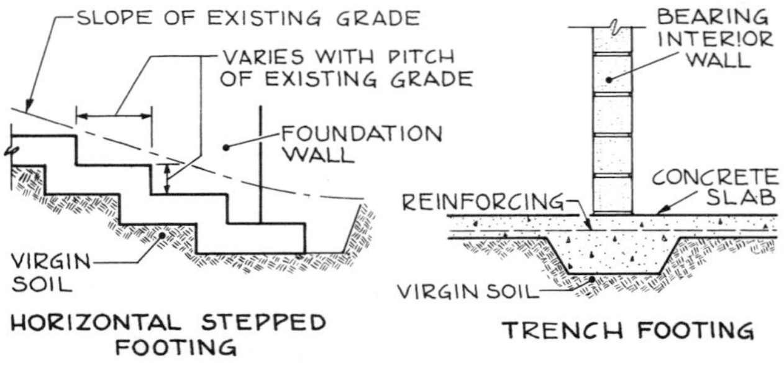

Figure F8 Horizontal stepped footing and trench footing.

TYPE OF FOOTING	MATERIAL USED	SHAPE	SIZE DETERMINED BY SOIL BEARING CAPACITY *	TYPE OF FOUNDATION WALL, PIER OR COLUMN	MAJOR USE
WALL	CONCRETE, REINFORCED CONCRETE	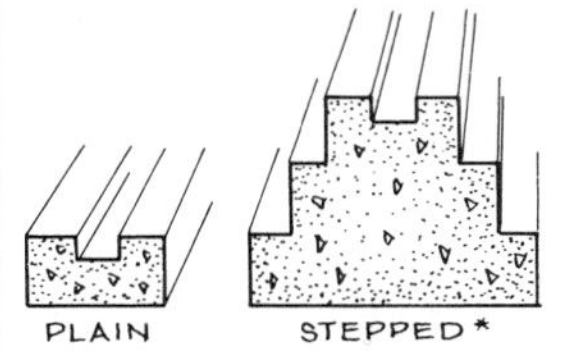 PLAIN; STEPPED*	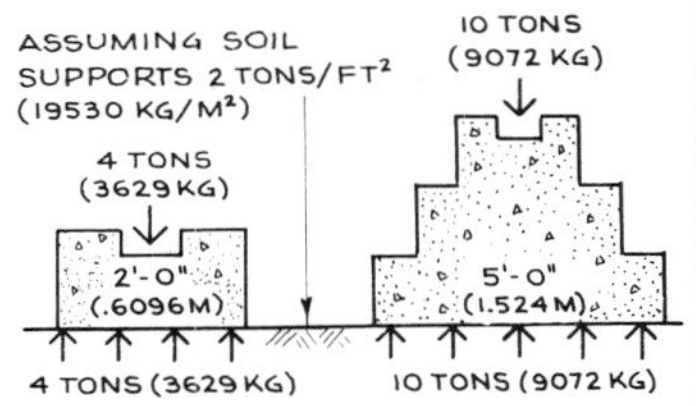 ASSUMING SOIL SUPPORTS 2 TONS/FT² (19530 KG/M²); 4 TONS (3629 KG); 2'-0" (.6096M); 4 TONS (3629 KG); 10 TONS (9072 KG); 5'-0" (1.524M); 10 TONS (9072 KG)	CONCRETE, REINFORCED CONCRETE, BRICK, CONCRETE BLOCK, STONE	FOUNDATION WALLS, OR ANY WALL THAT SUPPORTS A LOAD
COLUMN OR PIER	CONCRETE, REINFORCED CONCRETE	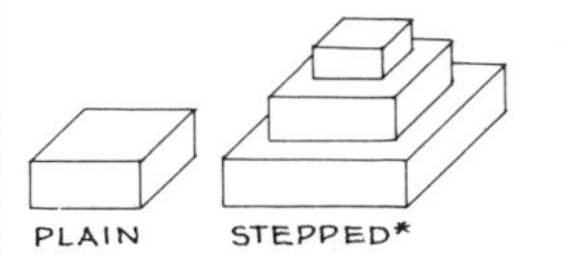 PLAIN; STEPPED*	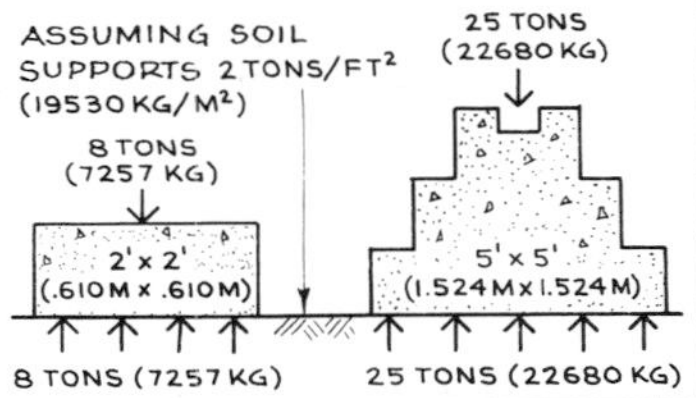 ASSUMING SOIL SUPPORTS 2 TONS/FT² (19530 KG/M²); 8 TONS (7257 KG); 2'x 2' (.610M x .610M); 8 TONS (7257 KG); 25 TONS (22680 KG); 5'x 5' (1.524M x 1.524M); 25 TONS (22680 KG)	WOOD, STEEL, ALUMINUM REINFORCED CONCRETE COLUMNS OR CONCRETE BLOCK, REINFORCED CONCRETE, BRICK, STONE PIERS	COLUMNS OR PIERS
GRILLAGE	STEEL AND CONCRETE	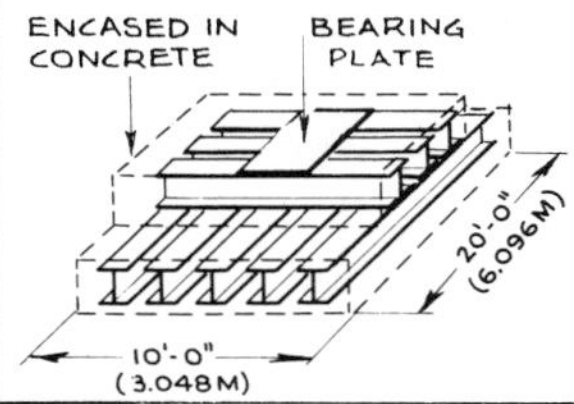 ENCASED IN CONCRETE; BEARING PLATE; 20'-0" (6.096M); 10'-0" (3.048M)	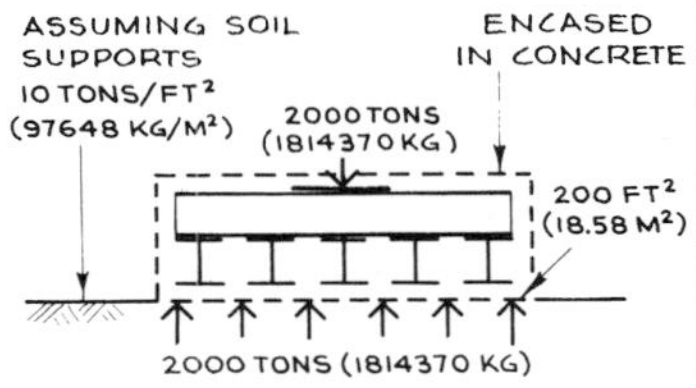 ASSUMING SOIL SUPPORTS 10 TONS/FT² (97648 KG/M²); ENCASED IN CONCRETE; 2000 TONS (1814370 KG); 200 FT² (18.58 M²); 2000 TONS (1814370 KG)	STRUCTURAL STEEL COLUMNS OR STEEL PIPE COLUMNS	COLUMNS WITH VERY HEAVY LOADS
MAT	REINFORCED CONCRETE	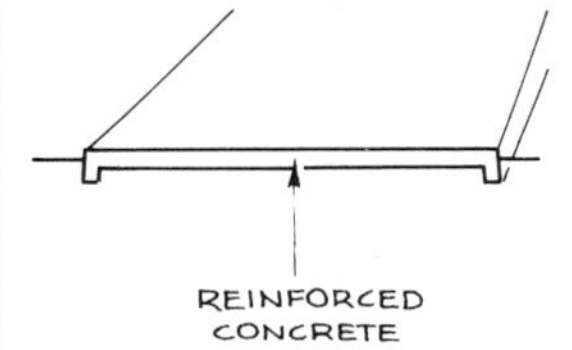 REINFORCED CONCRETE	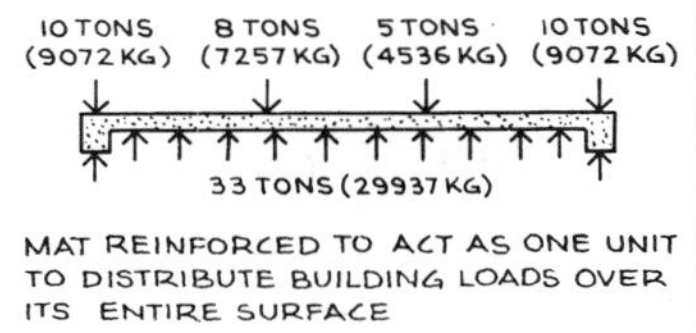 10 TONS (9072 KG); 8 TONS (7257 KG); 5 TONS (4536 KG); 10 TONS (9072 KG); 33 TONS (29937 KG); MAT REINFORCED TO ACT AS ONE UNIT TO DISTRIBUTE BUILDING LOADS OVER ITS ENTIRE SURFACE	NO FOUNDATION WALLS, MAT ACTS AS BOTH FOOTING AND FOUNDATION	AS FOUNDATION FOR ENTIRE BUILDING, USED GENERALLY IN AREAS WHERE THERE IS NO FREEZING
RAFT OR BOAT	REINFORCED CONCRETE WATER-PROOFED	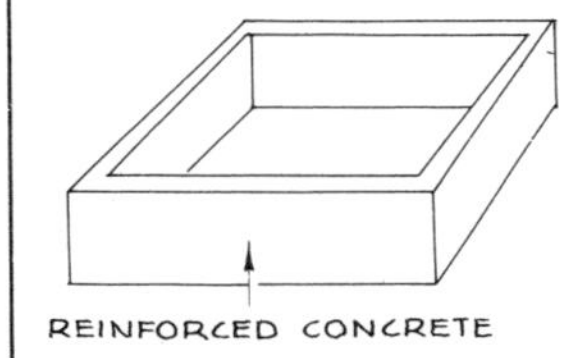 REINFORCED CONCRETE	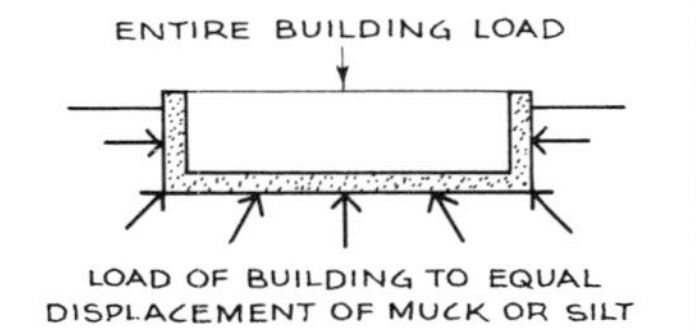 ENTIRE BUILDING LOAD; LOAD OF BUILDING TO EQUAL DISPLACEMENT OF MUCK OR SILT	ENTIRE BUILDING IS SUPPORTED ON RAFT OR BOAT	IN AREAS WHERE QUICK-SAND, MUCK, OR SILT ARE THE BEARING SOILS
CANTI-LEVER	REINFORCED CONCRETE OR STEEL GRILLAGE WITH CONCRETE	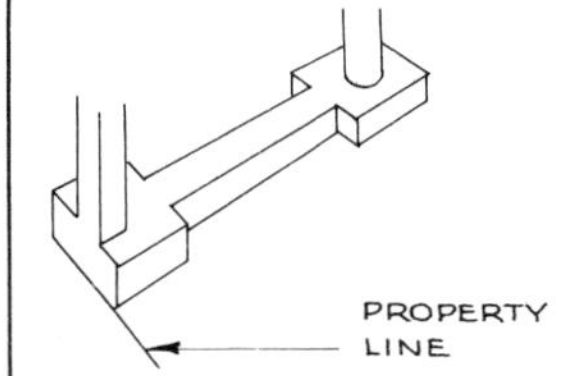 PROPERTY LINE	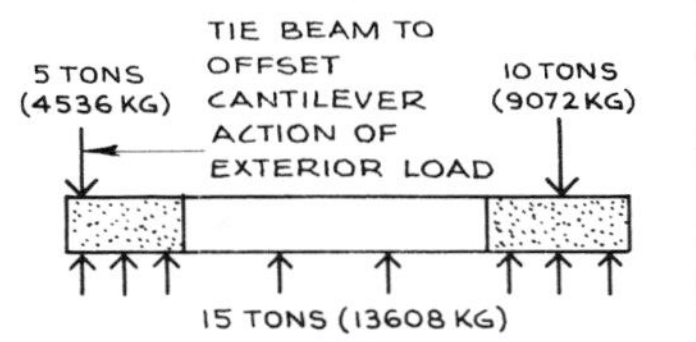 5 TONS (4536 KG); TIE BEAM TO OFFSET CANTILEVER ACTION OF EXTERIOR LOAD; 10 TONS (9072 KG); 15 TONS (13608 KG)	REINFORCED CONCRETE AND STEEL	USUALLY FOR LARGE BUILDINGS AT PROPERTY LINES
PILE	WOOD, STEEL, CONCRETE, REINFORCED CONCRETE OR VARIOUS COMBINATIONS	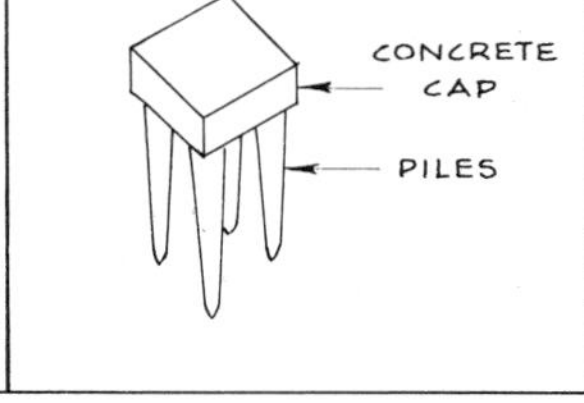 CONCRETE CAP; PILES	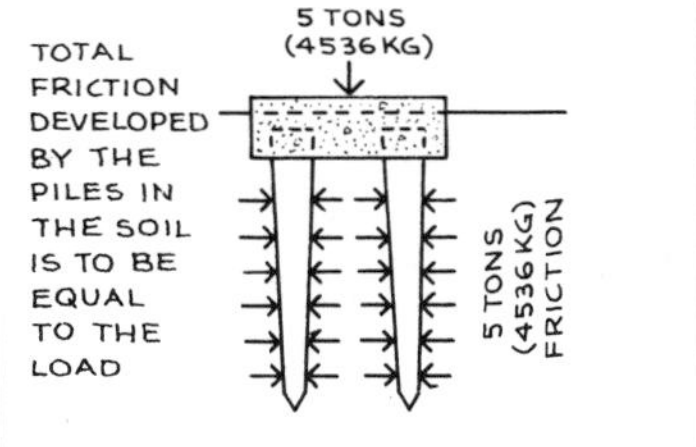 TOTAL FRICTION DEVELOPED BY THE PILES IN THE SOIL IS TO BE EQUAL TO THE LOAD; 5 TONS (4536 KG); 5 TONS (4536 KG) FRICTION	CONCRETE, REINFORCED CONCRETE, BRICK, CONCRETE BLOCK, STEEL, STONE WALLS, FOUNDATIONS, PIERS OR COLUMNS	IN AREAS WHERE SOIL IS OF LOW BEARING CAPACITY

Figure F9 Types of footings.

delta. Here it is necessary to design the basement and subbasements of the building as a boat that will float on muck and support the weight of the building by displacement. Third, if the soils do not have sufficient bearing strength and the boat-design system is not applicable, a method of support called piling is used. Such piling is always capped by footings on which the rest of the building (superstructure) is constructed.

FOSSIL FUELS

Because of the international situation developing from the oil crisis of the 1970s, fossil fuels will be playing a very important role in the future in meeting the needs for energy.

Fossil fuels consist of natural gas, petroleum, and coal, all of which are derived from organic materials. Petroleum may vary from a dark, syrupy liquid to a clear, colorless liquid, and coal varies from peat and lignite (brown coal) to hard coal (anthracite).

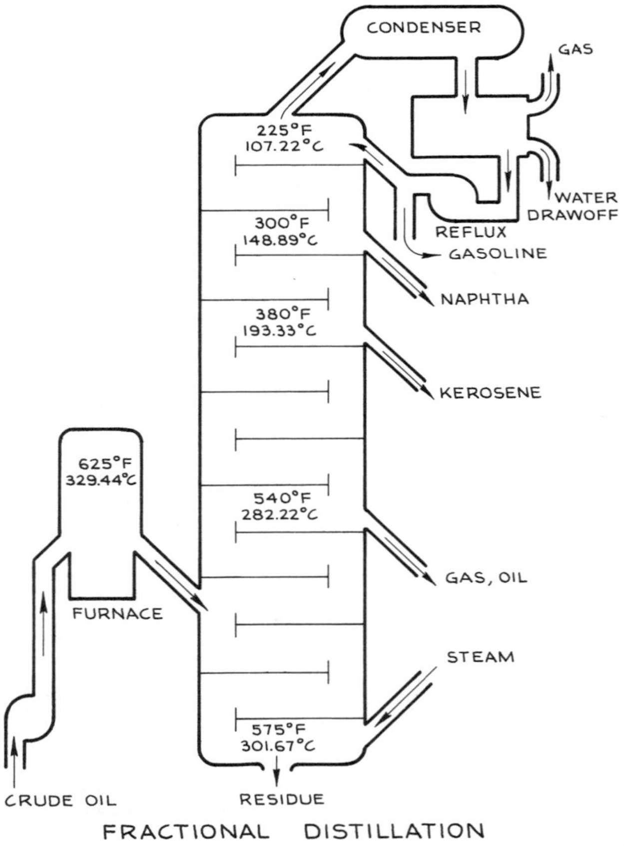

Figure F10 Fractional distillation of fossil fuel.

There are over 250,000 chemical by-products from bituminous coal alone, and more than 3000 products are manufactured from petroleum and natural gas. Figure *F10* shows an essential step in processing fossil fuels.

FOUNDATIONS

The word "foundations" is the inclusive term used to describe the parts of a building that rest upon the supporting earth (subsoil) and are designed to take and distribute the total load of the superstructure. Foundations are usually placed below the surface of the earth and may include, in addition to foundation walls and footings, retaining walls, piling, bulkheads, and caissons, each of which is discussed under its own heading.

The main problem with foundations is to determine the bearing capacity of the soil and to select the methods by which the loads should be distributed to the various types of soil. Water (ground water) and climate in relation to frost line are two other factors that influence the final design and materials chosen for the construction of foundations. *See* Caisson; Footings; Foundations: Walls; *and* Piles and Piling.

FOUNDATIONS: WALLS

The foundation wall is the part of a building's foundation that is located between the footings at the bottom and the supports for the floor nearest grade at the top. It has three functions: (1) to provide protection against penetration by water or frost; (2) to provide lateral support for the enclosed building area against the pressure exerted by the surrounding earth; and (3) to distribute vertical load from the building's superstructure to the footings.

In light construction or residential work or buildings where no basement area is required, the structure of foundation walls is controlled by climate, that is, by the frost line, as well as by prevailing water conditions at the building site and the area surrounding it.

Frost Line. The frost line along the northern belt of the United States, from the northern tip of Maine to Washington State, varies from 4 to 6 ft in depth. This means that, with or without a basement, the foundation walls must extend to this depth before the dangers of freezing are eliminated. In Florida, Texas, and any other region where freezing seldom occurs, foundation walls can be omitted unless basement areas are desired. In the majority of buildings, however, basement areas are required.

Water Conditions. Whenever basements are to be used for storage, mechanical equipment, and service areas, or must be habitable, the problem of overcoming water—whether from rain and snow, ground water (water table), or tides or underground springs—can become terribly complex. Good construction requires that this problem be solved completely.

Materials used for foundation walls must be strong and permanent in type, since they are buried in the earth and must outlast the building itself. The proper methods for their use and the methods and materials for waterproofing and dampproofing are discussed more fully under specific headings. The problem fundamentally is that water seeks its own level, and the architect must either get rid of it or build a swimming pool in reverse, actually a boat, where the inside is dry and the water is outside.

FRIABLE

A material or substance that crumbles easily can be described as friable.

FRIT

"Frit" refers to ground glass used as the basis for glazes and enamels (*see* Brick; Clay Tile; *and* Glazes and Porcelain Enamels).

FUSIBLE ALLOYS

Fusible alloys (*see* Table *F7*) are alloys that melt below 450°F (232.22°C) and as low as 158°F (70°C). The common metals that melt at low temperatures are zinc, lead, cadmium, bismuth, and tin.

Table F7 Common fusible alloys, including those generally used in fire-control systems

Name	Melting temperature °F (°C)	Composition (percent)			
		Bi	Pb	Sn	Cd
Rose's alloy	212.0 (100.0)	50	28	22	
Onion's alloy	212.0 (100.0)	50	30	20	
Newton's alloy	207.0 (91.67)	50	31	19	
D'Arcet's alloy	208.0 (92.22)	50	25	25	
Wood's metal	162.0 (72.22)	50	25	12.5	12.5

The fusible alloys have various habits on solidification, some expanding, some contracting, and others having a growth only after solidification, which may continue for approximately 21 to 42 days. Total growth may be as much as 0.008 in./in. (0.2032 mm/mm). All fusible alloys creep under light continuous loads.

Eutectic Alloys. A eutectic alloy is an alloy in which the melting point coincides with the freezing temperature, that is, it has no plastic range between liquid and solid (*see* Table *F8*). Generally it is whatever mixture of constituent metals that provides an alloy of the lowest

Table F8 Eutectic alloys

Melting temperature °F (°C)	Composition (percent)				
	Bi	Pb	Sn	Cd	Other elements
478.4 (248.0)		82		18	
430.0 (221.11)			96.5		Ag 3.5
390.0 (198.89)			92		Zn 8
361.0 (182.78)		38	62		
349.0 (176.11)			67	33	
293.0 (145.0)		31	51	18	
291.0 (143.33)	60			40	
281.0 (138.33)	58		42		
266.0 (130.0)	56		40		Zn 4
255.0 (123.89)	55.5	44.5			
217.0 (108.78)	54		26	20	
203.0 (95.0)	52.5	32	15.5		
197.0 (91.67)	52	40		8	
174.0 (78.89)	57		17.		In 26
158.0 (70.0)	50	26.7	13.3	10	
136.0 (57.78)	49	18	12		In 21
117.0 (47.22)	44.7	22.6	8.3	5.3	In 19.1

Table F9 Typical noneutectic alloys

Melting temperature °F (°C)	Composition (percent)				
	Bi	Pb	Sn	Cd	Other elements
293–349 (145.0–176.11)	12.6	47.5	39.9		
290–325 (143.33–162.78)	14	43	43		
266–343 (130.0–172.78)	20	50	30		
269–282 (131.67–138.89)	5	32	45	8	
255–266 (123.89–130.0)	56	2	40.9	0.7	In 0.4
248–306 (120.0–152.22)	21	42	37		
217–440 (102.78–226.67)	48	28.5	14.5		Sb 9
203–289 (95.0–142.78)	33.3	33.4	33.3		
203–237 (95.0–113.89)	59.4	14.8	25.8		
203–219 (95.0–103.89)	56	22	22		
181–198 (82.78–92.22)	52	31.7	15.3	1	
158–190 (70.0–87.76)	42.5	37.7	11.3	8.5	
142–149 (61.11–65.0)	48	25.6	12.8	9.6	In 4

possible melting point. Such alloys are used for electric fuses, automatic sprinkling devices, safety plugs in boilers, and the like.

Table F10 Uses of fusible alloys

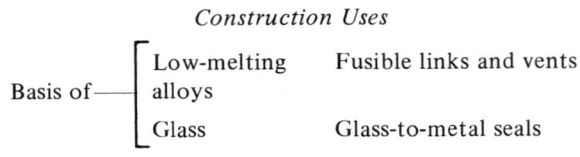

	Construction Uses	
Basis of	Low-melting alloys	Fusible links and vents
	Glass	Glass-to-metal seals
	Nonconstruction Uses	

Dies for "lost-wax" patterns, fusible cores in forming and electroforming, heat transfer medium, liquid seals, molds, spray coatings for patterns, etc.

Noneutectic alloys are too numerous to list completely (*see* Table *F9*). They remain pasty or liquid over a comparatively wide range of temperature.

Uses

Fusible alloys, because of their low melting points and high boiling points, find wide use in heat-treating and tempering baths. Their expansion characteristics when solidifying make them tremendously useful for installing tools, punches, bearings, etc., into handles, housings, presses, etc. The uses of fusible alloys are outlined in Table *F10*. The glass-to-metal seal is also described under Glass, Insulating.

FUSIBLE LINKS

These are metal links that melt and break at low temperatures (*see* Fusible Alloys) and are used for safety devices such as those for keeping fire doors open until the fusible link melts at a predetermined temperature.

G

GALVANIZING

Physical and Chemical Properties

Galvanizing is the process whereby a protective coat of zinc is applied to steel or iron sheet, strip, wire, castings, and formed shapes to protect them from corrosion. The advantage of a zinc coating is that, should any part of the steel or iron become exposed, galvanic reaction between the zinc and steel or iron causes the zinc to corrode and form compounds that cover and continue to protect the steel or iron as long as any zinc remains. *See also* Zinc Coatings; Steel, Galvanized; Metals, Coatings.

Galvanized zinc coatings are measured in oz/ft^2 (g/m^2) of surface, except on sheets. Sheets are coated on both sides, and the weight is given in oz/ft^2 (g/m^2) of sheet, or double the weight of coating by surface. Thus a sheet with a coating of 1 oz/ft^2 (305.15 g/m^2) actually has 0.5 oz/ft^2 (152.58 g/m^2) on each side, or a thickness roughly equivalent to 0.001 in. (0.0254 mm) per side.

Galvanized coatings are available on all types of sheet and iron products. For sheet and strip, coatings vary from 0.75 to 3.0 oz/ft^2 (228.75 to 915.45 g/m^2). For wire, castings, and prefabricated products, they vary from 0.3 to 1.5 oz/ft^2 (91.55 to 475.73 g/m^2).

Types and Uses

Zinc coatings are applied by several methods, namely, hot-dipping, electrogalvanizing, metallizing, and sheradizing. Most hot-dipped, galvanized zinc coatings freeze into a crystalline surface pattern known as spangles.

Zinc coatings are classified into five classes (*see* Table *G1*). Each class has a definite weight for the coating and therefore relates to the end use of the galvanized material. The generally available gauges and sizes of galvanized sheet and coils are given in Table *G2*.

Galvanized steel or iron is used for roofing, siding, decking, dockwork, and flat metalwork. Galvanized castings, machine parts, wire, and prefabricated products are used extensively in construction, industry and manufacturing.

History and Manufacture

In 1827 Crawford took out the first patent for hot-dip galvanizing. The great majority of galvanized sheet and strip is produced by a continuous galvanizing process (*see* Figure *G1*).

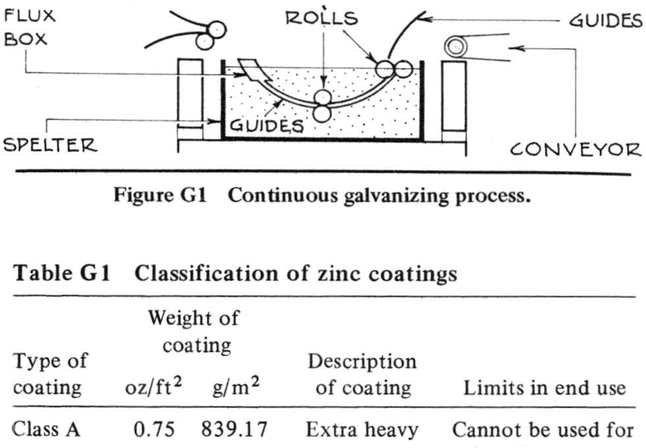

Figure G1 Continuous galvanizing process.

Table G1 Classification of zinc coatings

Type of coating	Weight of coating oz/ft^2	g/m^2	Description of coating	Limits in end use
Class A	0.75	839.17	Extra heavy	Cannot be used for forming, except for corrugating
Class B	2.75 2.50 2.00 1.75 1.50 1.25	839.17 762.88 610.30 534.01 457.73 367.44	Heavy	Cannot be used for forming, except for corrugating and curving for large radii
Class C	2.00 1.75 1.50 1.25	610.30 534.01 457.73 367.44	Moderately heavy	Used for moderate bending and forming
Class D	Not specified		Thin	For general utility purposes
Class E	Usually 0.75 1.25	 228.76 367.44	Tightly adherent	Used for severe forming

Table G2 Gauges and sizes of galvanized sheets and coils

Gauge	Nominal thickness		Weight[a]		Sheet size limits: Max. width		Sheet size limits: Max. length		Coils: Max. width	
	in.	mm	lb/ft^2	kg/m^2	in.	m	in.	m	in.	m
8	0.168	4.27	7.031	3.43						
9	0.1532	3.99	6.406	3.13						
10	0.1382	3.51	5.781	2.82	72	1.829	200	5.08	72	1.829
11	0.1233	3.13	5.156	2.52						
12	0.1084	2.75	4.531	2.21						
13	0.0934	2.37	3.906	1.91						
14	0.0785	1.99	3.281	1.66						
15	0.0710	1.80	2.969	1.45						
16	0.0635	1.61	2.656	1.30						
17	0.0575	1.46	2.406	1.18	72	1.829	200	5.08	60	1.524
18	0.0516	1.32	2.156	1.05						
19	0.0456	1.16	1.906	0.93						
20	0.0396	1.01	1.656	0.81						
21	0.0366	0.93	1.531	0.75	48	1.219	200	5.08	60	1.524
					60	1.524	168	4.267	48	1.219
22	0.0336	0.85	1.406	0.69	60	1.524	168	4.267	48	1.219
23	0.0306	0.75	1.281	0.63	51	1.295	144	3.658	48	1.219
24	0.0276	0.70	1.156	0.56						
25	0.0247	0.63	1.031	0.50	48	1.219	200	5.08	48	1.219
26	0.0217	0.55	0.906	0.44						
27	0.0202	0.51	0.844	0.42						
28	0.0187	0.48	0.781	0.38	45	1.143	200	5.08	45	1.143
29	0.0172	0.44	0.719	0.35						
30	0.0157	0.40	0.656	0.32						

[a]The galvanized sheet gauge established by custom is based on each of its gauge weights being 2.5 oz/ft^2 (662.88 g/m^2) heavier than the same U.S. Standard Gauge number.

GAUGES

In data on materials made of metal in sheet or wire form, the thickness was once (and still is in the construction field) given in gauges, or gauge numbers. The metal industry as a whole suffered confusion from these gauge designations, as each of the major metal industries set its own standards. Thus, a gauge number for steel products indicated a given thickness in fractions of an inch, but the same gauge number used for copper or other nonferrous metals indicated a different thickness in fractions of an inch. Moreover, certain metal industries (e.g., zinc) developed their own distinctive gauge systems which had nothing to do with other systems in existence. To complicate matters further, the British and some European countries each had its own gauge system, and other countries used the metric system.

To remedy this chaotic state of affairs, a general effort was made in the engineering professions, with almost universal success, to eliminate gauge numbers entirely. The thickness is now usually designated in decimals of an inch for sheet material and in circular mils or diameter in decimals of an inch for wire. However, so many tables and charts, particularly in materials for construction, still use the old gauge systems that, as a rule, both values are given. (*See* Table *G3*.)

Table G3 Sheet and wire gauge equivalents in decimals of an inch and in centimeters

Gauge Number	Ferrous products										Nonferrous products			
	American Steel Wire Gauge[a] for wire except music wire		Manufacturers Standard Gauge for sheet		British Imperial Gauge for wire		U.S. Standard Gauge for sheet		Birmingham or Stubbs Gauge[b] for strip and flat wire		American Wire Gauge[c] for sheet, strip, and wire		American Zinc Gauge for sheet	
	in.	cm	in.	cm	in.	cm	in.	cm	in.	cm	in.	cm	in.	cm
7/0	0.4900	1.245			0.500	1.27	0.500	1.27						
6/0	0.4615	1.172			0.464	1.18	0.46875	1.191			0.5800			
5/0	0.4305	1.093			0.432	1.10	0.4375	1.111			0.5165			
4/0	0.3938	1.000			0.400	1.02	0.40625	1.032	0.454	1.15	0.4600	1.168		
3/0	0.3625	0.9208			0.372	0.945	0.375	0.9525	0.425	1.08	0.4096	1.040		
2/0	0.3310	0.8407			0.348	0.884	0.34375	0.8731	0.380	0.965	0.3648	0.9266		
0	0.3065	0.7785			0.324	0.823	0.3125	0.7938	0.340	0.864	0.3249	0.8252		
1	0.2830	0.7188			0.300	0.762	0.28125	0.7144	0.300	0.762	0.2893	0.7348	0.002	0.00508
2	0.2625	0.6668			0.276	0.701	0.265625	0.6747	0.284	0.721	0.2576	0.6643	0.004	0.01016
3	0.2437	0.6190	0.2391	0.6073	0.252	0.640	0.25	0.6350	0.259	0.658	0.2294	0.5827	0.006	0.01524
4	0.2253	0.5723	0.2242	0.5695	0.232	0.589	0.234375	0.5953	0.238	0.605	0.2043	0.5189	0.008	0.02032
5	0.2070	0.5258	0.2092	0.5314	0.212	0.538	0.21875	0.5556	0.220	0.559	0.1819	0.4620	0.010	0.0254
6	0.1920	0.4877	0.1943	0.4935	0.192	0.488	0.203125	0.5159	0.203	0.516	0.1620	0.4115	0.012	0.03048
7	0.1770	0.4496	0.1793	0.4554	0.176	0.447	0.1875	0.4763	0.180	0.457	0.1443	0.3665	0.014	0.03556
8	0.1620	0.4115	0.1644	0.4176	0.160	0.406	0.171875	0.4366	0.165	0.419	0.1285	0.3264	0.016	0.04664
9	0.1483	0.3767	0.1495	0.3797	0.144	0.366	0.15625	0.3969	0.148	0.376	0.1144	0.2906	0.018	0.04572
10	0.1350	0.3429	0.1345	0.3416	0.128	0.325	0.140625	0.3572	0.134	0.340	0.1019	0.2588	0.020	0.0508
11	0.1205	0.3061	0.1196	0.3038	0.116	0.295	0.125	0.3175	0.120	0.305	0.09074	0.2305	0.024	0.06096
12	0.1055	0.2680	0.1046	0.2657	0.104	0.264	0.109375	0.2778	0.109	0.277	0.08081	0.2053	0.028	0.07112
13	0.0915	0.232	0.0897	0.2278	0.092	0.234	0.09375	0.2381	0.095	0.241	0.07196	0.1828	0.032	0.08128
14	0.0800	0.203	0.0747	0.1897	0.080	0.203	0.078125	0.1984	0.083	0.211	0.06408	0.1628	0.036	0.09144
15	0.0720	0.183	0.0673	0.1709	0.072	0.183	0.070312	0.1786	0.072	0.183	0.05707	0.1450	0.040	0.1016
16	0.0625	0.159	0.0598	0.1519	0.064	0.163	0.0625	0.1588	0.065	0.165	0.05082	0.1291	0.045	0.1143
17	0.0540	0.137	0.0538	0.1367	0.056	0.142	0.05625	0.1429	0.058	0.147	0.04526	0.1150	0.050	0.1270
18	0.0475	0.121	0.0478	0.1214	0.048	0.122	0.05	0.1270	0.049	0.124	0.04030	0.1024	0.055	0.1397
19	0.0410	0.104	0.0418	0.1060	0.040	0.102	0.04375	0.1111	0.042	0.107	0.03589	0.09116	0.060	0.1524
20	0.0348	0.0884	0.0359	0.0912	0.036	0.0914	0.0375	0.09525	0.035	0.089	0.03196	0.08118	0.070	0.1778
21	0.0318	0.0808	0.0329	0.0836	0.032	0.0813	0.034375	0.08731	0.032	0.081	0.02846	0.07229	0.080	0.2032
22	0.0286	0.0726	0.0299	0.0758	0.028	0.0711	0.03125	0.07938	0.028	0.071	0.02535	0.06439	0.090	0.2286
23	0.0258	0.0655	0.0269	0.0683	0.024	0.0610	0.028125	0.07144	0.025	0.064	0.02257	0.05733	0.100	0.254
24	0.0230	0.0584	0.0239	0.0607	0.022	0.0559	0.025	0.0635	0.022	0.056	0.02010	0.05105	0.125	0.3175
25	0.0204	0.0518	0.0209	0.0531	0.020	0.0508	0.021875	0.05556	0.020	0.051	0.01790	0.04547	0.250	0.635
26	0.0181	0.0460	0.0179	0.0455	0.018	0.0457	0.01875	0.04763	0.018	0.046	0.01594	0.04049	0.375	0.9525
27	0.0173	0.0439	0.0164	0.0417	0.0164	0.0417	0.0171875	0.04366	0.016	0.041	0.01419	0.03604	0.500	1.27
28	0.0162	0.0411	0.0149	0.0378	0.0149	0.0378	0.015625	0.03969	0.014	0.036	0.01264	0.03211	1.000	2.54
29	0.0150	0.0381	0.0135	0.0343	0.0136	0.0345	0.0140625	0.03572	0.013	0.033	0.01126	0.02860		
30	0.0140	0.0356	0.0120	0.0305	0.0124	0.0315	0.0125	0.03175	0.012	0.030	0.01003	0.02548		
31	0.0132	0.0335	0.0105	0.0267	0.0116	0.0295	0.0109375	0.02778	0.010	0.025	0.008928	0.02268		
32	0.0128	0.0325	0.0097	0.0246	0.0108	0.0274	0.0101563	0.0258	0.009	0.023	0.007950	0.02019		
33	0.0118	0.0300	0.0090	0.0229	0.0100	0.0254	0.009375	0.02381	0.008	0.020	0.007080	0.01798		
34	0.0104	0.0264	0.0082	0.0208	0.0092	0.0234	0.0085938	0.02183	0.007	0.018	0.006304	0.01601		
35	0.0095	0.024	0.0075	0.0191	0.0084	0.0213	0.0078125	0.01984	0.005	0.013	0.005614	0.01426		
36	0.0090	0.023	0.0067	0.0170	0.0076	0.0193	0.0070313	0.01786	0.004	0.010	0.005000	0.01270		
37	0.0085	0.022	0.0064	0.0163	0.0068	0.0173	0.0066406	0.01687			0.004453	0.01131		
38	0.0080	0.020	0.0060	0.0152	0.0060	0.0152	0.00625	0.01588			0.003965	0.01007		
39	0.0075	0.019			0.0052	0.0132					0.003531	0.008969		
40	0.0070	0.018			0.0048	0.0122					0.003145	0.007988		
41	0.0066	0.017			0.0044	0.0112								
42	0.0062	0.016			0.0040	0.0102								
43	0.0060	0.015			0.0036	0.0091								

Table G3 Sheet and wire gauge equivalents in decimals of an inch and in centimeters (*continued*)

	Ferrous products										Nonferrous products			
	American Steel Wire Gauge[a] for wire except music wire		Manufacturers Standard Gauge for sheet		British Imperial Gauge for wire		U.S. Standard Gauge for sheet		Birmingham or Stubbs Gauge[b] for strip and flat wire		American Wire Gauge[c] for sheet, strip, and wire		American Zinc Gauge for sheet	
Gauge Number	in.	cm	in.	cm	in.	cm	in.	cm	in.	cm	in.	cm	in.	cm
44	0.0058	0.015			0.0032	0.0081								
45	0.0055	0.014			0.0028	0.0071								
46	0.0052	0.013			0.0024	0.0061								
47	0.0050	0.013			0.0020	0.0051								
48	0.0048	0.012			0.0016	0.0041								
49	0.0046	0.012			0.0012	0.0030								
50	0.0044	0.011			0.0010	0.0025								

[a]Also known as the Washburn and Moen gauge.
[b]Used very little except for gauging imported wire.
[c]Also known as the Brown and Sharpe gauge.

GEM STONES

Physical and Chemical Properties

Gem stones are minerals of crystalline structure that have great beauty, durability, and value, depending on rarity and perfection of color, structure, and finished form. The specific gravity and index of refraction usually give positive identification of the type of gem.

Types and Uses

The common gem stones are agate, amethyst, californite, chrysoberyl, chrysocolla, diamond, emerald, garnet, jade, onyx, opal, rock crystal (quartz), ruby (corundum), sapphire, topaz, tourmaline, and turquoise. Formerly, gem stones were used primarily for jewelry and ornaments although there was some use of garnet for polishing and abrasives. Agate, topaz, turquoise, opal, and jade have been used in stained glass windows.

Since 1900 there has been an increasing demand for the following industrial gem stones:

1. Industrial diamonds for glass and metal cutting, drill bits, wire dies, and abrasive wheels.
2. Jewel bearings for timekeeping devices, meters, compasses, mechanical instruments, and other instrumentation.
3. Garnet, which has largely replaced sand as an abrasive.

Garnet. Garnet, next to the diamond in importance among the abrasive gem stones, is the general name of a group of minerals varying in color, hardness, toughness, and method of fracture. It is used for coating abrasive paper and cloth and for cutting glass and stone. Some garnet is made into wheels by the silicate or shellac processes, but vitrified wheels are not made because of the low melting point of garnet (2372°F, 260°C).

The best garnet abrasives come from red almandite, obtained in large crystals chiefly from New York State. Andradite is another variety of common garnet employed for coating abrasive paper and cloth, as is rhodolite, a pale rose to purple garnet.

Mined garnet is crushed, ground, and then separated and graded in settling tanks and sieves. Hornblende is a common impurity and is difficult to separate, but good-quality abrasive garnet should be free of this mineral.

GERMANIUM

Physical and Chemical Properties

Symbol: Ge
Atomic number: 32
Specific gravity: 5.323 (25°C, 77°F)
Melting point: 937.4°C, 1719.32°F
Boiling point: 2830°C, 5126°F

Germanium is a silvery-gray solid of metallic appearance which is similar chemically to carbon and silicon. It is not corroded by air or water and is available in lump, ingot, and powder form.

Types and Uses

Germanium is a semiconductor, and its major use is in making electronic transistors and rectifiers. Almost all uses of germanium are in its semimetallic form. Its uses in compound form are shown by asterisks in Table *G4*.

Table G4 Uses of germanium

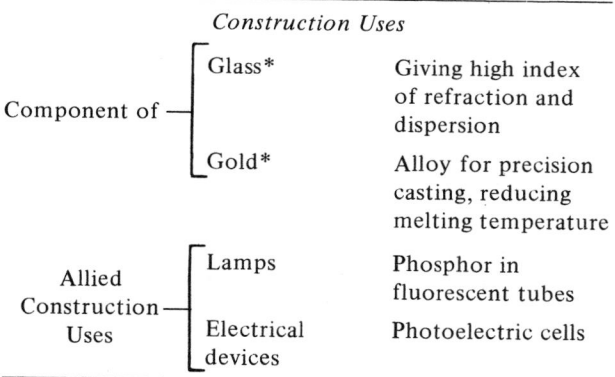

	Construction Uses	
Component of	Glass*	Giving high index of refraction and dispersion
Component of	Gold*	Alloy for precision casting, reducing melting temperature
Allied Construction Uses	Lamps	Phosphor in fluorescent tubes
Allied Construction Uses	Electrical devices	Photoelectric cells

History and Manufacture

In 1871 Mendeleyev predicted that a new element would be located between silicon and tin in the atomic table. Winkler discovered the new element in 1886 and named it germanium after his native country. Although it has been known for many years, its importance in electronics dates only from about 1950. Germanium is recovered as a by-product from zinc and other base metal ores. From these sources germanium tetrachloride is obtained. This is hydrolyzed in water to obtain a dioxide, which in turn is reduced by heat to a powdered metal. This powder is converted into ingots by heating it in an inert atmosphere.

GLASS AND GLAZING

Glass, which once was used in construction only in sheet form for windows and otherwise was familiar as tableware and objects of art, has now become an integral part of very many materials for construction. Foamed and cellular glass, curtain wall panels, reflective and insulating glass, and textiles are examples of these comparatively new products.

A major breakthrough in glass manufacturing has brought about a complete overhaul of the terminology and of the various types of glass materials used in construction. In this process, molten glass floats on a bath of molten tin and remains untouched until it hardens (*see* Glass, Float). The process produces a glass that has perfectly flat surfaces with an exceptionally brilliant finish and requires no further treatment except sizing and cutting. Float glass is suitable for window, heavy sheet and plate glass uses, and for the manufacturing of insulating, tempered, reflective, and other types of glass. Float glass was first manufactured in the United States in 1963, and most manufacturers in their promotion and publicity have treated float glass as a separate type of glass material. Until the construction field becomes familiar with float glass, there will be some confusion in terminology.

Where once, in the construction field, we could describe fully all installation of glass under one category of work, glazing, now only the setting of glass into metal or wood frames for doors and windows is the work of the glazier (*see* "Types and Uses" *and* "Application").

It is easier to understand the recent trends in the characteristics and uses of glass if one remembers that glass is by definition a ceramic material. In the United States only glasses of inorganic origin are recognized as such, to differentiate sharply the boundary between glass and plastics.

The necessity for reducing heat loss and heat gain in buildings, due to the slow depletion of fossil fuels, has resulted in extensive new research in glass. The existing heat-absorbing and glare-reducing glasses and the existing one-way mirrors have been elaborated upon and combined to introduce a whole new set of glass products. Tinted glasses (bronze, silver, gold, gray, blue, and green) have replaced to a great extent the heat-absorbing and glare-reducing glasses, but in terminology only. Reflective glass as a one-way mirror is still used for observation windows and panels. Both the above types, tinted and reflective, are now combined into an insulating glass to provide maximum control of heat loss and heat gain within structures.

Basic research in solar energy, generally halted in the United States in the 1930s, has now been revived because of the steady price increases in foreign crude oil and the emphasis on new energy sources. Usually glass is used for solar energy collectors; these consist of glass layers that transmit the sun's rays to a heat-absorbing coated metal plate, which in turn transmits the heat to water circulating in tubing. When considering the use of solar energy, one should always check the manufacturers of glass and solar collectors and current research projects for the latest data.

Physical and Chemical Properties

Scientifically, glass is defined as an inorganic product of fusion which has cooled to a rigid condition without crystallizing. Physicists consider glass to be an undercooled (or supercooled) liquid, though physically solid, since its structure is not crystalline but amorphous, as is characteristic of liquids. Glass, then, is a solid, supercooled, liquid ceramic material characterized by transparency, brittleness, hardness, and chemical inertness.

Glass differs completely from other ceramics in that most ceramics are shaped cold and then fired to produce the material, whereas glass is shaped at high temperatures and then allowed to cool. Also, glass can be reheated and shaped again. Glass therefore can be considered a thermoplastic material. In the molten state it is a solution of several different chemical compounds chemically bonded together; if allowed to cool slowly or held too long in the plastic state, these compounds tend to crystallize out of the solution. When this happens, the glass is said to be devitrified; actually it has frozen. To avoid this condition, the glass is carried through the crystallization temperature quickly so that it becomes solid as an undercooled liquid and the excessively high viscosity of the solid material stops the formation of crystals.

It is necessary that glass remain soft enough to yield to formative pressure over a considerable range of temperature, ending at a viscosity (stiffness) high enough to retain its final shape. The rate at which viscosity changes with temperature has an important effect on the forming of glass while hot and plastic. The viscosity can be altered and controlled by the composition of the glass.

Physical and mechanical properties vary with the composition of the glass and cover the following wide ranges: specific gravity, 2.125 to 8.12; annealing point, 662 to 1634°F (35 to 890°C); softening point, 932 to 2012°F (500 to 1100°C); coefficient of expansion, from 0.00000056 to 0.0000014; compressive strength, 90,000 to 180,000 lbf/in.^{2} (620.55 to 1241.10 MN/m^2); tensile strength, 4000 to 1,500,000 lbf/in.^{2} (27.58 to 10 342.5 MN/m^2).

Strength. The smaller the cross section of glass, the greater will be the unit strength. Glass fibers have a tensile strength of 1,500,000 lbf/in.^{2} (10 342.5 MN/m^2), whereas a 1-in.^{2} (6.45-cm^2) cross section of glass has a strength of only 5000 lbf/in.^{2} (34.48 MN/m^2). Actual values are even lower and depend on surface conditions and on the presence or absence of even microscopic imperfections. Glass shows little change in strength as the temperature is raised until the softening point, when it becomes ductile. When it ruptures, glass always fails in tension.

Light Transmission. Commonly used glasses transmit 85 to 95% of visible light. Most glasses are impervious to ultraviolet light.

Important Properties. For the construction field the important factors to know about glass are its thermal expansion (both from the internal stresses developed and the actual expansion of the glass), its thermal conductivity (relation to heat and cold losses within the building), its durability under conditions of end use, and occasionally its transmission of ultraviolet rays.

Chemical Properties. The chemical properties of a glass, like the physical and mechanical properties, are controlled by its composition and can vary over a wide range. In general, glass is attacked only by hydrofluoric acid, is very slowly dissolved by water, and is more rapidly soluble in alkaline solutions.

Constituents. The oxides of silicon, sodium, and calcium (sand, soda, and lime) are the most widely used ingredients for glass, and from this base glass virtually all other types have been derived (*see* Table *G5*).

In making glass the necessary oxides are not used directly; instead, the raw materials that will furnish them to the glass are used. Table *G6* shows this relation between raw materials and glass-forming oxides and lists the percent of raw materials remaining in the glass.

The glass used in architecture is usually a silicon-sodium-calcium type of composition (soda-lime glass) in which the difference in characteristics is controlled by the quantities of sodium and calcium used. In many of the special glasses other elements are used as additions or as substitutes for one or another of the basic components. For example, sodium, potassium, and lithium are mutually interchangeable; aluminum, magnesium, barium, boron, and lead may be substituted for calcium; in some cases aluminum, boron, or titanium may be used instead of silicon; and phosphorus may entirely replace silicon. Both sodium and potassium increase thermal expansion but decrease thermal conductivity. Lead also decreases thermal conductivity, whereas silicon and boron increase it. The electrical conductivity of glass is similarly controlled by the composition of the glass and increases as the temperature rises.

Any of the other elements introduced into the glass composition affects only color, viscosity, or permanence or adds some special desired physical property. Table *G7* lists all the other elements used in glass, their effects, and the types of glass produced.

Table G5 Typical chemical compositions of float, window, and plate glass

Type of glass	Silicon dioxide (SiO_2)	Sodium oxide (Na_2O)	Calcium oxide (CaO)	Magnesium oxide (MgO)	Aluminum oxide (Al_2O_3) and iron oxide ($FeO \cdot Fe_2O_3$)	Potassium oxide (K_2O)
			(percent of content)			
Window and heavy sheet glass (Colburn process)	71.7	13.6	9.7	4.3	0.7	
Window and heavy sheet glass (Fourcault process)	72.0	15.0	8.5	4.5	1.5	0.5
Plate glass	72.2	14.0	11.5	2.15	0.15	
Float glass	70.0–72.2	12.0–15.0	8.5–12.0	2.15–4.5	0.15–0.7	0–0.5

Table G6 Raw materials for glass

Major glass-forming oxides	Raw materials that furnish oxides to glass		Percentage of raw material remaining in glass
Aluminum oxide (Al_2O_3)	Hydrated alumina	$Al(OH)_3$	65.4
	Calcined alumina	Al_2O_3	100.0
	Feldspar	Al_2O_3	18.0
		$K_2O(Na_2O)$	13.0
		SiO_2	68.0
	Lepidolite	Al_2O_3	27.0
		K_2O	10.0
		Li_2O	4.5
		F_2	4.0
		SiO_2	53.0
	Cryolite	Na_3AlF_6	Some loss of F as SiF_4
Calcium oxide (CaO)	Limestone	$CaCO_3$	56.0
	Burned lime	CaO	100.0
	Hydrated lime	$Ca(OH)_2$	75.7
	Dolomite (limestone), raw	CaO	30.4
		MgO	21.8
	Dolomite (limestone), burned.	CaO	58.2
		Mg	41.8
Magnesium oxide (MgO)	Magnesium carbonate	$MgCO_3$	47.6
	Dolomite (see above)		
Potassium oxide (K_2O)	Potassium carbonate	K_2CO_3	61.8
	Calcined potassium nitrate	KNO_3	45.6
Silicon oxide (SiO_2)	Sand (quartz)	SiO_2	100.0 (less impurities)
Sodium oxide (Na_2O)	Sodium carbonate	Na_2CO_3	58.5
	Sodium sulfate	Na_2SO_4	43.7
	Sodium nitrate	$NaNO_3$	36.5
	Sodium fluosilicate	Na_2SiF_6	
	Borax, hydrated	B_2O_3	36.5
		Na_2O	16.3
	Borax, anhydrous	B_2O_3	69.2
		Na_2O	30.8
	Feldspar (see above)		

Table G7 Effect of various elements on glass characteristics and uses of the special glasses they produce

Element	Form used	Effect on glass composition	Advantages	Disadvantages	Type of glass and major use
Antimony	Sb_2O_3	Affects color	Counteracts the greenish color caused by iron impurities		Many types of glass, infrared transparent glass
Arsenic	As_2O_3	Affects color	Counteracts the greenish color caused by iron impurities		Many types of glass, infrared transparent glass
Barium	$BaCO_3$	Partially or completely replaces calcium (lime); barium carbonate changes to barium oxide (BaO) during melt	Increases index of refraction		Optical glass
Beryllium	BeO	Partially replaces either calcium or silicon (sand)	Increases index of refraction and resistance to weathering	Little used because of cost	Many types of glass
Boron	B_2O_3	Partially or completely replaces silicon	Increases index of refraction; reduces coefficient of expansion and deepens colors produced by other elements		Optical and ovenware glass
Cadmium	CdS	Produces color	Produces yellow glass and absorbs certain portions of the color spectrum		Special optical glass and filter-type glass
Carbon	C	Helps reduce materials in the melting; introduced into composition in form of anthracite coal	Increases boiling action by combining with oxygen to form CO_2 during reduction of other materials	Very hard to control because of gaseous CO_2	All types of glass
Cerium	Ce_2O_3	Produces color	Gives slight yellow tint and blue fluorescence in glass; absorbs certain portions of color spectrum; absorbs ultraviolet light		Special optical glasses, protection from X-rays
Cesium	CsO	Partially or completely replaces sodium	Increases index of refraction	Little used because of cost	Optical
Chlorine	NaCl	Helps agitate the melting process	Improves quality of glass (sodium becomes part of the glass)		All types of glass
Chromium	Cr_2O_2	Oxidizing agent; produces color	Produces yellow-green color; absorbs certain portions of color spectrum		Colored glass and special optical glass
Cobalt	Co_2O_3	Decolorizer; produces color	Produces blue color; with nickel produces a neutral tint; therefore used to neutralize colors due to impurities		All types of glass, colored glass, some optical glass
Copper	Cu_2O	Produces color	Produces red color		Colored glass
	CuO	Produces color	Produces blue color		Colored glass

Table G7 Effect of various elements on glass characteristics and uses of the special glasses they produce (*continued*)

Element	Form used	Effect on glass composition	Advantages	Disadvantages	Type of glass and major use
Fluorine	CaF_2	Opacifier and opalizer	Lowers index of refraction and viscosity at working temperatures (calcium becomes part of the glass)		Opalescent glass
Gold	$AuCl_3$	Produces color	Produces ruby color	Little used because of cost	Colored glass
Iron	Fe_2O_3	Produces color	Produces aquamarine color		Colored glass
	Fe	Usually an impurity	Lower viscosity at melting temperatures	Causes greenish color in glass	All types of glass
Lead	PbO	Increases density; produces color; introduced in the form of litharge	Increases brilliance, refractive index, and density; in large quantities produces canary yellow		Flint glass for tableware, optical, and X-ray glasses
Manganese	MnO_2	Affects and produces color	Counteracts the greenish color caused by iron impurities and produces amethyst purple to dark red color; also is transparent to infrared radiation	After a length of time glass decolorized with manganese turns a deep purple	All types of glass, colored glass
Neodymium and praseodymium	Oxides of both	Produces color	Produces a bluish pink color; reduces glare		Colored glass
Nickel	NiO	Produces color	Produces reddish brown in sodium-calcium glass, and violet in potassium-calcium glass; with cobalt produces a neutral tint		Colored glass
Nitrogen	$NaNO_3$	Oxidizing agent	Oxygen is released and nitrogen dissipated		All types of glass
Oxygen	(Always with other elements as oxide)	Oxidizing agent	Actual component of glass		All types of glass
Phosphorus	P_2O_5	Opacifier and opalizer in sodium-calcium glass			Opalescent glass
	$Al(PO_3)_3$	Completely replaces silicon			Scientific and optical glass
Potassium	K_2CO_3	Flux	Promotes fusing of the materials		Many types of glass
		Completely replaces silicon	Produces more brilliant glass; increases viscosity and resistance to weathering; lowers index of refraction	Not generally used, as it is more costly than sodium	Special glasses

Table G7 Effect of various elements on glass characteristics and uses of the special glasses they produce (*continued*)

Element	Form used	Effect on glass composition	Advantages	Disadvantages	Type of glass and major use
Selenium	Se	Produces color and in small percentages acts as a decolorizer	Produces pink color and with cadmium sulfide produces ruby red color		Colored glass
Silver	Ag_2O	Produces color	Produces yellow color		Colored glass
Strontium	SrO	Substitute for calcium or barium		Seldom used because of cost	Special glass
Sulfur	Na_2SO_4	Reducing agent and produces color	When present in glass as sodium sulfide (Na_2S), produces yellow color (sodium becomes part of the glass)		All types of glass, colored glass
Tin	SnO	Chemically reduces the iron in sodium-calcium glass	Helps to oxidize and reduce iron content		Many types of glass, glass for special ultraviolet transmission
		Combines with fluorides and phosphates	Produces milky opalescence		Opalescent glass
		Improves ruby colored glass of copper or gold type	Improves color of ruby glass		Colored glass
Titanium	TiO_2	Produces color	Produces violet color, and with iron or cerium a yellow color		Colored glass
Uranium	Na_2UO_4	Produces color	Produces greenish yellow fluorescent color; absorbs certain portions of the color spectrum		Colored glass, optical and scientific glass
Vanadium	V_2O_5	Produces color	Produces green color; absorbs certain portions of the color spectrum		Colored glass, optical glass
Zinc	ZnO	Flux substitute for calcium oxide	Increases resistance to weathering and decreases the coefficient of expansion		Many types of glass
Zirconium	ZrO_2	Increases viscosity	Increases index of refraction and resistance to weathering	Seldom used because of cost	Special glass

Coloring. Colorless glass employed in the architectural field should not contain more than 0.1% of iron oxide. Colored glass may be either translucent or opaque. The possible color range is indicated in Table *G8*, which lists coloring materials and their contents in glass. The colors are either solution colors or finely divided particles dispersed throughout the glass.

Workability. Glass can be fabricated to any desired shape. It can be cast, pressed, rolled, blown, polished, and ground.

Commercial Forms. Glass is available for use in construction in flat form as window, heavy sheet, float plate, tempered, heat-strengthened, patterned, wired, heat-absorbing, insulating (double glass), glare-reducing, laminated, structural, corrugated, and mirror glass; and in cellular, block (solid or hollow), and powdered form.

Types and Uses

In the construction field we think of glass primarily in relation to windows and other applications covered by

Table G8 Materials used for coloring glass

Coloring material	Percentage in glass	Color produced
Copper	0.03 to 0.1	Ruby
Copper oxide	0.2 to 2.0	Blue-green
Cadmium sulfo-selenide	0.03 to 0.1	Ruby and orange
Cadmium sulfide	0.03 to 0.1	Yellow
Ferric oxide	Up to 4.0	Yellow-green
Chromium oxide	0.05 to 0.2	Green to yellow-green
Ferrous oxide		Blue-green
Gold	0.01 to 0.03	Ruby
Carbon and sulfur compounds		Amber
Iron oxide and manganese oxide	1.0 to 2.0 2.0 to 4.0	Amber
Uranium oxide	0.1 to 1.0	Yellow with green fluorescence
Selenium		Pink
Selenides		Amber
Manganese oxide	0.5 to 3.0	Pink-purple
Nickel oxide	0.05 to 0.5	Brown and purple
Neodymium oxide	Up to 2.0	Pink
Cobalt oxide	0.001 to 0.1	Blue

the term "glazing." It is also used as a surfacing material, alone or as part of a system of construction (curtain walls, interior partitions), or as a purely decorative material (colored glass, mirrors). These are all glass in sheet form.

Glass breakage in many areas has brought about changes and additions to building codes involving safety in buildings. There are safety requirements regarding the use of tempered and wire glass in areas where breakage hazards exist; examples are storm doors, shower doors, entrance doors, and side lights. It is necessary to check all governing safety codes when selecting the types of glass to be installed in a building.

Glass block is treated more as a specialized form of masonry. Fibers, cellular glass, and powdered glass are fabricated into many products with widely diverse characteristics and architectural uses.

Each of the forms of glass that is important in construction is discussed separately under its own heading. Table *G9* (pages 302–309) summarizes data on these types of glass and gives their general uses.

Bent Glass. Window, heavy sheet, plate, patterned, structural, glare-reducing, heat-absorbing, and colored glass are available in bent form within definite limitations. The radii should be dimensioned to the outside surface of the glass and never be less than 6 in. (152.4 mm). When the design requires the bending of glass over $\frac{3}{8}$ in. (9.53 mm) thick, the manufacturers of the glass should be consulted. Bent glass is produced by heating flat glass and allowing it to follow by gravity the form of the mold, which is the exact shape of the bent form. Bent glass has greater distortion and wider tolerances than flat glass. (*See* Table *G10* and Figure *G2*.)

Optical Glasses. A few optical glasses are of interest to the architect for highly specialized applications. Specifically, these are (1) glass with a very low iron content that transmits ultraviolet light, (2) glass with a high ferrous iron content that cuts off both ultraviolet light and heat radiation, and (3) glass containing nickel that transmits ultraviolet light but no visible light.

Polarized Glass. Polarized glass consists of sheet glass coated with an organic crystalline chemical that polarizes transmitted light.

Application

Glazing. In construction specifications, "glazing" is the term used for the installation of glass. The materials used to make a weathertight joint between the glass and the frame into which the glass sets are called glazing materials. Glass is held in place with various types of wire clips, metal triangular points, shims, gaskets, tapes, special adhesives, foam rope, and glazing beads; it is kept away from the frame with various types of shims (*see* Figure *G3*).

Glazing materials in general use are as follows:

1. Wood sash putty is made with a color pigment and linseed oil, with or without drying oils. The wood should be treated with boiled linseed oil or a priming paint before putty is applied. Putty should never be painted before it is thoroughly dry.
2. Metal sash putty is made of materials that will adhere to nonporous materials. It should always be applied as recommended by its manufacturer. Generally, it should be painted 2 weeks after application. Metal sash putty is made in two types, exterior glazing and interior glazing.
3. Elastic glazing compound is made from selected processed oils and color pigments compounded so that it will remain plastic and resilient over a long period of time. Elastic glazing compounds are generally used where vibration and twisting may occur. They should be painted immediately after a thin skin has formed on the outside.

Table G9 Types of glass used in construction

Type of glass	Thickness[a] in.	Thickness[a] mm	Transparency and color	Transmittance[b] Visible (percent)	Transmittance[b] Total solar energy (percent)	Major uses	Other uses
Block, hollow	$3\frac{7}{8}$	98.43	Clear, patterned, translucent, and opaque	Depends on type, pattern, and texture		Exterior walls where insulation and specialized lighting are required	Interior panels, partitions, screens, etc.
Block, solid	$3\frac{7}{8}$	98.43	Slight greenish color	Translucent		Prisons and mental institutions	Interior panels, screens, etc.
Cellular	$1\frac{3}{4}$	44.45	Black	Opaque		Roof, panels, perimeter, and wall insulation	Industrial insulation
	2	50.80					
	$2\frac{1}{2}$	63.50					
	3	76.20					
	4	101.60					
Corrugated	$\frac{1}{2}$	12.70	Translucent, textured surface			Interior partitions, screens, etc.	Exterior glazing
	1	25.40					
Corrugated hexagonal wired	$1\frac{1}{8}$	28.58	Translucent, textured surface	Controlled by fire-resistance values		Roofs, skylights, exterior walls	Fire-resistant partitions
Fibers	5–16 μ; continuous length					Insulating materials, reinforcing plastics, and papers	
Flake	0.0003	0.0076				Filler for plastics	
	1–4 μ; $\frac{1}{32}$-in. (0.78-mm) diameter gives tilelike appearance; substitute for mica					Filler for paints and plastics	
Float window, single-strength, AA, A, B	0.085–0.101	2.159–2.565	Clear	91	87	Glazing windows	Glazing storm sash
Float window, double-strength AA, A, B	0.115–0.134	2.921–3.404		90	85	Glazing windows	Glazing storm sash

Float, clear	$\frac{3}{32}$	2.35	Clear	91	86	Glazing large windows, store fronts, etc.	Mirrors, shelves, furniture tops, showcases, insulation, tempered and laminated glass
	$\frac{1}{8}$	3.18	Clear	90	85		
	$\frac{3}{16}$	4.76		90	83		
	$1\frac{3}{64}$	5.09		90	82		
	$\frac{1}{4}$	6.35		89	79		
	$\frac{5}{16}$	7.94		88	77		
	$\frac{3}{8}$	9.53		87	82		
	$\frac{1}{2}$	12.70		86	62		
	$\frac{5}{8}$	15.88		84	62		
	$\frac{3}{4}$	19.05		83	58		
	1	25.40		80	51		
Float, clear mirror	$\frac{1}{8}$	3.18	Clear	90	85	Mirrors	One-way mirrors
	$\frac{1}{4}$	6.35		89	78		
Float, tinted bronze	$\frac{3}{16}$	4.76	Bronze	56	55	Glazing where glare elimination and better insulation is required	Insulating, tempered, and laminated glass
	$\frac{1}{4}$	6.35		50	48		
	$\frac{3}{8}$	9.53		37	33		
	$\frac{1}{2}$	12.70		28	24		
Float, tinted gray	$\frac{3}{16}$	4.76	Gray	50	52	Glazing where glare elimination and good insulation are required	Insulating, tempered, and laminated glass
	$\frac{1}{4}$	6.35		43	46		
	$\frac{3}{8}$	9.53		28	33		
	$\frac{1}{2}$	12.70		19	24		
Float, tinted blue-green	$\frac{3}{32}$	2.35	Blue-green	85	68	Glazing where glare elimination and heat absorbing are required	Insulating, tempered, and laminated glass
	$\frac{1}{8}$	3.18		83	65		
	$\frac{3}{16}$	4.76		79	53		
	$\frac{1}{4}$	6.35		75	47		
Glare reducing	Same as gray, bronze, and blue-green tinted float or plate glass						
Heat-absorbing	Same as blue-green tinted float, heavy sheet or plate glass						
Heat-strengthened	$\frac{1}{4}$	6.35	Black to white with ceramic glazing	Opaque with colored ceramic glaze on one side		Spandrels, curtain wall, panels, etc.	Decorative panels and screens
	$\frac{5}{16}$	7.94					
Heavy sheet, AA, A, B	$\frac{3}{16}$	4.76	Clear	90.0	83.8	Glazing windows, store fronts, etc.	Shelves, furniture tops, showcases; insulating, tempered, and laminated glass
	$\frac{7}{32}$	5.49		89.8	81.3		
	$\frac{1}{4}$	6.35		89.6	80.0		
	$\frac{3}{8}$	9.53		87.7	75.0		
	$\frac{7}{16}$	11.11		87.0	73.0		

Table G9 Types of glass used in construction (*continued*)

Type of glass	Thickness[a] in.	Thickness[a] mm	Transparency and color	Transmittance[b] Visible (percent)	Transmittance[b] Total solar energy (percent)	Major uses	Other uses
Heavy sheet, tinted, AA, A, B	$\frac{1}{8}$ $\frac{3}{16}$ $\frac{7}{32}$	3.18 4.76 5.49	Gray, bronze, and blue-green	43.5 43.7 13.8	64.0 65.7 40.9	Glazing where glare elimination and good insulation are required	Insulating, tempered, and laminated glass
Insulating, A quality, double strength window or heavy sheet, glass-to-glass seal	$\frac{3}{8}$ $\frac{7}{16}$	9.53 11.11	Clear	80 80	76 72	Glazing for small glass areas where insulation is required	
Insulating, $\frac{1}{8}$ in. (3.18-mm) clear sheet, float or plate, glass-to-metal seal	$\frac{9}{16}$ $\frac{13}{16}$	14.29 20.64	$\frac{1}{4}$-in. (6.35-mm) air space, clear $\frac{1}{2}$-in. (12.70-mm) air space, clear	81 81	69 69	Glazing for medium and large glass areas where insulation is required	
Insulating, $\frac{1}{4}$-in. (6.35-mm), clear, float or plate glass, glass-to-metal seal	$\frac{13}{16}$ $1\frac{1}{16}$	20.64 26.99	$\frac{1}{4}$-in. (6.35-mm) air space, clear $\frac{1}{2}$-in. (12.70-mm) air space, clear	77	59	Glazing for large glass areas where insulation is required	
Insulating, $\frac{1}{4}$ in. (6.35-mm), tinted or clear, float or plate glass, glass-to-metal seal	$\frac{13}{16}$ $1\frac{1}{16}$	20.64 26.99	$\frac{1}{4}$-in. (6.35-mm) air space, tinted $\frac{1}{2}$-in. (12.70-mm) air space, tinted	37–65 Depending on tint	35	Glazing for large glass areas where glare elimination, heat absorbing, and insulation are required	
Insulating, clear or tinted, reflective, combinations of float, plate, heat-processed or tempered	$\frac{13}{16}$	20.64	$\frac{1}{4}$-in. (6.35-mm) air space, clear, tinted, reflective $\frac{1}{2}$-in. (12.70-mm) air space, clear, tinted, reflective	10–35 Depending on combination of tinted, clear, and reflective glass	6–34	Glazing for large glass areas, doors, windows, side lights, etc., where glare elimination, heat absorbing, insulation, diminishing heat gain, and safety are required	

Insulating, $\frac{1}{4}$-in. (6.35-mm), float, plate, and patterned glass, glass-to-metal seal	$\frac{13}{16}$ $1\frac{1}{16}$	20.64 26.99	$\frac{1}{4}$-in. (6.35-mm) air space, translucent $\frac{1}{2}$-in. (12.70-mm) air space, translucent	Depends on type of patterned glass Depends on type of patterned glass		Glazing where translucence and insulation are required	
Laminated glass, safety	$\frac{5}{32}$ $\frac{7}{32}$ $\frac{15}{64}$ $\frac{1}{4}$ $\frac{17}{64}$ $\frac{3}{8}$ $\frac{1}{2}$ $\frac{5}{8}$ $\frac{3}{4}$ $\frac{7}{8}$ 1	3.92 5.49 5.88 6.35 6.66 9.53 12.70 15.88 19.05 22.23 25.40	Clear, tinted, or reflective	Depends on type of glass used for lamination and type of plastic		Glazing where protection against hazard of impact or breakage is required	
Laminated, bullet-resisting	$\frac{3}{4}$ $\frac{7}{8}$ 1 $1\frac{3}{16}$ $1\frac{9}{16}$ $1\frac{3}{4}$ 2 $2\frac{1}{2}$ 3	19.05 22.23 25.40 30.16 39.69 44.45 50.80 63.50 76.20	Clear, tinted, or reflective	Depends on type of glass used for lamination and type of plastic		Glazing where bullet resistance is required	Areas where robberies are possible
Mirror, AA or A window, or sheet, float, or plate glass	Thickness and size limitations depend on type of glass used		Silver or other metallic colors	Opaque or one-way mirror		Mirrors	Controlled vision by means of one-way mirrors
Patterned	$\frac{1}{8}$ $\frac{7}{32}$ $\frac{3}{8}$	3.18 5.49 9.53	Generally translucent but varies with each pattern	Depends on type of pattern and surface finish		Glazing where light and privacy are required	Decorative panels and screens
Plate, silvering, mirror glazing, and glazing qualities	$\frac{1}{8}$ $\frac{1}{4}$	3.18 6.35	Clear or tinted	90.6 89.1	86.1 79.9	Glazing for medium to large glass areas	Mirrors, shelves, furniture tops; insulating, tinted, reflective, and tempered glass

Table G9 Types of glass used in construction (*continued*)

Type of glass	Thickness[a] in.	Thickness[a] mm	Transparency and color	Transmittance[b] Visible (percent)	Transmittance[b] Total solar energy (percent)	Major uses	Other uses
Plate, heavy	$\frac{5}{16}$	7.94	Clear or tinted	88.0	76.5	Glazing for large glass areas, store fronts, shelves, furniture tops, etc.	Aquariums, shelves, screens, glass-top furniture, etc.
	$\frac{3}{8}$	9.53					
	$\frac{1}{2}$	12.70					
	$\frac{5}{8}$	15.88					
	$\frac{3}{4}$	19.05					
	1	25.40					
	$1\frac{1}{4}$	31.75					
Powdered	Ground to a very fine powder					Used as filler in resilient flooring	Used as filler in paints, plastics, and asphalt roofing and siding
Reflective, clear, float or plate	$\frac{1}{4}$	6.35	Clear	39	35	To reduce heat gain in structures and to reduce glare	
Reflective, gray float or plate	$\frac{1}{4}$	6.35	Tinted gray	18	35		
	$\frac{5}{16}$	7.95		16	28		
Reflective, bronze, float or plate	$\frac{1}{4}$	6.35	Tinted bronze	21	35		
	$\frac{5}{16}$	7.95		20	26		
Reflective, blue-green, float or plate	$\frac{1}{4}$	6.35	Tinted blue-green	50	18	To reduce heat gain in structures, absorb heat, and reduce glare	
Reflective, tempered, clear	$\frac{1}{4}$	6.35	Clear	50	34	To reduce heat gain and glare; used in areas where impact shock is present and safety is required	
Reflective, tempered, gray	$\frac{1}{4}$	6.35	Tinted gray	24	34		
	$\frac{5}{16}$	7.95		20	24		

Reflective,	$\frac{1}{4}$	6.35	Tinted bronze	30	24		
tempered,	$\frac{5}{16}$	7.95		20	20		
Structural	$\frac{1}{4}$	6.35	Black, white, and limited color range	Opaque		Curtain wall systems; facings for the exterior or interior of building	Toilet partitions, wall finishes in bathroom, tops for furniture
	$\frac{11}{32}$	8.62					
	$\frac{7}{16}$	11.11					
	$\frac{3}{4}$	19.05					
	1	25.40					
	$1\frac{1}{4}$	31.75					
	$\frac{7}{8}$	22.23					
	2 sheets $\frac{7}{16}$ thick	2 sheets 11.11 thick					
Tempered, float or plate glass	$\frac{7}{32}$	5.59	Clear, tinted, or reflective	Depends on type of glass		Glazing where impact shock is present and safety is required	Areas where thermal shock is present; furniture tops
	$\frac{1}{4}$	6.35					
	$\frac{3}{8}$	9.53					
	$\frac{1}{2}$	12.70					
	$\frac{3}{4}$	19.05					
	1	25.40					
	$1\frac{1}{4}$	31.75					
Tempered glass, patterned	$\frac{7}{32}$	5.59	Generally translucent but varies with each pattern	Depends on type of pattern and surface finish		Glazing where light, safety, and privacy are required	Areas where impact shock is present
	$\frac{3}{8}$	9.53					
Tinted, gray, float or plate	$\frac{1}{8}$	3.18	Tinted gray	62		To reduce heat gain and glare	
	$\frac{3}{16}$	4.76		50			
	$\frac{1}{4}$	6.35		41			
	$\frac{5}{16}$	7.94		38			
	$\frac{3}{8}$	9.53		28			
	$\frac{1}{2}$	12.70		19			
Tinted, bronze, float or plate	$\frac{1}{8}$	3.18	Tinted bronze	69			
	$\frac{3}{16}$	4.76		59			
	$\frac{1}{4}$	6.35		52			
	$\frac{3}{8}$	9.53		37			
	$\frac{1}{2}$	12.70		28			
Tinted, blue-green float or plate	$\frac{1}{8}$	3.18	Tinted blue-green	83		To reduce heat gain, absorb heat, and reduce glare	
	$\frac{3}{16}$	4.76		79			
	$\frac{1}{4}$	6.35		75			
Tinted, gray, sheet glass	$\frac{1}{8}$	3.18	Tinted gray	31		To reduce heat gain and glare	
	$\frac{7}{32}$	5.49		14			

Table G9 Types of glass used in construction (*continued*)

Type of glass	Thickness[a] in.	Thickness[a] mm	Transparency and color	Transmittance[b] Visible (percent)	Transmittance[b] Total solar energy (percent)	Major uses	Other uses
Tinted, gray, tempered or heat-processed, float or plate	$\frac{1}{8}$	3.18	Tinted gray	69		To reduce heat gain and glare; used where impact shock is present and safety is required	
	$\frac{3}{16}$	4.76		50			
	$\frac{1}{4}$	6.35		42			
	$\frac{5}{16}$	7.95		31			
	$\frac{3}{8}$	9.53		29			
	$\frac{1}{2}$	12.70		20			
Tinted, bronze, tempered or heat-processed, float or plate	$\frac{1}{8}$	3.18	Tinted bronze	15			
	$\frac{3}{16}$	4.76		59			
	$\frac{1}{4}$	6.35		51			
	$\frac{5}{16}$	7.95		50			
	$\frac{3}{8}$	9.53		38			
	$\frac{1}{2}$	12.70		29			
Tinted, blue-green, tempered or heat-processed, float or plate	$\frac{1}{8}$	3.18	Tinted blue-green	82		To reduce heat gain, absorb heat and reduce glare; used where impact shock is present and safety is required	
	$\frac{1}{4}$	6.35		75			
Tinted, gray and bronze, tempered or heat-processed, sheet glass	$\frac{3}{16}$	4.76	Tinted gray and bronze	59		To reduce heat gain and glare; used where impact shock is present and safety is required	
	$\frac{7}{32}$	5.49		50			
Window, single-strength AA, A, B, photo	$\frac{1}{16}$	1.59	Clear	91.3	89.0	Framing pictures, maps, documents, photographs, etc.	Instrument dials
Window, single strength AA, A, B, picture	$\frac{5}{64}$	1.96	Clear	91.3	89.0	Framing pictures, maps, documents, photographs, etc.	
Window, single-strength AA, A, B	$\frac{3}{32}$	2.35	Clear	90.3	87.3	Glazing for small glass areas	Mirrors, framing

Window, double-strength AA, A and B	$\frac{1}{8}$	3.18	Clear	90.3	85.8	Glazing for medium to large glass areas	Mirrors, framing
Window, double-strength, greenhouse	$\frac{1}{8}$	3.18	Green cast	90.3	85.8	Glazing for hothouses, greenhouses	
Wired, diamond welded mesh	$\frac{1}{4}$ $\frac{3}{8}$	6.35 9.53	Translucent	Generally translucent, depending on type of pattern		Skylights and similar areas where the hazard of breakage is present; glazing fire doors and windows	Areas where security and safety are required
Wired, hexagonal mesh	$\frac{1}{4}$ $\frac{3}{8}$	6.35 9.53					
Wired, narrow spaced, vertical wires	$\frac{1}{4}$ $\frac{3}{8}$	6.35 9.53					
Wired, patterned	$\frac{1}{4}$ $\frac{3}{8}$	6.35 9.53					

[a] All glass has a required minimum tolerance in its thickness; check with manufacturers for current data.
[b] These percentages vary because of the physical and chemical properties of the glass. Check with manufacturers on *U* factors, heat gains, transmittance, glare, etc., before selecting a particular type of glass.

Table G10 Size limitations for bent glass forms

Type of glass	Thickness		Length of arc or girth		Height of bent glass	
	in.	mm	ft	m	ft	m
Float or window, single strength[a]	0.085–0.100	2.16–2.54 mm	2	0.610	4	1.219
Float or window, double strength[a]	0.115–0.133	2.92–3.38 mm	4	1.219	5	1.524
			5	1.524	4	1.219
Float or heavy	$\frac{3}{16}$	4.76	5	1.524	7	2.134
sheet[a]	$\frac{7}{32}$	5.48	7	2.134	8	2.438
Plate[a] or float	$\frac{1}{8}$	3.18	4	1.219	5	1.524
			5	1.524	4	1.219
	$\frac{1}{4}$	6.34	10	3.048	12	3.658
			12	3.658	10	3.048
Patterned[a]	$\frac{1}{8}$	3.18	4	1.219	5	1.524
			5	1.524	4	1.219
	$\frac{7}{32}$	5.48	4	1.219	7	2.134
			7	2.134	4	1.219
Heat-absorbing[a]	$\frac{1}{8}$	3.18	4	1.219	5	1.524
			5	1.524	4	1.219
	$\frac{7}{32}$	5.48	4	1.219	7	2.134
			7	2.134	4	1.219
	$\frac{1}{4}$	6.35	8	1.438	11.5	3.505
			11.5	3.505	8	2.438
Structural[a]	$\frac{7}{16}$	11.12	5	1.524	7	2.134
	$\frac{11}{32}$	8.62	7	2.134	8	2.438
Colored[a], plate	$\frac{7}{32}$	4.48	7.50	2.286	9.67	2.948
or float	$\frac{1}{4}$[b]	6.34	9.17	2.794	12	3.658

[a] Available in bent form Types 1, 3, and 4, as shown in Figure *G2*.
[b] Available in bent form Types 2, 5, and 6, as shown in the same figure.

4. Polybutene tape, a nondrying mastic, is made in extruded ribbon shapes of various widths and thicknesses and must be applied with pressure for proper adhesion. It remains plastic over extremely long periods of time. It is used as a continuous bed material with polysulfide sealing compound.

5. Polysulfide elastomer sealing compound is a two-part synthetic rubber, based on a polysulfide polymer. The activator and the base compound are generally mixed at the job and applied with a caulking gun or spatula. It is also available premixed in a frozen condition, packed with dry ice. The use of this material requires that surrounding areas be protected by masking with tape and that any spillage be immediately removed because, once the compound has set, it is almost impossible to remove.

6. Compression materials are extruded or molded shapes made of rubber, neoprene, vinyl, or other plastics. To achieve a weathertight joint, the shape must be compressed not less than 15% (*see* Figure *G4*).

7. Glass-to-glass joints are made with special types of silicone adhesives (*see* Figure *G5*).

Correct preparation of the materials which are to receive the glass and glazing compound is of utmost importance. Priming material should be allowed to dry thoroughly before the glass and glazing material are applied. All projections in the glazing rabbets should be removed so that recommended minimum clearances for the glass are maintained. When two dissimilar metals are in contact, the joint surface should be treated with an insulating or sealing material to eliminate the possibility of galvanic action.

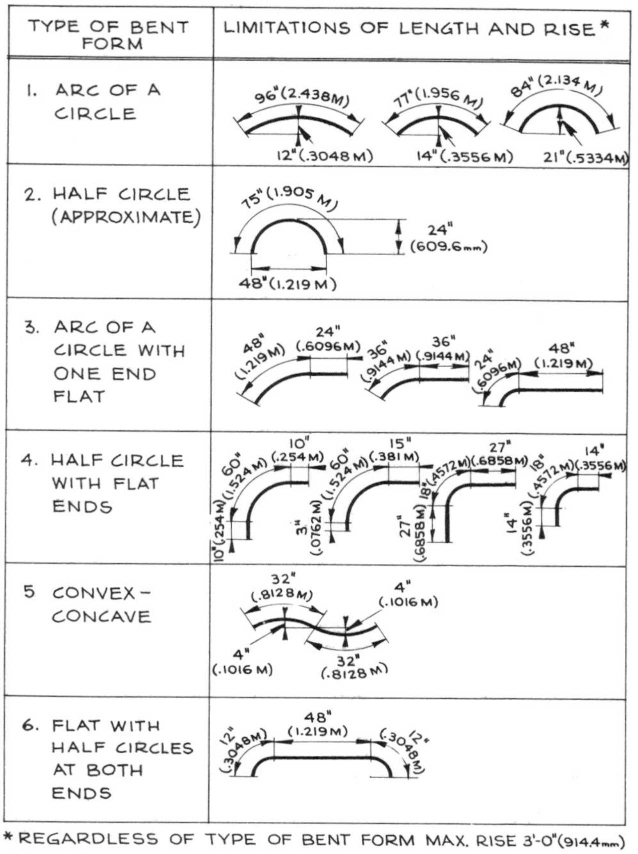
TYPE OF BENT FORM
LIMITATIONS OF LENGTH AND RISE *
1. ARC OF A CIRCLE
96"(2.438M) 12"(.3048 M)
77"(1.956 M) 14"(.3556 M)
84"(2.134 M) 21"(.5334M)
2. HALF CIRCLE (APPROXIMATE)
75"(1.905 M)
24" (609.6mm)
48"(1.219 M)
3. ARC OF A CIRCLE WITH ONE END FLAT
48" (1.219M) 24" (.6096M)
36" (.9144 M) 36" (.9144M)
24" (.6096M) 48" (1.219M)
4. HALF CIRCLE WITH FLAT ENDS
60" (1.524M) 10" (.254 M) 10" (.254M)
60" (1.524M) 15" (.381 M) 3" (.0762 M)
18" (.4572M) 27" (.6858M) 27" (.6858 M)
18" (.4572M) 14" (.3556M) 14" (.3556M)
5 CONVEX - CONCAVE
32" (.8128 M) 4" (.1016 M)
4" (.1016 M) 32" (.8128 M)
6. FLAT WITH HALF CIRCLES AT BOTH ENDS
12" (.3048M) 48" (1.219M) 12" (.3048M)
* REGARDLESS OF TYPE OF BENT FORM MAX. RISE 3'-0"(914.4mm)

Figure G2 Types of bent glass forms.

Wood should always be primed with a priming paint, never with shellac or varnish. Absorbent hard woods that are not to be painted may be primed with varnish if an elastic glazing compound is used. Nonabsorbent hard woods require the use of metal sash putty.

Steel, if not already given a shop-applied bonderized paint coat or zinc coating, should be treated to inhibit rust and then prime-painted with a paint that will furnish a surface to which the glazing material can adhere.

Aluminum should be similarly treated when oil-base glazing compounds are used. Any protective coating should be completely removed.

Gaskets, shims, tapes, rope, sealants, and all the various plastic accessories and frames for glazing should always be checked for method of installation, wind pressure, weathering, and color in relation to their end use. Glass and plastic manufacturers should always be consulted for data on the proper type of material for the glazing in the building to be constructed.

Installation of glass and glazing materials should not be done at temperatures below 40°F (4.44°C). When it

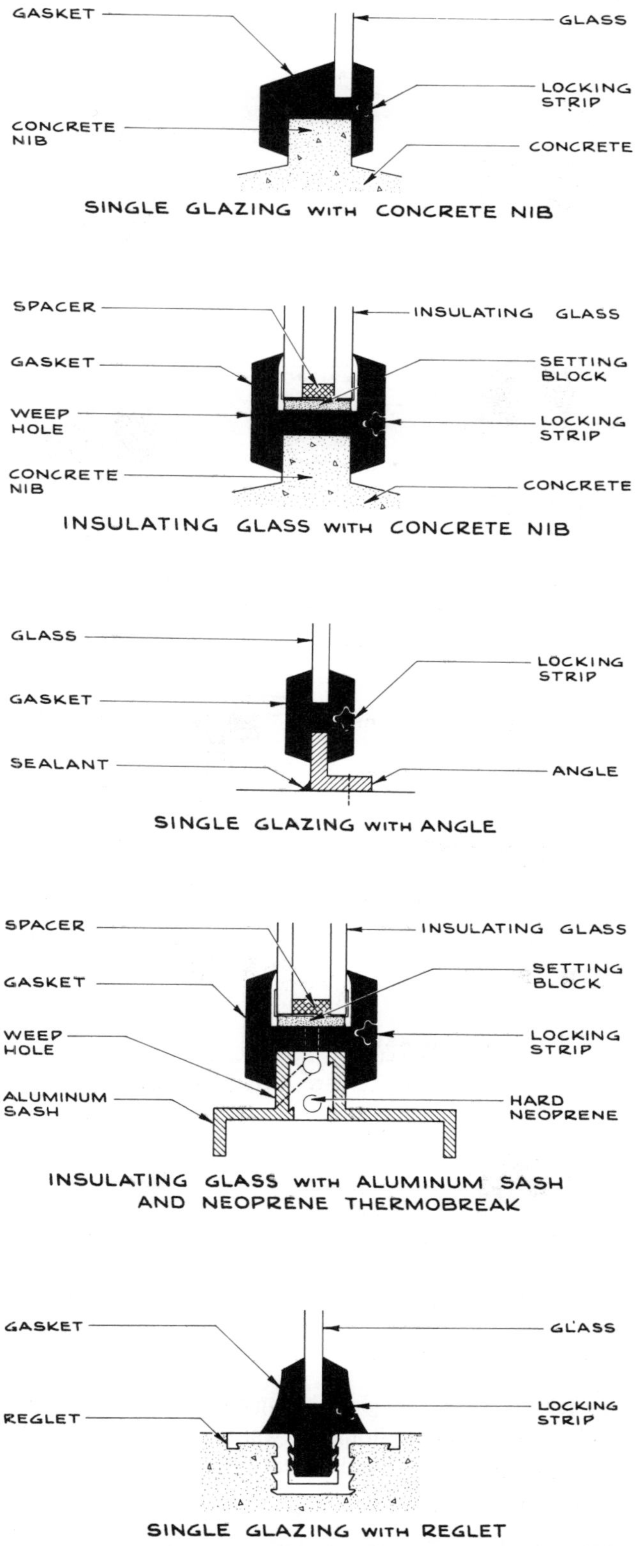

Figure G3 Typical examples of plastic gaskets for glazing.

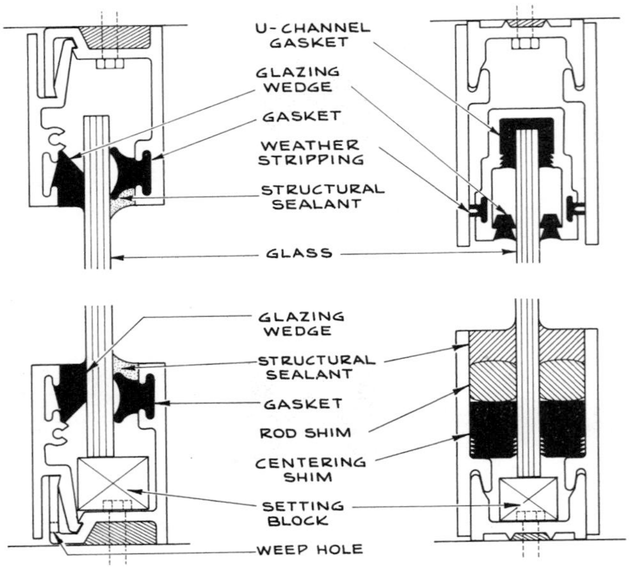

Figure G4 Plastic gaskets, shims, rope, etc., for glazing.

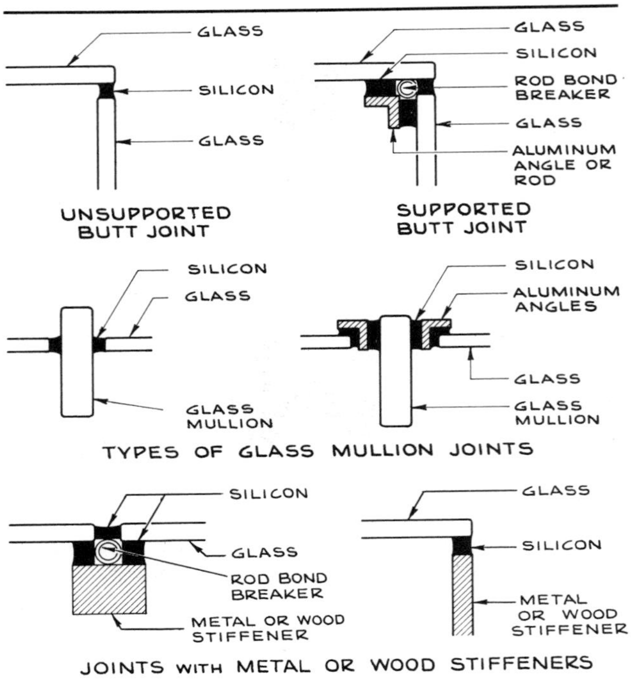

Figure G5 Types of glass-to-glass joints.

is necessary to install glass during cold weather, the glass and rabbet surfaces should be warmed so that no moisture can be trapped and cause failure of the weathertight seal.

Stained Glass. The use of plastics as a matrix in which colored glass is held together with the faceted glass surfaces projecting beyond the matrix has brought about the return of stained glass windows similar to the old-fashioned leaded kind. Panels are available in thicknesses of $\frac{5}{8}$ and $\frac{3}{4}$ in. (15.88 and 19.05 mm) and up to 16 ft^2 (1.486 m^2) in area, with a length-to-width ratio not to exceed 4 to 1 (*see* Plastics).

Condensed Checklist

These are general precautions in relation to glazing. For specific recommendations for each kind of flat glass and for other glass products, see the applicable glass headings.

1. Always select a type of glass that will meet design conditions and limitations. Then check the special installation instructions and glazing material requirements for this glass.
2. Always check the transparency of the glass in relation to its end use in the design.
3. Always establish what clearances are necessary for the type of frame into which the glass is being installed (*see* Table *G11*).
4. Always check what type of priming is necessary with the type of frame into which the glass is being installed.
5. Make sure that the glazing is such that under no conditions will the glass ever be put under tension.
6. Determine wind loads, and select the correct thickness and size of glass to meet these loads (*see* Table *G12*).
7. Check local, municipal, and state codes and also codes of the Fire Underwriters, insurance companies, labor departments, and federal government (Army, Navy, etc.) for limitations, restrictions, and fire resistance ratings.
8. Always check whether the glass requires special orientation, placement, handling, and cutting.
9. Shop drawings, field measurements, and detail drawings may be required for certain types of glass (*see* specific headings).

Conditions Favorable to the Use of Glass

Specific recommendations for various types of glass materials are listed under the applicable glass headings.

Conditions Unfavorable to the Use of Glass

1. In any area where it will be placed under tension.
2. Where it will receive constant strong impact shock.
3. Where very quick or localized quick extreme changes in temperature can occur.

Table G11 General clearances for various types of glass

Kind of glass[a]	Thickness of glass in.	Thickness of glass mm	Minimum clearance around edges or "minimum edge" in.	Minimum clearance around edges or "minimum edge" mm	Minimum clearance between glass and back of rabbet in.	Minimum clearance between glass and back of rabbet mm	Minimum depth of rabbet to hold glass in.	Minimum depth of rabbet to hold glass mm
Heavy sheet	$\frac{3}{16}$	4.76	$\frac{1}{8}$	3.18	$\frac{1}{8}$	3.18	$\frac{1}{2}$	12.70
	$\frac{7}{32}$	5.49	$\frac{1}{4}$	6.35	$\frac{1}{8}$	3.18	$\frac{5}{8}$	15.88
Insulating (double glass)	$\frac{1}{2}$	12.70	$\frac{1}{8}$	3.18	$\frac{1}{8}$	3.18	$\frac{1}{2}$	12.70
	$\frac{3}{4}$	19.05	$\frac{1}{4}$	6.35	$\frac{1}{8}$	3.18	$\frac{3}{4}$	19.05
	1	25.40	$\frac{1}{4}$	6.35	$\frac{1}{8}$	3.18	$\frac{3}{4}$	19.05
Patterned	$\frac{1}{8}$	3.18	$\frac{1}{8}$	3.18	$\frac{1}{8}$	3.18	$\frac{3}{8}$	9.53
	$\frac{3}{16}$	4.76	$\frac{1}{8}$	3.18	$\frac{1}{8}$	3.18	$\frac{1}{2}$	12.70
	$\frac{7}{32}$	5.49	$\frac{1}{4}$	6.35	$\frac{1}{8}$	3.18	$\frac{5}{8}$	15.88
Plate	$\frac{1}{8}$	3.18	$\frac{1}{8}$	3.18	$\frac{1}{16}$	1.59	$\frac{3}{8}$	9.53
	$\frac{1}{4}$ and over	6.35 and over	$\frac{1}{4}$	6.35	$\frac{1}{8}$	3.18	$\frac{5}{8}$	15.88
Tempered, heat-absorbing, glare-reducing	$\frac{1}{4}$ and over	6.35 and over	$\frac{1}{8}$	3.18	$\frac{1}{8}$	3.18	$\frac{5}{8}$	15.88
Window, single and double strength, and $\frac{1}{8}$ in.	0.087	2.21	$\frac{1}{8}$	3.18	$\frac{1}{16}$	1.59	$\frac{3}{8}$	9.53
	0.113	1.87	$\frac{1}{8}$	3.18	$\frac{1}{16}$	1.59	$\frac{3}{8}$	9.53
	$\frac{1}{8}$	3.18	$\frac{1}{8}$	3.18	$\frac{1}{16}$	1.59	$\frac{3}{8}$	9.53

[a]Float glass (glass produced by the float process) can be used to make any of the types listed in this table.

Table G12 Maximum size of glass glazed on four sides

Wind velocity mph (km/h)	Thickness of glass: 0.087 in. (2.21 mm) ft^2 (m^2)	0.113 in. (2.87 mm) ft^2 (m^2)	$\frac{1}{8}$ in. (3.18 mm) ft^2 (m^2)	$\frac{3}{16}$ in. (4.76 mm) ft^2 (m^2)	$\frac{1}{4}$ in. (6.35 mm) ft^2 (m^2)	$\frac{5}{16}$ in. (7.94 mm) ft^2 (m^2)	$\frac{3}{8}$ in. (9.53 mm) ft^2 (m^2)	$\frac{1}{2}$ in. (12.7 mm) ft^2 (m^2)	$\frac{5}{8}$-1 in. (15.88–25.4 mm) ft^2 (m^2)	$1\frac{1}{4}$ in. (31.75 mm) ft^2 (m^2)
30	35.0	64.50	72	162	288	244[a]				
(48.27)	(3.25)	(5.99)	(6.69)	(15.05)	(26.76)	(22.67)				
40	17.5	32.25	36	81	144	225	248[a]			
(64.36)	(1.63)	(3.01)	(3.34)	(7.53)	(13.38)	(20.90)	(23.04)			
55	11.6	21.50	24	54	96	150	216			
(88.50)	(1.08)	(2.00)	(2.23)	(5.02)	(8.92)	(13.94)	(20.07)			
65	8.7	16.10	18	41	72	112	162	244[a]		
(104.59)	(0.73)	(1.50)	(1.67)	(3.81)	(6.69)	(10.41)	(15.05)	(22.67)		
80	5.8	10.85	12	27	48	75	108	192		
(128.72)	(0.54)	(1.01)	(1.12)	(2.51)	(4.46)	(6.97)	(10.03)	(17.84)		
100	3.5	6.45	7	16	29	45	65	115		
(160.90)	(0.33)	(0.60)	(0.65)	(1.49)	(2.69)	(4.18)	(6.04)	(10.68)		
120	2.5	4.60	5	11	20	32	46	82	85[a]	81[a]
(193.08)	(0.23)	(0.43)	(0.47)	(1.02)	(1.86)	(2.94)	(4.27)	(7.62)	(7.90)	(7.53)

[a]Maximum sizes. These may increase by virtue of new manufacturing methods and processes.

History and Manufacture

The earliest glass artifacts are green glass rods made about 2600 B.C. found in Babylon in the near East and glass beads found in Egypt from about 2500 B.C. In these early times most glass was used for jewelry, ornaments, and containers and continued to be so employed until the time of the Roman Empire. It was sometime between 323 and 30 B.C., either in Egypt or perhaps in Syria (since Syrian soda glass was adaptable), that the blowing of glass was discovered. As a result, the Roman Empire used glass to a tremendous extent, in fact more widely than at any other time until the 20th century. The Romans were the first to use glass for windows. With the fall of their empire, their highly organized glassmaking industry was scattered into small localized centers.

The first revival of an important glass industry was in Venice about A.D. 1000, and from here the manufacture of glass spread throughout Europe. With the exception of the glass used for stained glass windows in Gothic architecture and the small panes used for other windows, the use of glass as we know it today did not exist. Glass used for windows was made by blowing a cylindrical shape, cutting it while still hot, and allowing this cut cylinder to flatten on a smooth surface. Naturally the size and the quality of such glass were limited.

What might be called the beginning of modern sheet glass was the mechanization of the blown cylindrical shape developed by J. H. Lubbers. He was able to make glass cylinders approximately 40 ft long and 30 in. in diameter, which were then similarly cut and flattened.

Fourcault Process. The next development was the method known as the Fourcault process (*see* Figure *G6*). It consisted of inserting into a heated pool of glass

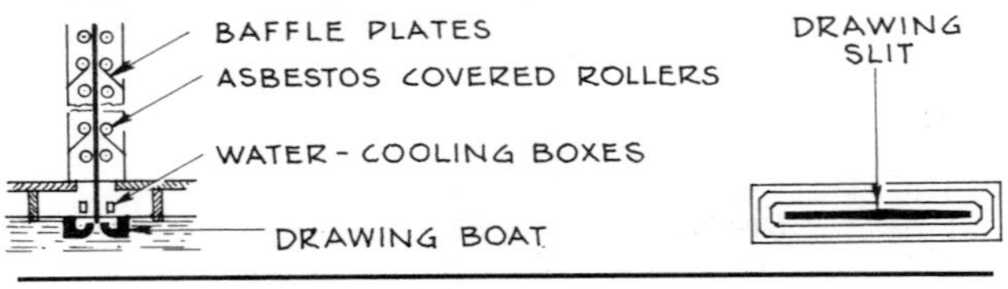

Figure G6 The Fourcault process for glass.

a small boat with a slot in the middle which was cooled with water. Glass from the molten batch pushed up through the slot and, being semicooled, was sufficiently hard to be taken hold of and pulled up vertically. When the resulting sheet was sufficiently cooled, it was cut off. This technique produced the first large sheets of glass, limited in size only by the width of the slot and the height that could be drawn.

Colburn or Libbey-Owens Process. Another method, the Colburn or Libbey-Owens process, was soon developed (*see* Figure *G7*). It is similar, except that at a certain height the glass is bent horizontally; it can then be rolled into a sheet over 200 ft long, from which the required sizes are cut.

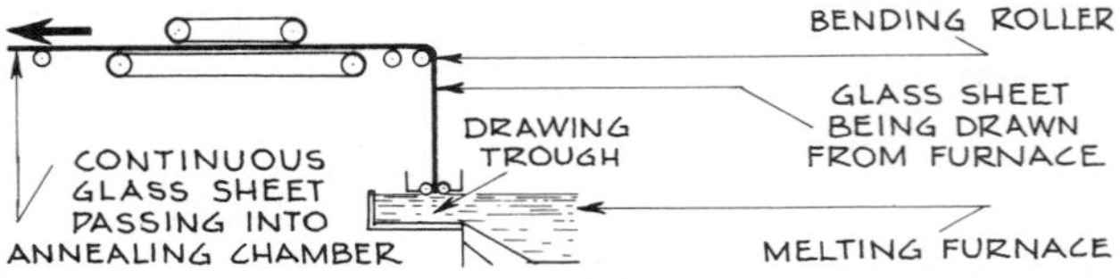

Figure G7 The Colburn or Libby-Owens process for glass.

Brothers and Chance Process. The difference between sheet and plate glass originally was that one was blown and the other was cast, rolled, and polished. But after 1890, when Brothers and Chance developed a process of casting molten glass onto an inclined table which fed to a pair of rollers, this differentiation became less clear-cut (*see* Glass, Plate).

Modern Applications. Since the discussion here is oriented toward construction, no mention is made of early researches on the composition and making of optical and special glasses, or of glass as an art form. In construction, until very recently, glass had few applications other than sheet and plate glass for simple glazing. These uses included mirrors for decorative interior effects and such highly developed art forms as stained glass windows. Although colored and various other kinds and forms of glass were known fairly early (*see* "History and Manufacture" under other glass headings), it is only since World War II that glass has become a major material in construction for a number of widely diverse applications, including insulation, textiles, curtain walls, and prefabricated panels.

GLASS BLOCK

Physical and Chemical Properties

Glass blocks are of two types: (1) hollow block, which consists of two pressed-glass shapes fused together into a single unit at elevated temperatures, in which the air in the hollow space has been dehydrated and partially evacuated; and (2) solid glass block. Both types are made of glass of the same general chemical composition as window, plate, or float glass (*see* Glass and Glazing).

The properties of glass block are as follows: compressive strength; 400 to 600 lbf/in.^{2} (2.758 to 4.137 MN/m^2) of gross bearing area (never use as load-bearing); weight; 20 lbf/in.^{2} (97.64 kg/m^2); maximum wind resistance when installed in panels; 20 lbf/ft^2 (47.88 N/m^2); coefficient of heat transmission; 0.44 to 0.60; light transmission; 50 to 85%, depending on the type of block.

Available Forms and Colors. Glass block is available clear, prismed for light diffusion, with or without glass fiber inserts, textured with various ribbings and patterns, and with or without a ceramic glaze on one face. The ceramic glazes are available in 12 standard colors.

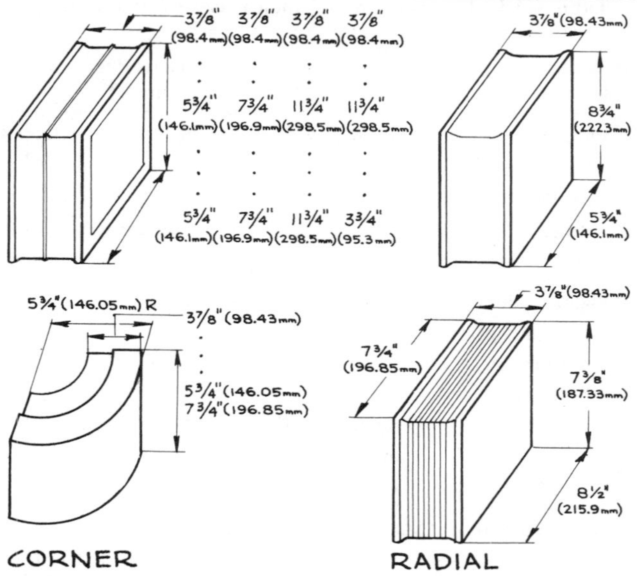

Figure G8 Standard sizes of hollow glass block (upper left), solid glass block (upper right), and special shapes (below).

Types and Uses

Glass blocks are made in various standard sizes on a 4-in. (101.6-mm) module, as shown in Figure *G8.* Various surface treatments of the interior and exterior faces (*see* Figure *G9*)—including filtering coatings, ceramic glazes, and the introduction of a glass fiber insert—permit a wide latitude of characteristics (translucence, diffusion, insulation) and uses for glass block. Table *G13* summarizes data on the stock types of glass block.

The major use of glass block is in areas where light transmission, insulation, and glare control are of major importance. Table *G14* gives the resistance, or *R* factor, for various sizes of glass blocks.

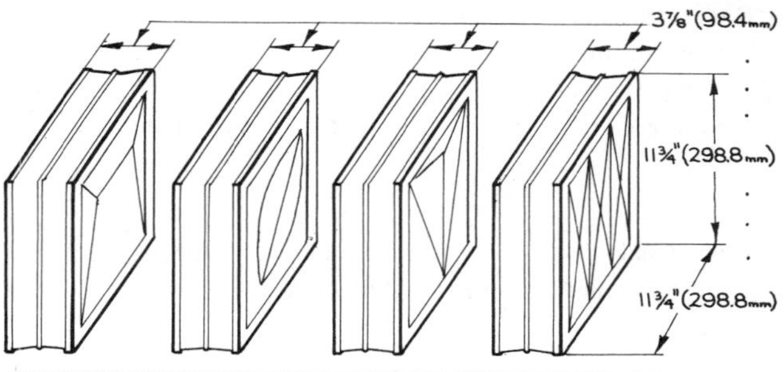

Figure G9 Sculptured-face glass block.

Table G13 Description and major uses of different types of glass block

Type of pattern		Glass fiber insert	Sizes[a]		Characteristics	Major use
Inside face	Outside face		in.	mm		
Prisms on both faces that are light-directing	Small corrugations on both faces	None	$7\frac{3}{4} \times 7\frac{3}{4}$	196.85 × 196.85	Maximum light transmission	In areas above eye level
		White insert	$7\frac{3}{4} \times 7\frac{3}{4}$ $11\frac{3}{4} \times 11\frac{3}{4}$	196.85 × 196.85 298.45 × 298.45	Reduction of glare and solar heat	
		Green insert	$7\frac{3}{4} \times 7\frac{3}{4}$ $11\frac{3}{4} \times 11\frac{3}{4}$	196.85 × 196.85 298.45 × 298.45	Better reduction of glare and solar heat	
Prisms on both faces that are light-diffusing	Small corrugations on both faces	None	$7\frac{3}{4} \times 7\frac{3}{4}$	196.85 × 196.85	Maximum light transmission	All areas; diffuses light in all directions
		White insert	$7\frac{3}{4} \times 7\frac{3}{4}$ $11\frac{3}{4} \times 11\frac{3}{4}$	196.85 × 196.85 298.45 × 298.45	Reduction of glare and solar heat	
		Green insert	$7\frac{3}{4} \times 7\frac{3}{4}$ $11\frac{3}{4} \times 11\frac{3}{4}$	196.85 × 196.85 298.45 × 298.45	Better reduction of glare and solar heat	

Table G13 Description and major uses of different types of glass block (*continued*)

Type of pattern: Inside face	Type of pattern: Outside face	Glass fiber insert	Sizes[a] in.	Sizes[a] mm	Characteristics	Major use
Flutes on both faces perpendicular to each other	Smooth on both faces	None	$5\frac{3}{4} \times 5\frac{3}{4}$ $7\frac{3}{4} \times 7\frac{3}{4}$ $11\frac{3}{4} \times 11\frac{3}{4}$	145.05 × 145.05 196.85 × 196.85 298.45 × 298.45	Flutes can be installed either vertically or horizontally	Decorative panels
		White insert	$7\frac{3}{4} \times 7\frac{3}{4}$	196.85 × 196.85	Reduction of glare and solar heat	
Slight random, uneven configurations on both faces	Smooth on both faces	None	$7\frac{3}{4} \times 7\frac{3}{4}$ $11\frac{3}{4} \times 11\frac{3}{4}$	196.85 × 196.85 298.45 × 298.45	Handmade appearance; should not be used for sun exposure	Decorative panels
	Ceramic glaze finish on one face	None	$5\frac{3}{4} \times 5\frac{3}{4}$ $7\frac{3}{4} \times 7\frac{3}{4}$	145.05 × 145.05 196.85 × 196.85	Low light transmission; limited color range	Used as color accents in glass block panels
Flutes on both faces running in same direction	Smooth on both faces	None	$5\frac{3}{4} \times 5\frac{3}{4}$ $7\frac{3}{4} \times 7\frac{3}{4}$ $11\frac{3}{4} \times 11\frac{3}{4}$	145.05 × 145.05 196.85 × 196.85 298.45 × 298.45	Flutes can be installed either vertically or horizontally	Decorative panels
		White insert	$7\frac{3}{4} \times 7\frac{3}{4}$	196.85 × 196.85	Reduction of glare and solar heat	
Concentric circle design or uneven pattern on both faces	Smooth on both faces	None	$5\frac{3}{4} \times 5\frac{3}{4}$ $7\frac{3}{4} \times 7\frac{3}{4}$ $11\frac{3}{4} \times 11\frac{3}{4}$	145.05 × 145.05 196.85 × 196.85 298.45 × 298.45	Almost transparent; should not be used for sun exposure	Decorative panels
		White insert	$11\frac{3}{4} \times 11\frac{3}{4}$	298.45 × 298.45	Reduction of glare, solar heat, and transparency	
Ribbed and slightly etched on both faces	Ribbed on both faces; either slightly etched or not etched	None	$7\frac{3}{4} \times 7\frac{3}{4}$ $11\frac{3}{4} \times 11\frac{3}{4}$	196.85 × 196.85 298.45 × 298.45	Ribs can be installed either vertically or horizontally	Decorative panels
Clear glass on both faces	Smooth clear glass on both faces	None	$7\frac{3}{4} \times 7\frac{3}{4}$ $11\frac{3}{4} \times 11\frac{3}{4}$	196.85 × 196.85 298.45 × 298.45	Clear vision	Used to give vision areas in glass block panels
Acid-etched on both faces	Smooth on both faces	None	$3\frac{3}{4} \times 11\frac{3}{4}$	95.25 × 298.45	Modular, maximum light transmission	Modular coordinated with other glass block sizes
		White	$3\frac{3}{4} \times 11\frac{3}{4}$	95.25 × 298.45	Reduction of glare and solar heat	
		Green insert	$3\frac{3}{4} \times 11\frac{3}{4}$	95.75 × 298.45	Better reduction of glare and solar heat	
	Ceramic glaze finish on one side	None	$3\frac{3}{4} \times 11\frac{3}{4}$	95.75 × 298.45		Used as accent color in glass block panels
Prisms parallel or at 45° on both faces (which are light-directing)	Smooth outside face; small corrugations on inside face	White insert	$11\frac{3}{4} \times 11\frac{3}{4}$	298.45 × 298.45	Maximum light transmission; stopping direct down rays of sun, thereby reducing glare and solar heat	Skylights in roofs
		Green insert	$11\frac{3}{4} \times 11\frac{3}{4}$	298.45 × 298.45	Better reduction of glare and solar heat	

Table G13 Description and major uses of different types of glass block (*continued*)

Type of pattern: Inside face	Type of pattern: Outside face	Glass fiber insert	Sizes[a] in.	Sizes[a] mm	Characteristics	Major use
Sculptural type of design pressed deep into face	Same sculptural face design	None	$11\frac{3}{4} \times 11\frac{3}{4}$	298.45 × 298.45	Flexible massive design effects	Decorative walls, panels, screens
	Ceramic glaze finish on one side	None	$11\frac{3}{4} \times 11\frac{3}{4}$	298.45 × 298.45		
Solid; no inside faces	Etched on one face, smooth on other	None	$3\frac{3}{4} \times 8\frac{3}{4}$	95.25 × 222.25	Maximum light transmissions; translucent and unbreakable	Prisons, mental institutions, bonded warehouses

[a]All blocks are $3\frac{7}{8}$ in. (98.43 mm) thick.

Table G14 Heat losses of various sizes and types of glass block

Size of block[a] in.	Size of block[a] mm	Glass fiber insert at middle	Resistance R, reciprocal of U: Btu/ft$^2 \cdot$h$\cdot$°F	Resistance R, reciprocal of U: 5.678 W/m$^2 \cdot$°C	Approximate weight lb	Approximate weight kg
$5\frac{3}{4} \times 5\frac{3}{4}$	145.05 × 145.05	without	1.667	0.293	3.5833	1.625
$7\frac{3}{4} \times 7\frac{3}{4}$	196.85 × 196.85	without	1.786	0.314	7.0675	3.206
$7\frac{3}{4} \times 7\frac{3}{4}$	196.85 × 196.85	with	2.083	0.366	7.0675	3.206
$11\frac{3}{4} \times 11\frac{3}{4}$	298.45 × 298.45	without	1.923	0.339	15.75	7.144
$11\frac{3}{4} \times 11\frac{3}{4}$	298.45 × 298.45	with	2.273	0.498	15.75	7.144
$3\frac{3}{4} \times 11\frac{3}{4}$	95.25 × 298.45	without	1.667	0.293	10.50	5.663
$3\frac{3}{4} \times 11\frac{3}{4}$	95.25 × 298.45	with	1.923	0.339	10.50	5.663

[a]All blocks are $3\frac{7}{8}$ in. (98.43 mm) thick.

Application

Glass block vertical walls are limited in length and height and in square foot area. When glass block panels are framed in a chase made of masonry or of steel, maximum length is 25 ft (7.62 m), maximum height 20 ft (6.096 m), and maximum panel area 144 ft^2 (13.378 m^2). When panels are not set in chases, maximum length and height are 10 ft (3.048 m), and maximum panel area is 100 ft^2 (9.29 m^2). Table *G15* gives data pertinent to the construction of curved glass block walls.

Mortar. Mortar for glass block is measured by volume and is compounded as follows: 1 part portland cement; $\frac{1}{4}$ to $1\frac{1}{4}$ parts lime, and sand equal to $2\frac{1}{4}$ to 3 times the amount of cement and lime combined. All sand must pass a No. 12 sieve. For exterior work, either waterproof portland cement or an integral waterproofing admixture should be used. Accelerators or antifreeze admixtures should never be used. All mortar joints are $\frac{1}{4}$ in. (6.35 mm) and tooled with a slightly concave surface. The joint used should be $\frac{1}{8}$ in. (3.175 mm) back from the face to expose the square shoulders of the glass block.

Condensed Checklist

1. The orientation should always be checked when glass blocks are to be installed. A correct type of glass block must be chosen to meet the conditions for light transmission, insulation, solar heat, decoration, and privacy on the basis of orientation and design.
2. Height, width, and square-foot area of glass block panels should fall within the maximum size limitations.
3. Allowance for an extra $\frac{1}{2}$-in. (12.7-mm) deflection of the steel or reinforced concrete at heads of glass block panels should always be made.
4. Glass block should never be used as a load-bearing wall or partition.
5. The treatment of head, jamb, and sill should always be checked when installing glass block (*see* Figures *G10* and *G11*).
6. For interior glass block panels, the treatment of head, jamb, and sill should always be checked.
7. Treatment at openings for doors and windows should always be checked.

Table G15 Minimum radius for glass block panels with not less than $\frac{1}{8}$ in. (3.18 mm) inside joint and not greater than $\frac{3}{4}$ in. (19.05 mm) outside joint

Size of glass block	Outside radius		Joint thickness			
			Inside		Outside	
	ft	m	in.	mm	in.	mm
$5\frac{3}{4}$ in. (145.05 mm)	4.25 min.	1.294	$\frac{1}{8}$	3.175	$\frac{5}{8}$	15.88
X $5\frac{3}{4}$ in. (145.05 mm)	4.67	1.423	$\frac{1}{8}$	3.175	$\frac{1}{2}$	12.7
X $3\frac{7}{8}$ in. (98.43 mm)	5.00	1.524	$\frac{1}{8}$	3.175	$\frac{1}{2}$	12.7
There is no maximum radius	5.33	1.625	$\frac{1}{8}$	3.175	$\frac{1}{2}$	12.7
beyond 7.33 ft (2.235 m)	5.67	1.728	$\frac{3}{16}$	4.763	$\frac{1}{2}$	12.7
	6.00	1.829	$\frac{3}{16}$	4.763	$\frac{1}{2}$	12.7
	6.33	1.929	$\frac{3}{16}$	4.763	$\frac{1}{2}$	12.7
	6.67	2.033	$\frac{1}{4}$	6.350	$\frac{1}{2}$	12.7
	7.00	2.134	$\frac{1}{4}$	6.350	$\frac{1}{2}$	12.7
	7.33	2.234	$\frac{1}{4}$	6.350	$\frac{1}{2}$	12.7
$7\frac{3}{4}$ in. (196.85 mm)	5.75 min.	1.753	$\frac{1}{8}$	3.175	$\frac{5}{8}$	15.875
X $7\frac{3}{4}$ in. (196.85 mm)	6.33	1.929	$\frac{3}{8}$	9.525	$\frac{3}{4}$	19.050
X $3\frac{7}{8}$ in. (98.43 mm)	6.67	2.033	$\frac{1}{4}$	6.350	$\frac{5}{8}$	15.875
There is no maximum radius	7.00	2.134	$\frac{1}{8}$	3.175	$\frac{1}{2}$	12.700
beyond 8.33 ft (2.539 m)	7.42	2.261	$\frac{1}{8}$	3.175	$\frac{7}{16}$	11.113
	7.67	2.338	$\frac{7}{16}$	11.113	$\frac{3}{4}$	19.050
	8.00	2.438	$\frac{5}{16}$	7.938	$\frac{5}{8}$	15.875
	8.33	2.539	$\frac{1}{4}$	6.350	$\frac{1}{2}$	12.700
$11\frac{3}{4}$ in. (298.45 mm)	8.50	2.591	$\frac{1}{8}$	3.175	$\frac{5}{8}$	15.875
X $11\frac{3}{4}$ in. (298.45 mm)						
X $3\frac{7}{8}$ in. (98.43 mm)						
There is no maximum radius						
beyond 8.5 ft (2.591 m)						

8. Enough clearance must always be allowed at head and jambs for installation of expansion material (*see* Figure *G12*).
9. Horizontal reinforcing should always be installed every 2 ft (0.6096 m) in height (*see* Figures *G13* to *G15*). Note the method of treating this reinforcing at jambs.
10. Emulsion-type asphalt paint should always be applied to sills so that the masonry bond is broken.
11. Oakum must be correctly installed and set $\frac{3}{8}$ in. back from finish surfaces at all areas where glass block is set into chases. A nonhardening and nonstaining caulking compound should always be used.
12. For prefabricated skylight panels made of glass block, the actual size of panel, method of installation, orientation of panel, and type of glass block best suited must always be checked. After the orientation of the panel has been established, the skylight must be installed so that the arrow marked on the prefabricated panel faces north. (*See* Figures *G16* and *G17*.)

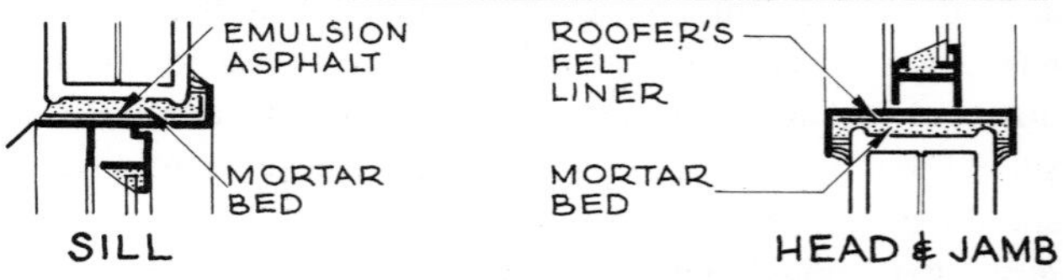

Figure G10 Head, sill, and jamb of metal window installed in a glass block wall.

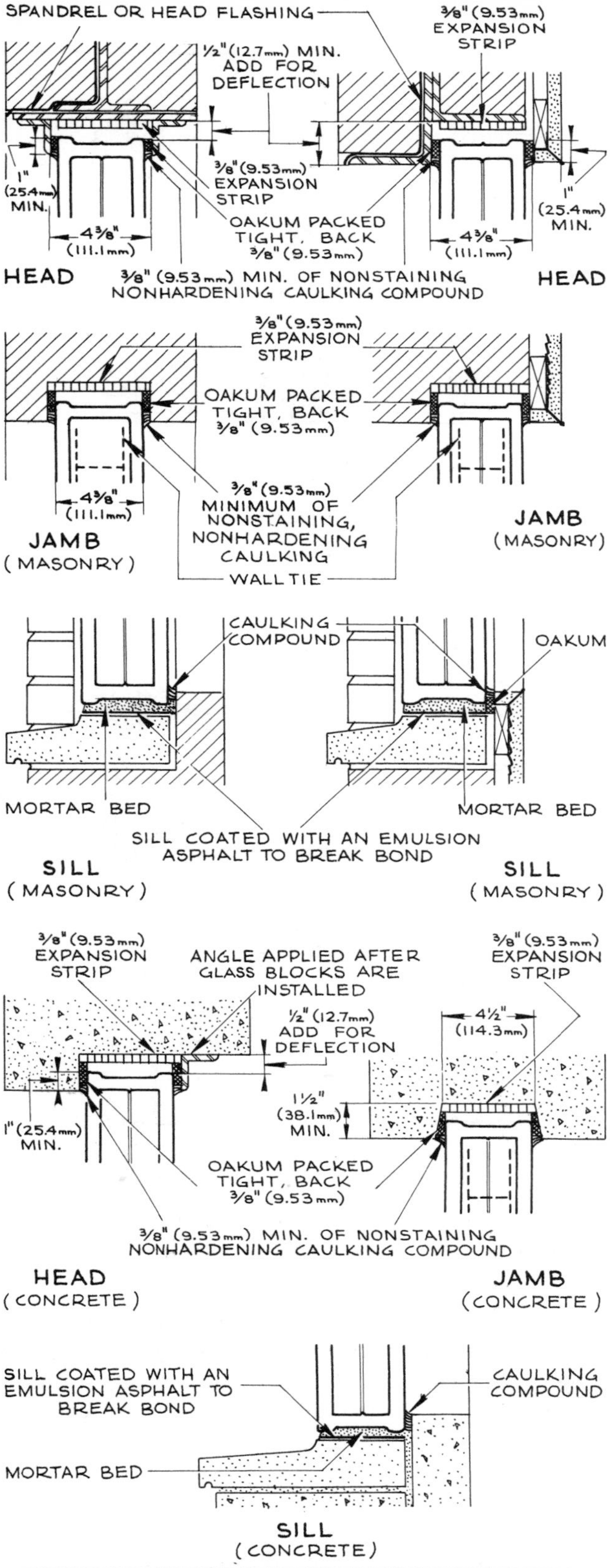

Figure G11 Detailing at head, jamb, and sill for glass block.

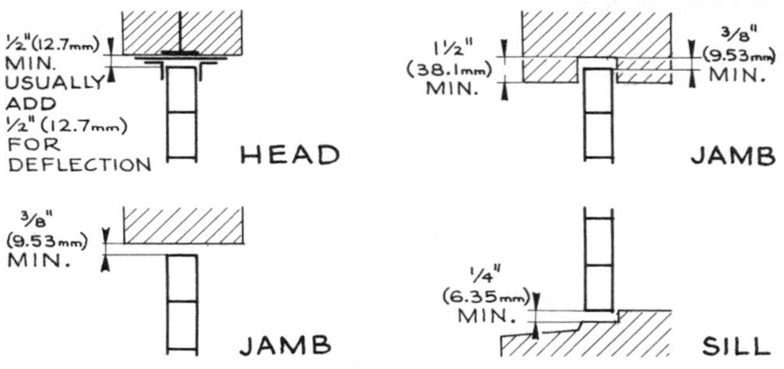

Figure G12 Minimum clearances at head, jamb, and sill for glass block.

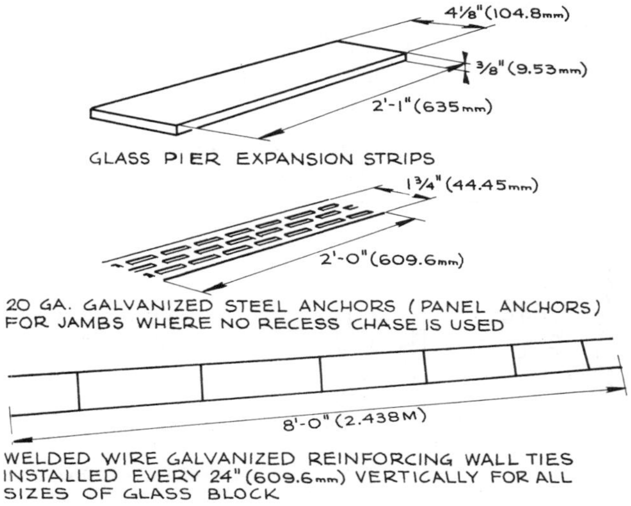

Figure G13 Accessories for glass block work.

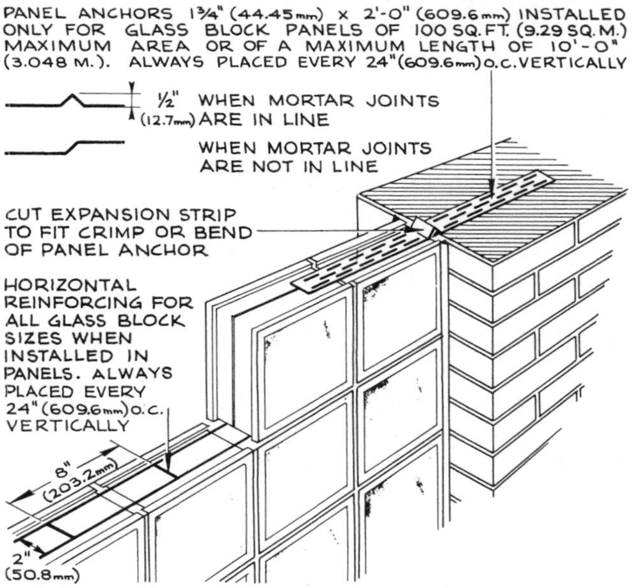

Figure G14 Panel anchors and wall ties for glass block panels.

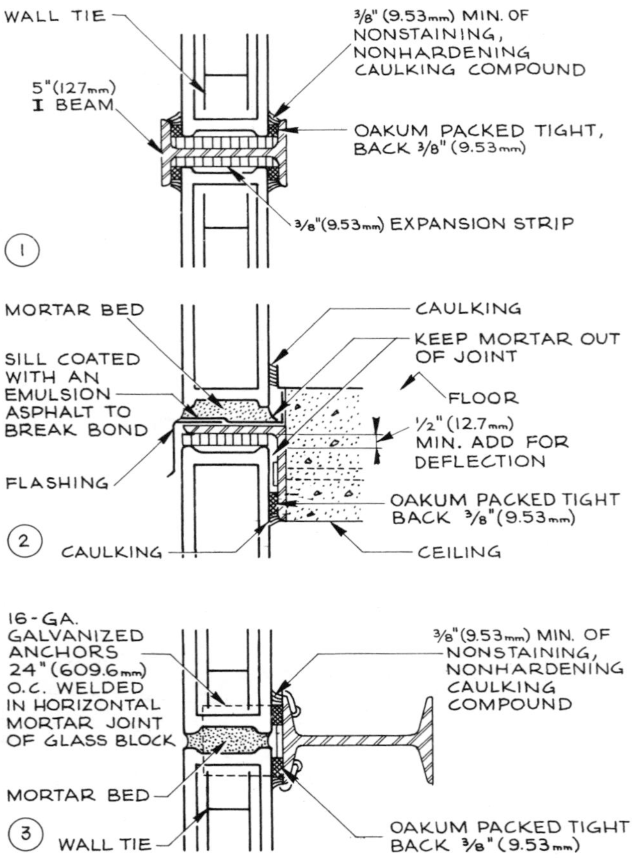

Figure G15 Methods of stiffening glass block work with steel.

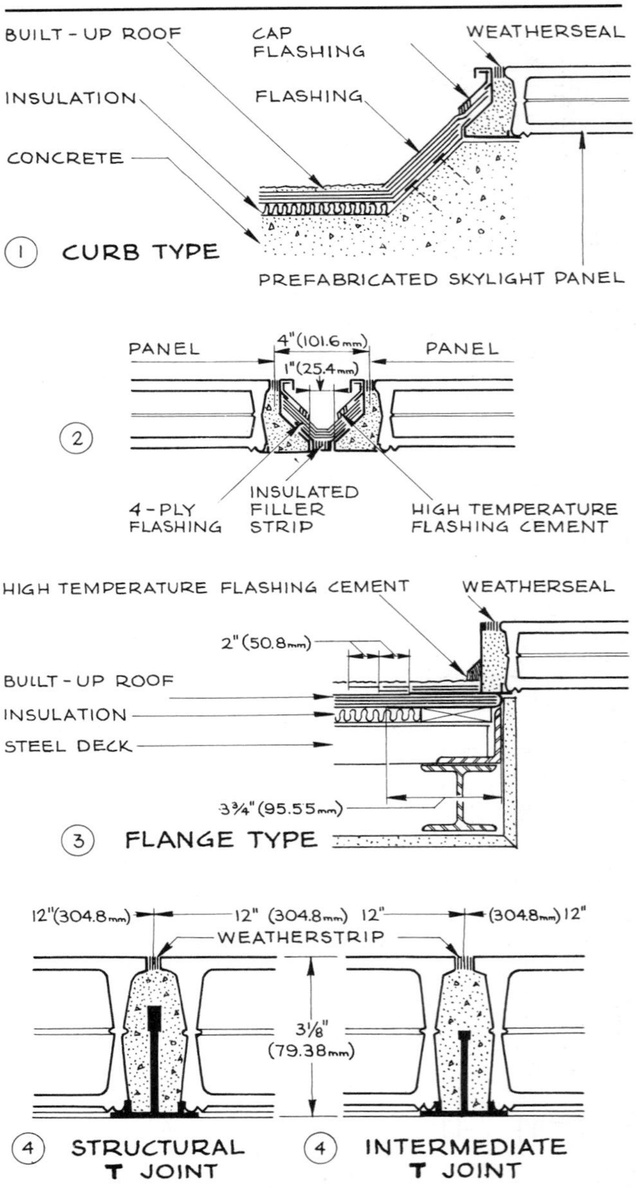

Figure G16 Details of installation methods for glass block skylight panels.

Conditions Favorable to the Use of Glass Block

1. Where light transmission, glare, and solar heat must be controlled.
2. Where heat loss is an important factor.
3. For light transmission and decorative effects.
4. For decorative translucent interior partitions.
5. For roof skylights where diffuse light and control of glare, direct overhead sun rays, and solar heat are necessary or desirable.

Conditions Unfavorable to the Use of Glass Block

1. Where large transparent glass areas are necessary because of area limitations.
2. Where direct impact will be encountered.

History and Manufacture

Glass blocks are a relatively new architectural material. They were developed during the 1929 depression and were first produced about 1933. The process is simple in principle but highly complicated in practice. First, two square or rectangular flat-bottomed dish shapes are made of pressed glass in the patterns and designs required. These two halves are then fused together at elevated temperatures, and the air in the hollow space is dehydrated and evacuated at a two-thirds vacuum to a controlled partial vacuum. Finally the edges are coated with a grit-bearing plastic to give them a good bonding surface. Solid glass block is a pressed glass shape.

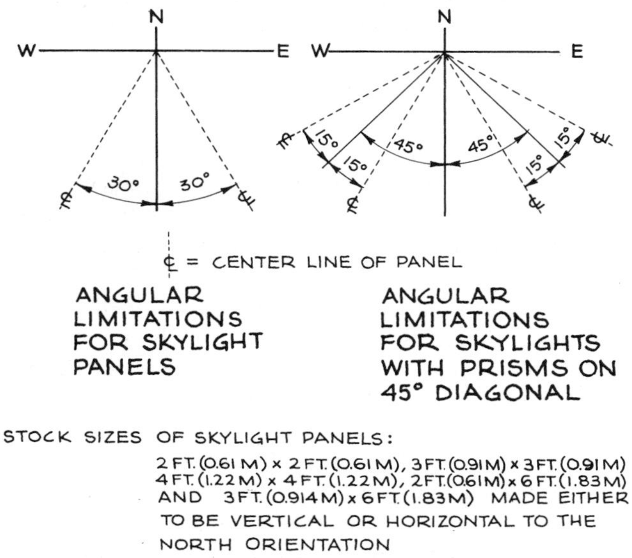

Figure G17 Orientation and angular limitations for glass block skylight panels.

GLASS, CELLULAR OR FOAMED

Physical and Chemical Properties

Cellular or foamed glass is glass containing millions of trapped air bubbles. It floats on water; it has an absorption of only 2% and very good thermal insulation properties. Its other properties and characteristics that are of interest in construction resemble those of glass wool (*see* Table *G16*).

Types and Uses

Cellular or foamed glass is available in block form in the following sizes: blocks 1 × 1.5 ft (0.3048 × 0.4572 m) and boards 2 × 4 ft (0.6096 × 1.2192 m) in thicknesses ranging from $1\frac{1}{2}$ through 4 in. (38.1 through 101.6 mm) in increments of $\frac{1}{2}$ in. (12.7 mm); and blocks 1.5 × 2 ft (0.4572 × 0.6096 m) in thicknesses ranging from 2 to 4 in. (50.8 through 101.6 mm) in increments of $\frac{1}{2}$ in. (12.7 mm).

For roof insulation it is available in blocks 1.5 × 2 ft (0.4572 × 0.6096 m) with a taper of $\frac{1}{8}$ in./ft (3.175 mm/0.3048 m) on the 2-ft dimension.

The major use of foamed cellular glass in construction is as an insulation material for roofs, building perimeters, and areas with controlled temperatures (*see* Insulation and Insulating Materials). Foam or cellular glass, when used for roof insulation, is generally installed by the roofing contractor.

Table G16 Chemical and physical properties of cellular or foamed glass

Combustibility	Noncombustible, will not burn
Compressive strength	100 lbf/in.^{2} (0.6895 MN/m^2)
Dimensional stability	Excellent
Maximum working temperature	940°F (504.44°C)
Thermal conductivity, U	50°F 0.36 Btu/ft^2 · h · °F (10°C 0.052 W/m^2 · °C)
Thermal resistance, R (reciprocal of U)	50°F 2.78 Btu/ft^2 · h · °F (10°C 19.32 W/m^2 · °C)
Water vapor permeability	0.00 perm-in. (0.00 perm-cm)

Foamed or cellular glass multiple sections are sandwiched between sheets of kraft paper to form rigid insulation boards.

Because cellular or foamed glass has the valuable property of bearing strength and is also vaporproof, it can be used to advantage in any area where control of temperature and humidity is necessary (especially for thermal insulating self-supporting partition walls and in applications where floor surfaces must support heavy loads).

History and Manufacture

Cellular glass is manufactured by placing into molds a mixture of crushed glass grains and carbonaceous material or limestone. The molds are passed through a furnace, and at the temperature at which the crushed glass grains become soft, the added material produces bubbles of gas. After coming out of the furnace, the cellular glass is removed from the molds and sawed into stock sizes.

GLASS, CORRUGATED

Physical and Chemical Properties

Corrugated glass is $\frac{3}{8}$-in. (9.53 mm)-thick rolled glass with a pattern on both sides, which has been formed into corrugations. The forming causes the pattern to be flattened on one side. The chemical composition of the glass is the same as that of float, plate, or window glass. Corrugated glass is available in standard corrugations in either wired or unwired types and in flatter corrugations, unwired only.

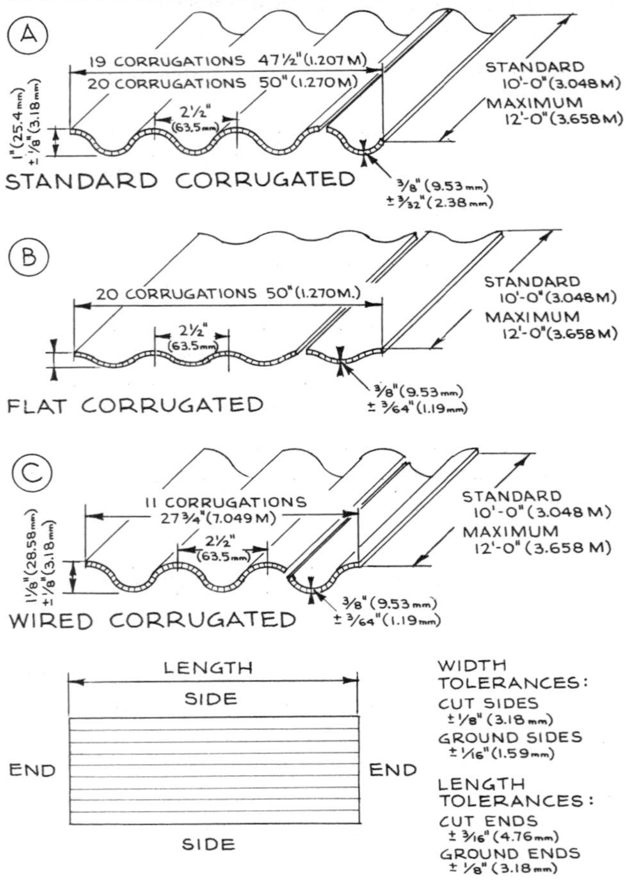

Figure G18 Types, dimensions, and tolerances of corrugated glass. (A) Available in widths less than 47.5 in. (120.65 cm) and 50 in. (127.00 cm) in 2.5-in. (6.35-cm) increments. Weight is approximately 16.3 lb/ft^2 (79.59 kg/m^2). (B) Available in widths less than 50 in. (127.00 cm) in 2.5-in. (6.35-cm) increments. Weight is approximately 5.3 lb/ft^2 (25.87 kg/m^2). (C) Available in widths less than 27.75 in. (70.48 cm) in 2.5-in. (6.35-cm) increments. Weight is approximately 6.62 lb/ft^2 (32.32 kg/m^2). Maximum span is 5 ft (1.524 m) for slopes up to 60°. For slopes greater than 60°, maximum span is 8 ft (2.438 m).

Types and Uses

Corrugated glass, wired and unwired, is manufactured in standard sizes with definitely established corrugations, thicknesses, and tolerances as shown in Figure *G18.* Generally, unwired corrugated glass is used for decorative purposes on exterior or interior wall partitions and in areas where obscurity is necessary, whereas wired corrugated glass is used for skylights, clerestories, and areas where fire resistance is required and where breakage must be avoided. Corrugated glass has a fire-polished finish but can be sandblasted or acid-etched to make it more obscure.

Wired corrugated glass is also made in an amber color which reduces glare and has heat-absorbing qualities.

Application

Condensed Checklist

1. Corrugated glass exposed to the sun or adjacent to heat sources should never be painted because this may cause heat traps to occur, and breakage may result.
2. In the installation of corrugated glass, methods of joining should always be checked. If butt joints or mitered joints are to be used, the edges should be ground; if the edges are held in a frame, they do not have to be ground, but moldings must be $\frac{1}{2}$ in. (12.7 mm) deep at minimum. (*See* Figure *G19.*)
3. Corrugated glass must never be set directly on masonry, concrete or any other unyielding material. For such conditions, setting blocks or a cushioning type of material should always be used.
4. Detail treatment at head and sill should be checked (*see* Figure *G20*).
5. Shop drawings must be required by specifications when fittings and moldings are not used.
6. Shapes other than rectangles, any holes to be drilled, and bending and notching of the glass are special conditions which require consultation with the manu-

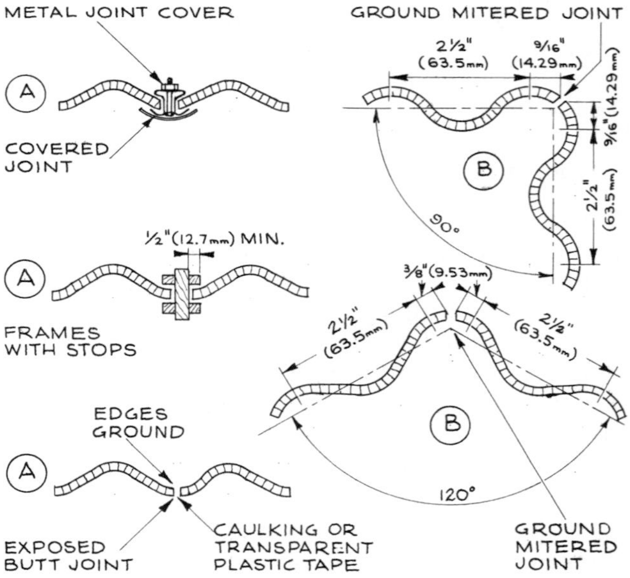

Figure G19 Typical vertical joints for corrugated glass. (A) Sides parallel to corrugations can be cut, ground, and mitered. Ends of corrugations can be cut and ground only. (B) As the angle decreases from 180° (straight), the length of the glass increases and has to be taken into account when detailing.

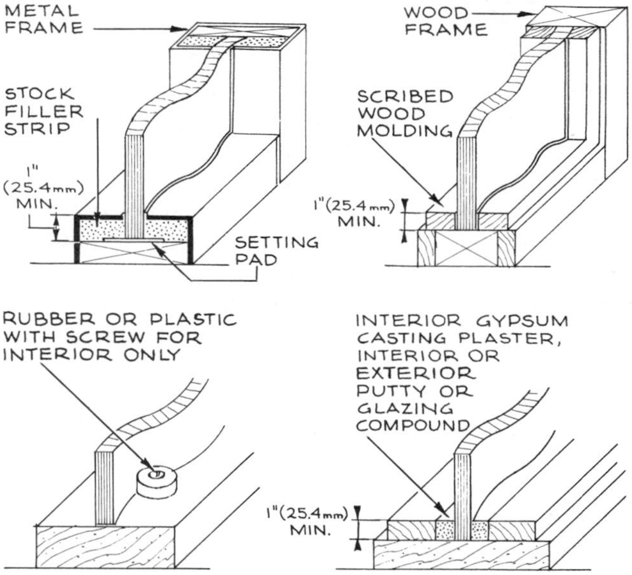

Figure G20 Typical sill and head details for corrugated glass.

facturer to establish the limitations and restrictions inherent in this type of glass.

7. When wired corrugated glass is used for roofing, skylights, clerestories, etc., spans must not be greater than 5 ft (1.524 m) for flat surfaces or slopes up to 60°; for slopes greater than 60°, the maximum span is 8 ft (2.438 m).
8. For roofing, the methods of lapping, joining, etc., must be correctly detailed (*see* Figure *G21*).
9. Local, municipal, and state codes and also the code of the Fire Underwriters, insurance companies, labor departments, and federal government (Army, Navy, etc.) should be checked for fire ratings and limitations and restrictions.

Conditions Favorable to the Use of Corrugated Glass

1. Wired or unwired corrugated glass may be used where a decorative, easily maintained, durable and translucent partition, wall, or screen is desired.
2. Wired corrugated glass may be used where a fire-retarding, strong, nonbreakable, translucent material is necessary.
3. In any area where borrowed light and privacy are desirable.

Conditions Unfavorable to the Use of Corrugated Glass

In areas where loads must be supported or where unwired corrugated glass will be under stress.

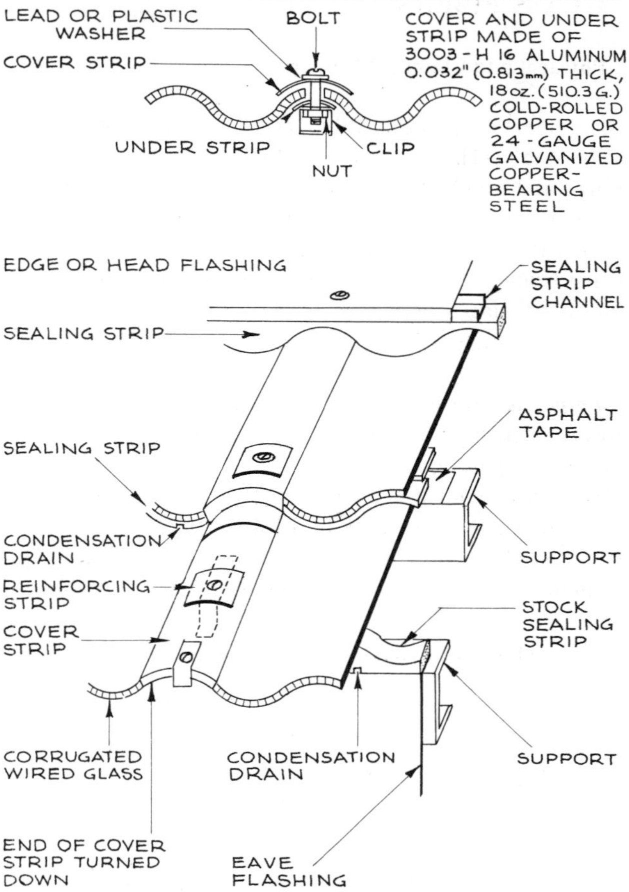

Figure G21 Details of roof construction with corrugated wire glass.

History and Manufacture

Wired corrugated glass was extensively used before 1932, but unwired corrugated glass is a relatively new material. The widespread acceptance of contemporary architectural design concepts and the search for new materials have led to a new outlook on wired corrugated glass, which until recently was used only for skylights and on industrial buildings. Its use on both the exterior and the interior of all types of buildings led to a demand for and the development of unwired corrugated glass. Today corrugated glass is an accepted material for all types of construction.

The manufacture of corrugated glass consists of first rolling glass in the same way as for patterned glass and then letting the still hot glass be formed into corrugations by its own weight on a corrugated surface in the manufacturing belt line.

GLASS FIBERS

Physical and Chemical Properties

Glass fibers are incombustible, nonabsorbent, and chemically stable. They withstand attack by insects, rodents, fungi, and molds. Glass fibers for thermal insulation and acoustic work are made from a low-alkali glass composition. Fibers with applications in the electrical field are made from glass with no alkali metal oxides. Laboratory-produced glass fibers have been made with a breaking strength as high as 500,000 lbf/in.^{2} (3447.5 MN/m^2).

Commercial Forms. Glass fibers are marketed in several forms, namely, fibers for textiles; fibers for reinforcing plastics, paper, concrete, and felts; glass wool; and spun glass. Glass wool has a density of $1\frac{1}{2}$ lb/ft^3 (24.03 kg/m^3); spun glass, a density of 14 lb/ft^3 (224.28 kg/m^3). Glass wool in construction is part of the large category of material known as mineral wool (*see* Mineral Wool).

Types and Uses

Glass fibers for textiles are used for decorative fabrics, tapes, fabrics for insulation and coverings in the electrical field, filters in the chemical field, and other applications where an incombustible, nonshrinking, nonstretching fabric is an important design requirement. Glass fibers, either as a cloth or felted from various types and lengths of glass fibers, are widely used for reinforcing paper, tapes, and plastics. Cut short lengths of glass fibers, known as chopped glass, are used as filler and reinforcing in molded plastic.

Glass wool finds many uses as thermal and acoustic insulating material in the construction field. It can be used in areas with temperatures ranging from subzero to 400°F (204.44°C). Tables *G17* to *G19* list the various types of glass wool, giving data on sizes, thermal and acoustic properties, and major uses.

Depletion of the earth's fossil fuels causes the installation of thermal insulation to become a very important

Table G17 Glass wool materials used in construction for acoustic treatments

	Standard dimensions								Sound absorption coefficient for frequencies of	
	Width[a]		Length		Thickness		Density			
Types of glass wool	ft	m	ft	m	in.	mm	lb/ft^3	kg/m^3	250–4000	Uses
Plain[b]	2, 3, 4,	0.610, 0.914,	200	60.96	$\frac{1}{2}$	12.70	$\frac{1}{2}$	8.01	0.14–0.73	Major use: acoustic
(resilient	and 6	1.219					$\frac{3}{4}$	12.02	–	absorption pads
blanket)		and 1.829					1	16.02	0.16–0.78	behind perforated
			100	30.48			$1\frac{1}{2}$	24.03	–	materials
							2	32.04	–	Other uses: insulation
			200	60.96	$\frac{3}{4}$	19.05	$1\frac{1}{2}$	24.03	–	and sound absorp-
							2	32.04	–	tion for appliances,
			100	30.48			1	16.02	–	ducts, pipes, heating
										units, etc.
			100	30.48	1	25.4	$\frac{1}{2}$	8.01	0.25–0.76	
							1	16.02	0.30–0.80	
			50	15.24			$\frac{3}{4}$	12.02	–	
							$1\frac{1}{2}$	24.03	–	
							2	32.04	–	
			100	30.48	$1\frac{1}{2}$	38.10	$\frac{3}{4}$	12.02	–	
							1	16.02	–	
					2	50.8	$1\frac{1}{2}$	24.03	–	
Plain[b]	2, 3,	0.610, 0.914,	200	60.96	$\frac{1}{2}$[c]	12.7	$1\frac{1}{2}$	24.03	0.54–0.75	
(semirigid	and 4	and 1.219	100	30.48			3	48.06	0.47–0.81	
blanket)							$1\frac{1}{2}$	24.03	0.61–0.83	
							3	48.06	0.66–0.85	
Acoustic rigid ceiling tiles	1.978	0.603	1.976	0.603	$1\frac{1}{4}$[c]	31.75	0.19	3.10	0.83–0.89	Suspended ceilings; can also be attached with adhesive
			3.978	1.212						

[a]Also made in 1.5- and 4.5-ft (0.457- and 1.372-m) widths on special order.
[b]This type of glass wool is used as thermal insulation for appliances, air conditioning, heating, ventilating, and plumbing installations. It is rarely used as thermal insulation in buildings except in prefabricated or preassembled components such as curtain walls. Both the resilient and semirigid blankets are available with various factory-applied surfaces such as paper, plastic, and aluminum foil.
[c]Also made in thicknesses of $1\frac{1}{2}$, 2, 3, and 4 in. (38.1, 50.8, 76.2 and 101.6 mm) on special order.

Table G18 Rigid glass wool materials used for thermal insulation of foundations, slabs, and roofs

Type of glass wool	Standard dimensions: Width (in.)	Width (mm)	Length (ft)	Length (m)	Thickness (in.)	Thickness (mm)	Thermal conductance[a] at 75°F (18.33°C) mean temperature: Btu/ft^2 · h · °F	W/m^2 · °C	Uses: Major	Uses: Other
Rigid	12	304.8	4	1.219	$\frac{3}{4}$	19.05	0.33	1.874	Interior edge of	Existing exterior
insulation	18	457.2			1	25.40	0.25	1.428	concrete slabs	face of founda-
(plain)	24	609.6			$1\frac{1}{2}$	38.10	0.17	0.965	on fill and in-	tion walls
					2	50.80	0.13	0.738	terior face of	
Rigid	4	101.6	3	0.914	1	25.40	0.28	1.590	foundation	
insulation	6	152.4			$1\frac{1}{2}$	38.10	0.19	1.079	walls	
(asphalt-	8	203.2			2	50.80	0.14	0.795		
coated)	12	304.8								
	24	457.2								
		609.6								
Rigid	20	508.0	4	1.219	$\frac{3}{4}$	19.05	0.33	1.874		
insulation					1	25.40	0.25	1.420		
(faced with					$1\frac{1}{2}$	38.10	0.17	0.968		
paper on					2	50.80	0.13	0.738		
one side)										
Rigid	24	609.6	4	1.219	$\frac{1}{2}$	12.70	0.50	2.839	For flat roofs	For pitched roofs
insulation					$\frac{3}{4}$	19.05	0.33	1.874		
(with paper					$\frac{7}{8}$	27.23	0.30	1.703		
asphalted					1	25.40	0.25	1.420		
to one					$1\frac{1}{4}$	31.75	0.20	1.136		
side)					$1\frac{1}{2}$	38.10	0.17	0.965		
					$1\frac{3}{4}$	44.45	0.15	0.852		
					2	50.80	0.13	0.738		

[a]Units are given in terms of the absolute joule per second or watt.

Table G19 Glass wool materials used in construction for thermal insulation

Type of glass	Standard dimensions: Width (in.)	Width (mm)	Length (ft)	Length (m)	Thickness (in.)	Thickness (mm)	Thermal conductance[a]: Btu/ft^2 · h · °F	W/m^2 · °C	Uses: Major	Uses: Other
Thermal	15	381.0	30–80	9.144–	$3\frac{1}{2}$	88.9	0.090	0.511	Side walls; be-	Sound-deadening
insulation	19	482.6		24.384	$3\frac{5}{8}$	92.1	0.077	0.437	tween roof	walls; floors
rolls[b]	23	584.2			4	101.6	0.074	0.420	rafters; in ceil-	and ceilings;
					5	127.0	0.071	0.403	ings with un-	crawl spaces
					6 and	152.4 and	0.053 and	0.301 and	insulated areas	
					$6\frac{1}{2}$	165.1	0.051	0.290	above	
Thermal	11	279.4	2 and 4	0.610 and	$3\frac{1}{2}$	88.9	0.090	0.511		
insulation	15	381.0		1.219	$3\frac{5}{8}$	92.1	0.077	0.437		
batts[b]	19	482.6			4	101.6	0.074	0.420		
	23	584.2			5	127.0	0.071	0.403		
					6 and	152.4 and	0.053 and	0.301 and		
					$6\frac{1}{2}$	165.1	0.051	0.290		

Table G19 Glass wool materials used in construction for thermal insulation (*continued*)

Type glass wool	Standard dimensions: Width in.	Width mm	Length ft	Length m	Thickness in.	Thickness mm	Thermal conductance[a] at 75°F (18.33°C) mean temperature: Btu/ft² · h · °F	W/m² · °C	Uses: Major	Uses: Other
Thermal insulation, not enclosed	15 23	381.0 584.2	3.813–8	1.162–2.438	$3\frac{1}{2}$ 4 $6\frac{1}{2}$ 7	88.9 101.6 165.0 177.8	0.090 0.074 0.051 0.045	0.511 0.420 0.290 0.257	Side walls; between roof rafters; in ceilings with uninsulated areas above	Sound-deadening walls; floors and ceilings; crawl spaces
Thermal insulation rolls with aluminum foil on one face[c]	15 23	381.0 584.2	40	12.192	$3\frac{1}{2}$ 4 6	88.9 101.6 152.4	0.090 0.073 0.050	0.511 0.414 0.283		
Thermal insulation batts with aluminum foil on one face[c]	15 23	381.0 584.2	4	1.219	$3\frac{1}{2}$ 4 6	88.9 101.6 152.4	0.090 0.073 0.050	0.511 0.414 0.283		
Loose glass wool pellets					4	101.6	0.060	0.341	Ceilings with uninsulated areas above	For packing in small crevices and cracks

[a]Units are given in terms of the absolute joule per second or volt.
[b]These blankets and batts are enclosed with vapor barrier or paper on one side and perforated or nonperforated paper on the other side. Both types have extensions for attachments.
[c]Aluminum foil gives thermal resistance because of the air space.

component in the design of buildings. Therefore it is always advisable to consult environmental engineers for the selection of thermal insulation. (*See* Insulation and Insulating Materials.)

Application

Condensed Checklist

1. When glass wool products are used as sound-absorbing and acoustic materials, specialists in the field of acoustics should always be consulted to determine the correct kind, thickness, and latest type of perforated or decorative material to be used to achieve the desired acoustic treatment.
2. When glass wool materials are used for thermal insulation, the temperature ranges of the locality where the building is to be built should always be checked so that the correct thickness and type of insulation material will be installed in the north, south, east, and west walls and in the roof areas (*see* Figure *G22*), to achieve maximum energy conservation.
3. Heating and air-conditioning engineers should be consulted for the type of thermal insulation best suited for the building. This is particularly important for the thermal insulation of areas requiring special humidity control and temperature control.
4. When glass wool materials are used for insulating flat or low-pitched roofs, it is necessary to check the correct methods of installation for the type of decking being used, including the treatment of decking before application, the need for a vapor barrier, and the method of terminating a day's work (*see* Figure *G23*).
5. When glass wool materials are used for perimeter insulation, proper installation depends on details at the joint of the concrete slab to the exterior wall and on selection of a suitable thickness and type of product for the climate of the locality. (Figure *G24* shows details for concrete slab and block.
6. When selecting glass wool for thermal insulation, the *R* factor should always be checked, as most building codes will require exterior walls and ceilings under roofs to meet a definite resistance factor. An example

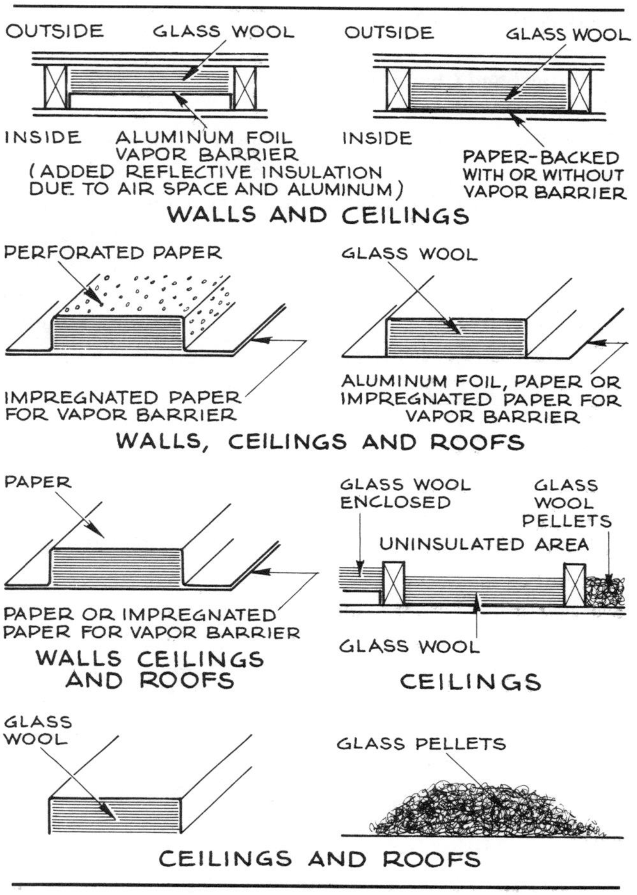

Figure G22 General types of glass wool thermal insulation, major uses, and details of installation.

would be that all exterior walls must have an R factor of 22.5.

7. Heating and air-conditioning engineers should always be consulted about the location where vapor barrier(s) should be installed.

Conditions Favorable to the Use of Glass Fiber Materials

1. Where a highly absorbent, inert, incombustible acoustical material is required to back up a perforated or decorative material in acoustic treatments.
2. Where an inert, incombustible, permanent, thermal insulation material that is unaffected by mold, rot, insects, and rodents is required.
3. Glass wool tile should be used where an incombustible, lightweight acoustic ceiling is required.

Conditions Unfavorable to the Use of Glass Fiber Materials

1. Where temperatures over 500°F (260°C) will be encountered.

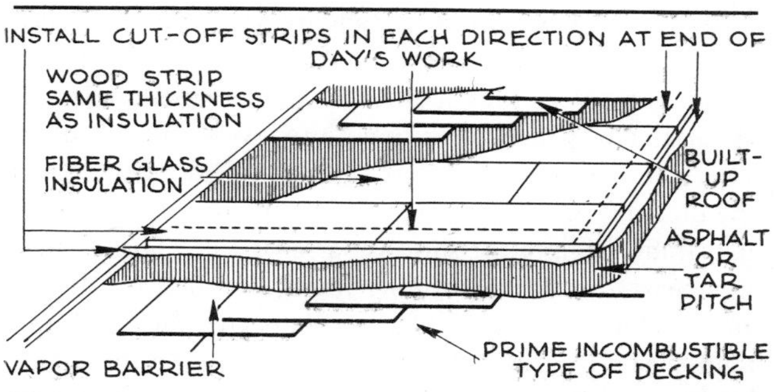

Figure G23 General method of installing glass wool roof insulation.

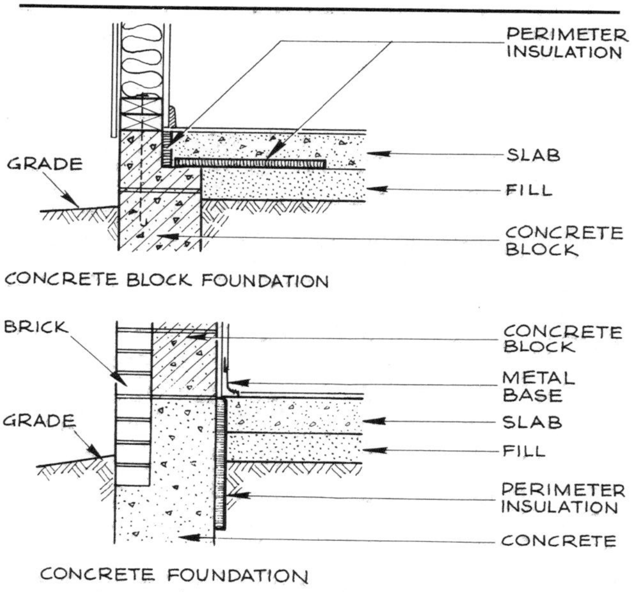

Figure G24 Details of glass wool perimeter insulation.

History and Manufacture

The first glass fibers were made by melting the end of a glass rod, attaching the droplet to a rotating wheel, and drawing or spinning a fiber. In 1713 Réaumur showed glass fabric to the Paris Academy of Science. In 1893, at the Columbian Exposition, Libby produced glass fibers and wove them with silk to produce fabrics. In 1929, Rosengarth invented a process in which molten glass flows onto the center of a rapidly rotating ceramic disk with radial serrations. By centrifugal force the streams of glass in the serrations are whipped off the edge into fibers. This is known as spun glass.

In 1938 the two major research efforts in this country to produce industrially valuable glass fiber products—those of Owens-Illinois and Corning-Glass—were merged. Their joint efforts developed first the manufacturing methods for glass wool and then those for fibers for textiles.

Glass Wool. Glass wool is made by passing molten glass through small holes or orifices. As the streams of molten glass pour down, they are caught in high-pressure jets of air or steam, which blow them into a wool-like product. The temperature of the glass, the size of the orifices, and the pressure of the jets control the type of fibers formed; they may be long or short, coarse or fine. These fibers then pass through a forming hood, which controls size and thickness. These glass wool fibers are fabricated into rigid sheets or boards by treatment with a thermosetting binder and by passing them through compression rolls and ovens.

Textile Fibers. The manufacture of textile fibers starts with glass that has been formed into $\frac{5}{8}$-in. (15.875-mm) spheres carefully inspected for impurities. There are two processes, the staple fiber process and the continuous filament process.

GLASS, FLOAT

Physical and Chemical Properties

Float glass is the same type of glass as plate, window, and heavy sheet (*see* Glass). The surface, rather than the composition or thickness, of the glass is the distinguishing feature. Float glass has perfectly flat surfaces and a brilliant surface finish equal to that of ground and polished plate glass.

Properties are as follows: specific gravity, 2.52; weight, from 1.20 to 6.55 lb/ft^2 (5.86 to 31.98 kg/m^2); softening point, 1337°F (725°C); coefficient of linear expansion, 0.00000445; tensile strength (modulus of rupture), 6000 lbf/in.^{2} (41.37 MN/m^2); total visible white light $\frac{1}{4}$ in. (6.35 mm) thickness, 78%; thermal conductance, or U value, 1.0 Btu/ft^2 h °F (0.1761 W/m^2).

Float glass is available in thicknesses ranging from $\frac{1}{8}$ to

Table G20 Types of float glass and their characteristics

Type of float glass	Thickness		Maximum size		Weight[a]		Transmittance[b]	
	in.	mm	ft	m	lb/ft^2	kg/m^2	Visible (percent)	Total solar energy (percent)
Window, single-strength AA, A, B	0.085–0.101	2.159–2.565	8 × 10.83	2.438 × 3.302	1.22	5.956	91	87
Window, double-strength AA, A, B	0.115–0.134	2.921–3.404	8 × 10.83	2.438 × 3.302	1.62	7.909	90	85
Clear	$\frac{3}{32}$	2.35	8 × 10.83	2.438 × 3.302	1.23	6.005	91	86
	$\frac{1}{8}$	3.18	8 × 10.83	2.438 × 3.302	1.64	8.006	90	85
	$\frac{3}{16}$	4.76	10 × 10.83	3.048 × 3.302	2.45	11.961	90	83
	$\frac{13}{64}$	5.09	10 × 10.83	3.048 × 3.302	2.66	12.986	90	82
	$\frac{1}{4}$	6.35	10.83 × 16.67	3.302 × 5.080	3.28	16.013	89	79
	$\frac{5}{16}$	7.94	10.00 × 15.00	3.048 × 4.572	4.10	20.016	88	77
	$\frac{3}{8}$	9.53	10.83 × 16.67	3.302 × 5.080	4.92	24.019	87	72
	$\frac{1}{2}$	12.70	10.83 × 16.67	3.302 × 5.080	6.56	32.026	86	62
	$\frac{5}{8}$	15.88	10.00 × 15.00	3.048 × 4.572	8.20	40.032	84	62
	$\frac{3}{4}$	19.05	10.00 × 15.00	3.048 × 4.572	9.85	48.088	83	58
	1	25.40	Subject to manufacturing operations		13.13	64.101	80	51
Clear, mirror	$\frac{1}{8}$	3.18	8 × 10.83	2.438 × 3.302	1.64	8.006	90	85
	$\frac{1}{4}$	6.35	7 × 10.00	2.134 × 3.048	3.27	15.964	89	78
Tinted, bronze	$\frac{3}{16}$	4.76	10 × 20	3.048 × 6.096	2.45	11.961	56	55
	$\frac{1}{4}$	6.36	10 × 20	3.048 × 6.096	3.28	16.013	50	48
	$\frac{3}{8}$	9.53	10 × 15	3.048 × 4.572	4.92	24.019	37	33
	$\frac{1}{2}$	12.70	10 × 15	3.048 × 4.572	6.56	32.026	28	24
Tinted, gray	$\frac{3}{16}$	4.76	10 × 20	3.048 × 6.096	2.45	11.961	50	52
	$\frac{1}{4}$	6.35	10 × 20	3.048 × 6.096	3.28	16.013	43	46

Table G20 Types of float glass and their characteristics (*continued*)

Type of float glass	Thickness		Maximum size		Weight[a]		Transmittance[b]	
	in.	mm	ft	m	lb/ft^2	kg/m^2	Visible (percent)	Total solar energy (percent)
Tinted, gray	$\frac{3}{8}$	9.53	10 × 15	3.048 × 4.572	4.92	24.019	28	33
	$\frac{1}{2}$	12.70	10 × 15	3.048 × 4.572	6.56	32.026	19	24
Tinted, green	$\frac{3}{32}$	2.35	8 × 10	2.438 × 3.048	1.23	6.005	85	68
	$\frac{1}{8}$	3.18	8 × 10	2.438 × 3.048	1.64	8.006	83	65
	$\frac{3}{16}$	4.76	10 × 10.83	3.048 × 3.302	2.45	11.961	79	53
	$\frac{1}{4}$	6.35	10 × 10.83	3.048 × 3.302	3.28	16.013	75	47
Tempered, clear	$\frac{1}{8}$	3.18	3.83 × 8	1.168 × 2.438	1.64	8.006	90	85
	$\frac{3}{16}$	4.76			2.45	11.961	90	82
	$\frac{1}{4}$	6.35			3.28	16.013	88	77
Tempered, tinted bronze	$\frac{3}{16}$	4.76	3.83 × 8	1.168 × 2.438	2.45	11.961	59	55
	$\frac{1}{4}$	6.35			3.28	16.013	52	48
Tempered, tinted gray	$\frac{3}{16}$	4.76	3.83 × 8	1.168 × 2.438	2.45	11.961	50	53
	$\frac{1}{4}$	6.35			3.28	16.013	41	48
Tempered, tinted green	$\frac{1}{8}$	3.18	3.83 × 8	1.168 × 2.438	1.64	8.006	83	65
	$\frac{3}{16}$	4.76			2.45	11.961	78	53
	$\frac{1}{4}$	6.35			3.28	16.013	74	48

[a]These weights vary slightly because of the manufacturing process and type of glass.
[b]These percentages vary slightly because of the manufacturing process and type of glass.

1 in. (3.18 to 25.4 mm), and in sizes up to a maximum of 10.33 ft (3.150 m) in width and 25 ft (7.620 m) in length.

Types and Uses

Float glass is available as window glass, heavy sheet glass, tinted glass, tempered glass, mirrors, insulating glass, heat-absorbing glass, laminated glass, and general glazing, all of which are fully covered under specific glass headings (*see also* Table *G20*).

Application

Condensed Checklist

1. Wind loads should always be checked to obtain the correct thickness of glass required for the size of opening; correct clearance dimensions for the thickness of glass selected should also be checked.
2. The glazing details are determined by the type of material into which float glass is to be installed. They include type of glazing material, type of glazing points or clips necessary for setting blocks, and clearances (*see* Figures *G25* to *G28*).
3. Operating windows or doors must not be handled or opened when glazing compounds are used until the compound has set.
4. Glazing must never be done at temperatures below 40°F (3.33°C) unless precautions are taken to prevent moisture from condensing in rabbets where glass is to be installed.
5. Local, municipal, and state codes and also the codes of Fire Underwriters, insurance companies, labor departments, and the federal government (Army, Navy, etc.) should always be checked for limitations and all currently applicable requirements.
6. The type of float glass chosen must meet all requirements in transparence, translucence, diffusion, and strength for its end use.
7. When using tinted float glass, it is necessary to check that all other types of glass to be used are of the same tint.

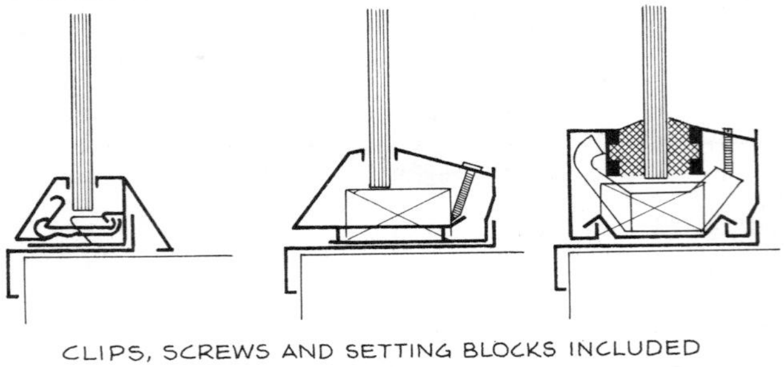

Figure G25 Float glass or heavy float glass installed in a typical store front of rolled or extruded stainless steel or aluminum.

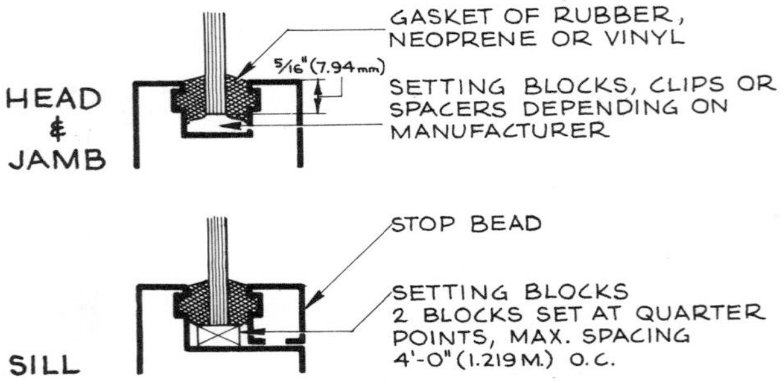

Figure G26 Float glass set in aluminum frame with extruded gaskets.

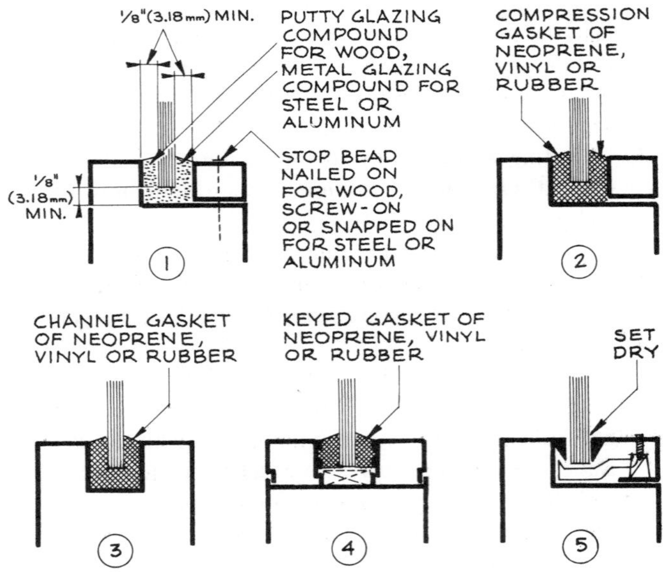

Figure G27 Float glass or heavy float glass set in frames of steel, stainless steel, or aluminum with glazing compound (1); steel, stainless steel, or aluminum with various types of gaskets (2, 3, 4); and stainless steel or aluminum set dry (5).

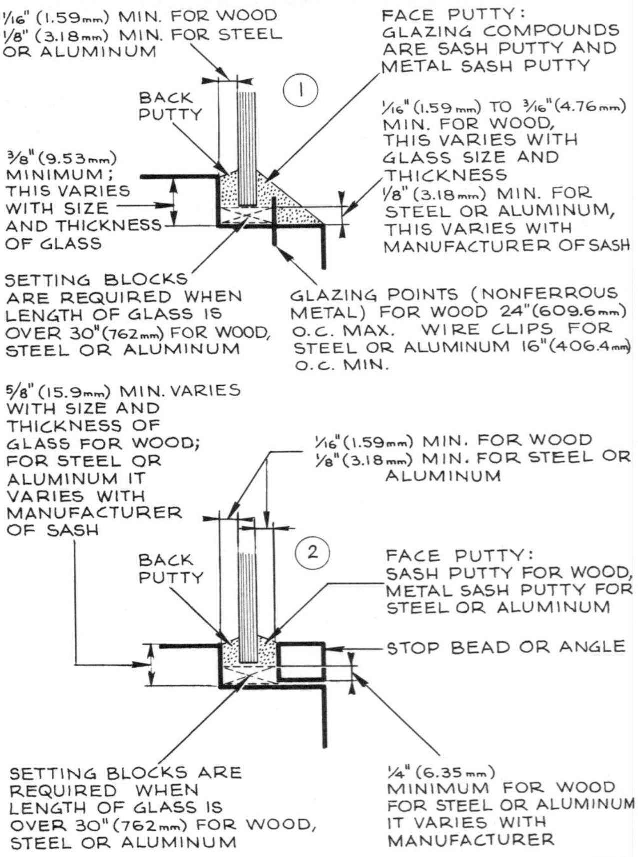

Figure G28 Float glass and heavy float glass set in wood, steel, or aluminum by face glazing (1) and with a stop bead (2).

8. All governing codes regarding the type of float glass to be installed in hazardous areas should always be checked.

Conditions Favorable to the Use of Float Glass

1. For shelves and tops for tables, desks, etc.
2. For any area where perfect transparence is important.
3. In general, float glass $\frac{1}{8}$ in. (3.18 mm) thick should be used for small windows, doors, and sidelights; and $\frac{1}{4}$-in. (6.35-mm) float glass for windows, doors, partitions, sidelights, store fronts, and display cases.
4. Heavy float glass should be used for large windows, partitions, store fronts, tops for tables and other articles of furniture, long shelving, aquariums, etc.
5. Tinted float glass should be used in areas where glare prevention and thermal insulation are important (*see* Glass: Insulating).
6. In general, float glass can be used in all areas where plate, window, and heavy sheet are installed.

Conditions Unfavorable to the Use of Float Glass

1. In areas where fire resistance is necessary.
2. In any area where float or heavy float glass will be placed in tension.

History and Manufacture

In the construction field the types of flat glass routinely used were plate and sheet, but in 1959 in England a new manufacturing process was installed by the Pilkington Brothers. In this manufacturing method a ribbon of molten glass floats on a bath of liquid tin and remains untouched until it hardens (*see* Figure *G29*). This float glass, as it is now termed, has parallel surfaces and the high optical quality of plate glass without grinding or polishing, as well as a fire-finished surface with the brilliance of the finest sheet glass.

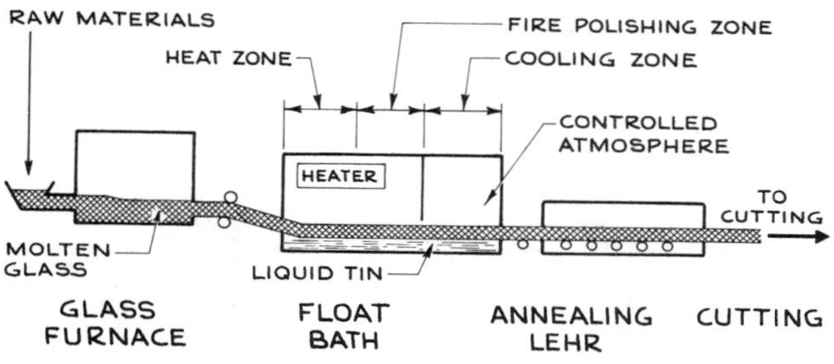

Figure G29 Flowchart of the manufacturing of float glass.

Float glass was introduced into the construction field in 1963, and, since it meets all the requirements of plate glass, it naturally has begun to replace plate glass. The thickness of float glass is controlled by the speed and flow of the molten glass; the width is controlled only by the maximum width of the molten tin, and since the general length of the float bath is more than 150 ft (45.288 m), there is no difficulty in producing glass lengths to meet currently required sizes.

GLASS, HEAT-STRENGTHENED

Physical and Chemical Properties

Heat-strengthened glass is float, plate, or patterned glass with a colored ceramic glaze fused to one side of it. The reheating of the glass, which is necessary to apply the ceramic glaze, and the subsequent cooling give the glass characteristics similar to those of tempered glass. It is about twice as strong as float or plate glass, and, like tempered glass, it cannot be drilled or cut. Exact sizes and details have to be supplied before it can be made. The color range is limited only by the colors available in ceramic glazes. Some heat-strengthened glass has an aluminum foil applied over the ceramic glaze to give added protection to the glaze and to supply reflective-type thermal insulation.

Heat-strengthened glass is available in thicknesses of $\frac{1}{4}$ and $\frac{5}{16}$ in. (6.35 and 7.94 mm) and in limited standard sizes, in colors to match tinted glass and reflective glass, but not as clear glass. (*See also* Glass: Tinted, Reflective, and Heat-Absorbing.)

Types and Uses

Heat-strengthened glass is opaque and comes in limited sizes, colors, and thicknesses, as shown in Table *G21*. This type of glass is now used as colored panels for spandrel glazing in curtain wall systems of construction.

Heat-strengthened glass is also available as a complete insulated unit consisting of heat-strengthened glass and insulation (urethane, foam, glass fiber, or other type of insulating material), with or without an interior finished panel (cement asbestos, plywood, laminates, steel, or aluminum). Because of the heat-absorbing characteristics of heat-strengthened glass, the methods of installation must be checked inasmuch as intense heat and expansion may cause heat-strengthened glass to break.

Application

Heat-strengthened glass must always rest on resilient setting blocks, and the frame into which the glass is installed must be strong enough so that no loads or stresses are transmitted to the glass.

Condensed Checklist

1. Heat-strengthened glass must be installed at temperatures above 40°F (4.44°C), and special attention should be given to ensuring that all bonding surfaces are free from moisture.
2. Local, municipal, and state codes and also the Fire Underwriters, insurance companies, labor departments and federal government (Army, Navy, etc.) should be checked for limitations and restrictions regarding heat-strengthened and structural glass.
3. The thickness, size, and color of the heat-strengthened glass chosen must meet the requirements for its end use.
4. Shop drawings and details of installation must be required by the specifications.

Table G21 Types of heat-strengthened glass

		Maximum sizes										
		Thickness		Tolerances		Width		Length		Weight		
Type	Color range	in.	mm	in.	mm	ft	m	ft	m	lb/in.^{2}	kg/m^2	Major use
Plate and float	Black, white, and 16 standard colors, plus colors to match tinted and reflective glass	$\frac{1}{4}$	6.35	$+\frac{1}{64}$ $-\frac{1}{32}$	+0.39 −0.78	4 6	1.219 1.829	7 12	2.134 3.658	3.27 3.29	229.04 2313.01	For curtain wall construction
Patterned	Black, white, and 12 standard colors	$\frac{5}{16}$	7.94	0.045	1.14	7.33	2.235	5	1.524	4.06	2854.46	For curtain wall construction

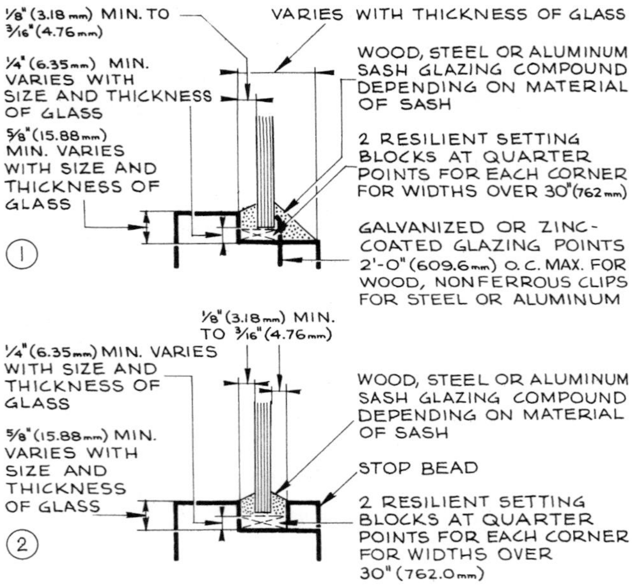

Figure G30 Glazing details for heat-strengthened glass set in wood, steel, or aluminum, showing face glazing (1) and glazing with a stop bead (2).

5. The material into which the glass is to be installed should be checked to make sure that correct clearances, setting blocks, clips, and type of glazing compound are used (*see* Figures *G30* and *G31*).
6. When heat-strengthened glass is used in operating windows or doors, these windows and doors must not be handled or opened until the glazing compound has set. This precaution should be clearly stated in the specifications.

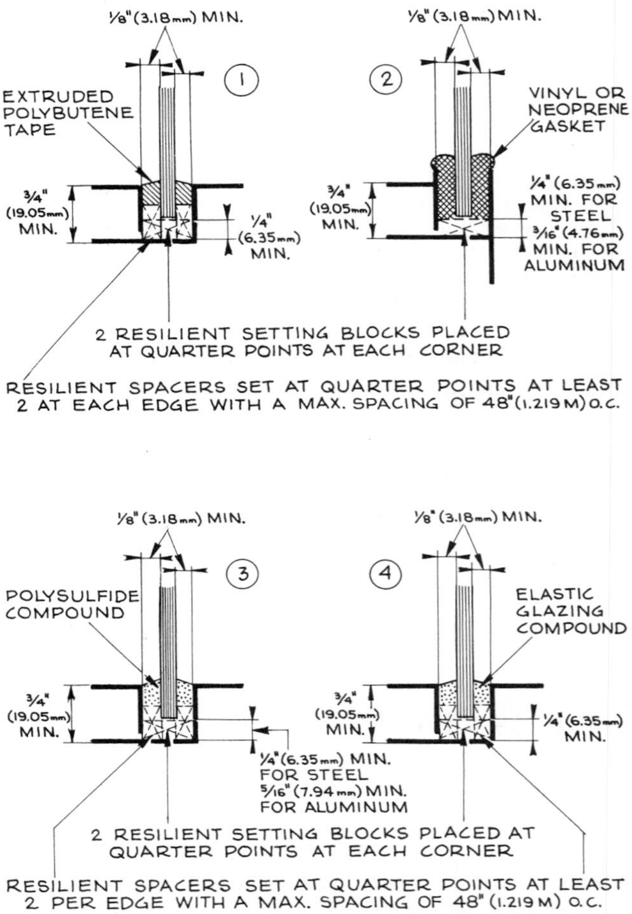

Figure G31 Heat-strengthened glass with tape (1), with a gasket (2), with polysulfide compound (3), and with elastic glazing compound (4).

7. When using heat-strengthened glass in a curtain wall system, it must be specified that a full-size mock-up be built to test the system for weather and climatic conditions. Also, a check should be made that there is no variation in colors between glasses which would destroy visual continuity.
8. When a curtain wall system incorporates plastic gaskets, the glass manufacturers should be consulted to make sure that the correct type of gasket is chosen and is installed properly.
9. Plastic gaskets should always be checked for color, stability, strength, and methods of application. A tested full-size mock-up should always be included in the specifications.

Conditions Favorable to the Use of Heat-Strengthened Glass

1. Where an opaque, colored, permanent unit is desired as a facing material within a curtain wall.

Conditions Unfavorable to the Use of Heat-Strengthened Glass

1. Where a fire-resistant material is necessary.
2. Where any permanent loads or stresses will be transmitted to the glass.

History and Manufacture

Heat-strengthened glass was developed parallel to the use of curtain wall construction soon after World War II. Heat-strengthened glass is manufactured by spraying a glaze on plate, float or patterned glass, heating until the glaze fuses, and then cooling the glass. This heat treatment causes the outside of the glass to be in compression and the inside to be in tension, giving it the characteristics of tempered glass.

GLASS, INSULATING

Physical and Chemical Properties

Insulating glass (double glass) may be defined as two or more sheets of glass of various types separated by a captured air space dehydrated at atmospheric pressure and usually $\frac{3}{16}$, $\frac{1}{4}$, or $\frac{1}{2}$ in. (4.76, 6.35, or 12.70 mm) thick. The types of glass used in various combinations are as follows: window "A" quality, plate glazing quality, heat-absorbing, float, tinted, heat-processed, patterned, tempered, and laminated glass. In all types the glass used for both sheets must be the same thickness, with a maximum variance of $\frac{1}{16}$ in. The specific compositions of the various types of glass used are fully described under Glass and Glazing; data on sizes, etc., are given under specific glass headings.

U Values. The important characteristic of insulating glass is its thermal insulating (U) value expressed in $Btu/ft^2 \cdot h \cdot °F$ ($W/m^2 \cdot °C$).

Table *G22* shows U values for single clear glass and types of insulating glass using clear glass. With the increase in air conditioning and the need to conserve energy, a twofold problem emerged: (1) better insulation was needed, and (2) methods to control solar heat gain had to be found. Tinted and reflective glass were developed and have proved successful in controlling solar heat gain for air conditioning, and combining these glasses to make insulation glass provided thermal insulation combined with solar heat-gain control. Table *G23* shows an example of calculations to determine the air-conditioning load for a structure.

Problem of Condensation. Insulating glass allows much higher inside relative humidity by overcoming the problem of condensation that usually occurs with a single sheet of glass.

Light Transmission. Light transmission varies with the combination of different types of glass (clear-clear, clear-tinted, reflective-clear, etc.), and each of these combinations has not only different light transmission but also different reflectance, heat gain, and U value, plus shading coefficients. All these various characteristics must be carefully evaluated when selecting a type of insulating glass.

Glass-to-Glass Seals. In a glass-to-glass seal two sheets of glass are hermetically sealed, that is, the edges of the glass are brought to a molten state and joined while molten. For this type of seal the two sheets of glass must have similar expansion characteristics in order to avoid the high stresses that can develop during the cooling of the sealed joint. This means that only similar types of glass can be made with a glass-to-glass seal.

Metal-to-Glass Seals. In a metal-to-glass seal the metal alloy used must have expansion characteristics similar to those of the glass; the thermal expansion must be uniform throughout the interval from room temperature to the melting point of glass; the metal must not soften or burn at sealing temperatures; adhesion between the metal and the glass must be good; and the metal must give off a minimum amount of gas (fumes) at sealing

Table G22 Difference in U value between single sheet of glass and various types of insulating glass

Type of glass	Thickness of glass		Air space thickness		U value[a]	
	in.	mm	in.	mm	Btu/ft² · h · °F	W/m² · °C
Single sheet of window, sheet, or plate glass	$\frac{1}{8}$	3.18	None		1.14	6.47
	$\frac{1}{4}$	6.35	None		1.12	6.36
Insulating glass with one air space, metal-to-glass seal	$\frac{1}{8}$	3.18	$\frac{1}{4}$	6.35	0.60	3.41
	$\frac{1}{8}$	3.18	$\frac{1}{2}$	12.70	0.55	3.12
	$\frac{1}{4}$	6.35	$\frac{1}{4}$	6.35	0.60	3.41
	$\frac{1}{4}$	6.35	$\frac{1}{2}$	12.70	0.55	3.12
Insulating glass with two air spaces, metal-to-glass seal	$\frac{1}{4}$	6.35	$\frac{1}{4}$	6.35	0.46	2.41
Insulating glass with one air space, glass-to-glass seal	$\frac{1}{8}$	3.18	$\frac{3}{16}$	4.77	0.54	3.07

[a]The U value is the amount of heat in Btu (W/m² · °C) which will pass through 1 ft² (0.0929 m²) of window area per hour per °F (°C) temperature difference between inside air and outside air.

temperature. The principle of metal-to-glass seals is based on the oxide film that develops when the metal is heated. Although the mechanism is not completely understood, in brief this is what happens: When the heated metal and the molten glass are brought together in intimate contact, the oxide film diffuses or dissolves to some degree in the glass and a hermetic seal is developed. The metals commonly used to make these special alloys for metal-to-glass seals are iron, chromium, copper, nickel, and cobalt. Platinum, tungsten, and molybdenum are also used.

Types and Uses

Insulating glass is available with either glass-to-glass or glass-to-metal seals in a large variety of standard sizes; other sizes can be manufactured to order. The use of various types of glass to make insulating glass results in a wide range of physical characteristics such as strength, transparency, type of light, and solar transmission from which to make a selection.

Sizes. Insulation glass is manufactured with a $\frac{3}{16}$-in. (4.76-mm) air space for glass-to-glass seals, and $\frac{1}{4}$-in. (6.35-mm) and $\frac{1}{2}$-in. (12.70-mm) air spaces for metal-to-glass seals. Total thickness varies with the thickness of the type of glass used in relation to the various widths of air space. Tables *G24* to *G26* give maximum sizes, weights, and characteristics of various types of insulating glass used in construction (*see* Figure *G32*).

Also available is insulating glass with a 2-in. (50.8-mm) air space. This type is used for a noise barrier (built into a special frame with sound-absorbent barriers), for installing remote-control venetian blinds within the air space, and for installing a metallic electric resistant film on the inside face of interior glass to produce a radiant heating unit.

Limitations in Patterns. Insulating glass can be manufactured only in straight-edge patterns. Units can be produced with four or five sides, providing that no angle is less than 45%. Triangles, circles, and bent shapes are not feasible. No cutouts, notches, holes, or finger pulls are possible.

Insulating glass using varying combinations of tinted, clear, reflective, heat-processed, and tempered glass is manufactured with the reflective surface located on various faces of the insulating glass: on the exterior glass face, on the interior face of the exterior glass, and on the face of the interior glass located in the air space. Gener-

Table G23 Example of calculating heat gain of a structure

Data for calculations

June 19

	A.M.					P.M.			
	8	9	10	11	12	1	2	3	4
N	29	33	35	37	38	37	35	33	29
S	29	45	69	88	95	88	69	45	29
E	215	192	145	80	41	37	35	31	26
W	26	31	35	37	41	80	145	192	215

September 19

	A.M.					P.M.			
	8	9	10	11	12	1	2	3	4
N	16	22	26	29	30	29	26	22	16
S	71	124	165	191	200	191	165	124	71
E	205	195	148	77	32	29	26	22	16
W	16	22	26	29	32	77	148	195	205

The building has the following glass areas: Outdoor air temperature is 95°F (35°C) and indoor air temperature is 75°F (23.88°C)

North	5,000 ft^2	(464.52 m^2)
South	10,000 ft^2	(929.03 m^2)
East	10,000 ft^2	(929.03 m^2)
West	5,000 ft^2	(464.52 m^2)
Total	30,000 ft^2	(2787.10 m^2)

Calculations are based on the fact that maximum heat gain occurs on September 19 at 2 P.M.

The following values are based on using double-strength, clear sheet glass with no shade:

Orientation	Glass area ft^2	Glass area m^2	Solar heat gain factor	Btu/h	W
North	5,000	464.52	X 26	= 130,000	11 724
South	10,000	929.03	X 165	= 1,650,000	483 615
East	10,000	929.03	X 26	= 260,000	72 206
West	5,000	464.52	X 148	= 740,000	216 894
Total solar heat gain[a]				= 2,780,000	784 439

Calculations for air-to-air heat gain due to indoor-outdoor temperature difference:
Heat gain in Btu/h (W) = Btu · h/ft^2 · °F (W/m^2 · °C) X total glass area in ft^2 (m^2) X [outdoor °F (°C) – indoor °F (°C)]

The following values are based on using 1-in. insulating glass, mirrored and bronze tinted:

Orientation	Glass area ft^2	Glass area m^2	Solar heat gain factor	Shading coefficient	Btu/h	W
North	5,000	464.52	X 26	X 0.16	= 20,800	6 096.48
South	10,000	929.03	X 165	X 0.16	= 168,000	49 240.80
East	10,000	929.03	X 26	X 0.16	= 41,600	12 192.96
West	5,000	464.52	X 148	X 0.16	= 118,400	34 703.04
Total solar heat gain					= 348,800	102 233.28

Air-to-air heat gain for 1-in. insulating glass with a Btu · h/ft^2 · °F U factor of 0.50 (2.839 W/m^2 · °C) from formula above:

$$0.50 \times 30{,}000 \times (95 - 75) = 300{,}000 \text{ Btu } (276\,930 \text{ W})$$

Total solar heat gain	= 348,800 Btu/h (102 233.28 W)
Air-to-air heat gain	= 300,000 Btu/h (276 930.00 W)
Air conditioning total load	= 648,800 Btu/h (379 163.28 W)

ally the interior glass is clear, and the exterior glass is clear or tinted.

A form of insulating safety glass is available without the use of an air space. This laminated glass uses clear glass with multiple layers of plastic painted with fine horizontal lines of metallic-type ink. By slightly offsetting the horizontally lined plastic sheets, a miniature-type louver is developed which blocks heat gain and glare. This type of glass is available in a wide variety of colors and also as clear or transparent.

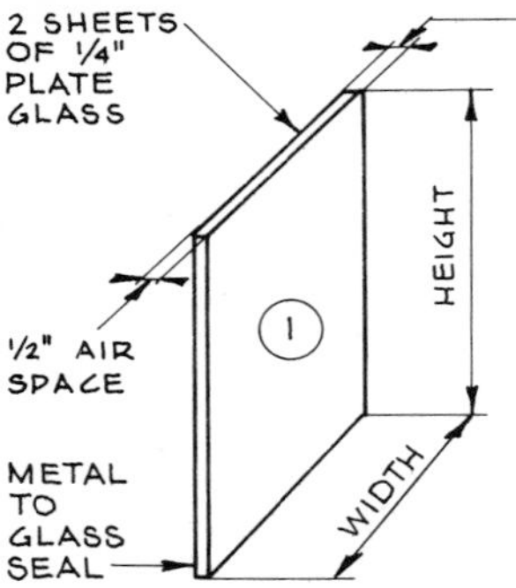

Thickness	1"	1"	1"	1"	1"	1"	1"	1"	1"	1"	1"	1"	1"
Height	76 3/4"	36" 48 1/8" 60 3/8"	44 1/2" 55 1/4" 68 3/4" 75" 93"	66" 72"	48 1/8" 60 3/8"	76 3/4"	52"	48 1/2" 64 1/2" 72 1/2"	52 1/2" 56 1/2"	50 3/8" 66 5/8"	48" 60" 72"	55 1/4" 68 3/4" 75" 93"	50" 58"
Width	33"**	35 1/2"	36"	42"	44 1/2"	45"**	45 3/8"*	46"	46 1/8"	47 3/4"	48"	48 1/8"	48 1/2"

CONTINUED →

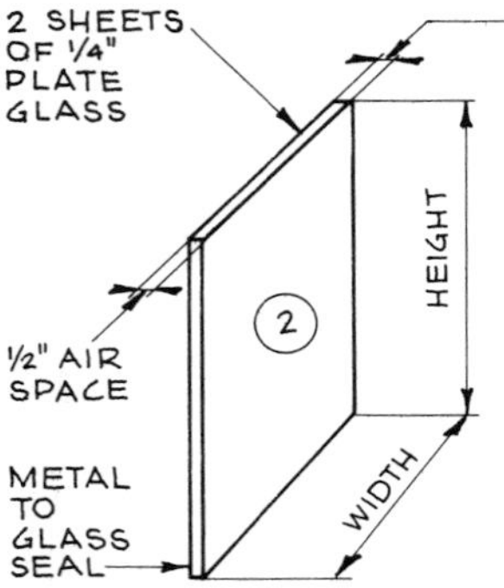

Thickness	1"	1"	1"	1"	1"	1"	1"	1"	1"	1"	1"	1"	1"
Height	56 1/2" 64 1/2" 72 1/2" 80 1/2" 96 1/2"	60 1/8"	58 1/8" 70 1/8"	60 3/8"	58 1/8" 66" 70 1/8"	76 3/4"	64 1/2" 72 1/2" 80 1/2" 96 1/2" 116 1/2"	72"	66 5/8"	68 3/4" 75" 93"	66"	72 1/2" 84" 96"	84" 96"
Width	50"	50 3/8"	52 1/2"	55 1/4"	56 1/2"	57"**	58"	60"	60 1/8"	60 3/8"	64 1/2"	66"	72"

** STOCK SIZES FOR SLIDING DOORS
* ALSO AVAILABLE MADE OF 3/16" SHEET GLASS WITH 1/2" AIR SPACE

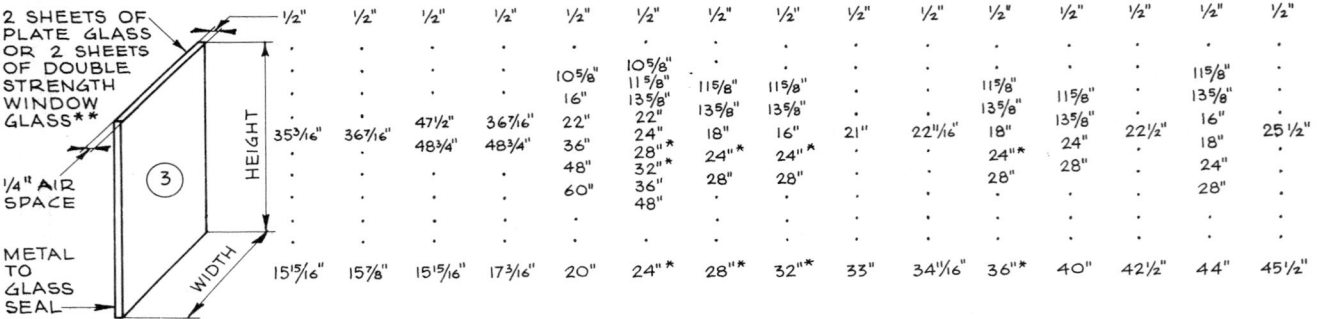

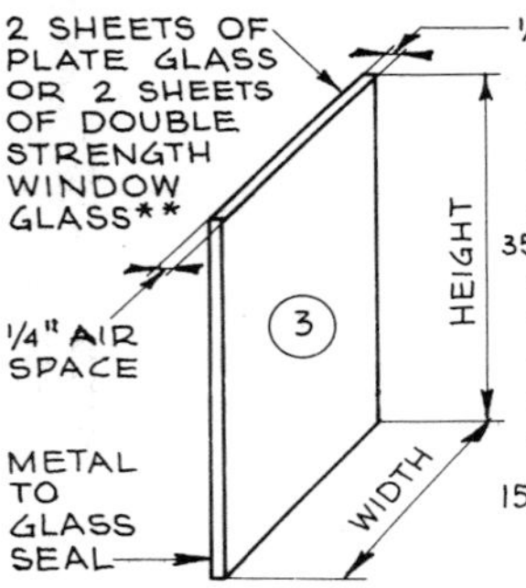

Thickness	1/2"	1/2"	1/2"	1/2"	1/2"	1/2"	1/2"	1/2"	1/2"	1/2"	1/2"	1/2"	1/2"	1/2"	1/2"
Height	35 3/16"	36 7/16"	47 1/2" 48 3/4"	36 7/16" 48 3/4"	10 5/8" 16" 22" 36" 48" 60"	10 5/8" 11 5/8" 13 5/8" 22" 24" 28"* 32"* 36" 48"	11 5/8" 13 5/8" 18" 24"* 28"	11 5/8" 13 5/8" 16" 24"* 28"	21"	22 11/16"	11 5/8" 13 5/8" 18" 24"* 28"	11 5/8" 13 5/8" 24" 28"	22 1/2"	11 5/8" 13 5/8" 16" 18" 24" 28"	25 1/2"
Width	15 15/16"	15 7/8"	15 15/16"	17 3/16"	20"	24"*	28"*	32"*	33"	34 11/16"	36"*	40"	42 1/2"	44"	45 1/2"

* CONSIDERED STANDARD SIZES AS LISTED EVEN THOUGH WIDTH AND HEIGHT DIMENSIONS ARE REVERSED
** SHOULD NOT BE TURNED. WIDTH DIMENSIONS ARE LISTED FIRST AND GLASS MUST BE INSTALLED WITH WIDTH PARALLEL TO FLOOR

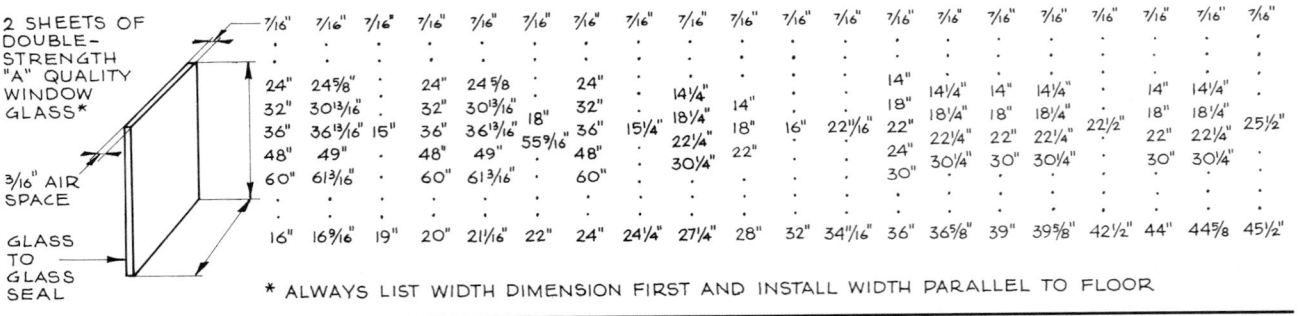

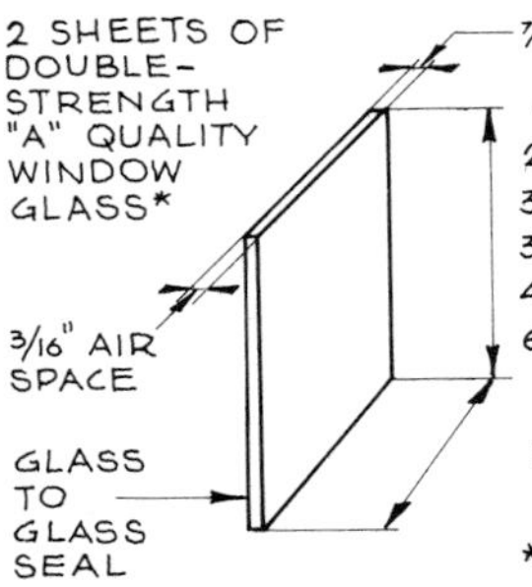

Thickness	7/16"	7/16"	7/16"	7/16"	7/16"	7/16"	7/16"	7/16"	7/16"	7/16"	7/16"	7/16"	7/16"	7/16"	7/16"	7/16"	7/16"	7/16"	7/16"	7/16"
Height	24" 32" 36" 48" 60"	24 5/8" 30 13/16" 36 13/16" 49" 61 3/16"	15"	24" 32" 36" 48" 60"	24 5/8 30 13/16" 36 13/16" 49" 61 3/16"	18" 55 9/16"	24" 32" 36" 48" 60"	15 1/4"	14 1/4" 18 1/4" 22 1/4" 30 1/4"	14" 18" 22"	16"	22 11/16"	14" 18" 22" 24" 30"	14 1/4" 18 1/4" 22 1/4" 30 1/4"	14" 18" 22" 30"	14 1/4" 18 1/4" 22 1/4" 30 1/4"	22 1/2"	14" 18" 22" 30"	14 1/4" 18 1/4" 22 1/4" 30 1/4"	25 1/2"
Width	16"	16 9/16"	19"	20"	21 1/16"	22"	24"	24 1/4"	27 1/4"	28"	32"	34 11/16"	36"	36 5/8"	39"	39 5/8"	42 1/2"	44"	44 5/8	45 1/2"

* ALWAYS LIST WIDTH DIMENSION FIRST AND INSTALL WIDTH PARALLEL TO FLOOR

Figure G32 Stock sizes of various types of clear insulating glass.

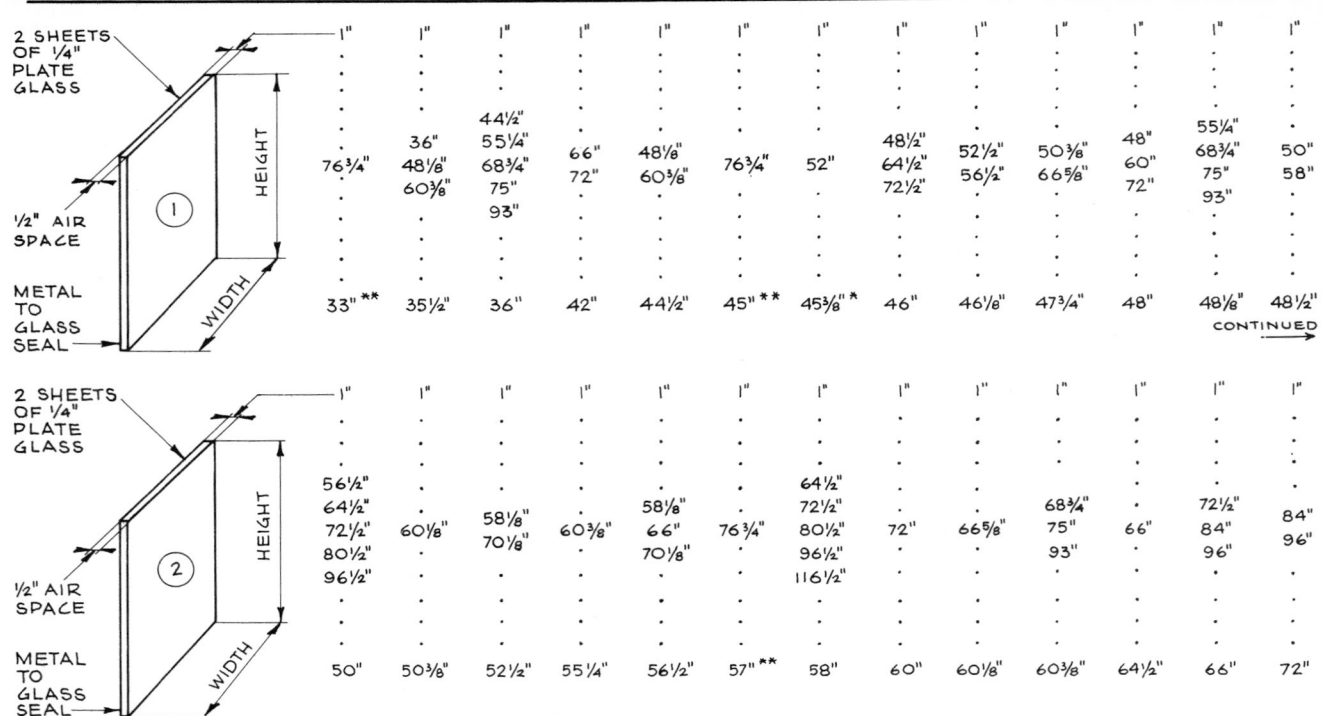

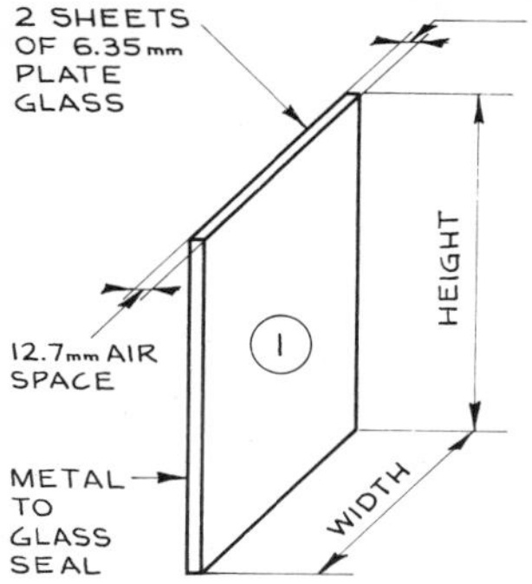

Thickness	Heights	Width
25.4mm	1.949 M	.8382 M**
25.4mm	.9144 M, 1.222 M, 1.534 M	.9017 M
25.4mm	1.130 M, 1.403 M, 1.746 M, 1.905 M, 2.362 M	.9144 M
25.4mm	1.676 M, 1.829 M	1.067 M
25.4mm	1.222 M, 1.534 M	1.130 M
25.4mm	1.949 M	1.143 M**
25.4mm	1.321 M	1.153 M*
25.4mm	1.232 M, 1.638 M, 1.842 M	1.168 M
25.4mm	1.334 M, 1.435 M	1.172 M
25.4mm	1.280 M, 1.692 M	1.213 M
25.4mm	1.219 M, 1.524 M, 1.829 M	1.219 M
25.4mm	1.403 M, 1.746 M, 1.905 M, 2.362 M	1.222 M
25.4mm	1.270 M, 1.473 M	1.232 M

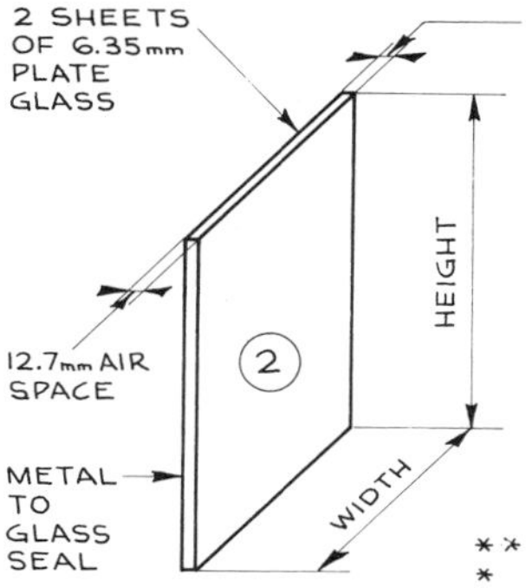

Thickness	Heights	Width
25.4mm	1.435 M, 1.638 M, 1.842 M, 2.045 M, 2.451 M	1.270 M
25.4mm	1.527 M	1.280 M
25.4mm	1.476 M, 1.782 M	1.334 M
25.4mm	1.534 M	1.403 M
25.4mm	1.476 M, 1.676 M, 1.781 M	1.435 M
25.4mm	1.949 M	1.448 M**
25.4mm	1.638 M, 1.842 M, 2.045 M, 2.451 M, 2.959 M	1.473 M
25.4mm	1.829 M	1.524 M
25.4mm	1.692 M	1.527 M
25.4mm	1.746 M, 1.905 M, 2.362 M	1.534 M
25.4mm	1.676 M	1.638 M
25.4mm	1.842 M, 2.134 M, 2.438 M	1.676 M
25.4mm	2.134 M, 2.438 M	1.829 M

** STOCK SIZES FOR SLIDING DOORS
* ALSO AVAILABLE - MADE OF 3/16" SHEET GLASS WITH 1/2" AIR SPACE

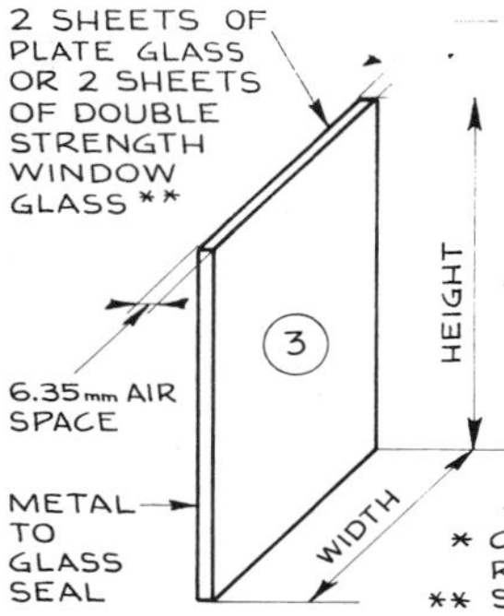

Thickness	Heights	Width
12.7mm	.894 M	.405 M
12.7mm	.926 M	.403 M
12.7mm	1.21 M, 1.24 M	.405 M
12.7mm	.926 M, 1.24 M	.437 M
12.7mm	.270 M, .406 M, .559 M, .914 M, 1.22 M, 1.52 M	.508 M
12.7mm	.270 M, .295 M, .346 M, .559 M, .610 M, .711 M*, .813 M*, .914 M, 1.22 M	.610 M*
12.7mm	.295 M, .346 M, .457 M, .610 M*, .711 M	.711 M*
12.7mm	.295 M, .346 M, .406 M, .610 M*, .711 M	.813 M*
12.7mm	.533 M	.838 M
12.7mm	.576 M	.881 M
12.7mm	.295 M, .346 M, .457 M, .610 M*, .711 M	.914 M*
12.7mm	.295 M, .346 M, .610 M, .711 M	1.02 M
12.7mm	.572 M	1.08 M
12.7mm	.295 M, .346 M, .406 M, .457 M, .610 M, .711 M	1.12 M
12.7mm	.648 M	1.16 M

* CONSIDERED STANDARD SIZES AS LISTED EVEN THOUGH WIDTH AND HEIGHT DIMENSIONS ARE REVERSED
** SHOULD NOT BE TURNED. WIDTH DIMENSIONS ARE LISTED FIRST AND GLASS MUST BE INSTALLED WITH WIDTH PARALLEL TO FLOOR

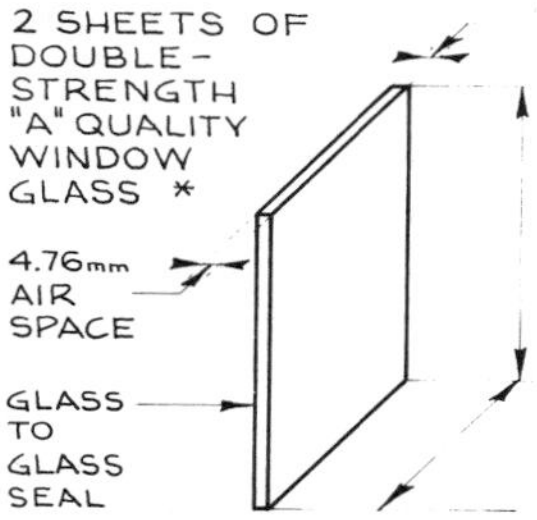

Thickness	Heights	Width
11.1mm	.6096 M, .8128 M, .9144 M, 1.219 M, 1.524 M	.4064 M
11.1mm	.6255 M, .7826 M, .9350 M, 1.245 M, 1.554 M	.4207 M
11.1mm	.3810 M	.4826 M
11.1mm	.6096 M, .8128 M, .9144 M, 1.219 M, 1.524 M	.5080 M
11.1mm	.6255 M, .7826 M, .9350 M, 1.245 M, 1.554 M	.5350 M
11.1mm	.4572 M, 1.411 M	.5588 M
11.1mm	.6096 M, .8128 M, .9144 M, 1.219 M, 1.524 M	.6096 M
11.1mm	.3874 M	.6160 M
11.1mm	.3620 M, .4636 M, .5652 M, .7684 M	.6922 M
11.1mm	.3556 M, .4572 M, .5588 M	.7112 M

CONTINUED →

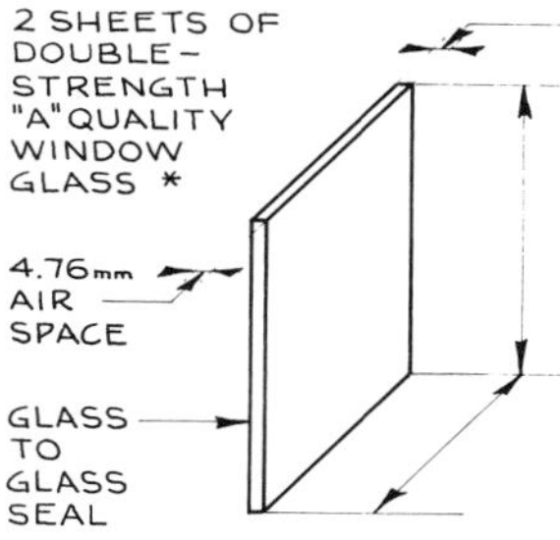

Thickness	Heights	Width
11.1mm	.4064 M	.8128 M
11.1mm	.5763 M	.8811 M
11.1mm	.3556 M, .4572 M, .5588 M, .6096 M, .7620 M	.9144 M
11.1mm	.3620 M, .4636 M, .5652 M, .7684 M	.9303 M
11.1mm	.3556 M, .4572 M, .5588 M, .7620 M	.9906 M
11.1mm	.3620 M, .4636 M, .5652 M, .7684 M	1.006 M
11.1mm	.5715 M	1.080 M
11.1mm	.3556 M, .4572 M, .5588 M, .7620 M	1.118 M
11.1mm	.3620 M, .4636 M, .5652 M, .7684 M	1.133 M
11.1mm	.6477 M	1.156 M

* ALWAYS LIST WIDTH DIMENSION FIRST AND INSTALL WIDTH PARALLEL TO FLOOR

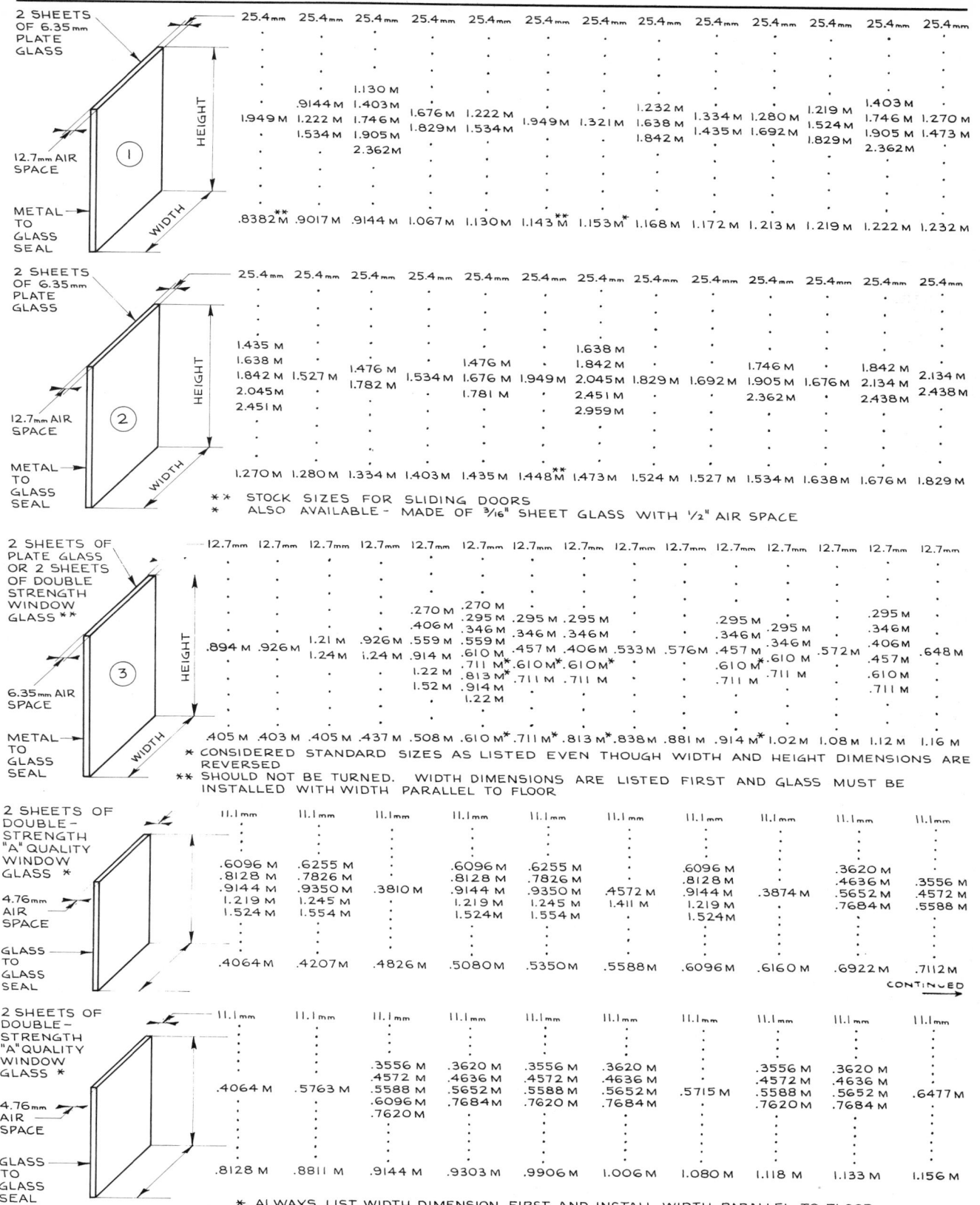

Figure G32 Stock sizes (metric) of various types of clear insulating glass (*continued*).

Table G24 Typical characteristics of insulating glass with glass-to-glass seal[a]

Type of clear glass	Total thickness in.	Total thickness mm	Maximum area ft^2	Maximum area m^2	Maximum dimension ft	Maximum dimension m	Weight lb/ft^2	Weight kg/m^2	Transmittance Visual (percent)	Transmittance Solar energy (percent)	*U* value Btu/h · ft^2 · °F	*U* value W/m^2 · °C	Heat gain Btu/h · ft	Heat gain W/m
Single-strength A window	$\frac{3}{8}$	9.53	10	0.929	5.83	1.778	2.4	11.72	80	76	0.54	3.07	194	612.07
Double-strength A window	$\frac{7}{16}$	11.11	24	2.23	5.83	1.778	3.2	15.62	80	72	0.54	3.07	188	1539.64

[a]Minimum size available is 1.33 × 1.17 ft (0.406 × 0.356 m).

Table G25 Typical characteristics of clear and tinted combinations of insulation glass with metal-to-glass seal[a]

Type of glass: Exterior	Type of glass: Interior	Total thickness $\frac{1}{4}$ in. (6.35 mm) in.	Total thickness $\frac{1}{4}$ in. (6.35 mm) mm	Total thickness $\frac{1}{2}$ in. (12.70 mm) in.	Total thickness $\frac{1}{2}$ in. (12.70 mm) mm	Maximum area[b] ft^2	Maximum area[b] m^2	Weight lb/ft^2	Weight kg/m^2	Transmittance Visual (percent)	Transmittance Solar energy (percent)	*U* value Btu/h · ft^2 · °F	*U* value W/m^2 · °C	Heat gain Btu/h · ft^2	Heat gain W/m^2
$\frac{1}{8}$-in. (3.18-mm) clear sheet	$\frac{1}{8}$-in. (3.18-mm) clear sheet	$\frac{9}{16}$	14.29	$\frac{13}{16}$	20.64	15	1.39	3.25	15.87	81	69	0.61	3.52	185	583.67
$\frac{1}{8}$-in. (3.18-mm) clear float or plate	$\frac{1}{8}$-in. (3.18-mm) clear float or plate	$\frac{9}{16}$	14.29	$\frac{13}{16}$	10.64	15	1.39	3.25	15.87	81	69	0.62	3.52	185	583.67
$\frac{3}{16}$-in. (4.76-mm) clear heavy sheet	$\frac{3}{16}$-in. (4.76-mm) clear heavy sheet	$\frac{11}{16}$	17.46	$\frac{15}{16}$	23.81	30	2.78	5.25	25.63	80	65	0.61	3.46	180	567.90
$\frac{7}{32}$-in. (5.49-mm) gray tinted heavy sheet	$\frac{3}{16}$-in. (4.76-mm) clear heavy sheet	$\frac{11}{16}$	17.46	$\frac{15}{16}$	23.81	30	2.78	5.25	25.63	12	34	0.54	3.07	113	356.52
$\frac{1}{8}$-in. (3.18-mm) gray tinted sheet	$\frac{1}{8}$-in. (3.18-mm) clear sheet	$\frac{9}{16}$	14.29	$\frac{13}{16}$	20.64	15	1.39	3.25	15.87	28	48	0.62	3.52	145	457.48
$\frac{1}{4}$-in. (6.35-mm) clear float or plate	$\frac{1}{4}$-in. (6.35-mm) clear float or plate	$\frac{13}{16}$	20.64	$1\frac{1}{16}$	26.99	70	6.50	6.50	31.74	77	59	0.54	3.07	170	537.35
$\frac{1}{4}$-in. (6.35 mm) tinted bronze float or plate	$\frac{1}{4}$-in. (6.35-mm) clear float or plate	$\frac{13}{16}$	20.64	$1\frac{1}{16}$	26.99	50[b]	4.65	6.50	31.74	45	35	0.54	3.07	115	362.83
$\frac{1}{4}$-in. (6.35-mm) tinted gray float or plate	$\frac{1}{4}$-in. (6.35-mm) clear float or plate	$\frac{13}{16}$	20.64	$1\frac{1}{16}$	26.99	50[b]	4.65	6.50	31.74	37	35	0.54	3.07	115	362.83
$\frac{1}{4}$-in. (6.35-mm) tinted blue-green float or plate	$\frac{1}{4}$-in. (6.35-mm) clear float or plate	$\frac{13}{16}$	20.64	$1\frac{1}{16}$	26.99	50[b]	4.65	6.50	31.74	65	35	0.54	3.07	115	362.83

[a]Insulating glass using tempered glass is limited to a maximum size of 40 ft^2 (3.72 m^2).
[b]Tinted insulating glass units over 50 ft^2 (4.65 m^2) must have tinted glass heat processed. Insulating glass using heat-processed glass is limited to a maximum size of 6 × 12 ft (1.829 × 3.658 m).

Table G26 **Typical types of insulating glass made with reflective glass, heat-processed glass, and tempered glass**[a]

Type of glass[b]	Location of clear, tinted, and reflective glass[c]	Transmittance[d] Visible (percent)	Transmittance[d] Solar energy (percent)	U value[d] Btu/h · ft² · °F	U value[d] W/m² · °C	Heat gain[d] Btu/h · ft²	Heat gain[d] W/m²
Clear float or plate	Clear with reflective coating on outside face of exterior glass, with clear inside glass	35	34	0.55	3.12	1.07	337.59
Bronze and clear float or plate	Bronze with reflective coating on outside face of exterior glass, with clear inside glass	20	20	0.55	3.12	75	236.63
Gray and clear float or plate	Gray with reflective coating on outside face of exterior glass, with clear inside glass	19	22	0.55	3.12	79	249.25
Clear float or plate	Clear with reflective coating on outside or inside face of exterior glass, with clear inside glass	20	10	0.35	1.99	40	126.20
Clear float or plate	Clear outside glass with reflective coating on face within air space of clear inside glass	20	10	0.35	1.99	65	205.08
Bronze and clear float or plate	Bronze outside glass with reflective coating on face within air space of clear inside glass	12	7	0.35	1.99	60	189.30
Gray with clear float or plate	Gray outside glass with reflective coating on face within air space of clear inside glass	10	6	0.35	1.99	60	189.30
Bronze and clear heat-processed or tempered float or plate	Bronze with reflective coating on exterior face of outside glass, with clear inside glass	25	26	0.38	2.16	85	268.18

Table G26 Typical types of insulating glass made with reflective glass, heat-processed glass, and tempered glass[a] (*continued*)

Type of glass[b]	Location of clear, tinted, and reflective glass[c]	Transmittance[d]: Visible (percent)	Transmittance[d]: Solar energy (percent)	U value[d]: Btu/h · ft² · °F	U value[d]: W/m² · °C	Heat gain[d]: Btu/h · ft²	Heat gain[d]: W/m²
Gray and clear heat-processed or tempered float or plate	Gray with reflective coating on exterior face of outside glass, with clear inside glass	21	24	0.37	2.01	82	258.71

[a]These types of insulating glass are all $1\frac{1}{16}$ in. (26.99 mm) thick, weigh 6.5 lb/ft² (31.74 kg/m²), and are available up to a maximum size of 6 X 12 ft (1.829 X 3.658 m).
[b]The thickness of all the various types of glass used to manufacture these types of insulating glass is $\frac{1}{4}$ in. (6.35 mm).
[c]The reflective surface is available in three locations: either the exterior or the interior face of the outside glass, or the face within the air space of the interior glass.
[d]These values vary according to the various types of reflective surface applied. Check with manufacturers on current types of reflective surfaces used in insulating glass, and obtain transmittance, *U*, and heat-gain values plus shading coefficients.

Table G27 Minimum clearances and rabbet depth for various types of insulating glass[a]

Type of insulating glass	Depth of protecting metal or wood frame: in.	Depth of protecting metal or wood frame: mm	Minimum clearance, glass to back of rabbet: in.	Minimum clearance, glass to back of rabbet: mm	Minimum clearance around edges: in.	Minimum clearance around edges: mm	Minimum depth of rabbet to hold glass: in.	Minimum depth of rabbet to hold glass: mm
Made of $\frac{1}{8}$-in. (3.175-mm) plate, float, or double-strength	$\frac{3}{8}$	9.53	$\frac{1}{8}$	3.18	$\frac{1}{8}$	3.18	$\frac{1}{2}$	12.70
Made of $\frac{3}{16}$-in. (4.763-mm) sheet	$\frac{3}{8}$	9.53	$\frac{1}{8}$	3.18	$\frac{1}{4}$	6.35	$\frac{5}{8}$	15.88
Made of $\frac{1}{4}$-in. (6.35-mm) float or plate	$\frac{3}{8}$	9.53	$\frac{1}{8}$	3.18	$\frac{1}{4}$	6.35	$\frac{3}{4}$	19.05
Made of $\frac{7}{32}$- or $\frac{1}{4}$-in. (5.56- or 6.35-mm) patterned	$\frac{1}{2}$	12.70	$\frac{1}{8}$	3.18	$\frac{1}{4}$	6.35	$\frac{3}{4}$	19.05

[a]Check with manufacturers on clearances, gaskets, and rabbet depth for the particular type of insulating glass selected in order to obtain a guarantee on percentage of breakage during installation.

Application

When designing with insulating glass, only the exterior sheet of glass should be used for calculating wind loads and strength. Required minimum clearances and rabbet depths are given in Table *G27*.

Condensed Checklist

1. The thickness of glass required for a given size of opening is controlled by wind loads.
2. Openings into which insulating glass is to be installed must be square and plumb. It is necessary to check that they are correct in size to meet the clearances necessary for the type of insulating glass being installed, because insulating glass cannot be changed in size once it has been manufactured.
3. Insulating glass must not have any areas covered with paint or paper because this can cause a heat trap that may result in breakage.
4. There can be no direct contact between the insulating glass and the frame into which it is installed.

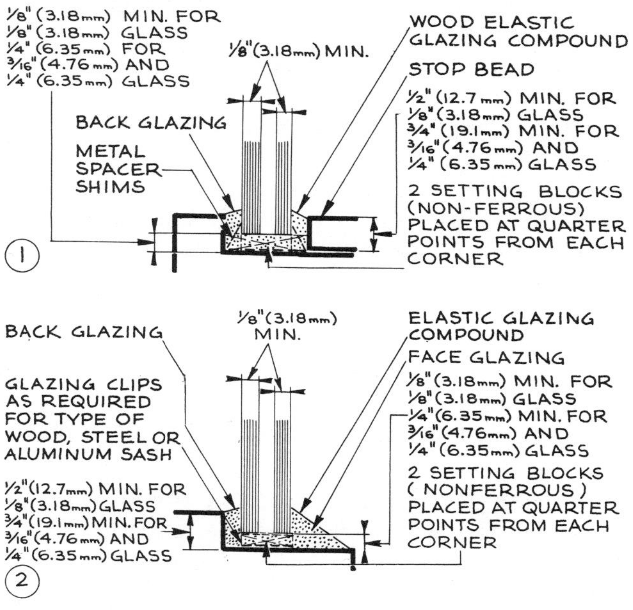

Figure G33 Insulating glass installed in wood, steel, or aluminum with glazing bead (1) and with face glazing (2).

5. Details of installation vary with the type of material into which insulating glass is to be set. This includes type of glazing material, gaskets, rope, tape, glazing clips and spacing shims, necessary setting blocks, and clearances for back-puttying, glazing compounds, or sealants. (*See* Figures *G33* to *G35*.)
6. The glazing compound must be a nonhardening type that does not contain any materials which will attack the metal-to-glass seal of the insulating glass. Putty should never be used.
7. Tinted and reflective glass and their various combinations should always be checked for the correct method of installation in reference to which face of the insulating glass is to face the exterior (*see* Plastics: Gaskets, Shims, Waterstops, etc.).
8. Insulating glass made with double-strength window (sheet) glass should be placed in the opening with the draw or wave distortions (manufacturer's width dimension) parallel with the floor.
9. Glazing must never be done at temperatures below 40°F (4.44°C) unless precautions are taken to prevent moisture from condensing in the rabbets where the glass is to be installed.
10. Operating windows or doors must not be handled or opened until the glazing compound, sealants, gaskets, tapes, and other setting accessories have set completely or been permanently installed.
11. Various codes (local, state, and also those of the Fire Underwriters, insurance companies, labor departments, and federal government) should be checked for requirements and limitations.
12. The insulating glass chosen must meet all other requirements in regard to transparency, translucency, light diffusion, thermal insulation, and strength for its end use in the design.
13. U values vary with the type of glass. Therefore one should always check the exact U values of the type of insulation glass chosen or contemplated for a building.
14. One should check all governing codes for safety glass requirements in relation to the size of glass areas that are to be constructed. Glass beyond a certain size must be safety glass, and all types of glass in hazardous areas must be safety glass.

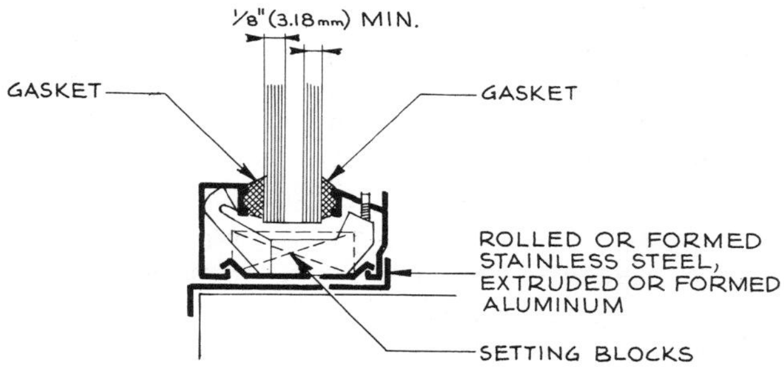

Figure G34 Insulating glass installed in typical store front.

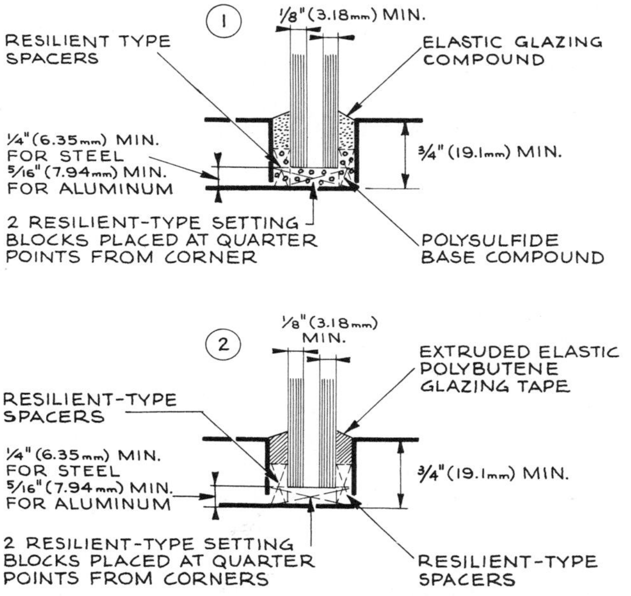

Figure G35 Insulating glass in steel or aluminum with multiple compound glazing (1) and with tape glazing (2).

15. It should be specified that a full-size mock-up of the curtain wall system selected, fabricated with the type of glass selected, must be submitted for checking the color harmony of the various glass types and for testing under the climatic conditions of the proposed building.
16. One should always check the manufacturer for installation details of the type of insulating glass selected, as many of these glasses are prone to breakage as a result of expansion, contraction, internal stresses, and extreme temperature changes within the insulating glass due to solar heat, wind, and type of frames.
17. It is always advisable to avoid a rabbet within masonry or concrete because the temperature between this type of rabbet and the heat absorbed by the glass will cause internal stresses within the insulating glass and cause breakage.

Conditions Favorable to the Use of Insulating Glass

1. Where thermal insulation for both heating and air conditioning are required.
2. Where relative humidity within the building is high and condensation must be avoided.

Conditions Unfavorable to the Use of Insulating Glass

1. Where fire resistance is necessary.
2. Where insulating glass might be placed in tension or other stress conducive to breakage.

History and Manufacture

The idea of insulating glass (double glass) was conceived by Haven, an engineer, in 1930. By 1937 the problem of the metal-to-glass seal was solved, and insulating glass became an accepted new architectural material. Further research led to a glass-to-glass seal, developed and manufactured in 1953. The manufacturing process for insulating glass is simple in principle but complex in actual production. Two sheets of glass are thoroughly washed, cleaned, and dried; then the two pieces of glass are hermetically sealed with either a glass-to-metal or a glass-to-glass seal, and the air which is captured between the sheets of glass is washed, cleaned, and dried. The entire process is done at atmospheric pressure; no vacuum is created.

With the great demand for air conditioning within buildings, the problem of solar heat gain through glass became an important factor in designing air-conditioning systems for buildings. The glass manufacturers first developed glare-reducing and heat-absorbing glass to combat this problem. Then, by adapting the one-way mirror for this purpose, they developed a whole new group of glasses, including tinted gray, bronze and blue-green glasses (glare-reducing and heat-absorbing) and the application of a metallic oxide reflective coating to one face of a glass (refractive glass). Now it is an easy matter to combine these various types of glass in the manufacture of insulating glass to obtain thermal insulation, glare reduction, and less solar heat gain.

GLASS, LAMINATED

Physical and Chemical Properties

Laminated glass is composed of two or more layers of either window or float (sheet) glass or polished plate or float glass, either clear, tinted or reflective, with a layer or more of a transparent or pigmented plastic sandwiched between the glass layers. A vinyl plastic, such as plasticized polyvinyl butyral resin, 0.015 to 0.025 in. (0.381 to 0.635 mm) thick, is generally used. The composition of the glass is the same as for plate, float, or window glass except that only the highest quality of each type of glass is made into laminated glass. When this type of glass receives a sharp impact and breaks, the adhesive character of the plastic holds the pieces of glass and prevents the sharp fragments from shattering and flying about. When four or more layers of glass are laminated with three or more layers of plastic, the product is a bullet-resisting glass.

Commercial Forms. Laminated glass is available in thicknesses ranging from $\frac{5}{32}$ to 3 in. (3.92 to 76.2 mm). It is made in various types of float, sheet, and plate glass and in a limited range of sizes.

Types and Uses

Two types of laminated glass are used in construction: (1) safety glass, composed of two layers of glass and one of plastic, and (2) bullet-resisting glass, composed of a minimum of four layers of glass and three layers of plastic.

Safety Glass. Safety glass is available with the plastic clear or pigmented, and with the glass either clear glass or heat-absorbing glass and tinted glass (these are fully covered under the specific headings). Safety glass is also manufactured of clear or tinted glass with translucent vinyl, which allows up to 63% light transmission but assures complete privacy. Table *G28* lists types, thick-

Table G28 Types, thicknesses, tolerances, weights, and maximum sizes of laminated safety glass

	Thickness							
	Actual		Tolerance		Weight		Maximum and standard sizes	
Type of glass	in.	mm	in.	mm	lb/ft²	kg/m²	ft and ft²	m and m²
Picture	$\frac{5}{32}$	3.92	$-\frac{1}{32}$	-0.78	1.92	7.37	7 ft²	0.650 m²
Single strength	$\frac{7}{32}$	5.49	$\pm\frac{1}{32}$	±0.78	2.42 to 2.49	11.81 to 12.16	15 ft²	1.394 m²
Combination of single and double strength	$\frac{15}{64}$	5.88	$\pm\frac{1}{32}$	±0.78	2.86 to 2.91	13.96 to 14.19	15 ft²	1.394 m²
Double-strength	$\frac{1}{4}$	6.35	$\pm\frac{1}{32}$	±0.78	3.24 to 3.32	15.82 to 16.21	15 ft²	1.394 m²
$\frac{1}{8}$-in. (3.18-mm) float or plate	$\frac{1}{4}$	6.35	$\pm\frac{1}{32}$	±0.78	3.16 to 3.19	15.43 to 15.57	5 × 7.5	1.524 × 2.286
Combination of $\frac{1}{8}$-in. (3.18-mm) and $\frac{1}{4}$-in. (6.35-mm) float or plate	$\frac{3}{8}$	9.53	$\pm\frac{1}{32}$	±0.78	4.88		Standard 2.5 × 6	Standard 0.762 × 1.829 Other sizes available
Combinations of $\frac{1}{4}$-in. (6.36-mm) float or plate glass with same or thicker glass	$\frac{1}{2}$	12.70	$\pm\frac{1}{16}$	±1.59	6.54	31.93		
	$\frac{5}{8}$	15.88	$\pm\frac{1}{16}$	±1.59	8.13	39.69		
	$\frac{3}{4}$	19.05	$\pm\frac{1}{16}$	±1.59	9.81	47.69		
	$\frac{7}{8}$	22.23	$\pm\frac{1}{16}$	±1.59	11.45	55.90		
	1	25.40	$\pm\frac{1}{16}$	±1.59	13.03	63.61		

nesses, weights, and maximum sizes of safety glass. The use of safety glass in construction is limited to transparent areas where strong impacts may be encountered and the hazard of flying glass must be avoided, for example, a press box in a baseball park.

Bullet-Resisting Glass. The types, thicknesses, weights, and maximum sizes of bullet-resisting glass are listed in Table *G29*. Bullet-resisting glass is finding new uses in places subject to robbery and for special observation windows in industrial operations, drive-in bank booths, and similar enclosed booths requiring protection.

Application

Condensed Checklist

1. Manufacturers of the glass and of the glazing compounds should always be consulted when installing safety or bullet-resisting glass, particularly on the exterior, for possible deteriorating effects of some glazing compounds on the layer or layers of plastic through chemical action.
2. Thicknesses and size limitations of safety glass or bullet-resisting glass should always be checked before finalized designs incorporating them are begun.
3. A check should always be made of insurance and underwriters' requirements for bullet-resisting glass to be installed in the design.
4. Local, municipal, and state codes and also the Fire Underwriters, insurance companies, labor departments, and federal government (Army, Navy, etc.) should also be checked for limitations and requirements.
5. The police departments and other security agencies should be consulted for the correct laminated glass to install for a particular end use.
6. The method of installing safety or bullet-resisting glass should be checked in relation to depth and size of rabbet and types of glazing materials.

Conditions Favorable to the Use of Safety and Bullet-Resisting Glass

1. Use safety glass in any area where the hazard of flying glass must be avoided.
2. Use bullet-resisting glass in places subject to robbery and in areas where exceptionally strong protection against possible flying objects at high speed or pressures is needed.

Table G29 Sizes, tolerances, and weights of laminated bullet-resisting glass

Thickness				Maximum sizes[a]							
Actual		Tolerance		Width		Length		Tolerances		Weight	
in.	mm	in.	mm	ft	m	ft	m	in.	mm	lb/in.^{2}	kg/m^2
$\frac{3}{4}$	19.05	$\pm\frac{1}{16}$	±1.59	3.83	1.168	7.00	2.134	0, $\frac{1}{8}$	0, 3.18	9.81	6897.12
$\frac{7}{8}$	22.23	−0, +$\frac{1}{8}$	−0, +3.18	3.83	1.168	7.00	2.134	0, $\frac{1}{8}$	0, 3.18	11.45	8050.15
1	25.40	−0, +$\frac{1}{8}$	−0, +3.18	3.83	1.168	7.00	2.134	0, $\frac{1}{8}$	0, 3.18	13.08	9196.16
$1\frac{1}{8}$ [b]	28.58	−0, +$\frac{1}{16}$	−0, +1.59	3.83	1.168	7.00	2.134	0, $\frac{1}{8}$	0, 3.18	14.72	10349.19
$1\frac{3}{16}$	29.16	$\pm\frac{1}{16}$	±1.59	3.83	1.168	7.00	2.134	0, $\frac{1}{8}$	0, 3.18	15.51	10904.62
$1\frac{1}{2}$ [b]	38.10	−0, +$\frac{3}{16}$	−0, +4.76	3.67	1.118	7.00	2.134	0, $\frac{1}{8}$	0, 3.18	19.62	13794.23
				3.83	1.168	6.67	2.032				
$1\frac{9}{16}$	38.69	−0, +$\frac{3}{16}$	−0, +4.76	3.50	1.067	7.00	2.134	0, $\frac{1}{8}$	0, 3.18	20.44	14370.75
				3.83	1.168	6.33	1.930				
2[b]	50.80	−0, +$\frac{3}{16}$	−0, +4.76	2.67	0.813	7.00	2.134	0, $\frac{1}{8}$	0, 3.18	26.16	18392.31
				3.83	1.168	4.83	1.473				
$2\frac{3}{32}$	53.15	−0, +$\frac{3}{16}$	−0, +4.76	2.50	0.762	7.00	2.134	0, $\frac{3}{16}$	0, 4.76	27.39	19257.09
				3.83	1.168	4.67	1.422				
$2\frac{1}{2}$	63.50	$-\frac{1}{16}$, +$\frac{3}{16}$	−1.58, +4.76	2.16	0.660	7.00	2.134	0, $\frac{1}{4}$	0, 6.35	32.70	22990.30
				3.83	1.168	3.83	1.168				
3	76.20	$\frac{1}{16}$, $\frac{3}{16}$	1.58, 4.76	1.67	0.508	7.00	2.134	0, $\frac{1}{4}$	0, 6.35	39.24	27488.47
				3.83	1.168	3.00	0.914				

[a] These are standard sizes; other sizes are available but must be specially manufactured.
[b] Meets Underwriters' specifications for indoor and outdoor installation in robber-resisting enclosures.

History and Manufacture

Laminated glass was invented by Benedictus of France, although Wood of England had anticipated it and received a patent in 1905. Laminated glass was evolved as a result of developments within the automobile industry and to a lesser extent the plastic industry. The tremendous demand for shatterproof glass for the closed automobile (today more than 90% of the total production of automobiles are closed cars) stimulated the glass industry into producing laminated glass. The first laminated glass failed in time as the available plastics discolored and weakened with age. It was not until the plastics industry developed a plastic which met all requirements that the laminated glass of today became possible.

In the process of manufacturing laminated glass, the glass is first carefully selected, ground, polished, and washed. Then it is formed and cut to exact size and shape. Next, it is assembled with the in-between layers of plastic, and finally the complete transparent sandwich is formed under heat and pressure.

GLASS, MIRRORS

Physical and Chemical Properties

Mirrors are made from polished plate, float, window, sheet, and picture glass. The reflecting surface is a thin coat of metal, generally silver, gold, copper, bronze, or chromium, applied to one side of the glass. For special mirrors, lead, aluminum, platinum, rhodium, or other metals may be used. The metal film can be semitransparent or opaque and left unprotected or protected with a coat of shellac, varnish, paint, or metal (usually copper). The silvering metals are deposited on the glass by various means—chemical deposition, amalgams, high-voltage discharge between electrodes in a semivacuum, evaporation in a semivacuum, or painting with organic sulfur compounds of metals.

Mirrors are available in all thicknesses and in sizes up to 6 × 12 ft (1.829 × 3.658 m). Larger sizes can be made to order. One-way mirrors are available made from clear, gray, or bronze glass, tempered glass, and also laminated safety glass.

Types and Uses

The two principal types of mirrors used in construction are float and polished plate glass mirrors and one-way mirrors. By special order, both types can be obtained in circular form to meet various radii. Float or plate glass for mirrors is usually one of the following three kinds: silvering quality, mirror glazing quality, and glazing quality float or polished plate glass, either clear or tinted. Tempered plate glass is also used. The quality of the mirror's reflection is controlled by the type of glass used.

Float and Polished Plate Glass Mirrors. Polished plate glass mirrors are made with a maximum advisable size of 6 X 12 ft (1.829 X 3.658 m), in thicknesses ranging from $\frac{1}{8}$ to $1\frac{1}{4}$ in (3.18 to 31.75 mm), depending on the end use of the mirror. They are silvered with silver, gold, or gun metal (bronze), and the silvering is protected by either copper plating or paint.

Door mirrors come in standard sizes, framed or unframed, 5.67 ft (1.727 m) high by 8 in. (0.204 m) less than door width for doors 2 to 2.67 ft (0.610 to 0.813 m).

One-Way Mirrors. The one-way mirror has a light transmission of approximately 5 to 11% and a light reflection of 45 to 55% on the reflective side. The metal film is a chromium alloy applied by evaporation in a semivacuum. When such a mirror is installed between two areas, one of which is brightly lighted (the reflective side) and the other dimly lighted (the opposite side), on the dimly lighted side it appears as a transparent piece of glass. Its major use is for areas where observation and research can be carried out without the observer being visible.

One-way mirrors are made $\frac{1}{8}$ in. (3.18 mm) thick, with a maximum size of 2.5 X 5 ft (0.762 X 1.524 m). Larger sizes are available on special order. All one-way mirrors have nonmirrored spots $\frac{3}{16}$ X 1 in. (4.76 X 25.4 mm) at top and bottom edges (where glass was held in position during manufacture). These should be concealed by the method of installation.

The one-way mirror concept led to the use of this type of glass for construction glazing for reflecting solar heat and at the same time reducing glare without sacrificing transparency to any extent. This use of one-way mirrors for glazing in buildings has led to the development of reflective glass for this construction purpose.

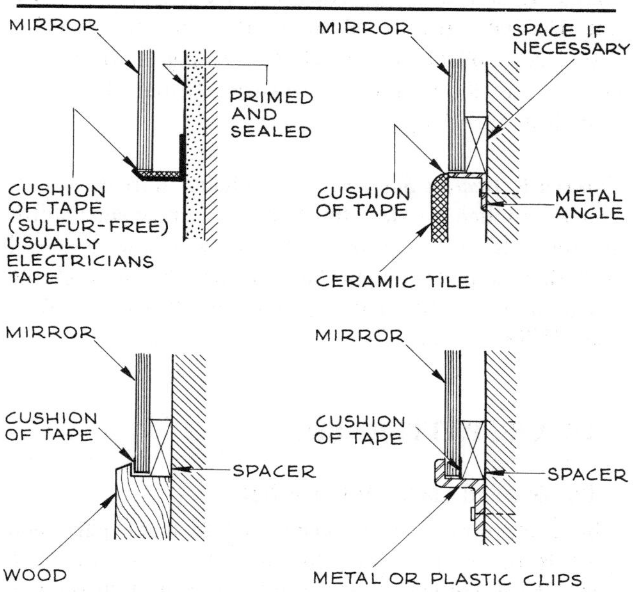

Figure G36 Common methods of supporting the weight of mirrors.

Application

Condensed Checklist

1. Mirrors should be made of a type of glass that meets the reflection requirements for its end use.
2. The type of silvering and the means for its protection must be considered in relation to the method of installation being used. Generally only mirrors with silvering that is electroplated with a protective layer of copper should be installed with mastic. Manufacturers of mirrors should be consulted for the correct mastic, methods, and requirements.
3. Proper installation of mirrors requires that in almost all cases the weight of the mirror be supported at the bottom (*see* Figure *G36*).
4. For mastic-type installation, 6 in. (12.7 mm) spots of mastic that do not cover more than 25% of the back surface of the mirror should be used.
5. Wall surfaces upon which mirrors are to be applied must be primed and sealed, smooth and firm, and thoroughly dry. There must be open space behind the mirror for ventilation.
6. To install mirrors with rosettes, a check should be made of the correct diameter of holes to be drilled in the mirror. Usually a $\frac{1}{4}$-in. (6.35-mm) diameter is used for rosette screws, and holes are drilled a minimum of 2 in. (50.8 mm) from the edge of the mirror. The mirror manufacturer should always be consulted for the quantity of rosettes necessary for the size of mirror to be installed. (*See* Figure *G37*.)

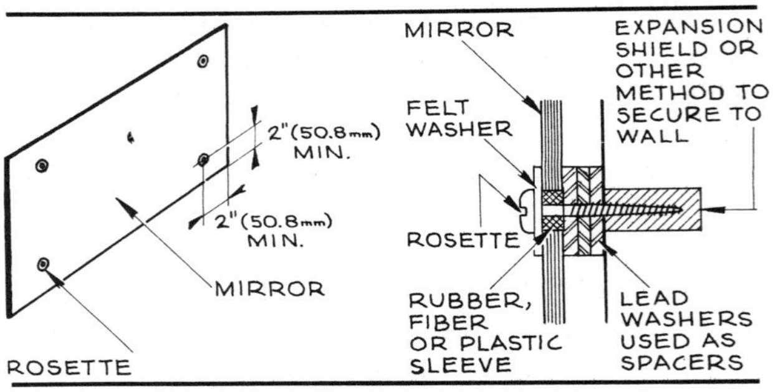

Figure G37 Mirrors installed with rosettes.

7. One-way mirrors can be installed in the same manner as plate glass mirrors or as window, heavy sheet, or plate glass. The metallic film is resistant to glazing compounds, handling, weathering and cleaning.

Conditions Favorable to the Use of Mirrors

1. In areas where reflection is important, either to increase the visual size of space or for reflective use by the occupants.
2. One-way mirrors are used for hidden observation, research, protection, etc.

Conditions Unfavorable to the Use of Mirrors

1. Where moisture may accumulate and be retained behind the mirror.
2. Where silvering is subject to chemical fumes.

History and Manufacture

In ancient times mirrors consisted of polished metal, usually bronze. Some glass mirrors backed with tin and silver have also been found. In the Middle Ages, although the method of backing glass with thin sheets of metal was well known, mirrors were almost exclusively made of steel, silver, and gold. Some glass mirrors were manufactured in Nuremberg in 1373, and small convex mirrors were made before 1500. Mirror manufacture as we know it today started in Venice. In 1507 two residents of Murano, an island in the Venetian Lagoon, obtained an exclusive privilege to manufacture mirrors for a period of 20 years. In 1564 the mirror makers of Venice allied themselves into a corporation, and soon thereafter glass mirrors replaced metal mirrors.

Early Amalgam Method. The mirrors were made by first blowing a cylinder of glass, slitting it, and flattening it on a piece of stone. The glass was then carefully polished. Adjacent, on a flat surface, a sheet of tin foil was laid on a blanket and a pool of mercury was poured on the tin foil; then a sheet of paper was laid over the mercury. Next the polished glass was allowed to fall slowly onto this; just before the glass reached the surface, the paper was drawn off so that a clean surface of mercury came in contact with the glass. Heavy weights were then placed on the glass, and the excess mercury was squeezed out, after which the amalgam (liquid alloy) of tin and mercury adhered to the glass.

The invention and production of plate glass in 1691 in France marked another advance in mirror manufacture. All mirrors were made in this manner until 1835, when Liebig discovered the chemical process of coating glass with metallic silver (now called silvering).

Silvering. In this process, silver-ammonia compounds are chemically reduced to metallic silver. Today most mirrors are made this way. Many are manufactured by an automatic process in which the cleaned glass passes along on a conveyor into an enclosed area where the necessary solutions meet in a spray and deposit the silver immediately on the glass. Many mirrors, however, are still silvered chemically by hand. The silvering can be protected with a coating of shellac, varnish, or paint. For almost permanent protection, an electroplated layer of copper can be applied on the silver.

One-Way Mirror Manufacture. What are known as one-way mirrors are manufactured in chambers where a partial vacuum is created (one ten-millionth of the air remains). Chromium alloy particles are suspended on a filament, and when an electric current passes through the filament, the metal evaporates and the metallic molecules collect on the glass, forming a strong, adherent, metallic film. (*See also* Glass: Tinted, Reflective, and Heat-Absorbing.)

Painted Reflective Surface. A metallic mirror surface can be applied to glass by painting with organic sulfur compounds of metals in a medium such as lavender oil and then applying slight heat, which decomposes the compounds and eliminates the medium, leaving a mirror-like metallic finish.

GLASS, PATTERNED

Physical and Chemical Properties

The composition of the glass used to make patterned glass is the same as that for float, window, and plate glass. Patterned glass is semitransparent, with distinctive geometric or linear designs on one or both sides; it is sometimes also sandblasted or etched. It diffuses transmitted light and gives varying degrees of privacy. It can be tempered and thereby becomes 3 to 5 times stronger than similar non-heat-treated patterned glass. It is available both tempered and untempered in a wide variety of patterns, textures, and degrees of transparency. Some patterns are also available as wired glass.

Types and Uses

Patterned glass is made in a wide variety of designs and finishes. These are summarized in Table *G30*, which lists available sizes, weights, and finishes, and Table *G31*, which describes the finishes.

Table G30 Types, thicknesses, size limitations, weights, and finishes for patterned glass

	Thickness		Maximum sizes				Weight		
			Width		Length				
Type of patterned glass	in.	mm	ft	m	ft	m	lb/in.^{2}	kg/m^2	Finishes available[a]
Corrugated (shallow corrugations on both sides but perpendicular	$\frac{7}{32}$	5.49	4.50	1.372	11.33	3.454	1.70–2.75	1898.29–1933.44	Fire, sandblasted, special
	$\frac{7}{32}$	5.49	5.00	1.524	12.00	3.658	3.60	2531.05	
Floral	$\frac{1}{8}$	3.18	4.00	1.219	11.00	3.353	1.75–2.00	1230.37–1406.14	Fire, special
Granular, fine	$\frac{1}{8}$	3.18	4.00	1.219	11.00	3.353	1.75–2.00	1230.37–1406.14	Fire, frosted, special
	$\frac{7}{32}$	5.49	5.00	1.524	11.33	3.454	2.75–2.80	1933.44–1968.60	Fire, frosted, special
Granular, medium	$\frac{1}{8}$	3.18	4.00	1.219	11.00	3.353	1.75–2.00	1230.37–1406.14	Fire, frosted, special
	$\frac{7}{32}$	5.49	5.00	1.524	11.33	3.454	2.75–2.80	1933.44–1968.60	
	$\frac{3}{8}$	9.53	5.00	1.524	12.00	3.658	5.00	3515.35	
Granular, heavy	$\frac{1}{8}$	3.18	4.00	1.219	11.00	3.353	1.75–2.00	1230.37–1406.14	Fire, frosted, special
	$\frac{7}{32}$	5.49	5.00	1.524	11.33	3.454	2.75–2.80	1933.44–1968.60	
Hammered	$\frac{1}{8}$	3.18	4.00	1.219	11.00	3.353	1.75–2.00	1230.37–1406.14	Fire, frosted, special
	$\frac{7}{32}$	5.49	5.00	1.524	11.33	3.454	2.75–2.80	1933.44–1968.60	
	$\frac{3}{8}$	9.53	4.00	1.219	8.44	2.540	5.00	3515.35	
Pebbled	$\frac{1}{8}$	3.18	4.00	1.219	11.00	3.353	1.75	1230.37	Fire, special
	$\frac{7}{32}$	5.49	4.50	1.372	11.33	3.454	2.75	1933.44	
Ribbed, small	$\frac{1}{8}$	3.18	4.00	1.219	11.00	3.353	1.75–2.00	1230.37–1406.14	Fire, frosted, special textured
	$\frac{7}{32}$	5.49	5.00	1.524	11.33	3.454	2.75–2.80	1933.44–1968.60	
	$\frac{3}{8}$	9.53	4.00	1.219	8.44	2.540	5.00	3515.35	
Ribbed, large	$\frac{1}{8}$	3.18	4.00	1.219	11.00	3.353	1.75–2.00	1230.37–1460.14	Fire, frosted, special
	$\frac{7}{32}$	5.49	5.00	1.524	11.33	3.454	2.75–2.80	1933.44–1968.60	
Squares, small	$\frac{1}{8}$	3.18	4.00	1.219	11.00	3.353	1.75–2.00	1230.37–1460.14	Fire, frosted, special
	$\frac{7}{32}$	5.49	5.00	1.524	11.33	3.454	2.75–2.80	1933.44–1968.60	
Squares, large	$\frac{1}{8}$	3.18	4.00	1.219	11.00	3.353	1.75	1230.37	Fire, special
	$\frac{7}{32}$	5.49	4.00	1.219	11.33	3.454	2.75	1983.44	
Striped, large, angular stripes	$\frac{1}{8}$	3.18	4.50	1.372	11.00	3.353	1.75–2.00	1230.37–1460.14	Fire, frosted, special
	$\frac{7}{32}$	5.49	4.50	1.372	11.33	3.454	2.75–2.80	1933.44–1968.60	
Striped, small, concave stripes	$\frac{1}{8}$	3.18	4.50	1.372	11.00	3.353	1.75–2.00	1230.37–1460.14	Fire, frosted, special
	$\frac{7}{32}$	5.49	4.50	1.372	11.33	3.454	2.75–2.80	1933.33–1968.60	
Striped, large, convex stripes	$\frac{1}{8}$	3.18	4.50	1.372	11.00	3.353	1.75–2.00	1230.37–1460.14	Fire, frosted, textured, special
	$\frac{7}{32}$	5.49	5.00	1.524	11.33	3.454	2.75–2.80	1933.44–1968.60	

[a] Special finishes are trademarked and patented.

Table G31 Types of finishes for patterned glass

Types of finish[a]	How finish is produced	Advantages	Disadvantages	Applications
Fire	Natural finish after rolling	Clear surface		Both sides
Frosted	Etched with hydrofluoric acid	Reduces glare; increases obscurity and diffusion	Reduces strength of glass to impact and mechanical stress	Both sides usually
Sandblasted	Fine-sand-blown with compressed air	Increases obscurity and diffusion	Hard to clean; reduces light transmission; makes glass very fragile	One side usually
Special	Trademarked and patented	Increases obscurity and gives more uniform diffusion	Reduces strength of glass to impact and mechanical stress	One side usually
Textured	Stippled surface rolled onto one side	Increases obscurity		One side only

[a]Weather finish is applied to one side or both sides, depending on end use of glass.

Tempered, heat-absorbing, tinted, and insulating (double glass) types of patterned glass are covered fully under separate headings for the various types of glass.

The major uses for patterned glass are for borrowing light but maintaining privacy; for decorative screens, windows, and walls; and for diffusing light. Patterned glass may be obtained bent for special purposes and decorative effects.

Application

Condensed Checklist

1. The type of patterned glass chosen should give the degree and type of privacy and light diffusion that the design requires.
2. The type of finish chosen should be one that does not weaken the glass or make it too fragile for the required end use.
3. Sizes should always be checked for availability in the pattern desired.
4. Thickness should always be checked in relation to size limitation, wind loads, and end use.
5. The pattern must be installed in the direction recommended by the manufacturer.
6. Ease and methods of maintenance should be checked for the type of patterned glass and the type of finish being used.
7. The smooth side of patterned glass must always be toward the face putty.
8. The governing codes should always be checked for the type of glass required in hazardous areas.
9. Patterned glass manufacturers should be consulted to obtain current patterns, methods of application, and glazing materials.
10. Before selecting a type of patterned glass to be used in an insulating glass, one should determine from glass manufacturers what insulating glass-patterned glass combinations are available.

Conditions Favorable to the Use of Patterned Glass

1. Where the design requires light but privacy.
2. Where the design requires decorative semitransparent walls, glass areas, or screens.
3. Where the light, either natural or artificial, is to be diffused.

Conditions Unfavorable to the Use of Patterned Glass

1. In any area where fire resistance is required, unless wired patterned glass that meets requirements is used. Local, municipal, and state codes of the Fire Underwriters, insurance companies, labor departments, and federal government (Army, Navy, etc.) should be checked for fire resistance requirements and limitations.
2. In any areas where tensional stresses may be imposed on the patterned glass.

History and Manufacture

This glass was first made in quantity in 1890 after the development of a process whereby molten glass was passed through a pair of rolls with designs on them which produced patterned sheets. The only limitations on the patterns are those imposed by a repeating type of design and by the diameter of the rolls. The natural finish is called a fire finish. Later, a method of etching or sandblasting the surface was developed, and finally, when tempered glass was invented, this process was adapted to patterned glass.

GLASS, PLATE

Physical and Chemical Properties

Plate glass is the same type of glass as window and heavy sheet (*see* Glass). The surface, rather than the composition or thickness, of the glass is the distinguishing feature. Plate glass is ground and polished on both sides to a perfectly flat plane.

Properties are as follows: specific gravity, 2.42; weight, 13.16 lb/ft^2/in. thickness (64.25 kg/m^2/25.4 mm thickness); softening point, 1350°F (732.22°C); coefficient of linear expansion, 0.00000445; compressive strength, 36,000 lbf/in.^{2} (248.22 MN/m^2); total visible white light, 90% for $\frac{1}{4}$-in. (6.35-mm) thickness; total radiant energy transmission ($\frac{1}{4}$-in. or 6.35-mm thickness), 77.4%; thermal conductance (U) value ($\frac{1}{4}$-in. or 6.35-mm thickness), 1.13 Btu/h·ft^2·°F (6.42 W/m^2·°C); thermal conductance, 100°F (37.78°C) differential.

The maximum uniformly distributed load, regardless of superficial area, which can be supported by any square piece of 1-in. (25.4-mm) thick plate glass supported on all four sides is 21,000 lb (9525.6 kg).

Plate glass is available in thicknesses ranging from $\frac{1}{8}$ to $1\frac{1}{4}$ in., and in sizes up to a maximum of 10.83 ft (3.25 m) in width and 25 ft (7.62 m) in length.

The manufacture of float glass with the same characteristics as plate glass but no need for grinding has caused plate glass to slowly take a minor role in glazing in construction and in the manufacture of tempered, tinted, laminated, insulating, and other types of glass.

Types and Uses

Plate glass is divided into two major types on the basis of thickness:

1. Plate glass in thicknesses up to $\frac{5}{16}$ in. (7.94 mm), available in three qualities: silvering, mirror glazing, and glazing.

2. Heavy plate glass in thicknesses from $\frac{5}{16}$ in. (7.94 mm) up to and including $1\frac{1}{4}$ in. (31.75 mm), generally in commercial quality only. Other data are in Table *G32*.

Plate glass is used for glazing, for mirrors, and also for making insulating, tinted, reflective, mirror, heat-strengthened, heat-absorbing, glare-reducing, laminated, and tempered glass, all of which are fully covered under specific glass headings.

Application

Condensed Checklist

1. One should always check wind loads to obtain the correct thickness of glass required for the size of opening, and also check correct clearance dimensions for the thickness of glass selected. (*See* Tables *G33* and *G34*.)
2. Glazing details are determined by the type of material into which the plate glass is to be installed. They include type of glazing materials, type of glazing points or clips, necessity for setting blocks, and clearances for back putty (*see* Figures *G38* to *G40*).
3. Operating windows or doors must not be handled or opened until the glazing compound has set.
4. Glazing must never be done at temperatures below 40°F (4.44°C) unless precautions are taken to prevent moisture from condensing in rabbets where glass is to be installed.
5. Local, municipal, and state codes and also the codes of Fire Underwriters, insurance companies, labor departments, and the federal government (Army, Navy etc.) should always be checked for limitations and all currently applicable requirements.
6. The type of plate or heavy plate glass chosen must meet all requirements in regard to transparency, translucency, diffusion, and strength for its end use.
7. When selecting gaskets for glazing, one should always check with the glass manufacturers regarding the correct size, type, and material of the gaskets to be used for the particular plate glass to be installed.
8. One should always specify that a test of at least two widths of a curtain wall system, using plate glass, gaskets, and spandrel panels, must be made under all weathering and climatic conditions that will be encountered where the building or buildings are to be constructed.

Conditions Favorable to the Use of Plate or Heavy Plate Glass

1. For shelves and tops for tables, desks, etc.
2. For any area where perfect transparency is important.
3. Use plate glass $\frac{1}{8}$ in. (3.18 mm) thick for small windows, doors, and sidelights; and $\frac{1}{4}$-in. (6.35-mm) plate glass for windows, doors, partitions, sidelights, store fronts, and display cases.

Type G32 Types, qualities, sizes, tolerances, and transparency of plate glass and heavy plate glass

Type of glass	Quality or finish	Maximum sizes: Thickness in.	Thickness mm	Tolerance in.	Tolerance mm	Width ft	Width m	Length ft	Length mm	Approximate weight lb/in.^{2}	Approximate weight kg/m^2	Transmittance: Total average daylight (percent)	Transmittance: Total solar radiation (percent)	Major uses
Plate glass (regular)	Silvering, mirror-glazing, glazing	$\frac{1}{8}$	3.18	$\pm\frac{1}{32}$	±0.784	6.33	1.930	10.67	3.251	1.64	1 153.30	90.6	86.1	General glazing
		$\frac{1}{4}$	6.35	$\pm\frac{1}{32}$	±0.784	10.83	3.302	20	6.096	3.27–3.29	2299.04–2313.10	89.1	79.9	
Plate glass (twin ground)	Silvering, mirror-glazing, glazing	$\frac{1}{4}$	6.35	$\pm\frac{1}{32}$	±0.784	10.83	3.302	20	6.096	3.27–3.29	2299.04–2313.10	89.1	79.9	Glazing where transparency is of paramount importance; for mirrors
Heavy plate glass (regular)	Commercial quality	$\frac{5}{16}$	7.94	$\pm\frac{1}{32}$	±0.784	10.83	3.307	18.17	5.537	4.06	2 854.46	88.0	76.5	Glazing of large areas where wind loads become a problem; showcase tops, shelves, interior partitions, furniture tops, etc.
		$\frac{3}{8}$	9.53	$\pm\frac{1}{32}$	±0.784	10.83	3.307	22	6.706	4.90–4.93	3445.04–3466.14	87.3	75.2	
		$\frac{1}{2}$	12.70	$\pm\frac{1}{32}$	±0.784	10.83	3.307	25	7.620	6.54–6.58	4598.08–4626.20	86.0	71.0	
		$\frac{5}{8}$	15.88	$\pm\frac{3}{64}$	±1.176	7.5	2.286	11	3.353	8.17	5 744.08	84.7	67.1	
		$\frac{3}{4}$	19.05	$\pm\frac{3}{64}$	±1.176	10.83	3.307	11.91	3.632	9.67–9.81	6798.69–6897.12	83.3	63.3	
		1	25.40	$\pm\frac{1}{16}$	±1.588	6.17	1.880	12.33	3.759	13.08–13.16	9196.16–9252.40	80.9	56.9	
		$1\frac{1}{4}$	31.75	$\pm\frac{1}{16}$	±1.588	6.17	1.880	12.33	3.759	16.45	11 565.50	78.1	50.1	
Heavy plate glass (twin ground)	Commercial quality	$\frac{3}{8}$	9.53	$\pm\frac{1}{32}$	±0.784	10	3.048	22	6.706	4.90	3 445.04	87.3	75.2	Glazing where transparency and wind loads are important
		$\frac{1}{2}$	12.70	$\pm\frac{1}{32}$	±0.784	9.75	2.972	25	7.620	6.54	4 598.08	86.0	71.0	
		$\frac{5}{8}$	15.88	$\pm\frac{3}{64}$	±1.176	7.5	2.286	11	3.353	8.17	5 744.08	84.7	67.1	
		$\frac{3}{4}$	19.05	$\pm\frac{3}{64}$	±1.176	7.5	2.286	11	3.353	9.81	6 897.12	83.3	63.3	
Heavy plate glass (unpolished)	As manufactured, with knurled pattern on the surface	$\frac{5}{32}$	3.92	$\pm\frac{1}{32}$	±0.784	6.33	1.930	10.67	3.251	1.64	1 153.30	Translucent		Interior partitions and glazing where privacy and borrowed light are important
		$\frac{21}{64}$	8.23	$\pm\frac{1}{32}$	±0.784	10.83	3.307	19.08	5.817	4.29	3 016.17			
		$\frac{1}{2}$	12.70	$\pm\frac{1}{32}$	±0.784	10.83	3.307	18.17	5.537	6.58	4 626.20			
		$\frac{5}{8}$	15.88	$\pm\frac{3}{64}$	±1.176	10.83	3.307	18.17	5.537	8.17	5 744.08			
		$\frac{7}{8}$	22.23	$\pm\frac{3}{64}$	±1.176	7.83	2.388	10.83	3.307	13.46	9 469.32			
		$1\frac{1}{8}$	28.58	$\pm\frac{1}{16}$	±1.588	6.17	1.880	12.33	3.759	14.80	10 405.44			
		$1\frac{3}{8}$	34.93	$\pm\frac{1}{16}$	±1.588	6.17	1.880	12.33	3.759	18.09	12 718.54			
Plate glass, colored (blue and peach)		$\frac{7}{32}$	5.49	$\pm\frac{1}{32}$	±0.784	6	1.829	10	3.048	2.86	2 010.78	58.5		Decorative partitions and glazing; mirrors
Plate glass, tapestry type	Plain with both sides rough; figured surface; polished on one side	$\frac{7}{32}$	5.49	$\pm\frac{1}{32}$	±0.784	5	1.524	12	3.658	2.86	2 010.78	Translucent		Interior partitions and glazing where privacy and borrowed light are important

Table G33 Maximum areas for plate glass glazed on four sides in relation to thickness

Wind velocity mph (km/h)	Thickness of glass									
	1/8 in. (3.18 mm) ft² (m²)	3/16 in. (4.76 mm) ft² (m²)	1/4 in. (6.35 mm) ft² (m²)	5/16 in. (7.94 mm) ft² (m²)	1/2 in. (9.53 mm) ft² (m²)	1/2 in. (12.70 mm) ft² (m²)	5/8, 3/4, 7/8, and 1 in. (15.88, 19.05, 22.23 and 25.44 mm) ft² (m²)	1 1/8 in. (28.58 mm) ft² (m²)	1 1/4 in. (44.45 mm) ft² (m²)	1 3/8 in. (34.93 mm) ft² (m²)
30	72	162	228	244[a]						
(48.27)	(6.69)	(15.05)	(21.18)	(22.67)						
40	36	81	144	225	248[a]					
(64.36)	(3.34)	(7.33)	(13.38)	(20.90)	(23.03)					
55	24	54	96	150	216					
(88.50)	(2.23)	(5.02)	(8.92)	(13.94)	(20.07)					
65	18	41	72	112	162	244[a]				
(104.59)	(1.67)	(3.81)	(6.69)	(10.41)	(15.05)	(22.67)				
80	12	27	48	75	108	192				
(128.72)	(1.12)	(2.51)	(4.46)	(6.97)	(10.03)	(17.84)				
100	7	16	29	45	65	115				
(160.90)	(0.65)	(1.49)	(2.69)	(4.18)	(6.04)	(10.68)				
120	5	11	20	32	46	82	85[a]	81[a]	81[a]	81[a]
(193.08)	(0.47)	(0.92)	(1.86)	(2.97)	(4.27)	(7.62)	(7.90)	(7.33)	(7.33)	(7.33)

[a] Maximum sizes may increase because of new manufacturing methods (float glass). Check with glass manufacturers.

Table G34 Minimum clearances for plate glass and heavy plate glass

Type of plate glass	Thickness		Minimum clearance: Around edges		Minimum clearance: Between glass and back of rabbet		Minimum clearance: Depth of rabbet	
	in.	mm	in.	mm	in.	mm	in.	mm
Plate	1/8	3.18	1/8	3.18	1/16	1.59	3/8	9.53
	1/4	6.35	1/4	6.35	1/8	3.18	5/8	15.88
Heavy plate	5/16 and over	7.94 and over	1/4	6.35	1/8	3.18	5/8	15.88

4. Use heavy plate glass for large windows, partitions, store fronts, tops for tables and other articles of furniture, long shelving, aquariums, etc.

Conditions Unfavorable to the Use of Plate or Heavy Plate Glass

1. In areas where fire resistance is necessary.
2. In any area where plate or heavy plate glass will be placed in tension.

History and Manufacture

Originally sheet glass and plate glass were clearly differentiated: sheet was blown, and plate was cast, rolled, ground, and polished. Today the demarcation is not so clear-cut (*see also* Glass, "History and Manufacturing").

The first cast and rolled plate glass was developed by Lucas de Nehou in 1688. Next, in 1890, Brothers and Chance developed the process of casting molten glass onto an inclined table which fed the glass to a pair of rollers, after which it was cut to the desired size, an-

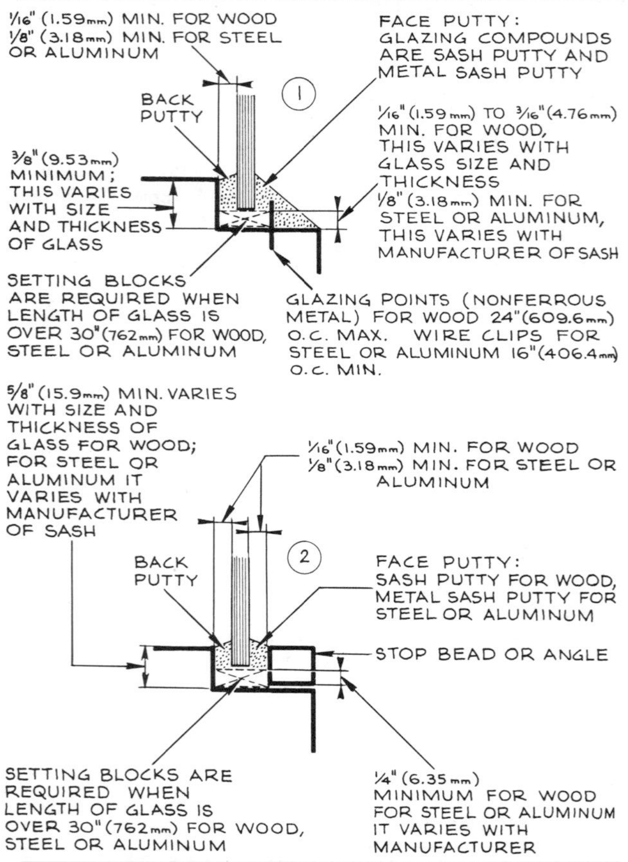

Figure G38 Plate glass or heavy plate glass set in wood, steel, or aluminum by face glazing (1) and with a stop bead (2).

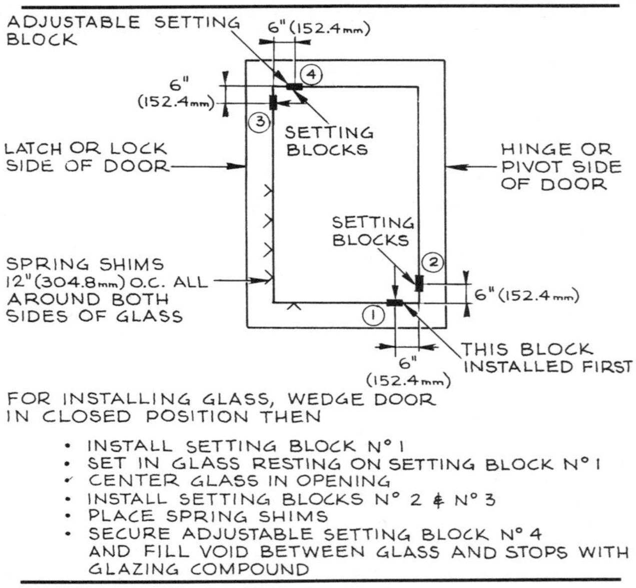

Figure G39 A typical method of installing plate glass or heavy plate glass in frames.

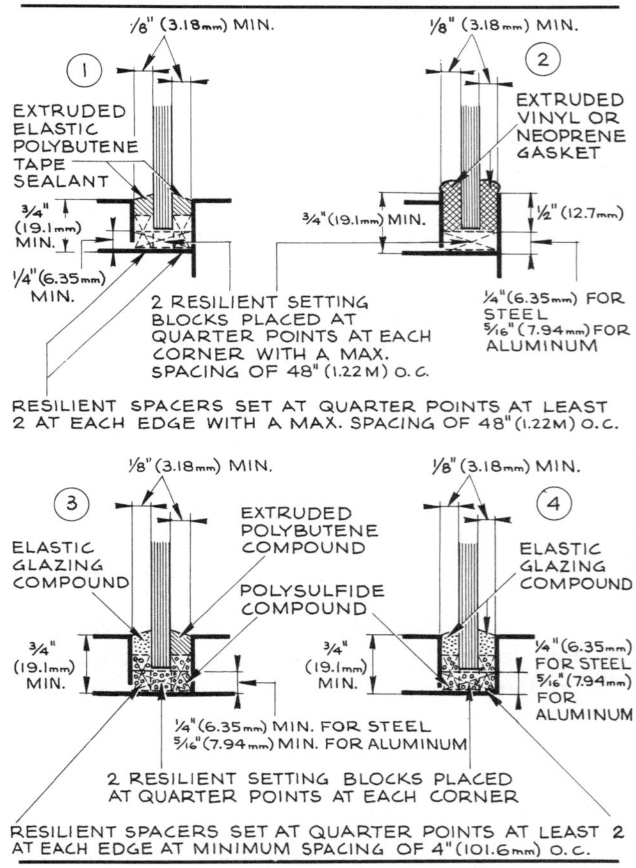

Figure G40 Plate glass or heavy plate glass set in steel or aluminum with tape (1), gasket (2), multiple seal (3), and multiple compound (4).

nealed, and finally ground and polished. The Bicheroux process developed in 1918 represents a peak of development of the discontinuous process. In it the glass rolled from the pot was received on traveling cars which moved past the inclined table. Shears cut between the cars so that each car with its load of glass could continue to the annealing furnace (leer).

The polishing and grinding were originally done on circular cast iron tables on which the glass was set in plaster. As the table revolved, a grinding mechanism, called a spider, with multiple polishing units, first ground down the glass, using sand and emery, and then polished it with soft felt pads, using jewelers' rouge (iron oxide) and water.

The next development was a continuous process in which the glass from the inclined rolling table passes in a continuous ribbon into an annealing furnace (leer) approximately 440 ft (134.112 m) long. In about $2\frac{1}{2}$ hours the glass passes through the leer and, upon emerging, is cut into lengths and ground and polished.

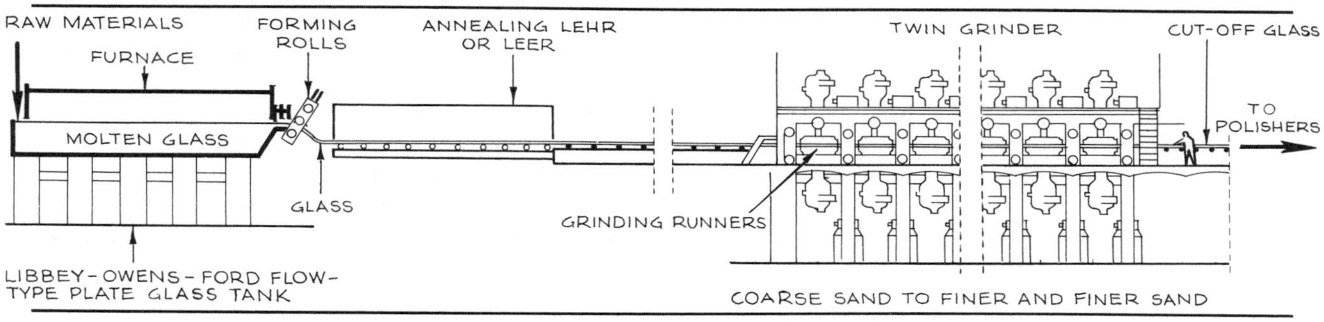

Figure G41 Flowchart for the twin grinding method of polishing plate glass.

Twin-Grinding Method. The most recent method is the twin grinding and polishing process for simultaneous and even continuous finishing of both surfaces, developed by Pilkington Brothers in England (*see* Figure *G41*). A version of this process was introduced into this country in 1953.

GLASS, STRUCTURAL

Physical and Chemical Properties

Structural glass is a float glass or a specially ground and polished glass that is opaque or colored. The color and opacity are obtained by adding metallic oxides. The original glass composition is a silicon, calcium, and sodium type. Structural glass may be heat-strengthened or tempered. Exact sizes must be specified and details have to be drawn so that this glass can be precut before installation. It is available in thicknesses from $\frac{1}{4}$ to $1\frac{1}{4}$ in. (6.35 to 31.75 mm) and in limited standard sizes.

Types and Uses

Structural glass is used as a facing material for vertical or horizontal surfaces of walls, partitions, and narrow flat surfaces. Other uses include toilet partitions, display fixtures, and table tops. It is available in various thicknesses and sizes and requires different methods of detailing for exterior and for interior installations. The surface can be decorated by sandblasting and grinding.

Table *G35* shows sizes, colors, thicknesses, weights, and major uses of structural glass.

Table G35 Types, sizes, colors, weights, and major uses of structural glass

Type of structural glass	Color range	Maximum sizes						Weight		Major uses
		Thickness		Width		Length				
		in.	mm	ft	m	ft	m	lb/ft^2	kg/m^2	
Ashlar	Opaque white,	$\frac{11}{32}$	8.62	0.67	0.204	1, 1.33	0.305, 0.406	4.4–4.5	21.48–21.97	Bathroom and
	black, and	$\frac{11}{32}$	8.62	1	0.305	1.33	0.406	4.4–4.5	21.48–21.97	kitchen wall
	8 standard	$\frac{11}{32}$	8.62	1.33	0.406	1.33	0.406	4.4–4.5	21.48–21.97	surfacing
	colors	$\frac{11}{32}$	8.62	2	0.610	2	0.610	4.4–4.5	21.48–21.97	
Sheets	Opaque white,	$\frac{1}{4}$[a]	6.35	6.33	1.930	10.83	3.302	3.29	16.06	Exterior and in-
	black, and	$\frac{11}{32}$	8.62	6.33	1.930	10.83	3.302	4.4–4.5	21.48–21.97	terior wall
	8 standard	$\frac{7}{16}$	11.11	6.33	1.930	10.83	3.302	5.6–5.76	27.34–28.19	surfacing
	colors	$\frac{3}{4}$	19.05	6.33	1.930	10.83	3.302	9.87–10.2	48.19–49.80	
Laminated	Opaque white,	$\frac{7}{8}$	22.23	5	1.524	5	1.524	11.2	54.68	Toilet partitions
	black, and	(2 sheets	(2 sheets							and solid par-
	8 standard	$\frac{7}{16}$ in.	11.11 mm							titions
	colors	thick)	thick)							
Heavy sheets	Opaque white	1	25.40	5	1.524	5	1.524	13.08	63.86	Countertops,
	black, and	$1\frac{1}{4}$	31.75	5	1.524	5	1.524	16.45	79.31	seats, and solid
	8 standard									partitions
	colors									

[a]In black color only.

Application

Structural glass should never be applied to a wood back-up material, but should be applied only to plaster or masonry. The back-up material must always be painted so that it is sealed and waterproofed.

Condensed Checklist

1. Local, municipal, and state codes and also codes of the Fire Underwriters, insurance companies, labor departments, and the federal government (Army, Navy, etc.), should be checked for limitations and restrictions.
2. The thickness, size, and color of the structural glass chosen must meet the design and engineering requirements for its end use.
3. When structural glass is installed as a surfacing material, the masonry or plaster back-up material must be sealed and made waterproof by a paint that meets the requirements of the manufacturers of the structural glass.
4. Sizes of structural glass used as a surfacing material must meet the following limitations. For wall areas over 15 ft (4.572 m) high, the maximum size is 6 ft^2 (0.557 m^2). For wall areas under 15 ft (4.572 m) in height, the maximum size is 10 ft^2 (3.048 m^2). The maximum width or length is 4 ft (1.219 m).
5. Structural glass must always be installed at temperatures above 40°F (4.44°C). Special attention should be given to ensuring that all bonding surfaces are free from moisture.
6. It is always advisable for the architect to draw large-scale elevations and sections of areas where structural glass is to be installed to ensure that horizontal and vertical joints, intersections, etc., are detailed correctly.
7. Treatment of horizontal and vertical joints, at internal and external corners, at the base, and at the cap (*see* Figures *G42* and *G43*) should be specified.
8. The joint compound must be one recommended by the manufacturer of the structural glass.
9. Setting cement for interior or exterior use must be of a type recommended by the manufacturer of the structural glass.
10. When using structural glass for toilet, shower, or solid partitions, one should check whether solid or laminated structural glass should be used. Lami-

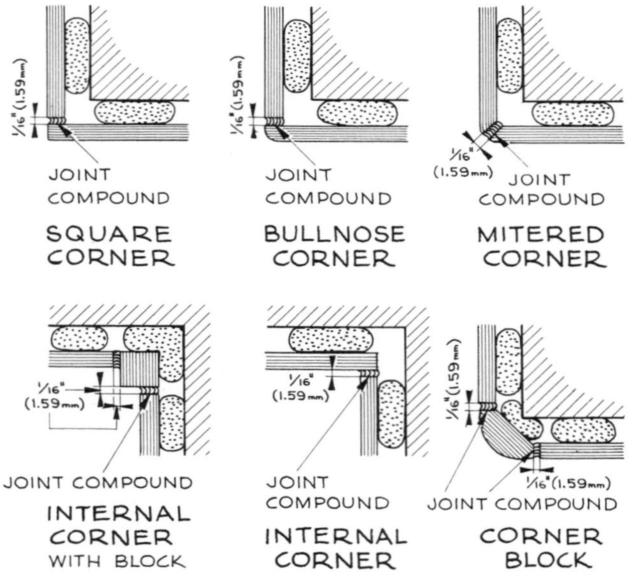

Figure G42 Details of typical exterior and interior corners for structural glass.

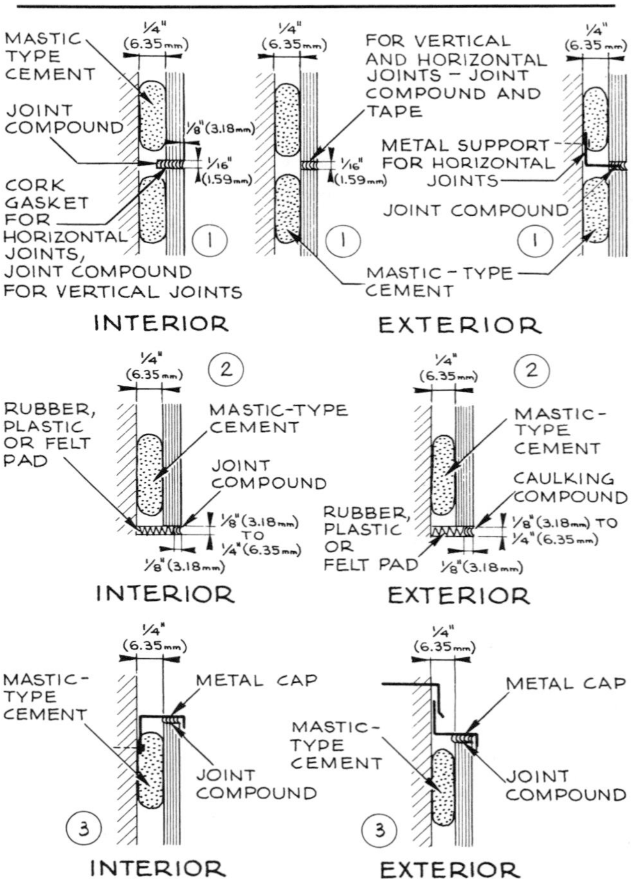

Figure G43 Details of typical joints for exterior and interior installations of structural glass.

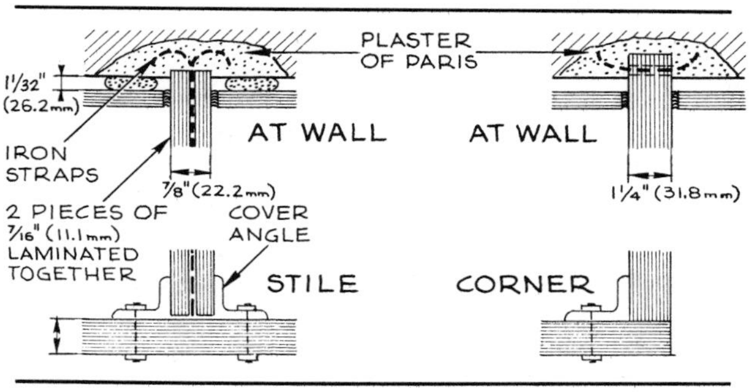

Figure G44 Details of joining structural glass together and at walls for toilet, shower, or solid partitions.

nated structural glass should never be used for shower stalls. (*See* Figure *G44*.)

11. When using structural glass for ceilings, one should always check with the manufacturer for methods of fastening and for installation details.
12. Specifications should always require that shop drawings and setting drawings be supplied for all structural glass installations by the structural glass contractor.

Conditions Favorable to the Use of Structural Glass

1. Where an opaque, colored, permanent facing material with low maintenance is required for the exterior or interior, particularly if it must withstand rough usage.
2. Where an opaque, colored, permanent, low-maintenance material is required for toilet, shower, or similar solid partitions, and for ceilings.

Conditions Unfavorable to the Use of Structural Glass

1. Where a fire-resistant material is necessary.
2. Where any permanent loads or stresses will be transmitted to the glass.

History and Manufacture

Structural glass was developed as an architectural material only about 1930. The manufacture of structural glass is similar to that of plate glass; the only difference lies in the raw materials used.

GLASS, TEMPERED

Physical and Chemical Properties

Tempered glass is float, heavy sheet, or plate glass or some suitable type of patterned glass that has been reheated to just below softening point and then suddenly cooled by subjecting both surfaces to jets of air. This causes the outside surfaces, which cool faster, to be in a state of compression and the inner portions of the glass to be in a state of tension, the two stresses being in balance. Properties include the following: modulus of rupture (ultimate tensile strength), 29,500 $lbf/in.^2$ (203.4 MN/m^2); maximum safe working temperature; 550°F (287.78°C); maximum thermal endurance; 400°F (204.44°C) temperature differential.

Strength. The greatest uniformly distributed load (regardless of superficial area) which can be supported by any square piece of 1-in. (25.4-mm)-thick tempered glass supported on all four edges is 100,000 lb (45 360 kg). Deformation under a given load is the same as for regular glass, but tempered glass may be deformed by twisting, bending, and tensional stresses much greater than the stress limits for regular glass before it fails. It is 3 to 5 times stronger in relation to impact forces and temperature variances.

Polarized Light. Sunlight from some angles provides polarized light that makes the strain patterns within tempered glass become visible as multicolored iridescence under certain conditions.

Frangibility. Tempered glass, if chipped or punctured, either at the edges or on any part of its flat surfaces, will shatter and disintegrate into small blunt crystals which have a tendency to fly apart if not held in a frame. For this reason it cannot be cut or drilled.

Sizes. Exact sizes and details must be supplied to the manufacturer before tempered glass can be made. It is available in thicknesses from $\frac{1}{4}$ to $1\frac{1}{4}$ in. (6.35 to 31.75 mm) and in sizes up to 6 × 12 ft (1.829 × 3.658 m).

Types and Uses

Tempered float or plate glass is made in $\frac{1}{2}$- and $\frac{3}{4}$-in. (12.7- and 19.05-mm) thicknesses for doors and sidelights, but thicknesses up to $1\frac{1}{4}$ in. (31.75 mm) are available for special design problems. Tempered patterned glass is made $\frac{3}{8}$ in. (9.53 mm) thick for doors, and $\frac{7}{32}$ and $\frac{3}{4}$ in. (5.49 and 19.05 mm) thick for general glazing and sidelights. A specially glazed tempered glass is made for use as blackboards. Table *G36* lists types, sizes, and weights of tempered glass.

Tempered glass is available clear, tinted, reflective, and insulating. (*see also* Glass: Laminated; Glass: Heat-Strengthened; Glass: Tinted, etc.)

Table G36 Types, sizes, and weights of tempered glass

Type of tempered glass	Thickness		Maximum sizes: Width		Maximum sizes: Length		Weight	
	in.	mm	ft	m	ft	m	lb/ft^2	kg/m^2
Float or plate glass for glazing	$\frac{1}{4}$	6.35	6	1.829	9	2.743	3.29	16.06
	$\frac{3}{8}$	9.53	6.42	1.957	9.17	2.794	4.39	21.43
	$\frac{1}{2}$	12.70	6.42	1.957	9.17	2.794	6.58	32.12
	$\frac{3}{4}$	16.05	6.42	1.957	9.17	2.794	9.67	47.21
	1	25.40	6.17	1.880	9.17	2.794	13.16	64.25
	$1\frac{1}{4}$	31.75	6.17	1.880	9.17	2.794	16.45	80.31
Float or plate glass for side-lights with doors	$\frac{1}{2}$	12.70	6	1.829	9	2.743	6.58	32.12
	$\frac{3}{4}$	19.05	6	1.829	9	2.743	9.67	47.21
Float or plate glass for doors	$\frac{1}{2}$	12.70	3.5	1.067	8	2.438	6.58 + hardware	32.12 + hardware
	$\frac{3}{4}$	19.05	4	1.219	9	2.743	9.67 + hardware	47.21 + hardware
Float or patterned glass[a]	$\frac{7}{32}$	5.59	5	1.524	11.33	3.454	2.76	13.43
	$\frac{3}{8}$	9.53	6	1.829	12	3.658	5.00	24.41
Float or patterned glass for doors	$\frac{3}{8}$	9.53	5	1.524	7.33	2.235	5.00 + hardware	24.41 + hardware

[a] Some types of patterned glass are not made in these ranges. *See* Glass, Patterned for size limitations.

Since tempered glass cannot be drilled or cut, all dimensions, holes, kick plates, hardware, and any decorative treatment by sandblasting must be designed and detailed before the glass is manufactured. The materials generally used for kick plates and hardware are aluminum, bronze, and stainless steel. Standard maximum and minimum dimensions and tolerances have been established for tempered plate glass doors (*see* Figure *G45*).

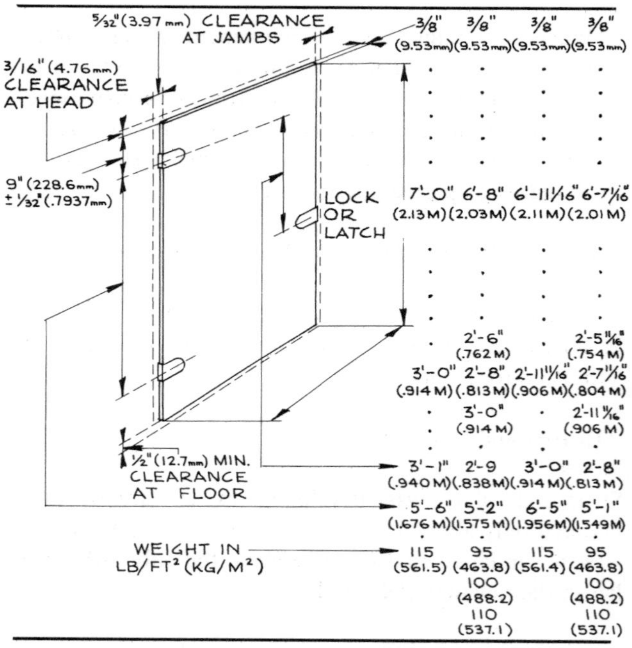

Figure G45 Tempered float or plate glass doors, showing limitations in dimensions, in sizes of holes, and in location of finish hardware.

Tolerances, Clearances. Tolerances for all tempered plate glass, including doors and sidelights, are as follows: for thickness, $+\frac{1}{32}$ in. (0.78 mm); for width and length (height), $-\frac{3}{32}$ in. (0.788 mm). Clearance for doors and sidelights are as follows: $\frac{1}{8}$ in. (3.18 mm) at top, sides, and between doors and sidelights; and $\frac{3}{16}$ in. (4.76 mm), preferably, and $\frac{1}{4}$ in. (6.35 mm), at maximum, at bottom of doors.

Stock Fittings. All tempered plate glass doors have stock fittings for hardware, pivots, locks, and latches, as shown in Figure *G46*. Fittings for pivot and floor check are an integral part of the door, installed during the manufacturing process and not on the site.

Tempered Patterned Glass. Designs and patterns can be made on tempered plate glass by sandblasting. Full-size details of the designs or patterns must be made and given to the manufacturer. The depth of sandblasting is limited to $\frac{1}{64}$ in. (0.39 mm).

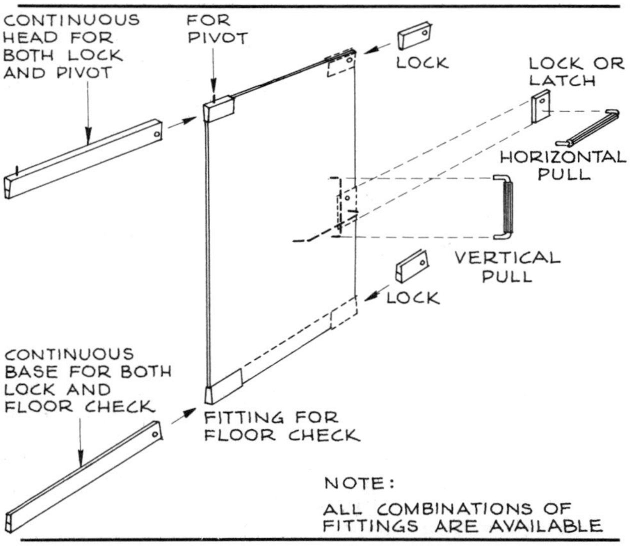

Figure G46 Various stock fittings for tempered float or plate glass doors.

Tempered patterned glass is available in a limited number of patterns, and the type of finish controls the maximum size available. The architect should always contact the manufacturer of patterned glass for data on current stock sizes, as this is a relatively new material, and the addition of new patterns and discontinuation of current ones must be expected. Table *G37* lists commonly available types, sizes, and finishes of tempered patterned glass.

The stock sizes and limitations for tempered patterned glass doors for interior use are shown in Figure *G47*. Tolerances in width and height are $+\frac{1}{16}$ in. (1.59 mm).

Chalkboards. Tempered glass chalkboards (blackboards) are made of float or plate glass covered with a colored, vitreous glaze that contains an abrasive usually aluminum oxide. This glaze is sprayed on, and the plate glass is heated to 1150°F (621.11°C), at which temperature the glaze fuses with the glass. Then the chalkboard is heated and cooled to make the glass tempered.

These chalkboards, $\frac{1}{4}$ in. (6.35 mm) thick and weighing 3.32 lb/ft^2 (16.21 kg/m^2), are available in a minimum size of 1.5 × 2 ft (0.457 × 0.610 m) and a maximum size of 6 × 12 ft (1.829 × 3.658 m). Standard sizes are as follows: 3.5 and 4 ft (1.067 and 1.219 m) in height; and 5 to 9 ft (1.524 to 2.743 m) in length in 1-ft (0.3048-m) increments. The standard colors are green, gray, blue, ivory, and tan. Tempered glass chalkboards can be installed in any of the stock metal frames used for this purpose.

Application

Condensed Checklist

1. All door sizes, sidelights, transoms, jambs, heads, hardware, locks, and any other special requirements must be detailed, and all types of hardware must be selected, before tempered glass is ordered.

Table G37 Tempered patterned glass

Type of pattern	Thickness		Recommended areas		Maximum size				Type of finishes available[a]
	in.	mm	ft^2	m^2	ft	mm	ft^2	m^2	
Concave stripes, small	$\frac{7}{32}$	5.59	20	1.86	5 × 7.33	1.524 × 2.235	36	3.34	Fire, textured, frosted, special[b]
Granular, fine	$\frac{7}{32}$	5.59	20	1.86	5 × 7.33	1.524 × 2.235	36	3.34	Fire, frosted
Granular, medium	$\frac{7}{32}$	5.59	20	1.86	5 × 7.33	1.524 × 2.235	36	3.34	Fire, frosted, special[b]
	$\frac{3}{8}$	9.53	36	3.34	5 × 7.33	1.524 × 2.235	36	3.34	Fire, frosted, special[b]
Hammered	$\frac{7}{32}$	5.59	20	1.86	5 × 7.33	1.524 × 2.235	36	3.34	Fire, frosted
	$\frac{3}{8}$	9.53	36	3.34	5 × 7.33	1.524 × 2.235	36	3.34	Fire
Pebbled	$\frac{7}{32}$	5.59	20	1.86	5 × 7.33	1.524 × 2.235	36	3.34	Fire, frosted
Ribbed, small	$\frac{7}{32}$	5.59	20	1.86	5 × 7.33	1.524 × 2.235	36	3.34	Fire, frosted
	$\frac{3}{8}$	9.53	36	3.34	5 × 7.33	1.524 × 2.235	36	3.34	Fire
Ribbed, large	$\frac{7}{32}$	5.59	15	1.39	4 × 5	1.219 × 1.524	20	1.86	Fire, frosted
Squares, small and large	$\frac{7}{32}$	5.59	20	1.86	5 × 7.33	1.524 × 2.235	36	3.34	Fire, frosted

[a] For description of finishes, *see* Glass, Patterned.
[b] Special finishes are trademarked or patented.

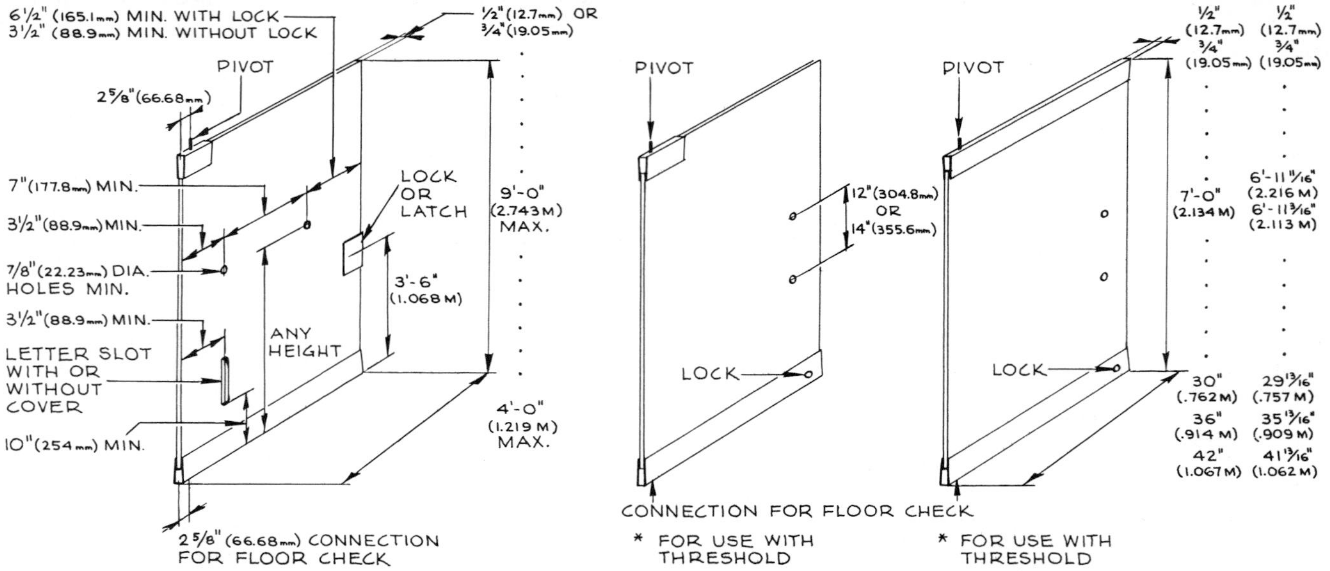

Figure G47 Tempered tinted, reflective, or patterned glass doors, showing stock sizes and finish hardware limitations.

2. Specifications should always require that shop drawings be supplied, that samples of the metal for fittings and their finish be submitted for approval, and that samples of the type of hardware (locks, door closers and checks, push and pull fittings, etc.) also be submitted for approval.
3. Tempered glass cannot be used as a fire door. Local, municipal, and state codes and also the codes of Fire Underwriters, insurance companies, labor departments, and the federal government (Army, Navy, etc.) should be checked for requirements.
4. The manufacturers of tempered glass should always be consulted for advice when there is a design problem that cannot be answered with stock doors and sidelights.
5. The owners should always be informed that tempered glass can be easily insured.
6. When using tempered patterned glass, one should always check with the manufacturer to learn which of the various surface finishes for light diffusion are obtainable in tempered form.
7. A check should always be made of the type of material into which tempered glass is to be installed, the necessary clearances, type of glazing material, type of glazing points or clips, and the necessity for setting blocks and spacers.
8. The manufacturers should be checked for the availability of standard sizes and for the maximum sizes that can be obtained.

Conditions Favorable to the Use of Tempered Glass

1. Where the design requires glass in areas where thermal shock, impact, and rough usage occur.
2. For doors where the design requires maximum transparency with a minimum or absence of vertical or horizontal mullions.

Conditions Unfavorable to the Use of Tempered Glass

1. Where fire doors are required.
2. In areas where impact is constantly present from heavy equipment moving on trucks, dollies, motorized lifts, etc., or where trucks or automobiles can come in direct contact with the glass.

History and Manufacture

Sometime during the period when Tiberius was Caesar of the Roman Empire, about A.D. 23 to 37, a master craftsman of the glass industry invented malleable glass which would bend, could be hammered like metal, and could be dropped without breaking. The inventor was brought before Tiberius and the remarkable qualities of the glass were demonstrated to him. Tiberius, for reasons unknown, ordered the inventor killed on the spot. Thus one of the great discoveries of man was lost for approximately 2000 years. We will never know exactly what this glass was, but the tempered glass that was developed around 1950 seems to possess almost all the qualities that the lost glass reputedly had.

The manufacture of tempered glass is a very highly controlled process which has already been described under "Physical and Chemical Properties." Because of the heating and sudden surface cooling, this type of glass is not as flat as plate glass and will have minor warpage, particularly along the exposed edges.

GLASS, TINTED, REFLECTIVE, AND HEAT-ABSORBING

Physical and Chemical Properties

Any type of glass that absorbs percentages of the total radiant energy of the sun or reflects light and heat from the sun's rays is known as tinted, reflective, and heat-absorbing glass. These types of glass are available annealed, heat-processed, and tempered. They are used not only for single glazing but also in the manufacture of a wide variety of insulating glass.

The demand for air conditioning in buildings brought about the development of these glasses, although heat-absorbing and glare-reducing glasses already were available, because of the tremendous heat gain through solar energy in the summer months. Later the energy problem accelerated the demand for better insulation with less heat gain and heat absorption.

Types and Uses

Tinted Glass. Tinted glass is made from float, heavy sheet, polished or rough plate, and some patterned glass. It is available in gray, bronze, and blue-green, either annealed, heat-processed, or tempered (*see* Table *G38*). These types of glass are heat absorbing and glare reducing, and, because of these characteristics, extra precautions are required in the installation, including the correct type of rabbet, gaskets, glazing material, and treatment of the edges of the glass. In general, tinted glass has a one-way effect which masks or semi-obscures interior colors, drapes, and blinds from the exterior during daylight.

Tinted Tempered Glass. Tinted tempered glass is made from float, plate, sheet, and certain patterned glass that is heated and then cooled (*see* Glass, Tempered). This makes the glass 4 to 5 times stronger than untempered

Table G38 Thicknesses, sizes, weights, visual characteristics, and relative heat gains of tinted glass[a]

Type of glass	Thickness		Maximum size		Weight		Transmittance[b] (percent)	Relative heat gain	
	in.	mm	ft	mm	lb/ft^2	kg/m^2		$Btu/ft^2 \cdot h$	W/m^2
Float or plate, gray	$\frac{1}{8}$	3.18	5.00 × 10.00	1.524 × 3.048	1.65	8.06	62	184	580.52
	$\frac{3}{16}$	4.76	10.50 × 20.00	3.200 × 6.096	2.45	11.96	50	165	520.58
	$\frac{1}{4}$	6.35	10.50 × 20.00	3.200 × 6.096	3.27	15.96	41	150	473.25
	$\frac{5}{16}$	7.94	10.00 × 16.00	3.048 × 4.877	4.08	19.92	38	148	466.94
	$\frac{3}{8}$	9.53	10.17 × 22.00	3.099 × 6.706	4.90	23.72	28	130	410.15
	$\frac{1}{2}$	12.70	10.00 × 25.00	3.048 × 7.620	6.54	31.93	19	115	362.83
Float or plate, bronze	$\frac{1}{8}$	3.18			1.65	8.06	69	185	583.68
	$\frac{3}{16}$	4.76	10.50 × 20.00	3.200 × 6.096	2.45	11.96	59	168	530.04
	$\frac{1}{4}$	6.35	10.50 × 20.00	3.200 × 6.096	3.27	15.96	52	156	492.18
	$\frac{3}{8}$	9.53	10.17 × 22.00	3.099 × 6.706	4.90	23.72	37	130	410.15
	$\frac{1}{2}$	12.70	10.00 × 25.00	3.048 × 7.620	6.54	31.93	28	115	362.83
Float or plate, blue-green,	$\frac{1}{8}$	3.18	6.17 × 10.00	1.88 × 3.048	1.65	8.06	83	179	564.75
	$\frac{3}{16}$	4.76	10.50 × 20.00	3.200 × 6.096	2.45	11.96	79	161	507.96
	$\frac{1}{4}$	6.35	10.50 × 20.00	3.200 × 6.096	3.27	15.96	75	155	489.03
Sheet glass, gray	$\frac{1}{8}$	3.18	5.00 × 7.00	1.524 × 2.134	1.65	8.06	31	170	536.35
	$\frac{7}{32}$	5.49	8.33 × 7.00	2.540 × 2.134	2.88	14.06	14	145	457.48

[a]Tolerance in thickness is $\pm\frac{1}{32}$ in. (0.78 mm) for all except sheet glass. Sheet glass tolerances are as follows: ±0.115 to 0.134 in. (2.92 to 3.40 mm) for $\frac{1}{8}$-in. (3.18-mm) thickness, ±0.212 to 0.230 in. (5.34 to 6.84 mm) for $\frac{7}{32}$-in. (5.49-mm) thickness.
[b]These figures vary with type of glass; always check with manufacturers for their transmittance (percent) and relative heat gain ($Btu/ft^2 \cdot h$, W/m^2) values.

Table G39 Thicknesses, sizes, weights, visual characteristics, and relative heat gains of tinted, tempered and heat-absorbing glass[a]

Type of glass	Thickness		Maximum size		Weight		Transmittance[b] (percent)	Relative heat gain[b]	
	in.	mm	ft	mm	lb/ft^2	kg/m^2		$Btu/ft^2 \cdot h$	W/m^2
Gray,[c]	$\frac{1}{8}$	3.18	2.83 × 5.50	0.864 × 1.676	1.66	8.06	69	190	599.45
float or plate	$\frac{3}{16}$	4.76	5.00 × 10.00	1.524 × 3.048	2.54	12.40	50	165	520.58
	$\frac{1}{4}$	6.35	12.00 × 10.00	3.658 × 3.048	3.28	16.01	42	150	473.25
	$\frac{5}{16}$	7.95	12.00 × 10.00	3.658 × 3.048	4.08	19.92	31	135	425.93
	$\frac{3}{8}$	9.53	12.00 × 10.00	3.658 × 3.048	4.92	34.02	29	130	410.15
	$\frac{1}{2}$	12.70	12.00 × 10.00	3.658 × 3.048	6.56	32.03	20	115	362.83
Bronze,[c]	$\frac{1}{8}$	3.18	2.83 × 5.50	0.813 × 1.676	1.65	8.06	15	184	580.52
float or plate	$\frac{3}{16}$	4.76	5.00 × 10.00	1.524 × 3.048	2.54	12.40	59	168	530.04
	$\frac{1}{4}$	6.35	12.00 × 10.00	3.658 × 3.048	3.28	16.01	51	150	473.25
	$\frac{5}{16}$	7.95	12.00 × 10.00	3.658 × 3.048	4.08	19.92	50	142	448.01
	$\frac{3}{8}$	9.53	12.00 × 10.00	3.658 × 3.048	4.92	24.02	38	130	410.15
	$\frac{1}{2}$	12.70	12.00 × 10.00	3.658 × 3.048	6.56	32.03	29	115	362.83
Blue-green,	$\frac{1}{8}$	3.18	2.83 × 5.67	0.864 × 1.727	1.65	8.06	82	180	567.90
float or plate, heat-absorbing	$\frac{1}{4}$	6.35	10.00 × 8.00	3.048 × 2.438	3.28	16.01	75	150	473.25
Gray and bronze,	$\frac{3}{16}$	4.76	5.00 × 8.00	1.524 × 2.438	2.51	12.25	59	168	530.04
sheet glass	$\frac{7}{32}$	5.49	5.00 × 8.00	1.524 × 2.438	2.82	13.78	50	145	457.48

[a]Check tolerances with manufacturers.
[b]Check with manufacturers for $Btu/ft^2 \cdot h$ (W/m^2) and transmittance values as the type of glass affects these figures.
[c]These types are available in $\frac{5}{8}$ in. (15.88 mm) and 1 in. (25.4 mm) for entrance doors and sliding doors.

glass. Tinted tempered glass is used mostly in areas where there is the hazard of glass breakage. It is also used in the manufacture of insulating glass (*see* Table *G39*). The demand for better insulation has increased the demand for tinted glasses and reflective glasses. The use of tinted tempered glass with a color tint and reflective qualities that match the type of glass used in the entire building is of great importance to the continuity of the total building design. (*See also* Glass, Insulating.)

Reflective Glass. Reflective glass is made from float, plate, sheet, and patterned glass that has been annealed, heat processed, or tempered. A metallic film is then applied to one side of the glass, similar to one-way mirrors (*see* Glass; Mirrors). This film reflects sunlight and therefore eliminates a large percent of heat gain to a building but does not greatly affect the transparency of the glass (*see* Table *G40*). One important advantage of this type of glass is that on the exterior it reflects like a mirror and therefore completely conceals drapes, blinds, colors, and activity on the inside of the building. It is advisable, however, to remember that at night the reverse happens; when the interior lights are on, a person inside the building cannot see out, whereas anyone on the outside has a clear view of the interior.

All reflective glass is manufactured with a metallic coating on one side of the clear or tinted glass. These metal coatings act as one-way mirrors allowing vision through the coating and present a mirrorlike reflective appearance on the outside. Metallic coatings that give the glass a gold or silver color are available. Thus, by the use of clear and tinted glasses, a wide variety of metallic colors are available in single glazing glass and insulating glass. It is advisable always to require a full-size mock-up of a building's curtain wall in order to test it for color harmony and climatic conditions.

Tinted, reflective and heat-absorbing glass is available in three tints—gray, bronze, and blue-green—in annealed, tempered, and heat-processed types (*see* Tables *G38* to *G40*).

These glasses are also used for some kinds of laminated glass and in great quantities for insulating glass, for various combinations of these types, and also for combinations with clear glass and heat-strengthened glass.

Manufacturers should be consulted for exact information on size and thickness limitations, and data on transmittance, insulation, reflectance, and heat-gain values. These values are of great importance in determining the types of heating and cooling systems to be installed and in the consumption of energy.

Table G40 Thicknesses, sizes, weights, visual characteristics, and relative heat gains of reflective glass

Type of glass	Thickness		Maximum size[a]		Weight		Transmittance[b] (percent)	Reflectance[b] (percent)	Relative heat gain[b]	
	in.	mm	ft	mm	lb/ft^2	kg/m^2			$Btu/ft^2 \cdot h$	W/m
Clear, float or plate	1/4	6.35	6.00 × 12.00	1.829 × 3.658	3.28	16.01	39	35	135	425.93
Gray, float or plate	1/4	6.35	6.00 × 12.00	1.829 × 3.658	3.28	16.01	18	35	100	315.50
	5/16	7.95			4.08	19.92	16	28	93	293.42
Bronze, float or plate	1/4	6.35	6.00 × 12.00	1.829 × 3.658	3.28	16.01	21	35	105	331.28
	5/16	7.95			4.08	19.92	20	26	82	258.71
Blue-green, float or plate	1/4	6.35	6.00 × 12.00	1.829 × 3.658	3.28	16.01	50	18	129	406.99
Tempered, clear	1/4	6.35	6.00 × 12.00	1.829 × 3.658	3.28	16.01	50	34	160	504.80
Tempered, gray	1/4	6.35	6.00 × 12.00	1.829 × 3.658	3.28	16.01	24	34	118	372.29
	5/16	7.95			4.08	19.92	20	24	135	425.93
Tempered, bronze	1/4	6.35	6.00 × 12.00	1.829 × 3.658	3.28	16.01	30	34	120	378.66
	5/16	7.95			4.08	19.92	20	20	99	312.34

[a]Check with manufacturers for minimum and maximum sizes available.
[b]These figures vary with the type of glass; always check with manufacturers for their transmittance, reflectance, and $Btu/ft^2 \cdot h$ (W/m^2) values.

Application

Tinted, reflective, and heat-absorbing glass requires careful detailing for installation because of the stresses due to expansion and contraction caused by the absorption of heat, and hence these types of glass require large clearances in installation. Because they absorb heat, it is also necessary to control the covering of the edges. Edge coverings must not be excessive in size because these covered areas will remain cold while the remainder of the glass heats up, and cracking may result. All edges must be clean-cut; nipping of the edges will cause stress concentration in the edges and consequent breakage.

When any of these types of glass is selected, the manufacturer should be consulted in relation to edge-of-glass treatment and type of rabbet, stops, gaskets, and other types of glazing materials to be used for application.

Condensed Checklist

1. Wind loads should always be checked for the thickness of glass required for a given size of opening.
2. The type of material into which the selected glass is to be installed should be correlated with the type of glazing material, gaskets, type of glazing chips or points, necessary setting blocks, and clearances. One should always check with the glass manufacturers. (*See* Figure *G48*.)
3. Operating windows and doors should not be handled or opened until the glazing compound gaskets, sealants, etc., all have set. This restriction should always be specified.
4. No glazing should be done at temperatures below 40°F (4.44°C) unless precautions are taken to prevent the condensation of moisture in rabbets and the like.
5. Local, municipal, and state codes and also the codes of the Fire Underwriters, insurance companies, labor departments, and federal government (Army, Navy, etc.) should be checked for limitations and requirements.
6. The type of tinted, reflective and heat-absorbing glass chosen must meet all requirements for transparency, translucency, diffusion, strength, solar radiation transmission, thermal characteristics, and strength related to its end use in the design.

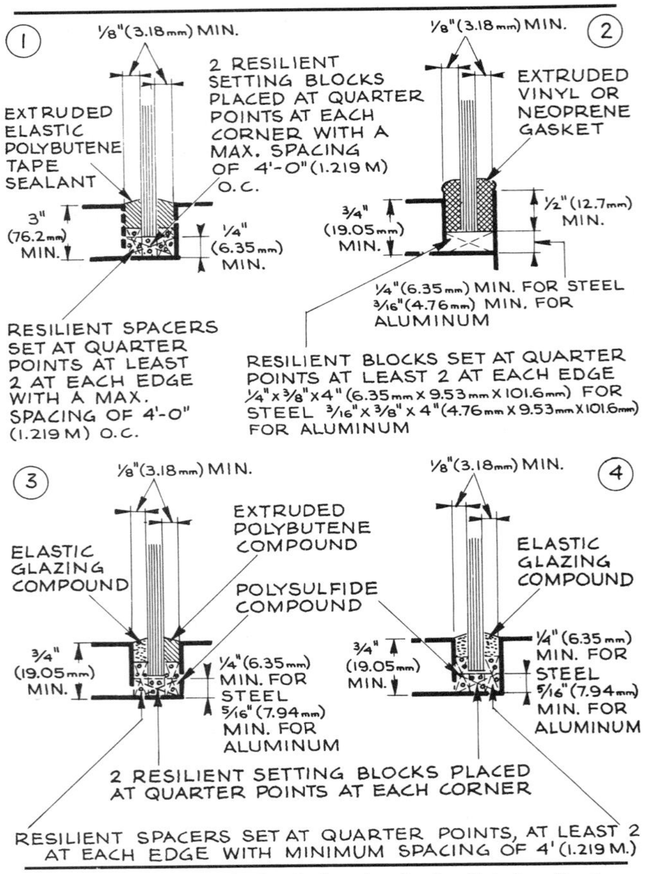

Figure G48 Typical single glazing details for tinted, reflective, and heat-absorbing glass.

7. When selecting tinted, reflective or heat-absorbing glass for areas where safety type glass is required, one should always check the governing safety codes for restrictions, limitations, and requirements.
8. When selecting tinted, reflective, or heat-absorbing glass for a curtain wall system, one should check that all other types of glass components match the glazing glass selected. It is advisable to require that full-size sections of the system be constructed for observation and testing.
9. When selecting insulating glass made with tinted, reflective, and heat-absorbing glass, the manufacturers should be consulted for current data on methods of application.

Conditions Favorable to the Use of Tinted, Reflective, and Heat-Absorbing Glass

1. In areas where solar radiation transmission will demand extra air conditioning and glare will destroy efficient human use of these areas.
2. Where glare must be controlled even though solar radiation transmission is not excessive.

Conditions Unfavorable to the Use of Tinted, Reflective, and Heat-Absorbing Glass

1. Where transparency is of major importance.
2. Where fire resistance is necessary unless a wired type of this glass is to be used.

History and Manufacture

The characteristics of tinted, reflecting, and heat-absorbing glass were known for many years through research in the field of optical glass, but it was not until fairly recently, actually after World War II and in the 1950s, that methods of applying this knowledge to the mass production of glass for construction were developed.

In tinted reflective and heat-absorbing glass, the problems were to reduce glare, reduce heat gain, and absorb heat without destroying true color vision. The colors that were finally developed to accomplish this are gray, bronze, and blue-green tints.

Heat-absorbing glass posed a similar problem—controlling the increased ferrous iron content while still retaining vision and light transmission. Today production methods have been developed that make these types of glass available and economically feasible for construction applications.

GLASS, WINDOW AND HEAVY SHEET

Physical and Chemical Properties

Window and heavy sheet glass is a soda-lime (silicon, calcium, and sodium) type of glass. Its composition and characteristics are fully covered under the main heading, Glass. Because of the manufacturing process, a wave or draw distortion runs in one direction through the sheet, and the degree of distortion controls the grading and usefulness of this category of glass. The surface has a brilliant fire-polished finish (the natural as-manufactured finish, also called fire finish).

Float glass now is available as window and heavy sheet glass. In the future float glass manufacturing methods may be used for all types of glass.

Window and heavy sheet glass are available in various types, strengths, and qualities, in thicknesses ranging from 0.043 to 0.256 in. (1.588 to 12.7 mm) and in sizes

Table G41 Data for window, heavy sheet, and picture glass

Type of glass	Strength or classification in. (mm)	Quality[a]	Thickness ft	Thickness mm	Weight lb/ft²	Weight kg/m²	Maximum sizes ft	Maximum sizes m	Transmittance: Total average daylight (percent)	Transmittance: Total solar radiation (percent)
Photo	1/16 (1.588)	AA, A, B	0.058–0.068	1.47–1.73	0.81	3.95	3 × 4.17	0.914 × 1.270	90.8–91.8	87.3–90.6
Picture[b]	5/64 (1.984)		0.070–0.080	1.778–2.032	1.01	4.93	3 × 4.17	0.914 × 1.270	90.8–91.8	87.3–90.6
Window	3/32 (2.381) single strength	AA, A, B	0.085–0.100	2.16–2.54	1.22	5.96	12 × 12	1.12 × 1.12	89.8–90.8	86.7–87.8
	1/8 (3.175) double strength	AA, A, B	0.115–0.134	2.92–3.40	1.65	8.06	5 × 6.67	1.524 × 2.032	89.8–90.8	84.7–86.8
	1/8 (3.175) double strength	Greenhouse	0.115–0.134	2.92–3.40	1.65	8.06	1.33 × 1.50 1.33 × 2.00 1.50 × 1.67 1.67 × 1.67	0.406 × 0.457 0.406 × 0.610 0.457 × 0.508 0.508 × 0.508	89.8–90.8	84.7–6.8
Heavy sheet	3/16 (4.763)	AA, A, B	0.182–0.205	4.62–5.21	2.45	11.96	10 × 7	3.048 × 2.134	89.6–90.5	82.9–84.6
	7/32 (5.556)		0.212–0.230	5.38–5.84	2.88	14.06	10 × 7	3.048 × 2.134	88.7–90.8	80.6–81.9
	1/4 (6.35)		0.240–0.260	6.10–6.60	3.24	15.82	10 × 7	3:048 × 2.134	88.7–90.6	79.0–81.0
	3/8 (9.525)		0.357–0.384	9.07–9.75	4.86	23.73	5 × 7	1.524 × 2.134	87.2–88.2	74.0–76.0
	7/16 (11.113)		0.400–0.430	10.16–10.92	5.67	27.68	5 × 7	1.524 × 2.134	86.2–87.8	72.0–74.0
Tinted	1/8 (3.175)	AA, A, B	0.115–0.134	2.92–3.40	1.65	8.06	5 × 7	1.524 × 2.134	31.0–56.0	56.0–73.0
	3/16 (4.763)		0.182–0.205	4.62–5.21	2.45	11.96	10 × 7	3.098 × 2.134	43.2–44.2	65.2–66.2
	7/32 (5.556)		0.212–0.230	5.38–5.84	2.88	14.06	8.33 × 7	2.540 × 2.134	13.5–14.2	40.5–41.2

[a]Quality AA is specially selected glass for highest grade work; quality A is select glass for superior glazing; quality B is suitable for general glazing; greenhouse grade is for general glazing but in limited sizes.
[b]This is thin window glass for framing purposes. Do not confuse with picture window glass.

up to a maximum of 10 × 7 ft (3.048 × 2.134 m). Both types are now available in float glass.

Types and Uses

Table *G41* lists the various types, qualities, sizes, weights, and solar radiation values of window and heavy sheet glass.

Single-strength window glass bears a red label, and double-strength glass a blue label. The quality for each type of window glass is marked clearly on the label. (*See* Table *G42*.)

Window and heavy sheet glass, clear and tinted, are used for general glazing. They are also used for mirrors, insulating glass, tempered glass, and laminated glass. They are available as tinted, glare-reducing, and heat-absorbing glass (*see* specific heading).

Application

Condensed Checklist

1. The width dimension should always be listed first so that the glass will be cut with the draw or wave distortions parallel to the bottom of the glazed opening.
2. Wind loads must always be checked to determine the correct thickness of glass for the size (ft² or m²

Table G42 Maximum sizes for window and heavy sheet glass glazed on four sides

Wind velocity		Average thickness of glass									
		0.087 in. (2.21 mm)		0.113 in. (2.87 mm)		0.125 in. (3.18 mm)		0.625 in. (4.76 mm)		0.250 in. (6.35 mm)	
mph	km/h	ft^2	m^2	ft^2	m^2	ft^2	m^2	ft^2	m^2	ft^2	m^2
30	48.27	35.00	3.25	64.50	5.99	72	6.69	162	15.05	288	26.77
40	64.36	17.50	1.63	32.25	3.00	36	3.34	81	7.53	144	13.38
55	88.50	11.60	1.08	21.50	2.00	24	2.23	54	5.02	96	8.82
65	104.60	8.70	0.81	16.10	1.50	18	1.67	41	3.81	72	6.69
80	128.72	5.80	0.54	10.85	1.01	12	1.12	27	2.51	48	4.46
100	160.90	3.50	0.33	6.45	0.60	7	0.65	16	1.49	29	2.69
120	193.08	2.50	0.23	4.60	0.43	5	0.46	11	1.02	20	1.86

Table G43 Minimum clearances for various types of window glass

Type of window glass	Average thickness of glass		Minimum clearance					
			Around edges		Between glass and back of rabbet		Depth of rabbet	
	in.	mm	in.	mm	in.	mm	in.	mm
Single strength	0.087	2.210	$\frac{1}{8}$	3.18	$\frac{1}{16}$	1.59	$\frac{3}{8}$	9.53
Double strength	0.113	2.870	$\frac{1}{8}$	3.18	$\frac{1}{16}$	1.59	$\frac{3}{8}$	9.53
Heavy sheet	0.125	3.175	$\frac{1}{8}$	3.18	$\frac{1}{16}$	1.59	$\frac{3}{8}$	9.53
	0.188	4.775	$\frac{1}{8}$	3.18	$\frac{1}{8}$	3.18	$\frac{1}{2}$	12.70
	0.219	5.563	$\frac{1}{4}$	6.35	$\frac{1}{8}$	3.18	$\frac{5}{8}$	15.88

area) of the opening. Clearance dimensions for the thickness of glass selected should also be checked. (*See* Tables *G41* and *G43*.)

3. Local, municipal, and state codes and also codes of the Fire Underwriters, insurance companies, labor departments, and the federal government (Army, Navy, etc.) should be checked for limitations and requirements.
4. The type of float, window, or heavy sheet glass chosen must meet all other requirements for its end use.
5. The type of material into which glass is to be installed and the thickness of the glass determine the type of glazing material, type of glazing points or clips, the necessity for setting blocks, and the clearances (*see* Figures *G49* to *G51*).
6. Rabbets into which glass is to be installed must be prepared to receive the glass. If of wood or steel, they must be primed; with aluminum rabbets any protective coating must be removed.
7. Glazing must never be done at temperatures below 40°F (4.44°C) unless precautions are taken to prevent moisture from condensing in rabbets where glass is to be installed.
8. Operating windows must not be handled or opened until after the glazing compound is set.
9. When using float glass, one should always check the thickness required for window or sheet glass and the rabbet requirements.
10. When using tinted float, window or heavy sheet, one should always check clearances, depth of rabbet, and type of glazing materials.

Conditions Favorable to the Use of Window or Heavy Sheet Glass

1. Use float and window glass for general glazing of all types of windows, doors, storm sash, etc.
2. Use picture glass for covering photographs, pictures, maps, and other framing purposes.
3. Use float heavy sheet glass for windows, doors, etc., where greater strength is required and also for shelves, display cases, furniture tops, etc.

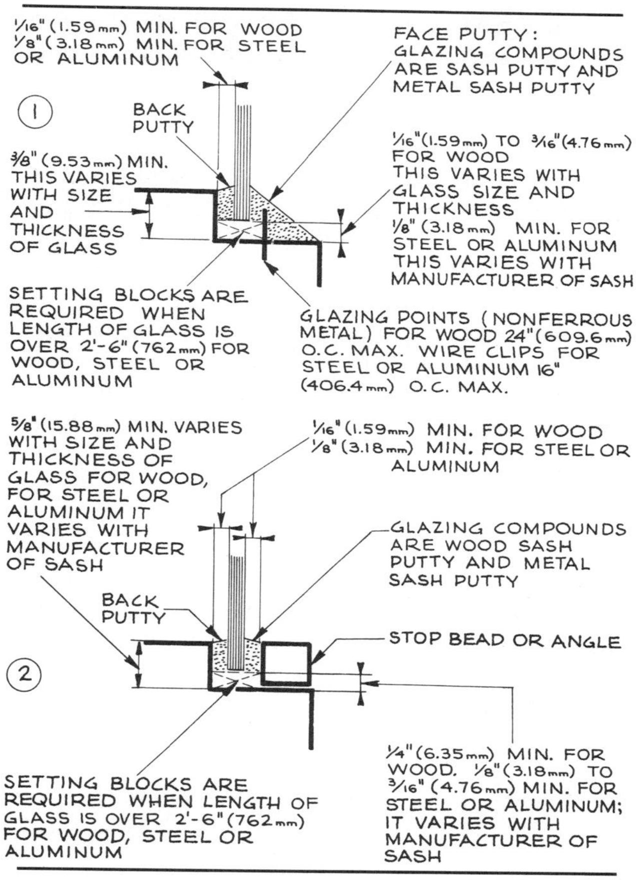

Figure G49 Window and heavy sheet glass glazed in wood, steel, or aluminum, showing face glazing (1) and glazing with a stop bead (2).

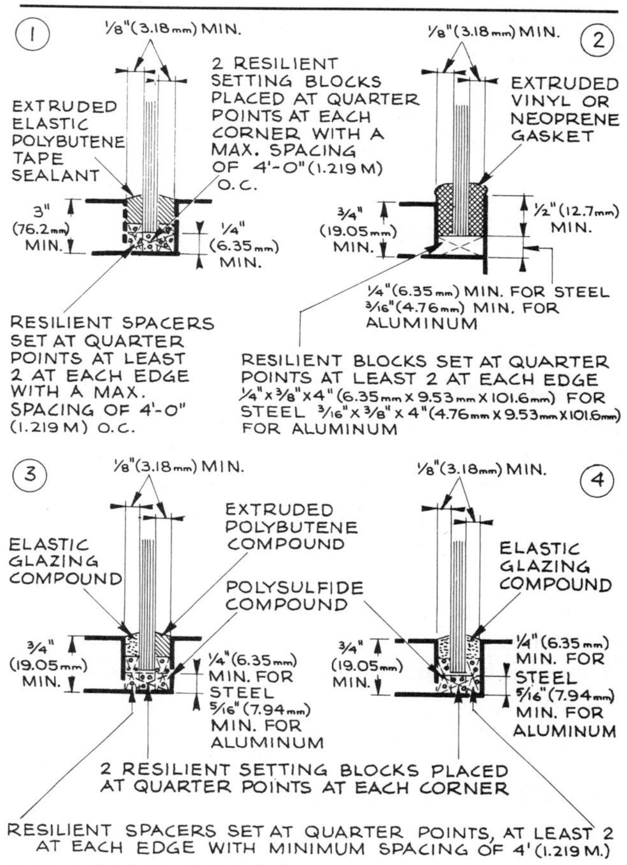

Figure G50 Heavy sheet glass set in steel or aluminum with tape glazing (1), gasket (2), multiple-seal (3), and multiple-compound glazing (4).

Conditions Unfavorable to the Use of Window and Heavy Sheet Glass

1. In areas where fire resistance is necessary.
2. In any area where glass will be placed in tension.

History and Manufacture

Modern manufacture of window and sheet glass is a continuous process. Raw materials consisting of silica sand (silicon), ground limestone (calcium) and soda ash (sodium) are stored in bins similar to farm silos. These raw materials, drawn by gravity, are individually weighed, proportioned, thoroughly mixed, and transported to the furnace. Broken glass (cullet) is added to the mixture to facilitate melting. This mixture is added to one end of the furnace, replenishing the exact amount of glass that is being drawn at the other end of the furnace. The furnaces have capacities up to 1700 tons (1727.2 metric tonnes) or more and are heated by gas to

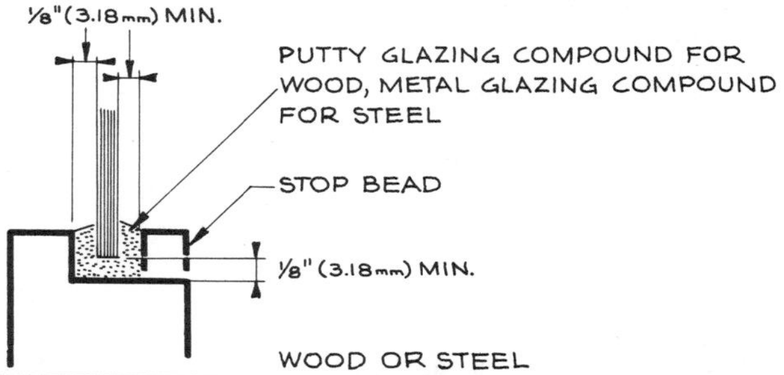

Figure G51 Heavy sheet glass set in wood or metal frames with stop bead.

temperatures as high as 2900°F (1593.33°C). The molten glass moves slowly from the feeding end to the drawing end. The glass is then either drawn vertically through a long roller furnace for annealing called a leer and at the top is cut into sheets (Fourcault process), or it is drawn vertically and then bent over a roller from which it passes into a horizontal leer and then is cut into sheets (Colburn or Libbey-Owens process). (*See also* Glass, "History and Manufacture.")

The most recent advance in manufacturing is the float glass process, now used to produce window and heavy sheet glass (*see* Glass; Float).

GLASS, WIRED

Physical and Chemical Properties

The composition of the glass for wired glass is the same as that used for window, heavy sheet, float, and plate glass. Wired glass is rolled flat glass in which hexagonal twisted wire mesh or diamond-shaped welded wire mesh is embedded at the middle of its thickness. The glass must be not less than $\frac{1}{4}$ in. (6.35 mm) thick, the mesh no larger than $\frac{7}{8}$ in. (22.23 mm), and the wire not less than gauge No. 24.

Wired glass is fire resistant and can be used in openings exposed to fire hazards. It will remain intact even after cracking or breakage and will not shatter when it receives a heavy impact. It is available with its surfaces polished or patterned and with various finishes. Float, plate, corrugated, and other types of glass are available with wire mesh and are covered fully under their particular headings.

Wired glass is available with various types of wire and wire mesh, among them hexagonal, diamond, square, and vertical. Wire sizes vary from 0.020 to 0.023 in. (0.508 to 0.584 mm). Diamond mesh generally is $\frac{7}{8}$ in. (22.23 mm) by $1\frac{1}{8}$ in. (28.58 mm); square mesh is generally $\frac{1}{2}$ in. (12.7 mm) on each side; vertical wires are spaced $\frac{1}{2}$ in. (12.7 mm) apart; and there is the standard hexagonal wire mesh. When selecting a type of wire glass, one should always check with the manufacturers for current data on the various types of wire and wire mesh available in the various types of glass (float, plate, patterned, etc.).

Types and Uses

Wired glass used as a fire-resistant material is subject to definite size limitations. These limitations are as follows: minimum thickness, $\frac{1}{4}$ in. (6.35 mm); maximum length and width, 4 ft (1.219 m); maximum area, not to exceed 5 ft^2 (0.4645 m^2). Also, the glass must be installed in nonflammable materials. Wired glass for other applica-

Table G44 Description, thickness, sizes, type of mesh, weight, and finishes of wired glass

Type of wired glass	Thickness		Maximum sizes: Width		Maximum sizes: Length		Weight		Available finishes[a,b]	Type of wire mesh
	in.	mm	ft	m	ft	m	lb/ft^2	kg/m^2		
Patterned, granular, fine	$\frac{1}{4}$	6.35	4	1.219	12	3.658	3.50	17.087	Fire, frosted, special	Hexagonal only
Patterned, granular, medium	$\frac{1}{4}$	6.35	4	1.219	12	3.658	3.50	17.087	Fire, frosted, special	Hexagonal or diamond
Patterned, hammered	$\frac{1}{4}$	6.35	4	1.219	12	3.658	3.50	17.087	Fire, frosted, special	Hexagonal or diamond
	$\frac{3}{8}$	9.53	4	1.219	8.33	2.540	5.00	24.410		Hexagonal or diamond
Float and polished, clear	$\frac{1}{4}$	6.35	4.83	1.473	11.83	3.607	3.40–3.50	16.599–17.087	Fire, frosted, special	Diamond
	$\frac{1}{4}$	6.35	5	1.524	12	3.658	3.40–3.50	16.599–17.087		Hexagonal
Patterned, ribbed, small	$\frac{1}{4}$	6.35	4	1.219	12	3.658	3.50	17.087	Fire, frosted, special	Hexagonal only
	$\frac{3}{8}$	9.53	4	1.219	8.33	2.540	5.00	24.410		Hexagonal only
Patterned, ribbed, large	$\frac{1}{4}$	6.35	4	1.219	12	3.658	3.50	17.087	Fire, frosted, special	Hexagonal only

[a]Special finishes are covered by trademark or patent.
[b]Frosted finish is on patterned surface for single-surface finish and on smooth side for special finishes. A frosted finish decreases light transmission by approximately 15 to 30% whereas the special finishes decrease it by 3% for one surface and by 6% for both surfaces.

tions can be thinner and used in larger sizes. Table *G44* shows types, sizes, and weights.

Wired glass in the role of a fire-resistant material is generally used in fire doors and fire door frames; it is also used for openings in corridors and room partitions, in exterior walls, in vertical shafts, and passageways leading to exterior fire escapes. It finds uses other than as a fire-resistant material in cases where there is a possibility of impact and abuse, and in any areas where flying glass would be dangerous, for example, skylights and overhead glazing. It also finds use as a burglar-proofing material.

Application

Condensed Checklist

1. Local, municipal, and state codes and also the Fire Underwriters, insurance companies, labor departments, and federal government (Army, Navy, etc.) should be checked for requirements and limitations for fire resistance.
2. The type of wired glass and the type of wire mesh should each be suited to its end use.
3. A finish should be chosen that gives the privacy and light diffusion required by the design.
4. Dimensions should be checked for availability in the type of wired glass desired.
5. Thickness should be checked in relation to size limitations, wind loads, and end use.
6. The materials into which wired glass is to be installed should be checked for fire resistance if exposed to fire hazard.

History and Manufacture

The search for fire-resistant materials led to the development of wired glass, and in 1899 wired glass was tested and approved by the Fire Underwriters.

Wired glass is manufactured by passing molten glass through a pair of rolls, with the wire mesh fed into the molten glass at approximately the center of its thickness.

GLAZES AND PORCELAIN ENAMELS

Physical and Chemical Properties

By definition a glaze is a transparent-to-opaque, fused, vitreous surface finish for ceramics (clay products). Porcelain enamel is a similar, fused vitreous surface for metals. Both provide a hard, smooth, dense surface that, like glass, is impervious to moisture and resistant to chemical action, high temperatures, mechanical and thermal shock, and scratching. If these are used on the exterior, resistance to weathering is an important characteristic.

A properly applied glaze also increases the mechanical strength of the clay body. Porcelain enamel, on the other hand, is applied to a basically strong material (usually a ferrous metal) mainly to protect the surface.

Coefficient of Expansion. A glaze or a porcelain enamel should have a coefficient of expansion compatible with that of the material to which it is applied; otherwise it may peel off or craze. In glazes, this effect has been largely overcome by the use of fillers such as talc and pyrophyllite in both body and surface finish.

Raw Materials. The main raw materials for both glazes and porcelain enamels are borax, feldspars, quartz, and china or ball clay. Other compounds are added for special properties, for example, cobalt for adherence, magnesium for hardness, and titanium for acid resistance and opacity.

Types and Uses

Glazes and porcelain enamels both provide a durable, attractive, colored surface that is particularly desirable where sanitary conditions and ease of maintenance are important.

Glazes in general are used for ceramic tile and baked clay products of all kinds—facing brick, ceramic veneer, finish surfaces of structural clay tile, decorative architectural applications, and sanitary ware. Salt glazes, because of their durability and low cost, are widely used for sewer and drain pipe, clay tile flue linings, structural clay tile, and facing brick.

Porcelain enamels find their greatest application in architecture as a surfacing finish material on sheet metal for the exterior and interior of buildings for walls and partitions, and in commercial and institutional equipment and built-in furniture.

History and Manufacture

Glazes. Glazes may be either salt glazes or sprayed glazes. The latter are applied wet or dry (in which case the method is called dipping).

Salt glazing is a method of applying a glaze to a heavy clay product by throwing salt on the fires near the end of the firing period. Salt glazes are possible only on ceramic products, as this method depends on direct chemical reaction between the sodium of the salt and the alumina or silica of the clay body. Salt glazes are especially resistant to chemical attack and mechanical shock but are limited in color by that of the natural clay body (white, tan, brown, mahogany color) and by the few metallic coloring oxides suitable for this glazing method (e.g., cobalt for blues, zinc powder for greens).

Spray glazing is a method of applying a glaze to ceramics in which the compounds for the finish, including fluxes, hardeners, opacifiers, and plasticizers, are sprayed over the material before or after drying. The whole is then heated to a high temperature so that the surface finish is fused to the body and becomes inseparable from it.

There are two methods of applying this type of glaze. In one the colored glaze is sprayed directly onto the clay product; in the other, a transparent glaze is sprayed over a white or colored coating (slip) so that the slip is between the glaze and the clay body.

These spray glazes are either raw (insoluble compounds) or fritted (i.e., mixtures containing soluble compounds that are made insoluble by being previously fused and then pulverized).

Polychrome glazed finishes require highly skilled craftmanship, as each different color glaze must have a separate burning at a lower fusing temperature than the preceding one.

Porcelain Enamels. This type of finish is an industrial outgrowth of enameling on precious metals, which originated as a decorative technique for jewelry and objects of art. The transition from art to industry came about rapidly in the early part of the nineteenth century.

Cast iron dry-process enamels, developed in Bohemia and England, were the first to be used on a large scale. Sheet iron enameling followed a few years later. In the United States a wet process on lightweight or thin castings was developed. Now sheet metal enameling is the most extensively used process and in this country is characterized by automatic equipment, technical control, and mass production. The term "porcelain enamel" came into common use about 1929. "Vitreous enamel" is the other term used to differentiate this fused type of coating from a paint type of surface finish.

Porcelain enamels consist basically of a transparent vitreous compound called flux or frit.

Flux is a fused and then pulverized combination of silica, alumina, soda, or potash (potassium carbonate), metallic oxides, and assorted compounds for imparting special properties. It is applied in a thin coat to the prepared metal and fired at a high temperature, ranging from 1365 to 1650°F (740.56 to 898.89°C), for a comparatively short time (2 to 8 minutes for sheet iron, 15 to 30 minutes for cast iron).

Several types of fluxes or frits are widely used: (1) titanium types, with titanium dioxide as opacifier, for white and for light pastel colors; (2) zirconium-opacified frits, which have less acid resistance but greater workability; (3) an acid-resistant antimony type for deep, bright colors (these have less resistance to abrasion and less covering power than the titanium type); and (4) lead-type frits for aluminum and aluminized steel.

Porcelain enamels can be applied by the spray method (wet process), by dipping (dry process), and by flow coating. One, two, or three layers of porcelain enamel may be applied to the base metal.

Base metals most commonly used are mild steel sheet and cast iron. Titanium enameling steel, aluminized steel, and stainless steel are newer, highly satisfactory types. On ferrous metals the first layer is usually a ground coat containing cobalt, nickel, or molybdenum oxide for good adhesion to the base metal.

The widespread use of aluminum has led to research on low-temperature porcelain enamels for aluminum. At present, aluminum must be pretreated, and lead-containing coatings are being used. Lithium oxide and lithium-cobalt compounds have been found to lower fusion temperatures, for example, 930°F (498.89°C) for aluminum, 1290 to 1380°F (698.89 to 748.89°C) for sheet iron; therefore, these lithium compounds will no doubt play an important role in enamels for this metal. (*See* Tables *G45* and *G46*.)

GOLD

Physical and Chemical Properties

Symbol: Au
Atomic number: 79
Specific gravity: 19.32 (20°C, 68°F)
Melting point: 1064.43°C, 1947.97°F
Boiling point: 2940°C, 5324°F

Gold is a soft, heavy, lustrous, bright yellow metal. It is corrosion resistant to air and is not attacked by acids.

Workability. Gold is the most malleable and ductile of all the metals and can be rolled, hammered, drawn, spun, cast, soldered, and welded but cannot be machined.

Commercial Forms. Gold is available in bar, rod, wire, tubing, sheet, foil, and plate form and in special shapes.

Table G45 Properties of various oxides in frits

Metallic oxide	Fusion temperature	Thermal expansion	Durability (resistance to mechanical and thermal shock, chemical resistance)
SiO_2	High melting point; high viscosity	Low	High chemical resistance
B_2O_3	Easy melting; low viscosity; quick setting	Very low	High surface hardness
Na_2O	Easy melting; low viscosity; slow setting	High	Low chemical resistance; low mechanical strength
K_2O	Easy melting; higher viscosity than Na_2O	High	Low chemical resistance; low mechanical strength
PbO	Easy melting; slow setting		High density; high refractive index
BaO	Assists melting; some fluxing action		Can replace lead oxide
CaO	Fluxing action at high temperatures; very quick setting		Excess causes devitrification
TiO_2	Fluxing action if replacing silica; reduces viscosity; slow setting		Good chemical resistance
ZnO	General fluxing action; high viscosity	Low	Decreases chemical resistance
MgO	Refractory; high viscosity; slow setting	Low	High mechanical strength
Al_2O_3	Refractory; very high viscosity; slow setting	Low	Prevents devitrification; high chemical resistance; high mechanical strength

Table G46 Refractive indices of opacifying metallic oxides

Tin oxide	2.04
Zirconium oxide	2.17
Antimony oxide	2.09
Titanium oxide (anatase)	2.52
Titanium dioxide (rutile)	2.76
Enamel glass	1.50

Table G47 Uses of gold

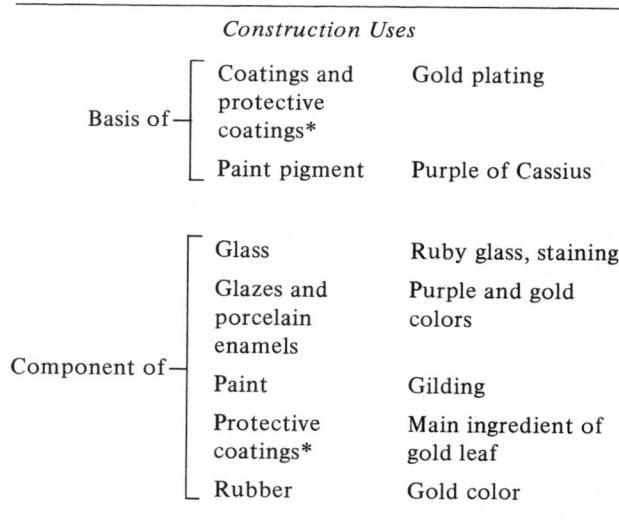

Construction Uses		
Basis of	Coatings and protective coatings*	Gold plating
	Paint pigment	Purple of Cassius
Component of	Glass	Ruby glass, staining
	Glazes and porcelain enamels	Purple and gold colors
	Paint	Gilding
	Protective coatings*	Main ingredient of gold leaf
	Rubber	Gold color
Nonconstruction Uses		
Coinage,* plating,* photography, special laboratory equipment*		

Types and Uses

Because most of the world's gold is government controlled and used to stabilize currency, gold remains high in cost and limited in availability for industrial applications. Most of the gold used industrially is in metallic form, as indicated by asterisks in Table *G47*.

Because of its color, luster, and chemical stability, gold is used for jewelry, objects of art, and coinage. Coinage gold is 90% gold and 10% copper. Table *G48* lists the gold alloys by color.

Gold is used to plate metals used under special conditions and as a decorative finish for items such as hard-

Table G48 Gold alloys by color

Color of gold	Composition (percent)						
	Au	Ag	Ni	Cu	Al	Zn	Fe
White	85		10			5	
Gray	86	8.6					5.4
Red	75			25			
Pink	50		6	42		2	
Pale yellow	92	0.8					7.2
Yellow	50	25.0		25			
Green	75	25.0					
Purple	79				21		
Blue	75						25.0

ware and plumbing fixtures. It can be applied to brass, bronze, copper, nickel, Monel, and silver.

Carat Designation. Gold content is expressed in carats, defined as parts of gold in 24 parts of the alloy: 24-carat gold is pure gold; 18-carat gold is $^{18}/_{24}$ or 75% gold.

Gold Leaf. Gold leaf is extensively used for lettering and sign work on doors and walls, for decorative purposes in buildings, and for art objects and furniture. It may also be used for wall or ceiling treatment when a special color or effect is desired and cost is not the only factor. Gold leaf, usually 18 to 22 carat, is applied with a coat of size or primer. One gram of gold will produce 6 ft^2 (0.5574 m^2) of gold leaf 0.000033 in. (0.00084 mm) thick, and 1 oz will produce 170 ft^2 (15.793 m^2). Table *G49* shows colors and types.

Table G49 Types and colors of gold leaf

Name of gold leaf	Composition (grains)		
	Gold	Silver	Copper
Red	456–460		20–24
Pale red	464		16
Extra deep	456	12	12
Deep	444	24	12
Citron	440	30	10
Yellow	408	72	
Pale yellow	384	96	
Lemon	360	120	
Green or pale	312	168	
White	240	240	

Gold Burnish. A liquid mixture called gold burnish, containing metallic gold or organic gold compounds, with or without silver or platinum, will whiten or give a green color to alloys.

Gold Substitutes. There are several substitutes for gold or a gold effect. A high-copper alloy called rich low brass is one of these. Another is gold shell, a duplex type of composite metal, also called doublé, in which a base of rich low brass is given a thin gold facing (less than 1% by weight) by rolling the two together at high temperature.

History and Manufacture

Because gold is widely found in its natural form, it was one of the first metals used by man and was highly prized by all the early civilizations, including the Egyptian, Minoan, Assyrian, Etruscan, Chinese, Mayan, and Inca. For the period from 1492, when Columbus discovered America, to approximately 1600, the amount of gold that came from America was equal to 35% of the world production for the same period. This had a profound effect on Europe and unbalanced its entire economic and political structure.

Naturally occurring metallic gold contains some silver and sometimes small amounts of copper, platinum, palladium, and other elements. Today most of the gold produced in the United States is refined by the electrolytic process, which produces a gold more than 99.99% pure.

In the electrolytic process the impure gold is cast into plates that serve as the anode. Thin, pure gold plates suspended between the anode plates act as the cathode. This assembly of anodes and cathodes is suspended in a solution of gold chloride and hydrochloric acid. As the electric current passes from anode to cathode, the gold is dissolved out of the impure plates and electrodeposited onto the cathode. The impurities either remain in solution or precipitate to the bottom as a slime or sludge. Gold is also obtained as a by-product in lead, zinc, and copper production and in the refining of many rare metals.

GRATINGS

There are three methods of fabrication for metal gratings: welded, pressure-locked, and riveted. The metals used are steel, galvanized steel, aluminum, and stainless steel. Bearing bars vary in thickness from $\frac{1}{8}$ to $\frac{3}{16}$ in. (3.18 to 4.76 mm) and in depth from $\frac{3}{4}$ to over $1\frac{3}{4}$ in. (19.05 to 44.45 mm), with cross bars from $\frac{7}{64}$ to $\frac{3}{16}$ in. (2.78 to 4.76 mm) in thickness and from $\frac{3}{16}$ to 1 in. (4.76 to 25.4 mm) in depth for steel, galvanized steel, aluminum, and stainless steel (*see* Figure *G52*).

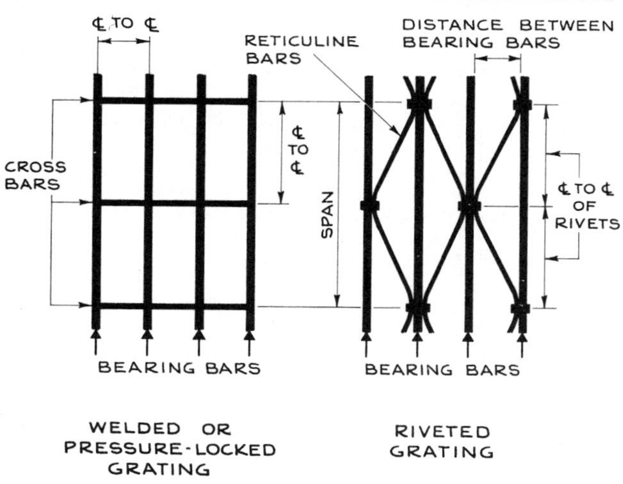

Figure G52 Standard types of gratings.

Types and Uses

Metal bar gratings are designated by a standard marking system, which identifies five characteristics of the grating, as follows:

W-19-4 (1 × $\frac{3}{16}$) steel: W means welded, *19* means bearing bars spaced $\frac{19}{16}$ or 1$\frac{3}{16}$ in. (30.16 mm) on center, *4* means cross bars spaced 4 in. (101.6 mm) on center, *(1 × $\frac{3}{16}$)* is the bearing bar size, and *steel* indicates the material.

R-18-7 (1$\frac{1}{4}$ × $\frac{1}{8}$) steel: R means riveted, *18* means bearing bars spaced $\frac{18}{16}$ or 1$\frac{1}{8}$ in. (29.58 mm) between faces, *7* means rivets spaced 7 in. (177.8 mm) on center, *(1$\frac{1}{4}$ × $\frac{1}{8}$)* is the bearing bar size, and *steel* is the material.

P-15-2 (1$\frac{1}{4}$ × $\frac{3}{16}$) aluminum: P means pressure-locked, *15* means bearing bars spaced $\frac{15}{16}$ in. (23.81 mm) on center, *2* means cross bars spaced 2 in. (50.8 mm) on center, *(1$\frac{1}{4}$ × $\frac{3}{16}$)* is the bearing bar size, and *aluminum* is the material.

Aluminum gratings will span from 2 to 8 ft (0.610 to 2.438 m), depending on bearing bar size, and steel gratings will span from 2 to 9 ft (0.610 to 2.743 m), again depending on bearing bar size.

Gratings are used in the construction field for flooring, stairs, vault covers, racks and shelving, trench and pit covers, areaways, bridge flooring, and concrete reinforcement. The type of material selected depends on the end use and the exposure—whether interior or exterior.

GRAVEL

Gravel is a coarse, granular material resulting from the erosion and disintegration of rocks by natural forces, both chemical and mechanical. It is composed of small, usually smooth, rounded stones or pebbles. It is distinguished from sand by the size of the grain, which is normally more than $\frac{1}{4}$ in. (6.35 mm) and may be as large as 3$\frac{1}{2}$ in. (88.9 mm). Larger fragments are called cobbles and boulders. Pea gravel is screened gravel between $\frac{1}{4}$ and $\frac{1}{2}$ in. (6.35 and 12.7 mm) in size.

Gravel deposits may occur along lakes, rivers (bank gravel), or beaches (marine gravel). Most gravel deposits are the result of water and glacial action. Water deposits are normally stratified. Glacial ones are not.

Gravel is made up of a quartz residue; it may contain some sand, shale, sandstone, and other rock materials. Bank or lake-bed gravel is intermixed with sand, silt, or clay.

Types and Uses

Commercial gravel is washed to remove the clay and organic materials, graded, and sold by the ton (1.016 tonne) or cubic yard, a cubic yard (0.765 m^3) weighing about 3000 lb (1360.77 kg). In contrast to sand, the weight of gravel is relatively constant regardless of moisture content.

Pit-run gravels are those in which no change is made in natural grading by screening or other means.

Crushed gravel is gravel that has been artificially crushed and is not to be confused with crushed stone, although it is used for many of the same purposes in construction work.

Gravel used as an aggregate in concrete may create problems because of the presence of hydrous or glassy silicates, the molecular water of which may react harmfully with soluble alkalis in cement.

GRIT

The word "grit" refers to granular abrasive material such as aluminum oxide or silicon carbide which may be coated on cloth, paper, or wheels for sanding, grinding, or polishing purposes. It may also be cast into a metal surface or used as a surface treatment of concrete to provide a nonslip surface. (*See also* Abrasives.)

GUTTERS AND LEADERS

Gutters may be defined as the conduits that catch rain water as it drains off a sloped roof. Leaders are the conduits that carry off water collected in the gutters or on the tops of flat roofs and conduct it to the ground. Gutters are commonly made of copper, aluminum, Monel, zinc, zinc-coated (galvanized) steel, stainless steel, or wood. (*See* Figures *G53* and *G54*.) They are either built in as part of the roof or suspended from the edge of the roof. Leaders are always made of metal; if exposed on the face of the building, they generally match the material of the gutter if it is made of metal.

Today we know the exact rainfall in any geographic area and can calculate accurately the amount of water to be handled by gutters and leaders. A widely diversified series of stock gutters made of the materials previously listed is available. These gutters are summarized in Table *G50*. Methods used to calculate the size of the gutter and of the leader are beyond the scope of this book.

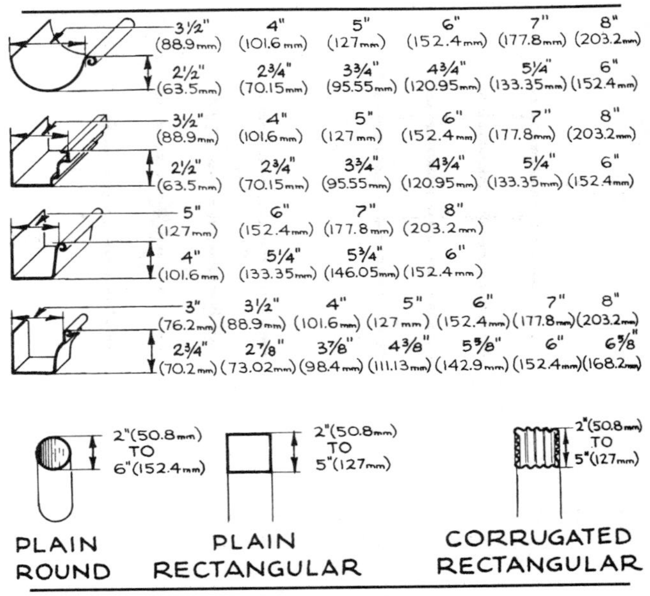

Figure G53 Typical types of gutters and leaders.

Table G50 Typical materials used for gutters and leaders

Material	Advantages	Disadvantages	Major use
Aluminum (*see* Aluminum Sheet and Strip)	Needs no painting; large selection of sizes and shapes; will not stain adjoining materials	Lightweight; must be well secured; difficult to solder on construction site; expansion joints require careful study and detailing	All types of buildings
Bronze (*see* Bronze)	Needs no painting; flexibility of design since these items are not stock ones but are fabricated for job	Will stain adjoining materials; requires careful study and detailing	Monumental architectural buildings where cost is not important
Copper (*see* Copper Sheet and Strip)	Needs no painting; large selection of sizes and shapes; easily soldered	Will stain adjoining materials; number of expansion joints requires careful study and detailing	All types of buildings
Monel (*see* Monel)	Needs no painting; not stock but fabricated for job; will not stain adjoining materials	Requires careful study and detailing as fabricated for job	All types of buildings
Stainless steel (*see* Stainless Steel Sheet and Strip)	Needs no painting; easily soldered; will not stain adjoining materials	Number of expansion joints requires careful study and detailing; limited selection of sizes and shapes	All types of buildings
Steel (*see* Steel, Galvanized; Steel Sheet and Strip)	Large selection of sizes and shapes; easily soldered	Requires paint; if galvanized, paint must be compatible with zinc	Residential, factory, and farm buildings
Terneplate (*see* Terneplate)	Not stock but fabricated for job; will not stain adjoining materials	Requires paint that is compatible with lead; requires careful study and detailing as fabricated for job	Residential
Wood (*see* Wood Gutters)	Easily joined and installed; will not stain adjoining materials	Requires staining or painting; limited selection of sizes and shapes	Residential
Zinc (*see* Zinc Sheet and Strip)	Will not stain adjoining materials; easily soldered; requires no painting	Limited selection of sizes and shapes	Residential

Application

Condensed Checklist

1. The width and shape of the gutter will control the distance between expansion joints and the distance from leader to first expansion joint.
2. The form of the gutter must allow for expansion of any water that freezes in it; that is, one side should be sloping, or the gutter should have a rounded cross section (*see* Figure *G54*).

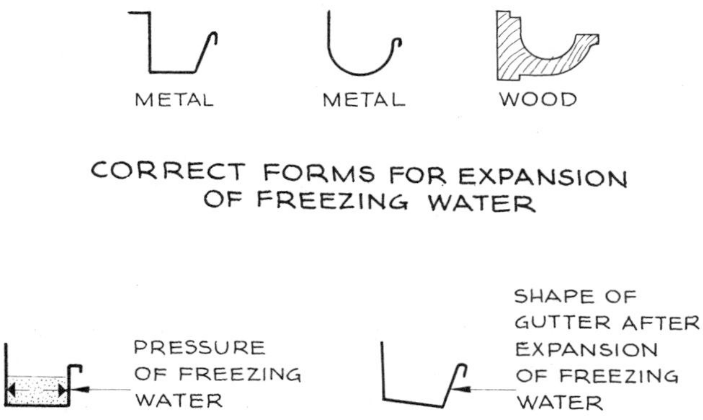

Figure G54 Typical shapes of gutters to avoid damage from the expansion of freezing water.

3. The gutter must be set with enough pitch or be sufficiently large for the water to drain to the leaders and not spill out of the gutter.
4. The outer edge of the gutter should always be at a lower level than the inside face.
5. The method of supporting a gutter and the number of supports should be such that the face of the gutter remains straight and the gutter does not sag even when filled.
6. Leaders should be sufficient in number and size to take the total water capacity of the gutter.
7. The method of connecting from gutter to leader to face of building should be checked.
8. If gutters are built into the roof and the connection of leader to gutter is partially concealed, some method must be provided to get at this joint should it freeze and break.
9. On interior leaders, some method of getting heat from the building to the connection at the roof must be devised so that freezing and breakage cannot occur.
10. The method of securing leaders to the face of the building should be checked (*see* Figure *G55*).

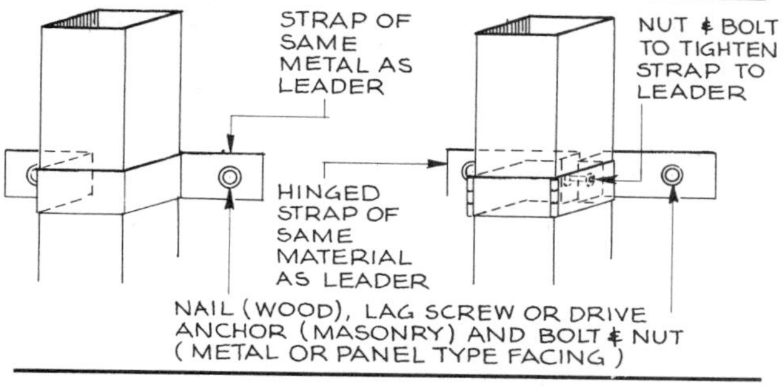

Figure G55 Various methods of attaching a leader to the face of a building.

11. Galvanic action. Leader and gutter must be made of the same metal or of compatible metals. All nails, supports, hangers, brackets, ties, etc., should always be checked; they should be made either of the same metal as the gutters and leaders or of compatible metals.

Conditions Favorable to the Use of Gutters and Leaders

1. Where there is heavy rainfall.
2. Where roof areas are large and the quantity of water collected, if allowed to spill off, would destroy surrounding landscape through erosion.
3. On flat roofs, internal leaders are usually necessary.

Conditions Unfavorable to the Use of Gutters and Leaders

1. In areas characterized by severe freezing and thawing or in very cold climates.

GYPSUM

Physical and Chemical Properties

Gypsum is a natural hydrated calcium sulfate (formula: $CaSO_4 \cdot 2H_2O$; specific gravity: about 2.3.). It is a common mineral, usually found combined with the oxides of iron and aluminum, calcium and magnesium carbonates, etc. If pure, it is white in color, but more often it is gray, red, or brown because of impurities.

Gypsum has a unique property that makes it valuable in construction. When it is heated (calcined) to 325 to 340°F (162.78 to 171.11°C), it loses about three-fourths of its combined water. Then, if water is added, it hydrates into felted crystals cemented together and resets in any desired shape to its original rocklike form:

$$CaSO_4 \cdot 2H_2O \xrightarrow{\text{calcined}} CaSO_4 \cdot \tfrac{1}{2}H_2O + \tfrac{1}{2}H_2O$$

$$\xrightarrow{\text{2nd settle}} CaSO_4 + \tfrac{1}{2}H_2O$$

Gypsum is available either calcined or uncalcined in multiwalled paper sacks or bags.

Types and Uses

The bulk of gypsum is used in construction, most of it in calcined form. Gypsum is important in the architectural field for making base coat and finishing plaster, acoustical plaster, fireproof and other special plasters, plaster boards, plaster lathing, building block (partition tile), roof tile, and reinforced plaster deckings.

Uncalcined gypsum is an effective, economical retarder in cement, used to control the setting of portland cement and concrete. It is also used as a filler in paints, paper, and textiles. The uses of the two calcined types of gypsum, plaster of paris and Keene's cement, are discussed further under Plaster and Plastering.

Compact, finely granular gypsum is called alabaster. In England gypsum is also called plaster stone and, because it is used in potteries for making molds, potter's stone.

History and Manufacture

Gypsum was used as a plaster by the Egyptians before 2100 B.C. The Greeks and Romans used it extensively, and our word "plaster" comes from the Greek word for both the raw material and the calcined product. Gypsum and lime have both been used in construction since ancient times and even today are competing materials to a limited extent. Each can be used to give a smooth, hard surface finish.

Calcination. Calcination takes place in rotary kilns (a continuous process) or in kettles (in batches). More recently, other types of continuous-flow machinery have been developed. The temperature of calcination controls the type of gypsum produced. At 325 to 390°F (162.78 to 198.89°C) the result is plaster of paris of various types. Calcination with alum, etc., at 900 to 1000°F (482.22 to 537.78°C) produces Keene's cement and other special types which contain small quantities of other materials such as borax or potassium carbonate.

Gypsum for wall plaster may have retardants, fiber, and other materials ground with it. Calcined gypsum for use in prefabricated products may have accelerators added to it.

H

HAFNIUM

Physical and Chemical Properties

Symbol: Hf
Atomic number: 72
Specific gravity: 13.31 (20°C, 68°F)
Melting point: 2150°C, 3902°F
Boiling point: 5400°C, 9752°F

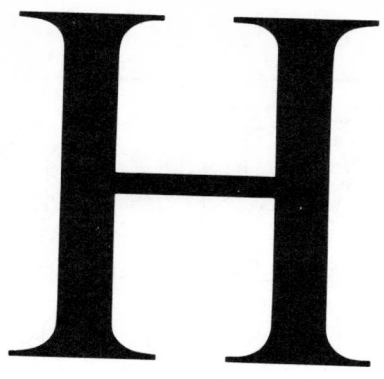

Hafnium is a strong, brilliantly lustrous metal which resembles zirconium very closely in physical and chemical properties and occurs together with it in ore minerals. Hafnium has corrosion resistance to almost all chemicals except fluorides and remains untarnished in almost all industrial atmospheres.

Machinability. In very pure form hafnium is soft and ductile, and can be hot- or cold-rolled, swaged, hammered, and drawn. Its machinability resembles that of 302 stainless steel.

Commercial Forms. Hafnium is available in ingot form.

Types and Uses

At present only a small quantity of hafnium in metal and compound form is available, and most of this is sold to the Atomic Energy Commission. Experimental work is in progress to increase both its production and its industrial applications.

Possible future uses of hafnium are as follows: X-ray tubes, jewelry, filaments in gas-filled devices, electronics, special glasses, high-temperature ceramics, abrasives, refractories, high-temperature insulation, special metal alloys.

Although zirconium and hafnium have many similar properties, there is one important difference between them. Hafnium has a high neutron cross section, that is, it absorbs neutrons and can in fact be used as a shield; zirconium, on the other hand, has a low neutron cross section, that is, it is transparent to neutrons. This property is resulting in a separate field of use for hafnium, more as a control than as a shield. Hafnium is the preferred metal for control rods in nuclear reactors since it is free from radiation damage.

History and Manufacture

In 1922 Coster and Hevesy discovered a new element and named it hafnium after the Latin name for Copenhagen. It was little more than a laboratory curiosity until 1951, when a process to produce hafnium-free zirconium was developed. Hafnium hydroxide, a by-product of this process, is calcined into the oxide, from which the metal is obtained by a process similar to that used for zirconium.

HANGERS

Hangers are a related group of devices for joining similar or different materials in the construction of a building (*see* Figures *H1*, *H2*, and *H3*). As a group they can be classed as rough hardware. There are four general types:

1. Hangers used in wood construction. Special hangers are made to attach one wood member to another piece of wood; other special types attach wood to masonry. Wood is also hung from various types of steel structural members.
2. Steel-to-steel hangers. These are specially designed to support one piece of the basic structural steel by suspension from another.
3. Various small hangers used to suspend acoustic materials and other types of ceiling material from basic construction made of steel, wood, or concrete.
4. Numerous hanger specialties used to support and hang wood forms for reinforced concrete work.

Hangers are widely used in wood frame construction, since one beam which intersects another at right angles cannot be adequately joined by nails or cleats, nor can structural joints be made between wood and another

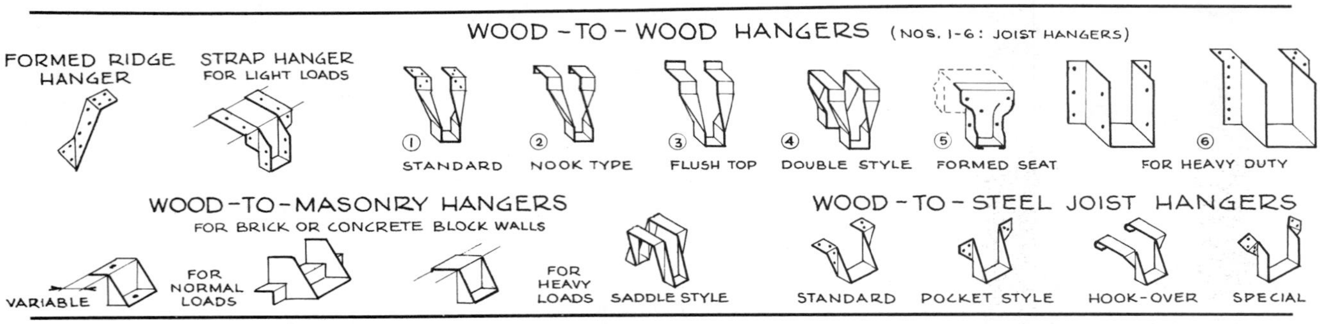

Figure H1 Wood-to-wood hangers, wood-to-masonry hangers, and wood-to-steel hangers.

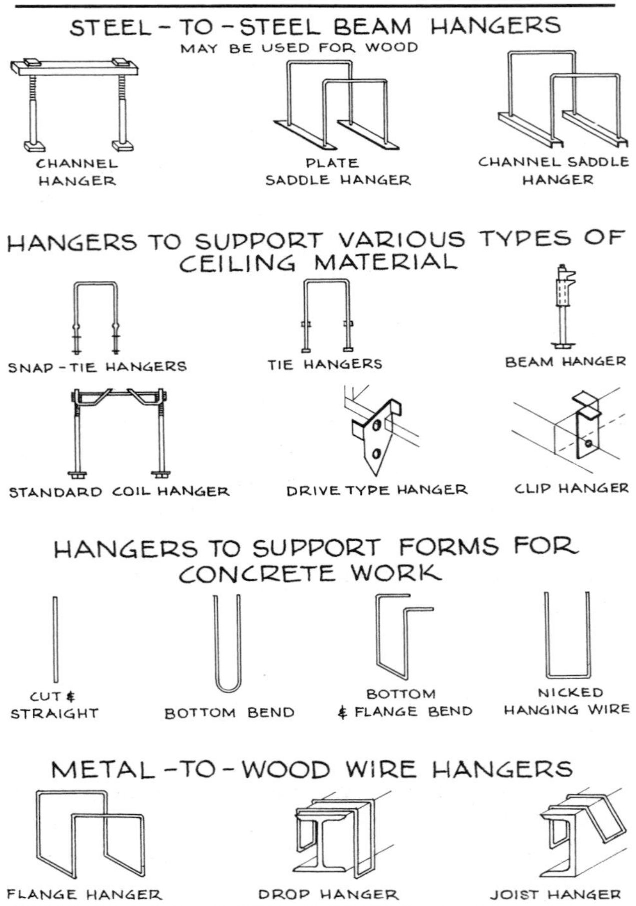

Figure H2 Steel-to-steel beam hangers (may be used for wood), hangers to support various types of ceiling materials, and metal-to-wood wire hangers to support forms for concrete work.

material as easily by any other means that is feasible for on-site work. Throughout the book, under specific headings, various types of hangers suited to the material being discussed are shown together with other pertinent accessories. *See also* Hardware, Rough.

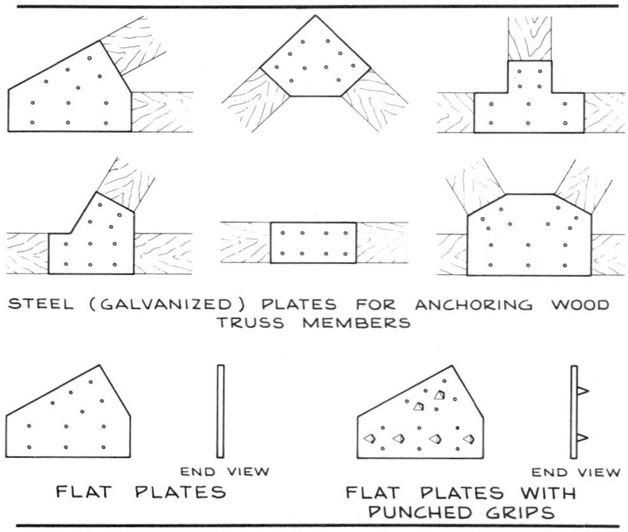

Figure H3 Types of wood-to-wood truss anchoring devices.

HARDNESS TESTS

There are four testing methods for obtaining the resistance of a material to plastic deformation.

1. *Brinnell Hardness Test:* a hard steel ball under a specific load is forced into the surface of a material. The Brinnell hardness number (HB) is derived by dividing the applied load by the surface area of the resulting impression.

2. *Rockwell Hardness Test:* a diamond spheroconical penetrator (Rockwell C scale, HRC) or a hard steel ball (Rockwell B scale, HRB) is forced into the surface of a material under sequential minor and major loads. The Rockwell hardness value is derived by applying directly to an arbitrarily calibrated dial the difference between the depths of impressions from two loads.

3. *Shore (Scleroscope) Hardness Test:* a diamond-tipped hammer is dropped by gravity from a fixed height, and a comparison of the effect of drop and rebound,

with the rebound applied to a graduated scale, measures the hardness of the material.

4. *Vickers Hardness Test:* This is similar to the Brinell test except a pyramid-shaped diamond penetrator is used instead of a ball to obtain the Vickers hardness (VH).

HARDWARE

In construction hardware is divided into two major categories, rough hardware and finish hardware.

Rough Hardware. Rough hardware covers all the miscellaneous items that go into a building for its basic structure (foundation, framing, roofing). It includes nails, anchors, inserts, nuts and bolts, screws, and ties (*see* specific headings). A word of caution is in order here. Rough hardware, although usually concealed, may cause staining of adjacent or covering materials; it may corrode and eventually even fail to perform its function if care is not taken in its selection and application. It is important to know or obtain information about the correct combination of type of rough hardware item and constituent metal in relation to the materials with which it is to be used, as well as the correct procedures of use. These same precautions apply to finish hardware, particularly that used on the exterior.

Finish Hardware. Finish hardware covers all the items which are used in the finishing of a building. It includes hinges, knobs, latches, window fasteners, and door closers, usually made of aluminum, steel, brass, or bronze. Complete coverage of all finish hardware is beyond the scope of this book. In the construction field it is advisable to either call in a specialist in this field or become acquainted with every item of detailed information necessary for the specification of finish hardware.

Two major types of finish hardware constantly encountered in the construction field are those for doors and windows. The tendency today is increasingly toward what is known as the package unit. For example, doors are now sold, as windows have been for years, together with all component parts (including hinges, locks, closers, and screws), delivered to the job ready for installation.

Hinges are the finish hardware elements on which doors and windows, cabinets, etc., turn, swing, or slide, and open and close. They may be of three types: concealed, exposed, and invisible. If made of steel, they can be obtained either with a prime coat ready to paint, or plated with brass, bronze, chrome, or other suitable metal.

Locks and latches are another finish hardware item of great importance. These fall into two categories, exposed and concealed locking and latching mechanisms.

Knobs can be obtained in many designs, shapes, and materials (including brass, bronze, aluminum, steel, glass, china, and plastic). The aluminum and steel can be finished or plated in a variety of ways (discussed under specific metal headings). Figures *H4* and *H5* show basic types of locks and hinges.

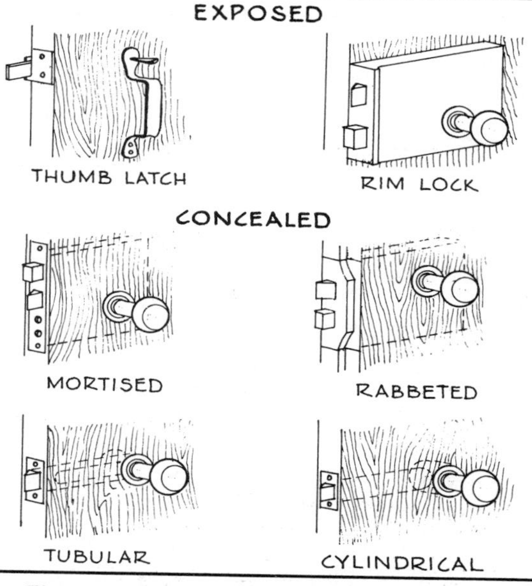

Figure H4 Basic types of locks in finish hardware.

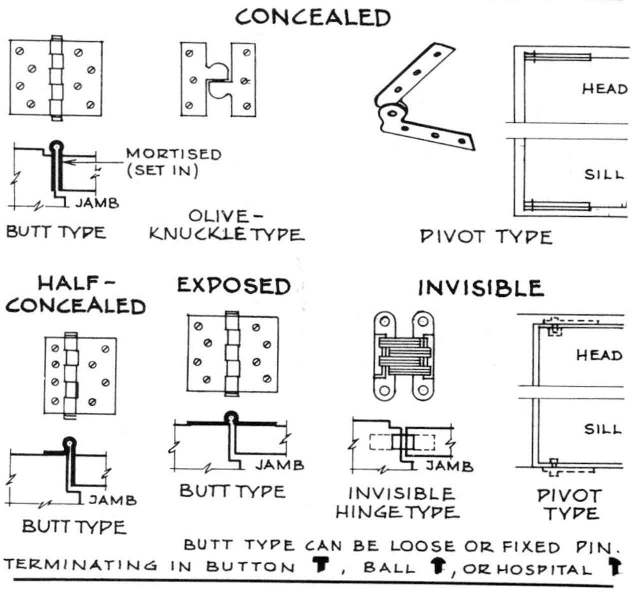

Figure H5 Basic types of hinges (butt) in finish hardware.

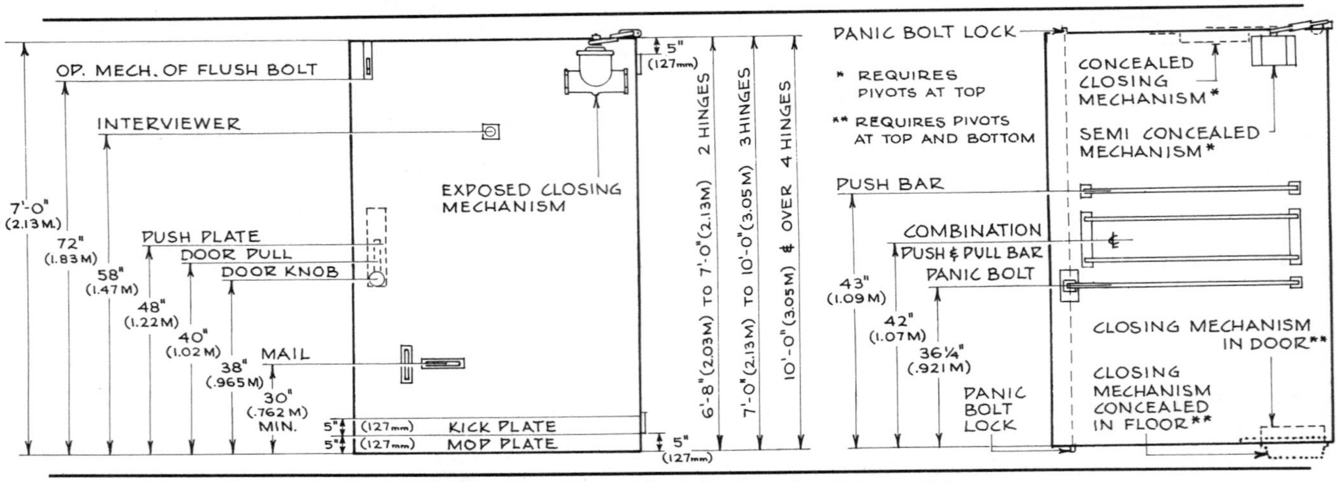

Figure H6 Location of typical finish hardware for all types of doors.

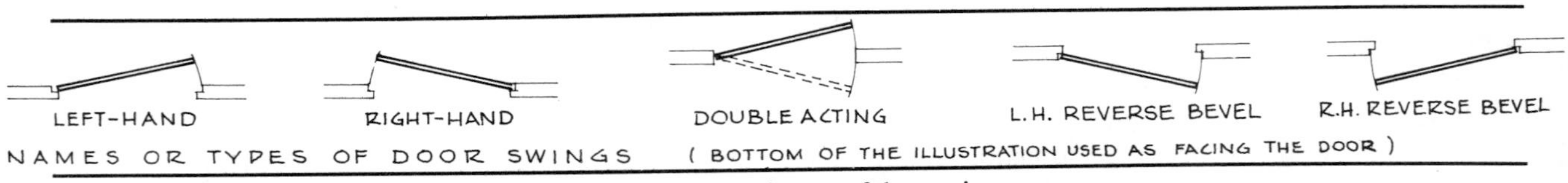

Figure H7 Names and types of door swings.

In listing hardware for doors there is a basic formula and series of names for the type of door swing and direction of swing (*see* Figure *H7*). Each component of finish hardware for doors has a definite optimal location that is governed either by the scale of the human body or, for components not controlled by the user, by whatever location results in optimal operational efficiency (*see* Figure *H6*).

HARDWARE, ROUGH

Rough hardware includes all the miscellaneous small items that are necessary to attach, join, anchor, and hold various materials together within building construction. Figure *H8* shows some typical kinds of rough hardware.

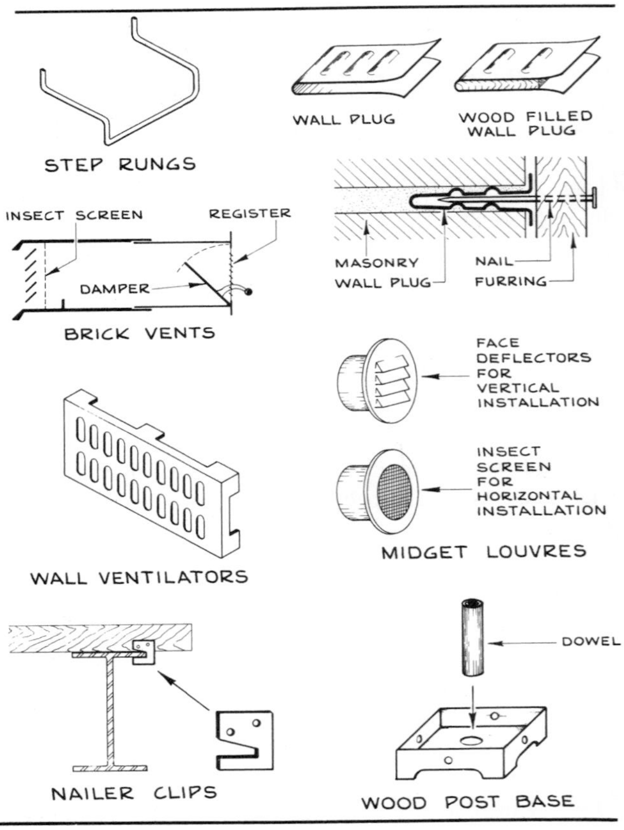

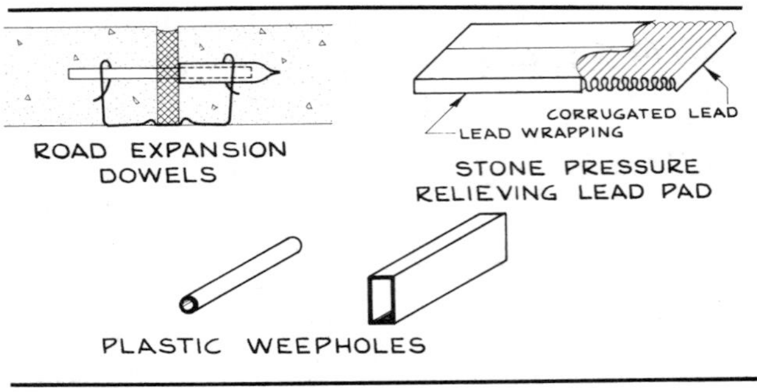

Figure H8 Typical examples of rough hardware.

See Anchors; Inserts; Reglets; Rivets; Screws, Nuts, Bolts, and Related Devices. Each unit on a specific material also describes the required rough hardware.

HELIUM

Physical and Chemical Properties

Symbol: He
Atomic number: 2
Melting point: below -272.2°C, -457.96°F
Boiling point: -268.6°C, -451.48°F

Helium is a colorless, odorless, tasteless gas that does not combine chemically with any other substance.

Types and Uses

Helium is used in construction under pressure and in caisson work as a substitute for nitrogen in air mixtures. It is also used as an inert gas shield in welding and in metallurgical procedures, and for filling luminescent (neon) light bulbs.

History and Manufacture

In 1868, on the basis of studies by Hanssen, a Frenchman, Sir Norman Lockyer recognized a new element and named it helium.

Commercial helium is produced by liquefaction of certain natural gases, but it can also be obtained as a by-product from the production of liquid air. Large-scale manufacture was first stimulated by military demands for lighter-than-air craft.

HYDROGEN

Physical and Chemical Properties

Symbol: H
Atomic number: 1
Melting point: -259.14°C, -434.45°F
Boiling point: -252.87°C, -423.17°F

Hydrogen is a colorless, odorless, tasteless, and highly flammable gas; it is the lightest of the elements. All acids contain hydrogen, which can be replaced by most metals to form salts of the acid. Hydrogen combines with oxygen and removes it from many compounds; this process is called reduction or deoxidation. Hydrogenation is the process of combining hydrogen chemically with various substances.

Types and Uses

Hydrogen has no direct uses in construction. It is an element in fuel gases (such as coal gas, water gas, and natural gas) and is important in welding. It might be of interest in industrial construction to know that steel in service at high temperatures can be embrittled by hydrogen and thus lose strength.

Unlike oxygen, which is plentiful in its uncombined state, hydrogen is chiefly found associated with carbon, oxygen, and nitrogen in compounds ranging from water (H_2O) to the most complex fats and proteins. Hundreds of products derived from petroleum are compounds of just two elements, carbon and hydrogen, and are known as hydrocarbons.

Deuterium ("heavy hydrogen") and tritium are forms of hydrogen used in atomic energy research and applications, especially in fusion research.

History and Manufacture

Henry Cavendish, the English scientist, is usually credited with the discovery of hydrogen in 1766. The name "hydrogen," meaning water producer, was given to the gas by Lavoisier in 1783. Hydrogen is produced commercially by several processes. In one it is obtained from water by electrolysis. Water that has been acidified with sulfuric acid to make it a conductor of electricity then has an electric current passed through it; hydrogen collects at the negative pole and oxygen at the positive pole. In another and far more commonly used process, a mixture of hydrogen and carbon monoxide called water gas is formed when steam is passed over hot coal or coke. This mixture is cooled and compressed so that the carbon monoxide is liquefied and the hydrogen can be collected.

I

INDIUM

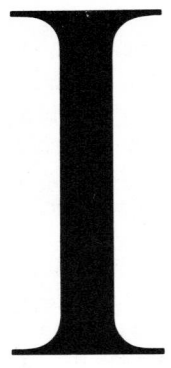

Physical and Chemical Properties

Symbol: In
Specific gravity: 7.31 (20°C, 68°F)
Melting point: 156.61°C, 313.9°F
Boiling point: 2080°C, 3776°F

Indium is a brilliantly lustrous, silver-white diamagnetic metal that is stable in air. When the pure metal is bent, it gives a high-pitched cry similar to that of tin. It is the softest of the metals and deforms under compression almost indefinitely. It alloys with other metals and increases their hardness, strength, corrosion resistance, and fluidity. For the composition of commercial indium, see Table *I 1*.

Table I 1 Composition of commercial indium

In	Sn, Pb, Cd	Cu, Bi, Ag, Fe, Tl
	(percent of content)	
99.97	0.01	0.001 or less

Commercial Forms. Indium is available in bar, rod or stick, ingot, pencil, foil, ribbon, shot, and wire form.

Types and Uses

Indium compounds are at present of minor importance. The uses of indium in metallic form, as noted by asterisks in Table *I 2*, are in some cases still in the experimental or development stage. Many are proving very valuable and have a high potential for much wider industrial application.

Actual commercial uses are mainly for plating and in special alloys such as fusible, glass- and ceramic-sealing, and soldering alloys. Indium can be used for plating onto lead, zinc, copper, cadmium, tin, gold, silver, and iron.

Table I 2 Uses of indium

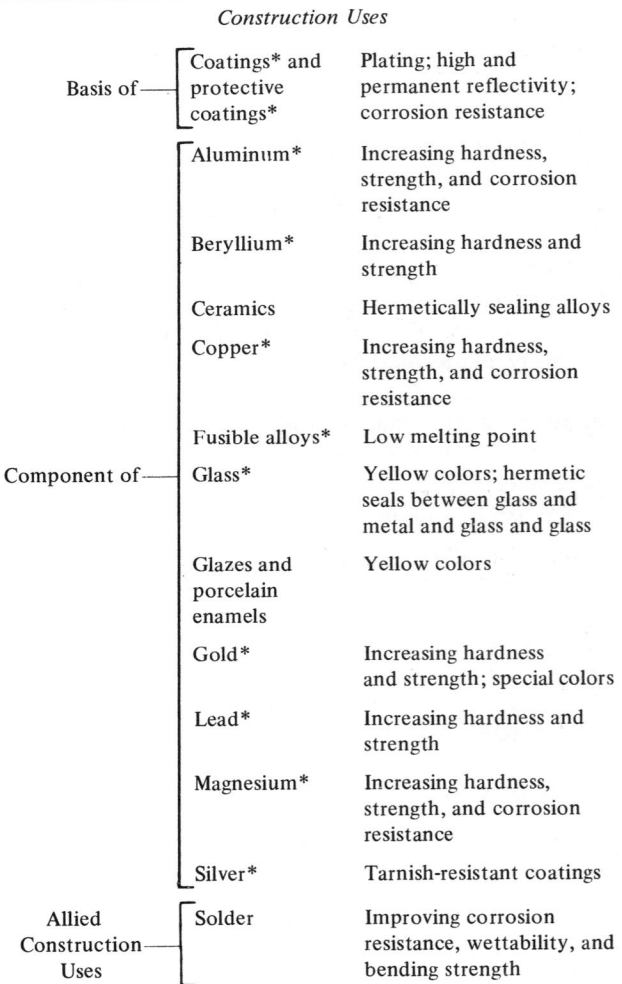

	Construction Uses	
Basis of	Coatings* and protective coatings*	Plating; high and permanent reflectivity; corrosion resistance
Component of	Aluminum*	Increasing hardness, strength, and corrosion resistance
	Beryllium*	Increasing hardness and strength
	Ceramics	Hermetically sealing alloys
	Copper*	Increasing hardness, strength, and corrosion resistance
	Fusible alloys*	Low melting point
	Glass*	Yellow colors; hermetic seals between glass and metal and glass and glass
	Glazes and porcelain enamels	Yellow colors
	Gold*	Increasing hardness and strength; special colors
	Lead*	Increasing hardness and strength
	Magnesium*	Increasing hardness, strength, and corrosion resistance
	Silver*	Tarnish-resistant coatings
Allied Construction Uses	Solder	Improving corrosion resistance, wettability, and bending strength
	Nonconstruction Uses	
Special reflectors, in chromium plating to reduce brittleness,* electrophotographic plates, special alloys,* magnets and permanent magnets,* titanium alloys,* yellow to red colors in fluorescent lamps		

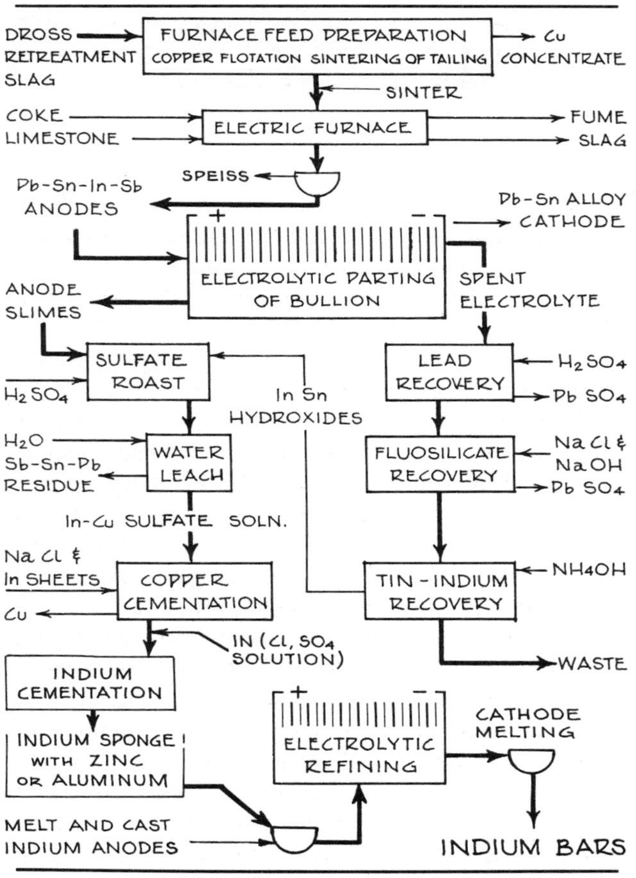

Figure I 1 Flowchart of indium recovery process.

History and Manufacture

In 1863 Reich and Richter discovered a new element and named it indium because of the indigo-blue line of its spectrum. In 1924 only one gram was available worldwide, and indium was listed at $311.04 per ounce. In 1959 indium cost $2.25 per ounce. Commercial indium is obtained from by-products of zinc and lead refining by several complex chemical processes; it is purified by electrolysis (*see* Figure *I 1*).

INSERTS

An insert is any device preset into concrete to provide permanent anchorage or a means of attachment for other materials. An insert may take the form of a notch, slot, hole, or specially shaped space; it may be a wire loop or a free-standing wire or a strip of material suitable for nailing, or a device onto which, for example, a wall could be attached or from which a ceiling could be hung.

The installation of inserts and related devices into forms for concrete prior to concrete pouring and setting, so that they become part of the basic construction, eliminates the necessity for expansion-type anchors or for subsequent drilling or chipping away of completed construction, with consequent saving of time, cost, and aggravation, if done accurately. *See* Anchors; Hardware, Rough; Hangers; *and* Ties.

Figure *I 2* shows the basic types of inserts and the materials of which they are commonly made. Inserts for masonry work (brick, stone, structural clay tile, etc.) are usually described or illustrated in this book as part of the discussion of accessories for the material in question, located under "Application."

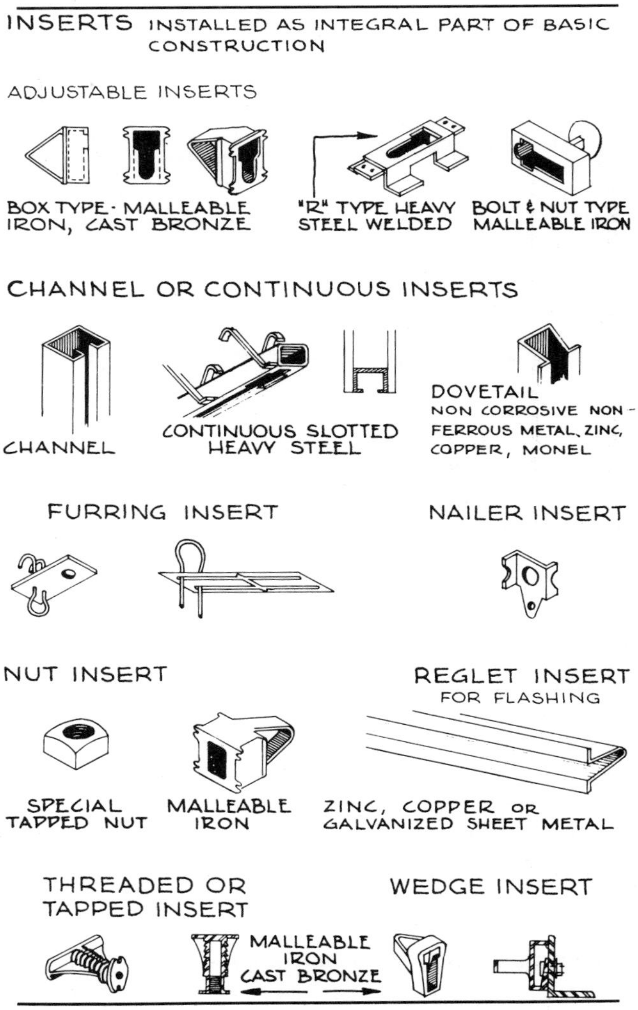

Figure I 2 Types of inserts.

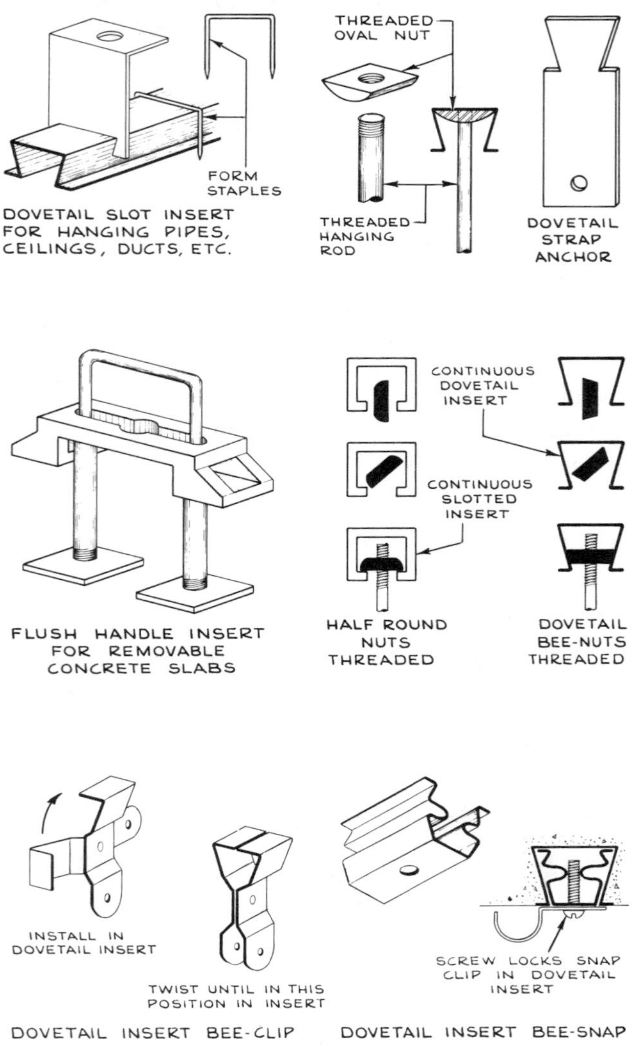

Figure I 2 Types of inserts (*continued*).

INSULATION AND INSULATING MATERIALS

Thermal insulation protects a building from excessive heat loss in winter and from heat gain in summer. In cold climates the problem of heat loss is a strong economic factor in any construction project and affects the design, choice of materials, and details of construction. Figure *I 3* shows winter design temperature zones; Figure *I 4* shows sun loads in July. (For noise control *see* Acoustics and Acoustical Materials.)

Heat transfer occurs in one of three ways: conduction, convection, and radiation. Heat is lost in all three ways from buildings. The main areas of heat loss are ceilings and roofs, walls, floors, windows and glass areas. Infiltration of air through cracks in the structure and from the opening of doors and windows adds to the overall heat loss.

The rate of heat loss varies with the material (*see* Table *I 3*). The denser a material, the better its conductivity. Conductivity also varies widely with the way in which any of the components are joined in a structure and may vary widely for different grades of the same material. Table *I 4* lists the commonly used thermal insulating materials.

Any material with a thermal conductivity of less than 0.5 Btu (0.5275 kJ) is arbitrarily considered an insulator. Most methods of controlling heat loss are based on the use of materials with low conductivity. The common exception is aluminum foil, the insulating value of which is based on its high reflectivity, often in combination with an air space. For example, 13 sheets of aluminum foil with twelve $\frac{3}{4}$-in. (19.05-mm) air spaces, for a total thickness of $9\frac{3}{4}$ in. (247.65 mm), develops an *R* value of 55.56. (*See* Table *I 3.*) The amount of heat flowing through any section or element of a building is represented by the factor *U*. The *U* factor depends on the thermal resistance of the section or element and is expressed in British thermal units per square foot of building section per hour per degree Fahrenheit ($Btu/ft^2 \cdot h \cdot {}^\circ F$), or watts per square meter per degree Celsius ($W/m^2 \cdot {}^\circ C$), of temperature difference between the air on the inside of the structure and the air on the outside. The *U* factor is almost always a fraction of 1 $Btu/ft^2 \cdot h \cdot {}^\circ F$ (or of 5.678 $W/m^2 \cdot {}^\circ C$), and obviously the higher the *U* value, the greater is the heat loss.

Definition of Terms Used in Insulation. Among the terms used to determine the insulation requirements for a building perhaps the most common is Btu (1.055 kJ), which is used to measure the amount of heat. For example, there are in 1 lb (0.4536 kg) of coal approximately 13,000 Btu (13 715 kJ), in 1 ft^3 (0.02832 m^3) of gas 1000 Btu (1055 kJ), and in 1 gal (4.546 liters) of fuel oil 141,000 Btu (148 755 kJ). In addition, Btu and the following terms are also defined under their own headings: *C* factor, *K* factor, *U* factor, and *R* factor.

Both *K* and *C* values (factors) are used when referring to an individual material, but for a composite of several materials (sandwich panel) only its *C* value is used. The *U* value is used when referring to the sum total of all materials comprising a completed wall, floor, ceiling, or roof. The *R* value is the reciprocal of the *K*, *C*, or *U* value. The *R* values are now the basic values used for insulation requirements as stated in building codes, specifications, etc.

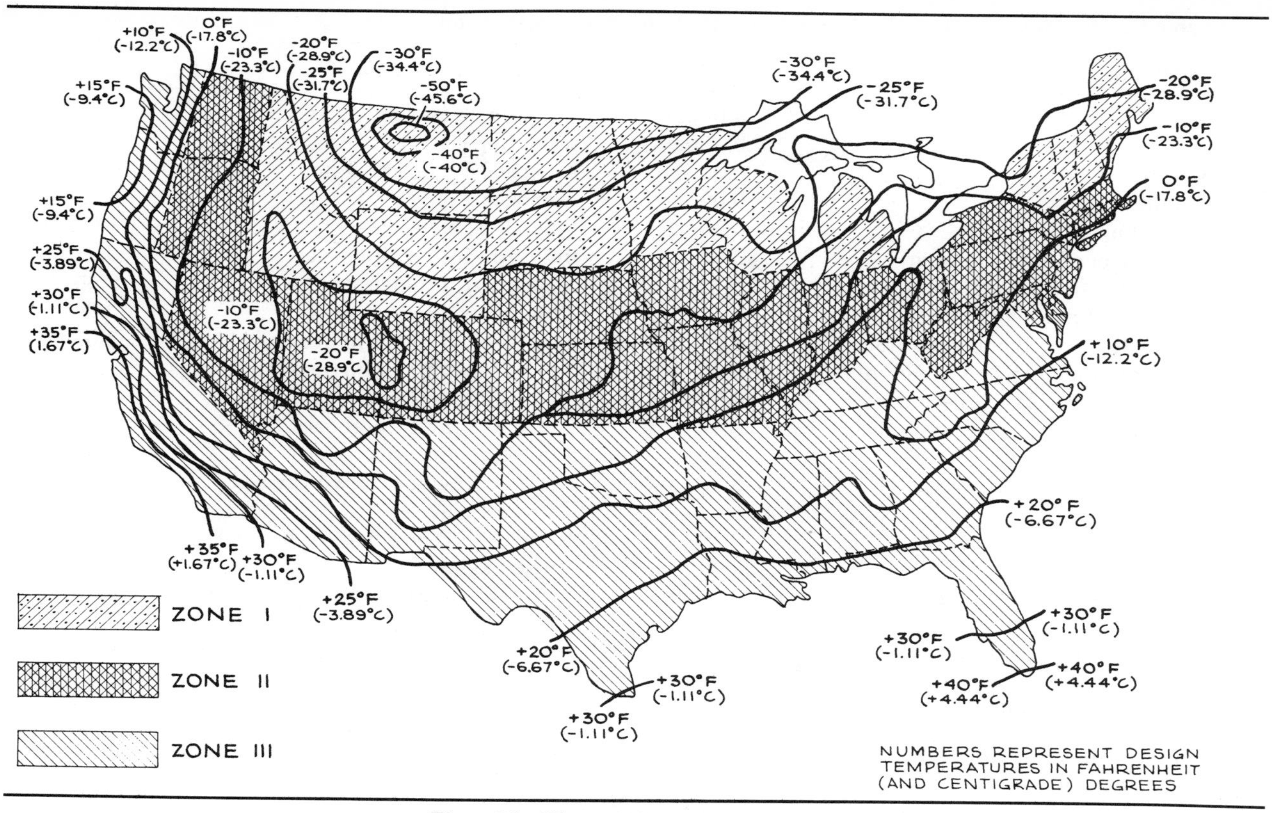

Figure I 3 Winter design temperature zones.

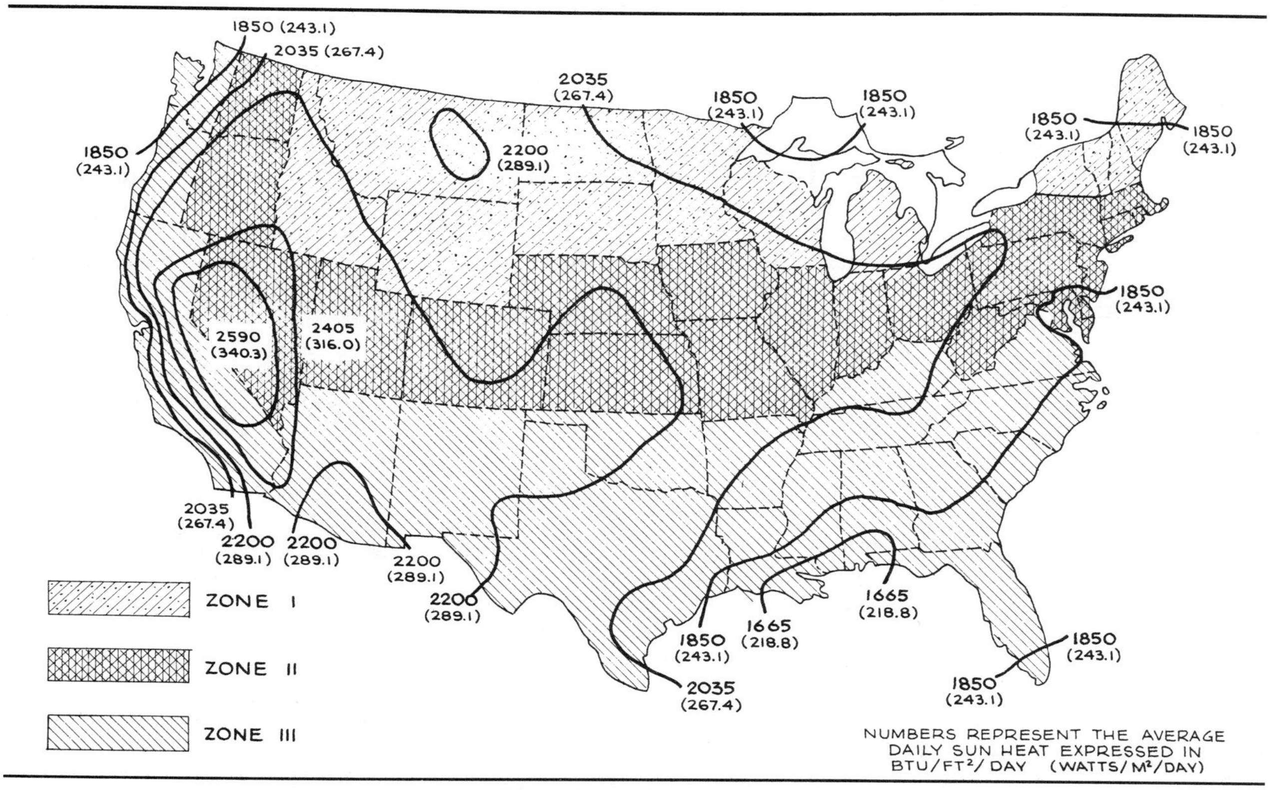

Figure I 4 Sun loads in July.

Table I 3 Resistance values of materials commonly used in construction[a]

Material	R value	Material	R value
Silver	0.00034	Insulating glass, one air space	1.61
Copper	0.00039	Insulating glass, two air spaces	2.13
Steel	0.0031	Air space $3\frac{1}{2}$ in. (88.9 mm)	0.91
Granite	0.051	Air space $\frac{3}{4}$ in. (19.05 mm)	0.92
Slate shingles	0.096	Fiberglass, paper-faced	11.0
Concrete per 1-in. (25.4 mm)	0.08	Built-up roofing, 4-ply slag	0.33
4-in. (101.6-mm) concrete block	0.71	Asphalt strip shingles	0.44
8-in. (203.2-mm) concrete block	1.11	White pine V-joint T & G	0.94
12-in. (304.8-mm) concrete block	1.28	2-in. (50.8-mm) wood decking	2.03
Common brick per 1-in. (25.4 mm)	0.20	3-in. (76.2-mm) wood decking	3.28
Face brick per 1-in. (25.4 mm)	0.11	1-in. (25.4-mm) fiberboard	2.78
Plate glass	0.18	2-in. (50.8-mm) fiberboard	5.26
Window glass	0.89	3-in. (76.2-mm) fiberboard	8.33
$\frac{1}{2}$-in. (12.7-mm) plaster board	0.45	8-in. (203.2-mm) lapped beveled wood siding	.0.81
$\frac{3}{8}$-in. (11.03-mm) plywood	0.59	10-in. (254.0-mm) lapped beveled wood siding	1.05
$\frac{3}{8}$-in. (11.03-mm) insulation board	2.06	1-in. (25.4-mm) fiberglass perimeter insulation	4.30
1-in. (25.4-mm) polyurethane foam	6.25	$\frac{1}{4}$-in. (6.35-mm) asbestos-cement siding	0.21
2-in. (50.8-mm) polyurethane foam	14.29	Wood shingles	0.87

[a]The resistance factor (R) is the reciprocal of conductivity (K), conductance (C), or overall heat transfer coefficient values (U):

$$R = \frac{1}{K}, \frac{1}{C}, \text{ or } \frac{1}{U}.$$

Table I 4 Thermal insulating materials used in construction

Physical form	Type of material	Major uses
Powders	Diatomaceous earth; sawdust; silica aerogel	Filler
Loose fibrous materials	Cork granules; glass wool; mineral wool; shredded bark; vermiculite; perlite; mica	Flat areas such as air spaces above ceilings, adjacent to roof, to reduce conduction and convection
Batt or blanket insulation[a]	Glass wool; mineral wool; wood fibers, etc., enclosed by paper, cloth, wire mesh, or aluminum	Air spaces, particularly in vertical walls and flat surfaces, to reduce conduction and convection
Board, sheet, and integrant insulation[a]	Cork; fiber; paper pulp; cellular glass; glass fiber	Sheathing for walls, to increase strength as well as reduce heat loss; rigid insulation on roofs; perimeter insulation along edges of slab floors
Reflective insulation	Aluminum foil, often combined in layers with one or more adjoining air spaces or combined with sheets of paper	Used principally for all types of refrigerated or controlled environmental spaces
Special types of block and brick insulation and refractories	Insulating block made of cork, expanded glass, 85% magnesia, or vermiculite; insulating refractory block or brick made of diatomaceous earth or kaolin (clay); heavy refractories made of fire clay, magnesite, or silica	Special controlled temperature and high-temperature insulation problems; for example, refrigerator rooms, pipes, ducts, boilers, fire chambers of boilers, and chimneys
Foam-type insulation	Rigid boards; 2-component on-job application with special applicators; polystyrene and polyurethane	Interior-applied insulation for walls above or below grade; roof insulation; air space and perimeter insulation

[a]These are available with vapor barrier as part of insulation.

Example of using Table I 3. Exterior wall of a building consists of face brick, 1-in. (25.4-mm) air space, $\frac{3}{8}$-in. (9.53-mm) plywood sheathing, metal studs, 3-in. (76.2-mm) mineral wool batts, and $\frac{1}{2}$-in. (12.7-mm) plaster board:

Outside air: 15 mph (24.24 km/h) wind	0.17*R*
Face brick	0.11*R*
1-in. (25.4-mm) air space	0.90*R*
$\frac{3}{8}$-in. (9.53-mm) plywood	0.59*R*
3-in. (88.9-mm) mineral wood batts	11.00*R*
$\frac{1}{2}$-in. (12.7-mm) plaster board	0.45*R*
Still inside air	0.68*R*
Total *R* =	13.90
U =	1/13.90 = 0.072

Much better resistance and *U* values can be obtained by changing various materials:

Outside air: 15 mph (24.14 km/h) wind	0.17*R*
Face brick	0.11*R*
1-in. (25.4-mm) air space with aluminum on sheathing	2.80*R*
$\frac{3}{8}$-in. (9.53-mm) insulating board with aluminum foil on one face	2.96*R*
2-in. (50.8-mm) polyurethane foam board	14.29*R*
$\frac{3}{4}$-in. (19.05-mm) vertical white pine paneling	0.94*R*
Still inside air	0.68*R*
Total *R* =	21.95
U =	1/21.95 = 0.046

Dead Air Space. Minute dead air spaces in a porous material are more effective than the air entrapped in a fibrous material. Air loses its value as an insulator and actually increases heat loss if the spaces are large enough for convection currents to aid heat transfer by conductivity. Therefore any increase in thickness of air space above $\frac{3}{4}$ in. (19.05 mm) has little value in reducing conduction.

Condensed Moisture Problems. In the highly insulated buildings constructed today in this country, inside temperatures are usually kept at approximately 70°F (21°C), and often the humidity is also maintained at a relatively high level. This creates another problem in relation to insulation, that of condensation of water vapor on the cooler parts of the insulating material or structural unit. Because a material conducts more heat when damp than when dry, this condensed moisture lowers the insulating values in general and can cause eventual deterioration of the structure. Therefore the warmer side of the wall should be sealed to be impervious to water vapor, and the vapor barrier on the colder side (usually outside) should be of a type that allows any vapor within the wall to escape to the outdoors. This vapor barrier may be a metal foil, an asphalt- or tar-impregnated paper, or the like (*see* Aluminum Foil; Paper and Paper Pulp Products; Plastic Sheet and Film). In every case it should be carefully joined and installed to present a continuous, unbroken moistureproof barrier.

INTEGRANT

The word "integrant" has been introduced in this book to eliminate confusion in terminology and to provide a clear definition and differentiation of the words "sheet," "board," "panel," and "decking."

In this book, an integrant is defined as a flat unit made of multiple thin, noncellular layers of a single material or of several different materials and forming an inseparable unit throughout. It is greater than $\frac{3}{16}$ in. (4.76 mm) in thickness, and it is rigid and semisupporting.

The sandwich panel is a structural system in which all the parts act together as a single structural unit. Integrants such as plywood and fiberboard decking are a type of sandwich panel and are limited only by their thickness (*see* Panels and Sandwich Panels).

Table *I 5* lists the different types of integrant and the materials from which they are made.

IODINE

Physical and Chemical Properties

Symbol: I
Atomic number: 53
Specific gravity: 4.93 (20°C, 68°F), solid
Melting point: 113.5°C, 236.3°F
Boiling point: 184.35°C, 363.83°F

Iodine, a nonmetallic element, is a solid at ordinary temperatures and consists of opaque orthorhombic crystals with a metallic luster.

Types and Uses

Elemental iodine has no construction uses. Its only application having any relation to construction is in the production of synthetic rubbers and high-purity titanium, silicon, hafnium, and zirconium. The major uses of iodine are in photography and the medical field.

Table I 5 Typical integrants used in construction

Material	Advantages	Disadvantages	Major uses
Asbestos (*see* Asbestos-Cement Panels and Integrants)	Insulation value; permanent hard finish on both faces	Difficult to treat joints	Decking, partitions, panels, exterior walls for residences
Glass insulating (*see* Glass, Insulating)	Good insulation value; can be either transparent or translucent	Limited stock sizes; other sizes must be made to order	Windows and fixed glass areas where insulation is important
Glass, laminated	Shatterproof, shockproof, and bulletproof also available	Limited stock sizes; limited to relatively small sizes; other sizes made to order	Areas where shatter-proof, shockproof, and bulletproof materials are required
Paper[a] (*see* Paper and Paper Pulp Products; Fibers)	Insulation value; available with one or both faces with finished surface; easily installed	Difficult to treat joints	Decking, partitions, sheathing, siding, wall and ceiling finishing materials
Plaster[a] (*see* Plaster Integrants and Sheets	Easily installed; available with one finished surface generally	Joints require special treatment to be concealed; relatively no insulation value	Sheathing, wall and ceiling finishing materials, partitions
Plastic (foam)[a] (*see* Plastics: Sheet; Plastics; Laminates; Plastics: Surfacing, Siding, and Paneling	Insulation value; permanent finish on both faces; easily joined	Limited width; unlimited length if faces are metal	Partitions, panels, exterior walls, roofs
Wood (plywood)[a] (see Wood Integrants)	Easily installed; available with one finished surface generally	Difficult to treat joints	Decking, sheathing, wall and ceiling finishing materials, siding, forms for concrete

[a]The available finish surfacing materials are aluminum, asbestos cement, baked enamels (paint), copper, paper, plastics, stainless steel, steel, galvanized steel, terneplate, textiles, etc.

History and Manufacture

Bernard Courtois discovered iodine in 1811 while engaged in the production of potassium nitrate. Because iodine gives off a violet-colored vapor when heated, it was named from *ioeides*, the Greek word for violet. Commercial quantities of iodine are extracted from seaweed, brine wells, and nitrate deposits in Chile.

IRIDIUM

Physical and Chemical Properties

Symbol: Ir
Atomic number: 77
Specific gravity: 22.42 (17°C, 62.6°F)
Melting point: 2410°C, 4370°F
Boiling point: 4130°C, 7466°F

Iridium is an extremely hard, grayish white metal, the most corrosion-resistant one known. It is available in powder and pellet form and sometimes in wire and sheet form.

Types and Uses

Iridium metal is used almost entirely for hardening platinum; it has also been used to make extrusion dies for high-melting glasses.

History and Manufacture

Tennant isolated iridium in 1804. Its name is derived from the Greek word *iris*, meaning rainbow, because of the varying colors of its salts. The process for obtaining iridium is explained under Platinum.

IRON

Physical and Chemical Properties

Symbol: Fe
Atomic number: 26
Specific gravity: 7.874 (20°C, 68°F)
Melting point: 1535°C, 2795°F
Boiling point: 2750°C, 4982°F
Tensile strength: 60,000 lbf/in.^{2}, 413.7 MN/m^2 (in annealed wire)

Iron is the second most abundant metal and the fourth most abundant element in the earth's crust. Pure iron is a tough, malleable, silvery white metal that is as soft and ductile as copper. It is easily magnetized and is the most magnetically permeable of the metals. It oxidizes rapidly in air and is readily attacked by most acids. Pure iron is difficult to obtain, however, and practically all commercial forms contain some carbon (*see* Iron Alloys).

Iron can be hardened by heating and sudden cooling and made pliable or more workable by heating and slow cooling. At very low temperatures iron is very brittle, at red heat it is soft, and at white heat it can be welded. As pure iron passes through these temperature ranges, it undergoes changes in its structure and properties that are vitally important in the preparation of steel.

The purest form of commercial iron is powdered iron, usually electrolytic, which is 99.965% pure. The commercial form in which iron is usually first prepared is crude or pig iron. This impure pig iron containing 3 to 4% carbon and varying amounts of phosphorus, silicon, sulfur, and manganese is the starting point from which all other kinds of iron and iron alloys (or steel) are produced.

Types and Uses

Iron is used as a metal in the form of cast and wrought iron but primarily for the production of the many kinds of steel, including stainless steel (*see* specific headings, including Iron Alloys).

Discussion of the compounds of iron is beyond the scope of this book. Table *I 6* summarizes uses in both

Table I 6 Uses of iron

Construction Uses		
Basis of	Abrasives*	Specialized sandblasting material
	Aggregate*	Hard and nonsparking concrete
	Cast iron*	Main ingredient
	Cement	Seawater-resistant cement (ferric oxide replaces alumina)
	Color pigments	Cream, buff, light red to maroon; cloudy blue to clear dark blue; yellow, brown, and red ochers
	Stainless steel*	Main ingredient
	Steel*	Main ingredient
	Wrought iron*	Main ingredient
Component of	Admixtures	Waterproofing ingredient; color pigments
	Aluminum*	Increasing strength and hardness
	Brass*	Increasing strength and hardness
Construction Uses (Continued)		
Component of	Bronze*	Increasing strength
	Ceramics	Surface decorative effects
	Concrete*	For hard and nonsparking surfaces
	Copper*	Strength and hardness
	Glazes and porcelain enamels	Yellow, red, brown, and blue colors
	Nickel*	Increasing strength and hardness
	Paint	Drier
	Paper	Yellow, red, brown, and blue colors
	Rubber	Yellow, red, brown, and blue colors
	Textiles	Yellow, red, brown, and blue colors; mordant
	Wood	Preservative
Allied Construction Uses	Water	Purification

Table I 6 Uses of iron (*continued*)

Nonconstruction Uses
Blue printing paper, engraving and lithography, manufacture of paper, metallurgy,* silver-tone photographic paper

compound and metallic form, the latter indicated by asterisks.

In commercial terminology the word "iron" is used to indicate wrought iron and cast or pig iron.

History and Manufacture

Meteoric iron containing some nickel was probably the first form in which man used this metal. The earliest Egyptian word for iron, *benipe*, is said to mean "metal from the sky."

The smelting of iron and the making of steel are skills which must have been discovered and rediscovered many times during the history of man. The primitive process of iron smelting probably involved building a fire against rich iron ore deposits or burning ore with fuel in a pit, using a natural draft of air. Archaeologists think that the first furnace was developed somewhere in the East or in Africa and consisted of a pot made of some sort of clay in which a fire was made. To this fire was periodically added a mixture of iron ore and charcoal from a trough, and a forced air draft was supplied by a crude bellows. As smelting and reduction took place, unmelted iron mixed with impurities collected at the bottom of the pot. When a sufficient quantity had collected, it was removed, hammered, and forged.

An iron blade dating back to about 3041 B.C. which was found in a pyramid indicates that methods of smelting and forging iron had already been developed. By about 1500 B.C. bellows for creating a forced draft were being used in Egypt. Methods for working and hardening steel which must have required centuries to develop were well known by 1041 B.C. in Greece. The Greeks obtained iron ore from the southern coast of the Black Sea; and the Romans, who continued the working and hardening of steel until the fall of the Roman Empire, obtained their ores from Elba and Spain.

Catalan Forge. With the Dark Ages, Western man reverted to primitive methods of making iron, although steelmaking continued in the Saracen world and the Far East. About A.D. 1293 the ironworkers of Spain developed the Catalan forge, a hearth type of furnace. It consisted of a crucible into which a charge of iron ore and charcoal was placed. A forced draft supplied by a water blower was injected into the furnace through holes in the bottom called tuyeres. This furnace could produce approximately 140 lb, or 63.504 kg, of iron in 5 hours.

A version of the Catalan forge was introduced into the southern American colonies, where it was further improved and became known as the American Bloomery. It was used until 1901, when the development of new processes caused it to be abandoned.

First Blast Furnace. Until 1350, apart from the lost ancient art of steelmaking, all methods of extracting iron from ore produced wrought iron. At this date a new type of furnace, the first blast furnace, which used charcoal as a fuel, was developed in an effort to decrease manufacturing costs. However, it produced not wrought iron, but a hard, brittle metal that could not be hammered, forged, or welded; because it could be easily cast, it became known as cast iron.

From this time, the bulk of iron and steel production was divided into two steps: first, the extraction of the metal into ingot or pig iron form, and, second, the remelting and processing of the crude iron into its final fabricated form. Technical development followed this sequence: bigger and bigger furnaces were built to produce greater quantities; then processes for converting pig iron into wrought iron were developed. Steel was produced at first from wrought iron, and then also directly from pig iron, but in limited quantities. In 1856 the Bessemer method was described by Sir Henry Bessemer, and the steel epoch began.

Blast Furnace Operation. The blast furnace remains the chief method of processing ore to iron in ingot form. At first untreated ores were charged directly, but since 1900 screening and washing of the ores and sintering of the fine ores and flue dusts have become general practice. (*See* Figure *I 5*.)

The blast furnace is a round shaft almost 200 ft (61 m) high, made of steel and lined with special refractory brick. The top is sealed off with an air lock so that gas cannot escape during the continuous feeding of charges into the furnace. About 8 ft (2.5 m) from the bottom is a series of water-cooled nozzles (tuyeres) through which preheated air is injected into the charge, and at the bottom are two tap holes, one for the slag and the other for molten metal. At a higher level there is another tap hole for slag (during operations iron and slag are drawn off 5 or 6 times a day). The present-day blast furnace is a continuous 24-hour operation. The charge consists of

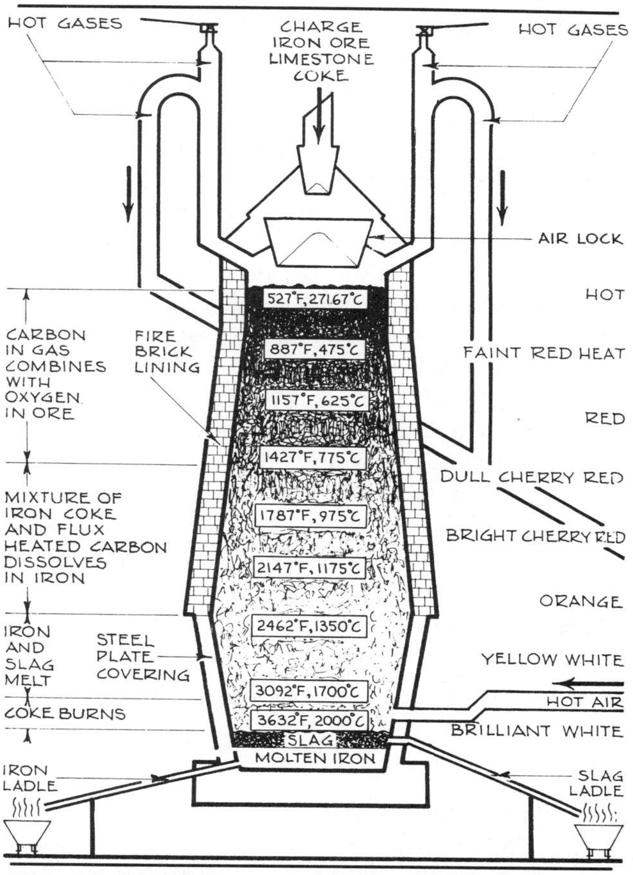

Figure I 5 Flowchart of blast furnace for production of pig iron.

iron ore, crushed limestone, and coke. At the air nozzles around the bottom of the furnace, the heat is so intense that the coke burns to form carbon monoxide. This gas rises through the charge toward the top, where it reduces the ore to iron. As the freshly reduced iron comes into contact with the hot carbon monoxide, it picks up carbon and is converted into the more fusible pig iron. Pure iron ordinarily melts at 2802.2°F (1539°C), but with a 4.3% carbon content it melts at 2066°F (1130°C). The pig iron collects at the bottom and is either cast into pigs or handled in large reservoirs, from which it is transferred directly to steelmaking processes.

A blast furnace that produces 600 tons of pig iron will consume 1160 tons of iron ore, 580 tons of coke, 290 tons of limestone, 2370 tons of air blast, and 17,000 tons of water (to wash dust from the gas and keep furnace parts cool). In producing the 600 tons of pig iron, the furnace also produces 330 tons of slag, 70 tons of dust, and 3400 tons of gas. (In metric equivalents, to produce 609.6 tonne of pig iron, 1178.6 tonne of iron ore, 589.3 tonne of coke, 294.6 tonne of limestone, 2407.9 tonne of air blast, and 17,272 tonne of water will be used; 335.3 tonne of slag, 71.1 tonne of dust, and 3454.4 tonne of gas will also be produced.)

Blast Furnace By-products. The slag from blast furnace operation is broken up into granules by a stream of air and is used as an aggregate for concrete and concrete block. Since slag is mainly calcium aluminum silicate, large quantities are also used for the manufacture of portland cement.

The gas is cleaned, washed, and filtered to remove dust and then used to generate steam for driving blower engines for the blast furnace and electricity for motors, cranes, etc., within the operation.

The dust is mixed with fine ore, sintered, and resmelted.

Coke and Its By-products. The coke fuel which is necessary for the blast furnace is generally made from bituminous coal. The coal is placed in large ovens and heated by gas flame to cause a distillation process called coking. The volatile matter within the coal is driven off as gas, and the residue is coke. The gaseous matter driven off is one of the most valuable by-products of the iron and steel industry. One ton of coked coal will yield the following by-products: 8 gal of tar, 20 lb of ammonium sulfate, 6000 ft^3 of surplus gas, and about 3 gal of light oils. (In metric equivalents, 1.016 tonne yields 36.37 liters of tar, 9.1 kg of ammonium sulfate, 16.99 m^3 of surplus gas, and about 13.64 liters of light oils.) These coal chemicals are processed into many other products (*see* Bitumens).

Pig Iron Characteristics. The characteristics of pig iron produced by the blast furnace depend on the type of ore and flux, on the proportion of coke to ore, and on the way the furnace is run, mainly the regulation of the temperature during furnace operation.

IRON ALLOYS

Physical and Chemical Properties

All iron contains some carbon. Even so-called pure iron, called ferrite, has a carbon content of 0.025%. The key, therefore, to the difference between iron, cast iron, wrought iron, and steel (by definition an iron-carbon alloy) is the carbon-iron relationship (*see* Table *I 7*).

The specific properties of the many kinds of iron and steel depend (1) on the percentage of carbon present; (2) on the presence of other elements, deliberately added or residual (*see* Tables *I 8*, *I 9* and *I 10*); (3) on what happens as the metal cools from liquid to solid (i.e., freezes); and (4) on the conditions of cooling after it freezes into the solid state. The properties are further controlled by mechanical means—hot or cold working, casting, forging, heat treatment, etc. (*see also* Alloys).

To introduce into the construction field some of the terminology used in technical literature on iron and steel (including stainless steel) and to give an inkling of the complexity of this subject, the chemistry of iron and the iron-carbon equilibrium diagram will be briefly described here.

Table I 7 General classification of the major iron alloys by carbon content

Iron	Wrought iron	Cast iron	Steel
Trace of carbon	Less than 0.1% carbon	More than 2.0% carbon	Less than 2.0% carbon

Allotropic Modifications. As iron is heated or cools from a molten state, it goes through a series of allotropic modifications which differ in their characteristics and relation to carbon. At ordinary temperatures pure iron (ferrite) is alpha iron, which is stable up to 1411°F (766°C). It is soft, ductile, and strongly magnetic. When pure it is almost as soft and ductile as copper. It holds almost no carbon in solid solution, at most 0.025%. Alpha iron becomes nonmagnetic from above about 1414°F (768°C) to about 1670°F (910°C). At these temperatures it becomes gamma iron, which will hold carbon in solid solutions up to 2% at 2066°F (1130°C). Gamma iron is described as hard, or tough, and this hardness is attributable to the carbon content. All solid solutions of carbon in gamma iron are known as

Table I 8 General effects of alloying elements on high-strength steels and other alloys of iron

Alloying element	Amount of alloying element (percent)	Effects on alloy
Aluminum		Deoxidizer; control of grain size; for high surface hardness and wear resistance in nitriding steels (with 0.95–1.30% aluminum)
Boron	0.0005–0.003	Increases hardenability (e.g., in aluminum-killed steels)
Carbon		Principal hardening element in steel
Chromium		Increases strength, hardness, toughness, and corrosion resistance
Cobalt		Resistance to softening at high temperatures; for high-speed tool steels, magnets
Columbium (niobium)		Fine grain size; increases ductility and impact strength
Copper	0.15 (max.)	Strength and increased resistance to corrosion
	Up to 0.70	Increases corrosion resistance; with phosphorus, gives even greater corrosion resistance
	0.75 or more	Increases strength when steel is reheated to 800 to 1000°F (426.67 to 537.78°C)
Lead	0.15–0.35	Increases machinability
Manganese	11.0–14.0	Deoxidizer; increases strength and hardness
Molybdenum	0.10–0.60	Steels retain tensile and creep strength at high temperatures; prevents temper brittleness
Nickel		Increases strength and toughness; improves corrosion resistance
Nitrogen	Above about 0.004	Increases hardness and tensile and yield strength
Phosphorus	0.07–0.15	Increases machinability, strength, hardness, and resistance to atmospheric corrosion
Silicon	Up to 0.35	Deoxidizer
	0.75 or more up to 2.50	Improves strength
Sulfur	0.05–0.06	Increases machinability
Titanium		Where high stress-rupture and hardening effects are required in steels at high temperatures; deoxidizer
Vanadium		Hardens, strengthens, and toughens
Zirconium		Deoxidizer; increases grain growth; improves hot-rolling properties

Table I 9 Common master alloys and their major uses in carbon steels and alloy steels

Master alloys	Major use in carbon steel or alloy steel
Calcium-aluminum-silicon	Deoxidizes; degasifies
Calcium-manganese-silicon	Deoxidizes; degasifies
Calcium molybdate	Adds molybdenum; adds strength
Calcium-silicon	Deoxidizes; eliminates sulfur
Ferroaluminum	Adds aluminum; deoxidizes
Ferroboron	Deoxidizes; desulfurizes; adds hardenability
Ferrocolumbium	Adds columbium (niobium); inhibits intergranular corrosion
Ferrochromium	Adds chromium; increases tensile strength and corrosion resistance; increases hardness and resistance to shock
Ferromanganese	Adds manganese; adds strength, utility, and hardness; controls sulfur; deoxidizes
Ferromanganese-silicon	Adds manganese; adds strength, ductility, and hardness; controls sulfur; deoxidizes
Ferromolybdenum	Adds molybdenum; adds hardness and corrosion resistance
Ferrophosphorus	Adds phosphorus; adds strength without brittleness
Ferroselenium	Adds selenium; aids machining qualities
Ferrosilicon	Adds silicon; improves strength for high-tensile-strength construction steel
Ferrosilicon-aluminum	Deoxidizes; controls grain size
Ferrotitanium	Eliminates oxygen and nitrogen
Ferrotungsten	Adds tungsten; adds hardness and heat resistance
Ferrovanadium	Adds vanadium; deoxidizes; imparts fine grain size; increases strength, ductility, and resiliency
Ferrozirconium	Zirconium combines with sulfur to give better surface
Manganese-silicon	Adds manganese and silicon without additional iron
Misch metal	Gives impact resistance; improves rolling properties
Silicon-manganese	Adds manganese, adds strength, ductility, and hardness; controls sulfur; deoxidizes and acts as scavenger
Silicon-spiegel	Adds manganese; adds strength, ductility, and hardness; controls sulfur; deoxidizes
Silicon-zirconium	Adds zirconium and some silicon; eliminates sulfur and nitrogen
Spiegeleisen	Adds manganese; adds strength, ductility, and hardness; controls sulfur; deoxidizes

Table I 10 Element limitations, minimums, and maximum percentages of incidental elements[a]

Manufacturing process	P (max. percent)	C (max. percent)	Si (min. percent)	Incidental elements Cu (max. percent)	Ni	Cr	Mo
Basic electric	0.025	0.025	0.20–0.35	0.35	0.25	0.20	0.06
Basic open-hearth or basic oxygen	0.035	0.040	0.20–0.35	0.35	0.25	0.20	0.06
Acid electric or acid open-hearth	0.050	0.050	0.15	0.035	0.25	0.20	0.06

[a]To improve machinability, 0.15–0.35% lead is added as an alloy.

austenite. Any carbon not in solid solution is in the form of iron carbide (Fe_3C), called cementite, which is hard and brittle. If large amounts of carbon are present, some of it may be in the form of free carbon called graphite.

Effects of Cooling. The iron-carbon diagram in Figure *I 6* shows what happens during very slow cooling of the molten metal. The properties of any slowly cooled steel are determined by the proportion and grain size (inherent size of the crystals) of alpha iron (ferrite) and cementite. Primary (large) grains of alpha will produce more ductility in a steel than the small grains of alpha in pearlite. Small grains, on the other hand, promote strength; for example, pearlite containing 99.20% iron and 0.80% carbon is the strongest slowly cooled iron-carbon alloy. By sudden cooling (quenching), or by

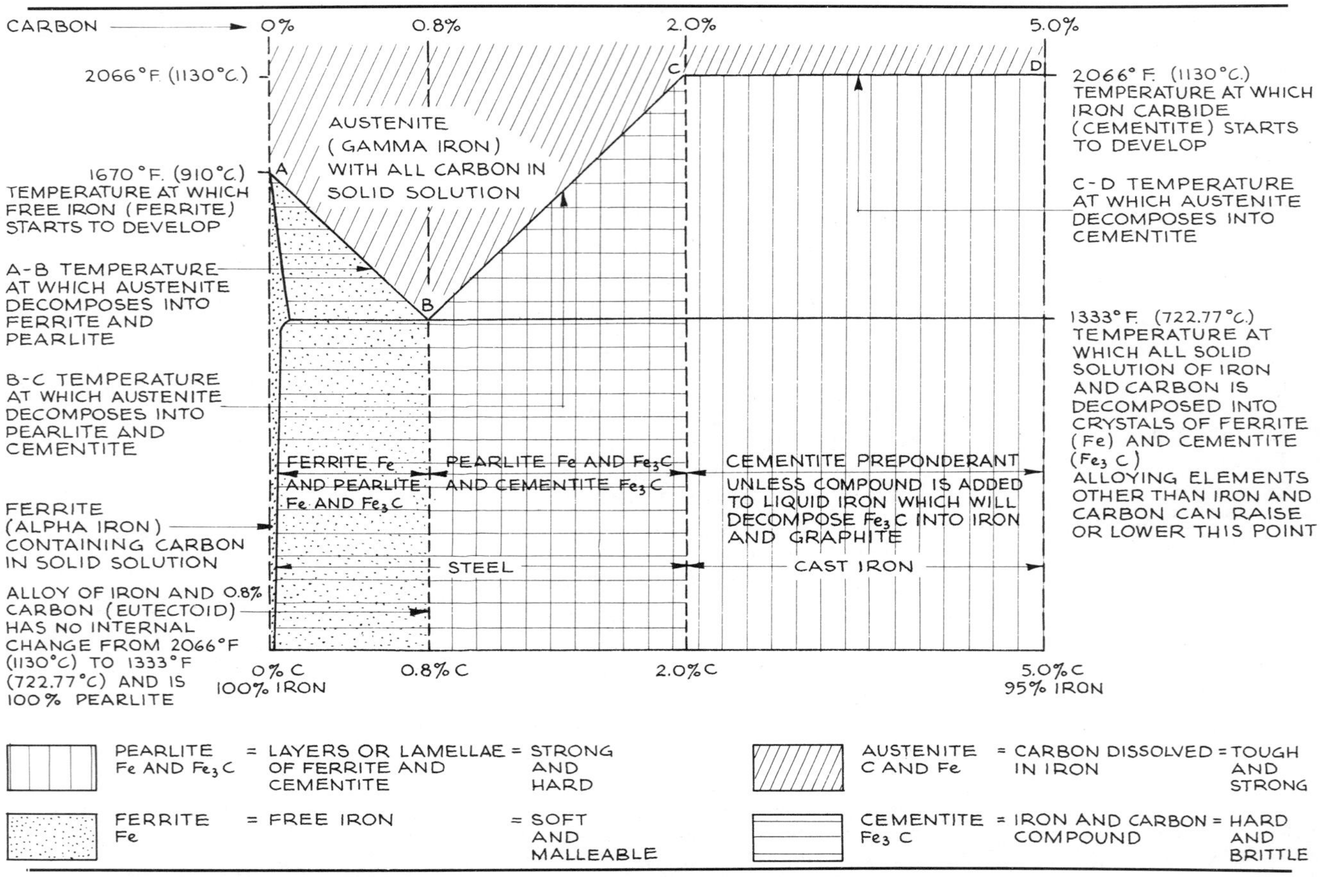

Figure I 6 Iron-carbon equilibrium.

adding alloying elements that permit austenite to exist at atmospheric temperatures, certain of the austenitic characteristics can be retained.

Martensite is the hardest constituent obtainable from the decomposition of austenite in unalloyed iron-carbon alloys. It is produced by the most rapid cooling that can be accomplished and represents a supersaturated solid solution of cementite in alpha iron.

Tempering. Tempering is the means of obtaining the exact type of transformation desired in a steel. The steel is heated to a predetermined temperature and then cooled at a controlled rate to produce one or more of the following: hardness, softness, toughness (strength plus useful ductility), desired grain size.

Numerical Designation of Grades

A system of nomenclature using numbers and symbols, roughly comparable to that used for identifying aluminum alloys, is used to identify the grades of standard steels.

Carbon Steels. A four-numeral series designates the chemical composition of carbon steel; the first two digits indicate the type of steel and alloy constituents, and the last two numbers indicate the approximate middle of the carbon range. The various grades of carbon steel and their meanings are as follows:

MT10xx	Nonresulfurized basic open-hearth and acid Bessemer carbon steel grades. "MT" refers to mechanical tubing grades.
11xx	Resulfurized basic open-hearth and acid Bessemer carbon steel grades.
12xx	Rephosphorized and resulfurized basic open-hearth carbon steel grades.

Alloy Steels. A four-numeral (and, for certain types of alloy steel, a five-numeral) designation is used to identify most of the alloy steels specified to chemical composition. The last two digits indicate the approximate middle of the carbon range. The first two digits are as follows:

13xx	Manganese 1.75%
23xx*	Nickel 3.50%
25xx*	Nickel 5.00%
31xx	Nickel 1.25%, chromium 0.65%
E33xx	Nickel 3.50%, chromium 1.55%, electric furnace
40xx	Molybdenum 0.75%
41xx	Chromium 0.50 or 0.95%, molybdenum 0.12 or 0.20%
43xx	Nickel 1.80%, chromium 0.50 or 0.80%, molybdenum 0.25%
E43xx	Same as 43xx but produced in electric furnace
44xx	Manganese 0.80%, molybdenum 0.40%
45xx	Manganese 0.55%, molybdenum 0.50%
46xx	Nickel 1.85%, molybdenum 0.75%
47xx	Nickel 1.05%, chromium 0.45%, molybdenum 0.20 or 0.35%
50xx	Chromium 0.28 or 0.40%
51xx	Chromium 0.80, 0.88, 0.93, 0.95, or 1.00%
E50100	Carbon 1.00%, chromium 0.50%
E51100	Carbon 1.00%, chromium 1.00%
E52100	Carbon 1.00%, chromium 1.45%
61xx	Chromium 0.60, 0.80, or 0.95%, vanadium 0.12 or 0.10% min. or 0.15% min.
7140	Carbon 0.40%, chromium 1.60%, molybdenum 0.35%, aluminum 1.15%
81xx	Nickel 0.30%, chromium 0.40%, molybdenum 0.12%
86xx	Nickel 0.55%, chromium 0.50%, molybdenum 0.20%
87xx	Nickel 0.55%, chromium 0.50%, molybdenum 0.25%
88xx	Nickel 0.55%, chromium 0.50%, molybdenum 0.35%
92xx	Manganese 0.85%, silicon 2.00% (9262, chromium 0.25 to 0.40%)
93xx	Nickel 3.25%, chromium 1.20%, molybdenum 0.12%
98xx	Nickel 1.00%, chromium 0.80%, molybdenum 0.25%
14Bxx	Boron
50Bxx	Chromium 0.50 or 0.28%, boron
51Bxx	Chromium 0.80%, boron
81Bxx	Nickel 0.33%, chromium 0.45%, molybdenum 0.12%, boron
86Bxx	Nickel 0.55%, chromium 0.50%, molybdenum 0.20%, boron
94Bxx	Nickel 0.45%, chromium 0.40%, molybdenum 0.12%, boron

*Nonstandard steel.

The suffix letter "H" is added to the conventional series number as a means of identifying steels specified to hardenability band limits.

The letter "B" between the second and third digits of the grade number indicates a boron steel (example: 94B17).

The letters "BV" between the second and third digits of the grade number indicate a boron-vanadium steel (example: 43BV14).

The following series designations apply only to alloy plates:

28xx	Nickel 9.00%
99x	Nickel 1.15%, chromium 0.50%, molybdenum 0.25%

In construction steels, the elements that occur normally are carbon, manganese, and silicon. A steel containing one additional alloying element such as nickel, chromium, or molybdenum is known as a single alloy steel. When two such alloy elements are added, the steel is termed a double (binary) alloy steel; when three alloy elements are added, it is termed a triple (ternary) alloy steel.

Stainless Steels. A system of numbers is used to identify stainless and heat-resisting steels by type and according to three general groups. In a three-numeral number, the first numeral indicates the group and the last two numerals indicate the type. Modifications of types are indicated by suffix letters. The meaning of these numbers is as follows:

3xx or TP-3xx	Chromium-nickel steels—nonhardenable by heat treatment, austenitic and nonmagnetic.
4xx or TP-4xx	Chromium steels—hardenable by heat treatment, martensitic and magnetic.
4xx or TP-4xx	Chromium steels—nonhardenable by heat treatment, ferritic and magnetic.
5xx	Chromium steels—low chromium heat-resisting.

Note. "TP" refers to tubular products grades.

Types and Uses

The major alloys of iron are discussed under the following headings: Iron, Cast; Iron, Wrought; Stainless Steel; Steel; Steel: Alloy Steels. Most of the alloys of iron that do not fall into the category of cast or wrought iron and steel are the products of open hearth methods with the alloying elements added to the molten iron. Table *I 11*

Table I 11 Iron alloys used in construction

Type of iron alloy	Composition or control of alloying element	Major use
Boiler iron	Low carbon	Heavy iron plate and sheet $\frac{1}{2}$ to $2\frac{1}{2}$ in. (12.7 to 63.5 mm) in thickness
Busheled iron	Scrap steel and iron	For use only under special conditions as it is not uniform in content
Ingot iron[a]	0.02% carbon	Sheet and strip for porcelain enameling plates or shapes and bars where rust resistance is required
Nickel-molybdenum	60% Ni, 20% Mo, 20% Fe	Chemical laboratory piping and equipment (acid resistant except nitric acid)
Stainless iron	12–20% chromium	Chemical laboratory piping and equipment (acid resistant)

[a]Ingot iron is considered a kind of low-carbon steel.

Table I 12 General physical and mechanical properties of gray and malleable cast iron

Type	Specific gravity	Tensile strength		Compressive strength		Coefficient of expansion
		lbf/in.^{2}	MN/m^2	lbf/in.^{2}	MN/m^2	
Gray cast iron	7.1	20,000–45,000	137.900–310.275	80,000–135,000	551.600–930.825	0.0000056
Malleable cast iron	7.3 to 7.4	50,000–53,000	344.750–365.435	60,000–120,000	431.700–827.400	0.0000066

Table I 13 Typical chemical composition of cast iron

Fe	C	S	Si	Mn	P
	(percent of content)				
Remainder	1.7–4.0	0.1–0.2	0.5–3.0	0.5–1.0	0.5–1.0

lists some of these iron alloys that occasionally appear in construction materials.

Iron and Nickel Alloys. Iron-nickel alloys include those that are entirely austenitic in character at room temperatures; otherwise the alloy is considered nickel-steel. The iron-nickel alloys show peculiar thermal expansion, stiffness, and magnetic properties and are not heat-hardenable. Iron with 36% Ni has a coefficient of expansion of 0 (zero). Within the range of 30 to 60% Ni, any desired coefficient of expansion can be obtained. Iron-nickel alloys with up to 30% Ni are nonmagnetic; with more than 30% Ni, varying magnetic properties are possible. With 27 to 44% Ni, the iron-nickel alloys show increased stiffness with increasing temperatures, with maximum stiffness at 36% Ni. If 8 to 12% chromium is substituted for iron, stiffness does not change with increasing temperatures. These peculiar characteristics permit magnets to be "tailor-made" to meet specific requirements; joints that are strainless are possible; and metal or bimetals that expand an exact, predetermined amount at a certain temperature can be made (e.g., thermostats). Combination metal springs that maintain a constant stiffness regardless of temperature represent another practical use of these alloys.

IRON, CAST

Physical and Chemical Properties

Cast iron is an iron-carbon alloy that contains more than 1.7% carbon and is poured while molten into forms. Usually it has a high carbon content and contains varying amounts of silicon, sulfur, manganese, and phosphorus. It can be easily cast into almost any shape, but it is too hard and brittle to be shaped by hammering, rolling, or pressing. General physical and mechanical properties of cast iron are given in Table *I 12*.

Strictly speaking, pig iron is cast iron as it comes from the furnace, and some pig irons and cast irons do have chemical compositions that are similar. The term "cast iron" usually refers to products made from pig iron which has been resmelted in a crucible or furnace and had its chemical composition adjusted before being cast into molds. A typical chemical composition of cast iron is given in Table *I 13*.

The characteristics of the different types of cast irons are determined by internal structure, which in turn can be varied by altering their compositions and by altering the melting, casting, and heat treatment procedures used.

Table I 14 Characteristics of cast irons used in the construction field

Type	Appearance when broken	Characteristics	Structural constituents
White cast iron	Silvery-white color	Hard and brittle; cannot be machined	Carbon chemically combined with iron, mostly as cementite
Gray cast iron[a]	Gray color	Soft and tough; easily melted and machined	Carbon in uncombined state; iron mostly ferrite
Malleable cast iron[b]	Light gray color	Strong, tough, and malleable	Made from white cast iron by annealing to reduce carbon content; iron mostly ferrite

[a]Graphite carbon in flakes.
[b]Graphite carbon in irregular, spherical-shaped particles.

Structural Constituents. Structural constituents of cast iron include ferrite, cementite, pearlite, and graphite carbon (*see* Iron Alloys). Graphite is one of the decomposition products of cementite when it breaks down into ferrite and graphite carbon. These structural constituents affect the characteristics of cast iron in the following manner: ferrite affects ductility; cementite controls the hardness (brittleness); pearlite, which is present in all cast irons except one (not used in construction), affects the toughness; and the graphite by its particle size and shape distribution within the cast iron largely controls its strength and ductility. Silicon tends to precipitate carbon in the form of flakes of graphite. Sulfur tends to keep carbon in solution or in combination with iron. Therefore cast irons high in silicon and low in sulfur make castings that are easy to machine and are not brittle under shock. Cast iron low in silicon and high in sulfur, on the other hand, has a hard surface and resists abrasion. Phosphorus is closely limited when density is important as it causes high porosity and may produce planes of weakness under shock.

Heat Treatment. Heat treatment usually consists of annealing, which relieves internal stresses or decreases the hardness to improve machinability.

Commercial Forms. Cast irons are obtainable in almost any form.

Types and Uses

Cast iron is used in the construction field mainly for piping and fittings, for ornamental ironwork, for hardware, as the base metal for porcelain enameled plumbing fixtures, and for miscellaneous castings such as floor and wall brackets for railings, vents, fireplace dampers, circular stairs, stair nosings, stadium seats, bench supports, manhole covers, and gratings.

The types of cast iron generally used are gray cast iron and malleable cast iron. The other cast irons are white cast iron, high-strength cast irons, and special cast iron alloys, which rarely appear in construction work. (*See* Table *I 14*.)

Cast irons find their largest consumption in the automotive and transportation fields, in heavy machinery, and in industry generally. Cast iron continues to play an important part in industry because it has significant compressive strength and the ability to absorb energy and stop vibration.

Nodular Cast Iron. After many years of neglect, research programs have developed methods for superior castings. One of these new developments is nodular cast iron, which has superior strength, ductility, toughness, machinability, and corrosion resistance; it also has better casting, finishing, and machining characteristics than cast iron. Here the graphite carbon is in small, rounded, well-dispersed particles instead of flakes or spheroids made up of tiny flakes. Nodular cast iron was developed simultaneously in England and the United States. In England, cerium was used; in the United States, magnesium.

Alloy Cast Irons. Any cast iron containing from 0.1 to 5.0% of alloying elements is known as alloy cast iron, a series of which is shown in Table *I 15*.

High-Strength Cast Iron. High-strength cast iron is any cast iron made with steel scrap or containing nickel, molybdenum, chromium, or other elements in amounts too small to classify it as alloy cast iron.

High Silicon Irons. These irons show high resistance to chemicals, particularly sulfuric acid, and find wide application in research, processing, and production within the entire chemical field (*see* Table *I 16*).

Table I 15 Alloying elements used in alloy cast irons

Alloying element	Amount of alloying element (percent)	Effects on characteristics of cast iron
Nickel	0.25–5.0	Promotes machinability; reduces size of graphite carbon flakes
	12.0–35.0	Increases resistance to heat and corrosion by common acids, alkalis, and seawater
Chromium	1.0–3.0	Increases tensile strength and resistance to wear and heat; reduces graphite carbon content and size of graphite carbon flakes
Molybdenum	0.25–1.25	Increases tensile strength; causes graphite carbon flakes to be more nodular in form
Vanadium	0.10–0.50	Increases tensile strength and hardness; causes graphite carbon flakes to be smaller and more uniformly distributed
Copper	Up to 1.0	Increases fluidity, corrosion resistance, toughness, and hardness
Aluminum, titanium, zirconium	Very small percentages and usually added with silicon	Increase mechanical properties; deoxidizers and scavengers
Silicon	Up to 2.5	Graphitizer and deoxidizer
	11.0–12.0	Develops resistance to all types of acids except hydrofluoric and hot hydrochloric
Boron	0.025–0.05	Substantially increases strength and hardness

Table I 16 Typical composition of high-silicon iron

Fe	Alloying elements			
	Si	C	Mn	Mo
	(percent of content)			
Remainder	14.5	0.85	0.65	0.0–3.0

Galvanizing and Plating. All cast irons can be galvanized or plated (*see* Zinc Coatings). When cast iron is used where it is subject to corrosion, it either should be a corrosion-resistant cast iron alloy or should be protected by galvanizing, other plating, or an asphalt-type coating.

History and Manufacture

Cast iron was the result of the development of the blast furnace in about 1350. Because it was cheaply produced in quantity, was easily cast, and had good compressive strength and adequate tensile strength, it was used extensively for structural purposes until 1850, when steel entered the picture. At present cast iron has been superseded by steel for structural purposes except for special applications (e.g., cast iron ring sections in tunnels). In the industrial field, many other applications for cast iron have been developed, particularly in the automotive and heavy machinery industries.

IRON, WROUGHT

Physical and Chemical Properties

Wrought iron is almost pure iron with less than 0.1% carbon, usually not more than 0.08%. It contains a small amount, approximately 2.0%, of slag (iron silicate) in purely physical association, not alloyed. For a typical chemical composition, see Table *I 17*. The iron silicate is in the form of microscopically fine threads or fibers distributed throughout the iron, generally about 250,000 per cross-sectional inch, in the direction of rolling as a rule.

Wrought iron is soft, malleable, tough, fatigue resistant, and resistant to progressive corrosion. Properties of wrought iron are as follows: specific gravity, 7.70; melting point, 2750°F (1510°C); coefficient of expansion, 0.00000741; tensile strength: 48,000 lbf/in.^{2} (330.960 MN/m^2) if unalloyed.

Workability. Wrought iron has good machinability and can be forged, bent, rolled, drawn, and spun. It can be welded by any of the commonly used processes.

Commercial Forms. Wrought iron is available in the form of pipes, plates, sheets, special shapes, and bars.

Table I 17 Typical chemical composition of wrought iron

Type	Fe	C	Mn	P	S	Si	Slag (by weight)
	(percent of content)						
Byers No. 1	Remainder	0.08	0.015	0.062	0.010	0.158	1.20
Mechanical	Remainder	0.08	0.029	0.115	0.015	0.185	2.85
Hand-puddled	Remainder	0.06	0.045	0.068	0.009	0.101	1.97

Types and Uses

Wrought iron is now used in the construction field primarily in the form of genuine wrought iron pipe, chain, sheet, and ornamental ironwork. Genuine wrought iron pipe is used extensively for plumbing, heating, and air conditioning, where a corrosion-resistance, tough, durable material is required.

Fine craftsmanship in ornamental ironwork is disappearing, and ironwork as an art form completely vanished with the last great ironworker, Edgar Brandt, a Frenchman. Brandt utilized modern techniques (pneumatic hammer, stamping, press, acetylene welding, etc.) and gave the last new creative impetus to the field of ornamental ironwork.

Because it is intrinsically related to classical architecture and requires highly skilled craftsmanship, and because today's fabricators have collected large quantities of the original casting molds and merely repeat them, the art has grown stagnant. It is possible, however, that a rebirth may occur, similar to what happened to stair railing design and other ornamental metalwork when the design possibilities of aluminum and expanded metal were fully realized by manufacturers, fabricators, designers, and architects.

Today wrought ironwork is used only in furniture, railings, fences, grilles, screen enclosures, and small decorative objects.

Application

The term "ornamental ironwork" refers either to cast iron made from foundry pig or to wrought iron. Usually any form that is repeated is made of cast iron, and forms such as bar, rod, pipe, or bent shapes are made of wrought iron.

Condensed Checklist

1. The iron alloys used should have the necessary corrosion resistance and strength.
2. Assembly of components and the size limitation of openings should be checked. Bolts, screws, etc., must be of a good grade of wrought iron.
3. The correct method of welding and type of welding electrode should be used.
4. Methods of anchoring ornamental ironwork at bottom and top should provide the strength and stiffness required by the design.
5. Anchoring attachments must be correctly installed and strongly supported by the structure.
6. If ornamental ironwork is to be used as structural support, its strength should be checked. Prefabricated column units are usually either 8 ft (2.438 m) high, although these can be cut to a minimum of 7.25 ft (2.21 m), or 7.33 ft (2.234 m) high, although these can be cut to a minimum of 6.5 ft (1.98 m).
7. Field dimensions must be accurate and must match shop drawings.
8. Local fabricators of ornamental ironwork should be checked to see whether they can handle the job.

History and Manufacture

Puddling. It was not until 1784 that Henry Cort, an Englishman, by using a reverberatory furnace with the hearth hollowed out to form a sand-lined puddle and by using coal instead of charcoal, was able to convert pig iron (cast iron) into wrought iron by stirring this puddle during refining. His process became known as dry puddling. Although this process was wasteful, with an iron loss of approximately 30%, it was used until about 1830, when Hall, by using an iron oxide material instead of sand for lining the furnace, made it possible to reduce iron losses to 10% and shortened the time of heating. He also found that all grades of pig iron could be used to make wrought iron. This process was called wet puddling.

In 1915 analysis of the physical and chemical properties of the wrought iron metal itself led to separation of the original process into three distinct steps: (1) refining the pure iron, (2) production of the proper slag, and (3) mixing the two in the correct amount to produce wrought iron.

K

KALAMEIN

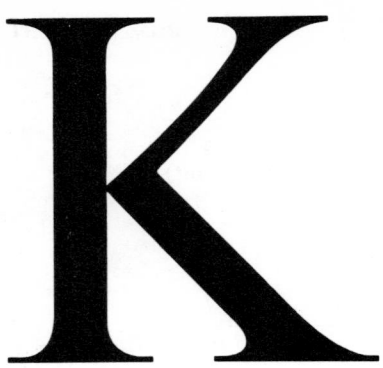

"Kalamein" is the term used for all fire-retarded (labeled) metal-covered woodwork. Originally, around 1880, in order to fire-retard wood interior trim, the first metal-covered woodwork was produced and introduced in New York City.

With the development of adhesives and incombustible materials such as plastic foams, particle board, asbestos cement, and plastics, kalamein work has become greatly diversified and includes a wide variety of construction materials such as doors, partitions, movable partitions, panels, and louvers.

Almost all building codes accept kalamein as an incombustible material, and all manufactured products of this type, including trim and connectors, have Fire Underwriters labels.

The metals generally used are aluminum, steel, galvanized steel, terne, and, for special installations, copper, brass, and bronze. Fire-retarded (labeled) metal-covered materials are available in a wide variety of finishes, among them baked enamel, anodized, painted, vinyl-coated, plastic laminates, and all the various types of metallic finishes. *See also* Asbestos Cement; Fireproofing; Paint; Plastics; Steel, Galvanized; Wood.

KAOLIN

Kaolin, or china clay, is a very pure white clay used with ball clays and fluxes in the manufacture of floor and wall tiles. *See* Clay Tile: Floor and Wall (Ceramic Tile).

KEENE'S CEMENT

Keene's cement is used for interior plaster wall and ceiling surfaces where moisture and/or heavy wear is present (*see* Cement and Gypsum).

KEROSENE

Kerosene is obtained from crude oil by distillation. Physical characteristics are as follows: a clear liquid that boils at 350 to 550°F (176.67 to 287.78°C) and freezes below -25°F (-31.67°C). In the construction field it is used as a fuel for temporary heating units and as a solvent for adhesives.

K FACTOR

K factor is the coefficient of thermal conductivity of an insulating material. In other words, it is a measurement of the rate of heat flow through a unit area of a homogeneous material of a set thickness under steady conditions, perpendicular to the temperature gradient (i.e., temperature difference surface to surface) per degree change in temperature. In the Imperial system the unit of measurement is British thermal unit per inch per square foot per hour per degree Fahrenheit. It is written thus: $Btu/in./ft^2h°F$.

The metric unit is expressed in watt per meter per degree Celsius, written thus: W/m°C. (A watt measures heat flow per hour.) The conversion factor (Imperial to metric) is

$$1\ Btu/in./ft^2h°F = 0.1442\ W/m°C.$$

See also Btu; *C* Factor; *R* Factor; *U* Factor; Insulation and Insulating Materials.

KRYPTON

Physical and Chemical Properties

Symbol: Kr
Atomic number: 36
Melting point: -156.6°C, -249.88°F
Boiling point: -152.3°C, -242.14°F

An inert, rare gaseous element of the helium family, solid krypton is a white crystalline substance with a face-centered cubic structure which is common to all the rare gases.

Types and Uses

Krypton is used in fluorescent lights. In 1960 the meter was defined in terms of the orange-red spectral line of ^{86}Kr. One meter equals 1,650,763.73 wavelengths (*in vacuo*) of the orange-red line of ^{86}Kr. This has replaced the standard meter bar of platinum-iridium alloy in Paris.

History and Manufacture

In 1899 krypton was discovered by Sir William Ramsay and M. W. Travers. Krypton is produced in air-separation plants.

L

LAMPS

Lamps are here defined as the medium whereby electricity can be turned into light. There are two types, the bulb and the tube. In construction it is important to know shape, connector, type of base, type of ballast, and size of lamps. When this information has been obtained, the space necessary for installing lamps and fixtures can be calculated. The commonly used shapes of bases, connectors, and lamps are illustrated in Figures *L1* to *L6*.

Lighting has generally been calculated in terms of the lumens required for work, display, or other activity or function carried on in a given area, without realization or consideration of the amount of energy (watts) being used. Now we think in terms of the amount of light (number of lumens) produced in relation to the number of watts (energy) expended. Table *L1* summarizes the characteristics of the various types of lamps. Table *L2* lists the lumens per watt for the various types of lamps.

	MINI-CAN SCREW (MC)
	CANDELABRA (CAND.)
	INTERMEDIATE (INTER.)
	DOUBLE CONTACT BAYONET CANDELABRA (D. C. BAY.)
	DISC (LUMILINE)
	MEDIUM (MED.)
	THREE CONTACT MEDIUM (3 C. MED.)
	MEDIUM PREFOCUS (MED. PF.)
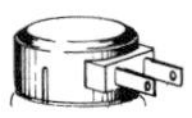	MEDIUM SIDE PRONG

Figure L1 Bases, connectors, etc., for incandescent lamps.

	MOGUL (MOG.)
	THREE CONTACT MOGUL (3 C. MOG.)
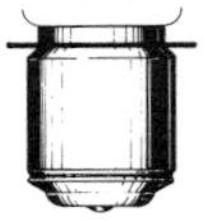	MOGUL PREFOCUS (MOG. PF.)
	MEDIUM SKIRTED (MED. SKT.) (MECHANICAL)
	MEDIUM SKIRTED (MED. SKT.) (CEMENT)

Figure L1 Bases, connectors, etc., for incandescent lamps (*continued*).

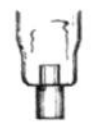	RECESSED SINGLE CONTACT (RSC)
	MOGLE (MOG.)

Figure L2 Bases, connectors, etc., for metal-halide lamps.

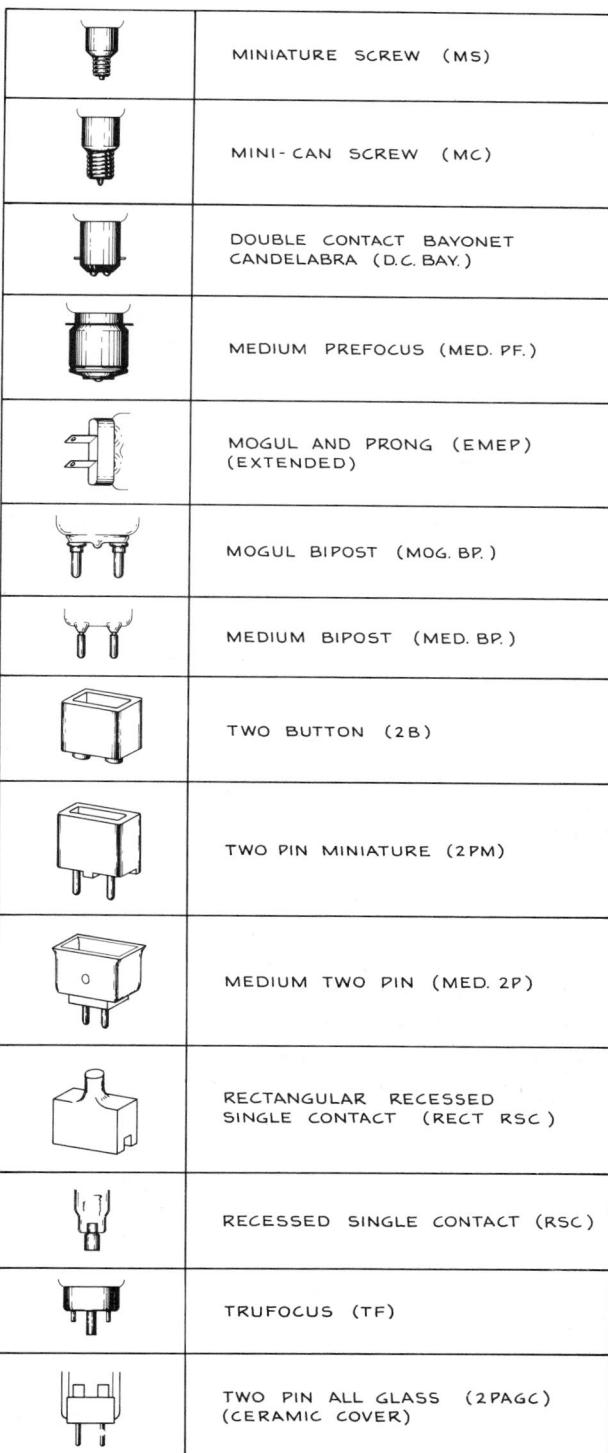

	MINIATURE SCREW (MS)
	MINI-CAN SCREW (MC)
	DOUBLE CONTACT BAYONET CANDELABRA (D.C. BAY.)
	MEDIUM PREFOCUS (MED. PF.)
	MOGUL AND PRONG (EMEP) (EXTENDED)
	MOGUL BIPOST (MOG. BP.)
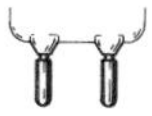	MEDIUM BIPOST (MED. BP.)
	TWO BUTTON (2B)
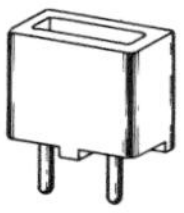	TWO PIN MINIATURE (2PM)
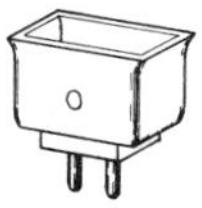	MEDIUM TWO PIN (MED. 2P)
	RECTANGULAR RECESSED SINGLE CONTACT (RECT RSC)
	RECESSED SINGLE CONTACT (RSC)
	TRUFOCUS (TF)
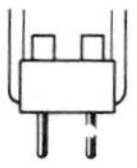	TWO PIN ALL GLASS (2PAGC) (CERAMIC COVER)

Figure L3 Bases, connectors, etc., for tungsten-halogen lamps.

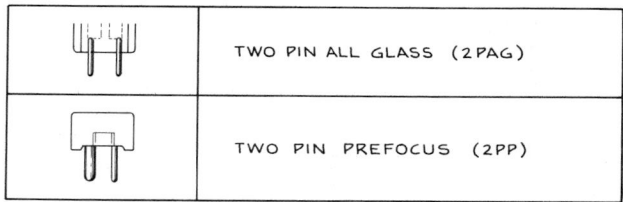

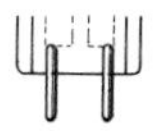	TWO PIN ALL GLASS (2PAG)
	TWO PIN PREFOCUS (2PP)

Figure L3 Bases, connectors, etc., for tungsten-halogen lamps (*continued*).

	SINGLE PIN (T-6 SLIMLINE)
	SINGLE PIN (T-8 SLIMLINE)
	SINGLE PIN (T-12 SLIMLINE)
	SHROUDED SINGLE PIN (SSP)
	MINIATURE BIPIN (MIN. BP.) (MINIATURE T-5 LAMP)
	MEDIUM BIPIN (MED. BP.) (MEDIUM T-8 LAMP)
	MEDIUM BIPIN (MED. BP.) (MEDIUM T-12 LAMP)
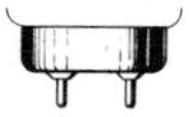	MOGUL BIPIN (MOG. BP.) (MOGUL T-17 LAMP)
	RECESSED DOUBLE CONTACT (T-12 LAMP)
	4-PIN (CIRCLINE)

Figure L4 Bases, connectors, etc., for fluorescent lamps.

	MEDIUM (MED.)
	ADMEDIUM (ADMIN.)
	MOGUL (MOG.)

Figure L5 Bases, connectors, etc., for mercury lamps.

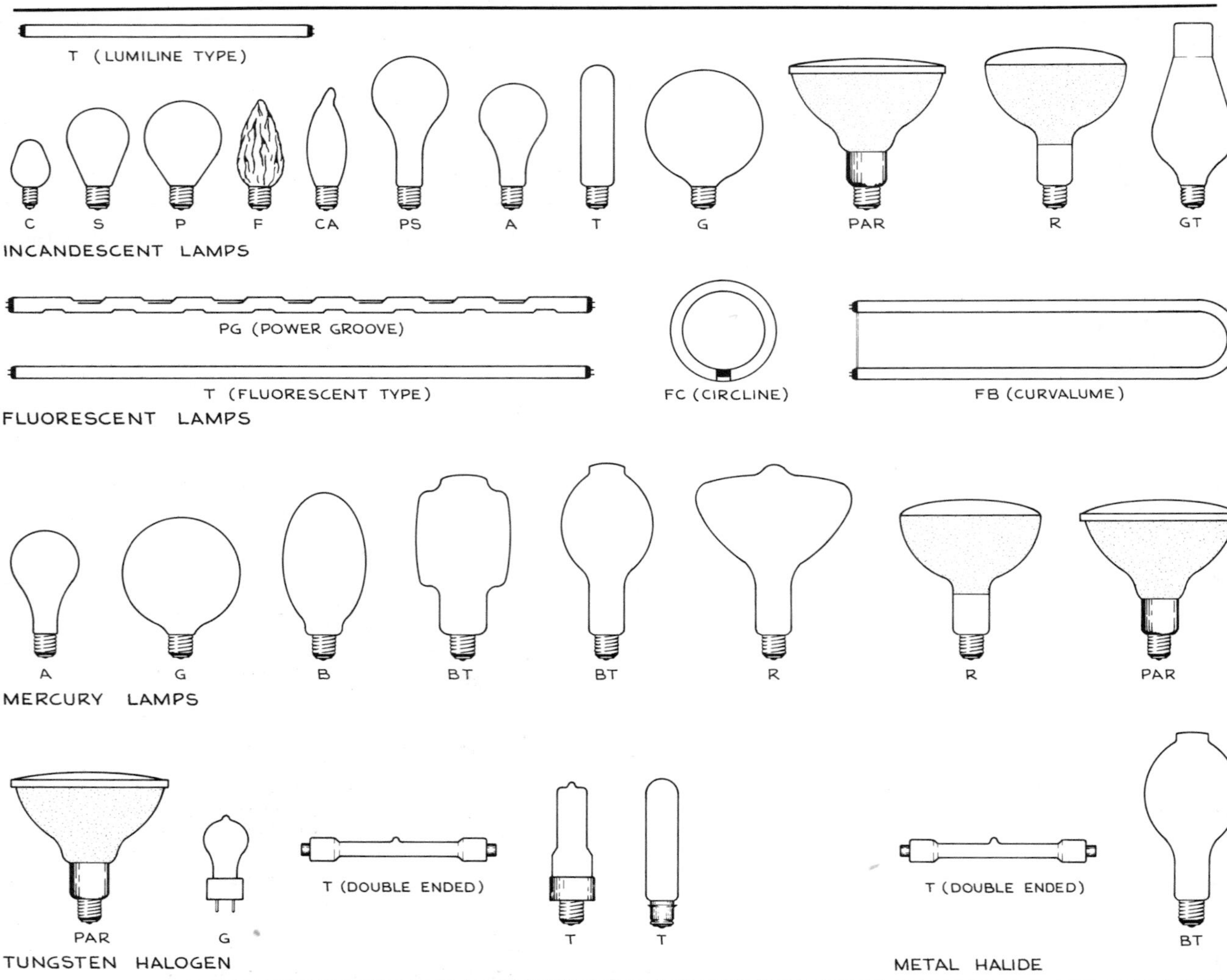

Figure L6 Types of lamps used in the construction field.

Table L1 Characteristics of various lamps

Type of lamp	Method of producing light	Type of lamp by wattage	Mean lumens (40% average life)	Average life (hours)
Incandescent	Tungsten filament in a gas or vacuum	6–2000 5000 and 10,000	40–59,000 143,000 and 290,500	300 to special lamps with 6000
Tungsten halogen	Tungsten filament with a halogen, usually iodine or bromine	12–2000	300–55,000	50–1000
Mercury	Mercury vapor and argon gas	40–1000	910–43,600	16,000–24,000
Fluorescent[a]	Mercury vapor in an inert gas such as argon, krypton, or neon, with a coating of fluorescent materials on the inner surface of the tube		65–100	300–32,000
Preheat		4–90		
Miniature		4–13		
Slimline		21–75		
Rapid-start		32–105		
Very high output		110–215		

Table L1 Characteristics of various lamps (*continued*)

Type of lamp	Method of producing light	Type of lamp by wattage	Mean lumens (40% average life)	Average life (hours)
Metal halide	Argon gas, mercury, and halide salts, usually iodides of sodium, thallium, and indium, or of sodium and scandium, or of sodium, thorium, and scandium; these iodides form an amalgam of mercury with selected iodides	175 400 1000 1500	10,200–10,800 24,600–27,200 79,000–82,500 134,900–142,500	6000–15,000
Sodium	Sodium and mercury amalgam and xenon gas	100 150 250 400 1000	8,550 14,400 23,200–27,300 45,000 127,400	12,000–20,000

[a]Fluorescent tubes are available for germicidal purposes, plant growth, and black light, which excites fluorescent paints.

Table L2 Types of lamps in relation to lumens per watt

Type of lamp	Lumens per watt
Incandescent carbon	0–4
Incandescent tungsten	6–36
Tungsten halogen	6–36
Mercury	34–63
Fluorescent	24–84
Metal halide	84–100
Sodium	32–140
Theoretical efficiency of white light	220

LEAD

Physical and Chemical Properties

Symbol: Pb
Atomic number: 82
Specific gravity: 11.35 (20°C, 68°F)
Melting point: 327.5°C, 621.5°F
Boiling point: 1740°C, 3164°F

Lead is a blue-gray, soft, very heavy metal (the heaviest of the common metals). It is extremely workable, has good corrosion resistance, is easily recovered from scrap materials, and is relatively impenetrable to radiation.

Corrosion Resistance. The corrosion resistance of lead is one reason for its wide use. This corrosion resistance arises from the fact that metallic lead does not react with many compounds or solutions, and with certain others it forms compounds that act as protective coatings against further action or corrosion. Lead embedded in cinders should be protected, however.

Toxicity. Lead (in fume, vapor, or dust form) and lead compounds are toxic if ingested in measurable quantity. Lead carbonate (white lead) and lead sulfate have been almost completely eliminated as the white paint solid (pigment) in interior paints because of the lead poisoning of children from lead-base paint. The only exception is basic lead silicochromate, which is nontoxic.

Commercial Forms. Lead is available (1) extruded, in the form of pipe, rod, wire, ribbon, cames, traps, bends, wedge lead, and special shapes; (2) rolled into sheet, foil, strip, blanks for drawing, and blanks for various shapes; (3) cast, as sand or die castings (gravity or pressure); and (4) in miscellaneous forms including metallic powder, wool, and shot.

Types and Uses

There are several grades of lead metal, of which corroding lead, chemical lead, and common desilverized lead are of interest to the construction field (*see* Table *L3*). Corroding lead is used for fine white lead paints, red lead, litharge, and orange mineral. Chemical lead and common desilverized lead are used for sheet, pipe, lead wool, powdered lead, ribbon lead, and alloys. In Table *L4* the uses of lead in metallic form are indicated by asterisks.

Table L3 Typical composition of chemical lead and corroding lead

Type of lead	Pb	Cu	Ag	Sb, Sn and As	Bi	Fe	Zn
	(percent of content)						
Chemical	99.90	0.04–0.08	0.002–0.02	0.002	0.005	0.002	0.001
Corroding	99.94	0.0025 combined		0.011	0.05	0.002	0.0015

Table L4 Uses of lead

Construction Uses

	Use	Purpose
Basis of	Adhesives	Applying textiles to surfaces; acid-resistant adhesive for metals, ceramics, glass, etc.
	Caulking	Lead wool
	Color pigments	Blue-gray and red; with other metals yellow, green, orange, and red
	Glazing compounds	Main ingredient of some putty
	Protective coatings*	Hot-dipped coatings on iron, steel, and copper; acid-resistant linings
	Vibration control*	Deadening vibrations under foundations and machinery
Component of	Asphalt	Filler; opacifier in asphalt tiles
	Brass*	Improving cutting qualities
	Bronze*	Aiding machinability
	Caulking	Ingredient
	Copper*	Aiding machinability
	Cork	Filler; opacifier
	Fusible alloys*	Ingredient
	Glass	Ingredient increasing luster, brilliance, toughness, and refractive index; protection against X-rays and gamma rays
	Glazes and porcelain enamels	Ingredients; frits; flux; gold color
	Glazing compounds	Ingredient

Construction Uses (Continued)

	Use	Purpose
Component of	Iron*	Aiding machinability
	Linoleum	Filler; opacifier
	Paint	Bronzing; drier; waterproofing; rust inhibitor; mildew control; paint solid (red lead, white lead, litharge)
	Plastics	Filler; opacifier; molds; stabilizer against light and heat
	Protective coatings	Prime red lead coatings for rust resistance on iron and steal
	Rubber	Filler; opacifier; accelerator; increasing toughness
	Steel*	Aiding machinability
	Terne*	A sheet iron or steel metal with a lead-tin coating
	Textiles	Waterproofing; fireproofing; mordant; dyeing; printing
	Tin*	Ingredient for making terne metal
	Wood	Preservative
Allied Construction Uses	Plumbing*	Pipes, traps, etc.; lead wool; caulking; sheets
	Electrical*	Cable coverings
	Solder*	Ingredient

Nonconstruction Uses

Cable coverings,* electronics, engraving, printing,* radiation barrier in nuclear energy plants,* radio, seals,* shield for radioactive materials,* toys,* type metal,* X-ray shields.*

Hardware. Lead finds many uses in rough hardware items such as expansion shields for securing bolts, screws, and other accessories in masonry, washers, lead-headed nails, etc.

Special Uses. There are also some specialized uses of lead in construction, for example, as an antispark material for floors in areas where explosions may occur; in chemical laboratories for sinks, pipes, and special equipment; and as a shield against radiation.

Of the many uses of lead in compound form, those in enamels and glazes are of most interest in the construction field.

The applications of lead more directly involved in construction are as a component of the various solders used for metal flashing and also in plumbing.

History and Manufacture

The name for lead is derived from the Anglo-Saxon *lead*, and the symbol, Pb, from the Roman name *plumbum* for lead and lead water pipes.

The discovery of lead is lost in antiquity. It is claimed that the Egyptians used lead about 7000 B.C. and later employed it for glazing pottery and making lead-bearing solder. They also used white lead as an ointment but not as a pigment. The Chinese used lead for money about 2000 B.C. In Babylon the famous handing gardens were floored with lead. Lead was added to ancient bronzes, and lead pipes were used extensively by the Romans and also in Asia. The Romans had 15 standard sizes in regular 10-ft lengths.

Lead was used in glass as early as 800 B.C. In the Middle Ages sheet lead was used for roofing, and the earliest bullets were made of lead and its alloys.

United States Deposits. In the United States, lead was mined in Virginia around 1621, and in 1690 it was discovered in the Mississippi Valley. The lead ores there and to the east are relatively free from precious metals, whereas in the western states the silver and gold content stimulated exploitation of the early lead discoveries. Zinc is commonly associated with lead and in many ores was a handicap because of the technical difficulties in separating the two metals. It was not until after 1925 that these difficulties were overcome and that the ores were mined commercially and processed for both lead and zinc on a scale far greater than was previously possible.

Modern Processing. Lead ores are first crushed and ground to separate physically most ore particles from waste rock particles; then they are treated, usually by flotation techniques, to concentrate the minerals. By this method even an ore containing lead, copper, and zinc minerals can be separated into component parts from which each metal can then be extracted separately.

The flotation process takes advantage of the fact that sulfide mineral particles will adhere to the surface of an air bubble. This affinity varies with different sulfides and can be controlled by flotation reagents that affect the surface properties of the mineral particle. Lead sulfide is readily floated using pine oil or cresylic acid as a frother and phosphorus pentasulfide as a frother-promoter. Lime and sodium cyanide prevent flotation of pyrite (FeS) and sphalerite (ZnS), which are commonly associated with lead ores.

Before lead is smelted, the products from the flotation process have to be roasted and sintered to eliminate sulfur and to give the desired characteristics to the feed for the blast furnace. The lead bullion produced in the blast furnace is then refined. The lead refining flowchart (Figure *L7*) shows the entire process, starting with the charge and including marketable products. On this flowchart, lead and the various by-products are underlined. It is of interest that tin, zinc, arsenic, sulfur, and bismuth are obtained in addition to the by-products shown.

LEAD CARBONATE, BASIC

Physical and Chemical Properties

Basic lead carbonate, commonly known as white lead, is a white, amorphous powder that is insoluble in water, soluble in acids, and darkened by hydrogen sulfide. The generally accepted formula is $2PbCO_3 \cdot Pb(OH)_2$, with a specific gravity of 6.14. White lead decomposes at 752°F (400°C).

Commercial white leads may contain from 62 to 75% lead carbonate ($PbCO_3$), the remainder being lead hydroxide. Superior pigment characteristics seem related to increased basicity, that is, a higher $Pb(OH)_2$ content. Basic lead carbonate is available as dry pigment, paste-in-oil, and semipaste. White lead as the white paint solid (pigment) for paints has been almost completely eliminated because of the many incidents of lead poisoning of children.

Hiding Power. Comparative measurement of the hiding power of white pigments (paint solids) is based on that of white lead. A pound of white lead is called one hiding unit. The chief faults of white lead are its relatively poor hiding power, a tendency to turn gray in the presence of sulfur fumes, and chalking.

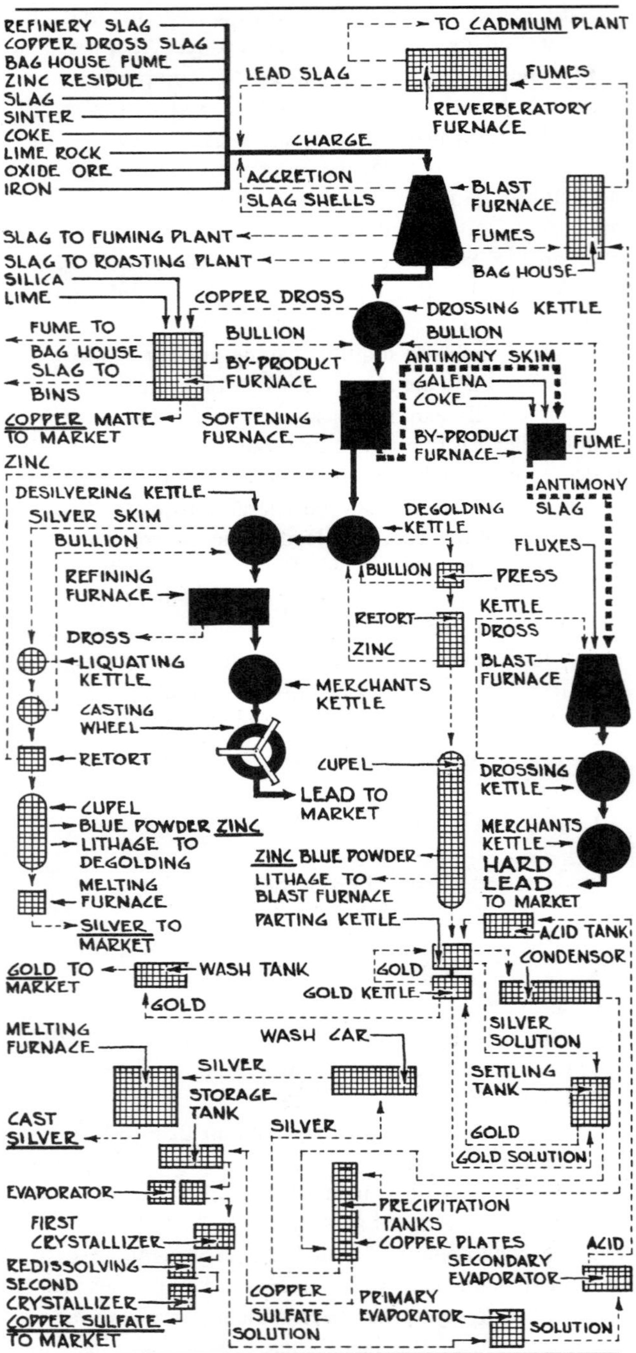

Figure L7 Flowchart of lead smelting and refining process.

Types and Uses

Today, white lead is used mainly for the white paint solid in some exterior paints. It is no longer used in interior paints. In exterior paints it imparts the adhesion, toughness, durability, and elasticity needed for exterior use.

White Lead Cement. A cement made by mixing paste white lead with a good grade of varnish to heavy brushing or troweling consistency is used for mounting canvas murals on walls, cementing canvas decking or roofing into place, applying metal letters to glass, and setting marble. A typical composition consists of $1\frac{1}{4}$ to $1\frac{1}{2}$ gal (5.68 to 6.82 liters) of varnish to 100 lb (45.36 kg) of heavy paste white lead.

History and Manufacture

White lead was known as an ointment by the ancient Egyptians. Its first use as a paint dates back to about the 4th century B.C. and is attributed to the Greeks and Romans. The oldest commercial manufacturing process is the Old Dutch or corroding process, which was abandoned some years ago in favor of faster methods, some of them, such as the Carter process, based on the same principle.

Carter Process. In the Carter process a small stream of molten lead is blown with steam or air to a fine powder. This powder, moistened with acetic acid and water, is placed in long, slowly revolving, wooden cylinders, through which air containing carbon dioxide is blown. The powder is thus oxidized and reacts with the acetic acid to form basic lead acetate, which in turn is decomposed by the carbon dioxide to form lead carbonate. This conversion occurs in about 12 to 14 days, as compared to about 100 days in the Old Dutch process.

Electrolytic Method. The speediest method, which obtains white lead almost instantly, is the electrolysis of a lead acetate solution between lead anodes and iron cathodes in a partitioned cell.

LEAD COATINGS

Physical and Chemical Properties

Lead and lead alloy coatings are thin, protective coatings that are alloyed with or mechanically bonded to the base metal. They are generally used on iron, steel, or copper. Lead, like zinc, forms a coating that is highly resistant to corrosion. It has a further advantage in that, if pinholes are present, as is normal in almost any metal coating, these holes become sealed by a corrosion product against further attack. Normal lead alloy coatings contain 90 to

95% lead and 2.5% or less of tin, with small amounts of antimony, zinc, and silver. Terneplate coatings are 80% lead and 20% tin (*see* Terneplate). The tin and antimony alloy readily with both the lead and the iron or steel and also harden the lead; zinc, silver, and other alloying elements tend to increase the fluidity of the molten lead.

Types and Uses

The lead-coated materials most commonly used in construction are terneplate for roofing and lead-coated copper for flashing. A lead coating makes an excellent surface for the reception of oil paints.

History and Manufacture

Lead coatings, although known earlier, came into widespread use toward the end of the 19th century and have increased steadily in importance to the present time. There are three processes for coating with lead: hot-dipping, electroplating, and spraying.

Hot-Dip Process. The hot-dip process, which is described fully under Terneplate, is similar to that used for zinc galvanizing.

Electroplating Process. Electroplating involves the use of fluoborate and sulfamate solutions with high current densities. The speed of electroplating is much greater with lead than with other metals. This process is usually limited to relatively small objects.

Spraying Process. In the spraying process the material to be coated is first sand-blasted or roughened and cleaned. Then, using special spray guns, it is covered with lead to the desired thickness, building up several layers if necessary. As the coating is merely anchored to the base material mechanically, a rough surface is necessary for secure bonding.

LEAD OXIDE: LITHARGE

Physical and Chemical Properties

Litharge is lead monoxide (PbO), a buff-yellow, odorless powder in its pure form. It also exists in yellow and red modifications that are very slightly soluble in water. The specific gravity is 9.53; the melting point is 1630.4°F (888°C).

Commercial Forms. Litharge is available in chemically pure (C.P.), fused, flake, and powder form in a large range of particle sizes.

Types and Uses

Litharge usually has less than 1% impurities. Its major uses are in rubber, storage batteries, glass, and glazes. In construction its uses include color pigments (the making of basic lead silicochromate and chrome pigments) and as an ingredient and filler in paints, linoleum, plastics, and cork. In linoleum and plastics it is a stabilizer, that is, it helps maintain the structure of the plastic formula, and prevents disintegration, chemical action, and temperature changes. It also acts as a binder.

Litharge Adhesive. Litharge combined with glycerin makes a quick-setting, acid-resistant adhesive with good tensile strength which is used for joining glass, metals, ceramics, etc. A typical composition—90 cm^3 glycerin to 1 lb (0.453 kg) freshly made litharge—sets in approximately 2 hours and develops a tensile strength of 668 $lbf/in.^2$ (4.74 MN/m^2) after 4 months.

History and Manufacture

Litharge was used first by the Egyptians in glazes and later (about 800 B.C.) in glass; these uses have continued to the present. With the development of the rubber industry and the discovery of the storage battery, litharge has found two important new applications: as an accelerator, toughener, and controlling ingredient in rubber and in the making of storage battery plates.

Preparation. Litharge is prepared by roasting molten lead in a reverberatory or cupel furnace in the presence of oxygen to form lead monoxide. The final product can be in lump, flake, or powder form.

LEAD OXIDE: RED LEAD

Physical and Chemical Properties

Red lead (Pb_3O_4) is a bright, orange-red powder, partially soluble in acids and insoluble in water. Its specific gravity is 8.32. It decomposes above 932°F (500°C).

"True red lead" is the actual Pb_3O_4 content in commercial pigment; "pure red lead" may contain unconverted litharge (lead monoxide, PbO) but is otherwise free from impurities. Red lead is available in dry, paste, or paint form.

Table L5 Types of red lead

Ingredients	Dry lead Grade A	Dry lead Grade B	Dry lead Grade C (% of content)	Paste-in-oil Grade B	Paste-in-oil Grade C
Red lead (Pb_3O_4)	85.0	95.0	97.0	95.0[a]	97.0[a]
Impurities, moisture, and insoluble materials	1.0	1.0	1.0		
Coarse particles retained on No. 325 sieve	1.0	1.0	1.0	1.5	1.5
Linseed oil				6.0[a]–8.00	6.0[a]–8.0
Moisture and other volatile materials				0.5	0.5
Paint pigment				92.0[a]–94.0	92.0[a]–94.0

[a]Minimum percentage allowed; other values are maximum or fixed.

Types and Uses

Red lead comes dry in three grades—A, B, and C—and in paste-in-oil in two grades—B and C (*see* Table *L5*). It finds its main construction use in heavy protective paint for iron and steel. Here the true lead oxide content is generally 95 to 97%, and the pigment is extremely fine. A large quantity of red lead is used in storage batteries, and it is an important ingredient in glazes, porcelain enamels, and glass.

History and Manufacture

Red lead was used in ancient India as a cosmetic and in medicinal preparations. Its use in glazes parallels the history of litharge. About the beginning of 1800 the use of red lead in protective paints for iron became important, and later it was even more widely used on steel. Today it still makes one of the best rust-inhibitive paints.

Preparation. Red lead is manufactured by heating litharge in a reverberatory furnace at a controlled temperature of 900 to 950°F (482.22 to 510°C). The litharge absorbs oxygen from the air and is converted into red lead (Pb_3O_4). In about 24 hours, 85% red lead is obtained, and the percentage increases with longer hours. Heating white lead in a similar furnace yields a red lead called orange mineral, which is used to make vermilion colors and inks.

LEAD SHEET AND STRIP

Physical and Chemical Properties

Lead sheet and strip are fabricated from either soft lead 99.9% pure or from lead alloyed with usually 6 to 7% antimony to add stiffness and strength. Lead sheet or strip is dull gray in color, heavy, strong, spark resistant, and corrosion resistant, especially to salt air. It will not stain adjoining materials. It reacts with uncured cement, concrete, and mortar and therefore cannot be used in direct contact with these materials unless isolated and protected by asphalt coatings or similar means during the curing stages. It makes an excellent base for oil paints but not for aluminum paints, in which a reaction between lead and the aluminum powder occurs.

Types and Uses

Weight. The weight of lead sheet and strip is usually calculated by pounds rather than by gauge or thickness. Lead sheet weighs approximately 1 lb/ft^2 for each $\frac{1}{64}$ in. (4.882 kg/m^2 for each 0.16 mm) in thickness. Above 16 lb (7.2576 kg) this weight-to-thickness ratio does not hold, and below 1 lb (0.4536 kg) it is usually given in thousandths of an inch (mm). Thus 4-lb lead is $\frac{1}{16}$ in. (1.8144-kg lead is 0.64 mm) thick, and 8-lb lead is $\frac{1}{8}$ in. (3.6288-kg lead is 1.28 mm) thick. For roofing and flashing in residential construction, the 2.5-lb (1.134-kg) weight is normally used; in industrial and commerical construction, the 3- or 4-lb (1.3608- or 1.8144-kg) weight of antimonial lead is generally used.

Size. Size is standardized as follows: standard sheets are 8 × 20 ft (2.4384 × 0.096 m), although most sheet lead is ordered cut to size. Standard strip comes in 20-ft (6.096-m) lengths and in 12-, 18-, and 24-in. (30.48-, 45.72-, and 60.96-cm) widths. Because of its expansion rate—that of antimonial lead is given as 0.000015 in./°F (0.000381 mm/0.5556°C)—care should be taken to limit the size of sheets used (*see* Item 3 *under* "Condensed Checklist").

Special Uses. Special uses of sheet lead include waterproof linings (e.g., pans in plumbing installations), chemical laboratory sinks or tabletops, and radiation protection. Lead has been used for protection against X-rays ever since the danger of exposure to these rays became known. Now it is used to shield against nuclear radiation (*see* Radiation Protection).

Application

Good practice dictates that, when lead is built into a masonry wall or otherwise rigidly held, continuous loose lock seams be used to join it to exposed sheets. In general it is advisable always to use holding cleats and never to nail the roofing or flashing directly.

The advantage of using lead for roofing and flashing is that the lead is so easily formed that most of the work can be done directly on the job without any preforming; however, close supervision of the work is necessary to assure the use of proper methods.

Condensed Checklist

1. All surfaces on which lead is to be applied should be smooth; rough projections should be eliminated, nails set, and screws countersunk.
2. Corrosion between materials is possible. Lead should not come into direct contact with uncured concrete, mortar, or cement. It should be protected with an asphalt coating or with building paper during the curing stages. The paper should be sized with rosin and should not contain acids. Although lead is, to all practical purposes, inert in contact with metals, a check should be made of the possibility of galvanic action.
3. Optimal sizes should be used. Because of expansion rates, sheet and strip lengths should not be greater than as follows: For roofing, the maximum should be 2 ft (0.6096 m) in width between seams and 4 ft (1.2192 m) in length. For flashing, the maximum is 6 to 8 in. (15.24 to 20.32 cm) in width without bending and 8 ft (2.4384 m) in length.
4. Fastening must be done correctly. Holding cleats or clips should always be used. Nails should never be driven directly through roofing or flashing. All clips should be copper or, if exposed, lead-coated copper (to match the roofing) and nailed with flat-headed copper nails. Aluminum nails should not be used.
5. Joints should not be soldered. Instead, loose lock, batten, or overlap seams should be used (*see* Figure *L8*). Loose lock and overlap seams should always be filled with a nonhardening compound. Overlap joints should be confined to batten caps, ridges, hips, cap and through-wall flashing, and valleys where the pitch is greater than 6 to 12. In all other conditions loose lock joints are generally used.
6. When soldering is necessary, a 50% lead- 50% tin solder with a rosin flux should generally be used, and soldering irons should not be too hot.
7. Types of cross seams for roofing and flashing (*see* Figure *L9*).
8. Treatment of standing and batten seams at roof ridges or hips (*see* Figure *L10*).
9. Eave and valley intersections for loose lock standing or batten seams (*see* Figure *L11*).

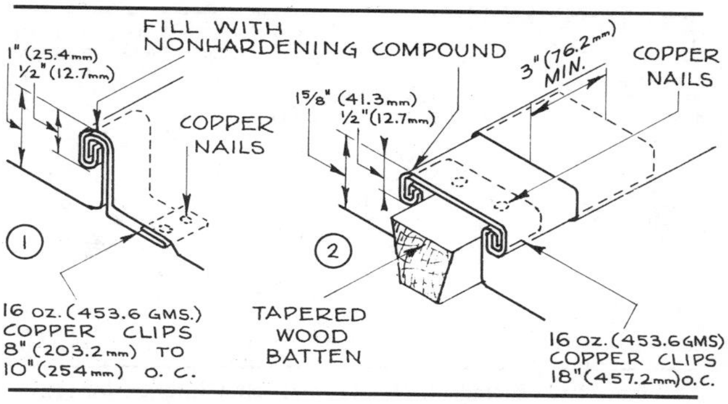

Figure L8 Details of loose lock standing (1) and batten (2) seams.

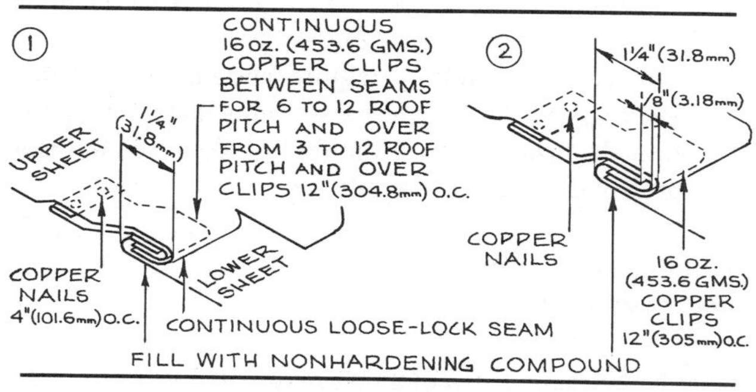

Figure L9 Cross seams (loose lock flat overlap) for roofing (1) and flashing (2). The first is used for horizontal seams between battens; the second as general purpose cross seams.

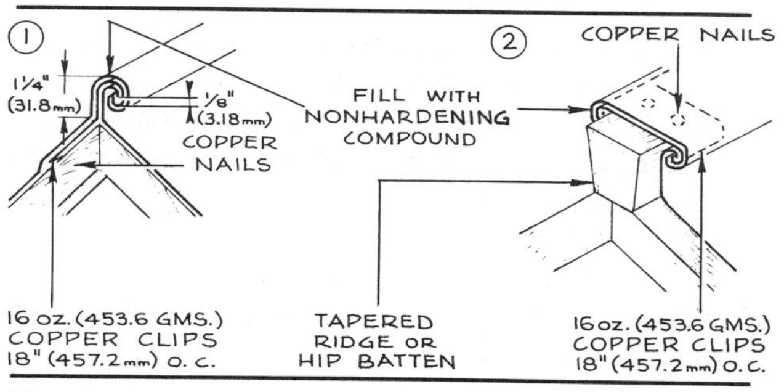

Figure L10 Details of standing (1) and batten (2) seams at roof ridge or hip.

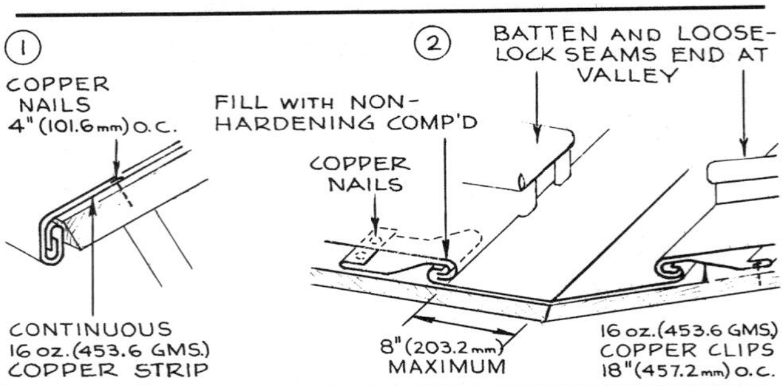

Figure L11 Details of eave (1) and valley intersections (2).

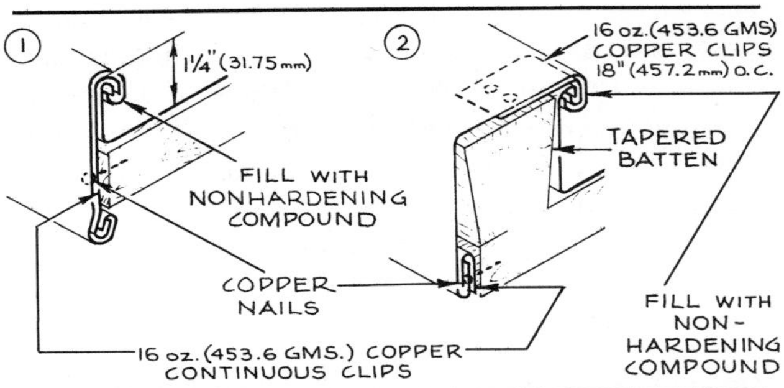

Figure L12 Details of loose lock standing (1) and batten (2) seams at roof edges.

10. Treatment of loose lock standing or batten seams at roof edge (*see* Figure *L12*).
11. Storage and handling require certain precautions. Lead sheet rolls should be stored in a standing position. In handling, care should be taken not to bang or damage ends of the rolls.

Conditions Favorable to the Use of Lead Sheet and Strip

1. Where staining of adjoining materials by the roofing and flashing material is to be avoided.
2. Where a strong, corrosion-resistant material is needed for roofing and flashing, particularly where exposure to salt water is a problem.
3. Where it is advantageous to have a fire-resistant roofing material. Municipal, state, and local codes should be checked and also the fire rating codes of the Fire Underwriters, insurance companies, labor departments, and federal government (Army, Navy, etc.).
4. Where strong vertical lines are desired on a roof.
5. Where a roofing material requiring no further maintenance is desired.

Conditions Unfavorable to the Use of Lead Sheet and Strip

1. Where a color other than dull gray is required.
2. In warm climates since the surface and color of lead are not highly reflective.
3. Where a lightweight roofing material is desired.

History and Manufacture

One of the most famous early uses of sheet lead was for the floors of the hanging gardens of Babylon. Later, in the Middle Ages, many famous cathedrals and buildings had lead roofing and flashing. These uses have continued to the present day. Commercial production of sheet and strip lead in the United States started at about the end of the 19th century. After 1925, as domestic demand increased and more ore became available, the United States become one of the largest producers of sheet and strip lead.

Cold-Rolling Process. Lead sheet and strip are made by cold-rolling methods. Molten lead is cast into rectangular slabs 4 to 6 in. (11.16 to 15.24 cm) thick and allowed to cool. The slabs are then rolled to 1- in. (2.54-cm) thickness on reversible-type rolling machinery in which the lower rollers are fixed and the upper rollers are movable and mounted one on top of the other. These 1-in. (2.54-cm)-thick sheets are then rolled to the thickness desired, taken to the cutting table, and marked, squared, trimmed, and cut to size.

LEAD SILICATES

Physical and Chemical Properties

Lead forms a number of compounds with silicon, including normal lead silicate ($PbSiO_3$), which is a white crystalline powder insoluble in most solvents, and also several basic lead silicates, which contain varying amounts of PbO. All of these are made by fusing pure silica sand with litharge in proper amounts at a high temperature to form the desired silicate compound.

One of the basic lead silicates used as a paint pigment is a complex salt of lead oxide and silicon which has a core of unconverted silica and a surface coating of the active material, the complex lead silicate salt. It contains about 45% lead and is available in powder form.

Basic lead silicates have many desirable pigment characteristics, especially for submerged exposure, where they increase the corrosion-inhibitive properties of mixed pigment paints.

Types and Uses

The use of lead silicates in ceramics and for fire-proofing fabrics is not new, but the use of basic lead silicates as a white pigment (paint solid) is very recent. These lead silicates have great potential; they possess good hiding power, form a stable elastic film, and are very economical. Eventually they may replace white lead in many paints because of their low cost.

Some of the complex lead silicates are used as stabilizers in vinyl plastics to arrest deterioration due to exposure.

LEAD SULFATES

Physical and Chemical Properties

Basic Lead Sulfate. Basic lead sulfate (also known as sublimed white lead) is an amorphous white powder, slightly soluble in hot water or acids. Its usual formula is given as $PbO \cdot PbSO_4$, but many types are possible. Basic lead sulfate has in general the same properties as basic lead carbonate, or white lead, and is available as powder or paste in a variety of combinations.

Blue Lead. Basic lead sulfate-blue, also referred to as blue lead, is actually a mixture of compounds Table *L6* shows its usual composition. It is a blue-gray powder that is insoluble in water or alcohol and makes an excellent rust-inhibitive constituent in paints. It is available in powder and paste form in various combinations, depending on use.

Table L6 Dry basic lead sulfate-blue

	Percent of content	
Ingredients	Maximum	Minimum
Lead sulfate ($PbSO_4$)		45
Lead oxide (Pb_3O_4)		30
Lead sulfide (PbS)	12	
Lead sulfite ($Pb_2SO_3 \cdot H_2O$)	5	
Carbon and impurities	5	
Residue retained on No. 325 sieve	1	

Types and Uses

Basic sulfate of white lead is used in glazes and porcelain enamels and in the rubber industry.

Basic lead sulfate-blue is used mainly in rust-inhibiting paints for iron and steel and excels in humid atmospheres and in fresh- and salt-water exposures. It has limited use as a blue-gray color pigment and also as the white paint solid (pigment) for some rust-inhibiting exterior paints, although it has less holding power than lead carbonate (white lead). In the future it is expected that basic lead sulfate-blue will be completely eliminated in paints.

History and Manufacture

The history of the use of lead sulfates as a paint pigment is similar to the histories of white lead and red lead (*see* Lead Carbonate; Lead Oxide).

Preparation of Basic Lead Sulfate. The basic sulfate of white lead is prepared by heating lead sulfide ores and molten lead in special furnaces to high temperatures in an oxidizing atmosphere and drawing off the lead sulfate fumes. In another method, metallic lead and lead oxide are mixed with water, and sulfuric acid is added under controlled conditions to form basic sulfate of white lead.

Preparation of Blue Lead. To prepare blue lead, lead ore and coal or coke are thoroughly mixed and charged into a special furnace, where they are burned by means of a blast of air at the bed of the furnace. The sulfide of the ore is partially oxidized to lead sulfate and lead oxide, which combine to form basic sulfate of blue lead.

LIME

Physical and Chemical Properties

Lime, also called quicklime, is calcium oxide (CaO), a white or grayish white, finely crystalline substance that sometimes has a yellow or brown tint because of iron impurities. Its specific gravity is 3.37; its melting point, 4676°F (2580°C); and its boiling point, 5162°F (2867.78°C). Lime reacts vigorously with water to form calcium hydroxide [$Ca(OH)_2$], known as slaked lime. This slaking must be carefully controlled because of the large quantities of heat developed during hydration.

Setting and Hardening. The setting of lime is largely due to drying out and to recrystallization (recarbonation) of calcium hydrate. The hardness is due to the lime combining with carbon dioxide of the air to form calcium carbonate, with some pozzolanic effect with the silica in the sand (*see* Cement).

Commercial Forms. Commercial limes consist primarily of calcium oxide or calcium-magnesium oxide, depending on whether the source is limestone or dolomite, a type of limestone containing 46% of magnesium (*see* Limestone). The hydrated limes are marketed in 50-lb (22.68-kg) paper bags, and quicklime is marketed in 80-lb (36.29-kg) multiwalled paper bags, in steel drums, and in bulk.

Types and Uses

Fifty years ago lime was used principally in construction and in agriculture. Today about three-fourths of all lime produced is used as a chemical in industry. It is used in the production of steel and magnesium; in the refining of copper, zinc, and other nonferrous metals; in the manufacture of paper, cement, and glass; and in many other diversified applications.

As a construction material, lime is available in the following types: (1) quicklime; (2) mason's hydrated lime; (3) special finishing hydrated lime; and (4) normal finishing hydrated lime.

Use of Quicklime. Quicklime, as used in construction, contains a high (20% minimum) percentage of magnesium oxide. It is never used unslaked. The major consumers are the chemical and process industries.

Use of Hydrated Limes. Hydrated lime is quicklime combined (slaked) with just enough water to satisfy its chemical affinity for water. It has the advantage of being dry and yet, when mixed with other ingredients, ready for use in plastering in a much shorter time than unslaked lime (quicklime). In order to develop the necessary plasticity in normal finishing hydrated lime, it must be made into putty and soaked from 12 to 15 hours before using, whereas the special finishing hydrated lime may be used $\frac{1}{2}$ hour after being made into a putty, as the plasticity will be developed within that time. The finishing hydrated limes produced in this country are all made from dolomite and contain approximately 60% of calcium and 40% of magnesium.

Hydraulic Lime. Hydraulic lime contains 10 to 17% each of aluminum oxide, silicon dioxide, and iron oxide with 40 to 45% lime and some magnesium oxide. It is treated similarly to hydrated lime. It, too, is in dry form and derives its name from its ability to set under water. It is used only in special cases where slow underwater setting is required. It has largely been replaced by portland cement.

History and Manufacture

Lime is one of the oldest manufactured building materials known to man and was used as mortar and plaster by all the early civilizations. The Egyptians used lime plaster before 2600 B.C. The Greeks used it extensively for mortars and plasters, and the Romans developed a mixture of lime putty and volcanic ash for the first real cement. The Chinese used lime in building the Great Wall, and the Aztecs and Incas used it extensively in the Americas before Columbus. The great murals and frescoes of Michelangelo, Raphael, and other artists of the Renaissance were made with lime putties. In the United States the early Spanish missions used lime stucco, and in 1662 the colonists in Rhode Island produced lime (quicklime).

Up to about 1900, practically all plaster and masonry mortar used in the United States was composed of lime. About 1900, the process of hydrating the oxide in the plant, under strict chemical control, was developed and resulted in the commercial hydrated limes of today.

Manufacture of Quicklime. Quicklime is manufactured by first crushing, grinding, and grading limestone (or dolomite) and then heating it to about 2000°F (1093°C) in horizontal rotary kilns similar to those used in making cement or in vertical kilns. The carbonates, thus heated, decompose into carbon dioxide and calcium oxide (or calcium-magnesium oxide).

Manufacture of Hydrated Lime. Hydrated lime is made by grinding quicklime, slaking the powder with a controlled amount of water, and then air-separating and sifting it to a fine dry powder, so that 98% will pass through a 200-mesh screen or sieve.

LINOLEUM

Physical and Chemical Properties

Linoleum is a resilient, waterproof floor covering that consists of a backing covered with a relatively thick layer of wearing surface. This wearing surface is a mixture that contains oxidized linseed oil processed in a special way, combined with wood or cork flour, various fillers, resin binders (wood or gum rosins), driers, and inert color pigments (*see* Table *L7*). In general it is durable and easy to maintain. It is resistant to alkalis, weak acids, and ordinary household and cooking oils. It is attacked by acids, petroleum solvents, and acetate solvents, and it is softened by prolonged exposure to high humidities.

Table L7 Typical compositions of wearing surface mixes, applied to backing to make linoleum

Material	Plain	Jaspé	Marble or spatter	Inlaid
	(percent of content)			
Oxidized mixture of bast linseed oil, driers, and resins	30	38.5	47	38
Cork flour	25	8.5	7	
Wood flour	6	20.0	29	38
Pigments and whiting	18	33.0	17	16
Scrap linoleum	21			
Clay				8

Fillers. Linoleum fillers may be stone dusts, whiting, diatomite, barium carbonate, or clay.

Driers. Linoleum driers are commonly litharge or red lead; lead resinate has also been used.

Color Pigments. Lithopone and titanium dioxide are standard linoleum pigments for white. Synthetic iron oxides provide yellows, reds, browns, and blues. For other colors *see* Color Pigments. For bright colors wood flour is considered better than cork.

Backings. Linoleum backings may be burlap (jute), a cotton fabric such as canvas, or felt made of rags, waste paper, and wood fiber. Linoleum is manufactured to rigid requirements for backing and ingredients, and tests for impact, deflection and indentation.

Types and Uses

Linoleum can be divided into five classifications: plain (battleship), jaspé, marbled, or spatter, straight-line inlaid, and molded inlaid. It is available in three gauges: service ($\frac{1}{16}$ in. or 1.6 mm), standard ($\frac{3}{32}$ in. or 2.4 mm), and heavy ($\frac{1}{8}$ in. or 3.2 mm), but not all classifications are manufactured in all three gauges. Usually plain and jaspé linoleum are made in heavy gauge; marbled in all gauges; and spatter, molded, and straight-line inlaid in standard gauge only. Linoleum comes in rolls 2 and 6 ft (60.96 cm and 1.8288 m) wide and up to 30 ft (9.144 m) in length for all classifications, and in 9-in.^{2} (22.86-cm^2) tiles for plain and marbled. Feature strips 1 in. (2.54 cm) wide and borders from 3 to 24 in. (7.62 to 60.96 cm) wide are available for all classifications except inlaid.

Application

Condensed Checklist

1. The latest colors, patterns, and textures should always be checked when using linoleum because many of these are quickly discontinued, and the thicknesses in which the various linoleums are made are often limited.
2. The floor surface upon which linoleum is to be applied should be smooth, as the linoleum will follow the contour of this floor.
3. The type of adhesive should be correct for the material and for the conditions of use. It is always best to apply the linoleum with adhesives recommended by the linoleum manufacturer.
4. For laying linoleum, the temperature of the rooms should be 70°F (21.11°C) minimum for 24 hours before and after laying, and the air should be relatively dry.
5. Sheet linoleum should be fitted to the room or area in a way that results in a minimum of seams. The seams should be at right angles to those of the subflooring, that is, perpendicular to the grain of the wood or boards.
6. Installation: Turning back half a sheet at a time, the installer should apply adhesive to the subflooring, replace the linoleum, and embed it in the adhesive. He should then roll it with a 150-lb (68.04-kg) roller in both directions to eliminate trapped air.
7. For linoleum tile, asphalt-saturated felt should first be applied with adhesive to the subflooring. Then linoleum tile is applied in the same manner as asphalt tile except that, after it is laid, it is rolled with a 150-lb (68.04-kg) roller in both directions. (*See* Figure *L13*.)

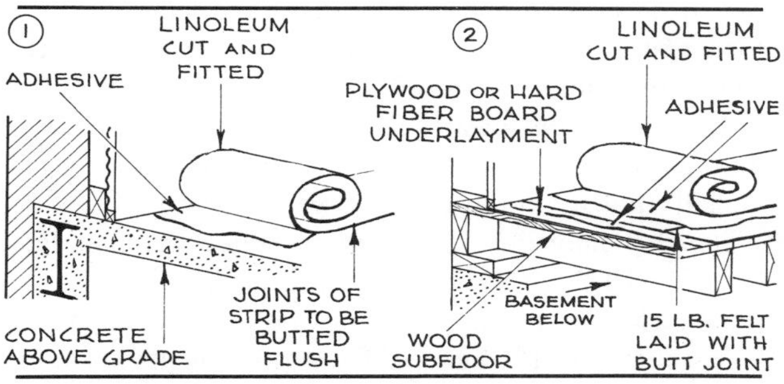

Figure L13 Methods of applying linoleum to concrete subfloor above grade (1) and to wood subfloor (2).

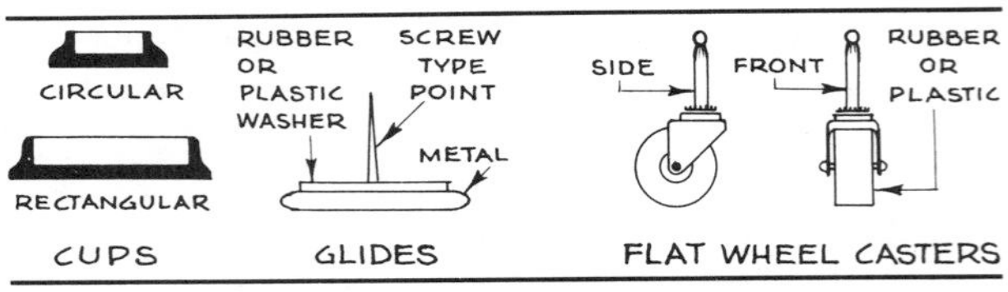

Figure L14 Typical accessories for legs of furniture resting on linoleum to prevent indentation of the surface.

8. All legs of radiators, built-in furniture, etc., should rest upon metal inserts, not the linoleum.
9. All legs of movable furniture should have large-size glides, cups, or flat roll casters (*see* Figure *L14*).
10. Linoleum should be cleaned with neutral soaps and cleaners and waxed with water-emulsion wax.

Conditions Favorable to the Use of Linoleum

1. In areas where a resilient, durable, colorful, grease-proof, waterproof type of flooring is needed. In areas of very heavy traffic such as corridors and hallways, it is best to use linoleum tile.
2. In areas where there is spillage of water, such as baths, toilets, and shower rooms.
3. In areas where there is spillage of grease, fruit juices, etc., such as kitchens and cafeterias.
4. For countertops where grease and cooking spillage occur, and for desk tops subject to heavy wear.

Conditions Unfavorable to the Use of Linoleum

1. On concrete slabs below grade and subflooring where dampness can penetrate from below.
2. On the exterior or in areas where any strong alkalis or acids may be present.

History and Manufacture

Linoleum was invented and named by Walton in 1860 in England. Walton became curious about the film that developed over an open can of paint and based his idea for a new material on the oxidization of linseed oil. He coined its name from two Roman words, *linum* meaning "flax" and *oleum* meaning "oil." It was from this idea that all the other resilient flooring materials such as asphalt, rubber, cork and plastic were developed.

The first linoleum in the United States was manufactured in 1872, in a plant that Walton built on Staten Island. The process used today is basically the same one he developed in 1864 except for improvements in manufacturing and the introduction of felts for backing about 1911. Originally only boiled linseed oil was used, but soybean oil has also proved to be satisfactory and is becoming more widely used.

Manufacturing Procedures. The first step in processing is oxidation of the combined linseed oil, driers, and resins in large tanks. Oxygen is forced through the tanks, and in 24 to 30 hours a sticky amber-colored mass develops. This mass is cut into chunks, dusted with wood flour so that it does not stick, and then cured from 5 to 30 days. After curing it is combined with color pigments (the same color pigments that are used for paints) and cork or wood flour, or both, plus mineral fillers (whiting, most commonly) and sometimes scrap linoleum and clay.

The prepared mix is now extruded and applied to the backing by several different methods, depending on the various color designs, patterns and inlaying to be made. In principle, the extruded mix and backing are passed through rollers, where heat and pressure combine the two together.

The maturing of finished linoleum continues from a few days to 7 weeks and takes place in vertical stoves, heated by steam radiators to 150 to 180°F (65.56 to 100°C), which can hold up to 5 linear miles of linoleum. The maturing is carefully controlled so that the finished product does not become too stiff.

Plain, Jaspé, Marbelized Linoleum. For plain linoleum, one heated roll is fed the mix and another cold roll is fed the backing; as the mix and backing pass through the rollers, heat and pressure consolidate the two into a smooth sheet. For jaspé, two or more colored mixes are combined in a horizontal striated pattern and then fed into the rollers. For marbelized linoleum one or more of the colors is supplied in multisized pellets to the background mix.

Inlaid Linoleum. For inlaid linoleum, sheets of different colored and textured mixes are fed individually to cylinders which cut out only the amounts of specific colors and textures needed for the final inlaid pattern and press them onto the backing in sequence until the final inlaid pattern is complete. The covered backing now passes through heated rollers which consolidate the two materials into the finished product.

Molded Linoleum. For molded linoleum the various colored mixes are applied to the backing in granular form through stencils. When the pattern is completed, the entire loose mix is compressed between the heated plates of a hydraulic press, which molds the pattern.

LITHIUM

Physical and Chemical Properties

Symbol: Li
Atomic number: 3
Specific gravity: 0.534 (20°C, 68°F)
Melting point: 179°C, 354.2°F
Boiling point: 1317°C, 2402.6°F

Lithium is a soft, silvery, highly reactive metal that must be kept under kerosene or gasoline or in inert gases, as it decomposes rapidly in air and reacts with water and acids, setting hydrogen free. Although it is the lightest metal known, it has no value as a structural material because of its low melting point and its chemical reactivity.

Commercial Forms. Lithium is available in ingot, rod, wire, ribbon, and shot form. Table *L8* shows the composition of lithium produced in the electrolytic cell.

Table L8 Composition of lithium metal

Li	K	Ca	Fe	N	Na
		(percent of content)			
99.21	0.009	0.02	0.001	0.06	0.70

Types and Uses

Lithium serves in a wide variety of organic chemical reactions of little interest to the construction field. Table *L9* gives in detail the main industrial applications (an asterisk indicates use of metallic lithium). The largest use is in the ceramics industry, where lithium compounds and minerals are important ingredients. Lithium carbonate and other compounds are used in producing glasses, glazes, and porcelain enamels with a high gloss and with scratch and chemical resistance.

Lithium is used in metallurgical procedures as a degasifier and desulfurizer. It is able to remove hydrogen, oxygen, sulfur, and nitrogen from molten metals since it combines with these elements at fairly low temperatures and extracts them from the metals as gaseous compounds. It also improves certain characteristics of the base metal being treated.

Lithium has certain advantages as an alloying agent in metals, but the problem of preventing vaporization of the lithium before it can alloy still exists. For example, 1% lithium is used in a recently marketed aluminum alloy of high strength. Magnesium alloys with up to 14% lithium are being made for certain military applications. Lithium and copper alloys, containing 0.005 to 0.008% lithium, have higher conductivity and tensile strength than phosphorized copper.

Table L9 Uses of lithium

	Construction Uses	
Component of	Aluminum*	Increasing hardness and tensile strength
	Cast iron*	Increasing density and impact resistance
	Copper*	High conductivity; desulfurizer; deoxidizer; grain refining
	Glass	Improving abrasion and chemical resistance
	Glazes and porcelain enamels	Flux; addition to frit; increasing acid and torsion resistance; improving bonding and adhesion; lowering firing temperatures
	Lead*	Increasing hardness and strength
	Magnesium*	Increasing tensile strength and corrosion resistance
	Nickel*	Desulfurizer
	Petroleum	Sulfur removal; catalysts
	Plastics	Stabilizer
	Steel*	Desulfurizer; deoxidizer; tensile strength; increasing fluidity
	Textiles	Bleaching; dyeing; drying agent
Allied Construction Uses	Air conditioning	Moisture absorption; dehumidification
	Solders	Fluxes
	Welding	Fluxes; coatings for electrodes
	Nonconstruction Uses	

Ceramic-bonded grinding wheels, drying agent, greases, infrared instruments, lubricants, lithium-silicon alloys,* refractory ceramics, special lightweight alloys*

In alloying with copper and for degasifying in nonferrous metals, lithium rod of the desired weight is encased in hermetically sealed metal tubes; for example, 0.32 oz (9g) of lithium per 100 lb (45.36 kg) of molten metal adds 0.2% lithium to the alloy.

Lithium chloride is the compound used to reduce humidity in air conditioning.

History and Manufacture

Lithium, discovered by Arfvedson in 1817, was named from the Greek *lithos*, meaning "stone." Lithium metal was first isolated in quantity in 1855 by Bunsen and Matthiessen, who electrolyzed the fused chloride. The industrial potentials of lithium were first realized during World War I in Germany when it was used in hardened lead alloys to replace lead-tin-antimony alloys and in a light, strong aluminum alloy where zinc was substituted for copper. In World War II its potentials were further developed by its use (including its compounds) in carbon dioxide absorption, hydrogen generation, magnesium-lithium alloys, organic syntheses, and dry batteries. Before World War I annual consumption was estimated at 400,000 lb (181 440 kg), but by the later 1950s it was estimated at over 30,000,000 lb (13 608 000 kg). A diagrammatic electrolytic cell for lithium production is shown in Figure *L15*.

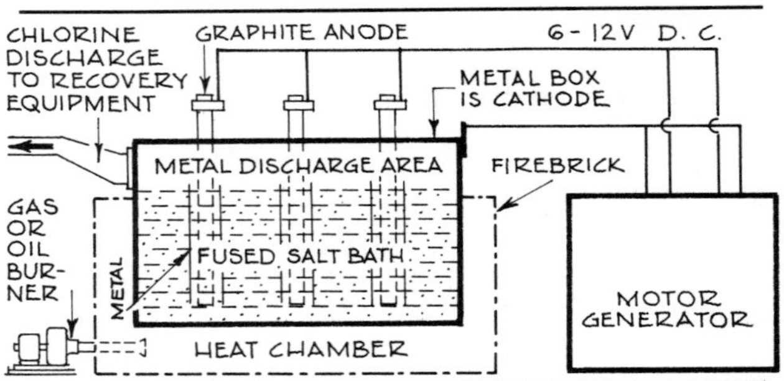

Figure L15 Diagrammatic electrolytic cell for lithium production.

LITHOPONE

Physical and Chemical Properties

Lithopone is a composite pigment containing barium sulfate and zinc sulfide in approximately equal molecular proportions. It owes its opacity to the zinc sulfide. Normal lithopone contains 28 to 30% zinc sulfide and 70 to 72% barium sulfate by weight. Other grades, less widely used, may contain 50 to 60% zinc sulfide. A larger proportion of zinc sulfide or the addition of titanium dioxide increases the hiding power of lithopone.

Types and Uses

Lithopone is used most widely in the manufacture of white and tinted interior paints and to a lesser extent in exterior paints, industrial enamels, and road-marking paints. Here lithopone contributes, in addition to white color, the following properties: improved brushing and washability, gloss retention, flow, and wear resistance. A large amount is also used in resilient floor materials (asphalt, plastic, cork, linoleum) as an opaque filler and lightener and for other pigment properties. It is also used in textiles, rubber, paper, plastics, and inks for its white pigment properties (*see* Barium; Paint; Painting; Zinc Sulfide).

History and Manufacture

About 1790 the Scottish chemist Orr discovered lithopone as a white pigment (paint solid). The manufacture of lithopone started in the United States in 1906. It is made by adding barium sulfide to a weak solution of zinc sulfate. A double decomposition takes place, precipitating a finely divided mixture of zinc sulfide and barium sulfate. The precipitate is filtered, washed, and dried; it is then heated, plunged into water, ground, and again washed and dried. The starting zinc sulfate solution consists of sulfuric acid and skimmings, together with fume dust, sludge, or other secondary sources of zinc; barium sulfide is obtained by roasting impure barium sulfate with coal or coke.

M

MAGNESIUM

Physical and Chemical Properties

Symbol: Mg
Atomic number: 12
Specific gravity: 1.738 (20°C, 68°F)
Melting point: 651°C, 1203.8°F
Boiling point: 1107°C, 2024.6°F
Tensile strength: 14,000 lbf/in.^{2}, 96.53 MN/m^2
Coefficient of expansion: 0.000025/°C

Magnesium, the sixth most abundant element in the earth's crust, is a chemically active, moderately hard, silvery metal. It oxidizes and tarnishes in moist air but not in dry air and is attacked by salt water. Magnesium is soluble in most acids, insoluble in water, nontoxic, and nonmagnetic. In finely divided form it will ignite easily, but solid magnesium will not burn unless heated above its melting point. It is the lightest structural metal, being one-third lighter than aluminum.

Workability. Magnesium can be worked by all the usual methods. It can be cast, extruded, rolled, drawn, spun, forged, blanked, and coined; it can be brazed with special care. It can be riveted and welded by gas, arc, and resistance welding.

Commercial Forms. Magnesium is available in ingot, bar, rod, ribbon, sheet, plate, tube, channel, powder, casting, and dust form. Metallic magnesium is 99.98% pure.

Types and Uses

Table *M1* shows the uses of magnesium in metallic form (indicated by asterisks) and indicates some of its many uses in compound form. Magnesia or magnesium oxide, being the most important, is marked by a double asterisk.

Table M1 Uses of magnesium

	Construction Uses	
Basis of	Brick**	Refractories
	Cement**	Oxychloride (Sorel)
	Insulation**	Main ingredient
Component of	Aluminum*	Lightening; increasing strength and corrosion resistance
	Brick**	Special surface markings
	Copper*	Deoxidizer; cleaner
	Glass**	Pyrex glass
	Glazes and porcelain enamels**	Colorless glazes; pink and yellow tones; frits
	Iron*	Increasing strength
	Monel*	Whitener; cleaner
	Nickel*	Deoxidizer; cleaner
	Paper**	Sizing; filler
	Paints**	Fireproof; filler; drier; waterproof; luminescent
	Plastics**	Lubricant; stabilizer
	Rubber**	Filler
	Textiles**	Sizing; dyes; printing; bleaching; fire-, water-, and mothproofing
	Titanium*	Reducing agent
	Wood	Fire retarding
	Zinc*	Increasing strength
	Zirconium*	Reducing agent
Allied Uses	Welding	Flux
	Refrigeration**	Brines

Table M1 Uses of magnesium (*continued*)

Nonconstruction Uses
Filters, flashlight powder,* fluorescent screens for X-ray photography, odor absorbent, printing.

Magnesia obtained from ground and burned magnesite (a natural mineral containing some iron carbonate and ferric oxide) is used to make refractory bricks for industry which can withstand temperatures of a high order, ranging from 3000° to 4000°F (1648.89 to 2204.44°C).

Recently, single crystals of magnesium oxide which exhibited ductility were reported. This promises ductility in a ceramic material, a truly important achievement.

Structural Uses. The major structural use of metallic magnesium at present is in alloy form for the aircraft, automotive, machine tool, and railroad industries, where its light weight is important. When small quantities of aluminum, manganese, zirconium, zinc, rare earth metals, and thorium are combined with magnesium, the resulting alloys have excellent physical properties.

Uses of these magnesium alloys in the construction field have been confined to household equipment and small accessories. But, since magnesium is so abundant and has so many characteristics valuable in a structural metal, it is possible to forecast a large development of its uses in construction in the future. For example, one alloy of magnesium has a specific gravity of 1.77 and an ultimate tensile strength of 42,000 lbf/in.^{2} (289.59 MN/m^2); thus it is lighter than many aluminum alloys and as strong.

Other Uses. Magnesium is used for expendable anodes with other metals, which then serve as cathodes, to protect them from corrosion. For example, underground pipelines, well casings, and tanks are protected by placing magnesium anodes, connected by wires, adjacent to them. Any corroding action takes place with the magnesium and not with the other metal.

Applications

Condensed Checklist

1. Magnesium should not be used in direct contact with iron, steel, copper, brass, or wood.
2. Generally, since magnesium is high in the galvanic series, great care should be taken in fabrication that the surfaces are completely cleaned before use or before being coated and painted.
3. In cutting and working the metal, the shavings, filings, and dust should be carefully collected and kept dry in clean, covered iron containers until disposed of by burning. If these cuttings and powders become ignited, the fire can be extinguished by covering with finely powdered graphite.

History and Manufacture

In ancient Greece a white mineral from Magnesia, a district of northern Greece, was called *magnesia lithos* (magnesian stone). When Sir Humphrey Davy, in 1808, isolated a new element from magnesian stone, he named it magnesium. In 1828 Bussy first prepared a nearly pure metal, and in 1833 Faraday produced pure metallic magnesium by electrolysis of magnesium chloride with a voltaic cell. The modern electrolytic process used for major industrial production is based on the electrolytic cell developed in 1852 by Bunsen. In 1941 the ferrosilicon thermal process was discovered and is used today for a small portion of the total production.

Electrolytic Process. The electrolytic process utilizes seawater and well or lake brines which contain magnesium chloride. The flowchart of the electrolytic process shown in Figure *M1* uses seawater. The seawater, containing 0.13% magnesium, is pumped into tanks and mixed with lime (CaO). Insoluble magnesium hydroxide [$Mg(OH)_2$] is formed, settles, and is drawn off and filtered. It is then treated with hydrochloric acid made from natural gas and the chlorine from the electrolytic cell. This treatment produces magnesium chloride solution, which is evaporated, partially dehydrated, and then

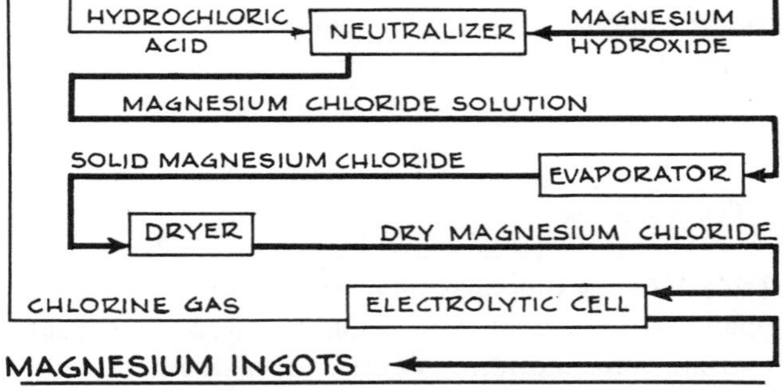

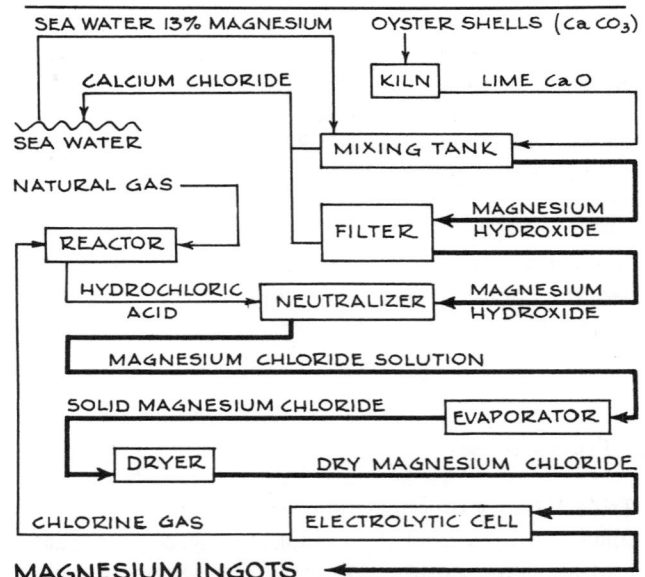

Figure M1 Electrolytic process for production of metallic magnesium.

decomposed in an electrolytic cell to 99.9% pure metallic magnesium and chlorine.

Thermal Processes. A thermal process may also be used for the production of magnesium, based on either ferrosilicon or carbon.

In the ferrosilicon method, dolomite (limestone) ore is heated to obtain magnesium oxide, which is then mixed with ferrosilicon. This mixture, pressed into small pellets and charged into a retort, is heated under a vacuum to 2200°F (1221°C). The silicon reacts with magnesium oxide to form silicon dioxide and magnesium vapor. One end of the retort is cooled, and the magnesium vapor condenses into fine crystals which are then removed, melted with a flux, and cast into ingots of 99.98% pure magnesium.

The carbothermic method follows a similar pattern, except that the magnesium vapor is chilled with hydrogen gas to produce a dust, which in turn is crystallized into the pure metal.

Melting, Alloying, Refining. The melting, alloying and refining of magnesium are similar in many respects to the same processes for other metals. Because of its high reactivity in the molten state, magnesium is alloyed with higher melting metals by the use of a nickel-magnesium alloy or alloys of magnesium with Monel, zinc, copper, and aluminum. Progress has been made in the development of protective surface treatments and in anodizing magnesium alloys.

MAGNETS AND MAGNETISM

The construction field is familiar with magnets as a recent innovation in cabinet door hardware, for example, magnetic catches. Throughout this book, however, substances have been described as ferromagnetic, paramagnetic, and diamagnetic. A short explanation seems in order.

Magnetism is the force of attraction between iron and certain substances known as magnets. Ferromagnetic substances are those that are highly magnetic, the most commonly known ones being iron, cobalt, and nickel; paramagnetic substances are those that are weakly magnetic; and diamagnetic substances may be described as nonmagnetic or actually repelled by the poles of an electromagnet. Magnets may be natural, for example, lodestone, the natural mineral form of iron oxide, also known as magnetite; or they may be artificial, that is, magnetized by special treatment.

Naturally magnetic substances were known long before the Christian era and have been used since A.D. 1000 for compasses in navigation. But what magnetism is cannot be explained definitively even today, although many theories have been developed. Practical knowledge of magnets and magnetism exists and is the basis of the production, distribution, and use of electricity as well as of communication by wire, radio, television, and other electronic means. Magnets are also essential for the operation of many modern industrial devices and machines.

MANGANESE

Physical and Chemical Properties

Symbol: Mn
Atomic number: 25
Specific gravity: 7.21 to 7.44
Melting point: 1244°C, 2271.2°F
Boiling point: 2097°C, 3806.6°F

Manganese is a reddish gray or silvery, brittle metallic element somewhat similar to iron in general chemical reactivity. It oxidizes superficially in air, rusts in moist air, reacts with sulfur, and, in the form of fused manganese, dissolves carbon. Manganese exists in various allotropic modifications of which only the alpha is stable and of interest to the construction field. The hard, brittle nature of alpha manganese has prevented fabrication of the pure metal for any practical purpose.

Commercial Forms. Manganese metal is available principally as electrolytic manganese in 500-lb (226.8-kg) net weight steel drums; it is also available as electric furnace metal and as metal produced by a thermic process. These products do not have the purity of the electrolytic metal, however (*see* Table *M2*).

Table M2 Composition of electrolytic manganese

Mn	C	S (percent of content)	Fe	Ca, Cu, Mg, Ag
99.97	0.004	0.0135	0.001	0.0115

Types and Uses

Approximately 95% of the world's production of manganese ores, manganiferous iron ores, and manganiferous zinc residuum is used in metallurgical processes, the bulk of it in the manufacture of iron and steel. Manganese in compound form also has many uses. Its uses in metallic form are noted by asterisks in Table *M3*.

Table M3 Uses of manganese

	Construction Uses	
Basis of	Color pigment	Brown and green
	Glazes and porcelain enamels	Purple and black colors
	Paint	Driers
	Textiles	Bleach; brown colors
Component of	Aluminum*	Increasing corrosion resistance, stiffness, and hardness
	Rubber	Increasing tackiness
	Copper*	Aiding vibration damping
	Glass	Neutralizing iron stains; pink through violet and purple to black colors
	Glazes and porcelain enamels	Color control
	Iron*	Increasing abrasive resistance; counteracting brittleness by sulfur control; deoxidizer; corrosion resistance
	Magnesium*	Corrosion resistance; increasing strength, stiffness, and hardness
	Nickel*	High strength
	Silver*	In alloys as substitute for nickel
	Stainless steel*	Substitute for nickel
	Steel*	Abrasive resistance; counteracting brittleness by sulfur control; deoxidizer; corrosion resistance
	Titanium*	Increasing strength while maintaining ductility
	Wood	Preservative
	Zinc*	In alloys as substitute for nickel

Use in Iron and Steel. For metallurgical purposes manganese ores are converted into standard, low-carbon and medium-carbon ferromanganese, spiegeleisen, silicomanganese, and silicospiegel, which are added during the production of iron and steel as a means of getting manganese into the steel.

Electrolytic manganese can replace ferromanganese of any grade in many applications and generally leads to better control of the carbon and phosphorus content.

All iron and steels contain some manganese. In high-strength steels a manganese content of 0.25 to 1.2% increases the tensile and yield strength by about 25%.

Use as Alloying Constituent. Manganese is not only used to control sulfur embrittlement and as a deoxidizer but is also essential as an alloying constituent in iron, steel, copper, zinc, aluminum, titanium and manganese-base alloys.

Practically all commercial aluminum and magnesium products contain manganese. Many copper alloys contain manganese to give special properties, such as vibration dampening and high expansion, which are particularly useful in instrumentation.

New Alloys. Because of the advent of pure manganese metal, all the binary systems of manganese need reviewing, and this will result in new alloys which will certainly play a role in the construction field. Some new alloys now in production are the following: 75% manganese and 20% copper (high-dampening); 50% manganese, 45% iron, and 5% aluminum (low expansion). Manganese stainless steels recently developed offer fine corrosion resistance to the atmosphere.

Other Uses. In the electric dry cell battery, the negative pole (the container) is of zinc and the positive pole of carbon; the depolarizing mix contains manganese dioxide. The manganese dioxide is responsible for the depolarizing action that helps to maintain the efficiency of the cell. Manganese is vital to plant and animal life and is essential to reproduction in animals.

History and Manufacture

Manganese was first recognized by Scheele, a Swedish chemist, in 1774. It was named for the magnetic properties of its mineral, pyrolusite, using the Latin word for magnet, *magnes.* In 1856 Mushet, by adding manganese, made the Bessemer steel-making process a practical success, and in 1888 Hadfield discovered the high-manganese steels that bear his name. Manganese has since become essential in the manufacture of steel;

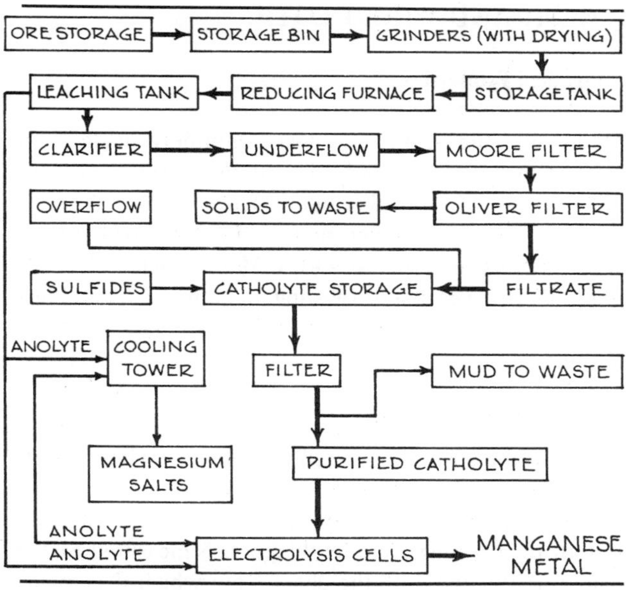

Figure M2 Flowchart of electrolytic production, or electrowinning, of manganese.

approximately 14 lb (6.34 kg) of manganese, chiefly in the form of ferromanganese, is used for each ton of steel produced, principally for the purpose of controlling the sulfur content.

Although manganese has been known and commonly used for a long time, it is only since 1930 that precise and extensive work on its potentialities as a metal has been possible, and it was only in the 1930s that the high-purity form was introduced to industry.

Electrolytic Production. Production of electrolytic manganese on a commercial scale began in the United States in 1941. This process is based on the electrolysis of sulfate or chloride solutions of manganese (*see* Figure *M2*). The careful purification of high-grade ores, the difficulties due to the resolubility of manganese in acids, and the necessity for special alloys for the anodes and cathodes, all combine to make this a highly technical process. Nevertheless, annual production of electrolytic manganese has risen from 43,700 lb (19822.32 kg) in 1939 to 7,000,000 lb (3175200 kg) in 1952. In 1958 production capacity was about 50 tons (50.8 metric tonnes), or 100,000 lb (45360 kg) per day.

MASONRY

"Masonry" originally referred to the art of building in stone. Today the term includes all types of building materials that consist of units held together with mortar, for example, stonework, brickwork, clay tile products, concrete block, gypsum block, and sometimes even glass block work. The characteristics of masonry work are a resultant of the properties of the masonry units, of the mortar, and of the methods of bonding, reinforcing, anchoring, tying, joining, etc., the units into a whole.

MERCURY

Physical and Chemical Properties

Symbol: Hg
Atomic number: 80
Specific gravity: 13.546 (20°C, 68°F)
Melting point: -38.87°C, -37.97°F
Boiling point: 356°C, 672.8°F

Mercury is a heavy, relatively inert, silvery metal. It is the only metal that is liquid at room temperature. Mercury vapor and dust are injurious to the human body.

Commercial Form. Mercury is available in liquid form in iron flasks containing 76 lb.

Types and Uses

Mercury has diversified uses based on its properties of liquidity at room temperature, high specific gravity and electrical conductivity, and its toxicity. These uses, both in compound form and in metallic form (indicated by asterisks), are shown in Table *M4.*

Amalgams. Mercury forms special alloys with one or more metals. These alloys are called amalgams and may be liquid or solid. A liquid amalgam of mercury and thallium is used for low temperature thermometers down to -76°F (-24.44°C). An amalgam of mercury and silver is used in producing mirrors (*see* Glass Mirrors).

Mercury Boiler for Electric Power. There is a mercury vapor boiler which uses mercury instead of water to produce electric power. In this boiler higher temperatures can be obtained at lower pressures. For instance, a pressure of 100 psi (20.307 kg/m^2) produces 928°F (497.78°C) mercury vapor, whereas a 3200 psi (2249.824 kg/m^2) pressure produces only 706°F (374.44°C) saturated water vapor; this is the highest pressure at which saturated water vapor can exist. Such a system is about 33% more efficient than a steam (water) system and is now actually being used by a public service company

Table M4 Uses of mercury

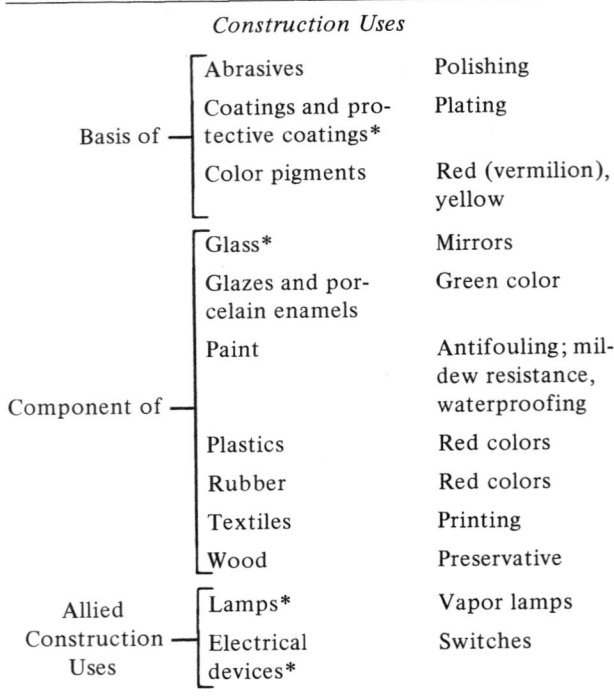

	Construction Uses	
Basis of	Abrasives	Polishing
	Coatings and protective coatings*	Plating
	Color pigments	Red (vermilion), yellow
Component of	Glass*	Mirrors
	Glazes and porcelain enamels	Green color
	Paint	Antifouling; mildew resistance, waterproofing
	Plastics	Red colors
	Rubber	Red colors
	Textiles	Printing
	Wood	Preservative
Allied Construction Uses	Lamps*	Vapor lamps
	Electrical devices*	Switches
	Nonconstruction Uses	
Barometers,* electronics,* mercury-vapor boilers,* photography, printing, special batteries,* thermometers*		

that supplies electrical power to a large section of a certain state. This indicates the possibility that very pure mercury used in some sort of sealed system would provide economically sound and efficient mercury boilers, even on a small scale.

History and Manufacture

Mercury has been found in Egyptian tombs dating back to 1500 B.C. and was known to the early Chinese and Hindus. The term "mercurius," together with the symbol for the planet Mercury, was used by alchemists about A.D. 700. The symbol Hg now used is from the Roman *hydrargyrum*, which in turn is derived from the Greek words meaning "liquid silver."

Production. Mercury occurs in its natural state, but the principal source is cinnabar ore, which is mercury sulfide (HgS). Mercury from this ore is obtained by roasting in air or with lime. In both methods the mercury is condensed from the gases, filtered, washed in nitric acid to remove other metals, and placed in flasks.

MESH AND WIRE CLOTH

Physical Properties

Mesh and wire cloth are most commonly made from steel (uncoated, coated, hot-dipped, galvanized or plated), copper, and aluminum. They are also available in brass, bronze, Monel metal, stainless steel, nickel silver, and various special alloys. Mesh is made by three methods: by weaving or twisting metal wires; by welding the wires; and by piercing sheet metal and then stretching or expanding it.

Measuring Systems. Woven wire mesh is always measured by the size of the openings between wires, and its thickness by the gauge or size of the wire. Expanded metal mesh is measured by the approximate or average size of the openings, and its thickness is measured by the gauge or thickness of the sheet from which it is made. All wire cloth is measured by the number of spaces or openings to the inch.

Types and Uses

Design Elements. The wire used to form mesh may be round, half-round, square, oval, half-oval, hexagonal, octagonal, flat, or other, more complex shapes. The patterns into which it can be woven or into which sheet can be shaped include diamond, crimp, herringbone, and Z-rib, to name only a few. The type of metal and finish will depend on the purpose and conditions of use. (*See* Figures *M3* and *M4*.

Use of Mesh. Mesh is used in construction for reinforcing concrete; for cement, plaster, and stucco lathing; for fencing, partitions, screens, guards for windows and openings, and ornamental metalwork. These are described under subsequent headings.

In industry mesh has very wide use for the size-grading of ores, minerals, sand, gravel, etc.

Use of Wire Cloth. Wire cloth is used in construction most frequently for insect screening of doors and windows. In vents and many other small openings in buildings, wire cloth serves also as a rodent barrier and flameproof covering, since wire cloth lets air through but will stop flames.

In industry wire cloth serves many of the same purposes as mesh, especially for all finer grading and for filtering. For such purposes it is available in mesh sizes as fine as 400 openings per inch (25.4 mm), made of wire only 0.001 in. (0.025 mm) in diameter.

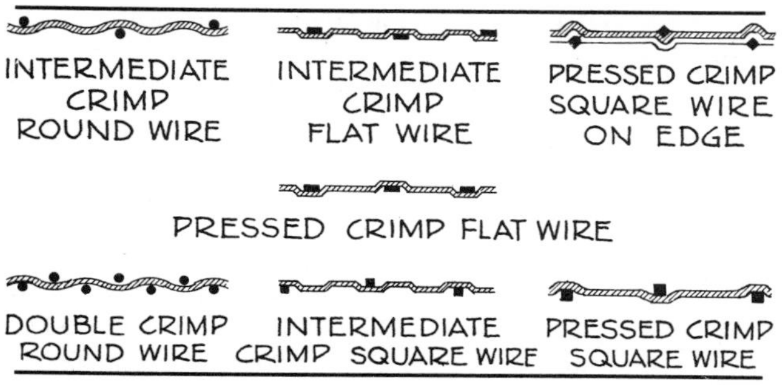

Figure M3 Various methods of forming wires and of pressing wires to form mesh.

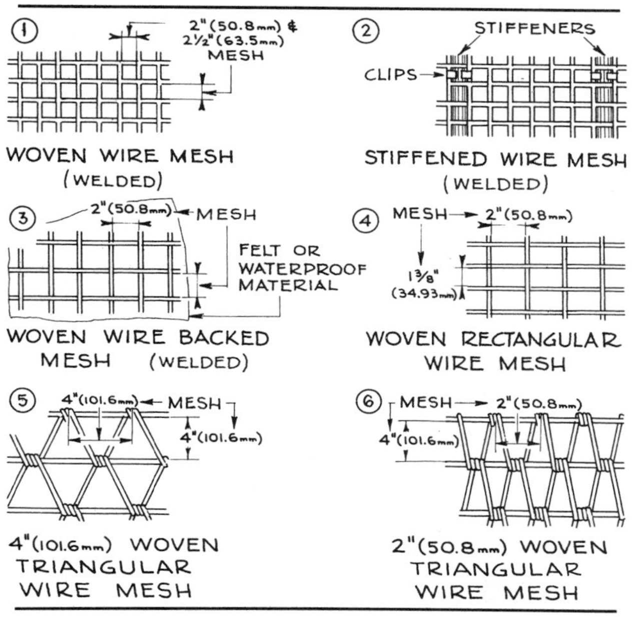

Figure M4 Welded and woven wire mesh.

History and Manufacture

The origin of mesh can be traced back to ancient times when metal pieces or strips were heated and hammered or crudely riveted together. Later, heated metal was pulled by hand through a rough die made of stone to form a wire that was used for chain mail armor and other types of woven wire products, all of which were made by hand.

Woven Mesh. In 1878 barbed wire first appeared, and the improved machines used for its fabrication led in 1890 to the development of a practical loom for weaving wire. After 1900 wire mesh and cloth as we now know them were developed. Weaving methods are constantly improving, and welding has been added to the methods of fabrication.

Expanded Mesh. Expanded mesh is a more recent type than woven or twisted mesh and cloth. To form the expanded mesh, the sheet is first cut or pierced in staggered slots or other patterns; then the sheet is held by the two sides parallel to the slots and stretched by pressure until the desired openings or form are obtained. Sheet may also be stamped, perforated, or deformed into an open mesh.

Welded Mesh. In welded wire mesh the wires are welded together.

MESH AND WIRE CLOTH FOR CONCRETE REINFORCING

Reinforcing mesh for concrete work is made in a variety of types, including triangular, welded, expanded, and chicken wire. Of these, welded wire mesh is the one generally used for reinforced concrete work. The other types are used as forms and for cement-gun work, residential concrete slabs, sidewalks, terraces, and miscellaneous other types of concrete work. For details see Figures *M5* to *M7* and Tables *M5* to *M8*. The mesh is generally made of steel or steel alloys and may be plain, galvanized, or painted. It is available in roll and sheet form.

See also Concrete, Reinforced; Gauges; Steel for Concrete; Wire and Wire Rope.

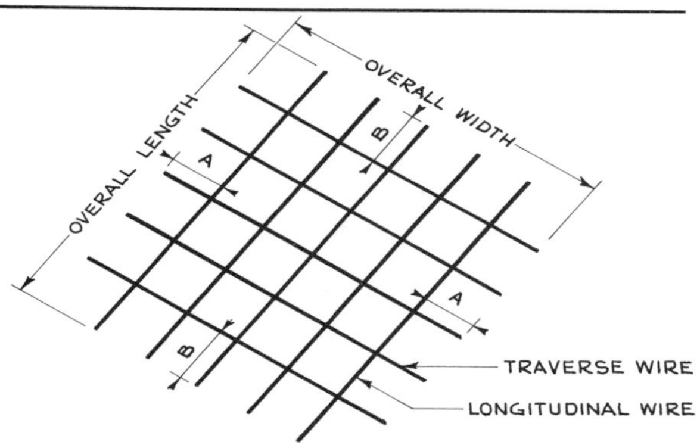

A. SIDE OVERHANGS MAY BE VARIED AS REQUIRED AND DO NOT HAVE TO BE EQUAL. OVERHANG LENGTHS LIMITED ONLY BY OVERALL WIDTH OF ROLL.

B. END OVERHANGS MAY BE VARIED AS REQUIRED. MORE ECONOMICAL FABRIC RESULTS, HOWEVER, IF THE SUM OF THE OVERHANGS EQUALS A TRANSVERSE WIRE SPACING

METHOD OF DESIGNATING:

WWF 6 x 12 - W16 x W26

LONGITUDINAL WIRE SPACING	=	6" (152.4mm)	LONGITUDINAL WIRE SIZE	=	W16
TRANSVERSE WIRE SPACING	=	12" (304.8mm)	TRANSVERSE WIRE SIZE	=	W26

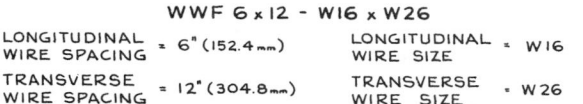

Figure M5 Welded wire mesh for concrete reinforcing.

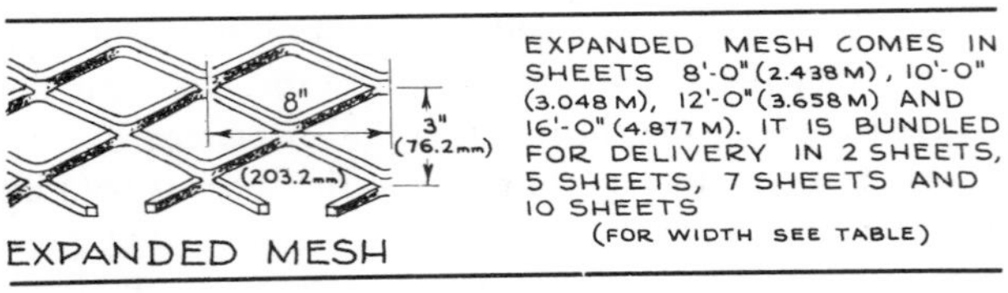

Figure M6 Typical example of expanded wire mesh.

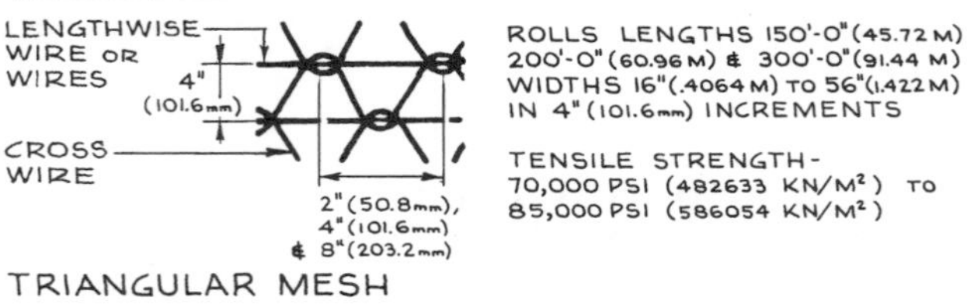

Figure M7 Triangular mesh.

Table M5 Triangular wire mesh for cement-gun type of work

Type number	Number of wires lengthwise spaced 4 in. (101.6 mm)	Gauge number of lengthwise wires	Gauge number of crosswise wires spaced 2 in. (50.8 mm)	Approximate weight per 100 ft² (9.29 m²)	
				lb	kg
7A	1	12	14	31	14.06
6A	1	10	14	37	16.78
5A	1	8	14	44	19.56
4A	1	6	14	53	24.04
29A	1	12	$12\frac{1}{2}$	42	19.05
28A	1	10	$12\frac{1}{2}$	48	21.77
27A	1	8	$12\frac{1}{2}$	55	24.95
26A	1	6	$12\frac{1}{2}$	64	29.03

Table M6 Gauge, welded wire designation, and dimensions of wire for welded wire fabric for concrete reinforcing

American Steel and Wire Gauge Number	W & D size numbers		Area		Nominal diameter	
	Smooth	Deformed	in.²	cm²	in.	mm
	W31	D31	0.310	2.000	0.628	15.95
	W30	D30	0.300	1.936	0.618	15.70
	W28	D28	0.280	1.807	0.597	15.16
	W26	D26	0.260	1.679	0.575	14.61
	W24	D24	0.240	1.549	0.553	14.05
	W22	D22	0.220	1.419	0.529	13.44
	W20	D20	0.200	1.290	0.504	12.80
0000000			0.189	1.219	0.490	12.45
	W18	D18	0.180	1.167	0.478	12.14
000000			0.167	1.076	0.4615	11.72
	W16	D16	0.160	1.032	0.451	11.46
00000			0.146	0.942	0.4305	10.94
	W14	D14	0.140	0.903	0.422	10.72
0000			0.122	0.787	0.394	10.01
	W12	D12	0.120	0.774	0.390	9.91
	W11	D11	0.110	0.710	0.374	9.54
	W10.5		0.105	0.678	0.366	9.30
000			0.103	0.665	0.3625	9.21
	W10	D10	0.100	0.645	0.356	9.04
	W9.5		0.095	0.613	0.348	8.84
	W9	D9	0.090	0.581	0.338	8.59
00			0.086	0.556	0.331	8.41
	W8.5		0.085	0.548	0.329	8.36
	W8	D8	0.080	0.516	0.319	8.10
	W7.5		0.075	0.484	0.309	7.85
0			0.074	0.478	0.3065	7.79
	W7	D7	0.070	0.452	0.298	7.57
	W6.5		0.065	0.419	0.288	7.32

Table M6 Gauge, welded wire designation, and dimensions of wire for welded wire fabric for concrete reinforcing (*continued*)

American Steel and Wire Gauge Number	W & D size numbers		Area		Nominal diameter	
	Smooth	Deformed	in.^{2}	cm^2	in.	mm
1			0.063	0.407	0.283	7.19
	W6	D6	0.060	0.387	0.276	7.01
	W5.5		0.055	0.355	0.264	6.71
2			0.054	0.348	0.2625	6.67
	W5	D5	0.050	0.323	0.252	6.40
3			0.047	0.303	0.244	6.20
	W4.5		0.045	0.290	0.240	6.10
4	W4	D4	0.040	0.258	0.225	5.72
	W3.5		0.035	0.226	0.211	5.36
5	W4		0.034	0.219	0.207	5.26
	W3		0.030	0.194	0.195	4.95
6	W2.9		0.029	0.187	0.192	4.88
7	W2.5		0.025	0.161	0.177	4.37
8	W2.1		0.021	0.136	0.162	4.12
	W2		0.020	0.129	0.159	4.04
9			0.017	0.111	0.148	3.76
	W1.5		0.015	0.097	0.138	3.51
10	W1.4		0.014	0.090	0.135	3.43

Table M7 General stock styles of welded wire fabric for reinforcing concrete[a]

Designation[b]	Spacing of wires				Diameter of wires				Sectional area				Weight	
	Longitudinal		Transverse		Longitudinal		Transverse		Longitudinal		Transverse			
	in.	mm	in.	mm	in.	mm	in.	mm	in.^{2}/ft	cm^2/0.31 m	in.^{2}/ft	cm^2/0.31 m	lb/100 ft^2	kg/9.29 m^2
6 × 6 W1.4 × W1.4	6	152.4	6	152.4	0.135	3.43	0.135	3.43	0.029	0.187	0.029	0.187	21	9.53
6 × 6 W2.1 × W2.1	6	152.4	6	152.4	0.162	4.12	0.162	4.12	0.041	0.265	0.041	0.265	30	13.61
6 × 6 W2.9 × W2.9	6	152.4	6	152.4	0.192	4.88	0.192	4.88	0.058	0.374	0.058	0.374	42	19.05
6 × 6 W4 × W4	6	152.4	6	152.4	0.225	5.72	0.225	5.72	0.080	0.516	0.080	0.516	58	26.31
4 × 4 W1.4 × W1.4	4	101.6	4	101.6	0.135	3.43	0.135	3.43	0.043	0.277	0.043	0.277	31	14.06
4 × 4 W2.1 × W2.1	4	101.6	4	101.6	0.162	4.12	0.162	4.12	0.062	0.400	0.062	0.400	44	19.96
4 × 4 W2.9 × W2.9	4	101.6	4	101.6	0.192	4.88	0.192	4.88	0.087	0.561	0.087	0.561	62	28.12
4 × 4 W4 × W4	4	101.6	4	101.6	0.225	5.72	0.225	5.72	0.120	0.774	0.120	0.774	85	38.06
4 × 12 W2.1 × W0.9	4	101.6	12	304.8	0.162	4.12	0.1055	2.67	0.062	0.400	0.009	0.058	25	11.30
4 × 12 W2.5 × W1.1	4	101.6	12	304.8	0.177	4.37	0.1205	3.06	0.074	0.478	0.011	0.071	31	14.06

[a] Welded smooth wire fabric with wires smaller than size W1.4 is manufactured from galvanized wire.
[b] Old designation was by steel wire gauge: 6 × 6 – 10 × 10. This now is 6 × 6 W1.4 × W1.4.

Table M8 Types of welded wire fabric most commonly used in construction

Size of square mesh		Wire designation	Sectional area		Weight	
in.	mm		in.^{2}/ft	cm^2/0.31 m	lb/100 ft^2	kg/9.29 m^2
6	152.4	W1.4	0.029	0.187	21	9.53
6	152.4	W2.1	0.041	0.265	30	13.61
6	152.4	W2.9	0.058	0.374	42	19.05

MESH AND WIRE CLOTH FOR FENCING, PARTITIONS, GRILLES, AND SCREENS

Mesh for fences, partitions, guards, grilles, screens, and ornamental work is usually fabricated from the following metals: steel, aluminum, brass, bronze, copper, Monel alloy, nickel silver, and stainless steel.

Choice of Pattern and Finish. Choice of pattern and finish may be made in any of the designs previously discussed and illustrated here in some detail (*see* Tables *M9* through *M13* and Figures *M8* and *M9*). Expanded metal is available in a wide variety of designs and finishes for use as decorative screens and area dividers. It is advisable to check with manufacturers to obtain the latest designs and finishes, as new designs are constantly being introduced and exploited for decorative purposes.

A wide variety of finishes is also possible. For example, steel wire or mesh may be plain, painted, coated, galvanized, hot-dipped, or plated.

Table M9 Typical woven round, flat, or square wire meshes used in construction

Round wire mesh (square or diamond)					Flat and square wire mesh				
	Size of wire		Openings between wires					Openings between wires for both square and flat wires	
	Diameter				Width of flat wire[a]		Square wire gauge number		
Gauge number	in.	mm	in.	mm	in.	mm		in.	mm
6	0.192	4.877	2	50.8					
8	0.162	4.115	$1\frac{3}{4}$	44.45					
9	0.148	3.759	$1\frac{3}{4}$	45.45					
10	0.135	3.429	$1\frac{1}{2}$	38.10	$\frac{3}{16}, \frac{1}{4}, \frac{5}{16}, \frac{3}{8}$	4.76, 6.35, 7.94, 9.53	10	$\frac{3}{4}, 1, 1\frac{1}{4}, 1\frac{1}{2}$	19.05, 25.4, 31.75, 38.1
12	0.105	2.667	$\frac{7}{8}$, 1	22.23 25.4	Same as above		12	Same as above	
14	0.080	2.032	$\frac{3}{4}$	19.05	Same as above		14	Same as above	
16	0.063	1.600	$\frac{3}{8}$	9.53	Same as above		16	Same as above	
18	0.047	1.194	$\frac{3}{8}$	9.53	Same as above		18	Same as above	

[a]Thickness same for round wire.

Table M10 Typical expanded metal mesh used in construction[a]

Type and gauge designation	Openings				Size of strand			
	Width		Length		Width		Length	
	in.	mm	in.	mm	in.	mm	in.	mm
$\frac{1}{2}$ in. (12.7 mm) No. 18	0.461	11.71	1.2	30.48	0.085	2.159	0.045	1.143
$\frac{3}{4}$ in. (19.05 mm) No. 16	0.923	23.44	2.0	50.8	0.086	2.184	0.059	1.499
$\frac{3}{4}$ in. (19.05 mm) No. 13	0.923	23.44	2.0	50.8	0.090	2.286	0.089	2.261
$\frac{3}{4}$ in. (19.05 mm) No. 10	0.923	23.44	2.0	50.8	0.135	3.429	0.089	2.261
$\frac{3}{4}$ in. (19.05 mm) No. 9	0.923	23.44	2.0	50.8	0.135	3.429	0.134	3.404
$1\frac{1}{2}$ in. (39.1 mm) No. 18	1.33	33.78	3.0	76.2	0.063	1.600	0.045	1.143
$1\frac{1}{2}$ in. (38.1 mm) No. 16	1.33	33.78	3.0	76.2	0.096	2.438	0.060	1.524
$1\frac{1}{2}$ in. (38.1 mm) No. 13	1.33	33.78	3.0	76.2	0.096	2.438	0.088	2.235
$1\frac{1}{2}$ in. (38.1 mm) No. 10	1.33	33.78	3.0	76.2	0.128	3.251	0.088	2.235
$1\frac{1}{2}$ in. (38.1 mm) No. 9	1.33	33.78	3.0	76.2	0.128	3.251	0.132	3.353

[a]Expanded metal mesh generally is available in the following sizes: widths of 3 ft (0.914 m), 4 ft (1.219 m), and 6 ft (1.829 m), and lengths of 6 ft (1.829 m), 8 ft (2.438 m), and 12 ft (3.658 m).

Table M11 Typical flattened expanded metal mesh used in construction[a]

Type and gauge designations		Openings				Size of strand			
		Width		Length		Width		Length	
Size	No.	in.	mm	in.	mm	in.	mm	in.	mm
$\frac{1}{2}$ in. (12.7 mm)	18–20	0.461	11.71	1.281	32.54	0.097	2.464	0.041	1.041
$\frac{1}{2}$ in. (12.7 mm)	16–18	0.461	11.71	1.281	32.54	0.106	2.692	0.052	1.321
$\frac{3}{4}$ in. (19.05 mm)	16–18	0.800	20.32	2.135	54.23	0.122	3.099	0.047	1.194
$\frac{3}{4}$ in. (19.05 mm)	14–16	0.800	20.32	2.135	54.23	0.129	3.277	0.059	1.499
$\frac{3}{4}$ in. (19.05 mm)	9–11	0.923	23.44	2.135	54.23	0.190	4.826	0.111	2.819
$1\frac{1}{2}$ in. (38.1 mm)	16–8	1.333	33.78	3.203	81.36	0.120	3.048	0.049	1.305
$1\frac{1}{2}$ in. (38.1 mm)	14–16	1.333	33.78	3.203	81.36	0.130	3.302	0.061	1.549
$1\frac{1}{2}$ in. (38.1 mm)	9–11	1.333	33.78	3.203	81.36	0.190	4.826	0.111	2.819

[a] Flattened expanded metal mesh generally is available in the following sizes: widths of 3 ft (0.914 m), 4 ft (1.219 m), and 6 ft (1.829 m), and lengths of 6 ft (1.829 m), 9 ft (2.438 m), and 12 ft (3.658 m).

Table M12 Types of hexagonal mesh (chicken wire) used for fencing, lathing, and reinforcing

Type of mesh	Size of mesh		Wire gauge number	Width of rolls		Length of rolls		Weight of 150-ft (176.17-m) roll	
	in.	mm		ft	m	ft	m	lb	kg
Lathing	1	25.4	18	3	0.914	150	176.17	88	39.92
	1	25.4	20	3	0.914	150	176.17	43.8	19.87
Reinforcing	1	25.4	18	3	0.914	150	176.17	88	39.92
	1	25.4	18	4	1.219	150	176.17	116	52.62
	$1\frac{1}{2}$	38.1	17	3	0.914	150	176.17	71	32.21
	$1\frac{1}{2}$	38.1	17	4	1.219	150	176.17	94	42.64

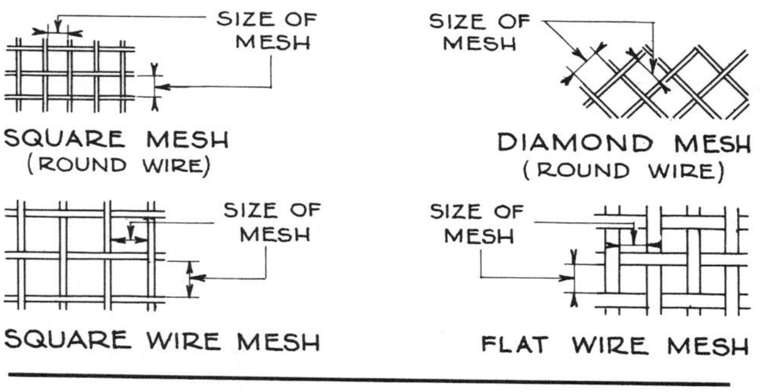

Figure M8 Woven round, flat, and square wire meshes.

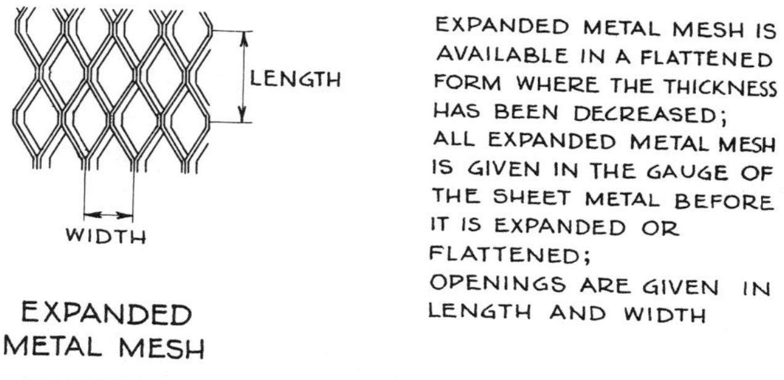

LENGTH
WIDTH
EXPANDED METAL MESH
EXPANDED METAL MESH IS AVAILABLE IN A FLATTENED FORM WHERE THE THICKNESS HAS BEEN DECREASED;
ALL EXPANDED METAL MESH IS GIVEN IN THE GAUGE OF THE SHEET METAL BEFORE IT IS EXPANDED OR FLATTENED;
OPENINGS ARE GIVEN IN LENGTH AND WIDTH

Figure M9 Expanded and flattened expanded meshes.

Table M13 Typical triangular wire mesh

Type number	Number of wires lengthwise per 4 in. (101.6 mm)	Gauge number of lengthwise wires	Effective sectional area per 1 ft (0.3048 m) of mesh by length in.^{2}	mm^2	Approximate weight per 100 ft^2 lb	9.2903 m^2 kg
		Crosswise wires, No. 14 gauge, spaced 4 in. (101.6) mm)[a]				
032	1	No. 12	0.032	0.206	22	9.98
040	1	No. 11	0.040	0.258	25	11.34
049	1	No. 10	0.049	0.316	28	12.70
058	1	No. 9	0.058	0.374	32	14.52
068	1	No. 8	0.068	0.439	35	15.80
080	1	No. 7	0.080	0.516	40	18.14
093	1	No. 6	0.093	0.600	45	20.41
107	1	No. 5	0.107	0.690	50	22.68
126	1	No. 4	0.126	0.813	57	25.86
146	1	No. 3	0.146	0.942	65	29.48
153	1	$\frac{1}{4}$ in. (6.35 mm)	0.153	0.987	68	30.84
168	1	No. 2	0.168	1.084	74	33.57
180	2	No. 6	0.180	1.167	78	35.38
208	2	No. 5	0.208	1.342	89	40.37
245	2	No. 4	0.245	1.599	103	46.72
267	3	No. 6	0.267	1.741	111	50.35
287	3	No. $5\frac{1}{2}$	0.287	1.870	119	53.93
309	3	No. 5	0.309	1.994	128	58.06
336	3	No. $4\frac{1}{2}$	0.336	2.168	138	62.60
365	3	No. 4	0.365	2.355	149	68.59
395	3	No. $3\frac{1}{2}$	0.395	2.549	160	72.58
		Crosswise wires, No. 14 gauge, spaced 8 in. (203.2 mm)[b]				
036P	1	No. 12	0.036	0.232	17	7.71
044P	1	No. 11	0.044	0.284	20	9.07
053P	1	No. 10	0.053	0.342	24	10.89
062P	1	No. 9	0.062	0.400	27	12.25
072P	1	No. 8	0.072	0.465	31	14.06
084P	1	No. 7	0.084	0.542	35	15.80
097P	1	No. 6	0.097	0.626	40	18.14
		Crosswise wires, No. $12\frac{1}{2}$ gauge, spaced 8 in. (203.2 mm)[c]				
041R	1	No. 12	0.041	0.265	21	9.53
049R	1	No. 11	0.049	0.316	24	10.89
058R	1	No. 10	0.058	0.374	28	12.70
067R	1	No. 9	0.067	0.432	31	14.06
077R	1	No. 8	0.077	0.497	35	15.80
089R	1	No. 7	0.089	0.574	40	18.14
102R	1	No. 6	0.102	0.658	44	19.96

[a] Effective sectional area by width = 0.022 in.^{2} (0.142 cm^2).
[b] Effective sectional area by width = 0.009 in.^{2} (0.058 cm^2).
[c] Effective sectional area by width = 0.014 in.^{2} (0.090 cm^2).

Choice of metal, pattern, and finish is therefore dictated by end use, design considerations, economic factors, and current availability.

Frames. Frames for supporting mesh screens, partitions, grilles, and the like not only must be strong enough to withstand the conditions of use; they are also an important design feature of the building. These frames are generally made of channels and round rods. Square and rectangular bars, pipe, tube, and angles also are used between frames and as frames, depending entirely on the design and end use. They are available in a wide assortment of sizes and shapes made from the same metal as the mesh.

Accessories. Special hardware and accessories for this type of metalwork are also readily available in the same metals as the mesh and framework. To avoid galvanic action, all components (mesh, frame, and accessories) should preferably be of one metal or, if different, of compatible metals.

Chain Link Fencing. Chain link fencing is a distinctive type of woven wire mesh with diamond-shaped openings, generally made of steel (hot-dipped, galvanized, or zinc coated) and aluminum wire. The most commonly used size is the 2-in. (50.8-mm) mesh made of wire gauge No. 6, 9, or 11, available in 100-ft (30.48-m) rolls in the following widths (heights): 3, 3.5, and 4 to 10 ft (0.914, 1.067, and 1.219 to 3.048 m), inclusive, in 12-in. (30.48-cm) increments. The ends of the wires at the top and bottom edges may be either barbed or knuckled (*see* Figure *M10*).

Supporting posts for chain link fencing should be spaced 10 ft (3.048 m) o.c. at maximum. They should be embedded in a concrete base to a depth of 2 ft (0.61 m) for fences up to 4 ft (1.219 m) high, and to a depth of 3 ft (0.914 m) for fences over 4 ft high (1.219 m). Top and bottom rails, usually of pipe, are optional for supporting purposes for heights of 6 ft (1.829 m) or less. For fences higher than 6 ft (1.829 m), top and bottom rails are necessary with pipe-type posts, optional for H-type column posts.

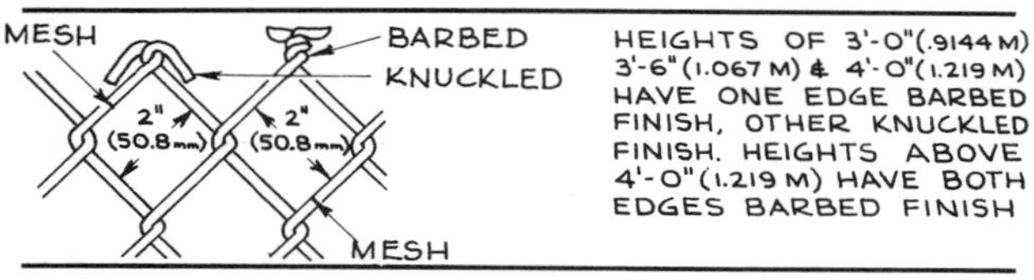

Figure M10 Barbed and knuckled finishes on edges of chain link fencing.

Gates require special supports, the size and design of which depend on the size of the opening and the type of gate (swinging or sliding).

Other Fencing. Many of the wire meshes shown in other tables and illustrations can also be used for fencing, and barbed wire can be added at top or bottom if desired. Chicken wire fencing, on the other hand, is seldom used in construction for anything other than plaster lathing and reinforcing of concrete (*see* Figure *M11* and Table *M12*).

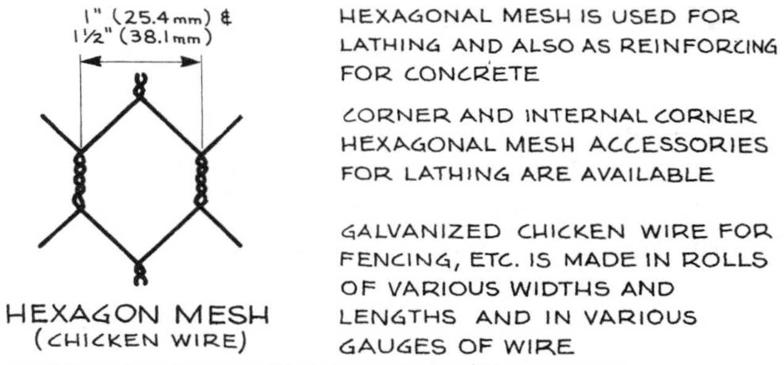

Figure M11 Hexagonal mesh (chicken wire).

MESH AND WIRE CLOTH FOR INSECT SCREENING

Wire cloth is generally used for insect screening. The most common size is woven of No. 22 gauge wire with 12 openings to the inch. Other types of wire cloth are used as rodent barriers and filters and for flameproofing vents.

A different kind of insect screening is made of sheet aluminum, slit and formed into tiny louvers that not only keep insects out but also control heat and glaring sunlight (*see* Figure *M12* and Table *M14*).

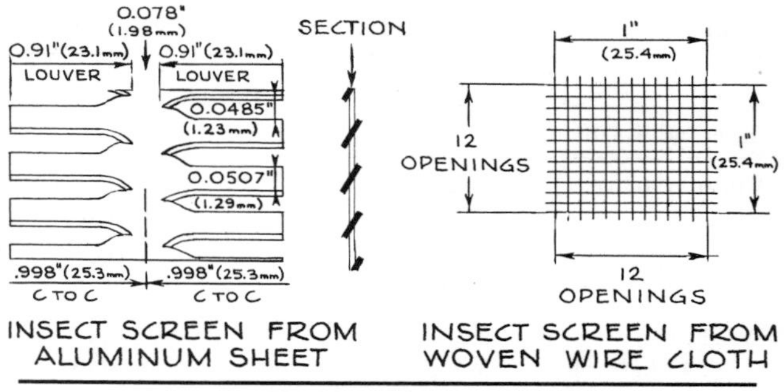

Figure M12 Insect screening.

Table M14 Types and sizes of insect screening

Type	Width of rolls	Length of rolls	Mesh size	Gauge of wire or sheet	Materials
Sheet aluminum	18–48 in. in 1-in. increments (45.72–121.92 cm in 2.54-cm increments)	50 ft (15.24 m)	0.0507 in. (1.288 mm)	0.029	Aluminum sheet
Woven wire mesh cloth	6 ft max. (1.829 m)	50 ft	No. 12	0.029	Steel, stainless steel, copper, bronze, and alclad aluminum

MESH AND WIRE CLOTH FOR LATHING

Lathing mesh for plaster, stucco, and cement work is generally made of steel or noncorrosive steel alloy wire or sheet. It may vary in type (expanded, ribbed, or wire mesh, or combinations of these), in finish (plain, galvanized, or painted), in size, and in weight (*see* Tables *M15* and *M16* and Figures *M13* and *M14*). The patented, trademarked types of mesh shown for concrete reinforcing are equally suitable for lathing. Chicken wire fencing is also used as plaster lath.

When using expanded or wire mesh lath, it is good practice always to check how the lathing is to be attached, supported, and anchored, since there will be considerable difference in the unsupported spanning length among the various gauges and patterns of lathing. (*See also* "Application" under Plaster and Plastering.)

Table M15 Types of wire mesh lath

Size	Welded woven wire (Type 1)	Welded stiffened woven wire (Type 2)	Back welded woven wire (Type 3)			Rectangular woven wire (Type 4)	Triangular woven wire (Type 5)	Triangular woven wire (Type 6)
Gauge number	18, 19, 20, 21	18, 19, 20, 21	12	14	16	14	14	14
Mesh in. (mm)	2, $2\frac{1}{2}$ (50.8, 63.5)	2, $2\frac{1}{2}$ (50.8, 63.5)	3, 4 (0.914, 1.219)	2 (50.8)	2 (50.8)	$1\frac{3}{8} \times 2$ (34.93 × 50.8)	4 × 4 (101.6 × 101.6)	2 × 4 (50.8 × 101.6)
Width of rolls ft (m)	3 (0.914)	3 to 10 (0.914 to 3.048)	4 (1.219)	4.167 (1.260)	2.375 (0.724)	2.063 (0.629)	3, 4 (0.914, 1.219)	3, 4 (0.914, 1.219)
Length of rolls ft (m)	150 (45.72)	150 (45.72)	50 (15.24)	108 (32.92)	50 (15.24)	219 (66.75)	150 (45.72)	150 (45.72)
Sheet sizes ft (m)				4.167 × 4.333 (1.260 × 1.321)				

Table M16 Types of expanded and ribbed mesh lath

Sizes and weight	Types 1	2	3	4	5	6	7	8
Width in. (cm)	27 (68.58)	24 (60.96)	24 (60.96)	16.875, 18 (42.86, 45.72)	24 (60.96)	24 (60.96)	27 (68.58)	18 (45.72)
Length ft (m)	8 (2.438)	8 (2.438)	8 (2.438)	9 (2.743)	8 (2.438)	8 (2.438)	8 (2.438)	8 (2.438)
Weight lb/yd² (0.542 kg/m²)	2.2, 2.5, 3.0, 3.4 (1.192, 1.355, 1.626, 1.843)	2.75, 3.0, 3.4 (1.491, 1.676, 1.843)	3.4 (1.843)	3.0, 3.4 (1.626, 1.491)	2.75, 3.0, 3.4 (1.491, 1.626, 1.843)	3.0, 3.4, 4.0 (1.626, 1.491, 2.168)	3.0, 3.4 (1.626, 1.491)	3.4 (1.491)

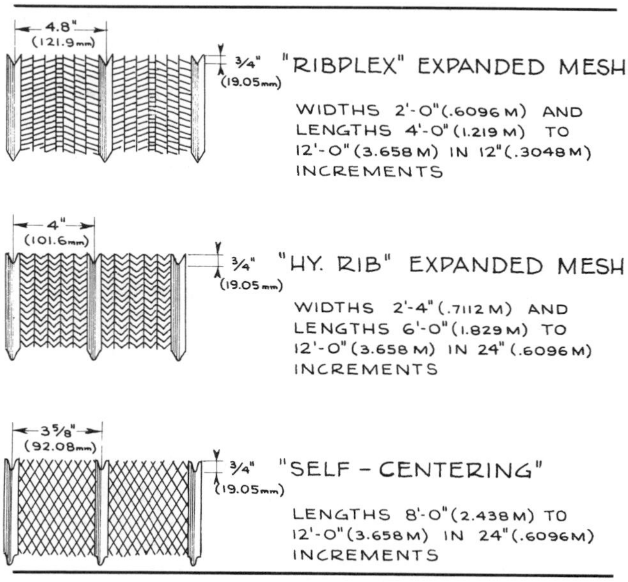

Figure M13 Ribbed expanded mesh.

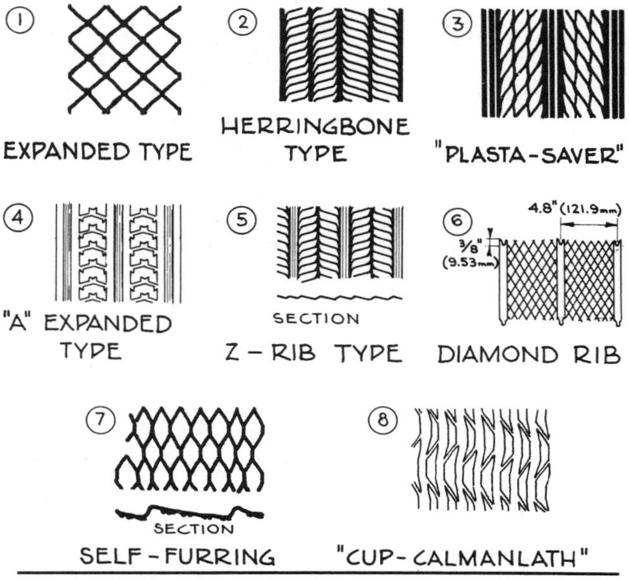

Figure M14 Expanded and ribbed mesh lath.

METALS

Physical and Chemical Properties

Metals are, in general, those substances that have a peculiar luster and hardness, can conduct heat and electricity, are opaque, and possess certain mechanical properties, the most remarkable being the power of resisting deformation (*see* Table *M17*).

Table M17 Characteristics of metals and nonmetals

Metals	Nonmetals
Crystalline structure	Crystalline or amorphous structure
Easily conduct heat or electricity	Do not easily conduct heat or electricity
Malleable and ductile	Solids are brittle and nonductile
Form positive ions in solution	Form negative ions in solution
Lose or lend electrons during chemical reactions	Gain or borrow electrons during chemical reactions
Set free at negative pole (cathode) in electrolysis	Set free at positive pole (anode) in electrolysis

Electron Theory. A metal is defined in chemistry as an element that yields positively charged ions in aqueous solution of its salts. Another definition would be that a metal is a lender of electrons, whereas a nonmetal is a borrower of electrons.

The electron theory is illustrated in its simplest form by Figure *M15*, which shows lithium, the metal, and fluorine, the nonmetal, combining by mutual lending and borrowing of electrons to form lithium fluoride. Note that lithium has three electrons and fluorine nine and that the atomic number for each is the same as the number of electrons in the respective atom.

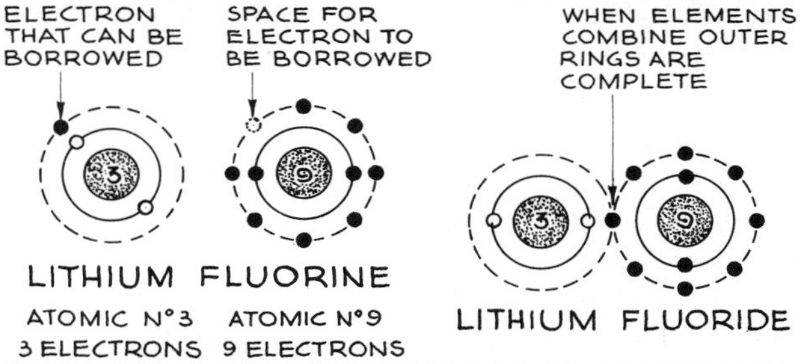

Figure M15 Schematic diagram of the electron theory.

Distinctions Between Metals and Nonmetals. Some elements, called amphoteric, act as metals under one set of conditions and as nonmetals, or negative ions, under others. In the light of recent advances in physics and chemistry, it is now known that the distinctions between metals and nonmetals do not always hold true. Many elements formerly considered nonmetals, liquids, or gases (e.g., hydrogen, boron, silicon) act as metals under special laboratory conditions which are now finding practical industrial applications. However, in this book we are considering metallic elements in relation to construction materials and avoiding technical and theoretical discussions, many of which can be controversial.

Data for Construction. What is important to know in the construction field is data on those physical, chemical, and mechanical properties that influence the use of a metallic material in construction. These data should include melting point, tensile strength, coefficient of expansion, corrosion resistance to various chemicals, workability or ease of handling in both fabrication and construction, and reaction to weathering.

Corrosion. Most metals deteriorate as a result of the slow action of oxygen, water vapor, carbon dioxide and other substances in the air, soil, water, or chemical agents to which they are exposed.

Tarnish is a surface film; while the resulting discoloration may be unsightly, no damage is done to the metal. When the interior of the metal is affected, the corrosion process may lead to mechanical failure of the metal part. The rusting of iron and steel is a familiar example of this.

Table M18 Galvanic or electropotential series of metals

Cesium	Cobalt
Lithium	Nickel (active)
Rubidium	Inconel (active)
Potassium	Tin
Calcium	Lead-tin
Sodium	Lead-antimony
Magnesium	Lead
Magnesium alloys	Brasses
Beryllium	Hydrogen
Aluminum	Antimony
Manganese	Bismuth
Zinc	Arsenic
Aluminum 1100	Copper
Chromium	Mercury
Gallium	Bronzes
Aluminum 2017T	Copper-nickel
Iron or steel	Monel
Cast iron	Silver solder
Chromium-iron (active)	Silver
Stainless steel 301, 302 (active)	Graphite
Cadmium	Palladium
Stainless steel 316 (active)	Platinum
Indium	Gold

Galvanic or Two-Metal Corrosion. One of the most important facts that should be known about a metal or an alloy is its reaction with other metals or alloys with which it may be in contact. These data are given in the galvanic or electropotential series, also called the electromotive force series. Here the metals are listed in a sequence in which each metal is corroded by all that follow it (*see* Table *M18*). In other words, when two different metals are in contact with each other in the presence of moisture, there will be a flow of current from one metal (the anode) to the other metal (the cathode), and one will be eaten away, or disintegrated, while the other (the cathode) will remain intact.

It is not always true that there is greater corrosion the further down the scale one goes; in certain cases one metal immediately following another may be very corrosive. The most active ones are noted in the series. An important point to remember in utilizing the galvanic scale is that moisture is the chief problem in this type of corrosion, and moisture depends to a great extent on climate. In the desert, electrochemical action will be at a minimum. On the seacoast the action will be much greater not only because of the ever-present moisture but also because of the salt. Some means of separating dissimilar metals must therefore always be found.

Prevention of Corrosion. Covering a metal with paint or other type of protective coating is an example of external means of preventing corrosion. Alloying to make a metal more noble (e.g., nickel alloys) or to cause formation of a protective surface film (e.g., stainless steel) is an example of an internal method of preventing corrosion.

The protective coating can be formed by chemical action (e.g., anodizing, phosphatizing) or by simple heating. A less corrosive metal may be applied by dipping (e.g., galvanizing, or dipping in zinc, and tinning), plating, cladding (e.g., alclad aluminum), spraying with molten metal (also called metallizing), or cementation (i.e., heating in a metal powder to form a coating). These various external means are described under the following headings: Aluminum Finishes; Lead Coatings; Metals: Sprayed Metal Coatings (Metallizing); Nickel Plating; Nitriding; Painting of Metal; Plating; Zinc Coatings.

Types and Uses

It is interesting to note that most of the elements are metals, actually 88 out of the 103 presently known elements. In the table of metallic elements, Table *M19*, all metals are listed alphabetically with their symbols and atomic numbers. In construction we need to know the symbol in order to read technical data; the atomic number, while not of immediate practical value, is the key to the atomic structure.

No metal is used in elemental, pure form in construction. Even iron, on which so much of our structure is

Table M19 List of metals

Element	Symbol	Atomic number	Element	Symbol	Atomic number
Actinium[a]	Ac	89	Mercury	Hg	80
Aluminum	Al	13	Molybdenum	Mo	42
Americium[a]	Am	95	Neodymium[b]	Nd	60
Antimony	Sb	51	Neptonium[a]	Np	93
Arsenic	As	33	Nickel	Ni	28
Barium	Ba	56	Niobium	Nb	41
Berkelium[a]	Bk	97	Nobelium[a]	No	102
Beryllium	Be	4	Osmium	Os	76
Bismuth[a]	Bi	83	Palladium	Pd	46
Boron	B	5	Platinum	Pt	78
Cadmium	Cd	48	Plutonium[a]	Pu	94
Calcium	Ca	20	Polonium[a]	Po	84
Californium[a]	Cf	98	Potassium	K	19
Carbon	C	6	Praseodymium[b]	Pr	59
Cerium[b]	Ce	58	Promethium[b]	Pm	61
Cesium	Cs	55	Protactinium[a]	Pa	91
Chromium	Cr	24	Radium[a]	Ra	88
Cobalt	Co	27	Rhenium	Re	75
Copper	Cu	29	Rhodium	Rh	45
Curium[a]	Cm	96	Rubidium	Rb	37
Dysprosium[b]	Dy	66	Ruthenium	Ru	44
Erbium[b]	Er	68	Samarium[b]	Sm	62
Einsteinium[a]	Es	99	Scandium	Sc	21
Europium[b]	Eu	63	Selenium	Se	34
Fermium[a]	Fm	100	Silicon	Si	14
Francium[a]	Fa	87	Silver	Ag	47
Gadolinium[b]	Gd	64	Sodium	Na	11
Gallium	Ga	31	Strontium	Sr	38
Germanium	Ge	32	Tantalum	Ta	73
Gold	Au	79	Technetium	Tc	43
Hafnium	Hf	72	Tellurium	Te	52
Holmium[b]	Ho	67	Terbium[b]	Tb	65
Hydrogen	H	1	Thallium[a]	Tl	81
Indium	In	49	Thorium[a]	Th	90
Iridium	Ir	77	Thulium[b]	Tm	69
Iron	Fe	26	Tin	Sn	50
Lanthanum[b]	La	57	Titanium	Ti	22
Lawrencium[a]	Lw	103	Tungsten	W	74
Lead[a]	Pb	82	Uranium[a]	U	92
Lithium	Li	3	Vanadium	V	23
Lutetium[b]	Lu	71	Ytterbium[b]	Yb	70
Magnesium	Mg	12	Yttrium[b]	Y	39
Manganese	Mn	25	Zinc	Zn	30
Mendelevium[a]	Md	101	Zirconium	Zr	40

[a]Radioactive metals.
[b]Rare earth metals.

based, is an alloy, strictly speaking, which differs in character depending on the amount of impurities, carbon or slag, present. Almost all metals are used as alloys or compounds. Alloys are therefore fully discussed as a separate heading.

Among the metals are two groups which at present have either limited or no use in materials for construction—the radioactive metals and the rare earth group of metals. Both are relatively new, but, because of the ever-growing activity in the fields of atomic research and space travel, it is almost certain that more of them will appear in construction materials. The rare earth metals are already being used in increasing amounts to produce steels and other alloys able to resist very high temperatures.

History and Manufacture

Comparison of the abundance of the various metals and the relative difficulties of obtaining them as metals gives an interesting and valuable insight into future technological developments. The table of relative abundance, Table *M20*, reveals why the metals commonly used by man have been his tools even though some are very scarce in nature. The phenomenal development of knowledge of the physical sciences and their industrial application which has taken place in the last few decades will surely lead to new materials and new structural methods in the construction of tomorrow.

Primary and Secondary Metals. All metals are categorized into two basic types from the production stand-

Table M20 Relative abundance of metals

	ABUNDANT METALS	
Easily converted	*Difficult to convert*	
Iron	Aluminum	Potassium
	Calcium	Silicon
	Magnesium	Sodium
	Manganese	Titanium
	SCARCE METALS	
Easily converted	*Difficult to convert*	
Antimony	Barium	Platinum
Arsenic	Beryllium	Rare earth metals
Bismuth	Boron	Rhenium
Cadmium	Cesium	Rhodium
Cobalt	Chromium	Ruthenium
Copper	Gallium	Strontium
Gold	Germanium	Tantalum
Lead	Hafnium	Thallium
Mercury	Indium	Tungsten
Nickel	Iridium	Vanadium
Selenium	Lithium	Zirconium
Silver	Molybdenum	
Tellurium	Niobium[a]	
Tin	Osmium	
Zinc	Palladium	

[a]Originally known as columbium.

point: primary metals, or those obtained directly from an ore or as a by-product in the extraction of other metals from an ore; and secondary metals, or those obtained from scrap and waste products.

METALS: ALUMINIZING

Aluminizing is the application of an aluminum coating by the hot-dip method to iron and steel sheet, strip, and coils. There are two grades: *Type 1* has an aluminum silicon coating that has excellent resistance to corrosion and can be used at high temperatures up to 1250°F (676.67°C); *Type 2*, known as architectural grade, has a pure aluminum coating that is resistant to atmospheric corrosion and can be used at temperatures up to 900°F (482.22°C) (*see* Table *M21*).

Aluminum coatings do not give any protection by anodic (galvanic) action, but do protect the iron and steel from moisture and other corrosive elements. The edges are also protected, even though there is corrosion of the iron and steel, because there is no rust bleeding or undercutting of the aluminum coating at the edge.

When using these aluminized iron and steel materials, one should always protect the aluminum or isolate it from other metals in order to stop galvanic action, except in the case of metals that are compatible with aluminum.

METALS: CHEMICAL FINISHES

Chemical finishes may be defined as the result of either a process that has no effect on the surface of the metal other than cleaning it, or a process that affects the surface of the metal in a specific way. These chemical finishes can be divided into four categories (*see also* Table *M22*).

1. A cleaning of the metal surface without affecting it in any other way.
2. An etched matte finish to the metal surface.
3. A brightening of the surface of the metal.
4. Conversion coatings, which are generally used to prepare the surface of the metal for painting or for receiving another type of finish. Conversion coatings can also be used to produce a patina or statuary finish.

In general, a standard designation for chemical finishes on metal utilizes the letter "C" followed by numerals. For example, an etched matte finish for aluminum is designated as follows: C20 for unspecified, C21 for fine matte, C22 for medium matte, C23 for coarse matte, and C2x for other types.

METALS: COATINGS FOR METALS

Coatings are used on metal not only to protect the metal from corrosion but also to add color, texture, and pat-

Table M21 Type 2 aluminized sheet and coils

Nominal thickness		Gauge number	Weight		Sheets				Coils	
					Width (max.)		Length (max.)		Width (max.)	
in.	mm		lb/ft²	kg/m²	in.	mm	in.	mm	in.	mm
0.0934	2.372	13	3.789	18.50	36	914.4	192	4876.8	36	914.4
0.0785	1.994	14	3.164	15.45						
0.0710	1.781	15	2.851	13.92						
0.0635	1.613	16	2.539	12.40						
0.0575	1.461	17	2.289	11.18						
0.0516	1.311	18	2.038	9.95						
0.0456	1.159	19	1.789	8.73	48	1219.2	192	4876.9	48	1219.2
0.0396	1.937	20	1.539	7.51						
0.0366	0.930	21	4.414	6.91						
0.0366	0.930	22	1.289	6.29						
0.0306	0.777	23	1.164	5.68						
0.0276	0.701	24	1.039	5.07						
0.0247	0.672	25	0.914	4.46	44	1117.6	192	4876.9	40	1016.0
0.0217	0.551	26	0.789	3.85						
0.0202	0.513	27	0.727	3.55	42	1066.8	144	3667.6	36	914.4
0.0187	0.475	28	0.664	3.24						

Table M22 Types of chemical finishes for metals and general processes used

Metal	Cleaning method	Method for matte etching	Method for brightening	Process for conversion coatings
Aluminum	Chlorinated and hydrocarbon solvents and inhibited chemical cleaners	Alkali and acid solutions	Chemical or electrolytic brightening	Alkaline chromate, acid cromate-fluoride, and acid chromate-fluoride-phosphate
Copper and copper alloys	Chlorinated and hydrocarbon solvents and inhibited chemical solvents	Acid solutions of sulfuric and nitric acids		Acid chloride; acid sulfate, carbonate, oxide, and sulfide treatments to produce patinas and statuary finishes
Stainless steel				Hot chemical oxidizing bath, and controlled heat treatment
Iron and steel	Pickling, chlorinated solutions, and alkaline solutions			Acid phosphate

tern. These coatings can be divided into four categories: anodic, vitreous (porcelain enamel), laminated, and organic.

Anodic Coatings. Anodic coatings are almost entirely restricted to aluminum (*see* Aluminum Finishes).

Vitreous Coatings. The vitreous coating generally used in construction is porcelain enamel. Porcelain enamel is an inorganic coating bonded to the metal by fusion at 800°F (426.67°C) or higher. In the construction field the metals commonly used are enameling iron or steel, decarbonized enameling sheet steel, cold-rolled sheet steel, and aluminum.

These coatings are nonporous, impermeable, and resistant to maintenance cleaning processes and chemical stain. Colors are permanent and coatings are obtainable in a variety of textures, patterns, and visual effects. The process of manufacturing porcelain enamel requires the following three steps: (1) design and fabrication of the metal, (2) preparation of the metal surface, and (3) formulation and application of the coating material, and treatment at the temperature required for fusion.

Certain basic requirements, principles, and standards must be followed to obtain a satisfactory porcelain enamel coating for iron, steel, and aluminum. These are, in brief, as follows:

1. Panels should be, in general, no larger than 4 by 8 ft (1.219 by 2.438 m).
2. Corner radii should never be less than $\frac{1}{16}$ in. (1.59 mm), $\frac{1}{8}$ in. (3.18 mm) being preferable.
3. Gauges of metal should be from No. 14 to 22 in order to control warping, although thinner gauges can be used when the metal is laminated to a structural core.
4. The metal must be cleaned and etched; for iron and steel, shot or sandblasting is sometimes used.
5. Porcelain enamel coats range between 2.0 and 4.0 mils (0.0508 and 0.1016 mm) with a maximum thickness of 20 mils (0.508 mm).
6. It is advisable to use semimatte to full matte finishes because they minimize reflections and thereby eliminate to a large degree visual distortions caused by even small deviations in the flat surface.
7. The type of texture and the method of obtaining the texture, either by embossing the metal or by stippling or graining the coating, should be specified.
8. The method of producing patterns should be specified as either by screening, printing, stippling, or graining or by decals when two or more colors are used.
9. The enameling industry has developed a 24 "Nature Tone" color pallet; and if more than three colors are desired, an increase in gauge thickness or more bracing is necessary.

Porcelain Enamel on Iron and Steel. All welding of corners and attachments must be done before porcelain enamel is applied on iron and steel. Brazing, soldering, or silver soldering cannot be used.

Usually two coats of porcelain enamel are applied to iron and steel. The first coat (ground coat) contains nickel and/or cobalt oxides to form a chemical-mechanical bond to the base metal; the firing temperatures range from 1450 to 1550°F (787.78 to 843.33°C). The second coat (cover coat) contains the ingredients to produce the final color, texture, pattern, and corrosion-resistant finish, and is applied at temperatures similar to those used for the first coat.

The type of alloy for both sheet and extrusion must be carefully checked; in general, 0 temper is recommended, as the firing process anneals the alloy. Currently, alloys 3003, 1100, and 6061 for sheet and strip, extrusion alloys 6061 and 7104, and casting alloys B443.0-F and 356.0-T6 are used.

Porcelain Enamel on Aluminum. For aluminum only one coat is generally required, and the coating is similar to those prepared for iron and steel except that the firing temperatures are lower, ranging from 950 to 1000°F (510 to 537.78°C).

Laminated Coatings. These coatings consist of adhesive-bonded plastic film laminated to a base metal. The films generally used are polyvinyl chloride (PVC, or vinyl) and polyvinyl fluoride (PVF) applied to steel, galvanized steel, and aluminum.

Polyvinyl chloride (PVC) is available in strips and sheets from 3 to 53 in. (76.2 to 1346.2 mm) wide on steel ranging in thickness from 0.008 to 0.075 in. (0.2032 to 1.905 mm), and on aluminum ranging in thickness from 0.010 to 0.125 in. (0.254 to 3.175 mm). The color selection for interior use is almost unlimited, but for the exterior there is generally a selection of approximately seven nonfading colors and a wide variety of patterns, textures, and simulated wood finishes. The vinyl (PVC) film varies in thickness from 0.005 to 0.012 in. (0.127 to 0.3048 mm) in increments of 0.001 in. (0.0254 mm).

Polyvinyl fluoride (PVC) is available on galvanized steel up to 0.03125 in. (0.7938 mm) thick and on aluminum ranging in thickness from 0.01 to 0.05 in. (0.254 to 1.27 mm) in coils 48 in. (1219.2 mm) wide and in sheets up to 50 in. (1270.0 mm) in width. The polyvinyl fluoride (PVF) film is only 0.002 in. (0.0508 mm) thick and is available in a medium-gloss smooth finish in six colors (the number of colors may vary). The same type of PVF film is also laminated onto wood, plywood, asbestos cement, plastics, and various other materials.

Organic Coatings. Organic coatings are used not only on metals but also on many other construction materials and are more fully covered under Paint *and* Painting, especially Painting on Metal. These coatings can be divided into three classifications for metal: (1) protective primers or undercoatings; (2) protective clear finish coatings; and (3) coatings that are pigmented (decorative) and also protective.

All metals must be cleaned before any organic coating is applied. In general these cleaning processes include chemical treatments, solvent cleaning and degreasing, and mechanical methods. This cleaning is of utmost importance, as the failure of any organic coating is usually due to lack of adhesion to the base metal.

There are various methods of applying organic coatings on metals. Paints, varnishes, and lacquers are applied by brushing, spraying, and rolling; baked enamels are applied by spraying, dipping, curtain coating, flow coating, and roller coating. *See also* Paint; Painting; Painting on Metal.

METALS: MECHANICAL FINISHES

A mechanical finish is defined here as a finish (surface texture) on a metal provided by mechanical means only (*see* Table *M23*). Mechanical finishes can be produced on all metals by grinding, polishing, or otherwise treating the surface to obtain the desired effects. On nonferrous metals and stainless steels, mechanical finishes are often applied to create final, permanent surface effects that need no additional protective coatings. For example, a polished bronze surface can be permitted to weather naturally through years of time without additional treatment or continuous polishing. On the other hand, any mechanical finish meant to provide the final surface effect on a ferrous metal such as steel, cast iron, or genuine wrought iron must always receive a protective coating to prevent corrosion.

Mechanical finishes may often be applied to provide the proper base for other treatment, for example, surface treatment of metals prior to plating processes.

Types and Uses

The mechanical finishes generally used in the construction field are shown in Table *M23*, and the types, names, and methods of producing them are given in Table *M24*.

Grinding. This is usually the first step in a series of operations to produce a higher polish. It is used principally for removing surface variations from castings and for deburring. Grinding is done by bonded abrasive wheels or disks or by a belt sander, using aluminum oxide or emery abrasive in sizes from 25 to 50 grit. This range of grit sizes produces a surface that has a definitely ground appearance.

Polishing. Polishing is used to remove any abrasions or marks on the metal surface resulting from previous operations, and to bring about a finish of greater smoothness and higher reflectivity than the ground finish. It is produced over a ground finish in a series of operations that may include roughing, greasing, buffing, and coloring. Increasingly finer grit sizes on high-speed wheels are employed, the final operation using canvas or felt wheels that bear a lubricant containing very fine abrasives.

Roughing is the preliminary operation used to prepare deeply scratched or rough surfaces for polishing.

Greasing or oiling refers to a refined roughing procedure, using 100 to 200 grit abrasive (in contrast to the 50 to 100 grit for roughing) plus a lubricant. It is necessary for finishing castings and other fabricated work marred by previous operations.

Buffing brings about a high luster finish on metals and is often the final polishing procedure. It is produced over the polished finish by the use of high-speed soft muslin disks sewed together or felt wheels, with a buffing agent such as tripoli powder mixed with grease binder.

Coloring refers to the high gloss and luster produced as a final finish over a buffed surface by the use of very soft high-speed wheels with a soft silica abrasive embedded in grease. The metal does not actually change color but takes on a higher gloss.

Burnishing. Burnishing or gentle tumbling with a large number of small parts in a barrel with suitable abrasive chips in a burnishing soap solution can be used to perform four basic types of mechanical finishing: (1) grinding, (2) polishing, (3) buffing, and (4) coloring. This finish can be used on parts where perfection of surface is not required, if the parts are not too complex in shape. A good luster and an even appearance which serves to minimize visible surface irregularities can be obtained.

Brush Finishing. Finishing methods employing various types of brushes are used to bring about a coarse or smooth lined or soft satiny sheen to metal surfaces. The size of the wires determines the degree of coarseness or fineness of the finish. The angle of contact can be changed to give various effects. Variations of these processes may be developed for the particular finish desired.

Scratch brush finishes are produced by using a rotating brush with coarse to fine wires to obtain a coarse or smooth lined texture over a polished finish.

Satin finish is the soft, smooth texture produced over a buffed or polished surface by brushing it with a fine wire wheel or by rubbing with a very fine abrasive such as an emery cloth belt.

Tampico brush finishing produces a somewhat duller smooth finish, free from scratches, streaks, or glare. This procedure employs a rotating brush of Tampico fiber and a brushing compound of fine abrasive and oil mixed to the consistency of a fine paste. When a satisfactory finish at low cost is desired and the buffing operation would be too costly and time-consuming, Tampico brushing can effectively accomplish the desired surface finishing after treatment of the surfaces with 120 grit plus a lubricant. This finishing procedure will be more satisfactory than buffing on surfaces from which scratches have not been completely removed.

Blasting-Type Finish. This is a finish having a slightly rough, uniform matt surface, produced by blasting silica sand (in varying degrees of particle size), steel shot, carborundum, rice hulls, etc., against the metal surface by air pressure. Careful control of nozzle pressure, distance, and angle is necessary to achieve certain effects and to prevent distortion. The resulting rough surface should be protected against accumulation of dirt by covering with a protective coating.

Hammered Finish. This is the simplest mechanical finish. It can be produced with a great many different effects on ferrous and nonferrous metals (mainly aluminum and copper) in colors that vary from the natural metal to smoky black shades in the deeper parts. A ball peen hammer or similar tool is used to produce irregular depressed patterns that simulate the appearance of an object hammered by hand out of a lump of rough metal. The raised portions of the surface may be brightened by polishing or highlighting. The final finished surface should be covered with a clear lacquer or varnish, followed by one or more coats of wax thoroughly rubbed on.

Etched Finish. An etched effect may be produced by mechanical means with abrasives, using a charged wheel, sandblasting, or a Tampico brush. The parts of the design that are to remain bright are covered with an adhesive paper or tape, and the entire surface is blasted or rubbed until the desired type and depth of etch have been attained. By again masking portions of the etched surface and re-etching, various depths or effects may be achieved. When the desired effect has been secured, the masking material is removed with a suitable solvent and the untreated surface may be processed by polishing or highlighting.

Highlighting. This is a process applied to hammered or etched surfaces having raised portions to give a two-tone effect. Usually, the higher surfaces are brightened by brushing, rubbing, or polishing, and the lower surfaces are left a darker color or different texture. In many types of ornamental surfaces, numerous corners and de-

Table M23 Types of mechanical finishes for steel, stainless steel, copper alloys, and aluminum

Type of finish	Method of obtaining finish: Steel or iron	Stainless steel	Copper alloys	Aluminum
Mill (as fabricated)	Hot or cold rolling and casting	Hot or cold rolling and casting	Hot or cold rolling, casting, and extruding	Hot or cold rolling, casting, and extruding
Bright rolled		Cold rolling	Cold rolling	Cold rolling
Grit texture (directional)	Seldom used	Polishing, buffing, hand rubbing, brushing, or cold rolling	Polishing, buffing, hand rubbing, brushing, or cold rolling	Polishing, buffing, hand rubbing, brushing, or cold rolling
Matte texture (nondirectional)	Seldom used	Sand or shot blasting	Sand or shot blasting	Sand or shot blasting
Bright polished		Polishing and buffing	Polishing and buffing	Polishing and buffing
Patterned	Rolling with matched design rolls or with a design and a smooth roll	Rolling with matched design rolls or with a design and a smooth roll	Rolling with matched design rolls or with a design and a smooth roll	Rolling with matched design rolls or with a design and a smooth roll

Table M24 Mechanical finishes

Type of finish	Metal	Name	Method of obtaining finish
Mill (as fabricated)	Aluminum and copper alloys	Unspecified	Natural finish produced by extrusion, casting, and hot or cold rolling with unpolished rolls
		Specular as fabricated	Mirrorlike cold-rolled finish produced with highly polished rolls
		Non-specular as fabricated	A more uniform finish than is provided under "unspecified"
	Copper alloys	Matte	Extruding, casting, and hot or cold rolling followed by annealing
	Steel	Hot rolled	Hot rolling, tight mill scale, and rust powder
		Cold rolled	Cold rolling, extremely smooth surface
	Stainless steel	Hot rolled No. 1	Hot rolling to a specified thickness followed by annealing and descaling
		Cold rolled No. 2D	Cold rolling, descaling or pickling operation, and a final pass through unpolished rolls
		No. 2B	Cold rolling, descaling or pickling operation, and a final pass through highly polished rolls
		Bright annealed	Annealing in a controlled-atmosphere furnace after cold-rolled finish No. 2B
Grit textured (directional)	Aluminum and copper alloys	Fine satin Medium satin Coarse satin	Wheel or belt polishing with grits of varying degrees of fineness
		Hand rubbed	Rubbing with abrasive cloths or stainless steel wool of increasing degrees of fineness, finishing with No. 0 or No. 00

Table M24 Mechanical finishes (*continued*)

Type of finish	Metal	Name	Method of obtaining finish
	Copper alloys	Uniform	A single pass of No. 80 grit belt moving at 5500–6000 surface ft/min (1.676.4–1.828.8 m/min)
		Brushed	Coarser directional finishes produced by power-driven wire wheel brushes, brush-backed sander heads, abrasive-impregnated foamed nylon disks, or abrasive cloth wheels
	Steel	Hand cleaning	Wire brushes, abrasive paper or cloth, scrapers, chisels, or chipping hammers
		Power-tool cleaning	Power-driven brushes, grinders, and sanders
	Stainless steel		Same as for steel
Buffed	Aluminum and copper alloys	Smooth specular (mirrorlike)	The brightest mechanical finish obtainable; ground and/or polished in one or more stages, with the final finishing prior to buffing done with 320-grit abrasive
		Specular	Buffing only, with no preliminary grinding or polishing for copper alloys ground and/or polished as above with some subsequent buffing
	Stainless steel	Polished	
		No. 3	Polishing with a 100-grit abrasive
		No. 4	Bright polished finish produced by finishing with a 120 to 150-grit mesh abrasive, following initial grinding with coarser abrasives
		No. 6	Soft satin finish produced by Tampico-brushing the No. 4 finish, using medium abrasives
		No. 7	A highly reflective finish produced by buffing a surface that has been finely ground but on which grit lines are visible
		No. 8	The most reflective finish, commonly produced by polishing with successively finer abrasives and then buffing; no grit lines visible
Nondirectional textured (sand-or shot-blast)	Aluminum and copper alloys	Fine matte Medium matte Coarse matte	Sand-blasting with silica sand or aluminum oxide of different degrees of fineness
		Fine shot blast Medium shot blast Coarse shot blast	Peened finish produced by shot-blasting, using steel shot of different sizes
	Aluminum	Extra-fine matte	Sand-blasting with very fine silica sand or aluminum oxide
	Steel	Cleaning for painting: blast cleaning to "white metal"; commercial blast cleaning; brush-off blast cleaning; "near-white" blast cleaning and flame cleaning	Shot- or sand-blasting with different sizes of sand and shot. For heavy steel parts; done with oxyacetylene torch and scraped and wire-brushed afterward
Patterned (for light-gauge material)	Aluminum, copper alloys, steel, and stainless steel	Embossing	Passing through matched-design rolls, embossing patterns on both sides
		Coining	Passing through a design roll and a smooth roll, coining one side only

pressions make the polishing of the lower surfaces a difficult and costly operation. Best results are obtained by polishing the higher surfaces and leaving the lower surfaces darker for a pleasing contrast. For even greater contrast between high and low surfaces, the metal can be painted dark before the final polishing operation. A final coat of clear lacquer or varnish is generally recommended to produce a finish of good appearance and reasonable permanence.

Mill Finishes. Metal fabricated at the mill may be given various finishes, depending on the type of alloy and method of fabrication.

"*Hot-rolled finish*" describes the surface produced on structural and mild steel shapes at the rolling mill. A thin scale remains over the smooth rolled surface, to which paints can be applied satisfactorily. (*See also* Steel Sheet and Strip.)

"*Cold finish*" is the finish produced on a metal bar or plate by the cold-drawing process. The surface is smooth and free of scale, has sharp corners and angles, and usually shows fine parallel lines in the direction of drawing.

"*Extruded finish*" is the mill finish produced in the process of extruding metal sections. It is a clean, pleasing commercial finish, usually with decided fine lines in the direction of extrusion.

See also Bonderizing; Nitriding; Parkerizing.

METALS: SPRAYED METAL COATINGS (METALLIZING)

Metals can be sprayed onto other metals or materials by a relatively simple procedure. A wire or powder of the metal to be sprayed is fed into the metallizing gun. As the wire or powder passes through the oxy-gas flame of the gun, it melts and is sprayed by compressed air onto the surface being metallized. The metal thus applied is made up of minute particles which are flattened by impact and interlock with the surface irregularities of the base metal. Such coatings tend to be porous. In construction, sprayed metal coatings of aluminum, zinc, and their alloys are used primarily on structural steel for protection against corrosion.

Sprayed metal coatings are used mainly for large assemblies, castings, forgings, etc., in place of painting or dipped metal coatings.

Metals that can be sprayed, and their chief functions or characteristics are as follows:

Aluminum: good corrosion and heat resistance
Brass: machine finishing properties
Copper: brazing applications and electrical uses
Lead: corrosion resistance; radiation shielding
Nickel: corrosion resistance; machine finishing
Zinc: all-around corrosion resistance; most commonly used metal for this purpose

Other procedures for metallizing by spraying or by special techniques are described under specific metal headings. *See also* Glass, Mirrors.

MICA

Physical and Chemical Properties

Mica is the name given to a group of minerals that crystallize in block or "book" form and can be easily split in one direction into very thin sheets. These sheets are very flexible, elastic, and tough. The mica minerals have similar physical properties but vary in chemical composition and structure. All are complex hydrous silicates with varying proportions of magnesium and potassium, usually some iron and fluorine, and sometimes sodium, lithium, chromium, titanium, barium, and manganese.

The color varies from black to colorless and transparent; the specific gravity ranges from 2.7 to 3.1. Micas are very poor conductors of heat and electricity. In most varieties, the water is an integral part of the molecule. A familiar exception is vermiculite, a ferromagnesium mica that is widely used in construction and is discussed separately.

Commercial Forms. The commercially important varieties of mica are muscovite (potassium mica) and phlogopite (magnesium mica). These micas are available in block and sheet form and as dry-ground and wet-ground flakes. Wet-ground mica differs from dry-ground mica in that it retains its sheen and has smooth rounded edges in contrast to the torn, hackled, and abraded flakes of the dry-ground form.

Types and Uses

Dry-ground mica is used in large tonnages as a surface finish ingredient in roofing materials and as an ingredient and bonding agent in wallboard joint cements. It is also used in paint, rubber plastics, and welding rods.

The fine wet-ground micas are used in paint, wallpapers, and tires. Mica spangles are used in wallpaper and for decorative effects.

Sheets of mica, usually muscovite, are invaluable in the electronic and electrical industries. Mica splittings are made into a built-up mica sheet material called micanite, used as a substitute for natural sheet mica.

MINERAL WOOL

"Mineral wool" is a term given to glass fibers and to the mass of fibrous material that results when air or steam is blown through molten rock or slag or a mixture of both by a special technique. Other terms used are "rock wool," "slag wool," and "glass wool." Mineral wool is made up of fine, pliant, intertwined, vitreous fibers. In general, it is incombustible and will not absorb or maintain moisture; it is insect- and verminproof; it has thermal resistance and sound-absorbing qualities; and it will not deteriorate or decompose. Its other properties such as stability, resilience, density, and uniformity can be closely controlled during manufacture.

Types and Uses

Mineral wool is available in bulk both as loose wool and in granulated or pellet form in graded sizes for various methods of application. It is also laminated with paper into batts, blankets, and quilted form (felted fibers stitched between layers of treated kraft paper).

Mineral wool is used mainly as an insulating material against both heat and sound in walls, ceilings, floors over unheated spaces, and partitions. It is also used for fireproofing, as a binder and filler for synthetic resin-bonded panels, and as a filtering medium. *See* Table *M25*.

Application

In batts, blankets, and pads with vapor barriers, the vapor barriers should always be located on the heated (warm-in-winter) side of the building section. In southern climates where mainly air conditioning (little or no heating) is required, the vapor barrier is generally located on the warm exterior side of the building section. In general, mineral wool should not be used in contact with earth or in floor areas below or near ground level, where moisture can collect in the mineral wool. The only exception to this precaution is mineral wool that has been specially treated to be waterproof, rotproof, insectproof (termites), and moldproof (fungus), and has been manufactured specifically for perimeter insulation.

See also Insulation and Insulating Materials; *C* Factor; *K* Factor; *R* Factor; *U* Factor.

MOLYBDENUM

Physical and Chemical Properties

Symbol: Mo
Atomic number: 42
Specific gravity: 10.22 (20°C, 68°F)
Melting point: 2610°C, 4730°F
Boiling point: 5560°C, 10,040°F
Tensile strength: 75,000 lbf/in.^{2}, 517.125 MN/m^2 (for unannealed rod)

Molybdenum is a silvery-white, ductile, mechanically strong metal that tarnishes when exposed to moist air. It is the most common refractory metal. In high-temperature service it can be protected from oxidation by siliconized coatings. Tensile strength may be as high as 250,000 lbf/in.^{2} (1723.75 MN/m^2) for fine wire.

Workability. Molybdenum may be forged, both hot- and cold-rolled, swaged, wire drawn, bent, flanged and spun.

Commercial Forms. Metallic molybdenum is available in the form of powder, ingot, rod, sheet, wire, ribbon, and various other shapes.

Types and Uses

The major use of molybdenum is in steel, to which it is added principally in the form of molybdic oxide and ferromolybdenum. This and other important uses of molybdenum are shown in Table *M26* (asterisks indicate metallic uses).

History and Manufacture

Pliny used the term *molybdaena* to designate substances containing lead. Lead sulfate and minerals of similar appearance were called molybdaena until 1778, when Scheele demonstrated their chemical differences. From his researches a new element was identified and named molybdenum.

Table M25 Types, sizes, and uses of mineral wool materials

Type and form of mineral wool	Composition of material	Finished shape of material	Method of application
Acoustic pads or blankets	Mixed with binders and formed into pads or blankets	Soft, spongy pad or blanket	Laid in perforated material or nailed or stapled to walls behind perforated material
Acoustic ceiling tiles	Mixed with binders and formed into tile shapes	Smooth or perforated; stonelike texture; lines and various other patterns	Nailed, stapled, applied with adhesive, or supported by suspended metal frame
Expansion joint fillers	Mixed with binders and materials that withstand water, rot, insects, mold, fungus, and termites	Narrow strips	With asphaltic-type cement
Fibers (loose)	Mixed with various binders	Soft, porous, smooth surface	Sprayed on surface
	Granular or pellet form	Loose and fluffy	Blown or poured
Insulation batts or blankets	Mineral wool fibers adhered to backing or completely enclosed[a]	Batt or blanket, sometimes quilted	Nailed, stapled, or stuffed
Rigid insulation	Mixed with binders and materials that withstand water, rot, insects, mold, fungus and termites	Narrow strips or boards	With asphaltic-type cement

[a]Insulation batts or blankets come with coverings that act as vapor barriers and reflective insulation or are perforated to

Size of material						
Thickness		Width		Length		
in.	mm	in.	mm	ft and in.	mm or m	Major use
1, 2, and 3	25.4, 50.8, and 76.2	22, 24, and 30	558.8, 609.6, and 762.0	2′-6″, 5′-0″	0.762 m, 1.524 m	Acoustic pad or blanket behind perforated ceiling materials
$1\frac{3}{16}$, $1\frac{7}{16}$	30.16, 36.51	12	304.8	1′-0″, 2′-0″	304.8 mm, 609.6 mm	Acoustic pad or blanket behind perforated tile ceiling material
$1\frac{3}{16}$, $1\frac{7}{16}$	30.16, 36.51	24	609.6	2′-0″, 4′-0″	609.6 mm, 1219.2 mm	
$\frac{3}{4}$, 1, $1\frac{1}{4}$, $1\frac{1}{2}$, $1\frac{3}{4}$, 2, $2\frac{1}{4}$, $2\frac{1}{2}$, and 3	19.05, 25.4, 31.75, 38.1, 44.5, 50.8, 57.15, 63.5, and 76.2	11, 15, 19, and 23	279.4, 381.0, 482.6, and 584.2	2′-0″, 4′-0″, and 8′-0″	0.610 m, 1.219 m, and 2.438 m	Acoustic pads or blankets for partition walls
$2\frac{1}{2}$	63.5	12	304.8	1′-0″, 2′-0″, and 3′-0″	304.8 mm, 609.6 mm, and 914.4 mm	Acoustic treatment of walls, usually above 6 ft, and of ceilings
$\frac{1}{2}$, $\frac{5}{8}$, $\frac{3}{4}$, $\frac{7}{8}$, and 1	12.7, 15.88, 19.05, 22.23, and 25.4	12	304.8	12″	304.8 mm	
$\frac{5}{8}$	15.88	12	304.8	24″	609.6 mm	
$\frac{5}{8}$	15.88	24	609.6	24″	609.6 mm	
$\frac{5}{8}$	15.88	$23\frac{3}{4}$	600.08	$23\frac{3}{4}$″ and $47\frac{3}{4}$″	600.08 mm and 1209.69 mm	
$\frac{1}{2}$, $\frac{3}{4}$, 1, and $2\frac{1}{2}$	12.7, 19.05, 25.4, and 63.5	4, 6, 8, 10, and 24	101.6, 152.4, 203.2, 304.8, and 609.6	8′-0″, 10′-0″	2.438 m, 3.048 m	Expansion joint fillers for concrete or masonry materials
$\frac{1}{2}$, $\frac{3}{4}$, and 1	12.7, 19.05, and 25.4					Acoustic treatment for ceilings
Usually 4	Usually 101.6					Blown in between wall studding or poured between ceiling beams for thermal insulation for exterior walls, ceilings, and roofs
$1\frac{1}{2}$, 2, 3, $3\frac{5}{8}$, 4, and 6	38.1, 50.8, 96.2, 92.08, 101.6, and 152.4	11, 15, 19, and 23	279.4, 381.0, 482.6, and 584.2	2′-0″, 4′-0″, and 8′-0″	0.610 m, 1.219 m, and 2.438 m	Thermal insulation for exterior walls, ceilings and roofs
$\frac{1}{2}$, $\frac{3}{4}$, 1, $1\frac{1}{4}$, $1\frac{1}{2}$, $1\frac{3}{4}$, 2, $2\frac{1}{4}$, $2\frac{1}{2}$, and 3	12.7, 19.05, 25.4, 31.75, 38.1, 44.5, 50.8, 57.15, 63.5, and 76.2	12, 24, and 48	304.8, 609.6, and 1219.2	4′-0″, 8′-0″, and 10′-0″	1.219 m, 2.438 m, and 3.048 m	Insulation for flat roofs and perimeter insulation above or below ground

relieve condensation, or in various combinations to fit all possible thermal insulation requirements.

Table M26 Uses of molybdenum

	Construction Uses	
Basis of	Coatings* and protective coatings*	On steel and other metals
	Color pigments	Orange, light red
	Cast iron*	Increasing strength, toughness, and wear resistance
Component of	Glass	Mirrors
	Glazes and porcelain enamels*	Yellows
	Iron*	Increasing strength, toughness, and wear resistance
	Stainless steel*	Increasing corrosion resistance
	Steel*	Increasing hardness, elastic limit, wear resistance, and machinability; substitute for tungsten
Allied Construction Uses	Zinc*	Black coatings
	Electrical devices*	Contacts in mercury switches
	Lamps*	Filaments

MONEL

Physical and Chemical Properties

The high nickel-copper metal marketed under the trade name Monel is a bright, strong, ductile silver-white alloy whose approximate composition is 66% nickel, 31.5% copper, and 2.5% combined iron, manganese, silicon, and other elements. This alloy is corrosion resistant, particularly to salt water and acids. It has a low coefficient of expansion comparable to that of concrete; high rigidity, which permits transfer of movement to the expansion joint without buckling; and good fatigue strength, which protects against fatigue cracks that result from flexing caused by expansion and contraction. Other physical properties are as follows: melting point, 2370 to 2460°F (1298.89° to 1348.89°C); tensile strength, 70,000 to 120,000 lb/in.^{2} (482.65 to 827.4 MN/m^2); density, 8.8 g/cm^3; weight, 0.319 lb/in.^{3}.

Workability. Monel alloy is readily workable and machinable by rolling mill techniques and can be forged, drawn, and cast. It lends itself to both welding and soldering.

Commercial Forms. Monel is available in rod, bar, wire, plate, sheet, strip, and tube form.

Table M27 Gauges of Monel alloy for roofing and flashing

Characteristics	Gauge number 22	24	25	26
Thickness				
in.	0.031	0.025	0.021	0.018
(mm)	(0.788)	(0.635)	(0.553)	(0.457)
Weight				
lb/ft^2	1.424	1.148	0.965	0.827
(kg/m^2)	(6.952)	(5.605)	(4.711)	(4.037)
Width of sheet				
ft	2, 2½, 3	2, 2½, 3	2, 2½, 3	2, 2½, 3
(m)	(0.6096, 0.762, 0.914)	(0.6096, 0.762, 0.914)	(0.6096, 0.762, 0.914)	(0.6096, 0.767, 0.914)
Length of sheet				
ft	8, 10	8, 10	8, 10	8
(m)	(2.438, 3.048)	(2.438, 3.048)	(2.438, 3.048)	(2.438)
Recommended width of pan for roofing				
ft			24	20
(m)			(7.315)	(6.096)
Recommended type of seam for roofing			Standing, batten, or flat	Standing or batten

Table M28 Weight and thickness of gauges of Monel alloy sheet other than roofing and flashing sheet

	Gauge number						
Characteristics	16	17	18	19	20	21	23
Thickness							
in.	0.062	0.056	0.05	0.043	0.037	0.034	0.028
(mm)	(1.575)	(1.422)	(1.27)	(1.092)	(0.94)	(0.864)	(0.711)
Weight							
lb/ft^2	2.848	2.572	2.297	1.975	1.700	1.562	1.286
(kg/m^2)	(13.904)	(12.557)	(11.214)	(9.642)	(8.3)	(7.626)	(6.278)

Types and Uses

Only the regular Monel alloy (two-thirds nickel and one-third copper) is used to any extent as a construction material. It is available in a number of tempers and in most fabricated forms.

***Roofing Sheet*.** Monel alloy roofing sheet is used for roofing, flashing, and other sheet metal building applications. It is available in the following gauges for these purposes:

Gauge No. 22: suggested for gutters 48 in. (1.2192 m) in girth and larger, and edge strips for stone coping or cornices.

Gauge No. 24: suggested for gutters 36 to 48 in. (0.9144 to 1.2192 m) in girth, edge strips for wood coping or cornices, eaves, gravel stops, and valley flashing for slate or tile roofing.

Gauge No. 25: suggested for gutters 36 in. (0.9144 m) or smaller in girth, molded gutters, gravel stops, flat seam coping, and base flashing over 10 in. (254 mm) in width.

Gauge No. 26: suggested for hung gutters, leaders, downspouts, heads, straps, cleats, through-wall flashing, counter and cap flashing, base flashing up to 10 in. (254 mm) in width, valley flashing for wood or asphalt roofing.

The table of gauges for roofing and flashing, Table *M27,* lists available sizes of Monel alloy sheet and gives other data pertinent to roofing applications.

***Heavier Gauges*.** For use in kitchen, hospital, laundry, and laboratory equipment, heavier gauges are available. Heavier gauges are also used for marine construction and in special processing and manufacturing equipment. *See* Table *M28*.

***Crimped Monel Alloy*.** Crimped Monel alloy has been used for various flashing applications, penthouse sidings, and parapet covers as well as for roofing. Here expansion joints are not necessary, thus eliminating a common maintenance problem. These crimps, or miniature corrugations, are about five per inch (25.4 mm) and 0.031 in. (0.787 mm) deep.

Application

Condensed Checklist

1. All bends should be made with a radius twice the thickness of the sheet.
2. Attachment: Nails should never be driven through the metal; one should always secure with Monel alloy cleats or clips, usually 12 in. (304.8 mm) o.c., using special roofing nails of Monel alloy, as shown in the illustrations.
3. When calculating center-to-center distance of standing seams, one should deduct $3\frac{1}{4}$ in. (82.55 mm) from the width of sheet; for batten seams, $2\frac{5}{8}$ in. (76.8 mm) should be deducted.
4. Joining and soldering: When soldering, pretinning of the sheet edges is required. A soldered joint should not be relied on for strength, which can best be obtained by lock seaming, riveting, etc. Solder should be used only as a sealing medium.
5. Methods of joining sheets for roofing are shown in Figure *M16*. Note the amount of taper in the

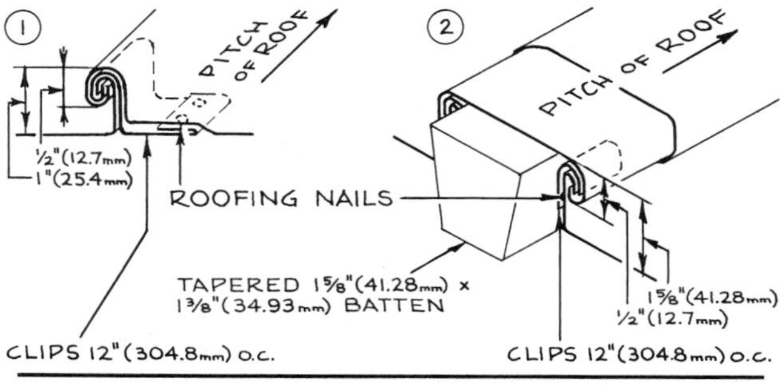

Figure M16 Details of standing seam (1) and batten seam (2).

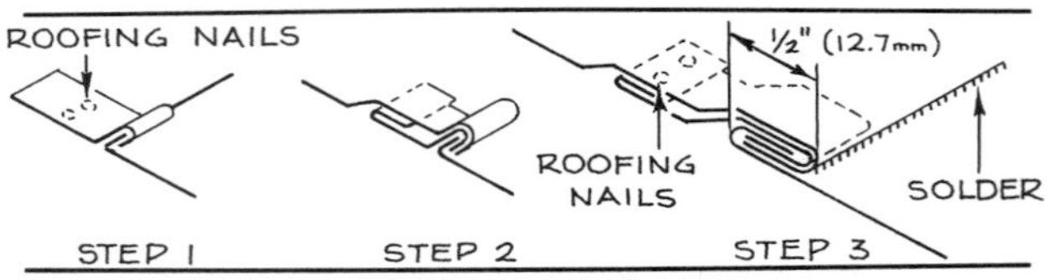

Figure M17 Details of soldered locked flat seam for cross seams and roofs with less than 3 to 12 pitch.

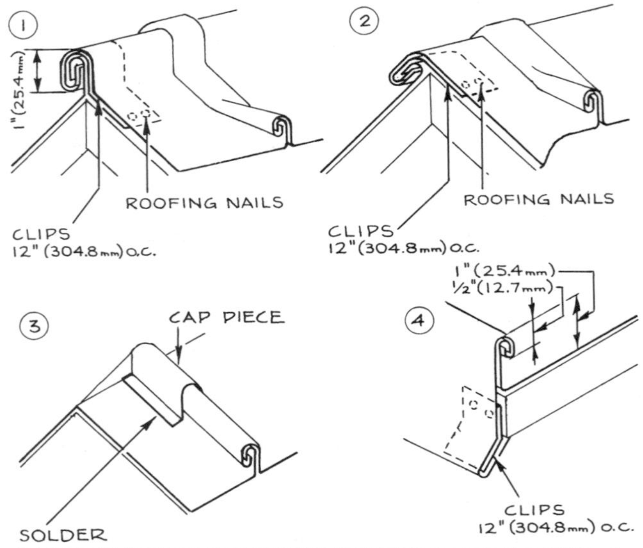

Figure M18 Details of standing (1), flat (2), and continuous standing (3) seams at ridge or hip and gable end (4).

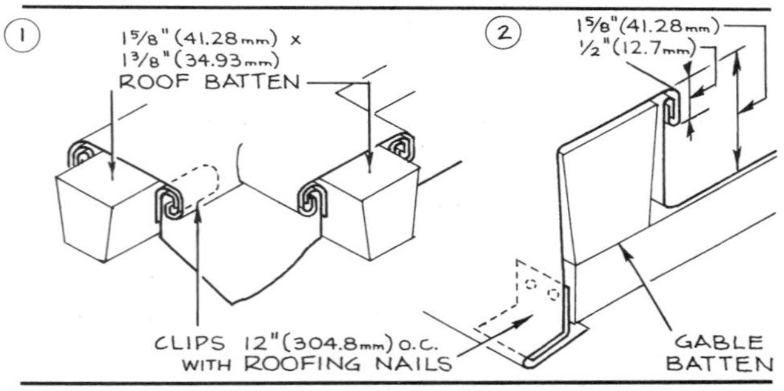

Figure M19 Details of batten seam at ridge or hip (1) and gable end (2).

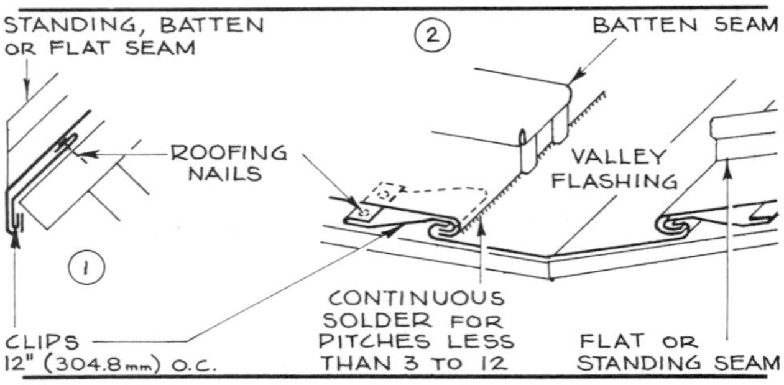

Figure M20 Details of standing, flat, or batten seams at eave (1) and valley (2).

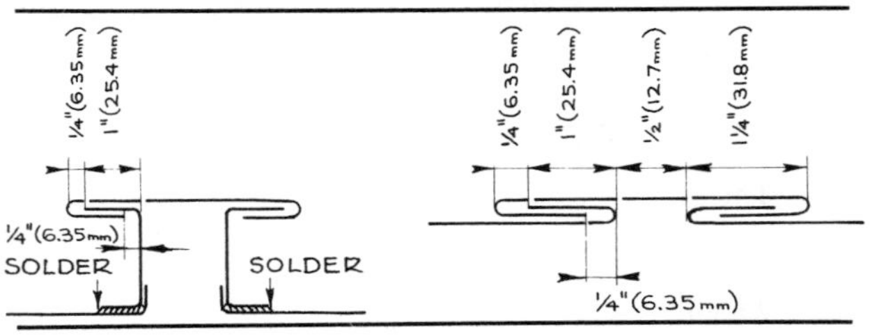

Figure M21 Details of expansion joints.

wooden batten. Consult Table *M27* for type of seam and recommended pan width in relation to gauge.

6. All cross seams should be locked flat seams which are soldered according to manufacturer's specifications (*see* Figure *M17*).
7. On roofs with pitches less than 3 to 12, locked flat or soldered seams should be used; with greater pitches, standing or batten seams.
8. Treatment of standing, flat, and continuous standing seams at roof ridges or hips and at gable ends (*see* Figure *M18*).
9. Treatment of batten seams at ridges or hips and gable ends of roof (*see* Figure *M19*).
10. Treatment of standing, flat or batten seams at eave and valley intersections (*see* Figure *M20*).
11. For expansion joints, allowance should be made for a free movement of 0.1104 or $\frac{7}{64}$ in. (2.8 mm) in an 8-ft (2.438 m) length through a 150°F (65.56°C) temperature change. A good rule of thumb is an expansion joint for every continuous length of over 24 ft (7.315 m) (*see* Figure *M21*).

Only the soft soldering fluxes known in the trade as acid fluxes are recommended for use with Monel alloy. Rosin has too mild a cleaning action, and silver solder is more costly than the commonly used 50-50 and 60-40 tin-lead solders. The proprietary (prepared) or "cut acid" flux for Monel alloy is made by adding pure zinc to commercial hydrochloric acid. A small amount of ammonium chloride is sometimes added to improve the fluxing action. All traces of flux should be removed after smoldering is completed.

Conditions Favorable to the Use of Monel Nickel-Copper Alloy

1. Where a strong, durable, corrosion-resistant material is needed for roofing or flashing.
2. Where strong horizontal or vertical lines are required on a roof.
3. Where staining of adjoining material must be avoided.
4. Where a fire-resistant roofing material is desired. Municipal, state, and local codes should be checked

and also the fire rating codes of the Fire Underwriters, insurance companies, labor departments, and federal government (Army, Navy, etc.).

5. Where a material with high reflectivity but low glare is required.

Conditions Unfavorable to the Use of Monel Nickel-Copper Alloy

1. Where a colored roof is desired.
2. Where the roofing or flashing might present an unfavorable galvanic couple (e.g., steel or iron).
3. Where corrosion resistance and great durability are not particularly necessary.

History and Manufacture

Monel nickel-copper alloy was introduced in 1905 by the International Nickel Company, Inc. The early version, containing about two-thirds nickel and one-third copper, was originally produced as a "natural" alloy directly from Canadian ore. At that time the manufacturing process consisted of oxidizing a matte of about 75% nickel and the remainder copper with small amounts of other elements. The oxides formed were then reduced first with carbon and finally with a controlled quantity of charcoal or tar coke in a special open-hearth furnace. The resulting molten metal was refined in an electric furnace, and the end product, Monel nickel-copper alloy, was drawn off into ingots.

Since then, the effects of adding other elements have been investigated. Also, as the original ore deposit was used up, it became necessary to control the composition of the resulting alloy. As a consequence, various modifications of the original alloy have been developed for specialized applications.

See also Nickel.

MORDANT

A mordant is a substance, usually in liquid form, that fixes a coloring dye on or in a textile, leather, or other material, so that the color becomes permanent. The term also may refer to a corrosive substance (acid) that is used to etch on metals.

MORTAR

Physical and Chemical Properties

Mortar may be defined as a mixture of cementitious materials with or without siliceous materials which, after being prepared in a plastic state with or without water, hardens into a stonelike mass.

Constituents. *Cementitious materials* include lime (hydrated and hydraulic) and cement (pozzolanic, natural, slag, masonry, and portland Types I, II, and III).

Siliceous materials (aggregate) include sand, crushed stone, burned shale, slag, minerals, and various types of stone dust.

Water for mortar must be clean, potable, and free from deleterious amounts of acids, alkalis, and organic materials, including dumped industrial wastes.

Sand is usually white pure silica sand. It is graded as shown in Table *M29* according to intended size of mortar joints in the masonry.

Color can be imparted to mortar by adding inorganic pigment compounds, which should not be used in quantities exceeding 10% of the weight of the cement (*see* Table *M30*).

Admixtures for giving special characteristics, as well as various plasticizers, emulsions, and hardeners, may be added (*see* Admixtures).

Table M29 Gradation of sand to size of mortar joint

Sieve number	Percentage of sand retained (by weight)			
	Joints over $\frac{1}{4}$ in. (6.25 mm)		Joints $\frac{1}{4}$ in. (6.25 mm) or less	
	Maximum	Minimum	Maximum	Minimum
4 (4.76 mm)	0	0	0	0
8 (2.38 mm)	10	0	0	0
16 (1.19 mm)	40	15	15	5
30 (595 microns)	65	35	60	40
50 (297 microns)	85	75	85	75
100 (149 microns)	98	95	99	95

Table M30 Inorganic color pigments

Inorganic color pigment	Color
Cobalt oxide	Blues
Oxides of iron	Grays, blacks, reds, browns, and buffs
Chromium oxide	Green
Carbon black	Grays to blacks

Composition. The composition of mortar varies in relation to end use. Mortars for general masonry construction and reinforced masonry consist of various types of portland cement or masonry cement plus lime (hydrated or lime putty), sand, and water (*see* Tables *M31* and *M32*).

Mortars for preassembled and prefabricated masonry units or for masonry that must meet special structural requirements must include additives that impart greater bonding, compressive, and tensile strength. As an alternative there also exists a different method using a completely packaged adhesive-type mortar applied with a special type of caulking gun.

Properties of Mixture. Only a brief discussion of how variations in the type of material used can affect the properties of the mortar mixture is possible here.

Table M31 Weights of materials for mortar for general masonry construction

	Weight	
Material[a]	lb/ft^3	kg/m^3
Portland cement	94	1505.88
Portland blast-furnace slag cement	94 (approx.)	1505.88 (approx.)
Masonry cement	Weight printed on bag	
Hydrated lime	40	640.80
Lime putty	80	1281.60
Sand	80 of dry sand	1281.60 of dry sand

[a]Water shall be clean and potable (fit to drink).

Lime affects mainly the volume stability (shrinkage and expansion) of the mortar, one of its most important characteristics. Lime putty by itself has a high degree of shrinkage but also the very desirable property of high water retentivity. A high-magnesia lime, on the other hand, expands excessively, and mortar containing it can have a disruptive effect on masonry. Cement itself varies in its coefficient of expansion depending on its chemical composition and the degree of hydration.

Aggregates also can have a marked effect on volume stability. Clay, shale, or any substance for which there is much difference between its coefficient of expansion and that of the cement paste is undesirable. So is any aggregate that shows a wide difference in the degree of expansion among the aggregate particles.

Cements can vary mortar characteristics, particularly the strength, as follows. Portland cement has a good compressive strength but poor plasticity. Hydraulic lime cement, natural cements, and pozzolana have a lower strength as a rule. Also, natural and pozzolan cements show wide variation in many of their characteristics and therefore are unpredictable in masonry work requiring strength. Slag and special portland cements do not stain adjoining materials and are used in nonstaining mortars.

Pulverized additives serve as fillers and plasticizers in mortar.

Resin or *polymer additives* give the mortar bonding, compressive, and tensile strength.

Control of Characteristics. Two methods of controlling the characteristics of mortar for general construction and for reinforced masonry have been developed:

1. Property specifications, which are based on mixing to meet laboratory tests for requirements (*see* Tables *M33* and *M34*).

2. Proportion specifications, which are based on predetermined proportions of dry mix (*see* Tables *M35*, *M36*, and *M37*).

Table M32 Materials for mortar for reinforced masonry

Cementitious materials	Lime	Admixtures[a]	Aggregates	Water
Portland cement Type I, IA, II, IIA, III, and IIIA	Hydrated	Air-entraining agents	Damp and loose sand	Clean and potable
Blended cements Types IS, IS(MS), IS-A, IS-A(MS), IP, and IP-A	Lime putty	Pure mineral oxide colors	Damp and loose sand	Clean and potable
Masonry cement	Correctly proportioned mix in bags		Damp and loose sand	Clean and potable

[a]Antifreeze compounds, accelerators, integral waterproofing compounds, or other admixtures shall not be used.

Table M33 Property specifications for mortar

Type of mortar	Average compressive strength in 28 days[a]		
	lb/in.^{2}	MN/m^2	kg/cm^2
M	2500	17.24	172.38
S	1800	12.41	124.11
N	750	5.17	51.71
O	350	2.40	24.03
K	75	0.52	5.17

[a] 1 lb/in.^{2} = 0.006895 MN/m^2 = 0.06895 kg/cm^2.

Table M34 Property or physical requirements for mortar for reinforced masonry

Compressive strength	
7 days[a]	1600 lb/in.^{2} (11.03 MN/m^2)
28 days	2500 lb/in.^{2} (17.24 MN/m^2)
Water retention	70% (flow after suction, minimum percent of original flow)
Air content	18% (volume, maximum percent)

[a] If mortar meets 28-day strength, but not 7-day strength, it is acceptable.

Table M35 Proportion specifications for mortar for general masonry construction by volume

Type of mortar	Parts by volume of portland cement, or portland blast-furnace slag cement	Parts by volume of masonry cement	Parts by volume of hydrated lime or lime putty	Aggregate, measured in a damp, loose condition
M	1	1	—	Not less than $2\frac{1}{4}$ and not more than 3 times the sum of the volumes of the cement and lime used
	1	—	$\frac{1}{4}$	
S	$\frac{1}{2}$	1	—	
	1	—	Over $\frac{1}{4}$ to $\frac{1}{2}$	
N	—	1	—	
	1	—	Over $\frac{1}{2}$ to $1\frac{1}{4}$	
O	—	1	—	
	1	—	Over $1\frac{1}{2}$ to $2\frac{1}{2}$	
K	1	—	Over $2\frac{1}{2}$ to 4	

Table M36 Mortar for reinforced masonry proportions by volume

Type of mortar	Parts by volume of portland cement or portland blast-furnace slag cement	Parts by volume of masonry cement	Parts by volume of hydrated lime or lime putty	Fine aggregate measured in a damp, loose condition
PM	1	1	$\frac{1}{4}$ to $\frac{1}{2}$	$2\frac{1}{4}$ to 3 times the sum of the volumes of cementitious materials
PL	1	—	—	

Table M37 Grout proportions by volume

Type of grout	Parts by volume of portland cement or portland blast-furnace slag cement	Parts by volume of hydrated lime or lime putty	Aggregate measured in a damp, loose condition	
			Fine	Coarse
Fine	1	0 to $\frac{1}{10}$	$2\frac{1}{4}$ to 3 times the sum of the volumes of the cementitious materials	
Coarse	1	0 to $\frac{1}{10}$		1 to 2 times the sum of the volumes of the cementitious materials

Table M38 Gradation of sand to size for high-bond mortar

Sieve size	Percentage passing
4 (4.76 mm)	100
8 (2.38 mm)	95–100
16 (1.19 mm)	70–100
30 (595 microns)	40–75
50 (297 microns)	20–35
100 (149 microns)	2–15
200 (75 microns)	0–2

Table M39 Gradation of pulverized filler

Sieve size	Percentage passing
100 (149 microns)	100
200 (75 microns)	94
325 (58 microns)	86
400 (38 microns)	82

Table M40 Material proportions for high-bond mortar

Liquid polymer	4 gal (18.18 liters)
Clean sand[a]	$3\frac{1}{4}$ ft^3 (0.092 m^3) with allowable range of $2\frac{1}{2}$–$3\frac{1}{2}$ ft^3 (0.071–0.099 m^3)
Portland cement Type I or II	94 lb/ft^3 (1505.88 kg/m^3)
Pulverized additive	50 lb/ft^3 (801.00 kg/m^3)
Water	Up to a maximum of 4 gal (18.18 liters)

[a]Volume of mix = volume of sand.

Table M41 Physical properties of high-bond mortar

Type of mortar	Compressive strength[a]		Tensile strength[a]	
	lb/in.^{2}	MN/m^2	lb/in.^{2}	MN/m^2
High-bond	8575	59.13	955	6.59

[a]1 lbf/in.^{2} = 0.006895 MN/m^2.

Mortars for preassembled and prefabricated masonry units and for special structural conditions must meet rigid requirements. These mortars consist of a high-bond additive, clean mason's sand, portland cement Type I or III, a plasticizer additive, and water that is potable and free from acids and impurities (*see* Tables *M38*, *M39*, *M40*, and *M41*).

Commercial Forms. Mortar is available ready-mixed, premixed, prehydrated, and job-mixed.

Ready-mixed mortar is obtainable in bags, compounded to the specification required. This type requires only the addition of water to make up the mortar.

Premixed mortar is mixed at the plant to the required job specification and transported to the site. In this case the final water ratio is controlled at the site by the use of the slump test, the concrete truck carrying water for this purpose. Premixed mortar may also be prepackaged for mixing at the job site.

Prehydrated mortar is an intermediate category made by mixing the dry ingredients with only sufficient water to produce a damp mass of a consistency such that it will retain its form when pressed into a ball with the hands and will not flow under the trowel. The mortar is then allowed to stand for a period of not less than 1 hour or more than 2 hours, after which time it is remixed with the addition of sufficient water to produce satisfactory workability.

Job-mixed mortar is mortar in which the materials are mixed on the site by hand or a mechanical mixer according to the job specifications. It is mixed in a drum-type batch mixer for a minimum of 3 minutes or by hand (generally for small jobs). The measurement of ingredients should at all times be controlled and accurately maintained, and the aggregate ratio should be not less than $2\frac{1}{4}$ times (in proportion-controlled mixes) and not more than $3\frac{1}{2}$ times (in property-controlled mixes) of the total separate volumes of cementitious materials used. Sufficient water should be added to the mix to give satisfactory workability, and slump tests should be made.

Types and Uses

General Mortars. The general types of mortar (*see* Table *M42*) are as follows:

Types M and *S*—these are high-strength mortars suitable for general use and recommended specifically for reinforced brick masonry and plain masonry below grade.

Type N—this is a medium-strength mortar suitable for general use in exposed masonry above grade.

Types O and *K*—these are low-strength mortars suitable for non-load-bearing walls of solid masonry units, partitions, and load-bearing walls in which the compressive stresses developed do not exceed 100 lbf/in.^{2} (0.6895 MN/m^2) and where exposures are not severe.

High-bond mortars—these are mortars used for preassembled and prefabricated units and for masonry requiring special structural end use.

Reinforced masonry and grout mortars—these types of mortars are used for reinforced masonry construction.

Table M42 Types of mortar and their major use[a]

Type of mortar	Variations	Proportions of constituents (parts): Cement, portland cement, or blast-furnace slag cement	Masonry cement	Hydrated lime	Lime putty	Sand as aggregate	Major uses
M	(1)	1	1	$\frac{1}{4}$		Not less than $2\frac{1}{4}$ and not more than 3 times the sum of the volume of cement and lime used	Heavy-loaded and below-grade masonry subjected to rigorous exposure
	(2)	1	–	–			
S	(1)	$\frac{1}{2}$	1	–			Load-bearing partitions and above-grade masonry subjected to severe exposure
	(2)	1	–	Over $\frac{1}{4}$ to $\frac{1}{2}$			
N	(1)	–	1	–			Load-bearing partitions and above-grade masonry subjected to less severe exposure
	(2)	1	–	Over $\frac{1}{2}$ to $1\frac{1}{4}$			
O	(1)	–	1	–			Interior masonry and above-grade, lightly loaded bearing masonry subjected to exposure in mild climates
	(2)	1	–	Over $1\frac{1}{2}$ to $2\frac{1}{2}$			
K	(1)	1	–	Over $2\frac{1}{2}$ to 4			Interior partitions; non-load-bearing or lightly load-bearing
Lime		–	–	1 of either hydrated lime or lime putty		3	Interior non-load-bearing work where shrinkage is not important
Nonstaining cement		Nonstaining portland, 1	–	$\frac{1}{5}$ of the cement and sand	–	3	Limestone, marble, ceramic veneer, glazed brick, and block masonry work
Nonstaining cement lime		Nonstaining portland, $\frac{1}{2}$	–	$\frac{1}{2}$	–	3	Cut stone and marble, and where nonstaining mortar is required
Nonstaining waterproof cement		Nonstaining waterproof portland, 1	–	$\frac{1}{5}$	–	3	Where waterproofing is required (caps, copings, sills, etc.)
Gypsum		Neat-setting unfibered gypsum (retarded 4-hour set), 1	–	–	–	3	Gypsum blocks, decking, etc.
Pointing	(1)	Nonstaining portland, 1	–	Sufficient amount to make as stiff a mixture as can be worked		2 white	Stonework facing
	(2)	White nonstaining	–	–	1	3	Stonework facing
Grout	(1)	Portland nonstaining, 1; portland, 1	–	–	–	1	Stone or brick facing; setting bearing plates and railings
	(2)	Portland or blast-furnace cement, 1	–	0 to $\frac{1}{10}$	–	$2\frac{1}{4}$ to 3 times the sum of the volumes of cementitious materials	
Coarse grout	(3)		–	0 to $\frac{1}{10}$	–	Coarse aggregate 1 to 2 times the sum of the volumes of cementitious materials	
For reinforced masonry	PM	Portland or blast-furnace slag cement, 1	1	–	–	$2\frac{1}{4}$ to 3 times the sum of the volumes of cementitious materials	Reinforced-type masonry
	PL		–	$\frac{1}{4}$ to $\frac{1}{2}$	–		

Table M42 Types of mortar and their major use[a] (*continued*)

Type of mortar	Variations	Proportions of constituents: Cement, portland cement, or blast-furnace slag cement	Masonry cement	Hydrated lime	Lime putty	Sand as aggregate	Major uses
High-bond mortar	(1)[b]	1 bag portland cement Type I or II	—	—	—	Volume of sand equals volume of mix	Preassembled, prefabricated masonry units and masonry units with special structural requirements
	(2)[c]	Portland cement, applied with caulking gun	—	—	—		
Parging		Same mortar as used for facing to be parged					

[a]Use the same weight per cubic foot as noted in the tables of proportion and property specifications.
[b]With liquid polymer, pulverized additive.
[c]With resin, hardener, and polymeric emulsion.

Chemically Resistant Mortars. Chemically resistant mortars for industrial and other special installations are available in several types:

Waterproofed portland cement, which is resistant to lactic acid, petroleum, vegetable oil, molasses, etc.

Sodium silicate cements, which are resistant to hot and cold acids (except hydrofluoric) and temperatures up to 1800°F (982.2°C).

Synthetic resin cements, which are impervious to liquids, resistant to acids, detergents, water, steam, fats, oils, and weak alkalis, and resistant to temperatures up to 300°F (148.9°C).

Sulfur-base cements, which are resistant to all acids up to 200°F (93.3°C) but not to weak alkalis and oils.

Bituminous mortars, which are resistant to acids and certain alkalis below 100°F (55.5°C) but not to oils, fats, greases, and some organic solvents.

These are but some of the many kinds of chemically resistant mortars available. An important precaution: for any installation where particular types of chemicals will be present or in use, it is advisable to consult with the manufacturers who specialize in chemically resistant mortars to select the correct type.

Refractory Mortars. Refractory mortars have a fire clay base to which may be added silica, chrome (chrome iron ore), silicon carbide, and alumina for special compositions (*see* Table *M43*). The field of refractory mortars is a highly specialized one, and in every instance the manufacturers of these mortars should be consulted.

The main types are heat-setting, air-setting, and plastic refractory. In the air-setting type there are three classes which should give a 200-lbf/in.^{2} (1.379 MN/m^2) modulus of rupture when used in a joint. Plastic refractory mortars are used to construct monolithic walls.

Table M43 Refractory mortars

Type of mortar	Pyrometric cone equivalent[a] (not lower than): Heat setting	Plastic refractories
Super duty	No. 31	No. $32\frac{1}{2}$
High-heat duty	No. 28	No. 31
Intermediate-heat duty	No. 26	
Low-heat duty	No. 16	

[a]These numbers refer to pyrometric cones, which are pyramids of established mixtures of oxides that melt at exact known temperatures and are used to evaluate temperatures of refractory materials. If cone No. 31 melts at *x temperature*, a refractory material that melts at this same temperature is therefore designated as a No. 31 refractory.

Refractory mortars for brick linings where high temperatures occur, as in incinerator and power plant chimneys, are Type I (wet mix) and Type II (dry mix). These are manufactured products that come ready for use.

In the case of very high temperatures, the highest temperature to be encountered should be determined, and the manufacturers of both brick and mortars should be consulted so that the correct mortar can be specified.

Application

Adhesion (technically known as bonding) between mortar and brick is affected by the water retentivity of the mortar, the rate of suction of the brick or other masonry unit when laid, the consistency (flow) of the mortar, and the technique of forming the mortar joint. For maximum bond strength, the water retentivity of the mortar should be high (flow after suction exceeding 70% of the initial flow), the flow of the mortar should be the maximum consistent with workability (using the maximum water possible), the suction rate of the brick when laid should be 20 g (0.7 oz)/min or less, and the forming of the mortar joint should be accompanied by pressure to ensure intimate contact between mortar and masonry unit. The strength of the bond is further increased by continuous pressure during the setting of the mortar.

High-bond mortars, including the packaged type, are used to construct large preassembled and prefabricated masonry units for the exterior of buildings and to construct masonry units that must meet special structural requirements. When high-bond mortars are used, rigid controls must be followed, and the mortars must be pretested before the units are manufactured.

When high-strength mortars are used, the metals to be used for dowels, anchors, ties, etc., should be checked. In general, steel with heavy zinc galvanizing or cadmium coating is used. With some of these mortars, stainless steel should not be used because of chemical reaction with certain ingredients in the propietary mortar mix. It is advisable to clarify this with the mortar manufacturer.

History and Manufacture

The history of mortar is closely linked to that of brick, ceramic tile, and other clay tile products since it was developed for use with these structural materials.

N

NAILS

Physical and Chemical Properties

The nail is the oldest and simplest mechanical fastener. It consists of a shaft that is pointed at one end and usually formed into or attached to a head at the other end. The head may be plain, colored, coated, plated, or enameled. The shaft may be plain, coated, plated, or enameled. It may be shiny and smooth or formed in a number of ways that increase its holding power, for example, serrated, saw-toothed, annular, helically threaded, or spirally fluted. Nails are manufactured not only with different thicknesses and lengths but also with a large variety of heads and points. (*See* Figure *N1*.)

Holding Power, Shearing Strength. In wood construction, where the majority of nails are used, the holding power and shearing strength are of major importance. These characteristics depend on the material used, the diameter of the shank, the length of the nail, and the treatment of the surface of the shaft.

With the great variety of materials used in construction for roofing, siding, underlayment, boards, and sheet material, a large variety of nails have been developed for the application of these materials to other materials or to each other. The holding power of the nail should be checked in relation to its end use.

Compatibility of Materials. The compatibility of the constituent metal with other materials in contact with the nail is equally important to prevent loss of holding power and staining or discoloration of the structure.

Identification. Nails are made of ferrous and nonferrous metals and are identified by gauge and length, and by penny. The word "penny" comes from the very early use of nails around 1690 and indicates the cost in pennies per hundred nails. Thus, a 10-penny nail meant that 100 nails of this type could be bought for 10 pennies. The term "penny" is still used today, though

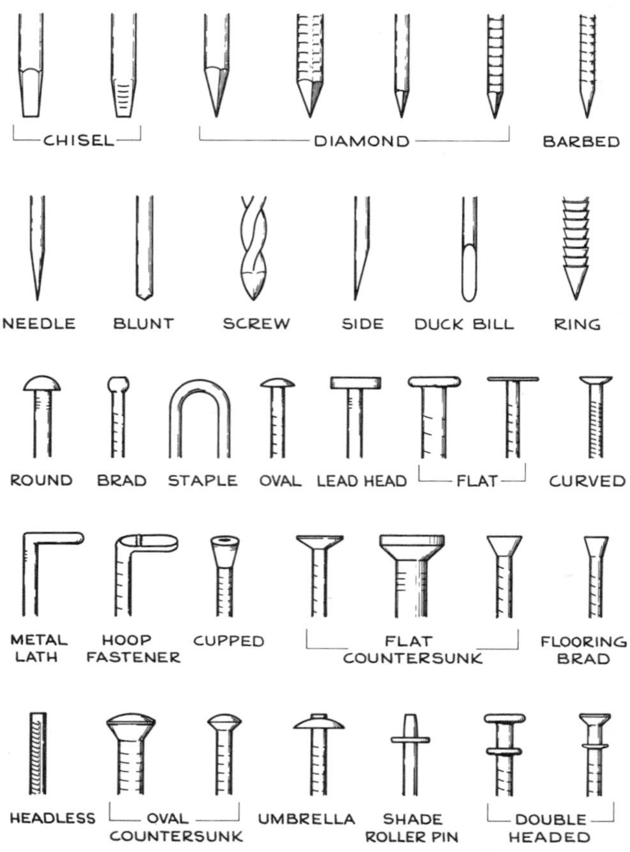

Figure N1 Various types of heads and points available in nails.

with no relation whatsoever to cost, but is slowly giving way to gauge, diameter, and length designations.

Penny sizes are designated with a small letter "d" after a number, for example, 6d = 6-penny nail. Equivalents in penny size to inches and millimeters are as follows:

Penny	*Inches*	*Millimeters*
2d	1	25.40
3d	$1\frac{1}{4}$	31.75
4d	$1\frac{1}{2}$	38.10

Penny	*Inches*	*Millimeters*
5d	$1\frac{3}{4}$	44.45
6d	2	50.80
7d	$2\frac{1}{4}$	57.15
8d	$2\frac{1}{2}$	63.50
9d	$2\frac{3}{4}$	69.85
10d	3	76.20
12d	$3\frac{1}{4}$	82.55
16d	$3\frac{1}{2}$	88.90
20d	$3\frac{3}{4}$	95.25
30d	$4\frac{1}{2}$	114.30
40d	5	127.00
50d	$5\frac{1}{2}$	139.70
60d	6	152.40

Types and Uses

The various types of nails may be divided into three major categories:

1. Nails for support and fastening in rough construction, especially of structural members, as shown in Figure *N2.*

2. Nails to support or fasten a finished material, as shown in Figure *N3.* These in many instances are exposed nails.

3. Concealed nails, for support and fastening of rough or finished materials (*see* Figure *N4*).

The first group includes all the large nails. Data are given on available metals, finishes, and sizes. These are the nails that must be checked for holding power and shear strength and the type of metals used in fabrication because the strength of the nail must be equal to its holding power and shear stress for the various types of wood framing.

The second group of nails is subdivided in the same manner. Because this type of nail is exposed, the material used for fabrication is especially important from the viewpoints of appearance and effect on contacting materials. Nails are available with pigmented coatings in a limited range of colors to match popular shingle and siding materials.

The third group, nails that are concealed, is most often made of steel or aluminum. Here the holding power is important because, once the material is set in place and the nail is concealed, it is almost impossible to improve its holding power or method of installation without tearing apart and perhaps partially destroying the installed material.

For miscellaneous nails, *see* Figures *N5*, *N6*, and *N7.*

Application

Condensed Checklist

1. The type of nail recommended by the manufacturer of the material to be fastened should be used.
2. A check should be made for galvanic action between the nail and the product to be held by the nail, particularly metals.

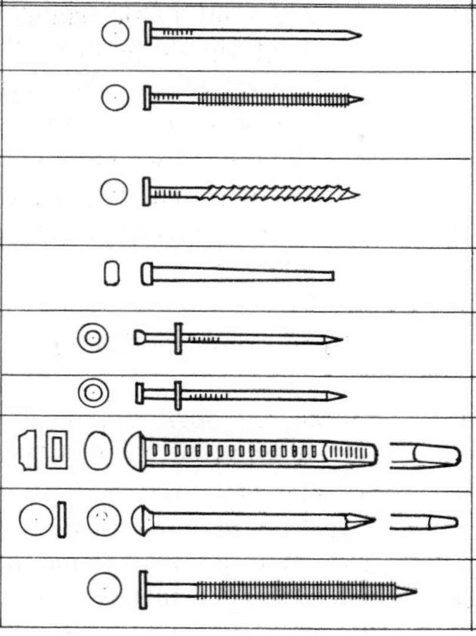

NAME	SHAPE	MATERIAL	FINISH	SIZES
COMMON		STEEL OR ALUMINUM	SMOOTH	PENNY 6 TO 60 INCHES 2"(50.8mm) TO 6" (152.4mm)
ANNULAR		STEEL, HARDENED STEEL, COPPER, BRASS, BRONZE, SILICON BRONZE, NICKEL SILVER, ALUMINUM, MONEL OR STAINLESS STEEL	BRIGHT, HARDENED	
HELICAL				
COMMON CUT STRIKE		STEEL OR IRON	PLAIN OR ZINC-COATED	PENNY 20 TO 100 INCHES 4"(101.6mm) TO 8" (203.2mm)
DOUBLE-HEADED		STEEL	BRIGHT OR ZINC-COATED	PENNY 5 TO 30 INCHES 1¾"(44.5mm) TO 4½"(114.3mm)
		ALUMINUM	BRIGHT	
SQUARE		STEEL	SMOOTH, BRIGHT, ZINC-COATED	INCHES 3"(76.2mm) TO 16" (406.4mm)
ROUND WIRE				
ANNULAR		ALUMINUM	BRIGHT OR HARD	

Figure N2 Nails for support and fastening in rough construction.

NAME	SHAPE	MATERIAL	FINISH	SIZES
SIDING & SHINGLE	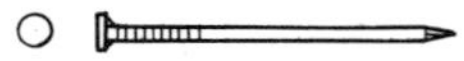	STEEL, COPPER OR ALUMINUM	SMOOTH, BRIGHT, ZINC-OR CEMENT-COATED	PENNY 2 TO 40 INCHES 1"(25.4mm) TO 5"(127mm)
ROOFING (BARBED)		STEEL OR ALUMINUM		INCHES ¾"(19.1mm) TO 2½"(63.5mm)
ROOFING		STEEL	BRIGHT OR ZINC-COATED	INCHES ¾"(19.1mm) TO 1¼"(31.75mm)
NONLEAKING ROOFING	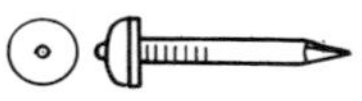			INCHES ¾"(19.1mm) TO 2"(50.8mm)
WALL BOARD	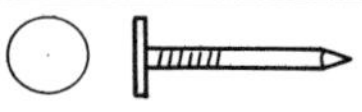	STEEL OR ALUMINUM	SMOOTH, BRIGHT, BLUED OR CEMENT-COATED	INCHES 1"(25.4mm) TO 1¾"(44.45mm)
ANNULAR AND HELICAL SQ. HEAD		STEEL	GALVANIZED	INCHES 1⅛"(28.6mm)
ANNULAR AND HELICAL WITH OR WITHOUT NEOPRENE WASHER		ALUMINUM OR GALVANIZED	BRIGHT OR HARDENED	PENNY 2½ TO 8 INCHES 1⅛"(28.6mm) TO 2½"(63.5mm)
HOOK NAIL	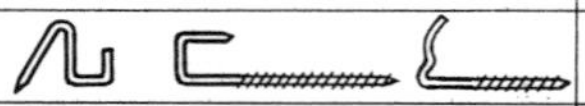	ALUMINUM OR STEEL	SMOOTH, BRIGHT	INCHES 1"(25.4mm) TO 1½"(38.1mm)
SHINGLE NAIL	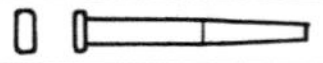	STEEL OR CUT IRON	PLAIN OR ZINC-COATED	PENNY 2 TO 6 1"(25.4mm) TO 2"(50.8mm)
CUT SLATING (NONFERROUS)		COPPER, MUNTZ METAL OR ZINC		INCHES 1¼"(31.8mm) TO 2"(50.8mm)
FLOORING		STEEL	SMOOTH, BRIGHT, CEMENT-COATED; DEEP, LONG, NARROW OR CUPPED HEADS	PENNY 3 TO 20 INCHES 1¼"(31.8mm) TO 4"(101.6mm)
GUTTER SPIKES ROUND	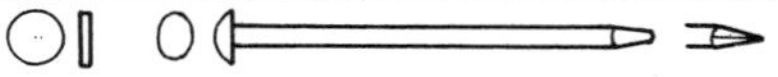	STEEL	BRIGHT OR ZINC-COATED	INCHES 5½"(140mm) TO 10½"(266.7mm)
ANNULAR		COPPER	BRIGHT	INCHES 7"(177.8mm) TO 8"(203.2mm)

Figure N3 Nails to support a finish material exposed or concealed by the material.

3. The size of the head in relation to the material to be supported by the nail should be checked, as the size of the head must increase in proportion to the area of material being held by the nail.
4. In wood framing and wood trusses, the correct number of nails must be used at any given joint so that their holding power is great enough to withstand the stresses imposed upon them.
5. One should always check whether the nail will be subject to moisture and to exterior wear (steel nails, for example, will rust and mark up finished material or show through paint).
6. Nails with shanks that are serrated, helically threaded, etc., not only have increased holding power, but also are permanent. Usually it is impossible to remove them without destroying the material about the shaft.
7. One should always check the holding power and the type of nail to be used when a particular type of material is to be connected to a nonwood material, for example, when the first ply of a built-up roof is to be applied to a poured gypsum decking.
8. The method of assembly should be checked to determine what type of nail should be used.
9. When automatic nailing equipment is to be used, a check should be made that the correct type of nail is selected for the particular type of automatic nailing equipment used.
10. The type of wood or woods to be assembled, joined, or connected should always be checked, as well as the type of nail best suited for this type of wood or woods.

History and Manufacture

Throughout their empire the Romans used a forged nail similar to those manufactured today. In the American colonies nails were so scarce that in 1646 the Virginia legislature passed a measure providing that, if a colonist decided to leave his home, he would be given the number of nails estimated to have been used in the structure so he would not burn his home down to obtain the nails and take them with him. Nails continued to be forged until about 1720, when a method of fabricating nails

NAME	SHAPE	MATERIAL	FINISH	SIZES
COMMON		STEEL	SMOOTH	PENNY 2 TO 20 INCHES 1"(25.4mm) TO 4"(101.6mm)
ANNULAR		ALUMINUM	BRIGHT, HARD, GALVANIZED, ENAMELED OR BLUED	PENNY 2 TO 8 INCHES 1"(25.4mm) TO 2½"(63.5mm)
HELICAL				
FINE NAILS		STEEL	BRIGHT	PENNY 2 TO 3 INCHES 1"(25.4mm) TO 1½"(38.1mm)
LATH			BLUED OR CEMENT-COATED	PENNY 2 TO 4 INCHES 1"(25.4mm) TO 1½"(38.1mm)
LATH		STEEL, ALUMINUM	SMOOTH, BRIGHT, BLUED, CEMENT-COATED	INCHES 1"(25.4mm) TO 1½"(38.1mm)
CASING OR BRAD			BRIGHT OR CEMENT-COATED	PENNY 2 TO 60 INCHES 1"(25.4mm) TO 6"(152.4mm)
FINISHING		STEEL	SMOOTH	PENNY 2 TO 20 INCHES 1"(25.4mm) TO 4"(101.6mm)
ANNULAR FINISHING		ALUMINUM	BRIGHT, HARD OR ENAMELED	PENNY 2 TO 4 INCHES 1"(25.4mm) TO 1½"(38.1mm)
HELICAL FINISHING				PENNY 4 TO 8 INCHES 1½(38.1mm) TO 2½"(63.5mm)
FINISHING CUT NAIL		STEEL OR CUT IRON	SMOOTH	STANDARD: PENNY 3 TO 20 1½"(38.1mm) TO 4"(102mm) FINE: PENNY 6 TO 10 2"(50.8mm) TO 3"(76.2mm)
CEMENT		STEEL	SMOOTH, BRIGHT OR OIL-QUENCHED	INCHES ½"(12.7mm) TO 3"(76.2mm)
CEMENT (FLUTED HELICAL)			HARDENED	INCHES 1½"(38.1mm) TO 3"(76.2mm)
OFFSET (LATH)	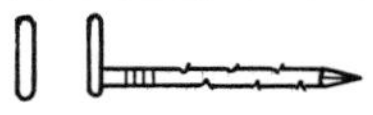		BRIGHT, BLUED, ZINC-COATED	INCHES 1⅛"(28.6mm)
HOOKED (LATH)				INCHES 1¼"(31.8mm) TO 1¾"(44.5mm)
STAPLE	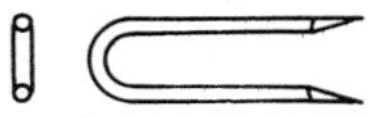			INCHES 1"(25.4mm) TO 1½"(38.1mm)

Figure N4 Concealed nails for support and fastening of rough or finish material.

from wire was developed. Since then almost all nails have been made in this way. A much smaller quantity is cut from plate or flat stock and from sheet metal.

Research and technical improvements, which developed various treatments to the shank of the nail, have increased the holding power of nails to as much as five times that of smooth-shank nails. These nails were developed in answer to the problems posed by (1) the increasing use of soft woods and short-length lumber from second growth trees, and (2) the poor holding power of nails in green or partially seasoned wood and in vacuum-processed woods in which oil carried the preservative.

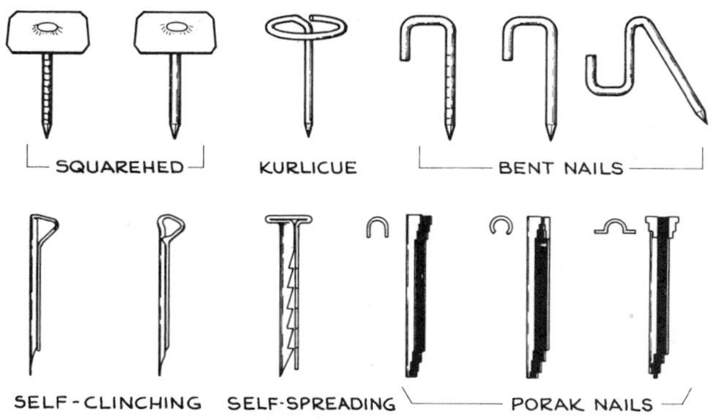

Figure N5 Sheet-metal self-clinching nails.

NAME	SHAPE	MATERIAL	FINISH	SIZES
DOWEL PINS	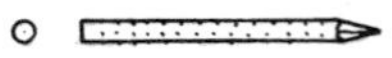	STEEL	BRIGHT	INCHES 5/8"(15.9mm) TO 2"(50.8mm)
SCUTCHEON	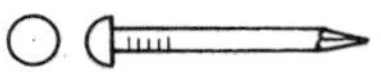	BRASS	BRIGHT	INCHES 1/4"(6.35mm) TO 1 1/4"(31.8mm)
BRADS		STEEL	BRIGHT	INCHES 3/8"(9.53mm) TO 1 1/2"(38.1mm)
FLAT-HEAD NAILS		STEEL	BRIGHT	INCHES 3/8"(9.53mm) TO 1 1/2"(38.1mm)
TACKS	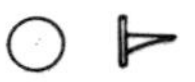	STEEL	BLUED	INCHES 3/8"(9.53mm) TO 3/4"(19.1mm)
BLIND STAPLES		STEEL	BLUED, GALVANIZED, BRIGHT	INCHES 3/8"(9.53mm) TO 5/8"(15.9mm)
DOUBLE POINTED TACKS		STEEL	BLUED	INCHES 1/4"(6.35mm) TO 15/32"(11.9mm)
MITER JOINT FASTENERS		STEEL	BRIGHT	

Figure N6 Miscellaneous nails.

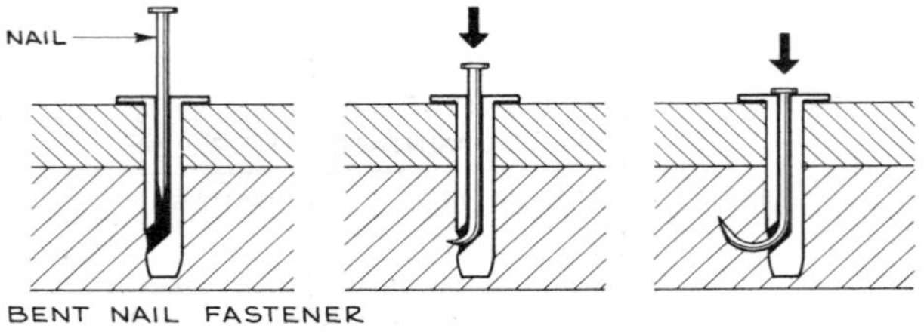

Figure N7 Nails for low-density materials.

NATURAL GAS

Natural gas is a mixture of gaseous hydrocarbons consisting primarily of methane and ethane with propane, butane, pentane, and small amounts of heavier hydrocarbons and other gases.

Around A.D. 800 in China natural gas wells were drilled and piped through bamboo tubes. Up until around 1920 the use of natural gas was localized because there was no way to transport natural gas over long distances. Most gas was obtained from the distillation of coal because this gas was easier to transport, store, and process.

In the 1920's pipelines for transporting natural gas long distances were developed and immediately expanded the use of natural gas, which up to this time had played a very small part in industry. By 1960 natural gas accounted for 14.6% of the energy consumed in the world and for over one-third of the energy used in the United States.

Natural gas cannot be used directly from the well. Gas processing serves two basic purposes: to make a gas that meets specific standards of use, and to recover the liquid and solid compounds. *See* Tables *N1* and *N2.* These liquid and solid compounds are raw materials for the chemical industry. Propane and butane are bottled under slight pressure, heavier hydrocarbons are made into gasoline, and carbon black, sulfur, helium, and argon are also produced.

See also Fossil Fuels; Petroleum; Coal; Petrochemicals.

Table N1 Characteristics of gases from two gas fields: Texas and the Netherlands

	Percentages of composition of the natural gas								
Location	N	O	CO_2	Methane	Ethane	Propane	Butane	Pentane	Hexane
Texas	12.8	2	0.5	76.2	4	2.6	1.3	0.6	
The Netherlands	14.32	0.01	0.87	81.3	2.84	0.43	0.14	0.08	0.02

Table N2 Example of the composition of a commercial gas

Percentages in composition of a commercial gas					
Methane	Ethane	Propane	Butane and heavy hydrocarbons	CO_2 + N	Heating power
83.5	7.0	2.1	1.4	6.0	9400[a]

[a]Expressed as 9400 kilocalories (39,330,000 joules) per cubic meter (60°F or 16°C, 14.51 pounds per square inch).

NEON

Physical and Chemical Properties

Symbol: Ne
Atomic number: 10
Melting point: -248.67°C, -415.61°F
Boiling point: -246.048°C, -410.89°F

Neon is a rare inert gas of the atmosphere, comprising 0.0018% by volume of topospheric air. It is widely distributed in nature, appearing also in gases trapped within the earth.

Neon is colorless, odorless, and tasteless, and has unusually high electrical conductivity and light emissive power, which make it valuable in lamps and other electrical devices.

Types and Uses

Neon is used to produce the familiar neon sign. Neon glow lamps are used for safety and night lights, and as indicator lights on environmental control panels, including more sophisticated usage in computer readout devices.

History and Manufacture

Neon was discovered in 1898 by Sir William Ramsay and M. W. Travers as a component of liquefied pure argon. It is produced on a commercial basis in air-separation plants by fractional distillation.

NICKEL

Physical and Chemical Properties

Symbol: Ni
Atomic number: 28
Specific gravity: 8.902 (25°C, 77°F)
Melting point: 1453°C, 2647°F
Boiling point: 2732°C, 4949.6°F
Tensile strength: 46,000 lbf/in.^{2}, (317.17 MN/m^2) for annealed 99.99% nickel
Coefficient of expansion: 0.000013/°C

Nickel is an inert, silvery metal that is resistant to strong alkalis and to most acids, but not to nitric acid. It resembles iron in strength and toughness and copper in resistance to oxidation and corrosion. Nickel is ferromagnetic, but above 680°F (360°C) it becomes paramagnetic.

Workability. Nickel takes a high polish and can be hot- and cold-rolled, forged, bent, extruded, spun, punched, and deep drawn. It can be welded by metallic arc, electric resistance, oxyacetyline, and atomic hydrogen processes; it can also be brazed and soldered.

Commercial Forms. The purified metal is available in many grades, in electrolytic cathode (*see* Table *N3* and *N5*), ingot, pellet, shot, sponge and powder form. The bulk of nickel sold is high purity grade. Much nickel is also sold as oxide powder and oxide sinter.

Types and Uses

When alloyed with other metals, nickel imparts its qualities of strength, hardness, toughness, ductility, corrosion resistance, and strength at high temperatures to the resulting material. When alloyed with nonferrous metals, it provides the added qualities of whiteness, magnetism, electrical resistance, and control of expansion. The major use of nickel is therefore in alloys–about 40% in ferrous alloys and 27% in various nonferrous alloys. A

Table N3 Composition of high-purity cathode nickel

Ni + Co	Fe	Cu	C	S
	(percent of content)			
99.9	Less than 0.01	Less than 0.01	0.01	0.001

Table N4 Uses of nickel

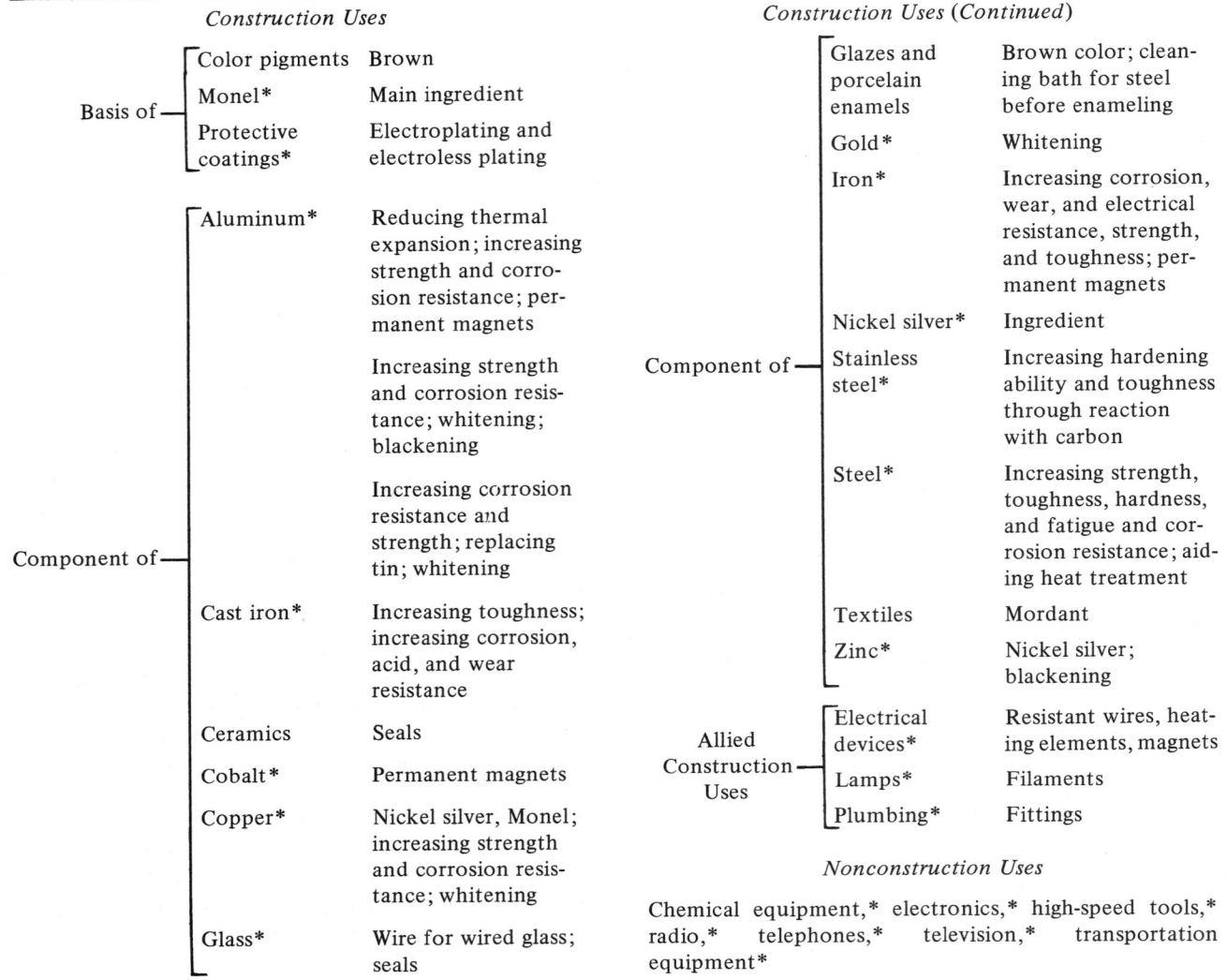

Construction Uses

Basis of	Color pigments	Brown
	Monel*	Main ingredient
	Protective coatings*	Electroplating and electroless plating
Component of	Aluminum*	Reducing thermal expansion; increasing strength and corrosion resistance; permanent magnets
		Increasing strength and corrosion resistance; whitening; blackening
		Increasing corrosion resistance and strength; replacing tin; whitening
	Cast iron*	Increasing toughness; increasing corrosion, acid, and wear resistance
	Ceramics	Seals
	Cobalt*	Permanent magnets
	Copper*	Nickel silver, Monel; increasing strength and corrosion resistance; whitening
	Glass*	Wire for wired glass; seals

Construction Uses (Continued)

Component of	Glazes and porcelain enamels	Brown color; cleaning bath for steel before enameling
	Gold*	Whitening
	Iron*	Increasing corrosion, wear, and electrical resistance, strength, and toughness; permanent magnets
	Nickel silver*	Ingredient
	Stainless steel*	Increasing hardening ability and toughness through reaction with carbon
	Steel*	Increasing strength, toughness, hardness, and fatigue and corrosion resistance; aiding heat treatment
	Textiles	Mordant
	Zinc*	Nickel silver; blackening
Allied Construction Uses	Electrical devices*	Resistant wires, heating elements, magnets
	Lamps*	Filaments
	Plumbing*	Fittings

Nonconstruction Uses

Chemical equipment,* electronics,* high-speed tools,* radio,* telephones,* television,* transportation equipment*

major use of nickel is in plating, both electroplating and electroless plating. These and other metallic uses are indicated with an asterisk in Table *N4*.

The nickel alloys of particular interest to the construction field are certain steels (low-nickel), stainless steels, Monel metals, aluminum, steel, and iron alloys, and nickel silvers. Each of these is important enough to merit discussion under a separate heading. Other alloys having important specialized applications but less used in the construction field are heat-resistant alloys and electrical resistance alloys; thermal expansion alloys; and high nickel-copper, copper-nickel, and coinage alloys. There are also special nickel alloys and super alloys for industrial, transportation, and chemical fields.

Table N5 Composition of high-purity nickel (carbonyl process)

Ni	Fe	Cu	Co	C	S
		(percent of content)			
99.97	0.01	0.0008	0.0005	0.01	0.002

Specialized Alloys. *Inconel* 600 is a tradename for an alloy which usually contains 76% nickel, 15% chromium, and 7% iron, with small, controlled amounts of copper, manganese, silicon, and carbon. It combines all the good qualities of nickel with the superior oxidation resistance of chromium. It is readily workable and machinable and can be drawn, cast, welded, and soldered. Because of its special corrosion resistance it finds use in food processing and chemical industries. Inconel is, in general, difficult to replace for high-temperature service. *Monel* is an alloy of approximately two-thirds nickel and one-

Table N6 Typical nickel alloys for heating elements

Ni	Fe	Cu	Cr	Maximum temperature	
(percent of content)				°F	°C
45		55		1000	537.78
60	24		16	1800	982.22
80			20	2000	1093.33

third copper that has many uses in construction and industry (*see* Monel).

Electrical resistance alloys are used for heating elements, thermometry, and control of electrical current and voltage (*see* Table *N6*).

Magnetic and *nonmagnetic alloys* consist of alloys of iron and nickel with percentages of other elements. The highest permeabilities are associated with alloys of iron with about 80% nickel plus several percentages of molybdenum or of copper and aluminum. Permanent magnets are manufactured by powder metallurgy and contain nickel with cobalt and aluminum, and sometimes copper and titanium. Nonmagnetic alloys are those with 4 to 15% nickel and at least 10% chromium, or with at least 45% copper or an iron-base material with controlled amounts of manganese and/or chromium.

Thermal expansion alloys are based on nickel and iron, frequently with controlled amounts of cobalt, chromium, silicon, manganese, and titanium. Some of these alloys have high coefficients of expansion similar to those of the fusible alloys. Since nickel-iron alloys can be manufactured to almost match the coefficient of expansion of another material, they are used for glass and ceramic seals, as wire for wired glass, for instrumentation, and for measuring tapes.

Copper-nickel alloys, 2 to 45% nickel, are used for marine service and are toxic to marine life.

Coinage is another use for nickel (25 to 100%) in various countries of the world. The 5-cent coin of the United States contains 25% nickel, and 75% copper, as do the outside layers of the sandwich metal used for United States 10-, 25-, and 50-cent pieces.

History and Manufacture

Nickel in alloy form has an ancient history. Prehistoric man used meteoric iron containing 5 to 15% nickel for his implements and jewelry. Early coins of about 235 B.C. from Asia Minor were found to contain nickel in almost the same proportions as today's United States nickel, and the early Chinese long used an alloy of nickel, copper, and zinc called *paktong*, which is much like the nickel silver in current use.

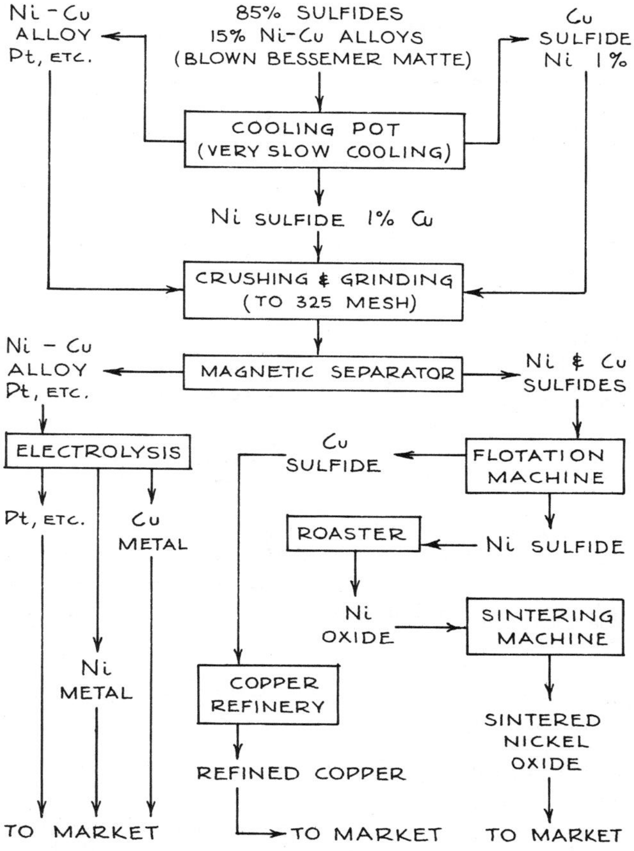

Figure N8 Flowchart of nickel production today.

Nickel itself was not isolated until 1751 when Cronstedt prepared an impure metallic nickel from arsenical ore that the Saxon miners called *kufpernickel*. Cronstedt decided to apply their name to his new metal. In 1775 Bergman confirmed Cronstedt's results, and the name "nickel" was accepted.

In 1838 Germany began the first production of refined metallic nickel from imported ores. New Caledonia was the world's source of nickel until 1905, when Canada became the largest producer. At present, New Caledonia is second in importance to Canada, which in 1959 supplied about 70% of the world's output (not counting Soviet bloc production). The United States is the largest producer of secondary nickel from scrap, as a result of the fact that we are the largest users of nickel but produce only about 4 to 5% of the world's supply.

Preparation of nickel is both costly and complex, the exact process depending on the type of ore used. *See* Figure *N8* for a flowchart of nickel production today.

Orford Process. The Orford process (Canadian), now only of historic interest, was the first continuous and

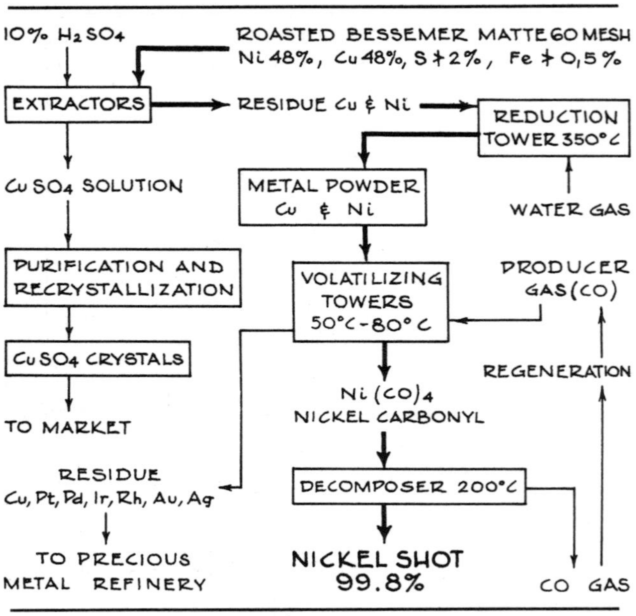

Figure N9 The original Mond process for the production of nickel shot.

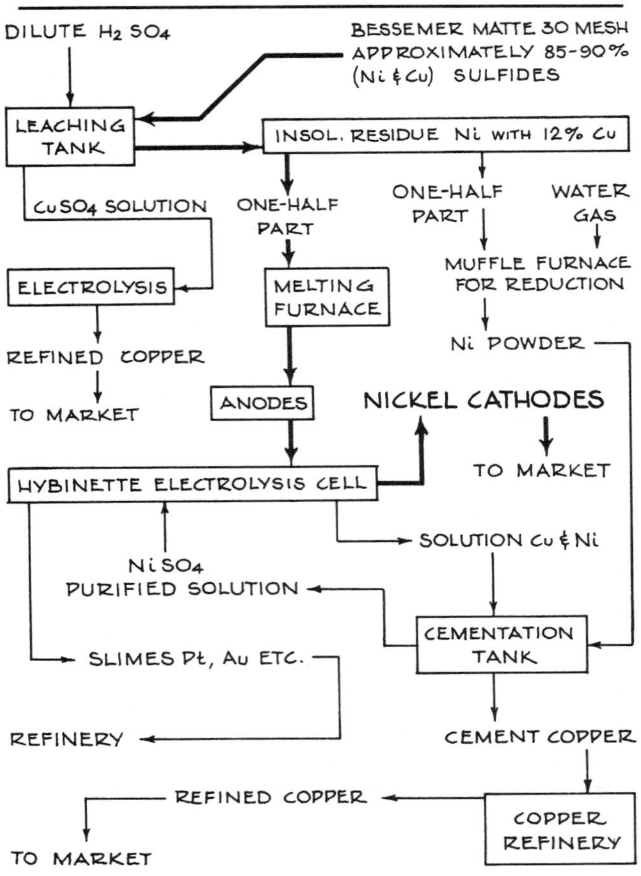

Figure N10 The Hybinette process for the electrolytic production of nickel.

therefore commercially feasible method. However, it was difficult and laborious to operate.

Mond Process. The Mond process (English), perfected in 1895, is also elaborate but produces a high-purity nickel and separates the precious metals without electrolysis. The nickel is separated as a volatile product and then deposited in pure metallic form. Copper is the other mjaor product in this method, which is still being used (*see* Figure *N9*).

Hybinette Process. The Hybinette process (also Canadian) represents electrolytic treatment of a nickel-copper residue. The residue is cast into anodes of impure metal, which are then dissolved electrolytically, and refined nickel is deposited on the cathodes. The remaining slimes are treated for the recovery of precious metals (*see* Figure *N10*).

By-products. The emphasis in much of the early research was on methods to convert the copper-nickel matte into usable alloys without separation of the copper and nickel. Monel alloy and its derivatives were results of these efforts. But as the demand for pure nickel and for controlled nickel alloys, as well as for various rare and precious metals, grew, intensive research was carried out in various countries for means of separating, concentrating, and refining nickel-bearing ores.

In 1973 the result of this research was the opening of a new type of nickel refining plant that produces nickel pellets and powders (*see* Table *N7*). This refinery is based on the reversibility of the carbonyl reaction. Carbon monoxide, at atmospheric pressure and at a temperature less than the boiling point of water, combines with and extracts nickel from impure metallic nickel as a vapor, leaving undesirable impurities behind. At a slightly higher temperature, the process reverses itself, releasing

Table N7 Carbonyl nickel pellets and powders

Type	Ni	Co	Cu	Fe	S	C	O
				(percent of content)			
Pellets	99.97	0.0005	0.0008	0.01	0.002	0.01	
Powder	Remainder			Less than 0.01	Less than 0.001	Less than 0.10	Less than 0.15
	Remainder			Less than 0.01	Less than 0.001	Less than 0.20	Less than 0.15

pure nickel and making carbon monoxide available to repeat the process. The impurities are processed to produce copper, cobalt, and platinum-group metals.

NICKEL PLATING

Physical and Chemical Properties

Electroplating is the most common method of applying a coating of nickel or nickel alloy to other metals for decorative or protective purposes. It can be applied to the following basis metals and their alloys: aluminum, beryllium-copper, brass, copper, iron, magnesium, Monel and other nickel alloys, lead-base alloys, steel, tin, and zinc. Nonconductors such as plastics are successfully plated by metallizing the nonconductor material.

Finish Characteristics. A pure nickel coating is off-white in color, hard, and corrosion resistant. It can be given a matte, satin, or bright polished finish but tends to tarnish in the presence of sulfur.

Undercoatings. Nickel electroplating is widely used as an undercoating for plating with chromium and other decorative metals such as gold, silver, and platinum. An undercoat of copper is frequently applied to create a better surface for nickel deposition.

Corrosion Resistance. Nickel plating is electropositive in character, and any breaks in the coating decrease its protective character, which otherwise is directly related to its thickness and the type of environment (marine, rural, industrial) to which it is exposed. Such breaks do not seem to occur in coat thicknesses of 0.0015 to 0.0020 in. (0.0381 to 0.0508 mm), and heavy coatings of 0.002 to 0.003 in. (0.0508 to 0.0762 mm) are about equal in corrosion resistance to wrought nickel sheet. Galvanic corrosion between nickel and other metals can ordinarily be avoided by using a metal or alloy that is close to nickel in the galvanic series and by having a sufficiently thick, nonporous, continuous coating on the basis metal.

Types and Uses

Coating Thickness. The thickness of the nickel coating is determined by its end use, as can be seen in Table *N8*; that of the copper undercoat varies similarly.

Color Range. A limited color range is possible in nickel plating and is used principally for decorative coatings. Various processes exist for codepositing with other metals such as cobalt, copper, and iron for special properties or decorative effects. For example, nickel-cobalt gives a pure white color resembling silver; nickel-copper combinations yield various shades of yellow.

Precautions. When nickel-chromium type of plating of materials is used in construction, precautions should be taken; these consist of checking with manufacturers to learn wearing characteristics, size limitations, best method of plating for end use in construction, and thickness of coating.

Electroless Plating. Electroless plating is a method based on chemical reaction to form a protective metal coating of nickel or nickel alloys.

Industrial Plating. Two other methods used industrially to apply protective nickel coatings are (1) cladding or pressure welding of nickel to one or both sides of a base plate of another metal by hot rolling, and (2) spraying or vapor deposition of nickel.

Table N8 Comparative thickness of nickel coatings in relation to type of use[a]

Base metal	Plating sequence	Minimum thickness for type of use: Very severe use, in.	mm	Severe use, in.	mm	General use, in.	mm	Mild use, in.	mm
Steel	Copper	0.001	0.0254	0.0006	0.01524	0.00035	0.00889	0.002	0.00508
	Nickel	0.001	0.0254	0.0006	0.01524	0.0004	0.01016	0.002	0.00508
	Chromium	0.00001	0.000254	0.00001	0.000254	0.00001	0.000254	0.00001	0.000254
Zinc and its alloys	Copper			0.0007	0.01778	0.00045	0.01143	0.0002	0.00508
	Nickel			0.005	0.1270	0.0003	0.00762	0.0003	0.00762
	Chromium			0.00001	0.000254	0.00001	0.000254	0.00001	0.000254
Copper and alloys with 50% copper	Nickel			0.005	0.0127	0.0003	0.00762	0.0001	0.00254
	Chromium			0.00001	0.000254	0.00001	0.000254	0.00001	0.000254

[a] In plating the mil is generally used: 1 mil = 0.001 in. = 25 microns.

Electroforming. Electroforming is a process by which nickel shapes are made by starting with a suitable core, then depositing on it a heavy layer of nickel, and finally removing the core. This method is used for making the dies which press the sound tracks on phonograph records.

Decorative Plating. As a rule, decorative nickel plating with a mirror-bright finish requires polishing and buffing, but such finishes can also be obtained by patented processes directly from the plating process without polishing or buffing. A recent innovation is plating that is a leveling process as well.

History and Manufacture

Nickel was first used in electroplating in 1843, but the high price and poor quality of the nickel plating kept it from becoming an established industry until 1870. Since then a growing percentage of the total annual consumption of nickel is used for electroplating.

The fundamentals of the process are very simple. First, the surface of the metal to be coated is prepared by polishing or buffing, as the better the surface before plating, the less finishing is required after plating. Next, immediately before plating, the metal surface is cleaned (pickled) and often given a thin or thick electroplating of copper to either clean or smooth out the surface. Finally, the object to be coated, which acts as the cathode, is immersed in a bath of electrolyte containing salts such as nickel sulfate, chloride, carbonate, and sulfamate in various combinations. The anode is either nickel, in which case it replenishes nickel taken out of the electrolyte, or an insoluble one that requires the addition of nickel salts to the bath to replace the nickel used up in plating.

As current passes through the electrolyte, the nickel is electrodeposited to the desired thickness on the object to be plated. Many of the plating solutions and techniques are patented.

NICKEL SILVERS

Physical and Chemical Properties

Nickel silvers are alloys of copper, nickel, and zinc. Lead is added in small quantities to give better machining characteristics, and both tin and lead are added in small quantities for casting alloys. A variety of colors is obtainable (*see* Figure *N11*). Nickel silvers have good corrosion resistance to marine atmosphere and salt water, but will tarnish in chemically polluted atmospheres. They are ductile, have good mechanical properties, and are easily cast, rolled, drawn, extruded and soldered, brazed, and welded.

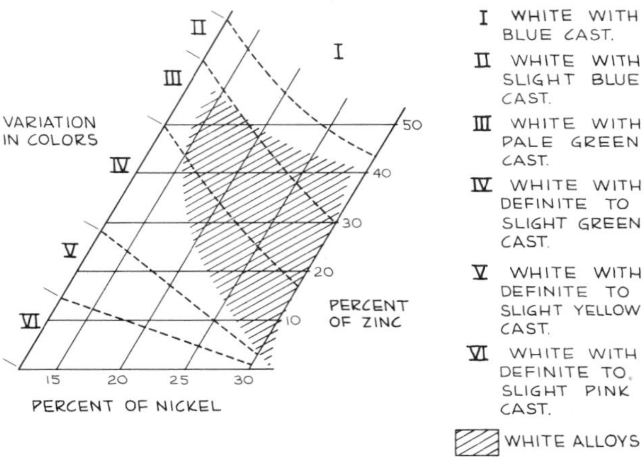

Figure N11 Color variations in nickel silvers.

Types and Uses

Nickel silvers are used more extensively in Europe than in the United States for finish hardware, stair railings, doors and door trim, etc. In the United States nickel silvers are used generally for screws, rivets, saddles, weatherstripping, and edging strips. *See* Table *N9*.

The higher the nickel content, the higher is the corrosion resistance of the alloys. For casting, 18% nickel content is generally used, and a high copper content is required for parts needing much fabricating.

Other uses for nickel silvers include marine hardware, ornamental metalwork, bathroom accessories and plumbing fittings, miscellaneous white metal hardware, and food equipment.

See also Brass; Bronze; Copper Alloys; Nickel.

History and Manufacture

The ancient Chinese smelted a natural alloy which they called *paktong*, meaning "white copper," and used for making gongs and other musical instruments. This was perhaps the first alloy worked by man. In A.D. 1600 it was imported into Europe by the East India Company, and in 1824 a similar alloy was developed in Germany. The English called it German silver and, in 1844, found it to be an excellent base for silver plating. It has been so used to the present day. These early alloys all contained small amounts of silver (from 2 to 20%).

Table N9 Nickel silvers used in the construction field

Name		Composition (percent maximum, unless shown as range)						
Copper alloy	Trade name	Cu	Ni	Zn	Pb	Fe	Mn	Total of other alloys
745	65-10	63.5–68.5	9.0–11.0	19.50–26.15	0.10	0.25	0.50	0.50
752	65-18	63.0–66.5	16.5–19.5	12.65–19.15	0.10	0.25	0.50	0.50
757	65-12	63.5–66.5	11.0–13.0	19.20–24.20	0.10	0.25	0.50	0.50

NIOBIUM (COLUMBIUM)

Physical and Chemical Properties

Symbol: Nb (Cb)
Atomic number: 41
Specific gravity: 8.57 (20°C, 68°F)
Melting point: 2468°C, 4474.4°F
Boiling point: 4927°C, 8900.6°F
Tensile strength: 35,000 to 40,000 lbf/in.^{2}, (241.33 to 275.8 MN/m^2) for the completely recrystallized form

Niobium is a steel-gray, lustrous metal that is malleable, ductile, and tough, and very similar to tantalum. It is corrosion resistant and generally acid resistant.

Workability. Niobium can be drawn, stamped, spun, formed into complicated shapes, welded by a special process, and readily machined, but not cast.

Commercial Forms. Niobium is available in powder, bar, and ingot form and in all normal mill-product shapes. For the composition of high-purity niobium, see Table *N10*.

Types and Uses

Niobium has only a few uses in its compound form. Its uses in metallic form are indicated by asterisks in Table *N11*.

Niobium is very important in making high-temperature alloys. Ferroniobium is a master alloy of niobium and iron, with some manganese and silicon, which is used to add niobium to stainless steels; ferroniobium-tantalum alloy is used similarly.

History and Manufacture

In 1801 an English chemist, Hatchett, in analyzing a mineral from New London, Connecticut, discovered a new element which he named columbium after the source country of the mineral. In 1844 Rose announced the discovery of a new element, closely akin to tantalum, which he named niobium after Niobe, the daughter of Tantalus. Later it was proved that columbium and niobium were the same element. (The American Chemical Society has adopted the name niobium, but the American metallurgical industries retain the name columbium.) It was not until 1929 that Balke isolated pure nobium metal. The method of obtaining niobium metal is described under Tantalum.

Table N10 Composition of high-purity niobium

Cb	Ta	Fe	Si	C	Ti
(percent of content)					
99.7	0.10[a]	0.01[a]	0.01[a]	0.05[a]	0.01[a]

[a] Maximum.

Table N11 Uses of niobium

Construction Uses		
Basis of	Abrasives	Main ingredient
Component of	Stainless steel*	Stabilizing by fixation of carbon
	Steel*	Increasing impact strength and resistance to oxidation; reducing creep; aiding nitriding
Allied Construction Uses	Welding*	Stainless steel electrodes
Nonconstruction Uses		
High-temperature alloys,* acidproof equipment,* with carbides for cutting tools, cermets		

NITRIDING

Nitriding is a process for the surface hardening (also called case hardening) of steel by the absorption of nitrogen. To form a nitrided case, special low-alloy steels containing nitride-forming elements (such as aluminum, chromium, or molybdenum) are heated in an ammonia (NH_3) atmosphere or in contact with a nitrogenous material to a temperature of 875°F (468.33°C) for 5 to 100 hours, the time depending on the depth of case desired, and then quenched. In this process the steels, also called nitralloys, may contain, in addition to aluminum, chromium, and molybdenum, small amounts of carbon, manganese, and silicon. Other alloys and stainless steels not containing aluminum may also be nitrided, but the surface is not as hard as in the nitralloys. Of the nitriding elements mentioned, aluminum seems most effective.

Nitrided steels are highly corrosion resistant (except to gasolines containing tetraethyl lead) and have the following advantages over other methods of surface hardening: a harder, more wear-resistant case; low distortion; and greater resistance to surface fatigue.

See also Metals: Mechanical Finishes; Metals: Sprayed Metal Coatings (Metallizing).

NITROGEN

Physical and Chemical Properties

Symbol: N
Atomic number: 7
Specific gravity: 0.88 for the liquid
Melting point: -209.86°C, -345.75°F
Boiling point: -195.8°C, -320.44°F

Nitrogen is a colorless, odorless, tasteless gas that constitutes about four-fifths of the earth's atmosphere. It is chemically rather inert and neither burns nor supports combustion. It is soluble in water and slightly soluble in alcohol. It constitutes one of the main building blocks of organic compounds.

Types and Uses

Nitrogen itself is a source of cold when liquefied and serves to produce conditioned atmospheres for industrial processes. A small amount is used in incandescent lamps to prevent arcing. It is a basic ingredient of nitric acid, ammonia, cyanamides, dyes, molding compounds or plastics derived from urea, cyanides and nitrides of case-hardening metals, and numerous synthetic organic and other nitrogen compounds. Nitric acid is used extensively in cleaning, preparing, and pickling metals and in the manufacture of plastics.

History and Manufacture

Nitrogen was first recognized in 1772 by Scheele and discovered independently about the same time by both Priestly and Rutherford. Lavoisier called it *azote*, from the Greek for "no life," and established that it was an element. Later, Chaptal introduced the name "nitrogen" to show that the element is a constituent of niter, but "azote" persists in organic chemistry as "azo" to indicate nitrogen-containing chemicals.

Nitrogen today is made commercially by fractional distillation of liquid air and sold in cylinders.

NONMETALS

The nonmetallic elements are of interest insofar as they affect the wood, metals, plastics, and other materials used in construction, either directly or by being present in the atmosphere and in other physical environmental conditions that affect the structure to be designed and constructed. Of the 103 known elements, only 15 are nonmetallic. These are listed in Table *N12*. They have certain characteristics that are the complete opposite of the characteristics of metallic elements, namely, they form negative ions in solution, gain or borrow electrons during chemical reactions, and set them free at the positive pole in electrolysis. *See also* Elements; Metals.

Table N12 List of nonmetals

Element	Symbol	Atomic number	Physical state
Argon	Ar	18	Gas
Astatine[a]	At	85	Gas
Bromine	Br	35	Liquid
Chlorine[b]	Cl	17	Gas
Fluorine	F	9	Gas
Helium	He	2	Gas
Iodine	I	53	Solid
Krypton	Kr	36	Gas
Neon	Ne	10	Gas
Nitrogen[b]	N	7	Gas
Oxygen[b]	O	8	Gas
Phosphorus[b]	P	15	Solid
Radon[a]	Rn	86	Gas
Sulfur[b]	S	16	Solid or liquid
Xenon	Xe	54	Gas

[a]Radioactive.
[b]Nonmetals most frequently encountered in construction.

O

OPALIZER

An opalizer is a material or substance which imparts to a material a milky iridescence similar to that of an opal (gemstone).

ORE

An ore is a mineral from which a metal or metallic compound can be extracted commercially. The valueless portion of the ore is known as gangue, which often consists of quartz, calcite, feldspar, and other rock. The valuable portion, or mineral ore, may contain oxides, sulfides, native metal, and single or double salts of the metal. Some minerals carry two or more metals, for example, stannite (Cu_2FeSnS_4). Others are mined for their nonmetallic elements, for example, pyrite (FeS_2) for sulfur compounds.

After the ore is mined, the valuable constituents must be separated and concentrated. The principal methods used are (1) hand picking, which depends on appearance; (2) gravity concentration, based on different densities of the constituents; (3) magnetic separation; (4) electrostatic separation, utilizing the attraction between unlike electric charges and the repulsion between like charges; and (5) the flotation method, which depends on the ability of the substance used to treat the ore to wet selectively some mineral particles, whereas others remain unwetted and adhere to air bubbles which float to the surface and are removed as a concentrate in the froth.

OSMIUM

Physical and Chemical Properties

Symbol: Os
Atomic number: 76
Specific gravity: 22.57 (highest of the metals)
Melting point: 3045°C, 5513°F
Boiling point: 5027°C, 9080.6°F

Osmium is a hard, brittle, white metal that is insoluble in most acids and gives off poisonous fumes when heated. It is available in powder and pellet form.

Types and Uses

The metal is principally used with ruthenium and platinum to produce alloys of great hardness. These alloys are used mainly for instrument pivots.

History and Manufacture

In 1804 Tennant isolated osmium. Its name is derived from the Greek word meaning "smell" because of the chlorine-like odor of osmium tetroxide. The process for obtaining osmium is described and shown under Platinum.

OXYGEN

Physical and Chemical Properties

Symbol: O
Atomic number: 8
Specific gravity: 1.14 (−182.96°C, −297.33°F)
Melting point: −218.4°C, −361.12°F
Boiling point: −182.96°C, −293.33°F

Oxygen is a colorless, odorless, tasteless gas. It is very active chemically, combining with nearly all elements to form oxides. Approximately 50% of the earth's crust is oxygen in combined form.

Types and Uses

Oxidation, the chemical combination of oxygen with another element, is one of the fundamental processes

Table O1 Oxidation

Type of oxidation	Necessary conditions	Method of stopping or preventing
Fire (rapid oxidation)	A combustible material plus air or oxygen	Shutting off the supply of air or oxygen by the use of an incombustible material
Decay or fermentation (slow aerobic oxidation)	Animal or vegetable matter plus air or oxygen, plus moisture, plus bacteria, plus warmth	Preventing growth of bacteria
Rusting (slow oxidation)	Metal plus air or oxygen, plus moisture, and sometimes aided by carbon dioxide	Excluding air and moisture from the metal by oiling, painting, varnishing, enameling, plating, galvanizing, or tinning

Table O2 Uses of oxygen

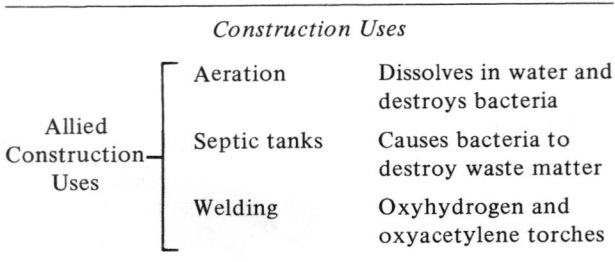

	Construction Uses	
Allied Construction Uses	Aeration	Dissolves in water and destroys bacteria
	Septic tanks	Causes bacteria to destroy waste matter
	Welding	Oxyhydrogen and oxyacetylene torches
	Nonconstruction Uses	
Combustion for smelting, melting, etc.		

in the chemical industry (*see* Table *O1*). The major uses of oxygen in the field of construction are shown in Table *O2*.

Vegetable and animal oils, heated and agitated by a current of oxygen, become partially oxidized, deodorized, and polymerized and are increased in density, viscosity, and drying power. This process is used in paints, varnishes, and plasticizers.

Other Compounds. Oxidized microcrystalline waxes are made from petroleum refining residues and used in floor polishes.

Siloxanes are compounds of silicon and oxygen and vary from relatively mobile fluids through oils, greases, and rubbers to resins and plastics (*see* Silicones).

Aldol is a carbon, hydrogen, and oxygen compound which is a solvent for organic substances and cellulose acetate. It is also used in cadmium plating, dyes, and synthetic resins.

Oxalic acid is a carbon, hydrogen, and oxygen compound which is used as a bleaching agent, in metal polishes, and in ink and rust removers.

History and Manufacture

Oxygen was discovered in 1774 by Priestley, an English minister and scientist. The discovery of oxygen, followed by the consequent explanation of the true nature of combustion by Lavoisier, marked the birth of chemistry as a science. Oxygen is obtained by the fractional distillation of liquid air. Liquid air is allowed to evaporate, and nitrogen evaporates first, followed by oxygen.

P

PAINT

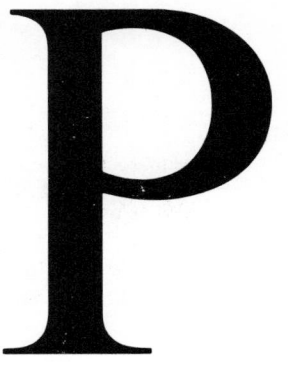

Physical and Chemical Properties

In today's terminology paint is only one part of what are known as organic coatings; these include paints, varnishes, enamels, lacquers, stains, sealers, and all the miscellaneous accessory products such as thinners. These coatings have the following properties in varying degrees, depending on the composition of the coating: good flowing and leveling; satisfactory spreading rate and film thickness; fast drying; high impermeability; good adhesion; flexibility; and hardness, abrasion-resistance, and durability. Such a coating may be a fluid, with or without the fluid suspension of finely divided solids, which, when applied to a suitably prepared surface, forms a solid film by one of five methods after the solvent evaporates:

1. Chemical conversion by absorption of oxygen. These are known as oxygen-convertible coatings.
2. Simple solidification after evaporation of the solvent. These are known as thermoplastic coatings or nonconvertible coatings.
3. Thermal polymerization by heating at elevated temperatures. These are known as thermosetting coatings.
4. The use of a catalyst which reacts chemically to produce the film.
5. Coalescence of latex particles with the evaporation of the water comprising the fluid.

The finely ground solids are technically known as the pigment, and the liquid portion is called the vehicle, or medium. Because in the construction field the architect ordinarily equates "pigment" with "color," in this book the "pigment" of the paint industry is called white paint solid. The white paint solid consists of white pigments (all paints other than white paints also contain color pigments), plasticizers, and extender pigments.

The vehicle consists of a binder which forms the film, driers to speed up formation of the film, plasticizers, and a volatile solvent known technically as the thinner. The type of solvent used in a paint determines whether it is a water-base paint (where the solvent is water), or an oil-base paint (where the solvent is an organic liquid). Water-base paints also contain emulsifiers, stabilizers, and antifoam agents. (*See* Metals: Coatings.)

White Paint Solids (or "Pigment"). Originally, white paint solids (pigments) were added only to give color to paints, but today they are known to affect many other properties of paint and are rarely used singly. White paint solids must be considered for their hiding power, settling, workability, stability after exposure, color, ability to protect organic vehicle binders from damaging rays of sunlight, rust-inhibiting properties, and resistance to mold, fungi, and chemicals.

Hiding power of a paint, that is, the ability of a paint to obscure the underlying material, varies greatly, depending on the chemical composition of the white paint solid and on the particular shape and size of the solid particles. The difference between the index of refraction of the vehicle and that of the white paint solid largely determines the hiding power of the paint. When the index difference between the white paint solid and the vehicle is large, the paint has good hiding power; as this difference decreases, the paint approaches transparency. For example, white lead has a refractive index of 1.95 and titanium dioxide, 2.70; the refractive index of most vehicles varies from 1.47 to 1.52.

The measure of hiding power is based on basic lead carbonate, better known as white lead, which has been completely eliminated from paints because of the toxicity of lead. Federal and state governments now require that there be less than 0.5% lead base on the dried film of paint intended for use in or upon homes or dwellings of any kind. The various lead oxides, sulfates, carbonates, and chromates are used only for rust-inhibiting paints for iron and steel. Titanium dioxide is the prime (and practically the only) white hiding pigment used today in non-rust-inhibiting paints. From Table *P1* it is clear that titanium dioxide has the best hiding power of these white paint solids.

Table P1 Characteristics and major use of white paint solids in organic coatings

White paint solids (white pigments)	Hiding power[a]	Major use
Basic lead carbonate (white lead)	1.00	No longer used
Lithopone (zinc sulfide and barium sulfate)	1.79	Interior
Titanium dioxide, anatase	6.53	Interior and exterior
Titanium dioxide, rutile	8.16	Interior and exterior
Zinc oxide	1.48	Exterior

[a]Hiding power is based on the hiding power of 1 lb of white lead, which equals 1 hiding unit. Thus 1 lb of white lead mixed to painting consistency with oil will completely obscure 15 ft^2 (1.39 m^2) of standard surface with white background and black stripes.

A high index of refraction not only imparts good hiding power to white paint solids but also minimizes the damaging effects of sunlight, since the ability to deflect sunlight depends on the ability of the white paint solid to refract light.

Colored paints differ from white paints in that the incident light is selectively absorbed. The light that is refracted (reflected) gives the color. The darker the color of a paint, the better is its hiding power. (*See also* Color Pigments.)

Particle shape and size not only affect the hiding power of a white paint solid, they also can give smoothness, reinforcement, density, and impermeability to the paint film. Particle shapes are classified as (1) nodular or rounded, a shape that imparts the greatest density; (2) acicular or needle-like, a shape that reinforces the film mechanically; and (3) lamellar or plate-like, a shape that makes the film less permeable to moisture and also contributes some mechanical reinforcement.

Aluminum, zinc, and some lead compounds are used alone or with white paint solids to impart rust-inhibiting qualities or impermeability (*see also* Painting of Metal).

The particle size of white paint solids, color pigments and extenders used for paint in the field of construction ranges in average diameter from 0.0001 to 0.0060 mm.

Chemical reactivity of the paint solids affects the characteristics of the paint film they form. Some paint solids are reactive, that is, they have the ability to react chemically with the small amounts of free acids in the vehicle to form stable reaction products (soaps) which develop special characteristics in the paint and film. Soaps from zinc oxide harden the film, and those from zinc chromate, lead silicochromate, and red lead develop rust-inhibiting characteristics. Paint solids that do not react, that is, nonreactive ones such as titanium dioxide, have a tendency to lose gloss and to chalk or flake off.

Chalking, which was once considered an undesirable characteristic, is now used and controlled. A paint as it ages collects dirt and changes color. But if chalking is correctly controlled, as the paint ages and begins to chalk slightly, the rain then washes off the chalking along with the dirt, permitting the surface color to remain unchanged and thus utilizing the chalking effect to keep the paint cleaner and brighter and to prolong its usefulness.

Extender pigments originally were used as cost-reducing materials in paints and were known as fillers. Generally they are white paint solids of low opacity that contribute very little to hiding power. Today, in good paints, extender pigments are used to develop beneficial characteristics. They can help to control gloss and to adjust consistency and workability; they can improve brushing properties and storage qualities. They minimize settling of the white paint solids when they have high oil-absorption properties. They are essential components of priming and sealing paints and flat finishes. (*See* Table *P2*.)

Plasticizers are nondrying materials that are combined with white paint solids and with binders forming hard films to soften and increase their flexibility.

The Vehicle, or Liquid Portion, of Paint. The vehicle contains volatile and nonvolatile constituents. The volatile solvent (thinner) facilitates application and contributes, through its evaporation, to the drying of the paint, but is is not a permanent part of the paint film. The nonvolatile constituents bind the white paint solid particles together and make the film adhere to the surface on which it is applied. Together with the white paint solid these nonvolatile constituents are responsible for the protective qualities and durability of the paint.

The vehicle may contain resins or drying oils alone, or a blend of drying oils and resins. In the latter case it is called oleoresinous.

The possible variations in the composition of vehicles are almost unlimited, and many formulations have become very complex in order to produce paint for specific purposes.

Table P2 Composition and characteristics of extender pigments used in organic coatings

Type of external pigment	Composition and description	How used in paints
Barite	Barium sulfate, finely ground; very heavy white powder	To improve brushing quality
Blanc fixe	Synthetic barium sulfate; more finely ground than barite	To improve brushing quality
Diatomite	Silica; finely ground white powder	To produce flatness and low gloss; to give flexibility to film; to improve brushing quality; to reduce cracking and chalking
Kaolin	Hydrated aluminum silicate, finely ground; white clay	To reduce settling; to strengthen film
Metallic soaps	Aluminum, zinc, and calcium stearates, finely ground; white powder	To thicken vehicles; to prevent settling; to increase water resistance; to harden film; to produce flatness.
Mica	Silicates of chemical composition; finely ground powder consisting of very fine, thin flakes	To reduce cracking, checking, and chalking; to reduce tarnishing and improve leafing in metallic paints; to increase the brilliance of color pigments
Silex	Naturally occurring amorphous silica; finely ground white powder	To give surface to paints which are often refinished (traffic line paint)
Whiting	Calcium carbonate, natural or chemically prepared, finely ground	To improve gloss, flow, abrasion resistance, and durability

Drying Oils. The word "drying" when used in connection with coatings refers not only to the evaporation of the volatile constituents but also to the setting or hardening process that takes place when thin films of drying oils or resins are exposed to the atmosphere. The drying in this case results from the absorption of oxygen (oxidation) or the combination of molecules within the oil (polymerization), or both. Linseed oil is the best known example of the drying oils, and lubricating oil of the nondrying oils. Linseed oil was the nonvolatile vehicle in most exterior wood paints, but now has generally been replaced by alkyds and latices (latex-based paints).

Drying oils are seldom used in their natural state but are refined and treated to eliminate, as far as possible, undesirable qualities and to increase and develop the best qualities. They are also used as a raw material for synthetic resins.

The natural oils used in coatings and as raw material for synthetic resins are as follows: castor (from castor plant seeds); fish (from tuna, shark, and menhaden); safflower (from safflower seeds); soybean (soya beans); and tung (from seeds of the tung tree *Aleurites fordii*). Linseed oil used in coatings is of four types: linseed oil, raw linseed oil, boiled linseed oil, and bodied linseed oil.

Resins. The resins used for the nonvolatile ingredients in vehicles may be either natural resins or synthetic resins. Natural resins have been used from ancient times in paint and varnish. Most interior paints use varnish as the vehicle because of its faster drying and better leveling qualities. (*See* Table *P3*.) Varnish is used as a clear finish for furniture, floors, woodwork, etc., but is not used in exterior paints other than special exterior-type varnishes.

Synthetic resins, which are constantly changing, being replaced, and being improved as a result of the continuing developments in plastics, adhesives, and rubber, are replacing the natural resins in organic coatings. They give superior properties such as durability, waterproofness, toughness, resistance to chemicals, and dielectric (nonconductor or insulator) strength. (*See* Table *P4*.)

Some synthetic resins are used singly, some in combination with others, some with catalysts, and some with drying oils and other vehicle ingredients to produce paints for specific purposes.

Solvents. The volatile portion of the vehicle, or solvent, includes both the liquids that actually dissolve the nonvolatile ingredients and the liquids that act more as suspending agents and might be classed as thinners or diluents. The primary reason for using solvents is to permit easier application or to make a solution of resin which then can dry to form a film. The solvents control the viscosity of the paint so that it can be applied by brushing, spraying, or dipping, as desired; they also

Table P3 Natural resins for organic coatings

Type of natural resin	How obtained	Solvents for paints	Major use in paints
Copal	A resinous exudation from trees found in the East Indies, the Philippines, Australia, and Africa, either fossil or of recent origin	Alcohol, turpentine	Varnishes, lacquers
Damar	A resinous exudation from trees found in the East Indies and Malaya, either of recent origin or semifossil	Alcohol, benzene, turpentine, and other oils	Spirit varnishes
Ester gum	Combination of glycerin and rosin		Varnishes
Mastic gum	A resinous exudation from the pistacia tree, found in the islands of the Mediterranean	Alcohol, ether	Lacquers
Rosin	Residue after distillation of turpentine oil from raw turpentine	Alcohol, benzene, ether, and other oils	Dark and clear varnishes; used as raw material for synthetic varnishes
Sandrac	A resinous exudation from the callitric tree, found in Morocco	Alcohol, turpentine, ether, acetone	Varnishes and lacquers
Shellac	Produced from the lac bug in India, Thailand, and Indochina	Alcohol	Shellac, varnish

Table P4 Synthetic resins for organic coatings

Type of synthetic resin	How obtained	Solvents for paints	Characteristics	Major uses
Acrylate resins	Polymers and copolymers of acrylic acid, methacrylic acid, and esters of these acids	Xylene	Thermoplastic films resistant to aging, ultraviolet rays, and oxidation; they are also colorless and can be produced in almost any degree of flexibility	Very durable protective paints; baked enamels
Alkyd resins	Union of dibasic acids or anhydrides with polybasic alcohol	Varies with type of alkyd resin: water, mineral spirits	Films that are tough and resist ultraviolet rays better than any other presently known resin; good color, retention, adhesion, flexibility, and weather resistance; some are nonyellowing	Nonvolatile ingredient in vehicles, baking enamels
Casein	Phosphoprotein in skimmed milk, by precipitation with acid or by lactic fermentation	Water	Powder type forms a fast-drying, hard, fairly durable film; paste type, a fast-drying, durable, washable film; both types are susceptible to mold	Interior water paints when dry form is used; both exterior and interior paints when paste form is used
Cellulosic polymers	Chemically treated natural cellulose, cellulose nitrate, cellulose acetate, ethyl cellulose	Varies with type of lacquer	Films that are flexible, durable, hard, and resistant to alkalis and dilute acids	Main ingredient of lacquers
Chlorinated rubber	Combination of rubber and chlorine	Linseed, oil, nitrobenzenes, esters, ketones	Films that are resistant to alkalis, some acids, gasoline, oil, and grease	Nonvolatile ingredient in vehicles; industrial maintenance

Table P4 Synthetic resins for organic coatings (*continued*)

Type of synthetic resin	How obtained	Solvents for paints	Characteristics	Major uses
Epoxy resins, catalyzed	Condensation of phenol, acetone and epychlorohydrin	Ketone, xylol, epoxy reducer	Two-compound pigmented enamel and catalyst; films are hard, tough, and resistant to chemicals, alkalis, abrasion, and cleaning	Very durable wall and floor coatings; the pigmented enamel or primer of the two-component coating; industrial maintenance; durable wall and floor coatings; nonvolatile ingredient in vehicles
Epoxy esters		Mineral spirits	Films that are hard and tough and have a limited degree of resistance to chemicals and alkalis	
Fluorocarbons	Compounds of carbon and fluorine with or without hydrogen	Varies with type of fluorocarbon	Extreme inertness and stability	Baked coatings
Latex acrylics and vinyl acetate-acrylics	Water emulsion of synthetic resins and rubbers	Water	Film dries rapidly, has nonpaint odor, and is washable	Nonvolatile ingredient in vehicles
Latex polyvinyl acetate	Same as above	Same as above	Same as above	Same as above
Latex styrene butadiene	Same as above	Same as above	Same as above	Same as above
Melamine resins	Melamine, formaldehyde, and an aliphatic alcohol	Varies with type of melamine-alkyd combination	Combined with alkyd resins for films with very fast curing, hardness, durability, and resistance to soaps	Nonvolatile ingredient in vehicles
Phenolic resins	Condensation of phenol with formaldehyde	Varies with type of phenolic resin	Film that dries rapidly and has great water and weather resistance	Varnish, baking enamels
Polystyrene resins	Polymerization of the hydrocarbon, styrene; reaction with linseed oil, tung oil, and dehydrated castor oil	Benzene	Film that is fast drying and water resistant; has a light color and good color retention	Nonvolatile ingredient in vehicles
Silicone alkyd	Minimum of 25% of silicone resins with alkyd	Mineral spirits, xylol	Films that have outstanding color and gloss retention; alkyd protects film until coated material is heated, when the silicone resin becomes heat-cured	Heat-resistant applications
Silicone resins	Organosiloxane polymers	Varies with type of silicone resin	Film that is resistant to weathering, oxidation, corrosive chemicals and heat	Exterior water-resistant paints for masonry surfaces
Urethanes, moisture-cured	Urethane polymer	Urethane reducer	Film that has outstanding abrasion resistance, excellent flexibility, and chemical, alkali, and water resistance	Industrial maintenance; wood and concrete floors

Table P4 Synthetic resins for organic coatings (*continued*)

Type of synthetic resin	How obtained	Solvents for paints	Characteristics	Major uses
Urethanes, oil-modified	Urethanes modified with drying oils and alkyds	Mineral spirits	Film that is tough, flexible, durable, and resistant to water, acid, alkali, grease, and cleaning materials	Areas where surface will receive excessive abrasive punishment
Vinyl resins	Plasticized copolymers of vinyl chloride and acetate	Ketones, esters	Film that has poor wetting and fast drying properties, excellent flexibility, and inertness to water and most chemicals	Exterior areas
Vinyl resins, catalyzed	Catalyzed vinyl chloride or acetate	Esters	Clear film that can have a dull sheen or dull or medium rubbed effect; resistant to water, household cleaners, alcohol, lipstick, and wax crayons	Interior wood, plywood, furniture, cabinets, etc.
Urea-formaldehyde resins	Raw materials heated and mixed with fillers to produce molding powders	Mixed with other ingredients for finishing process	Thermosetting, insoluble, strong, rigid surface	Baked enamels

Table P5 Paint solvents (thinners) for organic coatings

Type of solvent	Degree of danger	How obtained	Major solvent for the following
Mineral spirits VM and P naphtha	Relatively safe Low flash point, hazardous	Aliphatic hydrocarbon obtained from petroleum	Drying oils, natural resins, and most synthetic resins
Benzene Petroleum ether (benzine)	Toxic and hazardous Relatively safe	Aromatic hydrocarbon by-product from coke manufacture and synthetic production of aromatic hydrocarbons	Drying oils, natural resins, and most synthetic resins
Toluene Xylene	Hazardous	Same as above	Synthetic resins and lacquer resins
Dipentene Pine oil Turpentine	Relatively safe	Terpenes obtained from pine wood by distillation of the gum and destructive distillation of the wood	Drying oils, natural resins, and most synthetic resins
Denatured alcohol	Relatively safe	Alcohols obtained from fermentation of grain or synthetically	Shellac
Isopropanol alcohol Butanol alcohol Amyl alcohol	Relatively safe	Alcohols obtained from fermentation or synthetically	Varnish lacquers
Ethyl acetate Isopropyl acetate Butyl acetate Amyl acetate	Relatively safe	Esters (reaction products of acetic acid and ethyl, isopropyl, butyl, and amyl alcohols)	Lacquer formulations
Acetone Methyl ethyl ketone Methyl isobutyl ketone	Hazardous	Ketones (oxidation of alcohols and natural gas)	Varnish lacquers and synthetic resins
Higher boiling ethers	Hazardous	Esters (ethers obtained from ethylene glycol)	Paint solvents
Water	Safe	Clean and without chemical pollutants	Water-base paints and emulsion paints

affect the consistency, leveling, drying, adhesion, and durability of the paint. They must be carefully selected to meet the requirements of the drying oils and resins.

The quantity of solvent may vary from several percent by weight of the vehicle for raw linseed oil paints, and roughly 50% by weight of the vehicle for synthetic resin paints, to as high as 90% by weight of the vehicle for some lacquers. (*See* Table *P5*.)

Almost all solvents are highly flammable and should be stored and handled away from open flames, sparks, or sources of dangerously high temperature.

The solvent (thinner) most generally used in ordinary oil and varnish-vehicle paints is mineral spirits (petroleum distillation products). Some synthetic-resin formulations require solvent power and use toluol and xylol (coal-tar distillates), hydrogenated naphtha, and chlorinated hydrocarbons. Lacquers require solvents incorporating ketones, esters, or alcohols. Water is the solvent for latex-emulsion and portland cement paints.

Sealers. These are divided into three categories:

1. Clear sealers and water-repellent compounds. These consist of waxes, resins, silicone resins, and drying oils used singly or in combination with a water repellent such as aluminum stearate for the nonvolatile ingredient and a solvent such as petroleum naphtha or mineral spirits. Their compositions vary widely and basically consist of water-insoluble and water-repellent substances dissolved in a solvent. They are effective for shedding moisture and rain, but when they are to be used against aquastatic pressure it is advisable to consult with paint manufacturers in order to select the correct material to answer this specific problem.

2. Floor sealers. These are a special varnish or lacquer used in finishing wood floors. Lacquer types are used for oil-treating floors before varnishing and waxing, whereas varnish types (similar to a spar varnish) are thinned with a greater percentage of solvent to provide penetrating properties and consequently are used as a floor finish.

3. Knot sealers. These consist of a phenolic or vinyl resin in an alcohol solvent and are used to seal knots and pitch streaks in wood to prevent staining and peeling of the finish paint. Shellac is also used for this purpose.

Wax polish is seldom used as a clear coating by itself but is used to protect other finishes. The waxes generally are (1) carnauba wax, obtained from an exudation from the leaves of the wax palm found in Brazil; (2) beeswax, obtained from the honeycomb of the bee; and (3) ceresin wax, obtained from ozocerite, a natural waxlike hydrocarbon found in Utah.

There are two general types, one in which mineral spirits or similar organic solvents are used and another in which wax is emulsified with water. The waxes are often modified with resins and oils. Shellac resin is often added to impart nonslip qualities. In construction one should always check the type of flooring material to which wax polish is to be applied, to be sure that the correct type of wax polish is selected, that is, one that will not attack the finish floor. Asphalt tile, for example, is attacked by mineral spirits and organic solvents.

Bituminous Coatings. These are made from coal tar (*see* Tar) and asphalt (*see* Asphalt). Protective coatings of this type are used extensively for submerged or buried ferrous metal and for waterproofing masonry materials. They create a barrier against the permeation of moisture and oxygen.

Coal tar pitch must be melted to fluid consistency for application. It is used as a protective coating for metal and is usually applied by dipping. It becomes soft at 120 to 160°F (48.9 to 71°C) and therefore will sag where exposed to sunlight. Also, it becomes brittle at temperatures below freezing and is susceptible to cracking and disbonding. Because of these characteristics it has limited use.

Driers. Paint driers are added to the vehicle of oxygen-convertible paints that contain drying oils to speed up the drying of the paint or varnish film. Organic salts (e.g., naphthenates, resinates, octoates, linoleates) of heavy metals such as lead, cobalt, calcium, and manganese are the most commonly used ones. They are actually catalysts. When dissolved in the vehicle, they absorb oxygen from the air (since metals oxidize faster than oils) and pass it on to the oils in the vehicle, thus speeding up the oxygen conversion of the oils and consequently the drying process. Usually, two or three of the metal salts are combined to obtain the proper drier. For proper formulation of paints on the job, driers should be used with extreme caution and in limited amounts. Most paints are now manufactured with the correct amount of drier already added. (*See* Table *P6*.)

Types and Uses

In the construction field, organic coatings may be classified in two ways: one, according to use, or type of surface to which it is applied; and two, according to characteristics of the coating itself. Thus organic coatings may be divided into three categories according to use: (1) paint for wood, (2) paint for metal, and (3) paint for concrete, plaster, masonry, and similar materials. These are described in detail below under specific headings (*see* Painting of Concrete, Plaster, Masonry, and Mis-

Table P6 Driers used in organic coatings

Type of driers	Advantages	Disadvantages
Calcium salts	Fast and hard drying	Cannot be used alone but usually with cobalt
Cobalt salts	Fast and powerful	Cause surface drying and wrinkling
Iron salts	Used only in dark baking enamels	Have dark color
Lead salts	Completely hard drying throughout the film	Are too slow[a]
Manganese salts	Fast and hard drying	Cause brittleness and brown or pinkish discoloration to light colors
Zinc salts	Increase final film hardness	Delay initial setting time when used with other materials

[a]According to federal and state requirements, there must be less than 0.5% lead base on the dried film of paint intended for use in or upon people's living areas of any kind.

cellaneous Surfaces; Painting of Metal; Painting of Wood).

Organic coatings can be categorized according to their characteristics into (1) pigmented coatings, including paints and pigmented enamels; (2) clear coatings such as varnishes, lacquers, shellac, sealers, and wax polish; (3) bituminous coatings; (4) cement mortar coatings; (5) plastic and synthetic rubber coatings; (6) rust-preventing coatings; (7) heat- and fire-resistant coatings; and (8) various special categories such as coatings resistant to fungi, mold, insects, bacteria, or marine environments, to mention the most common ones. These organic coatings are applied for both exterior and interior use.

Pigmented Paints. These are available as paints, enamels, and baked enamels. They usually consist of white paint solid and a vehicle and include house paints, machinery paints, red lead paints, metallic paints, colored lacquers, and certain specialized products such as vinyl-resin paints. Today almost all pigmented paints are ready mixed for immediate application except for stirring, adding activator, or thinning. Almost all paint manufacturers have set up calibrated color systems with a very large selection of colors. Each color has an exact formula which can be repeated, thus ensuring indefinite reproduction of the same color. These developments have come about because today even normal use requires a carefully formulated type of paint to meet specific requirements, and a painting contractor cannot be expected or required to formulate paints from raw ingredients, nor is he equipped to do so.

Enamels in the strictest sense are pigmented paints that use varnish as the vehicle. The use of fortifying resins in oil-based paints has almost eliminated the difference between paints and enamels.

Baked enamels are always factory applied because they require controlled elevated temperatures to form the dry film. They are usually thermosetting and based on synthetic resins. The usual temperature range is from 200 to 300°F (93.33 to 148.89°C). The time required for baking usually ranges from 15 to 60 minutes, although some enamels may require only 1 minute and others several hours. The film formed is insoluble in the solvent from which it was applied, and it cannot be softened appreciably by heat. These enamels are available in a very wide range of colors, and the film they form is hard, durable, washable, and resistant to commonly used alkalis and acids. They can also be formulated to meet specialized end uses.

Clear Coatings. These generally do not have the durability of pigmented coatings in outside exposures and are used primarily to beautify and protect surfaces without obscuring their natural appearance.

Varnish is usually a combination of drying oils and fortified resins, either natural or synthetic, which dry as a result of chemical action induced by air drying or baking and by evaporation of the solvent, followed by oxidation and polymerization of the resins and drying oils. The common solvents for varnish are turpentine and mineral spirits. Varnish is used not only as a clear coating but also to a great extent as a vehicle of pigmented paints for quick drying and smooth leveling. The types of oil and resin and the ratio of oil to resin are the factors that control the properties of varnish. The oils contribute to the elasticity and the resins to the hardness in the finished film.

Varnish that is used as the vehicle for many of the pigmented paints, except exterior house paint, is formulated to meet the end use of the paint. It is used and handled almost entirely by the manufacturers of paints.

Lacquer is the term frequently applied to any coating that dries quickly and only by evaporation of the solvent.

A true lacquer has nitrocellulose as the basic nonvolatile ingredient and is characterized by fast drying and distinctive odor, with resins, plasticizers, and drying oils added only to improve the adhesion and elasticity (flexibility) of the film. The term "lacquer" also includes any air-drying or baking-type clear composition that is generally based on nitrocellulose or modified cellulose resins. Plasticizers only improve the flexibility of the lacquer film and are either specially treated oils or chemical compounds such as blown castor oil or dibutyl phthallate. The solvents for lacquers are certain acetates, ketones, and alcohols. They dry in 5 to 15 minutes and to a firm film in $\frac{1}{2}$ to 4 hours.

Lacquers are available clear or pigmented and are usually applied by dipping for small objects or by spraying. The film is tough and thin, but not as durable when exposed to sunlight and moisture as that formed by high-grade varnishes.

Shellac varnish is a solution of refined lac resin in denatured alcohol which acts more like a lacquer than a varnish, as it dries quickly by evaporation of the alcohol. It is available in white (bleached) and an orange color range. It is furnished in "cuts," that is, the ratio of pounds (kilograms) of resin to 1 gal (1 liter) of alcohol. Cuts of 4, 4.5, and 5, which cover the range of light, medium, and heavy-bodied shellacs, are the ones most commonly used. Shellac is used to provide a clear finish for woodwork, to seal knots and pitch stains in wood before painting, and to seal bituminous coatings before applying pigmented paints. Shellac should not be kept more than 6 months in liquid form.

Coal tar pitch (plain or plasticized) when combined with a mineral filler is known as coal tar enamel. It will not sag at 160°F (71.11°C) and will not crack or disbond down to -20°F (-28.89°C). It is applied hot.

Cold-applied coal tar paints are made by heating coal tar pitch to a liquid and then combining it with a solvent derived from coal tar distillation, generally xylol or coal tar naphtha. There are various types, some using coal tar pitch and others processed pitch, together with solvents and with or without fillers added. Those with common coal tar pitch give a coating similar to common hot-applied coal tar pitch. Those made with processed coal tar pitch, with or without fillers, are superior and give a coating similar to a coal tar enamel. Processed coal tar pitch paint is thixotropic and can be applied in heavy coats. ("Thixotropic" means the ability when undisturbed to become stiff or heavy-bodied or set and when stirred to become more fluid.)

Coal tar emulsion paints use water as the dispersing medium and common or processed coal tar pitch as the nonvolatile ingredient. They adhere to damp surfaces and are almost odorless. They are not as watertight as the organic solvent types but show better resistance to sunlight. For protecting iron, steel, and concrete that is to be installed under or in fresh or salt water, an epoxy-coal tar combination, two-compound enamel is available in various formulations. These epoxy-coal tar combinations can be applied by brush or roller, but the best method is spray application.

Asphalt coatings are available as enamels, cold-applied paints, and emulsions. In comparison to coal tar products they are generally more resistant to weathering, less susceptible to temperature extremes, and less resistant to moisture penetration. They are used in the same manner as coal tar products. Asphalt enamels are generally applied by hot-dipping to ferrous metal and are extensively used for waterproofing masonry surfaces. Emulsion and cold-applied asphalt paints are used for dampproofing and waterproofing masonry surfaces and for coating ferrous metals where hot dipping is not practicable.

Cement Mortar Coatings. These consist of portland cement and lime with water as the vehicle, and with color pigments, water repellents, and extender pigments as the white paint solid. In this case the cement and lime also act as white paint solid. Cement mortar coatings are used as dampproof coatings for masonry surfaces and also on steel as protective coatings. When portland cement hydrates in contact with steel, the calcium hydroxide which is liberated prevents rusting. Cement mortar coatings containing latex mixed with Type 1 portland cement, with or without sand, are used for areas subject to continual wetting by water or continued high humidity. (*See also* Mortars.)

Stains. Since stains are composed of a pigment (white paint solid) and a vehicle, they are also considered a type of pigmented paint. However, they have a low pigment content so that they will not obscure the natural grain of the wood and are further differentiated by their low viscosity and high penetrating properties. Practically all wood that is to be finished clear is stained to produce a particular coloring or to obtain uniformity in coloring (this is known as shade staining).

There are four general types of stains:

1. *Water stains*, consisting of dry ingredients with water as the vehicle. These are the most permanent but raise the grain of the wood.
2. *Non-grain-raising stains*, consisting of dry ingredients with various alcohols as the vehicle. These are not as permanent as water stains.
3. *Oil stains*, consisting of dry ingredients with drying oils and solvents as the vehicle. These have good color,

but fade in sunlight and tend to bleed into the finishing coat.

4. *Pigmented stains*. These are similar to oil stains but are applied and then wiped off to produce the effects desired.

Special Types of Organic Coatings. Today films are obtained not only by the five methods already mentioned but by other, new means such as (1) flame spraying, which is a means of depositing a finished film directly on the surface; (2) mixing the ingredients in a two-part combination where one part acts as a catalyst and a film is formed by chemical reaction; (3) powdered resins which are held in fluffed suspension by air, preheated materials receiving a finishing film by being placed in this fluffed suspension for a few seconds; and (4) the collodial dispersion of synthetic resins in a plasticizer, with or without solvents and other materials. These last are called plasticols and can form films on surfaces directly. Energy-curing and ultraviolet methods are also used.

A plastic (polyethylene) and certain synthetic rubbers (neoprene and Thiokol) are used where toughness, flexibility, and high resistance to the natural elements and to chemicals are of utmost importance, and where other types of coatings are ineffective. Polyethylene is so inert that it is applied by flame spraying. It is carried as a powder in a stream of air through an oxygen-propane flame, where it melts and is deposited as a continuous film on the surface to be finished. Thiokol is also applied in this manner, but both neoprene and Thiokol are also available in liquid form for brush or spray application.

Multicolored organic coatings (enamels) are interior paints in which droplets of paint of various colors are suspended by means of controlled surface tension in an aqueous phase (medium). When this special paint is applied by spraying, these multicolored droplets give accent colors on backgrounds of other colors. These enamels are generally nitrocellulose lacquer formulations.

Experimental nonlacquer vehicles with mineral spirits as solvents have been successfully used in multicolor formulations and are now available commercially. Their advantages lie in their great hiding power, that is, their ability to hide irregularities in surfaces with one coat. They are used successfully on masonry (especially concrete block), plaster, paper pulp and fiber materials, and wood.

Plastics, epoxy, urethane, latex, and polysulfide in various combinations and formations are available as sealants, vapor barriers, toppings, waterproof membranes, mortars, waterproofing floor sealers and finishes, and ceramic-type wall finishes. It is advisable always to check with the manufacturers for the best type of material for a special end use (*see also* Adhesives; Admixtures; Mortar; Plastics).

A peculiar competition now exists between two methods of achieving exactly the same finish: one is the lamination of a vinyl sheet onto a surface; and the other, the application of the same kind of coating by means of a plastisol formulation, that is, collodial dispersion of synthetic resin in a plasticizer.

Rust-Preventative Paints. Protective paint for ferrous metal must be characterized by rust-inhibitive properties, low permeability to corrosive agents, low absorption of water, and the ability to wet the surface to which it is applied (*see* Metals: Coatings). These paints are of two types: (1) primers which are a base for finished paints, and (2) finish coatings. (*See* Table *P7*.)

Fire-Resistant and Flameproof Paints. These paints are of two types: (1) those that will not support combustion and therefore will check flame spread, and (2) those that not only will not support combustion but in addition will swell up and stop heat transfer to combustible materials. The second type by definition is called intumescent paint. Materials which have been painted thus actually create a fireproof area when affected by heat.

Intumescent coatings are available with fire ratings approved by the Underwriters' Laboratories, Inc. The fire ratings are based on the number of coats of intumescent paint. These coatings will develop up to a 2-hour fire rating on structural steel and will give class A protection to wood and plywood. An intumescent spray-on coating with up to a 4-hour rating is available for fireproofing structural steel.

Mold-Resistant, Antibacterial, and Insecticidal Paints. Most paints can be made resistant to mold, rot, and insects by adding counteracting ingredients. Bactericidal agents for paint are available, but they have the same characteristics and properties as those used for mold, rot, and insects.

Application

Organic coatings are applied by various methods, depending on the material to be coated and the type of coating to be applied. Paints, varnishes, and lacquers are usually applied by brush, roller, or spray gun, whereas baking enamels are generally applied by mechanical methods.

These mechanical methods include not only spraying but also dipping, curtain flow, and roller coating. There

Table P7 Ingredients used for rust-inhibiting organic coatings

Type of pigment	Color	Rust-inhibiting qualities	How used in paints	Major use
Lead chromate	Red and yellow	Extremely good	60% lead chromate and 40% litharge with drying oils or resins	Priming paint for ferrous metals
Lead oxide: litharge	Orange-red	Extremely good	As part of red lead pigments or mixed with drying oils or resins	Priming paint for ferrous metals
Lead oxide: red lead	Orange-red	Alkaline nature neutralizes acidic corrosive agents which can penetrate the film or which result from deterioration of the film	Mixed with linseed oil or phenolic type varnish; ready mixed paints should have 97% to 98% grade red lead pigments; job-mixed uses 85% and 95% grade red lead pigments	Priming paint for ferrous metals
Basic lead silicochromate	Off-white	Very good	Used with vehicles that can be solvent- or water-based, air-dried, or baked	Primers and finish coats for iron and steel
Zinc chromate	Yellow	Slight water solubility of compound releases rust-inhibiting chromate ions	As part of nonvolatile ingredients of the vehicle; best result with zinc chromate with low concentration of chloride and sulfate	Priming paints for ferrous metals
Zinc dust[a]	Gray	Anodic to iron and therefore protects steel or iron by sacrificial action; quantity must be sufficient to maintain electrical contact between zinc and steel or iron	Mixture of 80% zinc dust and 20% zinc oxide added to nonvolatile ingredient of vehicle	Rust-inhibiting paints, underwater paints (for fresh water)
Zinc molybdate	White	Anodic to iron and steel	Part of nonvehicle ingredient of vehicle	Rust-inhibiting paints

[a]Zinc dust is used with chlorinated rubber, catalyzed epoxy resins, and other vehicle types for organic zinc-rich primers.

are four basic methods of spraying: (1) spraying using compressed air or other gases; (2) hot spraying, using heat to thin the coating to spraying consistency; (3) airless spraying using hydraulic pressure; and (4) electrostatic spraying, whereby the material to be sprayed is electrically charged to attract the sprayed material.

Lacquers and enamels are cured either by forced drying, using temperatures up to 200°F (93.33°C), or by baking, using temperatures over 200°F (93.33°C) and up to 600°F (315.56°C).

A paint or organic coating must be chosen to meet its end uses, be they decorative, protective, or both. Its proper use in any building is one of the major concerns of the architect. Painting therefore is discussed separately and in great detail (*see* Painting; Painting of Concrete, Plaster, Masonry, and Miscellaneous Surfaces; Painting of Metal; Painting of Wood).

Accessory Materials. In paint and painting the following accessory materials are used and therefore briefly described here:

Colors-in-oil, or concentrated pigments dispersed in drying oil, usually to a semipaste consistency. They have generally been replaced by universal colorants, but are still used when a particular color is desired and is hand-mixed. Because of their intense color only small quantities are needed. (*See also* Color Pigments.)

Driers such as "liquid drier," "oil drier," and "Japan drier," which are driers in liquid form to be added to hand-mixed paints to accelerate their drying.

Putty, composed of whiting and drying oil with a small percentage of solvent, drier, and sometimes white lead. This is used to fill holes and cracks before applying paint. (*See also* Putty.)

Glazing compounds, including putty, the most common one. In some regions the painter still does the glazing—hence the overlapping use and specification of materials in glazing and painting.

Caulking compounds, which are similar in composition and properties to glazing compounds and are used to seal joints where metal or wood meets masonry (*see also* Caulking).

Paste wood filler, composed generally of ground silica, quick-drying varnish, and solvent, with or without color pigments. This is used to fill the pores of open grain woods, especially wood floors, before applying finish coatings.

Paint and varnish removers, composed of either a strong caustic soda (lye) solution or a mixture of organic solvents such as alcohols, acetone, toluene, xylene, and methylene chloride with paraffin to slow evaporation. Both types loosen the paint or varnish to the point that it can be scraped off.

Condensed Checklist

1. One should always check the type of material upon which the paint is to be applied and consult with the paint manufacturers to select the type of organic coating that is best adapted for the material to be painted.
2. The type of organic coating to be used should always be specified either by official name, trade name, or the required composition of the white paint solids, vehicles, and thinners.
3. For specialized coating problems such as rust inhibition, flameproofing or fireproofing, dampproofing, or chemical resistance, one should always contact the paint manufacturers and other specialists in these fields in order to select the correct organic coating for a particular end use.
4. When selecting colored paints, particularly for the exterior, one should always determine the exact nature of the pigment used to make sure that colors will not fade or change in the sunlight. On the interior, the type of artificial light used for illumination should be determined, particularly in the case of fluorescent, mercury, and sodium-type lighting, to make sure that the colors chosen will harmonize and create the effect intended under whatever lighting conditions are encountered. The reflectance of the color should help give the desired overall effect. Paint manufacturers should be consulted so that the correct color pigments will be selected to meet any special lighting conditions.
5. When metallic (powdered metal) paints are used, the materials and especially the metals that are to be painted always should be checked to make sure that no galvanic action will be set up between the material to be painted and the metallic ingredient of the paint. The same precaution is needed when other paints are to be used with metallic paint.
6. In the case of materials that are to be prefinished and painted in the mill, factory, or shop before installation within the building, it is always advisable to consult with paint manufacturers to obtain the best system for the particular end use.
7. Local, municipal, and state codes and the codes of the Fire Underwriters, insurance companies, labor departments, and federal government (Army, Navy, etc.) should always be checked for restrictions on the use of highly volatile thinners and solvents. They should also be checked for fire resistance ratings (i.e., the advantages and limitations) of flameproof, fire-resistant, and fireproof paints.
8. For wood finishes that are to receive heavy usage, for example, wood floors, both the manufacturers of paints and the manufacturers of the woods should be consulted so that the correct varnish, lacquer, or wax can be selected.
9. For resilient flooring, the manufacturers of the flooring material should be consulted for the correct type of wax or clear finish.
10. When iron, steel, wood, and wood products are to be fire-retarded, local, municipal, and state codes should be checked for regulations concerning the use of intumescent organic coatings for fireretarding and flameproofing.
11. When concrete, iron, and steel are to be installed under fresh or salt water, a check should always be made with paint manufacturers for the type of organic coating that should be used.
12. When organic coatings that have highly volatile vehicles or catalyzing compounds are used, the work areas where the coatings are being applied should be exceptionally well ventilated.
13. When using organic coatings one should always check that installation temperature and relative humidity meet the requirements specified for the particular organic coatings involved.
14. A check should always be made that organic coatings meet federal, state, municipal, and county health codes.

Conditions Favoring the Use of Paint

1. Where colorful, decorative, durable, washable and wear-resistant finish surfaces are needed.
2. Where materials, particularly ferrous metals, need a corrosion-resistant and rust-inhibitive coating.
3. Where a dampproofing, water-repellent, or waterproof coating is needed (above grade).
4. Where a colorful, hard, abrasion-resistant, heat-resistant, protective coating that is also resistant to common alkalis, acids, and fumes is required for prefinished materials.

5. Where a clear finish is needed for materials, particularly wood, where the natural grain will be protected and visible.
6. Where fire-retarding and flameproofing of iron, steel, and wood and wood products is required.
7. Where a colorful, durable, abrasive-resistant ceramic-like finish is required on walls.
8. Where composition-type roofing, asphalt shingles, corrugated iron, terne, steel, and aluminum need revitalization, protection, and preservation.

Conditions Unfavorable to the Use of Paint

1. As a waterproof coating where there is aquastatic water pressure, unless a specific type of organic coating is recommended and proved to be effective for this particular end use.
2. Where the surfaces will receive extremely rough usage and are subject to impact, whether constant or intermittent.
3. Where a clear, maintenance-free finish is desired on exterior wood.
4. Where the floor surfaces will receive little or no maintenance, and traffic will be very heavy.
5. Where no repainting or refinishing of the surface is to be done.

History and Manufacture

The first use of paint antedates any written records, even those of cave paintings, painted designs on earthenware pots, and the use of paint for ceremonial dances and religious festivals. From ancient times until fairly recently the formulation of paint had changed only slightly in principle, but with today's discoveries and researches in the field of plastics, rubbers, and adhesives, the formulations now prepared are entirely different from any that were known before.

The early Egyptians used lime, ochres, lampblack, white lead, and secret formulated substances as the white paint solids, which they combined with natural resins (drying oils), pitches (asphalt), egg yolk and egg white (casein), and waxes (beeswax) for the binder (vehicle). Among most ancient cultures the manufacture of paint was a secret art, and the paints were used principally for decoration. The same general pattern of development, use, and manufacture could be traced in the Orient, among the Maya of Central America, the Inca of South America, and the Aztec of Mexico. Although the Greeks were perhaps the first to recognize the value of paint as protective coating, its decorative and artistic aspects were still more important to them.

Thus the artist's use of paint governed its history until about 1700, when the other functions of paint began to assume an equal or greater importance. From this time on, these other uses, such as protection against corrosion in metals, dampproofing of masonry, protection of wood against weathering, maintenance of hygenic conditions, and control of illumination, to give a few examples, became the major functions of paint.

The discovery and development of white pigments almost parallels the development of paint for protective coatings. White lead has been known since about 2500 B.C.; zinc oxide came into use in Europe in 1840, basic lead sulfate in 1855, and titanium dioxide in 1918.

Although the production of synthetics started in 1870 with the plastic, celluloid, it was not until after World War I that their use in paints was considered. Today the entire field of paints and painting is being constantly changed by the use of synthetics.

Manufacturing Process. No simple description of the manufacture of paints is possible because of the immense variety of paints that are being manufactured. Today the laboratory is the starting point in the manufacture of paint. Here formulas are developed and given severe tests for application, use, and manufacture. When a formula meets these tests successfully, it is in many cases tested further in a pilot plant before it actually goes into large-scale plant production.

Mixing Operation. The actual manufacturing process starts with the mixing of the raw materials. Let us follow the process for house paint. The laboratory formula gives the quantities of raw materials and the sequence in which the raw materials are to be placed in the mixer. The ingredients might include a vehicle consisting of several oils or varnishes, wetting agents and sometimes thinners, and a white paint solid consisting of dry white pigments, colored pigments, and extender pigments. The mixer is a symmetrically shaped steel vessel with a motorized vertical shaft to which are attached blades; to the sides of the vessel are attached stationary blades.

The fundamental purpose of the mixing operation is to distribute the ingredients into a uniform paste and to wet the white paint solids with the vehicle as thoroughly as possible. From the mixer the paste passes on to the grinding operation.

Grinding Operation. The grinding operation is actually a high-speed dispersion operation. This operation is necessary because the fine dry pigments of the white paint solid, although already ground fine enough, tend to gather together into larger particles. The grinding operation disperses these large particles and causes them to be wetted by the vehicle.

Thinning, Tinting, and Straining. The last steps are thinning, tinting, and straining before the paint is placed in containers, labeled, and packaged for use. The paste from the grinding operation has to be thinned to the final desired viscosity. As it arrives in containers from the grinding operation, a mixer similar to an egg beater is lowered into the container. The formula may call for additional quantities of varnishes, oils, thinners, driers, and other liquids, but never any dry ingredients. These liquids are added, and the mixer stirs all into a uniform thinned-down mass. Tinting is done at this stage. The formula describes approximately the right amount of tinting color to add; but rarely is the color exactly as desired, and from here on the paint passes through quality control checks for adjustments until the final shade is reached. When the paint has been thinned and tinted, it is strained and put into cans which are labeled and packaged for shipment.

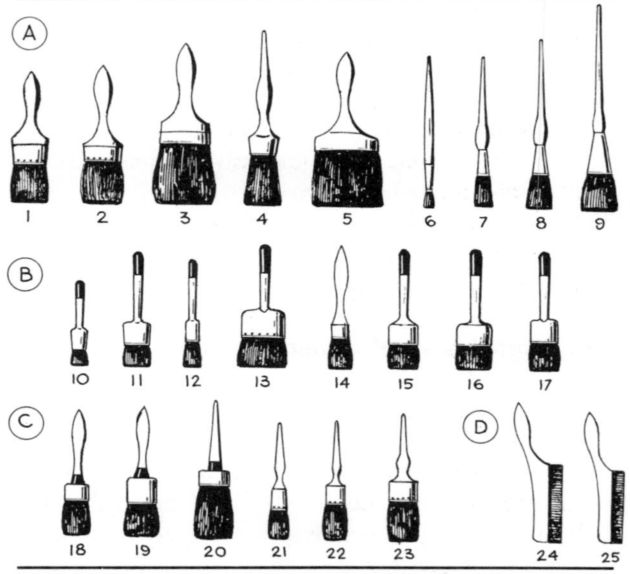

Figure P1 Brushes for applying paint.

PAINTING

Painting may be defined as the application of organic coatings (paints) to surfaces that are to be decorated or protected, or both.

Types and Uses

Painting, or the application of coatings, in construction has become so complicated that it is divided into four major types: (1) on-site painting; (2) mill, shop, or factory painting; (3) industrial painting; (4) maintenance painting.

Tools and Techniques. A brief review of the tools and techniques currently used in professional painting clearly illustrates this complexity.

On-site painting is generally done with brushes or rollers or by spraying the paint with a spray gun (*see* Figures *P1* to *P4* and Table *P8*).

Water-repellent, waterproofing, and dampproofing coatings are applied by any one of the following means: brush, spray, trowel, or mop, depending on the type of coating and its viscosity and on whether it is applied hot or cold.

Mill, shop, or factory painting is accomplished by a wide variety of methods. The type of method is controlled by the type of material to be coated and the end use of the material when installed in a building. Some of the more common methods are (1) dip application, in which the material is dipped into the coating; (2) roller coating, in which the coating is applied to the material by passing through rollers; (3) tumbling, in which small objects, along with the coatings, are placed in a metal barrel which then rotates; (4) centrifuging, in which small objects placed in a perforated container are dipped in the coating and then placed in a centrifuge; (5) spraying, done in special booths with spray guns; (6) silk screening, in which the coating is applied through silk screen stencils onto the material; (7) knife coating,

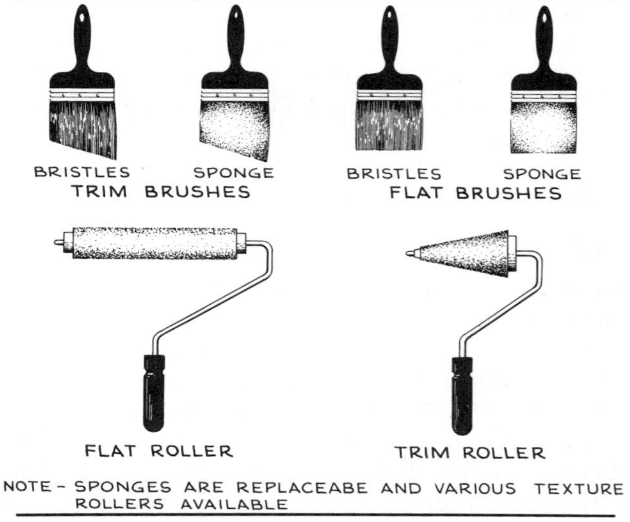

Figure P2 Typical types of rollers and special types of brushes.

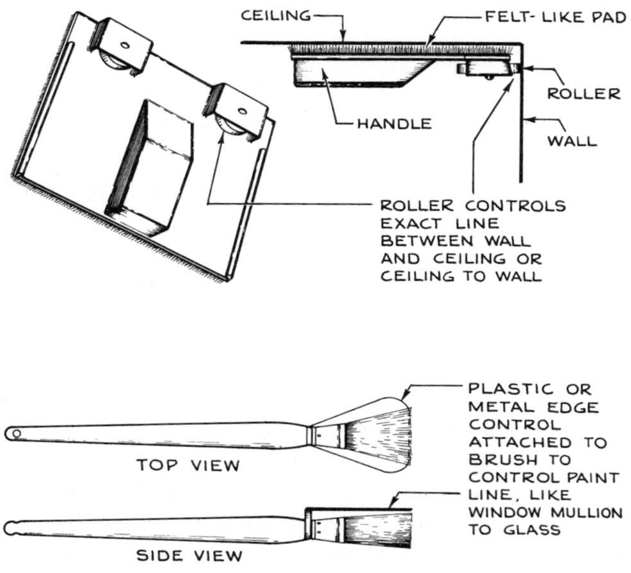

Figure P3 Types of roller and brush for controlling edge of paint.

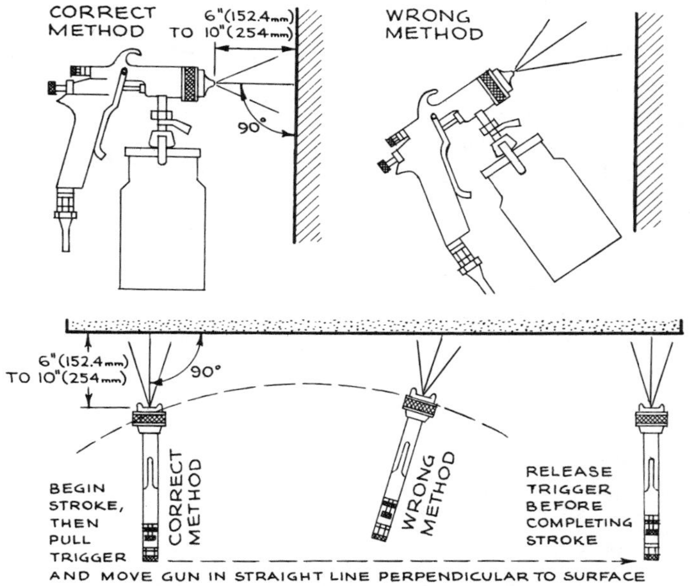

Figure P4 Methods of spraying paint.

Table P8 Types of organic coatings (paint) and methods for on-site painting

Type of paint	Where used	How applied
Asphaltic paints	Masonry and concrete walls; iron and steel	Brush, mop
Caulking compounds	Sealing exterior joints between materials	Caulking gun
Cement floor paint	Interior cement floors	Brush, spraying
Clear finishes	Exterior and interior wood surfaces	Brush, spraying, roller
Clear finishes for floors	Interior wood floors	Brush, spraying, roller
Dampproofing	Exterior and interior concrete or masonry above or below grade	Brush, spraying, trowel, mopping
Enamels	Interior trim, doors, windows, etc., and areas where there is cooking and moisture	Brush, spraying, roller
Filler	To fill holes, cracks, and pores of wood	Brush, hand, small trowel
Flat paints	Interior walls and ceilings	Brush, spraying, roller
House paints	Exterior of wood finished structures	Brush, spraying, roller
Lacquers	Wood cabinets; furniture	Brush, spraying
Masonry paints	Exterior and interior concrete, masonry, stucco, or cement surfaces	Brush, spraying, roller
Metallic paints	Exterior and interior as protective undercoating or finish or for solar reflection	Brush, spraying
Multicolored paints	Interior walls	Spraying, brush
Porch and trim paints	Exterior trim, doors, windows, railings, etc.	Brush, roller
Putty	Filling small holes in wood	Hand, small trowel
Rust-inhibiting paints	Exterior and interior ferrous metals	Brush, spraying, roller
Shingle stains	Exterior wood shingles and siding	Brush, cloth, sponge, spraying
Stains	Exterior and interior wood	Brush, cloth, sponge
Waterproofing	Exterior or interior walls below grade (not used when aquastatic head is present)	Brush, spraying, trowel, mopping
Water repellants	Exterior above-grade concrete, masonry, etc.	Spraying, brush, roller

Table P9 Application of finishes in mill, shop, and factory work

Product	Type of paints	How applied
Furniture, metal	Lacquers, enamels, baked enamels	Spraying, heating
Furniture, wood	Varnishes, lacquers, stains	Spraying, brush
Millwork and cabinetwork	Varnishes, enamels, lacquers, stains	Spraying
Panel	Baked enamels	Spraying, heating
Sheet, board, and integrant	Lacquers, varnishes, stains, enamels, baked enamels	Spraying, rolling, knifing, dipping, heating, brush
Shingles and siding	Shingle stains	Spraying, brush
Shop priming coats	Rust-inhibiting coating	Brush, spraying, dipping
Signs and decoration	Enamels, lacquers, baked enamels	Silk screen, heating
Specialized equipment	Lacquers, enamels, baked enamels	Spraying, heating

in which the paint is drawn under a knife or blade that is in contact with the material; (8) calender coating, in which a dry coat is applied with heat and rollers to the material. (*See* Table *P9*.)

In any case, good painting always depends on the proper preparation of the material to receive the paint, the selection of organic coatings best suited for this material, the use of the correct method or sequence of application of these organic coating materials, and the control of temperature, humidity, and general physical environment where the painting is to be done.

Industrial Painting. Industrial finishes and coatings include those for (1) factory maintenance; (2) painting for special conditions such as corrosion resistance and marine environments; (3) painting in the railroad and transportation industries; (4) painting of durable goods, wooden goods, insulation and paper, textiles, and leather; (5) tin decorating; and (6) other specialized types of painting in which other than the general types of site painting and shop or factory painting are necessary. In all such work one should consult with the paint manufacturers, the material manufacturers, and specialists in the particular fields to decide on the correct type of organic coatings and methods of painting to meet the conditions involved. Also included in this category are fire-resistant, flameproof, and fireproof paints and chemical-, bacterial-, and marine-resistant types of finishes.

Painting in Ordinary Construction. In the construction field, there are three different types of painting in normal construction: (1) on-site painting; (2) finish painting done completely in a mill, shop, or factory on materials that arrive ready to be installed as completed work; and (3) mill, shop, or factory prepainting on materials that arrive ready to receive on-site finish coatings.

Types of Materials Painted. Painting can be further broken down into the types of materials that are to receive the paint: painting of wood, painting of metal, and painting of concrete, plaster, and miscellaneous surfaces, including masonry (*see* specific headings under Painting).

Applications

Mill, shop, or factory painting is usually done under closely controlled conditions in which proper temperature, moisture, and humidity and correctly prepared material surfaces can be realized with comparative ease. The finish should be the best possible.

In on-site painting, surfaces have to be prepared; work areas have to be cleaned and controlled while painting is done and while the organic coating dries; and the organic coating has to be mixed, stored, tinted, and thinned. Painting is affected by temperature; best results are obtained at 60 to 80°F (15.56 to 26.67°C). Good ventilation is also necessary, especially when strong solvents are used. With so many variables and requirements, it is obviously very difficult to obtain good painted finishes with on-site painting.

Maintenance Painting. All painted surfaces have a relatively short life and therefore repainting and maintenance painting are always necessary. The important items that should be checked are as follows: (1) that all loose paint is removed; (2) that all blisters, cracks, etc., are removed and leveled; (3) that any surface irregularities in the existing painted surface are filled, sanded, and puttied so that the entire wall surface is as clean and smooth as possible; (4) that all stains are removed before new coats are applied; and (5) that the new organic coating is of a type that can be applied over the existing paint or coating. If it is necessary to return to the original surface, paint can be removed with paint or varnish remover, wire brushes, burning, sanding, sandblasting, and in the case of water paints, washing.

Condensed Checklist:

1. The type of painting required should always be specified. There are three types:

   Type I or "recommended painting"—this type can be recommended by a contractor; it has the usual life expectancy of a well-painted surface.
   Type II or "de luxe painting"—this type affords the best finish possible and provides the maximum durability that can be reasonably expected from a painted surface.
   Type III or "minimum painting"—this type is merely a marginal finish; the appearance, durability, and protective quality are not the maximum that can be obtained.

2. Local painting practices should always be checked so that the types of organic coatings to be used and the methods of application of the coatings are in accordance with the effective regulations and limitations. For example, certain cities and localities prohibit the use of any other painting method except by brush, and in some cases even the maximum size of the brush is controlled.
3. It is advisable to specify, if possible, the total number of colors that are to be used on the exterior and on the interior. The number of colors can be designated as follows: for exterior painting, 5 colors consisting of 3 deep colors and 2 tints; for interior painting, 10 colors consisting of 4 deep colors and 6 tints. An alternative method is to specify by rooms: all rooms, maximum of 3 colors; halls and stairs, maximum of 2 colors; closets and storage areas, maximum of 1 color. This information can also be incorporated into the finish schedule.
4. If possible, factory-tinted colored paints or tinted colors formulated by mixing standard factory-tinted colored paints by $\frac{1}{2}$-pint, pint, quart, gallon, or 5-gallon combinations should be used in preference to mixing colors by hand at the site of construction.
5. One should always make sure that painting for wood, painting for metal, and painting for masonry surfaces including plaster are clearly and separately specified so that the type of organic coating, preparation of the surface, number of coats, and method of application are completely covered for each category.
6. One should always make sure that all the surfaces and materials which do not require painting, and all millwork, cabinet work, equipment, etc., which are to be prefinished in mill, shop, or factory and require no further painting, are clearly shown or specified as "not to be painted."
7. One should always make sure that no exterior on-site painting is performed at temperatures below 50°F (10°C), or in rain or snow, or when surfaces are damp or covered with frost, unless the organic coatings are of a type that can be applied over damp or wet surfaces. If painting must be done in near-freezing weather (but never at or below 32°F, or 0°C), the paint must be warmed (never above 100°F, or 37.8°C) to improve application and wetting properties. The paint manufacturers should be consulted for their recommendations to meet this condition.
8. One should always make sure that interior on-site painting is done at temperatures above 50°F (10°C). If temporary heating is used, adequate ventilation must be provided. Open-flame heating (salamanders) should not be used because of the explosive qualities of the solvents and thinners for paints.
9. A check should always be made of organic coatings for lead content, as federal and state codes require less than 0.5% lead content, based on the dried film of paint, for use on residences and dwellings of any kind.
10. One should always specify the necessity for adequate ventilation for organic coatings having highly volatile ingredients.
11. A check should always be made that the organic coating to be used is not hazardous to the workman applying it; if it is hazardous, methods for protecting the workman should be specified.

Conditions Favorable to the Use of Painting

1. For both the exterior and interior surfaces of materials that cannot be left in their natural finished states and where color and decoration are of major importance.
2. For both the exterior and interior of buildings—on the exterior where protection from the elements is necessary, and on the interior for materials that can be damaged by moisture (not excessive) or by cooking, broiling, or toasting equipment.
3. For corrosion-resistant coatings on ferrous metal on the exterior or interior.
4. For all types of interior and exterior wood where the natural grain of the wood is to be visible.
5. For dampproofing and making water-repellent exterior surfaces both above and below grade. For similar interior surfaces, always consult the paint manufacturers in order to select the type of organic coating best suited for the existing condition.

6. For finishing interior wood or cement floors. Always consult with the manufacturers of both the flooring material and the paint in order to select the type of organic coating best suited for the condition.
7. For fire-retarding and retarding flame spread on combustible materials and for fire-retarding up to 4 hours on iron and steel.
8. For finishing interior masonry walls with a ceramic-type, hard, durable, abrasion-resistant surface.
9. For organic coatings that will require the minimum of maintenance. Always consult with paint manufacturers for this characteristic, as well as for recommended methods of application and maintenance.

Conditions Unfavorable to the Use of Painting

1. For waterproofing where any aquastatic pressure is present or may develop, unless a specific type of organic coating is recommended for this particular end use.
2. For fire-resisting, flameproofing, or fireproofing materials unless local, municipal, and state codes as well as the codes of the Fire Underwriters, insurance companies, and federal specifying agencies are thoroughly checked for limitations and requirements covering the use of intumescent types of organic coating.
3. For any specialized protective coatings for resistance against acids and alkalis, mold, rot, or insects and for bactericidal purposes unless specialists in these fields within the companies that manufacture paint or independent professional specialists are consulted.
4. For permanent dampproofing or waterproofing of exterior surfaces above grade, where transparency and color are of major importance.
5. For a relatively permanent stain on exterior natural woods, particularly on cypress, cedar, white fir, and redwood.

History and Manufacture

Painting other than creative painting by artists became a trade in architecture and construction sometime during the 15th century. Although the painting of buildings was fairly common in early times, its purpose was primarily artistic until about 1700. Since then, its protective function has grown in importance, and the manufacture of paint has become a major industry that now includes laboratories, research centers, pilot plants, and experimental testing centers in addition to the huge plants for mass-producing paints.

Painting, which originally consisted of no more than mixing the raw ingredients at the site, adding color pigments, thinners, etc., and then applying the paint, has also increased in complexity. As a result painting is now not only a highly specialized trade but also an important industry.

Glazing by the Painter. When window glass first came to be used in relatively large quantities, its installation was part of the painting trade (even today in certain areas glazing is the responsibility of the painting contractor). In time, glazing also became a highly specialized and eventually a separate trade.

PAINTING OF CONCRETE, PLASTER, MASONRY, AND MISCELLANEOUS SURFACES

Physical and Chemical Properties

The painting of concrete, plaster, masonry, and miscellaneous surfaces is done primarily for purposes of decoration, lighting, and sanitation and not for protecting and preserving the construction materials, except in cases where above-grade masonry surfaces require coating to minimize water seepage through the wall. The type of paint used for exterior concrete work is also suitable for exterior concrete and cinder block, asbestos-cement materials, brick, stone, stucco, and cement plaster. All these materials have an element of alkaline reactivity to a lesser or greater extent, and this reactivity plays an important role in the selection of the type of paint used, the time of painting, and the pretreatment of the surface. Paints with a portland cement base or an oil base are commonly used, but latex-base, resin-emulsion, epoxy, and urethane types are also widely used. Before any type of paint is applied, any peeling, chipping, flaking of existing paint, and efflorescence must be removed. Muriatic (hydrochloric) acid solutions are generally used to remove efflorescence.

Painting the interior of outside walls made of concrete, cinder and concrete block, brick, or stone is more difficult as dampness can penetrate through the wall and damage any tight-sealing, nonbreathing type of paint. For all other interior walls made of these materials, any of the varieties of interior paints described in other sections on paint and painting can be used.

Interior painting of plaster, plaster integrants, and paper pulp and fiber sheets, boards, and integrants requires surfaces that are clean and from which dirt, dust, excess mortar, oil, grease, mold, and mildew are removed or neutralized. In many cases no pretreatment is necessary. With porous fiber and paper pulp boards, two primer coats may be necessary, and the same holds true for dark-colored paper pulp boards. In almost all cases the primer coat not only is base for the finish coat, but also is sealer and neutralizer of the cleaned surface.

Types and Uses

There are two types of painting of concrete, plaster, and miscellaneous surfaces: (1) on-site painting of any of these materials, and (2) painting done in the mill, shop, or factory of asbestos-cement materials, plaster integrants, and paper pulp and fiber sheets, boards, and integrants. Table *P10* lists the types of paint and pretreatment as well as the number of coats for painting these materials under either set of conditions.

Painting done in the mill, shop, or factory of plaster, paper pulp, and fiber sheets, boards, and integrants is usually one of two types: (1) a pretreatment for on-site painting, consisting of a preservative and sealer or primer coat; or (2) a baked enamel finish, using a high-grade enamel, so that the material can be installed within the structure as a finished material.

Application

Condensed Checklist

1. For exterior painting with an organic coating on concrete work, cinder and concrete block, brick, stone, asbestos-cement materials, stucco, cement plaster, paper pulp, and fiber materials, the person in charge should always make sure that the surfaces are clean,

Table P10 Organic coatings (paints) used for concrete, plaster, masonry, and miscellaneous materials

Type of paint	Surface preparation	Priming coat	Intermediate coat	Finish coat	Major use
Asphalt-base paint (emulsion type)	Surface free of foreign material and not dripping wet	Asphalt-base paint, (emulsion type)		Asphalt-base paint, emulsion type	Exterior dampproofing of concrete and concrete block below grade
Cement-base paint	Surface clear and fairly smooth; efflorescence removed and surface thoroughly wetted	Same as finish coat; slightly moisten priming coat before applying finish coat		Cement-base paint, white or tinted	Exterior and interior concrete, cinder or concrete block, brick, stone, stucco, and cement plaster
Chlorinated rubber paint		No primer necessary		Silicone paint	Exterior concrete, masonry, stucco, and cement plaster
Coal-tar paint	Surface free of foreign matter and thoroughly dry	Coal-tar paint	Coal-tar paint	Coal-tar paint	
Concrete-floor paint (varnish base or rubber base)	Clean, smooth surface etched with 10–15% muriatic acid solution	Varnish- or rubber-base primer		Varnish- or rubber-base cement floor paint	Interior concrete floors
Epoxy paint	Surface cleaned of all foreign matter	Epoxy primer		Epoxy 2-package or oil-modified epoxy	Areas where chemical fumes and possible spills and splashes occur
Epoxy-tar combinations					Submerged concrete exposed to fresh and salt water
Oil-vehicle paint; alkyd, linseed oil, oil-modified phenolics, epoxies, and urethanes	Thoroughly clean, dry smooth surface; efflorescence removed and pretreatment with chemicals to inhibit alkaline reaction, using 2% zinc chloride and 3% phosphoric acid-water solution	Oil-base primer coat tinted same color as finish coat		Oil-base paint, white or tinted	Exterior and interior concrete, cinder or concrete block, asbestos-cement materials, brick, stone, stucco, and cement plaster

Table P10 Organic coatings (paints) used for concrete, plaster, masonry, and miscellaneous materials (*continued*)

Type of paint	Surface preparation	Priming coat	Intermediate coat	Finish coat	Major use
Silicone paint	Surface thoroughly clean; dry; all stains, efflorescence, oils, and grease removed; holes, cracks, or loose masonry repaired and patched	No primer necessary		Silicone paint	Exterior concrete, masonry, stucco and cement plaster
Urethanes	Surface cleaned of all foreign matter			Urethane	Finish for concrete floors and as an anti-dusting treatment
Varnish-base paint	Surface clean, smooth, and dry	Varnish-base primer sealer		Varnish-base paint, white or tinted (glossy, semiglossy, matte)	Interior concrete, cinder or concrete block, brick, stone, plaster, plaster integrants, paper pulp and fiber sheets, boards, and integrants
Vinyl paint	Surface cleaned of all foreign matter	Porous material: apply coating of latex-Type I portland cement mortar		Vinyl paint	Protection of concrete, brick, and masonry surfaces against acids, salts, alkalis, corrosive chemicals and fumes, and moisture
Water-base paint (latex, acrylic, and resin-emulsion)	Surface clean, smooth, and dry	Same as finish coat		Water-base paint, white or tinted	

relatively smooth, thoroughly dry, with efflorescence removed, and treated to be alkali-inhibited or sufficiently aged before painting. For cement-base paints, the same precautions hold true except that the surface has to be thoroughly wetted before paint is applied.

2. For interior painting with organic coatings on concrete, plaster, and miscellaneous surfaces, the surfaces must be clean, smooth, or relatively smooth (not a glossy surface), thoroughly dry, with efflorescence removed, and alkali-inhibited (or sufficiently aged) before painting. For cement-base paints, the same holds true except that the surface must be thoroughly wetted before paint is applied.
3. For painting cement floors, the cement must be treated chemically before organic coatings are applied in order to give good penetration and adhesion.
4. When dampproofing concrete and masonry walls below grade, a check should always be made that no aquastatic pressure can develop in the future, as these dampproof coatings are not waterproof. The surfaces must be free of foreign matter, relatively smooth, and free of any loose mortar or concrete before the dampproof coating is applied. For the asphalt-base type, the surface can be damp but never dripping wet; for the coal-tar base type, the surface must be thoroughly dry.
5. When using a clear dampproofing or water-repelling coating on the exterior, one should always ascertain from the manufacturer of the coating what its life will be under the actual conditions of use and when it will become ineffective and have to be reapplied.
6. In general, the method of application and limits in regard to temperature, drying time, etc., for painting concrete, plaster, and miscellaneous surfaces are the same as those for painting metal and wood.
7. When using prefinished materials of the types listed here, one should always learn from the manufacturers the specific types of painted finishes supplied and then choose the material that meets the requirements and conditions of the end use.

Conditions Favorable to the Painting of Concrete, Plaster, Masonry, and Miscellaneous Surfaces

1. Where colorful decoration, good sanitary conditions, and improved lighting conditions are important for both exterior and interior surfaces made of these materials.

2. Where the decoration, sealing, and finishing of the surface of concrete floors is important.
3. Where dampproofing is necessary on concrete or masonry walls below grade.
4. Where a transparent water-repelling coating is necessary for concrete or masonry walls above grade.
5. Where a ceramic-type finish for concrete or concrete block walls that is decorative, hard, durable, and abrasion resistant is required.
6. Where the refinishing of asphalt shingles, corrugated iron, terne, steel, or aluminum is necessary or desired.
7. Where walls and floors must be protected from chemical fumes and possible spills and splashes.
8. Where brick, concrete, and masonry surfaces must be protected against acids, salts, alkalis, chemicals, fumes, and moisture.

Conditions Unfavorable to the Painting of Concrete, Plaster, Masonry, and Miscellaneous Surfaces

1. Where waterproof coatings are required on either exterior or interior concrete and masonry surfaces, where aquastatic pressure is present.
2. Where surfaces will receive extremely rough usage and abuse, unless a ceramic type of organic coating is installed.

PAINTING OF METAL

Physical and Chemical Properties

All metals used in construction corrode to some extent in natural environments. Some take on a natural protective coating which prevents further corrosion, and others corrode so slowly that they will outlast the life of the building. Iron and steel, on the other hand, are actually man-made, and their existence as such is contrary to nature. They are created from natural stable oxides of iron, and they therefore tend to revert to the oxide form (common rust is ferric oxide and ferric hydroxide).

In order for iron and steel to rust (corrode), three factors must be present: oxygen, moisture, and a combination of anodic and cathodic areas. Corrosion is induced and accompanied by the flow of current resulting from potential differences between a positively charged (anode) and a negatively charged (cathode) area and involves certain chemical changes. Today the expression "electrochemical" is universally applied to the corrosion process (*see* Metals).

Oxygen and moisture (in the form of water vapor, mist, fog, rain, and snow) are both present in the atmosphere. Therefore only in areas where the relative humidity is always below 30% will corrosion not take place, and unfortunately there are few if any inhabitable areas with this atmosphere on the earth's surface. Salt water, a conductor of electricity, accelerates electrochemical action.

All iron and steel (except the stainless steels and weathering steels) used in construction have areas on their surfaces where chemical and physical differences exist that will create anodes and cathodes; corrosion is thus inevitable if oxygen and moisture are present. The most important use of organic coatings for metals, then, is as a protective coating for iron and steel. It is absolutely necessary that the iron and steel surfaces be properly cleaned for the organic coating to give optimal protection.

Types and Uses

There are three types of painting on metal: (1) on-site painting; (2) finished painting done in a mill, shop, or factory; and (3) mill, shop, or factory prepainting preparation of the metal for on-site finish painting. In construction, on-site painting and the prepainting preparation of metal for on-site finish painting are of utmost importance. Finish painting applied by mill, shop, or factory is of lesser importance, in the sense that once the organic coating is selected, only on-site checking or a check of the painted material or product by an independent laboratory is necessary.

On-Site Painting. On-site painting consists of cleaning the metal and then applying two or three coats of an organic coating: (1) a priming coat and (2) a finish coat, or (1) a priming coat, (2) an undercoat or intermediate coat, and (3) a finish coat. When the metal has prepainting preparation and is factory-primed, the on-site priming coat is eliminated. The priming coat should always be of the rust-inhibiting type; the type of painting for undercoat and finish coat depends on the requirements and conditions of the end use of the painted material.

Table *P11* lists the types of organic coating used for painting metal. The types listed for priming, intermediate, and finish coats are those that will give the most satisfactory painting for the end use as shown. It is generally advisable that all coats be of the same type of organic coating. When combinations of paints are used, paint manufacturers should always be consulted in order to select the system best suited to the end use.

Table P11 Types of organic coatings (paint) used for painting metal

Type of organic coating	Surface preparation and pretreatment	Priming coat	Intermediate coat[a] (undercoat)	Finish coat	Major use on iron and steel
Alkyd vehicle	Blast cleaning, pickling, flame cleaning; no pretreatment or solvent cleaning	Red-lead alkyd varnish primer	Same as priming coat except tinted with carbon black or lamp black to a color (that contrasts with priming coat)	Aluminum alkyd, black alkyd, white or tinted alkyd paint	Exterior exposed to severe weather conditions; interior where mild chemical exposure, high humidity, and infrequent condensation exist
Chlorinated rubber	Hand or power tool cleaning, blast cleaning; no pretreatment necessary	Rust-inhibiting chlorinated rubber	Same as primer or finish coat	Stabilized pigmented, chlorinated rubber coating	Interior and exterior iron and steel exposed to excessive moisture and chemicals
Coal-tar	Blast cleaning; surface to be cleaned and prime coat immediately applied	Coal-tar enamel primer applied hot	None	Coal-tar enamel applied hot	Exterior where iron and steel are to be installed underground or in and under water
Epoxy resins	Hand or power tool cleaning, blast cleaning; in some cases prime coat immediately applied	Epoxy resins with rust-inhibiting component	Same as primer or finish coat	Epoxy resin pigmented finish coat	Interior and exterior iron and steel where resistance to acids, alkalis, and chemicals is required
Oil-base vehicle	Solvent cleaning, wire brushing; no pretreatment necessary	Red-lead oil-base primer	Same as priming coat except tinted with carbon black or lamp black to a color that contrasts with priming coat	Aluminum varnish or black, white, or tinted oil-base paint	Exterior exposed to normal weather conditions; interior where moderately corrosive conditions exist
Phenolic vehicle	Blast cleaning, pickling, flame cleaning; no pretreatment necessary	Red-lead mixed-pigment phenolic varnish primer	None	Aluminum phenolic, black phenolic, white or tinted phenolic paint	Exterior where iron or steel is immersed in fresh water or exposed to high humidity and condensation; interior only where conditions are the same as the exterior ones
Polystyrene	Solvent cleaning, hand or power tool cleaning, blast cleaning; no pretreatment necessary	Zinc-rich polystyrene	Same as primer or finish coat	Finish coat recommended for use over polystyrene	Iron and steel subjected to fresh and salt water, brackish water, and chemical fumes
Urethane	Blast cleaning; no pretreatment necessary	Rust-inhibiting urethane	Same as primer or finish coat	Pigmented urethane finish coat	Industrial finishes for corrosion, chemical, and abrasion resistance

Table P11 Types of organic coatings (paint) used for painting metal (*continued*)

Type of organic coating	Surface preparation and pretreatment	Priming coat	Intermediate coat[a] (undercoat)	Finish coat	Major use on iron and steel
Vinyl vehicle	Blast cleaning, pickling; after cleaning surface to be pretreated with basic zinc chromate-vinyl butyral washcoat	Vinyl red-lead primer	Same as priming coat except tinted with lamp black to a contrasting color	Aluminum vinyl, black vinyl, or vinyl-alkyd paint in white, black, red, yellow, or orange	Exterior where iron or steel is immersed in salt or fresh water or is exposed to high humidity and condensation; interior where flame resistance, mildew resistance, corrosion resistance, and easy maintenance are necessary

[a]The intermediate coat can be the same as the finish coat but tinted to a contrasting color.

Prepainting Preparation. In general, most metals used in construction have received surface preparation and priming coats in the mill, shop, or factory before they arrive at the site of construction, and on-site painting involves primarily the intermediate and finish coats of paint. The prepainting preparation in mill, shop, or factory can be performed in many ways, as shown in Table *P12*, covering methods of cleaning iron and steel, and Table *P13*, covering methods for nonferrous metals. For all types of painting of metal, the methods of cleaning in relation to the type of organic coating applied are of utmost importance. The tables also list the characteristics of the resulting surface.

Finish Painting. Finish painting done in a mill, shop, or factory produces finishes under ideal condition; it thus provides the best type of painting that can be obtained to meet the requirements and conditions of end use.

Blast Cleaning. Blast cleaning is classified into four grades: (1) blast cleaning to "white" metal, in which all rust, paint, mill scale, and gray mill scale binder are removed, exposing the white metal; (2) commercial blast cleaning, which includes complete removal of all rust, paint, and mill scale but not of gray mill scale binder, and which will therefore appear rather streaky; (3) brush-off blast cleaning, which involves complete

Table P12 Methods for surface preparation of iron and steel for painting

Type of cleaning	How done	Characteristics of cleaned surface
Alkaline cleaning	Metal sprayed or immersed in various types of alkalis	Removes all oil and grease
Conversion cleaning	Converts the metal surface, acid phosphate solutions, and proprietary solutions	Metal surface has a slate gray color and a fine matte surface
Flame cleaning	Oxyacetylene flame consisting of a series of small, closely spaced flames that are very hot and are projected at high velocity	Reduces ordinary rust to iron oxide and pops off loose mill scale; after flame cleaning the surface should be wire-brushed
Hand cleaning	Wire brushes, abrasive paper or cloth, scrapers, chisels, and chipping hammers	Used primarily for spot-type cleaning
Pickling (phosphoric acid)	Metal is immersed in warmed dilute phosphoric acid with added rust inhibitors; does not need finishing	Removes all dirt, rust, and mill scale and gives the surface a protective film which retards rusting and is a good base for painting

Table P12 Methods for surface preparation of iron and steel for painting (*continued*)

Type of cleaning	How done	Characteristics of cleaned surface
Pickling (sulfuric acid)	Metal is immersed in warmed dilute sulfuric acid with other chemicals which confine the action largely to rust and scale and is then rinsed	Removes all dirt, rust, and mill scale and gives the surface a slight etching which helps adhesion of the paint
Power tool cleaning	Power brushes, grinders, and sanders	Same as above
Rust removers	Applied by brush or spraying, the phosphate type forms a film and retards rusting	Generally used in maintenance painting and with on-site painting where slight rusting has occurred
Shot and sand blasting[a]	Sand or steel grit (crushed shot) in a range of Nos. 16–45 screen sizes and dry compressed air at 80 to 100 $lbf/in.^2$	Removes all dirt, rust, tight mill scale, and other surface impurities; also roughens the surface, thus providing the best condition for adhesion of the paint
Solvent cleaning	Wiped with various types of solvents	Removes dirt, oil, and grease
Steam (vapor) cleaning	Exposes metal to chlorinated solvents	Leaves metal clean of all grease, oil, and dirt

[a]There are four methods of blast cleaning: (1) blast cleaning to "white metal," (2) commercial blast cleaning, (3) brush-off blast cleaning, and (4) "near-white" blast cleaning.

Table P13 Typical methods of cleaning nonferrous metals for painting

Type of metal	Method of cleaning surface	Type of paints used after this cleaning
Aluminum	Let weather for a month; wipe with mineral spirits, or special preparation fully described under Aluminum Finishes	Never use a lead base paint (organic coating) directly on an aluminum surface (*see* Aluminum Finishes)
Copper, bronze, and their alloys	Remove loose corrosion by sanding, and wipe with mineral spirits or apply dilute solution of hydrochloric or acetic acid and rinse	Use paints (organic coatings) recommended by the manufacturers of copper, bronze, and their alloys (*see* Bronze; Copper; Copper Alloys)
Galvanized iron	Three methods: (1) wipe with mineral spirits; (2) let surface weather for 6 months or until it turns dull; (3) apply dilute solution of hydrochloric, phosphoric, or acetic acid and then rinse	Use zinc-dust types of priming paints (*see* Zinc Coatings)
Terneplate	Wipe with mineral spirits; do not allow to rust before paintings	Use paints (organic coatings) recommended by the manufacturers of terneplate (*see* Terneplate)

removal of rust and paint as well as loose mill scale, but does not include removal of tight mill scale, and which requires such surfaces to be solvent-cleaned before painting; and (4) near-white blast cleaning, which is similar to brush-off blast cleaning but does not include the complete removal of all loose mill scale and tight mill scale, and which therefore requires solvent cleaning.

Most metals used in construction besides iron and steel are left in their natural states and colors. When they are to be painted, their surfaces must be similarly prepared. It is advisable to consult with both the manufacturers of these metals and the paint manufacturers to select the correct type of organic coating for a particular metal. The reason is that many organic coatings have a metallic base that will cause galvanic action between the coating and the metal to be covered, and thus the paint film will be destroyed. This is why aluminum paints should never be used on iron or steel on the exterior unless a prime coat isolates the aluminum from the iron or steel.

The other methods of cleaning metal are chemical treatments, solvent cleaning, hand and power-tool cleaning, and flame cleaning (*see* Metals: Aluminizing; Metals: Chemical Finishes; Metals: Coatings; Metals: Mechanical Finishes; Metals: Sprayed Metal Coatings).

To summarize, all metal surfaces should be clean, all oil or grease should be removed, and, most important, all moisture should be removed before painting is started. (*See* Tables *P12* and *P13*.)

Application

Condensed Checklist

1. For finish painting of metal done in a mill, shop, or factory, it is necessary to check all the conditions of end use for the painted material and then to specify the surface preparation, the prepainting treatment, and the type of organic coating to be used, as well as the number of coats to be applied. It is always advisable to consult with the paint manufacturers when other than normal weather conditions are to be encountered so that the correct surface preparation, pretreatment, and type of organic coating are selected.
2. For mill, shop, or factory prepainting preparation of the metal for on-site finish painting, it is also necessary to check the end use of the painted surface and then to specify the type of surface preparation, pretreatment, and type of priming paint (if required) that should be used. For other than normal weather conditions, the paint manufacturers should be consulted.
3. For on-site painting of metal other than prepainting-prepared metal, it is necessary to specify that rust, scale, and loose paint be removed with wire brushing and sanding and that all oil or grease be removed before painting is started.
4. For on-site painting of metal, the person in charge should make sure that no painting is done on the exterior when the temperature is below 50°F (10°C). For the interior he must make sure that the correct type of temporary heat is used and that adequate ventilation is supplied.
5. For on-site painting of metal, it is necessary to specify and make sure that metal surfaces are dry and free from moisture before painting is started. Moisture is one of the major causes of paint failure.
6. One should always check with paint manufacturers for types of rust-inhibiting organic coating that meet a special end use such as installation near salt water or immersion in fresh or salt water.
7. When metals will be exposed to chemicals and chemical fumes, paint manufacturers should always be consulted for the best painting system to meet this end use.
8. The correct method of applying organic coatings should always be determined, and the method for the coatings selected should be specified.

Conditions Favorable to the Painting of Metal

1. For all iron and steel on the exterior.
2. For all iron and steel on the interior where corrosive conditions will be encountered.
3. For any metal on the exterior or interior where staining of other surfaces may occur because of corrosion of the metal, where two metals will be in contact with each other and thus cause galvanic action, where the metal can be attacked by a nonmetallic material, and where the color of the metal is not in harmony with the color scheme of the building.
4. For iron and steel that are to be installed underground and in water.

Conditions Unfavorable to the Painting of Metal

1. For metals that are to be exposed to or in contact with strong chemical fumes, acids, and strong alkalis, except when an organic coating will meet the requirements and is resistant to the acids, alkalis, and fumes that will be encountered.
2. For flameproofing or fireproofing metals unless local, municipal, and state codes as well as the codes of the Fire Underwriters, insurance companies, and federal specifying agencies are thoroughly checked for their limitations and requirements covering fire resistance, flameproofing, and fireproofing of metals with intumescent-type organic coatings.

PAINTING OF WOOD

Physical and Chemical Properties

All wood used in construction requires painting or treatment of some type (*see also* Wood Finishes; Wood Preservatives). Both exterior and interior wood that is to be painted should always be finish lumber, and interior wood should always be sandpapered to obtain a satisfactory painted finish. For exterior and interior wood that is only to be stained, and in any situation where only texture and color control the finish, any type of finished or dimension lumber may be used.

Preparation of Surfaces. The proper preparation of wood surfaces is of utmost importance in obtaining an effective paint job. The moisture content of the wood must be considered. Wood that is not properly seasoned (green lumber) or wood that has become damp because of wet weather or through the application of plaster within the structure should be allowed to dry properly beforehand. Usually a week of clear dry weather is adequate to dry lumber that has become wet. Dampness in wood from plastering requires waiting until the plaster is dry for both exterior and interior wood. Once the prime coat has been applied, the moisture content of the wood becomes fairly well stabilized.

All exterior wood and trim that are to comprise the finished exterior surface should be back-primed. All interior wood that is to be stained, painted, or decoratively treated and become the finished wall surface should also be back-primed.

Great care should be taken in installing insulation materials because they will absorb and retain moisture or entrap water vapor if they are not sufficiently ventilated. This moisture can penetrate to the back of the paint film and destroy it. Water vapor can penetrate most materials. Therefore both exterior and interior water vapor penetrates exterior walls. To stop this penetration, it is necessary to install vapor barriers. It is always advisable to consult mechanical engineers to determine what type of insulation and vapor barriers should be installed.

All wood that is to be stained, painted, or given a transparent coating should have nailholes, small voids, etc., filled with putty (for best painting work, this should be done after the priming coat is applied); large pores should be filled with sealers, and knots and pitch streaks should be given a covering coat of a knot sealer.

Constituents of Paint for Wood. Until recently all exterior wood painting was done with organic coatings containing white lead as the white paint solid (including driers, extenders, and color pigments as part of the paint solids) in a vehicle consisting of linseed oil as the nonvolatile ingredient and turpentine as the volatile ingredient. Because of the lead poisoning of children, the use of lead compounds in paints as a white paint solid has been completely eliminated with the exception of certain lead compounds in rust-inhibiting paints for iron and steel. The white paint solid almost exclusively used in paints now is titanium dioxide.

Titanium dioxide is mixed in various proportions as the white paint solid; a combination of drying oils and synthetic resins comprises the nonvolatile ingredient, and mineral spirits the volatile ingredient, of the vehicle. These organic coatings also contain extender pigments, color pigments, and driers in closely controlled amounts so that the quantity of chalking, flowing, leveling, and spreading is also controlled. The total pigment is usually 37 to 36% by volume of the total nonvolatile ingredients, and the total nonvolatile ingredients are not more than 73% by volume of the paint.

Clear interior and exterior wood finishes are comprised of natural oils and natural and synthetic resins. For clear exterior finishes, paint manufacturers should always be consulted to obtain the correct system for a particular end use.

Methods of Application. In general, the painting of wood is done with brushes, rollers, and spraying. As spraying requires highly skilled craftsmen, it is generally used for painting done in a mill, shop, or factory.

Types and Uses

There are two types of painting of wood: (1) on-site painting, and (2) painting done in a mill, shop, or factory. Most of the doors, windows, trim, and various other wood items that are factory-assembled or prefabricated have received a prepainting treatment consisting generally of preservative and wood sealer. Such treatment permits the moisture content of the wood to be relatively stabilized, and the products can be stored and transported without being damaged by moisture before they are installed in construction. This pretreatment is not a prime coat; on-site painting is done as if the wood had not been treated. For the general types of paint used on wood in construction and the nature of the various coats applied, see Table *P14*.

Exterior On-Site Painting. Exterior on-site painting of wood today is done by either the three-coat or two-coat method. Both methods, if properly executed, give almost equal painting jobs. The two-coat method may require repainting sooner, but the cost of the additional coat at a later date does not greatly exceed the cost of applying the third coat initially. In some cases a primer or sealer coat is sufficient as a first coat, with a second coat serving as finish coat. This can be considered a one-coat method.

With either the two- or three-coat method the thickness of the coating should be approximately the same. Each coat should be allowed to dry thoroughly to a firm film before the next coat is applied. In clear dry weather, generally 48 hours is sufficient for oil-based vehicle paints, and 24 hours for synthetic-resin vehicle paints.

Exterior staining of wood is best applied by dipping,

Table P14 Types of organic coatings (paint) used on wood in construction

Type of paint	Surface preparation	Priming coat	Intermediate coat	Finish coat	Major use on wood
Alkyd (oil, latex, soya)	Puttying, applying filler, cleaning for exterior and interior work; sandpapering for interior only	Alkyd primer for 3-coat work can be white; for 3-coat work, tint to same color as finish coat	Same as finish coat	Alkyd paint and enamels, semiglossy or matte; clear, white, or tinted	Interior and exterior wood
Epoxy (catalyzed)	Puttying, applying filler, cleaning for exterior and interior work; sandpapering for interior only	Oil, alkyd, or epoxy primer	None needed	Epoxy catalyzed, clear, pigmented, glossy, semiglossy, or matte	Exterior wood
Lacquers[a]	Sanding, applying filler, puttying, staining if necessary	Lacquer primer, clear or tinted	Same as finish coat	Lacquer, clear or tinted	Mill, shop, or factory type of finish painting; limited use for on-site painting
Latex (acrylic emulsion chlorinated rubber, acrylic, styrene-butadiene)	Puttying, applying filler, cleaning for exterior and interior work; sandpapering for interior only	Latex primer	None needed	Latex, white or pigmented	Exterior and interior woods
Oil-base vehicle	Puttying, applying filler for exterior and interior work; sandpapering for interior only	Oil-base primer for 3-coat work can be white; for 2-coat work, tint to same color as finish coat	Same as oil-base finish coat	Oil-base vehicle for exterior, white or tinted; varnish vehicle for interior, white or tinted	Exterior wood
Phenolic (modified linseed-tung modified)	Puttying, applying filler, sanding, staining if necessary	Oil, alkyd, or phenolic primer	None needed	Phenolic, clear	Exterior and interior wood
Stains[b]	Puttying for exterior			Oil-base latex, alkyd stains	Exterior wood shingles and siding
Urethane (modified)	Puttying, applying filler, sanding, staining if necessary	None needed	None needed	Urethane, clear	Interior wood walls and floors
Varnish[a]	Puttying, applying filler, sanding, staining if necessary	Clear varnish primer	Same as finish coat	Varnish, clear, semiglossy, glossy, or matte for interior; high-grade spar varnish for exterior	Exterior and interior wood
Varnish for floors*	Sanding, applying filler	Lacquer floor sealer for floors that have been oil treated	Varnish sealer	Varnish sealer	Interior hard wood floors
Vinyl (catalyzed)	Clean with lacquer thinner	None needed	None needed	Vinyl (catalyzed)	Interior wood walls and floors

[a]Varnishes and lacquers are available with alkyd, epoxy, phenolic, urethane, and vinyl bases.
[b]*See* Wood Preservatives.

usually by the manufacturers of shingles, but on-site staining can be done with brushes, mops, or squeegees with a single coat. The stain should be applied liberally so that it will be well soaked up by the wood.

Exterior transparent paint for wood requires a three-coat method. The wood should be sanded, cleaned, and stained or filled, and after 24 hours three coats of a high-grade exterior-type synthetic-resin varnish are applied with 24-hour drying of the coats between successive applications.

Exterior or interior floors and steps made of softwood or hardwood require that the original coating be reasonably thin, as restaining will be necessary. The two-coat method is used, with each coat thinned to make a thin film.

Many stains contain wood preservatives as part of their formulation. Any exterior wood that has not received a mill or factory prefinish should first be back-primed and then given a coating of wood preservative if it is to be stained or left in its natural wood color (*see* Wood Finishes and Preservatives).

Interior On-Site Painting. Interior on-site painting of wood requires a three-coat method to obtain best appearance and durability. The two-coat method will often give a satisfactory job. In many cases a primer or sealer coat is sufficient as a first coat, and the second coat is then also the finish coat; this can be considered a one-coat method.

Water-base paints are generally not as good as oil- or synthetic-based paints for interior on-site painting as the water tends to raise the grain of the wood. However, if a good primer or sealer coat made specifically for use with water-base paints is used, a satisfactory job can be obtained.

Interior transparent paint for floors requires that the floor be first machine-sanded smooth and cleaned. A paste wood filler must then be applied to fill the pores of the wood.

Sealer for Floors (a special penetrating type of varnish) is applied by two methods:

1. Two coats with a brush. The first coat is well brushed into the surface to obtain penetration, and the second or finish coat is applied thin.

2. Flooding and squeegeeing. This method requires special machine applicators and equipment and is usually done by highly skilled craftsmen. After the floor sealer is dry (usually after 24 hours), burnishing with No. 0 steel wool on a power-driven type of sander is necessary to remove highlights and to smooth the raised grain. If waxing is desired, it can be applied any time after the floor sealer is dry.

Staining and Other Treatments of Interior Wood. When interior wood is to be stained, the type of stain and the species of wood to be stained are of utmost importance. It is therefore advisable to check with paint and stain manufacturers to determine the best method for a particular end use. Also, the type of finish required—waxed, lacquered, or varnished—must be compatible with the type of stain and the type of wood. In almost all cases the wood to be stained or bleached must be sanded smooth, dry, and cleaned of all stains and foreign matter before any stain or bleach is applied.

Special types of painting of wood surfaces for fire resistance, flameproofing, fireproofing, acid resistance, and resistance to mold, rot, insects, etc., require consultation with the organic coating manufacturers so that the correct type of organic coating can be selected.

Application

Condensed Checklist

1. For finish painting of wood done in a mill, shop, or factory, it is necessary to keep in mind the end-use requirements for this painted material before specifying the preparation of the wood, type of paint, and number of coats to be applied. It is always advisable to consult with the paint manufacturers for other than normal interior environments.
2. For on-site painting of wood, it is necessary to make sure that the wood is clean, smooth, dry, and free from stains before being painted (it is important that all wood be sufficiently dry before work is begun).
3. For on-site painting of wood, it is necessary to make sure that no painting is done on the exterior in temperatures below 50°F (10°C) and that adequate heat and ventilation are supplied for interior painting if the temperature is below 50°F (10°C).
4. For exterior on-site painting of wood, all wood should be back-primed and all edges, especially end grain edges, should be back-primed.
5. For all exterior on-site painting or staining of wood, one should make sure that all countersunk nail-head holes are filled and knots given sealer coats, and that the wood is smooth, dry, and clean of dust, foreign matter, and oil stains before any paint is applied.
6. For all exterior wood that is to be left natural in color, one should make sure that a clear type of wood preservative is applied (*see* Wood Preservatives).
7. For any interior wood that is to be fire-retarded, paint manufacturers should be consulted about the type of intumescent paint to use; federal, state, and municipal codes should be checked for their requirements in regard to fire-retarding wood.

8. Most factory-assembled wood products have been treated with a mold-, rot-, and insectproofing type of treatment. One should always check with the manufacturers as to what types of finish coating are available; an example would be white vinyl-finish organic coating.

Conditions Favorable to the Painting of Wood

1. For all wood that is exposed to the weather.
2. For all interior wood except that which has been prefinished in a mill, shop or factory.

Conditions Unfavorable to the Painting of Wood

1. For flameproofing, fire-retarding and fireproofing wood unless local, municipal, and state codes, insurance companies, Fire Underwriters, and government agencies are checked to make sure that they permit the use of organic coatings for these purposes.
2. For protecting wood against air- and earth-type termites, carpenter ants, mold, and fungus unless the treatment meets strict requirements and/or guarantees resistance against the above-mentioned insects, mold, and fungus.

PALLADIUM

Physical and Chemical Properties

Symbol: Pd
Atomic number: 46
Specific gravity: 12.02 (20°C, 68°F)
Melting point: 1552°C, 2825.6°F
Boiling point: 3140°C, 5652°F

Palladium is a soft, silvery-white, ductile metal that is less resistant to corrosion and oxidation than platinum. It has a very special characteristic in that hydrogen, but no other gas, will diffuse through the hot metal.

Workability. Palladium may be hot- or cold-worked, drawn, spun, bent, cast, welded, brazed, and hammered into foil.

Commercial Forms. Palladium is available in sheet, wire, foil, and leaf form.

Types and Uses

Palladium is often substituted for platinum because of its lower density and cost. Palladium has limited use in compound form; its uses in metallic form are indicated by asterisks in Table *P15*.

Table P15 Uses of palladium

	Construction Uses	
Component of	Textiles	Mordant
Allied Construction Uses	Electrical devices*	Electrical contacts, limiting fuses, resistance windings

History and Manufacture

Wollaston isolated palladium about 1803 and named it in honor of the asteroid Pallas. The process for obtaining palladium is shown under Platinum.

PANELS AND SANDWICH PANELS

A panel is defined in this book as a prefabricated unit that has a finished surface on one side and either a finished or unfinished surface on the other side, and that incorporates insulation, vapor barrier, soundproofing, etc., within its construction. Such panels are rigid and self-supporting to a greater or lesser extent.

Curtain-wall systems are based on the installation of these panels coordinated for use with modular and non-modular window units. Aluminum, stainless steel, steel, porcelain-enameled metal, bronze, asbestos-cement, brick, glass, mosaic tile, plastics, stone, concrete, glazed-face concrete block, and structural clay facing tile have all been used for panel curtain-wall construction.

Sandwich panels are a development from the construction methods used in the aviation industry. The sandwich panel is made of a material in the form of a cellular core, which has a skin of other materials applied and bonded to both sides of this core, and thereby becomes an entity in which all components work as one. The sandwich panel is actually a structure and should be designed for the particular condition of use. With future developments in the fields of adhesives, rubbers, and plastics, it will be possible to fabricate economical sandwich panels for use in construction as supporting floors, roofs, partitions, etc.

An integrant is a type of sandwich panel limited by its thickness. If it includes a vapor barrier, methods of eliminating or disposing of condensation, soundproofing, insulation, etc., it should meet the requirements of a curtain wall panel.

Application

Condensed Checklist

1. Method of joining the panels.
2. Method of attaching the panels to the frame of the building and at floors.
3. Treatment of electrical conduit and wiring, heating, air-conditioning, and other mechanical ducts, piping, etc., at the back (interior face) of the panel.
4. Weathertightness of the panel under normal general conditions and highly localized specific conditions that are often artificially created (e.g., leakage from strong updrafts created by the vast vertical area of a very tall, broad building).
5. Expansion and contraction of the panel unit in relation to the joining and to the structural framework.
6. Panels, including the system of installation, must be tested with full-scale sections for the wind velocities, rainfall, temperature changes, etc., that exist at the area of construction.
7. Samples of both exterior and interior finishes, colors, textures, etc., must be submitted for approval.

PAPER AND PAPER PULP PRODUCTS

Physical and Chemical Properties

Most paper and paper pulp products consist of fibers processed mechanically and chemically from cellulosic raw material. The only exceptions are special papers made from mineral fibers, for example, from asbestos and synthetic fibers. Plastics also have become a component in paper and paper pulp products.

The main source of this cellulosic raw material is the wood from both coniferous (softwood) and broadleaf (hardwood) trees, but predominantly from the coniferous group, which includes northern and temperate-zone woods such as spruce, balsam fir, and hemlock, and southern woods such as pine (*see* Table *P16*). There are other sources of vegetable cellulose for paper pulp, but

Table P16 Types of wood pulp used for making paper and paper pulp products

Type of pulp	Process	Characteristics	Major uses
	Wood pulp from trees		
Mechanical pulp: (coniferous and broadleaf trees)	Mechanical grinding of wood chips, washed and screened for size	Good absorption, bulk, and opacity; little strength or permanence	Paper pulp boards
Chemical pulps:			
1. Soda pulp (broadleaf or hardwood trees)	Chemical digestion by caustic process, using sodium hydroxide from soda ash plus lime	Bleached; combined with bleached sulfitic or sulfate pulp	Fine printing papers (wallpaper)
2. Sulfate (kraft) pulp (coniferous trees)	Chemical digestion by caustic process, using sodium hydroxide from sodium sulfide plus lime	Unbleached; good physical strength	Laminating papers, reinforced building paper, and paper pulp board
3. Sulfite pulp (coniferous trees)	Chemical digestion by acid process, using limestone plus sulfur dioxide to form sulfurous acid	Bleached	Paper grade and dissolving grade for plastics and fibers
		Unbleached	Strong papers for construction and electrical work
	Pulp from paper waste		
Waste paper shavings and cuttings, de-inked paper	Mechanically disintegrated; de-inking by alkalis	Little strength	Low-grade printing paper, paper boxes
Waste paper stock	Mechanically disintegrated	Low strength	Filler for boards, building paper
Screenings	Chemical digestion	Waste products from other processes of papermaking	Coarse paper, paper pulp board

Table P17 Sources of pulp other than wood for paper and paper pulp products

Type of pulp	Source	Process	Major uses
Bagasse	Sugarcane stalks	Mechanical disintegration and mild chemical digestion	Boards, insulation (mainly for acoustics)
Esparto	Esparto grass	Chemical digestion by sodium hydroxide (caustic soda) alone or with sodium sulfide	High-grade printing papers, wallpaper
Jute	Old sacking, burlap, and string	Chemical digestion by cooking with lime	Buff drawing paper, strong wrapping paper
Rag	Old and new cotton rags Old cotton and woolen rags	Removal of foreign matter; cleaning and washing; then chemical digestion by caustic soda, caustic lime, or caustic lime plus soda ash	Fine drawing papers Roofing felts
Rope	Manila hemp from old manila rope; waste from manufacturing	Chemical digestion by lime with soda ash ($CaO + Na_2CO_3$)	Abrasive papers, electrical insulation
Straw	Wheat, oat, barley, and rye straws	Chemical digestion by lime, soda ash, or caustic soda; unbleached	Corrugated paper boards (mainly for insulation)

these are used in much smaller quantity and for special types of paper (*see* Table *P17*).

Many new types of paper and paper materials are constantly becoming available because of the combination of various synthetic resins and plastic materials with the paper pulps or the finish paper materials. These papers are made by mixing plastics with the paper pulp, impregnating with plastics, laminating with various plastic or metallic films, and sizing and coating the papers with special surface finishes for particular end uses, such as the various papers for reproduction (*see* Plastics: Film; Wood: Hardboard and Particle Board).

Paper Pulp Materials. Wood consists of cellulose, lignin, carbohydrates, proteins, resins, and fats. Of these, cellulose is the preponderant ingredient and the basis of paper pulp; in some cases part of the carbohydrates and lignin may also be retained in the resulting pulp. Other materials used in making paper pulp include the following.

Colors: synthetic dyes, natural vegetable coloring materials, and mineral pigments.

Processing chemicals: alum (aluminum potassium sulfate), ferrous sulfate, casein, lime, limestone, salt, calcium sulfite, sodium sulfate, sulfur, chlorine, sodium carbonate.

Sizing: rosin, glue, and synthetic resins.

Fillers and coatings: clay, talc, gypsum, chalk, titanium pigments, synthetic resins, and others.

Important Properties. The important properties of paper and paper pulp products are water absorption, density, tensile strength, and porosity. These properties vary according to the kind of fiber, pulping process, and subsequent manufacturing and finishing operations.

Papers for drafting, sketching, drawing, and typing must have whiteness and transparency or opaqueness, be erasable and strong, and have a smooth or slightly rough texture. Papers for reproduction must be able to take all types of printing (black and white or color), and papers for copying and reproduction must have sensitized surfaces for some of the photocopying and reproduction systems. What differentiates paper from paper board is thickness: paper is no thicker than 0.012 in. (0.305 mm); anything thicker than this is considered board.

Chemical Treatments. Paper can be treated to be resistant to fire or flame, water and water vapor, insects, rodents, molds, fungi, and bacteria. It can also be made tearproof.

Commercial Forms. Paper is available in sheets, rolls, tubes, and boards in a wide diversity of types ranging from facial tissues and filter papers to thick, spongy insulating boards and from fine, thin printing papers to high-density building boards and decking. Paper may be smooth or creped, single or laminated, cored, corrugated, uncoated, or finished and sized in a wide variety of surfaces and textures.

Types and Uses

The direct uses of paper and paper pulp products in construction fall into four categories: (1) actual building materials; (2) containers and protective coverings for all sorts of materials used in construction; (3) forms for concrete or for creating voids in concrete slabs; and (4) all the various papers for reproduction, letters, envelopes, records, billing, and other business stationery used in the designing, administration, supervision, and construction of buildings (*see* Table *P18*).

Building Materials. Building materials include roofing felts, building paper, flashing, thermal and sound insulation, and a wide assortment of sheets and boards, both composite and integrant in character, which are usually marketed under trade names. A general breakdown of types is possible. Boards and integrants may be high

Table P18 Types of paper and paper pulp used in construction

Type of material	Type of pulp	Major uses	Process	Thickness or weight
Acoustic fiberboards	Wood, straw, bagasse, corn stalks	Acoustic treatment for walls and ceilings	Paper laminated or pulp pressed into boards with or without holes, slots, etc.	Usually $\frac{1}{2}$–2 in. in $\frac{1}{2}$-in. increments (12.7–50.8 mm in 12.7-mm increments)
Aluminum reflective insulation	Kraft plus aluminum foil	Thermal insulation; vapor barrier; concealed flashing	Two piles of kraft paper laminated with asphalt and reinforced with jute, sisal, or glass fibers, with aluminum foil laminated to one or both sides	Controlled by end use
Asphalt felt	Low-grade cotton and woolen rags, old paper stock	Sheathing paper under siding and shingles; built-up roofing	Saturated with asphalt	12–30 lb/100 ft^2 (5.44–13.6 kg/9.29 m^2)
Asphalt reinforced building paper with glass fibers	Kraft and glass fibers	Sheathing paper for roofing and siding; paper to cure or protect finish-type floors; vapor barrier; masonry wall waterproofing	Two layers or more of asphalt-treated papers laminated together with asphalt and reinforced with glass fibers	Controlled by end use
Asphalt sheathing paper	Sulfate and waste pulp	Sheathing paper for wood-frame buildings	Treated with asphalt	25–30 lb/500 ft^2 (13.34–13.61 kg/64.45 m^2)
Asphalt shingles and siding	Felts	Roofing and siding finish materials	Saturated with asphalt and surface treated with asphalt and special granules	Various sizes, shapes, forms, textures, and colors
Black waterproof paper	Jute, kraft	Vapor barrier; sound deadening under roofing; sheathing and flooring materials	Saturated and coated with asphalt or tar	35–50 lb/500 ft^2 (15.88 to 22.68 kg/64.45 m^2)
Carpet felt	Waste paper	Under carpets	Directly from paper machine	Controlled by end use
Chip board	Waste paper	Where strength and quality are not needed	Thin sizes on paper machine; thicker on rolling-type machine	From approximately $\frac{1}{16}$ in. (1.59 mm) up

Table P18 Types of paper and paper pulp used in construction (*continued*)

Type of material	Type of pulp	Major uses	Process	Thickness or weight
Concrete forms	Kraft	Cylindrical forms for concrete work; to create voids in concrete	Cylinders of 6-in. (152.4-mm)-wide strips laminated with waterproof adhesive, wax-coated, and plastic-, wax-, or silicone-lined	Inside diameter, 6 in. (152.4 mm) min. to 4 ft (1.219 m); min. thickness of ply, 0.015 in. (0.38 mm); thickness of tube, 0.125–0.5 in. (3.18–12.7 mm)
Concrete forms	Kraft	Domes for waffle-type concrete slabs	Formed with synthetic resins and plastic wax or silicone coating	1.58 × 1.58 and 2.5 × 2.5 ft (0.483 × 0.483 and 0.762 × 0.762 m) and generally 10 in. (254.0 mm) deep
Copper foil concealed flashing	Kraft plus copper foil	Vapor barrier; waterproofing; concealed flashing	Same process as aluminum reflective insulation except copper foil applied to one side only	1, 2, 3 oz (28.35, 56.70, 85.05 g); refers to copper
Deadening felt	Mixtures of rags and papers	In wall and floors for sound deadening	Directly from paper machine	Controlled by end use
Fiberboard	Waste paper, wood pulp	Inexpensive finish material for interiors	Layers of chip board laminated together and sized	Usually $\frac{3}{16}$, $\frac{1}{4}$, $\frac{1}{2}$, and $\frac{3}{4}$ in. (4.76, 6.35, 12.7, and 19.05 mm)
Insulation blanket	Wood pulp (or wheat, oat, barley, corn, or rye)	Thermal insulation	Fibers loosely held together to entrap maximum amount of air	Usually $\frac{1}{4}$, 1, 2, and 3 in. (6.35, 25.4, 50.8, and 76.2 mm)
Insulating fiberboards	Wood pulp (or wheat, oat, barley, corn, or rye), bagasse	Insulating sheathing board; core for integrants	Fibers sized and lightly pressed together to entrap air; coated or uncoated or impregnated with asphalt	Usually $\frac{1}{4}$, $\frac{3}{8}$, $\frac{1}{2}$, $\frac{3}{4}$, 1, 2, and 3 in. (6.35, 9.53, 12.7, 19.05, 25.4, 50.8, and 76.2 mm)
Insulating fiber lath	Mechanically shredded wood fibers	Lathing with thermal insulation; sound-deadening value for plaster	Fibers sized and lightly pressed together to entrap air; also with aluminum foil applied to one side	Usually $\frac{3}{8}$, $\frac{1}{2}$, $\frac{3}{4}$, and 1 in. (9.53, 12.7, 19.05, and 25.4 mm)
Jute or sisal reinforced building paper	Kraft and jute cords	Sheathing paper under roofing, siding, and flooring; to cure or protect finish-type floors; vapor barrier	Two plies of paper laminated with asphalt and reinforced with jute or sisal cords; plain or creped	33 lb/500 ft^2 (14.97 kg/46.45 m^2)
Kapok	Seed pods of kapok	Thermal and sound insulation	Fibers from seed pod treated and pasted to paper	Controlled by end use
Plaster lathing board	Wood, straw, waste paper	Lathing for plaster	Outside reinforcing for plaster core; also with aluminum foil applied to one side	Usually $\frac{3}{8}$ and $\frac{1}{2}$ in. (9.53 and 12.7 mm)

Table P18 Types of paper and paper pulp used in construction (*continued*)

Type of material	Type of pulp	Major uses	Process	Thickness or weight
Plaster board	Chip board or wood pulp	For dry-wall type of interior finish for walls and ceilings	Chip board or sized wood pulp paper applied as reinforcing onto plaster core; also with aluminum foil on one side	Usually $\frac{3}{8}$, $\frac{1}{2}$, and $\frac{3}{4}$ in. (9.53, 12.7, and 19.05 mm)
Plastic-coated paper	Kraft	Vapor barriers	Two plies of kraft paper laminated with asphalt, reinforced with jute, sisal, or glass fibers, and coated with plastic on one side	Controlled by end use
Plastic laminates	Kraft and sulfite	Durable, decorative surface finishes	Layers of kraft paper with plastic binders, decorative undersurface, and clear plastic covering, all compressed and heated into fused sheet	Usually $\frac{1}{16}$, $\frac{5}{32}$, and $\frac{1}{4}$ in. (1.59, 3.92, and 6.35 mm)
Reinforced building paper with glass fibers	Kraft and glass fibers	Sheathing paper for roofing, siding, flooring; to cure or protect finish-type floors	Two layers of paper laminated together with asphalt and reinforced with glass fibers	Controlled by end use
Roofing paper	Felts	Built-up roofs	Saturated with asphalt and coated with crushed slate	Controlled by end use
Roof sheathing paper	Waste paper	Sheathing paper under roofing	Saturated with asphalt	Approximately $\frac{3}{16}$ in. (4.76 mm)
Rosin-sized paper	Wood	Sheathing paper for roofing and siding; protecting variously finished surfaces in buildings	Paper sized with rosin; dense and nonporous	4, 5, 6, and 8 lb/ft^2 (19.53, 24.42, 29.29, and 55.06 kg/m^2)
Sheathing paper	Wood, waste paper	Sheathing paper against dust and wind only; general-use paper during construction	Directly from paper machine	Controlled by end use
Sheathing-paper felt	Dry felt	Waterproof sheathing paper	Saturated with asphalt and then coated on both sides with asphalt	50–60 lb/200 ft^2 (22.68–27.22 kg/18.58 m^2)
Tarred felt	Low-grade cotton rags or paper stock	Sheathing paper for roofing and siding; built-up roofs	Saturated with tar	12, 14, and 15 lb/100 ft^2 (5.44, 6.35, and 6.80 kg/9.29 m^2)

Table P18 Types of paper and paper pulp used in construction (*continued*)

Type of material	Type of pulp	Major uses	Process	Thickness or weight
Tarred threaded felt	Low-grade cotton rags or paper stock	Sheathing paper for roofing and siding where high tear strength is required (it is also rodent-proof); masonry wall waterproofing	Saturated with tar, threads running lengthwise	Controlled by end use
Wallpaper	Mechanical pulp, bleached chemical pulp	Plain or decorative wall coverings	Coated with clay or casein and clay and pigments; printed; embossed	Wide variation, controlled by end use
Wallpaper (synthetic resin coated)	Mechanical pulp, bleached mechanical paper	Plain or decorative water-resistant wall coverings	Coated with synthetic waterproof coatings	Wide variation, controlled by end use
Wallpaper (laminated plastic waterproof film)	Mechanical pulp, bleached chemical pulp	Plain or decorative waterproof wall coverings	Waterproof plastic film applied to one face	Wide variation, controlled by end use

density or medium density or very loosely held together. In the last case, they usually have good insulating values, which may be further increased for sound by perforations and for heat by lamination with a vapor barrier or reflective material, or both. They may be rigid enough to be self-supporting and even bear a light load, or they may be thin flexible sheets. They may be uniform throughout in composition or layered and combined with a wide variety of other materials such as plaster, foams, synthetic resins, and adhesives (*see* Wood: Hardboard and Particle Board). Mineral and glass wool for insulation is enclosed in paper or attached with adhesive to paper. These papers are available perforated or solid, waterproof, and with vapor barriers or other treatment, depending where thermal insulation is to be installed.

Containers and Protective Coatings. Paper also serves to package and protect other materials. It may be in the form of bags, boxes, cartons, wrappers, all sorts of protective paper for concrete work, and coverings and drop sheets to protect materials already installed.

Roll Identification and Sizes. Paper which comes in rolls for use in the architectural field is identified by the following:

Weight per 100 ft^2 (9.29 m^2).
Width and length of a roll in feet (meters).
Width and area (ft^2 or m^2) per roll.
Thickness in decimals of an inch (or in millimeters).

The standard sizes of rolls are 3, 3.33, 4, and 5 ft (0.914, 1.016, 1.219, and 1.524 m) in width and 10 ft (3.048 m) in length, or 250, 500, or 666 ft^2 (23.23, 46.45, or 61.87 m^2) per roll. Metal-coated papers are available 6 to 60 in. (15.24 to 152.4 cm) wide and 10 ft (3.048 m) long. Copper-coated papers are identified by ounces per square foot, and aluminum-type reflective foil by width and square feet per roll. Papers for curing and protecting flooring materials and for vapor barriers are available up to 8 ft (2.438 m) wide and containing as much as 2400 ft^2 (222.97 m^2).

Fireproofing and Flameproofing. Fireproof papers cannot in reality be absolutely fireproof, but paper pulp products can be made sufficiently fire resistant to meet many fire code requirements. The principle is to prevent the paper or paper pulp product from bursting into flame upon exposure to high temperatures, since cellulose when heated develops combustible gases. When these materials are made flame retardant, they are difficult to ignite, will not support combustion, and will self-extinguish after the source of heat has been removed. Table *P19*, which lists some typical flameproofing materials (many of which are marketed under trade names), gives a general idea of their composition and effects on paper pulp products.

Mineral-Fiber Papers. Paper made of asbestos is the only mineral-fiber paper known and used extensively in construction (*see* Asbestos Paper). Glass, silica, and ceramic fiber papers and mica papers are special types used where high temperatures are encountered, particularly in the electrical field for wire insulation.

Table P19 Typical flameproofing materials for paper and paper pulp products

Flame retardant	Application	Advantages	Disadvantages	Durability
Ammonium chloride and phosphate	Water solution	Prevents afterglow; good fire protection	Can be destroyed by severe wetting	Good if used in areas where wetting does not occur; with severe wetting the fire-protection properties are reduced or eliminated
Ammonium sulfate	Water solution	Excellent flame resistance	Fair afterglow	
Boric acid-sodium borate mixture; sodium phosphate	Water solution	Good fire protection	Does not prevent afterglow; can be destroyed by severe wetting	
Antimony oxide; chlorinated paraffin and antimony oxide; antimony trioxide	Pigment to paper	Excellent flame resistance; insoluble in water	Fair afterglow	Good, since compounds are insoluble in water
Zinc borate	Pigment to paper	Good fire retardant	Water resistant	Good in areas where severe wetting can occur

Miscellaneous Types. These include chemically resistant papers treated with chemically resistant synthetic resins, stretchable papers, and waterproof papers, which originally were treated with a copper-ammonium solution or coated with rubber latex, but now have synthetic resins mixed with the pulp to produce a waterproof paper. Wallpaper, which is a decorative rather than a construction material, can be included here. The washable paper types usually have a plastic film laminated to the exposed surface.

Other Papers. Not only is paper an important building material; it is also essential to the practice of architecture, engineering, and construction. Its applications within the various offices include tracing paper, drawing paper, colored papers, cardboard, laminated paper illustration boards, blueprint paper, and other types for reproduction, masking, and repair tapes, as well as all paper materials for office paperwork and records. (For a description of board for drawing, rendering, and mounting, *see* Paper Pulp Sheets.)

The sizes of flat paper used by industry and various government agencies for drawings, especially those on tracing paper, have been standardized for convenience in handling. These sizes, although used in some architectural, engineering, and construction offices, are not generally known throughout the profession. The dimensions are based on the $8\frac{1}{2}$ by 11 in. (21.59 by 27.94 cm) module (standard letterhead size) so that all drawings can be folded into an $8\frac{1}{2}$ by 11 in. (21.59 by 27.94 cm) unit that not only is suitable for mailing purposes but also permits all such material to fit into standard filing cabinets, manila folders, stock envelopes, etc. (*see* Tables *P20* and *P21*).

Metric conversion will cause all paper, envelope, folder, and other stationery sizes to change when the changeover is complete and standard for the United States. Figure *P5* shows the standard metric sizes for paper for drawing and writing materials. It is impossible to make any kind of direct transition from our existing paper sizes as shown in Tables *P20* and *P21* to their metric equivalents. An example to illustrate this is our

Table P20 Standardized sizes for paper based on the letterhead module of 8.5 × 11 in. (21.59 × 27.94 cm)

Size designation	Width		Length	
	in.	cm	in.	cm
A	8.5	21.59	11	27.94
B	11	27.94	17	41.18
C	17	41.18	22	55.88
D	22	55.88	34	86.36
E	34	86.36	44	111.76
F	28	71.12	40	101.60
T	11	27.94	34	86.36

Table P21 Standard paper sizes

Size designation	Dimensions	
	in.	cm
Folio note	5.5 × 8.5	13.97 × 21.59
Pocket note	6 × 9.5	15.24 × 24.13
U.S. government writing	8 × 10.5	20.32 × 26.67
Commercial writing	8.5 × 11	21.59 × 27.94
Legal cap	8.5 × 14	21.59 × 35.56
Foolscap	13 × 16	33.02 × 40.64
Denny	16 × 21	40.64 × 33.34
Folio	17 × 22	41.18 × 55.88
Royal	19 × 24	48.26 × 60.96
Super royal	20 × 28	50.80 × 71.12
Elephant	23 × 28	58.42 × 71.12
Imperial	23 × 31	58.42 × 78.74

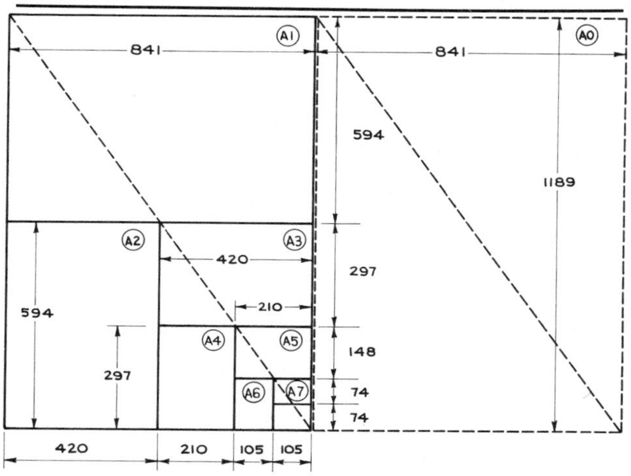

A SIZE	mm	A SIZE	mm
A0	841 x 1189	A6	105 x 148
A1	594 x 841	A7	74 x 105
A2	420 x 594	A8	52 x 74
A3	297 x 420	A9	37 x 52
A4	210 x 297	A10	26 x 37
A5	148 x 210		

MEASUREMENTS REPRESENT TRIMMED SIZES

Figure P5 Metric paper sizes for all drawing and writing material.

standard letterhead paper, which is $8\frac{1}{2}$ by 11 in. (215.9 by 279.4 mm); it has no relationship whatsoever to the metric standard letterhead size, which is 210 by 297 mm (8.27 by 11.69 in.). It is important to consider that even the sizes of existing drafting tables are not adaptable to the new metric sizes of paper (*see* Table *P22*).

Watermarks. Watermarks on paper originated in Italy about 1250 and consisted of simple shapes like circles or triangles and other uncomplicated designs that could be twisted in wire. The principle was also simple: as the wet paper passed or pressed over the wire form, the fibers were thinned out along the course of the wire and the impression remained as a thinner area in the finished paper. Until 1850 watermarks continued to be made of wire; then a period of elaboration of designs and techniques began. Portrait busts, scenes, and all sorts of fanciful figures were first molded in wax; next, dies were made from the wax model. Then fine brass wire gauze was pressed between the dies so that an exact replica was formed of the original wax model. As the paper was built up by hand, the water drained through the gauze and left the paper the exact thickness of the original wax model. Later, colored watermarking was invented. Today the watermark is again a simple affair made by a rubber form used on the dandy roll.

Papier Maché. Papier maché was invented in 1740 in France; the term refers to products made of paper pulp or paper saturated with water and made into pulp, plus a binder (animal or vegetable glue), which is then molded into any desired form. Papier maché can be shaped mechanically in molds or by hand over a supporting framework of some kind. After the form dries, it can be painted, coated, or dyed any color.

Papier maché has been used to make toys, dishes, novelties, and displays and also as a medium to make large-scale models of sculptural works. Papier maché was used extensively in the past but now has been largely replaced as a result of the wide variety of synthetic adhesives and resins available and the many combinations of sawdust, vermiculite, perlite, finely shredded or chopped paper, and other granular or powdered materials with a synthetic adhesive or resin to form a modeling material that has the same workability as papier maché and results in a permanent material (*see* Adhesives; Plastics; Rubber).

Application

Condensed Checklist

1. Action in contact with metals: If metal siding or roofing comes into contact with paper, the sheathing paper should be of the type recommended by the metal manufacturer (some metals are attacked by certain papers).
2. Types and degrees of resistance: Papers treated for resistance against insects, vermin, mold, fungi,

Table P22 Metric paper sizes in relation to standard existing drafting boards and new metric drafting boards

Metric paper size (mm)		Existing drafting boards in.	Existing drafting boards mm	Metric drafting board mm	
A0	1189 X 841	54 X 32	1372 X 813	A0	1270 X 920
A1	841 X 594	44 X 30	1092 X 737	A1	920 X 650
A2	595 X 420	32 X 24	813 X 584	A2	650 X 470

rodents, etc., are available, and it is important to choose the correct type for a particular end use.

3. Lapping and attachment: A check should always be made for the method best suited to the particular job.
4. Strength, weight, and type of pulp used: A better building paper will do a far superior job for a slight extra cost.
5. Code requirements: State, local, municipal, federal, insurance company, labor department, and Fire Underwriters' codes should be checked for fireproofing, dampproofing, waterproofing, vapor barrier, and other requirements.
6. Qualities as underlayment for roofing and siding: Sheathing paper so used should provide either some or all of the following: dampproofing and wind, dust, and vapor barrier protection, depending on the job requirements.
7. Qualities as underlayment for finish floors: These papers should provide sound deadening, vapor barrier protection, dampproofing, and waterproofing, as required by the job.
8. Qualities as vapor barriers: These include nontearing, strength (by reinforcement), waterproofing, and dampproofing.
9. Qualities as thermal insulation: These include, besides heat resistance, both moisture absorption and proofing against insects, rodents, mold, and fungi.
10. Treatment of termination point of concealed flashing: If a waterproof paper laminated with metal foil is used for concealed flashing, careful detailing of the transition point from concealed flashing to exposed metal flashing is necessary.
11. Qualities and effectiveness for protection and curing of finish floors, pavements, sidewalks, etc.: A check should always be made of the strength, reinforcement, moisture retention, and waterproofing of the paper itself and of the best method of lapping and attaching for the job.
12. Qualities in aluminum-laminated reflective building paper: Reflectivity of the aluminum finish, method of attachment, and depth and number of reflective areas should always be checked.
13. Waterproofing or dampproofing exterior foundation walls below grade: The paper used should always be waterproof and protected during backfilling. State, local, municipal, federal, insurance, and other codes should be checked for the required number of layers, for the type of asphalt or tar (hot or cold application), and for the conditions at top and bottom of foundations.
14. Built-up roofs: One should always check whether the paper is saturated with tar or asphalt, its thickness, type of adhesive asphalt or tar (hot or cold application), and method of protection for the finish surface.
15. Waterproof membranes: Papers for forming waterproof membranes for walls and floors below water level should be treated with tar, asphalt, or synthetic resin. The number of layers and particularly the type of joint connection at the exterior foundation wall and horizontal membrane under the basement floor, as well as the method of protecting the membrane during and after installation, must be carefully detailed and specified.
16. Papers used for drawing, drafting, and detailing: These should be of a type to produce clear, sharp, and clean reproductions.
17. Papers for sketching, renderings, and illustrations. These should be selected for suitability to the medium to be used, namely, pencil, ink, magic markers, pastels, water color, opaque colors (acrylic and tempera), or other.
18. Paper sizes: Sizes for construction-type drawings, including architectural, structural, mechanical, and topographical, should be selected for ease of handling in shops and offices and at construction sites.

Conditions Favorable to the Use of Paper and Paper Pulp Products

1. Where a thin material is needed directly behind a finish surface material on floor, walls, or roof to serve as a wind, dust, and vapor barrier, as dampproofing and waterproofing, and as sound-deadening (floor) and acoustic (wall and ceiling) materials.
2. For all types of built-up roofing.
3. Where a finish surface must be protected during construction.
4. Where it is necessary to protect and cure a finish floor surface during construction.
5. For dampproofing and waterproofing below-grade foundation walls.
6. For reflective insulation in walls, ceilings, and roofs.
7. For concealed flashing.
8. Where construction below water level requires a waterproof membrane.
9. For all types of designing, drawing, drafting, and detailing by architects, engineers, and the entire construction field.
10. For all types of administration, supervision, and management papers and forms.

Conditions Unfavorable to the Use of Paper and Paper Pulp Products

1. For any permanent exterior finish surface, unless adequately pretreated.
2. For exposed flashing.
3. For areas where there is excessive moisture.
4. For situations where fire resistance is essential.
5. For areas where there will be rough usage.

History and Manufacture

The Chinese are believed to have made paper from bamboo and rags as early as about 2000 B.C.

Until 1450 textile rags were still the only source of pulp. Wood finally became and has remained the major source of supply.

In the United States the manufacture of paper is a leading industry; in Canada, Sweden, and Finland, it is one of their largest industries.

Steps in Paper Making. In papermaking the fundamental steps are as follows.

1. The pulping process, or preparation of fibers for uniformly distributed suspension in water so that they form a sheet of interlacing fibers on a rolling screen through which the water can drain off.

2. Formation on rollers of the sheet of paper and, after this, removal by pressure and evaporation of the moisture to a point of equilibrium with the atmosphere.

3. The finishing of the surface to specific requirements, which may include additional steps to laminate layers of paper together or with other materials.

Mechanical Pulp. To make mechanical pulp, the wood is literally ground into pulp. After cleaning, screening, and thickening, it is ready for the final papermaking process.

Chemical Pulp. To produce chemical pulp, the wood is first chipped into small pieces and then digested by either the sulfite or the sulfate (Kraft) process. The soda process, the earliest method for converting wood into pulp, is used much less frequently now. In both the acid (sulfite) process and the caustic or alkaline process (sulfate or soda), the wood chips plus the chemical digester are placed in sealed chambers and heated, or "cooked." In the acid process, the temperature ranges from 212 to 276.8°F (100 to 135.56°C), and in the alkaline process from 644 to 662°F (340 to 350°C). After cooking, the pulp is blended, screened, and, if destined for higher grade papers, bleached; then it is washed and pumped to a storage chest.

Beating Process. The next process involves the beater, where the fiber and other materials are mixed, mashed, and beaten to make a mass of the right consistency so that on the paper machines the fibers will bond together and drain off water more slowly and uniformly. During this beating process, which is the main control point for

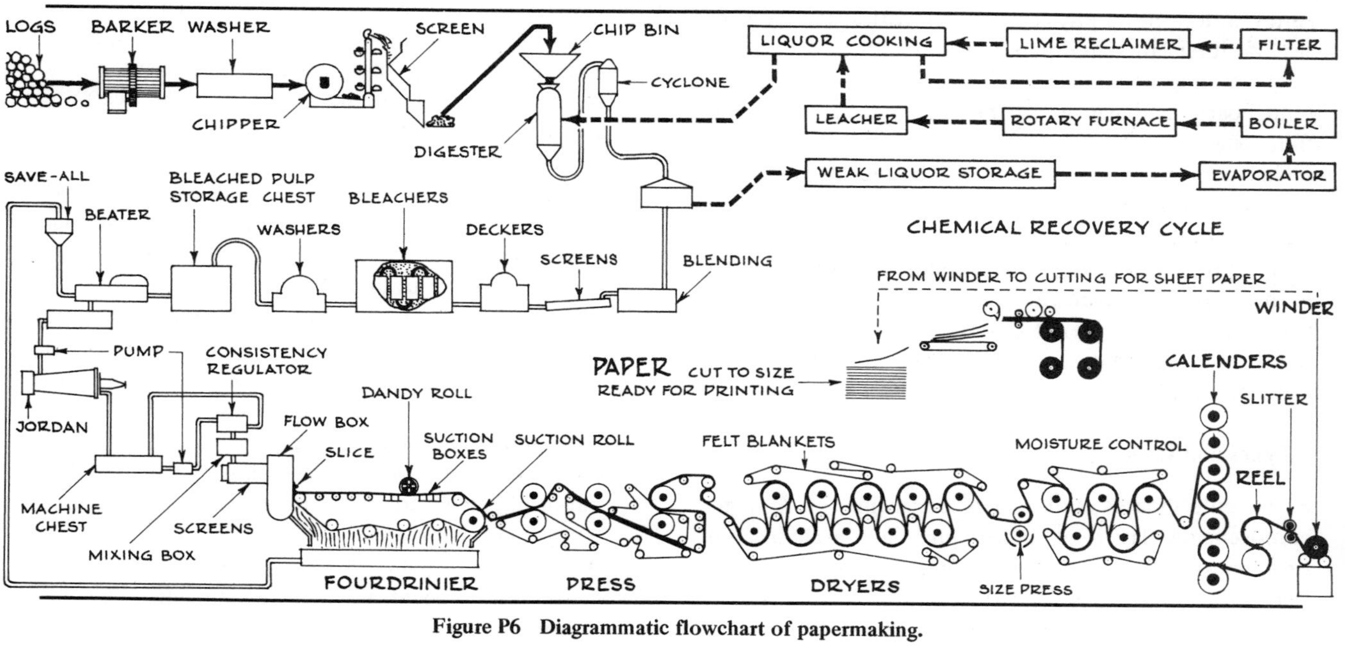

Figure P6 Diagrammatic flowchart of papermaking.

establishing the composition of the paper, other materials such as rosin, synthetic resins, alum, dyestuffs, clay, and chalk are added, depending on the type of paper to be made.

Formation on Rollers. From the beater the liquid mass of pulp flows down to the paper machine, where it first passes over an endless wire mesh on rollers through which the water drains off and a sheet of paper is formed. In this group of rollers, the top roller (dandy) for smoothing the surface may also contain the wire or rubber form which makes the watermark. From the wire rollers the newly formed sheet passes through a suction roller (at this point it contains about 4 lb of water/lb of paper) and then through a series of rolls which press the paper to remove more water, leaving a moisture content of about 65%. The next set of rollers consists of steam-heated dryers that reduce the moisture content to about 5 to 8%. The sheet then passes through a series of rolls which give it the final surface finish. (*See* Figure *P6*.)

In the case of many building materials, the paper in roll or sheet form goes through the laminator. As many as six sheets may be laminated for integrants (wall boards) and asphalted paper products.

PAPER PULP BOARDS AND PANELS

Physical and Chemical Properties

Paper pulp board and panel material for construction varies widely in its physical characteristics. It may consist of either laminated paper pulp boards or panels or a single homogeneous layer of wood paper pulp. Paper pulp boards and panels can be (1) surfaced with metal foil, plastic coating, or film; (2) impregnated with asphaltic materials or synthetic resins; and (3) laminated in combinations of metal foil or foils and plastic foams (*see* Paper Pulp Decking).

The type of board and panel should always be selected to meet its particular end use. Paper pulp boards and panels may be considered as thicker types or paper pulp sheets and rigidized sheets (*see* Paper and Paper Pulp Products, Table *P18*).

Types and Uses

Paper pulp boards and panels consisting of a single homogeneous layer of paper pulp are available for small-construction sheathing material with aluminum foil on one surface, impregnated with asphaltic material or plain (*see* Insulation and Insulating Materials).

Dimensions: usually 4 × 8 ft (1.219 × 2.438 m), and $\frac{1}{2}$, $\frac{5}{8}$, $\frac{3}{4}$, and 1 in. (12.70, 15.88, 19.05, and 25.44 mm) thick.

Paper pulp boards and panels consisting of a single homogeneous layer of paper pulp or laminated layers are available for roof insulation.

Dimensions: usually 2 × 4 ft (0.610 × 1.219 m), and $\frac{3}{4}$, 1, $1\frac{1}{2}$, and 2 in. (19.05, 25.4, 38.1 and 50.8 mm) thick.

Laminated paper pulp boards and panels with plastic foam or metal foil or various combinations are used for roof insulation and as part of roofing systems. Manufacturers should be consulted in regard to current data, sizes, thickness, U and R factors, and methods of application. Paper pulp acoustical tiles are available plain, perforated, textured, or patterned.

Dimensions: 1 × 1 ft (0.305 × 0.305 m) and 1.33 × 1.33 ft (0.406 × 0.406 m), and $\frac{1}{2}$ and $\frac{3}{4}$ in. (12.70 and 19.05 mm) thick.

Some types of paper pulp boards or panels are available up to a maximum size of 8 × 16 ft (2.438 × 4.877 m) and $\frac{1}{2}$ in. (12.7 mm) thick.

Paper pulp boards and panels are available with various textures, surface finishes, colors, and patterns for interior wall finishes (*see* Table *P23*).

Application

Condensed Checklist

1. Paper pulp board and panel materials change slightly in size in all directions with variations in humidity. Therefore, during installation, particularly in dry weather, they should be conditioned by wetting, always following the manufacturer's specifications for amount and method of wetting.
2. Boards and panels applied to walls and roofs as sheathing are secured at supports with sheathing-type nails spaced 8 in. (203.2 mm) o.c., and 4 in. (101.6 mm) o.c. around the edges.
3. Decorative boards and panels applied to interior walls are secured at supports with $1\frac{1}{2}$-in. (38.1-mm) finishing nails spaced 8 in. (203.2 mm) o.c., and 4 in. (101.6 mm) o.c. around the edges.
4. Nails should be placed a minimum of $\frac{3}{8}$ in. (9.525 mm) from the edges of the board or panel, and if flat-head screws are used, holes should be precountersunk in the material.
5. When board or panel material is to be applied with mastic, the type of adhesive recommended by the manufacturer of the paper pulp sheet material

Table P23 Types of paper pulp boards and panels used in construction

Type of paper pulp	Dimensions: Thickness in.	mm	Width ft	m	Length ft	mm	Major use
Acoustical tiles[a]	$\frac{1}{2}$	12.70	1	0.305	1	0.305	Acoustic ceiling treatment
	$\frac{3}{4}$	19.05	1.33	0.406	1.33	0.406	
Light construction sheathing	$\frac{1}{2}$	12.70	4	1.219	8	2.438	Wall and roof sheathing for residential and light-frame construction
	$\frac{5}{8}$	15.88					
	$\frac{3}{4}$	19.05					
	1	25.40					
Paper pulp boards or panels, laminated with plastic foam	1–4	25.4–101.6	2	0.610	4	1.219	Flat-roof insulation generally in combination with roofing systems
Paper pulp boards and panels with decorative finishes[b]	$\frac{1}{2}$	12.70	4	1.219	8	2.438	Interior wall finishes
	$\frac{3}{4}$	19.05					
Roof insulation[c]	$\frac{3}{4}$	19.05	2	0.610	4	1.219	Flat-roof insulation
	1	25.40					
	$1\frac{1}{2}$	38.10					
	2	50.80					

[a]Check codes for fire-resistance requirements for acoustical tile ceilings.
[b]These are available with T & G (tongue-and-groove) so nailing can be concealed.
[c]Always check for flame-spread and fire resistance.

should be used, following his directions for use and methods of supporting the sheet until the adhesive has set.

6. Joints must not be set completely tight but must allow for inherent movement caused by changes in humidity. Joints may be either exposed or covered.
7. When using paper pulp board or panel material, a check should be made of fire-resistant and flame-spread requirements as well as all governing codes and fire insurance company regulations.
8. When using paper pulp acoustical tiles, one should always check with manufacturers for current methods for application, including type of adhesive, suspension system, and furring.
9. When paper pulp board or panels are used for roof insulation, the manufacturers should always be consulted for the correct methods of attaching insulation to the roof substrate.
10. When paper pulp boards or panels laminated or in combination with plastic foam are used for roofing insulation, a check should always be made with roofing manufacturers for the correct type of roofing material or roofing system for use with such insulation, and for the correct method of application.

Conditions Favorable to the Use of Paper Pulp Boards and Panels

1. Where a strong, durable wall or ceiling finish with a panel effect is desired.
2. Where underlayment for resilient flooring is needed to bring various floors to the same elevation.
3. Where a colored, textured, and patterned prefinished material is desired for interior walls, ceilings, and built-in furniture.
4. Where a durable, strong, acoustical material of the perforated type is desired for walls and ceilings.
5. Where a perforated surface upon which various objects can be hung or supported is desired, particularly if flexibility of hanging is important.

Conditions Unfavorable to the Use of Paper Pulp Boards and Panels

1. In areas where atmospheric humidity is high or where moisture can collect. Plastic-surfaced prefinished

paper pulp boards and panels may be used, but all joints must be made waterproof.

2. Where a smooth unbroken wall or ceiling surface is desired.

History and Manufacture

Paper pulp boards and panels were developed soon after the invention of the papermaking machine in 1799, but it was not until about 1914 that paper board to serve as a construction material was produced. Research in plastics and adhesives and the rapid advances that were made in industrial paper processes during the years 1914 to 1930 resulted in the development of fiber and paper pulp boards and panels for both exterior and interior wall construction, insulation, sheathing, and acoustical tile-type materials. (*See also* Adhesives; Paints; Paper and Paper Pulp Products; Plastics; Rubbers.)

PAPER PULP DECKING

Physical and Chemical Properties

Roof decking consists of several thicknesses of paper pulp or fiber board laminated with waterproof adhesives and combined with various surface materials such as asbestos cement, plywood, metal foil, plastic film, plastic laminates, paper, and asphalt. Some of the laminations may be asphalt-impregnated. Thus the decking may incorporate materials that add to its inherent strength, increase its insulation and acoustic values, act as a vapor barrier, provide a nailing surface, and serve as a finish material for the interior ceiling (*see* Decking).

Types and Uses

Paper pulp decking is manufactured in various thicknesses, lengths, widths, and finishes. Three types of edge treatment are available: (1) tongue-and-groove on two sides and squared-off edges on two sides, (2) squared edges on all four sides, and (3) tongue-and-groove on all four sides. Table *P24* summarizes the pertinent data, including maximum span and conditions of use, for each type.

All types of paper pulp decking, because they serve simultaneously as structural roof decking, insulation, vapor barrier, and interior finish ceiling, must be carefully chosen to be suitable for the climatic conditions at the location of use. When selecting paper pulp decking, the following factor must also be carefully considered:

1. The range of interior relative humidity in the proposed building.

2. The range of outside temperatures in the area where the building is to be constructed, particularly when temperatures lower than 32°F (0°C) are normally encountered.

3. The type of roofing material to be installed over the decking. If it is to be nailed onto the decking, the thickness of decking required and the type of nail to be used must be determined (*see* Nails). If the decking is to be covered with built-up roofing with hot asphalt or tar, the method of joining the decking to prevent leakage of tar or asphalt must also be decided.

4. The thermal insulating value to be met by the decking and the location of the vapor barrier in connection with this value, so as to prevent condensation on the interior surface of the decking.

5. The type, color, texture, and acoustic value of the prefinished interior ceiling surface.

Table P24 Paper pulp roof decking

Type of decking		Span		Dimensions						
				Thickness		Width		Length		
Treatment of edges	Surface finishes and other characteristics	ft	m	in.	mm	ft	m	ft	m	Conditions of use
Tongue and groove on two sides, square edges at ends	Prefinished painted surface on one side; vapor barrier included	2.67	0.813	$1\frac{3}{8}$	34.93	2	0.610	8	2.438	All climates (interior relative humidity should be kept below 50%)
		4.00	1.219	$1\frac{7}{8}$	47.63	2	0.610	8	2.438	
				2	50.80					
				3	76.20					
	Prefinished acoustic paper pulp sheet on one side; vapor barrier included	4.00	1.219	$2\frac{1}{16}$	52.39	2	0.610	8	2.438	
				$3\frac{1}{16}$	77.79					
	Prefinished acoustic perforated paper pulp sheet on one side; vapor barrior included	4.00	1.219	2	50.80	2	0.610	8	2.438	
				3	76.20					

Table P24 Paper pulp roof decking (*continued*)

Type of decking: Treatment of edges	Type of decking: Surface finishes and other characteristics	Span ft	Span m	Thickness in.	Thickness mm	Width ft	Width m	Length ft	Length m	Conditions of use
Tongue and groove on two sides, square edges at ends	Prefinished plywood sheet on one side; vapor barrior included	4.00	1.219	$2\frac{1}{4}$	57.15	2.75	0.838	8	2.438	
				$2\frac{5}{8}$	66.68					
	Prefinished painted surface on one side; without vapor barrier	1.33	0.406	$\frac{5}{8}$	15.88	2	0.610	8	2.438	Climates where temperature does not fall below 40°F (4.44°C) (interior relative humidity should be kept below 50%)
		2.00	0.610	$\frac{15}{16}$	23.81	2	0.610	8	2.438	
		2.67	0.813	$1\frac{3}{8}$	34.93	2	0.610			
		4.00	1.219	$1\frac{7}{8}$	47.63	2	0.610	8	2.438	
				2	50.80					
				3	76.20					
	Prefinished acoustic paper pulp sheet on one side; without vapor barrier	4.00	1.219	$2\frac{1}{16}$	52.39	2	0.610	8	2.438	
				$3\frac{1}{16}$	77.79					
	Prefinished acoustic perforated paper pulp sheet on one side; without vapor barrier	4.00	1.219	2	50.80	2	0.610	8	2.438	
				3	76.20					
	Prefinished plywood sheet on one side; without vapor barrier	4.00	1.219	$2\frac{1}{4}$	57.15	2.75	0.838	8	2.438	
Square edges on four sides	Prefinished painted surface on one side; without vapor barrier	1.33	0.406	$\frac{5}{8}$	15.88	2	0.610	8	2.438	Climates where temperature does not fall below 40°F (4.44°C) (interior relative humidity should be kept below 50%)
						3	0.914			
		1.33	0.406	$\frac{5}{8}$	15.88	4	1.219	4–8 in 1-ft increments	1.219–2.438 m in 0.305-m increments	
						8	2.438			
		2.00	0.610	$\frac{15}{16}$	23.81	2	0.610	4–8 in 1-ft increments	1.219–2.438 m in 0.305-m increments	
						3	0.914			
	Both sides covered with asphalt coating (for vapor barrier) and then asbestos-cement sheets applied as finish surface	4.00	1.219	$\frac{9}{16}$	14.29	2	0.610	8	2.438	
				2	50.80	2				
Tongue and groove on four sides	Prefinished painted surface on one side; vapor barrier included	2.67	0.813	2	50.80	2	0.610	8	2.438	All climates (interior relative humidity should be kept below 50%)
		4.00	1.219	3	76.20	2	0.610	8	2.438	
	Prefinished painted surface on one side; without vapor barrier	2.67	0.813	2	50.80	2	0.610	8	24.38	Climates where temperature does not fall below 40°F (4.44°C) (interior relative humidity should be kept below 50%)
		4.00	1.219	3	76.20	2	0.610	8	24.38	

Application

Condensed Checklist

1. The maximum unsupported span of the decking selected should be checked.
2. Local, municipal, and state codes and also the codes of the Fire Underwriters, insurance companies, labor departments, and federal government (Army, Navy, etc.) should be checked for limitations and fire ratings.
3. Specifications should require that the material be carefully handled so that edges are not damaged and the interior finish is not soiled or stained during shipping, handling, and installation.
4. The method of joining at the sides and ends (treatment of edges) in relation to air leakage, moisture penetration, and effect on the design of the interior finished ceiling must be carefully chosen.
5. Termination of the decking at gable ends, fascias, hips, ridges, valleys, and vertical walls must be correctly detailed.
6. Methods of flashing and counterflashing of any piercing of the decking such as openings for pipes, skylights, vents, chimneys, ventilating fans, etc., should be carefully checked.
7. The type of roofing material selected should be checked in relation to the required size of nails to make sure that the decking is of sufficient thickness that nails will not penetrate through the finish interior surface.
8. Manufacturers should be checked for current sizes, thicknesses, spans, and finishes.

Conditions Favorable to the Use of Paper Pulp Decking

1. Where the design requires a decking that will supply strength for spans between roof rafters, beams, and purlins as well as thermal insulation and a wide variety of prefinished surfaces for interior ceilings.

Conditions Unfavorable to the Use of Paper Pulp Decking

1. In areas where the relative interior humidity will be greater than 50%.
2. Where the structural system requires unsupported spans greater than 4 ft (1.219 m).

History and Manufacture

Paper pulp or fiber roof decking is a relatively new material, being actually first produced just before World War II. This product is an outgrowth of laminated paper pulp materials for prefabricated interior partitions. The manufacturing process is relatively simple in that it requires only the laminating of paper pulp or fiber boards with waterproof adhesives under factory-controlled conditions.

PAPER PULP SHEETS AND RIGIDIZED SHEETS

Physical and Chemical Properties

Paper pulp sheet material for construction varies widely in its physical and chemical characteristics. It may consist of either laminated sheets of paper or a single homogeneous layer of wood paper pulp; it may be veneered with a plastic film or aluminum foil; and it may have plain, patterned, colored, or decorative surface treatment. The sheets may be reinforced with glass or plastic fibers. They may have a core of corrugated paper or plastic foam, and they may be made very dense and hard by applying heat and pressure. Some types are impregnated with various synthetic resins or asphalt that increase their durability and strength. Sheets and rigidized sheets so treated are also moisture resistant. (*See* Tables *P25 and P26*.)

Types and Uses

Sheet and rigidized sheet consisting of a single homogeneous layer of wood pulp, compressed with heat and pressure, are available in three types—plain, perforated, and corrugated—and in combinations of these. One or both faces may have a finish surface.

Dimensions: usually 4 × 8 ft (1.219 × 2.438 m) and $\frac{1}{8}$, $\frac{3}{16}$, or $\frac{1}{4}$ in. (3.18, 4.76, or 6.35 mm) thick.

Perforations in a square pattern: usually 1 in. (25.4 mm) o.c. in both directions, with a $\frac{3}{16}$ in. (4.76 mm) diameter for sheet $\frac{1}{8}$ or $\frac{3}{16}$ in. (3.18 or 4.76 mm) thick, and a $\frac{9}{32}$ in. (7.05 mm) diameter for $\frac{1}{4}$-in. (6.35-mm)-thick sheet.

Perforations in a diagonal pattern: usually $\frac{1}{2}$ in. (12.70 mm) o.c. in both directions, with a $\frac{3}{16}$-in. (4.76-mm) diameter.

The integral color of sheets is usually in the browns, but this type of sheet is also made with a factory-applied paint finish, commonly cream, gray, or green, and with a plastic finish on one side, in a wide range of colors and patterns including wood grain, marble, and

Table P25 Properties of paper pulp sheets

Type of sheet	Thickness in.	Thickness mm	Specific gravity	Modulus of rupture[a] lbf/in.^{2}	Modulus of rupture[a] MN/m^2	Maximum absorption (percent)
Wood paper pulp,	$\frac{1}{8}$	3.18	1.02	6,100	42.06	20
untreated	$\frac{1}{4}$	6.35	1.02	5,850	40.34	20
Wood paper pulp,	$\frac{1}{8}$	3.18	1.11	10,350	71.36	12
treated with resin	$\frac{1}{4}$	6.35	1.05	9,800	67.57	12

[a] 1 lbf/in.^{2} = 0.006895 MN/m^2.

Table P26 Properties of laminated paper sheets impregnated with resins

Type of sheet	Thickness in.	Thickness mm	Specific gravity	Modulus of rupture[a] lbf/in.^{2}	Modulus of rupture[a] MN/m^2	Absorption (percent of weight after 24-hour immersion)
Laminated paper with 30% content of resin impregnation	$\frac{1}{8}$; $\frac{3}{32}$ and smaller	3.18; 2.35 and smaller	1.41	31,300	215.81	2

[a] 1 lbf/in.^{2} = 0.006895 MN/m^2.

Table P27 Laminated paper sheet material used for interior finishes

Dimensions: Width ft	Width m	Length ft	Length m	Thickness in.	Thickness mm	Major use
4	1.219	4–16 in 1-ft increments	1.219–4.877 in 0.305-m increments	$\frac{3}{32}$, $\frac{1}{8}$, $\frac{3}{16}$, $\frac{1}{4}$, $\frac{5}{16}$	2.35, 3.18, 4.76, 6.35 7.94	For interior finishes of walls and ceilings

terrazzo. This category of sheet material is strong and moisture resistant and shows little change in dimensions with changes in humidity.

In addition to the types described, sheets are available with various surface textures, for example, grooved vertically, grooved into squares, and ribbed.

Laminated Sheet Material. Laminated sheet material is widely used in construction for the surface finish of interior walls and ceilings and as core material, in plain, corrugated, or cellular form, for sandwich panel construction, doors, partitions, and furniture (*see* Table *P27*).

Laminated Paper Sheets. Laminated paper sheets, an important material in architectural, structural and mechanical engineering, and construction offices, are used for all types of drawings, including detailing and presentation-type drawings. Paper pulp sheet for a rendering or presentation should preferably be one that has a 100% rag paper surface or is a specially treated paper pulp with synthetic resins, as these types can withstand rough erasures without damage to the finish surface. Various textures and finishes—smooth, slightly rough, and rough—are available to fit the medium used to make the presentation drawing. Table *P28* gives the standard types and sizes of sheet material.

Bending of Sheets. All paper pulp sheet can be bent to various minimum radii by two methods: (1) the cold bend method, in which supporting bands hold the sheet in curved position against a rigid strong backing;

Table P28 Types and sizes of laminated sheet material used in drawing, rendering, and mounting[a]

Standard types of sheet or illustration board	Size[b] Width in.	Width cm	Length in.	Length cm	Approximate thickness
Illustration board[c]	20 and 22	50.80 and 55.88	30	76.20	Made in single thickness of $\frac{1}{16}$ in. (1.59 mm) and in double thickness of $\frac{1}{8}$ in. (3.18 mm)
	27	68.58	40	101.60	
	28	71.12	42	106.68	
	30	76.20	40 and 52	101.60 and 132.08	
	40	101.60	60	152.40	
Illustration board,[c] extra large	38	96.52	60	152.40	Made in triple thickness only of $\frac{3}{16}$–$\frac{1}{4}$ in. (4.76–6.35 mm)
	48	121.92	72	182.88	
	50	127.00	62	157.48	
Thin sheet[c,d] in various numbers of laminations	22	55.88	30	76.20	Made in 1, 2, 3, and 4 ply
	23	58.42	29	73.66	Made in 1, 2, 3, 4, and 5 ply
	30	76.20	40	101.60	Made in 1, 2, 3, 4, and 5 ply
Plain mat or mounting board	22	55.88	28	71.12	Made in 6, 10, 12, 14, and 18 ply
	30	76.20	40	101.60	Made in 14, 18, and 28 ply
	40	101.60	60	152.40	Made in 14, 18, and 28 ply
Colored poster-type board	28	71.20	44	111.76	Made in single thickness of $\frac{1}{16}$ in. (1.59 mm), in full range of colors and black and white
Metallic poster-type board	25	63.50	44	111.76	Made in single thickness of $\frac{1}{16}$ in. (1.56 mm) in gold, red, blue, and green
	28	71.20	44	111.76	Made in gold and silver
Cardboard	30	76.20	40	101.60	Made in double thickness of $\frac{1}{8}$ in. (3.18 mm)
	40	101.60	60	152.40	Made in triple thickness of $\frac{3}{16}$–$\frac{1}{4}$ in. (4.76–6.35 mm)

[a]The finish of the surface can vary from very smooth to very rough. Finest finishes are available in hot-pressed (smooth) and cold-pressed (slightly rough to rougher) types.

[b]Smaller sizes cut from these sizes are available. For example, a sheet 30 × 40 in. (76.2 × 101.6 cm) yields two 15 × 20 in. (38.1 × 50.8 cm) or four 10 × 15 in. (25.4 × 38.1 cm) sheets.

[c]The surface finish can be either rag paper or wood pulp paper.

[d]Includes a thin sheet made from reprocessed paper called "chipboard."

and (2) the cold-moist bend method, in which the sheet material is moistened and then dried into a self-supporting fixed curve.

Application

Condensed Checklist

1. Paper pulp sheet materials, except those impregnated with resin, change slightly in size in all directions with variations in humidity. Therefore, during installation, particularly in dry weather, they should be conditioned by wetting, always following the manufacturer's specifications for amount and method of wetting.
2. Sheets applied to walls are secured at supports with $1\frac{1}{2}$-in. (38.1-mm) finishing nails spaced 8 or 4 in. (203.2 or 101.6 mm) o.c. around the edges. Sheets $\frac{1}{8}$ in. (3.18 mm) thick should not be used without solid backing.
3. Sheets applied to ceilings are secured at supports with $1\frac{1}{2}$-in. (38.1-mm) finishing nails spaced 6 or 4 in. (152.4 or 101.6 mm) o.c. around the edges. Nails should be countersunk. Sheet material $\frac{1}{8}$ or $\frac{3}{16}$ in. (3.18 or 4.76 mm) thick should not be used unless applied to a solid backing.
4. Nails should be placed a minimum of $\frac{3}{8}$ in. (9.53 mm) from the edges of the sheet, and if flat-head screws are used, holes should be precountersunk in the sheet material.

5. When sheet material is to be applied with mastic, the type of adhesive recommended by the manufacturer of the paper pulp sheet material should be used, following his directions for use and methods of supporting the sheet until the adhesive has set.
6. Joints must not be set completely tight but must allow for inherent movement caused by changes in humidity. This rule applies to all types except the resin-impregnated. Joints may be either exposed or covered, as shown in Figures *P7* and *P8*.

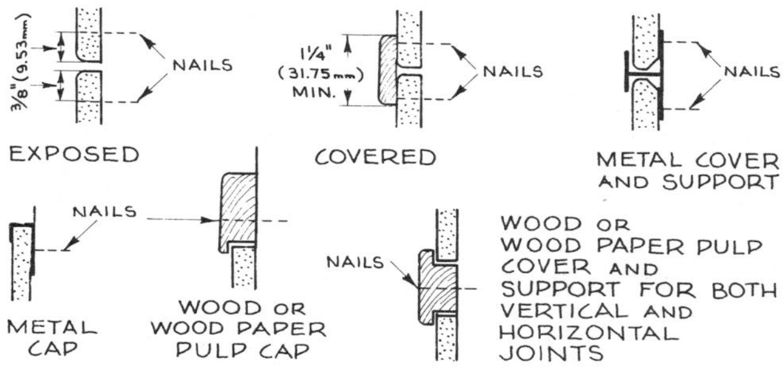

Figure P7 Details of horizontal and vertical joints.

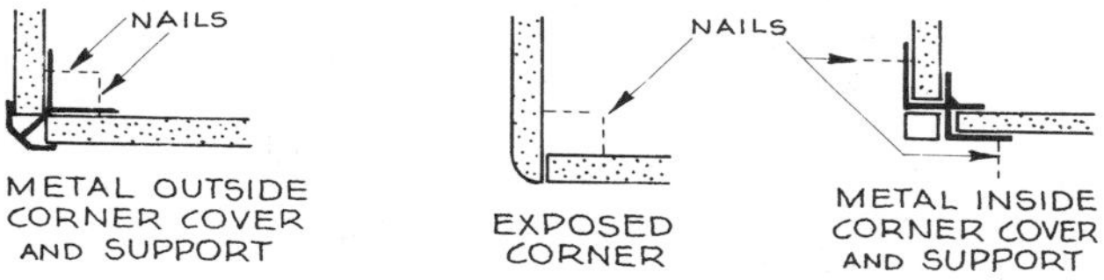

Figure P8 Details of corners.

7. When used as underlayment for resilient flooring, sheets should be nailed 4 in. (101.6 mm) o.c. in both directions.
8. Prefinished paper pulp sheets with a plastic surface should always be applied with adhesive and with cover supports of metal or other type of material.
9. When any prefabricated sandwich-type material that has a core of paper pulp sheet is used, the architect or other responsible person should learn from the manufacturer what its physical and mechanical properties are, how the paper pulp was treated, what type of adhesive was used, and what effects humidity might have. One should always consult with specialists on the proper structural and acoustic use of the material.
10. When perforated sheet is used, the type of perforation should be evaluated in terms of its acoustic value. If the sheet is to be used as pegboard, the correct method of support should be specified (*see* Wood: Hardwood and Particle Board).
11. A check should always be made with manufacturers for current types, sizes, finishes, and methods of application.
12. There are various methods for applying paper pulp sheets with high-strength adhesives to existing wall or ceiling surfaces and to new wood or metal stud or beam construction. A check should always be made with the manufacturers of adhesives, light-gauge metal, and paper pulp sheet to select the best materials and methods of application.

Conditions Favorable to the Use of Paper Pulp Sheets

1. Where a strong, durable wall or ceiling finish with a panel effect is desired.
2. Where a colored, textured, and patterned prefinished material is desired for interior walls, ceilings, and built-in furniture.

Conditions Unfavorable to the Use of Paper Pulp Sheets

1. In areas where atmospheric humidity is excessive or where moisture can collect. Plastic-surfaced prefinished paper pulp sheet or sheet with laminated plastic film may be used, but all joints must be made waterproof.
2. Where a smooth unbroken wall or ceiling surface is desired.

History and Manufacture

Paper pulp sheets for purposes other than writing and printing were developed soon after the invention of the papermaking machine in 1799 (e.g., wallpaper), but it was not until about 1914 that paper sheet to serve as a construction material was produced. From 1930 to the present, paper pulp sheets have undergone radical changes due to the development of synthetic resins and films. The most important change in construction sheet material was the development of waterproof, washable, and durable surface finishes and resin impregnation to produce exterior materials. Research in plastics and adhesives and the rapid advances that were made in industrial paper processes during the years 1914 to 1930 resulted in the development of fiber and paper pulp sheets for both exterior and interior wall construction. Such wood paper pulp and laminated paper products in constantly improved form undoubtedly will continue to appear because they are related directly to advances in plastics and adhesives.

PARKERIZING

Parkerizing is a trademarked process for producing a rust-inhibitive phosphate coating on iron or steel. It is achieved by dipping the metal in acid phosphate solution. *See also* Bonderizing; Metals: Mechanical Finishes; Metals: Sprayed Metal Coatings.

PARTITIONS

Partitions are defined in this book as the walls that divide interior space within a building. They may be either permanent or movable. Tables *P29* and *P30* list the materials commonly used for both types.

Table P29 Permanent built-in partitions

Material	Thickness	Finish	Generally accepted maximum unsupported height	Major use
Aluminum: expanded metal type	0.046–1.187 in. (1.17–30.15 mm)	Natural unfinished, anodized, or porcelain-enameled	10 ft (3.048 m)	Where decorative perforated, colored, or natural aluminum metallic partitions are desired
Aluminum mesh	0.035–1.00 in. (0.89–25.4 mm)	Natural unfinished, anodized, or plastic-coated	8 ft (2.438 m)	Where utility-type wire mesh partitions which can be colored are desired
Brick	$3\frac{1}{2}$–$3\frac{3}{4}$ in. (88.9–95.25 mm)	Requires no finish	9 ft (2.743 m)	Decorative partitions
Clay tile: non-load-bearing structural tile	2, 3, 4 in. (50.8, 76.2, 101.6 mm)	Requires plaster or other type of finish	9 ft (2.743 m) when not over 6 ft (1.829 m) long for 2 in. (50.8 mm) thickness, 12 ft (3.658 m) for 3 in. (76.2 mm), 15 ft (4.572 m) for 4 in. (101.6 mm)	General non-load-bearing partitions
Clay tile: structural facing tile	$3\frac{3}{4}$ in. (95.25 mm)	Colored, textured, or patterned finish on one or both sides	17 ft (5.182 m)	Partitions where durable, colored, texture- or pattern-finished surfaces are desired
Concrete block	$2\frac{5}{8}$, $3\frac{5}{8}$, $5\frac{5}{8}$ in. (66.68, 92.08, 142.88 mm)	Generally painted, plastered, or given special finishes applied by furring or adhesive	9 ft (2.743 m) for $2\frac{5}{8}$ in. (66.68 mm); 12 ft (3.658 m) for $3\frac{5}{8}$ and $5\frac{5}{8}$ in. (92.08 and 142.88 mm)	General nonload-bearing partitions
Concrete block with glazed finish	$1\frac{3}{4}$ in. (44.45 mm)	Glazed finish on one or both sides	9 ft (2.743 m), not over 6 ft (1.829 m) long	Where durable, colored, texture- or pattern-finished surfaces are desired
	$3\frac{3}{4}$ in. (95.25 mm)	Glazed finish on one or both sides	12 ft (3.658 m)	Where durable, colored, texture- or pattern-finished surfaces are desired
Glass block	$3\frac{7}{8}$ in. (98.43 mm)	Clear, translucent, or opaque (colored)	Up to 25 ft (7.620 m) high and 25 ft (7.620 m) wide	Where clear, translucent, or opaque (colored) decorative partitions are desired
Glass: corrugated	1 in. (25.4 mm)	Translucent, textured finish	12 ft (3.658 m)	Decorative translucent partitions

Table P29 Permanent built-in partitions (*continued*)

Material	Thickness	Finish	Generally accepted maximum unsupported height	Major use
Gypsum block	2, 3, 4, 6 in. (50.8, 76.2, 101.6 mm, 152.4 mm)	Requires plaster finish or other type of applied finish material	9 ft (2.743 m) for 2 in. (50.8 mm), 13 ft (3.962 m) for 3 in. (76.2 mm), 17 ft (5.182 m) for 4 in. (101.6 mm), 20 ft (6.096 m) for 6 in. (152.4 mm)	For fire-rated partitions and also where painted or other type of finish surface is desired
Plaster	2 in. (50.8 mm)	Plaster on both faces	9 ft (2.743 m) high, as wire mesh has to be anchored to floor and ceiling	Where thin partitions and painted or repairable finishes are desired
Stainless steel: expanded metal type	0.042–0.296 in. (1.07–7.52 mm)	High-polish, polished, brushed, or natural cold- or hot-rolled finish; limited color	8 ft (2.438 m)	Where decorative, perforated metallic or colored (minimum selection) partitions are desired
Steel: expanded metal type	0.022–1.187 in. (0.56–30.15 mm)	Painted, plated, or porcelain-enameled	10 ft (3.048 m)	Where decorative, perforated, colored, or various metallic colored partitions are desired
Steel mesh	0.035–1.00 in. (0.89–25.4 mm)	Galvanized, painted, or plastic-coated	8 ft (2.438 m)	Where utility-type wire mesh partitions that can be colored are desired
Stone: ashlar	4 in. (101.6 mm) (requires 4 in. or 101.6 mm of back-up material)	Wide variety of textures from smooth to very rough; limited range of colors	25 ft (7.620 m)	Where durable, decorative partitions (with stone color and texture) are desired
Stone: cut stone	$1\frac{3}{4}$–$2\frac{1}{2}$ in. (44.45–63.5 mm) (requires minimum of 7 in. or 177.8 mm, including 4 in. or 101.6 mm of back-up material)	Smooth or rough textured, or polished in a wide variety of colors and color combinations	25 ft (7.620 m)	Where durable, decorative partitions are desired
Wood	$1\frac{1}{2}$, $3\frac{1}{2}$ in. (38.1, 88.9 mm) rough	Plaster, plywood, paneling, plasterboard, etc., applied to wood studs	10 ft (3.048 m)	Residential and light construction

Table P30 Materials used for prefabricated and field-installed partitions

Material[a]	Advantages	Disadvantages	Major uses
Aluminum (*see* Aluminum Sheet and Strip)	Wide range of colors in porcelain enamel; lightweight; wide range of stock widths and sizes for toilet room partitions; fairly easily installed	Requires special erection and assembly systems; provisions for installing electrical wiring have to be a part of the partition; requires shop and assembly drawings	Toilet and shower stalls, toilet room partitions, office partitions
Asbestos (*see* Asbestos-Cement Board and Sheet; Asbestos Panels and Integrants; Fibers; Paper and Paper Pulp Products)	Has fire rating; fairly easily installed	Limited range of colors; doors require special bracing and framing; electrical wiring can be installed only in base; requires shop and installation drawings	Office partitions
Glass, corrugated (*see* Glass, Corrugated)	Rough texture; hard surface; translucent; easily installed	Limited sizes, in both width and length; doors require special bracing and framing; electrical wiring can be installed only in base	Decorative office partitions
Glass, structural (*see* Glass, Structural)	High gloss; hard finish; fairly wide range of colors	Requires shop and assembly drawings; wall surfaces are often finished in same material	Toilet and shower stalls, toilet room partitions
Marble (*see* Stone, Marble)	Wide range of colors; polished hard surface	Requires shop and installation drawings; in many cases wall surfaces are finished in same material	Toilet and shower stalls, toilet room partitions
Mesh (*see* Mesh and Wire Cloth)	Easily installed; wide variety of types of mesh from purely utilitarian to decorative	Provides no privacy	Storage and utility type of partitions, decorative screens
Plaster (*see* Plaster Integrants)	Easily installed; easily painted	No color selection; doors require special bracing and framing; electrical wiring can be installed only in base	Office partitions
Plastic (foam) (*see* Plastics: Sheet; Plastics: Laminates; Plastics: Surfacing, Sidings, and Panels)	Wide range of colors, textures, finishes	Requires special erection and assembly systems; provisions for installing electrical wiring as part of the partition	Office partitions
Plastic (*see* Plastics)	Wide range of colors and textures; finishes can be translucent or opaque; easily installed	Only decorative value; provides no privacy	Decorative panels and partitions
Slate (*see* Stone, Slate)	Polished hard surface	Requires shop and installation drawings; in many cases wall surfaces are finished in same material	Toilet and shower stalls, toilet room partitions

Table P30 Materials used for prefabricated and field-installed partitions (*continued*)

Material[a]	Advantages	Disadvantages	Major uses
Steel (*see* Steel Sheet and Strip)	Wide range of colors in porcelain enamel, baked enamel, and painted; wide range of stock widths and sizes for toilet room partitions; easily installed; has fire rating	Requires special erection and assembly systems; provisions for electrical wiring have to be a part of the partition; requires shop and installation drawings	Toilet and shower stalls, toilet room partitions, office partitions
Wood (*see* Wood Integrants; Panels and Sandwich Panels)	Wide range of types of wood for finishes; easily installed; easily painted	Requires shop and assembly drawings; provisions for electrical wiring have to be a part of the partition	Office partitions

[a] Aluminum, steel, and wood are the materials most widely used for prefabricated office partitions. These prefabricated partitions are available as solid full height or with glass above. They can also be acoustically treated.

PERLITE

Perlite is a glassy, volcanic rock, similar to obsidian. Its general composition is as follows: silicon dioxide 65 to 75%, aluminum oxide 10 to 20%, water 2 to 5%, and small amounts of soda ($NaCO_3$), potash (KCO_3), and lime (CaO). Its specific gravity is about 2.4 (unexpanded).

Perlite is often characterized by a minute globular structure and a high percentage of loosely combined water. As a result of these two characteristics, when perlite is heated quickly to softening (above 1500°F or 815.56°C), the trapped water, now converted into steam, forces it to expand many times its volume to form a light, fluffy, cellular material similar in structure to pumice. For example, a cubic foot of crude perlite weighs 85 lb (38.56 kg), and a cubic foot of expanded perlite, 7.5 to 12 lb (3.402 to 5.54 kg). Expanded perlite is made in processing furnaces which are patented by various companies, and many of the perlite products have also been patented.

Expanded perlite is widely used in construction as a lightweight, inert, fire-resistant aggregate with acoustic and thermal insulation value for concrete, plaster, stucco, etc. It is also used as loose fill insulation, as a filler in paints and plastics, and for refractory brick, roofing tile, and special filtering purposes.

PETROCHEMICALS

The petrochemical industry has grown from a very small start in the early 1940s to a major industry that is still developing. Petroleum, natural gas, and coal are the present raw materials from which the organic compounds known as petrochemicals are derived. At present, 175 substances are designated as petrochemicals, including paraffin, olefin, naphthene, methane, ethane, propane, ethylene, propylene, benzene, and toluene.

Ammonia is considered here because the hydrogen used to form ammonia is a product of petroleum refining. Synthetic fertilizers are also petrochemicals.

Because of the long-term outlook concerning fossil fuels, this industry will be researching and developing new types of raw materials for the production of petrochemicals.

PETROLEUM

Petroleum may vary from a dark, syrupy liquid to a clear, colorless liquid; it often contains hydrogen-carbon-sulfur compounds. Four main types of hydrocarbons are present in crude oil: normal paraffins, isoparaffins, cycloparaffins (also called naphthenes), and aromatics. The U.S. Bureau of Mines classifies crude oils as paraffin base, naphthene base, or mixed base by determining the properties of key fractions distilled from the oil. In addition to the hydrocarbons, compounds of sulfur, nitrogen, and oxygen are present in small amounts, and usually there are traces of vanadium, nickel, chlorine, and arsenic.

The word "petroleum" comes from the Greek and Latin words *petra* ("rock") and *oleum* ("oil"). The Dead Sea was known as Lake Asphaltites because of the semisolid petroleum washed up on its shores from underwater storages.

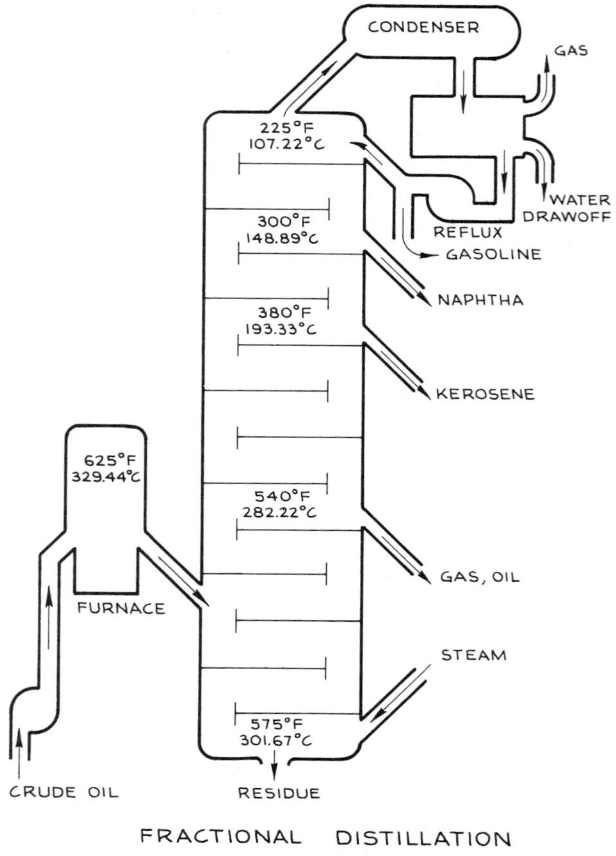

Figure P9 Fractional distillation of petroleum.

Table P31 Basic products from petroleum refining

Product	Use
Gases	Industrial heating
	Petrochemicals
	Liquid petroleum
Gasolines	Motor gasolines
	Aviation gasolines
Gas and diesel oils	Diesel engine fuels
Fuel oils	Residential, commercial, industrial heating, and electrical energy production
Lubricating oils	Automobile, bus, truck, etc., oils
	Gear oils
	Greases
Asphaltic bitumens	Roofing, paving, flooring, waterproofing and damp-proofing, coatings, protective coatings
Waxes	Surface and wood treatments
Thinners	Paints (organic coatings) and varnishes
Carbon black	Automotive tires, dyes and color pigment
Petroleum chemicals	Olefins
	Inorganic chemicals
	Organic chemicals
	Aromatic compounds
	Polymers
By-products	Paint dryers, emulsifying agents
	Cresylic acids

Until the beginning of the 19th century, illumination in the world was not much better than the illumination of the ancient Greeks and Romans. By the middle of the 19th century, kerosene or coal oil, from the distillation of coal, was in common use in America and Europe as an illuminating oil. It was found that crude oil yielded kerosene, and sources of crude oil were eagerly sought. The first well specifically drilled for oil was completed in August 1859 by Colonel Edwin L. Drake in Pennsylvania, and the petroleum industry was born. By 1969 there were over 600,000 producing oil wells in 60 different countries. Figure *P9* shows diagrammatically the fractional distillation of petroleum.

Over 80% of the world's organic chemicals are made from petroleum. Residential, commercial, and industrial heating is based to a great extent on petroleum or gas (natural gas mixed with petroleum gases).

The petroleum refining process produces the following major products: gases, gasolines, kerosene, gas and diesel oils, fuel oils, lubricating oils, thinners and waxes, asphaltic bitumens, organic chemicals, inorganic chemicals, aromatic compounds, polymers, and by-products. (*See* Table *P31*.)

In the construction field, perhaps the most important products from petroleum are the polymers: plastics, synthetic rubbers, and synthetic fibers. In plastics, the major products are polyethene, polyvinyl chloride, and polystyrene. In synthetic rubbers the major raw materials are butadiene, ethylene, benzene, and propylene; and in fibers, benzene (the raw material for nylon), ethylene (the raw material for polyesters), and the propylene derivative, acrylonitrile, for acrylics.

Coal is also a fossil fuel which contains the same basic constituents as crude oil. Fortunately, vegetable matter and trees contain the same basic ingredients as crude oil, but unfortunately not in a gaseous, solid, or liquid state such as our fossil fuels.

See also Coal; Fossil Fuels; Natural Gas; Petrochemicals; Plastics.

PHOSPHATIZING

The process of producing a phosphate conversion coating on ferrous metal by dipping it into a suitable aqueous solution of phosphoric acid is called phosphatizing. Its purpose is to improve paint adhesion and increase corrosion resistance.

PHOSPHOR

A phosphor is any substance which exhibits fluorescence or phosphorescence, that is, radiates light on impact of light of a different wavelength. The common phosphors used as coatings on fluorescent tubes are zinc sulfide, calcium tungstate, magnesium tungstate, zinc silicate, cadmium silicate, and cadmium borate.

PHOSPHORUS

Physical and Chemical Properties

Symbol: P
Atomic number: 15
Specific gravity: 1.82 (white), 2.20 (red), 2.25 to 2.69 (black)
Melting point: 44.1°C, 111.38°F
Boiling point: 280°C, 536°F

Phosphorus is a nonmetallic element that exists in several allotropic forms, among them white, red, and black.

White phosphorus (sometimes yellowish because of impurities) is a soft, waxy solid with a specific gravity of 1.83 and a melting point of 11.2°F (-11.55°C). It is insoluble in water and alcohol. At room temperature it exhibits phosphorescence (i.e., glows in the dark) and ignites spontaneously in air. It must therefore be handled submerged in water and also with care, as it is poisonous and can cause severe burns.

Red phosphorus is an amorphous powder with a specific gravity of 2.2 and a melting point of 1337°F (730.56°C). It is insoluble in all solvents and ignites in air at 500°F (260°C).

Black phosphorus consists of lustrous black crystals and is also insoluble in all solvents.

White and red phosphorus are commercially available but have few nonincendiary uses.

Phosphorus does not occur free in nature. Its compounds, however, are abundant and widely distributed, to be found in many rocks and mineral deposits.

Table P32 Uses of phosphorus

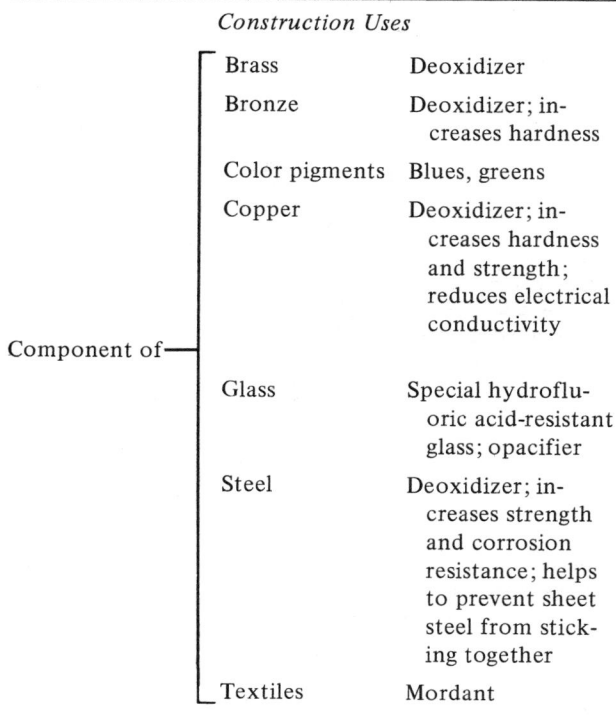

	Construction	Uses
Component of	Brass	Deoxidizer
	Bronze	Deoxidizer; increases hardness
	Color pigments	Blues, greens
	Copper	Deoxidizer; increases hardness and strength; reduces electrical conductivity
	Glass	Special hydrofluoric acid-resistant glass; opacifier
	Steel	Deoxidizer; increases strength and corrosion resistance; helps to prevent sheet steel from sticking together
	Textiles	Mordant

Phosphorus is one of the elements essential to the growth and development of plants and animals.

Types and Uses

Various phosphorus compounds find important uses in metallurgy as alloying elements and deoxidizing agents in iron, steel, bronze, and brass. Alloyed with tin and copper, phosphorus is used to deoxidize and add copper to brass and bronze; as ferrophosphorus, it is used to add phosphorus to iron and steel. Its other uses in compound form are shown in Table *P32.*

History and Manufacture

The Greek word *phosphorus*, "meaning light bearing," was the ancient name for the planet Venus when it appeared before sunrise. Phosphorus was discovered in 1669 by Brand, who prepared it from urine. About a hundred years later Karl Scheele developed a much easier method of preparing the element from bones. In modern industrial methods phosphate rock (tricalcium phosphate), when heated, produces elementary phosphorus vapor, which is collected under water.

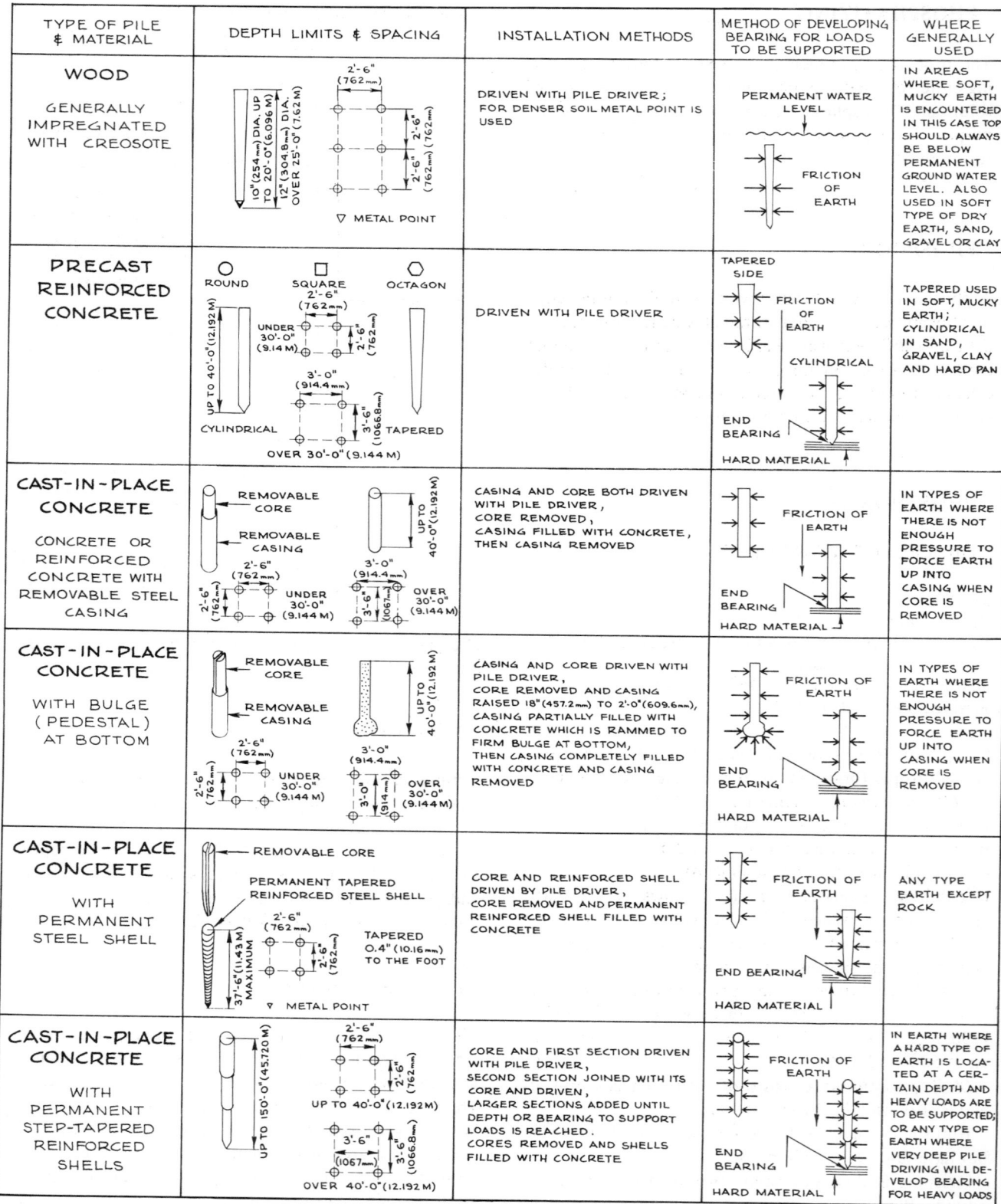

TYPE OF PILE & MATERIAL	DEPTH LIMITS & SPACING	INSTALLATION METHODS	METHOD OF DEVELOPING BEARING FOR LOADS TO BE SUPPORTED	WHERE GENERALLY USED
WOOD GENERALLY IMPREGNATED WITH CREOSOTE	10" (254 mm) DIA. UP TO 20'-0" (6.096 M) 12" (304.8 mm) DIA. OVER 25'-0" (7.62 M) 2'-6" (762 mm) 2'-6" (762 mm) 2'-6" (762 mm) ▽ METAL POINT	DRIVEN WITH PILE DRIVER; FOR DENSER SOIL METAL POINT IS USED	PERMANENT WATER LEVEL FRICTION OF EARTH	IN AREAS WHERE SOFT, MUCKY EARTH IS ENCOUNTERED; IN THIS CASE TOP SHOULD ALWAYS BE BELOW PERMANENT GROUND WATER LEVEL. ALSO USED IN SOFT TYPE OF DRY EARTH, SAND, GRAVEL OR CLAY
PRECAST REINFORCED CONCRETE	ROUND SQUARE OCTAGON UP TO 40'-0" (12.192 M) 2'-6" (762 mm) UNDER 30'-0" (9.14 M) 2'-6" (762 mm) 3'-0" (914.4 mm) 3'-6" (1066.8 mm) CYLINDRICAL TAPERED OVER 30'-0" (9.144 M)	DRIVEN WITH PILE DRIVER	TAPERED SIDE FRICTION OF EARTH CYLINDRICAL END BEARING HARD MATERIAL	TAPERED USED IN SOFT, MUCKY EARTH; CYLINDRICAL IN SAND, GRAVEL, CLAY AND HARD PAN
CAST-IN-PLACE CONCRETE CONCRETE OR REINFORCED CONCRETE WITH REMOVABLE STEEL CASING	REMOVABLE CORE REMOVABLE CASING UP TO 40'-0" (12.192 M) 2'-6" (762 mm) 2'-6" (762 mm) UNDER 30'-0" (9.144 M) 3'-0" (914.4 mm) 3'-6" (1067 mm) OVER 30'-0" (9.144 M)	CASING AND CORE BOTH DRIVEN WITH PILE DRIVER, CORE REMOVED, CASING FILLED WITH CONCRETE, THEN CASING REMOVED	FRICTION OF EARTH END BEARING HARD MATERIAL	IN TYPES OF EARTH WHERE THERE IS NOT ENOUGH PRESSURE TO FORCE EARTH UP INTO CASING WHEN CORE IS REMOVED
CAST-IN-PLACE CONCRETE WITH BULGE (PEDESTAL) AT BOTTOM	REMOVABLE CORE REMOVABLE CASING UP TO 40'-0" (12.192 M) 2'-6" (762 mm) 2'-6" (762 mm) UNDER 30'-0" (9.144 M) 3'-0" (914.4 mm) 3'-0" (914 mm) OVER 30'-0" (9.144 M)	CASING AND CORE DRIVEN WITH PILE DRIVER, CORE REMOVED AND CASING RAISED 18" (457.2 mm) TO 2'-0" (609.6 mm), CASING PARTIALLY FILLED WITH CONCRETE WHICH IS RAMMED TO FIRM BULGE AT BOTTOM, THEN CASING COMPLETELY FILLED WITH CONCRETE AND CASING REMOVED	FRICTION OF EARTH END BEARING HARD MATERIAL	IN TYPES OF EARTH WHERE THERE IS NOT ENOUGH PRESSURE TO FORCE EARTH UP INTO CASING WHEN CORE IS REMOVED
CAST-IN-PLACE CONCRETE WITH PERMANENT STEEL SHELL	REMOVABLE CORE PERMANENT TAPERED REINFORCED STEEL SHELL 37'-6" (11.43 M) MAXIMUM 2'-6" (762 mm) 2'-6" (762 mm) TAPERED 0.4" (10.16 mm) TO THE FOOT ▽ METAL POINT	CORE AND REINFORCED SHELL DRIVEN BY PILE DRIVER, CORE REMOVED AND PERMANENT REINFORCED SHELL FILLED WITH CONCRETE	FRICTION OF EARTH END BEARING HARD MATERIAL	ANY TYPE EARTH EXCEPT ROCK
CAST-IN-PLACE CONCRETE WITH PERMANENT STEP-TAPERED REINFORCED SHELLS	UP TO 150'-0" (45.720 M) 2'-6" (762 mm) 2'-6" (762 mm) UP TO 40'-0" (12.192 M) 3'-6" (1067 mm) 3'-6" (1066.8 mm) OVER 40'-0" (12.192 M)	CORE AND FIRST SECTION DRIVEN WITH PILE DRIVER, SECOND SECTION JOINED WITH ITS CORE AND DRIVEN, LARGER SECTIONS ADDED UNTIL DEPTH OR BEARING TO SUPPORT LOADS IS REACHED. CORES REMOVED AND SHELLS FILLED WITH CONCRETE	FRICTION OF EARTH END BEARING HARD MATERIAL	IN EARTH WHERE A HARD TYPE OF EARTH IS LOCATED AT A CERTAIN DEPTH AND HEAVY LOADS ARE TO BE SUPPORTED; OR ANY TYPE OF EARTH WHERE VERY DEEP PILE DRIVING WILL DEVELOP BEARING FOR HEAVY LOADS

Figure P10 Types of piling.

PICKLING

Pickling is the treatment of metal surfaces with a strong oxidizing agent, such as nitric acid, to make them chemically clean and provide a strong inert oxide film.

PILES AND PILING

A pile is a vertical member driven into the ground to help bear the vertical load of any structure resting upon it. The basic principle is that any length of material driven into any substance other than water develops friction along its length, and it is this friction that provides support. Thus piling is one method of increasing the load capacity of any type of soil to the degree required by a given architectural problem.

When construction is to be on soils of low-load capacity and piling becomes necessary, the architect automatically will have to turn to a specialist in this field. However, he should know certain fundamentals that will enable him to talk intelligently about this phase of the construction.

The material used for piling and the amount of moisture to which it will be exposed are the key factors. Wood piling that is partially submerged or subject to alternate wetting and drying will deteriorate very quickly, whereas wood completely submerged will last as long as the life of any building. It follows that, if wood is to be used, it must either remain dry or be totally submerged, that is, permanently below the ground water level.

Steel of any type in contact with water will rust out, and only in a few special cases can steel members be used for piling.

Concrete is the only material that can be used under any set of conditions–submerged, partially or alternately wet, or completely dry–and still remain intact throughout the life of the building. Figure *P10* shows the various types of piling, together with commonly used dimensions and depth limitations. A check should be made with all local, municipal, state, and federal agencies to be sure that all code requirements are met.

TYPE OF PILE & MATERIAL	DEPTH LIMITS & SPACING	INSTALLATION METHODS	METHOD OF DEVELOPING BEARING FOR LOADS TO BE SUPPORTED	WHERE GENERALLY USED
CAST-IN-PLACE CONCRETE CONCRETE OR REINFORCED CONCRETE STEEL PIPE	USED FOR ALL DEPTH WITH POINT AND CLOSED END USED WITH ALL TYPES OF SPACING OPEN END	STEEL PIPE DRIVEN WITH PILE DRIVER; IF OPEN END EARTH REMOVED BY COMPRESSED AIR, THEN CONCRETE OR REINFORCED CONCRETE INSTALLED	FRICTION OF EARTH CLOSED END END BEARING HARD MATERIAL	IN EARTH WHERE A HARD TYPE OF EARTH IS LOCATED AT A CERTAIN DEPTH AND HEAVY LOADS ARE TO BE SUPPORTED; OR ANY TYPE OF EARTH WHERE VERY DEEP PILE DRIVING WILL DEVELOP BEARING FOR HEAVY LOADS
COMBINATION WOOD PILE AND CAST-IN-PLACE CONCRETE PILE	REMOVABLE CORE WOOD PILE PERMANENT REINFORCED STEEL SHELL	WOOD PILE DRIVEN BY PILE DRIVER, PERMANENT REINFORCED STEEL SHELL WITH CORE PLACED IN TOP AND DRIVEN WITH PILE DRIVER, CORE REMOVED AND SHELL FILLED WITH CONCRETE	FRICTION OF EARTH PERMANENT GROUND WATER LEVEL	SPECIAL CONDITIONS WHERE BELOW DENSE EARTH IS A WATER CONDITION WITH OR WITHOUT SOFT, MUCKY EARTH
STRUCTURAL STEEL I BEAMS	UP TO AVAILABLE SIZE AND LENGTH OF STEEL 2'-6" (.762 M.) GENERAL SPACING 2'-0" (.610 M.) VERY DENSE SOIL	STRUCTURAL STEEL DRIVEN WITH PILE DRIVER; TOP OF STEEL INCASED IN CONCRETE OR COATED WITH RUST-INHIBITING PAINT 6"(152.4mm) ABOVE GRADE AND 2'-0"(.610 M.) BELOW GRADE	FRICTION OF EARTH END BEARING HARD MATERIAL	USED GENERALLY FOR DENSE HARD TYPE EARTH FOR HEAVY LOADS
SHEET PILING	VARIOUS DEPTHS TYPICAL TYPE	SHEET PILING DRIVEN WITH PILE DRIVER	SHEET PILING BOTTOM OF EXCAVATION WATER	USED WHERE A WATER CONDITION EXISTS AND EXCAVATIONS MUST BE KEPT DRY. ACTUALLY BUILDING A DAM TO KEEP WATER OUT

Figure P10 Types of piling (*continued*).

PIPE

Even though pipe had been used for thousands of years, it was only during the 20th century that a detailed understanding of velocity distribution and of energy losses was finally developed.

In the construction field, pipe is used for the transportation of hot and cold water, air-conditioning fluids, heating duct systems, natural gas, steam, storm water, sewage, chemical wastes, etc. Pipe is also used as lally columns, hollow or filled with concrete, as structural columns, and for many types of railings and hand railings.

Pipe is obtainable in sizes from a fraction of an inch or millimeter up to 30 ft (9.144 m) and larger. It is manufactured of steel, wrought and cast iron, concrete, asbestos cement, asphalt, clay tile, plastic, impregnated paper tubes, copper, brass, lead, and glass. In chemical research and manufacture, pipes made of gold, platinum, and silver and pipes with various types of interior and exterior plating are used.

See also Clay Tile; Concrete; Copper; Brass; Asbestos; Plastic; Lead; Glass; Steel; and Iron for pipe made of these materials.

PLASTER AND PLASTERING

Physical and Chemical Properties

In construction terminology the words "plaster" and "gypsum" are often used interchangeably, often incorrectly. In this book gypsum is treated as an ingredient for making plaster, and plaster is defined as the end product that results when various cementitious materials are mixed with water and other materials to make a pasty substance which has many other uses besides plastering. Products made of plaster include a wide variety of block, board, sheet, and decking, often marketed under trade names.

Ingredients. Plaster contains one or more of the following ingredients: lime (slaked quicklime or hydrated lime to make lime putty), gypsum (plaster of paris, Keene's cement), or gypsum-type plaster mixes, cement (portland or portland pozzolana), hair, fibers, sand, and water. It may contain air-entraining agents to give it acoustical properties, perlite or vermiculite to make it more fire resistant, and mineral aggregates for special properties. Keene's cement is manufactured from gypsum, which when heated to between 1000 and 14,000°F (537.78 and 760.00°C) produces a hard-burned gypsum. When the hard-burned gypsum is ground and a small amount of alum is added, a water-resistant plaster is produced. Plaster of paris mixed with various synthetic resins and a catalyst produces a high-strength casting plaster which also expands slightly and thus produces fine impressions from a mold. This is used for decorative plaster work. Plaster mixtures made with special types of plaster are applied over electrical radiant-heat wires or hot-water radiant copper tubing coils. These may be installed on the job site, or preassembled complete units of plaster-embedded coils or wires may be used.

Requirements for Cementitious Materials. All cementitious materials for making plaster are manufactured to specifications controlled to meet various code requirements, acoustic and fire-resistance requirements, and moisture absorption, setting time, strength, and other property requirements for their end use. They come to the job in paper bags, ready to be mixed and used, and are also available in barrels or bulk, but it is always advisable to check with manufacturers about availability in these forms.

Plastering. The plaster mixture is applied in coats (layers) to masonry surfaces, wire lath, wood lath, or various types of plaster board and sheet to give a hard finish surface to interior or exterior walls and ceilings. This surface finish may be smooth or rough and serves as a base for paint, wallpaper, plastic laminates, textiles, and the like. In construction terminology this is the procedure called plastering. The basic ingredients for making plaster for plastering are listed in Table *P33*.

Types and Uses

The different kinds of plaster are named on the basis of their main cementitious material or on the basis of the characteristics of the plaster mixtures. Those commonly used for plastering and for plaster products are summarized in Table *P34*.

High-Strength Gypsum Plaster. High-strength gypsum plaster mixes must meet definite compressive and tensile strength requirements (*see* Table *P35*).

Proportion of Fiber. The proportion of fiber and hair for plasters is also controlled to specific requirements (*see* Table *P36*).

Plastering Methods. Plastering is done according to two basic methods, two-coat and three-coat. A three-coat plastering job consists of a first, binding coat called the

Table P33 Ingredients for making plaster

Ingredient	Composition and preparation	Finished form for use in making plaster
Cement	Oxides of aluminum, calcium, iron, magnesium, and silicon; pulverized, burned (calcined), ground, and gypsum added	Portland cement
Epoxy	Epoxy resin and hardener with fine white sand and inorganic color pigments	Epoxy-type plaster
Fiber	Hemp, sisal, jute or wood fibers	Clean, long, and free from tannic acid
Gypsum	$2\ CaSO_4 \cdot H_2O$ burned (calcined) at 401°F (205°C)	Plaster of paris
	Plaster of paris mixed with clay, lime, and other materials in combinations covered by trademarks and patents	Gypsum plaster
	Same as above but mixed to meet established requirements	High-strength gypsum plaster
	Gypsum plaster with high strength; when blended with lime putty it produces a hard, strong finish with minimal or no shrinkage cracks	Gauging plaster
	Plaster of paris mixed with alum, borax or other materials and burned (calcined) at 932°F (500°C)	Keene's cement
	Gypsum plaster mixed with fibers	Fibered gypsum plaster
	Gypsum plaster mixed with fine white sand	Prepared gypsum plaster
	Gypsum plaster mixed with vermiculite, perlite, or other suitable mineral aggregate	Lightweight gypsum plaster, fire-resistant gypsum plaster
	Gypsum plaster mixed with ingredients to develop small air bubbles throughout the plaster in combinations covered by trademarks or patents	Acoustic plaster
	Gypsum plaster mixed with ingredients to develop more adhesive strength in combinations covered by trademarks or patents	Bonding plaster
	Gypsum plaster mixed with sand and other ingredients in combinations covered by trademarks and patents	Gypsum sand-float-finish plaster
Hair	Goat or cattle hair $\frac{1}{2}$ to 2 in. (12.7 to 50.8 mm)	Free from dust, knots, and balls
Lime	Lime (quicklime) which has been treated with just enough water to produce $CaO(OH)_2$ in dry form	Hydrated lime
	Lime (quicklime) with water, hydrated (slaked) on the job	Lime putty
Sand	Natural sands with 100% passing through a No. 4 (4.76 mm) sieve, washed and cleaned	White or light gray fine sand
Water		Clean, fresh water containing no salt, sulfur, or other harmful substances

Table P34 Cementitious materials and types of plaster used for plastering and plaster products

Type of cementitious material or plaster	Kinds of mixtures or on-the-job procedure	Major use
Acoustic plaster	Mixed with water	Interior acoustic treatment for walls and ceilings on gypsum plaster base
Bonding plaster	Mixed with water	Interior finish for smooth concrete walls or ceilings
Epoxy-type plaster	Mixed with hardener and white sand	Strong, durable, water-resistant interior and exterior finish; also used with hard, colored aggregates
Fibered gypsum plaster	Mixed with water and sand	For first scratch coat for 3-coat plastering jobs

Table P34 Cementitious materials and types of plaster used for plastering and plaster products (*continued*)

Type of cementitious material or plaster	Kinds of mixtures or on-the-job procedure	Major use
Fire-resistant plaster	Gypsum plaster, water, and perlite, vermiculite, or other mineral aggregate	For fireproofing other materials such as steel
Gypsum or high-strength gypsum plaster	Dry	Ingredient for hard or sand float finish with lime plaster
	With water, fiber, or hair and paper	Plaster sheets and integrants, plaster lathing
	With water, sand, lime putty, hair, or fiber	Two and 3-coat finish surfaces for interior walls and ceilings
	With water	Plaster block
	With water, hair, fiber, and sand	Mortar for plaster block
	With water, lightweight aggregate, and metal reinforcing	Plaster decking
Gypsum sand-float-finish plaster	With water	Sand float finish on 2- or 3-coat gypsum plaster
Hydrated lime	With water to make lime putty	Ingredient of hard-finish coat for gypsum and Keene's cement plaster; ingredient of sand float finish for Keene's cement and for 2- and 3-coat portland cement plasters
Keene's cement	With water, lime putty, and fine white sand	Hard finish coat for 2- or 3-coat gypsum plaster, and for areas where water and moisture are present (showers, bathrooms, etc.)
	With water, lime putty, and sand	Sand float finish coat for 2- or 3-coat gypsum plaster
Lightweight plaster	Gypsum plaster, water, and vermiculite or perlite aggregate	Lightweight plaster used where weight is important; *see also* Fire-resistant plaster
Lime putty	Lime putty mixed with sand and gypsum plaster	Two- and 3-coat finish surfaces for interior walls and ceilings
Molding plaster	Mixed with water per manufacturer's directions	Ornamental plaster work and castings
Plaster of paris	Dry	Ingredient of gypsum-type plasters
	With water	Making ornamental and sculptural casts
Portland cement or portland pozzolana	Mixed with water, sand, and lime putty	Two- and 3-coat finish surfaces for exterior and interior walls and ceilings, and for areas where water-resistant walls and ceilings are required
	Mixed with water, sand, and lime putty	Sand float finish for cement plasters
Prepared gypsum plaster	Mixed with water per manufacturers' directions	Two- and 3-coat finish surfaces for interior walls and ceilings

Table P35 Physical requirements for high-strength gypsum plaster

	Minimum strength			
	Compressive		Tensile	
Time elapsed	$lbf/in.^2$	MN/m^2	$lbf/in.^2$	MN/m^2
After 3 hours	1500	10.34	150	1.03
Fully dry	3500	24.13	450	3.10

Table P36 Proportions of hair or fiber in major types of plaster

Plaster made with:	Proportion of hair or fiber to type of plaster
Cement	1 lb (0.4536 kg) to 1 bag = 94 lb (42.638 kg)
Gypsum	Not added on job; usually premixed
Lime	1 bushel (36.384 liters or 35 245.4 cm^3) of hair or $\frac{1}{2}$ bushel (18.192 liters or 17 622.7 cm^3) of hair to 1 yd^3 (0.7646 m^3) of lime plaster

scratch coat ("pricking up," according to British terminology); a second coat, called the brown coat (straightening); and a final coat, called the hard finish (finishing). In two-coat work, the scratch and brown coats are combined into one coat; the second coat is the hard finish.

There is a one-coat method which consists of a high-strength plaster scratch coat followed immediately by doubling back with a high-strength finish plaster.

***Sgraffito*.** Sgraffito is a highly decorative type of plaster work developed in Italy during the Renaissance. This technique consists of applying two to three thin coats of plaster of different colors and then cutting away certain areas of one or two of the coats to produce a three-dimensional colored design.

***Epoxy-Type Plaster*.** An epoxy resin and hardener mixed with fine white silica sand provides a strong, durable, abrasion- and water-resistant plaster with strong bonding characteristics as finish for both exterior or interior surfaces. It is available in various colors, it can be troweled to a relatively smooth or irregular surface, and hard, colored aggregates can be applied by hand (seeded) to the surface to given an irregular, pebbled stone effect.

Application

The methods of applying plaster to walls and ceiling, the type of surface to which plaster can be applied, and the types and proportions of materials for the various plasters are outlined in Table *P37*.

***Lath*.** The surface, other than masonry, to which plaster is applied is called lath. Various types of wire and expanded metal mesh, wood lath, plaster board, or other material that will supply sufficient keying or bonding for plaster can serve as lath. Furring and running channels (*see* Tables *P38* and *P39*), corner beads, corner lath, casings, screeds, grounds, nails (*see* Table *P40*), wire and hangers are some of the necessary accessories that are also installed to form a base upon which plaster can be applied to walls and ceilings.

Metal lath for plastering includes the following commonly used types: (1) expanded flat or $\frac{3}{8}$ in. (9.35 mm) ribbed lath, made from copper alloy steel or galvanized steel sheet; (2) wire lath woven with a $\frac{1}{2}$ in. (12.7 mm) mesh size from cold-drawn wire made of copper alloy steel, welded and galvanized; (3) paper-backed self-furring lath with a 2 in. (50.8 mm) mesh, woven from 16-gauge wire, backed with an absorbent material (*see* Table *P41* and Figures *P11* and *P12*. (For a more complete listing of metal lathing, *see* Mesh and Wire Cloth for Lathing.)

Lathing board, categorized as plaster sheet in this book, is usually one of the following types: (1) solid; (2) solid insulating, with highly reflective aluminum foil on one side; and (3) perforated, with $\frac{3}{4}$-in. (19.05-mm) holes 4 in. (101.6 mm) apart in each direction. The most common size is 16 in. (406.4 mm) wide and 48 in. (1.219 m) long, $\frac{3}{8}$ or $\frac{1}{2}$ in. (9.53 or 12.7 mm) thick. The solid and insulation types are also available in various widths and floor-to-ceiling lengths. (For a complete list of dimensions, *see* Plaster Integrants and Sheets.)

Thickness of plaster coats depends on the type of material to which the plaster is applied. Generally, the total thickness is $\frac{5}{8}$ in. (15.88 mm) on metal lath and masonry, and $\frac{1}{2}$ in. (12.7 mm) on lathing board and gypsum block. In three-coat plastering, the scratch coat (pricking-up) is $\frac{1}{4}$ in. (6.35 mm) thick at minimum; the brown coat (straightening) is also at least $\frac{1}{4}$ in. (6.35 mm) thick; and the hard finish (finishing) is $\frac{3}{32}$ in. (2.38 mm) thick, with a minimum of $\frac{1}{16}$ in. (1.59 mm) at any point. For two-coat work, the base coat is $\frac{1}{2}$ in. (12.7 mm) thick and the hard finish the same as for three-coat work.

The lath for plastering has to be level, plumb, and well secured to the backing material. Since the plaster finish must be perfectly flat and plumb and have the correct thickness, it is necessary to have what are called leveling elements installed.

Table P37 Common methods of mixing, proportioning, and applying plaster

Types of plaster	Surface to which it is to be applied	First 2 coats for a 3-coat plastering job (proportions by weight unless otherwise noted)				First coat for a 2-coat plastering job (proportions by weight unless otherwise noted)		Final coat for both 2-coat and 3-coat plastering jobs (by volume unless otherwise noted)			
		Scratch coat		Brown coat				Hard finish		Sand float finish	
		Aggregate	Plaster[a]	Aggregate	Plaster[a]	Aggregate	Plaster	Lime putty[b]	Plaster	Aggregate	Plaster
Acoustic planter	On gypsum plaster base only; interior only							Mixed and applied as directed by manufacturer			
Bonding plaster	Smooth concrete; interior only					Mixed and applied as directed by manufacturer					
Fibered gypsum plaster		Used as the plaster for the mix instead of sand, gypsum, and added fiber or hair									
Gypsum fire-resistant or lightweight plaster	All types of lath; interior only	2 sand 2 ft^3 (56.64 liters) perlite or vermiculite	1 100 lb (45.36 kg)	3 sand 3 ft^3 (84.96 liters) perlite or vermiculite	1 100 lb (45.36 kg)	$2\frac{1}{2}$ sand $2\frac{1}{2}$ ft^3 (70.8 liters) perlite or vermiculite	1 100 lb (45.36 kg)	1	$\frac{1}{3}$ gypsum plaster	$1\frac{1}{2}$ sand	1
	On masonry; interior only	3 sand 2 ft^3 (56.64 liters) perlite or vermiculite	1 100 lb (45.36 kg)			3 sand 3 ft^3 (84.96 liters) perlite or vermiculite	1 100 lb (45.36 kg)				
Gypsum sand-float-finish plaster	On brown coat of gypsum plaster; interior only									Mixed and applied as directed by manufacturer	
High-strength gypsum	All types of lath and masonry; interior or exterior	2 sand, 1 high-strength gypsum plaster						$\frac{1}{4}$	1 high-strength gypsum		
Keene's cement	On brown coat of gypsum plaster; interior only							By weight $\frac{1}{4}$	1 Keene's cement, $\frac{1}{10}$ fine white sand	$4\frac{1}{2}$ sand	$1\frac{1}{2}$ Keene's cement, 2 lime putty[b]
Lime (quicklime) slaked on job	All types of lath and masonry; interior or exterior	$2\frac{3}{4}$ sand	1 lime putty, 1 portland or Keene's cement	3 sand	1 lime putty, 1 portland or Keene's cement			1	$\frac{1}{3}$ gypsum plaster	2 sand	1 lime putty, $\frac{1}{4}$ gypsum plaster
Molding plaster	Ornamental plaster work and castings	Mix with water per manufacturer's directions									
Prepared gypsum plaster	All types of lath or masonry; interior only	Mix with water per manufacturer's directions						Any type of final coat plaster (gypsum, Keene's, lime, or prepared finish coat)			
Portland cement plaster	All types of lath or masonry; interior or exterior	3 sand, $\frac{1}{4}$ lime putty,[b] 1 portland cement and water								3	1 portland cement, $\frac{1}{4}$ lime putty[b]
Portland pozzolana cement plaster	All types of lath or masonry; interior or exterior	3 sand, $\frac{1}{5}$ lime putty,[b] 1 portland pozzolana cement and water								3	1 portland pozzolana cement, $\frac{1}{5}$ lime putty[b]

[a]Neat gypsum plaster = gypsum plaster only, with water.
[b]Lime putty = hydrated lime and water.

Table P38 Spacing of supports and furring channels[a] for a suspended ceiling to be plastered

Hanger spacing center to center		Maximum channel spacing center to center			
		Main supporting channels[b]		Furring channels[c]	
ft	m	ft	m	in.	cm
Up to 3	Up to 0.314	4	1.219	16	40.64
$3\text{–}3\frac{1}{2}$	0.914–1.067	$3\frac{1}{2}$	1.067	19	48.26
$3\frac{1}{2}\text{–}4$	1.067–1.219	3	0.914	24	60.96

[a]Furring channels are sold by the 1000 ft.
[b]These are $1\frac{1}{2}$-in. (38.1-mm) cold-rolled channels 475 lb/1000 ft (215.46 kg/304.8 m).
[c]These are $\frac{3}{4}$-in. (19.05-mm) cold-rolled channels 300 lb/1000 ft (136.08 kg/304.8 mm).

Table P39 General maximum vertical heights for metal stud partitions

Spacing		Stud		Maximum height			
				Lath screwed		Lath clipped	
in.	cm	in.	mm	ft	m	ft	m
16	40.64	$1\frac{5}{8}$	41.28	10	3.048		
		$2\frac{1}{2}$	63.50	$13\frac{1}{2}$	4.115	$9\frac{1}{2}$	2.896
		$3\frac{5}{8}$	92.08	17	5.182	12	3.658
		4	101.60	18	5.486	13	3.962
		6	152.40	18	5.486	13	3.962
24	60.96	$1\frac{5}{8}$	41.28	10	3.048		
		$2\frac{1}{2}$	63.50	$13\frac{1}{2}$	4.115	$8\frac{1}{2}$	2.591
		$3\frac{5}{8}$	92.08	17	5.182	$10\frac{1}{2}$	3.200
		4	101.60	18	5.486	$12\frac{1}{2}$	3.810
		6	152.40	18	5.486	$12\frac{1}{2}$	3.810

Table P40 Types of nails used for lathing

Type of nail	Size of nails					Type of lath	Type of support
	Length			Head diameter			
	in.	mm	Gauge	in.	mm		
Barbed roofing nail	$1\frac{1}{2}$	38.10	11	$\frac{7}{16}$	11.11	Wire mesh or expanded metal	Horizontal wood
Barbed roofing nail	2	50.80	11	$\frac{7}{16}$	11.11	Ribbed expanded metal lath or self-furring lath	Vertical or horizontal wood
Blued nail	$1\frac{1}{8}$	28.58	13	$\frac{3}{8}$	9.53	Lathing board	Vertical or horizontal wood
Blued "U" head nail			12	"U" head		Paper-backed lath	Vertical or horizontal wood
Common nail	$1\frac{1}{2}$	38.10	4-penny			Wire mesh or expanded metal	Vertical wood
Wire staples	1	25.40	14			Wire mesh or expanded metal	Vertical wood
Zinc-coated furring nail	Allow $\frac{1}{4}$ in. (6.35 mm) between lath and backing					Wire mesh, expanded metal, or ribbed lath	Concrete or masonry

Table P41 General application of various types of metal lath

Type of lath[a]	Horizontal spacing of supports, in.	Horizontal spacing of supports, mm	Type of horizontal support	Vertical spacing of supports, in.	Vertical spacing of supports, mm	Type of vertical support	Laps, in.	Laps, mm
Expanded metal	Up to and including $13\frac{1}{2}$	342.9	Metal	Up to and including 16	406.4	Wood or metal	$\frac{1}{2}$	12.7
	Up to and including 16	406.4	Wood or concrete	Up to and including 16	406.4	Wood or metal	$\frac{1}{2}$	12.7
Paper-backed self-furring	Up to and including 16	406.4	Wood or concrete	Up to and including 16	406.4	Wood or metal	1	25.4
$\frac{3}{8}$- in. (9.53-mm) ribbed expanded metal	Up to and including 24	609.6	Wood, concrete, or metal	Up to and including 24	609.6	Wood, concrete, or metal	1	25.4
Wire	Up to and including $13\frac{1}{2}$	342.9	Wood, concrete, or metal	Up to and including 16	406.4	Wood	$\frac{1}{2}$	12.7

[a] Lath is secured to supports every 6 in. (152.4 mm); laps not secured to supports are tied with wire every 9 in. (228.6 mm) for walls and every 6 in. (152.4 mm) for ceilings. Long dimensions of lath are installed perpendicular to supports.

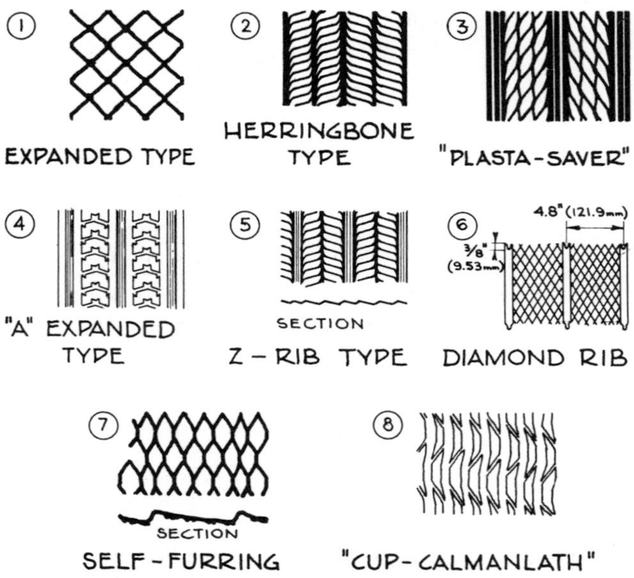

Figure P11 Commonly used types of metal lath for plastering.

Leveling elements include grounds and screeds. For walls, a screed is installed at the base of the wall with its top about 4 in. (101.6 mm) above the finish floor. This screed is run horizontally, leveled, and set at the exact thickness of the finished plaster. Usually it is covered by another finish material. Around all openings and at the intersection with the ceiling, grounds are installed. These grounds may be permanently installed, either to be covered by another finish material (trim) or made of a special type of metal against which a finish material can be installed. Temporary grounds are another type; these are removed after the brown coat is dry.

Grounds, like screeds, must be of the correct plaster thickness and installed level, straight, and plumb. A ground can be placed where ceilings intersect with walls if the joint is to be covered with some sort of trim piece. If the joint is not to be covered, the ceiling can extend on down the wall to some point where a joint is made. This can be accomplished with temporary grounds. The ceiling and wall intersection may also be made without any grounds, but it is very difficult to obtain a straight, true intersection by this method and to avoid cracks.

Accessories. Accessories made especially for plaster work include nails, casings, window stools, bases, and picture molds. Current literature on these types of accessories should always be checked to see whether new ones have been developed. (*See* Figure *P13*.)

Condensed Checklist

1. Lathing should be plumb, straight, and securely fastened to supports; laps should be well wired.
2. Base screeds should be plumb, level, and located correctly (*see* Figure *P14*).

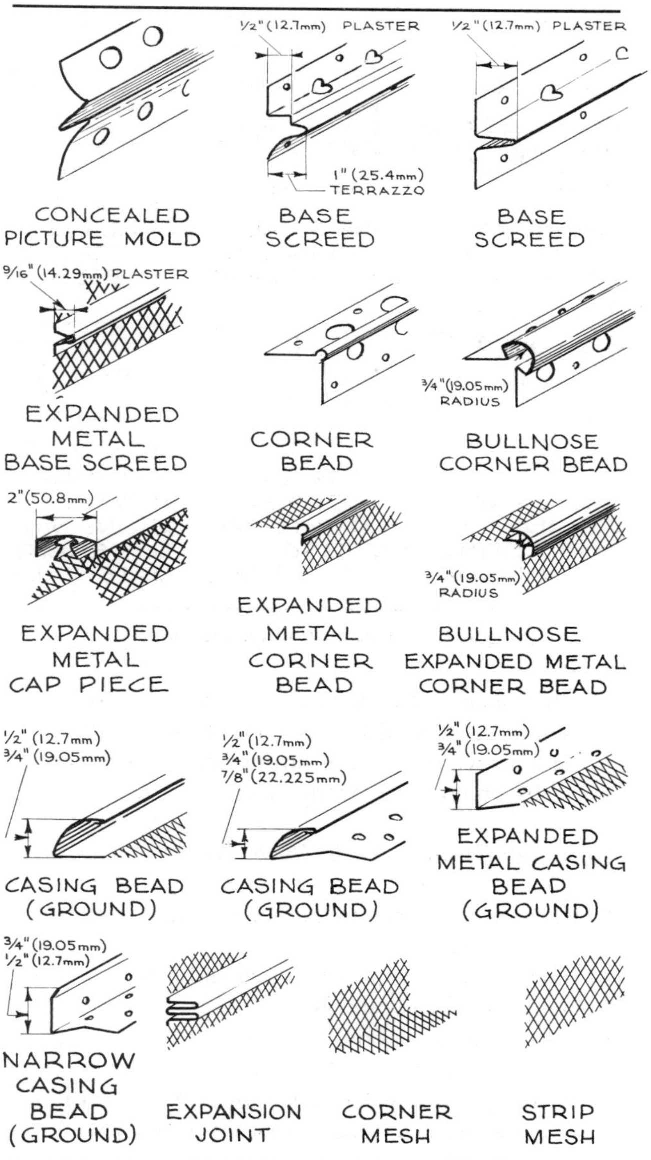

Figure P12 Accessories for lathing.

3. Grounds should be installed plumb and level and located correctly to meet all the various conditions at openings, etc. (*see* Figure *P15*).
4. All anchorage for cabinets, furniture, stair handrails, electrical outlets, etc., should be installed before plastering is started.
5. Corner beads should be installed at all external corners (*see* Figure *P16*).
6. All internal corners should be reinforced by lapping wire lath, or corner lath should be installed where other types of lath are used (*see* Figure *P17*).

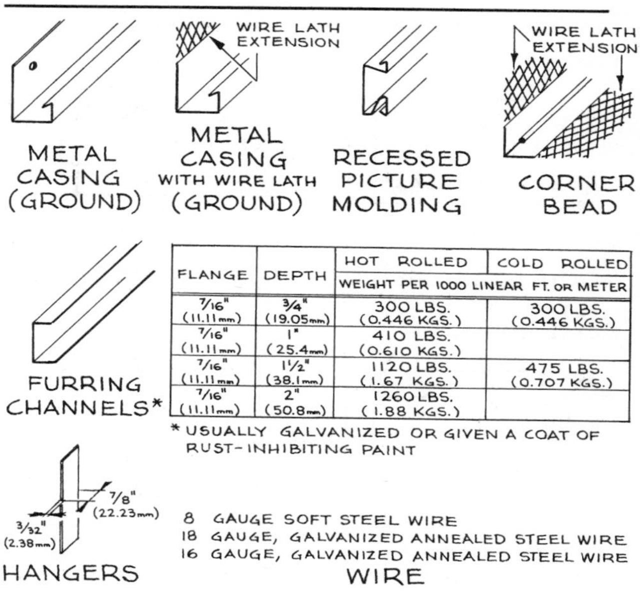

FLANGE	DEPTH	HOT ROLLED	COLD ROLLED
		WEIGHT PER 1000 LINEAR FT. OR METER	
7/16" (11.11mm)	¾" (19.05mm)	300 LBS. (0.446 KGS.)	300 LBS. (0.446 KGS.)
7/16" (11.11mm)	1" (25.4mm)	410 LBS. (0.610 KGS.)	
7/16" (11.11mm)	1½" (38.1mm)	1120 LBS. (1.67 KGS.)	475 LBS. (0.707 KGS.)
7/16" (11.11mm)	2" (50.8mm)	1260 LBS. (1.88 KGS.)	

Figure P13 Typical accessories for plastering.

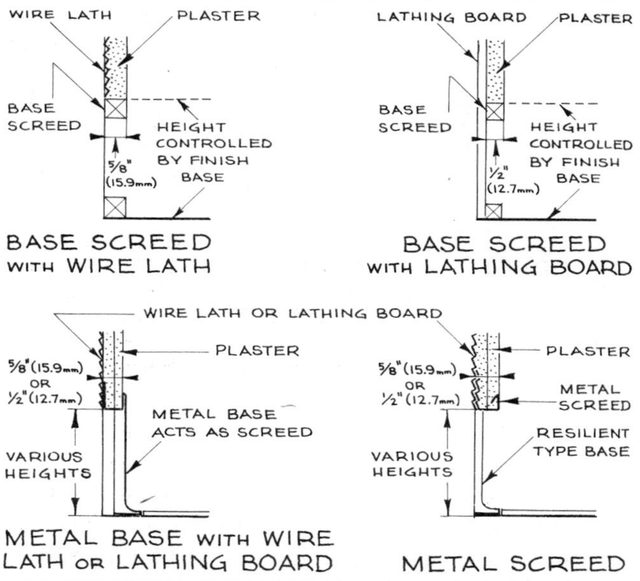

Figure P14 Typical methods for installing base screed.

7. Plaster work must never be installed when the temperature is below 40°F (4.44°C).
8. Mixtures for the various coats should be checked to see that proportions are correct.
9. Manufacturer's directions for applying the various types of plaster should be followed scrupulously. In construction one should always check that the

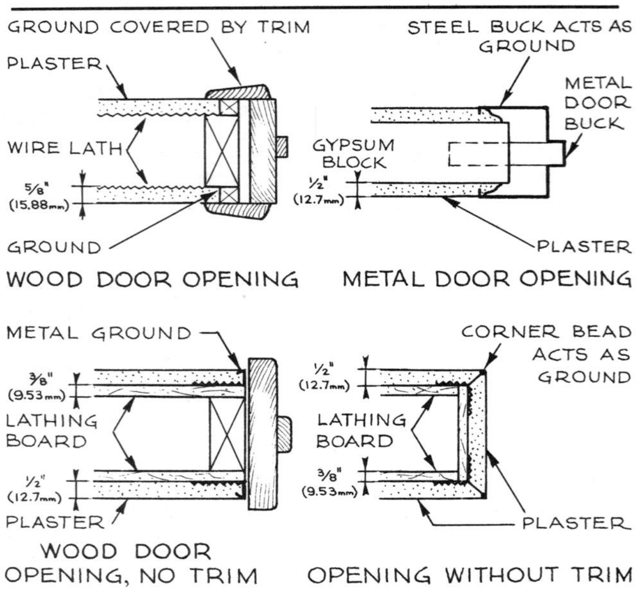

Figure P15 Typical installation of grounds at openings.

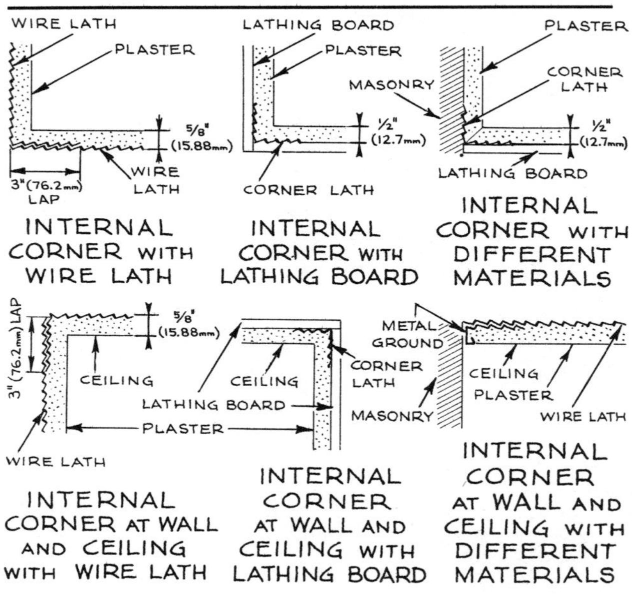

Figure P17 Treatment of internal corners.

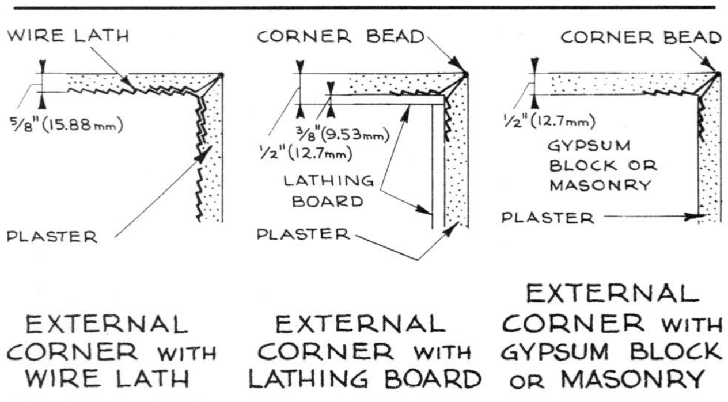

Figure P16 Typical external corner installation.

method or system of application is suitable and conforms to the end-use requirements of the plaster.

10. State, local, and municipal codes and also the fire rating codes of the Fire Underwriters, insurance companies, and federal government (Army, Navy) should be checked for the various types of plaster.
11. When using the one-coat method, information should be obtained from the manufacturer regarding the type of substrate necessary for this method and the correct application methods, as well as the thickness, methods of mixing, and type of finish.
12. When using epoxy-type plaster, one should always check with the manufacturer for mixing methods, pot-life, color, size of aggregates, and methods of application.
13. One should always check the types of nails and screws to be used for installing the various types of lath to wood or metal furring and to partitions on walls, on both the exterior and the interior.

Conditions Favorable to the Use of Plaster

1. Where a hard, smooth surface for interior walls and ceilings is desired.
2. Where a rough, sandy-textured finish is desired for interior walls and ceilings.
3. Where a hard, smooth, or rough and sandy type of finish is desired on exterior walls and ceilings (using plaster specifically designated for exterior work).
4. Where an existing rough surface is to be changed to a hard, smooth surface.
5. Where maximum fire resistance for the minimum thickness is to be obtained.

Conditions Unfavorable to the Use of Plaster

1. In areas that receive rough usage.
2. In areas where there is excessive moisture, except when Keene's cement or portland cement plaster is to to installed.
3. On ceilings where deflection may occur, as plaster has practically no tensile strength.
4. In very large areas where there may be expansion or contraction of the supporting materials.

PLASTER BLOCK

Physical and Chemical Properties

Plaster blocks, also known as gypsum partitions, are usually made of gypsum, asbestos, or vegetable fibers as binders, and reinforcing. They have a compressive strength of 75 lbf/in.^{2} (0515 MN/m^2) when dry and 25 lbf/in.^{2} (0.171 MN/m^2) when wet. They have a good fire resistance because gypsum ($CaSO_4 \cdot 2H_2O$) releases its water of crystallization very slowly when exposed to fire.

Types and Uses

Gypsum block is used for lightweight, fire-resistant interior partitions; solid plaster block is used only for furring and for fireproofing columns (*see* Table *P42* and Figure *P18*).

Gypsum block is set on a base course of clay tile as it is easily damaged by moisture and water.

Table P42 Types of plaster block

Type of block	Thickness		Weight		Unsupported height (maximum)	
	in.	mm	lb/ft^2	kg/m^2	ft	m
Solid	2	50.8	10	48.82	10	3.048
Solid	3	76.2	15	73.23	13	3.962
Hollow	3	76.2	10	48.82	13	3.962
Hollow	4	101.6	13	63.47	17	5.182
Hollow	6	152.4	22	107.43	30	9.144

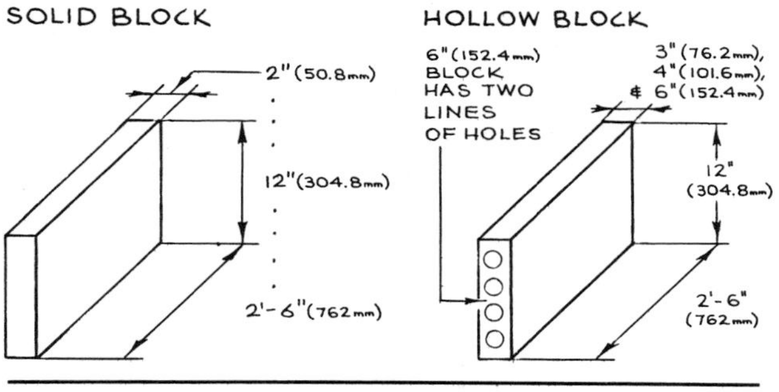

Figure P18 Types of plaster block.

Mortar. The mortar for gypsum block consists of 1 part neat gypsum plaster to 3 parts clean, sharp sand by weight.

Finish. All gypsum block is textured on its surfaces to provide a good bond for plastering; usually one side is printed with the manufacturer's name and the fire rating of the block. Plaster block work must thus be either plastered or furred to receive some other type of finish surface.

Application

Condensed Checklist

1. Plaster block must always have a base course of clay tile or other water-resistant material.
2. Wire lath should always be used to strengthen the upper corners of openings.
3. Lintels for large openings present a structural problem and have to be detailed and designed for each particular opening. Steel, reinforced gypsum, or reinforced door bucks may be used.
4. State, local, and municipal codes and also the fire rating codes of the Fire Underwriters, insurance companies, and federal government (Army, Navy, etc.) should be checked (*see* Table *P43*).

Conditions Favorable to the Use of Plaster Block

1. For lightweight interior partitions, particularly where the finish surface is to be plaster.
2. Where a soundproof partition wall is necessary (using hollow block).
3. Where fire-resistant partitions are required.
4. For enclosing vertical shafts with fire-resistant material.
5. For partitions around stairs, elevator shafts, columns, vertical ducts, and any other space that requires fire-resistant rating enclosure.

Conditions Unfavorable to the Use of Plaster Block

1. As load-bearing partitions.
2. In areas where there is moisture, spilling, or splashing of water, or where floors may be continually wet.
3. On the exterior.

History and Manufacture

The use of the present-day type of plaster block started about 1870. Except for minor changes to meet fire-resistance ratings, building codes, and standardization of sizes and compressive strengths, these blocks have not

Table P43 Fire-resistance ratings of gypsum block

Type of block	Thickness in.	Thickness mm	Gypsum plaster finish: ½ in. (12.7 mm) thick (1 to 3 mix)	Height limitations ft	Height limitations m	Fire resistance[a] (hours)
Solid	2	50.8	Plaster on one side	Used only for furring and		4
Hollow	3	76.2	Plaster on one side	fireproofing steel columns		4
Hollow	3	76.2	Plaster on one side	13		1½
Hollow	3	76.2	Plaster on both sides	13		3
Hollow	4	101.6	Plaster on one side	17		3
Hollow	4	101.6	Plaster on both sides	17		4
Hollow	6	152.4	Plaster on both sides	30		5[b]

[a]Based on gypsum block laid in gypsum-sand mortar.
[b]Estimated.

been changed from their original form. The process of manufacture is simple. Molds are filled with the correct mixture of gypsum, water, fibers, and sometimes sand; then, after the mixture has set, the molds are removed.

PLASTER DECKING

Physical and Chemical Properties

Gypsum plaster decking is available in two forms:

Precast gypsum decking: This consists of units 2 in. (50.8 mm) thick, 15 in. (381 mm) wide, and 10 ft (3.048 m) long, reinforced with wire mesh and having tongue-and-groove, galvanized steel edges on all four sides.

Poured gypsum decking: This consists of decking poured on the job and requires the setting up of metal T-sections as purlins, permanent forms made of various types of material, and wire mesh. Into these forms is poured a 2 or 2¼ in. (50.8 or 57.15 mm) slab of gypsum plaster composed of calcined gypsum with not more than 12½% of its weight consisting of wood chips, shavings, or fibers. When gypsum, which sets within 30 minutes, is poured on the job, it develops enough heat to be installed under even severe freezing conditions. Because gypsum dries from the underside toward the top, the type of permanent form should always allow for the passage of air to facilitate drying.

Types and Uses

Precast Gypsum Decking. Precast gypsum decking is manufactured in only one size. Its span limitations and thermal factors vary as shown in Table *P44*. This decking

Table P44 Insulation values, span limitations, and safe loads for precast gypsum decking

Size of decking	Insulation values[a] Btu/ft^2 · h · °F	Insulation values[a] W/m^2 · h · °C	Maximum span between supports ft	Maximum span between supports m	Uniform safe load[b] lb/ft^2	Uniform safe load[b] kg/m^2
2 in. (50.8 mm) thick, 15 in. (381 mm) wide, 10 ft (3.048 m) long, and 12 lb/ft^2 (58.58 kg/m^2) in weight	0.52 without insulation	2.95 without insulation	4	1.219	230	1122.86
	0.28 with ½ in. of insulation	1.58 with 12.7 mm of insulation	5	1.524	146	712.77
	0.19 with 1 in. of insulation	1.08 with 25.4 mm of insulation	6	1.829	105	512.61
			7	2.134	75	366.15

[a]Values include built-up roof.
[b]All uniform safe loads have a safety factor of 4.

may be cantilevered a maximum of 2.5 ft (726 mm) in length (the 10-ft, or 3.048-m dimension) and 6 in. (152.4 mm) in width (the 15-in., or 381-mm dimension). It has good fire resistance.

Poured Gypsum Decking. Poured gypsum decking, because of its light weight, can be used with light steel T-sections and various types of board or integrant as permanent forms and finish ceiling. Its greatest advantages are its fast setting time (30 minutes), its development of sufficient heat to allow it to be installed in severe freezing weather, and its fire resistance. (*See* Table *P45*.)

Both types of decking can be nailed into. The roof must be installed as soon as possible to cover the decking because it will deform when wet (but will not break). Table *P46* shows the holding power of nails installed in both types.

Application

Condensed Checklist

1. Spans of supporting members must not be greater than 7 ft (2.134 m) for precast gypsum decking.
2. For precast gypsum, detailing at wall intersections and ends of decking must be correctly treated, joints must be staggered, and anchoring clips must be installed correctly (*see* Figure *P19*).

Table P45 Components for poured gypsum decking

Sizes and weights	Gypsum integrant	Insulating-type integrant or board	Permanent form materials acoustic-type or board		Asbestos-cement integrant or sheet	Poured gypsum plaster	Wire mesh for all poured gypsum plaster
Thickness							Longitudinal wires No. 12 gauge, 4 in. (101.6 mm) o. c.; transverse wires No. 12 gauge, 8 in. (203.2 mm) o.c.; cross-sectional area = 0.026 in.^{2}/ft of width
in.	$\frac{1}{2}$	1	1	$1\frac{1}{2}$	$\frac{1}{4}, \frac{3}{8}, \frac{1}{2}$	2, $2\frac{1}{2}$	
(mm)	(12.7)	(25.4)	(25.4)	(38.1)	(6.35, 9.53, 12.7)	(50.8, 63.5)	
Width							
in.	32, 38	24, 32, 48	24, 32	24, 32, 48	24, 32		
(mm)	(812.8, 965.2)	(609.6, 812.8, 1219.2)	(609.6, 812.8)	(609.6, 812.8, 1219.2)	(609.6, 812.8)		
Length							
ft	Up to 12	Up to 12	Up to 12		4 only		
(m)	(Up to 3.658)	(Up to 3.658)	(Up to 3.658)		(1.219 only)		
Weight[a]							
lb/ft^2	2.1	1.35	1.1	1.75	2.4, 3.6, 4.8	8.33, 9.39	
(kg/m^2)	(10.25)	(6.59)	(5.37)	(8.54)	(11.72, 17.58, 23.43)	(40.67, 45.48)	

[a]Weights of permanent form material will vary with exact type used.

Table P46 Typical nails and their holding power in gypsum plaster decking

			Holding power[a]					
			Poured decking				Precast decking	
	Length of nail		Wet[b]		Dry			
Type of nail	in.	mm	lb	kg	lb	kg	lb	kg
Square head screw type	$1\frac{3}{4}$	44.45	24	10.89	186	84.37	46	20.87
Roofing type	$1\frac{3}{4}$	44.45	8	3.63	169	76.66		
for gypsum	$1\frac{1}{2}$	38.10	6	2.72	116	52.62		
Galvanized	$1\frac{3}{4}$	44.45	11	5.98	158	71.67	69	31.30
roofing	$1\frac{1}{2}$	38.10					25	11.34
Square cut	$1\frac{1}{2}$	38.10					150	68.04
	2	50.8	34	15.42	180	81.65		

[a]Direct pull resistance in pounds and kilograms.
[b]24 to 48 hours after pouring.

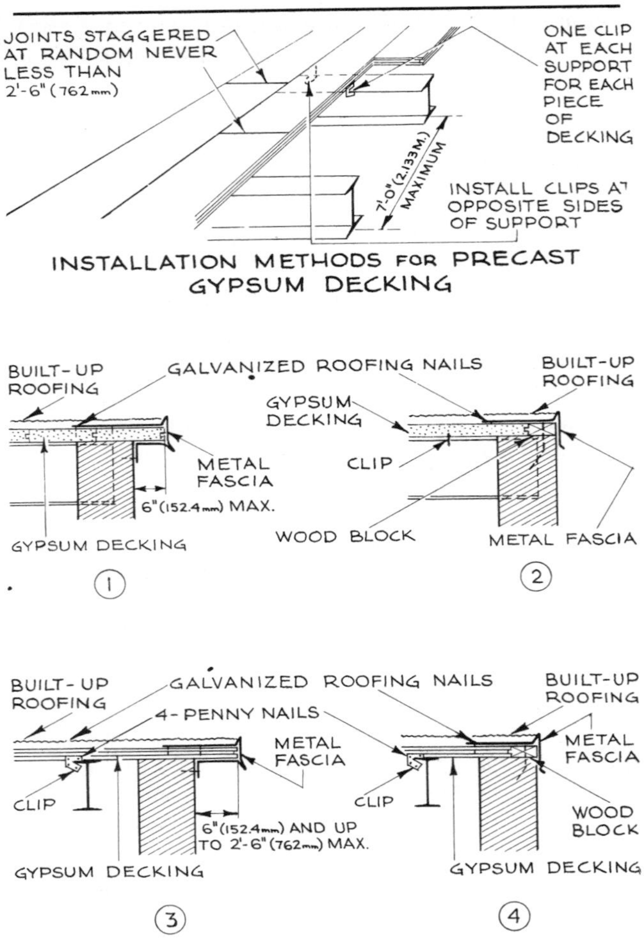

Figure P19 **Installation methods for precast gypsum decking.**

3. Permanent forms should meet all design requirements encountered in poured gypsum decking construction, particularly if they are to become the finish ceiling.
4. The type of purlins and their spacing should meet the floor or roof load requirements (*see* Figure *P20*).
5. Expansion joints should be installed every 200 ft (60.96 m) and also at intersections of various wings of a building and at any expansion joint which is part of the building.
6. Gypsum decking must be protected with wood plank runways for wheelbarrows, etc., that will pass across it during pouring and after setting.
7. Roofing material should be installed as soon as possible. If the gypsum decking is wet, it must be allowed to dry out before the roof is installed.
8. State, local, and municipal codes should be checked and also the fire rating codes of the Fire Underwriters, insurance companies, labor departments, and federal government (Army, Navy, etc.).

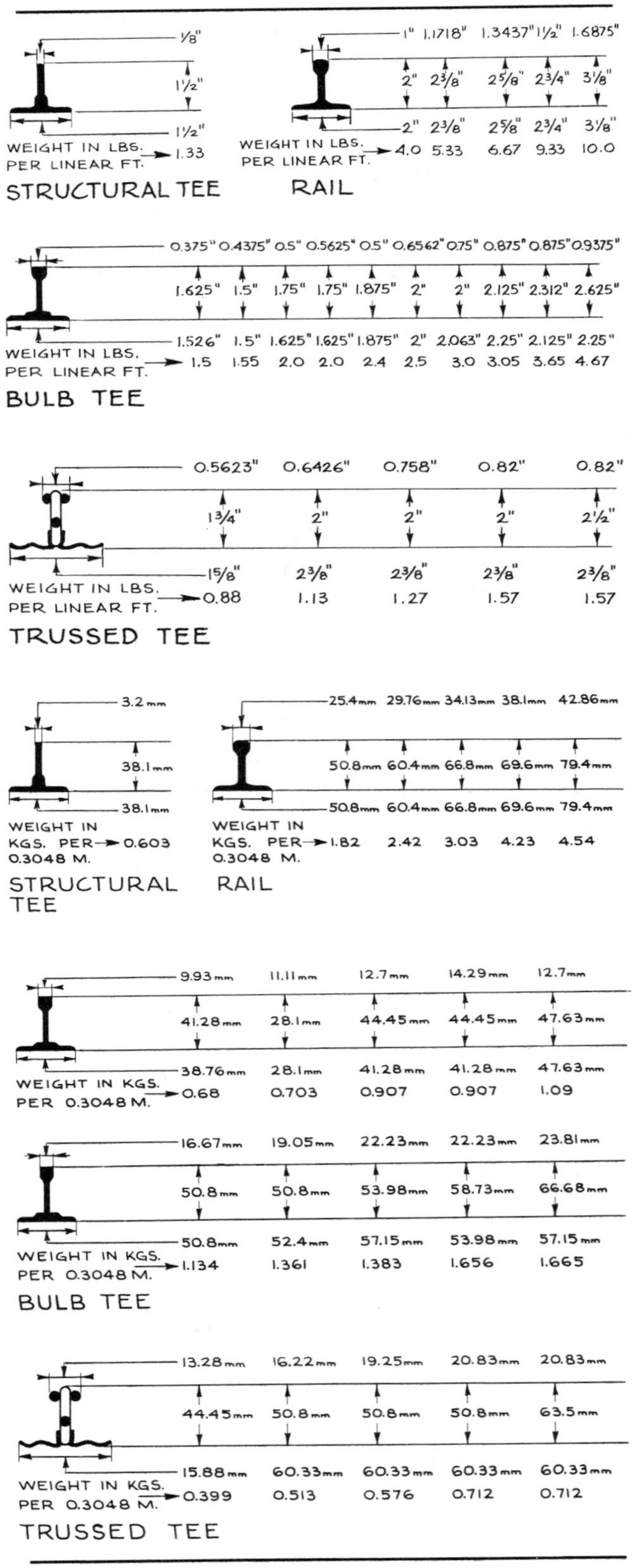

Figure P20 **Purlins for poured gypsum decking.**

Conditions Favorable to the Use of Gypsum Decking

1. Where a lightweight, fire-resistant roofing or flooring material is desired.
2. Where construction must proceed in freezing weather.
3. Where the design requires a simple, quickly and easily installed roofing or flooring material.

Conditions Unfavorable to the Use of Gypsum Decking

1. In areas where there is excessive moisture.
2. For floors on which there will be heavy storage or mechanical equipment or over which heavy loads will be transported.

History and Manufacture

The use of poured gypsum plaster dates back to the Greeks, who made decorative colored walks and floors of plaster. The technique of reinforcing gypsum plaster with metal started with the development of reinforced concrete in the 1860s. The search for lightweight materials for large roof areas led to the development of present-day poured reinforced gypsum and precast gypsum decking.

The manufacture of precast gypsum decking is simple. The exterior metal tongue-and-groove perimeter, with the wire mesh welded in place, is the form into which a correctly proportioned gypsum plaster is poured. After setting, it is ready for installation.

PLASTER INTEGRANTS AND SHEETS

Physical and Chemical Properties

Plaster integrants and sheets of all types, sold under a variety of trade names, generally consist of a core of gypsum plaster with fibers or hair, reinforced with paper laminated to both sides. Plaster integrants and sheets for interior walls and ceilings (dry-wall construction) have a sized paper on the side that is to be the finished surface of the walls and ceilings and an unsized paper on the back. They are available with a vinyl waterproof surface on one side for use in bathrooms as a base for ceramic wall tile or other type of finish surface material. They are also available with aluminum foil on one side to give thermal insulation. Plaster lath integrants are the same as plaster integrants except that low-grade paper is used on both sides and they are made in smaller sizes in either solid or perforated types. They are also obtainable with aluminum foil on one side for thermal insulation.

Types and Uses

Plaster integrants and sheets are available in various forms: regular, insulating, fire-resistant, waterproofed on one side, back-up, sheathing, lath, laminated, and decorative (*see* Figure *P21*). Table *P47* lists the common types, sizes, and major uses.

These sheets and integrants may be bent to radii which are controlled by the thickness (*see* Table *P48*).

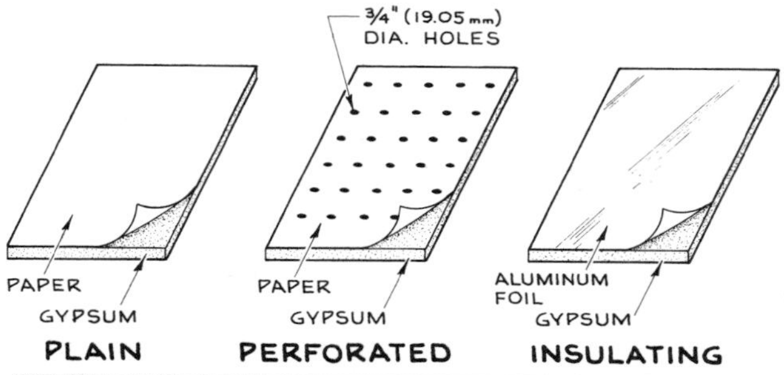

Figure P21 Typical types of plaster sheets.

Table P47 Types, sizes, surfaces, and major use of plaster boards, sheets, and integrants

Type of plaster board sheet, and integrant	Thickness		Width		Length		Edge treatment	Surface	Major use
	in.	mm	ft	m	ft	m			
Regular	$\frac{3}{4}$	6.35			6–14	1.829–4.267	Square	One side sized paper, other side unsized paper	Finish surface for walls and ceilings to be painted, papered, etc.
	$\frac{3}{8}$	9.53	4	1.219	in 1-ft (0.305-m) increments				
	$\frac{1}{2}$	12.7							
	$\frac{5}{8}$	15.88							
	$\frac{3}{8}$	9.53	4	1.219	6–14	1.829–4.267	Tapered	Same as above	Same as above
	$\frac{1}{2}$	12.7			in 1-ft (0.305-m) increments				
	$\frac{5}{8}$	15.88							

Table P47 Types, sizes, surfaces, and major use of plaster boards, sheets, and integrants (*continued*)

Type of plaster board, sheet, and integrant	Thickness		Width		Length		Edge treatment	Surface	Major use
	in.	mm	ft	m	ft	m			
Vinyl-coated waterproofed	$\frac{1}{4}$ $\frac{3}{8}$ $\frac{1}{2}$	6.35 9.53 12.7	4	1.219	6–14	1.829–4.267 in 1-ft (0.305-m) increments	Square with lapping	One side vinyl coated, other side unsized paper	Walls and ceilings in bathrooms and other areas where moisture is present
Insulating[a] (reflective)	$\frac{3}{8}$ $\frac{1}{2}$ $\frac{5}{8}$	9.53 12.7 15.88			6–14	1.829–4.267 in 1-ft (0.305-m) increments	Square or tapered	Same as regular with aluminum foil laminated to one side	Interior surface of exterior walls and ceilings and under roofs for thermal insulation
Backing	$\frac{3}{8}$ $\frac{1}{2}$	9.53 12.7	2, 4 2	0.61, 1.219 0.61	8	2.438	Square, tongue and groove	Both sides unsized paper	As base for laminating another layer of plaster board
Fire-resistant fire rating	$\frac{1}{2}$, $\frac{5}{8}$ $\frac{5}{8}$	12.7, 15.88 15.88	4 2	1.219 0.61	6–14	1.829–4.267 in 1-ft (0.305-m) increments	Tapered, tongue and groove	One side sized paper, other side unsized, or both sides unsized paper	Where fire ratings must be met in buildings
Decorative	$\frac{3}{8}$	9.53	4	1.219	6–10	1.829–3.048 in 1-ft (0.305-m) increments	Beveled	One side decorative paper and/or textured, other side unsized	Finished surfaces for walls and ceilings
Laminated	1	25.4	2	0.61	Mill-cut up to 12	3.658	Interlocking	Both sides unsized paper	Non-load-bearing partitions
Sheathing	$\frac{1}{2}$	12.7	2	0.61	8	2.439	Tongue and groove	Core of gypsum treated with asphalt, outside paper treated for water resistance	Sheathing for wood or metal stud frame buildings
Lath (solid)	$\frac{3}{8}$ $\frac{1}{2}$	9.53 12.7	$1\frac{1}{3}$	33.87	4, 8	1.219, 2.439	T & G	Both sides unsized paper	Same as above
Lath (perforated)	$\frac{3}{8}$	9.53	$1\frac{1}{3}$	33.87	4	1.219	Round	Both sides unsized paper; one $\frac{3}{4}$ in. (19.05-mm) hole for each 16 in.^{2} (103.23 cm^2) of surface	Lath for plastering

Table P47 Types, sizes, surfaces, and major use of plaster boards, sheets, and integrants (*continued*)

Type of plaster board, sheet, and integrant	Thickness		Width		Length		Edge treatment	Surface	Major use
	in.	mm	ft	m	ft	m			
Lath (reflective insulation)	$\frac{3}{8}, \frac{1}{2}$	9.53, 12.7	$1\frac{1}{3}$	33.87	4	1.219	Round	Both sides unsized paper, with aluminum foil laminated to one side	Lath for plastering
Lath (long-length)	$\frac{1}{2}$	12.7	2	0.61	Mill-cut up to 12	3.658	Tongue and groove	Both sides unsized paper	Lath for plastering
Lath (long-length reflective insulating)	$\frac{3}{8}$	9.53	2	0.61	Mill-cut up to 12	3.658	Tongue and groove	Both sides unsized paper with aluminum foil laminated to one side	Lath for plastering

[a]This lath should not be used as backing for ceramic wall tile.

Table P48 Bending radii for various thicknesses of plaster integrants and sheets

Thickness of sheet or integrant			
in.	$\frac{1}{4}$	$\frac{1}{2}$	$\frac{3}{8}$
(mm)	(6.35)	(12.7)	(9.53)
Minimum radius for plaster sheets and integrants bent lengthwise, 8 ft (2.438 m) minimum length			
ft	5	7.5	20
(m)	(1.524)	(2.286)	(6.096)
Minimum radius for plaster sheets and integrants bent along the 4-ft (1.219-m) width			
ft	15	25	
(m)	(4.572)	(7.620)	

Application

Condensed Checklist

1. When nails are used, they must be correctly installed for both the spacing and the method of nailing. For decorative sheet, the nail head should match the color of the surface and be driven in flush to the surface. For other types of sheet, the nail should be driven in such a way that a cuplike indentation appears on the surface. (*See* Table *P49*.)
2. In wood frame construction, all joints of plaster sheets and integrants must be backed with a supporting material in such a way that they are held rigid and cannot bend (*see* Figure *P22*). This applies to all types other than lathing (*see* Item 8).
3. All vertical joints should be staggered for integrants and sheets that are installed horizontally.
4. When plaster sheets and integrants are used for sheathing, all joints must be backed by supports that will permit wind and vapor barriers to be correctly installed and sealed.
5. In laminated (or layered) type of construction using plaster sheets and integrants, the base of back-up integrant (first layer) must be correctly secured and the finish-surface integrants must be laminated with their joints staggered in relation to the joints of the back-up material. In all laminated construction systems, only the adhesives recommended by the manufacturer of the sheet material should be used.
6. For partitions built by the laminated system, the base integrants must be correctly secured to floor and ceiling, and the surface-finish integrants must be laminated with their joints staggered in relation to to the back-up joints. (*See* Figure *P22*).

Table P49 Types of nails and spacing for plaster board sheets and integrants

Thickness and type of plaster board, sheet, or integrant in.	mm	Type of nail	Spacing of nails center to center
$\frac{5}{8}$	15.88	Barbed; $\frac{1}{4}$-in. (6.35-mm) head, $1\frac{3}{8}$ in. (34.93 mm) long, 0.098 gauge, bright or cement-coated	7 in. (177.8 mm) on ceilings; 8 in. (203.2 mm) on walls
$\frac{1}{2}$ $\frac{1}{2}$ (fire-resistant)	12.7 12.7	5-penny; $\frac{15}{64}$-in. (8.5-mm) flat head, $1\frac{5}{8}$ in. (41.28 mm) long, $13\frac{1}{2}$ gauge, bright or cement-coated	7 in. (177.8 mm) on ceilings; 8 in. (203.2 mm) on walls
$\frac{3}{8}$ $\frac{1}{4}$	9.53 6.35	4-penny; $\frac{7}{32}$-in. (5.56-mm) flat head, $1\frac{3}{8}$ in. (34.93 mm) long, 14 gauge, bright or cement-coated	7 in. (177.8 mm) on ceilings; 8 in. (203.2 mm) on walls
$\frac{1}{4}$ (applied over existing surfaces)	6.35	6-penny; $\frac{1}{4}$-in. (6.35-mm) flat head, $1\frac{7}{8}$ in. (47.63 mm) long, 13 gauge, bright or cement-coated	7 in. (177.8 mm) on ceilings; 8 in. (203.2 mm) on walls
$\frac{5}{8}$	15.88	6-penny; $\frac{1}{4}$-in. (6.35-mm) flat head, $1\frac{7}{8}$ in. (47.63 mm) long, 13 gauge, bright or cement-coated	6 in. (152.4 mm) on ceilings
$\frac{3}{8}$	9.53	Colored nail to match finish, $1\frac{1}{8}$ in. (28.58 mm) long	12 in. (304.8 mm) on walls
$\frac{1}{4}$ $\frac{3}{8}$ $\frac{1}{2}$	6.35	Barbed; $\frac{1}{4}$-in. (6.35-mm) head, $1\frac{1}{4}$ in. (31.75 mm) long, 0.098 gauge, bright or cement-coated	7 in. (177.8 mm) on ceilings; 8 in. (203.2 mm) on walls
$\frac{3}{8}$ (lath, all types)	9.53	Blued; $\frac{19}{64}$-in. (7.35-mm) flat head, $1\frac{1}{8}$ in. (28.58 mm) long, 13 gauge	5 in. (127 mm) for both ceilings and walls
$\frac{1}{2}$ (lath, all types)	12.7	Blued; $\frac{19}{64}$-in. (7.35-mm) flat head, $1\frac{1}{8}$ in. (28.58 mm) long, 13 gauge	4 in. (101.6 mm) for both ceilings and walls

7. When metal studs are used, the methods of attaching back-up sheet or integrant to studs at joints, corners, and ends must be correctly detailed as shown in Figure *P23*).

8. Plaster sheet and integrant lathing must be correctly attached and installed. The methods are similar to those for attaching to wood studs and metal studs. (*See* Figure *P24*.)

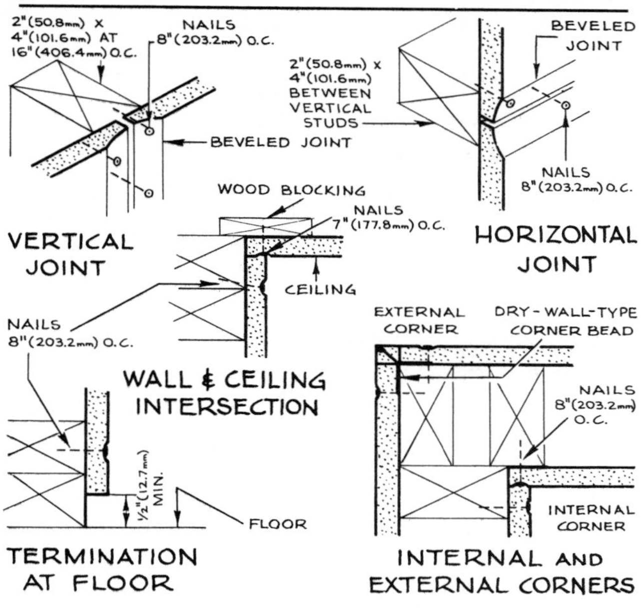

Figure P22 Methods of joining plaster sheets and integrants.

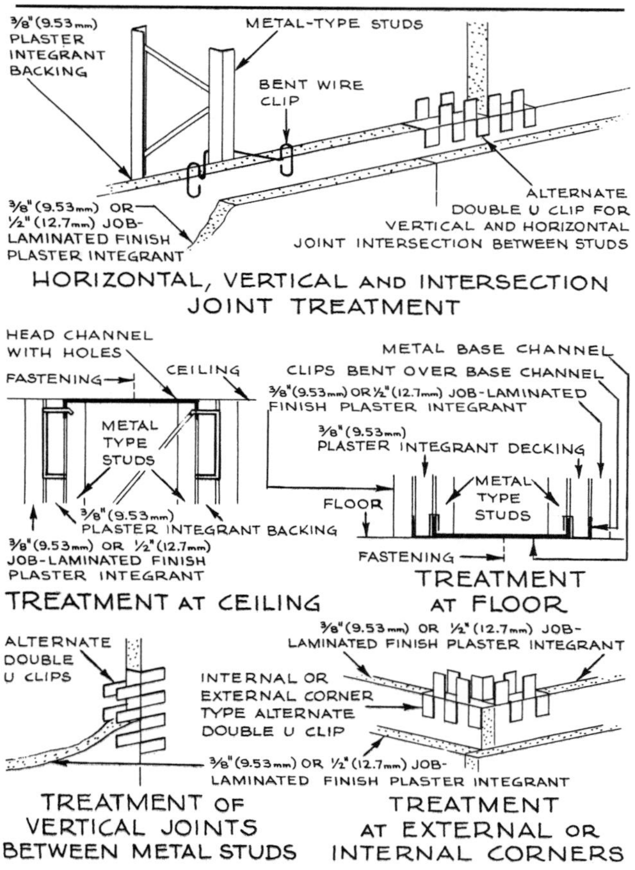

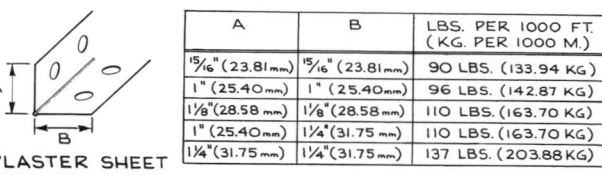

Figure P23 Detailing at joints and corners of laminated (layered) plaster sheets and integrants attached to steel or aluminum studs.

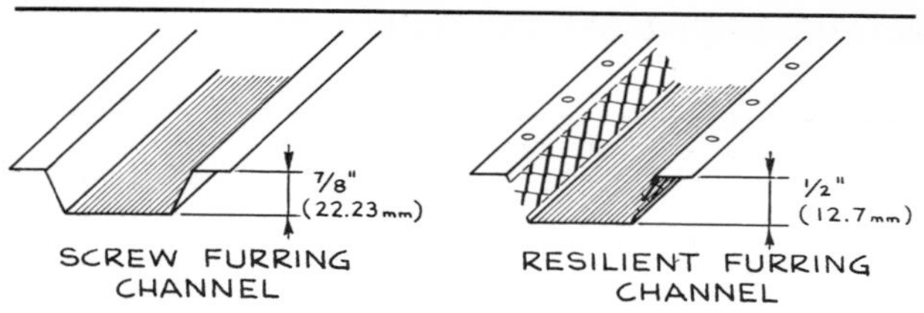

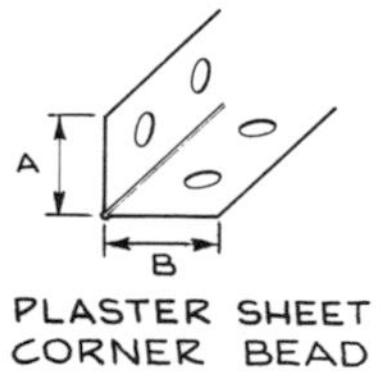

A	B	LBS. PER 1000 FT. (KG. PER 1000 M.)
15/16" (23.81mm)	15/16" (23.81mm)	90 LBS. (133.94 KG)
1" (25.40mm)	1" (25.40mm)	96 LBS. (142.87 KG)
1 1/8" (28.58mm)	1 1/8" (28.58mm)	110 LBS. (163.70 KG)
1" (25.40mm)	1 1/4" (31.75mm)	110 LBS. (163.70 KG)
1 1/4" (31.75mm)	1 1/4" (31.75mm)	137 LBS. (203.88 KG)

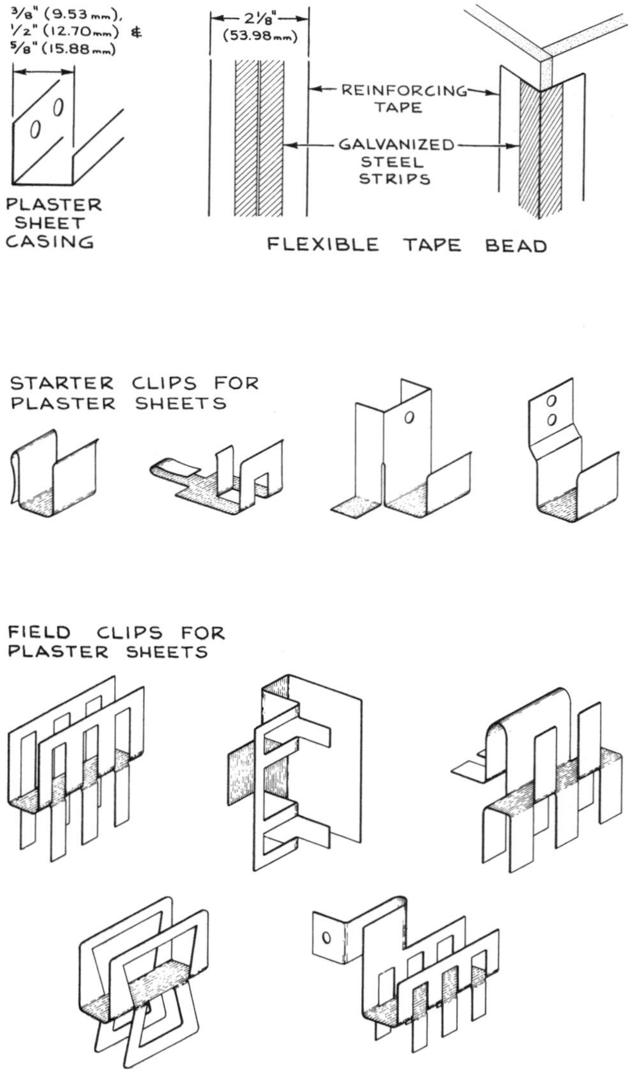

Figure P24 Typical accessories used with plaster sheets and integrants.

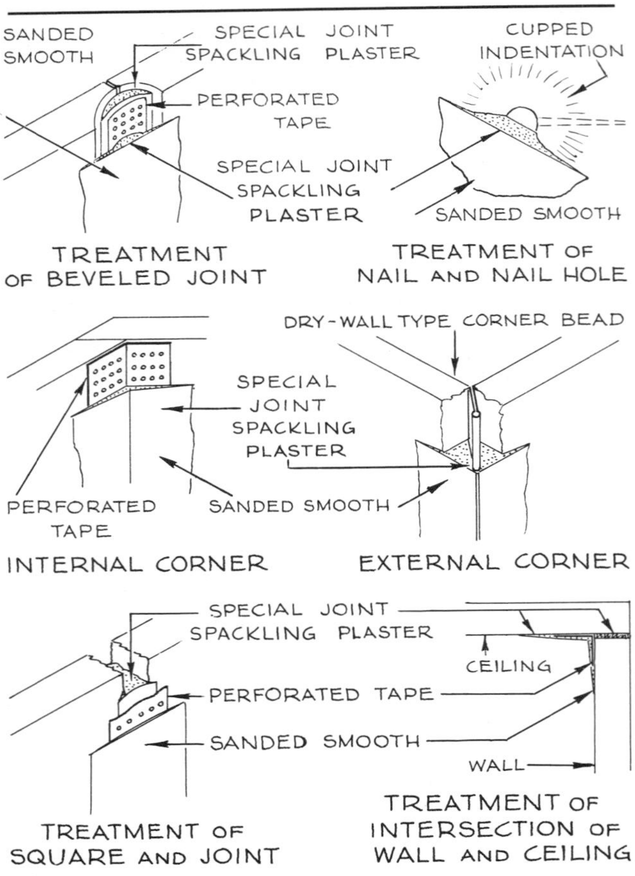

Figure P25 Finishing of joints, nails, and nail holes in plaster sheets and integrants.

9. Treatment of joints and nail holes with filler and subsequent taping and spackling should be done in strict accordance with manufacturers' directions. (*See* Figure *P25*.)
10. The treatment of joints and nails should always be rechecked after the prime coat of paint has been applied. Because of the whiteness of plaster material, many defects become apparent only at this time.
11. When installing plaster board, sheet, and integrant on metal furring, stud partitions, and walls, one should always use self-tapping, self-drilling, highly corrosion-resistant Phillips head steel screws (*see* Figure *P26*). These steel screws can be used with wood furring and studs.
12. When self-tapping and self-drilling steel screws are used, the manufacturer should always be consulted regarding the correct type of drill, type of special bit, and control of countersinking bit to be used for installation.

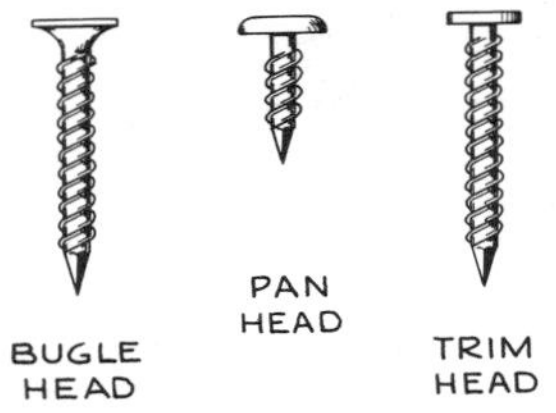

TYPE	SIZE
BUGLE	7/8"(22.23 mm), 1"(25.4 mm), 1 1/4"(31.75 mm), 1 5/16"(33.34 mm), 1 1/2"(38.1 mm), 1 5/8"(41.28 mm), 1 7/8"(47.63 mm) & 2 1/4"(57.15 mm)
PAN	3/8"(9.53 mm) & 3/4"(19.05 mm)
TRIM	1 5/8"(41.28 mm) & 2 1/4"(57.15 mm)

Figure P26 Types of self-tapping and self-drilling corrosion-resistant steel screws.

Conditions Favorable to the Use of Plaster Integrants and Sheets

1. As finish where a hard, smooth, fire-resistant surface finish that is quickly and easily installed is desired for walls and ceilings.
2. As sheathing where extra aluminum reflective insulation and vapor barrier are required on exterior walls.
3. Where lightweight, easily installed partitions are desired.
4. Where an easily and quickly installed lathing for walls and ceilings is required.
5. Where a waterproof base is needed for ceramic wall tile or other finish material.
6. Where an easily and rapidly installed base material is required for an acoustic tile type of ceiling finish.

Conditions Unfavorable to the Use of Plaster Integrants and Sheets

1. Where excessive moisture is present, except when vinyl waterproof plaster sheet is installed.
2. Where the finish will be subject to rough usage.

PLASTICS

Physical and Chemical Properties

The term "plastic" is essentially a commercial classification to which no strictly scientific definition can be applied. It is used to describe a product of synthetic origin which is capable of being shaped by flow in some stage of manufacture and which is not rubber, wood, glass, natural resin, leather, or metal.

The materials designated as rubber, fibers, resins, and plastics are of similar molecular structures, and by appropriate chemical and physical treatment it is possible to interconvert any one of them. Therefore certain structural features are common to all, and in turn are related to similarity in physical properties between materials not necessarily chemically related. In construction, plastics were first regarded as synthetic building materials which would provide answers to all the difficulties encountered in using existing materials. This belief resulted in great disillusionment, as the early plastics were not adequately tested and caused great embarrassment to all those involved in construction.

During World War II tremendous experimentation and development brought greater understanding of these materials, and the plastics industry turned to the industrial world for an outlet for its constantly improving products. Soon plastics became an ever-expanding, diversified material in construction as well.

Today the construction field faces a difficult problem because plastics, in the broad sense, not only are materials for flooring (including surfacing of exterior and interior athletic areas such as gyms and playing fields), counter tops, skylights, ceilings, roofing, siding, glazing, and textiles, but also have entered, via what one might call the back door, into the painting, lighting, and adhesives fields, each of which they are completely changing. Plastics are also entering the fields of plumbing, heating, ventilating, air conditioning, and electrical work. To complicate matters further, separate complete industries for synthetic rubbers, fibers, films, adhesives, and organic coatings exist and are considered as distinctly separate from the "plastics industry."

In the broad sense, there are three types of plastics: (1) cellulose plastics, (2) synthetic resin plastics, and (3) plastics derived from proteins and natural resins.

Synthetic resin plastics are subdivided into two classes: (1) thermosetting plastics (also called thermocuring plastics), which through heat are converted to a cured or infusible form that cannot be remelted and that have a molecular structure consisting of molecules linked in a three-dimensional network arrangement; and (2) thermoplastic resins, which may be softened by heat and then will regain their original properties upon cooling. The latter have a molecular structure consisting of molecules that are essentially linear or threadlike in form. They also have two major weaknesses, namely, they are soluble in a wide range of organic solvents and they cannot be used under high temperatures.

The use of plastics (using the word in its broadest sense) will steadily increase in construction, and it will be necessary to become familiar with their chemistry and physical properties in order to use and specify these materials. Fortunately, the trend in the plastics industry at present is to tailor-make plastics, under trade names as a rule, to meet the requirements and conditions of use. Then, after a product meets and answers these conditions, rigid tests and requirements are set up to standardize and control its quality and performance.

In the past the plastics industry generally directed its research, production, diversification, and expansion on the premise of an ever-available supply of low-cost fossil fuels. Figure *P27* illustrates the process from fossil fuels (hydrocarbons) to finished plastic materials. However, awareness of the inevitable, eventual depletion of the earth's fossil fuels is affecting the plastics industry and stimulating the search for other sources of raw materials from which to derive its products.

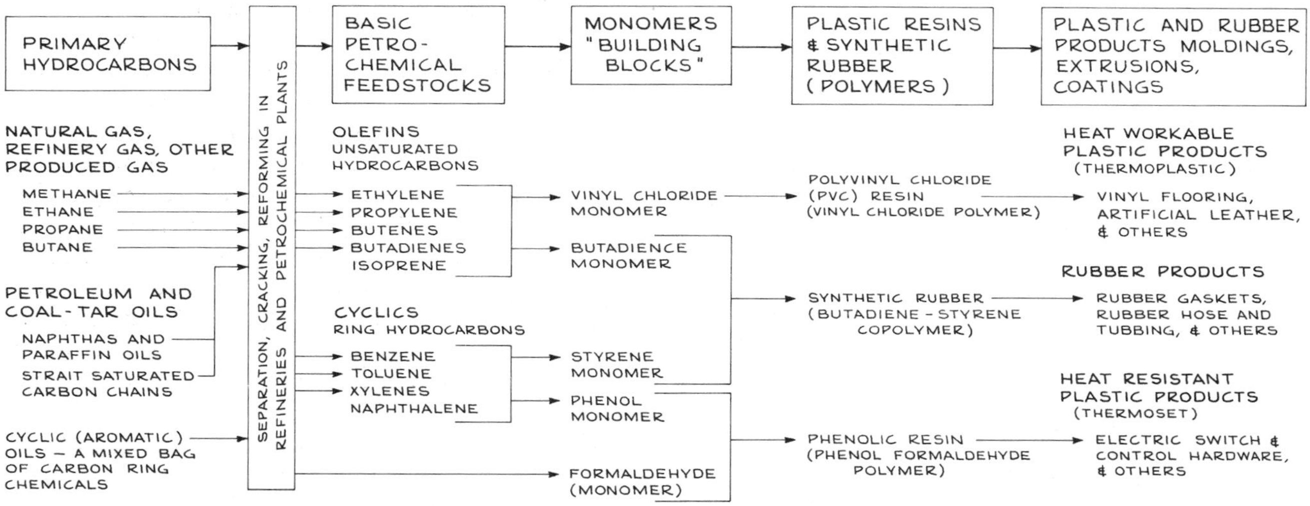

Figure P27 Flow diagram of source of plastics from fossil fuels to the various types of plastic materials.

Types and Uses

It is beyond the scope of this book to give a complete listing of the various types of plastics and their uses. However, Tables *P50* and *P51* list the plastics generally used in construction and those that are now used in other fields, as it is highly probable that in one form or another the latter will also find their way into structural materials.

In construction the words "plastic" and "resin" are often used indiscriminately in connection with synthetic products. In this book the word "resin" is used in relation to organic coatings (paints). Table *P52* gives the derivation of commonly used plastics and resins.

Materials made of plastic that are of direct interest in construction are discussed under separate headings. Because of the close relationship between plastics, rubbers, adhesives, fibers, textiles, and paints, it is advisable to read all these sections as a unit for a better understanding of today's materials for construction (*see* Table *P53*).

Application

Condensed Checklist

1. When any plastic material is selected, a check should always be made with the manufacturer to make sure that it meets the requirements and conditions of its end use. One should know its life, dimensional stability, durability, and color stability, as well as the effects of sunlight, weathering, heat and cold, and the normally used cleaning powders and solutions. One should learn how the material is applied and, particularly if by adhesive, the exact type of adhesive and method of application.
2. When plastics are used for structural joining or structural laminating, one should always check with the manufacturers of these structural members to make sure that the correct methods for this type of work are followed and are controlled by adequate testing.
3. When plastics are used for special joining in buildings (e.g., in panel wall construction), the architect, engineer, and/or other person responsible for the job should always consult with the manufacturers of these materials, and specify in complete detail how they are to be mixed and applied. It is most important to know, once they are mixed, how long they will remain usable (i.e., their pot-life).
4. For plastics applied with adhesive a good rule to follow is never to allow any type of adhesive to be substituted for the one recommended by the manufacturer of the plastic.
5. In general, one should always make sure that the plastic has been fully tested for the intended end use and, if possible, should examine a similar actual use of it in a building.

Conditions Favorable to the Use of Plastics

1. Where the plastic will do a better job than any other material now generally used.

Conditions Unfavorable to the Use of Plastics

1. Where other materials will do an equal or better job and serve the same purpose.

History and Manufacture

Plastics are an outgrowth of the integration of diversified and unrelated scientific knowledge. The basic investigations into the chemistry, physics, and biology of high-molecular-weight natural products such as cellulose, cotton, hair, natural resins, rubber, starch, and wool contributed one part of this knowledge. Another part came about from improvements in methods and instruments used in investigation. At first optical methods were used; then came ultramicroscopy, X-rays, colloidal chemistry techniques, and the ultracentrifuge. Now with the comparatively new science of nuclear physics, vast new areas of investigation are possible.

Early Plastics. The origin of plastics and the lacquers may date from 1833, when Branconnot, a professor of chemistry at Nancy, described the preparation of a "xyloidine" by treating starch, sawdust, and cotton with nitric acid. He found that this material was soluble in wood vinegar and attempted to make coatings, films, and shaped articles with it. In 1834 Von Liebig prepared melamine, but 100 years passed before it could be mass-produced. In 1846 Schoenbein, experimenting with nitrocellulose, predicted that it could be a substitute for gunpowder, with the result that militarists all over Europe immediately started extensive tests. Because crude preparatory methods were employed, disastrous explosions occurred in many places. These disasters stopped this type of production and allowed scientific research to proceed and to find means of stabilizing the product.

The art of molding plastics started with the simple hand-type hydraulic press, which was developed almost immediately after the discovery of the vulcanization of rubber in 1839 by Goodyear.

Table P50 Chemical and physical characteristics of plastics used in construction

Type of plastic	Derivation	Chemical and physical characteristics	Soluble or attacked by:
ABC *thermoplastic*	Acrylonitrile-butadiene-styrene copolymer; a thermoplastic polymer blend that may be produced from all three monomers or may be a mixture of copolymers	Dimensional stability over a temperature range from −40 to 160°F (−4.44 to 71.1°C); combustible but slow burning, low water absorption	Attacked by nitric and sulfuric acids, aldehydes, acetones, esters, and chlorinated hydrocarbons
Acrylic *thermoplastic*	A thermoplastic polymer or copolymer of acrylic acid, methacrylic acid, esters of these acids, or acrylonitrile	Clarity and transparency, low specific gravity, shock resistance, stability to weathering, slow burning. These can be converted to thermosetting resins by adding acrylic anhydride, acrylamide, or glycol esters of acrylic acid. They have the unusual property of transferring light through the body of the plastic, emitting it at the edge or end of the sheet, rod, bar, or object with little or no temperature change	Soluble in aromatic hydrocarbons, chlorinated hydrocarbons, esters, and ketones
Alkyd *thermosetting*	A thermosetting reaction product of a polyhydric alcohol and a polybasic acid in the presence of a drying oil which acts as a modifier; an alkyd is actually a polyester resin, which has a similar derivation but is not oil-modified	Resistance to severe exposure, weathering and heat; toughness, flexibility, good adhesion; is transparent	Soluble in mineral spirits
Allyl *thermosetting*	A special class of polyester resins derived from esters of allyl alcohol and dibasic acids	Highly resistant to chemicals, moisture, abrasion, and heat; curing even without application of heat or pressure; is transparent	Depends on type of allyl
Casein *thermosetting* or *thermoplastic*	Casein from milk acidified with dilute sulfuric, hydrochloric, or lactic acid	Nonflammable and tough, but has poor water resistance and dimensional stability; not suitable for rapid fabrication	Soluble in water
Cellulose acetate *thermoplastic*	Reaction of cellulose with acetic acid, acetic anhydride and sulfuric acid as catalyst	Flammable, not self-extinguishing; tough, high impact; subject to dimensional change due to cold flow, heat, or moisture absorption	Soluble in acetone
Cellulose acetate butyrate *thermoplastic*	Reaction of purified cellulose with acetic and butyric anhydrides, with sulfuric acid as catalyst	Similar to cellulose acetate	Soluble in acetone
Cellulose acetate propionate *thermoplastic*	Similar to cellulose acetate butyrate but made with propionic anhydride instead of butyric anhydride	Similar to cellulose acetate	Soluble in acetone
Cellulose nitrate (nitricellulose) *thermoplastic*	Cellulose treated with a mixture of concentrated nitric and sulfuric acids	Resistant to water, hydrocarbons, dilute acids, and alkalis; highly flammable; discolors on long exposure to sunlight	Soluble in acetone

Table P50 Chemical and physical characteristics of plastics used in construction (*continued*)

Type of plastic	Derivation	Chemical and physical characteristics	Soluble or attacked by:
Ethylcellulose *thermoplastic*	Alkali cellulose and ethyl chloride or sulfate, and cellulose and ethyl alcohol in the presence of dehydrating agents	Tough film which retains flexibility at low temperatures; lowest density of the commercial cellulose plastics	Soluble in most organic liquids
Epoxy *thermosetting*	Thermosetting resin based on the reactivity of the epoxide group	Toughness, high adhesion; corrosion resistant and resistant to common solvents, oils, and chemicals	Attacked by catalyzed epoxy reducer
Fluorocarbon plastics *thermoplastic*	Compound of carbon and fluorine with or without hydrogen	Chemically inert, nonflammable; resistant to high heat, chemicals, moisture, weathering, ozone, and ultraviolet radiation	Depends on type of fluorocarbon plastic
Furans *thermosetting*	Formaldehyde and acetylene which react to form butyl endole	Highly chemical resistant; resistant to high heat, acids, alkalis, alcohol and hydrocarbons; have high adhesion; black or dark in color	Depends on type of furan
Melamines *thermosetting*	Synthetic resin made from melamine and formaldehyde	Resistant to heat and water; and as organic coatings, exceptional curing speed and hardness; resistance to water, solvents, and soaps	Attacked by ketones; in paint (organic coatings) affected by mineral spirits and water
Phenol-formaldehyde resin *thermosetting*	Condensation of phenol or substitute phenols with aldehydes such as formaldehyde, acetaldehyde, and furfural	Gray to black color; hard; resistant to moisture, solvents, and heat up to 299°F (200°C); dimensionally stable; noncombustible	Decomposed by oxidizing acids
Polyamides *thermoplastic*	Polymerization of a dibasic acid and a diamine	Resistant to common solvents and alkalis; low water absorption; strong, tough, and elastic	Attacked by strong mineral acids
Polybutene *thermoplastic*	Any of the several thermoplastic isotactic polymers of isobutene of varying molecular weights	Combustible; resistant to abrasion, cracking, and weathering	Soluble in hydrocarbon solvents
Polycarbonate *thermoplastic*	Thermoplastic resin derived from bisphenol A and physgene A linear polyester of carbonic acid	Transparent; weather and ozone resistant; combustible but self-extinguishing; high impact strength; dimensionally stable; resistant to stains and heat	Soluble in chlorinated hydrocarbons; attacked by strong alkalis and aromatic hydrocarbons
Polyester *thermosetting*	Polycondensation of dicarboxylic acids with hydroxy alcohols; these are a special type of alkyd resin and are not modified with fatty acids or drying oils	Cure at room temperature when catalyzed and require little or no pressure; resistant to corrosive chemicals, solvents, etc.; are combustible and require addition of fire-retarding agents	Depends on type of polyester material
Polyethylene *thermosetting*	Polymerizing of ethylene with peroxide catalyst	Tough, white, flexible material resistant to solvents and corrosive solutions; good impact strength	Attacked by surfactants

Table P50 Chemical and physical characteristics of plastics used in construction (*continued*)

Type of plastic	Derivation	Chemical and physical characteristics	Soluble or attacked by:
Polystyrene *thermoplastic*	Polymerized styrene	Transparent; high strength and impact resistance; excellent electrical and thermal insulator; resists organic acids, alkalis, and alcohols; combustible and not self-extinguishing	Soluble in hydrocarbon solvents
Polyterpene *thermoplastic*	Polymerization of turpentine in the presence of catalysts such as aluminum chloride or mineral acids	A liquid soluble in most organic solvents	Soluble in organic solvents
Polyurethane *thermoplastic* or *thermosetting*	Condensation reaction of a polyisocyanate and a hydroxyl-containing material such as a polyol or drying oil	Fiber: high elastic modulus. Coatings: hardness, gloss, flexibility, abrasion resistance, and adhesion; resistant to impact, weathering, acids, and alkalis. Elastomers: resistance to abrasion, weathering, and organic solvents. Foams: flexible and rigid, excellent thermal conductivity, flame-resistant	Depends on type of polyurethane
Polyvinyl *thermoplastic*	Any of a series of polymers (resins) derived by polymerization or copolymerization of vinyl monomers, including vinyl chloride and acetate, acrylonitrile, styrene, the vinyl esters, and numerous others—specifically, polyvinyl chloride, acetate, alcohol, etc.	In general, resistant to weathering, oils, greases, acids, fungus, moisture, and petroleum hydrocarbons; combustible but self-extinguishing; abrasion resistant. Always check type of vinyl for its particular properties	Depends on type of vinyl
Silicone	Any of the large group of organic siloxane polymers; silicon is heated in methyl chloride to yield methychlorosilanes, which are separated and purified by distillation, the desired compound being mixed with water	Low surface tension; extreme water repellency; resistant to weathering, oxidation, and high temperature. Unhalogenated types are combustible	Soluble in organic solvents
Urea-formaldehyde *thermosetting*	Urea and formaldehyde are united in a two-stage process in the presence of pyridine, ammonia, or certain alcohols with heat and control of pH to form intermediates which are converted to thermosetting resins by heat and pressure in the presence of catalysts	Excellent light diffusion; strong, rigid; resistant to wear, household solvents, oils, and stain; good dimensional stability	Attacked by strong acids, alkalis, and solvents
Zien	Derived from corn processing	Resistant to dilute acids, anhydrous alcohols, turpentine, esters, and oils; combustible	Soluble in alcohol

Table P51 Plastic types, common and trade names, and uses in the construction field

Type of plastic	Common or trade name	Major use in construction	Minor construction uses
ABC *thermoplastic*	Lustran, Abson, Dylel, Cycolac, Kralastic	Pipe and fittings; shower stalls; bathtub enclosures; etc.	Telephones; building panels
Acrylic *thermoplastic*	Lucite, Plexiglas	Skydomes; skylights, organic coatings; adhesives; safety glass; finish hardware; lighting fixtures	Fibers (Acrilan, Orlon, Creslan, Zefran) for textiles, carpets, etc.; signs; textile finishes
Alkyd *thermosetting*	Alkyd	Vehicle for exterior and interior organic coatings and baked enamels	Textiles; rubber compounds; the entire electrical field
Allyl *thermosetting*	Contact adhesives	Adhesives for lamination and coatings	Varnishes; lacquers; heat-resistant furniture finishes; molding compositions
Casein *thermosetting* or *thermoplastic*	Casein glue	Household adhesives	Adhesives for interior wood laminations
Cellulose acetate *thermoplastic*	Tenite, Lumerith, Plastacele, cellophane	Organic coatings; sheet, rods, tubes, and film	Artificial leather and fibers; transparent film
Cellulose acetate butyrate *thermoplastic*	Cellophane	Similar to cellulose acetate	Similar to cellulose acetate
Cellulose acetate propionate *thermoplastic*	Cellophane	Similar to cellulose acetate	Similar to cellulose acetate
Cellulose nitrate (nitrocellulose) *thermoplastic*	Celluloid pyroxylin	Similar to cellulose acetate	Similar to cellulose acetate
Ethyl cellulose *thermoplastic*	Household glues	Hot-melt adhesives	Coating for cables, paper, textiles, etc.
Epoxy *thermosetting*	Epoxies	General adhesives; material-to-material adhesive; matrix for terrazzo flooring and seamless floors; surface coatings; special organic coatings	Special cements and mortars; organic coatings for floors
Fluorocarbon plastics *thermoplastic*	Teflon, Aclar, Halon, Fluorel	Air-, tent-, and cable-type structures; sheet, film, tubes, rods, tapes, and fibers; protective coatings; dark in color	High-temperature wire and cable insulation; refrigerants; gaskets
Furans *thermosetting*	Alkor cement, Tygon resins, furane plastics	Protective coatings; areas where resistance to acids, alkalis, alcohols, and hydrocarbons is required	Electrical coatings; adhesives; pipe and fittings
Phenol-formaldehyde *thermosetting*	Phenolics, Bakelite	Organic coatings; baked enamels	Adhesives; finish hardware; laminations and impregnations
Polyamides *thermoplastic*	Fiberthin, nylon, Facilon	Tarpaulins; air, tent, and cable structures; carpet	Laminated sheet facings; rollers; hardware; flexible floor coverings
Melamines *thermosetting*	Melamine	Plastic laminates	Organic coatings (combined with alkyds and acrylics); light fixtures

Table P51 Plastic types, common and trade names, and uses in the construction field (*continued*)

Type of plastic	Common or trade name	Major use in construction	Minor construction uses
Polybutene *thermoplastic*	Vistanex	Pressure-sensitive tapes; caulking and sealing compounds	Hot-melt adhesives; cable insulation and special sealants
Polycarbonate *thermoplastic*	Rowlex, Lexan	Nonbreakable glazing material	Electric light fixtures; molded products; film
Polyester *thermosetting*	Dacron, Mylar	Reinforced plastics; matrix for seamless and terrazzo flooring	Foams; protective coatings; textiles (Dacron); laminations; adhesives
Polyethylene *thermoplastic*	Polyethylene	Films for dampproofing; pipe for automatic exterior watering systems	Electric light fixtures; sheeting and films; the electrical field
Polystyrene *thermoplastic*	Styron, Styrofoam	Foam thermal insulation; transparent, translucent, opaque, and colored	Electrical equipment; electric light fixtures
Polyterpene *thermoplastic*	Polyterpene	Organic coatings	Wax polishes; curing concrete
Polyurethane	Polyurethane fiber, coatings, elastomers, and foams	Fibers: bristles for brushes. Coatings: baked coatings, organic coatings, two-component formulations, and roofing. Elastomers: sealants and caulking agents. Foams; thermal insulation	Fire-retardent wood coatings Adhesives; films; reinforced plastic Pipe and duct insulation; carpet underlayment
Polyvinyl *thermoplastic*	Vinyls, PVC, Hi-temp, Geon, PVA	Organic coatings, adhesives, films, cements, and mortars; piping, siding, gutters, flooring, athletic surfaces, synthetic turfs; pressure-sensitive tapes; hot-melt adhesives	Window and door frames; electrical insulation; film and sheeting; fibers; protective coatings; sealants and laminating agents
Silicone	RTV, silicones, siloxanes	Liquids: adhesives; water-resistant and protective coatings; weatherproofing concrete. Resin: sealants; gaskets; coatings	Sealing and caulking; electrical insulation; films
Urea-formaldehyde *thermosetting*	Plaskon	Baked enamels; laminating, protective coatings	Adhesives; housings for radio, TV, hi-fi, business machines, etc.
Zien	Zien	Adhesives; wood laminations	Imitation shellac; grease-resistant coatings

Table P52 Plastics and resins classified according to derivation

Natural polymers			
Hydrocarbons	Cellulose	Polymerization	Condensation
Chlorinated rubber, rubber hydrochloride	Cellulose acetate butyrate, cellulose acetate propionate, cellulose nitrate, ethylcellulose	Acrylate, allyl, halocarbons, polyethylene, polyterpene, vinyl	Allyl, epoxy, furan, phenol-formaldehyde, polyamides, polyester, polyurethane, silicones

Table P53 Types of synthetic rubbers[a]

Type of rubber	Common or trade name
Sodium polysulfide	Thiokol
Polychloroprene	Neoprene
Butadiene-styrene copolymer	SBR
Acrylonitrile-butadiene copolymer	Nitrile rubber
Ethylene-propylene-diene	EPDM rubbers
Synthetic polyisoprene	Coral, Natsyn
Isobutylene-isoprene copolymer	Butyl rubber
Polyacrylonitrile	Hycar
Polysiloxane	Silicone rubber
Epichlorohydrin polyurethane	Vulkolian

[a]*See* Rubber.

In 1865 Schutzinberger acetylated cellulose in an effort to produce a nonexplosive substance similar to nitrocellulose.

Celluloid. The first successful plastic was made by Hyatt, a printer in the United States; patents to the Hyatts were issued between 1870 and 1872. In 1871 the Celluloid Manufacturing Company was formed. The first use of Celluloid was for dental plate blanks and later in sheet form for automobile side curtains, as well as the now famous Celluloid collars. The Hyatts also were the first to attempt injection molding, which was later revived and used successfully by the Germans.

Cellulose Acetate. In 1879 Franchimont found that the esterification reaction could be catalyzed by sulfuric acid. In 1882 Stevens discovered that amyl acetate could be used as a solvent for nitrocellulose. His work led finally to the first transparent, flexible photographic films, developed by Goodwin of Newark, New Jersey, between 1887 and 1898. In 1894 Cross and Bevan in England patented a process for producing a type of cellulose acetate, and in 1911 Dreyfus of France perfected a manufacturing process for preparation of the acetylated compound and its hydrolysis. Cellulose acetate was first used as safety photographic film. During World War I it was used to coat airplane wings, and after the war served as acetate rayon fiber for safety glass. By injection molding, it was also used for tool handles, goggles, etc.

Experimentation with synthetic fibers started in 1884 with de Chardonnet's discovery of a method of producing fibers from nitrocellulose.

Casein Plastics. These had their beginning in 1898, when casein was condensed with formaldehyde by Spitteler and Krische. Production of casein plastics started in 1900 in France and Germany and in 1914 in England. The monopoly was broken during World War I, and in 1919 the United States also undertook production. The limitations of casein plastics in relation to humidity and dimensional stability soon became apparent, and today the use of casein plastics is limited almost entirely to the production of buttons.

Cellulose Esters. In 1905 Von Suida conceived the idea of ethylating cellulose as a method for dyeing cellulose. The cellulose esters were studied simultaneously by Lilienfeld of Austria, Leuchs of Germany, and Dreyfus of France. It was found that cellulose esters were soluble in organic liquids and had potentialities in plastics, in lacquer, and in fibers.

Bakelite. In 1909 Baekeland in the United States developed and produced the phenol-formaldehyde resins best known as Bakelite.

Developments from World War I to II. During and after World War I great strides were made in the field of plastics, and some peculiar shortages developed. For example, before and after World War I the motion picture and photographic film industry came to depend on amyl acetate as a solvent; then the two major suppliers of raw materials for amyl acetate, the United States and Russia, within a short interval of time experienced historical changes that made the supply of amyl acetate almost nonexistent. The United States went dry (Prohibition Era), and the Russian Revolution stopped the other source of materials. Fortunately, in the years 1920 to 1923 Weizmann perfected the butyl alcohol process, and the anhydrous ethyl acetate which was produced sufficed to end this shortage.

From World War I to World War II, new developments and processes appeared in the plastic field and some of the earlier basic problems were solved.

Polymers. Although as early as 1833 Berzelius had introduced the word "polymer"—which he used to indicate the presence of the same atoms in the same proportions in compounds having different molecular weights—it was not until this interwar period that a clear definition could be made.

A polymer is a substance (often synthetic) composed of giant molecules that have been formed by the union of a considerable number of simple molecules with one another. The number of molecules that unite to form a polymer molecule varies from 2 to 100 or even thousands. The simple molecule that will undergo such a change is known as a monomer, and the union of monomers to form polymers is called polymerization. The monomer molecules may be all alike, or two or more varieties of monomers may be involved in the formation of a particular polymer. Polyethylene plastic, a polymer, is formed by the union of ethylene molecules, whereas styrene synthetic rubber, a copolymer, requires the union of styrene and butadiene monomers. Condensation polymers are those in which the union of the molecules involves the formation and elimination of water or some simple substance in each step of the process.

Comparison of Natural and Synthetic Products. The comparative properties of natural and synthetic products were studied by Staudinger in Europe. In 1926 he demonstrated the interrelationship existing between the structure determined physically and the size and structure determined chemically by analysis. He showed that synthetic products have many properties in common with natural products and that the synthetics can be used as prototypes in their evaluation.

Resinification was found to be the result of two types of chemical reactions: (1) one involving a condensation reaction, where the polymer differs from the starting material by the elements eliminated in the condensation, and (2) one involving a polymerization reaction, where the polymer and starting material have the same chemical composition.

Research in the United States developed the functionability concepts that enabled a clear distinction to be made between the chemistry of the thermoplastic resins and the behavior of the thermosetting type of product.

Effect on Construction Materials. From World War II to the present, materials and design concepts in construction have been affected by plastics in all their forms. To mention a few examples, there are new paints, adhesives, and caulking compounds; new flooring, wall, partition, and roofing materials; new water repellents; hard high-gloss types of finish for walls; finish flooring that is almost painted on; and laminates. In the electrical field there are plastic wire coverings, switch boxes, and switch plates; and in plumbing and heating, there are plastic pipes, ducts, and the like.

The introduction of plastics into mortars, cements, and concrete has imparted to these materials tensile strengths that open up entirely new uses and applications for masonry products and concrete. The use of synthetic fibers with tensile strengths over 200,000 lbf/in.^{2} (1379 MN/m^2) and natural fibers in concrete has created a new structural material called fibered concrete. This will, of course, result in new types of reinforced concrete and even structural design.

The use of copolymer resin or other type of resin as matrix makes stained glass windows more readily available (*see* Glass and Glazing). There are also now plastic panels of a stained glass type usable on both the interior and the exterior. The United States (Federal) Pavilion of the 1967 New York World's Fair is an example of the use of this material.

PLASTICS: FIBERS, TEXTILES, AND CARPETS

Physical and Chemical Properties

Plastic fibers are extensively used for producing textiles and carpets. Examples are curtaining and drapes of all types; imitation leather wall and furniture coverings; all types of rugs and carpets; and twine, rope, and cables. Each is made from different plastics to tailor-fit the particular end product and usage. In general, the following are used: acrylics (Orlon, Acrilan), cellulose acetates, fluorocarbons (Teflon), polamides (nylon), polyesters (Dacron), polyurethanes, and polyvinals (vinyls). (*See also* Plastics; Rubbers; Fibers.)

The twine, rope, and cables have tensile strengths up to 50,000 lbf/in.^{2} (344.75 MN/m^2) and greater. The nylons have not only great tensile strength but also high elasticity. They are not affected by rot, vermin, insect, mold, or fungus. They also have good weathering characteristics and are resistant to salt water, common detergents, soaps, and other general household or construction site maintenance materials.

Types and Uses

In the construction field, plastic rope and cables are used for all types of hand or pulley hoists, and are available for mechanically operated hoists and cranes.

The textiles and fabrics are used extensively for drapes, curtains, wall coverings, and furniture. They also find extensive use in construction as temporary enclosure materials and protective coverings, and for tent, air, and cable structures (*see* Plastics: Film; Plastics: Sheet; Air Structures; Cable Structures; Tent Structures).

Plastic carpets are used extensively as floor covering and also as wall covering in areas where people can be hurt by falling (nursing homes) and/or for acoustical treatment of walls and ceilings. Plastic carpets are also used as exterior coverings for terraces and walks and as artificial grass. There is a similarity between plastic carpeting and plastic surfacing for playgrounds, athletic playing fields, and gymnasiums because the same plastic materials are used (*see* Plastics: Flooring).

There is available a large variety of plastic material combinations such as plastic foam (urethane or synthetic rubber) covered with a plastic fabric for manufacturing play and athletic equipment, matting, and covering for all types of furniture.

PLASTICS: FILM

Physical and Chemical Properties

Plastic films are manufactured from fluorocarbons, polyamides, polyesters, polyethylenes, polyurethanes, and polyvinyls. These films are available plain or reinforced with glass fibers. They find wide use in the construction field for dampproofing, waterproofing, temporary enclosures, covers for protecting materials and concrete that is curing, and vapor barriers. They are also used for air, cable, and tent structures. (*See* Table *P54*.) In these applications plastics either replace an existing material because they do a better job or represent new materials that require entirely new design concepts.

Table P54 Types of plastic films, common and trade names, and major uses

Type of plastic	Common or trade name	Major uses
Fluorocarbons[a]	Teflon	Air, cable, and tent structures
Polyamides[a]	Nylon	Air structures; protective coverings
Polyester	Mylar	Drawing paper and drafting paper
Polyethylene[a]	Polyethylene films	Protective coverings; dampproofing; temporary enclosures; air and tent structures
Polyurethane[a]	Polyurethane films	Protective coverings; dampproofing
Polyvinyls[a]	Vinyl films	Temporary enclosures: air, cable, and tent structures; dampproofing

[a]These plastics are also used for temporary structures such as collapsible greenhouses, swimming pool covers, and the like.

The plastic films used for dampproofing and similar applications are generally made of polyethylene and polyvinyl. They are available in rolls 6 ft (1.829 m) wide and in almost any length.

Types and Uses

All of these films have good resistance to weathering and moisture but vary in their resistance to acids, solvents, and alkalis. They vary in other characteristics also: whether they are flammable, self-extinguishing, or noncombustible; whether transparent, translucent, or opaque; and whether or not they can be colored. When a plastic film seems to meet the requirements for a particular end use, it is always advisable to check with the manufacturers of various plastic films to obtain the information necessary to select the correct plastic film for end use in question.

PLASTICS: FLOORING

Physical and Chemical Properties

Plastic flooring covers a wide variety of plastic materials used for interior and exterior flooring surfaces, ranging from the more familiar patterned and colored tiles and sheet, seamless colored flooring, and terrazzo seamless flooring to the newer and more extensively used exterior and interior play and utility surfaces for football, track, tennis, storage, parking garages, and even interior ice-skating surfaces (*see* Asphalt; Cement; Paint; Painting; Rubber; Terrazzo).

The flooring material most widely used in construction is vinyl flooring. Vinyl resilient flooring materials are divided into three major types: (1) solid vinyl, (2) vinyl and asbestos combined, and (3) a thin vinyl layer applied to other types of resilient flooring materials. These materials have good resilience; are durable, easy to maintain, and colorful; and have good resistance to grease and alkalis. They are made to meet rigid tests and strict federal specifications.

Commercial Forms. Vinyl resilient flooring materials are available in sheet and tile form. Inlays and feature strips are also available.

Types and Uses

Vinyl Sheet Flooring. Vinyl sheet flooring is manufactured in a wide range of thicknesses as follows: 0.069, 0.080, 0.100, 0.110, 0.180, and 0.224 in. (1.75, 2.03, 2.54, 2.79, 4.57, and 5.69 mm). It is available in rolls 6, 9, and 12 ft (1.829, 2.743, and 3.658 m) wide and up to 50 ft (15.740 m) long. The sheet flooring is available with smooth or embossed surfaces, with or without a thin or thick foam interlayer or underlayer. The thicker vinyl is used in areas where heavy traffic will be encountered. Feature strips are $\frac{1}{2}$ and 1 in. (12.7 and 25.4 mm) wide, and are usually available in only a limited color range. Decorative inlays are available from each manufacturer. The colors, textures, embossing, and patterns vary widely, and each manufacturer is constantly bringing out new types.

Vinyl Tiles. Vinyl tiles (all vinyl) are made in two thicknesses, $\frac{3}{32}$ and $\frac{1}{8}$ in. (2.35 and 3.18 mm); in squares 9 X 9, 12 X 12, 18 X 18, and 36 X 36 in. (22.86, 30.48, 45.72, and 91.44 cm), and in rectangles 18 X 36 in. (45.72 X 91.44 cm). Generally other sizes may also be available from various manufacturers. The colors, textures, patterns, inlays, and feature strips are similar to those for the sheet materials.

Vinyl Asbestos Tile. Vinyl asbestos tiles are made in three thicknesses, $\frac{1}{16}$, $\frac{3}{32}$, and $\frac{1}{8}$ in. (1.59, 2.35, and 3.18 mm); in squares 9 X 9 and 12 X 12 in. (22.86 and 30.48 cm). Some manufacturers make larger sizes in the thicker tile. The colors, textures, and patterns cover a wide range. Feature strips are generally limited to the $\frac{1}{2}$- and 1-in. (12.7- and 25.4-mm) widths, in 2-ft (0.610-m) lengths in all thicknesses; generally they are also limited to only four colors plus black and white. Inserts are available in large variety. For both vinyl tile and sheet-form vinyl, cove bases are generally available in six colors plus black and white, and are $\frac{1}{8}$ in. (3.18 mm) thick, 4 ft (1.219 m) long, and $2\frac{1}{2}$ or 4 in. (63.5 or 101.6 mm) high.

Other Vinyl Materials. Floorings that consist of a thin vinyl surface on another material are covered in this book under the heading for the particular base material. There are available for entrances to buildings vinyl matting and carpet runner. Other vinyl flooring accessories available are stair treads and nosings, corner guards, thresholds, and various edging strips.

Application

Vinyl sheet and tile flooring and vinyl asbestos tiles can be applied below grade, above grade, and on grade. In general, it is always advisable to follow the methods of application set forth by the manufacturers of resilient vinyl flooring materials and always to use the adhesive recommended for the particular end use of the flooring by the vinyl flooring manufacturer.

Seamless and Paint-Type Flooring. Seamless and thin-sprayed, rolled, or troweled, paint-type finish floorings are relatively new types of flooring, although special floor paints and oxychloric cement have been used for many years for finished flooring.

Seamless flooring consists of an epoxy, urethane, polyester, or other synthetic resin matrix with or without stone chips, plastic granules (chips), or other fillers to obtain a texture, color, or design (*see* Terrazzo; Paint; Painting). These seamless floorings vary in thickness from 20 mils (0.508 mm) to $\frac{3}{8}$ in. (9.53 mm). They can be applied to concrete, plywood, wood, asphalt, and other base materials, depending on the type of seamless flooring selected and its end use. A check with manufacturers of these types of flooring for methods of application is always advisable.

Athletic Surfacing. This type of flooring (surfacing) is generally restricted to recreation, play, and sports areas, both exterior and interior. One of the most common types is known as Astroturf, used for professional football and baseball. These surfaces are produced in two basic types, exterior and interior surfaces. They vary in thickness from 20 mils (0.508 mm) to 1 in. (25.4 mm) and thicker. They can be sprayed, broomed, troweled, and rolled, and applied as sheets, rolls, and squares. They are applied to concrete, asphalt, or wood, depending on the type of surfacing material and its end use. The plastics are made of vinyl, urethane, acrylic, or other type of resins and are available in a limited color selection. Straight, curved, and other markings required for particular sports can be installed with all the various systems for athletic surfacing. A check should be made with the manufacturers of these athletic surfacings to obtain the latest data, and an inspection should always be made of an existing installation of the surfacing material selected that has already been in use.

There is available a selective surfacing for interior use which can be used for multiple sports including ice skating and dancing.

Application

Condensed Checklist

1. The surface to which flooring or surfacing is to be applied must be dry, clean, and level. If underlayment is used, it should be the type recommended by the flooring manufacturer.
2. Flooring or surfacing should not be laid at temperatures below 55°F (12.78°C). In general, the areas should be heated to 70°F (21.11°C) for 48 hours before installation of the flooring, and temperatures

above 55°F (12.78°C) should be maintained thereafter.

3. When plastic-type finish flooring materials are used, the thickness of the various other finish flooring materials to be used within the structure should be carefully checked, and proper allowance made for any difference in thickness within the actual construction of the building, so that unsightly changes in level do not occur.
4. A check should always be made that only adhesives recommended by the flooring or surfacing manufacturer are used and that no substitutes are permitted unless approved in writing by the manufacturer of the flooring or surfacing.
5. Generally only neutral cleaners and emulsion-type waxes should be used for cleaning and finishing plastic flooring or surfacing. It is usually good practice to use the type of cleaner, finisher, and/or wax that is recommended by the flooring or surfacing manufacturer and to make the general contractor responsible for the protection, cleaning, and waxing of the flooring, so that a completed building is handed over to the owner.
6. One should always check with manufacturers of plastic flooring or surfacing for the correct type of substrate; for resurfacing or renovation, a check should be made regarding what should be done to the existing substrate before new flooring or surfacing is applied.
7. For exterior application, the manufacturer should be consulted for limitations of wetness and for restrictions and precautions if there is the possibility of rain during and after application.
8. One should always check with manufacturers for methods of application, finishing, cleaning, and maintaining the selected plastic flooring or surfacing.

Conditions Favorable to the Use of Plastic Flooring

1. Where a colorful, textured, tough, durable, easily maintained, grease-resistant type of finish flooring material is required for areas of both light and heavy human traffic.
2. As a finish flooring material for above-grade, on-grade, and below-grade floors.
3. Where a colorful, durable, easily maintained exterior or interior sports flooring or surfacing material is required.
4. Where soft, resilient, tough, colorful, finished interior flooring, which also requires simple maintenance, is required.
5. Where a flooring surface must move and shift with the substrate without cracking and chipping.

Conditions Unfavorable to the Use of Plastic Flooring

1. On exterior floor surfaces except for athletic, playground, and sports areas.
2. In areas where specific chemicals that can attack plastic-type flooring are used. For special areas of this type the flooring manufacturer should always be consulted to learn whether the plastic flooring will withstand the chemicals in question.

PLASTICS: FOAM

Physical and Chemical Properties

Plastic foams are generally of two basic types: (1) *polyurethane*, made either by reacting isocyanates with carboxylic compounds and adding a compound that produces carbon dioxide to make a urethane foam, or by reacting a diisocyanate with a compound containing an active hydrogen; and (2) *polystyrene*, a polymerized styrene with compounds added to produce carbon dioxide to make a polystyrene foam.

Polyurethane foam is available as rigid board in 3- and 4-ft (0.914- and 1.219-m) widths, in 4- to 12-ft (1.219- to 3.658-m) lengths, and in thicknesses ranging from $\frac{11}{16}$ in. (0.65 in. or 16.51 mm) to 2 ft (0.61 m). It is also available for on-site or in-plant application in the form of a two-component system to be either hand-mixed and poured by automatic equipment, frothing equipment, or spraying equipment. Table *P55* lists chemical and physical characteristics.

Table P55 Chemical and physical characteristics of polyurethane foam[a]

Btu/in./ft^2 · h · °F	0.11–0.14
W/m^2 · °C	0.0159–0.0202
Resistance, *R*	9.09–7.14
Reciprocal of W/m^2 · °C	62.89–49.50
Flame spread	25
Density	
lb/ft^3	1.8–4.0
kg/m^3	28.83–64.08

[a] Always check codes for requirements in regard to flame spread and fire resistance.

Polystyrene foam is available as rigid board in widths of 1.33, 2, and 4 ft (0.406, 0.610, and 1.219 m), in lengths of 3, 8, 9, and 12 ft (0.914, 2.438, 2.743, and 3.658 m), and in thicknesses of $\frac{1}{2}$, $\frac{3}{4}$, 1, $1\frac{1}{2}$, 2, 3, and 4 in. (12.7, 19.05, 25.4, 38.1, 50.8, 76.2, and 101.6 mm). It is also available with tongue-and-groove for use as sheathing on residential construction. Table *P56* lists chemical and physical characteristics.

Table P56 Chemical and physical characteristics of polystyrene foam[a]

Btu/in./ft^2 · h · °F	0.17–0.28
W/m^2 · °C	0.02451–0.04038
Resistance, *R*	5.88–3.85
Reciprocal of W/m^2 · °C	40.80–24.76
Flame spread	25
Density	
lb/ft^3	1.0–2.1
kg/m^3	16.02–33.64

[a]Polystyrene is flammable and has to be protected with nonflammable materials in order to meet fire-resistance ratings.

With regard to resistance to heat and fire, both foams are available in two types: those called slow-burning plastics, and those that will not support combustion, known as fire-resistant plastics.

Types and Uses

Polyurethane and polystyrene foams are used in construction generally as thermal insulation for roofs, exterior walls, and floors over open spaces or unheated areas, and for perimeter, pipe, and duct insulation. When used as thermal insulation, mechanical engineers should always be consulted for the correct thickness, resistance (*R* factor), and application methods. Code requirements in relation to fire resistance should also be checked.

Polyurethane foam is used to manufacture sandwich panels consisting of polyurethane foam combined with metals, asbestos cement, plywood, wood, and other suitable materials for use as curtain walls, panels, roofing, etc. Forms for complicated reinforced concrete, imitation heavy timber wood beams, and soundproofing interior partitions are also fabricated from polyurethane foam.

Both types of foam are used for floats and buoyancy units in boats, canoes, floating docks, and similar marine purposes.

In these applications plastics either replace an existing material because they do a better job or represent new materials that require entirely new design concepts.

PLASTICS: GASKETS, SHIMS, WATERSTOPS, ETC.

Physical and Chemical Properties

Plastic gaskets, shims, waterstops, and other miscellaneous plastic extruded shapes are made from vinyl, neoprene, nylon, urethane, and various other synthetic resins (*see* Plastics; Rubbers, Paints; Painting). These various small components are used with metals, concrete, wood, masonry, and other plastics as joint closures, weatherstripping, waterstops, glazing, control joints, and expansion joints. (*See also* Expansion and Expansion Joints; Flashing; Glass and Glazing.)

Types and Uses

Plastic gaskets are used generally for glazing and curtain wall systems. They are of three types: (1) zipper type, where, by inserting a solid unit within the gasket, it becomes rigid and stable; (2) interlocking friction type, where one unit interlocks into another to give a unified rigid whole; and (3) compression type, where the gaskets compress within a given space (*see* Figures *P28* and *P29*). Window, curtain wall, and gasket manufacturers should be consulted for current data on methods, materials, application, and systems for the particular end use required.

Plastic shims, joint closures, construction joint filler tapes, and various other plastic components can be broken down into two basic types: (1) those that are installed in a shop, factory, or mill, and (2) those that are installed in the field (*see* Figure *P30*). The manufacturers of these types of plastic components should be checked for current data in reference to the end use required (*see* Concrete Block; Brick, Burned; Stainless Steel; Steel; Aluminum; Curtain Walls.)

There is continuing research in this entire area and, as new synthetic materials are developed, there will be an increase in new plastic components of this type.

Plastic waterstops and expansion joint fillers and/or covers have to a great extent eliminated the use of metal bellows for expansion joints (*see* Figure *P31*). Plastic materials for expansion joints include not only waterstops but also various systems for roofing expansion joints, curtain walls, panel systems, etc.

Application

Condensed Checklist

1. When using plastic gaskets for glazing and curtain

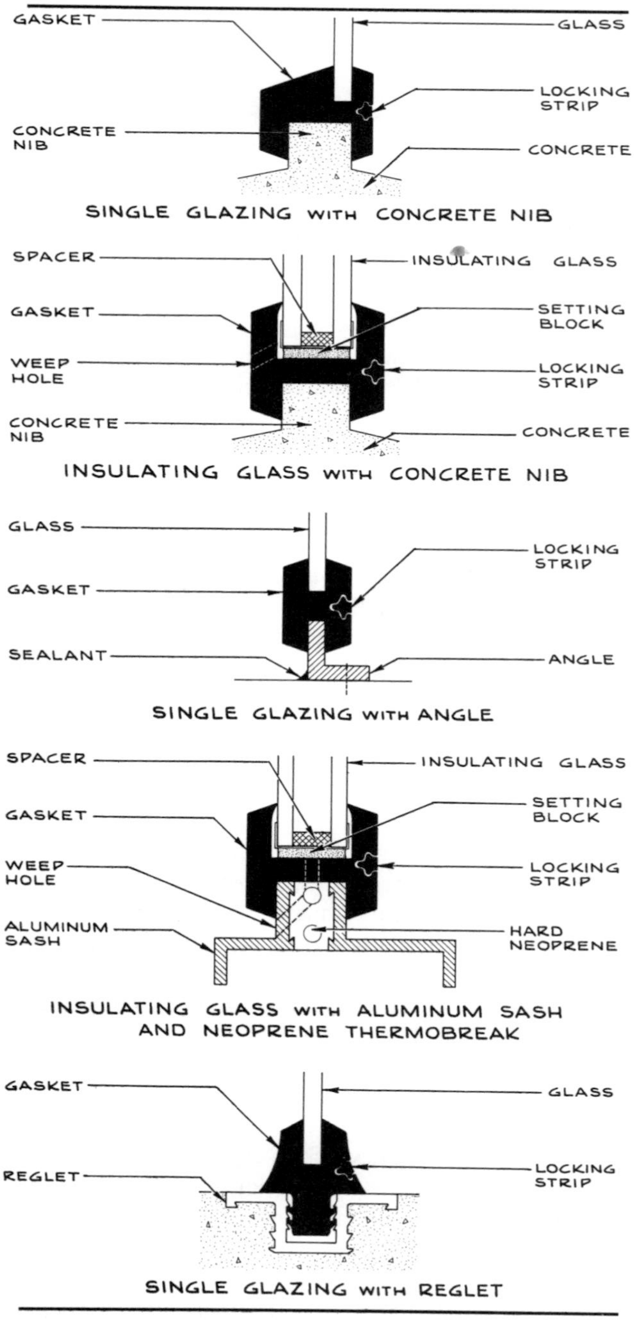

Figure P28 Types of plastic gaskets for glazing.

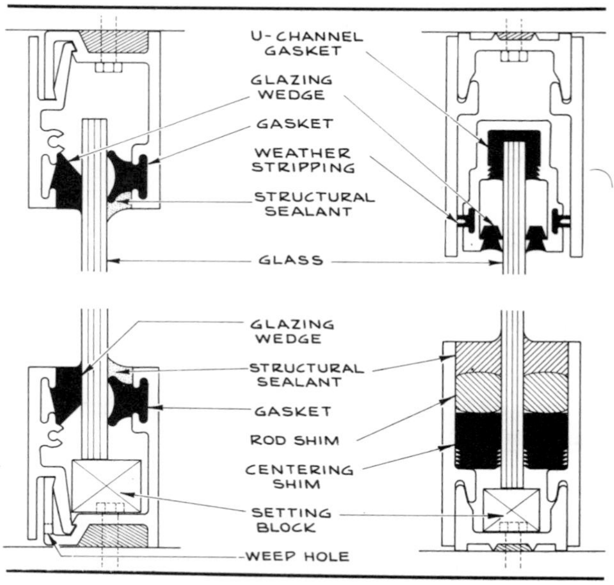

Figure P29 Typical types of extruded plastic wedges, gaskets, shims, and sealers, for glazing.

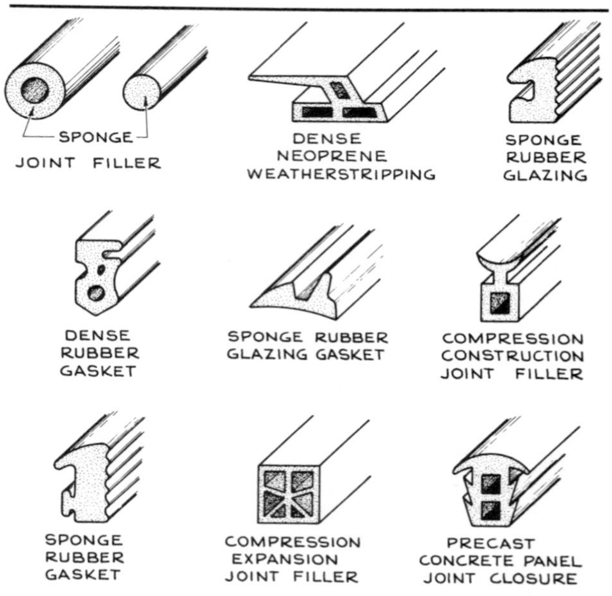

Figure P30 Types of extruded plastics for glazing, weatherstripping, joint sealers, etc.

wall systems, one should always check with manufacturers of gaskets, glass, and curtain wall systems for data regarding strength, stability, color, and weatherability.

2. A check should be made with window, curtain wall, and door manufacturers regarding the correct type of weatherstripping in relation to end use. An example would be a site near the ocean with heavy salt air, or an area with smog containing various chemicals that can attack the plastic weatherstripping.
3. A careful check should be made with manufacturers of precast concrete, curtain walls, tilt-up, or other structural systems for methods of sealing joints between units.

BULB TYPE

BULB TYPE SPLIT FLANGE

WIDTH		THICK.		BULB		WEB WIDTH		MAX. JOINT MOVEMENT		WEIGHT		STANDARD LENGTHS	
IN.	mm.	IN.	mm.	IN.	mm.	IN.	mm.	IN.	mm.	LBS. PER 100'	KGS. PER 30.5M	FT.	M.
4*	101.6	3/16	4.76	7/16	11.11	1 3/8	34.9	1/4	6.35	50	22.7	100	30.48
6*	152.4	3/16	4.76	7/16	11.11	2	50.8	1/2	12.70	74	33.6	100	30.48
6	152.4	3/16	4.76	1 1/8	28.6	2 3/4	69.9	1	25.40	98	44.5	50	15.24
6*	152.4	3/8	9.53	5/8	15.9	2	50.8	1/2	12.70	140	63.5	50	15.24
9	228.6	3/16	4.76	7/16	11.11	2 3/4	69.9	1	25.40	130	59.0	50	15.24
9*	228.6	3/8	9.53	5/8	15.9	2 3/4	69.9	1	25.40	220	99.8	50	15.24
9	228.6	1/2	12.70	3/4	19.1	2 3/4	69.9	1	25.40	252	114.3	50	15.24
9	228.6	3/8	9.53	1 1/2	38.1	3 1/4	82.6	1 3/4	44.45	244	110.7	50	15.24

DUMBBELL TYPE

4	101.6	3/16	4.76	3/8	9.53					47	21.3	100	30.48
5	127.0	3/16	4.76	1/2	12.7					70	31.8	100	30.48
6	152.4	3/16	4.76	3/4	19.1					108	49.0	100	30.48
6	152.4	3/8	9.53	3/4	19.1					150	68.0	50	15.24

* THESE BULB TYPES ARE ALSO AVAILABLE WITH SPLIT FLANGE

Figure P31 Typical types of water stops for expansion joints.

4. When using these types of plastic component, one should always check with the manufacturers for data on the type of plastic, its durability, its strength, etc., in relation to its end use.

PLASTICS: LAMINATES

Plastic laminates used in construction are almost all of one general type, made of several layers including (1) Kraft papers impregnated with phenolic resins, (2) aluminum foil to dissipate heat and make the laminates cigaretteproof, (3) a pattern sheet saturated with melamine, and (4) the final hard-wearing, transluscent surface sheet of melamine. These thin multiple layers are cured with intense 265 to 305° (128.44 to 151.67°C) heat and 800 to 1200 lbf/in.^{2} (5.52 to 827 MN/n^2) pressure for 60 to 90 minutes to form the plastic laminate in its finished state. These laminates are available in a large selection of colors, patterns, and finishes plus a variety of surface (textured) finishes.

In industry many other materials are incorporated into high-pressure plastic laminates, among them canvas, asbestos paper, asbestos textiles, glass textiles, combinations of cotton canvas, cotton line, nylon textile, and many others. These industrial plastic laminates are used for gears, pulleys, bobbins, and structural parts. Tremendous quantities are used in the radio, television, communications, electronics, and electrical power industries.

Commercial Forms. The laminates commonly used as a construction material are available in sheet and roll form. Plastic laminates are available in various grades.

1. *General Purpose* grade, $\frac{1}{16}$ in. (0.0635 mm) thick, used for vertical and horizontal surfaces. It has the greatest impact resistance, is the most dimensionally stable, and offers the widest selection of colors, finishes, and patterns.
2. *Vertical Surface* grade, $\frac{1}{32}$ in. (0.03625 mm) thick, used only for vertical surfaces such as wall panels, cabinets, and doors, where maximum durability and impact resistance are not prime requirements, as its dimensional stability is less than that of General Purpose grade. It is not recommended for use on surfaces wider than 2 ft (0.6096 m).
3. *Cabinet Liner* grade, 0.025 in. (0.635 mm) thick, used for the interior of cabinets, millwork, etc., and available in a limited range of colors, generally white, black, and a few neutral colors. It imparts structural balance and dimensional stability to cabinets, millwork, etc.
4. *Post-Forming* grade, 0.050 in. (1.270 mm) thick, used where small radii are required. It is similar to the General Purpose grade in use and properties.
5. *Balancing Sheet* grade, $\frac{1}{16}$ in. (0.0635 mm) thick, used to structurally balance and give dimensional stability to paneling, doors, etc.
6. *Back Sheet* grade, 0.050 in. (1.270 mm) thick, used for structural balancing with $\frac{1}{16}$-in. (0.0635-mm) General Purpose grade. This grade was developed by removing the melamine face of regular $\frac{1}{16}$-in. (0.0635-mm) General Purpose grade which was damaged during the manufacturing process.
7. *Backing Sheet* grade, 0.020 and 0.030 in. (0.508 and 0.762 mm) thick, used to stop moisture absorption in the core and give structural balancing to components constructed of $\frac{1}{32}$-in. (0.03625-mm) Vertical Surface grade.

Types and Uses

In this book only the finished plastic laminate itself is considered, and not the large variety of materials to which plastic laminates have been bonded to produce prefabricated units such as doors and the many different wall surfacing materials, for example, wood plies with laminate, and paper pulp or fiber sheets, boards, or panels with laminated plastic surfaces. In general, the most common core materials used in construction are particle board, hardwood-faced fir-core plywood, and fir plywood. Where moisture is present, the plywoods are of the exterior type, using waterproof adhesives.

Finishes. Plastic laminates are available in gloss or matte finish and in a wide variety of designs, patterns, stable colors, and surface textures. Table *P57* lists sizes, characteristics, and uses of the types of plastic laminates common in construction.

Application

Plastic laminates are applied in the field or in fabricators' shops, either with urea-formaldehyde, resorcinol, or another type of rigid water-resistant or waterproof glue, using pressure with or without heat, or with contact adhesives of the type recommended by the manufacturer of the plastic laminate in question.

Laminate edges are applied with hot-melt glues, rigid glues, or contact adhesives (*see* Figure *P32*).

There are three basic methods of applying plastic laminates to a core: (1) using a "plywood-type press," with cold or thermosetting adhesives (rigid glues); (2) using a "pinch roller," where laminates and core are passed

Table P57 Grades, sizes, characteristics, and major uses of plastic laminates in construction

	Dimensions							
	Thickness		Width		Length			
Grade of laminate[a]	in.	mm	ft	m	ft	m	Characteristics	Major uses
General purpose	$\frac{1}{16}$	0.0635	2, 2½, 3, 4	0.6096, 0.7620, 0.9144, 1.2192	5, 6, 7, 8 10	1.5240, 1.8288 2.1336, 2.4384 3.0480	For horizontal and vertical surfaces; good impact resistance, good dimensional stability; good resistance to household solvents, acids, and alkalis; resistant to heat up to 275°F (1135°C).	Countertops of all types; wall surfaces where rough usage will be encountered; furniture
Vertical surface	$\frac{1}{32}$	0.03625	2, 4	0.6096, 1.2192	6, 8, 10	1.8288, 2.4384, 3.0480	Less dimensional stability than general purpose grade; suitable for vertical surfaces only	Vertical wall paneling, doors, cabinets, etc.; greater widths than 2 ft (0.6096 mm) should be avoided.
Postforming[b]	0.050	1.270	2½, 3, 4	0.7620, 0.9144 1.2192	6, 8, 10	1.8288, 2.4384, 3.048	Used for surfaces with curves, circular shapes, and irregular shapes	Same as above
Cabinet liner	0.025	0.635	2, 2½, 3, 4	0.6098, 0.7620 0.9144, 1.2192	6, 8, 10	1.8288, 2.4384, 3.048	Color selection limited; good face durability and dimensional stability	Used for insides of cabinets, millwork, etc., to give structural balance and prevent moisture absorption
Balancing sheet	$\frac{1}{16}$	0.0625	2, 2½, 3, 4	0.6098, 0.7620 0.9144, 1.2192	6, 8 10	1.8288, 2.4384, 3.048	Equals in thickness and approximates the construction of general purpose grade; affords the maximum structural balance and dimensional stability	Paneling having large unsupported areas and wall paneling using $\frac{1}{16}$ in. (0.0635 mm) laminates
Back sheet	0.050	1.270	2, 2½, 3, 4	0.6096, 0.7620 0.9144, 1.2192	5, 6, 7, 8, 10	1.5240, 1.8288, 2.1336, 2.4384, 3.0480	General purpose grades whose melamine face surfaces were damaged during the manufacturing process and then removed	Same as balancing sheet
Backing sheet	0.020, 0.030	0.508, 0.762	2, 2½, 3, 4	0.6096, 0.7620 0.9144, 1.2192	6, 8, 10	1.8288, 2.4384, 3.048	Stops moisture from penetrating the core and give some structural balance	Used with vertical surface grade for wall paneling

[a]Fire-resistant plastic laminates are available to be applied to a fire-retardant core to obtain a Class 1 or "A" flame-spread rating.
[b]Both general purpose and post-forming grades are available cigarette-proofed.

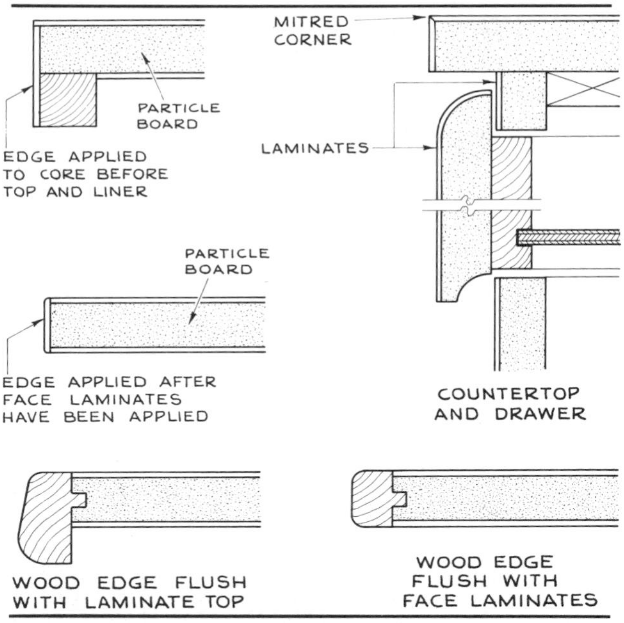

Figure P32 Typical methods of applying plastic laminate edges.

through two rollers, with contact adhesives; and (3) using contact adhesives and hand rollers or impact hammers. The third method is used for curved or irregular contoured surfaces.

Condensed Checklist

1. The laminate selected must meet the conditions of its end use in the building. It is advisable to consult with manufacturers of high-pressure laminates to make sure that the correct laminate is chosen.
2. Always check the type of construction required for the laminate and whether forming will be necessary. These factors control the type and thickness of the laminate used. For curved forms with small radii, a laminate having the proper thickness and suitable for curving (postforming) must be selected.
3. One should always make sure that the adhesives used are those recommended by the manufacturer of the laminate and that no substitutions are made.
4. It is advisable to check the locality in which the laminate-covered structure is to be built to make sure that local mills, shops, or factories have the equipment and are able to handle this type of work, particularly postforming.
5. Selection of the type of edge to be installed should be checked with the manufacturers of plastic laminates to determine the correct type for the end use.
6. When a wall finish of plastic laminate is selected, the type of laminate to be used should be checked in relation to impact resistance, rough usage, and moisture conditions.
7. When using plastic laminates one should always check whether liners, balancing sheets, or backing sheets are necessary for the particular end use.
8. One should always check to make sure that the colors, patterns, finishes, and surface textures selected will be available when the plastic laminate is installed, be it 6 months, 1 year, 2 years, or more in the future.
9. One should always check code requirements for the use of fire-retardant cores with fire-resistant plastic laminates (Class 1 or A flame-spread rating) or with standard plastic laminates (Class 2 or B flame-spread rating).

Conditions Favorable to the Use of Plastic Laminates

Where a hard, colorful, decorative, heat- and water-resistant, durable, easily maintained, cigaretteproof surface for walls, cabinets, furniture, horizontal counter tops, window sills, etc., is required on the interior of a building.

Conditions Unfavorable to the Use of Plastic Laminates

1. Where a fireproof material is needed, unless local, municipal, and state authorities, insurance companies, Fire Underwriters, and agencies of the federal government accept fire-resistant or standard-type plastic laminates on a fire-retardant core for this purpose.
2. On the exterior.

PLASTICS: PIPING

Physical and Chemical Properties

All types of plastic piping have a resistance to a broader range of chemicals than metal piping. Seven materials, all thermoplastic, are used in the manufacture of plastic pipe: acrylonitrile-butadiene-styrene (ABS), polyethylene (PE), polybutylene (PB), polypropylene (PP), polyvinyl chloride (PVC), chlorinated polyvinyl chloride (CPVC), and styrene rubber (SR) plastic.

Plastic piping will burn, and its ignition temperature is between 700 and 800°F (371.11 and 416.67°C). Because it collapses long before ignition, however, it does not provide a means for fire to travel from space to space.

The thermal expansion of plastic piping due to changes in temperature is quite considerable, as shown in Tables *P58* and *P59*, and care should be taken to provide for this expansion when dealing with both vertical and horizontal long runs of pipe.

Table P58 Thermal expansion for PVC piping

Length		Expansion with temperature change (in. for °F, mm for °C)													
		Temperature													
ft	m	40°F	4.45°C	50°F	10°C	60°F	10.56°C	70°F	21.11°C	80°F	26.67°C	90°F	32.22°C	100°	37.78°C
20	6.096	0.278	7.06	0.348	8.84	0.418	10.62	0.487	12.37	0.557	14.15	0.626	15.90	0.696	17.68
40	12.192	0.557	14.12	0.696	17.68	0.835	21.23	0.974	24.74	1.114	28.30	1.253	31.80	1.392	35.36
60	18.288	0.835	21.18	1.044	26.52	1.253	31.85	1.462	37.11	1.670	42.44	1.879	47.70	2.088	53.04
80	24.384	1.114	28.25	1.392	35.36	1.670	42.47	1.949	49.48	2.227	56.59	2.506	63.60	2.784	70.71
100	30.48	1.382	35.31	1.740	44.20	2.088	53.09	2.436	61.85	2.784	70.74	3.132	79.50	3.480	88.39

Table P59 Thermal expansion for ABS piping

Length		Expansion with temperature (in. for °F, mm for °C)													
		Temperature													
ft	m	40°F	4.45°C	50°F	10°C	60°F	10.56°C	70°F	32.22°C	80°F	26.67°C	90°F	32.22°C	100°F	37.78°C
20	6.096	0.536	13.61	0.67	17.02	0.804	20.42	0.938	23.83	1.072	27.23	1.206	30.63	1.34	34.04
40	12.192	1.072	27.23	1.34	34.04	1.608	40.84	1.876	47.65	2.144	54.46	2.412	61.27	2.68	68.07
60	18.288	1.608	40.84	2.01	51.05	2.412	61.27	2.814	71.48	3.216	81.69	3.618	91.90	4.02	102.11
80	24.384	2.144	54.46	2.68	68.07	3.216	81.69	3.752	95.30	4.288	108.52	4.824	122.53	5.36	136.14
100	30.48	2.680	68.07	3.35	85.09	4.020	102.11	4.690	179.13	5.360	137.14	6.030	153.16	6.70	170.18

Plastic piping is resistant to fungi, termites, and bacteria. Rodents and some insects will sometimes gnaw on buried plastic pipe to get at food or water.

Plastic piping is joined by several methods, which vary according to the physical characteristics of each type of plastic piping (*see* Table *P60*). The basic methods are: (1) solvent cementing, which involves using a solvent cement to chemically weld the pipes and fittings together; (2) insert fittings, which are inserted into the pipe and secured with stainless steel clamps; (3) elastomeric seals for bell-end piping and fittings; (4) threaded fittings, used for the types of plastic pipe that can be threaded; and (5) heat fusion, used for plastic pipe for which no solvent is available to chemically weld them.

There are available all types of fittings for the various types of plastic piping and a wide variety of transition fittings to connect plastic piping to other types of piping. (*See* Figures *P33* to *P36*.)

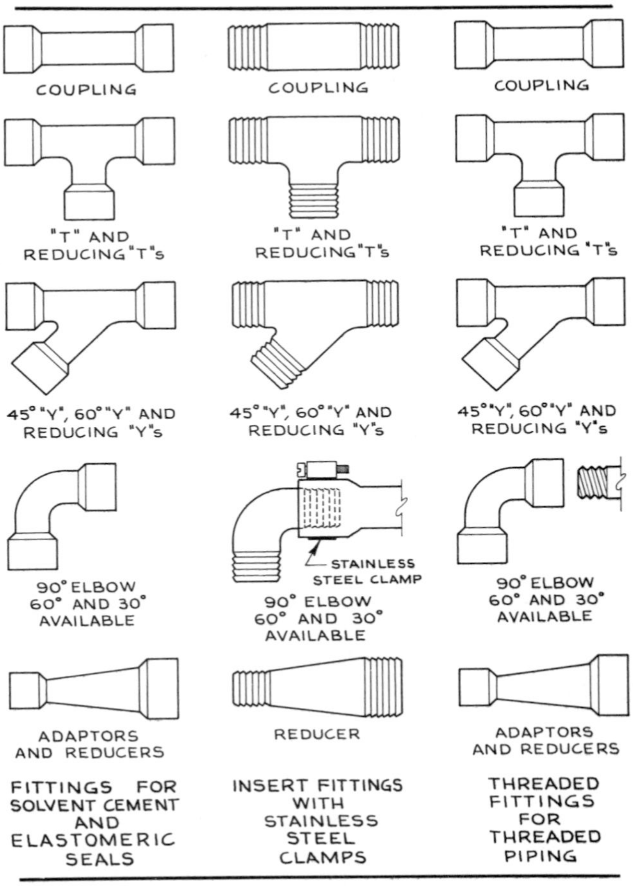

Figure P33 Typical types of fittings for plastic pipe.

Types and Uses

Plastic piping is used for water, sewage, drains, wastes and vents, gas distribution, industrial and chemical process piping, irrigation, and electrical and communications conduit.

In the construction field ABS, PVC, and CPVC are widely used, and PE, PB, PP, and SR find use in special areas of construction. Table *P60* summarizes data for each type.

Table P60 Types of plastic pipe

Material	Type	Maximum operating temperature[a] °F	°C	Joining methods	Major uses
Acrylonitrile-butadiene-styrene	ABS rigid	100 (pressure) 180 (nonpressure)	37.78 82.22	Solvent cement, threading, transition fittings	Drain, waste, vent, sewer, conduit, and water piping
Polyethylene	PE flexible	100 (pressure) 190 (nonpressure)	37.78 82.22	Insert fittings, socket fusion, butt fusion, transmission fittings	Water and gas service and mains, chemical waste, and irrigation systems
Polybutylene	PB flexible	180 (pressure) 200 (nonpressure)	82.22 93.33	Insert fittings, socket fusion, butt fusion, transmission fittings	Water and gas service and mains and irrigation systems
Polyvinyl chloride	PVC rigid	100 (pressure) 180 (nonpressure)	37.78 82.22	Solvent cement, elastomeric seal, threading, mechanical couplings, transmission fittings	Water and gas service and mains, drain, waste, vent, house sewer, and conduit piping, industrial process piping, and irrigation systems
Chlorinated polyvinyl chloride	CPVC rigid	180 (pressure)	82.22	Solvent cement, threading, mechanical couplings, transmission systems	Hot- and cold-water piping and chemical process piping
Polypropylene	PP rigid	100 (pressure) 180 (nonpressure)	37.78 82.22	Mechanical couplings, socket fusion, butt fusion, threading	Chemical waste and chemical process piping
Styrene rubber plastic	SR rigid	150 (nonpressure)	65.56	Solvent cement, transition fittings, elastomeric seal	Storm drains, subsoil dewatering systems, and sewage disposal fields

[a]Pressure rating is based on an operating temperature of 73°F (22.76°C); nonpressure limits may be exceeded if time of exposure is short.

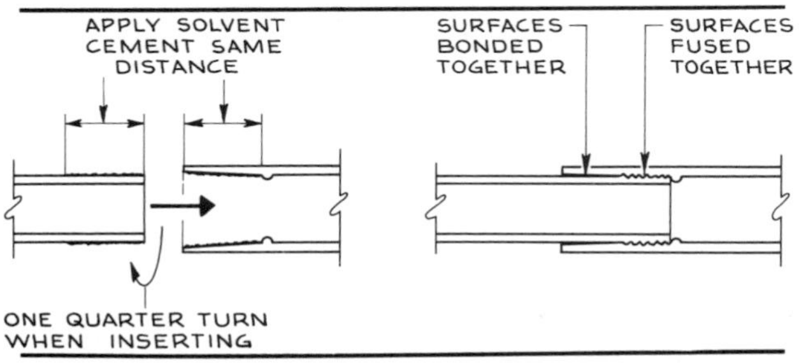

Figure P34 Typical solvent cement joining of plastic pipe couplings.

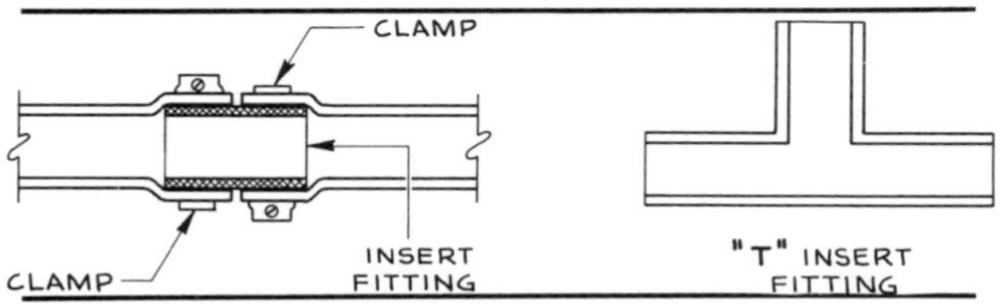

Figure P35 Typical plastic pipe insert couplings with clamps.

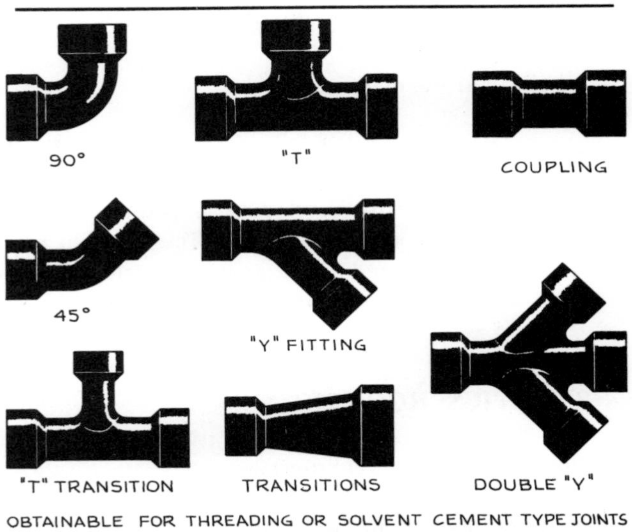

Figure P36 Typical bell-type plastic pipe fittings.

Application

Condensed Checklist

1. Check the type of fluid or gas to be conducted, and then check with plastic pipe manufacterers to ascertain the correct plastic pipe for the end use in question.
2. Check installation methods for meeting the thermal expansion requirements of the type of plastic piping used.
3. Check the width and depth of trench, the bedding, and the installation methods.
4. Check the type of joining and fittings to be used, to make sure that they are correct.
5. Check with municipal, state, and local codes and health departments for their restrictions and requirements regarding the installation of plastic piping.

Conditions Favorable to the Use of Plastic Pipe

1. For sewers, drains, waste and vents pipes, and storm drains in the construction of buildings.
2. For industrial and chemical piping.
3. For electrical and communications conduit.
4. For irrigation of lawns, athletic fields, golf courses, truck gardens, etc.
5. For sewers, pipes for industrial and chemical wastes, and exterior gas piping.

Conditions Unfavorable to the Use of Plastic Pipe

1. For industrial wastes that are corrosive to plastic piping.
2. For gas piping within buildings.

History and Manufacture

Plastic piping was used during and after World War II as a substitute for metal piping. From this beginning, it has grown and is still developing as perhaps the best piping for the construction field. The petrochemical industry developed and is continuing to develop thermoplastics which will meet all requirements for piping.

PLASTICS: ROOFING

Physical and Chemical Properties

Plastic roofing is made from urethanes, polyurethanes, polyester resins, vinyls, acrylics, epoxies, and silicones. The type of plastic varies with the type of material and the end use.

Types and Uses

Plastic roofing is available in two basic systems: (1) those that can be sprayed, rolled, brushed, or squeegeed, and (2) sheet material that can be floated onto or adhered to the substrate. In general, plastic roofing systems are applied by highly skilled and qualified roofers, as all these roof systems carry a guarantee. (*See also* Flashing; Expansion and Expansion Joints; Roofing.)

The roofings are available as a complete system to be applied to a substrate and include (1) foam insulation, (2) sheeting or coating, (3) protective topping, and (4) stone or granule-type finish surface.

The various plastic roofing methods should always be checked for the type of substrate upon which the plastic roofing will be applied.

Application

Condensed Checklist

1. Always check with plastic roofing manufacturers for current data and application methods.
2. When selecting the type of roofing to be used on a leaking, deteriorated roof, always check with the manufacturers of plastic roofing for the system or systems that can be directly applied to this particular type of existing roofing material.
3. Always check the guarantee or warranty for the length of time it is in effect, and check its coverage of any defects or shortcomings in the installation, flashing, roof drains, skylights, vents, or any equipment that will be mounted on the roof.
4. Check with the governing building codes, fire departments, and insurance companies for their restrictions and requirements concerning plastic roofing for the building or buildings that are to be constructed.
5. Always check the type of insulation, the *R* value required, and whether the insulation will be part of the plastic roof system or will act as substrate for the plastic roof system.
6. When selecting a plastic roofing system, always check all the flashing details, as each system has its own flashing methods and flashing details (*see* Figures *P37* and *P38*).
7. Before selecting a plastic roofing or plastic roofing system, check with the manufacturers of plastic roofing regarding the suitability of their roofing for application to the particular roofing substrate that has been selected for the structure.

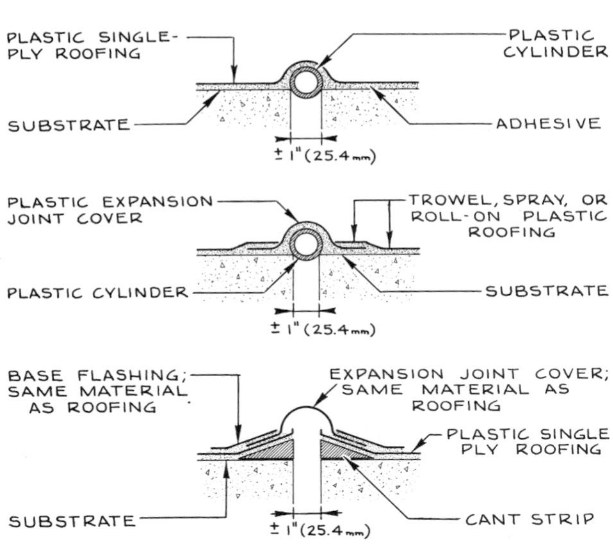

Figure P37 Typical types of flashing and expansion joint covers for plastic roofing.

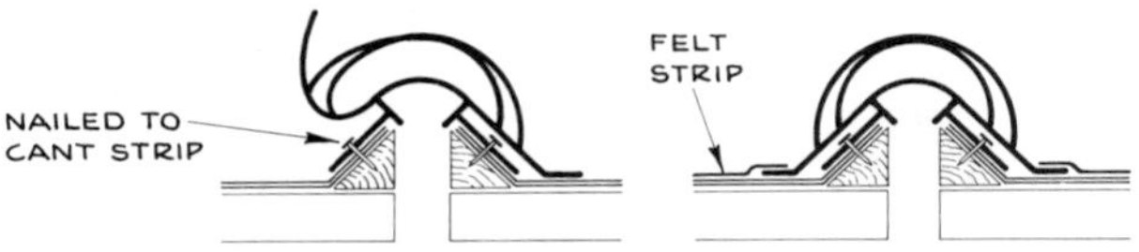

Figure P38 Extruded vinyl expansion joint cover for plastic roofing.

Conditions Favorable to the Use of Plastic Roofing

1. Where an old, existing, leaking roof requires reroofing.
2. Where an overall inclusive roofing system is required, including insulation, roofing membrane, finish wearing surface, flashing, expansion joints, cant strips, and fascias.

Conditions Unfavorable to the Use of Plastic Roofing

1. On roofs with pitches greater than 1 to 12.
2. In areas where there are rigid fire restrictions.

PLASTICS: SHEET

Physical and Chemical Properties

The construction use of plastics for skylights; for the diffusion of light; as fixed window areas, translucent roofs, etc.; and for interior partitions, doors, luminous ceilings, etc., is a comparatively recent and continuing development. In these applications plastics either replace an existing material because they do a better job or represent new materials that require entirely new design concepts.

Sheet. Plastic sheets today are usually made of acrylate, polycarbonate, polyester, or polystyrene plastics, either plain or reinforced with glass fibers. Plastic sheet materials are available flat or corrugated and in various deformed shapes.

Acrylic plastic sheet is available in thicknesses of $\frac{1}{32}$, $\frac{1}{16}$, and up to 4 in. (0.784, 1.588, and 101.6 mm) and even thicker ($4\frac{1}{4}$ in. or 107.95 mm); in sizes up to 120 X 144 in. (304.8 X 365.74 cm). It is available transparent, translucent, opaque, colored translucent, colored transparent, colored semiopaque, patterned and reflective (mirrored). It has better thermal values than glass.

Polycarbonate Sheet. Polycarbonate refers to synthetic thermoplastic resin derived from biphenol A and phosgene, a linear polyester of carbonic acid. It can be formed from any dihydroxy compound and any carbonate diester or by ester interchange. As for its physical and chemical properties, it is transparent, noncorrosive, combustible but self-extinguishing, dimensionally stable, and weather, abrasion, stain, and heat resistant; it also has very high impact strength. It is soluble in chlorinated hydrocarbons and attacked by strong alkalis and aromatic hydrocarbons. Depending on its thickness, it has from 20 to 30% less heat loss than plate or float glass; its light transmission ranges from 82 to 89%, and it is 43% lighter than aluminum and approximately half the weight of glass. Because of the large thermal expansion and contraction of polycarbonate sheet (approximately eight times greater than the values for plate or float glass), its installation requires that the method of glazing allows for equal or greater movement of the sheet than anticipated (*see* Table *P61*).

Polycarbonate sheet is available transparent, transparent bronze and gray, and obscure; in thicknesses from $\frac{1}{8}$ to $\frac{1}{2}$ in. (3.18 to 12.7 mm) and sheet sizes 24 X 48 in. (60.96 X 121.92 cm) up to 72 X 96 in. (1.829 X 2.438 m). There are also laminated types for areas requiring bullet resistance and security (*see* Table *P62*).

The major use for polycarbonate sheet is for glazing in areas subject to vandalism, where security and safety are required. The codes of the area where polycarbonate sheet is to be installed should always be checked to determine the limitations on the number of floors of a multifloored structure that can be glazed with this

Table P61 Types, sizes, and rabbet dimensions of polycarbonate sheet[a]

Type of sheet	Dimensions: Thickness in.	Thickness mm	Width ft	Width m	Length ft	Length m	Weight lb/ft²	Weight kg/m²	Rabbet depth in.	Rabbet depth mm
Clear[b]	0.010	0.254	2	0.610	4	1.219	0.063	0.308	0.385	9.779
	0.015	0.381					0.094	0.459	0.390	9.906
	0.020	0.508					0.130	0.635	0.395	10.033
	0.030	0.762					0.190	0.928	0.405	10.287
	0.040	1.016	2	0.610	4	1.219	0.250	1.221	0.415	10.541
			4	1.219	8	2.438				
	0.060	1.524	2	0.610	4	1.219	0.380	1.855	0.435	11.059
			4	1.219	6	1.829				
			4	1.219	8	2.438				
	0.080	2.032	4	1.219	8	2.438	0.500	2.441	0.455	11.557
	0.093	2.362	4	1.219	8	2.438	0.580	2.832	0.468	11.887
	0.125 $\frac{1}{8}$	3.175	4	1.219	6	1.829	0.780	3.808	0.50	12.70
			4	1.219	8	2.438				
			5	1.524	8	2.438				
			6	1.829	8	2.438				
	0.187 $\frac{3}{16}$	4.750	4	1.219	6	1.829	1.170	5.712	0.75	19.05
			4	1.219	8	2.438				
			5	1.524	8	2.438				
			6	1.829	8	2.438				
	0.250 $\frac{1}{4}$	6.350	2.833	0.864	6.500	1.981	1.560	7.616	0.75	19.05
			3.167	0.965	6.500	1.981				
			4	1.219	6	1.829				
			4	1.219	8	2.438				
			5	1.524	8	2.438				
			6	1.829	8	2.438				
	0.375 $\frac{3}{8}$	9.525	2	0.610	4	1.219	2.340	11.424	1.00	25.4
			4	1.219	8	2.438				
			5	1.524	8	2.438				
	0.50 $\frac{1}{2}$	12.700	4	1.219	8	2.438	3.140	15.329	1.00	25.4
Bronze[c]	0.125 $\frac{1}{8}$	3.175					0.780	3.808	0.50	12.70
	0.250 $\frac{1}{4}$	6.350					1.560	7.616	0.75	19.05
			4	1.219	8	2.438				
Gray[c]	0.125 $\frac{1}{8}$	3.175					0.780	3.808	0.50	12.70
	0.250 $\frac{1}{4}$	6.350					1.560	7.616	0.75	19.05

[a] All polycarbonate sheets have a tensile strength of 9500 lbf/in.² (65.505 MN/m²).
[b] Light transmission equals 82–89%.
[c] Light transmission equals 50%.

material. Generally, fire departments require that this type of glazing be limited to two floors.

All these plastic materials are available transparent, translucent, or opaque, and in a wide variety of stable colors. With regard to resistance to heat and fire, they are available in two types: those called slow-burning plastics and those that will not support combustion, known as fire-resistant plastics.

Types and Uses

Plastic sheet can be used to replace glass in any type of window or door, skylight, or shower and bathtub enclosure, and in any area where a transparent, translucent or opaque material in a wide variety of colors is needed on the interior. Polycarbonate sheet and laminated sheets are used in place of glass where security,

Table P62 Types, sizes, and weights of security and bullet-resistant laminated polycarbonate sheets[a]

		Thickness of sheet		Dimensions of sheet				Rabbet dimensions		
Type	Class	Security in. (mm)	Bullet-resistant[a] in. (mm)	Weight of sheet lb/ft² (kg/m²)	Width of sheet ft (m)	Length of sheet ft (m)	Width of sheet ft (m)	Rabbet depth in. (mm)	Edge clearance in. (mm)	Edge of rabbet to sheet in. (mm)
Security[a]		3/8 (9.525)		4.0 (19.53)	2 (0.6096)	3 (0.9144)	12 (0.3048)	3/8 (9.525)	1/4 (6.350)	1/8 (3.175)
		1/2 (12.700)		4.8 (21.83)	25 (0.7620)	5 (1.5240)	24 (0.6096)	9/16 (14.288)	3/8 (9.525)	3/16 (4.763)
		5/8 (15.875)		5.6 (27.34)	4 (1.2192)	7 (2.1336)	36 (0.9144)	3/4 (19.050)	1/2 (12.700)	1/4 (6.350)
							48 (1.2192)	1 (25.400)	3/4 (19.050)	1/4 (6.350)
Bullet-resistant[b]	Class I		1 3/16 (30.163)	6.5 (31.73)	3 (0.9144)	4 (1.2192)	Glazing must meet requirements for end use of bullet-resistant sheet			
	Class II		1 9/16 (39.688)	8.0 (39.06)						
	Class III		1 3/4 (44.45)	8.0 (39.06)						

[a]Security-resistant polycarbonate sheets are available in clear, bronze, and gray.
[b]Bullet-resistant polycarbonate sheets are available in clear, bronze, gray and green.

Table P63 Chemical and physical properties of clear plastic sheet[a]

Thickness in. (mm) of sheet	1/8 (3.18)	3/16 (4.76)	1/4 (6.35)	3/8 (9.35)	1/2 (12.7)
Btu/ft² · h · °F	1.06	1.01	0.96	0.88	0.81
(W/m² · °C)	(6.019)	(5.735)	(5.451)	(4.997)	(4.599)
Resistance, *R*	0.943	0.990	1.04	1.14	1.23
Reciprocal of W/m² · °C	0.166	0.174	0.183	0.200	0.217
Light transmittance (percent)	92	92	92	92	92
Recommended minimum radius					
in.	22.5	33.7	45.0	67.5	90.0
(cm)	(7.15)	(85.60)	(114.30)	(171.45)	(228.60)

[a]At approximately 325°F (162.78°C) clear plastic sheet becomes soft and pliable and can be formed to almost any shape.

protection against very rough usage, and the stopping of glass breakage are required (*see* Table *P62*).

In like manner, corrugated sheet can be used for roofing, skylights, light fixtures, decorative panels, and the other uses enumerated for sheet plastic.

Corrugated plastic sheet material is frequently used in conjunction with other corrugated materials such as corrugated aluminum, stainless steel and galvanized or coated steel sheet, and corrugated asbestos. In fact, the forms and corrugations of the plastic sheet are specially dimensioned for correlated use with these materials. Also, all the filler strips, accessories, and attaching devices are interchangeable. However, a softer grade of rubber or other filler material is recommended for plastics than for the harder metals and glass.

Tables *P63*, *P64*, and *P65* give typical stock sizes and weights for both plain and corrugated plastic sheet. Table *P64* includes shiplap and V-crimp materials.

Table P64 Typical sizes and weights of corrugated, shiplap, and V-crimp plastic sheet material

	Dimensions						Weight		Safe span for 100-lb (45.36-kg) load		Side lap		Corrugations				
	Thickness		Length		Width								Pitch		Depth		Number per width
Type of sheet	in.	mm	ft	m	ft	m	oz/ft²	g/m²	ft	m	in.	mm	in.	mm	in.	mm	
$\frac{1}{4}$ in.	$\frac{1}{16}$	1.59	8	2.438	2.17	0.661	8	2441.2	2.67	0.813	2	50.8	1.25	31.75	$\frac{1}{4}$	6.35	$22\frac{1}{2}$
(6.35 mm)			10	3.048	3[a]	0.914					$2\frac{1}{2}$	63.5					29
corrugated			12	3.658													
$2\frac{1}{2}$ in.	$\frac{1}{16}$	1.59	4	1.219	2.17[a]	0.661	8	2441.2	4	1.219	1	25.4	2.66	67.56	$\frac{1}{2}$	12.70	10
(63.5 mm)			5	1.524	2.83[a]	0.867											13
corrugated			6	1.829	3.33[a]	1.016											15
			8	2.438	4.17[a]	1.270											19
			10	3.048	2.29[a]	0.699					$1\frac{1}{2}$	37.2					$10\frac{1}{2}$
			12	3.658	2.75	0.838											$12\frac{1}{2}$
					4.29	1.308											$19\frac{1}{2}$
2.67 in.	$\frac{1}{16}$	1.59	8	2.438	2.91	0.889	8	2441.2	4.5	1.372	$1\frac{1}{2}$	37.2	2.67	67.82	$\frac{7}{8}$	24.23	$13\frac{1}{2}$
(67.82 mm)			10	3.048													
corrugated			12	3.658													
4.20 in.	$\frac{1}{16}$	1.59	8	2.438	3.50[a]	1.067	8	2441.2	4.5	1.372	1	25.4	4.20	106.68	$1\frac{1}{16}$	26.99	10
(106.68 mm)	$\frac{3}{32}$	2.35	10	3.048	3.50[a]	1.067	12	3661.8									
corrugated			12	3.658													
3 in.	$\frac{1}{16}$	1.59	8	2.438	2.575	0.785	6	1830.9	4.5	1.372	$1\frac{1}{2}$	37.2	3	76.20	$\frac{3}{4}$	19.05	10
(76.2 mm)	$\frac{3}{32}$	2.35	10	3.048	2.58	0.786	8	2441.2									12
corrugated			12	3.658													
2 in.	$\frac{1}{16}$	1.59	4	1.219	3.311	1.009	8	2441.2	4	1.219	1	25.4	2	50.8	$\frac{7}{16}$	11.11	20
(50.8 mm)			5	1.524	3.975	1.212											24
shiplap			6	1.829	4.477	1.364											27
			8	2.438													
			10	3.048													
			12	3.658													
3 in.	$\frac{1}{16}$	1.59	4	1.219	2.06	0.628	6	1830.9	4	1.219	1	25.4	3	76.2	$\frac{7}{16}$	11.11	8
(76.2 mm)			5	1.524													
shiplap			6	1.829													
			8	2.438													
			10	3.048													
			12	3.658													
V-crimp	$\frac{1}{16}$	1.59	8	2.438	2.17	0.661	6	1830.9	2.67	0.813	2 V's		1	25.4	$\frac{1}{2}$	12.7	5 V's
			10	3.048													
			12	3.658													

[a]These types have edges finished with both edges down. All other types of corrugated sheet have one edge down.

Table P65 Most common sizes and weights of plastic sheet[a]

Type of sheet[b]	Dimensions							Weight	
	Thickness			Width[c]		Length			
	in.	decimals	mm	ft	m	ft	m	lb/ft²	kg/m²
Transparent, translucent, opaque, colored (transparent, translucent, semiopaque), patterned and reflective (mirrored)	$\frac{1}{8}$	0.125	3.175	2–10	0.610–3.048	3–12	0.914–3.658	0.75	3.66
	$\frac{1}{16}$	0.187	4.763	2–10	0.610–3.048	3–12	0.914–3.658	1.10	5.37
	$\frac{1}{4}$	0.250	6.350	2–10	0.610–3.048	3–12	0.914–3.658	1.50	7.32

[a]Available for all sizes of window glass.
[b]Plastic sheet $\frac{3}{32}$ in. (0.10 in. or 2.45 mm) to $\frac{1}{4}$ in. (0.25 in. or 6.35 mm) thick is available on reels 250–600 ft (76.22–182.88 m) long.
[c]Available in 4-ft (1.219-m) width for all thicknesses.

Plastic sheet is also used for prefabricated skylights (skydomes), which are made as complete packaged units in a wide variety of shapes and sizes.

Sandwich-type skylights consist of the combination of plastic sheet with a cellular type of aluminum frame into a sandwich panel in which the frame and plastic sheets are joined into a structural unit. These panels are self-supporting, insulating, and highly glare resistant. They can be used as walls, partitions, and doors.

The laminating of plastic sheets to a plastic foam, which can vary in thickness, produces a sandwich panel that has high thermal values, strength, decorative value, and some supporting strength.

PLASTICS: SURFACING, SIDING, AND PANELS

Physical and Chemical Properties

Plastic surfacing is a method of applying plastic or plastic cementitious coatings to masonry, plywood, cement asbestos, stucco, etc., which produces a solid colored or mixed colored, pebbled, granular, or stone chip surface that is durable, weather resistant, decorative, and easily cleaned and maintained. Generally these coatings are made with epoxies, acrylics, polyesters, polyurethanes, vinyls, and portland cement, with various types of aggregates such as fine sand, or applied marble, granite, or stone granules, chips, and pebbles. These surfacing materials are available as a ceramic type of surface not only in a wide range of colors but also in mixed colors obtained from the marble, granite, or stone chips, granules or pebbles.

Types and Uses

Plastic Surfacing. These surfacing materials are available for on-site or factory application, and they can also be applied to existing surfaces. They are generally used for both exterior and interior surfaces including walls, ceilings, and the underside of balconies, porches, roofs, and canopies.

Paneling. Many prefabricated panel units, generally 4 by 8 ft (1.219 by 2.438 m), are available in a wide variety of colors and textures and are used for the finish surface on existing or new structures.

Also produced are a wide variety of plastic panel units used for exterior wall surfaces, form linings, bathroom walls, shower stalls, countertops, lavatories, etc., and they are available in a wide variety of colors, textures, and patterns, including synthetic marble and wood grain.

Plastics are now being used for plumbing fixtures; complete shower stalls including walls, ceiling, and base; bathtubs with walls, seats, and other design features; and counter tops with lavatory back splash. These constitute a relatively new product development in plastics.

Curtain wall systems consisting of numerous types of plastic polyester-type resins with or without insulation are also available. They vary in thickness from $\frac{5}{8}$ to 4 in. (15.88 to 101.6 mm), with surface textures of marble, granite and stone chips, granules, and pebbles. Also available are complete sandwich wall panels consisting of interior finish panel sheet glass, fiber, or foam-type insulation, and exterior finish material with exposed aggregate.

Application

Condensed Checklist

1. When selecting surfacing materials, always check with the manufacturer for current data on color, textures, and methods of application.
2. When applying surfacing material to existing surfaces, check that the type of surfacing matrix selected is adaptable for this type of material.
3. Always specify that samples of texture, finish, and color must be submitted for approval.
4. Always require that the methods of attachment, types of joints, and conditions at roof and grade be detailed and specified, and that shop drawings be submitted for approval, when prefabricated surfacing panels are selected for a structure.
5. When a panel design requires projections with 90° angle corners, always check the methods of obtaining these sharp corners and the restrictions on the size of aggregates due to these corner angles.
6. When selecting a plastic curtain wall system, always specify that full-size sample panels must be tested for the actual conditions that will be encountered when installed on the structure.
7. Always require shop drawings for prior approval when plastic curtain wall systems are to be installed on a structure.
8. When prefabricated shower stalls and bathtubs with enclosing walls are selected, always obtain installation drawings from the manufacturer that include the locations of all rough plumbing and fitting connections.
9. When selecting a complete countertop-lavatory-backsplash unit, always obtain installation drawings that include tolerances and the locations of all plumbing connections.

PLATING

"Plating" usually refers to the electroplating, or deposition by electrolysis, of a coating of metal on an object. Its purpose is to protect the basis metal from corrosion or to provide a decorative finish or both. It may also serve as the basis for further plating or other finishing.

One metal can also be plated on another by cladding, or rolling a thin sheet of the finish metal directly onto the basis sheet metal.

In electroplating the metal to be plated is pretreated by proper polishing, sandblasting, or pickling, and, if necessary or desirable, a chemical precoating or preplating with a metal such as copper for better adhesion of the finish. It is then placed as the cathode in a chemical solution which may contain salts of the metal to be plated, or the plating metal may serve as the anode. As electric current passes through the bath, the plating metal is deposited on the basis metal to whatever thickness is desired.

The following metals, listed in approximate order of commercial importance, are used for electroplating: nickel, silver, chromium, copper (including brass and bronze), zinc, gold, cadmium, tin, lead, cobalt, platinum, rhodium, indium.

Aluminum and stainless steel are plated, or clad, to other metals by pressing and rolling.

PLATINUM

Physical and Chemical Properties

Symbol: Pt
Atomic number: 78
Specific gravity: 21.45 (20°C, 68°F)
Melting point: 1772°C, 3221°F
Boiling point: 3827°C, 6920°F
Tensile strength: 54,000 lbf/in.^{2}, 372.33 MN/m^2

Platinum is a soft, silvery-white ductile metal that does not tarnish at any temperature. It is resistant to corrosion, oxidation, alkalis, and most acids.

Workability. Platinum can be hot- or cold-worked, drawn, rolled, welded, brazed, and hammered into foil. It has also been coined and worked similarly to gold and silver.

Commercial Forms. Platinum is available in sheet, wire, foil, and leaf forms. There are four grades of platinum, ranging from Grade 1, which is 99.99% pure, to Grade 4 (commercial), which is 99% pure.

Types and Uses

Because of its resistance to chemicals and oxidation, platinum finds many specialized uses in metallic form, as indicated by asterisks in Table *P66*.

Possible new applications may be developed for platinum, palladium, rhodium, and iridium as protective skin or layers on basis metals used in space vehicle construction. These applications are based on the superior resistance of the platinum-group metals and alloys to high-temperature oxidation. The protective metals could be applied by electroplating, cladding, or possibly spraying techniques.

Table P66 Uses of platinum

	Construction Uses	
Basis of	Color pigments	Violet toning
	Glass*	Mirrors
	Cobalt*	Permanent magnets
	Glass*	Dies and orifices for handling molten glass
Component of	Glazes and porcelain enamels	Silvery mirror film
	Textiles*	With gold in spinnerets for producing viscose rayon; with nickel in spinnerets for producing glass fibers
	Zinc	Etching
	Nonconstruction Uses	

Electroplating,* fountain-pen nibs,* indelible inks, jewelry,* medicine, photography, radio,* telephone,* television,* X-ray screens,* special plating finishes on metals

History and Manufacture

The first known reference to platinum bears the date 1557. In 1773 de l'Isle obtained malleable platinum, and the first pure platinum was obtained by Wollaston in 1803. The method of obtaining platinum is complex and involves the processing of several related elements grouped under the term "platinum metals;" these are platinum, palladium, rhodium, iridium, osmium, and ruthenium. It is interesting to note that gold, silver, nickel, and copper are also recovered in this processing. (*See* Figure *P39*.)

POINTING

"Pointing" refers in general to the troweling of mortar into a joint after the masonry unit is laid. It also refers to the repairing of masonry walls by filling in or "pointing" of vertical and horizontal joints that have weathered or areas where mortar has crumbled out.

Tuck pointing refers specifically to a decorative type of pointing done by cutting out a mortar joint to about a $\frac{1}{2}$ in. (12.75 mm) depth and refilling it with fresh mortar to a pattern.

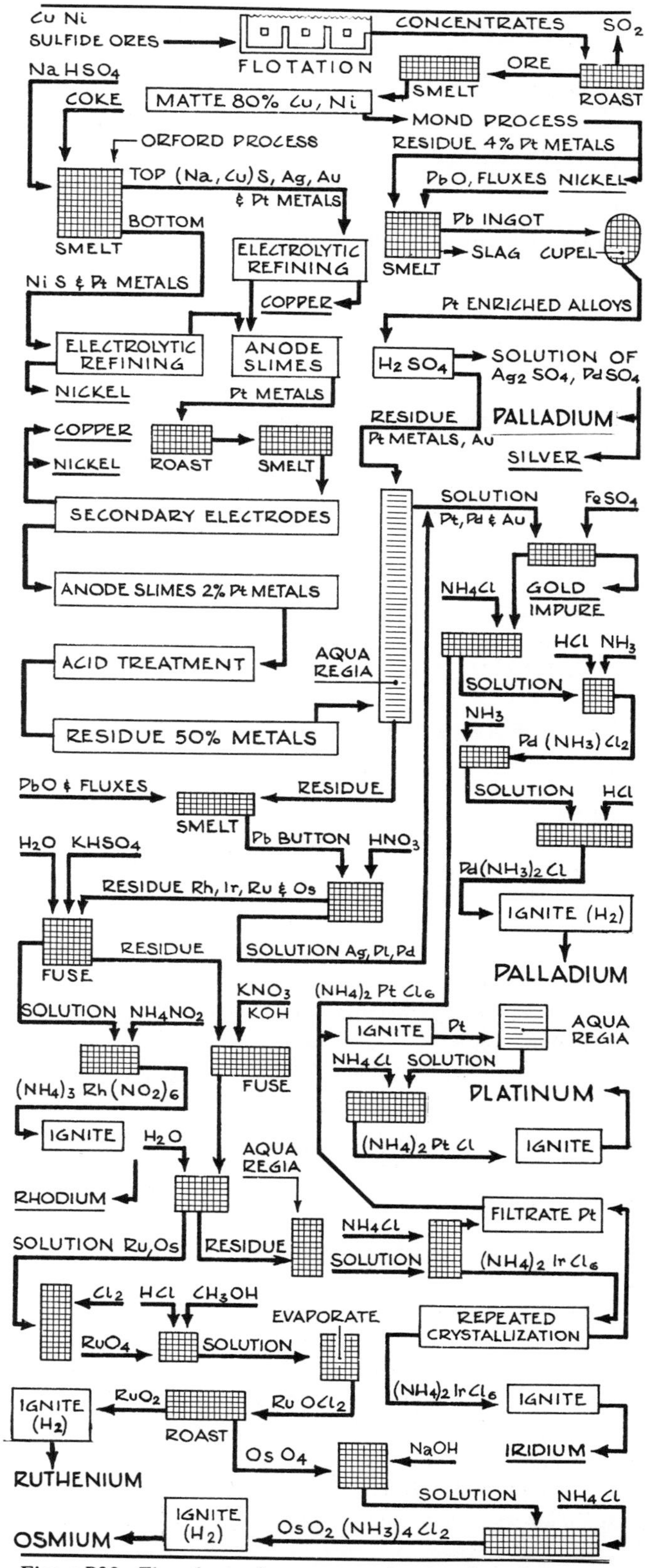

Figure P39 Flowchart of recovery and separation of the platinum metals.

PORCELAIN

Porcelain consists of a controlled mixture of refined plastic or nonplastic minerals shaped by one of many methods and then subjected to high temperatures (*see* Ceramics). In the construction field porcelain is used for all types of plumbing fixtures, electrical insulators, and equipment for chemical and biological laboratories.

PORCELAIN ENAMEL

Porcelain enamel (vitreous enamel) may be defined as an applied surface finish bonded (fused) to a metal body through high temperatures (*see* Glazes and Porcelain Enamels; *see also* Aluminum, Stainless Steel, Steel, *and* Casting).

POTASSIUM

Physical and Chemical Properties

Symbol: K
Atomic number: 19
Specific gravity: 0.826 (20°C, 68°F)
Melting point: 63.65°C, 172.17°F
Boiling point: 774°C, 1550°F

Potassium is a soft, silvery-white metal with a brilliant luster. It belongs to the alkali metals group and is one of the most reactive metallic elements and strongest reducing agents. It reacts violently with water and in air; it does not alloy with other metals. Potassium is never found free in nature; it always occurs in compound form. It is present in all plant life and is a constituent of many aluminum silicate minerals such as mica and feldspar.

Types and Uses

There are no industrial uses for potassium in metallic form, but its compounds find a wide variety of uses both in their own right and as constituents of construction materials, as shown in Table *P67*.

Potassium was originally used for the reduction of difficult-to-isolate metals like aluminum, magnesium, silicon, and boron.

Table P67 Uses of potassium

	Construction Uses	
Component of	Asphalt	Coating for granules for roofing and siding
	Color pigments	Black, red, brown, and yellow colors
	Glass	Fluxing ingredient; etching; silvering mirrors
	Glazes and porcelain enamels	Frits
	Gold	Electrogilding
	Paint	Driers
	Paper	Sizing; bleaching; flame retarding
	Steel	Tempering
	Textiles	Mordant; dyeing; bleaching; flame retarding
	Wood	Stains; preservative; cleaning
Allied Construction Uses	Refrigeration	Freezing mixtures
	Water	Purification
	Welding	Coatings on rods
	Nonconstruction Uses	
Blueprinting, electroplating, engraving and lithography, heat treatment of steel, manufacture of paper, glass, refractories, metallurgy, pharmaceuticals, photography		

History and Manufacture

The history of potassium is linked to that of sodium. Minerals containing compounds of both were used as early as 1700 B.C. for glazes and later by the Egyptians in making glass. Davy, an English chemist, isolated both elements in 1807. The one he isolated from potash he named potassium. The Germans, however, adopted *kalium*, a Latinized version of the Arabic word "alkali," and this is the source of the chemical symbol, K.

Today potassium is obtained chiefly by the electrolysis of potassium chloride, but there is no industrial demand for the element as such.

POT LIFE

Pot life is the period of time during which a sealant, adhesive, or coating, after being mixed with a curing agent, solvent, or other compounding ingredient, remains suitable for use.

POZZOLANA

Pozzolana is a siliceous or siliceous-aluminous material that in itself possesses little or no cementitious value but that will, in finely divided form and in the presence of moisture, react chemically with calcium hydroxide (slaked lime) at ordinary temperatures to form compounds possessing cementitious properties. Pozzolana originally referred to volcanic ash or tufa but now includes industrial materials (slags) which possess pozzolanic characteristics in varying degrees. Pozzolana is more widely used in Europe than in America in cement and concrete work. (*See also* Cement.)

PROTECTIVE COATINGS

Protective coatings may be defined in a broad sense as methods used to protect a material against corrosion, wear, microbes, insects, mold, color fading, atmospheric conditions, effect of climate, ultraviolet and other rays of the sun, etc., and to add color, texture, or patterns.

These coatings are subdivided into "film", less than 0.1 mil (0.00254 mm), and "coatings", greater than 0.1 mil (0.00254 mm). For further information, see the following sections: Aluminum Finishes; Galvanizing; Metals: Chemical Finishes; Metals: Coatings; Metals: Mechanical Finishes; Metals: Sprayed Metal Coatings; Paint; Painting; Painting of Concrete, Plaster, Masonry, and Miscellaneous Surfaces; Painting of Metal; Painting of Wood; Steel, Galvanized; Wire; Zinc Coatings.

PUMICE AND PUMICITE

Physical and Chemical Properties

Pumice is a porous, frothlike, volcanic glass that was molten at the moment of effusion and did not crystallize because of rapid cooling but instead frothed with the sudden release of dissolved water and gases as it solidified. Its color is usually grayish white; its true specific gravity (of the glass forming the pumice) is 2.3 to 2.4. But pumice itself is light in weight, about one-third as heavy as sand and gravel. It consists of silica 65 to 75%, alumina 12 to 15%, and soda and potash each 4 to 5%. The dead air cells that characterize its internal structure give pumice excellent insulating properties against heat and sound. Its composition makes it virtually fireproof. Artificial or slag pumice is manufactured for the same uses as the natural product.

Types and Uses

Pumice in powdered or ground form is used as a fine abrasive for polishing; as a heat-insulating, lightweight aggregate in concrete; as an ingredient of plaster and lightweight pozzolana cement; and in the manufacture of brick. It serves as a filler in paints and plastics. Minor uses include insecticides, filtration, absorbents, soil conditioning, and surfacing and ice control of roads.

Pumicite. Pumicite is volcanic dust similar in composition to pumice, from which it differs in that it is fine grained and has sharp edges suitable for abrasive purposes.

PUTTY

Putty is a cement composed of fine powdered chalk (whiting) or lead oxide (white lead) mixed with boiled or raw linseed oil. It may contain other drying oils such as soybean and perilla. As the oil oxidizes, the putty hardens; if rapid hardening is desired, litharge or special driers may be added. Putty is used in glazing to set sheets of glass into frames and to fill up holes, etc., in woodwork. Special putty mixtures (metal sash putty) are available for interior or exterior glazing of aluminum and steel window sash. (*See also* Glass and Glazing for an account of its use, and Lead Oxide *and* Paint for a description of its ingredients.)

R

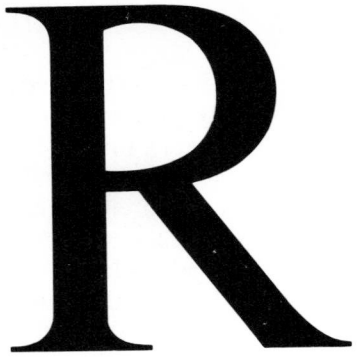

RADIATION PROTECTION

The various types of rays (alpha, beta, gamma, neutron) differ chiefly in their ability to penetrate and ionize matter. It is the latter characteristic that injures living tissue and must be guarded against. The alpha particle is a positively charged helium nucleus which is completely stopped by 3 or 4 in. (76.2 or 101.6 mm) of air or a piece of paper. Beta particles are high-speed electrons of varying energies. In general, they produce less ionization in matter than do alpha particles but are more penetrating. Most common substances—for example, 1 in. (25.4 mm) of wood—will completely absorb beta rays. Gamma radiation, on the other hand, has great powers of penetration. High-energy gamma rays will not be wholly absorbed by even a foot of lead, whereas lower energy gamma may be safely absorbed by $\frac{3}{16}$ in. (4.763 mm) or less of lead. Gamma rays produce large-scale ionization by a secondary process, that is, the gamma ray first produces a high-speed electron, and this, in turn, produces the ionization. Neutrons emitted by the cyclotron and the atomic pile are uncharged particles that will also ionize certain material indirectly. Neutrons are classified according to their energy levels as very fast, fast, slow, and thermal, the last having the lowest energy level. To shield against neutrons, metallic cadmium or a hydrogenous material such as water or paraffin is generally used. However, because gamma rays are emitted when the neutrons are absorbed, it is still necessary to provide a protective shield, usually of lead.

The use of protective shielding is identical in all cases, regardless of the type of ray, the thickness of the shielding depending on the penetrating power of the rays (*see* Table *R1*).

Lead has the advantage of being the densest of any commonly available material. Another advantage, particularly in shielding against neutrons and gamma rays, is that lead does not become contaminated. It may be used continuously without fear of its becoming radioactive and emitting its own harmful rays. For this reason, it is important that lead specified for shielding

Table R1 Comparative neutron cross section of possible structural materials (elements) in ascending order of absorption

Element	Absorption cross section[a]
Carbon	0.0045
Beryllium	0.0085
Bismuth	0.015
Lead	0.2
Silicon	0.2
Aluminum	0.22
Magnesium	0.3
Zirconium	0.4
Tin	0.6
Zinc	0.9
Niobium[b]	1.2
Chromium	2.5
Iron	2.5
Molybdenum	2.6
Germanium	2.8
Copper	3
Arsenic	4.3
Nickel	4.5
Vanadium	4.5
Antimony	5
Titanium	5
Palladium	6
Platinum	8
Manganese	13
Selenium	15
Tungsten	18
Cobalt	36
Silver	60
Gold	95
Cadmium	2500

[a] Values, which refer to thermal neutrons with a velocity of 2200 m/s, are expressed in barnes, an area of 10^{-24} cm^2.
[b] Formerly known as Columbium.

purposes be free of alloying elements, particularly those that may become radioactive upon exposure to high-energy radiation. In atomic reactors, lead and concrete are used in relation to a distance factor to protect personnel from exposure to all types of dangerous radiation, chiefly neutrons and gamma rays.

Neutrons escaping from the uranium section of reactors are absorbed by a thin cadmium shield. This absorption causes the cadmium to emit gamma rays, which in turn are stopped by a lead shield.

Some common applications of lead in the development of atomic energy and in the handling of radioactive materials include the following: lead bricks (interlocking or ordinary rectangular), lead plates and castings, "hot" laboratory table and sink tops, storage containers, carrying and shipping containers, lead rubber gloves and aprons, lead glass aprons, lead glass windows, Geiger counter tube shields, lead shot, lead sheet sheathing, lead-headed nails, and lead plywood.

RAMMED EARTH

Rammed earth construction is based on a method of making blocks or shapes for light construction in arid regions. It consists of simply filling a form with a mixture of earth and water, pounded into a compressed mass. These blocks, after drying, are used to build walls. Rammed earth is similar to adobe brick except that any loamy earth or clay earth can be used, whereas adobe brick is made of clay.

RARE EARTH METALS

Physical and Chemical Properties

The rare earths are a closely related family of 15 metals, listed in Table *R2* in related groups together with yttrium. Until recently they have been described generally as silver-white to gray in color; as a group, they tarnish in air, ignite in air at about 302°F (144.44°C), and in most preparations and alloys give off sparks when filed. They are soft and malleable and with some exceptions become harder as the atomic number increases. They are generally paramagnetic and good heat conductors but vary in their electrical conductivity. Recent research shows that, although these elements are similar chemically, they differ markedly with respect to nuclear cross sections and other properties.

Table R2 Rare earth metals

Element	Symbol	Atomic number
Lanthanum	La	57
Cerium	Ce	58
Praseodymium	Pr	59
Neodymium	Nd	60
Promethium[a]	Pm	61
Samarium	Sm	62
Europium	Eu	63
Gadolinium	Gd	64
Terbium	Tb	65
Dysprosium	Dy	66
Holmium	Ho	67
Erbium	Er	68
Thulium	Tm	69
Ytterbium	Yb	70
Lutetium	Lu	71
Yttrium	Y	39

[a]Promethium is an unstable fission product of uranium and is not found in nature.

Types and Uses

The rare earth metals are most frequently used in combination and are referred to as cerium, misch metal or ferrocerium, lanthanum-enriched misch metal, and didymium salts (*see* Table *R3*).

The rare earth metals find their greatest use in metallic form, as indicated by asterisks in Table *R4*, which summarizes the industrial applications of the group.

Cerium Compounds. Cerium, the most common of the rare earths, is now produced commercially, and both the metal and its compounds are finding increasing use in industry. In fact, misch metal is also being called cerium standard alloy and is marketed in a series of controlled mixtures, up to 99+% cerium. These are used for master alloys to improve the characteristics of steels and many light metals. Cerium compounds are used in drugs, as opacifiers in porcelain enamels, as abrasives (better than rouge) for polishing lenses and mirrors, and as blue and yellow colors in pigments and in the ceramic and textile industries.

Other Compounds. Some of the other rare earth metals also find uses in compound form. Lanthanum is used for glass with high refractive indices, in electronic devices, and in gas mantles. Praesodymium finds application as a decolorizer for glass, for greenish yellow colors in glass, glazes, and porcelain enamels, and with neodymium for

Table R3 Typical compositions of rare earth metals

Name	Ce	La	Nd (percent of content)	Pr	Sm	Other rare earth metals
Cerium (Grade AA)	99.9	None	0.01	0.01	0.01	Eu 0.01
Cerium standard alloy No. 1[a]	50.0–55.0	22.0–25.0	15.0–17.0			8.0–10.0
Cerium standard alloy No. 2[a]	30.0–38.0	32.0–35.0	22.0–25.0	7.0–9.0		
Lanthanum	0.025	Not known	0.025	0.025	0.025	99.90
Misch metal	52.0	Not known	18.0	5.0	1.0	24.0
Neodymium	1.0	Not known	78.0	15.0	2.0	4.0
Praseodymium		Not known	5.0	55.0	1.0	39.0

[a]Containing 0.5 to 1.5 Fe at maximum.

Table R4 Uses of rare earth metals

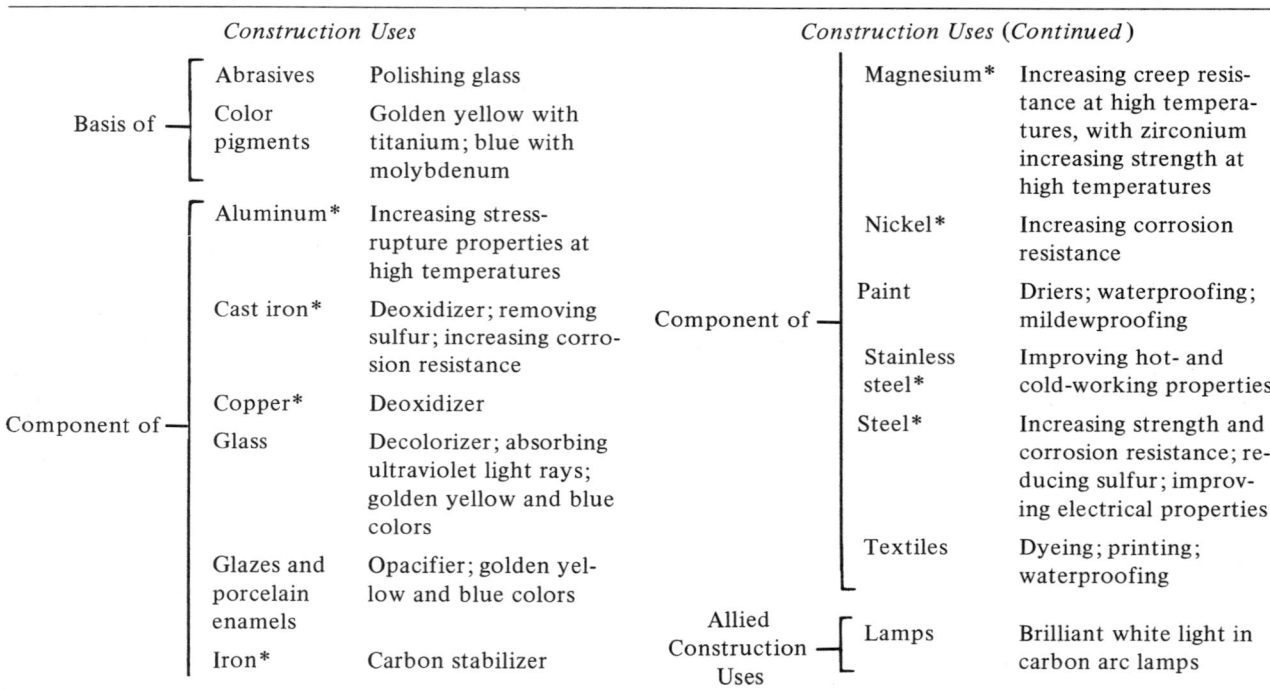

Construction Uses		
Basis of	Abrasives	Polishing glass
Basis of	Color pigments	Golden yellow with titanium; blue with molybdenum
Component of	Aluminum*	Increasing stress-rupture properties at high temperatures
Component of	Cast iron*	Deoxidizer; removing sulfur; increasing corrosion resistance
Component of	Copper*	Deoxidizer
Component of	Glass	Decolorizer; absorbing ultraviolet light rays; golden yellow and blue colors
Component of	Glazes and porcelain enamels	Opacifier; golden yellow and blue colors
Component of	Iron*	Carbon stabilizer
Construction Uses (Continued)		
Component of	Magnesium*	Increasing creep resistance at high temperatures, with zirconium increasing strength at high temperatures
Component of	Nickel*	Increasing corrosion resistance
Component of	Paint	Driers; waterproofing; mildewproofing
Component of	Stainless steel*	Improving hot- and cold-working properties
Component of	Steel*	Increasing strength and corrosion resistance; reducing sulfur; improving electrical properties
Component of	Textiles	Dyeing; printing; waterproofing
Allied Construction Uses	Lamps	Brilliant white light in carbon arc lamps

neutral-gray colored glass. Neodymium is used for reddish lilac colors in glass and porcelain enamels. Samarium and europium phosphors are used in tube lighting. Yttrium is used in gas mantles.

History and Manufacture

In 1794 Gadolin discovered in a mineral from Ytterby, Sweden, a new metallic oxide. Its existence was confirmed in 1797 by Ekeberg. They did not realize that their discovery was actually a mixture of 15 different elements.

Some of these new elements received their names in amusing and interesting ways. Yttrium, erbium, terbium, and ytterbium are all named after Ytterby, a small Swedish town. Holmium was derived from Holmia, a Latinized name for Stockholm; europium, from Europe; lutetium, from Lutetia, the Roman name for the original site of Paris; and thulium from Thule, the ancient Roman name for Scandinavia.

Rare earth metals are obtained from their ores by heating with sulfuric acid, thus converting them into sulfates which are then dissolved in water. By precipitation they are removed, and the compounds converted into commercial rare earth metal salts. Separation of cerium is relatively easy, but separation of the remaining rare earth metals is extremely difficult because of the great similarity of their properties. However, separation has been attained through ion-exchange techniques, and small lots of pure metals are commercially available.

REDUCING AGENT

A reducing agent is any element or substance that enables an atom or ion to add one or more electrons to its structure.

REGLETS

A reglet is a slot into which flashing or roofing material can terminate in either a vertical or horizontal wall (*see* Figure *R1*). This reglet or slot may be specially cut into stone or shaped in a clay tile structural unit by the extruding die, or it may be a metal accessory that forms the reglet or slot when inserted into cement, concrete, or stucco work. The detail drawings under Flashing show various types of reglets and their uses.

See also Inserts.

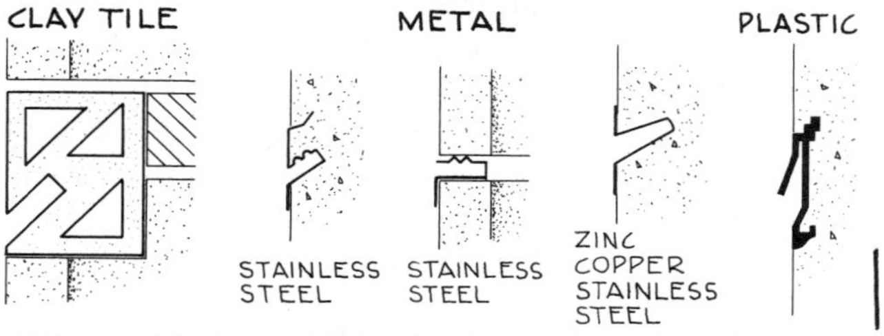

Figure R1 Typical reglets.

RETAINING WALLS

Retaining walls are independent walls built to hold back a bank of earth—at a change in grade, for example—and calculated to take the lateral load of the earth retained. The structural problem in designing these walls is to make sure that the lateral pressure from the banked earth does not tip the wall over. The three most common methods of providing adequate compensating pressure are (1) to make the retaining wall massive enough to withstand the pressure by weight alone; (2) to build a footing to extend under the retained earth; and (3) to build the footing to extend away from the retained earth. For earth embankments, a method called cribbing is used. (*See* Figures *R2 and R3 and* Table *R5*.)

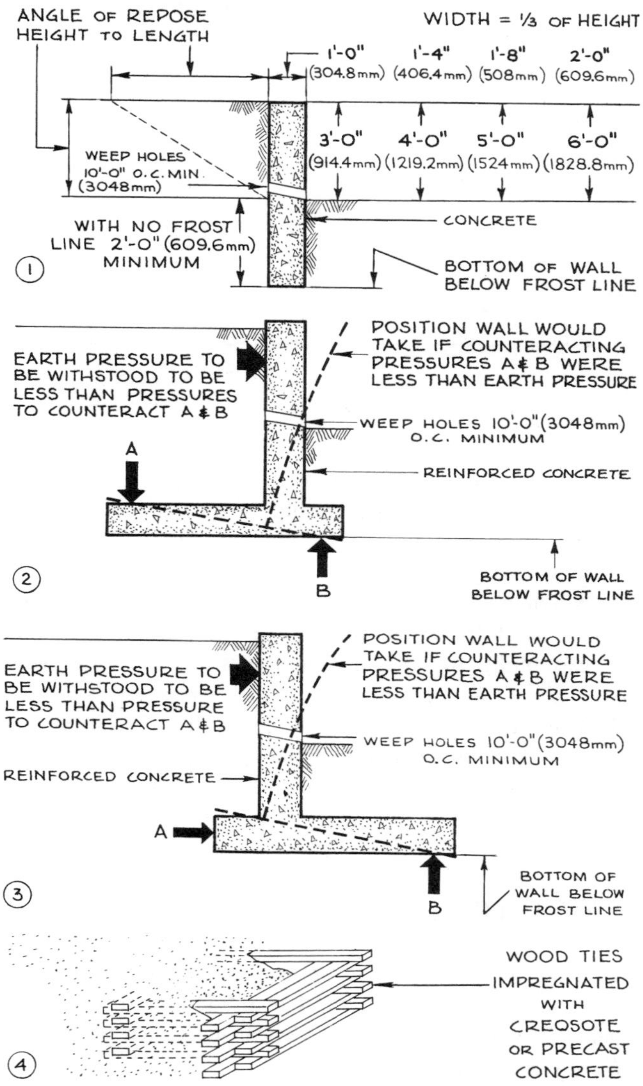

Figure R2 Type of construction for retaining walls.

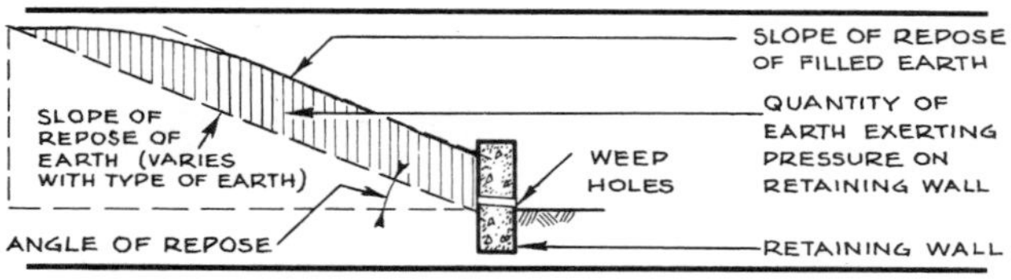

Figure R3 Use of angle and slope of repose in designing retaining walls and in banking earth.

Table R5 Natural slopes for various types of banked earth

Type of earth	Slope of repose: Length ft and in. (m)	Slope of repose: Height ft (m)	Angle of repose	Weight lb/ft^3 (kg/m^3)
Clean sand	1′6″ (0.4572)	1 (0.3048)	33°41′	90 (1441.8)
Sand and clay; dry sand; gravel and clay; gravel, sand, and clay; top soil	1′4″ (0.4063)	1 (0.3048)	36°53′	100 (1602)
Rotten rock, soft	1′4″ (0.4063)	1 (0.3048)	36°53′	110 (1762.2)
Rotten rock, hard	1′0″ (0.3048)	1 (0.3048)	45°	100 (1602)
Cinders	1′0″ (0.3048)	1 (0.3048)	45°	30–65 (480.6–1041.3)
Damp, soft clay	2′0″ (0.6096)	1 (0.3048)	26°34′	100 (1602)

Condensed Checklist

1. Design and material must be correlated. Masonry such as stone or brick should be used only for fairly low walls; otherwise concrete, usually reinforced, is used.
2. Adequate drainage from the banked earth must be provided by weep holes; otherwise, water pressure can build up, and, should the water freeze, it will destroy the retaining wall.
3. The position of the retaining wall in relation to property lines will affect the type of support below grade since one cannot build on another person's property.

R FACTOR

Thermal resistance (*R* factor) is a measure of the ability to retard heat flow, rather than the ability to transmit heat. The *R* value is the numerical reciprocal of the *U*, *K*, or *C* value. Thus

$$R = \frac{1}{C} \text{ or } \frac{1}{U} \text{ or } \frac{1}{K};$$

also

$$R = \frac{1}{K} \text{ for 1-in. material,}$$

and

$$R = \frac{2}{K} \text{ for 2 in. of the same material.}$$

A wall with a *U* value of 0.07 has 14.3 units of resistance.

The letter *R* is used in combination with numerals to designate thermal resistance values. *R*-11, also written as 11*R*, equals 11 resistance units. All insulating products having the same *R* are equal in insulating value regardless of material and thickness.

With the increasing use of energy and the resulting emphasis on energy conservation and the importance of insulating for both heat and cold, it became necessary to find some common factor that could be used in building codes, ordinances, and specifications. The *R* factor has become this common factor. Now all that is necessary is to state that, for example, the insulation for the X-Y-Zee Project shall be *R*-11 for walls and *R*-14 for roofs.

See also Btu; *C* Factor; *K* Factor; *U* Factor; Insulation and Insulating Materials; Mineral Wool.

RHODIUM

Physical and Chemical Properties

Symbol: Rh
Atomic number: 45
Specific gravity: 12.4 (20°C, 68°F)
Melting point: 1966°C, 3570°F
Boiling point: 3727°C, 5840.6°F

Rhodium is an extremely white, very hard metal that is highly resistant to tarnish and corrosion. It can be worked above 1500°F (815.56°C) and can be brazed and machined without difficulty. It is available in the form of rod, sheet, wire, and plating solutions.

Types and Uses

Because of its hardness and luster, rhodium is used as a coating on other materials, such as reflectors for motion picture projectors and aircraft searchlights, and with platinum metals. It has no uses in compound form at present. *See* Table *R6*.

Table R6 Uses of rhodium

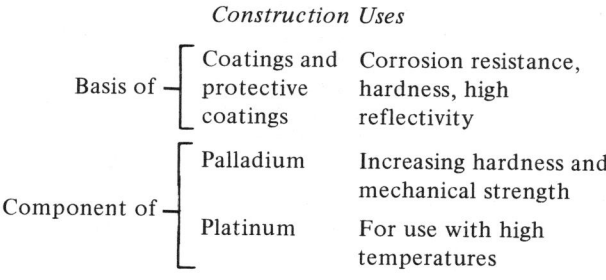

Construction Uses		
Basis of	Coatings and protective coatings	Corrosion resistance, hardness, high reflectivity
Component of	Palladium	Increasing hardness and mechanical strength
	Platinum	For use with high temperatures
Nonconstruction Uses		
Nozzles for extruding glass		
Possible Future Uses		
On glass for high-reflectivity mirrors		

History and Manufacture

About 1803 Wollaston, working with the platinum metals, isolated rhodium and gave it this name from the Greek *rhodon*, meaning "rose," because of the rose-red color of its salts. The process for obtaining rhodium is described under Platinum.

RIVETS

Rivets may be defined as devices used to join or fasten materials. The rivet, a metal cylinder or rod with a head at one end, is inserted through holes in the materials being joined, and then the protruding end is flattened to tie the two pieces of material together. Figure *R4* shows some of the various types of rivets used to join or fasten lightweight materials. There are two types of rivets: (1) those for holding thin metals or metal and another material together or for fastening them onto the basic structure, and (2) rivets used to join structural steel. The latter type must not only hold the two steel members together but also withstand the shear, bending, and other stresses developed by the structure. The design and application of structural steel rivets is beyond the scope of this book.

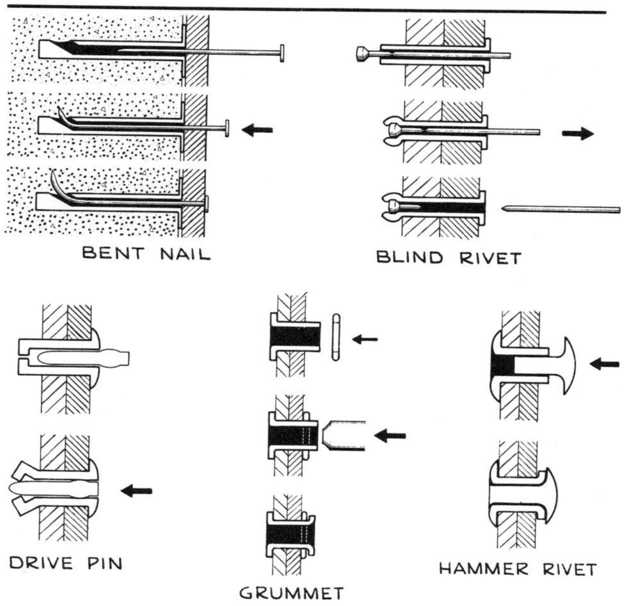

Figure R4 Types of rivets used to join or fasten lightweight materials.

ROOFING

"Roofing," in this book, refers to the materials that are installed and applied to weatherproof or waterproof the overhead cover of structures. Table *R7* lists the various types of roofing and the materials of which they are made.

ROSIN

Rosin is a natural resin obtained from coniferous trees (*see* Paint; Soldering).

ROUGE

Rouge is an oxide of iron that is obtained by heating yellow oxalate until it decomposes into a red powder, known as jeweler's rouge. Rouge is used as an abrasive for very fine polishing.

Table R7 Typical materials used for roofing

Material	Type of roof	Advantages	Disadvantages	Major uses
Aluminum (*see* Aluminum, Corrugated; Aluminum Sheet and Strip)	Pitched	Will not stain adjoining materials; requires no painting; reflective insulation value; lightweight; life of better than 20 years; fire resistant	Difficult to solder; has to be isolated from noncompatible materials to stop galvanic action; no color selection except in enameled type	Industrial and farm buildings
Aluminum shingles (*see* Aluminum Shingles and Siding)	Pitched	Wide range of colors and types; easily installed; fire resistant	Lightweight; must be insulated from noncompatible materials to stop galvanic or corrosive action	Residential and small commercial buildings
Asbestos (*see* Asbestos-Cement Shingles for Roofing and Siding; Asbestos, Corrugated)	Pitched	Will not stain adjoining materials; easily installed; life of better than 20 years; fire resistant	Limited selection of colors	Residential buildings
Asphalt, built-up (*see* Asphalt Roofing and Siding	Flat or slight pitch	Has life guaranteed by years with a maximum of 20 years; will not stain adjoining materials	No selection of color except for granule topping; requires skilled craftsman to install; fire resistance depends on material applied as topping; can be used as terrace with tile or concrete surface	All types of buildings
Asphalt shingles (*see* Asphalt Roofing and Siding	Pitched	Will not stain adjoining materials; easily installed; wide selection of colors	Limited fire resistance; life up to 20 years for only a few types	Residential buildings
Cement (*see* Clay Tile Roofing)	Pitched	Will not stain adjoining materials; fire resistant; life of better than 20 years	Very limited selection of colors; heavy; structure has to be designed to receive load; limited to climates where severe freezing is not encountered	Residential buildings
Clay tile (*see* Clay Tile Roofing)	Pitched	Will not stain adjoining materials; life of better than 20 years; fire resistant	Limited selection of colors; heavy; structure has to be designed to receive load	Residential buildings
Copper (*see* Copper Sheet and Strip)	Pitched, flat	Fire resistant; life of better than 20 years; easily soldered; requires no painting	Will stain adjoining materials; no color selection	Used on all types of buildings
Lead (*see* Lead Sheet and Strip)	Low pitched, flat	Fire resistant; life of better than 20 years; easily soldered; will not stain adjoining materials; requires no painting	Heavy; no color selection	Rarely used, but can be used on all types of buildings
Monel (*see* Monel)	Pitched	Fire resistant; life of better than 20 years; will not stain adjoining materials; requires no painting	No color selection; requires special solder; has to be isolated from noncompatible materials to stop galvanic action	Used on all types of buildings
Plastic (*see* Plastics: Roofing)	Flat or slight pitch	Easily installed; life of better than 20 years; can be applied to roof of almost any shape	Substrate must be of type to receive plastic-type roofing; requires skilled craftsman to install	All types of buildings
Stainless steel (*see* Stainless Steel Sheet, Strip, and Plate)	Pitched	Fire resistant; will not stain adjoining materials; life of better than 20 years; reflective insulation value	Difficult to solder; no color selection	Used on all types of buildings
Steel (*see* Steel, Corrugated; Steel, Galvanized; Steel Sheet, Strip, and Plate)	Pitched	Fire resistant; easily soldered; will not stain adjoining materials if galvanized and painted	Should be painted with paint compatible to zinc or paints that inhibit corrosion; life up to 20 years if well galvanized or protected with paint	Industrial and farm buildings

Table R7 Typical materials used for roofing (*continued*)

Material	Type of Roof	Advantages	Disadvantages	Major uses
Steel shingles (*see* Steel Sheet, Strip, and Plate)	Pitched	Wide range of colors and types; easily installed; fire resistant	Requires careful installation in order not to damage vinyl or baked enamel coating	Residential and small commercial buildings
Stone (*see* Stone, Slate)	Pitched	Fire resistant; easily installed; life of better than 20 years; will not stain adjoining materials	Limited selection of colors; heavy; structure must be designed to receive load	Residential buildings
Tar built-up (*see* Asphalt Roofing and Siding; Tar)	Flat or slight pitch	Same advantages as asphalt built-up roofing	Same disadvantages as asphalt built-up roofing	All types of buildings
Terneplate (*see* Terneplate)	Pitched	Fire resistant; easily painted; will not stain adjoining materials; easily soldered	Must be painted; life good for 20 years if paint kept in good repair	Residential buildings
Textiles (*see* Textiles)	Pitched or flat	Easily painted; easily installed; can be used as terrace (walked on); will not stain adjoining materials	Limited fire resistance; must be painted; life good for 20 years if paint kept in good repair	Residential buildings
Wood (*see* Wood Shingles)	Pitched	Easily applied; easily stained; life up to 15 years if not stained or painted; will not stain adjoining materials	No fire resistance; requires staining to increase life; will bleach to gray if not stained	Residential buildings
Zinc (*see* Zinc Sheet and Strip, Zinc Sheet, Corrugated)	Pitched	Fire resistant; will not stain adjoining materials; life of better than 20 years; easily soldered; easily painted	No color selection	Residential buildings

RUBBER

Physical and Chemical Properties

The properties of rubber that lead to its wide use are (1) its elasticity, more specifically, high elongation with rapid recovery over a wide range of temperatures and cohesive strength with the flexibility needed to cushion shocks and impacts; (2) its impermeability to gases and to water; and (3) its low specific gravity.

The water repellence of rubber is not absolute. Some water is absorbed, but the amount is slight compared to the water absorption of other materials and also slight compared to the absorption of oil. Rubber seems to be compressible, but actually there is no reduction in volume.

Rubber compositions are relatively unaffected by oxygen, acids, bases, many organic solvents, and other chemicals. They also show good electrical properties and outstanding performance against abrasive wear.

In designing with rubber, either natural or synthetic, it is important to know its low temperature flexibility and the compression set characteristics. Good compression recovery is usually desirable. Staining characteristics also are very important in architectural applications.

For the physical, mechanical, and chemical properties of common rubbers, see Tables *R8* and *R9*.

Natural Rubber (Caoutchouc). Natural rubber (caoutchouc) is a compound, polyisoprene, containing only carbon and hydrogen (C_5H_8). It belongs to the class of bodies known as terpenes and is related in chemical composition to the constituents of turpentine. Natural rubber is a coherent elastic solid that is insoluble in water and unaffected by alkalis or moderately concentrated acids. These chemicals can react, however, with nonrubber accessory substances.

Natural rubber has a low specific gravity (1.13), which decreases with increases of temperature. It is slowly affected by atmospheric oxygen. At 32 to 50°F (0 to 10°C) it is hard and opaque but reverts to a soft condition above 68°F (20°C). As the temperature increases, the rubber becomes softer, stickier, weaker, and less elastic. These changes are greatly accelerated at temperatures of 122 to 140°F (50 to 60°C). At a little below 392°F (200°C) rubber decomposes, yielding liquid hydrocarbons of the terpene series.

Latex. The source of natural rubber is latex, the milky fluid which occurs in a wide variety of trees growing in

Table R8 Physical and mechanical properties of commonly used rubbers

Kind of elastomer	Hardness[a] durometer	Tensile strength[b]: Pure gum lbf/in.^{2}	Pure gum MN/m^2	Black gum lbf/in.^{2}	Black gum MN/m^2	Rebound compression recovery	Abrasion resistance	Water swell resistance	Flame resistance	Elongation (percent): Pure gum	Black gum
Natural	A30–A90	2500 3500	17.24 24.13	3500 4500	24.13 31.03	Excellent	Excellent	Fair	Poor	750 550	850 650
Synthetic polyisoprene	A30–A90	2500 3500	17.24 24.13	3500 4500	24.13 31.03	Excellent	Excellent	Excellent	Poor	750 550	850 650
Butadiene-styrene	A40–A90	200 2500	1.38 17.24	300 3500	2.07 24.13	Good	Good to excellent	Excellent	Poor	400 500	600 600
Butadiene-acrylonitrine	A40–A95	200 2500	1.38 17.24	300 3500	2.07 24.13	Good	Good	Good	Fair	500	600
Neoprene (chloroprene)	A40–A95	3000 3000	20.69 20.69	4000 4000	27.58 27.58	Very good	Good	Fair	Good	800 500	900 600
Ethylene-propylene terpolymer	A40–A90	2000	13.79	3000	20.69	Good	Good	Good	Poor	400	600
Thiokol (polysulfide)	A40–A85	Less than 1000 lbf/in.^{2} (6.90 MN/m^2)				Good	Poor	Excellent	Poor	450	650
Butyl (isobutylene-isoprene)	A40–A75	2500 2500		3000 3500		Bad to cold; very good to hot	Fair	Excellent	Poor	750 650	950 850
Silicone (polysiloxane)	A40–A85	600	4.14	1000	6.90	Excellent	Poor	Excellent	Fair	60	400

[a]Hardness is measured on a durometer and expressed on a scale ranging from 0 to 100: very soft to very hard.

[b]1 lbf/in.^{2} = 0.006895 MN/m^2.

Table R9 Chemical properties of commonly used rubbers

Kind of elastomer	Weather resistance			Chemical resistance			
	Oxidation	Sunlight aging	Heat aging	Organic solvents including alcohol	Gasoline and petroleum oils	Animal and vegetable oils	Acids
Natural	Good	Poor	Good	Poor to most; good to alcohols	Poor	Poor to good	Fair to good
Synthetic polyisoprene	Good	Fair	Good	Poor to most; good to alcohols	Poor	Poor to good	Good
Butadiene-styrene	Good	Poor	Very good	Poor to most; good to alcohols	Poor	Poor to good	Fair to good
Butadiene-acrylonitrile	Good	Good	Very good	Excellent	Excellent	Excellent	Excellent
Neoprene (chloroprene)	Excellent	Very good	Excellent	Fair to most; poor to alcohols	Good	Good	Excellent
Ethylene-propylene terpolymer	Excellent	Excellent	Good	Fair	Poor	Fair	Good
Thiokol (polysulfide)	Very good	Very good	Fair	Excellent	Excellent	Excellent	Good
Butyl (isobutylene-isoprene)	Very good	Very good	Excellent	Poor to most; good to alcohols	Poor	Excellent	Excellent
Silicone (polysiloxane)	Very good	Excellent	Outstanding	Poor to most; fair to alcohols	Fair	Excellent	Excellent to dilute acids; fair to concentrated acids

the tropics. The principal commercial source of natural rubber is the tree *Hevea brasiliensis*, which yields the purest and best type of latex. Hevea latex contains 30 to 40% rubber and small quantities of compounds essential to the commercial applications of rubber products.

Synthetic Polyisoprene. Large quantities of synthetic polyisoprene, which duplicates natural rubber more closely than any other man-made polymer, are now produced in the United States, France, Japan, and the Soviet Union.

Vulcanizing Agents. When rubber is heated with sulfur at 248 to 320°F (120 to 160°C), it forms vulcanized rubber, which is stronger, more elastic, and less affected by changes of temperature than raw rubber. The amount of sulfur and heat applied and of other substances mixed with the rubber modify the changes throughout.

Some of the accessory substances in latex are particularly active in accelerating vulcanization. Acetone-soluble fatty acids (oleic, stearic) dissolve and disperse some of the mineral powders mixed with rubber during manufacture. Other acetone-soluble substances aid in preserving vulcanized rubber against atmospheric oxidation so that it remains supple and elastic longer.

Rubber compositions with high resistance to natural deterioration contain less than 1% sulfur; soft rubber contains 2 to 10% sulfur; and hard rubber contains 20 to 50%. Synthetic rubbers require less sulfur than does crude rubber. Other vulcanizing agents include organic compounds (some with and others without sulfur), selenium, and tellurium.

Accelerators usually are added to speed up the vulcanization process. The first accelerators were inorganic compounds (litharge white lead, lime, magnesium); now organic oxides are used alone or with zinc oxide, litharge, or magnesium, often assisted by additions of oleic acid, stearic acid, and pine tar.

Antioxidants, or age resisters, are chemicals that retard deterioration. Crude rubber contains natural antioxidants; synthetic rubbers must have additions of 1 to 2% of certain organic chemicals that have no direct effect on vulcanization. Other substances are added to rubber to modify its characteristics such as stiffness, strength, resistance to abrasion, or chemical action.

Reinforcing pigments such as carbon black, zinc oxide, certain clays, calcium silicate, and magnesium carbonate stiffen and strengthen rubber compositions so that the total energy needed to extend a strip of the composition to its breaking point is greater than that for unreinforced rubber. Fillers (whiting, barite) stiffen without increasing the total energy of rupture.

Coloring of Rubber. Colors are imparted to rubber by inorganic mineral pigments and a few of the organic dyes. Zinc oxide, lithopone, titanium dioxide, and zinc sulfide are used for white; ferric oxides, ultramarine blue, and zinc chromate for reds, blues, and yellows. Phthalocyanine colors and azo dyes are also used.

Synthetic Rubbers. The synthetic rubbers all resemble crude rubber to varying degrees. Some of them are of the nonvulcanizing (noncuring) type and are often listed as plastics. It is beyond the scope of this book to differentiate and describe the synthetic polymers (elastomers) which can now be compounded to replace or supplement natural rubbers and which are often a part of the many synthetic products replacing or supplementing cellulose, glass, animal glues, and other natural materials.

Natural rubber compounds have the best resistance to cutting, chipping, and crack growth. Butyl rubber has the lowest permeability to gases. Nitrile rubbers show the least swelling in oils and most solvents. Acrylate rubbers (polyethylacrylate) are remarkably resistant to dry heat and sunlight. Epon rubbers resist ozone.

The Buna series of synthetic rubbers developed in Germany closely approximate crude rubber. Buna S is made from butadiene and styrene; Buna N, from butadiene and acrylonitrile.

In general, a wide range of natural and synthetic rubber compositions is available, vulcanized and unvulcanized, in varying degrees of hardness or softness. Natural rubber and most United States synthetic rubbers are also available as latex (liquid form).

Types and Uses

The greatest consumption of rubber is in the electrical and transportation fields. Architectural and construction uses of the major types of rubber listed in the tables are as follows: (1) extrusions and moldings, (2) resilient flooring, (3) adhesives, (4) paints and protective coatings, (5) bituminous toppings and fillers, and (6) insulation materials. *See also* Adhesives; Glass and Glazing; Paint; Painting; Plastics.

Extrusions and Moldings. Extrusions and moldings made of foamed or unfoamed rubbers are used for sealing and glazing in general and in specialized applications such as compression seals for curtain wall construction. This group includes filler strips of all sorts for expansion and construction joints, closure strips for corrugated sheet materials, and products for vibration control.

Both natural and synthetic rubbers are used for these products. The synthetic polymers are usually neoprene, Buna N, and polyvinyl chloride, alone or compounded to meet various specifications of end use. The rubber may be solid, open cellular, or closed cellular.

Filler or closure strips for corrugated materials are manufactured in a very wide range of sizes to fit all types of corrugations and to seal all kinds of joints (horizontal, vertical, and diagonal to the corrugations).

The use of an adhesive to provide a better bond between closure and sheet material is usually recommended.

Rubber extrusions are currently being used in glazing to replace putty and glazing compounds. In quite a few instances, sheets of glass have been installed with only rubber extrusions, the glass being held in the frame primarily by the compression of the rubber, without any grooves or stopbeads as holding devices. This represents an entirely different approach to the installation of glass and similar materials in any sort of framework.

Resilient Flooring. Resilient flooring in tile or continuous roll form also includes minor items like foamed rubber undercushions for carpets, stair treads and nosings, and protective mats and pads.

As a rule rubber floor tiles are made of natural rubber for greatest resilience. Sizes, thicknesses, filler strips, and cove bases and other specialties, as well as methods for application and precautions to follow in installation, are much the same as those for plastic (vinyl) resilient flooring tiles.

Rubber flooring in roll or sheet form is applied according to the general rules given for linoleum.

The types of flooring vary widely in their specific resistance to different chemicals and types of environment. Each kind of flooring requires a special adhesive, type of underlayment (if any), methods of cleaning and maintenance, etc., as recommended by the flooring manufacturer. (*See also* Flooring; Linoleum; Plastics; Flooring.)

Adhesives. In the field of adhesives, it is often hard to differentiate between plastics and synthetic rubbers (*see* Adhesives; Plastics).

Paints and Protective Coatings. For the use of rubber in paints and protective coatings, *see* Paint.

Bituminous Toppings and Fillers. Compounds of rubber and bituminous materials are formulated to be spread as cement topping for industrial types of floors. Similar products are used as crack and joint fillers.

Insulation Materials. Insulation materials consist of boards, tubing used primarily for industrial and residential plumbing (cold water lines), and other shapes formed from foamed rubber filled with inert nitrogen.

Other Uses. Miscellaneous uses include adhesive and insulating tapes, washers, bumpers, and other hardware items, polishing wheels, belts, and resilient doors. Inflatable rubber balloons have been successfully used as forms in thin-shell concrete construction.

Application

Research has identified and synthesized the basic components that make up the various rubbers, resins, plastics, adhesives, paints, etc. Theoretically it is now possible to create an infinite variety of new molecular structures that can combine any desired grouping of favorable properties of the original natural materials without many of their drawbacks. This tailoring of a synthetic fabricated product to specific functions and conditions of use characterizes today's research programs and the new products appearing on the market. The difficulty for anyone not thoroughly versed in chemistry is to understand and interpret technical terminology and data and then to evaluate and use the materials correctly. In the construction field it is necessary to rely upon the manufacturer to guarantee his product on the basis of testing procedures that try to duplicate in accelerated form actual conditions of use.

History and Manufacture

Heavy black rubber balls made by the American Indians were first brought to Europe by Columbus and the later Spanish explorers. But three centuries passed before rubber became a commercial commodity in Europe, and then it was used as an eraser, not for its elastic or waterproof qualities. It was given its name by the English chemist Priestley, who observed that it rubbed out pencil marks.

Early Rubber Sources. Originally the wild trees of the Amazon were the only source of rubber. Eventually the English found ways of propagating the *Hevea* rubber trees elsewhere and established the first plantations in Ceylon. Malaya, the Dutch East Indies, and Ceylon became the largest producers of cultivated natural rubber.

Through the early 1800s rubber was used mainly for waterproof footgear but proved unsatisfactory for this purpose, as it hardened in cold weather, softened in hot weather, and was deteriorated by petroleum.

Discovery of Vulcanization. The successful commercial use of rubber is the result of the accidental discovery of vulcanization by Goodyear. In 1839, while following up on the earlier, unsuccessful efforts of Hayward in this country and Ludensdorff in Germany to reduce the tendency of rubber to become sticky by combining it with sulfur, Goodyear dropped a mixture of rubber with white lead and sulfur on a stove and inadvertently cooked it. Hancock in England worked on the same problem of vulcanizing (curing) rubber with sulfur and heat and patented his discovery in 1843. One year later Goodyear received his United States patent. These two inventors, Hancock and Goodyear, devising all sort of products and processing machinery, laid the groundwork for almost every important development of the rubber industry.

Rubber Manufacture. Rubber manufacture actually begins on the plantation. The collected latex is converted into smoked sheet or pale crepe crude rubber and shipped to processing plants.

Preparation of cements (solutions of rubber in organic solvents) is the oldest manufacturing operation in the rubber industry. The first manufactured products in this country were made from textiles coated with such liquid rubber. These early cement dipping and spreading techniques were gradually superseded by more complex procedures that included mastication, scientific compounding and mixing of rubber compositions, vulcanizing, fabrication into sheets, extrusions, spread coatings, and assembly of fabricated parts.

Mastication of the raw polymer is the first step for all but cements and latex products. When rubber is pressed between rollers or extruded under high pressure, it becomes plastic and sticky. This process accelerates with increased temperature. Chemical softeners (less than 1%) may be, but usually are not, added for most rubbers.

Mixing is the most important procedure, upon which the success of subsequent operations and of the finished product depends. While the rubber is in the plastic condition induced by mastication, powder and plastic solids are added either in roller-type mixing mills or, for large quantities, in Banbury or internal mixers. Banbury mixers three stories high can handle 1000-lb (453.6-kg) batches and complete the mixing in as little as 5 to 8 minutes. Large mills roll the mixed rubber composition into sheets.

Hot vulcanization or curing can be done in a number of ways, including hot water and steam curing under pressure, hot air curing, radio-frequency heating, and high-energy radiation. Often vulcanization and molding or extruding of the final product are done at the same time under heat and pressure.

Development of Synthetics. The most important development in the rubber industry has been the use of synthetic rubber compositions. Synthetic rubbers of various sorts were produced in the United States, Great Britain, France, Germany, and Russia during the latter part of the 19th century and the beginning of this century. By the early 1930s several commercial types were in use.

However, it was the loss of the sources of natural rubber during World War II that forced intensive research to produce a high-quality substitute in large amounts. The result was several highly successful types now being used in increasing tonnages. During the intervening years several new polymers were developed and have gained acceptance in the construction industry.

RUTHENIUM

Physical and Chemical Properties

Symbol: RU
Atomic number: 44
Specific gravity: 12.41 (20°C, 68°F)
Melting point: 2310°C, 4190°F
Boiling point: 3900°C, 7052°F

Ruthenium is a silvery-white, nonductile metal that is insoluble in most acids. It is available in powder and pellet form.

Types and Uses

Ruthenium metal is used principally as a hardener for platinum and palladium alloys. Hard alloys of osmium also contain some ruthenium.

History and Manufacture

In 1884 Claus discovered and named ruthenium after the Russian town Ruthenia. The process for obtaining ruthenium is described under Platinum.

S

SALT

A salt in chemical terminology is the compound produced by reaction between an acid and a base or any substance that yields ions other than hydrogen or hydroxyl ions. In another sense, a salt is obtained by displacing the hydrogen of an acid by a metal. (*See* Acids, Bases, and Salts.) Usually the word "salt" refers to common table salt, or sodium chloride (NaCl), which is widely distributed in nature and essential to life.

Uses

Salt is a mainstay of the chemical industry and has literally thousands of uses. Some of the most familiar are in commercial refrigeration brines, as a constituent in the manufacture of glass, for salt glazes on brick and clay products, as a flux in metallurgy, and for roadway and sidewalk control of snow and ice (salt lowers the freezing point of water in proportion to its concentration).

Salt is the raw ingredient for the manufacture of hydrochloric acid, chlorine, and a large group of very important alkali compounds such as soda ash (Na_2CO_3).

Manufacture

Salt is obtained by evaporation of seawater and other natural brines, by mining rock salt, and by pumping from underground salt beds.

SAND

Physical and Chemical Properties

Sand, gravel, silt, and clay are all products of the natural or artificial disintegration of rocks and minerals. Sand is obtained from glacial, river, lake, marine, residual, and wind-blown (very fine sand) deposits. It is distinguished from gravel and silt on the basis of particle size: grains from $\frac{1}{16}$ in. (159 mm) through $\frac{1}{4}$ in (6.35 mm) in diameter are classed as sand, those less than $\frac{1}{16}$ in. as silt, and those larger than $\frac{1}{4}$ in. as gravel.

Constituents. The composition and uniformity of sand vary with the type of deposit. In some localities feldspar, iron oxides, clay, and calcareous materials may be important constituents. Mica, garnet, zircon, and tourmaline are also usually present. Sometimes by-product recovery of ilmenites, gold, and other heavy minerals is achieved.

Silica Sand. Quartz or silica sands, consisting of over 98% silicon dioxide, are the most useful commercially as well as the most abundant type. Silica sand is strong, hard, a poor conductor of electricity, and chemically inert. It is insoluble in water and water solutions of the common mineral acids but is attacked (etched) by dilute alkaline solutions.

Commercial Units of Measurement. Sand is sold by the ton or cubic yard, a cubic yard weighing 2600 to 3100 lb (1179.36 to 1406.16 kg), depending on the composition and the particle size. The wetness of the sand can also markedly affect the weight.

Grades. The grade varies from bank-run (unprepared) sand to sand that is washed, screened for size, and otherwise processed for specific requirements.

The fineness modulus of sand is computed by adding the cumulative percentages retained on the six standard screens used for grading sand and then dividing the sum by 100 (*see* Table *S1*).

Types and Uses

Building and paving uses represent by far the major consumption of sand. Sand for concrete, cement, mortar, plaster, stucco, and terrazzo is classified as a fine aggregate and must meet certain requirements for each type of use (*see* Concrete; Mortar; etc.).

Table S1 Computation of fineness modulus from a typical sand analysis

Screen number	Metric equivalents	Individual percentages retained	Cumulative percentages retained
4	4.76 mm	1	1
8	2.38 mm	18	1 + 18 = 19
16	1.19 mm	20	19 + 20 = 39
30	595 microns	19	39 + 19 = 58
50	298 microns	18	58 + 18 = 76
100	149 microns	16	76 + 16 = 92
Left in pan		8	
Total		100%	285%
	Fineness modulus = 285 ÷ 100 = 2.85		

Clay Sand. Clay sand is used in road construction work and in metallurgy for making sand molds and furnace linings.

Pure Silica Sand. Pure silica sand is a basic raw ingredient of glass and other ceramics; it is also used in enamels, as a filler, as an abrasive, and for filtering purposes. A comparatively new use is for the manufacture of silicones.

Abrasive Sand. Abrasive sand is any kind of natural sand used for abrasive and grinding purposes, but the term does not include the sharp grains obtained by crushing quartz and used in carefully graded sizes for sandpaper.

Sandblast Sand. Sandblast sand is any sand employed in a high-pressure blast of air for cleaning metal, stone, or brick, and for giving a dull rough finish to glass or metal.

SANDPAPER

Sandpaper may be defined as a heavy paper coated with sand grains on one side and used as an abrasive, especially for finishing wood. In industry, sand has been largely replaced by crushed garnet and artificial aluminum oxide, but the abrasive papers are still commonly referred to as sandpaper. Where sand is still the abrasive material, crushed quartz sand is to be used rather than ordinary sand, which is too round to cut well.

Grades of sandpaper are based on the size of the crushed quartz grains as follows: Nos. $3\frac{1}{2}$, 3, $2\frac{1}{2}$, 2, $1\frac{1}{2}$, 0, 00 and 000 (*see* Abrasives *for correlation with grit sizes*).

SCREWS, NUTS AND BOLTS, AND RELATED DEVICES

Screws, nuts and bolts, small rivets, and washers are a group of related rough hardware items whose basic function is to hold pieces of material together.

Screws and nuts and bolts have certain similarities. Both are usually removable, and their strength lies in the type of metal used and in the number of threads per inch (the more threads, the greater their strength). It is possible to have as many as 64 threads per inch. The difference between a screw and a nut and bolt is clear if each is defined.

Bolts. A bolt is a round section of metal headed at one end, threaded at the other. The head design may be round, flat, oval, square, hexagonal, etc.

Nuts. A nut is a piece of metal drilled and threaded (tapped) to receive the threaded end of a bolt; it, too, is produced in various shapes.

Screws. A screw is also a round section of metal headed at one end and threaded at the other, but with the threaded end tapering to a point.

Rivets. Rivets are used for a permanent joint. A rivet also consists of a smooth, round section of metal headed at one end and long enough to extend beyond the materials being joined. This protruding end is spread to form a head by hammering. Structural steel rivets are not considered here, as they do not fall into the category of rough hardware. (*See also* Hardware: Rough; Rivets.)

Washers. Washers can accomplish several results: They can be used to lock the head or bolt into place; they may provide a seal or insulate incompatible materials from each other; they can widen or increase the area of force-exerting contact between head or nut and material being joined. In the last function, the washer both increases the holding power of the screw or nut and protects the material being held, should it be brittle or fragile enough to be injured by the pressure exerted against it.

Special washers with cylindrical-shaped protrusions on one face are made for high-strength bolts. When the bolt is tightened, the protrusions are squashed and the gap remaining is measured with flat metal gauges of various thicknesses. These varisized gauges can set the thickness to give the exact bolt tension that may be required.

Related Devices. Figures *S1* and *S2* show all general types of nuts and bolts, screws, rivets, washers, etc. Also

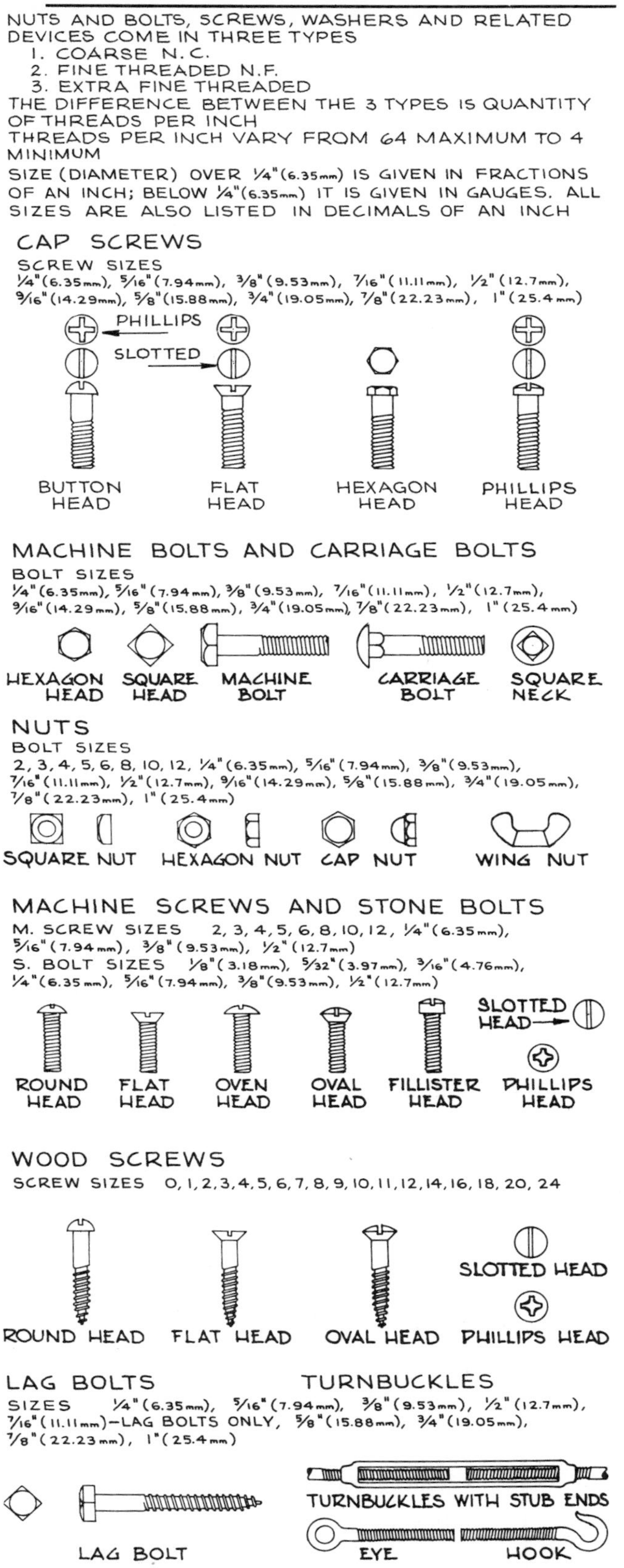

NUTS AND BOLTS, SCREWS, WASHERS AND RELATED DEVICES COME IN THREE TYPES
1. COARSE N.C.
2. FINE THREADED N.F.
3. EXTRA FINE THREADED
THE DIFFERENCE BETWEEN THE 3 TYPES IS QUANTITY OF THREADS PER INCH
THREADS PER INCH VARY FROM 64 MAXIMUM TO 4 MINIMUM
SIZE (DIAMETER) OVER 1/4"(6.35mm) IS GIVEN IN FRACTIONS OF AN INCH; BELOW 1/4"(6.35mm) IT IS GIVEN IN GAUGES. ALL SIZES ARE ALSO LISTED IN DECIMALS OF AN INCH
CAP SCREWS
SCREW SIZES
1/4"(6.35mm), 5/16"(7.94mm), 3/8"(9.53mm), 7/16"(11.11mm), 1/2"(12.7mm), 9/16"(14.29mm), 5/8"(15.88mm), 3/4"(19.05mm), 7/8"(22.23mm), 1"(25.4mm)
PHILLIPS
SLOTTED
BUTTON HEAD
FLAT HEAD
HEXAGON HEAD
PHILLIPS HEAD
MACHINE BOLTS AND CARRIAGE BOLTS
BOLT SIZES
1/4"(6.35mm), 5/16"(7.94mm), 3/8"(9.53mm), 7/16"(11.11mm), 1/2"(12.7mm), 9/16"(14.29mm), 5/8"(15.88mm), 3/4"(19.05mm), 7/8"(22.23mm), 1"(25.4mm)
HEXAGON HEAD
SQUARE HEAD
MACHINE BOLT
CARRIAGE BOLT
SQUARE NECK
NUTS
BOLT SIZES
2, 3, 4, 5, 6, 8, 10, 12, 1/4"(6.35mm), 5/16"(7.94mm), 3/8"(9.53mm), 7/16"(11.11mm), 1/2"(12.7mm), 9/16"(14.29mm), 5/8"(15.88mm), 3/4"(19.05mm), 7/8"(22.23mm), 1"(25.4mm)
SQUARE NUT
HEXAGON NUT
CAP NUT
WING NUT
MACHINE SCREWS AND STONE BOLTS
M. SCREW SIZES 2, 3, 4, 5, 6, 8, 10, 12, 1/4"(6.35mm), 5/16"(7.94mm), 3/8"(9.53mm), 1/2"(12.7mm)
S. BOLT SIZES 1/8"(3.18mm), 5/32"(3.97mm), 3/16"(4.76mm), 1/4"(6.35mm), 5/16"(7.94mm), 3/8"(9.53mm), 1/2"(12.7mm)
SLOTTED HEAD
ROUND HEAD
FLAT HEAD
OVEN HEAD
OVAL HEAD
FILLISTER HEAD
PHILLIPS HEAD
WOOD SCREWS
SCREW SIZES 0, 1, 2, 3, 4, 5, 6, 7, 8, 9, 10, 11, 12, 14, 16, 18, 20, 24
SLOTTED HEAD
ROUND HEAD
FLAT HEAD
OVAL HEAD
PHILLIPS HEAD
LAG BOLTS
TURNBUCKLES
SIZES 1/4"(6.35mm), 5/16"(7.94mm), 3/8"(9.53mm), 1/2"(12.7mm), 7/16"(11.11mm)—LAG BOLTS ONLY, 5/8"(15.88mm), 3/4"(19.05mm), 7/8"(22.23mm), 1"(25.4mm)
TURNBUCKLES WITH STUB ENDS
LAG BOLT
EYE
HOOK

Figure S1 Screws, nuts and bolts, washers, and related devices.

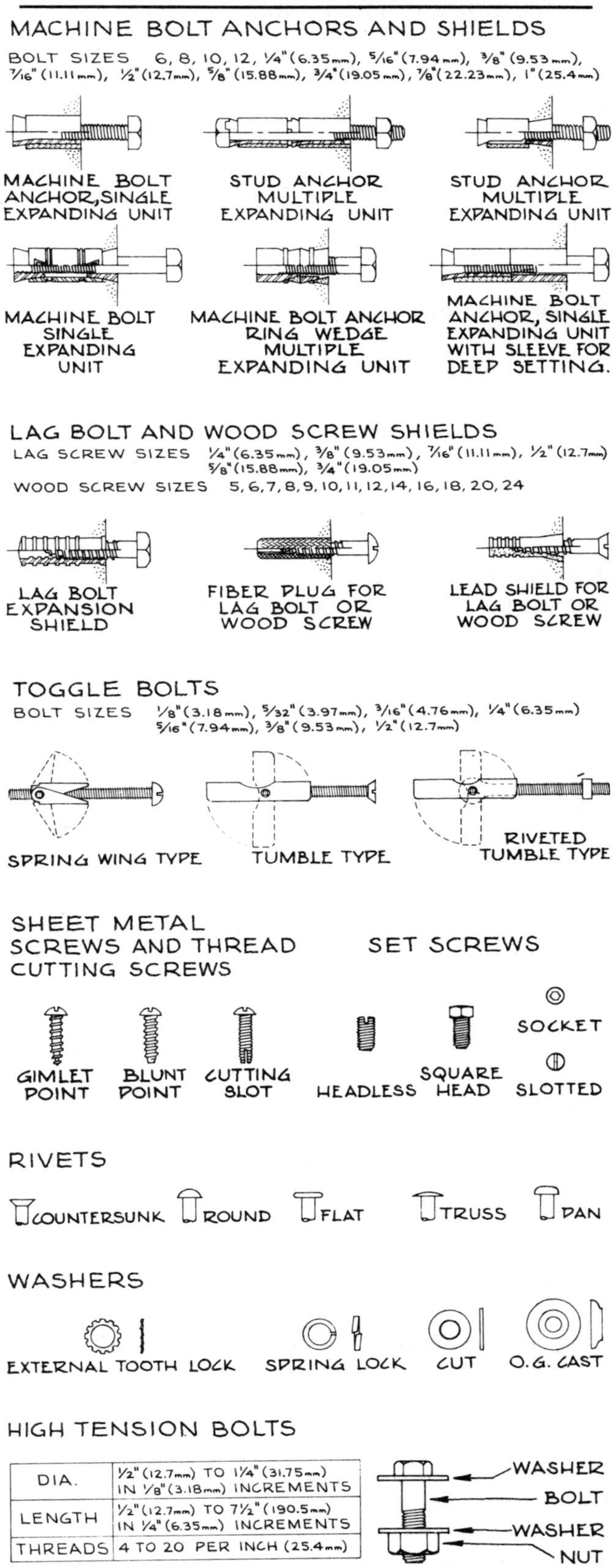

DIA.	1/2"(12.7mm) TO 1 1/4"(31.75mm) IN 1/8"(3.18mm) INCREMENTS
LENGTH	1/2"(12.7mm) TO 7 1/2"(190.5mm) IN 1/4"(6.35mm) INCREMENTS
THREADS	4 TO 20 PER INCH (25.4mm)

Figure S2 Screws, nuts and bolts, washers, and related devices.

shown are the lag bolt (which is a type of screw for masonry), the turnbuckle, the hook-and-eye, and a group of devices for anchoring a screw or a bolt into masonry by what is known as an expansion shield. The principle of the expansion devices is that, as the bolt or screw turns in the shield, the shield expands in size, thus exerting pressure against the adjoining material and becoming more securely engaged. The toggle bolt is a winged device used in hollow walls.

High-Tension Bolts. In construction the use of high-tension bolts instead of rivets or welding is common practice, as nuts and bolts can be made of high-strength steel alloys that can meet all the requirements for shear, compression, and strength that exist in the usual structures.

Materials and Finishes. Nuts, bolts, screws, washers, and related devices made of steel, brass, and aluminum are the ones more commonly used in construction. However, they are also available in copper, zinc, bronze, and many other metals. They are also made with many electroplated finishes such as chromium, brass, bronze, nickel, and zinc (galvanized), as well as plastic coated.

Countersinking. Countersinking the head is common practice and can be defined as beveling the edge of a hole for reception of the head so that it does not extend above the surface of the material.

See also Anchors; Hangers; Hardware: Rough; Nails; Rivets; *and* Ties.

SEALANT

A sealant is an adhesive type of material, either flowable or viscous, applied to seal joints or openings against the penetration of liquids, air or gases. *See* Asphalt; Paint; Painting, Plastics; Rubber; *and* Tar.

SELENIUM

Physical and Chemical Properties

Symbol: Se
Atomic number: 34
Specific gravity: 4.79 (gray)
Melting point: 217°C, 422.6°F
Boiling point: 684.9°C, 1264.82°F

Selenium is a brittle, semimetallic element that closely resembles tellurium but in all its compounds displays the properties of a nonmetal. Its compounds are highly toxic, although elemental selenium is not toxic. Its photosensitivity and asymmetric electrical conductivity are distinctive properties utilized in industrial applications. Molten selenium is partially or completely miscible with most metals and readily forms selenides by direct reaction with base metals.

Commercial Forms. Selenium is usually purified by distillation and is available as 99% selenium powder or as high-purity selenium shot.

Table S2 Uses of selenium

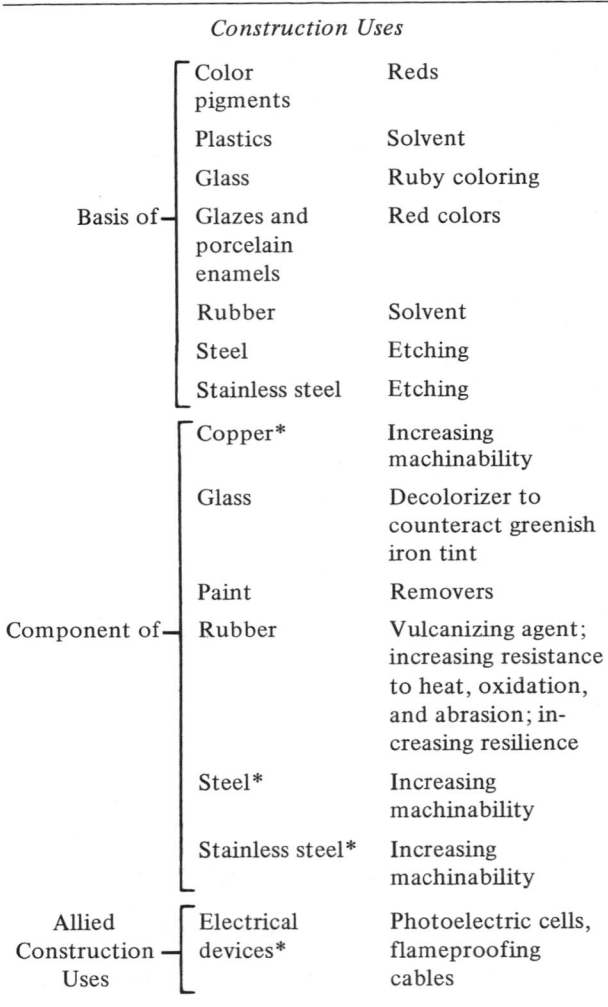

Construction Uses		
Basis of	Color pigments	Reds
	Plastics	Solvent
	Glass	Ruby coloring
	Glazes and porcelain enamels	Red colors
	Rubber	Solvent
	Steel	Etching
	Stainless steel	Etching
Component of	Copper*	Increasing machinability
	Glass	Decolorizer to counteract greenish iron tint
	Paint	Removers
	Rubber	Vulcanizing agent; increasing resistance to heat, oxidation, and abrasion; increasing resilience
	Steel*	Increasing machinability
	Stainless steel*	Increasing machinability
Allied Construction Uses	Electrical devices*	Photoelectric cells, flameproofing cables
Nonconstruction Uses		
Electronics,* printing, lubricating oils, medicine, oxidizing agent, coatings on steel before painting,* corrosion-resistant coatings*		

Types and Uses

The uses of selenium are shown in Table *S2*, an asterisk indicating those involving elemental selenium.

Not only visible light but also radium and X-rays affect the electrical conductivity of selenium. The selenium photocell used in exposure meters generates its own electromotive force (emf).

History and Manufacture

In 1817 Berzelius was studying a method for the production of sulfuric acid, which led to the discovery of a new element. He named it selenium from the Greek word *selene*, meaning "moon." It remained a laboratory curiosity until 1873, when Willoughby Smith discovered that the current resistance of selenium decreased as the intensity of illumination increased. This led to the development of the photoelectric cell and many applications important in our daily lives.

Methods of Recovery. Selenium is a by-product of the refining of lead, copper, and nickel. The three methods of recovery of selenium are roasting with soda, roasting with sulfuric acid, and smelting with soda and niter.

SHEET AND RIGIDIZED SHEET

A sheet is defined in this book as a flat unit that is normally very thin [approximately $\frac{1}{4}$ in. (6.35 mm) or less] in relation to its length and breadth and that is applied by adhesives, nails, or other attaching devices. It is nonrigid, that is, thin and flexible. Sheetlike materials (other than metals) that are thicker than $\frac{1}{4}$ in. (6.35 mm) are described under Board. Those made of layers and of more than one material are described under Integrant.

Rigidized sheet is defined as any sheet material that is deformed for the purpose of stiffening in any one direction or in all directions.

Table *S3* lists all the different types of sheet and the materials from which they are made.

Table S3 Typical kinds of sheet and rigidized sheet used in construction

Material	Advantages	Disadvantages	Major uses
Aluminum (*see* Aluminum, Corrugated; Aluminum Sheet and Strip)	Does not corrode; durable; easily formed; will not stain adjoining materials; reflective insulating value	Cannot be used where fire-retarding material is required; difficult to join and solder at job site	Exterior doors and frames, panel facing, roofing, gutters and leaders, flashing, expanded-type mesh, ceilings, insulation, etc.
Asbestos (*see* Asbestos-Cement, Board and Sheet)	Fire resistant; durable; available in limited colors and textures; will not stain adjoining materials	Brittle; nails and joints are exposed	Siding, wall and ceiling finishing material, fire-retarding material
Copper (*see* Copper Sheet and Strip)	Easily formed, joined, and soldered; durable	Corrodes; will stain adjoining materials unless lead coated; can corrode other metals by galvanic action	Gutters, leaders, flashing, roofing, decorative interior metalwork, etc.
Glass (*see* Glass and Glazing; *see also* Glass headings that follow)	Fire resistant if wired; can be transparent or translucent; can be shock resistant, glare or heat absorbing, or mirror	Generally limited to relatively small sizes except plate and heavy sheet; requires special setting and glazing if other than window, heavy sheet, plate, or patterned is used	All stock windows, fixed glass, sliding doors, decorative partitions, doors, (tempered), chalkboards, mirrors, skylights, etc.
Lead (*see* Lead Sheet and Strip)	Noncorroding; vibration deadening; resistant to radiation and X-ray; acid resistant; easily formed, joined, and soldered	Very soft; heavy, will sag and tear loose on vertical or steeply pitched surfaces if supports and attachments are not calculated to overcome this characteristic; not fire resistant	X-ray and radiation protection, vibration control, roofing, flashing, acid-resistant linings and piping, etc.

Table S3 Typical kinds of sheet and rigidized sheet used in construction (*continued*)

Material	Advantages	Disadvantages	Major uses
Monel metal (*see* Monel)	Noncorroding; easily formed; will not stain adjoining materials; durable; easily soldered; acid resistant	Has to be isolated from noncompatible materials to stop galvanic action	Roofing, flashing, sinks, decorative interior metalwork, etc.
Paper (*see* Paper Pulp Sheets and Rigidized Sheets; *see also* Fibers)	Easily applied; vapor barrier often included; insulating and acoustic values; flameproof; water resistant	Recommended that joints be exposed; not fire resistant	Sheathing, interior wall and ceiling finishing materials, acoustic treatments, etc.
Plaster (*see* Plaster Integrants and Sheets)	Easily installed; joints can be concealed; easily painted; available with prefinished surface on one side; fire resistant	Difficult to obtain smooth, even surface at joints and nail holes; no acoustic value; very limited insulating value	Wall and ceiling finishing material, sheathing, lathing for plaster, etc.
Plastic Laminates (*see* Plastics Laminates)	Very hard; durable; resistant surface; wide selection of colors and patterns; easily applied	Difficult to treat edges; joints visible in plain colors; limited to flat surfaces (can be formed to limited curved shapes in shops especially equipped to do this work)	All types of countertops, surfacing material for doors, walls, etc.
Plastics (*see* Plastics: Sheets, Films, and Foam)	Obtainable in transparent, translucent, or opaque forms; wide selection of colors	Distorts under relatively low temperatures; relatively high expansion; treatment of joints is difficult	Skylights, wall tile, fixed illuminating panels in exterior walls, hung luminous ceilings, etc.
Stainless steel (*see* Stainless Steel Sheet, Strip, and Plate)	Noncorroding; will not stain adjoining materials; durable; relatively easily soldered; acid resistant	Requires attachment with and to compatible materials so galvanic action will not occur; very rigid but fairly easily formed	Doors, windows, frames, trim, gutters and leaders, flashing, sinks, wall tile, etc.
Steel (*see* Steel, Galvanized; Steel Sheet, Strip, and Plate)	Strong; fire resistant; easily formed, soldered, and joined; corrosion resistant if galvanized	Will corrode unless painted on interior; rust-inhibiting paint necessary on exterior	Doors, window frames, door bucks, railings, gratings, expanded metal mesh, interior partitions and walls in offices
Steel, corrugated (*see* Steel, Galvanized; Steel, Corrugated)	Strong; easily installed, joined, and soldered; will not stain adjoining materials if galvanizing is protected	Will corrode as zinc is sacrificial metal; thus protective galvanizing will be penetrated unless it is painted and periodically repainted	Roofing, siding, and partitions for industrial, farming, and temporary buildings
Terneplate (*see* Terneplate)	Durable; easily joined and soldered; fire resistant	Must be painted to prevent corrosion; requires attaching devices that are compatible with lead and steel to stop galvanic action	Roofing, flashing, kalamine work
Wood (*see* Wood Integrants) Some plywoods are thinner than $\frac{1}{4}$ in. (6.35 mm)	Strong; easily stained and easily applied and installed; large selection of wood species and finishes and design effects	Not fire resistant; can be obtained flameproof; limited textures	Applied wall and ceiling surface finishing for interiors, furniture
Zinc (*see* Zinc Sheet and Strip; Zinc Sheet, Corrugated)	Noncorroding; will not stain adjoining materials; easily installed, joined, and soldered; easily painted	Requires attachments that are compatible to zinc to stop galvanic action	Roofing, siding, flashing

SHELLAC

Shellac is made from an insect that lives on various trees of southern Asia. When pure, shellac varies from pale orange to lemon yellow in color, and commercial shellac has a high content of common resin. Shellac is composed of polyhydric acids which, with loss of water, condense to form esters, thus forming polyester resins in the final coating. Shellac dissolved in alcohol with resin is used as a clear coating on hardwood floors. Shellac is also used in adhesives, varnishes and floor waxes, and some molding plastics.

SHINGLES

Shingles may be defined as small, thin pieces of material, usually thinner at one end than the other. The thicker end is called the butt. They are made of various materials and are used to cover roofs and exterior walls by applying so that one shingle overlaps the other. The size or exposure of shingles for roofs is controlled by the pitch of the roof; the size of the shingle when used for vertical exterior walls is controlled by the width between the horizontal lines forming a design and is important from an esthetic viewpoint only. *See* Roofing; Siding; specific material headings.

SIDING

Siding is defined in this book as any exterior wall covering that is installed either vertically or horizontally and fastened to the structure of the wall. The term "siding" here does not include materials that are prefabricated panel units, curtain wall components, or sandwich-type panels. The only exception is plywood, included under Wood Integrants. Table *S4* lists types of sidings and their constituent materials.

Table S4 Typical materials used for siding

Material	Advantages	Disadvantages	Major uses
Aluminum, corrugated (*see* Aluminum, Corrugated)	Does not corrode; will not stain adjoining materials; easily installed; fire resistant; weathertight	Has to be isolated from noncompatible metals to stop galvanic action; all fastening devices must be aluminum or compatible metals; no color selection	Commercial, industrial, farm, and similar buildings
Aluminum, corrugated plastic-coated	Does not corrode; will not stain adjoining materials; easily installed; wide color and texture selection; fire resistant; weathertight	All fastening devices must be aluminum or compatible metals, and prepainted to match corrugated color	Commercial, industrial, farm, and similar types of building
Aluminum siding, flat (*see* Aluminum Sheet and Strip)	Does not corrode; wide variety of colors in porcelain enameled finish; easily installed; does not stain adjoining materials; fire resistant; weathertight	All fastening devices must be aluminum or compatible metals; must be isolated from noncompatible metals to stop galvanic action	Residential buildings
Aluminum shingles and siding, flat, plastic-coated	Does not corrode; wide variety of colors and textures; easily installed; does not stain adjoining materials; fire resistant; weathertight	All fastening devices must be aluminum or compatible metals, and prepainted to match corrugated color	Residential buildings
Asbestos-cement sheet (*see* Asbestos-Cement Sheet and Board; Asbestos, Corrugated)	Easily installed; does not stain adjoining materials; fire resistant; durable	Brittle; limited color selection; vertical joints exposed and require flashing behind to make nonlapping types weathertight	Industrial, residential, farm, and other, similar buildings
Asbestos-cement sheet and corrugated sheet, plastic-coated	Easily installed; does not stain adjoining materials; wide range of colors; fire resistant; weathertight	Brittle; must be predrilled; vertical joints require special treatment to make weathertight corrugated laps	Industrial, residential, and farm buildings

Table S4 **Typical materials used for siding** ***(continued)***

Material	Advantages	Disadvantages	Major uses
Asbestos-cement shingles (*see* Asbestos-Cement Shingles for Roofs and Siding)	Easily installed; does not stain adjoining materials; fire resistant; weathertight; durable	Brittle; limited color selection	Residential buildings
Asbestos-cement shingles, plastic-coated	Easily installed; does not stain adjoining materials; wide range of colors; fire resistant; weathertight	Brittle; must be predrilled; all fastening devices must be prepainted to match shingle color	Residential buildings
Asphalt siding (*see* Asphalt Roofing and Siding)	Easily installed; does not stain adjoining materials; weathertight; large selection of colors and textures; durable	Only certain types have fire ratings; maximum life up to 20 years	Residential, farm storage, and similar buildings
Brick, nailable (*see* Brick, Burned)	Easily installed; does not stain adjoining materials; fire resistant; durable; weathertight	Heavy; limited color selection	Residential buildings
Concrete, artificial stone (*see* Concrete, Artificial Stone)	Easily installed using molds; fire resistant; large selection of colors and textures; weathertight	Heavy; unless carefully done has artificial character; can stain adjoining materials during application	Residential buildings
Fiber board (*see* Paper and Paper Pulp Products; Paper Pulp Sheets and Rigidized Sheets)	Easily installed; will not stain adjoining materials; thick types have insulation value	Joints have to be flashed or protected; requires painting; thick types are relatively soft	Low-cost residential construction, storage and temporary buildings
Glass, structural (*see* Glass, Structural) *Note:* This is not a siding in the true sense of the word.	Durable; wide range of colors; will not stain adjoining materials; weathertight	Heavy; requires special metal supporting and attaching devices	Stores, commercial and similar buildings
Plastic siding	Easily installed; does not stain adjoining materials; wide range of colors; weathertight	Fire ratings have to be checked; all fastening devices must be prepainted to match siding color	Residential buildings
Stainless steel siding (*see* Stainless Steel Sheet, Strip, and Plate; Stainless Steel, Corrugated)	Durable; easily installed; does not stain adjoining materials; fire resistant; weathertight	No color selection; attaching devices have to be stainless steel or compatible metal; problem of galvanic action has to be considered in connection with other metals	Stores, commercial and similar buildings
Stainless steel siding, plastic-coated	Easily installed; does not stain adjoining materials; fire resistant; wide range of colors; weathertight	All fastening devices must be stainless steel or compatible metal; problem of galvanic action has to be considered with other metals	Residential buildings, stores, commercial and similar types of building
Steel, corrugated (*see* Steel, Galvanized; Steel, Corrugated)	Easily installed; fire resistant; weathertight	Has to be protected by painting to stop corrosion; will stain adjoining materials if paint is not kept in good condition	Industrial, farm, storage, temporary, and similar buildings

Table S4 Typical materials used for siding (*continued*)

Material	Advantages	Disadvantages	Major uses
Steel, corrugated and coated (*see* Steel, Corrugated; Steel, Galvanized)	Easily installed; fire resistant; does not stain adjoining materials; weathertight	Limited selection of colors	Industrial and similar buildings
Steel shingles and siding plastic-coated	Easily installed; does not stain adjoining materials; wide range of colors; fire resistant; weathertight	All fastening devices should not break coating during installation, and should be prepainted to match shingle or siding color	Residential buildings, stores, commercial and similar types of buildings
Wood integrants (plywood) (*see* Wood Integrants)	Easily installed; large sizes; easily painted or stained	Joints have to be flashed or protected to be weathertight; limited to wood colors; must be stained or painted periodically	Residential buildings
Wood integrants (plywood), plastic-coated	Easily installed; wide selection of colors; weathertight	Joints have to be flashed or protected to be weathertight; all fastening devices must be prepainted to match plywood color	Residential buildings
Wood shingles (*see* Wood Shingles)	Easily installed; weathertight; easily stained or painted; large selection of color with prestained type	Must be stained or painted periodically; if left natural will bleach to gray	Residential buildings
Wood shingles, siding, and paneling, plastic-coated	Easily installed; wide selection of colors; weathertight	Fasteners and nails must be prepainted to match shingle or siding color	Residential buildings
Wood siding (*see* Wood Siding and Paneling)	Easily installed; weathertight; easily stained or painted; if left unfinished, will bleach to a gray color	Limited to wood colors which will fade to gray; if natural or other wood color are desired, periodic staining is required	Residential buildings
Zinc, corrugated (*see* Zinc Sheet, Corrugated)	Easily installed; will not stain adjoining materials; does not corrode	Attaching devices must be compatible metals to stop corrosion; must be isolated from noncompatible metals to stop galvanic action; no color selection	Industrial buildings

SILICON

Physical and Chemical Properties

Symbol: Si
Atomic number: 14
Specific gravity: 2.33 (25°C, 77°F)
Melting point: 1410°C, 2570°F
Boiling point: 2355°C, 4271°F

Silicon is next to oxygen in abundance in the earth's crust. Practically all the sand, clay, and rocks of the world are composed of silica (silicon dioxide) or silicates or mixtures of both. Many source books still describe silicon as a nonmetallic element, similar to carbon in many respects, and list two forms: a dark brown, amorphous powder, and a hard lustrous crystal that is a semiconductor and is identical in structure to the diamond. Recent research indicates that the crystalline form represents pure silicon and that silicon is a metal (or semimetal) closely related to germanium, tin, and lead. However, it is one of the least electropositive (or metallic) of the metals and completely lacks ductility.

Commercial Forms. Silicon is available in lump, briquet, powder, and crystalline forms.

Types and Uses

Elemental Silicon. Elemental silicon, indicated by asterisks in Table S5, has a limited yet important number of uses: (1) those that utilize its electrical properties, (2) metallurgical applications, and (3) combinations with carbon and ceramic materials to form corrosion- and heat-resistant materials. Recently it has found important commercial use in the preparation of silicones, a group of compounds valuable in construction for their effects on many commonly used building materials (*see* Silicones).

Silica and Silicates. Silica and silicates are among the fundamental materials used by man. One or the other or both in various combinations are major constituents of sand, quartz, sandstone, diatomite, asbestos, talc, mica, clay, sand, and granite. Silica and silicates are the foundation materials for the widely various ceramic industries, using the term in its broadest sense. Silicate slags, a by-product of the metallurgical industries, are a raw material for portland cement. Finely divided silicas are commonly used for producing flat varnishes.

Quartz consists of silica (silicon dioxide) and is the most common and widely distributed mineral in the earth's crust. It occurs in grains of sand, as a constituent of many rocks including sandstone and granite, in crystalline masses, and as large crystals.

Table S5 Uses of silicon

Construction Uses

Basis of	Abrasives	Main ingredient
	Adhesives	Main ingredient
	Protective coatings*	Iron, steel
	Glass	Main ingredient
	Quartz	Main ingredient
	Refractories	Main ingredient
	Rubber	Main ingredient of silicone rubber for high temperatures
	Sand and gravel	Main ingredient
Component of	Aluminum*	Hardening; corrosion and wear resistance; white color
	Asbestos	Ingredient
	Brasses*	Hardening; corrosion resistance
	Bronzes*	Hardening; corrosion resistance
	Cast iron*	Acid and corrosion resistance
	Cement	Ingredient
	Ceramics	Ingredient
	Clays	Ingredient
	Copper*	High strength; good electrical conductivity; acid, wear, and corrosion resistance; deoxidizer
	Glass	Coating for textiles
	Glazes and porcelain enamels	Ingredient

Construction Uses (Continued)

Component of	Iron*	Heat, corrosion, and wear resistance; deoxidizer
	Magnesium	Increasing strength
	Paper	Coatings; fireproofing; water repellent
	Plastics	Filler in floorings
	Coatings and protective coatings	Water repellents
	Rubber	Stabilizer; filler; reinforcing
	Steel*	Corrosion and heat resistance; deoxidizer; increasing electrical resistance
	Stone	Ingredient
	Textiles	Coatings; mordant; fireproofing; water repellent
	Varnish	Flatting agent
Allied Construction Uses	Air conditioning	Dehumidifying; dehydrating
	Electrical devices*	Insulation
	Lamps	Fluorescence

Nonconstruction Uses

Photocells,* solar batteries,* special dentist's mirrors

Possible Future Uses

Silicone is abundant, cheap, light, and resistant to heat and corrosion but has no ductility; ductile silicon is not now feasible, but when this is discovered, a whole new concept of structural materials will be possible

Quartz is widely used as an abrasive, as a refractory material, and as a flux in smelting ores. Fused quartz transmits ultraviolet light and is used for this purpose. The colorless pure variety of quartz exhibits peculiar optical and electrical properties which are utilized in radio, television, and electronics and for other specialized optical purposes.

History and Manufacture

Fused silica, or sand, was used for glass making as early as 2500 B.C. and was considered an elemental substance by the early chemists. Lavoisier, in 1787, was the first to suspect, and Davy later concluded, that silica is actually a compound. In 1823 Berzelius prepared an impure silicon and established it as an element. Crystalline silicon was first prepared by electrolysis in 1854 by Deville; later Potter developed a process that is the forerunner of the large-scale methods used today.

SILICONES

Silicones are a group of synthetic, semiorganic polymers (polysiloxanes) composed of silicon, carbon, hydrogen, and oxygen. They are produced in the basic forms of fluids, resins, and elastomers. These are compounded into greases, rubbers, and protective coatings characterized by a unique combination of properties: they are heat stable, water repellent, serviceable over a wide temperature span, and resistant to oxidation and weathering; they also retain good physical and dielectric properties in severe operating conditions. They are used for adhesives, protective coatings, surface treatments for glass and ceramics, textile finishes, electrical insulating materials, and extreme-temperature rubber.

Water-Repellent Action. The water-repellent action of silicones is of special interest in the construction field. Very thin films of silicone polymers are formed by reaction with the absorbed water on cellulose or glass; such a film is strongly bound and can be removed only by chemical action or by severe abrasion and may be readily cleaned by alcohol or by solutions of wetting agents. *See* Plastics.

SILVER

Physical and Chemical Properties

Symbol: Ag
Atomic number: 47
Specific gravity: 10.5 (20°C, 68°F)
Melting point: 961.93°C, 1763.47°F
Boiling point: 2212°C, 4013.6°F

Table S6 Grades of commercial silver

Fine	Sterling		Coinage	
Ag	Ag	Cu and others	Ag	Cu
99.9%	92.5%	7.5%	90%	10%

Silver is the whitest of all metals, soft, ductile and malleable. It has excellent resistance to oxidation up to its melting point but tarnishes readily in the presence of sulfur and its compounds. Silver is superior to all other metals in thermal and electrical conductivity. Taking 100 as the relative value, its thermal conductivity is 100 as opposed to 94 for copper, and its electrical conductivity rates 100 compared to 97.6 for copper.

Workability. Silver is next to gold in workability. It can be rolled, extruded, spun, coined, bent, drawn, hammered, brazed, and arc welded.

Commercial Forms. Silver is available in ingot, bar, rod, tube, sheet, foil, wire, shot, and powder form and as silver clad on base metals. For grades of commercial silver, see Table *S6*.

Types and Uses

Many applications of silver depend on its high thermal or electrical conductivity and corrosion resistance. Theoretically, it could to a great extent replace copper in electrical equipment. A major use of silver is in mirrors. Table *S7* shows the many uses of silver compounds and metallic silver, the latter indicated by asterisks.

History and Manufacture

Silver, found native and in ores, was one of the first metals used. As early as 3000 B.C. man had learned to separate silver from lead. Its use was first for jewelry and coinage, later for mirrors and tableware. Today silver is produced from the ore argentite and as a by-product in the production of lead, zinc, and copper and also from the refining of many rare metals (*see* Lead, Figure *L7*).

Today silver is extensively used in solder and brazing alloys, electrical contacts, printed circuits, cement for

Table S7 Uses of silver

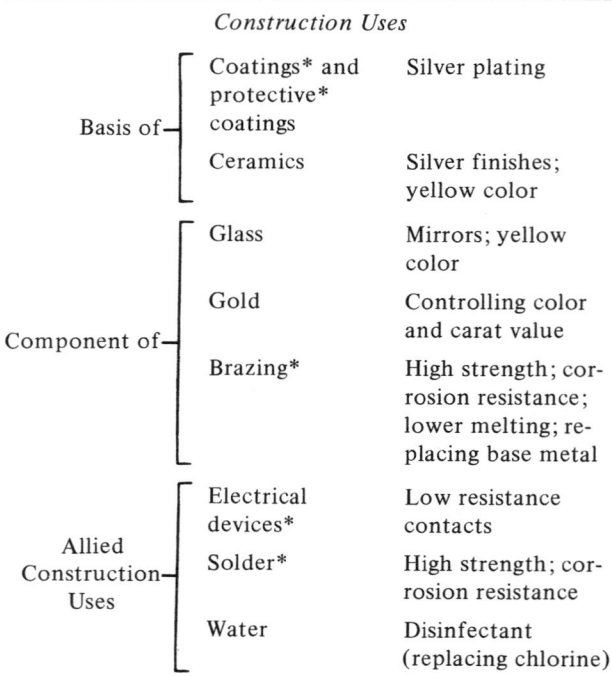

	Construction Uses	
Basis of	Coatings* and protective* coatings	Silver plating
	Ceramics	Silver finishes; yellow color
Component of	Glass	Mirrors; yellow color
	Gold	Controlling color and carat value
	Brazing*	High strength; corrosion resistance; lower melting; replacing base metal
Allied Construction Uses	Electrical devices*	Low resistance contacts
	Solder*	High strength; corrosion resistance
	Water	Disinfectant (replacing chlorine)
	Nonconstruction Uses	
Art objects,* coinage,* electronics,* laboratory equipment,* photography, silverware*		

glass, and seeding clouds for rain; a tremendous quantity is used in photography. Because of this enormous demand for silver, the coinage of the United States is now mostly copper with a silver-copper cladded front and back surface.

In the Western Hemisphere the principal producers of silver are Canada, Mexico, Peru, and the United States.

SIZING

"Sizing" refers to the material and/or the process that fills and seals up the pores in surfaces of materials before the application of paint, enamels, and other surface finishes. *See* Concrete; Concrete Block; Paper and Paper Pulp Products; Plaster and Plastering; Textiles; Wood.

SKYLIGHTS

Skylights may be defined simply as sources of natural light, or windows, in a ceiling. Many methods of installation are possible and commonly used. Skylights may be an integral part of the structural system for the roof. They may be installed in sidewalks to light basement areas. A skylight may consist of a large glass area in a pitched roof (studio lighting). In any case, five important requirements must be met:

Table S8 Transparent or translucent materials used for skylights

Material	Advantages	Disadvantages	Major uses
Glass block (*see* Glass Block)	Has good insulation value; condensation drainage is not necessary unless inside area has high moisture content; easily made weathertight	Requires special systems for support of small units; does not meet rigid fire code requirements; will shatter and break unless solid type of unit is used; light transmission and dispersion controlled by type of glass block selected	Sidewalk-type skylights and skylights for interior areas where the elimination of condensation drainage is important, and insulation is important
Glass, corrugated[a] (*see* Glass, Corrugated; Glass, Wired)	Because of corrugations it can span up to approximately 7 ft (2.134 m) without intermediate supports; generally meets fire code requirements; will withstand shock without shattering and falling; good light dispersion and transmission	Requires condensation drainage and special corrugated closures to make it weathertight	Utility-type skylights

Table S8 Transparent or translucent materials used for skylights (*continued*)

Material	Advantages	Disadvantages	Major uses
Glass, wired[a] (*see* Glass, Wired)	Generally meets fire code requirements; can withstand shock without shattering and falling	Limited span; generally requires intermediate supports; requires condensation drainage; light dispersion depends on type of glass selected; requires glazing compounds to make it weathertight	Utility-type skylights
Plastic, sandwich type[a] (*see* Plastics; Plastics, Sheet and Ridigized Sheet)	Can withstand shock with out shattering and falling; good light transmission and dispersion; easily made weathertight; good insulation value; does not require condensation drainage	Does not meet rigid fire code requirements; at present limited in size to 4 X 20 ft (1.219 X 6.096 m)	Skylights for interior areas where elimination of condensation drainage is important, and insulation is important
Plastic sheet[a] (*see* Plastics, Sheet and Ridigized Sheet)	Can withstand shock without shattering and falling; good light transmission and dispersion; easily made weathertight; wide selection of colors	Limited in size as plastic sheet has to be deformed to give strength; does not meet rigid fire code requirements; requires condensation drainage	Rectangular, square, or circular skylights in flat roofs
Plastic sheet, corrugated[a] (*see* Plastics, Sheet and Rigidized Sheet)	Because of corrugations it can span up to approximately 7 ft (2.134 m) without intermediate supports; can withstand shock without shattering and falling; good light dispersion and transmission; wide selection of colors	Requires condensation drainage; special corrugated closures needed to make it weathertight; does not meet rigid fire code requirements	Utility-type skylights

[a] Skylights using these materials are available as completely prefabricated units including flashing, supports, frame, etc.

1. The transparent or translucent material must be such that it cannot break, shatter, or fall.
2. The outside surface must be protected from the hazard of falling objects, or the material must be able to resist the impact of falling objects.
3. The structure must be weathertight.
4. Condensation drippings must be caught and drained off, or else the material must be such that condensation does not occur.
5. Fire protection requirements must be met.

Most skylights are available as prefabricated units that meet all the requirements just mentioned with the possible exception of fire code requirements.

Table *S8* lists the transparent and translucent materials commonly used for skylights.

SLAG

Physical and Chemical Properties

Slag is a nonmetallic product that separates in the smelting of metals. It is formed from the earthy materials in the ore and from the flux. The term usually refers to iron blast-furnace slag.

Composition. The chemical composition of American blast-furnace slag falls within the range of silica 30 to 42%, alumina 10 to 14%, lime 36 to 45%, sulfur 1 to 3%, magnesia about 2%, and some ferric and manganese oxide. The sulfur present is mainly in the form of sulfides which do not corrode metals to any appreciable extent. Slag is alkaline, and this alkalinity generally prevents corrosion.

Physical Properties. The physical properties of slag depend on the method used in treating or cooling the molten slag as it solidifies. The weight of unexpanded slag is 70 to 85 lb/ft^3 (1121.4 to 1361.7 kg/m^3); that of expanded slag is about 25 to 50 lb/ft^3 (400.5 to 810 kg/m^3) for coarse aggregate, and 45 to 70 lb/ft^3 (720.9 to 1121.4 kg/m^3) for fine aggregate.

In general, slag is strong and durable and one of the lightest aggregates for concrete construction. It has a rough surface, a porous structure with sizable internal voids, and, when crushed, is angular and roughly cubical in shape, providing good bonding.

Types and Uses

The three types of slag are ordinary air-cooled slag, expanded slag, and granulated slag. For most uses, slag is screened to size after being crushed.

Slag is the raw material for the manufacture of portland cement and mineral wool. Crushed slag is an all-purpose construction aggregate for (1) portland cement concrete, particularly lightweight structural concrete, and bituminous mixtures; (2) fill, base, and subgrade in the construction of roads, driveways, parking areas, and airfields; (3) concrete block and other masonry units; (4) concrete pipe; and (5) built-up roofing, as a granular cover material. Two comparatively recent uses for granulated slag are (1) in landscaping as a liming material and soil conditioner, and (2) in plaster consisting of a mixture of such slag, expanded vermiculite, and gypsum.

History and Manufacture

Slag processing grew from a minor industry in about 1900 to become a major supplier of raw ingredients and aggregate for the portland cement and highway construction industries.

To produce ordinary air-cooled slag, the molten slag is poured from the furnace into ladles that transport it to a slag bank or pit, where it solidifies under atmospheric conditions. After it cools, it may be sprayed with water and crushed to 24 sizes to meet the grading requirements of the various State Highway Departments.

Expanded slag is made more porous through the application of a limited quantity of water to molten slag or by any of several foaming methods utilizing streams of atomized air, steam, or water. Granulated slag is formed when molten slag is suddenly chilled by immersion in water.

See also Iron for description of blast furnace by-products.

SODIUM

Physical and Chemical Properties

Symbol: Na
Atomic number: 11 (20°C, 68°F)
Specific gravity: 0.9712
Melting point: 97.81°C, 208.06°F
Boiling point: 892°C, 1637.6°F

Sodium is a light, soft, ductile, malleable, silvery-white metal characterized by its intense chemical activity. It has high thermal and electrical conductivity (fourth after silver, copper, and aluminum). Sodium belongs to the alkali metals; like potassium, it is a very strong reducing agent. It combines violently with the oxygen in water and burns in air. It is never found free in nature but always occurs in compound form, frequently combined with potassium.

Commercial Forms. Sodium is available in brick, amalgam, and coated powder form. It is always kept either in a vacuum or immersed in an inert liquid, such as kerosene, which does not contain either free oxygen or water.

Types and Uses

Metallic sodium has very limited uses (indicated by asterisks in Table *S9*). Its compounds are so widely used that it is impossible even to begin listing them. The table gives only those applications in construction materials in which sodium compounds (usually carbonates, sulfates, or phosphates) are important components. (*See also* Salt.)

Natural sodium carbonates and sulfates are widely distributed and are commercially mined, as is salt.

History and Manufacture

Sir Humphrey Davy isolated sodium by electrolysis in 1807. Its symbol, Na, is derived from the Latin word *natrium*, meaning "metal of soda."

Sodium metal 99.9% pure is obtained by electrolysis of fused sodium hydroxide (caustic soda). Because it reacts with air and water, sodium metal is kept submerged in kerosene. Sodium is used in the refining process for the production of aluminum metal (*see* Aluminum).

Table S9 Uses of sodium

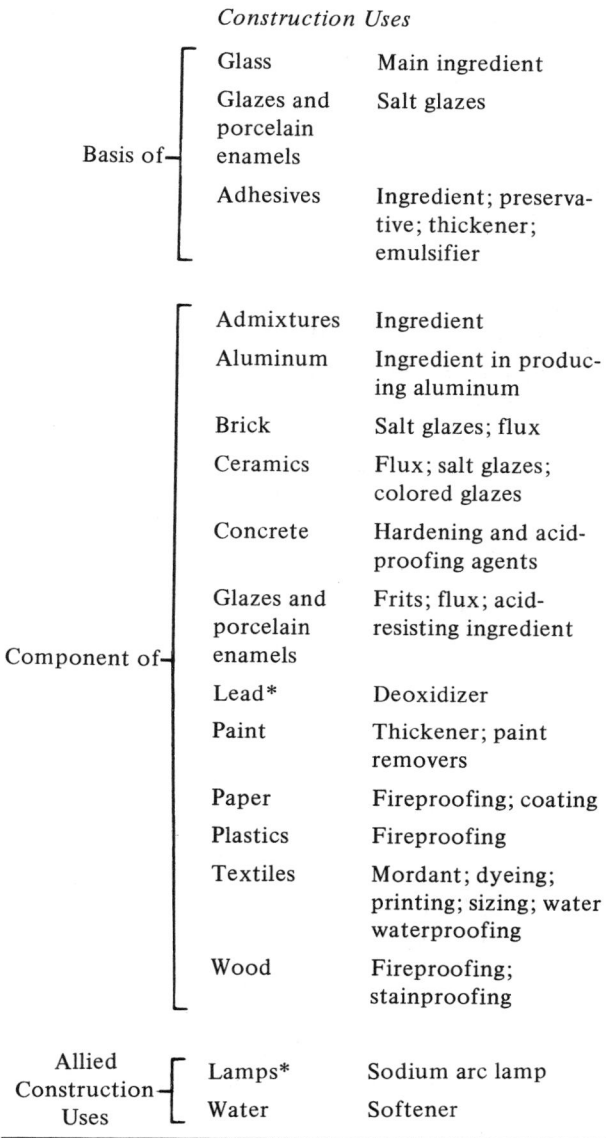

	Construction	*Uses*
Basis of	Glass	Main ingredient
	Glazes and porcelain enamels	Salt glazes
	Adhesives	Ingredient; preservative; thickener; emulsifier
Component of	Admixtures	Ingredient
	Aluminum	Ingredient in producing aluminum
	Brick	Salt glazes; flux
	Ceramics	Flux; salt glazes; colored glazes
	Concrete	Hardening and acid-proofing agents
	Glazes and porcelain enamels	Frits; flux; acid-resisting ingredient
	Lead*	Deoxidizer
	Paint	Thickener; paint removers
	Paper	Fireproofing; coating
	Plastics	Fireproofing
	Textiles	Mordant; dyeing; printing; sizing; water waterproofing
	Wood	Fireproofing; stainproofing
Allied Construction Uses	Lamps*	Sodium arc lamp
	Water	Softener

SOLDERING

Physical and Chemical Properties

Soldering is a method used to join metals, to make electrical connections, and to seal joints hermetically by filling them with another, lower melting metal or alloy called the solder. Since the temperatures used in soldering are comparatively low, there is no alloying action between the solder and the metals being joined, which are usually stronger than the solder itself.

Strength of Soldered Joints. Soldered joints have very little tensile, shear, or impact strength; therefore this method should not be used where a strong joint is required. Instead, other methods such as riveting, spot welding, or interlocking of seams should be utilized.

Composition. Solders are mostly alloys of tin and lead in various proportions with small percentages of other elements added to give special characteristics. The commonly used ones are described in detail under "Types and Uses."

Temperature Ranges. The melting points of solders generally used in construction range from 374 to 594°F (190° to 312°C) as shown in Figure *S3*, although there are some high-temperature solders with melting points as high as 744°F (395.5°C).

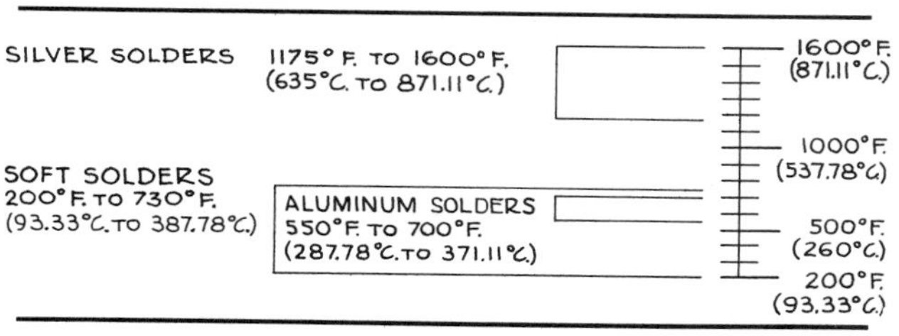

Figure S3 Temperature ranges for soldering.

Commercial Forms. Solders are available in ingot, slab, bar, wire, strip, tape, powder, paste, and sheet form and in pie-cut shape.

Types and Uses

Methods of Soldering. The usual methods of soldering are as follows:

1. *Soldering iron.* This is the oldest method. In it the iron piece is preheated or self-heated and applied to the joint along with the solder and the flux; the heat from the iron forms the soldered joint.
2. *Torch.* The parts to be soldered are heated by the torch flame, and then the solder and flux are applied. This method is limited to metals that can be heated without altering their characteristics.
3. *Sweat method.* Heating the metals to be joined causes the solder to run into the joint. This is the method used for joining copper tubing and fittings.

Other methods of soldering used in industry include the bath or dip-spray method, the induction method, and the electrical resistance method. An ultrasonic soldering method developed in England and used commercially there is now finding acceptance in the United States.

Types of Solder. Solders can be divided into four major groups: tin-lead type, tin-lead-antimony type, silver-lead type, and miscellaneous types with specialized uses. The composition, characteristics, and uses of each group are summarized in Tables *S10* to *S13* inclusive. Tin-lead, tin-lead-antimony, and silver-lead solders can contain only limited amounts of other elements, considered impurities, as shown in Table *S14*.

Tin-lead solder of the 50% tin-50% lead variety is the most commonly used general-purpose solder. Some tin-lead solders are used for coating the metals before soldering. This is known as pretinning. The first and the last four tin-lead solders listed in Table *S10* can be used for this purpose.

Tin-lead-antimony solders should not be used with galvanized iron. Only the last one listed in Table *S11* can be used for pretinning.

Silver-lead solders have low melting points and form comparatively strong and highly corrosion-resistant joints (*see* Table *S12*).

Fluxes. Fluxes for soldering are generally of three types: corrosive, neutral, and noncorrosive. They serve to remove oxides from the metals to be soldered and help in melting the solder.

Corrosive fluxes also are known as acid-type and salt-type fluxes and include chlorides of zinc, ammonium, calcium, magnesium, aluminum, and other metals. The chloride fluxes are highly effective in removing oxides from almost any kind of metal. Care should always be taken with the corrosive fluxes. The residue must be

Table S10 Tin-lead solders

Composition (percent)			Specific gravity	Melting range		Major uses
Sn	Pb	Sb (max.)		Solid °F (°C)	Liquid °F (°C)	
70	30	0.12	8.32	361 (182.8)	378 (192.2)	Coating metals
60	40	0.12	8.65	361 (182.8)	374 (190.0)	General purpose solder
50	50	0.12	8.85	361 (182.8)	421 (216.1)	Most common general purpose solder
45	55	0.12	8.97	361 (182.8)	441 (227.2)	Roofing seams
35	65	0.25	9.50	361 (182.8)	477 (230.5)	General purpose solder
30	70	0.25	9.70	361 (182.8)	491 (255.0)	Torch and machine soldering
25	75	0.25	10.00	361 (182.8)	511 (266.1)	Torch and machine soldering
20	80	0.50	10.20	361 (182.8)	531 (277.2)	Coating and joining metals
15	85	0.50	10.50	440 (232.9)	550 (287.8)	Coating and joining metals
10	90	0.50	10.80	514 (267.8)	570 (298.9)	Coating and joining metals
4.5–5.5	95.5–94.5	0.12	11.30	518 (270.0)	594 (312.2)	Coating and joining metals

Table S11 Tin-lead-antimony solders

Composition (percent)				Melting range		
				Solid	Liquid	
Sn	Pb	Sb (max.)	Specific gravity	°F (°C)	°F (°C)	Major uses
40	58.0	2.0	9.25	365 (185.0)	448 (231.1)	General purpose solder
35	63.2	1.8	9.44	365 (185.0)	470 (243.3)	General purpose solder
30	68.4	1.6	9.65	364 (184.4)	482 (250.0)	Torch and machine soldering
25	73.7	1.3	9.96	364 (184.4)	504 (256.7)	Torch and machine soldering
20	79.0	1.0	10.17	363 (183.9)	517 (269.4)	Coating and machine soldering

Table S12 Silver-lead solders

Composition (percent)					Melting range		
					Solid	Liquid	
Sn	Pb	Ag	Sb max.	Specific gravity	°F (°C)	°F (°C)	Major uses
0	97.5	2.5	0.4	11.35	579 (303)	579 (303)	Torch-heating type of soldering for copper, brass, and similar metals; not corrosion resistant in humid conditions
0.75 to 1.25	97.75 to 97.25	1.5	0.4	11.28	588 (309)	588 (309)	Torch-heating type of soldering for copper, brass, and similar metals

quickly removed, as it not only is corrosive to the metal being jointed but also is electrically conductive as a rule and therefore cannot be used for most electrical work.

Neutral fluxes, although not chemically neutral, are mild in type and are used for easily soldered metals such as copper, brass, lead, and tin plate. Stearic acid is a typical neutral solder. In many cases the residue will not affect the soldered metals and therefore does not have to be removed.

Table S13 Miscellaneous types of solder

Composition (percent)				Melting range		
				Solid	Liquid	
Pb	Ag	Zn	Cd	°F (°C)	°F (°C)	Major uses
85.4		0.3	14.3	470 (243)	477 (247)	Copper roofing
	5		95.0	639 (337)	744 (395)	High temperature
		50.0	50.0	508 (264)	619 (326)	High temperature

Table S14 Maximum permissible percentage of other elements (impurities) in tin-lead, tin-lead-antimony, and silver-lead solders

Bi	Cu	Fe	Al	Zn	Total of other elements
0.25	0.08	0.02	0.005	0.005	0.08

Noncorrosive fluxes leave residues that are noncorrosive and nonconductive and therefore need not be removed. Rosin is the principal flux of this type. Noncorrosive fluxes are weak in their fluxing action, and their use is limited to the easily soldered base metals.

Application

One should always check with the manufacturers of the metals to be soldered for the correct type of solder, flux, and method of soldering.

When pretinning is called for, the recommended thickness of the pretinning will differ with the metal to be soldered. For brass the thickness is 0.0005 in. (0.0127 mm), for copper 0.003 in. (0.0762 mm) and for steel 0.0002 in. (0.0508 mm); for other metals the manufacturers should be consulted.

Stainless steel requires a zinc-ammonium type of flux with some free hydrochloric acid, which will require almost 1 minute to function properly.

Aluminum and aluminum alloys can be soldered under controlled conditions. These soldered areas should be protected against moisture, which will start galvanic action. Glass, ceramics and some plastics, when coated with a 0.0005-in. (0.0127-mm) film of copper or silver, can also be soldered.

Condensed Checklist

1. Type and size of joint: When preparing joints for soldering, the area to be filled with solder should

range in width from 0.002 to 0.004 in. (0.0508 to 0.1016 mm). Below 0.002 in. (0.0508 mm) the opening is too small for the solder to enter, and above 0.004 in. (0.1016 mm) the force of gravity will begin to overcome the capillary action on which the joint depends, causing voids to appear.
2. Type of solder: This is particularly important when the necessary temperatures of soldering may affect the metals and their protective oxide films.
3. Flux: The type of flux to be used should be checked, and, if corrosive, cleaning methods should be specified.
4. Method of soldering: This should be suitable for the metals to be joined and for the strains to which the joint will be subjected.
5. Galvanic action: The metal to be joined and the solder should be compatible. Pretinning may help to prevent galvanic action, as also will sacrificial strips of metal acting as cathodes.

Conditions Favorable to the Use of Soldering

1. Where a watertight joint is required.
2. For electrical connections.

Conditions Unfavorable to the Use of Soldering

1. Where the joint will be subject to tension, shear, or impact forces.
2. Where galvanic action between the solder and the joined metals may occur.

History and Manufacture

The Egyptians are known to have made a lead-bearing solder in ancient times, and the lead linings of the hanging gardens of Babylon were soldered. Solders of tin-lead composition were used by the Romans, the two known types (both of which are still used) being ***argentarum***, 50% tin and 50% lead, and ***tertiarium***, 1 part tin and 2 parts lead.

Solders are made by the simple process of melting the ingredients, mixing, and allowing to cool. The oldest method of soldering consists of preheating an iron bar and applying it to the solder and the metal surfaces to be joined. The heat melts the solder onto the metal surfaces, where it hardens and forms a joint.

SPRAYING

Spraying is the process of coating metal with paint, another metal, or any other material by the use of air or hydraulic pressure. *See* Metals: Sprayed Metal Coatings; Paint; Painting.

Spraying is also the process used for applying plastic-type roofing, flooring, and surfacing.

STAINLESS STEEL

Physical and Chemical Properties

The stainless steels generally used in architecture are highly alloyed steels that contain more than 10% chromium. They are characterized by their resistance to heat, oxidation, and corrosion.

The stainless steels are hardened by cold working. They also have low heat conductivity but high thermal expansion; both these characteristics, particularly the latter, are important when stainless steel is used in buildings.

The physical and chemical properties of stainless steels are shown in Tables *S15* and *S16*.

Table S15 Chemical compositions of stainless steels used in construction

Type of stainless steel (AIAI number)	Chemical composition (percent)						
	Fe	Cr	Ni	Mn	C	Mo	Si
201[a]	67.51–73.51	16.0–18.0	3.5–5.5	5.5–7.5	0.15 max.		1.00 max.
202[a]	63.51–70.01	17.0–19.0	4.0–6.0	7.5–10.0	0.15 max.		1.00 max.
301[b]	70.84–74.84	16.0–18.0	6.0–8.0	2.0 max.	0.15 max.		1.00 max.
302[b]	67.84–70.84	17.0–19.0	8.0–10.0	2.0 max.	0.15 max.		1.00 max.
304[b]	65.84–67.84	18.0–20.0	8.0–12.0	2.0 max.	0.15 max.		1.00 max.
316[b]	61.83–68.83	16.0–18.0	10.0–14.0	2.0 max.	0.10 max.	2.0–3.0	1.00 max.
430[b]	79.81–83.81	14.0–18.0		1.0 max.	0.12 max.		1.00 max.

[a] Contains 1.0% maximum silicon, 0.06% phosphorus and sulfur.
[b] Contains 1.0% maximum silicon, 0.04% phosphorus, and 0.03% sulfur.

Table S16 Physical and mechanical properties of stainless steels used in construction

		Type of stainless steel						
		201	202	301[a]	302[b]	304	316[b]	430[c]
Property	Form	$lbf/in.^2$ (MN/m^2)	$lbf/in.^2$ (MN/m^2)	$lbf/in.^2$ (MN/m^2)	$lbf/in.^2$ (MN/m^2)	$lbf/in.^2$ (MN/m^2)	$lbf/in.^2$ (MN/m^2)	$lbf/in.^2$ (MN/m^2)
Yield strength	Sheet, Strip	55,000 (379.23)	50,000 (344.75)	40,000 (275.80)	40,000 (275.80)	42,000 (289.59)	42,000 (289.59)	50,000 (344.75)
	Bar				35,000 (241.33)		30,000 (206.85)	
	Tubing						35,000 (241.33)	
Tensile strength[d]	Sheet, Strip	115,000 (792.93)	105,000 (723.98)	110,000 (758.45)	90,000 (620.55)	84,000 (579.18)	84,000 (579.18)	75,000 (517.12)
	Bar				85,000 (586.08)		80,000 (551.60)	
	Tubing						85,000 (586.08)	
Weight $lb/in.^3$ (kg/m^2)		0.28 (196.86)	0.28 (196.86)	0.29 (203.89)	0.29 (203.89)	0.29 (203.89)	0.29 (203.89)	0.28 (196.86)
Coefficient of inner expansion in./in. · °F (cm/cm · °C)		0.000097 (0.000016)		0.0000094 (0.000017)	0.0000094 (0.000017)	0.0000094 (0.000017)	0.0000095 (0.000017)	0.0000151 (0.000027)

[a] Strength values are increased with less decrease in ductility than in 302 when 301 is cold-worked.
[b] Strength values are increased and ductility values decreased when 302 and 316 are cold-worked.
[c] Strength values are not increased by cold working to the same degree as in 301, 302, and 316.
[d] 1 $lbf/in.^2$ = 0.006895 MN/m^2.

Added Elements. Nickel and manganese are important added elements that produce special characteristics such as strength, toughness, and ease of fabrication in stainless steels. Columbium (niobium), molybdenum, phosphorus, selenium, silicon, sulfur, titanium, and zirconium are also used to give special characteristics.

Effects of Chromium. The chromium in these alloys is thought to be the element that gives the corrosion resistance. Part of the chromium combines with the carbon and some of the iron in steel to form chromium-iron carbides, and the remainder dissolves in the iron. The best corrosion resistance is obtained when as much of the chromium as possible is dissolved in the iron and as little as possible is combined in the carbide form. A thin, stable, hard, continuous, invisible film is formed on the surface, which acts as a barrier against progressive attack by corrosive agents, as long as oxygen in some form is present.

General Classification. Stainless steels can be grouped according to chemical composition and response to heat treatment as follows: (1) ferritic steels, which are nonhardenable steels with 15 to 30% chromium and a low carbon content of 0.08 to 0.20%; (2) martensitic steels, which are hardenable by quenching and contain 10 to 18% chromium and 0.08 to 1.10% carbon; and (3) austenitic steels, which are hardenable without quenching and contain 16 to 26% chromium and 6 to 22% nickel. Types 301, 302, 303, 304, and 316 are austenitic stainless steels, and Type 430 is a ferritic stainless steel, which is somewhat less resistant to corrosion than the austenitic stainless steels. (For an explanation of ferritic, martensitic, etc., *see* Iron Alloys.) Two austenitic stainless steels, Types 201 and 202, are available for construction applications. Type 201 is an austenitic alloy that is similar to Types 301 and 302 but is stronger, harder, and characterized by more spring-back in fabrication. Types 305 and 410 are also used, but primarily for bolts,

Table S17 Chemical compositions of types 305 and 410 stainless steel

Type of stainless steel	Chemical composition (percent)					
	Fe	Cr	Ni	Mn	C	Si
305[a]	64.88–76.88	10.00–19.00	10.00–13.00	2.0 max.	0.12 max.	1.00 max.
410[b]	83.85–85.85	11.50–13.50	0.50 max.	1.0 max.	0.15 max.	1.00 max.

[a] An austenitic stainless steel.
[b] A chromium stainless steel.

Table S18 Stainless steels commonly used in construction

Type of stainless steel	Properties and working characteristics	Corrosion-resistant characteristics	Available forms	Major uses	Other uses
301	Attains high strength by cold working and retains good ductility	Excellent corrosion resistance except to direct salt spray and seawater	Strip	Exterior flashing, gutters, leaders, trim, and interior work	Springs, appliances, household and dairy equipment
302 and 304 (considered equal 302/304)	Used in annealed condition generally; good for deep drawing and severe forming	Excellent corrosion resistance except to direct salt spray and seawater	Sheet, strip, bar, and tubing	Exterior work, windows, doors, panels, trim	In textile, paper, and chemical industries and for dairy and food handling
316	Easily formed	Superior corrosion resistance, especially to salt spray and seawater; also resistant to many chemicals	Sheet, strip, bar, and tubing	Exterior use in sea coast areas	All types of marine uses and in chemical, paper, and petroleum industries
430	Easily formed	Excellent corrosion resistance	Sheet, strip, bar, and tubing	Exterior use in sea coast areas	Automobile trim and chemical equipment
201	Higher strength, and better spinning and drawing characteristics than 301	Excellent corrosion resistance	Sheet	Exterior and interior work (similar to 301 uses)	Same applications as 301
202	Higher strength, and better spinning and drawing characteristics than 302	Excellent corrosion resistance	Sheet	Exterior work (similar to 302 uses)	Same applications as 302

nuts, screws, and other types of fasteners (*see* Table *S17*).

Workability. Stainless steels may be cast, forged, rolled, and drawn. They can be machined, bent, formed, riveted, and welded by regular welding processes except forge and hammer welding.

Commercial Forms. Stainless steels are available in structural sections, in sheet, strip, plate, bar, tubing, and wire form, and as castings.

Stainless steel is extruded in a great variety of shapes by forcing a heated billet through dies. This requires tremendous pressure, and molten glass is used as the lubricant. The chemical composition of the glass must be

Table S19 Finishes and textures for stainless steels

Sheet-rolled finishes	Polished finishes	Textured patterns	Etching	Color coatings
No. 1 a dull finish No. 2D a dull, non-reflective finish No. 2B a bright, moderately reflective finish Bright annealed: a bright, highly reflective finish Stainless steel strip No. 1 similar to No. 2D for sheet No. 2 similar to No. 2B for sheet Bright annealed: Similar to "bright, annealed" for sheet	No. 3 an intermediate polished finish No. 4 a bright machine polished finish with a visible "grain" No. 6 a dull satin finish No. 7 a bright, highly reflective finish No. 8 a bright, "mirror" finish	Both sheet and strip are available in: (1) one-directional ribbed or fluted (2) two-directional (3) nondirectional Pattern elments vary in size from about $\frac{1}{8}$ in. (3.18 mm) to 8 in. (203.2 mm), and in depth from 0.002 in. (0.051 mm) to 1 in. (25.4 mm) or more	A matte finish is produced by acid etching or abrasive blasting In both methods, areas not to be etched are protected with a coating	Various systems are used for color coatings: (1) acrylic (2) porcelain enamel (3) oxide conversion: (a) pale gold through bronze (b) gray to flat black

varied from grade to grade of stainless steel or steel. Commercial applications of this process are subject to certain limitations. A web thickness of $\frac{1}{8}$ in. (3.18 mm) is the minimum that can be produced satisfactorily, and the minimum cross-sectional area is 0.280 in.^{2} (1.81 cm^2).

Types and Uses

The stainless steels are divided into the following groups: (1) those with 12 to 30% chromium as the principal alloy; (2) those with 6 to 22% nickel with high percentages of chromium; and (3) the newer high-manganese, low-nickel stainless steels. In construction only a few of many stainless steels are employed, as shown in Table *S18*.

Finishes. Finishes differ for the various fabricated forms of stainless steel. Sheets are supplied in pickled, cold-rolled and polished finishes; strip is supplied pickled or cold-rolled, and with ground and polished finishes (*see* Table *S19*). Plate is supplied annealed and pickled. Round bars are supplied hot-rolled, annealed and pickled, cold-finished, centerless ground, and polished. Square, hexagonal, flat, and symmetrically shaped bars are supplied hot-rolled, annealed and pickled, and cold-drawn. Special sections and shapes and irregular shapes are supplied in pickled finish only. Tubular products are supplied in cold-drawn, annealed and pickled, and polished finishes. Not only can wire have a variety of finishes, but also the finishes are related to the size and end use of the wire.

Construction Uses. In construction stainless steel finds a wide variety of uses, for example, exterior and interior wall finishes, doors, windows, trim, grilles, louvers, screens, countertops, railings, flashing, gutters and leaders, signs and letters, and appliances. It also is used for areas such as kitchens, dairies, and laboratories. Figure *S4* shows the general thickness of stainless steel for each type of use.

Other Uses. The stainless steels also find wide use in the transportation, aviation, and marine fields, in heavy construction equipment, and in industry generally.

Application

Because of its strength, stainless steel requires less thickness than other materials to form stronger shapes. This is particularly true for stainless steel sheet.

Joining. Stainless steel can be welded and soldered, but brazing is not advisable. In some applications, structural adhesives are replacing mechanical fasteners and welding.

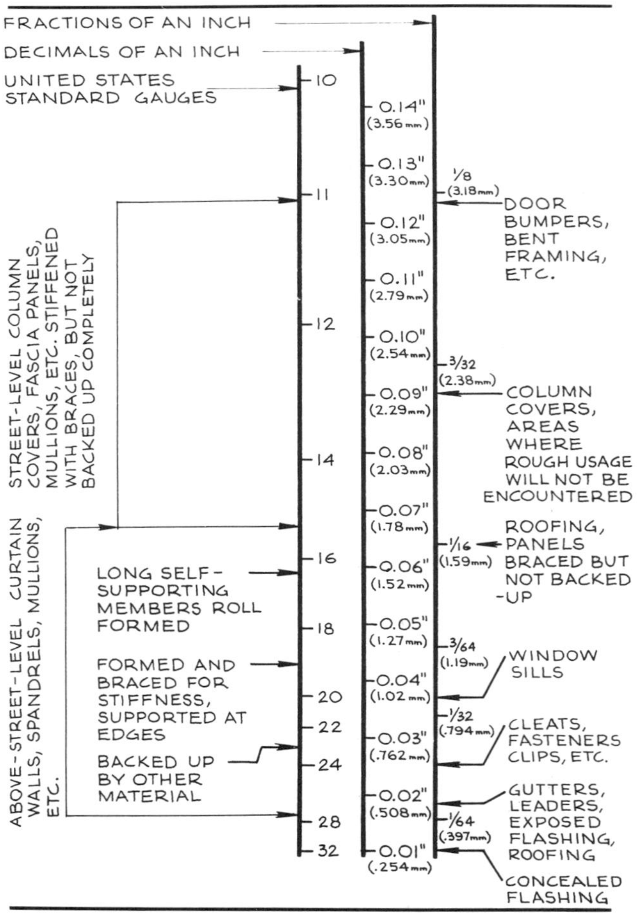

Figure S4 Recommended thicknesses of stainless steel sheet for various construction applications.

Welding of stainless steel varies according to the type of alloy. Generally, austenitic types can be fusion- and resistance-welded except that care must be taken, as the physical properties of these stainless steels (increased electrical resistance, reduced thermal conductivity, increased thermal expansion, and slightly lower melting point) differ from those of the carbon steels. Generally, ferritic types can be fusion- and resistance-welded, but the welded joint should not be subject to shock or bending stress. Ferritic types have almost the same physical characteristics as the austenitic types except that their ductility increases rapidly with heating. Neither type can be forged or hammer-welded. In general, the more rapid the method of welding, the less chance there is for warping and buckling.

Soldering of stainless steels can be accomplished with soft or hard solders. Soldering should be used only as a method of filling or sealing joints and not for obtaining mechanical strength. In general, care should be taken that the surfaces are prepared (tinning before soldering is advisable), that correct fluxes are used, and that the entire area is cleaned after soldering.

Cleaning. It is recommended that, after fabrication and after any surface finishing, the surface of the stainless steel be cleaned with acid or acid mixtures containing oxidizing agents. This is called passivating. It hastens the formation and improves the continuity of the transparent protective film which naturally forms when any clean surface of stainless steel is exposed to oxidizing conditions.

Condensed Checklist

1. The correct type of stainless steel (301, 302, 303, 316, 430F, 201, or 202) must be selected to meet the conditions and requirements of its end use.
2. One should always check local, municipal, and state codes and also the Fire Underwriters, insurance companies, labor departments, and federal government (Army, Navy, etc.) for requirements and restrictions.
3. Shop drawings must always be required by the specifications. One should make sure that all nuts, bolts, screws, and fastening devices are of stainless steel of the correct type and are compatible.
4. Stainless steel must not be in contact with any metal that might cause galvanic action or any other material that might affect it (*see* Stainless Steel Windows; Metals).
5. Sufficient expansion joints must always be installed so that no distortions can occur because of expansion or contraction.
6. When welding is necessary, particularly with large, flat surfaces, the manufacturers and fabricators of stainless steel should always be consulted to determine the correct type of welding to use and to learn whether there is any possibility that, because of the expansion caused by the heat of the welding, permanent distortions of the flat surface will occur even if sufficient stiffeners have been used. One should always specify that all welding flux, oxides, and discolorations must be removed by pickling, grinding, etc., so that these areas match the finish of the adjacent areas.
7. If soldering is used, it must be used only for filling and sealing, not for mechanical strength. The soldered area must be immediately cleaned with a neutralizer and washed with clean water. The stainless steel manufacturers and fabricators should always be consulted for the correct type of solder, method of soldering, and type of fluxes to be used.

8. When structural adhesives are specified for laminating and joining stainless steels, the manufacturer of the stainless steel should be consulted for the correct type and method.
9. Any damage to the surface caused by fabrication must be repaired by grinding, polishing, or buffing.
10. Any material used for stiffeners, anchors, etc., that is less corrosion resistant than stainless steel must be given a protective coating as recommended by the stainless steel manufacturer.
11. Stainless steel should be protected by a coating during fabrication, transportation, and installation. This coating should be removed after completion of the work.
12. As a general rule all fabrication should be done in the shop, where controlled work conditions are available.

Conditions Favorable to the Use of Stainless Steel

1. Where a very strong, durable, corrosion-resistant material requiring a minimum of maintenance is necessary.
2. Where a highly polished, mirrorlike surface material requiring minimal maintenance is desired.

Conditions Unfavorable to the Use of Stainless Steel

1. In areas where corrosion resistance and durability are not important.

History and Manufacture

The stainless steels appeared between 1903 and 1913 as a result of research conducted by Brealey in England, Becket in the United States, and Strauss and Maurer in Germany. These men shared in the initial development of the stainless steels.

Electric Furnace Operation. All stainless steel is produced in the electric furnace, where high temperatures, sometimes more than 3200°F (1760°C), and precise chemical controls can be obtained. The first electric steel furnace was built in 1906 in the United States, and it soon replaced the crucible process for making high-grade alloy steels and tool steels.

The furnace is completely lined with refractory brick. Large electrodes, 40 in. (101.6 cm) or larger in diameter, extend down into the furnace, and the temperature is controlled by raising and lowering these electrodes. The electric current arcs to the slag, creating heat between the electrode and the top of the slag, and then passes through the slag, which offers resistance and thus creates more heat; the current then passes in sequence through the metal and back up through the slag, and arcs to the other electrode. Because the metal is hotter at the electrodes, a constant motion or stirring of the molten metal

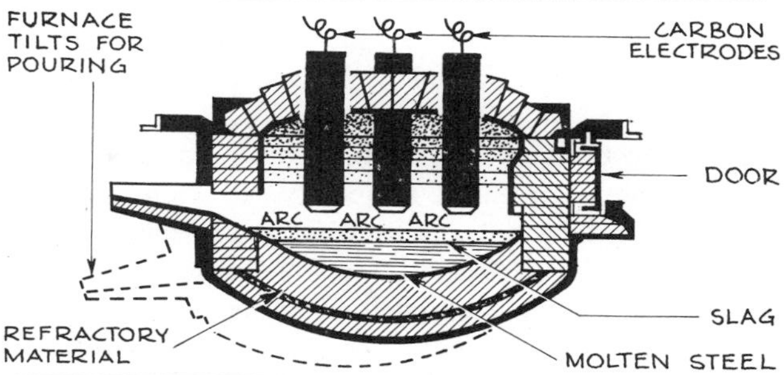

Figure S5 Schematic electric furnace.

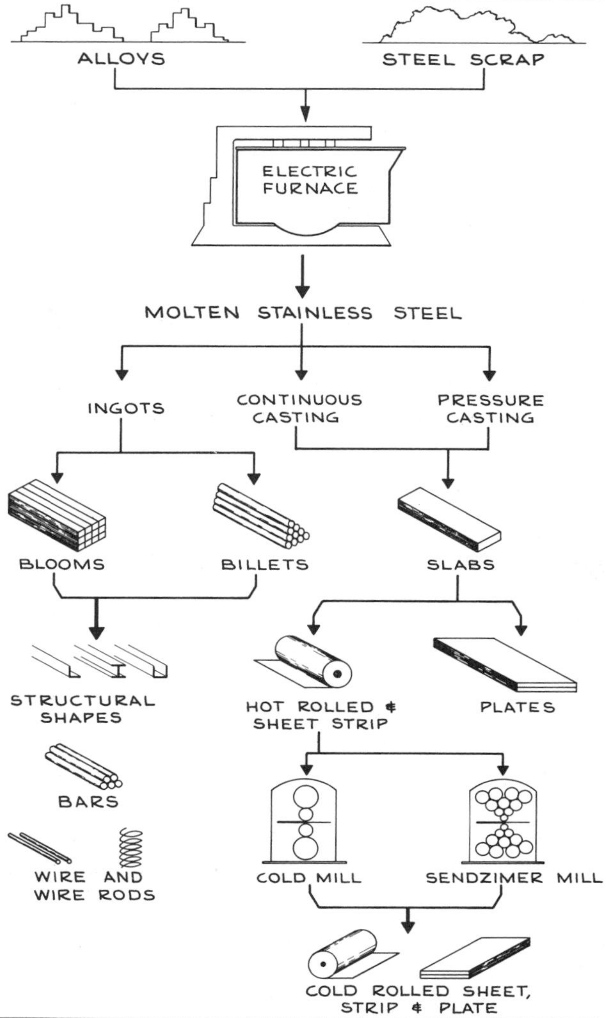

Figure S6 Making stainless steel products and materials.

occurs. In approximately $1\frac{1}{2}$ to 5 hours, depending on the quality of steel required, the process is finished and the molten metal is poured into ladles. (*See* Figure *S5*.)

Control Procedures. During the process, analyses of the molten metal are made frequently, additions such as various oxidizing or reducing slags may be made, alloying elements may be added, and the carbon and sulfur content is precisely controlled. When the required chemical composition is reached, the contents of the furnace are poured into ingots. The same precision and control is followed in the mill operations. All procedures are slower and more precise than those used for regular steel mill operations (*see* Figure *S6*).

STAINLESS STEEL BARS, ANGLES, CHANNELS, AND OTHER EXTRUDED SHAPES

Stainless steel bars and special shapes are made from AIAI Types 301, 302, 303, 304, 316, and 430. They are available as rounds, squares, flats, half ovals, hexagons, octagons, angles, channels, and tubes. Other special sections are available but depend on the various mills. Tubular shapes are described separately.

Types and Uses

Available Shapes. Stainless steel is available in virtually any form and size in which other metals are available. Tables *S20–S23* give common sizes, weights, and shapes available. When designing with stainless steel, however, it is advisable always to contact the manufacturers to find the current special and stock shapes that are available.

Stainless steel rounds, squares, octagons, and hexagons are available in sizes from $\frac{1}{8}$ to 8 in. (3.18 to 203.2 mm) in $\frac{1}{8}$-in. (3.18-mm) increments (the size refers to the diameter of the inscribed circle). Rounds are furnished hot-rolled, annealed and pickled, cold-drawn, and turned. They can be centerless ground and polished. When round bars are to be turned, allowance in diameter for turning should be specified (*see* Table *S20*).

Flat stainless steel bars are available up to 10 in. (254 mm) in width and $\frac{1}{8}$ in. (3.18 mm) or more in thickness. Squares, hexagons, flats, and symmetrical shapes are furnished hot-rolled, annealed and pickled, and cold-drawn. Common sizes of stainless steel angles are shown in Tables *S21* to *S23*, as are the common sizes of T's, I's, and channels.

Extruded stainless steel angles, channels, T's, +'s, and other shapes are available (*see* Figure *S7*).

Special and irregular sections and shapes are furnished with a pickled finish. For further finishing, they require first coarse and then increasingly finer abrasive grinding.

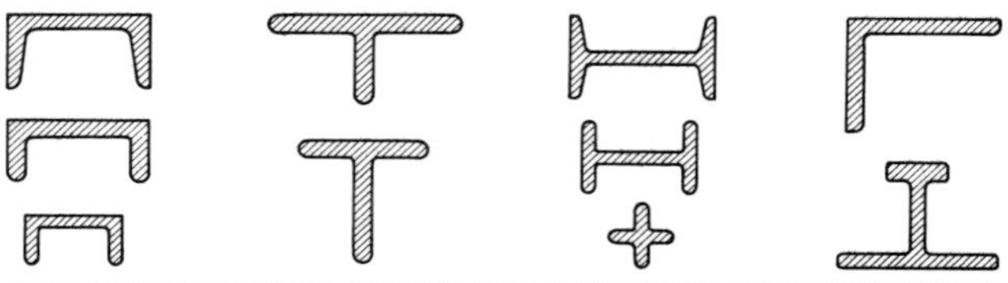

Figure S7 Typical extruded stainless steel shapes.

Construction Uses. Bars and special shapes are used in construction for railings, screens, grilles, louvers, fastening devices, store fronts, etc.

Application

Condensed Checklist

1. Before designing with stainless steel bars and shapes, it is always best to check the availability of the various sizes, shapes, and types of finish in which the required items can be furnished.
2. When stainless steel bars and shapes are to be welded, the manufacturers of stainless steel should always be consulted for the correct method of welding to produce the desired end result.

Table S20 Lathe turning allowances for round bars

	Diameter of hot-rolled bars					
	$1\frac{1}{2}$ to 3 in. (38.1 to 76.2 mm)		Over 3 to 6 in. (76.2 to 152.4 mm)		Over 6 to 8 in. (152.4 to 203.2 mm)	
	in.	mm	in.	mm	in.	mm
Minimum decrease in diameter for lathe turning	$\frac{1}{8}$	3.18	$\frac{1}{4}$	6.35	$\frac{5}{8}$	15.88

Table S21 Common sizes of extruded or cold-rolled stainless steel angles with unequal legs

	Size of legs				Thickness		Weight	
Shape	in.	mm	in.	mm	in.	mm	lb/ft	kg/m
Extruded, unequal legs 	2	50.8	1	25.40	$\frac{3}{16}$	4.76	1.75	2.60
	2	50.8	1	25.40	$\frac{1}{4}$	6.35	2.35	3.50
	2	50.8	$1\frac{1}{2}$	38.10	$\frac{3}{10}$	4.76	2.12	3.16
	2	50.8	$1\frac{1}{2}$	38.10	$\frac{1}{4}$	6.35	2.78	4.14
	$2\frac{1}{2}$	63.5	$1\frac{1}{4}$	31.75	$\frac{1}{4}$	6.35	2.85	4.24
	$2\frac{1}{2}$	63.5	$1\frac{1}{2}$	38.10	$\frac{1}{4}$	6.35	3.19	4.75
	$2\frac{1}{2}$	63.5	$1\frac{1}{2}$	38.10	$\frac{3}{8}$	9.53	4.70	7.00
	$2\frac{1}{2}$	63.5	2	50.8	$\frac{3}{16}$	4.76	2.75	4.09
	$2\frac{1}{2}$	63.5	2	50.8	$\frac{1}{4}$	6.35	3.62	5.39
	$2\frac{1}{2}$	63.5	2	50.8	$\frac{3}{8}$	9.53	5.30	7.89
	3	76.2	1	25.40	$\frac{1}{2}$	12.7	6.01	8.94
	3	76.2	$1\frac{1}{2}$	38.10	$\frac{1}{4}$	6.35	3.51	5.22
	3	76.2	$1\frac{1}{2}$	38.10	$\frac{3}{8}$	9.53	5.30	7.89
	3	76.2	2	50.8	$\frac{3}{16}$	4.76	3.07	4.57
	3	76.2	2	50.8	$\frac{1}{4}$	6.35	4.10	6.10
	3	76.2	2	50.8	$\frac{3}{8}$	9.53	5.90	8.78
	3	76.2	2	50.8	$\frac{1}{2}$	12.7	7.70	11.46
	$3\frac{1}{2}$	88.9	$1\frac{1}{2}$	38.10	$\frac{1}{2}$	12.7	7.64	11.37
	$3\frac{1}{2}$	88.9	$2\frac{1}{2}$	63.5	$\frac{1}{4}$	6.35	4.90	7.29
	$3\frac{1}{2}$	88.9	$2\frac{1}{2}$	63.5	$\frac{3}{8}$	9.53	7.23	10.76
	$3\frac{1}{2}$	88.9	3	76.2	$\frac{1}{4}$	6.35	5.40	8.04
	$3\frac{1}{2}$	88.9	3	76.2	$\frac{3}{8}$	9.53	7.90	11.76
	$3\frac{1}{2}$	88.9	3	76.2	$\frac{1}{2}$	12.7	10.20	15.18
	4	101.6	3	76.2	$\frac{1}{4}$	6.35	5.80	8.63
	4	101.6	3	76.2	$\frac{3}{8}$	9.53	8.50	12.65
	4	101.6	3	76.2	$\frac{1}{2}$	12.7	11.10	16.52
Cold-rolled, unequal legs	1	25.40	$\frac{5}{8}$	15.88	$\frac{1}{8}$	3.18	0.64	0.95
	$1\frac{3}{4}$	31.75	$\frac{3}{4}$	19.05	$\frac{1}{8}$	3.18	0.80	1.19
	$1\frac{1}{2}$	38.10	1	25.40	$\frac{1}{8}$	3.18	1.01	1.50
	2	50.80	1	25.40	$\frac{1}{8}$	3.18	1.23	1.83

3. All types of fastening devices should be of stainless steel. If they are attached to a metal other than stainless steel, they must be isolated so that no galvanic action can occur.
4. Specifications should always require that shop drawings be submitted and that all fastening devices be shown.
5. When stainless steel bars and shapes are to be adjacent to or a part of other stainless steel materials or construction, it is always necessary to check that finishes can be matched.

Conditions Favorable to the Use of Stainless Steel Bars and Shapes

1. For railings, grilles, screens, etc., where a strong, maintenance-free, permanent material is required.
2. In areas where other stainless steel is used so that the entire design is integrated.

Conditions Unfavorable to the Use of Stainless Steel Bars and Shapes

Where economy is paramount.

Table S22 Common sizes of extruded or cold-rolled stainless steel angles with equal legs

	Size of legs				Thickness		Weight	
Shape	in.	mm	in.	mm	in.	mm	lb/ft	kg/m
Extruded, equal legs	1	25.40	1	25.40	$\frac{3}{16}$	4.76	1.15	1.71
	1	25.40	1	25.40	$\frac{1}{4}$	6.35	1.49	2.22
	$1\frac{1}{4}$	31.75	$1\frac{1}{4}$	31.75	$\frac{3}{16}$	4.76	1.48	2.20
	$1\frac{1}{4}$	31.75	$1\frac{1}{4}$	31.75	$\frac{1}{4}$	6.35	1.92	2.88
	$1\frac{1}{2}$	38.10	$1\frac{1}{2}$	38.10	$\frac{3}{16}$	4.76	1.80	2.68
	$1\frac{1}{2}$	38.10	$1\frac{1}{2}$	38.10	$\frac{1}{4}$	6.35	2.39	3.56
	$1\frac{3}{4}$	44.45	$1\frac{3}{4}$	44.45	$\frac{3}{16}$	4.76	2.12	3.16
	$1\frac{3}{4}$	44.45	$1\frac{3}{4}$	44.45	$\frac{1}{4}$	6.35	2.75	4.09
	2	50.8	2	50.8	$\frac{3}{16}$	4.76	2.44	3.63
	2	50.8	2	50.8	$\frac{1}{4}$	6.35	3.19	4.75
	2	50.8	2	50.8	$\frac{5}{16}$	7.94	3.92	5.83
	2	50.8	2	50.8	$\frac{3}{8}$	9.53	4.70	7.00
	$2\frac{1}{2}$	63.5	$2\frac{1}{2}$	63.5	$\frac{3}{16}$	4.76	3.07	4.57
	$2\frac{1}{2}$	63.5	$2\frac{1}{2}$	63.5	$\frac{1}{4}$	6.35	4.10	6.10
	$2\frac{1}{2}$	63.5	$2\frac{1}{2}$	63.5	$\frac{5}{16}$	7.94	5.00	7.44
	$2\frac{1}{2}$	63.5	$2\frac{1}{2}$	63.5	$\frac{3}{8}$	9.53	5.90	8.78
	3	76.2	3	76.2	$\frac{3}{16}$	4.76	3.71	5.52
	3	76.2	3	76.2	$\frac{1}{4}$	6.35	4.90	7.29
	3	76.2	3	76.2	$\frac{5}{16}$	7.94	6.10	9.08
	3	76.2	3	76.2	$\frac{3}{8}$	9.53	7.20	10.72
	3	76.2	3	76.2	$\frac{1}{2}$	12.70	9.40	13.99
	$3\frac{1}{2}$	88.9	$3\frac{1}{2}$	88.9	$\frac{1}{4}$	6.35	5.80	8.36
	$3\frac{1}{2}$	88.9	$3\frac{1}{2}$	88.9	$\frac{3}{8}$	9.53	8.50	12.65
	$3\frac{1}{2}$	88.9	$3\frac{1}{2}$	88.9	$\frac{1}{2}$	12.70	11.10	16.52
Cold-rolled, equal legs	$\frac{1}{2}$	12.70	$\frac{1}{2}$	12.7	$\frac{1}{8}$	3.18	0.38	0.57
	$\frac{5}{8}$	15.88	$\frac{5}{8}$	15.88	$\frac{1}{8}$	3.18	0.48	0.71
	$\frac{3}{4}$	19.05	$\frac{3}{4}$	19.05	$\frac{1}{16}$	1.59	0.39	0.58
	$\frac{3}{4}$	19.05	$\frac{3}{4}$	19.05	$\frac{1}{8}$	3.18	0.59	0.88
	1	25.40	1	25.40	$\frac{1}{8}$	3.18	0.80	1.19
	1	25.40	1	25.40	$\frac{3}{16}$	4.76	1.16	1.73
	$1\frac{1}{4}$	31.75	$1\frac{1}{4}$	31.75	$\frac{1}{8}$	3.18	1.01	1.50
	$1\frac{1}{4}$	31.75	$1\frac{1}{4}$	31.75	$\frac{3}{16}$	4.76	1.48	2.20
	$1\frac{1}{2}$	38.10	$1\frac{1}{2}$	38.10	$\frac{1}{8}$	3.18	1.23	1.83
	$1\frac{1}{2}$	38.10	$1\frac{1}{2}$	38.10	$\frac{3}{16}$	4.76	1.80	2.68
	2	50.8	2	50.8	$\frac{1}{8}$	3.18	1.65	2.46
	2	50.8	2	50.8	$\frac{3}{16}$	4.76	2.44	3.63

History and Manufacture

The fabrication of stainless steel bars and special shapes is similar to the processes used for the same shapes in steel except the entire procedure in each case is slower and requires greater control. Generally, rounds, squares, flats, half ovals, hexagons, octagons, and special shapes are either hot-rolled or cold-drawn; angles, channels, and tubes are generally hot-rolled. Round bars may also be turned on a lathe.

Table S23 Common sizes of extruded or cold-rolled T's, I's, Z's, and channels

Shape	Depth		Depth of flange		Thickness		Weight	
	in.	mm	in.	mm	in.	mm	lb/ft	kg/m
Extruded "T"	$1\frac{1}{4}$	31.75	$1\frac{1}{4}$	31.75	$\frac{3}{16}$	4.76	1.48	2.20
	$1\frac{1}{2}$	38.10	$1\frac{1}{2}$	38.10	$\frac{3}{16}$	4.76	1.65	2.46
	2	50.80	2	50.80	$\frac{1}{4}$	6.35	3.19	4.75
	$2\frac{1}{2}$	63.50	$2\frac{1}{2}$	63.50	$\frac{1}{4}$	6.35	4.10	6.10
	$2\frac{1}{2}$	63.50	$2\frac{1}{2}$	63.50	$\frac{3}{8}$	9.53	5.90	8.78
	1	25.40	$1\frac{1}{2}$	38.10	$\frac{3}{16}$	4.76	1.48	2.20
	2	50.8	$1\frac{1}{4}$	31.75	$\frac{3}{16}$	4.76	1.96	2.87
	$2\frac{3}{8}$	60.33	$1\frac{1}{2}$	38.10	$\frac{3}{16}$	4.76	2.36	3.51
	$1\frac{1}{4}$	31.75	1	25.40	$\frac{1}{4}$	6.35	1.71	2.55
	$1\frac{1}{4}$	31.75	2	50.80	$\frac{1}{4}$	6.35	2.57	3.82
	$1\frac{1}{2}$	38.10	$2\frac{1}{2}$	63.50	$\frac{1}{4}$	6.35	3.19	4.75
	2	50.80	$1\frac{1}{2}$	38.10	$\frac{1}{4}$	6.35	2.75	4.09
	3	76.20	$2\frac{1}{2}$	63.50	$\frac{3}{8}$	9.53	6.70	9.97
	4	101.60	$3\frac{1}{2}$	88.90	$\frac{5}{16}$	7.94	7.66	11.40
Extruded channels	1	25.40	$\frac{1}{2}$	12.70	$\frac{1}{8}$	3.18	0.75	1.12
	$1\frac{1}{4}$	21.75	1	25.40	$\frac{1}{4}$	6.35	2.35	3.50
	$1\frac{1}{2}$	38.10	$\frac{1}{2}$	12.70	$\frac{1}{8}$	3.18	0.96	1.43
	$1\frac{1}{2}$	38.10	$\frac{3}{4}$	19.05	$\frac{3}{16}$	4.76	1.75	2.60
	$1\frac{1}{2}$	38.10	1	25.40	$\frac{1}{4}$	6.35	2.57	3.83
	2	50.80	$\frac{3}{4}$	19.05	$\frac{3}{16}$	4.76	2.02	3.01
	2	50.80	$\frac{5}{8}$	15.88	$\frac{1}{4}$	6.35	2.35	3.50
	2	50.80	1	25.40	$\frac{1}{4}$	6.35	3.15	4.69
	$2\frac{1}{2}$	63.50	$\frac{5}{8}$	15.88	$\frac{3}{16}$	4.76	2.28	3.39
	$2\frac{1}{2}$	63.50	1	25.40	$\frac{1}{4}$	6.35	3.57	5.31
	3	76.20	$1\frac{3}{8}$	34.93	$\frac{3}{16}$	4.76	4.10	6.10
	3	76.20	$1\frac{1}{2}$	38.10	$\frac{1}{4}$	6.35	4.75	7.07
	4	101.60	$1\frac{3}{4}$	44.45	$\frac{1}{4}$	6.35	6.69	9.96
Cold-rolled channels, equal sides	$\frac{1}{2}$	12.70	$\frac{1}{2}$	12.70	0.093	2.36	0.40	0.60
	$\frac{3}{4}$	19.05	$\frac{3}{4}$	19.05	0.093	2.36	0.57	0.85
	1	25.40	1	25.40	0.109	2.77	1.03	1.53
	$1\frac{1}{4}$	31.75	$1\frac{1}{4}$	31.75	0.109	2.77	1.32	1.96
	$1\frac{1}{2}$	38.10	$1\frac{1}{2}$	38.10	0.109	2.77	1.59	2.37
	2	50.80	2	50.80	0.125	3.18	2.41	3.59
Cold-rolled channels, unequal sides	$\frac{5}{8}$	15.88	$\frac{5}{16}$	7.94	0.078	1.98	0.29	0.43
	$\frac{3}{4}$	19.05	$\frac{3}{8}$	9.53	0.083	2.11	0.40	0.60
	$1\frac{1}{2}$	38.10	1	25.40	0.109	2.77	1.22	1.82
	$1\frac{3}{4}$	44.45	$1\frac{1}{8}$	28.58	0.109	2.77	1.40	2.08
	2	50.80	1	25.40	0.125	3.18	1.59	2.37
	$2\frac{3}{8}$	60.33	$2\frac{3}{16}$	55.56	0.156	3.96	3.41	5.08
Extruded I	$1\frac{1}{2}$	38.10	1	25.40	$\frac{3}{16}$	4.76	2.02	3.01
	2	50.8	$1\frac{1}{4}$	31.75	$\frac{1}{4}$	6.35	3.43	5.11
	$2\frac{1}{2}$	63.50	$1\frac{1}{2}$	38.10	$\frac{1}{4}$	6.35	4.22	6.29
	3	76.20	$2\frac{3}{8}$	60.33	$\frac{1}{4}$	6.35	6.60	9.82
	4	101.60	$2\frac{3}{4}$	69.85	$\frac{1}{4}$	6.35	8.32	12.38

STAINLESS STEEL SHEET, STRIP, AND PLATE

Physical Properties

Stainless steel strip and sheet are made from the alloy types generally used in construction applications, namely, 201, 202, 301, 302, 304, 316, and 430. Their chemical properties are fully discussed under Stainless Steel. Their physical properties differ only in types of finish and the fact that stainless steel sheet, strip, and plate are hardened by cold working. They have low heat conductivity and high thermal expansion.

Commercial Forms. Sheet is available in seven standard finishes (*see* Table *S24*). Strip is available in three

Table S24 Finishes for stainless steel sheet

Finish number	Type of finish	Fabricating method	Characteristics	Major use
1	Unfinished surface	Hot-rolled, annealed, and pickled	Unaffected by forming and annealing, whereas other finishes will be destroyed	In areas where appearance is unimportant
2D[a]	Dull, nonreflective surface	Cold-rolled before final annealing and pickling	Good for shapes and forms that require severe forming and then are polished and buffed	Roofing and other areas where reflectivity is unimportant
2B[a]	Bright surface with glossy satin sheen	Given a final light cold rolling	Good for shapes and forms that do not require severe drawing or forming; harder temper; fine surface for polishing and buffing	Exterior surfacing for walls and panels; not highly reflective
4[b]	Commercial type of uniform bright surface with good luster; all surface imperfections removed	Ground with successively finer abrasives	Good for shapes and forms that require only simple drawing and forming so finish will not be destroyed; welds and fabrication markings can be ground and polished to match surface	Most generally used finish in architectural panels, trim, fascias, etc.
6	Soft satin finish	A No. 4 finish brushed with Tampico fibers and fine abrasive	Low reflectivity with a uniform finish	Where reflectivity is undesirable and for contrast to highly polished finishes
7	Highly polished, semimirrorlike surface	Buffing	High reflectivity and high luster	Where high reflectivity is important and for contrast to other types of finish
8[c]	Mirrorlike finish	Buffing	Reflectivity same as plate glass mirror	Special decorative effects and as mirrors

[a] The chemical composition and thickness of the stainless steel influence these finishes. Care should be taken if exact matching of finishes is important.
[b] Shows finger marks less than the other finishes.
[c] Not standard. Available only on special order.

Table S25 Finishes for stainless steel strip

Number of finish	Type of finish	Method of obtaining	Characteristics	Major uses
1	Nonreflective type of surface	Cold-rolled, annealed, and pickled	Smoothness varies among light and heavy gauges	For mass production of trim, moldings, fascias, roof drainage sections, and panels
2	Bright with glossy satin sheen type of surface	Cold-rolled, annealed, pickled, and rerolled	Variations in smoothness can be caused by cold working	Same as above
Bright finish	Finishes similar to 4, 6, and 7 for sheet	Ground, polished, and buffed	Supplied in specially cut lengths	Same as above

Table S26 Edge classification for stainless steel strip

Type of edge	Description	Fabricating method
No. 1	Round or square	By edge rolling as produced
No. 3	Square	By slitting; not filed
No. 5	Square	Slitting, then rolling or filing

standard finishes and in three types of edge classification (*see* Tables *S25* and *S26*).

Types and Uses

Thicknesses. Stainless steel sheet and strip are available in various thicknesses, as shown in Table *S27*. It should be noted that sheet and strip thicknesses are expressed in different types of gauges and that, although the gauge numbers may be identical, the actual thicknesses vary. Also, many tables list thickness equivalents that are based on a different weight in pounds per square foot, and this practice, too, results in slight variations in actual thickness.

Types of Sheet. Stainless steel sheet is manufactured in widths of 24 in. (609.6 mm) and more, either in rolls or cut to size. Seven finishes are available. These are described in Table *S24*, which also lists the official designation for each type of finish, the method of obtaining it, its characteristics, and its major use.

Type 301 sheet is available in quarter ($\frac{1}{4}$) and half ($\frac{1}{2}$) hard temper.

Types of Strip. Stainless steel strip is available in coils or cut to length, with three finishes and three edge classifications. Table *S25* gives the official designation for each type of finish and describes its characteristics and major use. Edge classifications may be found in Table *S26*. Type 301 strip is available in $\frac{1}{4}$, $\frac{1}{2}$, and $\frac{3}{4}$ hard temper.

Texture Surfaces. Both sheet and strip are available in various textured surfaces that are produced by rolling between special rolls that emboss or imprint the texture on the sheet or strip.

Stainless Steel Plate. Stainless steel plate (*see* Table *S28*) is manufactured only with No. 1 finish (annealed and pickled). Any other type of finish requires coarse grinding, followed by finer grinding and polishing and then buffing. These finishes can be obtained by special order.

Application

When designing with stainless steel, it is a good general rule to remember that stainless steel is very strong and therefore much thinner gauges can be used than for other metals. But it must also be remembered that thinner materials have a greater tendency to show distortions

Table S27 Gauges, thicknesses, and weights of stainless steel sheet and strip

	Strip (U.S. Standard Gauge)						Strip (Birmingham Strip Gauge)					
	Thickness		Weight[a]				Thickness		Weight[a]			
			Types 201, 202, 301, 302, 304, 316 6		Type 430				Types 201, 202, 301, 302, 304, 316		Type 430	
Gauge number[b]	Decimals of an inch	mm	lb/ft^2	kg/m^2	lb/ft^2	kg/m^2	Decimals of an inch[b]	mm	lb/ft^2	kg/m^2	lb/ft^2	kg/m^2
8	0.17188	4.3658	7.2187	44.242	7.0813	34.571	0.6144	4.1758	6.930	33.832	6.798	33.188
10	0.14063	3.5720	5.9062	28.834	5.7937	28.285	0.1345	3.4163	5.628	27.476	5.521	26.954
11	0.1250	3.1750	5.2500	25.631	5.1500	25.142	0.1196	3.0378	5.040	24.605	4.944	24.137
12	0.10938	2.7783	4.5937	22.427	4.5063	22.000	0.1046	2.6568	4.578	22.350	4.491	21.925
14	0.07813	1.9845	3.2812	16.019	3.2187	15.714	0.0747	1.8974	3.486	17.019	3.420	16.694
16	0.06250	1.5875	2.6250	12.815	2.5750	12.571	0.0598	1.5189	2.730	13.328	2.678	11.121
18	0.05000	1.2700	2.1000	10.252	2.0600	10.057	0.0478	1.2141	2.058	10.047	2.019	9.857
20	0.03750	0.9525	1.5750	7.689	1.5450	7.543	0.0359	0.9119	1.470	7.177	1.442	7.040
22	0.03125	0.7938	1.3125	6.409	1.2875	6.287	0.0299	0.7595	1.176	5.741	1.154	5.634
24	0.02500	0.6350	1.0500	5.126	1.0300	5.029	0.0239	0.6071	0.924	4.511	0.906	4.423
26	0.01875	0.4763	0.7875	3.845	0.7725	3.771	0.0179	0.4537	0.756	3.691	0.742	3.622
28	0.01563	0.3970	0.6562	3.186	0.6438	3.143	0.0149	0.3785	0.588	2.871	0.577	2.817
30	0.01250	0.3175	0.5250	2.563	0.5150	2.514	0.0120	0.3048	0.504	2.461	0.494	2.412

[a] Chromium-nickel stainless steels are taken as 42 lb/ft · in. thickness (19.05 kg/0.3048 m · 25.4 mm thickness), and chromium stainless steels as 41.2 lb/ft · in. thickness (18.69 kg/0.3048 m · 25.4 mm thickness).

[b] Gauges of stainless steel sheet and strip commonly used in construction.

Table S28 Gauges, thicknesses, and weights of stainless steel plate

U.S. Standard Gauge	Thickness: Decimals of an inch	Thickness: mm	Thickness: Fractions of an inch	Weight[a]: Types 301, 302, 304, 316 lb/ft²	Weight[a]: Types 301, 302, 304, 316 kg/m²	Weight[a]: Type 430 lb/ft²	Weight[a]: Type 430 kg/m²
	1.000	25.4	1	41.342	201.83	40.478	197.61
	0.9375	23.81	15/16	38.759	189.22	37.949	185.27
	0.875	22.23	7/8	36.175	176.61	35.419	172.92
	0.8125	20.64	13/16	33.591	163.99	32.889	160.56
	0.75	19.05	3/4	31.007	151.38	30.359	148.21
	0.6875	17.46	11/16	28.432	138.81	27.829	135.87
	0.625	15.88	5/8	25.839	126.15	25.299	123.51
	0.5625	14.29	9/16	23.255	113.53	22.769	111.16
	0.5	12.70	1/2	20.671	100.92	20.239	98.81
	0.46875	11.91	15/32	19.379	94.61	18.974	72.63
	0.4375	11.11	7/16	18.087	88.30	17.709	86.46
	0.40625	10.32	13/32	16.795	81.99	16.444	80.28
	0.375	9.53	3/8	15.503	75.79	15.179	74.11
	0.34375	8.75	11/32	14.211	69.36	13.914	67.33
0	0.3125	7.94	5/16	12.920	63.08	12.050	58.83
1	0.28125	7.14	9/32	11.628	56.77	11.385	55.58
2	0.265625	6.75	17/64	10.981	53.61	10.752	52.49
3	0.25	6.35	1/4	10.336	50.46	10.120	49.41
4	0.234375	5.95	15/64	9.690	46.31	9.487	41.43
5	0.21875	5.55	7/32	9.044	44.15	8.855	43.23
6	0.203125	5.16	13/64	8.398	41.00	8.222	40.14
7	0.1875	4.76	3/16	7.752	37.85	7.590	37.05

[a]Weights are based on 0.2871 lb/in.^{3} (7.95 g/cm^3) for Types 301, 302, 304, and 316, and on 0.2811 lb/in.^{3} (7.78 g/cm^3) for Type 430.

when large, smooth, flat surfaces are desired. Therefore, even though stainless steel is very strong, one must not expect a large, thin sheet of metal—for example, a 26- or 28-gauge piece 2 ft (0.6096 m) high—not to have some distortions on its surface. It should also be borne in mind that stainless steel has a relatively high coefficient of expansion, and care must be taken when it is welded, especially in the case of fairly large, flat surfaces, that the welding does not cause permanent distortions.

Condensed Checklist

1. The correct type of stainless steel, gauge, and finish must be chosen to meet the requirements and limitations of its end use in the building.
2. Local and state codes and also the codes of the Fire Underwriters, insurance companies, labor departments, and federal government (Army, Navy, etc.) should be checked for fire-resistance requirements and other limitations.
3. All fastening devices must be of stainless steel or of other metals that will not cause galvanic action.
4. If stainless steel is to be used with other metals where galvanic action can occur, complete isolation between the metals must be achieved and these areas must be kept free from moisture.
5. Specifications must always require that shop draw- be furnished and that they show methods of fastening, type of welding, and type of finish.
6. Stainless steel must be protected with paper wrappings, adhesive paper, or plastic film coatings, etc., during fabrication, transportation, and erection and during construction until such time as the building is completed or no further damage is possible.
7. Finishes should be matchable even though various thicknesses and types of stainless steel are used.

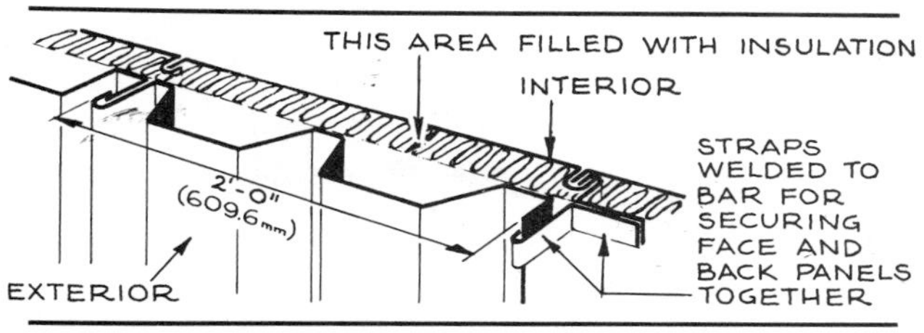

Figure S8 Typical stainless steel panel, using stock components.

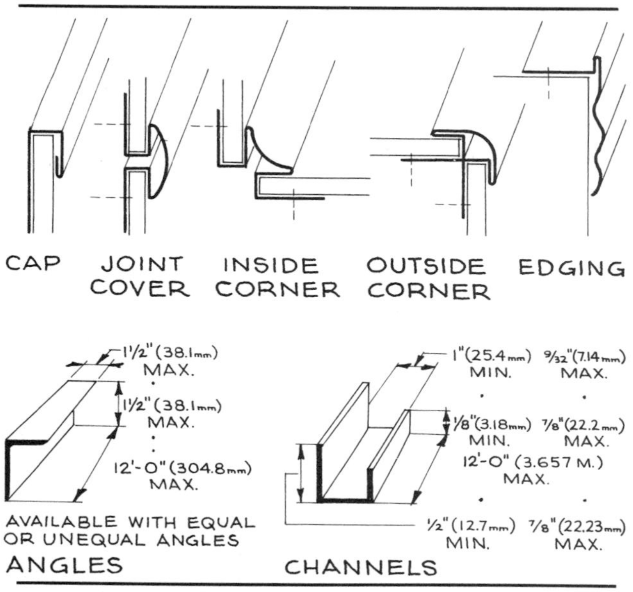

Figure S9 Typical stainless steel moldings and trim.

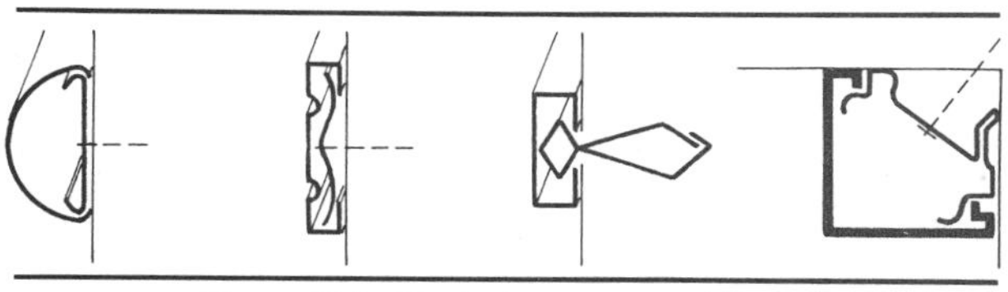

Figure S10 Typical snap-on moldings.

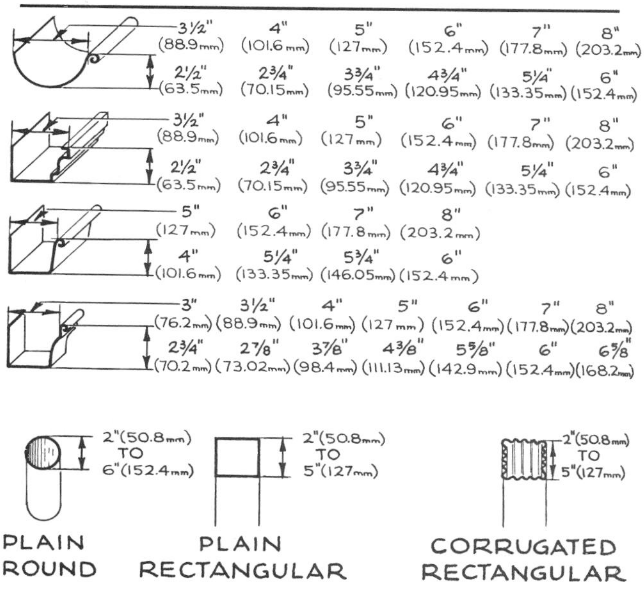

Figure S11 Typical stock stainless steel gutters and leaders.

8. Wall panels: When stainless steel is used as the wall finish on a building, either for panels or as a surfacing, the manufacturers of stainless steel should be consulted for stock shapes currently available or for limitations on fabricating special shapes. Stainless steel is available corrugated, ribbed, or fluted and in many other of the forms that are generally made in other metals. (*See* Figure *S8*.)

   One should always check with the manufacturers of stainless steel to determine the best sealing compounds for the various joints in the stainless steel wall surfacing or panels.
9. Moldings and trim: When stainless steel stock moldings and trim are used, it is advisable always to check with manufacturers for type of finish, gauge, and method of attachment. If the work is shop-fabricated, all corners must be mitered and welded. (*See* Figures *S9* and *S10*.)
10. Gutters and leaders: Stainless steel gutters and leaders must always be installed with all attachments such as straps, screws, bolts, and hangers made of stainless steel. Gutter lengths should not exceed 40 ft (12.192 m), and hangers should be 2.5 ft (1.662 m) o.c. at maximum. Always use 50-50 lead-tin solder. All accessories such as end closures, lead connections, 90-degree corners, hangers, and straps are available in stainless steel. Built-in gutters should always have sufficient expansion joints; the location and number of expansion joints are controlled by the width of the gutter. One should always check with manufacturers of stainless steel for the spacing of expansion joints in relation to the gutter design contemplated. Specifications should always call for shop drawings that show details of expansion joints and all methods of attachment. (*See* Figure *S11*.)
11. Store fronts: When stainless steel is being used for store fronts, one should always check with the manufacturers for the latest stock types of components such as moldings, awning boxes, mullions, glazing members, trim, and fascias, including type of finish and gauge of stainless steel (*see* Figure *S12*).
12. Roofing: When stainless steel sheet with No. 2D finish is used for batten roofing, sufficient allowance

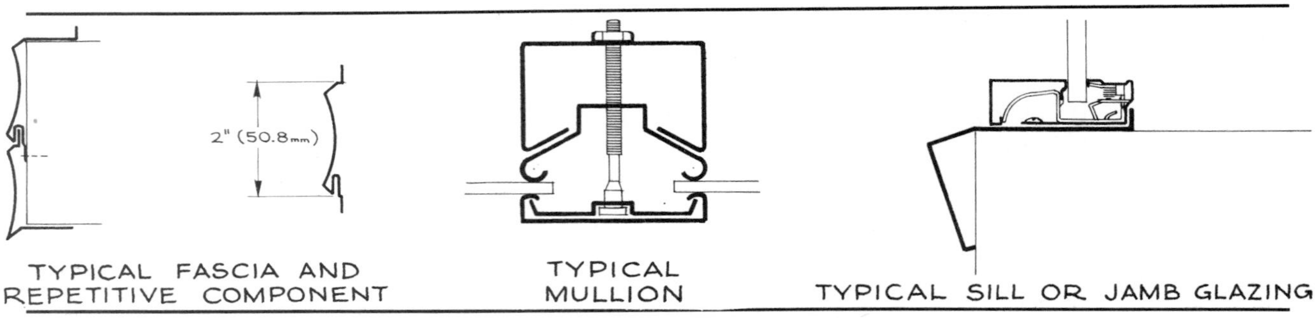

Figure S12 Typical components for stainless steel store fronts.

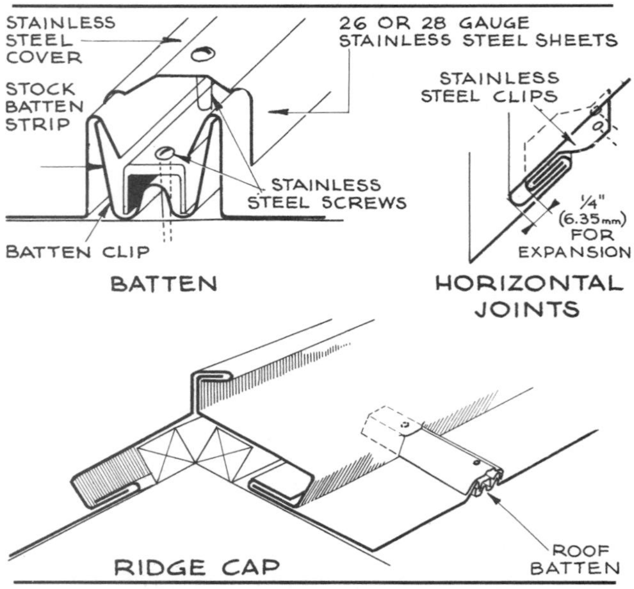

Figure S13 Details of stainless steel batten roofing.

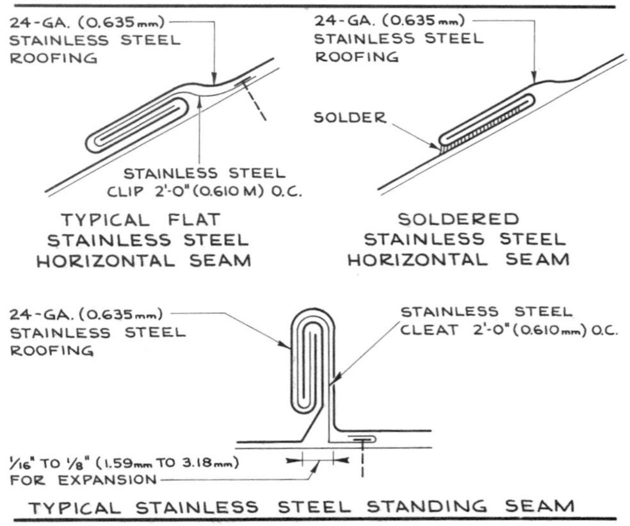

Figure S14 Typical examples of standing and flat stainless steel seams.

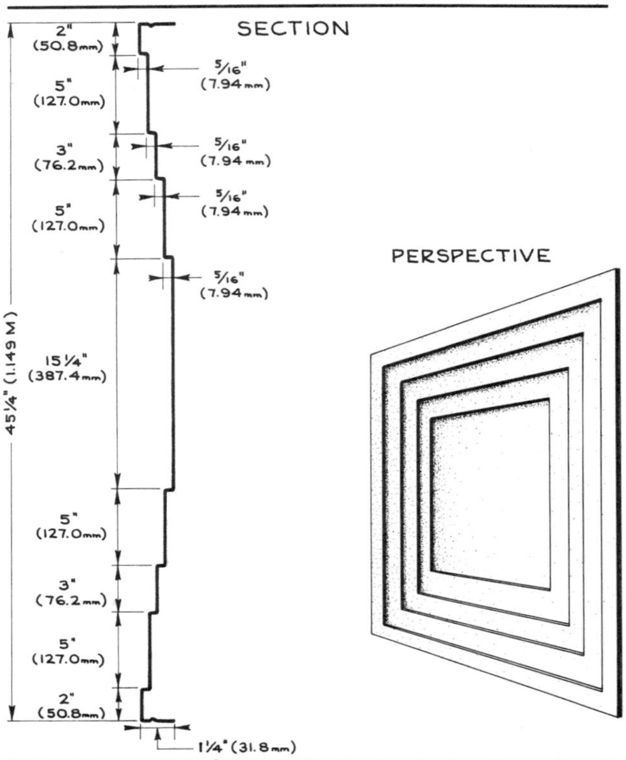

Figure S15 Pressed stainless steel spandrel panel for a curtain wall.

must be made for expansion, and stock stainless steel batten-type fasteners must be used. All other fastening devices should be stainless steel, and any roofing or flashing that is in contact with mortar should be protected with asphaltic coating. (*See* Figures *S13* and *S14*.) Stock batten strips, ridge caps, and closures are available. The general applications and details are similar to those for other types of metal roofing. (*See also* Aluminum Sheet and Strip; Monel; Terneplate.)

13. Panels and curtain walls: Stainless steel as shown in Figures *S15*, *S16*, and *S17* is used for building

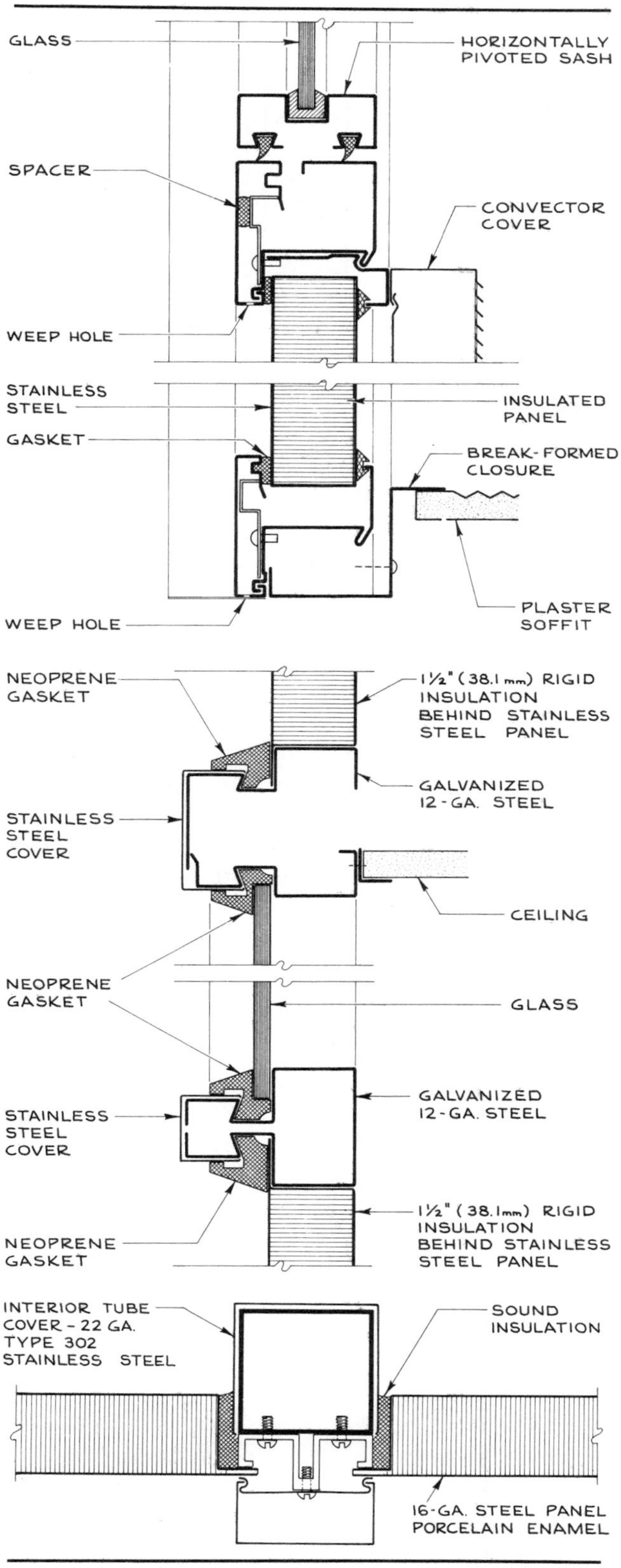

Figure S16 Typical stainless steel panels and curtain walls.

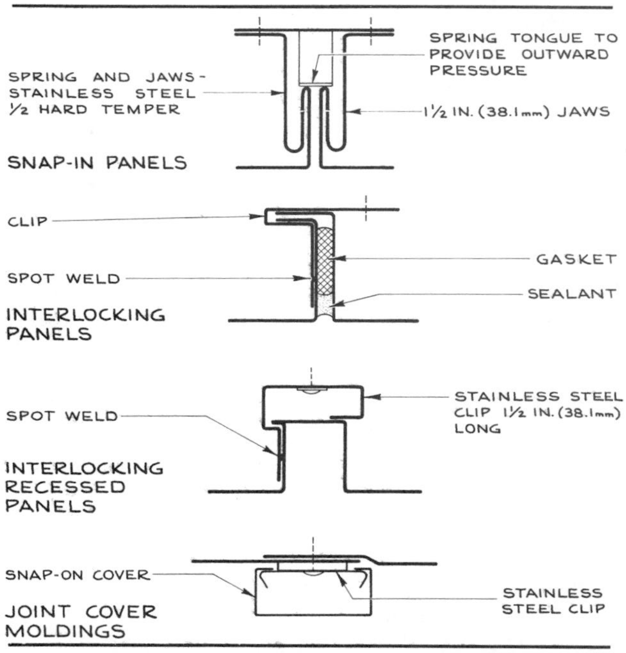

Figure S17 Methods of joining stainless steel panels and curtain walls.

panels and curtain walls. One should always check the type of stainless steel, the type of finish, and the methods of fastening panels or curtain walls to the structure.

Conditions Favorable to the Use of Stainless Steel Sheet, Strip and Plate

1. Where a strong, decorative, permanent, corrosion-resistant material that is relatively maintenance-free is required.
2. Where a surface with high reflectivity is needed.
3. Where a material that will not bend or distort and will permanently hold its shape is needed for cornices, fascias, moldings, gutters, and trim.

Conditions Unfavorable to the Use of Stainless Steel Sheet, Strip and Plate

1. Where economy is of major importance.
2. Where a fire-resistant material is needed.

History and Manufacture

Stainless steel sheet and strip were first used in industry. The first major use in construction was on the Empire State Building and the Chrysler Building. From then until World War II the use of stainless steel in the con-

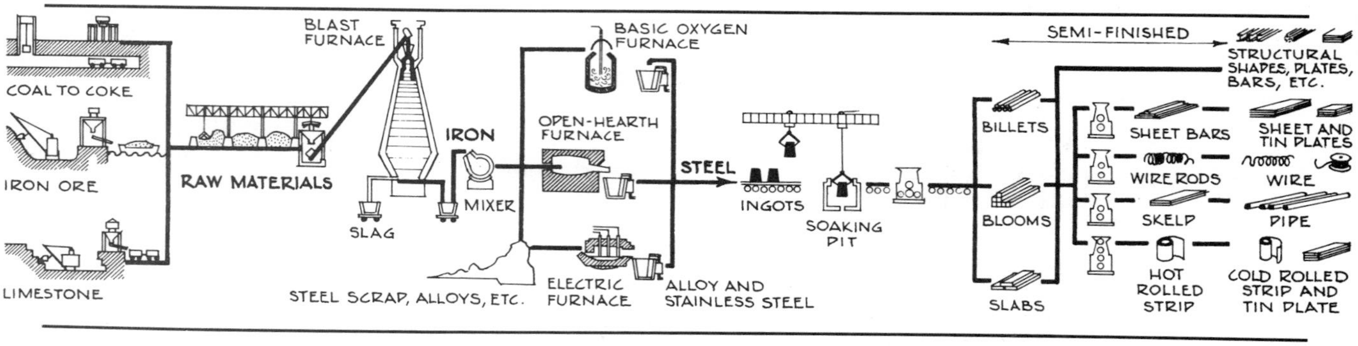

Figure S18 Process of manufacturing stainless steels.

struction field steadily increased. With the end of that war stainless steel again became available and continues to be one of the significant materials in contemporary construction.

The manufacture of sheet and strip is similar to the manufacture of steel sheet and strip with only one difference: the strength of stainless steel requires that the process be slower and more carefully controlled (*see* Figure *S18*).

STAINLESS STEEL TUBULAR SHAPES

Physical Properties

Stainless steel tubular shapes for construction are generally made from alloy Types 304, 316, and 430. The chemical and physical properties for all are fully covered under Stainless Steel.

Commercial Forms. Tubular products are available in pipe sizes ranging from $\frac{1}{8}$ to 12 in. (3.18 to 304.8 mm) and with outside diameters of 0.045 to 12.750 in. (1.14 to 323.85 mm). Available shapes include rounds, squares, rectangles, ovals, hexagons, and various others. All tubular shapes are supplied in six standard finishes.

Types and Uses

Stainless steel tubular products are fabricated by the following methods: (1) continuous welding of sheet material bent into the desired shape; (2) rotary piercing of solid billets; and (3) extrusion of solid billets, using glass as the lubricant (*see* Stainless Steel).

Table S29 Standard sizes and thicknesses of stainless steel pipe

Nominal pipe size[a]		Outside diameter		Nominal or average wall thickness per schedule number									
				No. 5		No. 10		No. 40		No. 80		No. 166	
in.	mm	in.	mm	in.	mm	in.	mm	in.	mm	in.	mm	in.	mm
$\frac{1}{2}$	12.70	0.840	21.336			0.083	2.108	0.109	2.769	0.147	3.734	0.187	4.750
$\frac{3}{4}$	19.05	1.050	26.670					0.113	2.870	0.154	3.912	0.218	5.537
1	25.40	1.315	33.401					0.133	3.378	0.179	4.547		
$1\frac{1}{4}$	31.75	1.660	42.164	0.065	1.654			0.140	3.556	0.191	4.851	0.250	6.350
$1\frac{1}{2}$	38.10	1.900	48.260			0.109	2.769	0.145	3.683	0.200	5.080	0.280	7.112
2	50.80	2.375	60.325					0.154	3.912	0.218	6.537	0.343	8.712
$2\frac{1}{2}$	63.5	2.875	73.025					0.203	5.156	0.276	7.010	0.375	9.525
3	76.2	3.500	88.900					0.216	5.508	0.300	7.620		
$3\frac{1}{2}$	88.9	4.000	101.600	0.083	2.108	0.120	3.048	0.226	5.740	0.318	8.077	0.438	11.125
4	101.6	4.500	114.300					0.237	6.020	0.337	8.560	0.531	13.030
5	127.0	5.563	141.300			0.134	3.404	0.258	6.533	0.375	9.525	0.625	15.875
6	152.4	6.625	169.275	0.109	2.769	0.135	3.429	0.280	7.112	0.432	10.973	0.718	18.237
8	203.2	8.625	219.075			0.148	3.759	0.322	8.179			0.906	23.102
10	254.0	10.750	273.050	0.134	3.404	0.165	4.191	0.365	9.271	0.500	12.700	1.125	28.575
12	304.8	12.750	323.850	0.165	4.191	0.180	4.572	0.375	9.525			1.312	33.325

[a] Also manufactured in $\frac{1}{8}$ in. (3.18 mm), $\frac{1}{4}$ in. (6.35 mm), and $\frac{3}{8}$ in. (9.53 mm). Always check with manufacturers for the availability of the sizes you are planning to use.

Table S30 Standard sizes and weights for square and rectangular tubular shapes[a]

Shape				Wall thickness gauges and standard weights															
				20		18		16		14		12		11		10		9	
in.	mm ×	in.	mm	lb/ft	kg/m	lb/ft	kg/m	lb/ft	kg/m	lb/ft	kg/m	lb/ft	kg/m	lb/ft	kg/m	lb/ft	kg/m	lb/ft	kg/m
1/2	12.70	1/2	12.70	0.221	0.329	0.301	0.448	0.385	0.573										
5/8	15.88	5/8	15.88	0.281	0.418	0.384	0.572	0.495	0.737	0.612	0.911								
3/4	19.05	3/4	19.05	0.340	0.506	0.467	0.695	0.606	0.902	0.753	1.121								
7/8	22.23	7/8	22.23	0.400	0.595	0.550	0.819	0.716	1.066	0.894	1.331								
1	25.40	1	25.40	0.459	0.683	0.634	0.944	0.827	1.231	1.035	1.540	1.321	1.966	1.436	2.137				
1 1/8	28.58	1 1/8	28.58	0.519	0.772	0.717	1.067	0.937	1.394	1.176	1.751	1.506	2.241	1.640	2.441				
1 1/4	31.75	1 1/4	31.75	0.578	0.860	0.800	1.191	1.048	1.560	1.317	1.960	1.691	2.517	1.844	2.744				
1 3/8	34.93	1 3/8	34.93	0.638	0.950	0.884	1.316	1.158	1.723	1.458	2.170	1.877	2.793	2.048	3.048				
1 1/2	38.10	1 1/2	38.10	0.697	1.037	0.967	1.439	1.269	1.889	1.600	2.381	2.062	3.069	2.252	3.351	2.489	3.704		
1 3/4	44.45	1 3/4	44.45	0.816	1.214	1.134	1.688	1.490	2.217	1.882	2.801	2.433	3.621	2.660	3.959	2.945	4.383		
1 7/8	47.63	1 7/8	47.63	0.876	1.304	1.217	1.811	1.600	2.381	2.023	3.011	2.618	3.896	2.864	4.262	3.173	4.722	3.476	5.173
2	50.80	2	50.80	0.935	1.392	1.300	1.934	1.711	2.546	2.164	3.230	2.803	4.171	3.068	4.566	3.401	5.061	3.728	5.548
2 1/4	57.15	2 1/4	57.15			1.467	2.183	1.932	2.875	2.446	3.641	3.174	4.724	3.476	5.173	3.856	5.739	4.231	6.297
2 1/2	63.50	2 1/2	63.50					2.153	3.204	2.728	4.060	3.544	5.274	3.884	5.780	4.312	6.417	4.734	7.045
2 5/8	66.68	2 5/8	66.68					2.263	3.368	2.869	4.270	3.729	5.550	4.088	6.084	4.539	6.755	4.985	7.419
3	76.20	3	76.20					2.595	3.862	3.293	4.901	4.286	6.378	4.700	6.995	5.223	7.773	5.471	8.142
3 1/2	88.90	3 1/2	88.90					3.036	4.518	3.857	5.740	5.017	7.481	5.516	8.209	6.134	9.129	6.747	10.041
4	101.60	4	101.60					3.479	5.177	4.422	6.581	5.768	8.584	6.332	9.423	7.045	10.484	7.753	11.538
3/8	9.53	1	25.40	0.310	0.461	0.425	0.633												
1/2	12.70	1	25.40	0.340	0.506	0.467	0.695	0.606	0.902	0.753	1.121								
1/2	12.70	1 1/4	31.75	0.400	0.595	0.550	0.819	0.716	1.066	0.894	1.331								
1/2	12.70	1 1/2	38.10	0.495	0.737	0.634	0.944	0.826	1.229	1.035	1.540								
5/8	15.88	3/4	19.05	0.311	0.463	0.426	0.634	0.550	0.819	0.682	1.015								
5/8	15.88	1 1/2	38.10	0.489	0.728	0.675	1.005	0.882	1.313	1.106	1.646	1.414	1.704						
5/8	15.88	2	50.80	0.608	0.905	0.842	1.253	1.103	1.642	1.388	2.066	1.784	2.655						
3/4	19.05	1	25.40	0.400	0.595	0.550	0.819	0.716	1.066	0.894	1.331								
3/4	19.05	1 1/4	31.75	0.459	0.683	0.634	0.944	0.827	1.231	1.035	1.540								
3/4	19.05	1 1/2	38.10	0.519	0.772	0.717	1.067	0.937	1.394	1.176	1.751								
3/4	19.05	1 3/4	44.45			0.800	1.191	1.047	1.560	1.317	1.960								
3/4	19.05	2	50.80	0.638	0.950	0.894	1.316	1.158	1.723	1.458	2.170	1.877	2.808	2.098	3.048				
7/8	22.23	1 1/4	31.75	0.489	0.728	0.675	1.004	0.882	1.313	1.106	1.646	1.414	2.104						
1	25.40	1 1/8	28.58	0.489	0.728	0.675	1.004	0.882	1.313	1.106	1.646								
1	25.40	1 1/4	31.75	0.519	0.772	0.717	1.067	0.937	1.394	1.176	1.751	1.506	2.241	1.640	2.441				
1	25.40	1 1/2	38.10	0.578	0.860	0.800	1.191	1.048	1.560	1.317	1.960	1.691	2.517	1.844	2.744				
1	25.40	1 3/4	44.45	0.638	0.950	0.884	1.316	1.158	1.723	1.458	2.170	1.877	2.793	2.048	3.048				

Table S30 Standard sizes and weights for square and rectangular tubular shapes[a] *(continued)*

Shape				Wall thickness gauges and standard weights															
				20		18		16		14		12		11		10		9	
in.	mm ×	in.	mm	lb/ft	kg/m	lb/ft	kg/m	lb/ft	kg/m	lb/ft	kg/m	lb/ft	kg/m	lb/ft	kg/m	lb/ft	kg/m	lb/ft	kg/m
1	25.40	2	50.80	0.697	1.037	0.967	1.439	1.269	1.889	1.600	2.381	2.062	3.069	2.252	3.351	2.489	3.718		
1	25.40	$2\frac{1}{2}$	63.50	0.816	1.214	1.134	1.688	1.490	2.217	1.882	2.801	2.433	3.621	2.660	3.959	2.945	4.382		
1	25.40	3	76.20			1.300	1.934	1.711	2.546	2.164	3.230	2.803	4.171	3.068	4.566	3.401	5.601		
1	25.40	$3\frac{1}{2}$	88.90			1.467	2.183	1.932	2.876	2.446	3.641	3.174	4.724	3.476	5.173	3.856	5.739		
$1\frac{1}{4}$	31.75	$1\frac{3}{4}$	44.45	0.697	1.037	0.967	1.439	1.269	1.889	1.600	2.381	2.062	3.069	2.252	3.351				
$1\frac{1}{4}$	31.75	2	50.80	0.757	1.217	1.050	1.563	1.379	2.052	1.741	2.591	2.247	3.434	2.456	3.655				
$1\frac{1}{4}$	31.75	3	76.20			1.383	2.058	1.821	2.710	2.305	3.430	2.989	4.299	3.272	4.859	3.628	5.399	3.979	5.922
$1\frac{1}{4}$	31.75	$3\frac{1}{2}$	88.90			1.550	2.307	2.042	3.039	2.587	3.850	3.359	5.000	3.680	5.477	4.084	6.078	4.483	6.677
$1\frac{1}{4}$	31.75	4	101.60			1.716	2.619	2.263	3.368	2.869	4.270	3.729	5.550	4.088	6.084	4.539	6.755	4.995	7.434
$1\frac{3}{8}$	34.93	$1\frac{1}{2}$	38.10			0.661	0.983	1.213	1.705	1.528	2.274	1.969	2.930	2.150	3.200				
$1\frac{3}{8}$	34.93	$6\frac{3}{8}$	161.93					3.368	5.012	4.280	6.370	5.582	8.307	6.128	9.120	6.817	9.207	7.501	11.163
$1\frac{1}{2}$	38.10	2	50.80	0.816	1.214	1.134	1.688	1.490	2.084	1.882	2.801	2.433	3.621	2.660	3.959				
$1\frac{1}{2}$	38.10	$2\frac{1}{4}$	57.15			1.217	1.811	1.600	2.381	2.023	3.011	2.618	3.896	2.864	4.261	3.173	4.722	3.476	5.173
$1\frac{1}{2}$	38.10	$2\frac{1}{2}$	63.50			1.300	1.934	1.711	2.546	2.164	3.221	2.803	4.172	3.068	4.566	3.401	5.601	3.728	5.548
$1\frac{1}{2}$	38.10	3	76.20			1.467	2.183	1.932	2.875	2.446	3.640	3.174	4.724	3.476	5.173	3.850	5.739	4.231	6.297
$1\frac{1}{2}$	38.10	$3\frac{1}{2}$	88.90			1.633	2.430	2.153	3.204	2.728	4.060	3.544	5.263	3.884	5.780	4.312	6.417		
$1\frac{3}{4}$	44.45	2	50.80			1.217	1.811	1.600	2.381	2.023	3.011	2.618	3.896	2.864	4.761	3.173	4.722	3.476	5.173
$1\frac{3}{4}$	44.45	4	101.60			1.883	2.802	2.484	3.697	3.151	4.689	4.100	6.008	4.496	6.691	4.995	7.434	5.488	8.167
2	50.80	3	76.20			1.633	2.430	2.153	3.204	2.728	4.060	3.544	5.263	3.884	5.780	4.312	6.417	4.734	7.045
2	50.80	4	101.60					2.595	3.862	3.293	4.901	4.286	6.378	4.700	6.995	5.223	7.773	5.741	8.544
2	50.80	5	127.00					3.037	4.520	3.857	5.740	5.027	7.481	5.516	8.209	6.134	9.129	6.747	10.041
$2\frac{1}{2}$	63.50	3	76.20					2.374	3.533	3.011	4.060	3.915	5.827	4.292	6.387	4.767	7.094	5.237	7.794
$2\frac{1}{2}$	63.50	4	101.60					2.816	4.191	3.575	5.387	4.656	6.929	5.108	7.601	5.679	8.452	6.244	9.292
$2\frac{1}{2}$	63.50	5	127.00					3.258	4.849	4.139	6.160	5.397	8.032	5.924	8.817	6.590	9.807	7.250	10.790
3	76.20	$3\frac{1}{2}$	88.90					2.816	4.191	3.575	5.387	4.056	6.929	5.108	7.601	5.679	8.452	6.244	9.292
3	76.20	4	101.60					3.037	4.520	3.857	5.740	5.027	7.481	5.516	8.209	6.134	9.129	6.747	10.041
3	76.20	5	127.00					3.479	5.177	4.422	6.581	5.768	8.584	6.332	9.423	7.045	10.484	7.763	11.538

[a]If no value is given in any column under gauge number, the tubular shape in question is not manufactured in that thickness.

Table S31 Finishes for stainless steel tubular shapes

Designation of finish	Type of finish	How obtained	Relation to finishes for sheet	Major use
No. 2 or P	Matte surface	Annealed and pickled	Similar to No. 1 finish for sheet	In concealed areas
No. 2 A	Matte surface	Pickled but not annealed	Similar to No. 2D finish for sheet	Where reflectivity is unimportant
No. 2B	Semireflective surface	Bright-annealed	Similar to No. 2B finish for sheet	Exposed areas where reflectivity is secondary
SWP	Smooth white	Grit-blasted and then pickled		In areas where contrast with other stainless steels is desired
80 grit	Smooth surface	Ground with No. 80 grit abrasive	Similar to No. 2B finish for sheet	Exposed areas where reflectivity is secondary
120 grit	Semipolished surface	Ground with No. 120 grit abrasive	Similar to No. 4 finish for sheet	Most generally used finish
180 grit	Satin polished surface	Ground with No. 180 grit abrasive	Similar to No. 6 finish for sheet	For contrast with high-polish finishes
240 grit	High-luster polished surface	Ground with No. 240 grit abrasive	Similar to No. 7 finish for sheet	Where high reflectivity is important
320 grit	Mirror finish	Ground with No. 320 grit abrasive	Similar to No. 8 mirror finish for sheet	Special decorative effect where mirror-like quality can be used to advantage

Generally, stainless steel tubular shapes are made from Types 316 and 430, and continuous welded tubing from Type 304 (because of its superior welding characteristics). Stainless steel pipe is made in stock sizes but varies widely in thickness for the different sizes of pipe, as shown in Table *S29*.

Stainless welded tubing is made in diameters up to 30 in. (762 mm) o.d., and seamless tubing up to $8\frac{5}{8}$ in. (219.1 mm) o.d. Squares, rectangles, ovals, and other shapes are available ranging from $\frac{1}{2}$-in. (12.7-mm) squares to rectangles 3 in. × 5 in. (76.2 mm × 127 mm) in size (*see* Table *S30*). Generally, these shapes are formed from $8\frac{5}{8}$-in. (219.1-mm) or smaller tubing.

Finishes. All round tubing can be polished; square, rectangular, and other shapes can be drawn from polished round tubing in some cases. The finishes for tubular shapes are described in Table *S31*, which gives the manufacturer's designation for each type of finish, the method whereby it is obtained, and the major use.

Construction Uses. Stainless steel tubular products are used in construction for doors, door frames, railings, screens, window frames, etc.

Application

Condensed Checklist

1. The current availability of the various stainless steel shapes and finishes should be checked.
2. Shop drawings that show all details and describe all methods of attachment should be required by the specifications.
3. Welding procedures must conform to the recommendations of the stainless steel manufacturers.
4. Fastening devices should be of stainless steel. If secured to a material other than stainless steel, they should be isolated to prevent the possibility of galvanic action.

History and Manufacture

Since 1931 the use of stainless steel tubular shapes has steadily increased until today they are an everyday material in the construction field.

Tubing is manufactured by two processes; the seamless and the continuous weld. With stainless steel, greater care and control are necessary than with steel; otherwise the processes are the same as those used for steel.

STAINLESS STEEL WINDOWS AND DOORS

Physical Properties

Stainless steel windows and doors are manufactured from stainless steel Types 202, 302, 304, and 430 in gauge Nos. 16, 18, 20, and 22, with a minimum tensile strength of 75,000 lbf/in.^{2} (517.13 MN/m^2). All are rolled, formed, or shaped, then shop-welded, and usually given a No. 2B or No. 4 finish. Stainless steel extrusions are generally used for doors. Hardware is of stainless steel or of a metal that is plated (generally with chromium or cadmium) so that no galvanic action can occur. Windows and doors are available in limited types and sizes.

Types and Uses

Stainless steel windows are manufactured in limited stock sizes and types as shown in Figures *S19* through *S23*. A study of the illustrations showing details of reversible and awning-type windows explains how great strength is obtained from very thin gauges of metal.

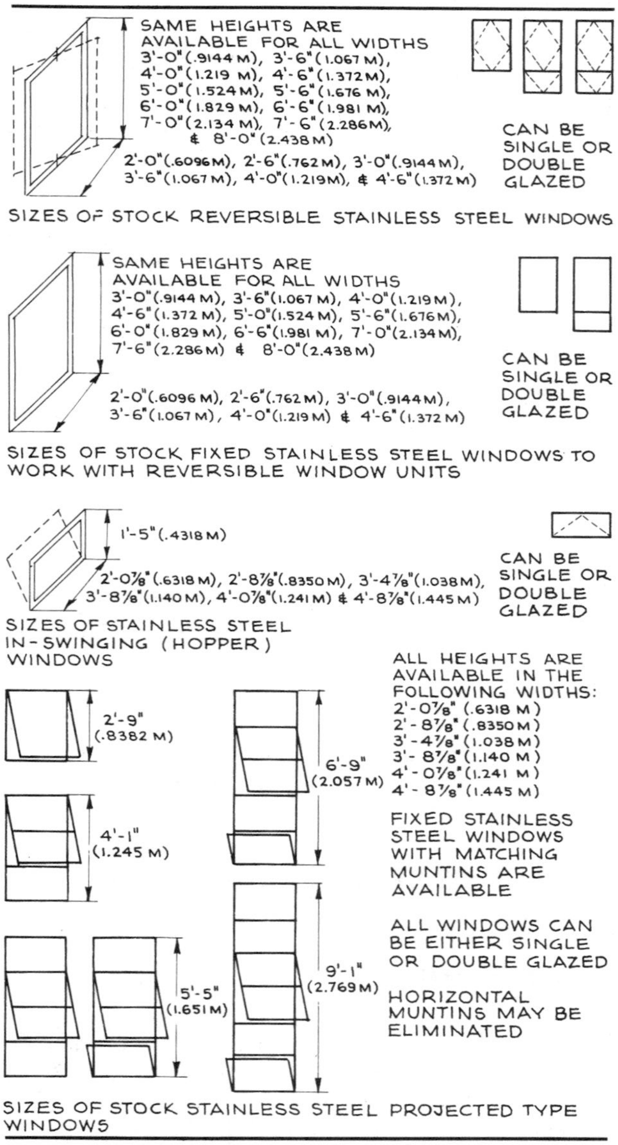

Figure S19 Sizes of typical stainless steel windows.

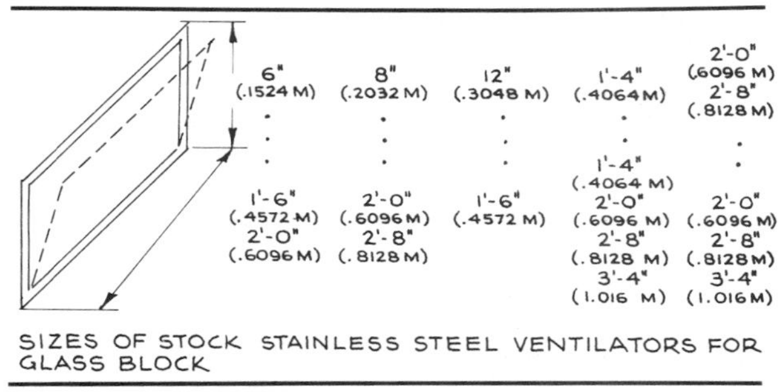

Figure S20 Sizes of stock stainless steel ventilators for glass block.

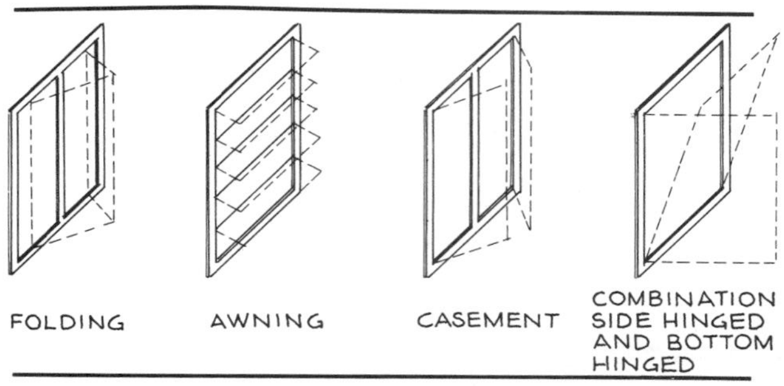

Figure S21 Miscellaneous types of stainless steel windows.

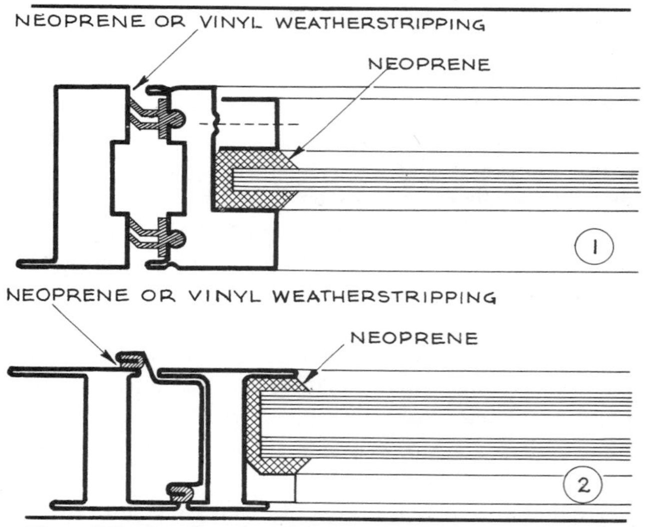

Figure S22 Details of stainless steel awning and pivot-type windows.

Stainless steel doors are manufactured from extruded shapes and/or rolled, formed, and shaped shop-welded sheet and strip in gauges Nos. 16, 18, 20, 22, and 24 (*see* Figures *S24* and *S25*). The doors are available with narrow stiles (1 to $2\frac{1}{4}$ in., or 25.4 to 57.15 mm), with medium stiles ($2\frac{1}{2}$ to $3\frac{1}{2}$ in., or 63.5 to 88.9 mm), and with wide stiles (4 to 8 in., or 101.6 to 203.2 mm), and in the following types: swing, balanced, revolving, and sliding.

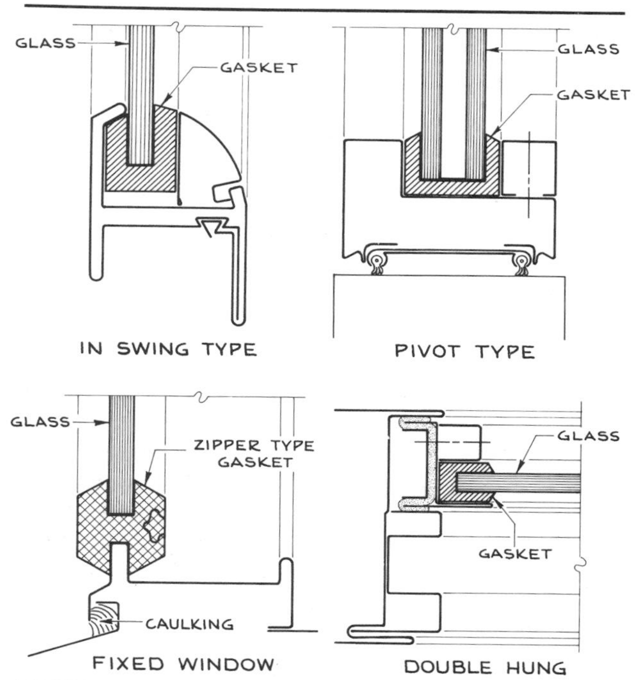

Figure S23 Typical examples of stainless steel windows.

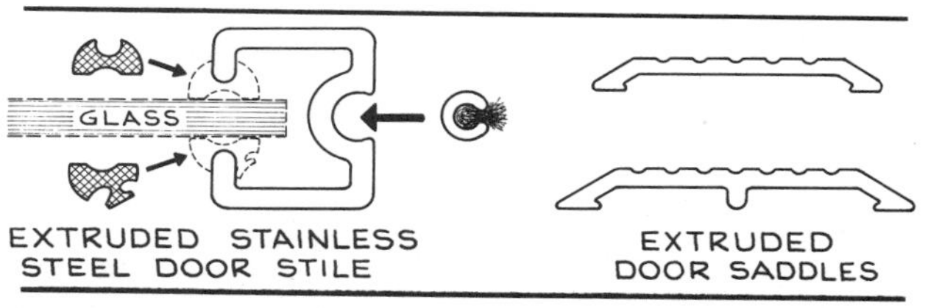

Figure S24 Typical examples of extruded door frame and saddle.

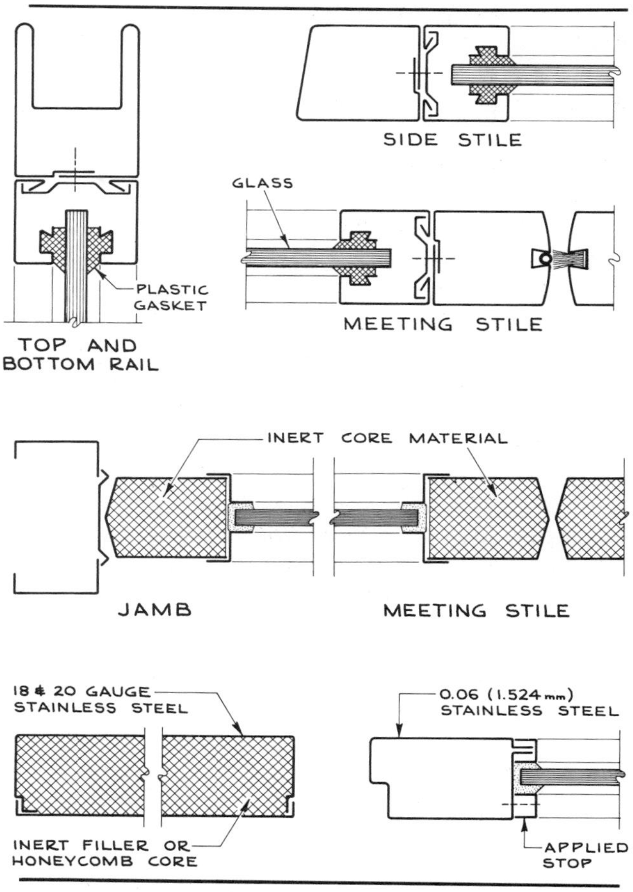

Figure S25 Typical stainless steel doors.

Application

Condensed Checklist

1. One should always check that the materials into which stainless steel windows and doors are to be installed are compatible, that is, that they should not cause galvanic action or affect the stainless steel. If the materials are masonry, the stainless steel should be protected with a heavy coat of rubber-base asphalt wherever it comes in contact with mortar. In the case of aluminum, complete isolation of the materials and fastenings must be achieved. Stainless steel and aluminum must never come into direct contact.
2. All hardware and fastening devices must be made of a metal or be plated with a metal that is compatible with stainless steel.
3. The type of finish available for the stainless steel window and doors is important wherever other stainless steel is adjacent so that the two either match or contrast.
4. Latest catalogues and manufacturers of stainless steel windows and doors should be consulted for the stock sizes currently available.
5. Shop drawings should always be specified, and all fastening devices and methods of attachment should be shown in these drawings.
6. The manufacturers should always be consulted for methods of glazing and correct glazing materials.

Conditions Favorable To the Use of Stainless Steel Windows and Doors

1. In areas where a strong, permanent, maintenance-free type of window is desired.
2. Where there is a maximum of operating sash and most windows are fixed glass because of complete air conditioning.
3. In areas where strong, permanent, maintenance-free doors are desired.

Conditions Unfavorable to the Use of Stainless Steel Windows and Doors

Where economy is important.

History and Manufacture

Stainless steel windows and doors as manufactured stock construction materials are a product of the 1950s. In an air-conditioned building, a minimum number of movable windows for natural ventilation is necessary, and the plastics and synthetic rubber industries have answered the problem of how to weatherseal the stainless steel rolled forms and shapes in relation to each other and other materials.

Fabrication begins when the stainless steel strip or sheet is formed into an appropriate curved or tubular section for window sash and frames and for doors and frames. These are cut to size with mitered corners and assembled. All corners are welded under carefully shop-controlled conditions so that strong, secure, and weather-tight joints are made. All welded surfaces are carefully ground and polished to match. Narrow-stile doors are manufactured from stainless steel extrusions.

STAINLESS STEEL WIRE

Physical Properties

Stainless steel wire for construction use is made from Types 301, 302, 304, 316, and 430. Its tensile strength varies from 95,00 to 350,000 lbf/in.^{2} (655.03 to 2413.25 MN/m^2). Only the chromium-nickel types develop high strengths. Stainless steel wire is available in coils in diameters ranging from 0.5 to 0.003 in. (12.7 to 0.076 mm), inclusive; in straightened and cut form it is available in diameters of 0.03 in. (0.762 mm) and larger. It is also available in various finishes including

Table S32 Stainless steel wire cloth

Type of mesh	Wire size in.	Wire size mm	Width of opening in.	Width of opening mm	Open area (percent)
4 × 4	0.047	1.194	0.203	5.156	65.9
4 × 4	0.035	0.889	0.215	5.461	74.0
5 × 5	0.035	0.889	0.165	4.191	68.0
5 × 5	0.023	0.584	0.177	4.496	78.3
8 × 8	0.032	0.813	0.093	2.362	54.4
8 × 8	0.028	0.711	0.097	2.464	60.2
8 × 8	0.025	0.635	0.100	2.540	64.0
10 × 10	0.032	0.813	0.068	1.727	46.2
10 × 10	0.025	0.635	0.075	1.905	56.3
10 × 10	0.023	0.584	0.077	1.956	59.3
12 × 12	0.023	0.584	0.060	1.524	51.8
12 × 12	0.018	0.457	0.065	1.651	60.8
16 × 16	0.018	0.457	0.045	1.143	50.7
20 × 20	0.016	0.406	0.034	0.864	43.2
24 × 24	0.014	0.356	0.028	0.711	44.2
30 × 30	0.017	0.432	0.016	0.406	23.9
30 × 30	0.012	0.305	0.021	0.533	40.8
40 × 40	0.010	0.254	0.015	0.381	36.0
50 × 40	0.009	0.229	0.011	0.279	31.7
60 × 50	0.0065	0.165	0.010	0.254	39.4
60 × 60	0.011	0.279	0.006	0.165	11.7
80 × 70	0.0055	0.1397	0.007	0.178	34.5
100 × 90	0.0045	0.1143	0.006	0.165	33.0
150 × 150	0.0026	0.0660	0.0041	0.104	37.4
180 × 180	0.0025	0.0635	0.0031	0.079	30.6
200 × 200	0.0021	0.0533	0.0029	0.0737	33.6
325 × 325	0.0014	0.0356	0.0017	0.0432	30.0
400 × 400	0.0010	0.0254	0.0015	0.0381	36.0

Table S33 Stainless steel wire commonly used in construction

Type of wire	Type of stainless steel used	Tensile strength lbf/in.^{2}	Tensile strength MN/m^2	Major uses
Cold drawn	302, 430	90,000–100,000	620.55–689.50	Nuts, bolts, screws, rivets, and similar products
Weaving	302	70,000–150,000	482.65–1034.25	Wire mesh and cloth
Rope	302	140,000–355,100	965.30–2448.42	Tension members and wire rope

pickled; metallic coated; oil, diamond, or soap drawn; and ground and polished.

Types and Uses

The types of stainless steel wire generally used in construction are shown in Tables *S32* and *S33*. Finishes are shown in Table *S34*.

Stainless steel wire is used for decorative screens and railings and for other decorative purposes. It is fabricated into nuts, bolts, screws, and rivets. It is also used in prestressed concrete work.

Application

Condensed Checklist

1. When designing with stainless steel wire tension members, it is always advisable to consult with structural engineers. The chemical composition of the wire, the method of manufacture, and the type of treatment (annealed, cold drawn, hot drawn) affect its strength in relation to its end use as a tension member.
2. When stainless steel wire is used for decorative effects, available sizes and types of finish should be checked. Stainless steel mesh and wire cloth are generally available in all sizes common to manufactured metal products, but the type of finish available in a desired mesh size may be limited.
3. Nuts, bolts, screws, rivets, and similar products should not be adjacent to metal that might cause galvanic action.

Conditions Favorable to the Use of Stainless Steel Wire

1. In structural areas for tension members where a small-sized, noncorrosive, easily maintained, and permanent material is necessary.

Table S34 Finishes for stainless steel wire

Finish	Description
Pickled	Nonreflective finish confined to annealed wire and wire from which the drawing lubricant has been removed
Metallic coated	Nonreflective finish produced by drawing with copper or lead, which acts as lubricant during manufacture
Oil drawn	Reflective finish produced by using oil as lubricant
Diamond drawn	Smooth, bright finish confined to 0.03-in. (0.762-mm) and finer wire
Soap drawn	Semibright finish with minimum luster
Ground and polished	Highly reflective polished finish, confined to straightened and cut wire $\frac{3}{32}$ in. (2.38 mm) and larger in diameter

2. In the form of stainless steel nuts, bolts, screws, rivets, and similar products for the assembly or attachment of stainless steel to other materials.
3. Where a noncorrosive, permanent, easily maintained mesh or wire cloth is required.

Condition Unfavorable to the Use of Stainless Steel Wire

Where design features are secondary and only tensile strength is important.

History and Manufacture

Stainless steel wire was first used industrially, but soon found its way into construction applications, first for nuts and bolts, screws, and rivets, and later, for wire mesh and cloth. When recently developed construction systems using exposed or covered tension members (similar to catenaries) were put into practice for contemporary buildings, stainless steel wire and wire rope, because of their strength and permanence, were preferred materials.

The process of wire drawing is generally the same as that used for other metals (*see* Wire and Wire Rope). The main differences are due to the strength of stainless steel, which requires different types of lubrication and more careful control.

STAIRS

In large buildings today, stairs are used mainly for fire exits, although in residences they still serve as the major means of travel from one floor to another. For the former, both the entire stair and the surrounding structure must be of fireproof materials such as concrete or steel, whereas for residences the stair can be made of any material (wood, aluminum, concrete, or any combination of materials).

The monumental or grand staircase is occasionally used in larger buildings but only for decorative purposes, and in most localities building codes permit only one flight of stairs, that is, from the main floor to a mezzanine or the second floor, but not beyond that point.

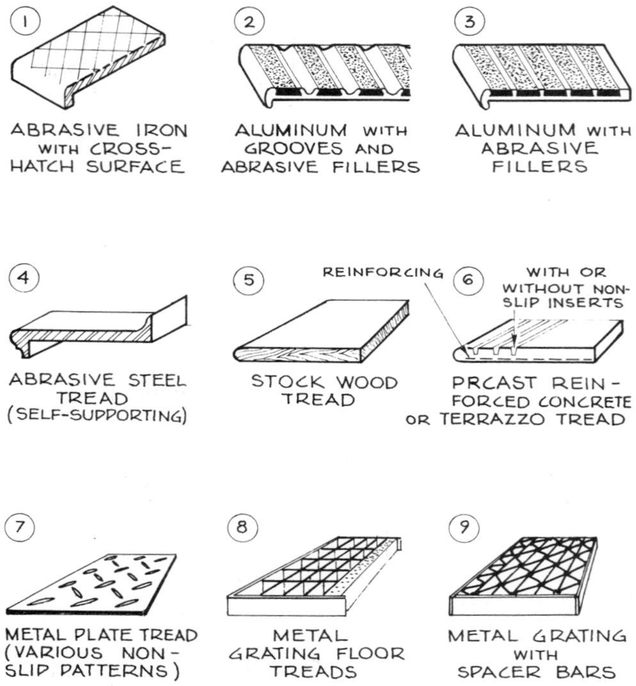

Figure S26 Typical nonslip nosings and prefabricated stair treads.

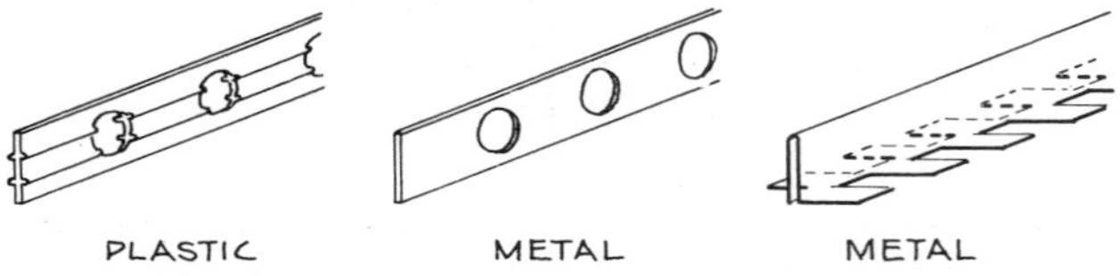

Figure S27 Typical abrasive insert stips for concrete or terrazzo stairs.

Most codes still limit the number of steps in one flight to 12 and the width to 1.83 ft (0.559 m) per person, setting a minimum width of 3.66 ft (1.118 m) and a maximum of 5.5 ft (1.676 m) without the introduction of a center railing. Stairs are designed to fit the human body, and the ratio of tread to riser is the controlling factor. A good rule to follow is that the height of the riser multiplied by the length of the tread should always be equal to a minimum of 70 and a maximum of 80 in.^{2}. For example, a good stair has a 7-in. (0.1778 m) riser and a 10-in. (0.254 m) tread; thus $7 \times 10 = 70$ in.^{2} ($0.1778 \times 0.254 = 0.0452$ m^2).

Many of the stairs installed in buildings are constructed wholly or in part of stock components, and in some cases are completely fabricated stairs made up of these components. These stock parts include treads, risers, and nonslip nosings for the stairs proper (*see* Figures *S26* and *S27*), as well as the completely prefabricated stairs.

Hand railings and their numerous component parts represent another large diversified group. All these stock prefabricated items must be considered from the viewpoint of proper material in relation to design and usage.

Completely prefabricated stairs are available in three main types: iron circular stairs, disappearing stairs (fold-down type), and various package units of prefabricated stairs, usually of wood (*see* Figures *S28* through *S30*).

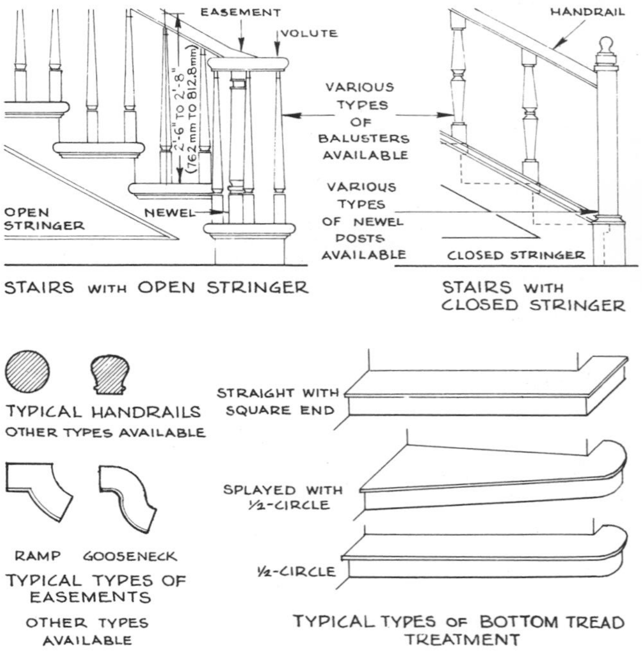

Figure S28 Prefabricated wood stairs, including available stock components.

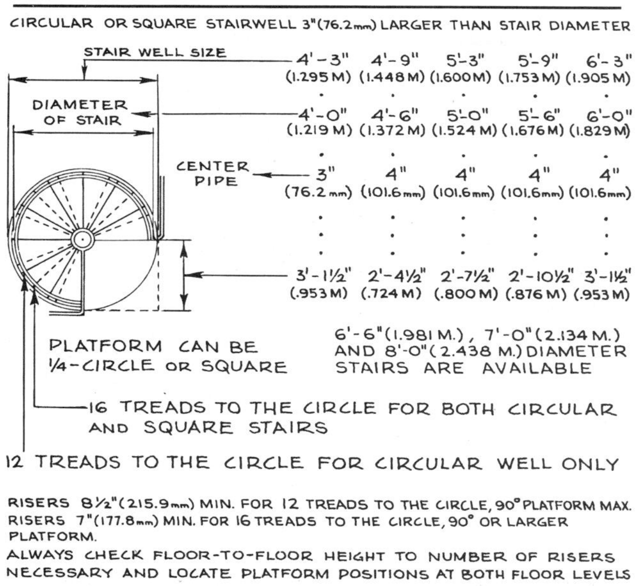

Figure S29 Typical prefabricated iron circular stairs.

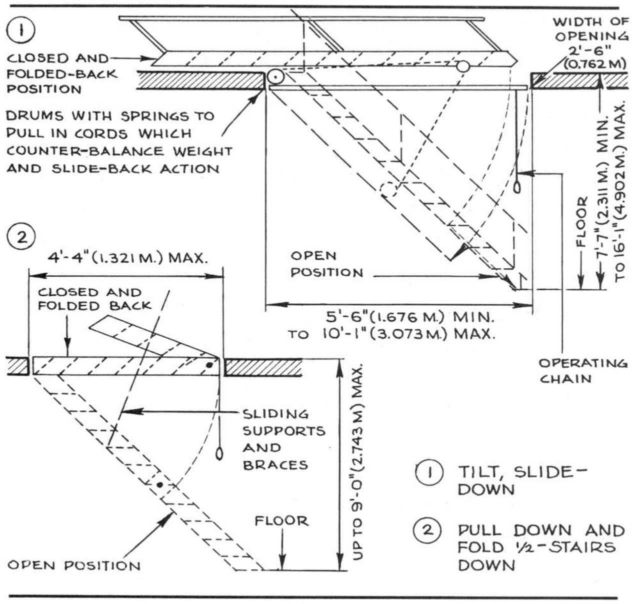

Figure S30 Typical prefabricated disappearing-type stairs.

STEEL

Physical and Chemical Properties

The word "steel" usually refers to plain carbon steels, defined as alloys of iron and carbon that do not contain more than 2% carbon and are malleable in block or ingot form. Stainless steels and alloy steels are described separately (*see* Stainless Steel; Steel: Alloy Steels).

In the plain or straight carbon steels the iron is always in excess of 95%. Phosphorus, sulfur, oxygen, and nitrogen are present, the last three always as impurities. Manganese, silicon, aluminum, copper, and nickel may be present either as residual impurities from the furnace method used or as elements deliberately added in small quantities to control the properties of the steel.

Official Definition of Carbon Steel. Officially a steel is classed as carbon steel when (1) no minimum content is specified or required for aluminum, boron, chromium, cobalt, columbium (niobium), molybdenum, nickel, titanium, tungsten, vanadium, zirconium, or any other element added to obtain a desired effect; or (2) the specified minimum for copper does not exceed 0.40%; or (3) the maximum content specified for any of the following elements does not exceed the stated percentage: manganese 1.65%, copper 0.60%, and silicon 0.60%.

Modification of Properties. The properties of a carbon steel vary greatly not only in relation to chemical composition, that is, the amount of carbon and other elements present; but they are also controlled to a great extent by the kind of heat treatment and mechanical work used during manufacture, that is, whether the carbon steel is cast, hot- or cold-rolled, slowly or rapidly cooled.

Classification and Designation. Because so many factors control the characteristics of carbon steel and other steels, there exists no simple system of designation or identification. Steel may be specified according to (1) chemical composition, (2) method of manufacture, (3) mechanical properties, (4) a system of numerical designations of grades of standard steels. The numerical designations are briefly summarized under Iron Alloys.

Classification According to Carbon Content. Both cast and wrought carbon steels have been generally classified into approximate grades as follows on the basis of carbon content:

Ingot iron: lowest possible content of carbon and other alloying elements; made in basic open-hearth furnaces; corrosion resistant; used largely for sheet.

Extra soft or dead soft steel: 0.08 to 0.18% carbon; this type of steel is characterized by ductility, toughness, weldability, and is used where cold-workability is desirable; generally it is used where strength or stiffness is not important (e.g. pipe, rivets, sheet, wire).

Table S35 Effect of carbon content on strength of typical carbon steels

Type of steel	Carbon percentage	Tensile strength of annealed steel[a] Ultimate[b] lbf/in.	Ultimate[b] MN/m	Yield[b] lbf/in.	Yield[b] MN/m
Very mild	0.05–0.15	40,000–55,000	275.80–379.23	24,000–30,000	165.48–206.85
Mild	0.15–0.25	48,000–65,000	303.96–448.18	30,000–36,000	206.85–248.22
Low-carbon	0.25–0.40	60,000–70,000	413.70–482.65	36,000–40,000	248.22–275.80
Medium-carbon	0.40–0.65	70,000–80,000	482.65–551.60	40,000–48,000	275.80–303.96
Higher-carbon	0.60–0.70	80,000–94,000	551.60–648.13	48,000–56,000	303.96–386.12
Spring	0.70–0.80	94,000–118,000	648.13–813.61	56,000–64,000	386.12–441.28
Pearlitic	0.75–0.85	120,000	827.40	70,000	482.65

[a] 1 lbf/in.^{2} = 0.006895 MN/m^2.
[b] Approximate values. The amount of working, annealing, etc., controls the exact figures.

Mild, structural grade: 0.15 to 0.29% carbon; characterized by strength combined with easy machinability; used for buildings, bridges, bolts, boilers, railroad rolling stock.

Medium grade: 0.25 to 0.35% carbon; harder and stronger than mild structural grade yet can be hot-forged; used for shipbuilding and machinery.

Medium hard grade: 0.35 to 0.65% carbon.

Spring grades: 0.85 to 1.05% carbon.

High-carbon tool steels: 1.05 to 1.20% carbon.

The role of the carbon–iron equilibrium is given in detail under Iron Alloys.

Effect of Carbon Content. Carbon in amounts up to 0.8% increases strength with each increment of carbon added and decreases ductility in the same proportion, as shown in Table *S35*. Hardness also increases with greater carbon content. For structural purposes, steel needs carbon to give strength but not in such quantity as to affect ductility. For example, steel bars or rods used for reinforcing concrete may contain as much carbon as will still allow them to be twisted and bent cold; cold-rolled steel must have enough toughness so it can be cold-deformed.

Effect of Other Elements. The elements generally added to carbon steels are manganese and silicon. Manganese minimizes hot shortness during the rolling of steel and is important for contributing to the strength, toughness, and hardness of the wrought product. It also reduces the oxygen and sulfur content. Silicon reduces the oxygen content and provides increased soundness and slightly higher strength.

The elements most harmful to steel are sulfur (a maximum of 0.05% for structural steels), as it affects weldability and can cause cracking during forging and rolling; and phosphorus, as it makes steel brittle, fragile under shock, and nonuniform in ductility. Sulfur, although kept at 0.05% maximum for structural steel and high-strength low-alloy steels, is added to many steels when good machinability is required. Phosphorus, although kept at 0.04% maximum for structural steel and high-strength low-alloy steels, is added to some steels to provide increased corrosion resistance. Cor-ten A has 0.07 to 0.15% of phosphorus, and Mayari R has 0.12% maximum of phosphorus (*see* Steel Alloys).

Lead is added to improve the machinability of steel. Lead and iron do not alloy; instead, the lead is distributed throughout the steel and held in suspension. When the steel is worked with cutting tools, the lead helps it break off rapidly and makes it easily machined. (*See also* Iron Alloys for specific effects of other elements.)

Grain Size. The grain size (size of crystals) influences both the strength and the toughness of the steel (*see* Table *S36*). Generally, the smaller the crystal, the greater will be the strength and toughness.

Table S36 Effect of grain size on basic characteristics of steel

Characteristic affected	Effect: Fine grain (small crystals)	Effect: Coarse grain (larger crystals)
Strength	Superior	Inferior
Ductility	Superior	Inferior
Toughness	Greatly superior	Inferior
Formability	Inferior	Superior
Machinability (rough)	Inferior	Superior
Machinability (finishing)	Superior	Inferior

The grain size of the steel is influenced by the following: the composition of the steel, the kind of processing and heat treatment if heat treatment is used, and, especially the range of temperature in which the steel is worked. Generally, the lower the processing temperatures, the finer is the grain size (*see* Figure *S31*). When steel is made to conform to specific fine-grain or coarse-grain numbers, the grain size referred to is that obtained upon heating the steel to an elevated temperature similar to that used for heat treatment, and this grain size is controlled by alloying elements in the steels, such as aluminum. For specification purposes, a steel is considered fine grained when the grain size numbers are from 5 to 8 inclusive, and coarse grained when the grain size numbers are from 1 to 5 inclusive. These requirements are accepted if 70% of the grains examined fall within these percentages.

Workability. Carbon steels can be wrought, rolled, cast, forged, and welded, but not, at present, extruded.

Experimental methods for extruding steel have advanced sufficiently to predict that extruded as well as cast and wrought steel shapes will in time be available for use in construction. When extrusions do become available, radical changes in handbooks, data, formulas, textbooks and in many of the now-standard construction materials can be expected.

Cold extrusions are available in many forms: bolts, rivets, nuts, and various small parts for gasoline engines and machines of all types. In the construction field cold extrusions at present are limited to nuts, bolts, rivets, and some small structural shapes (*see* Stainless Steel). Since the production of cold extrusions has almost doubled every 5 years, it will be necessary always to check with manufacturers of steel and stainless steel products to determine what is currently available.

Commercial Forms. Wrought carbon steels are available as sheet and strip, structural shapes, bars, rods, plate, pipe, tubing, and wire. The cast steels are available in almost any cast form. For the typical composition of a cast carbon steel commonly used in construction, *see* Table *S37*.

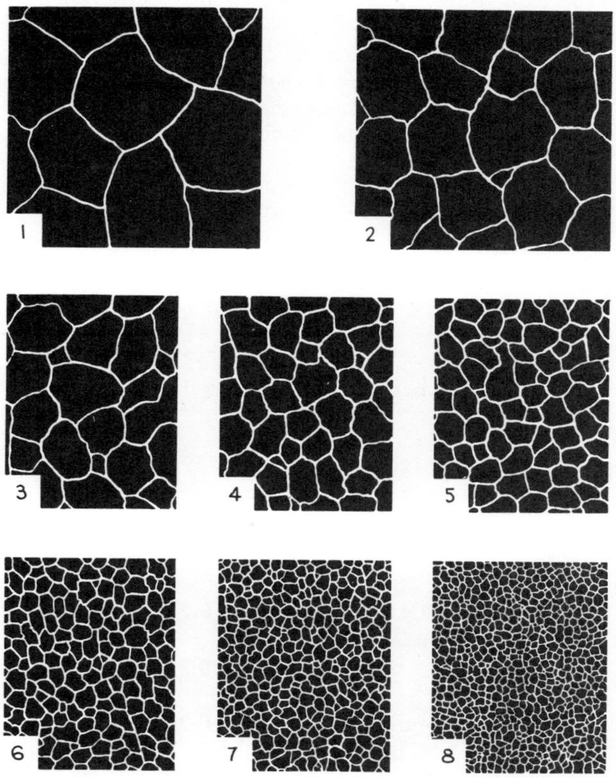

Figure S31 Standard grain-size numbers.

Types and Uses

Wrought Carbon Steel. Wrought carbon steels are used for structural steel, reinforcing rods for concrete, sheet and strip, corrugated steel sheet, mesh and wire cloth, ornamental and hollow metalwork, windows, doors, miscellaneous steel nails, screws, rivets, etc. (*see* Tables *S38* and *S39*). All major manufactured and fabricated products are covered under separate headings (e.g., *see* Steel for Concrete; Steel Sheet, Strip; Steel, Structural).

Carbon Steel Castings. Carbon steel castings used in construction are almost always medium-strength carbon steel and are treated by heat to relieve casting strains and to improve ductility and yield point.

Table S37 Typical composition of carbon steel used in construction (percent of content)

					S (max.)		
Type of steel	C	Mn	Si	P (max.)	Acid steel	Basic steel	Fe
Medium-carbon	0.2–0.5	0.5–1.0	0.25–0.75	0.5	0.06	0.05	Remainder

Table S38 Some typical steels used in construction

Major use	Manufacturing process	Major use	Manufacturing process
Steel castings	Basic open-hearth, basic oxygen, electric	Structural shapes	Basic oxygen, basic open-hearth, acid open-hearth
Cold-rolled steel	Basic oxygen, basic open-hearth		
Reinforcing for concrete	Basic oxygen, basic open-hearth	Wire for mesh and wire cloth	Basic oxygen
Forgings	Basic oxygen, basic open-hearth, acid open-hearth	Wire for prestressed concrete	Electric
Structural plates	Basic oxygen, basic open-hearth, acid open-hearth	Rivets	Basic oxygen, basic open-hearth
		Screws, nails, etc.	Basic oxygen
Small structural shapes	Basic oxygen, basic open-hearth	High-strength construction bolts	Basic oxygen, basic open-hearth, electric

Table S39 Chemical and physical requirements for structural steel

	Chemical requirements					
Type of steel product	C	Mn	P (max. percent)	S	Si	Cu[a] (min. percent)
Shapes[b]	0.26		0.04	0.05		0.20
Plates						
To $\frac{3}{4}$ in. incl. (19.05 mm)	0.25		0.04	0.05		0.20
Over $\frac{3}{4}$ to $1\frac{1}{2}$ in. incl. (19.05 to 38.1 mm)	0.25	0.80–1.20	0.04	0.05	0.15–0.30	0.20
Over $1\frac{1}{2}$ to $2\frac{1}{2}$ in. incl. (38.1 to 63.5 mm)	0.26	0.80–1.20	0.04	0.05	0.15–0.30	0.20
Over $2\frac{1}{2}$ to 4 in. incl. (63.5 to 101.6 mm)	0.27	0.85–1.20	0.04	0.05	0.15–0.30	0.20
Over 4 in. (101.6 mm)	0.29	0.85–1.20	0.04	0.05	0.15–0.30	0.20
Bars						
To $\frac{3}{4}$ in. incl. (19.05 mm)	0.26		0.04	0.05	0.15–0.30	0.20
Over $\frac{3}{4}$ to $1\frac{1}{2}$ in. incl. (19.05 to 38.1 mm)	0.27	0.60–0.90	0.04	0.05		0.20
Over $1\frac{1}{2}$ to 4 in. incl. (38.1 to 101.6 mm)	0.28	0.60–0.90	0.04	0.05		0.20
Over 4 in. (101.6 mm)	0.29	0.60–0.90	0.04	0.05		0.20

[a] For copper steel, this is the minimum percentage required.
[b] Manganese content of 0.85 to 1.35% and silicon content of 0.15 to 0.30% are required for shapes over 426 lb/ft (633.97 kg/m).

History and Manufacture

The earliest steels were probably made by accident, where or how we do not know. The famous Damascus and Toledo steels used for swords were actually molybdenum steel, as the ore used contained molybdenum. The original Wootz or Indian steel used for the same purpose contained small quantities of aluminum, added by unknown methods.

Blister Steel. Later, wrought iron bars were packed with charcoal in clay containers and heated for days to obtain a harder and stronger iron for tools and swords. Here the iron absorbed enough carbon to become what is known as blister steel. This was the only known method of making steel until the invention of the Bessemer process. Blister steel was a high-quality steel but could be produced only in small quantities.

Crucible Steel. Crucible steel was rediscovered in 1742 by Huntsman in England. The process consisted of placing scrap and other materials in barrel-like shapes (crucibles) made of clay or graphite (carbon) and lowering them into melting holes, where they were heated and melted for $2\frac{1}{2}$ to 4 hours by gas and air. This process is obsolete and has been replaced by the open-hearth and electric furnaces.

Bessemer Process. In approximately 1847, Henry Bessemer in England and William Kelly in the United States, working independently, realized that at 2300°F or 1260°C (the temperature of molten iron) silicon, manganese, and carbon would burn when exposed to oxygen in a blast of air and the heat given off would not only maintain the temperature of the molten metal but would raise it by 300 to 500°F (149 to 260°C). Kelly made his commercial converter in Johnstown, Pennsylvania, in 1857, whereas Bessemer was granted his original patents in 1855. A conflict arose, and Kelly, by proving that he used the process in 1847, was also granted patents. The two groups merged in 1866, and thereafter their method was called the Bessemer process. Quality steel could now be produced in quantity from the pig iron produced in blast furnaces.

Open-Hearth Furnace. About the time (1875) that Bessemer steel reached a peak of production in the United States, accounting for 86% of the ingot output, other methods for producing mild steels were being developed in Europe and England. By about 1868 the Siemens brothers in England and the Martin brothers in France had developed the principal features of the open-hearth furnace. In the open-hearth furnace, waste heat is used to preheat the air and gas for combustion, thus greatly increasing the temperature of the flame in the furnace.

The modern open-hearth furnace (*see* Figure *S32*) is built on a framework of stilts to create chambers for preheating the air and gas; this also allows the molten metal to be discharged from the furnace by gravity. The floor of the hearth is in the shape of an elongated saucer, in which the air and gas flames sweep across the molten metal. Thus the name "open-hearth."

The entire furnace is built of refractory brick and encased in steel. The loading floor is about level with the hearth, and loading is done through a series of water-cooled doors. The floor at which the refined molten metal is poured into large ladles (big enough to hold the entire molten steel capacity of the furnace) is approximately 15 ft (4.57 m) below the loading floor and at the opposite side of the furnace.

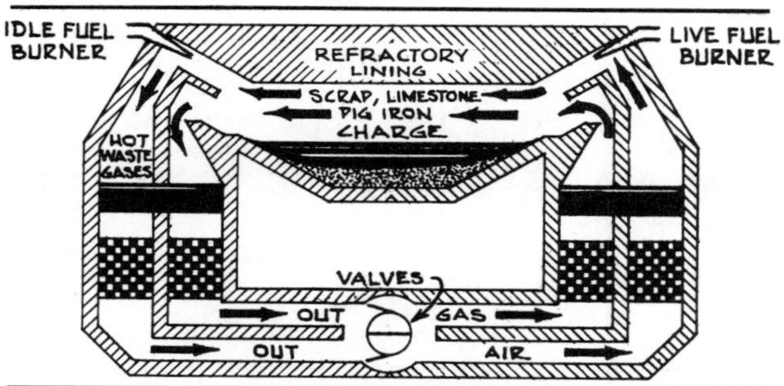

Figure S32 Open-hearth furnace.

Around the open-hearth furnace are all the components required for the process: gas producers; storage facilities for pig iron, scrap, iron ore, limestone, and fluorspar (flux); extra refractories for repairing the furnace (this may include a complete small plant for producing the refractories); a small electric furnace to produce spiegeleisen master alloy plus storage space for other master alloys. Besides these components there must be all the various types of equipment needed for moving these materials and for loading and unloading the furnace.

The furnace charge consists of steel scrap, pig iron, limestone, and fluxes. Throughout the entire refining process, which takes 6 to 14 hours, the direction of flow of the burning gases is reversed at frequent intervals. The iron on top of the charge melts first and starts to trickle down; the silicon and manganese react with the excess oxygen to form a slag. After 4 or 5 hours the molten metal is free of silicon and most of the manganese is in the slag. The limestone decomposes to produce lime and carbon dioxide. The carbon dioxide causes the melting metal to boil; the lime, as it floats up to become part of the slag, combines with some of the iron oxide, freeing the iron, and also dissolves and carries away any phosphorus oxide that may be present. When the limestone is gone, the molten metal continues to boil as the carbon combines with the oxygen of the remaining iron oxide to produce carbon dioxide.

At the latter part of the refining process the molten metal is tested for composition by taking liquid samples which are cast, cooled, and tested. Control of the composition near the end of the process, especially for percentages of phosphorus and carbon, is managed by various methods; adjusting temperatures, adding lime or iron ore to change the composition of the slag, adding fluorspar to increase fluidity, or adding low-phosphorus pig iron to increase the carbon. It is at this point that master alloys are added to deoxidize, desulfurize, or introduce the special elements for various alloy steels. When the desired composition is reached and confirmed by laboratory tests (there is always a testing laboratory adjacent to the open-hearth furnace), the temperature is raised sufficiently for casting the molten metal. The hole in the bottom of the furnace is unplugged, and the molten metal is poured into a ladle whose capacity is as nearly as possible equal to the quantity of molten metal in the furnace, so that the floating slag will run off into another ladle.

Oxygen Furnace. The basic oxygen furnace uses a stream of oxygen (in a water-cooled lance) forced down into the charge of molten pig iron and scrap with burnt lime and fluorspar, which form the slag, thus causing the rapid oxidation of carbon, manganese, and silicon. This reaction provides the heat necessary for melting the scrap, formation of the slag, and refining of the steel (*see* Figure *S33*).

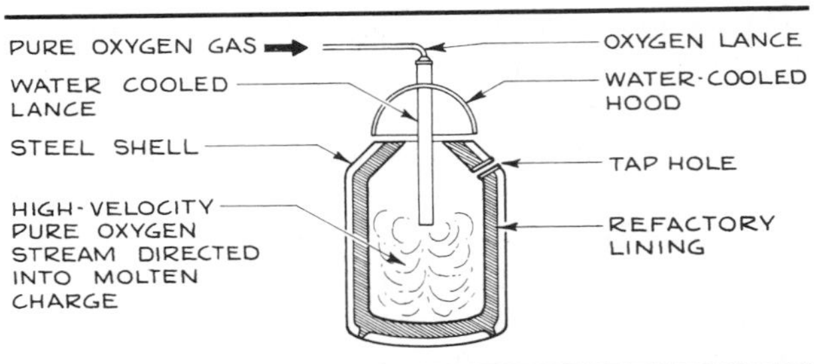

Figure S33 Basic oxygen furnace.

Other Methods. The electric arc furnace (*see* Figure *S34*) is fully described under Stainless Steel.

Liquid steel contains measurable amounts of dissolved oxygen, hydrogen, and nitrogen gases. Generally, the effect of these gases on the steel is insignificant. When a high degree of uniformity, soundness, and other qualities is required, supplementary vacuum treatment is used because of the uncontrolled amounts of dissolved gases. The various methods of vacuum treatment reduce hydrogen, oxygen, and carbon content, allow closer control of the composition of the steel, and improve microcleanliness. Highly specialized steels are produced by methods that melt and refine steel entirely under a vacuum. This type of manufacturing is very costly.

Vacuum-arc remelt steels are manufactured by forming an ingot of high-quality steel from an electric furnace; this ingot is then used as the electrode in a water-cooled copper crucible, which is put under a high vacuum. A negative charge is applied to the electrode and a positive charge to the crucible. The electrode melts, and trapped gases and vaporized nonmetallic impurities are drawn off by the vacuum. Steels made by this process are used when clean, uniform, and unique properties are required.

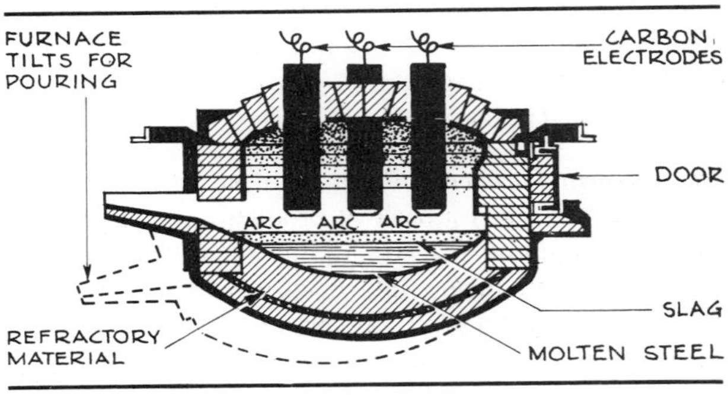

Figure S34 Electric furnace.

Acid and Basic Steel. Two types of steel are produced from basic oxygen and open-hearth furnaces: acid and basic steel.

1. The acid process requires ores and pig iron of low phosphorus content, as this process does not eliminate either sulfur or phosphorus; it also requires an acid-type slag and flux for the refining. The refractories used to line the furnace must contain similar acids, usually a mixture of iron, manganese, and silicon oxides, with at least 50% of silicon, so that foreign elements are not absorbed into the molten metal.

2. The basic process uses refractories of burned limestone, dolomite, or magnesite, which are not acid in character; burned lime is added in the charge to make a basic slag which absorbs and holds the phosphorus oxides developed during the refining. By this method the phosphorus content can be lowered, and ore or pig iron of higher phosphorus content can be used. The basic open-hearth furnace produces by far the most steel.

Blooming Mills. Steel ingots made by the various methods (open-hearth, or electric) are processed prior to fabrication in a blooming mill, where they are first heated to a uniform temperature of about 2200°F (1204°C) in underground furnaces (soaking pits) and then rolled into three forms: blooms, billets, and slabs.

The entire process from ore to finished fabricated steel products is summarized in the flowchart of Figure *S35*.

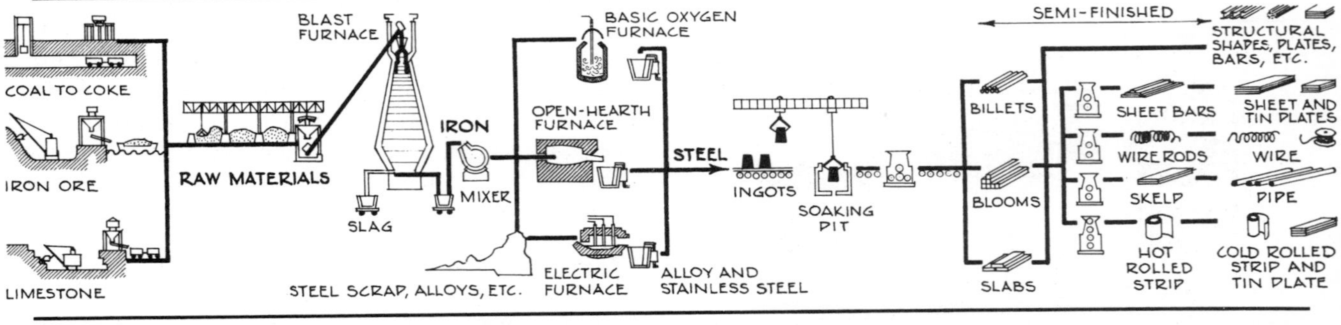

Figure S35 Flowchart of iron and steel manufacture.

STEEL: ALLOY STEELS

Physical and Chemical Properties

Alloy steels are steels to which various elements such as manganese, silicon, aluminum, titanium, and molybdenum have been added in sufficient quantity to produce properties unobtainable in carbon steels in cast, rolled, or heat-treated form.

Definition and Designation. Officially, a steel is considered to be alloy steel when (1) the maximum content of alloying elements exceeds one or more of the following limits: manganese 1.65%, silicon 0.60%, copper 0.60%; or (2) when a definite range or minimum quantity of any of the following is specified or required: aluminum, boron, chromium (to 3.99%), cobalt, columbium (niobium), molybdenum, nickel, titanium, tungsten, vanadium, zirconium, or any other element added to obtain a desired alloying effect. (*See* Table *S40.*)

Standard Construction Alloy Steels. The elements that occur normally are carbon, manganese, and silicon. A steel with one additional alloying element is known as a single-alloy steel. When two alloying elements are added, it is known as a double (binary) alloy steel; when three alloying elements are added, it is known as a triple (ternary) alloy steel.

Table S40 Compositions and numerical designations of commonly used alloy steels

Digit designation	Percentages of alloying elements other than iron[c]
13xx	1.75 Mn
23xx	3.50 Ni
25xx	5.00 Ni
31xx	1.25 Ni, 0.65 Cr
E33xx*	3.50 Ni, 1.55 Cr
40xx	0.25 Mo
41xx	0.05 or 0.95 Cr, 0.12 or 0.2 Mo
43xx	1.80 Ni, 0.5 or 0.8 Cr, 0.25 Mo
46xx	1.80 Ni, 0.25 Mo
47xx	1.05 Ni, 0.45 Cr, 0.20 to 0.35 Mo
50xx	0.28 or 0.40 Cr
51xx	0.80, 0.88, 0.93, 0.95 or 1.00 Cr
E5xxxx*	High C, High Cr
E50100*	100 C, 0.50 Cr
E51100*	1.00 C, 1.00 Cr
E52100*	1.00 C, 1.45 Cr
61xx	0.60, 0.80, or 0.95 Cr, or vanadium as noted, 0.12, 0.10 or 0.15 min vanadium
7140	0.40 C, 1.60 Cr, 0.35 Mo, 1.15 Al
81xx	0.30 Ni, 0.40 Cr, 0.12 Mo
86xx	0.55 Ni, 0.50 Cr, 0.20 Mo
87xx	0.55 Ni, 0.50 Cr, 0.25 Mo
88xx	0.55 Ni, 0.50 Cr, 0.35 Mo
92xx	0.85 Mn, 2.0 Si
93xx	3.25 Ni, 1.20 Cr, 0.12 Mo
98xx	1.00 Ni, 0.80 Cr, 0.25 Mo
14Bxx	Boron
50Bxx	0.50 or 0.28 Cr, Boron
51Bxx	0.80 Cr, Boron
81Bxx	0.33 Ni, 0.45 Cr, 0.12 Mo, Boron
86Bxx	0.55 Ni, 0.50 Cr, 0.20 Mo, Boron
94Bxx	0.45 Ni, 0.40 Cr, 0.12 Mo, Boron

*Electric furnace steel.

Manganese is present in all construction alloy steels, as it is essential to steel production not only in melting but also in rolling and other processing operations. A manganese content of 11.0 to 14.0% with a carbon content of 1.00 to 1.40% produces an austenitic alloy steel that is resistant to wear and abrasion under high impact stresses.

Silicon is not a carbide-forming element but enters into solution in the ferrite. Silicon is generally present in fully deoxidized construction alloy steels in amounts up to 0.35%. It increases hardenability and strengthens low-alloy steels.

Nickel is not a carbide-forming element but enters into solution in the ferrite. It increases internal strength and elastic limit; in heat-treated steel it increases strength and toughness. Nickel in combination with chromium produces alloy steels with higher elastic ratios, greater hardenability, and higher impact and fatigue resistance.

Chromium forms a solid solution with both the alpha and the gamma phases of iron. With carbon it forms a complex series of carbide compounds of chromium and iron. It is essentially a hardening element and increases wear resistance and cutting ability. Chromium in combination with nickel, a toughening element, produces alloy steels with superior mechanical properties.

Molybdenum forms a solid solution with the ferrite phase and can form a complex carbide, depending on the content of carbon and molybdenum. Mainly it enables alloy steels to retain their strength and resistance to creep at elevated temperatures.

Vanadium is a strong carbide-forming element. It dissolves to some degree in ferrite, adding strength and toughness. Vanadium steels show a much finer structure; grain growth tendencies are minimized for temperatures in the heat-treating range, thus allowing higher hardening and normalizing temperatures to be used.

Boron is usually added to steel to improve hardenability, that is, to increase the depth of hardening during quenching. Boron-treated steels usually have a 0.0005 to 0.003% range of boron content. This small amount of boron appears to be the optimum range for enhancing the hardenability of other alloys. Boron is an alloy intensifier. It has a relatively high nuclear cross section and can be used for neutron absorption.

Aluminum is used as a deoxidizer and for the control of inherent grain size. It is most effective in controlling grain growth and as an alloying addition in amounts of 0.95 to 1.30% in nitriding steel. The high surface hardness after nitriding is due to the formation of a hard, stable aluminum nitride compound.

Copper is used as an alloying element to increase the resistance to atmospheric corrosion and to increase yield strength.

Columbium (niobium) imparts a fine grain size and prevents grain coarsening at temperatures as high as 1875°F (1023.89°C); it prevents air hardening, retards softening during tempering, hastens nitriding reactions, and increases resistance to creep at elevated temperatures. It also increases the ductility of steels slightly and the impact strength markedly.

Titanium is used primarily as a deoxidizer and an effective grain growth inhibitor. It has the greatest carbide-forming tendency of any of the alloying elements used in steel alloys. In very low-carbon steels it has strong effect on strengthening the ferrite by solid solution effect and ranks next to silicon in this respect.

Zirconium is a grain growth inhibitor and is a more potent deoxidizer than boron, silicon, titanium, vanadium, or manganese.

Cobalt hardens or strengthens the ferrite when it is dissolved in the ferrite and thus resists softening under elevated temperatures.

Alloy steels are sometimes designated by the element or elements that have the greatest effect in bringing about the peculiar characteristics of the alloy concerned, regardless of the percentages of the element or elements contained within the alloy steel.

The alloying elements are added to increase the following properties: strength, hardness, ease and depth of hardenability, performance at high or low temperatures, electromagnetic properties, wear resistance, electrical conductivity, or resistivity. (For a detailed explanation of the effects of specific elements, *see also* Iron Alloys.)

In structural applications only the properties of strength, expansion, resistance to corrosion, ductility, and workability are of interest to the architect and engineer.

High-Strength Low-Alloy Steels. There are a group of trade name steels with improved mechanical properties and resistance to atmospheric corrosion (*see* Table *S41*). They are easily fabricated by shearing, gas cutting, hot and cold forming, punching, riveting, and welding. They are produced to mechanical property requirements and chemical composition limits and are available in the form of sheet and strip, plates, bars, structural shapes, pipes, tubes, and wire.

Super Alloys. Super alloys is the name given to a relatively new group of high-strength oxidation-resistant alloys for high-temperature service, that is, between 1000 and 2000°F (537.8° and 1098.3°C). At 1200°F

Table S41 Alloying elements in typical high-strength, low-alloy steels[a]

Common trade name[b]	Percentages of elements other than iron[c]										
	C	Mn	P	Si	Cu	Ni	Cr	Mo	Al	Zr	V
Cor-ten A[d]	0.12	0.20–0.50	0.07–0.15	0.25–0.75	0.25–0.55	0.65 max.	0.30–1.25				0.02–0.10
Cor-ten B[d]	0.10–0.19	0.90–1.25	0.04 max.	0.15–0.30	0.25–0.40		0.40–0.65				
Cor-ten C[d]	0.12–0.19	0.90–1.35	0.04 max.	0.15–0.30	0.25–0.40		0.40–0.70				0.04–0.10
Mayari R[d]	0.12	0.50–1.00	0.12	0.20–0.90	0.20–0.50	1.00 max.	0.40–1.00			0.10 max.	
Mayari R-50[d]	0.20	0.75–1.25	0.04	0.15–0.30	0.20–0.40	0.25–0.50	0.40–0.70				0.01–0.10
Mayari R-60[d]	0.20	0.75–1.35	0.04	0.15–0.30	0.20–0.40	0.25–0.50	0.40–0.70				0.01–0.10
Max High Tensil	0.08–0.15	0.50–0.75	0.04 max.	0.60–0.90			0.50–0.65				
Tri-ten[e]	0.22 max.	1.25 max.	0.04 max.	0.10–0.30	0.20–0.60	0.50–1.00					0.02 min.
Aldecor[e]	0.12 max.	0.15–0.40	0.08–0.15	0.35–0.75	0.35–0.60			0.16–0.28			
Double-strength	0.12 max.	0.50–1.00	0.04 max.		0.30–1.00	0.50–1.10		0.10 min.	0.12–0.27		
High-steel	0.12 max.	0.60–0.90	0.05–0.12	0.15 max.	0.95–1.30	0.45–0.75		0.08–0.18		0.05–0.15	
ASTM A-441[e]	0.22	0.85–1.25	0.04	0.30 max.	0.20 min.						0.02 min.
V steels[e]	0.22 and 0.25 max.	1.25 and 1.35 max.	0.04	0.30 max.	0.20 min.						0.02 min.

[a]Standard ASTM specifications cover all the trade name steels.
[b]Through license agreement these trade name alloys are made by more than one company.
[c]All these high-strength, low-alloy steels contain a maximum of 0.05% sulfur.
[d]These are known as weathering steels.
[e]All these high-strength, low-alloy steels have resistance to weathering, depending on copper and chromium content.

(648.9°C) they have a rupture strength of 30,000 to 75,000 lbf/in.^{2} (206.85 to 517.13 MN/m^2), whereas that of stainless steels is only 15,000 to 25,000 lbf/in.^{2} (103.43 to 172.39 MN/m^2) and that of carbon and low-alloy steels only 3000 to 6000 lbf/in.^{2} (20.69 to 41.37 MN/m^2). Super alloys include not only the elements found in the alloy steels such as nickel and chromium but also cobalt, tungsten, columbium (niobium), and boron. Also, the gas content for nitrogen and argon is controlled.

Types and Uses

In construction the high-alloy steels are used very little, if at all, except in specialized industrial and laboratory buildings. The high-strength low-alloy steels, however, are used in increasing quantity as sheet and strip, reinforcing for prestressed concrete, special structural steels, and cables for elevators, high-strength bolts, etc. For more information on structural uses, *see* Steel for Concrete; Steel Mesh and Wire Cloth; and other headings under Steel.

Super alloys have no construction uses. At present they are confined mainly to gas turbines, jet aircraft, and rocket applications. In this respect they are similar to many other materials which were originally developed to answer very highly specialized engineering and industrial problems. Eventually they find other uses and often gradually replace some material in the construction field that has been commonly used for a long time.

History and Manufacture

Although research on the specific effect of various elements on steel had begun in the early 1800s, nickel, manganese, tungsten, and silicon steels were not used until the latter part of the 19th century. The earliest commercial alloy steel made in the United States was the chromium steel used by Bauer for tools in 1865 and later incorporated into the Eads Bridge, built in 1874 across the Mississippi at St. Louis.

Molybdenum, vanadium, and copper steels are products of the early 20th century. After World War I and especially during and after World War II, formerly rare and entirely new metals became available to industry and were responsible for the development of countless alloy steels that answered steel consumers' special engineering requirements. Some alloy steels, many special steels such as tool and heat-resisting steels, and nearly all stainless steels are now made in electric furnaces (*see* Stainless Steel).

STEEL, CORRUGATED SHEET

Physical and Chemical Properties

Corrugated steel is rigidized sheet fabricated from low-carbon cold- or hot-rolled steel sheets which are either galvanized or covered with some type of bituminous coating. If galvanized, corrugated steel is silvery in color and has a glittering frosted surface (*see* Steel, Galvanized; Zinc Coatings). If coated with bituminous material, it may have a single-layered or a multiple, built-up surface and it may have a colored finish, depending on the manufacturer. It is available with vinyl and other types of plastic coating in various colors and textures. Corrugated steel of either type is strong, thin, and resistant to corrosion under normal atmospheric conditions and will not stain adjoining materials. It can be lapped, thus forming a weatherseal and making joining simple.

Contact with Other Materials. Galvanized sheet can be used in direct contact with wood, concrete, mortar, lead, tin, zinc, and aluminum. With other metals it should be insulated to prevent electrolytic action. With redwood or red cedar, both of which contain acids that attack zinc, it should be coated with asphalt-type paint or other similar coating for isolation. With a bituminous coating or built-up surface finish, it can be in direct contact with almost any material. Corrugated steel with vinyl or other types of plastic can also be in direct contact with almost any material. Manufacturers of both plastic and bituminous coated corrugated steel should be consulted when special chemical or electrical conditions are present.

Available Gauges. Corrugated steel is generally available in 18-, 20-, 22-, 24-, and 28-gauge sheet and strip.

Types and Uses

Corrugated steel is available in the commonly used industrial and standard types and in special curved and perforated types. For Applications, *see* Zinc, Corrugated.

Standard Type. Standard corrugated steel (*see* Figure *S36*) is available in the following dimensions:

Sheet thickness:	16 gauge	0.0598 in.	1.519 mm
	18	0.0478	1.214
	20	0.0359	0.912
	22	0.0299	0.760
	24	0.0239	0.607
	26	0.0179	0.455
	28	0.0149	0.379
	30	0.0135	0.343

Widths: 26 in. (660.4 mm)
Lengths: 7 ft (2.134 m), 8 ft (2.436 m), 10 ft (3.048 m), and 12 ft (3.658 m)
Depths of corrugations: $\frac{1}{4}$ in. (6.35 mm), $\frac{3}{8}$ in. (9.35 mm), $\frac{1}{2}$ in. (12.7 mm), $\frac{5}{8}$ in. (15.88 mm), and $\frac{3}{4}$ in. (19.05 mm)
Pitch of corrugations: $\frac{3}{8}$ in. (9.35 mm), $\frac{5}{8}$ in. (15.88 mm), $1\frac{1}{4}$ in. (31.75 mm), and 2 in. (50.8 mm)

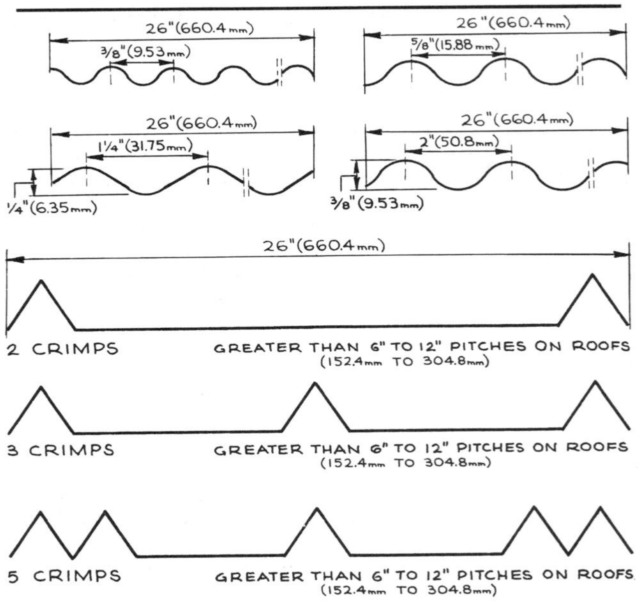

Figure S36 Standard corrugated steel sheet with galvanized or bituminous-coated surface. *Corrugated type:* Side lap, $1\frac{1}{2}$ corrugations for roofing, 1 corrugation for siding. End laps, 6 in. (152.4 mm) for roofing with 3 to 12 or greater pitch and 9 in. (228.6 mm) for roofing with up to 3 to 12 pitch. *Crimp Type:* Side lap, 1 V-crimp for both roofing and siding. End laps, 4 in. (101.6 mm) for siding and 6 in. (152.4 mm) for roofing with 6 to 12 or greater pitch.

Most commonly used size: 26 in. × 8 ft (660.4 mm × 2.436 m)

Industrial Type. This type is used principally in heavy construction for roofing and siding and to some extent for flashing (*see* Figure *S37*). Dimensions are as follows:

Sheet thickness:	18 gauge	0.0478 in.	1.213 mm
	20	0.0359	0.912
	24	0.0239	0.607

Widths: $27\frac{1}{2}$ in. (698.5 mm) and 33 in. (838.2 mm)
Lengths: 5 ft (1.524 m) to 12 ft (3.658 m) in 2-in. (50.8-mm) increments
Depth of corrugations: $\frac{1}{2}$ in. (12.7 mm), $\frac{3}{4}$ in. (19.04 mm), and $\frac{7}{8}$ in. (22.23 mm)
Pitch of corrugations: $2\frac{1}{2}$ in. (63.5 mm), 3 in. (76.2 mm), and 5 in. (127.0 mm)
Most commonly used size: $27\frac{1}{2}$ in. × 8 ft (698.5 mm × 2.436 m)

The industrial type includes ribbed and beaded sheet. Both the ribbed and beaded types are available in the same sheet thicknesses and lengths as other industrial corrugated sheet. Widths and depths of corrugations are:

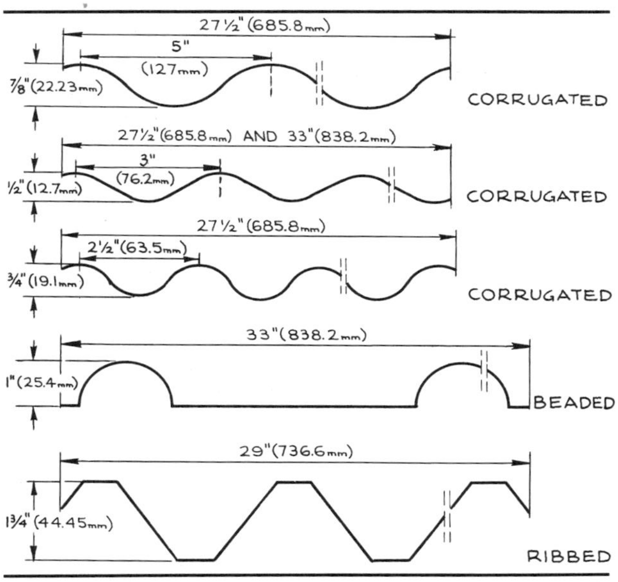

Figure S37 Industrial types of corrugated steel sheet with galvanized or bituminous-coated surface. *Corrugated type:* Side lap, $1\frac{1}{2}$ corrugations for roofing, 1 corrugation for siding. End laps, 4 in. (101.6 mm) for siding, 6 in. (152.4 mm) for roofing with 3 to 12 or greater pitch, and 9 in. (228.6 mm) for roofing with 2 to 12 up to 3 to 12 pitch. *Beaded and ribbed type:* Side laps, 1 bead or rib for both roofing and siding. End laps, 4 in. (101.6 mm) for siding and 6 in. (101.6 mm) for roofing with 3 to 12 or greater pitch. For lower pitches, end laps are greater and are sealed with bituminous joint.

Beaded: width 33 in. (838.2 mm); depth 1 in. (25.4 mm) with 6 beads to a 33-in. (838.2-mm) width

Ribbed: width 29 in. (736.6 mm); depth of bead $1\frac{3}{4}$ in. (44.5 mm), with 5 ribs to a 29-in. (736.6 mm) width

History and Manufacture

Corrugated iron was first manufactured about 1850 by passing a flat sheet through two corrugated rollers. Today the same principle is used with various types of roller designs to make the standard corrugated type, V type, pleated or ribbed type, and various other forms of patterned, rigidized steel sheet. The general shapes, whatever their design, are always planned so that one sheet overlapping the other makes a weathertight joint.

STEEL FOR CONCRETE

Physical and Chemical Properties

Steel is used as reinforcement for concrete work in the form of (1) hot-rolled reinforcing bars; (2) drawn wire woven into rope or made into reinforcing mesh; and (3) plain and corrugated sheet steel pans, decking, and other forms of concrete.

Reinforcing Bars. Reinforcing bars are made from carbon steel for new billets, from scrapped carbon steel axles of cars and locomotive tenders, or from standard section T-rails. The bars are hot-rolled and furnished to physical requirements only, as shown in Tables *S42* and *S43.* Reinforcing bars are available as rounds.

Reinforcing Mesh. For a description of the types and characteristics of steel used for reinforcing mesh, *see* Steel Mesh and Wire Cloth; *see also* Mesh and Wire Cloth for Concrete Reinforcing.

Pans, Decking. For steel sheet used to make pans, decking, and various steel sheet and strip forms for concrete, *see* Steel, Corrugated Sheet; Steel Sheet, Strip, and Plate.

Types and Uses

Reinforcing Bars. Round reinforcing bars are made both plain and deformed (*see* Figures *S38* and *S39*).

Spiral reinforcing rods for columns are available in four sizes. The spirals are prefabricated and are shipped to the site collapsed, together with special spacers for controlling the pitch of the spiral. Certain spirals, however, cannot be collapsed and must be shipped as a complete column unit.

Table S42 Physical requirements for steel reinforcing bars and spirals

	Tensile strength minimum[a]		Yield minimum[a]	
Type of steel	lbf/in.^{2}	MN/m^2	lbf/in.^{2}	MN/m^2
Billet steel	70,000–100,000	482.65–689.50	40,000–75,000	275.80–517.125
Rail steel	80,000– 90,000	551.6 –620.55	50,000–60,000	344.75–413.70
Axle steel	70,000– 90,000	482.65–620.55	40,000–60,000	275.80–413.70

[a] 1 lbf/in.^{2} = 0.006895 MN/m^2.

Table S43 Physical requirements for deformed bars for reinforcing concrete.

Type of steel	Size numbers	Grade	Tensile strength minimum[a] lbf/in.^{2}	Tensile strength minimum[a] MN/m^2	Yield minimum[a] lbf/in.^{2}	Yield minimum[a] MN/m^2	Minimal percentage of elongation in 8-in. (203.2 mm) length: 12	11	10	9	8	7	6	5	4.5	Cold bend test 180° unless otherwise noted (d = nominal diameter of size of deformed bar)	
Billet steel	#3 to and including #11	40	70,000	482.65	40,000	275.80	#4	#3	#8	#9	#10	#11				Under size #6	4*d*
							#5	#7								#6 and larger	5*d*
							#6										
	#3 to and including #11, #14 and #18	60	90,000	620.55	60,000	413.70		#9	#7	#3						Under size #6	4*d*
								#10	#8	#4						#6	5*d*
								#11		#5						#7 and #8	6*d*
								#14		#6						#9, #10 and #11	8*d*
								#18								#14 and #18	10*d* (90°)
	#11, #14 and #18	75	100,000	689.50	75,000	517.13								#11		#11	8*d* (90°)
														#14		#14 and #18	10*d* (90°)
														#18			
Rail steel	#3 to and including #11	50	80,000		50,000							#4	#3	#8		Under size #9	6*d*
												#5	#7	#9		#9 and #10	8*d*
												#6		#10		#11	8*d* (90°)
														#11			
	#3 to and including #11	60	90,000	620.55	60,000	413.70							#3	#7	#8	Under size #9	6*d*
													#4		#9	#9 and #10	8*d*
													#5		#10	#11	8*d* (90°)
													#6		#11		
Axle steel	#3 to and including #11	40	70,000		40,000	275.80	#4	#3	#8	#9	#10	#11				Under size #6	4*d*
							#5	#7								#6 and larger	5*d*
							#6										
	#3 to and including #11	60	90,000	620.55	60,000	413.70					#3	#8				Under size #6	4*d*
											#4	#9				#6	5*d*
											#5	#10				#7 and #8	6*d*
											#6	#11				#9, #10 and #11	8*d*
											#7						

[a] 1 lbf/in.^{2} = 0.006895 MN/m^2.

Rod Designation. Reinforcing rods are designated by number and by unit weight per foot. Numerical designations are based on the normal diameter of the bar expressed in eighths of an inch (*see* Table *S44*).

The nominal diameter of a deformed bar is equivalent to the diameter of a plain bar having the same weight per foot as the deformed bar.

Accessories. A series of accessories holds reinforcing bars in their correct positions both vertically and horizontally (*see* Table *S45* and Figure *S40*).

Wire Rope. In prestressed concrete a seven-wire, stress-relieved wire rope is used (*see* Table *S46*). The wire is made of cold-drawn steel (*see also* Wire and Wire Rope).

Reinforcing Mesh. Mesh used for reinforcing concrete is made from various gauges of cold-drawn steel wire similar to that used for wire rope (*see* Tables *S47* and *S48*). The most commonly used types of welded wire cloth are shown in Table *S49*.

Mesh in rolls is generally 5 or 7 ft (1.524 or 2.134 m) wide and 150 to 200 ft (45.72 to 60.96 m) long. Mesh with the longitudinal wire heavier than gauge No. 0 always comes in sheets.

On drawings, welded wire fabric is usually designated as follows:

WWF6 × 12–W16 × W26

Here the 6 × 12 after WWF indicates spacing of longitudinal wires *times* spacing of transverse wires, and the W16 × W26 after the dash indicates size of longitudinal wires *times* size of transverse wires (*see also* Figure *S41*).

"One-way mesh" is the term used for mesh with rectangular openings, which can be used in only one direction; "two-way mesh" denotes mesh with square openings, which can be used either lengthwise or crosswise.

Side laps should be one-half of a mesh opening for one-way mesh and one full mesh opening for two-way mesh. End laps are always one mesh opening for both types.

Steel Pans. Removable and permanent steel forms for tin pan construction (also called concrete joist construction) are made from hot-rolled low-carbon steel containing less than 0.25% carbon (*see* Figures *S42* and *S43*).

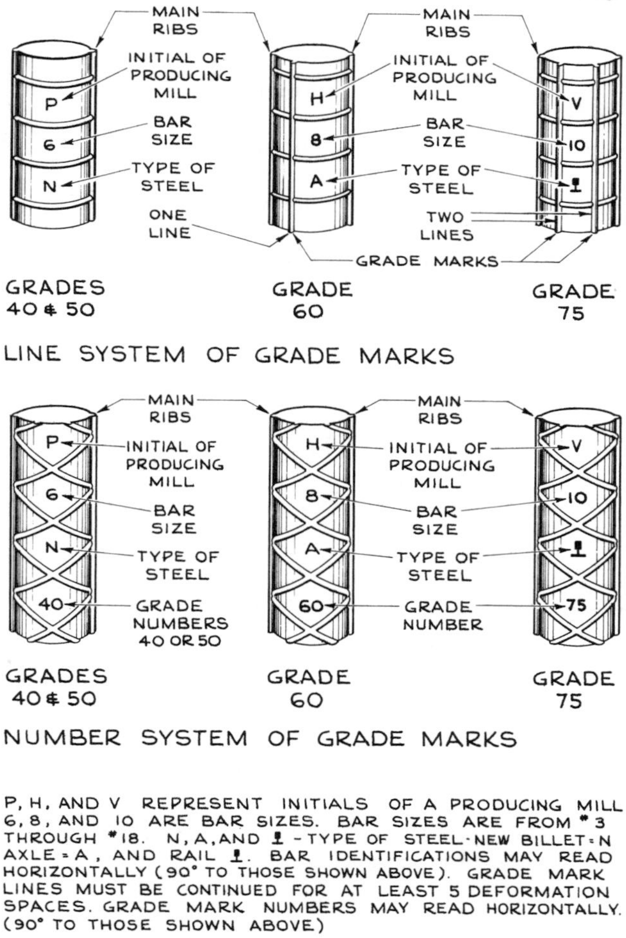

Figure S38 Identification marks for reinforcing bars and spirals.

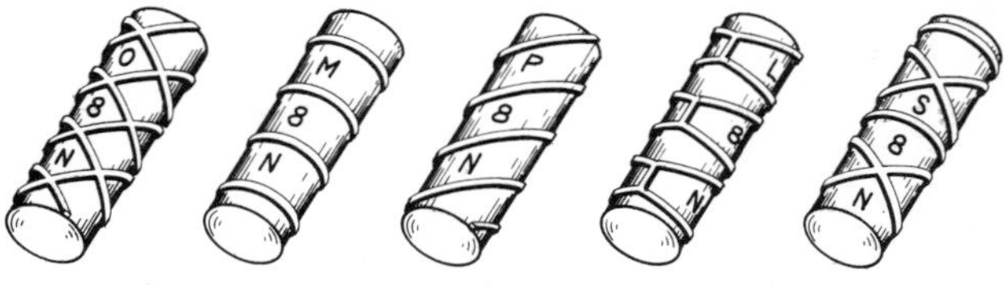

Figure S39 Types of steel reinforcing bars and spirals.

These forms are divided into two major categories: (1) removable forms made of 16-gauge or heavier smooth steel sheet; and (2) permanent forms, usually made of 24-gauge corrugated steel strip. These forms come in two standard widths of 20 and 30 in. (508.0 and 762.0 mm), and in standard lengths of 3 ft (0.914 m). Two types of fillers (intermediates) are available for meeting length conditions: 1 and 2 ft (0.3048 and 0.6096 m). There are also two types of filler for meeting width conditions: 10 in. by 3 ft (254 by 914 mm) and 15 in. by 3 ft (381 mm by 0.014 m). There are two types of special end forms, a 6-in. (152.4-mm) one for straight ends and a 3-ft (0.914-m) one for tapered ends. All forms are available in 6- to 14-in. (152.4- to 366.4-mm) heights in 2-in. (50.8-mm) increments. Special lengths can be ordered. (*See* Figures *S44* and *S45*.)

Forms for waffle-type slab systems are all of the removable type and are made from 16-gauge or heavier, smooth steel sheet. They are available in the following sizes: 18- and 30-in. (457.2 and 762.0 mm) squares, and in depths of 8 to 14 in. (203.2 to 355.6 mm). (*See* Figure *S46*.)

Several types of steel decking and corrugated steel serve as forms for concrete slabs and slab-and-joist

Table S44 Deformed bar designation numbers, nominal weights and dimensions, and deformation requirements

Bar designation number[b]	Nominal weight		Nominal dimensions[a]: Diameter		Cross sectional area		Perimeter		Deformation requirements: Maximum average spacing		Minimum average height		Maximum gap (cord of 12.5% of nominal perimeter)	
	lb/ft	kg/m	in.	mm	$in.^2$	cm^2	in.	mm	in.	mm	in.	mm	in.	mm
3	0.376	0.560	0.375	9.53	0.11	0.71	1.178	29.92	0.262	6.655	0.015	0.381	0.143	3.632
4	0.668	0.994	0.500	12.70	0.20	1.29	1.571	39.90	0.350	8.890	0.020	0.508	0.191	4.852
5	1.043	1.552	0.625	15.88	0.31	2.00	1.963	49.86	0.437	10.100	0.028	0.711	0.239	6.071
6	1.502	2.235	0.750	19.05	0.44	2.84	2.356	59.84	0.525	13.335	0.038	0.965	0.286	7.265
7	2.044	3.043	0.875	22.23	0.60	3.87	2.749	69.83	0.612	15.545	0.044	1.117	0.334	8.484
8	2.670	3.974	1.000	25.40	0.79	5.10	3.142	79.81	0.700	17.780	0.050	1.270	0.383	9.728
9	3.400	5.060	1.128	28.65	1.00	6.45	3.544	90.02	0.790	20.066	0.056	1.422	0.431	10.947
10	4.303	6.404	1.270	32.26	1.27	8.19	3.990	101.35	0.889	22.581	0.064	1.626	0.487	12.370
11	5.313	7.907	1.410	35.41	1.56	10.06	4.430	112.52	0.987	25.070	0.071	1.803	0.540	13.716
14	7.650	11.385	1.693	43.00	2.25	14.52	5.320	125.13	1.185	30.099	0.085	2.159	0.648	16.459
18	13.600	20.420	2.257	57.33	4.00	25.81	7.090	180.09	1.580	40.132	0.102	2.591		

[a]Nominal dimensions of a deformed bar are equivalent to those of a plain round bar with the same lb/ft (kg/m) as the deformed bar.
[b]Bar numbers are based on the number of $\frac{1}{8}$ in. (3.18 mm) included in the normal diameter of the bar.

Table S45 Standard types of reinforcing for concrete bar supports

Classification	Physical characteristics	Major use
A: Bright basic	Cold-drawn steel wire; no rust protection	Where concrete surfaces are concealed from view or in areas where rust spots and blemishes on the concrete surface are unimportant
B: Pregalvanized	Pregalvanized cold-drawn steel wire with minimum rust protection	Where nominal protection is required for a short period of time
C: Plastic-coated	Cold-drawn steel wire with plastic coating applied by dipping, or with premolded plastic tips added to the legs or with legs molded to the top wire	Where there is moderate exposure and/or the concrete surface will be subjected to light grinding or sandblasting
D: Stainless steel protected	Cold-drawn steel wire with a stainless steel tip permanently attached to the bottom of each leg	Same as C
E: Special stainless steel protected	No nonstainless steel wire of the bar support closer than $\frac{3}{4}$ in. (19.05 mm) from the form surface	Where there is moderately severe exposure and/or the concrete surface will be subjected to heavy grinding and severe sandblasting

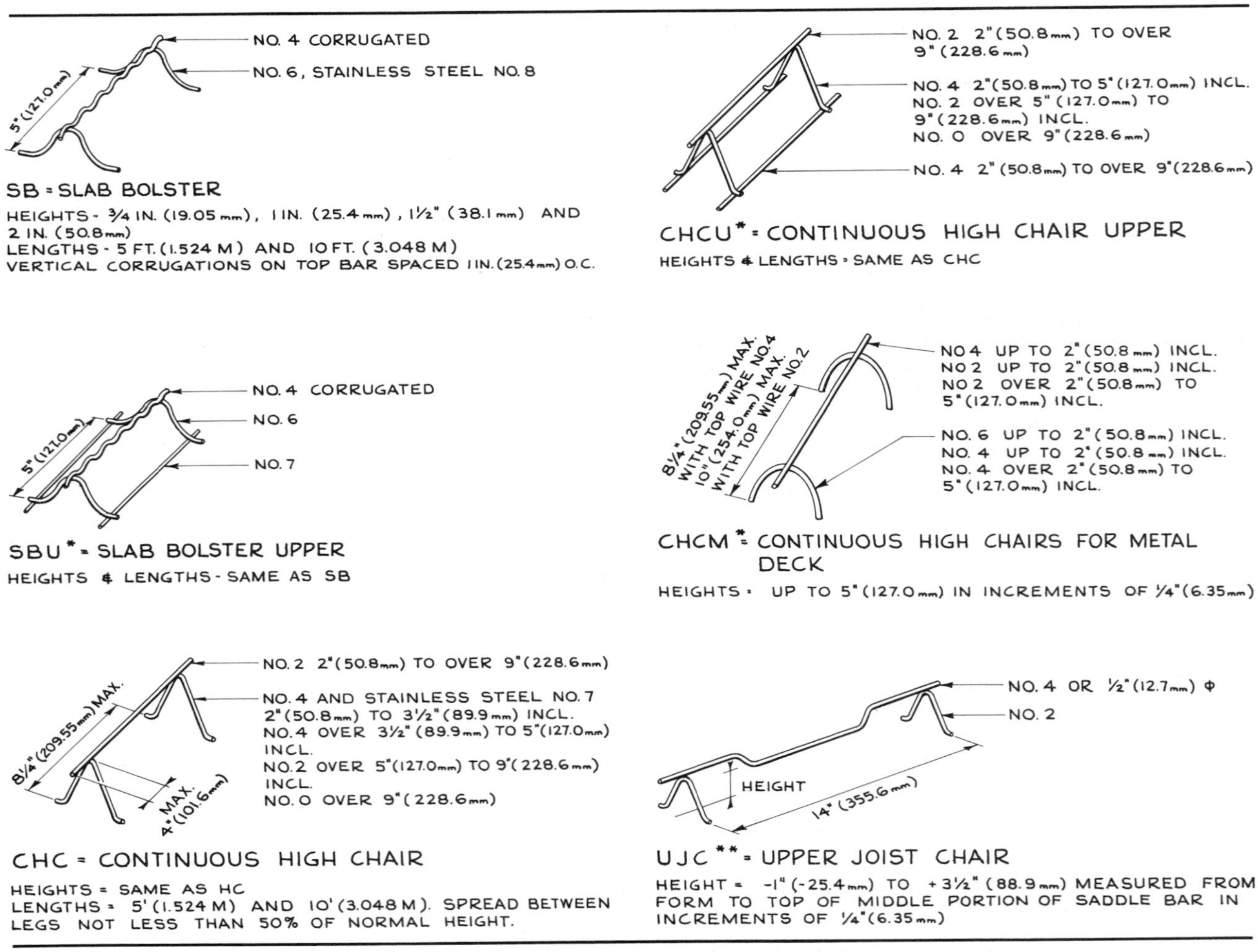

Figure S40 Standard types and dimensions of bar supports for reinforced concrete.

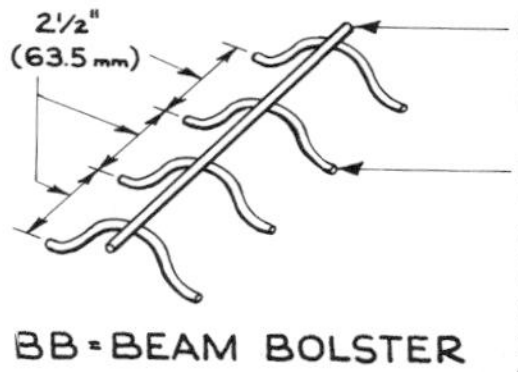

NO. 7 1"(25.4mm) TO 2"(50.8mm) INCL.
NO. 4 OVER 2"(50.8mm) TO OVER 3½"(88.9mm)

NO. 7 AND STAINLESS STEEL NO. 9 UP TO 1½"(38.1mm) INCL.
NO. 7 AND STAINLESS STEEL NO. 8 OVER 1½"(38.1mm) TO 2"(50.8mm) INCL.
NO. 4 AND STAINLESS STEEL NO. 7 OVER 2"(50.8mm) TO 3½"(88.9mm) INCL.
NO. 4 OVER 3½"(88.9mm)

2½" (63.5mm)

BB = BEAM BOLSTER

HEIGHTS = 1"(25.4mm), 1½"(38.1mm), 2"(50.8mm) AND OVER 2"(50.8mm) TO 5"(127.0mm) IN INCREMENTS OF ¼"(6.35mm)
LENGTH = 5 FT. (1.524 M)

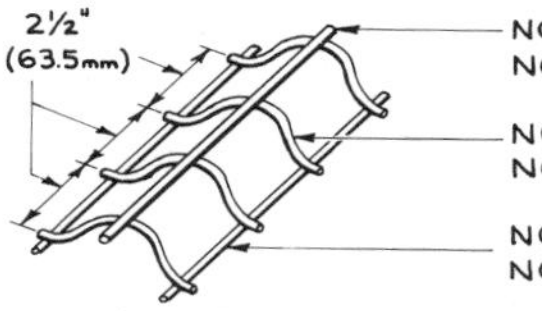

2½" (63.5mm)

NO. 7 UP TO 2"(50.8mm) INCL.
NO. 4 OVER 2"(50.8mm)

NO. 7 UP TO 2"(50.8mm) INCL.
NO. 4 OVER 2"(50.8mm)

NO. 7 UP TO 2"(50.8mm) INCL.
NO. 4 OVER 2"(50.8mm)

BBU* = BEAM BOLSTER UPPER

HEIGHTS & LENGTHS - SAME AS BB

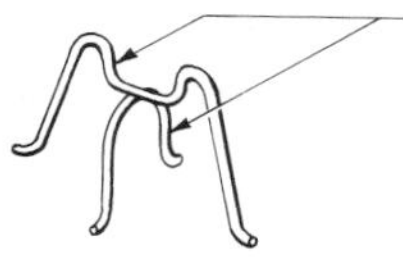

NO. 7 AND STAINLESS STEEL NO. 9

HEIGHTS = ¾"(19.05mm), 1"(25.4mm), 1½"(38.1mm) AND 1¾"(44.45mm)

BC = INDIVIDUAL BAR CHAIR

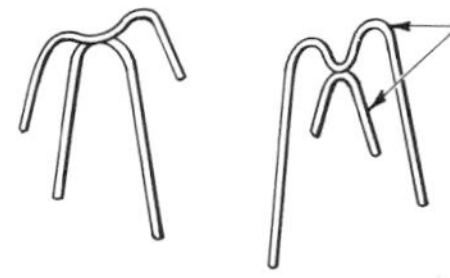

NO. 4 2"(50.8mm) TO 5"(127.0mm) INCL.
NO. 4 OVER 5"(127.0mm) TO 9"(228.6mm) INCL. AND OVER 9"(228.6mm)
THE LONGEST LEG WILL GOVERN THE SIZE OF WIRE TO BE USED.

HCM* = HIGH CHAIR FOR METAL DECK

HEIGHTS = SAME AS HC

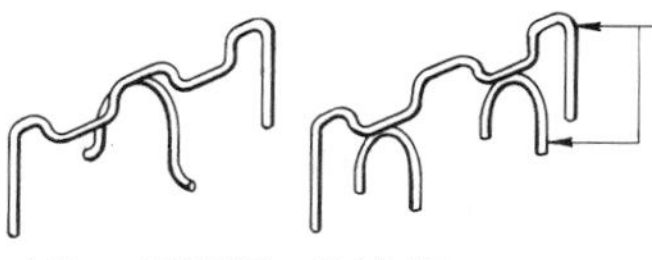

NO. 6 AND STAINLESS STEEL NO. 9

HEIGHTS - ¾"(19.05mm), 1"(25.4mm), & 1½"(38.1mm)
WIDTHS - 4"(101.6mm), 5"(127.0mm) & 6"(152.4mm)

JC = JOIST CHAIR

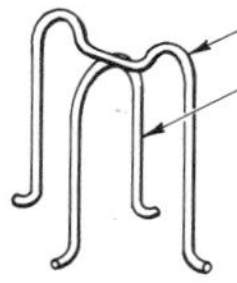

NO. 4 AND STAINLESS STEEL NO. 7 2"(50.8mm) TO 3½"(89.9mm) INCL.
NO. 4 OVER 3½"(89.9mm) TO 5"(127.0mm) INCL.
NO. 2 OVER 5"(127.0mm) TO 9"(228.6mm) INCL.
NO. 0 OVER 9"(228.6mm)

HC = INDIVIDUAL HIGH CHAIR

HEIGHTS - 2"(50.8mm) TO 15"(381.0mm) IN INCREMENTS OF ¼"(6.35mm)
LEGS AT 20° OR LESS WITH VERTICAL. OVER 12"(304.8mm) LEGS ARE REINFORCED WITH WELDED CROSS WIRES OR ENCIRCLING WIRES.

NOTES:
* AVAILABLE IN CLASS "A" ONLY, EXCEPT ON SPECIAL ORDER
** AVAILABLE IN CLASS "A" ONLY, WITH UPTURNED OR END BEARING LEGS.
SB, SBU, BB, BBU, BC, AND JC IN ORDER TO PROVIDE STABILITY, THE SPREAD BETWEEN LEG SUPPORTS SHALL BE A MINIMUM OF 70% OF THE NORMAL HEIGHT.
HC AND HCM IN ORDER TO PROVIDE STABILITY, THE LEG SPREAD OF THE SUPPORT ON THE MINOR AXIS OF THE SUPPORT SHALL BE A MINIMUM OF 55% OF THE NORMAL HEIGHT. CHC, CHCU, AND CHCM IN ORDER TO PROVIDE STABILITY, THE LEG SPREAD OF THE SUPPORT ON THE MINOR AXIS SHALL NOT EXCEED THE MIN. OR MAX. PERCENTAGES OF THE NORMAL HEIGHT SHOWN.

NORMAL HEIGHT		DISTANCE BETWEEN SUPPORTS % OF NORMAL HEIGHT	
IN.	mm	MAX.	MIN.
UNDER 4	UNDER 101.6	70	NO LIMIT
4	101.6	70	95
6	152.4	65	80
8	203.2	60	85
10	254.0	55	80
12 & OVER	304.8	50	75

Figure S40 Standard types and dimensions of bar supports for reinforced concrete (*continued*).

Table S46 Mechanical requirements for wire rope (strand) for prestressed concrete

Nominal diameter of wire rope (strand)		Minimum bearing strength		Nominal steel area of wire rope (strand)		Nominal weight of wire rope (strand)		Minimum difference in diameter between center wire and other wire		Approximate number of linear feet per reel, meter per reel		Requirements for yield strength: Initial load		Requirements for yield strength: Minimum load at 1% extension	
in.	mm	lbf	N	in.^{2}	cm^2	lb/100 ft	kg/304.8 m	in.	mm	ft	m	lbf	N	lbf	N
0.25	6.35	9,000	40 032	0.036	0.232	122	55.24	0.001	0.0254	25,000	7 620	900	4 003.2	7,650	34 027.2
0.31	7.94	14,500	64 496	0.056	0.361	198	89.81	0.0015	0.0381	15,000	4 572	1,450	6 449.6	12,300	54 710.4
0.375	9.53	20,000	88 960	0.080	0.516	274	124.29	0.002	0.0508	10,000 or 15,000	3 048 or 4 572	2,000	8 896.0	17,000	75 616.0
0.44	11.11	27,000	120 096	0.109	0.713	373	169.19	0.025	0.0635	8,000 or 9,000 or 12,000	2 438.4 or 2 735.2 or 3 657.6	2,700	12 009.6	23,000	102 309.6
0.50	12.70	36,000	160 128	0.144	0.929	494	224.08	0.003	0.0761	6,000 or 9,000	1 828.8 or 2 735.2	3,600	16 012.8	30,600	136 108.6

Table S47 Gauge, welded wire designation and dimensions of wire for welded wire fabric for concrete reinforcing

American Steel and Wire Gauge Number	W & D size numbers		Area		Nominal diameter	
	Smooth	Deformed	in.^{2}	cm^2	in.	mm
	W31	D31	0.310	2.000	0.628	15.95
	W30	D30	0.300	1.936	0.618	15.70
	W38	D28	0.280	1.807	0.597	15.16
	W26	D26	0.260	1.679	0.575	14.61
	W24	D24	0.240	1.549	0.553	14.05
	W22	D22	0.220	1.419	0.529	13.44
	W20	D20	0.200	1.290	0.504	12.80
0000000			0.189	1.219	0.490	12.45
	W18	D18	0.180	1.167	0.478	12.14
000000			0.167	1.076	0.4615	11.72
	W16	D16	0.160	1.032	0.451	11.46
00000			0.146	0.942	0.4305	10.94
	W14	D14	0.140	0.903	0.422	10.72
0000			0.122	0.787	0.394	10.01
	W12	D12	0.120	0.774	0.390	9.91
	W11	D11	0.110	0.710	0.374	9.54
	W10.5		0.105	0.678	0.366	9.30
000			0.103	0.665	0.3625	9.21
	W10	D10	0.100	0.645	0.356	9.04
	W9.5		0.095	0.613	0.348	8.84
	W9	D9	0.090	0.581	0.338	8.59
00			0.086	0.556	0.331	8.41
	W8.5		0.085	0.548	0.329	8.36
	W8	D8	0.080	0.516	0.319	8.10
	W7.5		0.075	0.484	0.309	7.85
0			0.074	0.478	0.3065	7.79
	W7	D7	0.070	0.452	0.298	7.57
	W6.5		0.065	0.419	0.288	7.32
1			0.063	0.407	0.283	7.19
	W6	D6	0.060	0.387	0.276	7.01
	W5.5		0.055	0.355	0.264	6.71
2			0.054	0.348	0.2625	6.67
	W5	D5	0.050	0.323	0.252	6.40
3			0.047	0.303	0.244	6.20
	W4.5		0.045	0.290	0.240	6.10
4	W4	D4	0.040	0.258	0.225	5.72
	W3.5		0.035	0.226	0.211	5.36
5			0.034	0.219	0.207	5.26
	W3		0.030	0.194	0.195	4.95
6	W2.9		0.029	0.187	0.192	4.88
7	W2.5		0.025	0.161	0.177	4.37
8	W2.1		0.021	0.136	0.162	4.12
	W2		0.020	0.129	0.159	4.04
9			0.017	0.111	0.148	3.76
	W1.5		0.015	0.097	0.138	3.51
10	W1.4		0.014	0.090	0.135	3.43

Table S48 General stock styles of welded wire fabric for reinforcing concrete[a]

	Spacing of wires				Diameter of wires				Sectional area					
	Longitudinal		Transverse		Longitudinal		Transverse		Longitudinal		Transverse		Weight	
Stock designation[b]	in.	mm	in.	mm	in.	mm	in.	mm	in.^{2}/ft	cm^2/0.31 m	in.^{2}/ft	cm^2/0.31 m	lb/100 ft^2	kg/9.29 m^2
6 × 6–W1.4 × W1.4	6	152.4	6	152.4	0.135	3.43	0.135	3.43	0.029	0.187	0.029	0.187	21	9.53
6 × 6–W2.1 × W2.1	6	152.4	6	152.4	0.162	4.12	0.162	4.12	0.041	0.265	0.041	0.265	30	13.61
6 × 6–W2.9 × W2.9	6	152.4	6	152.4	0.192	4.88	0.192	4.88	0.058	0.374	0.058	0.374	42	19.05
6 × 6–W4 × W4	6	152.4	6	152.4	0.225	5.72	0.225	5.72	0.080	0.516	0.080	0.516	58	26.31
4 × 4–W1.4 × W1.4	4	101.6	4	101.6	0.135	3.43	0.135	3.43	0.043	0.277	0.043	0.277	31	14.06
4 × 4–W2.1 × W2.1	4	101.6	4	101.6	0.162	4.12	0.162	4.12	0.062	0.400	0.062	0.400	44	19.96
4 × 4–W2.9 × W2.9	4	101.6	4	101.6	0.192	4.88	0.192	4.88	0.087	0.561	0.087	0.561	62	28.12
4 × 4–W4 × W4	4	101.6	4	101.6	0.225	5.72	0.225	5.72	0.120	0.774	0.120	0.774	85	38.06
4 × 12–W2.1 × W0.9	4	101.6	12	304.8	0.162	4.12	0.1055	2.67	0.062	0.400	0.009	0.058	25	11.30
4 × 12–W2.5 × W1.1	4	101.6	12	304.8	0.177	4.37	0.1205	3.06	0.074	0.478	0.011	0.071	31	14.06

[a]Welded smooth wire fabric with wires smaller than size W1.4 is manufactured from galvanized wire.
[b]Old designation was by steel wire gauge: 6 × 6 – 10 × 10. This now is 6 × 6–W1.4 × W1.4.

Table S49 Types of welded wire fabric most commonly used in construction

Size of square mesh		Wire designation	Sectional area		Weight	
in.	mm		in.^{2}/ft	cm^2/0.31 m	lb/100 ft^2	kg/9.29 m^2
6	152.4	W1.4	0.029	0.187	21	9.53
6	152.4	W2.1	0.041	0.265	30	13.61
6	152.4	W2.9	0.058	0.374	42	19.05

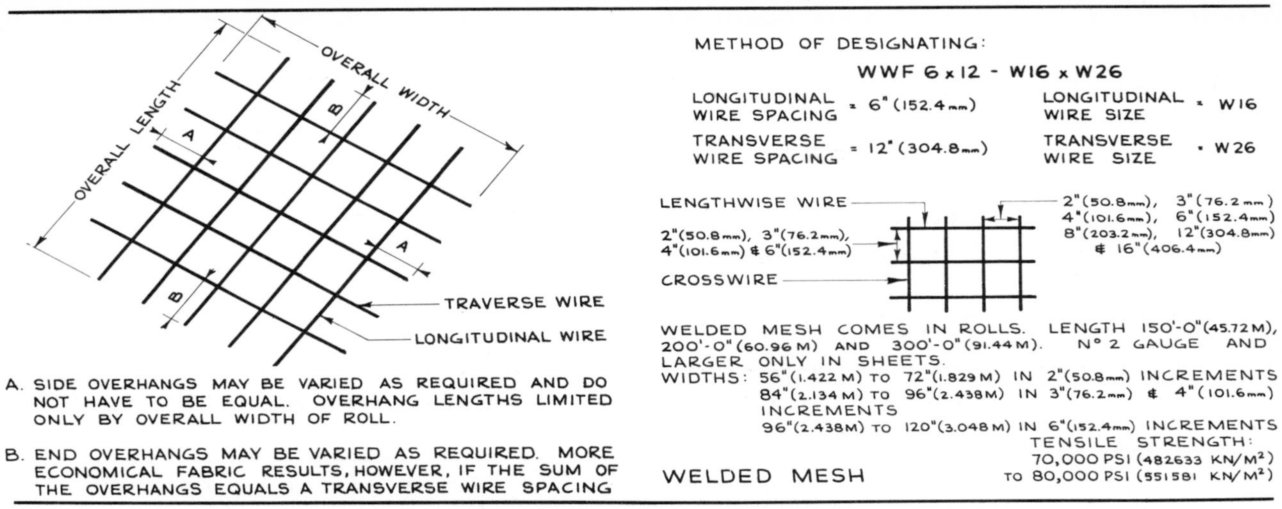

OVERALL WIDTH
OVERALL LENGTH
A
B
TRAVERSE WIRE
LONGITUDINAL WIRE
A. SIDE OVERHANGS MAY BE VARIED AS REQUIRED AND DO NOT HAVE TO BE EQUAL. OVERHANG LENGTHS LIMITED ONLY BY OVERALL WIDTH OF ROLL.
B. END OVERHANGS MAY BE VARIED AS REQUIRED. MORE ECONOMICAL FABRIC RESULTS, HOWEVER, IF THE SUM OF THE OVERHANGS EQUALS A TRANSVERSE WIRE SPACING
METHOD OF DESIGNATING:
WWF 6 x 12 - W16 x W26
LONGITUDINAL WIRE SPACING = 6"(152.4mm)
TRANSVERSE WIRE SPACING = 12"(304.8mm)
LONGITUDINAL WIRE SIZE = W16
TRANSVERSE WIRE SIZE = W26
LENGTHWISE WIRE
2"(50.8mm), 3"(76.2mm), 4"(101.6mm) & 6"(152.4mm)
CROSSWIRE
2"(50.8mm), 3"(76.2mm) 4"(101.6mm), 6"(152.4mm) 8"(203.2mm), 12"(304.8mm) & 16"(406.4mm)
WELDED MESH COMES IN ROLLS. LENGTH 150'-0"(45.72M), 200'-0"(60.96M) AND 300'-0"(91.44M). N° 2 GAUGE AND LARGER ONLY IN SHEETS.
WIDTHS: 56"(1.422M) TO 72"(1.829M) IN 2"(50.8mm) INCREMENTS
84"(2.134M) TO 96"(2.438M) IN 3"(76.2mm) & 4"(101.6mm) INCREMENTS
96"(2.438M) TO 120"(3.048M) IN 6"(152.4mm) INCREMENTS
TENSILE STRENGTH: 70,000 PSI (482633 KN/M²) TO 80,000 PSI (551581 KN/M²)
WELDED MESH

Figure S41 Steel welded wire fabric for reinforcing concrete.

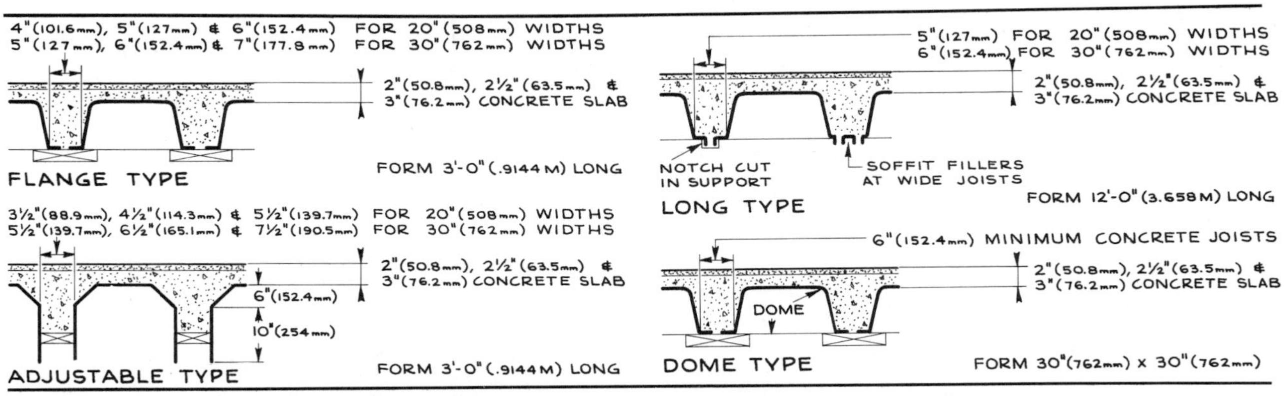

Figure S42 Typical removable steel forms for concrete joist construction.

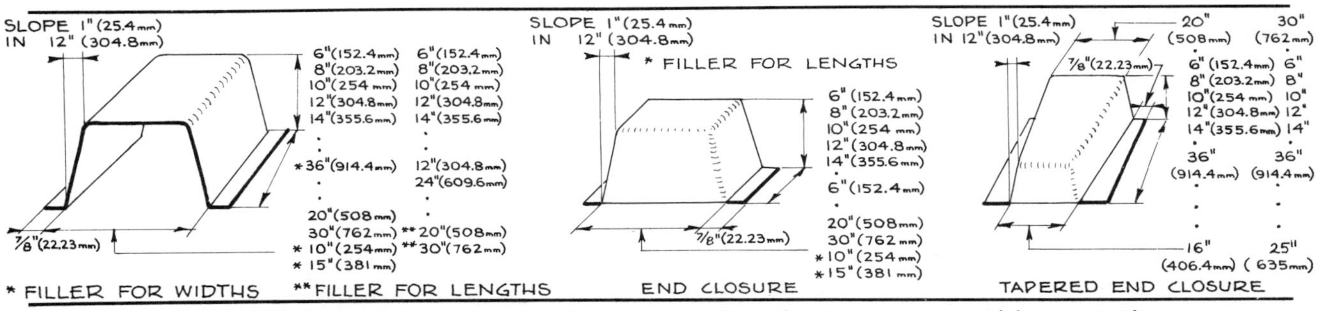

Figure S43 Removable and permanent flange-type steel forms for tin pan or concrete joist construction.

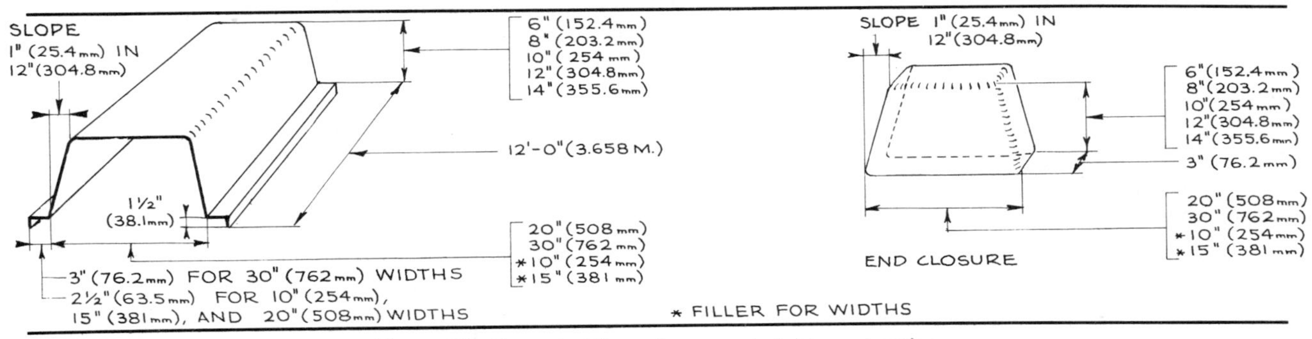

Figure S44 Long steel forms for concrete joist construction.

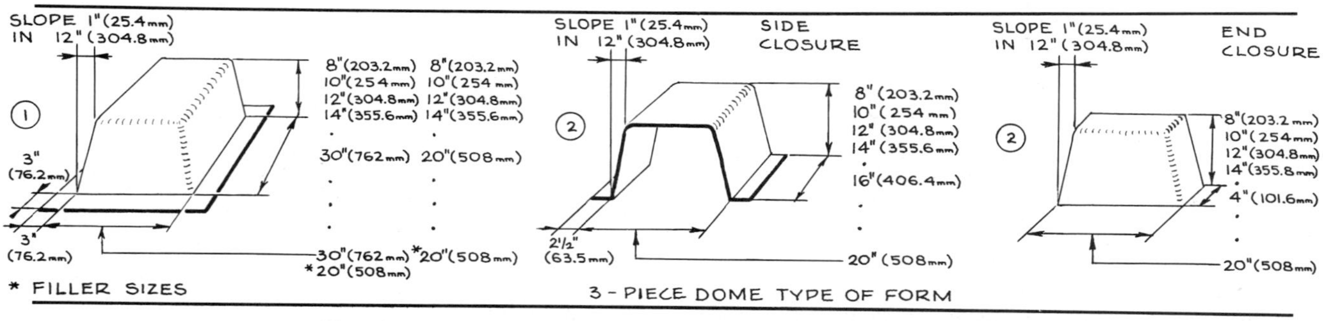

Figure S45 Removable dome forms for two-way concrete joist construction.

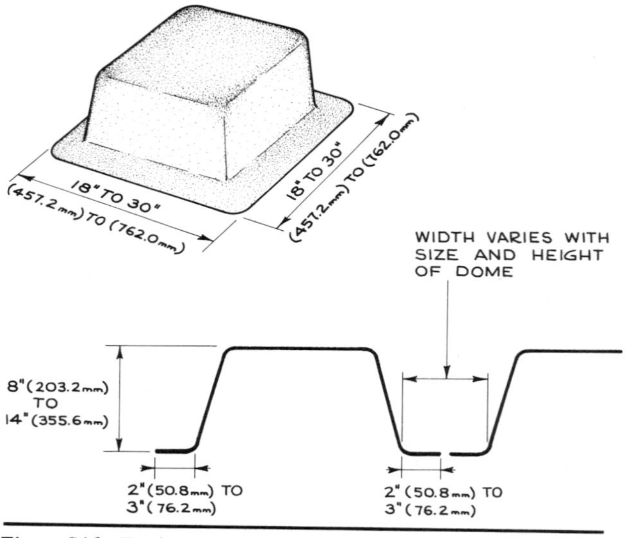

Figure S46 Typical types of removable dome forms for waffle concrete slab systems.

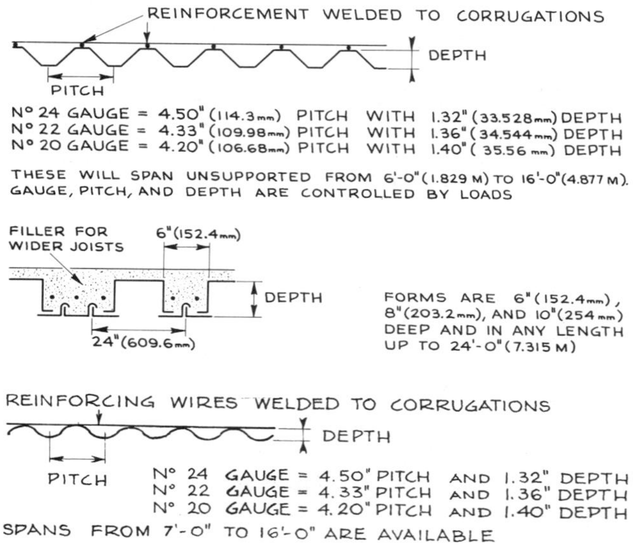

Figure S47 Typical steel decking and corrugated steel forms for concrete slab and joist construction.

construction (*see* Figure *S47*). They are made from heavy-gauge carbon steel sheet, either galvanized or plain. These types depend on the corrugations or the bent form to provide enough strength to span between the supporting beams. The reinforcing rods may be welded to the forms, or the forms may be shaped to receive reinforcing rods placed in the forms during construction.

Other Forms. There is continuous development of new forms and even systems of construction based on steel forms for concrete, especially permanent forms that act as part of the concrete reinforcement. Latest developments should always be checked for materials promising fast, inexpensive, fireproof construction.

Complete series of sheet steel shapes for forming all types of concrete foundations, retaining walls, walls, etc., are available for either leasing or buying. The shapes are generally 2 ft (0.6096 m) wide and 4 ft (1.219 m) long, although other lengths are available. There are also miscellaneous widths for fillers and special shapes for openings, external and internal corners, etc. Accessories for these sheet steel shapes include spacers, braces, ties, and clips.

Light steel framing for plywood forms includes angles, T-shapes, and bent shapes, which may be either leased or bought. They come in 2-ft (016096-m) widths and in 4-, 6-, 7-, and 8-ft (1.219-, 1.829-, 2.134-, and 2.438-m) lengths, together with fillers of the same lengths, in widths of 1, 2, 4, 6 through 20 in. (25.4, 50.8, 101.6, 152.4 through 508.0 mm) in 2-in. (50.8-mm) increments. There are also miscellaneous sizes for corners, openings, piers, etc., which can be assembled on the job. Accessories consist of spacers, braces, ties, clips, wedges, etc., which are part of these forms for concrete.

History and Manufacture

Until 1783, steel rolling mills produced only flat and square shapes. In that year Cort, an Englishman, built a rolling mill with grooved rolls which was the prototype of present-day rolling mills using steel in billet form.

The billets are made in a blooming mill, where ingots are first reduced to blooms and then to billets of appropriate size. At the rolling mill the billets are checked for imperfections, which are generally removed by actylene burning, and then passed through a horizontal furnace, which heats them to the correct rolling temperature. Each billet now passes through a series of grooved rollers at constantly increasing speed, each roller reducing the cross section and increasing the length. Each successive roller operates at a higher speed. Depending on the type of rolling mill, the rods either continue on through rollers in a straight line to the cooling area or are reversed in direction several times by what is called a transfer table before they go through finishing rollers. Some single-line rolling mills have the final (finish) roller delivering at the rate of 2600 ft/min.

One of the various possible methods of rolling a billet to its final finished form is shown in Figure *S48.*

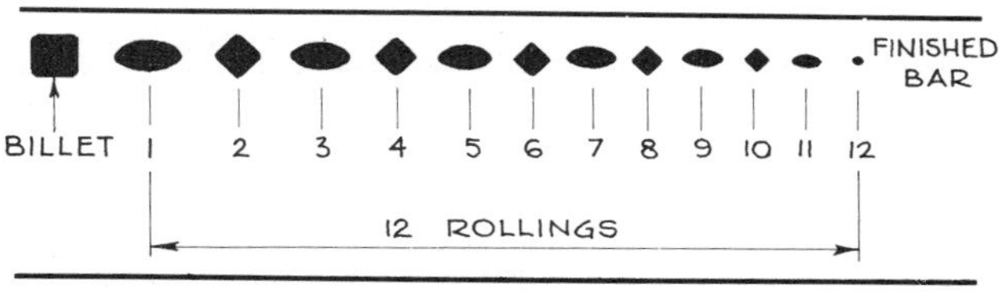

Figure S48 Oval and square method of rolling billets into round bars, showing progressive reduction through 12 passes.

STEEL, GALVANIZED

Physical and Chemical Properties

Galvanizing is the application of a coating of zinc to iron or steel sheet, strip, wire, and prefabricated metal products for protection against many types of corrosion. The various methods of applying galvanized coatings include the hot-dip process, spraying, electroplating, and sherardizing. (*See also* Galvanizing; Zinc Coatings.)

Weight of Coatings. Galvanized coatings are measured in ounces per square foot of surface except on sheets. Sheets are coated on both sides, and the weight is usually given in ounces per square foot (grams per square meter) of sheet, or double the weight of coating by surface. Designation by actual thickness in decimals of an inch (or in millimeters) is a preferable and less confusing method.

Galvanized coatings are available on all types of steel products, the thickness depending on the end use. Coatings vary from 0.75 to 3.0 oz/ft^2 (288.75 to 915.45 g/m^2) on sheet and sheet products and from 0.3 to 1.5 oz/ft^2 (91.55 to 475.73 g/m^2) of surface for wire and prefabricated metal products.

Table S50 Corrosion of zinc coatings in various environments

Type of environment	Location	Penetration per year (mils)
Industrial	Altoona, Pa.	0.26
Marine	Sandy Hook, N.J.	0.19
Rural	Lafayette, Ind.	0.09

The durability of zinc coatings on steel for corrosion resistance depends on the thickness of the coating and on the type of corroding environment (*see* Table *S50*).

Steel for Complicated Shapes. Steel sheet that is to be fabricated into complicated forms is of a soft and ductile type with low carbon content; if it is to be used in corrosive conditions, it is alloyed with 0.20% copper.

Types and Uses

Galvanized steel sheet is available in plain and corrugated form. The plain sheet is used directly in construction, mainly for ductwork and flat metalwork. The corrugated sheet is more commonly used for roofing, siding, and decking applications (*see* Steel, Corrugated). Galvanized steel is also available formed into various other rigid shapes, and perforated.

Roofing, Siding, Decking. In Table *S51*, which lists typical sizes of galvanized sheet and strip for roofing, siding, or decking, the dimensions are for plain sheet before corrugating or other forming. For roofing and siding the weight of coating is usually 2 oz/ft^2 (567 g/m^2) of sheet, and for decking it is 1.5 oz/ft^2 (42.53 g/m^2) of sheet.

Cladding. Galvanized steel sheet and strip are also used in construction for cladding, that is, covering wooden structures, primarily for fire resistance. This cladding of wood windows and doors with galvanized steel sheet is more specifically known as kalamein work.

Table S51 Typical dimensions of galvanized steel sheet and strip before forming[b]

Gauge of steel	Width available		Maximum length[a]		Thickness[a]	
	in.	mm	ft	m	in.	mm
8 and 9	2–30	50.8–0.762	15	4.572	0.1756–0.1458	4.45–3.703
10–13 incl.	2–48	50.8–1219.2	15	4.572	0.1457–0.0860	3.700–2.184
14–22 incl.	2–60	50.8–1524.0	16	4.877	0.0859–0.0322	2.182–0.819
23–28 incl.	2–48	50.8–1219.0	16	4.877	0.321–0.180	0.815–0.457
29–30 incl.	2–38	50.8–965.2	16	4.877	0.0179–0.0159	0.455–0.404

[a] Minimum length in all cases is 42 in. (1066.8 mm).
[b] Weight of zinc coating depends on class of coating desired.

Application

Galvanized sheet steel can be used in direct contact with wood, concrete, mortar, lead, tin, zinc, and aluminum. With other metals it should be insulated to prevent corrosion. When used with redwood or red cedar, both of which contain acids that attack zinc, it should be protected with coatings of asphalt-type paint or similar material. With a bituminous coating it can be in direct contact with almost any material. Galvanized sheets and coils should be stored in areas where they are not subject to moisture and extreme changes in temperature. (*See* Zinc Coatings.)

The heavier gauges of galvanized steel such as No. 22 will obviously produce a more durable roof and a better job of metalwork generally than the lighter gauges such as No. 28.

Seams. The sheets can be joined by standing seams (suitable for 26- and 28-gauge steels); batten seams (for 24-, 26-, and 28-gauge steels); and pressed seams (for 22- to 28-gauge steels).

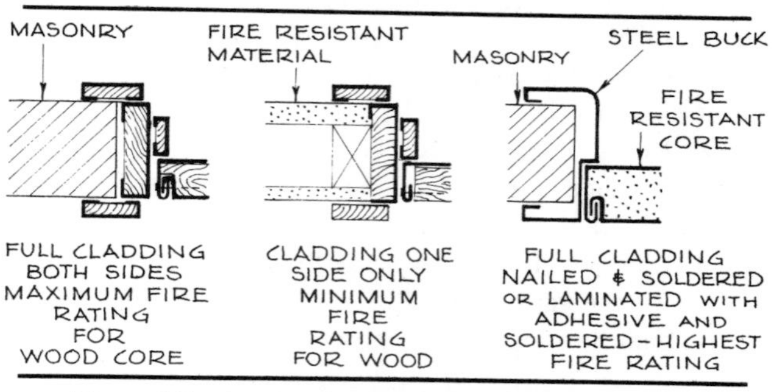

Figure S49 Details for cladding with galvanized steel.

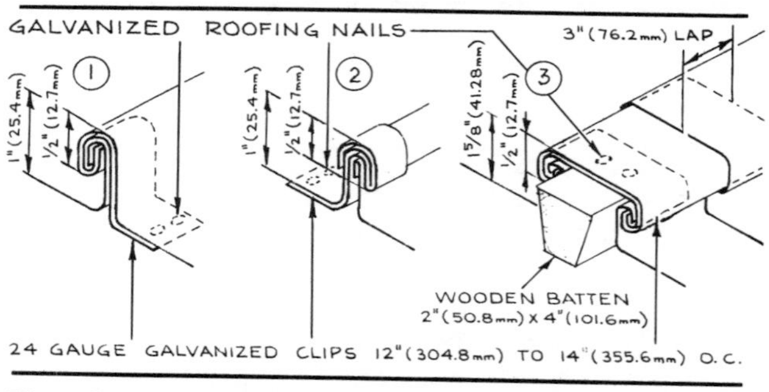

Figure S50 Details for standing (1), pressed (2), and batten (3), seams for galvanized steel roofing.

Condensed Checklist

1. The actual thickness of zinc coatings and the method of stating weight values—whether in terms of sheet or surface area—should be checked.
2. In the construction field, one should keep in mind the fact that weight, flexibility, and uniformity (tolerances) of zinc coatings vary in relation to the type of galvanizing process used, for example, hot dipping, electrogalvanizing, sherardizing, and spraying.
3. Special surface treatment may be necessary if the zinc coating is to be painted.
4. Fire ratings for cladding or kalamein work, particularly in relation to sheet gauge and thickness of coating, should be checked (*see* Figure *S49*).
5. For roofing, the building paper should be at least a 55-lb (24.95-kg) paper.
6. In roofing, the correct type of seam (standing, pressed, or batten) must be chosen in relation to sheet thickness. Figure *S50* shows details for each type of seam.
7. The correct type of cross seam, that is, flat lock or double lock, must be chosen in relation to roof pitch (*see* Figure *S51*).

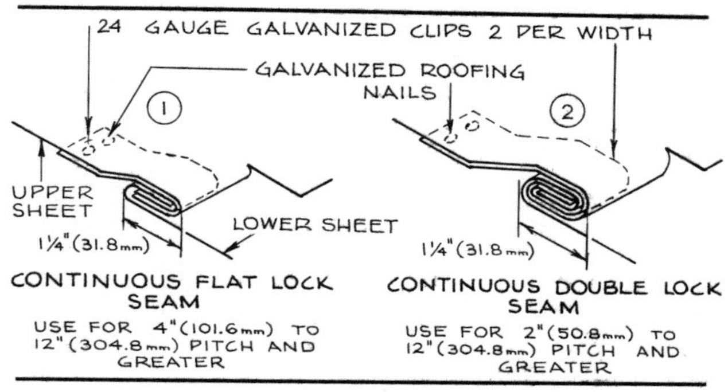

Figure S51 Flat-lock (1) and double-lock (2) seams for galvanized steel roofing.

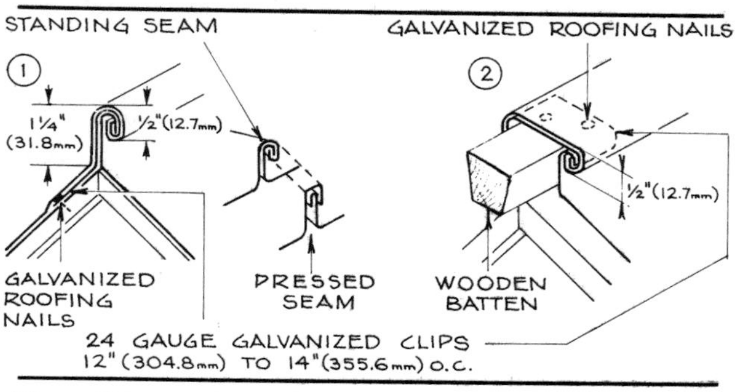

Figure S52 Details of ridge or hip for standing and pressed seams (1) and for batten seams (2) for galvanized steel roofing.

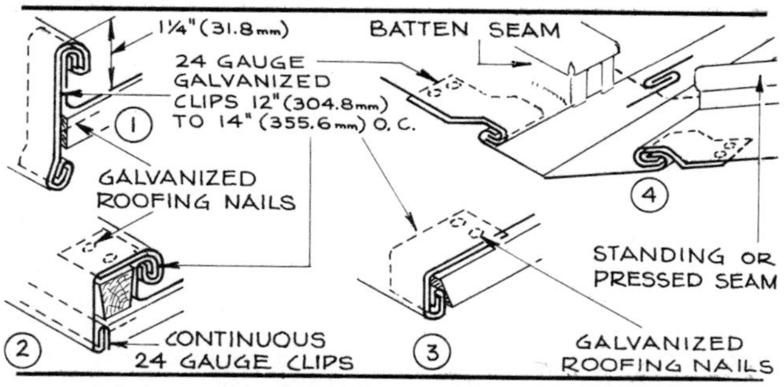

Figure S53 Details of gable end for standing and pressed seams (1) and batten seams (2), and treatment of roof edge (3) and valley (4) for standing, pressed, and batten seams for corrugated steel roofing.

8. Treatment of ridge or hip for standing, pressed, or batten seams should be as shown (*see* Figure *S52*).
9. Treatment of valley, roof edge, and gable ends for standing, pressed, or batten seams should be shown (*see* Figure *S53*).

Conditions Favorable to the Use of Galvanized Steel

1. Where a thin, strong, corrosion-resistant material is needed for roofing, cladding, duct work, and any bent or formed lightweight sections.
2. Where a corrosion-resistant surface is needed on a complicated steel shape which is to be formed.
3. When a high degree of fire resistance is required, particularly in cladding and kalamein work. State and municipal codes should be checked and also the fire rating codes of the Fire Underwriters, insurance companies, labor department, and federal government (Army, Navy, etc.).

Conditions Unfavorable to the Use of Galvanized Steel

1. In areas where there is a severely corrosive environment, particularly where acids, chemical fumes, or other materials may attack the zinc coatings.
2. Where neither the characteristic surface appearance of galvanized steel nor the painting and the maintenance problems it entails are desired.
3. On large surface areas that are to be painted.

History and Manufacture

Sheet and strip are the galvanized steel products of most interest to the architect. The sheet or strip in coils or cut lengths is first heated in a continuous furnace to burn off any lubricants left from the rolling mill. Then the coils or cut lengths are passed through an annealing furnace and a series of heating and cooling areas. After passage through pickling baths of hydrochloric or sulfuric acid to clean off scale, followed by washing and fluxing tanks, they are ready for application of the zinc coating.

Coils may contain more imperfections than cut lengths. In general, only one side of sheet material is expected to meet surface requirements.

For a description of the various methods of galvanizing, *see* Zinc Coatings.

STEEL: GALVANIZED CORRUGATED PIPE

Physical and Chemical Properties

Corrugated galvanized steel pipe is manufactured in the following shapes: circular and elliptical pipe, pipe-arches, arches, underpasses, and helically corrugated pipe (*see* Figure *S54*). All shapes are available galvanized with tin and galvanized with the following bituminous protective coatings: fully coated both inside and outside; fully coated, with the bottom quarter smooth with an

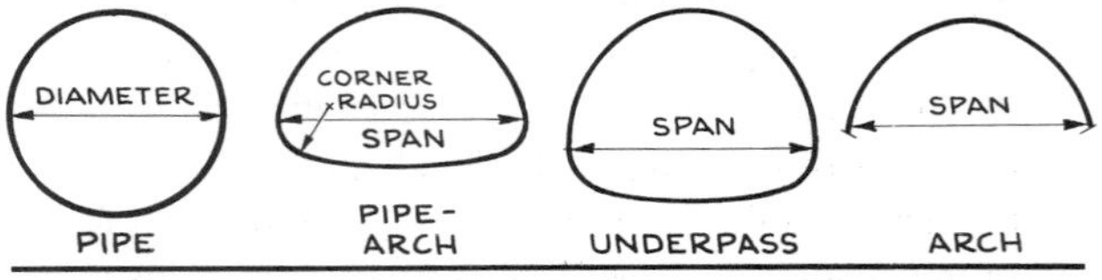

Figure S54 Forms and shapes of steel corrugated pipe.

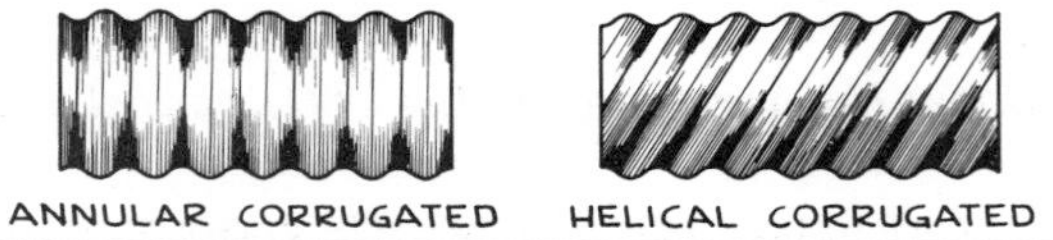

Figure S55 Types of corrugations.

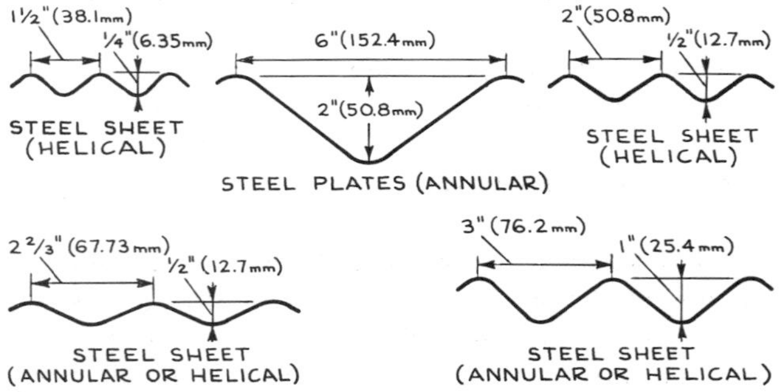

Figure S56 Dimensions of corrugations for steel sheets and plates.

asphalt pavement; fully coated, with the entire interior smooth with an asphalt lining. Corrugated pipe is also available with perforations for subdrainage.

Corrugated steel plate in gauges 1, 3, 5, 7, 8, 10, and 12 and corrugated steel sheet in gauges 8, 10, 12, 14, 16, and 18 are used to fabricate the various shapes of pipe, which range in size from 6 in. (152.4 mm) to 26 ft (7.93 m) in diameter for round pipe and come in a wide range of sizes for elliptical pipe, pipe-arches, arches, and underpasses. Corrugated pipe is also manufactured from 409 stainless steel.

Types and Uses

Steel corrugated pipe shapes are used for culverts, storm sewers, spillways, underpasses, service tunnels, and subdrainage. They are also used for industrial wastes if these wastes are noncorrosive to steel pipe.

Various types and sizes of corrugations in sheet steel and plate are used for producing the several shapes of corrugated pipe (*see* Figures *S55* and *S56*). These pipe shapes are joined with various types of couplings and fittings (*see* Figure *S57*).

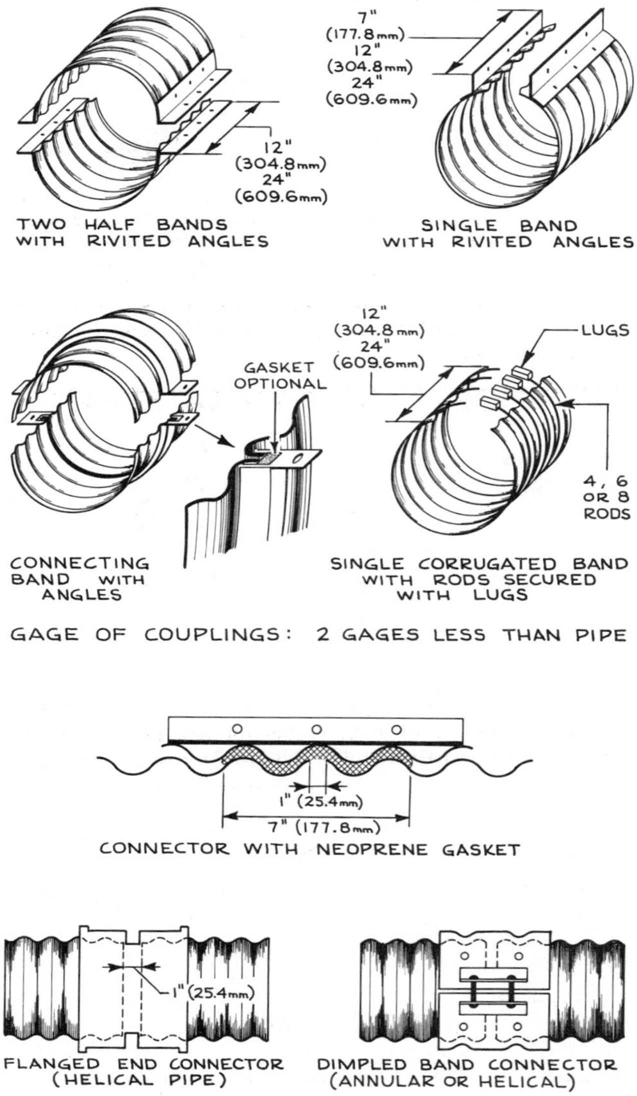

Figure S57 Types of connectors.

There is a very wide range of sizes for steel corrugated pipe and perforated pipe (*see* Table *S52*). For pipe-arches and part-circle culverts, *see* Table *S53*, and for structural plate pipe-arches and underpasses, *see* Table *S54*. The spans of arches vary from 5.667 to 20.33 ft (1.73 to 6.2 m), rises varying from 1.792 to 7.25 ft (0.55 to 2.21 m) with three different rises for each width of span.

Figure *S58* shows standard pipe fittings, and there are special fittings for inlets and outlets of steel corrugated pipe (*see* Figure *S59*). It is always necessary to protect outlets from scour and back- or under-cutting by water flow.

Table S52 Circular steel corrugated pipe sizes

Annular steel corrugated pipe		Perforated helical steel corrugated pipe
Diameters available in pipe with corrugations of $2\frac{2}{3}$ in. X $\frac{1}{2}$ in. (67.73 mm X 12.7 mm)	Diameters available in pipe with corrugations of 3 in. X 1 in. (76.2 mm X 25.4 mm)	Diameters available in pipe with corrugations of $1\frac{1}{2}$ in. X $\frac{1}{4}$ in. (38.1 mm X 6.35 mm) and 2 in. X $\frac{1}{2}$ in. (50.8 mm X 12.7 mm)
12 to 24 in. in 3-in. increments (304.8 to 609.6 mm in 76.2-mm increments) 24 to 96 in. in 6-in. increments (609.6 to 2456.4 mm in 152.4 mm increments)	36 to 120 in. in 6-in. increments (152.4 to 304.8 mm in 609.6-mm increments)	6 to 12 in. in 2-in. increments (152.4 to 304.8 mm in 50.8-mm increments) 12 to 24 in. in 3-in. increments (304.8 to 609.6 mm in 76.2-mm increments)

Table S53 Spans and rises for pipe-arches and part-circle culverts

Corrugated steel pipe-arches								Part-circle corrugated steel culverts			
$2\frac{2}{3}$ in. X $\frac{1}{2}$ in. corrugations (67.73 mm X 12.7 mm)				3 in. X 1 in. corrugations (76.2 mm X 25.4 mm)				Lengths = $25\frac{1}{2}$ in. (604.7 mm) = 24 in. (609.6 mm) when lapped			
Span		Rise		Span		Rise		Span		Rise[a]	
in.	mm	in.	mm	in.	mm	in.	mm	in.	mm	in.	mm
18	457.2	11	279.4	43	1092.2	27	685.8	12	304.8	$2 - 5\frac{5}{8}$	50.8 -143
22	558.8	13	303.2	50	1473.2	31	787.4	14	355.6	$2\frac{1}{4} - 7$	57.15-177.8
25	635	16	406.4	58	1473.2	36	914.4	16	406.4	$3\frac{1}{8} - 7\frac{3}{8}$	79.4 -187.4
29	736.6	18	457.2	65	1651	40	1016	18	457.2	$2\frac{1}{2} - 8\frac{3}{4}$	63.5 - 22.25
36	914.4	22	558.8	72	1828.8	44	1111.6	20	508.0	$2\frac{5}{8} - 9\frac{3}{8}$	66.8 -238.2
43	1092.2	27	685.8	73	1854.2	55	1397	22	558.8	$3\frac{1}{4} - 10\frac{3}{4}$	82.55-273.05
50	1270	31	787.4	81	2057.4	59	1498.6	24	609.6	$3\frac{5}{8} - 11\frac{1}{4}$	92.2 -285.75
58	1473.2	36	914.4	87	2209.8	63	1600.2	26	660.4	$3\frac{1}{4} - 12\frac{3}{8}$	82.55-314.4
65	1651	40	1016	95	2413	67	1701.8	28	771.2	$3\frac{1}{2} - 11\frac{3}{4}$	88.9 -298.45
72	1828.8	44	1111.6	103	2616.2	71	1803.4	30	762.0	$3\frac{5}{8} - 14\frac{5}{8}$	92.2 -371.6
79	2006.6	49	1244.6	112	2844.8	75	1905	32	812.8	$4\frac{1}{2} - 13\frac{7}{8}$	114.1 -325.6
85	2159	54	1371.6	117	2971.8	79	2006.6	34	883.6	$5 - 15\frac{1}{4}$	127 -387.35
				128	3251.2	83	2108.2	36	914.4	$5 - 16\frac{5}{8}$	127 -422.4
				137	3479.8	87	2209.8	42	1066.8	$5\frac{5}{8} - 17\frac{5}{8}$	133 -447.8
				142	3596.8	91	2311.4	48	1219.2	$6\frac{5}{8} - 14\frac{1}{2}$	168.4 -368.3

[a]Check with manufacturers for various rises and tolerances.

Application

When using steel corrugated pipe, pipe-arches, arches, underpasses, etc., one should always check for maximum and minimum height of cover, gauges, methods of installation, trench width, bedding, inlets, outlets, backfill, etc.

Condensed Checklist

1. The method of joining to be used.
2. Local, municipal, and state codes and highway codes for culverts, storm sewers, subdrainage, spillways, and underpasses.
3. The type of soil where the pipe is to be installed.

Table S54 Spans and rises for pipe-arches and underpasses

Pipe-arches								Underpasses			
6 in. X 2 in. (152 mm X 50.8 mm) corrugations for both pipe shapes											
18-in. (457.2-mm) corner radius, *R*				31-in. (787.4-mm) corner radius, *R*							
Span		Rise		Span		Rise		Span		Rise	
ft	m	ft	m	ft	m	ft	m	ft	m	ft	m
6.083	1.854	4.583	1.397	13.25	4.038	9.33	2.845	5.667	1.727	5.75	1.753
6.333	1.93	4.75	1.448	13.5	4.115	9.5	2.896	5.667	1.727	6.5	1.981
6.75	2.057	4.917	1.499	14.0	4.267	9.667	2.946	5.75	1.753	7.333	2.235
7.0	2.134	5.083	1.549	14.167	4.318	9.833	2.997	5.833	1.778	7.667	2.337
7.25	2.21	5.25	1.6	14.417	4.394	10.0	3.048	5.833	1.778	8.167	2.489
7.667	2.337	5.417	1.651	14.917	4.548	10.167	3.099	12.167	3.709	11.0	3.353
7.917	2.413	5.583	1.702	15.333	4.674	10.333	3.15	12.917	3.937	11.167	3.404
8.167	2.479	5.75	1.753	15.583	4.75	10.5	3.2	13.167	4.013	11.833	3.607
8.583	2.616	5.917	1.804	15.833	4.826	10.667	3.251	13.833	4.215	12.167	3.709
8.833	2.692	6.083	1.854	16.25	4.953	10.833	3.302	14.083	4.293	12.833	3.912
9.333	2.846	6.25	1.905	16.5	5.029	11.0	3.353	14.5	4.42	13.417	4.08
9.5	2.896	6.417	1.956	17.0	5.182	11.167	3.404	14.833	4.521	14.0	4.267
9.75	2.972	6.583	2.006	17.167	5.233	11.333	3.454	15.5	4.724	14.333	4.369
10.25	3.124	6.75	2.057	17.417	5.309	11.5	3.505	15.667	4.775	15.0	4.572
10.667	3.251	6.917	2.108	17.917	5.461	11.667	3.556	16.333	4.978	15.417	4.699
10.917	3.328	7.083	2.159	18.167	5.537	11.833	3.607	16.417	5.004	16.0	4.877
11.417	3.48	7.25	2.21	18.583	5.664	12.0	3.658	16.75	5.095	16.25	4.943
11.583	3.531	7.417	2.261	18.75	5.705	12.167	3.709	17.25	5.258	17.0	5.182
11.833	3.607	7.583	2.311	19.25	5.867	12.333	3.759	18.333	5.588	16.917	5.156
12.333	3.759	7.75	2.352	19.5	5.944	12.5	3.81	19.083	5.817	17.167	5.233
12.5	3.81	7.917	2.413	19.667	5.994	12.667	3.861	19.5	5.944	17.583	5.359
12.667	3.861	8.083	2.464	19.917	6.071	12.833	3.912	20.333	6.198	17.75	5.4
12.833	3.912	8.333	2.54	20.417	6.223	13.0	3.962				
13.417	4.08	8.417	2.566	20.583	6.274	13.167	4.013				
13.917	4.242	8.583	2.616								
14.083	4.293	8.75	2.657								
14.25	4.343	8.917	2.718								
14.833	4.521	9.083	2.769								
15.333	4.674	9.25	2.82								
15.5	4.724	9.417	2.87								
15.667	4.775	9.583	2.921								
15.833	4.826	9.833	2.997								
16.417	5.004	9.917	3.023								
16.583	5.055	10.083	3.073								

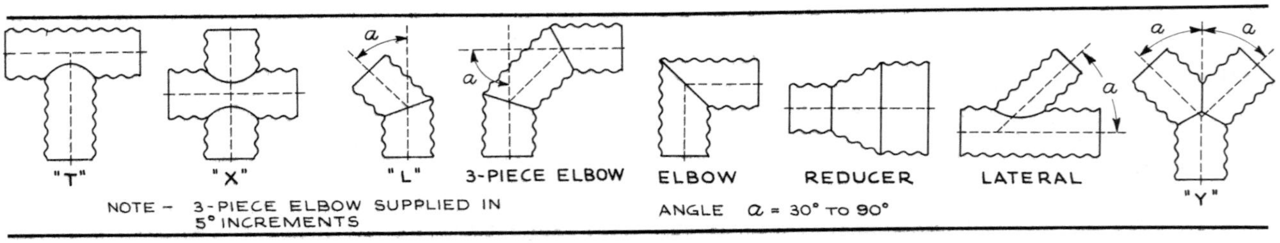

Figure S58 Standard fittings for steel corrugated pipe.

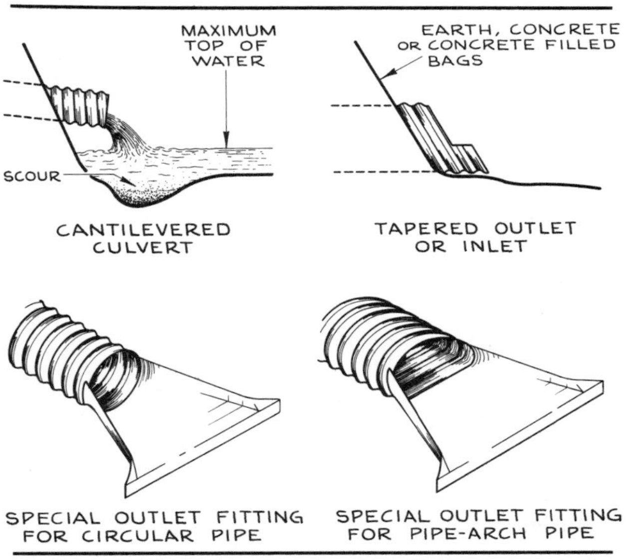

Figure S59 Fittings for inlets and outlets.

4. The waterway area to be handled.
5. The corrosive effect of water or liquid waste.
6. The type of corrugated pipe or perforated pipe to be used for solution of the design problem.
7. The treatment of inlets and outlets.

Conditions Favorable to the Use of Steel Corrugated Pipe

1. For culverts, storm sewers, spillways, underpasses, etc., where liquid is noncorrosive to steel corrugated pipe.
2. For subdrainage where liquid is noncorrosive to steel corrugated pipe.
3. For industrial wastes that are noncorrosive to steel corrugated pipe.
4. For flood control, levees, dams, etc.

Conditions Unfavorable to the Use of Steel Corrugated Pipe

1. For sewage pipes.
2. For industrial wastes that will attack steel corrugated pipe.

History and Manufacture

Stanley Simpson, a city engineer of Crawfordsville, Indiana, conceived the basic idea of applying the principle of corrugated paper to sheet metal and with James H. Watson, a sheet metal fabricator, produced the first corrugated pipe. In May 5, 1896, Watson obtained the original patent.

STEEL: HOLLOW METALWORK

Physical and Chemical Properties

The steel used for hollow metalwork is open-hearth sheet and strip steel of the furniture types. It is full pickled and cold-rolled, double-annealed and leveled. The gauges used are generally 10, 12, 14, 16, and 18 with $\frac{1}{8}$-in. (3.18-mm) and $\frac{3}{16}$-in. (4.76-mm) thick strips used as reinforcement for hardware.

Types and Uses

Steel hollow metalwork includes exterior doors and windows (double-hung windows), door bucks and frames, window trim, transoms, louvers, and many miscellaneous items such as moldings, bases, picture molds.

All these steel hollow metalwork products are available prime coated, galvanized, prepainted, baked enamel finished, vinyl coated, and with various types of overlaid finishes. One should always check with manufacturers of hollow metalwork regarding the availability of the type of finish required for a particular end use in a building. (*See* Figure *S60.*)

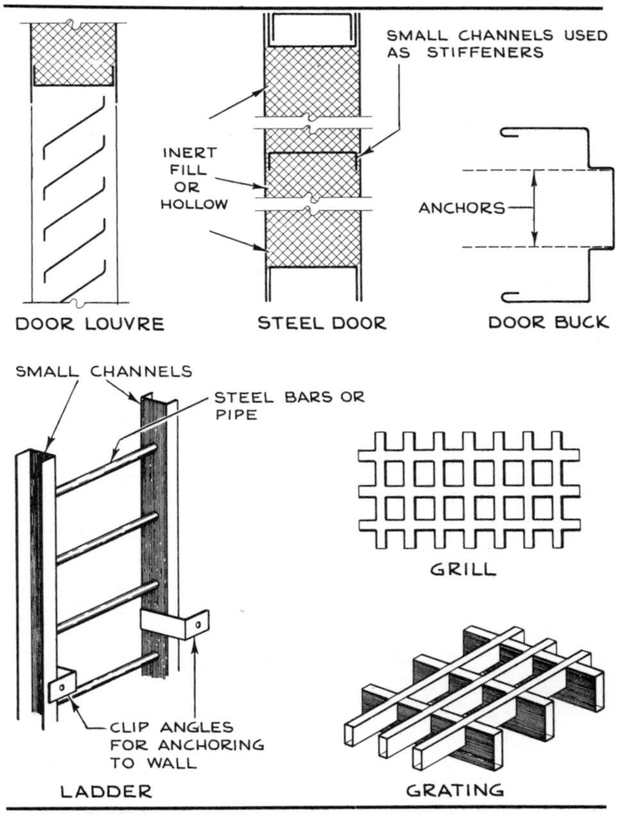

Figure S60 Typical hollow metalwork products.

The doors are made in 6.67-ft (2.032-m) and 7-ft (2.134-m) heights and in widths based on 2-in. (50.8-mm) increments up to 4 ft (1.219 m). Larger widths are available in special order.

Steel doors and bucks are available with vinyl coatings in a limited number of standard colors, and they can be obtained for interior installation with plastic laminates bonded to one or both sides of doors and also to door bucks. They are available in a wide variety of decorative colors, patterns, and natural woods.

The double-hung windows are made in modular and nonmodular sizes (*see* Figures *S61* and *S62*). Their construction consists of steel sheet and strip formed or drawn or joined by seams into hollow forms of the required shape; all joints are welded. They are manufactured as complete units and are given a suitable pretreatment for on-site painting or else a shop prime coat of paint. They are available with a vinyl coating in a limited number of standard colors.

(*See also* Doors; Windows; Steel Windows and Doors; Gratings).

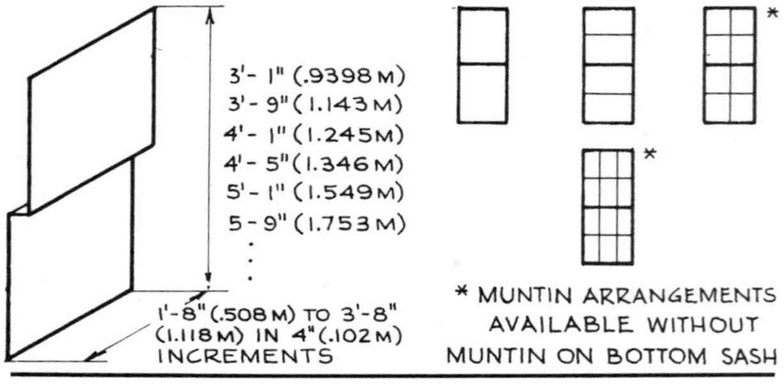

Figure S61 Steel hollow metal double-hung windows.

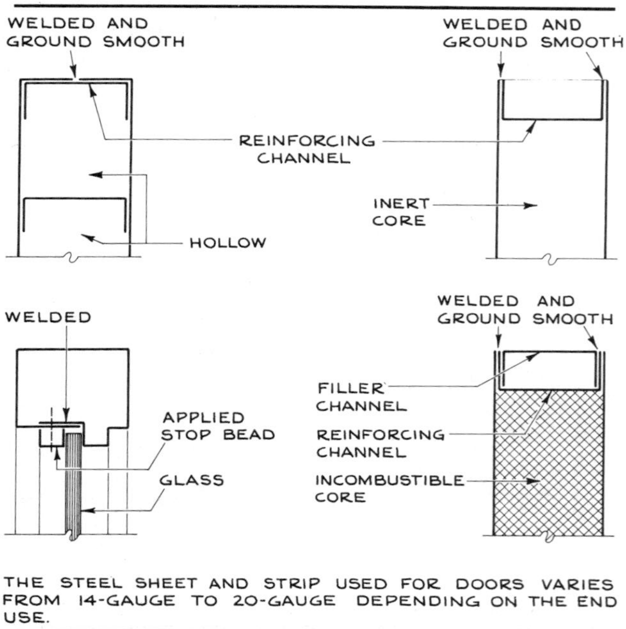

Figure S62 Typical hollow metal steel doors.

STEEL MESH AND WIRE CLOTH

Physical and Chemical Properties

The steel used for making mesh and wire cloth is commonly of two types: (1) low-carbon steel (0.08 to 0.15% carbon) for the various kinds of mesh used in fencing, partitions, and general wire work; and (2) a higher carbon steel (0.75 to 0.85% carbon) for mesh used in reinforcing concrete. The mesh and wire cloth may be plain or galvanized; mesh made of 11-gauge or finer wire is always galvanized.

Mesh and wire cloth are available with vinyl coatings in a limited number of standard colors, with other colors obtainable as special orders. The thickness of the vinyl coating depends on the requirements of the end use of the mesh and wire cloth. Manufacturers should be consulted as to the proper thickness of coating for the particular end use required in construction.

Special Types. Special kinds of mesh and wire cloth are made for use where excessively corrosive atmospheric conditions exist, where acid or other types of corrosive fumes are encountered, where special decorative effects are desired, and in many other situations where ordinary mesh and wire cloth are not suitable. The steels for these special purposes are alloy steels, stainless steel, high-temperature steels, or super alloys, depending on the end use. Many of these super alloys are not, by strict definition, really steel, as the iron content is low in comparison to that of other metallic elements also used.

Types and Uses

Large amounts of steel mesh and wire cloth are used in the construction field for all types of fencing; mesh partitions (wirework); grilles; insect screens; guards; lath for plaster, stucco, and cement; and all kinds of reinforcement for concrete. There are also innumerable minor uses such as vents, rodent barriers, screening for all types of ventilating and air-conditioning systems, and reinforcing for plastic, glass, fabrics, and paper. Perhaps the greatest single use of steel mesh and wire cloth is in industry for the grading of ores, paints, cements, etc.

The various types and sizes of steel mesh and wire cloth and their specific applications are described in the several headings under Mesh and Wire Cloth; Stainless Steel; Steel for Concrete; Wire.

History and Manufacture

The history and manufacture of steel mesh and wire cloth are described in general under Mesh and Wire Cloth; Wire.

STEEL SHEET, STRIP, AND PLATE

Physical and Chemical Properties

Steel sheet and strip are made from low-carbon, high-strength, low-alloy, and weathering steels generally containing about 0.15% and not exceeding 0.25% carbon except in the case of steels with a high manganese composition of approximately 1.25%, and steels with vanadium and columbium (niobium) combinations where the carbon content can be as high as 0.30%.

Strip by definition is sheet material that is 12 in. (304.8 mm) or less in width.

Plate is made from carbon and high-strength low-alloy steels that are graded by both physical and chemical requirements (*see* Tables *S60* through *S62*).

Thickness Designation. Steel sheet and strip are generally identified by gauge number on manufactured products for the construction field, whereas the manufacturers of sheet and strip designate thickness by decimals of an inch (or metric equivalent). Steel sheet and strip are available in thicknesses ranging from 0.012 to 0.2299 in. (0.3048 to 5.839 mm), in widths from 18 to 54 in. (0.457 to 1.372 m), in slit widths down to 2 in. (50.8 mm), and in coils with a maximum diameter of 60 in. (1.524 m) and a maximum weight of 20,000 lb (9072 kg). Flat sheets are also available with a maximum width of 48 in. (1.219 m) and a minimum length of 48 in. (1.219 m). (*See* Table *S55.*)

Quality Designation. Steel sheet and strip are available either hot-rolled or cold-rolled. The manufacturing process used affects the final temper, flatness, and finish. In general, hot-rolled sheet and strip should not be used where they are exposed, whereas cold-rolled sheet and strip are suitable for exposed parts.

Both hot- and cold-rolled sheet and strip are graded by both physical and chemical properties. They are available galvanized; electrolytically galvanized; with vinyl, alkyd, acrylic, and polyester paint coatings; with ferritic stainless steel coating; dull matte for cold-rolled, oiled or dry; with various edge treatments; and in coils of cut lengths. (*See* Tables *S56* to *S59.*)

Steel sheet and strip are also available in various tempers and qualities as follows:

1. Commercial quality, suitable for general use where special properties such as uniformity and good ductility are not necessary. It is made to meet bend test requirements only.

2. Drawing quality, produced for fabricating a specific identified part, where surface uniformity is not a primary consideration.

3. Physical quality, fabricated to meet clearly defined physical and mechanical requirements such as hardness, tensile strength, and uniformity of temper.

Table S55 Gauges and decimal requirements in pounds per square foot, kilograms per square meter, inches, and millimeters for typical steel sheet and strip

Gauge number	Manufacturers' gauge for sheet[a]			
	Weight		Thickness	
	lb/ft^2	kg/m^2	in.	mm
1	11.250	54.92	0.2813	0.7144
2	10.625	51.87	0.2656	0.6747
3	10.000	48.82	0.2500	0.6350
4	9.375	45.77	0.2243	0.5953
5	8.750	42.72	0.2187	0.5556
6	8.125	39.67	0.2031	0.5159
7	7.500	36.62	0.1875	0.4763
8	6.875	33.56	0.1719	0.4366
9	6.250	30.51	0.1563	0.3969
10	5.625	27.46	0.1406	0.3572
11	5.000	24.41	0.1250	0.3175
12	4.375	21.36	0.1094	0.2778
13	3.750	18.31	0.0938	0.2381
14	3.125	15.26	0.0781	0.1984
15	3.013	14.71	0.0703	0.1786
16	2.500	12.20	0.0625	0.1588
17	2.250	10.98	0.0563	0.1429
18	2.000	9.76	0.0500	0.1270
19	1.750	8.54	0.0438	0.1111
20	1.500	7.32	0.0375	0.0953
21	1.375	6.71	0.0344	0.0873
22	1.250	6.10	0.0313	0.0794
23	1.125	5.49	0.0281	0.0714
24	1.000	4.88	0.0250	0.0635
25	0.875	4.27	0.0219	0.0556
26	0.750	3.66	0.0188	0.0476
27	0.6875	3.36	0.0172	0.0437
28	0.625	3.05	0.0156	0.0397
29	0.5625	2.46	0.0141	0.0357
30	0.500	2.44	0.0125	0.0318

[a] This is also known as the U.S. Standards Revised.

Table S56 Physical and mechanical properties of structural-quality hot-rolled carbon steel sheet and strip

Grade	Tensile strength[a] minimum		Yield point[a] minimum		Chemical requirements					Ratio of bend diameter to thickness of sheet and strip
	lbf/in.^{2}	MN/m^2	lbf/in.^{2}	MN/m^2	C	Mn (max. percent)	P (max. percent)	S (max. percent)	Cu[b] (min. percent)	
A	45,000	310.28	25,000	172.38	0.25	0.25–0.60	0.04	0.20	0.20	0
B	49,000	337.85	30,000	206.85	0.25	0.25–0.60	0.04	0.20	0.20	1
C	52,000	358.54	33,000	227.54	0.25	0.25–0.60	0.04	0.20	0.20	$1\frac{1}{2}$
D	55,000	379.23	40,000	275.80	0.25	0.25–0.60	0.04	0.20	0.20	2
E	58,000	399.91	42,000	289.59	0.25	0.25–0.60	0.04	0.20	0.20	$2\frac{1}{2}$

[a]1 lbf/in.^{2} = 0.006895 MN/m^2.
[b]When copper steel is required.

Table S57 Structural-quality hot-dip zinc-coated steel sheet and strip

Grade	Tensile strength[a] minimum		Yield point[a] minimum		Chemical requirements				Ratio of bend diameter to thickness of sheet and strip
	lbf/in.^{2}	MN/m^2	lbf/in.^{2}	MN/m^2	C (max. percent)	P (max. percent)	S (max. percent)	Cu[b] (max. percent)	
A	45,000	310.28	33,000	227.54	0.20	0.04	0.04	0.20	$1\frac{1}{2}$
B	52,000	358.54	37,000	255.12	0.20	0.10	0.04	0.20	2
C	55,000	379.23	40,000	275.80	0.25	0.10	0.04	0.20	$2\frac{1}{2}$
D	65,000	448.18	50,000	344.75	0.40	0.20	0.04	0.20	[c]
E	82,000	565.39	80,000	551.60	0.20	0.04	0.04	0.20	[c]
F	70,000	482.65	50,000	344.75	0.50	0.04	0.04	0.20	[c]

[a]1 lbf/in.^{2} = 0.006895 MN/m^2.
[b]When copper steel is required.
[c]Bend test is not required for grades D, E, and F.

Table S58 Hot-rolled and cold-rolled high-strength, low-alloy steel sheet and strip[a]

Grade	Tensile strength[b] minimum		Yield point[b] minimum		Chemical requirements, ladle or cast analysis (percent)							Ratio of bend diameter to thickness of sheet and strip
	lbf/in.^{2}	Mn/m^2	lbf/in.^{2}	Mn/m^2	C max.	Mn max.	P max.	S max.	Cb[c] min.	V min.	N max.	
45	60,000	413.70	45,000	310.28	0.22 0.26[d]	1.35 1.40[d]	0.04 0.05[d]	0.05 0.06[d]	0.005 0.004[d]	0.01 0.005[d]		1
50	65,000	448.18	50,000	344.75	0.23 0.27[d]	1.35 1.40[d]	0.04 0.05[d]	0.05 0.06[d]	0.005 0.004[d]	0.01 0.005[d]		1
55	70,000	482.65	55,000	379.23	0.25 0.29[d]	1.35 1.40[d]	0.04 0.05[d]	0.05 0.06[d]	0.005 0.004[d]	0.01 0.005[d]		$1\frac{1}{2}$
60	75,000	517.13	60,000	413.70	0.26 0.30[d]	1.50 1.55[d]	0.04 0.05[d]	0.05 0.06[d]	0.005 0.004[d]	0.01 0.005[c]	0.012 0.015[d]	2
65	80,000	551.60	65,000	448.18	0.26 0.30[d]	1.50 1.55[d]	0.04 0.05[d]	0.05 0.06[d]	0.005 0.004[d]	0.01 0.005[d]	0.012 0.015[d]	$2\frac{1}{2}$
70	85,000	586.08	70,000	482.65	0.26 0.36[d]	1.65 1.70[d]	0.04 0.05[d]	0.05 0.06[d]	0.005 0.004[d]			3

[a]When copper is required, a minimum of 0.20% by ladle or cast analysis and of 0.18% by check analysis is required.
[b]1 lbf/in.^{2} = 0.006895 MN/m^2.
[c]Throughout the world the official new name of columbium (Cb) is niobium (Nb).
[d]Check analysis.

Table S59 Hot-rolled and cold-rolled high-strength, low-alloy steel sheet and strip with improved corrosion resistance[a]

Type of steel	Tensile strength minimum[b]				Yield point minimum[b]				Chemical requirements (max. percent, ladle or cast analysis)		
	Cut length		Coils		Cut length		Coils				
	lbf/in.^{2}	MN/m^2	lbf/in.^{2}	MN/m^2	lbf/in.^{2}	MN/m^2	lbf/in.^{2}	MN/m^2	C	M	S
Hot-rolled	70,000	482.65	65,000	448.18	50,000	344.75	45,000	310.28	0.22	0.25	0.05
Cold-rolled	65,000	448.18	65,000	448.18	45,000	310.28	45,000	310.28	0.26[c]	0.26[c]	0.06[c]

[a]Bend test requires that specimen shall be bent, at room temperature, 180° to an inside diameter of one thickness of the material without cracking on the outside of the bent portion.
[b]1 lbf/in.^{2} = 0.006895 MN/m^2.
[c]Check analysis.

Flatness. Both hot-rolled and cold-rolled sheet are produced in two standards of flatness: (1) commercial grade, which is roller leveled, and (2) stretcher-leveled, which represents sheet that has been actually stretched about 1 to 2%.

Optional Treatment. Pickling to remove the oxide scale present on all hot-rolled mill products is the buyer's option. Annealing or other heat treatment is the producer's prerogative and not often specified.

Steel Plate Quality Designations. Tables *S60* to *S62* inclusive outline physical and chemical requirements for steel plate.

Surface Finishes. Cold-rolled products usually have a dull surface texture suitable for the application of finishes such as paints, enamels, or lacquers. Depending on the end use, various degrees of freedom from surface disturbances or imperfections and special surfaces finishes can be specified.

Types and Uses

Steel sheet is used in construction for hollow metalwork and kalamein work. It is used in fabricated form as decking, galvanized sheet, expanded metal, panels and sandwich panels, and as a base metal for porcelain enamel. (*See* Figures *S63* and *S64.*)

Cold-rolled strip is available in five tempers, with six types of edges and three types of finishes (*see* Tables *S63* and *S64*).

Hot- and cold-rolled strip is used for hollow metalwork, decking, doors and windows, light bent sections, miscellaneous ironwork, roofing, siding, shingles, door bucks, window frames, partitions, and many miscellaneous items such as saddles, treads, small frames, and stairs. (*See* Figures *S39* and *S40.*)

Steel plate is widely used in the construction field, generally for bearing plates, column caps and bases, plates added to standard beams and columns to increase their strength, and trusses and built-up girders, as well as for increasing the strength of rolled sections and for joining structural sections. The major use of steel plate, however, is in shipbuilding, for heavy machinery, and in the transportation industries. Such steel plate is fabricated from steels that meet the special requirements of their end uses. In construction, steel plate is made from the same types of steel used for sheet and strip and for structural shapes. Therefore seldom is either steel plate or steel sheet and strip designated by property specifications.

Table S60 Structural-quality carbon steel plate[a]

Grade	Tensile strength[b]		Yield point[b]		Ratio of inside diameter of bend to thickness of plate						
	lbf/in.^{2}	MN/m^2	lbf/in.^{2}	MN/m^2	$\frac{3}{4}$ in. (19.05 mm) and under	Over $\frac{3}{4}$ to 1 in. (19.05 to 25.4 mm)	Over 1 to $1\frac{1}{2}$ in. (25.4 to 38.1 mm)	Over $1\frac{1}{2}$ to 2 in. (38.1 to 50.8 mm)	Over 2 to 3 in. (50.8 to 76.4 mm)	Over 3 to 4 in. (76.4 to 101.6 mm)	Over 4 in. 101.6 mm)
A	45,000–55,000	310.28–379.23	24,000	165.48	Flat on itself	Flat on itself	$\frac{1}{2}$	1	$1\frac{1}{2}$	2	$2\frac{1}{2}$
B	50,000–60,000	344.75–413.70	27,000	186.17	Flat on itself	Flat on itself	$\frac{3}{4}$	$1\frac{1}{2}$	2	$2\frac{1}{2}$	3
C	55,000–65,000	379.23–448.18	30,000	206.85	$\frac{1}{2}$	$\frac{1}{2}$	1	2	$2\frac{1}{2}$	3	$3\frac{1}{2}$
D	60,000–70,000	413.70–496.44	33,000	227.54		1	$1\frac{1}{2}$	$2\frac{1}{2}$	3	$3\frac{1}{2}$	4

[a]Chemical requirements: maximum of 0.04% of P, maximum of 0.05% of S, and when copper is required a minimum of 0.20%.
[b]1 lbf/in.^{2} = 0.006895 MN/m^2.

Table S61 Thicknesses and chemical requirements for high-yield-strength, tempered, and quenched alloy plate

Grade	Maximum thickness		Chemical requirements (percent)[a]										
	in.	mm	C	Mn	Si	Ni	Cr	Mo	V	Ti	Zr	Cu	B
A	$1\frac{1}{4}$	31.75	0.15– 0.21	0.08– 1.10	0.40– 0.80		0.50– 0.80	0.18– 0.28			0.05– 0.15		0.0025 max.
B	$1\frac{1}{4}$	31.75	0.12– 0.21	0.70– 1.00	0.20– 0.35		0.40– 0.65	0.15– 0.25	0.03– 0.08	0.01– 0.03			0.0005 700.005
C	$1\frac{1}{4}$	31.75	0.10– 0.20	1.10– 1.50	0.15– 0.30			0.20– 0.30					0.001– 0.005
D	$1\frac{1}{4}$	31.75	0.13– 0.20	0.40– 0.70	0.20– 0.35		0.85– 1.20	0.15– 0.25	[b]	0.04– 0.10		0.20– 0.40	0.0015– 0.005
E	4	101.6	0.12– 0.20	0.40– 0.70	0.20– 0.35		1.40– 2.00	0.40– 0.60	[b]	0.04– 0.10		0.20– 0.40	0.0015– 0.005
F	4	101.6	0.10– 0.20	0.60– 1.00	0.15– 0.35	0.70– 1.00	0.40– 0.65	0.40– 0.60	0.03– 0.08			0.15– 0.50	0.0005– 0.006
G	2	50.8	0.15– 0.21	0.80– 1.10	0.50 0.90		0.50– 0.90	0.40– 0.60			0.05– 0.15		0.0025 max.
H	2	50.8	0.12– 0.21	0.95– 1.30	0.20– 0.35	0.30– 0.70	0.40– 0.65	0.20– 0.30	0.03– 0.08				0.0005– 0.005
J	$1\frac{1}{4}$	31.75	0.12– 0.21	0.45– 0.70	0.20– 0.35			0.50– 0.65					0.001– 0.005
K	2	50.8	0.10– 0.20	1.10– 1.50	0.15– 0.30			0.45– 0.55					0.0015– 0.005
L	2	50.8	0.13– 0.20	0.40– 0.70	0.20– 0.35		1.15– 1.65	0.25– 0.40	[b]	0.04– 0.10		0.20 0.40	0.0015– 0.005
M	2	50.8	0.12– 0.21	0.45– 0.70	0.20– 0.35	1.20– 1.50		0.45– 0.60					0.001– 0.005
N	$\frac{3}{4}$	19.05	0.15– 0.21	0.80– 1.10	0.40– 0.90		0.50– 0.80	0.25 max.			0.05– 0.15		0.0005– 0.0025
P	4	101.6	0.12– 0.21	0.45– 0.70	0.20– 0.35	1.20– 1.50	0.85– 1.20	0.45– 0.60					0.001– 0.005

[a] All grades shall have a maximum of 0.035 percent of phosphorus and a maximum of 0.04 percent of sulfur.
[b] Vanadium may be substituted for part or all of titanium content on an equal control basis.

Table S62 A high-yield-strength, tempered, and quenched alloy plate

Thickness	Tensile strength[a] minimum		Yield strength[a] minimum		Ratio of bend diameter to thickness of plate[b]
	$lbf/in.^2$	MN/m^2	$lbf/in.^2$	MN/m^2	
To and including $\frac{3}{4}$ in. (19.05 mm)	110,000–130,000	758.45–896.35	100,000	689.50	
Over $\frac{3}{4}$ in. (19.05 mm) to and including $2\frac{1}{2}$ in. (63.5 mm)	110,000–130,000	758.45–896.35	100,000	689.50	
Over $2\frac{1}{2}$ in. (63.5 mm) to and including 4 in. (101.6 mm)	100,000–130,000	689.50–896.35	90,000	620.55	
1 in. (25.4 mm) and under					2
Over 1 in. (25.4 mm) to and including $2\frac{1}{2}$ in. (63.5 mm)					3
Over $2\frac{1}{2}$ in. (63.4 mm) to and including 4 in. (101. 6 mm)					4

[a] $1\ lbf/in.^2 = 0.006895\ MN/m^2$.
[b] These ratios are for a test specimen only; in fabrication, usually larger radii are used.

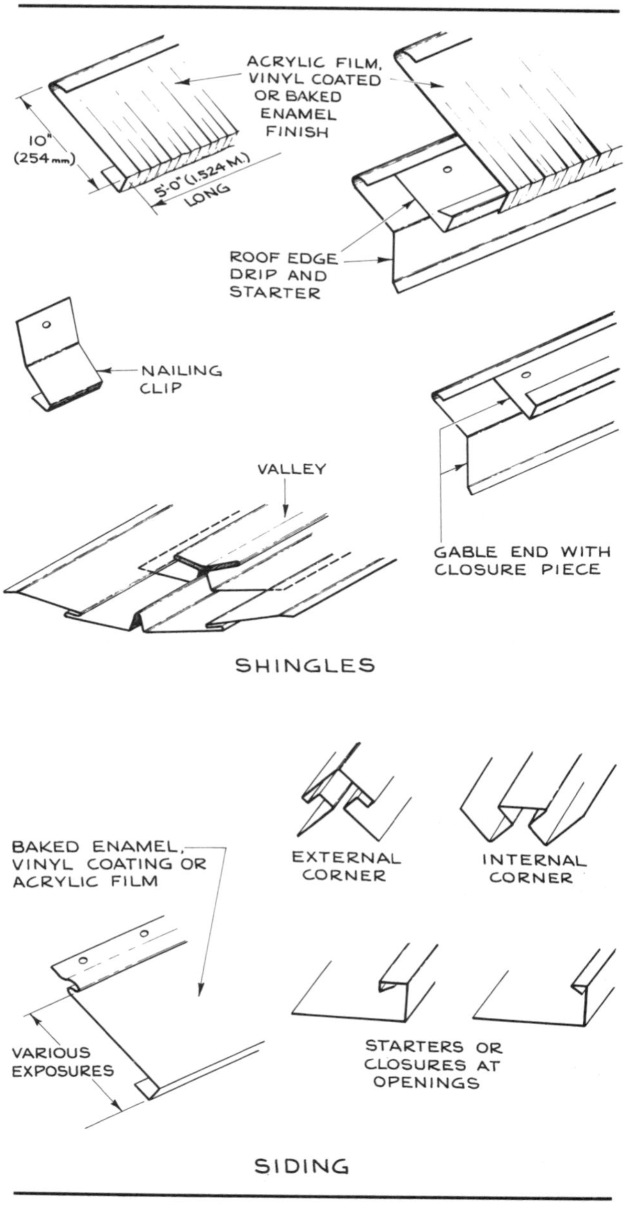

Figure S63 Typical steel shingles and siding.

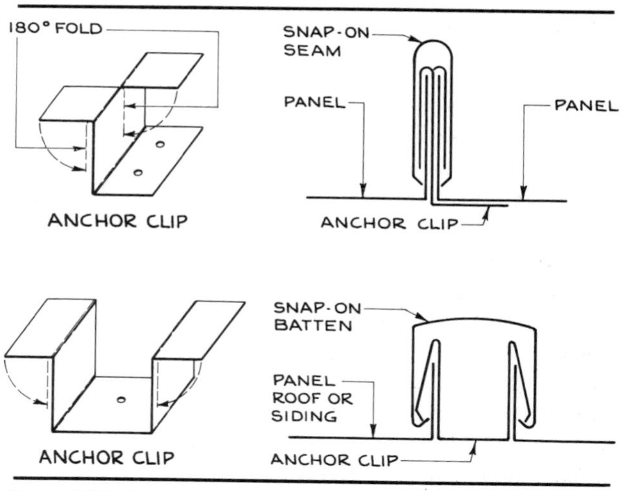

Figure S64 Snap-on type of standing or batten seams for steel roofing and siding.

Table S63 General uses of cold-rolled steel strip according to temper

Temper	Description	General use
No. 1 (Hard)	Stiff and springy	Flat work only
No. 2 (Half-hard)	Moderately stiff	Limited bending
No. 3 (Quarter-hard	Medium soft	Limited bending, drawing, and forming
No. 4 (Skin-rolled)	Soft and ductile	General bending, drawing, and forming
No. 5 (Dead soft)	Soft and ductile	Difficult bending, drawing, and forming

Instead, one deals with the countless products fabricated from steel sheet and strip, some of which have been enumerated and others of which are described under specific headings. (*See* Steel for Concrete; Steel, Galvanized; etc.)

Table S64 Types of edges and finishes for cold-rolled steel strip

Description	Edge and finish designation					
	No. 1	No. 2	No. 3	No. 4	No. 5	No. 6
Edge	Round, square or beveled	Natural mill edge	Square (made by slitting)	Rounded	Square (made by rolling or filing slit edge)	Square (not for exacting type of work)
Finish	Dull, no luster	Luster finish	High luster, good for electroplating	—	—	—

History and Manufacture

Hot-Rolling Procedures. The hot rolling of steel sheet and strip in quantity production began after the discovery of the Bessemer furnace, although some sheet and strip were rolled from the earlier crucible steels. The hot-rolling process was first used for rolling wrought iron and originally required much manpower to feed materials into the rolls, turn the rolls, and return the sheet to another set of rolls until the final thickness was reached. Today's mechanized continuous sheet and strip hot-rolling mills are so designed that almost before a slab from the reheating furnace is completely processed at the start of the mill, the other end seems to be being wound in fabricated form a quarter of a mile away.

Slabs from the blooming mill are heated to a temperature of 2250°F (1232°C) throughout; then the slab passes through a number of rough rollings which reduce thickness, increase length, and form the edges. Next it goes through several rolls for finishing. Each one of the rollings decreases thickness and increases length so that each successive rolling is accelerated in speed. From furnace to last finish rolling, the product is checked for temperature, thickness, width, and weight. At the end, if cut sheets are being produced, it is cut by shears and stacked; if rolls are required, it is coiled. The finished hot-rolled product is reduced to a thickness ranging from less than 0.05 in. (1.27 mm) to a maximum of 0.25 in (6.35 mm). Strip is never more than 12 in. (304.8 mm) wide.

Generally, all hot-rolled sheet and strip is given a cold flattening roll to improve the temper and surface of the steel.

Cold-Rolling Procedures. Cold rolling was discovered about 1867 when a workman by accident had his tongs drawn through a set of rolls. The foreman noticed that the destroyed tongs no longer had a blackened surface but instead were bright and smooth. From this incident was developed the cold rolling of steel.

Coils of steel from the hot-rolling mill are passed through an initial scalebreaker rolling; then they are pickled, rinsed, and dried. The subsequent cold rolling through several passes reduces the thickness by 40 to 85%.

Next, the coil is either directly annealed or cut into pieces and annealed. After annealing, it receives a final cold rolling which gives it final temper, flatness, and finish (called skin).

STEEL, STRUCTURAL

Physical and Chemical Properties

The steels used for structural steel are carbon steel; high-strength low-alloy steel; high-strength low-alloy steel with 50,000 lbf/in.^{2} (344.75 MN/m^2) yield point; high-strength low-alloy manganese-vanadium steel; the weathering type of steel; and high-strength low-alloy steel with columbium, vanadium, columbium-vanadium, and vanadium-nitrogen (*see* Table *S65*).

Carbon structural steel is fully covered under Steel, and the general characteristics of the high-strength low-alloy steels are also described under Steel: Alloy Steels; Steel Sheet, Strip, and Plate.

The same steels are used for high-strength bolts, washers, sheet, strip, plate, and bars.

Commercial Forms. Structural steel is available in angles, channels, I-beams, H-columns, T-shapes, Z-shapes, plates, round pipe columns, sheet piling, open web joists (bar joists), and light steel framing shapes. (*See also* subsequent headings under Steel, Structural.)

Types and Uses

Structural steel is fabricated in stock sizes and shapes as shown in Figures *S65* to *S92*. The diagrams give only dimensions and weights, since these are the important factors that can affect the architect's design. When special structural problems are encountered, it is possible to make built-up shapes from combinations of angles, plates, channels, etc. These built-up shapes fall into several categories, for example, plate girders, trusses, box beams, and rigid bends, to name only a few.

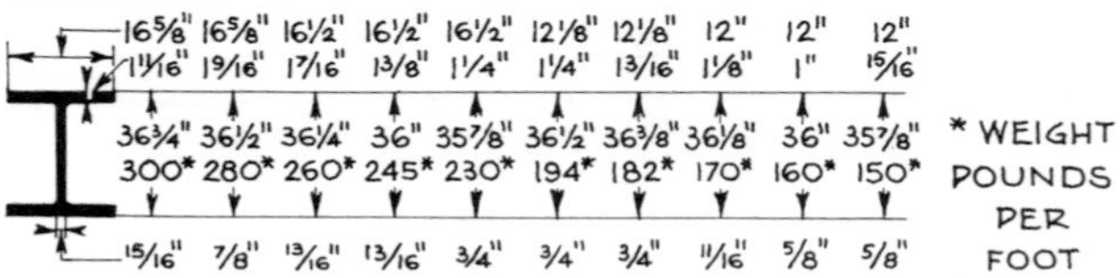

Figure S65 I-beams, 36-in. (91.44-cm)-wide flange.

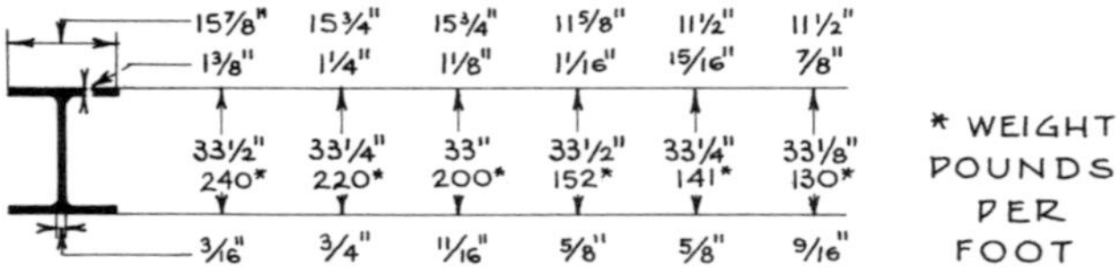

Figure S66 I-beams, 33-in. (83.82-cm)-wide flange.

Table S65 Physical and chemical requirements for carbon and high-strength, low-alloy steels for structural shapes

Type of high-strength low-alloy steel	Physical requirements: Tensile strength[c] minimum		Yield point[c] minimum		Chemical requirements (percent)												
	lbf/in.^{2}	MN/m^2	lbf/in.^{2}	MN/m^2	Ti	Mo	C	Mr	P	S	Si	Cu[a]	V	Cb	Zr	Cr	Ni
Structural steel	58,000–80,000	399.91–551.6	36,000	248.22			0.26 max.		0.04 max.	0.05 max.							
High-strength, low-alloy structural steel	63,000–70,000	434.39–482.65	42,000–50,000	289.59 344.75			0.15, 0.20	1.00, 1.35	0.15, 0.04	0.05		0.20 min.					
High-strength, low-alloy structural steel with 50,000 lbf/in.^{2} (344.75 MN/m^2) yield point	70,000	482.65	50,000	344.75	0.005–0.07 max.	0.10–0.25	0.10–0.20	0.60–1.25	0.035–0.04	0.04–0.05	0.15–0.90		0.01–0.10	0.04 max.	0.05–0.15	0.10–1.00	0.25–1.25
High-strength, low-alloy, manganese-vanadium structural steel	63,000–70,000	434.39–482,65	42,000–50,000	289.59–344.75			0.22	0.85–1.25	0.04	0.05	0.30	0.20 min.	0.02 min.				
High-strength, low-alloy structural steel	60,000–80,000	413.70–551.6	42,000–65,000	289.59–448.18			0.21–0.76	1.35–1.65	0.04	0.05	0.30		[b]	[b]			

[a]When copper is specified, a minimum content of 0.20% is required.
[b]There are four types: columbium (niobium) heat analysis, 0.005 to 0.05; vanadium heat analysis, 0.01 to 0.15; columbium (niobium) (0.05 max.) plus 0.02 to 0.15 vanadium heat analysis; and nitrogen 0.015 max. with vanadium, with a minimum ratio of 4 to 1 vanadium to nitrogen.
[c]1 lbf/in.^{2} = 0.006895 MN/m^2.

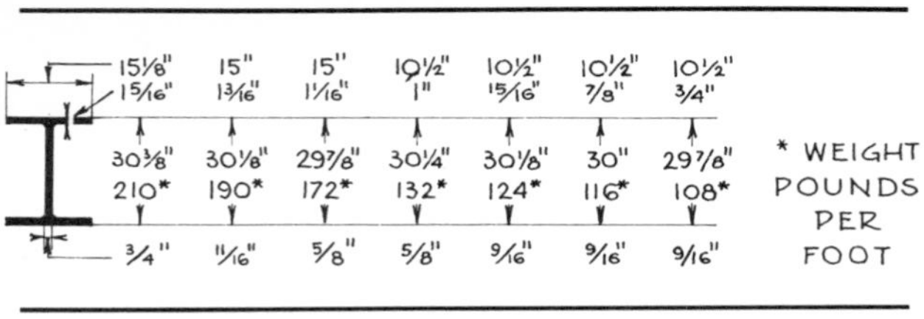

Figure S67 I-beams, 30-in. (76.20-cm)-wide flange.

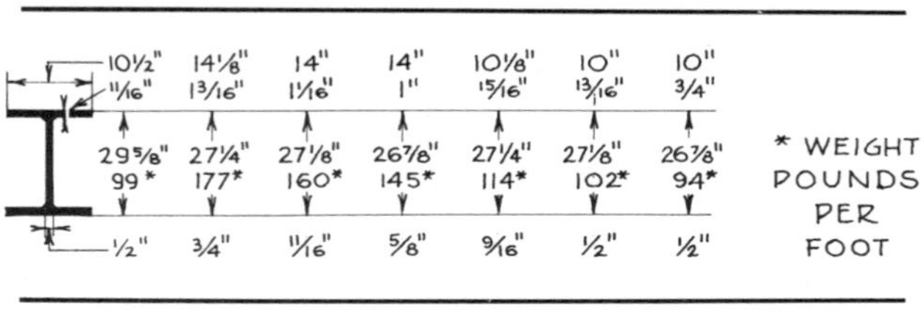

Figure S68 I-beams, 27-in. (68.58-cm)-wide flange.

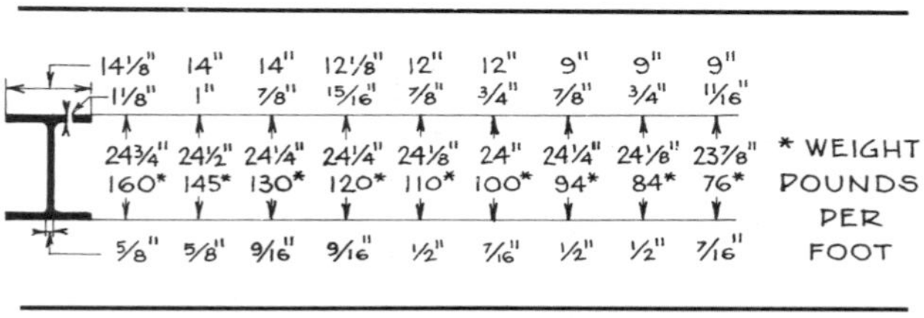

Figure S69 I-beams, 24-in. (60.96-cm)-wide flange.

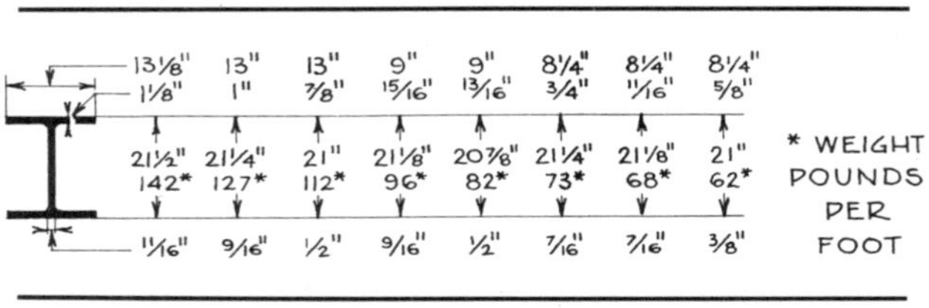

Figure S70 I-beams, 21-in. (53.34-cm)-wide flange.

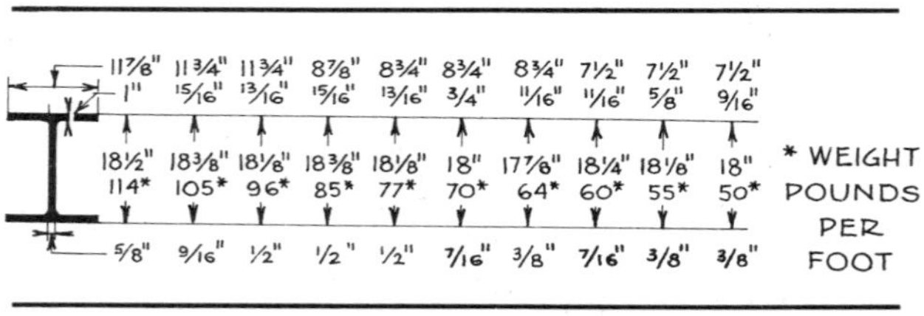

Figure S71 I-beams, 18-in. (45.72-cm)-wide flange.

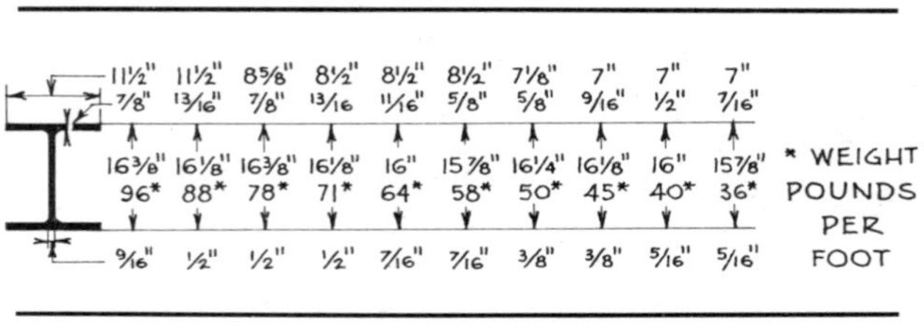

Figure S72 I-beams, 16-in. (40.64-cm)-wide flange.

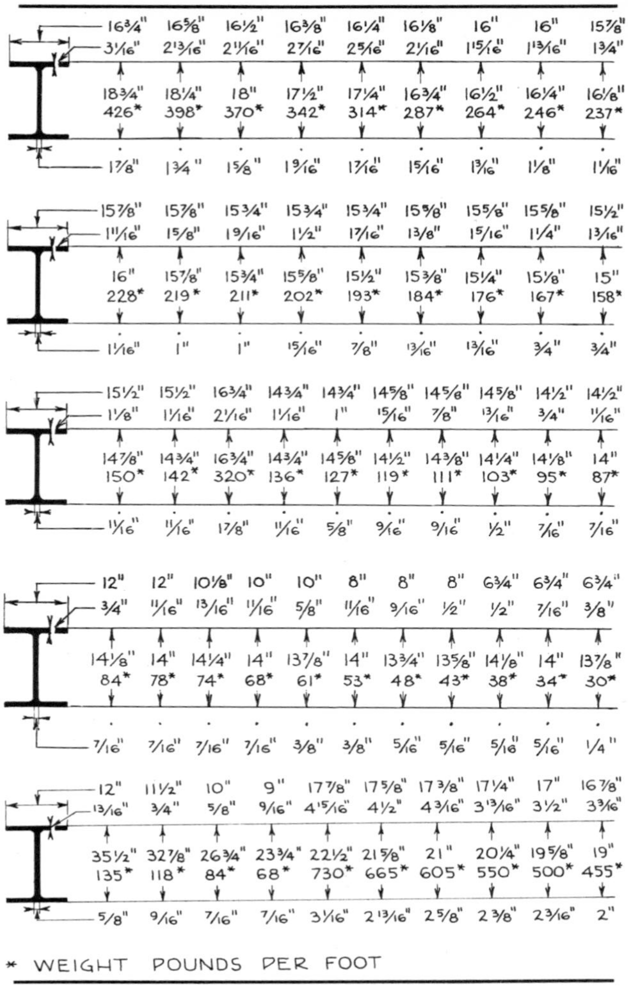

Figure S73 I-beams, 14-in. (35.56-cm)-wide flange.

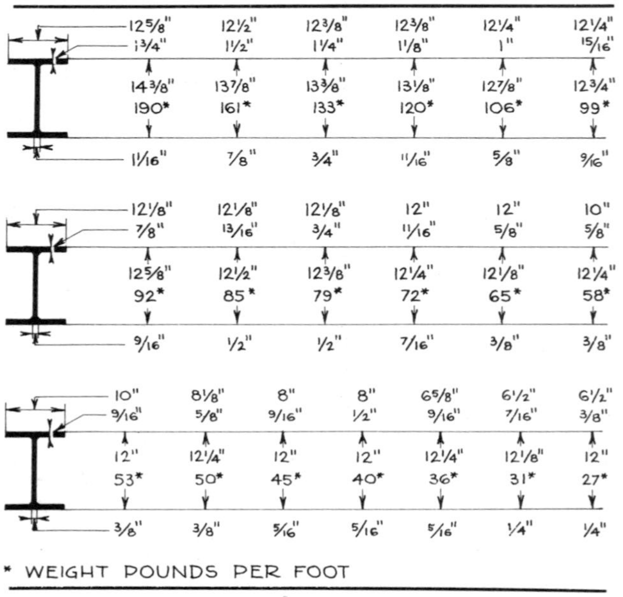

Figure S74 I-beams, 12-in. (30.48-cm)-wide flange.

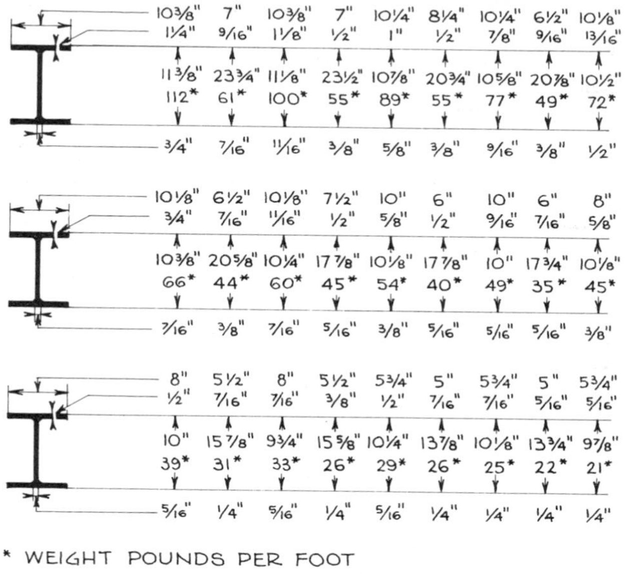

Figure S75 I-beams, 10-in. (25.40-cm)-wide flange.

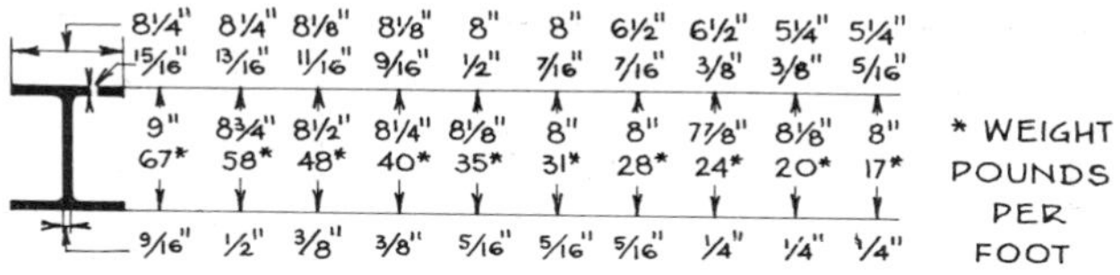

Figure S76 I-beams, 8-in. (20.32-cm)-wide flange.

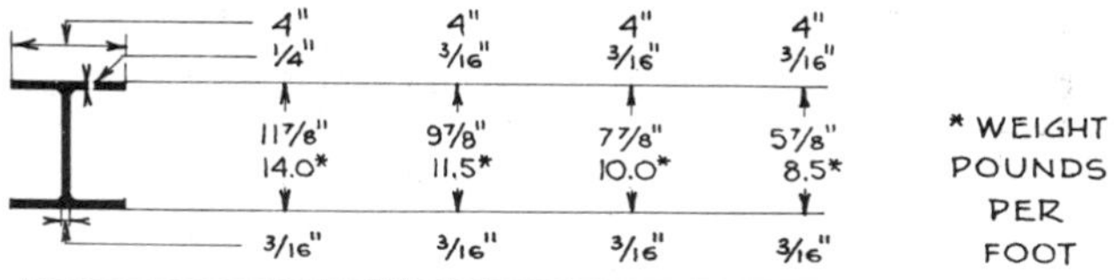

Figure S77 Joists.

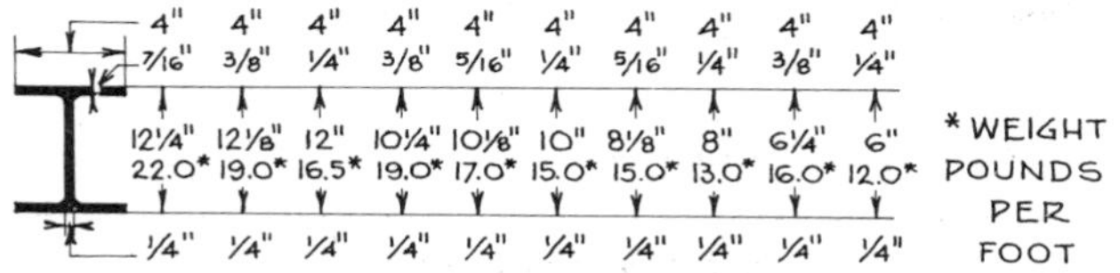

Figure S78 Light beams.

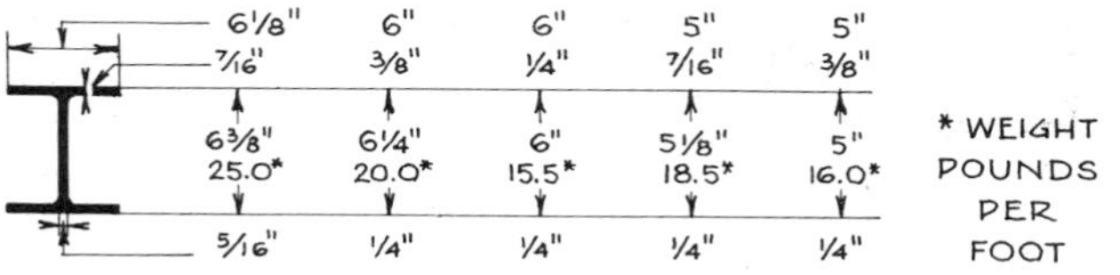

Figure S79 Stanchions.

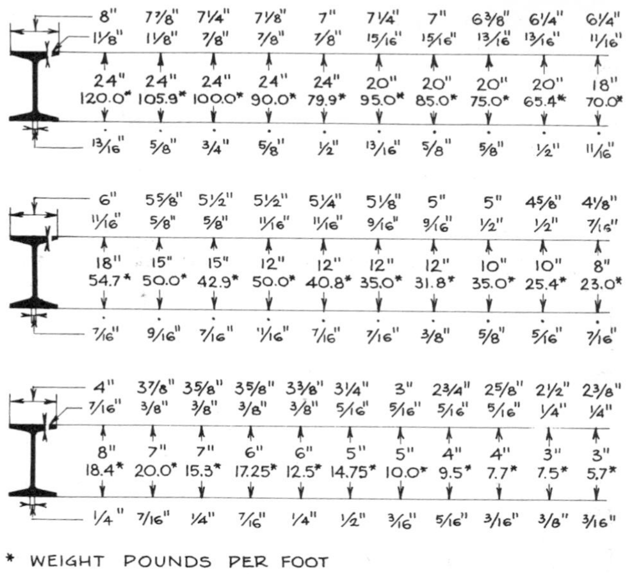

Figure S80 American standard beams.

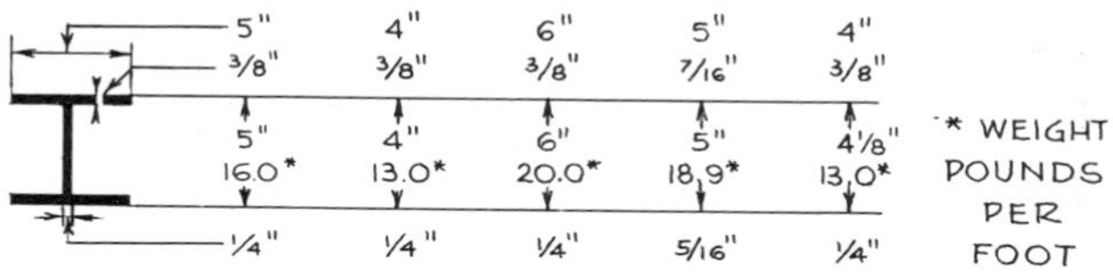

Figure S81 H-beams.

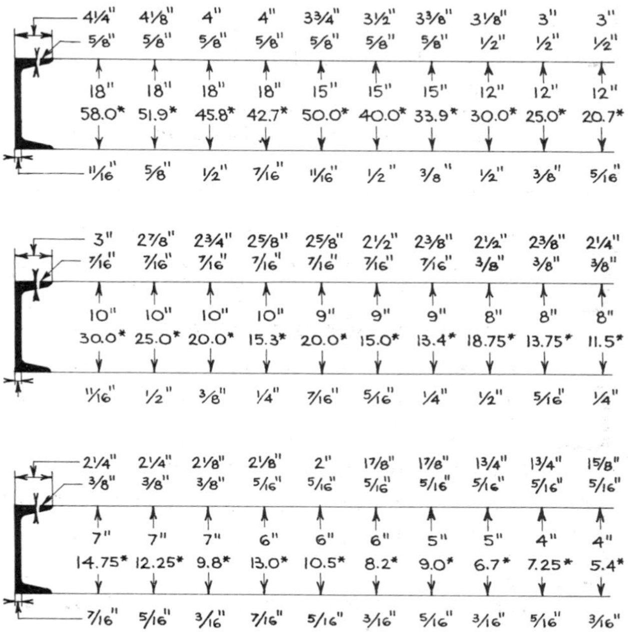

Figure S82 American standard channels.

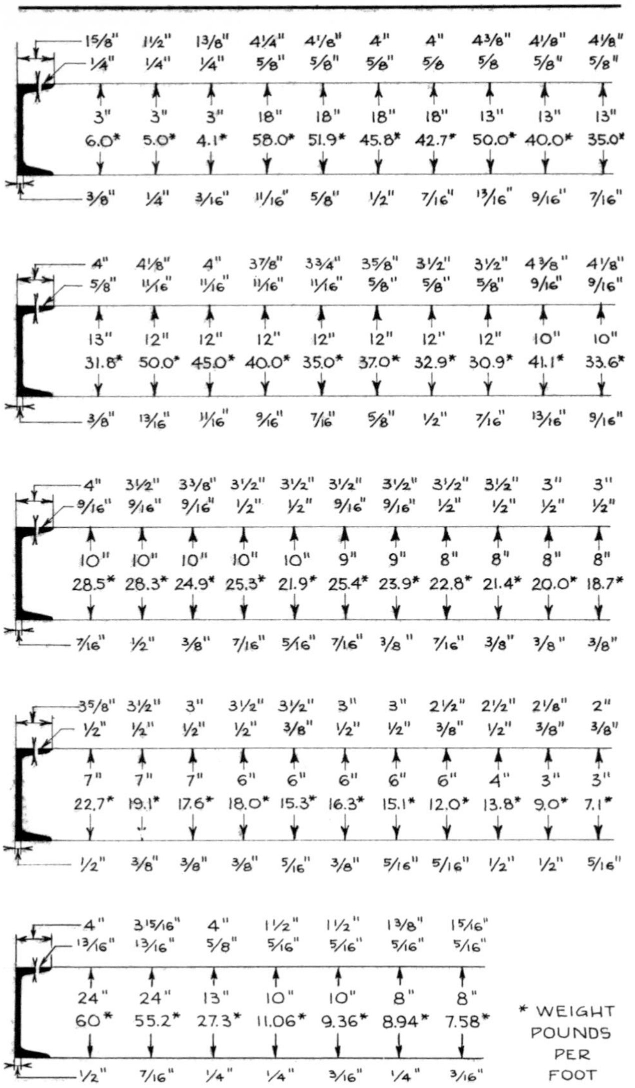

Figure S82 American standard channels (*continued*).

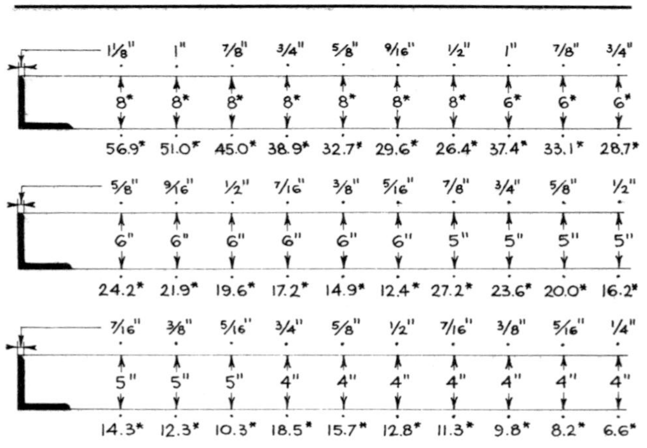

Figure S83 Equal angles.

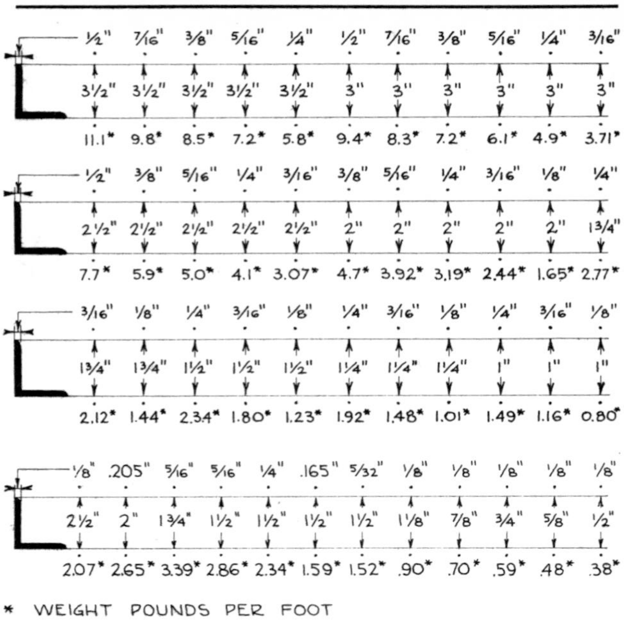

Figure S83 Equal angles (*continued*).

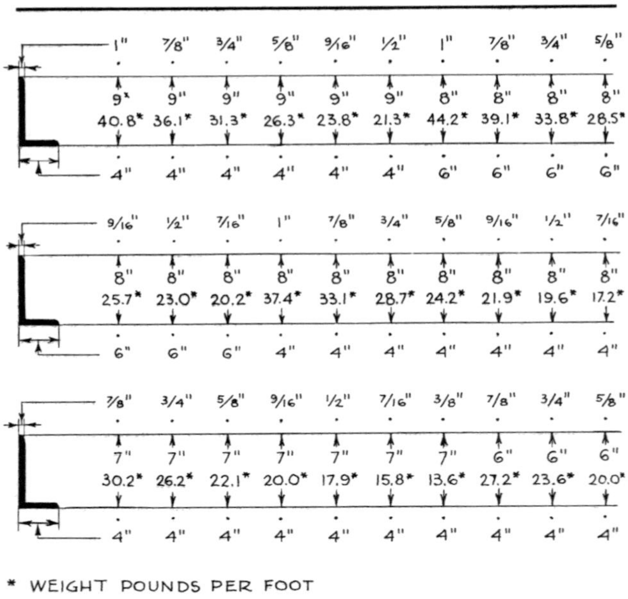

Figure S84 Unequal angles.

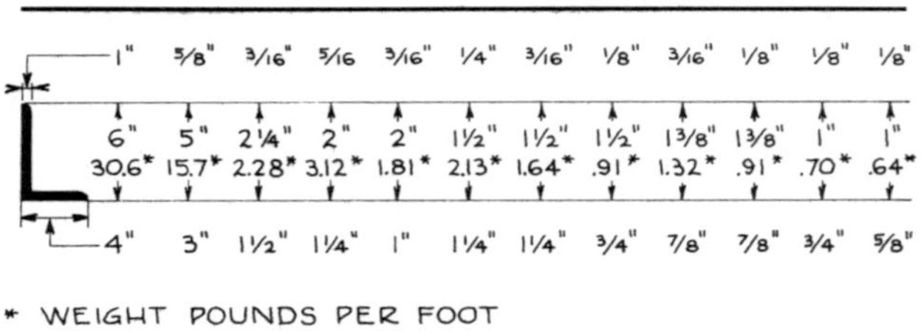

Figure S85 Unequal angles.

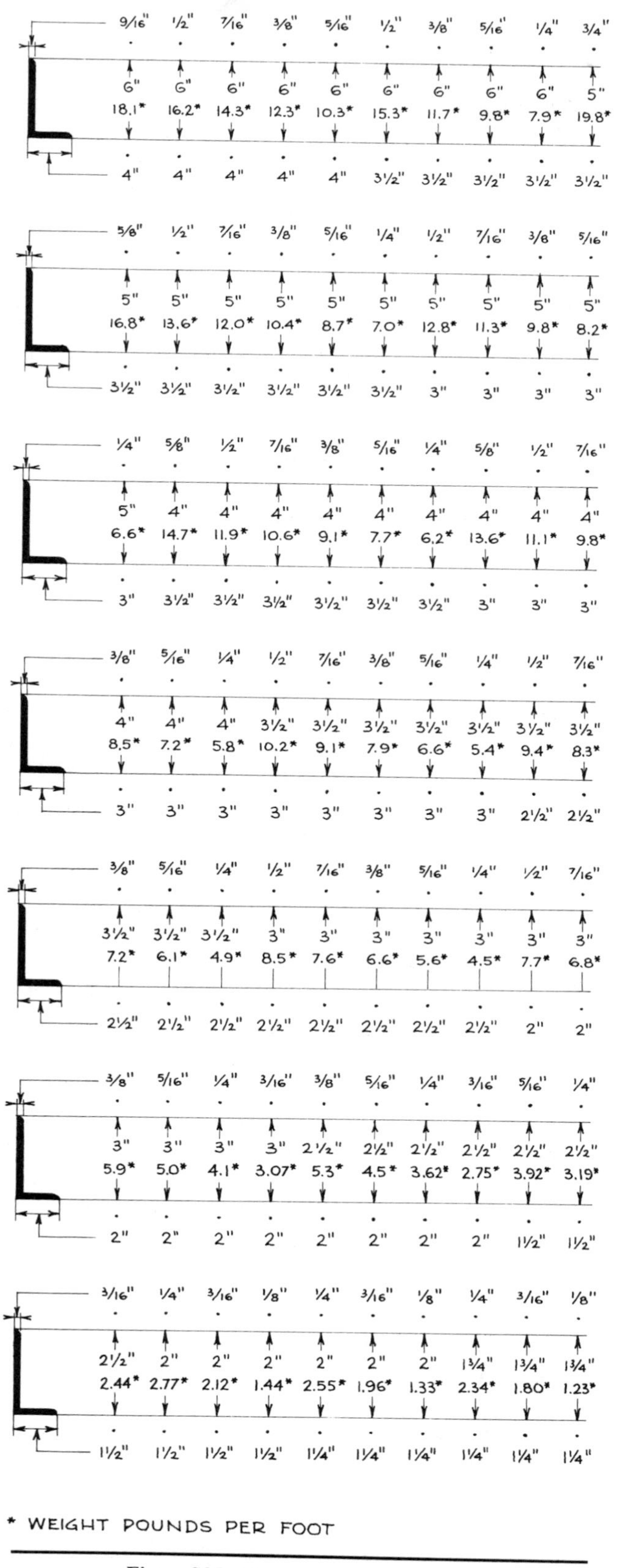

Figure S85 Unequal angles (*continued*).

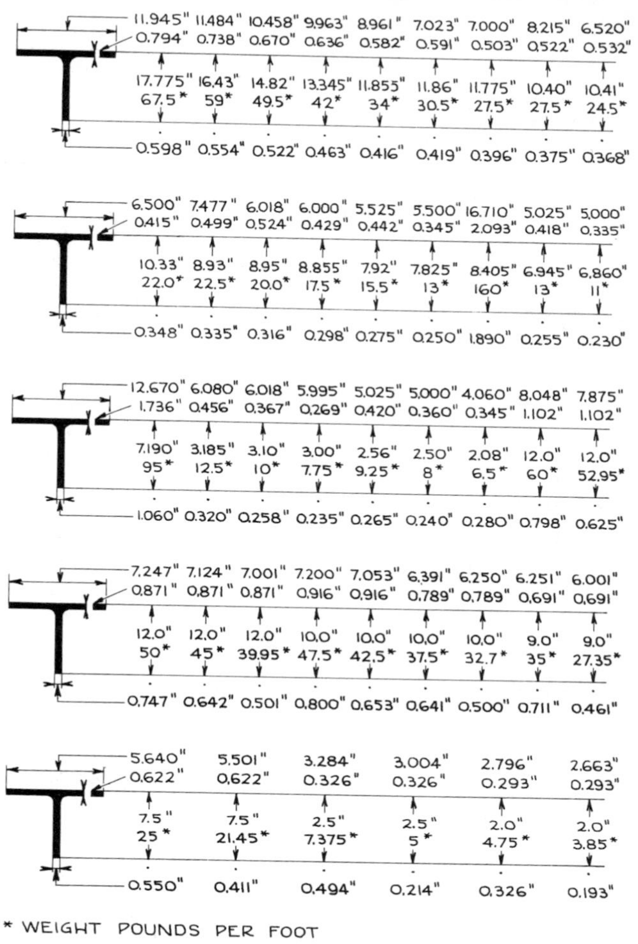

Figure S86 Structural T's.

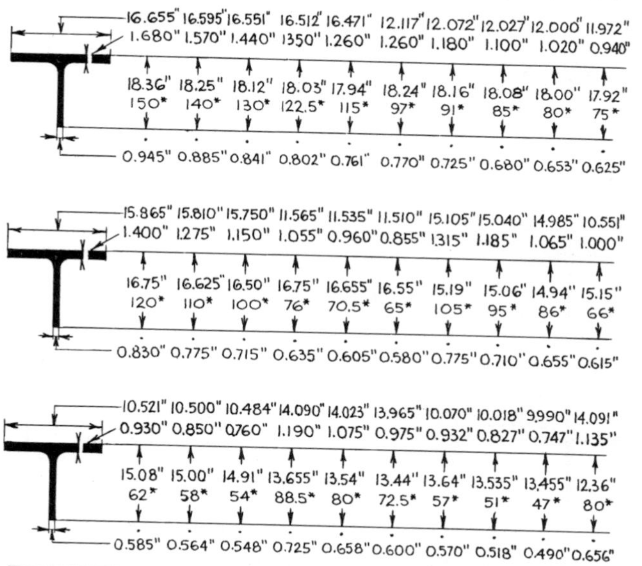

Figure S87 Structural T's.

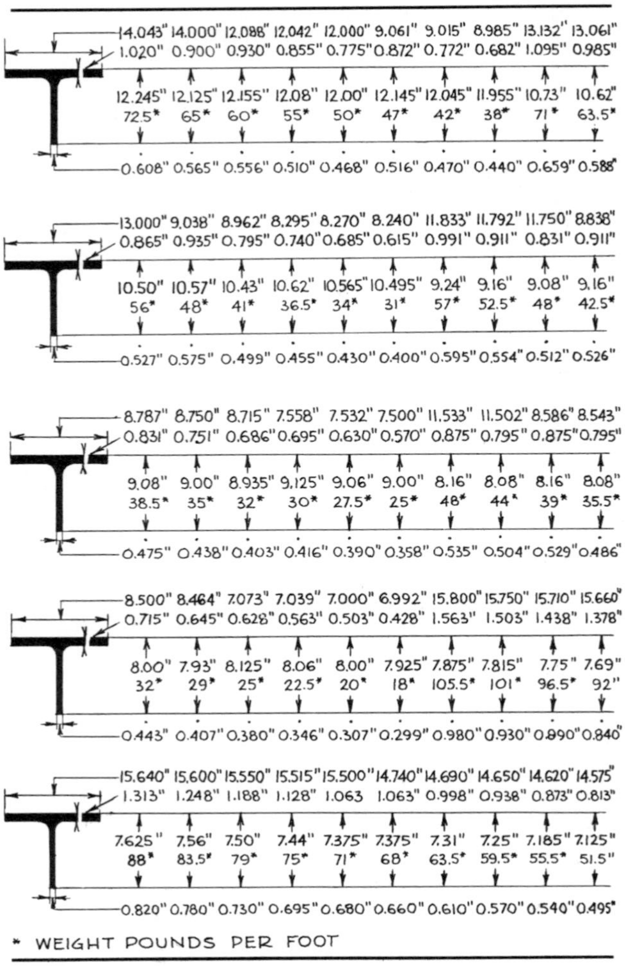

Figure S87 Structural T's (*continued*).

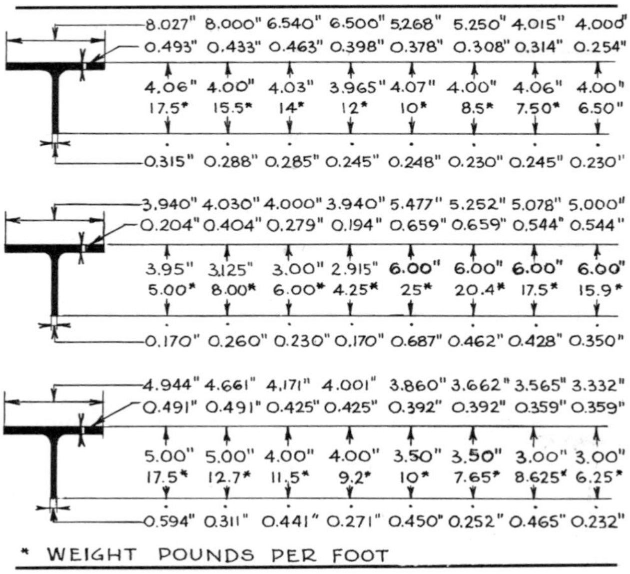

Figure S88 Structural T's (*continued*).

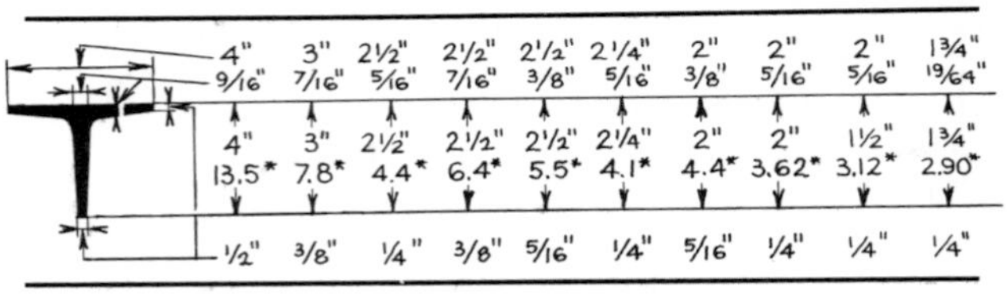

Figure S89 Equal and unequal T's.

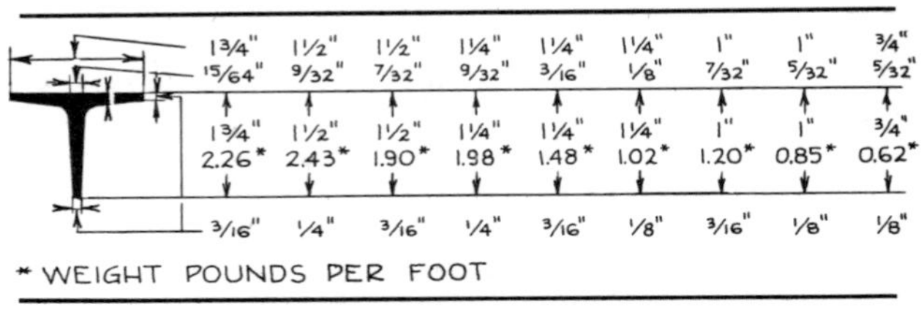

Figure S90 Equal and unequal T's.

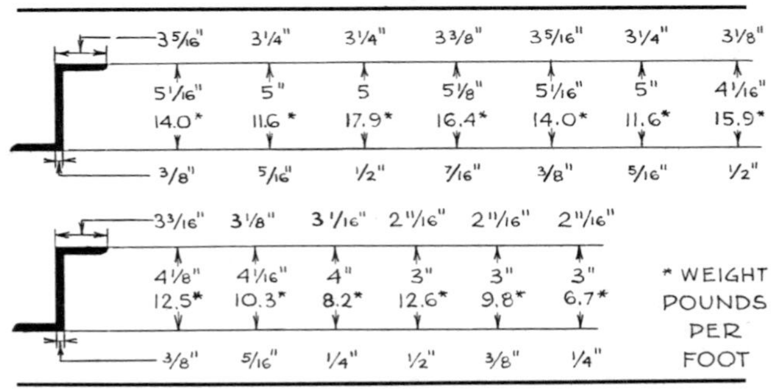

Figure S91 Z's.

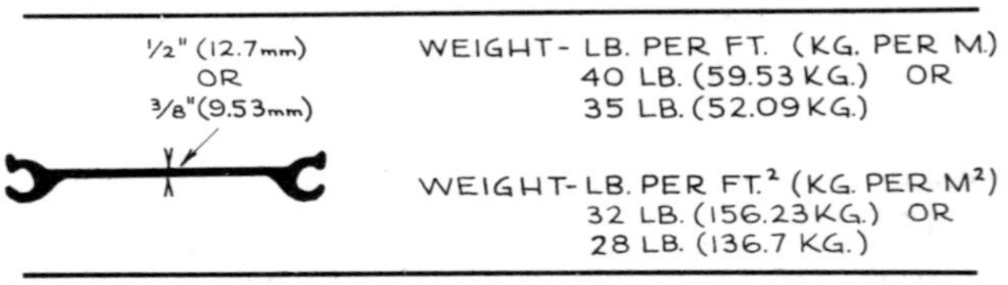

Figure S92 Steel sheet piling.

Stainless steel shapes are produced by extrusion, and many small steel parts now being cold-extruded include bolts, nuts, and special fasteners, as well as many parts for the automotive industry. It is expected that in the near future many parts for construction will also be extruded.

Application

Structural steel design is beyond the scope of this book. Innumerable textbooks, handbooks, catalogues, etc., give all the data and information pertinent to structural design. In construction, when working with structural steel, one should always check the exact actual sizes of the members to be used. For example, a 36-in.

(91.44-cm) flange is not always 36 in. high, and the flange width is not always 16.5 in. (41.9 cm), but varies according to the weight of the beam. These differences, although small, may affect the design and method of detailing in connection with windows, ceilings, partitions, and other parts of the building.

History and Manufacture

Before 1783 the only shapes made by rolling were flat or square bars. In that year Henry Cort, an Englishman, built grooved rolls that permitted the rolling of other shapes (*see* Steel for Concrete). Since then, new methods of rolling various shapes have been continuously developed. The general principles can be illustrated by following the rolling process for an angle.

Rolling Process for an Angle. First the rough ingot, 19 in. by 23 in. (48 cm by 58 cm) in section, is heated to rolling temperature, about 2200°F (1204°C). Then, by rolling through 17 different grooves, it is reduced to a rectangular billet 10 in. by 4 in. (25.4 cm by 10 cm) with a cross-sectional area of 40 in.^{2} (258 cm^2). At this point it is reheated and slowly reduced in cross-sectional area by further rolling, which begins to form the protrusions that will make the 90° final bend, to be completed in the last series of rollings. The finished product is an angle with a cross-section area of 5.75 in.^{2} (17.34 cm^2). Figure *S93* shows these various stages from rectangular billet to finished angle and also shows how an I-beam is produced and shaped from billet to finished section.

The reduction and shaping of the billet to the finished form causes the crystals within the metal to break down into smaller, more compact crystals, thus making the steel stronger and more malleable.

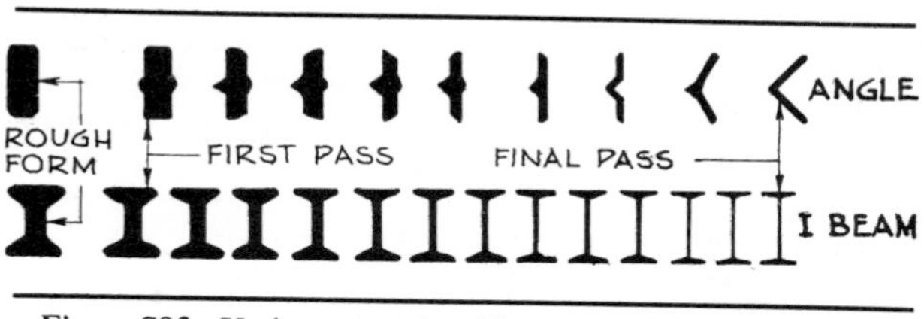

Figure S93 Various steps in rolling steel angles and I-beams.

STEEL, STRUCTURAL: LIGHT-GAUGE SHAPES AND SYSTEMS

Physical and Chemical Properties

Light-gauge steel shapes are formed from flat rolled carbon steel and usually should have a minimum yield point of 40,000 lbf/in.^{2} (275.8 MN/m^2). Carbon content is not specified, as mechanical property specifications are the controlling factor. Light-gauge steel shapes range from gauge No. 12 to No. 20 inclusive and are either single bent shapes or bent shapes welded together.

Types and Uses

These light-gauge steel shapes can be used for spans up to 20 ft (6.096 m), using the heaviest gauge with 12-in. (304.8-mm) o.c. spacing, when the total safe load does not exceed 148 lb/ft^2 (722.54 kg/m^2).

All shapes are available in a minimum of 6 ft (1.524 m) and a maximum of 40 ft (12.192 m) in length. There are two basic types, nailable and nonnailable. Both types are available punched or solid (*see* Figures *S94* and *S95*).

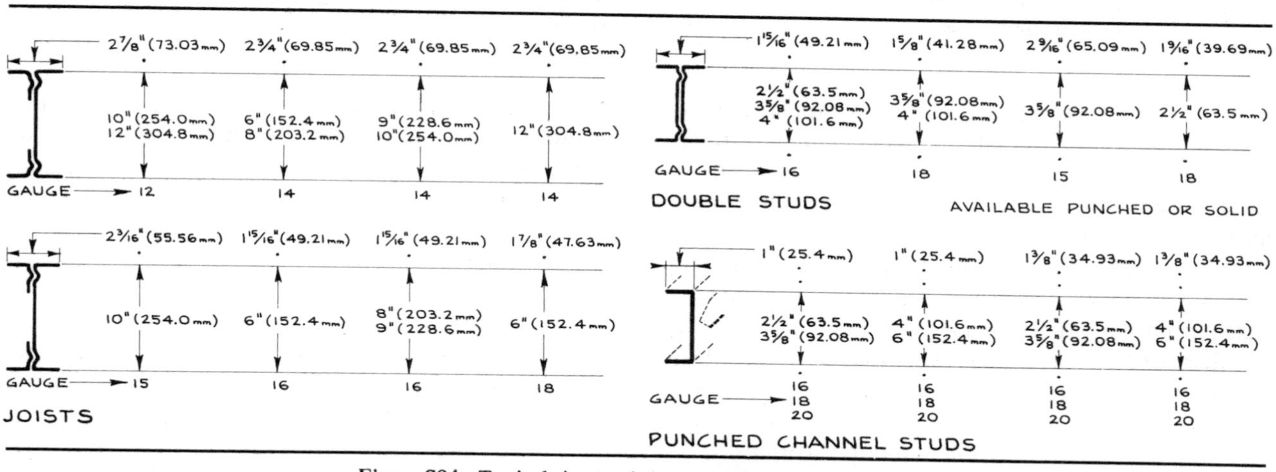

Figure S94 Typical sizes and shapes of light steel sections.

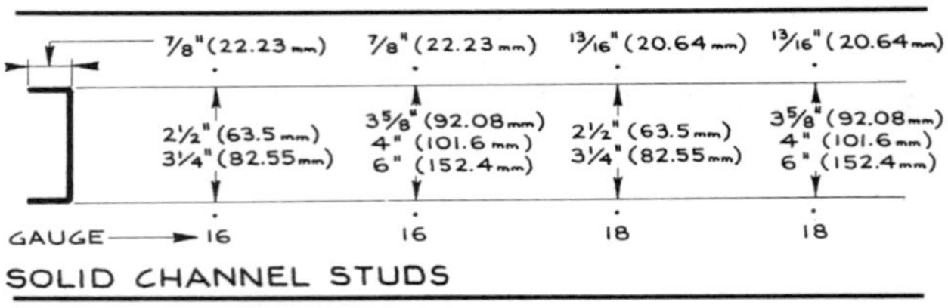

Figure S94 Typical sizes and shapes of light steel sections (*continued*).

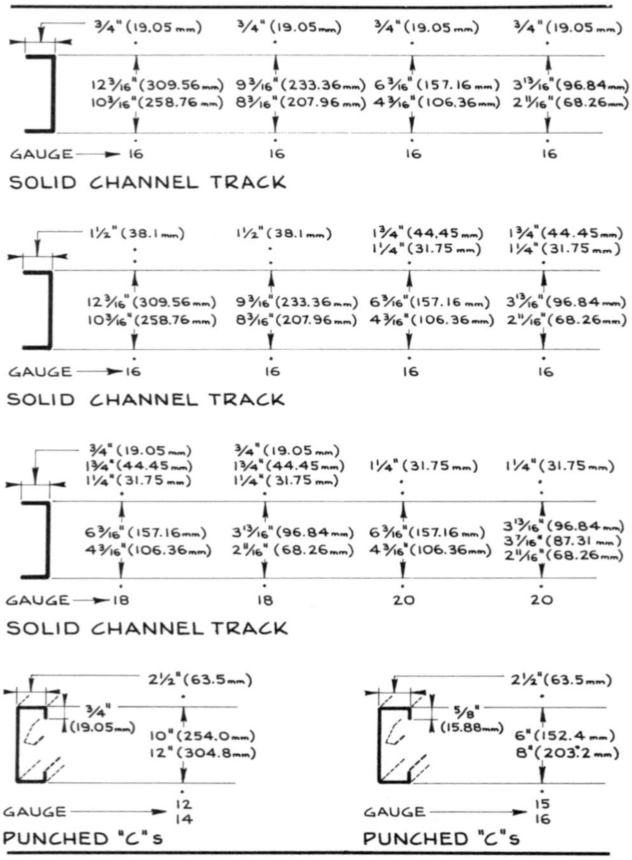

Figure S95 Typical sizes and shapes of light steel sections.

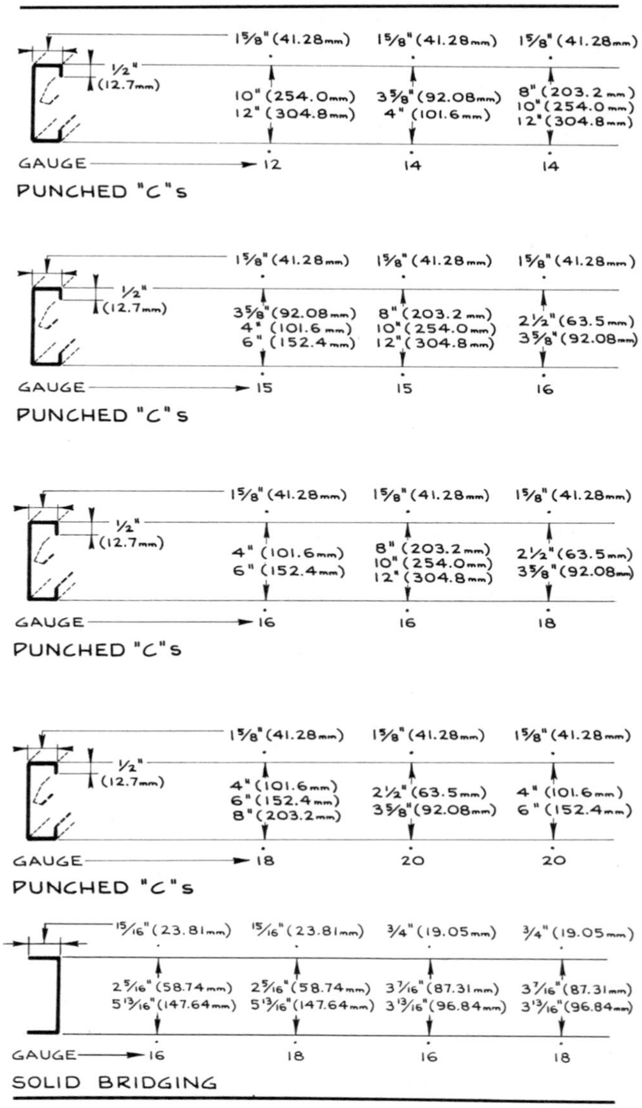

Figure S95 Typical sizes and shapes of light steel sections (*continued*).

Two types of finishes, zinc chromate primer and galvanized, are available. Studs are manufactured in thicknesses ranging from $2\frac{1}{2}$ to 6 in. (38.1 to 152.4 mm), and joists in depths from 4 to 12 in. (101.6 to 304.8 mm). Joists can be used for studs where a load or wall requires a thickness up to 12 in. (304.8 mm). Metal stud and joist systems have been adapted for residential, motel, multiple housing, and commercial types of structures, and these systems are used instead of wood and masonry construction. Because of their fire-resistant qualities and the ease of erection and installation of wiring and piping, they are used extensively for interior partitions in all types of structures.

Accessories. Accessories for this type of construction include bridging, bolts, nuts, screws, and anchors, together with devices for fastening units together, such as clips and nails.

Application

When choosing light-gauge steel framing it is always advisable to check with manufacturers for the latest data on this type of construction.

Light-gauge steel framing can be installed directly at the site of construction, or it can be prefabricated off site or on site in a temporary enclosure. When choosing this type of construction system, the selection of all interior and exterior materials, as well as the methods of installing them, should be checked.

Condensed Checklist

1. Complete shop drawings, details, and erection drawings, to be supplied by the manufacturer, must always be specified.
2. The type of floor and roofing materials must be determined, and the total safe load per square foot (0.0929 m^2) calculated, because these will control the size, depth, and shape, as well as the spacing, of the light-gauge steel shapes.
3. Type of finish material for floors, walls, and ceilings: Placing one finished material over another used as a submaterial can conceal the location of nailing grooves.
4. Type of shape used, whether solid or punched, in relation to piping, conduits, ducts, etc.
5. Types and locations of anchor bolts or drive anchors and method of installation, particularly at connections to foundation.
6. Locations of any openings for stairs, chimneys, vents, ducts, etc., in relation to the type of framing, particularly headers.
7. Roof overhangs, balconies, and cantilevers of any type: Light-gauge steel shapes can cantilever in one direction, but not in two directions in the same plane without difficulty.
8. Fire ratings: Local, municipal, and state codes should be checked and also the fire rating codes of the Fire Underwriters, labor departments, insurance companies, and federal government, including the Army and Navy.
9. Possibility of galvanic action with metals to be used in conjunction with light steel shapes.
10. Ability of the manufacturer in the locality to supply all necessary sizes, shapes, and accessories for the structure.
11. Types of self-tapping screws for connecting light-gauge steel units together and also for installing the materials to be applied to the light steel frame (*see also* Plaster Integrants and Sheets).
12. When off-site prefabrication is selected as the method of construction, one should check with manufacturers in the area where the proposed structure is to be erected.
13. All environmental control systems should be checked for their installation requirements in a light-gauge steel structure. Included are electric, plumbing, heating, ventilation, and air-conditioning systems.
14. The types of tools, machines, and equipment required for light-gauge steel construction should be checked.
15. For multistory buildings or complicated commercial structures, one should always consult with a structural engineer.

Conditions Favorable to the Use of Light-Gauge Steel Shapes

1. Where the design has clear spans not exceeding 30 ft (9.144 m), yet not so small that another construction system would be simpler.
2. Where the design requires a degree of fire resistance not obtainable with wood framing.
3. Where a strong, rigid, rotproof, termiteproof type of construction is required.
4. Where the type of construction requires a prefabricated construction system.
5. Where interior spaces must be subdivided with floor-to-ceiling partitions.
6. For interior partitions where electrical service and piping are required.

Conditions Unfavorable to the Use of Light-Gauge Steel Shapes

1. Where the locality is not near enough to manufacturers of this type of material.
2. Where the design requires a better than 4-hour fire-resistance rating.

STEEL, STRUCTURAL: OPEN WEB JOISTS, TRUSSES, AND SPACE FRAMES

Types and Uses

Open web steel joists, or bar joists, are prefabricated lightweight trusses composed of small T-angles and bars made of steel with a minimum yield strength between 36,000 and 50,000 lbf/in.^{2} (248.22 and 344.75 MN/m^2). They are now divided into six types which replace all previous designations: J and H for spans up to 60 ft (18.298 m); LJ and LH long-span joists for spans up to 96 ft (29.261 m); and DLJ and DLH deep long-span joists for spans up to 144 ft (43.981 m). Sizes, safe load capacities, bridging, method of welding, manufacture, etc., are standardized by the Steel Joist Institute (*see* Figure *S96*).

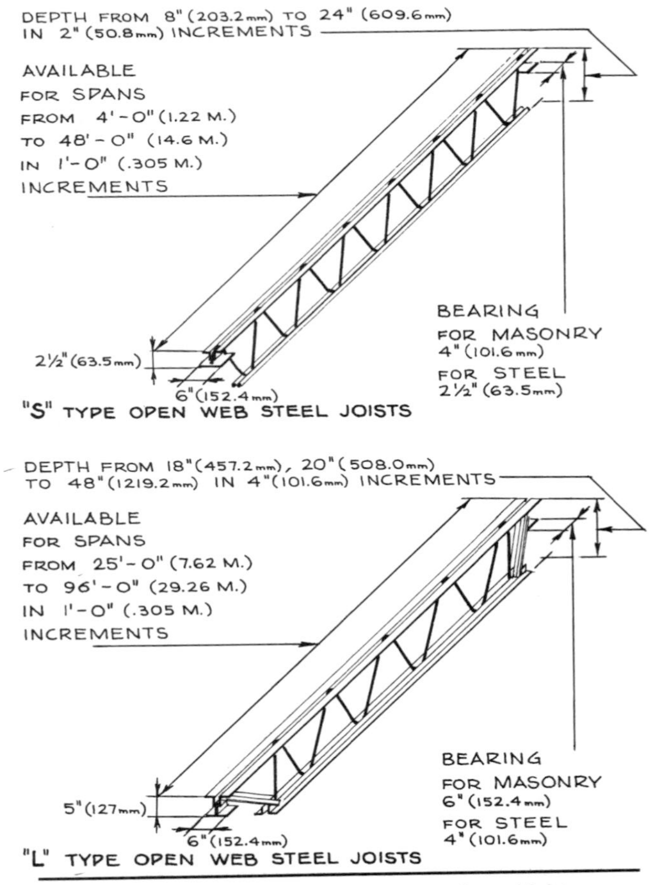

Figure S96 Typical sizes for open web steel joists.

See Steel, Structural for a description of physical and chemical properties and for history and manufacturing methods.

Steel trusses are available prefabricated or custom fabricated. When a structure requires trusses, manufacturers of light steel structural members and structural steel manufacturers should always be consulted as to what is available and, if a special truss is required, the length of time necessary for its manufacture and delivery. When trusses are required, one should always check with structural engineers.

The space frame was developed as a result of the search for a truss-type structural system that could span not only in one direction but also in many directions. Space frames can cover a large area with a minimum of supports. Figures *S97* and *S98* show various space frame components and typical kinds of space frames. When a space frame is required for a particular type of structure, one should always consult with a structural engineer and check with the manufacturers of space frames and space frame components.

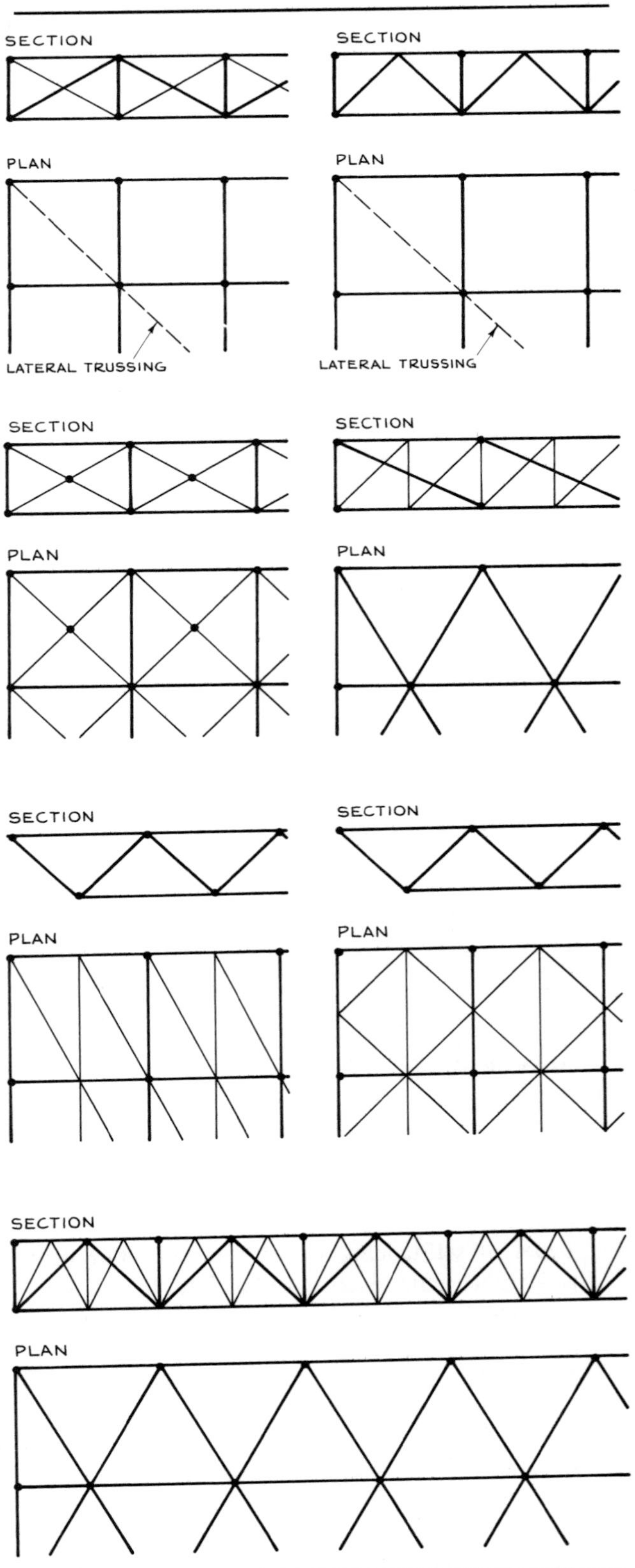

Figure S97 Typical types of space frames.

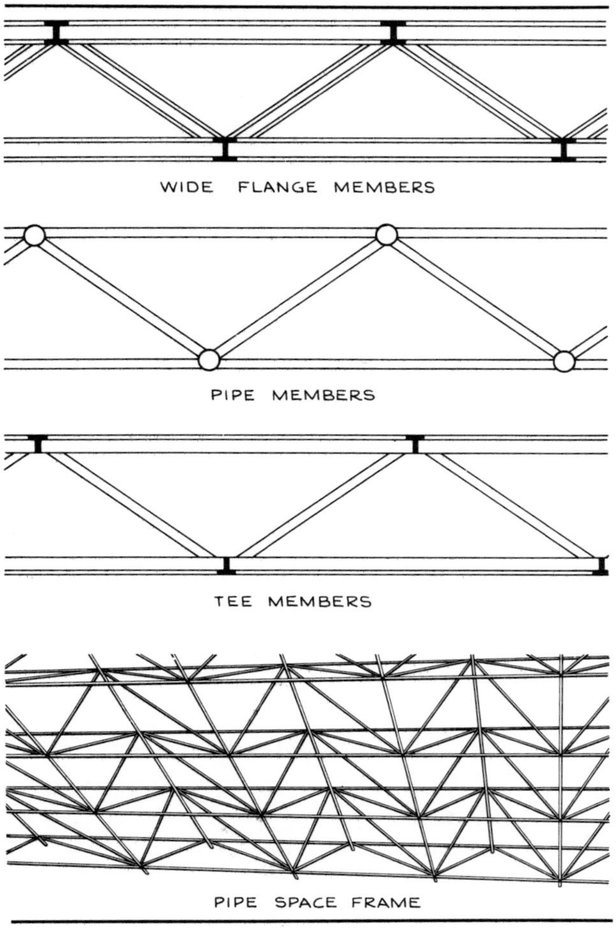

Figure S98 Typical types of space frame structural units.

Application

There are continual developments of new structural features in the various open web steel joist construction systems. Joists can be obtained (1) with the top member forming a raceway for conduits, pipes, etc., (2) with preinstalled wood nailer strips, and (3) bent to form nailing grooves.

Accessories. Accessories for open web steel joists include all kinds of anchors, clips, bridging, extensions for ceilings, headers for openings, and bearings. The local joist manufacturers should always be checked for available accessories and details of latest developments.

For J- and H-joists the bridging may be diagonal, as shown in Table *S66*, or horizontal and attached to the top and bottom chords (*see* Table *S67*).

Condensed Checklist

1. Type and weight of floor or roof construction: This will control spacing and therefore the size of the joists.
2. Type and spacing of bridging for various spans (*see* Table *S66*).
3. Type of bearing: On steel where two joists butt against one another and can be welded, a bearing of less than $2\frac{1}{2}$ in. (63.5 mm) can be used; otherwise, the necessary bearing must be provided (*see* Figure *S99*).

Table S66 Spacing of bridging for J, H, LJ, HJ, DLJ, and DLH series of open web steel joists

Type	Clear spans		Bridging locations	Depth for types LJ, LH, DLJ, and DLH		Joist spacing		Bridging angle size	
	ft	m		in.	m	ft	m	in.	mm
J3, H3	Up to 13	Up to 3.962	1 row at center	18	0.457	6.415	1.956	$1 \times 1 \times \frac{1}{8}$	25.4 × 25.4 × 3.18
	13–17	3.962–5.182	2 rows at $\frac{1}{3}$ points			8.166	2.489	$1\frac{1}{4} \times 1\frac{1}{4} \times \frac{1}{8}$	31.75 × 31.75 × 3.18
	17–28	" "	3 rows at $\frac{1}{4}$ points			9.830	2.997	$1\frac{1}{2} \times 1\frac{1}{2} \times \frac{1}{8}$	38.1 × 38.1 × 3.18
						11.500	3.505	$1\frac{3}{4} \times 1\frac{3}{4} \times \frac{1}{8}$	44.45 × 44.45 × 3.18
J4, H4	Up to 16	Up to 4.877	1 row at center						
	16–21	4.877–6.401	2 rows at $\frac{1}{3}$ points	20	0.508	6.415	1.956		
	21–32	6.401–9.754	3 rows at $\frac{1}{4}$ points			8.083	2.464	Same as above	
J5, H5	Up to 16	Up to 4.877	1 row at center			9.830	2.997		
	16–21	4.877–6.401	2 rows at $\frac{1}{3}$ points			11.500	3.505		
	21–33	6.401–10.058	3 rows at $\frac{1}{4}$ points	24	0.61	6.332	1.930		
	33–38	10.058–11.582	4 rows at $\frac{1}{5}$ points			8.083	2.464	Same as above	
	38–40	11.582–12.192	5 rows at $\frac{1}{6}$ points			9.750	2.972		
						11.415	3.480		

Table S66 Spacing of bridging for J, H, LJ, HJ, DLJ, and DLH series of open web steel joists (*continued*)

Type	Clear spans ft	Clear spans m	Bridging locations
J6, H6	Up to 18	Up to 5.486	
	18–22	5.486–6.706	
	22–36	6.706–10.973	Same as above
	36–40	10.973–12.192	
	40–48	12.192–14.630	
J7, H7	Up to 20	Up to 6.096	
	20–25	6.096–7.620	
	25–41	7.627–12.497	Same as above
	41–46	12.497–14.021	
	46–48	14.021–14.630	
J8, H8	Up to 21	Up to 6.401	
	21–27	6.401–8.230	
	27–43	8.230–13.106	Same as above
	43–48	13.106–14.630	
	48–60	14.630–18.288	
J9, H9	Up to 23	Up to 7.010	
	23–30	7.010–9.144	
	30–46	9.144–14.021	Same as above
	46–52	14.021–15.850	
	52–60	15.850–18.288	
J10, H10	Up to 24	Up to 7.315	
	24–30	7.315–9.144	
	30–47	9.144–14.326	Same as above
	47–53	14.326–16.154	
	53–60	16.154–18.288	
J11, H11	Up to 24	Up to 7.315	
	24–31	7.315–9.449	
	31–48	9.449–14.630	Same as above
	48–55	14.630–16.764	
	55–60	16.764–18.298	

Depth for types LJ, LH, DLJ, and DLH in.	m	Joist spacing ft	m	Bridging angle size in.	mm
28	0.711	6.166	1.880		
		8.000	2.438	Same as above	
		9.667	2.946		
		11.415	3.480		
32	0.813	6.083	1.854		
		7.830	2.388	Same as above	
		9.581	2.921		
		11.332	3.454		
36	0.914	7.750	2.362	$1\frac{1}{4} \times 1\frac{1}{4} \times \frac{1}{8}$	31.75 × 31.75 × 3.18
		9.500	2.986	$1\frac{1}{2} \times 1\frac{1}{2} \times \frac{1}{8}$	38.1 × 38.1 × 3.18
		11.250	3.429	$1\frac{3}{4} \times 1\frac{3}{4} \times \frac{1}{8}$	44.45 × 44.45 × 3.18
40	1.016	7.581	2.311		
		9.415	2.870	Same as above	
		11.166	3.404		
44	1.118	7.415	2.261		
		9.250	2.819	Same as above	
		11.000	3.353		
48	1.219	7.250	2.210		
		9.083	2.769	Same as above	
		10.913	3.327		
52	1.321	9.000	2.743	$1\frac{1}{2} \times 1\frac{1}{2} \times \frac{1}{8}$	38.1 × 38.1 × 3.18
		10.750	3.277	$1\frac{3}{4} \times 1\frac{3}{4} \times \frac{1}{8}$	44.45 × 44.45 × 3.18
		12.581	3.835	$2 \times 2 \times \frac{1}{8}$	50.8 × 50.8 × 3.18
56	1.422	8.830	2.692		
		10.667	3.251	Same as above	
		12.415	3.785		
60	1.524	8.581	2.616		
		10.500	3.200	Same as above	
		12.332	3.759		
64	1.626	8.415	2.565		
		10.332	3.150	Same as above	
		12.166	3.708		
68	1.727	8.166	2.489		
		10.166	3.099	Same as above	
		12.000	3.658		
72	1.829	8.000	2.438		
		10.000	3.048	Same as above	
		11.830	3.607		

Table S67 Horizontal bridging for J and H types of joists

Joist spacing		Sizes			
		Round		Angle	
ft	m	in.	mm	in.	mm
Up to and including 3	Up to and including 0.914	$\frac{1}{2}$	12.7	$\frac{3}{4} \times \frac{3}{4} \times \frac{1}{8}$	19.05 X 19.05 X 3.18
3–4	0.914–1.219	$\frac{5}{8}$	15.88	$\frac{3}{4} \times \frac{3}{4} \times \frac{1}{8}$	19.05 X 19.05 X 3.18
4–5	1.219–1.524			$1 \times 1 \times \frac{1}{8}$	50.8 X 50.8 X 3.18
5–6.25	1.524–1.905			$1\frac{1}{4} \times 1\frac{1}{4} \times \frac{1}{8}$	31.75 X 31.75 X 3.18
6.25–7.50	1.905–2.286			$1\frac{1}{2} \times 1\frac{1}{2} \times \frac{1}{8}$	38.1 X 38.1 X 3.18

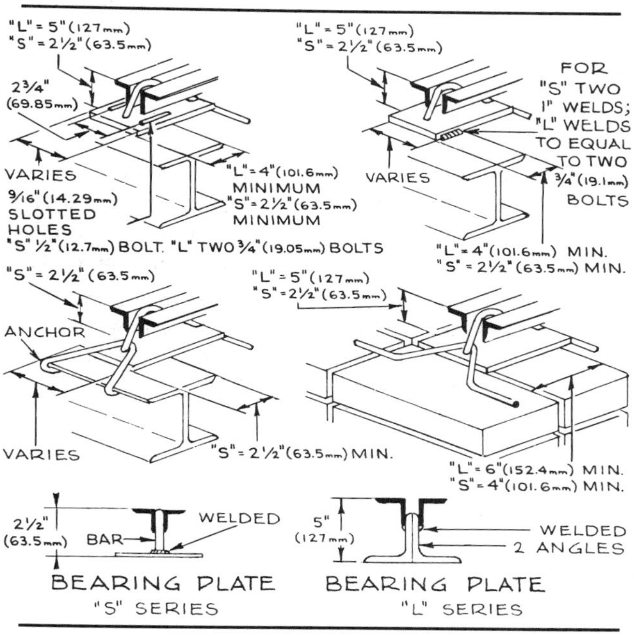

Figure S99 Typical types of bearing for open web steel joists.

4. Deflection: This should never exceed $\frac{1}{360}$ of the span.
5. Camber of joists: The top chord of the L series can be cambered to take deflection for dead flat roofs or to give the roof sufficient pitch for drainage. The top chord can also be pitched one way or two ways, and joists can be square end or underslung end. (*See* Figure *S100*.)
6. Methods of varying the fire resistance of open web steel joists, trusses, and space frames by type of flooring and ceiling materials and construction. State and local codes should be checked and also the fire rating codes of the Fire Underwriters, insurance companies, labor departments, and federal government (Army, Navy, etc.). (*See* Table *S68*).

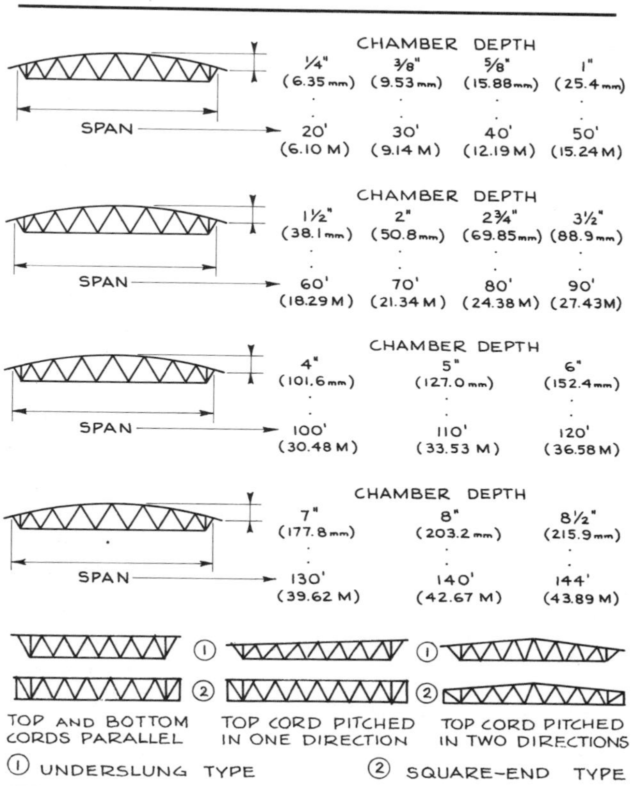

Figure S100 Standard camber (bowing up) for various lengths of open web steel joists and various standard treatments of top cord and ends of open web steel joists.

7. Joist extensions for overhangs, balconies, etc.: Before extending the top or bottom chord of open web steel joists to create a design with roof overhangs, balcony, etc., one should always consult local joist manufacturers. These extensions should never exceed the allowable total safe load in pounds per linear foot (lb/ft, or 1.4882 kg/m) of the joist used. Extensions from 2.5 to 5.5 ft (0.762 to 1.676 m)

Table S68 Fire-resistance ratings for various floor and roof systems

Fire rating	Floor	Ceiling[a]	Roof	Ceiling
1-hour	2-in. (50.8-mm) reinforced concrete slab	$\frac{1}{2}$-in. (12.7-mm) acoustical tile with concealed grid suspended from joists	2-in. (50.8-mm) structural wood fiber units	$\frac{3}{4}$-in. (19.05-mm) acoustical tile with concealed grid suspended from joists
	2-in. (50.8-mm) reinforced concrete slab	$\frac{1}{2}$-in. (12.7-mm) acoustical tile with exposed grid suspended from joists	Insulation on 26-gauge, 0.025-in. (0.635-mm), or heavy steel decking	$\frac{5}{8}$-in. (15.88-mm) acoustical tile with exposed grid suspended from joists
	2-in. (50.8-mm) reinforced concrete slab	$\frac{1}{2}$-in. (50.8-mm) gypsum board fastened to joists	2-in. (50.8-mm) vermiculite on centering on joists	$\frac{1}{2}$-in. (50.8-mm) acoustical tile with exposed grid suspended from joists
$1\frac{1}{2}$-hour	2-in. (50.8-mm) reinforced concrete slab	$\frac{5}{8}$-in. (15.88-mm) acoustical tile with concealed grid suspended from joists		
2-hour	$2\frac{1}{2}$-in. (63.5-mm) reinforced concrete slab	$\frac{5}{8}$-in. (15.88-mm) acoustical tile with concealed grid suspended from joists	2-in. (50.8-mm) gypsum planks	$\frac{5}{8}$-in. (15.88-mm) acoustical tile with exposed grid suspended from joists
	$2\frac{1}{2}$ in. (63.5 mm) reinforced concrete slab	$1\frac{1}{2}$-in. (50.8-mm) acoustical tile with exposed grid suspended from joists	Insulation on 22-gauge, 0.03125-in. (0.794-mm), or heavier steel decking	Suspended $\frac{7}{8}$-in. (22.23-mm) metal lath and plaster
	2-in. (50.8-mm) reinforced concrete slab	$\frac{5}{8}$-in. (15.88-mm) gypsum board fastened to joists		
	$2\frac{1}{2}$-in. (63.5-mm) reinforced concrete slab	$\frac{1}{2}$-in. (50.8-mm) gypsum board fastened to joists		
3-hour	$2\frac{1}{2}$-in. (63.5-mm) reinforced concrete slab	$\frac{3}{4}$ in. (19.05-mm) acoustical tile with concealed grid suspended from joists		
	$2\frac{1}{2}$-in. (63.5-mm) reinforced concrete slab	$\frac{5}{8}$-in. (15.88-mm) gypsum board fastened to joists		
4-hour	$2\frac{1}{2}$-in. (63.5-mm) reinforced concrete slab	$\frac{3}{4}$-in. (19.05-mm) metal lath and plaster		

[a] All acoustical tile, gypsum board, and grid should be checked against the building codes in force where the structure is to be erected.

are standard, and most joists manufacturers are equipped to supply such extensions.

8. Galvanic action: There must be no possibility of galvanic action between the joists and any materials with which they may come in direct contact.
9. The ability of the local joist manufacturer to supply all the design needs: In particular, the manufacturer's joists must meet standard specifications of the Steel Joist Institute.
10. One should always check the type, spacing, and

bearing of the trusses that are required so that they will meet the end-use conditions for the structure.

11. For large trusses one should always check whether the truss must be cambered to overcome the deflection due to the loads on the truss.
12. When a space frame is required, one should always check which is the best type of structural unit system to meet the design conditions of the structure.
13. The manufacturers of space frames and space frame components should always be consulted when the design requires a space frame.
14. One should always consult structural engineers when a space frame is required for a structure.

Conditions Favorable to the Use of Open Web Steel Joists, Trusses, and Space Frames

1. Where a lightweight, strong floor or roof construction system without the use of forms for pouring concrete is required.
2. Where space is needed between floor and ceiling to locate pipes, conduits, recessed light fixtures, ducts, and other accessories for ventilation and air-conditioning.
3. Where sound and thermal insulation between floors is required.
4. Where thermal insulation is required directly under the roof.
5. Where a large floor area or roof area with the minimum of supporting columns is required.
6. Where a large clear area is required on the lower floors of a multifloor structure.
7. Where, with the minimum of supports, single or multiple floors can be hung from trusses.

Conditions Unfavorable to the Use of Open Web Steel Joists, Trusses and Space Frames

1. Where a 4-hour fire rating is not sufficient to meet code requirements for the job.

STEEL, TUBULAR SHAPES

Physical and Chemical Properties

Tubular shapes may be made of several kinds of steel. The end use controls the exact chemical composition and mechanical properties of the steel (*see* Table *S69*).

Grades. There are three types of round structural tubing and pipe: standard weight, extra-strong, and double extra-strong. A number of grades, methods of designation, and types of steel are used for round, square, and rectangular pipe, based on the type of steel used in manufacture, as shown in Tables *S70*, *S71*, and *S72*.

Table S69 Chemical composition of steel used for structural piping and shaped tubing

Grade	C	Mn	P	S	V	Cb	Cu
			(percent of content)				
A, B	0.75–0.30	0.25–0.60	0.04	0.04			0.20[a]
I, II		1.35			0.01[b]		
III		1.35			0.01[b]	0.05[b]	

[a]When copper steel is required, 0.20% is the minimum.
[b]These are minimum values.

Table S70 Characteristics of structural steel tubing and pipe

Grade	Type	Method of manufacture	Tensile strength Minimum lbf/in.^{2}	Tensile strength Minimum MN/m^2	Yield strength Minimum lbf/in.^{2}	Yield strength Minimum MN/m^2
B	Pipe:[a] black or hot-dipped galvanized	Electric-resistance welded or seamless	60,000	413.70	35,000	241.33
		Continuous weld	45,000	310.28	25,000	172.38
A	Round tubing	Cold-formed welded and seamless round, square, rectangular, or special-shape structural tubing for welded, bolted, or riveted construction	45,000	310.28	33,000	227.54
B			58,000	400.11	42,000	789.59

Table S70 Characteristics of structural steel tubing and pipe (*continued*)

Grade	Type	Method of manufacture	Tensile strength Minimum lbf/in.²	Tensile strength Minimum MN/m²	Yield strength Minimum lbf/in.²	Yield strength Minimum MN/m²
A B	Shaped tubing	Cold-formed welded tubing in sizes with a maximum periphery of 48 in. (1.219 m) and a maximum wall thickness of 0.500 in. (12.7 mm); and seamless with a maximum periphery of 32 in. (0.813 m) and a maximum wall thickness of 0.500 in. (12.7 mm)	45,000 58,000	310.28 400.11	39,000 46,000	269.92 317.17
None	Round or shaped tubing	Hot-formed welded and seamless round, square, rectangular, and special-shape structural tubing for welded, bolted, or riveted structural purposes; square and rectangular tubing furnished in sizes 1–10 in. (25.4–254 mm) with wall thickness of 0.095–1.000 in. (2.413–15.4 mm); round tubing in nominal wall thickness of 0.109–1.000 in. (2.769–25.4 mm), depending on size	58,000	400.11	36,000	248.22
I, II[a] III[b]	Round or shaped tubing	Hot-formed welded or seamless high-strength, low-allow round, square, or special-shape structural tubing for welded, bolted, or riveted structural purposes	70,000 65,000	482.65 448.18	50,000 50,000	344.75 344.75

[a] Also available in a manganese-vanadium high-strength, low-alloy weathering steel.
[b] Also available in a columbium-vanadium high-strength, low-alloy weathering steel.

Table S71 Designation of structural tubing and pipe

Designation in inches and decimals of an inch (millimeters)	Type and shape	Description in inches and decimals of an inch (millimeters)
TS 5 × 5 × 0.500 (TS 127 × 127 × 12.7)	Structural tubing: square	5-in. (127-mm) outside dimensions with 0.500-in. (12.7-mm) wall thickness
TS 10 × 8 × 0.375 (TS 254 × 203.2 × 9.53)	Structural tubing: rectangular	10 by 8 in. (254 by 203.2 mm) outside dimensions with 0.375-in. (9.53-mm) wall thickness
TS 4 O.D. × 0.250 (TS 101.6 O.D. × 6.35)	Structural tubing: round pipe[a]	4-in. (101.6-mm) outside diameter with 0.250-in. (6.35-mm) wall thickness
Pipe 4 STD (Pipe 101.6 STD)	Pipe[a]	4-in. (101.6-mm) nominal pipe size, 4.5-in. (114.3-mm) outside diameter with 0.237-in. (6.02-mm) wall thickness

[a] Pipe is usually designated by its nominal diameter. In sizes $\frac{1}{2}$ through 12 in. (12.7 through 304.8 mm) the nominal diameter is smaller than the actual outside diameter (O.D.). In sizes over 12 in. (304.8 mm) the nominal and actual outside diameters are the same.

Table S72 Dimensions and weights of standard weight, extra-strong, and double extra-strong structural tubing and pipe

Standard weight, extra-strong, and double extra-strong		Standard weight			Extra-strong			Double extra-strong		
Nominal diameter in. (mm)	Outside diameter in. (mm)	Inside diameter in. (mm)	Wall thickness in. (mm)	Weight lb/ft (kg/m)	Inside diameter in. (mm)	Wall thickness in. (mm)	Weight lb/ft (kg/m)	Inside diameter in. (mm)	Wall thickness in. (mm)	Weight lb/ft (kg/m)
$\frac{1}{2}$	0.840	0.622	0.109	0.85	0.546	0.147	1.09			
(12.7)	(21.34)	(15.80)	(2.769)	(1.26)	(13.87)	(3.734)	(1.62)			
$\frac{3}{4}$	1.050	0.824	0.113	1.13	0.742	0.154	1.47	0.434	0.308	2.44
(19.05)	(26.67)	(20.93)	(2.870)	(1.68)	(18.85)	(3.912)	(2.19)	(11.02)	(7.813)	(3.63)
1	1.315	1.049	0.133	1.68	0.957	0.179	2.17	0.599	0.358	3.66
(25.40)	(33.40)	(26.64)	(3.378)	(2.50)	(24.31)	(4.547)	(3.23)	(15.21)	(9.193)	(5.45)
$1\frac{1}{4}$	1.660	1.380	0.140	2.27	1.278	0.191	3.00	0.896	0.382	5.21
(31.75)	(42.16)	(35.05)	(3.556)	(3.38)	(32.46)	(4.849)	(4.46)	(22.76)	(9.703)	(7.75)
$1\frac{1}{2}$	1.900	1.610	0.145	2.72	1.500	0.200	3.63	1.190	0.400	6.41
(38.10)	(48.26)	(40.89)	(3.683)	(4.05)	(38.10)	(5.080)	(5.40)	(30.23)	(10.16)	(9.54)
2	2.375	2.067	0.154	3.65	1.939	0.218	5.02	1.503	0.436	9.03
(50.80)	(60.33)	(52.50)	(3.921)	(5.43)	(49.25)	(5.537)	(7.47)	(38.18)	(11.074)	(13.44)
$2\frac{1}{2}$	2.875	2.469	0.203	5.79	2.323	0.276	7.66	1.771	0.552	13.69
(63.50)	(73.13)	(62.71)	(5.156)	(8.67)	(59.00)	(7.010)	(11.40)	(44.98)	(14.021)	(20.37)
3	3.500	3.068	0.216	7.58	2.900	0.300	10.25	2.300	0.600	18.58
(76.20)	(88.90)	(77.93)	(5.486)	(11.28)	(73.66)	(7.620)	(15.25)	(58.42)	(15.240)	(27.65)
$3\frac{1}{2}$	4.000	3.548	0.226	9.11	3.364	0.318	12.51			
(89.90)	(101.60)	(90.12)	(5.704)	(13.56)	(85.45)	(8.077)	(18.62)			
4	4.500	4.076	0.237	10.79	3.826	0.337	14.99	3.152	0.674	27.54
(101.60)	(114.30)	(102.36)	(6.020)	(16.06)	(97.18)	(8.560)	(22.29)	(80.06)	(17.12)	(40.99)
5	5.563	5.047	0.258	14.62	4.813	0.375	20.78	4.053	0.750	38.55
(127.00)	(143.30)	(128.19)	(6.553)	(21.76)	(122.70)	(9.525)	(30.92)	(102.95)	(19.05)	(57.37)
6	6.625	6.065	0.280	18.97	5.761	0.432	28.57	4.897	0.864	53.16
(152.40)	(168.28)	(154.05)	(7.112)	(28.23)	(146.33)	(10.973)	(42.52)	(124.38)	(21.946)	(79.11)
8	8.625	7.981	0.322	28.55	7.625	0.500	43.39	6.875	0.875	72.42
(203.20)	(219.08)	(202.72)	(8.179)	(32.49)	(193.68)	(12.700)	(64.57)	(174.63)	(22.225)	(107.78)
10	10.750	10.020	0.365	40.48	9.750	0.500	54.74			
(254.00)	(273.05)	(254.51)	(9.271)	(60.24)	(247.65)	(12.700)	(81.46)			
12	12.750	12.000	0.375	49.56	11.750	0.500	65.42			
(304.80)	(323.85)	(304.80)	(9.526)	(73.75)	(298.45)	(12.700)	(97.36)			
14[a]	14.000	13.250	0.375	54.57	13.000	0.500	72.09			
(355.60)	(355.60)	(333.55)	(9.526)	(81.20)	(330.20)	(12.700)	(107.27)			
16[a]	16.000	15.250	0.375	62.58	14.000	0.500	82.77			
(406.40)	(406.40)	(387.35)	(9.526)	(93.12)	(355.60)	(12.700)	(123.16)			
18[a]	18.000	17.250	0.375	70.59	17.000	0.500	93.45			
(254.2)	(254.2)	(438.15)	(9.526)	(105.04)	(431.80)	(12.700)	(139.05)			
20[a]	20.000	19.250	0.375	78.60	19.000	0.500	104.13			
(508.0)	(508.0)	(488.95)	(9.526)	(116.96)	(482.60)	(12.700)	(154.91)			
24[a]	24.000	23.250	0.375	94.02	23.000	0.500	125.49			
(609.6)	(609.6)	(590.55)	(9.526)	(140.79)	(584.20)	(12.700)	(186.73)			

[a] These sizes are considered round structural tubing.

Sizes. Sizes for round structural tubing range from $\frac{1}{2}$ to 40 in. (12.7 to 1016.0 mm) in diameter, in wall thicknesses of 0.109 to 0.875 in. (2.77 to 22.23 mm); and for square and rectangular tubing range from 1 to 40 in. (25.4 to 1016.0 mm) across flat sides, with wall thicknesses of 0.095 to 1.0 in. (2.413 to 25.4 mm). These tubular shapes are manufactured in random mill lengths of 16 to 22 ft (4.877 to 6.706 m), also 32 to 44 ft (9.754 to 13.411 m) in multiple and cut lengths. All are available in black, galvanized, and weathering steel.

Table S73 Dimensions, weights, and allowable concentrated safe loads for tubular columns (pipe)[a,b]

	Outside diameter							
	$\frac{1}{2}$ in. (12.7 mm)		$\frac{3}{4}$ in. (19.05 mm)			1 in. (25.4 mm)		
	Thickness of wall[c]							
	0.109 in. (2.769 mm)	0.147 in. (3.734 mm)	0.113 in. (2.870 mm)	0.154 in. (3.912 mm)	0.308 in. (7.823 mm)	0.133 in. (3.378 mm)	0.179 in. (4.547 mm)	0.358 in. (9.193 mm)
Unbraced length	Concentrated load in 1000 lb (453.6 kg) F_y = 25,000 lbf/in.^{2} (172.38 MN/m^2)							
ft. (m)	lb (kg)	lb (kg)	lb (kg)	lb (kg)	lb (kg)	lb (kg)	lb (kg)	lb (kg)
1	3.4	4.3	4.6	6.0	10.0	7.0	9.0	15.0
(0.3048)	(1.54)	(1.95)	(2.09)	(2.72)	(4.56)	(3.19)	(4.08)	(6.80)
2	2.7	3.4	4.0	5.0	8.0	6.0	8.0	13.0
(0.610)	(1.22)	(1.54)	(1.81)	(2.27)	(3.63)	(2.72)	(3.63)	(5.90)
3	1.9	2.3	3.3	4.1	6.0	6.0	7.0	11.00
(0.914)	(0.86)	(1.04)	(1.50)	(1.86)	(2.72)	(2.72)	(3.19)	(5.00)
4	1.1	1.3	2.4	2.9	3.8	5.0	6.0	9.0
(1.219)	(0.50)	(0.59)	(1.09)	(1.32)	(1.72)	(2.27)	(2.72)	(4.08)
5			1.5	1.9		3.6	4.4	3.6
(1.524)			(0.68)	(0.86)		(1.63)	(2.00)	(1.63)
6						2.5	3.0	4.1
(1.829)						(1.13)	(1.36)	(1.86)
7						1.9		
(2.134)						(0.86)		
8								
(2.438)								
9								
(2.743)								
10								
(3.048)								
Weight	0.085 lb/ft (1.26 kg/m)	1.09 lb/ft (1.62 kg/m)	1.13 lb/ft (1.68 kg/m)	1.47 lb/ft (2.19 kg/m)	2.44 lb/ft (3.63 kg/m)	1.68 lb/ft (2.50 kg/m)	2.17 lb/ft (3.23 kg/m)	3.66 lb/ft (5.45 kg/m)

[a]Heavy line indicates Kl/r = 120.
[b]Round tubular columns are available with F_y = 36,000 lbf/in.^{2} (248.22 MN/m^2). Check with manufacturers.
[c]First column is standard weight, second column is extra-strong, and third column is double extra-strong.

Types and Uses

Tubular shapes are used for structural columns, beams, and scaffolding and also as piles, foundations, deep water wells, etc. The smaller shapes are utilized for furniture, railings, bucks, etc. The great majority of tubing shapes are used in the construction field for columns, beams, scaffolding, and piles and pile casings, as well as for piping and electrical conduits. (*See* Tables *S73* to *S76*.)

Application

Although tubular shapes are available with fireproof exterior covering, connections at floors and ceilings can become very difficult and complicated.

State and local codes should be checked and also the fire rating codes of the Fire Underwriters, insurance companies, labor departments, and federal government (Army, Navy, etc.).

Condensed Checklist

1. Diameter specifications: Tubular sections should be specified by outside as well as inside diameters.
2. Unsupported height and load: The size may be decreased by adding reinforcement to the concrete core.
3. Wind bracing and lateral bracing: This should be done in a way that will not affect design.
4. Connections between tubular shapes and other structural shapes: Brackets, plates (projecting through the tubular shape), caps and bases, etc., can

Table S73 Dimensions, weights, and allowable concentrated safe loads for tubular columns (pipe)[a, b] *(continued)*

Unbraced length	Outside diameter: 1¼ in. (31.75 mm)			Outside diameter: 1½ in. (38.10 mm)		
Thickness of wall[c]	0.140 in. (3.556 mm)	0.191 in. (4.849 mm)	0.382 in. (9.703 mm)	0.145 in. (3.683 mm)	0.200 in. (5.080 mm)	0.400 in. (10.160 mm)
	Concentrated load in 1000 lb (453.6 kg) F_y = 25,000 lbf/in.^{2} (172.38 MN/m^2)					
ft. (m)	lb (kg)	lb (kg)	lb (kg)	lb (kg)	lb (kg)	lb (kg)
1 (0.3048)	10.0 (4.56)	13.0 (5.90)	22.0 (9.98)	12.0 (5.44)	15.0 (6.80)	27.0 (12.25)
2 (0.610)	9.0 (4.08)	12.0 (5.44)	20.0 (9.07)	11.0 (5.00)	15.0 (6.80)	25.0 (11.34)
3 (0.914)	8.0 (3.63)	11.0 (5.00)	18.0 (8.16)	10.0 (4.56)	14.0 (6.35)	23.0 (10.43)
4 (1.219)	7.0 (3.19)	10.0 (4.56)	16.0 (7.26)	9.0 (4.08)	13.0 (5.90)	21.0 (9.53)
5 (1.524)	6.0 (2.72)	8.0 (3.62)	13.0 (5.90)	9.0 (4.08)	11.0 (5.00)	18.0 (8.16)
6 (1.829)	5.0 (2.27)	7.0 (3.19)	10.0 (4.56)	7.0 (3.19)	10.0 (4.56)	15.0 (6.80)
7 (2.134)	4.1 (1.86)	5.0 (2.27)	7.0 (3.19)	6.0 (2.72)	8.0 (3.63)	12.0 (5.44)
8 (2.438)	3.2 (1.45)	3.9 (1.77)		5.0 (2.27)	6.0 (2.72)	9.0 (4.08)
9 (2.743)				4.0 (1.81)	5.0 (2.27)	7.0 (3.19)
10 (3.048)				3.2 (1.45)	4.1 (1.86)	
Weight	2.27 lb/ft (3.38 kg/m)	3.00 lb/ft (4.46 kg/m)	5.21 lb/ft (7.75 kg/m)	2.72 lb/ft (4.05 kg/m)	3.63 lb/ft (5.40 kg/m)	6.41 lb/ft (9.54 kg/m)

affect design as a result of projections, braces, and differences in levels.

5. Method of attachment at floor or foundation: Anchor bolts must be set accurately in concrete.
6. Electrical conduits, piping and duct work in relation to designing with tubular shapes: Tubular shapes do not allow for any vertical method of concealing these mechanical utilities.
7. Galvanic action: Materials, especially if other metals are in direct contact with the tubular shapes, should be checked for the possibility of galvanic action.
8. When using structural steel tubing for columns, beams, piles, etc., one should always consult a structural engineer.
9. When using structural steel tubing as a structural system, one should always check the codes and always consult with a structural engineer in regard to structural design and connections.
10. When designing a structure using structural steel tubing, manufacturers should always be consulted regarding the sizes and shapes as well as the grades, types, and lengths that are available.

Conditions Favorable to the Use of Tubular Shapes

1. Where the design calls for small, strong vertical supports and beams.
2. Where a continuous glass wall or window wall is desired with the minimum thickness of visible supports.
3. Where the design requires a simple exposed structural system.

Conditions Unfavorable to the Use of Tubular Shapes

1. Where passage of utilities such as piping, ducts, and conduits must be within the vertical support.
2. When the building must meet strict fire ratings.

Table S74 Dimensions, weights, and allowable concentrated safe loads for round tubular columns[a]

	Outside diameter				
	3 in. (76.2 mm)			$3\frac{1}{2}$ in. (89.9 mm)	
	Thickness of wall				
	0.216 in. 5.486 mm	0.300 in. 7.620 mm	0.600 in. 15.240 mm	0.226 in. 5.740 mm	0.318 in. 8.077 mm
Unbraced length	Concentrated load in 1000 lb (453.6 kg) F_y = 36,000 lbf/in.^{2} (248.22 MN/m^2)				
ft (m)	lb (kg)	lb (kg)	lb (kg)	lb (kg)	lb (kg)
6 (1.829)	38 (17.24)	52 (23.59)	91 (41.28)	48 (21.71)	66 (29.93)
7 (2.134)	36 (16.33)	48 (21.77)	84 (38.10)	46 (20.87)	63 (28.68)
8 (2.438)	34 (15.42)	45 (20.41)	77 (34.93)	44 (19.96)	59 (26.16)
9 (2.743)	31 (14.06)	41 (18.60)	69 (31.30)	41 (18.60)	55 (24.95)
10 (3.048)	28 (12.70)	37 (16.78)	60 (27.22)	38 (17.24)	51 (23.13)
11 (3.353)	25 (11.34)	33 (14.97)	51 (23.13)	35 (15.88)	47 (21.32)
12 (3.658)	22 (9.98)	28 (12.70)	43 (19.50)	32 (14.52)	43 (19.50)
13 (3.962)	19 (8.62)	24 (10.99)	37 (16.78)	29 (13.15)	38 (17.24)
14 (4.267)	16 (7.26)	21 (9.53)	32 (14.52)	25 (11.34)	33 (14.97)
15 (4.572)	14 (6.35)	18 (8.16)	28 (12.70)	22 (9.98)	29 (13.15)
16 (4.877)	12 (5.44)		24 (10.99)	19 (8.62)	25 (11.34)
17 (5.182)	11 (5.00)		22 (9.98)	17 (7.71)	23 (10.43)
18 (5.486)	10 (4.56)			15 (6.80)	20 (9.07)
19 (5.791)	9 (4.08)			14 (6.35)	18 (8.16)
20 (6.096)				12 (5.44)	16 (7.26)
22 (6.706)				10 (4.56)	
24 (7.315)					
26 (7.926)					
28 (8.534)					
30 (9.144)					
Weight	7.58 lb/ft (11.28 kg/m)	10.25 lb/ft (15.25 kg/m)	18.58 lb/ft (27.65 kg/m)	9.11 lb/ft (13.56 kg/m)	12.51 lb/ft (18.62 kg/m)

[a]Heavy line indicates Kl/r = 120.

Table S74 Dimensions, weights, and allowable concentrated safe loads for round tubular columns[a] (*continued*)

Unbraced length	Outside diameter 4 in. (101.6 mm)			Outside diameter 5 in. (127.0 mm)		
	Thickness of wall					
	0.237 in. 6.020 mm	0.337 in. 8.560 mm	0.674 in. 17.520 mm	0.258 in. 6.553 mm	0.375 in. 9.525 mm	0.750 in. 19.050 mm
	Concentrated load in 1000 lb (453.6 kg) F_y = 36,000 lbf/in.^{2} (248.22 MN/m^2)					
ft (m)	lb (kg)	lb (kg)	lb (kg)	lb (kg)	lb (kg)	lb (kg)
6 (1.829)	59 (26.16)	81 (36.74)	147 (66.68)	83 (37.65)	118 (53.52)	216 (97.98)
7 (2.134)	57 (25.85)	78 (35.38)	140 (63.50)	81 (36.74)	114 (51.71)	209 (94.80)
8 (2.438)	54 (24.49)	75 (34.02)	133 (60.33)	78 (35.38)	111 (50.35)	202 (91.63)
9 (2.743)	52 (23.58)	71 (32.21)	126 (57.42)	76 (34.47)	107 (48.53)	195 (88.45)
10 (3.048)	49 (22.23)	67 (30.39)	118 (53.52)	73 (31.11)	103 (46.72)	187 (84.82)
11 (3.353)	46 (20.87)	63 (28.68)	109 (49.44)	71 (32.21)	99 (44.91)	178 (80.74)
12 (3.658)	43 (19.50)	59 (26.16)	100 (45.36)	68 (30.84)	95 (43.09)	170 (77.11)
13 (3.962)	40 (18.14)	54 (24.49)	91 (41.28)	65 (29.48)	91 (41.28)	160 (72.58)
14 (4.267)	36 (16.33)	49 (22.23)	81 (36.74)	61 (27.67)	86 (39.01)	151 (68.49)
15 (4.572)	33 (14.97)	44 (19.96)	70 (31.75)	58 (26.31)	81 (36.74)	141 (63.96)
16 (4.877)	29 (13.15)	39 (17.69)	62 (28.12)	55 (24.95)	76 (34.47)	130 (58.96)
17 (5.182)	26 (11.34)	34 (15.42)	55 (24.95)	51 (23.13)	71 (32.21)	119 (54.98)
18 (5.486)	23 (10.43)	31 (14.06)	45 (20.41)	47 (21.32)	65 (29.48)	108 (48.99)
19 (5.791)	21 (9.53)	28 (12.70)	44 (19.96)	43 (19.50)	59 (26.16)	97 (44.00)
20 (6.096)	19 (8.62)	25 (11.34)	40 (18.14)	39 (17.69)	54 (24.49)	87 (39.46)
22 (6.706)	15 (6.80)		33 (14.97)	32 (14.52)	44 (19.95)	72 (32.66)
24 (7.315)	13 (5.90)			27 (12.25)	37 (16.78)	61 (27.67)
26 (7.926)				23 (10.43)	32 (14.52)	52 (23.59)
28 (8.534)				20 (9.07)	27 (12.25)	44 (19.96)
30 (9.144)				17 (7.71)	24 (10.99)	
Weight	10.79 lb/ft (16.06 kg/m)	14.98 lb/ft (22.29 kg/m)	27.54 lb/ft (40.98 kg/m)	14.62 lb/ft (21.76 kg/m)	20.78 lb/ft (30.92 kg/m)	38.55 lb/ft (57.38 kg/m)

Table S75 Dimensions, weights, and allowable concentrated safe loads for rectangular tubular columns[a, b]

Unbraced length	Outside dimensions: 12 × 6 in. (304.8 × 152.4 mm)			10 × 8 in. (254.0 × 203.8 mm)				10 × 6 in. (254.0 × 152.4 mm)	
Thickness of wall	0.3125 in. (7.938 mm)	0.375 in. (6.350 mm)	0.500 in. (12.700 mm)	0.250 in. (6.350 mm)	0.3125 in. (7.938 mm)	0.375 in. (9.525 mm)	0.500 in. (12.700 mm)	0.250 in. (6.350 mm)	0.3125 in. (7.938 mm)
	Concentrated load in 1000 lb (453.6 kg), F_y = 36,000 lbf/in.^{2} (248.22 MN/m^2)								
ft (m)	lb (kg)	lb (kg)	lb (kg)	lb (kg)	lb (kg)	lb (kg)	lb (kg)	lb (kg)	lb (kg)
6 (1.829)	208 (94.35)	247 (112.04)	318 (144.24)	173 (78.47)	213 (96.62)	251 (113.85)	324 (146.97)	150 (68.04)	184 (83.46)
7 (2.134)	205 (92.99)	242 (109.71)	312 (141.52)	171 (77.57)	210 (95.26)	248 (112.49)	320 (145.15)	147 (66.68)	180 (81.64)
8 (2.438)	201 (91.17)	238 (107.96)	306 (138.80)	169 (76.66)	207 (93.89)	245 (111.14)	316 (143.34)	144 (65.31)	177 (80.28)
9 (2.743)	197 (89.36)	233 (105.69)	299 (135.63)	167 (75.75)	205 (92.99)	242 (109.77)	311 (141.08)	141 (63.96)	173 (78.47)
10 (3.048)	193 (87.54)	228 (103.42)	292 (132.45)	165 (74.84)	202 (91.63)	238 (107.96)	307 (139.25)	138 (62.60)	169 (76.66)
11 (3.353)	188 (85.28)	222 (100.70)	285 (129.28)	162 (73.48)	199 (90.27)	234 (106.14)	302 (136.99)	135 (61.23)	165 (74.84)
12 (3.658)	184 (83.46)	217 (98.43)	278 (126.10)	160 (72.58)	195 (88.46)	230 (104.33)	297 (134.72)	132 (59.88)	161 (73.03)
13 (3.962)	179 (81.19)	211 (95.71)	270 (122.47)	157 (71.22)	192 (87.09)	226 (102.51)	291 (132.00)	128 (58.06)	157 (71.22)
14 (4.267)	174 (78.92)	205 (92.99)	262 (118.84)	154 (69.85)	189 (85.73)	222 (100.70)	286 (129.73)	125 (56.70)	152 (68.94)
15 (4.572)	169 (76.66)	199 (90.27)	253 (114.76)	151 (68.49)	185 (83.92)	218 (98.88)	280 (127.00)	121 (54.89)	148 (67.13)
16 (4.877)	163 (73.93)	192 (87.09)	245 (111.14)	148 (67.13)	181 (82.10)	213 (96.62)	274 (124.28)	117 (53.07)	143 (64.86)
17 (5.182)	159 (73.12)	186 (84.31)	236 (107.05)	145 (65.77)	177 (80.28)	209 (94.80)	268 (121.56)	113 (51.26)	138 (62.60)
18 (5.486)	152 (68.94)	179 (81.19)	226 (102.51)	142 (64.42)	174 (78.92)	204 (92.53)	262 (118.84)	109 (49.44)	133 (60.33)
19 (5.791)	146 (66.23)	172 (78.02)	217 (98.43)	139 (63.05)	170 (77.11)	199 (90.27)	255 (115.62)	105 (47.63)	127 (57.51)
20 (6.096)	140 (63.50)	164 (74.39)	207 (93.89)	135 (61.23)	165 (74.84)	194 (88.00)	249 (112.95)	100 (45.36)	122 (55.34)
22 (6.706)	127 (57.51)	149 (67.59)	186 (84.31)	129 (58.51)	157 (71.22)	184 (83.46)	235 (106.60)	91 (41.28)	110 (49.90)
24 (7.315)	114 (51.71)	133 (60.33)	164 (74.39)	121 (54.89)	148 (67.13)	173 (78.47)	220 (99.79)	81 (36.74)	98 (44.45)
26 (7.926)	100 (45.36)	115 (62.16)	141 (63.96)	114 (51.71)	138 (62.60)	162 (73.48)	205 (92.99)	71 (32.21)	85 (38.56)
28 (8.534)	86 (39.01)	100 (45.36)	121 (54.89)	106 (48.08)	128 (58.06)	150 (68.04)	189 (85.73)	61 (27.67)	73 (33.11)
30 (9.144)	75 (34.02)	87 (39.46)	106 (48.08)	97 (44.00)	118 (53.52)	137 (62.14)	172 (78.02)	53 (24.04)	64 (29.03)
32 (9.754)	66 (29.93)	76 (34.47)	93 (42.18)	89 (40.37)	107 (48.53)	124 (60.25)	155 (70.31)	47 (21.32)	56 (25.40)
34 (10.363)	58 (26.31)	67 (30.39)	87 (39.46)	79 (35.83)	96 (43.55)	110 (49.90)	137 (62.14)	41 (18.60)	49 (22.23)
36 (10.973)	52 (23.59)	60 (27.22)	73 (33.11)	71 (32.21)	85 (38.56)	98 (44.45)	122 (55.34)	37 (16.78)	44 (19.95)
38 (11.582)	47 (21.32)	54 (24.49)	66 (29.93)	64 (29.03)	76 (34.47)	88 (40.42)	110 (49.90)	33 (14.97)	40 (18.14)
40 (12.192)	42 (19.05)	49 (22.23)	59 (26.16)	57 (25.85)	69 (31.30)	80 (39.01)	99 (44.91	30 (13.61)	36 (16.33)
Weight	35.49 lb/ft (52.82 kg/m)	41.93 lb/ft (62.40 kg/m)	54.15 lb/ft (80.59 kg/m)	28.83 lb/ft (42.90 kg/m)	35.49 lb/ft (52.82 kg/m)	41.93 lb/ft (62.40 kg/m)	54.15 lb/ft (80.59 kg/m)	25.44 lb/ft (36.86 kg/m)	31.24 lb/ft (46.49 kg/m)

[a] Heavy line indicates Kl/r = 200.

[b] Rectangular tubular columns are available with F_y = 50,000 lbf/in.^{2} (344.75 MN/m^2).

Outside dimensions									
10 × 6 in. (254.0 × 152.4 mm)		8 × 6 in. (203.2 × 152.4 mm)				8 × 4 in. (101.6 mm)			
Thickness of wall									
0.375 in. (9.525 mm	0.500 in. (12.500 mm)	0.250 in. (6.350 mm)	0.3125 in. (7.938 mm)	0.375 in. (9.529 mm)	0.500 in. (12.700 mm)	0.250 in. (6.350 mm)	0.3125 in. (7.938 mm)	0.375 in. (9.529 mm)	0.500 in. (12.700 mm)
Concentrated load in 1000 lb (453.6 kg) F_y = 36,000 lbf/in.^{2} (248.22 MN/m^2)									
lb (kg)	lb (kg)	lb (kg)	lb (kg)	lb (kg)	lb (kg)	lb (kg)	lb (kg)	lb (kg)	lb (kg)
216 (97.98)	277 (125.65)	129 (58.51)	158 (71.67)	186 (84.31)	236 (107.05)	105 (47.63)	127 (57.51)	149 (67.59)	188 (85.28)
212 (96.17)	272 (123.38)	127 (57.51)	155 (70.31)	182 (82.55)	231 (104.78)	101 (45.81)	123 (55.79)	144 (65.31)	181 (82.10)
208 (94.35)	266 (120.66)	124 (56.25)	152 (68.94)	178 (80.74)	226 (102.51)	98 (44.45)	118 (53.52)	138 (62.60)	173 (78.47)
204 (92.53)	260 (117.94)	122 (55.34)	149 (67.59)	174 (78.92)	221 (100.25)	94 (42.64)	114 (51.71)	132 (59.88)	165 (74.84)
199 (90.27)	254 (115.21)	119 (54.98)	145 (65.77)	170 (77.11)	215 (97.54)	89 (40.37)	108 (48.99)	126 (57.42)	157 (71.21)
194 (88.00)	248 (112.94)	116 (52.62)	142 (64.41)	166 (75.30)	210 (95.26)	85 (38.56)	103 (46.72)	119 (54.98)	148 (67.13)
189 (85.73)	241 (109.32)	113 (51.66)	138 (62.60)	161 (73.03)	203 (92.08)	80 (36.30)	97 (44.00)	112 (51.10)	138 (62.60)
184 (83.46)	234 (106.14)	110 (49.90)	134 (60.78)	156 (70.76)	197 (89.36)	76 (34.47)	91 (41.28)	105 (47.63)	129 (58.51)
178 (80.74)	227 (102.97)	107 (48.53)	130 (58.96)	152 (68.94)	191 (86.73)	71 (32.21)	85 (38.56)	97 (44.00)	118 (53.52)
173 (78.47)	219 (99.34)	103 (46.72)	126 (57.42)	146 (66.23)	184 (83.46)	65 (29.48)	78 (35.38)	89 (40.73)	107 (48.53)
167 (75.75)	211 (95.71)	100 (45.36)	121 (54.89)	141 (63.96)	177 (80.28)	60 (27.22)	71 (32.21)	81 (36.74)	96 (43.55)
161 (73.03)	203 (92.08)	96 (43.55)	117 (53.07)	136 (59.24)	169 (76.66)	54 (24.49)	64 (29.03)	72 (32.66)	85 (38.56)
154 (69.85)	195 (88.46)	92 (41.73)	112 (51.10)	130 (58.96)	162 (73.48)	48 (21.77)	57 (25.85)	64 (29.03)	76 (34.47)
148 (67.13)	186 (84.31)	88 (39.92)	107 (48.53)	124 (56.25)	154 (69.85)	43 (19.50)	51 (25.85)	58 (26.31)	68 (30.84)
141 (63.96)	177 (80.28)	84 (38.10)	102 (46.27)	118 (53.52)	146 (66.23)	39 (17.69)	46 (20.87)	52 (23.59)	61 (27.67)
127 (57.51)	158 (71.67)	76 (34.47)	92 (41.73)	105 (47.63)	129 (58.51)	32 (14.52)	38 (17.24)	43 (19.50)	51 (23.13)
112 (51.10)	138 (62.60)	67 (30.39)	80 (36.30)	92 (41.73)	110 (49.90)	27 (12.25)	32 (14.52)	36 (16.33)	43 (19.50)
96 (43.55)	118 (53.52)	58 (26.31)	69 (31.30)	78 (35.38)	94 (42.64)	23 (10.43)	27 (12.25)	31 (14.06)	
83 (37.65)	102 (46.27)	50 (22.68)	59 (26.16)	68 (30.84)	81 (36.74)				
72 (32.66)	89 (40.37)	43 (19.50)	52 (23.59)	59 (26.16)	71 (32.21)				
64 (29.03)	78 (35.38)	38 (17.24)	46 (20.87)	52 (23.59)	62 (28.12)				
56 (25.40)	69 (31.30)	34 (15.42)	40 (18.14)	46 (20.87)	55 (24.25)				
50 (22.68)	62 (28.12)	30 (13.61)	36 (16.33)	41 (18.60)	49 (22.23)				
45 (20.41)	55 (24.95)	27 (12.25)	32 (14.52)	37 (16.78)					
41 (18.60)		24 (10.99)							
36.83 lb/ft (54.81 kg/m)	47.35 lb/ft (70.47 kg/m)	22.04 lb/ft (32.80 kg/m)	26.99 lb/ft (40.16 kg/m)	31.73 lb/ft (47.22 kg/m)	40.55 lb/ft (60.35 kg/m)	18.87 lb/ft (28.01 kg/m)	23.02 lb/ft (34.17 kg/m)	27.04 lb/ft (40.24 kg/m)	34.48 lb/ft (51.31 kg/m)

Table S76 Dimensions, weights, and allowable concentrated safe loads for structural square tubular columns [a, b]

Unbraced length	Outside dimensions: 12 × 12 in. (304.8 × 304.8 mm)		10 × 10 in. (254.0 × 254.0 mm)				
	Thickness of wall						
	0.375 in. (9.525 mm)	0.500 in. (12.700 mm)	0.250 in. (6.350 mm)	0.3125 in. (7.938 mm)	0.375 in. (9.525 mm)	0.500 in. (12.700 mm)	0.625 in. (15.875 mm)
	Concentrated load in 1000 lb (453.6 kg) F_y = 36,000 lbf/in.^{2} (248.22 MN/m^2)						
ft (m)	lb (kg)	lb (kg)	lb (kg)	lb (kg)	lb (kg)	lb (kg)	lb (kg)
6 (1.829)	357 (161.93)	467 (211.83)	196 (88.81)	242 (109.77)	285 (129.28)	370 (167.83)	471 (213.65)
7 (2.134)	354 (160.57)	464 (210.47)	195 (88.46)	240 (108.86)	283 (128.87)	366 (166.02)	467 (211.83)
8 (2.438)	352 (159.67)	461 (209.11)	193 (87.54)	238 (107.96)	280 (127.00)	363 (164.66)	462 (209.56)
9 (2.743)	349 (158.31)	457 (207.29)	191 (86.73)	235 (106.60)	277 (125.65)	359 (162.84)	457 (207.29)
10 (3.048)	346 (157.55)	453 (205.48)	189 (85.73)	233 (105.69)	274 (124.28)	355 (161.03)	452 (205.76)
11 (3.353)	343 (155.58)	449 (203.67)	187 (84.82)	230 (104.33)	271 (122.93)	351 (159.21)	447 (202.76)
12 (3.658)	340 (154.22)	445 (201.85)	185 (83.92)	228 (103.42)	268 (121.56)	347 (157.40)	441 (200.04)
13 (3.962)	337 (152.86)	441 (200.04)	182 (82.55)	225 (102.07)	265 (120.20)	342 (155.13)	436 (197.77)
14 (4.267)	334 (151.50)	437 (198.22)	180 (81.64)	222 (100.70)	261 (118.39)	338 (153.32)	430 (195.05)
15 (4.572)	331 (150.14)	432 (195.95)	178 (80.74)	219 (99.34)	258 (117.03)	333 (151.65)	424 (192.33)
16 (4.877)	327 (148.33)	428 (194.14)	175 (79.38)	216 (97.98)	254 (115.21)	328 (148.78)	417 (189.17)
17 (5.182)	324 (146.97)	423 (191.87)	173 (78.47)	213 (96.62)	250 (113.40)	323 (146.51)	411 (186.43)
18 (5.486)	320 (145.15)	418 (189.60)	170 (77.11)	209 (94.80)	246 (111.59)	318 (144.24)	404 (183.25)
19 (5.791)	316 (143.34)	413 (187.34)	167 (75.75)	206 (93.44)	242 (109.77)	312 (141.52)	397 (180.07)
20 (6.096)	313 (141.98)	408 (185.07)	165 (74.84)	203 (92.08)	238 (107.96)	307 (139.25)	390 (179.90)
22 (6.706)	305 (138.35)	398 (180.53)	159 (73.12)	195 (88.46)	230 (104.33)	295 (138.81)	376 (170.55)
24 (7.315)	296 (134.27)	387 (175.54)	153 (69.40)	189 (85.73)	221 (100.25)	284 (128.82)	361 (163.75)
26 (7.926)	288 (130.63)	375 (170.10)	147 (66.68)	180 (81.64)	211 (95.71)	271 (122.93)	345 (156.49)
28 (8.534)	278 (126.10)	363 (164.66)	140 (63.50)	172 (78.02)	202 (91.63)	258 (117.03)	328 (148.78)
30 (9.144)	270 (122.47)	351 (159.21)	133 (60.33)	164 (74.39)	191 (86.73)	245 (111.14)	311 (141.08)
32 (9.754)	260 (117.94)	338 (153.32)	126 (57.42)	155 (70.31)	181 (82.10)	231 (104.78)	293 (132.90)
34 (10.363)	250 (113.40)	325 (147.42)	199 (54.98)	146 (66.23)	170 (77.11)	216 (97.98)	274 (124.28)
36 (10.973)	240 (108.86)	311 (141.07)	111 (50.34)	136 (61.69)	159 (73.12)	201 (91.17)	255 (115.67)
38 (11.582)	229 (103.87)	297 (134.72)	104 (47.17)	126 (57.42)	147 (66.68)	185 (83.92)	234 (106.14)
40 (12.192)	218 (98.88)	282 (127.91)	95 (43.09)	116 (52.62)	134 (60.78)	168 (76.20)	213 (96.62)
Weight	58.05 lb/ft (86.39 kg/m)	76.00 lb/ft (113.10 kg/m)	32.23 lb/ft (47.96 kg/m)	39.74 lb/ft (59.14 kg/m)	47.03 lb/ft (69.99 kg/m)	60.95 lb/ft (90.72 kg/m)	77.40 lb/ft (115.19 kg/m)

[a] Heavy lines indicate Kl/r = 200.
[b] Square tubular columns are available with F_y = 50,000 lbf/in.^{2} (344.75 MN/m^2).

Table S76 Dimensions, weights, and allowable concentrated safe loads for structural square tubular columns [a, b] (*continued*)

Unbraced length	Outside dimensions: 8 × 8 in. (203.2 × 203.2 mm); Thickness of wall: 0.250 in. (6.350 mm)	0.3125 in. (7.938 mm)	0.375 in. (9.525 mm)	0.500 in. (12.700 mm)	0.625 in. (15.875 mm)
	Concentrated load in 1000 lb (453.6 kg) F_y = 36,000 lbf/in.^{2} (248.22 MN/m^2)				
ft (m)	lb (kg)	lb (kg)	lb (kg)	lb (kg)	lb (kg)
6 (1.829)	153 (69.40)	187 (84.82)	220 (99.79)	283 (128.87)	362 (164.20)
7 (2.134)	151 (68.49)	185 (83.92)	217 (98.43)	279 (126.55)	357 (161.93)
8 (2.439)	149 (67.59)	183 (83.01)	214 (97.07)	275 (123.38)	352 (159.67)
9 (2.743)	147 (66.68)	180 (81.64)	211 (95.71)	271 (122.93)	347 (157.40)
10 (3.048)	145 (65.77)	177 (80.28)	208 (94.35)	267 (121.11)	341 (154.68)
11 (3.353)	142 (64.41)	174 (78.92)	205 (92.99)	262 (118.84)	336 (152.41)
12 (3.658)	140 (63.50)	171 (77.57)	201 (91.17)	257 (116.57)	330 (149.69)
13 (3.962)	137 (62.14)	168 (76.20)	197 (89.36)	252 (114.31)	323 (146.51)
14 (4.267)	135 (61.23)	165 (74.84)	193 (87.54)	247 (112.04)	317 (143.79)
15 (4.572)	132 (59.88)	162 (73.48)	189 (85.73)	242 (109.77)	310 (140.62)
16 (4.877)	129 (58.51)	158 (71.67)	185 (83.92)	237 (107.50)	303 (137.44)
17 (5.182)	127 (57.51)	155 (70.31)	181 (82.10)	231 (104.78)	296 (134.27)
18 (5.486)	124 (56.25)	151 (68.49)	177 (80.28)	225 (102.07)	288 (130.63)
19 (5.791)	121 (54.89)	148 (67.13)	172 (78.02)	219 (99.34)	281 (127.46)
20 (6.096)	118 (53.52)	144 (65.31)	168 (76.20)	213 (96.62)	273 (123.83)
22 (6.706)	111 (50.35)	136 (59.24)	158 (71.67)	200 (90.72)	256 (116.12)
24 (7.315)	104 (47.17)	127 (57.51)	148 (67.13)	187 (84.82)	239 (108.41)
26 (7.926)	97 (44.00)	119 (54.98)	137 (62.14)	173 (78.47)	221 (100.25)
28 (8.534)	90 (40.82)	109 (49.44)	126 (57.42)	158 (71.67)	202 (91.63)
30 (9.144)	82 (37.19)	100 (45.36)	115 (52.16)	142 (64.41)	182 (82.55)
32 (9.754)	74 (33.57)	89 (40.37)	102 (46.27)	126 (57.42)	161 (73.03)
34 (10.363)	66 (29.93)	79 (35.83)	91 (41.28)	111 (50.35)	143 (64.86)
36 (10.973)	59 (26.16)	71 (32.21)	81 (36.74)	99 (44.91)	127 (57.51)
38 (11.582)	53 (24.04)	63 (28.58)	73 (33.11)	89 (40.37)	114 (51.71)
40 (12.192)	47 (21.32)	57 (25.85)	66 (29.93)	81 (36.74)	103 (46.72)
Weight	25.44 lb/ft (38.86 kg/m)	31.24 lb/ft (46.49 kg/m)	36.83 lb/ft (54.81 kg/m)	47.35 lb/ft (70.48 kg/m)	60.40 lb/ft (89.89 kg/m)

STEEL WINDOWS AND DOORS

Physical and Chemical Properties

Steel windows are made from new billets of structural grade steel which are hot-rolled into solid sections of a minimum metal thickness of $\frac{1}{8}$ in. (3.18 mm) and a minimum depth of $1\frac{3}{8}$ in. (28.58 mm). They are manufactured as complete units including hardware, weatherstripping, and operating mechanism and are pretreated for on-site painting, given a shop prime coat, vinyl-coated in various colors, or finished with a shop-applied baked enamel color finish (*see also* Windows).

Steel doors are made from structural quality hot-rolled or cold-rolled carbon steel sheet and strip in various thicknesses. They are manufactured by bending, forming, welding, and bracing into doors of all types, either as a complete unit including buck, door, and hardware or as doors only. They can be galvanized, primed, prefinished with baked-on color finishes, job-painted, or given special laminated finishes (*see also* Doors).

Types and Uses

Steel window units are made in modular and nonmodular sizes for residential, commercial, institutional, industrial, monumental, and specialized types of construction (*see* Figures *S101* to *S108*).

In addition to the windows illustrated, there are vertical pivoted, utility, top-hung, and continuous windows, as well as new types constantly being introduced by the various manufacturers. In selecting steel windows, a check should be made of the latest manufacturers' data.

Steel doors are made in standard sizes and thicknesses: 2 to 3 ft (0.610 to 0.914 m) wide in increments of 2 in. (50.8 mm), 6.667 and 7 ft (2.032 and 2.134 m) in height, and $1\frac{3}{8}$ and $1\frac{3}{4}$ in. (34.93 and 44.45 mm) thick. They are also manufactured in roll-down or side-coiling, grille, and shutter types of doors, as well as numerous kinds of fire doors. Many of the roll-down or side-coiling doors can be manually or mechanically operated. Door manufacturers should be consulted for the latest types available. (*See also* Steel Hollow Metalwork.)

Application

Condensed Checklist

1. The type of steel window or door chosen should be checked for the necessary strength and to see that it has received pretreatment or a corrosion-resistant prime coat.
2. When double glazing is used, one should always check that the type of window selected is available for double glazing.
3. All metals with which the steel windows or doors

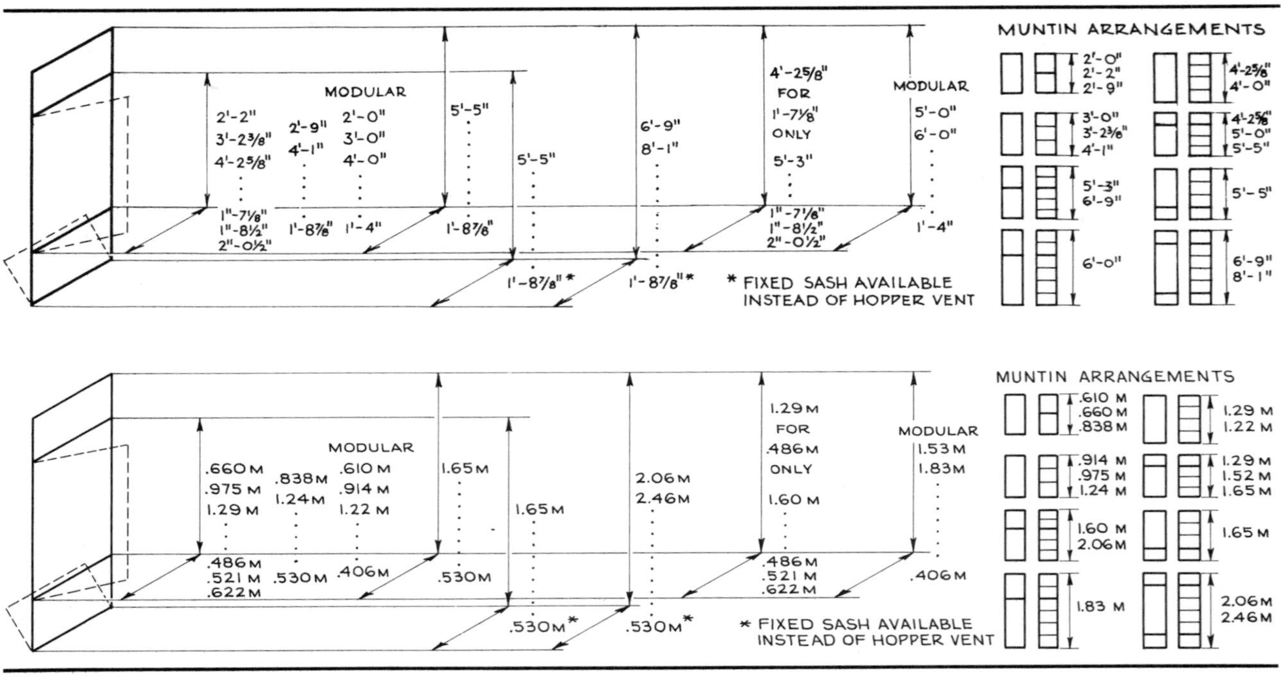

Figure S101 Sizes of single-casement windows.

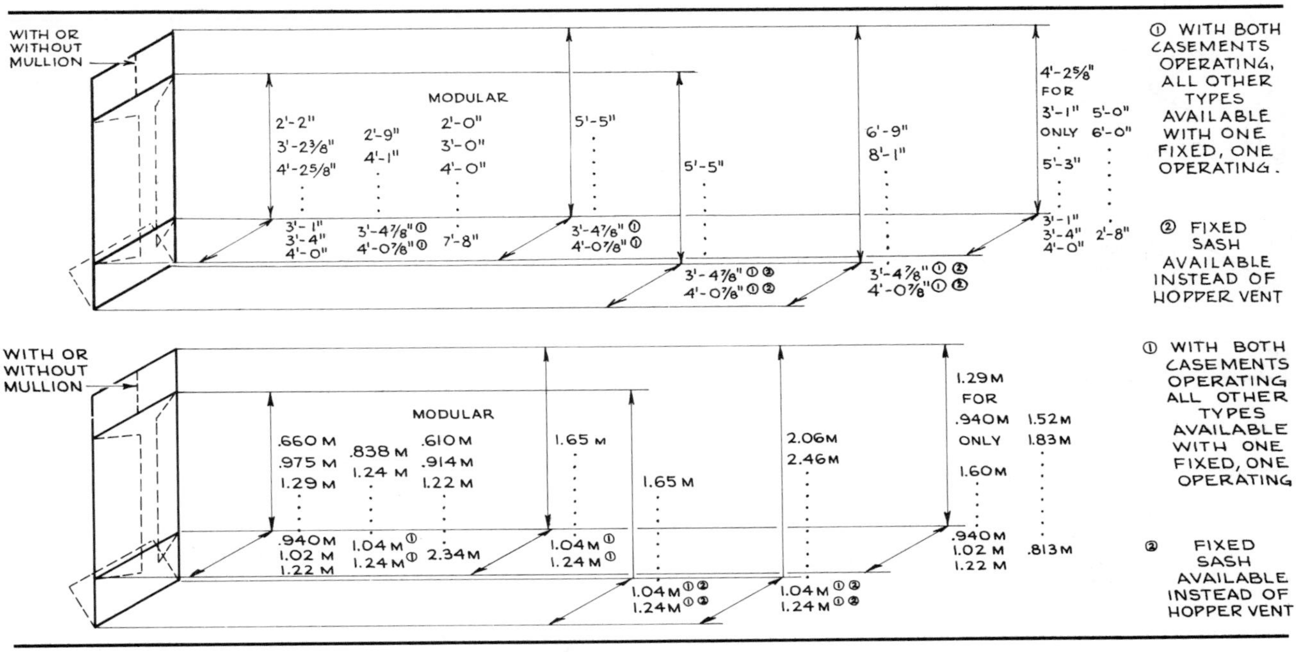

Figure S102 Sizes of double-casement windows.

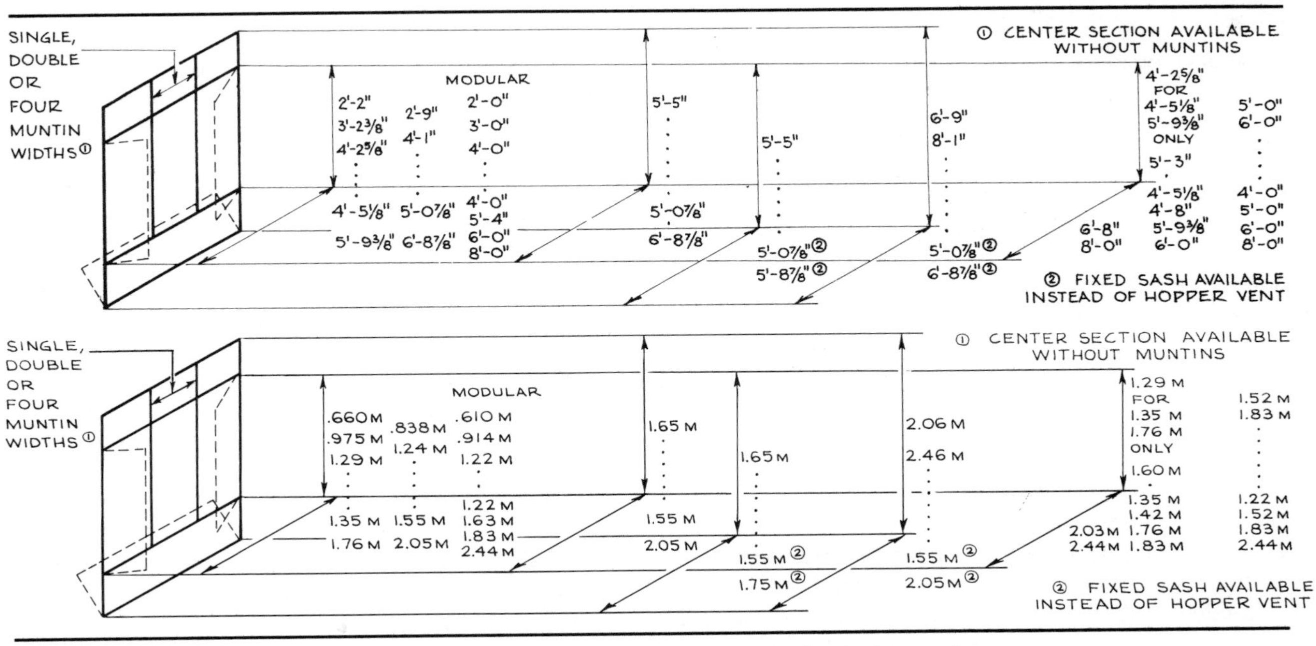

Figure S103 Sizes of two-casement windows with fixed center section.

are likely to come in contact should be checked for compatibility so that no galvanic action can start.

4. The caulking, glazing compounds, and gaskets used must be of the correct type for steel.
5. All hardware should be of a metal that will not cause galvanic action. It generally is made of brass or stainless steel, or is chrome plated.
6. One should always check with manufacturers of windows regarding the availability of the type of window selected.
7. When vinyl coatings or baked enamel finishes are selected for the type of window to be used, manufacturers should be consulted regarding the availability of these finishes and the range of colors.

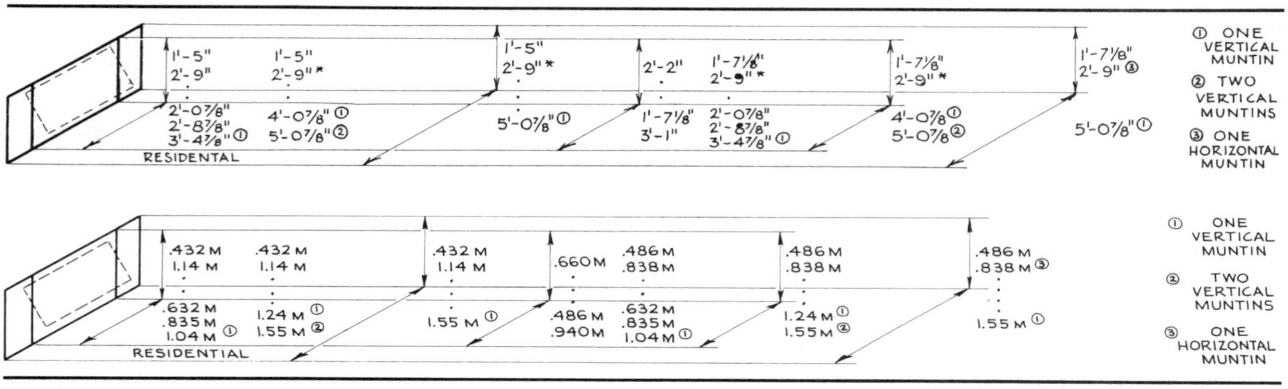

Figure S104 Sizes of out-projected and in-projected windows.

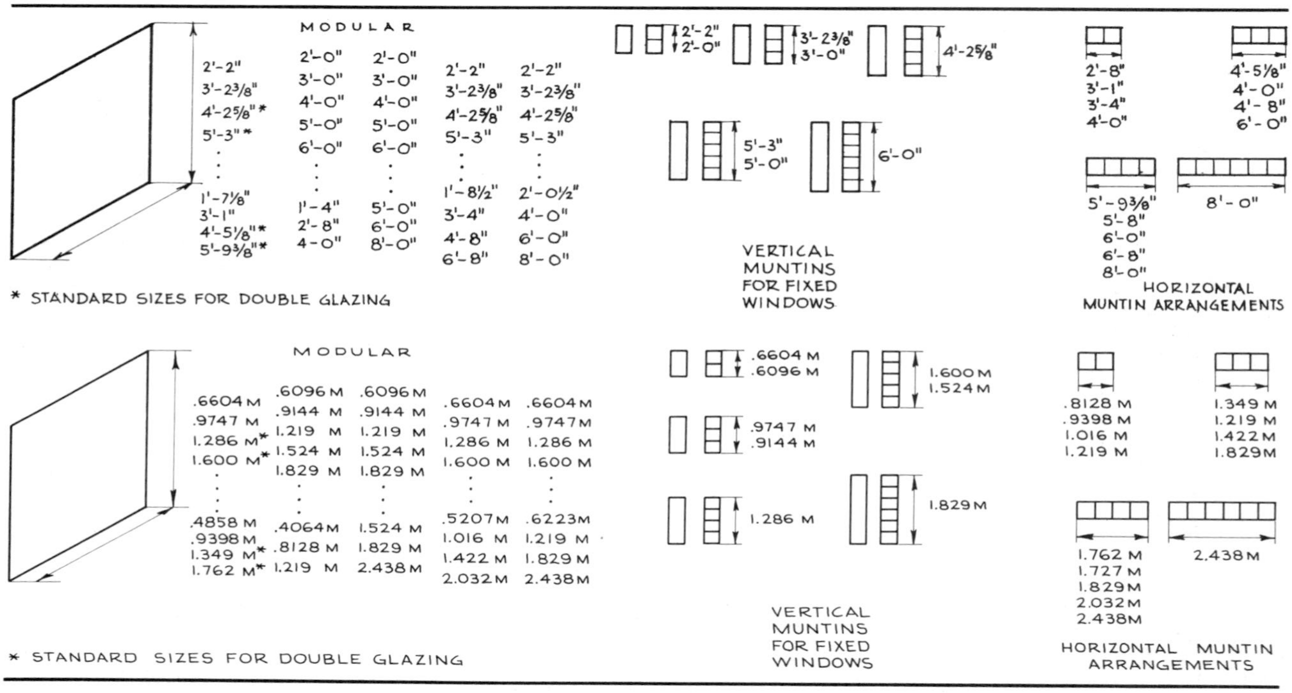

Figure S105 Sizes of fixed windows.

8. When using steel windows in curtain wall construction, one should always specify that a full-size section of the curtain wall must be tested under the same climatic conditions that prevail where the windows will be installed.
9. When selecting the type of steel door to be used, manufacturers should be consulted for the latest data on types of doors, finishes, and availability.
10. When fire doors are required, one should always check all building codes and fire codes of the locality where the doors are to be installed.
11. For roll-down or side-coiling types of doors, one should always check the size of enclosure for limitations on the head-of-door-to-ceiling dimension, and the size necessary for coiling at side walls. The weight of doors plus enclosures must be checked to be sure that adequate structural support is installed.
12. One should always check the type of hardware to be installed, and make sure that steel doors have necessary reinforcing and indentations for hinges, locks, closers, latches, etc.
13. When the type of finish has been selected, manufacturers should be consulted regarding its availability.

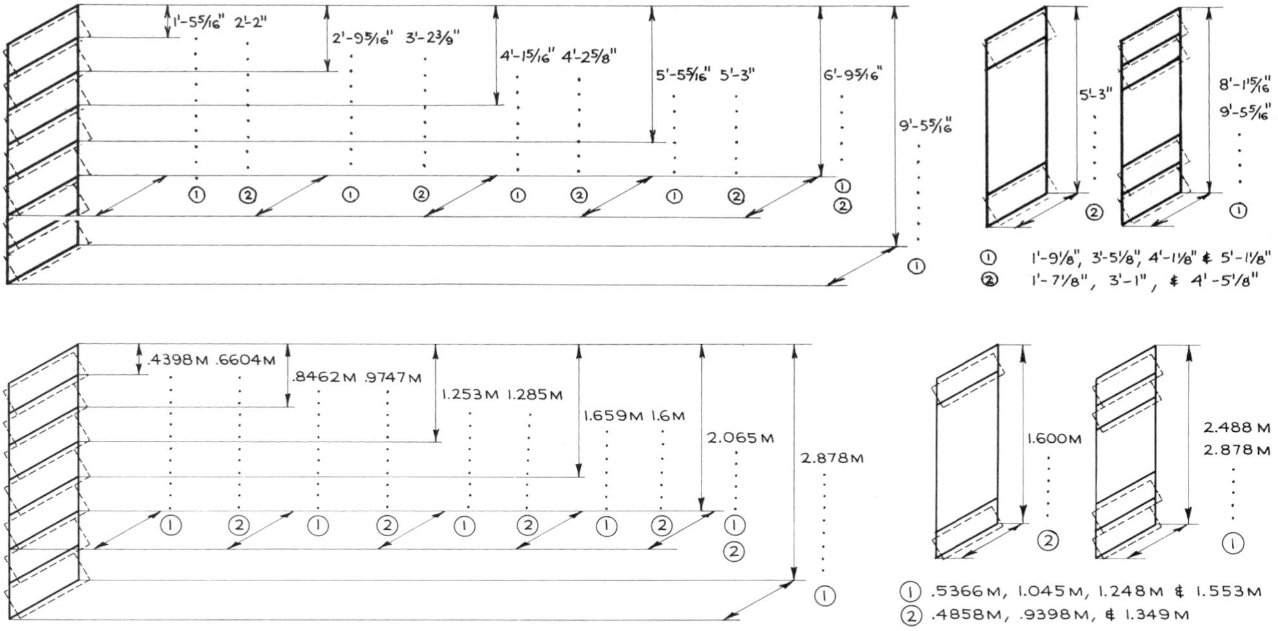

Figure S106 Sizes of awning-type windows.

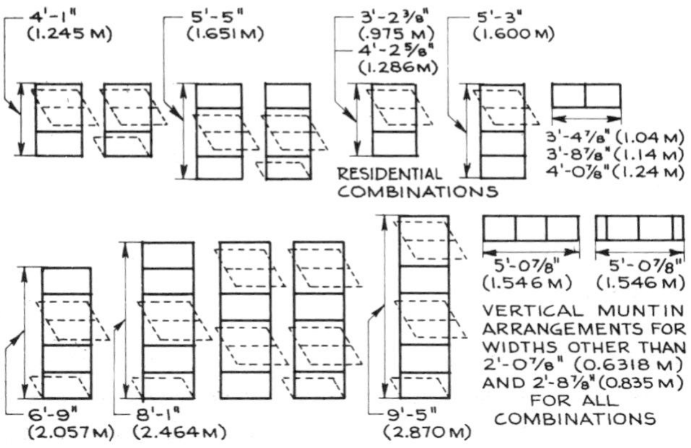

Figure S107 Heights of combination out-projected, in-projected, and fixed sections.

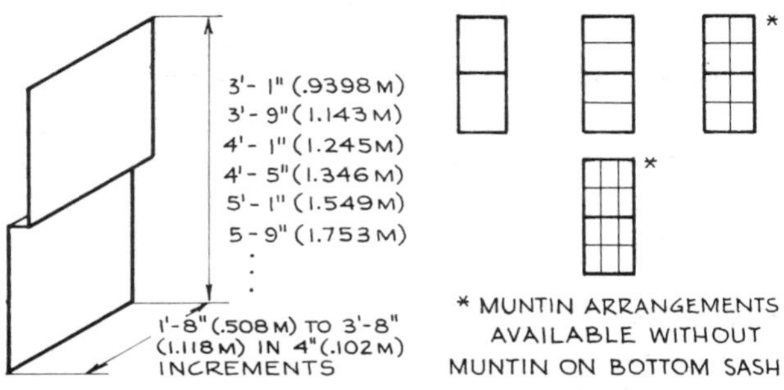

Figure S108 Typical sizes of steel double-hung windows.

14. When a complete packaged unit is selected, one should always check the type of hardware, finishes, colors, and shapes that are available from the manufacturers of packaged units.

Conditions Favorable to the Use of Steel Windows and Doors

1. Where a strong, rigid metal is required.
2. Where fire-resistant windows are required.
3. Where the windows are required to help support structural loads.
4. Where fire doors are required (*see* Figure *S109*).
5. Where strong, rigid, burglarproof doors are required.
6. Where large exterior openings are required.
7. For closing off areas and spaces within buildings.
8. For all types of exterior doors (*see* Figure *S110*) other than main entrance doors for which a maximum of glass is desired and a minimum thickness of frames is required.

Conditions Unfavorable to the Use of Steel Windows and Doors

1. Where maintenance must be kept to a minimum.
2. Where painting is not desired.
3. Where maximum visibility is required for entrance doors and sidelights.

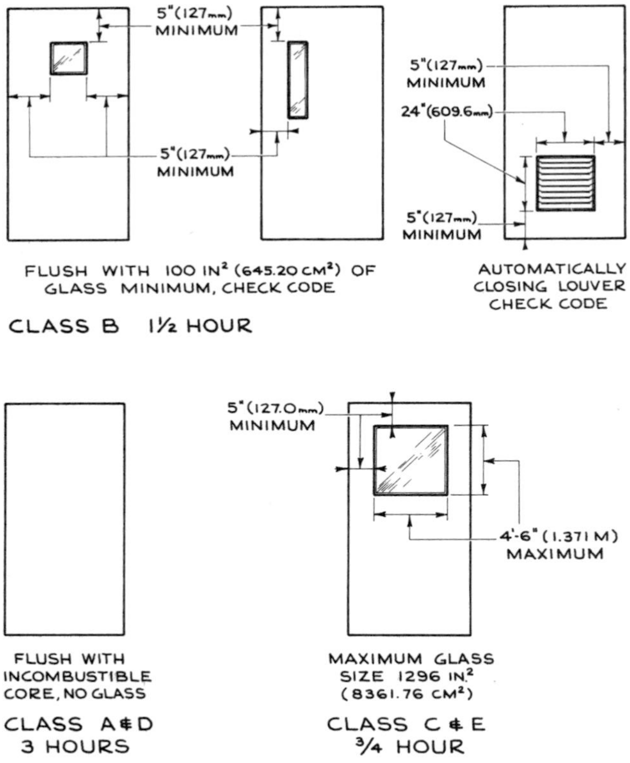

CLASS A & D - MAXIMUM SIZE

TYPE OF LOCK	SINGLE DOOR FT.	SINGLE DOOR M.	PAIR OF DOORS FT.	PAIR OF DOORS M
3-POINT LOCK	4 x 10	1.219 x 3.048	8 x 10	2.438 x 3.048
APPROVED SINGLE LATCH	4 x 10	1.219 x 3.048	8 x 8	2.438 x 2.438
FIRE EXIT HARDWARE	4 x 8	1.219 x 2.438	8 x 7.167	2.438 x 2.184

CLASS A, B & C - MAXIMUM SIZE

TYPE OF LOCK	SINGLE DOOR FT.	SINGLE DOOR M	PAIR OF DOORS FT.	PAIR OF DOORS M
3-POINT LOCK	4 x 8	1.219 x 2.438	8 x 10	2.438 x 3.048
APPROVED SINGLE LATCH	4 x 9	1.219 x 2.743	8 x 9	2.438 x 2.743
FIRE EXIT HARDWARE	4 x 8	1.219 x 2.438	8 x 7.167	2.438 x 2.184

Figure S109 Typical fire doors with maximum-size glass and louver openings.

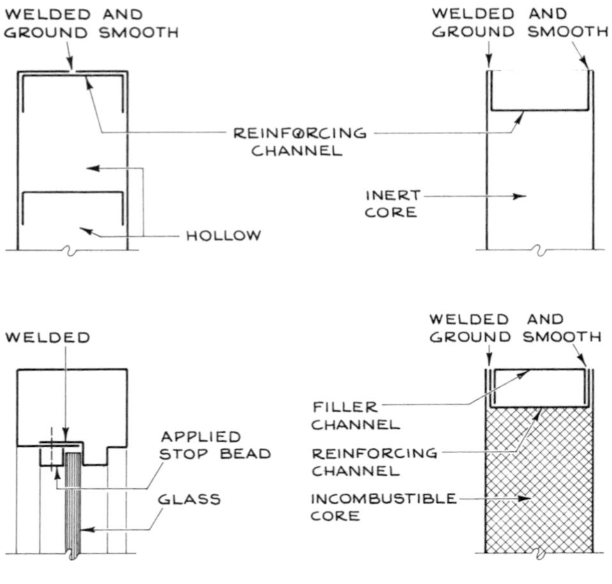

THE STEEL SHEET AND STRIP USED FOR DOORS VARIES FROM 14-GAUGE TO 20-GAUGE DEPENDING ON THE END USE.

Figure S110 Typical steel door types.

STONE

Physical and Chemical Properties

Stone or rock can be divided into three groups on the basis of geologic origin: igneous, sedimentary, and metamorphic. Igneous stone is the result of solidification from a molten state. Sedimentary stone is composed of sand, clay, and other substances derived from the breaking down of existent stone into small particles which are taken up and carried by water and then settled from the water into beds. These particles, together with the remains of plants and animals, are formed into stone or rock by mechanical pressure or are cemented together by chemical or organic action. Metamorphic stone is the ultimate product from both igneous and sedimentary stone formed either by pressure, heat, or moisture, or various combinations of these forces.

Tables *S77* and *S78* list data on physical properties, such as variations in weight, specific gravity, compressive strength, and shear strength, for stone commonly used in construction. (*See also* Stone, Granite, *and other stone headings*.)

Effect of Heat. Almost all natural stone is affected drastically by quick changes in temperature and therefore should not be used where very high temperatures will be encountered. Granite and slate will explode and completely disintegrate if exposed to severe heat such as temperatures above 2125°F (1162.78°C). Limestone and marble are ruined by relatively high temperatures. Only sandstone, which is fine grained and compact, can withstand heat up to very high temperatures. Most stones meet normal building code requirements for construction fire-resistance ratings. For example, a 4-in. (101.6-mm)-thick slab of limestone has a fire endurance of 1 hour and 12 minutes.

Commercial Forms. Stone is available in seven forms: rubble stone, dimension stone, monumental stone, flagstone, crushed and broken stone, and stone dust or powder (*see* "Types and Uses").

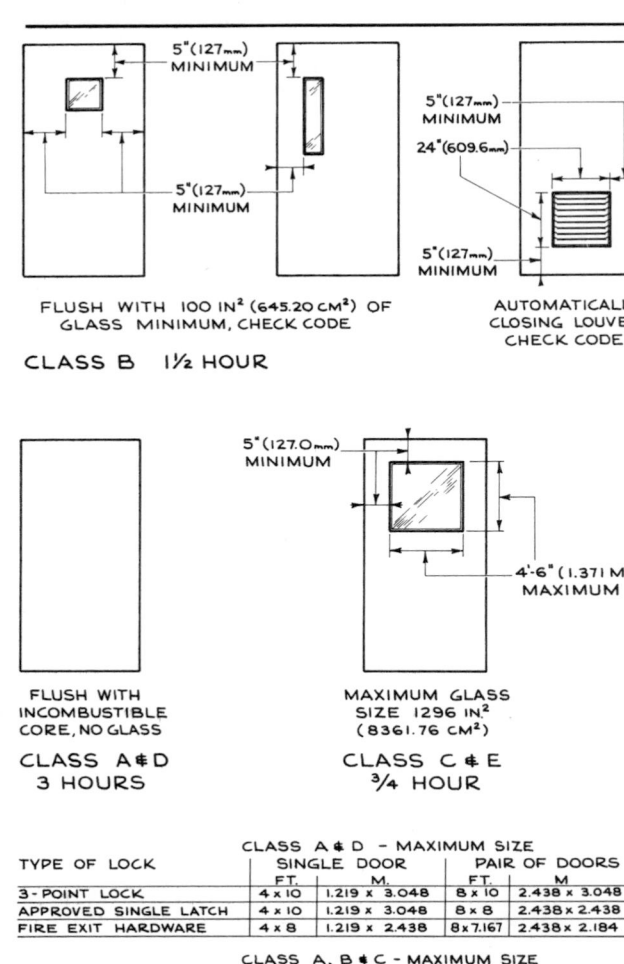

Table S77 Properties of stone used in construction

Type of stone	Name of stone	Principal ingredient	Weight lb/ft³	Weight kg/m³	Specific gravity	Compressive strength[a] lbf/in.²	Compressive strength[a] MN/m²	Modulus of rupture[a] lbf/in.²	Modulus of rupture[a] MN/m²
Igneous	Granite	Silica	163–169	242.58–251.51	2.61–2.70	15,600–30,800	107.56–212.37	1490–3060	10.27–21.10
Metamorphic	Marble	Calcium carbonate ($CaCO_3$)	165–170	245.55–252.99	2.64–2.72	9,058–17,787	62.46–123.32	1197–3442	8.24–23.73
	Slate		171–176	254.48–261.92	2.74–2.82	10,000–15,000	68.95–103.43	9000–5235	62.06–88.95
	Soapstone		185	275.32	2.73	9,000–14,000	62.06–69.53	3060–5235	21.10–36.00
Sedimentary	Limestone	Calcium carbonate ($CaCO_3$)	131–170	194.95–252.99	2.10–2.75	2,600–21,320	17.93–147.00	640–1995	4.41–13.76
	Sandstone		140–166	208.35–247.04	2.14–2.66	4,000–64,000	27.58–441.28	1500	10.34

[a]1 lbf/in.² = 0.006895 MN/m².

Table S78 Coefficients of expansion[a] of stone used in construction

Type of stone	Coefficient of expansion: Centigrade (Celsius)	Coefficient of expansion: Fahrenheit
Granite	0.0000084	0.0000047
Limestone	0.0000080	0.0000044
Marble	0.0000100	0.0000056
Sandstone	0.0000110	0.0000061
Slate	0.0000104	0.0000058

[a]*Example:* With a temperature change of 90°F, 100 ft of granite would change in length by 100 × 12 × 90 × 0.0000047 = 0.04 in. or approximately $\frac{1}{32}$ in.

Types and Uses

Many types of stone are used in construction, ranging from on-the-site rough fieldstone to stone that is quarried, cut, shaped, and polished to exact dimensions. Many localities have small quarries supplying limited amounts of stone of special colors, textures, and shapes. Table *S79* lists the commonly used stones together with their characteristics and major uses.

The following terminology is used to grade and classify stone for construction purposes:

1. *Rough building stone* (*fieldstone*)—rock-faced masses of various sizes and shapes.

2. *Rubble stone*—irregular stone fragments having at least one good face, delivered from the quarries. Irregular in shape and sized within fixed dimensions, usually 12 in. (304.8 mm) high and 2 ft (609.6 mm) long, rubble stone can be cut and made into blocks and pieces for building walls, veneers, copings, sills, etc. It is obtained by the ton or carload.

3. *Dimension stone* (*cut stone, ashlar*)—delivered from finished stone mills to a specific size, squared to dimensions each way and to a specific thickness. There are two types of surface finishes for dimension stone. In one the face is rough, or the natural split of the stone; in the other the surface is smooth, slightly textured, or polished. Dimension stone is used for exterior or interior surface veneers of buildings, prefabricated panels, preassembled systems, toilet partitions, flooring, copings, stair treads, sills, bearing walls, etc. It is obtained by the cubic foot (0.02833-m³). Ashlar is now included under dimension stone. Ashlar is smaller, rectangular stone with a flat-faced surface, generally square or rectangular, having sawed or dressed beds and joints.

4. *Monumental stone*—either rough or finished stone meeting the same requirements as dimension stone and used for gravestones, monuments, etc.

5. *Flagstone*—flat slabs of thin stone, generally from 1 to 2 in. (25.4 to 50.8 mm) thick, either irregular or squared, with the surface smooth, slightly rough, or polished. Flagstone is used on the exterior for paths, walks, and terraces, and on the interior as stair treads, flooring, blackboards, coping, sills, countertops, etc. It is obtained by the square foot.

6. *Crushed and broken stone*—chips, granules, or irregular shapes that have been graded and sized for construction work. Crushed stone usually begins at $\frac{1}{4}$ in. (6.35 mm) and runs by various stages to the $2\frac{1}{2}$ in. (63.5-mm) size. It differs from large-size gravel in being

Table S79 Common types of stone used in construction

Common name of	Type of stone	Characteristics: Texture	Characteristics: Appearance	Characteristics: Color	Major use	Other uses
Black granite (traprock)	Igneous	Fine to coarse grained, polished	Generally uniform	Black	Veneer for exterior of buildings	Crushed rock for concrete and asphaltic concrete
Bluestone	Sedimentary	Smooth to rough	Uniform	Blue-gray	Flagstones	Sills, countertops, flooring
Granite	Igneous	Fine to coarse grained, polished	Generally uniform; some colors have spots, veins, and variations in grain	Wide range of colors; white and black	Veneer for exterior of buildings	Paving blocks, curbing, monuments
Limestone	Sedimentary	Fine grained	Uniform in color	Buff-gray	Veneer for exterior of buildings	Copings, sills, flagstones, interior finishes
Marble	Metamorphic	Fine grained, polished	Uniform or with wide variations of veining and colors	Wide range of colors; white and black	Veneer for exterior and interior finishes of buildings	Flooring, monuments, countertops, table and desk tops, sills, and toilet partitions
Sandstone	Sedimentary	Rough	Generally uniform	White, gray, yellow, brown, red	Veneer for exterior of buildings	Flagstones, interior finishes
Slate	Metamorphic	Smooth to rough	Generally uniform	Blue, gray, green, reddish	Roof shingles, flagstones, veneer for exterior and interior of buildings	Blackboards, sills, countertops
Soapstone	Metamorphic	Smooth to rough	Generally uniform	Gray, green, blue	Flagstones	Chemical table tops and sinks, flooring, veneer for exterior of buildings
Travertine	Sedimentary	Smooth to rough, also polished	Irregularly shaped pores	Gray, white, buff	Veneer for interior finish	Countertops

usually composed of only one kind of rock. It is used as aggregate in concrete work and asphalt walks, roads, driveways, paths, etc.; as surfacing material for asphalt shingles, siding, and built-up roofing; and in terrazzo and artificial stonework. It is obtained by the ton.

7. *Stone dust or powder*—used for surfacing asphalt paving, as fill in paints, for resilient flooring, etc.

Application

The types of stonework are based on the shape of stone and the surface treatment or finish of the stone (*see* Figure *S111*). The term "ashlar" means only that the stone face showing on the finished surface has its beds and joints sawed or dressed. Ashlar stonework can have a rough, smooth, or polished finish, depending on the treatment of the face. Coursed ashlar has continuous horizontal joints, stacked ashlar has continuous vertical and horizontal joints, and random ashlar has neither continuous horizontal nor continuous vertical joints. Fieldstone always has a rough, irregular appearance, as the

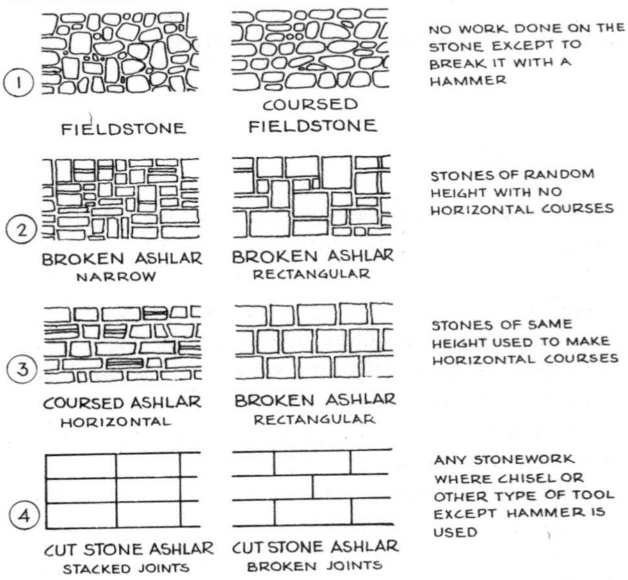

Figure S111 Common types of stonework.

natural surface or broken surface of the stone is exposed. Cut stone usually consists of large, thin slabs of stone with the face smooth, textured, slightly textured, or polished.

Mortar. Mortar for stonework other than fieldstone is always made with a nonstaining cement because the color, veining, and texture of stone can be destroyed if any staining occurs at joints.

Pointing mortar consists of 1 part nonstaining cement, 1 part hydrated lime, and 6 parts clean white sand that will pass a No. 16 sieve. Joint sealants are multicomponent polysulfide or polyurethane synthetic rubbers and silicones.

High-strength epoxy adhesives with or without expansion bolts are commonly used to construct preassembled stone units, panels, and decorative screens and fences (*see* Figure *S112*).

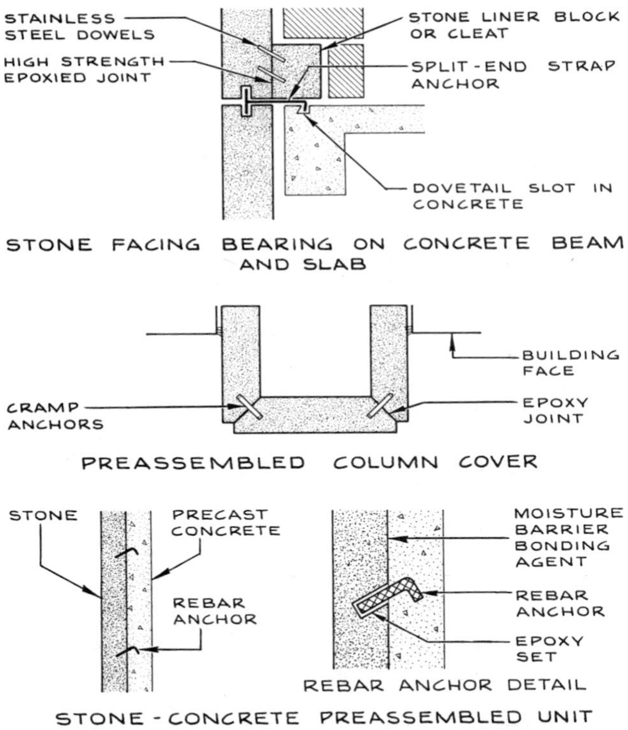

Figure S112 Typical application methods of high-strength epoxy adhesive for stone facing and prefabricated and preassembled units.

Condensed Checklist

1. The type of stone, its surface finish, its color, and its texture should be selected to meet the design requirements for its end use.
2. The stone should be checked for variations in grain, color, veining, and surface texture.
3. Availability of the stone selected and of skilled labor for this type of work in the locality in which construction is to take place should both be checked.
4. Samples of stone to be used should always be obtained. For cut stone, it is always necessary to specify that shop drawings be made, including setting drawings, and that each piece of stone be labeled for its exact location.
5. Local, municipal, and state codes and also those of the Fire Underwriters, insurance companies, labor departments, and federal government (Army, Navy, etc.) should be checked for limitations and fire ratings for stonework to be installed.
6. In veneer work, care should be taken to eliminate moisture from entering behind the veneer. The back-up material should always be coated with an asphalt type of dampproofing before stone is applied.
7. A check should be made whether the stone needs to be drenched with water before setting.
8. When heavy, large pieces of stone are used, small button-shaped pieces of lead, plastic, or aluminum should be used for aligning and supporting.
9. All anchors, ties, dowels, and other accessories used for supporting and aligning stone should be made of nonstaining, noncorroding metal, and care should be taken that these are tied, welded, or riveted to the supporting structure, especially structural steel, in such a way that no galvanic action can occur at the point of contact.
10. All flashing, whether exposed or concealed, should be of nonstaining metal.
11. The type of nonstaining cement, hydrated lime, sand, and water should be such that each mortar ingredient and the resulting mortar mix are nonstaining.
12. All cornices, copings, etc., should be set with vertical joints unfilled.
13. The ends of step and sill should be embedded in mortar at the sides; the rest of the joints should be left unfilled, to be filled later with nonstaining, nonhardening pointing compound.
14. The location of expansion joints (pressure joints) should always be established, and an expansion joint of corrugated lead with a lead cover should be installed. An alternative method is to fill joints with nonstaining, nonhardening pointing compound.
15. All stonework should be kept clean during and after installation. Specifications should require that all stonework be protected until the building is occupied.

16. All stonework should be cleaned only with soap powder, clean water, and fiber brushes or by a method specified by the stone manufacturer.
17. Temperature during installation: If mortar is used, stonework should not be installed in freezing weather unless adequate precautions are taken (protection by covering, heating, etc.).
18. When preassembled units and panels are used, a check should always be made that the high-strength epoxy adhesives, expansion bolts, etc., used in assembly meet the rigid requirements of end use of the panels and units.
19. All sealants should meet the requirements for their end use in the assembly of preassembled units or panels in the on-site construction operation.

Conditions Favorable to the Use of Stone

1. Where a permanent, strong, durable finishing material is required on exterior or interior surfaces of a building.
2. Where the color, texture, and pattern of stone, particularly veining, are desired for the design of exterior or interior finish surfaces.

Conditions Unfavorable to the Use of Stone

1. In areas where industrial atmospheres will attack stone.
2. Where materials adjacent to and serving as frame for stone will stain it, for example, copper, bronze, unprotected steel, and woods that contain tannic acid.
3. Where economy is of first importance.
4. In areas where intense heat is present, such as interiors and back hearths of fireplaces.

History and Manufacture

Man first built with what was at hand; the three basic materials available to him were wood, stone, and mud or clay. The history of architecture until as late as 1900 was largely the story of these materials as used in construction. Stone was the structural material, the exterior and interior finishing material, the flooring material, and in many cases the roofing material. It was also used for retaining walls, roads, walks, paths, steps, and all types of sculpture, statuary, and decorative and ornamental applications. Today stone is used almost entirely as a surface finishing material for both the exterior and interior of buildings.

Quarrying. Today stone is obtained from quarries in almost the same manner as in ancient times except that modern tools are used. In fact, some quarries used by the Greeks, Romans, and Carthaginians are still in operation.

The method is first to create a channel or slot to a certain depth (floor) and then to remove large blocks of stone from both sides or one side of the channel until a flat area (floor) is exposed. The size of this floor area is controlled by the characteristics of the natural stone deposit or by operational limitations.

To create the channels, holes are drilled close together horizontally to a certain established depth and the stone between the holes is removed. Then wedges are placed into this narrow horizontal slot, and the entire block of stone is broken free. (The depth of the holes is controlled by natural horizontal planes of separation in the rock itself or by horizontal separations, or fractures, made by drilling holes and discharging light powder charges. For softer stones a channeling machine is used which actually chops a horizontal slot.)

These large blocks are then subdivided into smaller units either by drilling and wedging or by the wire saws now used by most stone producers (*see* Figure *S113*). The wire saw cuts a smooth surface, thus reducing finishing in the mill; it can also subdivide large blocks of stone into sizes for finishing in the mill. In the quarry a new and deeper slot is made, and the entire quarrying sequence just described is repeated.

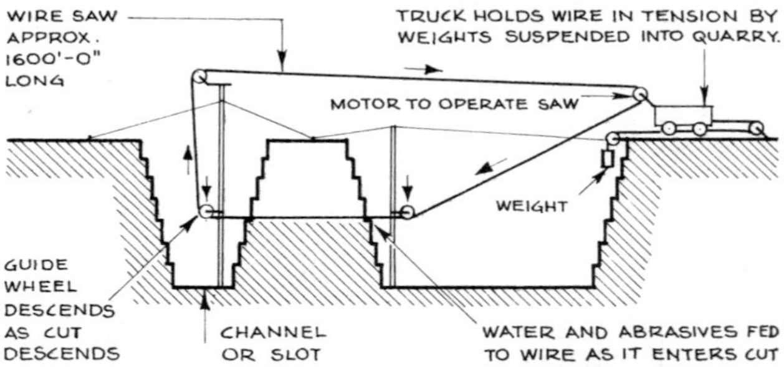

Figure S113 Cutting stone with wire saw.

Flame Drilling. A relatively new means of cutting long, narrow horizontal slots in the quarries is the flame drill. This device feeds liquid oxygen and fuel oil through a nozzle to develop a temperature of over 5000°F (2760°C). When a stream of water is added to this flame, the combination disintegrates the stone into fragments which are automatically blown out of the cut. This method produces a horizontal slot 8 in. (203.2 mm) wide. The noise developed is so loud that the operation is usually done only during the night.

Mill Operations. From the quarry the blocks of stone are transported to the finishing mill, which may be located near the quarry or far removed. Here the stone is subdivided and shaped into finished materials. Cutting is done by circular saws, gang saws that move horizontally back and forth, and wire saws. All types of saws use abrasives with water as the cutting materials. Stone that is not polished is shaped with pneumatic tools; stone to be polished has its surface ground down, first with coarse abrasives, and then with increasingly finer abrasives until the stage of final polish, which is done with a felt and a very fine polishing material.

STONE: GRANITE

Physical and Chemical Properties

Granite is igneous stone formed of feldspars and quartz with smaller amounts of mica and hornblende. Since all its ingredients except mica are hard, granite itself is hard, strong, durable, impervious to moisture, and the most difficult stone to cut and finish. Almost all granite contains water (up to 2%) and less than 1% of MnO, P_2O_3, TiO_3, ZrO_3, and BaO. (*See* Tables *S80* and *S81*.)

Grain Classification. Granite is classified as fine-grained, medium-grained, and coarse-grained. Medium-grained granite contains feldspar crystals about $\frac{1}{4}$ in. (6.35 mm) in diameter. In granite used as building stone, uniformity in texture and grain size is desirable.

Coloring. The color of granite depends on the color of the feldspar, hornblende, and mica. Feldspar produces red, pink, brown, buff, gray, and almost white granite; hornblende or black mica produces dark green and black granite.

Commercial Forms. Granite is available as dimension stone (rough or finished), including building stone, paving stones, curbing, and flagstones; and as broken or crushed stone graded to various sizes, including granite dust (*see* "Types and Uses").

Types and Uses

Granite comes in combinations of colored crystals which give an overall appearance of a white, gray, pink, red, brown, green, blue, or black stone. It can be divided into three categories on the basis of use: building stone (rough or finished); paving, curbing, and flagstones; and crushed or broken granite.

Building Stone. Building stone is divided into two general types, cut stone and ashlar-type stone:

1. *Cut stone*, which consists of large, thin slabs of sawed or polished granite veneer, $\frac{7}{8}$ in. (22.23 mm) thick at minimum and up to $2\frac{1}{2}$ in. (63.5 mm) thick and more, is applied to exterior surfaces of buildings and, rarely, for interior surfaces. With the development of new types of anchoring methods and the introduction of high-strength epoxy adhesives, thin granite veneers are now applied directly to the building structure (*see* Figure *S114*). Such a veneer stone is set with its back 1 in. (25.4 mm) at minimum from the back-up material.

Cut stone also includes molded sills, copings, lintels, window and door trim, columns, and stair treads, which are used with cut stone veneer or ashlar-type stonework. For cut stone, vertical and horizontal joints are generally $\frac{1}{4}$ in. (6.35 mm).

2. *Ashlar-type stone*, which consists of smaller pieces 6 to 24 in. (152.4 to 609.6 mm) long by 2 to 16 in. (50.8 to 406.4 mm) high and from $3\frac{1}{2}$ to $4\frac{1}{2}$ in. (88.9 to 114.3 mm) thick, is laid up in various ashlar patterns. This type is used as veneer on the exterior or interior of

Table S80 Chemical composition of granite

SiO_2	Al_2O_3	Fe_2O_3, FeO	K_2O (percent of content)	Na_2O	CaO	Water and other elements
68.1–76.0	13.6–17.4	0.9–4.6	2.4–5.7	2.0–6.6	1.0–3.0	1.04–2.0

Table S81 Physical properties of granite

Compressive strength[a]		Modulus of rupture[a]		Specific gravity	Weight	
lb/in.^{2}	MN/m^2	lb/in.^{2}	MN/m^2		lb/ft^3	kg/m^3
15,600–30,800	107.56–212.37	1490–3060	10.27–21.10	2.61–2.70	163–169	2611.26–2707.38

[a] 1 lbf/in.^{2} = 0.006895 MN/m^2.

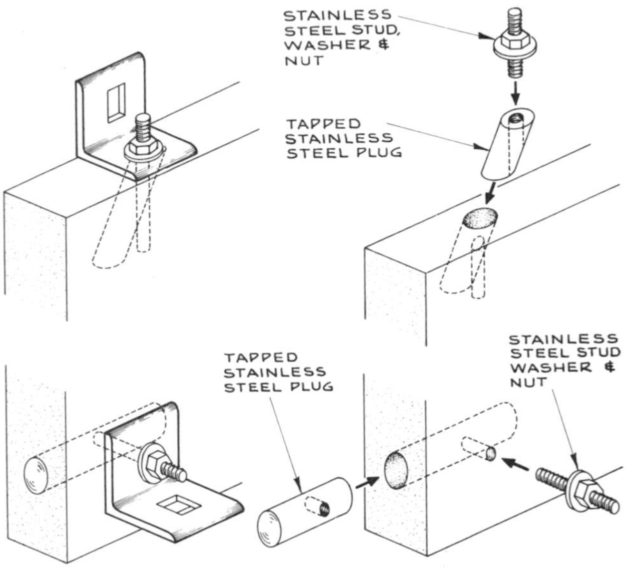

Figure S114 New methods of anchoring granite.

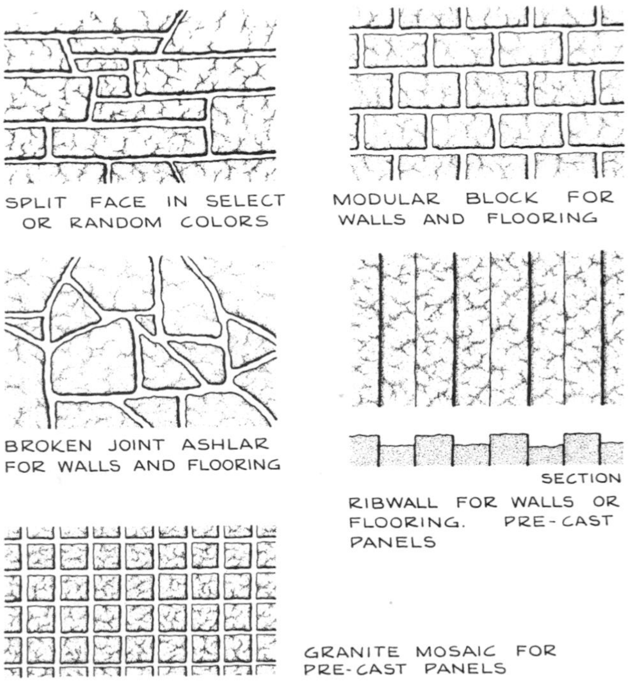

Figure S115 Ashlar and other types of granite units for walls and flooring.

buildings. Ashlar comes with various surface textures, usually designated as split, tooled, sawed, seam, and bold rock, and its edges are squared or angular, finished by sawing, tooling, or splitting. Vertical and horizontal joints vary from $\frac{1}{2}$ to $1\frac{1}{2}$ in. (12.7 to 38.1 mm). (*See* Figure *S115*.)

Paving, Curbing, and Flagstones. Granite paving, curbing, and flagstones are used generally for exterior paving, walls, and curbing for roadways and driveways because of their great durability. (*See* Figure *S115*).

Paving granite is rough-split and roughly squared, having one good face, and is 3 to 5 in. (76.2 to 127.0 mm) thick and wide by 4 to 12 in. (101.6 to 304.8 mm) long. It is generally laid with a 1-in. (25.4-mm) joint for paving, and $\frac{1}{2}$- to $1\frac{1}{2}$-in. (12.7- to 38.1-mm) joints for walls.

Curbing is made in two types, straight and sloped. Table *S82* gives the various stock sizes available. For straight curbing there are special cuts: (1) stock corner pieces with 2- and 3-ft (609.6- and 914.4-mm) radii in 4- to 7-in. (101.6- to 177.8-mm) depths, with dressed top and split or sawed face; (2) catchbasin inlets, 1.5 ft (457.2 mm) deep, 6 ft (1.83 m) long, and 4 to 7 in. (101.6 to 177.8 mm) wide, with sawed top and split face; and (3) stock curbing cut to radii of 10, 12, 15, and 20 ft (3.05, 3.66, 4.57, and 6.10 m) and from 25 to 50 ft (7.62 to 15.24 m) in 5-ft (1.27-m) increments.

Flagstones are sawed to $\frac{7}{8}$- and 1-in. (22.23- and 25.4-mm) thicknesses, with squared edges, usually 2 and 3 ft (609.6 and 914.4 mm) square. Other sizes are also available.

Retaining walls of granite have been used successfully for as much as a 24-ft (7.32-m)-high wall by prestressing to a reinforced concrete footing (*see* Figure *S116*).

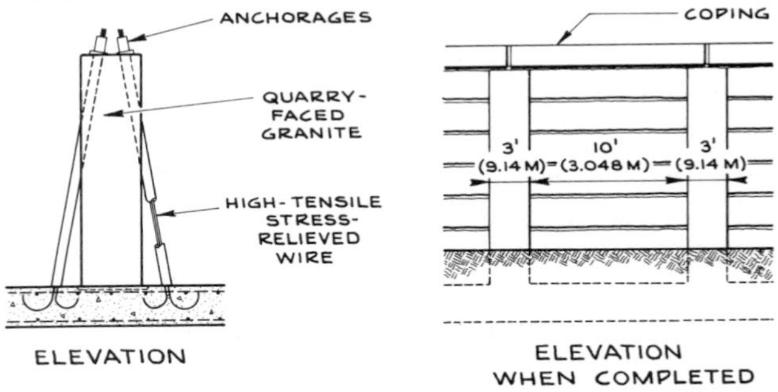

Figure S116 Prestressed granite retaining wall.

Crushed or Broken Granite. Crushed or broken granite and granite dust are produced by quarries using quite different operational methods. The larger, irregular blocks of granite (riprap) are used for river and harbor installations, railroad ballast, highway fills, or any other applications where large, irregular blocks are needed. Crushed granite, graded to size, is used for terrazzo and artificial stone. Granite dust is used as a filler and has the same applications as limestone dust.

Table S82 Types, sizes, and finishes of granite curbing

Type of curb	Dimensions: Width in.	Width mm	Depth in.	Depth mm	Length ft	Length m	Finish: Top	Finish: Face
Vertical	3	76.2	17	431.8	3	0.9144		
	$3\frac{1}{2}$–5	88.9–127	17	431.8	1	0.3048		
	4, 5	101.6, 127	16, 18	406.4, 457.2	3	0.9144	Sawed	Split
	6	152.4	12	304.8	4	1.2192		
	6, 7	152.4, 177.8	18, 20	406.4, 508.0	6	1.8288		
	4	101.6	16	406.4	1	0.3048	Tooled	Split
	5	127.0	16	406.4	3	0.9144		
	3	76.2	13	330.2	$\frac{3}{4}$–$1\frac{1}{2}$	0.0229–0.457		
	4	101.6	16	406.4	1–2	0.3048–0.6096	Rough quarry split	Rough quarry split
	6	152.4	19	482.6	$1\frac{1}{3}$–$2\frac{1}{2}$	0.406–0.762		
Sloped	3–6	76.2–152.4	12	304.8	3	0.9144		
	$4\frac{1}{2}$	114.3	11, 12	279.4, 304.8	1	0.3048		
					2	0.6096	Split	Split
					3	0.9144		

Application

Accessories. There are many accessories for installing granite, including anchors, dowels, and ties made of various metals (*see* Figure *S117*). Steel should never be used unless it is heavily galvanized or coated with asphalt-type paint. When using these accessories, one must always check that the metal will not corrode and thereby stain the granite and also that, when they are attached to another metal, the two metals are compatible. Corrosion can be sufficient to break the anchorage if incompatible materials are used.

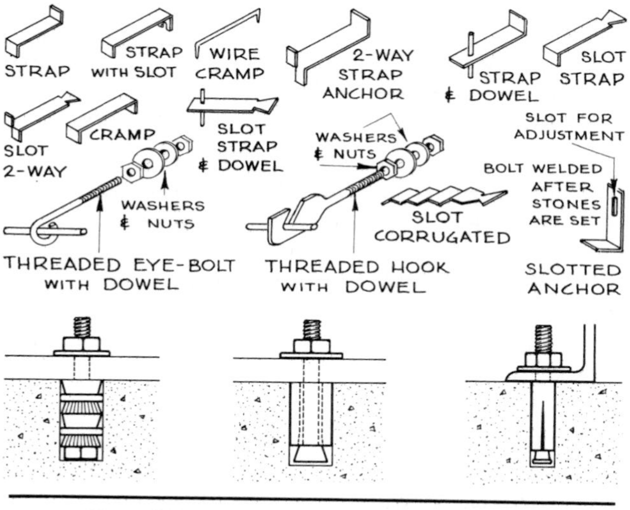

Figure S117 Accessories for granite stonework.

Condensed Checklist

1. The locality should always be checked for availability of the type of granite desired and of skilled labor to handle and set the stone.
2. Local, state, and municipal codes and also the codes of the Fire Underwriters, insurance companies, and federal government (Army, Navy, etc.) should be checked.
3. For cut stone, samples should always be obtained and shop drawings and setting drawings should be specified, including the requirement that all stones be numbered.
4. The granite must be examined for the type of finish, grade of stone, and variations in grade and color.
5. Back-up material, if concrete or masonry, should be coated with asphalt-type dampproofing before the granite is set.
6. All cement must be nonstaining cement or nonstaining waterproof cement; the sand must contain no impurities or salts that will stain the stone; the water must be fresh and clean.
7. Mortar and grout must be of the correct type, mixed exactly.
8. Installation should be such that no moisture can develop behind the granite. The space between the granite and the back-up material must be open except for spot attachments, and air must be able to circulate. Where solid grouting is necessary, nonstaining waterproof mortar should be used.

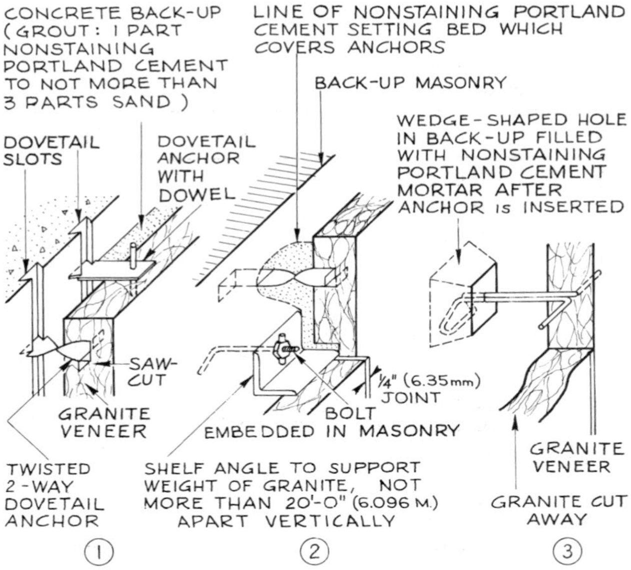

Figure S118 Details and methods of applying granite veneer, showing dovetail slots in concrete (1), shelf angle to support weight of granite (2), and wire anchors in vertical joint (3).

9. For exterior granite work, the type and material of anchors, ties, dowels, etc., and the type of flashing, both concealed and exposed, must be such that no staining can occur and no galvanic action can take place when these accessories are secured to the back-up support. For example, an aluminum dowel must not be used with a steel tie wire. (*See* Figures *S118* and *S119*.)
10. The number of anchors to support the pieces of granite must be correct in relation to the thickness of the granite (*see* Table *S83*).

Table S83 Number of anchors in relation to size of granite slabs

Size of granite slab[a]	Number of anchors[b]
Up to 2 ft^2 (0.158 m^2)	2
From 2 to 4 ft^2 (0.1858 to 0.3716 m^2)	3
From 4 to 12 ft^2 (0.3716 to 1.1148 m^2)	4
From 12 to 20 ft^2 (1.1148 to 1.8581 m^2)	6
Over 20 ft^2 (1.8581 m^2)	1 extra anchor for every additional 3 ft^2 (0.2787 m^2)

[a] The most practical large size is 8 ft (2.438 m) high and 12 ft (3.658 m) wide.
[b] For panel wall grid systems, check with the manufacturer regarding the number and types of anchor required.

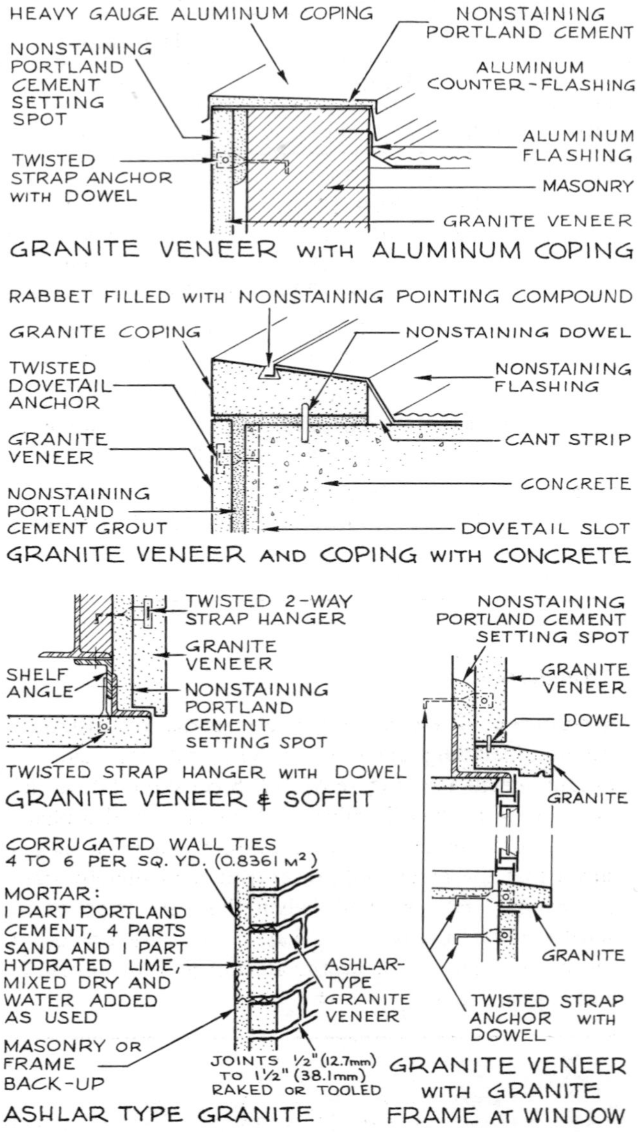

Figure S119 Details of granite veneer installations.

11. For cut stone the architect should always check the type of joint and especially the number and type of expansion (pressure-relieving) joints; usually one every 30 ft (9.144 m) is sufficient (*see* Figure *S120*).
12. When thin slabs of granite are used for curtain walls on buildings, one should always check the methods of fastening anchoring devices to prefabricated panels.
13. When preassembled or prefabricated panels are used, it is necessary to check that the high-strength epoxy adhesives, fastening devices, and sealants used in assembly meet the rigid requirements of the end use.

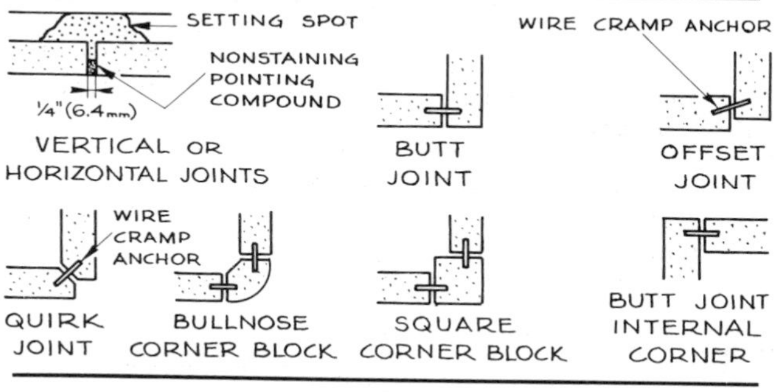

Figure S120 Typical joints for granite.

14. Oil, grease, smoke, or wash from concrete construction or scaffolding must not contact granite.
15. All unfinished granite work must be covered and protected when work stops; in case of impending heavy rains it must be covered and protected.
16. For cleaning granite only water and fiber brushes should be used, never acid and wire brushes.

Conditions Favorable to the Use of Granite

1. Where a highly permanent and durable exterior surface finish of a distinctive crystalline type and coloring is required.
2. Where a highly permanent ashlar stonework with a distinctive color is required.
3. Where colored paving, steps, and flagstones that are exceptionally durable and permanent are required.
4. Where a material that can withstand very rough usage and requires practically no maintenance is required.

Conditions Unfavorable to the Use of Granite

1. Where extremely high temperatures will be encountered, as in the interior and back hearth of a fireplace.

History and Manufacture

Granite was used as a building stone by the ancients, but because of its hardness was usually set with its hand-split texture exposed, although it was sometimes hand-finished and polished. With the development of steel and steam power, the cutting and finishing of granite become commercially possible and granite became a more common building material. With the development of electrical power and improved methods of sawing, finishing, and polishing the use of granite in construction steadily increased because of its durability and permanence. Today granite is readily available and is competitive costwise with all other types of stone as well as other construction materials.

Quarrying and Finishing. The quarrying and finishing of cut-stone or dimension granite is similar to the process used for other types of stone and is fully covered in the general discussion under Stone. Large quantities of granite are used for paving, curbing, ashlar-type building stone, and, to some extent, flagstones because of its durability and hardness. This type of granite is dressed with pneumatic tools to the required dimensions, and the surface is finished with surfacing machines which first chip off large irregularities by means of a heavy tool with blunt projections and then, with a lighter tool, work the surface down to an even plane with a tooled-surface finish. Ashlar-type granite is commonly left with a hand-split surface, its sides having previously been either saw-cut or tool-finished.

Carving and Lettering. Carving or lettering on granite, which was formerly done by hand or pneumatic tools, is now done by sand blasting. The surface of the granite is first covered with either a heavy adhesive tape or a heavy rubberlike compound (dope). The carving or lettering is transferred to this surface, and, just as in masking for airbrush work in architectural renderings, all parts that are to receive sand blasting are cut away. Uniformity in letters is achieved by using dies which, when heated and given a hammer blow, cut through to the granite. A stream of fine sand or powdered silicon carbide and air driven by compressed air at 80 to 100 lb/in.^{2} (0.0563 to 0.0703 kg/mm^2) quickly cuts away the exposed granite, but the sand or silicon carbide has no effect on the coating. Great precision is thus possible.

STONE: LIMESTONE

Physical and Chemical Properties

Limestone is sedimentary rock made up chiefly of calcareous shells of organisms that live in oceans and lakes. These shell deposite, supplemented by chemically precipitated calcium and magnesium carbonates and, in some localities, mixed or interbedded with sand, clay, iron oxide, and other detritus, eventually form the various kinds of limestone (*see* Table *S84*).

Classification by Carbonates. The chief constituent of all limestone is calcium carbonate ($CaCO_3$). If the magnesium carbonate content is below 10%, the stone is known as high-calcium limestone; if the content is above 10%, it is called dolomitic or magnesian limestone; and as the content approaches 45%, the stone is called dolomite, which is the double carbonate of calcium and magnesium ($CaCO_3 \cdot MgCO_3$).

Table S84 Composition of the common types of limestone

Type of limestone	CaO	MgO	CO_2	Fe_2O_3	Al_2O_3	SO_3	SiO_2 and insolubles
	(percent of content)						
Oolitic	54.54–54.80	0.59–0.72	42.90–43.30	0.08–0.32	0.28–0.68	0.05–0.07	
Dolomitic	27.40–27.76	15.40–18.90	38.72–42.10	1.10			10.40–13.18
Travertine	30.80	19.70	45.70	0.60			3.2

Table S85 Physical properties of the common types of limestone

Type of limestone	Compressive strength[a]		Modulus of rupture[a]		Specific gravity	Absorption (percent by weight)	Weight	
	lbf/in.^{2}	MN/m^2	lbf/in.^{2}	MN/m^2			lb/ft^3	Kg/m^3
Oolitic	2,600–9,700	17.93–66.88	640–1610	4.17–11.10	1.87–2.42	3.6–12.3	117–150	1874.34–2403.00
Dolomitic	13,500–21,320	93.08–147.00	1565–1995	10.79–13.76	2.45–2.85	3.4–9.0	155–170	2483.10–2723.40
Travertine	17,000	117.22	955	6.59	2.53	2.8	158	2531.16

[a]1 lbf/in.^{2} = 0.006895 MN/m^2.

Classification by Impurities. The prevailing impurities in limestones affect the color and are another basis of differentiation of the kinds of limestone. The iron oxides of ferruginous limestones produce reddish and yellowish colors; in carbonaceous or bituminous limestones, organic materials such as peat give a dark gray to black color. Argillaceous limestones contain clay, and cherty limestones contain silica.

Classification by Physical Characteristics. In some instances the names of limestones depend on the physical character of the stone. The range of limestone colors varies from light creamy buff to brownish buff and dark brown, from silvery gray to a slightly bluish cast, and variegated mixtures of these colors. In some limestones the shell structure remains intact and is predominantly of one kind, for example, coral, crinoid and coquina limestones, and chalk. Common compact limestone is a fine-grained, dense, homogeneous aggregate, light gray to dark brown in color. Oolitic limestone consists of small grains of calcium carbonate with a concentrically laminated structure. Travertine and Mexican onyx are made up of calcium carbonate deposited from solution. In travertine the deposits were made in successive layers that result in a banded structure characterized by many irregular cavities. Mexican onyx is stone deposited from cold-water solutions in limestone caves. (*See* Table *S85*.)

Limestone thermal resistance factors are shown in Table *S86*. Limestone is considered a fireproof material, as it calcines above 1500°F (815.56°C). Generally, limestone 4 in. (101.6 mm) thick has a fire endurance time of 1 hour and 12 minutes.

Table S86 Thermal resistance factors of limestone

Thickness		Resistance factor
in.	mm	
2	50.8	0.31
3	76.2	0.45
4	101.6	0.62
5	127.0	0.77
6	152.4	0.93
8	203.2	1.24
10	254.0	1.54
12	304.8	1.85

Commercial Forms. Limestones vary in texture porosity, hardness, strength, and color. To be suitable for use as dimension stone, a limestone must be compact, easily workable, uniform in texture, and attractive in color. Purity is not essential unless there are large amounts of silica, which may make the stone difficult to work, or of iron sulfides, which may cause iron stains on weathered surfaces.

Crushed or broken limestone is also available as aggregate for concrete and asphaltic concrete for construction purposes and as an ingredient for the chemical, metallurgical, and processing industries (*see* "Types and Uses").

Types and Uses

The uses of limestone fall into two major and completely separate categories:

1. Dimension stone, or building limestone—this includes oolitic, dolomitic and crinoid limestones, traver-

tine, coral, coquina, and Mexican onyx, which can be polished and is listed as onyx marble.

2. Crushed or broken limestone—this includes oolitic, dolomitic, siliceous, argillaceous, ferruginous, carbonaceous, oyster shell, and common compact limestones.

Building Limestone. Building limestone is available in gray and buff colors and in combinations of these two colors. The surface texture or fineness of grain is graded as statuary (A), select (B), standard (C), rustic (D), variegated (E), and Old Gothic (F). Grades A, B, C, and D come in buff or gray and vary only in fineness of grain from fine to coarse. Grade E is a mixture of buff and gray and is of unselected grain size. Grade F is a mixture of D and E and includes stone with seams and markings. The various finishes for limestone are shown in Table *S87*. Grade A has very limited use in construction.

Building limestone is divided into two general categories, cut stone and ashlar.

Table S87 Types of finish for limestone

Type of finish	How made	Appearance	Major use
Broached	Produced by planing; the planer cuts out smooth valleys with a rough surface between valleys	Rough-ribbed texture with smooth valleys	Cut stone, panels, and preassembled units
Carborundum	Cut by carborundum-type planer	Very smooth finish	Cut stone, molded stone surfaces, and preassembled units
Chat-sawed	By using a coarse abrasive during gang sawing	A coarse, pebbled surface	Cut stone, panels, preassembled units, and ashlar
Hand-tooled	By hand with various types of tools	Various types of finishes having handmade appearance	Confined to small, important areas because of cost
Honed	Machine-rubbed with fine sand and water	Superfine, smooth finish	Cut stone, molded stone surfaces, and preassembled units
Machine-tooled	Cut by machine in only one direction, by hand in the other direction	2, 4, 6, and 8 parallel concave or convex grooves to the inch (25.4 mm)	Types of ashlar only
Plucked	By rough-planing the surface and breaking or plucking out small particles	Rough texture	Cut stone, molded stone surfaces, and preassembled units
Rock face	Sawed top and bottom; exposed face, a natural split dressed by machine	Rough, irregular texture	Types of ashlar only
Rusticated	Rustication done by machine, but surfaces that cannot be made by machine are done by hand	Extra-deep joint effect	Types of ashlar, cut stone, and preassembled units
Sand-sawed	Gang saw only; no other finishing work	Granular surface containing saw marks, moderately smooth	Types of ashlar only
Sculptured	Done by hand by special craftsman	Reproduction of sculptor's original model	Important areas where design or building requires sculpture
Shot-sawed	Gang saw, using steel shot for abrasive	Medium-rough pebbled surface to a surface with irregular, rough, parallel lines; brown tones obtained from rust stains from ground steel shot	Types of ashlar only

Table S87 Types of finish for limestone (*continued*)

Type of finish	How made	Appearance	Major Use
Smooth machine	By planers	Relatively smooth with a certain amount of texture	Cut stone, molded stone, and preassembled units
Split face	Sawed top and bottom; exposed face a natural split when stone is broken	Rough, irregular texture	Types of ashlar only
Textured, light	Cut by machine	Light textures, varying from slight texture to various vertical ribbing, minimum thickness $3\frac{1}{2}$ in. (63.5 mm)	Cut stone, panels, and preassembled units
Textured, medium	Cut by machine in one direction	Various types and combinations of vertical ribbing, minimum thickness $3\frac{1}{2}$ in. (63.5 mm)	Cut stone, panels, and preassembled units
Textured, rough	Cut by machine in one direction.	Various types and combinaations of vertical ribbing, minimum thickness 4 in. (101.6 mm)	Cut stone, panels, and preassembled units
Wet-rubbed	Machine-rubbed with fine sand and water	Smooth finish	Cut stone, molded stone, and preassembled units

Table S88 Recommended panel sizes and thicknesses for efficient fabrication and handling

Thickness		Size			
		Width		Height	
in.	mm	ft	m	ft	m
6	152.4	5	1.524	14	4.267
5	127.0	5	1.524	13	3.962
4	101.6	5	1.524	10	3.048
3	76.2	4	1.219	8	2.438
2	50.8	3	0.914	5	1.524

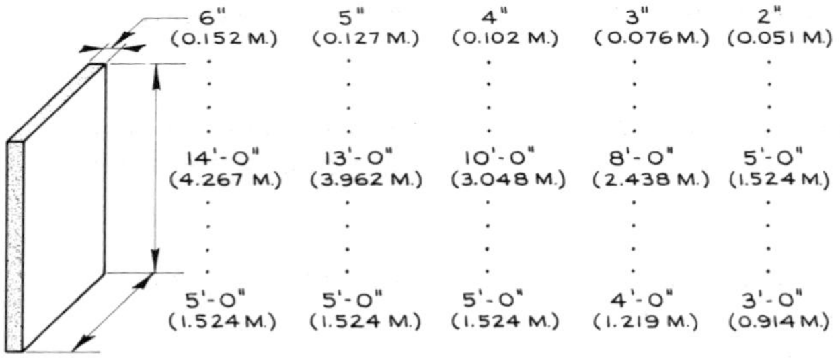

Figure S121 Typical limestone panels.

Cut stone includes veneer and stock shapes such as molded sills, copings, lintels, window and door trim, and columns, which are used with cut stone veneer or ashlar-type stonework. Veneer consists of large, thin slabs of limestone applied as exterior or interior surface finish to a building, set with the slab backs $1\frac{1}{2}$ in. (37.1 mm) from the back-up material (*see* Table *S88* and Figure *S121*). Veneer slabs also are formed into panels for curtain wall construction or used in the manufacture of preassembled and prefabricated panels for both interior and exterior surfaces of buildings. Cut stone joints are usually $\frac{1}{4}$ in. (6.35 mm) for both vertical and horizontal joints.

When limestone is used in construction with its grain running horizontally, it is on its "natural bed"; when it is laid with its grain running vertically, it is on its "edge." Copings, treads and risers, cornices, etc., are generally set on their edges.

Ashlar stone consists of smaller, thicker pieces of limestone which are laid in various ashlar patterns and used as veneer on the exterior or interior of buildings. Ashlar is available in two types: (1) stone sawed on four sides to any desired specific size, with the exposed face prepared in various finishes and set with $\frac{1}{4}$ in. (6.35 mm) joints for the larger sizes and with $\frac{1}{2}$ in. (12.7 mm) joints for smaller sizes (these joint sizes are applicable to strips of various lengths up to 5 ft); (2) split-face ashlar, sawed top and bottom, with the split face in $2\frac{1}{4}$, 5, and $7\frac{3}{4}$ in. (57.15, 127.0, and 196.85 mm) heights and in random lengths, set with $\frac{1}{2}$ in. (12.7-mm) joints.

Crushed and Broken Limestone. Crushed and broken limestone has many direct uses in construction, both as aggregate and chemical raw ingredient.

Aggregate uses include the following: (1) aggregate for concrete and reinforced concrete, processed to meet specifications; (2) aggregate for concrete and asphalt paving, also prepared and tested to meet requirements; (3) untested, crushed limestone, graded and sized, used as fill under concrete slabs and asphalt paving; (4) aggregate, crushed to the size of sand grains, washed and graded, used as sand in mortar, plaster, and concrete; (5) screened chips, used as surface finish for built-up roofs; (6) certain colored limestones granules, used in terrazzo work; and (7) outdoor surfacing material for playgrounds, walks, parking areas, and driveways.

Limestone aggregate for the first use is crushed, graded, sized, and then tested by the Los Angeles abrasion test. This method uses the sodium sulfate and magnesium sulfate tests. Each involves freezing or crystallization of the given substance in the pores of the aggregate to create heavy interior strain. The resistance of the stone to disruption is a measure of its soundness.

Crushed and broken limestone is used in the manufacture of concrete block. Other uses include surface aggregate for prefabricated panels, textures for pavements and sidewalks, and textures for the various types of surfacing techniques.

Limestone dust, fine enough so that 80% will pass a 200-mesh screen, is used as a filler in asphalt-type surface finishes and as a substitute for chalk whiting in cold water paints and calcimine. It is also used as a filler in rubber, paint, paper, linoleum, plastics, and certain textiles.

As a raw chemical ingredient, limestone finds its greatest consumption in the manufacture of portland cement, the preferred types being the high-calcium and argillaceous limestones. The next greatest use is for the manufacture of lime. The type used for lime must have a calcium carbonate content ranging from 97 to 99%.

Dolomite limestone with a high magnesium carbonate content is used to manufacture magnesia-type high-temperature insulation, as a source for the production of magnesium metal, and for certain limes used in prepared hard white finish plasters.

Limestone also serves as a raw ingredient in the manufacture of mineral wool, glass, and paper. Related metallurgical and processing applications include the use of limestone as a flux in the manufacture of pig iron, steel, and other metals, and as a refractory lining for smelting furnaces.

Heavy construction uses include aggregate for highway construction and for railroad ballast; large, irregular blocks of limestone, called riprap, are utilized for retaining walls and for river and harbor installations.

Application

Many accessories such as anchors (*see* Figure *S122*), dowels, and ties, made of various metals, are available for installing limestone. Any steel used should be galvanized after bending. Any metal item that is attached to or in contact with another metal should be compatible so that there is no possibility of galvanic action which will stain the limestone or may even corrode the anchorage sufficiently to break it. The combination of high-strength epoxy adhesives with expansion bolts, metal angles, etc., has made it possible to design and construct many intricate and complex preassembled and prefabricated units for exterior and interior components of buildings (*see* Figure *S123* and Table *S89*).

Limestone is one of the most widely used dimension or building stones. As with all cut stone, certain precautions (discussed subsequently) must be taken to ensure successful installation.

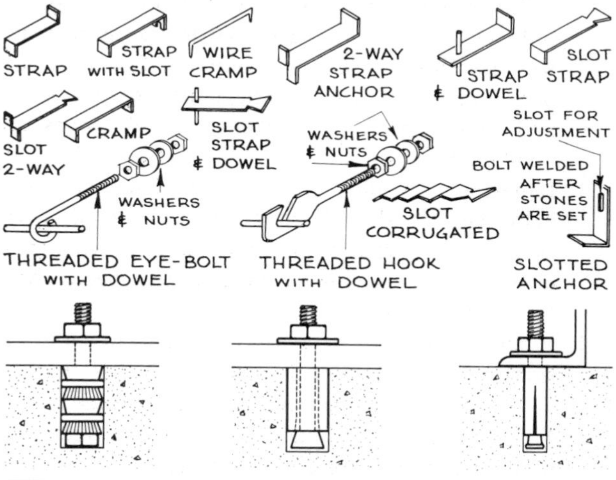

Figure S122 Typical anchors for limestone.

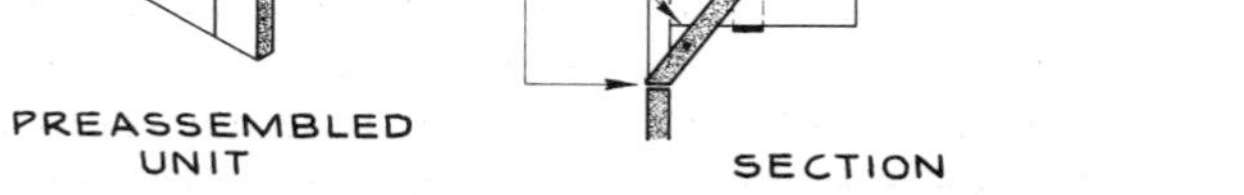

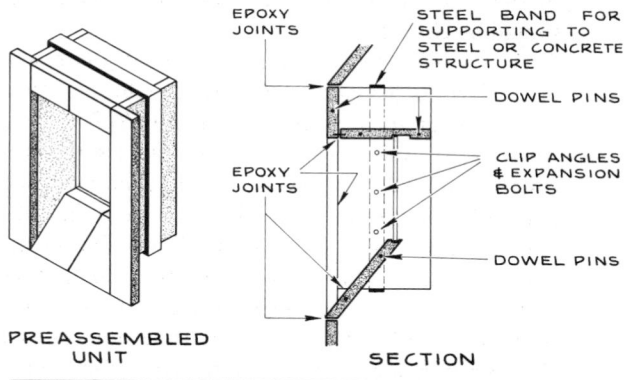

Figure S123 Preassembled and prefabricated limestone building components.

Table S89 Properties of high-strength adhesive for preformed limestone units

Type of adhesive[b]	Tensile strength[a]		Compressive strength[a]		Compressive double shear[a]		Water absorption in 24 hours (percent)
	$lbf/in.^2$	MN/m^2	$lbf/in.^2$	MN/m^2	$lbf/in.^2$	MN/m^2	
Two-component epoxy[b] consisting of epoxy resin, hardener, inert mineral filler[c], thixotropic agent	3500	24.13	6000	41.37	400	2.76	0.50

[a] 1 $lbf/in.^2$ = 0.006895 MN/m^2.
[b] Nonreactive color pigments are used to match exposed joints to limestone.
[c] Filler content shall not exceed 50% of total composition by weight.

Condensed Checklist

1. The locality should always be checked for availability of the type of limestone desired and of the skilled labor for handling and setting the stonework.
2. Samples should always be required for cut stone. Shop drawings and setting diagrams should be specified, and special care taken that all stones are numbered.
3. Limestone should be carefully handled when delivered. It should be stored on wood strips covered with waterproof paper, using a wood that does not contain tannic acid, and then covered with tarpaulins or waterproof paper.
4. If any stone arrives damaged, immediate notification should be sent to the manufacturer so that the damaged stone can be replaced without delay.
5. The stone should be examined for the correct type of finish, grade, and color, especially for variation in grade and color.
6. Local, state, and municipal codes and also the codes of the Fire Underwriters, and insurance companies should be checked.
7. For cut stone, the types and number of joints, and particularly of expansion or pressure-relieving joints should be carefully checked (*see* Figure *S124*).
8. Anchors, dowels, ties, and other accessories for installation and the flashing, both concealed and exposed, should be checked to be sure that the material used and the details of installation are correct. These items should be of a compatible material that will not react with the stone or the back-up material. For example, aluminum dowels should not be used with a steel tie wire. (*See* Figures *S125* and *S126*.)
9. All cement should be nonstaining waterproof cement, and the sand should not contain impurities or salts that will react with the stone. The water should be fresh and clean.
10. When high-strength epoxy adhesives are used, one should always check with the adhesive manufacturer and the stone supplier and manufacturer regarding the methods which should be used for this particular end use (*see* Figures *S127–129*).
11. Shop drawings and details, including types of adhesives to be used, must be submitted for approval.
12. Testing of a full-size section of preassembled units under the conditions that will be encountered when installed should be included in the specifications.
13. If back-up material is concrete or masonry, it

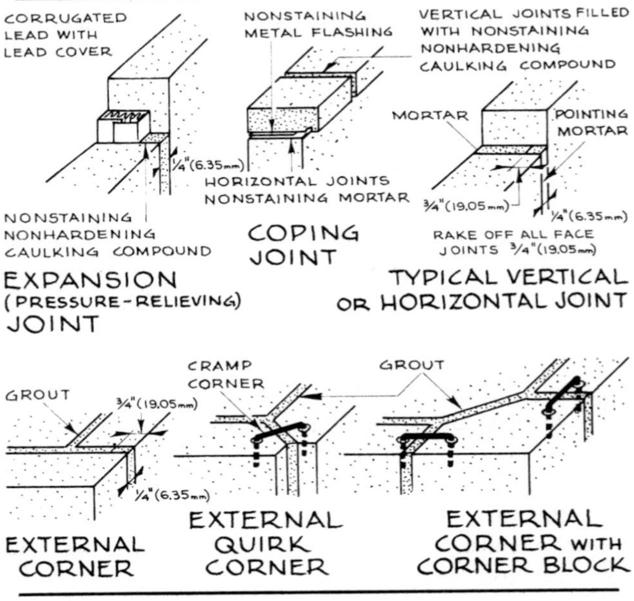

Figure S124 Details of typical vertical, horizontal, and corner joints for limestone.

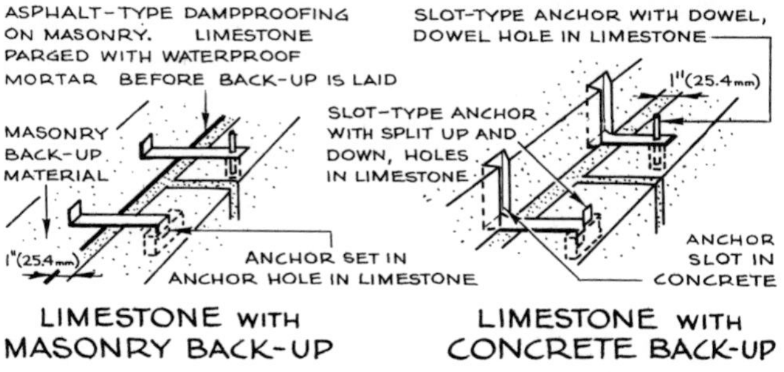

Figure S125 Details of dampproofing between limestone and back-up material.

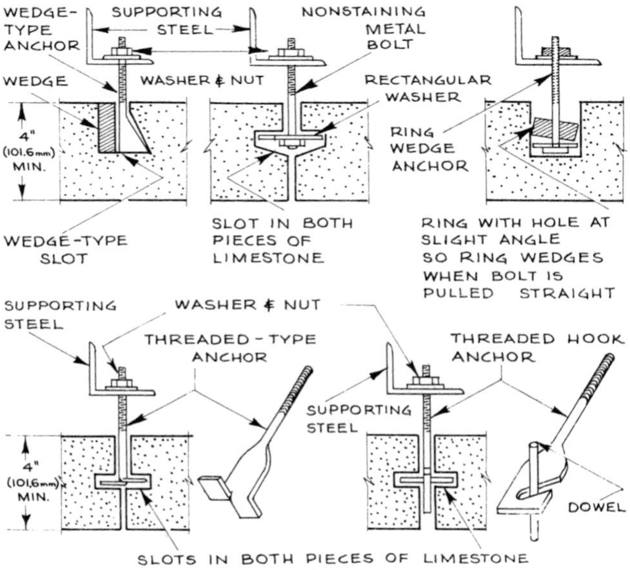

Figure S126 Details for hung ceilings and for soffits, including accessories.

should be coated with asphalt-type dampproofing before the limestone is set (*see* Figure *S130*).

14. Mortar and grout must be of the correct type of mix, prepared exactly (*see* Table *S90*).
15. Oil, grease, smoke, and wash from concrete construction should not come in contact with the limestone, nor should the scaffolding touch the stone.
16. All unfinished limestone work should be covered and protected when work stops; in case of impending heavy rains, it should also be covered and protected.
17. Finished limestone work should be protected with waterproof paper and wood where it is exposed to construction traffic.
18. Only soap powder, water, and fiber brushes should be used for cleaning limestone, never acid or wire brushes.

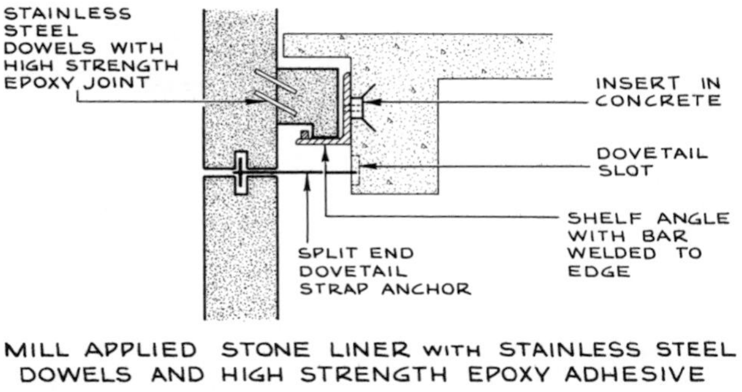

Figure S127 Mill-applied stone liner with stainless steel dowels and high-strength epoxy.

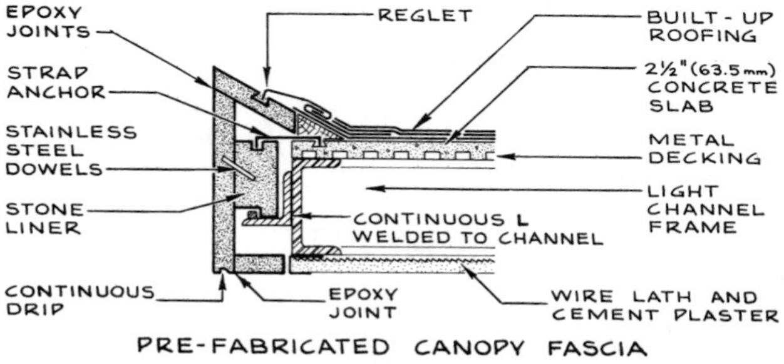

Figure S128 Prefabricated limestone canopy fascia.

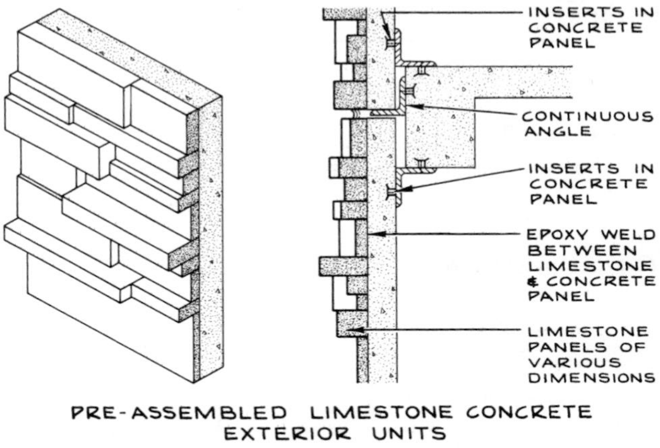

Figure S129 Preassembled limestone-faced concrete exterior units.

Conditions Favorable to the Use of Limestone

1. Where the design requires a permanent material with a durable red, gray, or buff (cut limestone) color and a permanent surface finish.
2. Where the design requires an ashlar stonework pattern which also has great flexibility in surface finish and texture.

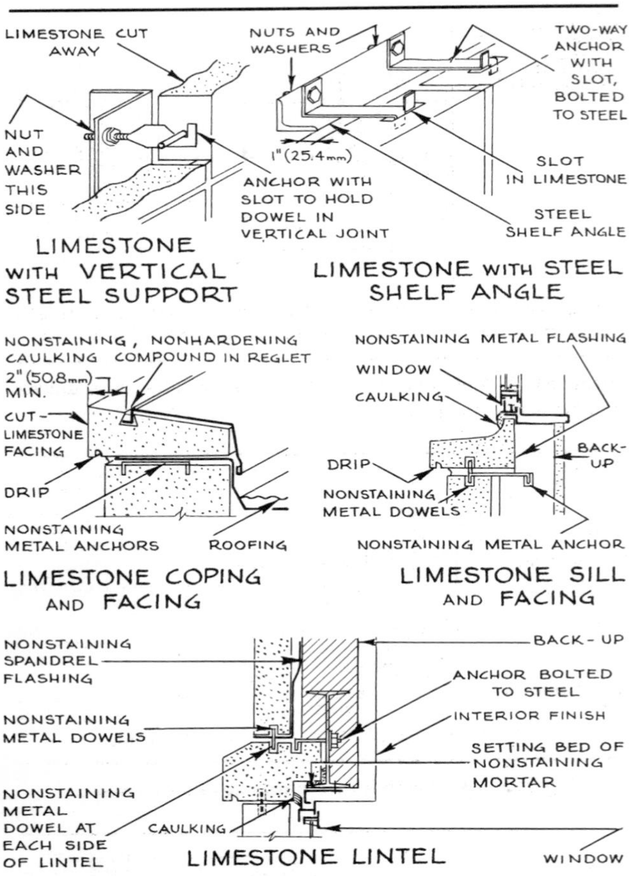

Figure S130 Typical details of anchorage and flashing.

3. Where a buff or gray, moderately smooth, durable floor surface in either a regular or irregular pattern is desired.

Conditions Unfavorable to the Use of Limestone

1. In areas where industrial fumes, smoke, and acids are prevalent.
2. In areas that receive rough usage, as limestone soils readily from oil or grease and can be easily damaged by hard bumps.
3. In areas where extremely high temperatures will be encountered, as in the interior and back hearth of a fireplace.

History and Manufacture

Because it is so widely distributed, easily worked, and durable, limestone may have been one of the first stones used by man, particularly in dry climates. The Egyptians, Etruscans, early Greeks, people of the Near and Far East, and the Romans all used limestone in their buildings. After the fall of Rome, throughout the Dark Ages, limestone continued to be used as an architectural material. In the Americas, the Incas, Mayas, and Aztecs also built with limestone. Today limestone is steadily growing in importance as an ingredient in the chemical industry, as well as for a building stone and raw material for other materials for architecture.

The quarrying and finishing of limestone is described in general under Stone. The limestone finishing mill also includes planers, which plane slabs of limestone smooth and to exact thickness. These machines can cut curved as well as straight shapes and can also be used to give various surface finishes. Limestone can be turned on a lathe. The finished surface of limestone can be given different textures by means of special tools.

Table S90 Types of mortar for limestone

Type of mortar	Nonstaining cement	Nonstaining waterproof cement[a]	Sand	Hydrated lime or quicklime putty
Setting	1		6	1
		1	6	1
Pointing	1		2	Sufficient quantity to make as stiff a mortar as can be worked
Grout	1		$1\frac{1}{2}$	
		1	$1\frac{1}{2}$	

[a]Waterproofers, if not ground with cement, can be added to nonstaining cement (no more than 2% by weight of cement and lime combined). Waterproofers used are ammonium and calcium soaps (stearates).

STONE: MARBLE

Physical and Chemical Properties

Probably all marbles are metamorphosed limestones. The word "marble" is derived from the Latin *marmor*, meaning shining stone, and is applied to crystalline and compact varieties of calcium carbonate ($CaCO_3$), sometimes combined with magnesium carbonate ($MgCO_3$), which are decorative when polished. By general definition, then, marble is any limestone, granular or compact in texture, that is capable of taking a polish or being used for fine construction work. Many types of stone are therefore classed as marbles, for example, Mexican onyx, alabaster (not a stalagmitic carbonate of lime, which is

Table S91 Typical compositions of United States marbles

Source	CaO	CO_2	MgO (percent of content)	Al_2O_3	Fe_2O_3
Vermont	55.86–55.90	43.78–43.80	0.27–0.34	0.06	0.02
Alabama	55.38–55.54	42.90–43.35	0.38–0.51	0.13	0.03–0.07
Georgia	53.23–54.48	42.17–43.22	0.90–0.92	0.35–0.94	0.04–0.42
Tennessee	55.38–55.80	42.65–43.52	0.06	0.14–0.45	0.06–0.16
North Carolina	53.70	53.54	0.80	0.06	0.005
Missouri	55.19	43.25	0.40	0.30	0.03

Table S92 Typical physical properties (maximums and minimums) of United States marbles

Compressive strength[a]		Shear strength[a]		Modulus of rupture[a]		Specific gravity	Weight	
$lbf/in.^2$	MN/m^2	$lbf/in.^2$	MN/m^2	$lbf/in.^2$	MN/m^2		lb/ft^3	kg/m^3
9058–17,787	62.46–122.64	11,000	75.85	1197–3442	8.25–23.73	2.64–23.73	165–170	2651.55–2731.90

[a] $1\ lbf/in.^2 = 0.006895\ MN/m^2$.

the alabaster of ancient times, but a stalagmitic sulfate of lime), certain gypsums, and granites. For the chemical composition and physical properties of marble, see Tables *S91* and *S92*.

Color. Marble varies in color from white to black and is found in innumerable variations of veining and color combinations. The crystalline structure of polished marble gives it a beautiful luster because light penetrates a short distance and then is reflected by the deeper-lying crystals.

Durability. Marble in a dry atmosphere or protected from rain is very durable but when exposed to severe weather or industrial fumes its surfaces will disintegrate and crumble.

Commercial Forms. Marble is available as dimension stone (rough or finished), including flooring, as crushed and broken marble graded to various sizes, and as marble dust.

Types and Uses

Commercial Classifications. Marbles are classified as A, B, C, and D on the basis of their natural qualities, group D being those with the maximum variations and defects but including the most beautifully colored examples.

Group A includes sound marbles with uniform and favorable working qualities.

Group B consists of marbles similar to group A but with less favorable working properties. They have occasional natural faults, and a limited amount of repair is needed. Terms used by manufacturers for such repair are "sticking" and "waxing."

Group C are marbles with uncertain variations in working qualities. They contain flaws, voids, veins, and lines of separation and always require repairing. Manufacturers' terms for repair are "sticking," "waxing," "filling," and "reinforcing."

Group D are marbles similar to group C but with a larger proportion of natural variations and with maximum variations in working qualities. The same methods of repair as used for group C are necessary.

Imported and Native Stone. Over a hundred different types of imported marbles are available, varying from monotone to highly decorative types (*see* Table *S93*). The quarries in the United States produce about 120 different marbles, which range from white to black and include almost every shade in the color spectrum (*see* Table *S94*).

Building Marble. Marble is used in construction for both the exterior and interior of buildings. Exterior marble is usually group A, but if marbles of groups B, C, and D are used, they should be treated for soundness and require spraying at least once a year with a cellulose acetate, plastic lacquer, or a clear plastic surface treatment. Interior marble can be of any group. The finishes of marble can range from rough to polished as shown in Table *S95*.

Marble for use in the exterior and interior of buildings comes in various sizes and thicknesses and requires various kinds of setting beds and types of joints. Tables *S96* and *S97* summarize these data, including sizes, joints, and setting beds in relation to where the marble is used. (*See also* Figures *S131* to *S*134.)

Table S93 Colors of common imported marbles and the countries of origin

Country of origin	Black	Black and white	Brown	Monotone: Light buff	Monotone: Dark buff	Monotone: Light gray	Green	Red	Rose	White	Yellow or gold	Veined and/or brecciated ground color: Bluish or gray	Veined and/or brecciated ground color: Cream or white	Veined and/or brecciated ground color: Tan or yellowish
Algeria								X						
Belgium	X							X				X		
England												X		
France	X	X	X	X	X	X	X	X			X	X	X	X
Greece							X							
Germany								X				X		
Ireland	X													
Israel				X										
Italy	X		X	X	X	X	X	X	X	X	X	X	X	X
Mexico													X	
Morocco	X											X	X	
Norway									X					
Peru											X			
Portugal							X		X				X	X
Spain	X			X		X		X	X		X	X	X	X
Sweden							X							
Switzerland														X
Trieste						X								X
Yugoslavia				X					X					

Table S94 Colors of marbles and the states that produce them[a]

Color of marble	States in United States producing marble: Alabama	California	Georgia	Minnesota	Missouri	New Jersey	Tennessee	Vermont
Black							X	X
Black and white			X					
Buff, brown, yellowish				X	X		X	X
Gray			X	X			X	X
Grayish brown					X			
Grayish pink							X	
Green						X		X
Light green								X
Pink			X	X	X		X	X
Red and reddish brown			X		X		X	X
Rose			X		X		X	
White, bluish			X					X
White, creamy	X	X						X
White, greenish								X

[a]Always check with marble suppliers for availability.

Monumental Marble. Marble finds great use as a material for all types of sculpture and memorial monuments.

Crushed and Broken Marble. Crushed or broken marble is graded and used as coloring aggregate in terrazzo and artificial stone and as the finish surface of built-up roofs. It is also utilized as surface aggregate for prefabricated panels, textures for pavements and sidewalks, and textures for the various types of surfacing techniques. Marble in this form is also used as aggregate for concrete and reinforced concrete work and for concrete and asphalt paving. Since marble is relatively pure calcium carbonate, it is interchangeable in certain uses with high-calcium limestone. Marble dust is used in the same manner as limestone dust.

Table S95 Common exterior and interior finishes for marbles

Type of finish	Method of manufacturing	Texture	Major use	Other uses
Grit	Rubbing with grit as abrasive	Smooth dull finish	Exterior facings for buildings	Exterior copings, sills, and molded forms; interior stair treads, saddles, and floors
Honed	Rubbing by machine or hand with special abrasives	Velvety, dull gloss on surface	Exterior facings for buildings and interior facings for walls	Exterior copings, sills, and molded forms; interior stair treads and floors
Natural	Any sawing, using sand and water as abrasive	Textured by sawing; moderately rough	Exterior facings for buildings	Exterior coping, sills, and molded forms; some interior uses
Polished	Polishing a honed surface with textile buffer with special fine-powdered polishing abrasives	Mirrorlike glossy surface which brings out color and veining of marble	Exterior facings for buildings, interior facings for walls	Exterior copings, sills, and molded forms; interior toilet partitions
Sand	Rubbing with sand alone as abrasive	Smooth dull surface	Exterior facings for buildings	Exterior copings, sills, and molded forms
Sand, blown	Sand blasting of face of marble to be exposed	Smooth mat surface	Exterior facings for buildings	Exterior copings, sills, and molded forms
Sand, wet	Rubbing with a revolving cast-iron surface with sand and water as abrasive	Smooth surface	Interior stair treads, saddles, and plat platforms	
Split face	Sawing top and bottom and machine-splitting exposed face	Rough, natural split of marble	Exterior ashlar stonework of small narrow stones	Interior ashlar stonework of small narrow stones

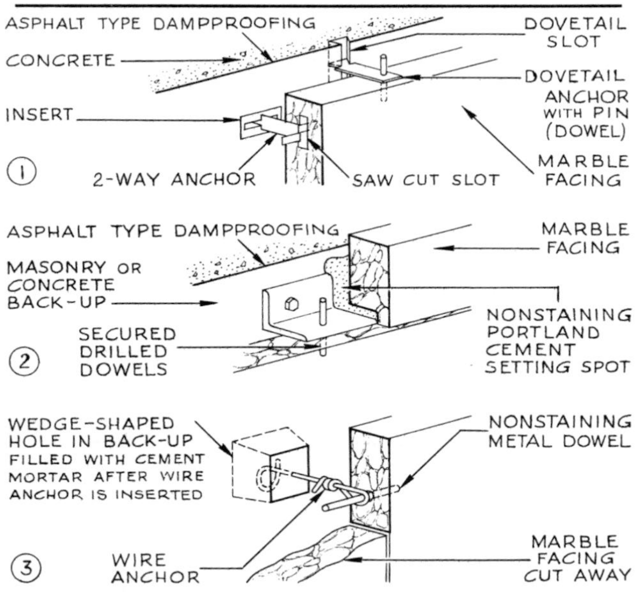

Figure S131 Details of methods of applying marble facings.

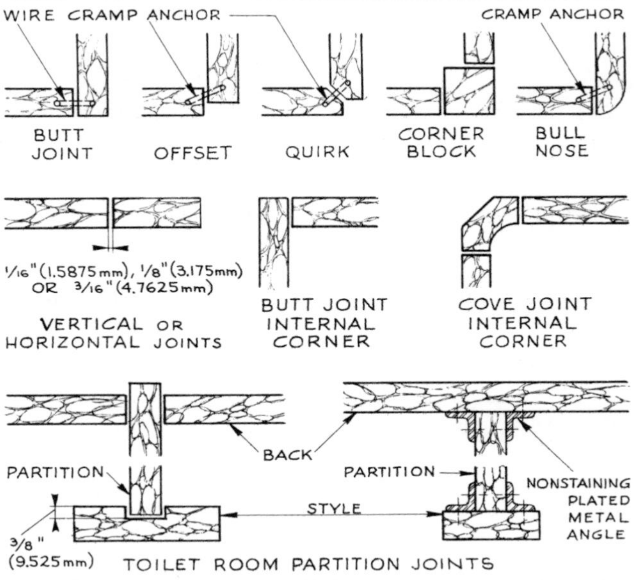

Figure S132 Typical joints for marble.

Application

There are many accessories for installing marble, such as anchors, dowels, and ties made of various metal, as well as methods for preassembling complex prefabricated units (*see* Figures *S135* to *S137*). These accessories should always be of a metal that will not corrode and thereby stain the marble when they are attached to another metal or any part of the back-up installation. Should galvanic action occur between two metals, one

Table S96 Exterior uses of marble, including sizes, joints, and setting beds

Type of marble	Thickness in.	Thickness mm	Length or size in. or ft^2	Length or size mm or m^2	Width in.	Width mm	Joint in.	Joint mm	Setting bed[a] in.	Setting bed[a] mm
Ashlar	$3\frac{1}{2}$	88.9	1-5 in.	0.3048–1.524 mm	$\frac{7}{8}$	5.56	$\frac{1}{2}$	12.7	1	25.4
					$1\frac{1}{4}$	31.75				
					$2\frac{1}{4}$	57.15				
					$3\frac{1}{2}$	89.90				
					5	127.00				
					$7\frac{3}{4}$	196.85				
					$10\frac{1}{2}$	226.70				
Copings, sills	No edge less than $\frac{7}{8}$	5.56	Of size required				Vertical joints left void $\frac{1}{16}$	1.588	$\frac{1}{2}$	12.4
Finish surfaces	$\frac{7}{8}$	5.56	2–20 ft^2	0.185–1.858 m^2			$\frac{1}{16}$	1.588	1	25.4
	$1\frac{1}{4}$	31.75					$\frac{1}{8}$	3.175		
	$1\frac{1}{2}$	38.0					$\frac{3}{16}$	4.763		
	2	50.8								
Flagstones	$1\frac{1}{4}$	31.75	Irregular or square 2–6 ft^2	0.185–0.557 m^2			$\frac{1}{16}$	1.580 if squared	$2\frac{1}{2}$	63.5 with fill and sand bed
	$1\frac{1}{2}$	38.1					$\frac{1}{2}$	12.70 if irregular	$\frac{1}{2}$	12.7 without fill
	2	50.8								
Molded forms	No edge less than $\frac{7}{8}$	5.56	Of size required				Same as for adjoining marble		Same as for adjoining marble	
Panels	$1\frac{1}{4}$	31.75	Of size required						Setting frame must allow for caulking, setting, and expansion	
Preassembled forms and shapes	$\frac{7}{8}$	5.56	Of size and shape required				Same as for adjoining marble		Same as for adjoining marble	

[a]Back of rough marble to face of back-up material.

Table S97 Interior uses of marble, including sizes, joints, and setting bed

Interior uses of marble	Thickness in.	Thickness mm	Length or size in. or ft^2	Length or size mm or m^2	Width in.	Width mm	Joint in.	Joint mm	Setting bed[a] in.	Setting bed[a] mm
Ceilings	Generally $\frac{7}{8}$	22.23	Generally 2–20 ft^2	0.1858–1.858 m^2			$\frac{1}{16}$	1.59	3	76.2
	For larger sizes $1\frac{1}{4}$	31.75								
	$1\frac{1}{2}$	38.10								
Tops of counters, tables, etc.	$\frac{7}{8}$	22.23	Of sizes required				$\frac{1}{16}$	1.59	$2\frac{1}{2}$	63.5 minimum with fill and sand bed
	$1\frac{1}{4}$	31.75							$\frac{1}{2}$	12.7 without fill
	$1\frac{1}{2}$	38.10								
	2	50.80 and thicker								
Floor tiles	$\frac{7}{8}$	22.23	16 in.	406.4 mm	8	203.2	Same as adjoining marble		Same as adjoining marble	
			12 in.	304.8 mm	12	304.8				
			20 in.	508.0 mm	10	254.0				

Table S97 Interior uses of marble, including sizes, joints, and setting bed (*continued*)

Interior uses of marble	Thickness		Length or size		Width		Joint		Setting bed[a]	
	in.	mm	in. or ft²	mm or m²	in.	mm	in.	mm	in.	mm
Molded and pre-assembled forms	No edge less than		Of size required				$\frac{1}{16}$	1.59	$\frac{1}{2}$	12.7
									minimum with fill and sand bed	
	$\frac{7}{8}$	22.23							$1\frac{1}{2}$	38.1
									without fill	
Risers and stringers for stairs	$\frac{7}{8}$	22.23	Of size required				$\frac{1}{16}$	1.59	$\frac{1}{2}$	12.7
									for risers	
									$1\frac{1}{2}$	38.1
									for stringers	
Saddles	$\frac{7}{8}$	22.23	Of size required				$\frac{1}{16}$	1.59	$2\frac{1}{2}$	63.5
	$1\frac{1}{4}$	31.75					if necessary		minimum with fill and sand bed	
	$1\frac{1}{2}$	38.10							$\frac{1}{2}$	12.7
									without fill	
Sills	$\frac{7}{8}$	22.23	Of size required				$\frac{1}{16}$	1.59	$\frac{1}{2}$	12.7
	$1\frac{1}{4}$	31.75					if necessary			
Stair treads and platforms	$\frac{7}{8}$	22.23	Of size required				$\frac{1}{16}$	1.59	$2\frac{1}{2}$	63.5
	$1\frac{1}{4}$	31.75					if necessary		minimum for masonry stairs	
	2	50.80					$\frac{1}{2}$	12.7	$\frac{1}{2}$	12.7
							for steel pan type			
Toilet room partitions	$\frac{7}{8}$	22.23	Of size required				If rabbeted		End partitions and stiles for toilet partitions to floor set into floor	
	$1\frac{1}{4}$	31.75					$\frac{3}{8}$	9.53		
							If into stiles			
							$\frac{1}{16}$	1.59	1	25.4
							$\frac{1}{8}$	3.18		
Wall finishes	$\frac{7}{8}$	22.23	Generally				$\frac{1}{16}$	1.59	Normally	
			2–20 ft²	0.1858–1.858 m²					$1\frac{1}{2}$	38.1
									If reinforcing is necessary	
									$1\frac{1}{2}$	63.5

[a]Back of marble to face of back-up material.

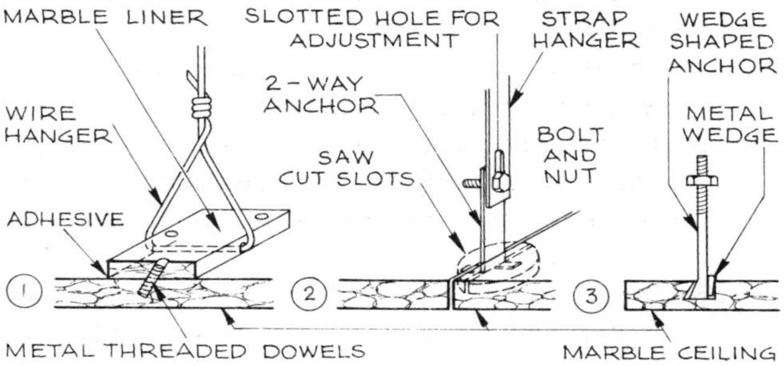

Figure S133 Three methods of hanging marble ceilings or soffits.

of the metals may corrode sufficiently to break the anchorage. Steel should never be used unless it is heavily galvanized or coated with asphalt-type paint.

The precautions given for the installation of limestone and granite usually apply equally to marble stonework. Figures *S131* to *S134* show exact installation details.

The combination of high-strength epoxy adhesives with expansion bolts, metal angles, etc., has made it possible to design and construct many intricate and complex preassembled and prefabricated units for exterior and interior components of buildings.

Condensed Checklist

1. One should always check the locality for the availability of the type of marble desired and of skilled labor for setting and handling. Some local quarry may be producing small quantities of a distinctive marble unavailable in other areas.
2. Local, municipal, state, and other codes should be checked for limitations and fire-resistant ratings.
3. For cut stone, samples should be obtained, and shop

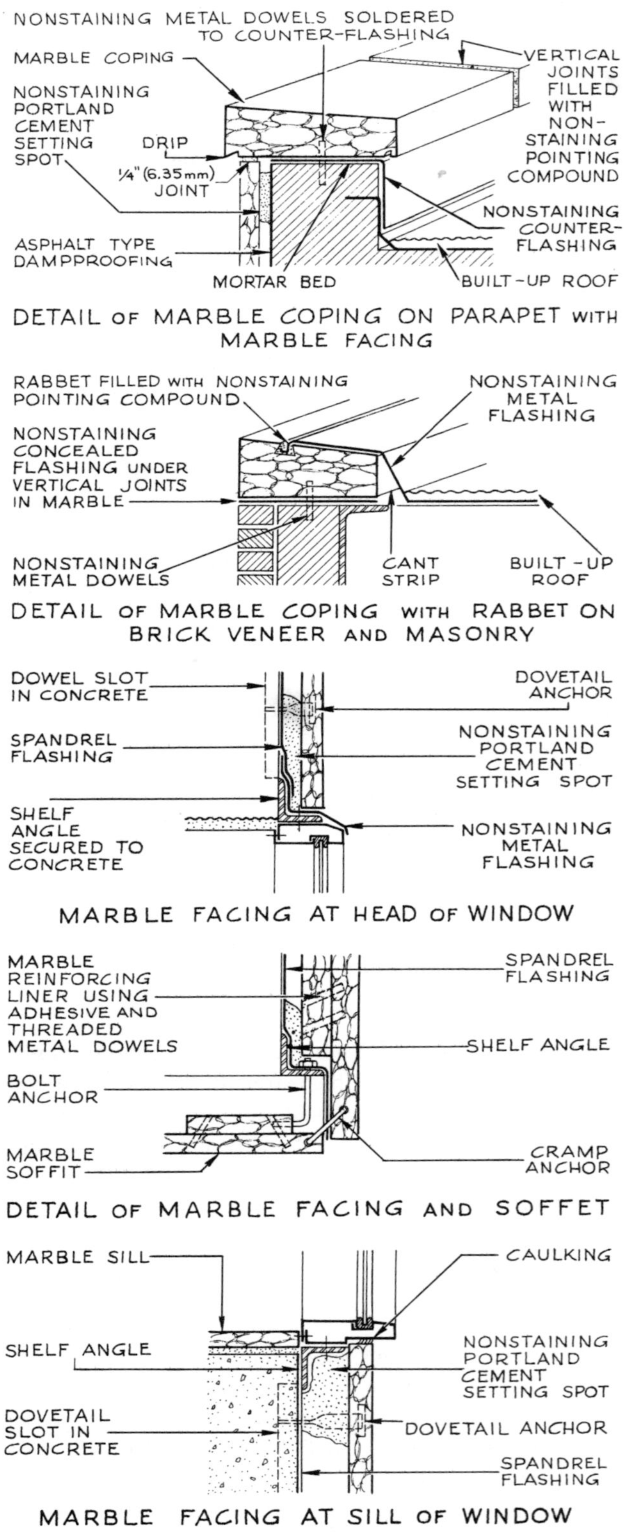

Figure S134 Details of marble coping, head and sill of a window, and marble facing and soffit.

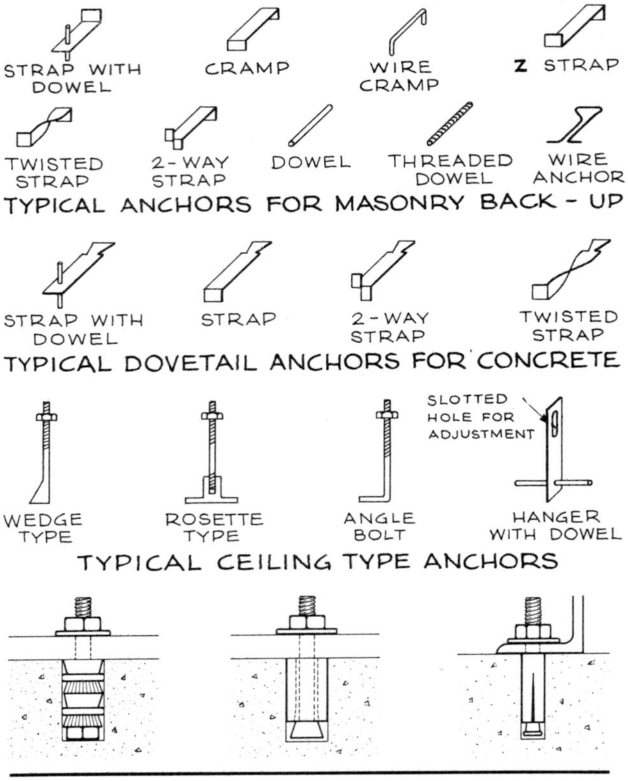

Figure S135 Accessories for marble work.

drawings and setting drawings are always required; all stones should be numbered and keyed into the setting drawings.

4. Marble must be carefully handled when delivered. It should be stored on wood strips covered with waterproof paper (the wood must not contain tannic acid) and kept covered with tarpaulins or waterproof paper.
5. The type of finish, grade, color, and variations in grade and color should be checked.
6. If any stone arrives damaged, care should be taken that immediate notification of the damaged stones reaches the manufacturer so that replacement can be made at once.
7. Back-up material, if concrete or masonry, should be coated with asphalt-type dampproofing before the marble is set.
8. The correct type of setting, pointing, and grouting material must be used (*see* Table *S98*).
9. Installation must be such that no moisture can develop behind the marble; the space between marble and back-up material must be open except for spot-attachments so that air can circulate. Where solid grouting is necessary, nonstaining waterproof mortar should be used.

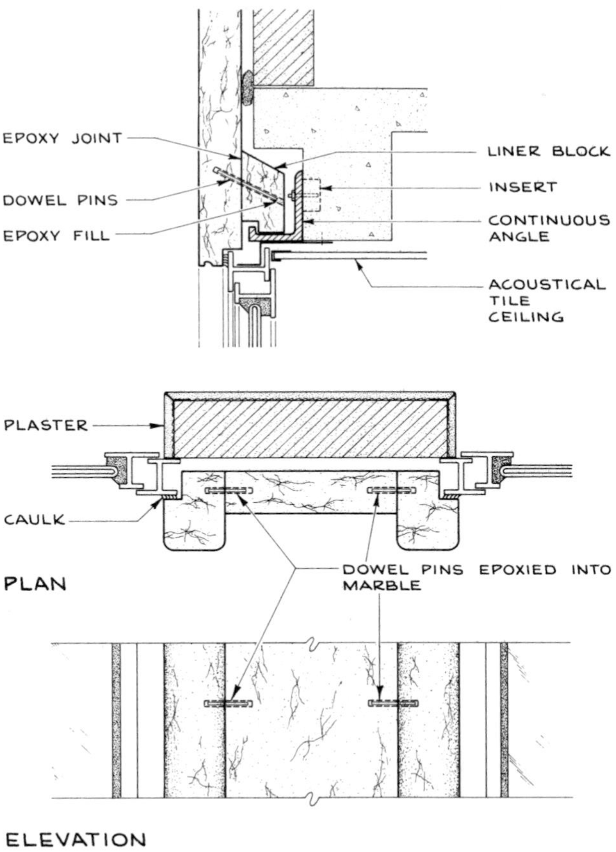

Figure S136 Preassembled marble units.

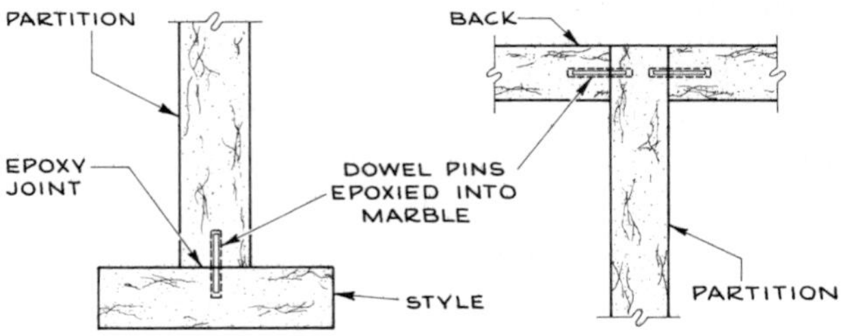

Figure S137 Toilet room partitions assembled with dowels and high-strength epoxy.

10. For exterior marble work, the type and material of anchors, ties, dowels, etc., and the type of flashing, both concealed and exposed, should be nonstaining and compatible with the back-up support so that no galvanic action can occur.
11. The number of anchors needed to support each piece of marble and the joint sizes must be correctly calculated in relation to the thickness of the marble (*see* Table *S99*).
12. The type of joint in general and particularly the number of expansion joints should be checked. Usually one expansion joint every 30 ft (9.144 m) is sufficient. These expansion joints are $\frac{3}{16}$ in. (4.76 mm) wide and filled with nonstaining pointing compound.
13. Neither oil, grease, smoke, or wash from concrete construction nor scaffolding should come in contact with the marble.

Table S98 Setting, pointing, and grouting materials for exterior and interior marble work

Location and type of marble	Setting material	Pointing material	Grout	Method of maintaining uniform joints
Interior wall finishes	Plaster of paris, nonstaining portland cement, litharge and glycerine, synthetic resin nonstaining adhesive[a]	Nonstaining, nonhardening pointing compound; nonstaining portland cement and water		When pointing compound is used, plastic or aluminum supporting points (cushions)[b]
Interior saddles, flooring, sills, and stain treads	1 part nonstaining portland cement to not more than 3–5 parts fine sand with water to a relatively dry mix	Nonstaining portland cement and water		For floors: marble tamped and bedded into place, then removed; marble wetted or bed wetted and marble replaced to exact finish level
Interior toilet room partitions		Litharge and glycerine or nonstaining pointing compound		Supported by plated angles, bolts, nuts, etc.
Exterior wall facings	Nonstaining quick-setting portland cement and water[a]	Nonstaining white portland cement and water; nonstaining pointing compound	1 part nonstaining portland cement to not more than 3 parts fine sand and water[c]	When pointing compound is used alone, plastic or aluminum supporting points (cushions)[b]
Exterior prefabricated or preassembled units	Stainless steel anchors, angles, clip angles, etc.	Gaskets, sealants, or pointing mortar	High-strength adhesives	Lead pads and plastic buttons

Table S98 Setting, pointing, and grouting materials for exterior and interior marble work (*continued*)

Location and type of marble	Setting material	Pointing material	Grout	Method of maintaining uniform joints
Exterior ashlar type	1 part nonstaining portland cement to 3 parts fine sand and water			
Exterior copings, stair treads, sills, and flagstones	1 part nonstaining portland cement to not more than 3 parts fine sand and water	Nonstaining portland cement and water		For flagstones: back of marble wetted and then tamped and bedded into place

[a]Exterior and interior marbles are set with a series of spots of the setting material instead of covering the entire surface.
[b]Supporting points (cushions) are available $\frac{1}{16}$ in. (1.588 mm), $\frac{1}{8}$ in. (3.175 mm), and $\frac{3}{16}$ in. (4.763 mm) thick.
[c]Exterior marble is sometimes set with the space between the back of the marble and the back-up material filled with mortar (grouting).

Table S99 Number of anchors and sizes of joints in relation to size and thickness of marble

Size of marble			Thickness of marble		Size of joints	
ft^2	m^2	Number of anchors	in.	mm	in.	mm
Up to 2	0.1858	2	$\frac{7}{8}$	5.56	$\frac{1}{16}$	1.59
From 2	0.1858	3	$\frac{7}{8}$	5.56	$\frac{1}{16}$	1.59
to 4	0.3716		$1\frac{1}{4}$	31.75	$\frac{1}{8}, \frac{3}{16}$	3.18, 4.763
From 4	0.3716	4	$\frac{7}{8}$	5.56	$\frac{1}{16}$	1.58
to 12	1.1148		$1\frac{1}{4}, 1\frac{1}{2}$	31.75, 38.1	$\frac{1}{8}, \frac{3}{16}$	3.18, 4.763
From 12	1.1148	6	$1\frac{1}{4}, 1\frac{1}{2}$	31.75, 38.1	$\frac{1}{8}, \frac{3}{16}$	3.18, 4.763
to 20	1.858		2	50.8		
Over 20	1.858	1 extra anchor for	$\frac{7}{8}$	5.56	$\frac{1}{16}$	1.59
		every additional	$1\frac{1}{4}, 1\frac{1}{2}$	31.75, 38.1	$\frac{1}{8}, \frac{3}{16}$	3.18, 4.763
		3 ft^2 (0.2787 m^2)	2	50.8		

14. All unfinished marble work must be covered and protected when work stops or if heavy rains are expected.
15. Finished marble work should be protected with waterproof paper and wood where it is exposed to construction traffic.
16. Only soap powder, water, and fiber brushes should be used for cleaning marble, never acid or wire brushes.
17. When high-strength epoxy adhesives are used the adhesive manufacturer and the stone supplier and manufacturer should be consulted regarding the methods which should be used for this particular end use.
18. Shop drawings and details, including types of adhesives to be used, must be submitted for approval.
19. Testing of a full-size section of preassembled unit, under the conditions which that will be encountered when installed, should be included in the specifications.

Conditions Favorable to the Use of Marble

1. Where the design requires a highly decorative, durable, permanent surface finish with a distinctive color and character.
2. Where the design requires an ashlar-patterned material that is durable, permanent, and distinctive in color.
3. Where a material with the characteristics mentioned is desired for flooring, sills, stair treads and risers, columns, and the like.
4. Where a material that can withstand rough usage and needs little maintenance is desired.

Conditions Unfavorable to the Use of Marble

1. In areas where there are industrial fumes or acids, or air pollution nearby.
2. In areas where extremely high temperatures will be encountered, for example, in the interior and back hearth of a fireplace.
3. Where economy is of great importance.

History and Manufacture

The first use of marble is attributed to the Egyptians of about 4751 B.C., who obtained it from Algeria in North Africa. However, the Babylonians, the early people of India, the Etruscans, the Mycenaeans, the Hebrews, and the Phoenecians, all used marble. The Greeks used marble from the very beginning for their buildings and sculpture and developed the working of marble to an art unsurpassed in history. Today we still use marbles obtained from Greece. The city of Carthage used marble from quarries in North Africa, which previously had been used by the Egyptians. Long before the Romans conquered Carthage in 146 B.C. they used large quantities of this marble, and even today marble is obtained from these same quarries, which were rediscovered only in 1949. The Romans used tremendous quantities of marble and had over 42 quarries developed and in operation. Two of these are still active. After the fall of the Roman Empire it was not until the 12th century that marble was rediscovered in Europe, although during this period the peoples of the Near and Far East never stopped building with marble. During the Renaissance the great sculptors and architects redeveloped the art of working marble, and in the Near and Far East building with marble continued, one of the most famous examples being the Taj Mahal.

American Quarries. In the United States the marble quarries of Vermont, New York, and New Jersey were the first discovered, and, as the United States expanded, new quarries were developed. Today the marbles of Georgia, Missouri, Tennessee, Alabama, Minnesota, and California, including those from the early quarries, are the best known and most used. The quarrying and finishing of marble is fully covered under Stone.

STONE: SLATE

Physical and Chemical Properties

Slate is a metamorphic rock formed from an argillaceous sedimentary stone consisting of a fine clay and sometimes sand or volcanic dust. This sediment was formed in layers and, as metamorphism occurred, resulted in a stone that splits readily into thin slabs, which are strong and durable and have high tensile strength (*see* Table *S100*). Slate also contains large quantities of colorless mica in small, irregularly shaped scales, about 0.002 by 0.006 in. (0.05 by 0.15 mm) in size, and minute, long grains of quartz.

Color. The small quantities of other ingredients usually give the various colors to the slate. Common slate colors are black, blue, purple, red, green, or gray. Carbonaceous materials or iron sulfides produce the dark colors such as black, blue, and gray; iron oxide, the red and purple; and chlorite, the green. Select slate is uniform in color, and ribbon slate contains stripes of darker colors.

Commercial Forms. Slate is available in the form of dimension stone, including roofing slate (shingles), blackboards, countertops, sills, facings, preassembled and prefabricated units, and flagstones; as slate dust or flour; and as granules.

Types and Uses

Slate flour and granules are the forms in which this stone is most used outside of the construction field. Slat roofing, structural and sanitary slate, flagstones, facings, preassembled and prefabricated units, and blackboards are the major types used in construction. These are all considered a form of dimension stone and quantitatively are much less important that slate flour and granules (*see* Table *S101*).

Roofing Slate. Roofing slate is available in standardized colors given the following specific names: unfading black, blue-black, gray, blue-gray, purple, variegated, green, and red.

The grading of roofing slate varies according to individual manufacturer. In general, a good slate should give a sharp metallic ring when struck with the knuckles. Slate may be either smooth or rough textured, with an overall uniform color or a uniform color with a darker colored ribbon or ribbons running crosswise (this type of slate

Table S100 Physical properties of slate

Compressive strength[a]		Modulus of rupture[a]		Specific gravity	Weight	
$lbf/in.^2$	MN/m^2	$lbf/in.^2$	MN/m^2		lb/ft^3	kg/m^3
10,000–23,000	68.95–158.59	11,000–18,000	75.85–124.11	2.74–2.80	171–176	2739.42–2815.52

[a] 1 $lbf/in.^2$ = 0.006895 MN/m^2.

Table S101 Slate commonly used in construction

Type of slate	Color of slate	Where quarried
Roofing slate or shingles, blackboards, countertops, sills	Blue-gray, dark blue	Maryland, Maine, Pennsylvania, Virginia, Vermont
Flagstones	Blue-gray, dark blue	Maryland, Maine, Pennsylvania, Virginia, Vermont
	Red, purple	New York, Vermont
	Green	Vermont
	Variegated	Vermont
Panels, preassembled and prefabricated	Blue-gray, dark blue	Maine, Vermont, Pennsylvania, Virginia
	Red, purple	Vermont
	Green	Vermont
	Variegated	Vermont

may be cut in such a manner that the ribbons are covered up in laying). A bent type of slate is available which is smooth, uniform in color, but naturally bent with an approximately 12-ft (3.65-m) radius.

All roofing slate is predrilled with two nail holes per slate and is split into a series of standard thicknesses ranging from $\frac{1}{4}$ to $\frac{1}{2}$ in. (6.35 to 12.7 mm). A thickness that falls between any two standard thicknesses is called a special thickness. Slate over $\frac{1}{2}$ in. (12.7 mm) thick is usually longer than 16 in. (406.4 mm). (*See* Figure *S138* and Table *S102*.)

The most commonly used roofing slate has a thickness specifically called commercial standard thickness, which is the quarry run of production and has variations in thickness above and below $\frac{3}{16}$ in. (4.76 mm). Thicker slates of this type have variations only greater than the specified thickness; for example, $\frac{1}{4}$-slate is $\frac{1}{4}$ in. (6.35-mm) or thicker, and each of the terms "$\frac{3}{16}$-slate," "full $\frac{3}{16}$-slate," and "not less than $\frac{3}{16}$-slate" refers to hand-selected slate all of the same thickness.

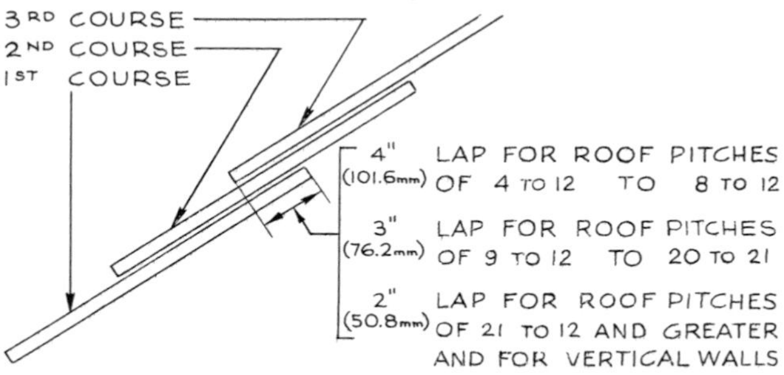

Figure S138 Relation of lapping of the three courses of shingles to roof pitch.

Slate for flat roofs, terraces, and promenades is generally smaller and thicker than roofing slate for pitched roofs (*see* Tables *S103* and *S104*).

Blackboard Slate. Blackboard slate is select, unfading black slate of uniform color and thickness, with all edges ground and accurately squared so that joints can be made tight, smooth, and on the same plane. The surface is always ground to a perfect plane with carborundum abrasive. The stock thicknesses are generally $\frac{1}{4}$ to $\frac{3}{8}$ in. (6.35 to 9.53 mm); stock heights are 3, $3\frac{1}{2}$, and 4 ft (0.9144, 1.0667, and 1.2192 m); and the standard length is 5 ft (1.524 m). Larger sizes and greater thicknesses are also available.

Countertop, Sill Slate. Countertops, sills, etc., are select slate, generally $\frac{7}{8}$ in. (22.23 mm) thick, unfading black, blue-black, gray, or blue-gray and uniform in color, with edges accurately ground and squared to sizes as required. Slate for these purposes is available up to 2 in. (50.8 mm) thick.

Flagstones. Flagstones are $\frac{7}{8}$ in. (22.23 mm) thick and available in all colors, with smooth or rough texture. They are made in three types: irregular with two edges sawed, regular with four edges sawed and squared, and random with no edges sawed.

Slate Granules and Flour. Slate granules and flour, made from waste slate or stone not suited for cutting, are used in large quantities—the flour for asphalt and tar papers and for asphalt siding and shingles (to prevent the material from sticking together) and the granules as a surfacing material. Powdered slate is also used as a filler, similarly to limestone dust.

Application

Condensed Checklist

1. The locality should always be checked for availability of slate of the type, color, and texture desired.
2. For roofing, the weight of the type of slate to be used should be checked against the weight permitted by the supporting members, as the weight of a slate roof may require larger or stronger supporting members.
3. Methods of laying slate for roofing that affect design and texture: The various thicknesses of slate

Table S102 Sizes, weights, thicknesses,[a] and exposures of roofing slate for roof pitches not less than 4 to 12

Size of slate shingle				Number of shingles in each square ($100\ ft^2 = 9.29\ m^2$)	Exposure with 3-in. (76.2-mm) lap		Nails required for each square ($100\ ft^2 = 9.29\ m^2$)			
Length		Width								
in.	mm	in.	mm		in.	mm	lb	oz	kg	g
10	254.0	6	152.4	686	$3\frac{1}{2}$	88.9	7	12	3.18	368.55
		7	177.8	588			7	4	3.18	113.40
		8	203.2	515			5	14	2.27	396.90
11	279.4	7	177.8	515	4	101.6	5	14	2.27	396.90
		8	203.2	450			5	2	2.27	56.70
12	304.8	6	152.4	533	$4\frac{1}{2}$	114.3	6	1	2.72	28.35
		7	177.8	457			5	3	2.27	85.05
		8	203.2	400			4	9	1.81	255.15
		9	228.6	355			4	1	1.81	28.35
		10	254.0	320			3	10	1.36	283.5
14	355.6	7	117.8	374	$5\frac{1}{2}$	139.7	4	4	1.81	113.40
		8	203.2	327			3	12	1.36	340.20
		9	228.6	291			3	5	1.36	141.75
		10	254.0	261			3	3	1.36	85.05
		11	279.0	238			2	11	0.91	311.85
		12	304.8	218			2	8	0.91	226.80
16	406.4	8	203.2	277	$6\frac{1}{2}$	164.1	3	2	1.36	56.70
		9	228.6	246			2	12	0.91	368.55
		10	254.0	222			2	8	0.91	226.80
		11	279.4	210			2	5	0.91	141.75
		12	304.8	184			2	2	0.91	56.70
		14	355.6	160			1	13	0.45	368.55
18	457.2	9	228.6	213	$7\frac{1}{2}$	189.5	2	7	0.91	198.45
		10	254.0	192			2	3	0.91	85.05
		11	279.4	175			2	0	0.91	0
		12	304.8	160			1	13	0.45	368.55
		13	330.2	148			1	11	0.45	311.85
		14	355.6	137			1	9	0.45	255.15
20	508.0	9	228.6	189	$8\frac{1}{2}$	214.9	2	3	0.91	85.05
		10	254.0	170			1	15	0.45	425.25
		11	279.4	154			1	12	0.45	340.20
		12	304.8	141			1	10	0.45	283.50
		13	330.2	132			1	8	0.45	226.80
		14	355.6	121			1	6	0.45	170.10
22	558.8	10	254.0	152	$9\frac{1}{2}$	240.3	1	12	0.45	340.20
		11	279.4	138			1	9	0.45	255.15
		12	304.8	126			1	7	0.45	198.45
		13	330.2	117			1	5	0.45	141.75
		14	355.6	108			1	4	0.45	113.40
24	609.6	11	279.4	125	$10\frac{1}{2}$	266.7	1	7	0.45	198.45
		12	304.8	114			1	5	0.45	141.75
		13	330.2	106			1	3	0.45	85.05
		14	255.6	98			1	2	0.45	56.70
		16	406.4	86			1	0	0.45	0
26	660.4	14	255.6	89	$11\frac{1}{2}$	292.1	1	0	0.45	0

[a] Standard thickness of $\frac{3}{16}$ in. ($\frac{4}{76}$ mm) weighs 800 lb (262188 kg) per square; $\frac{1}{4}$ in. (6.35 mm) weighs 900 lb (410.67 kg); $\frac{3}{8}$ in. (9.53 mm) weighs 1100 lb (498.96 kg); $\frac{1}{2}$ in. (12.7 mm) weighs 1700 lb (771.12 kg); and $\frac{3}{4}$ in. (19.05 mm) weighs 2600 lb (1179.36 kg). A square equals $100\ ft^2$, or $9.29\ m^2$.

Table S103 Sizes of slate for flat roofs, promenades, and walks

Size of slate				Thickness of slate					
Length		Width		Ordinary and light service		Promenade and heavy service		Terraces and walks	
in.	mm	in.	mm	in.	mm	in.	mm	in.	mm
6	152.4	6	152.4	$\frac{3}{16}$	4.76	$\frac{1}{4}$	6.35	$\frac{3}{4}$	19.05
		8	203.2			$\frac{3}{8}$	9.53	$1\frac{1}{4}$	31.75
		9	228.6						
10	254.0	6	152.4						
		7	177.8	Same as above		Same as above		Same as above	
		8	203.2						
12	304.8	6	152.4						
		7	177.8	Same as above		Same as above		Same as above	
		8	203.2						

Table S104 Face and back finishes for slate

Types of finish	Method of manufacture	Major use
Natural cleft face	Natural split	Exterior and interior installations
Gauged-rubbed	Face rubbed with diamond-head gauging machine	Tight setting conditions
Sand-rubbed	Face rubbed with diamond-head surfacing machine	Treads, risers, paving, and flooring in heavy-traffic areas
Honed	Face rubbed with diamond-head surfacing machine to make a satin-smooth finish	Stools, furniture, tops, facings, generally interior areas; also where both surfaces are exposed

give various specific design effects. Two of these effects are classified as "textured" and "graduated." "Textured" refers to the use of rough-textured slate up to $\frac{3}{8}$ in. (9.53 mm) thick with uneven butts and varying thicknesses or sizes. "Graduated" uses large-size slates with variations in thickness, size, and color. (*See* Figure *S139*.)

4. Method of applying the slate: Slate roofing may be applied onto lath which is spaced with its center-to-center distance equal to slate exposure, or on a nailable surface which must first be covered with 15-lb (6.804-kg) asphalt or tar-saturated felt for the commercial standard thickness of slate, 30-lb (13.608-kg) felt for slate up to $\frac{3}{4}$ in. (19.05 mm) thick, and 45-, 55- or 65-lb (20.412-, 24.548-, or 29.484-kg) rolled asphalt roofing for slate over $\frac{3}{4}$ in. (19.05 mm) thick.
5. Type and size of nails for applying slate roofing: Two nails are used for each slate. Nails are commonly large flat-head wire slate nails made of copper, copper alloy, or aluminum. In dry climates hot-dipped galvanized steel can be used. As a general

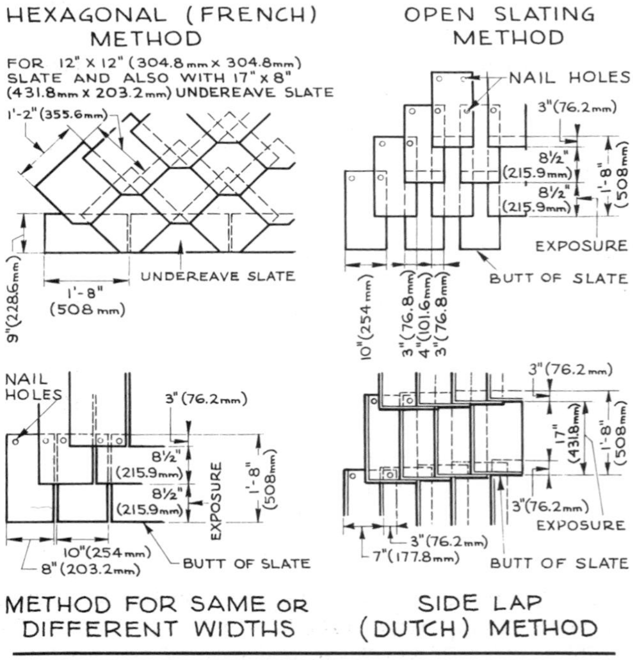

Figure S139 Methods of laying slate roofing.

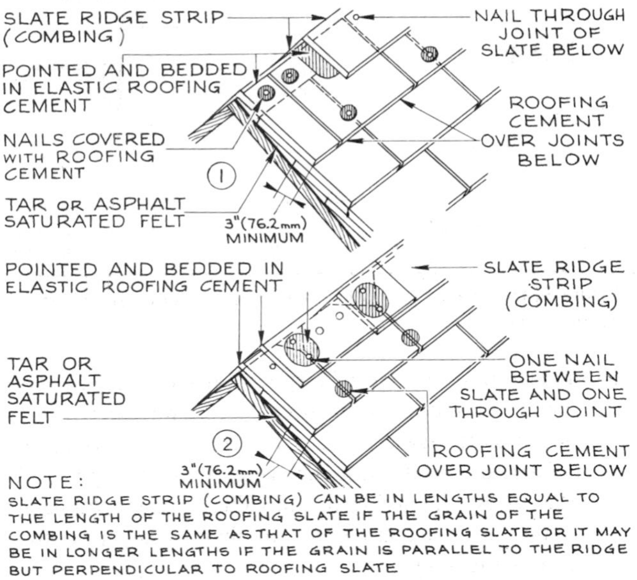

Figure S140 Details of ridge for strip saddle ridge (1) and saddle ridge (2).

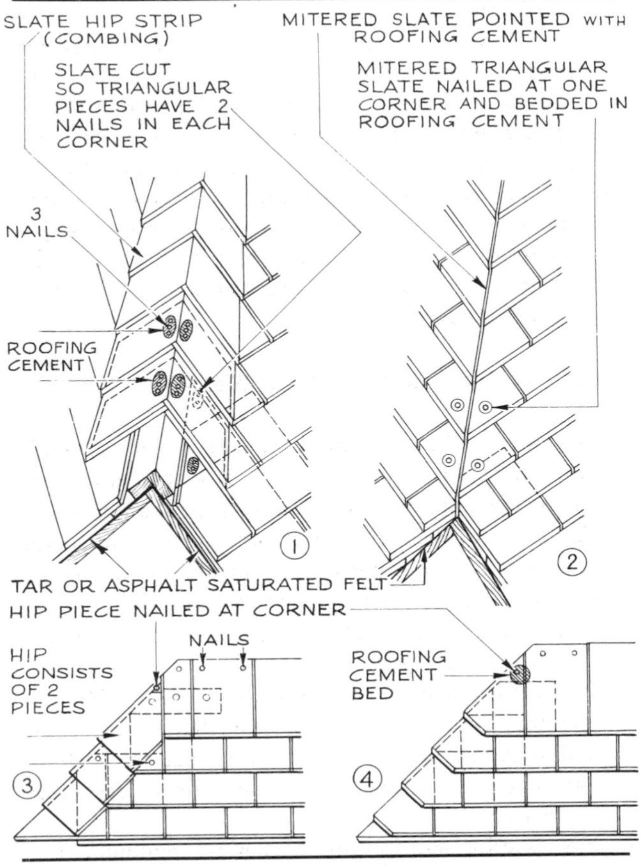

Figure S141 Details of various treatments of hips: saddle (1), mitered (2), Boston, made with two pieces (3), and fantail (4).

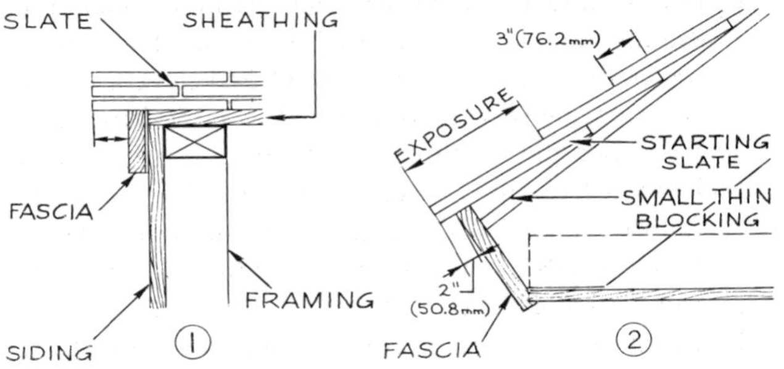

Figure S142 Details of eaves (1) and gable ends (2).

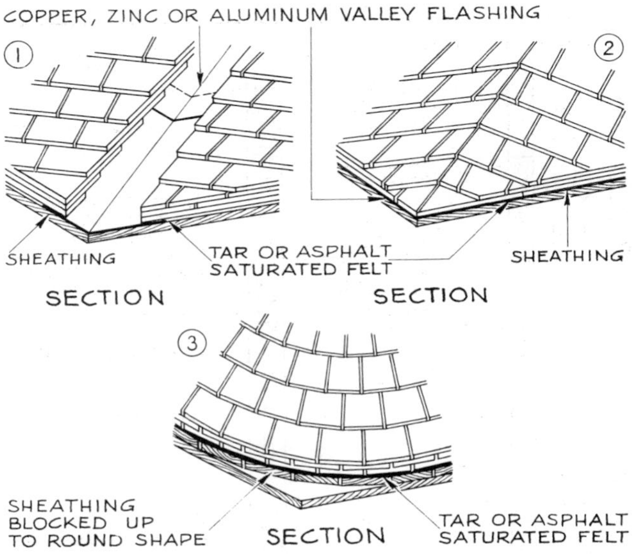

Figure S143 Details of types of valleys: open (1) and closed (2).

rule, the nail should be 1 in. (25.4 mm) longer than the thickness of the slate, usually three-penny nails for most slate roofs and four-penny for slates $1\frac{2}{3}$ ft × 10 in. (508 × 254 mm) and larger. Nails should never be driven in so tightly as to put a strain on the slate, and all exposed nails should be covered with roofing cement.

6. Treatment of slate roofing at ridges (*see* Figure *S140* for details).
7. Treatments of slate roofing at hips (*see* Figure *S141*).
8. Treatment of slate roofing at eaves and gable ends (*see* Figure *S142*).
9. Treatment of slate roofing at valleys (*see* Figure *S143* for details).
10. Local, municipal, and state codes and also the fire rating codes of the Fire Underwriters, insurance companies, labor departments, and federal government (Army, Navy, etc.) should be checked.

11. For blackboards, stock sizes and method of handling lengths should be investigated and correctly handled. Generally, lengths up to 5 ft (1.524 m) consist of one piece, lengths of over 5 to 9 ft (1.524 to 2.743 m) of two pieces, those over 9 to 13.5 ft (2.743 to 4.115 m) of three pieces, those over 13.5 to 18 ft (4.115 to 5.486 m) of four pieces, those over 18 to 22.5 ft (5.486 to 6.858 m) of five pieces, and those over 22.5 to 27 ft (6.858 to 8.23 m) of six pieces.
12. Blackboards should be correctly installed with stock chalk rails and trim of the type and material that meets design requirements (*see* Figure *S144*).

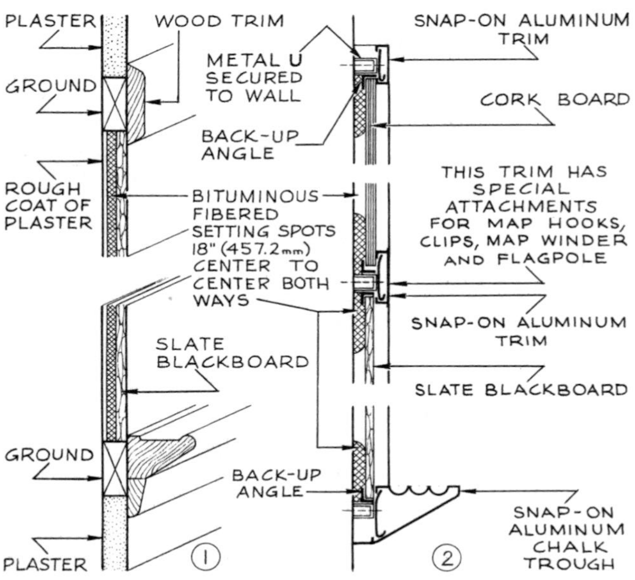

Figure S144 Details of blackboards installed with wood trim on plaster walls (1) and with metal trim on concrete block walls (2).

13. When epoxy adhesives are used for joints, one should always check with the adhesive manufacturer and suppliers and the manufacturers of panels or preassembled or prefabricated slate units to make sure that the correct joining methods are used for the particular end use.
14. Shop drawings and details, including types of adhesives, anchors, and dowels, must be submitted for approval.
15. When special climatic conditions prevail, a full-size unit must be tested under the conditions to which it will be exposed when installed.
16. Shop drawings should always be required for slate sills, exterior or interior copings, and countertops.
17. For sills, the method of treating window jambs must be correctly detailed (*see* Figure *S145*).

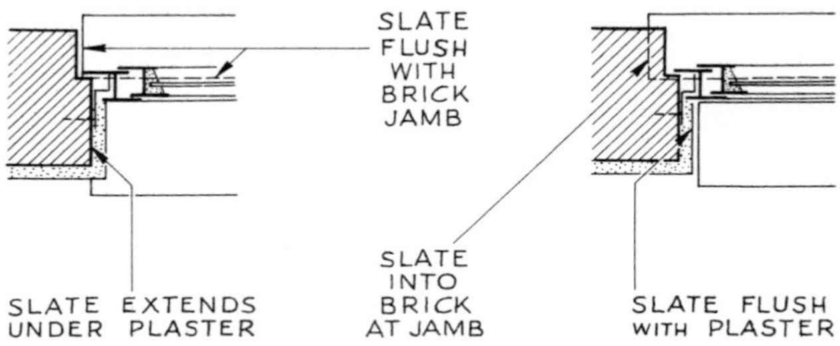

Figure S145 Details of both exterior and interior slate sills at window jambs.

18. Flagstones must be of a type that will meet the design requirements for both exterior and interior use. When slate flagstones cut to size and in variegated colors are used, shop drawings must be specified that show the laying of individual stones, and each slate must be marked by number to key into the setting drawing.
19. Installation of slate flooring: An allowance must always be made of $1\frac{3}{4}$ in. (44.45 mm) from top of slate to concrete slab on the interior and of 2 in. (50.8 mm) for exterior work. Mortar for setting slate is usually 1 part portland cement, 3 parts fine sand, and enough water to make a relatively dry mix. The slates are bedded and pressed into the setting mortar and brought to the finish level; any extra mortar squeezing up in the joints is removed. Before the mortar has set, the joints should be filled and troweled flush to the top of the slate, using a pointing mortar made of 1 part portland cement, $1\frac{1}{2}$ parts fine sand, and enough water to make a workable paste.

Conditions Favorable to the Use of Slate

1. As roofing when the design requires a material that is permanent, is fireproof, has a dark, uniform, nonfading color, and involves relatively no maintenance.
2. As flooring when a durable material is required for exterior or interior use and an irregular or rectangular pattern, a semirough or relatively smooth texture, and a uniform or varied dark color range are desirable from the design viewpoint.
3. For window sills, both interior or exterior, when the architect wishes to use a material that is permanent and requires relatively no maintenance, and whose dark color is utilized in the design or at least does not detract from it.
4. For exterior and interior facings for buildings. For such panels and preassembled and prefabricated units, *see* Figure *S146*.

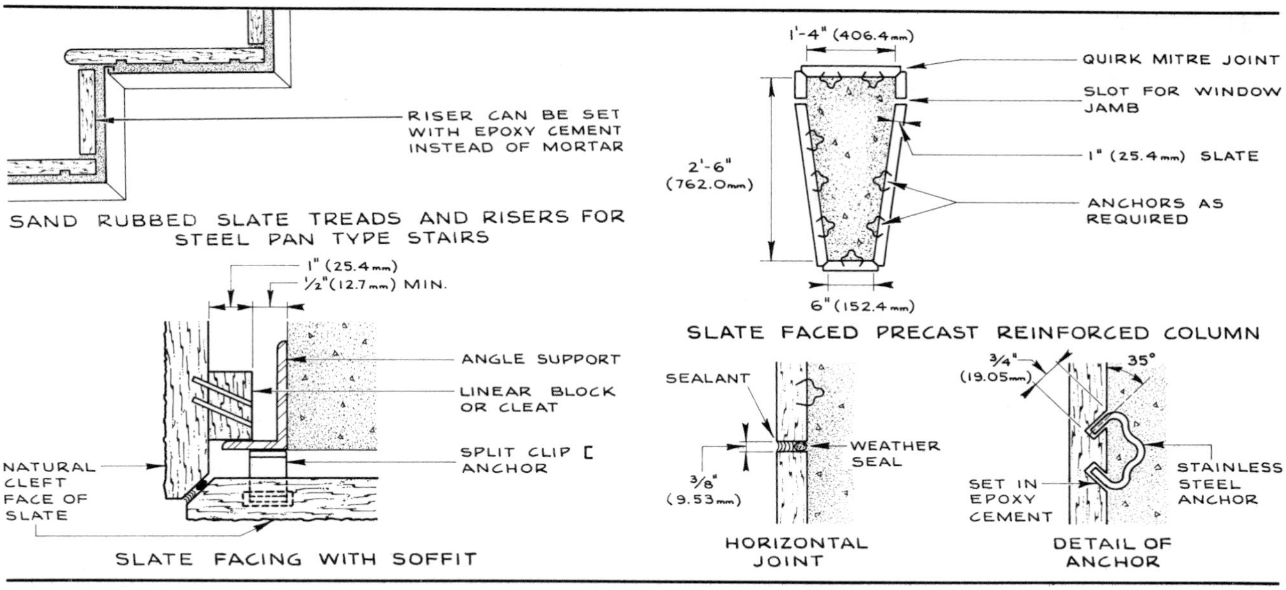

Figure S146 Details of slate with the use of high-strength epoxy adhesive.

Conditions Unfavorable to the Use of Slate

1. For a roofing material where economy is an important factor.
2. Where a highly smooth, polished type of flooring material is desired.

History and Manufacture

Slate was known to the ancients, who used it for flooring. Its first use as blackboards is unknown, but any civilization that has a written language has to find a material on which to mark and then erase in order to teach. In more northern climates, where protection from rain was important, slate was developed as a roofing material sometime before the Roman Empire. From the Gothic period to modern times slate has been extensively used as a roofing material because of its durability (it lasts well over a hundred years).

Quarrying and Finishing. Slate is quarried similarly to other stone, the details being described under the general heading Stone. A channel of stone is removed, and successive large blocks are broken off. These blocks are divided into smaller sizes, depending on the required size of slate, and are transported to the finishing and splitting sheds. The small blocks are then split into the required thickness, squared by hand or machine to finish sizes, and their edges ground if necessary.

Roofing Slate Manufacture. For roofing slate, a piece of slate usually 3 in. (76.2 mm) thick, with length and width sufficiently large to allow for squaring, is used. The splitter places a chisel against the edge of the block, as near to the middle as possible, and parallel to the cleavages in the slate. He then lightly taps it with a mallet. A crack appears, and, after slight wedging with the chisel, the block is split into 2 pieces with smooth and even surfaces. This is repeated until 16 to 18 separate pieces of slate have been obtained of the desired thickness. Originally slate was split to approximately $\frac{1}{6}$ in. (4.23 mm) thickness. To square the edges, the slate is held tightly top and bottom and is cut off either with a knife (by the same principle as a paper cutter) or by machine-driven rotating blades. Finally, two holes are made near the top for nailing.

Other Procedures. To grind edges for other types of slate, simple grinding wheels with abrasives are used. The split surface is smooth or rough, depending on the type of slate. For blackboards the surface is lightly finished by surface grinding with fine abrasives and water.

STRAPS

Straps may be defined as thin pieces of various types of metal which are used to hold materials together. In construction, straps refer to thin strips of metal that are called anchors or hangers if bent, and ties if not bent or shaped (*see* Anchors; Ties).

STRONTIUM

Physical and Chemical Properties

Symbol: Sr
Atomic number: 38
Specific gravity: 2.54
Melting point: 769°C, 1416.2°F
Boiling point: 1384°C, 2523.2°F

Strontium is a pale yellow, soft metal that decomposes water and must be kept immersed in naphtha. It is available in lump and rod form.

Types and Uses

There are at present few uses for strontium metal because of its similarity to calcium and barium, which are more readily available. Its compounds have several important applications (*see* Table *S105*).

Table S105 Uses of strontium

	Construction Uses	
Component of	Glass	Iridescence
	Paint	White paint solid; luminous agent

History and Manufacture

Strontium was named after Strontian, a town in Scotland. In 1808 Davy isolated strontium metal by electrolysis of the fused chlorides. Most strontium metal is now prepared by thermal reaction of strontium oxide and aluminum.

SULFUR

Physical and Chemical Properties

Symbol: S
Atomic number: 16
Specific gravity: 2.07 (rhombic) (20°C, 68°F)
Melting point: 112.8°C, 235.04°F
Boiling point: 444.674°C, 832.41°F

Sulfur is a nonmetallic element that exists in several forms and has an exceedingly complex molecular structure. It is most familiarly known as a pale yellow to yellowish brown, tasteless, odorless crystalline solid (its rhombic form). Sulfur is insoluble in water, slightly soluble in alcohol and ether, and soluble in benzene. It is a poor conductor of heat and electricity. It combines with all metals except gold and platinum to form sulfides, but not with equal ease. It is an active oxidizing agent. In air it burns readily to form sulfur dioxide, which dissolves in water to form sulfuric acid.

Commercial Forms. Sulfur is available in lump, roll, and powder form, 90 to 99.9% pure.

Types and Uses

Perhaps the most important form of sulfur is sulfuric acid, for this acid is used at some point in the manufacture of almost all industrial products.

In construction, pure sulfur has only one direct application; that is in sulfur cements, which consist of sulfur plus various inert mineral fillers. These are used as mortar in masonry where acid resistance is important, for pointing up acidproof brick, and for anchoring metal into masonry materials, for example, bolts, reglets, and pipe

Table S106 Uses of sulfur

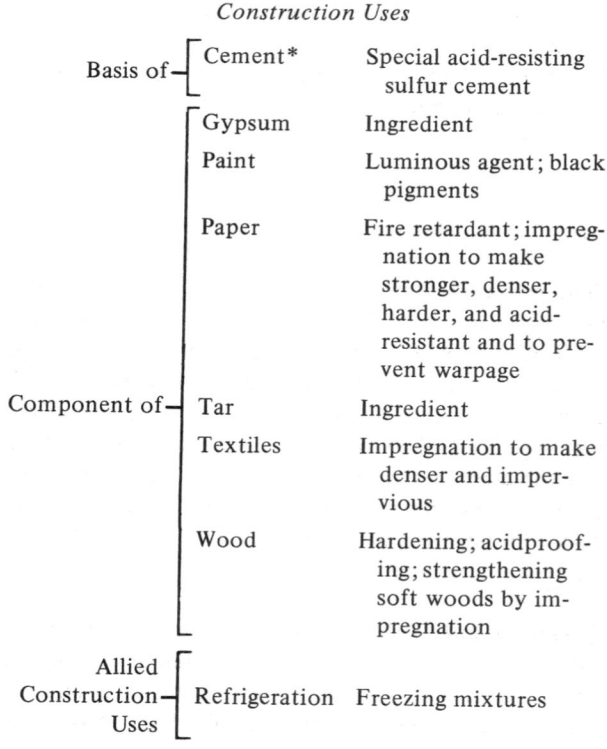

	Construction Uses	
Basis of	Cement*	Special acid-resisting sulfur cement
Component of	Gypsum	Ingredient
	Paint	Luminous agent; black pigments
	Paper	Fire retardant; impregnation to make stronger, denser, harder, and acid-resistant and to prevent warpage
	Tar	Ingredient
	Textiles	Impregnation to make denser and impervious
	Wood	Hardening; acidproofing; strengthening soft woods by impregnation
Allied Construction Uses	Refrigeration	Freezing mixtures
	Nonconstruction Uses	
Electroplating, industrial metal cleaning, pickling and processing, metallurgy.		

railing posts. Table *S106* shows the various other uses of sulfur compounds in relation to construction materials.

Preformed pulp and paper articles impregnated with sulfur become very strong, dense, hard, weather-resistant materials. Wood also is impregnated with sulfur, which, unlike creosote, does not change the color of the wood. Even soft woods become phenomenally hardened and strengthened, almost as if petrified. Various sulfur-soluble dyes can be added to impart color to the wood. Wood so treated is less moisture absorbent and resists acid elements. The sulfur also acts as a preservative against insects and bacterial growth in wood in contact with the ground.

Other types of fiber and even stone can likewise be sulfurized. Sulfur, moreover, is an important ingredient in rubbers. In metals, especially iron, sulfur is usually considered as impurity, but is specially added to steel used for screws (to improve machinability).

History and Manufacture

Sulfur, also called brimstone or burning stone, was used from ancient times in religious ceremonies, for purification (fumigation) of buildings, and as early as 2000 B.C. for bleaching cloth. The Romans also used it in medicine and for incendiaries in warfare. The Greeks called it *theion*, which we still use in the form of "thio" to indicate a sulfur-containing compound. It was first classified as an element by Lavoisier in 1777, and in 1809 Gay-Lussac and Thenard proved it to be one.

T

TALC

Physical and Chemical Properties

Talc is a mineral that occurs in several commercially useful forms. It is a hydrous magnesium silicate theoretically containing fixed amounts of silicon dioxide, magnesium oxide, and water. Most commercial talcs also contain varying amounts of quartz, calcite, dolomite, and iron oxide. Talc may be green, yellow, gray, or white in color and has a pearly or silvery luster. It is greasy to the touch and soft enough to cut with a knife, although it hardens under heat application. It is highly resistant to acids, alkalis, and heat. Its specific gravity is 2.6 to 2.8.

Talc deposits vary in structure; they may be soft, foliated or compact, massive aggregates but may also occur as platelike or crystalline aggregates or in fibrous or micaceous form. The compact form is called block talc, soapstone, or potstone. The very pure grades of block talc that meet specifications for electronic insulators are marketed as block steatite talc. Some of the less pure grades are sold as building stone. Most commercial grade talc, however, is marketed in the ground form.

Types and Uses

Talc enters into many materials used in construction, generally in powder form. It is used in tar and asphalt papers, asphalt shingles and siding, and roll roofing, mainly as a protective surface dusting. It serves as an extender, filler, or pigment in paints, paper, rubber, putty, plaster, linoleum, and textiles. The massive form and ground particles are molded and cut into ceramic insulators, into heater parts, and as a covering for steam pipes. The very pure grades of block talc (steatite) are used as electronic insulators. Miscellaneous uses include rope, slate pencils, and grease crayons.

History and Manufacture

Since talc is so easy to cut, it has been a popular stone for carving throughout the ages. In Egypt it was used for amulets; in Assyria it was covered with a vitreous glaze and used for signet seals, and in China for ornamental carvings.

TANTALUM

Physical and Chemical Properties

Symbol: Ta
Atomic number: 73
Specific gravity: 16.654
Melting point: 2996°C, 5424.8°F
Boiling point: 5225°C, 9437°F
Tensile strength: 35,000 to 40,000 lbf/in.^{2}, 241.337 to 275.8 MN/m^2

Tantalum is a strong, rather dull gray metal with high dielectric properties. It is one of the most acid-resistant metals and is similar to glass in its resistance to corrosion and chemical attack. It is soluble in caustic solutions, however. It forms very stable oxide films that allow alternating current to pass in one direction only and therefore are used in rectifiers. The composition of tantalum metal is given in Table *T1*.

Table T1 Composition of tantalum metal

Ta	C (percent of content)	Fe
99.9 min.	0.01 max.	0.001 max.

Workability. Tantalum is ductile, malleable, and tough, and therefore easily worked. However, it can absorb large volumes of gases at high temperatures (absorbing 740 times its own volume of hydrogen) and must be cold-worked. It can be drawn, stamped, spun, formed into complicated shapes, welded by special processes, and readily machined, but not cast.

Commercial Forms. It is available in powder, ingot, bar, swaged rod, sheet, ribbon, foil, tube, and wire form in powder metallurgy bars and special fabricated shapes.

Types and Uses

The uses of tantalum in metallic form are indicated by asterisks in Table *T2*. Tantalum compounds have very few uses to date.

Ferrotantalum-columbium (niobium) is used to add tantalum to stainless steels. The major uses of tantalum metal are for acid-resistant chemical equipment, corrosion-resistant tools, and electronic equipment (where its "getting" ability and dielectric properties can be utilized).

Tantalum can be made as hard as a diamond by the addition of silicon. Because of its corrosion resistance, tantalum is used extensively for surgical implants that remain permanently in the body.

Table T2 Uses of tantalum

	Construction Uses	
Basis of	Abrasives	Main ingredient
Component of	Stainless steel*	Stabilizing by fixation of the carbon
	Steel*	Resistance to scaling at high temperatures
	Titanium	High-temperature alloy
Allied Construction Uses	Lamps*	Filaments
	Nonconstruction Uses	
Chemical equipment,* corrosion-resistant tools, electronics,* optical glass		

History and Manufacture

In 1802 the Swedish chemist Ekeberg discovered a new element. It was so difficult to dissolve the mineral he was investigating that he named it tantalum, after the mythical god Tantalus, whose punishment in Hades gave us the word "tantalizing."

Tantalum was not isolated until 1903, when von Bolton prepared the pure, ductile metal. Tantalum and niobium (columbium) are almost always coexistent in their ores and are extracted together.

Extraction Methods. In one method the ores are fused with caustic soda to form sodium niobium tantalate. From this are obtained potassium tantalum fluoride and niobium potassium fluoride, which are then processed separately to yield their respective metals. Tantalum is obtained from potassium tantalum fluoride by electrolysis. Niobium potassium fluoride is treated with an alkali to change it to niobium oxide, which in turn is mixed with niobium carbide and heated in a vacuum to produce pure niobium (columbium) metal and carbon monoxide. A second method uses solvent organic separation of these two elements.

TAR

Physical and Chemical Properties

Tar is obtained either from the carbonization of bituminous coal or from the distillation of wood. The common names are "pitch" or "coal tar pitch," and "pine pitch."

Pitch is readily adhesive, highly waterproof, and durable; it is unaffected by many acids and alkalis. Pitch is available in forms ranging from hard solid to almost water-thin liquid (*see* Creosote).

Coal Tar Pitch. Coal tar pitch is the residue left after the other substances are distilled from coal tar. The main source is as a by-product from making coke. Coal tar is a sticky, black, viscous liquid containing hydrocarbons, phenols, nitrogen bases, sulfur compounds, nonphenolic oxygen compounds, and nonbasic nitrogen compounds. Its specific gravity ranges from 0.08 to 1.20, and its melting point from 60 to 250°F (15.56 to 121.11°C). It is one of the most stable bituminous materials.

Pine Pitch. Pine pitch is a black, viscous mass, soluble in benzol. It has a specific gravity of 1.06 to 1.10 and a minimum melting point of 148°F (64.44°C).

Types and Uses

Pine pitch is used to impregnate hemp fibers to make oakum, to impregnate and coat building papers and felt, and to preserve wood. In construction it has been almost completely replaced by coal tar pitch.

Coal tar pitch is available in three types, which differ in their consistency and uses as follows: (1) soft pitch, used for paving and for waterproof coatings; (2) medium pitch, used for built-up roofing, tar-base paints, protective metal coatings, waterproof coatings, expansion joints and joint fillers, and the impregnation and coating of paper, felt, paper pulp and fiber board, sheets, and integrants; and (3) hard pitch, used mainly for coatings on metals (*see also* Paints).

Coal tar pitch can withstand sunlight, water, cold, and heat better than asphalt and is therefore used more extensively than asphalt for built-up roofs. However, it does soften with direct radiant heat from sunlight in very hot weather.

Application

Since coal tar products are the most stable of the bitumens, they are used for purposes similar to those served by asphalt products but where greater durability and protection are desired. (*See* Asphalt; all other Asphalt headings, for details of use.)

Condensed Checklist

1. For built-up roofs, the roof pitch should be not greater than 2 to 12 (*see* Figure *T1*), and the type of bond (guarantee), based on 5 years, 10 years, etc., should be specified.
2. Local, municipal, and state codes and also the codes of the Fire Underwriters, insurance companies, and departments of the federal government (Labor, Army, Navy, etc.) should be checked for fire ratings, for types of built-up roof in relation to lengths of bond period, and other requirements.
3. Type of finish surfacing material: Slag, gravel, or stone chips may be used.
4. All tar pitch should be applied in a heated condition, and ladled and mopped according to the roofing manufacturer's directions.
5. The type of tar building paper should be compatible with the type of material with which it will be in direct contact. Certain materials are affected by tar.
6. Where protective coatings are applied on metal, the architect should determine what type of coating (tar or asphalt) is correct for the metal in question.
7. For use as expansion joints or as fillers in construction joints, especially for corrugated material, the filler material (tar or asphalt) should be determined according to the material with which it is being used.
8. For waterproof coatings and membranes on walls and floors below grade, a check should always be made of water conditions to decide what and how much is to be sealed out (*see also* Asphalt Coatings). If a hydrostatic condition exists, specialists should be consulted.

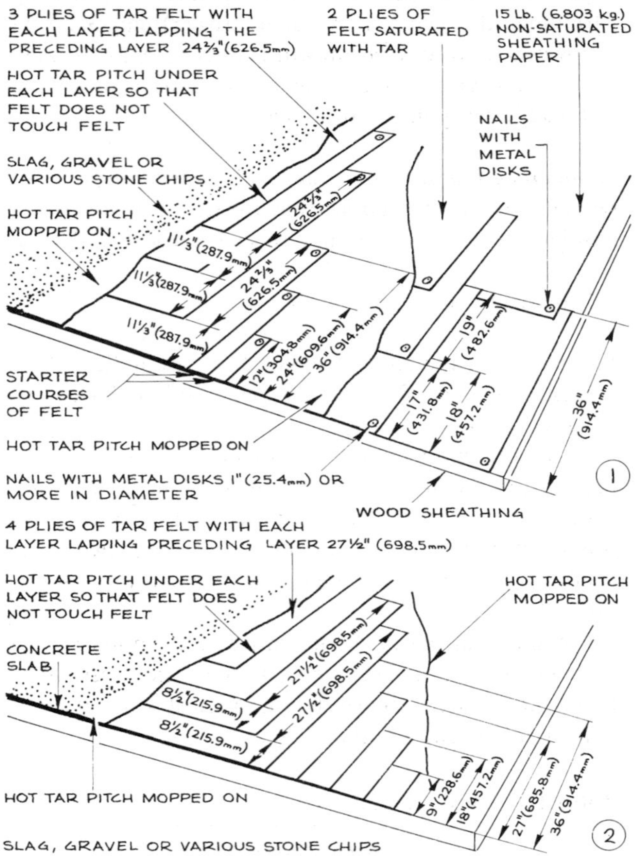

Figure T1 Typical built-up roofs using tar pitch.

9. For paving, the type of traffic, drainage, and correct installation techniques should be determined, as well as whether paving should be installed by the penetration method, mixed in place, or mixed in a central plant. (*See also* Asphalt Paving.)

Conditions Favorable to the Use of Tar

1. Where metals or materials need a durable protective coating.
2. For built-up roofs except in very hot climates.
3. For durable waterproofing and membranes for walls and slabs below grade.
4. For paving driveways, roads, paths, and walks, except in very hot climates.

Conditions Unfavorable to the Use of Tar

1. For built-up roofs with pitches over 2 to 12.

2. For paving, built-up roofs, and coatings in areas where solar heat and hot climates will cause tar to soften.

History and Manufacture

The first tar pitch was made from the distillation of wood and was used for waterproofing joints in wood ships. The exact date of its discovery is unknown. Coal tar was discovered about 1665 by Becher, a professor of medicine, in Germany. The introduction of coal gas lighting in 1792 in England was the next step in the development of the coal tar industry. In 1825 coal tar distillates were used to prevent decay in railroad ties, and in 1901 the first road was surfaced with coal tar. In 1907 Baekeland discovered the Bakelite resins in coal tars, and since then new coal tar products have continued to appear.

TELLURIUM

Physical and Chemical Properties

Symbol: Te
Specific gravity: 6.24 (20°C, 68°F)
Atomic number: 52
Melting point: 449.5°C, 841.1°F
Boiling point: 989.8°C, 1813.64°F

Tellurium is a brittle, silvery-white, semimetallic element that resembles sulfur and selenium in many respects but exhibits metallic properties not evidenced by these two elements. Both tellurium and its compounds are toxic if ingested. Molten tellurium is partially or completely miscible with most metals and forms tellurides by direct reaction with base metals.

Commercial Forms. Tellurium is usually cast, distilled, or electrodeposited and is available in ingot, stick, or powder form as 99% tellurium.

The composition of high-purity tellurium is given in Table *T3*.

Table T3 Composition of tellurium metal

Te	Se	Cu	Pb	SiO_2	Other elements
		(percent of content)			
99.7	0.02	0.015	0.003	0.04	0.222

Types and Uses

Tellurium has several uses in compound form. Its uses in metallic or elemental form are indicated by asterisks in Table *T4*. An important construction application involves the use of tellurium compounds, in particular sodium tellurate, for chemical finishes on steel, copper and its alloys, aluminum, and zinc.

Table T4 Types and uses of tellurium

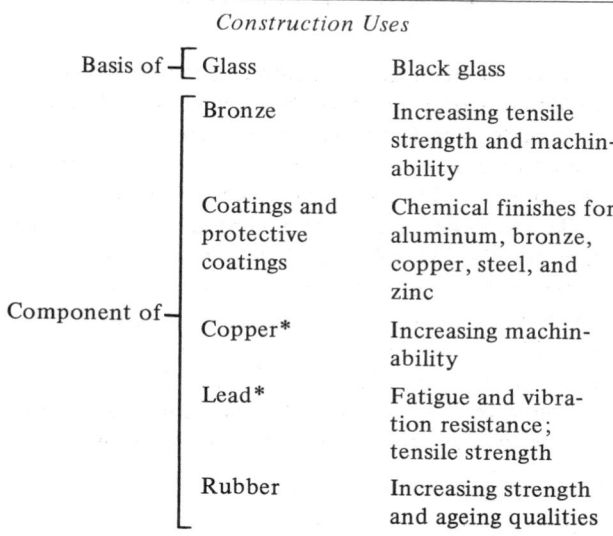

	Construction Uses	
Basis of	Glass	Black glass
Component of	Bronze	Increasing tensile strength and machinability
	Coatings and protective coatings	Chemical finishes for aluminum, bronze, copper, steel, and zinc
	Copper*	Increasing machinability
	Lead*	Fatigue and vibration resistance; tensile strength
	Rubber	Increasing strength and ageing qualities
	Nonconstruction Uses	
Corrosion inhibitor,* stainless steel,* thermoelectric cooling		

History and Manufacture

Although the peculiar properties of tellurium had been noted earlier, it was not until 1798 that Klaproth recognized it as a distinct element and named it after the Greek word *tellus*, meaning "earth."

Tellurium is recovered mainly from the slimes of copper or lead refining. First the precious metals are separated from these slimes; then the remaining slag is treated to precipitate tellurium dioxide, which is converted into tellurium metal by smelting or electrolytic processes.

TEMPERATURE

Temperature is the degree of hotness or coldness measured on a definite scale. According to the kinetic theory

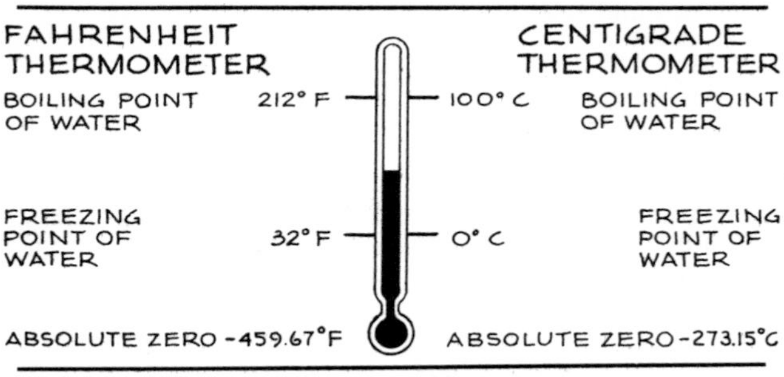

Figure T2 Typical mercury thermometer.

of heat, temperature is a measure of the average amount of kinetic energy possessed by each individual molecule of a body. The Kelvin (K) is the thermodynamic scale used to measure temperature in the metric system. At absolute zero, 0 K, all atomic vibration ceases.

The two scales commonly used for measuring temperature are the Fahrenheit and the Celsius, the preferred name for Centigrade. In most instruments for measuring temperature there are two fixed points: the temperature of boiling water (212° Fahrenheit, 100° Celsius) and the temperature of melting ice (32° Fahrenheit, 0° Celsius). Absolute zero is -459.4° Fahrenheit and -273° Celsius. To convert Fahrenheit into Celsius the equation is

$$\tfrac{5}{9}(°\text{F} - 32) = °\text{C}.$$

To convert Celsius into Fahrenheit the equation is

$$\tfrac{9}{5}(°\text{C}) + 32 = °\text{F}.$$

The most generally used instrument for measuring temperature is the mercury thermometer (*see* Figure *T2*). Its use is limited by the boiling point (166.33°F, 74.63°C) and the freezing point (-53.66°F, -47.59°C) of mercury. In this book temperatures are given in both Fahrenheit and Celsius to acquaint practitioners in the construction field with both scales.

TEMPERING

Tempering is the process of heating metal, glass, or other material to a temperature below the transformation stage and then cooling it at a controlled rate to change its hardness, strength, toughness, and/or other characteristics. *See* Glass; Stainless Steel; Steel; Plastics.

TENT STRUCTURES

The earliest tent used by prehistoric man consisted of supporting poles covered by animal hide, leaves, or straw. Perhaps the largest early tent structure was the one made for the Barnum & Bailey Circus. With the development of wire rope, stronger steel, and plastic films, sheeting, etc., the tent type of structure became more widely used in the building construction field (*see* Figure *T3*).

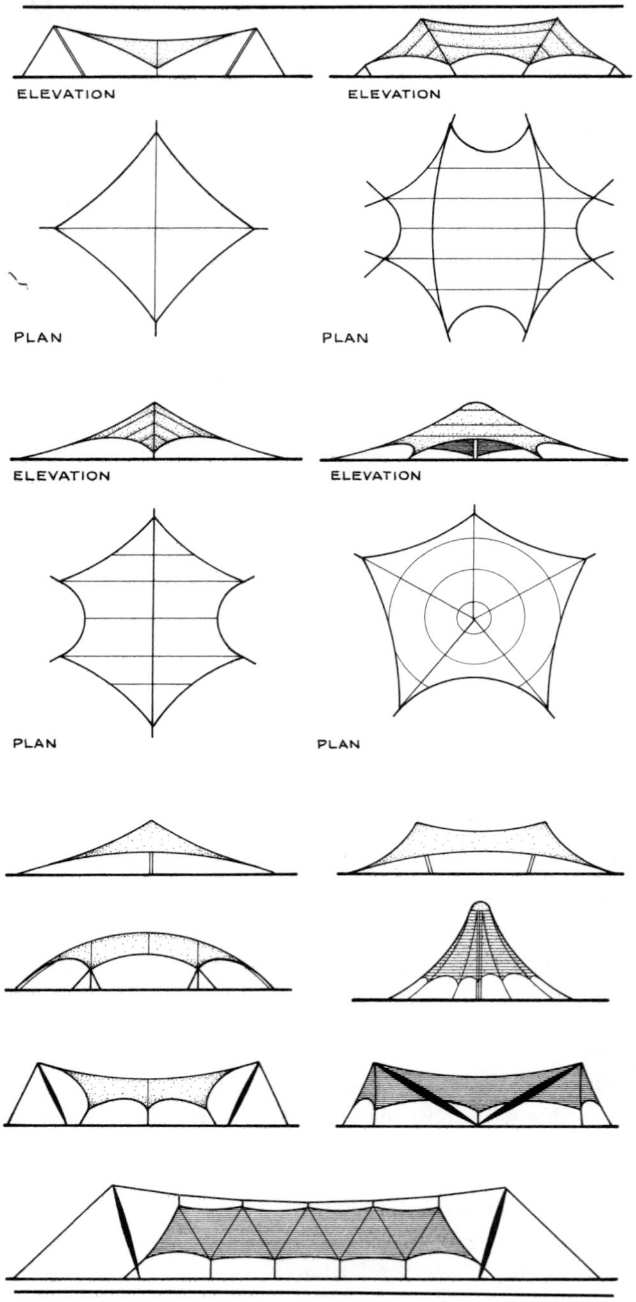

Figure T3 Various types of tent structures.

At present tent structures are generally built for limited time use (1 month, 1 year, or 2 years) for events such as Olympic Games, exhibitions of all types, and festivals.

It is beyond the scope of this book to deal with the engineering, particularly the aerodynamics, required for the design of tent structures.

See also Cable Structures; Air Structures; Plastics; Wire.

TERNEPLATE

Physical and Chemical Properties

Terneplate is dull gray in color, lightweight, and strong. It originally consisted of sheets or strips of iron in two grades, coke iron and charcoal iron, coated with a lead-tin alloy. Later steel was used. Today terneplate for the construction field (short terne or roofing terne) consists of two types—26- and 28-gauge 304 dead soft stainless steel (18% chromium, 8% nickel), and 26-, 28-, and 30-gauge copper-bearing steel (0.2% copper)—coated on both sides with hot-dipped alloy of 80% lead and 20% tin. Terneplate for industry (long terne) also consists of two types of sheet steel—14- to 30-gauge low-carbon steel with or without 0.2% copper, and rimmed, capped, or aluminum-killed sheet steel—with hot-dipped coatings of an alloy of lead and $12\frac{1}{2}$ to 25% tin.

Base Weight. Terneplate, like tinplate, is produced to a base weight expressed in pounds per base box. In other words, the weight of a coating refers to the total weight of terne metal distributed on both sides of 112 sheets, 14 in. X 20 in. (35.56 cm X 50.80 cm) in size, or an area equal to 217.78 ft^2 (20.28 m^2). Sometimes the weight is given per double base box which consists of 112 sheets 20 in. X 28 in. (50.80 cm X 72.12 cm) in size.

Expansion Rate. Terneplate has a slight expansion rate of only 0.825 in./100 ft/100°F (2.096 cm/30.40 m/55.56°C) change in temperature. This low rate of expansion means that runs of 30 ft (9.144 m) can be made without expansion joints where ends are free; where ends are secured, runs of 15 ft (4.572 m) are possible between expansion joints.

Chemical Reactivity. Terneplate reacts with acids and aluminum. In fact, its reactivity with contacting materials parallels the chemical reactivity of lead and tin. Lead and tin act as cathodes to iron and may accelerate corrosion at pinholes if they exist. Heavy-coated sheet has no pinholes but should be painted. Terneplate is usually given a shop coat of paint on both sides, but painting after installation is also necessary.

Types and Uses

The two grades of terneplate, short terne and long terne, have already been mentioned.

Short Terne (Roofing Terne). This type comes in sheets for terne stainless steel, and sheets and 50-ft (15.24-m) rolls for copper-bearing steel. It has a coating of 20 to 40 lb (9.072 to 18.144 kg) for 112 sheets of 20 in. X 28 in. (50.80 cm X 72.12 cm) on 26-, 28-, and 30-gauge copper-bearing steel and 26- and 28-gauge stainless steel (10 to 20 lb, or 4.536 to 9.072 kg, per single-box unit, as compared to the double-box unit).

The coating consists usually of 80% lead and 20% tin. This grade is generally used for roofing, flashing, gutters, and leaders (*see* Tables *T5* and *T6*).

Table T5 Physical characteristics of terne-coated copper-bearing steel sheet (roofing terne)

		Thickness		Weight of coatings per 112 sheets		Weight		50-ft (15.24-m) rolls		Rolled flat and cut into sheet, maximum size	
Type	Gauge	in.	mm	lb	kg	lb/ft^2	kg/0.0929 m^2	in.	mm	in.	mm
JC	30	0.120 plus coating	3.048 plus coating	20–40	9.072–18.144	0.54	0.245	4	101.6	28 × 144	711.2 × 3,657.6
								6	152.4		
								7	177.8		
1X	28	0.0148 plus coating	0.387 plus coating	20–40	9.072–18.144	0.65	0.295	8	203.2		
								10	254.0		
								12	304.8		
2X	26	0.0179 plus coating	0.455 plus coating	40	18.144	0.78	0.354	14	355.6		
								20	508.0		
								24	609.6		
								28	711.2		

Table T6 Physical characteristics of terne-coated stainless steel, short terne (roofing terne)

						Standard size sheets[a]			
		Thickness		Weight		Width		Length	
Type	Gauge	in.	mm	lb/ft^2	kg/0.0929 m^2	in.	mm	ft	m
TCS	28	0.0148 plus coating	0.376 plus coating	0.71	0.322	20 24	508 609.6	8–10	2.438–3.048
TCS	26	0.0179 plus coating	0.455 plus coating	0.82	0.372	28 36	711.2 914.4		

[a]Special widths available up to 3 × 12 ft (0.9144 × 3.6576 m).

Long Terne. This is a sheet mill product that comes in rolls up to 90 ft (27.432 m) in length and in cut lengths in large sizes up to 4 ft × 10 ft (1.219 m × 3.048 m). It has a coating of 8 to 40 lb (3.63 to 18.14 kg) per 112 sheets of 14 in. × 20 in. (35.56 cm × 50.8 cm) on 30- to 14-gauge (maximum) metal. The coating consists of $87\frac{1}{2}$ to 80% lead and $12\frac{1}{2}$ to 20% tin. Long terne is used for gasoline tanks, automobile body parts, doors and frames, and other fireproof construction, as well as for roofing. For doors and frames and for fireproof construction, the gauge and weight of coating are set by the National Board of Fire Underwriters.

Usage. Terne-coated copper-bearing steel should be painted after installation, and the paint should be well brushed in. The underside surfaces should have either a mill-applied shop coat or a coat or red iron oxide-linseed oil paint before installation. The exposed surface should receive a base coat of red iron oxide-linseed oil paint even if there is a mill-applied shop coat, and a second coat of any color, good-quality, long oil exterior paint.

Terne-coated stainless steel does not need to be painted. Upon exposure it weathers to a dark gray.

Application

Condensed Checklist

1. For flashing, all locked flat seams should be soldered.
2. Solder should be 50% lead and 50% tin, and only a rosin flux should be used (no acid).
3. Building paper must be coated with rosin and sized. Tar papers or other papers containing acids or aluminum should not be used.
4. Terneplate should always be held by clips, never nailed directly. Two $\frac{7}{8}$-in. (22.23-mm) zinc, lead, or rosin-coated roofing nails should be used for JC, 1X, and 2X, and $\frac{7}{8}$-in. (22.23-mm) stainless steel nails for TCS.
5. Expansion joints should be installed as shown in Figure *T4* every 15 ft (4.572 m) when the ends are secured, and at the end of runs of 30 ft (9.144 m) when the ends are free.
6. The sheets or strips for roofing should be joined by standing or batten seams (*see* Figure *T5*).
7. On roofs with 3 to 12 pitch or greater, standing or batten seams should be used; on roofs with lesser pitch, soldered locked flat seams should be used (*see* Figure *T6*).

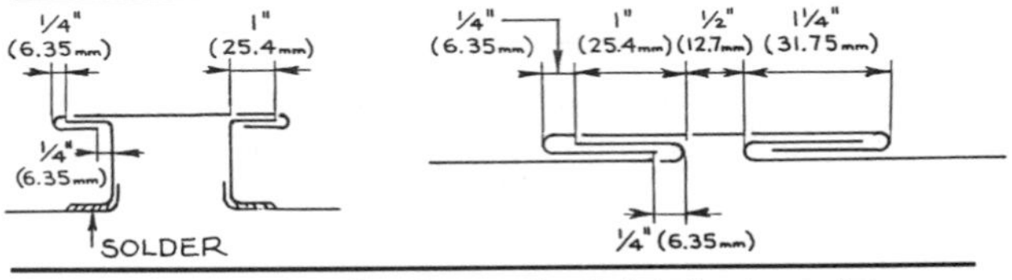

Figure T4 Details of expansion joints for terneplate.

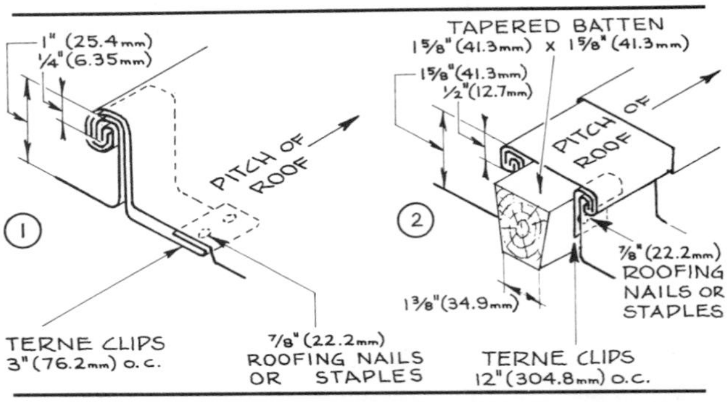

Figure T5 Standing seam (1) and batten seam (2) details.

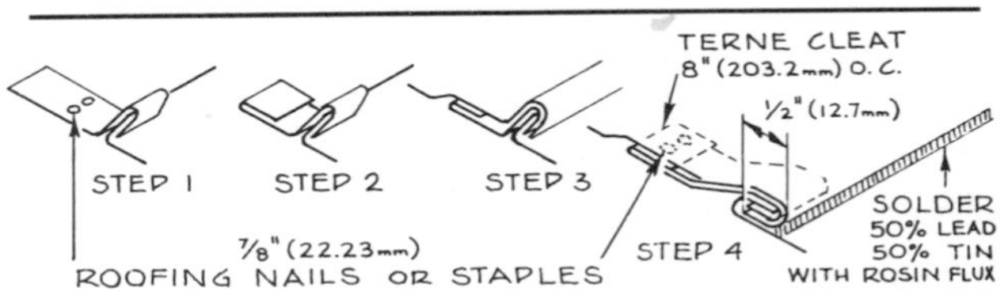

Figure T6 Details of half-inch soldered locked flat seam for roofs with less than 3 to 12 pitch.

Table T7 Finish widths of short terne after reduction of seam allowance

	Starting width							
	14 in. (35.56 cm)		20 in. (50.8 cm)		24 in. (60.96 cm)		28 in. (71.12 cm)	
	Finish width							
Type of seam	in.	mm	in.	mm	in.	mm	in.	mm
1 in. (25.4 mm) double-locked standing	$10\frac{3}{4}$	273.05	$16\frac{3}{4}$	425.45	$20\frac{3}{4}$	527.05	$24\frac{3}{4}$	628.65
2 in. (50.8 mm) X 2 in. (50.8 mm) batten,	9	228.6	15	381	19	482.6	23	584.2
center to center of battens	11	279.4	17	431.8	21	533.4	25	635
4 in. (101.6 mm) X 4 in. (101.6 mm) batten,			11	279.4	15	381	19	482.6
center to center of battens			15	381	19	482.6	23	584.2

Table T8 Coverage data for one square of roof

		Square feet (square meters) for one square (100 ft², 9.29 m²) of roof							
Terne width		1 in. (25.4 mm) double locked standing seam		2 in. X 2 in. (50.8 mm X 50.8 mm) batten seam		4 in. X 4 in. (101.6 mm X 101.6 mm) batten seam		Bermuda roof 2 in. (50.8 mm) riser	
in.	mm	ft²	m²	ft²	m²	ft²	m²	ft²	m²
14	355.6	130	12.08	164	15.24			134	12.45
20	508	119	11.06	142	13.19	174	16.17	122	11.33
24	609.6	116	10.78	134	12.45	158	14.68	118	10.96
28	711.2	113	10.50	128	11.89	148	13.75	115	10.68

8. All cross seams should be locked flat seams which have been soldered.
9. Calculations for roof pattern design should be based on finished widths, making correct allowance for seam deductions as shown in Tables *T7* and *T8*. The 20-in. (508-mm)-wide rolls are generally used.
10. Treatment of standing, flat, and continuous standing seams at roof ridges or hips and at gable ends of roof (*see* Figure *T7*).
11. Treatment of batten seams at ridges or hips and at gable ends of roof (*see* Figure *T8*).
12. Eave and valley intersections for standing, flat, or batten seams (*see* Figure *T9*).
13. Treatment of horizontal seams on Bermuda-type installations (*see* Figure *T10*).
14. The undersides of terne types JC, 1X, and 2X must have a mill-applied shop coat or an applied coat of red iron oxide-linseed oil paint before installation.
15. The exposed surfaces of terne types JC, 1X, and 2X must have a coat of red iron oxide-linseed oil

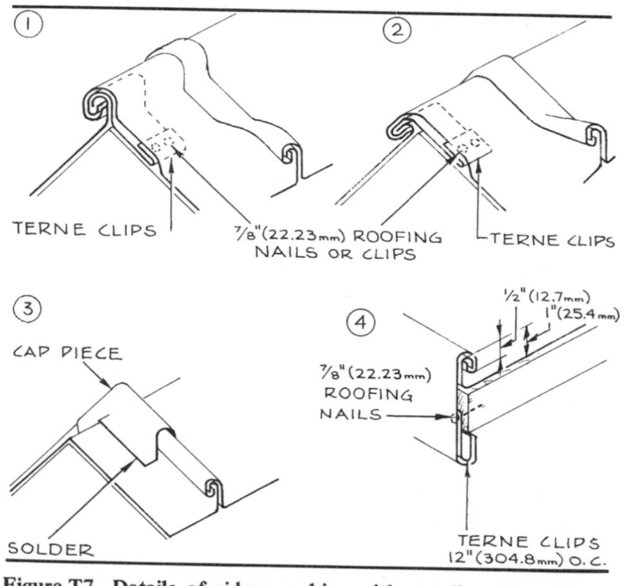

Figure T7 Details of ridges or hips with standing (1), flat (2), and continuous standing (3) seams and of gable end (4).

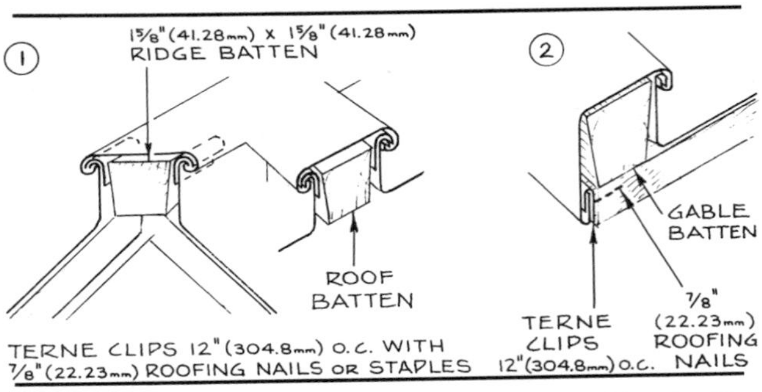

Figure T8 Details of batten seams at ridge or hip (1) and gable end (2).

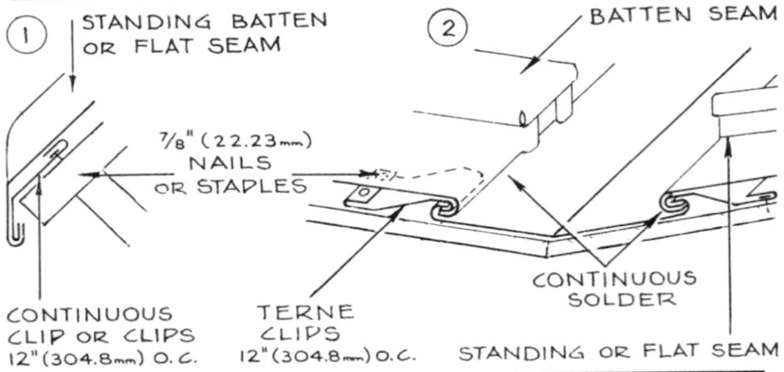

Figure T9 Details of standing, flat, and batten seams for eaves (1) and valleys (2).

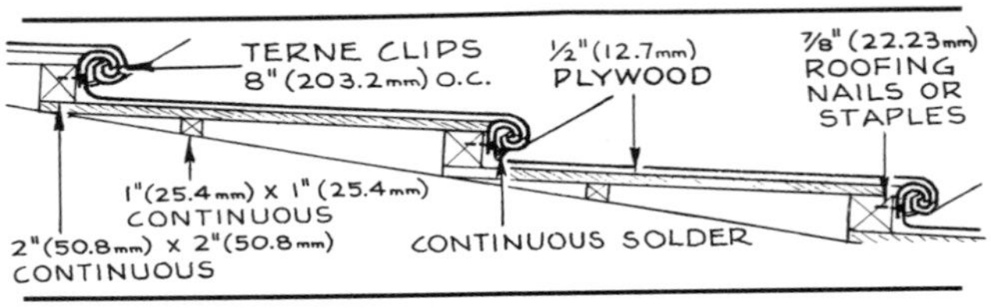

Figure T10 Details of Bermuda-type terne roof.

paint and a second coat of good-quality, long oil exterior paint in any color. The mill-applied shop coat is not to be considered the first coat.

16. Terne-coated stainless steel requires no painting and weathers to a uniform dark gray. It may be painted with a good-quality, long oil exterior paint if desired.

Conditions Favorable to the Use of Terneplate

1. Where a lightweight, low-expansion, strong roofing or flashing material is required.
2. Where strong horizontal or vertical lines are desired on a roof.
3. Where staining of materials adjoining the roofing or flashing must be avoided.
4. Where a fire-resistant roofing material is advantageous. Municipal, state, and local codes should be checked and also the fire rating codes of the Fire Underwriters, insurance companies, labor departments, and federal government (Army, Navy, etc.).
5. Where maintenance must be kept to a minimum, use terne-coated stainless steel.

Conditions Unfavorable to the Use of Terneplate

1. Where painting of the material is not desired.
2. Where maintenance must be kept to a minimum, don't use terne type JC, 1X, or 2X.
3. Where the terne will come into direct contact either with aluminum or with materials that contain acids.

History and Manufacture

Terneplate originated in Wales about 1700, although at that date it was known as tin roofing. Its use in the United States became widespread in the early 19th century. Andrew Jackson's home, The Hermitage, had a terne roof which was installed in 1835. Recently, however, The Historical Society had all the buildings in The Hermitage complex reroofed with terne-coated stainless steel, which was painted with red iron oxide so that the entire complex appears visually the same as it has since it was originally constructed.

Terneplate is manufactured by a process similar to the zinc galvanizing process. The metal to be coated is pickled and then passed first through a weak solution of hydrochloric (or sulfuric) acid and then through a heated solution of zinc chloride (or zinc chloride plus hydrochloric acid), which acts as a flux for better adhesion of the coating. Next the metal is passed into a molten lead-tin alloy which is kept at about 700 to 725°F (371.11 to 385°C) and covered with palm oil to prevent oxidation. As the metal sheet emerges, it passes between rollers which remove excess metal and oil and give a smooth coating. The sheet is then cleaned with flannel disk rolls and brushes, using sawdust, bran, or a similar absorbent to eliminate the oil.

TERRA COTTA

Terra cotta is a clay tile material that was used extensively before 1920. In the contemporary vernacular the term "terra cotta" has been replaced by the term "ceramic veneer" (*see* Clay Tile, Ceramic Veneer).

TERRAZZO

Physical and Chemical Properties

Terrazzo is a dense, durable floor and wall material consisting of a topping mixture of colored stone granules with either a matrix of portland cement (white or gray) and water or a matrix of synthetic resins. This topping is applied to an underbed of concrete fill or a concrete slab for a matrix of portland cement, and an underbed of wood, metal, or concrete fill or a concrete slab for a matrix of synthetic resins. Generally, the resin is an epoxy, a polyester, or a polyacrylate. Metal or plastic dividing strips must be installed for portland cement terrazzo to eliminate possible cracking, whereas for synthetic resin terrazzo (thin-set terrazzo) dividing strips are not necessary except in extremely large areas, 30 ft (9.144 m) or longer in length, and for decorative purposes. Both types, after the topping has set, require that it be ground and polished to a smooth surface finish. The range and variations of colors are controlled by the colors available in stones and in the pigments suitable for portland cement or the synthetic resin selected.

Mosaic. Another form of terrazzo, called mosaic, is made by combining small pieces of stone, pottery, glass, marble, metal, and other materials to create a design of some sort and placing them in a thin bed of cement.

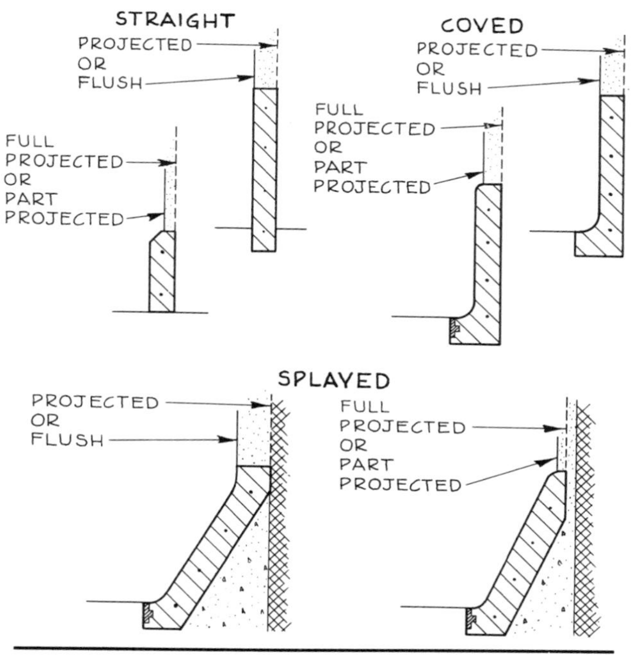

Figure T11 Precast terrazzo bases.

Commercial Forms. Terrazzo is available in precast form and made-in-place form, with either a smoothly polished, nonslip or sparkproof (conductive) surface.

Types and Uses

Both types of terrazzo are used for floors and bases where durability, resistance to wear, and minimal maintenance are necessary. It can be applied to walls, partitions, stair treads, and risers. Precast portland cement-matrix reinforced terrazzo stair treads and risers, caps, window stools, saddles, and other shapes are available (*see* Figures *T11* and *T12*). A limited number of precast shapes are available in the synthetic resin type of terrazzo. Both types of terrazzo floors may be made nonslip and nonsparking. The proportion of ingredients for portland cement-matrix terrazzo is shown in Table *T9*. The water used must be clean and fresh. *See also* Tables *T10*, *T11*, and *T12*.

Application

Setting Bed. Portland cement-matrix terrazzo for floors may be applied to concrete, steel decking, and wood. A setting bed must be installed in all cases. For applying to concrete there are two methods: bonding the terrazzo directly to the concrete, or separating it from the concrete. For wood or steel decking, separation is achieved by means of tar paper or sand. (*See* Figure *T13*.) When terrazzo is to be bonded to the concrete, the surface must be thoroughly cleaned and given a thin coating of cement and water (neat cement) to create good bonding. The underbed of concrete fill for all types of terrazzo is generally a mixture by volume of 1 part cement, 4 parts coarse screened sand, and water.

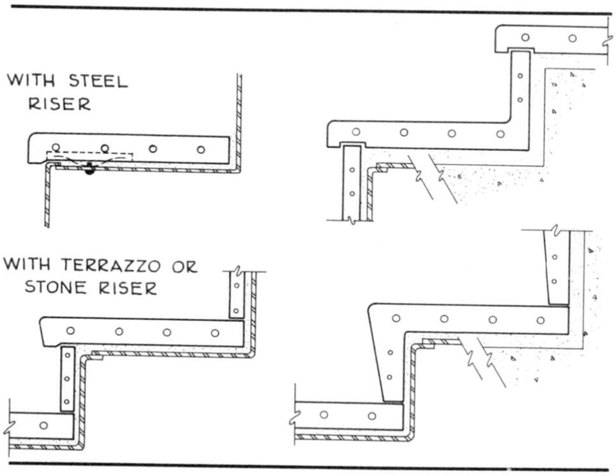

Figure T12 Precast terrazzo treads and risers.

Table T9 Ingredients for terrazzo

Type of terrazzo	Ingredients are mixed dry: Marble granules lb (kg)	Portland cement lb (kg)	Abrasive granules lb (kg)	Dividing strips*	Color pigments for cement
Terrazzo topping	200 (90.70)	94 (42.64)		Brass, zinc alloys, and plastic $1\frac{1}{4}$ in. (31.75 mm) deep	Nonfading, limeproof mineral pigments
Nonslip heavy duty	150 (68.04)	94 (42.64)	50 (26.68)		

*These are the generally used types. Any noncorrosive metal may be used, e.g., nickel silver, aluminum alloys, stainless steel, etc.

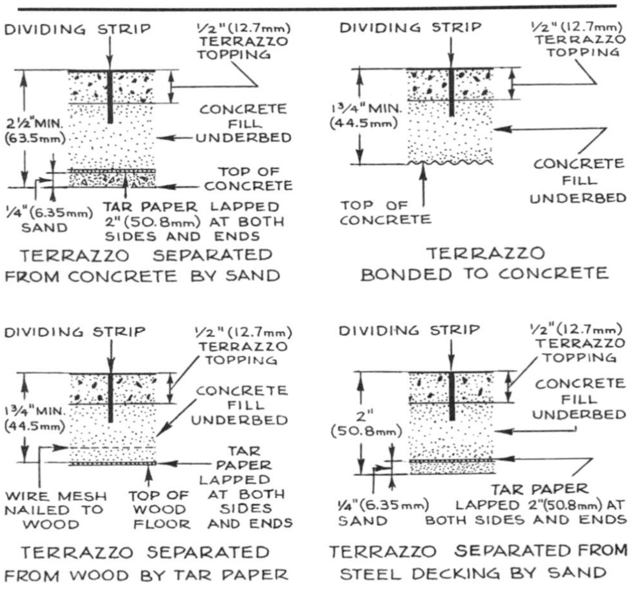

Figure T13 Methods of applying terrazzo to different types of subflooring materials.

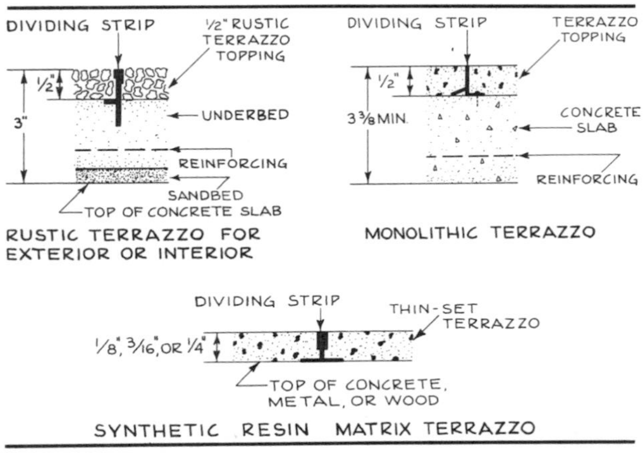

Figure T14 Dividing strips for rustic, monolithic, and synthetic resin types of terrazzo.

Table T10 Marble chip sizes

Number	Passes screen in.	Passes screen mm	Retained on screen in.	Retained on screen mm
0[a]	$\frac{1}{8}$	3.2	$\frac{1}{16}$	1.6
1[a]	$\frac{1}{4}$	6.4	$\frac{1}{8}$	3.2
2[a]	$\frac{3}{8}$	9.5	$\frac{1}{4}$	6.4
3[b,c]	$\frac{1}{2}$	12.7	$\frac{3}{8}$	9.5
4[b,c]	$\frac{5}{8}$	15.9	$\frac{1}{2}$	12.7
5[c]	$\frac{3}{4}$	19.2	$\frac{5}{8}$	15.9
6[c]	$\frac{7}{8}$	22.4	$\frac{3}{4}$	19.2
7[c]	1	25.4	$\frac{7}{8}$	22.4
8[c]	$1\frac{1}{8}$	28.6	1	25.4

[a]These sizes available separately.
[b]These sizes available separately from some quarries.
[c]Larger sizes are frequently available grouped, for example, 3–4 mixed, 7–8 mixed, and 4–7 mixed.

Synthetic resin-matrix terrazzo can be installed directly on concrete fill, a concrete slab, metal, or wood without any setting bed (*see* Figure *T14*).

Dividing Strips. Dividing strips are all $1\frac{1}{4}$ in. (31.75 mm) deep. Those less than $\frac{1}{8}$ in. (3.18 mm) thick are the same thickness along the entire $1\frac{1}{4}$ in. (31.75 mm) depth, whereas those $\frac{1}{8}$ in. (3.18 mm) or more in thickness have only the top $\frac{3}{8}$ in. (9.53 mm) of the strip the specified thickness. They are usually installed after the concrete fill underbed is spread and brought to a level about $\frac{5}{8}$ in. (15.88 mm) below finish floor level and while it is still not set. After the underbed has set, the terrazzo topping is placed in the areas between dividing strips and compacted with heavy rollers until excess water is extracted. Then it is troweled until the tops of the dividing strips are visible.

Table T11 Types, sizes, characteristics, and major uses of terrazzo with portland cement matrix

Type	Thickness in.	Thickness cm	Weight lb/ft²	Weight kg/m²	Chips	Advantages	Disadvantages	Major Use
Sand cushion	$2\frac{1}{2}$ min.	6.35 min.	27±	131.82±	Marble	Low transmission of sound to structure; easily maintained	Affected by acids and alkalis; will stain	Floors and areas where there will be heavy traffic
Bonded to concrete	$1\frac{3}{4}$ min.	4.92 min.	18±	87.88±	Marble	Easily maintained	Transmits sound to structure; affected by acids and alkalis; will stain	Floors, walls, and stairs
Monolithic	$\frac{1}{2}$ min.	1.27 min.	7±	34.17±	Marble	Easily maintained	Transmits sound to structure; affected by acids and alkalis; size of stone chips limited; will stain	Floors and areas where there will be heavy traffic
Rustic	$2\frac{3}{4}$ min.	6.74 min.	31±	151.34±	Marble, quartz, granite, or pebbles	Nonslip; resists weathering; exterior or interior paving or flooring	No color selection as it is not ground or polished	Exterior paving and sidewalks
Precast	Varies		Varies		Marble	Easily maintained	Affected by acids and alkalis; will stain	Stairs, bases, caps, window stools, etc.

Table T12 Types, sizes, characteristics, and major uses of terrazzo with synthetic resin matrix

Type	Thickness in.	Thickness mm	Weight lb/ft²	Weight kg/m²	Chips	Advantages	Disadvantages	Major use
Epoxy polyester	$\frac{1}{4}$–$\frac{3}{8}$	6.35–9.6	3–$4\frac{1}{2}$	14.65–21.97	Marble or other type of chips for special floors	Lightweight, nonstaining, rapidly installed, easily maintained, and nonslip	Not fireproof but self-extinguishing	Floors, walls, remodeling, areas where chemical resistance is required
Polyacrylate	Same as above		Same as above		Polyacrylate, polyester, marble, or other type of chips for special floors	Same as above	Not fireproof but fire retardant and self-extinguishing	
Polyester	Same as above		Same as above					
Precast	Varies		Varies		Binder and marble or other chips	Same as above	Not fireproof but fire retardant and self-extinguishing depending on binder	Floors, stairs, bases, caps, window stools, etc.

Dividing strips for synthetic resin-matrix terrazzo are $\frac{1}{8}$, $\frac{3}{16}$, and $\frac{1}{4}$ in. (3.18, 4.76, and 6.35 mm) deep and are secured to the underbed with an adhesive that is compatible with the matrix of the topping.

Topping. For terrazzo topping 70% of the marble granules must show; for heavy-duty nonslip topping a proportion of three marble granules to one abrasive must show; and for light-duty nonslip topping, where abrasive is sprinkled on the finish, a proportion of four marble granules to one abrasive granule must show. The surface must be kept moist for 6 days and then machine-ground, using first No. 24 and then No. 80 grit abrasive stones. Next, the entire surface is given a thin coating of neat cement to fill all voids. This coating should contain the same cement and color used for the topping. After not less than 72 hours, the terrazzo surface is finished with No. 80 grit abrasive stones and thoroughly washed.

The installation and finishing of synthetic resin-matrix terrazzo generally is in accordance with the strict specifications of the manufacturer or supplier of the synthetic resin material.

Condensed Checklist

1. Check that all necessary changes in level are installed in concrete that is to receive terrazzo finish (e.g., mat sinkages, steps, platforms).
2. Make sure all dividing strips and expansion joints are installed according to design (*see* Figure *T15*).
3. If a terrazzo base is part of the installation, check whether all necessary metal lath and temporary grounds have been used and whether the work is installed correctly. Dividing strips for portland cement-matrix terrazzo base should be installed not more than 5 ft (1.524 m) apart.
4. For stairs and stair platforms of concrete, when portland cement matrix is used, check whether the rough concrete for treads and platforms is 2 in. (50.8 mm) below the level of the finished terrazzo tread or platform and $1\frac{1}{2}$ in. (38.1 mm) back of the finished terrazzo riser. For other areas (stringers, curbs, etc.) there should also be a 2-in. (50.8-mm) setback from the finished terrazzo surface.

   For synthetic resin-matrix terrazzo, all of the above dimensions should be $\frac{1}{8}$, $\frac{3}{16}$, or $\frac{1}{4}$ in. (3.18, 4.76, or 6.35 mm), depending on the thickness of synthetic resin-matrix terrazzo selected.
5. Request a sample of the specified terrazzo finish. The finished product should be like the sample.
6. For walls, check whether the surface to which terrazzo is to be applied is at the correct depth behind the finish face and whether dividing strips for portland cement-matrix terrazzo are installed according to specified design.
7. For steel stairs where portland cement-matrix terrazzo is to be used, check whether metal lath is installed and secured to the steel of the stairs and whether 2 in. (50.8 mm) has been allowed for terrazzo treads, $1\frac{1}{2}$ in. (38.1 mm) for risers, and 2 to 3 in. (50.8 to 76.2 mm) for platforms.

   For synthetic resin-matrix terrazzo all of the above dimensions should be $\frac{1}{8}$, $\frac{3}{16}$, or $\frac{1}{4}$ in. (3.18, 4.76, or 6.35 mm), depending on the thickness of synthetic resin-matrix terrazzo selected.
8. For portland cement-matrix terrazzo always check whether the correct type and grit size of grinding and polishing abrasive are being used.

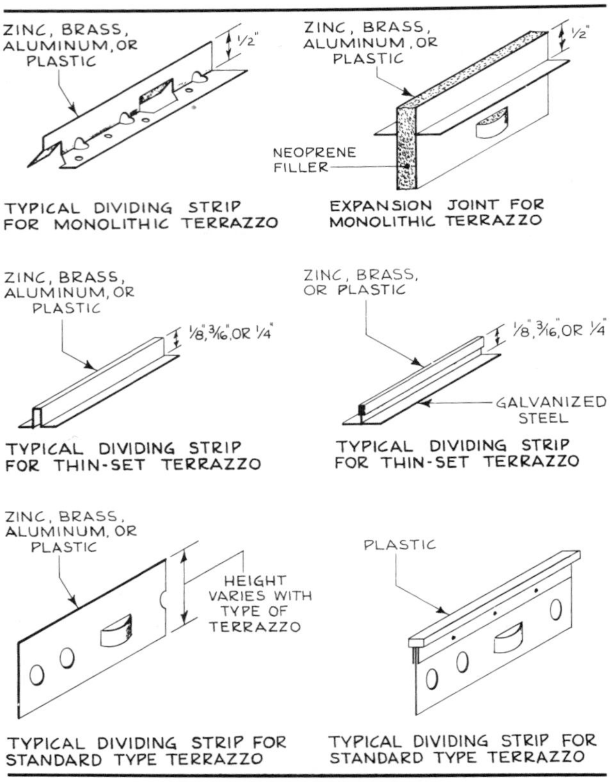

Figure T15 Typical dividing strips.

For synthetic resin-matrix terrazzo, always check the strict specifications of the manufacturer of the synthetic resin material for the method and equipment to be used in grinding and polishing the terrazzo to obtain a smooth polished finish.

9. Make sure that precast terrazzo units are reinforced and made with the same ingredients that are used for on-site work.
10. To make a terrazzo floor sparkproof (conductive), always consult specialists in this field.
11. Make sure that new portland cement-matrix terrazzo floors are cured properly. They should be scrubbed two or three times a week with neutral soap and mopped on alternate days. After 2 months of this treatment, the floors acquire a fine finish and require less upkeep.

    Synthetic resin-matrix terrazzo should be cured according to the strict specifications of the manufacturer of the synthetic resin material.
12. For nonslip portland cement-matrix terrazzo, always check the type and size of abrasive granules to be used.
13. For portland cement-matrix terrazzo, always check with The National Terrazzo and Mosaic Association, Inc., regarding the color selections that are available.

14. For synthetic resin-matrix terrazzo, always check that the colored stone chips selected will not be too large for the thickness of the thin-set terrazzo to be installed.

Conditions Favorable to the Use of Terrazzo

1. Where a hard, durable, colored floor, stair, or wall finish that is easy to maintain is required.
2. Where an area will receive very heavy wear.
3. Where a floor material that permits flexibility in color, design, and pattern is desired.

Conditions Unfavorable to the Use of Terrazzo

1. Where alkalis or acids will come in contact with floors, stairs, or walls.
2. Synthetic resin-matrix terrazzo, in areas where alkalis, acids, or solvents are used or stored unless the synthetic resin selected is not attacked by these substances.

History and Manufacture

The Greeks used regular-shaped pieces of marble, ceramics, and semiprecious stones (with gypsum as the cementitious material) to decorate their floors. The Romans, who developed the first concrete, made of decorative mosaic-type floors a very important art form, and after the fall of the Roman Empire, Byzantine architecture brought it to its highest point. Mosaics covered not only floors and walls but also columns and domed ceilings.

Terrazzo as we have known it was developed after the discovery of portland cement. With the research and experimentation in plastics after 1945 a new type of terrazzo flooring was developed in which synthetic resins serve as the binder (matrix) instead of portland cement.

Having these two types of terrazzo with the almost unlimited color range in plastics, the almost unlimited colors available in marbles, the many other types of stone and chips, and the wide range of cement pigments, it is possible to obtain a flooring material that will meet almost any type of use or design requirement.

Unless various other types of materials are used in addition to marble to form the design, a mosaic floor is made in the same manner as terrazzo floors. When using various materials other than marble, one should always consult with the manufacturers of portland cement and synthetic resins to obtain the correct materials, methods, and equipment.

TEXTILES

Physical and Chemical Properties

The yarns from which textiles are woven must be made from fibers that can resist the action of materials used in the manufacturing process and in all subsequent operations such as washing, cleaning, bleaching, and dyeing.

These materials may include alkalis, inorganic and organic acids (nitric, sulfuric, acetic), bleaches, hydrocarbons (dry-cleaning solvents), inorganic salts (seawater, soap solutions), dyes, and mordants. Textiles must also be able to resist extremes of heat and cold.

Textiles reflect the advances in the technology of plastics, adhesives, and synthetics in general. Many of today's textiles are nonfading, waterproof, weatherproof, fireproof, resistant to sunlight including ultraviolet rays, and unaffected by insects, fungi, and rodents.

Types and Uses

Textiles may be grouped as woven, knitted, or felted. Commonly, only felts made of wool or a combination of wool and other fibers are classified as textiles. Most of those used in architectural applications are considered paper pulp products. In woven textiles the yarns are interlaced at right angles to each other. In knit fabrics the yarns are looped so that one row of loops is caught in the preceding row of loops. Textiles are also usually identified on the basis of fiber content and count (*see* Fibers).

Construction uses of textiles are now largely limited to the field of interior decoration. Direct architectural applications include (1) canvas for roofing, detailed as shown in Figure *C2* (*see* Canvas), and (2) various types of cloth flashing, for example, cloth impregnated with asphalt and often reinforced with mesh, metal foil, or glass fibers, or cloth laminated with plastics.

Textiles have proved their practicability in exterior furniture, awnings, umbrellas, etc. It is entirely possible in the near furture for textiles to become a practical material for construction applications such as exterior finish not only for roofing but also side walls of buildings. For example, there has been a new development in using textiles for roofing in a two-part adhesive combination in which the textile has been treated with one component of an adhesive. To install the roofing, the activator of the adhesive is applied to the roof surface, and on contact the textile becomes adhered to the roof.

The introduction of plastics and plastic fibers into the textile field has completely changed the terminology in this field. Plastic fibers, foam, and sheet materials are now used in construction for both interior and exterior surfacing such as athletic fields and gyms; for exterior and interior carpets; as enclosure materials for air, cable, and tent structures; and for roofing. *See* Plastics; Fibers; Paper and Paper Pulp Products.

Fire Resistance. Fire resistance and flameproofing are of major importance wherever textiles are used in public buildings. Table *F3* (*see* Fibers) lists the fire retardants commonly applied to textiles. It can be used by the architect as a basis for specifying according to desired requirements or for identifying and comparing the relative merits of each type, many of which are sold under trade names.

All textiles should always be checked for flame spread and whether they are self-extinguishing if they do burn. Also, all building codes should be checked for acceptance of the textile materials selected for a building.

History and Manufacture

A history of textiles other than the brief summary under Fibers is beyond the scope of this book.

THALLIUM

Physical and Chemical Properties

Symbol: Tl
Atomic number: 81
Specific gravity: 11.85 (20°C, 68°F)
Melting point: 303.5°C, 578.3°F
Boiling point: 1457°C, 2654.6°F

Thallium is a bluish gray, leadlike metal which, when ignited, gives a green light. Its salts are highly poisonous. It can be cast, rolled, and extruded and is available in stick, ingot, wire and powder forms. The composition of electrolytic thallium is shown in Table *T13.*

Table T13 Composition of electrolytic thallium

Tl	Fe	Pb	Cu	Other elements
99.98	0.0003	0.008	0.003	0.0087

Types and Uses

Thallium has several uses in compound form, none of them related to construction. Its uses in metallic form are indicated by asterisks in Table *T14.*

Table T14 Uses of thallium

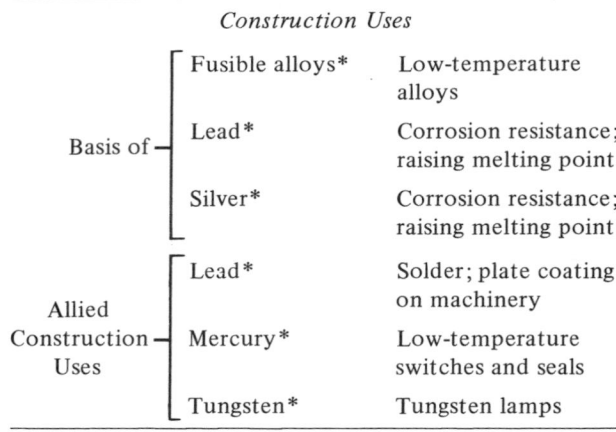

	Construction Uses	
Basis of	Fusible alloys*	Low-temperature alloys
	Lead*	Corrosion resistance; raising melting point
	Silver*	Corrosion resistance; raising melting point
Allied Construction Uses	Lead*	Solder; plate coating on machinery
	Mercury*	Low-temperature switches and seals
	Tungsten*	Tungsten lamps

History and Manufacture

Sir William Crookes discovered thallium in 1861. He compared its spectrum color to the bright green tint of new vegetation and named it after the Latin *thallus*, a budding twig.

Thallium is obtained as a by-product of the manufacture of sulfuric acid and the smelting of lead and zinc. Thallium metal is obtained by electrolysis, precipitation, and reduction.

TIES

Ties may be defined as all the small items that do not quite fit into the category of hangers or anchors but perform the same function, that is, they hold various metals and other materials together during construction. They may be small pieces of metal, mesh, or wire, or small strips of sheet metal. For example, small pieces of wire are used to join pieces of metal lathing for plasterwork. Where a concrete block wall or partition meets another concrete wall perpendicularly, small corrugated pieces of metal or small pieces of wire mesh are used to tie the two walls together. When wood or metal forms for concrete are built, ties join the wood or metal pieces comprising the form and hold the walls open and in position during the pouring of concrete (*see* Figure *T16*).

A general word of caution: Ties are among the small items in construction that are generally called for and expected to be installed; however, this is often a wrong assumption and they have been left out. The omission may be serious. One important tie easily forgotten is that which holds thin stone veneer to a steel structure.

When ties made of metal are used, a careful check

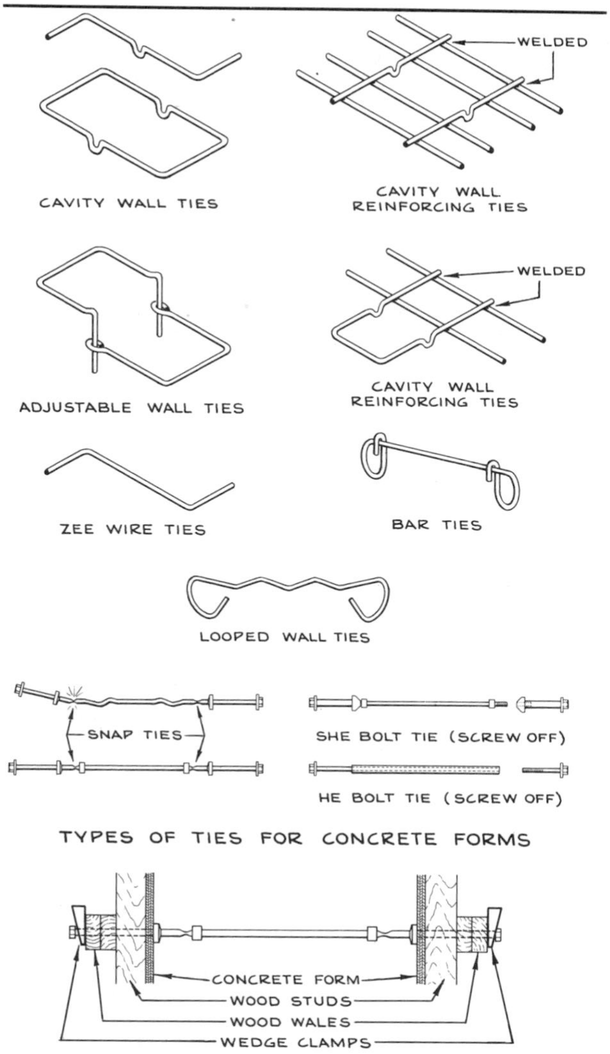

Figure T16 Various ties used in construction.

should always be made that no galvanic or chemical action can develop between them and the materials they connect. *See* Brick, Burned; Concrete; Stone.

TILE

Tile was originally defined as a thin solid slab made usually of burned clay, glazed or unglazed, and used structurally or decoratively in building. Gradually the word "tile" was extended to include (1) tile-shaped units, that is, square or rectangular, relatively thin, flat shapes made of materials other then burned clay; and (2) burned clay or "tile" products made in shapes other than flat solid slabs, for example, structural clay tile and drain and sewer tile.

In this book the first category of tile is described on the basis of the constituent materials. Take, for example, aluminum wall tile, which consists of aluminum sheet cut and formed to size and then covered with a wear-resistant, colored finish. The aluminum sheet is described under Aluminum Sheet and Strip; the finish under Aluminum Finishes or Paint and Painting, or Glazes and Porcelain Enamels, as the case may be.

The second category of tile is divided into the major types of clay products as follows: Clay Tile: Floor and Wall Tile; Clay Tile Drain and Vitrified Clay Pipe; Clay Tile Flue Linings; Clay Tile Roofing; Clay Tile, Structural; Clay Tile, Ceramic Veneer (*see* entries for each of these items).

TIN

Physical and Chemical Properties

Symbol: Sn
Atomic number: 50
Specific gravity: 7.31 for white tin
Melting point: 231.89°C, 449.4°F
Boiling point: 2270°C, 4118°F
Coefficient of expansion: 0.000023/°C

Tin is a soft, ductile, malleable, bluish white metal. Because it is normally covered with a thin film of stannic oxide, it resists corrosion by air, moisture, many acids found in various foods in the absence of air, sulfur dioxide, and hydrogen sulfide (which usually tarnishes and corrodes other metals). Tin will take a highly reflective polish and has the ability to wet other metals. Cast tin when bent emits a creaking noise known as the "cry" of tin. The composition of electrolytic tin of highest purity is shown in Table *T15*. At temperatures below 64.4°F (18°C) high-purity tin very slowly changes to a gray powder and eventually crumbles if kept long in a very cold climate.

Table T15 Composition of electrolytic tin of highest purity

Sn	Sb	Pb	Bi	Cu	Fe
		(percent of content)			
99.9908	0.0041	0.0009	0.0002	0.0002	0.0038

Workability. Tin flows readily under pressure and can be easily worked by spinning, rolling, and extrusion. It can be rolled as thin as 0.00015 in. (0.0038 mm).

Commercial Forms. It is commercially available as a high-purity metal (minimum of 99.80% tin) in pig, ingot, shot, anode, and powder form, and in various fabricated products such as sheet, wire, foil, tube, and pipe.

Types and Uses

The main use of tin is in metallic form as either pure tin or tin-containing alloys for protective coatings on stronger metals. The major use (up to one-half of the total tin consumption in the United States) is as "tin" cans for the food industries. The next most important use is as an alloying element. Tin compounds represent only 1% of the total consumption. The uses of tin in metallic form are indicated by asterisks in Table *T16*.

Construction uses of tin include bronzes, brasses, terneplate (*see* specific heading), mirrors, gilding, solders, hardware, and fusible alloys. The compounds are used in glass, glazes, and porcelain enamels.

Table T16 Uses of tin

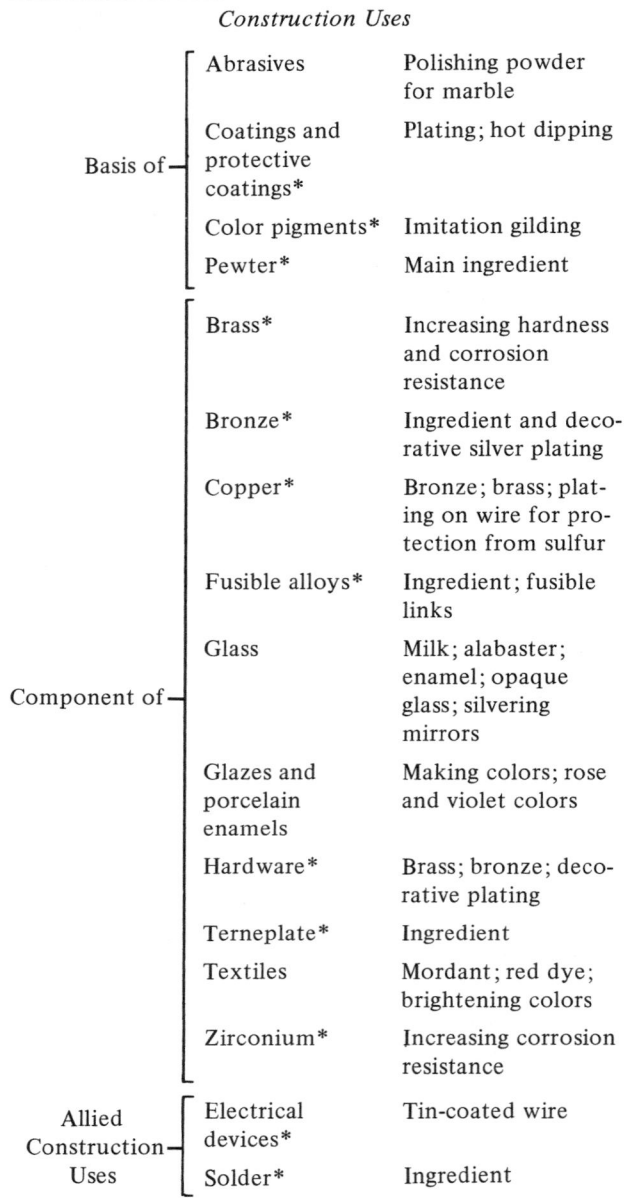

Construction Uses

Basis of	Abrasives	Polishing powder for marble
	Coatings and protective coatings*	Plating; hot dipping
	Color pigments*	Imitation gilding
	Pewter*	Main ingredient
Component of	Brass*	Increasing hardness and corrosion resistance
	Bronze*	Ingredient and decorative silver plating
	Copper*	Bronze; brass; plating on wire for protection from sulfur
	Fusible alloys*	Ingredient; fusible links
	Glass	Milk; alabaster; enamel; opaque glass; silvering mirrors
	Glazes and porcelain enamels	Making colors; rose and violet colors
	Hardware*	Brass; bronze; decorative plating
	Terneplate*	Ingredient
	Textiles	Mordant; red dye; brightening colors
	Zirconium*	Increasing corrosion resistance
Allied Construction Uses	Electrical devices*	Tin-coated wire
	Solder*	Ingredient

Nonconstruction Uses

Blueprint and sensitized papers, collapsible tubes,* electronics,* ornaments,* type metal for printing*

History and Manufacture

The earliest known use of tin dates back to about 3500 B.C., as evidenced by bronze implements discovered near the mouth of the Euphrates River. A ring and bottle of pure tin found in an Egyptian tomb indicate that even then tin was a valuable import commodity. Its source may have been the tin mined in South Africa by an ancient, unknown race whose workings were discovered in 1905. The ancient Chinese also used tin, and the Malay Peninsula, today's largest producer of tin, worked its ore deposits as far back as 800 B.C. When the Romans conquered Britain, Cornish tin, which has been worked continuously for nearly 3000 years, became available to them. Early Latin writings referred to tin as *plumbum candidum*, or white lead; later, tin became known as *stannum*, from which the symbol "Sn" is derived. (The word "tin" comes from the German *zinn*.) In America tin was known and used in Mexico and by the Incas in Peru before 1492. With the exception of Cornwall, these early sources are also today's largest producers of tin, namely, Southeastern Asia, Bolivia, and Central Africa.

Primitive Smelting. Tin smelting dates back to the Bronze Age. In the most primitive methods a raging fire was built in clay-lined trenches, and then tinstone in small pieces and more wood were thrown on alternately until a quantity of molten metal collected in the trench.

When the fire died, the molten metal was ladled into clay molds or holes in the ground. Tinstone, another name for the predominant tin mineral, cassiterite, could be panned out of streams, like gold, in small particles. Cornwall, before 1880, produced one-fifth to one-third of the world's output by methods that, apart from the substitution of a furnace for the trench or clay furnace, did not change much until after World War I.

Modern Tin Production. The first step in modern tin production is concentration of the ores. The three methods commonly used are gravity concentration flotation, and electrical concentration, the choice depending on the complexity of the ore, that is, whether it is almost pure cassiterite (SnO_2) or a sulfide combination like the Bolivian ores.

The flowchart for the Malayan dredge and cleaning shed concentrator (Figure *T17*) illustrates a relatively simple process yielding a very high tin concentrate (70 to 77% Sn).

The refining process includes primary smelting, retreatment of first-run slags to recover tin, and final removal of metallic impurities. All smelters (blast, reverberatory, or electric) use the same methods: smelting, liquation, poling, and casting.

Cornwall Process. The oldest, or Cornwall smelter, method (*see* Figure *T18*) starts with tin ore concentrates plus lime as a flux and coal slack as the reducing agent in a reverberatory furnace. After heating, stirring, and settling, four products result: crude tin, bottom slag, middle slag, and a top slag which is discarded. In the next step, called liquation, the crude tin is heated under careful temperature control so that only the impurities that have higher melting points than tin will remain. This removes most of the iron, arsenic, copper, and antimony. The semipure tin is drawn off and then heated above its boiling point and stirred by poles of green wood. This poling operation causes a boiling action, and impurities come to the surface, where they are taken off. The process is continued until tin of the required purity is obtained. The liquating and boiling process residues are hardhead (a crude mixture of tin and iron) and dross, both of which are retreated in a slag reverberatory furnace. The bottom slag is carried directly to the same slag furnace along with the middle slag, which has been crushed and stamped to obtain tin prills (crystals). This slag reverberatory furnace, charged with bottom slag and

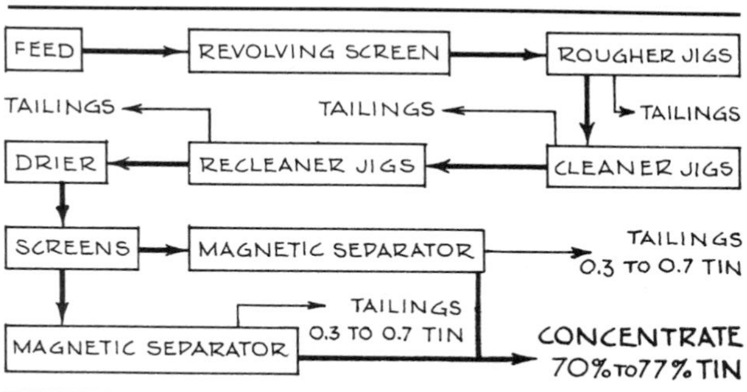

Figure T17 Malayan dredge and cleaning shed concentration.

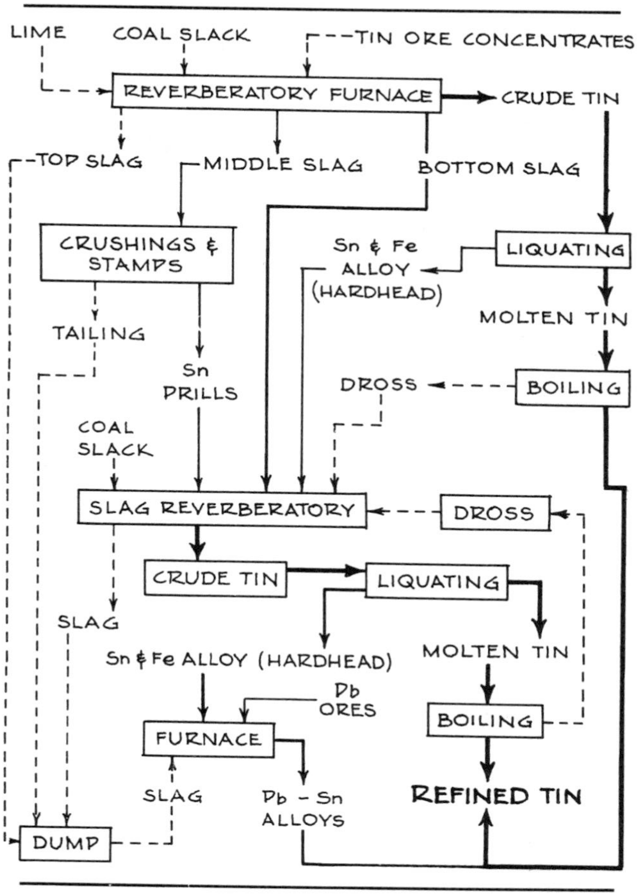

Figure T18 Cornwall smelter method of tin production.

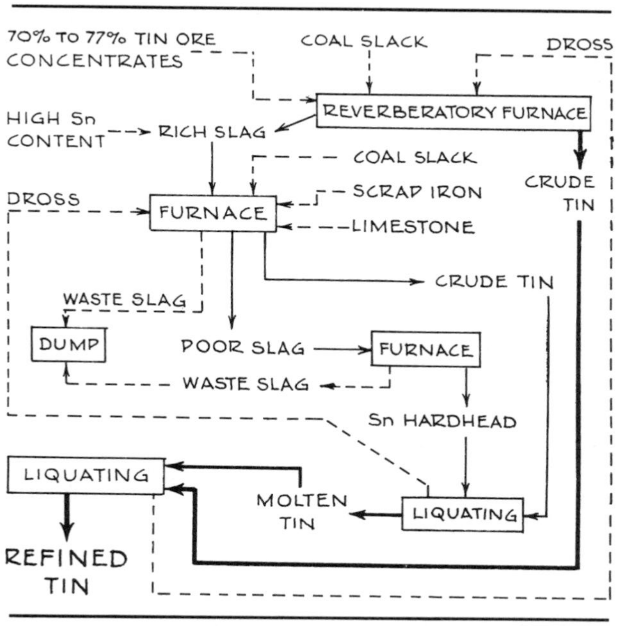

Figure T19 Process of refinement from high-tin concentrates.

prills from middle residues together with coal slack, produces crude tin and a slag which is discarded. This crude tin goes through a liquating process, as previously described, which yields refined tin and hardhead. This hardhead is treated separately to produce tin-lead alloys and a slag which is discarded. The final step is casting.

A more modern process for refining from high-tin concentrates follows the same general pattern as the Cornwall process (*see* Figure *T19*).

TITANIUM

Physical and Chemical Properties

Symbol: Ti
Atomic number: 22
Specific gravity: 4.54
Melting point: 1675°C, 3047°F
Boiling point: 3260°C, 5900°F
Coefficient of expansion: 0.0000085/°C
Tensile strength: 30,000 to 100,000 lbf/in.^{2}, 206.85 to 689.5 MN/m^2

Titanium is a light yet strong, ductile, silver-white metal, estimated to be the fourth most abundant structural metal in the earth's crust and the ninth most common element. It is paramagnetic and low in electrical conductivity (about 5% of that of aluminum) and thermal conductivity (similar to that of stainless steel). It also has a low coefficient of linear thermal expansion.

Its most valuable characteristics are its high strength-to-weight ratio and its superior resistance to corrosion by marine atmospheres and salt water. The strength depends on its purity: the higher its purity, the lower its strength, but the easier to fabricate it. Almost all metallic elements are soluble in titanium and form alloys that increase its strength up to 210,000 lbf/in.^{2} (1447.95 MN/m^2) for heat-treated alloys. Aluminum, vanadium, molybdenum, manganese, iron, and chromium are the elements most commonly used for this purpose. At high temperatures (about 1600°F, 871.11°C), titanium becomes excessively reactive in air and will absorb oxygen and nitrogen, which make it brittle.

Workability. Titanium can be worked by conventional methods if precautions are taken to avoid overheating the metal. It can be hot- or cold-rolled, forged, drawn, flanged, extruded, blanked, and bent; it can also be machined by standard shop operations similarly to 302 stainless steel, but special considerations are required in forming and machining operations. It can be punched, riveted, inert-gas arc welded, and resistance-welded by spot, seam, projection, or flash.

Commercial Forms. The sponge (or virgin) metal available from primary purification (*see* Table *T17*), is melted into ingots, and from ingots the following mill products are available: bar, billet, sheet, strip, plate, wire extrusions, tube, and pipe, as well as special forgings and castings.

Types and Uses

Titanium, which is one of the newest metals, has great possibilities as a future architectural material, although at present the bulk of it goes into military and highly specialized uses. Table *T18* shows both its metallic uses, indicated by asterisks, and the industrial applications of titanium compounds other than titanium dioxide. The compound titanium dioxide is widely used in the paint and other industries as a white pigment (solid) and for that reason is of special interest in construction (*see* Titanium Dioxide).

Alloys. Titanium alloys are used in both ferrous and nonferrous metallurgy to remove impurities and improve the characteristics of other metals. The more important of these are ferrocarbon-titanium, carbon-free ferrotitanium, ferromanganese-titanium and ferrosilicon-titanium, all used to deoxidize and denitrogenize steels, and cuprotitanium, used to deoxidize nonferrous metals.

Titanium base alloys per se are finding increasing uses wherever lightness, strength, and corrosion resistance are needed. They are much stronger and more corrosion

Table T17 Typical analysis of magnesium-reduced titanium sponge (Kroll method)

Ti	C	O	N	Al	Cu	Fe	Mg	Mn	Si	Sn	Cl	H
					(percent of content)							
99.33 min.	0.05 max.	0.15 max.	0.05 max.	0.005 max.	0.025[a]	0.2 max.	0.12[a]	0.06 max.	0.005	0.025	0.15 max.	0.05 max.

[a]The percentage must be less than this figure.

Table T18 Uses of titanium

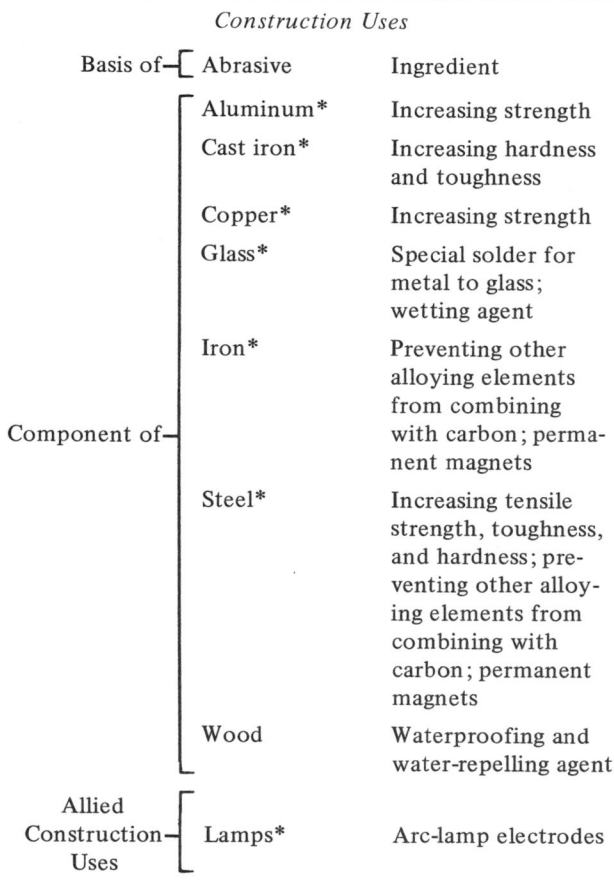

	Construction Uses	
Basis of	Abrasive	Ingredient
Component of	Aluminum*	Increasing strength
	Cast iron*	Increasing hardness and toughness
	Copper*	Increasing strength
	Glass*	Special solder for metal to glass; wetting agent
	Iron*	Preventing other alloying elements from combining with carbon; permanent magnets
	Steel*	Increasing tensile strength, toughness, and hardness; preventing other alloying elements from combining with carbon; permanent magnets
	Wood	Waterproofing and water-repelling agent
Allied Construction Uses	Lamps*	Arc-lamp electrodes

Nonconstruction Uses

Cermets,* corrosion-resistant chemical equipment,* electronics, X-ray*

Possible Future Uses

Doors, exterior and interior surface finishes, curtain walls, fascias, flashing, gutters, leaders, railings

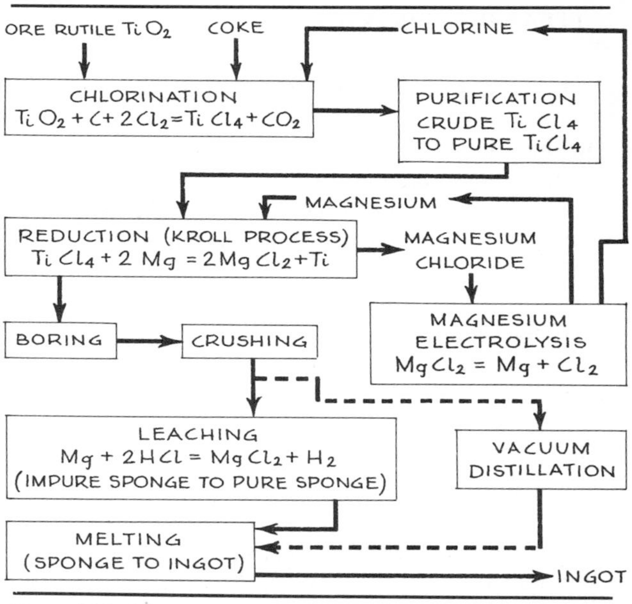

Figure T20 Flowchart of titanium extraction and refinement (Kroll process).

resistant than aluminum alloys, have great impact strength and unusual resistance to fatigue, and are more corrosion resistant to sea air and water than are stainless steels.

History and Manufacture

The element titanium was first discovered in England in 1790 by Gregor and again in 1795 by Klaproth, who named it after the Titans, the mythological first sons of the earth. Both men actually worked with titanium dioxide, which is found as a natural mineral. Titanium itself was not isolated until 1910, when Hunter prepared it in metallic form. It remained a laboratory curiosity until 1946, when the U.S. Bureau of Mines, using a process patented by Kroll, showed that it could be produced as a ductile metal on a commercial scale. Two years later, industrial titanium appeared and has proved so valuable that it is now one of the major modern metals.

Kroll Process. The Kroll refining process is based on the reduction of titanium tetrachloride by magnesium in an inert atmosphere of argon or helium and, with modifications, is still used today for large-scale production (*see* Figure *T20*). Use of the tetrachloride avoids oxygen and nitrogen contamination. Magnesium and titanium tetrachloride react spontaneously and can be readily controlled; the resulting magnesium chloride and residual magnesium can be reused after being distilled off and reprocessed. A ductile titanium is produced from the resulting sponge metal, which is equivalent at this point to pig iron. The sponge is first crushed and then arc-melted or induction-melted (not a commercial method) into ingots for further processing by various fabricators. Higher purity titanium can be produced from scrap or a crude electric-furnace reduction material such as titanium carbonitride by the iodide process or by a fused salt electrolytic process, but only in small quantities and at relatively high cost. Some commercial producers use a recently developed sodium reduction method similar to the Kroll process.

TITANIUM DIOXIDE

Physical and Chemical Properties

Titanium dioxide (TiO_2) is a markedly inert, stable compound found in nature in three crystalline forms (rutile, anastase, and brookite) which differ in their opacity and other physical characteristics. For example, the specific gravity is 3.9 for the anastase form and 4.2 for the rutile. The compound decomposes at about 3000°F (1648.89°C).

Titanium dioxide is manufactured as a white powder in two crystalline forms, anastase and rutile, from both ilmenite ore ($FeO \cdot TiO_2$) and titanium slag made from ilmenite. Both forms are insoluble in water, are attacked only by hot acids and alkalis, and are unaffected by sunlight, heat, or most gases that may be present in the atmosphere, such as hydrogen sulfide, ammonia, and sulfur dioxide.

The property that has made titanium dioxide so valuable in the paint industry (*see* Table *T19*) is the change in direction of a beam of visible light as it passes through TiO_2, measured as the refractive index. This is what gives the compound such great hiding or covering power in paints.

Table T19 Types of titanium dioxide for paints

Type	Form	Minimum percentage of TiO_2	Maximum percentage of $CaSO_4$
Titanium dioxide	Rutile or anastase	94	
Titanium calcium	Rutile or anastase	28	70
Titanium barium	Rutile base	28	70

Types and Uses

There are two general types of titanium pigment: it may be either pure titanium dioxide, or a composite in combinations with calcium sulfate or barium sulfate.

Table T20 Uses of titanium dioxide

Construction Uses		
Basis of	Color pigment	White
Component of	Asbestos	Giving whiteness and high reflection to asbestos-cement materials
	Asphalt	Giving whiteness and brighter colors to shingles, siding, and rolled roofing; overcoming dark colors of binders and extenders in floor tiles
	Cork	Whitening
	Fibers	Whitening
	Glass	Combined with other oxides to develop colors; increasing brilliancy, refractive index, and iridescence
	Glazes and porcelain enamels	Acid resistance; giving brightness, whiteness, opacity, and yellow color
	Linoleum	Whitening
Construction Uses (Continued)		
Component of	Paint	Opacity; hiding power; durability; heat resistance; white paint solid
	Paper	Filler; fire retardant; waterproofing and water-repelling agent; giving brightness, whiteness, and opacity; reducing transparency
	Plastics	Reducing transparency; giving brightness, whiteness, and opacity
	Protective coatings	Waterproofing and water repelling
	Rubber	Brightening, whitening
	Textiles	Mordant; fire retardant; waterproofing and water-repelling agent; any degree of hiding and translucency
	Wood	Waterproofing and water-repelling agent

The pure titanium dioxide may be either rutile or anastase in form. In general, both types are used for brightening, whitening, and opacifying materials such as cements, glass, paints, paper, plastics, rubber, textiles, asbestos, asphalt, fibers, glazes, and porcelain enamels. These and other uses are summarized in Table *T20*.

The value of titanium pigments is also dependent on particle size distribution. Modifications have been developed to produce certain desirable characteristics for various specific uses.

History and Manufacture

The early history of titanium dioxide parallels the story of the search for the element titanium until, in 1908, Rossi observed that impure titanium dioxide, when mixed with oil, has remarkable opacity. In 1912 Barton joined Rossi in research on the pigment possibilities of titanium. Working together, they finally found that adding finely divided calcium or barium sulfate to a solution of titanium in sulfuric acid facilitated its hydrolysis and yielded a composite pigment with a high opacity. Thus the first pigments used were composites. In about a decade, commercially feasible methods of producing pure titanium dioxide pigment were developed, and by 1932 both types were produced in the United States. The rutile form is more widely used because of its greater hiding power and resistance to chalking and fading. The anastase form has advantages in certain applications because of its higher dispersibility.

The process for producing titanium dioxide (*see* Figure *T21*) is based on the fact that ilmenite ore (TiO_2, FeO, Fe_2O_3) can be dissolved in sulfuric acid. The ferric iron in solution is reduced to ferrous by adding metallic iron, and the solution is evaporated to crystallize ferrous sulfate, which is separated by filtration. The filtrate is then boiled, and hydrous titanium oxide is precipitated. This precipitate is calcined to drive off moisture and develop opacity and then ground to achieve optimum particle size. The determination of size and distribution (by size) of the particles in a pigment depends on the medium and the method of dispersion.

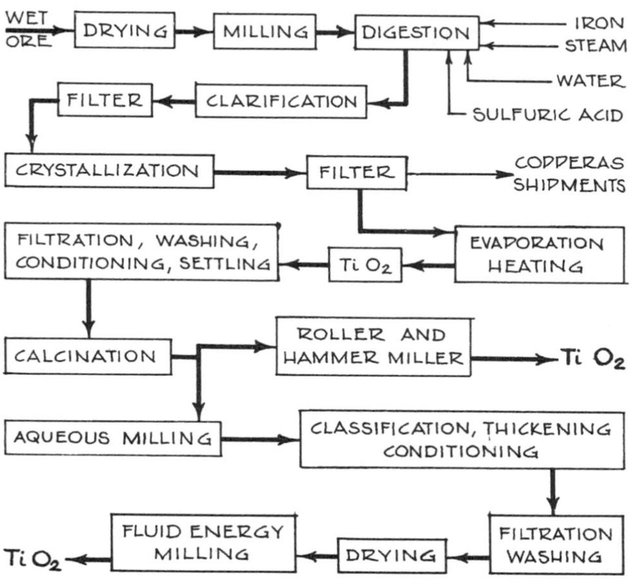

Figure T21 Flowchart showing production of titanium dioxide pigment (white paint solid).

TUNGSTEN

Physical and Chemical Properties

Symbol: W
Atomic number: 74
Specific gravity: 19.3 (20°C, 68°F)
Melting point: 3410°C, 6170°F
Boiling point: 5927°C, 10,700.6°F
Tensile strength: 18,000 to 590,000 lbf/in.^{2}, 124.11 to 4068.05 MN/m^2

Tungsten is a hard, brittle, gray metal that is relatively inert chemically. It is among the heaviest metals and is also very hard. It has the greatest tensile strength of all metals; for example, tungsten rod 0.05 in. in diameter has a tensile strength of 215,000 lbf/in.^{2} (1482.43 MN/m^2), and sheet 0.01 in. thick, 300,000 lbf/in. (2068.5 MN/m^2). The strength varies greatly depending on fabricating procedures, ranging from the low value given above for sintered ingot to 590,000 lbf/in.^{2} (4068.05 MN/m^2) for drawn wire 1.4 mils in diameter. An interesting feature is that strength increases with increased ductility—the exact opposite of what happens in all other metals. Tungsten has the lowest compressibility and coefficient of expansion of any metal.

Workability. Tungsten can be brazed readily but cannot be cast except by powder metallurgy; machining is not recommended.

Commercial Forms. The metal is available in powder, rod, wire, and sheet form and in powder metallurgical shapes and bars.

Table T21 Uses of tungsten

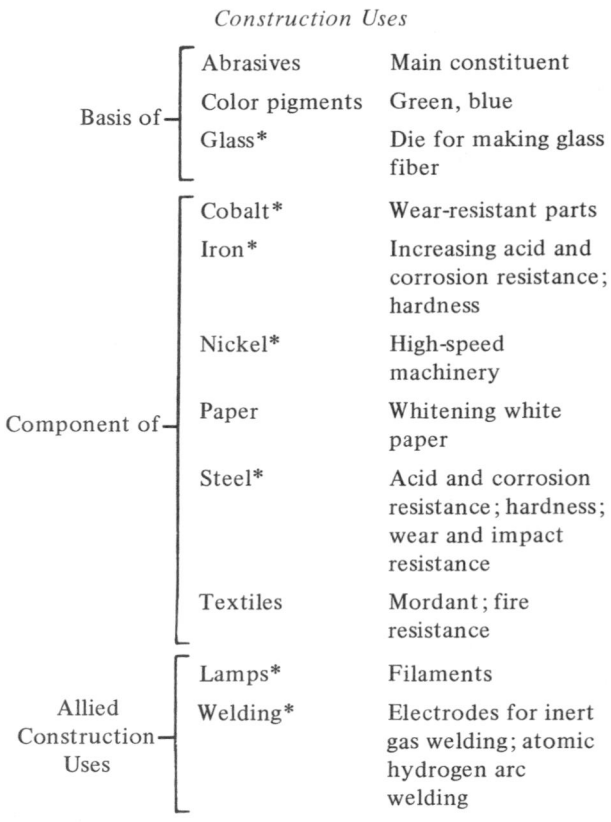

	Construction Uses	
Basis of	Abrasives	Main constituent
	Color pigments	Green, blue
	Glass*	Die for making glass fiber
Component of	Cobalt*	Wear-resistant parts
	Iron*	Increasing acid and corrosion resistance; hardness
	Nickel*	High-speed machinery
	Paper	Whitening white paper
	Steel*	Acid and corrosion resistance; hardness; wear and impact resistance
	Textiles	Mordant; fire resistance
Allied Construction Uses	Lamps*	Filaments
	Welding*	Electrodes for inert gas welding; atomic hydrogen arc welding
	Nonconstruction Uses	
Vibration damping devices,* X-ray tubes*		

Types and Uses

Tungsten compounds other than the carbides have only a few specialized uses; the uses of tungsten in metallic form are indicated by asterisks in Table *T21*.

About 30 to 40% of the world's production of tungsten goes into special steels. It can be added to steel in the form of ferrotungsten, scheelite, scrap, or metal powder. A chief property of tungsten in steel is its ability to impart or retain hardness at elevated temperatures. The greatest use for tungsten steels is not in construction but in all types of dies, punches, drills, and cutting and finishing tools.

Tungsten in the form of carbides often combined with cobalt represents an application as great as its use in steels. Tungsten carbides are important in cutting tools, wire-drawing dies, and abrasives.

History and Manufacture

Tungsten is the name commonly used in English-speaking countries, although wolfram is the official name in many foreign countries. The word "tungsten" means "heavy stone," but the origin of the term "wolfram" is obscure. In 1781 Scheele identified the acidic constituent of the mineral tungsten and named it tungstic acid. In 1783 the de Elhuyar brothers isolated tungsten by carbon reduction of the oxide.

Commercial Production. Most tungsten ore is converted to ferrotungsten, used for steel alloying.

Production of tungsten metal follows four stages: decomposition of the ore, preparation of a pure anhydrous or hydrated oxide, reduction to metal powder, and consolidation of the powder into solid metal.

TURPENTINE

Turpentine is obtained generally by steam distillation from the exudate of certain pine trees or from wood. Its main uses are as a thinner and solvent for paints, varnishes, and oil stains. *See* Paint; Painting.

U

U FACTOR

The *U* factor is the measurement of the overall coefficient of heat transfer (thermal transmission). It represents the combined thermal value of all the materials in a building section, including air spaces and surface air films. The *U* factor is expressed as the number of Btu's passing through 1 ft^2 (0.0929 m^2) in 1 hour, for each degree Fahrenheit (0.5556°C) in temperature difference. The symbol is $Btu/ft^2 \cdot h \cdot °F$. The conversion factor is 1 $Btu/ft^2 \cdot h \cdot °F = 5.678\ W/m^2 \cdot °C$.

See Btu; *C* Factor; *K* Factor; *R* Factor; Insulation and Insulating Materials.

URANIUM

Physical and Chemical Properties

Symbol: U
Atomic number: 92
Specific gravity: 18.95
Melting point: 1132°C, 2070.14°F
Boiling point: 3818°C, 6904.4°F

Uranium is a radioactive, heavy, nickel-white metal that is malleable and ductile. It can be welded, cast by powder metallurgy, and machined, but care must be taken as it ignites spontaneously.

Types and Uses

Uranium's most important application is in the field of nuclear energy, although there are other uses for it in compound and metallic form, the latter being indicated by asterisks in Table *U1*.

History and Manufacture

Klaproth discovered uranium in 1789. It is named after the planet Uranus. In 1842 Peligot found a method of producing uranium metal, and in 1896 Becquerel discovered the radioactive property of uranium. Uranium metal is now obtained by the reduction of uranium oxides and halides.

Table U1 Uses of uranium

	Construction Uses	
Basis of	Color pigment	Yellow; greenish
Component of	Glass	Yellow, greenish yellow, and black colors
	Glazes and porcelain enamels	Yellow, orange, brown, green, and black colors
	Iron*	Increasing strength and toughness
	Steel*	Increasing elastic limit and tensile strength
	Wood	Yellow to brown stains
	Nonconstruction Uses	
Photoelectric tubes,* X-ray photography*		

V

VANADIUM

Physical and Chemical Properties

Symbol: V
Atomic number: 23
Specific gravity: 6.11 (18.7°C, 65.66°F)
Melting point: 1890°C, 3434°F
Boiling point: 3338°C, 6036.4°F

Vanadium is a silvery-white metallic element that chemically can serve as either a metal or a nonmetal. It was originally considered brittle but is now known to be ductile in pure form. Pure ductile vanadium has only recently become available, but in limited quantities.

Workability. Vanadium can be forged, rolled, bent, stamped, pressed, machined, and welded.

Commercial Forms. Vanadium is obtainable in ingot, plate, bar, wire, sheet, strip, and chip form. The composition of commercial vanadium is shown in Table *V1*.

Types and Uses

Ferrovanadium, an alloy of vanadium and iron, is used for adding vanadium to iron and steel during their manufacture. Vanadium compounds have a few industrial uses, mostly as colors. The bulk of vanadium is used in metallic form, indicated by asterisks in Table *V2*.

History and Manufacture

About 1830, Sefström discovered a new element which he named vanadium after Vanadis, the Scandinavian goddess of beauty and youth, for the beautiful colors of its compounds in solution. By 1870 Roscoe had obtained metallic vanadium by hydrogen reduction of vanadous chloride, and in 1927 Marden and Rich succeeded in obtaining 99.9% pure metallic vanadium.

Table V1 Composition of commercial vanadium

V	O_2	H_2 (percent of content)	N_2	C
99.86	0.037	0.003	0.035	0.065

Table V2 Uses of vanadium

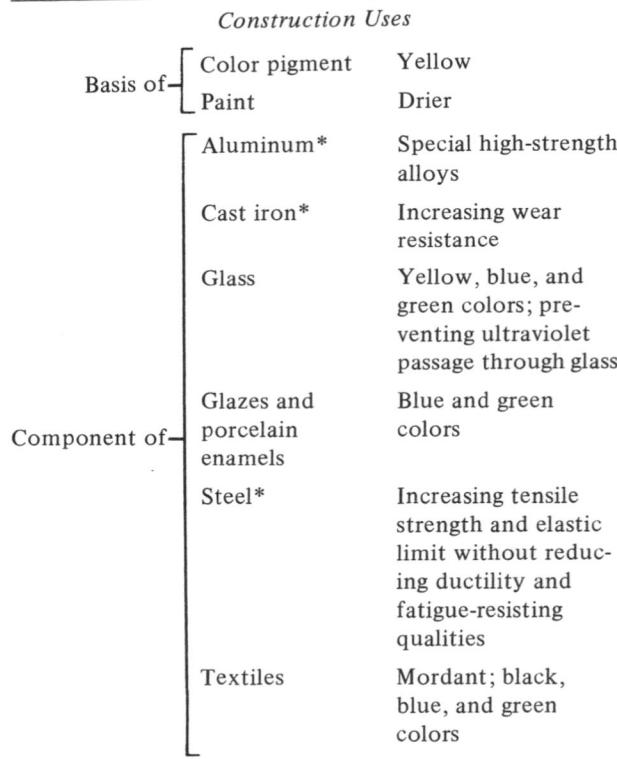

	Construction Uses	
Basis of	Color pigment	Yellow
	Paint	Drier
Component of	Aluminum*	Special high-strength alloys
	Cast iron*	Increasing wear resistance
	Glass	Yellow, blue, and green colors; preventing ultraviolet passage through glass
	Glazes and porcelain enamels	Blue and green colors
	Steel*	Increasing tensile strength and elastic limit without reducing ductility and fatigue-resisting qualities
	Textiles	Mordant; black, blue, and green colors
	Nonconstruction Uses	
Aniline black, X-ray photography		
	Possible Future Uses	
Because of its density and modulus of elasticity, vanadium alloys could become lightweight structural materials		

Commercial Production. In the early 1950s McKechnie and Seybolt developed the calcium iodide reduction process which produced vanadium on a commercial scale. In this process vanadium trioxide, calcium, and iodine are heated in a high-frequency furnace. The iodine combines with the calcium to form calcium iodide, the heat of reaction being sufficient to initiate reduction of the vanadium trioxide, and the vanadium collects as a solid metal. This process brought the price of vanadium down from $1800 per pound in 1950 to only $40 per pound in 1956. As a result vanadium, once considered a laboratory curiosity, is now a metal of high industrial potential.

VAPOR BARRIER

This is defined as a coating or material that is resistant to vapor transmission and will retard or stop the passage of water vapor. A material having a permeance of 1 perm or less is considered a vapor barrier. A perm is the unit of measure of the rate of water vapor transmission through a material; it is expressed in grains/ft^2 · h · in. of mercury pressure difference on the two faces. The metric equivalent is expressed in grams/m^2 · h · mm of mercury pressure difference on the two faces.

VERMICULITE

Physical and Chemical Properties

Vermiculite is a type of mica widely used in the construction industry. It consists of hydrated magnesium aluminum iron silicate characterized by a foliated structure. It occurs in yellowish to brown crystalline plates which can measure up to more than 9 in. (228.6 mm) across and 6 in. (152.4 mm) in thickness. The specific gravity of the unexpanded, crystalline form varies from about 2.3 to 2.7. Upon calcination at 1750°F (954.44°C) it expands in one direction only, at right angles to the line of cleavage, into threads with a vermicular motion (like a mass of small worms)—hence its name. During this process its volume increases as much as 16 times. In its exfoliated or expanded form vermiculite is characterized by lower density, low thermal conductivity, resistance to high temperatures, and chemical inertness.

Types and Uses

Vermiculite in expanded form is widely used in many types of thermal and acoustic insulation, as loose fill in buildings, as an aggregate in concrete, and as an ingredient in plaster, insulating concrete, and fireproofing compositions.

Unexpanded vermiculite is used as a filler in fire-resistant wallboard and in very fine powdered form as an extender in aluminum paint. It also has possibilities as a paint and calcimine pigment.

A new potential application is the removal of radioactive wastes in effluent solutions.

History and Manufacture

During the past several decades vermiculite has developed from a mineralogical curiosity to an important commercial product. It is mined with power shovels, concentrated by various wet methods, and otherwise processed. It is graded to commercial sizes, and the bulk of it is shipped to exfoliating plants, where it is expanded and then ground into pellet form, also screened to definite commercial sizes.

VIBRATION CONTROL

In the construction field, acoustics are considered only in relation to sound within an enclosed space, but another problem that plays an important part in connection with sound is the control of vibration. For example, a space can be treated acoustically, but next door to it, resting on the floor, there may be equipment with motors. Vibration from the motors will be transmitted to the floor area by pipes and ducts or by the structure itself no matter how much acoustic treatment exists, unless specific measures are taken to isolate the equipment or to absorb the vibration within the building (*see* Figure *V1*).

An extreme case is a building near railroad tracks. Here vibration is transmitted through the ground itself. Unless

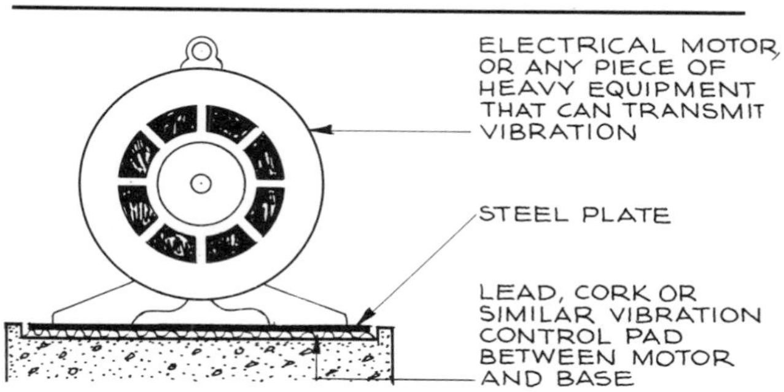

Figure V1 Vibration control of heavy mechanical equipment.

the footings are isolated from vibration, every time a train passes on the tracks, vibration will transmit the noise throughout the building. A well-known example is Park Avenue and the buildings built directly over the tracks of ConRail. All the buildings there have no basements, and every supporting column extends down below the level of the tracks and is completely vibration-isolated. In addition, all the streets, including Park Avenue and any sidestreets that are over the tracks, are completely isolated from the building to eliminate vibration (*see* Figure *V2*).

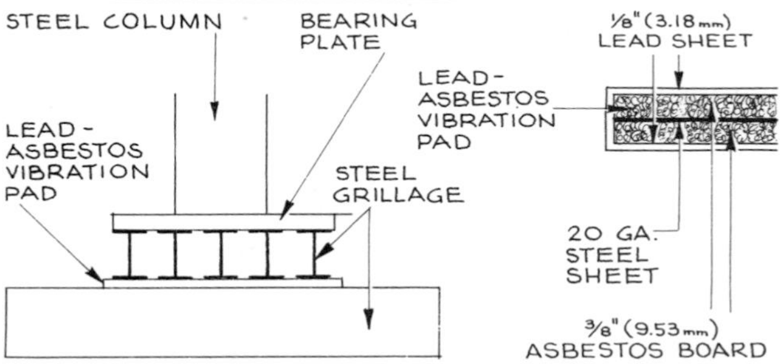

Figure V2 Vibration control of columnar footings.

Types and Uses

For foundation and footings of buildings, vibration control is usually achieved with lead pads. Machinery can be isolated by means of pads of cork, lead, rubber or synthetic rubber, plastic, and other types of special resilient materials. For low-frequency machines the supports should be damped. For high frequencies the supports should be elastic without damping. For more complicated and heavier machinery that creates greater vibration, floating platforms supported by springs or springs in oil are used.

Vibration that travels through pipes and ducts can be controlled by the special vibration joints that exist for piping and the special fabric connection made for ducts. Ducts may be lined with sound-absorbing material.

In construction when a problem of vibration arises from large equipment or from adjoining railroads, subways, heavy equipment, vehicular tunnels, etc., specialists in this field should be consulted, as this is a highly technical engineering problem beyond the scope of this book.

W

WATER

In planning any building one must be aware of water in relation to protection of the building from (1) the effects of water on the site, for example, ground water or subsoil water; and (2) water as a factor in humidity, for example, rain, snow, ice, water vapor, or humidity (*see* Table *W1*). One must also consider the effects of water containing various chemicals (salt water, sulfur-containing water, etc.) on the materials used in construction and on the mechanical aspects of the building such as plumbing (*see* Table *W2*). In the construction field the environmental control systems and equipment have to be carefully checked for water pollution and waste, as well as treatment of sewage. Industrial types of structures where manufacturing or chemical processes are in operation must be checked against surface water pollution and subsurface and water-table pollution.

Table W1 Water as a factor in site conditions and climate

Source of water	Type of damage	Corrective measures
Water vapor	Condensation from inside or outside air can cause staining, rotting, and molding in exterior walls; can freeze and damage exterior walls; can make crawl spaces, basements, and attics humid and damp; can collect on interior surface of glass and destroy window sills; can collect on ceilings if insulation is not calculated correctly; can start galvanic and chemical action between incompatible materials.	Installation of vapor barriers in correct position. Sufficient ventilation or use of correct type of insulation in various areas of building.
Rain and snow	Accumulation on roofs, if not drained off and away from building, can cause washouts and same damage as surface water. Poor drainage and construction may permit water to enter building through roof, vertical walls, doors, and windows, causing same type of damage as condensed water vapor, only more so.	Gutters and leaders of adequate size to carry off rain water from roofs. Adequate drainage of all walks, roads, parking areas, driveways, playfields, and large lawn areas. Installation of flashing, counterflashing, waterstops, and tables; dampproofing, waterproofing of exterior walls, parapets, etc.; weatherstripping and caulking of doors and windows.
Ice	Water in small holes, cracks, crevices, etc., can freeze, expand, and destroy materials used on exterior of buildings. If foundations are not deep enough, it can settle under them and upon freezing expand sufficiently to cause permanent destructive damage.	All joints in masonry and exterior walls made watertight and of a waterproof material. Vertical walls made of waterproof materials or treated to be water repellent. Location of foundations and footings below frost line.
Surface water	Rain, thawing ice, or snow can cause washouts and erosion, accumulate in large shallow ponds, create large areas of mud, saturate soil surrounding building, and cause same damage as ground water.	All leaders and gutters adequately drained into dry wells, storm sewers, or other provisions for water storage and dispersal. All parking lots, sidewalks, roadways, lawns, and playfields adequately drained.

Table W1 Water as a factor in site conditions and climate (*continued*)

Source of water	Type of damage	Corrective measures
Ground water	Basements and subbasement areas can become damp, wet, or completely flooded.	Water table for locality and specific site established and basement floor set above this level and highest point that it may reach. Basement floor and walls waterproofed with membrane type of waterproofing.
Subsoil water	Underground springs, brooks, and rivers can create hydrostatic head of water and flood basement or subbasement areas and make them permanently damp and wet.	Test borings made to check for subsoil water. Excavations for foundations examined for subsoil water. Any such water eliminated by piping, trenching, etc., to drain or divert it or by membrane waterproofing of basement or subbasement areas.

Table W2 Effect of chemical constituents in water

Description of water	Effect on materials	Protective or corrective measures
Soft water		
Water containing not more than 5 grains of calcium carbonate per gallon	This is best type of water for all uses, including drinking water and water for mixing cement, concrete, mortar, or plaster.	
Hard water		
Water containing salts of calcium and magnesium (degree of hardness given in grains of calcium carbonate per gallon of water)	Such water can eventually clog water pipes. In special buildings (factories, laboratories) it can cause production and equipment damage.	Water treated to eliminate salts by installation of water softeners.
Salt water		
Average salt content of 3.5%, consisting of sodium chloride 77.76% magnesium chloride 10.88% magnesium sulfate 4.74% calcium sulfate 3.60% potassium sulfate 2.46%	Salt (sea) water will attack and corrode most metals used in construction except certain stainless steels and aluminum alloys. It attacks paints, textiles, wood, and most plastics and rubbers unless correct type of product or correct protection is used. Such water is never to be used for mixing concrete, cement, plaster, or mortar.	Most metal must be protected with marine type of paint or asphaltic type of protective coating, or plated with special metals that resist the effects of salt water. Wood must be protected by special paints or stains that resist salt water.
Sulfur water		
Water containing a sulfur compound in solution (hydrogen sulfide usually)	Such water can corrode all piping, especially copper. It should never be used for mixing any materials used in construction.	Special purifiers used to eliminate sulfur in water. Special cements exist for foundations exposed to sulfur water.
Contaminated water		
Water that contains acids, alkalis, mineral or organic chemicals, or bacterial wastes	Such water can cause corrosion, disintegration and other damage to all piping. It should never be used for mixing materials such as concrete, cement, plaster, and mortar.	Special purifiers used to eliminate contaminating substances in water.

WATERPROOFING

Waterproofing may be defined as the method or means of sealing off against free water and moisture, usually under hydrostatic pressure, all parts of a building (exterior or foundation walls below grade and the lowermost floor) that come in direct contact with the earth.

Below-Grade Waterproofing. Below-grade waterproofing is necessary in many instances when buildings are built on sites where the basement or subbasement is below the prevailing water level. Thus, in a building near a river, it is likely that the basement or subbasement will be below the highest level reached by the river. Also, there are areas containing underground springs, brooks, rivers, or lakes where a basement or subbasement may have to be completely sealed off against moisture. This waterproofing must include the floor slab and the entire perimeter of all foundation or exterior walls which are in direct contact with the earth (*see* Figure *W1*).

There are several methods of waterproofing. In the one most generally used, called the membrane system, asphalt or tar coatings are applied hot or cold over the entire surface, alternating with several layers of saturated felt, the number of layers of felt depending on the seriousness of the condition.

The other method is to apply coats of special waterproof cement on the inside surfaces of the walls and lowermost floor, using pressure adhesion. Generally, the external application of a waterproof membrane gives the best building protection.

Above-Grade Waterproofing. On exterior walls above grade, where the basic problem is to keep out water which can penetrate the walls because of driving rains, many methods of waterproofing exist, as the problem is not difficult as below grade (*see* Figure *W2*).

Many ways of handling wall conditions above grade are described and illustrated in detail under Flashing, both concealed and exposed (*see also* Dampproofing). In general, it is always wise to assume that there will be trouble with water below grade and to prepare for it. This is equally true for walls above grade; it is wise to take precautions while constructing the building rather than after it has been completed.

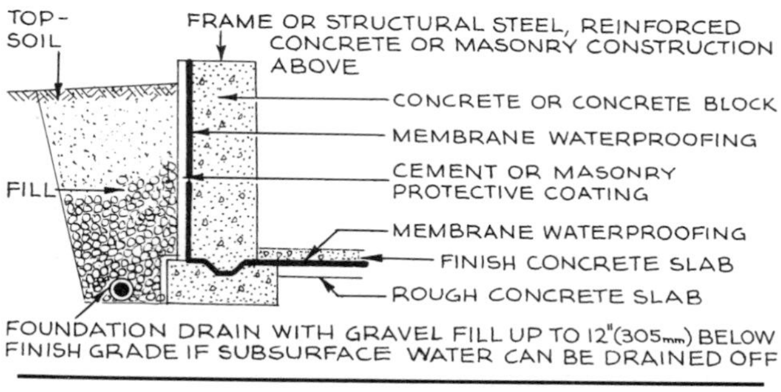

Figure W1 Waterproofing foundation walls below grade.

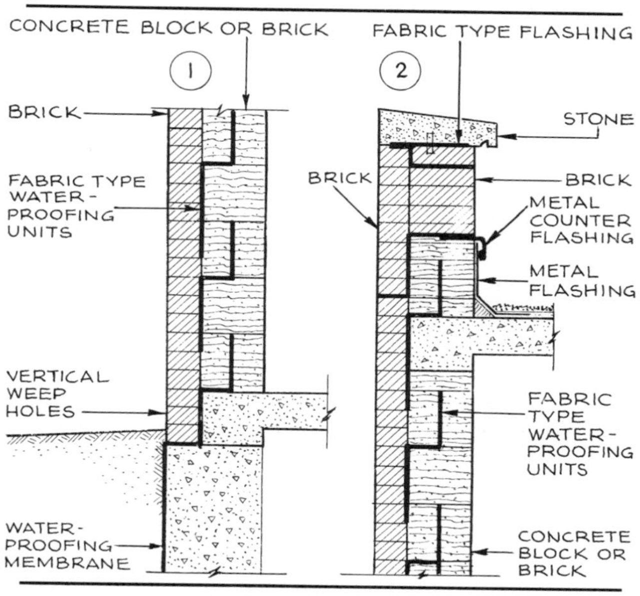

Figure W2 Waterproofing exterior walls above grade.

WAX

Waxes are a group of viscous to solid substances of various origins that consist of fatty acids in combination with alcohols. The animal waxes are beeswax from the honeybee, spermaceti from the sperm whale, and stearic acid from tallows. The main vegetable waxes are carnauba, bayberry, Japan wax, jojoba oil, and candelilla. There is also a group of mineral substances which, although not true waxes, are known as such. They are often used as substitutes for or blended with animal and vegetable waxes, which are in relatively short supply. These mineral waxes consist of mixtures of saturated hydrocarbons and include paraffin wax (a petroleum product), montan wax (derived from lignite, a kind of coal), ozocerite (a native, fossil paraffin), and ceresin (purified ozocerite). Many synthetic waxes are sold under trade names.

Waxes are harder, less greasy, and more brittle than fat. They have lower melting points than resins and are soluble in alcohol, mineral spirits, and most organic solvents but insoluble in water.

Construction Uses

The most familiar uses of waxes are as a fine finish on woods and for the maintenance and care of many types of flooring (*see* Wood Finishes). Waxes also have other, less well known but important uses in construction materials: as an impregnating and preserving agent; for waterproofing wood, cork, and textiles; in roof, antifouling, and waterproofing paints; as a flatting agent; for sizing, glossing, and impregnating textiles and paper; as a plasticizer in synthetic resins; and in wire coating and electrical insulation. (*See also* Paint; Petroleum.)

WEATHERSTRIPPING

Weatherstripping may be defined as the methods or devices used to make doors and windows that open to the exterior weathertight. The principle is to stop wind and water from coming into a building through cracks between doors or windows and their frames. This may be accomplished by interlocking one part with another and by friction. Weatherstripping is usually made of the following metals: aluminum, bronze, iron, nickel, silver, stainless steel, steel, and zinc. Pliable materials include felt, rubber, and plastic.

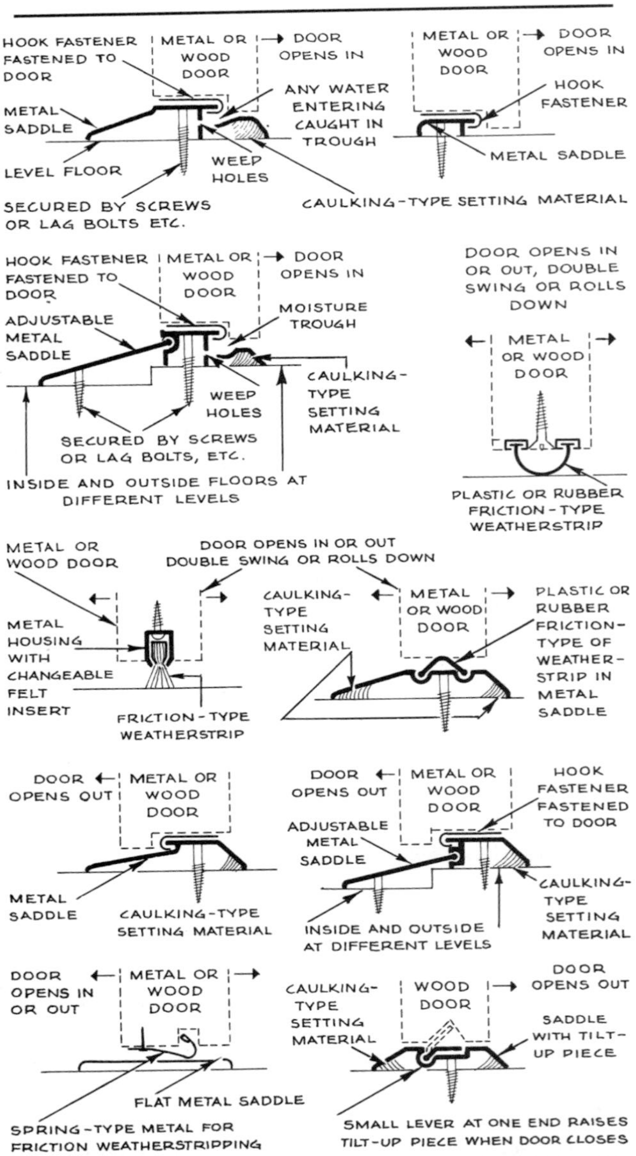

Figure W3 Typical details of weatherstripping at sills of doors.

For Doors. Typical details for weatherstripping doors are shown in Figures *W3* and *W4.* Metal doors for the exterior, whether obtained without their frames or as a package unit of door and frame, are generally manufactured without weatherstripping unless it is specified. Sliding metal doors, on the other hand, are always supplied as a unit with their frames, including built-in weatherstripping.

For Windows. Weatherstripping for windows is almost the same as for doors except in the case of double-hung windows (*see* Figure *W5*). Aluminum, bronze, steel, and stainless steel windows and many wood windows are manufactured with weatherstripping as part of the window-and-frame package unit. It is therefore advisable always to check whether the window selected includes weatherstripping.

See also Aluminum Windows and Doors; Stainless Steel Windows and Doors; Steel Windows and Doors; Wood Doors; Wood Windows.

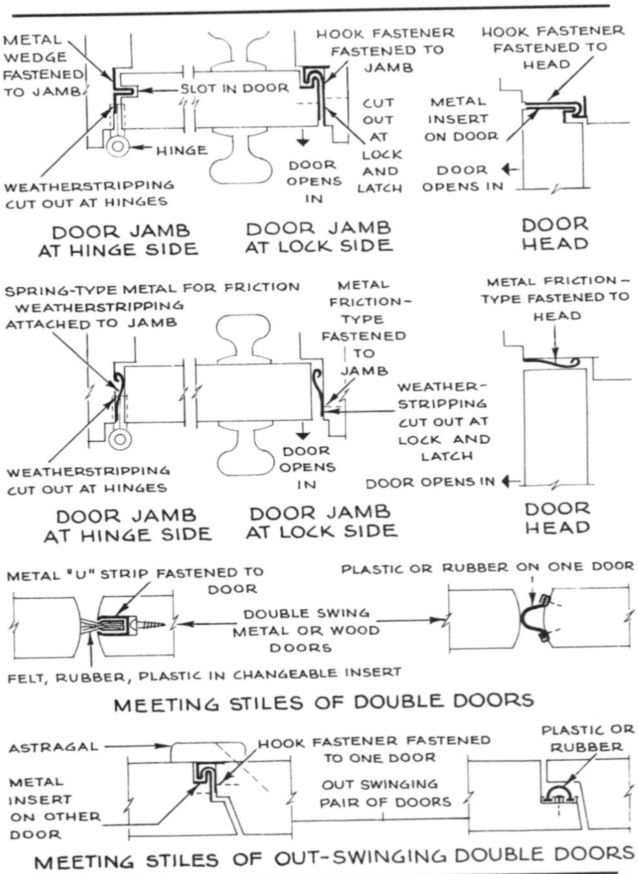

Figure W4 Typical details of weatherstripping at jambs, heads, and meeting stiles of doors.

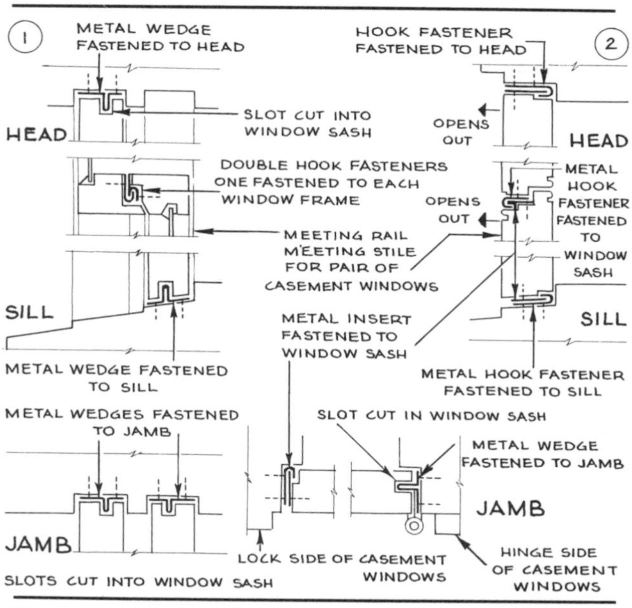

Figure W5 Typical details of weatherstripping for wood double-hung windows and casement windows.

WELDING

Welding may be defined as the process by which two metals are so joined that there is an actual union of the interatomic bonds. This may be brought about by close contact, heating, pressure, adding molten metal, or combinations of these methods. The resulting welded joints are as strong as the metals joined or stronger.

Welding may be divided into two broad categories: pressure welding, in which pressure and heat make the weld; and fusion welding, in which heat and added metal make the weld.

Effect of Welding on Metal Characteristics. In welding the critical factors are the effects of heating and cooling on the strength, ductility, and other characteristics of both the original metals and any metals added when the weld is made. This is the reason why cold working and annealing are often necessary after welding to restore the metal to its previous characteristics. The temperature ranges for welding are shown in Figure *W6*.

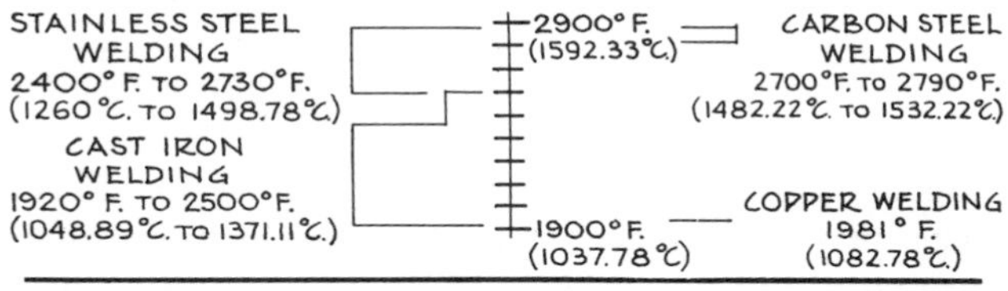

Figure W6 Temperature range for welding metals commonly used as construction materials.

When materials are to be welded, it is important that the process to be used, the type of welding rod, the effect of heating and cooling, the strength of the joint, and the type or types of metals to be joined, all are carefully checked with manufacturers, with welding handbooks, and with technical data on the particular metals to be joined. This is especially necessary if a welded joint is to be used under conditions of shear, tension, and compression.

Types and Uses

Various processes have been developed in each of the two broad categories of pressure and fusion welding:

Pressure Welding	*Fusion Welding*
1. Furnace or hammer	1. Fusion by gas flame
2. Electric welding without fusion	2. Fusion by electric arc
3. Pressure gas welding	3. Fusion by electric arc and exothermic reaction
4. Pressure after Thermit reaction	4. Fusion by Thermit reaction

In construction the most commonly used method is electric welding without fusion; and gas and electric-arc fusion welding are next. The other methods are used mostly by fabricators for shop work. Hammer welding is never used when the weld will be under stress.

Pressure Welding. Pressure welding using electrical resistance is generally termed resistance welding. The original process was invented by Thomson about 1908, and various types of processes now exist (*see* Figure *W7*).

Fusion Welding. In fusion welding the two methods of heating are gas flame and electric arc (*see* Figure *W8*). The gas flame now generally used is acetylene mixed with oxygen. It will deliver about 5500°F (3037.78°C) of heat, which is sufficient to melt the welding rod and the surrounding metal and fuse them together. In the electric-arc method, when the welding rod (or electrode) is brought near the joint of the metals to be welded, an electrical arc is formed which melts and fuses the metal and the welding rod. Either direct or alternating current may be used.

Electric Arc Improvements. Two recent developments in electric-arc welding are the atomic hydrogen-arc method and electrical tornado welding. In principle, the hydrogen-arc method consists of passing a stream of hydrogen through an electrical arc between tungsten

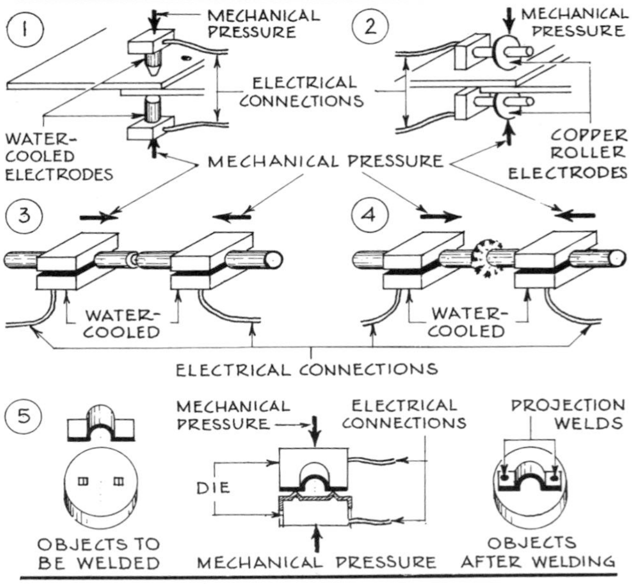

Figure W7 Methods of resistance (pressure) welding, including spot (1), seam (2), butt (3), flash (4), and projection (5).

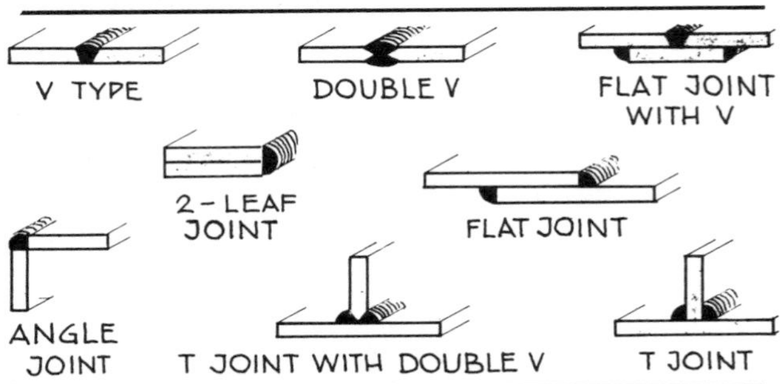

Figure W8 Types of fusion welding.

electrodes. The hydrogen molecules break up into atoms which recombine beyond the arc into molecules, releasing heat. This heat plus the heat of the arc produces a higher temperature than does any arc or gas flame. In electrical tornado welding, a magnetically controlled carbon arc is used with a suitable flux or shielding medium to prevent oxidation. This method permits great operating speeds.

Alumino-thermic Welding. Alumino-thermic welding utilizes Thermit, the trade name for a mixture of finely divided aluminum and iron oxide. This mixture, when ignited, reacts to produce a superheated (5000°F or 2760°C) liquid steel which pours into the joint area and forms a fusion weld. The metals to be joined must be preheated and otherwise prepared for this process by a rather complicated procedure. The Thermit method may be used with nonferrous as well as ferrous alloys by modifying the Thermit mixture to approximate the

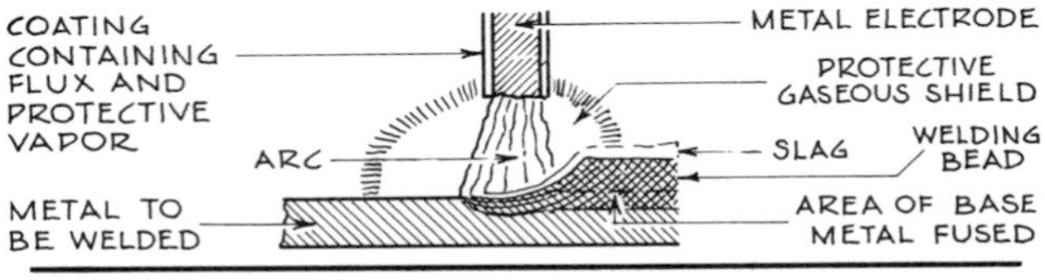

Figure W9 Process of arc welding where the coating of the welding rods is both flux and protective vapor.

composition of the metals to be joined. This method permits massive welds, utilizing up to 2 tons of metal in the total weld.

Fluxes. In almost all welding procedures a suitable flux must be used to keep oxide films from forming on the working surfaces of the metals. In arc welding the fluxes are usually applied as coatings on the welding rods; they not only provide a protective vapor shield about the arc but also may serve as a vehicle for transmitting desirable alloys into the deposited metal (*see* Figure *W9*).

In gas flame welding a suitable flux is also necessary, except when working with wrought iron and steel. It may be applied either as a coating on the welding rods, or as a paste or in strips to the joint area.

History

Heating two pieces of metal to red heat in a forge and then welding them together by hammering is as old a method as metalworking itself. Until 1890, although rolling and pressure welding techniques had been developed to supplement hammering, the basic method remained the same, and it is still used to a limited extent today. With the introduction of gas into industry a new group of welding processes evolved; later, with the advent of electricity, modern welding came into being. It was during World War I that welding as a method for joining metals became important. Since then, knowledge in this field has developed rapidly, and today welding is as essential to modern construction with metals as the nail is to wood construction.

WINDOWS

Windows may be broadly defined as the means of obtaining natural ventilation and light in buildings. With the development and increasing acceptance of air conditioning and the demand in the merchandising field for large display areas, the window is slowly losing its basic function of ventilation, together with all the attendant problems of hardware, insect screening, weatherstripping, and operating mechanisms. However, a new problem does arise, that of cleaning as part of general maintenance.

Table W3 Various types of stock windows

Type	Material	Operation	Operating hardware	Weatherstrip	Screen
Double-hung Standard	Aluminum, bronze, wood, stainless steel, steel	Manual Mechanical Electrical	Spiral, steel tape, wedge, weight balances	With or without	Installed outside; material same as for windows
Removable	Wood	Manual	Clock springs, spring balances, stainless steel bearings		
Reversible	Aluminum, bronze, steel, wood		Spiral, steel tape, wedge, weight balances with pivot		
Side hinge	Wood		Spiral, steel tape, wedge, weight balances with hinge on one side		
Casement	Aluminum, bronze, stainless steel, steel, wood	Manual Mechanical Electrical	Hinges	With or without	Installed inside; material same as for windows
Fixed	Aluminum, bronze, stainless steel, steel, wood	None	None	Without	None
Folding	Aluminum, stain-less steel, steel	Manual Mechanical Electrical	Hinge and slide	With or without	Installed inside; material same as for windows
Hinged (in-swinging hopper) Bottom-hinged	Aluminum, bronze, stainless steel, steel, wood	Manual	Hinges	With or without	Installed outside; mate-rial same as for windows
Top-hinged					Opposite of swing; material same as for windows
Horizontal sliding	Aluminum, bronze, stainless steel, steel, wood	Manual	Side	With or without	Installed outside; material same as for windows
Pivoted Horizontal Vertical	Aluminum, bronze, stainless steel, steel, wood	Manual Mechanical Electrical	Pivots	With or without	Specially made semiround screen; material same as for windows
Projected Typical Single	Aluminum, bronze, stainless steel, steel, wood	Manual Mechanical Electrical	Sliding lever-action hardware	With or without	Installed inside; material same as for windows
Austral			Sliding lever-action hardware with reversible-action arms		Inside and out; material same as for windows
Awning			Sliding lever-action hardware, interconnected		Installed inside; material same as for windows
Continuous			Sliding lever-action hardware	Without	None
Hopper		Manual		With or without	Installed outside; material same as for windows
Jalousie		Manual Mechanical Electrical	Sliding level action hardware	Without	Installed inside; material same as for windows
Louvered				With or without	

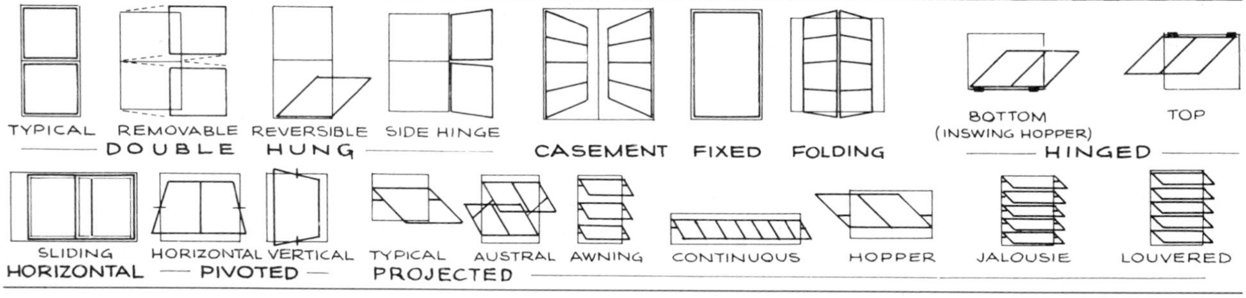

Windows are manufactured generally in classifications related to building types, for example, residential, commercial, monumental, institutional, or industrial, and for specialized types of construction such as prisons, detention houses, and mental institutions. All types of windows are also manufactured with plastic coatings in various colors which eliminate the maintenance problem of painting or staining.

Table *W3* lists as completely as possible the various types of stock windows currently available, together with a summary for each type of its general characteristics, method of operation, screening, and materials of which it is made.

Orientation, in relation to glare, radiant heat, cold, time of day, ultraviolet light, climate, variation of light due to weather, etc., is changing the concept of what constitutes a window. As a result, alternative means of obtaining natural light, such as light scoops, exterior baffles, and deep frames, are now accepted types of window. Research and experimentation will undoubtedly bring about changes in terminology and the uses of materials, as well as better natural lighting.

Once the type of window has been chosen that is the best suited for a building, the reader will find under the basic material headings complete data and details for each particular type of window. For example, if an awning type of window made of aluminum is chosen, under Aluminum Windows will be found data on sizes, finishes, type of glazing (single or double), and precautions in selection and application. Data on the metal as an element and on its commercial alloys may then be obtained under Aluminum and Aluminum Alloys.

WIRE AND WIRE ROPE

Physical and Chemical Properties

Wire is a thin rod or filament of metal, usually flexible. The material from which it is drawn is known as wire-rod, and any wire-rod that has had one cold drawing is considered wire, even though it may be as large as about 1 in. (25.4 mm) in cross section.

Metals Used. Both nonferrous and ferrous metals and their alloys are fabricated into wire. In construction the metals most generally used are iron, steel, stainless steel, aluminum, copper, brass, bronze, and nickel silver.

Strength. Wire can have a wide range of strengths, which are controlled by chemical composition, heat treatment, and amount of cold drawing of the alloy. Tensile strengths may be as high as 500,000 (344.75 kgf/mm^2) for a single wire, and for wire rope $\frac{3}{8}$ to $\frac{1}{4}$ in. (9.53 to 101.6 mm) in diameter, 20,000,000 lbf/in.^{2} (13 790 kgf/mm^2) or greater.

Shape. Wire is generally thought of as round, but it is produced in many shapes, including square, flat, oval, half-oval, half-round, triangular, hexagonal, and octagonal.

Size. Wire size, which used to be given in gauge numbers, is now almost universally designated in decimals of an inch or millimeter (diameter) or in circular mils (cross-sectional area), circular mil (0.0254 mm) being the area of a circle 1 mil (0.001 in. or 0.0254 mm) in diameter. In construction, however, the wire gauge numbers are still used. (*See* Table *W4*.)

There are two standard gauges in the United States: one for ferrous metals, called the American Steel Wire Gauge, and one for copper and other nonferrous metal wire, called the American Wire or B. & S. (Brown & Sharpe) Gauge.

There is also a third, special gauge system for piano wires, which are now used in construction for tension wires. In all these gauge systems, the larger the number, the finer is the wire. (For gauge equivalents of other countries, *see* Gauges.)

Finish. Wire is available in bright or plain finish, galvanized, plated with various metals, annealed or otherwise heat-treated, indented, bent, corrugated, and made into wire rope.

Requirements. The wire used for making wire rope, either bright or galvanized, must meet very rigid requirements which include testing for gauge, tensile strength, torsion, type of galvanized coatings, prestretched or nonprestretched, and stress at 0.7% extension under load, as shown in Table *W5*.

Types and Uses

Ferrous Wire. Ferrous wire not only is graded according to the American Steel Wire Gauge, but also is classified more simply as coarse round wire (20-gauge and coarser), fine round wire (16-gauge and finer), and shaped wire. The range from gauge No. 16 to No. 20 is not clearly defined because the same size of wire is considered fine for one end use and coarse for a different application. A 16-gauge wire is considered fine because it is the smallest size that can be drawn without intermediate heat treatment.

Table W4 Standard gauges used for wire correlated to diameters

Gauge number	American Steel Wire Gauge (Washburn and Muen) ferrous		American Wire Gauge (Brown and Sharpe) nonferrous		Music or piano wire gauge	
	in.	mm	in.	mm	in.	mm
0000000	0.4900	12.440				
000000	0.4615	11.722	0.5800	14.732	0.004	0.102
00000	0.4300	10.922	0.5165	13.119	0.005	0.127
0000	0.3938	10.003	0.4600	11.664	0.006	0.152
000	0.3625	9.208	0.4096	10.404	0.007	0.178
00	0.3310	8.407	0.3648	9.266	0.008	0.203
0	0.3065	7.785	0.3249	8.269	0.009	0.229
1	0.2830	7.188	0.2893	7.348	0.010	0.254
2	0.2625	6.668	0.2576	6.543	0.011	0.279
3	0.2437	6.543	0.2294	5.287	0.012	0.305
4	0.2253	5.723	0.2043	5.189	0.013	0.330
5	0.2070	5.258	0.1819	4.620	0.014	0.356
6	0.1920	4.877	0.1620	4.115	0.016	0.406
7	0.1770	4.496	0.1443	3.665	0.018	0.457
8	0.1620	4.115	0.1285	3.264	0.020	0.508
9	0.1483	3.767	0.1144	2.907	0.022	0.559
10	0.1350	3.329	0.1019	2.588	0.024	0.610
11	0.1205	3.060	0.0907	2.304	0.026	0.660
12	0.1055	2.680	0.0808	2.052	0.029	0.737
13	0.0915	2.324	0.0720	1.829	0.031	0.787
14	0.0800	2.032	0.0641	1.628	0.033	0.838
15	0.0720	1.826	0.0571	1.450	0.035	0.867
16	0.0625	1.588	0.0508	1.290	0.037	0.940
17	0.0540	1.372	0.0453	1.151	0.039	0.991
18	0.0475	1.207	0.0403	1.024	0.041	1.041
19	0.041	1.041	0.0359	0.912	0.043	1.092
20	0.0348	0.884	0.0320	0.813	0.045	1.143
21	0.0318	0.808	0.0285	0.724	0.047	1.194
22	0.0286	0.726	0.0253	0.643	0.049	1.245
23	0.0258	0.655	0.0226	0.574	0.051	1.295
24	0.0230	0.584	0.0201	0.511	0.055	1.397
25	0.0204	0.518	0.0179	0.455	0.059	1.499
26	0.0181	0.460	0.0159	0.404	0.063	1.600
27	0.0173	0.439	0.0142	0.361	0.067	1.702
28	0.0162	0.415	0.0126	0.320	0.071	1.803
29	0.0150	0.381	0.0113	0.287	0.075	1.905
30	0.0140	0.356	0.0100	0.254	0.080	2.032
31	0.0132	0.335	0.0089	0.226	0.085	2.159
32	0.0128	0.325	0.0079	0.201	0.090	2.286
33	0.018	0.300	0.0071	0.180	0.095	2.413
34	0.0104	0.264	0.0063	0.160	0.100	2.540
35	0.0095	0.241	0.0056	0.142	0.106	2.692
36	0.0090	0.229	0.0050	0.127	0.112	2.849
37	0.0085	0.216	0.0044	0.112	0.118	2.997
38	0.0080	0.203	0.0040	0.102	0.124	3.150
39	0.0075	0.191	0.0035	0.089	0.130	3.302
40	0.0070	0.179	0.0031	0.079	0.138	3.505
41	0.0066	0.168			0.146	3.708
42	0.0062	0.158			0.154	3.912
43	0.0060	0.152			0.162	4.115
44	0.0058	0.147			0.170	4.318
45	0.0055	0.140			0.180	4.572

Table W5 Mechanical properties of zinc-coated wire for the production of wire rope

Zinc coating	Nominal diameter		Ultimate tensile strength[a]		Stress at 0.7% extension under load[a]	
	in.	mm	lbf/in.^2	kgf/mm^2	lbf/in.^2	kgf/mm^2
Class A	0.040–0.110 incl.	1.016–2.794 incl.	220,000	151.69	150,000	103.43
	and	and				
	0.111 and larger[b]	2.819 and larger[b]	220,000	151.69	160,000	110.32
Class B	0.040 and larger[b]	1.016 and larger[b]	210,000	143.80	150,000	103.43
Class C	0.040 and larger[b]	1.016 and larger[b]	200,000	137.9	140,000	96.43

[a] $1\ \text{lbf/in.}^2 = 0.006895\ \text{MN/m}^2$.

[b] Implies that the wire sizes chosen by the wire rope manufacturers must meet these mechanical requirements.

Piano Wire. Piano wire, which is made of electric-furnace steel, was originally used only for musical instruments but now is also utilized for small steel springs, in prestressed concrete, for certain types of wire rope, and for other miscellaneous uses that demand a wire free from internal, physical, or surface defects.

Wire Rope. Wire rope has long been used extensively for prestressed concrete, structure components, elevators, bridges, and the like, but with the development of air structures, suspended and tent structures, and also air-inflated structures, it has become a very important construction material in combination with the plastic

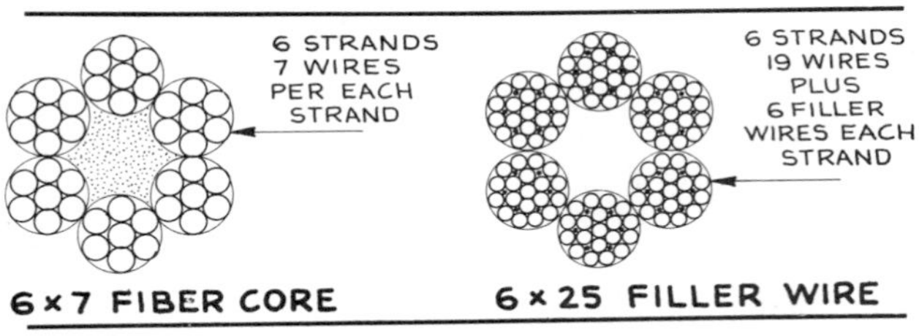

Figure W10 Typical types of wire rope.

sheets and textiles used with such structures. (*See* Figure *W10*; *see also* Air Structures; Tent Structures.)

The galvanized wire used for wire rope is manufactured with different classes of zinc coatings and is made into rope with different resistances to corrosion. For example, a rope in which all wires have Class A zinc coating has normal exterior corrosion resistance, whereas a rope with internal wires Class A zinc-coated but exterior wires Class B zinc-coated has a greater resistance to corrosion, and a rope in which all wires have Class C coating has the greatest resistance to corrosion. (*See* Tables *W6*, *W7*, and *W8*.)

When dealing with wire rope, it is always advisable to consult the manufacturers concerning suitability and requirements for the particular end use. Other tests may be required, depending on the use. In general, stiffness, tensile strength, and surface wear increase with an increase in carbon content and size of wire.

Wire rope is manufactured with specific designations such as 6 X 7, 6 X 19, 8 X 19, and 6 X 25 filler wire. These numerical designations mean, taking 6 X 7 as an example, that 6 strands of wire, each strand containing 7 wires, are used to make the rope (*see* Figure *W10*). Flexibility can be increased by increasing the number of wires in the strand, the number of strands in the wire rope, or both.

Other Wire Uses. Wire itself and wire products find many uses in the construction field, for example, as mesh and wire cloth, concrete reinforcing, lathing, fencing, wire rope, chain, bolts, screws, nails, ties, and countless similar accessory and hardware items.

History and Manufacture

Ancient Methods. Wire was first made by hammering metal and then cutting it into narrow strips. Although wire of this type must have been known in prehistoric times, the first written reference is found in the Bible.

Table W6 Minimum weights of zinc coating

Nominal diameter of coated wire		Minimum weight of coating					
in.	mm	oz/ft^2	g/m^2	oz/ft^2	g/m^2	oz/ft^2	g/m^2
0.040–0.061 incl.	1.016–1.549 incl.	0.40	122.06	0.80	244.12	1.20	366.18
0.062–0.079 incl.	1.575–1.907 incl.	0.50	152.58	1.00	305.15	1.50	457.73
0.080–0.092 incl.	2.032–2.337 incl.	0.60	183.09	1.20	366.18	1.80	549.27
0.093–0.103 incl.	2.362–2.616 incl.	0.70	213.61	1.40	427.21	2.10	640.92
0.104–0.199 incl.	2.642–3.023 incl.	0.80	244.12	1.60	488.24	2.40	732.36
0.120–0.142 incl.	3.048–3.607 incl.	0.85	259.79	1.70	518.76	2.55	778.13
0.143–0.187 incl.	3.632–4.750 incl.	0.90	274.64	1.80	529.27	2.70	823.91
0.188 and larger[a]	4.775 and larger[a]	1.00	3.05.15	2.00	610.30	3.00	915.45

[a]Implies that the wire sizes chosen by the wire rope manufacturers must meet these mechanical requirements.

Table W7 Minimum modulus of elasticity of prestretched wire rope

Nominal diameter of wire rope		Minimum modulus of elasticity[a]			
		Class A coatings[b]		Class B and Class C outer wire coatings	
in.	mm	$lbf/in.^2$	MN/mm^2	$lbf/in.^2$	MN/mm^2
$\frac{3}{8}$–4	9.53–101.6	20,000,000	137 900	19,000,000	130 705

[a]1 $lbf/in.^2 = 0.006895\ MN/m^2$.
[b]For Class B and Class C coatings throughout, consult manufacturers.

Table W8 Minimum breaking point of zinc-coated wire rope

Nominal diameter		Class A coating throughout		Class B coating throughout		Class C coating throughout		Class B coating outer wires, Class A coating inner wires		Class C coating outer wires, Class A coating inner wires		Gross metallic area		Weight	
in.	mm	2000-lb ton	metric tonne	2000-lb ton	metric tonne	2000-lb ton	metric tonne	2000-lb ton	metric tonne	2000-lb ton	metric tonne	in.^{2}	mm^2	ft-lb	kg-m
$\frac{3}{8}$	9.53	6.5	6.6	6.2	6.3	5.9	6.0	6.3	6.4	6.1	6.2	0.065	42.53	0.24	0.36
$\frac{7}{16}$	11.11	8.8	8.9	8.4	8.5	8.0	8.1	8.5	8.6	8.2	8.3	0.091	58.71	0.32	0.48
$\frac{1}{2}$	12.70	11.5	11.7	11.0	11.2	10.5	10.7	11.1	11.3	10.7	10.8	0.119	76.78	0.42	0.62
$\frac{9}{16}$	14.29	14.5	14.7	13.8	14.0	13.2	13.4	14.0	14.2	13.5	13.7	0.147	94.84	0.53	0.79
$\frac{5}{8}$	15.88	18.0	18.3	17.2	17.5	16.4	16.7	17.4	17.7	16.8	17.1	0.182	117.43	0.65	0.97
$\frac{11}{16}$	17.46	21.5	21.8	20.5	20.8	19.5	19.8	20.8	21.1	20.0	20.3	0.221	142.59	0.79	1.18
$\frac{3}{4}$	19.05	26.0	26.4	14.8	25.2	23.6	24.0	25.1	25.5	24.2	24.6	0.268	172.91	0.95	1.41
$\frac{13}{16}$	20.64	30.0	30.5	28.6	29.1	27.3	27.7	29.0	29.5	28.0	28.5	0.311	200.66	1.10	1.64
$\frac{7}{8}$	22.23	35.0	35.6	33.4	33.9	31.8	32.3	33.8	34.3	32.6	33.1	0.361	232.92	1.28	1.90
$\frac{15}{16}$	23.81	40.0	40.6	38.2	38.8	36.4	40.0	38.6	39.2	37.3	37.9	0.414	267.11	1.47	2.19
1	25.40	45.7	46.4	43.6	44.3	41.5	42.2	44.1	44.8	42.6	43.3	0.471	303.89	1.67	2.48
$1\frac{1}{8}$	28.58	57.8	58.7	55.1	56.0	52.5	53.3	55.8	56.7	53.9	54.8	0.596	384.54	2.11	3.14
$1\frac{1}{4}$	31.75	72.2	73.4	68.9	70.0	65.6	66.7	69.7	70.8	67.3	68.4	0.745	480.66	2.64	3.93
$1\frac{3}{8}$	34.95	87.8	89.2	83.8	85.1	79.8	81.1	84.8	86.2	81.8	83.1	0.906	584.55	3.21	4.78
$1\frac{1}{2}$	38.10	104.0	105.7	99.2	100.8	94.5	96.0	100.0	101.6	96.9	98.5	1.076	694.24	3.82	5.68
$1\frac{5}{8}$	41.28	123.0	125.0	117.0	118.9	112.0	113.8	120.0	132.9	117.0	118.9	1.270	819.40	4.51	6.71
$1\frac{3}{4}$	44.45	143.0	145.3	136.0	138.2	130.0	132.1	140.0	142.2	136.0	138.2	1.470	348.44	5.24	7.80
$1\frac{7}{8}$	47.63	164.0	166.6	156.0	158.5	149.0	151.4	160.0	162.5	156.0	158.5	1.690	1090.39	6.03	8.97
2	50.80	186.0	192.0	177.0	179.8	169.0	171.7	182.0	184.9	177.0	179.8	1.920	1238.78	6.85	10.19
$2\frac{1}{8}$	53.98	210.0	213.4	200.0	203.2	191.0	194.1	205.0	288.3	200.0	203.2	2.170	1400.08	7.73	11.50
$2\frac{1}{4}$	57.15	235.0	239.2	224.0	227.6	214.0	217.4	230.0	233.7	224.0	227.6	2.420	1561.38	8.66	12.89
$2\frac{3}{8}$	60.33	261.0	269.2	249.0	253.0	237.0	240.8	255.0	258.8	249.0	253.0	2.690	1735.59	9.61	14.30
$2\frac{1}{2}$	63.50	288.0	292.6	275.0	279.4	262.0	266.2	281.0	285.5	275.0	279.4	2.970	1916.24	10.60	15.77
$2\frac{5}{8}$	66.68	317.0	322.1	302.0	306.8	288.0	292.6	310.0	315.0	302.0	306.8	3.270	2109.80	11.62	17.29
$2\frac{3}{4}$	69.85	347.0	352.6	331.0	336.3	315.0	320.0	339.0	344.4	331.0	336.3	3.580	2309.82	12.74	18.96
$2\frac{7}{8}$	73.03	379.0	385.1	362.0	367.8	344.0	349.5	372.0	378.0	365.0	370.8	3.910	2522.73	13.90	20.69
3	76.20	412.0	418.6	393.0	399.3	374.0	380.0	405.0	411.5	397.0	403.4	4.250	2742.10	15.11	22.48
$3\frac{1}{4}$	82.55	475.0	482.6	453.0	460.3	432.0	438.9	466.0	473.5	457.0	464.3	5.040	3251.81	18.00	26.78
$3\frac{1}{2}$	88.90	555.0	563.9	529.0	537.5	504.0	512.1	545.0	533.7	534.0	542.5	5.830	3761.52	21.00	31.25
$3\frac{3}{4}$	95.25	640.0	650.3	611.0	620.8	582.0	591.3	628.0	639.1	616.0	625.9	6.670	4303.48	24.00	35.71
4	101.60	730.0	741.7	696.0	707.1	664.0	674.6	717.0	728.5	703.0	714.3	7.590	4897.07	27.00	40.18

To make round wire these strips were hammered into shape, and for long lengths several pieces were hammered together. Later these strips were pulled by hand through holes in a stone and still later through similar dies made of iron. This method of pulling limited the amount of reduction possible and the type of metal used to the strength of the man pulling, and the amount of wire to the length of his arms and legs.

Water power was used for the first time in Germany about 1350, in England in 1565, and in the United States in 1650. This and the invention of the steam engine in 1769 marked the beginning of modern wire-drawing techniques. The demand for wire rope about 1840 and then for wire for barbed wire and other fencing and for telegraph, telephone, and finally electric power transmission lines increased the demand for wire to tremendous tonnage. At present, the tonnage of wire produced in the steel industry alone is over 6% of total finished steel.

Modern Procedures. The manufacture of wire is similar for all metals. The description given here is for wire made from steel wire-rods. The rods are first cleaned, pickled in a hot, dilute acid bath to remove the iron oxide scale on their surfaces, and washed with water. At this point they are given a coating of lime emulsion, phosphate-base solution, or some other substance that will act as a lubricant in drawing and also neutralize any remaining acid. Depending on the continuous length of wire to be produced, a single length of wire-rod is prepared for drawing by electric-resistance butt-welding together the right number of wire-rod units. The wire drawing process consists of first pointing the wire-rod, then threading this point through the die, and connecting it to the drawing block. As wire is drawn, heat is created; therefore the die must be water-cooled and lubricants used on the wire-rods so that they will run through the die smoothly. The illustration of single-block wiredrawing in Figure *W11* shows the simple, single reduction from rod to wire.

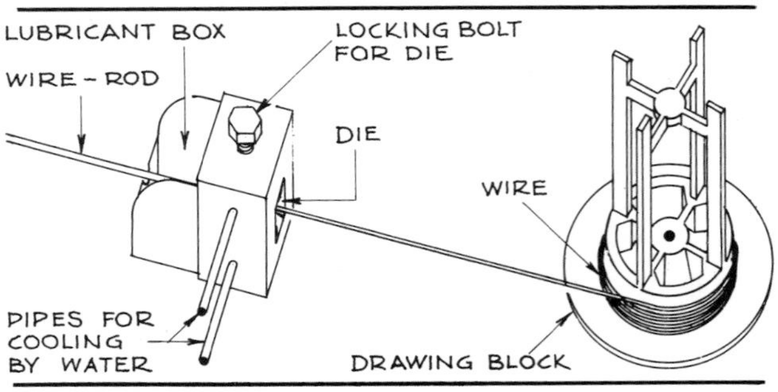

Figure W11 Single-block wire drawing.

When finer wires are required, a series of drawing dies is combined. The wire-rod starts through the first die, which reduces and lengthens it. It then passes through the second die, through which, because of its greater length, it must pass at higher speed; hence it passes through a third die at a still faster speed and through each successive die at an ever-increasing rate.

Wire cannot always be finished directly from wire-rod to wire, as the drawing can make the metal too hard and brittle. In this case, the wire is heated (annealed) between drawings. Although all wires are drawn in a similar fashion, each metal or alloy requires its own specific lubricant, cleaning solution, annealing technique, speed of drawing, and types and sequence of dies to meet the chemical and physical property requirements for the finished wire.

WOLLASTONITE

Wollastonite is a white, medium-hard, fibrous mineral that resembles a fibrous talc in its properties and major uses. It is a calcium silicate ($CaSiO_3$) containing 48.25% CaO and 52 to 55% SiO_2. It occurs in association with garnet in metamorphosed limestones.

Wollastonite is used in ceramic bodies of clay tile products such as floor and wall tile and as a filler and extender in paints, paper, etc. A fibrous cottony wollastonite aggregate prepared by a special milling method serves as filler in block and bulk insulation; as a diluent for asbestos fiber in magnesium oxychloride cement products; and as an additive to portland cement to increase strength. Chemically processed wollastonite yields a paint pigment suitable as a paint extender and paper filler. A strong, hard, dead white cement product that may be sawed, nailed, drilled, and cut without tearing or splitting can be made by mixing wollastonite with phosphoric acid, using boric acid as a set retarder, and firing the mixture cast in molds.

WOOD

Physical and Chemical Properties

Constituents. All wood consists of the following four components:

1. Cellulose comprises about 70% of the wood; it is subdivided into two types, alpha-cellulose and hemicellulose. Alpha-cellulose is the base of paper, paper pulp products, synthetic textiles, and plastics. At present, hemicellulose is little used; perhaps through

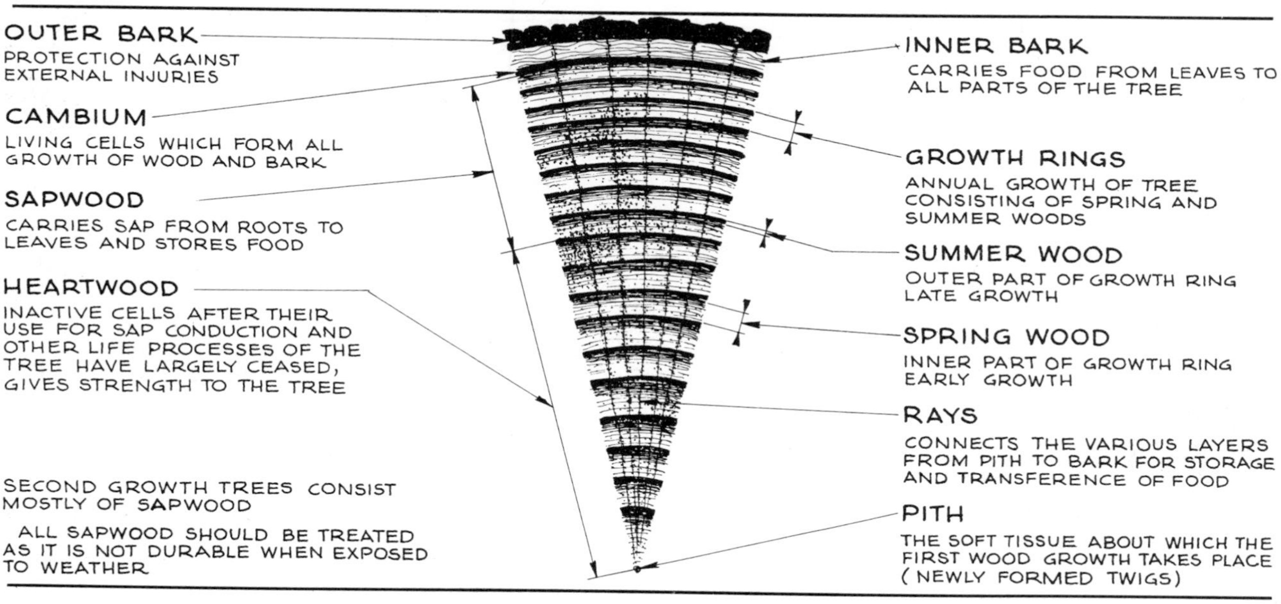
OUTER BARK
PROTECTION AGAINST EXTERNAL INJURIES
CAMBIUM
LIVING CELLS WHICH FORM ALL GROWTH OF WOOD AND BARK
SAPWOOD
CARRIES SAP FROM ROOTS TO LEAVES AND STORES FOOD
HEARTWOOD
INACTIVE CELLS AFTER THEIR USE FOR SAP CONDUCTION AND OTHER LIFE PROCESSES OF THE TREE HAVE LARGELY CEASED, GIVES STRENGTH TO THE TREE
SECOND GROWTH TREES CONSIST MOSTLY OF SAPWOOD
ALL SAPWOOD SHOULD BE TREATED AS IT IS NOT DURABLE WHEN EXPOSED TO WEATHER
INNER BARK
CARRIES FOOD FROM LEAVES TO ALL PARTS OF THE TREE
GROWTH RINGS
ANNUAL GROWTH OF TREE CONSISTING OF SPRING AND SUMMER WOODS
SUMMER WOOD
OUTER PART OF GROWTH RING LATE GROWTH
SPRING WOOD
INNER PART OF GROWTH RING EARLY GROWTH
RAYS
CONNECTS THE VARIOUS LAYERS FROM PITH TO BARK FOR STORAGE AND TRANSFERENCE OF FOOD
PITH
THE SOFT TISSUE ABOUT WHICH THE FIRST WOOD GROWTH TAKES PLACE (NEWLY FORMED TWIGS)

Figure W12 Section through tree trunk, showing internal structure of wood.

continued research its complete utilization can be realized.

2. Lignin comprises about 18 to 28% of the wood; it is the adhesive that gives strength and rigidity to the wood.

3. Extractives are not part of the wood structure but contribute such properties as color, odor, taste, and resistance to decay. They consist of tannin, starch, coloring matter, oils, resins, fats, and waxes. They can be removed from wood by neutral solvents, water, alcohol, acetone, benzene, and ether.

4. Ash-forming minerals make up from 0.2 to 1.0% of the wood and are part of the wood structure. They are the nutrient plant-food elements of the tree and are left as ash when the lignin and cellulose are burned.

The structure of wood (shown in Figure *W12*), its species, the way it is cut (its pattern of growth rings), and its moisture content are all factors that control the mechanical properties of a piece of lumber.

Linear Thermal Expansion. The coefficient of linear thermal expansion differs in any species according to the structural direction of the wood.

1. Expansion in a longitudinal direction (parallel to the fibers) is independent of the specific gravity of the wood and varies from 0.0000011 to 0.0000033 in./°F (0.0000020 to 0.0000059 mm/°C) for the different species.

2. Expansion across the fibers or in the radial and tangential direction varies directly with the specific gravity of the wood (when over-dry) from 0.0000146 to 0.0000341 in./°F (0.0000263 to 0.0000614 mm/°C) for different species.

Generally, the coefficient of thermal expansion can be overlooked because the movement (swelling and shrinking) of the wood caused by moisture under normal conditions of exposure is much greater. However, in buildings that must be kept dry and that are subject to a wide range of temperature changes, the coefficient of thermal expansion must be taken into consideration.

Moisture Content. Wood shrinks as it loses moisture and swells as it absorbs moisture. The water in wood can be divided into two categories: (1) free water in the cell cavities and intercellular spaces of the wood, and (2) absorbed water held in the capillaries of the walls of such wood elements as fibers and ray cells. The absorbed water is important in relation to shrinkage. When all free water is removed but all absorbed water remains, the so-called fiber saturation point (approximately 30% moisture content for all species) is reached. Shrinkage occurs at moisture content percentages below the fiber saturation point. Wood dried to 15% moisture has attained about one-half of the total shrinkage possible. For each 1% loss in moisture below fiber saturation point the wood shrinks about one-thirtieth of the total shrinkage; and likewise for each 1% increase in moisture content the wood swells about one-thirtieth of the total swelling possible. Wood shrinks the most in the direction of the annual growth rings, less across these rings, and generally very little parallel to the grain. Figures *W13* and *W14* show the shrinkage and distortions of various pieces of lumber in relation to their growth rings (as located in the original log) and the various kinds of warp.

Tables *W9* and *W10* show the shrinkage for various moisture contents of softwoods and hardwoods commonly used in construction.

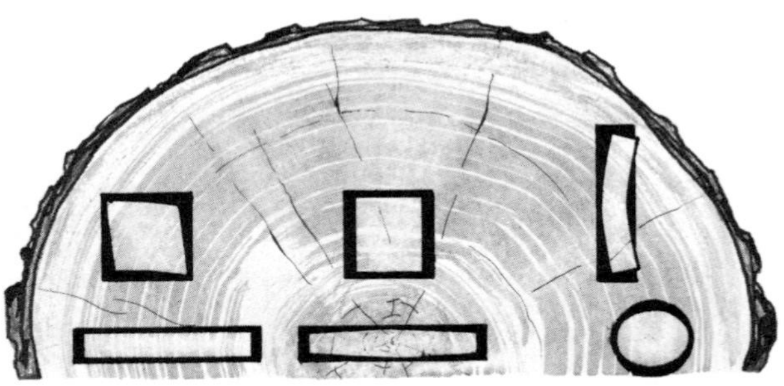

Figure W13 Shrinkage and distortions caused by the direction of annual rings.

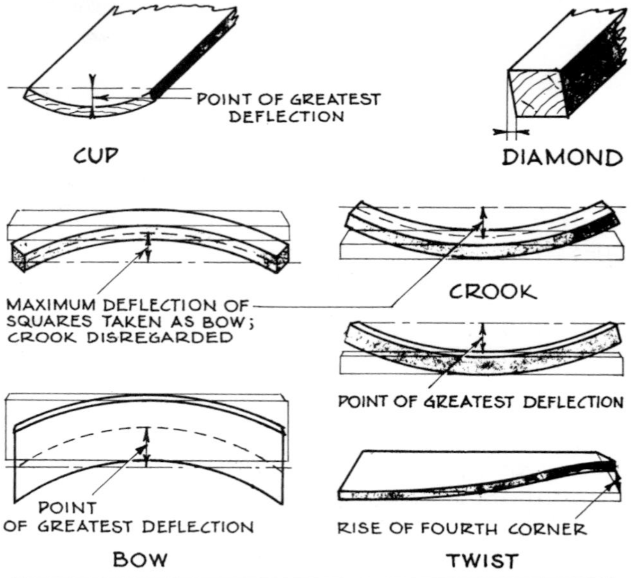

Figure W14 Various kinds of warpage in wood.

Table W9 Shrinkage of deciduous (hardwood) woods based on dimension of green wood

	Shrinkage values based on dimension of green wood (percent)								
	Dried to 20% moisture content[a]			Dried to 6% moisture content[a]			Dried to 0% moisture content		
Type of wood	In direction of growth rings (tangential)	Across growth rings (radial)	By volume	In direction of growth rings (tangential)	Across growth rings (radial)	By volume	In direction of growth rings (tangential)	Across growth rings (radial)	By volume
Alder, red	1.5	2.4	4.2	3.5	5.8	10.1	4.4	7.3	12.6
Ash, white	1.6	2.6	4.5	3.8	6.2	10.7	4.9	7.8	13.3
Ash, Oregon	1.4	2.7	4.4	3.3	6.5	10.6	4.1	8.1	13.2
Ash, black	1.7	2.6	5.1	4.0	6.2	12.2	5.0	7.8	15.2
Aspen, bigtooth quaking	1.1–1.2	2.2–2.6	3.8–3.9	2.6–2.8	5.4–6.3	9.2–9.4	3.3–3.5	6.7–7.9	11.5–11.8
Basswood	2.2	3.1	5.3	5.3	7.4	12.6	6.6	9.3	15.8
Beech, American	1.7	3.7	5.4	4.1	8.8	13.0	5.5	11.9	17.2
Birch, paper	2.1	2.9	5.4	5.0	6.9	13.0	6.3	8.6	16.2
Birch, sweet yellow	2.2–2.4	2.8–3.1	5.2–5.6	5.2–5.8	6.8–7.4	12.5–13.4	6.5–7.3	9.0–9.5	15.6–16.8
Butternut	1.1	2.1	3.5	2.7	5.1	8.5	3.4	6.4	10.6
Cherry, black	1.2	2.4	3.8	3.0	5.7	9.2	3.7	7.1	11.5
Chestnut, American	1.1	2.2	3.9	2.7	5.4	9.3	3.4	6.7	11.6
Cottonwood, eastern	1.3	3.1	4.7	3.1	7.4	11.3	3.9	9.2	13.9
Elm, American, cedar, rock	1.4–1.6	2.7–3.2	4.7–4.9	3.4–3.8	6.5–7.6	11.3–11.7	4.2–4.8	7.2–10.2	14.6–15.4
Elm, slippery	1.6	3.0	4.6	3.9	7.1	11.0	4.9	8.9	13.8
Hackberry	1.6	3.0	5.6	3.3	7.1	13.5	4.8	8.9	13.8
Hickory, pecan	1.6	3.0	4.5	3.9	7.1	10.9	4.9	8.9	13.6
Hickory, true	2.3–2.6	3.3–4.2	5.6–6.4	5.6–6.2	8.0–10.1	13.4–15.4	7.2–7.7	10.5–11.5	16.7–19.2

Table W9 Shrinkage of deciduous (hardwood) woods based on dimension of green wood (*continued*)

	Shrinkage values based on dimension of green wood (percent)								
	Dried to 20% moisture content[a]			Dried to 6% moisture content[a]			Dried to 0% moisture content		
Type of wood	In direction of growth rings (tangential)	Across growth rings (radial)	By volume	In direction of growth rings (tangential)	Across growth rings (radial)	By volume	In direction of growth rings (tangential)	Across growth rings (radial)	By volume
Locust, black	1.5	2.4	3.4	3.7	5.8	8.2	4.6	7.2	10.2
Magnolia, southern	1.8	2.2	4.1	4.3	5.3	9.8	5.4	6.6	12.3
Maple	1.2–1.6	2.4–3.2	3.9–5.0	2.4–3.9	5.7–7.6	9.3–11.9	3.0–4.8	7.1–9.9	11.6–14.7
Oak, red	1.3–1.8	2.7–3.5	4.5–6.3	3.2–4.4	6.6–8.5	10.8–15.2	4.0–5.0	8.6–11.3	14.5–19.0
Oak, white	1.4–1.8	2.4–3.6	4.2–5.6	3.3–5.3	5.8–8.6	10.0–13.4	4.4–6.6	8.8–12.7	12.7–16.4
Poplar, yellow	1.3	2.4	4.1	3.2	5.7	9.8	4.6	8.2	12.7
Sweetgum	1.7	3.3	5.0	4.2	7.9	12.0	5.3	10.2	15.8
Sycamore, American	1.7	2.5	4.7	4.1	6.1	11.4	5.0	8.4	14.1
Walnut, black	1.8	2.6	4.3	4.4	6.2	10.2	5.5	8.7	14.4

[a]The 20% moisture content values are one-third of the values for 0% moisture content, and the 6% moisture content values are four-fifths of the values for 0% moisture.

Table W10 Shrinkage of evergreen (softwood) woods based on dimension of green wood

	Shrinkage values based on dimension of green wood (percent)								
	Dried to 20% moisture content[a]			Dried to 6% moisture content[a]			Dried to 0% moisture content		
Type of wood	In direction of growth rings (tangential)	Across growth rings (radial)	By volume	In direction of growth rings (tangential)	Across growth rings (radial)	By volume	In direction of growth rings (tangential)	Across growth rings (radial)	By volume
Cedar, Alaska	0.9	2.0	3.1	2.2	4.8	7.4	2.8	6.0	9.2
Cedar, incense	1.1	1.7	2.5	2.6	4.2	6.1	3.3	5.2	7.7
Cedar, red eastern	1.0	1.6	2.6	2.5	3.8	6.2	3.1	4.7	7.8
Cedar, red western	0.8	1.7	2.3	1.9	4.0	5.4	2.4	5.0	6.3
Cedar, Port Orford	1.5	2.3	3.4	3.7	5.5	8.1	4.6	6.9	10.1
Cypress, bald	1.3	2.1	3.5	3.0	5.0	8.4	3.8	6.2	10.5
Fir, Douglas	1.2–1.7	2.1–2.6	3.5–3.9	2.9–4.0	5.0–6.2	8.5–9.4	3.8–4.8	6.9–7.6	10.7–12.4
Fir, balsam	1.0	2.3	3.7	2.3	5.5	9.0	2.9	6.9	11.2
Fir, western	1.1–1.5	2.4–3.3	3.3–4.6	2.6–3.7	5.7–6.6	7.8–11.0	3.3–4.5	7.0–9.2	9.8–13.0
Hemlock, eastern	1.0	2.3	3.2	2.4	5.4	7.8	3.0	6.8	9.7
Hemlock, western,	1.4	2.6	4.0	3.4	6.3	9.5	4.2	7.8	12.4
Larch, western	1.4	2.7	4.4	3.4	6.5	10.6	4.5	9.1	14.0
Pine, eastern white	0.8	2.0	2.7	1.8	4.8	6.6	2.1	6.1	8.2
Pine, lodgepole	1.5	2.2	3.8	3.6	5.4	9.2	4.5	7.4	12.3
Pine, pitch	1.3	2.4	3.6	3.2	5.7	8.7	4.0	7.1	10.9
Pine, pond	1.7	2.4	3.7	4.1	5.7	9.0	5.1	7.1	11.2
Pine, ponderosa	1.3	2.1	3.2	3.1	5.0	7.7	3.9	6.2	9.7
Pine, red	1.5	2.4	3.8	3.7	5.8	9.2	3.8	7.2	11.3
Pine, southern yellow short-leaf and long-leaf	1.5–1.8	2.5–2.6	4.1	3.5–4.4	5.9–6.2	9.8	4.2–5.4	7.2–7.7	11.9–12.3

Table W10 Shrinkage of evergreen (softwood) woods based on dimension of green wood (*continued*)

	Shrinkage values based on dimension of green wood (percent)								
	Dried to 20% moisture content[a]			Dried to 6% moisture content[a]			Dried to 0% moisture content		
Type of wood	In direction of growth rings (tangential)	Across growth rings (radial)	By volume	In direction of growth rings (tangential)	Across growth rings (radial)	By volume	In direction of growth rings (tangential)	Across growth rings (radial)	By volume
Pine, sugar	1.0	1.9	2.6	2.3	4.5	6.3	2.9	5.6	7.9
Pine, western white	1.4	2.5	3.9	3.3	5.9	9.4	4.1	7.4	11.8
Redwood	0.9	1.5	2.3	2.1	3.5	5.4	2.6	4.4	6.8
Spruce, eastern	1.3–1.4	2.3–2.6	3.8–3.9	3.0–3.3	5.4–6.2		3.8–4.1	6.8–7.8	11.3–11.8
Spruce, Engelmann	1.1	2.2	3.5	2.7	5.3	8.3	3.8	7.1	11.0
Spruce, Sitka	1.4	2.5	3.8	3.4	6.0	9.2	4.3	7.5	11.5
Tamarack	1.2	2.5	4.5	3.0	5.9	10.9	3.7	7.4	13.6

[a]The 20% moisture content values are one-third of the values for 0% moisture content, and the 6% moisture content values are four-fifths of the values for 0% moisture.

Generally, but not always, wood should be seasoned to the moisture content shown in Table *W11* and Figure *W15*; the values vary and are controlled by where the wood is to be used (exterior or interior).

Seasoning. Lumber when first cut is called green. Before it is used it should be seasoned either by air drying or by kiln drying.

The advantages of air-dried wood over green wood are reduction of weight; reduction of shrinkage, checking, honeycombing, and warping; increase in strength and nail-holding power; decrease in the attack of various fungi and insects; and improvement of the ability of the wood to hold paint and receive preservatives.

The advantages of kiln drying over air drying are greater reduction in weight; control of moisture content to any desired value; reduction in drying time; killing of any fungi or insects; setting the resins in resinous wood; and less degrade.

"Degrade" is the loss in quality during seasoning of the lumber through (1) unequal shrinkage, which includes checks, honeycomb, warp, loosening of knots, and collapse; and (2) action of fungi, that is, molds, stains and decay. Sometimes chemical stains occur in some softwoods.

Resistance to Decay. Some species of both hardwoods and softwoods have high resistance to decay (*see* Table *W12*) and can be used under conditions where decay hazards exist; other species of wood will need preservative treatment (*see* Wood Preservatives).

Table W11 Recommended moisture contents for wood materials at time of installation in various regions of the United States

	Average moisture content recommended at time of installation (percent of weight of oven-dry wood)[a]		
Where wood is used in building	Dry southwestern states	Damp southwestern states	Other states of the United States
Interior finish and softwood flooring	6	11	8
Siding, sheathing, framing, and exterior trim	9	12	12

[a]If 10% is tested and the average moisture content is within ±1%, the entire lot of this material will be satisfactory. Framing lumber of higher moisture content is often used because material of a specified moisture content may not be available except on special order.

Table W12 Common species of trees with high resistance to decay

Type of tree	
Deciduous (hardwoods)	Evergreen (softwoods)
Cherry, black	Bald cypress, old growth
Chestnut	Cedars
Locust, black	Cypress, Arizona
Oak	Redwood
Sassafras	Yew, Pacific
Walnut, black	

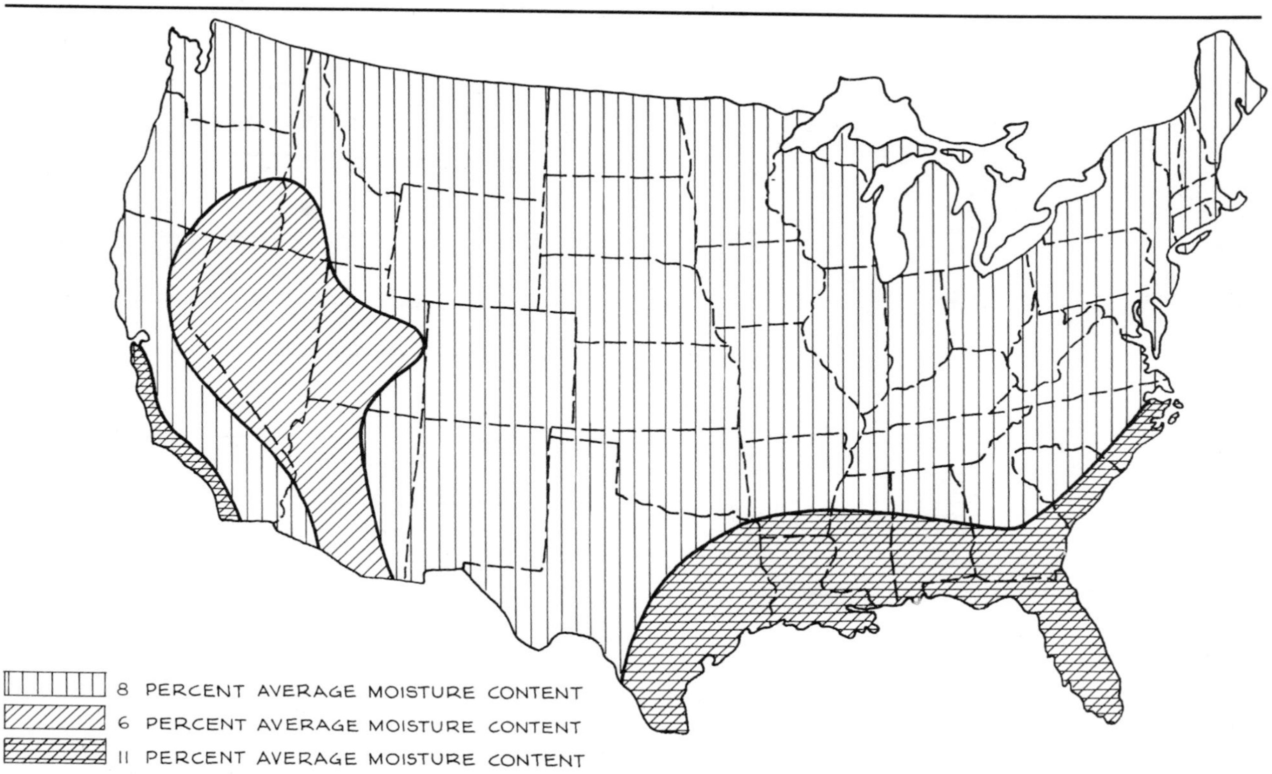

Figure W15 Recommended average moisture content for interior finish woodwork in the United States.

Commercial Forms. Wood is available as lumber, timber, piling, veneer, and plywood (considered a wood integrant in this book.

Quartersawing, Plainsawing, and Riftsawing. A log of wood can be cut in three different ways to make lumber: it can be plainsawed, quartersawed and riftsawed. "Plainsawed" refers to wood cut tangent to the annual growth rings or, in commercial practice, cut with the annual growth rings at an angle of 0 to 45°. "Quartersawed" refers to wood cut radially to the annual growth rings, or parallel to the rays, or, in commercial practice, cut with the annual growth rings at an angle of 45 to 90°. "Riftsawed" is similar to quartersawed except that the cuts strike the medullary rays at a slight angle to miss the flake effect (*see* Figure *W16*). Figures *W17* and *W18* show how a log is divided into lumber and the difference between plainsawed and quartersawed. Table *W13* lists the advantages and disadvantages of plainsawed and quartersawed methods of sawing wood.

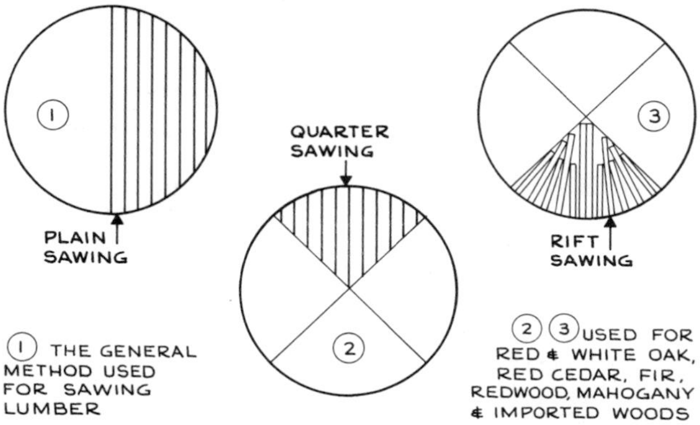

Figure W16 Methods of sawing lumber.

Types and Uses

There are two major classes of wood: (1) deciduous trees (hardwoods), which shed their leaves at the end of each growing season, except in the warmest regions; and (2) evergreens (softwoods), which do not shed their scalelike or needlelike leaves at the end of each growing season with the exception of cypress, tamarack, and larch. The terms "hardwood" and "softwood" as commonly used cause confusion because these terms have no direct relation to the actual physical hardness or softness of the wood.

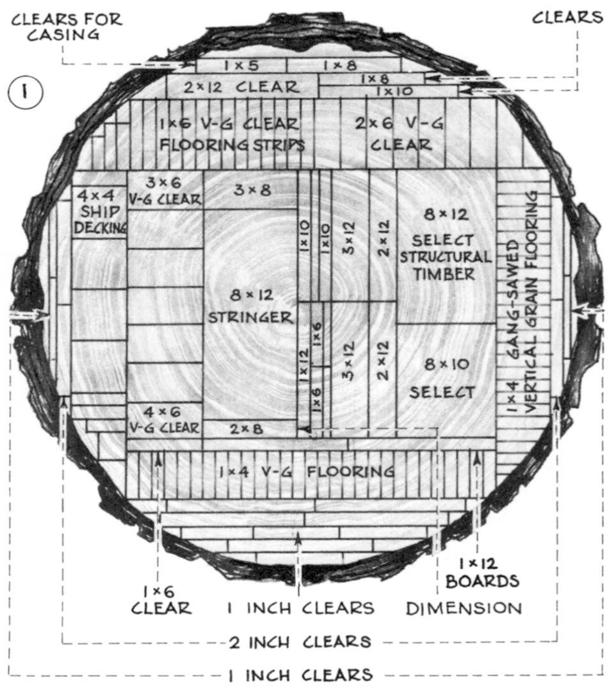

Figure W17 **Division of log into sizes before cutting.**

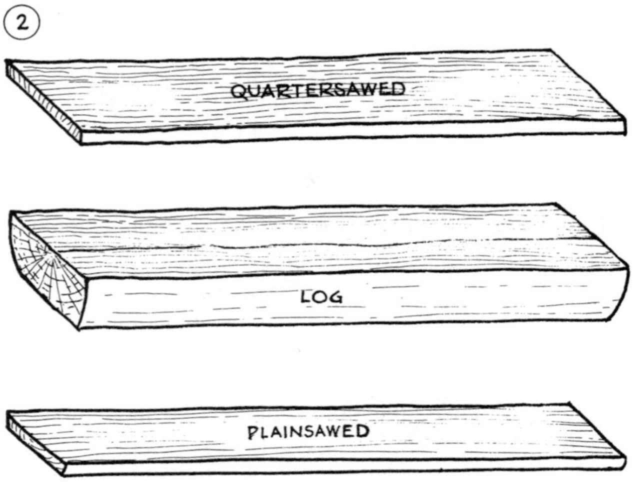

Figure W18 **Quartersawed and plainsawed lumber.**

Tables *W14* and *W15* list the deciduous and evergreen woods that are used in construction in the United States and describe them according to color, characteristics, weight, grain patterns, and uses. Imported woods in general construction use are described on page 782.

The general, overall uses of wood in architecture are shown in Table *W16*.

Table W13 **Advantages and disadvantages of plainsawed and quartersawed[a] lumber**

Plainsawed lumber		Quartersawed lumber	
Advantages	Disadvantages	Advantages	Disadvantages
1. Less waste 2. Growth ring patterns are distinct 3. Knots affect surface appearance and strength less 4. Shakes and pitch pockets affect only a few pieces 5. Shrinks and swells less in thickness 6. Does not collapse so easily in drying 7. Usually less costly as it takes less time to saw	1. The sapwood can extend far in at the surface edges of the pieces 2. Tends to have raised grain caused by separation of the growth rings 3. Pronounced (decorative) rays and grain patterns are less distinct 4. Tends to have surface checks and to split in seasoning and in use 5. Shrinks and swells in width 6. Tends to wear unevenly 7. Tends to twist and cup 8. In some species it does not hold paint well and allows liquids to pass into and through it	1. The sapwood appears at the edges and is limited only by the width of the sapwood in the logs 2. Has less raised grain caused by separation of the growth rings 3. Pronounced rays, and interlocked and wavy grain are more distinct 4. Less surface checks and splits in seasoning and in use 5. Shrinks and swells less in width 6. Wears evenly 7. Twists and cups less 8. In some species it holds paint better and does not allow liquids to pass into or through it	1. More waste 2. Growth patterns less distinct 3. Knots disfigure surface appearance and weaken strength 4. Shakes and pitch pockets extend through numerous pieces 5. Shrinks and swells in thickness 6. Tends to collapse in drying 7. More costly as it takes more time to saw

[a]Riftsawed lumber has the same characteristics as quartersawed lumber except that the medullary rays (thick-walled cells) are cut at a slight angle to eliminate "flake" effect.

Table W14 Characteristics of deciduous (hardwood) trees of the United States

Name of tree	Location	Color of heartwood[a]	Characteristics	Weight[b] lb/ft³: Moisture content 15%	Weight[b] lb/ft³: Moisture content 8%	Grain patterns: Plainsawed or rotary-cut veneer	Grain patterns: Quartersawed or quartersliced veneer	Major uses	Other uses
Alder, red	Pacific Coast between Alaska and California	Pale pinkish brown	Low shrinkage and shock resistance; intermediate strength	28.8	28.0	Faint growth rings	Generally scattered, large flakes	Sash, doors, and millwork paneling	Furniture
Ash, black	Eastern half of the United States	Dark grayish brown	Lighter weight than white or Oregon ash	35.3	34.3	Conspicuous growth rings and occasional burl	Growth rings not conspicuous; occasional burl		Furniture, sporting goods, handles, shipping containers
Ash, Oregon	Pacific Coast	Grayish brown	Lower strength than white ash	40.7	39.2			Veneer	
Ash, white	Eastern half of the United States	Grayish brown	Strong; hard; high resistance to shock; moderately high shrinkage	42.7	41.5				
Aspen	Northeastern and Great Lake states	Light brown	Lightweight; soft; low in strength and shock resistance; high shrinkage	27.0–27.3	26.1–27.3	Faint growth rings	Relatively clear	Veneer	Excelsior, pulpwood, boxes and turned articles
Basswood	Eastern half of the United States	Creamy white to creamy brown	Soft and light in weight; large shrinkage in width and thickness	26.0	25.5	Faint growth rings	Relatively clear	Sash, doors, moldings, veneer, paneling	Wooden ware, venetian blinds, excelsior, pulpwood
Beech, American	Eastern one-third of the United States	White with reddish to reddish brown tinge	Heavy; hard; strong; high resistance to shock and adaptable to steam bending; high shrinkage	44.3	43.2	Faint growth rings	Numerous small flakes up to $\frac{1}{8}$ in. (3.175 mm) in height	Flooring; veneer	Furniture, wooden ware, handles, containers
Birch, paper	Northeastern and Great Lake states and along Appalachian Mountains to northern Georgia	Light brown	Lower in weight; softer and less shock resistant than sweet birch	38.9	38.2	Faint growth rings	Relatively clear	Veneer, turned products	Spools, bobbins, toys
Birch, sweet or yellow		Dark reddish brown	Heavy; hard; strong; good shock resistance; considerable shrinkage	43.4–47.2	42.4–46.0	Occasionally wavy; growth rings not conspicuous	Occasionally wavy	Veneer, paneling, doors	Furniture, wooden ware, handles, containers
Butternut	Eastern half of the United States	Light chestnut brown	Lightweight; soft; low shock resistance	27.4	26.4	Faint growth rings	Faint growth rings	Veneer	
Cherry, black	Eastern half of the United States	Light to dark reddish brown	Strong; stiff; moderately hard; high shock resistance; moderately large shrinkage	36.1	34.8	Faint growth rings; occasional burl	Occasional burl	Veneer, paneling	Furniture, wooden ware

Table W14 Characteristics of deciduous (hardwood) trees of the United States (*continued*)

				Weight[b] lb/ft^3		Grain patterns			
Name of tree	Location	Color of heartwood[a]	Characteristics	Moisture content 15%	Moisture content 8%	Plainsawed or rotary-cut veneer	Quartersawed or quartersliced veneer	Major uses	Other uses
Chestnut, American	Practically all standing chestnut killed by blight	Grayish brown	Lightweight; moderately hard; low in strength; resistance to shock and shrinkage	30.5	29.5	Conspicuous growth rings	Relatively clear	Core stock for plywood panels	Furniture, boxes
Cottonwood	Eastern half of the United States	Grayish white to light grayish brown	Moderately soft; low shock resistance; large shrinkage	28.9	28.0	Faint growth rings	Relatively clear	Veneer	Pulpwood, excelsior, boxes
Elm, American	Eastern half of the United States	Light grayish brown with reddish tinge	Hard; heavy; high shock resistance and shrinkage; good bending qualities	36.3	35.5	Wavy pattern within each growth ring	Faint growth ring stripe	Veneer	Furniture, boxes, implements
Elm, slippery	Same as above but not not the Atlantic Coastal Plain, Florida, or the Gulf Coast	Dark brown with shades of red	Soft; moderately heavy; high shock resistance and shrinkage; good bending qualities	37.8–44.2	36.7–42.7	Conspicuous growth ring with fine pattern within each growth ring	Relatively clear	Veneer	Furniture, boxes, implements
Hackberry	East of the Great Plains	Light yellow or greenish gray	Moderately heavy; strong; high shock resistance; relatively large shrinkage	37.4	36.2	Conspicuous growth rings	Relatively clear	Veneer, paneling	Furniture
Hickory, pecan	Eastern half of the United States	Reddish brown	Heavy; hard; tough and strong; large shrinkage	46.5	45.0	Growth rings not conspicuous	Faint growth ring stripes	Flooring	Handles
Hickory, true	Eastern half of the United States	Reddish brown	Heavy; hard; tough and strong; large shrinkage	51.2–53.4	50.3–52.6	Growth rings not conspicuous	Faint growth ring stripes		Handles, sports apparatus, furniture
Locust, black	From Pennsylvania along the Appalachian Mountains to northern Georgia	Golden brown	Very heavy and very hard; very high resistance to shock; very high strength; high shrinkage; high decay resistance	49.0	46.7	Conspicuous growth rings	Relatively clear	Fence posts, poles	Rough construction
Magnolia	Atlantic and Gulf Coasts to Texas	Light to dark yellowish brown with greenish or purplish tinge	Moderately heavy; hard and stiff; low shrinkage; high shock resistance	35.5	34.4	Faint growth rings	Relatively clear	Sash, doors, veneer, paneling, millwork	Venetian blinds, furniture

Maple	Various types of maples grow in almost all parts of the United States	Light reddish brown	Moderately heavy; strong; stiff; hard; high resistance to shock and shrinkage	33.9–44.5	32.8–43.4	Faint growth rings; occasional bird's-eye; occasionally curly and wavy	Occasionally curly and wavy	Flooring, veneer, paneling, millwork	Furniture, wooden ware, pulpwood
Oak, red	Various types of red oaks grow in almost all parts of United States	Grayish brown with fleshy tinge	Strong; heavy; hard; high shrinkage and shock resistance	41.1–47.2	40.1–45.8	Conspicuous growth rings	Pronounced flake	Flooring, veneer, millwork	Furniture, wooden ware, handles, fence posts
Oak, white		Grayish brown, rarely with fleshy tinge	Strong; heavy; hard; good decay resistance; high shrinkage and shock resistance	45.0–48.5	43.5–47.9	Conspicuous growth rings	Pronounced flake	Flooring, doors, veneer, paneling	Furniture, fence posts, boats
Poplar, yellow	From Connecticut and New York southward to Florida and westward to Missouri	Light to yellowish brown with greenish or purplish tinge	Moderately strong; light in weight; soft; low shock resistance; large shrinkage	30.3	29.2	Faint growth rings	Relatively clear	Paneling, siding cores for plywood	Furniture, pulpwood
Sweetgum	From Connecticut westward to Missouri and southward to the Gulf of Mexico	Reddish brown	Moderately hard; heavy, strong, and stiff; good shock resistance; requires careful drying	36.4	35.5	Faint growth rings; occasional irregular dark streaks (figured gum)	Relatively clear; occasional irregular dark streaks (figured gum)	Veneer, trim, paneling, millwork	Furniture, pulpwood
Sycamore, American	From Maine to Nebraska, Texas, and northern Florida	Flesh brown	Moderately heavy; hard, strong, and stiff; good shock resistance; high shrinkage	35.7	34.7	Faint growth rings	Pronounced flakes up to $\frac{1}{4}$ in. (6.35 mm) in height	Flooring, veneer	Furniture, handles
Walnut, black	From Vermont to the Great Plains southward into Louisiana and Texas	Chocolate brown, occasionally with darker purplish streaks	Heavy; hard, strong, and stiff; little shrinkage; good shock resistance	38.6	37.0	Growth rings not conspicuous; occasionally wavy, curly with burl and other markings		Veneer, paneling, millwork	Furniture

[a]The sapwood is light in color or white unless discolored by fungus or chemical stains.
[b]For metric equivalents of these values, *see* Table *W14a*.

Table W14a Metric equivalents for moisture contents of deciduous (hardwood) trees[a]

	Weight			
	Moisture content 15%		Moisture content 8%	
	lb/ft³	kg/m³	lb/ft³	kg/m³
Alder, red	28.8	461.38	28.0	448.56
Ash, black	35.3	565.51	34.3	549.49
Ash, Oregon	40.7	652.01	39.2	627.98
Ash, white	42.7	684.05	41.5	664.83
Aspen	27.0–27.3	432.54–437.35	26.1–27.3	418.12–437.35
Basswood	26.0	416.52	25.5	408.51
Beech, American	44.3	709.69	43.2	692.06
Birch, paper	38.9	624.18	38.2	611.96
Birch, sweet or yellow	43.4–47.2	695.26–756.14	42.4–46.0	679.25–736.92
Butternut	27.4	438.94	26.4	422.93
Cherry, black	36.1	578.32	34.8	587.50
Chestnut, American	30.5	488.61	29.5	472.59
Cottonwood	28.9	462.98	28.0	488.56
Elm, American	26.3	480.53	35.5	658.71
Elm, slippery	37.8–44.2	605.56–708.08	36.7–42.7	587.93–684.05
Hackberry	37.4	599.15	36.2	579.92
Hickory, pecan	46.5	744.93	45.0	720.50
Hickory, true	51.2–53.4	820.22–855.47	50.3–52.6	805.81–842.65
Locust, black	49.0	784.98	46.7	748.13
Magnolia	35.5	658.71	34.4	551.09
Maple	33.9–44.5	543.08–712.89	32.8–43.4	525.45–695.26
Oak, red	41.1–47.2	658.42–756.14	40.1–45.8	642.40–738.72
Oak, white	45.0–48.5	720.50–776.97	43.5–47.9	696.87–767.36
Poplar, yellow	30.3	485.40	29.2	467.78
Sweetgum	36.4	583.13	35.5	658.71
Sycamore, American	35.7	571.91	34.7	555.89
Walnut, black	38.6	618.37	37.0	592.74

[a]For convenient reference, values in pounds per cubic foot have been repeated.

Classification and Grading of Wood

When a log is sawed into lumber, the various pieces of lumber differ greatly in quality. In order to classify or standardize these widely varying qualities, a series of lumber grades, each having a relatively narrow range of quality, has been established for both deciduous (hardwood) and evergreen (softwood) lumber. The grade of a piece of lumber is based on the number, character, and location of imperfections that may lower the strength, durability, and use of the lumber. Common imperfections are knots, checks, pitch pockets, shakes, and stains. The best grades of lumber are practically free from these. In each lower grade these imperfections increase in quantity. These defects do not prevent such lumber from finding widespread use, however.

Deciduous (Hardwood) Grades. The grade classification for hardwood is based on the amount of usable lumber in the piece. The standard lengths range from 4 to 16 ft in 1-ft increments (1.219 to 4.877 m in 0.3048-m increments), with not more than 50% of odd length to be admitted. The standard grades are described in Table *W17*. Table *W18* lists standard thicknesses in each grade.

In hardwood, Firsts, Seconds, and the face side of Selects meet specified requirements as to knots, holes, and other imperfections. Firsts and Seconds (generally written as "FAS") are nearly always combined into one grade. There is a grade called Sound Wormy, which has the same requirements as No. 1 Common except that imperfections are allowed in the cutting.

Dimension stock is cut from kiln-dried lumber into three classes; see finishes and grades in Table *W19*.

Table W15 Characteristics of evergreen (softwood) trees of the United States

Type of wood	Location	Color of heartwood[a]	Characteristics	Weight lb/ft³[b] Moisture content 15%	Moisture content 8%	Grain patterns Plainsawed or rotary-cut veneer	Quartersawed or quartersliced veneer	Major uses	Other uses
Cedar, Alaska	Pacific coast from Alaska to southern Oregon	Yellow	Moderately heavy; strong, stiff, hard; high resistance to shock; small heartwood shrinkage; good resistance to decay	31.6	30.4	Faint growth rings	Relatively clear	Paneling	Furniture
Cypress, bald	From Delaware to Florida along the Atlantic Coastal Plain, along the Gulf Coast and up the Mississippi Valley to Indiana	Light yellowish brown to reddish brown	Moderately heavy; strong, hard; small shrinkage; one of the most decay-resistant woods	32.6	31.4	Conspicuous, irregular growth rings	Growth ring stripe not conspicuous	Siding, paneling, sash, doors, construction, millwork	Piling, furniture, flooring
Fir, Douglas	From the Rocky Mountains to the Pacific Coast and from Mexico to Central British Columbia	Orange red to red; sometimes yellow	Varies widely in weight and strength; for high strength the density is determined by percentages of summerwood and rate of growth density rule	30.5–34.3	29.2–33.1	Conspicuous growth rings	Growth ring stripe not conspicuous	Construction, veneer, sash, doors, millwork	Piling, furniture, flooring
Fir, true eastern	New England, New York, Pennsylvania, and the Great Lake states	Nearly white to pale reddish brown	Light in weight; soft; higher strength than eastern fir; small to moderately large shrinkage	26.9	26.4	Growth rings not conspicuous	Growth rings not conspicuous	Paneling, siding	Pulpwood
Fir, true western	From the Rocky Mountains to the Pacific Coast	Nearly white to pale reddish brown	Light in weight; soft; higher strength than eastern fir; small to moderately large shrinkage	26.7–28.3	25.8–27.2	Conspicuous growth rings	Relatively clear	Siding, sheathing, paneling, construction, sash, doors, millwork	Venetian blinds, molding
Hemlock, eastern	From New England along the Appalachian Mountains to Alabama and Georgia and the Great Lake states	Light reddish brown	Coarse; moderately light in weight; hard, strong; low shock resistance	29.0	28.0	Growth rings not conspicuous	Faint growth ring stripe	Sheathing, construction	Pulpwood
Hemlock, western	Pacific Coast of Oregon and Washington, Rocky Mountains north to Canada and Alaska	Light reddish brown, brown	Moderately light in weight; low in strength, hardness, shock resistance; large shrinkage	29.6	28.7	Growth rings not conspicuous	Faint growth ring stripe	Construction, sheathing, siding, flooring	Furniture, pulpwood
Cedar, incense	California, Oregon, and Nevada	Reddish brown	Light in weight; soft; moderately low in strength, shock resistance, and stiffness; small shrinkage; spicy odor	25.5	24.2	Faint growth rings	Faint growth ring stripe	Shingles, paneling, construction	Pencils, venetian blinds, fencing

Table W15 Characteristics of evergreen (softwood) trees of the United States (*continued*)

Type of wood	Location	Color of heartwood[a]	Characteristics	Weight lb/ft³ [b]: Moisture content 15%	Weight lb/ft³ [b]: Moisture content 8%	Grain patterns: Plainsawed or rotary-cut veneer	Grain patterns: Quartersawed or quartersliced veneer	Major uses	Other uses
Larch, western	Montana, Idaho, Oregon, and Washington	Russet to reddish brown	Stiff; moderately strong; hard; high shock resistance; heavy; large shrinkage	39.4	38.2	Conspicuous growth rings	Relatively clear	Construction, paneling, flooring, sash, doors	Piling, poles
Pine, eastern white	From Maine along the Appalachian Mountains to Georgia and the Great Lake states	Cream to light reddish brown	Light in weight; small shrinkage; moderately soft; low in strength and shock resistance	25.4	24.2	Faint growth rings	Clear	Sash, doors, paneling, molding	Patterns
Pine, lodge pole (knotty pine)	Rocky Mountains and Pacific Coast to Alaska	Light reddish brown	Moderately light in weight; large shrinkage; soft; stiff; low in strength and shock resistance	29.2	28.2	Growth rings not conspicuous; faintly pocked	Growth rings not conspicuous; faintly pocked	Siding, paneling, flooring	Posts, poles
Pine, pitch	Maine and northern New York, south to Tennessee and Georgia	Brownish red	Medium strong; heavy, hard, stiff; high in shock resistance and shrinkage	34.9	33.8	Faint growth rings		Construction	Posts, poles
Pine, pond	New Jersey to Florida	Dark orange	Moderately strong; stiff; large shrinkage; medium hard; high in shock resistance	38.7	37.5	Faint growth rings		Construction	Posts, poles
Pine, ponderosa	From Arizona and New Mexico to South Dakota and westward to the Pacific Coast mountains	Orange to reddish brown	Moderately soft, stiff; light in weight; low in strength and shock resistance; little tendency to warp and twist	28.6	27.5	Growth rings not conspicuous	Faint growth ring stripe	Sash, doors, moldings, paneling, millwork, veneer	Blinds, sheathing, posts, poles, piling
Pine, red	New England states, New York, Pennsylvania, and the Great Lake states	Orange to reddish brown	Moderately heavy; strong, stiff, soft; high shock resistance; large shrinkage	31.4	30.4	Growth rings not conspicuous	Faint growth ring stripe	Construction, siding, flooring, sash, doors, millwork	Blinds, piling, posts, poles
Pine, southern yellow	New York and New Jersey, southward to Florida, westward to Texas and Oklahoma, and north to the Ohio Valley and Missouri	Orange to reddish brown	Longleaf are heavy, strong, stiff, hard, with moderately large shrinkage and high shock resistance	41.6–43.9	40.3–42.6	Conspicuous growth rings	Relatively clear	Construction (structural types follow density rule)	Bridges, docks, piling
		Orange to reddish brown	Shortleaf are lighter in weight than longleaf	35.7–36.3	34.6–35.2	Conspicuous growth rings	Relatively clear	Paneling, sheathing, construction	
Pine, sugar	From the Pacific Coast and Cascade Mountains of Oregon along coast ranges and Sierra Nevada of California	Light creamy brown	Small shrinkage; light in weight; moderately soft; low in strength, shock resistance, and stiffness; little tendency to warp and check	26.0	24.0	Faint growth rings	Relatively clear	Sash, doors, millwork, construction	Blinds, patterns, matches

Pine, western white	Western Montana, northern Idaho, Washington, Oregon, and California	Cream to light reddish brown	Moderately light in weight; soft, stiff; low in strength and shock resistance; large shrinkage	28,0	27.1	Faint growth rings	Relatively clear	Construction, sash, doors, millwork, sheathing, siding, paneling	Blinds, patterns, matches
Cedar, Port Orford	Pacific Coast from Coos Bay, Oregon, south to California; extends only 40 miles inland	Light yellow to pale brown	Moderately light in weight; stiff, strong, hard; resistant to shock; highly resistant to decay; moderate shrinkage; little tendency to warp	30.1	28.9	Faint growth rings	Clear	Sash, doors, flooring, paneling	Furniture, venetian blinds, mothproofing
Cedar red, eastern	Eastern half of the United States except Maine, Florida, and small area along Gulf Coast	Brick red to deep reddish brown	Moderately heavy; hard; low in strength and stiffness; high in shock resistance; very small shrinkage	33.5	32.2	Occasional streaks of white sapwood		Flooring, millwork	Mothproofing, pencils, fenceposts
Cedar, red, western	Pacific Coast north to Alaska; also in Idaho and Montana	Reddish brown	Light in weight; moderately soft; low in strength and shock resistance; very small shrinkage	23.4	22.4	Growth rings not conspicuous	Faint growth ring stripe	Shingles, siding, paneling, sash, doors, millwork	Poles, posts, piling
Redwood	In the Sierra Nevada of California	Cherry to deep reddish brown	Moderately strong; light in weight; stiff and hard; small shrinkage; high decay resistance	28.6	27.4	Growth rings not conspicuous; occasionally wavy and burled	Faint growth ring stripe; occasionally wavy and burled	Sash, doors, siding, paneling, millwork	Outdoor furniture, fencing, blinds, tanks, silos
Spruce, eastern	Great Lake states, New England, and the Appalachian Mountains	Nearly white	Moderately light in weight; low in shrinkage; strong, stiff, tough, and hard	29.4–28.4	28.7–27.2	Faint growth rings	Clear	Construction, millwork	Pulpwood, piano sounding boards
Spruce, Engelmann	High elevations of the Rocky Mountains	Nearly white	Light in weight; low in strength; limber, soft; moderately small shrinkage	24.1	23.2	Faint growth rings	Clear	Flooring, sheathing, construction	Poles, pulp and pulp making
Spruce, Sitka	Northwestern coast of the United States from California to Alaska	Light reddish brown	Moderately light in weight; low in bending and compressive strength; soft; low resistance to shock; small shrinkage	28.1	27.1	Growth rings not conspicuous	Faint growth ring stripe	Sash, doors, millwork paneling	Blinds, furniture, pulpwood, piano sounding boards
Tamarack	From Maine to Minnesota	Russet brown	Intermediate in weight and mechanical properties	37.6	36.3	Conspicuous growth rings	Relatively clear	Construction	Pulpwood, fencing, poles

[a]The sapwood is light in color or white unless discolored by fungus or chemical stains.
[b]For metric equivalents of these values, *see* Table *W15a*.

Table W15a Metric equivalents for moisture content of evergreen (softwood) trees[a]

	Weight			
	Moisture content 15%		Moisture content 8%	
	lb/ft³	kg/m³	lb/ft³	kg/m³
Cedar, Alaska	31.6	506.23	30.4	487.01
Cypress, bald	32.6	522.25	31.4	503.03
Fir, Douglas	30.5–34.3	488.61–549.49	29.2–33.1	467.78–530.26
Fir, true eastern	26.9	430.94	26.4	422.93
Fir, true western	26.7–28.3	427.73–453.37	25.8–27.2	413.32–435.74
Hemlock, eastern	29.0	464.58	28.0	448.56
Hemlock, western	29.6	474.19	28.7	459.77
Cedar, incense	25.5	408.51	24.2	387.68
Larch, western	39.4	613.19	38.2	611.96
Pine, eastern white	25.4	406.91	24.2	387.68
Pine, lodge pole (knotty pine)	29.2	467.78	28.2	451.76
Pine, pitch	34.9	551.09	33.8	541.48
Pine, pond	38.7	619.97	37.5	600.75
Pine, ponderosa	28.6	618.37	27.5	440.55
Pine, red	31.4	503.03	30.4	487.01
Pine, southern yellow	41.6–43.9	666.43–703.28	40.3–42.6	645.61–682.45
	35.7–36.3	571.91–581.53	34.6–25.3	554.29–565.51
Pine, sugar	26.0	416.52	24.0	384.48
Pine, western white	28.0	448.56	27.1	434.14
Cedar, Port Orford	30.1	482.20	28.9	462.98
Cedar, red eastern	33.5	536.67	32.2	515.84
Cedar, red western	23.4	347.87	22.4	358.85
Redwood	28.6	458.17	27.4	438.94
Spruce, eastern	29.4–28.4	470.99–459.97	28.7–27.2	459.77–435.74
Spruce, Engelmann	24.1	386.08	23.2	371.66
Spruce, Sitka	28.1	450.16	27.1	434.14
Tamarack	37.6	602.35	36.3	480.53

[a] For convenient reference, values in pounds per cubic foot have been repeated.

Table W16 Uses of wood in construction

Structural	Unfinished	Exterior finish	Interior finish	Milled (prefabricated finish)
1. Beams, joists, and rafters 2. Studs and posts 3. Girders 4. Sills, plates, and girts 5. Trusses 6. Structural laminated members 7. Decking 8. Piles	1. Subflooring 2. Wall and roof sheathing 3. Furring, blocking, and grounds 4. Sleepers and screens 5. Lathing 6. Cross bridging and bracing 7. Rough door and window bucks 8. Stair stringers and carriages	1. Exterior trim 2. Shingles, siding, and board-and-batten 3. Gutters 4. Fascias, caps, and moldings 5. Stairs and railings	1. Interior trim 2. Paneling 3. Flooring 4. Stairs, treads, and risers 5. Shelving 6. Built-ins (site construction)	1. Doors and screen doors 2. Windows, screens, and shutters 3. Moldings and trim 4. Stairs and railings 5. Fireplace mantels 6. Cabinets 7. Dressing tables 8. Built-in furniture

Table W17 Standard hardwood grades, inspected on the poorer side of the piece (except sheets)

Grade and lengths allowed	Width allowed in.	Width allowed mm	Surface measure of pieces ft²	Surface measure of pieces m²	Percentage of piece that must work into clear-face cuttings	Maximum number of cuttings allowed	Minimum size of cuttings required in.	by ft	mm	by m
Firsts[a] 8–16 ft (2.438–4.877 m) will admit 30% of 8–11 ft (2.438–3.353 m), one-half of which may be 8 ft and 9 ft (2.438 m and 2.743 m)	6+	152.4+	4-9 10-14 15+	0.37–0.84 0.93–1.30 1.39+	$91\frac{2}{3}$	1 2 3	4 or 3	5 7	101.6 76.2	1.524 2.134
Seconds[a] 8–16 ft (2.438–4.877 m) will admit 30% of 8–11 ft (2.438–3.353 m), one-half of which may be 8 ft and 9 ft (2.438 m and 2.743 m)	6+	152.4+	4, 5 6, 7 8–11 12–15 6+ 6, 7 8–11 12–15	0.37, 0.46 0.56, 0.65 0.75–1.02 1.11–1.39 0.56+ 0.56, 0.65 0.75–1.02 1.11–1.39	$83\frac{1}{3}$ $91\frac{2}{3}$	1 1 2 3 4 2 3 4	4 or 3	5 7	101.6 76.2	1.524 2.134
Selects 6–16 ft (1.829–4.877 m) will admit 30% of 6–11 ft (1.829–3.353 m), one-sixth of which may be 6 ft and 7 ft (1.829 m and 2.134 m)	4+	101.6+	2, 3 4+	0.19, 0.28 0.37+	$91\frac{2}{3}$ $83\frac{2}{3}$	1 1	4 or 3	5 7	101.6 76.2	1.524 3.134
No. 1 Common 4–16 ft (1.219–4.877 m) will admit 10% of 4–7 ft (1.219–2.134 m), one-half of which may be 4 ft and 5 ft (1.219 m and 1.524 m)	3	76.2	1 2 3, 4 5–7 3, 4 5–7 8–10 11–13 14+	0.10 0.20 0.28, 0.37 0.46–0.65 0.28, 0.37 0.46–0.65 0.75–0.93 1.02–1.21 1.3+	100 75 $63\frac{2}{3}$	0 1 2 3 1 2 3 4 5	4 or 3	2 3	101.6 76.2	0.610 0.914
No. 2 Common 4–16 ft (1.219–4.877 m) will admit 10% of 4–7 ft (1.219–2.134 m), one-third of which may be 4 ft and 5 ft (1.219 m and 1.524 m)	3	76.2	1 2, 3 4, 5 6, 7 2, 3 4, 5 6, 7 8, 9 10, 11 12, 13 14+	0.10 0.20, 0.28 0.37, 0.46 0.56, 0.65 0.20, 0.28 0.37, 0.46 0.56, 0.65 0.75, 0.84 0.93, 1.02 1.11, 1.21 1.3+	$66\frac{2}{3}$ 50	1 2 3 4 1 2 3 4 5 6 7	2	2	50.8	0.610
No. 3A Common[b] 4–16 ft (1.219–4.877 m) will admit 50% of 4–7 ft (1.219–2.134 m), one-half of which may be 4 ft and 5 ft (1.219 m and 1.524 m)	3	76.2	1+	0.10+	$33\frac{1}{3}$	Not specified	3	2	76.2	0.610
No. 3B Common[c] 4–16 ft (1.219–4.877 m) will admit 50% of 4–7 ft (1.219–2.134 m), one-half of which may be 4 ft and 5 ft (1.219 m and 1.524 m)	3	76.2	1+	0.10+	25	Not specified	$1\frac{1}{2}$	2	38.1	0.610

[a]Firsts and seconds are combined as one grade, FAS, in which the percentage of firsts varies from 20 to 40%, depending on the species of wood.
[b]This grade also admits pieces with one sound face and the other, the good one, graded not below No. 2 Common.
[c]Cuttings must be sound; clear face is not required.

Table W18 Standard thicknesses of hardwood lumber

Thickness							
Rough		Surfaced two sides (S2S)[a]		Rough		Surfaced two sides (S2S)[a]	
in.	mm	in.	mm	in.	mm	in.	mm
$\frac{3}{8}$	9.53	$\frac{3}{16}$	4.76	$2\frac{1}{2}$	63.50	$2\frac{1}{4}$	57.15
$\frac{1}{2}$	12.70	$\frac{5}{16}$	7.94	3	76.20	$2\frac{3}{4}$	69.85
$\frac{5}{8}$	15.88	$\frac{7}{16}$	11.11	$3\frac{1}{2}$	88.90	$3\frac{1}{4}$	82.55
$\frac{3}{4}$	19.05	$\frac{9}{16}$	14.29	4	101.60	$3\frac{3}{4}$	95.25
1	25.40	$\frac{13}{16}$	20.64	$4\frac{1}{2}$	124.30	[b]	[b]
$1\frac{1}{4}$	31.75	$1\frac{1}{16}$	26.95	5	127.00	[b]	[b]
$1\frac{1}{2}$	38.10	$1\frac{5}{16}$	33.34	$5\frac{1}{2}$	139.70	[b]	[b]
$1\frac{3}{4}$	44.45	$1\frac{1}{2}$	38.10	6	152.40	[b]	[b]
2	50.80	$1\frac{3}{4}$	44.45				

[a] Thickness for surfaced one side (S1S) requires arrangement with the lumber supplier.
[b] No rules cover these sizes, which require arrangement with the lumber supplier.

Table W19 Grading of hardwood dimension stock

Classes of dimension stock	Grades available in each class					Finishes
Solid dimension flat stock Kiln-dried dimension flat stock	Clear two faces	Clear one face	Paint	Core	Sound	Rough, semifinished, and finished
Solid dimension squares	Clear	Select	Sound			Rough
	Clear	Select	Paint	Sound		Surfaced

Table W20 Official nomenclature of commercial evergreens (softwoods)

Official U.S. Forest Service nomenclature	Name adopted by American Lumber Standards	Other names
Alaska cedar	Alaska cedar	Yellow cedar, Sitka cypress, yellow cypress
Eastern red cedar	Eastern red cedar	Red cedar, cedar, juniper
Incense cedar	Incense cedar	Cedar, white cedar
Port Orford cedar	Port Orford cedar	Lawson's cypress, Oregon cedar, white cedar
Atlantic white cedar	Southern white cedar	White cedar
Western red cedar	Western red cedar	Red cedar, cedar, western cedar
Bald cypress	Red cypress (coast type) or white cypress (inland type)	Cypress, tidewater cypress, red cypress, Gulf Coast red cypress, Louisiana red cypress, bald cypress, black cypress
Balsam fir	Balsam fir	Balsam, eastern fir
Fraser fir	Balsam fir	Balsam, eastern fir
California red fir	White fir	Red fir, fir, balsam, Colorado white fir, yellow fir
Grand fir	White fir	Red fir, fir, balsam, Colorado white fir, yellow fir
Pacific silver fir	White fir	Red fir, fir, balsam, Colorado white fir, yellow fir
Subalpine fir	White fir	Red fir, fir, balsam, Colorado white fir, yellow fir
White fir	White fir	Red fir, fir, balsam, Colorado white fir, yellow fir
Douglas fir	Douglas fir	Red fir, Oregon fir, Douglas spruce, yellow fir, Puget Sound pine

Table W20 Official nomenclature of commercial evergreens (softwoods) (*continued*)

Official U.S. Forest Service nomenclature	Name adopted by American Lumber Standards	Other names
Noble fir	Noble fir	Red fir
Eastern hemlock	Eastern hemlock	Hemlock, hemlock spruce, spruce pine
Mountain hemlock	Mountain hemlock	Weeping spruce, alpine spruce, hemlock spruce
Western hemlock	West Coast hemlock	Hemlock, hemlock spruce, Pacific hemlock, Alaska pine
Alligator juniper	Western juniper	Juniper, white cedar, cedar
Rocky Mountain juniper	Western juniper	Juniper, white cedar, cedar
Utah juniper	Western juniper	Juniper, white cedar, cedar
Western juniper	Western juniper	Juniper, white cedar, cedar
Western larch	Western larch	Tamarack, larch
Eastern white pine	Northern white pine	White, cork, soft, northern, and pumpkin pine
Jack pine	Jack pine	Scrub pine
Loblolly pine	Southern yellow pine	Old field, slash, shortleaf, Virginia, sap, yellow, and North Carolina pine
Pitch pine	Southern yellow pine	Old field, slash, shortleaf, Virginia, sap, yellow, and North Carolina pine
Shortleaf pine	Southern yellow pine	Old field, slash, shortleaf, Virginia, sap, yellow, and North Carolina pine
Virginia pine	Southern yellow pine	Old field, slash, shortleaf, Virginia, sap, yellow, and North Carolina pine
Lodgepole pine	Lodgepole pine	Scrub and spruce pine
Longleaf pine	Longleaf yellow and southern yellow pine	Southern, yellow, hard, Georgia, pitch, heart, and fat pine
Slash pine	Longleaf yellow and southern yellow pine	Southern, yellow, hard, Georgia, pitch, heart, and fat pine
Ponderosa pine	Ponderosa pine	Bull pine, Arizona white pine, western soft pine, western pine
Red pine	Norway pine	Hard and northern pine
Sugar pine	Sugar pine	Big pine
Western white pine	Idaho white pine	White and soft pine
Redwood	Redwood	Sequoia, coast redwood
Black spruce	Eastern spruce	
Red spruce	Eastern spruce	
White spruce	Eastern spruce	
Blue spruce	Engelmann spruce	White spruce, silver spruce, balsam, mountain spruce
Engelmann spruce	Engelmann spruce	White spruce, silver spruce, balsam, mountain spruce
Sitka spruce	Sitka spruce	Spruce, tideland spruce, western spruce, yellow spruce, silver spruce
Tamarack	Tamarack	Larch, hackmatack, red larch, black larch
Pacific yew	Pacific yew	Yew, western yew, mountain mahogany

Evergreen (Softwood) Grades. Softwoods are graded according to the American Lumber Standards. These standards cover both the nomenclature and the grading, as shown in Table *W20*.

Softwood lumber is divided into two major categories, *construction* and *remanufacturing*. Construction lumber is expected to be used in construction as graded and sized after sawing and planing. Remanufactured lumber will undergo further manufacturing steps and will have a completely different form in the finished product; examples are door trim, bases, and shiplap siding.

In general, softwood lumber is classified by use: *yard lumber*, lumber for ordinary construction and general building purposes (*see* Table *W21*); *structural lumber*, used where working stresses are required (*see* Table *W22*); and *factory and shop lumber*, used primarily for remanufacturing purposes. Softwood lumber is also classified according to the extent of finishing as (1) *rough lumber*, which is lumber that has not been dressed and shows saw marks on all four surfaces; and (2) *dressed lumber*, lumber that has been dressed by a planing machine and is further graded and designated as S1S, dressed one side; S2S, dressed two sides; S1E, dressed one edge; S2E, dressed two edges; and various combinations such as S1S1E, S1S2E, S2S1E, and S4S.

Lumber that in addition to being dressed has been matched, shiplapped, or patterned is divided into three types: (1) *matched lumber*, lumber made with tongue and groove, and when end-matched, the tongue and groove are at the ends also; (2) *shiplapped lumber*, lumber rabbeted or worked on both edges to make a close-lapped joint when placed together; and (3) *patterned lumber*, lumber that is shaped to a pattern or molded form, in addition to being dressed, matched, or shiplapped.

Table *W23* gives the American standard thicknesses and widths for softwood lumber. The lengths are in even numbers.

Table W21 Lumber classification and grade names for stress-graded lumber[a]

Lumber classification	Size				Grade name	Bending strength ratio (percent)
	Thickness		Width			
	in.	mm	in.	mm		
Light framing	2–4	50.8–101.6	4[b]	101.6[b]	Construction	34
					Standard	19
					Utility	9
Structural light framing	2–4	50.8–101.6	2–4	50.8–101.6	Select structural	67
					Select structural-1	55
					Select structural-2	45
					Select structural-3	26
Studs	2–4	50.8–101.6	2–4	50.8–101.6	Stud	26
Structural joists and planks	2–4	50.8–101.6	6	152.4 and wider	Select structural	65
					Select structural-1	55
					Select structural-2	45
					Select structural-3	26
Appearance framing	2–4	50.8–101.6	2–4	50.8–101.6	Appearance	55

[a] Sizes shown are nominal.
[b] Widths narrower than 4 in. (101.6 mm) may have different strength ratios from those shown.

Table W22 Sizes of stress-graded and non-stress-graded lumber for construction

Type	Thickness						Face width					
	Nominal[a]		Minimum dressed				Nominal[a]		Minimum dressed			
			Dry		Green				Dry		Green	
	in.	mm	in.	mm	in.	mm	in.	mm	in.	mm	in.	mm
Boards	1	25.40	$\frac{3}{4}$	19.05	$\frac{25}{32}$	19.84	2	50.80	$1\frac{1}{2}$	38.10	$1\frac{9}{16}$	39.69
	$1\frac{1}{4}$	31.75	1	25.40	$1\frac{1}{32}$	26.20	3	76.20	$2\frac{1}{2}$	63.50	$2\frac{9}{16}$	65.09
	$1\frac{1}{2}$	38.10	$1\frac{1}{4}$	31.75	$1\frac{9}{32}$	32.54	4	101.60	$3\frac{1}{2}$	88.90	$3\frac{9}{16}$	90.49
							5	127.00	$4\frac{1}{2}$	114.30	$4\frac{5}{8}$	117.48
							6	152.40	$5\frac{1}{2}$	139.70	$5\frac{5}{8}$	142.88
							7	177.80	$6\frac{1}{2}$	165.10	$6\frac{5}{8}$	168.28
							8	203.20	$7\frac{1}{4}$	184.15	$7\frac{1}{2}$	190.58
							9	228.60	$8\frac{1}{4}$	209.55	$8\frac{1}{2}$	215.90
							10	254.00	$9\frac{1}{4}$	234.95	$9\frac{1}{2}$	241.30
							11	279.40	$10\frac{1}{4}$	260.35	$10\frac{1}{2}$	262.70
							12	304.80	$11\frac{1}{4}$	285.75	$11\frac{1}{2}$	292.10
							14	355.60	$13\frac{1}{4}$	336.55	$13\frac{1}{2}$	315.90
							16	406.40	$15\frac{1}{4}$	387.35	$15\frac{1}{2}$	393.70
Dimension lumber	2	50.80	$1\frac{1}{2}$	38.10	$1\frac{9}{16}$	39.69	2	50.80	$1\frac{1}{2}$	30.10	$1\frac{9}{16}$	39.69
	$2\frac{1}{2}$	63.50	2	50.80	$2\frac{1}{16}$	52.39	3	76.20	$2\frac{1}{2}$	63.50	$2\frac{9}{16}$	65.09
	3	76.20	$2\frac{1}{2}$	63.50	$2\frac{9}{16}$	65.09	4	101.60	$3\frac{1}{2}$	88.90	$3\frac{9}{16}$	90.49
	$3\frac{1}{2}$	88.90	3	76.20	$3\frac{1}{16}$	78.79	5	127.00	$4\frac{1}{2}$	114.30	$4\frac{5}{8}$	117.48
	4	101.60	$3\frac{1}{2}$	88.90	$3\frac{9}{16}$	90.49	6	157.40	$5\frac{1}{2}$	139.70	$5\frac{5}{8}$	142.88
	$4\frac{1}{2}$	114.30	4	101.60	$4\frac{1}{16}$	103.19	8	203.20	$7\frac{1}{4}$	184.15	$7\frac{1}{2}$	190.50
							10	254.00	$9\frac{1}{4}$	234.95	$9\frac{1}{2}$	241.30
							12	304.80	$11\frac{1}{4}$	285.75	$11\frac{1}{6}$	292.10

Table W22 Sizes of stress-graded and non-stress-graded lumber for construction (*continued*)

	Thickness						Face width					
			Minimum dressed						Minimum dressed			
	Nominal[a]		Dry		Green		Nominal[a]		Dry		Green	
Type	in.	mm	in.	mm	in.	mm	in.	mm	in.	mm	in.	mm
Dimension Lumber	$4\frac{1}{2}$	114.30	4	101.60	$4\frac{1}{16}$	103.19	14	355.60	$13\frac{1}{4}$	336.55	$13\frac{1}{2}$	315.90
							16	406.40	$15\frac{1}{4}$	387.35	$15\frac{1}{2}$	393.70
Timbers	5 and greater	127.00			$\frac{1}{2}$ less than nominal	12.7	5 and greater	127.00			$\frac{1}{2}$ less than nominal	12.7

[a]Nominal sizes shown are not to be taken as actual sizes.

Table W23 Standard thicknesses and widths of softwood lumber

Type of lumber material	Thickness				Width			
	Nominal[a]		Dressed		Nominal[a]		Dressed[b]	
	in.	mm	in.	mm	in.	mm	in.	mm
Finish	$\frac{3}{8}$	9.53	$\frac{5}{16}$	7.94	2	50.8	$1\frac{1}{2}$	38.10
	$\frac{1}{2}$	12.70	$\frac{7}{16}$	11.11	3	76.2	$2\frac{1}{2}$	63.50
	$\frac{5}{8}$	15.88	$\frac{9}{16}$	14.29	4	101.6	$3\frac{1}{2}$	88.90
	$\frac{3}{4}$	19.05	$\frac{5}{8}$	15.88	5	127.0	$4\frac{1}{2}$	114.30
	1	25.40	$\frac{3}{4}$	19.05	6	152.4	$5\frac{1}{2}$	139.70
	$1\frac{1}{4}$	31.75	1	25.40	7	177.8	$6\frac{1}{2}$	165.10
	$1\frac{1}{2}$	38.10	$1\frac{1}{4}$	31.75	8	203.2	$7\frac{1}{4}$	184.15
	$1\frac{3}{4}$	44.45	$1\frac{3}{8}$	34.93	9	228.6	$8\frac{1}{4}$	209.55
	2	50.80	$1\frac{1}{2}$	38.10	10	254.0	$9\frac{1}{4}$	234.85
	$2\frac{1}{2}$	63.50	2	50.80	11	279.4	$10\frac{1}{4}$	260.35
	3	76.20	$2\frac{1}{2}$	63.50	12	304.8	$11\frac{1}{4}$	285.75
	$3\frac{1}{2}$	88.90	3	76.20	14	355.6	$13\frac{1}{4}$	336.55
	4	101.60	$3\frac{1}{2}$	88.90	16	406.4	$15\frac{1}{4}$	387.35
Flooring	$\frac{3}{8}$	9.53	$\frac{5}{16}$	7.94	2	50.8	$1\frac{1}{8}$	25.72
	$\frac{1}{2}$	12.70	$\frac{7}{16}$	11.11	3	76.2	$2\frac{1}{8}$	53.98
	$\frac{5}{8}$	15.88	$\frac{9}{16}$	14.29	4	101.6	$3\frac{1}{8}$	79.88
	1	25.40	$\frac{3}{4}$	19.05	5	127.0	$4\frac{1}{8}$	104.78
	$1\frac{1}{4}$	31.75	1	25.40	6	152.4	$5\frac{1}{8}$	130.18
	$1\frac{1}{2}$	38.10	$1\frac{1}{4}$	31.75				
Ceiling	$\frac{3}{8}$	9.53	$\frac{5}{16}$	7.94	3	76.2	$2\frac{1}{8}$	53.98
	$\frac{1}{2}$	12.70	$\frac{7}{16}$	11.11	4	101.6	$3\frac{1}{8}$	79.38
	$\frac{5}{8}$	15.88	$\frac{9}{16}$	14.29	5	127.0	$4\frac{1}{8}$	104.78
	$\frac{3}{4}$	19.05	$\frac{11}{16}$	17.46	6	152.4	$5\frac{1}{8}$	130.18
Partition	1	25.4	$\frac{25}{32}$	19.84	3	76.2	$2\frac{1}{8}$	53.93
					4	101.6	$3\frac{1}{8}$	79.38
					5	127.0	$4\frac{1}{8}$	104.78
					6	152.4	$5\frac{1}{8}$	130.18

Table W23 Standard thicknesses and widths of softwood lumber (*continued*)

Type of lumber material	Thickness				Width			
	Nominal[a]		Dressed		Nominal[a]		Dressed[b]	
	in.	mm	in.	mm	in.	mm	in.	mm
Stepping	1	25.4	$\frac{3}{4}$	19.05	8	203.2	$7\frac{1}{4}$	184.15
	$1\frac{1}{4}$	31.75	1	25.40	10	254.0	$9\frac{1}{4}$	234.85
	$1\frac{1}{2}$	38.10	$1\frac{1}{4}$	31.75	12	304.8	$11\frac{1}{4}$	285.75
	2	25.40	$1\frac{1}{2}$	38.10				
Bevel siding	$\frac{1}{2}$	12.7	$\frac{7}{16}$ butt,	11.11 butt,	4	101.6	$3\frac{1}{2}$	88.90
			$\frac{3}{16}$ tip	4.76 tip				
	$\frac{9}{16}$	14.29	$\frac{15}{32}$ butt,	11.91 butt,	5	127.0	$4\frac{1}{2}$	114.30
			$\frac{3}{16}$ tip	4.76 tip				
	$\frac{5}{8}$	15.88	$\frac{9}{16}$ butt,	14.29 butt,	6	152.4	$5\frac{1}{2}$	139.70
			$\frac{3}{16}$ tip	4.76 tip				
	$\frac{3}{4}$	19.05	$\frac{11}{16}$ butt,	17.46 butt,	8	203.2	$7\frac{1}{4}$	184.15
			$\frac{3}{16}$ tip	4.76 tip				
	1	25.40	$\frac{3}{4}$ butt,	19.05 butt,	10	254.0	$9\frac{1}{4}$	234.95
			$\frac{3}{16}$ tip	4.76 tip	12	304.8	$11\frac{1}{4}$	285.75
Bungalow siding	$\frac{3}{4}$	19.05	$\frac{11}{16}$ butt,	17.46 butt,	8	203.2	$7\frac{1}{4}$	184.15
			$\frac{3}{16}$ tip	4.76 tip	10	154.0	$8\frac{1}{4}$	234.95
					12	304.8	$11\frac{1}{4}$	285.75
Rustic and drop siding shiplapped, $\frac{3}{8}$-in. (9.55-mm) lap	$\frac{5}{8}$	15.88	$\frac{9}{16}$	14.29	4	101.6	3	76.2
	1	25.40	$\frac{23}{32}$	18.26	5	127.0	4	101.6
					6	152.4	5	127.0
Rustic and drop siding shiplapped, $\frac{1}{2}$-in. (12.7-mm) lap	$\frac{5}{8}$	15.88	$\frac{9}{16}$	14.29	4	101.6	$2\frac{7}{8}$	73.03
					5	127.0	$3\frac{7}{8}$	98.43
					6	152.4	$4\frac{7}{8}$	123.83
	1	25.40	$\frac{23}{32}$	18.26	8	203.2	$6\frac{5}{8}$	174.63
					10	254.0	$8\frac{5}{8}$	225.43
					12	304.8	$10\frac{5}{8}$	276.23
Rustic and drop siding dressed and matched	$\frac{5}{8}$	15.88	$\frac{9}{16}$	14.29	4	101.6	$3\frac{1}{8}$	73.38
					5	127.0	$4\frac{1}{8}$	104.78
	1	25.40	$\frac{23}{32}$	18.26	6	152.4	$5\frac{1}{8}$	130.18
					8	203.2	$6\frac{7}{8}$	174.63
					10	254.0	$8\frac{7}{8}$	225.43
Shiplap, $\frac{3}{8}$-in. (9.53-mm) lap	1	25.40	$\frac{3}{4}$	19.05	4	101.6	$3\frac{1}{8}$	98.43
					6	152.4	$5\frac{1}{8}$	130.18
					8	203.2	$6\frac{7}{8}$	174.63
					10	254.0	$8\frac{7}{8}$	225.43
					12	304.8	$10\frac{7}{8}$	276.23
					14	355.6	$12\frac{7}{8}$	327.03
					16	406.4	$14\frac{7}{8}$	377.83

Table W23 Standard thicknesses and widths of softwood lumber (*continued*)

Type of lumber material	Thickness				Width			
	Nominal[a]		Dressed		Nominal[a]		Dressed[b]	
	in.	mm	in.	mm	in.	mm	in.	mm
Shiplap, $\frac{1}{2}$-in. (12.7-mm) lap	1	25.40	$\frac{3}{4}$	19.05	4	101.6	3	76.2
					6	152.4	5	127.0
					8	203.2	$6\frac{3}{4}$	171.45
					10	254.0	$8\frac{3}{4}$	222.25
					12	304.8	$10\frac{3}{4}$	273.05
					14	355.6	$12\frac{3}{4}$	323.85
					16	406.4	$14\frac{3}{4}$	374.65
Centermatch, $\frac{1}{4}$-in. (6.35-mm) tongue	1	25.40	$\frac{3}{4}$	19.05	4	101.6	$3\frac{1}{8}$	98.43
	$1\frac{1}{4}$	31.75	1	25.40	5	127.0	$4\frac{1}{8}$	104.78
	$1\frac{1}{2}$	38.10	$1\frac{1}{4}$	31.75	6	152.4	$5\frac{1}{8}$	130.18
					8	203.2	$6\frac{7}{8}$	174.63
					10	254.0	$8\frac{7}{8}$	225.43
					12	304.8	$10\frac{7}{8}$	276.23
2-in. (50.8-mm) D & M, $\frac{3}{8}$-in. (9.53-mm) tongue	2	50.8	$1\frac{1}{2}$	38.10	4	101.6	3	76.20
					6	152.4	5	127.00
					8	203.2	$6\frac{3}{4}$	171.45
					10	254.0	$8\frac{3}{4}$	222.25
					12	304.8	$10\frac{3}{4}$	273.05
2-in. (50.8-mm) shiplap, $\frac{1}{2}$-in. (12.7-mm) lap	2	50.8	$1\frac{1}{2}$	38.10	4	101.6	3	76.20
					6	152.4	5	127.00
					8	203.0	$6\frac{3}{4}$	171.45
					10	254.0	$8\frac{3}{4}$	222.25
					12	304.8	$10\frac{3}{4}$	273.05

[a] For nominal thicknesses under 1 in. (25.4 mm), the board measure is based on the nominal surface dimension of width times length. With the exception of nominal thicknesses under 1 in. (25.4 mm), the nominal thicknesses and widths in this table are the same as the board measure or count sizes.
[b] Figures are for dry lumber with a moisture content of 19% or less. Green lumber has a moisture content above 19%.

Application

Finished woods depend on the color, figure, and luster of the particular species used and on the way in which it bleaches or takes fillers, stains, and transparent finishes (*see* Wood Finishes). Other decorative features depend on knotholes, pin or worm holes, bird pecks, decay in isolated pockets, configuration, bird's-eyes, mineral streaks, swirls in grain, and ingrown bark. "Texture" and "grain" of wood do not have any definite meanings, and when specifying woods, the exact types of texture, grain, and finish should be accurately defined, leaving no room for misinterpretation.

Wood Joining. Wood can be joined together with a wide variety of fastenings and glues, including nails, spikes, screws, bolts, lag screws, pins, and metal connectors of various shapes. It is beyond the scope of this book to cover the withdrawal resistance, effect of coatings, lead holes, etc., for these connections. When joints must meet exact structural requirements, it is advisable to consult structural engineers so that the joints can be designed correctly for the type of wood, direction of grain, and the changes in dimension that occur through changes in moisture content. In general, nails and screws are the most common fastening devices used; the type of metal they contain (because of possible

Table W24 Color effects of weathering on various woods

Type of wood	Changes to light gray color with silvery sheen	Changes to light gray color with moderate sheen	Changes to dark gray color with little or no sheen
Evergreen (softwood)	Bald cypress Cedar: Alaska Port Orford	Hemlock: eastern western Pine: eastern white ponderosa sugar western white Spruce: eastern Sitka	Douglas fir Fir: commercial white Larch: western Pine: southern yellow Cedar: western red Redwood
Deciduous (hardwood)		Aspen Basswood Birch Cottonwood Hickory Maple Sweet gum Yellow poplar	Ash Chestnut Oak: red white Walnut: black

Table W25 Checks, cupping, and twisting caused by weathering on various woods

Type of wood	Checks		Cupping and pull loose				
	Inconspicuous	Conspicuous	Slight	Distinct	Pronounced	Very pronounced	Twisting
Evergreen (softwood)	Bald cypress Cedar: Alaska Port Orford western red Redwood	Douglas fir Fir: commercial white Hemlock: eastern western Larch: western Pine: eastern white ponderosa southern yellow sugar western white Spruce: eastern Sitka	Bald cypress Cedar: Alaska Port Orford western red Redwood	Douglas fir Fir: commercial white Hemlock: eastern western Larch: western Pine: eastern white ponderosa southern yellow sugar western white Spruce: eastern Sitka			
Deciduous (hardwood)	Aspen Yellow poplar	Ash Basswood Birch Chestnut Cottonwood Hickory Maple Oak: red white Sweetgum Walnut: black		Aspen Basswood	Yellow poplar Chestnut Walnut: black	Ash Birch Cottonwood Hickory Maple Oak: red white Sweetgum	Birch Cottonwood Elm Sweetgum Sycamore Tupelo Tropical hardwoods

rust stains) and their concealment are the major problems in architectural design with wood. (*See also* Wood Joining.)

Weathering. Weathering of wood causes changes in color, roughening and checking of the surface, and cupping and tearing loose of the wood from its fastening. Wood does not erode rapidly; it takes approximately 100 years for $\frac{1}{4}$ in. (6.35 mm) to waste away. Weathering changes all wood to a gray color (darker in some woods). Tables *W24* and *W25* list the specific effects of weathering on the various types of wood.

Tooling. The various species of woods have different characteristics for working with handtools and machine tools, as shown by Tables *W26* and *W27*.

Other Uses of Wood. Wood pulp is used for manufacturing paper, paper pulp boards, decking, acoustical tile, etc. (*see* Paper and Paper Pulp Products). Particle board and hardboard are completely covered under Wood Hardboard and Particleboard.

Condensed Checklist

1. Availability of the type of lumber desired should always be checked in the locality.
2. Nominal and dressed dimensions, including available lengths of the type of wood to be used, should be carefully checked.
3. The species of wood to be used should always be specified either by its official commercial name or, if some imported wood, by botanical name.
4. The end use of the type of lumber being specified must be determined in order to specify the correct grade.
5. All lumber should be air-dried or kiln-dried.
6. Local, municipal, and state codes and also the codes of the Fire Underwriters, insurance companies, labor department, and federal government (Army, Navy, etc.) should be checked for strength limitations, thickness and width limitations, and fire-retarding and fire-resistance requirements and treatments.
7. All hardwood or softwood should be checked for its classification for painting if that type of finish is desired (*see* Wood Finishes; Wood Preservatives).

Conditions Favorable to the Use of Wood

1. Where a strong, lightweight, durable, weatherproof, easily worked, joined, installed, and finished material is needed.
2. Where a material that offers a large variety of colors, grain patterns, and textures is desired.

Table W26 Hardwoods and softwoods classified for ease of working with hand tools

Type of wood	Easy to work	Relatively easy to work	Difficult to work
Deciduous (hardwood)	Alder (red) Basswood Butternut Chestnut Yellow poplar	Birch (paper) Cottonwood Magnolia Sweetgum Sycamore Tupelo Walnut	Ash, white Beech Birch Cherry Elm Hackberry Hickory Locust Maple Oak
Evergreen (softwood)	Idaho white pine Incense cedar Northern white cedar Northern white pine Ponderosa pine Port Orford cedar Southern white pine Southern white cedar Sugar pine Western red cedar	Bald cypress Balsam fir Eastern hemlock Eastern red cedar Eastern spruce Lodgepole pine Redwood Sitka spruce White fir Western hemlock	Douglas fir Southern yellow pine Western larch

Table W27 Comparative machinability of hardwoods

Species of wood	Planing	Shaping	Turning	Boring	Mortising	Sanding	Steam bending	Nailing	Screwing
	Comparisons based on quality produced in percentages of pieces produced								
	Perfect and excellent surface (percent)						Unbroken pieces (percent)	Free from complete splits (percent)	
Ash	75	55	79	94	58	75	67	65	71
Basswood	64	10	68	76	51	17	2	79	68
Beech		24	90	99	92	49	75	42	58
Birch	63	57	80	97	97	34	72	32	48
Chestnut	74	28	87	91	70	64	56	66	60
Cottonwood	21	3	70	70	52	19	44	82	78
Elm	33	13	65	94	75	66	74	80	74
Hackberry	74	10	77	99	72		94	63	63
Hickory	76	20	84	100	98	80	76	35	63
Magnolia	65	27	79	71	32	37	85	73	76
Mahogany	90	68	39	100	100		41	68	78
Maple, hard	54	72	82	99	95	38	57	27	52
Maple, soft	41	25	76	80	34	37	59	58	61
Oak, red	91	28	84	99	95	81	86	66	78
Oak, white	87	35	85	95	99	83	91	69	74
Poplar, yellow	70	13	81	87	63	19	58	77	67
Sweetgum	51	28	86	92	58	23	67	69	69
Sycamore	22	12	85	98	96	21	29	79	74
Tupelo, black	48	32	75	82	24	21	42	65	63
Tupelo, water		52	79	62	33	34	46	64	63
Walnut, black	62	34	91	100	98		78	50	59

3. Where a strong, lightweight, easily worked structural framing material is required.

Conditions Unfavorable to the Use of Wood

1. As a fire-resistant material unless it has been especially treated for fire resistance and is accepted by codes governing its use.
2. Where the design calls for large, unbroken, smooth areas.

Descriptions of Imported Woods

Adodire. This wood grows extensively from Sierra Leone westward to the Cameroons and south to Zaire. It has a creamy yellow color, generally a wavy grain, and, when quartersawed, a mottled figure. Its strength properties are almost identical to those of English oak. It is used for decorative veneers for plywood, furniture, and cabinetwork.

Andioba. This wood, widely distributed in tropical America, is similar to mahogany, a reddish brown to dark reddish brown in color, and easy to work, paint, and glue. Because it is very durable against decay and insects, it is suited for frame construction and flooring in tropical climates and is also used for furniture, cabinetwork, decorative veneer, and plywood.

Angelique. This wood, considered an African mahogany, is found in French Guiana and Surinam. There are two types: "angelique gris," which has heartwood that is russet brown when cut and then turns to dark brown with a purplish cast; and "angelique rouge," which has heartwood more distinctly reddish and generally has wide bands of purplish color. Both types have strength superior to that of teak or white oak and are highly resistant to decay and marine borers. Angelique therefore can be used for harbor installations, piers, dock fenders, dock flooring, and underwater members.

Apamate. This wood, golden to dark brown in color, is distributed from southern Mexico through Central America to Venezuela and Equador. It weighs approximately 38 lb/ft^3 (608.76 kg/m^3), is easily machined, and takes finishes well. It is lighter than oak but equal to it in bending and compression parallel to the grain. It is used for furniture, interior moldings, doors, flooring, and decorative veneers.

Balsa Wood. This is the wood of large and fast-growing trees found from southern Mexico to Ecuador and northern Brazil. It is the lightest of the commercial woods, weighing 10 to 20 lb/ft^3 (160.20 to 192.24 kg/m^3), and also has the qualities of strength, stiffness, and workability. It is used for life preservers, buoys, floats, and vibration insulators. A similar lightweight Japanese wood, used for instruments, floats, and applications where lightness is required, is named kiri.

Bamboo. A genus of gigantic treelike grasses, bamboo weighs 22 lb/ft^3 (352.44 kg/m^3) and is widely used for making blinds and furniture. The giant bamboo of Ceylon grows to a height of 120 ft (67.056 m).

The larger sizes are extensively used in the East as timber wood.

Boxwood. The common species is native to Europe and Asia but also grows in America. It is used for rulers, instruments, inlay work, and small carvings. The wood is light yellow in color, hard, fine-grained, and dense. It does not warp easily. The weight is about 65 lb/ft^3 (1041.3 kg/m^3).

Courbaril. This wood is found in the West Indies, southern Mexico, Central America, and the Amazon basin. The sapwood is gray white, and the heartwood varies from brown to a purplish cast. Its strength is similar to that of shagbark hickory. The wood machines well, finishes smoothly, and glues well. It weighs 50 lb/ft^3 (811 kg/m^3), has moderate decay resistance, and is used for decorative veneer and furniture; the sapwood is utilized where blond wood is required.

East India Walnut. This refers to the wood of trees found in tropical Asia and Africa (but not belonging to the walnut family). The wood is hard, dense, and close-grained. Its color is dark brown with gray streaks. It is used for furniture, flooring, paneling, and decorative work. Its weight averages about 50 lb/ft^3 (801 kg/m^3).

Ebony. The hard, black woods of the genus *Diospyros*, found in western Africa and India, represent the true ebonies. Black ebony has a black heartwood and brownish white sapwood. It is very heavy, durable, and characterized by a fine open grain and the ability to take a polish. It weighs 78 lb/ft^3 (1249 kg/m^3). Ebony wood is shipped in short billets and is graded according to color and source.

Greenheart. The wood of this tree is found in British Guiana and is especially valued in marine construction because of its resistance to fungi and termites. The wood is very strong and hard, with good wearing qualities. It weighs 62 lb/ft^3 (993.24 kg/m^3). Woods sometimes marketed as greenheart include itauba, from the lower Amazon Valley, and Surinam greenheart, from the Dutch and French Guianas. The tonka bean tree of Panama is very resistant to marine borers and is used as a substitute for greenheart.

Jarrah. This wood comes from southwestern Australia. It has dark red mahogany-colored heartwood and a pale-colored sapwood, with grain that is generally straight. It weighs 44 lb/ft^3 (704.88 kg/m^3), has high strength properties, and is resistant to termites and fungi. Jarrah is used for piers, jetties, piling, bridges, harbor installations, and flooring, but under heavy traffic it will splinter.

Kakrodua. This wood grows in western, Africa and is becoming a substitute for teak. Its weight is 44 lb/ft^3 (704.88 kg/m^3). Although it strongly resembles teak, it does not have the oily nature of teak and is finer textured. Kakrodua is used for the same purposes as teak, namely, decorative veneer, flooring, cabinetwork, furniture, and boats.

Kapur. Nine species that grow over parts of Malaya, Sumatra, Borneo, North Borneo, and Sarawak are all combined under the name "kapur." The wood has light reddish brown heartwood and pale-colored sapwood. Its weight is 48 lb/ft^3 (769.96 kg/m^3).

Khaya. A class of woods from trees of the genus *Khaya*, growing chiefly in tropical west Africa, is known commercially as African mahogany. The woods closely resemble mahogany, but they are more strongly figured, slightly lighter in weight and softer, and have greater shrinkage. Much African mahogany is cut into veneer; the standard thickness of face veneer is $\frac{1}{28}$ in. (0.0357 in. or 0.9068 mm).

Lauan. The wood from various trees growing in the Philippines, Malaya, and Sarawak is known in the American market as Philippines mahogany. The woods resemble mahogany in general appearance, weight, and strength, but the shrinkage and swelling with changes in moisture are greater than in the true mahoganies. Much of the so-called mahogany normally shipped from the

Philippines is the wood from the apitong tree, which also grows in Borneo and Malaya. The wood is reddish brown and weighs approximately 44 lb/ft^3 (704.88 kg/m^3).

Limba. This wood grows in west-central Africa and the Congo. Its color varies from gray-white to creamy brown, and it may contain dark streaks. It is not resistant to decay, insects, and termites. Generally straight-grained, it is easily worked, glues easily, and is easy to season. It is used for interior trim and paneling, furniture, and as a veneer for plywood known as korina.

Mahoe. The wood of a tropical American tree (*Hibiscus elatus*), also called blue mahoe and majagua, is employed as a substitute for walnut. The wood has a gray-blue color, or sometimes brownish gray with streaks, and an aromatic odor. It is hard, with a coarse, open grain.

Mahogany. Mahogany is available in large logs. The wood is hard, durable, and uniformly fine in grade. It seasons well and has a rich reddish color after exposure; when freshly cut it is pinkish to golden brown. It has a high luster when polished. The grain is often figured; the beautifully curled grain specimens are from selected forks or crotches of the trees. Mahogany weighs 32 to 42 lb/ft^3 (512.44 to 672.84 kg/m^3), depending on the species. The hardness and closeness of grain of the different mahoganies also may vary considerably.

All true mahogany comes from trees of the genus *Swietenia*, family Meliaceae, native to the Americas. The tropical cedars, Spanish cedar, and Paraguayan cedar also belong to the same family. Swietenia mahogany is obtained from Mexico, Central America, the West Indies, and as far south as northern Argentina.

Today the word "mahogany" is applied to a variety of woods (*see* Khaya; Lauan; Mahoe; Sapele; Toon). Other mahoganies include (1) Australian red mahogany, which is hard, durable, and dark red in color, with a coarse, open grain; (2) Cameroon mahogany, from western Africa; and (3) crabwood, which is found in Brazil and the Guianas and is used as a substitute for mahogany. The crabwood has a deep reddish brown color and a coarse open grain and weighs 40 lb/ft^3 (640.8 kg/m^3). In general, African mahoganies are marketed under the names of shipping ports and the shipments vary depending on the species cut in the region.

Marblewood. A variety of ebony that comes from India, the wood is black with yellowish stripes. It has a close, hard, firm texture, takes a fine polish and weighs approximately 65 lb/ft^3 (1047.3 kg/m^3).

Meranti. This wood comes from Malaysia and Indonesia and is known as light red and dark red meranti. These two color varieties come from several species of *Shorea*, meranti being the trade name. The weight varies from 25 to 44 lb/ft^3 (400.5 to 704.88 kg/m^3). The wood is used for the same purposes as Philippine lauan.

Obeche. This wood comes from west-central Africa and grows to a height of 150 ft (45.72 m) and a diameter up to 5 ft (1.524 m). The trunk is free from branches for a considerable height, thus permitting clear lumber of considerable size to be obtained. The wood is creamy white to pale yellow with no difference in color between heartwood and sapwood. It has straight grain, machines easily, glues well, and takes nails and screws without splitting. Its weight is about 24 lb/ft^3 (384.48 kg/m^3). The major use is for veneers and core stock. The wood is also known as samba and wawa.

Okoume. This wood, which comes from west-central Africa and Guinea, has a salmon pink color and a texture coarser than that of birch. It offers unusual flexibility in working and finishing. It permits staining to either darker or lighter shades and is used for decorative plywoods, general utility plywoods, and doors.

Rosewood. This term refers chiefly to the wood of several species of *Dalbergia* found in northern South America. The color of rosewood is dark brown to purple. It takes a beautiful polish and has a characteristic fragrance. It is very hard and has a coarse, even grain. It weighs 54 lb/ft^3 (865.08 kg/m^3). Rosewood from India, also called Sissoo, is a beautiful brown hardwood. Rosewood from Borneo comes from other species of trees. It has a deep red color with light and dark streaks and a close texture.

Sapele. This term refers to the figured woods of various species of trees found in tropical Africa which are mixed with Khaya and exported from western Africa as African mahogany. Sapele weighs 39 lb/ft^3 (624.78 kg/m^3), is easily worked, glues well, and is used for veneers for decorative plywood.

Satinee. The wood of this tree, which is native to tropical America, particularly to the Guianas, is reddish brown, has a fine grain, is fairly hard, and takes a lustrous polish. The weight is 54 lb/ft^3 (865.08 kg/m^3).

Teak. True teak is the wood of a tree of southern Asia that resembles oak in appearance and is called Indian oak in England. It is strong and firm and has small shrinkage. It contains an essential oil that gives it a pleasant odor and makes it immune to the attacks of ants and other insects. The color is golden yellow, the grain is coarse and open, and the surface is greasy to the touch. It is one of the most durable woods. Its weight is 40 lb/ft^3 (640.8 kg/m^3).

Toon. The aromatic wood from certain trees in India, Burma, Java, and Australia is known as Indian mahogany. It is almost indistinguishable from the Spanish cedar or tropical American andioba. If seasoned well, the wood does not warp and is durable. It is easily worked and takes a fine polish. The color is a deep red, and the grain has a beautiful pattern. The weight is about 35 lb/ft^3 (567.7 kg/m^3).

Walnut. This wood grows not only in Turkey, Italy, France, and Yugoslavia but also in western and central Asia, extending to China and northern India, yet it is referred to as European walnut. The wood of any one locality may vary considerably in color, figure, and texture, but selected export timber generally shows typical characteristics. Color varies with a grayish brown background, marked with irregular dark-colored streaks accentuated by the natural wavy grain. The wood weighs about 40 lb/ft^3 (640.8 kg/m^3). It is easily machined, finished, and glued, and highly decorative veneers are obtained from stumps, burls, and crotches of a relatively small percentage of the trees. The major uses for walnut are for decorative veneers and as solid wood for cabinetwork, paneling, and furniture.

Zebrawood. This is the wood of a relatively small tree, rarely taller than 65 ft (19.81 m), which grows in western Africa. The yellow and dark brown alternating stripes which are characteristic of the wood give it its name. The main use is as a veneer for paneling and furniture; solid wood is also obtainable.

History and Manufacture

Wood was one of the first materials used by man to construct shelter for himself. The uses of wood continued to increase in quantity and complexity through the Stone Age, the Bronze Age, the Iron Age, and right up to the present time.

The greatest changes in the use of wood as a building material came about with the development of power other than manpower—first water, then steam, next oil, and finally electricity. Power made it possible to cut trees into lumber in large quantities and led to the development of standard sizes of lumber for building purposes. Later, through research, the strength, weight, moisture content, resistance to decay, shock resistance, effect of weathering, seasoning, etc., of the lumber from the different types of trees were established, and the sizes of lumber were graded in relation to the end use of the lumber.

Veneer (thin slices of wood) was known from ancient times and used for decoration as inlay. Rotary cutting, a process made possible by new power methods, led to the development of plywood.

Today, with the continuing developments in rubbers, plastics, and adhesives, wood is undergoing another change as a building material because wood can now be impregnated, laminated, bonded together permanently with sheet metals and other materials, permanently formed into shapes, and tailor-made to meet a wide range of structural requirements.

WOOD DOORS

Physical Properties

Wood doors are made in several ways: (1) from solid lumber, (2) from solid lumber combined with plywood or glass, (3) built up from solid lumber with laminated wood veneer surfaces, (4) built up on a hollow core of small pieces of wood or other materials with wood veneer surfaces, (5) reinforced with metal angles, plates, or channels and coated with a plastic-type finish generally a vinyl, and (6) flush, with either a hollow or solid core and plastic laminated finish.

Solid lumber doors are usually made from the same species of wood that is used for windows. The lumber is kiln-dried.

Laminated veneer surface doors are made from a wide range of deciduous (hardwood) and evergreen (softwood) trees. When solid wood is used in combination with plywood, the plywood is usually Douglas fir unless the species of wood for both solid and plywood surfaces must be matched. The plywoods used may vary widely in grade.

All doors are mill-made under controlled conditions. Exterior doors are joined with waterproof adhesives, and interior doors with water-resistant adhesives.

Types and Uses

Wood doors may be divided into two major categories according to use, exterior and interior doors.

Exterior Doors. Exterior doors, in addition to meeting appearance requirements, must be able to withstand the rigors of weather and therefore are made, using waterproof adhesives, from the more durable species of wood and are treated against rot, fungi, and insects.

Interior Doors. Interior doors must meet appearance requirements, act as sound barriers, and give privacy to the areas they separate.

Table *W28* and Figures *W19* to *W21* show the commonly available stock sizes, types (method of operation), forms (decorative styles), and methods of construction of wood doors.

Special-Purpose Doors. For fire resistance there are special flush doors made with fire-resistant cores which are

Table W28 Types, sizes, operation, and use of wood doors

Type of door	Method of opening	Forms used[a]	Standard sizes for interior use: Thickness	Width	Height
	Swing[b]	All forms, Nos. 1–8 incl.	$1\frac{3}{8}$, $1\frac{3}{4}$ in. (34.93, 44.45 mm)	$1\frac{5}{6}$–3 ft (0.559–0.914 m) in 2-in. (50.8-mm) increments[d]	$6\frac{2}{3}$, 7 ft (2.032, 2.134 m)
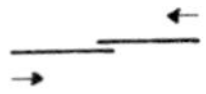	Sliding[b]	Nos. 1, 2, 3, 4, 6, 7	$1\frac{3}{8}$, $1\frac{3}{4}$ in. (34.93, 44.45 mm)	$1\frac{5}{6}$–3 ft (0.559–0.914 m) in 2-in. (50.8-mm) increments[d]	$6\frac{2}{3}$, 7 ft (2.032, 2.134 m)
	Folding	No. 4 only	$\frac{3}{8}$ in. (9.53 mm)	$3\frac{1}{2}$, $3\frac{5}{8}$, $4\frac{1}{4}$, 5 in. (88.9, 92.08, 107.95, 127.0 mm)	$6\frac{2}{3}$–12 ft (2.032–3.658 m)[e]
	Folding: 2 to 4 panels	Nos. 3, 4, 6, 8	$1\frac{1}{8}$, $1\frac{3}{8}$ in. (28.58, 34. 93mm)	4, 5, 6, 8 ft (1.219, 1.524, 1.829, 2.438 m)	$6\frac{2}{3}$–8 ft (2.032–2.438 m)[e]
	Folding: multiple panels[b]	No. 4 only	$1\frac{3}{8}$–$2\frac{1}{2}$ in. (34.93–63.5 mm) and thicker	1–4 ft (0.3048–1.219 m)[d]	8–20 ft (2.438–6.096 m)[e]
	Folding: louver type with multiple slats[b]	No. 4 solid or in special tapered form	$\frac{3}{8}$, 1 in. (9.53, 25.4 mm)	$8\frac{3}{8}$ in. (212.73 mm) and larger	$6\frac{2}{3}$, 8 ft (2.032, 2.438 m)[e]
	Coiled: vertical, narrow boards[b]	No. 4 only	$\frac{5}{8}$ in. (15.88 mm) thinner and thicker available	2 in. (50.8 mm) and less	8–20 ft (2.438–6.096 m)[e]
	Bi-parting	Nos. 1, 3, 4	$1\frac{1}{8}$–$1\frac{3}{8}$ in. (28.58–34.93 mm)	$1\frac{1}{2}$–3 ft (0.457–0.914 m)[d]	$2\frac{1}{2}$–8 ft (0.762–2.438 m)
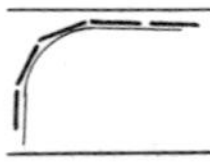	Roll-up	Nos. 1, 3, 4			
	Fold-up	Nos. 1, 3, 4			
	Roll-down[b]	Nos. 1, 3, 4	$1\frac{3}{8}$ in. (34.93 mm) larger thicknesses available	2–20 ft (0.610–6.096 m)[d]	

[a]See Figure W19 for door forms. [b]These types can be mechanically operated. [c]These doors are always mechanically operated.

rated as $\frac{3}{4}$-hour and 1-hour doors. Highly soundproof doors are manufactured with special sound-absorbing core materials, surfaces, and plastic gaskets. Certain manufacturers are equipped to design and fabricate doors to meet special purposes such as radiation protection and other highly specialized types of protection.

Decorative Doors. A large variety is available for both exterior and interior use. Designs range from block-type textures and elaborate paneling to sculptured surfaces.

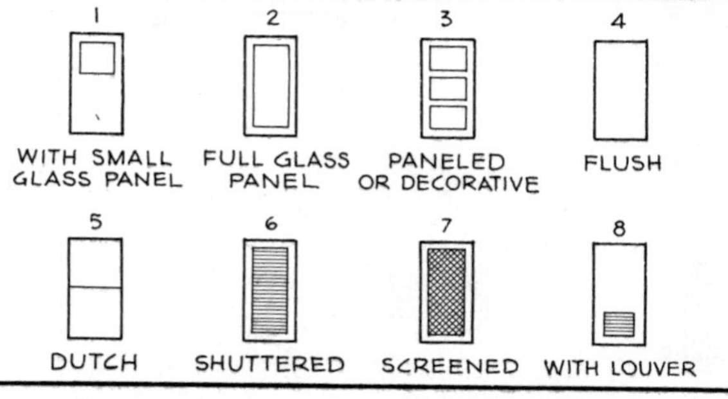

Figure W19 Basic general forms of wood doors.

Where used	Standard sizes for exterior use		
	Thickness	Width	Height
Exterior and interior	$1\frac{1}{2}$ in. (44.45 mm)	$2\frac{1}{2}$–3 ft (0.762–0.914 m) in 2-in. (50.8-mm) increments	7′–0″ (2.134 m)
Interior and exterior	$1\frac{1}{2}$ in. (44.45 mm)	$2\frac{1}{2}$–3 ft (0.762–0.914) in 2-in. (50.8-mm) increments	7′–0″ (2.134 m)
Interior openings	Sizes as required by the job		
Interior openings	Sizes as required by the job		
Area dividers	Sizes as required by the job		
Area dividers	Sizes as required by the job		
Area dividers	Sizes as required by the job		
By-pass doors and dumbwaiter doors	Sizes as required by the job		
Garage doors	$1\frac{3}{4}$ in. (44.45 mm) thicker doors available	7–24 ft (2.134–7.315 m)	$6\frac{1}{2}$–$12\frac{1}{2}$ ft (1.981–3.810 m)[e]
Garage doors	$1\frac{3}{4}$ in. (44.45 mm) thicker doors available	7–18 ft (2.134–5.486 m)	$6\frac{1}{2}$–$12\frac{1}{2}$ ft (1.981–3.810 m)
Small openings and area dividers[c]	Sizes as required by the job		

[d]Larger sizes are available. [e]Greater heights are available.

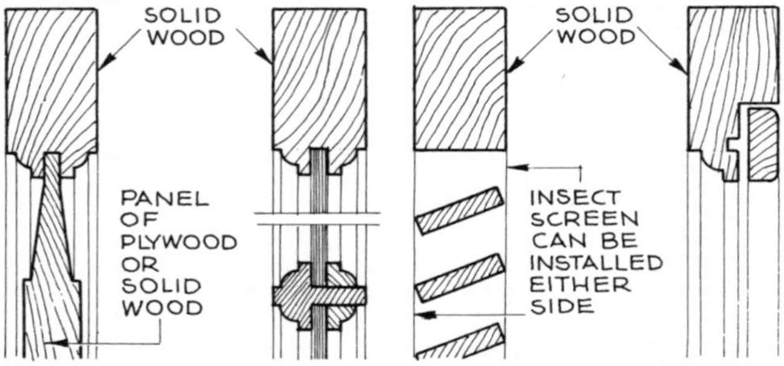

Figure W20 Construction details of door forms 1, 2, 3, 4, 5, 6, 7, and 8 when solid lumber is used.

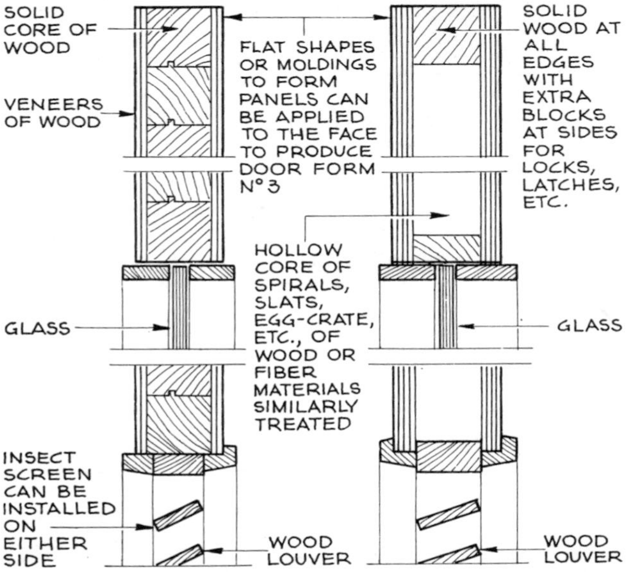

Figure W21 Construction details of door forms 1, 2, 3, 4, 5, and 8 when laminated lumber construction (hollow or solid core) is used.

Application

Condensed Checklist

1. The type, form, method of opening, and size of door selected must meet the requirements and conditions of design and use in the location in which it is to be installed.
2. Local, municipal, and state codes and also codes of the Fire Underwriters, insurance companies, labor departments, and federal government (Army, Navy, etc.) must be checked for limitations on size and height, method of opening, and type of hardware, and fire-resistance requirements for exterior doors and interior fire doors.

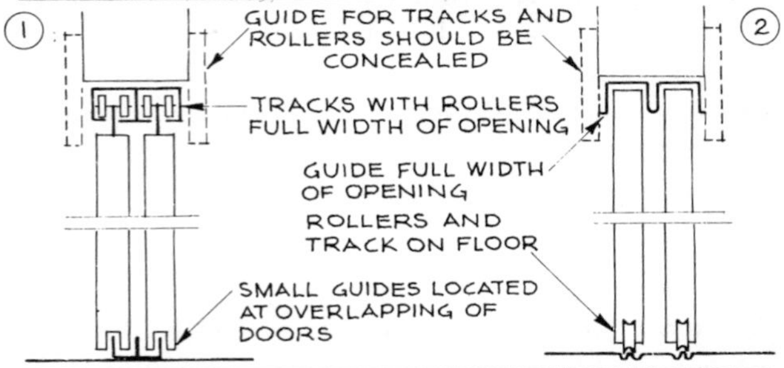

Figure W22 Details for top-hung sliding doors with floor guides (1) and for sliding doors with bottom rollers (2).

3. The method of weatherstripping of exterior doors and particularly the treatment of latching and locking hardware must be carefully determined and detailed.
4. Each swinging door should be checked to determine whether the door is to be a right-hand, left-hand, right-hand reverse, or left-hand reverse swing so that the correct hardware and door buck or frame can be specified.
5. Sliding doors should always be checked for treatment at head to conceal sliding hardware and checked for treatment at floor for guiding hardware (*see* Figure *W22*).
6. For folding doors, the size of the opening must always be calculated, taking into account the space needed for the stacked folding door so that the actual opening is sufficiently large when the door is completely pushed to one side, and also determining whether a pocket is necessary or desirable. One should always check with manufacturers of folding doors for available sizes, sound-deadening characteristics, and type and size of tracks (width and depth), to make sure that the door selected will meet the requirements of design and end use.
7. For roll-up and fold-up doors, one should always check the necessary clearances above the head and at the sides of the opening.
8. For roll-up and fold-up doors, it is always best to use types that are manufactured as complete package units (including hardware) because manufacturers of such units are best equipped to calculate the weight of the door against the balancing systems so it will open and close easily.
9. For roll-down (biparting) doors, one should always check the necessary clearances above the head and at the sides of openings. As in the case of roll-up and fold-up doors, the packaged unit should be used.
10. For large area dividers (folding doors), it is necessary to consult manufacturers for data on sound-

deadening qualities, mechanical operation, sizes, and clearances and heights necessary above the head of the door to make sure that the door selected will meet all requirements.

11. When using decorative doors, one should always check with manufacturers for available widths, heights, thicknesses, and type of hardware to be used. It is also necessary to determine whether the door is prefinished or requires job-site finishing and, in the latter case, the type of paint to be applied.
12. For both swing and sliding types of door, one should check whether they are available prefinished with various types of plastic coatings or laminates.
13. For coil-type room dividers, it is necessary to check the size of coil enclosure and whether two coils are necessary for soundproofing. The manufacturer should be consulted for availability, size, motor control, manual control, and other data pertinent to the selected end use.
14. For by-parting doors, one should always check the necessary clearances above head, below sill, and at the sides of the opening.

Conditions Favorable to the Use of Wood Doors

1. Where a durable, easily installed, economical exterior door available in a very wide range of designs, sizes, and types is desired.
2. Where a lightweight, easily installed interior door available in a very wide variety of designs, sizes, and types is desired.

Conditions Unfavorable to the Use of Wood Doors

1. Where a fire-resistance rating higher than Class B is necessary.

History and Manufacture

The first use of wood for a door is lost in history, but we know leather was used for hinges even before bronze was discovered so wood doors must have been common before the Bronze Age. Highly decorative doors have always been components of all architectural styles and continue to be so now. The development of new types of construction for doors follows the development of adhesives, from the animal adhesives known in ancient times to the present-day synthetic adhesives (*see* Wood Joining). Thus, as the casein and vegetable protein adhesives developed before World War I were improved and synthetic adhesives became available, wood doors in quantity production changed in form from solid wood to solid wood with plywood panels or veneer surfaces, then to hollow-core, and finally to sandwich-panel doors.

WOOD FINISHES

The sawmill finishes of wood obtainable in the yard are rough-sawn (rough lumber), rough-planed (semi-finish lumber), and planed (finish lumber). Lumber or wood in any of the three yard finishes can be stained directly without prior treatment, as only the texture and color are finally important. For a painted surface, finish lumber should always be used, and for fine smooth paint finishes the wood surface should first be sanded. For enamel and lacquer, finish lumber should always be used and must be sanded.

Hardwood Finishes. The deciduous woods (hardwoods) can be classified on the basis of pore structure as large and small pored (*see* Table *W29*). Those with small pores may be painted directly, but those with large pores require fillers.

Table W29 Classification of typical deciduous woods (hardwoods) by sizes of pores

Pore structure	Species of hardwood
Large pores	African mahogany (khaya), ash, butternut, chestnut, elm, hackberry, hickory, mahogany, and oak
Small pores	Alder (red), aspen, basswood, beech, cherry, cottonwood, sweetgum, magnolia, maple, yellow poplar, and sycamore

Softwood Finishes. The evergreens (softwoods) can be classified, particularly for exterior uses, into four groups on the basis of finishing requirements (*see* Table *W30*).

Plywood Finishes. Plywood is available in two classifications for the hardwoods, natural finish and paint finish. In softwoods, because of the wide bands of summer wood, the wood must always be treated with special primers to stop checking, raising of the grain, and cracking of the face ply, unless it has been pretreated in the manufacturing process.

Table W30 Classification of typical evergreens (softwoods) for exterior painting

Softwoods that can be painted with a wide range of types and qualitites of paint	Softwoods for which a suitable priming paint is necessary	Softwoods that require careful selection of priming paints and a high-quality finish paint	Softwoods that require a most careful selection of priming paints and finish paints[a]
Alaska cedar	Eastern white pine	True western firs	Douglas fir
Incense cedar	Sugar pine	Eastern hemlock	Western larch
Port Orford cedar	Western white pine	Western hemlock	Red pine
Eastern red cedar		Lodgepole pine	Southern yellow pine
Western red cedar		Ponderosa pine	Tamarack
Bald cypress		Eastern spruce	
Redwood		Engelmann spruce	
		Sitka spruce	

[a]Aluminum paint is one of the best prime paints for these woods.

Types and Uses

Wood finishes are divided into exterior finishes and interior finishes, each with markedly different requirements. The exterior finishes must withstand weathering and give protection against moisture and deterioration from various causes, whereas interior finishes must provide for a wide variety of decorative effects. Durability, resistance to moisture, etc., are comparatively minor factors.

See Paint, Painting, *and* Painting of Wood for description of methods and for types of paint suitable for wood finishing.

Exterior Finishes. In general, exterior paints should be applied in three coats, namely, a water-repellent preservative coat, a prime coat, and a finish coat. Two finish coats will last 8 to 10 years, compared to 4 to 5 years for a single finish coat. Exterior paints usually are latex, alkyd, or oil-base type. High-quality acrylic-latex, epoxy, acrylic, and vinyl paints are generally applied without a primer.

The white paint solids generally used are white lead, zinc oxide, titanium dioxide, and, to a lesser extent, zinc sulfide, lead titanate, and antimony oxide.

Water-repellent preservatives contain a fungicide, a small amount of resin, and a very small amount of water repellent, usually a wax or waxlike substance. The fungicide, usually pentachlorophenol, is toxic and care should be used to avoid excessive contact with the solution or vapor (if applied by spraying). The preservatives retard the growth of fungi, almost completely eliminate water staining at the ends of boards, reduce warping, and protect against decay woods with low natural resistance. Durability and protection can be best obtained by repeatedly brushing or spraying the wood to the point of refusal (where the wood will no longer absorb the preservative).

Pigmented penetrating stains are semitransparent. They are manufactured from solvent resin, latex, or linseed oil systems with inorganic pigments; some types have a water repellent and fungicide added.

Transparent coatings form a continuous film over the wood and should be applied over a water-repellent preservative. Both varnish and synthetics such as silicones, acrylic polymers, and vinyl fluorides or similar products can be used only in areas protected from sunlight.

Interior Finishes. Interior wood finishes are also available in opaque and transparent types.

Opaque interior finishes have generally a latex, alkyd, linseed oil, varnish, or other type of synthetic resin as the vehicle, with white paint solids, color pigments, and drying oils added.

Transparent finishes must be applied only to a wood surface that is completely free of blemishes. Such finishing consists of all or some combination of the following operations: staining, filling, sealing, surface coating, rubbing, and polishing. Gloss or semigloss varnish and nitrocellulose lacquer, shellac, flat oils, and waxes are generally used for transparent coatings.

Application

Condensed Checklist

1. For exterior and interior finishing the wood must be dry (15 to 20% moisture content). No exterior painting should ever be done in cold or freezing weather or if there is danger of dew, frost, or a temperature drop of more than 20°F (-6.66°C).
2. For exterior and interior finishing, sufficient time must elapse before applying a second coat. In warm, dry weather 24 hours is sufficient but 2 to 3 days is better. Never allow more than 1 or 2 weeks at maximum between successive coats.

3. For exterior and interior finishing, the priming coat must contain, in addition to the white paint solids with or without color pigments, enough volatile thinner to wet the wood quickly and enough bodied linseed oil to prevent undue penetration.
4. For exterior and interior finishing, generally three-coat work should always be specified to obtain best results (including preservative or sealer, primer, and finish coat).
5. For exterior siding, shingles, and millwork, back priming should always be specified to retard any change in moisture content of the wood. This will help stop curling, warping, and shrinkage.
6. For exterior natural finishes, paint manufacturers should always be consulted for the best method of obtaining the type of natural finish desired.
7. For exterior painting, staining, or natural finish, when wood preservatives are to be specified, the paint manufacturer should always be consulted to make sure that these preservatives will not affect the type of finish to be applied.
8. For interior finishes the type of wood and its classification should be known before the type of finish is selected.
9. The exact method for treatment of knots and resins in the wood should always be specified. Generally, after the prime coat is dry, the knots should be shellacked or sealed with a knot sealer of the type recommended by the paint manufacturer. For resins, particularly in the white and ponderosa pines, a coat of knot sealer should be used after the prime coat is dry.
10. When acrylic latex, epoxy, vinyl, or other synthetic resin coatings are selected, it is always necessary to check with the manufacturers of these coatings regarding the correct method of application.

WOOD FLOORINGS

Physical Properties

The woods used for floorings are the dense, durable, hard, wear-resistant and close-grained species of both deciduous and evergreen woods. Because their durability, color, and texture are of major importance in the construction field, the various flooring manufacturers' associations require that flooring meet rigid rules and regulations for kiln drying, grading, and moisture control (for both quartersawed and plainsawed wood flooring). Wood flooring is available as strip, edge-grain solid block, and preassembled into plywood or into thin, small solid strips forming thin square blocks.

Types and Uses

Wood flooring is made in three types: (1) strip flooring, (2) thin square blocks, and (3) solid end-grain blocks (*see* Figure *W23*).

Strip Flooring. Strip flooring is made from both deciduous (hardwood) and evergreen (softwood) trees. Tables *W31* and *W32* give the standard sizes, types, and species of wood for evergreen (softwood) and deciduous (hardwood) strip flooring.

The grading of hardwood strip flooring differs for unfinished and finished flooring, as shown in Tables *W33* and *W34*. Grading for hardwood strip floorings of all other species follows the grading rules set forth by the Maple and National Oak Flooring Manufacturers' Associations.

Strip flooring is available prefinished, ready for installation without any further treatment.

Thin Block Flooring. Thin square blocks are prefabricated flooring units made either from solid wood or from veneers similar to plywood. They are available either prefinished or unfinished, with either tongue-and-groove or square edges. These blocks are made from deciduous (hardwood) trees such as oak, maple, walnut, cherry, teak, hickory (true), hickory (pecan), and birch. Other woods are also available. Table *W35* and Figure *W24* show sizes, nailing methods, thicknesses, and patterns of the various types of thin block flooring. All types of wood block flooring are available prefinished, ready for installation without further treatment.

In general, all prefinished strip and grain block types of flooring have been treated against mold, fungi, and

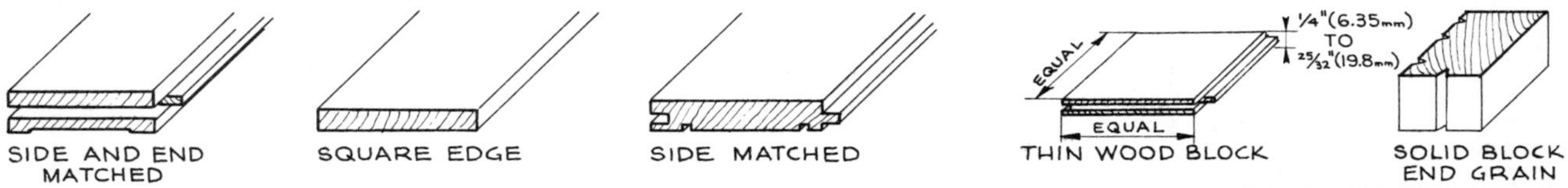

Figure W23 Types of wood flooring.

Table W31 Evergreen (softwood) strip flooring

Types of wood	Type of strip flooring	Nominal size: Thickness in.	Nominal size: Thickness mm	Nominal size: Width in.	Nominal size: Width mm	Actual size: Thickness in.	Actual size: Thickness mm	Actual size: Width in.	Actual size: Width mm	Nailing: Size of nail d[a]	Nailing: Size of nail in.	Nailing: Size of nail mm	Nailing: Nail spacing in.	Nailing: Nail spacing mm	Dimensions for figuring board-feet 144 in^3 (0.00236 m^3)
Douglas fir, western larch,	Tongue-and-grooved, side-	$\frac{3}{8}$	9.53	2	50.8	$\frac{5}{16}$	7.94	$1\frac{1}{2}$	38.1	6d	2	50.8	10	254	1 × $2\frac{3}{4}$ in. (25.4 × 69.85 mm)
										Wire nail, coated					
bald cypress, eastern hemlock,	matched; tongue-and-grooved end-	$\frac{1}{2}$	12.70	3	76.2	$\frac{7}{16}$	11.11	$2\frac{3}{8}$	60.33	7d or 9d	$2\frac{1}{4}$ or $2\frac{3}{4}$	57.15 or 69.85	12	304.8	1 × 3 in. (25.4 × 76.2 mm)
engelmann and spruce; and east-	matched also available	$\frac{5}{8}$	15.88	4	101.6	$\frac{9}{16}$	14.29	$3\frac{1}{4}$	82.55	Screw type or cut nail 8d	$2\frac{1}{2}$	63.5	10	254	1 × 4 in. (25.4 × 101.6 mm)
ern white, western white, western		1	25.40	5	127.0	$\frac{25}{32}$	19.84	$4\frac{1}{4}$	107.95	Wire nail, coated			10	254	1 × 5 in. (25.4 × 127.0 mm)
red, southern, and ponderosa pine		$1\frac{1}{4}$	31.75	6	152.4	$\frac{11}{16}$	29.99	$5\frac{3}{16}$	131.76	8d or 9d	$2\frac{1}{4}$ or $2\frac{3}{4}$	57.15 or 69.85	12	304.8	1 × 6 in. (25.4 × 152.4 mm)
		$1\frac{1}{2}$	38.1			$\frac{15}{16}$	33.34			Screw type or cut nail 9d	$2\frac{3}{4}$	69.85			
										Wire nail, coated					
	Shiplap[b]	2	50.8	4	101.6	$1\frac{5}{8}$	41.28	3	76.2	Sizes as recommended by manufacturer			10–12	254–304.8	2 × 4 in. (50.8 × 101.6 mm)
		$2\frac{1}{2}$	63.5	6	152.4	$2\frac{1}{8}$	53.98	5	127.0						$2\frac{1}{2}$ × 6 in. (63.5 × 152.4 mm)
				8	203.2										3 × 8 in. (76.2 × 203.2 mm)
		3	76.2	10	254.0	$2\frac{5}{8}$	66.68	7	177.8						4 × 10 in. (101.6 × 254.0 mm)
		4	101.6	12	304.8	$3\frac{5}{8}$	92.08	9	228.6						4 × 12 in. (101.6 × 304.8 mm)
								11	279.4						
	Tongue-and-grooved, side-matched[b]	2	50.8	4	101.6	$1\frac{5}{8}$	41.28	$3\frac{1}{8}$	79.38						
		$2\frac{1}{2}$	63.5	6	152.4	$2\frac{1}{8}$	53.98	$5\frac{1}{8}$	130.18						
				8	203.2			$7\frac{1}{8}$	180.98						
		3	76.2	10	254.0	$2\frac{5}{8}$	66.68	$9\frac{1}{8}$	231.78						
		4	101.6	12	304.8	$3\frac{5}{8}$	92.08	$11\frac{1}{8}$	282.58						
	For spline[b]	2	50.8	4	101.6	$1\frac{5}{8}$	41.28	$3\frac{1}{2}$	88.9						
		$2\frac{1}{2}$	63.5	6	152.4	$2\frac{1}{8}$	53.98	$5\frac{1}{2}$	139.7						
				8	203.2			$7\frac{1}{2}$	190.5						
		3	76.2	10	254.0	$2\frac{5}{8}$	66.68	$9\frac{1}{2}$	246.3						
		4	101.6	12	304.8	$3\frac{5}{8}$	92.08	$11\frac{1}{2}$	292.1						

[a]d is the abbreviation for penny.
[b]Not standard; used for very heavy-duty flooring.

Table W32 Deciduous (hardwood) strip flooring[a]

Type of strip flooring	Nominal size[b] Thickness in.	Nominal size Thickness mm	Nominal size Width in.	Nominal size Width mm	Actual size Thickness in.	Actual size Thickness mm	Actual size Width in.	Actual size Width mm	Weight lb/1000 board-ft	Weight kg/2.36 m^3	Nailing: Size of nail d[c]	Size of nail in.	Size of nail mm	Nail spacing in.	Nail spacing mm	Dimensions for figuring board-feet 144 in.^{3} (0.00236 m^3)
Tongue-and-grooved, end-matched; tongue-and-grooved side-matched also available	$\frac{25}{32}$	19.84	$1\frac{1}{2}$	38.1	$\frac{3}{4}$	19.05	$1\frac{1}{2}$	38.1	1820	825.55	7d 9d	$2\frac{1}{4}$ $2\frac{3}{4}$	57.15 69.85	10–12	254–304.8	1 × $2\frac{1}{2}$ in. (25.4 × 63.5 mm)
			2	50.8			2	50.8	1920	870.91	Screw-type or cut nail 8d	$2\frac{1}{2}$	63.5			1 × $2\frac{3}{4}$ in. (25.4 × 69.85 mm)
			$2\frac{1}{4}$	57.15			$2\frac{1}{4}$	57.15	2020	916.17						1 × 3 in. (25.4 × 76.2 mm)
			$3\frac{1}{4}$	82.55			$3\frac{1}{4}$	82.55	2110	957.10	Wire nail, coated					1 × 4 in. (25.4 × 101.6 mm)
	$\frac{3}{8}$[d]	9.53	$1\frac{1}{2}$	38.1	$\frac{11}{32}$	8.73	$1\frac{1}{2}$	38.1	1000	453.60	4d	$1\frac{1}{2}$	38.1	8	203.2	1 × 2 in. (25.4 × 50.8 mm)
			2	50.8			2	50.8			Screw-type or cut nail					1 × $2\frac{1}{2}$ in. (25.4 × 63.5 mm)
	$\frac{1}{2}$[d]	12.7	$1\frac{1}{2}$	38.1	$\frac{15}{32}$	11.91	$1\frac{1}{2}$	38.1	1300	589.68	5d	$1\frac{3}{4}$	44.45	10	254.0	1 × 2 in. (25.4 × 50.8 mm)
			2	50.8			2	50.8	13500	612.36	Screw-type, cut nail, or wire casing nail					1 × $2\frac{1}{2}$ in. (25.4 × 63.5 mm)
	$\frac{17}{16}$	26.99	2	50.8	$\frac{33}{32}$	26.20	2	50.8	2250	1020.60	7d 8d	$2\frac{1}{4}$ $2\frac{3}{4}$	57.15 69.85	10–12	254–304.8	$\frac{5}{4}$ × $2\frac{3}{4}$ in. (31.75 × 69.95 mm)
			$2\frac{1}{4}$	57.15			$2\frac{1}{4}$	57.15	2400	1088.64	Screw-type, cut nail, or 8d wire nail, coated					$\frac{5}{4}$ × 3 in. (31.75 × 76.2 mm)
			$3\frac{1}{4}$	82.55			$3\frac{1}{4}$	82.55								$\frac{5}{4}$ × 4 in. (31.75 × 101.6 mm)
Square edge	$\frac{25}{32}$	19.84	$2\frac{1}{2}$	63.5	$\frac{3}{4}$	19.05	$2\frac{1}{2}$	63.5	2250	1020.60	11-gauge barbed flooring nail	$2\frac{1}{4}$	57.15	10–12	254–304.8	1 × $3\frac{1}{4}$ in. (25.4 × 82.5 mm)
			$3\frac{1}{4}$	82.55			$3\frac{1}{4}$	82.55	2400	1088.64						1 × 4 in. (25.4 × 101.6 mm)
			$3\frac{1}{2}$	88.9			$3\frac{1}{2}$	88.9	2500	1134.00						1 × $4\frac{1}{4}$ in. (25.4 × 107.95 mm)
	$\frac{5}{16}$	7.94	$1\frac{1}{2}$ $2\frac{1}{2}$	38.1 63.5	$\frac{5}{16}$	7.94	$1\frac{1}{2}$ $2\frac{1}{2}$	38.1 63.5	1200	544.32	15-gauge barbed flooring nail	1	25.4	$3\frac{1}{2}$	88.9	Face count
	$\frac{17}{16}$	26.99	$2\frac{1}{2}$	63.5	$\frac{33}{32}$	26.20	$2\frac{1}{2}$	63.5	2500	1134.00	11-gauge barbed flooring nail	$2\frac{1}{4}$	57.15	10–12	254–304.8	$\frac{5}{4}$ × $3\frac{1}{4}$ in. (31.75 × 82.55 mm)
			$3\frac{1}{2}$	88.9			$3\frac{1}{2}$	88.9	2600	1179.36						$\frac{5}{4}$ × $4\frac{1}{4}$ in. (31.75 × 107.95 mm)

[a]The following woods are covered here: oak, beech, maple (hard), pecan (hickory), cherry, teak, ash and walnut.
[b]Some thicker sizes are available, e.g., $\frac{41}{32}$ in. (32.54 mm), $\frac{53}{32}$ in. (42.07 mm) and thicker.
[c]*d* is the abbreviation for penny.
[d]Always should be installed on subflooring.

Table W33 Grades of unfinished deciduous (hardwood) strip flooring

Species of wood	Grades	Allowable imperfections	How bundled: Tongue-and-grooved, side- and end-matched or side-matched	Major uses
Oak[a] (color not considered)	Clear (quartersawed or plainsawed)	Face practically clear, admitting an average of $\frac{3}{8}$ in. (9.53 mm) of bright sap	$1\frac{1}{4}$ (381.0 mm) and up; average length, $3\frac{3}{4}$ ft (1.431 m) [average length of square edge, 5 ft (1.524 m)]	For fine type of flooring
	Select (quartersawed or plainsawed)	Face may contain sap, small streaks, pin wormholes, slight imperfections in warping, and small tight knots that do not average more than one every 3 ft (0.914 m)	$1\frac{1}{4}$ ft (381.0 mm) and up; average length, $3\frac{1}{4}$ ft (0.991 m) [average length of square edge, (4 ft (1.219 m)]	Medium type of flooring
	Number 1 Common (plainsawed)	Face may contain flags, heavy streaks and checks, wormholes, knots, and minor imperfections in working	$1\frac{1}{4}$ ft (381.0 mm) and up; average length, $2\frac{3}{4}$ ft (0.838 mm) [average length of square edge, 3 ft (0.914 mm)]	Good type of flooring
	Number 2 Common (plainsawed)	May contain defects of all types	$1\frac{1}{4}$ ft (381.0 mm) and up; average length, $2\frac{1}{4}$ ft (0.686 m)	Good economical flooring
	$1\frac{1}{4}$-ft (381.0 mm) shorts, No. 1 Common and better	Made up of No. 1 Common, Select, and Clear	$\frac{3}{4}$–$1\frac{1}{2}$ ft (228.6 mm–457.2 mm); average length, $1\frac{1}{4}$ ft (381.0 mm)	Parquetry-type flooring and also used as fillers for small areas for other grades
	$1\frac{1}{4}$-ft (381.0 mm) shorts, No. 2 Common	Made up of No. 2 Common only	$\frac{3}{4}$–$1\frac{1}{2}$ ft (228.6 mm–457.2 mm); average length, $1\frac{1}{4}$ ft (381.0 mm)	
Beech, birch and hard maple[a]	First Grade	Face practically clear, with varying natural color of the wood	2 ft (609.6 mm) and up as the stock will produce, the proportion of 2 ft (609.6 mm) and 3 ft (0.914 m) bundles shall be what the stock will produce up to 33% in first and 45% in second in footage	Highest standard made; fine type of flooring
	Second Grade	Will admit tight sound knots and slight imperfections in dressing but must lay without waste		Fine type of flooring where slight imperfections are not objectionable
	Third Grade	Of such a character as will lay and give a good serviceable floor	$1\frac{1}{4}$ ft (381.0 mm) and up as the stock will produce; the proportion of $1\frac{1}{4}$–3 ft (381.0 mm–0.914 m) bundles shall be what stock will produce up to 65% in footage	Good economical flooring
Hard white maple[a]	Special grade: First Grade White Hard Maple	Special stock selected for uniformity of ivory white color	Lengths same as for First Grade hard white maple	Finest type of flooring
Beech[a]	Special grade: First Grade Red	Special stock selected for uniformity of rich red color	Lengths are the same as for First Grade Hard white maple	Finest type of flooring
Pecan (hickory)[a]	First Grade (unselected for color)	Practically free of defects with varying natural color of the wood	2 ft (609.6 mm) and up; not over 25% of the footage shall be 2 ft (609.6 mm) and 3 ft (0.914 m)	Fine type of flooring

Table W33 Grades of unfinished deciduous (hardwood) strip flooring (*continued*)

Species of wood	Grades	Allowable imperfections	How bundled: Tongue-and-grooved, side- and end-matched or side-matched	Major uses
Pecan (hickory)[a] (*continued*)	First Grade Red	Same as First Grade except face shall be heartwood	Same as First Grade	Fine type of flooring
	Second Grade	Will admit tight sound knots, pin wormholes, streaks, light stain, and slight imperfections in working	$1\frac{1}{4}$ ft (381.0 mm) and up as the stock will produce; the proportion of $1\frac{1}{4}$–3 ft (381.0 mm–0.914 m) bundles shall be what stock will produce up to 40% in footage	Will lay a sound floor without cutting
	Second Grade Red	Same as Second Grade except face shall be heartwood		
	Third Grade	Of such a character as will give a good serviceable floor	$1\frac{1}{4}$ ft (381.0 mm) and up as the stock will produce; the proportion of $1\frac{1}{4}$–3 ft (381.0 mm–0.914 m) bundles shall be what the stock will produce up to 60% in footage	Good economical flooring
Northern hard maple[b] Northern beech[b] Northern birch[b]	First Grade	Face practically free from defects but will admit occasional pin knots not over $\frac{1}{8}$ in (318 mm) diameter, spots and streaks not not over $\frac{1}{4}$ in. (6.35 mm) wide and 3 in. (76.2 mm) long, and other very minor imperfections	2 ft (0.610 m) and up as the stock will produce; not over 30% of total footage shall be in bundles under 4 ft (1.219 m) length	Highest standard grade for fine type of flooring
	Second Grade	Will admit sound knots and slight imperfections in dressing, prominent discolorations, numerous streaks and spots, but must lay without waste	2 ft (0.610 m) and up as the stock will produce; not over 45% of total footage shall be in bundles under 4 ft (1.219 m) in length	Fine type of flooring where slight imperfections are not objectionable
	Third Grade	Of such character as will give a good serviceable floor	$1\frac{1}{4}$ in. (381 mm) and up as the stock will produce; not over 65% of total footage shall be in bundles under 4 ft (1.219 m) in length	Good economical flooring
	Second and better	Combination of First and Second Grades developing in the strip without cross-cutting for each grade, where lowest grade pieces shall be not less than Second Grade	2 ft (0.610 m) and up as the stock will produce; not over 40% of total footage shall be in bundles under 4 ft (1.219 m) in length	Medium type of flooring
	Third and better	Combination of First, Second, and Third Grades developing in the strip without cross-cutting, where lowest grade pieces shall be not less than Third Grade	$1\frac{1}{4}$ (381 mm) and up as the stock will produce; not over 50% of total footage shall be in bundles under 4 ft (1.219 m) in length	Good economical flooring

Table W33 Grades of unfinished deciduous (hardwood) strip flooring (*continued*)

Species of wood	Grades	Allowable imperfections	How bundled: Tongue-and-grooved, side- and end-matched or side-matched	Major uses
Northern hard maple[b]	Special grade: Selected First Grade Light (or White) Northern Hard Maple	Selected for uniformity of light color	Lengths are the same as for First Grade beech, birch, and hard maple	Finest type of flooring
	Special grade: Selected First Grade Amber (or Brown) Northern Hard Maple	Selected for uniformity of amber color	Lengths are the same as for First Grade beech, birch, and hard maple	Finest type of flooring
Beech[b] Birch[b]	Special grade: Selected First Grade Red	Selected for uniformity of color	Lengths are the same as for First Grade	Finest type of flooring

[a]Grading rules of the National Oak Flooring Manufacturers' Association.
[b]Grading rules of the Maple Flooring Manufacturers' Association.

Table W34 Grades of prefinished deciduous (hardwood) strip flooring

Species of wood	Grade	Allowable imperfections	How bundled: Tongue-and-grooved, side- and end-matched, or side-matched	Major use
Oak[a]	Prime Grade	Face shall be selected after finishing, but sapwood and the natural variations in color are permitted	1.25 ft (381 mm) and up Minimum average length is 3.5 ft (1.067 m)	Finest type of flooring
	Standard Grade	Will contain sound wood characteristics after filling and finishing; will lay sound floor without cutting	1.25 ft (381 mm) and up Minimum average length is 2.75 ft (0.838 mm)	Medium type of flooring
	Tavern Grade	Shall be of such a nature as will lay and give a serviceable floor; purposely contains typical wood characteristics	1.25 ft (381 mm) and up Minimum average length is 2.25 ft (0.686 m)	Good economical flooring
	Standard and better	Combination of Prime and Standard Grades where no piece shall be lower than Standard Grade	1.25 ft (381 mm) and up Minimum average length is 3 ft (0.914 m)	Fine type of flooring
Beech and pecan	Tavern and better	Combination of Prime, Standard, and Tavern Grades where no piece shall be lower than Tavern Grade	1.25 ft (381 mm) and up Minimum average length is 3 ft (0.914 m)	Medium type of flooring

[a]Grading rules of the National Oak Manufacturers' Association.

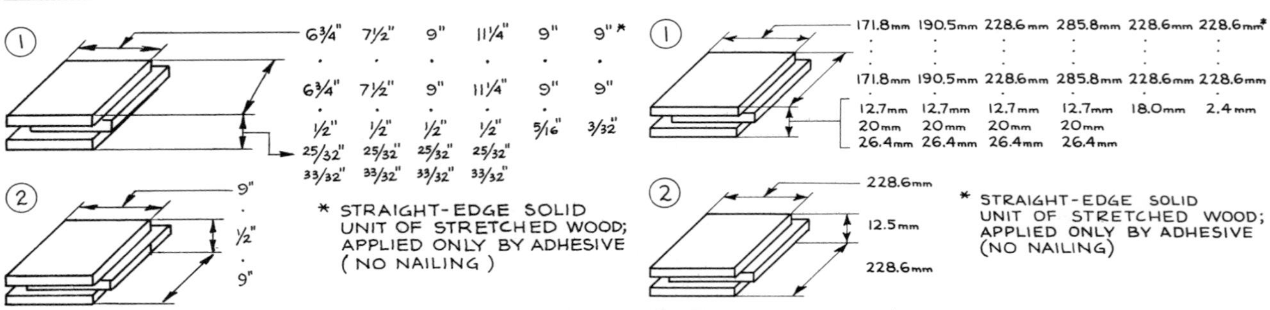

Figure W24 Thin wood block flooring. Sizes of tongue-and-groove type (1) and of plywood type (2).

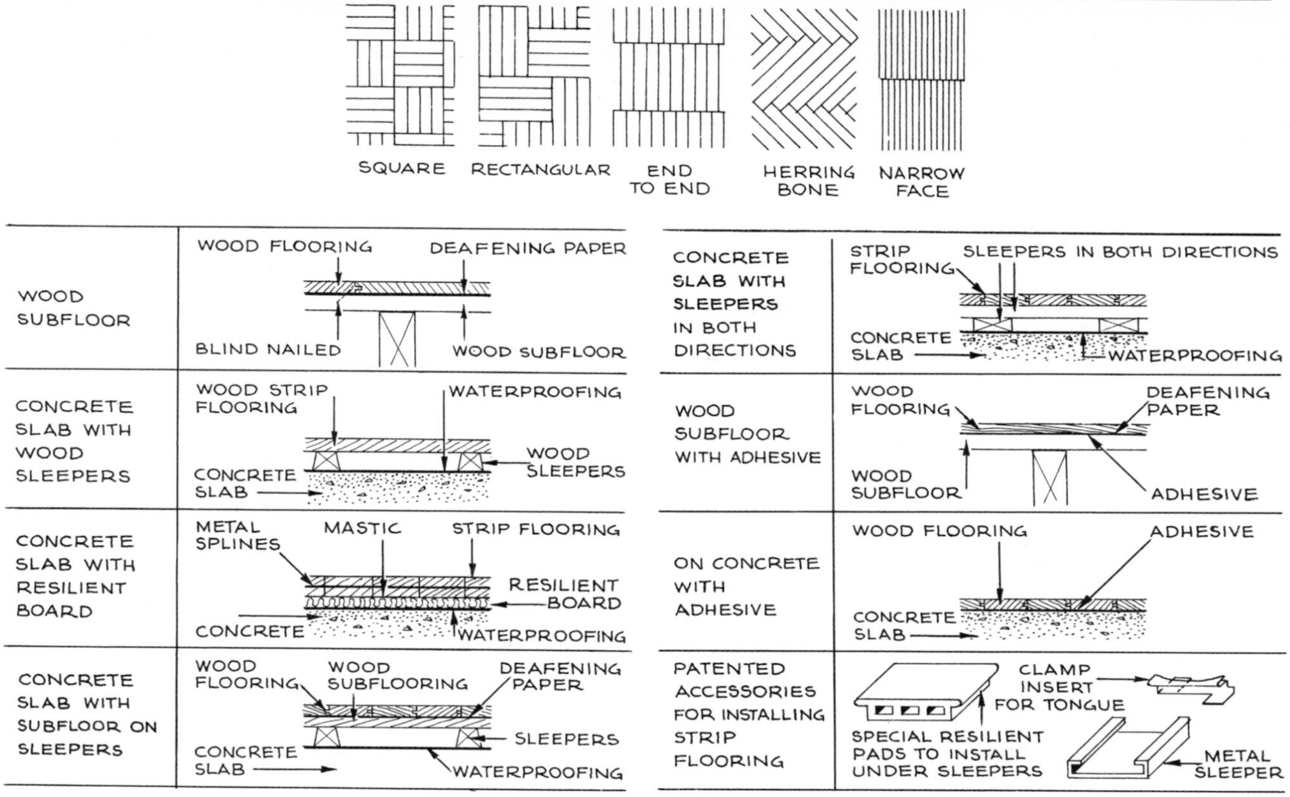

Figure W24 Thin wood block flooring (continued). Patterns for: Laying block flooring (top) and details of installation on various types of subflooring.

Table W35 Types, sizes, nailing data and pattern of solid deciduous (hardwood) thin block flooring

Type of block	How assembled	Sizes: Thickness, Nominal in.	mm	Actual in.	mm	Width in.	mm	Length in.	mm	Nailing: Size of nail	Nail spacing in.	mm	Type of pattern
Single-piece block	Side and end tongue-and-grooved, one-half each right and left hand matching	$\frac{33}{32}$	26.20	$\frac{3}{4}$	19.05	$1\frac{1}{2}$ 2 $2\frac{1}{4}$	38.1 50.8 57.15	$6\frac{3}{4}$ to $13\frac{1}{2}$	171.45 to 342.9	$2\frac{1}{4}$-in. (57.15-mm) spiral floor nail or 6d[b] 2-in. (50.8-mm) cut nail	10	254	Square, herring-bone
Prefabricated blocks[a]	Side and end tongue-and-grooved to interlock on all sides	$\frac{33}{32}$	26.20	$\frac{3}{4}$	19.05	$1\frac{1}{6}$ 2 $2\frac{1}{4}$	26.62 50.8 57.15	$7\frac{1}{2}$ in. (189.5 mm) and 9 in. (228.6 mm) squares 8 in. (203.2 mm) and 9 in. (228.6 mm) squares 6 ft $\frac{3}{4}$ in. (171.45 mm) and 9 in. (228.6 mm) squares		$2\frac{1}{4}$-in. (57.15-mm) spiral floor nail or 6d[b] 2-in. (50.8-mm) cut nail	10	254	Square
Prefabricated plywood blocks				$\frac{1}{2}$ $\frac{3}{4}$	12.7 19.05			9 in. (228.6 mm) squares					
Single slats	Side tongue-and-grooved	$\frac{53}{32}$ $\frac{33}{32}$	42.33 26.20	$\frac{41}{32}$ $\frac{25}{32}$	32.75 18.37	$1\frac{1}{2}$ 2 $2\frac{1}{4}$ $3\frac{1}{4}$	38.1 50.8 57.15 82.55	8–16 Generally 12	203.2–406.4 304.8	Size recommended by flooring	10	250	End to end

[a]This type can be installed with adhesive.
[b]d is the abbreviation for penny.

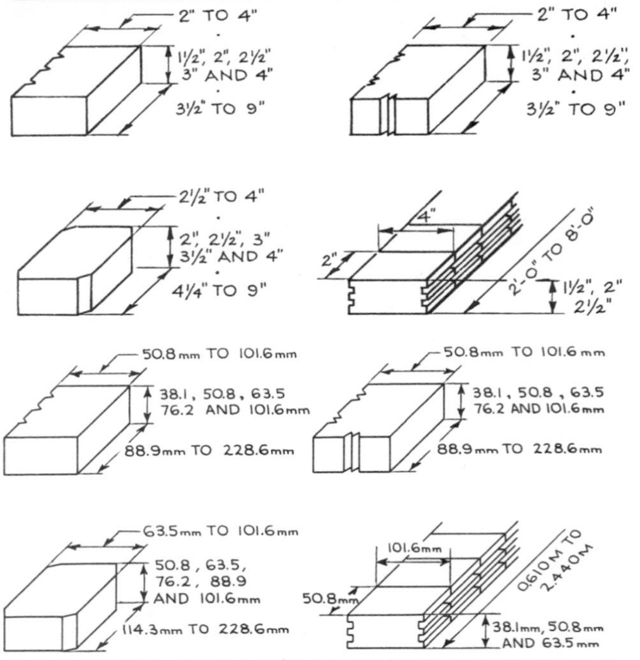

Figure W25 Common types and sizes of solid block flooring.

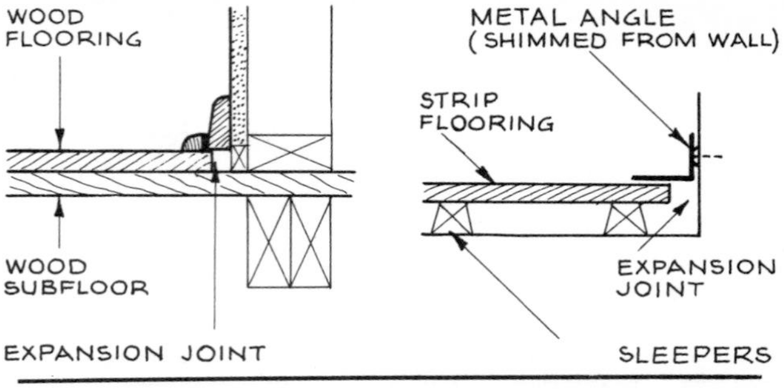

Figure W26 Details of allowances for expansion at perimeters of wood strip flooring.

insect attack. Unfinished flooring of these types is usually not treated except on special order. Prefinished strip and thin block types of flooring are impregnated with synthetic resins and given a finish that is very hard and durable, with a waxlike sheen; therefore they require very little maintenance.

Solid Block Flooring. Solid block flooring is generally made from southern yellow pine. The face is end-grained, and the blocks have been treated with creosote against decay, moisture, mold, and insects. Figure *W25* shows sizes and types.

Application

Condensed Checklist

1. For all types of wood flooring, one should always check that the species of wood and the type and grade of flooring selected meet the requirements and limitations of the areas where the flooring is to be installed.
2. For all types, the flooring manufacturers should be consulted for the best method of installing the flooring selected and particularly for the type of subflooring, sound-deafening material, sleepers, and adhesive to be used.
3. If wooden sleepers are used, they must be treated to be resistant against decay, mold, fungi, and insects.
4. For all types of wood flooring, sufficient expansion joints must be installed at the perimeter (*see* Figure *W26*).
5. It is always necessary to specify that the flooring be stored in the building for 3 or more days in an area with good ventilation. Both the storage area and the areas where the flooring is to be installed should have temperatures of not more than 70°F (21.11°C) or less than 50°F (10°C).
6. When wood flooring is used, it is always necessary to check whether there is a high relative humidity during spring, summer, and early fall in the area of construction. When there is high relative humidity, it is good practice to give the flooring 2 to 3 weeks of acclimatization before sanding and finishing.
7. Wood flooring that is to be used in a region within the termite belt or in a hot, humid region should always have been treated against insects, mold, fungi, and decay, especially if it is to be installed on concrete slabs on grade or below grade.
8. For all types of wood flooring, flooring manufacturers should be consulted for the best method of finishing the type, grade, and species of wood selected. In general, the use of a penetrating type of sealer with a finish coat of wax is recommended, and varnish and shellac are not recommended.
9. Edge-grain solid wood block flooring should always be treated with coal tar creosote to a final retention of 6 lb/ft^3 (96.12 kg/m^3).
10. Solid wood blocks should usually be laid at right angles to the line of traffic, on a coating of coal tar pitch $\frac{1}{8}$ in. (3.18 mm) thick. Two coats of hot coal tar pitch, applied to fill the voids between the blocks and to act as a binder, must be worked into the joints uniformly. The remaining thin film of coal tar pitch remaining on the surface soon wears off with use.

11. For all types of prefinished flooring, the flooring manufacturers should be consulted regarding the methods of application and protection and the correct types of substrate upon which the flooring should be installed.
12. When strip flooring is to be installed in gymnasiums, the treatment of the substrate against moisture penetration, the method of attachment to wood or metal sleepers, and the expansion joint around the entire perimeter should be carefully checked, and the manufacturers of this type of strip flooring should be consulted.

Conditions Favorable to the Use of Wood Flooring

1. Use solid block flooring where a durable, wear-resistant, easily maintained, and dustless type of flooring is needed. Solid block flooring is also nonsparking and has good resistance against oils, greases, and mild chemicals.
2. Use thin wood block flooring where the design requires a decorative flooring that is durable, wear-resistant, and easily maintained.
3. Use short length strip flooring of different species of wood where the design requires highly decorative patterns.
4. Use strip flooring where a generally decorative, durable, resilient, wear-resistant, and easily maintained flooring is desired.

Conditions Unfavorable to the Use of Wood Flooring

1. In areas where acids and strong alkalis will be used.
2. In areas where high heat and high humidity will be encountered.
3. In areas where water will be constantly spilled and remain on the flooring material.
4. In areas below grade unless special precautions are taken.

History and Manufacture

Stone, baked clay tiles, and wood were the earliest materials used for flooring. With the invention of the iron nail, wood became the most common flooring material until the invention of linoleum and the development of resilient flooring materials. Today, wood flooring is used principally in residential construction and in institutional and manufacturing buildings where a durable, resilient, nonsparking floor is required.

WOOD GUTTERS

Physical and Chemical Properties

Wood gutters are usually made from redwood or Douglas fir, the only trees from which extralong lengths of clear wood are easily available. Gutters are also made from cypress, white fir, spruce, red cedar, and hemlock. Only lumber that is free from knots, knotholes, and any other imperfections is used. Gutters are available in limited sizes and lengths; a few manufacturers make leaders.

Types and Uses

Wood gutters are made in two stock shapes, the Boston and the Ohio. Both are available in lengths up to 20 ft (6.096 m). Those made of Douglas fir can be obtained up to 40 ft (12.192 m) long. (*See* Figure *W27*.)

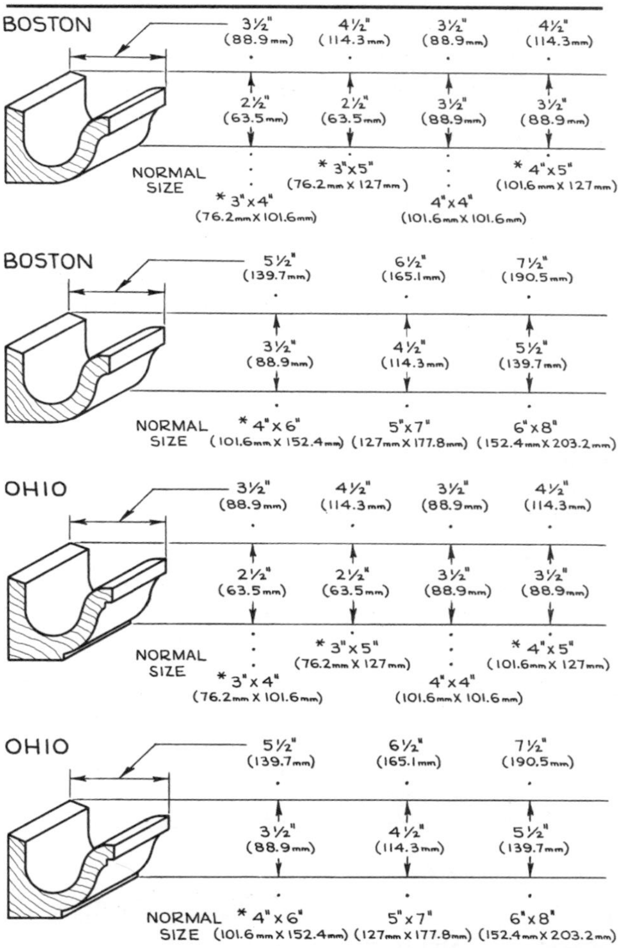

* STOCK SIZES FOR ALL SPECIES OF WOODS USED FOR GUTTERS AND 4"(101.6mm) x 4"(101.6mm) A STOCK SIZE FOR REDWOOD

Figure W27 Stock shapes and sizes of wood gutters.

Rectangular wood leaders are limited to two stock sizes: one has inside dimensions of $3\frac{1}{4} \times 3\frac{1}{8}$ (82.55 × 79.98 mm), and the other has inside dimensions of 3 × 4 in. (76.2 × 101.6 mm).

Wood gutters can be obtained with a vinyl coat in limited colors.

Application

Condensed Checklist

1. Rainfall data in the region of construction should always be checked so that adequate sizes of gutters and leaders can be selected. Gutters should be set at a slope of $\frac{1}{16}$ in. to 1 ft (1.59 mm to 0.3048 m). They may be set level if enough leaders of adequate size are installed.
2. The availability of the shape of gutter and species of wood desired should always be checked in the locality of construction.
3. The joining of wood gutters both horizontally and at corners should be correctly detailed (*see* Figure *W28*).
4. Metal leader outlets should always be used where connections to leaders are made, and basket strainers should be installed.
5. Methods of connection of leaders should be checked, especially with large overhangs, as unsightly angular leader connections may have a detrimental effect on the overall design.

Conditions Favorable to the Use of Wood Gutters

1. Where gutters that match the wood character of the building are desired.
2. Where the gutter is to be built on as part of the building.
3. For residential roofs where roof areas are not too large for the stock sizes of wood gutters.

Condition Unfavorable to the Use of Wood Gutters

For roof areas that require gutters larger than the sizes available in wood.

History and Manufacture

The earliest examples of wood gutters were V-type ones made by nailing together two boards which were then attached to the edge of the roofs to carry off the water, spilling it either directly onto the ground or into wooden containers. When metal in sheet form became available, it almost completely replaced wood for this purpose. Today the wood gutter is generally limited to residential buildings.

WOOD HARDBOARD AND PARTICLEBOARD

Types and Uses of Hardboard

Hardboard is defined as a panel manufactured primarily from interfelted lingo-cellulose fibers consolidated under heat and pressure in a hot press to a density of at least 31 lb/ft^3 (4.96.62 kg/m^3). Other materials may be added during manufacture to improve certain properties such as stiffness, hardness, finishing properties, and resistance to abrasion and moisture, as well as to increase strength, durability, and utility.

Hardboard is divided into two types: (1) basic hardboard, and (2) prefinished hardboard panels and siding. Basic hardboard is again divided into five types: tempered, standard, service-tempered, service, and industrialite. These five types are divided into two classes of surface finish. Basic hardboard is available smooth on one side (S1S) and smooth on two sides (S2S), whereas

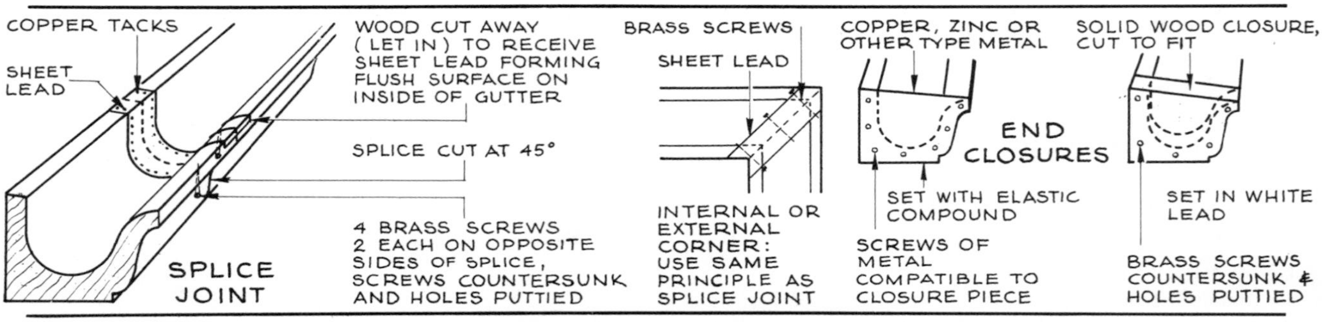

Figure W28 Recommended methods for joining wood gutters and for end closures.

prefinished hardboard is divided into two types: *Class I*, a finish with resistance to wear, stain, heat, fading, steam, abrasion, and moisture; and *Class II*, a finish with resistance to household stains and moderate resistance to wear and abrasion but with no requirements of resistance to heat, moisture, and steam. (*See* Tables *W36*, *W37*, and *W38*.)

Basic hardboard is available perforated (pegboard) or punched into designs for decorative grilles, screens, room dividers, etc. Prefinished hardboard is available striated and scored in one direction or in two directions, in a wide variety of finishes. The various applied finishes are plastic laminates, vinyl films, baked enamels, paints, stains, lacquers, and enamels. There are also numerous special finishes. It is therefore always advisable to check with the manufacturers of hardboard regarding the available types of basic hardboard and prefinished hardboard and their suitability for the required end use.

Prefinished hardboard is used on the exterior for siding, paneling, ceilings, overhangs, and doors. On the interior, it is used for walls, ceilings, underlayment, cabinetwork, displays, pegboard, acoustical treatment and the like. Standard sizes are 4 × 8 ft (1.219 × 2.436 m), in thicknesses of $\frac{1}{8}$, $\frac{3}{16}$, and $\frac{1}{4}$ in. (3.175, 4.763, and 6.350 mm).

Condensed Checklist

1. Always check with manufacturers regarding the correct type of adhesive for laminating or for bonding hardboard to substrates and other surfaces for either interior or exterior end use.
2. When selecting prefinished hardboard, always check with manufacturers for the types of finishes, textures, etc., that are currently available and will not be withdrawn from manufacture within a year's time.

Table W36 Types, sizes, markings, and tensile strengths of hardboard[a]

Class of hardboard	Surface finishes[a]	Thickness		Number and color of stains	Width		Length[b]		Tensile strength			
									Parallel to surface		Perpendicular to surface	
		in.	mm		ft	m	ft	m	lb/in.^{2}	MN/m^2	lb/in.^{2}	MN/m^2
1 Tempered	S1S	$\frac{1}{12}$	2.12	1 red	All can be obtained in 4 and 5 ft (1.219 and 1.524 m)		Classes 1, 2, 3 and 4 can be obtained in 3–10	0.914–3.048 in 1-ft (0.3048-m) increments	3500	24.133	150	1.034
	S1S, S2S	$\frac{1}{10}$, $\frac{1}{8}$, $\frac{3}{16}$, $\frac{1}{4}$, $\frac{5}{16}$, $\frac{3}{8}$	2.54, 3.18, 4.76, 6.35, 7.94, 9.53									
2 Standard	S1S, S2S	$\frac{1}{12}$, $\frac{1}{10}$, $\frac{1}{8}$, $\frac{3}{16}$, $\frac{1}{4}$, $\frac{5}{16}$, $\frac{3}{8}$	2.12, 2.54, 3.18, 4.76, 6.35, 7.94, 9.53	1 green					2500	17.238	100	0.690
3 Service-tempered	S1S, S2S	$\frac{1}{8}$, $\frac{3}{16}$, $\frac{1}{4}$, $\frac{3}{8}$	3.18, 4.76, 6.35, 9.53	2 red					2000	13.790	100	0.690
4 Service	S1S, S2S	$\frac{1}{8}$, $\frac{3}{16}$, $\frac{1}{4}$, $\frac{3}{8}$, $\frac{7}{16}$, $\frac{1}{2}$	3.18, 4.76, 6.35, 9.53, 11.11, 12.70	2 green					1500	10.343	75	0.517
	S2S	$\frac{5}{8}$, $\frac{11}{16}$, $\frac{3}{4}$, $\frac{13}{16}$, $\frac{7}{8}$, 1, $1\frac{1}{8}$	15.88, 17.46, 19.05, 20.64, 22.23, 25.40, 28.58									
5 Industrialite	S1S, S2S	$\frac{3}{8}$, $\frac{7}{16}$, $\frac{1}{2}$	9.53, 11.11, 12.70	1 blue			10–16	3.048–4.877 in 2-ft (0.610-m) increments	1000	6.895	35	0.241
	S2S	$\frac{5}{8}$, $\frac{11}{16}$, $\frac{3}{4}$, $\frac{13}{16}$, $\frac{7}{8}$, 1, $1\frac{1}{8}$	15.88, 17.46, 19.05, 20.64, 22.23, 25.40, 28.58									

[a] Hardboard is marked by class with different stripes and colors, as shown in the table.
[b] The lengths listed are the sizes generally carried by lumber suppliers. Lengths can be obtained up to 20 ft (18.192 m), and almost any length can be obtained in large quantities by consulting the manufacturers.

Table W37 Types, water resistances, flame spreads, sizes, and tensile strengths of prefinished hardboard substrate

Class	Thickness		Water absorption based on weight (percent)	Thickness swelling (percent)	Flame spread index	Width[a]		Length[a]		Tensile strength			
										Parallel to surface		Perpendicular to surface	
	in.	mm				ft	m	ft	m	lbf/in.^{2}	MN/m^2	lbf/in.^{2}	MN/m^2
1						All can be		All classes can be					
Tempered	$\frac{1}{8}$	3.18	20	16		obtained		obtained in					
	$\frac{3}{16}$	4.76	18	15	0–25	in 4 and		4–10	1.219–3.048	3500	24.133	150	1.034
	$\frac{1}{4}$	6.35	12	11		5 ft		in					
						(1.219 and		1-ft (0.3048-m)					
2						1.524 m)		increments					
Standard	$\frac{1}{8}$	3.18	25	18		4,	1.219,	and					
	$\frac{3}{16}$	4.76	25	18	26–75	5	1.524	10–20	3.048–6.096	2500	17.238	100	0.690
	$\frac{1}{4}$	6.35	20	14				in					
3								2-ft (0.6096-m)					
Service	$\frac{1}{8}$	3.18	25	22				increments					
tempered	$\frac{3}{16}$	4.76	20	18	76–200					2000	13.790	100	0.690
	$\frac{1}{4}$	6.35	20	14									
4													
Service	$\frac{1}{8}$	3.18	30	25									
	$\frac{3}{16}$	4.76	27	22	over 200					1500	10.344	75	0.517
	$\frac{1}{4}$	6.35	27	22									

[a]Widths and lengths as listed ahove are the standard sizes generally available. For special shapes and dimensions, contact the manufacturers of hardboard.

Table W38 Types, sizes, surface treatments, and densities of hardboard exterior siding

Type	Nominal dimensions						Surface-treatment		Density	
	Width[a]		Length		Thickness		Surface texture	Protective coating		
	in.	mm	ft	m	in.	mm			lb/ft^3	kg/m^3
Lap siding	4, 6, 8, 9, 10, 12	101.6, 152.4, 203.2, 228.6, 254.0, 344.8	4–16	1.219–4.877 in 2-ft (0.610-m) increments	$\frac{3}{8}$, $\frac{7}{16}$	9.53, 11.11	Embossed or smooth	Primed or unprimed	31	496.62
Panel siding	48 (4 ft)	219.2 (2.19 m)	4, 6, 7, 8, 9, 10, 12	1.219, 1.829, 2.134, 2.438, 2.743, 3.048, 3.658	$\frac{1}{4}$, $\frac{3}{8}$, $\frac{7}{16}$	6.35, 9.53, 11.11	Embossed, grooved, or smooth			

[a]Refers to exposed width. Actual width may be greater, depending on type of edge finish and/or lap.

3. For forming or bending hardboard, always check the minimum bending radii for the selected type of hardboard (*see* Table *W39*).
4. When hardboard is to be factory-primed or undercoated and then finished in a mill or factory or on the job site, always check with the manufacturer of the hardboard and the paint manufacturers regarding the painting system to be used.

Types and Uses of Particleboard

Particleboard is defined as a panel manufactured of ligno-cellulose materials, primarily in the form of discrete pieces or particles of wood as distinguished from fibers, combined with synthetic resins or other binders and bonded together under heat and pressure in a hot press by a process in which the entire interparticle bond

Table W39 Minimum radii for bending the different types of hardboard by either dry or moist cold method

Type of sheet	Thickness		Minimum radii for bending: Cold dry method, Smooth side in		Cold dry method, Smooth side out		Cold moist method, Smooth side in		Cold moist method, Smooth side out	
	in.	mm	in.	cm	in.	cm	in.	cm	in.	cm
Plain with one side smooth and the other side textile type of texture	$\frac{1}{8}$	3.18	12	30.48	10	25.40	7	17.78	5	12.70
	$\frac{3}{16}$	4.76	18	45.72	16	35.54	10	25.40	8	20.32
	$\frac{1}{4}$	6.35	27	68.58	24	60.96	15	38.10	12	30.48
	$\frac{5}{16}$	7.94	35	88.90	30	76.20	22	55.88	18	45.72
Plain with both sides smooth	$\frac{1}{8}$	3.18	10	25.40	10	25.40	7	17.78	7	17.78
	$\frac{1}{4}$	6.35	16	35.54	16	35.54	12	30.48	12	30.48
Plain, less dense, with one side smooth and the other side textile type of texture	$\frac{3}{16}$	4.76	20	50.80	18	45.72	12	30.48	10	25.40
	$\frac{1}{4}$	6.35	30	76.20	27	68.58	18	45.72	15	38.10
Plain (tempered) with one side smooth and the other side textile type of texture	$\frac{1}{8}$	3.18	9	22.86	7	17.78	6	15.24	4	10.16
	$\frac{3}{16}$	4.76	16	35.54	14	35.56	9	22.86	6	15.24
	$\frac{1}{4}$	6.35	25	63.50	22	55.88	14	35.56	10	25.40
	$\frac{5}{16}$	7.94	36	91.44	30	76.20	20	50.80	16	35.54
Plain (tempered) with both sides smooth	$\frac{1}{8}$	3.18	10	25.40	10	25.40	7	17.78	7	17.78
	$\frac{3}{16}$	4.76	16	35.54	16	35.54	12	50.48	12	50.48
	$\frac{1}{4}$	6.35	30	76.20	30	76.20	25	63.50	25	63.50

is created by the added binder. Other materials may be added before bonding to obtain special properties such as finishing properties, resistance to abrasion, and increased strength, durability, and utility.

Particleboard is divided into two types, interior and exterior, according to use and into three density grades and two strength classes (*see* Table *W40*). Particleboard has a uniformity of thickness and density, dimensional stability, and excellent glue-bonding characteristics. It does not warp, and it has no surface defects. Its density is about 40 lb/ft^3 (642 kg/m^3). Its major use is as a core for metals, wood veneers, hardboard, or plastic laminates and also for countertops, cabinets, shelves, wall paneling, and all types of doors. It may have its surfaces treated to receive paint or clear or stained finishes. The raw edges of particleboard should not be left exposed. They can be protected with wood veneer, metal, wood strips, plastic laminates, or painted, clear, or stained finishes. *See* Wood Integrants (Plywood).

Wood particles with synthetic resins or other binders are used to make various moldings and molded units. These are formed with heat and pressure and have densities of 45 to 85 lb/ft^3 (720.9 to 1361.7 kg/m^3).

WOOD INTEGRANTS (PLYWOOD)

Physical Properties

Plywood is the general term used to define layers (plies) of wood bonded together permanently, with the grain of one or more layers at 90° to the grain of the intervening layers.

Plywood is usually composed of an odd number of plies. The outside plies are called the faces or the face and the back, and the center ply or plies are called the core. The plies immediately below the faces or the face and the back, with grain laid 90° to them, are called crossbands. The core may be a veneer, solid lumber, or particle board, with the total panel thickness not less than $\frac{1}{16}$ in. (1.59 mm) or more than 3 in. (76.2 mm).

Another type of plywood is based on an even number of plies (four-ply and six-ply). When thicker layers or plies are used, often two adjacent layers have the grain running in the same direction. Thus four-ply would consist of 1-2-1 layers, and six-ply would have a 2-2-2 pattern of plies.

Table W40 Types, densities, strengths, physical properties, and sizes of particleboard

Type of particleboard	Density (grade), minimum average	Strength class[a]	Modulus of rupture, minimum average		Modulus of elasticity, minimum average		Internal bond, minimum average		Size					
									Thickness		Width		Length	
			lbf/in.^{2}	MN/m^2	lbf/in.^{2}	MN/m^2	lbf/in.^{2}	MN/m^2	in.	mm	ft	m	ft	m
Type 1[b]	A. High density: 50 lb/ft^3 (801 kg/m^3) and over	1	2400	16.55	350,000	2413.25	200	2.379						
		2	3400	23.44	350,000	2413.25	140	0.965						
	B. Medium density: between 37 and 50 lb/ft^3 (592.74 and 801 kg/m^3)	1	1600	11.03	250,000	1723.75	70	0.483	$\frac{1}{8}$–2 in $\frac{1}{16}$-in. or $\frac{1}{8}$-in. increments	3.18–50.8 (1.59-mm) (3.18-mm)	4	1.219	8	2.438
		2	2400	16.55	400,000	2758.00	60	0.414						
	C. Low density: 37 lb/ft^3 (592.74 kg/m^3) and under	1	800	5.52	150,000	1034.25	20	0.138						
		2	1400	9.65	250,000	1723.75	30	0.207						
Type 2[c]	A. Same as A above	1	2400	16.55	350,000	2413.25	125	0.862						
		2	3400	23.44	500,000	2447.50	400	2.758						
	B. Same as B above	1	1800	12.41	250,000	1723.75	65	0.448						
		2	2500	17.24	450,000	3102.75	60	0.414						

[a]Strength classification properties based on the panels currently produced.
[b]Mat-formed particle board, generally made with urea-formaldehyde resin binders, suitable for interior application.
[c]Mat-formed particle board made with durable and highly moisture- and heat-resistant binders, generally phenolic resins, suitable for exterior and certain exterior application.

Balanced Construction. The arrangement of plies in pairs on each side of the center ply, or core, so that for each ply there is an opposite, similar, and parallel ply, is called balanced construction. Matching the plies involves considerations of thickness, the kind of wood with reference to shrinkage and density, the moisture content when glued, and the angle or relative direction of the grain. In thin plywood the quality of each of the plies affects the shape and permanence of the final form. In five-ply lumber-core plywood the cross bands affect the quality and stability. (*See* Figure *W29* for construction details.)

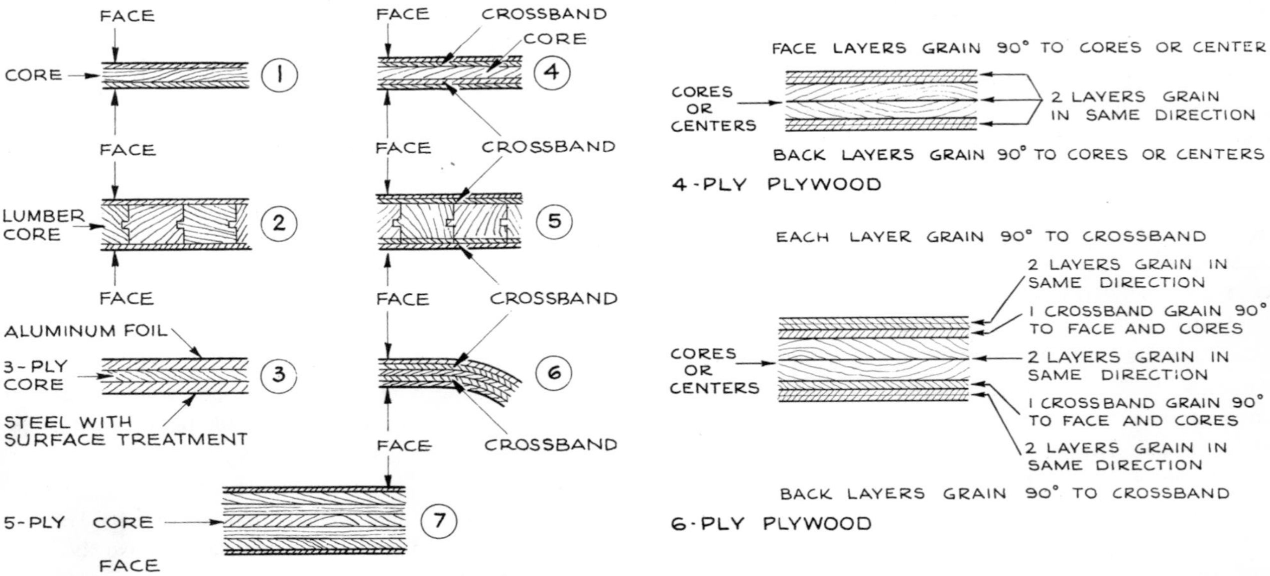

Figure W29 Details of the construction of various types of plywood.

Advantages. The advantages of plywood are (1) the approximate equalization of strength properties along its length and width; (2) greater resistance to checking and splitting; and (3) less change in dimension with change in moisture content. Assuming all plies are of equal thickness, the greater the number of plies for a given thickness, the more nearly equal are the strength and shrinkage properties along the length and width of plywood, and the greater the resistance to splitting.

Commercial Forms. Plywood is available in various types for exterior and interior use, some with metal foil surfaces or with special surfaces for chalkboards, and others with a variety of surface materials such as plastics, textile, or paper applied to one face.

Exterior plywood is made with 100% waterproof adhesives, whereas interior plywood is made with highly moisture-resistant adhesives that are not permanently waterproof and can be installed in any area that is not exposed to continuous moisture or extreme humidity.

Types and Uses

Plywood may be classified into two large categories as (1) Construction and Industrial plywoods, and (2) Decorative Construction and Industrial Hardwood plywoods.

Construction and Industrial Plywood. This category is controlled by U.S. Product Standard PS1 and includes panels made from some 70 or more species of wood, of which approximately 15 or more are hardwoods.

The grade of plywood is determined by the appearance and quality of the veneer, and the type of plywood by the moisture resistance of the glued joints. The designation and construction of each type of Construction and Industrial plywood, including the species of wood, sizes, and major uses, are given in Table *W41*, and the grades of softwood plywood and characteristics are described in Table *W42*.

Table W41 Product standards for exterior and interior plywoods

Type of plywood	Type of adhesive	Characteristics	Grade	Construction: Face[a]	Construction: Back[a]	Construction: Inner layers	Size: Thickness in	Size: Thickness m	Size: Width ft	Size: Width m	Size: Length ft	Size: Length m	Major uses
Product Standard Exterior	Hot-pressed phenol resin	Will retain form and strength when repeatedly wet and dried; for permanent exterior use	A–A	A	A	C							When appearance of both faces is important
			A–B	A	B	C							Same as above, but appearance of one side less important
			A–C	A	C	C	$\frac{1}{4}$ $\frac{3}{8}$ $\frac{1}{2}$ $\frac{5}{8}$ $\frac{3}{4}$	6.35 9.53 12.70 15.88 19.05	4	1.219	8	2.438	Where appearance of one side is important
			B–C	B	C	C							General utility
			C plugged C	C plugged	C	C							Used as underlayment where conditions exist; baths
			C–C	C	C	C	$\frac{5}{16}$ $\frac{3}{8}$ $\frac{1}{2}$ $\frac{5}{8}$ $\frac{3}{4}$	7.94 9.53 12.70 15.88 19.05	4	1.219	8	2.438	Unsanded grade as backing for rough construction
			B–B plyform	B	B	C	$\frac{5}{8}$ $\frac{3}{4}$	15.88 19.05	4	1.219	8	2.438	For concrete forms

Table W41 Product standards for exterior and interior plywoods (*continued*)

Type of plywood	Type of adhesive	Characteristics	Grade	Construction: Face[a]	Construction: Back[a]	Construction: Inner layers	Size: Thickness in.	Size: Thickness m	Size: Width ft	Size: Width m	Size: Length ft	Size: Length m	Major uses
Product Standard Interior	Casein or extended resin adhesive of phenol type	Will retain form and strength when occasionally subjected to a thorough wetting and subsequent normal drying	A–A	A	A	D	$\frac{1}{4}$ $\frac{3}{8}$ $\frac{1}{2}$ $\frac{5}{8}$ $\frac{3}{4}$	6.35 9.53 12.70 15.88 19.05	4	1.219	8	2.438	Where appearance of both faces is important
			A–B	A	B	D							Same as above, but appearance of one side less important
			A–D	A	D	D							Where appearance of one side is important
			B–D	B	D	D							Utility where one smooth side is required
			C plugged D	C plugged	D	C^b or D	$\frac{5}{16}$ $\frac{3}{8}$ $\frac{1}{2}$ $\frac{5}{8}$ $\frac{3}{4}$	7.94 9.53 12.70 15.89 19.05	4	1.219	8	2.438	Built-ins and tile backing
			C–D	C	D	D							Unsanded general utility standard
			B–B	B	B	D	$\frac{1}{4}$ $\frac{3}{8}$ $\frac{1}{2}$ $\frac{5}{8}$ $\frac{3}{4}$	6.35 9.53 12.70 15.88 19.05					General utility
			N–D	N	D	D							Built-in furniture where appearance of one face is important
			Underlayment	C plugged	D	C^b or D							Underlayment free of any moisture
			N–N N–A N–B	N	N A or B	C	$\frac{3}{4}$	19.05	4	1.219	8	2.438	Furniture, built-ins, cabinet doors, etc., with natural finish
			2–4–1	C plugged	D	C or D	$1\frac{1}{8}$ $1\frac{1}{4}$	28.58 31.75	4	1.219	8	2.438	Subflooring or underlayment, 4-ft (1.219-m) spans

[a]See Table W42 for classification of grades for Construction and Industrial plywoods.
[b]Veneer immediately adjacent to face shall be Grade C or better.

The species of woods are grouped on the basis of stiffness and U.S. Product Standard PS1 into five classifications: Groups 1, 2, 3, 4, and 5. The strongest woods are in Group 1, the weakest in Group 5 (see Table *W43*).

All plywood has a grade trademark on back and edge. A typical back-stamp and edge-mark are shown in Figure *W30*. Structural plywoods are made for construction where nail-bearing, shear, compression, tension, etc., are of maximum importance. Plywood types Exterior C-C, Structural I and II, and Standard Engineering Grades are given two numbers with a slash between them on the grade trademarks; the number on the left indicates spacing in inches for supports when used as roof decking with grain across supports, and the number on the right indicates spacing in inches for supports when the panel is used for subflooring with face grain across supports.

Table W42 Classification of grades[a] for softwood plywoods

A	B	C plugged	C	D	N
Smooth and paintable; may be more than one piece; neatly made repairs permitted	Solid surface veneer; circular repair plugs and tight knots permitted	Improved C veneer underlayment grade	Minimum veneer permitted in exterior type; knotholes to 1 in. (25.4 mm), splits and plugs, or other repairs permitted	Used only for inner plys and the back of interior type	Natural finish veneer; select all-heartwood; free of open defects

[a]These grades are further limited in size of knots, open splits, pitch pockets, knotholes, sander skips, patching defects, and wood discoloration.

Table W43 Classification of wood species for plywood

Classification of plywood	Species of wood
Group 1	Birch (sweet and yellow), Douglas fir 1,[a] larch (western), maple (sugar), pine (Caribbean), pine (southern, loblolly, longleaf, shortleaf, slash), and tanoak
Group 2	Cedar (Port Orford), Douglas fir 2,[b] fir (California red, grand, noble, Pacific silver, white), hemlock (western), lavan (almon, bagtikan, red lavan, tan gile, white lavan), maple (black), mengkulang, meranti, pine, (pond, red, western white), spruce (Sitka), sweetgum, and tamarack
Group 3	Alder (red), cedar (Alaska), pine, (jack, lodgepole, ponderosa, spruce), redwood, and spruce (black, red, white)
Group 4	Aspen (big tooth, quaking), birch (paper), cedar (incense, western red), fir (subalpine), hemlock (eastern), pine (eastern white, sugar), poplar (western[c]), and spruce (Engelmann)
Group 5	Fir (balsam) and poplar (balsam)

[a]Douglas fir 1 (Washington, Oregon, California, Idaho, Montana, Wyoming, British Columbia, Alberta).
[b]Douglas fir 2 (Nevada, Utah, Colorado, Arizona, New Mexico).
[c]Black cottonwood.

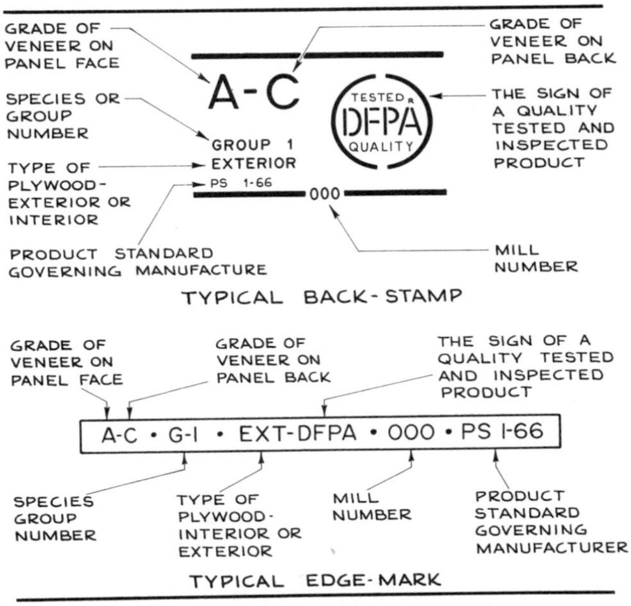

Figure W30 Examples of grade trademarks for plywood.

Overlay Plywoods. Plywoods made with overlays of paper pulp sheets or of paper or fabric impregnated with synthetic resins are graded as shown in Table *W44*, which also lists the species of wood, characteristics, sizes, and major uses of overlay plywoods. When these overlay plywoods are used, a full description should follow the grade designation by name, for example, Medium Density BB with B grade inner plies, Surfaced 1 Side only.

Several other types of plywood are also available with paper pulp sheets, cork, hardboards, plastic laminates, metal sheets, and linoleum on one or both surfaces. These are used for cabinets, doors, furniture tops, chalk boards, display boards, etc.

Overlay plywoods must meet the specific needs of their use in construction; therefore the manufacturers should be contacted when a particular special type of surface or finish for concrete forms is required.

Hardwood Plywoods. Four types of hardwood plywood are specified in the Commercial Standard grade: Technical, Type I, Type II, and Type III. The difference between the types is based on the resistance of the glue bond to severe service conditions, except for Technical, which differs from Type I in the permissible thickness and arrangement of the plies. Technical and Type I are similar to Exterior Type Construction and Industrial

Table W44 Overlay and special types of plywoods

Type of plywood	Source	Type of adhesive	Characteristics	Grade	Construction			Size						Major use
					Face[a]	Back[a]	Inner layers	Thickness in.	Thickness mm	Width ft	Width m	Length ft	Length m	
High-Density Overlay Exterior	Manufacturers of construction and industrial plywood	Hot-pressed phenol resin	Hard, glossy surface, high resistance to chemical attack, permanent exterior use	HDO	A	A	C							No further finishing with paint or varnish necessary
					B	B	C							Same as above except finish not as good
High-Density Overlay Ex-Exterior for concrete forms					B	B	C or C plugged	$\frac{5}{16}$ $\frac{3}{8}$ $\frac{1}{2}$ $\frac{5}{8}$ $\frac{3}{4}$	7.94 19.53 12.70 15.89 19.05	4	1.219	8	2.438	For concrete forms
Medium-Density Overlay Exterior				MDD	B	B or C	C							Smooth, uniform surface suitable for high-quality paint finishes
Overlay of hardboard, metal sheets, reinforced Fiberglas, plastic laminates, resin-treated papers, cork, or linoleum	Produced by remanufacturers	Depends on type of overlay	Overlay adds strength, stiffness, abrasion resistance, and decorative finishes	Various types of exterior or interior plywoods				Plywood thickness including overlay thickness		4	1.219	8	2.438	Doors, cabinets, chalkboards, built-ins; where a permanent surface or surfaces are important
Plywood, surface cut with grooves, wire-brushed and sandblasted	Manufacturers of construction and industrial plywoods	Hot-pressed phenol resin	Embossed, grooved, and other types of textures	Both exterior and exterior types of plywoods used as base for overlayment				$\frac{3}{4}$ $\frac{7}{8}$	19.05 22.23	4	1.219	8	2.438	Interior paneling and exterior siding

[a]Face and back grade refer to the veneer directly underlying the overlaid surface.

plywood in durability. Type II is similar to Interior Type Construction and Industrial plywood, and Type III has good dry strength but no water resistance. The grades for veneer are 1, 2, 3, and 4 in order of descending quality. Grade 1 is usualt described separately for each species, as grain pattern, small knots, burl, etc., are the governing factors in relation to end use. Grades 2, 3, and 4 are the same for all species of hardwoods. Grades 2 and 3 are usually, and Grade 4 is always, used for inner plies. The plywood is grade-designated by grade of veneer used on the face and back, for example, Grade 1-1 or Grade 1-2.

Plywood, particularly hardwood, is available in a wide variety of prefinishes, which may be clear or pigmented. It is also available fire-retarded and treated against decay, insects, and vermin.

Special Plywoods. Plywoods are made for special purposes such as marine vessels and aircraft. These require special glues, types of wood, and mechanical properties. Plywood is also made with various surface patterns such as small grooves, evenly spaced grooves, V-joints, and blasted surface. Prefinished plywood is supplied in various hardwood and softwood veneers, either stained, colored, or natural. Prefinished hardwood plywoods are available in vertical panels of varying widths. These panels have various types of side interlocking systems for joining that permit simple installation. (*See* Table *W44*).

Application

Condensed Checklist

1. The following should be checked to make sure that the plywood in question meets the requirements and limitations of the end use: official designation, interior or exterior use, and the grade of the plywood and of the veneers comprising it.
2. Local availability of the type and grade of plywood to be used should be checked.
3. Manufacturers of plywood should always be consulted to obtain current information on any new textures, species of wood, and new types of plywood available.
4. Manufacturers of plywood should be consulted for the best method of fastening and securing the plywood to meet design and end-use requirements and limitations.
5. Methods of joining plywood when used on the exterior (*see* Figures *W31* to *W34*).

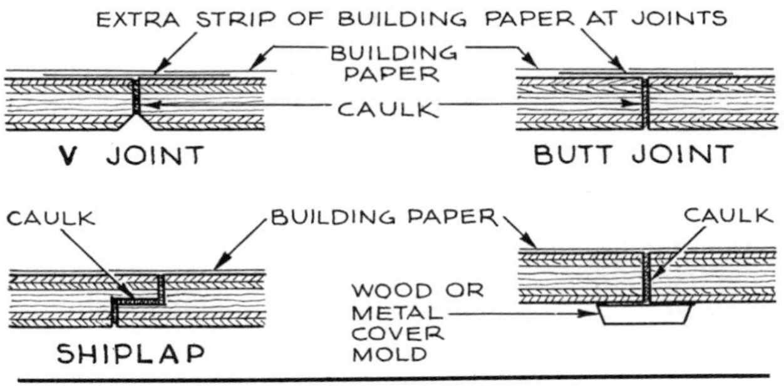

Figure W31 Vertical exterior joints.

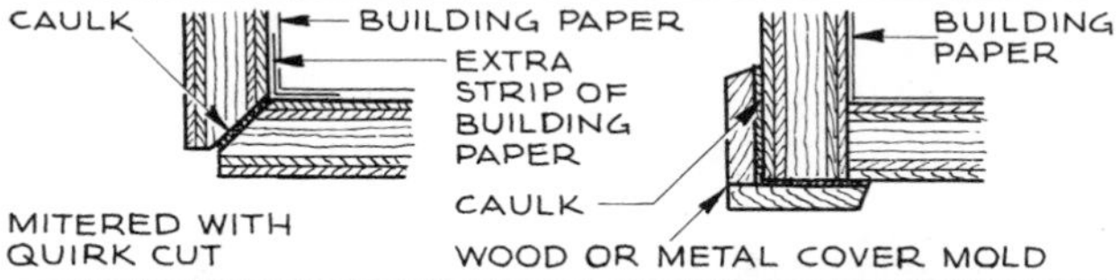

Figure W32 External exterior corners.

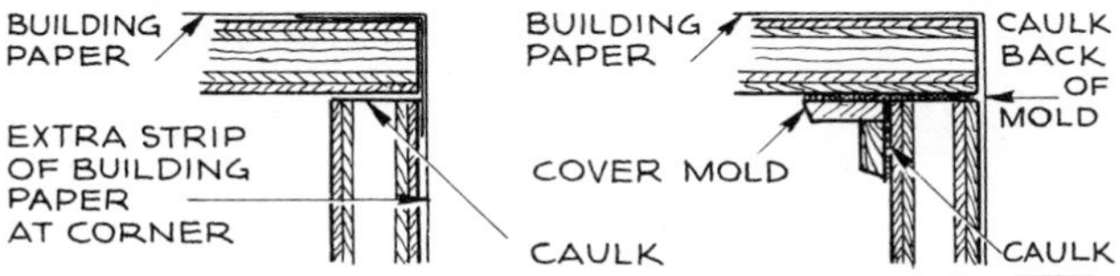

Figure W33 Internal exterior corners.

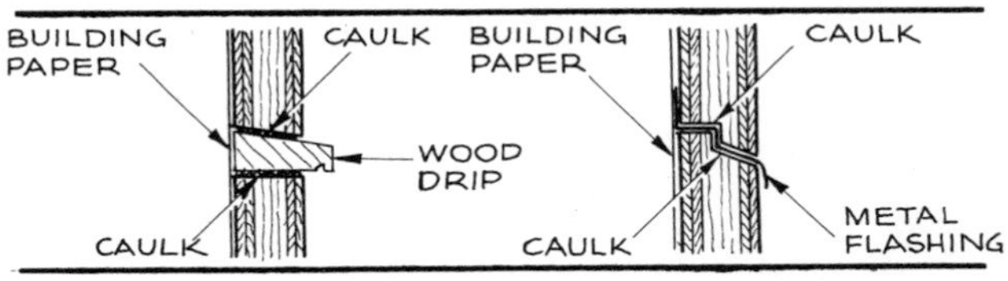

Figure W34 Horizontal exterior joints.

6. Methods of joining plywood when used on the interior for wall surfacing, cabinet, and millwork (*see* Figures *W35* to *W41*).
7. When prefinished plywood is used for interiors, the architect should always check what methods of fastening (e.g., blind nailing or metal clips) can be used and what types of finishes and species of hardwood are available and in what lengths. Generally, interior finish plywood is available $\frac{1}{4}$ in. (6.35 mm) thick, $16\frac{1}{4}$ in. (412.75 mm) wide, and in 6-, 7-, and 8-ft (1.829-, 2.134-, and 2.438-m) lengths.
8. The dimensions of areas where plywood is to be used should always be checked in drawings, and during construction to see that supports meet the sizes of the plywood, that all joints of plywood fall on supports, and that the areas are, as far as is practicable, multiples of stock plywood sizes.

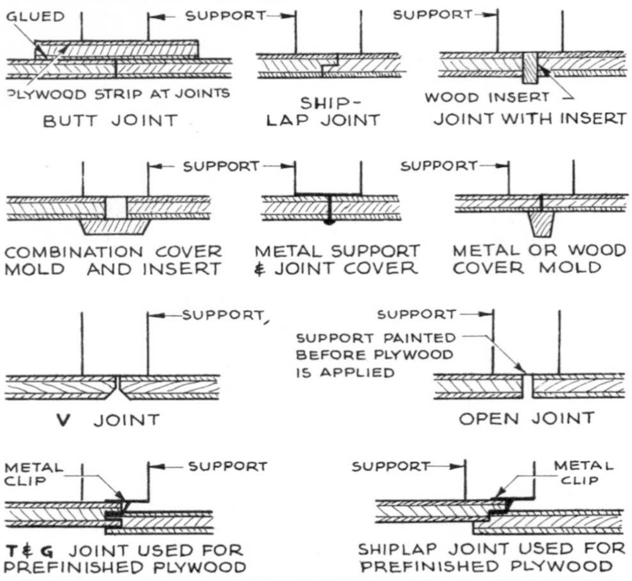

Figure W35 Typical details of interior joints.

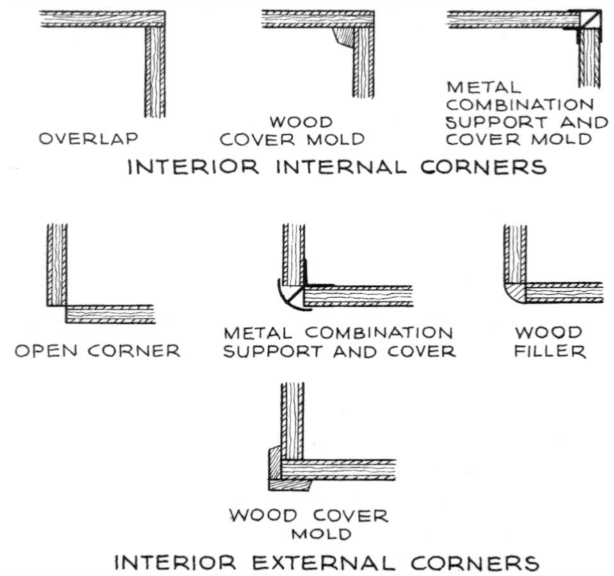

Figure W36 Details of interior corners.

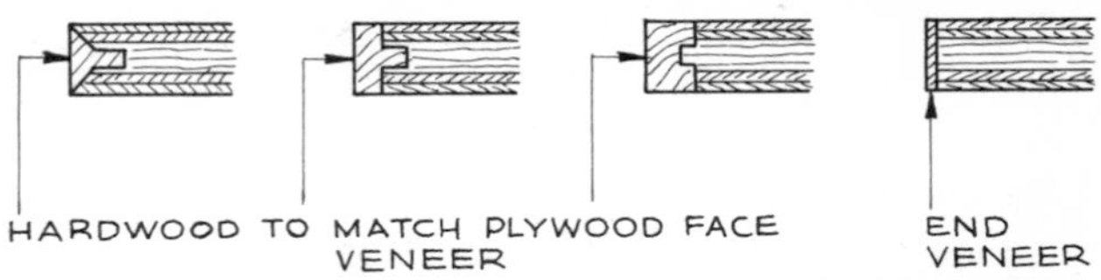

Figure W37 Treatment of end grains for cabinet and millwork.

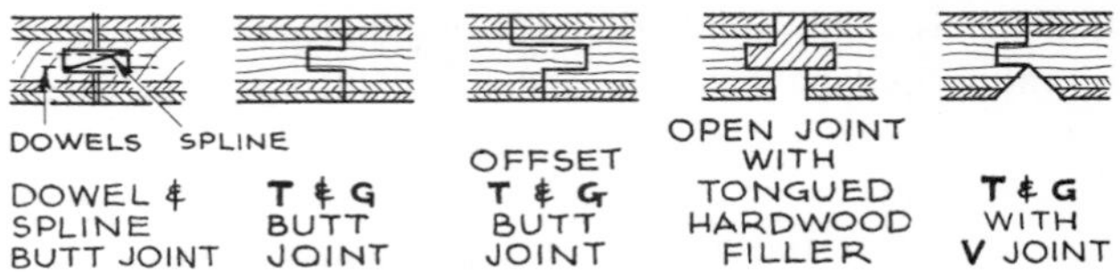

Figure W38 Vertical and horizontal joints for cabinet and millwork.

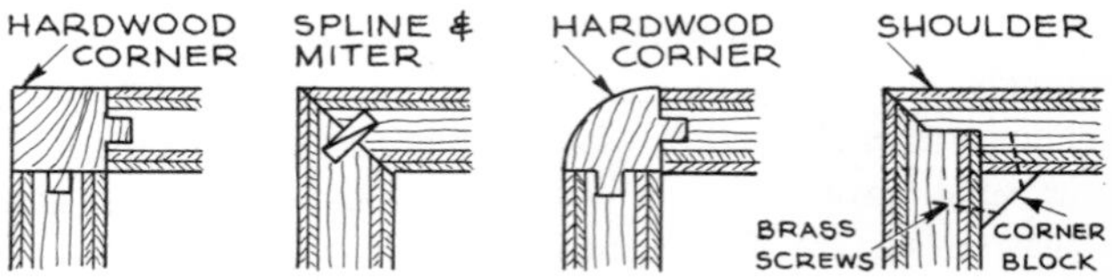

Figure W39 Treatment of external corners for cabinet and millwork.

Figure W40 Treatment of internal corners for cabinet and millwork.

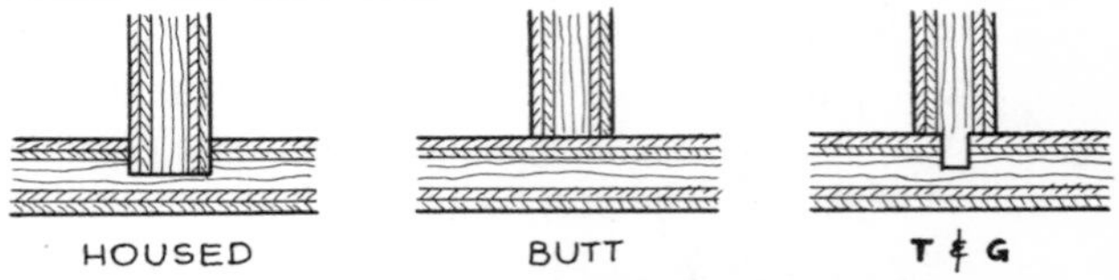

Figure W41 Treatment of perpendicular joints in cabinet and millwork.

9. When plywood is used to construct built-up or laminated structural forms, structural engineers should always be consulted.
10. When using overlay types of plywood, one should check with the manufacturer regarding availability and methods of installation.
11. Special plywoods for interior paneling and exterior siding in construction should be checked for type of plywood (exterior or interior), texture, embossing and grooves, and methods of installation.
12. When prefinished plywood is used in construction, the manufacturers should be consulted as to types of finishes, types of woods, and method of application.
13. When using fire-retarded plywoods, one should check municipal, city, state, and any other codes that have jurisdiction over fire-resistance ratings.

14. When using plywood with special characteristics such as chemical resistance and resistance to decay, insects, and vermin, the manufacturers should be consulted for availability and methods of application.

Conditions Favorable to the Use of Plywood

1. Where a strong, thin, nailable, stable, durable material is needed for enclosing (sheathing) the exterior walls, floors, and roofs of a frame building.
2. Where a weatherproof, durable material available in a variety of textures is desired for the exterior finish of a frame building.
3. Where a strong, durable, easily installed finish material for interior walls and ceilings is desired in a wood-grained decorative texture or plain surface.
4. Where a strong, decorative, durable material is desired for cabinetwork and millwork.
5. Where a flat, smooth, even surface with a minimum of joints is desired as underlayment for resilient flooring, carpets, etc.
6. Where natural or pigmented prefinished interior wall paneling is to be installed in construction.
7. Where a special permanent exterior or interior finish is required in construction for decoration, performance, chalkboard, etc., and for concrete forms.

Conditions Unfavorable to the Use of Plywood

1. Where it will be in contact with the ground or with water in a building.
2. Where strong acids, alkalis, or chemical fumes are present.
3. Where fire resistance is of major importance, and fire retardance is not sufficient.

History and Manufacture

Veneer (thin slices of wood) was known from ancient times and used for decorative inlay work and later as veneer on furniture. The only adhesives then available were animal glues. Rotary cutting, a process made possible by the use of power in combination with improved adhesives and the discovery of new synthetic ones, led to the development of plywood as we know it.

In the early 1900s starch and casein glues were developed, and shortly before World War I these were supplemented by vegetable protein glues. With these new adhesives plywood could be produced in quantity. When synthetic adhesives were introduced, waterproof plywood became possible. Today adhesives can be tailor-made to meet any condition of lamination for various materials, and with the continuing advances in the field of plastics and synthetic rubbers, many new laminated materials can be anticipated.

How Veneer Is Cut. Veneer can be cut from the log by various methods, as shown in Figures *W42* and *W43*. The rotary method is used for virtually all construction plywood; in general, all the other methods shown are for hardwoods.

METHOD OF CUTTING	NAME OF TYPE OR CUTTING	CHARACTERISTICS
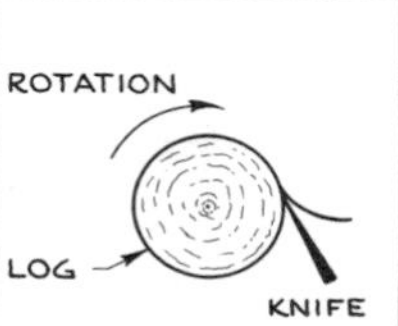	ROTARY METHOD	DIFFICULT TO MATCH GRAIN. SOME DECORATIVE HARDWOOD. METHOD USED FOR VIRTUALLY ALL CONSTRUCTION PLYWOOD
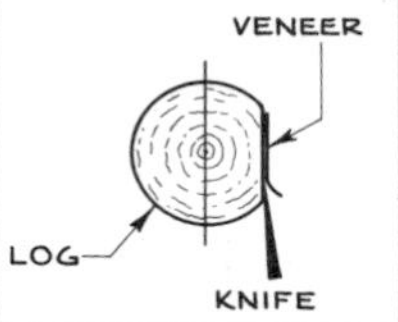	PLAIN SLICING METHOD	MATCHING OF GRAIN RELATIVELY EASY. FAIRLY NARROW PIECES. DECORATIVE HARDWOOD.
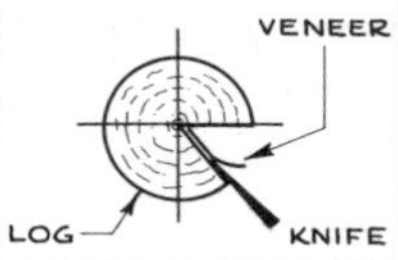	QUARTER SLICING METHOD	NARROW PIECES WITH GRAIN RUNNING IN ONE DIRECTION, DECORATIVE HARDWOOD
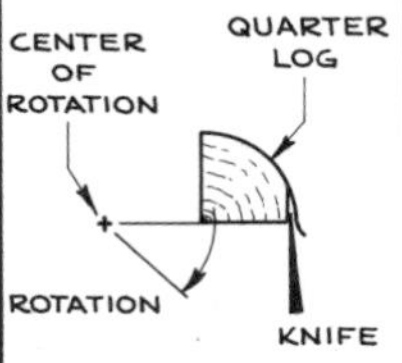	RIFT CUT (GENERALLY LIMITED TO OAK)	NARROW PIECES WITH FAIRLY UNIFORM GRAIN. HARDWOODS
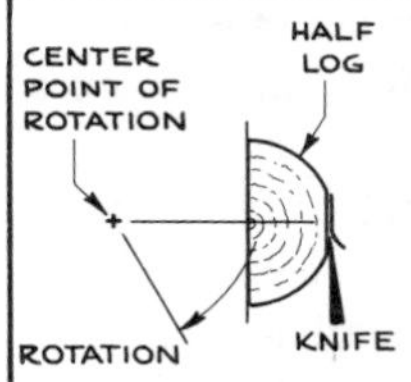	HALF-ROUND SLICING (USED FOR RED OAK)	SIMILAR TO ROTARY. CUT GRAIN CHARACTERISTICS, HARDWOODS
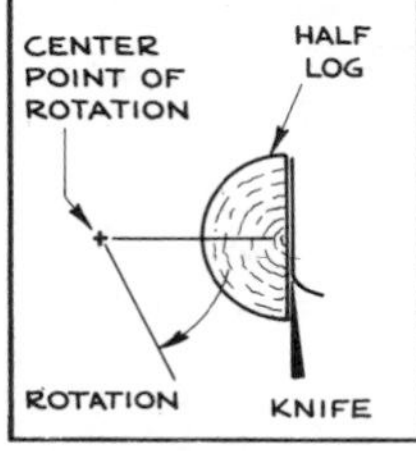	BACK-CUT	GRAIN CHARACTERISTICS SMALLER THAN PLAIN-SLICING METHOD DECORATIVE HARDWOODS

Figure W42 Methods of cutting veneers.

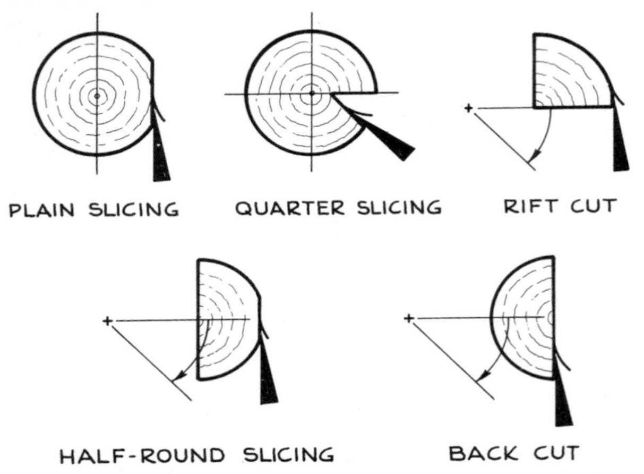

Figure W43 Methods of cutting veneers other than rotary cutting.

WOOD JOINING

Physical Properties

The species of wood, specific gravity, moisture content, direction of grain, thickness and depth of the pieces to be joined, imperfections, and character of the surface of the wood—all influence the method used for joining. Location of the joint on the exterior or interior of a building affects not only the actual shape of the joint but also the type of metal, adhesive, caulking, and plastering used to make it.

Types and Uses

In general, the joining of wood can be divided into two large categories: joints for the exterior and joints for the interior. In either location, wood may be joined with a wide variety of fastenings, among them nails, spikes, screws, bolts, lag screws, pins, metal connectors of various shapes, and adhesives. Each type of fastening requires that the design of the joint be adapted to the varying strength properties of the wood along and across the grain and to the dimensional changes that occur with changes in moisture content.

Design Factors

Not only are the structural aspects of joining wood important in construction, but the design aspects of wood joints are also important. Figures *W44* through *W47* show the basic wood joints in which design is a factor. (For joining plywood, *see* Wood Integrants.) These

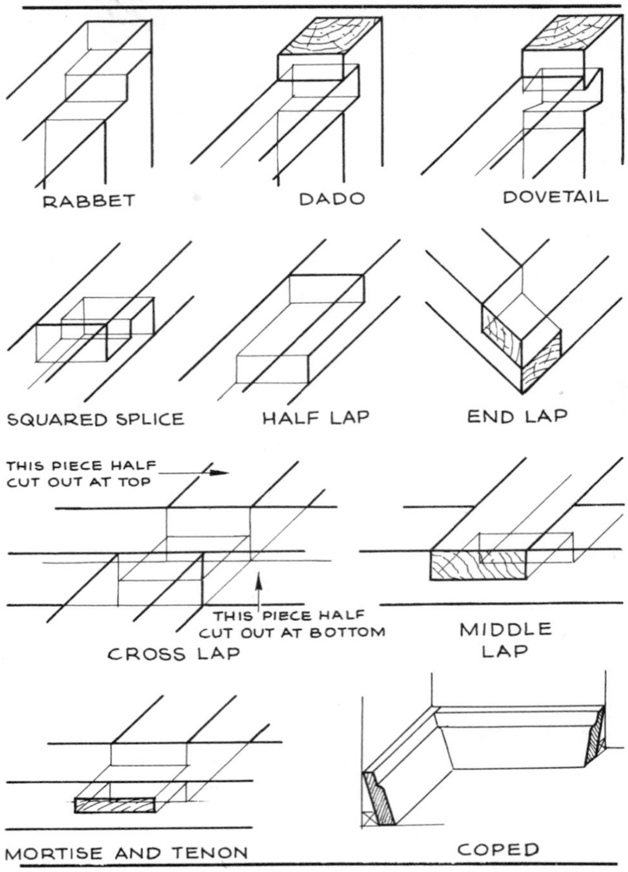

Figure W44 Typical wood joints.

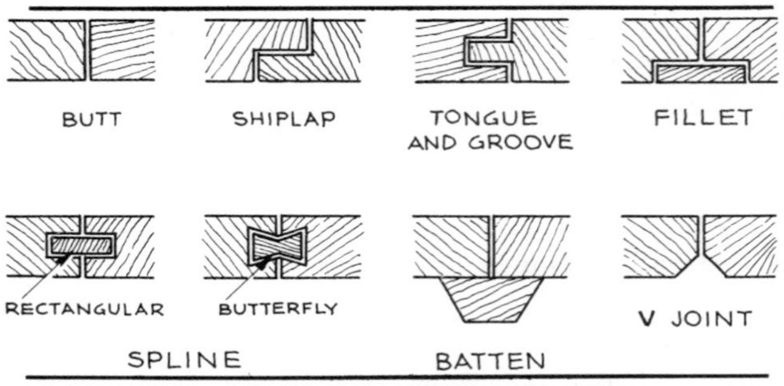

Figure W45 Horizontal and vertical joints.

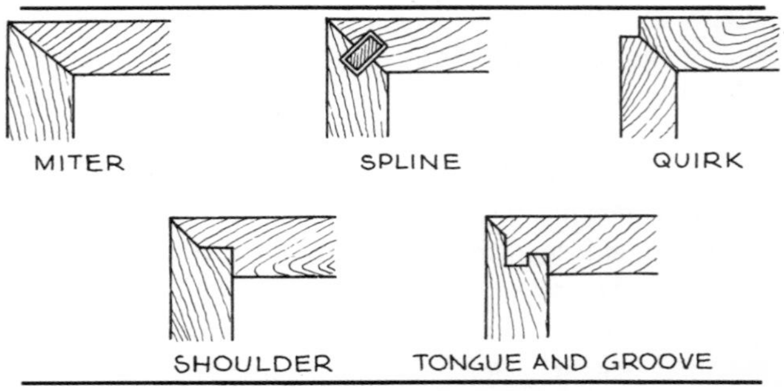

Figure W46 Miter joints.

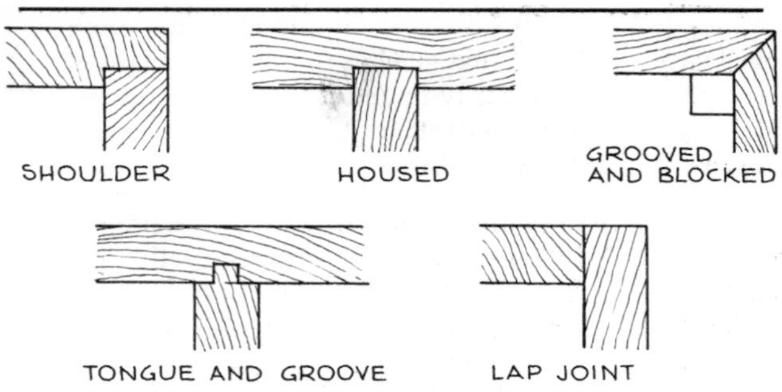

Figure W47 Miscellaneous joints.

joints can be glued, nailed, screwed, and doweled. The method of fastening is determined by the finish and also by the use of the joint. In general, the fastening device should be concealed by having the head covered by the next piece of wood or the head countersunk below the finished surface of the wood and the hole filled with a plug, putty, or a filler compound.

The various fastening devices for wood, other than those used for structural joints where stresses must be met, are discussed in the following paragraphs.

Nails. Nails usually give more strength to joints when driven into the side grain (perpendicular to the wood fibers) than when driven into the end grain (parallel to the wood fibers). They should be used so that their lateral resistance is utilized rather than their withdrawal resistance. The withdrawal resistance is affected by the type of head and point, type of shank, length of time that the nail remains in the wood, type of surface coating, density of the wood, species of wood, diameter of the nail, moisture content changes in the wood resulting from its end use, direction of driving, and depth of penetration. To increase withdrawal resistance beyond that of a plain nail with a smooth shank, some technique such as coating the surface or roughening and deforming the shank of the nail is used. Nail heads give larger bearing surfaces against the wood fibers. The nail points should be checked in relation to species of woods. Long, sharp points are usually good for soft woods but cause splitting in hard woods. The blunt or flat point reduces splitting but tears the wood fibers and thereby reduces withdrawal resistance. In fine finish woodwork, predrilling to make holes of a smaller size than the nail is recommended. The metal of which the nail is made is an important consideration on the exterior as corrosion will cause staining. (*See also* Nails.)

Screws. Screws have stronger withdrawal resistance than nails. The effective length of the screw is limited by the length at which the screw fails in tension. This length decreases as the density of the wood increases. Screws placed in the side grain of wood vary in their resistance to withdrawal in direct relation to the square of the density (specific gravity) of the wood. As with nails, predrilling and lubricating the surface of the screws are recommended.

For structural joints using spikes, lag screws, bolts, and metal connectors, the strength of the joint depends on the size and type of joining device, species of wood, thickness and width of lumber, distances from the ends, spacing, and direction of the stresses with respect to the grain direction of the wood. For any structural joining it is advisable to consult structural engineers to determine the correct method of joining.

For other types of fastening devices for wood, *see* Anchors; Hangers; Straps: Ties.

Adhesives. The joining of wood with adhesives depends on (1) the species of wood and its preparation, (2) the kind and quality of the adhesive and its preparation for use, (3) the process of application, (4) the types of joints, (5) the conditioning of the joints, and (6) the end use or function of the joints. Table *W45* lists the various species of woods and their properties for joining with adhesives (gluing properties).

The moisture content of the wood at the time of joining affects the final strength of the joint, the development of checks in the wood, and the warping of joined members. The most satisfactory moisture content of wood at joining is usually that which, when increased by the moisture content of the adhesive, approximately equals the average moisture content the joined members will have in their end use.

Wood surfaces that are to be joined with adhesives should be smooth and true, free from machine marks, chipped or loose grain, and any other surface irregularities. In joining, a film of adhesive unbroken by air bubbles or foreign particles should be in contact with the wood surfaces over the entire joining area.

Table *W46* lists and describes the types of adhesives in common use for joining wood.

Adhesives for joining wood and metal or other materials are usually combinations of a thermosetting resin (generally the phenolic type) and a thermoplastic resin or elastomer (generally a polyvinyl resin or synthetic rubber). The vehicle may be water, alcohol, ethyl acetate, or some other solvent.

Several methods of application are possible. One or both surfaces to be joined may be covered with the adhesive; when the solvent has evaporated, the joining is accomplished with heat and pressure. Or joining may be

Table W45 Species of woods classified by properties for joining with adhesives

Type of wood	Wood that glues very easily with different adhesives under a wide range of gluing conditions	Wood that glues well with different adhesives under a moderately wide range of gluing conditions	Wood that glues satisfactorily under well controlled gluing conditions	Wood that is difficult to glue and requires close control of gluing conditions or special treatment to obtain best results
Hardwoods (deciduous)	Aspen, chestnut, cottonwood, yellow poplar, black willow	Alder (red), basswood, butternut[a], elm (American[a], rock[a]), hackberry, magnolia,[a] mahogany,[a] sweet gum	Ash (white)[a], cherry (black)[a], maple (soft)[a], oak (red,[a] white), hickory (pecan), sycamore,[a] tupelo (black, water) walnut (black)	Beech (American), birch (yellow)[a], hickory (true)[a], maple (hard)[a]
Softwoods (evergreen)	Bald cypress, red cedar[b], fir (white, grand, noble, Pacific, silver, California red), larch	Red cedar (eastern),[a] Douglas fir, hemlock (western) pine (eastern white)[b], pine (southern yellow, (ponderosa)	Alaska cedar[a]	

[a]Glued more easily with resin type than with nonresin type of adhesive.
[b]Glued more easily with non resin type than with resin type of adhesive.

Table W46 Adhesives used in joining wood

Type of adhesive	Form	Preparation	Application	Properties: Advantages	Properties: Disadvantages	Major use
Animal	Dry or liquid	Dry: mixed with water, soaked, and melted. Liquid: applied from container	Pressed at room temperature	High dry strength; little organic staining of the wood	Low resistance to moisture or damp conditions	Furniture, millwork cabinetwork
Blood protein	Dry	Mixed with cold water, lime, caustic soda, and other chemicals, sometimes combined with soybean protein	Pressed at room temperature or in hot presses at 240°F (115.56°C) or higher	High dry strength; moderate resistance to water and damp conditions	Not suitable for exterior	Interior-type softwood plywoods
Casein	Dry	Mixed with water	Pressed at room temperature	Moderately high dry strength; moderate resistance to water and damp conditions	Not suitable for exterior	Laminated lumber for interior use
Vegetable protein (usually soybean)	Dry	Mixed with cold water, lime, caustic soda, and other chemicals, sometimes combined with blood protein	Pressed at room temperature; frequently hot-pressed	Moderate to low dry strength; moderate to low resistance to water and damp conditions	Not suitable for exterior	Interior-type softwood plywoods
Urea resin	Dry or liquid; may be blended with melamine or resins	Dry: mixed with water; hardeners, fillers, and extenders may be added to either dry or liquid form	Applied at room temperature or hot pressed at about 250°F (121.11°C)	High in both dry and wet strength; moderately durable under damp conditions	Not suitable for exterior use; low resistance to temperatures above 120°F (48.89°C)	Interior types of hardwood plywood, particleboard, and interior flush doors

Table W46 Adhesives used in joining wood (*continued*)

Type of adhesive	Form	Preparation	Application	Properties: Advantages	Properties: Disadvantages	Major use
Melamine resin	Powder with or without catalyst	Dry or with catalyst	Mixed with water and applied at room temperature; curing requires temperature of 250–300°F (121.11–148.89°C)	High in both dry and wet strength; very resistant to moisture and damp conditions	Not suitable for exterior	Primarily as fortifier for urea resins for interior types of hardwood plywood
Phenol resins	Dry, liquid, or film	Dry: with alcohol or water, film form: used as received	Dry: at room temperature, hot-press at 260–300°F (126.67–148.89°C)	High in both dry and wet strength; very resistant to moisture and damp conditions; resistant to high temperatures	Can discolor wood	Exterior softwood plywood and particleboard
Resorcinol resin and phenol-resorcinol resins	Liquid with hardener	Mixed with hardener	Applied at room temperature, cures at 70–150°F (21.11–65.56°C)	High in both dry and wet strength; very resistant to moisture and damp conditions; resistant to high temperatures	Can discolor wood	Laminated lumber and assembly joints that must withstand severe service
Polyvinyl acetate resin emulsions[a]	Liquid	Used from container	Applied and pressed at room temperature	High dry strength; low resistance to moisture and high temperatures	Not suitable for exterior	Furniture, flush doors, assembly of interior panel systems
Rubber-base contact adhesives	Liquid (neoprene rubber base in solvent or water emulsion)	Used as received	Applied to both surfaces and dried before pressing; instantaneous bonding	Initial strength immediately upon pressing; strength increases over period of weeks; water resistant	Not suitable for exterior	For nonstructural bonds for job-site bonding applications
Rubber-base elastometric construction adhesives	Puttylike (synthetic or natural rubber in a solvent)	Used as received	Extruded in beads and ribbons with or without supplemental nailing	Develops strength slowly; water resistant	Not suitable for exterior	Lumber and plywood to joists or wall studs; laminating plaster board, insulation foams, hardboard, etc.; assembly of panel systems
Thermoplastic synthetic resins	Solid chunks, pellets, ribbons, rods, or films; solvent-free	Melted for spreading	Requires special equipment for controlling bonding conditions	Rapid bonding; moisture resistant	Lower strength than conventional adhesives; minimal penetration	Edge bonding, films and paper overlays, laminations
Epoxy resins	Two-part liquids	Resin and curing agent liquids; pot life and cure conditions vary widely with composition	Applied at room temperature or higher temperatures	Good adhesion to metals, glass, plastics, wood, and wood products	Requires careful control of pot life and cure	Bonding wood to wood; bonding metals, masonry, and plastics to themselves or to wood

[a]Modified vinyl-resin emulsions are available which involve addition of a curing agent at time of use which greatly improves resistance to heat and moisture.

performed in two steps with different adhesives. The first step here is to spread the material on the metal or other material and to cure at high temperatures, and the second step is to join the metal or other material to the wood, using a resin-type adhesive that sets at room temperatures.

Casein and rubber latex combinations are used where high strength and durability are not essential. Their advantage is that they are applied and cured at room temperature.

In general, for all joining of wood with adhesives, the manufacturers of both the adhesives and the materials to be joined should be consulted to learn the best procedure and the correct adhesive to meet the design and end-use conditions and requirements.

Applications

In construction the joining of wood can be divided into two major categories: (1) joining done at the site, and (2) joining done under mill, shop, or factory conditions. These two groups can then be classed, as already described, into (1) exterior joints and (2) interior joints.

In general, the exterior joints made at the site should always be detailed from the viewpoint of maintenance, watertightness, effects of weathering, and overall design. Structural joints must meet structural requirements first and design aspects second. Interior joints, on the other hand, are detailed almost entirely from the viewpoint of design, that is, how they will look when finished and what effect they will have on the interior design.

All mill, shop, or factory joining can be very closely controlled, and joints can be obtained to meet almost any requirements of both strength and design. The closer a joint approaches invisibility, however, the more difficult and costly it is to obtain. This is equally true of joining done on site.

Condensed Checklist

1. The species of wood, type of fastening device, adhesive, and method of joining must always be checked to see that the joint meets design and end-use requirements.
2. Shop drawings should always be required for millwork, cabinetwork, or built-in furniture, and all fastening devices, adhesives, and types of joints must be shown and described.
3. The adhesives used must not stain the wood.
4. Structural engineers should always be consulted regarding joints that are required to meet structural loads and stresses.
5. For good joints, the lumber must be well seasoned and its moisture content not too great. For very fine work, the moisture content should be specified.

History and Manufacture

The earliest joining of wood was accomplished by using pieces of very hard wood as pins (nails or dowels) and cutting or notching the wood to make interlocking members into which the pin could be inserted. With the discovery of bronze, crude nails and clamps were used instead of wood pins. With the discovery of iron and therefore of the nail as we know it today, the techniques of joining wood were completely changed. A very limited number of adhesives such as asphalt and crude animal and vegetable glues were known to the ancients and were used primarily for wood inlay in furniture. Until the discovery of steam power, wood was joined by various combinations of nailing, notching, and limited gluing. Once the use of mechanical power permitted sawing lumber in quantity, there was a growing and continuous need for improved and diversified methods of joining lumber.

Today, because of the synthetic adhesives, the more recent types of fasteners with special coatings, the use of new shapes and new metals, and the improvements in lamination and treatment of wood, the joining of wood has become highly specialized into particular fields such as structural joints, furniture and millwork joints, and joints of dissimilar materials such as glass and wood or metal and wood.

WOOD MOLDINGS, TRIM, AND ORNAMENTAL SHAPES

Physical Properties

The wood used for moldings, trim, and ornamental shapes is chosen from the better grades of both deciduous and evergreen species. The species of wood should have the following properties: it should be easily worked both with hand and power tools; it should not warp, expand, or check easily; it should be easily nailed, screwed, doweled, and joined with adhesives; and, most important, it should be easily finished, painted, stained, etc. There are available a large variety of stock shapes, turnings, etc., for moldings, trim, and other ornamental forms of wood.

Synthetic resins combined with wood fibers, sawdust, or wood flour are formed with heat and pressure into plain and decorative moldings, trim, and miscellaneous shapes.

Types and Uses

In today's construction, moldings, trim, and ornamental shapes are delegated to a minor role. Their principal use is to cover the joint at the intersections of materials. Where two different materials or the same materials meet or intersect, expansion and contraction caused by changes in humidity and temperature and the settling that normally occurs in a building cause these intersections to open and close. The resulting cracks, even if not unsightly, allow the infiltration of air, water, etc., and must be covered or closed in some way to eliminate these difficulties. Because of the ease with which wood can be installed, it is commonly employed for this purpose.

Only a limited number of the stock shapes for moldings, trim, and ornamental woodwork are illustrated in Figures *W48* and *W49*; they include the principal types from the viewpoint of application or end use.

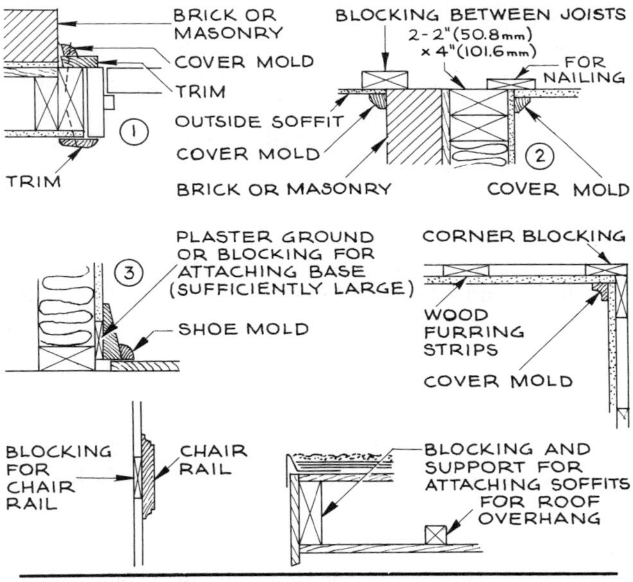

Figure W48 Typical details for applying various wood moldings and trim.

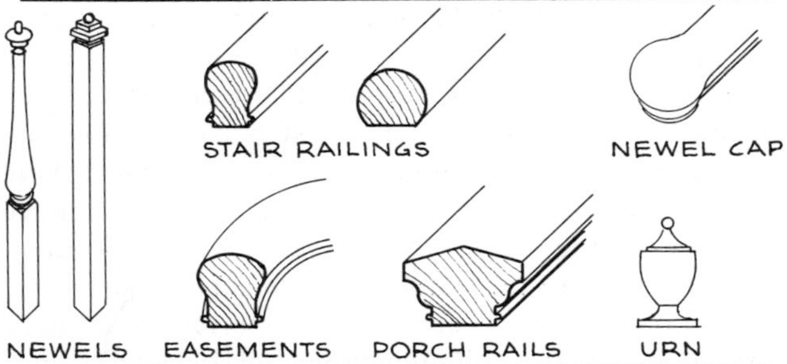

Figure W49 Wood ornamental shapes.

Classical columns and caps and designs of various periods in brackets, fireplace mantels, ornamental entrance doors and frames, and other shapes are also available as stock items. (*See also* Stairs; Wood Gutters.)

Application

Condensed Checklist

1. Moldings always require blocking or some other method by which they can be applied securely (*see* Figure *W50*).
2. Local lumber yards and manufacturers of moldings, trim, and ornamental shapes should always be checked for the types and shapes carried in stock.
3. When designing special shapes, one should always check that the shape in question can be but from a stock size of lumber and that there are millworking establishments in the locality where these shapes can be made.
4. The species and grade of lumber should always be specified for the moldings, trim, and ornamental shapes being specially made.
5. When prefabricated ornamental shapes are being used, the grade and species of lumber used in manufacture should be checked and dimensional drawings of prefabricated ornamental shapes should be obtained so that adequate blocking and clearance can be allowed.
6. One should always check that the moldings and trim selected will meet the requirements and limitations of the end use.

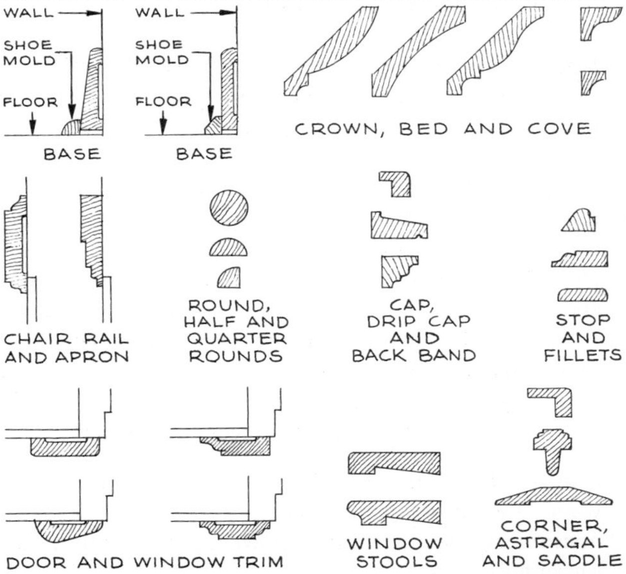

Figure W50 Some stock wood moldings and trim.

7. When using compressed wood moldings, it is necessary to check whether they have to be predrilled before application or applied with an adhesive.

Conditions Favorable to the Use of Moldings, Trim, and Ornamental Shapes

1. Where a method of covering a joint between two materials that will develop a crack or opening unless covered is needed.
2. Where decorative treatment that cannot be obtained economically by special manufacture is required.

Conditions Unfavorable to the Use of Moldings, Trim, and Ornamental Shapes

1. Where joints need to be covered because continual wetting can occur.

History and Manufacture

Wood by its very nature is easily carved, and throughout history man has carved and shaped it into furniture and decorations. Until power-driven machines were invented and developed, all moldings, trim, and ornamental shapes continued to be handmade. Today, except for wood stair railing easements, classical columns and column caps, and other decorative shapes that can be done only by hand, all moldings and trim are machine-made.

WOOD PRESERVATIVES

Wood can be treated to prevent or at least delay destruction by fire, insects, bacteria, fungi, and marine organisms.

Decay. Mold, stains, and decay in wood are caused by fungi. Their growth depends on mild temperatures ranging from 50 to 90°F (10 to 32.22°C) and dampness. Most decay occurs in wood that has a moisture content above the fiber saturation point. Usually wood maintained at 20% moisture content or less is safe from fungus damage. Wood that is under water or continuously dry will not decay. Care should be taken not to use wood where moisture can collect and remain.

Insects. There are several types of insects that attack wood, the most common being subterranean termites, nonsubterranean termites, powder-post beetles, and carpenter ants.

Subterranean termites live in the ground and build earthen tubes to reach their food—cellulose—which includes wood primarily but also paper and paper pulp products. The channels formed as they eat all the wood substance tend to follow the grain. In general, a complete barrier must be made between the wood and possible earthen tubes of the termites by (1) installing a shield made of metal or special termiteproof materials; (2) termiteproofing the wood used for construction close to earth; or (3) poisoning the soil adjacent to the building. It is good practice to make a periodic check around the building and under it for evidence of termites such as earthen tubes. If signs of termites are noted, immediate destruction of the tubes and poisoning of the surrounding earth are necessary.

Nonsubterranean termites, fortunately, are limited to a small portion of the United States (*see* map in Figure *W51*). The only relatively permanent method of arresting attack is to use lumber that has been given full-length termiteproofing treatment with wood preservatives. Attention to structural features and sanitation are also highly important.

Carpenter ants and powder-pest beetles use wood for shelter rather than for food, but if they are not found and are left undisturbed they can do extensive damage. They convert wood to powder, shredded fibers, or pellets. Their channels are likely to cut across the grain. Once located, the colony can be killed by dusting with an approved insecticide.

The damage to wood by the various marine organisms, mollusks, and crustaceans in salt water or brackish water is best arrested by heavy, thorough treatment with creosote or creosote solutions.

Where wood is exposed to severe marine borer hazard, it should be impregnated with a waterborne preservative or a combination of both creosote or creosote solutions and a waterborne preservative, as shown in Table *W47*.

Wood Preservatives. Wood preservatives used to forestall attack by decay, fungi, harmful insects, and marine borers are divided into two general groups: (1) oils such as creosote and petroleum solutions of pentachlorophenol (*see* Table *W48*), and (2) waterborne salts that are applied as water solutions (*see* Table *W49*).

Oil-type preservatives are ordinarily used for wood in contact with soil and water. Wood treated with waterborne preservatives is sometimes used in contact with the ground or water where paint is desired. Ammoniacal copper arsenite, chromated copper arsenate, or chromated zinc arsenate seems to be preferable for such uses. Some of the copper-containing preservatives at high retentions do provide limited protection to wood against marine borers.

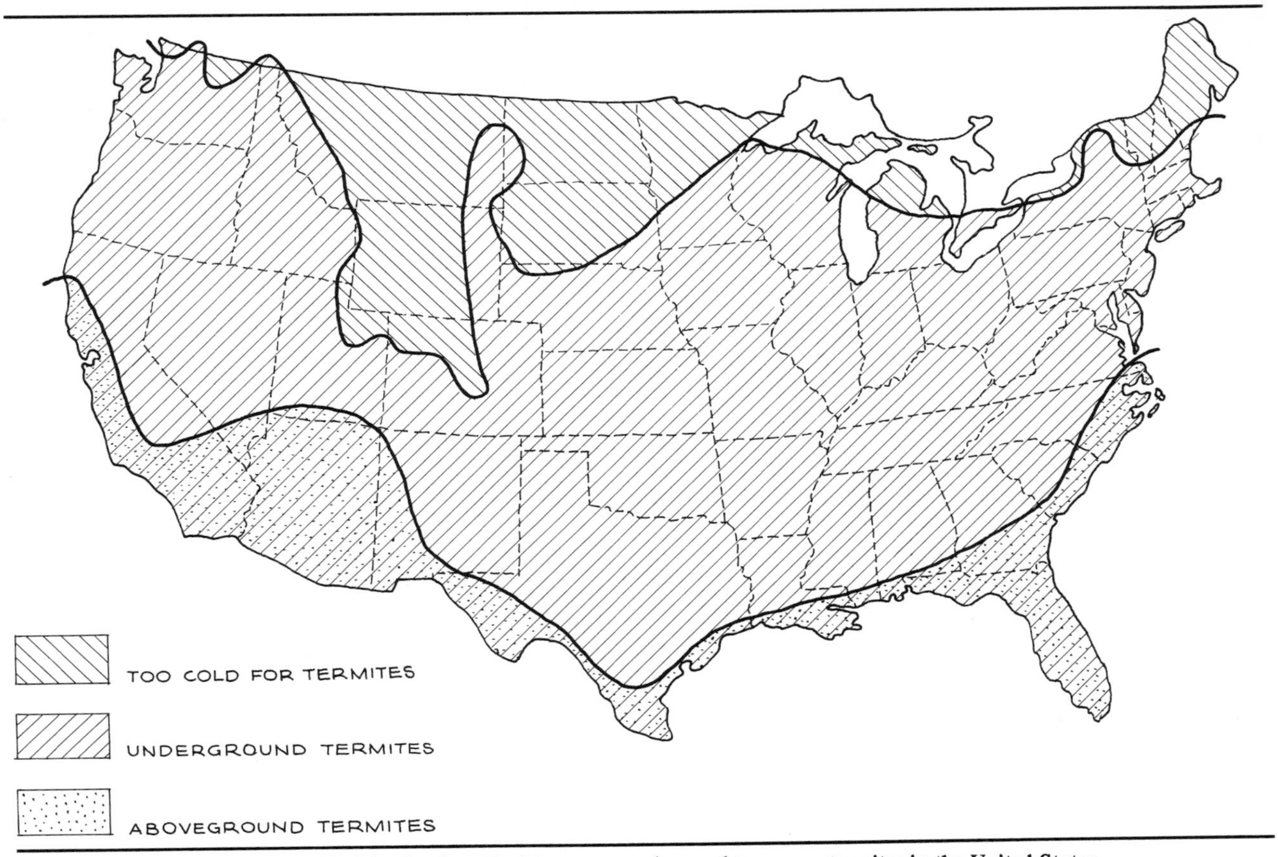

Figure W51 Distribution of subterranean and nonsubterranean termites in the United States.

Table W47 Treatment for wood piles exposed to severe marine borer attack

		Type of wood			
		Southern pine, Red pine		Coastal Douglas fir	
Type of borer	Type of treatment	lb/ft^3	kg/m^3	lb/ft^3	kg/m^3
Limnoria tripunctata only	Ammonical copper arsenite	2.5	40.05	2.5	40.05
	Chromated copper arsenate	2.5	40.05	2.5	40.05
Limnoria tripunctata and pholads	First treatment:				
	Ammonical copper arsenite	1.0	16.02	1.0	16.02
	Chromated copper arsenite	1.0	16.02	1.0	16.02
	Second treatment:				
	Creosote or creosote solutions	20.0	320.4	20.0	320.4
		20.0	320.4	Not recommended	

As a result of the great advances in methods of combatting decay, molds, fungi, and insects in the field of plant diseases and pest control, certain new compounds are now being effectively adapted to the preserving of wood. A malathion-impregnated asphalt felt paper is used for termite shields.

Fireproofing. When fire resistance is important, wood can be treated to meet the fire-resistance ratings of local and state codes and also the codes of the Fire Underwriters, insurance companies, and departments of the federal government (Labor, Army, Navy, etc.). The two methods are (1) covering the wood with a compound or

Table W48 Oil-type wood preservatives

Type of preservative	Composition	Advantages	Disadvantages
Coal-tar creosote	Black or brownish oil made from distilling coal tar	High toxicity to wood-destroying organisms; insolubility in water; low volatility; ease of application; permanence and depth of penetration can be determined	Dark brown color, cannot be painted; strong, unpleasant odor; easily ignited when first applied
Crystal-free coal-tar creosote	Coal-tar creosote from which some crystal-forming materials have been removed	Same advantages as coal-tar creosote; can be brushed and sprayed on more easily	Same disadvantages as coal-tar creosote
Anthracene oils	Coat-tar distillates of higher specific gravity and higher boiling ranges than coal-tar creosote	Same advantages as coal-tar creosote, plus less loss through evaporation during heating for application	Same disadvantages as coal-tar creosote
Creosotes derived from wood, oil, and water gas	Creosotes distilled from wood, oil, and water gas	Same advantages as coal-tar creosote	Less effective than coal-tar creosote
Creosote solutions	Mixture of coal tar or petroleum oils and 50–80% by volume of coal-tar creosote	Same advantages as coal-tar creosote	Less effective than coal-tar creosote
Pentachlorophenol	Mixture of petroleum oils and 5% of pentachlorophenol, also 2% penta in creosote	High protection against decay fungi and termites; can be painted; no unpleasant odor; less easily ignited than coal-tar creosotes (fire hazard compares to that of untreated wood if volatile solvents are used and allowed to evaporate)	May alter color if dark-colored petroleum oils are used; provides less protection against marine borers
Copper naphthenate	Mixture of petroleum oils and 0.3–0.5% copper metal equivalent (5–30% copper naphthenate)		Gives wood greenish or dark color and provides less protection against marine borers than creosote
Water-repellent preservatives[a]	Mineral spirits, and 10–25% maximum of nonvolative matter including the preservatives; preservatives shall be not less than 5% pentachlorophenol; copper naphthenate varies from 1 to 2%	Retards moisture changes in wood; good protection against decay and insects	Cannot be used in contact with ground or areas where continual dampness can occur unless preservative is thoroughly applied (water repellent has little value)

[a]Not less than 0.045% copper 8 quinolinolate for areas where foodstuffs will be in contact with treated woods.

Table W49 Waterborne wood preservatives

Type of preservative	Composition	Advantages	Disadvantages
Acid copper chromate[a] (Celcure)	31.8% copper oxide and 68.2% chromic acid; copper sulfate, potassium dichromate, or sodium dichromate may be substituted for copper oxide	Good protection against insects and decay; can be painted; no objectionable odor; impregnated with 0.5 lb/ft^3 (8.10 kg/m^3), the wood has same resistance to marine borer attack	Wood can be in contact with ground or water, but degree of impregnation depends on end use of wood

Table W49 Waterborne wood preservatives (*continued*)

Type of preservative	Composition	Advantages	Disadvantages
Chromated copper arsenate	Parts by weight: Chromium trioxide / Copper oxide / Arsenic pentoxide		
Type I[a] (Erdalith, Greensalt, Tanalith, and CCA)	61 / 17 / 22	Excellent protection against decay, fungi, and termites; good resistance to marine borer attack when only *Limnoria* and *Terdeo* borers are present	Wood can be in contact with ground; used in salt water when only *Limnoria* and *Terdeo* borers are present
Type II[a] (Boliden, K33)	35.3 / 19.6 / 45.1	Good protection against decay and insect attack; can be painted; no objectionable odor	Should be in contact with ground or water
Type III[a] (Wolman, CCA)	47 / 19 / 34 The following substitutions are permitted: sodium or potassium dichromate for chromium trioxide; copper sulfate, basic copper carbonate, or copper trioxide for copper oxide; and arsenic acid or sodium arsenate for arsenic pentoxide		
Chromated zinc chloride	80% zinc oxide and 20% chromium trioxide; the following substitutions are permitted: zinc chloride for zinc oxide, and sodium dichromate for chromium trioxide	Moderately effective in contact with ground or in installations; good protection under somewhat drier conditions	Should not be used in contact with ground or water; has some leaching action (applies to both chromated zinc chloride entries)
Chromated zinc chloride (FR)[b]	80% chromated zinc chloride, 10% boric acid, and 10% ammonium sulfate	Retention of $1\frac{1}{2}$–3 lb/ft^3 (24.03–48.06 kg/m^3) provides protection from decay and insect attack and has good fire-retardant characteristics	
Ammonical copper arsenite[a] (Chemonite)	49.8% copper oxide, 50.2% arsenic pentoxide, 1.7% acetic acid; the following substitutions are permitted: copper hydroxide for copper oxide, and arsenic trioxide for arsenic pentoxide	Good protection against decay and termite attack; protection against marine environment, provided pholad-type borers are not present	Wood can be in contact with ground or water, but degree of impregnation and high retention of preservatives depend on end use of wood
Chromated zinc arsenate[a]	20% arsenic acid, 21% sodium arsenate, 10% sodium dichromate, and 43% zinc sulfate	Good protection against decay and termites; can be painted; no objectionable odor	Wood can be in contact with ground but not water
Fluor chrome Arsenate phenol[a] (Wolman salts, Osmosalts)	22% fluoride, 37% chromium trioxide, 25% arsenic pentoxide, and 10% dinitrophenol; the following substitutions are permitted: sodium pentachlorophenate for dinitrophenol; sodium or potassium fluoride for fluoride, sodium chromate or dichromate for chromium trioxide, and sodium arsenate for arsenic pentoxide	Good protection against decay, fungi, and insects in aboveground wood construction, and moderate protection when in contact with ground	Wood cannot be used in contact with ground or water when good protection is required under these conditions
Zinc metal arsenite	60 parts arsenious acid and 40 parts zinc oxide with sufficient acetic acid to maintain preservations in solution	Good protection against decay and insects; can be painted; no objectionable odor	Wood can be used in contact with ground but generally not recommended for contact with water

[a] Many of these preservatives are trademark products covered by patents and should not be used without the specific consent of the patentee. [b] Designation for fire retardant.

material, and (2) impregnating the wood. Superficial coatings or layers of protective materials covering the surface of the wood retard the normal increases in temperature under fire conditions and thereby decrease the rate of flame spread. This in turn lessens the rate of flame penetration and therefore the destruction of wood in contact with fire.

Coatings or layers of protective material over the surface of the wood are good only for interior purposes because they are not durable when exposed to weather. These water-soluble, fire-retardant chemicals generally are trademarked formulations of ammonium phosphate, borax, or sodium silicate, combined with other materials to provide adherence to the wood, brushability, appearance, and color. There are also chemicals with low water solubility such as zinc borate, chlorinated paraffin, and chlorinated rubber which are used for fire-retardant coatings.

Complete impregnation (extending completely through the piece) is accomplished with a chemical that makes the wood itself not support combustion. Partial impregnation in many cases serves adequately, but if the lumber is to be cut or milled, complete impregnation is necessary. The chemicals commonly used for impregnation are monobasic ammonium phosphate, dibasic ammonium phosphate, ammonium sulfate, sodium tetraborate (borax), boric acid, and zinc chloride. The ammonium phosphates check both flaming and glowing. Borax checks flaming but is not a good glow retardant, whereas boric acid checks glowing but is not as good a flame retardant.

To obtain the correct treatment for a given enduse, it is always advisable to consult with the fire-retardant manufacturers regarding (1) permanence of the fire retardant, (2) its effect on the strength of the wood, (3) possible corrosive effect on metals which must be used to install the wood, (4) effect on paint and glue, (5) hygroscopicity (water absorption), and (6) toxicity. All governing codes for the use of treated wood should also be checked. These same precautions hold true for treated woods in general.

WOOD SHINGLES

Physical Properties

Ninety-eight percent of all wood shingles are made from western red cedar, and the rest from eastern white cedar, tidewater cypress, and California redwood. They are available in stock sizes, sawed or handsplit, with smooth surfaces or grooved, either quartersawed or plainsawed, and with thick or thin butts. They may be made of all heartwood or heartwood plus sapwood. Those made of all heartwood are more resistant to decay. Quartersawed and thick-butt shingles have greater resistance to warping. The difference between a shingle and a shake is that a shingle is sawed on both faces, whereas a shake is split on one or both faces.

Types and Uses

Wood shingles are graded separately for the different species of wood. Western red cedar is graded as No. 1, No. 2, and No. 3; bald cypress as No. 1, Bests, Primas, Economy, and Clippers; and redwood as No. 1 and No. 2. The No. 1 grade for all species of wood is all clear, all heartwood, and all quartersawed. The lower grades are quartersawed or plainsawed, have heartwood and sapwood, and have imperfections increasing proportionately in the lower grades.

The thickness of both random and dimension shingles is designated as follows:

4/2 equals 4 shingles to 2 in. (50.8 mm) of butt.
$5/2\frac{1}{4}$ equals 5 shingles to $2\frac{1}{4}$ in. (57.15 mm) of butt.
5/2 equals 5 shingles to 2 in. (50.8 mm) of butt.

Handsplit shingles (shakes) are similarly designated with more permissible variations.

Handsplit shakes are manufactured in three types, handsplit-resawn, tapersplit, and straightsplit. These are divided into two basic butt sizes as follows:

$\frac{1}{2}$ to 3/4-in. (12.7 to 19.05-mm) butts.
3/4 to $1\frac{1}{4}$-in. (19.05 to 31.75 mm) butts.

Shingles made in random widths are usually packed by the square (100 ft^2 or 9.29 m^2). Dimensioned widths are usually packed or delivered in 1000-shingle lots. Figure *W52* and Table *W50* show common sizes and types of wood shingles, including data on nailing and

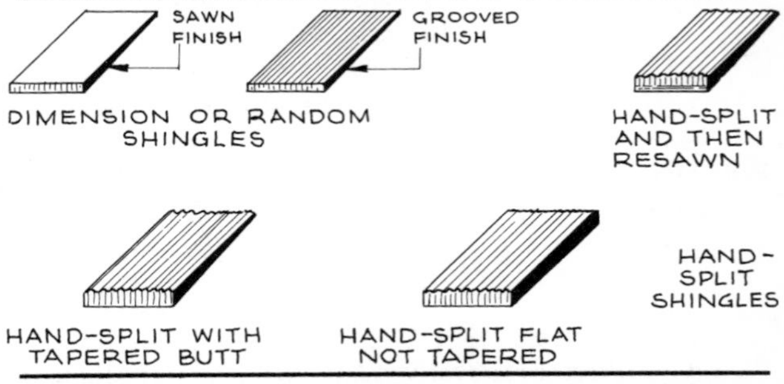

Figure W52 Types and surface treatments of wood shingles.

Table W50 Roof shingles for roofing and siding

Type of shingle	Size: Width in.	Width mm	Length in.	Length mm	Thickness designation	Nail size in.	Nail size mm	How used: Roofing, Maximum exposure: Pitches of 4 to 12 and steeper in.	mm	Pitches of 3 to 12 and 4 to 12 in.	mm	Siding: Maximum exposure for single course in.	mm	Maximum exposure for double course in.	mm
Dimension	5 or 6	127 or 152.4	16	406.4	5/2	$1\frac{1}{4}$	31.75	5	127.0	$3\frac{3}{4}$	95.25	$7\frac{1}{2}$	190.5	12	304.8
	5 or 6	127 or 152.4	18	457.2	$5/2\frac{1}{2}$	$1\frac{1}{4}$	31.75	$5\frac{1}{2}$	139.7	$4\frac{1}{4}$	107.95	$8\frac{1}{2}$	215.9	14	355.6
	6	152.4	24	609.6	4/2	$1\frac{1}{2}$	38.10	$7\frac{1}{2}$	190.5	$5\frac{3}{4}$	146.05	$11\frac{1}{2}$	292.1	16	406.4
Random	3 min.	76.2 min.	16	406.4	5/2	$1\frac{1}{4}$	31.75	5	127.0	$3\frac{3}{4}$	95.25	$7\frac{1}{2}$	190.5	12	304.8
	14 max.	355.6 max.	18	457.2	$5/2\frac{1}{4}$	$1\frac{1}{4}$	31.75	$5\frac{1}{2}$	139.7	$4\frac{1}{4}$	107.95	$8\frac{1}{2}$	215.9	14	355.6
			24	609.6	4/2	$1\frac{1}{2}$	38.10	$7\frac{1}{2}$	190.5	$5\frac{3}{4}$	146.05	$11\frac{1}{2}$	292.1	16	406.4
Handsplit	Random	Random	18	457.2	$\frac{1}{2}$–$\frac{3}{4}$ in. (12.7–29.05 mm)	2–$2\frac{1}{2}$	50.8–63.5	10	254.0			$11\frac{1}{2}$	292.1	16–$16\frac{1}{2}$	406.4–419.1
			24	609.6	$\frac{3}{4}$–$1\frac{1}{4}$ in. (19.05–31.75 mm)									16–$16\frac{1}{2}$	406.4–419.1

exposure according to use. Only approximately 0.1% of the total production of wood shingles consists of dimension width shingles; the remainder are random width shingles.

Wood shingles for siding are available in a wide variety of factory-applied colors.

Fire-protected wood shakes and shingles for roofing and siding are available to meet U.L. class B rating and U.L. class C rating. Fire-protected shingles installed by standard methods obtain a class C rating. By applying these types of shingle over an underlayment of plastic-coated steel foil on top of $\frac{1}{2}$-in. (12.7-mm) plywood or 2-in. (50.8-mm) tongue-and-groove sheathing, a class B rating is obtained. Fire-protected wood shingles and shakes are available in handsplit and resawn shingles and shakes. The fire-protecting impregnation improves weathering and resistance to decay without affecting the application of decorative stains.

Shingles and shakes are available on special order in an unlimited variety of butt designs (e.g., half-round).

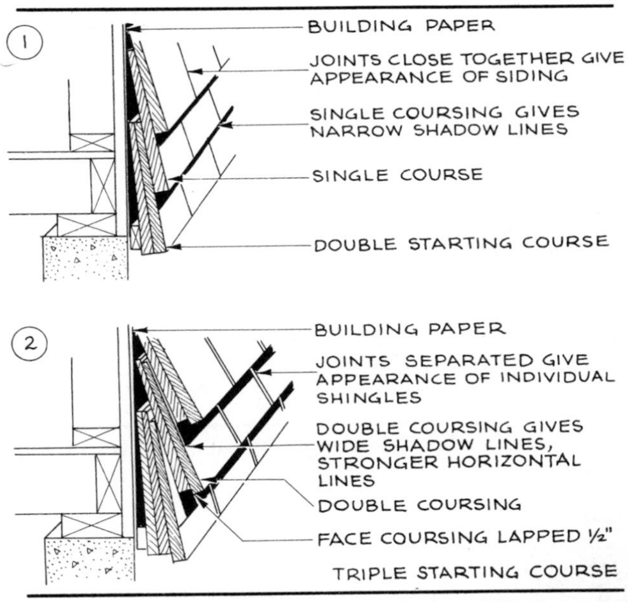

Figure W53 Details of single and double coursing for wood shingles used for siding.

Application

Condensed Checklist

1. The type, size, and grade of wood shingles must meet the requirements and limitations for their end use in the design.
2. Local, municipal, and state codes and also codes of the Fire Underwriters, insurance companies, labor departments, and federal government (Army, Navy, etc.) should always be checked for fire resistance and other limitations on the use of wood shingles.
3. For siding, a method of application that meets design requirement must be used, for example, single or double coursing, and joints close together or apart, as shown in Figure *W53*.
4. Treatment of external corners of wood shingles used as siding (*see* Figure *W54*).
5. For roofing, exposure of shingles in relation to roof pitch should always be checked, and treatment of ridges, hips, and valleys correctly detailed (*see* Figure *W55*).

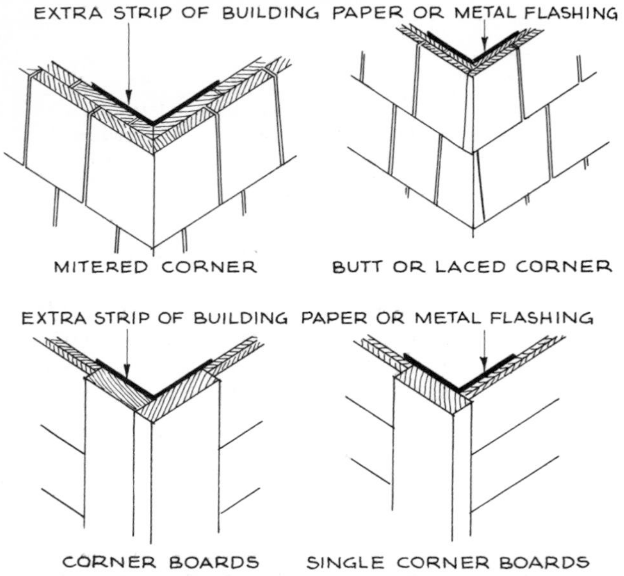

Figure W54 Details of exterior corners for wood shingle siding.

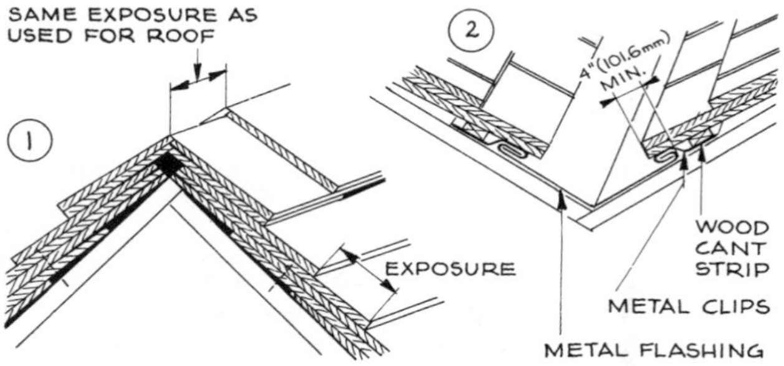

Figure W55 Details of hip or ridge (1) and open valley (2) for wood shingle roof.

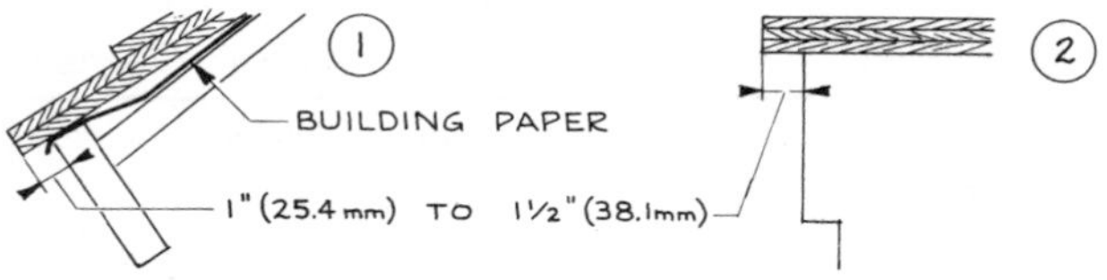

Figure W56 Details of eave and gable end for wood shingle roof.

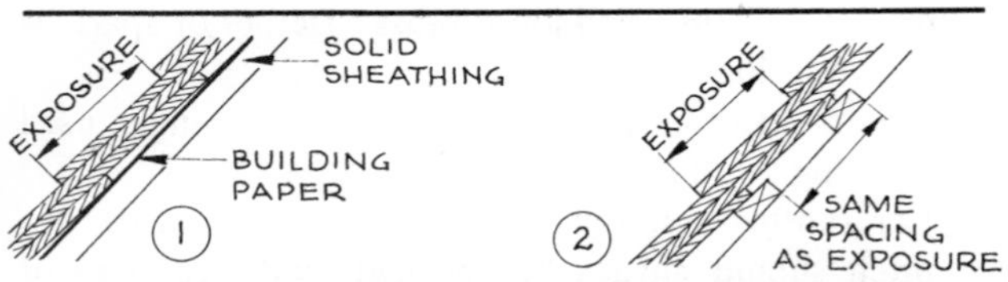

Figure W57 Details of solid sheathing (1) and strip sheathing (2) for wood shingle roof.

6. Treatment at gable ends and eaves for a wood shingle roof (*see* Figure *W56*).
7. For roofing, one should always check whether to use solid sheathing or strip sheathing. Solid sheathing on low-pitched roofs (pitches of 3 to 12 and 4 to 12) allows moisture to remain in the shingles longer and thus shortens the durability of the roof. On steeper pitches, both methods are recommended. (*See* Figure *W57*.)

Condition Favorable to the Use of Wood Shingles

As roofing and siding wherever a durable, smooth or textured, easily maintained, and easily colored covering material with the characteristic appearance of wood shingles is desired.

Condition Unfavorable to the Use of Wood Shingles

In localities where nonsubterranean termites exist unless the wood shingles have been specially treated against mold, fungi, and insect attack.

History and Manufacture

Wood logs are first cut into specific lengths, which are then cut into blocks of more or less triangular shape (quartersawed blocks). These blocks should have a true edge-grain face. These blocks are next sawed into shingles, which are graded for size, thickness, and any defects, and then kiln-dried. The surface may be hand-split, grooved or striated, or left sawed smooth.

WOOD SIDING AND PANELING

Physical Properties

Almost all exterior siding is made from evergreen trees (softwoods). It is usually No. 1 grade and sometimes No. 2 grade and is either kiln-dried or air-dried. It is available in a large variety of shapes and sizes, with the shapes controlled to provide weatherproof covering to the exterior of a building.

Paneling, on the other hand, is manufactured from both deciduous (hardwood) and evergreen (softwood) trees (*see* Wood for the various grades and sizes of both hardwood and softwood). Paneling, like siding, is available in various stock sizes and shapes. The grain, color, and texture of the wood, however, are the determining factors in its use.

Grading. All siding and paneling are made from common boards and common dimension lumber, cut and dressed to standard shapes and sizes. Paneling and siding are graded according to the lumber from which they are cut—usually A or B and Better, C, and D for softwoods, and Firsts, Seconds, Select, and No. 1 Common for hardwoods.

Types and Uses

Siding finds its greatest use in residential construction, whereas paneling is used in all types of buildings. Figures *W58* and *W59* show the sizes and shapes of siding that is usually installed horizontally on the exterior of buildings.

Siding designed to be installed vertically on the exterior of buildings can also be used for paneling on the interior. Figure *W60* shows the sizes and shapes available in this type of siding.

The difference between paneling and siding lies in the thickness and, more importantly, in the fact that pan-

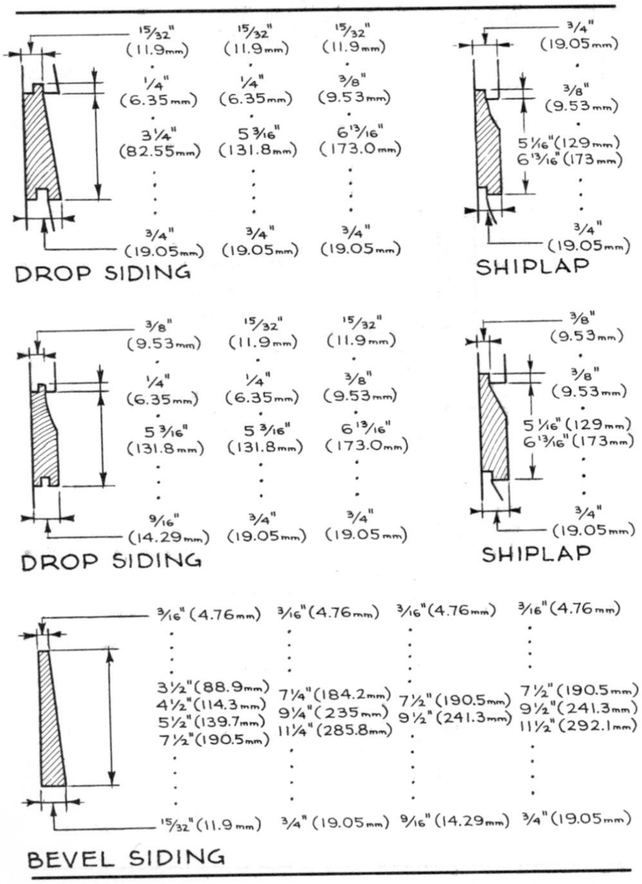

Figure W58 Typical shapes and sizes of wood siding to be used horizontally.

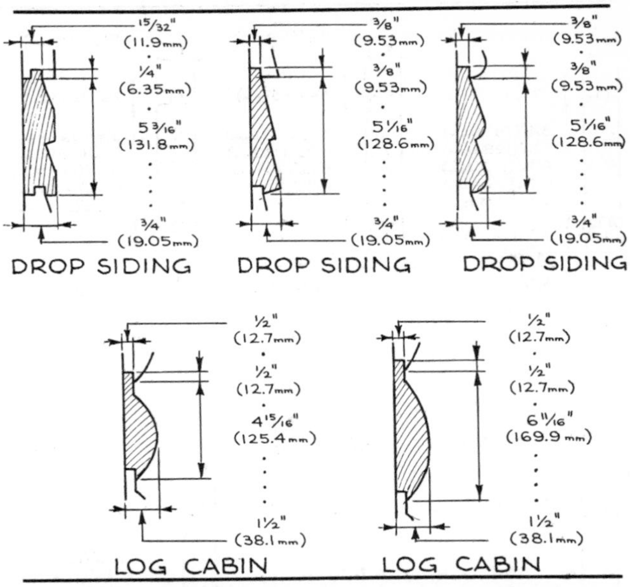

Figure W59 Typical shapes and sizes of wood siding that can be used both vertically and horizontally.

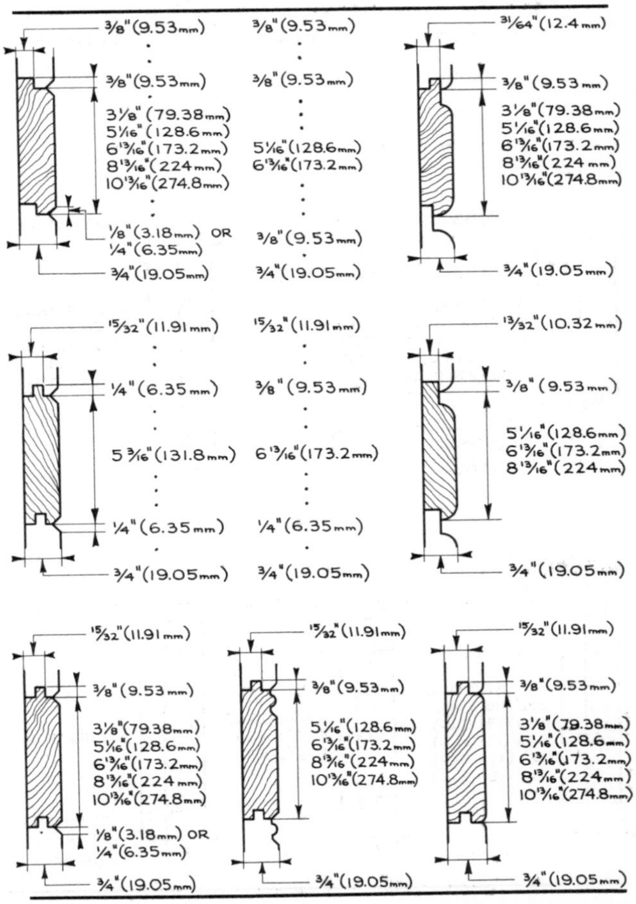

Figure W60 Exterior vertical wood siding that can also be used for interior paneling.

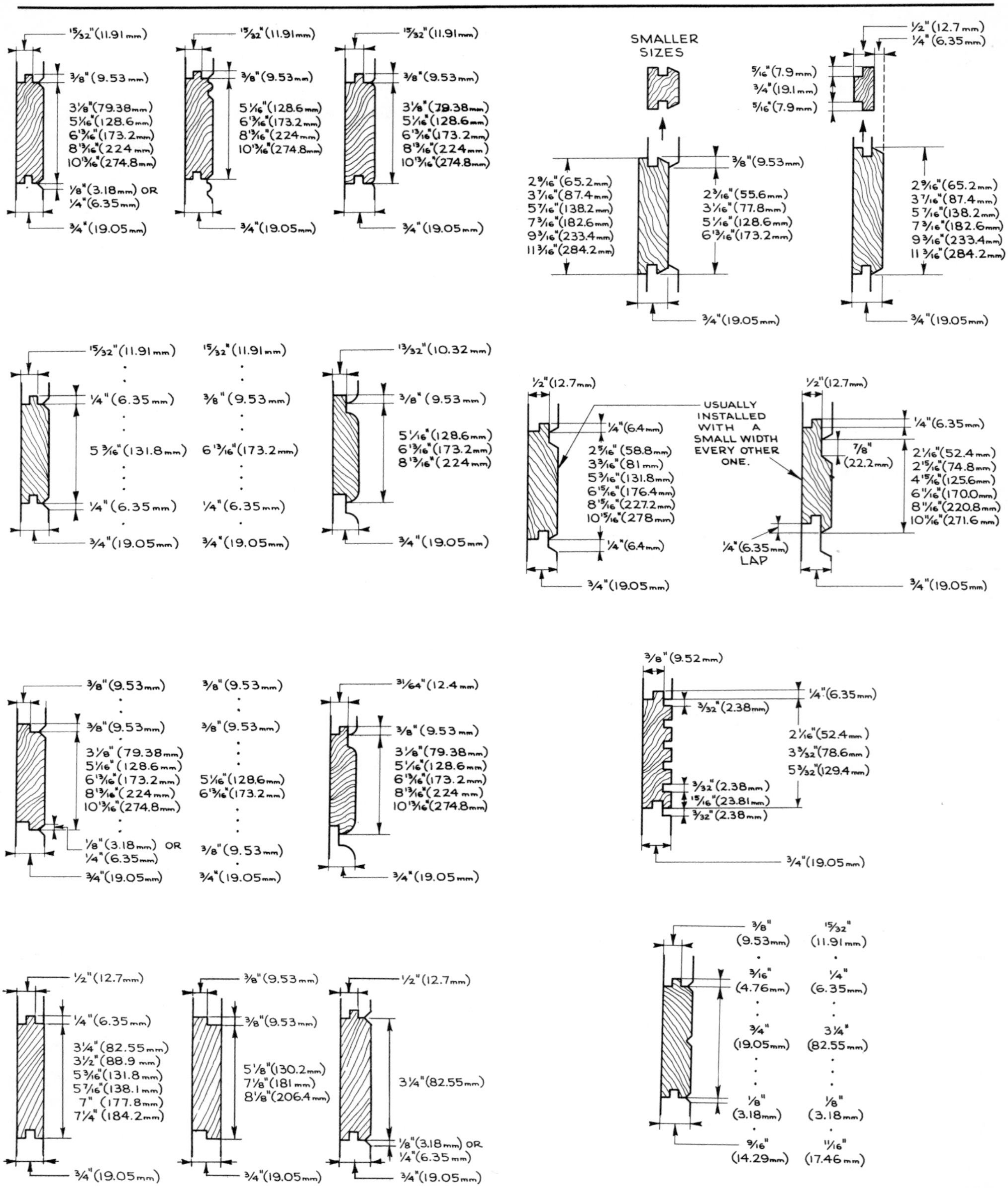

Figure W61 Shapes and sizes of wood paneling.

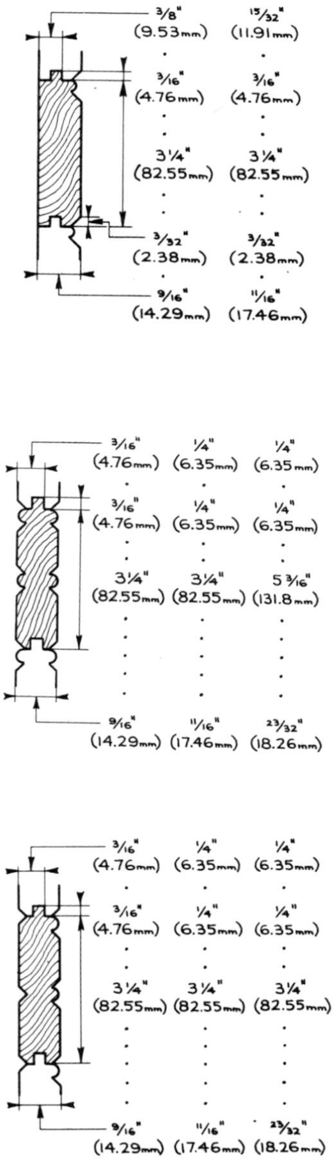

Figure W61 Shapes and sizes of wood paneling (*continued*).

eling shapes need not be weathertight. Figure *W61* shows the commonly used sizes and shapes of paneling. The distinction of lumber manufacturers between two types called ceiling and paneling has been ignored here because today this distinction is no longer valid.

There are miscellaneous shapes and sizes of wood that can be used for paneling, louvers, partitions, and decking. These are shown in Figure *W62*.

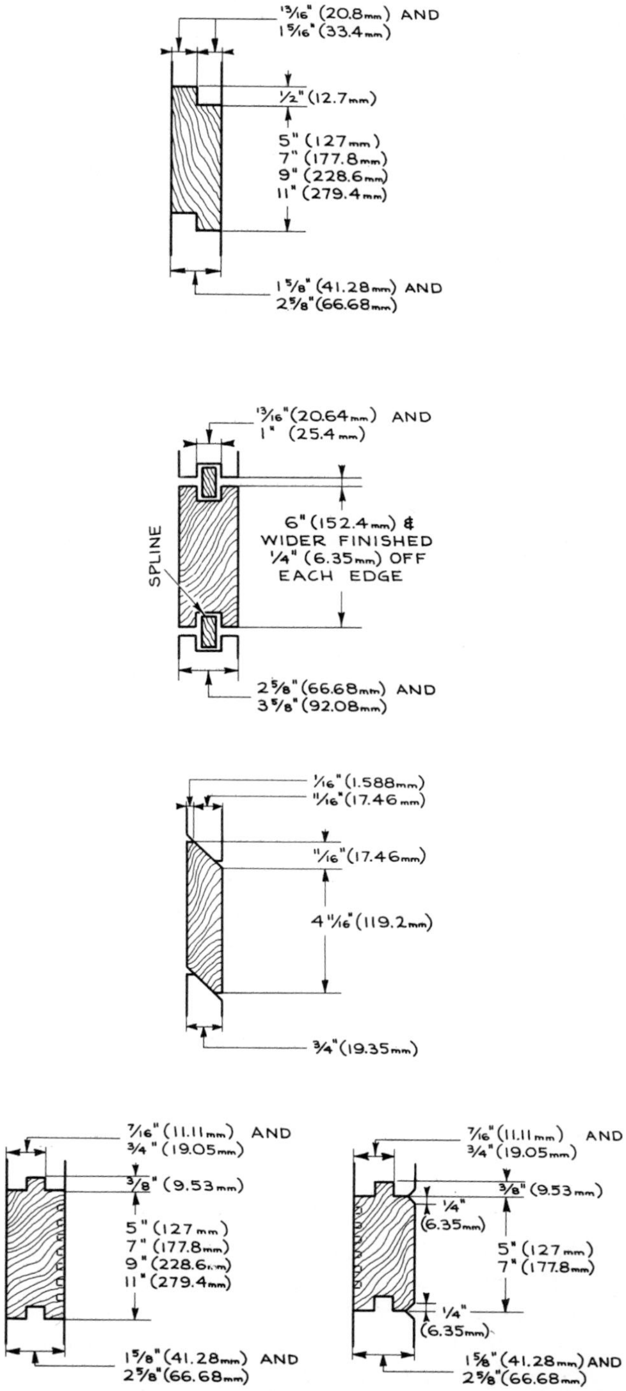

ALSO MADE WITH 1/8" (3.18mm) x 1/4" (6.35mm) GROOVES SPACED 1/2" (12.7mm) O. C. FOR ACOUSTIC TREATMENTS IN 5" (127mm) AND 6 7/8" (174.62mm) WIDTHS.

Figure W62 Miscellaneous shapes and sizes of paneling, decking, and partitions.

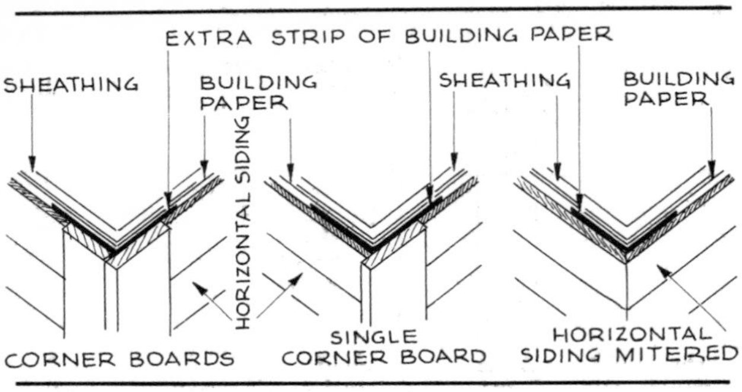

Figure W63 Details of exterior corners for wood siding.

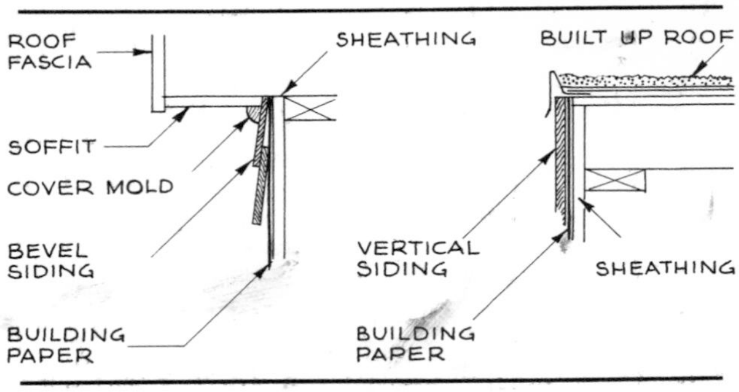

Figure W64 Details at grade for wood siding.

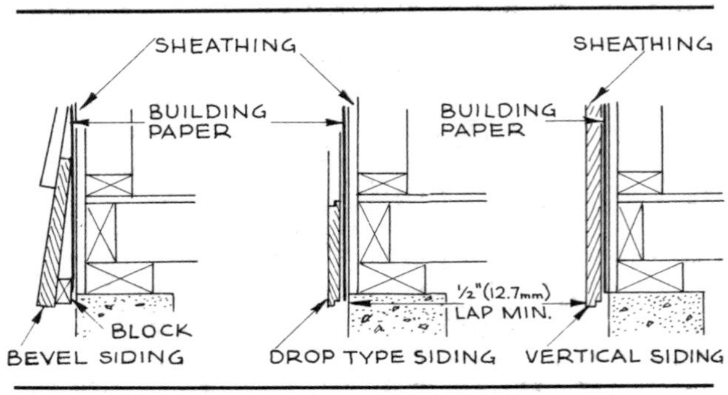

Figure W65 Details at roof for wood siding.

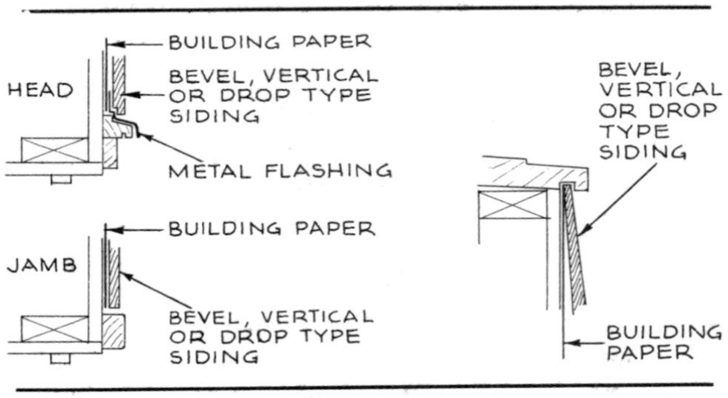

Figure W66 Details at door and window openings for wood siding.

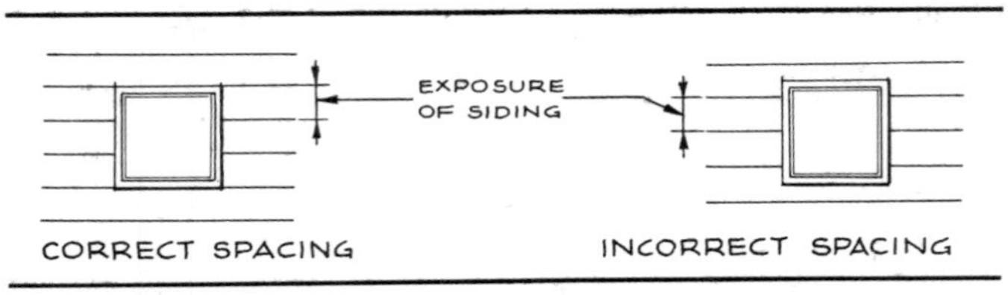

Figure W67 Examples of correct and incorrect spacing (exposure).

Application

Condensed Checklist

1. Siding or paneling must meet all the requirements and limitations of its end use in the design.
2. Treatment of siding at exterior corners, at intersection with roof, at openings, and at grade must be correctly detailed (*see* Figures *W63* to *W66*).
3. For all siding one should always check what type of nail to use, its corrosion resistance, and the method of installation (whether blind-nailed or countersunk with holes puttied or covered by the lap of the siding). The type of sheathing also controls the type of nail to be used.
4. Vertical siding should always be checked for the lengths available in the type, grade, and species of wood chosen so that no or very few horizontal joints will occur.
5. For horizontal siding the height of heads and sills of windows and heads of doors must be correlated with the exposure height of the siding; these must be calculated correctly in relation to each other so that unsightly small pieces do not occur above and below windows and doors (*see* Figure *W67*).
6. For all types of paneling, treatment at corners, at floor, at ceiling, and at openings must be correctly detailed (*see* Figure *W68*).
7. For all types of paneling, one should check the type of nails and installation (whether blind-nailed or

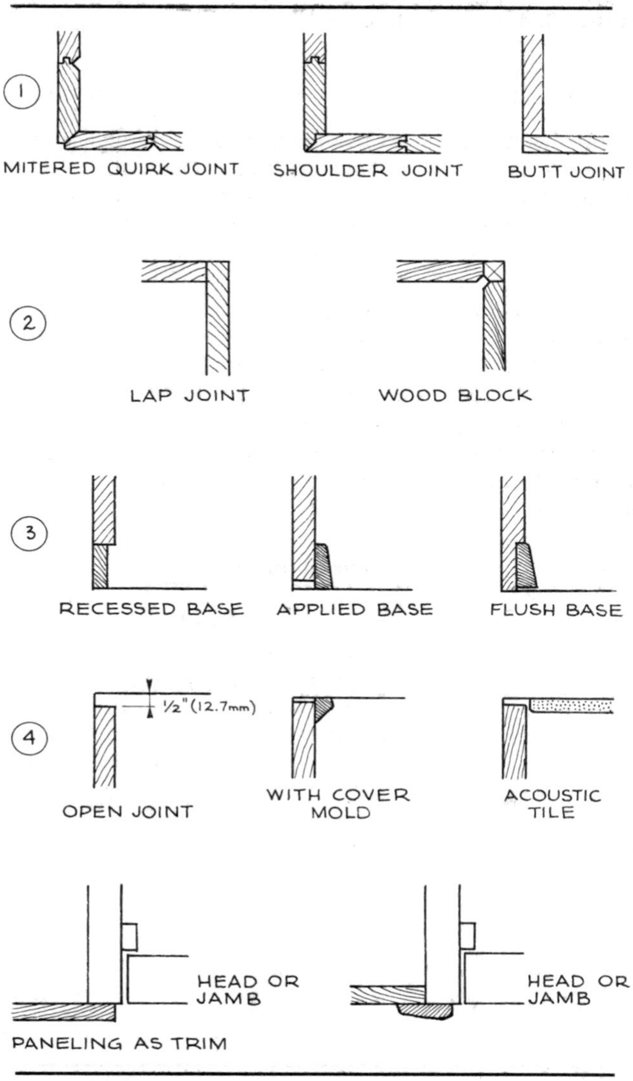

Figure W68 Detailing of paneling for various conditions.

countersunk with holes puttied or covered by the paneling lap).

8. For all types of paneling one should make sure that blocking, stripping, and backing are correctly installed to receive the type of paneling chosen.

Conditions Favorable to the Use of Wood Siding and Paneling

1. Wood siding for exterior of buildings where a strong, durable, easily installed, weathertight material that can be painted, stained, or left the natural wood color is required.
2. Wood paneling for interior wall surfacing where a strong, durable, easily installed material that can be painted, stained, or left the natural wood color and that is available in a large variety of stock sizes, shapes, and species of wood is required.

Condition Unfavorable to the Use of Wood Siding or Paneling

Where fire resistance is a controlling factor.

WOOD, STRUCTURAL

Physical Properties

Strength Factors. The strength of structural lumber is influenced by the strength and variability of the clear wood, the moisture content, the duration or the load, and the size, number, and locations of strength-reducing characteristics such as knots, cross grain, and splits. Decay in any form is severely restricted or prohibited in structural lumber. Table *W51* describes common types of characteristics and their effects on the grading of structural lumber.

Where combinations of strength-reducing characteristics are present, their combined effects are considered in structural grading. Generally, grading rules prohibit any serious combinations.

Density. Strength is closely related to the weight or density of the wood. Higher working stresses can be assigned to lumber by using the rate of growth (number of annual rings per inch) and the percentage of summerwood (to springwood) to select pieces of superior strength from certain species (Douglas fir, southern yellow pine). The rate of growth (number of annual growth rings per inch) must be within a specific range, and selection for density requires, in addition to a specified rate of growth, a minimum percentage of summerwood to springwood. Wood meeting these density requirements can be assigned a basic strength value higher by one-sixth.

Effect of Seasoning. When wood is seasoned, the direct effect of the loss of moisture is stiffening and strengthening of the wood fibers. In the seasoning of large pieces, splitting or checking may occur which partially offsets the increase in strength.

Overloading. Both the elastic limit and the ultimate strength of wood are higher under short-time loading

Table W51 Common imperfections that affect grading of stress-graded structural lumber

Type of imperfection	Description of imperfection	Effects on strength	Effect on grading structural lumber
Slope of grain	Areas where the direction of the wood fiber is not parallel to the edges of the piece of lumber	Tends to twist with changes in moisture content, the components of longitudinal, tensile and compressive stresses acting across the grain where wood is the least strong	Cross-grained pieces are undesirable; reduction of strength due to cross grain in structure is taken as twice the reduction observed in tests of small clear specimens
Knots	Knots interrupt the direction of grain and cause localized cross grain with steep slopes	Knots reduce tensile strength more than compressive and shear strength and affect stiffness slightly	The size, number and location of knots is restricted for structural lumber; cluster knots are prohibited
Shakes	A separation of the wood between the annual growth rings	In lumber subjected to bending, shakes reduce the resistance to shear; they do not affect the strength for longitudinal compression	Shakes are restricted in those parts of a bending member where shearing stresses are highest
Checks and splits	Actual split in the wood	Same as for shakes	Same as for shakes
Wane	Bark or lack of wood on the edge or corner of the piece of lumber	Affects nailing and bearing	Limited in structural lumber requirements for fabrication, bearing, nailing and appearance and not for effect on strength
Pitch pockets	Openings between annual growth rings containing pitch or bark	Have little or no effect on strength	Usually disregarded except if a large number occur; shake may be present or bond between annual growth rings may be weakened
Holes	Either a knothole or a hole caused by some other means	Same as for knots	Same as for knots

than under long-time loading. Wood is thus able to withstand considerable overloads for short periods or smaller overloads for longer periods.

Basic Stress. A system of stress-grading lumber has been set up in the United States for the development, manufacturing, and merchandising of structural wood (*see* Figure *W69*). Most commercial softwoods are stress-graded under standard practice, and the principle of stress grading is applied to hardwoods.

Stressed-grade structural lumber is available in various grades, depending on the type of wood, designated in general as Select Structural, Industrial, Construction, and Standard. Douglas fir and western larch have the following grades: Dense Select Structural, Dense Construction, and three grades of Industrial based on stress in lbf/in.^{2} (MN/m^2). Tables *W52* and *W53* show the characteristics of stress-graded structural lumber.

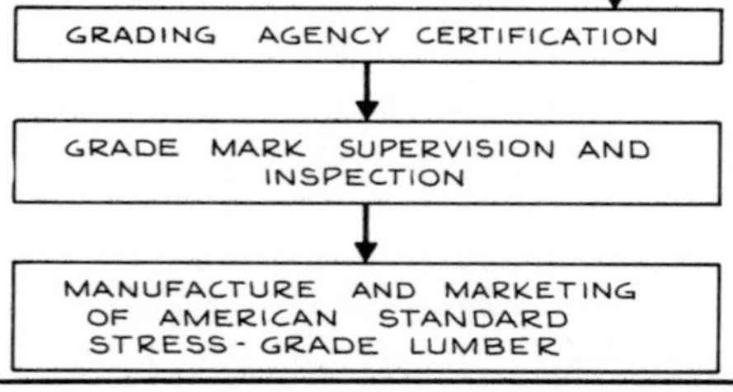

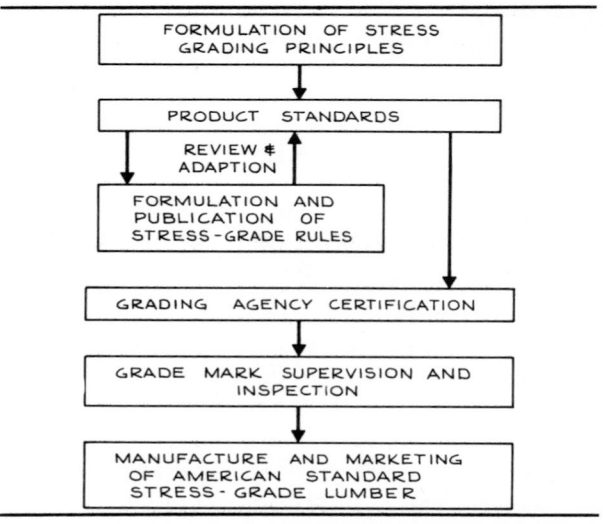

Figure W69 Diagram of system for stress-grading lumber.

Table W52 Characteristics of generally used stress-graded structural softwood

			Compression				Shear		Tension	
	Modulus of elasticity		Parallel to grain, maximum crushing strength		Perpendicular to grain, fiber stress at proportional limit		Parallel to grain, maximum shearing strength		Perpendicular to grain, maximum tensile strength	
Type of wood	million lbf/in.^{2}	MN/m^2	lbf/in.^{2}	MN/m^2	lbf/in.^{2}	MN/m^2	lbf/in.^{2}	MN/m^2	lbf/in.^{2}	MN/m^2
Cedar, Alaska	1.14	7 860	3050	21.09	350	2.41	840	5.79	330	2.28
	1.42	9 791	6310	43.51	620	4.28	1130	7.79	360	2.48
Cedar,	0.75	5 171	2390	16.48	240	1.66	690	4.76	180	1.24
Atlantic white	0.83	5 723	4700	32.41	410	2.83	800	5.52	220	1.52
Cedar,	0.64	4 413	1990	13.72	230	1.60	620	4.28	240	1.66
northern white	0.80	5 516	3960	27.30	310	2.14	850	5.86	240	1.66
Cedar,	1.90	13 101	3140	21.65	300	2.07	840	5.79	180	1.24
Port Orford	1.70	11 722	6250	43.09	720	4.96	1370	9.45	400	2.76
Cedar,	0.94	6 481	2770	19.10	240	1.66	770	5.31	230	1.60
red western	1.11	7 654	4560	31.44	460	3.17	990	6.82	220	1.52
Cypress, bald	1.18	8 136	3580	24.68	400	2.76	810	5.59	300	2.07
	1.44	9 929	6360	43.85	730	5.03	1000	6.90	270	1.87
Douglas fir,	1.56	10 756	3780	26.06	380	2.62	900	6.21	300	2.07
coast	1.95	13 445	7240	49.92	800	5.52	1330	9.17	340	2.34
Douglas fir,	1.51	10 411	3870	26.68	420	2.90	940	6.48	290	2.00
interior west	1.82	12 500	7440	51.30	760	5.24	1290	8.89	300	2.41
Douglas fir,	1.41	9 722	3470	23.93	300	2.48	950	6.55	340	2.34
interior north	1.79	12 342	6900	47.58	770	5.31	1400	9.65	390	2.69
Douglas fir,	1.16	7 998	3110	21.44	340	2.34	950	6.55	250	1.68
interior south	1.49	10 274	6220	42.89	740	5.10	1510	10.41	330	2.28
Hemlock,	1.07	7 378	3080	21.24	360	2.48	850	5.86	230	1.60
eastern	1.20	8 274	5410	37.30	650	4.48	1060	7.31		
Hemlock,	1.04	7 171	2880	19.86	370	2.55	930	6.41	330	2.28
mountain	1.33	9 170	6440	44.40	860	5.93	1540	10.62		
Hemlock,	1.31	9 033	3360	23.17	280	1.93	860	5.93	290	2.00
western	1.64	11 308	7110	49.02	550	3.79	1250	8.62	340	2.34
Larch,	0.96	6 619	3760	25.90	400	2.76	870	6.00	330	2.28
western	1.87	12 893	1640	52.68	930	6.41	1360	9.38	430	2.97
Pine, red	1.28	9 826	2730	18.82	260	1.79	690	4.76	300	2.07
	1.63	11 239	6070	41.85	600	4.13	1210	8.34	460	3.17
Pine,	0.99	6 826	2440	16.82	220	1.52	680	4.69	250	1.68
eastern white	1.24	8 560	4800	33.10	440	3.03	900	6.21	310	2.14
Pine,	1.19	8 205	2430	16.75	190	1.31	680	4.69	260	1.79
western white	1.46	10 067	5040	34.55	470	3.24	1040	7.17		
Pine,	1.08	7 447	2610	18.00	250	1.68	680	4.69	220	1.52
lodgepole	1.34	9 239	5370	37.03	610	4.21	880		290	2.00
Pine,	1.00	6 895	2450	15.89	280	1.93	700	4.83	310	2.14
ponderosa	1.29	8 895	5320	36.62	580	4.00	1130	7.79	420	2.30
Pine, slash	1.53	10 549	3820	26.34	530	3.65	960	6.62		
	1.98	13 652	8140	53.13	1020	7.03	1680	11.58		
Pine,	1.39	9 584	3530	24.34	350	2.41	910	6.27	320	2.21
shortleaf	1.75	12 066	7270	50.13	820	5.65	1390	9.58	470	3.24
Redwood,	1.18	8 136	4200	28.96	420	2.90	800	5.52	260	1.79
old-growth	1.34	9 239	6150	42.40	700	4.83	940	6.48	240	1.66
Redwood,	0.96	6 619	3110	21.44	270	1.87	990	6.82	300	2.07
young-growth	1.10	7 585	5220	35.99	520	3.59	1110	7.65	250	1.68
Spruce,	1.03	7 102	2180	15.03	200	1.38	640	4.41	240	1.66
Engelmann	1.30	8 964	4480	29.89	410	2.83	1200	8.27	350	2.41
Spruce, Sitka	1.23	8 481	2670	18.41	280	1.93	760	5.24	250	1.68
	1.57	9 825	5610	38.68	580	4.00	1150	7.93	370	2.55
Spruce, black	1.06	7 309	2570	17.72	140	0.97	660	4.55	100	0.69
	1.53	10 549	5320	36.68	530	3.65	1030	7.10		
Tamarack	1.24	9 550	3480	23.99	390	2.69	860	5.93	260	1.79
	1.64	11 308	7160	49.37	800	5.52	1280	8.81	400	2.76

Table W53 Characteristics of hardwood structural lumber

	Modulus of elasticity		Compression: Parallel to grain, maximum crushing strength		Compression: Perpendicular to grain, fiber stress at proportional limit		Shear: Parallel to grain, maximum shearing strength		Tension: Perpendicular to grain, maximum tensile strength	
Type of wood	million lbf/in.^{2}	MN/m^2	lbf/in.^{2}	MN/m^2	lbf/in.^{2}	MN/m^2	lbf/in.^{2}	MN/m^2	lbf/in.^{2}	MN/m^2
Ash, black	1.04	7 171	2300	15.86	350	2.41	860	5.93	490	3.38
	1.60	11 032	5970	41.16	760	5.24	1570	10.82	700	4.83
Ash, white	1.44	9 929	3990	27.51	670	4.62	1380	9.51	590	4.07
	1.74	11 997	7410	51.09	1160	8.00	1950	13.44	940	6.48
Ash, Oregon	1.13	7 791	3510	24.20	530	3.65	1190	8.20	590	4.67
	1.36	9 377	6040	41.65	1250	8.62	1790	12.34	720	4.96
Aspen, big tooth	1.12	7 722	2500	17.24	210	1.45	730	5.03		
	1.43	9 860	5300	36.54	450	3.10	1080	7.45		
Aspen, quaking	0.86	5 930	2140	14.75	180	1.24	660	4.55	230	1.59
	1.18	8 136	4250	29.30	370	2.55	850	5.86	260	1.79
Beech, American	1.38	9 515	3350	23.10	540	3.72	1290	8.89	720	4.96
	1.72	11 859	7300	50.33	1010	6.96	2010	13.86	1010	6.96
Birch, paper	1.17	8 067	2360	16.27	270	1.86	840	5.79	380	2.62
	1.59	10 963	5690	39.23	600	4.41	1210	8.34		
Birch, sweet	1.65	11 377	3740	25.79	470	3.24	1240	8.55	430	2.96
	2.17	14 962	8540	58.88	1080	7.45	2240	15.44	950	6.55
Birch, yellow	1.50	10 342	3380	23.30	430	2.96	1110	7.65	430	2.96
	2.01	13 859	8170	56.33	970	6.69	1880	12.96	920	6.55
Cottonwood,	1.08	7 447	2200	15.17	160	1.10	610	4.21	270	1.86
black	1.27	8 757	4500	31.03	300	3.47	1040	7.17	330	2.28
Cottonwood,	1.01	6 965	2280	15.72	200	1.38	680	4.69	410	2.83
eastern	1.37	9 446	4910	33.85	380	2.62	930	6.41	580	4.00
Elm, American	1.11	7 653	2910	20.06	360	2.48	1000	6.90	590	4.07
	1.34	9 239	5520	38.06	690	4.76	1510	10.41	660	4.55
Elm, slippery	1.23	8 481	3320	22.89	420	2.90	1110	7.65	640	4.41
	1.49	10 274	6360	43.85	820	5.65	1630	11.24	530	3.65
Locust, black	1.85	12 756	6800	46.89	1160	8.00	1760	12.14	770	5.31
	2.05	14 235	10,180	70.19	1830	12.62	2480	17.10	640	4.41
Maple, black	1.33	9 170	3270	22.55	600	4.14	1130	7.79	720	4.96
	1.62	11 170	6680	46.06	1020	7.03	1920	13.24	670	4.62
Maple, sugar	1.55	10 687	4020	27.72	640	4.41	1460	10.07		
	1.83	12 618	7830	53.99	1470	10.14	2330	16.06		
Oak, red northern	1.35	9 308	3440	23.72	610	4.21			750	5.17
	1.82	12 549	6760	47.30	1010	6.96			800	5.52
Oak, scarlet	1.48	10 205	4090	28.20	830	5.72	1410	9.72	700	4.83
	1.91	13 169	8330	57.43	1120	7.72	1890	13.13	870	
Oak, white live	1.58	10 894	5430	37.44	2040	14.07	2210	15.24		
	1.98	13 652	8900	61.37	2840	19.58	2660	18.34		
Oak, white	1.25	8 619	3560	24.55	670	4.62	1250	8.62	770	5.31
	1.78	12 273	7440	51.30	1070	7.38	2000	13.79	800	5.52
Poplar, yellow	1.22	8 412	2660	18.34	270	1.86	790	5.45	510	3.51
	1.58	10 894	5540	38.20	500	3.45	1190	8.21	540	3.72
Poplar, balsam	0.75	5 171	1690	11.65	140	0.96	500	3.45		
	1.10	7 585	4020	27.72	300	3.45	790	5.45		
Sweetgum	1.20	8 274	3040	20.96	370	2.55	990	6.83	540	3.72
	1.64	11 308	6320	43.58	620	4.27	1600	11.03	760	5.24
Tupelo, black	1.03	7 102	3040	20.96	480	3.31	1100	7.58	570	3.93
	1.20	8 274	5520	38.06	830	5.72	1340	9.24	500	3.45
Walnut, black	1.42	9 791	4300	29.65	490	3.38	1220	8.41	571	3.93
	1.68	11 584	7580	52.26	1010	6.96	1370	9.45	690	4.76

Types and Uses

It is not within the scope of this book to cover structural wood design; the aim is only to give the basic information that is important and must be considered when designing with wood structurally. The various manufacturers of lumber give standard stress grades for various species of lumber according to its size. The following items are available stress-graded (all the dimensions given are nominal):

Light framing: 2 to 4 in. (50.8 to 101.6 mm) thick and 4 in. wide; available in three grades: Construction, Standard, and Utility.

Structural light framing: 2 to 4 in. thick and 2 to 4 in. wide; available in four grades: Select Structural, Select Structural 1, Select Structural 2, and Select Structural 3.

Studs: 2 to 4 in. thick and 2 to 4 in. wide; available in one grade, Stud.

Structural joists and planks: 2 to 4 in. thick and 6 in. (152.4 mm) and wider; available in the same four grades as structural light framing.

Appearance framing: 2 to 4 in. thick and 2 to 4 in. wide; available in one grade, Appearance.

Solid wood decking: available in a wide variety of softwoods in Select and Commercial grades, in thicknesses ranging from 2 to 4 in. and in widths of 4 in. and greater (*see* Figure *W70*).

The working stresses for the various structural grades can be obtained from the lumber manufacturers. It is also good practice to specify the allowable unit working stress he requires.

For any structural wood design other than light framing for residential work, structural engineers should be consulted.

Solid structural wood is usually joined with nails; but for trusses, headers, and trimmers, and for obtaining flush ceilings, a very large variety of rough hardware, hangers, ties, bolts, split-rings, gusset plates, and the like is available (*see* Figures *W71* and *W72*).

(*See also* Hardware, Rough; Hangers; Ties, Anchors; Nails; Wood, Structural Laminated.)

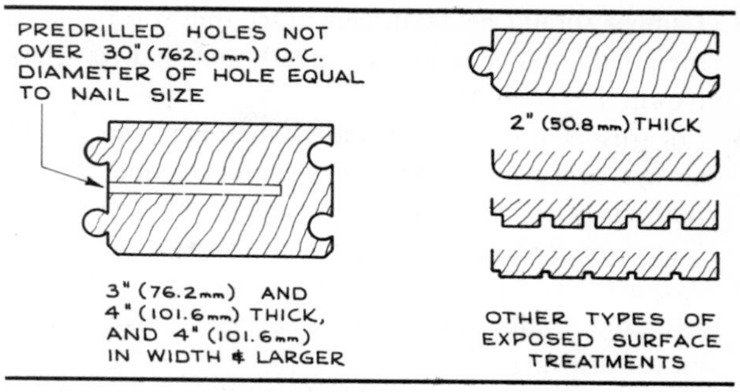

Figure W70 Typical solid wood decking.

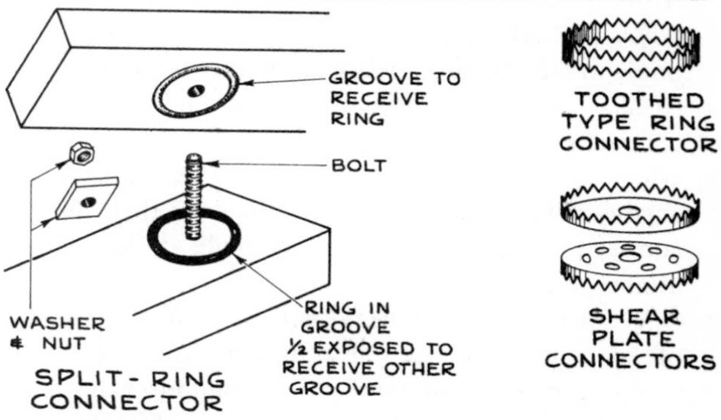

Figure W71 Typical methods of connecting wood trusses.

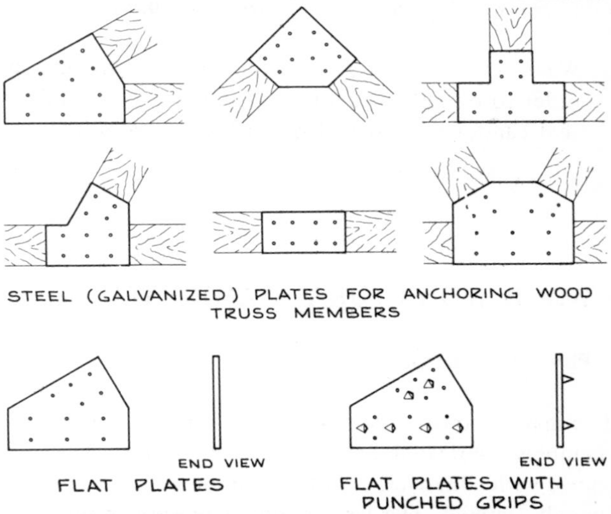

Figure W72 Types of wood-to-wood truss anchoring devices.

Application

Condensed Checklist

1. The lumber yards in the locality should be checked to see whether the types, sizes, and stress grades of structural lumber needed to meet the design requirements are available.
2. Local, municipal, and state codes and the codes of the Fire Underwriters, insurance companies, and federal government (Army, Navy, etc.) should be checked for regulations and limitations on stress characteristics for structural lumber and also for fire-resistance limitations and requirements for wood-framed buildings.

3. Lumber should be examined to see that it is the type specified, that it is grade-marked, and that its stress requirements are indicated.
4. For trusses or built-up structural members one should always consult with structural engineers, and specify that shop or fabrication drawings be submitted and that all methods of fastening be indicated.
5. Required working stresses that must be met by the structural lumber must always be specified and noted on the drawings.
6. Structural engineers should always be consulted for the best methods of joining and attaching wood structural members.
7. When prefabricated solid wood trusses are required, one should always check with the manufacturers as to size, design, method of joining, and method of connecting to the structure.
8. When solid wood trusses are required, a structural engineer should always be consulted for their design connections, and the type and grade of structural lumber required.
9. One should always check with the manufacturers of solid wood decking for limitations on support spacings, methods of connecting to supports and connecting decking together, and finishes for exposed surface.

Conditions Favorable to the Use of Structural Lumber

1. Where specific stresses are required in the framing in order to obtain a structurally sound and safe building that cannot be obtained with other types of lumber.

WOOD, STRUCTURAL LAMINATED

Physical Properties

Laminated construction may be defined as structural members fabricated of two or more layers of wood, joined with adhesives, with the grain of all layers approximately parallel. The laminations may vary in number, size, shape, and thickness, as well as in species of wood used. Laminated members are made from short and narrow lengths of lumber, and normal imperfections such as knots can be dispersed by selective placement of the laminations so that no loss of strength is incurred.

Avoiding Internal Stresses. Internal stresses caused by shrinking and swelling can be avoided by selecting

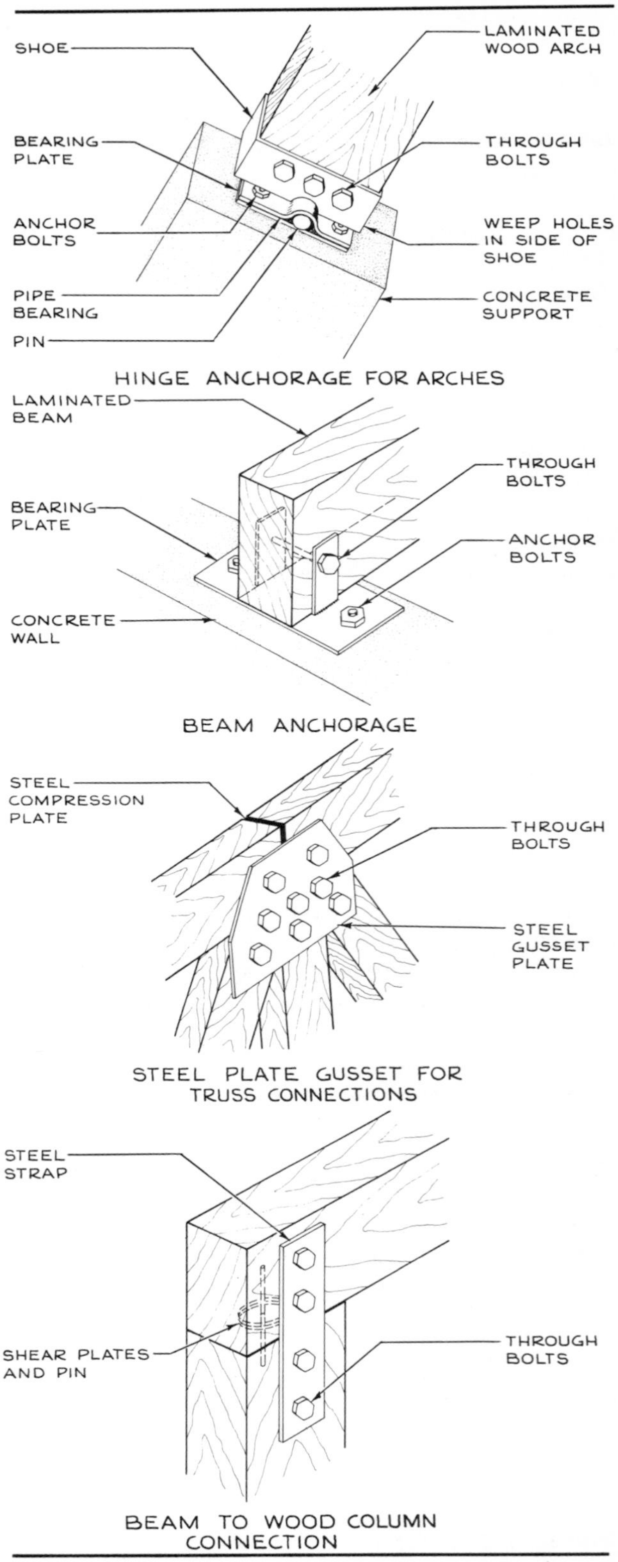

Figure W73 Various types of rough hardware for connecting laminated structural members.

laminations that will shrink or swell in equal amounts in the same direction and by selecting of lumber that is well seasoned.

Moisture Content. The range in moisture content among the laminations in the same assembly should be no greater than 5%. If the laminated member is to be used in an area where its moisture content may be raised to more than about 20%, heartwood of a durable species of wood should be used or the wood should be treated with preservatives and joined by highly moisture-resistant adhesives.

Evaluation of Joints. The quality of the joints made with adhesives for dry conditions is evaluated by the block shear test. Joints that are to be exposed to severe conditions should be capable of withstanding without significant delamination the high internal stresses that develop as a result of rapid wetting and drying. For these conditions an exterior type of adhesive must be used. (*See* Wood Joining; *see also* Adhesives, Plastics, Wood Finishes, *and* Wood Preservatives.)

Rough Hardware. For all types of structural laminated beams, girders, trusses, arches, etc., there are available a very large variety of rough hardware devices (*see* Figure *W73*).

Types and Uses

Laminated structural arches, trusses, rigid frames, girders, beams, purlins, and decking are made under controlled manufacturing conditions in the specific sizes, loading strengths, and shapes needed to meet design requirements and service conditions. As jigs, lumber, and manufacturing processes have become standardized, three-hinged arches and beams and purlins calculated for various spans, roof pitches, and loadings have become available as stock items. All structural laminated members meet rigid control and inspection standards and have product-quality marks (*see* Figure *W74*).

AITC P-143
QUALITY INSPECTED ® PS 56-73

Figure W74 A typical custom product quality mark.

Possible Shapes. Technical advances in adhesives have opened a relatively new field for the use of wood. Structural forms of almost any shape can now be built of laminated wood members. Dome shapes with interlocking triangular or radial patterns of small laminated wood members can span up to 300 ft (91.44 m) or more.

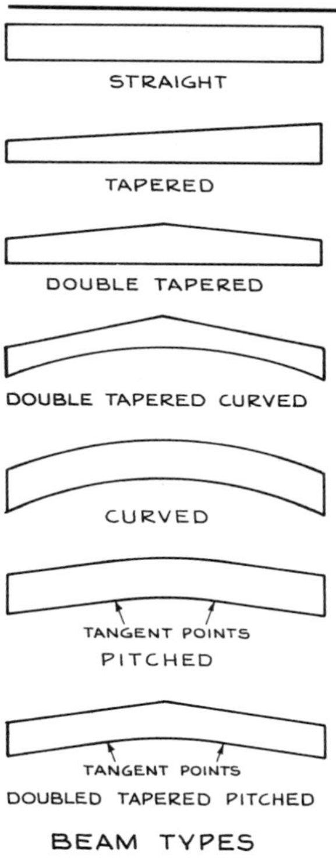

STANDARD SIZES			
NORMAL WIDTH		NET FINISHED WIDTH	
IN.	mm	IN.	mm
3	76.2	2¼	57.15
4	101.6	3⅛	79.38
6	152.4	5⅛	130.18
8	203.2	6¾	171.65
10	254.0	8¾	222.25
12	304.8	10¾	273.05
14	355.6	12¼	311.15
16	406.4	14¼	361.95

DEPTH DIMENSIONS CONTROLLED BY DESIGN REQUIREMENTS.

CONSTANT CURVATURE

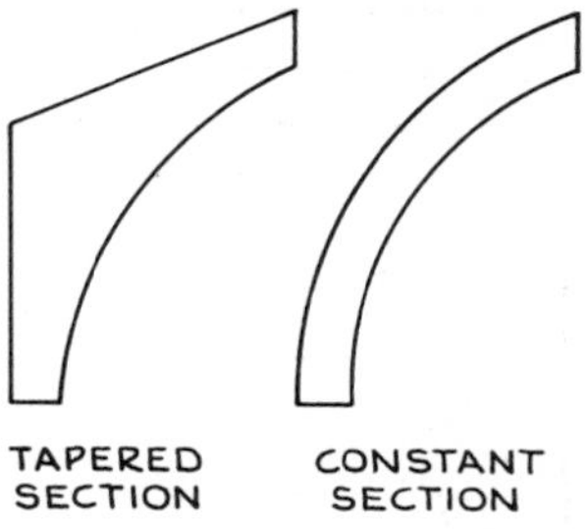

NET THICKNESS OF LAMINATIONS	MINIMUM RADIUS OF CURVATURE
⅜" (9.53 mm)	4'-8" (1.422 M)
½" (12.70 mm)	7'-2" (2.184 M)
¾" (19.05 mm)	12'-6" (3.180 M)
1⅝" (41.28 mm)	40'-0" (12.192 M)

TANGENT ENDS

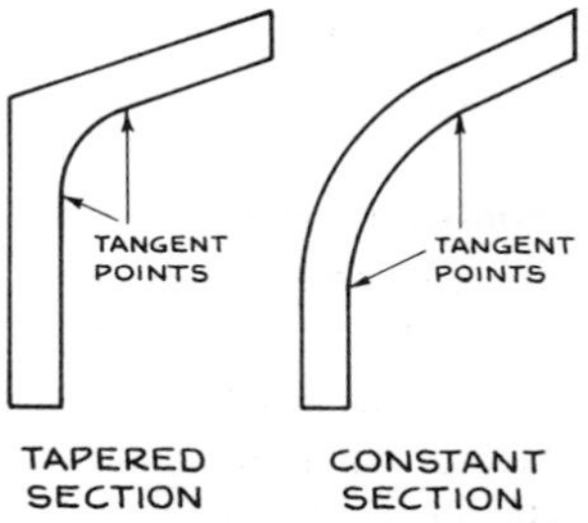

NET THICKNESS OF LAMINATIONS	MINIMUM RADIUS OF CURVATURE
⅜" (9.53 mm)	4'-0" (1.219 M)
½" (12.70 mm)	6'-0" (1.829 M)
¾" (19.05 mm)	9'-4" (2.845 M)
1⅝" (41.28 mm)	32'-0" (9.754 M)

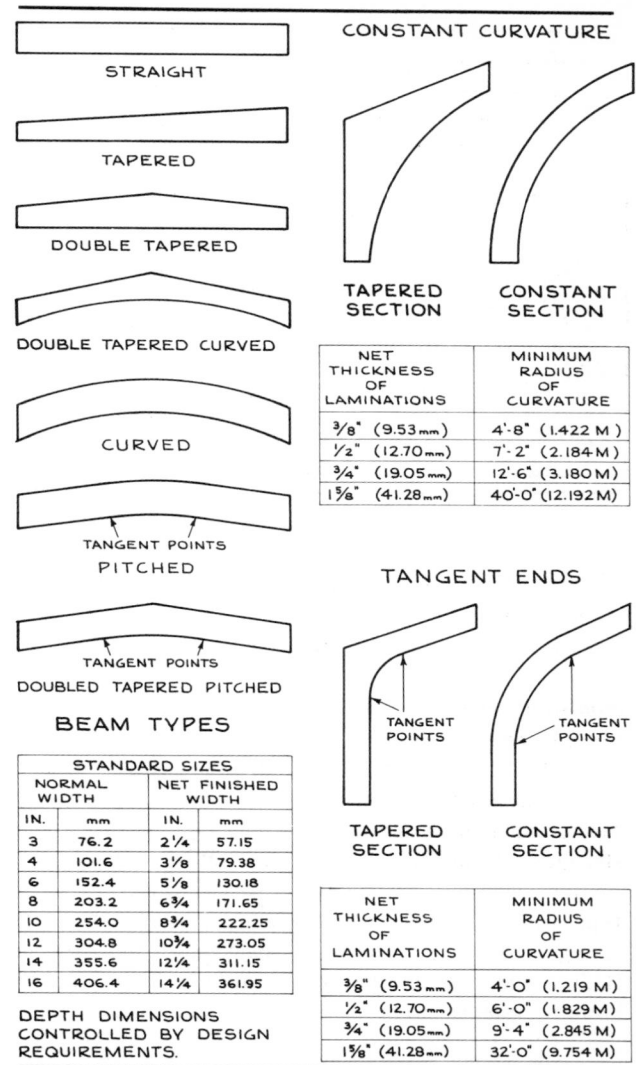

Figure W75 Typical structural laminated beams and arch types.

Figure *W75* shows typical structural laminated beams and arches with sizes, curvature limitations, and basic shapes, and Figure *W76* shows typical laminated decking types and sizes. In general, most laminated decking is made from Douglas fir, larch, southern pine, hemlock, fir, Idaho white pine, inland white fir, ponderosa pine, and inland red cedar. The exposed interior surface is available in the following finishes: smooth, grooved, striated, wire-brushed, prefinished, and finished in a wide variety of stains.

Laminated members made of combinations of plywood and solid wood can span 25 to 60 ft (7.620 to

SIZE			
NORMAL		ACTUAL	
IN.	mm	IN.	mm
3 x 6	76.2 x 152.4	2¼ x 5⅜	57.15 x 136.53
3 x 6 H AND 3 x 6 S	76.2 x 152.4	2 21/32 x 5⅜	67.47 x 136.53
4 x 6	101.6 x 152.4	3 x 5⅜	76.2 x 136.53
5 x 6	127.0 x 152.4	3¾ x 5⅜ 3 21/32 x 5⅜	92.25 x 136.53 92.82 x 136.53

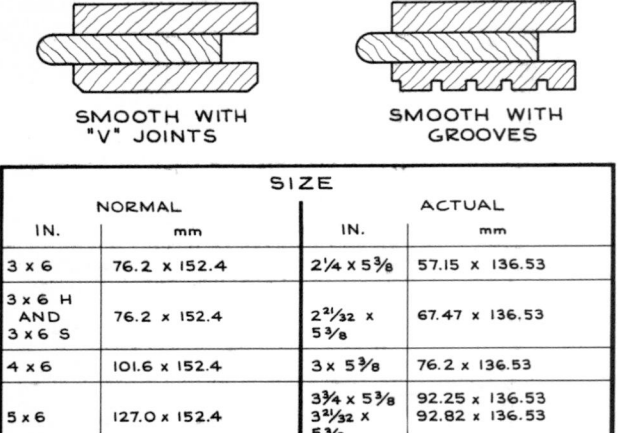

Figure W76 Typical laminated decking.

18.288 m) or more. Combinations of metal and wood using the wood for compression and the steel for tension are also available. (*See* Figure *W77.*)

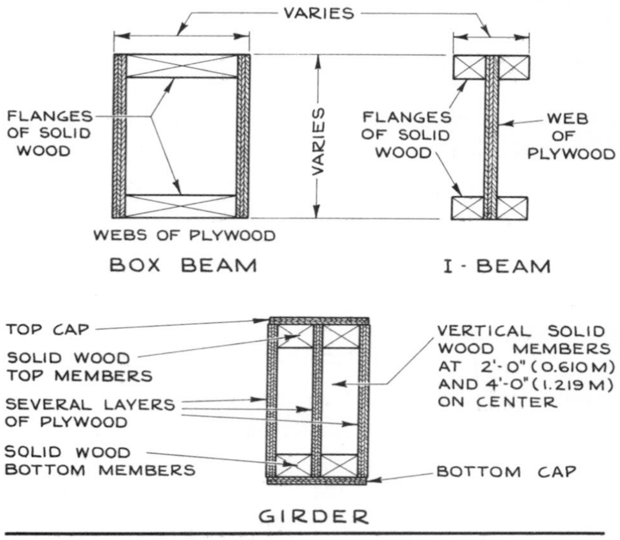

Figure W77 Examples of laminated members of plywood and solid wood.

Sandwich Panels. Although hollow-core doors have been known for about 30 years, the structural application of this principle is a development of the 1950s. Sandwich construction is formed by bonding two thin facings to a thick core. The thin facings are made of a strong, dense material and are the load-bearing elements of the construction. The core is a lightweight cellular-type material that separates and stabilizes the thin facings and carries shearing loads. Sandwich panels are finding wide use in the construction of prefabricated housing systems and in residential construction.

Stressed-skin Panels. Stressed-skin panels consist of wood stringers and plywood skins. Figure *W78* shows a typical stressed-skin panel. The stringers supply shear resistance, and the plywood skins resist bending. This type of panel is used efficiently for floors, walls, and roof components.

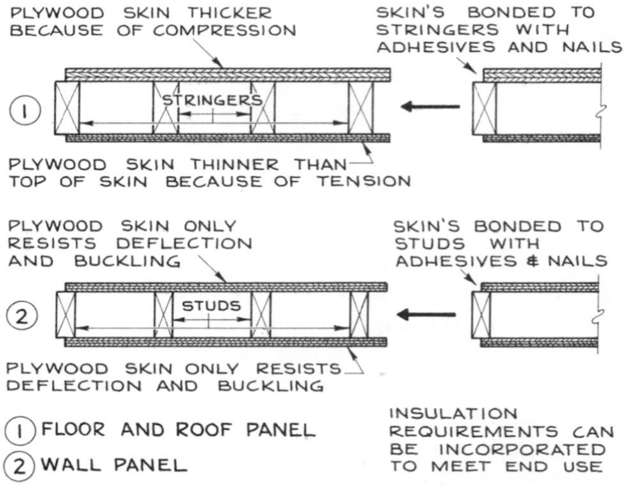

Figure W78 Typical examples of stressed-skin panels.

Application

All laminated structural members follow rigid rules and standards. It is beyond the scope of this book to cover the designing of laminated wood structural members.

Condensed Checklist

1. One should always check, with the aid of specialists in this field if necessary, the type of adhesive used, as well as the grade and stress requirements of the design (*see* Table *W54*), and one should make sure that the manufacturer is following the American Institute of Timber Construction Standards.
2. Shop drawings should always be called for by specifications and must show methods of attachment, grade, species of lumber, type of adhesive, and finish (if required).
3. Local, municipal, and state codes and also the codes of the Fire Underwriters, insurance companies, labor departments, and federal government (Army, Navy, etc.) should be checked for requirements.

Table W54 Grade classification of wood laminated structural members[a]

Grading criteria	Premium[b] Appearance grade[c]	Architectural[b] Appearance grade[c]	Industrial[b] Appearance grade[c]
Characteristics and requirements for appearance	Used where the finest appearance is demanded. In exposed surfaces, knotholes and other voids shall be replaced with clear wood inserts or a neutral colored filler. Inserts shall be selected with a similarity of color and grain to the adjacent wood. The wide face of exposed laminates shall be selected for appearance free of loose knots or voids and with a similarity of color and grain of the laminations at end and edge joints. Knot size is limited to 20% of the net face width of the lamination. No more than two maximum knots shall occur in a 6-ft (1.829-m) length. Exposed surfaces shall be surfaced smooth, misses not permitted. The corners of the wide face of exposed laminations in the final member shall be eased.	Used where appearance is an important requirement. In exposed surfaces knotholes and other voids in excess of $\frac{3}{4}$ in. (19.05 mm) shall be replaced with clear wood inserts or filler. The wide face of exposed laminations shall be free of loose knots and open knotholes. The material shall be selected for similarity of color and grain of the laminations at end and edge joints. Exposed surfaces shall be surfaced smooth, misses not permissed. The corners of the wide face of exposed laminations in the final member shall be eased.	Used where appearance is not of primary importance. Inserts and wood fillers are not required. The wide face of exposed laminations shall be free from loose knots and open knotholes, and shall be surfaced on two sides only, an occasional miss being permitted along individual laminations.

[a] Unless otherwise specified, laminated timber truss members shall be industrial grade or better, except that the wide face of exposed laminates is permitted to have loose knots and open knotholes.
[b] In all grades the laminations may possess the natural growth characteristics of the lumber grades.
[c] When opaque finishes are specified, similarity of color and grain is not required.

4. The relative humidity of the locality should be established; also, the relative humidity range in the actual building area where structural laminated members are to be installed should be determined so that the correct type of wood and adhesives, and preservative treatment if necessary, will be used.
5. Structural engineers should always be consulted when structural laminated members are to be used.
6. When laminated decking is selected, the size, span limitation, types of finishes, and type of lumber should always be checked.
7. For stressed-skin panels, a structural engineer should always be consulted to check design limitations for their end use.
8. Laminated members of plywood and solid wood, when selected, should be designed by a structural engineer, and the manufacturers of these types of components should be consulted.
9. When selecting stressed-skin panels or laminated plywood and solid wood beams, girders, and trusses, one should always check that a manufacturer of these types of construction components is available within economic distance.
10. Laminated wood structural elements are resistant to fire and in almost all conditions fall under the building codes for heavy timber construction. Therefore one should always make sure that the building code of the area where the structure is to be erected accepts laminated-wood types of structures as fire resistant.

Conditions Favorable to the Use of Structural Laminated Members

1. Where a strong, durable, lightweight, economical structural system that permits large open areas free from intermediate supports is desired.

2. Where wood structural members are required to support spans and loads that cannot be met by solid wood members.
3. Where a natural wood appearance is desired.
4. Where fire resistance is important.
5. For preassembled and prefabricated systems in the field of housing construction.

Conditions Unfavorable to the Use of Structural Laminated Members

1. As structural members for large unsupported spans where they must also support multiple stories above.

History and Manufacture

Laminated construction has been used for furniture for many decades but only during the past 25 years in this country for structural members. The first laminated structural members were arches made of softwood (evergreen) and casein glue near the beginning of this century in Europe. During World War I experiments were made with laminated wood structural members for aircraft and soon after were extended into the building field. In the United States the first use of laminated arches was in a building erected in the middle thirties by the Forest Products Laboratory. With the development of highly moisture-resistant synthetic adhesives, the use of laminated construction expanded steadily, and today laminated wood structural members are a recognized and distinctive structural material in the field of construction.

WOOD WINDOWS

Wood windows are generally made from ponderosa pine. They are also made from a wide variety of evergreen (softwood) trees but rarely from deciduous (hardwood) trees (*see* Table *W55*). Wood windows are mill-made under controlled conditions with the wood kiln-dried and treated against attack by mold, fungi, and insects. They are available in a large variety of sizes and types and with sash, frames, weatherstripping, trim, hardware, and glass as separate items or in various combinations or as a complete packaged unit.

Types and Uses

Generally, wood windows increase in size in 4-in. (101.6-mm) multiples for both height and width. Figures *W79* through *W84* show the types and sizes of stock wood windows commonly available.

Table W55 Species of wood used for windows

Deciduous trees (hardwoods)	Evergreen trees (softwoods)
Basswood, magnolia, red alder	Douglas fir, eastern hemlock, eastern spruce, Engelmann spruce, Idaho white pine, Norway pine, ponderosa pine, red cypress, redwood, Sitka spruce, southern yellow pine, sugar pine, West Coast hemlock, western larch, western red cedar, white fir

SIZE OF SINGLE CASEMENT				
1'-6"	3'-1¾"	3'-1¾"	6'-5¼"	8'-1"
1'-6¾"	3'-2½"	3'-2¼"	6'-6"	8'-1¾"
1'-7"	3'-9⅝"	3'-9⅝"	7'-0⅞"	8'-8½"
1'-7"	3'-3"	3'-3"	6'-7"	8'-3"
1'-7³⁄₁₆"	3'-3¼"	3'-3¼"	6'-7⅜"	8'-3⁷⁄₁₆"
1'-10"	3'-10¾"	3'-10¾"	7'-11¼"	9'-11½"
1'-11"	3'-11"	3'-11"	7'-11"	9'-11"
1'-11¼"	3'-11½"	3'-11½"	8'-0"	10'-0¼"
2'-3"	4'-7"	4'-7"	9'-3"	11'-7"

SIZE OF SINGLE CASEMENT				
457.2mm	959.2mm	959.2mm	1962.2mm	2429.8mm
476.6mm	977.9mm	977.9mm	1981.2mm	2449.2mm
482.6mm	1158.9mm	1158.9mm	2155.8mm	2620.3mm
482.6mm	990.6mm	990.6mm	2006.6mm	2480.6mm
487.4mm	997mm	997mm	2016.1mm	2492.8mm
558.8mm	1187.8mm	1187.8mm	2419.4mm	3035.3mm
584.2mm	1193.8mm	1193.8mm	2413mm	3022.6mm
590.6mm	1215.5mm	1215.5mm	2404.4mm	3054.4mm
685.8mm	1397mm	1397mm	2819.4mm	3530.6mm

Figure W79 Casement windows.

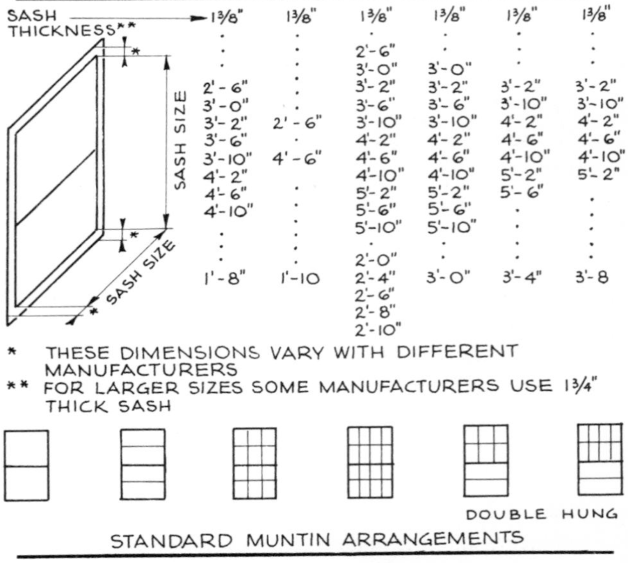

Figure W80 Double-hung windows.

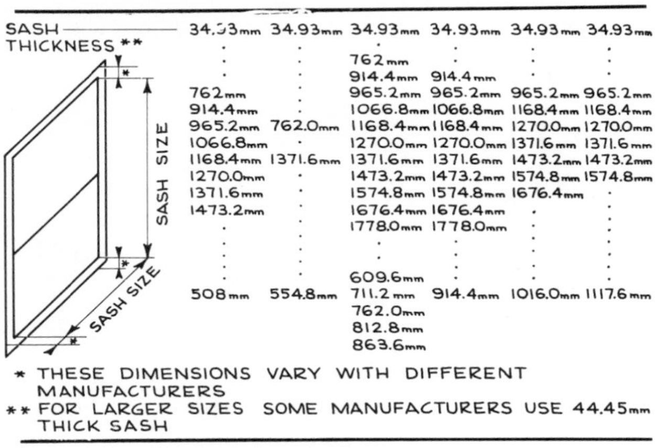

SASH THICKNESS **
34.93mm 34.93mm 34.93mm 34.93mm 34.93mm 34.93mm
SASH SIZE
SASH SIZE
762mm 914.4mm 965.2mm 1066.8mm 1168.4mm 1270.0mm 1371.6mm 1473.2mm
762.0mm
1371.6mm
762mm 914.4mm 965.2mm 1066.8mm 1168.4mm 1270.0mm 1371.6mm 1473.2mm 1574.8mm 1676.4mm 1778.0mm
914.4mm 965.2mm 1066.8mm 1168.4mm 1270.0mm 1371.6mm 1473.2mm 1574.8mm 1676.4mm 1778.0mm
965.2mm 1168.4mm 1270.0mm 1371.6mm 1473.2mm 1574.8mm 1676.4mm
965.2mm 1168.4mm 1270.0mm 1371.6mm 1473.2mm 1574.8mm
508mm 554.8mm
609.6mm 711.2mm 762.0mm 812.8mm 863.6mm
914.4mm 1016.0mm 1117.6mm
* THESE DIMENSIONS VARY WITH DIFFERENT MANUFACTURERS
** FOR LARGER SIZES SOME MANUFACTURERS USE 44.45mm THICK SASH

Figure W80 Double-hung windows (*continued*).

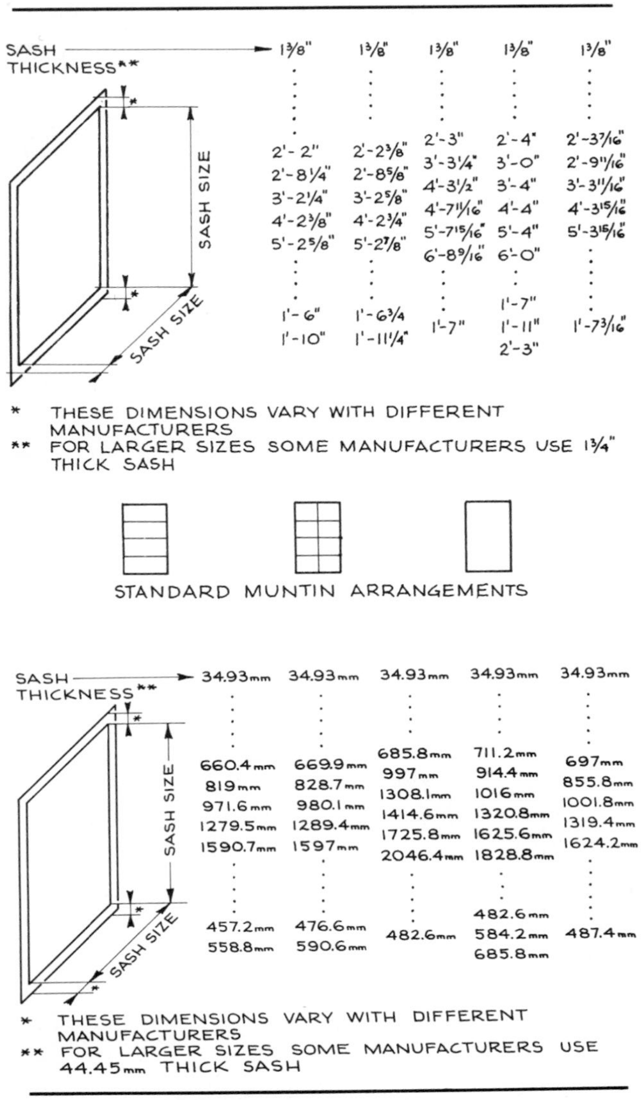

Figure W81 Fixed sash windows.

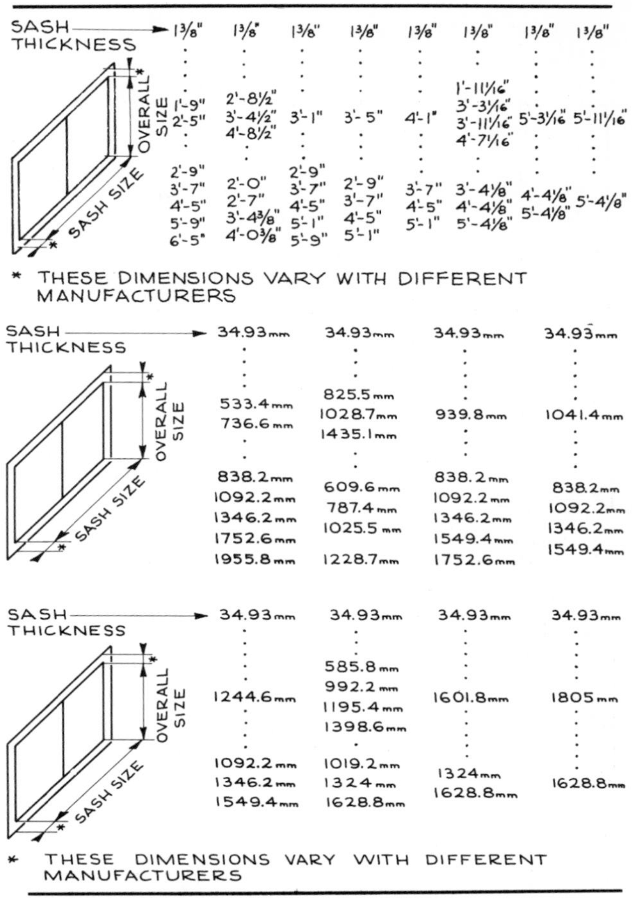

SASH THICKNESS
1⅜" 1⅜" 1⅜" 1⅜" 1⅜" 1⅜" 1⅜" 1⅜"
OVERALL SIZE
SASH SIZE
1'-9" 2'-5"
2'-8½" 3'-4½" 4'-8½"
3'-1"
3'-5"
4'-1"
1'-11 1/16" 3'-3 1/16" 3'-11 1/16" 4'-7 1/16"
5'-3 1/16
5'-11 1/16"
2'-9" 3'-7" 4'-5" 5'-9" 6'-5"
2'-0" 2'-7" 3'-4⅜" 4'-0⅜"
2'-9" 3'-7" 4'-5" 5'-1" 5'-9"
2'-9" 3'-7" 4'-5" 5'-1"
3'-7" 4'-5" 5'-1"
3'-4⅛" 4'-4⅛" 5'-4⅛"
4'-4⅛" 5'-4⅛"
5'-4⅛"
* THESE DIMENSIONS VARY WITH DIFFERENT MANUFACTURERS
SASH THICKNESS
34.93mm 34.93mm 34.93mm 34.93mm
OVERALL SIZE
SASH SIZE
533.4mm 736.6mm
825.5mm 1028.7mm 1435.1mm
939.8mm
1041.4mm
838.2mm 1092.2mm 1346.2mm 1752.6mm 1955.8mm
609.6mm 787.4mm 1025.5mm 1228.7mm
838.2mm 1092.2mm 1346.2mm 1549.4mm 1752.6mm
838.2mm 1092.2mm 1346.2mm 1549.4mm
SASH THICKNESS
34.93mm 34.93mm 34.93mm 34.93mm
OVERALL SIZE
SASH SIZE
1244.6mm
585.8mm 992.2mm 1195.4mm 1398.6mm
1601.8mm
1805mm
1092.2mm 1346.2mm 1549.4mm
1019.2mm 1324mm 1628.8mm
1324mm 1628.8mm
1628.8mm
* THESE DIMENSIONS VARY WITH DIFFERENT MANUFACTURERS

Figure W82 Sliding windows.

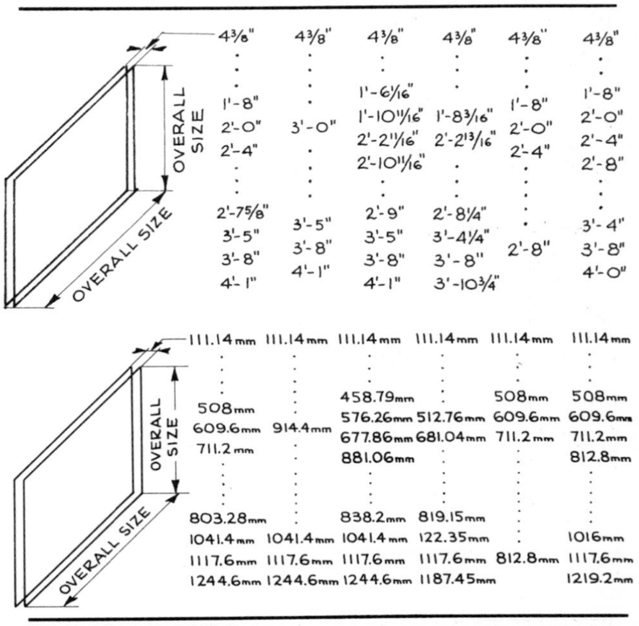

Figure W83 Single-awning hopper or casement-type windows.

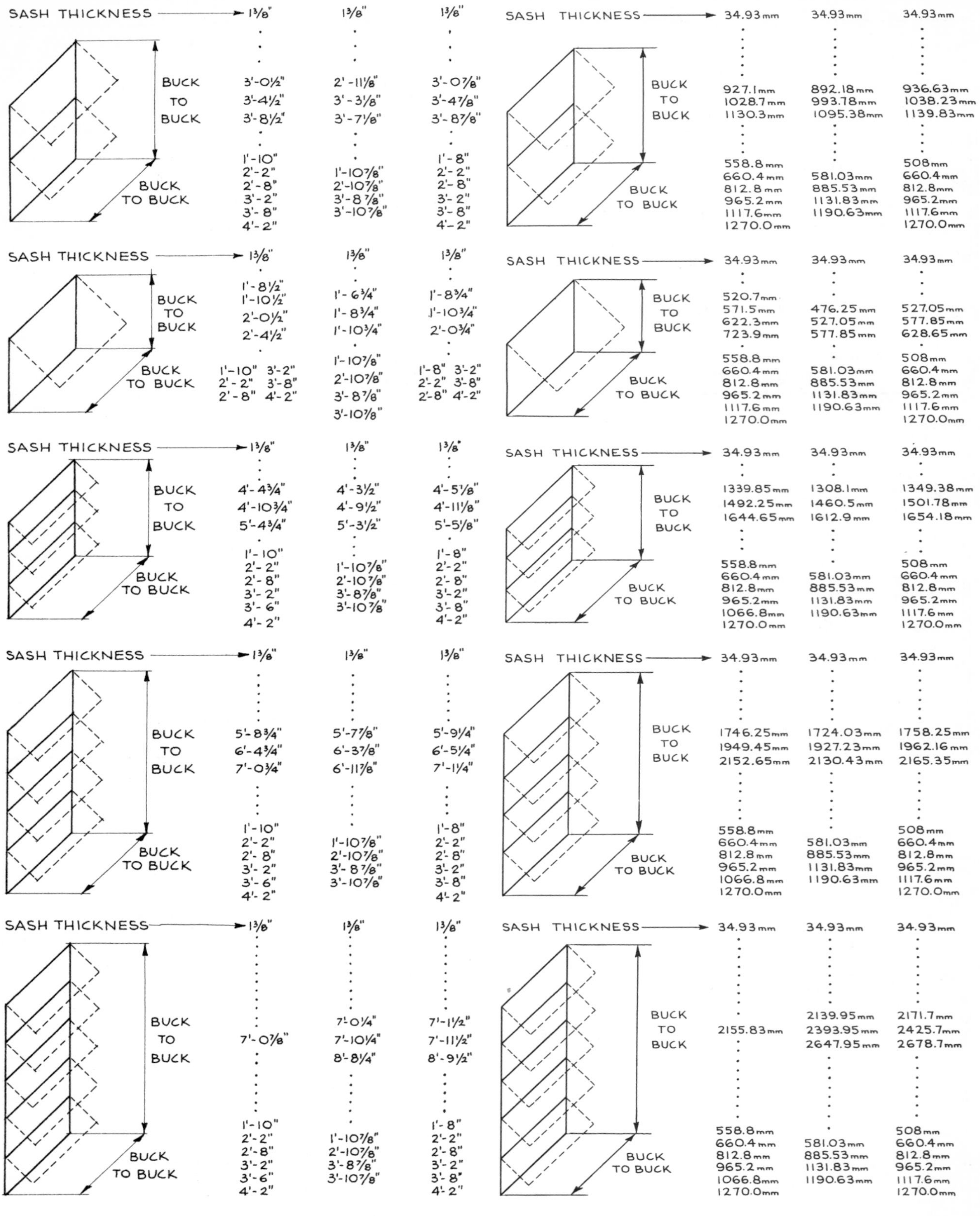

Figure W84 Multiple-awning windows.

There are fixed windows that can be combined with these various types or used as individual fixed windows. For all types of complete window units, insect screen and storm sashes can be obtained as part of the package. Usually the windows are glazed with window glass, but almost all manufacturers make these windows with insulating glass, plate glass, and all other types of glass except those that require special glazing treatment.

Various special window units such as quarter-round, half-round, circular, and many types of bay or bow windows are available. Some manufacturers make complete modular systems of windows which can act as a structural support. The overall trend is towards prefabricated units.

All window types are now available with vinyl coatings in a variety of colors. This vinyl coating makes the window completely maintenance-free for a considerably long period of time.

Application

Condensed Checklist

1. All dimensions of wood windows should be checked with the materials used for exterior and interior finishes so that they will fit correctly; for example, brick coursing, shingle or siding exposure and coursing, etc., must be calculated in relation to the size and design of window.
2. Wood windows must be of the correct thickness and have the right trim for the type of wall construction into which they are to be installed.
3. Available types of hardware, weatherstripping, and glazing should be checked to see that they meet the requirements of their end use.
4. Available species of wood from which windows are manufactured should be checked if the design requires that the windows match the exterior wood.
5. The lumber used for the windows should be treated against mold, fungi, and insects, especially if the windows are to be installed in a hot, humid location or climate.
6. One should always check the method of operating windows when screens or storm sashes are to be installed.
7. When installation is completed, all areas around the window must be made weathertight and have no air leakage.
8. Windows must never support any structural loads, and lintels must be calculated so that no deflection will cause stresses to be transferred to the windows.

Condition Favorable to the Use of Wood Windows

Where an economical, weathertight, airtight, durable, easily installed window is required.

Conditions Unfavorable to the Use of Wood Windows

1. Where a fire-resistant window is required.
2. Where repainting and maintenance should be avoided, unless vinyl-coated window units are used.

X

XENON

Physical and Chemical Properties

Symbol: Xe
Atomic number: 54
Melting point: −111.9°C, −169.42°F
Boiling point: −107.1°C, −160.78°F
Specific gravity: −3.52 (−103°C, −217.4°C) liquid

Xenon is a colorless, odorless, and tasteless gas used in xenon-filled arc lamps.

History and Manufacture

In 1898 Sir William Ramsay and M. W. Travers discovered xenon. It is produced in air-separation plants.

Z

ZINC

Physical and Chemical Properties

Symbol: Zn
Atomic number: 30
Specific gravity: 7.133 (25°C, 77°F)
Melting point: 419.58°C, 787.24°F
Boiling point: 907.0°C, 1664.6°F
Tensile strength: 4130 lbf/in.^{2} (28.48 MN/m^2) for pure zinc at room temperature; up to 45,000 lbf/in.^{2} (310.38 MN/m^2) in cast-alloy form

Zinc is a medium-hard, bluish white metal that is characterized by brittleness and low strength and is subject to creep.

Corrosion Resistance. Zinc is readily attacked by acids and alkalis. It is resistant to corrosion by water. On exposure to air, a film of zinc carbonate or oxide forms which protects zinc from further oxidation. Zinc can come into contact with lead, tin, aluminum, and wood, but in the presence of moisture it should be insulated from all other metals and from redwood and cedar, both of which contain zinc-corroding acids.

Workability. Zinc can be hot- and cold-rolled, drawn, formed, extruded, spun, punched, cast, and machined by ordinary methods; it can be riveted, soldered, and welded. It should preferably be bent against the grain and not heated above 150°F (65.56°C). At the temperature of boiling water (212°F, 100°C) zinc sheet becomes annealed.

Commercial Forms. Zinc is marketed as slab zinc, available in various grades of purity (*see* Table *Z1*), bar, plate, stick, sheet, shot, strip, foil, wire, tubing, and powder form and in special shapes.

Types and Uses

The most important uses of zinc are (1) as protective coatings (galvanizing) on iron and steel, (2) as die-casting metal, and (3) as an alloying element in brasses.

Zinc Compounds. Zinc compounds are essential ingredients in the paint, ceramic, rubber, paper, plastics, and textile industries, where they are used as pigments and fillers. These and other uses of zinc are summarized in

Table Z1 Grades of slab zinc and their major uses

Grade	Zn (min.)	Pb (max.)	Fe (max.)	Cd (max.)	Total Pb-Fe-Cd allowed	Major uses
	(percent of content)					
Special high grade	99.99	0.006	0.005	0.004	0.01	Die-casting alloys, electrogalvanizing
High grade	99.90	0.070	0.02	0.07	0.10	Brass, rolled zinc
Intermediate	99.50	0.20	0.03	0.5	0.5	Divided among rolled zinc, galvanizing, and brass
Brass special	99.00	0.60	0.03	0.5	1.0	
Selected	98.75	0.80	0.04	0.75	1.25	
Prime Western	98.32	1.60	0.08	[a]	[a]	Hot-dip galvanizing

[a] Not specified.

Table Z2 Uses of zinc

Construction Uses

Basis of–	Abrasives*	Polishing ingredient
	Admixtures	Ingredient of concrete hardener
	Protective coatings*	Galvanizing; plating; sherardizing
Component of–	Adhesives	Clarifying; bleaching; fungicide; white color; opacity
	Aluminum	Bonderizing; lubricant in dies
	Antimony*	Lowering melting point
	Asphalt	Filler; white coloring
	Bismuth*	Lowering melting point
	Brass*	Ingredient
	Bronze*	Minor ingredient
	Cement	Concrete hardener
	Ceramics	White, yellow and green colors; flux
	Copper*	Making brasses and some bronzes
	Cork	Whitening
	Fibers	Bleaching; waterproofing
	Glass	Etching; zinc glass; opacity; white color
	Glazes and porcelain enamels	White, yellow, and green colors; flux
	Gold*	Changing color
	Iron*	Galvanizing; plating; cathodic protection
	Linoleum	Filler and whitening
	Magnesium*	Increasing resistance to salt water; increasing strength and toughness
	Paint	White paint solid; lithopone; flattening agent; rust-inhibiting; yellow and green colors; fluorescence; drier; wetting; dispersing and hardening agent; fungi and mildew inhibiting; neutralizing lime and alkalinity of concrete and cement surfaces
Component of– (Cont.)	Paper	White coatings; brightening; bleaching pulp; filler
	Steel*	Galvanizing; sherardizing; bonderizing
	Plastics	Lubricant; stabilizer; white, yellow, and green colors; filler
	Protective coatings	Zinc dust paint for galvanized iron and steel
	Rubber	Vulcanizing; accelerator; drying; lubricant; white color; filler
	Textiles	Mothproofing; printing; dyeing; mordant; bleaching; fireproofing; stripping agent; coatings; whitening; filler
	Tin*	Promoting fluidity
	Titanium*	Hardening
	Wood	Preservative
Allied Construction Uses–	Lamps	Fluorescence
	Plumbing*	Fittings
	Solder*	Ingredient of aluminum solders

Nonconstruction Uses

Bearings, coinage, household appliances, lithography, luminous dials, metallurgy, ornaments, photoengraving, television, wet and dry batteries, X-ray

Table *Z2*; uses in metallic (or alloy) form are indicated by asterisks.

Zinc Die-Castings and Alloys. Zinc die-castings find many uses in the construction field, for instance, in hardware, electrical devices, light fixtures, and bathroom accessories. Zinc die-casting alloys have been well received because of low cost, ease of casting at low temperatures, adequate strength for many purposes, dimensional accuracy, minimum finishing requirements, and corrosion resistance. Die-casting alloys contain about 95% zinc with about 4% aluminum and magnesium in varying proportions, with or without copper. The castings can be finished by buffing, polishing, and brushing; they can be plated with copper, nickel, chromium, brass, silver, or black nickel; they can be chemically treated with chromates, phosphates, or molybdates; the surface can be painted, enameled, varnished, lacquered, or coated with plastic.

Rough Hardware. Rough hardware of many types made of zinc and zinc alloys or galvanized with zinc is manufactured for use with zinc installations, but it is also used for many other types of installations where corrosive attack, electrolytic action, and other special conditions must be met. Some of the more common rough hardware items are nails, screws, nuts, bolts, and washers.

Other Specialized Uses. Other specialized uses of zinc where corrosion resistance and nonstaining properties are required are in construction with stone, brick, concrete, cement, plaster, terrazzo, and wood. In these cases zinc can be used in accessories such as anchors, cavity wall ties, flashing, reglets, sash reglets, corner beads, screws, expansion joints, terrazzo strips, and weatherstripping.

Zinc or zinc and aluminum in varying proportions provide the best soldering material for aluminum.

History and Manufacture

Paracelsus is given the credit for the discovery in 1520 of zinc as metal, but again we do not really know who first isolated it in metallic form because Portuguese traders had brought zinc back from China before Paracelsus' discovery. The art of zinc smelting developed about 1730 in England. The modern zinc industry dates from 1806, when zinc smelting started in Liege, Belgium, as the result of a decree by Napoleon. It was there that the horizontal retort process was developed. The United States began production in 1858 and since 1909 has been the world's leading producer and consumer.

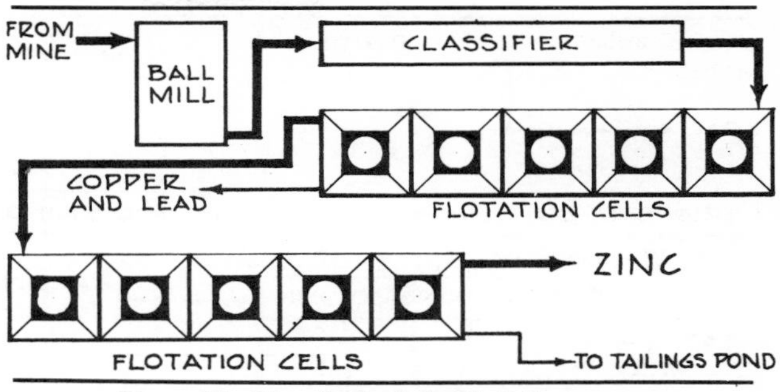

Figure Z1 Flowchart of zinc flotation process.

Horizontal Retort Process. The horizontal retort process developed in Belgium was for many years the only commercial method for recovering zinc. Now it is used only where gas for heat, ores with a minimum zinc content of 60%, and retort clays for making retorts are available.

Ore Treatment by Flotation. In modern zinc-smelting procedures, the ores must be concentrated prior to treatment for recovery of the metal. First, crushing and grinding the ore physically separates most ore particles from waste rock particles; then the ore is processed by flotation (*see* Figure *Z1*). In this method, advantage is taken of the fact that sulfide mineral particles will adhere to the surface of an air bubble. This affinity varies with different sulfides and can be controlled by flotation reagents that affect the surface properties of the mineral particles. The common zinc mineral sphalerite is frequently associated with lead, copper, and iron sulfides, from which it must be separated. The lead and copper sulfides are floated free of the ore pulp with pine oil, cresylic acid, and phosphorus pentasulfide as reagents, while the sphalerite (ZnS) and pyrite (FeS_2) are depressed with sodium cyanide and lime. After removal of the lead-copper minerals, a reagent such as copper sulfate is added to reactivate the sphalerite, and new increments of pine oil and other reagents are added to promote frothing and flotation of the zinc sulfide. The zinc concentrates, after being thickened, filtered, and dried, are then ready for reduction to metal.

Vertical Retort Process. The vertical retort process for carbon reduction (developed in 1929) was the first continuous operation (*see* Figure *Z2*). Here the roasted concentrates are mixed with coking coal, briquetted, dried, and then coked by heating with waste gases from the reduction furnace. After entering the top of the reduction furnace, the descending column of briquettes is heated to reduction temperature; metallic zinc vapor and carbon monoxide rise to the top and are removed. The

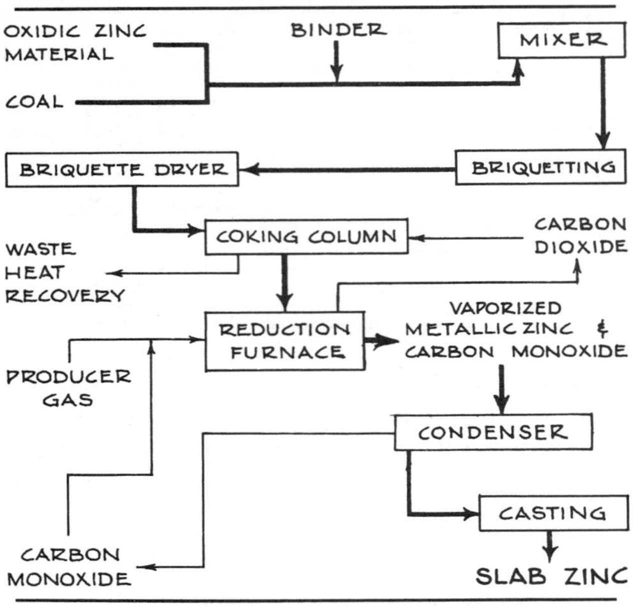

Figure Z2 Vertical retort zinc smelter.

zinc vapor is condensed and cast into slab zinc, and the hot carbon monoxide is reused for heating the reduction furnace.

Electrolytic Process. The electrolytic process of zinc production is based on electrolytic deposition of zinc from zinc sulfate solutions (*see* Figure *Z3*). It is the preferred method in areas of cheap power and is used throughout the world. Roasted zinc concentrate is leached with a weak solution of sulfuric acid. The leach solution of zinc sulfate that is filtered from the pulp also contains copper, cadmium, antimony, manganese, and other metal ion impurities. These are precipitated by controlled additions of zinc dust and filtered from the pregnant zinc solutions. The purified zinc solution is pumped to electrolytic cells in which the cathodes are pure aluminum and the anodes either chemical lead or a lead-silver alloy. Pure zinc is deposited on the cathodes, and the barren electrolyte is circulated back to the leaching process and begins the cycle again. The zinc is stripped from the cathodes, melted, skimmed, and cast into slabs.

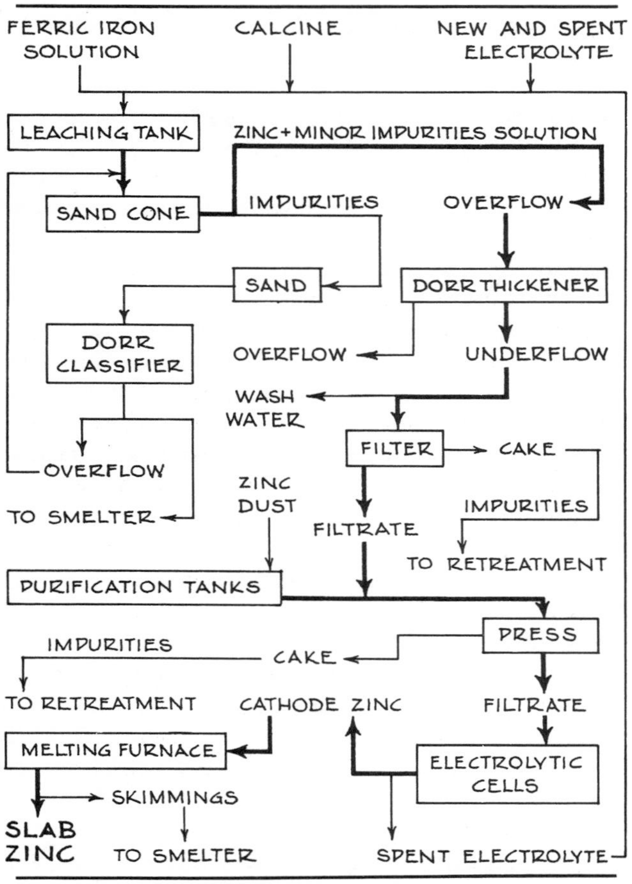

Figure Z3 Electrolytic process for recovery of zinc.

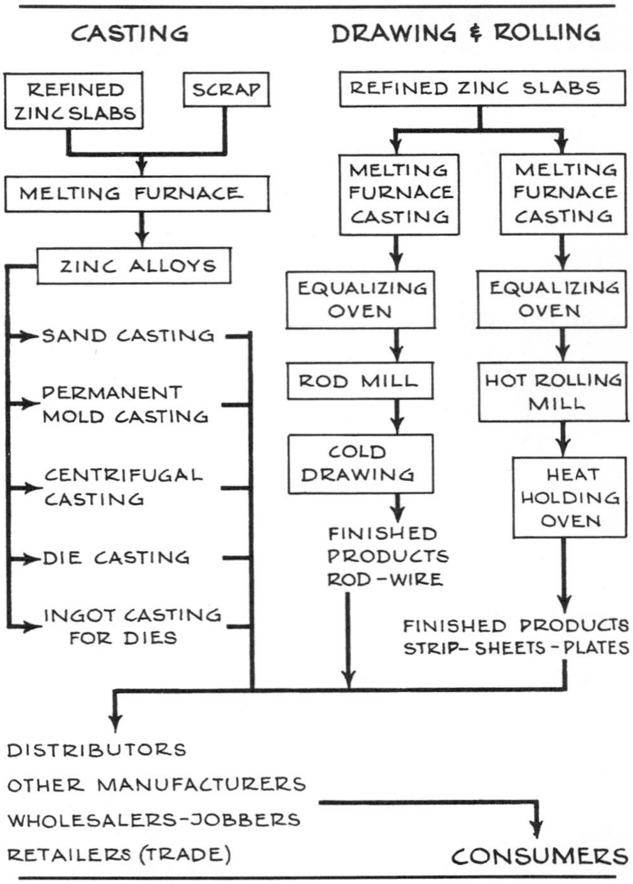

Figure Z4 Flowchart of zinc fabrication.

Fabrication. Slab zinc is fabricated into finished products by casting, drawing, and rolling as shown on the flowchart of zinc fabrication (*see* Figure *Z4*).

Zinc residues from metallurgical procedures are important sources of zinc compounds.

Secondary Zinc. Secondary zinc, new and old, is recovered from copper-base and zinc-base scrap. Zinc differs from most metals in that few zinc products are returned for reprocessing, and therefore the main source of secondary zinc is plant (new) scrap.

ZINC COATINGS

Physical and Chemical Properties

Hot-dipping, electrogalvanizing, spraying, and sherardizing are all processes whereby a protective coat of zinc is applied to steel and iron to seal them against corrosion. The advantage of coating with zinc is that, should the iron or steel become exposed through wear, aging, or discontinuities such as pinholes in the coating, galvanic reaction between the coating and base metal causes the zinc to corrode and form compounds that cover and continue to protect the iron and steel for as long as any zinc remains.

Most hot-dipped galvanized zinc coatings freeze into a crystalline surface pattern, known as *spangles*. These are not a criterion of quality, but for appearance's sake an effort is made to produce the largest and best form of spangles.

In older galvanizing methods for sheet and other shapes of metal, *flaking* of the zinc coat was a problem. During the hot-dipping process a multiple-layered structure of iron-zinc alloys is built up between the base metal and the outer surface of zinc coating. The middle layer tends to be hard and brittle and forms an irregular boundary with the surface finish layer, which usually has the composition of the zinc bath. Additions of aluminum and the development of continuous galvanizing processes have led to thinner but more ductile, or flexible, coatings.

In the electrolytic, sherardizing, spraying, and fused-coating processes of zinc coating, pure zinc or a zinc alloy is deposited as a single layer.

Thick versus Thin Coatings. Although the thickness of the zinc coating controls the rust-free life of the base metals, there are applications where a thick layer is not necessarily desirable inasmuch as a thick layer of zinc is brittle and will crack when bent sharply.

Types and Uses

The most common galvanized material used in construction is galvanized iron (steel) sheet and strip. It is available flat or corrugated, with the surface plain or refinished with various other surface materials.

There are several methods of preparing galvanized sheets for the application of paint or other organic coatings. Hot-dipped sheets where the zinc coating is wiped down when leaving the hot zinc bath have a zinc-iron alloy coating of 0.1 to 0.5 oz/ft^2 (30.52 to 152.50 g/m^2) which does not have spangles and is suitable for painting after cleaning.

Galvanized sheets that are heat treated after coating produce a zinc-iron alloy surface free of spangles and, with normal cleaning, providing a suitable surface for paint. The most widely used process is a hot phosphate treatment in which a chemical solution or a series of chemical solutions react with the zinc to produce a largely zinc phosphate surface film that, with normal cleaning, is a suitable surface for paint.

Galvanized sheet which is allowed to weather will produce a surface that is suitable for paint.

Paints manufactured specifically for painting galvanized surfaces (particularly paints with a latex vehicle) can be applied to galvanized surfaces without pretreatment.

Painting. Where severe corrosive conditions exist, it is advisable to protect galvanized iron with paint. Care should be taken as most paints will not adhere to spangled galvanized iron unless it is completely cleaned of grease and prepared for painting.

Discoloration During Storage. Galvanized sheets or rigidized sheets can become defaced and discolored when subjected to dampness and extremes of temperature. If the sheets are piled flat in the open or tightly bundled in a warehouse, the zinc coating can also be damaged by the consequent absence of oxygen and carbon dioxide between the sheets. This absence prevents the formation of a protective film of zinc carbonate; instead, zinc hydroxide forms and destroys the galvanizing. The discoloration, commonly called white storage stain, can be avoided by dipping in hydrochloric acid, rinsing, and then dipping in a solution of dichromate, after which final rinsing and drying are necessary.

History and Manufacture

Zinc coatings appeared during the first half of the 18th century, and in 1837 Crawford took out the first patent for hot-dip galvanizing. Currently, four methods are used to apply metallic zinc coatings: hot-dip galvanizing, electrogalvanizing, metallic spraying, and sherardizing. The most recent process is continuous galvanizing of steel strip and sheet from large coils instead of individual large sheets. Continuous galvanizing processes for zinc-coated wire and steel sheet and strip now account for 90% of the output of these items.

Hot-Dip Galvanizing. To date hot-dip galvanizing of steel represents the largest consumption of zinc in this country. The process may be semiautomatic or manual (for fabricated items) or mechanical (dipping of sheet, wire, and pipe).

Electrogalvanizing. In the electrolytic process, iron or steel, after thorough cleaning, is immersed in an electrolyte solution of zinc sulfate or cyanide which may contain small amounts of other substances. By electrolytic action a coat of pure zinc is deposited on the iron or steel. This process allows for complete control of the thickness of the coat, but it does not usually provide the thick coatings that give maximum protection.

Sherardizing. The sherardizing process consists of placing the thoroughly cleaned iron or steel objects to be coated into an enclosure in which they are surrounded with metallic zinc dust and then heating them. In this way a clear, thin zinc-alloy coating is obtained which molds itself to any surface design. This process is limited to objects of relatively small size or items that can be coiled, such as electrical conduits.

Metallic Spraying. Metallic spraying consists of applying a fine spray of molten zinc to the thoroughly cleaned iron or steel. A combination process exists in which the coating is heated and fused with the iron or steel, with which it forms an alloy. This produces a less brittle zinc coating which will not peel or flake on bending.

ZINC OXIDE

Physical and Chemical Properties

Zinc oxide (ZnO) is a fine, white powder that is used both as a pigment and as a chemical. It does not melt (volatilizes at 3092°F, 1700°C), is insoluble in water or alcohol, and is unaffected by sulfur. It is soluble in dilute acids, ammonium hydroxide, and strong alkalis. It is nontoxic. Its refractive index of 2.0 is largely responsible for its whitening and hiding power properties. The pigment is not affected by ultraviolet light but completely absorbs such light and converts it into heat. Zinc oxide has low heat conductivity and high electrical resistance.

Commercial Forms. Zinc oxide pigment is available in two types and in a wide range of grades based on specific use, chemical purity, and particle size and shape. Major differences in the two types are shown in Table *Z3*.

Table Z3 Minimum requirements of zinc oxides

Type	ZnO	S	Impurities	Moisture and other volatile matter	Retained on 325-mesh sieve
		(maximum percent of content)			
American process	98.0	0.2	2.0	0.5	1.0
French process	99.0	0.1	1.0	0.5	1.0

Types and Uses

The principal construction uses of zinc oxide are in plastics, rubber, paint, ceramics, resilient floor coverings, asbestos-cement shingles, and roofing granules. Other important uses include lubricating oils and greases, rayon, adhesive tapes, and phosphate solutions for treatment of steel prior to painting.

Zinc oxide is a major constituent of exterior house paints as it imparts whiteness, opacity, and tint retention; it also toughens paint films, reduces chalking, improves cleaning properties and durability, and provides mildew resistance.

History and Manufacture

The ancient Egyptians and Chinese used zinc oxide as a pigment. The modern history of zinc pigments (white paint solid) dates back to about 1840, when LeClair, in France, first produced zinc oxide in quantity. It was first manufactured in the United States in 1852, and its production has paralleled the growth of the rubber and paint industries, in which it plays major roles.

The two processes for manufacturing zinc oxide are the American process and the French process. In the American process, zinc ore is mixed with carbon (coal) and then heated, liberating zinc vapor, which is immediately oxidized to zinc oxide and collected in muslin bags. In the French process, metallic zinc is converted to zinc oxide by vaporizing and burning in air.

Table Z4 Zinc gauges commonly used for roofing, siding, and flashing

Gauge number	Weight		Thickness		Sheet maximum				Strip maximum			
					Length		Width		Length		Width	
	lb/ft²	kg/m²	in.	mm	ft	m	ft	m	ft	m	in.	mm
9	0.67	2.37	0.018	0.46					12	3.658		
10	0.75	3.66	0.020	0.51	8	2.438	5	1.524	for all gauges		20	508
11	0.90	4.39	0.024	0.61	for all		for all		in flat sheet		for all	
12	1.05	5.13	0.028	0.71	gauges		gauges		600	182.88	gauges	
									in rolls			

ZINC SHEET AND STRIP

Physical and Chemical Properties

Zinc sheet and strip are fabricated from either zinc or an alloy of zinc containing copper (for hardening) and small quantities of lead, cadmium, chromium, magnesium, and other nonferrous metals. The alloys are used for a wider range of stiffness and for special characteristics.

Sheet and strip are dull gray in color, strong, spark resistant, and corrosion resistant (especially to salt air) and will not stain adjoining materials. They can be used in direct contact with wood, concrete, mortar, lead, tin, galvanized iron or steel, and aluminum. With other metals, zinc sheet and strip should be isolated to stop electrolytic action, and with redwood or cedar the zinc should be coated with asphalt paint or other similar coating to avoid reaction with the acids of the wood.

Types and Uses

Rolled zinc in sheet or strip form is used in construction for roofing, flashing, weatherstripping, leaders, gutters, roof ridges and hips, termite shields, and terrazzo stripping. Sheet for roofing and strip for flashing are available in two classifications, soft and medium hard. Corrugated zinc sheet for roofing and siding is discussed separately. (*See* Zinc Sheet, Corrugated.)

Zinc Gauge System. The gauge system for zinc is distinctive for that metal. Rolled zinc gauges range from Nos. 3 to 28 inclusive. Gauge No. 3 weighs 0.22 lb/ft² (0.10 kg/m²) and is 0.006 in. (0.15 mm) thick; gauge No. 28 weighs 37.5 lb/ft² (17.01 kg/m²) and is 1 in. (25.4 mm) thick. The gauges commonly used for roofing, siding, and flashing are shown in Table *Z4*. To give a comparative picture, Gauge No. 10 is equal in thickness to 16-oz. (453.6-g) copper, whereas No. 11 is slightly thinner than 20-oz. (567.0-g) copper.

The section on gauges gives the equivalent thickness in decimals of an inch for all the commonly used gauge numbers.

Application

Sheet and strip zinc can be bent, but only at rounded angles, and can be soldered using a 50% tin and 50% lead solder with a hydrochloric acid flux. To allow for expansion, sheet and strip should always be held with clips, not nailed. If sheet or strip is soldered, expansion joints should be placed approximately every 15 ft (4572 m).

Condensed Checklist

1. Zinc should not come into direct contact with metals other than lead, galvanized iron or steel, tin, or aluminum.
2. All nails should be aluminum, solid zinc, or zinc-coated (galvanized). They should have large, flat heads and sharp points. Expansion joints should be placed every 15 ft (4572 m) if sections are soldered.
3. Methods of joining the sheet or strip for roofing and flashing by the use of standing or batten seams (*see* Figure *Z5*).

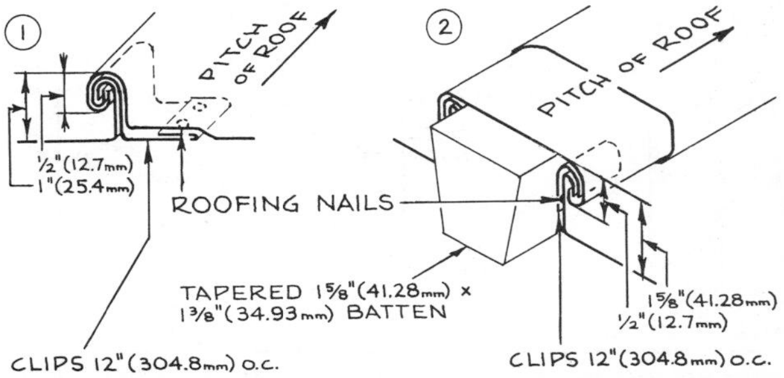

Figure Z5 Details of standing seam (1) and batten seam (2).

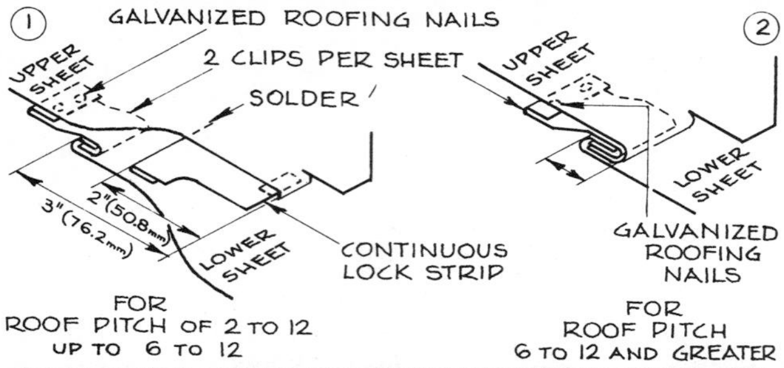

Figure Z6 Details of cross seams for standing and batten seam roof.

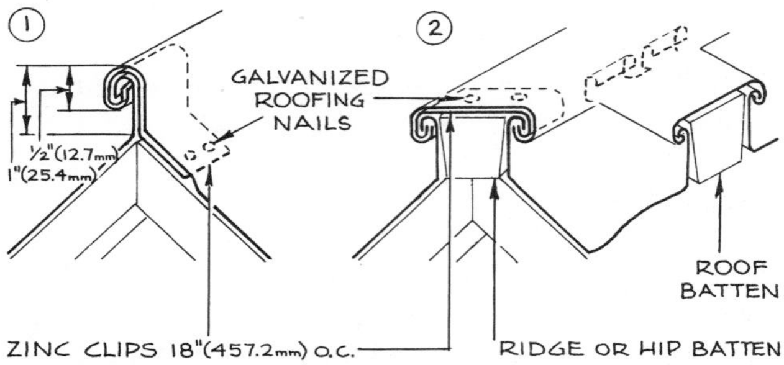

Figure Z7 Details of standing seam (1) and batten seam (2) at ridge and hip.

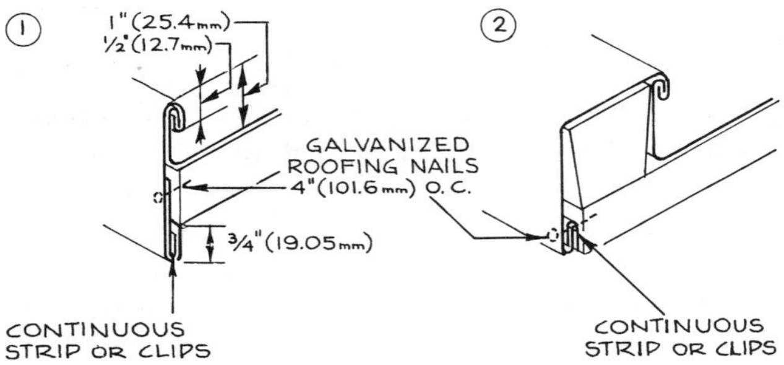

Figure Z8 Details of standing seam (1) and batten seam (2) at roof edge.

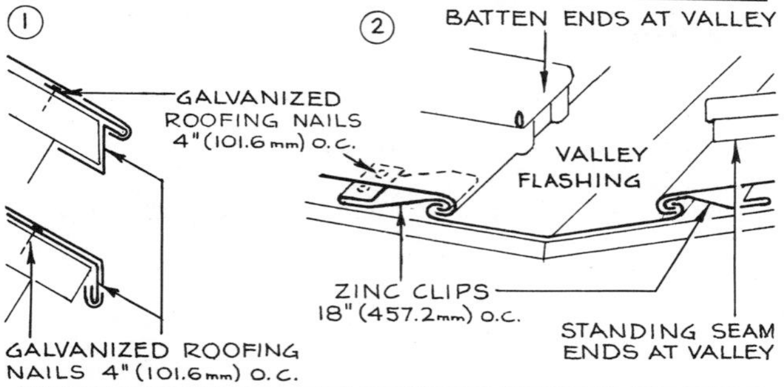

Figure Z9 Details of eave drip (1) and valley (2) for standing and batten seams.

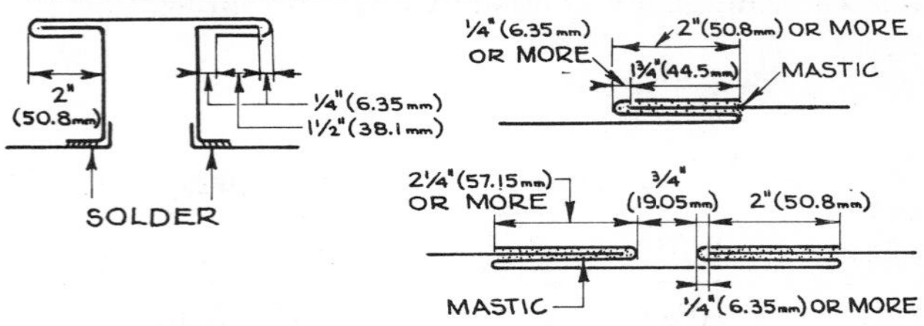

Figure Z10 Details of expansion joints.

4. Type of cross seam for roofs with low pitch and with steep pitch (*see* Figure *Z6*).
5. Treatment of seams at roof ridge and hip (*see* Figure *Z7*).
6. Treatment of seams at gable ends of roof (*see* Figure *Z8*).
7. Treatment of seams at eave drip and valley (*see* Figure *Z9*).
8. Treatment of expansion joints (*see* Figure *Z10*).
9. Procedures to prevent discoloration of zinc sheet and strip are the same as those for galvanized steel sheet. They should not be left open or tightly bundled but should be stored at constant temperature and humidity. (*See* Zinc Coatings.)

Conditions Favorable to the Use of Zinc Sheet and Strip

1. Where staining of adjoining materials by the roofing and flashing is to be avoided.
2. Where a strong, corrosion-resistant material is needed for roofing and flashing, especially near the sea.
3. Where a material that will adhere to concrete, mortar, and plaster and that will not be corroded by them is needed.
4. Where it is advantageous to have a fire-resistant roofing material. Municipal, state, and local codes should be checked and also the fire rating codes of the Fire Underwriters, insurance companies, labor departments, and federal government (Army, Navy, etc.).
5. Where strong vertical lines are desired on a roof.

Conditions Unfavorable to the Use of Zinc Sheet and Strip

1. Where the zinc may come into direct contact with metals other than lead, tin, aluminum, and galvanized iron or steel.
2. Where the zinc may come into direct contact with metals other than lead, tin, aluminum, and galvanized iron or steel.
3. Where a color other than dull gray is desired.
4. Where a highly reflective roofing material is desired or necessary.

History and Manufacture

Zinc in some sort of sheet or strip form was first produced in China. It was first manufactured in England about 1750 and in the United States after 1859. The present-day production of sheet and strip is by hot rolling. In this method the zinc slab is heated to about 350 to 400°F (176.67 to 204.44°C) and rolled in one direction on steel rollers. A 3 in. (76.2 mm) slab is reduced to a thickness of 0.125 to 0.14 in. (3.18 to 3.56 mm), allowed to cool to about 250°F (121.11°C), and then finished by cold rolling. For finish sheet the rough sheets are stacked in groups of 8 to 18 and rolled simultaneously. During the rolling the stacks are split frequently so that all sheets receive equal treatment. Zinc strip is finished by rotary-gauge slitters which cut the strip to size; zinc sheet is finished by shearing.

ZINC SHEET, CORRUGATED

Physical and Chemical Properties

Corrugated zinc is rigidized sheet zinc ranging in thickness from gauge No. 11 through No. 16 (of the zinc gauge system). It is usually fabricated from an alloy of zinc with a small percentage of copper for hardening and very small amounts of chromium and magnesium. It is dull gray in color, strong, spark resistant, and corrosion resistant, especially to salt air; it will not stain adjoining materials. It has another advantage in that it can be lapped, thus making a weatherseal and eliminating many joint problems. It can be used in direct contact with wood, concrete, mortar, lead, tin, galvanized iron, and aluminum. With other metals, it should be insulated to prevent electrolytic action. With redwood and red cedar, both of which contain acids that attack zinc, it should be coated with asphalt paint or similar coating.

Types and Uses

Corrugated zinc is available in two types that are especially suited for construction use: roofing and siding.

Stock widths: $27\frac{1}{2}$ and $40\frac{1}{2}$ in. (69.85 and 102.87 cm) for roofing; $26\frac{1}{4}$ and $39\frac{1}{2}$ in. (66.68 and 100.33 cm) for siding. Corrugations end downwards at one end and upwards at the other.

Stock lengths: 6, 7, 8, and 10 ft (1.829, 2.134, 2.438, and 3.048 m) for both types.

Overall depth (from one side of sheet to other): $\frac{7}{8}$ in. (22.23 mm) for roofing; $\frac{5}{8}$ in. (15.88 mm) for siding.

Center to center of corrugations: $2\frac{1}{2}$ in. (63.5 mm) for both types.

Minimum roof pitch: 3 to 12.

Stock gauges: 11, 12, 13, 14, 15, and 16 for both types (according to the American zinc gauge system).

Maximum roof spacing without support: based on gauge according to Table *Z5*.

Side lap: $1\frac{1}{2}$ corrugations for both types.

Weight: based on gauge.

Bending: Corrugated zinc can be bent parallel to the corrugations but not perpendicular to them.

Application

Accessories for corrugated zinc consist of various types of supports (clinch nails, rivets, etc.), washers, corner covers, filler strips, preformed covers, and ridge rolls (*see* Figure *Z11*).

Condensed Checklist

1. Lead heads, lead washers, and soft iron tin-coated rivets with zinc washers, also called burrs, should be used for attachment.

Table Z5 Maximum roof spacing and weight of corrugated zinc

Gauge number	Thickness		Maximum roof spacing		Weight			
					Roofing		Siding	
	in.	mm	in.	mm	lb/ft^2	kg/m^2	lb/ft^2	kg/m^2
11	0.024	0.61	42	1066.8	1.12	5.47	0.97	4.74
12	0.028	0.71	45	1143.0	1.30	6.35	1.13	5.52
13	0.032	0.81	48	1219.2	1.49	7.27	1.29	6.30
14	0.036	0.91	51	1295.4	1.68	8.20	1.45	7.08
15	0.040	1.02	54	1371.6	1.87	9.13	1.61	7.86
16	0.045	1.14	57	1447.8	2.06	10.06	1.77	8.64

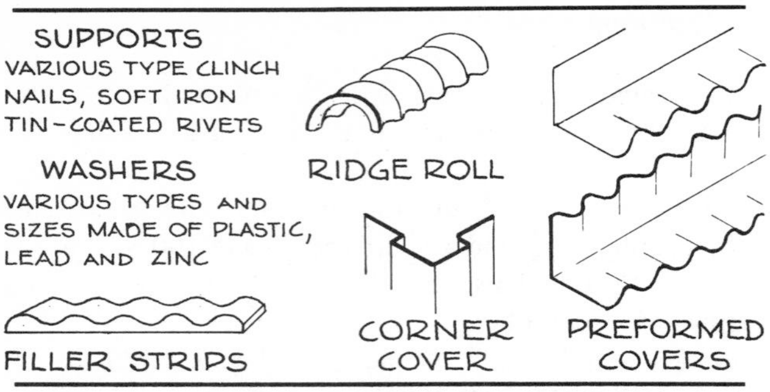

Figure Z11 Typical accessories for corrugated zinc.

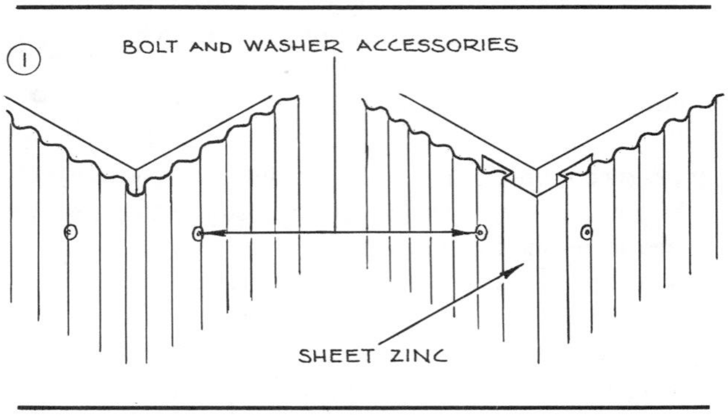

Figure Z12 External corner details.

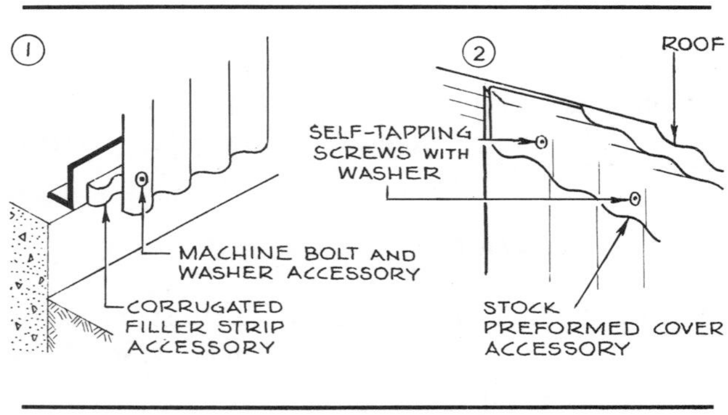

Figure Z13 Details of sill (1) and roof edge (2) joints.

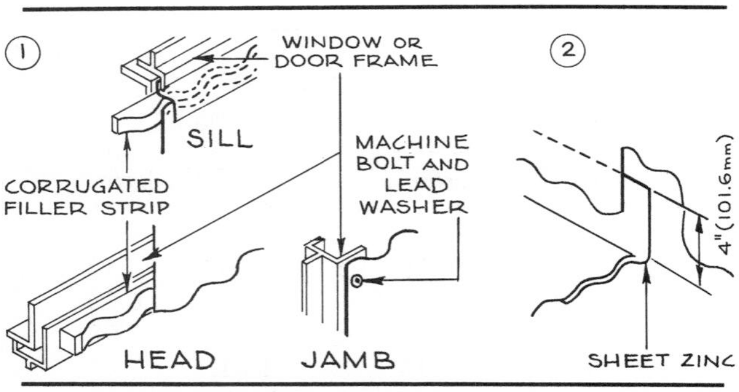

Figure Z14 Window and door (1) and roof (2) openings.

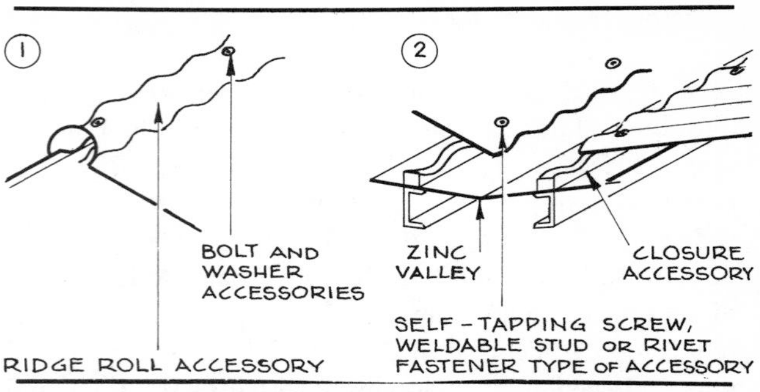

Figure Z15 Ridge and hip (1) and valley (2) details.

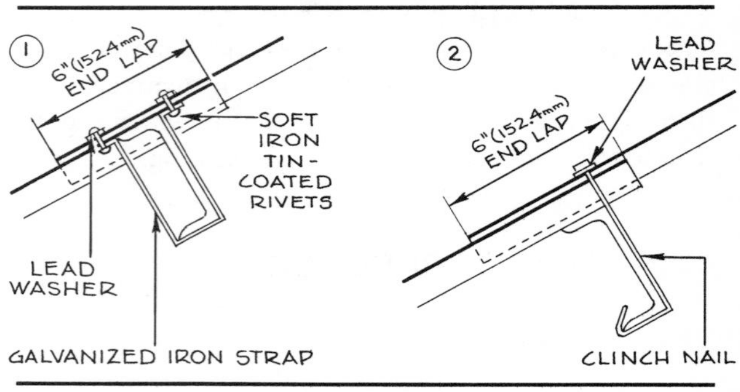

Figure Z16 Typical roof fasteners.

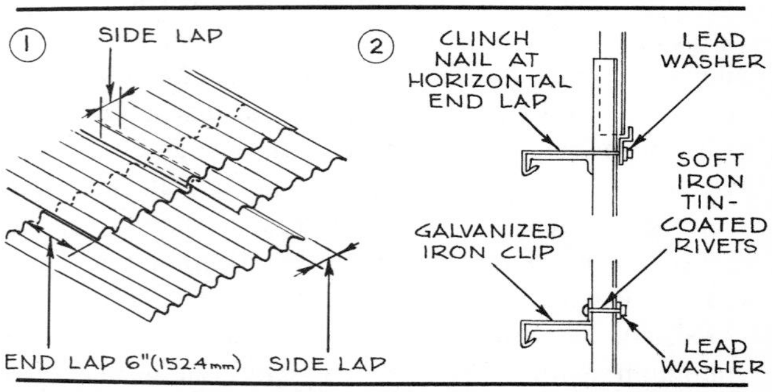

Figure Z17 Details of lapping (1) and side wall fasteners (2).

2. Treatment of external corners (*see* Figure *Z12*).
3. Treatment of joints at roof edge and sill at grade (*see* Figure *Z13*).
4. Treatment of openings for windows and doors: At head and sill, stock filler strips are generally used; at jambs, the corrugated zinc is riveted or otherwise secured to the window frame (*see* Figure *Z14*).
5. Treatment of roof intersections at ridge, hip, and valley (*see* Figure *Z15*).
6. Type of roof fastener: A large variety is available for different conditions (*see* Figure *Z16*).
7. Various types of lapping details for roofs and siding (*see* Figure *Z17*).
8. Types of side wall fasteners: A large variety is available for different conditions (*see* Figure *Z17*).

9. Storage and handling: Storage should be at constant temperature and humidity. Dampness can cause deterioration of surface finish, as will the presence of salts, acids, and alkalis.

Conditions Favorable to the Use of Corrugated Zinc

1. Where the surface is likely to receive slightly rough treatment.
2. Where maintenance must be kept to a minimum.
3. Where fire resistance is required. Municipal, state, and local codes should be checked and also the fire rating codes of the Fire Underwriters, etc.

Conditions Unfavorable to the Use of Corrugated Zinc

1. On curved surfaces where the curve is perpendicular to the corrugations.
2. Where a span larger than 4.75 ft (1.448 m) is demanded.
3. Where a color other than dull gray is desired.

History and Manufacture

Corrugated zinc was first manufactured about 1900 by rolling zinc sheet through two corrugated rollers, using a process similar to that for making corrugated aluminum.

ZINC SULFIDE

Physical and Chemical Properties

Zinc sulfide occurs in nature as the mineral sphalerite (blende) and is also artificially produced as a yellowish white powder, ZnS. It has the highest opacity of the zinc pigments, and when it contains minute quantities of metallic impurities it is phosphorescent.

Commercial Forms. Zinc sulfide is available in three grades: technical, chemically pure (C.P.), and phosphorescent.

Types and Uses

Zinc sulfide was used in the past as a pigment in paints and as a whitening agent and filler in linoleum, glass, rubber, plaster, glues, paper, etc., and for special luminous paints. Today zinc sulfide has been largely replaced by lithopone, a combination of zinc sulfide and barium sulfate.

ZIRCONIUM

Physical and Chemical Properties

Symbol: Zr
Atomic number: 40
Specific gravity: 6.506 (20°C, 68°F)
Melting point: 1852°C, 3365.6°F
Boiling point: 4377°C, 7910.6°F

Zirconium is a silvery-white metal that is soft and ductile in pure form. It is one of the more abundant metals in the earth's crust and is always found in combination with hafnium. Like hafnium, it is highly corrosion resistant to almost all chemicals except fluorides and remains untarnished in almost all industrial atmospheres.

Workability. Zirconium may be bent, drawn, formed, and cast. It may also be welded with an inert-gas shield and can be machined, but its shavings are pyrophoric (spontaneously explosive).

Commercial Forms. Zirconium is now available in two grades: commercial, containing hafnium, and pure or reactor grade. It can be obtained in ingot, billet, wire, pipe, tube, strip, sheet, plate, foil, rod, and powder form.

Types and Uses

Most zirconium metal is diverted into nuclear energy applications. Its general commercial uses are increasing, however, because of its excellent resistance to corrosion in certain chemical applications. Table *Z6* lists current applications, an asterisk indicating use of the metal.

Zirconium is usually added to nonferrous alloys, steel, and cast iron in the form of an alloy with other metals. Zirconium alloys show excellent strength up to 600°F (315.56°C) and higher.

Zirconium compounds are also being more widely used. Zirconium oxide, if stabilized, can be made into containers that are very inert and strong at high temperatures.

Nuclear Energy Uses. In the nuclear energy field, the low neutron-capture cross section and outstanding corrosion properties of zirconium make it the ideal "canning" material for fuel in reactor applications. For this purpose it must be as nearly hafnium-free as possible. Radioactive zirconium 95 is used in the plastics and petroleum industries for radioactive tracing (*see* Radiation Protection).

Table Z6 Uses of zirconium

	Construction Uses	
Basis of	Abrasive	Main ingredient
	Brick	Fire brick; refractory material
	Color pigment	White
Component of	Aluminum*	Permitting stronger, larger castings
	Brass*	Increasing hardness and strength
	Bronze	Increasing hardness and strength
	Cast iron*	Increasing strength
	Chromium*	Alloying agent
	Copper*	Increasing strength
	Enamel	Acidproofing
	Glass	Opacifier
	Construction Uses (Continued)	
Component of	Glazes and porcelain enamels	Opacifier
	Magnesium*	Alloying agent
	Nickel*	Degasifying; increasing strength
	Paint	Antichalking
	Stainless steel*	Aiding machinability
	Steel*	Deoxidizer; substitute for manganese in steel manufacture; ductility; shock and fatigue resistance
	Textiles	Water repellent
	Nonconstruction Uses	
Powder metallurgy, preservatives, and tanning		

History and Manufacture

In 1789 Klaproth discovered a new metal oxide in the mineral zircon. In 1824 Berzelius produced a metallic zirconium in the form of an impure black powder. Later, Lely and Hamburger produced zirconium metal sufficiently pure to show ductility. Starting in 1944, Kroll and his co-workers in the U.S. Bureau of Mines developed a method for large-scale production of high-purity zirconium in answer to the needs of the nuclear energy program. This process is based on the reduction of zirconium tetrachloride with magnesium.

INDEX

INDEX